AF535062

The Opacity Project
Volume 2

The Opacity Project
Volume 2

Compiled by the Opacity Project Team

With an introduction by K A Berrington, Queen's University of Belfast

Institute of Physics Publishing
Bristol and Philadelphia

British Library Cataloguing-in-Publication Data

A catalogue record for this book is available from the British Library.

ISBN 0 7503 0174 0

Library of Congress Cataloging-in-Publication Data are available.

Published by Institute of Physics Publishing, wholly owned by The Institute of Physics, London

Institute of Physics Publishing, Techno House, Redcliffe Way, Bristol BS1 6NX, UK

US Editorial Office: Institute of Physics Publishing, The Public Leger Building, Suite 1035, 150 South Independence Mall West, Philadelphia, PA 19106, USA

Printed in the UK by J W Arrowsmith Ltd, Bristol

Contents

INTRODUCTION

This is the second volume in 'The Opacity Project' series by the Opacity Project (OP) team and published by the Institute of Physics.

Volume 1 was published in 1995 (ISBN 0 7503 0288 7), and provided an excellent introduction to the aims and results of the Project, including 32 key papers reprinted from the literature, together with selected atomic data tables for the elements He through Si.

In this second volume we aim to complete these tables by including data for S, Ar, Ca and Fe. We also include reprints of related papers published since Volume 1 appeared.

It is not therefore necessary to give in this introduction to Volume 2 extensive details of the science of stellar opacities and of the Project. We therefore confine ourselves here to a summary only, and note that full details are to be found in Volume 1.

The *opacity* κ_ν of a medium is the reciprocal of the mean free path of frequency ν. The *intensity of radiation* at position $\mathbf{r}$ and in a direction $\hat{\mathbf{s}}$ is described by a function $I_\nu(\mathbf{r}, \hat{\mathbf{s}})$ such that $I_\nu \mathrm{d}a\mathrm{d}\omega$ is the amount of radiant energy crossing an area $\mathrm{d}a$, normal to $\hat{\mathbf{s}}$, in a solid angle $\mathrm{d}\omega$ and per unit frequency and unit time. If $\mathrm{d}s$ is an element of distance in the direction $\hat{\mathbf{s}}$, then $\kappa_\nu \mathrm{d}s$ is the probability of radiation being absorbed or scattered in traversing that distance. The equation of radiative transfer is

$$\frac{\mathrm{d}I_\nu}{\mathrm{d}s} = \kappa_\nu I_\nu + j_\nu \qquad (1)$$

where j_ν is the *emissivity*.

A medium of sufficiently low density may be considered to be an assembly of 'particles': electrons, nuclei, atoms and molecules. For such a medium the opacity is given by

$$\kappa_\nu = \sum_n N(n)\sigma_\nu(n) \qquad (2)$$

where $N(n)$ is the number of particles of type n per unit volume and $\sigma_\nu(n)$ is the cross section for absorption or scattering. In order to calculate κ_ν using (2) one needs to determine the populations $N(n)$ (the problem of the *equation of state*) and the cross sections $\sigma_\nu(n)$ (a problem of atomic or molecular physics). Full details concerning the equation of state, and the atomic physics, is given in 'The Opacity Project Volume 1'.

The following atomic processes contribute to opacities in stellar interiors.

1. Bound-free transitions,

$$A + h\nu \rightarrow A^+ + \mathrm{e}, \tag{3}$$

for photo-detachment from negative ions and photo-ionization of atoms and atomic positive ions.

2. Bound-bound transitions,

$$A(i) + h\nu \rightarrow A(f), \tag{4}$$

in atoms and atomic positive ions; and, at low temperatures, in molecules.

3. Free-free transitions,

$$A + \mathrm{e}_i + h\nu \rightarrow A^+ + \mathrm{e}_f, \tag{5}$$

where A may be a neutral atom or an atomic positive ion.

4. Scattering processes. The most important is electron scattering,

$$\mathrm{e}_i + h\nu \rightarrow \mathrm{e}_f + h\nu, \tag{6}$$

in which there is a change in the direction of photon but (at least for the case of Thompson scattering) no change in the photon frequency. At low temperatures Rayleigh scattering can also contribute.

We define *stellar envelopes* to be those parts of interiors for which the initial levels giving significant contributions to opacities are not markedly perturbed by the plasma environment, and for which equation (2) can be used. The OP work is concerned with opacities for envelopes. In equation (2) we use accurate cross sections for free atoms and we allow for plasma perturbations only in calculating the line profiles and the occupation probabilities. We do not include any molecular contributions. A large team has been involved, particularly in the atomic physics work (see Table 1 for details). The main emphasis of the project has been in obtaining accurate atomic data which, incidentally, may be of value in a wide range of other problems. The methods used in the OP work may not give accurate results if employed in the deepest layers of stellar interiors.

The OP atomic data were computed using collisional techniques for the relevant bound and free states. This involved representing the electron–ion collision quantum-mechanically in a configuration interaction and close-coupling approach, and solving Schrödinger's equation using the R-matrix method. For further details on the

R-matrix method, the interested reader should consult 'Atomic and molecular processes: an R-matrix approach' by P G Burke and K A Berrington, published by the Institute of Physics in 1993 (ISBN 0-7503-0199-6).

The method has some nice features. All of the important physics of a low energy collision are incorporated in an ab initio way. For example, the resonance structures which can so dominate a cross section at low energies arise naturally from such an approach. Also, the use of an energy-independent basis in the inner region makes the R-matrix method highly efficient for treating large numbers of energies, an important feature for calculating large numbers of bound states for radiative calculations, or for calculating cross sections over a fine energy grid in order to represent complicated resonance structures. Autoionizing states can radically alter the low temperature behaviour of collision rates, and are a major contributor to opacity.

A comprehensive and well-tested suite of computer programs has been developed to calculate atomic data using the R-matrix method. These programs have been published as 'RMATRX1' by K A Berrington, W B Eissner and P H Norrington, *Computer Physics Communications* **92** 290 (1995). The programs have a long history, and have involved many people in their development and use. The programs have been designed to deal with the general atom or ion, to calculate both collisional and radiative data. There are many options, eg. to be able to include relativistic effects, and extensions, eg. to molecules. These programs are used worldwide, and are the standard tools for the OP and other applications.

At this point, it is appropriate to mention the international IRON Project (IP), which grew out of the Opacity Project, and which involves many of the OP team. The aim is to systematically compute electron excitation cross sections for the iron group of elements, using R-matrix methods. The techniques are similar to those used in the OP; the main difference being the introduction of relativistic effect via the Breit-Pauli Hamiltonian in order to calculate fine-structure cross sections. Particular attention is given to requirements for the interpretation of data from specific space observations. Dissemination is also via data bases, with a series of papers 'Atomic data from the IRON Project' in the journal Astronomy and Astrophysics. The first such paper, 'The IRON Project I. Goals and methods' was published by D G Hummer, K A Berrington, W Eissner, A K Pradhan, H E Saraph and J A Tully, *A&A* **279** 298 (1993). In view of the close links between the OP and the IP, it is planned to produce 'The Opacity Project Volume 3: The Iron Project' in due course, to contain the key papers and data from the IP.

The contents of the present volume 'The Opacity Project Volume 2' are as follows.

- **Section 1. Review.** The paper here was from a talk given at a Joint Discussion meeting on 'Astrophysical applications of powerful new databases' at the 22nd General Assembly of the International Astronomical Union in The Hague, 22–23 August 1994. This summarises the atomic data produced by the OP and by the new Iron Project.

- **Section 2. Atomic data calculations.** The main OP atomic data work is contained in a series of papers 'Atomic Data for Opacity Calculations' referred to as ADOC and published in the Journal of Physics B. The first 20 papers in the ADOC series were reprinted in 'The Opacity Project Volume 1'. This section contains reprints of 2 further ADOC papers and another 2 papers concerned with the atomic data work associated with Fe ions. In addition, we reprint the paper from Astronomy and Astrophysics on TOPBASE, the central database for all the OP atomic data at the Centre des Données Astronomiques (CDS) in Strasbourg.

- **Section 3. Tables of selected atomic data.** All OP atomic data are available in the database system TOPBASE. 'The Opacity Project Volume 1' contained tabulations of selected energy levels and gf-values for the elements He–Si. In this volume we give selected printed data for energy levels and gf-values for the remaining elements S, Ar, Ca and Fe.

As the OP was a team effort, this introduction finishes with a Table listing the OP team members. On a personal note, it is a pleasure to acknowledge the leadership and dedication of Mike Seaton in bringing this Project to fruition.

Keith Berrington, 1996 September.

Table 1: The Opacity Project Team

Name	Affiliation	Contribution
K A Berrington	QUB	atomic data
P G Burke	QUB	atomic data
V M Burke	QUB	atomic data
K Butler	MUO	atomic data
W Däppen	USCA	equations of state
W Eissner	QUB	atomic data
J A Fernley	UCL and VILSPA	atomic data
A Hibbert	QUB	atomic data
D G Hummer	JILA and MUO	atomic data and equations of state
A E Kingston	QUB	atomic data
M Le Dourneuf	UR	atomic data
D J Lennon	QUB and MUO	atomic data
Ding Luo	JILA	atomic data
C Mendoza	IBMV and IVIC	atomic data
D Mihalas	UI	equations of state and opacities
S N Nahar	OSU	atomic data
G Peach	UCL	atomic data
A K Pradhan	OSU	atomic data and opacities
H E Saraph	UCL	atomic data
P M J Sawey	QUB	atomic data
M P Scott	QUB	atomic data
M J Seaton	UCL	atomic data and opacities
P J Storey	UCL	atomic data
K T Taylor	RHL and QUB	atomic data
J A Tully	OCA	atomic data
Yu Yan	UI and NCSA	atomic data and opacities
C J Zeippen	OPM	atomic data

Affiliations: IBMV=IBM Venezuela; IVIC=Instituto Venezolano de Investigaciones Cientificas; JILA=Joint Institute for Laboratory Astrophysics; MUO=Munich University Observatory; NCSA=National Center for Supercomputer Applications; OCA= Observatoire de la Côte d'Azur; OPM=Observatoire de Paris (Meudon); OSU=Ohio State University; QUB=Queen's University Belfast; RHL=Royal Holloway University of London; UCL=University College London; UI=University of Illinois; UR=University of Rennes; VILSPA=Villafranca Satellite Tracking Station.

Astrophysical Applications of Powerful New Databases
ASP Conference Series, Vol. 78, 1995
S. J. Adelman and W. L. Wiese (eds.)

Summary of the Opacity and IRON Projects

KEITH A. BERRINGTON
Department of Applied Mathematics and Theoretical Physics, Queen's University, Belfast BT7 1NN, U.K.
e-mail: K.Berrington@Queens-Belfast.AC.UK

Abstract. The recently-completed Opacity Project and the current IRON Project are international collaborations to calculate atomic data for primarily astrophysical applications, using state-of-the-art theoretical and computational methods. As the name implies, the Opacity Project was to calculate radiative opacity data for atoms and atomic ions; the IRON Project is to calculate electron collisional data mainly for the iron-group elements. The scope and accuracy of the atomic data are discussed. A summary of publications is given.

1. Introduction

This paper discusses two international collaborative projects, the Opacity Project and the IRON Project, with an emphasis on the atomic data produced. A companion paper by Professor M. J. Seaton describes the results of the Opacity Project in more detail, together with astrophysical implications. Although the motivation and areas of application are different, the two projects have much in common with their atomic physics. Both projects involve the calculation of large amounts of good atomic data. It is these aspects which will be discussed here.

1.1 Scientific Background

It is obvious that a knowledge of elemental atomic, molecular and optical processes is necessary to understand distant matter. At one level, the interpretation of spectra from astronomical objects certainly requires a knowledge of the processes that populate atomic and molecular energy levels that give rise to emission and absorption. But at a more fundamental level, the construction of theoretical and predictive models of, say, a star, requires a secure knowledge of the basic atomic processes in the stellar plasma.

The most important atomic processes are easily listed, namely:

- transition frequencies
- oscillator strengths
- photoionization cross sections
- recombination rate coefficients
- electron impact excitation rates
- ionization cross sections, etc.

Surprisingly, although the study of these processes can in principle be done in the laboratory, it has proven difficult to obtain the necessary range of atomic data of sufficient accuracy in controlled laboratory conditions. This has given tremendous impetus to theoretical atomic physicists to attempt to calculate the required atomic data from quantum-mechanical first principles.

A feature of recent times is that both atomic physics and observational astronomy are refining their respective databases to place strict limits on the models, and so advance our understanding of the universe and its objects. This decade is witnessing the deployment of powerful new instruments for astronomical spectroscopy. For example ESA's programme, including the Solar and Heliospheric Observatory (SoHO), and NASA's Great Observatories programme (HST, GRO, AXAF, and SIRTF), covering the infrared, optical, X-ray, and γ-ray regions of the spectrum, will provide several significant new windows on the universe. The increasing sophistication, precision and range of observational techniques and of theoretical models create new demands for more and better data on atomic processes.

In the past, the basic atomic data used by astronomers were provided by research activity in atomic, molecular, and optical physics. The data were often of variable quality: quantum-mechanical calculations are difficult and compromises have to be made to obtain tractable solutions. Uncertainties in the atomic data have always been a problem for the astrophysics user. Atomic physicists have not always focussed on the real needs. Yet when accurate atomic data are produced systematically for astrophysical applications, as in the recently-completed Opacity Project and the new IRON Project, the results can have important implications for astronomy.

1.2 Some Atomic Theory

It is appropriate at this stage to introduce some atomic theory to facilitate the discussion on accuracy.

Both the Opacity and the IRON Projects use similar theoretical techniques for calculating atomic data. These involve representing the electron-atom (or -ion) collision quantum-mechanically as a wavefunction Ψ, and solving Schrödinger's equation using the R-matrix method (Burke & Berrington 1993).

In the R-matrix method, Ψ is considered in two regions separated by a sphere of radius $r = a$, where r is the radial coordinate of the colliding electron.

- In the inner region $r < a$, the colliding electron must be considered indistinguishable from the atomic electrons, and a many-electron problem has to be solved. This is done in the most efficient way possible by expanding Ψ in terms of an energy-independent basis satisfying R-matrix boundary conditions. The equations are solved by a matrix diagonalization for all energies.

• In the outer region $r > a$, the colliding electron merely 'sees' a long range potential due to the atom. Ψ can be written straightforwardly as a no-exchange close-coupling expansion. The resultant coupled differential equations are solved subject to open-channel (energetically allowed) or closed-channel boundary conditions as appropriate, for each value of the total energy E.

• Matching the wavefunctions on the boundary $r = a$ yields the complete Ψ for all space.

The form of Ψ at infinity yields the observables, such as the cross section or collision strength $\Omega(E)$. The energy integral of $\Omega(E)$ over a Maxwellian distribution gives the collisional excitation rates normally used in astrophysics applications.

Ψ can also be used to obtain radiative data by calculating the appropriate dipole matrix elements. Bound states and their energies can be obtained from the discrete solutions Ψ given when all channels are closed. Thus, oscillator strengths (f-values, bound-bound data), or photoionization cross sections (bound-free data) can be calculated from the same theoretical approach.

The method has some nice features. All of the important physics of a low energy collision are incorporated in an *ab initio* way. For example, the resonance structures which can so dominate a cross section at low energies arise naturally from such an approach. Note that the use of an energy-independent basis in the inner region makes the R-matrix method highly efficient for calculating data for large numbers of E values, an important feature for calculating large numbers of bound states for radiative calculations, or for calculating cross sections and Ω(E) over a fine E grid to represent complicated resonance structures.

A comprehensive and well-tested suite of computer programs has been developed to calculate atomic data using the R-matrix method (Berrington et al. 1995). These programs have a long history, and have involved many people in their development and use. They have been designed to deal with the general atom or ion, to calculate both collisional and radiative data. There are many options, the most important of which is to be able to incorporate relativistic effects via terms of the Breit-Pauli Hamiltonian. These programs are used worldwide, and are the standard tools used in the Opacity and IRON Projects. They represent the state-of-the-art in computational atomic physics.

2. The Opacity Project

2.1 Motivation

Accurate calculations of stellar envelope opacities are vital if realistic structure, evolution, and pulsation models are to be quantitatively established for stars of all spectral types. Opacities from the Los Alamos Astrophysical Opacity Library (Huebner et al. 1977) have been almost universally applied, although the atomic data used as a basic ingredient have themselves been computed, in many cases, with the crude methods in use (necessarily) before the advent of modern supercomputers and *ab initio* calculations.

Nonetheless, with a few notable exceptions, Los Alamos opacities resulted in models that appeared to give a satisfactory agreement with observation. The long-standing difference between Cepheid masses inferred from pulsation models and evolution tracks - both computed with Los Alamos opacities - is one such exception. It was clearly important to check the suggestion by Simon (1982), that increasing the metal contribution to the total opacity by a factor of 2 to 3 would remove this discrepancy, as there were implications for stellar evolution.

The origin and magnitude of any increased metal contribution to the total opacity had to be identified through accurate atomic physics calculations and detailed consideration of the equation of state. Seaton (1987) proposed the calculation of the required atomic data by *ab initio* methods, and launched an international collaboration to do this, the Opacity Project (OP). An independent approach (OPAL) was taken at the Lawrence Livermore National Laboratory (Iglesias et al. 1987).

The OP and OPAL calculations indicate that Simon's prediction is broadly correct. The new opacities are now being widely used in stellar structure, evolution, and pulsation models.

2.2 Scope of the Atomic Data

The Opacity Project computed atomic data for opacity calculations for

- **H, He, Li, Be, B, C, N, O, F, Ne, Na, Mg, Al, Si, S, Ar, Ca,** and **Fe;**

- **energies** of terms having effective quantum numbers $\nu \leq 10$ and total angular momentum $L \leq 3$ or 4, all spin and parity combinations;

- **gf-values** for all dipole transitions between these bound terms;

- **photoionization cross sections** from all calculated bound terms, tabulated on a grid of photon energies suitable to describe the resonance structure in sufficient detail to calculate reliable opacities;

- **line broadening parameters.**

2.3 Accuracy of Atomic Data

The quality of the atomic data produced by the Opacity Project is discussed in several places, for example: the original papers in the J. Phys. B series *'Atomic Data for Opacity Calculations'* (for a collection see the book by the Opacity

Project Team 1995); the proceedings of the Caracas workshop on Astrophysical Opacities (Lynas-Gray et al. 1992); Seaton et al. (1994). It is therefore sufficient to give a summary of conclusions here.

The calculations are optimized for positive ions. For all ions up to the iso-electronic sequence of aluminum-like ions, the R-matrix calculations include all states belonging to ground complexes explicitly as target states in the expansion of the total wavefunction. For all positive ions in these systems the oscillator strengths and photoionization cross sections should be accurate to about 10 per cent, except for sensitive cases involving a lot of cancellation. The accuracy may be less good for some neutral atoms that have low-lying target states not belonging to ground complexes. The accuracy may also be less good for systems belonging to higher iso-electronic sequences (Seaton et al. 1994).

Further work is continuing, particularly to improve the data for the first few ionization stages of iron.

2.4 Publications and Database

Twenty-eight key research papers arising from the Project, including the J. Phys. B series *'Atomic Data for Opacity Calculations'* and the ApJ series *'Equations of State for Stellar Envelopes'*, together with calculated energies and oscillator strengths for light ions to Si, are reprinted in *'The Opacity Project Volume 1'* (Opacity Project Team 1995).

All numeric data are available from **TOPbase**, an on-line database at the Centre de Donnèes Astronomiques de Strasbourg (CDS) (Cunto et al. 1993). To use TOPbase through internet:

IP address: 130.79.128.5, **account:** topbase, **password:** seaton+

Inexperienced users are advised first to obtain the user-guide through anonymous ftp (same number) before accessing the database.

3. The IRON Project

3.1 Motivation

An indication has already been given of the wealth and precision of new astronomical observations likely to arise in the near future, which, together with the increasing sophistication of quantitative theories, require more and better atomic data.

The primary aim of the IRON Project is to systematically compute electron excitation cross sections for the iron group of elements for astrophysical applications. Particular attention is given to cross sections required for the interpretation of data from specific space observations, for example SoHO and ISO.

Atomic data for the SoHO Coronal Diagnostic Spectrometer (CDS) and Solar Ultraviolet Measurements of Emitted Radiation (SUMER) experiments were assessed at an international meeting at Cosiners House 26-27 March 1992, hosted by the SERC Rutherford Appleton Laboratory (Lang 1994). The reviews revealed many deficiencies in electron impact excitation data currently available, including a lack of reliable data for n = 2 to 4 transitions giving rise to UV lines, and the general inadequacy of Fe ion data. In some cases, an excitation rate is deduced from a cross section which has been calculated at just one impact energy, i. e., ignoring the energy variation required to compute a thermally averaged rate, and that cross section may have a large error! For example, of the Fe ions to be observed in SoHO, Fe IX - XVI, only the excitation rates for Fe IX are known to sufficient accuracy for a sufficient number of transitions over a sufficient temperature range (Mason 1994).

Predicted IR lines detectable by ISO include more than 100 IR lines from positive ions. Tully (1991, private communication) reviewed the atomic data requirements for these lines, and identified the need for much more accurate electron impact excitation rate coefficients for direct excitation, particularly of fine-structure levels, at low temperatures. Reliable excitation data hardly exist at all for low ionized stages of Ti, Cr, Mn, Fe, Co, and Ni.

Table 1. IRON Project participants

N. Badnell	Strathclyde University UK
K. A. Berrington	Queen's University Belfast UK
P. G. Burke	Queen's University Belfast UK
V. M. Burke	Daresbury Laboratory UK
K. Butler	Institut fur A & A. Munchen Germany
W. Eissner	Ruhr Universitat Bochum Germany
M. Galavis	IBM Venezuela Caracas Venezuela
A. Hibbert	Queen's University Belfast UK
D. G. Hummer	Boulder Colorado USA
M. Le Dourneuf	Rennes Universite France
D. J. Lennon	Institut fur A & A Munchen Germany
C. Mendoza	IBM Venezuela Caracas Venezuela
H. E. Mason	Cambridge University UK
P. Norrington	Queen's University Belfast UK
J. Pelan	Queen's University Belfast UK
A. K. Pradhan	Ohio State University Columbus Ohio USA
H. E. Saraph	University College London UK
M. J. Seaton	University College London UK
P. J. Storey	University College London UK
K. T. Taylor	Queen's University Belfast UK
J. A. Tully	Obs. de la Cote d'Azur Nice France
C. J. Zeippen	Obs. de Paris Meudon France
H. Zhang	Ohio State University Columbus Ohio USA

The IRON Project, like the Opacity Project, is an international collaboration. For information, a list of participants is given in Table 1.

3.2 Scope of the Atomic Data

The IRON Project is a significant organisational task. To manage a project of this size, and to provide a steady stream of results (a not insignificant consideration given research support constraints), the Project is divided into a series of short-term and long-term goals. These goals can be summarised as follows:

Goal 1. Fine-structure transitions in ground state configurations.
Excitation cross sections are being calculated for fine-structure transitions in the ground configuration of all ions of astrophysical interest in the

B, C, O, F, Al, Si, S, Cl iso-electronic sequences, also Fe I - VII

(i. e., ions with open p and d shells). These data are essential for the interpretation of IR lines to be observed by ISO, as well as for coronal spectra. The calculations are largely completed and are being published.

Goal 2. Fe ions.
Excitation cross sections are being calculated for all transitions involving states with principal quantum number up to and including $n = 4$. This is now underway, and will provide collisional rates for interpretation of observations from SoHO. In particular reliable collisional and radiative data will be available for Fe II, which appears in spectra of a wide variety of objects.

Further goals include:

- the calculation of collisional data for nearby elements, e. g., Ti, V, Cr, Mn, Co, and Ni;

- extending the Opacity Project calculations to radiative transitions beyond the electric dipole, in intermediate coupling, e. g., E2 and M1 transitions;

- the calculation of radiative properties and energies for heavy ions of relevance to current beam-foil and other experimental investigations.

3.3 Accuracy of the Atomic Data

Care is being taken to ensure that the new data calculated in the IRON Project are accurate. This involves a number of checks, both of self-consistency within the calculation, and by comparison with other data where available. An important feature of the Project is the twice-yearly meeting, which acts as an important forum for exchanging ideas, criticising results, etc., as well as for managing the Project.

Papers already produced by the project give an indication of the accuracy obtainable with modern methods. Rather than giving specific details (these in any case can be found in the literature), it is worth making some general remarks on the expected sources of inaccuracy - and on techniques to minimize their effects.

A. Target State Expansions

The collisional wavefunction involves the coupling of all the required initial and final target states, together with any other states that couple strongly to these. This could be a lot of states (!), and a compromise must be reached between what is necessary, and what is practicable. In the IRON Project, truncation is normally at a given principal quantum number n, i. e., an $n \leq 4$ calculation includes all states involving the outer electron up to $n = 4$. For ions, the omission of more highly excited states will lead to missing resonance structures in the calculation; such resonance structures could give important contributions to $\Omega(E)$ in the energy range of interest. A further consequence of omitting highly excited states is apparent at collisional energies above their threshold, where loss of flux into their channels would not be adequately accounted for. This is especially true in the so-called intermediate energy region, where ionization channels are also open. In practice the truncation of the target state expansion imposes an upper limit to the collision energy at which the calculation will be reliable, corresponding to the energy of the first omitted resonance or threshold. In the IRON Project, care is taken to ensure that the resultant energy range corresponds to the appropriate temperature range at which the ion is likely to be abundant.

B. Configuration-Interaction (CI) Effects in the Target

A time-consuming (for the worker) part of a calculation on a given atomic system is the job of developing a good representation of the target wavefunction, before the collision calculation can proceed. Again compromises have to be made; each target state could be constructed from a very elaborate CI basis, but it is often not easy to carry such a complicated wavefunction through to a collision calculation. A particular restriction of the collision method is that all target states have to be formed out of the same set of one-electron orbitals. Pseudo-orbitals can be introduced to provide further correlation. High quality experimental energy levels are used in the calculation if available. The IRON Project puts great effort into the atomic structure aspect of the calculation, as the final data will depend on the accuracy of the target wavefunctions.

C. Relativistic Effects

Fine-structure and other relativistic effects can be treated using the Breit-Pauli Hamiltonian or by a non-relativistic calculation in LS coupling followed by an algebraic transformation to more appropriate coupling schemes. The Breit-Pauli approach treats the relativistic effects more accurately, while the recoupling procedure allows more accurate wavefunctions for the target ion to be obtained. For most systems the latter is sufficient, at least for the iron-group elements.

D. Near-threshold Resonances

This is related to the other points mentioned so far, in that to obtain highly accurate $\Omega(E)$ at near-threshold energies requires attention to both the target and the collisional calculations. A particular difficulty relates to near-threshold resonances; these are difficult to pin down accurately as small changes in the calculation method (the introduction of fine-structure effects for example), can change the position of any near-threshold resonance and hence the threshold behaviour of $\Omega(E)$. Hardly any of these resonance positions have been experimentally determined, so there is little scope for independent verification. In practice this places a lower limit on the temperature at which a calculation can be trusted; the IRON Project is checking this point, and alerting users where appropriate.

E. High Angular Momenta and High-Energy Contributions

This is a more tractable problem. R-matrix calculations are normally, for practical reasons, carried out over a restricted range of collisional angular momentum and E. Facilities have been developed to 'top-up' $\Omega(E)$ for high angular momentum contributions, and for ensuring that electron excitation rates are calculated over a temperature range consistent with that of the calculation. These points may seem obvious, but have sometimes been overlooked in the past. The IRON Project is committed to publishing results which are within the range of validity of the methods used. For this reason, it may be unsafe to extrapolate IRON Project results outside their range of validity, for example to lower or higher temperatures than those given.

3.4 Publications and Database

The IRON Project is publishing its work in a series of papers in A&A (normally the Supplement Series) under the generic title 'Atomic data from the IRON Project'. Table 2 outlines the publications, and hence the data available, to date. Astronomy and Astrophysics makes published tables available electronically via anonymous ftp **130.79.128.5** at the CDS in Strasbourg.

Clearly, not all data can be published in this way; the final output will be very extensive, including energy-dependent collision strengths as well

Table 2. Papers published so far in the A&A series: Atomic data from the IRON Project

---

I. Goals and methods (Hummer et al. 1993)

II. Effective collision strengths for infrared transitions in carbon-like ions (Lennon & Burke 1994)

III. Rate coefficients for electron impact excitation of Boron-like ions: Na VI, Mg VIII, Al IX, Si X, S XII, Ar XIV, Ca XVI and Fe XXII (Zhang et al. 1994)

IV. Electron excitation of the $^2P^o(3/2)$ - $^2P^o(1/2)$ fine-structure transition in fluorine-like ions (Saraph & Tully 1994)

V. Effective collision strengths for transitions in the ground configuration of oxygen-like ions (Butler & Zeippen 1994)

VI. Collision Strengths and Rate Coefficients for Fe II (Zhang & Pradhan 1995)

VII. Radiative Transition Probabilities for Fe II (Nahar 1995)

VIII. Electron excitation of the $3d^4\,{}^5D_J$ ground state fine-structure transition in Ti-like ions V II, Cr III, Mn IV, Fe V, Co VI and Ni VII (Berrington 1995)

IX. Electron excitation of the $^2P^o$ 3/2 - 1/2 fine-structure transitions in chlorine-like ions from Ar II to Ni XII (Pelan & Berrington 1995)

X. Effective collision strengths for infrared transitions in silicon- and sulphur-like ions (Galavis et al. 1995)

XI. The $^2P^o_{1/2-3/2}$ fine-structure lines of Ar VI, K VII and Ca VIII (Saraph & Storey 1995)

---

as the thermally-averaged collision rates. All these data are being archived, and will eventually be accessible on-line via a system currently being developed called TIPbase, to be compatible with the Opacity Project's TOPbase.

4. Finally

This paper has described two collaborative projects on the interface of atomic physics and astrophysics. Both the Opacity Project and the IRON Project will produce an immense amount of good atomic data of use in many applications. Both Projects are committed to releasing the data into the public domain, and, perhaps even more important, to explaining the methods used and approximations made in obtaining the data, and the expected reliability of the results.

It is clear that the conception and execution of these projects was due to a synergism between astrophysicists and atomic physicists, harnessing expertise on both sides. It is hoped that this will continue in the future, and that other collaborative projects, or extensions to the present ones, will arise. Any suggestions?

References

Berrington, K. A. 1995, A&AS, 109, 193

Berrington, K. A., Eissner, W., and Norrington, P. 1995, Comput. Phys. Comm., submitted

Burke, P. G., and Berrington, K. A. (eds.) 1994, Atomic and Molecular Processes: an R-matrix Approach (Bristol, Institute of Physics)

Butler, K., and Zeippen, C. J. 1994, A&AS, 108, 1

Cunto, W., Mendoza, C., Ochsenbein, F., and Zeippen, C. J. 1993, A&A, 275, L5

Galavis, M. E., Mendoza, C., and Zeippen, C. J. 1995, A&AS, in press

Huebner W. F., Merts, A. L., Magee, N. H., and Argo, M. F. 1977, Los Alamos Sci. Lab. Rep. LA-6760-M

Hummer, D. G., Berrington, K. A., Eissner, W., Pradhan, A. K., Saraph, H. E., and Tully, J. A. 1993, A&A, 279, 298

Iglesias, C., Rogers, F. J., and Wilson, B. G. 1987, ApJ, 322, L45

Lang, J. (ed.) 1994, Atom. Data Nucl. Data, 57, 1ff

Lennon, D. J., and Burke, V. M. 1994, A&AS, 103, 273-277

Lynas-Gray A., Mendoza, C., and Zeippen, C. J. (eds.) 1992, Rev. Mex. Astron. Astrof., vol. 23

Mason, H. E. 1994, Atom. Data Nucl. Data, 57, 305

Nahar, S. N. 1995, A&A, 293, 967

Opacity Project Team (eds.) 1995, The Opacity Project Volume 1 (Bristol, Institute of Physics) ISBN 0-7503-0288-7

Pelan, J. C., and Berrington, K. A. 1995, A&AS, in press

Saraph, H. E., and Storey, P. J. 1995, A&AS, in press

Saraph, H. E., and Tully, J. A. 1994, A&AS, 107, 29

Seaton, M. J. 1987, J. Phys. B, 20, 6363

Seaton, M. J., Yan, Yu, Mihalas, D., and Pradhan, A. K. 1994, MNRAS, 266, 805

Simon, N. 1982, ApJ, 260, L87

Zhang, H., Graziani, M., and Pradhan, A. K. 1995, A&A, 283, 319-330

Zhang, H., and Pradhan, A. K. 1995, A&A, 293, 953

J. Phys. B: At. Mol. Opt. Phys. 27 (1994) 1315–1323. Printed in the UK

Atomic data for opacity calculations: XXI. The neon sequence

A Hibbert and M P Scott

Department of Applied Mathematics and Theoretical Physics, The Queen's University of Belfast, Belfast BT7 1NN, UK

Received 31 January 1992, in final form 24 December 1993

Abstract. Photoabsorption processes are studied for the ground and excited terms of the members of the neon isoelectronic sequence considered to be of significant astrophysical abundance. A two-state, *LS* close-coupling calculation is carried out. Oscillator strengths and total photoionization cross section data are calculated for even and odd *LS* terms with $S=0, 1$; $L \leqslant 4$, and an effective principal quantum number $\nu \leqslant 10$.

1. Introduction

This paper is part of a series of publications on atomic data required by the Opacity Project. This project is an international collaboration whose aim is to calculate, *ab initio*, a complete set of atomic photoabsorption data sufficient for the determination of accurate radiative opacities for stellar envelopes. The atomic data required for these calculations are discussed by Seaton in the first paper in this series (atomic data for opacity calculations (ADOC) I, Seaton 1987).

The members of the neon isoelectronic sequence which were considered to be of significant abundance to be included were Ne I, Na II, Mg III, Al IV, Si V, S VII, Ar IX, Ca XI and Fe XVII.

The calculations were carried out using the non-relativistic *R*-matrix method using computer codes described in the second paper in this series, ADOC II (Berrington *et al* 1987). A two-state F-like core is used i.e. $1s^22s^22p^5\ ^2P^o$ and $1s^22s2p^6\ ^2S$. The electron–ion scattering problem is solved to compute:

(a) energy levels for the Ne-like bound terms of even and odd *LS* where $S=0, 1$ and $L \leqslant 5$, and $\nu \leqslant 10$, where ν is the effective principal quantum number relative to the lowest coupled F-like state

(b) *gf*-values for all dipole allowed transitions between these bound terms

(c) the photon-energy-dependent total cross section for photoionization of all terms with $L \leqslant 4$.

All calculations have been carried out using *LS* coupling. No attempt has been made to include fine structure effects since the photoabsorption data obtained will be integrated over photon energy to obtain mean opacities where redistribution of oscillator strengths among fine structure levels may not be important when line broadening effects are considered.

Several theoretical studies have been carried out for photoabsorption by the ground states of the lower members of the neon isoelectronic series. These include the RPAE

0953-4075/94/071315+09$19.50

calculations of Lamoureux and Radojevic (1982). In particular, the photoionization of neutral neon from the ground state has been the subject of study by several authors including Burke and Taylor (1975), Amusia *et al* (1976, 1977), Luke (1973) and references therein. Photoionization from excited states of neon has been studied by Luke (1982), but little, if any, work has been carried out on excited states of the other members of the sequence. Cacelli *et al* (1983) have calculated partial photoionization cross sections for neon using Dyson amplitudes for photon energies up to 300 eV. These compare favourably with the experimental data of Wuilleumier and Krause (1974, 1979). Recent experimental work by Schartner *et al* (1990) looks at the Ne 2s photoionization cross section between threshold and 51 eV.

It is clear that while some data are available for the ground states of the members of the neon isoelectronic sequence, there is a need for a systematic calculation of photoabsorption data for a wider range of states of these ions.

2. The calculation

We include the two target states $2s^22p^5\,{}^2P^o$ and $2s2p^6\,{}^2S$ in the *R*-matrix calculation, each being described by a configuration interaction wavefunction in which we have used the single correlation function $\bar{3}d$ in addition to the Hartree–Fock orbitals of the ground state (Clementi and Roetti 1974). The most significant correlation effect (Mohan and Hibbert 1991) is the ${}^2S$ mixing of the $(2s2p^6+2s^22p^4\bar{3}d)^2S$. We represent the radial $\bar{3}d$ function by a single Slater type orbital with exponent optimized using the code CIV3 (Hibbert 1975) on the lowest ${}^2S$ energy arising from the two-configuration mixing. In the ground state we allow for $2s\rightarrow\bar{3}d$ replacement although, as may be seen from the eigenvector components of the wavefunctions listed in table 1, this correlation effect is less strong than that of the ${}^2S$ state. Burke and Taylor (1975) also included $\bar{3}s$ and $\bar{3}p$ correlation orbitals but Mohan and Hibbert (1991) found these to be of less importance than the $\bar{3}d$ orbital in describing the principal correlation effects. Our calculated ${}^2P^o$–${}^2S$ oscillator strengths are shown in table 1. It may be seen that the oscillator strengths obtained with the simple target wavefunctions used in the present work agree well with those obtained in much more extensive structure calculations (Blackford and Hibbert 1993, Johnson and Kingston 1987).

For the collision calculation we have used the *R*-matrix code (Berrington *et al* 1987) with 15 continuum basis functions per channel in the inner region.

3. Results

3.1. Bound states

From our collision wavefunctions we have calculated all Ne-like bound states for even and odd *LS* terms such that $S=0$, 1 and $L\leqslant 5$ and an effective quantum number up to 10.

Energy levels for some of the lower lying ${}^1S$ and ${}^1P^o$ bound terms of Ne are given in table 2 where they are compared with other experimental and theoretical data. The bound states found are all of the form $1s^22s^22p^5nl$ associated with the ${}^2P^o$ core. No bound states associated with the ${}^2S$ excited core were found for Ne in the energy range

Table 1. Target state data.

Z		10	11	12	13	14	16	18	20	26
3d exponent		3.38095	3.98516	4.53243	5.07859	5.62379	6.71191	7.79789	8.88251	12.13069
Eigenvectors										
$^2P^o$: $2s^22p^5$		0.99784	0.99827	0.99856	0.99879	0.99897	0.99923	0.99940	0.99952	0.99973
$2s2p^5(^1P)\ \overline{3d}$		−0.05571	−0.05005	−0.04564	−0.04189	−0.03868	−0.03351	−0.02954	−0.02640	−0.02001
$2s2p^5(^3P)\ \overline{3d}$		0.03477	0.03101	0.02818	0.02579	0.02377	0.02054	0.01807	0.01613	0.01220
^2S: $2s2p^6$		0.98503	0.98965	0.99224	0.99401	0.99526	0.99683	0.99775	0.99832	0.99916
$2s^22p^4\ \overline{3d}$		0.17237	0.14353	0.12432	0.10929	0.09729	0.07950	0.06703	0.05786	0.04086
Energies										
$E(^2P^o - {}^2S)$ (Ryd)		2.04347	2.46255	2.87956	3.29598	3.71155	4.12626	5.36593	6.18959	8.65183
Oscillator strengths ($^2P^o - {}^2S$)										
This work	f_l	0.084	0.085	0.082	0.079	0.076	0.069	0.063	0.057	0.045
	f_v	0.095	0.092	0.087	0.083	0.078	0.071	0.064	0.059	0.046
Blackford and Hibbert (1993)	f_l				0.086	0.081	0.073	0.066	0.061	0.049
Johnson and Kingston (1987)	f_l	0.095		0.088						
	f_v	0.096		0.089						

Table 2. Energy levels for some of the lower lying bound terms of Ne. The energies are given in Ryd.

Term	Experiment	Theory		
Energy (rel. to $1s^22s^22p^5\ ^2P^o$ Ne^+ threshold)				
$1s^22s^22p^6$	−1.585[a]	−1.6612[b]	−1.571[c]	
$1s^22s^22p^53s\ ^1P^o$	−0.348[a]	−0.3371[b]	−0.361[c]	
$1s^22s^22p^53p\ ^1S$	−0.191[a]	−0.1885[b]	−0.188[c]	
$1s^22s^22p^54p\ ^1S$	−0.0879[a]	−0.0903[b]	−0.0912[c]	
$1s^22s^22p^55p\ ^1S$	−0.0487[a]	−0.0531[b]	−0.0539[c]	
Energy (rel. to $1s^22s^22p^6$ Ne ground state)				
$1s^22s^22p^53s\ ^1P^o$	1.237[a]	1.3241[b]	1.349[d]	1.21[c]
$1s^22s^22p^54s\ ^1P^o$		1.5275[b]	1.564[d]	
$1s^22s^22p^55s\ ^1P^o$		1.5897[b]	1.628[d]	
$1s^22s^22p^56s\ ^1P^o$		1.6167[b]	1.656[d]	

[a] Moore (1949).
[b] This work.
[c] Luke (1982).
[d] Lamoureux and Radojevic (1982).

studied. Indeed, the first such state ($1s^22s2p^63s\ ^3S$) lies well above the ionization limit (Hibbert *et al* 1993).

For higher members of the isoelectronic sequence a number of the 2s-hole states become bound. We compare in table 3 our calculated energy positions of the $2s2p^6np\ ^1P^o$ states in Fe XVII with experimental values. Since our results were obtained using *LS* coupling, we have also calculated, using CIV3, some corrections to allow for the relativistic shifts of these energies. These include the overall *LS* shifts (mass correction, Darwin) and the correction due to the spin–orbit mixing of the $^3P^o_1$ and $^1P^o_1$ levels as well as the $2s^22p^5\ ^2P^o$ fine structure. When these corrections are added to the non-relativistic values, the energy positions relative to the ionization limit agree with the experimental values to within a few hundredths of a Rydberg. The experimental value for the ionization limit itself is quoted only to 0.02 Ryd (Corliss and Sugar, 1982) and is 3 Ryd lower than the previously quoted value of Reader and Sugar (1975).

Table 3. Energy positions (in Ryd) of $2s2p^6np\ ^1P^o$ states of Fe XVII relative to the ionization limit.

		Corrections			
np	This work	Mass corr. + Darwin	Spin–orbit mixings	Total	Expt[a]
3p	−27.7420	0.6374	0.3386	−26.7660	−26.8432
4p	−11.0887	0.7113	0.3233	−10.0541	−10.0968
5p	−3.7368	0.7414	0.3177	−2.6795	−2.7520

[a] Corliss and Sugar (1982).

3.2. *gf values*

We calculate *gf* values in the dipole approximation for all allowed transitions between our calculated Ne-like *LS* terms. As with other Opacity Project atomic data calculations (e.g. Berrington *et al* 1994) we tend to favour the length form of the results in preference to the velocity form since our calculation is based on a collisional approximation. However, in general, agreement between length and velocity forms is good.

In table 4 we give results of oscillator strengths for various 2p→$ns(^1P^o)$ transitions for members of the sequence and compare with other experimental and theoretical data.

Table 4. Oscillator strengths for various $2p^6\ ^1S \rightarrow 2p^5(^2P)ns\ ^1P^o$ transitions for the lower members of the Ne isoelectronic sequence. The excitation energies, ω, are given in Ryd.

Ion	*ns*	ω_{OP}[a]	ω_{RPAE}[b]	f_{OP}[a]	f_{CIV3}[c]	f_{RPAE}[a]	f_{others}
Ne	3s	1.3241	1.349	0.1696	0.176	0.160	0.161[d], 0.159[e] 0.13[f], 0.16[g]
	4s	1.5275	1.564	0.0348		0.0275	0.013[f]
	5s	1.5900	1.628	0.0127		0.00989	0.0042[f]
	6s	1.6167	1.656	0.0061		0.00490	
	7s	1.6309		0.0034			
	8s	1.6293		0.0021			
Na^+	3s	2.5473	2.560	0.2268	0.230	0.208	0.213[e]
	4s	3.1160	3.145				
	5s	3.3092	3.343				
	6s	3.3985	3.434				
	7s	3.4470					
	8s	3.4763					
Mg^{2+}	3s	4.0365	4.043	0.2450	0.252	0.225	0.231[e], 0.228[h] 0.246[i]
	4s	5.0801	5.108	0.0356		0.0403	
	5s	5.4592	5.492	0.0132		0.0141	
	6s	5.6490		0.0064		0.00653	
	7s	5.7397		0.0036			
	8s	5.8008		0.0023			
Al^{3+}	3s	5.8000	5.804	0.2501	0.258	0.229	0.238[e], 0.229[h] 0.251[i], 0.224[j]
	4s	7.4236	7.450	0.0405		0.0410	
	5s	8.0371	8.070	0.0151		0.0145	
	6s	8.3352	8.371	0.0074		0.00674	
	7s	8.5026		0.0042			
	8s	8.6058		0.0026			
Si^{4+}	3s	7.8405	7.845	0.2501	0.252	0.231	0.240[e]
	4s	10.1453	10.172	0.0422		0.0415	
	5s	11.0407	11.074	0.0159		0.0148	
	6s	11.4824	11.518	0.0079		0.00690	
	7s	11.7326		0.0045			
	8s	11.8880		0.0029			

[a] This work.
[b] Lamoureux and Radojevic (1982).
[c] Hibbert *et al* (1993).
[d] Semi-empirical model, Aymar *et al* (1970).
[e] Time-dependent HF calculation, Stewart (1975).
[f] Experiment, Lawrence and Lizst (1969).
[g] Experiment, Lewis (1967).
[h] Relativistic RPAE, Shorer (1979).
[i] Second-order-many body calculation of Ceyzeriat (1979).
[j] Multiconfiguration HF calculation of Guennou (1980).

We note that in the case of Na^+ the energies of some of the $1s^22s^22p^5ns\ ^1P^o$ states are very close to the $1s^22s^22p^5nd\ ^1P^o$ states, e.g. the excitation energy of the 3s state (−0.4485 Ryd) is close to the excitation energy of the 4s state (−0.4415 Ryd); the 4d (−0.2516 Ryd) is close to the 5s (−0.2483 Ryd); and the 5d (−0.1608 Ryd) to the 6s (−0.1590 Ryd). This strong mixing gives oscillator strengths which are not as reliable as for the other ions and so we do not give them in table 4. A similar problem is also experienced by Lamoureux and Radojevic (1982) which is why there is only one RPAE

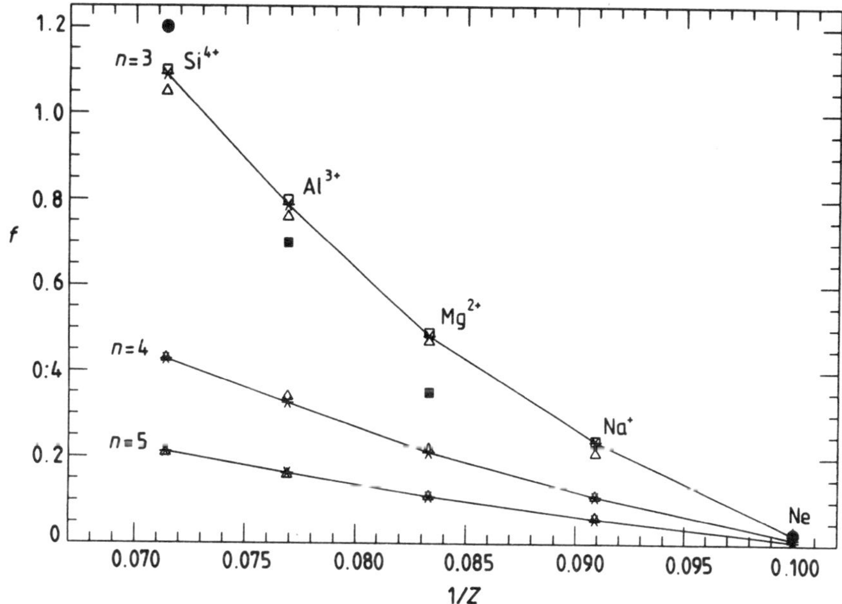

Figure 1. Oscillator strength for the $2p^6\ ^1S \rightarrow 2p^5nd\ ^1P^o$ transitions, plotted against $1/Z$. *, current work; △, RPAE, Lamoureux and Radojevic (1982); □, time-dependent HF calculation, Stewart (1975); ■, relativistic RPAE, Shorer (1979); ●, relativistic central field calculation, Crance (1973).

result for the Na^+ oscillator strengths. In figure 1 we plot against $1/Z$ oscillator strengths for several transitions of the form $2p^6\ ^1S \rightarrow 2p^5nd\ ^1P^o$. Again, our results are in reasonable agreement with the other theoretical and experimental data, if perhaps a little high in some cases.

3.3. Photoionization cross sections

Photoionization cross sections are calculated for each Ne-like LS term for $L \leqslant 4$. The photon energy mesh chosen was sufficiently fine to highlight low energy resonance structures. In figures 2 and 3 we present photoionization data for the ground state and the first three excited states of neon. In figure 2 we observe the first three members of the $1s^22s2p^6np\ ^1P^o$ resonance series converging onto the $1s^22s2p^6\ ^2S$ Ne^+ threshold. The positions of these resonances are in reasonable agreement with those reported by Burke and Taylor (1975). These authors report quantum defect values of 0.824 and 0.808 for the $1s^22s2p^63p\ ^1P^o$ and $1s^22s2p^64p\ ^1P^o$ resonances respectively which compare with values of 0.816 and 0.792 from figure 2. Given that we are using a less sophisticated CI type wavefunction than Burke and Taylor we would expect the quantum defect values to be lower. In figure 3 we give cross section data for photoionization from the first four states of neon and compare our results with data from Luke (1982). This data has been obtained from Luke's figure 2 by summing the εs and εd partial cross section

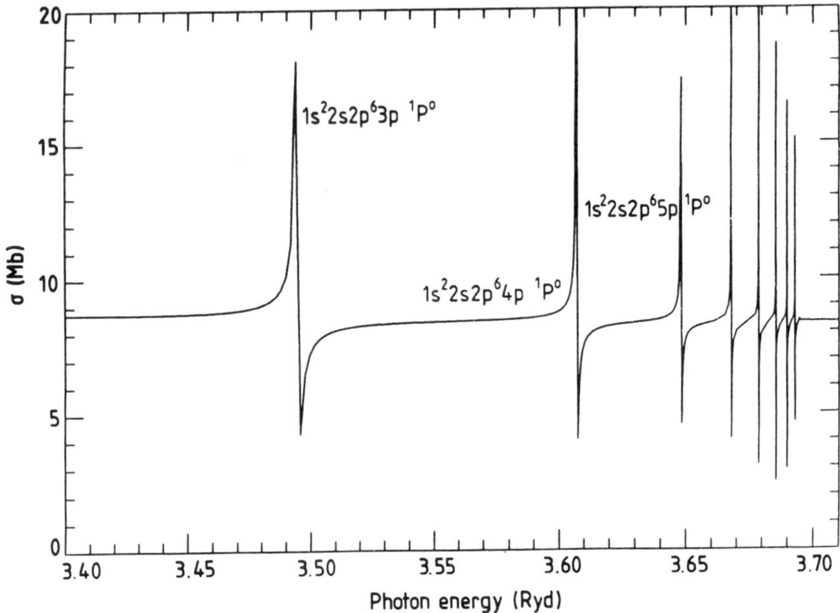

Figure 2. Total cross section for the photoionization of neutral neon from the $1s^2 2s^2 2p^6\ {}^1S$ ground state just above the $1s^2 2s^2 2p^5\ {}^1P^o$ threshold of Ne^+.

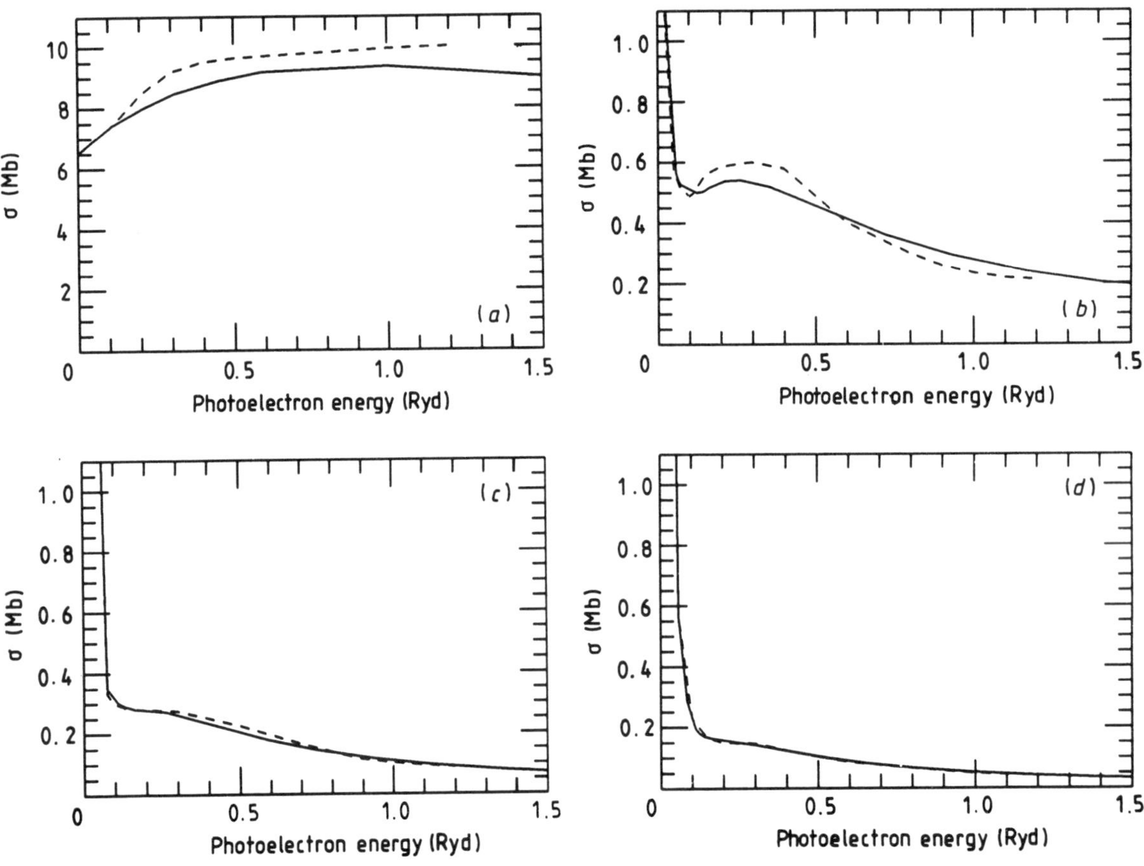

Figure 3. Total photoionization cross section (Mb) from the first ${}^1S$ states of neutral neon: (*a*) $1s^2 2s^2 2p^6$; (*b*) $1s^2 2s^2 2p^5 3p$; (*c*) $1s^2 2s^2 2p^5 4p$; (*d*) $1s^2 2s^2 2p^5 5p$. Full curves, this work; broken curves, Luke (1982) (see text).

results. We see that overall the agreement is fairly good which gives us confidence in using the data obtained to calculate accurate stellar opacities.

4. Conclusion

The photoabsorption data described here has been produced for the specific needs of the Opacity Project. As such the amount of data is vast and there is a limit to what can be discussed here. However, all data will be available for release when the project is finished. Further details are available from the authors.

Although this is a fairly simple calculation (e.g. we have ignored relativistic effects and have only included a limited amount of configuration interaction), comparison with other data indicate that our calculations are sufficiently accurate to enable reliable stellar opacities to be calculated. We note that this is the first attempt at a systematic study of the photoabsorption data for a wide range of excited terms for the lower members of the neon isoelectronic sequence which are considered to be of significant astrophysical abundance.

Acknowledgments

The computing was carried out on the CRAY/XMP-48 at the Atlas Supercomputing Centre at the Rutherford Appleton Laboratory on a UK SERC grant. We are grateful to the referees for constructive comments.

References

Amusia M Ya, Cherepkov N A, Pavlin I, Radojevic V and Zivanovic Dj 1977 *J. Phys. B: At. Mol. Phys.* **10** 1413–23

Amusia M Ya, Cherepkov N A, Zivanovic Dj and Radojevic V 1976 *Phys. Rev.* A **13** 1466–74

Aymar M, Feneuille S and Klapisch M 1970 *Nucl. Instrum. Methods* **90** 137–43

Berrington K A, Burke P G, Seaton M J, Storey P J, Taylor K T and Yu Yan 1987 *J. Phys. B: At. Mol. Phys.* **20** 6379–98

Berrington K A, Hibbert A and Scott M P 1994 *J. Phys. B: At. Mol. Opt. Phys.* to be submitted

Blackford H M S and Hibbert A 1993 *At. Data Nucl. Data Tables* to be published

Burke P G and Taylor K T 1975 *J. Phys. B: At. Mol. Phys.* **8** 2620–39

Cacelli I, Caravetta V and Moccia R 1983 *J. Phys. B: At. Mol. Phys.* **16** 1895–906

Ceyzeriat P 1979 *Doctoral thesis* Lyon, France

Clementi E and Roetti C 1974 *At. Data Nucl. Data Tables* **14** 177–478

Corliss C and Sugar J 1982 *J. Phys. Chem. Ref. Data* **11** 135–241

Crance M 1973 *At. Data* **5** 185–200

Guennou H 1980 Private communication in Lamoureux and Radojevic (1982)

Hibbert A 1975 *Comput. Phys. Commun.* **9** 141–72

Hibbert A, LeDourneuf M and Mohan M 1993 *At. Data Nucl. Data Tables* **53** 23–112

Johnson C J and Kingston A E 1987 *J. Phys. B: At. Mol. Phys.* **20** 5663–73

Lamoureux M and Radojevic V 1982 *J. Phys. B: At. Mol. Phys.* **15** 1341–51

Lawrence G M and Liszt H S 1969 *Phys. Rev.* A **178** 122–5

Lewis E L 1967 *Proc. Phys. Soc.* **92** 817–25

Luke T M 1973 *J. Phys. B: At. Mol. Phys.* **6** 30–41

—— 1982 *J. Phys. B: At. Mol. Phys.* **15** 1217–27

Mohan M and Hibbert A 1991 *Phys. Scr.* **44** 158–63

Moore C E 1949 in *Atomic Energy Levels* vol 1 NBS Circular No (Washington, DC: US Govt Printing Office)
Reader J and Sugar J 1975 *J. Phys. Chem. Ref. Data* **4** 353–440
Seaton M J 1987 *J. Phys. B: At. Mol. Phys.* **20** 6363–78
Schartner K-H, Mogel B, Mobus B, Schmoranzer H and Wildberger M 1990 *J. Phys. B: At. Mol. Opt. Phys.* **23** L527–32
Shorer P 1979 *Phys. Rev.* A **20** 642–55
Stewart R F 1975 *Mol. Phys.* **29** 1575–83
Wuilleumier F and Krause M O 1974 *Phys. Rev.* A **10** 242–58
—— 1979 *J. Electron Spectrosc.* **15** 15–20

J. Phys. B: At. Mol. Opt. Phys. **28** (1995) 2817–2827. Printed in the UK

Atomic data for opacity calculations: XXII. Computations for 2 472 790 multiplet gf-values in Fe VIII to Fe XIII

A E Lynas-Gray†, M J Seaton‡ and P J Storey‡

† Department of Physics, Nuclear Physics Building, Keble Road, Oxford, OX1 3RH, UK
‡ Department of Physics and Astronomy, University College London, Gower Street, London WC1E 6BT, UK

Received 5 January 1995

Abstract. In the calculation of opacities for stellar interiors it is found that very large numbers of spectrum lines of iron ions make important contributions. In the work of the Opacity Project the main computations of atomic data have been made using R-matrix methods, referred to as RMAT. For the important ionization stages of Fe VIII to Fe XIII further atomic data are required. For those stages calculations have been made using the code SUPERSTRUCTURE for a further 2 472 790 spectral multiplets in LS coupling. These are referred to as PLUS data. Results are compared for cases in which both RMAT and PLUS calculations have been made. For many individual transitions the differences can be quite large, which is a consequence of the larger amount of configuration-interaction included in the RMAT work. For statistical purposes, such as the calculation of Rosseland-mean opacities, the accuracy of the PLUS data should be adequate.

1. Introduction

The work of the Opacity Project (OP) has been concerned with the calculation of opacities for stellar envelopes (Seaton 1987, Seaton *et al* 1994). For a mixture of chemical elements the monochromatic opacity per unit length is

$$\kappa_\nu = \sum_{ijk} N(ijk)\sigma_\nu(ijk) + N_e\sigma_\nu(e) + \kappa_\nu(ff) \tag{1}$$

where $N(ijk)$ is the number-density for level i of element k in ionization stage j and $\sigma_\nu(ijk)$ the cross section for absorption of radiation by that level; N_e is the electron density and $\sigma_\nu(e)$ the cross section for photon scattering by free electrons; and $\kappa_\nu(ff)$ is a contribution from free–free transitions. The $\sigma_\nu(ijk)$ are determined by processes of photoionization and transitions in spectrum lines and include information about line profiles. Both $\sigma_\nu(ijk)$ and $\kappa_\nu(ff)$ include a correction factor for stimulated emission.

The net flux of radiant energy in a stellar interior is determined by the Rosseland-mean opacity, κ_R, defined as

$$\frac{1}{\kappa_R} = \int_0^\infty \frac{1}{\kappa_\nu} f(u)\,du \tag{2}$$

where

$$u = h\nu/kT \qquad f(u) = \frac{15}{4\pi^4}u^4\exp(-u)/[1-\exp(-u)]^2 \qquad \int_0^\infty f(u)\,du = 1\,. \tag{3}$$

The main OP atomic physics calculations have been made using R-matrix methods (Berrington *et al* 1987) and the results obtained will be referred to as RMAT data. They are

0953-4075/95/142817+11$19.50

available in the database system TOPbase (Cunto *et al* 1993). For an ion containing $(N+1)$ electrons the RMAT work uses close-coupling expansions

$$\Psi = \mathcal{A}\sum_i \psi_i\theta_i + \sum_j \Phi_j c_j \qquad (4)$$

where the ψ_i are functions for the N-electron 'target', the θ_i functions for an added electron, $\mathcal{A}$ is an anti-symmetrization operator, and the Φ_j are functions of bound-state type for the $(N+1)$-electron system. If one can include all the low-lying target states ψ_i, the expansion (2) gives good results for all low-lying states of the whole system and all states with one electron more highly excited or in the continuum (final states after photoionization).

For highly ionized systems, of importance for the opacity work, one would like to be able to include all target states belonging to the *ground complex* of the target, that is to say all states with the same set of principal quantum numbers as the ground state. That has been possible for all ions with $N \leqslant 12$ but for $N > 12$ the number of ground-complex states becomes large (mainly due to states with 3d electrons being included) and some compromises have to be made. Further details are given by Seaton *et al* (1992).

The work of the OPAL project (Iglesias *et al* 1990) was the first to show that very large numbers of lines of iron ions must be included in stellar opacity calculations, particularly in the temperature range of 10^5–10^6 K. In that range the important ions are Fe XIII (ground configuration $2p^63s^23p^2$, $N = 13$) to Fe VIII ($2p^63s^23p^63d$, $N = 18$). For those stages the RMAT calculations gave some 250 000 lines in LS coupling, and many more when fine structures are taken into account. Convergency checks showed that it was desirable to include more iron lines in the OP work and we have therefore made further calculations using the configuration-interaction code SUPERSTRUCTURE (Eissner *et al* 1974, Nussbaumer and Storey 1978). The results obtained, some 2 500 000 multiplets in LS coupling for Fe XIII to Fe VIII, are referred to as PLUS data.

2. Description of the PLUS calculations

The basic expansion is

$$\Psi = \sum_j \Phi_j c_j \qquad (5)$$

where the Φ_j are antisymmetric vector-coupled functions for a single configuration. For many cases in the present work one has a number of functions Φ_j all with the same configuration and total angular momenta SL, but different internal arrangements of angular momentum coupling.

The code SUPERSTRUCTURE calculates radial functions $P_{nl}(r)$ in statistical-model potentials $V_{nl}(r)$ containing scale parameters ζ_{nl} which are adjusted so as to minimize energies for specified levels or groups of levels. The coefficients c_j in (5) are obtained on diagonalising the matrix with elements $(\Phi_i|H|\Phi_j)$. We consider states of two types:

(i) C states, with configurations $3s^x3p^y3d^z$;
(ii) $T\ nl$ states with configurations $3s^{x'}3p^{y'}3d^{z'}nl$, $n = 4$ or 5 and $l = 0, 1, 2$ or 3.

Particular configurations for C and T states are denoted by Cj and Ti. We make calculations for two types of transitions:

(i) $3 \rightarrow 3$. Transitions between two C states. For these calculations we allow for configuration-interaction (CI) by including in (5) all C states of the ground complex.

Table 1. Number of levels and multiplets for RMAT and PLUS calculations. These figures do not include results obtained by extrapolation (a further 878 199 transitions to $n = 6$ levels and 878 199 photoionization cross sections).

Fe	RMAT		PLUS	
	Levels	Multiplets	Levels	Multiplets
XIII	817	38 286	3 288	27 553
XII	903	58 983	8 348	142 451
XI	1156	77 187	18 022	420 167
X	765	45 369	22 736	707 754
IX	1006	47 670	24 488	703 853
VIII	242	3 821	15 601	471 012
Total	4889	271 316	92 483	2 472 790

Table 2. Authors of the RMAT calculation.

Fe	Authors
XIII	C Mendoza
XII	C Mendoza
XI	K Butler, C Mendoza, C J Zeippen
X	K Butler, C Mendoza, C J Zeippen
IX	C Mendoza
VIII	P J Storey, K T Taylor

(ii) $3 \rightarrow n$ with $n = 4$ and 5. Transitions $C \rightarrow T\ nl$. Here we include only as many configurations as our computer resources permitted (we used machines with 32-bit wordlength and maximum main memory of 50 Mb). In many cases this meant that we included only one configuration but still had a number of states included in (5) with different internal arrangements of angular momenta: for such cases matrix diagonalizations were still required.

We do not make PLUS calculations for transitions with $n > 3$ for the initial level. Check calculations show that such transitions (other than those already included in the RMAT work) are not important because the occupation numbers $N(ijk)$ are small and the line frequencies are low. Extrapolation procedures used to obtain data for $3 \rightarrow n'$ with $n' > 5$, and for photoionization, will be discussed in section 4.

Table 1 gives details on the numbers of levels and multiplets included in the RMAT and PLUS calculations for Fe XIII to VIII and table 2 gives a list of the authors for the RMAT work.

3. Comparisons of RMAT and PLUS results

The RMAT data are more accurate than the PLUS data, firstly because more CI are included and secondly because the RMAT functions θ_i in (2) are fully optimized.

Comparisons of RMAT and PLUS data provide an interesting study of the accuracy which can be obtained, for complex systems, using simpler approximations such as those used in the PLUS work.

The first task, of course, is to make a one-to-one correspondence for the levels from the two calculations. That is not always easy for cases in which there is a lot of mixing. We use the RMAT level identifications as given in TOPbase.

3.1. Energy levels

The RMAT energies are given relative to ionization limits while those for PLUS are on arbitrary scales (but can be put on scales of absolute total energies). It is simplest to give PLUS energies as excitation energies relative to ground states.

We consider in some detail results for the fairly simple case of Fe XIII. The RMAT work included all target states belonging to the configurations $3s^23p$, $3s3p^2$, $3p^3$ and $3s^23d$ (but not 3s3p3d, $3s3d^2$ and $3d^3$ which also belong to the ground complex). Figure 1 shows RMAT levels for Fe XIII ^3P states. It includes a list of target states Ti, ground-complex configurations Cj giving ^3P terms, and Rydberg series Sk which can be identified as $Tj\ nl$. We also show the positions of series limits. The RMAT work does not give any levels above the first series limit $C1$ (those levels are included as resonances in the photoionization cross sections). In principle there can be CI interactions between all of the levels shown on figure 1. Some levels come close together and CI effects can then be particularly large.

Figure 2 shows similar results for Fe XIII $^3D^o$ levels. Note, from figures 1 and 2, how some configurations give a number of levels of the same SL (six levels for $C4$ ^3P (two being very close together) and 6 for $C11$ $^3D^o$).

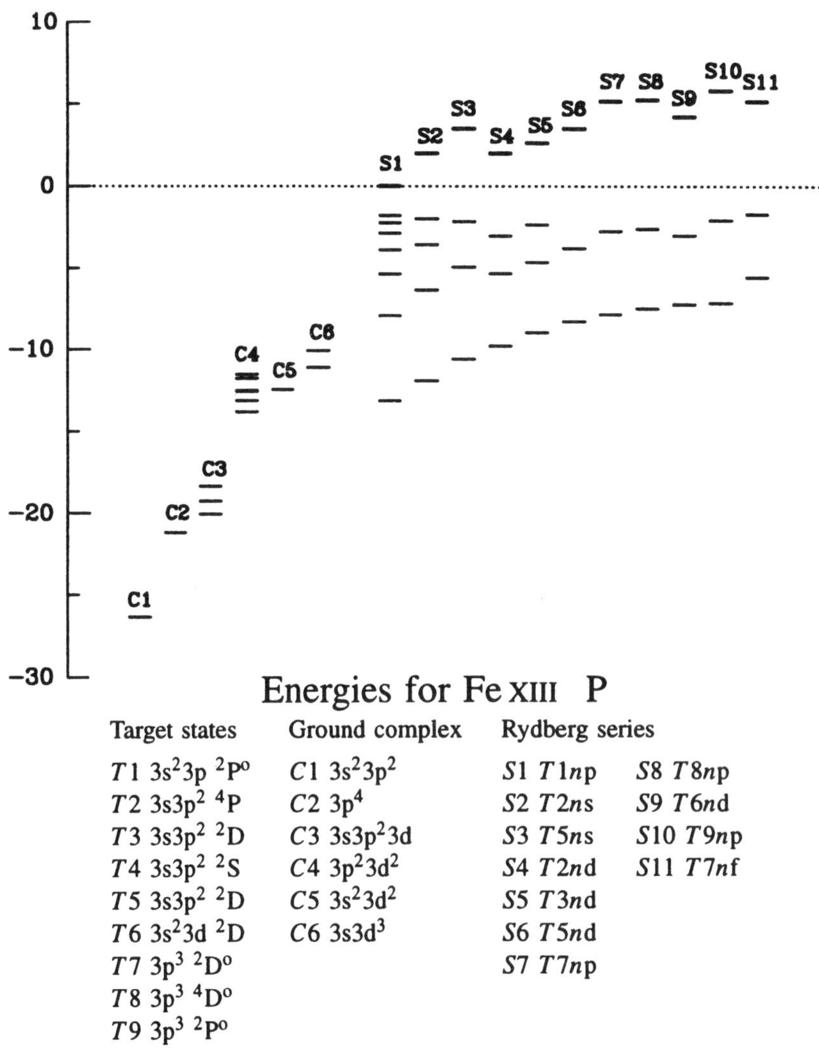

Figure 1. Energies for Fe XIII ^3P states, calculated using RMAT, relative to the energy for the ground state of Fe XIV. The Rydberg series may converge to the ground state or to excited states of Fe XIV. The thick lines ($S1$, $S2$ etc) shows positions of series limits.

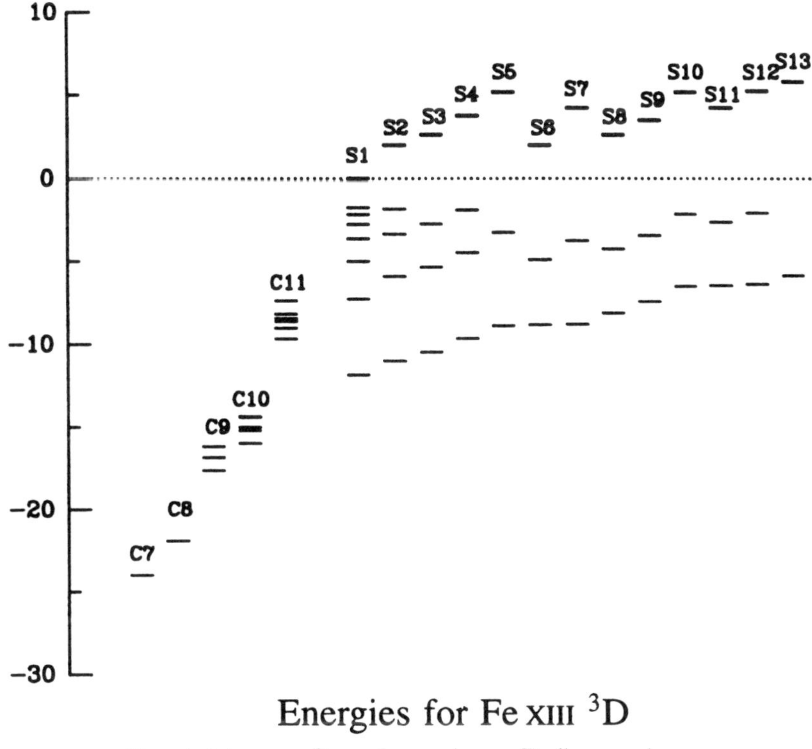

Target states	Ground complex	Rydberg series	
*T*1 $3s^2 3p\ ^2P^o$	*C*7 $3s3p^3$	*S*1 *T*1*n*d	*S*8 *T*3*n*f
*T*2 $3s3p^2\ ^4P$	*C*8 $3s^2 3p3d$	*S*2 *T*2*n*p	*S*9 *T*5*n*f
*T*3 $3s3p^2\ ^2D$	*C*9 $3p^3 3d$	*S*3 *T*3*n*p	*S*10 *T*7*n*d
*T*4 $3s3p^2\ ^2S$	*C*10 3s3p3d	*S*4 *T*5*n*p	*S*11 *T*6*n*f
*T*5 $3s3p^2\ ^2D$	*C*11 $3p3d^3$	*S*5 *T*7*n*s	*S*12 *T*8*n*d
*T*6 $3s^2 3d\ ^2D$		*S*6 *T*2*n*f	*S*13 *T*9*n*d
*T*7 $3p^3\ ^2D^o$		*S*7 *T*6*n*p	
*T*8 $3p^3\ ^4D^o$			
*T*9 $3p^3\ ^2P^o$			

Figure 2. As figure 1, for Fe XIII $^3D^o$.

3.2. *gf*-values

Figure 3 compares *gf*-values for all transitions for which both RMAT and PLUS calculations were made. The amount of scatter is, at first sight, quite surprising—maybe even rather alarming. Closer scrutiny of individual cases shows that the differences are real and due to the improved accuracy of the RMAT work.

We consider in some detail the case of Fe XIII transitions from the ground state, $3s^2 3p^2$ ^3P, to final $^3D^o$ states (see figures 1 and 2). The top plot in figure 4 shows quantum defects μ for $3s^2 3pnd$ $^3D^o$ levels from RMAT, against energy ϵ relative to the ionization limit. For a simple unperturbed series μ would vary smoothly as a function of ϵ. The irregular variations shown on the plot are due to series perturbations (CI effects). The lower plot of figure 4 shows the quantity $(n - \mu)^3 gf(n)$ where n is the principal quantum number for the final state. That quantity would also vary smoothly for an unperturbed series. For the RMAT data the variation is quite irregular (the value for $n = 10$ is way off scale for the figure). The plot also shows the PLUS results for $n = 4$ and 5. For $n = 5$ there is a factor of 2.4 between the PLUS and RMAT *gf*-values due to the perturbation of the RMAT level.

Figure 5 shows all *gf*-values obtained in both calculations for the transitions from Fe XIII $3s^2 3p^2$ ^3P to $^3D^o$ levels. For the larger *gf*-values the agreement is generally good

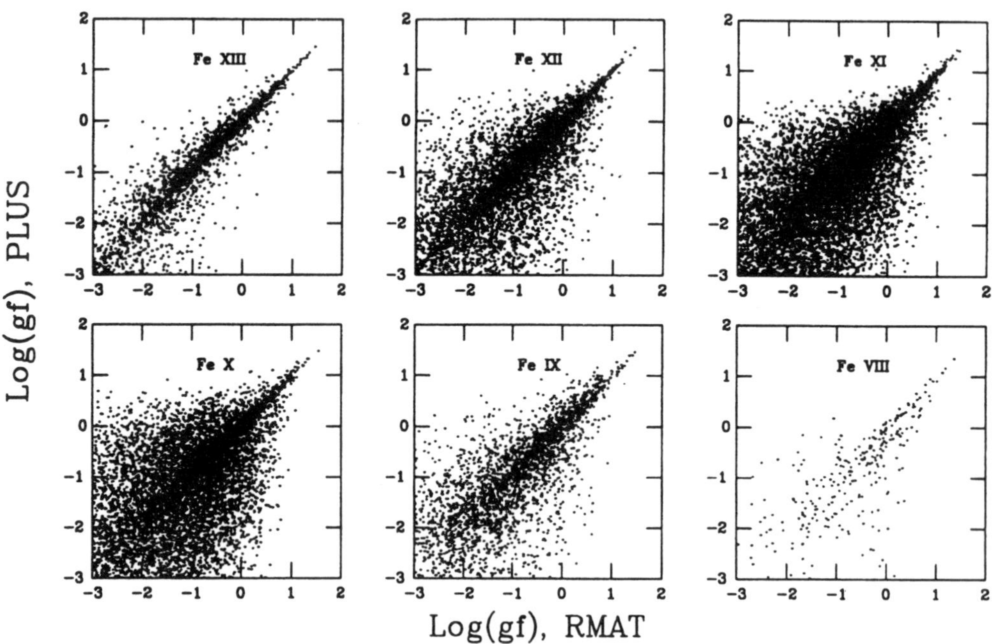

Figure 3. Comparison of $\log(gf)$ values for all transitions for which both RMAT and PLUS calculations have been made.

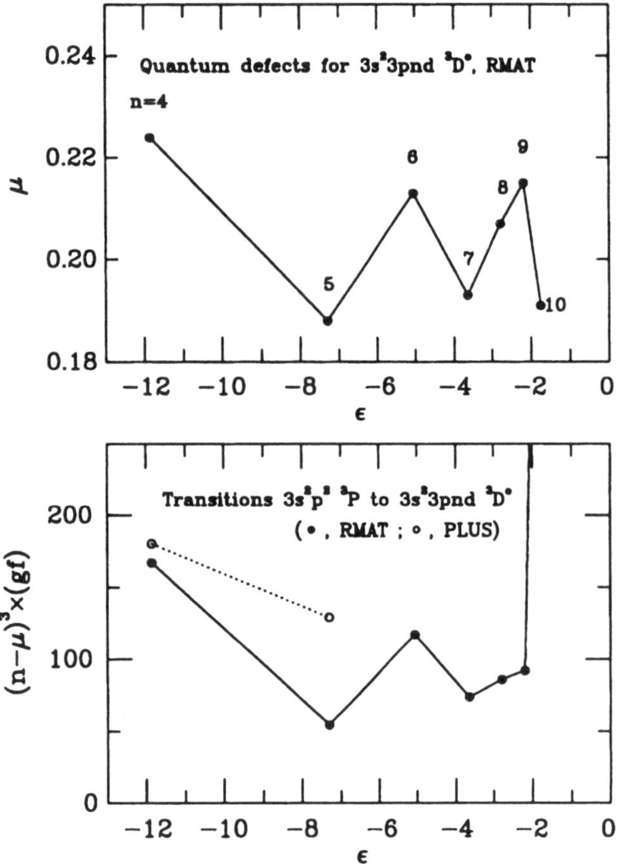

Figure 4. Top figure: quantum defects μ for $3s^2 3pnd$ $^3D^o$ levels in Fe XIII, RMAT calculations, against ϵ (the energy in Rydberg units relative to ionization limits). The irregular variations of μ are due to configuration-interaction effects. Lower figure: values of $(n - \mu)^3 (gf)$ for transitions $3s^2 3p^2$ ^3P to $3s^2 3pnd$ $^3D^o$ in Fe XIII. Full circles and full line, RMAT. Open circles and broken line, PLUS ($n = 4$ and 5 only). In the PLUS calculations we do not allow for interaction of $3s^2 3pnd$ with other configurations.

(but note that here we are using a very extended gf-scale). For the small gf-values there are differences by as much as several orders of magnitude. One general trend is that PLUS results tend to be larger for large values, smaller for small ones. That is a consequence of the additional mixing in the RMAT work.

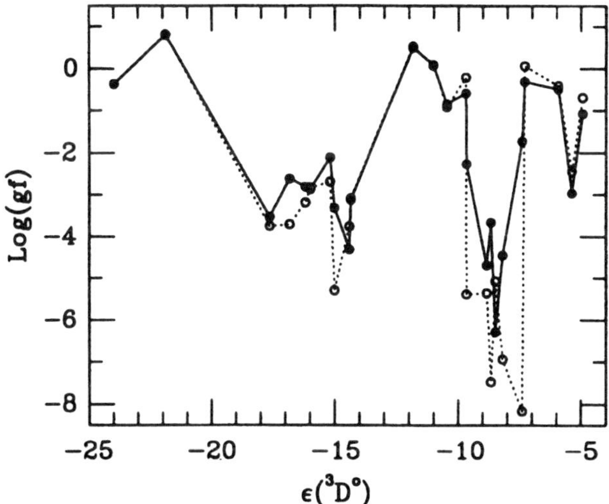

Figure 5. Values of $\log(gf)$ against $\epsilon(^3\mathrm{D}^o)$, the energy in Rydberg of the $^3\mathrm{D}^o$ levels relative to that of $2s^2 3p\ ^3\mathrm{P}^o$. Results for all transitions for which both RMAT and PLUS calculations have been made. RMAT: full circles and full line. PLUS: open circles and broken line.

4. Extrapolations of the PLUS data

For some high-density cases we find it desirable in the opacity work to include transitions $3 \rightarrow n$ with $n > 5$. To a good enough accuracy they can be included on making extrapolations of the PLUS data.

For a level $T\ nl$ we can put

$$E = E_T + \epsilon_{nl} \tag{6}$$

where E_T is the energy of the series limit and $\epsilon_{nl} = -z^2/(n - \mu_{nl})^2$ where μ_{nl} is the quantum defect and $z = (Z - n)$ is the charge seen by an outer electron. For the RMAT data the values of E_T are known and we can calculate the μ_{nl}. For the PLUS data the E_T are not known but we can obtain approximate values of μ_{nl}, assumed independent of n, using values of E for $4l$ and $5l$. The results obtained are in satisfactory agreement with values of μ_{nl} obtained from RMAT levels which are not markedly perturbed. Table 3 gives approximated values of μ_l.

Table 3. Mean values of quantum defects μ_l.

Fe	$l = 0$	1	2	3
XIII	0.530	0.406	0.196	0.036
XII	0.616	0.474	0.250	0.049
XI	0.708	0.556	0.300	0.054
X	0.802	0.648	0.354	0.057
IX	0.906	0.741	0.410	0.054
VIII	1.023	0.843	0.471	0.048

Consider transitions from an inital level i to higher levels nl. The quantity

$$F(\epsilon_{nl}) = (n - \mu_{nl})^3 f(nl, i) \tag{7}$$

tends to a finite limit for $n \rightarrow \infty$. Extrapolating to $\epsilon > 0$, the photoionization cross section in atomic units is

$$\sigma_{PI}(\epsilon) = (2\pi^2\alpha/z^2)F(\epsilon) \tag{8}$$

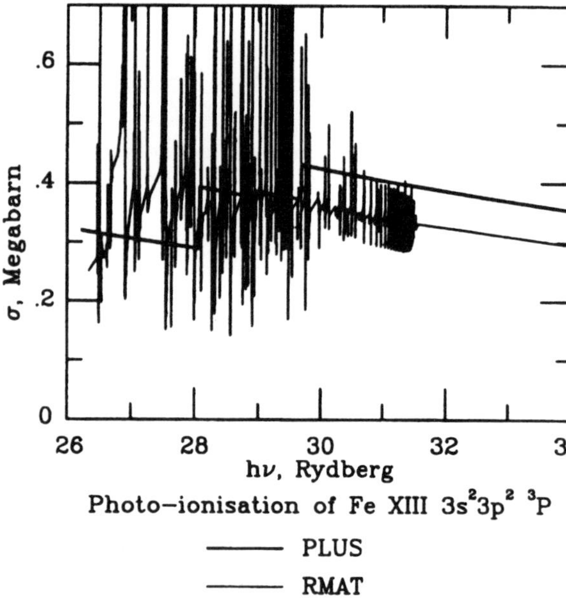

Figure 6. Cross section for photoionization of Fe XIII $3s^23p^2\ ^3P$. Thin line from RMAT calculations. Thick line from extrapolations of PLUS data.

where α is the fine-structure constant and ϵ is the energy of the ejected electron in Rydberg units. The photon energy is $\omega = (\epsilon - E_i)$ where E_i is the initial energy relative to the series limit ($E_i < 0$). For series without perturbations and hence photoionization cross sections without resonances it is usually a good approximation to take $\sigma \propto \omega^{-\beta}$ ($\beta = 3$ in the Kramers approximation). In regions of stellar interiors where high lines and further photoionization are important, the lines with ($n > 6$) will be blended to give a photoabsorption cross section which is a below-threshold extrapolation of $\sigma_{PI}(\epsilon)$ to $\epsilon < 0$. We use extrapolated $f(nl, i)$ for $n = 6$ and photoabsorption cross sections starting at $\epsilon = -z^2/(6.5 - \mu_l)^2$.

We take each photoionization cross section to be given by an analytical expression

$$\sigma(\omega) = S\omega^{-\beta} \qquad \omega \geqslant \omega_0 \tag{9}$$

where ω_0 is the energy for the photoionization threshold. The quantities S, β and ω_0 are archived for 878 199 transitions. Figure 6 compares the RMAT and PLUS cross sections for photoionization from the Fe XIII $3s^23p^2\ ^3P$ ground state. The threshold energies are in good agreement, 26.34 Rydberg for RMAT, 26.23 for PLUS. The RMAT cross section has a mass of resonances completely absent in the PLUS cross section. The widths of the resonances are determined by CI effects. There are some particularly broad resonances in the region just above the threshold. The agreement in the back ground cross sections is reasonably good. At the higher energies, beyond the region of resonances, the PLUS cross section is larger than the RMAT one by about 25%.

5. The importance of the PLUS data for the opacity work

All final OP opacity results (Seaton *et al* 1994) are calculated with inclusion of PLUS data whenever the corresponding RMAT data are not available. Despite the large scatter it is reassuring to see from figure 3 that there is little evidence for systematic differences between PLUS and RMAT data. When millions of lines have to be included the accuracy of the PLUS data is probably entirely adequate.

OP opacities are calculated as functions of temperature, T, and electron density in cm^{-3}, N_e (effects of fine structure are allowed for using methods described by Seaton *et al* 1994). We consider results for pure iron and take N to be the number of iron nuclei per unit volume.

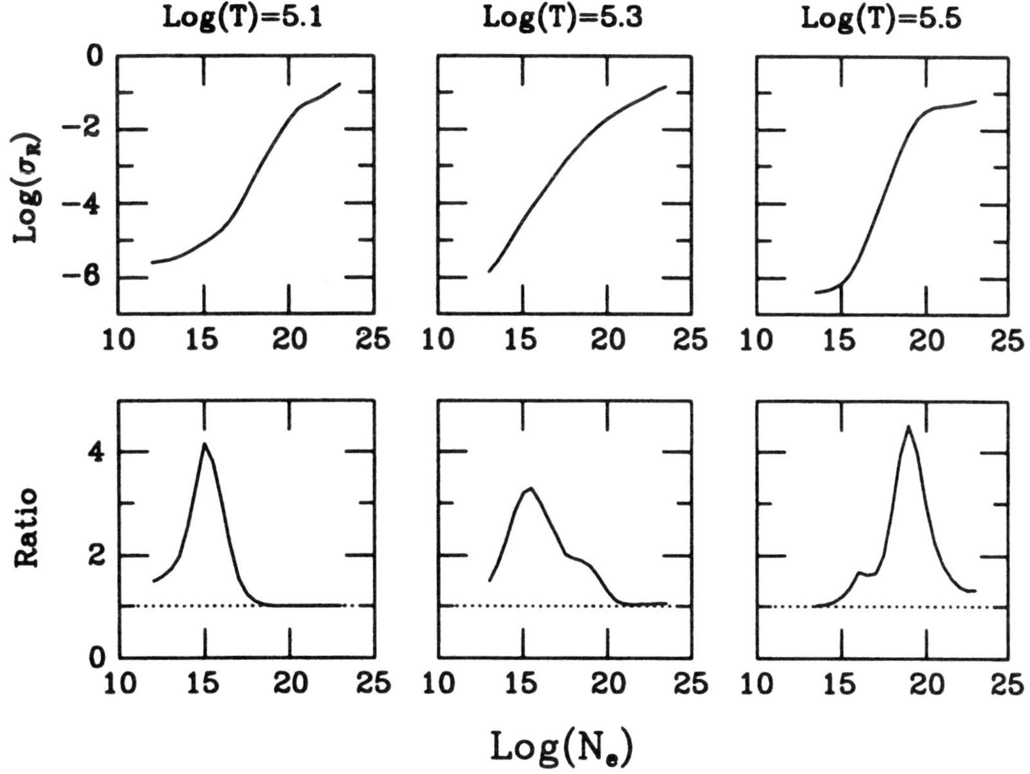

Figure 7. Opacity results for $\log(T) = 5.1$, 5.3 and 5.5 (T in K) against $\log(N_e)$ (N_e in cm^{-3}). Upper plots, results for $\log(\sigma_R)$ (σ_R in atomic units, a_0^2), calculated with inclusion of PLUS data. Lower plots, ratios of σ_R with and without PLUS data.

We define the monochromatic cross section to be $\sigma_\nu = \kappa_\nu/N$ and the Rosseland-mean cross section to be $\sigma_R = \kappa_R/N$.

Figure 7 shows values of σ_R calculated with inclusion of PLUS data, as functions of $\log(N_e)$ for $\log(T) = 5.1$, 5.3 and 5.5, together with the ratios of σ_R calculated with and without PLUS data. Those ratios have maximum values in the regions for which the ionization stages Fe VIII to XIII are giving dominant contributions.

Figure 8 shows results for the monochromatic cross sections σ_ν as functions of $u = h\nu/kT$ for the case of $\log(T) = 5.3$ and $\log(N_e) = 15.5$, calculated both without and with PLUS data, and for $0 \leqslant u \leqslant 13$ (the weighting function $f(u)$ is very small for $u > 13$). The main features are: (i) a maximum at $u \simeq 4$ mainly due to $3 \to 3$ transitions (in that region the calculations without and with PLUS data give similar results, since large numbers of $3 \to 3$ transitions are already included in the RMAT data): (ii) a minimum in the region of $u \simeq 5$ which occurs before the contributions from $3 \to n$ transitions with $n > 3$ become important; (iii) the region of $u \gtrsim 8$ where the opacity is largely dominated by contributions from $3 \to n$ transitions. On the plots of figure 8 there are large areas which are completely black; this is a consequence of there being very many lines. Figure 9 shows detailed result for the more restricted range of $5.4 \leqslant u \leqslant 5.6$. It is seen from figure 9 that the inclusion of more lines, from PLUS data, leads to increases in the minima of the cross sections which occur between the lines; and hence to a raising of the lower boundary of the black areas shown in figure 8. Since σ_R is a weighted harmonic mean, it is sensitive to the location of the minima. The importance of including PLUS data is that they fill in the gaps which occur when only RMAT data are included, which can give increases in Rosseland mean opacities by factors of up to about 4.5.

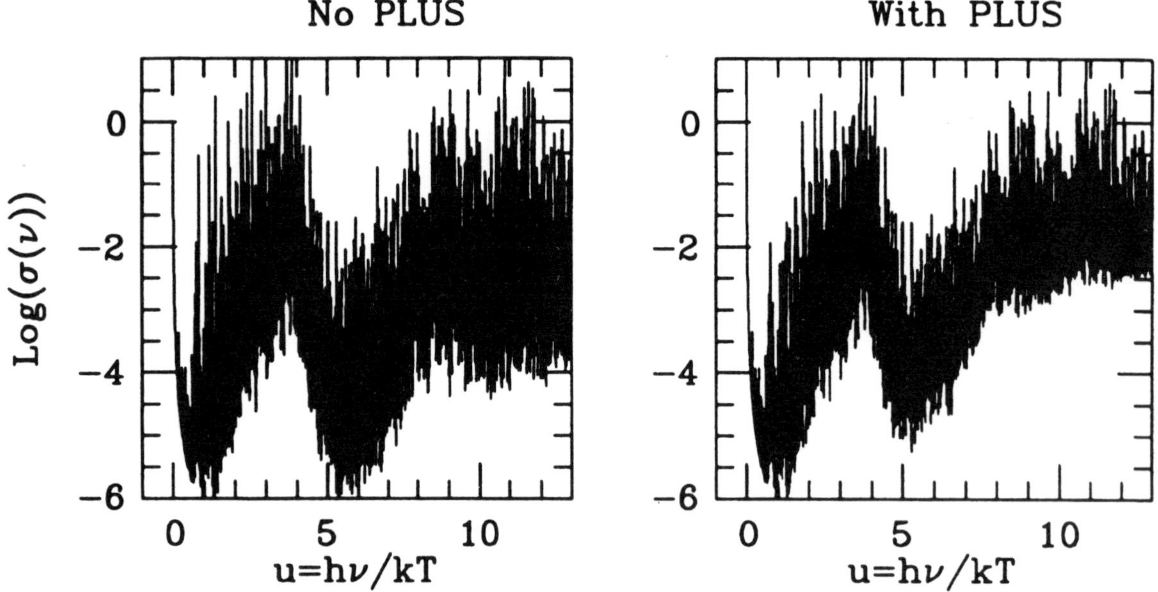

Figure 8. Monochromatic opacities for pure iron in atomic units for $\log(T) = 5.3$ and $\log(N_e) = 15.5$, calculated without and with PLUS data.

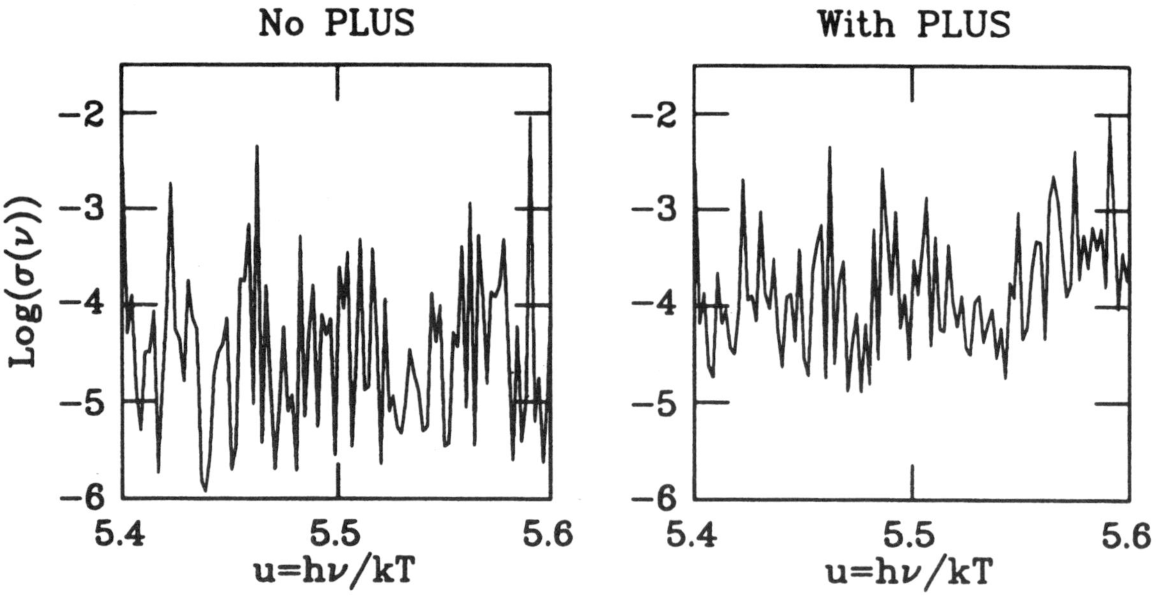

Figure 9. As figure 8, showing detailed results for a small range of $u = h\nu/kT$.

Multiplet oscillator strengths produced by calculations reported in this paper can be obtained by sending an electronic mail request to aelg@astro.ox.ac.uk (INTERNET) specifying the ion for which the data are required. It is intended that the multiplet oscillator strengths will eventually be provided through anonymous file transfer and TOPbase (Cunto *et al* 1993).

References

Berrington K A, Burke P G, Butler K, Seaton M J, Storey P J, Taylor K T and Yu Yan 1987 *J. Phys. B: At. Mol. Phys.* **20** 6379

Cunto W, Mendoza C, Ochsenbein F and Zeippen C J 1993 *Astron. Astrophys.* **275** L5

Eissner W, Jones M and Nussbaumer H 1974 *Comput. Phys. Commun.* **8** 270
Iglesias C A, Rogers F J and Wilson B G 1990 *Astrophys. J* **360** 221
Nussbaumer H and Storey P J 1978 *Astron. Astrophys.* **64** 139
Seaton M J 1987 *J. Phys. B: At. Mol. Phys.* **20** 6363
Seaton M J, Yu Yan, Mihalas D and Pradhan A K 1994 *Mon. Not. R. Astron. Soc.* **266** 805
Seaton M J, Zeippen C J, Tully J A, Pradhan A K, Mendoza C, Hibbert A and Berrington K A 1992 *Rev. Mexicana Astron. Astrof.* **23** 19

J. Phys. B: At. Mol. Opt. Phys. **26** (1993) L1–L6. Printed in the UK

LETTER TO THE EDITOR

Photoionization of Fe^+

Maryvonne Le Dourneuf†, Sultana N Nahar‡ and Anil K Pradhan‡

† Laboratoire de Théorie des Processus Atomiques et Moléculaires, UPR 261 du CNRS, Observatoire de Paris, 92195 Meudon, France
‡ Department of Astronomy, The Ohio State University, Columbus, OH 43210, USA

Received 10 November 1992

Abstract. New results are presented for the photoionization of Fe^+ in the close-coupling approximation using the R-matrix method. Cross sections are reported for the ^{6}D symmetry, including the ground state $3d^64s(^6D)$. The effect of photoionization of the 4s and the 3d shells is investigated through a series of calculations and it is shown that the latter makes the dominant contribution. The final 21-state expansion for the core ion Fe^{2+} includes all coupled states from the $3d^6$, $3d^54s$ and $3d^54p$ configurations. Photoionization of bound states along a Rydberg series is studied and the influence of strong photoexcitation-of-core type of resonances is pointed out. Partial cross sections for photoionization into individual final states of the core ion are also obtained. The new ground state cross section is up to two orders of magnitude higher than recently reported values.

Although Fe^+ is an ion of enormous importance in astrophysics, its complex electronic structure with a large number of closely spaced energy levels makes it difficult to carry out *ab initio* atomic calculations and the little previous work that has been reported is not accurate. The primary effect to take into account is the strong coupling between the various photoionization channels belonging to a number of excited states of the residual ion and orbital angular momenta of the continuum electron. The close-coupling method enables a detailed consideration of the coupling and related effects due to autoionization structures. However, the first step in the close-coupling calculations, that is to determine an adequately large and accurate wavefunction expansion for the states of the residual or the 'target' ion, involves extensive atomic structure calculations. Furthermore, the convergence of the expansion as it relates to the contribution to the total photoionization cross section from the excited states is not apparent from the term energy diagram. In particular, the relative contributions from the 4s and 3d outer shells need to be studied as one expects that the open 3d shell will play an important role in the photoionization of the ground state and other low lying states.

The factors outlined above require that several sets of calculations, with successively more extensive eigenfunction expansions, are needed to properly study the effect of convergence and resonances. In the present report we focus on the ^{6}D states, particularly the ground state $3d^6(^5D)4s\ ^6D$ and discuss some of the calculations that have been carried out using the R-matrix method. The computer codes employed were developed for the Opacity Project (Seaton 1987) and are described by Berrington *et al* (1987).

We consider the ground state photoionization of Fe^+,

$$3d^6(^5D)4s\ ^6D + h\nu \rightarrow (3d^6, 3d^54s \text{ or } 3d^54p)[S_iL_i] + e$$

0953-4075/93/010001+06$07.50

where the right-hand side represents the final states S_iL_i of the residual ion Fe^{2+} dominated by any one of the three configurations listed, which altogether give rise to 136 *LS* terms of septet, quintet, triplet and singlet symmetries. Thus for the photoionization of Fe^+ the total symmetries for the (electron+ion) system to be considered are: octets, sextets, quartets and doublets.

As we confine ourselves to ^6D states of Fe^+, only the septet and the quintet states of Fe^{2+} are coupled in the calculations. The target ion expansion for Fe^{2+} was developed using the SUPERSTRUCTURE program of Eissner *et al* (1974). The configuration basis, with the three primary configurations and five correlation configurations, is given in table 1. It was chosen to optimize primarily the 3d shell correlation. The 21 states of Fe^{2+} so obtained are also listed in table 1 along with their experimental and calculated energies.

Table 1. Energies (in Rydbergs) of the 21 target terms of Fe^{2+}. Note: the experimental energies are from Sugar and Corliss (1985). The Fe^{2+} eigenfunction expansion consists of the following: primary configurations: $3s^23p^63d^6$, $3s^23p^63d^54s$, $3s^23p^63d^54p$; correlation configurations: $3s^23p^43d^8$, $3p^63d^8$, $3s3p^63d^7$, $3s3p^53d^8$, $3s^23p^63d^54d$.

	Term		*E* (expt)	*E* (calc.)		Term		*E* (expt)	*E* (calc.)
1	$3d^6$	^5D	0.0000	0.0000	12	$3d^5(^4G)4p$	$^5F^o$	1.0619	1.0717
2	$3d^5(^6S)4s$	^7S	0.2742	0.2729	13	$3d^5(^4P)4p$	$^5D^o$	1.0661	1.0825
3	$3d^5(^6S)4s$	^5S	0.3736	0.4167	14	$3d^5(^4P)4p$	$^5S^o$	1.0653	1.0847
4	$3d^5(^4G)4s$	^5G	0.5783	0.5995	15	$3d^5(^4P)4p$	$^5P^o$	1.0810	1.1083
5	$3d^5(^4P)4s$	^5P	0.6061	0.6466	16	$3d^5(^4P)4p$	$^5F^o$	1.1041	1.1220
6	$3d^5(^4D)4s$	^5D	0.6359	0.6730	17	$3d^5(^4P)4p$	$^5D^o$	1.1208	1.1449
7	$3d^5(^6S)4p$	$^7P^o$	0.7516	0.7338	18	$3d^5(^4D)4p$	$^5P^o$	1.1272	1.1509
8	$3d^5(^4F)4s$	^5F	0.7585	0.8038	19	$3d^5(^4F)4p$	$^5G^o$	1.2339	1.2602
9	$3d^5(^6S)4p$	$^5P^o$	0.8133	0.8295	20	$3d^5(^4F)4p$	$^5F^o$	1.2402	1.2716
10	$3d^5(^4G)4p$	$^5G^o$	1.0358	1.0353	21	$3d^5(^4F)4p$	$^5D^o$	1.2520	1.2848
11	$3d^5(^4G)4p$	$^5H^o$	1.0512	1.0507					

In addition to the energies listed in table 1, a careful examination was made of the agreement between the length and the velocity oscillator strengths for transitions among the $3d^6$, $3d^54s$ states of even parity and the $3d^54p$ states of odd parity. The level of such an agreement is usually a good indicator of the accuracy of the target wavefunctions. We find that for most of over 300 transitions the agreement between length and velocity was around 15%. While this is somewhat larger than the usual aim of within 10%, given the complexity of the atomic structure calculations we consider this to be sufficient, as any further improvement would require a significantly larger basis set of correlation functions that might make the whole calculation impractical.

In order to study the effect of coupling of the final ion states on the photoionization cross sections, several sets of calculations were carried out using 3, 6, 9 and 21 states, respectively, of Fe^{2+}. Henceforth we refer to these as the 3CC, 6CC etc. The 3CC calculation included only the first three states, $3d^6\ ^5D$ and $3d^5(^6S)4s\ ^7S$ and ^5S, the 6CC included the next three additional states, the 9CC included the first nine states that contain two odd parity $3d^54p$ states, and finally the 21CC calculation included all of the septet and the quintet states of table 1.

The final continuum state wavefunctions for the $^6P^o$, $^6D^o$ and the $^6F^o$ symmetries were obtained at a fine mesh of the photoelectron energies, chosen to delineate the resonance structures in detail. The mesh was chosen to be in terms of the effective quantum number, ν, relative to successive target thresholds, such that an interval in ν of 1.0 is delineated by 100 points; often this corresponds to the same number of points for an energy region containing one or two resonances. All resonances with $\nu \leqslant 10.0$ are delineated fully.

Figure 1 presents the total cross sections for the ground state. As more states of the residual ion Fe^{2+} are included, the open channel region fills up with resonance structures belonging to the higher thresholds. In each figure, the last of the resonance structures corresponds to the highest threshold, as shown by arrows, in that particular approximation. For example, the 3CC results in figure 1(*a*) include the $3d^6\,^5D$, $3d^5(^6S)4s\,^7S$ and $3d^5(^6S)4s\,^5S$ states, and thus there are no resonances above the highest ^5S state. In figure 1 the first arrow corresponds to the ionization threshold in that particular approximation.

A very large increment in the cross section is obtained over the 3CC results when a few other $3d^54s$ states are included, as may be seen from figure 1(*b*) with the 6CC values. The *background* cross section increases by approximately a factor of two due to the coupling of the additional states. This points to the importance of the 3d shell in the photoionization process; the 6CC calculations include most of the contribution to ground state photoionization from the 4s and the 3d shells. The 9CC calculations in figure 1(*c*), with a few of the odd parity states dominated by the $3d^54p$ configuration, contain more resonances but no significant increase in the background cross section over the 6CC results. The 21CC results show yet more resonance structure but these are rather weak, reflecting the weakening effect of coupling to higher states.

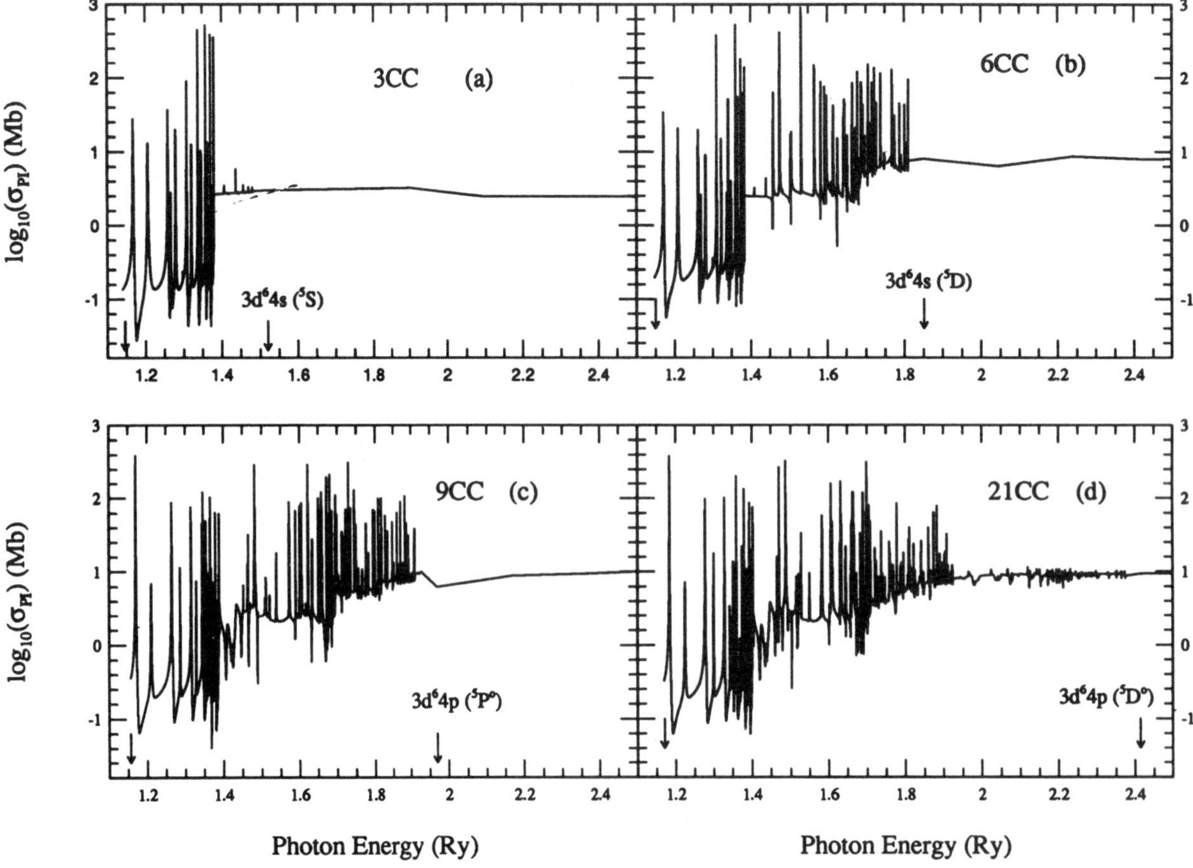

Figure 1. Ground state photoionization in the three-state, six-state, nine-state and 21-state close-coupling approximation.

The final 21CC cross sections for the ground state of Fe^+ are up to two orders of magnitude higher than the recently reported values by Sawey and Berrington (1992) who employed only a single state expansion for Fe^{2+} and obtained a nearly constant value of 0.1 Mb, compared to the present maximum value of approximately 10 Mb for the background cross section (figure 1(d)). The cross sections reported by Reilman and Manson (1979) in the central field approximation (which does not account for coupling or resonance effects) are more than a factor of two lower in the near-threshold region than the present values.

Figure 2 shows the partial cross sections for photoionization of the ground state of Fe^+ into several states of Fe^{2+}. As the incident photon energy increases to enable higher thresholds of the residual ion to be attained, the photoelectron flux redistributes itself into all the open channels at each threshold and one may calculate the partial photoionization cross sections for ionization into specific excited states of the residual ion (e.g. Nahar and Pradhan 1991). Under conditions of photoionization equilibrium and non-LTE stellar atmospheres the level population in excited states of ions is often determined by such partial photoionization from the previous ionization stage and it is useful to compute these cross sections explicitly. The arrow in each panel of figure 2 corresponds to the excited threshold of the residual ion. While we have calculated

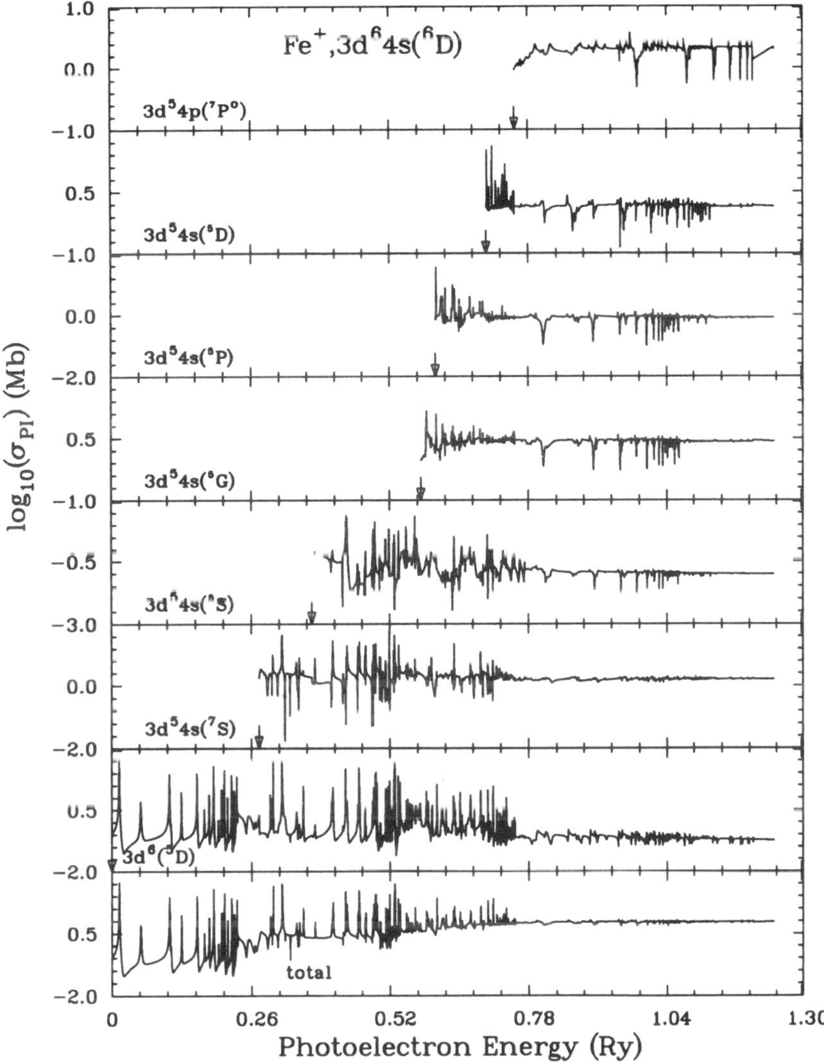

Figure 2. Cross sections for partial photoionization of the Fe^+ ground state into the ground and the excited states of the residual ion Fe^{2+}.

the partial cross sections for ionization into all coupled states included in the Fe^{2+} expansion, the results are given only for a few selected states in the figure.

Yu and Seaton (1987) have described strong resonance features in the photoionization of excited Rydberg states associated with dipole allowed transitions *within* the residual ion core, and have labelled these as *photoexcitation-of-core* (PEC) resonances. The PEC resonances are related to the dielectronic recombination process in that the free electron may recombine with the ion through these resonances if they undergo radiative stabilization by emitting a photon—the inverse process. For photoionization of bound states of Fe^{+} in the Rydberg series $3d^6(^5D)nd$, for example, the PEC features correspond to dipole transitions of the type

$$3d^6(^5D) \rightarrow 3d^54p(^5P^o, ^5D^o, ^5F^o)$$

in the Fe^{2+} core. As the outer electron is essentially a 'spectator' during the process, the PEC features are most easily discernible in the photoionization of bound states along a Rydberg series with weakly bound electrons. In these cases the PEC resonances modify considerably what one would otherwise expect to be a hydrogenic cross section.

Figure 3 shows the photoionization cross sections of bound states of the $3d^6(^5D)nd\ ^6D$ series of Fe^{+}. The PEC resonances are the broad features lying in between

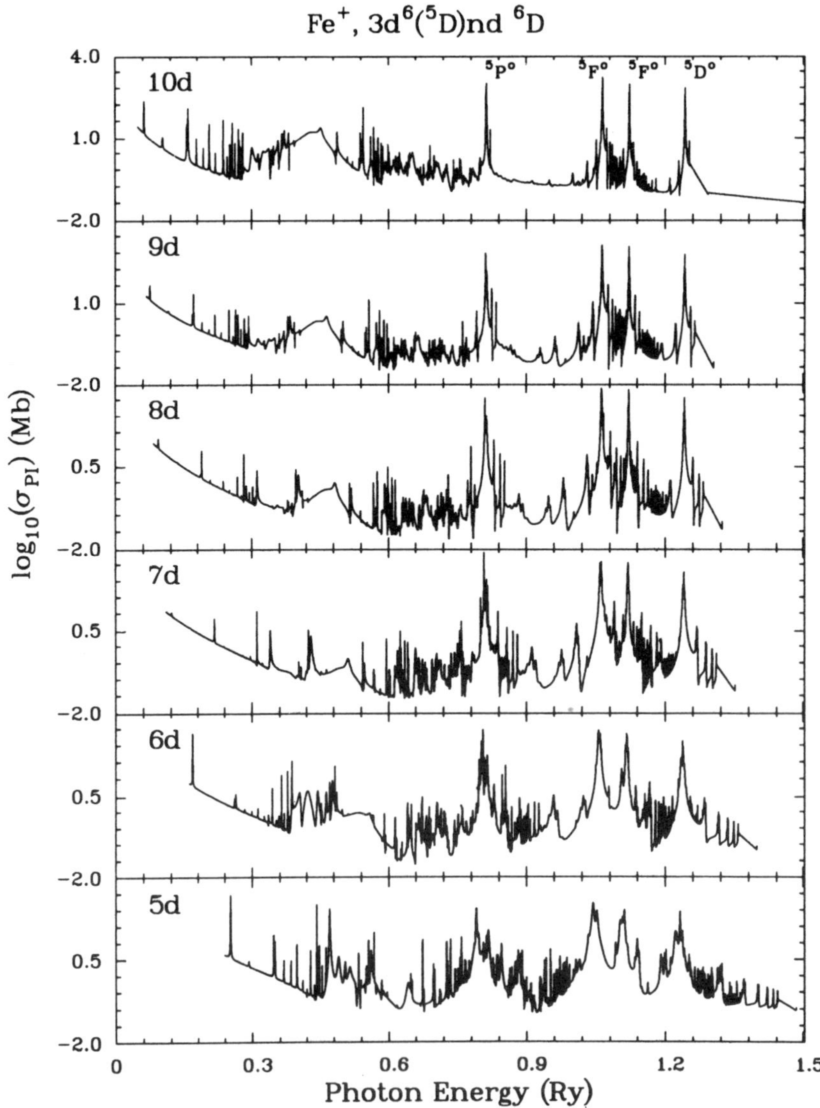

Figure 3. Photoionization cross sections of bound states of Fe^{+} in a Rydberg series showing the PEC resonance features.

the usual, narrower series of resonances converging onto the excited thresholds. As seen on looking up the panels of figure 3, the cross sections for all states along the Rydberg series of bound states show that the PEC resonances are at precisely the same photon energies corresponding to the dipole transitions within the Fe^{2+} core states. The prominent PEC resonances are identified in the figure.

Further calculations are in progress for the Opacity Project for photoionization cross sections and transition probabilities involving a large number of bound states of Fe^{+}, including the quartet and the doublet symmetries and up to 83 target terms in the Fe^{2+} eigenfunction expansion (perhaps the largest close-coupling calculations to date).

MLD would like to acknowledge support of the CNRS, France. This work was supported in part by grants from the US National Science Foundation (AST-8996215 and PHY-9115057). SNN acknowledges support from the College of Mathematical and Physical Sciences at the Ohio State University. The computational work was carried out at the Centre de Calcul Vectorial pour la Recherche at Palaiseau (France) under grant 1607, and the Ohio Supercomputer Center in Columbus, Ohio (USA).

References

Berrington K A, Burke P G, Butler K, Seaton M J, Storey P J, Taylor K T and Yu Yan 1987 *J. Phys. B: At. Mol. Phys.* **20** 6379

Eissner W, Jones M and Nussbaumer H 1974 *Comput. Phys. Commun.* **8** 270

Nahar S N and Pradhan A K 1991 *Phys. Rev.* A **44** 2935

Reilman R F and Manson S T 1979 *Astrophys. J. Suppl.* **40** 815

Sawey P M J and Berrington K A 1992 *J. Phys. B: At. Mol. Opt. Phys.* **25** 1451

Seaton M J 1987 *J. Phys. B: At. Mol. Phys.* **20** 6363

Sugar J and Corliss C 1985 *J. Phys. Chem. Ref. Data* **14** Suppl. 2

Yu Yan and Seaton M J 1987 *J. Phys. B: At. Mol. Phys.* **20** 6409

Photoionization cross sections and oscillator strengths for Fe III

Sultana N. Nahar
Department of Astronomy, The Ohio State University, Columbus, Ohio 43210
(Received 12 October 1995)

Ab initio calculations are carried out for the photoionization cross sections and oscillator strengths for Fe III in the close-coupling approximation employing the R-matrix method. A 49-state eigenfunction expansion for Fe IV, with states dominated by the ground configuration $3d^5$ and the excited configurations $3d\,^4 4s$ and $3d\,^4 4p$, is employed to ensure an extensive treatment of autoionizing resonances and other electron correlation effects related to channel couplings that considerably affect the effective cross sections. Of particular interest are the wide resonances due to photoexcitation within the core states that enhance the background cross sections by orders of magnitude at high energies. Coupled wave functions are obtained for 805 bound states of Fe III below the first ionization threshold; all of the 199 LS terms observed experimentally have been identified. Photoionization cross sections, with detailed autoionizing resonances, are obtained for the computed bound states as a function of photoelectron energy. The present results also include the oscillator strengths for 11 979 transitions between the bound states of Fe III that lie below the ionization threshold. In addition, the partial photoionization cross sections for 239 bound states, with the residual ion in the ground state, are also obtained. These cross sections are computed for applications to the determination of electron-ion recombination rate coefficients and population of levels in collisional-radiative models that do not assume local thermodynamic equilibrium. The present radiative data are compared with the observed energy values, lifetimes, and the photoionization cross sections calculated by others.

PACS number(s): 32.80.Fb

I. INTRODUCTION

Fe III is a very important ion in astrophysical sources but its detailed theoretical study is difficult because of its complexity. The strong electron correlation effects in this 24-electron ion require a large eigenfunction expansion for an accurate wave-function representation. In a previous calculation under the Opacity Project (OP) [1], the radiative data for energy levels, photoionization cross sections, and oscillator strengths were obtained for Fe III [2] employing a relatively limited eigenfunction expansion with the 16 terms dominated by only the ground configuration $3d^5$. As we show in this work, this expansion is not quite adequate to achieve the desired accuracy or completeness of the set of radiative quantities for Fe III. Consequently the OP data set requires substantial improvement in terms of missing states, corresponding oscillator strengths for bound-bound transitions, and absence of important resonances in photoionization cross sections, etc. The aim of the present work is to study Fe III in much greater detail and obtain more accurate radiative data. The present work forms part of the new Iron Project (IP) [3], which, in addition to collisional calculations (which is its primary aim), also involves improvement of some radiative data for complex atomic systems not adequately treated in the OP work. Recent studies of large-scale computations of Fe II [4] and of Fe I [5] show that the photoionization cross sections in the new calculations can differ considerably, even by orders of magnitude, from earlier calculations in either simpler atomic physics approximations, such as central field, Hartree-Slater, single configuration Hartree-Fock, or even limited close-coupling calculations that do not include all contributing channels. On the other hand, for practical applications it is important to obtain these parameters to high accuracy. For example, the abundances and ionization fractions of iron in astrophysical sources depend critically on photoionization and recombination parameters that might possibly explain the observed spectra of stars, gaseous nebulae, quasars, etc.

In this paper we present selected results for the energies, photoionization cross sections, and oscillator strengths of Fe III, illustrating the physical effects that manifest themselves in a large coupled-channel calculation of radiative processes. Of particular interest are the two main types of resonances in the cross sections, one due to the Rydberg series of quasibound states, above the ionization threshold, converging onto the excited target thresholds of Fe IV, and the other type due to the photoexcitation of a dipole transition from the ground state of Fe IV, i.e., $3d^5(^6S)\rightarrow 3d\,^4 4p(^6P^o)$. The second type of resonance is the photoexcitation-of-core (PEC) resonance and is intimately connected with the inverse dielectronic recombination process [6,7]. The PEC resonances are not included in any earlier work (the previous OP calculation did not include the excited $3d\,^5 4p$ configuration). The total number of bound states and oscillator strengths obtained are much larger than the previous OP data. The present work also provides the partial photoionization cross sections for ionization into the ground state of the residual ion. The partial cross sections are directly applicable to ongoing work on total, unified election-ion recombination rates [7], and astrophysical diagnostics.

II. COMPUTATIONS

In the close-coupling approximation (CC), the core ion, termed the "target," is represented by an N-electron system and the wave-function expansion $\Psi(E)$ for any symmetry, $SL\pi$, of the total $(N+1)$-electron system is represented in

1050-2947/96/53(3)/1545(8)/$10.00

terms of the core or the target ion states as

$$\Psi(E) = A\sum_i \chi_i \theta_i + \sum_j c_j \Phi_j, \tag{2.1}$$

where χ_i is the target ion wave function in a specific state $S_i L_i \pi_i$ and θ_i is the wave function for the $(N+1)$th electron in a channel labeled as $S_i L_i \pi_i k_i^2 \ell_i (SL\pi)$; k_i^2 is its incident kinetic energy. Φ_j's are the correlation wave functions of the $(N+1)$-electron system that compensates the orthogonality condition of the total wave function. The target state wave functions for Fe IV is obtained from the atomic structure computer code SUPERSTRUCTURE [8], prior to the R-matrix calculations. The 49-state eigenfunction expansion is chosen such that it includes the dominant dipole transitions in the core and important correlation effects. Comparison of the calculated energies of the 49 states with the observed energies is presented in Table I. The calculated energies agree with the observed values usually within a few percent, with a few exceptions; the largest difference is 17.5% for the state $3d^5(^2I)$. This indicates an overall reasonably good agreement of the calculated values with the observed ones in light of the complex atomic structure involved. The table also includes the list of principal and correlation configurations employed for the target ion and the values of the scaling parameters $\lambda_{n\ell}$, for the set of target orbitals $n\ell$, in the scaled Thomas-Fermi-Dirac potential in SUPERSTRUCTURE.

One crucial factor for proper representation of $\Psi(E)$ is the second sum in the wave function expansion that accounts for orthogonality constraints between the bound target and the free continuum orbitals, as well as for additional short-range electron correlation. For smaller atomic systems the sum can include all possible $(N+1)$-electron states with the chosen number of the target orbital set $n\ell$. For large atomic systems, such as Fe III, however, such a sum may correspond to a very large number of states that may not all be important but result in a huge Hamiltonian matrix to be diagonalized. Hence a judicious choice needs to be made for the set of $(N+1)$-electron correlation functions that are significant. We employ a different, suitably optimized, set of such functions for each total spin symmetry of the electron-ion system; we consider $(2S+1) = 1,3,5,7$. For a given total spin, only those target states coupled to that spin value are included in the CC expansion, thus significantly reducing the size of the calculations.

The CC calculations for Fe III are carried out in a manner similar to the OP [1] calculations. The theory is explained in Ref. [1]. The computations are carried out using the R-matrix codes developed for the OP [9], but extended for the IP work [3]. The calculations are, essentially, as in the earlier large-scale computations for the radiative data of Fe II [4]. Some modifications were made to the asymptotic code of the R-matrix package that computes the bound-free transition amplitudes to yield partial, state-specific (with given initial and final states), photoionization cross sections.

Calculations of Fe III are carried out in LS coupling, with the assumption that the relativistic effects may not be considerable because of the low ion charge. The present work considers all bound states of $S_t L_t \pi_t n\ell$, where $S_t L_t \pi_t$ are the 49 target states and $n \leqslant 10$ and $0 \leqslant \ell \leqslant n-1$. It is nontrivial to carry out the spectroscopic identification of the calculated

TABLE I. The Fe IV target states in the 49-state CC expansion for Fe III. E(obs) corresponds to the observed [12] and E(calc) to the calculated energies in Ry. Principal configurations: $3s^2 3p^6 3d^5$, $3s^2 3p^6 3d^4 4s$, and $3s^2 3p^6 3d^4 4p$. Correlation configurations: $3s^2 3p^5 3d^6$, $3s^2 3p^4 3d^7$, $3s^2 3p^6 3d^3 4s 4p$, $2p^6 3p^6 3d^7$, $3s 3p^6 3d^5 4s$, $3s^2 3p^6 3d^3 4s 4d$, and $3s 3p^6 3d^4 4s^2$. Thomas-Fermi scaling parameters: 1.40467($1s$), 1.11279($2s$), 1.27320($2p$), 1.27320($3s$), 1.07320($3p$), 1.05832($3d$), 1.07320 ($4s$), 1.05832($4p$), and 1.40794($4d$).

Configuration	Term	E(obs)	E(calc)
$3d^5$	$^6S^e$	0.000000	0.000000
$3d^5$	$^4G^e$	0.294169	0.343447
$3d^5$	$^4P^e$	0.321730	0.349341
$3d^5$	$^4D^e$	0.354206	0.406702
$3d^5$	$^2I^e$	0.429070	0.504217
$3d^5$	$^2D^e$	0.453314	0.504982
$3d^5$	$^2F^e$	0.471355	0.517330
$3d^5$	$^4F^e$	0.480416	0.538692
$3d^5$	$^2H^e$	0.512384	0.576651
$3d^5$	$^2G^e$	0.524726	0.603606
$3d^5$	$^2F^e$	0.557809	0.639729
$3d^5$	$^2S^e$	0.607998	0.694126
$3d^5$	$^2D^e$	0.675418	0.755665
$3d^5$	$^2G^e$	0.755404	0.856408
$3d^5$	$^2P^e$	0.912367	1.031666
$3d^5$	$^2D^e$	0.986434	1.095135
$3d^4(^5D)4s$	$^6D^e$	1.170799	1.159025
$3d^4(^5D)4s$	$^4D^e$	1.261186	1.264731
$3d^4(^3H)4s$	$^4H^e$	1.407729	1.420342
$3d^4(^3P_2)4s$	$^4P^e$	1.412215	1.439247
$3d^4(^3F_2)4s$	$^4F^e$	1.422743	1.445444
$3d^4(^3G)4s$	$^4G^e$	1.450157	1.470624
$3d^4(^3H)4s$	$^2H^e$	1.463188	1.482829
$3d^4(^3P_2)4s$	$^2P^e$	1.467623	1.501715
$3d^4(^3F_2)4s$	$^2F^e$	1.477001	1.508756
$3d^4(^3G)4s$	$^2G^e$	1.505378	1.533146
$3d^4(^3D)4s$	$^4D^e$	1.509078	1.541497
$3d^4(^1G_2)4s$	$^2G^e$	1.528647	1.558940
$3d^4(^1I)4s$	$^2I^e$	1.535894	1.559382
$3d^4(^1S_2)4s$	$^2S^e$	1.555802	1.598304
$3d^4(^3D)4s$	$^2D^e$	1.561891	1.605603
$3d^4(^1D_2)4s$	$^2D^e$	1.613219	1.664325
$3d^4(^1F)4s$	$^2F^e$	1.669099	1.723453
$3d^4(^5D)4p$	$^6F^o$	1.724550	1.818157
$3d^4(^5D)4p$	$^6P^o$	1.732119	1.831152
$3d^4(^3F_1)4s$	$^4F^e$	1.734927	1.807848
$3d^4(^3P_1)4s$	$^4P^e$	1.735791	1.810057
$3d^4(^5D)4p$	$^4P^o$	1.754276	1.862131
$3d^4(^5D)4p$	$^6D^o$	1.761746	1.861259
$3d^4(^3F_1)4s$	$^2F^e$	1.787629	1.871164
$3d^4(^3P_1)4s$	$^2P^e$	1.787918	1.874124
$3d^4(^5D)4p$	$^4F^o$	1.791167	1.904812
$3d^4(^1G_1)4s$	$^2G^e$	1.833408	1.915998
$3d^4(^5D)4p$	$^4D^o$	1.843957	1.968479
$3d^4(^3H)4p$	$^4H^o$	1.938032	2.040010
$3d^4(^3P_2)4p$	$^4D^o$	1.953335	2.073489
$3d^4(^3F_2)4p$	$^4G^o$	1.963043	2.077942
$3d^4(^3H)4p$	$^4I^o$	1.969819	2.070261
$3d^4(^3H)4p$	$^2G^o$	1.970958	2.081440

TABLE II. Comparison of the calculated energies, E(calc), of Fe III with the observed ones E(obs) [12,13]. "Ek" indicates energy measured by Ekberg [13]. All energies are in Rydberg units. The ionization potential is 2.2528 Ry.

Configuration	$SL\pi$	E(obs)	E(calc)	Configuration	$SL\pi$	E(obs)	E(calc)
$3d^5(^6S)4s$	$^7S^e$	1.97864	1.928	$3d^5(^6S)4p$	$^7P^o$	1.50127	1.485
$3d^5(^6S)4d$	$^7D^e$	0.91033	0.893	$3d^5(^6S)5s$	$^7S^e$	0.89244	0.884
$3d^5(^6S)5p$	$^7P^o$	0.73740	0.726	$3d^5(^6S)4f$	$^7F^{oa}$	0.57272	0.575
$3d^4(^5D)4s4p^3P^o$	$^7P^o$	0.54930	0.560	$3d^5(^6S)5d$	$^7D^e$	0.51767	0.511
$3d^5(^6S)6s$	$^7S^e$	0.51306	0.510	$3d^5(^6S)6p$	$^7P^o$	0.44233	0.437
$3d^5(^6S)5f$	$^7F^{oa}$	0.36542	0.364	$3d^5(^6S)5g$	$^7G^{ea}$	0.36067	0.360
$3d^5(^6S)6d$	$^7D^e$	0.33549	0.332	$3d^5(^6S)7s$	$^7S^e$	0.33356	0.332
$3d^5(^6S)6g$	$^7G^{ea}$	0.25041	0.250	$3d^5(^6S)6h$	$^7H^{oa}$	0.25005	0.250
$3d^6$	$^5D^e$	2.24898	2.127	$3d^5(^6S)4s$	$^5S^e$	1.87921	1.828
$3d^5(^4G)4s$	$^5G^e$	1.67448	1.586	$3d^5(^4P)4s$	$^5P^e$	1.64675	1.582
$3d^5(^4D)4s$	$^5D^e$	1.61698	1.532	$3d^5(^4F)4s$	$^5F^e$	1.49430	1.407
$3d^5(^6S)4p$	$^5P^o$	1.43953	1.421	$3d^5(^4G)4p$	$^5G^o$	1.21704	1.148
$3d^5(^4G)4p$	$^5H^o$	1.20162	1.137	$3d^5(^4G)4p$	$^5F^o$	1.19089	1.122
$3d^5(^4P)4p$	$^5S^o$	1.18758	1.137	$3d^5(^4P)4p$	$^5D^o$	1.18673	1.136
$3d^5(^4P)4p$	$^5P^o$	1.17188	1.119	$3d^5(^4D)4p$	$^5F^o$	1.14874	1.075
$3d^5(^4D)4p$	$^5D^o$	1.13378Ek	1.060	$3d^5(^4D)4p$	$^5P^o$	1.12566	1.056
$3d^5(^4F)4p$	$^5G^o$	1.01894	0.940	$3d^5(^4F)4p$	$^5F^o$	1.01264	0.930
$3d^5(^4F)4p$	$^5D^o$	1.00081	0.919	$3d^5(^6S)4d$	$^5D^e$	0.87193	0.853
$3d^5(^6S)5s$	$^5S^e$	0.86991	0.860	$3d^5(^6S)5p$	$^5P^o$	0.71835	0.708
$3d^5(^4G)4d$	$^5H^e$	0.61977	0.574	$3d^5(^4G)4d$	$^5F^e$	0.61590	0.571
$3d^5(^4G)4d$	$^5G^e$	0.61486	0.569	$3d^5(^4G)4d$	$^5I^e$	0.61351	0.570
$3d^5(^4G)5s$	$^5G^e$	0.59608	0.586	$3d^5(^4P)4d$	$^5F^e$	0.59045	0.555
$3d^5(^6S)4f$	$^5F^o$	0.56900	0.565	$3d^5(^4P)5s$	$^5P^e$	0.56707	0.559
$3d^5(^4D)4d$	$^5G^e$	0.55390	0.510	$3d^5(^4D)4d$	$^5D^{ea}$	0.54980	0.506
$3d^5(^4D)5s$	$^5D^e$	0.53888	0.526	$3d^5(^6S)6s$	$^5S^e$	0.50313	0.498
$3d^5(^6S)5d$	$^5D^e$	0.48854	0.462	$3d^5(^4G)5p$	$^5G^o$	0.44546	0.431
$3d^5(^4G)5p$	$^5H^o$	0.44140	0.428	$3d^5(^4G)5p$	$^5F^o$	0.43738	0.424
$3d^5(^4F)4d$	$^5H^e$	0.43134	0.379	$3d^5(^4F)4d$	$^5G^e$	0.42611	0.378
$3d^5(^4P)5p$	$^5D^o$	0.41952	0.402	$3d^5(^4P)5p$	$^5S^o$	0.41851	0.404
$3d^5(^4F)5s$	$^5F^e$	0.41197	0.400	$3d^5(^4P)5p$	$^5P^o$	0.40975	0.399
$3d^5(^4D)5p$	$^5F^o$	0.38410	0.369	$3d^5(^4D)5p$	$^5D^{oa}$	0.37830	0.366
$3d^5(^6S)5f$	$^5F^o$	0.36408	0.362	$3d^5(^6S)5g$	$^5G^e$	0.36065	0.360
$3d^5(^4F)5p$	$^5G^o$	0.25676	0.243	$3d^5(^4F)5p$	$^5F^o$	0.25239	0.243
$3d^5(^6S)6g$	$^5G^e$	0.25039	0.250	$3d^5(^6S)6h$	$^5H^{oa}$	0.25004	0.250
$3d^5(^4G)5d$	$^5H^e$	0.22433	0.210	$3d^5(^4G)5d$	$^5F^e$	0.22311	0.209
$3d^5(^4G)5d$	$^5G^e$	0.22303	0.209	$3d^5(^4G)5d$	$^5I^e$	0.22234	0.209
$3d^5(^4G)6s$	$^5G^e$	0.21790	0.214	$3d^5(^4P)6s$	$^5P^e$	0.18946	0.187
$3d^5(^4D)6s$	$^5D^{ea}$	0.16155	0.154	$3d^6$	$^3P^e2$	2.07027	1.952
$3d^6$	$^3H^e$	2.06828	1.920	$3d^6$	$^3F^e2$	2.05567	1.926
$3d^6$	$^3G^e$	2.02649	1.885	$3d^6$	$^3D^e$	1.97230	1.840
$3d^6$	$^3P^e1$	1.79726	1.665	$3d^6$	$^3F^e1$	1.79482	1.656
$3d^5(^4G)4s$	$^3G^e$	1.60844	1.511	$3d^5(^4P)4s$	$^3P^e$	1.58039	1.508
$3d^5(^4D)4s$	$^3D^e$	1.55089	1.460	$3d^5(^2I)4s$	$^3I^e$	1.52521	1.405
$3d^5(^2D_3)4s$	$^3D^e$	1.50181	1.416	$3d^5(^2F_2)4s$	$^3F^e$	1.48390	1.400
$3d^5(^2H)4s$	$^3H^e$	1.44384	1.339	$3d^5(^2G_2)4s$	$^3G^e$	1.43440	1.322
$3d^5(^4F)4s$	$^3F^e$	1.42845	1.334	$3d^5(^2F_1)4s$	$^3F^e$	1.40174	1.297
$3d^5(^2S)4s$	$^3S^e$	1.35375	1.250	$3d^5(^2D_2)4s$	$^3D^e$	1.28766	1.186
$3d^5(^2G_1)4s$	$^3G^e$	1.21091	1.090	$3d^5(^4G)4p$	$^3F^o$	1.17506	1.127
$3d^5(^4G)4p$	$^3H^o$	1.17286	1.127	$3d^5(^4P)4p$	$^3P^o$	1.16071	1.132
$3d^5(^4G)4p$	$^3G^o$	1.14164	1.090	$3d^5(^4P)4p$	$^3D^o$	1.13616	1.108
$3d^5(^4D)4p$	$^3D^o$	1.11474	1.070	$3d^5(^4D)4p$	$^3F^o$	1.10862	1.061
$3d^5(^4P)4p$	$^3S^o$	1.10107	1.069	$3d^5(^4D)4p$	$^3P^o$	1.07955	1.028

TABLE II. (*Continued*).

Configuration	$SL\pi$	E(obs)	E(calc)	Configuration	$SL\pi$	E(obs)	E(calc)
$3d^5(^2I)4p$	$^3K^o$	1.06551	0.998	$3d^5(^2I)4p$	$^3I^o$	1.06160	0.993
$3d^5(^2P)4s$	$^3P^e$	1.06035Ek	0.943	$3d^5(^2D)4p$	$^3F^o$	1.04642	1.007
$3d^5(^2I)4p$	$^3H^o$	1.04566	0.973	$3d^5(a^2D)4p$	$^3P^o$	1.02762	0.981
$3d^5(a^2F)4p$	$^3G^o$	1.02083	0.978	$3d^5(a^2D)4p$	$^3D^o$	1.02148	0.974
$3d^5(a^2F)4p$	$^3D^o$	1.01150	0.963	$3d^5(a^2F)4p$	$^3F^o$	1.00754	0.967
$3d^5(^2H)4p$	$^3H^o$	0.99622	0.938	$3d^5(^2H)4p$	$^3G^o$	0.99432	0.945
$3d^5(c^2D)4s$	$^3D^e$	0.98731Ek	0.872	$3d^5(^4F)4p$	$^3G^o$	0.98094	0.923
$3d^5(^2H)4p$	$^3I^o$	0.97810	0.927	$3d^5(a^2G)4p$	$^3F^o$	0.96940	0.913
$3d^5(^4F)4p$	$^3D^o$	0.96389	0.912	$3d^5(^4F)4p$	$^3F^o$	0.95654	0.892
$3d^5(a^2G)4p$	$^3H^o$	0.94920	0.885	$3d^5(a^2G)4p$	$^3G^o$	0.94050	0.874
$3d^5(b^2F)4p$	$^3F^o$	0.93400	0.861	$3d^5(b^2F)4p$	$^3G^o$	0.91153	0.846
$3d^5(b^2F)4p$	$^3D^o$	0.90768	0.841	$3d^5(^2S)4p$	$^3P^o$	0.89299	0.825
$3d^5(b^2D)4p$	$^3F^o$	0.81157	0.746	$3d^5(b^2D)4p$	$^3D^o$	0.80819	0.748
$3d^5(b^2D)4p$	$^3P^o$	0.79434Ek	0.723	$3d^5(b^2G)4p$	$^3H^o$	0.74034	0.664
$3d^5(b^2G)4p$	$^3F^o$	0.73557	0.657	$3d^5(b^2G)4p$	$^3G^o$	0.72907	0.652
$3d^5(^2P)4p$	$^3P^o$	0.62122Ek	0.544	$3d^5(^4G)4d$	$^3F^e$	0.59060	0.520
$3d^5(^4G)4d$	$^3I^e$	0.58673	0.537	$3d^5(^2P)4p$	$^3D^o$	0.58526Ek	0.500
$3d^5(^4G)5s$	$^3G^e$	0.58114	0.568	$3d^5(^2P)4p$	$^3S^o$	0.56596Ek	0.476
$3d^5(^4D)4d$	$^3G^e$	0.53051	0.526	$3d^5(^4D)5s$	$^3D^e$	0.52381	0.507
$3d^5(c^2D)4p$	$^3F^o$	0.50945Ek	0.433	$3d^5(c^2D)4p$	$^3D^o$	0.49426Ek	0.420
$3d^5(^2I)5s$	$^3I^e$	0.45865	0.446	$3d^5(^4G)5p$	$^3F^o$	0.43399	0.423
$3d^5(^4G)5p$	$^3H^o$	0.43338	0.422	$3d^5(^4G)5p$	$^3G^o$	0.42559	0.412
$3d^5(^4D)5p$	$^3D^o$	0.37360	0.362	$3d^5(^4D)5p$	$^3F^o$	0.37290	0.358
$3d^5(^2I)5p$	$^3I^o$	0.30724	0.294	$3d^5(^2I)5p$	$^3H^o$	0.30266	0.289
$3d^5(^4G)6s$	$^3G^e$	0.21115	0.207	$3d^5(^4D)6s$	$^3D^e$	0.15486	0.147
$3d^6$	$^1I^e$	1.97620	1.815	$3d^6$	$^1G^e2$	1.97137	1.836
$3d^6$	$^1S^e2$	1.93560	1.820	$3d^6$	$^1D^e2$	1.92656	1.793
$3d^6$	$^1F^e$	1.86192	1.717	$3d^6$	$^1G^e1$	1.73139	1.581
$3d^6$	b^1D^e	1.55075Ek	1.376	$3d^5(^2I)4s$	$^1I^e$	1.49256	1.365
$3d^5(^2D_3)4s$	$^1D^e$	1.46142	1.400	$3d^5(^2F_2)4s$	$^1F^e$	1.45181	1.360
$3d^5(^2H)4s$	$^1H^e$	1.40969	1.298	$3d^5(^2G_2)4s$	$^1G^e$	1.40068	1.288
$3d^5(^2F_1)4s$	$^1F^e$	1.36852	1.258	$3d^5(^2S)4s$	$^1S^e$	1.31960Ek	1.201
$3d^5(^2D_2)4s$	$^1D^e$	1.25435	1.151	$3d^5(a^2D)4p$	$^1F^o$	1.24115	0.976
$3d^5(a^2D)4p$	$^1D^o$	1.19781	1.007	$3d^5(^2G_1)4s$	$^1G^e$	1.17799	1.053
$3d^5(^2I)4p$	$^1H^o$	1.05259	0.983	$3d^5(^2I)4p$	$^1K^o$	1.05003	0.982
$3d^5(a^2F)4p$	$^1G^o$	1.02845	0.979	$3d^5(^2P)4s$	$^1P^e$	1.02769Ek	0.907
$3d^5(^2I)4p$	$^1I^o$	1.01588	0.942	$3d^5(a^2D)4p$	$^1P^o$	0.98898	0.947
$3d^5(a^2F)4p$	$^1D^o$	0.97920	0.931	$3d^5(a^2G)4p$	$^1G^o$	0.97863	0.917
$3d^5(a^2F)4p$	$^1F^o$	0.97293	0.924	$3d^5(^2H)4p$	$^1I^o$	0.96303	0.908
$3d^5(^2D)4s$	$^1D^e$	0.95464Ek	0.836	$3d^5(a^2G)4p$	$^1H^o$	0.93526	0.874
$3d^5(^2H)4p$	$^1H^o$	0.93292	0.862	$3d^5(a^2G)4p$	$^1F^o$	0.93114	0.858
$3d^5(b^2F)4p$	$^1D^o$	0.92586	0.857	$3d^5(b^2F)4p$	$^1G^o$	0.89492	0.825
$3d^5(b^2F)4p$	$^1F^o$	0.87996	0.800	$3d^5(^2S)4p$	$^1P^o$	0.87101	0.796
$3d^5(b^2D)4p$	$^1F^o$	0.79942	0.735	$3d^5(b^2D)4p$	$^1P^o$	0.78552Ek	0.718
$3d^5(b^2D)4p$	$^1D^o$	0.77580	0.710	$3d^5(b^2G)4p$	$^1H^o$	0.71479	0.637
$3d^5(b^2G)4p$	$^1G^o$	0.71026	0.630	$3d^5(b^2G)4p$	$^1F^o$	0.70085	0.618
$3d^5(^2P)4p$	$^1D^o$	0.57862	0.499	$3d^5(c^2D)4p$	$^1D^o$	0.48902	0.404
$3d^5(c^2D)4p$	$^1F^o$	0.47730	0.398				

[a]The term energy is from an incomplete set of observed fine-structure levels.

levels. Identification is carried out through detailed examination of the effective quantum number of the states employing the code ELEVID [10]. Oscillator strengths for transitions among all computed bound states are obtained. The photoionization cross sections are calculated for all bound states below the ionization threshold. The resonances in the cross sections are delineated for the Rydberg series corresponding to each target state of Fe IV, characterized by the effective quantum number ν, up to $\nu = 10$ *below* each target state. The resonances in the small energy region between $\nu > 10$ and the corresponding target threshold are averaged in the Gailitis averaging procedure [11].

The computations were carried on the Cray Y-MP computer. As for Fe II, it was necessary to divide the calculations into small groups of few symmetries, $SL\pi$, at a time because of the large amount of memory, disk space, and CPU time required. The largest cases were for the singlet and the triplet spin symmetries, $(2S+1) = 1,3$. The largest size of the Hamiltonian matrix was about 2000, with 150 channels, requiring about 1.3 Gbytes of computer disk space for each single continuum wave function and 32 Mbytes of RAM memory. The total CPU time required for the computations was 500 h with another 400 h for code development, trials, technical problems, etc.

III. RESULTS AND DISCUSSIONS

We present the results for calculated energies, oscillator strengths, and photoionization cross sections of Fe III in the three subsections below.

A. Energy levels

The first set of calculations is for the bound states of the system $e+$ Fe IV $\rightarrow$ Fe III. We obtain 805 LS bound states of Fe III, about four times more than observed in the laboratory [12,13], which lie below the first ionization threshold $3d^5(^6S)$. The total number of LS bound states is 2381, corresponding to $S_tL_t\pi_tn\iota$, where $S_tL_t\pi_t$ are the 49 target states and n goes from the ground state to 10 (11 for a few symmetries), and $0\leq\iota\leq9$. The states that lie above the first ionization threshold are usually quasibound autoionizing states, but may be bound states in pure LS coupling; we do not consider such states in the present work. All the observed 199 LS terms [12,13] are identified, indicating an adequate and reasonably complete representation for the wavefunction expansion Ψ. The National Institute of Standards and Technology table [12] contains 184 observed LS terms, while the recent measurements at Lund [13] find 15 additional LS terms. Of the 15 new observed terms, two states, $3d^6(^1D)$ and $3d\ ^44s(^1S)$, were predicted by our earlier calculations [4]. The calculated energy values are compared with the observed ones in Table II. The table lists the observed terms statistically averaged over the fine-structure levels. The calculated binding energy of the ground state of Fe III is 2.127 Ry, compared to the observed value of 2.2528 Ry, a difference of 5.6%. For most of the calculated energies the difference with the observed ones is within 7%; the largest difference is 21.4% for the $3d^5(a^2D)4p(^1F^o)$ state, which has a large percentage of admixture with other states. Inclusion of relativistic effects and higher configuration interactions should further improve the calculated energies, but considerably more computational effort would be needed for such *ab initio* calculations [4].

B. Oscillator strengths

The oscillator strengths, or the f values, of Fe III are obtained for the bound-bound transitions between all bound LS terms considered in this work. The total number of f values calculated is 249 584, corresponding to all states $S_tL_t\pi_tn\ell$, with $n\leq10$, and $0\leq\ell\leq9$. Of these 11 979 correspond to the states that lie below the ionization threshold. Currently available f values in the literature are from calculations by Sawey and Berrington [2], Kurucz and Peytremann [14], Fawcett [16], Biemont [15], and Ekberg [13]. Except for the first, which used the CC approximation for the OP work, all the rest of the references used variations of the Cowan's code [17] and agree well with each other. The present work will be compared with two of them through lifetime values. Although no measured f values are available for comparison, the lifetimes of four states of Fe III are available that were measured by Andersen, Petersen, and Biemont [18] The lifetime of a state j is the reciprocal of the sum of the transition probabilities, A_{ji}, to all lower states i. The transition probability A_{ji} between states i and j is related to the oscillator strength, f_{ij}, as $A_{ji} = 1/2\alpha^3(g_i/g_j)E_{ji}^2f_{ij}$, where α is the fine-structure constant, g_i and g_j are statistical weight factors, and E_{ji} is the transition energy. Table III presents the calculated lifetimes of the four states and compares with the measured ones. The present lifetimes are obtained from the calculated A values where the observed energy differences have been used for the transition energy instead of the calculated ones. In Table III, the Biemont-Kurucz-Peytremann (BKP) values correspond to the calculated lifetimes of Andersen, Petersen, and Biemont [18] where they used the f values for strong spectral lines calculated by Biemont [15], and weak lines calculated by Kurucz and Peytremann [14]. The present lifetimes of states z^5H^o and z^5G^o agree within experimental uncertainties of Andersen, Petersen, and Biemont, and for the state z^3H^o we agree with the BKP value, which is lower than the measured value. The present lifetime of state y^1I^o is lower than the measured value and the BKP value (the latter two agree with each other). The reason for the present lower value could be because of the relatively high angular momentum of the state $(L=6)$ for which it is often difficult to (i) include sufficient correlation from other states, and (ii) employ a sufficiently large R-matrix boundary to represent the bound-state wave function in the inner region for the high-L states.

C. Photoionization cross sections

1. Total photoionization cross sections

Total photoionization cross sections $\sigma_{\rm PI}$, including detailed autoionizing resonances, for the 805 bound states are obtained. By "total" we mean the sum of all individual cross sections leaving the residual ion in all of the coupled states of Fe IV (out of the 49 target states considered). Some specific results are presented that describe the important features in the photoionization cross sections.

As the energy of the photons increases, crossing various excited thresholds of the target, the photoionization cross

TABLE III. Lifetimes, τ (ns), of Fe III. The experimental values are from Andersen, Petersen, and Biemont [18].

State	Lifetime		
	Present	Expt	BKP
z^5H^o	1.9	1.6(0.3)	1.7
z^5G^o	2.1	2.1(0.3)	2.0
z^3H^o	1.4	1.7(0.2)	1.4
y^1I^o	0.82	1.2(0.2)	1.2

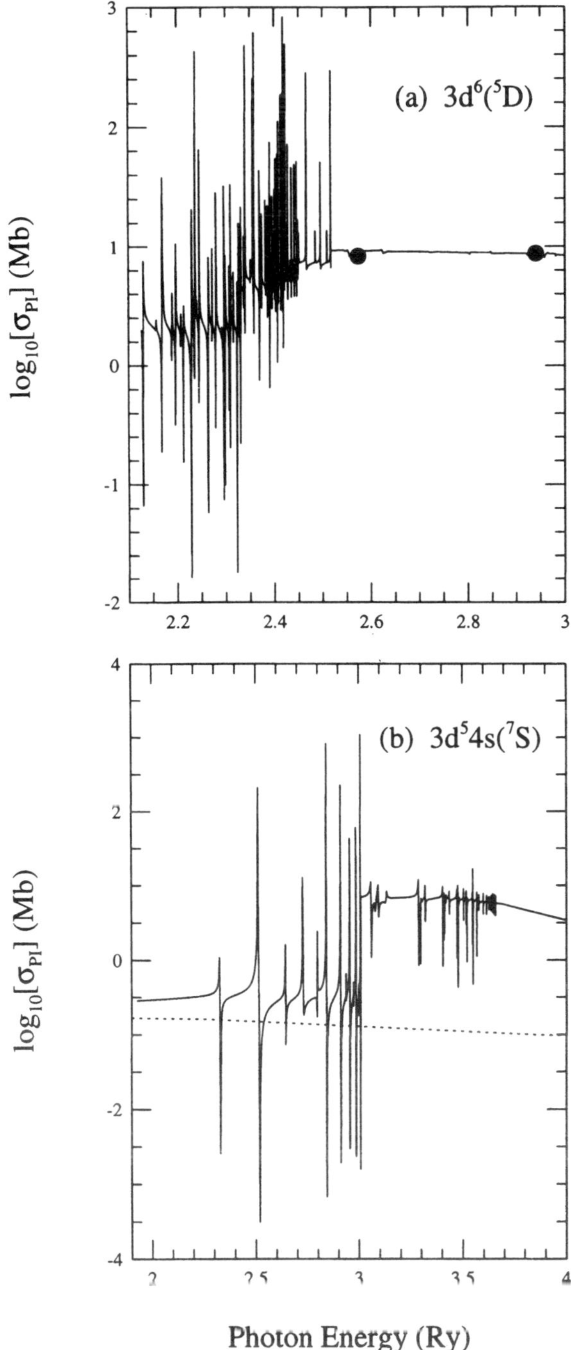

FIG. 1. Photoionization cross section σ_{PI} (solid lines) of (a) the ground state $3d^6(^5D)$ and (b) excited state $3d\ ^54s(^7S)$ of Fe III. The filled circles in (a) are from Ref. [19] and the dotted curve in (b) is from the 16-state CC calculations of Ref. [2].

sections show resonances due to the autoionizing Rydberg series of states belonging to each target state. The background cross sections due to direct ionization by absorption of photons can be affected considerably by these autoionizing resonances. Figure 1 presents such examples where the

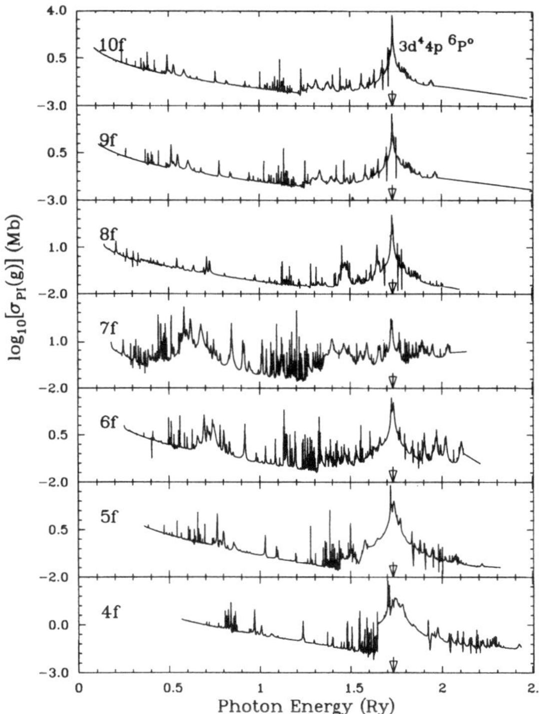

FIG. 2. Photoionization cross sections σ_{PI} of the Rydberg series of bound states of Fe III: $3d^5\,{}^6Snf(^5F^o)$, $4 \leq n \leq 10$. The PEC resonance, indicated by the arrow in each panel, corresponds to the dipole core transition $3d^5(^6S) \rightarrow 3d^4\,{}^5D4p(^6P^o)$ in Fe IV. (With the exception of the top panel, the upper limit is not shown on the y axis.)

photoionization cross sections for (a) the $3d^6(^5D)$ ground state, and (b) the $3d\ ^54s(^7S)$ state of Fe III are shown. In Fig. 1(a) the present ground-state cross sections show a large number of resonances in the low-energy region, whereas the higher-energy region is comparatively smooth since the coupling effects from the high-lying $3d\ ^44s$ and $3d\ ^44p$ states are weaker. These cross sections show patterns similar to the earlier OP data [2], except for the high-energy background, which falls slightly faster in the 49-state CC calculations (not shown here). The filled circles in the figure correspond to central-field cross sections [19], which, when extrapolated to the first ionization threshold, would give a threshold value 5 times higher than the present one, thus overestimating the cross sections in the near-threshold region considerably. In Fig. 1(b) the present cross sections (solid curve) for the $3d\ ^54s(^7S)$ state are compared with the earlier 16-state OP calculations of Sawey and Berrington [2] (dotted curve), and it is seen that their work misses all of the resonances because their CC eigenfunction expansion did not include higher states that couple to $3d\ ^54s(^7S)$.

The narrow resonances in the cross sections usually correspond to the Rydberg series of states $S_tL_t\pi_t\nu\ell$ belonging to the target state $S_tL_t\pi_t$, where ν is the effective quantum number. But, as mentioned earlier, there may be wide resonances in the cross sections due to PEC [6,11]. For this type of resonance the target goes through an allowed dipole transition from the ground state to a higher state, while the outer

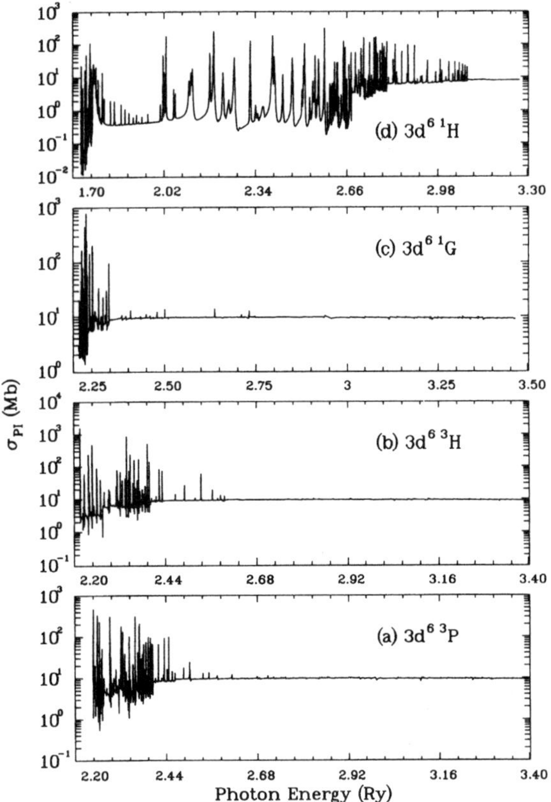

FIG. 3. Photoionization cross sections σ_{PI} for the metastable states (a) $3d^6(^3P)$, (b) $3d^6(^3H)$, (c) $3d^6(^1G)$, and (d) $3d^6(^1H)$ of Fe III.

electron remains a "spectator." This effect enhances the cross sections considerably, even by orders of magnitude. PEC resonances are more pronounced in the photoionization cross sections of excited bound states since the outer electron is loosely bound to the ion core. The large PEC resonance due to dipole-allowed core excitation of ground state $3d^5(^6S)$ of Fe IV to the excited $3d\,^44p(^6P^o)$ state is seen in Fig. 2. The figure shows the photoionization cross sections of seven excited states of Fe III belonging to the Rydberg series $3d^5\,^6Snf(^5F^o)$, where $n=4-10$. The arrow in each panel of this figure points to the position of the excited $^6P^o$ state of the core Fe IV, which remains at the same photon energy for each photoionization state of Fe III. This figure also exhibits the very nonhydrogenic behavior of the excited-state cross sections due to the presence of the resonances, especially due to the PEC resonance.

Figure 3 presents σ_{PI} for several metastable states of Fe III, (a) $3d^6(^3P)$, (b) $3d^6(^3H)$, (c) $3d^6(^1G)$, and (d) $3d^6(^1H)$. All four states show a large number of resonances near the threshold, indicating higher rates of ionization in this region while the high-energy regions show smooth background except for the $3d^5(^1H)$ state. Owing to their comparatively long lifetimes, photoionization and recombination from the metastable states are likely to be significant in the calculation of level populations and spectral line intensities from collisional-radiative models.

2. Partial photoionization cross sections and electron-ion recombination

This work also reports partial photoionization cross sections, leaving the residul ion in the ground state, of a large number of bound states of Fe III. These are the 239 septet and quintet states of Fe III that can couple to target ground state $3d^5(^6S)$. The partial photoionization cross sections are important in the determination of state-specific populations, recombination rates, etc. The present cross sections will be employed for electron-ion recombination of Fe III in a unified treatment for total electron-ion recombination [7]. The recombination cross sections are related to the photoionization cross sections through detailed balance (the Milne relation). Inclusion of autoionizing resonances in the cross sections thus accounts for both the radiative recombination and the dielectronic recombination (this unified treatment is complemented by calculations for the purely dielectronic recombination from high-n resonances). In the ionization balance equations under plasma equilibrium, one usually employs the total photoionization of the ground state on the one hand, and the inverse process of recombination into all bound states of the ion on the other; thus the partial photoionization cross sections into specific core states of the residual ion are needed.

The partial photoionization cross sections for Fe III, leaving the residual ion Fe IV in various excited states, are obtained for the ^5D ground state (Fig. 4). The bottom panel in

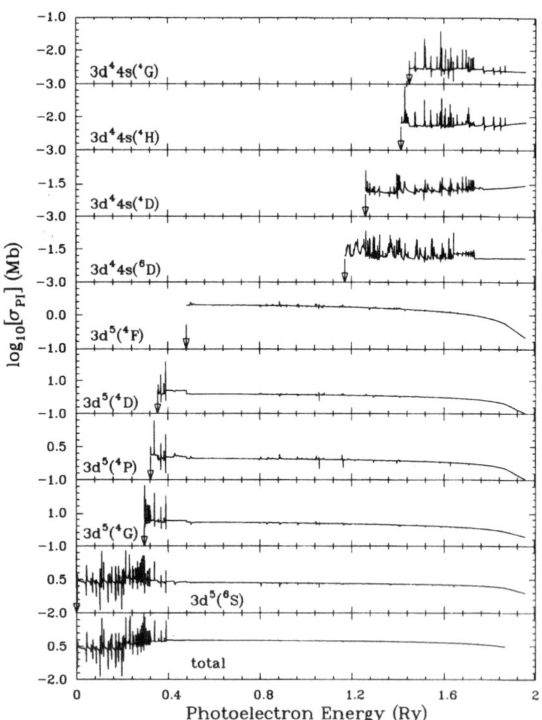

FIG. 4. Partial photoionization cross sections for the ground $3d^6(^5D)$ state of Fe III, leaving the Fe IV core in various states, $3d^5(^6S)$, $3d^5(^4G)$, $3d^5(^4P)$, $3d^5(^4D)$, and so on, as specified in the panels. The bottommost panel presents the total photoionization cross section. The arrows point to threshold energies of the target states.

Fig. 4 presents the *total* photoionization cross section for the state, whereas the upper ones presents the *partial* photoionization cross sections leaving the target Fe IV in states $3d^5(^6S)$, $3d^5(^4G)$, $3d^5(^4P)$, $3d^5(^4D)$ and so on, as specified in the panels. The arrows point to the threshold energies of these target states. It is evident from the figure that the partial cross section that dominates the total corresponds to that of the first ionization threshold of Fe IV, $3d^5(^6S)$. Nonetheless, the resonances in the higher partial cross sections might also contribute significantly to the probability for ionization into those levels.

IV. CONCLUSION

A comprehensive study of radiative processes in a complex atomic system, such as the ions of the iron group elements, reveals important physical phenomena characteristic of strongly correlated systems. Large-scale computations are reported for the detailed study of radiative processes in Fe III, with computed energy levels, oscillator strengths, and photoionization cross sections, in an *ab initio* manner. We consider over four times the number of bound states than have been observed in the laboratory. Good agreement is found between the calculated energies and all those that are observed. The accuracy of the oscillator strengths is inferred from the reasonably good agreement with the measured lifetimes of four bound states of Fe III. Total and partial photoionization cross sections (with final states specified) are computed for all the bound states and the resonance structures are delineated in detail. The parameters reported in the present work should be of higher accuracy than the previous OP data for Fe III. It is expected that the new photoionization data in particular will yield more accurate photoionization models than present models, which employ earlier photoionization data. The partial photoionization cross sections will be employed for the calculation of total election-ion recombination rate coefficients of Fe III.

The complete radiative data of the present work for energy levels, oscillator strengths, and photoinization cross sections of Fe III will be available through the Opacity Project electronic database, TOPbase [20].

ACKNOWLEDGMENTS

I would like to thank Professor Anil K. Pradhan for contributions (supported by NSF Grant No. PHY-9421898), Manuel Bautista for providing the Fe IV target wave functions, A. F. Robey and Dr. J. R. Fuhr from NIST for providing their most recent compilation of observed energy values of Fe III and Fe IV electronically, and Dr. J. O. Ekberg for his recently measured energy values of some Fe III states. This work was supported by NASA Grants No. NAGW-3315 and No. NAS-32643. The computational work was carried out on the Cray YMP computer at the Ohio Supercomputer Center.

[1] M.J. Seaton, J. Phys. B **20**, 6363 (1987).
[2] P.M.J. Sawey and K.A. Berrington, J. Phys. B **25**, 1451 (1992).
[3] D.G. Hummer, K.A. Berrington, W. Eissner, A.K. Pradhan, H.E. Saraph, and J.A. Tully, Astron. Astrophys. **279**, 298 (1993).
[4] S.N. Nahar and A.K. Pradhan, J. Phys. B **27**, 429 (1994).
[5] A.M. Bautista and A.K. Pradhan, J. Phys. B **28**, L173 (1995).
[6] Yu Yan and M.J. Seaton, J. Phys. B **20**, 6409 (1987).
[7] S.N. Nahar and A.K. Pradhan, Phys. Rev. A **49**, 1816 (1994); Astrophys. J. **447**, 966 (1995).
[8] W. Eissner, M. Jones, and N. Nussbaumer, Comput. Phys. Commun. **8**, 270 (1974).
[9] K.A. Berrington, P.G. Burke, K. Butler, M.J. Seaton, P.J. Storey, K.T. Taylor, and Yu Yan, J. Phys. B **20**, 6379 (1987).
[10] S.N. Nahar, Phys. Scr. **293**, 967 (1995).
[11] S.N. Nahar and A.K. Pradhan, Phys. Rev. A **44**, 2935 (1991).
[12] J. Sugar and C. Corliss, J. Phys. Chem. Ref. Data **14**, Suppl. 2 (1985).
[13] J.O. Ekberg, Astron. Astrophys. Suppl. **101**, 1 (1993).
[14] R.L. Kurucz and E. Peytremann, Smithsonian Astrophysical Observatory Special Report No. 362, 1975 (unpublished).
[15] E. Biemont, J. Quantum Spectrosc. Radiat. Transfer. **16**, 137 (1976).
[16] B.C. Fawcett, At. Data Nucl. Data Tables **41**, 181 (1989).
[17] R.D. Cowan, *The Theory of Atomic Structure and Spectra* (University of California Press, Berkeley, California, 1981).
[18] T. Andersen, P. Petersen, and E. Biemont, J. Quantum. Spectrosc. Radiat. Transfer. **17**, 389 (1977).
[19] R.F. Reilman and S.T. Manson, Astrophys. J. Suppl. **40**, 815 (1979).
[20] W. Cunto, C. Mendoza, F. Ochsenbein, and C.J. Zeippen, Astron. Astrophys. **275**, L5 (1993).

J. Phys. B: At. Mol. Opt. Phys. 28 (1995) L173–L179. Printed in the UK

LETTER TO THE EDITOR

Photoionization of neutral iron

Manuel A Bautista and Anil K Pradhan
Department of Astronomy, The Ohio State University, Columbus, Ohio 43210, USA

Received 16 December 1994

Abstract. Accurate calculations are reported for the photoionization of the ground and the excited states of Fe I with an extensive coupled channel expansion using the *R*-matrix method. A total of 52 *LS* terms of the core ion Fe II dominated by configurations $3d^64s$, $3d^7$, $3d^64p$, $3d^54s^2$, $3d^54s4p$ are included; these eigenstates thereby account for the photoionization of the outer 4s subshell as well as the open inner 3d subshell of the ground state $3d^64s^2(^5D)$ of Fe I. The coupling of all relevant photoionization channels give rise to extensive autoionization structures and ensure that no large and sharp discontinuity at the 3d inner-shell ionization edge is present. The new results differ by up to several orders of magnitude from earlier data in central field type approximations that are currently employed in plasma modelling of astrophysical and laboratory sources.

While the low ionization stages of iron are of considerable interest in laboratory and astrophysical plasmas, accurate atomic calculations for these species are particularly difficult primarily owing to the complex electron–electron correlation effects involved. A wavefunction representation of these systems requires a large configuration expansion for the bound or the free (continuum) states. Furthermore, a large number of atomic states needs to be considered in large-scale practical applications such as the calculation of opacities in astrophysical sources (Seaton 1987, Seaton *et al* 1994). For example, photoionization of Fe I is of interest in, and is directly applicable to, the analysis solar UV opacity (e.g. Bell *et al* 1994). It might be expected that the presence of autoionizing resonances in particular might be a major contributor to observed features in the absorption spectra. Previous calculations have been carried out in central field type approximations (e.g. Reilman and Manson 1979, Verner *et al* 1993); however, these do not incorporate the coupling effects and resonances that are known to be of considerable importance in the photoionization of strongly correlated systems such as the low ionization stages of iron (e.g. Nahar and Pradhan 1994). Earlier *R*-matrix calculations were carried out for Fe I by Sawey and Berrington (1992); however, their calculations included only the terms dominated by the ground configuration of Fe II and therefore do not accurately represent the coupling effects (for example, their calculations did not obtain the ground state of Fe I).

It is necessary to allow for channel couplings among many states of the target (or the core) ion Fe II for at least three reasons. First, the configuration interaction (CI) between the numerous low-lying *LS* terms is important. Second, it is known that strong dipole couplings between opposite parity terms within the target ion give rise to large photoexcitation-of-core (PEC) resonances at corresponding incident photon frequencies (Yu and Seaton 1987). This implies that terms dominated by even-parity configurations $3d^64s$ and $3d^7$ and the odd-parity configuration $3d^64p$ of Fe II should be included in the eigenfunction expansion. Third, the photoionization channels corresponding to the ionization of the open inner 3d shell are likely

0953-4075/95/060173+07$19.50

to contribute considerably to the total cross section. This requires that terms dominated by the configurations $3d^54s^2$ and $3d^54s4p$ of Fe II should also be included. Taking these points into account, we consider the photoionization of the ground state of Fe I as

$$h\nu + \text{Fe I}(3d^64s^2;\ ^5D) \rightarrow e + \text{Fe II}(3d^64s + 3d^7 + 3d^64p + 3d^54s^2 + 3d^54s4p). \quad (1)$$

The complexity and the extent of the eigenfunction expansion may be inferred from tables 1 and 2. Table 1 lists all the coupled LS terms in the present calculations. The atomic structure program SUPERSTRUCTURE (Eissner *et al* 1974) was used to optimize the target wavefunctions. In general, the target energies obtained were within 5–10% of the experimentally observed energies for the 52 terms, with few exceptions. Table 2 gives the calculated energies for the 30 LS terms of Fe II that couple with the ground-state symmetry ^5D of Fe I. A further indication of the accuracy of the target wavefunctions was the good agreement (< 10%) between the length and the velocity dipole oscillator strengths between the $3d^4s$, $3d^7$ terms and the $3d^64p$ terms of Fe II. These terms constitute the first sum in the usual close-coupling (CC) expression for the total wavefunction for the e + ion system

$$\Psi(E; SL\pi) = A\sum_i \chi_i\theta_i + \sum_j c_j\Phi_j \quad (2)$$

where χ_i is the target ion wavefunction in a specific state S_iL_i and θ_i is the wavefunction for the free electron. The second summation on the right-hand side represents short-range correlation functions that assumes considerable importance in order to obtain accurate e+ion wavefunctions, but which may also be problematic in that pseudoresonances may result if the two summations are not consistent (Berrington *et al* 1987). In the present work it was necessary to optimize the set of Φ_j functions for Fe I in a very careful manner so as to obtain accurate bound-state energies and bound–bound oscillator strengths for Fe I. In order that the calculations be computationally tractable one tries to keep the bound channel set as small as possible. However, for Fe I we found it essential to include rather a large number of $(N+1)$-electron functions and, furthermore, it was necessary to divide the overall calculations into groups of total (e + ion) symmetries $Sl\pi$ according to their multiplicities, i.e. $(2S+1) = 1, 3, 5$ and 7. For each set of $(2S+1)$ multiplicity we consider total angular momenta $L = 0$–7 for both parities; a total of 52 $SL\pi$s corresponding to the target states in table 1 and partial waves $\ell = 0$–7. The atomic structure program SUPERSTRUCTURE (Eissner and Nussbaumer 1974) was used to optimize the target wavefunctions computed using a one-electron orbital set obtained in a scaled Thomas–Fermi–Dirac potential. The data and the details related to the optimization of the N-electron target, and the $(N+1)$-electron functions, are omitted from this letter for brevity (these will be presented elsewhere).

Table 1. Fe II target terms in the eigenfunction expansion.

$3s^23p^63d^64s$:	$a^6D, a^4D, b^4P, a^4H, b^4F, a^4G, b^2P, b^2H, a^2F, b^2G, b^4D, a^2I, c^2G, b^2D, a^2S, c^2D$
$3s^23p^63d^7$:	$a^4F, a^4P, a^2G, a^2P, a^2H, a^2D, b^2F$
$3s^23p^63d^54s^2$:	$a^6S, b^4G, d^4P, c^4D, {}^4F$
$3s^23p^63d^64p$:	$z^2D^o, z^6F^o, z^6P^o, z^4F^o, z^4D^o, z^4P^o, z^4S^o, y^4P^o, z^4G^o, z^4H^o, z^2D^o, z^4I^o, y^4D^o, z^2G^o, y^4F^o, z^2I^o, x^4D^o, y^4G^o, z^2F^o, y^2G^o, z^2P^o, z^2H^o$
$3s^23p^63d^54s4p$:	z^8P^o, y^6P^o

Table 2. Energies (in Rydberg units) of terms coupled to photoionization of the ground state $3d^6 4s^2 (^5D)$ of Fe I. The experimental energies are from Sugar and Corliss (1985).

	Term		E (expt)	E (calc)		Term		E (expt)	E (calc)
1	$(^5D)4s$	$a\,^6D$	0.0	0.0	16	$(^5D)4p$	$z\,^4P^o$	0.4012	0.4265
2	$3d^7$	$a\,^4F$	0.0186	0.0182	17	$3d^5 4s^2$	$b\,^4G$	0.5462	0.4907
3	$(^5D)4s$	$a\,^4D$	0.0953	0.0720	18	$3d^5 4s^2$	$d\,^4P$	0.5961	0.5199
4	$3d^7$	$a\,^4P$	0.1302	0.1203	19	$(^3P)4p$	$z\,^4S^o$	0.5762	0.5399
5	$(^3P)4s$	$b\,^4P$	0.2303	0.1914	20	$3d^5 4s^2$	$c\,^4D$	0.6237	0.5463
6	$(^3H)4s$	$a\,^4H$	0.2242	0.1918	21	$(^3P)4p$	$y\,^4P^o$	0.5805	0.5504
7	$(^3F)4s$	$b\,^4F$	0.2452	0.2040	22	$(^3H)4p$	$z\,^4G^o$	0.5729	0.5504
8	$3d^5 4s^2$	$a\,^6S$	0.2461	0.2087	23	$(^3H)4p$	$z\,^4H^o$	0.5760	0.5516
9	$(^3G)4s$	$a\,^4G$	0.2761	0.2310	24	$(^3H)4p$	$z\,^4I^o$	0.5756	0.5565
10	$(^3D)4s$	$b\,^4D$	0.3480	0.2825	25	$(^3P)4p$	$y\,^4D^o$	0.5849	0.5646
11	$(^5D)4p$	$z\,^6D^o$	0.3515	0.3490	26	$(^3F)4p$	$y\,^4F^o$	0.5955	0.5625
12	$(^5D)4p$	$z\,^6F^o$	0.3732	0.3805	27	$(^6S)4s4p$	$y\,^6P^o$	0.6379	0.5620
13	$(^5D)4p$	$z\,^6P^o$	0.3798	0.3886	28	$(^3F)4p$	$x\,^4D^o$	0.6002	0.5722
14	$(^5D)4p$	$z\,^4F^o$	0.3872	0.4037	29	$(^3F)4p$	$y\,^4G^o$	0.6020	0.5791
15	$(^5D)4p$	$z\,^4D^o$	0.3807	0.4039	30	$3d^5 4s^2$	^4F	0.7699	0.6656

The first step in the photoionization calculations is for the bound-state energies and wavefunctions of Fe I. Table 3 gives the computed energies for a few of these states compared to experimental values. The general agreement is of the order of 1–5%. Figure 1(a) shows the photoionization cross section for the $3d^6 4s^2 (^5D)$ ground state up to incident photon energies above all ionization thresholds of Fe II included in the CC expansion. Figure 1(b) presents the same cross section from another set of calculations excluding the $3d^5 4s^2 (^6S$, ^4G, ^4D and $^4F)$ terms of Fe II that correspond to the ionization of the inner 3d shell. In figure 1(a) the thresholds corresponding to the first term dominated by configurations $3d^6 4p(^6D^o)$, $3d^5 4s^2 (^4G)$ and $3d^5 4s4p(^6P^o)$ are marked. Comparing with figure 1(b) that excludes the $3d^5 4s^2$ terms it is seen that there is strong enhancement in both the background and the resonances structures with the onset of additional inner-shell ionization channels into $3d^5 4s^2$ terms of Fe II. In particular the presence of large resonances just above the $3d^6 4p(^6D^o)$ term is noted; this is clearly due to strong coupling to the terms corresponding to the ionization of the 3d shell. As expected there is no resonance structure above the highest term included in the eigenfunction expansion $3d^5 4s^2 (^4F)$ (table 3).

Table 3. Bound-state energies of Fe I. The experimental energies are from Sugar and Corliss (1985).

Term		E (expt)	E (calc)	Term		E (expt)	E (calc)
$3d^6 4s^2$	$a\,^5D$	0.5771	0.5697	$3d^7 4p$	$z\,^5G^o$	0.2606	0.2431
$3d^7 4s$	$a\,^5F$	0.5128	0.4988	$3d^6 4s4p$	$y\,^5P^o$	0.2434	0.2204
$3d^7 4s$	$a\,^5P$	0.4197	0.3915	$3d^6 4s4p$	$x\,^5D^o$	0.2166	0.2100
$3d^6 4s4p$	$z\,^7P^o$	0.3616	0.3736	$3d^6 4s4p$	$x\,^5F^o$	0.2104	0.2119
$3d^7 4s$	$a\,^1G$	0.3569	0.3519	$3d^6 4s4p$	$e\,^7D$	0.1869	0.1851
$3d^6 4s4p$	$z\,^5D^o$	0.3425	0.3581	$3d^7 4p$	$y\,^5S^o$	0.1752	0.1846
$3d^6 4s4p$	$z\,^5F^o$	0.3328	0.3431	$3d^6 4s4p$	$x\,^5G^o$	0.1637	0.1568
$3d^7 4s$	$a\,^1P$	0.3298	0.3354	$3d^7 4p$	$w\,^5P^o$	0.1593	0.1748
$3d^7 4s$	$a\,^1D$	0.3201	0.3166	$3d^7 4p$	$^5D^o$	0.1536	0.1469
$3d^6 4s4p$	$z\,^5P^o$	0.3135	0.3292	$3d^7 5s$	$e\,^5F$	0.1479	0.1442

Figure 1(a) also shows a comparison with the results of Reilman and Manson (1979) in the central field approximation, and the more recent calculations of Verner *et al* (1993) in

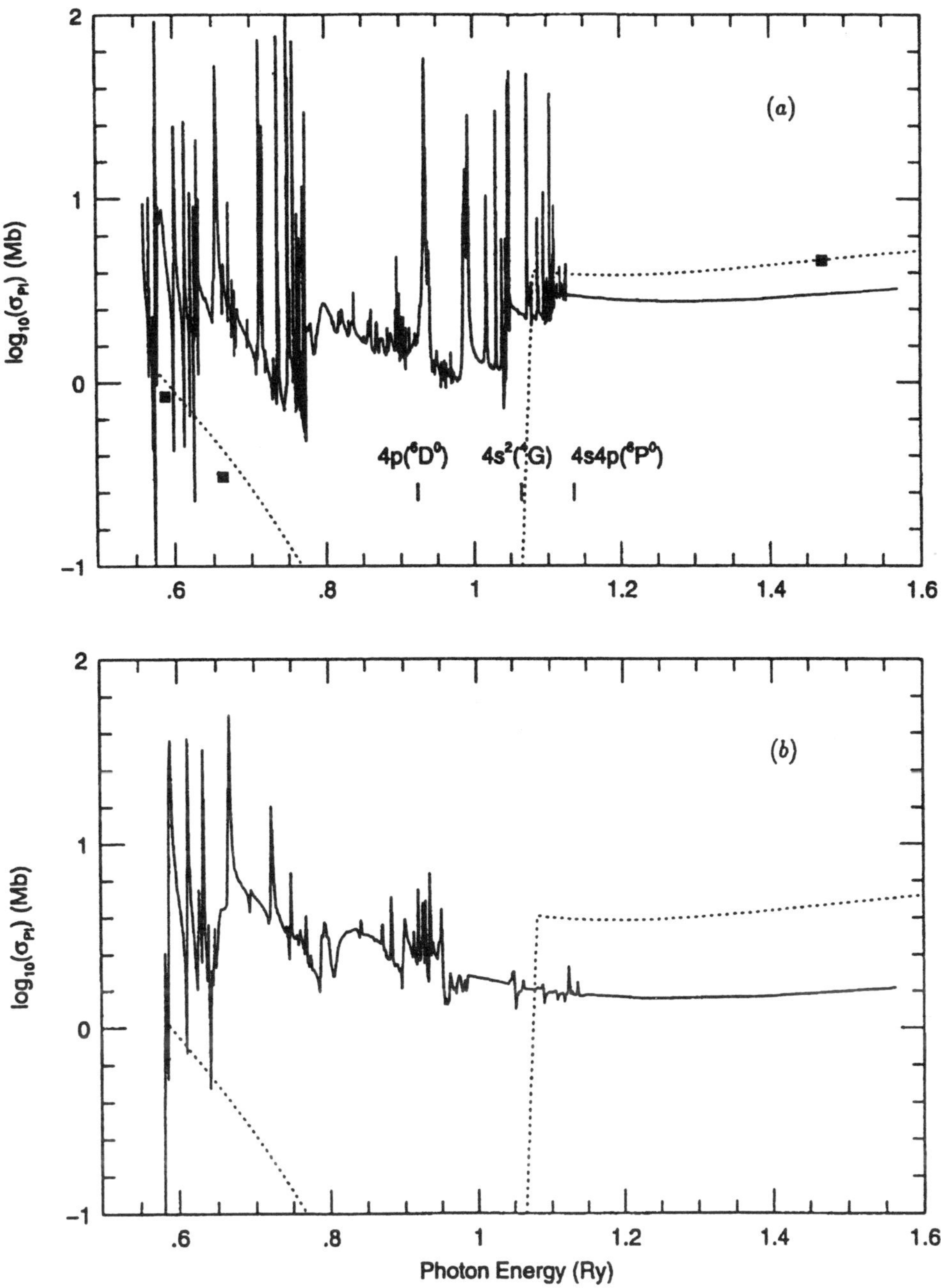

Figure 1. (a) Photoionization cross section of the ground state $3d^64s^2(^5D)$ of Fe I: full curve, present result; broken curve, Verner *et al* (1993); filled squares, Reilman and Manson (1979). (b) Cross section without coupling of the $3d^54s^2$ target Fe II terms contributing to the ionization of the inner 3d subshell.

the Dirac–Hartree–Slater approximation. Both of these approximations neglect the complex correlation effects, such as inter-channel coupling, that are responsible for the large cross sections in the near-threshold region. The two earlier results agree very well with each other. The present coupled channel results are *up to several orders of magnitude* higher in the energy region below about 1 Rydberg. A remarkable feature of the comparison in figure 1(a) is the large discontinuous jump in the earlier results, over three orders of magnitude, due to the onset of inner-shell ionization of the 3d subshell. The present results exhibit no such discontinuity since the ionization thresholds up to inner-shell ionization are explicitly coupled together. The total photoionization cross section is therefore continuous across all

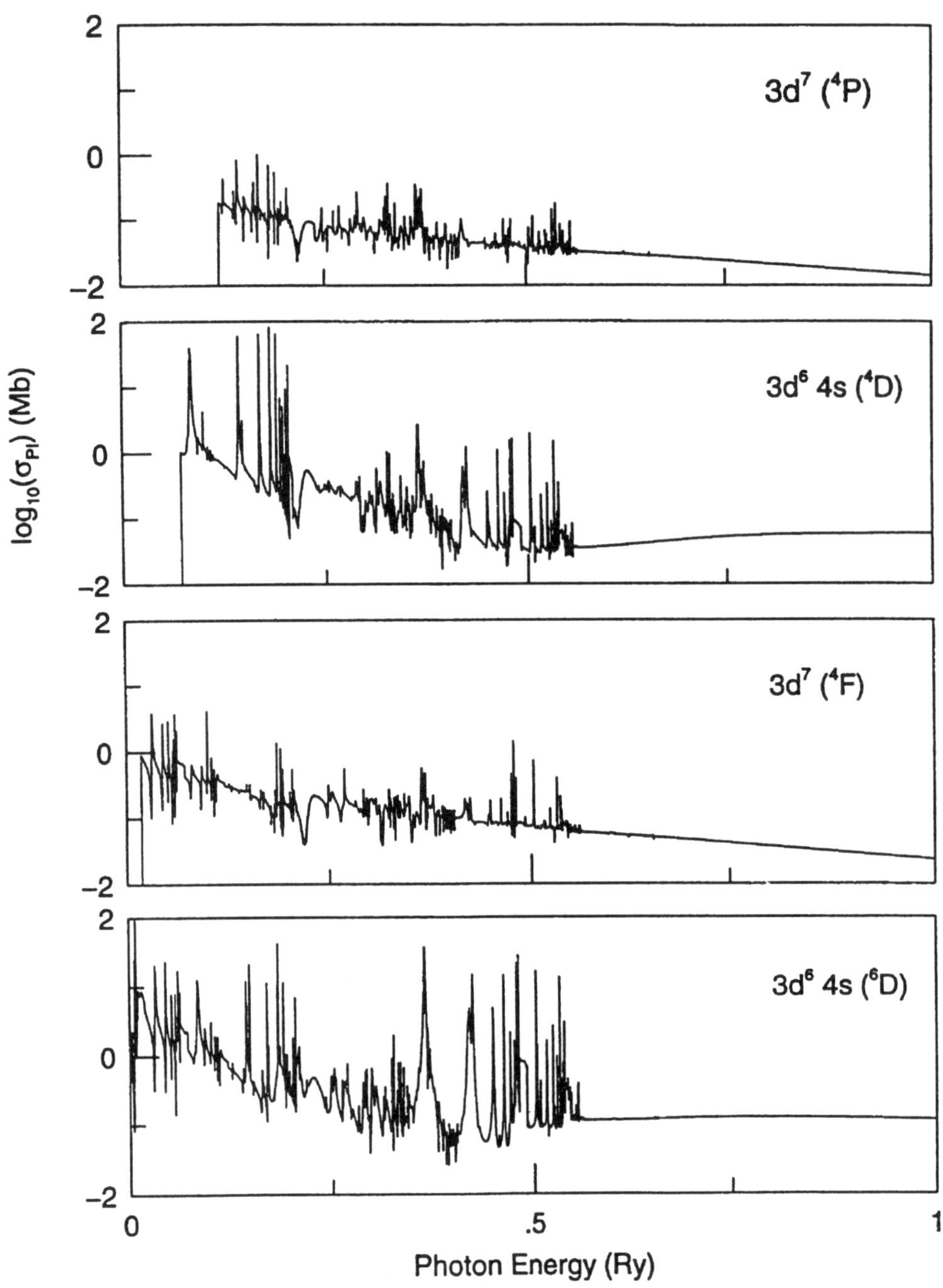

Figure 2. Partial photoionization cross sections into the ground and the excited states of Fe II.

thresholds in the eigenfunction expansion (tables 1 and 2). This might be regarded as a general feature of photoionization cross sections in the high-energy regime, in that the inner-shell photoionization 'edges' often seen in the central field type cross sections should not, in fact, be sharp discontinuities, but should consist of a relatively smoothly varying background, superimposed with autoionizing resonances, if the appropriate thresholds belonging to the high-lying states are considered. The sharp ionization edges are thus artifacts introduced in the cross sections owing to the neglect of coupled states in the high-energy region. It is clear from figure 1(a) that the amount of photoabsorption missing from the earlier works is considerable, and the integrated photoionization rate (as inferred from the area under the full and broken curves) would be in substantial error. In practice, however, it is computationally very expensive to couple a very large number of excited states for heavy atoms so as to

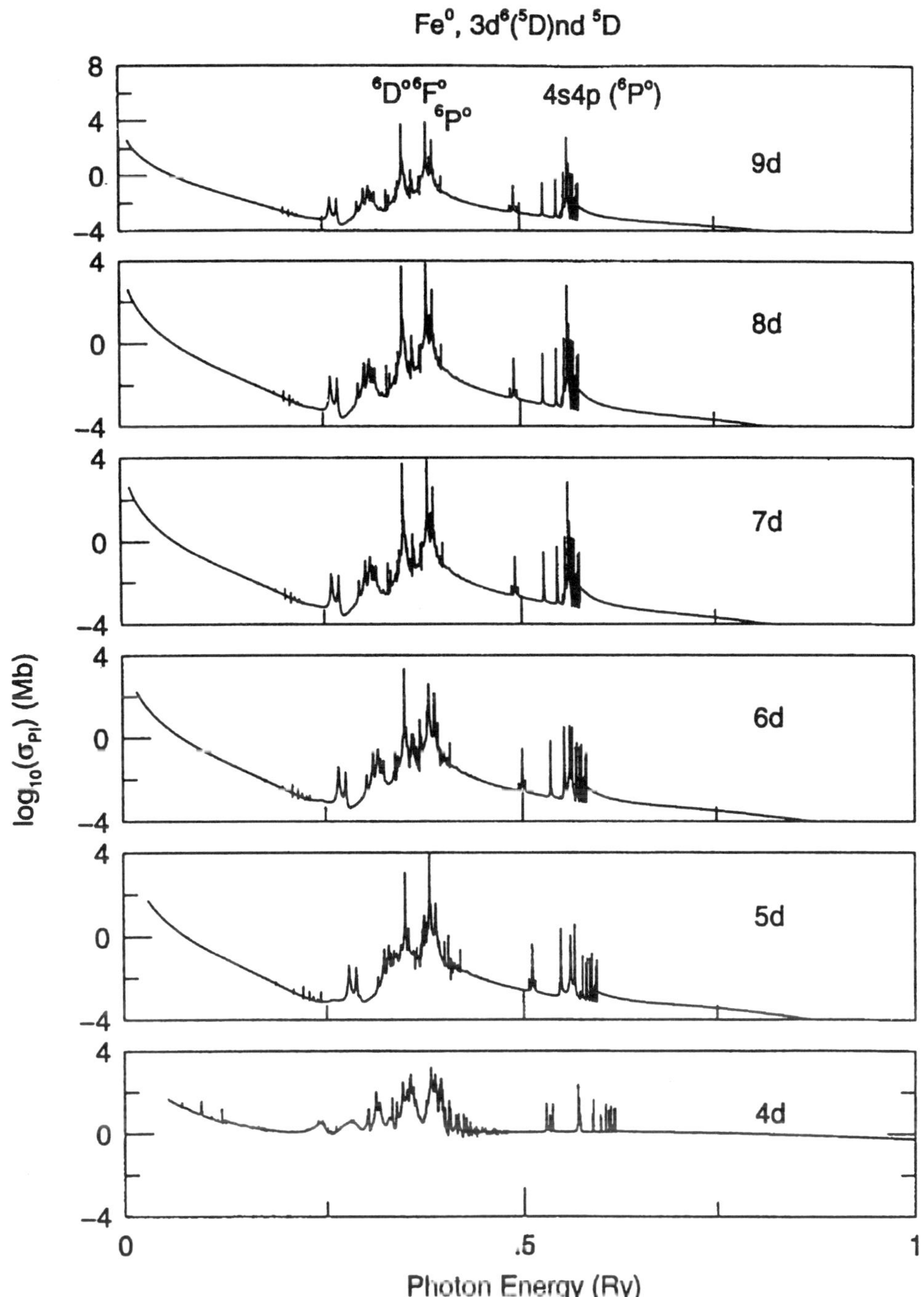

Figure 3. Photoionization of Fe I bound states in a Rydberg series showing the PEC resonance features.

avoid these inner-shell discontinuities. On the other hand, the magnitude of the jumps decreases with energy (i.e. successive inner-shell ionizations) and is usually much less than the large one seen for Fe I in figure 1(*a*) (Verner *et al* 1993). Furthermore, at higher photon energies (> 1 Rydberg), when the coupling of the ejected electron with the core ion is weak, the central field results approach the present cross sections. The convergence of all results in the high-energy region, where one would expect the central field approximation to yield reasonably accurate cross sections, also indicates that the CC results have converged in this energy region (this is not the case in figure 1(*b*) where the discrepancy with previous works at high energies is considerably larger).

In practical applications regarding non-LTE spectral models, it is often important to determine accurately the level population in the *excited* levels of the residual ion following photoionization. This requires that not only should the total but also the *partial* photoionization cross sections into the excited states of the ionized ion be calculated. Therefore, in the present case it is of interest to obtain the partial cross sections for photoionization of the ground state of Fe I into at least the lowest few terms of Fe II. Figure 2 gives these partial cross sections for

$$h\nu + \mathrm{Fe\ I}[3\mathrm{d}^6 4\mathrm{s}^2(^5\mathrm{D})] \rightarrow \mathrm{e} + \mathrm{Fe\ II}[3\mathrm{d}^6 4\mathrm{s}(^6\mathrm{D}), 3\mathrm{d}^7(^4\mathrm{F}), 3\mathrm{d}^6 4\mathrm{s}(^4\mathrm{D}), 3\mathrm{d}^7(^4\mathrm{P})]. \qquad (3)$$

The four metastable terms of Fe II considered above are important in astrophysical sources as these give rise to the prominent infrared lines that provide extremely useful diagnostics of electron density and iron abundance (Bautista *et al* 1994).

The PEC process referred to earlier is demonstrated in figure 3, which displays the photoionization cross sections of Fe I bound states in a Rydberg series, $3\mathrm{d}^6(^5\mathrm{D})4\mathrm{s}n\mathrm{d}(^5\mathrm{D})$ with $n = 4$–9. At the Fe II target thresholds $3\mathrm{d}^6 4\mathrm{p}(\mathrm{z}\,^6\mathrm{D}^\mathrm{o}, \mathrm{z}\,^6\mathrm{F}^\mathrm{o}, \mathrm{z}\,^6\mathrm{P}^\mathrm{o})$ and $3\mathrm{d}^5(^6\mathrm{S})4\mathrm{s}4\mathrm{p}(\mathrm{y}\,^6\mathrm{P}^\mathrm{o})$, the incident photon energies equal the energies of the strong dipole transitions from the ground state $3\mathrm{d}^6 4\mathrm{s}(\mathrm{a}\,^6\mathrm{D})$ and large autoionizing resonances are formed, enhancing the effective cross section by several orders of magnitude above the otherwise decreasing background. The prominent peaks shown in figure 3 correspond to these dipole transition energies

$$\mathrm{a}\,^6\mathrm{D} \rightarrow \mathrm{z}\,^6\mathrm{D}^\mathrm{o}, \mathrm{z}\,^6\mathrm{F}^\mathrm{o}, \mathrm{z}\,^6\mathrm{P}^\mathrm{o}, \mathrm{y}\,^6\mathrm{P}^\mathrm{o}.$$

While this letter presents results for the photoionization of a few states, calculations are in progress for all bound states of Fe I with effective quantum number $\leqslant 10.0$, concurrently with calculations for all resulting dipole length and velocity oscillator strengths. The dataset will be employed in improved calculations of low-temperature astrophysical opacities ($\log(T) < 4.5$) dominated by low-charge iron ions (Seaton *et al* 1994).

This work was supported in part by a grant for the Iron Project from the US National Science Foundation (PHY-9115057). The computational work was carried out on the Cray Y-MP at the Ohio Supercomputer Center in Columbus, Ohio.

References

Bautista M A, Pradhan A K and Osterbrock D O 1994 *Astrophys. J. Lett.* **432** L135
Bell R A, Paltoglou G and Tripicco M J 1994 *Mon. Not. R. Astron. Soc.* **268** 771
Berrington K A, Burke P G, Butler K, Seaton M J, Storey P J, Taylor K T and Yu Yan 1987 *J. Phys. B: At. Mol. Phys.* **20** 6379
Eissner W, Jones M and Nussbaumer H 1974 *Comput. Phys. Commun.* **8** 270
Nahar S N and Pradhan A K 1994 *J. Phys. B: At. Mol. Opt. Phys.* **27** 429
Reilman R F and Manson S T 1979 *Astrophys. J. Supp. Ser.* **40** 815
Sawey P M J and Berrington K A 1992 *J. Phys. B: At. Mol. Opt. Phys.* **25** 1451
Seaton M J 1987 *J. Phys. B: At. Mol. Phys.* **20** 6363
Seaton M J, Yu Y, Mihalas D and Pradhan A K 1994 *Mon. Not. R. Astron. Soc.* **266** 805
Sugar J and Corliss C 1985 *J. Phys. Chem. Ref. Data* **14** Suppl. No 2
Verner D A, Yakovlev D G, Band I M and Trzhaskovskaya M B 1993 *At. Data Nucl. Data Tables* **55** 233
Yu Yan and Seaton M J 1987 *J. Phys. B: At. Mol. Phys.* **20** 6409

Astron. Astrophys. 275, L5–L8 (1993)

Letter to the Editor

TOPbase at the CDS

W. Cunto[1], C. Mendoza[1,*], F. Ochsenbein[2], and C.J. Zeippen[3]

[1] IBM Venezuela Scientific Center, P.O. Box 64778, Caracas 1060A, Venezuela
[2] CDS – Observatoire Astronomique, 11 Rue de l'Université, F-67000 Strasbourg, France
[3] UPR 261 du CNRS et DAMAp, Observatoire de Paris, F-92195 Meudon, France

Received April 2, accepted April 27, 1993

Abstract. TOPbase, the Opacity Project atomic database, has been set up at the Centre de Données Astronomiques de Strasbourg (CDS), France, for general access via Internet. This database contains accurately calculated energy levels, f-values and photoionisation cross sections for astrophysically abundant ions. We briefly describe the physical model and computational method used, the atomic data specifications, the database management system and the network access arrangements. The new facility should be of value to the astronomical community.

Key words: Atomic data

1. Introduction

The *Opacity Project* (OP) refers to an international collaboration that was formed in 1984 to re-estimate stellar envelope opacities in terms of atomic data computed by *ab initio* methods (Seaton 1987). This effort was a response to a plea by Simon (1982) for a revision of the standard metal opacities (see Cox & Tabor 1976, Huebner 1985, Weiss et al. 1990) in order to solve long-standing problems related to pulsating stars. It has involved research groups from France, Germany, the United Kingdom, the United States and Venezuela. The massive sets of atomic data required in such opacity calculations, namely, energy levels, oscillator strengths, photoionisation cross sections, line-broadening parameters, etc., must fulfill two important requisites: (i) completeness, in the sense that all contributing atomic processes and astrophysically abundant ions must be considered and (ii) accuracy. With respect to the latter, one of the main challenges taken up by the OP was to employ state-of-the-art atomic computational methods. The new metal opacities resulting from the OP work are indeed showing increases as large as a factor of 3 in some astrophysically interesting regimes (Yu Yan 1992, Seaton 1992, 1993), in agreement with a recent independent calculation (Rogers & Iglesias 1992 and references therein). Furthermore, the statistical accuracy and completeness of the OP atomic data sets are superior to those ever reached in the field (Seaton et al. 1992, Mendoza 1992).

Send offprint requests to: C.J. Zeippen
*Visiting Fellow, IVIC, Venezuela

One of the main concerns of the OP team is that of data accessibility. An efficient database management system (DBMS) was therefore developed by Cunto & Mendoza (1992a), paying particular attention to time/space performance, query capabilities, portability and low cost. The complete package, OP data and DBMS, is usually referred to as TOPbase, and has been recently installed at the Centre de Données Astronomiques de Strasbourg (CDS) for general Internet use. The CDS, in cooperation with other world data centers, is committed to the implementation and maintenance of extensive databases for astronomical studies.

The present letter briefly describes the physical model and computational methods used, the specifications of the atomic data, the main features of the TOPbase DBMS and the network access arrangements available at the CDS, so as to encourage the use of this new facility.

2. Method and physical model

OP calculations are based on the close-coupling method of scattering theory (Burke & Seaton 1971), where an N-electron ionic system is described in terms of an $(N-1)$-electron parent ion—usually referred to as the *target* or *core*—and an active electron. Wavefunctions for the total system are thus expanded in terms of the target wavefunctions

$$\Psi^{SL\pi} = \mathcal{A}\sum_i \chi_i\theta_i + \sum_j c_j\Phi_j \ ,$$

where $\mathcal{A}$ is the antisymmetrisation operator, χ_i are the target wavefunctions and θ_i is the active electron function. The Φ_j are bound-state type functions for the total system introduced to compensate for orthogonality conditions imposed on the θ_i and to render short-range correlations. The application of the Kohn variational principle with the functions θ_i and the coefficients c_j as variational parameters leads to a set of integro-differential equations that are solved numerically. In the present calculations relativistic effects are neglected, and therefore, LS coupling is assumed throughout where states of the total N-electron ionic system are assigned the quantum numbers $SL\pi$ (total spin, orbital angular momentum and parity). The close-coupling approximation allows a formal treatment of two very important properties of atomic systems: electron correlations (second expansion) and threshold effects (first expansion). Most other contemporary atomic computational methods emphasise either of these two features but not both.

Moreover, the close-coupling approach can be used to treat the active electron in both bound and free states, and hence, allows the study of bound–bound and bound–free transitions in a consistent framework.

The integro-differential equations obtained from the Kohn variational principle are solved with codes based on the R-matrix method (Burke et al. 1971) and on new asymptotic techniques developed by Seaton (1985). The highly optimised computational package, described by Berrington et al. (1987), provides an efficient tool for computing large quantities of accurate data for atomic systems.

3. Atomic data

The OP team has computed atomic data for all stages of ionisation of astrophysically abundant species: $1 \leq Z \leq 14$; $Z = 16$, $Z = 18$, $Z = 20$ and $Z = 26$. In Table 1 we summarise for each isonuclear sequence the number of states and transitions calculated. The version of TOPbase (0.5) installed at the CDS contains data only for ions with electron number $N \leq 13$. The rest of the data will be progressively included after thorough checks. TOPbase consists of:

1. Term energies and configuration assignments for states of the "T type", labelled with the code $(Tnl; SL\pi)$ where T is a target state. The principal and orbital angular momentum quantum numbers of the active electron are respectively limited to the range $n \leq 10$ and $l \leq 4$. States with equivalent-electron configurations, where the active electron cannot be associated to a particular target state, have also been considered; they are usually referred to as of the "C type" and labelled $(Ci; SL\pi)$ where i is an index in a configuration list.
2. Oscillator strengths (f-values) for optically allowed transitions between all states considered in 1.
3. Photoionisation cross sections from states in 1. that lie below the ionisation threshold. The cross sections have been computed with energy meshes fine enough to resolve the resonance structure.

For each isoelectronic sequence, the atomic data are being reviewed in detail in the series of papers "Atomic data for opacity calculations" (ADOC) in the *Journal of Physics B*. For the version of TOPbase currently available at the CDS the source references are

- He sequence: Fernley et al. (1987) (ADOC-VII)
- Li sequence: Peach et al. (1988) (ADOC-IX)
- Be sequence: Tully et al. (1990) (ADOC-XIV)
- B sequence: Yu Yan et al. (1987) (ADOC-III); Yu Yan & Seaton (1987) (ADOC-IV); Fernley et al. (1993)
- C sequence: Luo et al. (1989) (ADOC-X); Luo and Pradhan (1989) (ADOC-XI)
- N sequence: Burke and Lennon (1993)
- O sequence: Butler and Zeippen (1993a)
- F sequence: Butler and Zeippen (1993b)
- Ne sequence: Scott (1993)
- Na sequence: Taylor (1993)
- Mg sequence: Butler et al. (1993)
- Al sequence: Mendoza et al. (1993)

Further extensive comparisons of the OP data with experiment and theory (Seaton et al. 1992, Mendoza 1992 and references therein) have ensured that the initial objectives of the Project have been reached.

Table 1. Number of states and transitions for atoms considered in the OP work

Z	States	Trans.
1	55	330
2	108	645
3	134	798
4	232	1624
5	238	1744
6	815	9160
7	1083	13 122
8	1391	18 037
9	1716	18 375
10	2009	32 172
11	2141	37 142
12	2459	46 175
13	2774	54 325
14	3251	65 188
16	4109	88 422
18	5289	126 950
20	6450	182 144
26	18 632	911 580
	52 886	1 607 933

4. The TOPbase DBMS

TOPbase is concerned with the definition, development, distribution and maintenance of a specific, portable and low-cost DBMS for intensive use of the OP atomic data. It emphasises powerful querying, time/space efficiency and portability by exploiting the specific properties of the data and by adopting efficient computational methods. The currently available prototype has been implemented on UNIX[1] based platforms for on-line interactive use, but a batch access mode will also be considered in the near future. The installation and use of the package are well documented (Cunto & Mendoza 1992b, 1992c).

The TOPbase DBMS manipulates data related to two atomic properties: bound states and dipole allowed transitions between bound states. The OP data volume and these two atomic properties condition TOPbase to be structured in three entities: term energies (e), f-values (f) and photoionisation cross sections (p). The latter is a bound-state attribute, and is therefore functionally dependent on e, but it is handled separately due to its large data volume.

The design framework takes into consideration the types of queries most likely to be performed by users: along isonuclear and isoelectronic sequences, within ionic systems, spectrocopic series and energy or wavelength ranges. Since data access is on a read-only basis, specific compact and efficient data structures have been devised to implement indexes and tables. Data loading from secondary storage is really the main bottle-neck in system performance due to the volume of data that may be involved. Hence, the TOPbase data manipulation scheme has been implemented at two levels: (i) searches in and time consuming block-data retrievals from secondary storage and (ii) fast and versatile functions performed iteratively on the data in main storage in order to satisfy the user's ultimate

[1] UNIX is a trademark of AT & T

needs. Two data structures have been implemented in main memory for this purpose: the *view* and the *table*. A search in the database is performed according to user selected criteria that generate a subset of highly cohesive data sharing a common meaning. This subset is loaded into special buffers located in main memory to allow further manipulation; the data structure implemented in main memory to store the loaded subset is referred to as the *view*. Views can be directed to different output devices, namely, the monitor, printer and disk files. Logical reorganisations of the data stored in a view are possible through the concept of *tables*. A table is a vector array that enables or disables data within the view according to selection criteria, inclusion/exclusion facilities and sorting. Tables can be output on the mentioned devices, and graphic displays of table columns and cross sections can also be rendered. Graphic processing is managed by interfacing TOPbase with standard graphics software; the CDS version is interfaced with the PLPLOT[2] graphics package.

4.1. Example 1

Consider the following set of typical TOPbase commands

```
cv e (nz 8 ne 6
dv
so e
dt 1 9 islp iconf te
```

The first command creates a view of the *e* entity selecting all the levels of O III ($Z = 8$ and $N = 6$). The 370 levels stored in main memory can be displayed with the Display View command (**dv**), the levels being listed in ascending energy order within each $SL\pi$ series. The third command creates a table with the levels sorted in ascending energy order, and by means of the fourth command only the lowest 9 are listed (see Table 2) showing three attributes for each level: its total quantum numbers $SL\pi$, where $\mathtt{islp} = 100(2S+1)+10L+\pi$, its configuration assignment **iconf** and its term energy **te** (the latter is expressed in this case with respect to the ground state while the usual default, e, is given relative to the ionisation limit).

Table 2. Term energies and configuration assignments for the lowest 9 states of O III, obtained by the TOPbase commands of Example 1

ISLP	ICONF	TE(RYD)
310	2s2 2p2	0.00000E+00
120	2s2 2p2	1.89210E-01
100	2s2 2p2	4.11000E-01
501	2s 2p3	5.28950E-01
321	2s 2p3	1.09202E+00
311	2s 2p3	1.30583E+00
121	2s 2p3	1.72423E+00
301	2s 2p3	1.82113E+00
111	2s 2p3	1.94945E+00
311	2s2 2p 3s	2.44021E+00

4.2. Example 2

The commands

```
cv f (ne 4 islp 100 ilv 1 jlv 1
dt nz islp jslp ilv jlv gf wl
```

will create a view and display a table containing the gf-values and wavelengths for the resonance transition of ions in the Be isoelectronic sequence as shown in Table 3.

Table 3. gf-values and wavelengths for the resonance transition of ions in the Be isoelectronic sequence, obtained by the TOPbase commands of Example 2

NZ	ISLP	JSLP	ILV	JLV	GF	WL(A)
4	111	100	1	1	1.40E+00	2.331E+03
5	111	100	1	1	1.03E+00	1.340E+03
6	111	100	1	1	7.00E-01	9.602E+02
7	111	100	1	1	6.26E-01	7.532E+02
8	111	100	1	1	5.22E-01	6.215E+02
9	111	100	1	1	4.48E-01	5.297E+02
10	111	100	1	1	3.92E-01	4.620E+02
11	111	100	1	1	3.49E-01	4.098E+02
12	111	100	1	1	3.14E-01	3.683E+02
13	111	100	1	1	2.86E-01	3.345E+02
14	111	100	1	1	2.62E-01	3.065E+02
16	111	100	1	1	2.25E-01	2.625E+02
18	111	100	1	1	1.97E-01	2.296E+02
20	111	100	1	1	1.75E-01	2.041E+02
26	111	100	1	1	1.32E-01	1.530E+02

4.3. Example 3

In order to plot the low-energy behaviour of the photoionisation cross section of the ground state of Si II, the following commands are issued

```
cv p (nz 14 ne 13 islp 211
se (ilv 1
px (x 0.0 1.5 y 0.01 10
qq
```

The first command creates a view of the photoionisation entity, *p*, containing the cross sections for all the levels of the $^2P^o$ (**islp=211**) series of singly ionised Si ($Z = 14$, $N = 13$). Since this view allows data manipulation beyond our initial request, a table is created by selecting (**se**) only the lowest level, i.e., the ground state. Its cross section is then plotted with the third command for specific abscissa and ordinate ranges (see Fig. 1). To exit TOPbase, the **qq** command must be typed.

5. TOPbase at the CDS

Since the TOPbase DBMS and atomic data are periodically updated, the installation of the package for network access at a data center largely simplifies and reduces the cost of program support. Above all, it ensures that the whole user community, which can grow very rapidly, is accessing a common version

[2]PLPLOT has been developed by T.J. Pearson (Caltech), M.J. LeBrun and G. Furnish (Texas) and T. Richardson (Duke)

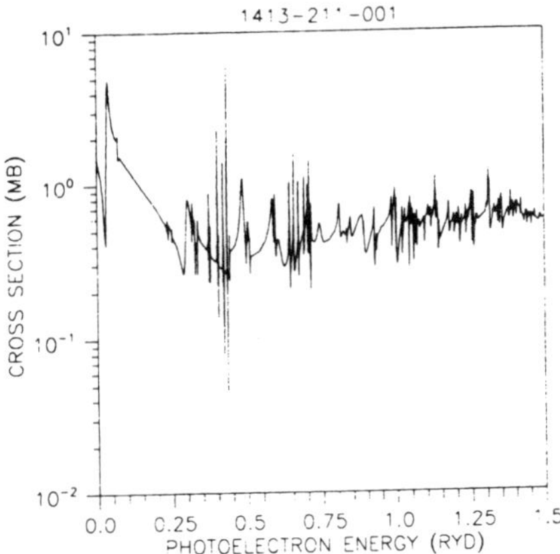

Fig. 1. Plot of the photoionisation cross section of the ground state of Si II in the low-energy region resulting from the TOPbase commands of Example 3.

of the package. Although Internet remote access can still be troublesome, it is set to improve in the short term. Within this network environment, the TOPbase implementation at the CDS is at a prototype stage; however, the many aspects that imply an efficient on-line service are being carefully considered and will be incorporated in future versions.

Documentation (LaTeX files) can be obtained via Internet from the CDS (IP `130.79.128.5`) with anonymous `ftp`. With such information, a user will be able to access and fully interact with the program through `telnet` at the following address

- IP: `130.79.128.5`
- account: `topbase`
- password: `Seaton+`

Graphic processing is limited to the X environment[3], and for this purpose, the user is asked for the screen identifier. Arrangements for file transfer with `ftp` are given in the documentation or when the program is invoked.

The examples given in Sect. 4 can be tried straightforwardly as an introduction to the package before transferring the documentation and getting familiar with the commands.

Acknowledgements. We would like to thank Drs. J. Dubau and G. Massacrier from the Observatoire de Paris, Meudon, France, for developing the interface for the PLPLOT graphics package and for assistance with its installation. The participation of W. Cunto and C. Mendoza in the Opacity Project and the development of TOPbase are supported by IBM Venezuela. Exchange visits of C.J. Zeippen and C. Mendoza to the IBM Venezuela Scientific Center and the Observatoire de Paris are funded by the CNRS and the CONICIT in the framework of a bilateral scientific agreement. Further support from the Observatoire de Paris and from the GdR 131 du CNRS "Structure Interne" is gratefully acknowledged.

[3]The X Window System is a trademark of MIT

References

Berrington, K.A., Burke, P.G., Butler, K., Seaton, M.J., Storey, P.J., Taylor, K.T. & Yu Yan 1987, J. Phys. B, 20, 6379

Burke, P.G., Hibbert, A. & Robb, W.D. 1971, J. Phys. B, 4, 153

Burke, P.G. & Seaton, M.J. 1971, Meth. Comp. Phys., 10, 1

Burke, V.M. & Lennon, D.J. 1993, J. Phys. B, to be published

Butler, K., Mendoza, C. & Zeippen, C.J. 1993, J. Phys. B, in press

Butler, K. & Zeippen, C.J. 1993a, J. Phys. B, to be published

Butler, K. & Zeippen, C.J. 1993b, J. Phys. B to be published

Cox, A.N. & Tabor, J.E. 1976, ApJS, 31, 271

Cunto, W. & Mendoza, C. 1992a, Rev. Mex. Astron. Astrof., 23, 107

Cunto, W. & Mendoza, C. 1992b, TOPbase User's Manual, Report CSC-01-92, IBM Caracas Scientific Center, Venezuela

Cunto, W. & Mendoza, C. 1992c, How to install TOPbase, Report CSC-02-92, IBM Caracas Scientific Center, Venezuela

Fernley, J.A., Hibbert, A., Kingston, A.E. & Seaton M.J. 1993, J. Phys. B, to be published

Fernley, J.A., Taylor, K.T. & Seaton M.J. 1987, J. Phys. B, 20, 6457

Huebner, W.F. 1985, in Physics of the Sun, P. Sturrock, T. Holzer, D. Mihalas & R. Ulrich (eds.). Reidel, Dordrecht, p. 33

Luo, D. & Pradhan, A.K. 1989, J. Phys. B, 22, 3377

Luo, D., Pradhan, A.K., Saraph, H.E., Storey, P.J. and Yu Yan 1989, J. Phys. B, 22, 389

Mendoza, C. 1992, in Atomic and Molecular Data for Space Astronomy: Needs, Analysis and Availability, P.L. Smith & W.L. Wiese (eds.). Lecture Notes in Physics 407, Springer-Verlag, Berlin, p. 85

Mendoza, C., Eissner, W., Le Dourneuf, M. & Zeippen, C.J. 1993, J. Phys. B, to be published

Peach, G., Saraph, H.E. & Seaton, M.J. 1988, J. Phys. B, 21, 3669

Rogers, F.J. & Iglesias, C.A. 1992, ApJS, 79, 507

Scott, M.P. 1993, J. Phys. B, to be published

Seaton, M.J. 1985, J. Phys. B, 18, 2111

Seaton, M.J. 1987, J. Phys. B, 20, 6363

Seaton, M.J. 1992, Rev. Mex. Astron. Astrofis., 23, 180

Seaton, M.J. 1993, in Inside the Stars, IAU Colloqium 137, in press

Seaton, M.J., Zeippen, C.J., Tully, J.A., Pradhan, A.K., Mendoza, C., Hibbert, A. & Berrington, K.A. 1992, Rev. Mexicana Astron. Astrof., 23, 19

Simon, N.R. 1982, ApJ, 260, L87

Taylor, K.T. 1993, J. Phys. B, to be published

Tully, J.A., Seaton, M.J. and Berrington, K.A. 1990, J. Phys. B, 23, 3811

Weiss, A., Keady, J.J. & Magee, H.H. 1990, At. Data Nucl. Data Tables, 45, 209

Yu Yan 1992, Rev. Mex. Astron. Astrofis., 23, 171

Yu Yan & Seaton, M.J. 1987, J. Phys. B, 20, 6409

Yu Yan, Taylor, K.T. & Seaton, M.J. 1987, J. Phys. B, 20, 6399

This article was processed by the author using Springer-Verlag LaTeX A&A style file 1990.

The Opacity Project – Atomic Data Tables

In this volume we tabulate all energy levels and a subset of gf-values as calculated by the OP for atoms and ions of the elements S, Ar, Ca and Fe. The subset of gf-values is restricted to transitions involving $n \leq 4$, or in some cases $n \leq 5$, where n is the effective principal quantum number relative to the ionisation threshold.

The arrangement of the tables follows the same format as in Volume 1. The ions are ordered in iso-electronic sequencies, He-like to Fe-like. The data are arranged in columns in the tables, and each page is headed with the ion name for easing searches. For each ion there are three tables:

- The first table lists all the calculated energy levels, grouped according to ${}^{2S+1}L^{\pi}$ symmetry, and numbered within each symmetry in ascending order (the index i). The states are further identified by a configuration label, but this identification is not to be taken as definitive; a colon (:) indicates an uncertainty in this identification, usually as a result of strong configuration mixing.
- The second table lists a subset of the calculated energies, i.e. those bound states with $n \leq 4$, or in some cases $n \leq 5$, as described above. These energies are placed in ascending order, with the energies in Rydberg units relative to the ground state. Since states with high angular momentum l were not calculated by the OP (for the purpose of opacity calculations such states were assumed to be hydrogenic), the table heading indicates the highest l of the calculated bound states.
- The third table lists a subset of the calculated gf-values. Again the table heading indicates the highest effective n and l considered. The states involved in the transition are identified by their ${}^{2S+1}L^{\pi}$ symmetry and index i and i', as defined in the first table. The *sign* of the gf-value is defined here to be positive if the energy of state i lies above that of i' in the transition labelling, otherwise a negative sign is printed. $|gf|$ is of course symmetric with respect to reversing the initial and final states: the oscillator strength f for a given transition can be obtained by simply dividing by the statistical weight $g = (2S+1)(2L+1)$ of the initial state.

Keith Berrington, 1996 September.

He-like S (S^{14+})

Term energies relative to 1s ^{2}S ionization threshold for each symmetry

i E(Ryds) Description	i E(Ryds) Description	i E(Ryds) Description	i E(Ryds) Description
^{1}S^e	5 −6.31147 1s6s	3 −9.00289 1s5d	^{3}P^o
1 −236.273 1s^2	6 −4.63042 1s7s	4 −6.25170 1s6d	1 −56.9238 1s2p
2 −56.8003 1s2s	7 −3.54141 1s8s	5 −4.59290 1s7d	2 −25.1909 1s3p
3 −25.1719 1s3s	8 −2.79586 1s9s	6 −3.51634 1s8d	3 −14.1418 1s4p
4 −14.1362 1s4s	^{1}D^e	7 −2.77828 1s9d	4 −9.04054 1s5p
5 −9.03800 1s5s	1 −24.9980 1s3d	^{1}P^o	5 −6.27339 1s6p
6 −6.27207 1s6s	2 −14.0614 1s4d	1 −55.9918 1s2p	6 −4.60652 1s7p
7 −4.60576 1s7s	3 −8.99960 1s5d	2 −24.9375 1s3p	7 −3.52543 1s8p
8 −3.52496 1s8s	4 −6.24971 1s6d	3 −14.0381 1s4p	8 −2.78466 1s9p
9 −2.78434 1s9s	5 −4.59162 1s7d	4 −8.98814 1s5p	
^{3}S^e	6 −3.51546 1s8d	5 −6.24329 1s6p	
1 −58.0815 1s2s	7 −2.77766 1s9d	6 −4.58766 1s7p	
2 −25.5103 1s3s	^{3}D^e	7 −3.51284 1s8p	
3 −14.2733 1s4s	1 −25.0089 1s3d	8 −2.77583 1s9p	
4 −9.10683 1s5s	2 −14.0673 1s4d		

Energies in ascending order from ground state for terms with effective $n \leq 5.0$, $L \leq 2$

Term	i	E(Ryds)	Term	i	E(Ryds)	Term	i	E(Ryds)	Term	i	E(Ryds)	Term	i	E(Ryds)
^{1}S^e	1	0.00000	^{3}S^e	2	210.762	^{1}P^o	2	211.335	^{1}D^e	2	222.211	^{3}D^e	3	227.270
^{3}S^e	1	178.191	^{3}P^o	2	211.082	^{3}S^e	3	221.999	^{1}P^o	3	222.234	^{1}D^e	3	227.273
^{3}P^o	1	179.349	^{1}S^e	3	211.101	^{3}P^o	3	222.131	^{3}S^e	4	227.166	^{1}P^o	4	227.284
^{1}S^e	2	179.472	^{3}D^e	1	211.264	^{1}S^e	4	222.136	^{3}P^o	4	227.232			
^{1}P^o	1	180.281	^{1}D^e	1	211.274	^{3}D^e	2	222.205	^{1}S^e	5	227.235			

gf-values for transitions involving terms with effective $n \leq 5.0$, $L \leq 2$

i i′	gf$_L$	i i′	gf$_L$	i i′	gf$_L$	i i′	gf$_L$	i i′	gf$_L$	i i′	gf$_L$
	^{1}S^e–^{1}P^o	3 3	−4.47E−1		^{1}D^e–^{1}P^o	3 3	1.85E+0	3 1	3.21E−2	1 3	−1.88E−1
1 1	−7.65E−1	3 4	−1.15E−1	1 1	2.10E+0	3 4	−4.01E−2	3 2	3.45E−1	1 4	−3.74E−2
1 2	−1.53E−1	4 1	1.07E−2	1 2	−1.81E−2		^{3}S^e–^{3}P^o	3 3	−3.12E−1	2 1	1.10E+0
1 3	−5.69E−2	4 2	1.12E−1	1 3	−5.26E−2	1 1	−1.35E−1	3 4	−1.45E+0	2 2	5.35E+0
1 4	−2.75E−2	4 3	−7.90E−2	1 4	−1.06E−2	1 2	−1.18E+0	4 1	1.27E−2	2 3	2.91E−1
2 1	−3.24E−2	4 4	−4.98E−1	2 1	3.62E−1	1 3	−2.89E−1	4 2	7.71E−2	2 4	−4.69E−1
2 2	−4.05E−1	5 1	4.24E−3	2 2	1.88E+0	1 4	−1.19E−1	4 3	5.61E−1	3 1	4.05E−1
2 3	−9.85E−2	5 2	2.52E−2	2 3	−2.98E−2	2 1	1.47E−1	4 4	−3.94E−1	3 2	1.23E+0
2 4	−4.05E−2	5 3	1.81E−1	2 4	−1.34E−1	2 2	−2.26E−1		^{3}D^e–^{3}P^o	3 3	5.22E+0
3 1	4.83E−2	5 4	−9.99E−2	3 1	1.31E−1	2 3	−1.30E+0	1 1	6.13E+0	3 4	3.98E−1
3 2	−5.64E−2			3 2	4.19E−1	2 4	−3.37E−1	1 2	1.64E−1		

He-like Ar (Ar^{16+})

Term energies relative to 1s ^{2}S ionization threshold for each symmetry

i	E(Ryds)	Description	i	E(Ryds)	Description	i	E(Ryds)	Description	i	E(Ryds)	Description
		^{1}S^e	5	−8.09768	1s6s	3	−11.5633	1s5d			^{3}P^o
1	−301.772	1s^2	6	−5.94183	1s7s	4	−8.02967	1s6d	1	−73.0206	1s2p
2	−72.8728	1s2s	7	−4.54495	1s8s	5	−5.89915	1s7d	2	−32.3292	1s3p
3	−32.3064	1s3s	8	−3.58847	1s9s	6	−4.51643	1s8d	3	−18.1532	1s4p
4	−18.1463	1s4s			^{1}D^e	7	−3.56848	1s9d	4	−11.6063	1s5p
5	−11.6032	1s5s	1	−32.1086	1s3d			^{1}P^o	5	−8.05446	1s6p
6	−8.05288	1s6s	2	−18.0611	1s4d	1	−71.9526	1s2p	6	−5.91469	1s7p
7	−5.91379	1s7s	3	−11.5595	1s5d	2	−32.0396	1s3p	7	−4.52680	1s8p
8	−4.52625	1s8s	4	−8.02735	1s6d	3	−18.0346	1s4p	8	−3.57574	1s9p
9	−3.57536	1s9s	5	−5.89765	1s7d	4	−11.5465	1s5p			
		^{3}S^e	6	−4.51541	1s8d	5	−8.02011	1s6p			
1	−74.3297	1s2s	7	−3.56775	1s9d	6	−5.89316	1s7p			
2	−32.6909	1s3s			^{3}D^e	7	−4.51242	1s8p			
3	−18.3021	1s4s	1	−32.1214	1s3d	8	−3.56566	1s9p			
4	−11.6814	1s5s	2	−18.0680	1s4d						

Energies in ascending order from ground state for terms with effective $n \leq 5.0$, $L \leq 2$

Term	i	E(Ryds)	Term	i	E(Ryds)	Term	i	E(Ryds)	Term	i	E(Ryds)	Term	i	E(Ryds)
^{1}S^e	1	0.00000	^{3}S^e	2	269.081	^{1}P^o	2	269.732	^{1}D^e	2	283.710	^{3}D^e	3	290.208
^{3}S^e	1	227.442	^{3}P^o	2	269.442	^{3}S^e	3	283.469	^{1}P^o	3	283.737	^{1}D^e	3	290.212
^{3}P^o	1	228.751	^{1}S^e	3	269.465	^{3}P^o	3	283.618	^{3}S^e	4	290.090	^{1}P^o	4	290.225
^{1}S^e	2	228.899	^{3}D^e	1	269.650	^{1}S^e	4	283.625	^{3}P^o	4	290.165			
^{1}P^o	1	229.819	^{1}D^e	1	269.663	^{3}D^e	2	283.704	^{1}S^e	5	290.168			

gf-values for transitions involving terms with effective $n \leq 5.0$, $L \leq 2$

i i'	gf$_L$	i i'	gf$_L$	i i'	gf$_L$	i i'	gf$_L$	i i'	gf$_L$	i i'	gf$_L$
	^{1}S^e–^{1}P^o	3 3	−4.51E−1		^{1}D^e–^{1}P^o	3 3	1.85E+0	3 1	3.15E−2	1 3	−1.86E−1
1 1	−7.72E−1	3 4	−1.15E−1	1 1	2.10E+0	3 4	−3.53E−2	3 2	3.38E−1	1 4	−3.69E−2
1 2	−1.54E−1	4 1	1.05E−2	1 2	−1.61E−2		^{3}S^e–^{3}P^o	3 3	−2.77E−1	2 1	1.10E+0
1 3	−5.71E−2	4 2	1.10E−1	1 3	−5.29E−2	1 1	−1.19E−1	3 4	−1.47E+0	2 2	5.37E+0
1 4	−2.75E−2	4 3	−7.03E−2	1 4	−1.07E−2	1 2	−1.19E+0	4 1	1.25E−2	2 3	2.55E−1
2 1	−2.87E−2	4 4	−5.03E−1	2 1	3.62E−1	1 3	−2.91E−1	4 2	7.59E−2	2 4	−4.64E−1
2 2	−4.09E−1	5 1	4.17E−3	2 2	1.88E+0	1 4	−1.20E−1	4 3	5.50E−1	3 1	4.05E−1
2 3	−9.90E−2	5 2	2.48E−2	2 3	−2.64E−2	2 1	1.44E−1	4 4	−3.47E−1	3 2	1.24E+0
2 4	−4.07E−2	5 3	1.78E−1	2 4	−1.35E−1	2 2	−2.00E−1		^{3}D^e–^{3}P^o	3 3	5.25E+0
3 1	4.75E−2	5 4	−8.84E−2	3 1	1.31E−1	2 3	−1.31E+0	1 1	6.14E+0	3 4	3.52E−1
3 2	−4.99E−2			3 2	4.19E−1	2 4	−3.40E−1	1 2	1.46E−1		

He-like Ca (Ca^{18+})

Term energies relative to 1s ^{2}S ionization threshold for each symmetry

i	E(Ryds)	Description	i	E(Ryds)	Description	i	E(Ryds)	Description	i	E(Ryds)	Description
		^{1}S^e	5	−10.1061	1s6s	3	−14.4438	1s5d			^{3}P^o
1	−375.272	1s^2	6	−7.41651	1s7s	4	−10.0300	1s6d	1	−91.1175	1s2p
2	−90.9455	1s2s	7	−5.67349	1s8s	5	−7.36871	1s7d	2	−40.3565	1s3p
3	−40.3298	1s3s	8	−4.47984	1s9s	6	−5.64154	1s8d	3	−22.6644	1s4p
4	−22.6563	1s4s			^{1}D^e	7	−4.45744	1s9d	4	−14.4921	1s5p
5	−14.4884	1s5s	1	−40.1080	1s3d			^{1}P^o	5	−10.0578	1s6p
6	−10.0559	1s6s	2	−22.5608	1s4d	1	−89.9133	1s2p	6	−7.38619	1s7p
7	−7.38510	1s7s	3	−14.4394	1s5d	2	−40.0306	1s3p	7	−5.65321	1s8p
8	−5.65253	1s8s	4	−10.0273	1s6d	3	−22.5312	1s4p	8	−4.46561	1s9p
9	−4.46516	1s9s	5	−7.36699	1s7d	4	−14.4249	1s5p			
		^{3}S^e	6	−5.64037	1s8d	5	−10.0192	1s6p			
1	−92.5779	1s2s	7	−4.45661	1s9d	6	−7.36199	1s7p			
2	−40.7605	1s3s			^{3}D^e	7	−5.63705	1s8p			
3	−22.8309	1s4s	1	−40.1227	1s3d	8	−4.45428	1s9p			
4	−14.5761	1s5s	2	−22.5688	1s4d						

Energies in ascending order from ground state for terms with effective $n \leq 5.0$, $L \leq 2$

Term	i	E(Ryds)	Term	i	E(Ryds)	Term	i	E(Ryds)	Term	i	E(Ryds)	Term	i	E(Ryds)
^{1}S^e	1	0.00000	^{3}S^e	2	334.511	^{1}P^o	2	335.241	^{1}D^e	2	352.711	^{3}D^e	3	360.828
^{3}S^e	1	282.694	^{3}P^o	2	334.915	^{3}S^e	3	352.441	^{1}P^o	3	352.740	^{1}D^e	3	360.832
^{3}P^o	1	284.154	^{1}S^e	3	334.942	^{3}P^o	3	352.607	^{3}S^e	4	360.695	^{1}P^o	4	360.847
^{1}S^e	2	284.326	^{3}D^e	1	335.149	^{1}S^e	4	352.615	^{3}P^o	4	360.779			
^{1}P^o	1	285.358	^{1}D^e	1	335.164	^{3}D^e	2	352.703	^{1}S^e	5	360.783			

gf-values for transitions involving terms with effective $n \leq 5.0$, $L \leq 2$

i i'	gf$_L$	i i'	gf$_L$	i i'	gf$_L$	i i'	gf$_L$	i i'	gf$_L$	i i'	gf$_L$
	^{1}S^e–^{1}P^o	3 3	−4.54E−1		^{1}D^e–^{1}P^o	3 3	1.85E+0	3 1	3.11E−2	1 3	−1.83E−1
1 1	−7.78E−1	3 4	−1.16E−1	1 1	2.09E+0	3 4	−3.16E−2	3 2	3.33E−1	1 4	−3.65E−2
1 2	−1.54E−1	4 1	1.04E−2	1 2	−1.44E−2		^{3}S^e–^{3}P^o	3 3	−2.48E−1	2 1	1.10E+0
1 3	−5.72E−2	4 2	1.08E−1	1 3	−5.31E−2	1 1	−1.07E−1	3 4	−1.48E+0	2 2	5.39E+0
1 4	−2.76E−2	4 3	−6.33E−2	1 4	−1.07E−2	1 2	−1.20E+0	4 1	1.23E−2	2 3	2.29E−1
2 1	−2.58E−2	4 4	−5.07E−1	2 1	3.63E−1	1 3	−2.93E−1	4 2	7.49E−2	2 4	−4.59E−1
2 2	−4.11E−1	5 1	4.12E−3	2 2	1.87E+0	1 4	−1.21E−1	4 3	5.43E−1	3 1	4.04E−1
2 3	−9.93E−2	5 2	2.46E−2	2 3	−2.36E−2	2 1	1.42E−1	4 4	−3.11E−1	3 2	1.24E+0
2 4	−4.08E−2	5 3	1.76E−1	2 4	−1.35E−1	2 2	−1.79E−1		^{3}D^e–^{3}P^o	3 3	5.27E+0
3 1	4.68E−2	5 4	−7.92E−2	3 1	1.32E−1	2 3	−1.33E+0	1 1	6.15E+0	3 4	3.17E−1
3 2	−4.48E−2			3 2	4.19E−1	2 4	−3.42E−1	1 2	1.32E−1		

He-like Fe (Fe^{24+})

Term energies relative to $1s\ ^2S$ ionization threshold for each symmetry

i	E(Ryds)	Description	i	E(Ryds)	Description	i	E(Ryds)	Description	i	E(Ryds)	Description
	$^1S^e$		5	−17.4647	$1s6s$	3	−25.0055	$1s5d$		$^3P^o$	
1	−643.772	$1s^2$	6	−12.8201	$1s7s$	4	−17.3641	$1s6d$	1	−157.408	$1s2p$
2	−157.164	$1s2s$	7	−9.80911	$1s8s$	5	−12.7569	$1s7d$	2	−69.7715	$1s3p$
3	−69.7333	$1s3s$	8	−7.74655	$1s9s$	6	−9.76686	$1s8d$	3	−39.1984	$1s4p$
4	−39.1866	$1s4s$		$^1D^e$		7	−7.71694	$1s9d$	4	−25.0696	$1s5p$
5	−25.0641	$1s5s$	1	−69.4396	$1s3d$		$^1P^o$		5	−17.4011	$1s6p$
6	−17.3983	$1s6s$	2	−39.0601	$1s4d$	1	−155.795	$1s2p$	6	−12.7802	$1s7p$
7	−12.7786	$1s7s$	3	−24.9993	$1s5d$	2	−69.3370	$1s3p$	7	−9.78237	$1s8p$
8	−9.78139	$1s8s$	4	−17.3604	$1s6d$	3	−39.0209	$1s4p$	8	−7.72779	$1s9p$
9	−7.72713	$1s9s$	5	−12.7545	$1s7d$	4	−24.9800	$1s5p$			
	$^3S^e$		6	−9.76524	$1s8d$	5	−17.3497	$1s6p$			
1	−159.323	$1s2s$	7	−7.71578	$1s9d$	6	−12.7480	$1s7p$			
2	−70.3024	$1s3s$		$^3D^e$		7	−9.76086	$1s8p$			
3	−39.4173	$1s4s$	1	−69.4600	$1s3d$	8	−7.71271	$1s9p$			
4	−25.1799	$1s5s$	2	−39.0711	$1s4d$						

Energies in ascending order from ground state for terms with effective $n \leq 5.0$, $L \leq 2$

Term	i	E(Ryds)	Term	i	E(Ryds)	Term	i	E(Ryds)	Term	i	E(Ryds)	Term	i	E(Ryds)
$^1S^e$	1	0.00000	$^3S^e$	2	573.469	$^1P^o$	2	574.435	$^1D^e$	2	604.711	$^3D^e$	3	618.766
$^3S^e$	1	484.448	$^3P^o$	2	574.000	$^3S^e$	3	604.354	$^1P^o$	3	604.751	$^1D^e$	3	618.772
$^3P^o$	1	486.363	$^1S^e$	3	574.038	$^3P^o$	3	604.573	$^3S^e$	4	618.592	$^1P^o$	4	618.791
$^1S^e$	2	486.607	$^3D^e$	1	574.311	$^1S^e$	4	604.585	$^3P^o$	4	618.702			
$^1P^o$	1	487.976	$^1D^e$	1	574.332	$^3D^e$	2	604.700	$^1S^e$	5	618.707			

gf-values for transitions involving terms with effective $n \leq 5.0$, $L \leq 2$

$i\ i'$	gf_L	$i\ i'$	gf_L	$i\ i'$	gf_L	$i\ i'$	gf_L	$i\ i'$	gf_L	$i\ i'$	gf_L
	$^1S^e$–$^1P^o$	3 3	−4.61E−1		$^1D^e$–$^1P^o$	3 3	1.84E+0	3 1	3.02E−2	1 3	−1.79E−1
1 1	−7.91E−1	3 4	−1.17E−1	1 1	2.09E+0	3 4	−2.43E−2	3 2	3.22E−1	1 4	−3.57E−2
1 2	−1.55E−1	4 1	1.01E−2	1 2	−1.11E−2		$^3S^e$–$^3P^o$	3 3	−1.90E−1	2 1	1.10E+0
1 3	−5.74E−2	4 2	1.06E−1	1 3	−5.35E−2	1 1	−8.12E−2	3 4	−1.52E+0	2 2	5.43E+0
1 4	−2.76E−2	4 3	−4.78E−2	1 4	−1.08E−2	1 2	−1.23E+0	4 1	1.20E−2	2 3	1.76E−1
2 1	−1.97E−2	4 4	−5.16E−1	2 1	3.63E−1	1 3	−2.97E−1	4 2	7.30E−2	2 4	−4.49E−1
2 2	−4.17E−1	5 1	4.00E−3	2 2	1.87E+0	1 4	−1.22E−1	4 3	5.26E−1	3 1	4.03E−1
2 3	−1.00E−1	5 2	2.40E−2	2 3	−1.81E−2	2 1	1.37E−1	4 4	−2.37E−1	3 2	1.24E+0
2 4	−4.11E−2	5 3	1.72E−1	2 4	−1.36E−1	2 2	−1.36E−1		$^3D^e$–$^3P^o$	3 3	5.32E+0
3 1	4.54E−2	5 4	−6.06E−2	3 1	1.32E−1	2 3	−1.36E+0	1 1	6.18E+0	3 4	2.43E−1
3 2	−3.43E−2			3 2	4.18E−1	2 4	−3.47E−1	1 2	1.01E−1		

Li-like S (S^{13+})

Term energies relative to $1s^2\ ^1$S ionization threshold for each symmetry

i	Energy(Ryds)	Description	i	Energy(Ryds)	Description	i	Energy(Ryds)	Description
	^{2}S^e			^{2}P^o			^{2}D^e	
1	−51.7704	$1s^22s$	1	−49.7546	$1s^22p$	1	−21.7875	$1s^23d$
2	−22.5554	$1s^23s$	2	−21.9986	$1s^23p$	2	−12.2552	$1s^24d$
3	−12.5719	$1s^24s$	3	−12.3431	$1s^24p$	3	−7.84286	$1s^25d$
4	−8.00332	$1s^25s$	4	−7.88754	$1s^25p$	4	−5.44617	$1s^26d$
5	−5.53842	$1s^26s$	5	−5.47190	$1s^26p$	5	−4.00112	$1s^27d$
6	−4.05896	$1s^27s$	6	−4.01726	$1s^27p$	6	−3.06326	$1s^28d$
7	−3.10189	$1s^28s$	7	−3.07405	$1s^28p$	7	−2.42029	$1s^29d$
8	−2.44737	$1s^29s$	8	−2.42785	$1s^29p$			

Energies in ascending order from ground state for terms with effective $n \leq 5.0$, $L \leq 2$

Term	i	E(Ryds)	Term	i	E(Ryds)	Term	i	E(Ryds)	Term	i	E(Ryds)
^{2}S^e	1	0.00000	^{2}P^o	2	29.7718	^{2}P^o	3	39.4273	^{2}P^o	4	43.8828
^{2}P^o	1	2.01580	^{2}D^e	1	29.9829	^{2}D^e	2	39.5152	^{2}D^e	3	43.9275
^{2}S^e	2	29.2150	^{2}S^e	3	39.1985	^{2}S^e	4	43.7670			

gf-values for transitions involving terms with effective $n \leq 5.0$, $L \leq 2$

i i'	gf$_L$	i i'	gf$_L$	i i'	gf$_L$	i i'	gf$_L$
	^{2}S^e–^{2}P^o	3 1	2.50E−2		^{2}P^o–^{2}D^e	3 3	−3.42E+0
1 1	−1.77E−1	3 2	2.70E−1	1 1	−4.05E+0	4 1	2.59E−2
1 2	−7.07E−1	3 3	−4.11E−1	1 2	−7.33E−1	4 2	3.26E−1
1 3	−1.80E−1	3 4	−8.51E−1	1 3	−2.70E−1	4 3	−3.61E−1
1 4	−7.54E−2	4 1	9.80E−3	2 1	−1.46E−1		
2 1	1.18E−1	4 2	5.88E−2	2 2	−3.52E+0		
2 2	−2.98E−1	4 3	4.35E−1	2 3	−8.19E−1		
2 3	−7.69E−1	4 4	−5.22E−1	3 1	1.31E−1		
2 4	−2.07E−1			3 2	−2.59E−1		

Li-like Ar (Ar^{15+})

Term energies relative to $1s^2\ ^1$S ionization threshold for each symmetry

i	Energy(Ryds)	Description	i	Energy(Ryds)	Description	i	Energy(Ryds)	Description
	^{2}S^e			^{2}P^o			^{2}D^e	
1	−67.1785	$1s^22s$	1	−64.8810	$1s^22p$	1	−28.4559	$1s^23d$
2	−29.3378	$1s^23s$	2	−28.7018	$1s^23p$	2	−16.0061	$1s^24d$
3	−16.3701	$1s^24s$	3	−16.1085	$1s^24p$	3	−10.2434	$1s^25d$
4	−10.4279	$1s^25s$	4	−10.2954	$1s^25p$	4	−7.11315	$1s^26d$
5	−7.21923	$1s^26s$	5	−7.14309	$1s^26p$	5	−5.22581	$1s^27d$
6	−5.29233	$1s^27s$	6	−5.24459	$1s^27p$	6	−4.00090	$1s^28d$
7	−4.04533	$1s^28s$	7	−4.01344	$1s^28p$	7	−3.16113	$1s^29d$
8	−3.19228	$1s^29s$	8	−3.16992	$1s^29p$			

Energies in ascending order from ground state for terms with effective $n \leq 5.0$, $L \leq 2$

Term	i	E(Ryds)	Term	i	E(Ryds)	Term	i	E(Ryds)	Term	i	E(Ryds)
^{2}S^e	1	0.00000	^{2}P^o	2	38.4767	^{2}P^o	3	51.0700	^{2}P^o	4	56.8831
^{2}P^o	1	2.29750	^{2}D^e	1	38.7226	^{2}D^e	2	51.1724	^{2}D^e	3	56.9351
^{2}S^e	2	37.8407	^{2}S^e	3	50.8084	^{2}S^e	4	56.7506			

gf-values for transitions involving terms with effective $n \leq 5.0$, $L \leq 2$

i i'	gf$_L$	i i'	gf$_L$	i i'	gf$_L$	i i'	gf$_L$
	^{2}S^e–^{2}P^o	3 1	2.42E−2		^{2}P^o–^{2}D^e	3 3	−3.44E+0
1 1	−1.55E−1	3 2	2.60E−1	1 1	−4.06E+0	4 1	2.55E−2
1 2	−7.26E−1	3 3	−3.61E−1	1 2	−7.33E−1	4 2	3.21E−1
1 3	−1.83E−1	3 4	−8.79E−1	1 3	−2.70E−1	4 3	−3.22E−1
1 4	−7.64E−2	4 1	9.48E−3	2 1	−1.30E−1		
2 1	1.13E−1	4 2	5.70E−2	2 2	−3.54E+0		
2 2	−2.62E−1	4 3	4.20E−1	2 3	−8.21E−1		
2 3	−7.92E−1	4 4	−4.59E−1	3 1	1.29E−1		
2 4	−2.11E−1			3 2	−2.31E−1		

Li-like Ca (Ca^{17+})

Term energies relative to $1s^2\ ^1$S ionization threshold for each symmetry

i	Energy(Ryds)	Description	i	Energy(Ryds)	Description	i	Energy(Ryds)	Description
		^{2}S^e			^{2}P^o			^{2}D^e
1	−84.5866	$1s^22s$	1	−82.0079	$1s^22p$	1	−36.0132	$1s^23d$
2	−37.0092	$1s^23s$	2	−36.2941	$1s^23p$	2	−20.2571	$1s^24d$
3	−20.6684	$1s^24s$	3	−20.3740	$1s^24p$	3	−12.9639	$1s^25d$
4	−13.1724	$1s^25s$	4	−13.0232	$1s^25p$	4	−9.00235	$1s^26d$
5	−9.12227	$1s^26s$	5	−9.03652	$1s^26p$	5	−6.61377	$1s^27d$
6	−6.68899	$1s^27s$	6	−6.63520	$1s^27p$	6	−5.06354	$1s^28d$
7	−5.11377	$1s^28s$	7	−5.07784	$1s^28p$	7	−4.00073	$1s^29d$
8	−4.03596	$1s^29s$	8	−4.01076	$1s^29p$			

Energies in ascending order from ground state for terms with effective $n \leq 5.0$, $L \leq 2$

Term	i	E(Ryds)	Term	i	E(Ryds)	Term	i	E(Ryds)	Term	i	E(Ryds)
^{2}S^e	1	0.00000	^{2}P^o	2	48.2925	^{2}P^o	3	64.2126	^{2}P^o	4	71.5634
^{2}P^o	1	2.57870	^{2}D^e	1	48.5734	^{2}D^e	2	64.3295	^{2}D^e	3	71.6227
^{2}S^e	2	47.5774	^{2}S^e	3	63.9182	^{2}S^e	4	71.4142			

gf-values for transitions involving terms with effective $n \leq 5.0$, $L \leq 2$

i i'	gf$_L$	i i'	gf$_L$	i i'	gf$_L$	i i'	gf$_L$
	^{2}S^e–^{2}P^o	3 1	2.35E−2		^{2}P^o–^{2}D^e	3 3	−3.46E+0
1 1	−1.38E−1	3 2	2.52E−1	1 1	−4.07E+0	4 1	2.52E−2
1 2	−7.41E−1	3 3	−3.22E−1	1 2	−7.33E−1	4 2	3.16E−1
1 3	−1.86E−1	3 4	−9.00E−1	1 3	−2.69E−1	4 3	−2.90E−1
1 4	−7.73E−2	4 1	9.24E−3	2 1	−1.18E−1		
2 1	1.10E−1	4 2	5.56E−2	2 2	−3.55E+0		
2 2	−2.33E−1	4 3	4.08E−1	2 3	−8.22E−1		
2 3	−8.11E−1	4 4	−4.09E−1	3 1	1.27E−1		
2 4	−2.15E−1			3 2	−2.09E−1		

Li-like Fe (Fe^{23+})

Term energies relative to $1s^2\ ^1S$ ionization threshold for each symmetry

i	Energy(Ryds)	Description	i	Energy(Ryds)	Description	i	Energy(Ryds)	Description
		$^2S^e$			$^2P^o$			$^2D^e$
1	−148.931	$1s^22s$	1	−145.493	$1s^22p$	1	−64.1359	$1s^23d$
2	−65.4616	$1s^23s$	2	−64.5190	$1s^23p$	2	−36.0099	$1s^24d$
3	−36.5651	$1s^24s$	3	−36.1664	$1s^24p$	3	−23.0454	$1s^25d$
4	−23.3268	$1s^25s$	4	−23.1248	$1s^25p$	4	−16.0033	$1s^26d$
5	−16.1652	$1s^26s$	5	−16.0490	$1s^26p$	5	−11.7572	$1s^27d$
6	−11.8588	$1s^27s$	6	−11.7859	$1s^27p$	6	−9.00142	$1s^28d$
7	−9.06932	$1s^28s$	7	−9.02062	$1s^28p$	7	−7.11212	$1s^29d$
8	−7.15971	$1s^29s$	8	−7.12558	$1s^29p$			

Energies in ascending order from ground state for terms with effective $n \leq 5.0$, $L \leq 2$

Term	i	E(Ryds)	Term	i	E(Ryds)	Term	i	E(Ryds)	Term	i	E(Ryds)
$^2S^e$	1	0.00000	$^2P^o$	2	84.4120	$^2P^o$	3	112.764	$^2P^o$	4	125.806
$^2P^o$	1	3.43800	$^2D^e$	1	84.7951	$^2D^e$	2	112.921	$^2D^e$	3	125.885
$^2S^e$	2	83.4694	$^2S^e$	3	112.365	$^2S^e$	4	125.604			

gf-values for transitions involving terms with effective $n \leq 5.0$, $L \leq 2$

i i'	gf_L	i i'	gf_L	i i'	gf_L	i i'	gf_L
	$^2S^e$–$^2P^o$	3 1	2.22E−2		$^2P^o$–$^2D^e$	3 3	−3.51E+0
1 1	−1.04E−1	3 2	2.37E−1	1 1	−4.10E+0	4 1	2.45E−2
1 2	−7.71E−1	3 3	−2.46E−1	1 2	−7.33E−1	4 2	3.06E−1
1 3	−1.91E−1	3 4	−9.43E−1	1 3	−2.69E−1	4 3	−2.18E−1
1 4	−7.90E−2	4 1	8.74E−3	2 1	−9.01E−2		
2 1	1.02E−1	4 2	5.30E−2	2 2	−3.60E+0		
2 2	−1.74E−1	4 3	3.86E−1	2 3	−8.28E−1		
2 3	−8.52E−1	4 4	−3.12E−1	3 1	1.23E−1		
2 4	−2.22E−1			3 2	−1.57E−1		

Be-like S (S^{12+})

Term energies relative to 2*s* ^{2}S ionization threshold for each symmetry

i	E(Ryds)	Description	*i*	E(Ryds)	Description	*i*	E(Ryds)	Description	*i*	E(Ryds)	Description
		$^1S^e$	11	−3.45727	2s7d	7	−4.80353	2p5p :	4	−2.69135	2p6d
1	−47.7638	$2s^2$	12	−2.72034	2p6p	8	−3.56137	2s7s	5	−1.43781	2p7d
2	−41.3314	$2p^2$	13	−2.65663	2p6f	9	−2.76590	2p6p	6	−.62490	2p8d
3	−20.0962	2s3s	14	−2.64417	2s8d	10	−2.69698	2s8s	7	−.06793	2p9d
4	−16.6489	2p3p	15	−2.09047	2s9d	11	−2.13828	2s9s			$^3F^e$
5	−11.0876	2s4s			$^1D^o$			$^3P^e$	1	−8.58452	2p4f
6	−8.55619	2p4p	1	−17.1802	2p3d	1	−43.0975	$2p^2$	2	−4.75829	2p5f
7	−7.01126	2s5s	2	−8.69922	2p4d	2	−17.3015	2p3p	3	−2.68226	2p6f
8	−4.90229	2s6s	3	−4.81382	2p5d	3	−8.77574	2p4p	4	−1.43202	2p7f
9	−4.69784	2p5p	4	−2.71362	2p6d	4	−4.85695	2p5p	5	−.62099	2p8f
10	−3.54202	2s7s	5	−1.45139	2p7d	5	−2.73971	2p6p	6	−.06516	2p9f
11	−2.73632	2s8s	6	−.63382	2p8d	6	−1.46815	2p7p			$^3F^o$
12	−2.65110	2p6p	7	−.07411	2p9d	7	−.64519	2p8p	1	−17.1899	2p3d
13	−2.12901	2s9s			$^1F^e$	8	−.08216	2p9p	2	−10.5803	2s4f
		$^1P^e$	1	−8.59005	2p4f			$^3P^o$	3	−8.69697	2p4d
1	−17.7078	2p3p	2	−4.76177	2p5f	1	−45.9622	2s2p	4	−6.77158	2s5f
2	−8.90327	2p4p	3	−2.68441	2p6f	2	−19.5923	2s3p	5	−4.81284	2p5d
3	−4.91445	2p5p	4	−1.43342	2p7f	3	−18.1723	2p3s	6	−4.72511	2p5g
4	−2.77073	2p6p	5	−.62195	2p8f	4	−16.8188	2p3d	7	−4.70175	2s6f
5	−1.48692	2p7p	6	−.06584	2p9f	5	−10.8809	2s4p	8	−3.45392	2s7f
6	−.65745	2p8p			$^1F^o$	6	−9.11061	2p4s	9	−2.71359	2p6d
7	−.09062	2p9p	1	−16.5208	2p3d	7	−8.58329	2p4d	10	−2.66421	2p6g
		$^1P^o$	2	−10.5546	2s4f	8	−6.91491	2s5p	11	−2.64391	2s8f
1	−44.2925	2s2p	3	−8.47133	2p4d	9	−5.02457	2p5s	12	−2.08887	2s9f
2	−19.6407	2s3p	4	−6.75442	2s5f	10	−4.81776	2s6p :			$^3G^e$
3	−17.8371	2p3s	5	−4.72893	2s6f	11	−4.72659	2p5d :	1	−8.55430	2p4f
4	−16.4589	2p3d	6	−4.72472	2p5g	12	−3.50441	2s7p	2	−6.75660	2s5g
5	−10.8580	2s4p	7	−4.67048	2p5d	13	−2.83451	2p6s	3	−4.75996	2p5f
6	−9.03665	2p4s	8	−3.44794	2s7f	14	−2.70084	2p6d	4	−4.67962	2s6g
7	−8.44879	2p4d	9	−2.66560	2p6g :	15	−2.66279	2s8p	5	−3.44883	2s7g
8	−6.89530	2s5p	10	−2.66252	2p6d :	16	−2.11208	2s9p	6	−2.68005	2p6f
9	−5.01389	2p5s	11	−2.63052	2s8f			$^3D^e$	7	−2.66666	2p6h
10	−4.77404	2s6p	12	−2.08583	2s9f	1	−19.1474	2s3d	8	−2.63638	2s8g
11	−4.67684	2p5d			$^1G^e$	2	−17.5925	2p3p	9	−2.08636	2s9g
12	−3.49878	2s7p	1	−8.53115	2p4f	3	−10.7005	2s4d			
13	−2.82448	2p6s	2	−6.75579	2s5g	4	−8.86478	2p4p			
14	−2.67740	2s8p	3	−4.75035	2p5f	5	−8.50785	2p4f			
15	−2.63977	2p6d	4	−4.67395	2s6g	6	−6.82679	2s5d			
16	−2.10900	2s9p	5	−3.44843	2s7g	7	−4.90009	2p5p			
		$^1D^e$	6	−2.67293	2p6f	8	−4.74167	2s6d			
1	−42.5856	$2p^2$	7	−2.66666	2p6h	9	−4.71149	2p5f			
2	−18.8200	2s3d	8	−2.63399	2s8g	10	−3.47251	2s7d			
3	−17.0639	2p3p	9	−2.08612	2s9g	11	−2.76245	2p6p			
4	−10.5964	2s4d			$^3S^e$	12	−2.66603	2p6f			
5	−8.69203	2p4p	1	−20.4327	2s3s	13	−2.65380	2s8d			
6	−8.48861	2p4f	2	−17.4146	2p3p	14	−2.09734	2s9d			
7	−6.77984	2s5d	3	−11.2082	2s4s			$^3D^o$			
8	−4.82371	2p5p	4	−8.80355	2p4p	1	−16.9138	2p3d			
9	−4.71056	2p5f	5	−7.07384	2s5s	2	−8.61067	2p4d			
10	−4.70187	2s6d	6	−4.94220	2s6s :	3	−4.77302	2p5d			

Be-like S (S^{12+})

Energies in ascending order from ground state for terms with effective $n \leq 4.0$, $L \leq 3$

Term	i	E(Ryds)	Term	i	E(Ryds)	Term	i	E(Ryds)	Term	i	E(Ryds)	Term	i	E(Ryds)
$^1S^e$	1	0.00000	$^1S^e$	3	27.6676	$^1P^e$	1	30.0560	$^3D^o$	1	30.8500	$^3P^o$	5	36.8829
$^3P^o$	1	1.80160	$^1P^o$	2	28.1231	$^3D^e$	2	30.1713	$^3P^o$	4	30.9450	$^1P^o$	5	36.9058
$^1P^o$	1	3.47130	$^3P^o$	2	28.1715	$^3S^e$	2	30.3492	$^1S^e$	4	31.1149	$^3D^e$	3	37.0633
$^3P^e$	1	4.66630	$^3D^e$	1	28.6164	$^3P^e$	2	30.4623	$^1F^o$	1	31.2430	$^1D^e$	4	37.1674
$^1D^e$	1	5.17820	$^1D^e$	2	28.9438	$^3F^o$	1	30.5739	$^1P^o$	4	31.3049	$^3F^o$	2	37.1835
$^1S^e$	2	6.43240	$^3P^o$	3	29.5915	$^1D^o$	1	30.5836	$^3S^e$	3	36.5556			
$^3S^e$	1	27.3311	$^1P^o$	3	29.9267	$^1D^e$	3	30.6999	$^1S^e$	5	36.6762			

gf-values for transitions involving terms with effective $n \leq 4.0$, $L \leq 3$

i i'	gf_L	i i'	gf_L	i i'	gf_L	i i'	gf_L	i i'	gf_L	i i'	gf_L
	$^1P^o$–$^1S^e$	5 1	1.60E−1	2 5	−1.01E−1	1 3	3.37E−1	2 1	1.52E−2	3 3	1.99E−3
1 1	2.25E−1	5 2	9.34E−3	3 1	8.22E−1	1 4	−3.88E−3	2 2	−3.07E−1	3 4	6.96E−2
1 2	−1.64E−1	5 3	3.46E−1	3 2	2.79E−2		$^3P^o$–$^3S^e$	3 1	6.70E−1	3 5	8.69E−1
1 3	−3.01E−2	5 4	9.78E−3	3 3	4.59E−1	1 1	−2.74E−1	3 2	−8.25E−1		$^3D^o$–$^3P^e$
1 4	−9.72E−2	5 5	2.11E−1	3 4	−8.06E−3	1 2	−2.69E−1	4 1	2.66E+0	1 1	8.59E+0
1 5	−1.04E−2		$^1P^o$–$^1P^e$	3 5	−5.53E−3	1 3	−4.54E−2	4 2	1.06E−1	1 2	3.43E−1
2 1	6.10E−1	1 1	−3.77E−1	4 1	3.43E−1	2 1	6.37E−1	5 1	8.33E−5		$^3D^o$–$^3D^e$
2 2	1.87E−3	2 1	−2.14E−1	4 2	1.36E+0	2 2	−1.34E−1	5 2	1.90E−3	1 1	6.20E−1
2 3	8.75E−2	3 1	−2.63E−2	4 3	1.63E−1	2 3	−4.48E−1		$^3D^e$–$^3P^o$	1 2	1.60E−1
2 4	−1.67E−2	4 1	1.56E−1	4 4	2.89E−2	3 1	2.55E−1	1 1	6.58E+0	1 3	−9.93E−5
2 5	−1.29E−1	5 1	4.18E−4	4 5	4.25E−1	3 2	−2.14E−1	1 2	4.21E−1		$^3F^o$–$^3D^e$
3 1	2.58E−2		$^1D^e$–$^1P^o$		$^1D^o$–$^1P^e$	3 3	−2.34E−3	1 3	−1.81E−2	1 1	2.10E−1
3 2	8.07E−2	1 1	2.48E−1	1 1	1.53E−1	4 1	7.30E−3	1 4	−3.96E−1	1 2	7.06E−1
3 3	2.10E−1	1 2	−4.10E−2		$^1D^o$–$^1D^e$	4 2	3.13E−1	1 5	−2.26E−1	1 3	−1.07E−2
3 4	−1.92E−1	1 3	−1.76E−1	1 1	1.05E+0	4 3	−3.23E−3	2 1	8.54E−1	2 1	1.46E+1
3 5	−7.70E−3	1 4	−5.94E−2	1 2	1.44E−1	5 1	9.95E−1	2 2	5.06E−1	2 2	2.39E−1
4 1	3.30E−2	1 5	−8.36E−4	1 3	−8.72E−3	5 2	1.43E−2	2 3	8.26E−1	2 3	5.33E−1
4 2	1.23E+0	2 1	1.60E+0	1 4	−9.62E−5	5 3	9.34E−1	2 4	−1.10E−1		
4 3	2.23E−3	2 2	2.67E−1		$^1F^o$–$^1D^e$		$^3P^o$–$^3P^e$	2 5	−2.96E−6		
4 4	2.66E−2	2 3	−9.81E−3	1 1	5.00E+0	1 1	−7.79E−1	3 1	1.19E+0		
4 5	−1.13E−4	2 4	−1.36E−1	1 2	7.43E−2	1 2	−8.91E−1	3 2	4.83E+0		

Be-like Ar (Ar^{14+})

Term energies relative to 2s ^{2}S ionization threshold for each symmetry

i	E(Ryds)	Description	i	E(Ryds)	Description	i	E(Ryds)	Description	i	E(Ryds)	Description
		$^1S^e$	9	−6.66234	2p5f			$^3S^e$	10	−4.61955	2s7d
1	−62.6170	$2s^2$	10	−6.26348	2s6d	1	−26.9187	2s3s	11	−4.04882	2p6p
2	−55.2536	$2p^2$	11	−4.60202	2s7d	2	−23.4621	2p3p	12	−3.93503	2p6f
3	−26.5305	2s3s	12	−3.99967	2p6p	3	−14.8134	2s4s	13	−3.53376	2s8d
4	−22.5717	2p3p	13	−3.92490	2p6f	4	−12.0616	2p4p	14	−2.79065	2s9d
5	−14.6758	2s4s	14	−3.52221	2s8d	5	−9.36680	2s5s	15	−2.35032	2p7p
6	−11.7725	2p4p	15	−2.78276	2s9d	6	−6.86478	2p5p	16	−2.27970	2p7f
7	−9.29758	2s5s	16	−2.32073	2p7p	7	−6.44376	2s6s			$^3D^o$
8	−6.75068	2p5p	17	−2.27307	2p7f	8	−4.72415	2s7s	1	−22.8796	2p3d
9	−6.38896	2s6s			$^1D^o$	9	−4.03326	2p6p	2	−11.8396	2p4d
10	−4.70307	2s7s	1	−23.1907	2p3d	10	−3.60005	2s8s	3	−6.73583	2p5d
11	−3.96234	2p6p	2	−11.9424	2p4d	11	−2.83918	2s9s	4	−3.96683	2p6d
12	−3.58263	2s8s	3	−6.78307	2p5d	12	−2.34735	2p7p	5	−2.29915	2p7d
13	−2.82942	2s9s	4	−3.99260	2p6d			$^3P^e$	6	−1.21758	2p8d
14	−2.31836	2p7p :	5	−2.31486	2p7d	1	−57.2793	$2p^2$	7	−.47647	2p9d
		$^1P^e$	6	−1.22789	2p8d	2	−23.3335	2p3p			$^3F^e$
1	−23.8013	2p3p	7	−.48361	2p9d	3	−12.0328	2p4p	1	−11.8086	2p4f
2	−12.1793	2p4p			$^1F^e$	4	−6.83405	2p5p	2	−6.71828	2p5f
3	−6.90009	2p5p	1	−11.8153	2p4f	5	−4.02342	2p6p	3	−3.95601	2p6f
4	−4.05906	2p6p	2	−6.72252	2p5f	6	−2.33466	2p7p	4	−2.29225	2p7f
5	−2.35624	2p7p	3	−3.95862	2p6f	7	−1.24132	2p8p	5	−1.21292	2p8f
6	−1.25541	2p8p	4	−2.29396	2p7f	8	−.49312	2p9p	6	−.47317	2p9f
7	−.50285	2p9p	5	−1.21408	2p8f			$^3P^o$			$^3F^o$
		$^1P^o$	6	−.47400	2p9f	1	−60.5542	2s2p	1	−23.2001	2p3d
1	−58.6488	2s2p			$^1F^o$	2	−25.9510	2s3p	2	−14.0846	2s4f
2	−26.0055	2s3p	1	−22.4104	2p3d	3	−24.3373	2p3s	3	−11.9388	2p4d
3	−23.9466	2p3s	2	−14.0549	2s4f	4	−22.7720	2p3d	4	−9.01427	2s5f
4	−22.3493	2p3d	3	−11.6734	2p4d	5	−14.4353	2s4p	5	−6.78140	2p5d
5	−14.4088	2s4p	4	−8.99504	2s5f	6	−12.4194	2p4s	6	−6.67944	2p5g
6	−12.3319	2p4s	5	−6.67907	2p5g	7	−11.8069	2p4d	7	−6.25906	2s6f
7	−11.6508	2p4d	6	−6.65951	2p5d	8	−9.18216	2s5p	8	−4.59785	2s7f
8	−9.16059	2s5p	7	−6.24596	2s6f	9	−7.02669	2p5s	9	−3.99225	2p6d
9	−7.00108	2p5s	8	−4.59118	2s7f	10	−6.73023	2p5d	10	−3.93488	2p6g
10	−6.65450	2p5d	9	−3.93462	2p6g	11	−6.34625	2s6p	11	−3.51976	2s8f
11	−6.32623	2s6p	10	−3.92381	2p6d	12	−4.65743	2s7p	12	−2.78075	2s9f
12	−4.65231	2s7p	11	−3.51439	2s8f	13	−4.13264	2p6s	13	−2.31465	2p7d
13	−4.11449	2p6s	12	−2.77746	2s9f	14	−3.96159	2p6d	14	−2.27902	2p7g
14	−3.91934	2p6d	13	−2.27912	2p7g	15	−3.55773	2s8p			$^3G^e$
15	−3.55118	2s8p	14	−2.27540	2p7d	16	−2.80828	2s9p	1	−11.7707	2p4f
16	−2.80555	2s9p			$^1G^e$	17	−2.40297	2p7s	2	−8.99757	2s5g
17	−2.39487	2p7s	1	−11.7420	2p4f	18	−2.30053	2p7d	3	−6.70494	2p5f
18	−2.27876	2p7d :	2	−8.99667	2s5g			$^3D^e$	4	−6.24745	2s6g
		$^1D^e$	3	−6.68723	2p5f	1	−25.4337	2s3d	5	−4.59210	2s7g
1	−56.6822	$2p^2$	4	−6.24622	2s6g	2	−23.6682	2p3p	6	−3.94849	2p6f
2	−25.0520	2s3d	5	−4.59167	2s7g	3	−14.2242	2s4d	7	−3.93766	2p6h
3	−23.0531	2p3p	6	−3.93812	2p6h	4	−12.1346	2p4p	8	−3.51533	2s8g
4	−14.1038	2s4d	7	−3.93669	2p6f	5	−11.7196	2p4f	9	−2.77797	2s9g
5	−11.9348	2p4p	8	−3.51484	2s8g	6	−9.07834	2s5d	10	−2.28913	2p7f
6	−11.6966	2p4f	9	−2.77771	2s9g	7	−6.88136	2p5p	11	−2.28091	2p7h
7	−9.02411	2s5d	10	−2.28270	2p7f	8	−6.67863	2p5f			
8	−6.79112	2p5p	11	−2.28072	2p7h	9	−6.29251	2s6d			

Be-like Ar (Ar^{14+})

Energies in ascending order from ground state for terms with effective $n \leq 4.0$, $L \leq 3$

Term	i	E(Ryds)	Term	i	E(Ryds)	Term	i	E(Ryds)	Term	i	E(Ryds)	Term	i	E(Ryds)
$^1\mathbf{S}^e$	1	0.00000	$^1\mathbf{S}^e$	3	36.0865	$^1\mathbf{P}^e$	1	38.8157	$^3\mathbf{D}^o$	1	39.7374	$^3\mathbf{P}^o$	5	48.1817
$^3\mathbf{P}^o$	1	2.06280	$^1\mathbf{P}^o$	2	36.6115	$^3\mathbf{D}^e$	2	38.9488	$^3\mathbf{P}^o$	4	39.8450	$^1\mathbf{P}^o$	5	48.2082
$^1\mathbf{P}^o$	1	3.96820	$^3\mathbf{P}^o$	2	36.6660	$^3\mathbf{S}^e$	2	39.1549	$^1\mathbf{S}^e$	4	40.0453	$^3\mathbf{D}^e$	3	48.3928
$^3\mathbf{P}^e$	1	5.33770	$^3\mathbf{D}^e$	1	37.1833	$^3\mathbf{P}^e$	2	39.2835	$^1\mathbf{F}^o$	1	40.2066	$^1\mathbf{D}^e$	4	48.5132
$^1\mathbf{D}^e$	1	5.93480	$^1\mathbf{D}^e$	2	37.5650	$^3\mathbf{F}^o$	1	39.4169	$^1\mathbf{P}^o$	4	40.2677	$^3\mathbf{F}^o$	2	48.5324
$^1\mathbf{S}^e$	2	7.36340	$^3\mathbf{P}^o$	3	38.2797	$^1\mathbf{D}^o$	1	39.4263	$^3\mathbf{S}^e$	3	47.8036			
$^3\mathbf{S}^e$	1	35.6983	$^1\mathbf{P}^o$	3	38.6704	$^1\mathbf{D}^e$	3	39.5639	$^1\mathbf{S}^e$	5	47.9412			

gf-values for transitions involving terms with effective $n \leq 4.0$, $L \leq 3$

i i'	gf$_L$	i i'	gf$_L$	i i'	gf$_L$	i i'	gf$_L$	i i'	gf$_L$	i i'	gf$_L$
	$^1\mathbf{P}^o$–$^1\mathbf{S}^e$	5 1	1.65E−1	2 5	−9.53E−2	1 3	3.01E−1	2 1	1.64E−2	3 3	2.12E−3
1 1	1.97E−1	5 2	8.27E−3	3 1	8.61E−1	1 4	−2.95E−3	2 2	−2.70E−1	3 4	6.41E−2
1 2	−1.44E−1	5 3	3.62E−1	3 2	2.52E−2		$^3\mathbf{P}^o$–$^3\mathbf{S}^e$	3 1	6.45E−1	3 5	7.65E−1
1 3	−2.82E−2	5 4	8.47E−3	3 3	4.04E−1	1 1	−2.64E−1	3 2	−7.22E−1		$^3\mathbf{D}^o$–$^3\mathbf{P}^e$
1 4	−1.06E−1	5 5	1.86E−1	3 4	−7.27E−3	1 2	−2.78E−1	4 1	2.69E+0	1 1	8.69E+0
1 5	−9.66E−3		$^1\mathbf{P}^o$–$^1\mathbf{P}^e$	3 5	−4.28E−3	1 3	−4.47E−2	4 2	9.35E−2	1 2	3.02E−1
2 1	6.32E−1	1 1	−3.84E−1	4 1	3.41E−1	2 1	5.58E−1	5 1	5.34E−6		$^3\mathbf{D}^o$–$^3\mathbf{D}^e$
2 2	1.65E−3	2 1	−1.88E−1	4 2	1.40E+0	2 2	−1.17E−1	5 2	1.85E−3	1 1	5.47E−1
2 3	7.62E−2	3 1	−2.23E−2	4 3	1.59E−1	2 3	−4.27E−1		$^3\mathbf{D}^e$–$^3\mathbf{P}^o$	1 2	1.39E−1
2 4	−1.37E−2	4 1	1.37E−1	4 4	2.68E−2	3 1	2.23E−1	1 1	6.66E+0	1 3	−4.78E−5
2 5	−1.25E−1	5 1	3.10E−4	4 5	3.72E−1	3 2	−1.89E−1	1 2	3.68E−1		$^3\mathbf{F}^o$–$^3\mathbf{D}^e$
3 1	3.04E−2		$^1\mathbf{D}^e$–$^1\mathbf{P}^o$		$^1\mathbf{D}^o$–$^1\mathbf{P}^e$	3 3	−2.08E−3	1 3	−1.59E−2	1 1	1.96E−1
3 2	7.67E−2	1 1	2.20E−1	1 1	1.33E−1	4 1	6.12E−3	1 4	3.50E 1	1 2	6.22E 1
3 3	1.84E−1	1 2	−3.82E−2		$^1\mathbf{D}^o$–$^1\mathbf{D}^e$	4 2	2.73E−1	1 5	−2.20E−1	1 3	−8.27E−3
3 4	−1.69E−1	1 3	−1.69E−1	1 1	1.05E+0	4 3	−2.82E−3	2 1	8.99E−1	2 1	1.47E+1
3 5	−7.35E−3	1 4	−6.02E−2	1 2	1.28E−1	5 1	1.05E+0	2 2	4.44E−1	2 2	2.31E−1
4 1	3.27E−2	1 5	−8.76E−4	1 3	−7.66E−3	5 2	1.34E−2	2 3	7.26E−1	2 3	4.64E−1
4 2	1.25E+0	2 1	1.61E+0	1 4	−6.53E−5	5 3	8.18E−1	2 4	−9.72E−2		
4 3	1.94E−3	2 2	2.32E−1		$^1\mathbf{F}^o$–$^1\mathbf{D}^e$		$^3\mathbf{P}^o$–$^3\mathbf{P}^e$	2 5	−1.68E−7		
4 4	2.32E−2	2 3	−7.78E−3	1 1	5.10E+0	1 1	−6.83E−1	3 1	1.19E+0		
4 5	−1.42E−4	2 4	−1.20E−1	1 2	6.89E−2	1 2	−9.38E−1	3 2	4.91E+0		

Be-like Ca (Ca^{16+})

Term energies relative to 2s ^{2}S ionization threshold for each symmetry

i	E(Ryds)	Description	*i*	E(Ryds)	Description	*i*	E(Ryds)	Description	*i*	E(Ryds)	Description
		$^1S^e$	9	-8.93466	2p5f			$^3S^e$	10	-5.92997	2s7d
1	-79.4706	$2s^2$	10	-8.04430	2s6d	1	-34.2939	2s3s	11	-5.55797	2p6p
2	-71.1774	$2p^2$	11	-5.91027	2s7d	2	-30.3991	2p3p	12	-5.42842	2p6f
3	-33.8538	2s3s	12	-5.50213	2p6p	3	-18.9186	2s4s	13	-4.53654	2s8d
4	-29.3845	2p3p	13	-5.41670	2p6f	4	-15.8205	2p4p	14	-3.58282	2s9d
5	-18.7638	2s4s	14	-4.52351	2s8d	5	-11.9792	2s5s	15	-3.38195	2p7p
6	-15.4906	2p4p	15	-3.57399	2s9d	6	-9.15713	2p5p	16	-3.30142	2p7f
7	-11.9025	2s5s	16	-3.34808	2p7p	7	-8.25782	2s6s			$^3D^o$
8	-9.01482	2p5p	17	-3.29374	2p7f	8	-6.05107	2s7s	1	-29.7346	2p3d
9	-8.20831	2s6s			$^1D^o$	9	-5.53672	2p6p	2	-15.5690	2p4d
10	-6.02847	2s7s	1	-30.0904	2p3d	10	-4.61375	2s8s	3	-9.01882	2p5d
11	-5.45298	2p6p	2	-15.6858	2p4d	11	-3.63959	2s9s	4	-5.46483	2p6d
12	-4.59619	2s8s	3	-9.07250	2p5d	12	-3.36957	2p7p	5	-3.32404	2p7d
13	-3.62946	2s9s	4	-5.49410	2p6d			$^3P^e$	6	-1.93554	2p8d
14	-3.31906	2p7p	5	-3.34187	2p7d	1	-73.4623	$2p^2$	7	-.98406	2p9d
		$^1P^e$	6	-1.94723	2p8d	2	-30.2550	2p3p			$^3F^e$
1	-30.7842	2p3p	7	-.99216	2p9d	3	-15.7903	2p4p	1	-15.5329	2p4f
2	-15.9558	2p4p			$^1F^e$	4	-9.13135	2p5p	2	-8.99846	2p5f
3	-9.20596	2p5p	1	-15.5409	2p4f	5	-5.52969	2p6p	3	-5.45223	2p6f
4	-5.56995	2p6p	2	-9.00347	2p5f	6	-3.36474	2p7p	4	-3.31601	2p7f
5	-3.38910	2p7p	3	-5.45532	2p6f	7	-1.96274	2p8p	5	-1.93011	2p8f
6	-1.97865	2p8p	4	-3.31801	2p7f	8	-1.00313	2p9p	6	-.98022	2p9f
7	-1.01412	2p9p	5	-1.93148	2p8f			$^3P^o$			$^3F^o$
		$^1P^o$	6	-.98119	2p9f	1	-77.1470	2s2p	1	-30.0998	2p3d
1	-75.0063	2s2p			$^1F^o$	2	-33.1990	2s3p	2	-18.0889	2s4f
2	-33.2593	2s3p	1	-29.1893	2p3d	3	-31.3917	2p3s	3	-15.6810	2p4d
3	-30.9456	2p3s	2	-18.0549	2s4f	4	-29.6145	2p3d	4	-11.5769	2s5f
4	-29.1292	2p3d	3	-15.3763	2p4d	5	-18.4897	2s4p	5	-9.07041	2p5d
5	-18.4595	2s4p	4	-11.5553	2s5f	6	-16.2285	2p4s	6	-8.95419	2p5g
6	-16.1279	2p4s	5	-8.95373	2p5g	7	-15.5312	2p4d	7	-8.03846	2s6f
7	-15.3535	2p4d	6	-8.92916	2p5d	8	-11.7691	2s5p	8	-5.90506	2s7f
8	-11.7451	2s5p	7	-8.02500	2s6f	9	-9.34986	2p5s	9	-5.49353	2p6d
9	-9.31681	2p5s	8	-5.89789	2s7f	10	-9.00848	2p5d	10	-5.42816	2p6g
10	-8.92230	2p5d	9	-5.42780	2p6g	11	-8.14259	2s6p	11	-4.52049	2s8f
11	-8.12362	2s6p	10	-5.41443	2p6d	12	-5.97416	2s7p	12	-3.57141	2s9f
12	-5.97142	2s7p	11	-4.51480	2s8f	13	-5.65338	2p6s	13	-3.34153	2p7d
13	-5.62831	2p6s	12	-3.56799	2s9f	14	-5.45737	2p6d	14	-3.30087	2p7g
14	-5.40963	2p6d	13	-3.30057	2p7g	15	-4.56468	2s8p			$^3G^e$
15	-4.55829	2s8p	14	-3.29323	2p7d	16	-3.60358	2s9p	1	-15.4878	2p4f
16	-3.60281	2s9p			$^1G^e$	17	-3.44192	2p7s	2	-11.5582	2s5g
17	-3.42735	2p7s	1	-15.4533	2p4f	18	-3.31990	2p7d	3	-8.98112	2p5f
18	-3.29043	2p7d	2	-11.5572	2s5g			$^3D^e$	4	-8.02670	2s6g
		$^1D^e$	3	-8.95953	2p5f	1	-32.6090	2s3d	5	-5.89879	2s7g
1	-72.7804	$2p^2$	4	-8.02562	2s6g	2	-30.6334	2p3p	6	-5.44286	2p6f
2	-32.1731	2s3d	5	-5.89835	2s7g	3	-18.2480	2s4d	7	-5.43130	2p6h
3	-29.9319	2p3p	6	-5.43163	2p6h	4	-15.9048	2p4p	8	-4.51569	2s8g
4	-18.1112	2s4d	7	-5.42883	2p6f	5	-15.4317	2p4f	9	-3.56843	2s9g
5	-15.6781	2p4p	8	-4.51519	2s8g	6	-11.6498	2s5d	10	-3.31053	2p7f
6	-15.4050	2p4f	9	-3.56818	2s9g	7	-9.18443	2p5p	11	-3.30299	2p7h
7	-11.5883	2s5d	10	-3.30300	2p7h	8	-8.95318	2p5f			
8	-9.08195	2p5p	11	-3.30161	2p7f	9	-8.07735	2s6d			

Be-like Ca (Ca^{16+})

Energies in ascending order from ground state for terms with effective $n \leq 4.0$, $L \leq 3$

Term	i	E(Ryds)	Term	i	E(Ryds)	Term	i	E(Ryds)	Term	i	E(Ryds)	Term	i	E(Ryds)
$^{1}S^{e}$	1	0.00000	$^{1}S^{e}$	3	45.6167	$^{1}P^{e}$	1	48.6863	$^{3}D^{o}$	1	49.7360	$^{3}P^{o}$	5	60.9809
$^{3}P^{o}$	1	2.32359	$^{1}P^{o}$	2	46.2113	$^{3}D^{e}$	2	48.8372	$^{3}P^{o}$	4	49.8560	$^{1}P^{o}$	5	61.0111
$^{1}P^{o}$	1	4.46429	$^{3}P^{o}$	2	46.2716	$^{3}S^{e}$	2	49.0715	$^{1}S^{e}$	4	50.0861	$^{3}D^{e}$	3	61.2226
$^{3}P^{e}$	1	6.00829	$^{3}D^{e}$	1	46.8616	$^{3}P^{e}$	2	49.2156	$^{1}F^{o}$	1	50.2813	$^{1}D^{e}$	4	61.3594
$^{1}D^{e}$	1	6.69019	$^{1}D^{e}$	2	47.2975	$^{3}F^{o}$	1	49.3708	$^{1}P^{o}$	4	50.3414	$^{3}F^{o}$	2	61.3817
$^{1}S^{e}$	2	8.29320	$^{3}P^{o}$	3	48.0789	$^{1}D^{o}$	1	49.3802	$^{3}S^{e}$	3	60.5519			
$^{3}S^{e}$	1	45.1767	$^{1}P^{o}$	3	48.5249	$^{1}D^{e}$	3	49.5387	$^{1}S^{e}$	5	60.7067			

gf-values for transitions involving terms with effective $n \leq 4.0$, $L \leq 3$

$i\ i'$	gf_L	$i\ i'$	gf_L	$i\ i'$	gf_L	$i\ i'$	gf_L	$i\ i'$	gf_L	$i\ i'$	gf_L
	$^{1}P^{o}$–$^{1}S^{e}$	5 1	1.69E−1	2 5	−9.11E−2	1 3	2.71E−1	2 1	1.73E−2	3 3	2.20E−3
1 1	1.75E−1	5 2	7.55E−3	3 1	8.93E−1	1 4	−2.38E−3	2 2	−2.40E−1	3 4	6.02E−2
1 2	−1.29E−1	5 3	3.74E−1	3 2	2.29E−2		$^{3}P^{o}$–$^{3}S^{e}$	3 1	6.25E−1	3 5	6.83E−1
1 3	−2.67E−2	5 4	7.63E−3	3 3	3.61E−1	1 1	−2.57E−1	3 2	−6.41E−1		$^{3}D^{o}$–$^{3}P^{e}$
1 4	−1.12E−1	5 5	1.65E−1	3 4	−6.62E−3	1 2	−2.86E−1	4 1	2.71E+0	1 1	8.77E+0
1 5	−9.12E−3		$^{1}P^{o}$–$^{1}P^{e}$	3 5	−3.46E−3	1 3	−4.41E−2	4 2	8.34E−2	1 2	2.70E−1
2 1	6.49E−1	1 1	−3.90E−1	4 1	3.40E−1	2 1	4.96E−1	5 1	6.76E−6		$^{3}D^{o}$–$^{3}D^{e}$
2 2	1.49E−3	2 1	−1.68E−1	4 2	1.43E+0	2 2	−1.04E−1	5 2	1.84E−3	1 1	4.89E−1
2 3	6.75E−2	3 1	−1.93E−2	4 3	1.57E−1	2 3	−4.11E−1		$^{3}D^{e}$–$^{3}P^{o}$	1 2	1.23E−1
2 4	−1.16E−2	4 1	1.22E−1	4 4	2.53E−2	3 1	1.98E−1	1 1	6.72E+0	1 3	−1.94E−5
2 5	−1.22E−1	5 1	2.36E−4	4 5	3.31E−1	3 2	−1.69E−1	1 2	3.27E 1		$^{3}F^{o}$ $^{3}D^{e}$
3 1	3.42E−2		$^{1}D^{e}$–$^{1}P^{o}$		$^{1}D^{o}$–$^{1}P^{e}$	3 3	−1.92E−3	1 3	−1.42E−2	1 1	1.82E−1
3 2	7.36E−2	1 1	1.98E−1	1 1	1.18E−1	4 1	5.27E−3	1 4	−3.14E−1	1 2	5.56E−1
3 3	1.64E−1	1 2	−3.62E−2		$^{1}D^{o}$–$^{1}D^{e}$	4 2	2.41E−1	1 5	−2.14E−1	1 3	−6.73E−3
3 4	−1.51E−1	1 3	−1.63E−1	1 1	1.05E+0	4 3	−2.54E−3	2 1	9.36E−1	2 1	1.47E+1
3 5	−7.03E−3	1 4	−6.07E−2	1 2	1.15E−1	5 1	1.09E+0	2 2	3.95E−1	2 2	2.26E−1
4 1	3.25E−2	1 5	−8.95E−4	1 3	−6.81E−3	5 2	1.27E−2	2 3	6.47E−1	2 3	4.11E−1
4 2	1.26E+0	2 1	1.61E+0	1 4	−4.50E−5	5 3	7.27E−1	2 4	−8.71E−2		
4 3	1.72E−3	2 2	2.05E−1		$^{1}F^{o}$–$^{1}D^{e}$		$^{3}P^{o}$–$^{3}P^{e}$	2 5	−3.62E−7		
4 4	2.07E−2	2 3	−6.38E−3	1 1	5.19E+0	1 1	−6.08E−1	3 1	1.19E+0		
4 5	−1.64E−4	2 4	−1.08E−1	1 2	6.36E−2	1 2	−9.76E−1	3 2	4.98E+0		

Be-like Fe (Fe^{22+})

Term energies relative to 2s ^{2}S ionization threshold for each symmetry

i E(Ryds) Description	i E(Ryds) Description	i E(Ryds) Description	i E(Ryds) Description
$^1S^e$	9 −17.6658 2p5f	$^3S^e$	10 −11.4148 2p6p
1 −142.031 $2s^2$	10 −14.7192 2s6d	1 −61.7532 2s3s	11 −11.2363 2p6f
2 −130.942 $2p^2$	11 −11.3404 2p6p	2 −56.5398 2p3p	12 −10.8386 2s7d
3 −61.1571 2s3s	12 −11.2198 2p6f	3 −34.2344 2s4s	13 −8.29487 2s8d
4 −55.1533 2p3p	13 −10.8111 2s7d	4 −30.0915 2p4p	14 −7.45183 2p7p
5 −34.0273 2s4s	14 −8.27738 2s8d	5 −21.7374 2s5s	15 −7.34112 2p7f
6 −29.6416 2p4p	15 −7.40614 2p7p	6 −17.9594 2p5p	16 −6.55108 2s9d
7 −21.6371 2s5s	16 −7.33049 2p7f	7 −15.0180 2s6s	$^3D^o$
8 −17.7575 2p5p	17 −6.53890 2s9d	8 −11.4013 2p6p	1 −55.6289 2p3d
9 −14.9596 2s6s	$^1D^o$	9 −10.9881 2s7s	2 −29.7512 2p4d
10 −11.3032 2p6p	1 −56.1190 2p3d	10 −8.40328 2s8s	3 −17.7827 2p5d
11 −10.9413 2s7s	2 −29.9105 2p4d	11 −7.43671 2p7p	4 −11.2863 2p6d
12 −8.38173 2s8s	3 −17.8557 2p5d	12 −6.62492 2s9s	5 −7.37252 2p7d
13 −7.36938 2p7p	4 −11.3261 2p6d	$^3P^e$	6 −4.83355 2p8d
14 −6.60829 2s9s	5 −7.39678 2p7d	1 −134.006 $2p^2$	7 −3.09352 2p9d
$^1P^e$	6 −4.84949 2p8d	2 −56.3497 2p3p	$^3F^e$
1 −57.0624 2p3p	7 −3.10456 2p9d	3 −30.0574 2p4p	1 −29.7000 2p4f
2 −30.2795 2p4p	$^1F^e$	4 −17.9385 2p5p	2 −17.7536 2p5f
3 −18.0384 2p5p	1 −29.7117 2p4f	5 −11.3761 2p6p	3 −11.2686 2p6f
4 −11.4301 2p6p	2 −17.7609 2p5f	6 −7.42890 2p7p	4 −7.36116 2p7f
5 −7.46162 2p7p	3 −11.2731 2p6f	7 −4.87129 2p8p	5 −4.82584 2p8f
6 −4.89269 2p8p	4 −7.36411 2p7f	8 −3.12000 2p9p	6 −3.08806 2p9f
7 −3.13479 2p9p	5 −4.82786 2p8f	$^3P^o$	$^3F^o$
$^1P^o$	6 −3.08949 2p9f	1 −138.923 2s2p	1 −56.1285 2p3d
1 −136.075 2s2p	$^1F^o$	2 −60.2770 2s3p	2 −33.1014 2s4f
2 −60.3537 2s3p	1 −54.8543 2p3d	3 −57.8847 2p3s	3 −29.9017 2p4d
3 −57.2723 2p3s	2 −33.0541 2s4f	4 −55.4719 2p3d	4 −21.1848 2s5f
4 −54.7986 2p3d	3 −29.4795 2p4d	5 −33.6533 2s4p	5 −17.8512 2p5d
5 −33.6112 2s4p	4 −21.1555 2s5f	6 −30.6504 2p4s	6 −17.6924 2p5g
6 −30.5110 2p4s	5 −17.6917 2p5g	7 −29.6980 2p4d	7 −14.7099 2s6f
7 −29.4559 2p4d	6 −17.6556 2p5d	8 −21.4509 2s5p	8 −11.3247 2p6d
8 −21.4196 2s5p	7 −14.6921 2s6f	9 −18.2336 2p5s	9 −11.2351 2p6g
9 −18.1823 2p5s	8 −11.2347 2p6g	10 −17.7625 2p5d	10 −10.8061 2s7f
10 −17.6462 2p5d	9 −11.2179 2p6d	11 −14.8573 2s6p	11 −8.27268 2s8f
11 −14.8354 2s6p	10 −10.7934 2s7f	12 −11.5453 2p6s	12 −7.39585 2p7d
12 −11.5259 2p6s	11 −8.26520 2s8f	13 −11.2818 2p6d	13 −7.34022 2p7g
13 −11.2162 2p6d	12 −7.33979 2p7g	14 −10.8938 2s7p	14 −6.53589 2s9f
14 −10.8751 2s7p	13 −7.32965 2p7d	15 −8.33490 2s8p	$^3G^e$
15 −8.32820 2s8p	14 −6.53030 2s9f	16 −7.53413 2p7s	1 −29.6339 2p4f
16 −7.51785 2p7s	$^1G^e$	17 −7.36703 2p7d	2 −21.1595 2s5g
17 −7.32691 2p7d	1 −29.5818 2p4f	18 −6.57839 2s9p	3 −17.7270 2p5f
18 −6.57183 2s9p	2 −21.1580 2s5g	$^3D^e$	4 −14.6944 2s6g
$^1D^e$	3 −17.6943 2p5f	1 −59.4689 2s3d	5 −11.2565 2p6f
1 −133.068 $2p^2$	4 −14.6930 2s6g	2 −56.8584 2p3p	6 −11.2393 2p6h
2 −58.8693 2s3d	5 −11.2363 2p6h	3 −33.3194 2s4d	7 −10.7956 2s7g
3 −55.8971 2p3p	6 −11.2363 2p6f	4 −30.2095 2p4p	8 −8.26626 2s8g
4 −33.1333 2s4d	7 −10.7942 2s7g	5 −29.5621 2p4f	9 −7.35351 2p7f
5 −29.9024 2p4p	8 −8.26562 2s8g	6 −21.2848 2s5d	10 −7.34307 2p7h
6 −29.5245 2p4f	9 −7.34313 2p7h	7 −18.0072 2p5p	11 −6.53112 2s9g
7 −21.2014 2s5d	10 −7.34014 2p7f	8 −17.6910 2p5f	
8 −17.8684 2p5p	11 −6.53054 2s9g	9 −14.7641 2s6d	

Be-like Fe (Fe^{22+})

Energies in ascending order from ground state for terms with effective $n \leq 4.0$, $L \leq 3$

Term	i	E(Ryds)	Term	i	E(Ryds)	Term	i	E(Ryds)	Term	i	E(Ryds)	Term	i	E(Ryds)
$^1S^e$	1	0.00000	$^1S^e$	3	80.8739	$^1P^e$	1	84.9686	$^3D^o$	1	86.4021	$^3P^o$	5	108.377
$^3P^o$	1	3.10800	$^1P^o$	2	81.6773	$^3D^e$	2	85.1726	$^3P^o$	4	86.5591	$^1P^o$	5	108.419
$^1P^o$	1	5.95601	$^3P^o$	2	81.7540	$^3S^e$	2	85.4912	$^1S^e$	4	86.8777	$^3D^e$	3	108.711
$^3P^e$	1	8.02501	$^3D^e$	1	82.5621	$^3P^e$	2	85.6813	$^1F^o$	1	87.1767	$^1D^e$	4	108.897
$^1D^e$	1	8.96301	$^1D^e$	2	83.1617	$^3F^o$	1	85.9025	$^1P^o$	4	87.2324	$^3F^o$	2	108.929
$^1S^e$	2	11.0890	$^3P^o$	3	84.1463	$^1D^o$	1	85.9120	$^3S^e$	3	107.796			
$^3S^e$	1	80.2778	$^1P^o$	3	84.7587	$^1D^e$	3	86.1339	$^1S^e$	5	108.003			

gf-values for transitions involving terms with effective $n \leq 4.0$, $L \leq 3$

i i'	gf_L	i i'	gf_L	i i'	gf_L	i i'	gf_L	i i'	gf_L	i i'	gf_L
	$^1P^o$–$^1S^e$	5 1	1.77E−1	2 5	−8.25E−2	1 3	2.09E−1	2 1	1.91E−2	3 3	2.36E−3
1 1	1.32E−1	5 2	6.35E−3	3 1	9.60E−1	1 4	−1.54E−3	2 2	−1.81E−1	3 4	5.31E−2
1 2	−9.77E−2	5 3	3.99E−1	3 2	1.81E−2		$^3P^o$–$^3S^e$	3 1	5.87E−1	3 5	5.17E−1
1 3	−2.38E−2	5 4	6.31E−3	3 3	2.73E−1	1 1	−2.43E−1	3 2	−4.81E−1		$^3D^o$–$^3P^e$
1 4	−1.26E−1	5 5	1.25E−1	3 4	−5.18E−3	1 2	−3.02E−1	4 1	2.75E+0	1 1	8.93E+0
1 5	−8.08E−3		$^1P^o$–$^1P^e$	3 5	−2.20E−3	1 3	−4.26E−2	4 2	6.30E−2	1 2	2.05E−1
2 1	6.84E−1	1 1	−4.01E−1	4 1	3.38E−1	2 1	3.73E−1	5 1	1.35E−4		$^3D^o$–$^3D^e$
2 2	1.20E−3	2 1	−1.28E−1	4 2	1.49E+0	2 2	−7.84E−2	5 2	1.85E−3	1 1	3.72E−1
2 3	5.03E−2	3 1	−1.37E−2	4 3	1.54E−1	2 3	−3.79E−1		$^3D^e$–$^3P^o$	1 2	9.08E−2
2 4	−7.97E−3	4 1	9.16E−2	4 4	2.25E−2	3 1	1.49E−1	1 1	6.83E+0	1 3	−9.07E−7
2 5	−1.15E−1	5 1	1.17E−4	4 5	2.48E−1	3 2	−1.28E−1	1 2	2.45E−1		$^3F^o$–$^3D^e$
3 1	4.24E−2		$^1D^e$–$^1P^o$		$^1D^o$–$^1P^e$	3 3	−1.65E−3	1 3	−1.07E−2	1 1	1.49E−1
3 2	6.78E−2	1 1	1.52E−1	1 1	8.86E−2	4 1	3.73E−3	1 4	−2.41E−1	1 2	4.20E−1
3 3	1.24E−1	1 2	−3.23E−2		$^1D^o$–$^1D^e$	4 2	1.80E−1	1 5	−2.02E−1	1 3	−4.37E−3
3 4	−1.14E−1	1 3	−1.52E−1	1 1	1.05E+0	4 3	−2.06E−3	2 1	1.01E+0	2 1	1.48E+1
3 5	−6.29E−3	1 4	−6.14E−2	1 2	8.86E−2	5 1	1.17E+0	2 2	2.98E−1	2 2	2.15E−1
4 1	3.21E−2	1 5	−9.06E−4	1 3	−5.11E−3	5 2	1.16E−2	2 3	4.88E−1	2 3	3.06E−1
4 2	1.28E+0	2 1	1.61E+0	1 4	−1.58E−5	5 3	5.46E−1	2 4	−6.63E−2		
4 3	1.27E−3	2 2	1.51E−1		$^1F^o$–$^1D^e$		$^3P^o$–$^3P^e$	2 5	−5.27E−6		
4 4	1.55E−2	2 3	−4.06E−3	1 1	5.35E+0	1 1	−4.57E−1	3 1	1.19E+0		
4 5	−2.02E−4	2 4	−8.26E−2	1 2	5.07E−2	1 2	−1.05E+0	3 2	5.11E+0		

B-like S (S^{11+})

Term energies relative to $2s^2$ ^{1}S ionization threshold for each symmetry

i	E(Ryds)	Description
		^{2}S^e
1	−37.4693	$2s2p^2$
2	−18.1206	$2s^2$ ^{1}S $3s$
3	−14.8361	$2s2p$ ^{3}P $3p$
4	−13.5339	$2s2p$ ^{1}P $3p$
5	−11.4259	$2p^2$ ^{1}S $3s$
6	−10.7212	$2p^2$ ^{1}D $3d$
7	−9.81517	$2s^2$ ^{1}S $4s$
8	−7.44281	$2s2p$ ^{3}P $4p$
9	−6.22414	$2s^2$ ^{1}S $5s$
10	−5.88105	$2s2p$ ^{1}P $4p$
11	−4.23393	$2s^2$ ^{1}S $6s$
12	−4.08293	$2s2p$ ^{3}P $5p$
13	−3.85967	$2p^2$ ^{1}D $4d$
14	−3.32996	$2p^2$ ^{1}S $4s$
15	−3.08755	$2s^2$ ^{1}S $7s$
16	−2.49416	$2s2p$ ^{1}P $5p$
17	−2.33685	$2s^2$ ^{1}S $8s$
18	−2.26479	$2s2p$ ^{3}P $6p$
19	−1.84332	$2s^2$ ^{1}S $9s$
		^{4}S^e
1	−15.3644	$2s2p$ ^{3}P $3p$
2	−7.64749	$2s2p$ ^{3}P $4p$
3	−4.18106	$2s2p$ ^{3}P $5p$
4	−2.32601	$2s2p$ ^{3}P $6p$
5	−1.21724	$2s2p$ ^{3}P $7p$
6	−.50187	$2s2p$ ^{3}P $8p$
7	−.01367	$2s2p$ ^{3}P $9p$
		^{2}P^e
1	−37.2301	$2s2p^2$
2	−15.5669	$2s2p$ ^{3}P $3p$
3	−13.7455	$2s2p$ ^{1}P $3p$
4	−12.8147	$2p^2$ ^{3}P $3s$
5	−11.7691	$2p^2$ ^{3}P $3d$
6	−10.8364	$2p^2$ ^{1}D $3d$
7	−7.69402	$2s2p$ ^{3}P $4p$
8	−5.97932	$2s2p$ ^{1}P $4p$
9	−5.02448	$2p^2$ ^{3}P $4s$
10	−4.51363	$2p^2$ ^{3}P $4d$
11	−4.18144	$2s2p$ ^{3}P $5p$
12	−3.87653	$2p^2$ ^{1}D $4d$
13	−2.51258	$2s2p$ ^{1}P $5p$
14	−2.33684	$2s2p$ ^{3}P $6p$
15	−1.45511	$2p^2$ ^{3}P $5s$
16	−1.22143	$2s2p$ ^{3}P $7p$
17	−1.17654	$2p^2$ ^{3}P $5d$
18	−.68837	$2s2p$ ^{1}P $6p$:
19	−.59394	$2p^2$ ^{1}D $5d$:
20	−.50356	$2s2p$ ^{3}P $8p$
21	−.01699	$2s2p$ ^{3}P $9p$
		^{4}P^e
1	−39.6851	$2s2p^2$
2	−15.2288	$2s2p$ ^{3}P $3p$
3	−13.2587	$2p^2$ ^{3}P $3s$
4	−11.5726	$2p^2$ ^{3}P $3d$
5	−7.61742	$2s2p$ ^{3}P $4p$
6	−5.11266	$2p^2$ ^{3}P $4s$
7	−4.45996	$2p^2$ ^{3}P $4d$
8	−4.16270	$2s2p$ ^{3}P $5p$
9	−2.31960	$2s2p$ ^{3}P $6p$
10	−1.48977	$2p^2$ ^{3}P $5s$
11	−1.21872	$2s2p$ ^{3}P $7p$
12	−1.15646	$2p^2$ ^{3}P $5d$
13	−.49856	$2s2p$ ^{3}P $8p$
14	−.01178	$2s2p$ ^{3}P $9p$
		^{2}D^e
1	−38.3443	$2s2p^2$
2	−16.4272	$2s^2$ ^{1}S $3d$
3	−15.1131	$2s2p$ ^{3}P $3p$
4	−13.7404	$2s2p$ ^{1}P $3p$
5	−12.6086	$2p^2$ ^{1}D $3s$
6	−11.2284	$2p^2$ ^{3}P $3d$
7	−11.0344	$2p^2$ ^{1}D $3d$
8	−9.91289	$2p^2$ ^{1}S $3d$
9	−9.17372	$2s^2$ ^{1}S $4d$
10	−7.54920	$2s2p$ ^{3}P $4p$
11	−7.15476	$2s2p$ ^{3}P $4f$
12	−6.00365	$2s2p$ ^{1}P $4p$
13	−5.84612	$2s^2$ ^{1}S $5d$
14	−5.50547	$2s2p$ ^{1}P $4f$
15	−4.57832	$2p^2$ ^{1}D $4s$
16	−4.32667	$2p^2$ ^{3}P $4d$
17	−4.11617	$2s2p$ ^{3}P $5p$
18	−4.05173	$2s^2$ ^{1}S $6d$
19	−3.94242	$2p^2$ ^{1}D $4d$
20	−3.92868	$2s2p$ ^{3}P $5f$
21	−2.97119	$2s^2$ ^{1}S $7d$
22	−2.75824	$2p^2$ ^{1}S $4d$
23	−2.51535	$2s2p$ ^{1}P $5p$
24	−2.30033	$2s2p$ ^{3}P $6p$
25	−2.27916	$2s2p$ ^{1}P $5f$
26	−2.26883	$2s^2$ ^{1}S $8d$
27	−2.18814	$2s2p$ ^{3}P $6f$
28	−1.79268	$2s^2$ ^{1}S $9d$
		^{4}D^e
1	−15.5235	$2s2p$ ^{3}P $3p$
2	−11.7756	$2p^2$ ^{3}P $3d$
3	−7.70164	$2s2p$ ^{3}P $4p$
4	−7.18141	$2s2p$ ^{3}P $4f$
5	−4.53244	$2p^2$ ^{3}P $4d$
6	−4.19749	$2s2p$ ^{3}P $5p$
7	−3.95349	$2s2p$ ^{3}P $5f$
8	−2.33996	$2s2p$ ^{3}P $6p$
9	−2.19863	$2s2p$ ^{3}P $6f$
10	−1.24027	$2s2p$ ^{3}P $7p$
11	−1.18103	$2p^2$ ^{3}P $5d$
12	−1.13655	$2s2p$ ^{3}P $7f$
13	−.50666	$2s2p$ ^{3}P $8p$
14	−.44911	$2s2p$ ^{3}P $8f$
15	−.01735	$2s2p$ ^{3}P $9p$
		^{2}F^e
1	−11.7104	$2p^2$ ^{3}P $3d$
2	−11.0195	$2p^2$ ^{1}D $3d$
3	−7.22805	$2s2p$ ^{3}P $4f$
4	−5.58107	$2s2p$ ^{1}P $4f$
5	−4.44711	$2p^2$ ^{3}P $4d$
6	−4.00609	$2p^2$ ^{1}D $4d$
7	−3.93580	$2s2p$ ^{3}P $5f$
8	−2.31937	$2s2p$ ^{1}P $5f$
9	−2.20889	$2s2p$ ^{3}P $6f$
10	−1.17112	$2p^2$ ^{3}P $5d$:
11	−1.11887	$2s2p$ ^{3}P $7f$:
12	−1.11148	$2p^2$ ^{3}P $5g$
13	−.67492	$2p^2$ ^{1}D $5d$
14	−.59070	$2p^2$ ^{1}D $5g$
15	−.54823	$2s2p$ ^{1}P $6f$
16	−.45264	$2s2p$ ^{3}P $8f$
		^{4}F^e
1	−11.9352	$2p^2$ ^{3}P $3d$
2	−7.24442	$2s2p$ ^{3}P $4f$
3	−4.57452	$2p^2$ ^{3}P $4d$
4	−3.98269	$2s2p$ ^{3}P $5f$
5	−2.21406	$2s2p$ ^{3}P $6f$
6	−1.21942	$2p^2$ ^{3}P $5d$
7	−1.14581	$2s2p$ ^{3}P $7f$
8	−1.11158	$2p^2$ ^{3}P $5g$
9	−.45555	$2s2p$ ^{3}P $8f$
		^{2}G^e
1	−11.2599	$2p^2$ ^{1}D $3d$
2	−7.19040	$2s2p$ ^{3}P $4f$
3	−5.80629	$2s^2$ ^{1}S $5g$
4	−5.49954	$2s2p$ ^{1}P $4f$
5	−4.01429	$2p^2$ ^{1}D $4d$
6	−3.99834	$2s^2$ ^{1}S $6g$
7	−3.94671	$2s2p$ ^{3}P $5f$
8	−2.94043	$2s^2$ ^{1}S $7g$
9	−2.31471	$2s2p$ ^{1}P $5f$
10	−2.23637	$2s^2$ ^{1}S $8g$
11	−2.19818	$2s2p$ ^{3}P $6h$:
12	−2.19745	$2s2p$ ^{3}P $6f$:
13	−1.77681	$2s^2$ ^{1}S $9g$
		^{4}G^e
1	−7.21653	$2s2p$ ^{3}P $4f$
2	−3.97241	$2s2p$ ^{3}P $5f$
3	−2.20827	$2s2p$ ^{3}P $6f$
4	−2.19836	$2s2p$ ^{3}P $6h$
5	−1.15202	$2s2p$ ^{3}P $7h$
6	−1.14400	$2s2p$ ^{3}P $7f$
7	−1.08022	$2p^2$ ^{3}P $5g$
8	−.45324	$2s2p$ ^{3}P $8f$
9	−.44756	$2s2p$ ^{3}P $8h$
		^{2}S^o
1	−12.7164	$2p^2$ ^{3}P $3p$
2	−4.87433	$2p^2$ ^{3}P $4p$
3	−1.36592	$2p^2$ ^{3}P $5p$
		^{4}S^o
1	−36.0535	$2p^3$
2	−12.1551	$2p^2$ ^{3}P $3p$
3	−4.69466	$2p^2$ ^{3}P $4p$
4	−1.28436	$2p^2$ ^{3}P $5p$
		^{2}P^o
1	−41.3332	$2s^22p$
2	−34.5353	$2p^3$
3	−17.2502	$2s^2$ ^{1}S $3p$
4	−15.8944	$2s2p$ ^{3}P $3s$
5	−14.5779	$2s2p$ ^{1}P $3s$
6	−14.0950	$2s2p$ ^{3}P $3d$
7	−12.7849	$2s2p$ ^{1}P $3d$
8	−12.1657	$2p^2$ ^{3}P $3p$
9	−11.4466	$2p^2$ ^{1}D $3p$
10	−10.5889	$2p^2$ ^{1}S $3p$
11	−9.46384	$2s^2$ ^{1}S $4p$
12	−7.89255	$2s2p$ ^{3}P $4s$
13	−7.18820	$2s2p$ ^{3}P $4d$
14	−6.33538	$2s2p$ ^{1}P $4s$
15	−5.99869	$2s^2$ ^{1}S $5p$
16	−5.63220	$2s2p$ ^{1}P $4d$
17	−4.66061	$2p^2$ ^{3}P $4p$
18	−4.30058	$2s2p$ ^{3}P $5s$
19	−4.14453	$2p^2$ ^{1}D $4p$
20	−4.12325	$2s^2$ ^{1}S $6p$
21	−3.95877	$2s2p$ ^{3}P $5d$
22	−3.77096	$2p^2$ ^{1}D $4f$
23	−3.02908	$2s^2$ ^{1}S $7p$:
24	−3.01348	$2p^2$ ^{1}S $4p$:
25	−2.67357	$2s2p$ ^{1}P $5s$
26	−2.40611	$2s2p$ ^{3}P $6s$
27	−2.35699	$2s2p$ ^{1}P $5d$
28	−2.29179	$2s^2$ ^{1}S $8p$
29	−2.20088	$2s2p$ ^{3}P $6d$
30	−1.81500	$2s^2$ ^{1}S $9p$

B-like S (S^{11+})

i	Energy(Ryds)	Description	i	Energy(Ryds)	Description	i	Energy(Ryds)	Description
	^{4}P°		20	−.57744	$2s2p$ ^{1}P $6d$:		^{4}F°	
1	−16.3164	$2s2p$ ^{3}P $3s$	21	−.46746	$2s2p$ ^{3}P $8d$	1	−14.8450	$2s2p$ ^{3}P $3d$
2	−14.5561	$2s2p$ ^{3}P $3d$		^{4}D°		2	−7.44585	$2s2p$ ^{3}P $4d$
3	−12.4606	$2p^2$ ^{3}P $3p$	1	−14.6445	$2s2p$ ^{3}P $3d$	3	−4.35006	$2p^2$ ^{3}P $4f$
4	−8.02672	$2s2p$ ^{3}P $4s$	2	−12.5631	$2p^2$ ^{3}P $3p$	4	−4.07964	$2s2p$ ^{3}P $5d$
5	−7.34825	$2s2p$ ^{3}P $4d$	3	−7.37539	$2s2p$ ^{3}P $4d$	5	−3.93284	$2s2p$ ^{3}P $5g$
6	−4.79648	$2p^2$ ^{3}P $4p$	4	−4.82232	$2p^2$ ^{3}P $4p$	6	−2.26816	$2s2p$ ^{3}P $6d$
7	−4.35777	$2s2p$ ^{3}P $5s$	5	−4.37849	$2p^2$ ^{3}P $4f$	7	−2.19458	$2s2p$ ^{3}P $6g$
8	−4.03491	$2s2p$ ^{3}P $5d$	6	−4.04629	$2s2p$ ^{3}P $5d$	8	−1.18196	$2s2p$ ^{3}P $7d$
9	−2.43354	$2s2p$ ^{3}P $6s$	7	−2.24944	$2s2p$ ^{3}P $6d$	9	−1.15051	$2s2p$ ^{3}P $7g$
10	−2.24365	$2s2p$ ^{3}P $6d$	8	−1.34244	$2p^2$ ^{3}P $5p$	10	−1.08649	$2p^2$ ^{3}P $5f$
11	−1.34425	$2p^2$ ^{3}P $5p$	9	−1.16884	$2s2p$ ^{3}P $7d$	11	−.47750	$2s2p$ ^{3}P $8d$
12	−1.26329	$2s2p$ ^{3}P $7s$	10	−1.12313	$2p^2$ ^{3}P $5f$	12	−.44636	$2s2p$ ^{3}P $8g$
13	−1.16558	$2s2p$ ^{3}P $7d$	11	−.46983	$2s2p$ ^{3}P $8d$		^{2}G°	
14	−.54508	$2s2p$ ^{3}P $8s$		^{2}F°		1	−4.39260	$2p^2$ ^{3}P $4f$
15	−.46777	$2s2p$ ^{3}P $8d$	1	−14.1932	$2s2p$ ^{3}P $3d$	2	−3.94509	$2s2p$ ^{3}P $5g$
16	−.04447	$2s2p$ ^{3}P $9s$	2	−13.0488	$2s2p$ ^{1}P $3d$	3	−3.88768	$2p^2$ ^{1}D $4f$
	^{2}D°		3	−11.9064	$2p^2$ ^{1}D $3p$	4	−2.30099	$2s2p$ ^{1}P $5g$
1	−35.3134	$2p^3$	4	−9.00790	$2s^2$ ^{1}S $4f$	5	−2.20363	$2s2p$ ^{3}P $6g$
2	−14.6261	$2s2p$ ^{3}P $3d$	5	−7.21679	$2s2p$ ^{3}P $4d$	6	−1.16516	$2s2p$ ^{3}P $7g$:
3	−12.9308	$2s2p$ ^{1}P $3d$	6	−5.80901	$2s^2$ ^{1}S $5f$:	7	−1.09382	$2p^2$ ^{3}P $5f$:
4	−12.3617	$2p^2$ ^{3}P $3p$	7	−5.68679	$2s2p$ ^{1}P $4d$:	8	−.63002	$2p^2$ ^{1}D $5f$
5	−11.6559	$2p^2$ ^{1}D $3p$	8	−4.32840	$2p^2$ ^{3}P $4f$	9	−.53091	$2s2p$ ^{1}P $6g$
6	−7.36445	$2s2p$ ^{3}P $4d$	9	−4.26662	$2p^3$ ^{1}D $4p$	10	−.45007	$2s2p$ ^{3}P $8g$
7	−5.70452	$2s2p$ ^{1}P $4d$	10	−4.01075	$2s^2$ ^{1}S $6f$		^{4}G°	
8	−4.70852	$2p^2$ ^{3}P $4p$	11	−3.96815	$2s2p$ ^{3}P $5d$	1	−4.40728	$2p^2$ ^{3}P $4f$
9	−4.37498	$2p^2$ ^{3}P $4f$	12	−3.93046	$2s2p$ ^{3}P $5g$	2	−3.94687	$2s2p$ ^{3}P $5g$
10	−4.22653	$2p^2$ ^{1}D $4p$	13	−3.86750	$2p^2$ ^{1}D $4f$	3	−2.20385	$2s2p$ ^{3}P $6g$
11	−4.04084	$2s2p$ ^{3}P $5d$	14	−2.94366	$2s^2$ ^{1}S $7f$	4	−1.16783	$2s2p$ ^{3}P $7g$:
12	−3.81741	$2p^2$ ^{1}D $4f$	15	−2.64935	$2p^2$ ^{1}S $4f$	5	−1.10209	$2p^2$ ^{3}P $5f$:
13	−2.38504	$2s2p$ ^{1}P $5d$	16	−2.39106	$2s2p$ ^{1}P $5d$	6	−.45031	$2s2p$ ^{3}P $8g$
14	−2.24168	$2s2p$ ^{3}P $6d$	17	−2.25285	$2s2p$ ^{1}P $5g$:			
15	−1.28341	$2p^2$ ^{3}P $5p$	18	−2.25283	$2s^2$ ^{1}S $8f$:			
16	−1.16623	$2s2p$ ^{3}P $7d$	19	−2.20551	$2s2p$ ^{3}P $6d$			
17	−1.12132	$2p^2$ ^{3}P $5f$	20	−2.19409	$2s2p$ ^{3}P $6g$			
18	−.80022	$2p^2$ ^{1}D $5p$	21	−1.77970	$2s^2$ ^{1}S $9f$			
19	−.59492	$2p^2$ ^{1}D $5f$:						

B-like S (S^{11+})

Energies in ascending order from ground state for terms with effective $n \leq 4.0$, $L \leq 4$

Term	i	E(Ryds)	Term	i	E(Ryds)	Term	i	E(Ryds)	Term	i	E(Ryds)	Term	i	E(Ryds)
$^2P^o$	1	0.00000	$^2P^e$	2	25.7663	$^2P^e$	3	27.5877	$^2P^o$	8	29.1675	$^2D^e$	7	30.2988
$^4P^e$	1	1.64810	$^4D^e$	1	25.8097	$^2D^e$	4	27.5928	$^4S^o$	2	29.1781	$^2F^e$	2	30.3137
$^2D^e$	1	2.98890	$^4S^e$	1	25.9688	$^2S^e$	4	27.7993	$^4F^e$	1	29.3980	$^2P^e$	6	30.4968
$^2S^e$	1	3.86390	$^4P^e$	2	26.1044	$^4P^e$	3	28.0745	$^2F^o$	3	29.4268	$^2S^e$	6	30.6120
$^2P^e$	1	4.10310	$^2D^e$	3	26.2201	$^2F^o$	2	28.2844	$^4D^e$	2	29.5576	$^2P^o$	10	30.7443
$^4S^o$	1	5.27970	$^4F^o$	1	26.4882	$^2D^o$	3	28.4024	$^2P^e$	5	29.5641	$^2D^e$	8	31.4203
$^2D^o$	1	6.01980	$^2S^e$	3	26.4971	$^2P^e$	4	28.5185	$^2F^e$	1	29.6228	$^2S^e$	7	31.5180
$^2P^o$	2	6.79790	$^4D^o$	1	26.6887	$^2P^o$	7	28.5483	$^2D^o$	5	29.6773	$^2P^o$	11	31.8693
$^2S^e$	2	23.2126	$^2D^o$	2	26.7071	$^2S^o$	1	28.6168	$^4P^e$	4	29.7606	$^2D^e$	9	32.1594
$^2P^o$	3	24.0830	$^2P^o$	5	26.7553	$^2D^e$	5	28.7246	$^2P^o$	9	29.8866	$^2F^o$	4	32.3253
$^2D^e$	2	24.9060	$^4P^o$	2	26.7771	$^4D^o$	2	28.7701	$^2S^e$	5	29.9073			
$^4P^o$	1	25.0168	$^2F^o$	1	27.1400	$^4P^o$	3	28.8726	$^2G^e$	1	30.0733			
$^2P^o$	4	25.4388	$^2P^o$	6	27.2382	$^2D^o$	4	28.9715	$^2D^e$	6	30.1048			

gf-values for transitions involving terms with effective $n \leq 4.0$, $L \leq 4$

i	i'	gf_L	i	i'	gf_L	i	i'	gf_L	i	i'	gf_L	i	i'	gf_L	i	i'	gf_L
		$^2P^e$–$^2S^o$	2	7	−1.18E−3	5	8	8.66E−3	1	8	−4.54E−1	4	9	−4.50E−1	1	9	−2.95E−1
1	1	−2.36E−1	2	8	−8.32E−4	5	9	6.25E−6	1	9	−1.18E−1	4	10	−9.37E−2	1	10	−1.16E−4
2	1	−2.65E−1	2	9	−2.12E−3	5	10	−5.20E−1	1	10	−5.53E−2	4	11	−8.46E−6	1	11	−3.62E−4
3	1	−2.85E−2	2	10	−1.06E−4	5	11	−7.84E−5	1	11	−7.08E−4	5	1	1.45E−2	2	1	3.84E+0
4	1	−1.46E−2	2	11	−5.78E−1	6	1	6.48E−3	2	1	8.35E−1	5	2	7.89E−1	2	2	2.76E−5
5	1	2.68E−1	3	1	3.16E−1	6	2	4.26E−1	2	2	2.74E−4	5	3	2.30E−4	2	3	6.04E−1
6	1	5.37E−2	3	2	2.70E−2	6	3	3.14E−4	2	3	4.31E−2	5	4	8.65E−6	2	4	−1.90E−2
		$^4P^e$–$^4S^o$	3	3	2.95E−2	6	4	1.38E−3	2	4	1.48E−1	5	5	6.91E−3	2	5	−6.94E−4
1	1	−6.76E−1	3	4	2.16E−1	6	5	2.74E−2	2	5	−1.14E−3	5	6	1.07E−1	2	6	−3.65E−6
1	2	−3.27E−1	3	5	−4.78E−6	6	6	2.27E−4	2	6	−3.16E−1	5	7	7.26E−3	2	7	−5.95E−1
2	1	5.86E−3	3	6	−2.22E−1	6	7	1.32E−1	2	7	−8.95E−3	5	8	9.65E−2	2	8	−3.83E−5
2	2	−3.81E−1	3	7	−3.53E−2	6	8	7.02E−2	2	8	−7.06E−2	5	9	−5.20E−3	2	9	−8.20E−4
3	1	5.69E−1	3	8	−1.50E−1	6	9	1.37E−1	2	9	−1.32E−2	5	10	−1.85E−3	2	10	−8.41E−4
3	2	−5.35E−1	3	9	−2.99E−3	6	10	−1.53E−2	2	10	−9.81E−4	5	11	−3.96E−5	2	11	−1.84E−1
4	1	6.89E+0	3	10	−1.97E−3	6	11	−1.07E−4	2	11	−1.96E−3	6	1	2.71E−2	3	1	1.55E+0
4	2	3.20E−1	3	11	−1.49E−4	7	1	2.38E−2	3	1	2.63E−1	6	2	1.85E+0	3	2	1.58E−2
		$^2S^e$–$^2P^o$	4	1	4.11E−2	7	2	3.00E−6	3	2	4.10E−2	6	3	2.30E−3	3	3	1.48E−1
1	1	1.76E−1	4	2	2.48E−4	7	3	2.89E−1	3	3	5.07E−1	6	4	2.70E−3	3	4	7.02E−1
1	2	−1.32E−1	4	3	1.75E−1	7	4	1.33E−2	3	4	7.67E−3	6	5	8.95E−5	3	5	−4.47E−3
1	3	−6.00E−3	4	4	8.54E−2	7	5	2.66E−4	3	5	4.13E−1	6	6	2.95E−3	3	6	−4.32E−2
1	4	−1.06E−1	4	5	1.62E−1	7	6	6.29E−4	3	6	1.74E−3	6	7	3.21E−1	3	7	−3.14E−3
1	5	−1.63E−1	4	6	2.67E−3	7	7	2.40E−3	3	7	−1.28E−1	6	8	1.22E−1	3	8	−2.80E−1
1	6	−1.63E+0	4	7	−1.89E−1	7	8	1.48E−4	3	8	−2.79E−1	6	9	2.10E−1	3	9	−7.05E−2
1	7	−8.03E−1	4	8	−1.12E−2	7	9	4.57E−6	3	9	−1.39E−2	6	10	−7.97E−4	3	10	−2.52E−3
1	8	−5.17E−4	4	9	−2.10E−1	7	10	6.13E−6	3	10	−2.05E−1	6	11	−4.85E−5	3	11	−5.25E−3
1	9	−5.98E−3	4	10	−7.22E−2	7	11	−7.50E−1	3	11	−6.02E−4			$^2D^e$–$^2P^o$	4	1	1.14E−1
1	10	−3.61E−1	4	11	−2.12E−2			$^2P^e$–$^2P^o$	4	1	5.80E−2	1	1	3.34E−1	4	2	7.97E−4
1	11	−1.51E−2	5	1	3.29E−3	1	1	8.70E−1	4	2	2.64E−1	1	2	−4.37E−1	4	3	5.52E−1
2	1	1.38E−1	5	2	2.57E−1	1	2	−5.38E−1	4	3	5.81E−3	1	3	−3.90E−2	4	4	1.74E−1
2	2	2.57E−5	5	3	5.48E−4	1	3	−3.66E−3	4	4	7.74E−1	1	4	−2.95E−1	4	5	6.42E−1
2	3	−4.53E−1	5	4	8.33E−4	1	4	−2.47E−2	4	5	1.28E−1	1	5	−2.24E−1	4	6	4.25E−3
2	4	−4.63E−2	5	5	3.38E−1	1	5	−3.35E−1	4	6	1.29E−7	1	6	−4.43E−2	4	7	−1.20E−2
2	5	−4.72E−1	5	6	1.79E−2	1	6	−2.80E−1	4	7	−8.42E−4	1	7	−4.20E−2	4	8	−3.28E−5
2	6	−2.57E−2	5	7	1.10E−2	1	7	−1.34E+0	4	8	−2.52E−1	1	8	−1.78E−1	4	9	−4.22E−6

B-like S (S^{11+})

i i'		gf$_L$	i i'		gf$_L$	i i'		gf$_L$	i i'		gf$_L$	i i'		gf$_L$	i i'		gf$_L$
4	10	−6.06E−1	8	7	6.63E−2			$^{2}\mathbf{P}^{e}$–$^{2}\mathbf{D}^{o}$	2	5	−4.62E−3	1	5	−2.96E−5	5	2	7.17E−3
4	11	−1.27E−3	8	8	3.44E−3	1	1	−4.18E−1	3	1	2.73E−2	2	1	1.08E+1	5	3	−1.20E+0
5	1	1.65E−2	8	9	4.64E−3	1	2	−5.52E−1	3	2	−7.70E−2	2	2	1.31E−2	5	4	−2.39E−3
5	2	2.21E−1	8	10	6.99E−1	1	3	−4.47E+0	3	3	−4.48E−2	2	3	1.69E−1	6	1	4.50E−1
5	3	2.24E−2	8	11	−1.51E−3	1	4	−7.21E−1	3	4	−6.26E−1	2	4	4.41E−2	6	2	2.83E−5
5	4	6.17E−3	9	1	7.49E−1	1	5	−5.88E−1	3	5	−5.69E−2	2	5	6.95E−1	6	3	3.06E−2
5	5	7.40E−1	9	2	4.67E−3	2	1	2.44E−2	4	1	6.18E−4			$^{4}\mathbf{P}^{e}$–$^{4}\mathbf{D}^{o}$	6	4	−7.20E−4
5	6	8.36E−3	9	3	2.70E+0	2	2	−5.62E−1	4	2	1.43E−2	1	1	−1.18E+1	7	1	1.13E−1
5	7	1.14E−4	9	4	1.60E−1	2	3	−1.02E−2	4	3	−1.58E−1	1	2	−8.93E−1	7	2	1.25E−2
5	8	−5.94E−2	9	5	1.61E−2	2	4	−6.35E−2	4	4	−2.11E−2	2	1	−6.89E−1	7	3	1.16E−1
5	9	−7.14E−1	9	6	1.69E−4	2	5	−3.13E−3	4	5	−2.59E−1	2	2	−4.75E−1	7	4	−6.25E−6
5	10	−3.39E−2	9	7	5.56E−2	3	1	1.16E−1	5	1	6.29E−1	3	1	2.46E−2	8	1	8.10E−6
5	11	−8.97E−6	9	8	1.63E−5	3	2	1.45E−3	5	2	6.64E−3	3	2	−1.53E+0	8	2	8.22E−1
6	1	4.39E−2	9	9	6.50E−4	3	3	−3.84E−1	5	3	4.09E−4	4	1	9.78E−1	8	3	1.14E−1
6	2	1.11E+0	9	10	2.13E−3	3	4	−4.15E−1	5	4	−5.67E−2	4	2	5.90E−4	8	4	−3.50E−3
6	3	4.32E−5	9	11	1.09E+0	3	5	−8.64E−2	5	5	−7.70E−1			$^{4}\mathbf{D}^{e}$–$^{4}\mathbf{D}^{o}$	9	1	3.04E−4
6	4	5.53E−4			$^{4}\mathbf{S}^{e}$–$^{4}\mathbf{P}^{o}$	4	1	2.84E−1	6	1	2.89E+0	1	1	−2.98E−1	9	2	8.16E−3
6	5	6.81E−4	1	1	3.64E−1	4	2	3.90E−3	6	2	6.20E−2	1	2	−1.42E+0	9	3	1.60E−4
6	6	1.08E−1	1	2	−5.29E−1	4	3	8.78E−3	6	3	9.71E−2	2	1	6.56E−2	9	4	−5.90E−1
6	7	1.43E−2	1	3	−3.43E−1	4	4	−2.66E−1	6	4	1.09E−1	2	2	4.29E−1			$^{2}\mathbf{F}^{e}$–$^{2}\mathbf{F}^{o}$
6	8	6.87E−1			$^{4}\mathbf{P}^{e}$–$^{4}\mathbf{P}^{o}$	4	5	−5.98E−1	6	5	6.87E−2			$^{4}\mathbf{F}^{e}$–$^{4}\mathbf{D}^{o}$	1	1	4.35E−1
6	9	6.15E−3	1	1	−1.01E+0	5	1	2.83E−1	7	1	1.33E+0	1	1	1.66E−1	1	2	8.69E−2
6	10	−7.60E−5	1	2	−3.74E+0	5	2	4.43E−1	7	2	3.83E−3	1	2	1.65E+0	1	3	1.95E−2
6	11	−3.37E−4	1	3	−7.24E−1	5	3	1.52E−4	7	3	4.45E−1			$^{2}\mathbf{D}^{e}$–$^{2}\mathbf{F}^{o}$	1	4	−5.50E−5
7	1	4.41E−2	2	1	1.35E+0	5	4	4.24E−4	7	4	3.36E−1	1	1	−5.83E+0	2	1	1.79E−1
7	2	3.11E+0	2	2	−2.12E−1	5	5	−1.51E−3	7	5	9.10E−2	1	2	−4.38E+0	2	2	4.20E−1
7	3	2.22E−4	2	3	−3.23E−1	6	1	6.05E−1	8	1	5.65E−2	1	3	−5.02E−1	2	3	1.75E−1
7	4	3.04E−3	3	1	1.02E+0	6	2	2.05E−2	8	2	6.02E−3	1	4	−4.10E−3	2	4	−1.14E−4
7	5	1.91E−3	3	2	6.03E−3	6	3	1.16E−1	8	3	2.94E−1	2	1	−3.19E−3			$^{2}\mathbf{G}^{e}$–$^{2}\mathbf{F}^{o}$
7	6	3.07E−2	3	3	−1.06E+0	6	4	1.38E−1	8	4	1.28E−3	2	2	−4.38E−1	1	1	2.89E−3
7	7	9.83E−2	4	1	3.42E−3	6	5	8.74E−2	8	5	1.99E−2	2	3	−1.23E−2	1	2	1.44E−1
7	8	1.89E−4	4	2	3.17E−1			$^{2}\mathbf{D}^{e}$–$^{2}\mathbf{D}^{o}$	9	1	4.03E−4	2	4	−9.88E+0	1	3	1.27E+0
7	9	1.97E−1	4	3	5.37E−1	1	1	−6.87E−1	9	2	1.10E−5	3	1	−1.16E+0	1	4	−6.86E−4
7	10	−4.12E−4			$^{4}\mathbf{D}^{e}$–$^{4}\mathbf{P}^{o}$	1	2	−1.76E+0	9	3	3.53E−4	3	2	−1.51E−3			$^{4}\mathbf{D}^{e}$–$^{4}\mathbf{F}^{o}$
7	11	−2.49E−4	1	1	1.52E+0	1	3	−4.23E−1	9	4	2.03E−5	3	3	−2.15E−3	1	1	−1.57E+0
8	1	1.84E−2	1	2	−1.22E−1	1	4	−1.47E−1	9	5	3.45E−4	3	4	−6.85E−4	2	1	1.18E+0
8	2	2.72E+0	1	3	−5.43E−1	1	5	−6.14E−1			$^{2}\mathbf{F}^{e}$–$^{2}\mathbf{D}^{o}$	4	1	2.38E−3			$^{4}\mathbf{F}^{e}$–$^{4}\mathbf{F}^{o}$
8	3	2.29E−3	2	1	1.11E−2	2	1	5.98E−4	1	1	6.94E−1	4	2	−6.50E−1	1	1	1.44E+0
8	4	1.25E−4	2	2	3.17E−1	2	2	−5.79E−3	1	2	8.98E−2	4	3	−6.58E−1			
8	5	4.91E−5	2	3	1.04E+0	2	3	−8.26E−1	1	3	4.50E−2	4	4	−1.17E−1			
8	6	2.69E−3				2	4	−6.23E−4	1	4	7.72E−1	5	1	2.14E−2			

B-like Ar (Ar^{13+})

Term energies relative to $2s^2$ ^{1}S ionization threshold for each symmetry

i	E(Ryds)	Description	i	E(Ryds)	Description	i	E(Ryds)	Description	i	E(Ryds)	Description
	$^2S^e$		23	−.17355	$2p^2$ ^{3}P $6d$	32	−1.99184	$2p^2$ ^{1}D $5d$	11	−.19505	$2p^2$ ^{3}P $6d$
1	−50.8111	$2s2p^2$	24	−.12765	$2s2p$ ^{1}P $7p$		**$^4D^e$**		12	−.12345	$2p^2$ ^{3}P $6g$
2	−24.2702	$2s^2$ ^{1}S $3s$		**$^4P^e$**		1	−21.2785	$2s2p$ ^{3}P $3p$		**$^2G^e$**	
3	−20.4779	$2s2p$ ^{3}P $3p$	1	−53.3576	$2s2p^2$	2	−16.9681	$2p^2$ ^{3}P $3d$	1	−16.3627	$2p^2$ ^{1}D $3d$
4	−18.9857	$2s2p$ ^{1}P $3p$	2	−20.9359	$2s2p$ ^{3}P $3p$	3	−10.7831	$2s2p$ ^{3}P $4p$	2	−10.1773	$2s2p$ ^{3}P $4f$
5	−16.5872	$2p^2$ ^{1}S $3s$	3	−18.7000	$2p^2$ ^{3}P $3s$	4	−10.1703	$2s2p$ ^{3}P $4f$	3	−8.33167	$2s2p$ ^{1}P $4f$
6	−15.7369	$2p^2$ ^{1}D $3d$	4	−16.7261	$2p^2$ ^{3}P $3d$	5	−7.13846	$2p^2$ ^{3}P $4d$	4	−7.80891	$2s^2$ ^{1}S $5g$
7	−13.2153	$2s^2$ ^{1}S $4s$	5	−10.6855	$2s2p$ ^{3}P $4p$	6	−6.06756	$2s2p$ ^{3}P $5p$	5	−6.53153	$2p^2$ ^{1}D $4d$
8	−10.4816	$2s2p$ ^{3}P $4p$	6	−7.82796	$2p^2$ ^{3}P $4s$	7	−5.77431	$2s2p$ ^{3}P $5f$	6	−5.77242	$2s2p$ ^{3}P $5f$
9	−8.80816	$2s2p$ ^{1}P $4p$	7	−7.05780	$2p^2$ ^{3}P $4d$	8	−3.55029	$2s2p$ ^{3}P $6p$	7	−5.44454	$2s^2$ ^{1}S $6g$
10	−8.27538	$2s^2$ ^{1}S $5s$	8	−6.02545	$2s2p$ ^{3}P $5p$	9	−3.38289	$2s2p$ ^{3}P $6f$	8	−4.01419	$2s^2$ ^{1}S $7g$
11	−6.36011	$2p^2$ ^{1}D $4d$	9	−3.52678	$2s2p$ ^{3}P $6p$	10	−2.62095	$2p^2$ ^{3}P $5d$	9	−3.87351	$2s2p$ ^{1}P $5f$
12	−5.94379	$2s2p$ ^{3}P $5p$	10	−2.96653	$2p^2$ ^{3}P $5s$	11	−2.04023	$2s2p$ ^{3}P $7p$	10	−3.38251	$2s2p$ ^{3}P $6h$
13	−5.76616	$2p^2$ ^{1}S $4s$	11	−2.58478	$2p^2$ ^{3}P $5d$	12	−1.93835	$2s2p$ ^{3}P $7f$	11	−3.38051	$2s2p$ ^{3}P $6f$
14	−5.71340	$2s^2$ ^{1}S $6s$	12	−2.02604	$2s2p$ ^{3}P $7p$	13	−1.06933	$2s2p$ ^{3}P $8p$	12	−3.06135	$2s^2$ ^{1}S $8g$
15	−4.21224	$2s^2$ ^{1}S $7s$	13	−1.06009	$2s2p$ ^{3}P $8p$	14	−1.00091	$2s2p$ ^{3}P $8f$	13	−2.50240	$2p^2$ ^{3}P $5g$
16	−4.06501	$2s2p$ ^{1}P $5p$	14	−.40309	$2s2p$ ^{3}P $9p$	15	−.40663	$2s2p$ ^{3}P $9p$	14	−2.41966	$2s^2$ ^{1}S $9g$
17	−3.46935	$2s2p$ ^{3}P $6p$	15	−.37484	$2p^2$ ^{3}P $6s$	16	−.35818	$2s2p$ ^{3}P $9f$	15	−2.02198	$2p^2$ ^{1}D $5d$
18	−3.17369	$2s^2$ ^{1}S $8s$	16	−.16026	$2p^2$ ^{3}P $6d$	17	−.18094	$2p^2$ ^{3}P $6d$		**$^4G^e$**	
19	−2.49962	$2s^2$ ^{1}S $9s$		**$^2D^e$**			**$^2F^e$**		1	−10.2103	$2s2p$ ^{3}P $4f$
	$^4S^e$		1	−51.8136	$2s2p^2$	1	−16.8875	$2p^2$ ^{3}P $3d$	2	−5.79495	$2s2p$ ^{3}P $5f$
1	−21.0950	$2s2p$ ^{3}P $3p$	2	−22.2958	$2s^2$ ^{1}S $3d$	2	−16.0620	$2p^2$ ^{1}D $3d$	3	−3.39400	$2s2p$ ^{3}P $6f$
2	−10.7194	$2s2p$ ^{3}P $4p$	3	−20.7995	$2s2p$ ^{3}P $3p$	3	−10.2278	$2s2p$ ^{3}P $4f$	4	−3.38264	$2s2p$ ^{3}P $6h$
3	−6.04117	$2s2p$ ^{3}P $5p$	4	−19.2348	$2s2p$ ^{1}P $3p$	4	−8.34467	$2s2p$ ^{1}P $4f$	5	−2.50319	$2p^2$ ^{3}P $5g$
4	−3.53356	$2s2p$ ^{3}P $6p$	5	−17.9398	$2p^2$ ^{1}D $3s$	5	−7.03695	$2p^2$ ^{3}P $4d$	6	−1.94560	$2s2p$ ^{3}P $7f$
5	−2.03220	$2s2p$ ^{3}P $7p$	6	−16.3149	$2p^2$ ^{3}P $3d$	6	−6.49543	$2p^2$ ^{1}D $4d$	7	−1.93455	$2s2p$ ^{3}P $7h$
6	−1.06300	$2s2p$ ^{3}P $8p$	7	−16.0957	$2p^2$ ^{1}D $3d$	7	−5.79451	$2s2p$ ^{3}P $5f$	8	−1.00571	$2s2p$ ^{3}P $8f$
7	−.40132	$2s2p$ ^{3}P $9p$	8	−14.8114	$2p^2$ ^{1}S $3d$	8	−3.90788	$2s2p$ ^{1}P $5f$	9	−.99937	$2s2p$ ^{3}P $8h$
	$^2P^e$		9	−12.4605	$2s^2$ ^{1}S $4d$	9	−3.39657	$2s2p$ ^{3}P $6f$	10	−.36148	$2s2p$ ^{3}P $9f$
1	−50.5484	$2s2p^2$	10	−10.6061	$2s2p$ ^{3}P $4p$	10	−2.56212	$2p^2$ ^{3}P $5d$	11	−.35769	$2s2p$ ^{3}P $9h$
2	−21.3286	$2s2p$ ^{3}P $3p$	11	−10.1386	$2s2p$ ^{3}P $4f$	11	−2.52197	$2p^2$ ^{3}P $5g$	12	−.11375	$2p^2$ ^{3}P $6g$
3	−19.2485	$2s2p$ ^{1}P $3p$	12	−8.84006	$2s2p$ ^{1}P $4p$	12	−2.01486	$2p^2$ ^{1}D $5d$		**$^2S^o$**	
4	−18.1744	$2p^2$ ^{3}P $3s$	13	−8.25772	$2s2p$ ^{1}P $4f$	13	−1.94332	$2s2p$ ^{3}P $7f$	1	−18.0667	$2p^2$ ^{3}P $3p$
5	−16.9610	$2p^2$ ^{3}P $3d$	14	−7.94548	$2s^2$ ^{1}S $5d$	14	−1.91424	$2p^2$ ^{1}D $5g$	2	−7.54879	$2p^2$ ^{3}P $4p$
6	−15.8642	$2p^2$ ^{1}D $3d$	15	−7.19850	$2p^2$ ^{1}D $4s$	15	−1.49888	$2s2p$ ^{1}P $6f$	3	−2.82263	$2p^2$ ^{3}P $5p$
7	−10.7756	$2s2p$ ^{3}P $4p$	16	−6.88361	$2p^2$ ^{3}P $4d$	16	−1.00616	$2s2p$ ^{3}P $8f$	4	−.29300	$2p^2$ ^{3}P $6p$
8	−8.81568	$2s2p$ ^{1}P $4p$	17	−6.45872	$2p^2$ ^{1}D $4d$	17	−.36269	$2s2p$ ^{3}P $9f$		**$^4S^o$**	
9	−7.71604	$2p^2$ ^{3}P $4s$	18	−5.98113	$2s2p$ ^{3}P $5p$	18	−.14520	$2p^2$ ^{3}P $6d$	1	−49.1898	$2p^3$
10	−7.12225	$2p^2$ ^{3}P $4d$	19	−5.75261	$2s2p$ ^{3}P $5f$	19	−.12332	$2p^2$ ^{3}P $6g$	2	−17.4126	$2p^2$ ^{3}P $3p$
11	−6.37947	$2p^2$ ^{1}D $4d$	20	−5.50574	$2s^2$ ^{1}S $6d$	20	−.04784	$2s2p$ ^{1}P $7f$	3	−7.34039	$2p^2$ ^{3}P $4p$
12	−6.05351	$2s2p$ ^{3}P $5p$	21	−5.09232	$2p^2$ ^{1}S $4d$		**$^4F^e$**		4	−2.72804	$2p^2$ ^{3}P $5p$
13	−4.13580	$2s2p$ ^{1}P $5p$	22	−4.14594	$2s2p$ ^{1}P $5p$	1	−17.1535	$2p^2$ ^{3}P $3d$	5	−.24166	$2p^2$ ^{3}P $6p$
14	−3.55257	$2s2p$ ^{3}P $6p$	23	−4.03747	$2s^2$ ^{1}S $7d$	2	−10.2458	$2s2p$ ^{3}P $4f$		**$^2P^o$**	
15	−2.91228	$2p^2$ ^{3}P $5s$	24	−3.86416	$2s2p$ ^{1}P $5f$	3	−7.19435	$2p^2$ ^{3}P $4d$	1	−55.2554	$2s^22p$
16	−2.61116	$2p^2$ ^{3}P $5d$	25	−3.50473	$2s2p$ ^{3}P $6p$	4	−5.80957	$2s2p$ ^{3}P $5f$	2	−47.4308	$2p^3$
17	−2.03544	$2s2p$ ^{3}P $7p$	26	−3.37110	$2s2p$ ^{3}P $6f$	5	−3.40151	$2s2p$ ^{3}P $6f$	3	−23.2576	$2s^2$ ^{1}S $3p$
18	−1.96187	$2p^2$ ^{1}D $5d$	27	−3.08804	$2s^2$ ^{1}S $8d$	6	−2.64777	$2p^2$ ^{3}P $5d$	4	−21.7086	$2s2p$ ^{3}P $3s$
19	−1.61937	$2s2p$ ^{1}P $6p$	28	−2.50503	$2p^2$ ^{3}P $5d$	7	−2.52209	$2p^2$ ^{3}P $5g$	5	−20.2143	$2s2p$ ^{1}P $3s$
20	−1.06813	$2s2p$ ^{3}P $8p$	29	−2.43767	$2s^2$ ^{1}S $9d$	8	−1.95003	$2s2p$ ^{3}P $7f$	6	−19.6017	$2s2p$ ^{3}P $3d$
21	−.41775	$2s2p$ ^{3}P $9p$	30	−2.35308	$2p^2$ ^{1}D $5s$	9	−1.00876	$2s2p$ ^{3}P $8f$	7	−18.1164	$2s2p$ ^{1}P $3d$
22	−.34069	$2p^2$ ^{3}P $6s$	31	−2.01135	$2s2p$ ^{3}P $7p$	10	−.36371	$2s2p$ ^{3}P $9f$	8	−17.4210	$2p^2$ ^{3}P $3p$

B-like Ar (Ar^{13+})

i	E(Ryds)	Description	i	E(Ryds)	Description	i	E(Ryds)	Description	i	E(Ryds)	Description
9	−16.5778	$2p^2$ ^{1}D $3p$		**^{2}D$^\circ$**		4	−12.2645	$2s^2$ ^{1}S $4f$	9	−1.48510	$2s2p$ ^{1}P $6g$
10	−15.6107	$2p^2$ ^{1}S $3p$	1	−48.3215	$2p^3$	5	−10.2116	$2s2p$ ^{3}P $4d$	10	−1.00265	$2s2p$ ^{3}P $8g$
11	−12.8025	$2s^2$ ^{1}S $4p$	2	−20.2319	$2s2p$ ^{3}P $3d$	6	−8.51942	$2s2p$ ^{1}P $4d$	11	−.36029	$2s2p$ ^{3}P $9g$
12	−11.0087	$2s2p$ ^{3}P $4s$	3	−18.2883	$2s2p$ ^{1}P $3d$	7	−7.84717	$2s^2$ ^{1}S $5f$	12	−.12747	$2p^2$ ^{3}P $6f$
13	−10.1802	$2s2p$ ^{3}P $4d$	4	−17.6507	$2p^2$ ^{3}P $3p$	8	−6.88860	$2p^2$ ^{3}P $4f$	13	−.12162	$2p^2$ ^{3}P $6h$
14	−9.22792	$2s2p$ ^{1}P $4s$	5	−16.8243	$2p^2$ ^{1}D $3p$	9	−6.83739	$2p^2$ ^{1}D $4p$	14	−.04086	$2s2p$ ^{1}P $7g$
15	−8.44219	$2s2p$ ^{1}P $4d$	6	−10.3880	$2s2p$ ^{3}P $4d$	10	−6.36939	$2p^2$ ^{1}D $4f$		**^{4}G$^\circ$**	
16	−8.08990	$2s^2$ ^{1}S $5p$	7	−8.49041	$2s2p$ ^{1}P $4d$	11	−5.79631	$2s2p$ ^{3}P $5d$	1	−6.98189	$2p^2$ ^{3}P $4f$
17	−7.27835	$2p^2$ ^{3}P $4p$	8	−7.35409	$2p^2$ ^{3}P $4p$	12	−5.75918	$2s2p$ ^{3}P $5g$	2	−5.77945	$2s2p$ ^{3}P $5g$
18	−6.70817	$2p^2$ ^{1}D $4p$	9	−6.96115	$2p^2$ ^{3}P $4f$	13	−5.45229	$2s^2$ ^{1}S $6f$	3	−3.39021	$2s2p$ ^{3}P $6g$
19	−6.25538	$2p^2$ ^{1}D $4f$	10	−6.79261	$2p^2$ ^{1}D $4p$	14	−4.95048	$2p^2$ ^{1}S $4f$	4	−2.54309	$2p^2$ ^{3}P $5f$
20	−6.17919	$2s2p$ ^{3}P $5s$	11	−6.31029	$2p^2$ ^{1}D $4f$	15	−4.02021	$2s^2$ ^{1}S $7f$:	5	−1.93971	$2s2p$ ^{3}P $7g$
21	−5.77959	$2s2p$ ^{3}P $5d$	12	−5.87683	$2s2p$ ^{3}P $5d$	16	−3.97452	$2s2p$ ^{1}P $5d$:	6	−1.00286	$2s2p$ ^{3}P $8g$
22	−5.59981	$2s^2$ ^{1}S $6p$	13	−3.98163	$2s2p$ ^{1}P $5d$	17	−3.85282	$2s2p$ ^{1}P $5g$	7	−.36033	$2s2p$ ^{3}P $9g$
23	−5.40536	$2p^2$ ^{1}S $4p$	14	−3.43859	$2s2p$ ^{3}P $6d$	18	−3.39443	$2s2p$ ^{3}P $6d$	8	−.13603	$2p^2$ ^{3}P $6f$
24	−4.33684	$2s2p$ ^{1}P $5s$	15	−2.72404	$2p^2$ ^{3}P $5p$	19	−3.37770	$2s2p$ ^{3}P $6g$	9	−.12162	$2p^2$ ^{3}P $6h$
25	−4.10440	$2s^2$ ^{1}S $7p$	16	−2.53416	$2p^2$ ^{3}P $5f$	20	−3.06604	$2s^2$ ^{1}S $8f$			
26	−3.93729	$2s2p$ ^{1}P $5d$	17	−2.16151	$2p^2$ ^{1}D $5p$	21	−2.49327	$2p^2$ ^{3}P $5f$			
27	−3.62346	$2s2p$ ^{3}P $6s$	18	−1.97484	$2s2p$ ^{3}P $7d$	22	−2.42264	$2s^2$ ^{1}S $9f$			
28	−3.38698	$2s2p$ ^{3}P $6d$	19	−1.91202	$2p^2$ ^{1}D $5f$	23	−2.17595	$2p^2$ ^{1}D $5p$			
29	−3.12590	$2s^2$ ^{1}S $8p$	20	−1.53929	$2s2p$ ^{1}P $6d$		**^{4}F$^\circ$**				
30	−2.68357	$2p^2$ ^{3}P $5p$	21	−1.02426	$2s2p$ ^{3}P $8d$	1	−20.4914	$2s2p$ ^{3}P $3d$			
31	−2.46522	$2s^2$ ^{1}S $9p$	22	−.37518	$2s2p$ ^{3}P $9d$	2	−10.4828	$2s2p$ ^{3}P $4d$			
32	−2.13837	$2p^2$ ^{1}D $5p$	23	−.23832	$2p^2$ ^{3}P $6p$	3	−6.91607	$2p^2$ ^{3}P $4f$			
33	2.08051	$2s2p$ ^{3}P $7s$	24	−.13065	$2p^2$ ^{3}P $6f$	4	−5.92249	$2s2p$ ^{3}P $5d$			
	^{4}P$^\circ$		25	−.07377	$2s2p$ ^{1}P $7d$	5	−5.76047	$2s2p$ ^{3}P $5g$			
1	−22.2025	$2s2p$ ^{3}P $3s$		**^{4}D$^\circ$**		6	−3.46509	$2s2p$ ^{3}P $6d$			
2	−20.1523	$2s2p$ ^{3}P $3d$	1	−20.2505	$2s2p$ ^{3}P $3d$	7	−3.37823	$2s2p$ ^{3}P $6g$			
3	−17.7681	$2p^2$ ^{3}P $3p$	2	−17.8878	$2p^2$ ^{3}P $3p$	8	−2.51293	$2p^2$ ^{3}P $5f$			
4	−11.1639	$2s2p$ ^{3}P $4s$	3	−10.3997	$2s2p$ ^{3}P $4d$	9	−1.98905	$2s2p$ ^{3}P $7d$			
5	−10.3686	$2s2p$ ^{3}P $4d$	4	−7.48736	$2p^2$ ^{3}P $4p$	10	−1.93275	$2s2p$ ^{3}P $7g$			
6	−7.44863	$2p^2$ ^{3}P $4p$	5	−6.96505	$2p^2$ ^{3}P $4f$	11	−1.03440	$2s2p$ ^{3}P $8d$			
7	−6.25700	$2s2p$ ^{3}P $5s$	6	−5.88341	$2s2p$ ^{3}P $5d$	12	−.99808	$2s2p$ ^{3}P $8g$			
8	−5.86969	$2s2p$ ^{3}P $5d$	7	−3.44326	$2s2p$ ^{3}P $6d$	13	−.38143	$2s2p$ ^{3}P $9d$			
9	−3.66090	$2s2p$ ^{3}P $6s$	8	−2.79349	$2p^2$ ^{3}P $5p$	14	−.35678	$2s2p$ ^{3}P $9g$			
10	−3.43659	$2s2p$ ^{3}P $6d$	9	−2.53713	$2p^2$ ^{3}P $5f$	15	−.12038	$2p^2$ ^{3}P $6f$			
11	−2.77302	$2p^2$ ^{3}P $5p$	10	−1.97543	$2s2p$ ^{3}P $7d$		**^{2}G$^\circ$**				
12	−2.10765	$2s2p$ ^{3}P $7s$	11	−1.02568	$2s2p$ ^{3}P $8d$	1	−6.96187	$2p^2$ ^{3}P $4f$			
13	−1.97169	$2s2p$ ^{3}P $7d$	12	−.37582	$2s2p$ ^{3}P $9d$	2	−6.39105	$2p^2$ ^{1}D $4f$			
14	−1.11496	$2s2p$ ^{3}P $8s$	13	−.27798	$2p^2$ ^{3}P $6p$	3	−5.77841	$2s2p$ ^{3}P $5g$			
15	−1.02291	$2s2p$ ^{3}P $8d$	14	−.13253	$2p^2$ ^{3}P $6f$	4	−3.88633	$2s2p$ ^{1}P $5g$			
16	−.43989	$2s2p$ ^{3}P $9s$		**^{2}F$^\circ$**		5	−3.39003	$2s2p$ ^{3}P $6g$			
17	−.37349	$2s2p$ ^{3}P $9d$	1	−19.7113	$2s2p$ ^{3}P $3d$	6	−2.52968	$2p^2$ ^{3}P $5f$			
18	−.26616	$2p^2$ ^{3}P $6p$	2	−18.4207	$2s2p$ ^{1}P $3d$	7	−1.95360	$2p^2$ ^{1}D $5f$			
			3	−17.1188	$2p^2$ ^{1}D $3p$	8	−1.93902	$2s2p$ ^{3}P $7g$			

B-like Ar (Ar^{13+})

Energies in ascending order from ground state for terms with effective $n \leq 4.0$, $L \leq 4$

Term	i	E(Ryds)	Term	i	E(Ryds)	Term	i	E(Ryds)	Term	i	E(Ryds)	Term	i	E(Ryds)
$^2P^o$	1	0.00000	$^2P^e$	2	33.9268	$^2P^e$	3	36.0069	$^2P^o$	8	37.8344	$^2D^e$	7	39.1597
$^4P^e$	1	1.89780	$^4D^e$	1	33.9769	$^2D^e$	4	36.0206	$^4S^o$	2	37.8428	$^2F^e$	2	39.1934
$^2D^e$	1	3.44180	$^4S^e$	1	34.1604	$^2S^e$	4	36.2697	$^4F^e$	1	38.1019	$^2P^e$	6	39.3912
$^2S^e$	1	4.44430	$^4P^e$	2	34.3195	$^4P^e$	3	36.5554	$^2F^o$	3	38.1366	$^2S^e$	6	39.5185
$^2P^e$	1	4.70700	$^2D^e$	3	34.4559	$^2F^o$	2	36.8347	$^4D^e$	2	38.2873	$^2P^o$	10	39.6447
$^4S^o$	1	6.06560	$^4F^o$	1	34.7640	$^2D^o$	3	36.9671	$^2P^e$	5	38.2944	$^2D^e$	8	40.4440
$^2D^o$	1	6.93390	$^2S^e$	3	34.7775	$^2P^e$	4	37.0810	$^2F^e$	1	38.3679	$^2S^e$	7	42.0401
$^2P^o$	2	7.82460	$^4D^o$	1	35.0049	$^2P^o$	7	37.1390	$^2D^o$	5	38.4311	$^2P^o$	11	42.4529
$^2S^e$	2	30.9852	$^2D^o$	2	35.0235	$^2S^o$	1	37.1887	$^4P^e$	4	38.5293	$^2D^e$	9	42.7949
$^2P^o$	3	31.9978	$^2P^o$	5	35.0411	$^2D^e$	5	37.3156	$^2S^e$	5	38.6682	$^2F^o$	4	42.9909
$^2D^e$	2	32.9596	$^4P^o$	2	35.1031	$^4D^o$	2	37.3676	$^2P^o$	9	38.6776			
$^4P^o$	1	33.0529	$^2F^o$	1	35.5441	$^4P^o$	3	37.4873	$^2G^e$	1	38.8927			
$^2P^o$	4	33.5468	$^2P^o$	6	35.6537	$^2D^o$	4	37.6047	$^2D^e$	6	38.9405			

gf-values for transitions involving terms with effective $n \leq 4.0$, $L \leq 4$

i i'	gf_L	i i'	gf_L	i i'	gf_L	i i'	gf_L	i i'	gf_L	i i'	gf_L
	$^2P^e$–$^2S^o$	2 7	−7.06E−4	5 8	6.97E−3	1 8	−4.81E−1	4 9	−3.98E−1	1 9	−3.24E−1
1 1	−2.43E−1	2 8	−8.22E−4	5 9	−4.64E−7	1 9	−1.37E−1	4 10	−8.62E−2	1 10	−1.53E−4
2 1	−2.33E−1	2 9	−1.61E−3	5 10	−4.61E−1	1 10	−5.94E−2	4 11	−3.19E−5	1 11	−4.49E−4
3 1	−2.52E−2	2 10	−1.12E−4	5 11	−5.30E−5	1 11	−2.75E−4	5 1	1.40E−2	2 1	3.90E+0
4 1	−1.17E−2	2 11	−6.24E−1	6 1	6.18E−3	2 1	8.70E−1	5 2	7.89E−1	2 2	3.00E−5
5 1	2.32E−1	3 1	3.41E−1	6 2	4.47E−1	2 2	2.95E−4	5 3	1.99E−4	2 3	5.19E−1
6 1	4.72E−2	3 2	2.53E−2	6 3	3.12E−4	2 3	3.86E−2	5 4	2.01E−6	2 4	−1.53E−2
	$^4P^e$–$^4S^o$	3 3	2.81E−2	6 4	1.30E−3	2 4	1.27E−1	5 5	5.65E−3	2 5	−7.32E−4
1 1	−5.93E−1	3 4	1.88E−1	6 5	2.15E−2	2 5	−7.79E−4	5 6	9.63E−2	2 6	−1.08E−5
1 2	−3.55E−1	3 5	−1.29E−6	6 6	1.09E−4	2 6	−2.75E−1	5 7	6.78E−3	2 7	−5.21E−1
2 1	6.86E−3	3 6	−1.89E−1	6 7	1.18E−1	2 7	−7.94E−3	5 8	8.20E−2	2 8	−2.04E−5
2 2	−3.35E−1	3 7	−3.12E−2	6 8	6.19E−2	2 8	−6.05E−2	5 9	−4.48E−3	2 9	−6.82E−4
3 1	5.48E−1	3 8	−1.36E−1	6 9	1.18E−1	2 9	−1.05E−2	5 10	−1.56E−3	2 10	−8.35E−4
3 2	−4.67E−1	3 9	−3.05E−3	6 10	−1.04E−2	2 10	−8.00E−4	5 11	−1.97E−5	2 11	−1.78E−1
4 1	7.05E+0	3 10	−2.17E−3	6 11	−4.01E−5	2 11	−1.76E−3	6 1	2.71E−2	3 1	1.66E+0
4 2	2.79E−1	3 11	−6.48E−5	7 1	2.39E−2	3 1	2.80E−1	6 2	1.91E+0	3 2	1.40E−2
	$^2S^e$–$^2P^o$	4 1	3.78E−2	7 2	1.19E−8	3 2	4.08E−2	6 3	1.94E−3	3 3	1.31E−1
1 1	1.55E−1	4 2	7.26E−5	7 3	2.79E−1	3 3	4.46E−1	6 4	2.34E−3	3 4	6.09E−1
1 2	−1.18E−1	4 3	1.46E−1	7 4	1.31E−2	3 4	7.27E−3	6 5	8.79E−5	3 5	−4.06E−3
1 3	−5.46E−3	4 4	7.69E−2	7 5	1.23E−4	3 5	3.57E−1	6 6	2.68E−3	3 6	−3.72E−2
1 4	−1.01E−1	4 5	1.42E−1	7 6	4.33E−4	3 6	1.11E−3	6 7	2.86E−1	3 7	−2.59E−3
1 5	−1.52E−1	4 6	2.43E−3	7 7	1.69E−3	3 7	−1.10E−1	6 8	1.07E−1	3 8	−2.42E−1
1 6	−1.67E+0	4 7	−1.59E−1	7 8	8.62E−5	3 8	−2.52E−1	6 9	1.79E−1	3 9	−5.85E−2
1 7	−8.33E−1	4 8	−9.95E−3	7 9	2.20E−7	3 9	−1.07E−2	6 10	−7.03E−4	3 10	−2.42E−3
1 8	−4.47E−4	4 9	−1.86E−1	7 10	5.47E−6	3 10	−1.81E−1	6 11	−4.78E−5	3 11	−4.40E−3
1 9	−5.83E−3	4 10	−6.18E−2	7 11	−6.56E−1	3 11	−4.46E−4		$^2D^e$–$^2P^o$	4 1	1.15E−1
1 10	−3.86E−1	4 11	−1.68E−2		$^2P^e$–$^2P^o$	4 1	7.43E−2	1 1	2.96E−1	4 2	9.58E−4
1 11	−1.23E−2	5 1	4.06E−3	1 1	7.62E−1	4 2	2.50E−1	1 2	−3.84E−1	4 3	4.81E−1
2 1	1.33E−1	5 2	2.43E−1	1 2	−4.78E−1	4 3	5.37E−3	1 3	−3.68E−2	4 4	1.58E−1
2 2	2.77E−5	5 3	4.96E−4	1 3	−4.04E−3	4 4	6.72E−1	1 4	−2.82E−1	4 5	5.63E−1
2 3	−3.94E−1	5 4	6.22E−4	1 4	−2.19E−2	4 5	1.15E−1	1 5	−2.17E−1	4 6	2.93E−3
2 4	−4.15E−2	5 5	3.00E−1	1 5	−3.17E−1	4 6	1.80E−6	1 6	−4.71E−2	4 7	−1.03E−2
2 5	−4.14E−1	5 6	1.41E−2	1 6	−2.88E−1	4 7	−1.22E−3	1 7	−4.34E−2	4 8	−3.50E−5
2 6	−2.07E−2	5 7	9.43E−3	1 7	−1.36E+0	4 8	−2.15E−1	1 8	−1.84E−1	4 9	−8.57E−6

B-like Ar (Ar^{13+})

i i'	gf_L	i i'	gf_L	i i'	gf_L	i i'	gf_L	i i'	gf_L	i i'	gf_L
4 10	−5.30E−1	8 7	7.25E−2		**$^2P^e$–$^2D^o$**	2 5	−4.20E−3	1 5	−7.44E−6	5 2	6.69E−3
4 11	−9.68E−4	8 8	3.54E−3	1 1	−3.75E−1	3 1	2.63E−2	2 1	1.13E+1	5 3	−1.06E+0
5 1	2.16E−2	8 9	3.86E−3	1 2	−5.40E−1	3 2	−6.46E−2	2 2	1.61E−2	5 4	−1.77E−3
5 2	2.14E−1	8 10	6.15E−1	1 3	−4.55E+0	3 3	−3.87E−2	2 3	1.56E−1	6 1	3.80E−1
5 3	1.94E−2	8 11	−4.86E−4	1 4	−7.61E−1	3 4	−5.54E−1	2 4	4.24E−2	6 2	2.09E−5
5 4	5.02E−3	9 1	7.52E−1	1 5	−6.40E−1	3 5	−5.12E−2	2 5	6.17E−1	6 3	3.01E−2
5 5	6.56E−1	9 2	1.03E−3	2 1	2.14E−2	4 1	2.23E−4		**$^4P^e$–$^4D^o$**	6 4	−5.17E−4
5 6	6.15E−3	9 3	2.80E+0	2 2	−4.81E−1	4 2	1.25E−2	1 1	−1.21E+1	7 1	1.14E−1
5 7	9.60E−5	9 4	1.60E−1	2 3	−8.60E−3	4 3	−1.36E−1	1 2	−9.61E−1	7 2	1.10E−2
5 8	−5.29E−2	9 5	1.54E−2	2 4	−5.61E−2	4 4	−1.91E−2	2 1	−5.95E−1	7 3	9.72E−2
5 9	−6.27E−1	9 6	2.74E−4	2 5	−3.13E−3	4 5	−2.30E−1	2 2	−4.13E−1	7 4	−3.65E−5
5 10	−3.02E−2	9 7	4.31E−2	3 1	1.14E−1	5 1	6.06E−1	3 1	2.17E−2	8 1	2.67E−5
5 11	−1.94E−5	9 8	1.80E−5	3 2	1.39E−3	5 2	5.44E−3	3 2	−1.34E+0	8 2	7.25E−1
6 1	4.26E−2	9 9	3.37E−4	3 3	−3.32E−1	5 3	3.73E−4	4 1	8.63E−1	8 3	1.05E−1
6 2	9.86E−1	9 10	6.76E−4	3 4	−3.68E−1	5 4	−4.98E−2	4 2	6.99E−4	8 4	−1.00E−3
6 3	7.09E−7	9 11	9.47E−1	3 5	−7.54E−2	5 5	−6.71E−1		**$^4D^e$–$^4D^o$**	9 1	1.98E−4
6 4	5.18E−4		**$^4S^e$–$^4P^o$**	4 1	2.68E−1	6 1	3.09E+0	1 1	−2.54E−1	9 2	7.37E−3
6 5	6.71E−4	1 1	3.19E−1	4 2	2.52E−3	6 2	5.98E−2	1 2	−1.25E+0	9 3	1.98E−4
6 6	1.02E−1	1 2	−4.54E−1	4 3	6.23E−3	6 3	1.05E−1	2 1	6.22E−2	9 4	−5.09E−1
6 7	1.42E−2	1 3	−3.02E−1	4 4	−2.27E−1	6 4	8.29E−2	2 2	3.73E−1		**$^2F^e$–$^2F^o$**
6 8	5.97E−1		**$^4P^e$–$^4P^o$**	4 5	−5.33E−1	6 5	6.43E−2		**$^4F^e$–$^4D^o$**	1 1	3.92E−1
6 9	4.01E−3	1 1	−9.77E−1	5 1	2.85E−1	7 1	1.21E+0	1 1	1.71E−1	1 2	7.71E−2
6 10	−1.33E−4	1 2	−3.82E+0	5 2	3.91E−1	7 2	2.25E−3	1 2	1.43E+0	1 3	1.73E−2
6 11	−2.24E−4	1 3	−7.71E−1	5 3	9.23E−5	7 3	3.78E−1		**$^2D^e$–$^2F^o$**	1 4	−4.37E−5
7 1	1.65E−2	2 1	1.17E+0	5 4	4.67E−4	7 4	3.08E−1	1 1	−5.92E+0	2 1	1.67E−1
7 2	3.32E+0	2 2	−1.83E−1	5 5	−1.34E−3	7 5	7.11E−2	1 2	−4.64E+0	2 2	3.73E−1
7 3	2.56E−4	2 3	−2.79E−1	6 1	6.14E−1	8 1	6.87E−2	1 3	−5.26E−1	2 3	1.48E−1
7 4	3.05E−3	3 1	8.93E−1	6 2	1.77E−2	8 2	4.99E−3	1 4	−2.03E−3	2 4	−9.85E−5
7 5	1.44E−3	3 2	5.22E−3	6 3	1.05E−1	8 3	2.70E−1	2 1	−4.24E−3		**$^2G^e$–$^2F^o$**
7 6	3.40E−2	3 3	−9.29E−1	6 4	1.21E−1	8 4	1.45E−3	2 2	−4.28E−1	1 1	3.82E−3
7 7	8.84E−2	4 1	2.67E−3	6 5	7.62E−2	8 5	1.89E−2	2 3	−1.40E−2	1 2	1.41E−1
7 8	1.80E−3	4 2	2.79E−1		**$^2D^e$–$^2D^o$**	9 1	6.54E−5	2 4	−9.86E+0	1 3	1.11E+0
7 9	1.72E−1	4 3	4.63E−1	1 1	−6.07E−1	9 2	4.69E−6	3 1	−1.00E+0	1 4	−3.44E−4
7 10	−4.33E−4		**$^4D^e$–$^4P^o$**	1 2	−1.77E+0	9 3	6.09E−5	3 2	−8.93E−4		**$^4D^e$–$^4F^o$**
7 11	−1.07E−4	1 1	1.33E+0	1 3	−4.21E−1	9 4	1.23E−5	3 3	−1.44E−3	1 1	−1.35E+0
8 1	1.26E−2	1 2	−1.07E−1	1 4	−1.59E−1	9 5	1.64E−4	3 4	−1.68E−4	2 1	1.04E+0
8 2	2.85E+0	1 3	−4.73E−1	1 5	−6.56E−1		**$^2F^e$–$^2D^o$**	4 1	1.62E−3		**$^4F^e$–$^4F^o$**
8 3	1.28E−4	2 1	8.93E−3	2 1	6.19E−4	1 1	6.54E−1	4 2	−5.63E−1	1 1	1.30E+0
8 4	2.57E−4	2 2	3.05E−1	2 2	−6.04E−3	1 2	9.07E−2	4 3	−5.88E−1		
8 5	4.31E−5	2 3	8.97E−1	2 3	−7.28E−1	1 3	4.42E−2	4 4	−1.06E−1		
8 6	3.72E−3			2 4	−5.05E−4	1 4	6.69E−1	5 1	1.73E−2		

B-like Ca (Ca^{15+})

Term energies relative to $2s^2$ ^{1}S ionization threshold for each symmetry

i	E(Ryds)	Description	i	E(Ryds)	Description	i	E(Ryds)	Description	i	E(Ryds)	Description
	^{2}S^e		20	−1.75919	$2s2p$ ^{3}P $8p$	28	−4.06145	$2p^2$ ^{1}D $5s$		**^{4}F^e**	
1	−66.1549	$2s2p^2$	21	−1.38306	$2p^2$ ^{3}P $6s$	29	−4.02989	$2s^2$ ^{1}S $8d$	1	−23.2622	$2p^2$ ^{3}P $3d$
2	−31.3091	$2s^2$ ^{1}S $3s$	22	−1.17987	$2p^2$ ^{3}P $6d$	30	−3.64707	$2p^2$ ^{1}D $5d$	2	−13.7472	$2s2p$ ^{3}P $4f$
3	−27.0097	$2s2p$ ^{3}P $3p$	23	−.89435	$2s2p$ ^{3}P $9p$	31	−3.53969	$2p^2$ ^{1}D $5g$	3	−10.3162	$2p^2$ ^{3}P $4d$
4	−25.3288	$2s2p$ ^{1}P $3p$	24	−.86778	$2s2p$ ^{1}P $7p$	32	−3.18173	$2s^2$ ^{1}S $9d$	4	−7.95576	$2s2p$ ^{3}P $5f$
5	−22.6397	$2p^2$ ^{1}S $3s$	25	−.46215	$2p^2$ ^{1}D $6d$	33	−2.98978	$2s2p$ ^{3}P $7p$	5	−4.81173	$2s2p$ ^{3}P $6f$
6	−21.6432	$2p^2$ ^{1}D $3d$		**^{4}P^e**		34	−2.89306	$2s2p$ ^{3}P $7f$	6	−4.39829	$2p^2$ ^{3}P $5d$
7	−17.1146	$2s^2$ ^{1}S $4s$	1	−69.0317	$2s2p^2$	35	−2.82562	$2s2p$ ^{1}P $6p$	7	−4.25352	$2p^2$ ^{3}P $5g$
8	−14.0208	$2s2p$ ^{3}P $4p$	2	−27.5331	$2s2p$ ^{3}P $3p$	36	−2.64900	$2s2p$ ^{1}P $6f$	8	−2.91644	$2s2p$ ^{3}P $7f$
9	−12.1100	$2s2p$ ^{1}P $4p$	3	−25.0316	$2p^2$ ^{3}P $3s$		**^{4}D^e**		9	−1.68729	$2s2p$ ^{3}P $8f$
10	−10.7735	$2s^2$ ^{1}S $5s$	4	−22.7698	$2p^2$ ^{3}P $3d$	1	−27.9234	$2s2p$ ^{3}P $3p$	10	−1.20385	$2p^2$ ^{3}P $6d$
11	−9.36183	$2p^2$ ^{1}D $4d$	5	−14.2542	$2s2p$ ^{3}P $4p$	2	−23.0512	$2p^2$ ^{3}P $3d$	11	−1.12107	$2p^2$ ^{3}P $6g$
12	−8.72715	$2p^2$ ^{1}S $4s$	6	−11.0440	$2p^2$ ^{3}P $4s$	3	−14.3650	$2s2p$ ^{3}P $4p$	12	−.84475	$2s2p$ ^{3}P $9f$
13	−8.08866	$2s2p$ ^{3}P $5p$	7	−10.1596	$2p^2$ ^{3}P $4d$	4	−13.6594	$2s2p$ ^{3}P $4f$		**^{2}G^e**	
14	−7.43166	$2s^2$ ^{1}S $6s$	8	−8.20565	$2s2p$ ^{3}P $5p$	5	−10.2505	$2p^2$ ^{3}P $4d$	1	−22.3561	$2p^2$ ^{1}D $3d$
15	−6.05174	$2s2p$ ^{1}P $5p$	9	−4.95976	$2s2p$ ^{3}P $6p$	6	−8.25387	$2s2p$ ^{3}P $5p$	2	−13.6642	$2s2p$ ^{3}P $4f$
16	−5.41370	$2s^2$ ^{1}S $7s$	10	−4.76213	$2p^2$ ^{3}P $5s$	7	−7.91476	$2s2p$ ^{3}P $5f$	3	−11.5705	$2s2p$ ^{1}P $4f$
17	−4.89068	$2s2p$ ^{3}P $6p$	11	−4.32538	$2p^2$ ^{3}P $5d$	8	−4.98409	$2s2p$ ^{3}P $6p$	4	−10.2262	$2s^2$ ^{1}S $5g$
18	−4.13121	$2s^2$ ^{1}S $8s$	12	−3.00504	$2s2p$ ^{3}P $7p$	9	−4.78972	$2s2p$ ^{3}P $6f$	5	−9.55753	$2p^2$ ^{1}D $4d$
19	−3.59347	$2p^2$ ^{1}D $5d$	13	−1.74733	$2s2p$ ^{3}P $8p$	10	−4.36512	$2p^2$ ^{3}P $5d$	6	−7.91113	$2s2p$ ^{3}P $5f$
20	−3.25546	$2s^2$ ^{1}S $9s$	14	−1.41430	$2p^2$ ^{3}P $6s$	11	−3.02109	$2s2p$ ^{3}P $7p$	7	−7.11277	$2s^2$ ^{1}S $6g$
21	−2.97218	$2s2p$ ^{3}P $7p$	15	−1.16464	$2p^2$ ^{3}P $6d$	12	−2.90271	$2s2p$ ^{3}P $7f$	8	−5.79565	$2s2p$ ^{1}P $5f$
22	−2.81656	$2s2p$ ^{1}P $6p$	16	−.88548	$2s2p$ ^{3}P $9p$	13	−1.75749	$2s2p$ ^{3}P $8p$	9	−5.22068	$2s^2$ ^{1}S $7g$
	^{4}S^e			**^{2}D^e**		14	−1.67822	$2s2p$ ^{3}P $8f$	10	−4.79130	$2s2p$ ^{3}P $6h$
1	−27.7155	$2s2p$ ^{3}P $3p$	1	−67.2848	$2s2p^2$	15	−1.18844	$2p^2$ ^{3}P $6d$	11	−4.78605	$2s2p$ ^{3}P $6f$
2	−14.2918	$2s2p$ ^{3}P $4p$	2	−29.0541	$2s^2$ ^{1}S $3d$	16	−.89242	$2s2p$ ^{3}P $9p$	12	−4.22595	$2p^2$ ^{3}P $5g$
3	−8.22185	$2s2p$ ^{3}P $5p$	3	−27.3756	$2s2p$ ^{3}P $3p$	17	−.83855	$2s2p$ ^{3}P $9f$	13	−3.99974	$2s^2$ ^{1}S $8g$
4	−4.96341	$2s2p$ ^{3}P $6p$	4	−25.6195	$2s2p$ ^{1}P $3p$		**^{2}F^e**		14	−3.68112	$2p^2$ ^{1}D $5d$
5	−3.01061	$2s2p$ ^{3}P $7p$	5	−24.1616	$2p^2$ ^{1}D $3s$	1	−22.9552	$2p^2$ ^{3}P $3d$	15	−3.57543	$2p^2$ ^{1}D $5g$
6	−1.74950	$2s2p$ ^{3}P $8p$	6	−22.2926	$2p^2$ ^{3}P $3d$	2	−21.9945	$2p^2$ ^{1}D $3d$	16	−3.16113	$2s^2$ ^{1}S $9g$
7	−.88819	$2s2p$ ^{3}P $9p$	7	−22.0470	$2p^2$ ^{1}D $3d$	3	−13.7273	$2s2p$ ^{3}P $4f$	17	−2.89986	$2s2p$ ^{3}P $7f$
	^{2}P^e		8	−20.6015	$2p^2$ ^{1}S $3d$	4	−11.6091	$2s2p$ ^{1}P $4f$	18	−2.89918	$2s2p$ ^{3}P $7h$
1	−65.8687	$2s2p^2$	9	−16.2471	$2s^2$ ^{1}S $4d$	5	−10.1321	$2p^2$ ^{3}P $4d$	19	−2.66283	$2s2p$ ^{1}P $6f$
2	−27.9797	$2s2p$ ^{3}P $3p$	10	−14.1634	$2s2p$ ^{3}P $4p$	6	−9.51421	$2p^2$ ^{1}D $4d$	20	−2.65590	$2s2p$ ^{1}P $6h$
3	−25.6419	$2s2p$ ^{1}P $3p$	11	−13.6222	$2s2p$ ^{3}P $4f$	7	−7.94145	$2s2p$ ^{3}P $5f$		**^{4}G^e**	
4	−24.4248	$2p^2$ ^{3}P $3s$	12	−12.1791	$2s2p$ ^{1}P $4p$	8	−5.81728	$2s2p$ ^{1}P $5f$	1	−13.7043	$2s2p$ ^{3}P $4f$
5	−23.0433	$2p^2$ ^{3}P $3d$	13	−11.5094	$2s2p$ ^{1}P $4f$	9	−4.80755	$2s2p$ ^{3}P $6f$	2	−7.93780	$2s2p$ ^{3}P $5f$
6	−21.7825	$2p^2$ ^{1}D $3d$	14	−10.3648	$2s^2$ ^{1}S $5d$	10	−4.29697	$2p^2$ ^{3}P $5d$	3	−4.80228	$2s2p$ ^{3}P $6f$
7	−14.3572	$2s2p$ ^{3}P $4p$	15	−10.3227	$2p^2$ ^{1}D $4s$	11	−4.25337	$2p^2$ ^{3}P $5g$	4	−4.79136	$2s2p$ ^{3}P $6h$
8	−12.1531	$2s2p$ ^{1}P $4p$	16	−9.95336	$2p^2$ ^{3}P $4d$	12	−3.67268	$2p^2$ ^{1}D $5d$	5	−4.22695	$2p^2$ ^{3}P $5g$
9	−10.9118	$2p^2$ ^{3}P $4s$	17	−9.47587	$2p^2$ ^{1}D $4d$	13	−3.55884	$2p^2$ ^{1}D $5g$	6	−2.91093	$2s2p$ ^{3}P $7f$
10	−10.2341	$2p^2$ ^{3}P $4d$	18	−8.15723	$2s2p$ ^{3}P $5p$	14	−2.91155	$2s2p$ ^{3}P $7f$	7	−2.89918	$2s2p$ ^{3}P $7h$
11	−9.38386	$2p^2$ ^{1}D $4d$	19	−7.94412	$2p^2$ ^{1}S $4d$	15	−2.67257	$2s2p$ ^{1}P $6f$	8	−1.68367	$2s2p$ ^{3}P $8f$
12	−8.24069	$2s2p$ ^{3}P $5p$	20	−7.88234	$2s2p$ ^{3}P $5f$	16	−1.68473	$2s2p$ ^{3}P $8f$	9	−1.67653	$2s2p$ ^{3}P $8h$
13	−6.08093	$2s2p$ ^{1}P $5p$	21	−7.18300	$2s^2$ ^{1}S $6d$	17	−1.14645	$2p^2$ ^{3}P $6d$	10	−1.11012	$2p^2$ ^{3}P $6g$
14	−5.00296	$2s2p$ ^{3}P $6p$	22	−6.09150	$2s2p$ ^{1}P $5p$	18	−1.12092	$2p^2$ ^{3}P $6g$	11	−.84241	$2s2p$ ^{3}P $9f$
15	−4.68402	$2p^2$ ^{3}P $5s$	23	−5.76893	$2s2p$ ^{1}P $5f$	19	−.84219	$2s2p$ ^{3}P $9f$	12	−.83572	$2s2p$ ^{3}P $9h$
16	−4.35598	$2p^2$ ^{3}P $5d$	24	−5.26929	$2s^2$ ^{1}S $7d$	20	−.77832	$2s2p$ ^{1}P $7f$		**^{2}S^o**	
17	−3.60925	$2p^2$ ^{1}D $5d$	25	−4.93301	$2s2p$ ^{3}P $6p$	21	−.49962	$2p^2$ ^{1}D $6d$	1	−24.3073	$2p^2$ ^{3}P $3p$
18	−3.01844	$2s2p$ ^{3}P $7p$	26	−4.77651	$2s2p$ ^{3}P $6f$	22	−.43213	$2p^2$ ^{1}D $6g$	2	−10.7244	$2p^2$ ^{3}P $4p$
19	−2.81596	$2s2p$ ^{1}P $6p$	27	−4.22910	$2p^2$ ^{3}P $5d$				3	−4.60018	$2p^2$ ^{3}P $5p$

B-like Ca (Ca^{15+})

i	E(Ryds)	Description	i	E(Ryds)	Description	i	E(Ryds)	Description	i	E(Ryds)	Description
4	−1.31665	$2p^2$ ^{3}P $6p$	7	−8.47196	$2s2p$ ^{3}P $5s$	12	−1.29977	$2p^2$ ^{3}P $6p$	4	−5.79224	$2s2p$ ^{1}P $5g$
		$^4\mathrm{S}^\circ$	8	−8.02487	$2s2p$ ^{3}P $5d$	13	−1.13183	$2p^2$ ^{3}P $6f$	5	−4.80055	$2s2p$ ^{3}P $6g$
1	−64.3283	$2p^3$	9	−5.11296	$2s2p$ ^{3}P $6s$	14	−.85822	$2s2p$ ^{3}P $9d$	6	−4.25672	$2p^2$ ^{3}P $5f$
2	−23.5607	$2p^2$ ^{3}P $3p$	10	−4.85214	$2s2p$ ^{3}P $6d$			**$^2\mathrm{F}^\circ$**	7	−3.60292	$2p^2$ ^{1}D $5f$
3	−10.4873	$2p^2$ ^{3}P $4p$	11	−4.53763	$2p^2$ ^{3}P $5p$	1	−26.1189	$2s2p$ ^{3}P $3d$	8	−2.90498	$2s2p$ ^{3}P $7g$
4	−4.49268	$2p^2$ ^{3}P $5p$	12	−3.09930	$2s2p$ ^{3}P $7s$	2	−24.6833	$2s2p$ ^{1}P $3d$	9	−2.65749	$2s2p$ ^{1}P $6g$
5	−1.25826	$2p^2$ ^{3}P $6p$	13	−2.94109	$2s2p$ ^{3}P $7d$	3	−23.2218	$2p^2$ ^{1}D $3p$	10	−1.68052	$2s2p$ ^{3}P $8g$
		$^2\mathrm{P}^\circ$	14	−1.81066	$2s2p$ ^{3}P $8s$	4	−16.0207	$2s^2$ ^{1}S $4f$	11	−1.12616	$2p^2$ ^{3}P $6f$
1	−71.1788	$2s^22p$	15	−1.70350	$2s2p$ ^{3}P $8d$	5	−13.7069	$2s2p$ ^{3}P $4d$	12	−1.11875	$2p^2$ ^{3}P $6h$
2	−62.3289	$2p^3$	16	−1.28824	$2p^2$ ^{3}P $6p$	6	−11.8049	$2s2p$ ^{1}P $4d$	13	−.83849	$2s2p$ ^{3}P $9g$
3	−30.1547	$2s^2$ ^{1}S $3p$	17	−.92850	$2s2p$ ^{3}P $9s$	7	−10.2538	$2s^2$ ^{1}S $5f$	14	−.76994	$2s2p$ ^{1}P $7g$
4	−28.4127	$2s2p$ ^{3}P $3s$	18	−.85602	$2s2p$ ^{3}P $9d$	8	−9.95830	$2p^2$ ^{3}P $4f$	15	−.45620	$2p^2$ ^{1}D $6f$
5	−26.7410	$2s2p$ ^{1}P $3s$			**$^2\mathrm{D}^\circ$**	9	−9.90961	$2p^2$ ^{1}D $4p$	16	−.43385	$2p^2$ ^{1}D $6h$
6	−25.9982	$2s2p$ ^{3}P $3d$	1	−63.3322	$2p^3$	10	−9.37195	$2p^2$ ^{1}D $4f$			**$^4\mathrm{G}^\circ$**
7	−24.3384	$2s2p$ ^{1}P $3d$	2	−26.7274	$2s2p$ ^{3}P $3d$	11	−7.93967	$2s2p$ ^{3}P $5d$	1	−10.0659	$2p^2$ ^{3}P $4f$
8	−23.5669	$2p^2$ ^{3}P $3p$	3	−24.5359	$2s2p$ ^{1}P $3d$	12	−7.89968	$2s2p$ ^{3}P $5g$	2	−7.92402	$2s2p$ ^{3}P $5g$
9	−22.6005	$2p^2$ ^{1}D $3p$	4	−23.8301	$2p^2$ ^{3}P $3p$	13	−7.76779	$2p^2$ ^{1}S $4f$	3	−4.80061	$2s2p$ ^{3}P $6g$
10	−21.5234	$2p^2$ ^{1}S $3p$	5	−22.8835	$2p^2$ ^{1}D $3p$	14	−7.12185	$2s^2$ ^{1}S $6f$	4	−4.27333	$2p^2$ ^{3}P $5f$
11	−16.6413	$2s^2$ ^{1}S $4p$	6	−13.9120	$2s2p$ ^{3}P $4d$	15	−5.90990	$2s2p$ ^{1}P $5d$	5	−2.90544	$2s2p$ ^{3}P $7g$
12	−14.6246	$2s2p$ ^{3}P $4s$	7	−11.7774	$2s2p$ ^{1}P $4d$	16	−5.75964	$2s2p$ ^{1}P $5g$	6	−1.68070	$2s2p$ ^{3}P $8g$
13	−13.6728	$2s2p$ ^{3}P $4d$	8	−10.5010	$2p^2$ ^{3}P $4p$	17	−5.23112	$2s^2$ ^{1}S $7f$	7	−1.13638	$2p^2$ ^{3}P $6f$
14	−12.6233	$2s2p$ ^{1}P $4s$	9	−10.0483	$2p^2$ ^{3}P $4f$	18	−4.80329	$2s2p$ ^{3}P $6d$	8	−1.11875	$2p^2$ ^{3}P $6h$
15	−11.7045	$2s2p$ ^{1}P $4d$	10	−9.86027	$2p^2$ ^{1}D $4p$	19	−4.78504	$2s2p$ ^{3}P $6g$	9	−.83889	$2s2p$ ^{3}P $9g$
16	−10.5485	$2s^2$ ^{1}S $5p$	11	−9.30355	$2p^2$ ^{1}D $4f$	20	−4.21589	$2p^2$ ^{3}P $5f$			
17	−10.4080	$2p^2$ ^{3}P $4p$	12	−8.03390	$2s2p$ ^{3}P $5d$	21	−4.00487	$2s^2$ ^{1}S $8f$			
18	−9.75802	$2p^2$ ^{1}D $4p$	13	−5.90174	$2s2p$ ^{1}P $5d$	22	−3.85984	$2p^2$ ^{1}D $5p$			
19	−9.23991	$2p^2$ ^{1}D $4f$	14	−4.85633	$2s2p$ ^{3}P $6d$	23	−3.58929	$2p^2$ ^{1}D $5f$			
20	−8.38830	$2s2p$ ^{3}P $5s$	15	−4.48592	$2p^2$ ^{3}P $5p$	24	−3.16436	$2s^2$ ^{1}S $9f$			
21	−8.29808	$2p^2$ ^{1}S $4p$	16	−4.26799	$2p^2$ ^{3}P $5f$	25	−2.91317	$2s2p$ ^{3}P $7d$			
22	−7.92188	$2s2p$ ^{3}P $5d$	17	−3.84449	$2p^2$ ^{1}D $5p$	26	−2.89638	$2s2p$ ^{3}P $7g$			
23	−7.29325	$2s^2$ ^{1}S $6p$	18	−3.55739	$2p^2$ ^{1}D $5f$	27	−2.72299	$2s2p$ ^{1}P $6d$			
24	−6.31103	$2s2p$ ^{1}P $5s$	19	−2.94556	$2s2p$ ^{3}P $7d$	28	−2.64937	$2s2p$ ^{1}P $6g$			
25	−5.86365	$2s2p$ ^{1}P $5d$	20	−2.71835	$2s2p$ ^{1}P $6d$			**$^4\mathrm{F}^\circ$**			
26	−5.33440	$2s^2$ ^{1}S $7p$	21	−1.70575	$2s2p$ ^{3}P $8d$	1	−27.0276	$2s2p$ ^{3}P $3d$			
27	−5.07088	$2s2p$ ^{3}P $6s$	22	−1.25445	$2p^2$ ^{3}P $6p$	2	−14.0201	$2s2p$ ^{3}P $4d$			
28	−4.79600	$2s2p$ ^{3}P $6d$	23	−1.12946	$2p^2$ ^{3}P $6f$	3	−9.99098	$2p^2$ ^{3}P $4f$			
29	−4.43221	$2p^2$ ^{3}P $5p$	24	−.85962	$2s2p$ ^{3}P $9d$	4	−8.08541	$2s2p$ ^{3}P $5d$			
30	−4.07484	$2s^2$ ^{1}S $8p$	25	−.80622	$2s2p$ ^{1}P $7d$	5	−7.90092	$2s2p$ ^{3}P $5g$			
31	−3.80422	$2p^2$ ^{1}D $5p$	26	−.59674	$2p^2$ ^{1}D $6p$	6	−4.88461	$2s2p$ ^{3}P $6d$			
32	−3.52690	$2p^2$ ^{1}D $5f$	27	−.43279	$2p^2$ ^{1}D $6f$	7	−4.78557	$2s2p$ ^{3}P $6g$			
33	−3.21533	$2s^2$ ^{1}S $9p$			**$^4\mathrm{D}^\circ$**	8	−4.23953	$2p^2$ ^{3}P $5f$			
34	−3.07179	$2s2p$ ^{3}P $7s$	1	−26.7462	$2s2p$ ^{3}P $3d$	9	−2.96113	$2s2p$ ^{3}P $7d$			
35	−2.95109	$2s2p$ ^{1}P $6s$	2	−24.1029	$2p^2$ ^{3}P $3p$	10	2.89713	$2s2p$ ^{3}P $7g$			
36	−2.90459	$2s2p$ ^{3}P $7d$	3	−13.9244	$2s2p$ ^{3}P $4d$	11	−1.71672	$2s2p$ ^{3}P $8d$			
37	−2.70408	$2s2p$ ^{1}P $6d$	4	−10.6536	$2p^2$ ^{3}P $4p$	12	−1.67512	$2s2p$ ^{3}P $8g$			
		$^4\mathrm{P}^\circ$	5	−10.0527	$2p^2$ ^{3}P $4f$	13	−1.11801	$2p^2$ ^{3}P $6f$			
1	−28.9783	$2s2p$ ^{3}P $3s$	6	−8.04061	$2s2p$ ^{3}P $5d$	14	−.86513	$2s2p$ ^{3}P $9d$			
2	−26.6383	$2s2p$ ^{3}P $3d$	7	−4.86069	$2s2p$ ^{3}P $6d$	15	−.83531	$2s2p$ ^{3}P $9g$			
3	−23.9660	$2p^2$ ^{3}P $3p$	8	−4.56532	$2p^2$ ^{3}P $5p$			**$^2\mathrm{G}^\circ$**			
4	−14.8009	$2s2p$ ^{3}P $4s$	9	−4.27125	$2p^2$ ^{3}P $5f$	1	−10.0412	$2p^2$ ^{3}P $4f$			
5	−13.8893	$2s2p$ ^{3}P $4d$	10	−2.94583	$2s2p$ ^{3}P $7d$	2	−9.39570	$2p^2$ ^{1}D $4f$			
6	−10.6065	$2p^2$ ^{3}P $4p$	11	−1.70702	$2s2p$ ^{3}P $8d$	3	−7.92312	$2s2p$ ^{3}P $5g$			

B-like Ca (Ca^{15+})

Energies in ascending order from ground state for terms with effective $n \leq 4.0$, $L \leq 4$

Term	i	E(Ryds)	Term	i	E(Ryds)	Term	i	E(Ryds)	Term	i	E(Ryds)	Term	i	E(Ryds)
$^2P^o$	1	0.00000	$^2P^e$	2	43.1991	$^2P^e$	3	45.5369	$^2P^o$	8	47.6119	$^2D^e$	7	49.1318
$^4P^e$	1	2.14710	$^4D^e$	1	43.2554	$^2D^e$	4	45.5593	$^4S^o$	2	47.6181	$^2F^e$	2	49.1843
$^2D^e$	1	3.89400	$^4S^e$	1	43.4633	$^2S^e$	4	45.8500	$^4F^e$	1	47.9166	$^2P^e$	6	49.3963
$^2S^e$	1	5.02390	$^4P^e$	2	43.6457	$^4P^e$	3	46.1472	$^2F^o$	3	47.9570	$^2S^e$	6	49.5356
$^2P^e$	1	5.31010	$^2D^e$	3	43.8032	$^2F^o$	2	46.4955	$^4D^e$	2	48.1276	$^2P^o$	10	49.6554
$^4S^o$	1	6.85050	$^4F^o$	1	44.1512	$^2D^o$	3	46.6429	$^2P^e$	5	48.1355	$^2D^e$	8	50.5773
$^2D^o$	1	7.84660	$^2S^e$	3	44.1691	$^2P^e$	4	46.7540	$^2F^e$	1	48.2236	$^2S^e$	7	54.0642
$^2P^o$	2	8.84990	$^4D^o$	1	44.4326	$^2P^o$	7	46.8404	$^2D^o$	5	48.2953	$^2P^o$	11	54.5375
$^2S^e$	2	39.8697	$^2P^o$	5	44.4378	$^2S^o$	1	46.8715	$^4P^e$	4	48.4090	$^2D^e$	9	54.9317
$^2P^o$	3	41.0241	$^2D^o$	2	44.4514	$^2D^e$	5	47.0172	$^2S^e$	5	48.5391	$^2F^o$	4	55.1581
$^2D^e$	2	42.1247	$^4P^o$	2	44.5405	$^4D^o$	2	47.0759	$^2P^o$	9	48.5783			
$^4P^o$	1	42.2005	$^2F^o$	1	45.0599	$^4P^o$	3	47.2128	$^2G^e$	1	48.8227			
$^2P^o$	4	42.7661	$^2P^o$	6	45.1806	$^2D^o$	4	47.3487	$^2D^e$	6	48.8862			

gf-values for transitions involving terms with effective $n \leq 4.0$, $L \leq 4$

i i'	gf_L	i i'	gf_L	i i'	gf_L	i i'	gf_L	i i'	gf_L	i i'	gf_L
	$^2P^e$–$^2S^o$	2 7	−4.84E−4	5 8	5.82E−3	1 8	−5.03E−1	4 9	−3.57E−1	1 9	−3.47E−1
1 1	−2.49E−1	2 8	−7.77E−4	5 9	−2.74E−8	1 9	−1.52E−1	4 10	−7.95E−2	1 10	−1.74E−4
2 1	−2.08E−1	2 9	−1.32E−3	5 10	−4.13E−1	1 10	−6.29E−2	4 11	−4.47E−5	1 11	−5.03E−4
3 1	−2.26E−2	2 10	−1.09E−4	5 11	−4.31E−5	1 11	−1.07E−4	5 1	1.36E−2	2 1	3.94E+0
4 1	−9.75E−3	2 11	−6.60E−1	6 1	6.03E−3	2 1	8.97E−1	5 2	7.88E−1	2 2	3.24E−5
5 1	2.04E−1	3 1	3.61E−1	6 2	4.62E−1	2 2	3.10E−4	5 3	1.76E−4	2 3	4.55E−1
6 1	4.20E−2	3 2	2.40E−2	6 3	2.83E−4	2 3	3.49E−2	5 4	3.29E−7	2 4	−1.27E−2
	$^4P^e$–$^4S^o$	3 3	2.65E−2	6 4	1.20E−3	2 4	1.11E−1	5 5	4.79E−3	2 5	−7.32E−4
1 1	−5.28E−1	3 4	1.66E−1	6 5	1.75E−2	2 5	−5.77E−4	5 6	8.73E−2	2 6	−1.77E−5
1 2	−3.78E−1	3 5	−8.32E−6	6 6	5.68E−5	2 6	−2.43E−1	5 7	6.35E−3	2 7	−4.66E−1
2 1	7.72E−3	3 6	−1.65E−1	6 7	1.07E−1	2 7	−7.18E−3	5 8	7.12E−2	2 8	−1.18E−5
2 2	−2.99E−1	3 7	−2.78E−2	6 8	5.52E−2	2 8	−5.31E−2	5 9	−3.90E−3	2 9	−6.00E−4
3 1	5.32E−1	3 8	−1.23E−1	6 9	1.03E−1	2 9	−8.75E−3	5 10	−1.35E−3	2 10	−8.08E−4
3 2	−4.14E−1	3 9	−2.96E−3	6 10	−7.34E−3	2 10	−6.73E−4	5 11	−1.18E−5	2 11	−1.72E−1
4 1	7.18E+0	3 10	−2.25E−3	6 11	−2.22E−5	2 11	−1.58E−3	6 1	2.71E−2	3 1	1.75E+0
4 2	2.46E−1	3 11	−3.00E−5	7 1	2.37E−2	3 1	2.93E−1	6 2	1.95E+0	3 2	1.27E−2
	$^2S^e$–$^2P^o$	4 1	3.56E−2	7 2	1.25E−7	3 2	4.06E−2	6 3	1.68E−3	3 3	1.17E−1
1 1	1.38E−1	4 2	8.98E−6	7 3	2.70E−1	3 3	3.98E−1	6 4	2.05E−3	3 4	5.39E−1
1 2	−1.05E−1	4 3	1.25E−1	7 4	1.28E−2	3 4	6.83E−3	6 5	8.37E−5	3 5	−3.68E−3
1 3	−5.07E−3	4 4	7.00E−2	7 5	6.91E−5	3 5	3.14E−1	6 6	2.45E−3	3 6	−3.27E−2
1 4	−9.82E−2	4 5	1.26E−1	7 6	3.18E−4	3 6	7.70E−4	6 7	2.58E−1	3 7	−2.23E−3
1 5	−1.45E−1	4 6	2.19E−3	7 7	1.36E−3	3 7	−9.61E−2	6 8	9.47E−2	3 8	−2.13E−1
1 6	−1.70E+0	4 7	−1.38E−1	7 8	5.50E−5	3 8	−2.29E−1	6 9	1.56E−1	3 9	−5.03E−2
1 7	−8.55E−1	4 8	−8.96E−3	7 9	1.01E−6	3 9	−8.64E−3	6 10	−6.16E−4	3 10	−2.29E−3
1 8	−3.94E−4	4 9	−1.66E−1	7 10	4.36E−6	3 10	−1.61E−1	6 11	−4.19E−5	3 11	−3.72E−3
1 9	−5.74E−3	4 10	−5.42E−2	7 11	−5.83E−1	3 11	−3.87E−4		$^2D^e$–$^2P^o$	4 1	1.15E−1
1 10	−4.06E−1	4 11	−1.44E−2		$^2P^e$–$^2P^o$	4 1	8.80E−2	1 1	2.65E−1	4 2	1.10E−3
1 11	−1.08E−2	5 1	4.68E−3	1 1	6.77E−1	4 2	2.39E−1	1 2	−3.43E−1	4 3	4.26E−1
2 1	1.28E−1	5 2	2.32E−1	1 2	−4.30E−1	4 3	4.97E−3	1 3	−3.52E−2	4 4	1.43E−1
2 2	2.96E−5	5 3	4.60E−4	1 3	−4.34E−3	4 4	5.95E−1	1 4	−2.71E−1	4 5	5.01E−1
2 3	−3.49E−1	5 4	4.90E−4	1 4	−1.98E−2	4 5	1.04E−1	1 5	−2.12E−1	4 6	2.16E−3
2 4	−3.75E−2	5 5	2.70E−1	1 5	−3.04E−1	4 6	6.58E−6	1 6	−4.91E−2	4 7	−8.94E−3
2 5	−3.68E−1	5 6	1.16E−2	1 6	−2.94E−1	4 7	−1.39E−3	1 7	−4.45E−2	4 8	−3.29E−5
2 6	−1.73E−2	5 7	8.24E−3	1 7	−1.37E+0	4 8	−1.88E−1	1 8	−1.88E−1	4 9	−1.29E−5

B-like Ca (Ca^{15+})

i	i'	gf$_L$	i	i'	gf$_L$	i	i'	gf$_L$	i	i'	gf$_L$	i	i'	gf$_L$	i	i'	gf$_L$
4	10	−4.72E−1	8	7	7.27E−2			$^{2}\mathbf{P}^{e}-{}^{2}\mathbf{D}^{o}$	2	5	−3.85E−3	1	5	−9.74E−7	5	2	6.20E−3
4	11	−7.62E−4	8	8	3.54E−3	1	1	−3.40E−1	3	1	2.56E−2	2	1	1.16E+1	5	3	−9.43E−1
5	1	2.59E−2	8	9	3.24E−3	1	2	−5.30E−1	3	2	−5.55E−2	2	2	1.71E−2	5	4	−1.47E−3
5	2	2.09E−1	8	10	5.48E−1	1	3	−4.61E+0	3	3	−3.40E−2	2	3	1.43E−1	6	1	3.28E−1
5	3	1.71E−2	8	11	−2.34E−4	1	4	−7.96E−1	3	4	−4.97E−1	2	4	4.01E−2	6	2	1.33E−5
5	4	4.23E−3	9	1	7.51E−1	1	5	−6.83E−1	3	5	−4.65E−2	2	5	5.54E−1	6	3	2.91E−2
5	5	5.89E−1	9	2	5.73E−4	2	1	1.93E−2	4	1	5.22E−5			$^{4}\mathbf{P}^{e}-{}^{4}\mathbf{D}^{o}$	6	4	−4.37E−4
5	6	4.82E−3	9	3	2.88E+0	2	2	−4.21E−1	4	2	1.10E−2	1	1	−1.23E+1	7	1	1.14E−1
5	7	8.14E−5	9	4	1.60E−1	2	3	−7.43E−3	4	3	−1.19E−1	1	2	−1.02E+0	7	2	9.98E−3
5	8	−4.76E−2	9	5	1.50E−2	2	4	−5.03E−2	4	4	−1.74E−2	2	1	−5.23E−1	7	3	8.33E−2
5	9	−5.59E−1	9	6	3.41E−4	2	5	−3.05E−3	4	5	−2.06E−1	2	2	−3.66E−1	7	4	−6.22E−5
5	10	−2.72E−2	9	7	3.72E−2	3	1	1.12E−1	5	1	5.89E−1	3	1	1.93E−2	8	1	4.46E−5
5	11	−2.26E−5	9	8	1.40E−5	3	2	1.32E−3	5	2	4.60E−3	3	2	−1.20E+0	8	2	6.51E−1
6	1	4.13E−2	9	9	2.43E−4	3	3	−2.92E−1	5	3	3.37E−4	4	1	7.72E−1	8	3	9.65E−2
6	2	8.87E−1	9	10	4.42E−4	3	4	−3.31E−1	5	4	−4.45E−2	4	2	7.64E−4	8	4	−6.05E−4
6	3	3.12E−6	9	11	8.37E−1	3	5	−6.70E−2	5	5	−5.94E−1			$^{4}\mathbf{D}^{e}-{}^{4}\mathbf{D}^{o}$	9	1	1.47E−4
6	4	4.72E−4			$^{4}\mathbf{S}^{e}-{}^{4}\mathbf{P}^{o}$	4	1	2.57E−1	6	1	3.25E+0	1	1	−2.22E−1	9	2	5.56E−3
6	5	6.53E−4	1	1	2.84E−1	4	2	1.74E−3	6	2	5.66E−2	1	2	−1.11E+0	9	3	1.46E−4
6	6	9.32E−2	1	2	−3.98E−1	4	3	4.58E−3	6	3	1.08E−1	2	1	5.84E−2	9	4	−4.48E−1
6	7	1.38E−2	1	3	−2.70E−1	4	4	−1.98E−1	6	4	6.52E−2	2	2	3.30E−1			$^{2}\mathbf{F}^{e}-{}^{2}\mathbf{F}^{o}$
6	8	5.26E−1			$^{4}\mathbf{P}^{e}-{}^{4}\mathbf{P}^{o}$	4	5	−4.80E−1	6	5	5.98E−2			$^{4}\mathbf{F}^{e}-{}^{4}\mathbf{D}^{o}$	1	1	3.55E−1
6	9	2.71E−3	1	1	−9.51E−1	5	1	2.87E−1	7	1	1.11E+0	1	1	1.68E−1	1	2	6.91E−2
6	10	−1.71E−4	1	2	−3.88E+0	5	2	3.50E−1	7	2	1.37E−3	1	2	1.27E+0	1	3	1.55E−2
6	11	−1.67E−4	1	3	−8.10E−1	5	3	5.82E−5	7	3	3.27E−1			$^{2}\mathbf{D}^{e}-{}^{2}\mathbf{F}^{o}$	1	4	−3.50E−5
7	1	4.88E−2	2	1	1.04E+0	5	4	4.84E−4	7	4	2.83E−1	1	1	−5.98E+0	2	1	1.55E−1
7	2	3.47E+0	2	2	−1.60E−1	5	5	−1.21E−3	7	5	5.76E−2	1	2	−4.85E+0	2	2	3.35E−1
7	3	2.45E−4	2	3	−2.46E−1	6	1	6.21E−1	8	1	7.85E−2	1	3	−5.47E−1	2	3	1.28E−1
7	4	2.95E−3	3	1	7.94E−1	6	2	1.56E−2	8	2	4.36E−3	1	4	−1.10E−3	2	4	−8.30E−5
7	5	1.12E−3	3	2	4.59E−3	6	3	9.53E−2	8	3	2.47E−1	2	1	−5.03E−3			$^{2}\mathbf{G}^{e}-{}^{2}\mathbf{F}^{o}$
7	6	3.53E−2	3	3	−8.28E−1	6	4	1.07E−1	8	4	1.57E−3	2	2	−4.06E−1	1	1	4.33E−3
7	7	8.02E−2	4	1	2.20E−3	6	5	6.75E−2	8	5	1.78E−2	2	3	−1.42E−2	1	2	1.37E−1
7	8	3.81E−3	4	2	2.49E−1			$^{2}\mathbf{D}^{e}-{}^{2}\mathbf{D}^{o}$	9	1	1.59E−5	2	4	−9.85E+0	1	3	9.83E−1
7	9	1.53E−1	4	3	4.08E−1	1	1	−5.44E−1	9	2	2.22E−7	3	1	−8.81E−1	1	4	−1.92E−4
7	10	−4.45E−4			$^{4}\mathbf{D}^{e}-{}^{4}\mathbf{P}^{o}$	1	2	−1.78E+0	9	3	2.62E−5	3	2	−5.15E−4			$^{4}\mathbf{D}^{e}-{}^{4}\mathbf{F}^{o}$
7	11	−5.52E−5	1	1	1.19E+0	1	3	−4.20E−1	9	4	9.59E−6	3	3	−1.02E−3	1	1	−1.18E+0
8	1	1.10E−2	1	2	−9.50E−2	1	4	−1.69E−1	9	5	1.15E−4	3	4	−5.46E−6	2	1	9.27E−1
8	2	2.96E+0	1	3	−4.20E−1	1	5	−6.89E−1			$^{2}\mathbf{F}^{e}-{}^{2}\mathbf{D}^{o}$	4	1	1.15E−3			$^{4}\mathbf{F}^{e}-{}^{4}\mathbf{F}^{o}$
8	3	1.13E−5	2	1	7.51E−3	2	1	6.36E−4	1	1	6.25E−1	4	2	−4.96E−1	1	1	1.18E+0
8	4	2.37E−4	2	2	2.89E−1	2	2	−6.11E−3	1	2	8.78E−2	4	3	−5.30E−1			
8	5	3.13E−5	2	3	7.91E−1	2	3	−6.50E−1	1	3	4.30E−2	4	4	−9.99E−2			
8	6	1.17E−3				2	4	−4.10E−4	1	4	5.89E−1	5	1	1.44E−2			

B-like Fe (Fe^{21+})

Term energies relative to $2s^2$ ^{1}S ionization threshold for each symmetry

i	E(Ryds)	Description
^{2}S^e		
1	−124.181	$2s2p^2$
2	−57.7737	$2s^2$ ^{1}S $3s$
3	−51.9670	$2s2p$ ^{3}P $3p$
4	−49.7220	$2s2p$ ^{1}P $3p$
5	−46.1173	$2p^2$ ^{1}S $3s$
6	−44.7037	$2p^2$ ^{1}D $3d$
7	−31.8156	$2s^2$ ^{1}S $4s$
8	−27.6466	$2s2p$ ^{3}P $4p$
9	−25.0401	$2s2p$ ^{1}P $4p$
10	−21.3212	$2p^2$ ^{1}D $4d$
11	−20.5171	$2p^2$ ^{1}S $4s$
12	−20.1283	$2s^2$ ^{1}S $5s$
13	−16.5125	$2s2p$ ^{3}P $5p$
14	−13.9890	$2s^2$ ^{1}S $6s$:
15	−13.6504	$2s2p$ ^{1}P $5p$:
16	−10.4932	$2s2p$ ^{3}P $6p$
17	−10.4125	$2p^2$ ^{1}D $5d$
18	−10.1524	$2s^2$ ^{1}S $7s$
19	−8.96557	$2p^2$ ^{1}S $5s$
20	−7.78431	$2s^2$ ^{1}S $8s$:
21	−7.62285	$2s2p$ ^{1}P $6p$:
22	−6.86793	$2s2p$ ^{3}P $7p$
23	−6.10281	$2s^2$ ^{1}S $9s$
^{4}S^e		
1	−52.9331	$2s2p$ ^{3}P $3p$
2	−28.0159	$2s2p$ ^{3}P $4p$
3	−16.6887	$2s2p$ ^{3}P $5p$
4	−10.5884	$2s2p$ ^{3}P $6p$
5	−6.92846	$2s2p$ ^{3}P $7p$
6	−4.56175	$2s2p$ ^{3}P $8p$
7	−2.94361	$2s2p$ ^{3}P $9p$
^{2}P^e		
1	−123.841	$2s2p^2$
2	−53.2837	$2s2p$ ^{3}P $3p$
3	−50.1731	$2s2p$ ^{1}P $3p$
4	−48.5356	$2p^2$ ^{3}P $3s$
5	−46.6302	$2p^2$ ^{3}P $3d$
6	−44.8705	$2p^2$ ^{1}D $3d$
7	−28.1047	$2s2p$ ^{3}P $4p$
8	−25.1287	$2s2p$ ^{1}P $4p$
9	−23.5046	$2p^2$ ^{3}P $4s$
10	−22.5620	$2p^2$ ^{3}P $4d$
11	−21.3438	$2p^2$ ^{1}D $4d$
12	−16.7172	$2s2p$ ^{3}P $5p$
13	−13.7952	$2s2p$ ^{1}P $5p$
14	−12.0145	$2p^2$ ^{3}P $5s$
15	−11.5092	$2p^2$ ^{3}P $5d$
16	−10.5878	$2s2p$ ^{3}P $6p$
17	−10.4309	$2p^2$ ^{1}D $5d$
18	−7.69252	$2s2p$ ^{1}P $6p$
19	−6.93999	$2s2p$ ^{3}P $7p$
20	−5.80979	$2p^2$ ^{3}P $6s$
21	−5.52154	$2p^2$ ^{3}P $6d$
22	−4.56514	$2s2p$ ^{3}P $8p$
23	−4.49792	$2p^2$ ^{1}D $6d$
24	−4.03114	$2s2p$ ^{1}P $7p$
25	−2.94871	$2s2p$ ^{3}P $9p$
26	−2.10590	$2p^2$ ^{3}P $7s$
27	−1.92148	$2p^2$ ^{3}P $7d$
^{4}P^e		
1	−128.065	$2s2p^2$
2	−52.6771	$2s2p$ ^{3}P $3p$
3	−49.3703	$2p^2$ ^{3}P $3s$
4	−46.2330	$2p^2$ ^{3}P $3d$
5	−27.9625	$2s2p$ ^{3}P $4p$
6	−23.6901	$2p^2$ ^{3}P $4s$
7	−22.4580	$2p^2$ ^{3}P $4d$
8	−16.6652	$2s2p$ ^{3}P $5p$
9	−12.0796	$2p^2$ ^{3}P $5s$
10	−11.4678	$2p^2$ ^{3}P $5d$
11	−10.5699	$2s2p$ ^{3}P $6p$
12	−6.92115	$2s2p$ ^{3}P $7p$
13	−5.85006	$2p^2$ ^{3}P $6s$
14	−5.50038	$2p^2$ ^{3}P $6d$
15	−4.55591	$2s2p$ ^{3}P $8p$
16	−2.94047	$2s2p$ ^{3}P $9p$
17	−2.12689	$2p^2$ ^{3}P $7s$
18	−1.90952	$2p^2$ ^{3}P $7d$
^{2}D^e		
1	−125.710	$2s2p^2$
2	−54.6783	$2s^2$ ^{1}S $3d$
3	−52.4592	$2s2p$ ^{3}P $3p$
4	−50.1300	$2s2p$ ^{1}P $3p$
5	−48.1740	$2p^2$ ^{1}D $3s$
6	−45.5625	$2p^2$ ^{3}P $3d$
7	−45.2375	$2p^2$ ^{1}D $3d$
8	−43.2800	$2p^2$ ^{1}S $3d$
9	−30.6082	$2s^2$ ^{1}S $4d$
10	−27.8395	$2s2p$ ^{3}P $4p$
11	−27.0802	$2s2p$ ^{3}P $4f$
12	−25.1596	$2s2p$ ^{1}P $4p$
13	−24.2261	$2s2p$ ^{1}P $4f$
14	−22.6552	$2p^2$ ^{1}D $4s$
15	−22.1619	$2p^2$ ^{3}P $4d$
16	−21.4810	$2p^2$ ^{1}D $4d$
17	−19.5423	$2s^2$ ^{1}S $5d$
18	−19.4163	$2p^2$ ^{1}S $4d$
19	−16.6025	$2s2p$ ^{3}P $5p$
20	−16.2251	$2s2p$ ^{3}P $5f$
21	−13.8151	$2s2p$ ^{1}P $5p$
22	−13.5438	$2s^2$ ^{1}S $6d$
23	−13.3566	$2s2p$ ^{1}P $5f$
24	−11.3416	$2p^2$ ^{3}P $5d$
25	−11.0693	$2p^2$ ^{1}D $5s$
26	−10.5313	$2p^2$ ^{1}D $5d$
27	−10.4880	$2s2p$ ^{3}P $6p$
28	−10.3354	$2p^2$ ^{1}D $5g$
29	−10.3239	$2s2p$ ^{3}P $6f$
30	−9.94183	$2s^2$ ^{1}S $7d$
31	−8.40189	$2p^2$ ^{1}S $5d$
32	−7.69767	$2s2p$ ^{1}P $6p$
33	−7.60337	$2s^2$ ^{1}S $8d$
34	−7.44955	$2s2p$ ^{1}P $6f$
35	−6.89974	$2s2p$ ^{3}P $7p$
36	−6.76439	$2s2p$ ^{3}P $7f$
37	−6.00553	$2s^2$ ^{1}S $9d$
38	−5.42569	$2p^2$ ^{3}P $6d$
39	−4.86239	$2p^2$ ^{1}D $6s$
^{4}D^e		
1	−53.2102	$2s2p$ ^{3}P $3p$
2	−46.6442	$2p^2$ ^{3}P $3d$
3	−28.1158	$2s2p$ ^{3}P $4p$
4	−27.1319	$2s2p$ ^{3}P $4f$
5	−22.5823	$2p^2$ ^{3}P $4d$
6	−16.7331	$2s2p$ ^{3}P $5p$
7	−16.2586	$2s2p$ ^{3}P $5f$
8	−11.5268	$2p^2$ ^{3}P $5d$
9	−10.6068	$2s2p$ ^{3}P $6p$
10	−10.3453	$2s2p$ ^{3}P $6f$
11	−6.94348	$2s2p$ ^{3}P $7p$
12	−6.77761	$2s2p$ ^{3}P $7f$
13	−5.53273	$2p^2$ ^{3}P $6d$
14	−4.57049	$2s2p$ ^{3}P $8p$
15	−4.46169	$2s2p$ ^{3}P $8f$
16	−2.95045	$2s2p$ ^{3}P $9p$
17	−2.87382	$2s2p$ ^{3}P $9f$
18	−1.92973	$2p^2$ ^{3}P $7d$
^{2}F^e		
1	−46.5018	$2p^2$ ^{3}P $3d$
2	−45.1239	$2p^2$ ^{1}D $3d$
3	−27.2289	$2s2p$ ^{3}P $4f$
4	−24.3652	$2s2p$ ^{1}P $4f$
5	−22.4103	$2p^2$ ^{3}P $4d$
6	−21.5248	$2p^2$ ^{1}D $4d$
7	−16.2987	$2s2p$ ^{3}P $5f$
8	−13.4249	$2s2p$ ^{1}P $5f$
9	−11.4244	$2p^2$ ^{3}P $5d$
10	−11.3600	$2p^2$ ^{3}P $5g$
11	−10.5229	$2p^2$ ^{1}D $5d$
12	−10.3632	$2p^2$ ^{1}D $5g$
13	−10.3632	$2s2p$ ^{3}P $6f$
14	−7.48512	$2s2p$ ^{1}P $6f$
15	−6.79008	$2s2p$ ^{3}P $7f$
16	−5.47349	$2p^2$ ^{3}P $6d$
17	−5.43836	$2p^2$ ^{3}P $6g$
18	−4.54364	$2p^2$ ^{1}D $6d$
19	−4.46832	$2p^2$ ^{1}D $6g$
20	−4.44918	$2s2p$ ^{3}P $8f$
21	−3.90620	$2s2p$ ^{1}P $7f$
22	−2.87938	$2s2p$ ^{3}P $9f$
23	−1.89147	$2p^2$ ^{3}P $7d$
24	−1.86904	$2p^2$ ^{3}P $7g$
^{4}F^e		
1	−46.9293	$2p^2$ ^{3}P $3d$
2	−27.2544	$2s2p$ ^{3}P $4f$
3	−22.6743	$2p^2$ ^{3}P $4d$
4	−16.3151	$2s2p$ ^{3}P $5f$
5	−11.5627	$2p^2$ ^{3}P $5d$
6	−11.3602	$2p^2$ ^{3}P $5g$
7	−10.3749	$2s2p$ ^{3}P $6f$
8	−6.79583	$2s2p$ ^{3}P $7f$
9	−5.55394	$2p^2$ ^{3}P $6d$
10	−5.43855	$2p^2$ ^{3}P $6g$
11	−4.47371	$2s2p$ ^{3}P $8f$
12	−2.88218	$2s2p$ ^{3}P $9f$
13	−1.94088	$2p^2$ ^{3}P $7d$
14	−1.86918	$2p^2$ ^{3}P $7g$
^{2}G^e		
1	−45.6786	$2p^2$ ^{1}D $3d$
2	−27.1328	$2s2p$ ^{3}P $4f$
3	−24.2936	$2s2p$ ^{1}P $4f$
4	−21.5887	$2p^2$ ^{1}D $4d$
5	−19.3554	$2s^2$ ^{1}S $5g$
6	−16.2509	$2s2p$ ^{3}P $5f$
7	−13.4913	$2s^2$ ^{1}S $6g$:
8	−13.3448	$2s2p$ ^{1}P $5f$:
9	−11.3341	$2p^2$ ^{3}P $5g$
10	−10.5364	$2p^2$ ^{1}D $5d$
11	−10.3848	$2p^2$ ^{1}D $5g$
12	−10.3363	$2s2p$ ^{3}P $6f$
13	−10.3363	$2s2p$ ^{3}P $6h$
14	−9.87758	$2s^2$ ^{1}S $7g$
15	−8.26473	$2p^2$ ^{1}S $5g$
16	−7.56984	$2s^2$ ^{1}S $8g$
17	−7.46066	$2s2p$ ^{1}P $6f$
18	−7.43681	$2s2p$ ^{1}P $6h$
19	−6.77298	$2s2p$ ^{3}P $7h$:
20	−6.77163	$2s2p$ ^{3}P $7f$:
21	−5.97541	$2s^2$ ^{1}S $9g$
22	−5.42183	$2p^2$ ^{3}P $6g$
^{4}G^e		
1	−27.1915	$2s2p$ ^{3}P $4f$
2	−16.2891	$2s2p$ ^{3}P $5f$

B-like Fe (Fe^{21+})

i	E(Ryds)	Description	*i*	E(Ryds)	Description	*i*	E(Ryds)	Description	*i*	E(Ryds)	Description
3	−11.3354	$2p^2$ ^{3}P $5g$	28	−10.6801	$2s2p$ ^{3}P $6s$	20	−6.83338	$2s2p$ ^{3}P $7d$	25	−7.55021	$2s^2$ ^{1}S $8f$
4	−10.3625	$2s2p$ ^{3}P $6f$	29	−10.3568	$2p^2$ ^{1}D $5f$	21	−5.62758	$2p^2$ ^{3}P $6p$	26	−7.43791	$2s2p$ ^{1}P $6g$
5	−10.3310	$2s2p$ ^{3}P $6h$	30	−10.3167	$2s2p$ ^{3}P $6d$	22	−5.45173	$2p^2$ ^{3}P $6f$	27	−6.78997	$2s2p$ ^{3}P $7d$
6	−6.78842	$2s2p$ ^{3}P $7f$	31	−10.0354	$2s^2$ ^{1}S $7p$	23	−4.68125	$2p^2$ ^{1}D $6p$	28	−6.76855	$2s2p$ ^{3}P $7g$
7	−6.77144	$2s2p$ ^{3}P $7h$	32	−8.66146	$2p^2$ ^{1}S $5p$	24	−4.50012	$2p^2$ ^{1}D $6f$:	29	−5.98118	$2s^2$ ^{1}S $9f$
8	−5.42314	$2p^2$ ^{3}P $6g$	33	−7.87060	$2s2p$ ^{1}P $6s$	25	−4.45115	$2s2p$ ^{3}P $8d$:	30	−5.41372	$2p^2$ ^{3}P $6f$
9	−4.46893	$2s2p$ ^{3}P $8f$	34	−7.67507	$2s^2$ ^{1}S $8p$	26	−3.94648	$2s2p$ ^{1}P $7d$			**^{4}F°**
10	−4.45660	$2s2p$ ^{3}P $8h$	35	−7.51001	$2s2p$ ^{1}P $6d$	27	−2.89995	$2s2p$ ^{3}P $9d$	1	−51.9867	$2s2p$ ^{3}P $3d$
11	−2.87890	$2s2p$ ^{3}P $9f$	36	−7.01754	$2s2p$ ^{3}P $7s$	28	−1.98766	$2p^2$ ^{3}P $7p$	2	−27.6346	$2s2p$ ^{3}P $4d$
12	−2.87053	$2s2p$ ^{3}P $9h$	37	−6.78303	$2s2p$ ^{3}P $7d$	29	−1.87757	$2p^2$ ^{3}P $7f$	3	−22.2184	$2p^2$ ^{3}P $4f$
13	−1.86073	$2p^2$ ^{3}P $7g$	38	−6.04884	$2s^2$ ^{1}S $9p$			**^{4}D°**	4	−16.4954	$2s2p$ ^{3}P $5d$
		^{2}S°	39	−5.59484	$2p^2$ ^{3}P $6p$	1	−51.5812	$2s2p$ ^{3}P $3d$	5	−16.2392	$2s2p$ ^{3}P $5g$
1	−48.3700	$2p^2$ ^{3}P $3p$			**^{4}P°**	2	−48.0906	$2p^2$ ^{3}P $3p$	6	−11.3505	$2p^2$ ^{3}P $5f$
2	−23.2452	$2p^2$ ^{3}P $4p$	1	−54.6565	$2s2p$ ^{3}P $3s$	3	−27.5004	$2s2p$ ^{3}P $4d$	7	−10.4755	$2s2p$ ^{3}P $6d$
3	−11.8444	$2p^2$ ^{3}P $5p$	2	−51.4463	$2s2p$ ^{3}P $3d$	4	−23.1483	$2p^2$ ^{3}P $4p$	8	−10.3307	$2s2p$ ^{3}P $6g$
4	−5.71268	$2p^2$ ^{3}P $6p$	3	−47.9050	$2p^2$ ^{3}P $3p$	5	−22.3113	$2p^2$ ^{3}P $4f$	9	−6.85821	$2s2p$ ^{3}P $7d$
5	−2.03965	$2p^2$ ^{3}P $7p$	4	−28.7181	$2s2p$ ^{3}P $4s$	6	−16.4325	$2s2p$ ^{3}P $5d$	10	−6.76942	$2s2p$ ^{3}P $7g$
		^{4}S°	5	−27.4543	$2s2p$ ^{3}P $4d$	7	−11.8016	$2p^2$ ^{3}P $5p$	11	−5.43516	$2p^2$ ^{3}P $6f$
1	−121.746	$2p^3$	6	−23.0799	$2p^2$ ^{3}P $4p$	8	−11.3877	$2p^2$ ^{3}P $5f$	12	−4.51475	$2s2p$ ^{3}P $8d$
2	−47.3436	$2p^2$ ^{3}P $3p$	7	−17.0362	$2s2p$ ^{3}P $5s$	9	−10.4391	$2s2p$ ^{3}P $6d$	13	−4.45553	$2s2p$ ^{3}P $8g$
3	−22.9187	$2p^2$ ^{3}P $4p$	8	−16.4124	$2s2p$ ^{3}P $5d$	10	−6.83691	$2s2p$ ^{3}P $7d$	14	−2.91079	$2s2p$ ^{3}P $9d$
4	−11.6966	$2p^2$ ^{3}P $5p$	9	−11.7794	$2p^2$ ^{3}P $5p$	11	−5.68976	$2p^2$ ^{3}P $6p$	15	−2.86970	$2s2p$ ^{3}P $9g$
5	−5.63261	$2p^2$ ^{3}P $6p$	10	−10.7769	$2s2p$ ^{3}P $6s$	12	−5.45483	$2p^2$ ^{3}P $6f$	16	−1.86873	$2p^2$ ^{3}P $7f$
6	−1.99115	$2p^2$ ^{3}P $7p$	11	−10.4308	$2s2p$ ^{3}P $6d$	13	−4.50052	$2s2p$ ^{3}P $8d$			**^{2}G°**
		^{2}P°	12	−7.05325	$2s2p$ ^{3}P $7s$	14	−2.90115	$2s2p$ ^{3}P $9d$	1	−22.2843	$2p^2$ ^{3}P $4f$
1	−130.964	$2s^22p$	13	−6.83089	$2s2p$ ^{3}P $7d$	15	−2.02600	$2p^2$ ^{3}P $7p$	2	−21.3646	$2p^2$ ^{1}D $4f$
2	−119.014	$2p^3$	14	−5.67413	$2p^2$ ^{3}P $6p$	16	−1.87968	$2p^2$ ^{3}P $7f$	3	−16.2720	$2s2p$ ^{3}P $5g$
3	−56.1966	$2s^2$ ^{1}S $3p$	15	−4.64251	$2s2p$ ^{3}P $8s$			**^{2}F°**	4	−13.3875	$2s2p$ ^{1}P $5g$
4	−53.8825	$2s2p$ ^{3}P $3s$	16	−4.49678	$2s2p$ ^{3}P $8d$	1	−50.6953	$2s2p$ ^{3}P $3d$	5	−11.3748	$2p^2$ ^{3}P $5f$
5	−51.6738	$2s2p$ ^{1}P $3s$	17	−3.00167	$2s2p$ ^{3}P $9s$	2	−48.8247	$2s2p$ ^{1}P $3d$	6	−10.4242	$2p^2$ ^{1}D $5f$
6	−50.5401	$2s2p$ ^{3}P $3d$	18	−2.89834	$2s2p$ ^{3}P $9d$	3	−46.8753	$2p^2$ ^{1}D $3p$	7	−10.3468	$2s2p$ ^{3}P $6g$
7	−48.3598	$2s2p$ ^{1}P $3d$	19	−2.01903	$2p^2$ ^{3}P $7p$	4	−30.2911	$2s^2$ ^{1}S $4f$	8	−7.46298	$2s2p$ ^{1}P $6g$
8	−47.3508	$2p^2$ ^{3}P $3p$			**^{2}D°**	5	−27.1964	$2s2p$ ^{3}P $4d$	9	−6.78045	$2s2p$ ^{3}P $7g$
9	−46.0172	$2p^2$ ^{1}D $3p$	1	−120.369	$2p^3$	6	−24.6318	$2s2p$ ^{1}P $4d$	10	−5.44527	$2p^2$ ^{3}P $6f$
10	−44.5729	$2p^2$ ^{1}S $3p$	2	−51.5637	$2s2p$ ^{3}P $3d$	7	−22.1698	$2p^2$ ^{3}P $4f$:	11	−5.43477	$2p^2$ ^{3}P $6h$
11	−31.1599	$2s^2$ ^{1}S $4p$	3	−48.6307	$2s2p$ ^{1}P $3d$	8	−22.0838	$2p^2$ ^{1}D $4p$:	12	−4.48460	$2p^2$ ^{1}D $6f$
12	−28.4811	$2s2p$ ^{3}P $4s$	4	−47.7113	$2p^2$ ^{3}P $3p$	9	−21.3342	$2p^2$ ^{1}D $4f$	13	−4.46244	$2p^2$ ^{1}D $6h$
13	−27.1538	$2s2p$ ^{3}P $4d$	5	−46.4007	$2p^2$ ^{1}D $3p$	10	−19.3872	$2s^2$ ^{1}S $5f$	14	−4.45080	$2s2p$ ^{3}P $8g$
14	−25.7760	$2s2p$ ^{1}P $4s$	6	−27.4843	$2s2p$ ^{3}P $4d$	11	−19.1794	$2p^2$ ^{1}S $4f$	15	−3.89201	$2s2p$ ^{1}P $7g$
15	−24.4871	$2s2p$ ^{1}P $4d$	7	−24.5985	$2s2p$ ^{1}P $4d$	12	−16.2927	$2s2p$ ^{3}P $5d$	16	−2.87477	$2s2p$ ^{3}P $9g$
16	−22.8030	$2p^2$ ^{3}P $4p$	8	−22.9349	$2p^2$ ^{3}P $4p$	13	−16.2378	$2s2p$ ^{3}P $5g$	17	−1.87468	$2p^2$ ^{3}P $7f$
17	−21.8700	$2p^2$ ^{1}D $4p$	9	−22.3060	$2p^2$ ^{3}P $4f$	14	−13.5553	$2s2p$ ^{1}P $5d$	18	−1.80049	$2p^2$ ^{3}P $7h$
18	−21.1494	$2p^2$ ^{1}D $4f$	10	−22.0150	$2p^2$ ^{1}D $4p$	15	−13.4577	$2s^2$ ^{1}S $6f$			**^{4}G°**
19	−19.9301	$2p^2$ ^{1}S $4p$	11	−21.2390	$2p^2$ ^{1}D $4f$	16	−13.3478	$2s2p$ ^{1}P $5g$	1	−22.3210	$2p^2$ ^{3}P $4f$
20	−19.8038	$2s^2$ ^{1}S $5p$	12	−16.4240	$2s2p$ ^{3}P $5d$	17	−11.3164	$2p^2$ ^{3}P $5f$	2	−16.2729	$2s2p$ ^{3}P $5g$
21	−16.9268	$2s2p$ ^{3}P $5s$	13	−13.5408	$2s2p$ ^{1}P $5d$	18	−10.7862	$2p^2$ ^{1}D $5p$	3	−11.3978	$2p^2$ ^{3}P $5f$
22	−16.2706	$2s2p$ ^{3}P $5d$	14	−11.6926	$2p^2$ ^{3}P $5p$	19	−10.4085	$2p^2$ ^{1}D $5f$	4	−10.3481	$2s2p$ ^{3}P $6g$
23	−14.1263	$2s2p$ ^{1}P $5s$	15	−11.3829	$2p^2$ ^{3}P $5f$	20	−10.3643	$2s2p$ ^{3}P $6d$	5	−6.78088	$2s2p$ ^{3}P $7g$
24	−13.7237	$2s^2$ ^{1}S $6p$	16	−10.7637	$2p^2$ ^{1}D $5p$	21	−10.3201	$2s2p$ ^{3}P $6g$	6	−5.45999	$2p^2$ ^{3}P $6f$
25	−13.4531	$2s2p$ ^{1}P $5d$	17	−10.4361	$2s2p$ ^{3}P $6d$	22	−9.88891	$2s^2$ ^{1}S $7f$	7	−5.43481	$2p^2$ ^{3}P $6h$
26	−11.6453	$2p^2$ ^{3}P $5p$	18	−10.3018	$2p^2$ ^{1}D $5f$	23	−8.28191	$2p^2$ ^{1}S $5f$	8	−4.46300	$2s2p$ ^{3}P $8g$
27	−10.7377	$2p^2$ ^{1}D $5p$	19	−7.55141	$2s2p$ ^{1}P $6d$	24	−7.57563	$2s2p$ ^{1}P $6d$	9	−2.87503	$2s2p$ ^{3}P $9g$

B-like Fe (Fe^{21+})

Energies in ascending order from ground state for terms with effective $n \leq 4.0$, $L \leq 4$

Term	i	E(Ryds)	Term	i	E(Ryds)	Term	i	E(Ryds)	Term	i	E(Ryds)	Term	i	E(Ryds)
$^2P^o$	1	0.00000	$^2P^e$	2	77.6803	$^2P^e$	3	80.7909	$^2P^o$	8	83.6132	$^2D^e$	7	85.7265
$^4P^e$	1	2.89900	$^4D^e$	1	77.7538	$^2D^e$	4	80.8340	$^4S^o$	2	83.6204	$^2F^e$	2	85.8401
$^2D^e$	1	5.25401	$^4S^e$	1	78.0309	$^2S^e$	4	81.2420	$^4F^e$	1	84.0347	$^2P^e$	6	86.0935
$^2S^e$	1	6.78300	$^4P^e$	2	78.2869	$^4P^e$	3	81.5937	$^2F^o$	3	84.0887	$^2S^e$	6	86.2603
$^2P^e$	1	7.12300	$^2D^e$	3	78.5048	$^2F^o$	2	82.1393	$^4D^e$	2	84.3198	$^2P^o$	10	86.3911
$^4S^o$	1	9.21800	$^4F^o$	1	78.9773	$^2D^o$	3	82.3333	$^2P^e$	5	84.3338	$^2D^e$	8	87.6840
$^2D^o$	1	10.5950	$^2S^e$	3	78.9970	$^2P^e$	4	82.4284	$^2F^e$	1	84.4622	$^2S^e$	7	99.1484
$^2P^o$	2	11.9500	$^2P^o$	5	79.2902	$^2S^o$	1	82.5940	$^2D^o$	5	84.5633	$^2P^o$	11	99.8041
$^2S^e$	2	73.1903	$^4D^o$	1	79.3828	$^2P^o$	7	82.6042	$^4P^e$	4	84.7310	$^2D^e$	9	100.355
$^2P^o$	3	74.7674	$^2D^o$	2	79.4003	$^2D^e$	5	82.7900	$^2S^e$	5	84.8467	$^2F^o$	4	100.672
$^2D^e$	2	76.2857	$^4P^o$	2	79.5177	$^4D^o$	2	82.8734	$^2P^o$	9	84.9468			
$^4P^o$	1	76.3075	$^2F^o$	1	80.2687	$^4P^o$	3	83.0590	$^2G^e$	1	85.2854			
$^2P^o$	4	77.0815	$^2P^o$	6	80.4239	$^2D^o$	4	83.2527	$^2D^e$	6	85.4015			

gf-values for transitions involving terms with effective $n \leq 4.0$, $L \leq 4$

i	i'	gf_L	i	i'	gf_L	i	i'	gf_L	i	i'	gf_L	i	i'	gf_L	i	i'	gf_L
		$^2S^e$–$^2P^o$	3	2	2.19E−2	5	4	2.88E−4	7	6	1.35E−4	1	11	−2.47E−6	4	2	2.19E−1
1	1	1.05E−1	3	3	2.21E−2	5	5	2.10E−1	7	7	9.98E−4	2	1	9.55E−1	4	3	3.91E−3
1	2	−8.06E−2	3	4	1.23E−1	5	6	7.73E−3	7	8	2.75E−5	2	2	3.55E−4	4	4	4.44E−1
1	3	−4.49E−3	3	5	−1.84E−5	5	7	6.13E−3	7	9	9.64E−7	2	3	2.68E−2	4	5	7.86E−2
1	4	−9.25E−2	3	6	−1.17E−1	5	8	3.80E−3	7	10	1.23E−6	2	4	8.11E−2	4	6	2.64E−5
1	5	−1.36E−1	3	7	−2.04E−2	5	9	−4.57E−6	7	11	−4.37E−1	2	5	−3.36E−4	4	7	−1.45E−3
1	6	−1.76E+0	3	8	−9.74E−2	5	10	−3.14E−1			$^2S^o$–$^2P^e$	2	6	−1.79E−1	4	8	−1.38E−1
1	7	−8.98E−1	3	9	−2.60E−3	5	11	−3.87E−5	1	1	2.62E−1	2	7	−5.70E−3	4	9	−2.71E−1
1	8	−2.52E−4	3	10	−2.48E−3	6	1	5.96E−3	1	2	1.58E−1	2	8	−3.97E−2	4	10	−6.43E−2
1	9	−5.19E−3	3	11	−1.22E−5	6	2	4.92E−1	1	3	1.69E−2	2	9	−5.86E−3	4	11	−5.53E−5
1	10	−4.55E−1	4	1	3.23E−2	6	3	2.07E−4	1	4	7.14E−3	2	10	−4.92E−4	5	1	1.32E−2
1	11	−8.89E−3	4	2	6.86E−5	6	4	1.04E−3	1	5	−1.49E−1	2	11	−1.13E−3	5	2	7.84E−1
2	1	1.20E−1	4	3	8.90E−2	6	5	1.16E−2	1	6	−3.21E−2	3	1	3.21E−1	5	3	1.24E−4
2	2	3.55E−5	4	4	5.39E−2	6	6	8.83E−6			$^2P^e$–$^2P^o$	3	2	4.00E−2	5	4	3.92E−6
2	3	−2.59E−1	4	5	9.30E−2	6	7	8.37E−2	1	1	5.09E−1	3	3	3.03E−1	5	5	3.47E−3
2	4	−2.88E−2	4	6	1.49E−3	6	8	4.11E−2	1	2	−3.31E−1	3	4	5.19E−3	5	6	6.86E−2
2	5	−2.77E−1	4	7	−9.75E−2	6	9	7.46E−2	1	3	−5.14E−3	3	5	2.30E−1	5	7	5.63E−3
2	6	−1.24E−2	4	8	−7.23E−3	6	10	−3.98E−3	1	4	−1.53E−2	3	6	3.66E−4	5	8	5.06E−2
2	7	−1.85E−4	4	9	−1.28E−1	6	11	−4.85E−6	1	5	−2.79E−1	3	7	−7.01E−2	5	9	−2.58E−3
2	8	−6.41E−4	4	10	−3.94E−2	7	1	2.32E−2	1	6	−3.04E−1	3	8	−1.76E−1	5	10	−8.74E−4
2	9	−8.37E−4	4	11	−1.10E−2	7	2	1.03E−6	1	7	−1.40E+0	3	9	−6.09E−3	5	11	−4.59E−6
2	10	−7.79E−5	5	1	5.85E−3	7	3	2.51E−1	1	8	−5.55E−1	3	10	−1.26E−1	6	1	2.76E−2
2	11	−7.35E−1	5	2	2.14E−1	7	4	1.16E−2	1	9	−1.84E−1	3	11	−3.81E−4	6	2	2.04E+0
3	1	4.01E−1	5	3	3.63E−4	7	5	2.87E−5	1	10	−6.87E−2	4	1	1.15E−1	6	3	1.20E−3

B-like Fe (Fe^{21+})

i i'	gf_L	i i'	gf_L	i i'	gf_L	i i'	gf_L	i i'	gf_L	i i'	gf_L
6 4	1.57E−3	2 3	−1.04E−2	7 9	−2.89E−2	3 4	−3.81E−1	5 3	−7.13E−1	1 4	−7.44E+0
6 5	9.24E−5	2 4	−1.55E−3	8 1	1.99E−1	3 5	−3.74E−2	5 4	−1.05E−3	2 1	4.25E−1
6 6	2.06E−3	2 5	−1.98E−1	8 2	1.94E−8	4 1	8.49E−5	6 1	2.37E−1	2 2	2.29E−1
6 7	2.01E−1	2 6	−7.44E−1	8 3	1.57E−1	4 2	8.06E−3	6 2	9.08E−7	2 3	3.06E−1
6 8	6.98E−2	2 7	−3.70E+0	8 4	1.62E−5	4 3	−8.62E−2	6 3	2.47E−2	2 4	−1.84E−1
6 9	1.12E−1	2 8	−3.22E+0	8 5	3.59E−2	4 4	−1.43E−2	6 4	−3.54E−4		$^4P^e$–$^4P^o$
6 10	−4.82E−4	2 9	−3.12E−4	8 6	−3.85E−1	4 5	−1.60E−1	7 1	9.95E−2	1 1	−8.99E−1
6 11	−3.12E−5	3 1	3.23E−2	8 7	−6.64E−3	5 1	5.54E−1	7 2	8.55E−3	1 2	−4.00E+0
	$^2P^e$–$^2D^o$	3 2	−3.31E−1	8 8	−3.68E−3	5 2	2.97E−3	7 3	5.79E−2	1 3	−9.00E−1
1 1	−2.64E−1	3 3	−8.76E−2	8 9	−7.89E−6	5 3	2.43E−4	7 4	−8.33E−5	2 1	7.70E−1
1 2	−5.14E−1	3 4	−3.19E−1	9 1	3.97E−1	5 4	−3.34E−2	8 1	6.76E−5	2 2	−1.17E−1
1 3	−4.72E+0	3 5	−1.26E−2	9 2	4.26E−4	5 5	−4.41E−1	8 2	5.11E−1	2 3	−1.81E−1
1 4	−8.83E−1	3 6	−2.29E−5	9 3	3.64E−2	6 1	3.53E+0	8 3	7.53E−2	3 1	5.98E−1
1 5	−7.67E−1	3 7	−1.70E−4	9 4	1.62E−5	6 2	4.68E−2	8 4	−3.23E−4	3 2	3.32E−3
2 1	1.57E−2	3 8	−6.23E−6	9 5	4.19E−1	6 3	1.04E−1	9 1	9.73E−5	3 3	−6.19E−1
2 2	−3.04E−1	3 9	−3.04E+0	9 6	−1.16E−3	6 4	3.88E−2	9 2	2.92E−3	4 1	1.39E−3
2 3	−5.17E−3	4 1	2.51E−1	9 7	−1.13E−1	6 5	4.71E−2	9 3	6.72E−5	4 2	1.88E−1
2 4	−3.90E−2	4 2	8.10E−3	9 8	−1.63E−3	7 1	9.70E−1	9 4	−3.21E−1	4 3	2.98E−1
2 5	−2.73E−3	4 3	−4.00E−1	9 9	−1.68E−4	7 2	4.56E−4		$^2D^o$–$^2F^e$		$^4P^e$–$^4D^o$
3 1	1.08E−1	4 4	−1.11E−1	10 1	2.23E−4	7 3	2.31E−1	1 1	−5.69E−1	1 1	−1.27E+1
3 2	1.11E−3	4 5	−2.84E−3	10 2	6.50E−4	7 4	2.20E−1	1 2	−1.23E+1	1 2	−1.15E+0
3 3	−2.16E−1	4 6	−4.56E−4	10 3	2.12E−3	7 5	3.73E−2	2 1	−7.65E−2	2 1	−3.84E−1
3 4	−2.49E−1	4 7	−2.78E−3	10 4	3.65E−1	8 1	9.61E−2	2 2	−1.59E−2	2 2	−2.72E−1
3 5	−5.35E−2	4 8	−1.88E−4	10 5	2.08E−2	8 2	3.44E−3	3 1	−4.00E−2	3 1	1.40E−2
4 1	2.34E−1	4 9	−1.61E−1	10 6	2.33E−4	8 3	2.03E−1	3 2	−1.15E−1	3 2	−8.96E−1
4 2	7.13E−4	5 1	2.04E−1	10 7	4.86E−4	8 4	2.02E−3	4 1	−4.28E−1	4 1	5.92E−1
4 3	1.91E−3	5 2	6.47E−4	10 8	−4.07E−1	8 5	1.45E−2	4 2	−3.43E−2	4 2	7.48E−4
4 4	−1.46E−1	5 3	2.80E−3	10 9	−2.62E−4	9 1	5.10E−8	5 1	4.95E−6		$^4P^o$–$^4D^e$
4 5	−3.71E−1	5 4	−3.73E−1	11 1	5.33E−4	9 2	5.07E−6	5 2	−4.24E−1	1 1	−8.87E−1
5 1	2.95E−1	5 5	−4.54E−1	11 2	1.57E−1	9 3	1.27E−6		$^2F^e$–$^2F^o$	1 2	−5.16E−3
5 2	2.68E−1	5 6	−5.49E−4	11 3	2.40E−3	9 4	6.49E−6	1 1	2.78E−1	2 1	7.08E−2
5 3	2.30E−5	5 7	−6.18E−4	11 4	4.54E−4	9 5	6.54E−5	1 2	5.22E−2	2 2	−2.43E−1
5 4	4.19E−4	5 8	−1.08E−5	11 5	1.63E−5		$^2D^e$–$^2F^o$	1 3	1.17E−2	3 1	3.17E−1
5 5	−9.59E−4	5 9	−1.46E−2	11 6	9.43E−5	1 1	−6.07E+0	1 4	−2.14E−5	3 2	−5.77E−1
6 1	6.36E−1	6 1	5.19E−2	11 7	1.74E−5	1 2	−5.28E+0	2 1	1.24E−1		$^4D^e$–$^4D^o$
6 2	1.16E−2	6 2	5.01E−5	11 8	7.89E−5	1 3	−6.06E−1	2 2	2.58E−1	1 1	−1.60E−1
6 3	7.62E−2	6 3	2.32E−2	11 9	−6.13E−1	1 4	−2.15E−4	2 3	9.08E−2	1 2	−8.42E−1
6 4	7.97E−2	6 4	−1.18E−3		$^2D^e$–$^2D^o$	2 1	−6.40E−3	2 4	−5.52E−5	2 1	4.90E−2
6 5	4.98E−2	6 5	−3.04E−3	1 1	−4.15E−1	2 2	−3.38E−1		$^2F^o$–$^2G^e$	2 2	2.42E−1
	$^2P^o$–$^2D^e$	6 6	−7.38E−2	1 2	−1.81E+0	2 3	−1.27E−2	1 1	−4.72E−3		$^4D^e$–$^4F^o$
1 1	2.03E+1	6 7	−3.33E−2	1 3	−4.17E−1	2 4	−9.85E+0	2 1	−1.26E−1	1 1	−8.60E−1
1 2	−4.02E+0	6 8	−4.23E−3	1 4	−1.93E−1	3 1	−6.44E−1	3 1	−7.29E−1	2 1	7.07E−1
1 3	−1.94E+0	6 9	−3.93E−4	1 5	−7.61E−1	3 2	−3.68E−5	4 1	6.36E−5		$^4D^o$–$^4F^e$
1 4	−1.19E−1	7 1	4.69E−2	2 1	6.74E−4	3 3	−3.44E−4		$^4S^e$–$^4P^o$	1 1	−1.53E−1
1 5	−3.47E−2	7 2	3.54E−1	2 2	−5.92E−3	3 4	−2.14E−4	1 1	2.10E−1	2 1	−9.32E−1
1 6	−4.00E−2	7 3	1.68E−3	2 3	−4.94E−1	4 1	4.95E−4	1 2	−2.90E−1		$^4F^e$–$^4F^o$
1 7	−5.39E−2	7 4	6.18E−3	2 4	−1.85E−4	4 2	−3.63E−1	1 3	−2.05E−1	1 1	9.19E−1
1 8	−9.17E−3	7 5	−5.45E−5	2 5	−3.05E−3	4 3	−4.08E−1		$^4S^o$–$^4P^e$		
1 9	−7.49E−1	7 6	−1.12E−2	3 1	2.45E−2	4 4	−8.98E−2	1 1	3.98E−1		
2 1	2.60E−1	7 7	−6.54E−2	3 2	−3.89E−2	5 1	9.22E−3	1 2	−9.90E−3		
2 2	−5.21E−5	7 8	−6.78E−2	3 3	−2.39E−2	5 2	4.78E−3	1 3	−4.96E−1		

C-like S (S^{10+})

Term energies relative to $2s^22p$ ^{2}P ionization threshold for each symmetry

i	E(Ryds)	Description	i	E(Ryds)	Description	i	E(Ryds)	Description	i	E(Ryds)	Description
		^{1}S°	15	−2.53564	$2s^22p$ ^{2}P $7d$	10	−1.48678	$2s^22p$ ^{2}P $9g$	31	−1.60133	$2p^3$ ^{2}P $4s$
1	−11.0571	$2s2p^2$ ^{2}P $3p$	16	−2.41384	$2p^3$ ^{2}D $4s$			**^{1}I°**	32	−1.56679	$2s^22p$ ^{2}P $9s$
2	−7.95941	$2p^3$ ^{2}D $3d$	17	−2.02011	$2s2p^2$ ^{2}D $5p$	1	−2.46617	$2s^22p$ ^{2}P $7i$	33	−1.51367	$2s^22p$ ^{2}P $9d$
3	−4.12238	$2s2p^2$ ^{2}P $4p$	18	−1.92888	$2s^22p$ ^{2}P $8d$	2	−1.88720	$2s^22p$ ^{2}P $8i$			**^{3}D°**
4	−1.80512	$2p^3$ ^{2}D $4d$	19	−1.80646	$2s2p^2$ ^{2}D $5f$	3	−1.49099	$2s^22p$ ^{2}P $9i$	1	−33.9918	$2s2p^3$
5	−.99249	$2s2p^2$ ^{2}P $5p$	20	−1.58955	$2p^3$ ^{2}D $4d$			**^{3}S°**	2	−13.9043	$2s^22p$ ^{2}P $3d$
		^{1}P°	21	−1.51828	$2s^22p$ ^{2}P $9d$	1	−32.3214	$2s2p^3$	3	−13.0525	$2s2p^2$ ^{4}P $3p$
1	−31.8080	$2s2p^3$			**^{1}F°**	2	−13.4914	$2s2p^2$ ^{4}P $3p$	4	−11.7881	$2s2p^2$ ^{2}D $3p$
2	−15.7610	$2s^22p$ ^{2}P $3s$	1	−13.5734	$2s^22p$ ^{2}P $3d$	3	−10.7237	$2s2p^2$ ^{2}P $3p$	5	−10.8813	$2s2p^2$ ^{2}P $3p$
3	−13.5628	$2s^22p$ ^{2}P $3d$	2	−11.8125	$2s2p^2$ ^{2}D $3p$	4	−9.91213	$2p^3$ ^{4}S $3s$	6	−9.57213	$2p^3$ ^{2}D $3s$
4	−11.7439	$2s2p^2$ ^{2}D $3p$	3	−7.68095	$2s^22p$ ^{2}P $4d$	5	−7.45844	$2p^3$ ^{2}D $3d$	7	−8.42130	$2p^3$ ^{4}S $3d$
5	−10.9059	$2s2p^2$ ^{2}S $3p$	4	−7.39177	$2p^3$ ^{2}D $3d$	6	−6.55901	$2s2p^2$ ^{4}P $4p$	8	−7.80519	$2s^22p$ ^{2}P $4d$
6	−10.2723	$2s2p^2$ ^{2}P $3p$	5	−6.81350	$2p^3$ ^{2}P $3d$	7	−3.97194	$2s2p^2$ ^{2}P $4p$	9	−7.70722	$2p^3$ ^{2}D $3d$
7	−8.60915	$2p^3$ ^{2}P $3s$	6	−5.12610	$2s2p^2$ ^{2}D $4p$	8	−3.49909	$2s2p^2$ ^{4}P $5p$	10	−6.93707	$2p^3$ ^{2}P $3d$
8	−8.48088	$2s^22p$ ^{2}P $4s$	7	−5.04333	$2s^22p$ ^{2}P $5d$	9	−3.05355	$2p^3$ ^{4}S $4s$	11	−6.41348	$2s2p^2$ ^{4}P $4p$
9	−7.72668	$2p^3$ ^{2}D $3d$	8	−4.84359	$2s^22p$ ^{2}P $5g$	10	−1.85857	$2s2p^2$ ^{4}P $6p$	12	−5.91990	$2s2p^2$ ^{4}P $4f$
10	−7.66990	$2s^22p$ ^{2}P $4d$	9	−4.56330	$2s2p^2$ ^{2}D $4f$	11	−1.62193	$2p^3$ ^{2}D $4d$	13	−5.13386	$2s2p^2$ ^{2}D $4p$
11	−6.45880	$2p^3$ ^{2}P $3d$	10	−3.65190	$2s2p^2$ ^{2}S $4f$	12	−.93840	$2s2p^2$ ^{2}P $5p$	14	−5.10200	$2s^22p$ ^{2}P $5d$
12	−5.26608	$2s^22p$ ^{2}P $5s$	11	−3.45711	$2s^22p$ ^{2}P $6d$	13	−.90480	$2s2p^2$ ^{4}P $7p$	15	−4.54571	$2s2p^2$ ^{2}D $4f$
13	−5.05391	$2s^22p$ ^{2}P $5d$	12	−3.44213	$2s2p^2$ ^{2}P $4f$	14	−.31453	$2s2p^2$ ^{4}P $8p$	16	−4.05858	$2s2p^2$ ^{2}P $4p$
14	−5.04139	$2s2p^2$ ^{2}D $4p$	13	−3.32069	$2s^22p$ ^{2}P $6g$			**^{3}P°**	17	−3.48328	$2s^22p$ ^{2}P $6d$
15	−4.48697	$2s2p^2$ ^{2}D $4f$	14	−2.51271	$2s^22p$ ^{2}P $7d$	1	−33.4280	$2s2p^3$	18	−3.46378	$2s2p^2$ ^{2}P $4f$
16	−4.20212	$2s2p^2$ ^{2}S $4p$	15	−2.45810	$2s^22p$ ^{2}P $7g$	2	−15.9245	$2s^22p$ ^{2}P $3s$	19	−3.39889	$2s2p^2$ ^{4}P $5p$
17	−3.89162	$2s2p^2$ ^{2}P $4p$	16	−2.07342	$2s2p^2$ ^{2}D $5p$	3	−13.8523	$2s^22p$ ^{2}P $3d$	20	−3.17675	$2s2p^2$ ^{4}P $5f$
18	−3.56974	$2s^22p$ ^{2}P $6s$	17	−1.91298	$2s^22p$ ^{2}P $8d$	4	−12.8880	$2s2p^2$ ^{4}P $3p$	21	−2.52708	$2s^22p$ ^{2}P $7d$
19	−3.45407	$2s^22p$ ^{2}P $6d$	18	−1.88154	$2s^22p$ ^{2}P $8g$	5	−11.6253	$2s2p^2$ ^{2}D $3p$	22	−2.43791	$2p^3$ ^{2}D $4s$
20	−2.61356	$2s^22p$ ^{2}P $7s$	19	−1.82418	$2s2p^2$ ^{2}D $5f$	6	−10.9607	$2s2p^2$ ^{2}S $3p$	23	−2.38565	$2p^3$ ^{4}S $4d$
21	−2.51188	$2s^22p$ ^{2}P $7d$	20	−1.55509	$2p^3$ ^{2}D $4d$	7	−10.7688	$2s2p^2$ ^{2}P $3p$	24	−2.07812	$2s2p^2$ ^{2}D $5p$
22	−2.05638	$2s2p^2$ ^{2}D $5p$	21	−1.50690	$2s^22p$ ^{2}P $9d$	8	−8.80077	$2p^3$ ^{2}P $3s$	25	−1.92293	$2s^22p$ ^{2}P $8d$
23	−1.98835	$2s^22p$ ^{2}P $8s$	22	−1.48665	$2s^22p$ ^{2}P $9g$	9	−8.53724	$2s^22p$ ^{2}P $4s$	26	−1.82177	$2s2p^2$ ^{4}P $6p$
24	−1.91275	$2s^22p$ ^{2}P $8d$			**^{1}G°**	10	−7.79120	$2s^22p$ ^{2}P $4d$	27	−1.80741	$2s2p^2$ ^{2}D $5f$
25	−1.79157	$2s2p^2$ ^{2}D $5f$	1	−7.89699	$2p^3$ ^{2}D $3d$	11	−7.58219	$2p^3$ ^{2}D $3d$	28	−1.73013	$2p^3$ ^{2}D $4d$
26	−1.71538	$2p^3$ ^{2}D $4d$	2	−4.85627	$2s^22p$ ^{2}P $5g$	12	−7.04353	$2p^3$ ^{2}P $3d$	29	−1.70717	$2s2p^2$ ^{4}P $6f$
27	−1.56561	$2s^22p$ ^{2}P $9s$	3	−4.58022	$2s2p^2$ ^{2}D $4f$	13	−6.35536	$2s2p^2$ ^{4}P $4p$	30	−1.51419	$2s^22p$ ^{2}P $9d$
28	−1.55717	$2p^3$ ^{2}P $4s$	4	−3.49221	$2s2p^2$ ^{2}P $4f$	14	−5.30420	$2s^22p$ ^{2}P $5s$			**^{3}F°**
29	−1.50644	$2s^22p$ ^{2}P $9d$	5	−3.33307	$2s^22p$ ^{2}P $6g$	15	−5.10340	$2s^22p$ ^{2}P $5d$	1	−14.1017	$2s^22p$ ^{2}P $3d$
		^{1}D°	6	−2.46209	$2s^22p$ ^{2}P $7g$	16	−5.04809	$2s2p^2$ ^{2}D $4p$	2	−11.9297	$2s2p^2$ ^{2}D $3p$
1	−32.3771	$2s2p^3$	7	−1.88532	$2s^22p$ ^{2}P $8g$	17	−4.50859	$2s2p^2$ ^{2}D $4f$	3	−7.99588	$2p^3$ ^{2}D $3d$
2	−14.1123	$2s^22p$ ^{2}P $3d$	8	−1.84801	$2s2p^2$ ^{2}D $5f$	18	−4.23002	$2s2p^2$ ^{2}S $4p$	4	−7.87765	$2s^22p$ ^{2}P $4d$
3	−11.9389	$2s2p^2$ ^{2}D $3p$	9	−1.73703	$2p^3$ ^{2}D $4d$	19	−4.03065	$2s2p^2$ ^{2}P $4p$	5	−7.13202	$2p^3$ ^{2}P $3d$
4	−10.4767	$2s2p^2$ ^{2}P $3p$	10	−1.48833	$2s^22p$ ^{2}P $9g$	20	−3.59229	$2s^22p$ ^{2}P $6s$	6	−5.88724	$2s2p^2$ ^{4}P $4f$
5	−9.37530	$2p^3$ ^{2}D $3s$			**^{1}H°**	21	−3.47814	$2s^22p$ ^{2}P $6d$	7	−5.17468	$2s2p^2$ ^{2}D $4p$
6	−7.88076	$2s^22p$ ^{2}P $4d$	1	−4.84978	$2s^22p$ ^{2}P $5g$	22	−3.36636	$2s2p^2$ ^{4}P $5p$	8	−5.11923	$2s^22p$ ^{2}P $5d$
7	−7.54820	$2p^3$ ^{2}D $3d$	2	−4.54683	$2s2p^2$ ^{2}D $4f$	23	−2.62269	$2s^22p$ ^{2}P $7s$	9	−4.84397	$2s^22p$ ^{2}P $5g$
8	−6.93558	$2p^3$ ^{2}P $3d$	3	−3.35683	$2s^22p$ ^{2}P $6g$	24	−2.52605	$2s^22p$ ^{2}P $7d$	10	−4.58536	$2s2p^2$ ^{2}D $4f$
9	−5.14673	$2s^22p$ ^{2}P $5d$	4	−2.46611	$2s^22p$ ^{2}P $7i$	25	−2.06445	$2s2p^2$ ^{2}D $5p$	11	−3.66637	$2s2p^2$ ^{2}S $4f$
10	−5.10220	$2s2p^2$ ^{2}D $4p$	5	−2.46112	$2s^22p$ ^{2}P $7g$	26	−1.98755	$2s^22p$ ^{2}P $8s$	12	−3.49711	$2s^22p$ ^{2}P $6d$
11	−4.52523	$2s2p^2$ ^{2}D $4f$	6	−1.88711	$2s^22p$ ^{2}P $8i$	27	−1.92223	$2s^22p$ ^{2}P $8d$	13	−3.44804	$2s2p^2$ ^{2}P $4f$
12	−3.94798	$2s2p^2$ ^{2}P $4p$	7	−1.88429	$2s^22p$ ^{2}P $8g$	28	−1.81027	$2s2p^2$ ^{4}P $6p$	14	−3.32325	$2s^22p$ ^{2}P $6g$
13	−3.49462	$2s^22p$ ^{2}P $6d$	8	−1.80593	$2s2p^2$ ^{2}D $5f$	29	−1.79131	$2s2p^2$ ^{2}D $5f$	15	−3.15217	$2s2p^2$ ^{4}P $5f$
14	−3.46943	$2s2p^2$ ^{2}P $4f$	9	−1.49089	$2s^22p$ ^{2}P $9i$	30	−1.68987	$2p^3$ ^{2}D $4d$	16	−2.53591	$2s^22p$ ^{2}P $7d$

C-like S (S^{10+})

i E(Ryds) Description	i E(Ryds) Description	i E(Ryds) Description	i E(Ryds) Description
17 −2.45827 $2s^22p$ ^{2}P $7g$	$^5\mathbf{P}^o$	16 −1.34001 $2s2p^2$ ^{2}S $5s$	$^1\mathbf{F}^e$
18 −2.09229 $2s2p^2$ ^{2}D $5p$	1 −13.2875 $2s2p^2$ ^{4}P $3p$	$^1\mathbf{P}^e$	1 −10.8857 $2s2p^2$ ^{2}D $3d$
19 −1.92882 $2s^22p$ ^{2}P $8d$	2 −6.52032 $2s2p^2$ ^{4}P $4p$	1 −15.1675 $2s^22p$ ^{2}P $3p$	2 −9.62721 $2s2p^2$ ^{2}P $3d$
20 −1.88173 $2s^22p$ ^{2}P $8g$	3 −3.44296 $2s2p^2$ ^{4}P $5p$	2 −11.5883 $2s2p^2$ ^{2}P $3s$	3 −8.71758 $2p^3$ ^{2}D $3p$
21 −1.83072 $2s2p^2$ ^{2}D $5f$	4 −1.85334 $2s2p^2$ ^{4}P $6p$	3 −10.5677 $2s2p^2$ ^{2}D $3d$	4 −7.58124 $2s^22p$ ^{2}P $4f$
22 −1.78296 $2p^3$ ^{2}D $4d$	5 −.91339 $2s2p^2$ ^{4}P $7p$	4 −9.84312 $2s2p^2$ ^{2}P $3d$	5 −4.83103 $2s^22p$ ^{2}P $5f$
23 −1.70622 $2s2p^2$ ^{4}P $6f$	6 −.30728 $2s2p^2$ ^{4}P $8p$	5 −8.85776 $2p^3$ ^{2}D $3p$	6 −4.76760 $2s2p^2$ ^{2}D $4d$
24 −1.51854 $2s^22p$ ^{2}P $9d$	$^5\mathbf{D}^o$	6 −8.22355 $2s^22p$ ^{2}P $4p$	7 −3.63921 $2s2p^2$ ^{2}P $4d$
25 −1.48675 $2s^22p$ ^{2}P $9g$	1 −13.3829 $2s2p^2$ ^{4}P $3p$	7 −7.91466 $2p^3$ ^{2}P $3p$	8 −3.36264 $2s^22p$ ^{2}P $6f$
$^3\mathbf{G}^o$	2 −8.74412 $2p^3$ ^{4}S $3d$	8 −5.13089 $2s^22p$ ^{2}P $5p$	9 −2.47893 $2s^22p$ ^{2}P $7f$
1 −7.93614 $2p^3$ ^{2}D $3d$	3 −6.54924 $2s2p^2$ ^{4}P $4p$	9 −4.66093 $2s2p^2$ ^{2}D $4d$	10 −2.10708 $2s2p^2$ ^{2}D $5d$
2 −5.93232 $2s2p^2$ ^{4}P $4f$	4 −5.94053 $2s2p^2$ ^{4}P $4f$	10 −4.35488 $2s2p^2$ ^{2}P $4s$	11 −2.04111 $2p^3$ ^{2}D $4p$
3 −4.85660 $2s^22p$ ^{2}P $5g$	5 −3.45824 $2s2p^2$ ^{4}P $5p$	11 −3.72395 $2s2p^2$ ^{2}P $4d$	12 −1.89771 $2s^22p$ ^{2}P $8f$
4 −4.59678 $2s2p^2$ ^{2}D $4f$	6 −3.18251 $2s2p^2$ ^{4}P $5f$	12 −3.50325 $2s^22p$ ^{2}P $6p$	13 −1.84675 $2s2p^2$ ^{2}D $5g$
5 −3.49554 $2s2p^2$ ^{2}P $4f$	7 −2.55650 $2p^3$ ^{4}S $4d$	13 −2.56738 $2s^22p$ ^{2}P $7p$	14 −1.49895 $2s^22p$ ^{2}P $9f$
6 −3.33432 $2s^22p$ ^{2}P $6g$	8 −1.85906 $2s2p^2$ ^{4}P $6p$	14 −2.10825 $2p^3$ ^{2}D $4p$	15 −1.49798 $2p^3$ ^{2}D $4f$
7 −3.17335 $2s2p^2$ ^{4}P $5f$	9 −1.72414 $2s2p^2$ ^{4}P $6f$	15 −2.03449 $2s2p^2$ ^{2}D $5d$	$^1\mathbf{G}^e$
8 −2.46210 $2s^22p$ ^{2}P $7g$	10 −.91792 $2s2p^2$ ^{4}P $7p$	16 −1.95718 $2s^22p$ ^{2}P $8p$	1 −10.7705 $2s2p^2$ ^{2}D $3d$
9 −1.88524 $2s^22p$ ^{2}P $8g$	11 −.83396 $2s2p^2$ ^{4}P $7f$	17 −1.54383 $2s^22p$ ^{2}P $9p$	2 −7.56085 $2s^22p$ ^{2}P $4f$
10 −1.83624 $2s2p^2$ ^{2}D $5f$	12 −.31065 $2s2p^2$ ^{4}P $8p$	18 −1.51218 $2p^3$ ^{2}D $4f$	3 −4.81140 $2s^22p$ ^{2}P $5f$
11 −1.77628 $2p^3$ ^{2}D $4d$	13 −.25258 $2s2p^2$ ^{4}P $8f$	19 −1.31390 $2s2p^2$ ^{2}P $5s$	4 −4.75727 $2s2p^2$ ^{2}D $4d$
12 −1.71581 $2s2p^2$ ^{4}P $6f$	$^5\mathbf{F}^o$	$^1\mathbf{D}^e$	5 −3.36034 $2s^22p$ ^{2}P $6f$
13 −1.70657 $2s2p^2$ ^{4}P $6h$	1 −5.90925 $2s2p^2$ ^{4}P $4f$	1 −36.5212 $2s^22p^2$	6 −3.35013 $2s^22p$ ^{2}P $6h$
14 −1.48828 $2s^22p$ ^{2}P $9g$	2 −3.16731 $2s2p^2$ ^{4}P $5f$	2 −29.4052 $2p^4$	7 −2.47292 $2s^22p$ ^{2}P $7f$
$^3\mathbf{H}^o$	3 −1.71681 $2s2p^2$ ^{4}P $6f$	3 −14.5875 $2s^22p$ ^{2}P $3p$	8 −2.46554 $2s^22p$ ^{2}P $7h$
1 −4.85123 $2s^22p$ ^{2}P $5g$	4 −.82925 $2s2p^2$ ^{4}P $7f$	4 −12.5364 $2s2p^2$ ^{2}D $3s$	9 −2.08309 $2s2p^2$ ^{2}D $5d$
2 −4.55615 $2s2p^2$ ^{2}D $4f$	5 −.24940 $2s2p^2$ ^{4}P $8f$	5 −10.6328 $2s2p^2$ ^{2}D $3d$	10 −1.89327 $2s^22p$ ^{2}P $8f$
3 −3.35705 $2s^22p$ ^{2}P $6g$	$^5\mathbf{G}^o$	6 −9.87633 $2s2p^2$ ^{2}S $3d$	11 −1.89032 $2s^22p$ ^{2}P $8h$
4 −2.46611 $2s^22p$ ^{2}P $7i$	1 −5.95070 $2s2p^2$ ^{4}P $4f$	7 −9.38227 $2s2p^2$ ^{2}P $3d$	12 −1.85545 $2s2p^2$ ^{2}D $5g$
5 −2.46138 $2s^22p$ ^{2}P $7g$	2 −3.18568 $2s2p^2$ ^{4}P $5f$	8 −8.16288 $2p^3$ ^{2}D $3p$	13 −1.50246 $2p^3$ ^{2}D $4f$
6 −1.88712 $2s^22p$ ^{2}P $8i$	3 −1.72751 $2s2p^2$ ^{4}P $6f$	9 −8.03334 $2s^22p$ ^{2}P $4p$	14 −1.49520 $2s^22p$ ^{2}P $9f$
7 −1.88506 $2s^22p$ ^{2}P $8g$	4 −1.70659 $2s2p^2$ ^{4}P $6h$	10 −7.66544 $2p^3$ ^{2}P $3p$	15 −1.49400 $2s^22p$ ^{2}P $9h$
8 −1.81261 $2s2p^2$ ^{2}D $5f$	5 −.83571 $2s2p^2$ ^{4}P $7f$	11 −7.52122 $2s^22p$ ^{2}P $4f$	$^1\mathbf{H}^e$
9 −1.70508 $2s2p^2$ ^{4}P $6h$	6 −.82197 $2s2p^2$ ^{4}P $7h$	12 −5.41712 $2s2p^2$ ^{2}D $4s$	1 −3.35141 $2s^22p$ ^{2}P $6h$
10 −1.49089 $2s^22p$ ^{2}P $9i$	7 −.25356 $2s2p^2$ ^{4}P $8f$	13 −5.04014 $2s^22p$ ^{2}P $5p$	2 −2.46684 $2s^22p$ ^{2}P $7h$
11 −1.48686 $2s^22p$ ^{2}P $9g$	8 −.24623 $2s2p^2$ ^{4}P $8h$	14 −4.80096 $2s^22p$ ^{2}P $5f$	3 −1.89246 $2s^22p$ ^{2}P $8h$
$^3\mathbf{I}^o$	$^1\mathbf{S}^e$	15 −4.70186 $2s2p^2$ ^{2}D $4d$	4 −1.85302 $2s2p^2$ ^{2}D $5g$
1 −2.46617 $2s^22p$ ^{2}P $7i$	1 −35.9233 $2s^22p^2$	16 −3.83855 $2s2p^2$ ^{2}S $4d$	5 −1.52842 $2p^3$ ^{2}D $4f$
2 −1.88720 $2s^22p$ ^{2}P $8i$	2 −28.3077 $2p^4$	17 −3.56798 $2s2p^2$ ^{2}P $4d$	6 −1.49457 $2s^22p$ ^{2}P $9h$
3 −1.70623 $2s2p^2$ ^{4}P $6h$	3 −14.2586 $2s^22p$ ^{2}P $3p$	18 −3.44107 $2s^22p$ ^{2}P $6p$	$^1\mathbf{I}^e$
4 −1.49099 $2s^22p$ ^{2}P $9i$	4 −11.6691 $2s2p^2$ ^{2}S $3s$	19 −3.34584 $2s^22p$ ^{2}P $6f$	1 −3.35030 $2s^22p$ ^{2}P $6h$
$^5\mathbf{S}^o$	5 −10.3753 $2s2p^2$ ^{2}D $3d$	20 −2.53835 $2s^22p$ ^{2}P $7p$	2 −2.46574 $2s^22p$ ^{2}P $7h$
1 −35.5585 $2s2p^3$	6 −7.93137 $2s^22p$ ^{2}P $4p$	21 −2.40819 $2s^22p$ ^{2}P $7f$	3 −1.89201 $2s^22p$ ^{2}P $8h$
2 −12.9839 $2s2p^2$ ^{4}P $3p$	7 −7.12087 $2p^3$ ^{2}P $3p$	22 −2.24428 $2s2p^2$ ^{2}D $5s$	4 −1.89033 $2s^22p$ ^{2}P $8j$
3 −10.4443 $2p^3$ ^{4}S $3s$	8 −4.99281 $2s^22p$ ^{2}P $5p$	23 −2.05985 $2s2p^2$ ^{2}D $5d$	5 −1.83255 $2s2p^2$ ^{2}D $5g$
4 −6.42944 $2s2p^2$ ^{4}P $4p$	9 −4.64782 $2s2p^2$ ^{2}D $4d$	24 −1.93843 $2s^22p$ ^{2}P $8p$	6 −1.49371 $2s^22p$ ^{2}P $9h$
5 3.41316 $2s2p^2$ ^{4}P $5p$	10 −4.48013 $2s2p^2$ ^{2}S $4s$	25 −1.89061 $2s^22p$ ^{2}P $8f$	7 −1.49336 $2s^22p$ ^{2}P $9j$
6 −3.19659 $2p^3$ ^{4}S $4s$	11 −3.43788 $2s^22p$ ^{2}P $6p$	26 −1.83625 $2p^3$ ^{2}D $4p$	$^3\mathbf{S}^e$
7 −1.83003 $2s2p^2$ ^{4}P $6p$	12 −2.52054 $2s^22p$ ^{2}P $7p$	27 −1.83520 $2s2p^2$ ^{2}D $5g$	1 −14.9450 $2s^22p$ ^{2}P $3p$
8 −.89920 $2s2p^2$ ^{4}P $7p$	13 −2.03071 $2s2p^2$ ^{2}D $5d$	28 −1.52931 $2s^22p$ ^{2}P $9p$	2 −11.9331 $2s2p^2$ ^{2}S $3s$
9 −.29828 $2s2p^2$ ^{4}P $8p$	14 −1.92457 $2s^22p$ ^{2}P $8p$	29 −1.50701 $2p^3$ ^{2}D $4f$	3 −10.6296 $2s2p^2$ ^{2}D $3d$
	15 −1.51994 $2s^22p$ ^{2}P $9p$	30 −1.49365 $2s^22p$ ^{2}P $9f$	4 −8.15571 $2s^22p$ ^{2}P $4p$

C-like S (S^{10+})

i	E(Ryds)	Description	i	E(Ryds)	Description	i	E(Ryds)	Description	i	E(Ryds)	Description
5	−8.05817	$2p^3$ ^{2}P $3p$	9	−7.99120	$2p^3$ ^{2}P $3p$	4	−4.79932	$2s^22p$ ^{2}P $5f$	6	−.28267	$2s2p^2$ ^{4}P $8d$
6	−5.10128	$2s^22p$ ^{2}P $5p$	10	−7.53350	$2s^22p$ ^{2}P $4f$	5	−3.36563	$2s^22p$ ^{2}P $6f$		^{5}**F**^{e}	
7	−4.73738	$2s2p^2$ ^{2}D $4d$	11	−5.97065	$2s2p^2$ ^{4}P $4d$	6	−3.35015	$2s^22p$ ^{2}P $6h$	1	−12.5160	$2s2p^2$ ^{4}P $3d$
8	−4.58094	$2s2p^2$ ^{2}S $4s$	12	−5.52839	$2s2p^2$ ^{2}D $4s$	7	−3.19019	$2s2p^2$ ^{4}P $5g$	2	−6.25399	$2s2p^2$ ^{4}P $4d$
9	−3.49341	$2s^22p$ ^{2}P $6p$	13	−5.10153	$2s^22p$ ^{2}P $5p$	8	−2.47686	$2s^22p$ ^{2}P $7f$	3	−3.49929	$2s2p^2$ ^{4}P $5d$
10	−2.55726	$2s^22p$ ^{2}P $7p$	14	−4.81129	$2s^22p$ ^{2}P $5f$	9	−2.46555	$2s^22p$ ^{2}P $7h$	4	−3.20708	$2s2p^2$ ^{4}P $5g$
11	−2.07048	$2s2p^2$ ^{2}D $5d$	15	−4.79056	$2s2p^2$ ^{2}D $4d$	10	−2.12093	$2s2p^2$ ^{2}D $5d$	5	−2.28472	$2p^3$ ^{4}S $4f$
12	−1.94980	$2s^22p$ ^{2}P $8p$	16	−3.91747	$2s2p^2$ ^{2}S $4d$	11	−1.89515	$2s^22p$ ^{2}P $8f$	6	−1.85362	$2s2p^2$ ^{4}P $6d$
13	−1.53897	$2s^22p$ ^{2}P $9p$	17	−3.71406	$2s2p^2$ ^{2}P $4d$	12	−1.89033	$2s^22p$ ^{2}P $8h$	7	−1.71124	$2s2p^2$ ^{4}P $6g$
14	−1.39044	$2s2p^2$ ^{2}S $5s$	18	−3.49414	$2s^22p$ ^{2}P $6p$	13	−1.85574	$2s2p^2$ ^{2}D $5g$	8	−.89347	$2s2p^2$ ^{4}P $7d$
15	−1.26173	$2p^3$ ^{2}P $4p$	19	−3.38027	$2s2p^2$ ^{4}P $5d$	14	−1.71051	$2s2p^2$ ^{4}P $6g$	9	−.81781	$2s2p^2$ ^{4}P $7g$
	^{3}**P**^{e}		20	−3.34726	$2s^22p$ ^{2}P $6f$	15	−1.52043	$2p^3$ ^{2}D $4f$	10	−.28593	$2s2p^2$ ^{4}P $8d$
1	−37.0543	$2s^22p^2$	21	−2.56352	$2s^22p$ ^{2}P $7p$	16	−1.49682	$2s^22p$ ^{2}P $9f$	11	−.23936	$2s2p^2$ ^{4}P $8g$
2	−29.8816	$2p^4$	22	−2.47124	$2s^22p$ ^{2}P $7f$	17	−1.49414	$2s^22p$ ^{2}P $9h$		^{5}**G**^{e}	
3	−14.8745	$2s^22p$ ^{2}P $3p$	23	−2.28474	$2s2p^2$ ^{2}D $5s$		^{3}**H**^{e}		1	−3.19070	$2s2p^2$ ^{4}P $5g$
4	−13.7604	$2s2p^2$ ^{4}P $3s$	24	−2.11683	$2s2p^2$ ^{2}D $5d$	1	−3.35141	$2s^22p$ ^{2}P $6h$	2	−1.71097	$2s2p^2$ ^{4}P $6g$
5	−12.2078	$2s2p^2$ ^{4}P $3d$	25	−2.06422	$2p^3$ ^{2}D $4p$	2	−3.19909	$2s2p^2$ ^{4}P $5g$	3	−.81522	$2s2p^2$ ^{4}P $7g$
6	−11.7159	$2s2p^2$ ^{2}P $3s$	26	−1.95361	$2s^22p$ ^{2}P $8p$	3	−2.46684	$2s^22p$ ^{2}P $7h$	4	−.23710	$2s2p^2$ ^{4}P $8g$
7	−10.7122	$2s2p^2$ ^{2}D $3d$	27	−1.89277	$2s^22p$ ^{2}P $8f$	4	−1.89245	$2s^22p$ ^{2}P $8h$		^{5}**H**^{e}	
8	−9.69920	$2s2p^2$ ^{2}P $3d$	28	−1.83535	$2s2p^2$ ^{2}D $5g$	5	−1.85271	$2s2p^2$ ^{2}D $5g$	1	−3.19950	$2s2p^2$ ^{4}P $5g$
9	−9.35306	$2p^3$ ^{4}S $3p$	29	−1.79857	$2s2p^2$ ^{4}P $6d$	6	−1.71580	$2s2p^2$ ^{4}P $6g$	2	−1.71617	$2s2p^2$ ^{4}P $6g$
10	−8.34881	$2p^3$ ^{2}D $3p$	30	−1.54025	$2s^22p$ ^{2}P $9p$	7	−1.53019	$2p^3$ ^{2}D $4f$	3	−.82172	$2s2p^2$ ^{4}P $7i$
11	−8.11536	$2s^22p$ ^{2}P $4p$	31	−1.51043	$2p^3$ ^{2}D $4f$	8	−1.49457	$2s^22p$ ^{2}P $9h$	4	−.81841	$2s2p^2$ ^{4}P $7g$
12	−7.77622	$2p^3$ ^{2}P $3p$	32	−1.49508	$2s^22p$ ^{2}P $9f$		^{3}**I**^{e}		5	−.24273	$2s2p^2$ ^{4}P $8i$
13	−6.75956	$2s2p^2$ ^{4}P $4s$	33	−1.29505	$2p^3$ ^{2}P $4p$	1	−3.35031	$2s^22p$ ^{2}P $6h$	6	−.23925	$2s2p^2$ ^{4}P $8g$
14	−6.12823	$2s2p^2$ ^{4}P $4d$		^{3}**F**^{e}		2	−2.46575	$2s^22p$ ^{2}P $7h$			
15	−5.07690	$2s^22p$ ^{2}P $5p$	1	−12.0512	$2s2p^2$ ^{4}P $3d$	3	−1.89205	$2s^22p$ ^{2}P $8h$			
16	−4.73772	$2s2p^2$ ^{2}D $4d$	2	−10.9845	$2s2p^2$ ^{2}D $3d$	4	−1.89033	$2s^22p$ ^{2}P $8j$			
17	−4.39433	$2s2p^2$ ^{2}P $4s$	3	−10.0017	$2s2p^2$ ^{2}P $3d$	5	−1.83277	$2s2p^2$ ^{2}D $5g$			
18	−3.67060	$2s2p^2$ ^{2}P $4d$	4	−8.76756	$2p^3$ ^{2}D $3p$	6	−1.49372	$2s^22p$ ^{2}P $9h$			
19	−3.57998	$2s2p^2$ ^{4}P $5s$	5	−7.58353	$2s^22p$ ^{2}P $4f$	7	−1.49342	$2s^22p$ ^{2}P $9j$			
20	−3.46977	$2s^22p$ ^{2}P $6p$	6	−6.05974	$2s2p^2$ ^{4}P $4d$		^{5}**P**^{e}				
21	−3.43822	$2s2p^2$ ^{4}P $5d$	7	−4.85795	$2s2p^2$ ^{2}D $4d$	1	−14.2599	$2s2p^2$ ^{4}P $3s$			
22	−2.73682	$2p^3$ ^{4}S $4p$	8	−4.81569	$2s^22p$ ^{2}P $5f$	2	−12.2364	$2s2p^2$ ^{4}P $3d$			
23	−2.54993	$2s^22p$ ^{2}P $7p$	9	−3.75248	$2s2p^2$ ^{2}P $4d$	3	−9.60750	$2p^3$ ^{4}S $3p$			
24	−2.08466	$2s2p^2$ ^{2}D $5d$	10	−3.41522	$2s2p^2$ ^{4}P $5d$	4	−6.90542	$2s2p^2$ ^{4}P $4s$			
25	−1.98389	$2p^3$ ^{2}D $4p$	11	−3.36331	$2s^22p$ ^{2}P $6f$	5	−6.15079	$2s2p^2$ ^{4}P $4d$			
26	−1.94922	$2s^22p$ ^{2}P $8p$	12	−3.20698	$2s2p^2$ ^{4}P $5g$	6	−3.63326	$2s2p^2$ ^{4}P $5s$			
27	−1.90822	$2s2p^2$ ^{4}P $6s$	13	−2.47867	$2s^22p$ ^{2}P $7f$	7	−3.45602	$2s2p^2$ ^{4}P $5d$			
28	−1.82703	$2s2p^2$ ^{4}P $6d$	14	−2.26945	$2p^3$ ^{4}S $4f$	8	−2.85961	$2p^3$ ^{4}S $4p$			
29	−1.53427	$2s^22p$ ^{2}P $9p$	15	−2.13821	$2s2p^2$ ^{2}D $5d$	9	−1.95733	$2s2p^2$ ^{4}P $6s$			
30	−1.50294	$2p^3$ ^{2}D $4f$	16	−2.07136	$2p^3$ ^{2}D $4p$	10	−1.83341	$2s2p^2$ ^{4}P $6d$			
31	−1.27136	$2s2p^2$ ^{2}P $5s$	17	−1.89744	$2s^22p$ ^{2}P $8f$	11	−.97876	$2s2p^2$ ^{4}P $7s$			
	^{3}**D**^{e}		18	−1.84687	$2s2p^2$ ^{2}D $5g$	12	−.88130	$2s2p^2$ ^{4}P $7d$			
1	−15.0637	$2s^22p$ ^{2}P $3p$	19	−1.81172	$2s2p^2$ ^{4}P $6d$	13	−.35099	$2s2p^2$ ^{4}P $8s$			
2	−12.8301	$2s2p^2$ ^{2}D $3s$	20	−1.71071	$2s2p^2$ ^{4}P $6g$	14	−.27762	$2s2p^2$ ^{4}P $8d$			
3	−11.7588	$2s2p^2$ ^{4}P $3d$	21	−1.51277	$2p^3$ ^{2}D $4f$		^{5}**D**^{e}				
4	−10.8957	$2s2p^2$ ^{2}D $3d$	22	−1.49810	$2s^22p$ ^{2}P $9f$	1	−12.3830	$2s2p^2$ ^{4}P $3d$			
5	−10.0379	$2s2p^2$ ^{2}S $3d$		^{3}**G**^{e}		2	−6.20996	$2s2p^2$ ^{4}P $4d$			
6	−9.84949	$2s2p^2$ ^{2}P $3d$	1	−11.0209	$2s2p^2$ ^{2}D $3d$	3	−3.48206	$2s2p^2$ ^{4}P $5d$			
7	−8.84587	$2p^3$ ^{2}D $3p$	2	−7.56986	$2s^22p$ ^{2}P $4f$	4	−1.84621	$2s2p^2$ ^{4}P $6d$			
8	−8.19181	$2s^22p$ ^{2}P $4p$	3	−4.86745	$2s2p^2$ ^{2}D $4d$	5	−.88882	$2s2p^2$ ^{4}P $7d$			

C-like S (S^{10+})

Energies in ascending order from ground state for terms with effective $n \leq 4.0$, $L \leq 4$

Term	i	E(Ryds)	Term	i	E(Ryds)	Term	i	E(Ryds)	Term	i	E(Ryds)	Term	i	E(Ryds)
$^3P^e$	1	0.00000	$^3P^e$	4	23.2939	$^3P^o$	5	25.4290	$^3D^e$	6	27.2048	$^3S^e$	5	28.9961
$^1D^e$	1	0.53310	$^1F^o$	1	23.4809	$^1P^e$	2	25.4660	$^1P^e$	4	27.2111	$^1D^e$	9	29.0209
$^1S^e$	1	1.13100	$^1P^o$	3	23.4915	$^1S^o$	1	25.9972	$^3P^e$	8	27.3551	$^3F^o$	3	29.0584
$^5S^o$	1	1.49580	$^3S^o$	2	23.5629	$^3G^e$	1	26.0334	$^1F^e$	2	27.4270	$^3D^e$	9	29.0631
$^3D^o$	1	3.06250	$^5D^o$	1	23.6714	$^3F^e$	2	26.0698	$^5P^e$	3	27.4468	$^1S^o$	2	29.0948
$^3P^o$	1	3.62630	$^5P^o$	1	23.7668	$^3P^o$	6	26.0936	$^3D^o$	6	27.4821	$^3G^o$	1	29.1181
$^1D^o$	1	4.67720	$^3D^o$	3	24.0018	$^1P^o$	5	26.1484	$^1D^e$	7	27.6720	$^1S^e$	6	29.1229
$^3S^o$	1	4.73290	$^5S^o$	2	24.0704	$^3D^e$	4	26.1586	$^1D^o$	5	27.6790	$^1P^e$	7	29.1396
$^1P^o$	1	5.24630	$^3P^o$	4	24.1663	$^1F^e$	1	26.1686	$^3P^e$	9	27.7012	$^1G^o$	1	29.1573
$^3P^e$	2	7.17270	$^3D^e$	2	24.2242	$^3D^o$	5	26.1730	$^1P^e$	5	28.1965	$^1D^o$	6	29.1735
$^1D^e$	2	7.64910	$^1D^e$	4	24.5179	$^1G^e$	1	26.2838	$^3D^e$	7	28.2084	$^3F^o$	4	29.1766
$^1S^e$	2	8.74660	$^5F^e$	1	24.5383	$^3P^o$	7	26.2855	$^3P^o$	8	28.2535	$^3D^o$	8	29.2491
$^3P^o$	2	21.1298	$^5D^e$	1	24.6713	$^3S^o$	3	26.3306	$^3F^e$	4	28.2867	$^3P^o$	10	29.2631
$^1P^o$	2	21.2933	$^5P^e$	2	24.8179	$^3P^e$	7	26.3421	$^5D^o$	2	28.3101	$^3P^e$	12	29.2780
$^1P^e$	1	21.8868	$^3P^e$	5	24.8465	$^1D^e$	5	26.4215	$^1F^e$	3	28.3367	$^1P^o$	9	29.3276
$^3D^e$	1	21.9906	$^3F^e$	1	25.0031	$^3S^e$	3	26.4247	$^1P^o$	7	28.4451	$^3D^o$	9	29.3470
$^3S^e$	1	22.1093	$^1D^o$	3	25.1154	$^1P^e$	3	26.4866	$^3P^o$	9	28.5170	$^1F^o$	3	29.3733
$^3P^e$	3	22.1798	$^3S^e$	2	25.1212	$^1D^o$	4	26.5776	$^1P^o$	8	28.5734	$^1P^o$	10	29.3844
$^1D^e$	3	22.4668	$^3F^o$	2	25.1246	$^5S^o$	3	26.6100	$^3D^o$	7	28.6330	$^1D^e$	10	29.3888
$^5P^e$	1	22.7944	$^1F^o$	2	25.2418	$^1S^e$	5	26.6790	$^3P^e$	10	28.7054	$^3F^e$	5	29.4707
$^1S^e$	3	22.7957	$^3D^o$	4	25.2662	$^1P^o$	6	26.7820	$^1P^e$	6	28.8307	$^3P^o$	11	29.4721
$^1D^o$	2	22.9420	$^3D^e$	3	25.2955	$^3D^e$	5	27.0164	$^3D^e$	8	28.8624	$^1F^e$	4	29.4730
$^3F^o$	1	22.9526	$^1P^o$	4	25.3104	$^3F^e$	3	27.0526	$^1D^e$	8	28.8914	$^3G^e$	2	29.4844
$^3D^o$	2	23.1500	$^3P^e$	6	25.3384	$^3S^o$	4	27.1421	$^3S^e$	4	28.8985			
$^3P^o$	3	23.2020	$^1S^e$	4	25.3852	$^1D^e$	6	27.1779	$^3P^e$	11	28.9389			

C-like S (S^{10+})

gf-values for transitions involving terms with effective $n \leq 4.0$, $L \leq 4$

i i'	gf_L	i i'	gf_L	i i'	gf_L	i i'	gf_L	i i'	gf_L	i i'	gf_L
	$^1P^o$–$^1S^e$	9 3	4.11E-2	4 5	-1.75E-1	1 6	-6.62E-4	2 4	-8.08E-2	7 5	9.97E-4
1 1	1.52E-1	9 4	7.26E-4	4 6	-1.83E-4	1 7	-3.29E-2	2 5	-3.57E-3	7 6	5.92E-4
1 2	-2.99E-1	9 5	2.12E-2	4 7	-8.10E-2	2 1	3.50E-1	2 6	-1.48E-3	7 7	1.15E-2
1 3	-9.70E-3	9 6	1.49E-2	5 1	1.00E-1	2 2	-1.34E-3	2 7	-5.14E-4	7 8	-2.15E-2
1 4	-6.08E-2	10 1	2.41E-1	5 2	8.15E-3	2 3	-1.24E-1	2 8	-1.42E-2	7 9	-2.02E-2
1 5	-2.35E-1	10 2	1.58E-2	5 3	-1.02E-3	2 4	-2.91E-1	2 9	-3.97E-1	7 10	-5.47E-1
1 6	-1.86E-4	10 3	4.09E-1	5 4	-1.02E-2	2 5	-1.07E-3	2 10	-1.47E-5	8 1	4.75E-2
2 1	8.03E-2	10 4	2.27E-3	5 5	-1.26E-3	2 6	-8.33E-2	3 1	4.61E-2	8 2	2.12E-3
2 2	1.17E-7	10 5	1.15E-2	5 6	-1.09E-4	2 7	-4.72E-4	3 2	4.85E-6	8 3	2.64E-1
2 3	-1.88E-1	10 6	1.73E-1	5 7	-1.09E-1	3 1	2.14E-1	3 3	9.39E-3	8 4	3.15E-4
2 4	-6.49E-2		**$^1S^o$–$^1P^e$**	6 1	4.81E-2	3 2	-3.82E-3	3 4	-1.33E-4	8 5	1.71E-3
2 5	-5.64E-3	1 1	1.81E-1	6 2	4.47E-1	3 3	-1.01E-1	3 5	-3.87E-2	8 6	2.41E-4
2 6	-5.74E-2	1 2	5.14E-2	6 3	4.41E-5	3 4	-4.01E-2	3 6	-6.84E-2	8 7	1.51E-3
3 1	1.23E+0	1 3	-3.66E-3	6 4	-4.95E-2	3 5	-1.97E-1	3 7	-5.58E-2	8 8	-2.01E-2
3 2	1.50E-4	1 4	-1.58E-1	6 5	-2.41E-3	3 6	-4.97E-4	3 8	-7.93E-4	8 9	-8.92E-1
3 3	1.15E-1	1 5	-1.62E-1	6 6	-1.10E-2	3 7	-1.42E-1	3 9	-9.19E-3	8 10	-4.08E-3
3 4	-6.57E-3	1 6	-4.51E-4	6 7	-1.41E-1	4 1	5.61E-2	3 10	-3.40E-2	9 1	1.69E-3
3 5	-4.71E-2	1 7	-4.12E-2	7 1	7.44E-5	4 2	6.28E-1	4 1	4.56E-1	9 2	5.90E-1
3 6	-3.87E-2	2 1	3.43E-4	7 2	2.52E-1	4 3	3.36E-4	4 2	2.99E-2	9 3	3.54E-3
4 1	9.02E-3	2 2	6.88E-4	7 3	4.55E-4	4 4	-2.32E-3	4 3	3.21E-2	9 4	5.71E-4
4 2	3.80E-3	2 3	1.11E-1	7 4	6.09E-3	4 5	-1.51E-2	4 4	2.03E-1	9 5	1.62E-1
4 3	7.77E-3	2 4	1.67E-2	7 5	5.36E-3	4 6	-2.97E-3	4 5	-2.25E-1	9 6	4.10E-3
4 4	-7.75E-4	2 5	1.06E-1	7 6	-8.14E-3	4 7	-1.10E-1	4 6	-1.72E-2	9 7	8.50E-3
4 5	-1.28E-1	2 6	1.52E-5	7 7	-2.26E-1	5 1	2.58E-3	4 7	-2.78E-2	9 8	5.00E-2
4 6	-1.26E-4	2 7	-1.39E-4	8 1	1.27E-1	5 2	1.74E-1	4 8	-9.34E-4	9 9	3.88E-9
5 1	3.31E-1		**$^1P^o$–$^1P^e$**	8 2	1.25E-3	5 3	1.82E-3	4 9	-2.40E-3	9 10	-5.82E-4
5 2	1.16E-2	1 1	-2.77E-3	8 3	2.52E-3	5 4	3.57E-4	4 10	-8.69E-3	10 1	1.74E-2
5 3	1.18E-1	1 2	-2.21E-1	8 4	2.94E-5	5 5	-1.48E-1	5 1	1.22E-2	10 2	6.23E-2
5 4	2.09E-1	1 3	-1.66E-1	8 5	4.38E-4	5 6	-7.62E-8	5 2	1.49E-3	10 3	1.57E-2
5 5	-1.79E-2	1 4	-1.07E+0	8 6	-2.96E-1	5 7	-6.57E-2	5 3	1.02E-1	10 4	3.50E-4
5 6	-6.84E-4	1 5	-1.84E-3	8 7	-5.33E-3	6 1	9.07E-1	5 4	4.07E-2	10 5	1.60E-2
6 1	4.51E-2	1 6	-3.55E-3	9 1	2.00E-2	6 2	2.87E-2	5 5	-5.92E-4	10 6	7.03E-3
6 2	5.34E-4	1 7	-2.45E-1	9 2	2.80E-3	6 3	9.50E-3	5 6	-4.88E-1	10 7	5.17E-3
6 3	9.53E-2	2 1	-1.68E-1	9 3	2.19E-2	6 4	6.71E-3	5 7	-6.36E-2	10 8	7.42E-4
6 4	2.76E-2	2 2	-5.27E-1	9 4	1.41E-1	6 5	7.54E-4	5 8	-4.43E-2	10 9	1.22E-2
6 5	1.07E-4	2 3	-6.86E-3	9 5	2.14E-1	6 6	5.43E-1	5 9	-5.44E-4	10 10	-4.47E-5
6 6	-1.19E-2	2 4	-1.42E-4	9 6	3.07E-2	6 7	4.39E-5	5 10	-3.58E-2		**$^1D^o$–$^1D^e$**
7 1	2.74E-3	2 5	-6.02E-4	9 7	6.13E-3		**$^1P^o$–$^1D^e$**	6 1	3.05E-3	1 1	7.62E-1
7 2	1.47E-1	2 6	-2.80E-1	10 1	2.09E-1	1 1	4.57E-1	6 2	1.00E-2	1 2	-8.13E-1
7 3	4.35E-3	2 7	-2.47E-3	10 2	1.13E-2	1 2	-1.38E-1	6 3	3.45E-1	1 3	-1.22E-2
7 4	1.28E-1	3 1	1.94E-1	10 3	1.31E-2	1 3	-5.69E-3	6 4	6.60E-2	1 4	-1.76E-1
7 5	4.45E-4	3 2	-7.30E-5	10 4	3.17E-2	1 4	-5.66E-2	6 5	2.38E-3	1 5	-1.46E+0
7 6	-5.47E-3	3 3	-3.54E-2	10 5	2.19E-2	1 5	-4.99E-1	6 6	-2.05E-3	1 6	-4.80E-2
8 1	1.09E-2	3 4	-1.51E-1	10 6	2.89E-1	1 6	-4.93E-2	6 7	-3.00E-1	1 7	-5.41E-1
8 2	2.06E-3	3 5	-7.99E-4	10 7	1.81E-3	1 7	-2.78E+0	6 8	-3.12E-1	1 8	-2.56E-1
8 3	1.15E-1	3 6	-6.31E-2		**$^1D^o$–$^1P^e$**	1 8	-1.60E-3	6 9	-2.20E-3	1 9	-5.37E-4
8 4	9.22E-3	3 7	-1.11E-4	1 1	-4.56E-3	1 9	-3.67E-2	6 10	-1.56E-2	1 10	-1.16E-1
8 5	9.31E-4	4 1	1.26E-1	1 2	-2.77E-1	1 10	-3.17E-1	7 1	4.18E-5	2 1	1.02E+0
8 6	-2.37E-1	4 2	-7.05E-4	1 3	-3.61E-1	2 1	2.49E-1	7 2	1.96E-1	2 2	2.19E-4
9 1	3.39E-2	4 3	-2.61E-1	1 4	-9.99E-3	2 2	1.46E-6	7 3	2.20E-3	2 3	5.12E-2
9 2	1.00E-1	4 4	-6.24E-3	1 5	-2.66E-1	2 3	-6.70E-1	7 4	3.34E-1	2 4	-6.81E-3

C-like S (S^{10+})

i	i'	gf$_L$	i	i'	gf$_L$	i	i'	gf$_L$	i	i'	gf$_L$	i	i'	gf$_L$	i	i'	gf$_L$
2	5	−2.06E−1	1	6	−3.02E−2	1	2	−6.12E−1	6	5	−2.15E−1	3	2	1.41E−1	3	5	−2.45E−3
2	6	−9.83E−2	1	7	−6.70E−1	1	3	−2.95E−3	7	1	3.82E−1	3	3	4.71E−1	3	6	−5.65E−5
2	7	−8.92E−5	1	8	−1.37E−2	1	4	−7.09E−1	7	2	1.48E−1	3	4	1.97E−2	3	7	−3.11E−1
2	8	−3.60E−5	1	9	−1.42E−1	2	1	−1.68E−1	7	3	−1.31E−5	3	5	1.14E−4	3	8	−3.13E−1
2	9	−1.79E−2	1	10	−1.38E−4	2	2	−7.58E−2	7	4	−8.10E−3	3	6	2.68E−1	3	9	−1.32E−4
2	10	−2.69E−3	2	1	8.79E−1	2	3	−4.37E−1	7	5	−1.84E−1	3	7	−1.10E−8	3	10	−1.26E−3
3	1	7.17E−1	2	2	1.88E−2	2	4	−4.60E−3	8	1	1.83E−3	3	8	−3.19E−1	3	11	−6.27E−2
3	2	1.41E−2	2	3	4.45E−1	3	1	6.96E−4	8	2	1.93E−1	3	9	−5.07E−1	3	12	−4.69E−4
3	3	5.33E−3	2	4	5.08E−1	3	2	2.19E−3	8	3	6.53E−4	3	10	−2.06E−2	4	1	9.42E−1
3	4	2.65E−1	2	5	−7.34E−2	3	3	2.42E−5	8	4	−1.64E−4	3	11	−1.58E−3	4	2	4.24E−2
3	5	−2.84E−1	2	6	−2.39E−2	3	4	−2.28E−2	8	5	−2.28E−1	3	12	−1.41E−1	4	3	2.49E−1
3	6	−2.18E−2	2	7	−3.00E−6			$^{1}\mathbf{G}^{o}$–$^{1}\mathbf{F}^{e}$	9	1	1.35E−1	4	1	2.72E−2	4	4	7.70E−1
3	7	−2.06E−2	2	8	−2.91E−1	1	1	2.97E−2	9	2	1.76E−4	4	2	2.62E−1	4	5	−2.72E−1
3	8	−7.11E−2	2	9	−4.11E−4	1	2	4.24E−2	9	3	1.94E−4	4	3	8.04E−3	4	6	−4.62E−3
3	9	−1.11E−5	2	10	−2.15E−1	1	3	8.82E−1	9	4	−4.06E−1	4	4	6.60E−1	4	7	−4.58E−2
3	10	−1.46E−2	3	1	1.07E+0	1	4	−3.84E−4	9	5	−4.55E−5	4	5	5.11E−3	4	8	−3.43E−2
4	1	4.86E−2	3	2	6.93E−4			$^{1}\mathbf{F}^{o}$–$^{1}\mathbf{G}^{e}$	10	1	1.19E+0	4	6	1.33E−1	4	9	−4.23E−1
4	2	1.59E−2	3	3	1.67E+0	1	1	−9.87E−2	10	2	1.85E−2	4	7	6.16E−5	4	10	−2.22E−3
4	3	5.45E−1	3	4	3.60E−2	2	1	−1.01E+0	10	3	1.58E−2	4	8	−2.40E−2	4	11	−2.60E−3
4	4	8.52E−2	3	5	3.38E−3	3	1	4.68E−3	10	4	8.60E−1	4	9	−3.72E−1	4	12	−8.67E−3
4	5	1.90E−3	3	6	3.21E−5			$^{1}\mathbf{G}^{o}$–$^{1}\mathbf{G}^{e}$	10	5	1.05E−4	4	10	−4.40E−1	5	1	4.45E−2
4	6	−7.69E−3	3	7	5.60E−2	1	1	3.88E−1	11	1	3.71E−3	4	11	−7.95E−4	5	2	6.87E−3
4	7	−9.45E−2	3	8	4.96E−2			$^{3}\mathbf{P}^{o}$–$^{3}\mathbf{S}^{e}$	11	2	3.16E−4	4	12	−1.67E−1	5	3	5.48E−2
4	8	−1.84E−3	3	9	1.06E+0	1	1	−2.19E−2	11	3	1.38E−1			$^{3}\mathbf{P}^{o}$–$^{3}\mathbf{P}^{e}$	5	4	2.24E−1
4	9	−1.74E−3	3	10	−1.44E−6	1	2	−2.71E−1	11	4	1.46E−3	1	1	6.06E−1	5	5	1.27E−3
4	10	−4.72E−1			$^{1}\mathbf{D}^{o}$–$^{1}\mathbf{F}^{e}$	1	3	−8.21E−1	11	5	6.67E−3	1	2	−3.84E−1	5	6	6.90E−4
5	1	6.10E−3	1	1	−6.75E−1	1	4	−8.33E−3			$^{3}\mathbf{S}^{o}$–$^{3}\mathbf{P}^{e}$	1	3	−1.31E−2	5	7	−6.15E−1
5	2	5.46E−1	1	2	−4.95E+0	1	5	−1.65E−1	1	1	7.20E−1	1	4	−3.54E−1	5	8	−9.76E−3
5	3	1.65E−3	1	3	−2.98E−1	2	1	−2.65E−1	1	2	−5.26E−1	1	5	−1.45E+0	5	9	−1.03E−3
5	4	6.40E−1	1	4	−1.92E−2	2	2	−3.33E−1	1	3	−2.99E−3	1	6	−1.07E−1	5	10	−4.09E−1
5	5	7.76E−3	2	1	−1.33E−1	2	3	−2.19E−3	1	4	−7.43E−3	1	7	−2.26E+0	5	11	−1.79E−3
5	6	4.35E−4	2	2	−2.39E−2	2	4	−2.27E−1	1	5	−2.79E−1	1	8	−2.87E−1	5	12	−9.19E−2
5	7	4.30E−5	2	3	−1.34E−4	2	5	−5.11E−4	1	6	−2.71E−1	1	9	−7.65E−4	6	1	1.25E−1
5	8	−8.29E−1	2	4	−4.28E+0	3	1	5.65E−1	1	7	−9.50E−3	1	10	−5.02E−3	6	2	6.87E−3
5	9	−2.32E−2	3	1	−4.88E−1	3	2	−1.77E−4	1	8	−4.49E+0	1	11	−1.62E−2	6	3	2.75E−2
5	10	−1.70E−1	3	2	−1.28E−3	3	3	−1.30E−1	1	9	−7.00E−1	1	12	−4.45E−1	6	4	1.03E−2
6	1	1.98E−1	3	3	−2.62E−2	3	4	−1.82E−1	1	10	−9.37E−2	2	1	5.71E−1	6	5	1.83E−5
6	2	4.52E−3	3	4	−2.72E−2	3	5	−4.01E−5	1	11	−3.47E−3	2	2	6.86E−5	6	6	7.01E−2
6	3	3.78E−1	4	1	1.69E−4	4	1	5.89E−7	1	12	−9.45E−2	2	3	−9.41E−1	6	7	−6.27E−6
6	4	9.07E−3	4	2	−5.00E−1	4	2	−6.88E−3	2	1	3.70E−1	2	4	−1.73E−1	6	8	−6.21E−2
6	5	8.95E−4	4	3	−4.44E−1	4	3	−3.06E−2	2	2	3.94E−3	2	5	−5.80E−6	6	9	−5.02E−4
6	6	1.30E−3	4	4	−5.57E−2	4	4	−1.90E−3	2	3	3.97E−2	2	6	−1.37E+0	6	10	−6.08E−2
6	7	1.86E−3	5	1	4.74E−3	4	5	−5.56E−3	2	4	6.88E−2	2	7	−5.93E−3	6	11	−1.44E−2
6	8	3.78E−3	5	2	9.77E−3	5	1	3.06E−1	2	5	−6.01E−1	2	8	−2.88E−3	6	12	−3.52E−1
6	9	7.68E−2	5	3	−5.47E−1	5	2	4.53E−4	2	6	−2.75E−3	2	9	−7.39E−5	7	1	2.24E−1
6	10	−1.14E−3	5	4	−8.05E−4	5	3	−3.01E−1	2	7	−4.02E−2	2	10	−9.68E−3	7	2	7.73E−3
		$^{1}\mathbf{F}^{o}$–$^{1}\mathbf{D}^{e}$	6	1	3.56E−3	5	4	−5.61E−3	2	8	−1.03E−3	2	11	−6.89E−1	7	3	4.94E−1
1	1	4.62E+0	6	2	1.82E−2	5	5	−3.19E−1	2	9	−7.99E−2	2	12	−7.90E−4	7	4	1.97E−1
1	2	9.57E−5	6	3	1.97E−3	6	1	1.13E−3	2	10	−5.27E−3	3	1	2.61E+0	7	5	2.39E−2
1	3	6.88E−1	6	4	−5.62E−1	6	2	8.73E−1	2	11	−1.95E−3	3	2	4.32E−4	7	6	6.59E−1
1	4	−2.05E−2			$^{1}\mathbf{F}^{o}$–$^{1}\mathbf{F}^{e}$	6	3	−1.81E−2	2	12	−9.83E−4	3	3	2.62E−1	7	7	−4.37E−6
1	5	−8.01E−3	1	1	−2.69E−2	6	4	−1.41E−3	3	1	1.23E−1	3	4	−3.13E−3	7	8	−3.66E−1

C-like S (S^{10+})

i i'	gf_L	i i'	gf_L	i i'	gf_L	i i'	gf_L	i i'	gf_L	i i'	gf_L
7 9	−3.25E−3		^{3}D^o–^{3}P^e	5 4	1.65E−1	9 8	1.40E−1	6 2	2.98E−2	11 9	4.61E−2
7 10	−2.78E−1	1 1	5.39E−1	5 5	1.28E−3	9 9	9.57E−2	6 3	5.65E−4		^{3}D^o–^{3}D^e
7 11	−4.90E−3	1 2	−1.17E+0	5 6	1.13E+0	9 10	4.57E−1	6 4	−1.07E−5	1 1	−1.10E−2
7 12	−4.87E−2	1 3	−1.07E−1	5 7	−4.80E−4	9 11	2.22E−4	6 5	−1.34E+0	1 2	−1.06E+0
8 1	3.68E−3	1 4	−4.21E−1	5 8	−1.33E−2	9 12	1.41E−6	6 6	−4.66E−2	1 3	−1.76E+0
8 2	5.26E−1	1 5	−2.60E−1	5 9	−2.00E−3		^{3}P^o–^{3}D^e	6 7	−8.45E−2	1 4	−4.42E+0
8 3	6.91E−3	1 6	−1.94E−1	5 10	−2.24E−3	1 1	−2.37E−2	6 8	−1.52E−2	1 5	−4.17E−2
8 4	2.80E−3	1 7	−1.05E+0	5 11	−1.84E−5	1 2	−4.32E−1	6 9	−1.46E−1	1 6	−3.24E−1
8 5	1.34E−4	1 8	−6.53E−3	5 12	−5.50E−1	1 3	−3.27E+0	7 1	3.06E−1	1 7	−7.46E−1
8 6	1.09E+0	1 9	−9.61E−2	6 1	1.04E−2	1 4	−2.62E+0	7 2	5.45E−3	1 8	−1.84E−4
8 7	1.13E−2	1 10	−4.21E−1	6 2	7.14E−1	1 5	−2.85E+0	7 3	2.39E−3	1 9	−2.12E−2
8 8	1.09E−3	1 11	−2.13E−3	6 3	2.34E−2	1 6	−1.47E+0	7 4	1.06E−3	2 1	3.85E−1
8 9	7.23E−3	1 12	−6.45E−3	6 4	4.74E−3	1 7	−8.50E−3	7 5	−6.35E−2	2 2	−5.55E−4
8 10	−3.85E−3	2 1	7.89E+0	6 5	8.47E−3	1 8	−1.79E−2	7 6	−8.59E−1	2 3	−3.66E−3
8 11	−1.49E−4	2 2	3.11E−7	6 6	1.06E+0	1 9	−6.49E−1	7 7	−3.58E−1	2 4	−4.37E−1
8 12	−1.28E+0	2 3	9.04E−1	6 7	5.29E−3	2 1	−1.25E+0	7 8	−1.28E−2	2 5	−1.62E−1
9 1	8.43E−2	2 4	−1.58E−2	6 8	2.24E−5	2 2	−7.20E−1	7 9	−5.69E−1	2 6	−1.39E−1
9 2	7.60E−5	2 5	−3.11E−5	6 9	−2.04E−2	2 3	−1.75E−2	8 1	7.16E−4	2 7	−3.18E−4
9 3	5.91E−1	2 6	−1.94E−3	6 10	−1.43E+0	2 4	−1.51E−6	8 2	6.73E−1	2 8	−1.10E−1
9 4	6.25E−2	2 7	−9.19E−2	6 11	−3.13E−3	2 5	−1.61E−3	8 3	1.32E−2	2 9	−2.18E−4
9 5	8.65E−4	2 8	−1.19E+0	6 12	−1.10E−2	2 6	−8.33E−3	8 4	8.32E−3	3 1	1.98E−3
9 6	2.13E−3	2 9	−3.85E−4	7 1	2.97E−2	2 7	−3.68E−4	8 5	5.03E−3	3 2	−1.15E−2
9 7	9.59E−4	2 10	−1.47E−3	7 2	1.85E+0	2 8	−1.34E+0	8 6	2.15E−3	3 3	−5.45E−1
9 8	3.68E−3	2 11	−2.35E−1	7 3	1.29E−3	2 9	−7.29E−4	8 7	9.75E−4	3 4	−1.29E−1
9 9	5.58E−4	2 12	−2.65E−3	7 4	1.03E−6	3 1	6.91E−2	8 8	−4.87E−5	3 5	−4.27E−3
9 10	−1.26E−3	3 1	1.60E+0	7 5	8.69E−2	3 2	−5.46E−6	8 9	−1.52E+0	3 6	−3.80E−4
9 11	−1.63E+0	3 2	4.10E−3	7 6	6.74E−3	3 3	−1.19E−4	9 1	6.77E−1	3 7	−1.09E−2
9 12	−1.73E−4	3 3	4.14E−1	7 7	1.75E−2	3 4	−1.11E−1	9 2	8.16E−4	3 8	−3.16E−4
10 1	5.79E−1	3 4	9.82E−1	7 8	1.89E−2	3 5	−2.83E−2	9 3	1.51E−3	3 9	−3.82E−3
10 2	6.94E−3	3 5	−2.65E−3	7 9	1.44E+0	3 6	−3.84E−1	9 4	1.85E−4	4 1	4.44E−1
10 3	8.05E−1	3 6	−2.39E−3	7 10	−1.42E−3	3 7	−1.31E−3	9 5	3.17E−3	4 2	1.66E+0
10 4	1.20E−1	3 7	−6.66E−2	7 11	−9.26E−5	3 8	−5.64E−2	9 6	7.66E−3	4 3	−4.37E−4
10 5	1.69E−4	3 8	−3.47E−5	7 12	−3.51E−4	3 9	−6.42E−3	9 7	8.85E−5	4 4	−5.31E−1
10 6	3.77E−3	3 9	−1.05E+0	8 1	1.76E+0	4 1	5.43E−3	9 8	−2.05E+0	4 5	−2.67E−2
10 7	7.26E−3	3 10	−7.14E−2	8 2	4.13E−5	4 2	−9.98E−4	9 9	−6.53E−4	4 6	−2.17E−2
10 8	4.57E−3	3 11	−1.02E−2	8 3	2.54E+0	4 3	−1.41E+0	10 1	6.32E−2	4 7	−1.16E+0
10 9	1.23E−3	3 12	−9.67E−3	8 4	3.73E−1	4 4	−1.90E−2	10 2	1.09E−3	4 8	−5.94E−6
10 10	5.08E−4	4 1	1.71E−1	8 5	1.07E−4	4 5	−2.82E−3	10 3	1.23E−4	4 9	−1.29E−1
10 11	4.41E−1	4 2	1.68E−3	8 6	1.12E−2	4 6	−6.26E−4	10 4	6.92E−4	5 1	1.54E+0
10 12	−6.56E−7	4 3	1.19E−1	8 7	2.18E−2	4 7	−2.72E−3	10 5	4.48E−3	5 2	2.35E−2
11 1	1.23E−2	4 4	3.09E−1	8 8	2.97E−2	4 8	−4.51E−3	10 6	4.42E−2	5 3	8.17E−4
11 2	3.87E+0	4 5	3.01E−3	8 9	4.40E−5	4 9	−8.00E−5	10 7	1.06E−3	5 4	3.35E−4
11 3	2.16E−3	4 6	−9.42E−4	8 10	6.08E−3	5 1	8.97E−4	10 8	8.47E−2	5 5	−3.76E−3
11 4	3.95E−3	4 7	−2.51E−1	8 11	1.49E+0	5 2	1.22E+0	10 9	2.48E−3	5 6	−3.28E−1
11 5	6.44E−3	4 8	−1.23E−6	8 12	−1.19E−5	5 3	3.24E−3	11 1	1.76E−4	5 7	−4.10E−1
11 6	8.94E−3	4 9	−1.68E−5	9 1	4.74E−2	5 4	−4.25E−1	11 2	8.19E−3	5 8	−4.03E−3
11 7	8.83E−3	4 10	−1.81E−1	9 2	5.19E+0	5 5	−3.85E−3	11 3	6.16E−3	5 9	−8.62E−1
11 8	4.31E−1	4 11	−1.76E−3	9 3	1.84E−5	5 6	−2.24E−2	11 4	5.75E−1	6 1	1.81E−2
11 9	1.51E−1	4 12	−4.12E−1	9 4	2.55E−3	5 7	−2.42E−1	11 5	9.03E−3	6 2	1.05E+0
11 10	5.57E−1	5 1	3.59E−1	9 5	7.90E−5	5 8	−2.24E−2	11 6	6.55E−2	6 3	7.28E−3
11 11	7.39E−6	5 2	5.46E−4	9 6	3.62E−4	5 9	−2.66E−3	11 7	2.89E−1	6 4	6.24E−3
11 12	4.58E−3	5 3	4.81E−1	9 7	1.53E−1	6 1	9.76E−1	11 8	2.76E−4	6 5	7.82E−5

C-like S (S^{10+})

i i'	gf$_L$	*i i'*	gf$_L$	*i i'*	gf$_L$	*i i'*	gf$_L$	*i i'*	gf$_L$	*i i'*	gf$_L$
6 6	1.50E−4		$^3\mathbf{F}^o$–$^3\mathbf{D}^e$	4 4	1.09E−2	5 5	−2.23E−1	2 5	−2.37E−2	1 2	−8.65E+0
6 7	−1.19E+0	1 1	1.83E+0	4 5	4.65E−2	6 1	1.77E−2	3 1	6.13E−2	1 3	−6.41E−1
6 8	−8.91E−6	1 2	−2.04E−3	4 6	1.18E−1	6 2	2.07E−2	3 2	6.09E−1	2 1	7.70E−1
6 9	−1.30E−1	1 3	−8.13E−4	4 7	1.53E−2	6 3	1.89E−3	3 3	8.03E−1	2 2	−5.53E−1
7 1	2.69E−5	1 4	−1.03E−2	4 8	2.77E+0	6 4	−2.33E+0	3 4	4.78E−1	2 3	−3.39E−1
7 2	2.14E−3	1 5	−9.97E−1	4 9	8.73E−5	6 5	−7.06E−3	3 5	−3.43E−3	3 1	8.56E−1
7 3	4.23E−1	1 6	−1.12E+0		$^3\mathbf{D}^o$–$^3\mathbf{F}^e$	7 1	1.27E+0	4 1	8.31E−4	3 2	1.27E−2
7 4	2.64E−3	1 7	−2.90E−3	1 1	−4.38E+0	7 2	1.18E−2	4 2	3.94E−3	3 3	−1.56E+0
7 5	4.14E−4	1 8	−3.98E−1	1 2	−1.05E+1	7 3	2.07E−4	4 3	3.35E−2		$^5\mathbf{P}^o$–$^5\mathbf{P}^e$
7 6	4.54E−2	1 9	−2.50E−3	1 3	−2.13E+0	7 4	8.23E−3	4 4	2.64E−3	1 1	1.62E+0
7 7	4.46E−2	2 1	5.35E−1	1 4	−7.17E−1	7 5	−1.51E−3	4 5	−2.06E−1	1 2	−7.83E−1
7 8	−1.89E−4	2 2	2.16E+0	1 5	−1.23E−2	8 1	1.67E−3		$^3\mathbf{G}^o$–$^3\mathbf{F}^e$	1 3	−8.68E−1
7 9	−2.35E−3	2 3	−2.00E−3	2 1	−6.74E−3	8 2	4.07E−3	1 1	2.25E−6		$^5\mathbf{D}^o$–$^5\mathbf{P}^e$
8 1	8.45E−1	2 4	−1.83E−1	2 2	−2.29E−1	8 3	5.58E−2	1 2	1.27E−2	1 1	2.45E+0
8 2	1.57E−2	2 5	−4.78E−2	2 3	−2.32E−1	8 4	4.13E−3	1 3	1.91E−1	1 2	−9.81E−4
8 3	6.65E−5	2 6	−2.47E−2	2 4	−8.83E−4	8 5	−1.24E+0	1 4	2.67E+0	1 3	−1.65E+0
8 4	3.31E−3	2 7	−9.76E−2	2 5	−1.31E+1	9 1	6.11E−2	1 5	−9.95E−4	2 1	1.01E−2
8 5	1.24E−2	2 8	−3.79E−3	3 1	−2.14E+0	9 2	6.47E−1		$^3\mathbf{F}^o$–$^3\mathbf{G}^e$	2 2	1.43E−1
8 6	2.95E−3	2 9	−1.18E+0	3 2	−5.07E−2	9 3	5.25E−2	1 1	−3.13E−1	2 3	2.40E+0
8 7	4.52E−5	3 1	5.61E−2	3 3	−7.50E−3	9 4	9.62E−2	1 2	−1.77E+1		$^5\mathbf{P}^o$–$^5\mathbf{D}^e$
8 8	6.27E−1	3 2	1.60E−3	3 4	−1.25E−3	9 5	−1.01E−4	2 1	−2.50E+0	1 1	−1.75E+0
8 9	1.48E−4	3 3	1.79E−3	3 5	−2.48E−3		$^3\mathbf{F}^o$–$^3\mathbf{F}^e$	2 2	−1.89E−1		$^5\mathbf{D}^o$–$^5\mathbf{D}^e$
9 1	2.33E−3	3 4	1.01E−1	4 1	6.05E−4	1 1	−1.46E−2	3 1	2.55E−1	1 1	−6.78E−1
9 2	1.19E−3	3 5	6.12E−3	4 2	−1.16E+0	1 2	−4.38E−1	3 2	−3.53E−2	2 1	9.84E−1
9 3	5.07E−2	3 6	1.52E−1	4 3	−2.60E−2	1 3	−1.51E+0	4 1	1.44E−2		$^5\mathbf{D}^o$–$^5\mathbf{F}^c$
9 4	1.09E−1	3 7	1.36E+0	4 4	−1.22E−1	1 4	−1.24E−2	4 2	−2.53E+0	1 1	−2.96E+0
9 5	3.00E−2	3 8	1.86E−2	4 5	−1.04E−2	1 5	−1.86E+0		$^3\mathbf{G}^o$–$^3\mathbf{G}^e$	2 1	2.35E+0
9 6	5.75E−1	3 9	−4.91E−6	5 1	6.32E−3	2 1	4.60E−3	1 1	1.34E+0		
9 7	1.04E+0	4 1	5.21E+0	5 2	3.42E−4	2 2	−5.19E−1	1 2	−7.86E−5		
9 8	6.87E−5	4 2	8.89E−2	5 3	−1.51E+0	2 3	−2.38E−3		$^5\mathbf{S}^o$–$^5\mathbf{P}^e$		
9 9	8.28E−5	4 3	1.09E−3	5 4	−8.34E−1	2 4	−1.44E+0	1 1	−7.46E−1		

C-like Ar (Ar^{12+})

Term energies relative to $2s^22p$ ^{2}P ionization threshold for each symmetry

i	E(Ryds)	Description	i	E(Ryds)	Description	i	E(Ryds)	Description	i	E(Ryds)	Description
	$^1S^o$		11	−6.91737	$2s^22p$ ^{2}P $5d$	11	−2.04564	$2s2p^2$ ^{2}P $5f$	17	−6.88242	$2s^22p$ ^{2}P $5d$
1	−16.1384	$2s2p^2$ ^{2}P $3p$	12	−6.42098	$2s2p^2$ ^{2}P $4p$		**$^1H^o$**		18	−6.73080	$2s2p^2$ ^{2}S $4p$
2	−12.5042	$2p^3$ ^{2}D $3d$	13	−5.87947	$2s2p^2$ ^{2}P $4f$	1	−7.11764	$2s2p^2$ ^{2}D $4f$	19	−6.50454	$2s2p^2$ ^{2}P $4p$
3	−6.63329	$2s2p^2$ ^{2}P $4p$	14	−4.78548	$2s^22p$ ^{2}P $6d$	2	−6.74537	$2s^22p$ ^{2}P $5g$	20	−5.12358	$2s2p^2$ ^{4}P $5p$
4	−3.88610	$2p^3$ ^{2}D $4d$	15	−4.62057	$2p^3$ ^{2}D $4s$	3	−4.69290	$2s^22p$ ^{2}P $6g$	21	−5.01290	$2s^22p$ ^{2}P $6s$
5	−2.41841	$2s2p^2$ ^{2}P $5p$	16	−3.68508	$2p^3$ ^{2}D $4d$	4	−3.45042	$2s^22p$ ^{2}P $7g$	22	−4.76069	$2s^22p$ ^{2}P $6d$
6	−.19516	$2s2p^2$ ^{2}P $6p$	17	−3.56982	$2s2p^2$ ^{2}D $5p$	5	−3.44882	$2s^22p$ ^{2}P $7i$	23	−3.78944	$2p^3$ ^{2}P $4s$
	$^1P^o$		18	−3.50550	$2s^22p$ ^{2}P $7d$	6	−3.30477	$2s2p^2$ ^{2}D $5f$	24	−3.72812	$2p^3$ ^{2}D $4d$
1	−44.2657	$2s2p^3$	19	−3.29800	$2s2p^2$ ^{2}D $5f$	7	−2.64074	$2s^22p$ ^{2}P $8i$	25	−3.65468	$2s^22p$ ^{2}P $7s$
2	−21.5560	$2s^22p$ ^{2}P $3s$	20	−2.90629	$2p^3$ ^{2}P $4d$	8	−2.63977	$2s^22p$ ^{2}P $8g$	26	−3.57045	$2s2p^2$ ^{2}D $5p$
3	−18.9337	$2s^22p$ ^{2}P $3d$	21	−2.67821	$2s^22p$ ^{2}P $8d$	9	−2.08655	$2s^22p$ ^{2}P $9i$	27	−3.49057	$2s^22p$ ^{2}P $7d$
4	−16.9045	$2s2p^2$ ^{2}D $3p$	22	−2.32209	$2s2p^2$ ^{2}P $5p$	10	−2.08594	$2s^22p$ ^{2}P $9g$	28	−3.28908	$2s2p^2$ ^{2}D $5f$
5	−15.9442	$2s2p^2$ ^{2}S $3p$	23	−2.11276	$2s^22p$ ^{2}P $9d$		**$^1I^o$**		29	−2.94569	$2s2p^2$ ^{4}P $6p$
6	−15.1991	$2s2p^2$ ^{2}P $3p$	24	−2.05933	$2s2p^2$ ^{2}P $5f$	1	−3.44909	$2s^22p$ ^{2}P $7i$	30	−2.92349	$2p^3$ ^{2}P $4d$
7	−13.3075	$2p^3$ ^{2}P $3s$		**$^1F^o$**		2	−2.64077	$2s^22p$ ^{2}P $8i$	31	−2.77583	$2s^22p$ ^{2}P $8s$
8	−12.2224	$2p^3$ ^{2}D $3d$	1	−18.9399	$2s^22p$ ^{2}P $3d$	3	−2.08657	$2s^22p$ ^{2}P $9i$	32	−2.66938	$2s^22p$ ^{2}P $8d$
9	−11.6534	$2s^22p$ ^{2}P $4s$	2	−16.9822	$2s2p^2$ ^{2}D $3p$		**$^3S^o$**		33	−2.58880	$2s2p^2$ ^{2}S $5p$
10	−10.7322	$2p^3$ ^{2}P $3d$	3	−11.8033	$2p^3$ ^{2}D $3d$	1	−44.8732	$2s2p^3$	34	−2.35832	$2s2p^2$ ^{2}P $5p$
11	−10.6352	$2s^22p$ ^{2}P $4d$	4	−11.1503	$2p^3$ ^{2}P $3d$	2	−18.9209	$2s2p^2$ ^{4}P $3p$	35	−2.17568	$2s^22p$ ^{2}P $9s$
12	−7.68766	$2s2p^2$ ^{2}D $4p$	5	−10.6527	$2s^22p$ ^{2}P $4d$	3	−15.7538	$2s2p^2$ ^{2}P $3p$	36	−2.10587	$2s^22p$ ^{2}P $9d$
13	−7.31740	$2s^22p$ ^{2}P $5s$	6	−7.75685	$2s2p^2$ ^{2}D $4p$	4	−14.8056	$2p^3$ ^{4}S $3s$		**$^3D^o$**	
14	−7.03481	$2s2p^2$ ^{2}D $4f$	7	−7.12505	$2s2p^2$ ^{2}D $4f$	5	−11.8986	$2p^3$ ^{2}D $3d$	1	−46.7720	$2s2p^3$
15	−6.81794	$2s^22p$ ^{2}P $5d$	8	−6.81043	$2s^22p$ ^{2}P $5d$	6	−9.42179	$2s2p^2$ ^{4}P $4p$	2	−19.3480	$2s^22p$ ^{2}P $3d$
16	−6.69967	$2s2p^2$ ^{2}S $4p$	9	−6.76897	$2s^22p$ ^{2}P $5g$	7	−6.45511	$2s2p^2$ ^{2}P $4p$	3	−18.4074	$2s2p^2$ ^{4}P $3p$
17	−6.32348	$2s2p^2$ ^{2}P $4p$	10	−6.06664	$2s2p^2$ ^{2}S $4f$	8	−5.52645	$2p^3$ ^{4}S $4s$	4	−16.9518	$2s2p^2$ ^{2}D $3p$
18	−4.99953	$2s^22p$ ^{2}P $6s$	11	−5.80462	$2s2p^2$ ^{2}P $4f$	9	−5.14857	$2s2p^2$ ^{4}P $5p$	5	−15.9293	$2s2p^2$ ^{2}P $3p$
19	−4.72356	$2s^22p$ ^{2}P $6d$	12	−4.72541	$2s^22p$ ^{2}P $6d$	10	−3.67385	$2p^3$ ^{2}D $4d$	6	−14.4188	$2p^3$ ^{2}D $3s$
20	−3.80035	$2p^3$ ^{2}D $4d$	13	−4.68591	$2s^22p$ ^{2}P $6g$	11	−3.00803	$2s2p^2$ ^{4}P $6p$	7	−13.0256	$2p^3$ ^{4}S $3d$
21	−3.76670	$2p^3$ ^{2}P $4s$	14	−3.69464	$2s2p^2$ ^{2}D $5p$	12	−2.33250	$2s2p^2$ ^{2}P $5p$	8	−12.1849	$2p^3$ ^{2}D $3d$
22	−3.64153	$2s^22p$ ^{2}P $7s$	15	−3.56046	$2p^3$ ^{2}D $4d$	13	−1.68918	$2s2p^2$ ^{4}P $7p$	9	−11.3090	$2p^3$ ^{2}P $3d$
23	−3.49963	$2s2p^2$ ^{2}D $5p$	16	−3.46864	$2s^22p$ ^{2}P $7d$	14	−1.16766	$2p^3$ ^{4}S $5s$	10	−10.7983	$2s^22p$ ^{2}P $4d$
24	−3.46823	$2s^22p$ ^{2}P $7d$	17	−3.44635	$2s^22p$ ^{2}P $7g$	15	−.82647	$2s2p^2$ ^{4}P $8p$	11	−9.24529	$2s2p^2$ ^{4}P $4p$
25	−3.28180	$2s2p^2$ ^{2}D $5f$	18	−3.30651	$2s2p^2$ ^{2}D $5f$	16	−.25227	$2s2p^2$ ^{4}P $9p$	12	−8.68457	$2s2p^2$ ^{4}P $4f$
26	−2.77310	$2s^22p$ ^{2}P $8s$	19	−2.80418	$2p^3$ ^{2}P $4d$	17	−.14554	$2s2p^2$ ^{2}P $6p$	13	−7.77165	$2s2p^2$ ^{2}D $4p$
27	−2.68254	$2p^3$ ^{2}P $4d$	20	−2.65415	$2s^22p$ ^{2}P $8d$		**$^3P^o$**		14	−7.09825	$2s2p^2$ ^{2}D $4f$
28	−2.65371	$2s^22p$ ^{2}P $8d$	21	−2.63997	$2s^22p$ ^{2}P $8g$	1	−46.1227	$2s2p^3$	15	−6.88486	$2s^22p$ ^{2}P $5d$
29	−2.55020	$2s2p^2$ ^{2}S $5p$	22	−2.26777	$2s2p^2$ ^{2}S $5f$	2	−21.7458	$2s^22p$ ^{2}P $3s$	16	−6.55177	$2s2p^2$ ^{2}P $4p$
30	−2.28762	$2s2p^2$ ^{2}P $5p$	23	−2.09601	$2s^22p$ ^{2}P $9d$	3	−19.2900	$2s^22p$ ^{2}P $3d$	17	−5.87579	$2s2p^2$ ^{2}P $4f$
31	−2.16620	$2s^22p$ ^{2}P $9s$	24	−2.08951	$2s^22p$ ^{2}P $9g$	4	−18.2154	$2s2p^2$ ^{4}P $3p$	18	−5.16637	$2s2p^2$ ^{4}P $5p$
32	−2.09511	$2s^22p$ ^{2}P $9d$	25	−2.02192	$2s2p^2$ ^{2}P $5f$	5	−16.7590	$2s2p^2$ ^{2}D $3p$	19	−4.89094	$2s2p^2$ ^{4}P $5f$
	$^1D^o$			**$^1G^o$**		6	−16.0076	$2s2p^2$ ^{2}S $3p$	20	−4.76313	$2s^22p$ ^{2}P $6d$
1	−44.9195	$2s2p^3$	1	−12.4205	$2p^3$ ^{2}D $3d$	7	−15.7910	$2s2p^2$ ^{2}P $3p$	21	−4.66360	$2p^3$ ^{2}D $4s$
2	−19.6015	$2s^22p$ ^{2}P $3d$	2	−7.14898	$2s2p^2$ ^{2}D $4f$	8	−13.5365	$2p^3$ ^{2}P $3s$	22	−4.56090	$2p^3$ ^{4}S $4d$
3	−17.1337	$2s2p^2$ ^{2}D $3p$	3	−6.77564	$2s^22p$ ^{2}P $5g$	9	−12.0516	$2p^3$ ^{2}D $3d$	23	−3.81612	$2p^3$ ^{2}D $4d$
4	−15.4513	$2s2p^2$ ^{2}P $3p$	4	−5.86504	$2s2p^2$ ^{2}P $4f$	10	−11.7151	$2s^22p$ ^{2}P $4s$	24	−3.62012	$2s2p^2$ ^{2}D $5p$
5	−14.1846	$2p^3$ ^{2}D $3s$	5	−4.69461	$2s^22p$ ^{2}P $6g$	11	−11.4483	$2p^3$ ^{2}P $3d$	25	−3.49236	$2s^22p$ ^{2}P $7d$
6	−12.0145	$2p^3$ ^{2}D $3d$	6	−3.84765	$2p^3$ ^{2}D $4d$	12	−10.7828	$2s^22p$ ^{2}P $4d$	26	−3.30767	$2s2p^2$ ^{2}D $5f$
7	−11.2959	$2p^3$ ^{2}P $3d$	7	−3.45221	$2s^22p$ ^{2}P $7g$	13	−9.17473	$2s2p^2$ ^{4}P $4p$	27	−2.96171	$2s2p^2$ ^{4}P $6p$
8	−10.8825	$2s^22p$ ^{2}P $4d$	8	−3.31709	$2s2p^2$ ^{2}D $5f$	14	−7.72538	$2s2p^2$ ^{2}D $4p$	28	−2.87979	$2p^3$ ^{2}P $4d$
9	−7.75637	$2s2p^2$ ^{2}D $4p$	9	−2.64319	$2s^22p$ ^{2}P $8g$	15	−7.32853	$2s^22p$ ^{2}P $5s$	29	−2.80508	$2s2p^2$ ^{4}P $6f$
10	−7.08047	$2s2p^2$ ^{2}D $4f$	10	−2.09788	$2s^22p$ ^{2}P $9g$	16	−7.05246	$2s2p^2$ ^{2}D $4f$	30	−2.66959	$2s^22p$ ^{2}P $8d$

C-like Ar (Ar^{12+})

i	E(Ryds)	Description
31	−2.37488	$2s2p^2$ ^{2}P $5p$
32	−2.10670	$2s^22p$ ^{2}P $9d$
33	−2.05662	$2s2p^2$ ^{2}P $5f$
	^{3}F^o	
1	−19.5895	$2s^22p$ ^{2}P $3d$
2	−17.1189	$2s2p^2$ ^{2}D $3p$
3	−12.5442	$2p^3$ ^{2}D $3d$
4	−11.5513	$2p^3$ ^{2}P $3d$
5	−10.8835	$2s^22p$ ^{2}P $4d$
6	−8.63190	$2s2p^2$ ^{4}P $4f$
7	−7.80418	$2s2p^2$ ^{2}D $4p$
8	−7.14849	$2s2p^2$ ^{2}D $4f$
9	−6.92041	$2s^22p$ ^{2}P $5d$
10	−6.76951	$2s^22p$ ^{2}P $5g$
11	−6.08458	$2s2p^2$ ^{2}S $4f$
12	−5.81767	$2s2p^2$ ^{2}P $4f$
13	−4.85211	$2s2p^2$ ^{4}P $5f$
14	−4.78520	$2s^22p$ ^{2}P $6d$
15	−4.68622	$2s^22p$ ^{2}P $6g$
16	−3.89350	$2p^3$ ^{2}D $4d$
17	−3.64150	$2s2p^2$ ^{2}D $5p$
18	−3.50552	$2s^22p$ ^{2}P $7d$
19	−3.44668	$2s^22p$ ^{2}P $7g$
20	−3.32083	$2s2p^2$ ^{2}D $5f$
21	−2.94663	$2p^3$ ^{2}P $4d$
22	−2.79118	$2s2p^2$ ^{4}P $6f$
23	−2.67850	$2s^22p$ ^{2}P $8d$
24	−2.64024	$2s^22p$ ^{2}P $8g$
25	−2.28245	$2s2p^2$ ^{2}S $5f$
26	−2.11335	$2s^22p$ ^{2}P $9d$
27	−2.09008	$2s^22p$ ^{2}P $9g$
28	−2.03061	$2s2p^2$ ^{2}P $5f$
	^{3}G^o	
1	−12.4702	$2p^3$ ^{2}D $3d$
2	−8.68881	$2s2p^2$ ^{4}P $4f$
3	−7.16939	$2s2p^2$ ^{2}D $4f$
4	−6.77556	$2s^22p$ ^{2}P $5g$
5	−5.87186	$2s2p^2$ ^{2}P $4f$
6	−4.87728	$2s2p^2$ ^{4}P $5f$
7	−4.69465	$2s^22p$ ^{2}P $6g$
8	−3.86018	$2p^3$ ^{2}D $4d$
9	−3.45250	$2s^22p$ ^{2}P $7g$
10	−3.33363	$2s2p^2$ ^{2}D $5f$
11	−2.80692	$2s2p^2$ ^{4}P $6f$
12	−2.80156	$2s2p^2$ ^{4}P $6h$
13	−2.64307	$2s^22p$ ^{2}P $8g$
14	−2.09841	$2s^22p$ ^{2}P $9g$
15	−2.05020	$2s2p^2$ ^{2}P $5f$
	^{3}H^o	
1	−7.13223	$2s2p^2$ ^{2}D $4f$
2	−6.74532	$2s^22p$ ^{2}P $5g$
3	−4.69331	$2s^22p$ ^{2}P $6g$
4	−3.45133	$2s^22p$ ^{2}P $7g$
5	−3.44887	$2s^22p$ ^{2}P $7i$
6	−3.31360	$2s2p^2$ ^{2}D $5f$
7	−2.79923	$2s2p^2$ ^{4}P $6h$
8	−2.64077	$2s^22p$ ^{2}P $8i$
9	−2.63993	$2s^22p$ ^{2}P $8g$
10	−2.08661	$2s^22p$ ^{2}P $9g$:
11	−2.08607	$2s^22p$ ^{2}P $9i$
	^{3}I^o	
1	−3.44909	$2s^22p$ ^{2}P $7i$
2	−2.80095	$2s2p^2$ ^{4}P $6h$
3	−2.64077	$2s^22p$ ^{2}P $8i$
4	−2.08657	$2s^22p$ ^{2}P $9i$
	^{5}S^o	
1	−48.5846	$2s2p^3$
2	−18.3266	$2s2p^2$ ^{4}P $3p$
3	−15.4418	$2p^3$ ^{4}S $3s$
4	−9.26319	$2s2p^2$ ^{4}P $4p$
5	−5.58840	$2p^3$ ^{4}S $4s$
6	−5.15659	$2s2p^2$ ^{4}P $5p$
7	−2.97062	$2s2p^2$ ^{4}P $6p$
8	−1.66206	$2s2p^2$ ^{4}P $7p$
9	−1.22538	$2p^3$ ^{4}S $5s$
10	−.81694	$2s2p^2$ ^{4}P $8p$
11	−.24187	$2s2p^2$ ^{4}P $9p$
	^{5}P^o	
1	−18.6831	$2s2p^2$ ^{4}P $3p$
2	−9.36779	$2s2p^2$ ^{4}P $4p$
3	−5.21324	$2s2p^2$ ^{4}P $5p$
4	−2.99708	$2s2p^2$ ^{4}P $6p$
5	−1.67704	$2s2p^2$ ^{4}P $7p$
6	−.82760	$2s2p^2$ ^{4}P $8p$
7	−.24890	$2s2p^2$ ^{4}P $9p$
	^{5}D^o	
1	−18.7941	$2s2p^2$ ^{4}P $3p$
2	−13.4181	$2p^3$ ^{4}S $3d$
3	−9.40418	$2s2p^2$ ^{4}P $4p$
4	−8.70536	$2s2p^2$ ^{4}P $4f$
5	−5.23690	$2s2p^2$ ^{4}P $5p$
6	−4.89529	$2s2p^2$ ^{4}P $5f$
7	−4.75621	$2p^3$ ^{4}S $4d$
8	−3.00515	$2s2p^2$ ^{4}P $6p$
9	−2.81442	$2s2p^2$ ^{4}P $6f$
10	−1.68265	$2s2p^2$ ^{4}P $7p$
11	−1.56444	$2s2p^2$ ^{4}P $7f$
12	−.84419	$2s2p^2$ ^{4}P $8p$
13	−.81081	$2p^3$ ^{4}S $5d$
14	−.75171	$2s2p^2$ ^{4}P $8f$
15	−.25105	$2s2p^2$ ^{4}P $9p$
16	−.19692	$2s2p^2$ ^{4}P $9f$
	^{5}F^o	
1	−8.66117	$2s2p^2$ ^{4}P $4f$
2	−4.87063	$2s2p^2$ ^{4}P $5f$
3	−2.80472	$2s2p^2$ ^{4}P $6f$
4	−1.55824	$2s2p^2$ ^{4}P $7f$
5	−.74905	$2s2p^2$ ^{4}P $8f$
6	−.19423	$2s2p^2$ ^{4}P $9f$
	^{5}G^o	
1	−8.71319	$2s2p^2$ ^{4}P $4f$
2	−4.89295	$2s2p^2$ ^{4}P $5f$
3	−2.81748	$2s2p^2$ ^{4}P $6f$
4	−2.80158	$2s2p^2$ ^{4}P $6h$
5	−1.56622	$2s2p^2$ ^{4}P $7f$
6	−1.55645	$2s2p^2$ ^{4}P $7h$
7	−.75436	$2s2p^2$ ^{4}P $8f$
8	−.75282	$2s2p^2$ ^{4}P $8h$
9	−.67232	$2p^3$ ^{4}S $5g$
10	−.19795	$2s2p^2$ ^{4}P $9f$
11	−.19234	$2s2p^2$ ^{4}P $9h$
	^{1}S^e	
1	−49.0145	$2s^22p^2$
2	−40.1957	$2p^4$
3	−19.7865	$2s^22p$ ^{2}P $3p$
4	−16.8425	$2s2p^2$ ^{2}S $3s$
5	−15.2566	$2s2p^2$ ^{2}D $3d$
6	−11.5513	$2p^3$ ^{2}P $3p$
7	−10.9855	$2s^22p$ ^{2}P $4p$
8	−7.19530	$2s2p^2$ ^{2}D $4d$
9	−7.04976	$2s2p^2$ ^{2}S $4s$
10	−6.95155	$2s^22p$ ^{2}P $5p$
11	−4.82145	$2s^22p$ ^{2}P $6p$
12	−3.53097	$2s^22p$ ^{2}P $7p$
13	−3.33582	$2s2p^2$ ^{2}D $5d$
14	−3.04964	$2p^3$ ^{2}P $4p$
15	−2.70867	$2s2p^2$ ^{2}S $5s$
16	−2.69305	$2s^22p$ ^{2}P $8p$
17	−2.12367	$2s^22p$ ^{2}P $9p$
	^{1}P^e	
1	−20.8543	$2s^22p$ ^{2}P $3p$
2	−16.7639	$2s2p^2$ ^{2}P $3s$
3	−15.4907	$2s2p^2$ ^{2}D $3d$
4	−14.6799	$2s2p^2$ ^{2}P $3d$
5	−13.5784	$2p^3$ ^{2}D $3p$
6	−12.4893	$2p^3$ ^{2}P $3p$
7	−11.3389	$2s^22p$ ^{2}P $4p$
8	−7.22930	$2s^22p$ ^{2}P $5p$
9	−7.15309	$2s2p^2$ ^{2}D $4d$
10	−6.86397	$2s2p^2$ ^{2}P $4s$
11	−6.11197	$2s2p^2$ ^{2}P $4d$
12	−4.90693	$2s^22p$ ^{2}P $6p$
13	−4.27014	$2p^3$ ^{2}D $4p$
14	−3.60285	$2p^3$ ^{2}D $4f$
15	−3.58548	$2s^22p$ ^{2}P $7p$
16	−3.38460	$2p^3$ ^{2}P $4p$
17	−3.30599	$2s2p^2$ ^{2}D $5d$
18	−2.73499	$2s^22p$ ^{2}P $8p$
19	−2.54958	$2s2p^2$ ^{2}P $5s$
20	−2.18003	$2s2p^2$ ^{2}P $5d$
21	−2.14116	$2s^22p$ ^{2}P $9p$
	^{1}D^e	
1	−49.7047	$2s^22p^2$
2	−41.4568	$2p^4$
3	−20.1740	$2s^22p$ ^{2}P $3p$
4	−17.8342	$2s2p^2$ ^{2}D $3s$
5	−15.5664	$2s2p^2$ ^{2}D $3d$
6	−14.7041	$2s2p^2$ ^{2}S $3d$
7	−14.1098	$2s2p^2$ ^{2}P $3d$
8	−12.7498	$2p^3$ ^{2}D $3p$
9	−12.1923	$2p^3$ ^{2}P $3p$
10	−11.1178	$2s^22p$ ^{2}P $4p$
11	−10.5267	$2s^22p$ ^{2}P $4f$
12	−8.11528	$2s2p^2$ ^{2}D $4s$
13	−7.23636	$2s2p^2$ ^{2}D $4d$
14	−7.04291	$2s^22p$ ^{2}P $5p$
15	−6.75395	$2s^22p$ ^{2}P $5f$
16	−6.23816	$2s2p^2$ ^{2}S $4d$
17	−5.89336	$2s2p^2$ ^{2}P $4d$
18	−4.85337	$2s^22p$ ^{2}P $6p$
19	−4.68848	$2s^22p$ ^{2}P $6f$
20	−4.05776	$2p^3$ ^{2}D $4p$
21	−3.75398	$2s2p^2$ ^{2}D $5s$
22	−3.60534	$2p^3$ ^{2}D $4f$
23	−3.55031	$2s^22p$ ^{2}P $7p$
24	−3.44626	$2s^22p$ ^{2}P $7f$
25	−3.37303	$2s2p^2$ ^{2}D $5d$
26	−3.29781	$2s2p^2$ ^{2}D $5g$
27	−3.22589	$2p^3$ ^{2}P $4p$
28	−2.70875	$2s^22p$ ^{2}P $8p$
29	−2.65423	$2p^3$ ^{2}P $4f$
30	−2.63952	$2s^22p$ ^{2}P $8f$
31	−2.35154	$2s2p^2$ ^{2}S $5d$
32	−2.13990	$2s^22p$ ^{2}P $9p$
33	−2.09192	$2s^22p$ ^{2}P $9f$
34	−2.06093	$2s2p^2$ ^{2}P $5d$
	^{1}F^e	
1	−15.0691	$2s2p^2$ ^{2}D $3d$
2	−14.3993	$2s2p^2$ ^{2}P $3d$
3	−13.4094	$2p^3$ ^{2}D $3p$
4	−10.6023	$2s^22p$ ^{2}P $4f$
5	−7.30282	$2s2p^2$ ^{2}D $4d$
6	−6.79197	$2s^22p$ ^{2}P $5f$
7	−6.00375	$2s2p^2$ ^{2}P $4d$
8	−4.70960	$2s^22p$ ^{2}P $6f$
9	−4.24134	$2p^3$ ^{2}D $4p$
10	−3.60737	$2p^3$ ^{2}D $4f$
11	−3.46019	$2s^22p$ ^{2}P $7f$

C-like Ar (Ar^{12+})

i	E(Ryds)	Description	i	E(Ryds)	Description	i	E(Ryds)	Description	i	E(Ryds)	Description
12	−3.39708	$2s2p^2$ ^{2}D $5d$	11	−3.42225	$2p^3$ ^{2}P $4p$	12	−8.23992	$2s2p^2$ ^{2}D $4s$			$^3\mathbf{G}^e$
13	−3.29538	$2s2p^2$ ^{2}D $5g$	12	−3.33622	$2s2p^2$ ^{2}D $5d$	13	−7.34486	$2s2p^2$ ^{2}D $4d$	1	−16.0340	$2s2p^2$ ^{2}D $3d$
14	−2.69700	$2p^3$ ^{2}P $4f$	13	−2.82269	$2s2p^2$ ^{2}S $5s$	14	−7.12157	$2s^22p$ ^{2}P $5p$	2	−10.5777	$2s^22p$ ^{2}P $4f$
15	−2.64897	$2s^22p$ ^{2}P $8f$	14	−2.71381	$2s^22p$ ^{2}P $8p$	15	−6.76347	$2s^22p$ ^{2}P $5f$	3	−7.39548	$2s2p^2$ ^{2}D $4d$
16	−2.13315	$2s2p^2$ ^{2}P $5d$	15	−2.14180	$2s^22p$ ^{2}P $9p$	16	−6.32486	$2s2p^2$ ^{2}S $4d$	4	−6.77262	$2s^22p$ ^{2}P $5f$
17	−2.08405	$2s^22p$ ^{2}P $9f$			$^3\mathbf{P}^e$	17	−6.08836	$2s2p^2$ ^{2}P $4d$	5	−4.85784	$2s2p^2$ ^{4}P $5g$
18	−2.03219	$2s2p^2$ ^{2}P $5g$	1	−50.3224	$2s^22p^2$	18	−4.89922	$2s^22p$ ^{2}P $6p$	6	−4.70442	$2s^22p$ ^{2}P $6f$
		$^1\mathbf{G}^e$	2	−42.0182	$2p^4$	19	−4.87987	$2s2p^2$ ^{4}P $5d$	7	−4.69424	$2s^22p$ ^{2}P $6h$
1	−15.7259	$2s2p^2$ ^{2}D $3d$	3	−20.5150	$2s^22p$ ^{2}P $3p$	20	−4.69336	$2s^22p$ ^{2}P $6f$	8	−3.64658	$2p^3$ ^{2}D $4f$
2	−10.5638	$2s^22p$ ^{2}P $4f$	4	−19.2375	$2s2p^2$ ^{4}P $3s$	21	−4.29807	$2p^3$ ^{2}D $4p$	9	−3.46702	$2s^22p$ ^{2}P $7f$
3	−7.28823	$2s2p^2$ ^{2}D $4d$	5	−17.3925	$2s2p^2$ ^{4}P $3d$	22	−3.86075	$2s2p^2$ ^{2}D $5s$	10	−3.44919	$2s^22p$ ^{2}P $7h$
4	−6.76349	$2s^22p$ ^{2}P $5f$	6	−16.9134	$2s2p^2$ ^{2}P $3s$	23	−3.61124	$2p^3$ ^{2}D $4f$	11	−3.43611	$2s2p^2$ ^{2}D $5d$
5	−4.69721	$2s^22p$ ^{2}P $6f$	7	−15.6610	$2s2p^2$ ^{2}D $3d$	24	−3.57606	$2s^22p$ ^{2}P $7p$	12	−3.29460	$2s2p^2$ ^{2}D $5g$
6	−4.69721	$2s^22p$ ^{2}P $6h$	8	−14.5110	$2s2p^2$ ^{2}P $3d$	25	−3.45053	$2s^22p$ ^{2}P $7f$	13	−2.79558	$2s2p^2$ ^{4}P $6g$
7	−3.62273	$2p^3$ ^{2}D $4f$	9	−14.1538	$2p^3$ ^{4}S $3p$	26	−3.42214	$2s2p^2$ ^{2}D $5d$	14	−2.72074	$2p^3$ ^{2}P $4f$
8	−3.45105	$2s^22p$ ^{2}P $7f$	10	−12.9719	$2p^3$ ^{2}D $3p$	27	−3.34054	$2p^3$ ^{2}P $4p$	15	−2.64528	$2s^22p$ ^{2}P $8f$
9	−3.44916	$2s^22p$ ^{2}P $7h$	11	−12.3292	$2p^3$ ^{2}P $3p$	28	−3.29812	$2s2p^2$ ^{2}D $5g$	16	−2.64068	$2s^22p$ ^{2}P $8h$
10	−3.40045	$2s2p^2$ ^{2}D $5d$	12	−11.2114	$2s^22p$ ^{2}P $4p$	29	−2.81216	$2s2p^2$ ^{4}P $6d$	17	−2.26049	$2s2p^2$ ^{2}S $5g$
11	−3.29204	$2s2p^2$ ^{2}D $5g$	13	−9.65744	$2s2p^2$ ^{4}P $4s$	30	−2.72622	$2s^22p$ ^{2}P $8p$	18	−2.08995	$2s^22p$ ^{2}P $9f$
12	−2.70727	$2p^3$ ^{2}P $4f$	14	−8.86884	$2s2p^2$ ^{4}P $4d$	31	−2.67638	$2p^3$ ^{2}P $4f$	19	−2.08909	$2s^22p$ ^{2}P $9h$
13	−2.64208	$2s^22p$ ^{2}P $8f$	15	−7.29452	$2s2p^2$ ^{2}D $4d$	32	−2.64218	$2s^22p$ ^{2}P $8f$	20	−2.01237	$2s2p^2$ ^{2}P $5g$
14	−2.64051	$2s^22p$ ^{2}P $8h$	16	−7.08840	$2s^22p$ ^{2}P $5p$	33	−2.39489	$2s2p^2$ ^{2}S $5d$			$^3\mathbf{H}^e$
15	−2.25940	$2s2p^2$ ^{2}S $5g$	17	−6.93638	$2s2p^2$ ^{2}P $4s$	34	−2.17482	$2s2p^2$ ^{2}P $5d$	1	−4.86969	$2s2p^2$ ^{4}P $5g$
16	−2.08911	$2s^22p$ ^{2}P $9h$	18	−6.03986	$2s2p^2$ ^{2}P $4d$	35	−2.13466	$2s^22p$ ^{2}P $9p$	2	−4.69615	$2s^22p$ ^{2}P $6h$
17	−2.08751	$2s^22p$ ^{2}P $9f$	19	−5.38805	$2s2p^2$ ^{4}P $5s$	36	−2.08553	$2s^22p$ ^{2}P $9f$	3	−3.64740	$2p^3$ ^{2}D $4f$
18	−2.01199	$2s2p^2$ ^{2}P $5g$	20	−5.03540	$2p^3$ ^{4}S $4p$			$^3\mathbf{F}^e$	4	−3.45112	$2s^22p$ ^{2}P $7h$
		$^1\mathbf{H}^e$	21	−4.93769	$2s2p^2$ ^{4}P $5d$	1	−17.1991	$2s2p^2$ ^{4}P $3d$	5	−3.29675	$2s2p^2$ ^{2}D $5g$
1	−4.69615	$2s^22p$ ^{2}P $6h$	22	−4.87257	$2s^22p$ ^{2}P $6p$	2	−15.9736	$2s2p^2$ ^{2}D $3d$	6	−2.80272	$2s2p^2$ ^{4}P $6g$
2	−3.64669	$2p^3$ ^{2}D $4f$	23	−4.12827	$2p^3$ ^{2}D $4p$	3	−14.8699	$2s2p^2$ ^{2}P $3d$	7	−2.64231	$2s^22p$ ^{2}P $8h$
3	−3.45111	$2s^22p$ ^{2}P $7h$	24	−3.59195	$2p^3$ ^{2}D $4f$	4	−13.4691	$2p^3$ ^{2}D $3p$	8	−2.09334	$2s^22p$ ^{2}P $9h$
4	−3.29543	$2s2p^2$ ^{2}D $5g$	25	−3.56630	$2s^22p$ ^{2}P $7p$	5	−10.6018	$2s^22p$ ^{2}P $4f$	9	−2.03179	$2s2p^2$ ^{2}P $5g$
5	−2.64231	$2s^22p$ ^{2}P $8h$	26	−3.40066	$2s2p^2$ ^{2}D $5d$	6	−8.78574	$2s2p^2$ ^{4}P $4d$			$^3\mathbf{I}^e$
6	−2.09334	$2s^22p$ ^{2}P $9h$	27	−3.27809	$2p^3$ ^{2}P $4p$	7	−7.39117	$2s2p^2$ ^{2}D $4d$	1	−4.69447	$2s^22p$ ^{2}P $6h$
7	−2.03167	$2s2p^2$ ^{2}P $5g$	28	−3.08369	$2s2p^2$ ^{4}P $6s$	8	−6.78649	$2s^22p$ ^{2}P $5f$	2	−3.45090	$2s^22p$ ^{2}P $7h$
		$^1\mathbf{I}^e$	29	−2.85740	$2s2p^2$ ^{4}P $6d$	9	−6.14599	$2s2p^2$ ^{2}P $4d$	3	−3.29882	$2s2p^2$ ^{2}D $5g$
1	−4.69446	$2s^22p$ ^{2}P $6h$	30	−2.71727	$2s^22p$ ^{2}P $8p$	10	−4.92749	$2s2p^2$ ^{4}P $5d$	4	−2.64019	$2s^22p$ ^{2}P $8h$:
2	−3.45087	$2s^22p$ ^{2}P $7h$	31	−2.56796	$2s2p^2$ ^{2}P $5s$	11	−4.89470	$2s2p^2$ ^{4}P $5g$	5	−2.64019	$2s^22p$ ^{2}P $8j$
3	−3.29849	$2s2p^2$ ^{2}D $5g$	32	−2.15261	$2s2p^2$ ^{2}P $5d$	12	−4.70795	$2s^22p$ ^{2}P $6f$	6	−2.08625	$2s^22p$ ^{2}P $9h$
4	−2.64019	$2s^22p$ ^{2}P $8h$	33	−2.12529	$2s^22p$ ^{2}P $9p$	13	−4.46488	$2p^3$ ^{4}S $4f$			$^5\mathbf{P}^e$
5	−2.64019	$2s^22p$ ^{2}P $8j$	34	−1.74092	$2s2p^2$ ^{4}P $7s$	14	−4.26420	$2p^3$ ^{2}D $4p$	1	−19.8258	$2s2p^2$ ^{4}P $3s$
6	−2.08624	$2s^22p$ ^{2}P $9h$			$^3\mathbf{D}^e$	15	−3.62784	$2p^3$ ^{2}D $4f$	2	−17.4228	$2s2p^2$ ^{4}P $3d$
		$^3\mathbf{S}^e$	1	−20.7333	$2s^22p$ ^{2}P $3p$	16	−3.46297	$2s^22p$ ^{2}P $7f$	3	−14.4552	$2p^3$ ^{4}S $3p$
1	−20.5942	$2s^22p$ ^{2}P $3p$	2	−18.1792	$2s2p^2$ ^{2}D $3s$	17	−3.44142	$2s2p^2$ ^{2}D $5d$	4	−9.82865	$2s2p^2$ ^{4}P $4s$
2	−17.1517	$2s2p^2$ ^{2}S $3s$	3	−16.8403	$2s2p^2$ ^{4}P $3d$	18	−3.29661	$2s2p^2$ ^{2}D $5g$	5	−8.89313	$2s2p^2$ ^{4}P $4d$
3	−15.5673	$2s2p^2$ ^{2}D $3d$	4	−15.8830	$2s2p^2$ ^{2}D $3d$	19	−2.83388	$2s2p^2$ ^{4}P $6d$	6	−5.45637	$2s2p^2$ ^{4}P $5s$
4	−12.6643	$2p^3$ ^{2}P $3p$	5	−14.9009	$2s2p^2$ ^{2}S $3d$	20	−2.80089	$2s2p^2$ ^{4}P $6g$	7	−5.16592	$2p^3$ ^{4}S $4p$
5	−11.2540	$2s^22p$ ^{2}P $4p$	6	−14.6858	$2s2p^2$ ^{2}P $3d$	21	−2.69490	$2p^3$ ^{2}P $4f$	8	−4.97054	$2s2p^2$ ^{4}P $5d$
6	−7.28540	$2s2p^2$ ^{2}D $4d$	7	−13.5637	$2p^3$ ^{2}D $3p$	22	−2.64746	$2s^22p$ ^{2}P $8f$	9	−3.12528	$2s2p^2$ ^{4}P $6s$
7	−7.24600	$2s2p^2$ ^{2}S $4s$	8	−12.5814	$2p^3$ ^{2}P $3p$	23	−2.18878	$2s2p^2$ ^{2}P $5d$	10	−2.86583	$2s2p^2$ ^{4}P $6d$
8	−7.01713	$2s^22p$ ^{2}P $5p$	9	−11.3007	$2s^22p$ ^{2}P $4p$	24	−2.08996	$2s^22p$ ^{2}P $9f$	11	−1.75761	$2s2p^2$ ^{4}P $7s$
9	−4.88837	$2s^22p$ ^{2}P $6p$	10	−10.5401	$2s^22p$ ^{2}P $4f$	25	−2.03196	$2s2p^2$ ^{2}P $5g$	12	−1.59636	$2s2p^2$ ^{4}P $7d$
10	−3.57347	$2s^22p$ ^{2}P $7p$	11	−8.68250	$2s2p^2$ ^{4}P $4d$				13	−1.02320	$2p^3$ ^{4}S $5p$

C-like Ar (Ar^{12+})

i	Energy(Ryds)	Description	i	Energy(Ryds)	Description
14	−.87800	$2s2p^2$ ^{4}P $8s$	11	−.75544	$2s2p^2$ ^{4}P $8g$
15	−.77380	$2s2p^2$ ^{4}P $8d$	12	−.68743	$2p^3$ ^{4}S $5f$
16	−.28561	$2s2p^2$ ^{4}P $9s$	13	−.22042	$2s2p^2$ ^{4}P $9d$
17	−.21171	$2s2p^2$ ^{4}P $9d$	14	−.19264	$2s2p^2$ ^{4}P $9g$
		^{5}D^e			**^{5}G^e**
1	−17.6071	$2s2p^2$ ^{4}P $3d$	1	−4.85855	$2s2p^2$ ^{4}P $5g$
2	−8.96143	$2s2p^2$ ^{4}P $4d$	2	−2.79626	$2s2p^2$ ^{4}P $6g$
3	−5.01461	$2s2p^2$ ^{4}P $5d$	3	−1.55248	$2s2p^2$ ^{4}P $7g$
4	−2.88520	$2s2p^2$ ^{4}P $6d$	4	−.74501	$2s2p^2$ ^{4}P $8g$
5	−1.60783	$2s2p^2$ ^{4}P $7d$	5	−.19130	$2s2p^2$ ^{4}P $9g$
6	−.78181	$2s2p^2$ ^{4}P $8d$			**^{5}H^e**
7	−.21703	$2s2p^2$ ^{4}P $9d$	1	−4.87026	$2s2p^2$ ^{4}P $5g$
		^{5}F^e	2	−2.80328	$2s2p^2$ ^{4}P $6g$
1	−17.7635	$2s2p^2$ ^{4}P $3d$	3	−1.55699	$2s2p^2$ ^{4}P $7g$
2	−9.01298	$2s2p^2$ ^{4}P $4d$	4	−1.55491	$2s2p^2$ ^{4}P $7i$
3	−5.03777	$2s2p^2$ ^{4}P $5d$	5	−.74806	$2s2p^2$ ^{4}P $8g$
4	−4.89537	$2s2p^2$ ^{4}P $5g$	6	−.74658	$2s2p^2$ ^{4}P $8i$
5	−4.48451	$2p^3$ ^{4}S $4f$	7	−.19347	$2s2p^2$ ^{4}P $9g$
6	−2.89720	$2s2p^2$ ^{4}P $6d$	8	−.19239	$2s2p^2$ ^{4}P $9i$
7	−2.80128	$2s2p^2$ ^{4}P $6g$			
8	−1.61528	$2s2p^2$ ^{4}P $7d$			
9	−1.55702	$2s2p^2$ ^{4}P $7g$			
10	−.78699	$2s2p^2$ ^{4}P $8d$			

C-like Ar (Ar^{12+})

Energies in ascending order from ground state for terms with effective $n \leq 4.0$, $L \leq 4$

Term	i	E(Ryds)	Term	i	E(Ryds)	Term	i	E(Ryds)	Term	i	E(Ryds)	Term	i	E(Ryds)
$^3P^e$	1	0.00000	$^1P^o$	3	31.3887	$^3P^o$	6	34.3148	$^1D^o$	5	36.1378	$^1F^o$	3	38.5191
$^1D^e$	1	0.61770	$^3S^o$	2	31.4015	$^3F^e$	2	34.3488	$^3P^e$	9	36.1686	$^3P^o$	10	38.6073
$^1S^e$	1	1.30790	$^5D^o$	1	31.5283	$^1P^o$	5	34.3782	$^1D^e$	7	36.2126	$^1P^o$	9	38.6690
$^5S^o$	1	1.73780	$^5P^o$	1	31.6393	$^3D^o$	5	34.3931	$^1P^e$	5	36.7440	$^1S^e$	6	38.7711
$^3D^o$	1	3.55040	$^3D^o$	3	31.9150	$^3D^e$	4	34.4394	$^3D^e$	7	36.7587	$^3F^o$	4	38.7711
$^3P^o$	1	4.19970	$^5S^o$	2	31.9958	$^1F^e$	1	34.4533	$^3P^o$	8	36.7859	$^3P^o$	11	38.8741
$^1D^o$	1	5.40290	$^3P^o$	4	32.1070	$^3P^o$	7	34.5314	$^3F^e$	4	36.8533	$^1P^e$	7	38.9835
$^3S^o$	1	5.44920	$^3D^e$	2	32.1432	$^3S^o$	3	34.5686	$^5D^o$	2	36.9043	$^3D^o$	9	39.0134
$^1P^o$	1	6.05670	$^1D^e$	4	32.4882	$^1G^e$	1	34.5965	$^1F^e$	3	36.9130	$^3D^e$	9	39.0217
$^3P^e$	2	8.30420	$^5F^e$	1	32.5589	$^3P^e$	7	34.6614	$^1P^o$	7	37.0149	$^1D^o$	7	39.0265
$^1D^e$	2	8.86560	$^5D^e$	1	32.7153	$^3S^e$	3	34.7551	$^3D^o$	7	37.2968	$^3S^e$	5	39.0684
$^1S^e$	2	10.1267	$^5P^e$	2	32.8996	$^1D^e$	5	34.7560	$^3P^e$	10	37.3505	$^3P^e$	12	39.1110
$^3P^o$	2	28.5766	$^3P^e$	5	32.9299	$^1P^e$	3	34.8317	$^1D^e$	8	37.5726	$^1F^o$	4	39.1721
$^1P^o$	2	28.7664	$^3F^e$	1	33.1233	$^1D^o$	4	34.8711	$^3S^e$	4	37.6581	$^1D^e$	10	39.2046
$^1P^e$	1	29.4681	$^3S^e$	2	33.1707	$^5S^o$	3	34.8806	$^3D^e$	8	37.7410	$^1S^e$	7	39.3369
$^3D^e$	1	29.5891	$^1D^o$	3	33.1887	$^1S^e$	5	35.0658	$^3F^o$	3	37.7782	$^3F^o$	5	39.4389
$^3S^e$	1	29.7282	$^3F^o$	2	33.2035	$^1P^o$	6	35.1233	$^1S^o$	2	37.8182	$^1D^o$	8	39.4399
$^3P^e$	3	29.8074	$^1F^o$	2	33.3402	$^3D^e$	5	35.4215	$^1P^e$	6	37.8331	$^3D^o$	10	39.5241
$^1D^e$	3	30.1484	$^3D^o$	4	33.3706	$^3F^e$	3	35.4525	$^3G^o$	1	37.8522	$^3P^o$	12	39.5396
$^5P^e$	1	30.4966	$^3P^e$	6	33.4090	$^3S^o$	4	35.5168	$^1G^o$	1	37.9019	$^1P^o$	10	39.5902
$^1S^e$	3	30.5359	$^1P^o$	4	33.4179	$^1D^e$	6	35.6183	$^3P^e$	11	37.9932	$^1F^o$	5	39.6697
$^1D^o$	2	30.7209	$^1S^e$	4	33.4799	$^3D^e$	6	35.6366	$^1P^o$	8	38.1000	$^1P^o$	11	39.6872
$^3F^o$	1	30.7329	$^3D^e$	3	33.4821	$^1P^e$	4	35.6425	$^1D^e$	9	38.1301	$^1F^e$	4	39.7201
$^3D^o$	2	30.9744	$^1P^e$	2	33.5585	$^3P^e$	8	35.8114	$^3D^o$	8	38.1375	$^3F^e$	5	39.7206
$^3P^o$	3	31.0324	$^3P^o$	5	33.5634	$^5P^e$	3	35.8672	$^3P^o$	9	38.2708	$^3G^e$	2	39.7447
$^3P^e$	4	31.0849	$^1S^o$	1	34.1840	$^3D^o$	6	35.9036	$^1D^o$	6	38.3079	$^1G^e$	2	39.7586
$^1F^o$	1	31.3825	$^3G^e$	1	34.2884	$^1F^e$	2	35.9231	$^3S^o$	5	38.4238			

gf-values for transitions involving terms with effective $n \leq 4.0$, $L \leq 4$

i i'	gf_L	i i'	gf_L	i i'	gf_L	i i'	gf_L	i i'	gf_L	i i'	gf_L
	$^1P^o$–$^1S^e$	3 6	−4.33E−5	6 5	4.57E−5	9 4	2.17E−3	1 2	4.46E−2	1 7	−5.06E−3
1 1	1.35E−1	3 7	−3.43E−2	6 6	−1.32E−1	9 5	5.25E−4	1 3	−4.27E−3	2 1	−1.46E−1
1 2	−2.66E−1	4 1	9.79E−3	6 7	−3.35E−3	9 6	−1.44E−5	1 4	−1.34E−1	2 2	−4.56E−1
1 3	−9.42E−3	4 2	3.21E−3	7 1	2.61E−3	9 7	−2.18E−1	1 5	−1.47E−1	2 3	−5.15E−3
1 4	−5.57E−2	4 3	6.49E−3	7 2	1.42E−1	10 1	8.12E−3	1 6	−3.58E−2	2 4	−3.33E−4
1 5	−2.43E−1	4 4	−4.61E−4	7 3	6.57E−4	10 2	1.75E+0	1 7	−2.92E−6	2 5	−7.15E−4
1 6	−8.86E−2	4 5	−1.10E−1	7 4	1.17E−1	10 3	5.71E−2	2 1	2.86E−4	2 6	−3.44E−4
1 7	−1.36E−3	4 6	−1.53E−2	7 5	4.34E−4	10 4	1.52E−3	2 2	5.85E−4	2 7	−3.13E−1
2 1	7.68E−2	4 7	−1.85E−9	7 6	−2.09E−1	10 5	1.38E−2	2 3	1.01E−1	3 1	1.68E−1
2 2	5.79E−7	5 1	3.53E−1	7 7	−6.23E−5	10 6	9.94E−2	2 4	1.25E−2	3 2	−1.14E−5
2 3	−1.67E−1	5 2	1.08E−2	8 1	5.60E−4	10 7	1.53E−2	2 5	9.15E−2	3 3	−2.70E−2
2 4	−5.73E−2	5 3	1.03E−1	8 2	1.21E−1	11 1	2.72E−1	2 6	−3.24E−5	3 4	−1.41E−1
2 5	−4.36E−3	5 4	1.83E−1	8 3	3.19E−6	11 2	2.51E−1	2 7	−2.24E−5	3 5	−9.41E−4
2 6	−2.41E−4	5 5	−1.65E−2	8 4	7.28E−5	11 3	4.16E−1		$^1P^o$–$^1P^e$	3 6	−1.98E−6
2 7	−6.76E−2	5 6	−1.34E−3	8 5	2.77E−2	11 4	1.11E−3	1 1	−2.74E−3	3 7	−5.36E−2
3 1	1.27E+0	5 7	−3.51E−4	8 6	−5.87E−3	11 5	1.00E−2	1 2	−2.12E−1	4 1	1.13E−1
3 2	1.45E−4	6 1	4.80E−2	8 7	−1.45E−4	11 6	9.28E−3	1 3	−1.48E−1	4 2	−4.67E−4
3 3	9.93E−2	6 2	7.36E−4	9 1	1.35E−2	11 7	1.67E−1	1 4	−1.13E+0	4 3	−2.21E−1
3 4	−5.45E−3	6 3	9.19E−2	9 2	1.14E−4		$^1S^o$–$^1P^e$	1 5	−1.51E−3	4 4	−6.96E−3
3 5	−4.16E−2	6 4	2.45E−2	9 3	1.00E−1	1 1	1.59E−1	1 6	−2.56E−1	4 5	−1.55E−1

C-like Ar (Ar^{12+})

i i'	gf$_L$	i i'	gf$_L$	i i'	gf$_L$	i i'	gf$_L$	i i'	gf$_L$	i i'	gf$_L$
4 6	−7.15E−2		^{1}D^o–^{1}P^e	8 3	1.14E−2	5 7	−5.64E−2	10 9	1.71E−3	4 10	−1.61E−3
4 7	−7.71E−4	1 1	−3.60E−3	8 4	4.57E−3	5 8	−4.38E−2	10 10	1.26E−3	5 1	1.05E−2
5 1	8.51E−2	1 2	−2.67E−1	8 5	1.22E−4	5 9	−3.84E−2	11 1	1.04E−2	5 2	5.19E−1
5 2	7.30E−3	1 3	−3.70E−1	8 6	3.04E−3	5 10	−8.50E−6	11 2	1.19E−2	5 3	1.19E−3
5 3	−8.59E−4	1 4	−6.91E−3	8 7	5.34E−1	6 1	6.81E−3	11 3	1.60E−2	5 4	5.61E−1
5 4	−8.83E−3	1 5	−2.79E−1		^{1}P^o–^{1}D^e	6 2	1.06E−2	11 4	2.51E−4	5 5	5.64E−3
5 5	−1.09E−3	1 6	−3.66E−2	1 1	4.00E−1	6 3	2.79E−1	11 5	3.49E−3	5 6	3.64E−4
5 6	−9.83E−2	1 7	−2.39E−4	1 2	−1.25E−1	6 4	5.64E−2	11 6	5.77E−3	5 7	−4.20E−4
5 7	−6.43E−5	2 1	2.99E−1	1 3	−4.73E−3	6 5	1.71E−3	11 7	2.17E−2	5 8	−7.38E−1
6 1	3.53E−2	2 2	−7.01E−4	1 4	−5.34E−2	6 6	−2.05E−3	11 8	1.05E−3	5 9	−1.64E−1
6 2	3.97E−1	2 3	−1.17E−1	1 5	−5.12E−1	6 7	−2.56E−1	11 9	8.76E−4	5 10	−7.18E−6
6 3	8.57E−5	2 4	−2.42E−1	1 6	−4.14E−2	6 8	−2.77E−1	11 10	1.15E−2	6 1	2.77E−2
6 4	−4.22E−2	2 5	−8.69E−4	1 7	−2.96E+0	6 9	−1.60E−2		^{1}D^o–^{1}D^e	6 2	1.26E+0
6 5	−1.92E−3	2 6	−4.14E−5	1 8	−2.25E−5	6 10	−9.40E−3	1 1	6.73E−1	6 3	9.09E−5
6 6	−1.19E−1	2 7	−7.98E−2	1 9	−3.33E−1	7 1	1.91E−3	1 2	−7.28E−1	6 4	1.50E−3
6 7	−8.11E−3	3 1	1.88E−1	1 10	−1.64E−2	7 2	1.86E−1	1 3	−1.21E−2	6 5	1.58E−2
7 1	6.57E−4	3 2	−3.04E−3	2 1	2.39E−1	7 3	3.93E−5	1 4	−1.64E−1	6 6	7.07E−2
7 2	2.21E−1	3 3	−8.55E−2	2 2	2.46E−6	7 4	2.85E−1	1 5	−1.50E+0	6 7	4.94E−2
7 3	3.10E−4	3 4	−3.88E−2	2 3	−5.87E−1	7 5	8.00E−4	1 6	−4.57E−2	6 8	1.21E−1
7 4	5.05E−3	3 5	−1.75E−1	2 4	−7.17E−2	7 6	2.10E−4	1 7	−5.75E−1	6 9	1.08E−4
7 5	4.41E−3	3 6	−1.24E−1	2 5	−2.70E−3	7 7	8.96E−3	1 8	−2.75E−1	6 10	−2.71E−6
7 6	−2.04E−1	3 7	−7.16E−7	2 6	−1.33E−3	7 8	−2.81E−2	1 9	−1.39E−1	7 1	1.15E−3
7 7	−6.81E−6	4 1	4.88E−2	2 7	−2.51E−6	7 9	−4.88E−1	1 10	−6.23E−4	7 2	1.82E+0
8 1	4.40E−6	4 2	5.54E−1	2 8	−2.77E−4	7 10	−2.88E−4	2 1	1.04E+0	7 3	2.02E−3
8 2	5.16E−3	4 3	1.05E−4	2 9	−1.48E−6	8 1	6.06E−3	2 2	2.15E−4	7 4	4.50E−4
8 3	3.07E−2	4 4	−1.82E−3	2 10	−4.56E−1	8 2	6.70E−1	2 3	4.40E−2	7 5	8.79E−2
8 4	1.51E−1	4 5	−1.34E−2	3 1	4.79E−2	8 3	2.22E−4	2 4	−5.53E−3	7 6	2.30E−2
8 5	2.01E−1	4 6	−9.75E−2	3 2	7.66E−6	8 4	4.80E−4	2 5	−1.84E−1	7 7	1.53E−1
8 6	7.50E−3	4 7	−9.78E−4	3 3	7.82E−3	8 5	1.57E−1	2 6	−9.48E−2	7 8	9.90E−3
8 7	−8.74E−5	5 1	2.02E−3	3 4	−1.06E−4	8 6	8.15E−4	2 7	−3.02E−4	7 9	1.23E−1
9 1	1.25E−1	5 2	1.53E−1	3 5	−3.31E−2	8 7	3.02E−3	2 8	−7.30E−6	7 10	−7.40E−4
9 2	1.06E−5	5 3	1.28E−3	3 6	−8.41E−2	8 8	4.33E−2	2 9	−6.67E−4	8 1	1.96E−1
9 3	1.44E−3	5 4	3.40E−4	3 7	−6.79E−2	8 9	−6.84E−5	2 10	−1.47E−2	8 2	1.96E−2
9 4	8.22E−6	5 5	−1.27E−1	3 8	−8.20E−4	8 10	−5.36E−5	3 1	7.63E−1	8 3	3.91E−1
9 5	4.82E−5	5 6	−5.95E−2	3 9	−1.57E−5	9 1	4.68E−2	3 2	1.22E−2	8 4	7.07E−3
9 6	1.57E−4	5 7	−7.93E−5	3 10	−9.62E−3	9 2	7.88E−8	3 3	4.76E−3	8 5	4.57E−3
9 7	−2.69E−1	6 1	9.43E−5	4 1	4.87E−1	9 3	2.71E−1	3 4	2.27E−1	8 6	1.85E−3
10 1	3.22E−2	6 2	7.80E−5	4 2	2.64E−2	9 4	6.33E−3	3 5	−2.43E−1	8 7	8.34E−3
10 2	3.40E−4	6 3	2.17E−4	4 3	2.80E−2	9 5	9.98E−4	3 6	−2.17E−2	8 8	6.08E−5
10 3	5.50E−3	6 4	1.12E−1	4 4	1.76E−1	9 6	2.18E−4	3 7	−1.78E−2	8 9	1.94E−3
10 4	6.41E−2	6 5	2.92E−1	4 5	−1.93E−1	9 7	1.05E−3	3 8	−5.66E−2	8 10	8.85E−2
10 5	1.17E−4	6 6	1.40E−2	4 6	−1.61E−2	9 8	1.22E−4	3 9	−1.22E−2		^{1}F^o–^{1}D^e
10 6	2.21E−1	6 7	−1.32E−5	4 7	−2.39E−2	9 9	2.55E−5	3 10	−1.79E−3	1 1	4.85E+0
10 7	2.93E−2	7 1	9.53E−3	4 8	−3.15E−4	9 10	−8.19E−1	4 1	5.74E−2	1 2	9.40E−5
11 1	2.28E−1	7 2	8.22E−6	4 9	−9.87E−3	10 1	1.92E−3	4 2	1.74E−2	1 3	5.96E−1
11 2	6.22E−3	7 3	2.54E−2	4 10	−2.75E−3	10 2	7.04E−2	4 3	4.81E−1	1 4	−1.66E−2
11 3	7.69E−3	7 4	4.92E−3	5 1	1.53E−2	10 3	4.34E−3	4 4	7.42E−2	1 5	−6.70E−3
11 4	3.04E−3	7 5	2.07E−2	5 2	1.17E−3	10 4	8.20E−5	4 5	1.02E−3	1 6	−2.62E−2
11 5	8.20E−5	7 6	3.51E−1	5 3	9.03E−2	10 5	2.07E−2	4 6	−6.88E−3	1 7	−5.65E−1
11 6	3.03E−2	7 7	4.21E−4	5 4	3.75E−2	10 6	2.71E−5	4 7	−8.27E−2	1 8	−2.96E−3
11 7	2.71E−1	8 1	9.05E−1	5 5	−5.01E−4	10 7	3.05E−1	4 8	−2.43E−4	1 9	−4.58E−5
		8 2	2.46E−2	5 6	−4.24E−1	10 8	1.44E−2	4 9	−4.37E−1	1 10	−1.57E−1

C-like Ar (Ar^{12+})

i i′	gf$_L$	*i i′*	gf$_L$	*i i′*	gf$_L$	*i i′*	gf$_L$	*i i′*	gf$_L$	*i i′*	gf$_L$
2 1	9.31E−1	3 4	−2.30E−2	3 1	2.59E−1	9 3	1.24E−1	3 10	−1.57E−2	3 1	2.68E+0
2 2	1.71E−2	4 1	2.51E−4	3 2	−1.97E−4	9 4	6.34E−3	3 11	−1.25E−1	3 2	4.56E−4
2 3	3.90E−1	4 2	−4.40E−1	4 1	2.75E−1	9 5	−3.46E−6	3 12	−7.30E−4	3 3	2.24E−1
2 4	4.44E−1	4 3	−4.08E−1	4 2	−6.89E−5	10 1	1.22E−1	4 1	3.86E−2	3 4	−1.28E−3
2 5	−6.24E−2	4 4	−3.90E−2	5 1	3.48E−3	10 2	1.61E−4	4 2	2.44E−1	3 5	−3.09E−3
2 6	−2.24E−2	5 1	2.76E−3	5 2	−1.71E−1	10 3	4.08E−5	4 3	7.46E−3	3 6	−8.59E−5
2 7	−2.81E−5	5 2	6.90E−3		^{1}G^o–^{1}G^e	10 4	1.57E−4	4 4	5.67E−1	3 7	−2.76E−1
2 8	−2.48E−1	5 3	−4.76E−1	1 1	3.60E−1	10 5	−3.62E−1	4 5	3.45E−3	3 8	−2.79E−1
2 9	−1.83E−1	5 4	−1.96E−4	1 2	−2.97E−5	11 1	1.82E−5	4 6	1.20E−1	3 9	−1.35E−4
2 10	−2.70E−3	6 1	3.53E−1		^{3}P^o–^{3}S^e	11 2	3.41E−3	4 7	7.73E−5	3 10	−5.12E−4
3 1	4.53E−2	6 2	4.02E−5	1 1	−1.96E−2	11 3	3.57E−2	4 8	−2.33E−2	3 11	−4.25E−4
3 2	2.93E+0	6 3	2.46E−2	1 2	−2.58E−1	11 4	5.21E−1	4 9	−3.16E−1	3 12	−5.06E−2
3 3	2.63E−4	6 4	−5.47E−5	1 3	−8.61E−1	11 5	−5.42E−5	4 10	−3.91E−1	4 1	1.03E+0
3 4	9.41E−4	7 1	6.95E−2	1 4	−1.77E−1	12 1	1.29E+0	4 11	−1.55E−1	4 2	3.80E−2
3 5	2.47E−2	7 2	1.69E−1	1 5	−5.87E−3	12 2	1.69E−2	4 12	−1.56E−5	4 3	2.27E−1
3 6	4.03E−2	7 3	1.04E−1	2 1	−2.32E−1	12 3	1.61E−2	5 1	1.03E−2	4 4	6.66E−1
3 7	6.19E−2	7 4	−9.03E−3	2 2	−2.91E−1	12 4	7.35E−5	5 2	1.63E+0	4 5	−2.32E−1
3 8	4.33E−1	8 1	4.58E−4	2 3	−1.40E−3	12 5	8.18E−1	5 3	1.12E−3	4 6	−4.63E−3
3 9	4.94E−4	8 2	1.57E−3	2 4	−2.00E−4		^{3}S^o–^{3}P^e	5 4	6.88E−4	4 7	−4.08E−2
3 10	−6.68E−7	8 3	2.56E−4	2 5	−2.57E−1	1 1	6.30E−1	5 5	3.43E−3	4 8	−2.77E−2
4 1	1.75E−2	8 4	−3.74E−1	3 1	4.83E−1	1 2	−4.74E−1	5 6	8.83E−4	4 9	−3.85E−1
4 2	3.60E+0		^{1}F^o–^{1}F^e	3 2	−6.11E−5	1 3	−3.28E−3	5 7	2.00E−1	4 10	−4.38E−4
4 3	2.20E−5	1 1	−1.91E−2	3 3	−1.12E−1	1 4	−6.16E−3	5 8	1.84E−1	4 11	−8.34E−3
4 4	3.30E−6	1 2	−5.84E−1	3 4	−3.79E−7	1 5	−2.77E−1	5 9	1.39E−1	4 12	−2.63E−3
4 5	1.15E−2	1 3	−4.70E−3	3 5	−1.62E−1	1 6	−2.51E−1	5 10	2.20E−1	5 1	6.17E−2
4 6	5.51E−2	1 4	−6.60E−1	4 1	7.64E−5	1 7	−1.21E−2	5 11	1.40E−3	5 2	7.57E−3
4 7	3.49E−4	2 1	−1.41E−1	4 2	−5.77E−3	1 8	−4.64E+0	5 12	−8.61E−6	5 3	4.15E−2
4 8	3.56E−2	2 2	−6.91E−2	4 3	−2.76E−2	1 9	−7.65E−1		^{3}P^o–^{3}P^e	5 4	1.97E−1
4 9	5.90E−1	2 3	−3.88E−1	4 4	−4.95E−3	1 10	−1.10E−1	1 1	5.36E−1	5 5	6.87E−4
4 10	−3.12E−6	2 4	−3.82E−3	4 5	−1.28E−3	1 11	−1.06E−1	1 2	−3.42E−1	5 6	1.18E−3
5 1	1.06E+0	3 1	4.50E−2	5 1	2.71E−1	1 12	−1.37E−3	1 3	−1.30E−2	5 7	−5.28E−1
5 2	1.05E−3	3 2	1.72E−1	5 2	3.54E−4	2 1	3.87E−1	1 4	−3.35E−1	5 8	−9.98E−3
5 3	1.78E+0	3 3	2.84E−1	5 3	−2.56E−1	2 2	3.01E−3	1 5	−1.46E+0	5 9	−5.34E−4
5 4	3.17E−2	3 4	−1.90E−6	5 4	−2.86E−1	2 3	3.55E−2	1 6	−9.11E−2	5 10	−3.60E−1
5 5	2.98E−3	4 1	7.09E−3	5 5	−3.34E−3	2 4	5.89E−2	1 7	−2.39E+0	5 11	−7.46E−2
5 6	2.71E−5	4 2	5.13E−1	6 1	4.50E−4	2 5	−5.16E−1	1 8	−3.00E−1	5 12	−3.31E−3
5 7	2.96E−2	4 3	1.61E−3	6 2	7.60E−1	2 6	−3.31E−3	1 9	−5.47E−4	6 1	1.35E−1
5 8	1.08E−3	4 4	−7.27E−6	6 3	−1.78E−2	2 7	−3.60E−2	1 10	−4.87E−3	6 2	6.73E−3
5 9	1.87E−4	5 1	6.65E−4	6 4	−1.90E−1	2 8	−7.50E−4	1 11	−4.82E−1	6 3	2.49E−2
5 10	1.07E+0	5 2	2.34E−4	6 5	−1.30E−3	2 9	−6.77E−2	1 12	−9.86E−3	6 4	8.59E−3
	^{1}D^o–^{1}F^e	5 3	1.72E−4	7 1	3.34E−1	2 10	−3.62E−3	2 1	5.46E−1	6 5	1.07E−5
1 1	−6.46E−1	5 4	−8.27E−3	7 2	1.38E−1	2 11	−5.06E−4	2 2	6.54E−5	6 6	6.59E−2
1 2	−5.31E+0		^{1}G^o–^{1}F^e	7 3	−2.83E−5	2 12	−2.15E−3	2 3	−8.14E−1	6 7	−7.62E−6
1 3	−2.90E−1	1 1	3.78E−2	7 4	−1.56E−1	3 1	1.32E−1	2 4	−1.56E−1	6 8	−5.45E−2
1 4	−7.04E−3	1 2	3.78E−2	7 5	−6.19E−3	3 2	1.37E−1	2 5	−3.03E−4	6 9	−6.91E−4
2 1	−1.35E−1	1 3	7.83E−1	8 1	1.67E−3	3 3	4.14E−1	2 6	−1.20E+0	6 10	−6.33E−2
2 2	−4.10E−2	1 4	−2.88E−4	8 2	1.70E−1	3 4	1.78E−2	2 7	−5.08E−3	6 11	−3.13E−1
2 3	−2.00E−3		^{1}F^o–^{1}G^e	8 3	6.29E−4	3 5	2.39E−4	2 8	−1.35E−3	6 12	−1.22E−2
2 4	−4.30E+0	1 1	−9.85E−2	8 4	−1.99E−1	3 6	2.30E−1	2 9	−2.10E−4	7 1	2.17E−1
3 1	−4.24E−1	1 2	−6.27E+0	8 5	−4.87E−6	3 7	−2.36E−6	2 10	−2.85E−3	7 2	5.84E−3
3 2	−1.20E−3	2 1	−8.78E−1	9 1	2.08E−4	3 8	−2.78E−1	2 11	−5.81E−4	7 3	4.14E−1
3 3	−2.49E−2	2 2	−4.60E−2	9 2	8.95E−5	3 9	−4.51E−1	2 12	−7.89E−1	7 4	1.79E−1

C-like Ar (Ar^{12+})

i i'	gf_L	i i'	gf_L	i i'	gf_L	i i'	gf_L	i i'	gf_L	i i'	gf_L
7 5	2.33E−2	11 9	1.98E−3	3 12	−1.22E−2	8 4	2.66E−3	3 1	6.01E−2	8 8	−1.34E+0
7 6	5.71E−1	11 10	2.80E−2	4 1	2.05E−1	8 5	1.46E−3	3 2	−6.95E−7	8 9	−1.47E−4
7 7	−1.91E−8	11 11	2.49E−1	4 2	2.09E−3	8 6	1.09E−4	3 3	−2.84E−6	9 1	6.73E−5
7 8	−3.09E−1	11 12	−5.33E−5	4 3	1.02E−1	8 7	1.56E−1	3 4	−1.03E−1	9 2	6.08E−3
7 9	−3.58E−3	12 1	5.54E−1	4 4	2.75E−1	8 8	1.22E−1	3 5	−2.89E−2	9 3	5.59E−3
7 10	−2.48E−1	12 2	7.85E−6	4 5	2.60E−3	8 9	1.02E−1	3 6	−4.42E−1	9 4	5.07E−1
7 11	−3.84E−2	12 3	8.63E−1	4 6	−4.30E−4	8 10	4.08E−1	3 7	−2.11E−3	9 5	8.54E−3
7 12	−5.35E−3	12 4	1.24E−1	4 7	−2.18E−1	8 11	8.79E−6	3 8	−9.82E−4	9 6	5.67E−2
8 1	6.22E−3	12 5	2.06E−4	4 8	0.00E+0	8 12	−6.85E−6	3 9	−5.08E−2	9 7	2.47E−1
8 2	4.96E−1	12 6	4.25E−3	4 9	−3.60E−9	9 1	4.98E−2	4 1	4.19E−3	9 8	4.52E−2
8 3	5.07E−3	12 7	6.98E−3	4 10	−1.52E−1	9 2	5.11E+0	4 2	−4.98E−4	9 9	−6.93E−4
8 4	1.78E−3	12 8	5.49E−3	4 11	−3.58E−1	9 3	6.18E−3	4 3	−1.23E+0	10 1	6.79E−1
8 5	7.17E−5	12 9	1.05E−4	4 12	−2.83E−3	9 4	7.77E−6	4 4	−1.48E−2	10 2	1.76E−3
8 6	9.59E−1	12 10	4.71E−4	5 1	3.71E−1	9 5	4.18E−5	4 5	−3.19E−3	10 3	1.14E−3
8 7	9.84E−3	12 11	3.38E−5	5 2	2.82E−4	9 6	3.28E−4	4 6	−4.38E−4	10 4	7.84E−4
8 8	1.21E−3	12 12	4.38E−1	5 3	4.14E−1	9 7	2.53E−3	4 7	−2.34E−3	10 5	1.58E−3
8 9	5.78E−3		**$^3D^o$–$^3P^e$**	5 4	1.50E−1	9 8	2.03E−1	4 8	−7.84E−8	10 6	3.34E−3
8 10	−4.25E−3	1 1	4.80E−1	5 5	2.11E−3	9 9	9.68E−3	4 9	−3.09E−3	10 7	2.79E−6
8 11	−1.13E+0	1 2	−1.03E+0	5 6	9.89E−1	9 10	1.27E−3	5 1	2.42E−3	10 8	9.56E−6
8 12	−1.06E−4	1 3	−1.02E−1	5 7	−4.52E−4	9 11	1.03E+0	5 2	1.07E+0	10 9	−1.80E+0
9 1	2.13E−2	1 4	−3.97E−1	5 8	−1.12E−2	9 12	−7.89E−4	5 3	1.32E−3	11 1	9.58E−4
9 2	4.07E+0	1 5	−2.64E−1	5 9	−1.99E−3	10 1	1.67E+0	5 4	−3.65E−1	11 2	2.43E−4
9 3	1.14E−4	1 6	−1.93E−1	5 10	−2.33E−3	10 2	1.68E−2	5 5	−2.27E−3	11 3	3.69E−3
9 4	1.79E−3	1 7	1.08E+0	5 11	4.85E−1	10 3	2.70E+0	5 6	−2.30E−2	11 4	1.60E−1
9 5	6.11E−3	1 8	−5.48E−3	5 12	−8.38E−6	10 4	3.82E−1	5 7	−2.22E−1	11 5	3.47E−1
9 6	5.96E−3	1 9	−9.99E−2	6 1	1.60E−2	10 5	3.64E−4	5 8	−4.87E−4	11 6	5.77E−1
9 7	7.33E−3	1 10	−4.83E−1	6 2	6.86E−1	10 6	1.12E−2	5 9	−1.57E−2	11 7	4.19E−2
9 8	3.80E−1	1 11	−7.40E−3	6 3	1.95E−2	10 7	1.88E−2	6 1	8.45E−1	11 8	9.01E−3
9 9	1.37E−1	1 12	−7.60E−4	6 4	3.24E−3	10 8	2.79E−2	6 2	2.92E−2	11 9	−4.24E−5
9 10	4.78E−1	2 1	8.15E+0	6 5	9.62E−3	10 9	8.85E−6	6 3	8.41E−4	12 1	6.81E−2
9 11	4.42E−3	2 2	6.39E−8	6 6	9.42E−1	10 10	4.00E−4	6 4	−7.61E−6	12 2	8.08E−4
9 12	−1.26E−3	2 3	7.79E−1	6 7	4.34E−3	10 11	4.50E−3	6 5	−1.16E+0	12 3	3.92E−5
10 1	8.83E−2	2 4	−8.69E−3	6 8	1.52E−5	10 12	1.47E+0	6 6	−2.56E−2	12 4	4.20E−6
10 2	1.32E−3	2 5	−4.62E−5	6 9	−1.90E−2		**$^3P^o$–$^3D^e$**	6 7	−7.76E−2	12 5	2.66E−3
10 3	5.61E−1	2 6	−2.22E−3	6 10	−1.26E+0	1 1	−2.23E−2	6 8	−1.28E−1	12 6	1.85E−2
10 4	6.29E−2	2 7	−7.84E−2	6 11	−9.68E−3	1 2	−4.17E−1	6 9	−1.05E−2	12 7	2.14E−6
10 5	4.24E−4	2 8	−1.03E+0	6 12	−3.07E−8	1 3	−3.32E+0	7 1	2.43E−1	12 8	3.76E−4
10 6	5.21E−4	2 9	−1.22E−4	7 1	3.30E−2	1 4	−2.78E+0	7 2	6.02E−3	12 9	8.17E−2
10 7	5.69E−4	2 10	−2.32E−4	7 2	1.77E+0	1 5	−2.85E+0	7 3	1.55E−3		**$^3D^o$–$^3D^e$**
10 8	3.48E−3	2 11	−2.48E−3	7 3	2.09E−4	1 6	−1.78E+0	7 4	6.22E−4	1 1	−9.53E−3
10 9	7.17E−5	2 12	−2.30E−1	7 4	5.83E−6	1 7	−1.02E−2	7 5	−4.87E−2	1 2	−1.02E+0
10 10	1.54E−4	3 1	1.73E+0	7 5	1.14E−1	1 8	−6.64E−1	7 6	−7.55E−1	1 3	−1.07E+0
10 11	6.29E−5	3 2	3.54E−3	7 6	7.04E−3	1 9	−1.88E−2	7 7	−3.10E−1	1 4	−4.52E+0
10 12	−1.42E+0	3 3	3.65E−1	7 7	1.77E−2	2 1	−1.10E+0	7 8	−5.22E−1	1 5	−4.67E−2
11 1	1.73E−2	3 4	8.47E−1	7 8	2.17E−2	2 2	−6.35E−1	7 9	−1.20E−2	1 6	−3.55E−1
11 2	8.99E−1	3 5	−2.13E−3	7 9	1.26E+0	2 3	−1.49E−2	8 1	4.69E−4	1 7	−7.88E−1
11 3	3.55E−5	3 6	−1.93E−3	7 10	−5.85E−4	2 4	−1.66E−5	8 2	5.87E−1	1 8	−2.14E−2
11 4	8.49E−4	3 7	−5.75E−2	7 11	−1.01E−4	2 5	−6.69E−4	8 3	9.31E−3	1 9	−2.25E−4
11 5	1.06E−2	3 8	−1.68E−5	7 12	−5.64E−5	2 6	−4.01E−3	8 4	7.37E−3	2 1	3.28E−1
11 6	1.43E−3	3 9	−9.01E−1	8 1	5.33E−2	2 7	−4.13E−4	8 5	3.99E−3	2 2	−3.41E−4
11 7	1.33E−1	3 10	−5.12E−2	8 2	5.47E+0	2 8	−2.70E−5	8 6	2.73E−3	2 3	−5.17E−3
11 8	1.42E−1	3 11	−7.73E−3	8 3	8.25E−5	2 9	−1.47E+0	8 7	4.44E−4	2 4	−3.86E−1

C-like Ar (Ar^{12+})

i i′	gf_L	*i i′*	gf_L	*i i′*	gf_L	*i i′*	gf_L	*i i′*	gf_L	*i i′*	gf_L
2 5	−1.47E−1	7 2	1.74E−3	1 7	−2.69E−3	1 3	−2.38E+0	9 5	−6.32E−3	2 2	−1.75E−1
2 6	−1.52E−1	7 3	4.12E−1	1 8	−8.88E−4	1 4	−7.33E−1	10 1	3.06E−4	3 1	1.99E−1
2 7	−1.19E−4	7 4	2.10E−3	1 9	−3.96E−1	1 5	−5.92E−3	10 2	1.13E−5	3 2	−5.84E−4
2 8	−6.14E−4	7 5	1.24E−3	2 1	4.72E−1	2 1	−9.61E−3	10 3	1.14E−2	4 1	1.45E+0
2 9	−9.62E−2	7 6	4.43E−2	2 2	1.90E+0	2 2	−2.07E−1	10 4	3.07E−4	4 2	−8.50E−3
3 1	2.39E−3	7 7	4.61E−2	2 3	−2.53E−3	2 3	−2.98E−1	10 5	−7.85E−1	5 1	3.13E−3
3 2	−9.11E−3	7 8	−1.98E−3	2 4	−1.58E−1	2 4	−6.46E−3		**$^3F^o$–$^3F^e$**	5 2	−1.82E+0
3 3	−4.70E−1	7 9	−1.66E−4	2 5	−4.27E−2	2 5	−1.31E+1	1 1	−1.59E−2		**$^3G^o$–$^3G^e$**
3 4	−1.19E−1	8 1	1.43E−3	2 6	−2.48E−2	3 1	−1.85E+0	1 2	−3.96E−1	1 1	1.23E+0
3 5	−4.03E−3	8 2	7.36E−4	2 7	−8.14E−2	3 2	−5.46E−2	1 3	−1.39E+0	1 2	−1.26E−4
3 6	−6.00E−4	8 3	6.06E−2	2 8	−1.02E+0	3 3	−6.00E−3	1 4	−1.62E−2		**$^5S^o$–$^5P^e$**
3 7	−1.03E−2	8 4	1.10E−1	2 9	−1.06E−3	3 4	−5.77E−4	1 5	−1.79E+0	1 1	−7.16E−1
3 8	−3.10E−3	8 5	3.93E−2	3 1	3.99E−4	3 5	−7.47E−4	2 1	2.54E−3	1 2	−8.99E+0
3 9	−1.32E−4	8 6	5.02E−1	3 2	3.27E−3	4 1	1.10E−4	2 2	−4.43E−1	1 3	−6.87E−1
4 1	4.03E−1	8 7	8.95E−1	3 3	1.64E−3	4 2	−1.02E+0	2 3	−1.75E−3	2 1	6.70E−1
4 2	1.45E+0	8 8	3.38E−4	3 4	1.05E−1	4 3	−2.16E−2	2 4	−1.27E+0	2 2	−4.80E−1
4 3	−1.30E−3	8 9	−6.35E−5	3 5	1.25E−2	4 4	−1.02E−1	2 5	−1.80E−2	2 3	−2.85E−1
4 4	−4.51E−1	9 1	8.38E−4	3 6	1.59E−1	4 5	−1.01E−2	3 1	6.73E−2	3 1	7.49E−1
4 5	−2.60E−2	9 2	9.66E−5	3 7	1.20E+0	5 1	4.88E−3	3 2	5.78E−1	3 2	1.14E−2
4 6	−1.83E−2	9 3	4.97E−3	3 8	2.64E−5	5 2	9.13E−5	3 3	7.38E−1	3 3	−1.37E+0
4 7	−1.02E+0	9 4	1.01E−1	3 9	−6.35E−4	5 3	−1.32E+0	3 4	4.11E−1		**$^5P^o$–$^5P^e$**
4 8	−1.19E−1	9 5	4.50E−1	4 1	7.98E−3	5 4	−7.64E−1	3 5	−2.54E−4	1 1	1.42E+0
4 9	−1.55E−4	9 6	2.21E−3	4 2	1.87E−3	5 5	−1.69E−1	4 1	1.28E−2	1 2	−6.74E−1
5 1	1.33E+0	9 7	1.87E−4	4 3	1.44E−3	6 1	1.23E−2	4 2	1.10E−1	1 3	−7.66E−1
5 2	2.30E−2	9 8	4.24E−1	4 4	7.05E−6	6 2	1.85E−2	4 3	6.13E−1		**$^5D^o$–$^5P^e$**
5 3	6.53E−4	9 9	−2.34E−5	4 5	1.03E−1	6 3	2.36E−3	4 4	1.19E−1	1 1	2.15E+0
5 4	−8.66E−4	10 1	9.25E−1	4 6	2.02E−1	6 4	−2.05E+0	4 5	−3.59E−4	1 2	−5.62E−4
5 5	−5.46E−3	10 2	1.56E−2	4 7	7.84E−3	6 5	−3.18E−3	5 1	1.53E−5	1 3	−1.43E+0
5 6	−2.84E−1	10 3	1.53E−4	4 8	1.85E+0	7 1	1.11E+0	5 2	7.14E−4	2 1	6.74E−3
5 7	−3.65E−1	10 4	5.68E−3	4 9	−1.71E−3	7 2	1.05E−2	5 3	6.56E−4	2 2	1.77E−1
5 8	−7.67E−1	10 5	1.96E−2	5 1	5.60E+0	7 3	3.79E−4	5 4	2.02E−4	2 3	2.10E+0
5 9	−7.63E−4	10 6	2.43E−3	5 2	8.50E−2	7 4	7.82E−3	5 5	−1.43E−1		**$^5P^o$–$^5D^e$**
6 1	1.55E−2	10 7	6.69E−5	5 3	8.71E−4	7 5	−4.72E−4		**$^3G^o$–$^3F^e$**	1 1	−1.50E+0
6 2	9.20E−1	10 8	7.05E−4	5 4	1.41E−2	8 1	5.79E−2	1 1	1.14E−3		**$^5D^o$–$^5D^e$**
6 3	5.52E−3	10 9	5.98E−1	5 5	2.59E−2	8 2	5.53E−1	1 2	1.52E−2	1 1	−5.85E−1
6 4	5.60E−3		**$^3F^o$–$^3D^e$**	5 6	6.80E−2	8 3	4.94E−2	1 3	1.94E−1	2 1	9.34E−1
6 5	5.27E−5	1 1	1.57E+0	5 7	4.15E−4	8 4	8.50E−2	1 4	2.36E+0		**$^5D^o$–$^5F^e$**
6 6	1.74E−4	1 2	−1.20E−3	5 8	4.03E−3	8 5	−4.33E−5	1 5	−7.12E−4	1 1	−2.55E+0
6 7	−1.04E+0	1 3	−9.33E−4	5 9	2.77E+0	9 1	3.07E−3		**$^3F^o$–$^3G^e$**	2 1	2.06E+0
6 8	−1.21E−1	1 4	−7.65E−3		**$^3D^o$–$^3F^e$**	9 2	4.70E−1	1 1	−3.20E−1		
6 9	−7.68E−6	1 5	−9.39E−1	1 1	−4.29E+0	9 3	7.83E−1	1 2	−1.80E+1		
7 1	2.98E−5	1 6	−8.79E−1	1 2	−1.12E+1	9 4	1.02E−1	2 1	−2.17E+0		

C-like Ca (Ca^{14+})

Term energies relative to $2s^22p$ ^{2}P ionization threshold for each symmetry

i E(Ryds) Description	i E(Ryds) Description	i E(Ryds) Description	i E(Ryds) Description
$^1S^o$	8 −14.4490 $2s^22p$ ^{2}P $4d$	3 −9.06251 $2s^22p$ ^{2}P $5g$	3 −25.6243 $2s^22p$ ^{2}P $3d$
1 −22.1059 $2s2p^2$ ^{2}P $3p$	9 −10.9259 $2s2p^2$ ^{2}D $4p$	4 −8.72861 $2s2p^2$ ^{2}P $4f$	4 −24.4211 $2s2p^2$ ^{4}P $3p$
2 −17.9425 $2p^3$ ^{2}D $3d$	10 −10.1264 $2s2p^2$ ^{2}D $4f$	5 −6.46753 $2p^3$ ^{2}D $4d$	5 −22.7706 $2s2p^2$ ^{2}D $3p$
3 −9.67581 $2s2p^2$ ^{2}P $4p$	11 −9.43731 $2s2p^2$ ^{2}P $4p$	6 −6.25346 $2s^22p$ ^{2}P $6g$	6 −21.9543 $2s2p^2$ ^{2}S $3p$
4 −6.51577 $2p^3$ ^{2}D $4d$	12 −9.18221 $2s^22p$ ^{2}P $5d$	7 −5.11554 $2s2p^2$ ^{2}D $5f$	7 −21.6990 $2s2p^2$ ^{2}P $3p$
5 −4.12811 $2s2p^2$ ^{2}P $5p$	13 −8.79138 $2s2p^2$ ^{2}P $4f$	8 −4.59571 $2s^22p$ ^{2}P $7g$	8 −19.1544 $2p^3$ ^{2}P $3s$
6 −1.29973 $2p^3$ ^{2}D $5d$	14 −7.36092 $2p^3$ ^{2}D $4s$	9 −3.70474 $2s2p^2$ ^{2}P $5f$	9 −17.4128 $2p^3$ ^{2}D $3d$
7 −1.16554 $2s2p^2$ ^{2}P $6p$	15 −6.35929 $2s^22p$ ^{2}P $6d$	10 −3.51376 $2s^22p$ ^{2}P $8g$	10 −16.7486 $2p^3$ ^{2}P $3d$
$^1P^o$	16 −6.24814 $2p^3$ ^{2}D $4d$	11 −2.77905 $2s^22p$ ^{2}P $9g$	11 −15.4158 $2s^22p$ ^{2}P $4s$
1 −58.7245 $2s2p^3$	17 −5.45883 $2s2p^2$ ^{2}D $5p$	12 −2.35930 $2s2p^2$ ^{2}D $6f$	12 −14.3288 $2s^22p$ ^{2}P $4d$
2 −28.2321 $2s^22p$ ^{2}P $3s$	18 −5.41350 $2p^3$ ^{2}P $4d$	13 −2.34435 $2s2p^2$ ^{2}D $6h$	13 −12.5234 $2s2p^2$ ^{4}P $4p$
3 −25.2009 $2s^22p$ ^{2}P $3d$	19 −5.07153 $2s2p^2$ ^{2}D $5f$	**$^1H^o$**	14 −10.8675 $2s2p^2$ ^{2}D $4p$
4 −22.9448 $2s2p^2$ ^{2}D $3p$	20 −4.65925 $2s^22p$ ^{2}P $7d$	1 −10.1587 $2s2p^2$ ^{2}D $4f$	15 −10.0962 $2s2p^2$ ^{2}D $4f$
5 −21.8813 $2s2p^2$ ^{2}S $3p$	21 −4.02057 $2s2p^2$ ^{2}P $5p$	2 −8.99235 $2s^22p$ ^{2}P $5g$	16 −9.85006 $2s2p^2$ ^{2}S $4p$
6 −21.0081 $2s2p^2$ ^{2}P $3p$	22 −3.70667 $2s2p^2$ ^{2}P $5f$	3 −6.24861 $2s^22p$ ^{2}P $6g$	17 −9.70891 $2s^22p$ ^{2}P $5s$
7 −18.8897 $2p^3$ ^{2}P $3s$	23 −3.56051 $2s^22p$ ^{2}P $8d$	4 −5.09739 $2s2p^2$ ^{2}D $5f$	18 −9.46080 $2s2p^2$ ^{2}P $4p$
8 −17.6162 $2p^3$ ^{2}D $3d$	24 −2.80942 $2s^22p$ ^{2}P $9d$	5 −4.59194 $2s^22p$ ^{2}P $7i$	19 −9.13351 $2s^22p$ ^{2}P $5d$
9 −15.8763 $2p^3$ ^{2}P $3d$	25 −2.55925 $2s2p^2$ ^{2}D $6p$	6 −4.59012 $2s^22p$ ^{2}P $7g$	20 −7.16541 $2s2p^2$ ^{4}P $5p$
10 −15.3458 $2s^22p$ ^{2}P $4s$	26 −2.33977 $2s2p^2$ ^{2}D $6f$	7 −3.51581 $2s^22p$ ^{2}P $8i$	21 −6.62833 $2s^22p$ ^{2}P $6s$
11 −14.1691 $2s^22p$ ^{2}P $4d$	**$^1F^o$**	8 −3.51484 $2s^22p$ ^{2}P $8g$	22 −6.42547 $2p^3$ ^{2}P $4s$
12 −10.8392 $2s2p^2$ ^{2}D $4p$	1 −25.2018 $2s^22p$ ^{2}P $3d$	9 −2.77800 $2s^22p$ ^{2}P $9g$:	23 −6.33021 $2s^22p$ ^{2}P $6d$
13 −10.0732 $2s2p^2$ ^{2}D $4f$	2 −23.0327 $2s2p^2$ ^{2}D $3p$	10 −2.77736 $2s^22p$ ^{2}P $9i$	24 −6.30715 $2p^3$ ^{2}D $4d$
14 −9.78926 $2s2p^2$ ^{2}S $4p$	3 −17.1081 $2p^3$ ^{2}D $3d$	11 −2.34724 $2s2p^2$ ^{2}D $6f$	25 −5.45185 $2s2p^2$ ^{2}D $5p$
15 −9.69438 $2s^22p$ ^{2}P $5s$	4 −16.3798 $2p^3$ ^{2}P $3d$	12 −2.34586 $2s2p^2$ ^{2}D $6h$	26 −5.42585 $2p^3$ ^{2}P $4d$
16 −9.27334 $2s2p^2$ ^{2}P $4p$	5 −14.1747 $2s^22p$ ^{2}P $4d$	**$^1I^o$**	27 −5.07143 $2s2p^2$ ^{2}D $5f$
17 −9.05166 $2s^22p$ ^{2}P $5d$	6 −10.9233 $2s2p^2$ ^{2}D $4p$	1 −4.59199 $2s^22p$ ^{2}P $7i$	28 −4.82736 $2s^22p$ ^{2}P $7s$
18 −6.61097 $2s^22p$ ^{2}P $6s$	7 −10.1827 $2s2p^2$ ^{2}D $4f$	2 −3.51584 $2s^22p$ ^{2}P $8i$	29 −4.64243 $2s^22p$ ^{2}P $7d$
19 −6.42245 $2p^3$ ^{2}D $4d$	8 −9.10937 $2s2p^2$ ^{2}S $4f$	3 −2.77800 $2s^22p$ ^{2}P $9i$	30 −4.35713 $2s2p^2$ ^{2}S $5p$
20 −6.32345 $2p^3$ ^{2}P $4s$	9 −9.06361 $2s^22p$ ^{2}P $5d$	4 −2.34570 $2s2p^2$ ^{2}D $6h$	31 −4.27476 $2s2p^2$ ^{4}P $6p$
21 −6.28531 $2s^22p$ ^{2}P $6d$	10 −8.95219 $2s^22p$ ^{2}P $5g$	**$^3S^o$**	32 −4.06238 $2s2p^2$ ^{2}P $5p$
22 −5.40741 $2s2p^2$ ^{2}D $5p$	11 −8.67170 $2s2p^2$ ^{2}P $4f$	1 −59.4271 $2s2p^3$	33 −3.66729 $2s^22p$ ^{2}P $8s$
23 −5.17398 $2p^3$ ^{2}P $4d$	12 −6.28831 $2s^22p$ ^{2}P $6d$	2 −25.2307 $2s2p^2$ ^{4}P $3p$	34 −3.54890 $2s^22p$ ^{2}P $8d$
24 −5.01386 $2s2p^2$ ^{2}D $5f$	13 −6.24325 $2s^22p$ ^{2}P $6g$	3 −21.6638 $2s2p^2$ ^{2}P $3p$	35 −2.88693 $2s^22p$ ^{2}P $9s$
25 −4.81968 $2s^22p$ ^{2}P $7s$	14 −6.20754 $2p^3$ ^{2}D $4d$	4 −20.5776 $2p^3$ ^{4}S $3s$	36 −2.80126 $2s^22p$ ^{2}P $9d$
26 −4.61535 $2s^22p$ ^{2}P $7d$	15 −5.48005 $2s2p^2$ ^{2}D $5p$	5 −17.2311 $2p^3$ ^{2}D $3d$	37 −2.57800 $2s2p^2$ ^{4}P $7p$
27 −4.32448 $2s2p^2$ ^{2}S $5p$	16 −5.29165 $2p^3$ ^{2}P $4d$	6 −12.8120 $2s2p^2$ ^{4}P $4p$	38 −2.52714 $2s2p^2$ ^{2}D $6p$
28 −3.98051 $2s2p^2$ ^{2}P $5p$	17 −5.09377 $2s2p^2$ ^{2}D $5f$	7 −9.46957 $2s2p^2$ ^{2}P $4p$	39 −2.33411 $2s2p^2$ ^{2}D $6f$
29 −3.65803 $2s^22p$ ^{2}P $8s$	18 −4.61625 $2s^22p$ ^{2}P $7d$	8 −8.34025 $2p^3$ ^{4}S $4s$	**$^3D^o$**
30 −3.53113 $2s^22p$ ^{2}P $8d$	19 −4.58991 $2s^22p$ ^{2}P $7g$	9 −7.26118 $2s2p^2$ ^{4}P $5p$	1 −61.5530 $2s2p^3$
31 −2.88122 $2s^22p$ ^{2}P $9s$	20 −3.96868 $2s2p^2$ ^{2}S $5f$	10 −6.26919 $2p^3$ ^{2}D $4d$	2 −25.6875 $2s^22p$ ^{2}P $3d$
32 −2.78003 $2s^22p$ ^{2}P $9d$	21 −3.67100 $2s2p^2$ ^{2}P $5f$	11 −4.35688 $2s2p^2$ ^{4}P $6p$	3 −24.6415 $2s2p^2$ ^{4}P $3p$
33 −2.53337 $2s2p^2$ ^{2}D $6p$	22 −3.53226 $2s^22p$ ^{2}P $8d$	12 −4.02806 $2s2p^2$ ^{2}P $5p$	4 −22.9934 $2s2p^2$ ^{2}D $3p$
34 −2.32651 $2s2p^2$ ^{2}D $6f$	23 −3.50937 $2s^22p$ ^{2}P $8g$	13 −2.73001 $2p^3$ ^{4}S $5s$	5 −21.8018 $2s2p^2$ ^{2}P $3p$
$^1D^o$	24 −2.78967 $2s^22p$ ^{2}P $9d$	14 −2.56793 $2s2p^2$ ^{4}P $7p$	6 −20.1448 $2p^3$ ^{2}D $3s$
1 −59.4636 $2s2p^3$	25 −2.77601 $2s^22p$ ^{2}P $9g$	15 −1.47440 $2s2p^2$ ^{4}P $8p$	7 −18.5239 $2p^3$ ^{4}S $3d$
2 −25.9872 $2s^22p$ ^{2}P $3d$	26 −2.55869 $2s2p^2$ ^{2}D $6p$	16 −1.19682 $2p^3$ ^{2}D $5d$	8 −17.5553 $2p^3$ ^{2}D $3d$
3 −23.2081 $2s2p^2$ ^{2}D $3p$	27 −2.35308 $2s2p^2$ ^{2}D $6f$	17 −1.09920 $2s2p^2$ ^{2}P $6p$	9 −16.5726 $2p^3$ ^{2}P $3d$
4 −21.3063 $2s2p^2$ ^{2}P $3p$	28 −2.34210 $2s2p^2$ ^{2}D $6h$	18 −.70727 $2s2p^2$ ^{4}P $9p$	10 −14.3479 $2s^22p$ ^{2}P $4d$
5 −19.8730 $2p^3$ ^{2}D $3s$	**$^1G^o$**	**$^3P^o$**	11 −12.6060 $2s2p^2$ ^{4}P $4p$
6 −17.3755 $2p^3$ ^{2}D $3d$	1 −17.8373 $2p^3$ ^{2}D $3d$	1 −60.8179 $2s2p^3$	12 −11.9487 $2s2p^2$ ^{4}P $4f$
7 −16.5407 $2p^3$ ^{2}P $3d$	2 −10.2058 $2s2p^2$ ^{2}D $4f$	2 −28.4488 $2s^22p$ ^{2}P $3s$	13 −10.9404 $2s2p^2$ ^{2}D $4p$

C-like Ca (Ca^{14+})

i	E(Ryds)	Description	i	E(Ryds)	Description	i	E(Ryds)	Description	i	E(Ryds)	Description
14	−10.1506	$2s2p^2$ ^{2}D $4f$	28	−2.77619	$2s^22p$ ^{2}P $9g$	5	−8.47627	$2p^3$ ^{4}S $4s$		**^{1}S^e**	
15	−9.59169	$2s2p^2$ ^{2}P $4p$	29	−2.57265	$2s2p^2$ ^{2}D $6p$	6	−7.20872	$2s2p^2$ ^{4}P $5p$	1	−64.1053	$2s^22p^2$
16	−9.13915	$2s^22p$ ^{2}P $5d$	30	−2.44606	$2s2p^2$ ^{4}P $7f$	7	−4.31038	$2s2p^2$ ^{4}P $6p$	2	−54.0850	$2p^4$
17	−8.78647	$2s2p^2$ ^{2}P $4f$	31	−2.36167	$2s2p^2$ ^{2}D $6f$	8	−2.76583	$2p^3$ ^{4}S $5s$	3	−26.1900	$2s^22p$ ^{2}P $3p$
18	−7.45546	$2p^3$ ^{2}D $4s$	32	−2.34211	$2s2p^2$ ^{2}D $6h$	9	−2.57397	$2s2p^2$ ^{4}P $7p$	4	−22.9192	$2s2p^2$ ^{2}S $3s$
19	−7.32834	$2p^3$ ^{4}S $4d$		**^{3}G^o**		10	−1.45842	$2s2p^2$ ^{4}P $8p$	5	−21.0381	$2s2p^2$ ^{2}D $3d$
20	−7.16027	$2s2p^2$ ^{4}P $5p$	1	−17.8975	$2p^3$ ^{2}D $3d$	11	−.69582	$2s2p^2$ ^{4}P $9p$	6	−16.8406	$2p^3$ ^{2}P $3p$
21	−6.86671	$2s2p^2$ ^{4}P $5f$	2	−11.9472	$2s2p^2$ ^{4}P $4f$		**^{5}P^o**		7	−14.5688	$2s^22p$ ^{2}P $4p$
22	−6.41758	$2p^3$ ^{2}D $4d$	3	−10.2301	$2s2p^2$ ^{2}D $4f$	1	−24.9592	$2s2p^2$ ^{4}P $3p$	8	−10.2834	$2s2p^2$ ^{2}S $4s$
23	−6.33259	$2s^22p$ ^{2}P $6d$	4	−9.06313	$2s^22p$ ^{2}P $5g$	2	−12.7463	$2s2p^2$ ^{4}P $4p$	9	−10.1306	$2s2p^2$ ^{2}D $4d$
24	−5.48645	$2s2p^2$ ^{2}D $5p$	5	−8.73666	$2s2p^2$ ^{2}P $4f$	3	−7.26830	$2s2p^2$ ^{4}P $5p$	10	−9.25753	$2s^22p$ ^{2}P $5p$
25	−5.36946	$2p^3$ ^{2}P $4d$	6	−6.87514	$2s2p^2$ ^{4}P $5f$	4	−4.34013	$2s2p^2$ ^{4}P $6p$	11	−6.40115	$2s^22p$ ^{2}P $6p$
26	−5.09528	$2s2p^2$ ^{2}D $5f$	7	−6.48487	$2p^3$ ^{2}D $4d$	5	−2.59431	$2s2p^2$ ^{4}P $7p$	12	−5.54028	$2p^3$ ^{2}P $4p$
27	−4.64410	$2s^22p$ ^{2}P $7d$	8	−6.25344	$2s^22p$ ^{2}P $6g$	6	−1.47003	$2s2p^2$ ^{4}P $8p$	13	−5.12649	$2s2p^2$ ^{2}D $5d$
28	−4.30165	$2s2p^2$ ^{4}P $6p$	9	−5.13234	$2s2p^2$ ^{2}D $5f$	7	−.70361	$2s2p^2$ ^{4}P $9p$	14	−4.68472	$2s^22p$ ^{2}P $7p$
29	−4.12024	$2s2p^2$ ^{4}P $6f$	10	−4.59553	$2s^22p$ ^{2}P $7g$		**^{5}D^o**		15	−4.53097	$2s2p^2$ ^{2}S $5s$
30	−4.08303	$2s2p^2$ ^{2}P $5p$	11	−4.11826	$2s2p^2$ ^{4}P $6f$	1	−25.0865	$2s2p^2$ ^{4}P $3p$	16	−3.57818	$2s^22p$ ^{2}P $8p$
31	−3.70395	$2s2p^2$ ^{2}P $5f$	12	−4.11178	$2s2p^2$ ^{4}P $6h$	2	−18.9848	$2p^3$ ^{4}S $3d$	17	−2.82181	$2s^22p$ ^{2}P $9p$
32	−3.55010	$2s^22p$ ^{2}P $8d$	13	−3.71101	$2s2p^2$ ^{2}P $5f$	3	−12.7893	$2s2p^2$ ^{4}P $4p$	18	−2.36827	$2s2p^2$ ^{2}D $6d$
33	−2.80223	$2s^22p$ ^{2}P $9d$	14	−3.51409	$2s^22p$ ^{2}P $8g$	4	−11.9714	$2s2p^2$ ^{4}P $4f$		**^{1}P^e**	
34	−2.59458	$2s2p^2$ ^{4}P $7p$	15	−2.77906	$2s^22p$ ^{2}P $9g$	5	−7.54925	$2p^3$ ^{4}S $4d$	1	−27.4242	$2s^22p$ ^{2}P $3p$
35	−2.54444	$2s2p^2$ ^{2}D $6p$	16	−2.45767	$2s2p^2$ ^{4}P $7f$	6	−7.26888	$2s2p^2$ ^{4}P $5p$	2	−22.8236	$2s2p^2$ ^{2}P $3s$
36	−2.46153	$2s2p^2$ ^{4}P $7f$	17	−2.45566	$2s2p^2$ ^{4}P $7h$	7	−6.88702	$2s2p^2$ ^{4}P $5f$	3	−21.3126	$2s2p^2$ ^{2}D $3d$
37	−2.34770	$2s2p^2$ ^{2}D $6f$	18	−2.36767	$2s2p^2$ ^{2}D $6f$	8	−4.34998	$2s2p^2$ ^{4}P $6p$	4	−20.4130	$2s2p^2$ ^{2}P $3d$
	^{3}F^o		19	−2.34436	$2s2p^2$ ^{2}D $6h$	9	−4.12800	$2s2p^2$ ^{4}P $6f$	5	−19.1754	$2p^3$ ^{2}D $3p$
1	−25.9739	$2s^22p$ ^{2}P $3d$		**^{3}H^o**		10	−2.60219	$2s2p^2$ ^{4}P $7p$	6	−17.9408	$2p^3$ ^{2}P $3p$
2	−23.1900	$2s2p^2$ ^{2}D $3p$	1	−10.1768	$2s2p^2$ ^{2}D $4f$	11	−2.46443	$2s2p^2$ ^{4}P $7f$	7	−14.9804	$2s^22p$ ^{2}P $4p$
3	−17.9873	$2p^3$ ^{2}D $3d$	2	−8.99255	$2s^22p$ ^{2}P $5g$	12	−2.29274	$2p^3$ ^{4}S $5d$	8	−10.2587	$2s2p^2$ ^{2}D $4d$
4	−16.8636	$2p^3$ ^{2}P $3d$	3	−6.24920	$2s^22p$ ^{2}P $6g$	13	−1.47371	$2s2p^2$ ^{4}P $8p$	9	−10.0357	$2s2p^2$ ^{2}P $4s$
5	−14.4465	$2s^22p$ ^{2}P $4d$	4	−5.10929	$2s2p^2$ ^{2}D $5f$	14	−1.38377	$2s2p^2$ ^{4}P $8f$	10	−9.42139	$2s^22p$ ^{2}P $5p$
6	−11.8793	$2s2p^2$ ^{4}P $4f$	5	−4.59035	$2s^22p$ ^{2}P $7g$:	15	−.70640	$2s2p^2$ ^{4}P $9p$	11	−9.05672	$2s2p^2$ ^{2}P $4d$
7	−10.9781	$2s2p^2$ ^{2}D $4p$	6	−4.59035	$2s^22p$ ^{2}P $7i$	16	−.64368	$2s2p^2$ ^{4}P $9f$	12	−6.97540	$2p^3$ ^{2}D $4p$
8	−10.2096	$2s2p^2$ ^{2}D $4f$	7	−4.10855	$2s2p^2$ ^{4}P $6h$		**^{5}F^o**		13	−6.50135	$2s^22p$ ^{2}P $6p$
9	−9.19307	$2s^22p$ ^{2}P $5d$	8	−3.51562	$2s^22p$ ^{2}P $8g$	1	−11.9158	$2s2p^2$ ^{4}P $4f$	14	−6.18317	$2p^3$ ^{2}D $4f$
10	−9.12059	$2s2p^2$ ^{2}S $4f$	9	−3.51506	$2s^22p$ ^{2}P $8i$	2	−6.86682	$2s2p^2$ ^{4}P $5f$	15	−5.92086	$2p^3$ ^{2}P $4p$
11	−8.96393	$2s^22p$ ^{2}P $5g$	10	−2.77778	$2s^22p$ ^{2}P $9g$	3	−4.11598	$2s2p^2$ ^{4}P $6f$	16	−5.14488	$2s2p^2$ ^{2}D $5d$
12	−8.68594	$2s2p^2$ ^{2}P $4f$	11	−2.77751	$2s^22p$ ^{2}P $9i$	4	−2.45638	$2s2p^2$ ^{4}P $7f$	17	−4.75174	$2s^22p$ ^{2}P $7p$
13	−6.84346	$2s2p^2$ ^{4}P $5f$	12	−2.45054	$2s2p^2$ ^{4}P $7h$	5	−1.37903	$2s2p^2$ ^{4}P $8f$	18	−4.29037	$2s2p^2$ ^{2}P $5s$
14	−6.51471	$2p^3$ ^{2}S $4d$	13	−2.35488	$2s2p^2$ ^{2}D $6f$	6	−.64036	$2s2p^2$ ^{4}P $9f$	19	−3.84317	$2s2p^2$ ^{2}P $5d$
15	−6.35864	$2s^22p$ ^{2}P $6d$	14	−2.34587	$2s2p^2$ ^{2}D $6h$		**^{5}G^o**		20	−3.61511	$2s^22p$ ^{2}P $8p$
16	−6.24355	$2s^22p$ ^{2}P $6g$		**^{3}I^o**		1	−11.9775	$2s2p^2$ ^{4}P $4f$	21	−2.84928	$2s^22p$ ^{2}P $9p$
17	−5.54281	$2s2p^2$ ^{2}D $5p$	1	−4.59199	$2s^22p$ ^{2}P $7i$	2	−6.89332	$2s2p^2$ ^{4}P $5f$	22	−2.37659	$2s2p^2$ ^{2}D $6d$
18	−5.41762	$2p^3$ ^{2}P $4d$	2	−4.11084	$2s2p^2$ ^{4}P $6h$	3	−4.13113	$2s2p^2$ ^{4}P $6f$		**^{1}D^e**	
19	−5.12147	$2s2p^2$ ^{2}D $5f$	3	−3.51584	$2s^22p$ ^{2}P $8i$	4	−4.11179	$2s2p^2$ ^{4}P $6h$	1	−64.8885	$2s^22p^2$
20	−4.65964	$2s^22p$ ^{2}P $7d$	4	−2.77800	$2s^22p$ ^{2}P $9i$	5	−2.46582	$2s2p^2$ ^{4}P $7f$	2	−55.5101	$2p^4$
21	−4.59021	$2s^22p$ ^{2}P $7g$	5	−2.45263	$2s2p^2$ ^{4}P $7h$	6	−2.45569	$2s2p^2$ ^{4}P $7h$	3	−26.6395	$2s^22p$ ^{2}P $3p$
22	−4.10025	$2s2p^2$ ^{4}P $6f$	6	−2.34570	$2s2p^2$ ^{2}D $6h$	7	−2.12079	$2p^3$ ^{4}S $5g$	4	−24.0140	$2s2p^2$ ^{2}D $3s$
23	−3.98380	$2s2p^2$ ^{2}S $5f$		**^{5}S^o**		8	−1.38529	$2s2p^2$ ^{4}P $8f$	5	−21.3948	$2s2p^2$ ^{2}D $3d$
24	−3.68283	$2s2p^2$ ^{2}P $5f$	1	−63.6116	$2s2p^3$	9	−1.37584	$2s2p^2$ ^{4}P $8h$	6	−20.4492	$2s2p^2$ ^{2}S $3d$
25	−3.56054	$2s^22p$ ^{2}P $8d$	2	−24.5465	$2s2p^2$ ^{4}P $3p$	10	−.64473	$2s2p^2$ ^{4}P $9f$	7	−19.7362	$2s2p^2$ ^{2}P $3d$
26	−3.50994	$2s^22p$ ^{2}P $8g$	3	−21.3165	$2p^3$ ^{4}S $3s$	11	−.63838	$2s2p^2$ ^{4}P $9h$	8	−18.2110	$2p^3$ ^{2}D $3p$
27	−2.80957	$2s^22p$ ^{2}P $9d$	4	−12.6261	$2s2p^2$ ^{4}P $4p$				9	−17.5885	$2p^3$ ^{2}P $3p$

C-like Ca (Ca^{14+})

i	E(Ryds)	Description	i	E(Ryds)	Description	i	E(Ryds)	Description	i	E(Ryds)	Description
10	−14.7245	$2s^22p$ ^{2}P $4p$	3	−10.3746	$2s2p^2$ ^{2}D $4d$	16	−2.39920	$2s2p^2$ ^{2}D $6d$	13	−10.4349	$2s2p^2$ ^{2}D $4d$
11	−14.0270	$2s^22p$ ^{2}P $4f$	4	−9.00473	$2s^22p$ ^{2}P $5f$			**^{3}P^e**	14	−9.46326	$2s^22p$ ^{2}P $5p$
12	−11.3406	$2s2p^2$ ^{2}D $4s$	5	−6.25381	$2s^22p$ ^{2}P $6f$	1	−65.5914	$2s^22p^2$	15	−9.31727	$2s2p^2$ ^{2}S $4d$
13	−10.3100	$2s2p^2$ ^{2}D $4d$	6	−6.24996	$2s^22p$ ^{2}P $6h$	2	−56.1568	$2p^4$	16	−9.05001	$2s2p^2$ ^{2}P $4d$
14	−9.35779	$2s^22p$ ^{2}P $5p$	7	−6.18591	$2p^3$ ^{2}D $4f$	3	−27.0364	$2s^22p$ ^{2}P $3p$	17	−8.96856	$2s^22p$ ^{2}P $5f$
15	−9.22638	$2s2p^2$ ^{2}S $4d$	8	−5.21130	$2s2p^2$ ^{2}D $5d$	4	−25.5943	$2s2p^2$ ^{4}P $3s$	18	−6.98834	$2p^3$ ^{2}D $4p$
16	−8.99773	$2s^22p$ ^{2}P $5f$	9	−5.18663	$2p^3$ ^{2}P $4f$	5	−23.4748	$2s2p^2$ ^{4}P $3d$	19	−6.88037	$2s2p^2$ ^{4}P $5d$
17	−8.78296	$2s2p^2$ ^{2}P $4d$	10	−5.06877	$2s2p^2$ ^{2}D $5g$	6	−22.9955	$2s2p^2$ ^{2}P $3s$	20	−6.49190	$2s^22p$ ^{2}P $6p$
18	−6.66501	$2p^3$ ^{2}D $4p$	11	−4.59251	$2s^22p$ ^{2}P $7f$	7	−21.5038	$2s2p^2$ ^{2}D $3d$	21	−6.25193	$2s^22p$ ^{2}P $6f$
19	−6.43948	$2s^22p$ ^{2}P $6p$	12	−4.59251	$2s^22p$ ^{2}P $7h$	8	−20.2199	$2s2p^2$ ^{2}P $3d$	22	−6.18720	$2p^3$ ^{2}D $4f$
20	−6.24528	$2s^22p$ ^{2}P $6f$	13	−3.96038	$2s2p^2$ ^{2}S $5g$	9	−19.8292	$2p^3$ ^{4}S $3p$	23	−5.95728	$2p^3$ ^{2}P $4p$
21	−6.18098	$2p^3$ ^{2}D $4f$	14	−3.66407	$2s2p^2$ ^{2}P $5g$	10	−18.4616	$2p^3$ ^{2}D $3p$	24	−5.72025	$2s2p^2$ ^{2}D $5s$
22	−5.82108	$2p^3$ ^{2}P $4p$	15	−3.51757	$2s^22p$ ^{2}P $8f$	11	−17.7557	$2p^3$ ^{2}P $3p$	25	−5.23228	$2s2p^2$ ^{2}D $5d$
23	−5.64210	$2s2p^2$ ^{2}D $5s$	16	−3.51071	$2s^22p$ ^{2}P $8h$	12	−14.8331	$2s^22p$ ^{2}P $4p$	26	−5.22175	$2p^3$ ^{2}P $4f$
24	−5.20608	$2p^3$ ^{2}P $4f$	17	−2.77924	$2s^22p$ ^{2}P $9f$	13	−13.0830	$2s2p^2$ ^{4}P $4s$	27	−5.00599	$2s2p^2$ ^{2}D $5g$
25	−5.17433	$2s2p^2$ ^{2}D $5d$	18	−2.77691	$2s^22p$ ^{2}P $9h$	14	−12.1654	$2s2p^2$ ^{4}P $4d$	28	−4.74105	$2s^22p$ ^{2}P $7p$
26	−4.99467	$2s2p^2$ ^{2}D $5g$	19	−2.41095	$2s2p^2$ ^{2}D $6d$	15	−10.3649	$2s2p^2$ ^{2}D $4d$	29	−4.59455	$2s^22p$ ^{2}P $7f$
27	−4.71100	$2s^22p$ ^{2}P $7p$	20	−2.34761	$2s2p^2$ ^{2}D $6g$	16	−10.0589	$2s2p^2$ ^{2}P $4s$	30	−4.14824	$2s2p^2$ ^{4}P $6d$
28	−4.58976	$2s^22p$ ^{2}P $7f$			**^{1}H^e**	17	−9.37962	$2s^22p$ ^{2}P $5p$	31	−4.09364	$2s2p^2$ ^{2}S $5d$
29	−4.06533	$2s2p^2$ ^{2}S $5d$	1	−6.25337	$2s^22p$ ^{2}P $6h$	18	−8.97082	$2s2p^2$ ^{2}P $4d$	32	−3.83151	$2s2p^2$ ^{2}P $5d$
30	−3.73149	$2s2p^2$ ^{2}P $5d$	2	−6.21649	$2p^3$ ^{2}D $4f$	19	−7.85587	$2p^3$ ^{4}S $4p$	33	−3.61131	$2s^22p$ ^{2}P $8p$
31	−3.58841	$2s^22p$ ^{2}P $8p$	3	−5.09730	$2s2p^2$ ^{2}D $5g$	20	−7.42764	$2s2p^2$ ^{4}P $5s$	34	−3.51647	$2s^22p$ ^{2}P $8f$
32	−3.51254	$2s^22p$ ^{2}P $8f$	4	−4.59419	$2s^22p$ ^{2}P $7h$	21	−6.97697	$2s2p^2$ ^{4}P $5d$	35	−2.84786	$2s^22p$ ^{2}P $9p$
33	−2.83257	$2s^22p$ ^{2}P $9p$	5	−3.68453	$2s2p^2$ ^{2}P $5g$	22	−6.79426	$2p^3$ ^{2}D $4p$	36	−2.77904	$2s^22p$ ^{2}P $9f$
34	−2.77663	$2s^22p$ ^{2}P $9f$	6	−3.51263	$2s^22p$ ^{2}P $8h$	23	−6.46273	$2s^22p$ ^{2}P $6p$	37	−2.71516	$2s2p^2$ ^{2}D $6s$
35	−2.68140	$2s2p^2$ ^{2}D $6s$	7	−2.77832	$2s^22p$ ^{2}P $9h$	24	−6.16945	$2p^3$ ^{2}D $4f$	38	−2.46211	$2s2p^2$ ^{4}P $7d$
36	−2.39301	$2s2p^2$ ^{2}D $6d$	8	−2.34947	$2s2p^2$ ^{2}D $6g$	25	−5.84998	$2p^3$ ^{2}P $4p$	39	−2.42353	$2s2p^2$ ^{2}D $6d$
37	−2.33281	$2s2p^2$ ^{2}D $6g$			**^{1}I^e**	26	−5.20136	$2s2p^2$ ^{2}D $5d$	40	−2.33326	$2s2p^2$ ^{2}D $6g$
		^{1}F^e	1	−6.25016	$2s^22p$ ^{2}P $6h$	27	−4.72777	$2s^22p$ ^{2}P $7p$			**^{3}F^e**
1	−21.7523	$2s2p^2$ ^{2}D $3d$	2	−5.08960	$2s2p^2$ ^{2}D $5g$	28	−4.44209	$2s2p^2$ ^{4}P $6s$	1	−23.2446	$2s2p^2$ ^{4}P $3d$
2	−20.0681	$2s2p^2$ ^{2}P $3d$	3	−4.59064	$2s^22p$ ^{2}P $7h$	29	−4.31067	$2s2p^2$ ^{2}P $5s$	2	−21.8580	$2s2p^2$ ^{2}D $3d$
3	−18.9792	$2p^3$ ^{2}D $3p$	4	−3.51530	$2s^22p$ ^{2}P $8h$	30	−4.18043	$2s2p^2$ ^{4}P $6d$	3	−20.6382	$2s2p^2$ ^{2}P $3d$
4	−14.1186	$2s^22p$ ^{2}P $4f$	5	−2.77790	$2s^22p$ ^{2}P $9h$	31	−3.80119	$2s2p^2$ ^{2}P $5d$	4	−19.0483	$2p^3$ ^{2}D $3p$
5	−10.3934	$2s2p^2$ ^{2}D $4d$	6	−2.77768	$2s^22p$ ^{2}P $8j$	32	−3.60142	$2s^22p$ ^{2}P $8p$	5	−14.1163	$2s^22p$ ^{2}P $4f$
6	−9.07324	$2s^22p$ ^{2}P $5f$	7	−2.34191	$2s2p^2$ ^{2}D $6g$	33	−2.84042	$2s^22p$ ^{2}P $9p$	6	−12.0656	$2s2p^2$ ^{4}P $4d$
7	−8.89915	$2s2p^2$ ^{2}P $4d$	8	−2.25000	$2s^22p$ ^{2}P $9j$	34	−2.66518	$2s2p^2$ ^{4}P $7s$	7	−10.4923	$2s2p^2$ ^{2}D $4d$
8	−6.93164	$2p^3$ ^{2}D $4p$			**^{3}S^e**	35	−2.50368	$2s2p^2$ ^{4}P $7d$	8	−9.12995	$2s2p^2$ ^{2}P $4d$
9	−6.27088	$2s^22p$ ^{2}P $6f$	1	−27.1249	$2s^22p$ ^{2}P $3p$	36	−2.43220	$2p^3$ ^{4}S $5p$	9	−9.00704	$2s^22p$ ^{2}P $5f$
10	−6.17466	$2p^3$ ^{2}D $4f$	2	−23.2731	$2s2p^2$ ^{2}S $3s$	37	−2.40162	$2s2p^2$ ^{2}D $6d$	10	−7.23032	$2p^3$ ^{4}S $4f$
11	−5.21806	$2p^3$ ^{2}P $4f$	3	−21.4054	$2s2p^2$ ^{2}D $3d$			**^{3}D^e**	11	−6.95886	$2p^3$ ^{2}D $4p$
12	−5.20658	$2s2p^2$ ^{2}D $5d$	4	−18.1492	$2p^3$ ^{2}P $3p$	1	−27.2852	$2s^22p$ ^{2}P $3p$	12	−6.93364	$2s2p^2$ ^{4}P $5d$
13	−5.03426	$2s2p^2$ ^{2}D $5g$	5	−14.8797	$2s^22p$ ^{2}P $4p$	2	−24.4108	$2s2p^2$ ^{2}D $3s$	13	−6.82976	$2s2p^2$ ^{4}P $5g$
14	−4.60658	$2s^22p$ ^{2}P $7f$	6	−10.4045	$2s2p^2$ ^{2}S $4s$	3	−22.8178	$2s2p^2$ ^{4}P $3d$	14	−6.20880	$2s^22p$ ^{2}P $6f$
15	−3.78398	$2s2p^2$ ^{2}P $5d$	7	−10.2833	$2s2p^2$ ^{2}D $4d$	4	−21.7652	$2s2p^2$ ^{2}D $3d$	15	−6.19899	$2p^3$ ^{2}D $4f$
16	−3.67900	$2s2p^2$ ^{2}P $5g$	8	−9.38661	$2s^22p$ ^{2}P $5p$	5	−20.6792	$2s2p^2$ ^{2}S $3d$	16	−5.26285	$2s2p^2$ ^{2}D $5d$
17	−3.52322	$2s^22p$ ^{2}P $8f$	9	−6.47855	$2s^22p$ ^{2}P $6p$	6	−20.4247	$2s2p^2$ ^{2}P $3d$	17	−5.21255	$2p^3$ ^{2}P $4f$
18	−2.78430	$2s^22p$ ^{2}P $9f$	10	−5.06286	$2p^3$ ^{2}P $4p$	7	−19.1570	$2p^3$ ^{2}D $3p$	18	−5.03604	$2s2p^2$ ^{2}D $5g$
19	−2.40901	$2s2p^2$ ^{2}D $6d$	11	−5.18547	$2s2p^2$ ^{2}D $5d$	8	−18.0515	$2p^3$ ^{2}P $3p$	19	−4.60489	$2s^22p$ ^{2}P $7f$
20	−2.34088	$2s2p^2$ ^{2}D $6g$	12	−4.74267	$2s^22p$ ^{2}P $7p$	9	−14.9354	$2s^22p$ ^{2}P $4p$	20	−4.15183	$2s2p^2$ ^{4}P $6d$
		^{1}G^e	13	−4.60163	$2s2p^2$ ^{2}S $5s$	10	−14.0431	$2s^22p$ ^{2}P $4f$	21	−4.11228	$2s2p^2$ ^{4}P $6g$
1	−21.5793	$2s2p^2$ ^{2}D $3d$	14	−3.60826	$2s^22p$ ^{2}P $8p$	11	−11.9455	$2s2p^2$ ^{4}P $4d$	22	−3.85716	$2s2p^2$ ^{2}P $5d$
2	−14.0654	$2s^22p$ ^{2}P $4f$	15	−2.84316	$2s^22p$ ^{2}P $9p$	12	−11.4804	$2s2p^2$ ^{2}D $4s$	23	−3.67880	$2s2p^2$ ^{2}P $5g$

C-like Ca (Ca^{14+})

i	Energy(Ryds)	Description	i	Energy(Ryds)	Description	i	Energy(Ryds)	Description
24	−3.52362	$2s^22p\ ^2\mathrm{P}\ 8f$	9	−2.77833	$2s^22p\ ^2\mathrm{P}\ 9h$	7	−.66710	$2s2p^2\ ^4\mathrm{P}\ 9d$
25	−2.78402	$2s^22p\ ^2\mathrm{P}\ 9f$	10	−2.45391	$2s2p^2\ ^4\mathrm{P}\ 7g$		$^5\mathbf{F}^e$	
26	−2.49507	$2s2p^2\ ^4\mathrm{P}\ 7d$	11	−2.45189	$2s2p^2\ ^4\mathrm{P}\ 7i$	1	−23.9078	$2s2p^2\ ^4\mathrm{P}\ 3d$
27	−2.45654	$2s2p^2\ ^4\mathrm{P}\ 7g$	12	−2.34990	$2s2p^2\ ^2\mathrm{D}\ 6g$	2	−12.3315	$2s2p^2\ ^4\mathrm{P}\ 4d$
28	−2.42497	$2s2p^2\ ^2\mathrm{D}\ 6d$		$^3\mathbf{I}^e$		3	−7.25509	$2p^3\ ^4\mathrm{S}\ 4f$
29	−2.34133	$2s2p^2\ ^2\mathrm{D}\ 6g$	1	−6.25017	$2s^22p\ ^2\mathrm{P}\ 6h$	4	−7.05939	$2s2p^2\ ^4\mathrm{P}\ 5d$
	$^3\mathbf{G}^e$		2	−5.09008	$2s2p^2\ ^2\mathrm{D}\ 5g$	5	−6.83355	$2s2p^2\ ^4\mathrm{P}\ 5g$
1	−21.9450	$2s2p^2\ ^2\mathrm{D}\ 3d$	3	−4.59065	$2s^22p\ ^2\mathrm{P}\ 7h$	6	−4.22467	$2s2p^2\ ^4\mathrm{P}\ 6d$
2	−14.0838	$2s^22p\ ^2\mathrm{P}\ 4f$	4	−3.51531	$2s^22p\ ^2\mathrm{P}\ 8h$	7	−4.11270	$2s2p^2\ ^4\mathrm{P}\ 6g$
3	−10.4984	$2s2p^2\ ^2\mathrm{D}\ 4d$	5	−2.77791	$2s^22p\ ^2\mathrm{P}\ 9h$	8	−2.52340	$2s2p^2\ ^4\mathrm{P}\ 7d$
4	−9.01788	$2s^22p\ ^2\mathrm{P}\ 5f$	6	−2.77769	$2s^22p\ ^2\mathrm{P}\ 8j$	9	−2.45667	$2s2p^2\ ^4\mathrm{P}\ 7g$
5	−6.85005	$2s2p^2\ ^4\mathrm{P}\ 5g$	7	−2.45184	$2s2p^2\ ^4\mathrm{P}\ 7i$	10	−2.14211	$2p^3\ ^4\mathrm{S}\ 5f$
6	−6.26325	$2s^22p\ ^2\mathrm{P}\ 6f$	8	−2.34237	$2s2p^2\ ^2\mathrm{D}\ 6g$	11	−1.42317	$2s2p^2\ ^4\mathrm{P}\ 8d$
7	−6.25011	$2s^22p\ ^2\mathrm{P}\ 6h$	9	−2.25000	$2s^22p\ ^2\mathrm{P}\ 9j$	12	−1.37656	$2s2p^2\ ^4\mathrm{P}\ 8g$
8	−6.21596	$2p^3\ ^2\mathrm{D}\ 4f$		$^5\mathbf{P}^e$		13	−.67109	$2s2p^2\ ^4\mathrm{P}\ 9d$
9	−5.26894	$2s2p^2\ ^2\mathrm{D}\ 5d$	1	−26.2736	$2s2p^2\ ^4\mathrm{P}\ 3s$	14	−.63890	$2s2p^2\ ^4\mathrm{P}\ 9g$
10	−5.19767	$2p^3\ ^2\mathrm{P}\ 4f$	2	−23.5024	$2s2p^2\ ^4\mathrm{P}\ 3d$		$^5\mathbf{G}^e$	
11	−5.07400	$2s2p^2\ ^2\mathrm{D}\ 5g$	3	−20.1784	$2p^3\ ^4\mathrm{S}\ 3p$	1	−6.85097	$2s2p^2\ ^4\mathrm{P}\ 5g$
12	−4.60021	$2s^22p\ ^2\mathrm{P}\ 7f$	4	−13.2799	$2s2p^2\ ^4\mathrm{P}\ 4s$	2	−4.10518	$2s2p^2\ ^4\mathrm{P}\ 6g$
13	−4.59252	$2s^22p\ ^2\mathrm{P}\ 7h$	5	−12.1917	$2s2p^2\ ^4\mathrm{P}\ 4d$	3	−2.44905	$2s2p^2\ ^4\mathrm{P}\ 7g$
14	−4.10428	$2s2p^2\ ^4\mathrm{P}\ 6g$	6	−8.01456	$2p^3\ ^4\mathrm{S}\ 4p$	4	−1.37388	$2s2p^2\ ^4\mathrm{P}\ 8g$
15	−3.96119	$2s2p^2\ ^2\mathrm{S}\ 5g$	7	−7.51894	$2s2p^2\ ^4\mathrm{P}\ 5s$	5	−.63663	$2s2p^2\ ^4\mathrm{P}\ 9g$
16	−3.66447	$2s2p^2\ ^2\mathrm{P}\ 5g$	8	−6.99382	$2s2p^2\ ^4\mathrm{P}\ 5d$		$^5\mathbf{H}^e$	
17	−3.52176	$2s^22p\ ^2\mathrm{P}\ 8f$	9	−4.48957	$2s2p^2\ ^4\mathrm{P}\ 6s$	1	−6.86543	$2s2p^2\ ^4\mathrm{P}\ 5g$
18	−3.51073	$2s^22p\ ^2\mathrm{P}\ 8h$	10	−4.18847	$2s2p^2\ ^4\mathrm{P}\ 6d$	2	−4.11360	$2s2p^2\ ^4\mathrm{P}\ 6g$
19	−2.78239	$2s^22p\ ^2\mathrm{P}\ 9f$	11	−2.69223	$2s2p^2\ ^4\mathrm{P}\ 7s$	3	−2.45443	$2s2p^2\ ^4\mathrm{P}\ 7g$
20	−2.77692	$2s^22p\ ^2\mathrm{P}\ 9h$	12	−2.52862	$2p^3\ ^4\mathrm{S}\ 5p$	4	−2.45194	$2s2p^2\ ^4\mathrm{P}\ 7i$
21	−2.44852	$2s2p^2\ ^4\mathrm{P}\ 7g$	13	−2.49383	$2s2p^2\ ^4\mathrm{P}\ 7d$	5	−1.37753	$2s2p^2\ ^4\mathrm{P}\ 8g$
22	−2.44107	$2s2p^2\ ^2\mathrm{D}\ 6d$	14	−1.53147	$2s2p^2\ ^4\mathrm{P}\ 8s$	6	−1.37576	$2s2p^2\ ^4\mathrm{P}\ 8i$
23	−2.34807	$2s2p^2\ ^2\mathrm{D}\ 6g$	15	−1.40865	$2s2p^2\ ^4\mathrm{P}\ 8d$	7	−.63922	$2s2p^2\ ^4\mathrm{P}\ 9g$
	$^3\mathbf{H}^e$		16	−.74679	$2s2p^2\ ^4\mathrm{P}\ 9s$	8	−.63794	$2s2p^2\ ^4\mathrm{P}\ 9i$
1	−6.86469	$2s2p^2\ ^4\mathrm{P}\ 5g$	17	−.66108	$2s2p^2\ ^4\mathrm{P}\ 9d$			
2	−6.25341	$2s^22p\ ^2\mathrm{P}\ 6h$		$^5\mathbf{D}^e$				
3	−6.21811	$2p^3\ ^2\mathrm{D}\ 4f$	1	−23.7305	$2s2p^2\ ^4\mathrm{P}\ 3d$			
4	−5.09820	$2s2p^2\ ^2\mathrm{D}\ 5g$	2	−12.2719	$2s2p^2\ ^4\mathrm{P}\ 4d$			
5	−4.59419	$2s^22p\ ^2\mathrm{P}\ 7h$	3	−7.03681	$2s2p^2\ ^4\mathrm{P}\ 5d$			
6	−4.11288	$2s2p^2\ ^4\mathrm{P}\ 6g$	4	−4.21065	$2s2p^2\ ^4\mathrm{P}\ 6d$			
7	−3.68468	$2s2p^2\ ^2\mathrm{P}\ 5g$	5	−2.51462	$2s2p^2\ ^4\mathrm{P}\ 7d$			
8	−3.51264	$2s^22p\ ^2\mathrm{P}\ 8h$	6	−1.41747	$2s2p^2\ ^4\mathrm{P}\ 8d$			

C-like Ca (Ca^{14+})

Energies in ascending order from ground state for terms with effective $n \leq 4.0$, $L \leq 4$

Term	i	E(Ryds)	Term	i	E(Ryds)	Term	i	E(Ryds)	Term	i	E(Ryds)	Term	i	E(Ryds)
$^3\mathbf{P}^e$	1	0.00000	$^1\mathbf{F}^o$	1	40.3896	$^3\mathbf{G}^e$	1	43.6464	$^1\mathbf{D}^o$	5	45.7184	$^1\mathbf{F}^o$	3	48.4833
$^1\mathbf{D}^e$	1	0.70290	$^1\mathbf{P}^o$	3	40.3905	$^1\mathbf{P}^o$	5	43.7101	$^3\mathbf{P}^e$	9	45.7622	$^3\mathbf{F}^o$	4	48.7278
$^1\mathbf{S}^e$	1	1.48610	$^5\mathbf{D}^o$	1	40.5049	$^3\mathbf{D}^o$	5	43.7296	$^1\mathbf{D}^e$	7	45.8552	$^1\mathbf{S}^e$	6	48.7508
$^5\mathbf{S}^o$	1	1.97980	$^5\mathbf{P}^o$	1	40.6322	$^3\mathbf{F}^e$	2	43.7334	$^1\mathbf{P}^e$	5	46.4160	$^3\mathbf{P}^o$	10	48.8428
$^3\mathbf{D}^o$	1	4.03840	$^3\mathbf{D}^o$	3	40.9499	$^3\mathbf{D}^e$	4	43.8262	$^3\mathbf{D}^e$	7	46.4335	$^3\mathbf{D}^o$	9	49.0188
$^3\mathbf{P}^o$	1	4.77350	$^5\mathbf{S}^o$	2	41.0449	$^1\mathbf{F}^e$	1	43.8391	$^3\mathbf{P}^o$	8	46.4370	$^1\mathbf{D}^o$	7	49.0507
$^1\mathbf{D}^o$	1	6.12780	$^3\mathbf{P}^o$	4	41.1703	$^3\mathbf{P}^o$	7	43.8924	$^3\mathbf{F}^e$	4	46.5431	$^1\mathbf{F}^o$	4	49.2116
$^3\mathbf{S}^o$	1	6.16430	$^3\mathbf{D}^e$	2	41.1806	$^3\mathbf{S}^o$	3	43.9276	$^5\mathbf{D}^o$	2	46.6066	$^1\mathbf{P}^o$	9	49.7151
$^1\mathbf{P}^o$	1	6.86690	$^1\mathbf{D}^e$	4	41.5774	$^1\mathbf{G}^e$	1	44.0121	$^1\mathbf{F}^e$	3	46.6122	$^3\mathbf{P}^o$	11	50.1756
$^3\mathbf{P}^e$	2	9.43460	$^5\mathbf{F}^e$	1	41.6836	$^3\mathbf{P}^e$	7	44.0876	$^1\mathbf{P}^o$	7	46.7017	$^1\mathbf{P}^o$	10	50.2456
$^1\mathbf{D}^e$	2	10.0813	$^5\mathbf{D}^e$	1	41.8609	$^3\mathbf{S}^e$	3	44.1860	$^3\mathbf{D}^o$	7	47.0675	$^1\mathbf{P}^e$	7	50.6110
$^1\mathbf{S}^e$	2	11.5064	$^5\mathbf{P}^e$	2	42.0890	$^1\mathbf{D}^e$	5	44.1966	$^3\mathbf{P}^e$	10	47.1298	$^3\mathbf{D}^e$	9	50.6560
$^3\mathbf{P}^o$	2	37.1426	$^3\mathbf{P}^e$	5	42.1166	$^5\mathbf{S}^o$	3	44.2749	$^1\mathbf{D}^e$	8	47.3804	$^3\mathbf{S}^e$	5	50.7117
$^1\mathbf{P}^o$	2	37.3593	$^3\mathbf{S}^e$	2	42.3183	$^1\mathbf{P}^e$	3	44.2788	$^3\mathbf{S}^e$	4	47.4422	$^3\mathbf{P}^e$	12	50.7583
$^1\mathbf{P}^e$	1	38.1672	$^3\mathbf{F}^e$	1	42.3468	$^1\mathbf{D}^o$	4	44.2851	$^3\mathbf{D}^e$	8	47.5399	$^1\mathbf{D}^e$	10	50.8669
$^3\mathbf{D}^e$	1	38.3062	$^1\mathbf{D}^o$	3	42.3833	$^1\mathbf{S}^e$	5	44.5533	$^3\mathbf{F}^o$	3	47.6041	$^1\mathbf{S}^e$	7	51.0226
$^3\mathbf{S}^e$	1	38.4665	$^3\mathbf{F}^o$	2	42.4014	$^1\mathbf{P}^o$	6	44.5833	$^1\mathbf{S}^o$	2	47.6489	$^1\mathbf{D}^o$	8	51.1424
$^3\mathbf{P}^e$	3	38.5550	$^1\mathbf{F}^o$	2	42.5587	$^3\mathbf{D}^e$	5	44.9122	$^1\mathbf{P}^e$	6	47.6506	$^3\mathbf{F}^o$	5	51.1449
$^1\mathbf{D}^e$	3	38.9519	$^3\mathbf{P}^e$	6	42.5959	$^3\mathbf{F}^e$	3	44.9532	$^3\mathbf{G}^o$	1	47.6939	$^3\mathbf{D}^o$	10	51.2435
$^5\mathbf{P}^e$	1	39.3178	$^3\mathbf{D}^o$	4	42.5980	$^3\mathbf{S}^o$	4	45.0138	$^1\mathbf{G}^o$	1	47.7541	$^3\mathbf{P}^o$	12	51.2626
$^1\mathbf{S}^e$	3	39.4014	$^1\mathbf{P}^o$	4	42.6466	$^1\mathbf{D}^e$	6	45.1422	$^3\mathbf{P}^e$	11	47.8357	$^1\mathbf{F}^o$	5	51.4167
$^1\mathbf{D}^o$	2	39.6042	$^1\mathbf{S}^e$	4	42.6722	$^3\mathbf{D}^e$	6	45.1667	$^1\mathbf{P}^o$	8	47.9752	$^1\mathbf{P}^o$	11	51.4223
$^3\mathbf{F}^o$	1	39.6175	$^1\mathbf{P}^e$	2	42.7678	$^1\mathbf{P}^e$	4	45.1784	$^1\mathbf{D}^e$	9	48.0029	$^1\mathbf{F}^e$	4	51.4728
$^3\mathbf{D}^o$	2	39.9039	$^3\mathbf{D}^e$	3	42.7736	$^3\mathbf{P}^e$	8	45.3715	$^3\mathbf{D}^o$	8	48.0361	$^3\mathbf{F}^e$	5	51.4751
$^3\mathbf{P}^o$	3	39.9671	$^3\mathbf{P}^o$	5	42.8208	$^5\mathbf{P}^e$	3	45.4130	$^3\mathbf{P}^o$	9	48.1786	$^3\mathbf{G}^e$	2	51.5076
$^3\mathbf{P}^e$	4	39.9971	$^1\mathbf{S}^o$	1	43.4855	$^3\mathbf{D}^o$	6	45.4466	$^1\mathbf{D}^o$	6	48.2159	$^1\mathbf{G}^e$	2	51.5260
$^3\mathbf{S}^o$	2	40.3607	$^3\mathbf{P}^o$	6	43.6371	$^1\mathbf{F}^e$	2	45.5233	$^3\mathbf{S}^o$	5	48.3603			

gf-values for transitions involving terms with effective $n \leq 4.0$, $L \leq 4$

i i'	gf_L	i i'	gf_L	i i'	gf_L	i i'	gf_L	i i'	gf_L	i i'	gf_L
	$^1\mathbf{P}^o$–$^1\mathbf{S}^e$	3 6	−2.79E−6	6 5	1.96E−5	9 4	5.34E−4	1 2	3.90E−2	1 7	−3.74E−3
1 1	1.21E−1	3 7	−3.17E−2	6 6	−1.17E−1	9 5	2.28E−2	1 3	−4.75E−3	2 1	−1.28E−1
1 2	−2.40E−1	4 1	1.09E−2	6 7	−3.12E−3	9 6	9.45E−2	1 4	−1.16E−1	2 2	−4.03E−1
1 3	−9.36E−3	4 2	2.81E−3	7 1	3.38E−3	9 7	−4.57E−5	1 5	−1.34E−1	2 3	−4.19E−3
1 4	−5.21E−2	4 3	5.36E−3	7 2	1.38E−1	10 1	1.35E−2	1 6	−3.11E−2	2 4	−4.27E−4
1 5	−2.48E−1	4 4	−1.48E−4	7 3	4.59E−4	10 2	5.75E−5	1 7	−2.73E−6	2 5	−6.64E−4
1 6	−9.85E−2	4 5	−9.44E−2	7 4	1.05E−1	10 3	9.16E−2	2 1	2.42E−4	2 6	−4.29E−4
1 7	−1.01E−3	4 6	−1.32E−2	7 5	3.72E−4	10 4	1.52E−3	2 2	5.13E−4	2 7	−3.35E−1
2 1	7.44E−2	4 7	−2.42E−6	7 6	−1.88E−1	10 5	3.50E−4	2 3	9.39E−2	3 1	1.47E−1
2 2	1.22E−6	5 1	3.74E−1	7 7	−1.30E−6	10 6	2.15E−5	2 4	9.70E−3	3 2	−5.49E−7
2 3	1.40E−1	5 2	1.02E−2	8 1	3.92E−4	10 7	−1.95E−1	2 5	7.94E−2	3 3	−2.13E−2
2 4	−5.12E−2	5 3	9.07E−2	8 2	1.25E−1	11 1	2.73E−1	2 6	−2.80E−6	3 4	−1.30E−1
2 5	−3.54E−3	5 4	1.63E−1	8 3	1.80E−5	11 2	1.49E−3	2 7	−1.69E−5	3 5	−9.09E−4
2 6	−7.05E−5	5 5	−1.52E−2	8 4	5.92E−5	11 3	4.92E−1		$^1\mathbf{P}^o$–$^1\mathbf{P}^e$	3 6	−2.68E−6
2 7	−7.70E−2	5 6	−8.19E−4	8 5	2.58E−2	11 4	1.54E−3	1 1	−2.71E−3	3 7	−4.85E−2
3 1	1.31E+0	5 7	−3.03E−4	8 6	−5.18E−3	11 5	2.54E−3	1 2	−2.06E−1	4 1	1.03E−1
3 2	1.35E−4	6 1	5.03E−2	8 7	6.38E−5	11 6	1.02E−6	1 3	1.30E−1	4 2	3.20E−4
3 3	8.51E−2	6 2	9.76E−4	9 1	8.08E−3	11 7	1.60E−1	1 4	−1.18E+0	4 3	−1.89E−1
3 4	−4.63E−3	6 3	8.63E−2	9 2	2.06E+0		$^1\mathbf{S}^o$–$^1\mathbf{P}^e$	1 5	−1.61E−3	4 4	−7.52E−3
3 5	−3.77E−2	6 4	2.17E−2	9 3	8.78E−5	1 1	1.42E−1	1 6	−2.71E−1	4 5	−1.40E−1

C-like Ca (Ca^{14+})

i i'	gf_L	i i'	gf_L	i i'	gf_L	i i'	gf_L	i i'	gf_L	i i'	gf_L
4 6	−6.47E−2		**$^1D^o$–$^1P^e$**	8 3	7.57E−3	5 7	−5.03E−2	10 9	4.97E−6	4 10	−1.09E−3
4 7	−7.01E−4	1 1	−3.01E−3	8 4	2.80E−3	5 8	−4.19E−2	10 10	−7.25E−1	5 1	1.48E−2
5 1	7.29E−2	1 2	−2.61E−1	8 5	1.17E−7	5 9	−3.30E−2	11 1	1.21E−2	5 2	5.01E−1
5 2	6.26E−3	1 3	−3.75E−1	8 6	1.81E−4	5 10	−2.61E−5	11 2	1.13E−4	5 3	9.51E−4
5 3	−7.64E−4	1 4	−4.41E−3	8 7	4.74E−1	6 1	1.04E−2	11 3	2.10E−2	5 4	4.99E−1
5 4	−7.68E−3	1 5	−2.90E−1		**$^1P^o$–$^1D^e$**	6 2	1.11E−2	11 4	1.36E−4	5 5	4.45E−3
5 5	−9.08E−4	1 6	−3.94E−2	1 1	3.56E−1	6 3	2.38E−1	11 5	9.32E−5	5 6	3.02E−4
5 6	−9.05E−2	1 7	−2.19E−4	1 2	−1.14E−1	6 4	5.06E−2	11 6	3.43E−3	5 7	−6.46E−4
5 7	−3.90E−5	2 1	2.59E−1	1 3	−3.93E−3	6 5	1.36E−3	11 7	1.05E−3	5 8	−6.39E−1
6 1	2.87E−2	2 2	−4.29E−4	1 4	−5.07E−2	6 6	−1.80E−3	11 8	3.28E−5	5 9	−1.69E−1
6 2	3.54E−1	2 3	−1.14E−1	1 5	−5.21E−1	6 7	−2.21E−1	11 9	9.25E−5	5 10	−4.30E−6
6 3	1.29E−4	2 4	−2.07E−1	1 6	−3.94E−2	6 8	−2.55E−1	11 10	1.15E−2	6 1	2.76E−2
6 4	−3.60E−2	2 5	−7.45E−4	1 7	−3.07E+0	6 9	−1.05E−2		**$^1D^o$–$^1D^e$**	6 2	1.21E+0
6 5	−1.53E−3	2 6	−7.45E−5	1 8	−3.47E−4	6 10	−7.13E−3	1 1	6.02E−1	6 3	9.61E−5
6 6	−1.05E−1	2 7	−7.64E−2	1 9	−3.52E−1	7 1	3.07E−3	1 2	−6.58E−1	6 4	1.46E−3
6 7	−5.44E−3	3 1	1.68E−1	1 10	−1.12E−2	7 2	1.79E−1	1 3	−1.24E−2	6 5	1.25E−2
7 1	6.64E−4	3 2	−2.53E−3	2 1	2.33E−1	7 3	9.35E−5	1 4	−1.55E−1	6 6	6.46E−2
7 2	1.98E−1	3 3	−7.24E−2	2 2	3.47E−6	7 4	2.52E−1	1 5	−1.53E+0	6 7	3.96E−2
7 3	1.63E−4	3 4	−3.73E−2	2 3	−5.19E−1	7 5	5.64E−4	1 6	−4.25E−2	6 8	1.00E−1
7 4	4.27E−3	3 5	−1.58E−1	2 4	−6.47E−2	7 6	1.13E−4	1 7	−5.98E−1	6 9	1.22E−4
7 5	3.59E−3	3 6	−1.10E−1	2 5	−2.16E−3	7 7	7.59E−3	1 8	−2.88E−1	6 10	−3.31E−6
7 6	−1.81E−1	3 7	−3.48E−6	2 6	−1.29E−3	7 8	−3.12E−2	1 9	−1.63E−1	7 1	6.70E−5
7 7	−5.08E−6	4 1	4.28E−2	2 7	−2.55E−5	7 9	−4.28E−1	1 10	−1.85E−4	7 2	2.00E+0
8 1	2.83E−5	4 2	4.92E−1	2 8	−1.06E−4	7 10	−6.26E−5	2 1	1.05E+0	7 3	1.36E−5
8 2	4.23E−3	4 3	1.30E−5	2 9	−2.08E−6	8 1	5.86E−3	2 2	2.12E−4	7 4	1.81E−4
8 3	3.12E−2	4 4	−1.42E−3	2 10	−4.95E−1	8 2	6.80E−1	2 3	3.75E−2	7 5	9.00E−2
8 4	1.35E−1	4 5	−1.20E−2	3 1	4.89E−2	8 3	2.00E−4	2 4	−4.61E−3	7 6	2.08E−2
8 5	1.73E−1	4 6	−8.66E−2	3 2	9.55E−6	8 4	3.19E−4	2 5	−1.66E−1	7 7	1.49E−1
8 6	6.84E−3	4 7	−7.97E−4	3 3	6.63E−3	8 5	1.42E−1	2 6	−8.85E−2	7 8	8.08E−3
8 7	−4.72E−5	5 1	1.69E−3	3 4	−8.76E−5	8 6	6.07E−4	2 7	−4.75E−4	7 9	1.12E−1
9 1	2.33E−4	5 2	1.37E−1	3 5	−2.91E−2	8 7	2.38E−3	2 8	−4.02E−5	7 10	−2.98E−4
9 2	3.16E−5	5 3	1.01E−3	3 6	−8.81E−2	8 8	3.68E−2	2 9	−7.60E−4	8 1	1.96E−1
9 3	1.03E−2	5 4	3.01E−4	3 7	−6.92E−2	8 9	−3.18E−5	2 10	−1.36E−2	8 2	9.85E−4
9 4	6.22E−2	5 5	−1.11E−1	3 8	−1.35E−3	8 10	−1.56E−5	3 1	8.04E−1	8 3	4.01E−1
9 5	1.73E−4	5 6	−5.44E−2	3 9	−2.38E−5	9 1	2.49E−5	3 2	1.12E−2	8 4	6.48E−3
9 6	2.24E−1	5 7	−3.15E−5	3 10	−9.02E−3	9 2	8.91E−2	3 3	4.41E−3	8 5	1.56E−3
9 7	−1.84E−4	6 1	4.99E−5	4 1	5.12E−1	9 3	1.87E−4	3 4	1.98E−1	8 6	6.12E−4
10 1	1.23E−1	6 2	6.87E−5	4 2	2.43E−2	9 4	1.77E−4	3 5	−2.09E−1	8 7	3.02E−3
10 2	5.14E−5	6 3	3.79E−7	4 3	2.49E−2	9 5	2.26E−2	3 6	−2.03E−2	8 8	0.00E+0
10 3	1.07E−3	6 4	1.07E−1	4 4	1.54E−1	9 6	2.89E−4	3 7	−1.61E−2	8 9	2.24E−4
10 4	2.04E−6	6 5	2.55E−1	4 5	−1.67E−1	9 7	2.92E−1	3 8	−4.71E−2	8 10	7.85E−2
10 5	2.88E−5	6 6	1.46E−2	4 6	−1.56E−2	9 8	1.45E−2	3 9	−1.18E−2		**$^1F^o$–$^1D^e$**
10 6	2.53E−5	6 7	−4.14E−6	4 7	−2.15E−2	9 9	2.23E−3	3 10	−1.59E−3	1 1	5.02E+0
10 7	−2.38E−1	7 1	3.97E−4	4 8	−3.16E−4	9 10	−6.41E−5	4 1	6.59E−2	1 2	8.87E−5
11 1	2.85E−1	7 2	1.09E−4	4 9	−9.39E−3	10 1	4.60E−2	4 2	1.91E−2	1 3	5.18E−1
11 2	5.98E−3	7 3	3.20E−2	4 10	−2.68E−3	10 2	6.66E−6	4 3	4.31E−1	1 4	−1.39E−2
11 3	2.73E−3	7 4	3.56E−3	5 1	1.89E−2	10 3	2.66E−1	4 4	6.66E−2	1 5	−5.93E−3
11 4	5.88E−4	7 5	2.11E−2	5 2	9.37E−4	10 4	6.34E−3	4 5	6.04E−4	1 6	−2.42E−2
11 5	9.32E−6	7 6	3.09E−1	5 3	8.21E−2	10 5	5.83E−4	4 6	−5.75E−3	1 7	−4.99E−1
11 6	1.44E−4	7 7	−4.08E−4	5 4	3.45E−2	10 6	1.67E−4	4 7	−7.21E−2	1 8	−2.45E−3
11 7	2.67E−1	8 1	1.02E+0	5 5	−4.70E−4	10 7	1.02E−3	4 8	−2.38E−5	1 9	−1.64E−5
		8 2	2.20E−2	5 6	−3.69E−1	10 8	4.70E−5	4 9	−3.96E−1	1 10	−1.52E−1

C-like Ca (Ca^{14+})

i	i'	gf$_L$	i	i'	gf$_L$	i	i'	gf$_L$	i	i'	gf$_L$	i	i'	gf$_L$	i	i'	gf$_L$
2	1	9.84E−1	3	4	−2.05E−2	3	1	2.42E−1	9	3	1.14E−1	3	10	−1.22E−2	3	1	2.73E+0
2	2	1.60E−2	4	1	3.03E−4	3	2	−1.62E−4	9	4	5.93E−3	3	11	−1.12E−1	3	2	4.75E−4
2	3	3.47E−1	4	2	−3.87E−1	4	1	2.35E−1	9	5	−1.76E−5	3	12	−4.68E−4	3	3	1.94E−1
2	4	3.92E−1	4	3	−3.67E−1	4	2	−1.95E−5	10	1	1.63E−4	4	1	4.82E−2	3	4	−5.42E−4
2	5	−5.38E−2	4	4	−3.30E−2	5	1	2.62E−3	10	2	2.81E−3	4	2	2.33E−1	3	5	−3.76E−3
2	6	−2.03E−2	5	1	1.88E−3	5	2	−1.57E−1	10	3	3.98E−2	4	3	6.66E−3	3	6	−7.80E−5
2	7	−3.35E−5	5	2	5.01E−3			$^1\mathbf{G}^o{-}^1\mathbf{G}^e$	10	4	4.55E−1	4	4	5.00E−1	3	7	−2.49E−1
2	8	−2.27E−1	5	3	−4.20E−1	1	1	3.34E−1	10	5	−7.44E−7	4	5	2.76E−3	3	8	−2.48E−1
2	9	−1.52E−1	5	4	−8.14E−5	1	2	−2.26E−5	11	1	1.15E−1	4	6	1.08E−1	3	9	−1.98E−4
2	10	−2.51E−3	6	1	3.23E−1			$^3\mathbf{P}^o{-}^3\mathbf{S}^e$	11	2	2.13E−4	4	7	7.60E−5	3	10	−4.24E−4
3	1	4.80E−2	6	2	3.52E−6	1	1	−1.80E−2	11	3	5.27E−5	4	8	−2.02E−2	3	11	−3.68E−4
3	2	2.87E+0	6	3	1.90E−2	1	2	−2.49E−1	11	4	3.55E−6	4	9	−2.75E−1	3	12	−4.48E−2
3	3	1.54E−4	6	4	−2.74E−5	1	3	−8.89E−1	11	5	−3.23E−1	4	10	−3.50E−1	4	1	1.10E+0
3	4	1.09E−3	7	1	5.80E−2	1	4	−1.87E−1	12	1	1.36E+0	4	11	−1.49E−1	4	2	3.52E−2
3	5	2.39E−2	7	2	1.45E−1	1	5	−5.14E−3	12	2	1.64E−2	4	12	−3.92E−5	4	3	2.08E−1
3	6	4.33E−2	7	3	9.65E−2	2	1	−2.05E−1	12	3	1.32E−2	5	1	1.03E−2	4	4	5.81E−1
3	7	5.90E−2	7	4	−1.18E−3	2	2	−2.58E−1	12	4	1.65E−5	5	2	1.67E+0	4	5	−1.98E−1
3	8	3.85E−1	8	1	1.15E−3	2	3	−9.61E−4	12	5	7.20E−1	5	3	9.64E−4	4	6	−4.64E−3
3	9	4.06E−4	8	2	2.81E−3	2	4	−1.87E−4			$^3\mathbf{S}^o{-}^3\mathbf{P}^e$	5	4	6.03E−4	4	7	−3.66E−2
3	10	−1.19E−7	8	3	1.99E−5	2	5	−2.80E−1	1	1	5.61E−1	5	5	2.95E−3	4	8	−2.26E−2
4	1	1.42E−2	8	4	−3.34E−1	3	1	4.18E−1	1	2	−4.31E−1	5	6	7.98E−4	4	9	−3.51E−1
4	2	3.93E+0			$^1\mathbf{F}^o{-}^1\mathbf{F}^e$	3	2	−2.88E−5	1	3	−3.62E−3	5	7	1.77E−1	4	10	−1.47E−4
4	3	4.20E−6	1	1	−1.40E−2	3	3	−1.00E−1	1	4	−5.16E−3	5	8	1.70E−1	4	11	−8.34E−3
4	4	1.99E−5	1	2	−5.44E−1	3	4	−4.87E−9	1	5	−2.76E−1	5	9	1.22E−1	4	12	−2.36E−3
4	5	1.51E−2	1	3	−4.81E−3	3	5	−1.51E−1	1	6	−2.37E−1	5	10	1.87E−1	5	1	7.90E−2
4	6	5.35E−2	1	4	−6.36E−1	4	1	2.59E−4	1	7	−1.48E−2	5	11	1.68E−3	5	2	8.30E−3
4	7	3.44E−4	2	1	−1.19E−1	4	2	−5.03E−3	1	8	−4.72E+0	5	12	−1.48E−5	5	3	3.24E−2
4	8	3.32E−2	2	2	−6.24E−2	4	3	−2.54E−2	1	9	−8.44E−1			$^3\mathbf{P}^o{-}^3\mathbf{P}^e$	5	4	1.77E−1
4	9	5.13E−1	2	3	−3.50E−1	4	4	−4.77E−3	1	10	−1.24E−1	1	1	4.81E−1	5	5	3.81E−4
4	10	−7.54E−7	2	4	−3.47E−3	4	5	−8.57E−4	1	11	−1.19E−1	1	2	−3.08E−1	5	6	1.65E−3
5	1	1.06E+0	3	1	4.15E−2	5	1	2.46E−1	1	12	−5.51E−4	1	3	−1.34E−2	5	7	−4.56E−1
5	2	2.98E−4	3	2	1.43E−1	5	2	2.66E−4	2	1	4.01E−1	1	4	−3.20E−1	5	8	−9.87E−3
5	3	1.88E+0	3	3	2.48E−1	5	3	−2.20E−1	2	2	2.47E−3	1	5	−1.47E+0	5	9	−2.64E−4
5	4	2.92E−2	3	4	−4.03E−7	5	4	−2.58E−1	2	3	3.21E−2	1	6	−8.03E−2	5	10	−3.21E−1
5	5	2.38E−3	4	1	7.58E−3	5	5	−2.78E−3	2	4	5.09E−2	1	7	−2.49E+0	5	11	−5.87E−2
5	6	2.73E−5	4	2	4.91E−1	6	1	5.05E−4	2	5	−4.47E−1	1	8	−3.10E−1	5	12	−2.67E−3
5	7	2.25E−2	4	3	2.33E−3	6	2	6.80E−1	2	6	−3.76E−3	1	9	−1.98E−4	6	1	1.39E−1
5	8	5.42E−4	4	4	−2.13E−5	6	3	−1.69E−2	2	7	−3.30E−2	1	10	−3.37E−3	6	2	6.51E−3
5	9	3.06E−5	5	1	4.31E−4	6	4	−1.66E−1	2	8	−4.50E−4	1	11	−5.21E−1	6	3	2.49E−2
5	10	9.56E−1	5	2	1.98E−4	6	5	−1.06E−3	2	9	−5.95E−2	1	12	−7.44E−3	6	4	8.76E−3
		$^1\mathbf{D}^o{-}^1\mathbf{F}^e$	5	3	8.33E−5	7	1	2.94E−1	2	10	−3.06E−3	2	1	5.30E−1	6	5	3.11E−5
1	1	6.10E 1	5	4	6.01E 3	7	2	1.18E 1	2	11	3.05E 4	2	2	6.46E 5	6	6	5.52E 2
1	2	−5.56E+0			$^1\mathbf{G}^o{-}^1\mathbf{F}^e$	7	3	−7.23E−5	2	12	−1.95E−3	2	3	−7.13E−1	6	7	−8.23E−6
1	3	−3.01E−1	1	1	4.30E−2	7	4	−1.41E−1	3	1	1.41E−1	2	4	−1.43E−1	6	8	−4.48E−2
1	4	−3.35E−3	1	2	3.66E−2	7	5	−4.88E−3	3	2	1.34E−1	2	5	−8.50E−4	6	9	−8.25E−4
2	1	−1.33E−1	1	3	6.89E−1	8	1	1.42E−3	3	3	3.71E−1	2	6	−1.06E+0	6	10	−5.84E−2
2	2	−4.60E−2	1	4	−1.48E−4	8	2	1.53E−1	3	4	1.55E−2	2	7	−4.48E−3	6	11	−2.78E−1
2	3	−2.89E−3			$^1\mathbf{F}^o{-}^1\mathbf{G}^e$	8	3	5.46E−4	3	5	3.48E−4	2	8	−8.51E−4	6	12	−9.66E−3
2	4	−4.30E+0	1	1	−9.65E−2	8	4	−1.76E−1	3	6	2.01E−1	2	9	−2.63E−4	7	1	2.20E−1
3	1	−3.68E−1	1	2	−6.27E+0	8	5	−6.73E−6	3	7	−8.90E−6	2	10	−2.12E−3	7	2	4.72E−3
3	2	−1.06E−3	2	1	−7.65E−1	9	1	1.64E−4	3	8	−2.46E−1	2	11	−4.50E−4	7	3	3.58E−1
3	3	−2.38E−2	2	2	−4.13E−2	9	2	8.26E−5	3	9	−4.00E−1	2	12	−8.65E−1	7	4	1.64E−1

C-like Ca (Ca^{14+})

i	i'	gf_L	i	i'	gf_L	i	i'	gf_L	i	i'	gf_L	i	i'	gf_L	i	i'	gf_L
7	5	2.31E−2	11	9	9.28E−5	3	12	−1.05E−2	8	4	2.74E−3	3	1	5.28E−2	8	8	−1.19E+0
7	6	5.05E−1	11	10	5.70E−6	4	1	2.38E−1	8	5	3.02E−3	3	2	−1.15E−8	8	9	−8.42E−5
7	7	−6.01E−6	11	11	1.27E−5	4	2	2.59E−3	8	6	3.31E−5	3	3	−7.58E−5	9	1	2.59E−4
7	8	−2.68E−1	11	12	−1.25E+0	4	3	8.82E−2	8	7	1.53E−1	3	4	−9.60E−2	9	2	4.89E−3
7	9	−4.25E−3	12	1	5.54E−1	4	4	2.49E−1	8	8	1.03E−1	3	5	−2.58E−2	9	3	5.42E−3
7	10	−2.31E−1	12	2	3.61E−5	4	5	2.41E−3	8	9	1.01E−1	3	6	−4.50E−1	9	4	4.54E−1
7	11	−3.30E−2	12	3	9.04E−1	4	6	1.99E−5	8	10	3.56E−1	3	7	−2.19E−3	9	5	8.47E−3
7	12	−4.48E−3	12	4	1.26E−1	4	7	−1.90E−1	8	11	1.74E−5	3	8	−2.06E−3	9	6	4.96E−2
8	1	8.71E−3	12	5	2.03E−4	4	8	−2.12E−6	8	12	−6.79E−7	3	9	−4.76E−2	9	7	2.12E−1
8	2	4.78E−1	12	6	4.19E−3	4	9	−3.77E−6	9	1	2.82E−2	4	1	3.39E−3	9	8	4.17E−2
8	3	4.43E−3	12	7	5.76E−3	4	10	−1.31E−1	9	2	5.52E+0	4	2	−1.16E−4	9	9	−6.98E−5
8	4	1.15E−3	12	8	4.32E−3	4	11	−3.18E−1	9	3	2.85E−4	4	3	−1.08E+0	10	1	3.13E−4
8	5	5.13E−4	12	9	7.16E−5	4	12	−2.49E−3	9	4	3.89E−4	4	4	−1.28E−2	10	2	2.14E−4
8	6	8.59E−1	12	10	2.05E−4	5	1	3.83E−1	9	5	2.16E−4	4	5	−3.61E−3	10	3	3.88E−3
8	7	8.59E−3	12	11	8.09E−6	5	2	1.18E−4	9	6	3.31E−4	4	6	−1.81E−4	10	4	1.46E−1
8	8	1.14E−3	12	12	3.90E−1	5	3	3.66E−1	9	7	3.96E−3	4	7	−2.16E−3	10	5	3.33E−1
8	9	4.96E−3			**$^3D^o$–$^3P^e$**	5	4	1.38E−1	9	8	2.07E−1	4	8	−3.60E−5	10	6	4.99E−1
8	10	−5.69E−3	1	1	4.33E−1	5	5	2.88E−3	9	9	1.11E−2	4	9	−2.28E−3	10	7	3.90E−2
8	11	−1.00E+0	1	2	−9.26E−1	5	6	8.70E−1	9	10	1.29E−3	5	1	3.80E−3	10	8	7.13E−3
8	12	−9.40E−5	1	3	−9.99E−2	5	7	−3.72E−4	9	11	9.05E−1	5	2	9.47E−1	10	9	−1.13E−4
9	1	1.87E−2	1	4	−3.80E−1	5	8	−9.43E−3	9	12	−5.40E−4	5	3	5.61E−4	11	1	6.69E−1
9	2	4.20E+0	1	5	−2.67E−1	5	9	−2.05E−3	10	1	1.69E+0	5	4	−3.14E−1	11	2	1.97E−3
9	3	2.06E−6	1	6	−1.95E−1	5	10	−1.85E−3	10	2	1.84E−3	5	5	−1.28E−3	11	3	7.43E−4
9	4	1.99E−3	1	7	−1.10E+0	5	11	−4.38E−1	10	3	2.83E+0	5	6	−2.41E−2	11	4	1.65E−4
9	5	5.45E−3	1	8	−4.37E−3	5	12	−1.83E−5	10	4	3.89E−1	5	7	−2.05E−1	11	5	1.59E−3
9	6	4.58E−3	1	9	−1.06E−1	6	1	2.11E−2	10	5	4.82E−4	5	8	−3.51E−5	11	6	2.71E−3
9	7	6.90E−3	1	10	−5.29E−1	6	2	6.68E−1	10	6	1.11E−2	5	9	−1.27E−2	11	7	1.28E−5
9	8	3.45E−1	1	11	−9.33E−3	6	3	1.69E−2	10	7	1.60E−2	6	1	7.37E−1	11	8	2.46E−5
9	9	1.27E−1	1	12	−9.37E−4	6	4	2.39E−3	10	8	1.78E−2	6	2	2.76E−2	11	9	−1.60E+0
9	10	4.08E−1	2	1	8.31E+0	6	5	1.06E−2	10	9	1.50E−5	6	3	1.04E−3	12	1	7.33E−2
9	11	3.47E−3	2	2	1.04E−6	6	6	8.46E−1	10	10	1.59E−4	6	4	−3.81E−6	12	2	8.47E−4
9	12	−2.55E−4	2	3	6.75E−1	6	7	3.63E−3	10	11	6.89E−4	6	5	−1.02E+0	12	3	2.73E−5
10	1	1.76E−2	2	4	−5.45E−3	6	8	1.32E−5	10	12	1.31E+0	6	6	−1.23E−2	12	4	2.44E−8
10	2	8.91E−1	2	5	−5.81E−5	6	9	−1.79E−2			**$^3P^o$–$^3D^e$**	6	7	−6.73E−2	12	5	1.14E−3
10	3	1.22E−4	2	6	−2.23E−3	6	10	−1.12E+0	1	1	−2.13E−2	6	8	−1.18E−1	12	6	1.27E−2
10	4	7.86E−4	2	7	−6.91E−2	6	11	−6.57E−3	1	2	−4.08E−1	6	9	−9.05E−3	12	7	5.61E−6
10	5	1.03E−2	2	8	−9.22E−1	6	12	−1.08E−5	1	3	−3.33E+0	7	1	2.17E−1	12	8	1.65E−4
10	6	1.27E−3	2	9	−8.27E−6	7	1	3.29E−2	1	4	−2.88E+0	7	2	7.17E−3	12	9	7.27E−2
10	7	1.32E−1	2	10	−1.99E−4	7	2	1.69E+0	1	5	−2.87E+0	7	3	1.09E−3			**$^3D^o$–$^3D^e$**
10	8	1.25E−1	2	11	−2.26E−3	7	3	8.97E−5	1	6	−2.02E+0	7	4	3.72E−4	1	1	−8.60E−3
10	9	2.04E−3	2	12	−2.21E−1	7	4	3.44E−5	1	7	−1.17E−2	7	5	−3.27E−2	1	2	−9.93E−1
10	10	2.80E−2	3	1	1.86E+0	7	5	1.23E−1	1	8	−7.04E−1	7	6	−6.66E−1	1	3	−1.94E+0
10	11	2.14E−1	3	2	3.10E−3	7	6	7.21E−3	1	9	−1.63E−2	7	7	−2.75E−1	1	4	−4.59E+0
10	12	−1.61E−5	3	3	3.26E−1	7	7	1.74E−2	2	1	−9.65E−1	7	8	−4.64E−1	1	5	−5.08E−2
11	1	8.89E−2	3	4	7.38E−1	7	8	2.61E−2	2	2	−5.69E−1	7	9	−9.71E−3	1	6	−3.78E−1
11	2	8.10E−9	3	5	−1.72E−3	7	9	1.09E+0	2	3	−1.32E−2	8	1	4.62E−4	1	7	−8.39E−1
11	3	5.36E−1	3	6	−1.66E−3	7	10	−4.32E−4	2	4	−4.23E−5	8	2	5.22E−1	1	8	−2.27E−2
11	4	6.15E−2	3	7	−5.09E−2	7	11	−4.30E−5	2	5	−3.28E−4	8	3	7.28E−3	1	9	−2.33E−4
11	5	3.60E−4	3	8	−2.62E−6	7	12	−3.89E−5	2	6	−2.67E−3	8	4	6.48E−3	2	1	2.84E−1
11	6	2.45E−4	3	9	−7.96E−1	8	1	5.56E−2	2	7	−3.92E−4	8	5	3.33E−3	2	2	−2.50E−4
11	7	6.28E−4	3	10	−4.35E−2	8	2	5.64E+0	2	8	−5.84E−5	8	6	2.80E−3	2	3	−6.69E−3
11	8	1.51E−3	3	11	−6.30E−3	8	3	8.06E−5	2	9	−1.58E+0	8	7	4.53E−5	2	4	−3.46E−1

C-like Ca (Ca^{14+})

i i′	gf$_L$	*i i′*	gf$_L$	*i i′*	gf$_L$	*i i′*	gf$_L$	*i i′*	gf$_L$	*i i′*	gf$_L$
2 5	−1.30E−1	7 2	1.53E−3	1 7	−2.34E−3	1 3	−2.59E+0	9 5	−9.66E−4	2 2	−1.68E−1
2 6	−1.55E−1	7 3	3.85E−1	1 8	−8.72E−4	1 4	−7.74E−1	10 1	2.24E−4	3 1	1.71E−1
2 7	−5.68E−5	7 4	1.91E−3	1 9	−3.87E−1	1 5	−2.98E−3	10 2	4.81E−4	3 2	−2.35E−4
2 8	−9.72E−4	7 5	2.00E−3	2 1	4.24E−1	2 1	−1.27E−2	10 3	1.19E−2	4 1	1.30E+0
2 9	−8.82E−2	7 6	4.21E−2	2 2	1.69E+0	2 2	−1.86E−1	10 4	4.15E−4	4 2	−2.13E−3
3 1	2.88E−3	7 7	4.48E−2	2 3	−2.66E−3	2 3	−3.21E−1	10 5	−6.94E−1	5 1	5.22E−3
3 2	−7.42E−3	7 8	−1.75E−3	2 4	−1.38E−1	2 4	−8.76E−3		**$^{3}F^{o}$–$^{3}F^{e}$**	5 2	−1.61E+0
3 3	−4.09E−1	7 9	−1.06E−4	2 5	−3.77E−2	2 5	−1.31E+1	1 1	−1.68E−2		**$^{3}G^{o}$–$^{3}G^{e}$**
3 4	−1.11E−1	8 1	1.24E−3	2 6	−2.35E−2	3 1	−1.60E+0	1 2	−3.63E−1	1 1	1.13E+0
3 5	−4.26E−3	8 2	4.92E−4	2 7	−7.05E−2	3 2	−5.65E−2	1 3	−1.27E+0	1 2	−1.04E−4
3 6	−8.96E−4	8 3	6.22E−2	2 8	−9.07E−1	3 3	−4.46E−3	1 4	−1.58E−2		**$^{5}S^{o}$–$^{5}P^{e}$**
3 7	−1.02E−2	8 4	1.09E−1	2 9	−8.91E−4	3 4	−2.61E−4	1 5	−1.75E+0	1 1	−6.97E−1
3 8	−2.65E−3	8 5	4.53E−2	3 1	1.00E−4	3 5	−2.68E−6	2 1	1.48E−3	1 2	−9.24E+0
3 9	−6.74E−5	8 6	4.45E−1	3 2	2.67E−3	4 1	4.83E−6	2 2	−3.81E−1	1 3	−7.37E−1
4 1	3.72E−1	8 7	7.80E−1	3 3	1.49E−3	4 2	−8.93E−1	2 3	−1.11E−3	2 1	5.91E−1
4 2	1.28E+0	8 8	5.93E−4	3 4	1.09E−1	4 3	−1.83E−2	2 4	−1.14E+0	2 2	−4.18E−1
4 3	−1.64E−3	8 9	−5.29E−5	3 5	1.45E−2	4 4	−8.51E−2	2 5	−1.62E−2	2 3	−2.47E−1
4 4	−3.85E−1	9 1	3.07E−5	3 6	1.46E−1	4 5	−1.02E−2	3 1	7.05E−2	3 1	6.66E−1
4 5	−2.41E−2	9 2	1.04E−4	3 7	1.05E+0	5 1	3.99E−3	3 2	5.45E−1	3 2	1.01E−2
4 6	−1.58E−2	9 3	6.62E−3	3 8	4.11E−5	5 2	−5.40E−6	3 3	6.60E−1	3 3	−1.22E+0
4 7	−9.14E−1	9 4	9.83E−2	3 9	−3.32E−4	5 3	−1.15E+0	3 4	3.53E−1		**$^{5}P^{o}$–$^{5}P^{e}$**
4 8	−1.11E−1	9 5	4.18E−1	4 1	7.28E−4	5 4	−6.93E−1	3 5	−1.80E−4	1 1	1.26E+0
4 9	−1.29E−4	9 6	8.98E−4	4 2	1.64E−3	5 5	−1.47E−1	4 1	1.20E−2	1 2	−5.86E−1
5 1	1.18E+0	9 7	4.14E−5	4 3	2.97E−3	6 1	9.35E−3	4 2	1.06E−1	1 3	−6.84E−1
5 2	2.30E−2	9 8	3.73E−1	4 4	4.89E−5	6 2	1.64E−2	4 3	5.54E−1		**$^{5}D^{o}$–$^{5}P^{e}$**
5 3	5.44E−4	9 9	−2.03E−4	4 5	8.84E−2	6 3	2.35E−3	4 4	1.10E−1	1 1	1.90E+0
5 4	−1.49E−3	10 1	9.84E−1	4 6	2.24E−1	6 4	−1.83E+0	4 5	−4.38E−5	1 2	−3.48E−4
5 5	−6.23E−3	10 2	1.53E−2	4 7	7.02E−3	6 5	−2.16E−3	5 1	9.42E−6	1 3	−1.26E+0
5 6	−2.45E−1	10 3	2.45E−4	4 8	1.63E+0	7 1	9.93E−1	5 2	2.98E−4	2 1	5.21E−3
5 7	−3.29E−1	10 4	3.37E−3	4 9	−6.84E−4	7 2	9.02E−3	5 3	1.06E−3	2 2	1.89E−1
5 8	−6.82E−1	10 5	1.02E−2	5 1	5.87E+0	7 3	4.36E−4	5 4	2.51E−5	2 3	1.84E+0
5 9	−5.79E−4	10 6	1.83E−3	5 2	8.32E−2	7 4	7.27E−3	5 5	−1.25E−1		**$^{5}P^{o}$–$^{5}D^{e}$**
6 1	1.35E−2	10 7	2.43E−5	5 3	9.91E−4	7 5	−3.43E−4		**$^{3}G^{o}$–$^{3}F^{e}$**	1 1	−1.30E+0
6 2	8.22E−1	10 8	3.38E−5	5 4	1.26E−2	8 1	5.41E−2	1 1	2.66E−3		**$^{5}D^{o}$–$^{5}D^{e}$**
6 3	4.50E−3	10 9	5.27E−1	5 5	2.47E−2	8 2	4.86E−1	1 2	1.62E−2	1 1	−5.08E−1
6 4	4.92E−3		**$^{3}F^{o}$–$^{3}D^{e}$**	5 6	4.54E−2	8 3	4.76E−2	1 3	2.00E−1	2 1	8.69E−1
6 5	4.05E−5	1 1	1.36E+0	5 7	1.71E−4	8 4	7.49E−2	1 4	2.08E+0		**$^{5}D^{o}$–$^{5}F^{e}$**
6 6	1.57E−4	1 2	−8.27E−4	5 8	8.37E−4	8 5	−6.11E−5	1 5	−3.66E−4	1 1	−2.21E+0
6 7	−9.16E−1	1 3	−1.11E−3	5 9	2.45E+0	9 1	3.03E−3		**$^{3}F^{o}$–$^{3}G^{e}$**	2 1	1.85E+0
6 8	−1.11E−1	1 4	−6.44E−3		**$^{3}D^{o}$–$^{3}F^{e}$**	9 2	4.25E−1	1 1	−3.15E−1		
6 9	−1.63E−6	1 5	−8.81E−1	1 1	−4.17E+0	9 3	7.00E−1	1 2	−1.81E+1		
7 1	5.21E−5	1 6	−7.35E−1	1 2	−1.18E+1	9 4	9.14E−2	2 1	−1.89E+0		

C-like Fe (Fe^{20+})

Term energies relative to $2s^22p$ ^{2}P ionization threshold for each symmetry

i	E(Ryds)	Description
	^{1}S^o	
1	−45.3358	$2s2p^2$ ^{2}P $3p$
2	−39.6053	$2p^3$ ^{2}D $3d$
3	−21.7811	$2s2p^2$ ^{2}P $4p$
4	−17.4332	$2p^3$ ^{2}D $4d$
5	−11.1533	$2s2p^2$ ^{2}P $5p$
6	−7.26868	$2p^3$ ^{2}D $5d$
7	−5.47726	$2s2p^2$ ^{2}P $6p$
8	−2.09799	$2s2p^2$ ^{2}P $7p$
9	−1.76258	$2p^3$ ^{2}D $6d$
	^{1}P^o	
1	−114.110	$2s2p^3$
2	−53.5904	$2s^22p$ ^{2}P $3s$
3	−49.3397	$2s^22p$ ^{2}P $3d$
4	−46.4017	$2s2p^2$ ^{2}D $3p$
5	−44.9733	$2s2p^2$ ^{2}S $3p$
6	−43.7704	$2s2p^2$ ^{2}P $3p$
7	−40.9840	$2p^3$ ^{2}D $3d$
8	−39.1498	$2p^3$ ^{2}P $3s$
9	−36.7015	$2p^3$ ^{2}P $3d$
10	−29.4049	$2s^22p$ ^{2}P $4s$
11	−27.7378	$2s^22p$ ^{2}P $4d$
12	−23.2903	$2s2p^2$ ^{2}D $4p$
13	−22.1885	$2s2p^2$ ^{2}D $4f$
14	−21.8819	$2s2p^2$ ^{2}S $4p$
15	−21.3027	$2s2p^2$ ^{2}P $4p$
16	−18.5290	$2s^22p$ ^{2}P $5s$
17	−17.7368	$2s^22p$ ^{2}P $5d$
18	−17.3282	$2p^3$ ^{2}D $4d$
19	−17.1945	$2p^3$ ^{2}P $4s$
20	−15.5468	$2p^3$ ^{2}P $4d$
21	−12.8694	$2s2p^2$ ^{2}D $5p$
22	−12.7470	$2s^22p$ ^{2}P $6s$
23	−12.3507	$2s2p^2$ ^{2}D $5f$
24	−12.2891	$2s^22p$ ^{2}P $6d$
25	−11.3926	$2s2p^2$ ^{2}S $5p$
26	−10.9425	$2s2p^2$ ^{2}P $5p$
27	−9.31281	$2s^22p$ ^{2}P $7s$
28	−9.03754	$2s^22p$ ^{2}P $7d$
29	−7.27804	$2s2p^2$ ^{2}D $6p$
30	−7.18208	$2p^3$ ^{2}D $5d$
31	−7.09946	$2s^22p$ ^{2}P $8s$
32	−6.96675	$2s2p^2$ ^{2}D $6f$
33	−6.91352	$2s^22p$ ^{2}P $8d$
34	−6.48367	$2p^3$ ^{2}P $5s$
35	−5.78066	$2s2p^2$ ^{2}S $6p$
36	−5.64659	$2p^3$ ^{2}P $5d$
37	−5.59748	$2s^22p$ ^{2}P $9s$
38	−5.46307	$2s^22p$ ^{2}P $9d$
39	−5.34797	$2s2p^2$ ^{2}P $6p$
	^{1}D^o	
1	−115.103	$2s2p^3$
2	−50.4753	$2s^22p$ ^{2}P $3d$
3	−46.7703	$2s2p^2$ ^{2}D $3p$
4	−44.2005	$2s2p^2$ ^{2}P $3p$
5	−42.2842	$2p^3$ ^{2}D $3s$
6	−38.8137	$2p^3$ ^{2}D $3d$
7	−37.6349	$2p^3$ ^{2}P $3d$
8	−28.1382	$2s^22p$ ^{2}P $4d$
9	−23.4221	$2s2p^2$ ^{2}D $4p$
10	−22.2673	$2s2p^2$ ^{2}D $4f$
11	−21.4320	$2s2p^2$ ^{2}P $4p$
12	−20.5172	$2s2p^2$ ^{2}P $4f$
13	−18.6133	$2p^3$ ^{2}D $4s$
14	−17.9220	$2s^22p$ ^{2}P $5d$
15	−17.0533	$2p^3$ ^{2}D $4d$
16	−15.9554	$2p^3$ ^{2}P $4d$
17	−12.9031	$2s2p^2$ ^{2}D $5p$
18	−12.4229	$2s^22p$ ^{2}P $6d$
19	−12.3548	$2s2p^2$ ^{2}D $5f$
20	−11.0031	$2s2p^2$ ^{2}P $5p$
21	−10.5575	$2s2p^2$ ^{2}P $5f$
22	−9.10082	$2s^22p$ ^{2}P $7d$
23	−7.88045	$2p^3$ ^{2}D $5s$
24	−7.26665	$2s2p^2$ ^{2}D $6p$
25	−7.07851	$2p^3$ ^{2}D $5d$
26	−7.00841	$2p^3$ ^{2}D $5g$
27	−6.98081	$2s2p^2$ ^{2}D $6f$
28	−6.95187	$2s^22p$ ^{2}P $8d$
29	−5.85834	$2p^3$ ^{2}P $5d$
30	−5.49301	$2s^22p$ ^{2}P $9d$
31	−5.38901	$2s2p^2$ ^{2}P $6p$
32	−5.14987	$2s2p^2$ ^{2}P $6f$
	^{1}F^o	
1	−49.3214	$2s^22p$ ^{2}P $3d$
2	−46.5166	$2s2p^2$ ^{2}D $3p$
3	−38.3794	$2p^3$ ^{2}D $3d$
4	−37.4183	$2p^3$ ^{2}P $3d$
5	−27.7408	$2s^22p$ ^{2}P $4d$
6	−23.4074	$2s2p^2$ ^{2}D $4p$
7	−22.3515	$2s2p^2$ ^{2}D $4f$
8	−20.8067	$2s2p^2$ ^{2}S $4f$
9	−20.3959	$2s2p^2$ ^{2}P $4f$
10	−17.7396	$2s^22p$ ^{2}P $5d$
11	−17.6152	$2s^22p$ ^{2}P $5g$
12	−16.9787	$2p^3$ ^{2}D $4d$
13	−15.7939	$2p^3$ ^{2}P $4d$
14	−12.9160	$2s2p^2$ ^{2}D $5p$
15	−12.4046	$2s2p^2$ ^{2}D $5f$
16	−12.3078	$2s^22p$ ^{2}P $6d$
17	−12.2436	$2s^22p$ ^{2}P $6g$
18	−10.8800	$2s2p^2$ ^{2}S $5f$
19	−10.4958	$2s2p^2$ ^{2}P $5f$
20	−9.03918	$2s^22p$ ^{2}P $7d$
21	−8.99439	$2s^22p$ ^{2}P $7g$
22	−7.30191	$2s2p^2$ ^{2}D $6p$
23	−7.06096	$2p^3$ ^{2}D $5d$
24	−7.03236	$2p^3$ ^{2}D $5g$
25	−6.97949	$2s2p^2$ ^{2}D $6f$
26	−6.96335	$2s2p^2$ ^{2}D $6h$
27	−6.91648	$2s^22p$ ^{2}P $8d$
28	−6.88733	$2s^22p$ ^{2}P $8g$
29	−5.76791	$2p^3$ ^{2}P $5d$
30	−5.63701	$2p^3$ ^{2}P $5g$
31	−5.48923	$2s2p^2$ ^{2}S $6f$
32	−5.46309	$2s^22p$ ^{2}P $9d$
33	−5.43886	$2s^22p$ ^{2}P $9g$
34	−5.11337	$2s2p^2$ ^{2}P $6f$
	^{1}G^o	
1	−39.4379	$2p^3$ ^{2}D $3d$
2	−22.3792	$2s2p^2$ ^{2}D $4f$
3	−20.5053	$2s2p^2$ ^{2}P $4f$
4	−17.6448	$2s^22p$ ^{2}P $5g$
5	−17.3630	$2p^3$ ^{2}D $4d$
6	−12.4195	$2s2p^2$ ^{2}D $5f$
7	−12.2569	$2s^22p$ ^{2}P $6g$
8	−10.5461	$2s2p^2$ ^{2}P $5f$
9	−9.00416	$2s^22p$ ^{2}P $7g$
10	−7.23340	$2p^3$ ^{2}D $5d$
11	−7.04804	$2p^3$ ^{2}D $5g$
12	−7.00726	$2s2p^2$ ^{2}D $6f$
13	−6.95341	$2s2p^2$ ^{2}D $6h$
14	−6.89339	$2s^22p$ ^{2}P $8g$
15	−5.64783	$2p^3$ ^{2}P $5g$
16	−5.44911	$2s^22p$ ^{2}P $9g$
17	−5.14151	$2s2p^2$ ^{2}P $6f$
18	−5.11716	$2s2p^2$ ^{2}P $6h$
	^{1}H^o	
1	−22.2975	$2s2p^2$ ^{2}D $4f$
2	−17.6333	$2s2p^2$ ^{2}P $5g$
3	−12.3894	$2s2p^2$ ^{2}D $5f$
4	−12.2414	$2s^22p$ ^{2}P $6g$
5	−8.99852	$2s^22p$ ^{2}P $7g$
6	−8.99852	$2s^22p$ ^{2}P $7i$
7	−7.05727	$2p^3$ ^{2}D $5g$
8	−6.99564	$2s2p^2$ ^{2}D $6f$
9	−6.95028	$2s2p^2$ ^{2}D $6h$
10	−6.89053	$2s^22p$ ^{2}P $8i$
11	−6.88718	$2s^22p$ ^{2}P $8g$
12	−5.65954	$2p^3$ ^{2}P $5g$
13	−5.47365	$2s2p^2$ ^{2}S $6h$
14	−5.44437	$2s^22p$ ^{2}P $9g$
15	−5.44160	$2s^22p$ ^{2}P $9i$
16	−5.10963	$2s2p^2$ ^{2}P $6h$
	^{1}I^o	
1	−9.00058	$2s^22p$ ^{2}P $7i$
2	−7.05445	$2p^3$ ^{2}D $5g$
3	−6.95732	$2s2p^2$ ^{2}D $6h$
4	−6.89087	$2s^22p$ ^{2}P $8i$
5	−5.44559	$2s^22p$ ^{2}P $9i$
6	−5.12111	$2s2p^2$ ^{2}P $6h$
	^{3}S^o	
1	−115.096	$2s2p^3$
2	−49.4961	$2s2p^2$ ^{4}P $3p$
3	−44.7354	$2s2p^2$ ^{2}P $3p$
4	−43.2330	$2p^3$ ^{4}S $3s$
5	−38.5831	$2p^3$ ^{2}D $3d$
6	−25.9638	$2s2p^2$ ^{4}P $4p$
7	−21.4973	$2s2p^2$ ^{2}P $4p$
8	−19.9422	$2p^3$ ^{4}S $4s$
9	−17.0869	$2p^3$ ^{2}D $4d$
10	−15.3569	$2s2p^2$ ^{4}P $5p$
11	−11.0209	$2s2p^2$ ^{2}P $5p$
12	−9.73645	$2s2p^2$ ^{4}P $6p$
13	−9.19347	$2p^3$ ^{4}S $5s$
14	−7.10831	$2p^3$ ^{2}D $5d$
15	−6.32477	$2s2p^2$ ^{4}P $7p$
16	−5.40325	$2s2p^2$ ^{2}P $6p$
17	−4.14998	$2s2p^2$ ^{4}P $8p$
18	−3.47814	$2p^3$ ^{4}S $6s$
19	−2.65429	$2s2p^2$ ^{4}P $9p$
20	−2.05202	$2s2p^2$ ^{2}P $7p$
21	−1.67538	$2p^3$ ^{2}D $6d$
	^{3}P^o	
1	−116.911	$2s2p^3$
2	−53.8903	$2s^22p$ ^{2}P $3s$
3	−49.9609	$2s^22p$ ^{2}P $3d$
4	−48.3724	$2s2p^2$ ^{4}P $3p$
5	−46.1458	$2s2p^2$ ^{2}D $3p$
6	−45.0804	$2s2p^2$ ^{2}S $3p$
7	−44.7449	$2s2p^2$ ^{2}P $3p$
8	−41.3600	$2p^3$ ^{2}P $3s$
9	−38.8497	$2p^3$ ^{2}D $3d$
10	−38.0011	$2p^3$ ^{2}P $3d$
11	−29.5012	$2s^22p$ ^{2}P $4s$
12	−27.9646	$2s^22p$ ^{2}P $4d$
13	−25.5543	$2s2p^2$ ^{4}P $4p$
14	−23.3165	$2s2p^2$ ^{2}D $4p$
15	−22.2238	$2s2p^2$ ^{2}D $4f$
16	−21.9300	$2s2p^2$ ^{2}S $4p$
17	−21.5918	$2s2p^2$ ^{2}P $4p$
18	−18.5720	$2s^22p$ ^{2}P $5s$
19	−17.8458	$2s^22p$ ^{2}P $5d$

C-like Fe (Fe^{20+})

i	E(Ryds)	Description	i	E(Ryds)	Description	i	E(Ryds)	Description	i	E(Ryds)	Description
20	−17.3815	$2p^3$ ^{2}P $4s$	27	−11.0933	$2s2p^2$ ^{2}P $5p$	34	−5.63818	$2p^3$ ^{2}P $5g$	3	−7.05441	$2p^3$ ^{2}D $5g$
21	−17.1527	$2p^3$ ^{2}D $4d$	28	−10.5535	$2s2p^2$ ^{2}P $5f$	35	−5.50529	$2s2p^2$ ^{2}S $6f$	4	−6.95746	$2s2p^2$ ^{2}D $6h$
22	−15.9852	$2p^3$ ^{2}P $4d$	29	−9.63783	$2s2p^2$ ^{4}P $6p$	36	−5.49145	$2s^22p$ ^{2}P $9d$	5	−6.89087	$2s^22p$ ^{2}P $8i$
23	−15.1843	$2s2p^2$ ^{4}P $5p$	30	−9.38377	$2s2p^2$ ^{4}P $6f$	37	−5.43987	$2s^22p$ ^{2}P $9g$	6	−6.11563	$2s2p^2$ ^{4}P $7h$
24	−12.9286	$2s2p^2$ ^{2}D $5p$	31	−9.07850	$2s^22p$ ^{2}P $7d$	38	−5.12585	$2s2p^2$ ^{2}P $6f$	7	−5.44559	$2s^22p$ ^{2}P $9i$
25	−12.7427	$2s^22p$ ^{2}P $6s$	32	−8.50138	$2p^3$ ^{4}S $5d$			**^{3}G$^\circ$**	8	−5.12112	$2s2p^2$ ^{2}P $6h$
26	−12.3720	$2s^22p$ ^{2}P $6d$	33	−7.89781	$2p^3$ ^{2}D $5s$	1	−39.5296	$2p^3$ ^{2}D $3d$			**^{5}S$^\circ$**
27	−12.3514	$2s2p^2$ ^{2}D $5f$	34	−7.31752	$2s2p^2$ ^{2}D $6p$	2	−24.7138	$2s2p^2$ ^{4}P $4f$	1	−120.697	$2s2p^3$
28	−11.4234	$2s2p^2$ ^{2}S $5p$	35	−7.18860	$2p^3$ ^{2}D $5d$	3	−22.4155	$2s2p^2$ ^{2}D $4f$	2	−48.5421	$2s2p^2$ ^{4}P $3p$
29	−11.0636	$2s2p^2$ ^{2}P $5p$	36	−7.00804	$2p^3$ ^{2}D $5g$	4	−20.5211	$2s2p^2$ ^{2}P $4f$	3	−44.2950	$2p^3$ ^{4}S $3s$
30	−9.60580	$2s2p^2$ ^{4}P $6p$	37	−6.99755	$2s2p^2$ ^{2}D $6f$	5	−17.6449	$2s^22p$ ^{2}P $5g$	4	−25.7034	$2s2p^2$ ^{4}P $4p$
31	−9.32686	$2s^22p$ ^{2}P $7s$	38	−6.94365	$2s^22p$ ^{2}P $8d$	6	−17.3950	$2p^3$ ^{2}D $4d$	5	−20.1651	$2p^3$ ^{4}S $4s$
32	−9.07518	$2s^22p$ ^{2}P $7d$	39	−6.28053	$2s2p^2$ ^{4}P $7p$	7	−14.7742	$2s2p^2$ ^{4}P $5f$	6	−15.2578	$2s2p^2$ ^{4}P $5p$
33	−7.28860	$2s2p^2$ ^{2}D $6p$	40	−6.12619	$2s2p^2$ ^{4}P $7f$	8	−12.4431	$2s2p^2$ ^{2}D $5f$	7	−9.65480	$2s2p^2$ ^{4}P $6p$
34	−7.14769	$2p^3$ ^{2}D $5d$	41	−5.82829	$2p^3$ ^{2}P $5d$	9	−12.2565	$2s^22p$ ^{2}P $6g$	8	−9.30832	$2p^3$ ^{4}S $5s$
35	−7.10235	$2s^22p$ ^{2}P $8s$	42	−5.48473	$2s^22p$ ^{2}P $9d$	10	−10.5576	$2s2p^2$ ^{2}P $5f$	9	−6.29061	$2s2p^2$ ^{4}P $7p$
36	−6.97714	$2s2p^2$ ^{2}D $6f$	43	−5.43317	$2s2p^2$ ^{2}P $6p$	11	−9.37738	$2s2p^2$ ^{4}P $6f$	10	−4.12334	$2s2p^2$ ^{4}P $8p$
37	−6.94136	$2s^22p$ ^{2}P $8d$	44	−5.14701	$2s2p^2$ ^{2}P $6f$	12	−9.37074	$2s2p^2$ ^{4}P $6h$	11	−3.52881	$2p^3$ ^{4}S $6s$
38	−6.53359	$2p^3$ ^{2}P $5s$			**^{3}F$^\circ$**	13	−9.00413	$2s^22p$ ^{2}P $7g$	12	−2.64070	$2s2p^2$ ^{4}P $9p$
39	−6.26082	$2s2p^2$ ^{4}P $7p$	1	−50.4581	$2s^22p$ ^{2}P $3d$	14	−8.40687	$2p^3$ ^{4}S $5g$			**^{5}P$^\circ$**
40	−5.88214	$2p^3$ ^{2}P $5d$	2	−46.7367	$2s2p^2$ ^{2}D $3p$	15	−7.24271	$2p^3$ ^{2}D $5d$	1	−49.1239	$2s2p^2$ ^{4}P $3p$
41	−5.76969	$2s2p^2$ ^{2}S $6p$	3	−39.6664	$2p^3$ ^{2}D $3d$	16	−7.04897	$2p^3$ ^{2}D $5g$	2	−25.8627	$2s2p^2$ ^{4}P $4p$
42	−5.60307	$2s^22p$ ^{2}P $9s$	4	−38.1531	$2p^3$ ^{2}P $3d$	17	−7.02709	$2s2p^2$ ^{2}D $6f$	3	−15.3333	$2s2p^2$ ^{4}P $5p$
43	−5.48061	$2s^22p$ ^{2}P $9d$	5	−28.1314	$2s^22p$ ^{2}P $4d$	18	−6.95405	$2s2p^2$ ^{2}D $6h$	4	−9.68715	$2s2p^2$ ^{4}P $6p$
44	−5.41260	$2s2p^2$ ^{2}P $6p$	6	−24.6126	$2s2p^2$ ^{4}P $4f$	19	−6.89313	$2s^22p$ ^{2}P $8g$	5	−6.31410	$2s2p^2$ ^{4}P $7p$
		^{3}D$^\circ$	7	−23.4840	$2s2p^2$ ^{2}D $4p$	20	−6.12246	$2s2p^2$ ^{4}P $7f$	6	−4.13763	$2s2p^2$ ^{4}P $8p$
1	−117.902	$2s2p^3$	8	−22.3904	$2s2p^2$ ^{2}D $4f$	21	−6.11520	$2s2p^2$ ^{4}P $7h$	7	−2.65152	$2s2p^2$ ^{4}P $9p$
2	−50.0410	$2s^22p$ ^{2}P $3d$	9	−20.8400	$2s2p^2$ ^{2}S $4f$	22	−5.64773	$2p^3$ ^{2}P $5g$			**^{5}D$^\circ$**
3	−48.6777	$2s2p^2$ ^{4}P $3p$	10	−20.4246	$2s2p^2$ ^{2}P $4f$	23	−5.44891	$2s^22p$ ^{2}P $9g$	1	−49.2992	$2s2p^2$ ^{4}P $3p$
4	−46.4581	$2s2p^2$ ^{2}D $3p$	11	−17.9219	$2s^22p$ ^{2}P $5d$	24	−5.14899	$2s2p^2$ ^{2}P $6f$	2	−41.0401	$2p^3$ ^{4}S $3d$
5	−44.9855	$2s2p^2$ ^{2}P $3p$	12	−17.6158	$2s^22p$ ^{2}P $5g$	25	−5.11715	$2s2p^2$ ^{2}P $6h$	3	−25.9254	$2s2p^2$ ^{4}P $4p$
6	−42.6727	$2p^3$ ^{2}D $3s$	13	−17.4323	$2p^3$ ^{2}D $4d$			**^{3}H$^\circ$**	4	−24.7608	$2s2p^2$ ^{4}P $4f$
7	−40.3763	$2p^3$ ^{4}S $3d$	14	−16.0406	$2p^3$ ^{2}P $4d$	1	−22.3274	$2s2p^2$ ^{2}D $4f$	5	−18.8375	$2p^3$ ^{4}S $4d$
8	−39.0197	$2p^3$ ^{2}D $3d$	15	−14.7265	$2s2p^2$ ^{4}P $5f$	2	−17.6340	$2s^22p$ ^{2}P $5g$	6	−15.3575	$2s2p^2$ ^{4}P $5p$
9	−37.7174	$2p^3$ ^{2}P $3d$	16	−12.9530	$2s2p^2$ ^{2}D $5p$	3	−12.4098	$2s2p^2$ ^{2}D $5f$	7	−14.8020	$2s2p^2$ ^{4}P $5f$
10	−27.9941	$2s^22p$ ^{2}P $4d$	17	−12.4276	$2s2p^2$ ^{2}D $5f$	4	−12.2413	$2s^22p$ ^{2}P $6g$	8	−9.70376	$2s2p^2$ ^{4}P $6p$
11	−25.6717	$2s2p^2$ ^{4}P $4p$	18	−12.4094	$2s^22p$ ^{2}P $6d$	5	−9.36078	$2s2p^2$ ^{4}P $6h$	9	−9.39401	$2s2p^2$ ^{4}P $6f$
12	−24.7297	$2s2p^2$ ^{4}P $4f$	19	−12.2441	$2s^22p$ ^{2}P $6g$	6	−9.00028	$2s^22p$ ^{2}P $7g$	10	−8.66028	$2p^3$ ^{4}S $5d$
13	−23.4362	$2s2p^2$ ^{2}D $4p$	20	−10.9018	$2s2p^2$ ^{2}S $5f$	7	−7.05799	$2p^3$ ^{2}D $5g$	11	−6.32273	$2s2p^2$ ^{4}P $7p$
14	−22.3037	$2s2p^2$ ^{2}D $4f$	21	−10.5156	$2s2p^2$ ^{2}P $5f$	8	−7.00825	$2s2p^2$ ^{2}D $6f$	12	−6.13361	$2s2p^2$ ^{4}P $7f$
15	−21.6532	$2s2p^2$ ^{2}P $4p$	22	−9.35036	$2s2p^2$ ^{4}P $6f$	9	−6.95096	$2s2p^2$ ^{2}D $6h$	13	−4.14396	$2s2p^2$ ^{4}P $8p$
16	−20.5100	$2s2p^2$ ^{2}P $4f$	23	−9.10040	$2s^22p$ ^{2}P $7d$	10	−6.89052	$2s^22p$ ^{2}P $7i$	14	−4.01817	$2s2p^2$ ^{4}P $8f$
17	−18.7236	$2p^3$ ^{2}D $4s$	24	−8.99485	$2s^22p$ ^{2}P $7g$	11	−6.88716	$2s^22p$ ^{2}P $8g$	15	−3.16371	$2p^3$ ^{4}S $6d$
18	−18.5139	$2p^3$ ^{4}S $4d$	25	−7.33771	$2s2p^2$ ^{2}D $6p$	12	−6.11161	$2s2p^2$ ^{4}P $7h$	16	−2.65499	$2s2p^2$ ^{4}P $9p$
19	−17.8552	$2s^22p$ ^{2}P $5d$	26	−7.23317	$2p^3$ ^{2}D $5d$	13	−5.66028	$2p^3$ ^{2}P $5g$	17	−2.56816	$2s2p^2$ ^{4}P $9f$
20	−17.2907	$2p^3$ ^{2}D $4d$	27	−7.03297	$2p^3$ ^{2}D $5g$	14	−5.47374	$2s2p^2$ ^{2}S $6h$			**^{5}F$^\circ$**
21	−15.9036	$2p^3$ ^{2}P $4d$	28	−7.01820	$2s2p^2$ ^{2}D $6f$	15	−5.44466	$2s^22p$ ^{2}P $9g$	1	−24.6729	$2s2p^2$ ^{4}P $4f$
22	−15.2356	$2s2p^2$ ^{4}P $5p$	29	−6.96391	$2s2p^2$ ^{2}D $6h$	16	−5.44170	$2s^22p$ ^{2}P $8i$	2	−14.7671	$2s2p^2$ ^{4}P $5f$
23	−14.7807	$2s2p^2$ ^{4}P $5f$	30	−6.95744	$2s^22p$ ^{2}P $8d$	17	−5.10964	$2s2p^2$ ^{2}P $6h$	3	−9.37526	$2s2p^2$ ^{4}P $6f$
24	−12.9332	$2s2p^2$ ^{2}D $5p$	31	−6.88764	$2s^22p$ ^{2}P $8g$			**^{3}I$^\circ$**	4	−6.12253	$2s2p^2$ ^{4}P $7f$
25	−12.3903	$2s2p^2$ ^{2}D $5f$	32	−6.10646	$2s2p^2$ ^{4}P $7f$	1	−9.36596	$2s2p^2$ ^{4}P $6h$	5	−4.01099	$2s2p^2$ ^{4}P $8f$
26	−12.3747	$2s^22p$ ^{2}P $6d$	33	−5.88021	$2p^3$ ^{2}P $5d$	2	−9.00058	$2s^22p$ ^{2}P $7i$	6	−2.56326	$2s2p^2$ ^{4}P $9f$

C-like Fe (Fe^{20+})

i E(Ryds) Description	*i* E(Ryds) Description	*i* E(Ryds) Description	*i* E(Ryds) Description
^{5}G^o	18 −10.7521 $2s2p^2$ ^{2}P $5d$	42 −5.44430 $2s^22p$ ^{2}P $9f$	24 −5.10750 $2s2p^2$ ^{2}P $6g$
1 −24.7636 $2s2p^2$ ^{4}P $4f$	19 −9.21678 $2s^22p$ ^{2}P $7p$	43 −5.16409 $2s2p^2$ ^{2}P $6d$	**^{1}H^e**
2 −14.8074 $2s2p^2$ ^{4}P $5f$	20 −7.57725 $2p^3$ ^{2}D $5p$	**^{1}F^e**	1 −16.9933 $2p^3$ ^{2}D $4f$
3 −9.39781 $2s2p^2$ ^{4}P $6f$	21 −7.05929 $2s2p^2$ ^{2}D $6d$	1 −44.7292 $2s2p^2$ ^{2}D $3d$	2 −12.3982 $2s2p^2$ ^{2}D $5g$
4 −9.37085 $2s2p^2$ ^{4}P $6h$	22 −7.03478 $2p^3$ ^{2}D $5f$	2 −42.4124 $2s2p^2$ ^{2}P $3d$	3 −12.2500 $2s^22p$ ^{2}P $6h$
5 −8.40803 $2p^3$ ^{4}S $5g$	23 −7.01941 $2s^22p$ ^{2}P $8p$	3 −41.0434 $2p^3$ ^{2}D $3p$	4 −10.5169 $2s2p^2$ ^{2}P $5g$
6 −6.13629 $2s2p^2$ ^{4}P $7f$	24 −6.20460 $2p^3$ ^{2}P $5p$	4 −27.6635 $2s^22p$ ^{2}P $4f$	5 −9.00087 $2s^22p$ ^{2}P $7h$
7 −6.11522 $2s2p^2$ ^{4}P $7h$	25 −5.62172 $2s2p^2$ ^{2}P $6s$	5 −22.6604 $2s2p^2$ ^{2}D $4d$	6 −7.06649 $2p^3$ ^{2}D $5f$
8 −4.02006 $2s2p^2$ ^{4}P $8f$	26 −5.52911 $2s^22p$ ^{2}P $9p$	6 −20.7226 $2s2p^2$ ^{2}P $4d$	7 −6.97829 $2s2p^2$ ^{2}D $6g$
9 −4.00653 $2s2p^2$ ^{4}P $8h$	27 −5.25872 $2s2p^2$ ^{2}P $6d$	7 −18.0231 $2p^3$ ^{2}D $4p$	8 −6.89103 $2s^22p$ ^{2}P $8h$
10 −3.02082 $2p^3$ ^{4}S $6g$	**^{1}D^e**	8 −17.6950 $2s^22p$ ^{2}P $5f$	9 −5.44721 $2s^22p$ ^{2}P $9h$
11 −2.56953 $2s2p^2$ ^{4}P $9f$	1 −122.443 $2s^22p^2$	9 −16.9362 $2p^3$ ^{2}D $4f$	10 −5.12363 $2s2p^2$ ^{2}P $6g$
12 −2.55860 $2s2p^2$ ^{4}P $9h$	2 −109.680 $2p^4$	10 −15.6084 $2p^3$ ^{2}P $4f$	**^{1}I^e**
^{1}S^e	3 −51.3682 $2s^22p$ ^{2}P $3p$	11 −12.5535 $2s2p^2$ ^{2}D $5d$	1 −12.3868 $2s2p^2$ ^{2}D $5g$
1 −121.384 $2s^22p^2$	4 −47.8787 $2s2p^2$ ^{2}D $3s$	12 −12.3671 $2s2p^2$ ^{2}D $5g$	2 −12.2327 $2s^22p$ ^{2}P $6h$
2 −107.765 $2p^4$	5 −44.2211 $2s2p^2$ ^{2}D $3d$	13 −12.2854 $2s^22p$ ^{2}P $6f$	3 −8.99910 $2s^22p$ ^{2}P $7h$
3 −50.7367 $2s^22p$ ^{2}P $3p$	6 −42.9685 $2s2p^2$ ^{2}S $3d$	14 −10.6629 $2s2p^2$ ^{2}P $5d$	4 −6.98794 $2s2p^2$ ^{2}D $6g$
4 −46.4169 $2s2p^2$ ^{2}S $3s$	7 −41.9521 $2s2p^2$ ^{2}P $3d$	15 −10.5187 $2s2p^2$ ^{2}P $5g$	5 −6.89063 $2s^22p$ ^{2}P $8j$
5 −43.7086 $2s2p^2$ ^{2}D $3d$	8 −39.9676 $2p^3$ ^{2}D $3p$	16 −9.02259 $2s^22p$ ^{2}P $7f$	6 −6.88638 $2s^22p$ ^{2}P $8h$
6 −38.1068 $2p^3$ ^{2}P $3p$	9 −39.1526 $2p^3$ ^{2}P $3p$	17 −7.55213 $2p^3$ ^{2}D $5p$	7 −5.44454 $2s^22p$ ^{2}P $9j$
7 −28.3100 $2s^22p$ ^{2}P $4p$	10 −28.5322 $2s^22p$ ^{2}P $4p$	18 −7.08604 $2s2p^2$ ^{2}D $6d$	8 −5.44384 $2s^22p$ ^{2}P $9h$
8 −22.5498 $2s2p^2$ ^{2}S $4s$	11 −27.5233 $2s^22p$ ^{2}P $4f$	19 −7.03606 $2p^3$ ^{2}D $5f$	**^{3}S^e**
9 −22.3196 $2s2p^2$ ^{2}D $4d$	12 −23.9991 $2s2p^2$ ^{2}D $4s$	20 −6.96282 $2s2p^2$ ^{2}D $6g$	1 −52.0494 $2s^22p$ ^{2}P $3p$
10 −18.0179 $2s^22p$ ^{2}P $5p$	13 −22.5415 $2s2p^2$ ^{2}D $4d$	21 −6.90613 $2s^22p$ ^{2}P $8f$	2 −46.9133 $2s2p^2$ ^{2}S $3s$
11 −16.1250 $2p^3$ ^{2}P $4p$	14 −21.0987 $2s2p^2$ ^{2}S $4d$	22 −5.67156 $2p^3$ ^{2}P $5f$	3 −44.2464 $2s2p^2$ ^{2}D $3d$
12 −12.5009 $2s^22p$ ^{2}P $6p$	15 −20.5505 $2s2p^2$ ^{2}P $4d$	23 −5.45669 $2s^22p$ ^{2}P $9f$	4 −39.9679 $2p^3$ ^{2}P $3p$
13 −12.4024 $2s2p^2$ ^{2}D $5d$	16 −18.1125 $2s^22p$ ^{2}P $5p$	24 −5.21036 $2s2p^2$ ^{2}P $6d$	5 −28.7430 $2s^22p$ ^{2}P $4p$
14 −11.6980 $2s2p^2$ ^{2}S $5s$	17 −17.6301 $2p^3$ ^{2}D $4p$	25 −5.12420 $2s2p^2$ ^{2}P $6g$	6 −22.7262 $2s2p^2$ ^{2}S $4s$
15 −9.13255 $2s^22p$ ^{2}P $7p$	18 −17.6269 $2s^22p$ ^{2}P $5f$	**^{1}G^e**	7 −22.5070 $2s2p^2$ ^{2}D $4d$
16 −7.03769 $2s2p^2$ ^{2}D $6d$	19 −16.9545 $2p^3$ ^{2}D $4f$	1 −44.4700 $2s2p^2$ ^{2}D $3d$	8 −18.2087 $2s^22p$ ^{2}P $5p$
17 −6.97013 $2s^22p$ ^{2}P $8p$	20 −16.5121 $2p^3$ ^{2}P $4p$	2 −27.5688 $2s^22p$ ^{2}P $4f$	9 −16.7386 $2p^3$ ^{2}P $4p$
18 −6.05505 $2s2p^2$ ^{2}S $6s$	21 −15.5536 $2p^3$ ^{2}P $4f$	3 −22.6275 $2s2p^2$ ^{2}D $4d$	10 −12.6065 $2s^22p$ ^{2}P $6p$
19 −5.83396 $2p^3$ ^{2}P $5p$	22 −13.2094 $2s2p^2$ ^{2}D $5s$	4 −17.6479 $2s^22p$ ^{2}P $5f$	11 −12.4894 $2s2p^2$ ^{2}D $5d$
20 −5.50620 $2s^22p$ ^{2}P $9p$	23 −12.5435 $2s^22p$ ^{2}P $6p$	5 −16.9470 $2p^3$ ^{2}D $4f$	12 −11.7863 $2s2p^2$ ^{2}S $5s$
^{1}P^e	24 −12.4845 $2s2p^2$ ^{2}D $5d$	6 −15.5928 $2p^3$ ^{2}P $4f$	13 −9.19706 $2s^22p$ ^{2}P $7p$
1 −52.4649 $2s^22p$ ^{2}P $3p$	25 −12.3439 $2s2p^2$ ^{2}D $5g$	7 −12.5520 $2s2p^2$ ^{2}D $5d$	14 −7.08215 $2s2p^2$ ^{2}D $6d$
2 −46.3303 $2s2p^2$ ^{2}D $3d$	26 −12.2455 $2s^22p$ ^{2}P $6f$	8 −12.3883 $2s2p^2$ ^{2}D $5g$	15 −7.01257 $2s^22p$ ^{2}P $8p$
3 −44.1156 $2s2p^2$ ^{2}P $3s$	27 −11.0243 $2s2p^2$ ^{2}S $5d$	9 −12.2554 $2s^22p$ ^{2}P $6f$	16 −6.23344 $2p^3$ ^{2}P $5p$
4 −42.9447 $2s2p^2$ ^{2}P $3d$	28 −10.5834 $2s2p^2$ ^{2}P $5d$	10 −12.2496 $2s^22p$ ^{2}P $6h$	17 −5.98250 $2s2p^2$ ^{2}S $6s$
5 −41.3284 $2p^3$ ^{2}D $3p$	29 −9.16522 $2s^22p$ ^{2}P $7p$	11 −10.8672 $2s2p^2$ ^{2}S $5g$	18 −5.53426 $2s^22p$ ^{2}P $9p$
6 −39.6546 $2p^3$ ^{2}P $3p$	30 −8.99766 $2s^22p$ ^{2}P $7f$	12 −10.4898 $2s2p^2$ ^{2}P $5g$	**^{3}P^e**
7 −28.8903 $2s^22p$ ^{2}P $4p$	31 −7.52556 $2s2p^2$ ^{2}D $6s$	13 −9.00450 $2s^22p$ ^{2}P $7f$	1 −123.400 $2s^22p^2$
8 −22.4758 $2s2p^2$ ^{2}D $4d$	32 −7.30965 $2p^3$ ^{2}D $5p$	14 −8.99617 $2s^22p$ ^{2}P $7h$	2 −110.582 $2p^4$
9 −22.2311 $2s2p^2$ ^{2}P $4s$	33 −7.06571 $2s2p^2$ ^{2}D $6d$	15 −7.08951 $2s2p^2$ ^{2}D $6d$	3 −51.9342 $2s^22p$ ^{2}P $3p$
10 −20.9160 $2s2p^2$ ^{2}P $4d$	34 −7.03287 $2p^3$ ^{2}D $5f$	16 −7.04780 $2p^3$ ^{2}D $5f$	4 −49.9902 $2s2p^2$ ^{4}P $3s$
11 −18.2662 $2s^22p$ ^{2}P $5p$	35 −6.99927 $2s^22p$ ^{2}P $8p$	17 −6.96438 $2s2p^2$ ^{2}D $6g$	5 −47.0548 $2s2p^2$ ^{4}P $3d$
12 −18.0845 $2p^3$ ^{2}D $4p$	36 −6.96616 $2s2p^2$ ^{2}D $6g$	18 −6.89343 $2s^22p$ ^{2}P $8f$	6 −46.5676 $2s2p^2$ ^{2}D $3d$
13 −16.9612 $2p^3$ ^{2}D $4f$	37 −6.88930 $2s^22p$ ^{2}P $8f$	19 −6.88902 $2s^22p$ ^{2}P $8h$	7 −44.3772 $2s2p^2$ ^{2}P $3s$
14 −16.6689 $2p^3$ ^{2}P $4p$	38 −6.11744 $2p^3$ ^{2}P $5p$	20 −5.66566 $2p^3$ ^{2}P $5f$	8 −42.6832 $2s2p^2$ ^{2}P $3d$
15 −12.6230 $2s^22p$ ^{2}P $6p$	39 −5.63853 $2p^3$ ^{2}P $5f$	21 −5.48032 $2s2p^2$ ^{2}S $6g$	9 −42.2145 $2p^3$ ^{4}S $3p$
16 −12.4552 $2s2p^2$ ^{2}D $5d$	40 −5.56918 $2s2p^2$ ^{2}S $6d$	22 −5.44695 $2s^22p$ ^{2}P $9f$	10 −40.3099 $2p^3$ ^{2}D $3p$
17 −11.3883 $2s2p^2$ ^{2}P $5s$	41 −5.52134 $2s^22p$ ^{2}P $9p$	23 −5.43862 $2s^22p$ ^{2}P $9h$	11 −39.4048 $2p^3$ ^{2}P $3p$

C-like Fe (Fe^{20+})

i	E(Ryds)	Description	i	E(Ryds)	Description	i	E(Ryds)	Description	i	E(Ryds)	Description
12	−28.6868	$2s^22p$ ^{2}P $4p$	21	−15.5968	$2p^3$ ^{2}P $4f$	26	−7.13371	$2s2p^2$ ^{2}D $6d$	14	−5.12397	$2s2p^2$ ^{2}P $6g$
13	−26.3389	$2s2p^2$ ^{4}P $4s$	22	−14.7927	$2s2p^2$ ^{4}P $5d$	27	−7.05099	$2p^3$ ^{2}D $5f$			^{3}I^e
14	−25.0480	$2s2p^2$ ^{4}P $4d$	23	−13.3058	$2s2p^2$ ^{2}D $5s$	28	−6.97284	$2s2p^2$ ^{2}D $6g$	1	−12.3876	$2s2p^2$ ^{2}D $5g$
15	−22.6123	$2s2p^2$ ^{2}D $4d$	24	−12.6199	$2s^22p$ ^{2}P $6p$	29	−6.90449	$2s^22p$ ^{2}P $8f$	2	−12.2327	$2s^22p$ ^{2}P $6h$
16	−22.3065	$2s2p^2$ ^{2}P $4s$	25	−12.5545	$2s2p^2$ ^{2}D $5d$	30	−6.15638	$2s2p^2$ ^{4}P $7d$	3	−8.99912	$2s^22p$ ^{2}P $7h$
17	−20.7980	$2s2p^2$ ^{2}P $4d$	26	−12.3450	$2s2p^2$ ^{2}D $5g$	31	−6.11689	$2s2p^2$ ^{4}P $7g$	4	−6.98879	$2s2p^2$ ^{2}D $6g$
18	−19.2683	$2p^3$ ^{4}S $4p$	27	−12.2563	$2s^22p$ ^{2}P $6f$	32	−5.66829	$2p^3$ ^{2}P $5f$	5	−6.89063	$2s^22p$ ^{2}P $8j$
19	−18.1769	$2s^22p$ ^{2}P $5p$	28	−11.0970	$2s2p^2$ ^{2}S $5d$	33	−5.45481	$2s^22p$ ^{2}P $9f$	6	−6.88637	$2s^22p$ ^{2}P $8h$
20	−17.8328	$2p^3$ ^{2}D $4p$	29	−10.7333	$2s2p^2$ ^{2}P $5d$	34	−5.26948	$2s2p^2$ ^{2}P $6d$	7	−6.11390	$2s2p^2$ ^{4}P $7i$
21	−16.9399	$2p^3$ ^{2}D $4f$	30	−9.38971	$2s2p^2$ ^{4}P $6d$	35	−5.12386	$2s2p^2$ ^{2}P $6g$	8	−5.44454	$2s^22p$ ^{2}P $9j$
22	−16.5753	$2p^3$ ^{2}P $4p$	31	−9.20853	$2s^22p$ ^{2}P $7p$			^{3}G^e	9	−5.44385	$2s^22p$ ^{2}P $9h$
23	−15.5791	$2s2p^2$ ^{4}P $5s$	32	−9.00455	$2s^22p$ ^{2}P $7f$	1	−45.0094	$2s2p^2$ ^{2}D $3d$			^{5}P^e
24	−14.9372	$2s2p^2$ ^{4}P $5d$	33	−7.62473	$2p^3$ ^{2}D $5p$	2	−27.6024	$2s^22p$ ^{2}P $4f$	1	−50.9504	$2s2p^2$ ^{4}P $3s$
25	−12.5969	$2s^22p$ ^{2}P $6p$	34	−7.47225	$2s2p^2$ ^{2}D $6s$	3	−22.8033	$2s2p^2$ ^{2}D $4d$	2	−47.0857	$2s2p^2$ ^{4}P $3d$
26	−12.5006	$2s2p^2$ ^{2}D $5d$	35	−7.11250	$2s2p^2$ ^{2}D $6d$	4	−17.6714	$2s^22p$ ^{2}P $5f$	3	−42.7089	$2p^3$ ^{4}S $3p$
27	−11.4191	$2s2p^2$ ^{2}P $5s$	36	−7.03638	$2p^3$ ^{2}D $5f$	5	−16.9963	$2p^3$ ^{2}D $4f$	4	−26.6120	$2s2p^2$ ^{4}P $4s$
28	−10.6972	$2s2p^2$ ^{2}P $5d$	37	−7.02544	$2s^22p$ ^{2}P $8p$	6	−15.6212	$2p^3$ ^{2}P $4f$	5	−25.0819	$2s2p^2$ ^{4}P $4d$
29	−9.83939	$2s2p^2$ ^{4}P $6s$	38	−6.96905	$2s2p^2$ ^{2}D $6g$	7	−14.7374	$2s2p^2$ ^{4}P $5g$	6	−19.5035	$2p^3$ ^{4}S $4p$
30	−9.47284	$2s2p^2$ ^{4}P $6d$	39	−6.89409	$2s^22p$ ^{2}P $8f$	8	−12.6303	$2s2p^2$ ^{2}D $5d$	7	−15.7002	$2s2p^2$ ^{4}P $5s$
31	−9.18530	$2s^22p$ ^{2}P $7p$	40	−6.20335	$2p^3$ ^{2}P $5p$	9	−12.3896	$2s2p^2$ ^{2}D $5g$	8	−14.9577	$2s2p^2$ ^{4}P $5d$
32	−8.85676	$2p^3$ ^{4}S $5p$	41	−6.13271	$2s2p^2$ ^{4}P $7d$	10	−12.2706	$2s^22p$ ^{2}P $6f$	9	−9.90132	$2s2p^2$ ^{4}P $6s$
33	−7.47946	$2p^3$ ^{2}D $5p$	42	−5.66902	$2p^3$ ^{2}P $5f$	11	−12.2498	$2s^22p$ ^{2}P $6h$	10	−9.48457	$2s2p^2$ ^{4}P $6d$
34	−7.09225	$2s2p^2$ ^{2}D $6d$	43	−5.60820	$2s2p^2$ ^{2}S $6d$	12	−10.8682	$2s2p^2$ ^{2}S $5g$	11	−8.97638	$2p^3$ ^{4}S $5p$
35	−7.02015	$2p^3$ ^{2}D $5f$	44	−5.54124	$2s^22p$ ^{2}P $9p$	13	−10.4905	$2s2p^2$ ^{2}P $5g$	12	−6.44520	$2s2p^2$ ^{4}P $7s$
36	−7.00949	$2s^22p$ ^{2}P $8p$	45	−5.44705	$2s^22p$ ^{2}P $9f$	14	−9.35551	$2s2p^2$ ^{4}P $6g$	13	−6.18880	$2s2p^2$ ^{4}P $7d$
37	−6.40483	$2s2p^2$ ^{4}P $7s$	46	−5.24496	$2s2p^2$ ^{2}P $6d$	15	−9.01473	$2s^22p$ ^{2}P $7f$	14	−4.22590	$2s2p^2$ ^{4}P $8s$
38	−6.18128	$2s2p^2$ ^{4}P $7d$			^{3}F^e	16	−8.99617	$2s^22p$ ^{2}P $7h$	15	−4.05530	$2s2p^2$ ^{4}P $8d$
39	−6.16105	$2p^3$ ^{2}P $5p$	1	−46.7154	$2s2p^2$ ^{4}P $3d$	17	−7.13523	$2s2p^2$ ^{2}D $6d$	16	−3.33811	$2p^3$ ^{4}S $6p$
40	−5.62153	$2s2p^2$ ^{2}P $6s$	2	−44.8416	$2s2p^2$ ^{2}D $3d$	18	−7.06455	$2p^3$ ^{2}D $5f$	17	−2.71191	$2s2p^2$ ^{4}P $9s$
41	−5.52817	$2s^22p$ ^{2}P $9p$	3	−43.2731	$2s2p^2$ ^{2}P $3d$	19	−6.97792	$2s2p^2$ ^{2}D $6g$	18	−2.59378	$2s2p^2$ ^{4}P $9d$
42	−5.22695	$2s2p^2$ ^{2}P $6d$	4	−41.1436	$2p^3$ ^{2}D $3p$	20	−6.90024	$2s^22p$ ^{2}P $8f$			^{5}D^e
		^{3}D^e	5	−27.6573	$2s^22p$ ^{2}P $4f$	21	−6.88905	$2s^22p$ ^{2}P $8h$	1	−47.4280	$2s2p^2$ ^{4}P $3d$
1	−52.2728	$2s^22p$ ^{2}P $3p$	6	−24.8989	$2s2p^2$ ^{4}P $4d$	22	−6.10884	$2s2p^2$ ^{4}P $7g$	2	−25.1936	$2s2p^2$ ^{4}P $4d$
2	−48.4375	$2s2p^2$ ^{4}P $3d$	7	−22.7930	$2s2p^2$ ^{2}D $4d$	23	−5.68371	$2p^3$ ^{2}P $5f$	3	−15.0140	$2s2p^2$ ^{4}P $5d$
3	−46.0898	$2s2p^2$ ^{2}D $3s$	8	−20.9686	$2s2p^2$ ^{2}P $4d$	24	−5.48259	$2s2p^2$ ^{2}S $6g$	4	−9.51240	$2s2p^2$ ^{4}P $6d$
4	−44.7510	$2s2p^2$ ^{2}D $3d$	9	−18.3501	$2p^3$ ^{4}S $4f$	25	−5.45179	$2s^22p$ ^{2}P $9f$	5	−6.20713	$2s2p^2$ ^{4}P $7d$
5	−43.3004	$2s2p^2$ ^{2}S $3d$	10	−18.0614	$2p^3$ ^{2}D $4p$	26	−5.43892	$2s^22p$ ^{2}P $9h$	6	−4.06696	$2s2p^2$ ^{4}P $8d$
6	−42.9669	$2s2p^2$ ^{2}P $3d$	11	−17.6914	$2s^22p$ ^{2}P $5f$	27	−5.10817	$2s2p^2$ ^{2}P $6g$	7	−2.60224	$2s2p^2$ ^{4}P $9d$
7	−41.3025	$2p^3$ ^{2}D $3p$	12	−16.9755	$2p^3$ ^{2}D $4f$			^{3}H^e			^{5}F^e
8	−39.8231	$2p^3$ ^{2}P $3p$	13	−15.6039	$2p^3$ ^{2}P $4f$	1	−16.9966	$2p^3$ ^{2}D $4f$	1	−47.6732	$2s2p^2$ ^{4}P $3d$
9	−28.8251	$2s^22p$ ^{2}P $4p$	14	−14.8690	$2s2p^2$ ^{4}P $5d$	2	−14.7619	$2s2p^2$ ^{4}P $5g$	2	−25.2785	$2s2p^2$ ^{4}P $4d$
10	−27.5400	$2s^22p$ ^{2}P $4f$	15	−14.7574	$2s2p^2$ ^{4}P $5g$	3	−12.3993	$2s2p^2$ ^{2}D $5g$	3	−18.3912	$2p^3$ ^{4}S $4f$
11	−24.7296	$2s2p^2$ ^{4}P $4d$	16	−12.6262	$2s2p^2$ ^{2}D $5d$	4	−12.2500	$2s^22p$ ^{2}P $6h$	4	−15.0506	$2s2p^2$ ^{4}P $5d$
12	−24.1849	$2s2p^2$ ^{2}D $4s$	17	−12.3683	$2s2p^2$ ^{2}D $5g$	5	−10.5173	$2s2p^2$ ^{2}P $5g$	5	−14.7586	$2s2p^2$ ^{4}P $5g$
13	−22.7159	$2s2p^2$ ^{2}D $4d$	18	−12.2816	$2s^22p$ ^{2}P $6f$	6	−9.36839	$2s2p^2$ ^{4}P $6g$	6	−9.53326	$2s2p^2$ ^{4}P $6d$
14	−21.2434	$2s2p^2$ ^{2}S $4d$	19	−10.7724	$2s2p^2$ ^{2}P $5d$	7	−9.00087	$2s^22p$ ^{2}P $7h$	7	−9.37105	$2s2p^2$ ^{4}P $6g$
15	−20.8789	$2s2p^2$ ^{2}P $4d$	20	−10.5184	$2s2p^2$ ^{2}P $5g$	8	−7.06544	$2p^3$ ^{2}D $5f$	8	−8.44361	$2p^3$ ^{4}S $5f$
16	−18.2415	$2s^22p$ ^{2}P $5p$	21	−9.43199	$2s2p^2$ ^{4}P $6d$	9	−6.98254	$2s2p^2$ ^{2}D $6g$	9	−6.21966	$2s2p^2$ ^{4}P $7d$
17	−18.0984	$2p^3$ ^{2}D $4p$	22	−9.37118	$2s2p^2$ ^{4}P $6g$	10	−6.89104	$2s^22p$ ^{2}P $8h$	10	−6.11754	$2s2p^2$ ^{4}P $7g$
18	−17.6442	$2s^22p$ ^{2}P $5f$	23	−9.02016	$2s^22p$ ^{2}P $7f$	11	−6.11715	$2s2p^2$ ^{4}P $7g$	11	−4.07526	$2s2p^2$ ^{4}P $8d$
19	−16.9647	$2p^3$ ^{2}D $4f$	24	−8.41583	$2p^3$ ^{4}S $5f$	12	−6.11456	$2s2p^2$ ^{4}P $7i$	12	−4.00805	$2s2p^2$ ^{4}P $8g$
20	−16.7046	$2p^3$ ^{2}P $4p$	25	−7.56907	$2p^3$ ^{2}D $5p$	13	−5.44721	$2s^22p$ ^{2}P $9h$	13	−3.04288	$2p^3$ ^{4}S $6f$

C-like Fe (Fe^{20+})

i	Energy(Ryds)	Description	i	Energy(Ryds)	Description
14	−2.60790	$2s2p^2$ ^{4}P $9d$	7	−3.01934	$2p^3$ ^{4}S $6h$
15	−2.55979	$2s2p^2$ ^{4}P $9g$	8	−2.55882	$2s2p^2$ ^{4}P $9i$
	^{5}G^e		9	−2.55882	$2s2p^2$ ^{4}P $9g$:
1	−14.7391	$2s2p^2$ ^{4}P $5g$			
2	−9.35709	$2s2p^2$ ^{4}P $6g$			
3	−6.11009	$2s2p^2$ ^{4}P $7g$			
4	−4.00221	$2s2p^2$ ^{4}P $8g$			
5	−2.55689	$2s2p^2$ ^{4}P $9g$			
	^{5}H^e				
1	−14.7632	$2s2p^2$ ^{4}P $5g$			
2	−9.36967	$2s2p^2$ ^{4}P $6g$			
3	−6.11811	$2s2p^2$ ^{4}P $7g$			
4	−6.11811	$2s2p^2$ ^{4}P $7i$			
5	−4.00547	$2s2p^2$ ^{4}P $8g$:			
6	−4.00547	$2s2p^2$ ^{4}P $8i$			

Energies in ascending order from ground state for terms with effective $n \leq 4.0$, $L \leq 4$

Term	i	E(Ryds)	Term	i	E(Ryds)	Term	i	E(Ryds)	Term	i	E(Ryds)	Term	i	E(Ryds)
^{3}P^e	1	0.00000	^{1}P^o	3	74.0603	^{3}G^e	1	78.3906	^{1}D^o	5	81.1158	^{1}F^o	3	85.0206
^{1}D^e	1	0.95700	^{1}F^o	1	74.0786	^{3}D^o	5	78.4145	^{3}P^e	9	81.1855	^{3}F^o	4	85.2469
^{1}S^e	1	2.01600	^{5}D^o	1	74.1008	^{1}P^o	5	78.4267	^{1}D^e	7	81.4479	^{1}S^e	6	85.2932
^{5}S^o	1	2.70300	^{5}P^o	1	74.2761	^{3}F^e	2	78.5584	^{3}P^o	8	82.0400	^{3}P^o	10	85.3989
^{3}D^o	1	5.49800	^{3}D^o	3	74.7223	^{3}D^e	4	78.6490	^{1}P^e	5	82.0716	^{3}D^o	9	85.6826
^{3}P^o	1	6.48900	^{5}S^o	2	74.8579	^{3}P^o	7	78.6551	^{3}D^e	7	82.0975	^{1}D^o	7	85.7651
^{1}D^o	1	8.29700	^{3}D^e	2	74.9625	^{3}S^o	3	78.6646	^{3}F^e	4	82.2564	^{1}F^o	4	85.9817
^{3}S^o	1	8.30400	^{3}P^o	4	75.0276	^{1}F^e	1	78.6708	^{1}F^e	3	82.3566	^{1}P^o	9	86.6985
^{1}P^o	1	9.29000	^{1}D^e	4	75.5213	^{1}G^e	1	78.9300	^{5}D^o	2	82.3599	^{3}P^o	11	93.8988
^{3}P^e	2	12.8180	^{5}F^e	1	75.7268	^{3}P^e	7	79.0228	^{1}P^o	7	82.4160	^{1}P^o	10	93.9951
^{1}D^e	2	13.7200	^{5}D^e	1	75.9720	^{5}S^o	3	79.1050	^{3}D^o	7	83.0237	^{1}P^e	7	94.5097
^{1}S^e	2	15.6350	^{5}P^e	2	76.3143	^{3}S^e	3	79.1536	^{3}P^e	10	83.0901	^{3}D^e	9	94.5749
^{3}P^o	2	69.5097	^{3}P^e	5	76.3452	^{1}D^e	5	79.1789	^{3}S^e	4	83.4321	^{3}S^e	5	94.6570
^{1}P^o	2	69.8096	^{3}S^e	2	76.4867	^{1}D^o	4	79.1995	^{1}D^e	8	83.4324	^{3}P^e	12	94.7132
^{1}P^e	1	70.9351	^{1}D^o	3	76.6297	^{1}P^e	3	79.2844	^{3}D^e	8	83.5769	^{1}D^e	10	94.8678
^{3}D^e	1	71.1272	^{3}F^o	2	76.6633	^{1}P^o	6	79.6296	^{3}F^o	3	83.7336	^{1}S^e	7	95.0900
^{3}S^e	1	71.3506	^{3}F^e	1	76.6846	^{1}S^e	5	79.6914	^{1}P^e	6	83.7454	^{1}D^o	8	95.2618
^{3}P^e	3	71.4658	^{3}P^e	6	76.8324	^{3}D^e	5	80.0996	^{1}S^o	2	83.7947	^{3}F^o	5	95.2686
^{1}D^e	3	72.0318	^{1}F^o	2	76.8834	^{3}F^e	3	80.1269	^{3}G^o	1	83.8704	^{3}D^o	10	95.4059
^{5}P^e	1	72.4496	^{3}D^o	4	76.9419	^{3}S^o	4	80.1670	^{1}G^o	1	83.9621	^{3}P^o	12	95.4354
^{1}S^e	3	72.6633	^{1}S^e	4	76.9831	^{1}D^e	6	80.4315	^{3}P^e	11	83.9952	^{1}F^o	5	95.6592
^{1}D^o	2	72.9247	^{1}P^o	4	76.9983	^{3}D^e	6	80.4331	^{1}D^e	9	84.2474	^{1}P^o	11	95.6622
^{3}F^o	1	72.9419	^{1}P^e	2	77.0697	^{1}P^e	4	80.4553	^{1}P^o	8	84.2502	^{1}F^e	4	95.7365
^{3}D^o	2	73.3590	^{3}P^o	5	77.2542	^{5}P^e	3	80.6911	^{3}D^o	8	84.3803	^{3}F^e	5	95.7427
^{3}P^e	4	73.4098	^{3}D^e	3	77.3102	^{3}P^e	8	80.7168	^{3}P^o	9	84.5503	^{3}G^e	2	95.7976
^{3}P^o	3	73.4391	^{1}S^o	1	78.0642	^{3}D^o	6	80.7273	^{1}D^o	6	84.5863	^{1}G^e	2	95.8312
^{3}S^o	2	73.9039	^{3}P^o	6	78.3196	^{1}F^e	2	80.9876	^{3}S^o	5	84.8169			

C-like Fe (Fe^{20+})

gf-values for transitions involving terms with effective $n \leq 4.0$, $L \leq 4$

$i\ i'$	gf_L	$i\ i'$	gf_L	$i\ i'$	gf_L	$i\ i'$	gf_L	$i\ i'$	gf_L	$i\ i'$	gf_L
	$^1P^o$–$^1S^e$	8 2	1.34E−1	2 2	−3.00E−1	9 4	4.99E−2	5 5	−7.82E−2	3 7	−5.77E−2
1 1	9.20E−2	8 3	2.96E−5	2 3	−2.69E−3	9 5	1.32E−4	5 6	−4.31E−2	3 8	−1.68E−3
1 2	−1.84E−1	8 4	2.91E−5	2 4	−4.50E−4	9 6	1.68E−1	5 7	−1.04E−5	3 9	−6.97E−5
1 3	−9.31E−3	8 5	2.06E−2	2 5	−5.18E−4	9 7	−2.10E−5	6 1	3.03E−5	3 10	−7.46E−3
1 4	−4.61E−2	8 6	−3.59E−3	2 6	−3.86E−4	10 1	1.14E−1	6 2	6.04E−5	4 1	5.63E−1
1 5	−2.58E−1	8 7	−1.98E−5	2 7	−3.74E−1	10 2	9.76E−5	6 3	6.46E−4	4 2	2.03E−2
1 6	−1.20E−1	9 1	9.75E−3	3 1	1.07E−1	10 3	5.53E−4	6 4	8.97E−2	4 3	1.91E−2
1 7	−1.07E−3	9 2	2.20E+0	3 2	−6.86E−6	10 4	2.12E−8	6 5	1.87E−1	4 4	1.11E−1
2 1	7.02E−2	9 3	1.54E−6	3 3	−1.29E−2	10 5	9.86E−6	6 6	1.40E−2	4 5	−1.19E−1
2 2	2.79E−6	9 4	3.95E−4	3 4	−1.04E−1	10 6	2.38E−5	6 7	−6.25E−7	4 6	−1.29E−2
2 3	−1.12E−1	9 5	2.21E−2	3 5	−8.30E−4	10 7	−1.75E−1	7 1	1.51E−5	4 7	−1.55E−2
2 4	−3.89E−2	9 6	7.01E−2	3 6	−1.66E−5	11 1	3.33E−1	7 2	3.87E−5	4 8	−2.60E−4
2 5	−2.37E−3	9 7	−1.84E−5	3 7	−3.97E−2	11 2	5.38E−3	7 3	3.17E−2	4 9	−7.63E−3
2 6	−3.23E−5	10 1	1.35E−2	4 1	8.16E−2	11 3	1.76E−3	7 4	1.14E−3	4 10	−2.31E−3
2 7	−9.68E−2	10 2	4.78E−7	4 2	−1.00E−4	11 4	4.66E−4	7 5	1.90E−2	5 1	2.67E−2
3 1	1.38E+0	10 3	7.67E−2	4 3	−1.31E−1	11 5	2.60E−6	7 6	2.26E−1	5 2	6.13E−4
3 2	1.27E−4	10 4	9.53E−4	4 4	−7.15E−3	11 6	3.02E−5	7 7	−8.11E−5	5 3	6.38E−2
3 3	5.95E−2	10 5	2.35E−4	4 5	−1.06E−1	11 7	1.97E−1	8 1	1.12E+0	5 4	2.79E−2
3 4	−3.32E−3	10 6	2.07E−7	4 6	−5.07E−2		$^1D^o$–$^1P^e$	8 2	1.90E−2	5 5	−3.09E−4
3 5	−2.94E−2	10 7	−1.46E−1	4 7	−6.57E−4	1 1	−2.02E−3	8 3	5.39E−3	5 6	−2.61E−1
3 6	−2.08E−7	11 1	2.70E−1	5 1	5.24E−2	1 2	−2.50E−1	8 4	1.66E−3	5 7	−3.99E−2
3 7	−2.68E−2	11 2	2.99E−4	5 2	4.80E−3	1 3	−3.83E−1	8 5	6.51E−7	5 8	−3.54E−2
4 1	1.29E−2	11 3	5.23E 1	5 3	1.82E 1	1 4	1.64E 3	8 6	1.70E 5	5 9	2.20E 2
4 2	2.18E−3	11 4	1.20E−3	5 4	−5.51E−3	1 5	−3.14E−1	8 7	3.46E−1	5 10	−5.03E−5
4 3	3.66E−3	11 5	1.61E−3	5 5	−6.83E−4	1 6	−4.67E−2		$^1P^o$–$^1D^e$	6 1	1.83E−2
4 4	4.71E−5	11 6	2.99E−6	5 6	−7.09E−2	1 7	−1.30E−4	1 1	2.68E−1	6 2	1.25E−2
4 5	−6.68E−2	11 7	1.16E−1	5 7	−3.59E−5	2 1	1.85E−1	1 2	−8.92E−2	6 3	1.72E−1
4 6	−9.44E−3		$^1S^o$–$^1P^e$	6 1	1.88E−2	2 2	−1.44E−4	1 3	−2.80E−3	6 4	3.80E−2
4 7	−8.08E−8	1 1	1.08E−1	6 2	2.67E−1	2 3	−9.81E−2	1 4	−4.59E−2	6 5	7.72E−4
5 1	4.19E−1	1 2	2.83E−2	6 3	1.60E−4	2 4	−1.46E−1	1 5	−5.37E−1	6 6	−1.53E−3
5 2	9.16E−3	1 3	−4.89E−3	6 4	−2.48E−2	2 5	−6.04E−4	1 6	−2.54E−2	6 7	−1.55E−1
5 3	6.89E−2	1 4	−8.13E−2	6 5	−8.59E−4	2 6	−8.84E−5	1 7	−3.31E+0	6 8	−2.03E−1
5 4	1.20E−1	1 5	−1.05E−1	6 6	−8.02E−2	2 7	−6.74E−2	1 8	−1.24E−3	6 9	−4.96E−3
5 5	−1.18E−2	1 6	−2.23E−2	6 7	−3.28E−3	3 1	1.27E−1	1 9	−3.94E−1	6 10	−4.27E−3
5 6	−2.21E−4	1 7	−6.61E−6	7 1	5.89E−4	3 2	−1.58E−3	1 10	−6.43E−3	7 1	5.90E−3
5 7	−1.92E−4	2 1	1.69E−4	7 2	1.49E−1	3 3	−4.97E−2	2 1	2.20E−1	7 2	1.65E−1
6 1	5.79E−2	2 2	3.68E−4	7 3	2.86E−5	3 4	−3.15E−2	2 2	5.74E−6	7 3	9.73E−5
6 2	1.51E−3	2 3	7.54E−2	7 4	3.00E−3	3 5	−1.22E−1	2 3	−3.86E−1	7 4	1.88E−1
6 3	6.86E−2	2 4	5.59E−3	7 5	2.31E−3	3 6	−8.18E−2	2 4	−4.94E−2	7 5	2.80E−4
6 4	1.60E−2	2 5	5.75E−2	7 6	−1.35E−1	3 7	−1.21E−5	2 5	−1.38E−3	7 6	8.52E−6
6 5	−2.10E−5	2 6	4.24E−5	7 7	−3.53E−6	4 1	3.10E−2	2 6	−1.14E−3	7 7	5.25E−3
6 6	−8.84E−2	2 7	−8.49E−6	8 1	4.16E−5	4 2	3.69E−1	2 7	−9.76E−5	7 8	−3.06E−2
6 7	−2.00E−3		$^1P^o$–$^1P^e$	8 2	2.70E−3	4 3	−9.57E−5	2 8	−2.30E−5	7 9	−3.11E−1
7 1	5.81E−3	1 1	−2.73E−3	8 3	2.83E−2	4 4	−8.55E−4	2 9	−7.62E−6	7 10	−1.01E−5
7 2	1.29E−1	1 2	−1.96E−1	8 4	1.03E−1	4 5	−9.30E−3	2 10	−5.72E−1	8 1	5.56E−3
7 3	2.73E−4	1 3	−1.03E−1	8 5	1.22E−1	4 6	−6.62E−2	3 1	5.10E−2	8 2	7.00E−1
7 4	7.90E−2	1 4	−1.26E+0	8 6	5.52E−3	4 7	−3.71E−4	3 2	1.50E−5	8 3	1.40E−4
7 5	2.28E−4	1 5	1.60E 3	8 7	1.80E 5	5 1	1.20E 3	3 3	1.47E 3	8 4	1.20E 4
7 6	−1.43E−1	1 6	−3.00E−1	9 1	1.55E−5	5 2	1.02E−1	3 4	−6.20E−5	8 5	1.10E−1
7 7	−7.49E−6	1 7	−2.73E−3	9 2	9.31E−6	5 3	6.16E−4	3 5	−2.10E−2	8 6	2.88E−4
8 1	2.62E−4	2 1	−9.37E−2	9 3	8.83E−3	5 4	2.32E−4	3 6	−8.36E−2	8 7	1.40E−3

C-like Fe (Fe^{20+})

i	i'	gf_L	i	i'	gf_L	i	i'	gf_L	i	i'	gf_L	i	i'	gf_L	i	i'	gf_L
8	8	2.59E−2	2	9	−7.28E−4	8	1	1.94E−1	5	2	1.46E−4	3	2	9.17E−2	5	2	1.43E−4
8	9	8.44E−7	2	10	−1.20E−2	8	2	8.03E−5	5	3	2.05E+0	3	3	1.79E−1	5	3	−1.55E−1
8	10	−3.54E−6	3	1	8.88E−1	8	3	4.16E−1	5	4	2.72E−2	3	4	−8.37E−7	5	4	−2.02E−1
9	1	1.32E−5	3	2	9.26E−3	8	4	5.71E−3	5	5	1.76E−3	4	1	6.77E−3	5	5	−1.82E−3
9	2	1.03E−1	3	3	3.63E−3	8	5	7.12E−4	5	6	4.72E−6	4	2	4.17E−1	6	1	2.98E−5
9	3	1.67E−4	3	4	1.41E−1	8	6	2.12E−4	5	7	1.49E−2	4	3	3.19E−3	6	2	4.93E−1
9	4	9.55E−5	3	5	−1.47E−1	8	7	1.89E−3	5	8	2.63E−4	4	4	−1.31E−5	6	3	−1.41E−2
9	5	1.78E−2	3	6	−1.76E−2	8	8	3.01E−6	5	9	8.73E−6	5	1	1.67E−4	6	4	−1.31E−1
9	6	1.87E−5	3	7	−1.19E−2	8	9	6.90E−5	5	10	7.03E−1	5	2	7.72E−5	6	5	−1.06E−3
9	7	2.25E−1	3	8	−3.20E−2	8	10	5.74E−2			**^{1}D^{o}–^{1}F^{e}**	5	3	3.29E−5	7	1	2.17E−1
9	8	1.26E−2	3	9	−9.83E−3			**^{1}F^{o}–^{1}D^{e}**	1	1	−5.69E−1	5	4	−4.86E−3	7	2	1.01E−1
9	9	1.73E−3	3	10	−1.15E−3	1	1	5.36E+0	1	2	−6.07E+0			**^{1}G^{o}–^{1}F^{e}**	7	3	−8.42E−5
9	10	−4.88E−6	4	1	8.35E−2	1	2	8.76E−5	1	3	−3.18E−1	1	1	4.39E−2	7	4	−9.80E−2
10	1	4.37E−2	4	2	2.27E−2	1	3	3.72E−1	1	4	−7.88E−4	1	2	3.05E−2	7	5	−3.13E−3
10	2	1.88E−8	4	3	3.27E−1	1	4	−9.34E−3	2	1	−1.16E−1	1	3	5.11E−1	8	1	1.07E−3
10	3	2.43E−1	4	4	4.94E−2	1	5	−4.54E−3	2	2	−4.29E−2	1	4	−3.34E−5	8	2	1.15E−1
10	4	5.77E−3	4	5	7.35E−5	1	6	−1.71E−2	2	3	−3.30E−3			**^{1}F^{o}–^{1}G^{e}**	8	3	4.25E−4
10	5	3.32E−4	4	6	−4.24E−3	1	7	−3.79E−1	2	4	−4.32E+0	1	1	−8.39E−2	8	4	−1.29E−1
10	6	8.66E−5	4	7	−5.08E−2	1	8	−1.90E−3	3	1	−2.65E−1	1	2	−6.30E+0	8	5	−5.72E−6
10	7	4.38E−4	4	8	−9.62E−4	1	9	−5.21E−6	3	2	−1.01E−3	2	1	−5.53E−1	9	1	1.35E−4
10	8	4.72E−6	4	9	−3.03E−1	1	10	−1.35E−1	3	3	−1.96E−2	2	2	−3.47E−2	9	2	4.86E−5
10	9	6.11E−7	4	10	−6.74E−4	2	1	1.10E+0	3	4	−1.74E−2	3	1	2.03E−1	9	3	8.96E−2
10	10	−5.37E−1	5	1	2.62E−2	2	2	1.39E−2	4	1	3.21E−4	3	2	−1.57E−4	9	4	4.96E−3
11	1	1.19E−2	5	2	4.65E−1	2	3	2.63E−1	4	2	−2.79E−1	4	1	1.62E−1	9	5	−9.51E−6
11	2	9.00E−6	5	3	5.49E−4	2	4	2.88E−1	4	3	−2.87E−1	4	2	−1.19E−5	10	1	1.28E−4
11	3	2.26E−2	5	4	3.78E−1	2	5	−3.78E−2	4	4	−2.77E−2	5	1	1.39E−3	10	2	1.87E−3
11	4	1.41E−4	5	5	2.68E−3	2	6	−1.65E−2	5	1	7.43E−4	5	2	−1.26E−1	10	3	3.95E−2
11	5	4.37E−5	5	6	2.17E−4	2	7	−1.94E−5	5	2	1.98E−3			**^{1}G^{o}–^{1}G^{e}**	10	4	3.32E−1
11	6	1.51E−3	5	7	−1.00E−3	2	8	−1.77E−1	5	3	−3.07E−1	1	1	2.66E−1	10	5	−3.55E−7
11	7	6.91E−4	5	8	−4.58E−1	2	9	−1.04E−1	5	4	−1.67E−5	1	2	−9.65E−6	11	1	1.00E−1
11	8	1.88E−5	5	9	−1.51E−1	2	10	−1.64E−3	6	1	2.57E−1			**^{3}P^{o}–^{3}S^{e}**	11	2	1.86E−4
11	9	2.81E−5	5	10	−2.30E−5	3	1	5.21E−2	6	2	8.32E−6	1	1	−1.51E−2	11	3	4.85E−5
11	10	8.26E−3	6	1	2.69E−2	3	2	2.71E+0	6	3	1.05E−2	1	2	−2.35E−1	11	4	3.19E−6
		^{1}D^{o}–^{1}D^{e}	6	2	1.12E+0	3	3	9.36E−5	6	4	−1.51E−5	1	3	−9.40E−1	11	5	−2.41E−1
1	1	4.57E−1	6	3	6.07E−5	3	4	1.27E−3	7	1	3.88E−2	1	4	−2.10E−1	12	1	1.50E+0
1	2	−5.10E−1	6	4	1.21E−3	3	5	1.84E−2	7	2	1.07E−1	1	5	−4.01E−3	12	2	1.56E−2
1	3	−1.28E−2	6	5	7.73E−3	3	6	4.46E−2	7	3	7.89E−2	2	1	−1.51E−1	12	3	9.93E−3
1	4	−1.38E−1	6	6	5.24E−2	3	7	4.84E−2	7	4	−3.69E−4	2	2	−1.94E−1	12	4	4.20E−6
1	5	−1.58E+0	6	7	2.35E−2	3	8	2.95E−1	8	1	7.35E−4	2	3	−4.96E−4	12	5	5.21E−1
1	6	−4.02E−2	6	8	6.76E−2	3	9	8.26E−5	8	2	1.21E−3	2	4	−1.57E−4			**^{3}S^{o}–^{3}P^{e}**
1	7	−6.39E−1	6	9	9.91E−5	3	10	−1.21E−6	8	3	1.70E−5	2	5	−3.24E−1	1	1	4.21E−1
1	8	−3.22E−1	6	10	−3.25E−6	4	1	1.11E−2	8	4	−2.42E−1	3	1	2.97E−1	1	2	−3.37E−1
1	9	−2.11E−1	7	1	6.98E−4	4	2	4.60E+0			**^{1}F^{o}–^{1}F^{e}**	3	2	−1.34E−6	1	3	−4.23E−3
1	10	−3.33E−7	7	2	2.31E+0	4	3	5.57E−6	1	1	−7.26E−3	3	3	−7.67E−2	1	4	−3.45E−3
2	1	1.07E+0	7	3	1.33E−4	4	4	8.30E−5	1	2	−4.33E−1	3	4	−5.74E−8	1	5	−2.80E−1
2	2	2.05E−4	7	4	1.13E−4	4	5	1.86E−2	1	3	−4.44E−3	3	5	−1.29E−1	1	6	−2.05E−1
2	3	2.59E−2	7	5	7.89E−2	4	6	4.24E−2	1	4	−6.05E−1	4	1	6.42E−4	1	7	−1.89E−2
2	4	−3.19E−3	7	6	1.36E−2	4	7	6.35E−4	2	1	−8.09E−2	4	2	−3.60E−3	1	8	−4.83E+0
2	5	−1.30E−1	7	7	1.26E−1	4	8	2.28E−2	2	2	−4.86E−2	4	3	−2.00E−2	1	9	−1.04E+0
2	6	−7.01E−2	7	8	5.53E−3	4	9	3.77E−1	2	3	−2.69E−1	4	4	−3.99E−3	1	10	−1.61E−1
2	7	−4.09E−4	7	9	8.82E−2	4	10	−6.16E−7	2	4	−3.00E−3	4	5	−3.88E−4	1	11	−1.45E−1
2	8	−7.31E−5	7	10	−7.78E−5	5	1	1.07E+0	3	1	3.13E−2	5	1	1.91E−1	1	12	−5.25E−5

C-like Fe (Fe^{20+})

i	*i'*	gf_L	*i*	*i'*	gf_L	*i*	*i'*	gf_L	*i*	*i'*	gf_L	*i*	*i'*	gf_L	*i*	*i'*	gf_L
2	1	4.31E-1	1	4	-2.95E-1	5	8	-8.95E-3	9	12	-6.34E-5	2	3	4.83E-1	6	7	2.50E-3
2	2	1.58E-3	1	5	-1.49E+0	5	9	-1.59E-5	10	1	1.76E-2	2	4	-1.50E-3	6	8	1.81E-6
2	3	2.48E-2	1	6	-5.52E-2	5	10	-2.38E-1	10	2	8.69E-1	2	5	-7.16E-5	6	9	-1.43E-2
2	4	3.53E-2	1	7	-2.68E+0	5	11	-3.77E-2	10	3	6.57E-5	2	6	-2.00E-3	6	10	-8.41E-1
2	5	-3.18E-1	1	8	-3.31E-1	5	12	-1.69E-3	10	4	7.89E-4	2	7	-5.25E-2	6	11	-4.45E-3
2	6	-4.90E-3	1	9	-1.62E-5	6	1	1.64E-1	10	5	8.75E-3	2	8	-7.01E-1	6	12	-2.30E-5
2	7	-2.52E-2	1	10	-2.25E-3	6	2	6.47E-3	10	6	1.00E-3	2	9	-1.97E-4	7	1	3.25E-2
2	8	-1.19E-4	1	11	-6.03E-1	6	3	1.58E-2	10	7	1.15E-1	2	10	-1.26E-4	7	2	1.56E+0
2	9	-4.39E-2	1	12	-4.82E-3	6	4	5.11E-3	10	8	9.15E-2	2	11	-1.96E-3	7	3	2.00E-5
2	10	-2.13E-3	2	1	4.97E-1	6	5	3.17E-4	10	9	2.04E-3	2	12	-1.97E-1	7	4	1.00E-4
2	11	-1.13E-4	2	2	6.49E-5	6	6	4.82E-2	10	10	2.39E-2	3	1	2.14E+0	7	5	1.14E-1
2	12	-1.51E-3	2	3	-5.19E-1	6	7	-6.30E-6	10	11	1.53E-1	3	2	2.42E-3	7	6	7.54E-3
3	1	1.56E-1	2	4	-1.13E-1	6	8	-3.60E-2	10	12	-6.33E-8	3	3	2.48E-1	7	7	1.55E-2
3	2	1.31E-1	2	5	-3.07E-3	6	9	-1.18E-3	11	1	8.76E-2	3	4	5.29E-1	7	8	3.30E-2
3	3	2.81E-1	2	6	-7.95E-1	6	10	-5.36E-2	11	2	2.68E-6	3	5	-1.07E-3	7	9	7.76E-1
3	4	1.22E-2	2	7	-3.22E-3	6	11	-2.14E-1	11	3	4.79E-1	3	6	-1.15E-3	7	10	-1.67E-4
3	5	7.05E-4	2	8	-3.90E-4	6	12	-7.02E-3	11	4	5.74E-2	3	7	-3.76E-2	7	11	-2.16E-9
3	6	1.43E-1	2	9	-2.82E-4	7	1	2.10E-1	11	5	1.81E-4	3	8	-2.32E-4	7	12	-2.23E-5
3	7	-2.86E-5	2	10	-1.33E-3	7	2	2.44E-3	11	6	8.27E-5	3	9	-5.94E-1	8	1	6.03E-2
3	8	-1.84E-1	2	11	-3.23E-4	7	3	2.64E-1	11	7	4.42E-4	3	10	-3.02E-2	8	2	5.89E+0
3	9	-2.99E-1	2	12	-1.01E+0	7	4	1.36E-1	11	8	8.19E-4	3	11	-4.48E-3	8	3	7.34E-5
3	10	-7.36E-3	3	1	2.82E+0	7	5	2.25E-2	11	9	2.90E-5	3	12	-7.14E-3	8	4	2.50E-3
3	11	-8.42E-2	3	2	5.17E-4	7	6	3.58E-1	11	10	1.08E-6	4	1	3.15E-1	8	5	4.60E-3
3	12	-2.87E-4	3	3	1.38E-1	7	7	-5.02E-5	11	11	9.07E-6	4	2	3.64E-3	8	6	8.26E-6
4	1	7.18E-2	3	4	2.69E-4	7	8	-1.83E-1	11	12	-9.22E-1	4	3	6.25E-2	8	7	1.31E-1
4	2	2.08E-1	3	5	-4.26E-3	7	9	-5.02E-3	12	1	5.50E-1	4	4	1.92E-1	8	8	6.77E-2
4	3	5.48E-3	3	6	-4.14E-5	7	10	-1.77E-1	12	2	7.17E-5	4	5	1.92E-3	8	9	8.99E-2
4	4	3.69E-1	3	7	-1.95E-1	7	11	-1.92E-2	12	3	9.73E-1	4	6	7.08E-4	8	10	2.63E-1
4	5	2.08E-3	3	8	-1.86E-1	7	12	-2.92E-3	12	4	1.33E-1	4	7	-1.37E-1	8	11	1.16E-4
4	6	8.42E-2	3	9	-4.38E-4	8	1	1.47E-2	12	5	1.74E-4	4	8	-2.41E-6	8	12	-9.59E-7
4	7	6.64E-5	3	10	-3.22E-4	8	2	4.43E-1	12	6	3.92E-3	4	9	-4.39E-6	9	1	2.06E-2
4	8	-1.43E-2	3	11	-2.93E-4	8	3	3.36E-3	12	7	4.23E-3	4	10	-9.64E-2	9	2	6.34E+0
4	9	-1.92E-1	3	12	-3.58E-2	8	4	5.14E-4	12	8	3.12E-3	4	11	-2.34E-1	9	3	7.59E-6
4	10	-2.67E-1	4	1	1.27E+0	8	5	2.57E-3	12	9	2.66E-5	4	12	-1.84E-3	9	4	5.30E-4
4	11	-1.24E-1	4	2	3.04E-2	8	6	6.54E-1	12	10	8.53E-5	5	1	4.09E-1	9	5	1.08E-3
4	12	-6.50E-5	4	3	1.67E-1	8	7	6.35E-3	12	11	5.51E-6	5	2	9.74E-8	9	6	1.87E-4
5	1	1.02E-2	4	4	4.18E-1	8	8	8.95E-4	12	12	2.87E-1	5	3	2.72E-1	9	7	4.16E-3
5	2	1.76E+0	4	5	-1.36E-1	8	9	3.43E-3			$^3D^o$–$^3P^e$	5	4	1.13E-1	9	8	1.83E-1
5	3	6.66E-4	4	6	-4.74E-3	8	10	-6.59E-3	1	1	3.34E-1	5	5	5.07E-3	9	9	1.31E-2
5	4	4.04E-4	4	7	-2.86E-2	8	11	-7.49E-1	1	2	-7.03E-1	5	6	6.37E-1	9	10	3.46E-4
5	5	2.17E-3	4	8	-1.35E-2	8	12	-6.08E-5	1	3	-9.49E-2	5	7	-2.26E-4	9	11	6.74E-1
5	6	6.40E-4	4	9	-2.74E-1	9	1	1.70E-2	1	4	-3.48E-1	5	8	-6.40E-3	9	12	-1.00E-4
5	7	1.34E-1	4	10	1.05E-5	9	2	4.18E+0	1	5	-2.66E-1	5	9	-1.89E-3	10	1	1.69E+0
5	8	1.40E-1	4	11	-7.39E-3	9	3	1.52E-5	1	6	-2.05E-1	5	10	-1.63E-3	10	2	4.02E-4
5	9	8.72E-2	4	12	-1.75E-3	9	4	1.65E-3	1	7	-1.13E+0	5	11	-3.40E-1	10	3	3.05E+0
5	10	1.32E-1	5	1	1.20E-1	9	5	4.56E-3	1	8	-2.66E-3	5	12	-2.10E-5	10	4	4.10E-1
5	11	1.86E-3	5	2	9.69E-3	9	6	2.40E-3	1	9	-1.18E-1	6	1	3.36E-2	10	5	8.35E-4
5	12	-1.36E-5	5	3	1.90E-2	9	7	5.46E-3	1	10	-6.28E-1	6	2	6.33E-1	10	6	1.06E-2
		$^3P^o$–$^3P^e$	5	4	1.34E-1	9	8	2.67E-1	1	11	1.22E-2	6	3	1.26E-2	10	7	1.24E-2
1	1	3.68E-1	5	5	8.13E-5	9	9	1.03E-1	1	12	-1.21E-3	6	4	1.30E-3	10	8	1.13E-2
1	2	-2.36E-1	5	6	2.34E-3	9	10	2.89E-1	2	1	8.63E+0	6	5	1.32E-2	10	9	6.81E-7
1	3	-1.38E-2	5	7	-3.22E-1	9	11	2.44E-3	2	2	4.19E-6	6	6	6.46E-1	10	10	6.12E-5

C-like Fe (Fe^{20+})

i i'	gf_L	i i'	gf_L	i i'	gf_L	i i'	gf_L	i i'	gf_L	i i'	gf_L
10 11	2.44E−4	6 5	−7.35E−1	12 3	1.06E−5	5 9	−3.94E−4	1 6	−4.86E−1	3 2	−5.24E−2
10 12	9.59E−1	6 6	−4.22E−3	12 4	5.12E−6	6 1	1.02E−2	1 7	−1.81E−3	3 3	−2.24E−3
	$^3P^o$–$^3D^e$	6 7	−5.80E−2	12 5	2.05E−4	6 2	6.20E−1	1 8	−8.43E−4	3 4	−3.57E−5
1 1	−1.96E−2	6 8	−7.75E−2	12 6	6.06E−3	6 3	2.98E−3	1 9	−3.52E−1	3 5	−1.96E−3
1 2	−3.91E−1	6 9	−7.16E−3	12 7	2.76E−6	6 4	3.61E−3	2 1	3.25E−1	4 1	5.03E−5
1 3	−3.35E+0	7 1	1.38E−1	12 8	4.84E−5	6 5	1.80E−5	2 2	1.26E+0	4 2	−6.57E−1
1 4	−3.13E+0	7 2	6.92E−3	12 9	5.40E−2	6 6	1.24E−4	2 3	−2.74E−3	4 3	−1.32E−2
1 5	−2.83E+0	7 3	4.93E−4		$^3D^o$–$^3D^e$	6 7	−6.70E−1	2 4	−9.90E−2	4 4	−5.76E−2
1 6	−2.53E+0	7 4	2.09E−5	1 1	−6.88E−3	6 8	−9.06E−2	2 5	−2.84E−2	4 5	−1.13E−2
1 7	−1.52E−2	7 5	−2.18E−2	1 2	−9.46E−1	6 9	−4.11E−9	2 6	−2.10E−2	5 1	2.64E−3
1 8	−7.87E−1	7 6	−4.81E−1	1 3	−2.11E+0	7 1	6.57E−5	2 7	−5.01E−2	5 2	−8.97E−5
1 9	−1.33E−2	7 7	−1.96E−1	1 4	−4.70E+0	7 2	1.12E−3	2 8	−6.82E−1	5 3	−8.22E−1
2 1	−7.13E−1	7 8	−3.66E−1	1 5	−5.63E−2	7 3	3.07E−1	2 9	−5.22E−4	5 4	−5.42E−1
2 2	−4.34E−1	7 9	−6.34E−3	1 6	−4.27E−1	7 4	1.21E−3	3 1	8.05E−6	5 5	−1.25E−1
2 3	−9.81E−3	8 1	4.34E−4	1 7	−9.42E−1	7 5	3.23E−3	3 2	1.69E−3	6 1	5.28E−3
2 4	−9.98E−5	8 2	3.92E−1	1 8	−2.66E−2	7 6	3.50E−2	3 3	1.52E−3	6 2	1.24E−2
2 5	−8.06E−5	8 3	4.45E−3	1 9	−1.73E−4	7 7	3.85E−2	3 4	1.02E−1	6 3	2.22E−3
2 6	−1.32E−3	8 4	4.85E−3	2 1	2.01E−1	7 8	−1.21E−3	3 5	1.66E−2	6 4	−1.37E+0
2 7	−3.24E−4	8 5	2.11E−3	2 2	−1.20E−4	7 9	−4.62E−5	3 6	1.13E−1	6 5	−1.38E−3
2 8	−7.27E−5	8 6	2.80E−3	2 3	−8.88E−3	8 1	8.93E−4	3 7	7.65E−1	7 1	7.56E−1
2 9	−1.78E+0	8 7	−4.12E−4	2 4	−2.63E−1	8 2	2.18E−4	3 8	5.86E−5	7 2	6.08E−3
3 1	3.89E−2	8 8	−8.94E−1	2 5	−9.12E−2	8 3	5.40E−2	3 9	−1.12E−4	7 3	4.94E−4
3 2	−7.29E−7	8 9	−3.96E−5	2 6	−1.42E−1	8 4	9.73E−2	4 1	3.90E−5	7 4	5.87E−3
3 3	−4.04E−4	9 1	2.15E−4	2 7	−2.58E−5	8 5	5.37E−2	4 2	8.44E−4	7 5	−2.50E−4
3 4	−7.82E−2	9 2	3.03E−3	2 8	−1.16E−3	8 6	3.31E−1	4 3	4.34E−3	8 1	4.39E−2
3 5	−1.61E−2	9 3	5.11E−3	2 9	−7.30E−2	8 7	5.67E−1	4 4	1.48E−4	8 2	3.60E−1
3 6	−4.01E−1	9 4	3.48E−1	3 1	3.34E−3	8 8	1.18E−3	4 5	5.93E−2	8 3	4.17E−2
3 7	−1.97E−3	9 5	7.61E−3	3 2	−4.44E−3	8 9	−2.40E−5	4 6	2.22E−1	8 4	5.61E−2
3 8	−2.69E−3	9 6	3.50E−2	3 3	−2.91E−1	9 1	1.31E−4	4 7	5.13E−3	8 5	−9.04E−5
3 9	−4.19E−2	9 7	1.50E−1	3 4	−9.08E−2	9 2	4.33E−5	4 8	1.20E+0	9 1	2.61E−3
4 1	2.10E−3	9 8	3.41E−2	3 5	−3.64E−3	9 3	8.25E−3	4 9	−1.35E−4	9 2	3.32E−1
4 2	4.34E−4	9 9	−1.80E−5	3 6	−1.33E−3	9 4	7.81E−2	5 1	6.34E+0	9 3	5.42E−1
4 3	−7.77E−1	10 1	2.70E−4	3 7	−8.88E−3	9 5	3.26E−1	5 2	8.19E−2	9 4	7.24E−2
4 4	−7.99E−3	10 2	1.94E−4	3 8	−1.87E−3	9 6	5.47E−5	5 3	1.10E−3	9 5	−2.70E−4
4 5	−3.56E−3	10 3	3.81E−3	3 9	−1.07E−5	9 7	4.29E−5	5 4	1.01E−2	10 1	8.05E−5
4 6	−3.42E−6	10 4	1.14E−1	4 1	2.99E−1	9 8	2.75E−1	5 5	2.05E−2	10 2	3.62E−4
4 7	−1.68E−3	10 5	2.89E−1	4 2	9.37E−1	9 9	−5.03E−5	5 6	2.66E−2	10 3	5.60E−3
4 8	−1.20E−4	10 6	3.55E−1	4 3	−2.05E−3	10 1	1.09E+0	5 7	7.86E−5	10 4	1.76E−4
4 9	−1.23E−3	10 7	3.27E−2	4 4	−2.66E−1	10 2	1.50E−2	5 8	3.18E−4	10 5	−5.13E−1
5 1	5.90E−3	10 8	4.01E−3	4 5	−2.07E−2	10 3	2.46E−4	5 9	1.79E+0		$^3F^o$–$^3F^e$
5 2	7.03E−1	10 9	−3.85E−5	4 6	−1.14E−2	10 4	2.24E−3		$^3D^o$–$^3F^e$	1 1	−1.68E−2
5 3	−3.01E−4	11 1	6.27E−1	4 7	−7.04E−1	10 5	6.23E−3	1 1	−3.97E+0	1 2	−2.90E−1
5 4	−2.22E−1	11 2	2.01E−3	4 8	−8.80E−2	10 6	1.41E−3	1 2	−1.29E+1	1 3	−9.88E−1
5 5	−2.71E−4	11 3	3.74E−4	4 9	−3.94E−5	10 7	4.98E−6	1 3	−3.03E+0	1 4	−1.38E−2
5 6	−2.15E−2	11 4	1.06E−4	5 1	8.72E−1	10 8	1.02E−5	1 4	−8.56E−1	1 5	−1.71E+0
5 7	−1.65E−1	11 5	9.61E−4	5 2	2.10E−2	10 9	3.82E−1	1 5	−5.71E−4	2 1	−3.61E−4
5 8	−2.18E−4	11 6	1.23E−3	5 3	4.02E−4		$^3F^o$–$^3D^e$	2 1	−1.65E−2	2 2	−2.68E−1
5 9	−8.94E−3	11 7	6.23E−6	5 4	−2.24E−3	1 1	9.67E−1	2 2	−1.39E−1	2 3	−5.75E−4
6 1	5.67E−1	11 8	6.91E−6	5 5	−7.22E−3	1 2	−3.62E−4	2 3	−3.04E−1	2 4	−8.68E−1
6 2	2.49E−2	11 9	−1.19E+0	5 6	−1.72E−1	1 3	−1.22E−3	2 4	−9.70E−3	2 5	−1.46E−2
6 3	1.16E−3	12 1	8.34E−2	5 7	−2.52E−1	1 4	−4.14E−3	2 5	−1.31E+1	3 1	6.78E−2
6 4	−1.58E−6	12 2	9.10E−4	5 8	−5.10E−1	1 5	−7.40E−1	3 1	−1.13E+0	3 2	4.49E−1

C-like Fe (Fe^{20+})

$i\ i'$	gf_L	$i\ i'$	gf_L	$i\ i'$	gf_L	$i\ i'$	gf_L	$i\ i'$	gf_L	$i\ i'$	gf_L
3 3	5.05E−1	5 4	7.66E−6	2 1	−1.36E+0		$^5\mathbf{S}^o$–$^5\mathbf{P}^e$		$^5\mathbf{P}^o$–$^5\mathbf{P}^e$		$^5\mathbf{P}^o$–$^5\mathbf{D}^e$
3 4	2.49E−1	5 5	−9.07E−2	2 2	−1.61E−1	1 1	−6.60E−1	1 1	9.40E−1	1 1	−9.29E−1
3 5	−9.98E−5		$^3\mathbf{G}^o$–$^3\mathbf{F}^e$	3 1	1.21E−1	1 2	−9.73E+0	1 2	−4.20E−1		$^5\mathbf{D}^o$–$^5\mathbf{D}^e$
4 1	1.02E−2	1 1	4.49E−3	3 2	−8.11E−5	1 3	−8.45E−1	1 3	−5.16E−1	1 1	−3.66E−1
4 2	8.83E−2	1 2	1.47E−2	4 1	1.01E+0	2 1	4.37E−1		$^5\mathbf{D}^o$–$^5\mathbf{P}^e$	2 1	6.94E−1
4 3	4.26E−1	1 3	1.85E−1	4 2	−8.19E−4	2 2	−2.99E−1	1 1	1.42E+0		$^5\mathbf{D}^o$–$^5\mathbf{F}^e$
4 4	8.88E−2	1 4	1.54E+0	5 1	3.34E−3	2 3	−1.79E−1	1 2	−7.59E−5	1 1	−1.58E+0
4 5	−1.98E−5	1 5	−7.96E−5	5 2	−1.19E+0	3 1	4.98E−1	1 3	−9.44E−1	2 1	1.41E+0
5 1	2.99E−5		$^3\mathbf{F}^o$–$^3\mathbf{G}^e$		$^3\mathbf{G}^o$–$^3\mathbf{G}^e$	3 2	7.58E−3	2 1	3.16E−3		
5 2	9.81E−5	1 1	−2.75E−1	1 1	8.97E−1	3 3	−9.10E−1	2 2	1.79E−1		
5 3	5.71E−4	1 2	−1.84E+1	1 2	−5.52E−5			2 3	1.34E+0		

N-like S (S^{9+})

Term energies relative to $2s^22p^2$ ^{3}P ionization threshold for each symmetry

i	E(Ryds)	Description	i	E(Ryds)	Description	i	E(Ryds)	Description	i	E(Ryds)	Description
		$^2S^e$	9	−4.27648	$2s^22p^2$ ^{3}P $5p$	35	−1.24911	$2s^22p^2$ ^{3}P $9d$	10	−6.25828	$2p^4$ ^{3}P $3s$
1	−27.4095	$2s2p^4$	10	−3.14374	$2s2p^3$ ^{3}D $4d$			**$^2P^o$**	11	−5.34933	$2s2p^3$ ^{5}S $4p$
2	−12.6621	$2s^22p^2$ ^{1}S $3s$	11	−3.03306	$2s2p^3$ ^{5}S $5s$	1	−31.7870	$2s^22p^3$	12	−4.52304	$2s^22p^2$ ^{3}P $5s$
3	−10.7455	$2s^22p^2$ ^{1}D $3d$	12	−2.92127	$2s^22p^2$ ^{3}P $6p$	2	−23.9887	$2p^5$	13	−4.20930	$2p^4$ ^{3}P $3d$
4	−8.56915	$2s2p^3$ ^{3}P $3p$	13	−2.33803	$2s2p^3$ ^{3}S $4s$	3	−12.5560	$2s^22p^2$ ^{3}P $3p$	14	−4.09130	$2s^22p^2$ ^{3}P $5d$
5	−7.22455	$2s2p^3$ ^{1}P $3p$	14	−2.13488	$2s^22p^2$ ^{3}P $7p$	4	−11.8822	$2s^22p^2$ ^{1}D $3p$	15	−3.54191	$2s2p^3$ ^{3}D $4p$
6	−6.31414	$2s^22p^2$ ^{1}S $4s$	15	−1.64385	$2s^22p^2$ ^{3}P $8p$	5	−11.6048	$2s^22p^2$ ^{1}S $3p$	16	−3.37125	$2s2p^3$ ^{3}P $4p$
7	−5.65724	$2s^22p^2$ ^{1}D $4d$	16	−1.60931	$2s2p^3$ ^{5}S $6s$	6	−9.77647	$2s2p^3$ ^{3}P $3s$	17	−3.06013	$2s^22p^2$ ^{3}P $6s$
8	−4.55492	$2p^4$ ^{1}S $3s$	17	−1.27967	$2s^22p^2$ ^{3}P $9p$	7	−8.32632	$2s2p^3$ ^{1}P $3s$	18	−3.04701	$2s2p^3$ ^{3}D $4f$
9	−3.63978	$2s^22p^2$ ^{1}S $5s$			**$^6S^o$**	8	−8.27276	$2s2p^3$ ^{3}D $3d$	19	−2.83950	$2s2p^3$ ^{5}S $5p$
10	−3.61293	$2p^4$ ^{1}D $3d$	1	−12.3922	$2s2p^3$ ^{5}S $3s$	9	−7.48287	$2s2p^3$ ^{3}P $3d$	20	−2.83053	$2s^22p^2$ ^{3}P $6d$
11	−3.29268	$2s^22p^2$ ^{1}D $5d$	2	−5.88833	$2s2p^3$ ^{5}S $4s$	10	−6.75329	$2s^22p^2$ ^{3}P $4p$	21	−2.22374	$2s^22p^2$ ^{3}P $7s$
12	−3.18279	$2s2p^3$ ^{3}P $4p$	3	−3.10864	$2s2p^3$ ^{5}S $5s$	11	−6.63341	$2s2p^3$ ^{1}D $3d$	22	−2.07630	$2s^22p^2$ ^{3}P $7d$
13	−2.22556	$2s^22p^2$ ^{1}S $6s$	4	−1.66310	$2s2p^3$ ^{5}S $6s$	12	−6.26416	$2s2p^3$ ^{1}P $3d$	23	−1.96214	$2s2p^3$ ^{3}S $4p$
14	−2.01974	$2s^22p^2$ ^{1}D $6d$	5	−.81569	$2s2p^3$ ^{5}S $7s$	13	−6.08474	$2s^22p^2$ ^{1}D $4p$	24	−1.68102	$2s^22p^2$ ^{3}P $8s$
15	−1.70159	$2s2p^3$ ^{1}P $4p$	6	−.27651	$2s2p^3$ ^{5}S $8s$	14	−5.91717	$2s^22p^2$ ^{1}S $4p$	25	−1.58458	$2s^22p^2$ ^{3}P $8d$
16	−1.37783	$2s^22p^2$ ^{1}S $7s$			**$^2P^e$**	15	−5.51201	$2s^22p^2$ ^{1}D $4f$	26	−1.51959	$2s2p^3$ ^{5}S $6p$
17	−1.25788	$2s^22p^2$ ^{1}D $7d$	1	−27.1800	$2s2p^4$	16	−5.20767	$2p^4$ ^{3}P $3p$:	27	−1.31882	$2s^22p^2$ ^{3}P $9s$
18	−.85275	$2s^22p^2$ ^{1}S $8s$	2	−13.6203	$2s^22p^2$ ^{3}P $3s$	17	−4.33263	$2p^4$ ^{1}D $3p$	28	−1.25027	$2s^22p^2$ ^{3}P $9d$
19	−.78714	$2s2p^3$ ^{3}P $5p$	3	−11.6546	$2s^22p^2$ ^{3}P $3d$	18	−4.25069	$2s^22p^2$ ^{3}P $5p$			**$^4P^o$**
20	−.76676	$2s^22p^2$ ^{1}D $8d$	4	−10.8352	$2s^22p^2$ ^{1}D $3d$	19	−3.69456	$2s2p^3$ ^{3}P $4s$	1	−12.8289	$2s^22p^2$ ^{3}P $3p$
21	−.48989	$2s^22p^2$ ^{1}S $9s$	5	−9.69480	$2s2p^3$ ^{3}D $3p$	20	−3.60436	$2p^4$ ^{1}S $3p$	2	−10.0932	$2s2p^3$ ^{3}P $3s$
22	−.43179	$2s^22p^2$ ^{1}D $9d$	6	−8.98510	$2s2p^3$ ^{3}P $3p$	21	−3.50397	$2s^22p^2$ ^{1}D $5p$	3	−8.37692	$2s2p^3$ ^{3}D $3d$
		$^2S^o$	7	−8.03359	$2s2p^3$ ^{1}D $3p$:	22	−3.43230	$2s^22p^2$ ^{1}S $5p$	4	−7.99061	$2s2p^3$ ^{3}P $3d$
1	−13.0342	$2s^22p^2$ ^{3}P $3p$	8	−7.48026	$2s2p^3$ ^{3}S $3p$	23	−3.21279	$2s^22p^2$ ^{1}D $5f$	5	−6.89096	$2s^22p^2$ ^{3}P $4p$
2	−8.76297	$2s2p^3$ ^{3}S $3s$	9	−7.23618	$2s2p^3$ ^{1}P $3p$	24	−3.12912	$2s2p^3$ ^{3}D $4d$	6	−5.49640	$2p^4$ ^{3}P $3p$
3	−8.45297	$2s2p^3$ ^{3}D $3d$	10	−7.14119	$2s^22p^2$ ^{3}P $4s$	25	−2.91902	$2s^22p^2$ ^{3}P $6p$	7	−4.32854	$2s^22p^2$ ^{3}P $5p$
4	−6.94315	$2s^22p^2$ ^{3}P $4p$	11	−6.46310	$2s^22p^2$ ^{3}P $4d$	26	−2.79772	$2s2p^3$ ^{3}P $4d$	8	−3.79606	$2s2p^3$ ^{3}P $4s$
5	−6.58197	$2s2p^3$ ^{1}D $3d$	12	−6.00707	$2p^4$ ^{3}P $3s$	27	−2.18332	$2s^22p^2$ ^{1}D $6p$	9	−3.17397	$2s2p^3$ ^{3}D $4d$
6	−5.14852	$2p^4$ ^{3}P $3p$	13	−5.68788	$2s^22p^2$ ^{1}D $4d$	28	−2.11962	$2s^22p^2$ ^{3}P $7p$	10	−3.00212	$2s2p^3$ ^{3}P $4d$
7	−4.35350	$2s^22p^2$ ^{3}P $5p$	14	−4.48774	$2s^22p^2$ ^{3}P $5s$	29	−2.10914	$2s^22p^2$ ^{1}S $6p$	11	−2.93933	$2s^22p^2$ ^{3}P $6p$
8	−3.19289	$2s2p^3$ ^{3}D $4d$	15	−4.13170	$2p^4$ ^{3}P $3d$	30	−2.07399	$2s2p^3$ ^{1}P $4s$	12	−2.15245	$2s^22p^2$ ^{3}P $7p$
9	−2.97683	$2s^22p^2$ ^{3}P $6p$	16	−4.09940	$2s^22p^2$ ^{3}P $5d$	31	−1.97518	$2s^22p^2$ ^{1}D $6f$	13	−1.63756	$2s^22p^2$ ^{3}P $8p$
10	−2.30394	$2s2p^3$ ^{3}S $4s$	17	−3.60078	$2s2p^3$ ^{3}D $4p$	32	−1.62113	$2s^22p^2$ ^{3}P $8p$	14	−1.28808	$2s^22p^2$ ^{3}P $9p$
11	−2.14071	$2s^22p^2$ ^{3}P $7p$	18	−3.49635	$2p^4$ ^{1}D $3d$	33	−1.49811	$2s2p^3$ ^{1}D $4d$	15	−1.09671	$2s2p^3$ ^{3}P $5s$
12	−1.64076	$2s^22p^2$ ^{3}P $8p$	19	−3.36664	$2s2p^3$ ^{3}P $4p$	34	−1.33739	$2s^22p^2$ ^{1}S $7p$			**$^6P^e$**
13	−1.48480	$2s2p^3$ ^{1}D $4d$	20	−3.30709	$2s^22p^2$ ^{1}D $5d$	35	−1.33122	$2s^22p^2$ ^{1}D $7p$	1	−11.4149	$2s2p^3$ ^{5}S $3p$
14	−1.29024	$2s^22p^2$ ^{3}P $9p$	21	−3.04801	$2s^22p^2$ ^{3}P $6s$	36	−1.27794	$2s2p^3$ ^{1}P $4d$	2	−5.50074	$2s2p^3$ ^{5}S $4p$
		$^4S^e$	22	−3.01312	$2s2p^3$ ^{3}D $4f$	37	−1.27262	$2s^22p^2$ ^{3}P $9p$	3	−2.91728	$2s2p^3$ ^{5}S $5p$
1	−9.18897	$2s2p^3$ ^{3}P $3p$	23	−2.83692	$2s^22p^2$ ^{3}P $6d$	38	−1.21684	$2s^22p^2$ ^{1}D $7f$	4	−1.55495	$2s2p^3$ ^{5}S $6p$
2	−3.43826	$2s2p^3$ ^{3}P $4p$	24	−2.21524	$2s^22p^2$ ^{3}P $7s$	39	−1.04613	$2s2p^3$ ^{3}P $5s$	5	−.74872	$2s2p^3$ ^{5}S $7p$
3	−.91231	$2s2p^3$ ^{3}P $5p$	25	−2.08337	$2s^22p^2$ ^{3}P $7d$			**$^4P^e$**	6	−.23221	$2s2p^3$ ^{5}S $8p$
		$^4S^o$	26	−2.02678	$2s^22p^2$ ^{1}D $6d$	1	−29.5622	$2s2p^4$			**$^2D^e$**
1	−32.9909	$2s^22p^3$	27	−1.97780	$2s2p^3$ ^{1}D $4p$	2	−13.8358	$2s^22p^2$ ^{3}P $3s$	1	−28.2764	$2s2p^4$
2	−12.6804	$2s^22p^2$ ^{3}P $3p$	28	−1.79024	$2s2p^3$ ^{3}S $4p$	3	−11.5274	$2s^22p^2$ ^{3}P $3d$	2	−13.2057	$2s^22p^2$ ^{1}D $3s$
3	−11.7974	$2s2p^3$ ^{5}S $3s$	29	−1.69734	$2s2p^3$ ^{1}P $4p$	4	−11.0689	$2s2p^3$ ^{5}S $3p$	3	−11.2476	$2s^22p^2$ ^{3}P $3d$
4	−8.92169	$2s2p^3$ ^{3}S $3s$	30	−1.66571	$2s^22p^2$ ^{3}P $8s$	5	−9.33977	$2s2p^3$ ^{3}D $3p$	4	−10.9618	$2s^22p^2$ ^{1}D $3d$
5	−8.25162	$2s2p^3$ ^{3}D $3d$	31	−1.58794	$2s^22p^2$ ^{3}P $8d$	6	−8.94839	$2s2p^3$ ^{3}P $3p$	5	−10.4445	$2s^22p^2$ ^{1}S $3d$
6	−6.80460	$2s^22p^2$ ^{3}P $4p$	32	−1.39526	$2s2p^3$ ^{1}D $4f$	7	−8.00738	$2s2p^3$ ^{3}S $3p$	6	−9.22262	$2s2p^3$ ^{3}D $3p$
7	−5.73762	$2s2p^3$ ^{5}S $4s$	33	−1.31304	$2s^22p^2$ ^{3}P $9s$	8	−7.27395	$2s^22p^2$ ^{3}P $4s$	7	−8.89406	$2s2p^3$ ^{3}P $3p$
8	−5.13522	$2p^4$ ^{3}P $3p$	34	−1.26562	$2s^22p^2$ ^{1}D $7d$	9	−6.42432	$2s^22p^2$ ^{3}P $4d$	8	−7.80909	$2s2p^3$ ^{1}D $3p$

N-like S (S^{9+})

i	E(Ryds)	Description
9	−7.37334	$2s2p^3$ ^{1}P $3p$
10	−6.53964	$2s^22p^2$ ^{1}D $4s$
11	−6.31027	$2s^22p^2$ ^{3}P $4d$
12	−5.72939	$2s^22p^2$ ^{1}D $4d$
13	−5.63971	$2p^4$ ^{1}D $3s$
14	−5.50497	$2s^22p^2$ ^{1}S $4d$
15	−4.07427	$2p^4$ ^{3}P $3d$
16	−4.03711	$2s^22p^2$ ^{3}P $5d$
17	−3.73787	$2s^22p^2$ ^{1}D $5s$
18	−3.53396	$2p^4$ ^{1}D $3d$
19	−3.45166	$2s2p^3$ ^{3}D $4p$
20	−3.32961	$2s^22p^2$ ^{1}D $5d$
21	−3.30735	$2s2p^3$ ^{3}P $4p$
22	−3.23327	$2s^22p^2$ ^{1}D $5g$
23	−3.22643	$2s^22p^2$ ^{1}S $5d$
24	−3.03245	$2s2p^3$ ^{3}D $4f$
25	−2.79796	$2s^22p^2$ ^{3}P $6d$
26	−2.70780	$2s2p^3$ ^{3}P $4f$
27	−2.60174	$2p^4$ ^{1}S $3d$
28	−2.28025	$2s^22p^2$ ^{1}D $6s$
29	−2.05743	$2s^22p^2$ ^{3}P $7d$
30	−2.03792	$2s^22p^2$ ^{1}D $6d$
31	−1.99336	$2s^22p^2$ ^{1}D $6g$
32	−1.98758	$2s^22p^2$ ^{1}S $6d$
33	−1.91492	$2s2p^3$ ^{1}D $4p$
34	−1.73051	$2s2p^3$ ^{1}P $4p$
35	−1.57210	$2s^22p^2$ ^{3}P $8d$
36	−1.40508	$2s^22p^2$ ^{1}D $7s$
37	−1.39309	$2s2p^3$ ^{1}D $4f$
38	−1.26867	$2s^22p^2$ ^{1}D $7d$
39	−1.24832	$2s^22p^2$ ^{1}S $7d$:
40	−1.24339	$2s^22p^2$ ^{1}D $7g$
41	−1.23616	$2s^22p^2$ ^{3}P $9d$:
		^{2}**D**$^\text{o}$
1	−32.2055	$2s^22p^3$
2	−12.6869	$2s^22p^2$ ^{3}P $3p$
3	−12.0329	$2s^22p^2$ ^{1}D $3p$
4	−10.3044	$2s2p^3$ ^{3}D $3s$
5	−8.86886	$2s2p^3$ ^{1}D $3s$
6	−8.23847	$2s2p^3$ ^{3}D $3d$
7	−7.76528	$2s2p^3$ ^{3}P $3d$
8	−6.93570	$2s2p^3$ ^{1}D $3d$.
9	−6.78515	$2s^22p^2$ ^{3}P $4p$
10	−6.55244	$2s2p^3$ ^{3}S $3d$
11	−6.25980	$2s^22p^2$ ^{3}P $4f$
12	−6.15170	$2s^22p^2$ ^{1}D $4p$
13	−6.08763	$2s2p^3$ ^{1}P $3d$
14	−5.54783	$2s^22p^2$ ^{1}D $4f$
15	−5.23442	$2p^4$ ^{3}P $3p$
16	−4.65294	$2p^4$ ^{1}D $3p$
17	−4.27596	$2s^22p^2$ ^{3}P $5p$
18	−4.00678	$2s^22p^2$ ^{3}P $5f$
19	−3.91123	$2s2p^3$ ^{3}D $4s$
20	−3.52510	$2s^22p^2$ ^{1}D $5p$
21	−3.22628	$2s^22p^2$ ^{1}D $5f$
22	−3.09151	$2s2p^3$ ^{3}D $4d$
23	−2.95711	$2s^22p^2$ ^{3}P $6p$
24	−2.88593	$2s2p^3$ ^{3}P $4d$
25	−2.78362	$2s^22p^2$ ^{3}P $6f$
26	−2.33732	$2s2p^3$ ^{1}D $4s$
27	−2.15663	$2s^22p^2$ ^{1}D $6p$
28	−2.12660	$2s^22p^2$ ^{3}P $7p$
29	−2.04685	$2s^22p^2$ ^{3}P $7f$
30	−1.98271	$2s^22p^2$ ^{1}D $6f$
31	−1.63215	$2s^22p^2$ ^{3}P $8p$
32	−1.58546	$2s2p^3$ ^{1}D $4d$
33	−1.55993	$2s^22p^2$ ^{3}P $8f$
34	−1.41915	$2s2p^3$ ^{3}S $4d$
35	−1.34248	$2s^22p^2$ ^{1}D $7p$
36	−1.29282	$2s2p^3$ ^{1}P $4d$
37	−1.27508	$2s^22p^2$ ^{3}P $9p$
38	−1.23338	$2s^22p^2$ ^{3}P $9f$
39	−1.22460	$2s^22p^2$ ^{1}D $7f$
40	−1.14155	$2s2p^3$ ^{3}D $5s$
		^{4}**D**$^\text{e}$
1	−11.6640	$2s^22p^2$ ^{3}P $3d$
2	−9.73266	$2s2p^3$ ^{3}D $3p$
3	−9.13091	$2s2p^3$ ^{3}P $3p$
4	−6.48407	$2s^22p^2$ ^{3}P $4d$
5	−4.48053	$2p^4$ ^{3}P $3d$
6	−4.12102	$2s^22p^2$ ^{3}P $5d$
7	−3.65461	$2s2p^3$ ^{3}D $4p$
8	−3.41978	$2s2p^3$ ^{3}P $4p$
9	−3.06971	$2s2p^3$ ^{3}D $4f$
10	−2.84640	$2s^22p^2$ ^{3}P $6d$
11	−2.71464	$2s2p^3$ ^{3}P $4f$
12	−2.08407	$2s^22p^2$ ^{3}P $7d$
13	−1.59152	$2s^22p^2$ ^{3}P $8d$
14	−1.25500	$2s^22p^2$ ^{3}P $9d$
		^{4}**D**$^\text{o}$
1	−12.9017	$2s^22p^2$ ^{3}P $3p$
2	−10.6382	$2s2p^3$ ^{3}D $3s$
3	−9.87684	$2s2p^3$ ^{5}S $3d$
4	−8.49441	$2s2p^3$ ^{3}D $3d$
5	−7.92139	$2s2p^3$ ^{3}P $3d$
6	−7.01083	$2s^22p^2$ ^{3}P $4p$:
7	−6.84348	$2s2p^3$ ^{3}S $3d$
8	−6.22431	$2s^22p^2$ ^{3}P $4f$
9	−5.34080	$2p^4$ ^{3}P $3p$
10	−4.91744	$2s2p^3$ ^{5}S $4d$
11	−4.34093	$2s^22p^2$ ^{3}P $5p$
12	−4.01484	$2s2p^3$ ^{3}D $4s$
13	−4.00902	$2s^22p^2$ ^{3}P $5f$
14	−3.23219	$2s2p^3$ ^{3}D $4d$
15	−2.99514	$2s2p^3$ ^{3}P $4d$
16	−2.93389	$2s^22p^2$ ^{3}P $6p$
17	−2.78414	$2s^22p^2$ ^{3}P $6f$
18	−2.62539	$2s2p^3$ ^{5}S $5d$
19	−2.15741	$2s^22p^2$ ^{3}P $7p$
20	−2.04659	$2s^22p^2$ ^{3}P $7f$
21	−1.64744	$2s^22p^2$ ^{3}P $8p$
22	−1.58818	$2s2p^3$ ^{3}S $4d$
23	−1.55102	$2s^22p^2$ ^{3}P $8f$
24	−1.39775	$2s2p^3$ ^{5}S $6d$
25	−1.28911	$2s^22p^2$ ^{3}P $9p$
26	−1.23638	$2s^22p^2$ ^{3}P $9f$
27	−1.19189	$2s2p^3$ ^{3}D $5s$
		^{6}**D**$^\text{o}$
1	−10.3278	$2s2p^3$ ^{5}S $3d$
2	−5.10324	$2s2p^3$ ^{5}S $4d$
3	−2.72520	$2s2p^3$ ^{5}S $5d$
4	−1.44723	$2s2p^3$ ^{5}S $6d$
5	−.68220	$2s2p^3$ ^{5}S $7d$
6	−.18823	$2s2p^3$ ^{5}S $8d$
		^{2}**F**$^\text{e}$
1	−11.5502	$2s^22p^2$ ^{3}P $3d$
2	−10.9728	$2s^22p^2$ ^{1}D $3d$
3	−9.56392	$2s2p^3$ ^{3}D $3p$
4	−7.94835	$2s2p^3$ ^{1}D $3p$
5	−6.39467	$2s^22p^2$ ^{3}P $4d$
6	−5.76454	$2s^22p^2$ ^{1}D $4d$
7	−4.24312	$2p^4$ ^{3}P $3d$
8	−4.07171	$2s^22p^2$ ^{3}P $5d$
9	−4.00438	$2s^22p^2$ ^{3}P $5g$
10	−3.64808	$2p^4$ ^{1}D $3d$
11	−3.58129	$2s2p^3$ ^{3}D $4p$
12	−3.34354	$2s^22p^2$ ^{1}D $5d$
13	−3.23524	$2s^22p^2$ ^{1}D $5g$
14	−3.06356	$2s2p^3$ ^{3}D $4f$
15	−2.81744	$2s^22p^2$ ^{3}P $6d$
16	−2.79072	$2s^22p^2$ ^{3}P $6g$
17	−2.69681	$2s2p^3$ ^{3}P $4f$
18	−2.07955	$2s^22p^2$ ^{3}P $7d$
19	−2.04392	$2s^22p^2$ ^{3}P $7g$
20	−2.03995	$2s^22p^2$ ^{1}D $6d$
21	−1.99646	$2s^22p^2$ ^{1}D $6g$
22	−1.95641	$2s2p^3$ ^{1}D $4p$
23	−1.58003	$2s^22p^2$ ^{3}P $8d$
24	−1.56872	$2s^22p^2$ ^{3}P $8g$
25	−1.39194	$2s2p^3$ ^{1}D $4f$
26	−1.37454	$2s2p^3$ ^{3}S $4f$
27	−1.27784	$2s^22p^2$ ^{1}D $7d$
28	−1.24358	$2s^22p^2$ ^{3}P $9d$
29	−1.24078	$2s^22p^2$ ^{1}D $7g$
30	−1.23045	$2s^22p^2$ ^{3}P $9g$
31	−1.03879	$2s2p^3$ ^{1}P $4f$
		^{2}**F**$^\text{o}$
1	−12.2308	$2s^22p^2$ ^{1}D $3p$
2	−8.12011	$2s2p^3$ ^{3}D $3d$
3	−7.73141	$2s2p^3$ ^{3}P $3d$
4	−6.86374	$2s2p^3$ ^{1}D $3d$
5	−6.34965	$2s2p^3$ ^{1}P $3d$
6	−6.22213	$2s^22p^2$ ^{3}P $4f$
7	−6.10650	$2s^22p^2$ ^{1}D $4p$
8	−5.60259	$2s^22p^2$ ^{1}D $4f$
9	−5.25129	$2s^22p^2$ ^{1}S $4f$
10	−4.81257	$2p^4$ ^{1}D $3p$
11	−3.99273	$2s^22p^2$ ^{3}P $5f$
12	−3.54474	$2s^22p^2$ ^{1}D $5p$
13	−3.24151	$2s^22p^2$ ^{1}D $5f$
14	−3.13508	$2s^22p^2$ ^{1}S $5f$
15	−3.05948	$2s2p^3$ ^{3}D $4d$
16	−2.88723	$2s2p^3$ ^{3}P $4d$
17	−2.77263	$2s^22p^2$ ^{3}P $6f$
18	−2.16488	$2s^22p^2$ ^{1}D $6p$
19	−2.03885	$2s^22p^2$ ^{3}P $7f$
20	−1.99135	$2s^22p^2$ ^{1}D $6f$
21	−1.96307	$2s^22p^2$ ^{1}D $6h$
22	−1.93893	$2s^22p^2$ ^{1}S $6f$
23	−1.56402	$2s^22p^2$ ^{3}P $8f$
24	−1.55833	$2s2p^3$ ^{1}D $4d$
25	−1.36358	$2s^22p^2$ ^{1}D $7p$
26	−1.30242	$2s2p^3$ ^{1}P $4d$
27	−1.23583	$2s^22p^2$ ^{1}D $7f$
28	−1.23365	$2s^22p^2$ ^{3}P $9f$
29	−1.22650	$2s^22p^2$ ^{1}D $7h$
30	−1.20826	$2s^22p^2$ ^{1}S $7f$
		^{4}**F**$^\text{e}$
1	−11.7749	$2s^22p^2$ ^{3}P $3d$
2	−9.67845	$2s2p^3$ ^{3}D $3p$
3	−6.51848	$2s^22p^2$ ^{3}P $4d$
4	−4.84810	$2s2p^3$ ^{5}S $4f$
5	−4.34271	$2p^4$ ^{3}P $3d$
6	−4.13570	$2s^22p^2$ ^{3}P $5d$
7	−4.00470	$2s^22p^2$ ^{3}P $5g$
8	−3.62977	$2s2p^3$ ^{3}D $4p$
9	−3.08350	$2s2p^3$ ^{3}D $4f$
10	−2.85542	$2s^22p^2$ ^{3}P $6d$
11	−2.79235	$2s^22p^2$ ^{3}P $6g$
12	−2.70405	$2s2p^3$ ^{3}P $4f$
13	−2.59619	$2s2p^3$ ^{5}S $5f$
14	−2.08946	$2s^22p^2$ ^{3}P $7d$
15	−2.04387	$2s^22p^2$ ^{3}P $7g$
16	−1.59531	$2s^22p^2$ ^{3}P $8d$
17	−1.56872	$2s^22p^2$ ^{3}P $8g$
18	−1.38204	$2s2p^3$ ^{3}S $4f$
19	−1.37332	$2s2p^3$ ^{5}S $6f$
20	−1.25708	$2s^22p^2$ ^{3}P $9d$

N-like S (S^{9+})

i	Energy(Ryds)	Description	i	Energy(Ryds)	Description	i	Energy(Ryds)	Description
21	−1.23081	$2s^22p^2$ ^{3}P $9g$	15	−1.56128	$2s^22p^2$ ^{3}P $8g$	4	−2.70805	$2s2p^3$ ^{3}P $4f$
	^{4}F^o		16	−1.40112	$2s2p^3$ ^{1}D $4f$	5	−2.03847	$2s^22p^2$ ^{3}P $7g$
1	−8.66283	$2s2p^3$ ^{3}D $3d$	17	−1.27629	$2s^22p^2$ ^{1}D $7d$	6	−1.56142	$2s^22p^2$ ^{3}P $8g$
2	−8.06494	$2s2p^3$ ^{3}P $3d$	18	−1.23578	$2s^22p^2$ ^{1}D $7g$	7	−1.23402	$2s^22p^2$ ^{3}P $9g$
3	−6.23223	$2s^22p^2$ ^{3}P $4f$	19	−1.23389	$2s^22p^2$ ^{3}P $9g$		**^{4}G^o**	
4	−4.00055	$2s^22p^2$ ^{3}P $5f$	20	−1.22115	$2s^22p^2$ ^{1}D $7i$	1	−8.61656	$2s2p^3$ ^{3}D $3d$
5	−3.28058	$2s2p^3$ ^{3}D $4d$	21	−1.20466	$2s^22p^2$ ^{1}S $7g$	2	−6.26839	$2s^22p^2$ ^{3}P $4f$
6	−3.01282	$2s2p^3$ ^{3}P $4d$	22	−1.03794	$2s2p^3$ ^{1}P $4f$	3	−4.01542	$2s^22p^2$ ^{3}P $5f$
7	−2.77681	$2s^22p^2$ ^{3}P $6f$		**^{2}G^o**		4	−3.26479	$2s2p^3$ ^{3}D $4d$
8	−2.04201	$2s^22p^2$ ^{3}P $7f$	1	−8.42296	$2s2p^3$ ^{3}D $3d$	5	−2.78654	$2s^22p^2$ ^{3}P $6f$
9	−1.56370	$2s^22p^2$ ^{3}P $8f$	2	−6.87801	$2s2p^3$ ^{1}D $3d$	6	−2.77799	$2s^22p^2$ ^{3}P $6h$
10	−1.23564	$2s^22p^2$ ^{3}P $9f$	3	−6.26243	$2s^22p^2$ ^{3}P $4f$	7	−2.59436	$2s2p^3$ ^{5}S $5g$
	^{6}F^e		4	−5.62648	$2s^22p^2$ ^{1}D $4f$	8	−2.04720	$2s^22p^2$ ^{3}P $7f$
1	−4.86150	$2s2p^3$ ^{5}S $4f$	5	−4.00953	$2s^22p^2$ ^{3}P $5f$	9	−2.04109	$2s^22p^2$ ^{3}P $7h$
2	−2.60671	$2s2p^3$ ^{5}S $5f$	6	−3.24752	$2s^22p^2$ ^{1}D $5f$	10	−1.56708	$2s^22p^2$ ^{3}P $8f$
3	−1.38033	$2s2p^3$ ^{5}S $6f$	7	−3.18604	$2s2p^3$ ^{3}D $4d$	11	−1.56277	$2s^22p^2$ ^{3}P $8h$
4	−.64071	$2s2p^3$ ^{5}S $7f$	8	−2.78382	$2s^22p^2$ ^{3}P $6f$	12	−1.37215	$2s2p^3$ ^{5}S $6g$
5	−.16072	$2s2p^3$ ^{5}S $8f$	9	−2.77795	$2s^22p^2$ ^{3}P $6h$	13	−1.23797	$2s^22p^2$ ^{3}P $9f$
	^{2}G^e		10	−2.04511	$2s^22p^2$ ^{3}P $7f$	14	−1.23481	$2s^22p^2$ ^{3}P $9h$
1	−11.0723	$2s^22p^2$ ^{1}D $3d$	11	−2.04106	$2s^22p^2$ ^{3}P $7h$		**^{6}G^o**	
2	−5.77870	$2s^22p^2$ ^{1}D $4d$	12	−1.99310	$2s^22p^2$ ^{1}D $6f$	1	−2.59444	$2s2p^3$ ^{5}S $5g$
3	−3.99911	$2s^22p^2$ ^{3}P $5g$	13	−1.96340	$2s^22p^2$ ^{1}D $6h$	2	−1.37242	$2s2p^3$ ^{5}S $6g$
4	−3.85021	$2p^4$ ^{1}D $3d$	14	−1.58281	$2s2p^3$ ^{1}D $4d$	3	−.63541	$2s2p^3$ ^{5}S $7g$
5	−3.34519	$2s^22p^2$ ^{1}D $5d$	15	−1.56360	$2s^22p^2$ ^{3}P $8f$	4	−.15704	$2s2p^3$ ^{5}S $8g$
6	−3.23947	$2s^22p^2$ ^{1}D $5g$	16	−1.56265	$2s^22p^2$ ^{3}P $8h$			
7	−3.13455	$2s^22p^2$ ^{1}S $5g$	17	−1.24026	$2s^22p^2$ ^{1}D $7f$			
8	−3.08301	$2s2p^3$ ^{3}D $4f$	18	−1.23645	$2s^22p^2$ ^{3}P $9f$			
9	−2.79067	$2s^22p^2$ ^{3}P $6g$	19	−1.23479	$2s^22p^2$ ^{3}P $9h$			
10	−2.70058	$2s2p^3$ ^{3}P $4f$	20	−1.22690	$2s^22p^2$ ^{1}D $7h$			
11	−2.04861	$2s^22p^2$ ^{1}D $6d$		**^{4}G^e**				
12	−2.03830	$2s^22p^2$ ^{3}P $7g$	1	−3.99944	$2s^22p^2$ ^{3}P $5g$			
13	−1.99982	$2s^22p^2$ ^{1}D $6g$	2	−3.10039	$2s2p^3$ ^{3}D $4f$			
14	−1.92775	$2s^22p^2$ ^{1}S $6g$	3	−2.79364	$2s^22p^2$ ^{3}P $6g$			

N-like S (S^{9+})

Energies in ascending order from ground state for terms with effective $n \leq 4.0$, $L \leq 4$

Term	i	E(Ryds)	Term	i	E(Ryds)	Term	i	E(Ryds)	Term	i	E(Ryds)	Term	i	E(Ryds)
$^4S^o$	1	0.00000	$^4F^e$	1	21.2160	$^4F^e$	2	23.3124	$^2F^o$	2	24.8707	$^4D^o$	7	26.1474
$^2D^o$	1	0.78540	$^4D^e$	1	21.3269	$^2F^e$	3	23.4269	$^4F^o$	2	24.9259	$^4S^o$	6	26.1863
$^2P^o$	1	1.20390	$^2P^e$	3	21.3363	$^4P^e$	5	23.6511	$^2P^e$	7	24.9573	$^2D^o$	9	26.2057
$^4P^e$	1	3.42870	$^2P^o$	5	21.3861	$^2D^e$	6	23.7682	$^4P^e$	7	24.9835	$^2P^o$	10	26.2376
$^2D^e$	1	4.71450	$^2F^e$	1	21.4407	$^4S^e$	1	23.8019	$^4P^o$	4	25.0002	$^2P^o$	11	26.3574
$^2S^e$	1	5.58140	$^4P^e$	3	21.4635	$^4D^e$	3	23.8599	$^2F^e$	4	25.0425	$^2S^o$	5	26.4089
$^2P^e$	1	5.81090	$^6P^e$	1	21.5760	$^2P^e$	6	24.0058	$^4D^o$	5	25.0695	$^2D^o$	10	26.4384
$^2P^o$	2	9.00220	$^2D^e$	3	21.7433	$^4P^e$	6	24.0425	$^2D^e$	8	25.1818	$^2D^e$	10	26.4512
$^4P^e$	2	19.1551	$^2G^e$	1	21.9186	$^4S^o$	4	24.0692	$^2D^o$	7	25.2256	$^4F^e$	3	26.4724
$^2P^e$	2	19.3706	$^4P^e$	4	21.9220	$^2D^e$	7	24.0968	$^2F^o$	3	25.2594	$^4D^e$	4	26.5068
$^2D^e$	2	19.7852	$^2F^e$	2	22.0181	$^2D^o$	5	24.1220	$^2P^o$	9	25.5080	$^2P^e$	11	26.5278
$^2S^o$	1	19.9567	$^2D^e$	4	22.0291	$^2S^o$	2	24.2279	$^2P^e$	8	25.5106	$^4P^e$	9	26.5665
$^4D^o$	1	20.0892	$^2P^e$	4	22.1557	$^4F^o$	1	24.3280	$^2D^e$	9	25.6175	$^2F^e$	5	26.5962
$^4P^o$	1	20.1620	$^2S^e$	3	22.2454	$^4G^o$	1	24.3743	$^4P^e$	8	25.7169	$^2F^o$	5	26.6412
$^2D^o$	2	20.3040	$^4D^o$	2	22.3527	$^2S^e$	4	24.4217	$^2P^e$	9	25.7547	$^2S^e$	6	26.6767
$^4S^o$	2	20.3105	$^2D^e$	5	22.5464	$^4D^o$	4	24.4964	$^2S^e$	5	25.7663	$^2D^e$	11	26.6806
$^2S^e$	2	20.3288	$^6D^o$	1	22.6631	$^2S^o$	3	24.5379	$^2P^e$	10	25.8497	$^4G^o$	2	26.7225
$^2P^o$	3	20.4349	$^2D^o$	4	22.6865	$^2G^o$	1	24.5679	$^4D^o$	6	25.9800	$^2P^o$	12	26.7267
$^6S^o$	1	20.5987	$^4P^o$	2	22.8977	$^4P^o$	3	24.6139	$^2S^o$	4	26.0477	$^2G^o$	3	26.7284
$^2F^o$	1	20.7601	$^4D^o$	3	23.1140	$^2P^o$	7	24.6645	$^2D^o$	8	26.0552	$^2D^o$	11	26.7311
$^2D^o$	3	20.9580	$^2P^o$	6	23.2144	$^2P^o$	8	24.7181	$^4P^o$	5	26.0999	$^4P^e$	10	26.7326
$^2P^o$	4	21.1087	$^4D^e$	2	23.2582	$^4S^o$	5	24.7392	$^2G^o$	2	26.1128			
$^4S^o$	3	21.1935	$^2P^e$	5	23.2961	$^2D^o$	6	24.7524	$^2F^o$	4	26.1271			

N-like S (S^{9+})

gf-values for transitions involving terms with effective $n \leq 4.0$, $L \leq 4$

i	i'	gf$_L$	i	i'	gf$_L$	i	i'	gf$_L$	i	i'	gf$_L$	i	i'	gf$_L$	i	i'	gf$_L$
		$^2S^o$–$^2P^e$	5	7	2.26E−1	4	10	−2.48E−3	2	12	−2.55E−5	7	3	4.33E−1	11	6	3.53E−4
1	1	2.69E−4	5	8	7.97E−2	4	11	−1.37E−2	3	1	9.50E−1	7	4	6.70E−2	11	7	1.28E−2
1	2	1.34E−1	5	9	1.79E−5	4	12	−2.69E−2	3	2	8.32E−6	7	5	1.70E−3	11	8	5.73E−3
1	3	−4.65E−1	5	10	6.89E−3	5	1	1.76E−2	3	3	2.47E−1	7	6	6.97E−5	11	9	8.75E−3
1	4	−4.04E−2	5	11	−1.38E−5	5	2	7.25E−4	3	4	4.40E−3	7	7	9.89E−3	11	10	3.79E−1
1	5	−3.08E−1			$^2S^e$–$^2P^o$	5	3	4.61E−2	3	5	−5.82E−4	7	8	8.51E−4	11	11	4.26E−3
1	6	−6.85E−2	1	1	3.26E−1	5	4	2.16E−1	3	6	−4.99E−3	7	9	−2.82E−4	11	12	−4.26E−5
1	7	−6.58E−2	1	2	−6.32E−2	5	5	2.13E−1	3	7	−2.11E−2	7	10	−2.14E−4			$^2P^o$–$^2D^e$
1	8	−8.49E−2	1	3	−1.22E−4	5	6	1.37E−1	3	8	−3.06E−1	7	11	−3.51E−1	1	1	−1.66E−1
1	9	−2.67E−2	1	4	−8.55E−3	5	7	2.25E−1	3	9	−7.05E−2	7	12	−1.70E−3	1	2	−2.11E−1
1	10	−2.23E−2	1	5	−1.01E−2	5	8	2.85E−6	3	10	−7.34E−2	8	1	3.93E−3	1	3	−1.71E+0
1	11	−5.91E−1	1	6	−2.27E−1	5	9	3.37E−3	3	11	−1.66E−2	8	2	1.19E−2	1	4	−2.53E+0
2	1	3.38E−1	1	7	−1.82E−1	5	10	−3.19E−4	3	12	−1.01E−3	8	3	4.30E−2	1	5	−2.26E+0
2	2	6.18E−1	1	8	−1.65E−1	5	11	−7.76E−3	4	1	1.66E+0	8	4	4.66E−1	1	6	−1.52E−2
2	3	2.92E−3	1	9	−2.36E+0	5	12	−3.41E−1	4	2	1.60E−4	8	5	3.77E−2	1	7	−7.80E−1
2	4	2.93E−5	1	10	−1.54E−2	6	1	1.58E−2	4	3	1.33E−1	8	6	1.67E−3	1	8	−4.46E−2
2	5	4.47E−4	1	11	−1.67E−2	6	2	1.28E−5	4	4	4.54E−1	8	7	9.70E−2	1	9	−2.07E−1
2	6	1.48E−4	1	12	−1.15E+0	6	3	7.72E−3	4	5	7.16E−4	8	8	1.02E−3	1	10	−4.35E−2
2	7	−1.96E−1	2	1	1.30E−1	6	4	7.02E−3	4	6	−6.01E−3	8	9	3.24E−6	1	11	−7.43E−1
2	8	−4.05E−1	2	2	5.29E−6	6	5	2.84E−1	4	7	−1.11E−3	8	10	−3.75E−1	2	1	7.93E−1
2	9	−1.01E−1	2	3	−2.71E−3	6	6	5.08E−4	4	8	−9.72E−3	8	11	−1.14E−1	2	2	−6.39E−6
2	10	−2.50E−1	2	4	−5.94E−3	6	7	4.11E−3	4	9	−1.79E−4	8	12	−7.48E−2	2	3	−5.94E−5
2	11	−5.54E−2	2	5	−7.49E−1	6	8	5.28E−6	4	10	−5.21E−3	9	1	2.02E−1	2	4	−6.25E−6
3	1	1.17E−1	2	6	−1.97E−2	6	9	2.14E−3	4	11	−1.33E−2	9	2	5.67E−3	2	5	−1.49E−6
3	2	3.15E−3	2	7	−3.47E−1	6	10	6.23E−3	4	12	−1.16E−2	9	3	3.34E−3	2	6	−1.60E−2
3	3	9.76E−2	2	8	−8.14E−3	6	11	4.68E−5	5	1	3.39E−2	9	4	3.74E−1	2	7	−1.29E−2
3	4	1.23E−4	2	9	−8.02E−3	6	12	−5.11E−3	5	2	1.46E−2	9	5	1.86E−1	2	8	−4.48E−3
3	5	2.85E−1	2	10	−2.45E−3			$^2P^e$–$^2P^o$	5	3	2.28E−2	9	6	2.64E−3	2	9	−3.82E−3
3	6	8.45E−3	2	11	−4.38E−4	1	1	4.13E−1	5	4	1.90E−4	9	7	5.78E−1	2	10	−2.51E−4
3	7	−6.80E−4	2	12	−1.88E−2	1	2	−1.02E+0	5	5	5.35E−4	9	8	1.44E−2	2	11	−4.28E−6
3	8	−1.91E−4	3	1	5.70E−1	1	3	−1.62E−2	5	6	5.46E−3	9	9	1.90E−3	3	1	7.35E−3
3	9	−1.51E−5	3	2	6.33E−5	1	4	−5.67E−3	5	7	−8.88E−3	9	10	−3.41E−1	3	2	8.39E−2
3	10	−2.59E−3	3	3	8.30E−2	1	5	−3.14E−3	5	8	−6.79E−1	9	11	−2.20E−2	3	3	−1.05E+0
3	11	−1.44E−2	3	4	2.26E−1	1	6	−4.01E−2	5	9	−3.77E−4	9	12	−1.54E−1	3	4	−2.79E−2
4	1	2.37E−3	3	5	9.52E−3	1	7	−2.87E−1	5	10	−6.97E−4	10	1	3.94E−4	3	5	−2.81E−4
4	2	1.31E−1	3	6	−2.53E−3	1	8	−3.14E−1	5	11	−3.53E−3	10	2	9.41E−3	3	6	−4.55E−1
4	3	1.01E−1	3	7	−1.33E−5	1	9	−7.68E−2	5	12	−1.36E−4	10	3	3.23E−1	3	7	−5.38E−2
4	4	1.69E−2	3	8	−8.03E−5	1	10	−2.58E−2	6	1	5.01E−1	10	4	2.34E−1	3	8	−1.25E−2
4	5	8.24E−5	3	9	−1.71E−4	1	11	−2.25E+0	6	2	7.30E−3	10	5	1.47E−1	3	9	−1.74E−2
4	6	1.85E−3	3	10	−3.65E−4	1	12	−2.88E−1	6	3	1.62E−1	10	6	1.29E−2	3	10	−1.33E−2
4	7	1.58E−3	3	11	−1.31E−1	2	1	3.31E−1	6	4	1.05E−1	10	7	1.53E−1	3	11	−1.08E+0
4	8	6.55E−2	3	12	−5.36E−3	2	2	3.69E−7	6	5	6.17E−2	10	8	5.74E−5	4	1	2.56E−2
4	9	5.29E−2	4	1	2.18E−1	2	3	−5.65E−1	6	6	5.29E−1	10	9	1.74E−3	4	2	7.66E−1
4	10	1.24E−1	4	2	2.10E−2	2	4	−3.73E−1	6	7	−5.14E−4	10	10	−5.87E−1	4	3	−6.17E−2
4	11	−8.23E−1	4	3	2.05E−1	2	5	−1.03E−1	6	8	−4.33E−2	10	11	−6.32E−2	4	4	−3.98E−1
5	1	8.09E−1	4	4	3.21E−3	2	6	−3.60E−1	6	9	−6.10E−1	10	12	−2.54E−2	4	5	−4.49E−2
5	2	7.31E−4	4	5	2.01E−3	2	7	−4.83E−3	6	10	−1.70E−3	11	1	2.69E−1	4	6	−3.53E−2
5	3	1.10E−2	4	6	2.98E−1	2	8	−8.71E−3	6	11	−2.52E−3	11	2	1.78E−3	4	7	−1.58E−1
5	4	3.53E−1	4	7	−1.47E−3	2	9	−6.29E−3	6	12	−2.69E−2	11	3	4.97E−1	4	8	−1.03E−1
5	5	4.26E−3	4	8	−7.51E−3	2	10	−3.26E−1	7	1	3.57E−3	11	4	9.69E−2	4	9	−1.26E−2
5	6	1.51E−3	4	9	−4.36E−1	2	11	−1.33E−2	7	2	1.03E−2	11	5	4.80E−2	4	10	−4.89E−1

N-like S (S^{9+})

i	i'	gf_L	i	i'	gf_L	i	i'	gf_L	i	i'	gf_L	i	i'	gf_L	i	i'	gf_L
4	11	−3.28E−1	9	8	1.28E−4	2	4	−2.57E−1	7	1	1.07E−1	11	9	2.61E−2	5	5	−2.10E−4
5	1	7.31E−3	9	9	−1.58E−4	2	5	−5.01E−3	7	2	8.27E−1	11	10	3.91E−2	5	6	−2.93E−4
5	2	2.06E−5	9	10	−1.63E−4	2	6	−1.02E−3	7	3	2.90E−2	11	11	−2.15E−1	5	7	−4.94E−3
5	3	−2.57E−6	9	11	−4.36E−2	2	7	−1.27E−2	7	4	9.89E−3			$^2D^e$–$^2D^o$	5	8	−3.61E−3
5	4	−1.38E−2	10	1	5.11E−5	2	8	−1.27E−1	7	5	3.12E−1	1	1	9.10E−1	5	9	−3.11E−3
5	5	−1.35E+0	10	2	1.38E−2	2	9	−5.29E−1	7	6	9.43E−9	1	2	−7.29E−5	5	10	−3.23E−2
5	6	−1.41E−2	10	3	2.49E−1	2	10	−4.40E−2	7	7	−2.75E−4	1	3	−1.82E−2	5	11	−2.06E−1
5	7	−1.00E−1	10	4	8.04E−3	2	11	−7.73E−5	7	8	−6.74E−1	1	4	−8.01E−1	6	1	9.42E−1
5	8	−3.12E−3	10	5	5.83E−3	3	1	2.68E−1	7	9	−3.44E−2	1	5	−4.72E−1	6	2	6.11E−2
5	9	−5.35E−1	10	6	5.15E−4	3	2	2.57E−3	7	10	−2.54E−2	1	6	−1.83E+0	6	3	1.56E−1
5	10	−2.08E−4	10	7	5.83E−3	3	3	4.94E−3	7	11	−5.29E−2	1	7	−1.89E+0	6	4	1.33E+0
5	11	−2.92E−2	10	8	6.29E−3	3	4	−2.07E−3	8	1	1.84E−2	1	8	−8.54E−1	6	5	−6.17E−3
6	1	2.41E−1	10	9	1.37E−4	3	5	−8.27E−5	8	2	5.30E−1	1	9	−2.75E−1	6	6	−5.98E−1
6	2	2.47E−2	10	10	−9.90E−5	3	6	−1.96E−1	8	3	2.36E−1	1	10	−7.94E−1	6	7	−8.01E−2
6	3	4.80E−3	10	11	−1.80E+0	3	7	−1.05E−2	8	4	2.72E−2	1	11	−1.76E−1	6	8	−4.22E−4
6	4	1.45E−2	11	1	3.02E−1	3	8	−6.26E−3	8	5	5.63E−1	2	1	5.33E−1	6	9	−2.40E−2
6	5	9.06E−3	11	2	1.01E−2	3	9	−3.87E−4	8	6	6.51E−6	2	2	−1.37E−1	6	10	−6.92E−2
6	6	−2.99E−2	11	3	2.48E−2	3	10	−4.23E−1	8	7	4.10E−4	2	3	−1.09E+0	6	11	−3.14E−3
6	7	−1.06E+0	11	4	8.88E−1	3	11	−2.13E+0	8	8	−1.35E−1	2	4	−8.88E−2	7	1	1.03E−1
6	8	−6.46E−2	11	5	3.14E−3	4	1	6.04E−1	8	9	−5.11E−1	2	5	−1.82E+0	7	2	7.57E−1
6	9	−6.17E−2	11	6	3.90E−2	4	2	1.28E−1	8	10	−2.17E−1	2	6	−1.47E−2	7	3	1.11E−2
6	10	−8.23E−3	11	7	6.17E−3	4	3	1.29E−1	8	11	−1.05E−1	2	7	−2.04E−3	7	4	4.86E−2
6	11	−3.54E−3	11	8	2.26E−1	4	4	−1.31E−3	9	1	4.76E−3	2	8	−1.22E−2	7	5	−7.23E−4
7	1	3.01E−2	11	9	2.16E−2	4	5	−1.13E−4	9	2	1.36E−2	2	9	−7.89E−3	7	6	−2.22E−3
7	2	1.09E+0	11	10	−1.56E−3	4	6	−1.36E−3	9	3	9.13E−1	2	10	−1.13E−2	7	7	−3.90E−1
7	3	6.57E−3	11	11	−2.24E−2	4	7	−6.37E−3	9	4	1.80E−2	2	11	−5.28E−2	7	8	−9.91E−3
7	4	4.66E−4	12	1	4.71E−2	4	8	−4.50E−2	9	5	9.49E−4	3	1	2.00E+0	7	9	−1.14E−2
7	5	1.02E−3	12	2	1.25E−3	4	9	−6.86E−2	9	6	5.18E−4	3	2	2.30E−1	7	10	−1.35E−3
7	6	1.36E−2	12	3	1.37E−2	4	10	−1.21E−3	9	7	1.89E−3	3	3	9.82E−2	7	11	−1.97E−2
7	7	2.17E−3	12	4	1.07E−1	4	11	−2.36E−2	9	8	−6.04E−2	3	4	−9.62E−3	8	1	1.19E−1
7	8	−1.68E−3	12	5	4.57E−1	5	1	5.84E−1	9	9	−3.87E−1	3	5	−7.26E−4	8	2	1.13E−1
7	9	−1.15E+0	12	6	3.97E−6	5	2	4.92E−7	9	10	−3.95E−2	3	6	−1.60E−1	8	3	1.05E+0
7	10	−1.12E−2	12	7	3.20E−3	5	3	6.27E−2	9	11	−4.05E−1	3	7	−1.52E−1	8	4	4.28E−1
7	11	−2.37E−2	12	8	1.22E−2	5	4	3.53E−1	10	1	1.05E−1	3	8	−8.24E−2	8	5	1.11E+0
8	1	1.25E+0	12	9	1.41E−3	5	5	−2.40E−5	10	2	8.88E−2	3	9	−1.49E−1	8	6	1.14E−3
8	2	2.96E−2	12	10	2.30E−4	5	6	−7.39E−1	10	3	5.20E−2	3	10	−9.40E−3	8	7	−1.39E−4
8	3	1.92E−2	12	11	1.99E−5	5	7	−1.03E−2	10	4	1.30E−4	3	11	−5.33E−1	8	8	−1.96E−1
8	4	4.18E−4			$^2P^e$–$^2D^o$	5	8	−4.77E−3	10	5	5.13E−2	4	1	2.09E+0	8	9	−1.15E−1
8	5	2.02E−4	1	1	1.23E+0	5	9	−1.61E−2	10	6	2.45E−4	4	2	3.86E−1	8	10	−2.57E−1
8	6	1.64E−1	1	2	−5.28E−3	5	10	−1.89E−3	10	7	9.08E−3	4	3	3.36E−1	8	11	−5.36E−3
8	7	8.30E−3	1	3	−7.20E−8	5	11	−1.27E−2	10	8	−6.42E−2	4	4	−1.27E−2	9	1	2.94E−2
8	8	−4.98E−3	1	4	−4.99E−2	6	1	7.08E−2	10	9	−7.61E−1	4	5	−5.32E−6	9	2	2.01E−2
8	9	−2.68E−2	1	5	−3.39E−1	6	2	1.86E−1	10	10	−2.41E−1	4	6	−3.90E−3	9	3	1.03E−1
8	10	−2.00E−4	1	6	3.03E−1	6	3	7.69E−2	10	11	−1.80E−2	4	7	−6.62E−3	9	4	4.46E−2
8	11	−1.40E−2	1	7	−1.70E−2	6	4	1.11E−1	11	1	3.19E−2	4	8	−4.80E−2	9	5	1.09E−3
9	1	9.02E−2	1	8	−1.33E−1	6	5	−3.25E−5	11	2	1.50E−2	4	9	−2.58E−4	9	6	3.64E−4
9	2	6.41E−4	1	9	−3.41E−2	6	6	−3.47E−2	11	3	3.72E−2	4	10	−1.08E−1	9	7	2.63E−3
9	3	4.87E−1	1	10	−3.07E+0	6	7	−1.15E+0	11	4	4.48E−5	4	11	−2.57E−2	9	8	−2.13E−3
9	4	2.26E−2	1	11	−6.30E−1	6	8	−7.16E−4	11	5	3.95E−4	5	1	1.53E−1	9	9	0.00E+0
9	5	3.77E−3	2	1	4.17E−1	6	9	−9.68E−3	11	6	4.54E−3	5	2	1.66E−3	9	10	−1.24E−2
9	6	2.04E−2	2	2	−8.48E−1	6	10	−2.66E−2	11	7	1.17E−2	5	3	3.01E−2	9	11	−2.58E−1
9	7	1.65E−2	2	3	−6.27E−1	6	11	−2.65E−2	11	8	1.43E−2	5	4	−7.38E−5	10	1	9.19E−2

N-like S (S^{9+})

i i'	gf_L	i i'	gf_L	i i'	gf_L	i i'	gf_L	i i'	gf_L	i i'	gf_L
10 2	8.14E−2	7 1	4.49E−1	6 2	−1.36E+0	5 3	9.64E−4	3 3	−5.25E−2	2 3	−3.56E−3
10 3	6.09E−1	7 2	4.75E−2	6 3	−7.46E−2	5 4	2.43E−6	3 4	−8.42E−1	2 4	−3.05E−3
10 4	1.96E−2	7 3	1.28E−1	6 4	−2.43E−2	5 5	−1.04E−4	3 5	−3.26E−1	2 5	−7.83E−1
10 5	8.75E−3	7 4	7.62E−6	6 5	−1.48E−5		**$^{2}F^{o}$–$^{2}G^{e}$**	3 6	−1.58E−1	3 1	8.52E−1
10 6	7.05E−4	7 5	−9.76E−2	7 1	8.03E−4	1 1	−2.46E+0	3 7	−1.92E−1	3 2	−7.45E−4
10 7	1.40E−4	8 1	1.32E+0	7 2	−7.26E−3	2 1	3.13E−5	3 8	−6.26E−2	3 3	−5.01E−1
10 8	9.40E−3	8 2	7.12E−2	7 3	−2.08E+0	3 1	2.60E−3	3 9	−4.16E−1	3 4	−1.06E−1
10 9	1.26E−3	8 3	1.40E−2	7 4	−7.89E−4	4 1	4.16E−1	3 10	−1.27E−2	3 5	−2.72E−1
10 10	7.62E−4	8 4	5.89E−2	7 5	−5.19E−3	5 1	2.55E+0	4 1	1.35E−1	4 1	5.54E−3
10 11	−7.19E−2	8 5	−7.39E−1	8 1	4.87E−1		**$^{2}F^{e}$–$^{2}G^{o}$**	4 2	1.04E+0	4 2	−1.18E−2
11 1	2.93E−1	9 1	1.90E−2	8 2	6.34E−3	1 1	−1.45E−1	4 3	5.57E−4	4 3	−8.55E−2
11 2	2.98E−1	9 2	4.16E−1	8 3	−7.73E−4	1 2	−1.07E−2	4 4	1.63E−3	4 4	−1.33E−3
11 3	1.38E−1	9 3	1.25E−2	8 4	−9.61E−1	1 3	−9.23E+0	4 5	3.34E−4	4 5	−8.88E−3
11 4	6.62E−3	9 4	1.05E−2	8 5	−1.55E−1	2 1	−3.58E−2	4 6	1.05E−4	5 1	4.74E−1
11 5	2.97E−3	9 5	−2.26E+0	9 1	1.76E+0	2 2	−1.90E−3	4 7	−1.16E+0	5 2	1.18E−2
11 6	4.71E−3	10 1	6.97E−2	9 2	6.02E−4	2 3	−3.13E+0	4 8	−1.13E−2	5 3	−1.01E+0
11 7	9.25E−3	10 2	4.84E−1	9 3	1.10E−3	3 1	−2.65E+0	4 9	−1.23E−2	5 4	−1.22E−1
11 8	2.37E−1	10 3	1.12E−3	9 4	−1.41E−2	3 2	−1.90E−2	4 10	−1.24E+0	5 5	−2.67E−2
11 9	4.98E−1	10 4	8.06E−2	9 5	−1.50E+0	3 3	−1.67E−1	5 1	2.11E+0	6 1	1.21E−1
11 10	2.38E−4	10 5	−5.16E−2	10 1	4.53E−1	4 1	8.77E−4	5 2	6.01E−3	6 2	1.94E+0
11 11	−1.01E−2	11 1	1.62E−1	10 2	9.53E−4	4 2	−2.26E+0	5 3	2.52E−1	6 3	−2.91E−4
	$^{2}D^{o}$–$^{2}F^{e}$	11 2	2.66E−1	10 3	3.43E−4	4 3	−5.06E−4	5 4	1.21E−1	6 4	−5.55E−1
1 1	−1.47E+0	11 3	1.71E−3	10 4	1.19E−2	5 1	2.12E−2	5 5	4.98E−1	6 5	−7.43E−3
1 2	−8.94E+0	11 4	2.34E−2	10 5	−2.41E−1	5 2	3.38E−3	5 6	3.60E−3	7 1	8.36E−1
1 3	−1.15E+0	11 5	3.64E−4	11 1	5.04E−3	5 3	−8.54E−1	5 7	−2.34E−3	7 2	8.54E−3
1 4	−3.33E−1		**$^{2}D^{e}$–$^{2}F^{o}$**	11 2	7.15E−3		**$^{2}G^{e}$–$^{2}G^{o}$**	5 8	−1.70E−3	7 3	3.33E−3
1 5	−1.18E+0	1 1	−2.23E−2	11 3	2.53E−2	1 1	−1.56E−3	5 9	−2.12E−2	7 4	−8.31E−6
2 1	−1.63E+0	1 2	−4.29E+0	11 4	4.65E−4	1 2	−1.68E+0	5 10	−1.52E−2	7 5	−5.48E−2
2 2	−2.03E−2	1 3	−2.52E+0	11 5	5.54E−4	1 3	−2.16E−3	6 1	8.13E−3	8 1	7.50E−1
2 3	−5.07E−1	1 4	−4.39E+0		**$^{2}F^{e}$–$^{2}F^{o}$**		**$^{4}S^{o}$–$^{4}P^{e}$**	6 2	2.14E−1	8 2	3.64E−4
2 4	−1.83E−2	1 5	−6.88E−1	1 1	7.41E−2	1 1	−6.25E−1	6 3	1.48E−1	8 3	1.92E−3
2 5	−2.42E+0	2 1	−1.64E+0	1 2	−2.95E−1	1 2	−4.58E−1	6 4	2.28E−2	8 4	4.87E−3
3 1	−7.73E−3	2 2	−6.53E−3	1 3	−5.44E−1	1 3	−6.63E+0	6 5	1.40E−3	8 5	−2.01E+0
3 2	−1.26E+0	2 3	−9.64E−5	1 4	−6.99E−2	1 4	−8.38E−1	6 6	1.70E−2	9 1	8.95E−1
3 3	−5.30E−1	2 4	−1.53E−2	1 5	−1.43E−2	1 5	−1.57E−2	6 7	3.37E−2	9 2	2.76E−2
3 4	−2.35E−1	2 5	−3.23E−1	2 1	4.88E−1	1 6	−6.25E−2	6 8	9.46E−1	9 3	2.23E−2
3 5	−4.55E−1	3 1	8.78E−3	2 2	−2.41E−2	1 7	−4.04E−1	6 9	−1.26E+0	9 4	2.04E−2
4 1	4.89E−3	3 2	−1.61E−1	2 3	−6.25E−2	1 8	−3.88E−2	6 10	−5.35E−4	9 5	1.40E+0
4 2	3.44E−2	3 3	−1.18E−2	2 4	−5.97E−1	1 9	−1.51E+0		**$^{4}S^{e}$–$^{4}P^{o}$**	10 1	3.13E−3
4 3	−1.25E+0	3 4	−4.06E−3	2 5	−4.08E−3	1 10	−9.28E−4	1 1	4.10E−1	10 2	5.68E−1
4 4	−1.09E−2	3 5	−8.55E−2	3 1	1.97E−5	2 1	1.03E−1	1 2	4.32E−1	10 3	5.10E−3
4 5	−1.34E−2	4 1	1.33E−1	3 2	−8.81E−1	2 2	4.98E−1	1 3	−3.59E−2	10 4	1.49E−3
5 1	6.64E−3	4 2	−1.35E−2	3 3	−2.48E−9	2 3	−5.77E−1	1 4	−9.86E−1	10 5	1.27E−3
5 2	5.67E−3	4 3	−4.16E−4	3 4	−2.69E−2	2 4	−6.82E−1	1 5	−1.18E−3		**$^{6}S^{o}$–$^{6}P^{e}$**
5 3	4.58E−4	4 4	−4.22E−2	3 5	−6.32E−3	2 5	−6.99E−2		**$^{4}P^{e}$–$^{4}P^{o}$**	1 1	−2.32E+0
5 4	−1.64E+0	4 5	−1.35E−3	4 1	2.23E+0	2 6	−1.28E−1	1 1	−4.13E−2		**$^{4}P^{o}$–$^{4}D^{e}$**
5 5	−4.62E−2	5 1	3.11E−2	4 2	7.96E−4	2 7	−3.72E−1	1 2	−5.91E−1	1 1	−1.97E+0
6 1	1.32E−3	5 2	−2.93E−3	4 3	−1.06E−3	2 8	−2.63E−1	1 3	−5.97E+0	1 2	−5.41E−1
6 2	2.28E−4	5 3	−9.19E−4	4 4	−3.89E−1	2 9	−1.10E+0	1 4	−7.89E−1	1 3	−4.10E−1
6 3	6.51E−2	5 4	−3.73E−3	4 5	−4.66E−2	2 10	−9.81E−3	1 5	−1.64E−2	1 4	−3.09E+0
6 4	−1.42E−3	5 5	−8.71E−2	5 1	5.65E−2	3 1	3.67E−1	2 1	−1.32E+0	2 1	9.51E−4
6 5	−1.32E−3	6 1	3.86E−3	5 2	1.98E−3	3 2	1.85E−1	2 2	−1.23E+0	2 2	−2.08E−2

N-like S (S^{9+})

$i\ i'$	gf_L	$i\ i'$	gf_L	$i\ i'$	gf_L	$i\ i'$	gf_L	$i\ i'$	gf_L	$i\ i'$	gf_L
2 3	−2.54E+0	2 6	−6.50E−1	6 5	−1.98E+0	10 4	1.18E−2	4 2	1.95E−2	7 1	1.15E−1
2 4	−6.08E−2	2 7	−8.42E−1	6 6	−5.74E−2	10 5	8.02E−3	4 3	5.98E−5	7 2	4.01E−3
3 1	5.19E−2	3 1	1.53E−2	6 7	−2.51E−3	10 6	7.86E−4	4 4	1.50E−2	7 3	−2.73E+0
3 2	3.88E−1	3 2	−4.19E−4	7 1	2.05E+0	10 7	8.00E−5	4 5	9.33E−3		$^4\mathbf{D}^e$–$^4\mathbf{F}^o$
3 3	8.03E−2	3 3	−3.09E−3	7 2	6.47E−4		$^4\mathbf{D}^e$–$^4\mathbf{D}^o$	4 6	7.97E−1	1 1	−2.62E−1
3 4	−5.50E−2	3 4	−2.57E−1	7 3	4.80E−3	1 1	6.95E−1	4 7	4.06E−1	1 2	−1.24E−1
4 1	9.60E−1	3 5	−1.78E−1	7 4	7.67E−5	1 2	−3.49E−4		$^6\mathbf{P}^e$–$^6\mathbf{D}^o$	2 1	−2.44E+0
4 2	6.51E−2	3 6	−6.00E−3	7 5	−2.28E−4	1 3	−4.16E−3	1 1	−4.01E+0	2 2	−4.53E−3
4 3	2.42E−2	3 7	−2.55E−2	7 6	−1.57E+0	1 4	−6.74E−1		$^4\mathbf{D}^o$–$^4\mathbf{F}^e$	3 1	−4.18E−3
4 4	−6.88E−2	4 1	3.02E−3	7 7	−4.26E−1	1 5	−4.41E−2	1 1	−3.45E+0	3 2	−3.71E+0
5 1	3.96E−1	4 2	−7.16E−3	8 1	8.12E−1	1 6	−1.20E−1	1 2	−9.43E−1	4 1	2.67E−3
5 2	9.99E−4	4 3	−3.13E+0	8 2	1.51E−3	1 7	−6.00E−1	1 3	−5.92E+0	4 2	5.11E−2
5 3	7.33E−4	4 4	−1.92E−2	8 3	2.05E−3	2 1	6.27E−1	2 1	1.60E−3		$^4\mathbf{F}^e$–$^4\mathbf{F}^o$
5 4	−3.22E+0	4 5	−1.53E−3	8 4	2.08E−3	2 2	2.21E+0	2 2	−3.65E+0	1 1	−9.80E−1
	$^4\mathbf{P}^e$–$^4\mathbf{D}^o$	4 6	−4.03E−4	8 5	1.34E−2	2 3	9.55E−4	2 3	−1.11E−1	1 2	−6.86E−1
1 1	−3.38E−3	4 7	−5.99E−3	8 6	−8.90E−1	2 4	−1.82E+0	3 1	4.32E−4	2 1	−1.05E+0
1 2	−1.15E+0	5 1	1.14E−2	8 7	−2.51E+0	2 5	−2.07E−2	3 2	−3.22E−3	2 2	−1.36E−1
1 3	−4.61E+0	5 2	2.13E+0	9 1	5.06E−2	2 6	−2.98E−4	3 3	−4.75E−4	3 1	7.13E−4
1 4	−5.05E+0	5 3	8.08E−3	9 2	7.75E−3	2 7	−3.50E−4	4 1	5.23E−2	3 2	4.09E−3
1 5	−4.77E+0	5 4	−7.23E−1	9 3	2.51E−4	3 1	1.01E+0	4 2	1.40E−1		$^4\mathbf{F}^e$–$^4\mathbf{G}^o$
1 6	−3.99E−1	5 5	−5.57E−2	9 4	2.28E−4	3 2	1.50E−1	4 3	−4.90E−2	1 1	−3.51E−1
1 7	−2.98E−1	5 6	−1.41E−3	9 5	3.57E−2	3 3	2.34E−3	5 1	9.97E−1	1 2	−2.34E+1
2 1	−2.13E+0	5 7	−5.38E−2	9 6	2.50E−2	3 4	−1.30E−2	5 2	6.98E−2	2 1	−4.71E+0
2 2	−1.18E+0	6 1	6.84E−1	9 7	1.54E−1	3 5	−7.62E−1	5 3	−1.65E−1	2 2	−3.58E−1
2 3	−1.45E−2	6 2	2.40E−3	10 1	3.21E−3	3 6	−5.50E−4	6 1	2.86E+0	3 1	4.53E−2
2 4	−1.33E−8	6 3	2.80E−3	10 2	1.82E+0	3 7	−8.53E−3	6 2	2.16E−2	3 2	−3.39E+0
2 5	−8.95E−3	6 4	−3.90E−3	10 3	4.00E−2	4 1	9.84E−1	6 3	−3.05E+0		

N-like Ar (Ar^{11+})

Term energies relative to $2s^22p^2$ ^{3}P ionization threshold for each symmetry

i	E(Ryds)	Description	i	E(Ryds)	Description	i	E(Ryds)	Description	i	E(Ryds)	Description
		^{2}S^e	12	−9.27503	$2s^22p^2$ ^{3}P $4d$	24	−5.03707	$2s2p^3$ ^{3}P $4d$	33	−3.11345	$2s^22p^2$ ^{1}D $6g$
1	−39.1162	$2s2p^4$	13	−8.35457	$2s^22p^2$ ^{1}D $4d$	25	−4.84981	$2s^22p^2$ ^{1}D $5f$	34	−3.09359	$2s^22p^2$ ^{1}S $6d$
2	−17.9491	$2s^22p^2$ ^{1}S $3s$	14	−7.95495	$2p^4$ ^{3}P $3d$	26	−4.33922	$2s2p^3$ ^{1}P $4s$	35	−2.96585	$2s2p^3$ ^{1}P $4f$
3	−15.6322	$2s^22p^2$ ^{1}D $3d$	15	−7.18637	$2p^4$ ^{1}D $3d$	27	−4.17067	$2s^22p^2$ ^{3}P $6p$	36	−2.95627	$2s^22p^2$ ^{3}P $7d$
4	−13.1613	$2s2p^3$ ^{3}P $3p$	16	−6.36547	$2s^22p^2$ ^{3}P $5s$	28	−3.54307	$2s2p^3$ ^{1}D $4d$	37	−2.26488	$2s^22p^2$ ^{3}P $8d$
5	−11.6013	$2s2p^3$ ^{1}P $3p$	17	−5.96531	$2s2p^3$ ^{3}D $4p$	29	−3.34711	$2s2p^3$ ^{1}P $4d$	38	−2.22917	$2s^22p^2$ ^{1}D $7s$
6	−9.18118	$2s^22p^2$ ^{1}S $4s$	18	−5.89661	$2s^22p^2$ ^{3}P $5d$	30	−3.27299	$2s^22p^2$ ^{1}D $6p$	39	−2.20258	$2s2p^3$ ^{3}D $5p$
7	−8.57056	$2p^4$ ^{1}S $3s$	19	−5.72188	$2s2p^3$ ^{3}P $4p$	31	−3.21978	$2s^22p^2$ ^{1}S $6p$	40	−2.15246	$2s2p^3$ ^{3}P $5p$
8	−8.32130	$2s^22p^2$ ^{1}D $4d$	20	−5.28386	$2s2p^3$ ^{3}D $4f$	32	−3.06353	$2s^22p^2$ ^{1}D $6f$	41	−2.04890	$2s^22p^2$ ^{1}D $7d$
9	−7.39108	$2p^4$ ^{1}D $3d$	21	−4.96612	$2s^22p^2$ ^{1}D $5d$	33	−3.04606	$2s^22p^2$ ^{3}P $7p$	42	−2.01832	$2s^22p^2$ ^{1}S $7d$
10	−5.52738	$2s2p^3$ ^{3}P $4p$	22	−4.34558	$2s^22p^2$ ^{3}P $6s$	34	−2.38469	$2s2p^3$ ^{3}P $5s$	43	−2.00522	$2s^22p^2$ ^{1}D $7g$
11	−5.37362	$2s^22p^2$ ^{1}S $5s$	23	−4.13025	$2s2p^3$ ^{1}D $4p$	35	−2.32504	$2s^22p^2$ ^{3}P $8p$	44	−1.93000	$2s2p^3$ ^{3}D $5f$
12	−4.94807	$2s^22p^2$ ^{1}D $5d$	24	−4.07368	$2s^22p^2$ ^{3}P $6d$	36	−2.12650	$2s^22p^2$ ^{1}D $7p$	45	−1.80536	$2s2p^3$ ^{3}P $5f$
13	−3.84656	$2s2p^3$ ^{1}P $4p$	25	−3.90324	$2s2p^3$ ^{3}S $4p$	37	−2.10347	$2s^22p^2$ ^{1}S $7p$	46	−1.78681	$2s^22p^2$ ^{3}P $9d$
14	−3.36197	$2s^22p^2$ ^{1}S $6s$	26	−3.81790	$2s2p^3$ ^{1}P $4p$	38	−2.01202	$2s2p^3$ ^{3}D $5d$	47	−1.45550	$2s^22p^2$ ^{1}D $8s$
15	−3.12624	$2s^22p^2$ ^{1}D $6d$	27	−3.40658	$2s2p^3$ ^{1}D $4f$	39	−1.99893	$2s^22p^2$ ^{1}D $7f$			^{2}D^o
16	−2.19169	$2s^22p^2$ ^{1}S $7s$	28	−3.15187	$2s^22p^2$ ^{3}P $7s$	40	−1.84404	$2s2p^3$ ^{3}P $5d$	1	−44.6885	$2s^22p^3$
17	−2.07195	$2s2p^3$ ^{3}P $5p$	29	−3.13156	$2s^22p^2$ ^{1}D $6d$	41	−1.82139	$2s^22p^2$ ^{3}P $9p$	2	−17.9418	$2s^22p^2$ ^{3}P $3p$
18	−2.03566	$2s^22p^2$ ^{1}D $7d$	30	−2.98668	$2s^22p^2$ ^{3}P $7d$			^{2}D^e	3	−17.1687	$2s^22p^2$ ^{1}D $3p$
19	−1.43403	$2s^22p^2$ ^{1}S $8s$	31	−2.39253	$2s^22p^2$ ^{3}P $8s$	1	−40.1119	$2s2p^4$	4	−15.1933	$2s2p^3$ ^{3}D $3s$
20	−1.33112	$2s^22p^2$ ^{1}D $8d$	32	−2.31441	$2s2p^3$ ^{3}D $5p$	2	−18.5762	$2s^22p^2$ ^{1}D $3s$	5	−13.5759	$2s2p^3$ ^{1}D $3s$
21	−.92228	$2s^22p^2$ ^{1}S $9s$	33	−2.28302	$2s^22p^2$ ^{3}P $8d$	3	−16.2004	$2s^22p^2$ ^{3}P $3d$	6	−12.7208	$2s2p^3$ ^{3}D $3d$
22	−.84939	$2s^22p^2$ ^{1}D $9d$	34	−2.17748	$2s2p^3$ ^{3}P $5p$	4	−15.8893	$2s^22p^2$ ^{1}D $3d$	7	−12.1592	$2s2p^3$ ^{3}P $3d$
		^{2}S^o	35	−2.04146	$2s^22p^2$ ^{1}D $7d$	5	−15.2877	$2s^22p^2$ ^{1}S $3d$	8	−11.2122	$2s2p^3$ ^{1}D $3d$
1	−18.3527	$2s^22p^2$ ^{3}P $3p$	36	−1.92797	$2s2p^3$ ^{3}D $5f$	6	−13.9075	$2s2p^3$ ^{3}D $3p$	9	−10.7334	$2s2p^3$ ^{3}S $3d$
2	−13.4555	$2s2p^3$ ^{3}S $3s$	37	−1.87369	$2s^22p^2$ ^{3}P $9s$	7	−13.5427	$2s2p^3$ ^{3}P $3p$	10	−10.3436	$2s2p^3$ ^{1}P $3d$
3	−12.9856	$2s2p^3$ ^{3}D $3d$	38	−1.80080	$2s^22p^2$ ^{3}P $9d$	8	−12.2847	$2s2p^3$ ^{1}D $3p$	11	−9.68357	$2s^22p^2$ ^{3}P $4p$
4	−10.8268	$2s2p^3$ ^{1}D $3d$			^{2}P^o	9	−11.8085	$2s2p^3$ ^{1}P $3p$	12	−9.30888	$2p^4$ ^{3}P $3p$
5	−9.86010	$2s^22p^2$ ^{3}P $4p$	1	−44.2063	$2s^22p^3$	10	−9.82513	$2p^4$ ^{1}D $3s$	13	−8.99548	$2s^22p^2$ ^{3}P $4f$
6	−9.20146	$2p^4$ ^{3}P $3p$	2	−35.1169	$2p^5$	11	−9.39156	$2s^22p^2$ ^{1}D $4s$	14	−8.88773	$2s^22p^2$ ^{1}D $4p$
7	−6.19949	$2s^22p^2$ ^{3}P $5p$	3	−17.7859	$2s^22p^2$ ^{3}P $3p$	12	−9.08689	$2s^22p^2$ ^{3}P $4d$	15	−8.64066	$2p^4$ ^{1}D $3p$
8	−5.45997	$2s2p^3$ ^{3}D $4d$	4	−16.9857	$2s^22p^2$ ^{1}D $3p$	13	−8.40483	$2s^22p^2$ ^{1}D $4d$	16	−8.21981	$2s^22p^2$ ^{1}D $4f$
9	−4.52340	$2s2p^3$ ^{3}S $4s$	5	−16.6819	$2s^22p^2$ ^{1}S $3p$	14	−8.17772	$2s^22p^2$ ^{1}S $4d$	17	−6.32846	$2s2p^3$ ^{3}D $4s$
10	−4.21613	$2s^22p^2$ ^{3}P $6p$	6	−14.5951	$2s2p^3$ ^{3}P $3s$	15	−7.88818	$2p^4$ ^{3}P $3d$	18	−6.10538	$2s^22p^2$ ^{3}P $5p$
11	−3.52145	$2s2p^3$ ^{1}D $4d$	7	−12.9567	$2s2p^3$ ^{1}P $3s$	16	−7.23403	$2p^4$ ^{1}D $3d$	19	−5.77496	$2s^22p^2$ ^{3}P $5f$
12	−3.08732	$2s^22p^2$ ^{3}P $7p$	8	−12.7709	$2s2p^3$ ^{3}D $3d$	17	−6.21668	$2p^4$ ^{1}S $3d$	20	−5.35142	$2s2p^3$ ^{3}D $4d$
13	−2.35051	$2s^22p^2$ ^{3}P $8p$	9	−11.8148	$2s2p^3$ ^{3}P $3d$	18	−5.82971	$2s^22p^2$ ^{3}P $5d$	21	−5.23779	$2s^22p^2$ ^{1}D $5p$
14	−2.04189	$2s2p^3$ ^{3}D $5d$	10	−10.8905	$2s2p^3$ ^{1}D $3d$	19	−5.77840	$2s2p^3$ ^{3}D $4p$	22	−5.16292	$2s2p^3$ ^{3}P $4d$
15	−1.84674	$2s^22p^2$ ^{3}P $9p$	11	−10.3959	$2s2p^3$ ^{1}P $3d$	20	−5.66422	$2s2p^3$ ^{3}P $4p$	23	−4.86628	$2s^22p^2$ ^{1}D $5f$
		^{2}P^e	12	−9.63045	$2s^22p^2$ ^{3}P $4p$	21	−5.48364	$2s^22p^2$ ^{1}D $5s$	24	−4.53951	$2s2p^3$ ^{1}D $4s$
1	−38.8619	$2s2p^4$	13	−9.29109	$2p^4$ ^{3}P $3p$	22	−5.29395	$2s2p^3$ ^{3}D $4f$	25	−4.19184	$2s^22p^2$ ^{3}P $6p$
2	−19.0539	$2s^22p^2$ ^{3}P $3s$	14	−8.83916	$2s^22p^2$ ^{1}D $4p$	23	−4.98733	$2s^22p^2$ ^{1}D $5d$	26	−4.01318	$2s^22p^2$ ^{3}P $6f$
3	−16.7105	$2s^22p^2$ ^{3}P $3d$	15	−8.69977	$2s^22p^2$ ^{1}S $4p$	24	−4.91033	$2s2p^3$ ^{3}P $4f$	27	−3.64425	$2s2p^3$ ^{1}D $4d$
4	−15.7312	$2s^22p^2$ ^{1}D $3d$	16	−8.22376	$2p^4$ ^{1}D $3p$	25	−4.87843	$2s^22p^2$ ^{1}S $5d$	28	−3.45076	$2s2p^3$ ^{3}S $4d$
5	−14.4738	$2s2p^3$ ^{3}D $3p$	17	−8.18128	$2s^22p^2$ ^{1}D $4f$	26	−4.85755	$2s^22p^2$ ^{1}D $5g$	29	−3.33250	$2s2p^3$ ^{1}P $4d$
6	−13.6449	$2s2p^3$ ^{3}P $3p$	18	−7.43362	$2p^4$ ^{1}S $3p$	27	−4.05893	$2s2p^3$ ^{1}D $4p$	30	−3.27675	$2s^22p^2$ ^{1}D $6p$
7	−12.5847	$2s2p^3$ ^{1}D $3p$:	19	−6.13009	$2s2p^3$ ^{3}P $4s$	28	−4.02928	$2s^22p^2$ ^{3}P $6d$	31	−3.07322	$2s^22p^2$ ^{1}D $6f$
8	−11.8704	$2s2p^3$ ^{3}S $3p$	20	−6.07489	$2s^22p^2$ ^{3}P $5p$	29	−3.88713	$2s2p^3$ ^{1}P $4p$	32	−3.05510	$2s^22p^2$ ^{3}P $7p$
9	−11.6175	$2s2p^3$ ^{1}P $3p$	21	−5.39044	$2s2p^3$ ^{3}D $4d$	30	−3.41490	$2s^22p^2$ ^{1}D $6s$	33	−2.94307	$2s^22p^2$ ^{3}P $7f$
10	−10.2268	$2p^4$ ^{3}P $3s$	22	−5.21254	$2s^22p^2$ ^{1}D $5p$	31	−3.40821	$2s2p^3$ ^{1}D $4f$	34	−2.48344	$2s2p^3$ ^{3}D $5s$
11	−10.1743	$2s^22p^2$ ^{3}P $4s$	23	−5.13534	$2s^22p^2$ ^{1}S $5p$	32	−3.14752	$2s^22p^2$ ^{1}D $6d$	35	−2.33134	$2s^22p^2$ ^{3}P $8p$

N-like Ar (Ar^{11+})

i	E(Ryds)	Description	i	E(Ryds)	Description	i	E(Ryds)	Description	i	E(Ryds)	Description
36	−2.25436	$2s^22p^2$ ^{3}P $8f$	9	−8.26377	$2s^22p^2$ ^{1}D $4f$			**^{2}G^o**	3	−4.74428	$2s2p^3$ ^{5}S $5s$
37	−2.13394	$2s^22p^2$ ^{1}D $7p$	10	−7.86530	$2s^22p^2$ ^{1}S $4f$	1	−12.9413	$2s2p^3$ ^{3}D $3d$	4	−2.71107	$2s2p^3$ ^{5}S $6s$
38	−2.00701	$2s^22p^2$ ^{1}D $7f$	11	−5.75265	$2s^22p^2$ ^{3}P $5f$	2	−11.1809	$2s2p^3$ ^{1}D $3d$	5	−1.51440	$2s2p^3$ ^{5}S $7s$
39	−1.98994	$2s2p^3$ ^{3}D $5d$	12	−5.30688	$2s2p^3$ ^{3}D $4d$	3	−9.01974	$2s^22p^2$ ^{3}P $4f$	6	−.75086	$2s2p^3$ ^{5}S $8s$
40	−1.92417	$2s2p^3$ ^{3}D $5g$	13	−5.25330	$2s^22p^2$ ^{1}D $5p$	4	−8.28395	$2s^22p^2$ ^{1}D $4f$	7	−.23400	$2s2p^3$ ^{5}S $9s$
41	−1.90355	$2s2p^3$ ^{3}P $5d$	14	−5.14030	$2s2p^3$ ^{3}P $4d$	5	−5.77361	$2s^22p^2$ ^{3}P $5f$			**^{4}P^e**
42	−1.83073	$2s^22p^2$ ^{3}P $9p$	15	−4.88409	$2s^22p^2$ ^{1}D $5f$	6	−5.46039	$2s2p^3$ ^{3}D $4d$	1	−41.6027	$2s2p^4$
43	−1.78058	$2s^22p^2$ ^{3}P $9f$	16	−4.75916	$2s^22p^2$ ^{1}S $5f$	7	−4.88997	$2s^22p^2$ ^{1}D $5f$	2	−19.3098	$2s^22p^2$ ^{3}P $3s$
		^{2}F^e	17	−3.99663	$2s^22p^2$ ^{3}P $6f$	8	−4.00898	$2s^22p^2$ ^{3}P $6f$	3	−16.5452	$2s^22p^2$ ^{3}P $3d$
1	−16.5858	$2s^22p^2$ ^{3}P $3d$	18	−3.61652	$2s2p^3$ ^{1}D $4d$	9	−4.00017	$2s^22p^2$ ^{3}P $6h$	4	−16.0368	$2s2p^3$ ^{5}S $3p$
2	−15.8795	$2s^22p^2$ ^{1}D $3d$	19	−3.39538	$2s2p^3$ ^{1}P $4d$	10	−3.64062	$2s2p^3$ ^{1}D $4d$	5	−14.0333	$2s2p^3$ ^{3}D $3p$
3	−14.3110	$2s2p^3$ ^{3}D $3p$	20	−3.27163	$2s^22p^2$ ^{1}D $6p$	11	−3.08922	$2s^22p^2$ ^{1}D $6f$	6	−13.6010	$2s2p^3$ ^{3}P $3p$
4	−12.4770	$2s2p^3$ ^{1}D $3p$	21	−3.08492	$2s^22p^2$ ^{1}D $6f$	12	−3.05428	$2s^22p^2$ ^{1}D $6h$	7	−12.5515	$2s2p^3$ ^{3}S $3p$
5	−9.19000	$2s^22p^2$ ^{3}P $4d$	22	−3.05397	$2s^22p^2$ ^{1}D $6h$	13	−2.94448	$2s^22p^2$ ^{3}P $7f$	8	−10.5324	$2p^4$ ^{3}P $3s$
6	−8.44359	$2s^22p^2$ ^{1}D $4d$	23	−3.02799	$2s^22p^2$ ^{1}S $6f$	14	−2.93903	$2s^22p^2$ ^{3}P $7h$	9	−10.2645	$2s^22p^2$ ^{3}P $4s$
7	−8.08467	$2p^4$ ^{3}P $3d$	24	−2.93690	$2s^22p^2$ ^{3}P $7f$	15	−2.25414	$2s^22p^2$ ^{3}P $8f$	10	−9.22724	$2s^22p^2$ ^{3}P $4d$
8	−7.41459	$2p^4$ ^{1}D $3d$	25	−2.24917	$2s^22p^2$ ^{3}P $8f$	16	−2.25031	$2s^22p^2$ ^{3}P $8h$	11	−8.05749	$2p^4$ ^{3}P $3d$
9	−5.95278	$2s2p^3$ ^{3}D $4p$	26	−2.14026	$2s^22p^2$ ^{1}D $7p$	17	−2.04416	$2s2p^3$ ^{3}D $5d$	12	−7.98690	$2s2p^3$ ^{5}S $4p$
10	−5.84623	$2s^22p^2$ ^{3}P $5d$	27	−2.01365	$2s^22p^2$ ^{1}D $7f$	18	−2.01149	$2s^22p^2$ ^{1}D $7f$	13	−6.41301	$2s^22p^2$ ^{3}P $5s$
11	−5.76750	$2s^22p^2$ ^{3}P $5g$	28	−1.99321	$2s^22p^2$ ^{1}D $7h$	19	−1.99358	$2s^22p^2$ ^{1}D $7h$	14	−5.88323	$2s^22p^2$ ^{3}P $5d$
12	−5.30469	$2s2p^3$ ^{3}D $4f$	29	−1.97804	$2s^22p^2$ ^{1}S $7f$	20	−1.92159	$2s2p^3$ ^{3}D $5g$	15	−5.86671	$2s2p^3$ ^{3}D $4p$
13	−5.00952	$2s^22p^2$ ^{1}D $5d$	30	−1.96733	$2s2p^3$ ^{3}D $5d$	21	−1.82430	$2s2p^3$ ^{3}P $5g$	16	−5.72046	$2s2p^3$ ^{3}P $4p$
14	−4.91875	$2s2p^3$ ^{3}P $4f$	31	−1.92358	$2s2p^3$ ^{3}D $5g$	22	−1.78024	$2s^22p^2$ ^{3}P $9f$	17	−5.29868	$2s2p^3$ ^{3}D $4f$
15	−4.86456	$2s^22p^2$ ^{1}D $5g$	32	−1.88869	$2s2p^3$ ^{3}P $5d$	23	−1.77734	$2s^22p^2$ ^{3}P $9h$	18	−4.42724	$2s2p^3$ ^{5}S $5p$
16	−4.10253	$2s2p^3$ ^{1}D $4p$	33	−1.82269	$2s2p^3$ ^{3}P $5g$			**^{4}S^e**	19	−4.36182	$2s^22p^2$ ^{3}P $6s$
17	−4.05245	$2s^22p^2$ ^{3}P $6d$	34	−1.77623	$2s^22p^2$ ^{3}P $9f$	1	−13.8972	$2s2p^3$ ^{3}P $3p$	20	−4.11709	$2s2p^3$ ^{3}S $4p$
18	−4.00669	$2s^22p^2$ ^{3}P $6g$			**^{2}G^e**	2	−5.81533	$2s2p^3$ ^{3}P $4p$	21	−4.05704	$2s^22p^2$ ^{3}P $6d$
19	−3.41071	$2s2p^3$ ^{1}D $4f$	1	−16.0325	$2s^22p^2$ ^{1}D $3d$	3	−2.21586	$2s2p^3$ ^{3}P $5p$	22	−3.16098	$2s^22p^2$ ^{3}P $7s$
20	−3.39280	$2s2p^3$ ^{3}S $4f$	2	−8.46432	$2s^22p^2$ ^{1}D $4d$	4	−.30719	$2s2p^3$ ^{3}P $6p$	23	−2.98206	$2s^22p^2$ ^{3}P $7d$
21	−3.16237	$2s^22p^2$ ^{1}D $6d$	3	−7.66965	$2p^4$ ^{1}D $3d$			**^{4}S^o**	24	−2.54337	$2s2p^3$ ^{5}S $6p$
22	−3.10790	$2s^22p^2$ ^{1}D $6g$	4	−5.76083	$2s^22p^2$ ^{3}P $5g$	1	−45.6013	$2s^22p^3$	25	−2.40461	$2s^22p^2$ ^{3}P $8s$
23	−2.99749	$2s2p^3$ ^{1}P $4f$	5	−5.31777	$2s2p^3$ ^{3}D $4f$	2	−17.9350	$2s^22p^2$ ^{3}P $3p$	26	−2.27950	$2s^22p^2$ ^{3}P $8d$
24	−2.96956	$2s^22p^2$ ^{3}P $7d$	6	−5.01301	$2s^22p^2$ ^{1}D $5d$	3	−16.8967	$2s2p^3$ ^{5}S $3s$	27	−2.24647	$2s2p^3$ ^{3}D $5p$
25	−2.93623	$2s^22p^2$ ^{3}P $7g$	7	−4.92383	$2s2p^3$ ^{3}P $4f$	4	−13.6530	$2s2p^3$ ^{3}S $3s$	28	−2.17339	$2s2p^3$ ^{3}P $5p$
26	−2.29043	$2s2p^3$ ^{3}D $5p$	8	−4.86947	$2s^22p^2$ ^{1}D $5g$	5	−12.7323	$2s2p^3$ ^{3}D $3d$	29	−1.92937	$2s2p^3$ ^{3}D $5f$
27	−2.27050	$2s^22p^2$ ^{3}P $8d$	9	−4.76486	$2s^22p^2$ ^{1}S $5g$	6	−9.68982	$2s^22p^2$ ^{3}P $4p$	30	−1.87805	$2s^22p^2$ ^{3}P $9s$
28	−2.25067	$2s^22p^2$ ^{3}P $8g$	10	−3.99573	$2s^22p^2$ ^{3}P $6g$	7	−9.19512	$2p^4$ ^{3}P $3p$	31	−1.79816	$2s^22p^2$ ^{3}P $9d$
29	−2.05619	$2s^22p^2$ ^{1}D $7d$	11	−3.42025	$2s2p^3$ ^{1}D $4f$	8	−8.43923	$2s2p^3$ ^{5}S $4s$	32	−1.57218	$2p^4$ ^{3}P $4s$
30	−2.00838	$2s^22p^2$ ^{1}D $7g$	12	−3.16350	$2s^22p^2$ ^{1}D $6d$	9	−6.10694	$2s^22p^2$ ^{3}P $5p$			**^{4}P^o**
31	−1.92798	$2s2p^3$ ^{3}D $5f$	13	−3.09314	$2s^22p^2$ ^{1}D $6g$	10	−5.38120	$2s2p^3$ ^{3}D $4d$	1	−18.1080	$2s^22p^2$ ^{3}P $3p$
32	−1.81844	$2s2p^3$ ^{3}P $5f$	14	−3.04588	$2s2p^3$ ^{1}P $4f$	11	−4.66758	$2s2p^3$ ^{5}S $5s$	2	−14.9761	$2s2p^3$ ^{3}P $3s$
33	−1.79324	$2s^22p^2$ ^{3}P $9d$	15	−2.90215	$2s^22p^2$ ^{1}S $6g$	12	−4.55660	$2s2p^3$ ^{3}S $4s$	3	−12.0009	$2s2p^3$ ^{3}D $3d$
34	−1.77474	$2s^22p^2$ ^{3}P $9g$	16	−2.93688	$2s^22p^2$ ^{3}P $7g$	13	−4.18699	$2s^22p^2$ ^{3}P $6p$	4	−12.4653	$2s2p^3$ ^{3}P $3d$
		^{2}F^o	17	−2.24938	$2s^22p^2$ ^{3}P $8g$	14	−3.06162	$2s^22p^2$ ^{3}P $7p$	5	−9.79550	$2s^22p^2$ ^{3}P $4p$
1	−17.4108	$2s^22p^2$ ^{1}D $3p$	18	−2.05770	$2s^22p^2$ ^{1}D $7d$	15	−2.66709	$2s2p^3$ ^{5}S $6s$	6	−9.62586	$2p^4$ ^{3}P $3p$
2	−12.5604	$2s2p^3$ ^{3}D $3d$	19	−2.01121	$2s^22p^2$ ^{1}D $7g$	16	−2.33071	$2s^22p^2$ ^{3}P $8p$	7	−6.29375	$2s2p^3$ ^{3}P $4s$
3	−12.1276	$2s2p^3$ ^{3}P $3d$	20	−1.98848	$2s^22p^2$ ^{1}D $7i$	17	−2.00862	$2s2p^3$ ^{3}D $5d$	8	−6.12420	$2s^22p^2$ ^{3}P $5p$
4	−11.1351	$2s2p^3$ ^{1}D $3d$	21	−1.96674	$2s^22p^2$ ^{1}S $7g$	18	−1.83376	$2s^22p^2$ ^{3}P $9p$	9	−5.43654	$2s2p^3$ ^{3}D $4d$
5	−10.5015	$2s2p^3$ ^{1}P $3d$	22	−1.93153	$2s2p^3$ ^{3}D $5f$	19	−1.40574	$2s2p^3$ ^{5}S $7s$	10	−5.27238	$2s2p^3$ ^{3}P $4d$
6	−8.97445	$2s^22p^2$ ^{3}P $4f$	23	−1.81944	$2s2p^3$ ^{3}P $5f$			**^{6}S^o**	11	−4.22144	$2s^22p^2$ ^{3}P $6p$
7	−8.92974	$2s^22p^2$ ^{1}D $4p$	24	−1.77174	$2s^22p^2$ ^{3}P $9g$	1	−17.6174	$2s2p^3$ ^{5}S $3s$	12	−3.07877	$2s^22p^2$ ^{3}P $7p$
8	−8.84129	$2p^4$ ^{1}D $3p$				2	−8.62992	$2s2p^3$ ^{5}S $4s$	13	−2.44380	$2s2p^3$ ^3P $5s$

N-like Ar (Ar^{11+})

i	E(Ryds)	Description
14	−2.33828	$2s^22p^2$ ^{3}P $8p$
15	−2.03491	$2s2p^3$ ^{3}D $5d$
16	−1.95778	$2s2p^3$ ^{3}P $5d$
17	−1.83988	$2s^22p^2$ ^{3}P $9p$
	^{6}**P**^{e}	
1	−16.4493	$2s2p^3$ ^{5}S $3p$
2	−8.16127	$2s2p^3$ ^{5}S $4p$
3	−4.51156	$2s2p^3$ ^{5}S $5p$
4	−2.57908	$2s2p^3$ ^{5}S $6p$
5	−1.43246	$2s2p^3$ ^{5}S $7p$
6	−.69655	$2s2p^3$ ^{5}S $8p$
7	−.19618	$2s2p^3$ ^{5}S $9p$
	^{4}**D**^{e}	
1	−16.7231	$2s^22p^2$ ^{3}P $3d$
2	−14.5171	$2s2p^3$ ^{3}D $3p$
3	−13.8252	$2s2p^3$ ^{3}P $3p$
4	−9.30203	$2s^22p^2$ ^{3}P $4d$
5	−8.40072	$2p^4$ ^{3}P $3d$
6	−6.02281	$2s2p^3$ ^{3}D $4p$
7	−5.92073	$2s^22p^2$ ^{3}P $5d$
8	−5.78616	$2s2p^3$ ^{3}P $4p$
9	−5.31698	$2s2p^3$ ^{3}D $4f$
10	−4.92674	$2s2p^3$ ^{3}P $4f$
11	−4.08809	$2s^22p^2$ ^{3}P $6d$
12	−2.99432	$2s^22p^2$ ^{3}P $7d$
13	−2.31976	$2s2p^3$ ^{3}D $5p$
14	−2.28787	$2s^22p^2$ ^{3}P $8d$
15	−2.20620	$2s2p^3$ ^{3}P $5p$
16	−1.93752	$2s2p^3$ ^{3}D $5f$
17	−1.81869	$2s2p^3$ ^{3}P $5f$
18	−1.80358	$2s^22p^2$ ^{3}P $9d$
	^{4}**D**^{o}	
1	−18.1947	$2s^22p^2$ ^{3}P $3p$
2	−15.5973	$2s2p^3$ ^{3}D $3s$
3	−14.5921	$2s2p^3$ ^{5}S $3d$
4	−13.0295	$2s2p^3$ ^{3}D $3d$
5	−12.3636	$2s2p^3$ ^{3}P $3d$
6	−11.2534	$2s2p^3$ ^{3}S $3d$
7	−9.81595	$2s^22p^2$ ^{3}P $4p$
8	−9.44017	$2p^4$ ^{3}P $3p$
9	−9.00851	$2s^22p^2$ ^{3}P $4f$
10	−7.45550	$2s2p^3$ ^{5}S $4d$
11	−6.47464	$2s2p^3$ ^{3}D $4s$
12	−6.15937	$2s^22p^2$ ^{3}P $5p$
13	−5.77786	$2s^22p^2$ ^{3}P $5f$
14	−5.50666	$2s2p^3$ ^{3}D $4d$
15	−5.23442	$2s2p^3$ ^{3}P $4d$
16	−4.23159	$2s^22p^2$ ^{3}P $6p$
17	−4.16555	$2s2p^3$ ^{5}S $5d$
18	−4.01159	$2s^22p^2$ ^{3}P $6f$
19	−3.64498	$2s2p^3$ ^{3}S $4d$
20	−3.08127	$2s^22p^2$ ^{3}P $7p$
21	−2.94564	$2s^22p^2$ ^{3}P $7f$
22	−2.54336	$2s2p^3$ ^{3}D $5s$
23	−2.39115	$2s2p^3$ ^{5}S $6d$
24	−2.34366	$2s^22p^2$ ^{3}P $8p$
25	−2.25540	$2s^22p^2$ ^{3}P $8f$
26	−2.06203	$2s2p^3$ ^{3}D $5d$
27	−1.94044	$2s2p^3$ ^{3}P $5d$
28	−1.92287	$2s2p^3$ ^{3}D $5g$
29	−1.84181	$2s^22p^2$ ^{3}P $9p$
30	−1.78058	$2s^22p^2$ ^{3}P $9f$
	^{6}**D**^{o}	
1	−15.1615	$2s2p^3$ ^{5}S $3d$
2	−7.68484	$2s2p^3$ ^{5}S $4d$
3	−4.28024	$2s2p^3$ ^{5}S $5d$
4	−2.44899	$2s2p^3$ ^{5}S $6d$
5	−1.35199	$2s2p^3$ ^{5}S $7d$
6	−.64329	$2s2p^3$ ^{5}S $8d$
7	−.15910	$2s2p^3$ ^{5}S $9d$
	^{4}**F**^{e}	
1	−16.8564	$2s^22p^2$ ^{3}P $3d$
2	−14.4498	$2s2p^3$ ^{3}D $3p$
3	−9.34269	$2s^22p^2$ ^{3}P $4d$
4	−8.22287	$2p^4$ ^{3}P $3d$
5	−7.35604	$2s2p^3$ ^{5}S $4f$
6	−6.00938	$2s2p^3$ ^{3}D $4p$
7	−5.92100	$2s^22p^2$ ^{3}P $5d$
8	−5.76767	$2s^22p^2$ ^{3}P $5g$
9	−5.32732	$2s2p^3$ ^{3}D $4f$
10	−4.92774	$2s2p^3$ ^{3}P $4f$
11	−4.11784	$2s2p^3$ ^{5}S $5f$
12	−4.09747	$2s^22p^2$ ^{3}P $6d$
13	−4.00658	$2s^22p^2$ ^{3}P $6g$
14	−3.40487	$2s2p^3$ ^{3}S $4f$
15	−3.00034	$2s^22p^2$ ^{3}P $7d$
16	−2.93650	$2s^22p^2$ ^{3}P $7g$
17	−2.35593	$2s2p^3$ ^{5}S $6f$
18	−2.31395	$2s2p^3$ ^{3}D $5p$
19	−2.28775	$2s^22p^2$ ^{3}P $8d$
20	−2.25079	$2s^22p^2$ ^{3}P $8g$
21	−1.94064	$2s2p^3$ ^{3}D $5f$
22	−1.82317	$2s2p^3$ ^{3}P $5f$
23	−1.80649	$2s^22p^2$ ^{3}P $9d$
24	−1.77487	$2s^22p^2$ ^{3}P $9g$
	^{4}**F**^{o}	
1	−13.2513	$2s2p^3$ ^{3}D $3d$
2	−12.5472	$2s2p^3$ ^{3}P $3d$
3	−8.98787	$2s^22p^2$ ^{3}P $4f$
4	−5.76542	$2s^22p^2$ ^{3}P $5f$
5	−5.57333	$2s2p^3$ ^{3}D $4d$
6	−5.29442	$2s2p^3$ ^{3}P $4d$
7	−4.00324	$2s^22p^2$ ^{3}P $6f$
8	−2.94188	$2s^22p^2$ ^{3}P $7f$
9	−2.25303	$2s^22p^2$ ^{3}P $8f$
10	−2.09469	$2s2p^3$ ^{3}D $5d$
11	−1.96855	$2s2p^3$ ^{3}P $5d$
12	−1.92333	$2s2p^3$ ^{3}D $5g$
13	−1.82307	$2s2p^3$ ^{3}P $5g$
14	−1.77818	$2s^22p^2$ ^{3}P $9f$
	^{6}**F**^{e}	
1	−7.37805	$2s2p^3$ ^{5}S $4f$
2	−4.13098	$2s2p^3$ ^{5}S $5f$
3	−2.36498	$2s2p^3$ ^{5}S $6f$
4	−1.29997	$2s2p^3$ ^{5}S $7f$
5	−.60883	$2s2p^3$ ^{5}S $8f$
6	−.13508	$2s2p^3$ ^{5}S $9f$
	^{4}**G**^{e}	
1	−5.76153	$2s^22p^2$ ^{3}P $5g$
2	−5.34106	$2s2p^3$ ^{3}D $4f$
3	−4.93832	$2s2p^3$ ^{3}P $4f$
4	−3.99603	$2s^22p^2$ ^{3}P $6g$
5	−2.93718	$2s^22p^2$ ^{3}P $7g$
6	−2.24968	$2s^22p^2$ ^{3}P $8g$
7	−1.94679	$2s2p^3$ ^{3}D $5f$
8	−1.82911	$2s2p^3$ ^{3}P $5f$
9	−1.77238	$2s^22p^2$ ^{3}P $9g$
	^{4}**G**^{o}	
1	−13.1848	$2s2p^3$ ^{3}D $3d$
2	−9.02934	$2s^22p^2$ ^{3}P $4f$
3	−5.78447	$2s^22p^2$ ^{3}P $5f$
4	−5.54822	$2s2p^3$ ^{3}D $4d$
5	−4.11343	$2s2p^3$ ^{5}S $5g$
6	−4.01352	$2s^22p^2$ ^{3}P $6f$
7	−4.00022	$2s^22p^2$ ^{3}P $6h$
8	−2.94812	$2s^22p^2$ ^{3}P $7f$
9	−2.93908	$2s^22p^2$ ^{3}P $7h$
10	−2.35344	$2s2p^3$ ^{5}S $6g$
11	−2.25719	$2s^22p^2$ ^{3}P $8f$
12	−2.25035	$2s^22p^2$ ^{3}P $8h$
13	−2.08378	$2s2p^3$ ^{3}D $5d$
14	−1.92203	$2s2p^3$ ^{3}D $5g$
15	−1.82440	$2s2p^3$ ^{3}P $5g$
16	−1.78158	$2s^22p^2$ ^{3}P $9f$
17	−1.77743	$2s^22p^2$ ^{3}P $9h$
	^{6}**G**^{o}	
1	−4.11387	$2s2p^3$ ^{5}S $5g$
2	−2.35387	$2s2p^3$ ^{5}S $6g$
3	−1.29255	$2s2p^3$ ^{5}S $7g$
4	−.60368	$2s2p^3$ ^{5}S $8g$
5	−.13138	$2s2p^3$ ^{5}S $9g$

N-like Ar (Ar^{11+})

Energies in ascending order from ground state for terms with effective $n \leq 4.0$, $L \leq 4$

Term	i	E(Ryds)	Term	i	E(Ryds)	Term	i	E(Ryds)	Term	i	E(Ryds)	Term	i	E(Ryds)
$^4S^o$	1	0.00000	$^2P^e$	3	28.8908	$^4S^e$	1	31.7041	$^4D^o$	5	33.2377	$^4D^o$	7	35.7853
$^2D^o$	1	0.91280	$^2P^o$	5	28.9194	$^4D^e$	3	31.7761	$^2D^e$	8	33.3166	$^4P^o$	5	35.8058
$^2P^o$	1	1.39500	$^2F^e$	1	29.0155	$^4S^o$	4	31.9483	$^2D^o$	7	33.4421	$^4S^o$	6	35.9114
$^4P^e$	1	3.99860	$^4P^e$	3	29.0561	$^2P^e$	6	31.9564	$^2F^o$	3	33.4737	$^2D^o$	11	35.9177
$^2D^e$	1	5.48940	$^6P^e$	1	29.1520	$^4P^e$	6	32.0003	$^2P^e$	8	33.7309	$^2P^o$	12	35.9708
$^2S^e$	1	6.48510	$^2D^e$	3	29.4009	$^2D^o$	5	32.0254	$^2P^o$	9	33.7865	$^4P^o$	6	35.9754
$^2P^e$	1	6.73940	$^4P^e$	4	29.5645	$^2D^e$	7	32.0586	$^2D^e$	9	33.7928	$^4D^o$	8	36.1611
$^2P^o$	2	10.4844	$^2G^e$	1	29.5688	$^2S^o$	2	32.1458	$^2P^e$	9	33.9838	$^2D^e$	11	36.2097
$^4P^e$	2	26.2915	$^2D^e$	4	29.7120	$^4F^o$	1	32.3500	$^2S^e$	5	34.0000	$^4F^e$	3	36.2586
$^2P^e$	2	26.5474	$^2F^e$	2	29.7218	$^4G^o$	1	32.4165	$^4D^o$	6	34.3479	$^2D^o$	12	36.2924
$^2D^e$	2	27.0251	$^2P^e$	4	29.8701	$^2S^e$	4	32.4400	$^2D^o$	8	34.3891	$^4D^e$	4	36.2992
$^2S^o$	1	27.2486	$^2S^e$	3	29.9691	$^4D^o$	4	32.5718	$^2G^o$	2	34.4204	$^2P^o$	13	36.3102
$^4D^o$	1	27.4066	$^4D^o$	2	30.0040	$^2S^o$	3	32.6157	$^2F^o$	4	34.4662	$^2P^e$	12	36.3262
$^4P^o$	1	27.4933	$^2D^e$	5	30.3136	$^2P^o$	7	32.6446	$^2P^o$	10	34.7108	$^4P^e$	10	36.3740
$^2S^e$	2	27.6522	$^2D^o$	4	30.4080	$^2G^o$	1	32.6600	$^2S^o$	4	34.7745	$^2S^o$	6	36.3998
$^2D^o$	2	27.6595	$^6D^o$	1	30.4398	$^4P^o$	3	32.7124	$^2D^o$	9	34.8679	$^4S^o$	7	36.4061
$^4S^o$	2	27.6663	$^4P^o$	2	30.6252	$^2P^o$	8	32.8304	$^4P^e$	8	35.0689	$^2F^e$	5	36.4113
$^2P^o$	3	27.8154	$^2P^o$	6	31.0062	$^4S^o$	5	32.8690	$^2F^o$	5	35.0998	$^2S^e$	6	36.4201
$^6S^o$	1	27.9839	$^4D^o$	3	31.0092	$^2D^o$	6	32.8805	$^2P^o$	11	35.2054	$^2D^e$	12	36.5144
$^2F^o$	1	28.1905	$^4D^e$	2	31.0842	$^2P^e$	7	33.0166	$^2D^o$	10	35.2577	$^4G^o$	2	36.5719
$^2D^o$	3	28.4326	$^2P^e$	5	31.1275	$^2F^o$	2	33.0409	$^4P^e$	9	35.3368	$^2G^o$	3	36.5815
$^2P^o$	4	28.6156	$^4F^e$	2	31.1515	$^4P^e$	7	33.0498	$^2P^e$	10	35.3745	$^4D^o$	9	36.5927
$^4S^o$	3	28.7046	$^2F^e$	3	31.2903	$^4F^o$	2	33.0541	$^2P^e$	11	35.4270			
$^4F^e$	1	28.7449	$^4P^e$	5	31.5680	$^2F^e$	4	33.1243	$^2S^o$	5	35.7412			
$^4D^e$	1	28.8782	$^2D^e$	6	31.6938	$^4P^o$	4	33.1360	$^2D^e$	10	35.7761			

gf-values for transitions involving terms with effective $n \leq 4.0$, $L \leq 4$

i	i'	gf_L	i	i'	gf_L	i	i'	gf_L	i	i'	gf_L	i	i'	gf_L	i	i'	gf_L
		$^2S^o$–$^2P^e$	2	10	−2.86E−1	4	8	6.54E−2	6	6	2.25E−1	2	2	5.94E−6	3	11	−5.77E−2
1	1	1.20E−4	2	11	−2.10E−2	4	9	4.26E−3	6	7	5.20E−2	2	3	−3.14E−3	3	12	−3.65E−3
1	2	1.14E−1	2	12	−2.48E−2	4	10	−6.06E−4	6	8	5.29E−2	2	4	−8.03E−3	3	13	−1.64E−5
1	3	−3.87E−1	3	1	1.05E−1	4	11	−6.40E−4	6	9	8.88E−2	2	5	−6.47E−1	4	1	2.49E−1
1	4	−3.76E−2	3	2	3.07E−3	4	12	−1.40E−4	6	10	2.07E−1	2	6	−1.80E−2	4	2	1.93E−2
1	5	−2.71E−1	3	3	8.44E−2	5	1	3.49E−4	6	11	1.97E−2	2	7	−3.06E−1	4	3	1.82E−1
1	6	−5.74E−2	3	4	3.20E−4	5	2	1.70E−1	6	12	6.14E−7	2	8	−7.92E−4	4	4	2.79E−3
1	7	−4.93E−2	3	5	2.37E−1	5	3	9.27E−2			$^2S^e$–$^2P^o$	2	9	−6.38E−3	4	5	3.04E−3
1	8	−3.17E−2	3	6	7.36E−3	5	4	1.35E−2	1	1	2.88E−1	2	10	−2.86E−4	4	6	2.52E−1
1	9	−8.97E−3	3	7	−4.32E−4	5	5	2.20E−4	1	2	−5.79E−2	2	11	−1.33E−3	4	7	−1.52E−3
1	10	−2.00E−3	3	8	−1.59E−4	5	6	8.86E−4	1	3	−5.21E−5	2	12	−3.93E−3	4	8	−6.32E−3
1	11	−8.55E−2	3	9	−7.51E−5	5	7	9.97E−5	1	4	−8.70E−3	2	13	−3.33E−6	4	9	−3.74E−1
1	12	−6.76E−1	3	10	−5.42E−4	5	8	4.21E−3	1	5	−8.41E−3	3	1	5.99E−1	4	10	−1.06E−2
2	1	3.33E−1	3	11	−1.94E−3	5	9	9.48E−5	1	6	−2.17E−1	3	2	7.32E−5	4	11	−3.36E−2
2	2	5.21E−1	3	12	−8.75E−3	5	10	2.30E−2	1	7	−1.43E−1	3	3	7.21E−2	4	12	−8.50E−4
2	3	9.69E−4	4	1	8.62E−1	5	11	1.87E−1	1	8	−2.21E−1	3	4	1.86E−1	4	13	−6.63E−2
2	4	1.34E−7	4	2	9.87E−4	5	12	−7.19E−1	1	9	−2.45E+0	3	5	7.15E−3	5	1	1.08E−2
2	5	1.53E−4	4	3	8.80E−3	6	1	2.30E−1	1	10	−2.95E−2	3	6	−1.91E−3	5	2	3.08E−4
2	6	1.31E−4	4	4	3.10E−1	6	2	1.52E−3	1	11	1.54E+0	3	7	4.30E−7	5	3	3.55E−2
2	7	−1.64E−1	4	5	4.12E−3	6	3	1.80E−5	1	12	−6.59E−4	3	8	−2.14E−4	5	4	1.81E−1
2	8	−3.45E−1	4	6	1.18E−3	6	4	3.18E−4	1	13	−2.00E−6	3	9	−1.37E−5	5	5	1.46E−1
2	9	−3.33E−1	4	7	1.95E−1	6	5	1.32E−1	2	1	1.26E−1	3	10	−1.52E−1	5	6	1.27E−1

N-like Ar (Ar^{11+})

i	*i′*	gf$_L$	*i*	*i′*	gf$_L$	*i*	*i′*	gf$_L$	*i*	*i′*	gf$_L$	*i*	*i′*	gf$_L$	*i*	*i′*	gf$_L$
5	7	1.99E−1	3	6	−4.02E−3	7	6	1.43E−4	11	6	7.72E−2	3	7	−4.15E−2	7	11	−7.10E−4
5	8	2.28E−3	3	7	−5.23E−3	7	7	9.56E−3	11	7	1.02E−2	3	8	−1.32E−2	7	12	−1.07E−2
5	9	2.01E−3	3	8	−2.77E−1	7	8	2.67E−4	11	8	3.82E−3	3	9	−1.28E−2	8	1	1.24E+0
5	10	−7.79E−3	3	9	−6.53E−2	7	9	−1.88E−4	11	9	1.27E−3	3	10	−8.44E−4	8	2	3.62E−3
5	11	−2.88E−1	3	10	−7.00E−3	7	10	−2.92E−1	11	10	1.87E−3	3	11	−1.43E−2	8	3	1.60E−2
5	12	−4.72E−5	3	11	−5.29E−3	7	11	−2.24E−4	11	11	1.45E−5	3	12	−1.21E+0	8	4	5.14E−4
5	13	−7.37E−3	3	12	−7.11E−2	7	12	−3.11E−3	11	12	−9.66E−1	4	1	2.39E−2	8	5	2.76E−4
6	1	1.76E−2	3	13	−7.49E−4	7	13	−8.22E−1	11	13	−2.99E−2	4	2	6.64E−1	8	6	1.39E−1
6	2	1.43E−9	4	1	1.80E+0	8	1	3.89E−3	12	1	2.67E−1	4	3	−4.56E−2	8	7	6.35E−3
6	3	1.15E−2	4	2	1.43E−4	8	2	1.25E−2	12	2	1.16E−4	4	4	−3.37E−1	8	8	−4.65E−3
6	4	2.92E−3	4	3	1.15E−1	8	3	2.15E−3	12	3	5.28E−1	4	5	−4.06E−2	8	9	−2.15E−3
6	5	3.08E−1	4	4	3.70E−1	8	4	7.21E−1	12	4	1.03E−1	4	6	−3.43E−2	8	10	−1.54E−3
6	6	3.06E−3	4	5	1.25E−3	8	5	3.40E−2	12	5	4.93E−2	4	7	−1.39E−1	8	11	−4.70E−4
6	7	5.09E−4	4	6	−4.81E−3	8	6	5.22E−3	12	6	1.75E−3	4	8	−1.11E−1	8	12	−1.10E−2
6	8	1.99E−5	4	7	−9.09E−4	8	7	1.52E−1	12	7	6.06E−3	4	9	−1.59E−3	9	1	1.06E−1
6	9	8.74E−4	4	8	−1.07E−2	8	8	2.22E−4	12	8	4.84E−3	4	10	−2.12E−3	9	2	3.23E−4
6	10	1.73E−4	4	9	−5.38E−4	8	9	−1.00E−5	12	9	2.39E−3	4	11	−4.28E−1	9	3	3.93E−1
6	11	2.80E−3	4	10	−1.34E−2	8	10	−1.61E−1	12	10	3.77E−5	4	12	−3.70E−1	9	4	2.65E−2
6	12	4.69E−3	4	11	−1.36E−1	8	11	−1.27E−1	12	11	8.86E−6	5	1	7.52E−3	9	5	3.72E−3
6	13	4.27E−5	4	12	−2.78E−3	8	12	−2.35E−2	12	12	3.30E−1	5	2	4.23E−5	9	6	1.99E−2
		$^2P^e$–$^2P^o$	4	13	−1.67E−6	8	13	−6.61E−7	12	13	1.64E−7	5	3	−1.52E−4	9	7	1.39E−2
1	1	3.68E−1	5	1	3.86E−2	9	1	1.70E−1			**$^2P^o$–$^2D^e$**	5	4	−1.21E−2	9	8	1.24E−4
1	2	−9.19E−1	5	2	1.18E−2	9	2	1.89E−2	1	1	−1.50E−1	5	5	−1.13E+0	9	9	−7.90E−6
1	3	−1.40E−2	5	3	1.97E−2	9	3	5.08E−2	1	2	−2.03E−1	5	6	−1.46E−2	9	10	−2.62E−4
1	4	−5.99E−3	5	4	3.32E−4	9	4	2.62E−2	1	3	−1.54E+0	5	7	−9.43E−2	9	11	−1.25E−4
1	5	−3.53E−3	5	5	6.17E−4	9	5	2.61E−1	1	4	−2.85E+0	5	8	−1.74E−3	9	12	−1.78E−2
1	6	−3.52E−2	5	6	6.02E−3	9	6	3.20E−4	1	5	−2.55E+0	5	9	−4.58E−1	10	1	2.97E−1
1	7	−2.15E−1	5	7	−5.95E−5	9	7	5.74E−1	1	6	−1.67E−2	5	10	−2.17E−3	10	2	4.24E−3
1	8	−3.76E−1	5	8	−5.70E−1	9	8	1.03E−4	1	7	−8.56E−1	5	11	−2.03E−3	10	3	6.03E−2
1	9	−7.90E−2	5	9	−3.22E−4	9	9	2.11E−4	1	8	−4.88E−2	5	12	−3.45E−2	10	4	7.21E−1
1	10	−2.50E+0	5	10	−3.66E−3	9	10	−4.27E−3	1	9	−2.07E−1	6	1	2.28E−1	10	5	1.91E−3
1	11	−2.73E−1	5	11	−5.67E−5	9	11	−1.56E−1	1	10	−7.03E−4	6	2	2.23E−2	10	6	3.86E−2
1	12	−3.10E−3	5	12	−3.15E−5	9	12	−1.15E−2	1	11	−5.73E−2	6	3	3.40E−3	10	7	7.97E−3
1	13	−2.40E−1	5	13	−8.84E−2	9	13	−3.25E−2	1	12	−7.40E−1	6	4	1.18E−2	10	8	1.77E−1
2	1	3.18E−1	6	1	5.45E−1	10	1	1.74E−2	2	1	7.07E−1	6	5	7.21E−3	10	9	1.64E−2
2	2	2.73E−8	6	2	6.92E−3	10	2	6.03E−1	2	2	−7.43E−6	6	6	−2.84E−2	10	10	−7.23E−5
2	3	−4.78E−1	6	3	1.32E−1	10	3	4.11E−2	2	3	−6.56E−5	6	7	−9.06E−1	10	11	−1.21E−3
2	4	−3.35E−1	6	4	8.75E−2	10	4	2.92E−2	2	4	−4.12E−6	6	8	−6.27E−2	10	12	−8.77E−4
2	5	−1.00E−1	6	5	5.80E−2	10	5	3.02E−3	2	5	−2.30E−6	6	9	−5.38E−2	11	1	7.70E−2
2	6	−3.09E−1	6	6	4.56E−1	10	6	2.99E−1	2	6	−1.51E−2	6	10	−2.83E−3	11	2	5.71E−4
2	7	−1.87E−3	6	7	−1.83E−9	10	7	2.23E−2	2	7	−1.10E−2	6	11	−7.84E−3	11	3	2.15E−2
2	8	−6.67E−3	6	8	−3.65E−2	10	8	1.38E−3	2	8	−3.01E−3	6	12	−2.07E−3	11	4	1.46E−1
2	9	−2.63E−3	6	9	−5.15E−1	10	9	1.59E−3	2	9	−4.32E−3	7	1	8.04E−2	11	5	3.03E−1
2	10	−1.02E−3	6	10	−1.49E−3	10	10	5.75E−3	2	10	−4.82E−1	7	2	9.52E−1	11	6	2.12E−4
2	11	−2.62E−4	6	11	−2.87E−2	10	11	5.85E−4	2	11	−1.36E−4	7	3	2.14E−3	11	7	2.46E−3
2	12	−4.08E−1	6	12	−4.52E−3	10	12	−1.38E−1	2	12	−5.53E−5	7	4	8.43E−4	11	8	1.85E−2
2	13	−7.96E−6	6	13	−7.43E−3	10	13	−2.65E−1	3	1	6.36E−3	7	5	9.32E−4	11	9	9.80E−4
3	1	9.44E−1	7	1	5.14E−3	11	1	3.51E−2	3	2	7.68E−2	7	6	4.38E−3	11	10	−1.82E−4
3	2	1.46E−5	7	2	9.36E−3	11	2	6.73E−2	3	3	−8.89E−1	7	7	2.58E−3	11	11	−5.50E−6
3	3	2.05E−1	7	3	3.98E−1	11	3	2.29E−1	3	4	−1.29E−2	7	8	−8.80E−4	11	12	−7.14E−5
3	4	3.32E−3	7	4	4.90E−2	11	4	1.71E−1	3	5	−9.43E−4	7	9	−1.03E+0	12	1	2.55E−3
3	5	−2.47E−4	7	5	1.38E−3	11	5	2.90E−2	3	6	−3.88E−1	7	10	−3.72E−1	12	2	1.59E−2

N-like Ar (Ar^{11+})

i	i'	gf_L	i	i'	gf_L	i	i'	gf_L	i	i'	gf_L	i	i'	gf_L	i	i'	gf_L
12	3	2.20E−1	3	6	−2.16E−1	7	10	−2.05E−3	12	2	1.77E−2	4	5	−8.18E−5	8	9	−2.35E−1
12	4	5.32E−3	3	7	−1.63E−2	7	11	−2.07E−3	12	3	3.35E−2	4	6	−5.77E−3	8	10	−2.01E−2
12	5	7.21E−3	3	8	−4.58E−2	7	12	−1.92E−1	12	4	3.15E−3	4	7	−1.01E−2	8	11	−5.26E−5
12	6	1.02E−4	3	9	−3.25E−3	8	1	6.21E−2	12	5	3.93E−4	4	8	−2.07E−2	8	12	−3.41E−3
12	7	3.06E−3	3	10	−2.05E−5	8	2	2.88E−1	12	6	2.57E−3	4	9	−2.56E−1	9	1	2.89E−2
12	8	1.20E−4	3	11	−3.49E−4	8	3	5.74E−2	12	7	3.47E−3	4	10	−5.94E−2	9	2	2.17E−2
12	9	7.34E−4	3	12	−6.53E−2	8	4	1.05E−2	12	8	9.53E−3	4	11	−3.92E−2	9	3	1.13E−1
12	10	7.80E−6	4	1	6.17E−1	8	5	5.05E−1	12	9	9.33E−3	4	12	−1.00E−4	9	4	4.32E−2
12	11	−3.11E−5	4	2	1.13E−1	8	6	1.10E−4	12	10	4.32E−3	5	1	1.92E−1	9	5	2.52E−3
12	12	−1.58E+0	4	3	1.04E−1	8	7	3.23E−5	12	11	2.20E−2	5	2	2.44E−3	9	6	9.31E−5
13	1	1.28E−1	4	4	−8.65E−4	8	8	−8.99E−3	12	12	1.37E−3	5	3	2.68E−2	9	7	1.82E−3
13	2	1.25E−6	4	5	−2.04E−4	8	9	−5.35E−1			**$^2D^e$–$^2D^o$**	5	4	−4.90E−5	9	8	−1.81E−3
13	3	5.54E−6	4	6	−2.02E−3	8	10	−4.70E−2	1	1	8.11E−1	5	5	−1.60E−4	9	9	−3.27E−2
13	4	9.93E−4	4	7	−9.32E−3	8	11	−2.79E−2	1	2	−1.21E−5	5	6	−1.86E−4	9	10	−3.04E−1
13	5	8.84E−6	4	8	−8.81E−2	8	12	−2.37E−1	1	3	−1.90E−2	5	7	−6.46E−3	9	11	−2.59E−4
13	6	3.84E−1	4	9	−1.04E−1	9	1	2.02E−2	1	4	−7.64E−1	5	8	−1.23E−4	9	12	−4.94E−5
13	7	2.55E−1	4	10	−4.03E−4	9	2	4.04E−4	1	5	−4.55E−1	5	9	−4.28E−2	10	1	2.28E−3
13	8	1.92E−2	4	11	−1.14E−2	9	3	8.28E−1	1	6	−1.76E+0	5	10	−5.68E−1	10	2	2.09E−3
13	9	1.42E−5	4	12	−5.49E−3	9	4	1.27E−2	1	7	−2.12E+0	5	11	−7.65E−3	10	3	9.90E−3
13	10	1.06E−1	5	1	6.35E−1	9	5	2.64E−2	1	8	−1.27E+0	5	12	−3.68E−4	10	4	3.96E−3
13	11	8.33E−5	5	2	6.15E−5	9	6	2.25E−4	1	9	−1.03E+0	6	1	1.07E+0	10	5	2.02E+0
13	12	−2.65E−5	5	3	5.49E−2	9	7	6.32E−5	1	10	−1.43E−1	6	2	5.80E−2	10	6	1.14E−2
		$^2P^e$–$^2D^o$	5	4	2.94E−1	9	8	−1.51E−3	1	11	−2.14E−3	6	3	1.49E−1	10	7	5.56E−3
1	1	1.08E+0	5	5	8.83E−6	9	9	1.56E−2	1	12	2.72E−2	6	4	1.13E+0	10	8	1.86E−3
1	2	−4.56E−3	5	6	−6.30E−1	9	10	−7.82E−1	2	1	5.15E−1	6	5	−5.17E−3	10	9	2.67E−3
1	3	−5.61E−5	5	7	−1.05E−2	9	11	−3.71E−3	2	2	−1.25E−1	6	6	−4.82E−1	10	10	1.26E−4
1	4	−4.51E−2	5	8	−7.70E−3	9	12	−4.01E−1	2	3	−9.31E−1	6	7	−8.53E−2	10	11	−4.04E−5
1	5	−3.24E−1	5	9	−1.36E−2	10	1	6.16E−4	2	4	−8.21E−2	6	8	−2.09E−3	10	12	−2.30E−3
1	6	−3.04E−1	5	10	−1.21E−4	10	2	2.19E−2	2	5	−1.57E+0	6	9	−7.43E−2	11	1	1.02E−1
1	7	−1.10E−2	5	11	−4.70E−3	10	3	1.54E−2	2	6	−1.14E−2	6	10	−1.24E−2	11	2	8.46E−2
1	8	−6.62E−2	5	12	−2.92E−2	10	4	9.89E−1	2	7	−1.58E−3	6	11	−9.98E−4	11	3	5.87E−1
1	9	−3.46E+0	6	1	8.34E−2	10	5	3.41E−2	2	8	−8.84E−4	6	12	−2.10E−1	11	4	2.32E−2
1	10	−4.37E+0	6	2	1.52E−1	10	6	5.75E−4	2	9	−2.83E−6	7	1	1.17E−1	11	5	8.05E−4
1	11	−9.64E−2	6	3	6.75E−2	10	7	1.39E−3	2	10	−1.58E−3	7	2	6.55E−1	11	6	4.60E−4
1	12	−4.04E−1	6	4	1.02E−1	10	8	1.04E−3	2	11	−2.78E−2	7	3	1.04E−2	11	7	5.69E−5
2	1	4.04E−1	6	5	−2.11E−5	10	9	1.08E−2	2	12	−1.95E−3	7	4	3.99E−2	11	8	2.51E−3
2	2	−7.17E−1	6	6	−4.01E−2	10	10	4.02E−3	3	1	2.29E+0	7	5	7.47E−4	11	9	1.09E−3
2	3	−5.65E−1	6	7	−9.54E−1	10	11	−1.50E−1	3	2	1.67E−1	7	6	−3.55E−3	11	10	1.83E−3
2	4	−2.21E−1	6	8	−1.00E−2	10	12	−8.74E−1	3	3	9.55E−2	7	7	−3.37E−1	11	11	6.89E−3
2	5	−3.16E−3	6	9	−1.37E−2	11	1	7.50E−2	3	4	−8.55E−3	7	8	−9.89E−3	11	12	−7.24E−4
2	6	−5.60E−4	6	10	−8.58E−3	11	2	2.90E−1	3	5	−5.39E−4	7	9	−8.31E−3	12	1	2.96E−1
2	7	−5.69E−3	6	11	−2.89E−3	11	3	1.52E−1	3	6	−1.10E−1	7	10	−1.54E−2	12	2	3.44E−1
2	8	−1.66E−3	6	12	−3.76E−2	11	4	5.27E−2	3	7	−1.42E−1	7	11	−1.10E−3	12	3	1.56E−1
2	9	−7.67E−5	7	1	1.12E−1	11	5	2.15E−3	3	8	−2.91E−1	7	12	−2.27E−1	12	4	5.14E−3
2	10	−1.12E−3	7	2	6.93E−1	11	6	5.77E−4	3	9	−1.08E−1	8	1	9.71E−2	12	5	2.15E−3
2	11	−8.04E−1	7	3	2.99E−2	11	7	4.69E−3	3	10	−2.07E−2	8	2	8.71E−2	12	6	2.38E−3
2	12	−6.14E−4	7	4	7.52E−3	11	8	1.08E−3	3	11	−1.02E−1	8	3	8.23E−1	12	7	4.77E−3
3	1	2.81E−1	7	5	2.73E−1	11	9	3.18E−7	3	12	−7.03E−3	8	4	3.97E−1	12	8	1.07E−2
3	2	1.77E−3	7	6	1.32E−6	11	10	9.78E−6	4	1	1.07E+0	8	5	9.77E−1	12	9	1.80E−2
3	3	4.60E−3	7	7	−3.64E−4	11	11	−1.49E+0	4	2	3.56E−1	8	6	1.00E−3	12	10	7.18E−3
3	4	−1.72E−3	7	8	−6.31E−1	11	12	−4.64E−2	4	3	2.56E−1	8	7	−2.41E−4	12	11	5.78E−1
3	5	−6.34E−5	7	9	−6.73E−2	12	1	3.24E−2	4	4	−8.60E−3	8	8	−2.34E−1	12	12	2.63E−4

N-like Ar (Ar^{11+})

i i'	gf_L	i i'	gf_L	i i'	gf_L	i i'	gf_L	i i'	gf_L	i i'	gf_L
	$^2\mathbf{D}^o$–$^2\mathbf{F}^e$	11 2	5.27E−2	9 3	6.13E−4	1 2	−3.52E−3	4 4	1.00E−3	2 2	−1.06E+0
1 1	−1.30E+0	11 3	1.21E−2	9 4	−1.76E−2	1 3	−8.93E+0	4 5	5.81E−4	2 3	−1.51E−3
1 2	−9.86E+0	11 4	3.55E−3	9 5	−1.48E+0	2 1	−4.55E−2	4 6	−1.65E−4	2 4	−7.03E−4
1 3	−1.25E+0	11 5	−2.57E+0	10 1	5.62E−4	2 2	−4.11E−2	4 7	−1.01E+0	2 5	−9.19E−1
1 4	−3.59E−1	12 1	4.36E−5	10 2	9.73E−3	2 3	−3.44E+0	4 8	−1.09E+0	2 6	−1.03E−3
1 5	−1.18E+0	12 2	6.17E−4	10 3	6.86E−3	3 1	−2.24E+0	4 9	−3.42E−3	3 1	7.13E−1
2 1	−1.35E+0	12 3	1.28E+0	10 4	9.46E−3	3 2	−1.53E−2	4 10	−5.64E−3	3 2	−3.54E−4
2 2	−2.60E−2	12 4	4.01E−4	10 5	8.95E−4	3 3	−1.39E−1	5 1	2.26E+0	3 3	−4.38E−1
2 3	−4.28E−1	12 5	−1.32E−3	11 1	5.99E−1	4 1	8.38E−4	5 2	4.44E−3	3 4	−9.70E−2
2 4	−1.73E−2		$^2\mathbf{D}^e$–$^2\mathbf{F}^o$	11 2	7.24E−4	4 2	−1.98E+0	5 3	2.17E−1	3 5	−2.42E−1
2 5	−2.67E+0	1 1	−2.05E−2	11 3	1.04E−4	4 3	−2.77E−5	5 4	1.03E−1	3 6	−3.83E−4
3 1	−3.82E−3	1 2	−4.32E+0	11 4	4.24E−3	5 1	1.11E−2	5 5	4.11E−1	4 1	2.63E−3
3 2	−1.06E+0	1 3	−2.63E+0	11 5	1.29E−2	5 2	1.28E−3	5 6	4.15E−3	4 2	−9.58E−3
3 3	−4.80E−1	1 4	−5.25E+0	12 1	6.02E−3	5 3	−7.63E−1	5 7	−8.96E−4	4 3	−8.03E−2
3 4	−1.94E−1	1 5	−7.70E−1	12 2	3.59E−3		$^2\mathbf{G}^e$–$^2\mathbf{G}^o$	5 8	−1.19E−2	4 4	−8.31E−4
3 5	−5.13E−1	2 1	−1.42E+0	12 3	1.02E−2	1 1	−2.29E−3	5 9	−9.53E−4	4 5	−5.28E−3
4 1	3.40E−3	2 2	−6.04E−3	12 4	1.18E−4	1 2	−1.76E+0	5 10	−1.41E−2	4 6	−5.87E−3
4 2	2.70E−2	2 3	−2.49E−4	12 5	1.63E−4	1 3	−9.86E−4	6 1	4.63E−3	5 1	4.01E−1
4 3	−1.07E+0	2 4	−2.43E−3		$^2\mathbf{F}^e$–$^2\mathbf{F}^o$		$^4\mathbf{S}^o$–$^4\mathbf{P}^e$	6 2	2.69E−1	5 2	1.33E−2
4 4	−1.01E−2	2 5	−1.09E−2	1 1	7.13E−2	1 1	−5.58E−1	6 3	1.37E−1	5 3	−8.16E−1
4 5	−1.00E−2	3 1	1.02E−2	1 2	−2.43E−1	1 2	−4.40E−1	6 4	2.08E−2	5 4	−1.10E−1
5 1	3.62E−3	3 2	−1.74E−1	1 3	−5.29E−1	1 3	−6.95E+0	6 5	2.70E−3	5 5	−2.57E−2
5 2	5.43E−3	3 3	−3.18E−2	1 4	−9.89E−2	1 4	−9.73E−1	6 6	1.16E−2	5 6	−4.58E−1
5 3	3.53E−4	3 4	−1.48E−2	1 5	−1.91E−2	1 5	−2.77E−2	6 7	1.18E−2	6 1	8.78E−2
5 4	−1.43E+0	3 5	−2.01E−3	2 1	3.93E−1	1 6	−7.31E−2	6 8	4.39E−3	6 2	1.70E+0
5 5	−2.97E−2	4 1	1.08E−1	2 2	−1.85E−2	1 7	−4.02E−1	6 9	8.39E−1	6 3	−7.64E−4
6 1	2.89E−3	4 2	−1.44E−2	2 3	−6.76E−2	1 8	−4.83E−3	6 10	−1.08E+0	6 4	−4.54E−1
6 2	1.31E−4	4 3	−4.56E−3	2 4	−6.76E−1	1 9	−5.54E−2	7 1	1.19E−1	6 5	−4.48E−3
6 3	4.85E−2	4 4	−1.61E−1	2 5	−5.53E−2	1 10	−1.53E+0	7 2	3.62E−4	6 6	−2.85E−1
6 4	−7.57E−4	4 5	−5.34E−5	3 1	4.77E−5	2 1	1.01E−1	7 3	6.39E−6	7 1	8.03E−1
6 5	−4.07E−4	5 1	2.73E−2	3 2	−7.40E−1	2 2	4.23E−1	7 4	2.56E−2	7 2	8.07E−3
7 1	3.71E−1	5 2	−3.67E−3	3 3	−8.16E−7	2 3	−4.84E−1	7 5	4.96E−1	7 3	2.49E−3
7 2	3.93E−2	5 3	−2.09E−3	3 4	−2.91E−2	2 4	−6.02E−1	7 6	1.63E−1	7 4	−4.51E−5
7 3	1.14E−1	5 4	−2.07E−3	3 5	−3.47E−3	2 5	−4.94E−2	7 7	2.38E−1	7 5	−1.08E−2
7 4	1.94E−6	5 5	−2.29E−1	4 1	1.94E+0	2 6	−1.17E−1	7 8	7.42E−1	7 6	−1.01E+0
7 5	−4.75E−2	6 1	2.19E−3	4 2	3.49E−4	2 7	−2.72E−1	7 9	7.90E−3	8 1	3.44E−3
8 1	9.32E−1	6 2	−1.17E+0	4 3	−1.63E−3	2 8	−1.53E−2	7 10	1.97E−4	8 2	4.96E−1
8 2	1.66E−1	6 3	−5.87E−2	4 4	−3.41E−1	2 9	−2.69E−1		$^4\mathbf{S}^e$–$^4\mathbf{P}^o$	8 3	6.17E−3
8 3	2.13E−2	6 4	−2.57E−2	4 5	−6.72E−2	2 10	−1.21E+0	1 1	3.61E−1	8 4	1.06E−3
8 4	5.82E−2	6 5	−7.72E−5	5 1	5.60E−2	3 1	3.43E−1	1 2	3.74E−1	8 5	−2.59E−3
8 5	−2.40E−2	7 1	6.76E−4	5 2	1.14E−3	3 2	1.69E−1	1 3	−3.46E−2	8 6	−9.78E−1
9 1	2.56E−1	7 2	−3.91E−3	5 3	1.03E−4	3 3	−4.70E−2	1 4	−8.24E−1	9 1	6.32E−1
9 2	4.89E−1	7 3	−1.79E+0	5 4	9.23E−6	3 4	−7.05E−1	1 5	−1.45E−4	9 2	3.70E−3
9 3	1.36E−3	7 4	−2.87E−4	5 5	2.57E−4	3 5	−2.84E−1	1 6	−1.73E−1	9 3	1.26E−3
9 4	8.16E−2	7 5	−3.05E−3		$^2\mathbf{F}^o$–$^2\mathbf{G}^e$	3 6	−1.48E−1		$^4\mathbf{P}^e$–$^4\mathbf{P}^o$	9 4	1.84E−3
9 5	−3.29E−2	8 1	3.65E−1	1 1	−2.05E+0	3 7	−1.58E−1	1 1	−3.74E−2	9 5	−1.79E+0
10 1	5.47E−3	8 2	3.75E−3	2 1	4.27E−5	3 8	−1.17E−3	1 2	−5.69E−1	9 6	−1.22E−3
10 2	9.24E−1	8 3	−9.08E−4	3 1	2.61E−3	3 9	−7.41E−2	1 3	−6.37E+0	10 1	1.04E+0
10 3	2.62E−3	8 4	−8.16E−1	4 1	2.77E−1	3 10	−4.41E−1	1 4	−7.41E−1	10 2	1.82E−2
10 4	1.74E−2	8 5	−9.15E−2	5 1	2.12E+0	4 1	1.23E−1	1 5	−1.49E−2	10 3	1.33E−2
10 5	−1.14E−2	9 1	1.42E+0		$^2\mathbf{F}^e$–$^2\mathbf{G}^o$	4 2	9.04E−1	1 6	−5.10E−1	10 4	9.28E−3
11 1	2.20E−1	9 2	1.48E−4	1 1	−1.67E−1	4 3	1.05E−3	2 1	−1.14E+0	10 5	1.20E+0

N-like Ar (Ar^{11+})

i i′	gf$_L$	*i i′*	gf$_L$	*i i′*	gf$_L$	*i i′*	gf$_L$	*i i′*	gf$_L$	*i i′*	gf$_L$
10 6	3.55E−4	1 7	−1.55E−2	5 6	−1.43E−2	9 5	4.93E−3	3 3	2.36E−3	6 2	2.25E−2
	$^6\mathbf{S}^o$–$^6\mathbf{P}^e$	1 8	−7.68E−1	5 7	−3.32E−2	9 6	1.39E−2	3 4	−9.63E−3	6 3	−1.08E−1
1 1	−2.02E+0	1 9	−5.96E−3	5 8	−1.37E−2	9 7	−3.01E+0	3 5	−6.54E−1	7 1	5.22E−1
	$^4\mathbf{P}^o$–$^4\mathbf{D}^e$	2 1	−1.85E+0	5 9	−2.26E−3	9 8	−3.87E−3	3 6	−1.28E−3	7 2	2.73E−5
1 1	−1.63E+0	2 2	−1.03E+0	6 1	5.68E−1	9 9	−8.93E−3	3 7	−5.58E−4	7 3	−4.85E+0
1 2	−4.71E−1	2 3	−1.22E−2	6 2	2.76E−3	10 1	6.20E−2	3 8	−8.22E−1	8 1	2.66E−3
1 3	−3.53E−1	2 4	−2.02E−4	6 3	1.19E−3	10 2	1.45E−3	3 9	−6.31E−2	8 2	2.42E+0
1 4	−3.43E+0	2 5	−1.93E−3	6 4	−2.67E−3	10 3	3.07E−4	4 1	1.10E+0	8 3	−3.97E−4
2 1	4.29E−4	2 6	−1.30E−2	6 5	−1.67E+0	10 4	5.64E−5	4 2	1.95E−2	9 1	1.28E−1
2 2	−2.05E−2	2 7	−1.69E+0	6 6	−2.65E−2	10 5	1.36E−2	4 3	4.24E−5	9 2	1.78E−2
2 3	−2.21E+0	2 8	−5.35E−5	6 7	−1.18E−2	10 6	3.53E−2	4 4	9.85E−3	9 3	8.69E−3
2 4	−5.72E−2	2 9	−2.21E−2	6 8	−1.98E−1	10 7	3.75E−2	4 5	4.99E−3		$^4\mathbf{D}^e$–$^4\mathbf{F}^o$
3 1	3.60E−2	3 1	1.19E−2	6 9	−6.85E−2	10 8	6.08E−3	4 6	5.06E−2	1 1	−2.49E−1
3 2	3.17E−1	3 2	−3.02E−4	7 1	1.61E+0	10 9	−9.28E−1	4 7	9.51E−1	1 2	−2.19E−1
3 3	7.31E−2	3 3	−2.00E−4	7 2	1.42E−3		$^4\mathbf{D}^e$–$^4\mathbf{D}^o$	4 8	5.25E−4	2 1	−2.03E+0
3 4	−3.43E−2	3 4	−2.28E−1	7 3	4.10E−3	1 1	5.80E−1	4 9	−2.33E−1	2 2	−5.30E−3
4 1	8.14E−1	3 5	−2.41E−1	7 4	1.90E−5	1 2	−1.95E−4		$^6\mathbf{P}^e$–$^6\mathbf{D}^o$	3 1	−4.64E−3
4 2	6.10E−2	3 6	−1.58E−1	7 5	−5.40E−4	1 3	−5.25E−3	1 1	−3.34E+0	3 2	−3.17E+0
4 3	1.80E−2	3 7	−1.31E−3	7 6	−1.85E+0	1 4	−5.89E−1		$^4\mathbf{D}^o$–$^4\mathbf{F}^e$	4 1	1.07E−3
4 4	−3.76E−2	3 8	−4.50E−2	7 7	−5.25E−3	1 5	−4.24E−2	1 1	−2.87E+0	4 2	2.15E−2
5 1	3.83E−1	3 9	−1.20E+1	7 8	−1.10E+0	1 6	−8.93E−1	1 2	−8.15E−1		$^4\mathbf{F}^e$–$^4\mathbf{F}^o$
5 2	3.40E−5	4 1	2.69E−3	7 9	−2.73E−1	1 7	−2.00E−1	1 3	−6.53E+0	1 1	−8.67E−1
5 3	5.94E−4	4 2	−6.07E−3	8 1	7.79E−3	1 8	−1.34E−3	2 1	8.75E−4	1 2	−6.54E−1
5 4	−2.75E+0	4 3	−2.63E+0	8 2	1.57E+0	1 9	−2.44E+0	2 2	−3.18E+0	2 1	−8.58E−1
6 1	5.86E−1	4 4	−2.30E−2	8 3	1.89E−2	2 1	5.68E−1	2 3	1.10E−1	2 2	−1.30E−1
6 2	1.41E+0	4 5	−2.18E−3	8 4	9.18E−3	2 2	1.90E+0	3 1	5.84E−4	3 1	2.16E−4
6 3	9.46E−2	4 6	−7.11E−3	8 5	8.27E−3	2 3	5.68E−4	3 2	−1.78E−3	3 2	6.16E−4
6 4	−4.03E−4	4 7	−2.33E−3	8 6	1.27E−3	2 4	−1.54E+0	3 3	−5.03E−4		$^4\mathbf{F}^e$–$^4\mathbf{G}^o$
	$^4\mathbf{P}^e$–$^4\mathbf{D}^o$	4 8	−9.37E−4	8 7	−4.08E−3	2 5	−1.38E−2	4 1	4.52E−2	1 1	−4.21E−1
1 1	−3.41E−3	4 9	−2.48E−2	8 8	−2.53E+0	2 6	−1.39E−4	4 2	1.17E−1	1 2	−2.36E+1
1 2	−1.12E+0	5 1	4.82E−3	8 9	−3.04E−2	2 7	−1.28E−5	4 3	−3.38E−2	2 1	−3.97E+0
1 3	−4.62E+0	5 2	1.86E+0	9 1	9.70E−1	2 8	−1.93E−1	5 1	8.23E−1	2 2	−3.08E−1
1 4	−5.48E+0	5 3	4.81E−3	9 2	6.67E−4	2 9	−8.59E−3	5 2	6.04E−2	3 1	2.91E−2
1 5	−5.40E+0	5 4	−6.00E−1	9 3	9.31E−4	3 1	8.63E−1	5 3	−8.41E−2	3 2	−2.94E+0
1 6	−9.05E−1	5 5	−4.09E−2	9 4	1.23E−3	3 2	1.40E−1	6 1	2.32E+0		

N-like Ca (Ca^{13+})

Term energies relative to $2s^22p^2$ ^{3}P ionization threshold for each symmetry

i	E(Ryds)	Description	*i*	E(Ryds)	Description	*i*	E(Ryds)	Description	*i*	E(Ryds)	Description
		^{2}S^e	6	−19.1847	$2s2p^3$ ^{3}P $3p$	15	−12.1762	$2s^22p^2$ ^{1}D $4p$	19	−8.45254	$2s2p^3$ ^{3}P $4p$
1	−52.8263	$2s2p^4$	7	−18.0363	$2s2p^3$ ^{1}D $3p$:	16	−12.1582	$2p^4$ ^{1}S $3p$	20	−8.10879	$2s2p^3$ ^{3}D $4f$
2	−24.1153	$2s^22p^2$ ^{1}S $3s$	8	−17.1941	$2s2p^3$ ^{3}S $3p$	17	−11.8835	$2s^22p^2$ ^{1}S $4p$	21	−7.90216	$2s^22p^2$ ^{3}P $5d$
3	−21.4198	$2s^22p^2$ ^{1}D $3d$	9	−16.9073	$2s2p^3$ ^{1}P $3p$	18	−11.3895	$2s^22p^2$ ^{1}D $4f$	22	−7.58503	$2s^22p^2$ ^{1}D $5s$
4	−18.6324	$2s2p^3$ ^{3}P $3p$	10	−15.3401	$2p^4$ ^{3}P $3s$	19	−8.98042	$2s2p^3$ ^{3}P $4s$	23	−7.53998	$2s2p^3$ ^{3}P $4f$
5	−16.8617	$2s2p^3$ ^{1}P $3p$	11	−13.6602	$2s^22p^2$ ^{3}P $4s$	20	−8.29315	$2s2p^3$ ^{3}D $4d$	24	−6.98778	$2s^22p^2$ ^{1}D $5d$
6	−13.4777	$2p^4$ ^{1}S $3s$	12	−12.7097	$2p^4$ ^{3}P $3d$	21	−8.17097	$2s^22p^2$ ^{3}P $5p$	25	−6.87185	$2s^22p^2$ ^{1}D $5g$
7	−12.4539	$2s^22p^2$ ^{1}S $4s$	13	−12.5855	$2s^22p^2$ ^{3}P $4d$	22	−7.68441	$2s2p^3$ ^{3}P $4d$	26	−6.84357	$2s^22p^2$ ^{1}S $5d$
8	−12.0536	$2p^4$ ^{1}D $3d$	14	−11.7827	$2p^4$ ^{1}D $3d$	23	−7.26798	$2s^22p^2$ ^{1}D $5p$	27	−6.73301	$2s2p^3$ ^{1}D $4p$
9	−11.5695	$2s^22p^2$ ^{1}D $4d$	15	−11.6087	$2s^22p^2$ ^{1}D $4d$	24	−7.14426	$2s^22p^2$ ^{1}S $5p$	28	−6.41304	$2s2p^3$ ^{1}P $4p$
10	−8.26886	$2s2p^3$ ^{3}P $4p$	16	−8.92564	$2s2p^3$ ^{3}D $4p$	25	−6.93044	$2s2p^3$ ^{1}P $4s$	29	−5.96776	$2s2p^3$ ^{1}D $4f$
11	−7.43411	$2s^22p^2$ ^{1}S $5s$	17	−8.61171	$2s^22p^2$ ^{3}P $5s$	26	−6.81200	$2s^22p^2$ ^{1}D $5f$	30	−5.48334	$2s^22p^2$ ^{3}P $6d$
12	−6.93700	$2s^22p^2$ ^{1}D $5d$	18	−8.44162	$2s2p^3$ ^{3}P $4p$	27	−6.14282	$2s2p^3$ ^{1}D $4d$	31	−5.38973	$2s2p^3$ ^{1}P $4f$
13	−6.35545	$2s2p^3$ ^{1}P $4p$	19	−8.10777	$2s2p^3$ ^{3}D $4f$	28	−5.75917	$2s2p^3$ ^{1}P $4d$	32	−4.80831	$2s^22p^2$ ^{1}D $6s$
14	−4.74196	$2s^22p^2$ ^{1}S $6s$	20	−8.00138	$2s^22p^2$ ^{3}P $5d$	29	−5.65466	$2s^22p^2$ ^{3}P $6p$	33	−4.48356	$2s^22p^2$ ^{1}D $6d$
15	−4.45853	$2s^22p^2$ ^{1}D $6d$	21	−6.96043	$2s^22p^2$ ^{1}D $5d$	30	−4.63608	$2s^22p^2$ ^{1}D $6p$	34	−4.42130	$2s^22p^2$ ^{1}S $6d$
16	−3.65836	$2s2p^3$ ^{3}P $5p$	22	−6.83581	$2s2p^3$ ^{1}D $4p$	31	−4.58293	$2s^22p^2$ ^{1}S $6p$	35	−4.40859	$2s^22p^2$ ^{1}D $6g$
17	−3.15874	$2s^22p^2$ ^{1}S $7s$	23	−6.55630	$2s2p^3$ ^{3}S $4p$	32	−4.39115	$2s^22p^2$ ^{1}D $6f$	36	−4.02518	$2s^22p^2$ ^{3}P $7d$
18	−2.97804	$2s^22p^2$ ^{1}D $7d$	24	−6.36478	$2s2p^3$ ^{1}P $4p$	33	−4.13180	$2s^22p^2$ ^{3}P $7p$	37	−3.84928	$2s2p^3$ ^{3}D $5p$
19	−2.40550	$2p^4$ ^{1}D $4d$	25	−5.96720	$2s2p^3$ ^{1}D $4f$	34	−4.02387	$2s2p^3$ ^{3}P $5s$	38	−3.76451	$2s2p^3$ ^{3}P $5p$
20	−2.14639	$2s^22p^2$ ^{1}S $8s$	26	−5.83814	$2s^22p^2$ ^{3}P $6s$	35	−3.60361	$2s2p^3$ ^{3}D $5d$	39	−3.52606	$2p^4$ ^{1}D $4s$
21	−2.02184	$2s^22p^2$ ^{1}D $8d$	27	−5.54080	$2s^22p^2$ ^{3}P $6d$	36	−3.38390	$2s2p^3$ ^{3}P $5d$	40	−3.49674	$2s2p^3$ ^{3}D $5f$
22	−1.99395	$2p^4$ ^{1}S $4s$	28	−4.46990	$2s^22p^2$ ^{1}D $6d$	37	−3.28452	$2p^4$ ^{3}P $4p$	41	−3.34793	$2s2p^3$ ^{3}P $5f$
23	−1.60765	$2s2p^3$ ^{1}P $5p$	29	−4.25193	$2s^22p^2$ ^{3}P $7s$	38	−3.15425	$2s^22p^2$ ^{3}P $8p$	42	−3.19757	$2s^22p^2$ ^{1}D $7s$
24	−1.44616	$2s^22p^2$ ^{1}S $9s$	30	−4.07107	$2s2p^3$ ^{3}D $5p$	39	−3.08308	$2s^22p^2$ ^{1}D $7p$	43	−3.08024	$2s^22p^2$ ^{3}P $8d$
25	−1.36607	$2s^22p^2$ ^{1}D $9d$	31	−4.05848	$2s^22p^2$ ^{3}P $7d$	40	−3.05759	$2s^22p^2$ ^{1}S $7p$	44	−2.99318	$2s^22p^2$ ^{1}D $7d$
26	−1.15830	$2s2p^3$ ^{3}P $6p$	32	−3.80231	$2s2p^3$ ^{3}P $5p$	41	−2.93799	$2s^22p^2$ ^{1}D $7f$	45	−2.95648	$2s^22p^2$ ^{1}S $7d$
		^{2}S^o	33	−3.68452	$2p^4$ ^{3}P $4s$	42	−2.84530	$2p^4$ ^{1}D $4p$	46	−2.94721	$2s^22p^2$ ^{1}D $7g$
1	−24.5638	$2s^22p^2$ ^{3}P $3p$	34	−3.49564	$2s2p^3$ ^{3}D $5f$	43	−2.48143	$2s^22p^2$ ^{3}P $9p$	47	−2.69095	$2p^4$ ^{3}P $4d$
2	−19.0496	$2s2p^3$ ^{3}S $3s$	35	−3.22749	$2s^22p^2$ ^{3}P $8s$	44	−2.12940	$2p^4$ ^{1}D $4f$	48	−2.43207	$2s^22p^2$ ^{3}P $9d$
3	−18.4194	$2s2p^3$ ^{3}D $3d$	36	−3.10463	$2s^22p^2$ ^{3}P $8d$	45	−2.10044	$2s^22p^2$ ^{1}D $8p$	49	−2.35150	$2p^4$ ^{1}D $4d$
4	−15.9702	$2s2p^3$ ^{1}D $3d$	37	−2.98340	$2s^22p^2$ ^{1}D $7d$	46	−2.07662	$2s^22p^2$ ^{1}S $8p$	50	−2.16947	$2s^22p^2$ ^{1}D $8s$
5	−14.1552	$2p^4$ ^{3}P $3p$	38	−2.75640	$2p^4$ ^{3}P $4d$			**^{2}D^e**	51	−2.03159	$2s^22p^2$ ^{1}D $8d$
6	−13.2863	$2s^22p^2$ ^{3}P $4p$	39	−2.53608	$2s^22p^2$ ^{3}P $9s$	1	−53.9500	$2s2p^4$	52	−2.00841	$2s^22p^2$ ^{1}S $8d$
7	−8.42343	$2s^22p^2$ ^{3}P $5p$	40	−2.44836	$2s^22p^2$ ^{3}P $9d$	2	−24.8495	$2s^22p^2$ ^{1}D $3s$	53	−2.00000	$2s^22p^2$ ^{1}D $8g$
8	−8.27694	$2s2p^3$ ^{3}D $4d$	41	−2.32342	$2p^4$ ^{1}D $4d$	3	−22.0481	$2s^22p^2$ ^{3}P $3d$			**^{2}D^o**
9	−7.26826	$2s2p^3$ ^{3}S $4s$			**^{2}P^o**	4	−21.7157	$2s^22p^2$ ^{1}D $3d$	1	−59.1725	$2s^22p^3$
10	−6.12286	$2s2p^3$ ^{1}D $4d$	1	−58.6269	$2s^22p^3$	5	−21.0160	$2s^22p^2$ ^{1}S $3d$	2	−24.0915	$2s^22p^2$ ^{3}P $3p$
11	−5.72693	$2s^22p^2$ ^{3}P $6p$	2	−48.2498	$2p^5$	6	−19.4969	$2s2p^3$ ^{3}D $3p$	3	−23.2063	$2s^22p^2$ ^{1}D $3p$
12	−4.18018	$2s^22p^2$ ^{3}P $7p$	3	−23.9098	$2s^22p^2$ ^{3}P $3p$	7	−19.0715	$2s2p^3$ ^{3}P $3p$	4	−20.9882	$2s2p^3$ ^{3}D $3s$
13	−3.64636	$2s2p^3$ ^{3}D $5d$	4	−22.9904	$2s^22p^2$ ^{1}D $3p$	8	−17.6661	$2s2p^3$ ^{1}D $3p$	5	−19.1835	$2s2p^3$ ^{1}D $3s$
14	−3.33676	$2p^4$ ^{3}P $4p$	5	−22.6391	$2s^22p^2$ ^{1}S $3p$	9	−17.1239	$2s2p^3$ ^{1}P $3p$	6	−18.1031	$2s2p^3$ ^{3}D $3d$
15	−3.18121	$2s^22p^2$ ^{3}P $8p$	6	−20.2922	$2s2p^3$ ^{3}P $3s$	10	−14.8877	$2p^4$ ^{1}D $3s$	7	−17.4436	$2s2p^3$ ^{3}P $3d$
16	−2.50516	$2s^22p^2$ ^{3}P $9p$	7	−18.4640	$2s2p^3$ ^{1}P $3s$	11	−12.8349	$2s^22p^2$ ^{1}D $4s$	8	−16.4265	$2s2p^3$ ^{1}D $3d$
17	−2.08326	$2s2p^3$ ^{3}S $5s$	8	−18.1693	$2s2p^3$ ^{3}D $3d$	12	−12.6170	$2p^4$ ^{3}P $3d$	9	−15.8472	$2s2p^3$ ^{3}S $3d$
		^{2}P^e	9	−17.0317	$2s2p^3$ ^{3}P $3d$	13	−12.3646	$2s^22p^2$ ^{3}P $4d$	10	−15.3972	$2s2p^3$ ^{1}P $3d$
1	−52.5474	$2s2p^4$	10	−16.0494	$2s2p^3$ ^{1}D $3d$	14	−11.8374	$2p^4$ ^{1}D $3d$	11	−14.2770	$2p^4$ ^{3}P $3p$
2	−25.3785	$2s^22p^2$ ^{3}P $3s$	11	−15.4592	$2s2p^3$ ^{1}P $3d$	15	−11.6720	$2s^22p^2$ ^{1}D $4d$	12	−13.5039	$2p^4$ ^{1}D $3p$
3	−22.6602	$2s^22p^2$ ^{3}P $3d$	12	−14.2694	$2p^4$ ^{3}P $3p$	16	−11.2667	$2s^22p^2$ ^{1}S $4d$	13	−13.0782	$2s^22p^2$ ^{3}P $4p$
4	−21.5274	$2s^22p^2$ ^{1}D $3d$	13	−13.0223	$2s^22p^2$ ^{3}P $4p$	17	−10.7220	$2p^4$ ^{1}S $3d$	14	−12.2716	$2s^22p^2$ ^{3}P $4f$
5	−20.1544	$2s2p^3$ ^{3}D $3p$	14	−12.9952	$2p^4$ ^{1}D $3p$	18	−8.70092	$2s2p^3$ ^{3}D $4p$	15	−12.2435	$2s^22p^2$ ^{1}D $4p$

N-like Ca (Ca^{13+})

i	E(Ryds)	Description	i	E(Ryds)	Description	i	E(Ryds)	Description	i	E(Ryds)	Description
16	−11.4365	$2s^22p^2$ ^{1}D $4f$	18	−5.95620	$2s2p^3$ ^{1}D $4f$	31	−3.06107	$2s^22p^2$ ^{3}P $8f$	11	−4.41935	$2s^22p^2$ ^{1}D $6f$
17	−9.34222	$2s2p^3$ ^{3}D $4s$	19	−5.51050	$2s^22p^2$ ^{3}P $6d$	32	−2.95323	$2s^22p^2$ ^{1}D $7f$	12	−4.37951	$2s^22p^2$ ^{1}D $6h$
18	−8.26969	$2s^22p^2$ ^{3}P $5p$	20	−5.43621	$2s^22p^2$ ^{3}P $6g$	33	−2.93417	$2s^22p^2$ ^{1}D $7h$	13	−4.00747	$2s^22p^2$ ^{3}P $7f$
19	−8.17720	$2s2p^3$ ^{3}D $4d$	21	−5.41744	$2s2p^3$ ^{1}P $4f$	34	−2.90257	$2s^22p^2$ ^{1}S $7f$	14	−4.00096	$2s^22p^2$ ^{3}P $7h$
20	−7.86987	$2s^22p^2$ ^{3}P $5f$	22	−4.49870	$2s^22p^2$ ^{1}D $6d$	35	−2.66642	$2p^4$ ^{3}P $4f$	15	−3.63456	$2s2p^3$ ^{3}D $5d$
21	−7.83471	$2s2p^3$ ^{3}P $4d$	23	−4.41598	$2s^22p^2$ ^{1}D $6g$	36	−2.41895	$2s^22p^2$ ^{3}P $9f$	16	−3.53302	$2s2p^3$ ^{3}D $5g$
22	−7.38612	$2s2p^3$ ^{1}D $4s$	24	−4.04144	$2s^22p^2$ ^{3}P $7d$	37	−2.10387	$2s^22p^2$ ^{1}D $8p$	17	−3.32493	$2s2p^3$ ^{3}P $5g$
23	−7.21650	$2s^22p^2$ ^{1}D $5p$	25	−4.00124	$2s^22p^2$ ^{3}P $7g$	38	−2.08061	$2p^4$ ^{1}D $4f$	18	−3.06760	$2s^22p^2$ ^{3}P $8f$
24	−6.83223	$2s^22p^2$ ^{1}D $5f$	26	−3.92951	$2s2p^3$ ^{3}D $5p$	39	−2.00588	$2s^22p^2$ ^{1}D $8f$	19	−3.06275	$2s^22p^2$ ^{3}P $8h$
25	−6.26400	$2s2p^3$ ^{1}D $4d$	27	−3.49231	$2s2p^3$ ^{3}D $5f$	40	−1.99389	$2s^22p^2$ ^{1}D $8h$	20	−2.95546	$2s^22p^2$ ^{1}D $7f$
26	−6.03554	$2s2p^3$ ^{3}S $4d$	28	−3.36169	$2s2p^3$ ^{3}P $5f$	41	−1.97081	$2s^22p^2$ ^{1}S $8f$	21	−2.93521	$2s^22p^2$ ^{1}D $7h$
27	−5.76449	$2s2p^3$ ^{1}P $4d$	29	−3.09419	$2s^22p^2$ ^{3}P $8d$			**^{2}G^e**	22	−2.65185	$2p^4$ ^{3}P $4f$
28	−5.66851	$2s^22p^2$ ^{3}P $6p$	30	−3.06269	$2s^22p^2$ ^{3}P $8g$	1	−21.8921	$2s^22p^2$ ^{1}D $3d$	23	−2.42344	$2s^22p^2$ ^{3}P $9f$
29	−5.45290	$2s^22p^2$ ^{3}P $6f$	31	−2.99901	$2s^22p^2$ ^{1}D $7d$	2	−12.3795	$2p^4$ ^{1}D $3d$	24	−2.42026	$2s^22p^2$ ^{3}P $9h$
30	−4.65252	$2s^22p^2$ ^{1}D $6p$	32	−2.95140	$2s^22p^2$ ^{1}D $7g$	3	−11.7404	$2s^22p^2$ ^{1}D $4d$	25	−2.08587	$2p^4$ ^{1}D $4f$
31	−4.40252	$2s^22p^2$ ^{1}D $6f$	33	−2.82006	$2p^4$ ^{3}P $4d$	4	−8.13115	$2s2p^3$ ^{3}D $4f$	26	−2.00709	$2s^22p^2$ ^{1}D $8f$
32	−4.16526	$2s2p^3$ ^{3}D $5s$	34	−2.44016	$2s^22p^2$ ^{3}P $9d$	5	−7.84700	$2s^22p^2$ ^{3}P $5g$	27	−1.99454	$2s^22p^2$ ^{1}D $8h$
33	−4.14274	$2s^22p^2$ ^{3}P $7p$	35	−2.42074	$2s^22p^2$ ^{3}P $9g$	6	−7.54599	$2s2p^3$ ^{3}P $4f$			**^{4}S^e**
34	−4.00865	$2s^22p^2$ ^{3}P $7f$	36	−2.41716	$2p^4$ ^{1}D $4d$	7	−7.00994	$2s^22p^2$ ^{1}D $5d$	1	−19.4850	$2s2p^3$ ^{3}P $3p$
35	−3.58514	$2s2p^3$ ^{3}D $5d$	37	−2.03858	$2s^22p^2$ ^{1}D $8d$	8	−6.88959	$2s^22p^2$ ^{1}D $5g$	2	−8.60420	$2s2p^3$ ^{3}P $4p$
36	−3.54134	$2s2p^3$ ^{3}D $5g$	38	−2.00275	$2s^22p^2$ ^{1}D $8g$	9	−6.66344	$2s^22p^2$ ^{1}S $5g$	3	−3.82239	$2s2p^3$ ^{3}P $5p$
37	−3.46197	$2s2p^3$ ^{3}P $5d$			**^{2}F^o**	10	−5.96791	$2s2p^3$ ^{1}D $4f$	4	−1.25931	$2s2p^3$ ^{3}P $6p$
38	−3.34370	$2p^4$ ^{3}P $4p$	1	−23.4937	$2s^22p^2$ ^{1}D $3p$	11	−5.43931	$2s^22p^2$ ^{3}P $6g$			**^{4}S^o**
39	−3.16188	$2s^22p^2$ ^{3}P $8p$	2	−17.9029	$2s2p^3$ ^{3}D $3d$	12	−5.41781	$2s2p^3$ ^{1}P $4f$	1	−60.2127	$2s^22p^3$
40	−3.09336	$2s^22p^2$ ^{1}D $7p$	3	−17.4109	$2s2p^3$ ^{3}P $3d$	13	−4.49991	$2s^22p^2$ ^{1}D $6d$	2	−24.0823	$2s^22p^2$ ^{3}P $3p$
41	−3.06860	$2s^22p^2$ ^{3}P $8f$	4	−16.3177	$2s2p^3$ ^{1}D $3d$	14	−4.42258	$2s^22p^2$ ^{1}D $6g$	3	−22.8878	$2s2p^3$ ^{5}S $3s$
42	−3.00969	$2p^4$ ^{1}D $4p$	5	−15.5991	$2s2p^3$ ^{1}P $3d$	15	−4.30368	$2s^22p^2$ ^{1}S $6g$	4	−19.2843	$2s2p^3$ ^{3}S $3s$
43	−2.94515	$2s^22p^2$ ^{1}D $7f$	6	−13.7428	$2p^4$ ^{1}D $3p$	16	−3.99804	$2s^22p^2$ ^{3}P $7g$	5	−18.1145	$2s2p^3$ ^{3}D $3d$
44	−2.60490	$2p^4$ ^{3}P $4f$	7	−12.2942	$2s^22p^2$ ^{1}D $4p$	17	−3.49436	$2s2p^3$ ^{3}D $5f$	6	−14.1482	$2p^4$ ^{3}P $3p$
45	−2.48738	$2s^22p^2$ ^{3}P $9p$	8	−12.2202	$2s^22p^2$ ^{3}P $4f$	18	−3.35947	$2s2p^3$ ^{3}P $5f$	7	−13.0833	$2s^22p^2$ ^{3}P $4p$
46	−2.42451	$2s^22p^2$ ^{3}P $9f$	9	−11.4885	$2s^22p^2$ ^{1}D $4f$	19	−3.05917	$2s^22p^2$ ^{3}P $8g$	8	−11.6519	$2s2p^3$ ^{5}S $4s$
47	−2.12114	$2s^22p^2$ ^{1}D $8p$	10	−10.9037	$2s^22p^2$ ^{1}S $4f$	20	−3.00328	$2s^22p^2$ ^{1}D $7d$	9	−8.32081	$2s^22p^2$ ^{3}P $5p$
48	−2.10132	$2p^4$ ^{1}D $4f$	11	−8.13898	$2s2p^3$ ^{3}D $4d$	21	−2.95486	$2s^22p^2$ ^{1}D $7g$	10	−8.17115	$2s2p^3$ ^{3}D $4d$
49	−2.07212	$2s2p^3$ ^{1}D $5s$	12	−7.84071	$2s^22p^2$ ^{3}P $5f$	22	−2.92002	$2s^22p^2$ ^{1}D $7i$	11	−7.34584	$2s2p^3$ ^{3}S $4s$
		^{2}F^e	13	−7.80278	$2s2p^3$ ^{3}P $4d$	23	−2.88527	$2s^22p^2$ ^{1}S $7g$	12	−6.60420	$2s2p^3$ ^{5}S $5s$
1	−22.5173	$2s^22p^2$ ^{3}P $3d$	14	−7.30240	$2s^22p^2$ ^{1}D $5p$	24	−2.48722	$2p^4$ ^{1}D $4d$	13	−5.67490	$2s^22p^2$ ^{3}P $6p$
2	−21.6828	$2s^22p^2$ ^{1}D $3d$	15	−6.85504	$2s^22p^2$ ^{1}D $5f$	25	−2.41820	$2s^22p^2$ ^{3}P $9g$	14	−4.15123	$2s^22p^2$ ^{3}P $7p$
3	−19.9613	$2s2p^3$ ^{3}D $3p$	16	−6.70370	$2s^22p^2$ ^{1}S $5f$	26	−2.03757	$2s^22p^2$ ^{1}D $8d$	15	−3.92803	$2s2p^3$ ^{5}S $6s$
4	−17.9057	$2s2p^3$ ^{1}D $3p$	17	−6.22213	$2s2p^3$ ^{1}D $4d$	27	−2.00500	$2s^22p^2$ ^{1}D $8g$	16	−3.60447	$2s2p^3$ ^{3}D $5d$
5	−12.8441	$2p^4$ ^{3}P $3d$	18	−5.81537	$2s2p^3$ ^{1}P $4d$	28	−1.98287	$2s^22p^2$ ^{1}D $8i$	17	−3.33801	$2p^4$ ^{3}P $4p$
6	−12.4928	$2s^22p^2$ ^{3}P $4d$	19	−5.44042	$2s^22p^2$ ^{3}P $6f$	29	−1.96097	$2s^22p^2$ ^{1}S $8g$	18	−3.15914	$2s^22p^2$ ^{3}P $8p$
7	−12.0748	$2p^4$ ^{1}D $3d$	20	−4.65598	$2s^22p^2$ ^{1}D $6p$			**^{2}G^o**	19	−2.49010	$2s^22p^2$ ^{3}P $9p$
8	−11.7126	$2s^22p^2$ ^{1}D $4d$	21	−4.41572	$2s^22p^2$ ^{1}D $6f$	1	−18.3608	$2s2p^3$ ^{3}D $3d$	20	−2.34383	$2s2p^3$ ^{5}S $7s$
9	−8.89647	$2s2p^3$ ^{3}D $4p$	22	−4.37855	$2s^22p^2$ ^{1}D $6h$	2	−16.3904	$2s2p^3$ ^{1}D $3d$	21	−2.10034	$2s2p^3$ ^{3}S $5s$
10	−8.11626	$2s2p^3$ ^{3}D $4f$	23	−4.33668	$2s^22p^2$ ^{1}S $6f$	3	−12.2781	$2s^22p^2$ ^{3}P $4f$			**^{6}S^o**
11	−7.95506	$2s^22p^2$ ^{3}P $5d$	24	−3.99823	$2s^22p^2$ ^{3}P $7f$	4	−11.5072	$2s^22p^2$ ^{1}D $4f$	1	−23.7329	$2s2p^3$ ^{5}S $3s$
12	−7.85958	$2s^22p^2$ ^{3}P $5g$	25	−3.54987	$2s2p^3$ ^{3}D $5d$	5	−8.32441	$2s2p^3$ ^{3}D $4d$	2	−11.8716	$2s2p^3$ ^{5}S $4s$
13	−7.54633	$2s2p^3$ ^{3}P $4f$	26	−3.53873	$2s2p^3$ ^{3}D $5g$	6	−7.85754	$2s^22p^2$ ^{3}P $5f$	3	−6.70010	$2s2p^3$ ^{5}S $5s$
14	−7.01567	$2s^22p^2$ ^{1}D $5d$	27	−3.44330	$2s2p^3$ ^{3}P $5d$	7	−6.85994	$2s^22p^2$ ^{1}D $5f$	4	−3.98141	$2s2p^3$ ^{5}S $6s$
15	−6.88081	$2s^22p^2$ ^{1}D $5g$	28	−3.32222	$2s2p^3$ ^{3}P $5g$	8	−6.26292	$2s2p^3$ ^{1}D $4d$	5	−2.37647	$2s2p^3$ ^{5}S $7s$
16	−6.79814	$2s2p^3$ ^{1}D $4p$	29	−3.10178	$2s^22p^2$ ^{1}D $7p$	9	−5.45541	$2s^22p^2$ ^{3}P $6f$	6	−1.35025	$2s2p^3$ ^{5}S $8s$
17	−5.97416	$2s2p^3$ ^{3}S $4f$	30	−3.06107	$2p^4$ ^{1}D $4p$	10	−5.44510	$2s^22p^2$ ^{3}P $6h$	7	−.65449	$2s2p^3$ ^{5}S $9s$

N-like Ca (Ca^{13+})

i	E(Ryds)	Description
	^{4}P^e	
1	−55.6452	$2s2p^4$
2	−25.6752	$2s^22p^2$ ^{3}P $3s$
3	−22.4564	$2s^22p^2$ ^{3}P $3d$
4	−21.8963	$2s2p^3$ ^{5}S $3p$
5	−19.6321	$2s2p^3$ ^{3}D $3p$
6	−19.1333	$2s2p^3$ ^{3}P $3p$
7	−17.9996	$2s2p^3$ ^{3}S $3p$
8	−15.7113	$2p^4$ ^{3}P $3s$
9	−13.7561	$2s^22p^2$ ^{3}P $4s$
10	−12.8192	$2p^4$ ^{3}P $3d$
11	−12.5352	$2s^22p^2$ ^{3}P $4d$
12	−11.1231	$2s2p^3$ ^{5}S $4p$
13	−8.86225	$2s2p^3$ ^{3}D $4p$
14	−8.60672	$2s^22p^2$ ^{3}P $5s$
15	−8.44125	$2s2p^3$ ^{3}P $4p$
16	−8.11196	$2s2p^3$ ^{3}D $4f$
17	−7.98637	$2s^22p^2$ ^{3}P $5d$
18	−6.82731	$2s2p^3$ ^{3}S $4p$
19	−6.32843	$2s2p^3$ ^{5}S $5p$
20	−5.86460	$2s^22p^2$ ^{3}P $6s$
21	−5.52906	$2s^22p^2$ ^{3}P $6d$
22	−4.27002	$2s^22p^2$ ^{3}P $7s$
23	−4.05406	$2s^22p^2$ ^{3}P $7d$
24	−4.00047	$2p^4$ ^{3}P $4s$
25	−3.86128	$2s2p^3$ ^{3}D $5p$
26	−3.78622	$2s2p^3$ ^{3}P $5p$
27	−3.73652	$2s2p^3$ ^{5}S $6p$
28	−3.49471	$2s2p^3$ ^{3}D $5f$
29	−3.23511	$2s^22p^2$ ^{3}P $8s$
30	−3.09858	$2s^22p^2$ ^{3}P $8d$
31	−2.84426	$2p^4$ ^{3}P $4d$
32	−2.54204	$2s^22p^2$ ^{3}P $9s$
33	−2.44515	$2s^22p^2$ ^{3}P $9d$
34	−2.24706	$2s2p^3$ ^{5}S $7p$
	^{4}P^o	
1	−24.2785	$2s^22p^2$ ^{3}P $3p$
2	−20.7374	$2s2p^3$ ^{3}P $3s$
3	−18.3009	$2s2p^3$ ^{3}D $3d$
4	−17.8273	$2s2p^3$ ^{3}P $3d$
5	−14.6586	$2p^4$ ^{3}P $3p$
6	−13.1995	$2s^22p^2$ ^{3}P $4p$
7	−9.13390	$2s2p^3$ ^{3}P $4s$
8	−8.37239	$2s^22p^2$ ^{3}P $5p$
9	−8.23497	$2s2p^3$ ^{3}D $4d$
10	−7.96244	$2s2p^3$ ^{3}P $4d$
11	−5.71021	$2s^22p^2$ ^{3}P $6p$
12	−4.17903	$2s^22p^2$ ^{3}P $7p$
13	−4.07776	$2s2p^3$ ^{3}P $5s$
14	−3.63993	$2s2p^3$ ^{3}D $5d$
15	−3.52886	$2s2p^3$ ^{3}P $5d$
16	−3.46301	$2p^4$ ^{3}P $4p$
17	−3.17063	$2s^22p^2$ ^{3}P $8p$
18	−2.49633	$2s^22p^2$ ^{3}P $9p$
	^{6}P^e	
1	−22.3745	$2s2p^3$ ^{5}S $3p$
2	−11.3225	$2s2p^3$ ^{5}S $4p$
3	−6.42620	$2s2p^3$ ^{5}S $5p$
4	−3.82562	$2s2p^3$ ^{5}S $6p$
5	−2.27956	$2s2p^3$ ^{5}S $7p$
6	−1.28593	$2s2p^3$ ^{5}S $8p$
7	−.60964	$2s2p^3$ ^{5}S $9p$
	^{4}D^e	
1	−22.6752	$2s^22p^2$ ^{3}P $3d$
2	−20.2039	$2s2p^3$ ^{3}D $3p$
3	−19.3989	$2s2p^3$ ^{3}P $3p$
4	−13.2289	$2p^4$ ^{3}P $3d$
5	−12.6195	$2s^22p^2$ ^{3}P $4d$
6	−8.98112	$2s2p^3$ ^{3}D $4p$
7	−8.58464	$2s2p^3$ ^{3}P $4p$
8	−8.13111	$2s2p^3$ ^{3}D $4f$
9	−8.02377	$2s^22p^2$ ^{3}P $5d$
10	−7.56658	$2s2p^3$ ^{3}P $4f$
11	−5.55232	$2s^22p^2$ ^{3}P $6d$
12	−4.06859	$2s^22p^2$ ^{3}P $7d$
13	−3.96499	$2s2p^3$ ^{3}D $5p$
14	−3.81338	$2s2p^3$ ^{3}P $5p$
15	−3.50453	$2s2p^3$ ^{3}D $5f$
16	−3.36463	$2s2p^3$ ^{3}P $5f$
17	−3.10755	$2s^22p^2$ ^{3}P $8d$
18	−2.96515	$2p^4$ ^{3}P $4d$
19	−2.45142	$2s^22p^2$ ^{3}P $9d$
	^{4}D^o	
1	−24.3796	$2s^22p^2$ ^{3}P $3p$
2	−21.4585	$2s2p^3$ ^{3}D $3s$
3	−20.2005	$2s2p^3$ ^{5}S $3d$
4	−18.4637	$2s2p^3$ ^{3}D $3d$
5	−17.6938	$2s2p^3$ ^{3}P $3d$
6	−16.4784	$2s2p^3$ ^{3}S $3d$
7	−14.4368	$2p^4$ ^{3}P $3p$
8	−13.2304	$2s^22p^2$ ^{3}P $4p$
9	−12.2788	$2s^22p^2$ ^{3}P $4f$
10	−10.4948	$2s2p^3$ ^{5}S $4d$
11	−9.50302	$2s2p^3$ ^{3}D $4s$
12	−8.40673	$2s2p^3$ ^{3}D $4d$
13	−8.30405	$2s^22p^2$ ^{3}P $5p$
14	−7.92907	$2s2p^3$ ^{3}P $4d$
15	−7.84884	$2s^22p^2$ ^{3}P $5f$
16	−6.28168	$2s2p^3$ ^{3}S $4d$
17	−6.02094	$2s2p^3$ ^{5}S $5d$
18	−5.71431	$2s^22p^2$ ^{3}P $6p$
19	−5.45748	$2s^22p^2$ ^{3}P $6f$
20	−4.24260	$2s2p^3$ ^{3}D $5s$
21	−4.16011	$2s^22p^2$ ^{3}P $7p$
22	−4.01047	$2s^22p^2$ ^{3}P $7f$
23	−3.68509	$2s2p^3$ ^{3}D $5d$
24	−3.58778	$2s2p^3$ ^{5}S $6d$
25	−3.54133	$2s2p^3$ ^{3}D $5g$
26	−3.50255	$2s2p^3$ ^{3}P $5d$
27	−3.40564	$2p^4$ ^{3}P $4p$
28	−3.17425	$2s^22p^2$ ^{3}P $8p$
29	−3.06880	$2s^22p^2$ ^{3}P $8f$
30	−2.61777	$2p^4$ ^{3}P $4f$
31	−2.49884	$2s^22p^2$ ^{3}P $9p$
32	−2.42480	$2s^22p^2$ ^{3}P $9f$
33	−2.13963	$2s2p^3$ ^{5}S $7d$
	^{6}D^o	
1	−20.8877	$2s2p^3$ ^{5}S $3d$
2	−10.7672	$2s2p^3$ ^{5}S $4d$
3	−6.15561	$2s2p^3$ ^{5}S $5d$
4	−3.67313	$2s2p^3$ ^{5}S $6d$
5	−2.18510	$2s2p^3$ ^{5}S $7d$
6	−1.22335	$2s2p^3$ ^{5}S $8d$
7	−.56604	$2s2p^3$ ^{5}S $9d$
	^{4}F^e	
1	−22.8309	$2s^22p^2$ ^{3}P $3d$
2	−20.1238	$2s2p^3$ ^{3}D $3p$
3	−13.0144	$2p^4$ ^{3}P $3d$
4	−12.6676	$2s^22p^2$ ^{3}P $4d$
5	−10.3670	$2s2p^3$ ^{5}S $4f$
6	−8.95736	$2s2p^3$ ^{3}D $4p$
7	−8.14347	$2s2p^3$ ^{3}D $4f$
8	−8.04831	$2s^22p^2$ ^{3}P $5d$
9	−7.86066	$2s^22p^2$ ^{3}P $5g$
10	−7.55721	$2s2p^3$ ^{3}P $4f$
11	−5.98251	$2s2p^3$ ^{3}S $4f$
12	−5.95648	$2s2p^3$ ^{5}S $5f$
13	−5.56375	$2s^22p^2$ ^{3}P $6d$
14	−5.43676	$2s^22p^2$ ^{3}P $6g$
15	−4.07639	$2s^22p^2$ ^{3}P $7d$
16	−4.00141	$2s^22p^2$ ^{3}P $7g$
17	−3.95588	$2s2p^3$ ^{3}D $5p$
18	−3.56068	$2s2p^3$ ^{5}S $6f$
19	−3.50682	$2s2p^3$ ^{3}D $5f$
20	−3.36686	$2s2p^3$ ^{3}P $5f$
21	−3.11255	$2s^22p^2$ ^{3}P $8d$
22	−3.06268	$2s^22p^2$ ^{3}P $8g$
23	−2.88979	$2p^4$ ^{3}P $4d$
24	−2.45477	$2s^22p^2$ ^{3}P $9d$
25	−2.42075	$2s^22p^2$ ^{3}P $9g$
26	−2.11427	$2s2p^3$ ^{5}S $7f$
	^{4}F^o	
1	−18.7389	$2s2p^3$ ^{3}D $3d$
2	−17.9186	$2s2p^3$ ^{3}P $3d$
3	−12.2394	$2s^22p^2$ ^{3}P $4f$
4	−8.45749	$2s2p^3$ ^{3}D $4d$
5	−8.00636	$2s2p^3$ ^{3}P $4d$
6	−7.83776	$2s^22p^2$ ^{3}P $5f$
7	−5.44992	$2s^22p^2$ ^{3}P $6f$
8	−4.00509	$2s^22p^2$ ^{3}P $7f$
9	−3.69545	$2s2p^3$ ^{3}D $5d$
10	−3.54568	$2s2p^3$ ^{3}D $5g$
11	−3.53120	$2s2p^3$ ^{3}P $5d$
12	−3.32269	$2s2p^3$ ^{3}P $5g$
13	−3.06498	$2s^22p^2$ ^{3}P $8f$
14	−2.69450	$2p^4$ ^{3}P $4f$
15	−2.42194	$2s^22p^2$ ^{3}P $9f$
	^{6}F^e	
1	−10.3952	$2s2p^3$ ^{5}S $4f$
2	−5.97551	$2s2p^3$ ^{5}S $5f$
3	−3.57199	$2s2p^3$ ^{5}S $6f$
4	−2.12254	$2s2p^3$ ^{5}S $7f$
5	−1.18193	$2s2p^3$ ^{5}S $8f$
6	−.53719	$2s2p^3$ ^{5}S $9f$
	^{4}G^e	
1	−8.16068	$2s2p^3$ ^{3}D $4f$
2	−7.84993	$2s^22p^2$ ^{3}P $5g$
3	−7.56350	$2s2p^3$ ^{3}P $4f$
4	−5.43974	$2s^22p^2$ ^{3}P $6g$
5	−3.99852	$2s^22p^2$ ^{3}P $7g$
6	−3.51409	$2s2p^3$ ^{3}D $5f$
7	−3.37251	$2s2p^3$ ^{3}P $5f$
8	−3.05938	$2s^22p^2$ ^{3}P $8g$
9	−2.41840	$2s^22p^2$ ^{3}P $9g$
	^{4}G^o	
1	−18.6538	$2s2p^3$ ^{3}D $3d$
2	−12.2914	$2s^22p^2$ ^{3}P $4f$
3	−8.43420	$2s2p^3$ ^{3}D $4d$
4	−7.86576	$2s^22p^2$ ^{3}P $5f$
5	−5.95238	$2s2p^3$ ^{5}S $5g$
6	−5.46225	$2s^22p^2$ ^{3}P $6f$
7	−5.44514	$2s^22p^2$ ^{3}P $6h$
8	−4.01268	$2s^22p^2$ ^{3}P $7f$
9	−4.00099	$2s^22p^2$ ^{3}P $7h$
10	−3.68261	$2s2p^3$ ^{3}D $5d$
11	−3.55683	$2s2p^3$ ^{5}S $6g$
12	−3.53353	$2s2p^3$ ^{3}D $5g$
13	−3.32495	$2s2p^3$ ^{3}P $5g$
14	−3.07038	$2s^22p^2$ ^{3}P $8f$
15	−3.06278	$2s^22p^2$ ^{3}P $8h$
16	−2.66055	$2p^4$ ^{3}P $4f$
17	−2.42558	$2s^22p^2$ ^{3}P $9f$
18	−2.42028	$2s^22p^2$ ^{3}P $9h$
19	−2.11237	$2s2p^3$ ^{5}S $7g$
	^{6}G^o	
1	−5.95300	$2s2p^3$ ^{5}S $5g$
2	−3.55744	$2s2p^3$ ^{5}S $6g$
3	−2.11286	$2s2p^3$ ^{5}S $7g$

N-like Ca (Ca^{13+})

Energies in ascending order from ground state for terms with effective $n \leq 4.0$, $L \leq 4$

Term	i	E(Ryds)	Term	i	E(Ryds)	Term	i	E(Ryds)	Term	i	E(Ryds)	Term	i	E(Ryds)
$^4S^o$	1	0.00000	$^2F^e$	1	37.6954	$^2D^o$	5	41.0292	$^2P^o$	9	43.1810	$^2S^o$	6	46.9264
$^2D^o$	1	1.04020	$^4P^e$	3	37.7563	$^4P^e$	6	41.0794	$^2P^e$	9	43.3054	$^4D^o$	8	46.9823
$^2P^o$	1	1.58580	$^6P^e$	1	37.8382	$^2D^e$	7	41.1412	$^2S^e$	5	43.3510	$^4D^e$	4	46.9838
$^4P^e$	1	4.56750	$^2D^e$	3	38.1646	$^2S^o$	2	41.1631	$^4D^o$	6	43.7343	$^4P^o$	6	47.0132
$^2D^e$	1	6.26270	$^4P^e$	4	38.3164	$^4F^o$	1	41.4738	$^2D^o$	8	43.7862	$^4S^o$	7	47.1294
$^2S^e$	1	7.38640	$^2G^e$	1	38.3206	$^4G^o$	1	41.5589	$^2G^o$	2	43.8223	$^2D^o$	13	47.1345
$^2P^e$	1	7.66530	$^2D^e$	4	38.4970	$^2S^e$	4	41.5803	$^2F^o$	4	43.8950	$^2P^o$	13	47.1904
$^2P^o$	2	11.9629	$^2F^e$	2	38.5299	$^2P^o$	7	41.7487	$^2P^o$	10	44.1633	$^4F^e$	3	47.1983
$^4P^e$	2	34.5375	$^2P^e$	4	38.6853	$^4D^o$	4	41.7490	$^2S^o$	4	44.2425	$^2P^o$	14	47.2175
$^2P^e$	2	34.8342	$^4D^o$	2	38.7542	$^2S^o$	3	41.7933	$^2D^o$	9	44.3655	$^2F^e$	5	47.3686
$^2D^e$	2	35.3632	$^2S^e$	3	38.7929	$^2G^o$	1	41.8519	$^4P^e$	8	44.5014	$^2D^e$	11	47.3778
$^2S^o$	1	35.6489	$^2D^e$	5	39.1967	$^4P^o$	3	41.9118	$^2F^o$	5	44.6136	$^4P^e$	10	47.3935
$^4D^o$	1	35.8331	$^2D^o$	4	39.2245	$^2P^o$	8	42.0434	$^2P^o$	11	44.7535	$^2P^e$	12	47.5030
$^4P^o$	1	35.9342	$^6D^o$	1	39.3250	$^4S^o$	5	42.0982	$^2D^o$	10	44.8155	$^4F^e$	4	47.5451
$^2S^e$	2	36.0974	$^4P^o$	2	39.4753	$^2D^o$	6	42.1096	$^2P^e$	10	44.8726	$^4D^e$	5	47.5932
$^2D^o$	2	36.1212	$^2P^o$	6	39.9205	$^2P^e$	7	42.1764	$^2D^e$	10	45.3250	$^2D^e$	12	47.5957
$^4S^o$	2	36.1304	$^4D^e$	2	40.0088	$^4P^e$	7	42.2131	$^4P^o$	5	45.5541	$^2P^e$	13	47.6272
$^2P^o$	3	36.3029	$^4D^o$	3	40.0122	$^4F^o$	2	42.2941	$^4D^o$	7	45.7759	$^4P^e$	11	47.6775
$^6S^o$	1	36.4798	$^2P^e$	5	40.0583	$^2F^e$	4	42.3070	$^2D^o$	11	45.9357	$^2F^e$	6	47.7199
$^2F^o$	1	36.7190	$^4F^e$	2	40.0889	$^2F^o$	2	42.3098	$^2P^o$	12	45.9433	$^2S^e$	7	47.7588
$^2D^o$	3	37.0064	$^2F^e$	3	40.2514	$^4P^o$	4	42.3854	$^2S^o$	5	46.0575	$^2G^e$	2	47.8332
$^2P^o$	4	37.2223	$^4P^e$	5	40.5806	$^4D^o$	5	42.5189	$^4S^o$	6	46.0645	$^2D^e$	13	47.8481
$^4S^o$	3	37.3249	$^2D^e$	6	40.7158	$^2D^e$	8	42.5466	$^4P^e$	9	46.4566	$^2F^o$	7	47.9185
$^4F^e$	1	37.3818	$^4S^e$	1	40.7277	$^2D^o$	7	42.7691	$^2F^o$	6	46.4699	$^4G^o$	2	47.9213
$^4D^e$	1	37.5375	$^4D^e$	3	40.8138	$^2F^o$	3	42.8018	$^2P^e$	11	46.5525	$^4D^o$	9	47.9339
$^2P^e$	3	37.5525	$^4S^o$	4	40.9284	$^2P^e$	8	43.0186	$^2D^o$	12	46.7088	$^2G^o$	3	47.9346
$^2P^o$	5	37.5736	$^2P^e$	6	41.0280	$^2D^e$	9	43.0888	$^2S^e$	6	46.7350	$^2D^o$	14	47.9411

gf-values for transitions involving terms with effective $n \leq 4.0$, $L \leq 4$

i i'	gf_L	i i'	gf_L	i i'	gf_L	i i'	gf_L	i i'	gf_L	i i'	gf_L
$^2S^o$–$^2P^e$		1 3	−3.30E−1	1 6	−5.03E−2	1 9	−7.02E−3	1 12	−5.03E−2	2 2	4.52E−1
1 1	5.03E−5	1 4	−3.49E−2	1 7	−4.08E−2	1 10	−1.33E−3	1 13	−6.94E−1	2 3	4.06E−4
1 2	9.93E−2	1 5	−2.40E−1	1 8	−2.36E−2	1 11	−8.76E−2	2 1	3.26E−1	2 4	4.09E−6

N-like Ca (Ca^{13+})

i	i'	gf_L	i	i'	gf_L	i	i'	gf_L	i	i'	gf_L	i	i'	gf_L	i	i'	gf_L
2	5	7.84E−5	6	5	5.07E−4	4	1	2.75E−1	7	11	9.53E−4	4	6	−4.06E−3	8	2	1.36E−2
2	6	8.44E−5	6	6	1.29E−3	4	2	1.82E−2	7	12	6.45E−5	4	7	−9.96E−4	8	3	1.23E−3
2	7	−1.44E−1	6	7	1.14E−4	4	3	1.64E−1	7	13	1.95E−3	4	8	−1.10E−2	8	4	6.61E−1
2	8	−3.06E−1	6	8	1.74E−3	4	4	2.22E−3	7	14	2.53E−3	4	9	−8.34E−4	8	5	2.76E−2
2	9	−3.03E−1	6	9	1.19E−5	4	5	3.60E−3			$^2P^e$–$^2P^o$	4	10	−1.23E−2	8	6	4.87E−3
2	10	−2.62E−1	6	10	1.08E−5	4	6	2.17E−1	1	1	3.32E−1	4	11	−1.64E−1	8	7	1.37E−1
2	11	−2.12E−7	6	11	1.85E−1	4	7	−1.09E−3	1	2	−8.36E−1	4	12	−1.40E−5	8	8	5.64E−5
2	12	−1.43E−5	6	12	−3.33E−2	4	8	−5.40E−3	1	3	−1.24E−2	4	13	−1.06E−3	8	9	−4.41E−5
2	13	−2.17E−2	6	13	−5.94E−1	4	9	−3.23E−1	1	4	−6.38E−3	4	14	−1.78E−3	8	10	−1.37E−1
3	1	9.85E−2			$^2S^e$–$^2P^o$	4	10	−9.65E−3	1	5	−3.77E−3	5	1	4.04E−2	8	11	−1.16E−1
3	2	2.63E−3	1	1	2.59E−1	4	11	−3.04E−2	1	6	−3.21E−2	5	2	1.00E−2	8	12	−4.36E−6
3	3	7.52E−2	1	2	−5.31E−2	4	12	−6.16E−2	1	7	−2.01E−1	5	3	1.79E−2	8	13	−2.14E−1
3	4	4.69E−4	1	3	−2.32E−5	4	13	−8.54E−8	1	8	−3.79E−1	5	4	3.47E−4	8	14	−1.12E−1
3	5	2.04E−1	1	4	−8.66E−3	4	14	−8.73E−4	1	9	−8.18E−2	5	5	5.91E−4	9	1	1.82E−1
3	6	6.11E−3	1	5	−7.31E−3	5	1	7.44E−3	1	10	−2.61E+0	5	6	5.10E−3	9	2	2.18E−2
3	7	−3.28E−4	1	6	−2.10E−1	5	2	1.10E−4	1	11	−2.63E−1	5	7	−7.09E−6	9	3	4.98E−2
3	8	−2.47E−4	1	7	−1.36E−1	5	3	3.01E−2	1	12	−2.56E−1	5	8	−4.89E−1	9	4	1.86E−2
3	9	−1.97E−5	1	8	−2.38E−1	5	4	1.56E−1	1	13	−9.44E−2	5	9	−2.00E−4	9	5	2.37E−1
3	10	−8.72E−4	1	9	−2.50E+0	5	5	1.17E−1	1	14	−1.59E−1	5	10	−3.37E−3	9	6	2.73E−4
3	11	−7.07E−4	1	10	−3.14E−2	5	6	1.18E−1	2	1	3.08E−1	5	11	−2.96E−5	9	7	5.02E−1
3	12	−2.59E−2	1	11	−1.70E+0	5	7	1.75E−1	2	2	3.83E−7	5	12	−7.60E−2	9	8	1.28E−7
3	13	−1.59E−2	1	12	−2.86E−6	5	8	2.59E−3	2	3	−4.11E−1	5	13	−3.50E−3	9	9	1.16E−4
4	1	8.94E−1	1	13	−2.07E−6	5	9	1.17E−3	2	4	−3.07E−1	5	14	−4.11E−3	9	10	−3.89E−3
4	2	7.73E−4	1	14	−6.66E−4	5	10	−6.70E−3	2	5	−9.34E−2	6	1	5.85E−1	9	11	−1.35E−1
4	3	8.21E−3	2	1	1.22E−1	5	11	−2.45E−1	2	6	−2.72E−1	6	2	6.42E−3	9	12	−3.06E−2
4	4	2.75E−1	2	2	6.56E−6	5	12	−6.00E−3	2	7	−1.21E−3	6	3	1.11E−1	9	13	−7.32E−2
4	5	4.00E−3	2	3	−3.01E−3	5	13	−5.83E−2	2	8	−5.12E−3	6	4	7.67E−2	9	14	−3.24E−2
4	6	8.97E−4	2	4	−8.45E−3	5	14	−5.85E−2	2	9	−1.79E−3	6	5	5.34E−2	10	1	6.93E−3
4	7	1.70E−1	2	5	−5.67E−1	6	1	8.60E−4	2	10	−1.07E−3	6	6	3.99E−1	10	2	6.40E−1
4	8	5.59E−2	2	6	−1.65E−2	6	2	1.33E−1	2	11	−2.69E−4	6	7	−5.47E−6	10	3	1.00E−3
4	9	4.03E−3	2	7	−2.70E−1	6	3	1.98E−4	2	12	−7.09E−5	6	8	−3.07E−2	10	4	7.84E−4
4	10	−6.96E−4	2	8	−3.81E−4	6	4	7.76E−4	2	13	−2.40E−1	6	9	−4.43E−1	10	5	2.46E−5
4	11	−1.27E−4	2	9	−5.34E−3	6	5	2.70E−4	2	14	−2.22E−1	6	10	−1.30E−3	10	6	3.21E−1
4	12	−5.91E−2	2	10	−1.71E−4	6	6	8.26E−4	3	1	9.35E−1	6	11	−2.86E−2	10	7	2.82E−2
4	13	−4.76E−3	2	11	−3.88E−4	6	7	6.20E−1	3	2	2.04E−5	6	12	−5.52E−3	10	8	1.99E−3
5	1	2.48E−1	2	12	−7.01E−5	6	8	3.05E−3	3	3	1.74E−1	6	13	−6.18E−3	10	9	4.36E−4
5	2	7.41E−4	2	13	−1.17E−3	6	9	6.34E−3	3	4	2.81E−3	6	14	−1.62E−6	10	10	5.76E−3
5	3	7.97E−5	2	14	−3.67E−3	6	10	8.69E−4	3	5	−1.37E−4	7	1	5.67E−3	10	11	3.10E−4
5	4	2.57E−4	3	1	6.16E−1	6	11	3.14E−3	3	6	−3.43E−3	7	2	8.61E−3	10	12	−2.54E−1
5	5	1.13E−1	3	2	8.12E−5	6	12	3.90E−3	3	7	−3.25E−3	7	3	3.59E−1	10	13	−3.40E−1
5	6	1.99E−1	3	3	6.50E−2	6	13	−1.06E−3	3	8	−2.46E−1	7	4	3.90E−2	10	14	−3.18E−1
5	7	4.60E−2	3	4	1.57E−1	6	14	−8.75E−4	3	9	−5.88E−2	7	5	1.14E−3	11	1	5.13E−2
5	8	4.46E−2	3	5	5.95E−3	7	1	2.01E−2	3	10	−6.31E−3	7	6	1.69E−4	11	2	1.63E−4
5	9	8.02E−2	3	6	−1.51E−3	7	2	8.46E−5	3	11	−6.52E−3	7	7	8.20E−3	11	3	2.38E−1
5	10	2.00E−1	3	7	−8.07E−6	7	3	1.39E−2	3	12	−5.10E−4	7	8	1.41E−4	11	4	1.68E−1
5	11	−6.84E−5	3	8	−3.31E−4	7	4	1.20E−3	3	13	−2.95E−2	7	9	−1.84E−4	11	5	2.89E−2
5	12	−2.68E−1	3	9	−3.06E−5	7	5	2.97E−1	3	14	−3.59E−2	7	10	−2.53E−1	11	6	6.92E−3
5	13	−1.54E−2	3	10	−1.44E−1	7	6	4.12E−3	4	1	1.89E+0	7	11	−5.98E−5	11	7	1.49E−3
6	1	8.11E−4	3	11	−8.46E−2	7	7	1.88E−5	4	2	1.33E−4	7	12	−7.35E−1	11	8	1.38E−3
6	2	1.96E−1	3	12	−9.93E−5	7	8	1.53E−5	4	3	1.03E−1	7	13	−8.88E−3	11	9	1.20E−3
6	3	8.39E−2	3	13	−1.21E−3	7	9	5.22E−4	4	4	3.13E−1	7	14	−2.15E−2	11	10	2.72E−4
6	4	1.34E−2	3	14	−2.43E−3	7	10	4.89E−5	4	5	1.39E−3	8	1	5.55E−3	11	11	3.58E−5

N-like Ca (Ca^{13+})

i	i'	gf_L	i	i'	gf_L	i	i'	gf_L	i	i'	gf_L	i	i'	gf_L	i	i'	gf_L
11	12	5.59E−7	2	8	−2.08E−3	6	8	−5.95E−2	10	8	1.48E−1	14	8	1.83E−1	4	4	−6.00E−4
11	13	−4.91E−1	2	9	−4.55E−3	6	9	−4.87E−2	10	9	1.44E−2	14	9	4.26E−3	4	5	−2.48E−4
11	14	−4.91E−1	2	10	−4.69E−1	6	10	−2.35E−3	10	10	−9.79E−5	14	10	3.00E−1	4	6	−2.32E−3
12	1	1.26E−2	2	11	−5.55E−4	6	11	−6.72E−3	10	11	−5.67E−4	14	11	−4.84E−4	4	7	−1.07E−2
12	2	1.96E−1	2	12	−2.14E+0	6	12	−4.05E−5	10	12	−2.37E−2	14	12	−7.66E−3	4	8	−8.17E−2
12	3	3.18E−2	2	13	−5.05E−3	6	13	−2.04E−3	10	13	−1.05E−3	14	13	−6.32E−1	4	9	−1.36E−1
12	4	8.91E−3	3	1	5.73E−3	7	1	8.86E−2	11	1	8.92E−2			$^2P^e$–$^2D^o$	4	10	−2.68E−3
12	5	4.02E−3	3	2	7.30E−2	7	2	8.37E−1	11	2	4.93E−4	1	1	9.65E−1	4	11	−1.36E−5
12	6	3.90E−6	3	3	−7.66E−1	7	3	1.17E−3	11	3	2.45E−2	1	2	−4.07E−3	4	12	−1.84E−4
12	7	1.85E−3	3	4	−5.25E−3	7	4	1.05E−3	11	4	1.33E−1	1	3	−1.64E−4	4	13	−1.13E−2
12	8	6.42E−2	3	5	−1.72E−3	7	5	9.15E−4	11	5	2.57E−1	1	4	−4.12E−2	4	14	−3.14E−1
12	9	6.37E−2	3	6	−3.33E−1	7	6	3.23E−3	11	6	2.25E−4	1	5	−3.11E−1	5	1	6.73E−1
12	10	6.73E−3	3	7	−3.62E−2	7	7	2.20E−3	11	7	2.18E−3	1	6	−2.97E−1	5	2	1.66E−4
12	11	1.51E−2	3	8	−1.31E−2	7	8	−7.35E−4	11	8	1.83E−2	1	7	−7.28E−3	5	3	4.86E−2
12	12	3.14E−1	3	9	−1.09E−2	7	9	−9.07E−1	11	9	5.91E−4	1	8	−6.38E−2	5	4	2.56E−1
12	13	7.05E−3	3	10	−6.52E−4	7	10	−3.29E−1	11	10	−1.53E−4	1	9	−3.42E+0	5	5	−3.99E−6
12	14	6.64E−3	3	11	−2.14E−2	7	11	−1.93E−4	11	11	−4.89E−7	1	10	−5.01E+0	5	6	−5.50E−1
13	1	2.49E−1	3	12	−2.09E−5	7	12	−6.05E−4	11	12	−1.40E−2	1	11	−3.91E−1	5	7	−1.09E−2
13	2	5.48E−3	3	13	−1.30E+0	7	13	−7.55E−3	11	13	−7.85E−5	1	12	−5.61E−3	5	8	−7.05E−3
13	3	5.31E−1	4	1	2.27E−2	8	1	1.25E+0	12	1	1.36E−1	1	13	−2.00E−2	5	9	−1.33E−2
13	4	1.00E−1	4	2	5.86E−1	8	2	1.97E−3	12	2	1.37E−4	1	14	−1.64E−2	5	10	−1.10E−4
13	5	4.72E−2	4	3	−3.48E−2	8	3	1.38E−2	12	3	1.31E−5	2	1	3.93E−1	5	11	−2.89E−2
13	6	1.19E−3	4	4	−2.99E−1	8	4	5.85E−4	12	4	6.27E−4	2	2	−6.14E−1	5	12	−3.13E−3
13	7	2.86E−3	4	5	−3.35E−2	8	5	2.08E−4	12	5	2.36E−6	2	3	5.17E−1	5	13	2.51E−3
13	8	1.67E−2	4	6	−3.18E−2	8	6	1.18E−1	12	6	3.49E−1	2	4	−1.94E−1	5	14	−2.37E−2
13	9	1.20E−2	4	7	−1.26E−1	8	7	6.20E−3	12	7	2.14E−1	2	5	−2.38E−3	6	1	9.13E−2
13	10	1.08E−3	4	8	−1.04E−1	8	8	−3.90E−3	12	8	1.70E−2	2	6	−3.54E−4	6	2	1.29E−1
13	11	1.76E−3	4	9	−2.81E−4	8	9	−8.65E−4	12	9	1.52E−5	2	7	−3.67E−3	6	3	6.10E−2
13	12	2.62E−2	4	10	−1.54E−3	8	10	−1.75E−3	12	10	9.25E−2	2	8	−4.33E−4	6	4	9.39E−2
13	13	1.50E−1	4	11	−3.69E−1	8	11	−3.16E−4	12	11	−2.55E−7	2	9	−4.69E−4	6	5	−3.17E−7
13	14	1.24E−1	4	12	−1.68E−5	8	12	−4.01E−2	12	12	−7.17E−1	2	10	−1.12E−4	6	6	−3.98E−2
		$^2P^o$–$^2D^e$	4	13	−4.22E−1	8	13	−7.27E−3	12	13	−2.42E−5	2	11	−4.15E−4	6	7	−8.07E−1
1	1	−1.36E−1	5	1	7.55E−3	9	1	1.13E−1	13	1	1.37E−1	2	12	−6.95E−4	6	8	−1.30E−2
1	2	−1.97E−1	5	2	6.13E−5	9	2	2.19E−4	13	2	7.00E−3	2	13	−8.86E−1	6	9	−1.09E−2
1	3	−1.36E+0	5	3	−5.50E−4	9	3	3.37E−1	13	3	1.06E−1	2	14	−2.91E−4	6	10	−6.69E−3
1	4	−3.11E+0	5	4	−9.04E−3	9	4	3.04E−2	13	4	2.67E−3	3	1	2.91E−1	6	11	−3.90E−2
1	5	−2.80E+0	5	5	−9.60E−1	9	5	3.70E−3	13	5	5.57E−3	3	2	1.21E−3	6	12	−2.31E−4
1	6	−1.37E−2	5	6	−1.31E−2	9	6	1.97E−2	13	6	2.06E−2	3	3	4.36E−3	6	13	−3.80E−3
1	7	−9.24E−1	5	7	−8.73E−2	9	7	1.13E−2	13	7	2.76E−2	3	4	−1.46E−3	6	14	−1.97E−2
1	8	−5.72E−2	5	8	−1.33E−3	9	8	1.23E−4	13	8	1.54E−1	3	5	−4.77E−5	7	1	1.18E−1
1	9	−2.13E−1	5	9	−3.98E−1	9	9	9.09E−5	13	9	9.11E−3	3	6	−2.14E−1	7	2	6.13E−1
1	10	−1.34E−3	5	10	−1.81E−3	9	10	−2.09E−4	13	10	2.06E−1	3	7	−1.57E−2	7	3	2.50E−2
1	11	5.13E−2	5	11	−3.08E−3	9	11	−3.32E−5	13	11	−2.99E−4	3	8	−6.94E−2	7	4	6.39E−3
1	12	−9.62E−3	5	12	−4.60E−5	9	12	−4.72E−2	13	12	−7.58E−3	3	9	−1.41E−2	7	5	2.40E−1
1	13	−7.47E−1	5	13	−4.09E−2	9	13	−1.05E−2	13	13	−7.49E−1	3	10	−6.09E−3	7	6	1.79E−6
2	1	6.36E−1	6	1	2.20E−1	10	1	2.96E−1	14	1	1.10E−1	3	11	−3.31E−7	7	7	−6.19E−4
2	2	−8.20E−6	6	2	2.01E−2	10	2	2.73E−3	14	2	1.50E−2	3	12	−5.98E−5	7	8	−5.51E−1
2	3	−6.95E−5	6	3	2.53E−3	10	3	7.01E−2	14	3	9.71E−2	3	13	−4.01E−4	7	9	−5.87E−2
2	4	−3.19E−6	6	4	1.00E−2	10	4	6.10E−1	14	4	3.87E−3	3	14	5.40E+0	7	10	4.10E−3
2	5	−3.21E−6	6	5	6.04E−3	10	5	1.10E−3	14	5	2.71E−3	4	1	6.22E−1	7	11	−1.92E−1
2	6	−1.41E−2	6	6	−2.22E−2	10	6	3.58E−2	14	6	2.92E−2	4	2	1.02E−1	7	12	−3.51E−2
2	7	−1.01E−2	6	7	−7.91E−1	10	7	6.98E−3	14	7	1.36E−2	4	3	8.74E−2	7	13	−1.28E−3

N-like Ca (Ca^{13+})

i i'	gf_L	i i'	gf_L	i i'	gf_L	i i'	gf_L	i i'	gf_L	i i'	gf_L
7 14	−3.43E−2	11 10	4.32E−4	2 5	−1.39E+0	6 1	1.18E+0	9 11	−1.19E−4	13 7	2.55E−3
8 1	7.92E−2	11 11	1.65E−4	2 6	−9.41E−3	6 2	6.10E−2	9 12	−1.00E−1	13 8	4.16E−3
8 2	2.30E−1	11 12	−3.44E−4	2 7	−1.31E−3	6 3	1.36E−1	9 13	−4.19E−4	13 9	8.51E−3
8 3	4.69E−2	11 13	−1.45E+0	2 8	−2.62E−4	6 4	9.94E−1	9 14	−5.54E−7	13 10	3.74E−3
8 4	8.13E−3	11 14	−1.12E−3	2 9	−8.12E−6	6 5	−4.30E−3	10 1	4.29E−3	13 11	3.58E−4
8 5	4.48E−1	12 1	1.07E−2	2 10	−5.92E−4	6 6	−4.05E−1	10 2	1.84E−3	13 12	3.31E−3
8 6	9.58E−5	12 2	2.36E−4	2 11	−2.81E−7	6 7	−8.86E−2	10 3	8.03E−3	13 13	5.09E−1
8 7	3.30E−5	12 3	1.65E−3	2 12	−1.44E−3	6 8	−1.87E−3	10 4	3.02E−3	13 14	−3.48E−2
8 8	−7.93E−3	12 4	2.53E−4	2 13	−3.70E−2	6 9	−6.51E−2	10 5	1.78E+0		**$^2D^o$–$^2F^e$**
8 9	−4.71E−1	12 5	2.51E−4	2 14	−3.34E−2	6 10	−1.17E−2	10 6	8.65E−3	1 1	−1.16E+0
8 10	−3.37E−2	12 6	3.06E−2	3 1	2.54E+0	6 11	−1.96E−1	10 7	4.60E−3	1 2	−1.06E+1
8 11	−2.37E−1	12 7	2.81E−1	3 2	1.22E−1	6 12	−1.08E−2	10 8	1.86E−3	1 3	−1.32E+0
8 12	−3.55E−1	12 8	3.79E−1	3 3	9.51E−2	6 13	−2.35E−3	10 9	2.70E−3	1 4	−3.81E−1
8 13	−5.13E−3	12 9	2.38E−2	3 4	−7.71E−3	6 14	−5.08E−3	10 10	6.49E−5	1 5	−4.35E−2
8 14	−2.15E−2	12 10	5.14E−3	3 5	−4.72E−4	7 1	1.17E−1	10 11	−2.49E−3	1 6	−1.09E+0
9 1	2.56E−2	12 11	5.47E−2	3 6	−8.40E−2	7 2	5.74E−1	10 12	−1.30E+0	2 1	−1.14E+0
9 2	2.79E−6	12 12	1.91E−2	3 7	−1.24E−1	7 3	1.14E−2	10 13	−1.20E−3	2 2	−2.72E−2
9 3	7.18E−1	12 13	1.37E−3	3 8	−3.07E−1	7 4	2.78E−2	10 14	−4.12E−5	2 3	−3.64E−1
9 4	9.95E−3	12 14	−4.38E−2	3 9	−1.02E−1	7 5	2.02E−3	11 1	9.96E−2	2 4	−1.63E−2
9 5	2.31E−2	13 1	2.59E−2	3 10	−4.31E−2	7 6	−4.79E−3	11 2	9.84E−2	2 5	−1.12E−3
9 6	3.08E−4	13 2	1.88E−2	3 11	−3.67E−4	7 7	−2.89E−1	11 3	5.52E−1	2 6	−2.88E+0
9 7	7.83E−5	13 3	3.34E−2	3 12	−1.62E−7	7 8	−6.86E−3	11 4	2.37E−2	3 1	−2.16E−3
9 8	−9.87E−4	13 4	1.76E−3	3 13	−7.65E−2	7 9	−8.19E−3	11 5	2.40E−4	3 2	−9.29E−1
9 9	−8.37E−3	13 5	1.08E−4	3 14	−1.08E+0	7 10	−1.41E−2	11 6	2.51E−4	3 3	−4.39E−1
9 10	−6.87E−1	13 6	2.69E−5	4 1	1.82E+0	7 11	−2.15E−1	11 7	4.15E−5	3 4	−1.65E−1
9 11	−4.02E−1	13 7	3.81E−2	4 2	3.35E−1	7 12	−1.83E−2	11 8	1.80E−3	3 5	−9.29E−4
9 12	−7.19E−2	13 8	1.25E−2	4 3	2.01E−1	7 13	−3.87E−5	11 9	2.73E−4	3 6	−5.27E−1
9 13	−1.17E−3	13 9	1.02E−3	4 4	−6.00E−3	7 14	−9.31E−3	11 10	7.44E−4	4 1	2.52E−3
9 14	−1.37E−2	13 10	1.18E−3	4 5	−1.52E−4	8 1	8.48E−2	11 11	3.31E−4	4 2	2.17E−2
10 1	7.24E−3	13 11	9.27E−4	4 6	−7.59E−3	8 2	7.44E−2	11 12	1.56E−7	4 3	−9.39E−1
10 2	5.63E−4	13 12	2.09E−3	4 7	−1.41E−2	8 3	6.90E−1	11 13	7.97E−3	4 4	−9.18E−3
10 3	1.05E−5	13 13	1.53E−2	4 8	−1.03E−2	8 4	3.67E−1	11 14	−4.96E−2	4 5	−5.43E−5
10 4	9.02E−1	13 14	−4.54E−1	4 9	−3.06E−1	8 5	8.67E−1	12 1	2.73E−2	4 6	−8.27E−3
10 5	3.36E−2		**$^2D^e$–$^2D^o$**	4 10	−6.19E−2	8 6	8.28E−4	12 2	2.73E−5	5 1	2.31E−3
10 6	3.88E−4	1 1	7.31E−1	4 11	−4.41E−4	8 7	−2.68E−4	12 3	1.14E−4	5 2	5.18E−3
10 7	1.43E−4	1 2	−1.05E−7	4 12	−1.50E−4	8 8	−2.00E−1	12 4	6.51E−4	5 3	2.74E−4
10 8	3.67E−4	1 3	−1.96E−2	4 13	−3.99E−2	8 9	−2.00E−1	12 5	8.68E−4	5 4	−1.26E+0
10 9	7.65E−3	1 4	−7.32E−1	4 14	−8.59E−2	8 10	−1.82E−2	12 6	9.56E−2	5 5	−7.00E−4
10 10	1.47E−3	1 5	−4.44E−1	5 1	2.20E−1	8 11	−3.13E−3	12 7	7.13E−4	5 6	−2.37E−2
10 11	−8.13E−1	1 6	−1.72E+0	5 2	3.03E−3	8 12	−1.05E+0	12 8	5.42E−1	6 1	3.45E−3
10 12	−1.61E−2	1 7	−2.24E+0	5 3	2.38E−2	8 13	−1.52E−3	12 9	2.38E−1	6 2	1.16E−4
10 13	−1.11E−3	1 8	−1.38E+0	5 4	−1.33E−5	8 14	−1.45E−5	12 10	4.21E−5	6 3	3.89E−2
10 14	−1.56E−3	1 9	−1.15E+0	5 5	−1.28E−4	9 1	3.15E−2	12 11	1.31E−1	6 4	−4.26E−4
11 1	7.03E−2	1 10	−1.38E−1	5 6	−1.40E−4	9 2	2.06E−2	12 12	1.13E−1	6 5	−3.09E−2
11 2	3.22E−1	1 11	−2.87E−2	5 7	−7.41E−3	9 3	1.05E−1	12 13	1.18E−5	6 6	−2.45E−4
11 3	1.84E−1	1 12	−4.32E−1	5 8	−1.87E−9	9 4	4.02E−2	12 14	−2.45E−6	7 1	3.28E−1
11 4	8.07E−3	1 13	−3.53E−6	5 9	−4.31E−2	9 5	2.65E−3	13 1	3.09E−1	7 2	3.50E−2
11 5	8.42E−4	1 14	−3.61E−5	5 10	−5.29E−1	9 6	2.82E−5	13 2	3.65E−1	7 3	1.03E−1
11 6	9.10E−5	2 1	4.98E−1	5 11	−4.30E−5	9 7	1.35E−3	13 3	1.91E−1	7 4	4.42E−6
11 7	3.27E−3	2 2	−1.19E−1	5 12	−6.02E−4	9 8	−1.62E−3	13 4	5.50E−3	7 5	−2.54E−2
11 8	6.16E−4	2 3	−8.10E−1	5 13	−5.59E−3	9 9	−3.56E−2	13 5	1.61E−3	7 6	−2.72E−2
11 9	1.23E−4	2 4	−7.52E−2	5 14	−6.30E−2	9 10	−2.57E−1	13 6	1.17E−3	8 1	8.20E−1

N-like Ca (Ca^{13+})

$i\ i'$	gf_L	$i\ i'$	gf_L	$i\ i'$	gf_L	$i\ i'$	gf_L	$i\ i'$	gf_L	$i\ i'$	gf_L
8 2	1.32E−1	2 4	−9.41E−4	9 7	−3.76E−3	4 2	−1.04E−5	6 2	4.35E−3	4 10	−1.51E−3
8 3	2.05E−2	2 5	−3.44E−3	10 1	4.19E−4	4 3	−2.21E−3	6 3	−6.94E−1	4 11	−5.84E−3
8 4	4.95E−2	2 6	−4.00E−6	10 2	7.38E−3	4 4	−2.96E−1		$^2G^e$–$^2G^o$	5 1	2.35E+0
8 5	−1.04E−1	2 7	−1.36E+0	10 3	5.23E−3	4 5	−5.89E−2	1 1	−2.77E−3	5 2	3.59E−3
8 6	−1.07E−2	3 1	1.16E−2	10 4	9.41E−3	4 6	−2.09E+0	1 2	−1.65E+0	5 3	1.92E−1
9 1	2.13E−1	3 2	−1.68E−1	10 5	8.90E−4	4 7	−6.75E−4	1 3	−9.15E−4	5 4	8.93E−2
9 2	3.93E−1	3 3	−3.44E−2	10 6	−1.35E+0	5 1	8.83E−4	2 1	1.55E−2	5 5	3.53E−1
9 3	1.34E−3	3 4	−2.25E−2	10 7	−1.11E−5	5 2	2.13E−1	2 2	1.94E+0	5 6	3.82E−3
9 4	7.60E−2	3 5	−7.12E−3	11 1	6.26E−1	5 3	2.28E−1	2 3	−7.42E−8	5 7	−3.29E−4
9 5	−4.44E−2	3 6	−3.77E−5	11 2	8.44E−4	5 4	6.56E−3		$^4S^o$–$^4P^e$	5 8	−1.02E−2
9 6	−1.64E−2	3 7	−1.20E−2	11 3	7.65E−5	5 5	4.02E−3	1 1	−5.05E−1	5 9	−4.32E−4
10 1	6.25E−3	4 1	9.07E−2	11 4	2.07E−3	5 6	4.22E−3	1 2	−4.27E−1	5 10	−6.21E−2
10 2	8.59E−1	4 2	−1.51E−2	11 5	5.30E−3	5 7	−3.06E−4	1 3	−7.16E+0	5 11	−1.24E−2
10 3	3.32E−3	4 3	−9.08E−3	11 6	6.04E−7	6 1	6.67E−2	1 4	−1.10E+0	6 1	1.44E−1
10 4	1.47E−2	4 4	−1.87E−1	11 7	−1.85E+0	6 2	1.32E−3	1 5	−4.08E−2	6 2	2.01E−3
10 5	−2.32E−2	4 5	−6.77E−4	12 1	2.52E−5	6 3	6.37E−4	1 6	−8.40E−2	6 3	4.45E−4
10 6	−1.78E−3	4 6	−1.05E−5	12 2	4.58E−1	6 4	1.05E−3	1 7	−4.09E−1	6 4	1.56E−2
11 1	2.35E−3	4 7	−1.00E−1	12 3	2.66E−1	6 5	1.58E−4	1 8	−7.46E−3	6 5	4.47E−1
11 2	2.32E−3	5 1	2.43E−2	12 4	8.85E−2	6 6	1.85E−3	1 9	−6.33E−2	6 6	1.47E−1
11 3	1.13E+0	5 2	−3.99E−3	12 5	3.47E−3	6 7	−4.63E−3	1 10	−6.81E−2	6 7	2.14E−1
11 4	3.95E−4	5 3	−2.72E−3	12 6	5.45E−3		$^2F^o$–$^2G^e$	1 11	−1.51E+0	6 8	6.65E−1
11 5	−1.60E+0	5 4	−1.63E−3	12 7	−8.23E−5	1 1	−1.76E+0	2 1	9.96E−2	6 9	−1.59E−5
11 6	−3.55E−3	5 5	−2.53E−1	13 1	4.12E−3	1 2	−6.26E−4	2 2	3.67E−1	6 10	−7.16E−1
12 1	2.69E−5	5 6	−5.71E−4	13 2	3.15E−3	2 1	4.82E−5	2 3	−4.19E−1	6 11	−5.50E−3
12 2	2.46E−3	5 7	−8.33E−3	13 3	5.79E−3	2 2	−2.37E−3	2 4	−5.33E−1	7 1	6.95E−4
12 3	1.47E−2	6 1	1.37E−3	13 4	2.30E−4	3 1	2.87E−3	2 5	−3.87E−2	7 2	3.12E−1
12 4	3.08E−1	6 2	−1.03E+0	13 5	5.81E−4	3 2	−1.02E−3	2 6	−1.04E−1	7 3	1.27E−1
12 5	−6.95E−6	6 3	−3.81E−2	13 6	2.46E−5	4 1	2.32E−1	2 7	−2.24E−1	7 4	1.55E−2
12 6	−1.23E−2	6 4	−2.23E−2	13 7	−2.91E−4	4 2	−6.08E−2	2 8	−9.77E−3	7 5	3.00E−5
13 1	2.34E−1	6 5	−1.11E−4		$^2F^e$–$^2F^o$	5 1	1.85E+0	2 9	−2.61E−1	7 6	5.09E−3
13 2	6.92E−2	6 6	−2.43E−4	1 1	6.84E−2	5 2	−1.04E−1	2 10	−3.88E−3	7 7	2.49E−3
13 3	1.19E−3	6 7	−2.22E−3	1 2	−2.10E−1	6 1	5.38E−4	2 11	−1.29E+0	7 8	3.02E−6
13 4	6.90E−4	7 1	7.02E−4	1 3	−4.92E−1	6 2	−1.80E+0	3 1	3.25E−1	7 9	7.44E−1
13 5	−3.28E−4	7 2	−1.18E−3	1 4	−1.03E−1	7 1	3.61E−1	3 2	1.53E−1	7 10	−1.28E−3
13 6	−2.24E+0	7 3	−1.55E+0	1 5	−2.18E−2	7 2	4.25E−5	3 3	−4.14E−2	7 11	−9.34E−1
14 1	5.01E−2	7 4	−1.93E−6	1 6	−1.19E−3		$^2F^e$–$^2G^o$	3 4	−6.08E−1		$^4S^e$–$^4P^o$
14 2	7.26E−3	7 5	−2.68E−3	1 7	−5.81E−3	1 1	−1.63E−1	3 5	−2.54E−1	1 1	3.23E−1
14 3	3.79E−4	7 6	−1.49E−4	2 1	3.29E−1	1 2	−1.11E−2	3 6	−1.33E−1	1 2	3.27E−1
14 4	1.51E−6	7 7	−1.57E−5	2 2	−1.55E−2	1 3	−8.66E+0	3 7	−1.39E−1	1 3	−3.09E−2
14 5	2.12E−4	8 1	3.05E−1	2 3	−6.57E−2	2 1	−4.84E−2	3 8	−7.08E−4	1 4	−7.02E−1
14 6	2.94E−3	8 2	2.22E−3	2 4	−6.47E−1	2 2	−5.72E−2	3 9	−7.44E−2	1 5	−1.54E−1
	$^2D^e$–$^2F^o$	8 3	−9.03E−4	2 5	−6.62E−2	2 3	−3.72E+0	3 10	−4.11E−5	1 6	−2.19E−8
1 1	−1.91E−2	8 4	−7.15E−1	2 6	−1.88E−3	3 1	−1.95E+0	3 11	−4.58E−1		$^1P^o$–$^1P^o$
1 2	−4.31E+0	8 5	−7.20E−2	2 7	−1.15E−1	3 2	−1.39E−2	4 1	1.16E−1	1 1	−3.45E−2
1 3	−2.68E+0	8 6	−1.85E−1	3 1	6.78E−5	3 3	−1.22E−1	4 2	7.97E−1	1 2	−5.58E−1
1 4	−5.86E+0	8 7	−1.68E−2	3 2	−6.39E−1	4 1	7.75E−4	4 3	1.33E−3	1 3	−6.61E+0
1 5	8.48E−1	9 1	1.23E+0	3 3	8.70E−7	4 2	1.73E+0	4 4	7.25E−1	1 4	7.31E−1
1 6	−5.07E−1	9 2	3.43E−5	3 4	−2.91E−2	4 3	−1.27E−5	4 5	6.29E−4	1 5	−5.46E−1
1 7	−8.86E−3	9 3	3.84E−4	3 5	−3.26E−3	5 1	1.33E+0	4 6	−4.15E−4	1 6	−9.29E−3
2 1	−1.26E+0	9 4	−1.54E−2	3 6	−1.93E−3	5 2	7.96E−4	4 7	−8.97E−1	2 1	−1.01E+0
2 2	−5.47E−3	9 5	−1.27E+0	3 7	−4.28E−4	5 3	−2.00E−4	4 8	−9.72E−1	2 2	−9.39E−1
2 3	−3.23E−4	9 6	−3.86E−1	4 1	1.72E+0	6 1	3.45E−3	4 9	−4.45E−4	2 3	−8.60E−4

N-like Ca (Ca^{13+})

i i'	gf$_L$	i i'	gf$_L$	i i'	gf$_L$	i i'	gf$_L$	i i'	gf$_L$	i i'	gf$_L$
2 4	−2.23E−4	11 1	1.16E+0	2 3	−1.03E−2	7 9	−1.44E−1	2 5	−8.94E−3	5 1	7.23E−1
2 5	−2.21E−4	11 2	1.67E−2	2 4	−4.43E−4	8 1	4.08E−3	2 6	−2.70E−4	5 2	5.40E−2
2 6	−1.03E+0	11 3	1.48E−2	2 5	−6.43E−4	8 2	1.38E+0	2 7	−1.81E−1	5 3	−9.30E−2
3 1	6.12E−1	11 4	1.12E−2	2 6	−4.84E−3	8 3	1.28E−2	2 8	−2.08E−7	5 4	−4.10E−2
3 2	−1.72E−4	11 5	2.07E−3	2 7	−2.10E−4	8 4	8.25E−3	2 9	−3.76E−3	6 1	2.02E+0
3 3	−3.91E−1	11 6	1.03E+0	2 8	−1.86E+0	8 5	7.26E−3	3 1	7.61E−1	6 2	1.98E−2
3 4	−8.77E−2		$^6\mathbf{S}^o$–$^6\mathbf{P}^e$	2 9	−2.34E−2	8 6	1.63E−3	3 2	1.27E−1	6 3	−3.75E−1
3 5	−6.51E−4	1 1	−1.78E+0	3 1	9.53E−3	8 7	−2.28E+0	3 3	2.27E−3	6 4	−7.61E−2
3 6	−2.19E−1		$^4\mathbf{P}^o$–$^4\mathbf{D}^e$	3 2	−2.26E−4	8 8	−8.84E−5	3 4	−6.60E−3	7 1	6.55E−3
4 1	1.12E−3	1 1	−1.39E+0	3 3	−3.09E−4	8 9	−3.51E−3	3 5	−5.64E−1	7 2	2.11E+0
4 2	−8.07E−3	1 2	−4.19E−1	3 4	−1.97E−1	9 1	9.87E−1	3 6	−6.56E−4	7 3	−3.11E+0
4 3	−7.55E−2	1 3	−3.06E−1	3 5	−2.48E−1	9 2	3.50E−3	3 7	−7.51E−1	7 4	−2.54E−2
4 4	−6.75E−4	1 4	−1.01E−4	3 6	−2.43E−1	9 3	7.28E−4	3 8	−1.03E−4	8 1	5.63E−1
4 5	−5.37E−3	1 5	−3.69E+0	3 7	−1.78E−4	9 4	5.86E−4	3 9	−3.14E−2	8 2	4.91E−4
4 6	−3.65E−3	2 1	2.10E−4	3 8	−1.09E−3	9 5	2.53E−3	4 1	1.54E−4	8 3	−8.72E−3
5 1	3.47E−1	2 2	−1.79E−2	3 9	−1.19E+1	9 6	5.34E−3	4 2	8.84E−4	8 4	−4.26E+0
5 2	1.17E−2	2 3	−1.95E+0	4 1	2.47E−3	9 7	7.00E−6	4 3	3.23E−2	9 1	9.57E−2
5 3	−6.97E−1	2 4	−3.50E−3	4 2	−5.14E−3	9 8	−2.65E+0	4 4	4.51E−1	9 2	1.50E−3
5 4	−9.45E−2	2 5	−5.30E−2	4 3	−2.27E+0	9 9	−1.04E−2	4 5	3.69E−2	9 3	2.25E−4
5 5	−4.09E−1	3 1	2.92E−2	4 4	−2.76E−2	10 1	5.36E−4	4 6	1.35E+0	9 4	8.10E−3
5 6	−1.64E−2	3 2	2.71E−1	4 5	−3.24E−3	10 2	1.13E−3	4 7	2.95E−1		$^4\mathbf{D}^e$–$^4\mathbf{F}^o$
6 1	7.14E−2	3 3	6.50E−2	4 6	−6.39E−3	10 3	2.52E−3	4 8	1.80E−7	1 1	−2.30E−1
6 2	1.50E+0	3 4	−4.98E−2	4 7	−1.66E−4	10 4	3.49E−1	4 9	−3.01E−5	1 2	−2.47E−1
6 3	−5.81E−4	3 5	−2.39E−2	4 8	−1.64E−3	10 5	4.38E−1	5 1	1.20E+0	2 1	−1.75E+0
6 4	−3.83E−1	4 1	7.18E−1	4 9	−1.34E−2	10 6	8.17E−1	5 2	1.88E−2	2 2	−5.93E−3
6 5	−2.44E−1	4 2	5.55E−2	5 1	1.65E−3	10 7	1.56E−1	5 3	4.61E−5	3 1	−4.45E−3
6 6	−3.76E−3	4 3	1.45E−2	5 2	1.65E+0	10 8	8.79E−5	5 4	6.65E−3	3 2	−2.72E+0
7 1	7.22E−1	4 4	−1.27E−1	5 3	3.09E−3	10 9	−3.15E−3	5 5	3.30E−3	4 1	1.57E+0
7 2	7.91E−3	4 5	−2.45E−2	5 4	−5.20E−1	11 1	6.97E−2	5 6	2.39E−2	4 2	1.60E−1
7 3	1.89E−3	5 1	1.72E−4	5 5	−2.81E−2	11 2	1.68E−3	5 7	3.67E−4	5 1	8.39E−4
7 4	−7.40E−5	5 2	1.24E+0	5 6	−1.55E−2	11 3	1.42E−4	5 8	8.42E−1	5 2	1.28E−2
7 5	−9.00E−1	5 3	8.03E−2	5 7	−7.31E−3	11 4	8.35E−4	5 9	−1.98E−1		$^4\mathbf{F}^e$–$^4\mathbf{F}^o$
7 6	−8.16E−3	5 4	−1.49E+0	5 8	−2.14E−2	11 5	1.76E−2		$^6\mathbf{P}^e$–$^6\mathbf{D}^o$	1 1	−7.77E−1
8 1	4.60E−3	5 5	−1.40E−5	5 9	−2.09E−3	11 6	6.54E−3	1 1	−2.86E+0	1 2	−6.02E−1
8 2	4.36E−1	6 1	3.58E−1	6 1	4.98E−1	11 7	2.18E−4		$^4\mathbf{D}^o$–$^4\mathbf{F}^e$	2 1	−7.32E−1
8 3	6.07E−3	6 2	6.99E−5	6 2	1.67E−3	11 8	2.89E−2	1 1	−2.45E+0	2 2	−1.20E−1
8 4	8.88E−4	6 3	4.36E−4	6 3	5.15E−4	11 9	−8.01E−1	1 2	−7.12E−1	3 1	1.91E−1
8 5	−8.61E−1	6 4	8.61E−7	6 4	−1.43E−3		$^4\mathbf{D}^e$–$^4\mathbf{D}^o$	1 3	−4.13E−2	3 2	8.21E−1
8 6	−3.49E−5	6 5	−2.39E+0	6 5	−1.44E+0	1 1	4.97E−1	1 4	−6.95E+0	4 1	2.52E−3
9 1	5.79E−1		$^4\mathbf{P}^e$–$^4\mathbf{D}^o$	6 6	−1.93E−2	1 2	−1.16E−4	2 1	5.01E−4	4 2	1.11E−2
9 2	2.18E−3	1 1	−3.34E−3	6 7	−1.77E−1	1 3	−6.12E−3	2 2	−2.82E+0		$^4\mathbf{F}^e$–$^4\mathbf{G}^o$
9 3	6.48E−4	1 2	−1.08E+0	6 8	−1.05E−2	1 4	−5.28E−1	2 3	−2.65E−3	1 1	−4.27E−1
9 4	9.66E−4	1 3	−4.59E+0	6 9	−4.77E−2	1 5	−3.74E−2	2 4	−1.09E−1	1 2	−2.38E+1
9 5	3.48E−5	1 4	−5.79E+0	7 1	1.38E+0	1 6	−9.11E−1	3 1	6.97E−4	2 1	−3.45E+0
9 6	−1.60E+0	1 5	−5.93E+0	7 2	1.96E−3	1 7	−3.43E−3	3 2	−7.61E−4	2 2	−2.76E−1
10 1	7.71E−4	1 6	−1.02E+0	7 3	3.53E−3	1 8	−1.60E−1	3 3	−5.54E−4	3 1	2.76E+0
10 2	1.83E−3	1 7	−7.83E−1	7 4	1.01E−5	1 9	−2.34E+0	3 4	−7.53E−4	3 2	−2.71E−2
10 3	1.82E−1	1 8	−8.56E−3	7 5	−6.93E−4	2 1	5.14E−1	4 1	4.14E−2	4 1	2.83E−4
10 4	2.05E−1	1 9	−6.76E−3	7 6	−1.61E+0	2 2	1.68E+0	4 2	1.00E−1	4 2	−2.56E+0
10 5	6.74E−1	2 1	−1.62E+0	7 7	−1.03E+0	2 3	−2.70E−5	4 3	−1.98E−4		
10 6	1.64E−3	2 2	−9.16E−1	7 8	−4.07E−3	2 4	−1.34E+0	4 4	−2.58E−2		

N-like Fe (Fe^{19+})

Term energies relative to $2s^22p^2$ ^{3}P ionization threshold for each symmetry

i	E(Ryds)	Description	i	E(Ryds)	Description	i	E(Ryds)	Description	i	E(Ryds)	Description
	^{2}S^e		3	−45.8544	$2s^22p^2$ ^{3}P $3d$	4	−46.3326	$2s^22p^2$ ^{1}D $3p$		**^{2}D^e**	
1	−105.977	$2s2p^4$	4	−44.2478	$2s^22p^2$ ^{1}D $3d$	5	−45.8735	$2s^22p^2$ ^{1}S $3p$	1	−107.484	$2s2p^4$
2	−47.9730	$2s^22p^2$ ^{1}S $3s$	5	−42.5204	$2s2p^3$ ^{3}D $3p$	6	−42.7495	$2s2p^3$ ^{3}P $3s$	2	−48.9884	$2s^22p^2$ ^{1}D $3s$
3	−44.1184	$2s^22p^2$ ^{1}D $3d$	6	−41.1699	$2s2p^3$ ^{3}P $3p$	7	−40.3613	$2s2p^3$ ^{1}P $3s$	3	−44.9441	$2s^22p^2$ ^{3}P $3d$
4	−40.4196	$2s2p^3$ ^{3}P $3p$	7	−39.7253	$2s2p^3$ ^{3}S $3p$	8	−39.7027	$2s2p^3$ ^{3}D $3d$	4	−44.5232	$2s^22p^2$ ^{1}D $3d$
5	−38.0357	$2s2p^3$ ^{1}P $3p$	8	−38.5213	$2s2p^3$ ^{1}D $3p$	9	−38.0515	$2s2p^3$ ^{3}P $3d$	5	−43.5566	$2s^22p^2$ ^{1}S $3d$
6	−33.5724	$2p^4$ ^{1}S $3s$	9	−38.1456	$2s2p^3$ ^{1}P $3p$	10	−36.8715	$2s2p^3$ ^{1}D $3d$	6	−41.5903	$2s2p^3$ ^{3}D $3p$
7	−31.4083	$2p^4$ ^{1}D $3d$	10	−36.0282	$2p^4$ ^{3}P $3s$	11	−36.0391	$2s2p^3$ ^{1}P $3d$	7	−41.0248	$2s2p^3$ ^{3}P $3p$
8	−25.4094	$2s^22p^2$ ^{1}S $4s$	11	−32.2984	$2p^4$ ^{3}P $3d$	12	−34.5459	$2p^4$ ^{3}P $3p$:	8	−39.1476	$2s2p^3$ ^{1}D $3p$
9	−24.1792	$2s^22p^2$ ^{1}D $4d$	12	−30.9306	$2p^4$ ^{1}D $3d$	13	−32.7335	$2p^4$ ^{1}D $3p$	9	−38.4490	$2s2p^3$ ^{1}P $3p$
10	−19.6539	$2s2p^3$ ^{3}P $4p$	13	−27.0835	$2s^22p^2$ ^{3}P $4s$	14	−31.6919	$2p^4$ ^{1}S $3p$	10	−35.4470	$2p^4$ ^{1}D $3s$
11	−17.1705	$2s2p^3$ ^{1}P $4p$	14	−25.5446	$2s^22p^2$ ^{3}P $4d$	15	−26.1438	$2s^22p^2$ ^{3}P $4p$	11	−32.1608	$2p^4$ ^{3}P $3d$
12	−15.4837	$2s^22p^2$ ^{1}S $5s$	15	−24.2316	$2s^22p^2$ ^{1}D $4d$	16	−25.0469	$2s^22p^2$ ^{1}D $4p$	12	−31.0005	$2p^4$ ^{1}D $3d$
13	−14.7972	$2s^22p^2$ ^{1}D $5d$	16	−20.6256	$2s2p^3$ ^{3}D $4p$	17	−24.5762	$2s^22p^2$ ^{1}S $4p$	13	−29.6156	$2p^4$ ^{1}S $3d$
14	−11.9497	$2p^4$ ^{1}S $4s$	17	−19.9962	$2s2p^3$ ^{3}P $4p$	18	−23.8790	$2s^22p^2$ ^{1}D $4f$	14	−25.9946	$2s^22p^2$ ^{1}D $4s$
15	−11.3988	$2p^4$ ^{1}D $4d$	18	−19.4108	$2s2p^3$ ^{3}D $4f$	19	−20.6794	$2s2p^3$ ^{3}P $4s$	15	−25.2003	$2s^22p^2$ ^{3}P $4d$
16	−10.3645	$2s2p^3$ ^{3}P $5p$	19	−17.8797	$2s2p^3$ ^{3}S $4p$:	20	−19.6646	$2s2p^3$ ^{3}D $4d$	16	−24.3306	$2s^22p^2$ ^{1}D $4d$
17	−10.2015	$2s^22p^2$ ^{1}S $6s$	20	−17.5076	$2s2p^3$ ^{1}D $4p$	21	−18.7955	$2s2p^3$ ^{3}P $4d$	17	−23.6791	$2s^22p^2$ ^{1}S $4d$
18	−9.78947	$2s^22p^2$ ^{1}D $6d$	21	−17.1544	$2s2p^3$ ^{1}P $4p$	22	−18.0156	$2s2p^3$ ^{1}P $4s$	18	−20.3150	$2s2p^3$ ^{3}D $4p$
19	−7.76249	$2s2p^3$ ^{1}P $5p$	22	−16.9935	$2s^22p^2$ ^{3}P $5s$	23	−16.8724	$2s2p^3$ ^{1}D $4d$	19	−19.9130	$2s2p^3$ ^{3}P $4p$
20	−7.04069	$2s^22p^2$ ^{1}S $7s$	23	−16.5691	$2s2p^3$ ^{1}D $4f$	24	−16.5544	$2s^22p^2$ ^{3}P $5p$	20	−19.4174	$2s2p^3$ ^{3}D $4f$
21	−6.78508	$2s^22p^2$ ^{1}D $7d$	24	−16.2627	$2s^22p^2$ ^{3}P $5d$	25	−16.2721	$2s2p^3$ ^{1}P $4d$	21	−18.5748	$2s2p^3$ ^{3}P $4f$
22	−5.32955	$2s2p^3$ ^{3}P $6p$	25	−14.8209	$2s^22p^2$ ^{1}D $5d$	26	−15.2399	$2s^22p^2$ ^{1}D $5p$	22	−17.7610	$2s2p^3$ ^{1}D $4p$
23	−5.01473	$2s^22p^2$ ^{1}S $8s$	26	−13.4894	$2p^4$ ^{3}P $4s$	27	−15.0826	$2s^22p^2$ ^{1}S $5p$	23	−17.2350	$2s2p^3$ ^{1}P $4p$
24	−4.84141	$2s^22p^2$ ^{1}D $8d$	27	−11.9794	$2p^4$ ^{3}P $4d$	28	−14.6091	$2s^22p^2$ ^{1}D $5f$	24	−16.5811	$2s2p^3$ ^{1}D $4f$
25	−3.63390	$2s^22p^2$ ^{1}S $9s$	28	−11.7076	$2s^22p^2$ ^{3}P $6s$	29	−12.7007	$2p^4$ ^{3}P $4p$	25	−16.1136	$2s^22p^2$ ^{3}P $5d$
26	−3.51117	$2s^22p^2$ ^{1}D $9d$	29	−11.2887	$2p^4$ ^{1}D $4d$	30	−12.0050	$2p^4$ ^{1}D $4p$	26	−15.7637	$2s2p^3$ ^{1}P $4f$
27	−2.73807	$2s2p^3$ ^{1}P $6p$	30	−11.2610	$2s^22p^2$ ^{3}P $6d$	31	−11.4325	$2s^22p^2$ ^{3}P $6p$	27	−15.6789	$2s^22p^2$ ^{1}D $5s$
	^{2}S^o		31	−10.7306	$2s2p^3$ ^{3}D $5p$	32	−11.1868	$2p^4$ ^{1}D $4f$	28	−14.8625	$2s^22p^2$ ^{1}D $5d$
1	−48.5342	$2s^22p^2$ ^{3}P $3p$	32	−10.5221	$2s2p^3$ ^{3}P $5p$	33	−11.1165	$2p^4$ ^{1}S $4p$	29	−14.6787	$2s^22p^2$ ^{1}S $5d$
2	−41.1724	$2s2p^3$ ^{3}S $3s$	33	−10.0992	$2s2p^3$ ^{3}D $5f$	34	−10.8676	$2s2p^3$ ^{3}P $5s$	30	−14.6656	$2s^22p^2$ ^{1}D $5g$
3	−40.0591	$2s2p^3$ ^{3}D $3d$	34	−9.80375	$2s^22p^2$ ^{1}D $6d$	35	−10.2698	$2s2p^3$ ^{3}D $5d$	31	−13.0019	$2p^4$ ^{1}D $4s$
4	−36.7371	$2s2p^3$ ^{1}D $3d$	35	−8.53661	$2s^22p^2$ ^{3}P $7s$	36	−10.0425	$2s^22p^2$ ^{1}D $6p$	32	−11.8989	$2p^4$ ^{3}P $4d$
5	−34.3464	$2p^4$ ^{3}P $3p$	36	−8.26055	$2s^22p^2$ ^{3}P $7d$	37	−9.97297	$2s^22p^2$ ^{1}S $6p$	33	−11.3116	$2p^4$ ^{1}D $4d$
6	−26.5427	$2s^22p^2$ ^{3}P $4p$	37	−8.02262	$2s2p^3$ ^{3}S $5p$:	38	−9.96194	$2s2p^3$ ^{3}P $5d$	34	−11.1788	$2s^22p^2$ ^{3}P $6d$
7	−19.7818	$2s2p^3$ ^{3}D $4d$	38	−7.85142	$2s2p^3$ ^{1}D $5p$	39	−9.69070	$2s^22p^2$ ^{1}D $6f$	35	−10.6147	$2s2p^3$ ^{3}D $5p$
8	−18.5454	$2s2p^3$ ^{3}S $4s$	39	−7.76032	$2s2p^3$ ^{1}P $5p$	40	−8.36462	$2s^22p^2$ ^{3}P $7p$	36	−10.4814	$2s2p^3$ ^{3}P $5p$
9	−16.8330	$2s2p^3$ ^{1}D $4d$	40	−7.35702	$2s2p^3$ ^{1}D $5f$	41	−8.17748	$2s2p^3$ ^{1}P $5s$	37	−10.3012	$2s^22p^2$ ^{1}D $6s$
10	−16.7269	$2s^22p^2$ ^{3}P $5p$	41	−6.79464	$2s^22p^2$ ^{1}D $7d$	42	−7.53820	$2s2p^3$ ^{1}D $5d$	38	−10.1889	$2p^4$ ^{1}S $4d$
11	−12.7531	$2p^4$ ^{3}P $4p$	42	−6.49560	$2s^22p^2$ ^{3}P $8s$	43	−7.35583	$2s2p^3$ ^{1}P $5d$	39	−10.1009	$2s2p^3$ ^{3}D $5f$
12	−11.5407	$2s^22p^2$ ^{3}P $6p$	43	−6.31351	$2s^22p^2$ ^{3}P $8d$	44	−6.94032	$2s^22p^2$ ^{1}D $7p$	40	−9.88978	$2s2p^3$ ^{3}P $5f$
13	−10.3138	$2s2p^3$ ^{3}D $5d$	44	−5.52973	$2s2p^3$ ^{3}D $6p$	45	6.80810	$2s^22p^2$ ^{1}S $7p$	41	0.82550	$2s^22p^2$ ^{1}D $6d$
14	−8.47561	$2s^22p^2$ ^{3}P $7p$	45	−5.42206	$2s2p^3$ ^{3}P $6p$	46	−6.72514	$2s^22p^2$ ^{1}D $7f$	42	−9.73481	$2s^22p^2$ ^{1}S $6d$
15	−8.30408	$2s2p^3$ ^{3}S $5s$	46	−5.17354	$2s2p^3$ ^{3}D $6f$	47	−6.38225	$2s^22p^2$ ^{3}P $8p$	43	−9.71784	$2s^22p^2$ ^{1}D $6g$
16	−7.52199	$2s2p^3$ ^{1}D $5d$	47	−5.10849	$2s^22p^2$ ^{3}P $9s$	48	−5.62303	$2s2p^3$ ^{3}P $6s$	44	−8.20636	$2s^22p^2$ ^{3}P $7d$
17	6.42570	$2s^22p^2$ ^{3}P $8p$	48	−4.98415	$2s^22p^2$ ^{3}P $9d$	49	−5.26003	$2s2p^3$ ^{3}D $6d$	45	−7.97075	$2s2p^3$ ^{1}D $5p$
18	−5.28414	$2s2p^3$ ^{3}D $6d$	49	−4.84615	$2s^22p^2$ ^{1}D $8d$	50	−5.10338	$2s2p^3$ ^{3}P $6d$	46	−7.79652	$2s2p^3$ ^{1}P $5p$
19	−5.06084	$2s^22p^2$ ^{3}P $9p$		**^{2}P^o**		51	−5.03191	$2s^22p^2$ ^{3}P $9p$	47	−7.30515	$2s2p^3$ ^{1}D $5f$
	^{2}P^e		1	−113.903	$2s^22p^3$	52	−4.94142	$2s^22p^2$ ^{1}D $8p$	48	−7.15896	$2s2p^3$ ^{1}P $5f$
1	−105.626	$2s2p^4$	2	−99.6753	$2p^5$	53	−4.91753	$2s^22p^2$ ^{1}S $8p$	49	−7.09380	$2s^22p^2$ ^{1}D $7s$
2	−49.6912	$2s^22p^2$ ^{3}P $3s$	3	−47.6199	$2s^22p^2$ ^{3}P $3p$	54	−4.80202	$2s^22p^2$ ^{1}D $8f$	50	−6.80777	$2s^22p^2$ ^{1}D $7d$

N-like Fe (Fe^{19+})

i	E(Ryds)	Description
51	−6.75332	$2s^22p^2$ ^{1}S $7d$
52	−6.72632	$2s^22p^2$ ^{1}D $7g$
53	−6.27891	$2s^22p^2$ ^{3}P $8d$
54	−5.45429	$2s2p^3$ ^{3}D $6p$
55	−5.39814	$2s2p^3$ ^{3}P $6p$
56	−5.17422	$2s2p^3$ ^{3}D $6f$
57	−5.05044	$2s^22p^2$ ^{1}D $8s$
58	−5.04790	$2s2p^3$ ^{3}P $6f$
59	−4.95942	$2s^22p^2$ ^{3}P $9d$
60	−4.85609	$2s^22p^2$ ^{1}D $8d$
61	−4.81892	$2s^22p^2$ ^{1}S $8d$
62	−4.80218	$2s^22p^2$ ^{1}D $8g$
	^{2}D^o	
1	−114.637	$2s^22p^3$
2	−47.8765	$2s^22p^2$ ^{3}P $3p$
3	−46.6458	$2s^22p^2$ ^{1}D $3p$
4	−43.6914	$2s2p^3$ ^{3}D $3s$
5	−41.3382	$2s2p^3$ ^{1}D $3s$
6	−39.5928	$2s2p^3$ ^{3}D $3d$
7	−38.6559	$2s2p^3$ ^{3}P $3d$
8	−37.4202	$2s2p^3$ ^{3}S $3d$:
9	−36.5579	$2s2p^3$ ^{1}D $3d$
10	−35.9341	$2s2p^3$ ^{1}P $3d$
11	−34.5234	$2p^4$ ^{3}P $3p$
12	−33.4709	$2p^4$ ^{1}D $3p$
13	−26.2478	$2s^22p^2$ ^{3}P $4p$
14	−25.1517	$2s^22p^2$ ^{1}D $4p$
15	−25.0629	$2s^22p^2$ ^{3}P $4f$
16	−23.9498	$2s^22p^2$ ^{1}D $4f$
17	−21.2302	$2s2p^3$ ^{3}D $4s$
18	−19.5624	$2s2p^3$ ^{3}D $4d$
19	−19.0449	$2s2p^3$ ^{3}P $4d$
20	−18.5709	$2s2p^3$ ^{1}D $4s$
21	−17.0599	$2s2p^3$ ^{3}S $4d$:
22	−16.7503	$2s2p^3$ ^{1}D $4d$
23	−16.5630	$2s^22p^2$ ^{3}P $5p$
24	−16.2709	$2s2p^3$ ^{1}P $4d$
25	−16.0281	$2s^22p^2$ ^{3}P $5f$
26	−15.2786	$2s^22p^2$ ^{1}D $5p$
27	−14.6395	$2s^22p^2$ ^{1}D $5f$
28	−12.8115	$2p^4$ ^{3}P $4p$
29	−12.1931	$2p^4$ ^{1}D $4p$
30	−11.8800	$2p^4$ ^{3}P $4f$
31	−11.4606	$2s^22p^2$ ^{3}P $6p$
32	−11.1364	$2s^22p^2$ ^{3}P $6f$
33	−11.1058	$2p^4$ ^{1}D $4f$
34	−11.0663	$2s2p^3$ ^{3}D $5s$
35	−10.2306	$2s2p^3$ ^{3}D $5d$
36	−10.1621	$2s2p^3$ ^{3}D $5g$
37	−10.0756	$2s2p^3$ ^{3}P $5d$
38	−10.0654	$2s^22p^2$ ^{1}D $6p$
39	−9.70857	$2s^22p^2$ ^{1}D $6f$
40	−8.40302	$2s^22p^2$ ^{3}P $7p$
41	−8.35414	$2s2p^3$ ^{1}D $5s$
42	−8.18165	$2s^22p^2$ ^{3}P $7f$
43	−7.61525	$2s2p^3$ ^{3}S $5d$:
44	−7.46060	$2s2p^3$ ^{1}D $5d$
45	−7.39072	$2s2p^3$ ^{1}D $5g$
46	−7.35165	$2s2p^3$ ^{1}P $5d$
47	−6.95374	$2s^22p^2$ ^{1}D $7p$
48	−6.73622	$2s^22p^2$ ^{1}D $7f$
49	−6.39366	$2s^22p^2$ ^{3}P $8p$
50	−6.26082	$2s^22p^2$ ^{3}P $8f$
51	−5.71480	$2s2p^3$ ^{3}D $6s$
52	−5.23880	$2s2p^3$ ^{3}D $6d$
53	−5.19407	$2s2p^3$ ^{3}D $6g$
54	−5.16330	$2s2p^3$ ^{3}P $6d$
55	−5.03999	$2s^22p^2$ ^{3}P $9p$
56	−4.95120	$2s^22p^2$ ^{1}D $8p$
57	−4.94603	$2s^22p^2$ ^{3}P $9f$
58	−4.80938	$2s^22p^2$ ^{1}D $8f$
	^{2}F^e	
1	−45.6565	$2s^22p^2$ ^{3}P $3d$
2	−44.4202	$2s^22p^2$ ^{1}D $3d$
3	−42.2356	$2s2p^3$ ^{3}D $3p$
4	−39.5237	$2s2p^3$ ^{1}D $3p$
5	−32.4632	$2p^4$ ^{3}P $3d$
6	−31.4271	$2p^4$ ^{1}D $3d$
7	−25.3927	$2s^22p^2$ ^{3}P $4d$
8	−24.3767	$2s^22p^2$ ^{1}D $4d$
9	−20.5883	$2s2p^3$ ^{3}D $4p$
10	−19.4135	$2s2p^3$ ^{3}D $4f$
11	−18.6093	$2s2p^3$ ^{3}P $4f$
12	−17.8260	$2s2p^3$ ^{1}D $4p$
13	−16.6267	$2s2p^3$ ^{3}S $4f$
14	−16.5785	$2s2p^3$ ^{1}D $4f$
15	−16.1867	$2s^22p^2$ ^{3}P $5d$
16	−15.9698	$2s^22p^2$ ^{3}P $5g$
17	−15.7814	$2s2p^3$ ^{1}P $4f$
18	−14.8936	$2s^22p^2$ ^{1}D $5d$
19	−14.6904	$2s^22p^2$ ^{1}D $5g$
20	−12.0995	$2p^4$ ^{3}P $4d$
21	−11.4083	$2p^4$ ^{1}D $4d$
22	−11.2190	$2s^22p^2$ ^{3}P $6d$
23	−11.1156	$2s^22p^2$ ^{3}P $6g$
24	−10.7296	$2s2p^3$ ^{3}D $5p$
25	−10.0993	$2s2p^3$ ^{3}D $5f$
26	−9.91139	$2s2p^3$ ^{3}P $5f$
27	−9.84418	$2s^22p^2$ ^{1}D $6d$
28	−9.72910	$2s^22p^2$ ^{1}D $6g$
29	−8.23073	$2s^22p^2$ ^{3}P $7d$
30	−8.16808	$2s^22p^2$ ^{3}P $7g$
31	−8.00449	$2s2p^3$ ^{1}D $5p$
32	−7.37158	$2s2p^3$ ^{3}S $5f$:
33	−7.35168	$2s2p^3$ ^{1}D $5f$
34	−7.17126	$2s2p^3$ ^{1}P $5f$
35	−6.82069	$2s^22p^2$ ^{1}D $7d$
36	−6.73369	$2s^22p^2$ ^{1}D $7g$
37	−6.29443	$2s^22p^2$ ^{3}P $8d$
38	−6.25119	$2s^22p^2$ ^{3}P $8g$
39	−5.52035	$2s2p^3$ ^{3}D $6p$
40	−5.17014	$2s2p^3$ ^{3}D $6f$
41	−5.13945	$2s2p^3$ ^{3}D $6h$
42	−5.06158	$2s2p^3$ ^{3}P $6f$
43	−4.97312	$2s^22p^2$ ^{3}P $9d$
44	−4.93811	$2s^22p^2$ ^{3}P $9g$
45	−4.86140	$2s^22p^2$ ^{1}D $8d$
46	−4.80650	$2s^22p^2$ ^{1}D $8g$
	^{2}F^o	
1	−47.0639	$2s^22p^2$ ^{1}D $3p$
2	−39.2641	$2s2p^3$ ^{3}D $3d$
3	−38.6272	$2s2p^3$ ^{3}P $3d$
4	−37.2083	$2s2p^3$ ^{1}D $3d$
5	−36.2749	$2s2p^3$ ^{1}P $3d$
6	−33.8202	$2p^4$ ^{1}D $3p$
7	−25.2197	$2s^22p^2$ ^{1}D $4p$
8	−24.9617	$2s^22p^2$ ^{3}P $4f$
9	−24.0275	$2s^22p^2$ ^{1}D $4f$
10	−23.1448	$2s^22p^2$ ^{1}S $4f$
11	−19.4973	$2s2p^3$ ^{3}D $4d$
12	−18.9757	$2s2p^3$ ^{3}P $4d$
13	−16.9869	$2s2p^3$ ^{1}D $4d$
14	−16.3574	$2s2p^3$ ^{1}P $4d$
15	−15.9892	$2s^22p^2$ ^{3}P $5f$
16	−15.2974	$2s^22p^2$ ^{1}D $5p$
17	−14.6746	$2s^22p^2$ ^{1}D $5f$
18	−14.4529	$2s^22p^2$ ^{1}S $5f$
19	−12.2975	$2p^4$ ^{1}D $4p$
20	−11.9714	$2p^4$ ^{3}P $4f$
21	−11.1076	$2s^22p^2$ ^{3}P $6f$
22	−11.0591	$2p^4$ ^{1}D $4f$
23	−10.1963	$2s2p^3$ ^{3}D $5d$
24	−10.1696	$2s2p^3$ ^{3}D $5g$
25	−10.0783	$2s^22p^2$ ^{1}D $6p$
26	−10.0540	$2s2p^3$ ^{3}P $5d$
27	−9.81793	$2s2p^3$ ^{3}P $5g$
28	−9.72811	$2s^22p^2$ ^{1}D $6f$
29	−9.67148	$2s^22p^2$ ^{1}D $6h$
30	−9.59814	$2s^22p^2$ ^{1}S $6f$
31	−9.50179	$2p^4$ ^{1}S $4f$
32	−8.16150	$2s^22p^2$ ^{3}P $7f$
33	−7.58533	$2s2p^3$ ^{1}D $5d$
34	−7.40438	$2s2p^3$ ^{1}P $5d$
35	−7.38329	$2s2p^3$ ^{1}D $5g$
36	−7.03416	$2s2p^3$ ^{1}P $5g$
37	−6.95748	$2s^22p^2$ ^{1}D $7p$
38	−6.74880	$2s^22p^2$ ^{1}D $7f$
39	−6.71669	$2s^22p^2$ ^{1}D $7h$
40	−6.66510	$2s^22p^2$ ^{1}S $7f$
41	−6.24899	$2s^22p^2$ ^{3}P $8f$
42	−5.21637	$2s2p^3$ ^{3}D $6d$
43	−5.19544	$2s2p^3$ ^{3}D $6g$
44	−5.15085	$2s2p^3$ ^{3}P $6d$
45	−5.01526	$2s2p^3$ ^{3}P $6g$
46	−4.95816	$2s^22p^2$ ^{1}D $8p$
47	−4.93725	$2s^22p^2$ ^{3}P $9f$
48	−4.81755	$2s^22p^2$ ^{1}D $8f$
49	−4.80057	$2s^22p^2$ ^{1}D $8h$
50	−4.76075	$2s^22p^2$ ^{1}S $8f$
	^{2}G^e	
1	−44.8039	$2s^22p^2$ ^{1}D $3d$
2	−31.8726	$2p^4$ ^{1}D $3d$
3	−24.4216	$2s^22p^2$ ^{1}D $4d$
4	−19.4273	$2s2p^3$ ^{3}D $4f$
5	−18.6096	$2s2p^3$ ^{3}P $4f$
6	−16.6020	$2s2p^3$ ^{1}D $4f$
7	−15.9844	$2s^22p^2$ ^{3}P $5g$
8	−15.7794	$2s2p^3$ ^{1}P $4f$
9	−14.9049	$2s^22p^2$ ^{1}D $5d$
10	−14.7105	$2s^22p^2$ ^{1}D $5g$
11	−14.3452	$2s^22p^2$ ^{1}S $5g$
12	−11.5421	$2p^4$ ^{1}D $4d$
13	−11.1052	$2s^22p^2$ ^{3}P $6g$
14	−10.1054	$2s2p^3$ ^{3}D $5f$
15	−9.90960	$2s2p^3$ ^{3}P $5f$
16	−9.84739	$2s^22p^2$ ^{1}D $6d$
17	−9.73760	$2s^22p^2$ ^{1}D $6g$
18	−9.54820	$2s^22p^2$ ^{1}S $6g$
19	−8.15831	$2s^22p^2$ ^{3}P $7g$
20	−7.37320	$2s2p^3$ ^{1}D $5f$
21	−7.17170	$2s2p^3$ ^{1}P $5f$
22	−6.82194	$2s^22p^2$ ^{1}D $7d$
23	−6.74052	$2s^22p^2$ ^{1}D $7g$
24	−6.69399	$2s^22p^2$ ^{1}D $7i$
25	−6.64016	$2s^22p^2$ ^{1}S $7g$
26	−6.24732	$2s^22p^2$ ^{3}P $8g$
27	−5.17282	$2s2p^3$ ^{3}D $6f$
28	−5.13922	$2s2p^3$ ^{3}D $6h$
29	−5.07664	$2s2p^3$ ^{3}P $6h$
30	−5.06001	$2s2p^3$ ^{3}P $6f$
31	−4.93401	$2s^22p^2$ ^{3}P $9g$
32	−4.86532	$2s^22p^2$ ^{1}D $8d$
33	−4.81021	$2s^22p^2$ ^{1}D $8g$
34	−4.78154	$2s^22p^2$ ^{1}D $8i$
35	−4.74839	$2s^22p^2$ ^{1}S $8g$
	^{2}G^o	
1	−39.9529	$2s2p^3$ ^{3}D $3d$
2	−37.3641	$2s2p^3$ ^{1}D $3d$

N-like Fe (Fe^{19+})

i	E(Ryds)	Description	i	E(Ryds)	Description	i	E(Ryds)	Description	i	E(Ryds)	Description
3	−25.0510	$2s^22p^2$ ^{3}P $4f$	13	−12.7499	$2p^4$ ^{3}P $4p$	34	−5.49163	$2s2p^3$ ^{3}D $6p$	17	−8.26820	$2s^22p^2$ ^{3}P $7d$
4	−24.0483	$2s^22p^2$ ^{1}D $4f$	14	−11.4646	$2s^22p^2$ ^{3}P $6p$	35	−5.42251	$2s2p^3$ ^{3}P $6p$	18	−6.32027	$2s^22p^2$ ^{3}P $8d$
5	−19.7663	$2s2p^3$ ^{3}D $4d$	15	−10.2579	$2s2p^3$ ^{3}D $5d$	36	−5.17671	$2s2p^3$ ^{3}D $6f$	19	−5.55166	$2s2p^3$ ^{3}D $6p$
6	−17.0383	$2s2p^3$ ^{1}D $4d$	16	−9.05510	$2s2p^3$ ^{5}S $6s$	37	−5.11568	$2s^22p^2$ ^{3}P $9s$	20	−5.45064	$2s2p^3$ ^{3}P $6p$
7	−16.0298	$2s^22p^2$ ^{3}P $5f$	17	−8.42224	$2s2p^3$ ^{3}S $5s$	38	−4.97834	$2s^22p^2$ ^{3}P $9d$	21	−5.18575	$2s2p^3$ ^{3}D $6f$
8	−14.6831	$2s^22p^2$ ^{1}D $5f$	18	−8.35072	$2s^22p^2$ ^{3}P $7p$			**^{4}P^o**	22	−5.06728	$2s2p^3$ ^{3}P $6f$
9	−11.9298	$2p^4$ ^{3}P $4f$	19	−6.39818	$2s^22p^2$ ^{3}P $8p$	1	−48.1300	$2s^22p^2$ ^{3}P $3p$	23	−4.98711	$2s^22p^2$ ^{3}P $9d$
10	−11.1306	$2s^22p^2$ ^{3}P $6f$	20	−5.89882	$2s2p^3$ ^{5}S $7s$	2	−43.3860	$2s2p^3$ ^{3}P $3s$			**^{4}D^o**
11	−11.1128	$2s^22p^2$ ^{3}P $6h$	21	−5.25272	$2s2p^3$ ^{3}D $6d$	3	−39.8733	$2s2p^3$ ^{3}D $3d$	1	−48.2733	$2s^22p^2$ ^{3}P $3p$
12	−11.0552	$2p^4$ ^{1}D $4f$	22	−5.03986	$2s^22p^2$ ^{3}P $9p$	4	−39.2764	$2s2p^3$ ^{3}P $3d$	2	−44.3655	$2s2p^3$ ^{3}D $3s$
13	−10.3182	$2s2p^3$ ^{3}D $5d$			**^{6}S^o**	5	−35.0822	$2p^4$ ^{3}P $3p$	3	−42.3750	$2s2p^3$ ^{5}S $3d$
14	−10.1755	$2s2p^3$ ^{3}D $5g$	1	−47.4188	$2s2p^3$ ^{5}S $3s$	6	−26.4111	$2s^22p^2$ ^{3}P $4p$	4	−40.0990	$2s2p^3$ ^{3}D $3d$
15	−9.81854	$2s2p^3$ ^{3}P $5g$	2	−24.6019	$2s2p^3$ ^{5}S $4s$	7	−20.8878	$2s2p^3$ ^{3}P $4s$	5	−39.0479	$2s2p^3$ ^{3}P $3d$
16	−9.73342	$2s^22p^2$ ^{1}D $6f$	3	−14.4908	$2s2p^3$ ^{5}S $5s$	8	−19.7345	$2s2p^3$ ^{3}D $4d$	6	−37.5127	$2s2p^3$ ^{3}S $3d$
17	−9.67334	$2s^22p^2$ ^{1}D $6h$	4	−9.12798	$2s2p^3$ ^{5}S $6s$	9	−19.2179	$2s2p^3$ ^{3}P $4d$	7	−34.7617	$2p^4$ ^{3}P $3p$
18	−8.17638	$2s^22p^2$ ^{3}P $7f$	5	−5.94388	$2s2p^3$ ^{5}S $7s$	10	−16.6892	$2s^22p^2$ ^{3}P $5p$	8	−26.4594	$2s^22p^2$ ^{3}P $4p$
19	−8.16547	$2s^22p^2$ ^{3}P $7h$	6	−3.89969	$2s2p^3$ ^{5}S $8s$	11	−12.9928	$2p^4$ ^{3}P $4p$	9	−25.0711	$2s^22p^2$ ^{3}P $4f$
20	−7.61162	$2s2p^3$ ^{1}D $5d$	7	−2.50959	$2s2p^3$ ^{5}S $9s$	12	−11.5154	$2s^22p^2$ ^{3}P $6p$	10	−22.6176	$2s2p^3$ ^{5}S $4d$
21	−7.39524	$2s2p^3$ ^{3}S $5g$			**^{4}P^e**	13	−10.9672	$2s2p^3$ ^{3}P $5s$	11	−21.4495	$2s2p^3$ ^{3}D $4s$
22	−7.38654	$2s2p^3$ ^{1}D $5g$	1	−109.789	$2s2p^4$	14	−10.2999	$2s2p^3$ ^{3}D $5d$	12	−19.8448	$2s2p^3$ ^{3}D $4d$
23	−7.04538	$2s2p^3$ ^{1}P $5g$	2	−50.1086	$2s^22p^2$ ^{3}P $3s$	15	−10.1729	$2s2p^3$ ^{3}P $5d$	13	−19.1444	$2s2p^3$ ^{3}P $4d$
24	−6.75224	$2s^22p^2$ ^{1}D $7f$	3	−45.5351	$2s^22p^2$ ^{3}P $3d$	16	−8.40885	$2s^22p^2$ ^{3}P $7p$	14	−17.1026	$2s2p^3$ ^{3}S $4d$
25	−6.71948	$2s^22p^2$ ^{1}D $7h$	4	−44.8166	$2s2p^3$ ^{5}S $3p$	17	−6.41524	$2s^22p^2$ ^{3}P $8p$	15	−16.6879	$2s^22p^2$ ^{3}P $5p$
26	−6.25875	$2s^22p^2$ ^{3}P $8f$	5	−41.7579	$2s2p^3$ ^{3}D $3p$	18	−5.67646	$2s2p^3$ ^{3}P $6s$	16	−16.0411	$2s^22p^2$ ^{3}P $5f$
27	−6.25042	$2s^22p^2$ ^{3}P $8h$	6	−41.1017	$2s2p^3$ ^{3}P $3p$	19	−5.27702	$2s2p^3$ ^{3}D $6d$	17	−13.5095	$2s2p^3$ ^{5}S $5d$
28	−5.28470	$2s2p^3$ ^{3}D $6d$	7	−39.6795	$2s2p^3$ ^{3}S $3p$	20	−5.21845	$2s2p^3$ ^{3}P $6d$	18	−12.8994	$2p^4$ ^{3}P $4p$
29	−5.19565	$2s2p^3$ ^{3}D $6g$	8	−36.5755	$2p^4$ ^{3}P $3s$	21	−5.05039	$2s^22p^2$ ^{3}P $9p$	19	−11.9032	$2p^4$ ^{3}P $4f$
30	−5.01853	$2s2p^3$ ^{3}P $6g$	9	−32.4516	$2p^4$ ^{3}P $3d$			**^{6}P^e**	20	−11.5270	$2s^22p^2$ ^{3}P $6p$
31	−4.94447	$2s^22p^2$ ^{3}P $9f$	10	−27.2180	$2s^22p^2$ ^{3}P $4s$	1	−45.4917	$2s2p^3$ ^{5}S $3p$	21	−11.1575	$2s2p^3$ ^{3}D $5s$
32	−4.93687	$2s^22p^2$ ^{3}P $9h$	11	−25.4586	$2s^22p^2$ ^{3}P $4d$	2	−23.8105	$2s2p^3$ ^{5}S $4p$	22	−11.1390	$2s^22p^2$ ^{3}P $6f$
33	−4.81978	$2s^22p^2$ ^{1}D $8f$	12	−23.5321	$2s2p^3$ ^{5}S $4p$	3	−14.0928	$2s2p^3$ ^{5}S $5p$	23	−10.3551	$2s2p^3$ ^{3}D $5d$
34	−4.80220	$2s^22p^2$ ^{1}D $8h$	13	−20.4748	$2s2p^3$ ^{3}D $4p$	4	−8.90040	$2s2p^3$ ^{5}S $6p$	24	−10.1650	$2s2p^3$ ^{3}D $5g$
		^{4}S^e	14	−19.9855	$2s2p^3$ ^{3}P $4p$	5	−5.80180	$2s2p^3$ ^{5}S $7p$	25	−10.1348	$2s2p^3$ ^{3}P $5d$
1	−41.6129	$2s2p^3$ ^{3}P $3p$	15	−19.4230	$2s2p^3$ ^{3}D $4f$	6	−3.80514	$2s2p^3$ ^{5}S $8p$	26	−8.57028	$2s2p^3$ ^{5}S $6d$
2	−20.1279	$2s2p^3$ ^{3}P $4p$	16	−17.8883	$2s2p^3$ ^{3}S $4p$	7	−2.44352	$2s2p^3$ ^{5}S $9p$	27	−8.41765	$2s^22p^2$ ^{3}P $7p$
3	−10.5998	$2s2p^3$ ^{3}P $5p$	17	−17.0727	$2s^22p^2$ ^{3}P $5s$			**^{4}D^e**	28	−8.18135	$2s^22p^2$ ^{3}P $7f$
4	−5.45974	$2s2p^3$ ^{3}P $6p$	18	−16.2305	$2s^22p^2$ ^{3}P $5d$	1	−45.8768	$2s^22p^2$ ^{3}P $3d$	29	−7.62263	$2s2p^3$ ^{3}S $5d$
5	−2.38070	$2s2p^3$ ^{3}P $7p$	19	−13.9739	$2s2p^3$ ^{5}S $5p$	2	−42.5887	$2s2p^3$ ^{3}D $3p$	30	−6.41936	$2s^22p^2$ ^{3}P $8p$
6	−.39353	$2s2p^3$ ^{3}P $8p$	20	−13.6060	$2p^4$ ^{3}P $4s$	3	−41.4848	$2s2p^3$ ^{3}P $3p$	31	−6.26248	$2s^22p^2$ ^{3}P $8f$
		^{4}S^o	21	−12.1125	$2p^4$ ^{3}P $4d$	4	−33.0573	$2p^4$ ^{3}P $3d$	32	−5.76814	$2s2p^3$ ^{3}D $6s$
1	−116.059	$2s^22p^3$	22	−11.7445	$2s^22p^2$ ^{3}P $6s$	5	−25.5778	$2s^22p^2$ ^{3}P $4d$	33	−5.60400	$2s2p^3$ ^{5}S $7d$
2	−47.8642	$2s^22p^2$ ^{3}P $3p$	23	−11.2471	$2s^22p^2$ ^{3}P $6d$	6	−20.7084	$2s2p^3$ ^{3}D $4p$	34	−5.30202	$2s2p^3$ ^{3}D $6d$
3	−40.2040	$2s2p^3$ ^{5}S $3s$	24	−10.6876	$2s2p^3$ ^{3}D $5p$	7	−20.0956	$2s2p^3$ ^{3}P $4p$	35	−5.20133	$2s2p^3$ ^{3}P $6d$
4	−41.5101	$2s2p^3$ ^{3}S $3s$	25	−10.5270	$2s2p^3$ ^{3}P $5p$	8	−19.4491	$2s2p^3$ ^{3}D $4f$	36	−5.19106	$2s2p^3$ ^{3}D $6g$
5	−39.5967	$2s2p^3$ ^{3}D $3d$	26	−10.1095	$2s2p^3$ ^{3}D $5f$	9	−18.6213	$2s2p^3$ ^{3}P $4f$	37	−5.05405	$2s^22p^2$ ^{3}P $9p$
6	−34.3434	$2p^4$ ^{3}P $3p$	27	−8.82266	$2s2p^3$ ^{5}S $6p$	10	−16.2908	$2s^22p^2$ ^{3}P $5d$	38	−4.94650	$2s^22p^2$ ^{3}P $9f$
7	−26.2548	$2s^22p^2$ ^{3}P $4p$	28	−8.55457	$2s^22p^2$ ^{3}P $7s$	11	−12.3174	$2p^4$ ^{3}P $4d$			**^{6}D^o**
8	−24.2899	$2s2p^3$ ^{5}S $4s$	29	−8.25007	$2s^22p^2$ ^{3}P $7d$	12	−11.2799	$2s^22p^2$ ^{3}P $6d$	1	−43.4120	$2s2p^3$ ^{5}S $3d$
9	−19.6450	$2s2p^3$ ^{3}D $4d$	30	−8.01280	$2s2p^3$ ^{3}S $5p$	13	−10.7922	$2s2p^3$ ^{3}D $5p$	2	−23.0183	$2s2p^3$ ^{5}S $4d$
10	−18.6227	$2s2p^3$ ^{3}S $4s$	31	−6.51046	$2s^22p^2$ ^{3}P $8s$	14	−10.5801	$2s2p^3$ ^{3}P $5p$	3	−13.7038	$2s2p^3$ ^{5}S $5d$
11	−16.6102	$2s^22p^2$ ^{3}P $5p$	32	−6.30730	$2s^22p^2$ ^{3}P $8d$	15	−10.1256	$2s2p^3$ ^{3}D $5f$	4	−8.68027	$2s2p^3$ ^{5}S $6d$
12	−14.3539	$2s2p^3$ ^{5}S $5s$	33	−5.76347	$2s2p^3$ ^{5}S $7p$	16	−9.92285	$2s2p^3$ ^{3}P $5f$	5	−5.66507	$2s2p^3$ ^{5}S $7d$

N-like Fe (Fe^{19+})

i	E(Ryds)	Description	i	E(Ryds)	Description	i	E(Ryds)	Description	i	E(Ryds)	Description
6	−3.71438	$2s2p^3\ ^5\mathrm{S}\ 8d$			$^4\mathbf{F}^o$	7	−8.15898	$2s^22p^2\ ^3\mathrm{P}\ 7g$	3	−5.55394	$2s2p^3\ ^5\mathrm{S}\ 7g$
7	−2.38020	$2s2p^3\ ^5\mathrm{S}\ 9d$	1	−40.5381	$2s2p^3\ ^3\mathrm{D}\ 3d$	8	−6.24791	$2s^22p^2\ ^3\mathrm{P}\ 8g$	4	−3.64041	$2s2p^3\ ^5\mathrm{S}\ 8g$
		$^4\mathbf{F}^e$	2	−39.3946	$2s2p^3\ ^3\mathrm{P}\ 3d$	9	−5.19539	$2s2p^3\ ^3\mathrm{D}\ 6f$	5	−2.32848	$2s2p^3\ ^5\mathrm{S}\ 9g$
1	−46.0978	$2s^22p^2\ ^3\mathrm{P}\ 3d$	3	−24.9980	$2s^22p^2\ ^3\mathrm{P}\ 4f$	10	−5.13925	$2s2p^3\ ^3\mathrm{D}\ 6h$			
2	−42.4695	$2s2p^3\ ^3\mathrm{D}\ 3p$	4	−19.9609	$2s2p^3\ ^3\mathrm{D}\ 4d$	11	−5.07678	$2s2p^3\ ^3\mathrm{P}\ 6h$			
3	−32.7256	$2p^4\ ^3\mathrm{P}\ 3d$	5	−19.2567	$2s2p^3\ ^3\mathrm{P}\ 4d$	12	−5.07446	$2s2p^3\ ^3\mathrm{P}\ 6f$			
4	−25.6510	$2s^22p^2\ ^3\mathrm{P}\ 4d$	6	−16.0154	$2s^22p^2\ ^3\mathrm{P}\ 5f$	13	−4.93432	$2s^22p^2\ ^3\mathrm{P}\ 9g$			
5	−22.3997	$2s2p^3\ ^5\mathrm{S}\ 4f$	7	−12.0222	$2p^4\ ^3\mathrm{P}\ 4f$			$^4\mathbf{G}^o$			
6	−20.6740	$2s2p^3\ ^3\mathrm{D}\ 4p$	8	−11.1249	$2s^22p^2\ ^3\mathrm{P}\ 6f$	1	−40.3958	$2s2p^3\ ^3\mathrm{D}\ 3d$			
7	−19.4600	$2s2p^3\ ^3\mathrm{D}\ 4f$	9	−10.4073	$2s2p^3\ ^3\mathrm{D}\ 5d$	2	−25.0766	$2s^22p^2\ ^3\mathrm{P}\ 4f$			
8	−18.6280	$2s2p^3\ ^3\mathrm{P}\ 4f$	10	−10.1923	$2s2p^3\ ^3\mathrm{P}\ 5d$	3	−19.9220	$2s2p^3\ ^3\mathrm{D}\ 4d$			
9	−16.6460	$2s2p^3\ ^3\mathrm{S}\ 4f$	11	−10.1683	$2s2p^3\ ^3\mathrm{D}\ 5g$	4	−16.0486	$2s^22p^2\ ^3\mathrm{P}\ 5f$			
10	−16.3198	$2s^22p^2\ ^3\mathrm{P}\ 5d$	12	−9.81911	$2s2p^3\ ^3\mathrm{P}\ 5g$	5	−13.3897	$2s2p^3\ ^5\mathrm{S}\ 5g$			
11	−15.9719	$2s^22p^2\ ^3\mathrm{P}\ 5g$	13	−8.17204	$2s^22p^2\ ^3\mathrm{P}\ 7f$	6	−11.9453	$2p^4\ ^3\mathrm{P}\ 4f$			
12	−13.3988	$2s2p^3\ ^5\mathrm{S}\ 5f$	14	−6.25670	$2s^22p^2\ ^3\mathrm{P}\ 8f$	7	−11.1433	$2s^22p^2\ ^3\mathrm{P}\ 6f$			
13	−12.1995	$2p^4\ ^3\mathrm{P}\ 4d$	15	−5.33470	$2s2p^3\ ^3\mathrm{D}\ 6d$	8	−11.1129	$2s^22p^2\ ^3\mathrm{P}\ 6h$			
14	−11.2970	$2s^22p^2\ ^3\mathrm{P}\ 6d$	16	−5.22744	$2s2p^3\ ^3\mathrm{P}\ 6d$	9	−10.3858	$2s2p^3\ ^3\mathrm{D}\ 5d$			
15	−11.1159	$2s^22p^2\ ^3\mathrm{P}\ 6g$	17	−5.19552	$2s2p^3\ ^3\mathrm{D}\ 6g$	10	−10.1770	$2s2p^3\ ^3\mathrm{D}\ 5g$			
16	−10.7696	$2s2p^3\ ^3\mathrm{D}\ 5p$	18	−5.01636	$2s2p^3\ ^3\mathrm{P}\ 6g$	11	−9.81920	$2s2p^3\ ^3\mathrm{P}\ 5g$			
17	−10.1300	$2s2p^3\ ^3\mathrm{D}\ 5f$	19	−4.94221	$2s^22p^2\ ^3\mathrm{P}\ 9f$	12	−8.50087	$2s2p^3\ ^5\mathrm{S}\ 6g$			
18	−9.92466	$2s2p^3\ ^3\mathrm{P}\ 5f$			$^6\mathbf{F}^e$	13	−8.18386	$2s^22p^2\ ^3\mathrm{P}\ 7f$			
19	−8.50663	$2s2p^3\ ^5\mathrm{S}\ 6f$	1	−22.4494	$2s2p^3\ ^5\mathrm{S}\ 4f$	14	−8.16552	$2s^22p^2\ ^3\mathrm{P}\ 7h$			
20	−8.27881	$2s^22p^2\ ^3\mathrm{P}\ 7d$	2	−13.4311	$2s2p^3\ ^5\mathrm{S}\ 5f$	15	−7.39582	$2s2p^3\ ^3\mathrm{S}\ 5g$			
21	−8.16802	$2s^22p^2\ ^3\mathrm{P}\ 7g$	3	−8.52770	$2s2p^3\ ^5\mathrm{S}\ 6f$	16	−6.26442	$2s^22p^2\ ^3\mathrm{P}\ 8f$			
22	−7.37518	$2s2p^3\ ^3\mathrm{S}\ 5f$	4	−5.57090	$2s2p^3\ ^5\mathrm{S}\ 7f$	17	−6.25044	$2s^22p^2\ ^3\mathrm{P}\ 8h$			
23	−6.32734	$2s^22p^2\ ^3\mathrm{P}\ 8d$	5	−3.65211	$2s2p^3\ ^5\mathrm{S}\ 8f$	18	−5.55300	$2s2p^3\ ^5\mathrm{S}\ 7g$			
24	−6.25140	$2s^22p^2\ ^3\mathrm{P}\ 8g$	6	−2.33685	$2s2p^3\ ^5\mathrm{S}\ 9f$	19	−5.32288	$2s2p^3\ ^3\mathrm{D}\ 6d$			
25	−5.55754	$2s2p^3\ ^5\mathrm{S}\ 7f$			$^4\mathbf{G}^e$	20	−5.19684	$2s2p^3\ ^3\mathrm{D}\ 6g$			
26	−5.54174	$2s2p^3\ ^3\mathrm{D}\ 6p$	1	−19.4796	$2s2p^3\ ^3\mathrm{D}\ 4f$	21	−5.01902	$2s2p^3\ ^3\mathrm{P}\ 6g$			
27	−5.18886	$2s2p^3\ ^3\mathrm{D}\ 6f$	2	−18.6454	$2s2p^3\ ^3\mathrm{P}\ 4f$	22	−4.94794	$2s^22p^2\ ^3\mathrm{P}\ 9f$			
28	−5.13946	$2s2p^3\ ^3\mathrm{D}\ 6h$	3	−15.9852	$2s^22p^2\ ^3\mathrm{P}\ 5g$	23	−4.93691	$2s^22p^2\ ^3\mathrm{P}\ 9h$			
29	−5.06928	$2s2p^3\ ^3\mathrm{P}\ 6f$	4	−11.1063	$2s^22p^2\ ^3\mathrm{P}\ 6g$			$^6\mathbf{G}^o$			
30	−4.99212	$2s^22p^2\ ^3\mathrm{P}\ 9d$	5	−10.1411	$2s2p^3\ ^3\mathrm{D}\ 5f$	1	−13.3910	$2s2p^3\ ^5\mathrm{S}\ 5g$			
31	−4.93808	$2s^22p^2\ ^3\mathrm{P}\ 9g$	6	−9.93211	$2s2p^3\ ^3\mathrm{P}\ 5f$	2	−8.50208	$2s2p^3\ ^5\mathrm{S}\ 6g$			

N-like Fe (Fe^{19+})

Energies in ascending order from ground state for terms with effective $n \leq 4.0$, $L \leq 4$

Term	i	E(Ryds)	Term	i	E(Ryds)	Term	i	E(Ryds)	Term	i	E(Ryds)	Term	i	E(Ryds)
$^4S^o$	1	0.00000	$^6P^e$	1	70.5673	$^4F^o$	1	75.5209	$^2F^o$	4	78.8506	$^2S^e$	7	84.6507
$^2D^o$	1	1.42200	$^2D^e$	3	71.1149	$^2S^e$	4	75.6394	$^2P^o$	10	79.1875	$^2D^e$	12	85.0585
$^2P^o$	1	2.15600	$^4P^e$	4	71.2424	$^4G^o$	1	75.6631	$^2S^o$	4	79.3219	$^2P^e$	12	85.1284
$^4P^e$	1	6.27000	$^2G^e$	1	71.2551	$^2P^o$	7	75.6976	$^4P^e$	8	79.4835	$^2D^e$	13	86.4434
$^2D^e$	1	8.57500	$^2D^e$	4	71.5358	$^4D^o$	4	75.9600	$^2D^o$	9	79.5011	$^4P^e$	10	88.8410
$^2S^e$	1	10.0820	$^2F^e$	2	71.6387	$^2S^o$	3	75.9998	$^2F^o$	5	79.7841	$^2P^e$	13	88.9754
$^2P^e$	1	10.4330	$^4D^o$	2	71.6935	$^2G^o$	1	76.1060	$^2P^o$	11	80.0199	$^2S^o$	6	89.5163
$^2P^o$	2	16.3837	$^2P^e$	4	71.8112	$^4P^o$	3	76.1857	$^2P^e$	10	80.0308	$^4D^o$	8	89.5995
$^4P^e$	2	65.9503	$^2S^e$	3	71.9406	$^2P^e$	7	76.3336	$^2D^o$	10	80.1248	$^4P^o$	6	89.6479
$^2P^e$	2	66.3678	$^2D^o$	4	72.3676	$^2P^o$	8	76.3562	$^2D^e$	10	80.6120	$^4S^o$	7	89.8042
$^2D^e$	2	67.0706	$^2D^e$	5	72.5024	$^4P^e$	7	76.3795	$^4P^o$	5	80.9768	$^2D^o$	13	89.8112
$^2S^o$	1	67.5248	$^6D^o$	1	72.6470	$^4S^o$	5	76.4623	$^4D^o$	7	81.2973	$^2P^o$	15	89.9152
$^4D^o$	1	67.7857	$^4P^o$	2	72.6730	$^2D^o$	6	76.4662	$^2P^o$	12	81.5131	$^2D^e$	14	90.0644
$^4P^o$	1	67.9290	$^2P^o$	6	73.3094	$^2F^e$	4	76.5352	$^2D^o$	11	81.5356	$^4F^e$	4	90.4080
$^2S^e$	2	68.0860	$^4D^e$	2	73.4703	$^4F^o$	2	76.6644	$^2S^o$	5	81.7126	$^4D^e$	5	90.4812
$^2D^o$	2	68.1825	$^2P^e$	5	73.5386	$^4P^o$	4	76.7825	$^4S^o$	6	81.7156	$^2P^e$	14	90.5144
$^4S^o$	2	68.1947	$^4F^e$	2	73.5894	$^2F^o$	2	76.7949	$^2F^o$	6	82.2388	$^4P^e$	11	90.6004
$^2P^o$	3	68.4391	$^4D^o$	3	73.6840	$^2D^e$	8	76.9114	$^2S^e$	6	82.4866	$^2S^e$	8	90.6496
$^6S^o$	1	68.6402	$^2F^e$	3	73.8233	$^4D^o$	5	77.0110	$^2D^o$	12	82.5881	$^2F^e$	7	90.6663
$^2F^o$	1	68.9951	$^4P^e$	5	74.3011	$^2D^o$	7	77.4031	$^4D^e$	4	83.0016	$^2F^o$	7	90.8392
$^2D^o$	3	69.4131	$^4S^e$	1	74.4461	$^2F^o$	3	77.4317	$^2P^o$	13	83.3255	$^2D^e$	15	90.8587
$^2P^o$	4	69.7263	$^2D^e$	6	74.4687	$^2P^e$	8	77.5377	$^4F^e$	3	83.3334	$^2D^o$	14	90.9073
$^4S^o$	3	69.8550	$^4S^o$	4	74.5489	$^2D^e$	9	77.6100	$^2F^e$	5	83.5957	$^4G^o$	2	90.9824
$^4F^e$	1	69.9612	$^4D^e$	3	74.5742	$^2P^e$	9	77.9134	$^4P^e$	9	83.6074	$^4D^o$	9	90.9879
$^4D^e$	1	70.1822	$^2D^o$	5	74.7207	$^2P^o$	9	78.0075	$^2P^e$	11	83.7606	$^2D^o$	15	90.9960
$^2P^o$	5	70.1855	$^2S^o$	2	74.8866	$^2S^e$	5	78.0233	$^2D^e$	11	83.8981	$^2G^o$	3	91.0080
$^2P^e$	3	70.2046	$^2P^e$	6	74.8891	$^4D^o$	6	78.5463	$^2G^e$	2	84.1864	$^2P^o$	16	91.0121
$^2F^e$	1	70.4025	$^4P^e$	6	74.9573	$^2D^o$	8	78.6387	$^2P^o$	14	84.3671			
$^4P^e$	3	70.5239	$^2D^e$	7	75.0341	$^2G^o$	2	78.6949	$^2F^e$	6	84.6319			

N-like Fe (Fe^{19+})

gf-values for transitions involving terms with effective $n \leq 4.0$, $L \leq 4$

i i'	gf_L	i i'	gf_L	i i'	gf_L	i i'	gf_L	i i'	gf_L	i i'	gf_L
	$^2S^o$–$^2P^e$	4 9	3.98E−3	2 1	1.15E−1	5 4	1.09E−1	8 7	5.84E−8	3 9	−4.35E−2
1 1	1.56E−7	4 10	−3.33E−4	2 2	8.44E−6	5 5	8.08E−2	8 8	1.26E−6	3 10	−4.62E−3
1 2	7.15E−2	4 11	−5.67E−2	2 3	−2.77E−3	5 6	9.69E−2	8 9	2.22E−4	3 11	−6.29E−3
1 3	−2.29E−1	4 12	−1.07E−1	2 4	−9.40E−3	5 7	1.29E−1	8 10	7.19E−5	3 12	−3.64E−4
1 4	−2.76E−2	4 13	−8.33E−4	2 5	−4.14E−1	5 8	2.22E−3	8 11	4.06E−4	3 13	−1.20E−4
1 5	−1.82E−1	4 14	−4.83E−5	2 6	−1.33E−2	5 9	4.99E−5	8 12	6.71E−6	3 14	−1.33E−4
1 6	−3.47E−2	5 1	2.89E−1	2 7	−2.02E−1	5 10	−5.08E−3	8 13	2.56E−6	3 15	−5.47E−2
1 7	−2.83E−2	5 2	6.23E−4	2 8	−1.53E−4	5 11	−1.71E−1	8 14	1.62E−6	3 16	−8.88E−4
1 8	−1.44E−2	5 3	1.06E−5	2 9	−3.56E−3	5 12	−4.20E−3	8 15	3.13E−3	4 1	2.07E+0
1 9	−5.67E−3	5 4	3.47E−4	2 10	−8.15E−5	5 13	−9.06E−2	8 16	−1.20E−3	4 2	1.21E−4
1 10	−8.59E−4	5 5	8.08E−2	2 11	−8.33E−5	5 14	−9.11E−2		$^2P^e$–$^2P^o$	4 3	7.50E−2
1 11	−2.55E−5	5 6	1.53E−1	2 12	−7.43E−5	5 15	−2.91E−6	1 1	2.54E−1	4 4	2.12E−1
1 12	−2.47E−5	5 7	3.56E−2	2 13	−1.18E−4	5 16	−4.90E−5	1 2	−6.55E−1	4 5	1.73E−3
1 13	−8.23E−2	5 8	2.73E−2	2 14	−4.53E−5	6 1	2.24E−3	1 3	−1.00E−2	4 6	−2.69E−3
1 14	−8.72E−1	5 9	6.41E−2	2 15	−5.00E−3	6 2	1.18E−1	1 4	−6.97E−3	4 7	−9.44E−4
2 1	3.08E−1	5 10	1.46E−1	2 16	−2.01E−3	6 3	1.68E−4	1 5	−4.38E−3	4 8	−1.02E−2
2 2	3.27E−1	5 11	−1.82E−1	3 1	6.50E−1	6 4	4.77E−4	1 6	−2.60E−2	4 9	−1.50E−3
2 3	8.39E−5	5 12	−2.01E−1	3 2	9.92E−5	6 5	3.05E−4	1 7	−1.81E−1	4 10	−1.00E−2
2 4	1.36E−5	5 13	−5.74E−5	3 3	4.82E−2	6 6	3.80E−4	1 8	−3.77E−1	4 11	−1.57E−1
2 5	4.11E−5	5 14	−1.21E−4	3 4	1.08E−1	6 7	4.57E−1	1 9	−8.87E−2	4 12	−1.56E−5
2 6	−9.14E−7	6 1	4.59E−4	3 5	3.69E−3	6 8	1.38E−3	1 10	−2.81E+0	4 13	−8.84E−5
2 7	−1.08E−1	6 2	2.35E−1	3 6	−8.87E−4	6 9	3.62E−3	1 11	−2.37E−1	4 14	−8.58E−5
2 8	−2.03E−1	6 3	6.80E−2	3 7	−2.86E−5	6 10	5.94E−4	1 12	−2.88E−1	4 15	−2.25E−3
2 9	−2.46E−1	6 4	1.14E−2	3 8	−5.34E−4	6 11	2.76E−3	1 13	−3.21E−1	4 16	−9.63E−2
2 10	−1.87E−1	6 5	3.27E−4	3 9	−4.82E−4	6 12	2.36E−3	1 14	−5.48E−2	5 1	4.58E−2
2 11	−4.85E−4	6 6	7.85E−4	3 10	−1.13E−1	6 13	−3.08E−3	1 15	−2.85E−3	5 2	7.03E−3
2 12	−5.17E−6	6 7	1.67E−5	3 11	−9.26E−2	6 14	−4.87E−1	1 16	−8.05E−6	5 3	1.38E−2
2 13	−5.34E−4	6 8	3.90E−4	3 12	−1.00E−4	6 15	−1.15E−6	2 1	2.88E−1	5 4	3.96E−4
2 14	−1.58E−2	6 9	2.73E−5	3 13	−2.80E−4	6 16	−1.11E−6	2 2	2.29E−6	5 5	5.63E−4
3 1	8.84E−2	6 10	2.60E−6	3 14	−7.24E−6	7 1	2.07E−3	2 3	−2.92E−1	5 6	4.42E−3
3 2	1.80E−3	6 11	5.87E−5	3 15	−2.69E−3	7 2	1.19E+0	2 4	−2.38E−1	5 7	−6.74E−5
3 3	5.84E−2	6 12	8.96E−5	3 16	−4.92E−2	7 3	1.46E−5	2 5	−7.85E−2	5 8	−3.38E−1
3 4	6.94E−4	6 13	1.37E−1	4 1	3.29E−1	7 4	2.75E−5	2 6	−1.99E−1	5 9	−1.67E−4
3 5	1.42E−1	6 14	−4.52E−1	4 2	1.60E−2	7 5	2.42E−5	2 7	−5.77E−4	5 10	−2.70E−3
3 6	4.33E−3		$^2S^e$–$^2P^o$	4 3	1.26E−1	7 6	3.16E−4	2 8	−3.03E−3	5 11	−2.48E−5
3 7	−1.87E−4	1 1	1.97E−1	4 4	1.50E−3	7 7	3.33E−3	2 9	−1.11E−3	5 12	−5.53E−2
3 8	−2.82E−4	1 2	−4.19E−2	4 5	4.11E−3	7 8	1.21E−2	2 10	−8.03E−4	5 13	−6.50E−3
3 9	−3.81E−7	1 3	−5.30E−7	4 6	1.54E−1	7 9	2.97E−6	2 11	−2.79E−4	5 14	−2.91E−6
3 10	−4.13E−4	1 4	−8.88E−3	4 7	−2.47E−4	7 10	2.93E−1	2 12	−4.19E−5	5 15	−9.21E−5
3 11	−4.01E−2	1 5	−5.36E−3	4 8	−4.03E−3	7 11	9.03E−3	2 13	−1.84E−5	5 16	−1.39E−4
3 12	−3.11E−4	1 6	−1.94E−1	4 9	−2.32E−1	7 12	7.09E−2	2 14	−2.99E−6	6 1	6.74E−1
3 13	−1.96E−4	1 7	−1.26E−1	4 10	−7.39E−3	7 13	5.41E−2	2 15	−5.69E−1	6 2	5.73E−3
3 14	−4.41E−3	1 8	−2.66E−1	4 11	−2.11E−2	7 14	2.56E−3	2 16	−1.03E−2	6 3	7.91E−2
4 1	9.51E−1	1 9	−2.56E+0	4 12	−4.92E−2	7 15	−1.62E−8	3 1	9.14E−1	6 4	5.54E−2
4 2	5.56E−4	1 10	−3.74E−2	4 13	−5.87E−4	7 16	−1.70E−5	3 2	3.37E−5	6 5	4.40E−2
4 3	6.06E−3	1 11	−2.01E+0	4 14	−5.54E−3	8 1	1.99E−2	3 3	1.21E−1	6 6	2.91E−1
4 4	2.04E−1	1 12	−2.89E−6	4 15	−1.45E−4	8 2	6.64E−8	3 4	1.74E−3	6 7	−9.31E−6
4 5	3.52E−3	1 13	−4.00E−5	4 16	−5.72E−4	8 3	1.52E−2	3 5	6.40E−5	6 8	−2.15E−2
4 6	5.13E−4	1 14	−3.26E−1	5 1	3.42E−3	8 4	1.38E−3	3 6	−2.30E−3	6 9	−3.12E−1
4 7	1.19E−1	1 15	−7.98E−6	5 2	4.36E−6	8 5	2.67E−1	3 7	−1.56E−3	6 10	−1.07E−3
4 8	3.77E−2	1 16	−1.41E−6	5 3	2.09E−2	8 6	4.42E−3	3 8	−1.85E−1	6 11	−2.69E−2

N-like Fe (Fe^{19+})

i	i'	gf_L	i	i'	gf_L	i	i'	gf_L	i	i'	gf_L	i	i'	gf_L	i	i'	gf_L
6	12	−3.31E−3	9	16	−1.41E−3	13	4	1.33E−1	2	8	−7.16E−4	5	15	−5.34E−2	9	7	7.28E−3
6	13	−1.37E−3	10	1	1.36E−2	13	5	2.63E−2	2	9	−5.10E−3	6	1	2.02E−1	9	8	1.21E−4
6	14	−1.22E−3	10	2	5.84E−1	13	6	4.06E−3	2	10	−4.40E−1	6	2	1.59E−2	9	9	1.97E−4
6	15	−2.28E−3	10	3	3.02E−4	13	7	7.35E−4	2	11	−1.77E+0	6	3	1.18E−3	9	10	−1.21E−4
6	16	−6.03E−4	10	4	3.49E−4	13	8	6.71E−4	2	12	−7.09E+0	6	4	6.74E−3	9	11	−3.79E−2
7	1	7.04E−3	10	5	6.37E−5	13	9	5.69E−4	2	13	−1.74E+0	6	5	3.88E−3	9	12	−7.46E−3
7	2	7.06E−3	10	6	2.37E−1	13	10	5.40E−5	2	14	−2.38E−9	6	6	−1.59E−2	9	13	−7.31E−3
7	3	2.71E−1	10	7	2.28E−2	13	11	2.68E−5	2	15	−2.55E−4	6	7	−5.73E−1	9	14	−4.52E−5
7	4	2.41E−2	10	8	8.33E−4	13	12	1.66E−6	3	1	4.46E−3	6	8	−4.78E−2	9	15	−5.73E−3
7	5	9.55E−4	10	9	1.93E−4	13	13	2.39E−5	3	2	5.86E−2	6	9	−4.01E−2	10	1	2.90E−1
7	6	1.63E−4	10	10	3.25E−3	13	14	1.31E−6	3	3	−5.40E−1	6	10	−1.57E−3	10	2	1.34E−3
7	7	6.39E−3	10	11	1.40E−5	13	15	−7.24E−1	3	4	−2.09E−9	6	11	−3.73E−5	10	3	8.48E−2
7	8	−1.34E−5	10	12	−1.72E−1	13	16	−7.15E−2	3	5	−3.17E−3	6	12	−1.23E−3	10	4	4.36E−1
7	9	−1.20E−4	10	13	−5.02E−1	14	1	2.53E−1	3	6	−2.39E−1	6	13	−2.43E−4	10	5	2.41E−4
7	10	−1.73E−1	10	14	−6.25E−2	14	2	3.00E−4	3	7	−2.47E−2	6	14	−5.20E−3	10	6	2.82E−2
7	11	−9.36E−6	10	15	−9.85E−5	14	3	6.10E−1	3	8	−1.02E−2	6	15	−1.22E−3	10	7	5.10E−3
7	12	−5.52E−1	10	16	−4.28E−5	14	4	1.16E−1	3	9	−8.27E−3	7	1	9.19E−2	10	8	1.00E−1
7	13	−1.78E−2	11	1	4.13E−4	14	5	5.68E−2	3	10	−4.47E−4	7	2	6.13E−1	10	9	1.02E−2
7	14	−6.77E−3	11	2	1.31E−1	14	6	5.72E−4	3	11	−1.66E−5	7	3	3.75E−4	10	10	−1.18E−4
7	15	−2.01E−3	11	3	3.62E−5	14	7	2.44E−3	3	12	−1.34E−4	7	4	1.05E−3	10	11	−2.92E−2
7	16	−9.01E−6	11	4	1.37E−5	14	8	2.76E−3	3	13	−1.22E−4	7	5	7.86E−4	10	12	−1.50E−1
8	1	1.20E−2	11	5	3.37E−5	14	9	1.29E−3	3	14	−2.24E−2	7	6	1.97E−3	10	13	−4.82E−3
8	2	1.49E−2	11	6	1.00E−4	14	10	1.06E−5	3	15	−1.50E+0	7	7	1.48E−3	10	14	−2.42E−4
8	3	1.39E−3	11	7	6.30E−4	14	11	1.01E−5	4	1	2.07E−2	7	8	1.03E−3	10	15	−9.46E−5
8	4	5.07E−1	11	8	6.85E−2	14	12	1.56E−5	4	2	4.32E−1	7	9	−6.73E−1	11	1	1.14E−1
8	5	2.12E−2	11	9	6.16E−2	14	13	1.88E−6	4	3	−1.71E−2	7	10	−2.46E−1	11	2	3.31E−4
8	6	3.52E−3	11	10	8.34E−3	14	14	7.36E−6	4	4	−2.20E−1	7	11	−4.78E−4	11	3	2.77E−2
8	7	1.20E−1	11	11	1.32E−2	14	15	2.02E−1	4	5	−2.32E−2	7	12	−1.51E−5	11	4	9.14E−2
8	8	3.39E−7	11	12	2.55E−1	14	16	−4.28E−4	4	6	−2.58E−2	7	13	−2.76E−4	11	5	1.91E−1
8	9	−1.51E−4	11	13	1.45E−4			**$^2P^o$–$^2D^e$**	4	7	−9.60E−2	7	14	−9.41E−6	11	6	1.75E−4
8	10	−8.75E−2	11	14	−1.32E−4	1	1	−1.06E−1	4	8	−8.29E−2	7	15	−5.27E−3	11	7	1.72E−3
8	11	−9.28E−2	11	15	−1.03E−4	1	2	−1.86E−1	4	9	−7.68E−5	8	1	1.29E+0	11	8	1.55E−2
8	12	−4.83E−6	11	16	−4.79E−5	1	3	−9.74E−1	4	10	−9.29E−4	8	2	9.25E−4	11	9	2.22E−4
8	13	−2.26E−1	12	1	1.02E−2	1	4	−3.59E+0	4	11	−8.21E−6	8	3	1.07E−2	11	10	−9.21E−5
8	14	−8.09E−2	12	2	4.95E+0	1	5	−3.31E+0	4	12	−1.63E−7	8	4	6.53E−4	11	11	−1.22E−2
8	15	−4.41E−3	12	3	2.65E−6	1	6	−1.21E−2	4	13	−6.09E−6	8	5	2.63E−4	11	12	−8.16E−2
8	16	−3.88E−3	12	4	4.12E−6	1	7	−1.08E+0	4	14	−3.00E−1	8	6	7.85E−2	11	13	−1.20E−1
9	1	2.14E−1	12	5	7.75E−5	1	8	−7.94E−2	4	15	−4.84E−1	8	7	5.02E−3	11	14	−2.03E−7
9	2	2.83E−2	12	6	7.51E−4	1	9	−2.31E−1	5	1	7.70E−3	8	8	−2.50E−3	11	15	−1.76E−5
9	3	3.88E−2	12	7	4.48E−5	1	10	−3.28E−3	5	2	2.18E−4	8	9	−2.66E−4	12	1	1.62E−1
9	4	7.93E−3	12	8	3.57E−2	1	11	−8.72E−3	5	3	−1.68E−3	8	10	−1.32E−3	12	2	1.39E−4
9	5	1.75E−1	12	9	1.24E−2	1	12	−1.53E−2	5	4	−4.85E−3	8	11	−4.37E−2	12	3	3.05E−6
9	6	1.65E−4	12	10	5.69E−2	1	13	1.17E−2	5	5	−6.69E−1	8	12	−8.92E−3	12	4	4.14E−4
9	7	3.52E−1	12	11	8.08E−2	1	14	−4.82E−2	5	6	−1.09E−2	8	13	−1.05E−3	12	5	2.03E−6
9	8	2.37E−5	12	12	4.76E−2	1	15	−7.47E−1	5	7	−7.10E−2	8	14	−1.19E−4	12	6	2.67E−1
9	9	−6.89E−5	12	13	2.06E−1	2	1	4.88E−1	5	8	−9.25E−4	8	15	−5.26E−3	12	7	1.53E−1
9	10	−3.65E−3	12	14	1.95E−3	2	2	−1.04E−5	5	9	−2.91E−1	9	1	1.27E−1	12	8	1.23E−2
9	11	−8.83E−2	12	15	−1.27E−5	2	3	−8.47E−5	5	10	−1.29E−3	9	2	9.05E−5	12	9	8.44E−6
9	12	−2.52E−2	12	16	−5.79E−5	2	4	−2.39E−6	5	11	−8.57E−7	9	3	2.41E−1	12	10	7.15E−2
9	13	−7.18E−2	13	1	5.14E−2	2	5	−8.10E−6	5	12	−2.01E−5	9	4	3.42E−2	12	11	−5.12E−1
9	14	−2.90E−1	13	2	5.64E−6	2	6	−1.27E−2	5	13	−3.52E−6	9	5	3.61E−3	12	12	−4.92E−3
9	15	−3.39E−3	13	3	1.98E−1	2	7	−8.15E−3	5	14	−4.57E−3	9	6	1.69E−2	12	13	−1.03E−3

N-like Fe (Fe^{19+})

i	i′	gf$_L$	i	i′	gf$_L$	i	i′	gf$_L$	i	i′	gf$_L$	i	i′	gf$_L$	i	i′	gf$_L$
12	14	−1.43E−5	16	6	7.83E−4	3	12	−2.63E−6	7	4	4.57E−3	10	11	−5.84E−1	14	3	3.81E−2
12	15	−4.06E−6	16	7	4.67E−4	3	13	−5.59E−4	7	5	1.72E−1	10	12	−1.33E−2	14	4	1.48E−3
13	1	2.95E−1	16	8	7.11E−3	3	14	−3.71E−2	7	6	−6.74E−6	10	13	−7.48E−5	14	5	6.46E−5
13	2	2.33E−4	16	9	1.54E−3	3	15	−5.55E+0	7	7	−1.49E−3	10	14	−7.27E−6	14	6	6.37E−4
13	3	3.06E−5	16	10	2.06E−7	4	1	6.27E−1	7	8	−3.96E−1	10	15	−2.26E−4	14	7	1.39E−3
13	4	5.21E−5	16	11	6.47E−6	4	2	7.61E−2	7	9	−3.70E−2	11	1	3.99E−3	14	8	1.87E−3
13	5	1.19E−4	16	12	1.70E−5	4	3	5.89E−2	7	10	−6.26E−3	11	2	3.23E−4	14	9	8.61E−4
13	6	3.94E−2	16	13	8.77E−7	4	4	−2.64E−4	7	11	−1.44E−1	11	3	5.59E−7	14	10	7.61E−4
13	7	2.39E−2	16	14	6.99E−1	4	5	−2.25E−4	7	12	−2.50E−2	11	4	2.39E−7	14	11	3.25E−5
13	8	2.52E−1	16	15	1.01E−3	4	6	−2.67E−3	7	13	−4.94E−5	11	5	9.62E−5	14	12	1.16E−6
13	9	1.06E−2			$^2P^e$–$^2D^o$	4	7	−1.18E−2	7	14	−1.32E−5	11	6	1.77E−2	14	13	1.18E−2
13	10	4.29E−1	1	1	7.28E−1	4	8	−5.79E−2	7	15	−2.34E−2	11	7	2.62E−1	14	14	−1.02E−4
13	11	−8.10E−3	1	2	−3.31E−3	4	9	−1.46E−1	8	1	1.10E−1	11	8	2.89E−1	14	15	−3.68E−1
13	12	−2.83E−1	1	3	−6.12E−4	4	10	−3.34E−3	8	2	1.53E−1	11	9	7.23E−3			$^2D^e$–$^2D^o$
13	13	−1.33E−2	1	4	−3.37E−2	4	11	−4.44E−5	8	3	2.36E−2	11	10	2.68E−3	1	1	5.62E−1
13	14	−4.62E−6	1	5	−2.86E−1	4	12	−9.64E−4	8	4	4.46E−3	11	11	3.55E−2	1	2	−5.16E−5
13	15	−7.22E−6	1	6	−2.80E−1	4	13	−9.05E−3	8	5	3.18E−1	11	12	1.74E−2	1	3	−2.10E−2
14	1	2.64E−3	1	7	−1.81E−3	4	14	−6.88E−2	8	6	7.96E−5	11	13	−6.92E−5	1	4	−6.74E−1
14	2	1.13E−4	1	8	−6.80E−2	4	15	−2.85E−1	8	7	1.13E−5	11	14	−1.04E−6	1	5	−4.20E−1
14	3	2.06E−4	1	9	−3.09E+0	5	1	7.52E−1	8	8	−6.19E−3	11	15	−2.19E−4	1	6	−1.58E+0
14	4	3.21E−4	1	10	−6.24E+0	5	2	3.21E−4	8	9	−3.38E−1	12	1	2.27E−3	1	7	−2.47E+0
14	5	2.37E−4	1	11	−4.34E−1	5	3	3.64E−2	8	10	−1.96E−2	12	2	3.67E−6	1	8	−1.62E+0
14	6	1.44E−4	1	12	−5.86E−3	5	4	1.80E−1	8	11	−1.57E−1	12	3	2.89E−4	1	9	−1.41E+0
14	7	5.70E−4	1	13	−7.39E−3	5	5	−1.39E−6	8	12	−2.89E−1	12	4	3.72E−4	1	10	−1.15E−1
14	8	3.26E−4	1	14	−1.27E−3	5	6	−3.93E−1	8	13	−1.88E−3	12	5	1.91E−4	1	11	−3.19E−2
14	9	7.62E−1	1	15	−1.70E−3	5	7	−1.07E−2	8	14	−2.49E−3	12	6	9.79E−3	1	12	−5.13E−1
14	10	8.53E−4	2	1	3.70E−1	5	8	−6.06E−3	8	15	−1.15E−2	12	7	4.01E−2	1	13	−6.64E−7
14	11	3.21E−5	2	2	−4.34E−1	5	9	−1.14E−2	9	1	3.82E−2	12	8	1.14E−1	1	14	−2.85E−3
14	12	−1.21E−4	2	3	−4.02E−1	5	10	−7.93E−5	9	2	1.16E−6	12	9	2.93E−1	1	15	−6.12E−5
14	13	−8.00E−1	2	4	−1.42E−1	5	11	−2.03E−2	9	3	5.30E−1	12	10	1.24E−1	2	1	4.68E−1
14	14	−7.63E−6	2	5	−1.44E−3	5	12	−1.93E−3	9	4	6.98E−3	12	11	7.65E−2	2	2	−9.55E−2
14	15	−4.29E−5	2	6	−1.38E−4	5	13	−2.08E−3	9	5	1.97E−2	12	12	5.17E−2	2	3	−5.82E−1
15	1	3.62E−4	2	7	−2.11E−3	5	14	−2.65E−4	9	6	4.12E−4	12	13	−1.03E−4	2	4	−6.04E−2
15	2	2.54E−2	2	8	−4.31E−5	5	15	−2.33E−2	9	7	5.88E−5	12	14	−1.76E−5	2	5	−1.02E+0
15	3	1.59E−1	2	9	−1.01E−3	6	1	1.15E−1	9	8	−5.18E−4	12	15	−1.98E−4	2	6	−6.16E−3
15	4	1.10E−2	2	10	−2.00E−5	6	2	9.22E−2	9	9	−2.37E−3	13	1	6.68E−2	2	7	−9.55E−4
15	5	9.00E−3	2	11	−6.26E−5	6	3	4.83E−2	9	10	−4.93E−1	13	2	2.98E−1	2	8	−3.33E−5
15	6	1.52E−4	2	12	−5.19E−6	6	4	7.39E−2	9	11	−3.33E−1	13	3	1.74E−1	2	9	−2.23E−5
15	7	8.42E−4	2	13	−1.05E+0	6	5	2.63E−5	9	12	−4.20E−2	13	4	7.71E−3	2	10	−1.40E−4
15	8	2.87E−4	2	14	−2.02E−2	6	6	−3.82E−2	9	13	−5.51E−5	13	5	1.25E−3	2	11	−2.14E−5
15	9	4.62E−5	2	15	−1.98E−3	6	7	−5.48E−1	9	14	−1.42E−3	13	6	2.88E−5	2	12	−9.01E−4
15	10	1.11E−6	3	1	3.11E−1	6	8	−1.66E−2	9	15	−8.44E−3	13	7	1.66E−3	2	13	−4.39E−2
15	11	3.23E−6	3	2	5.42E−4	6	9	−8.69E−3	10	1	1.73E−2	13	8	1.19E−4	2	14	−1.05E+0
15	12	5.22E−5	3	3	3.50E−3	6	10	−5.25E−3	10	2	5.38E−4	13	9	4.42E−5	2	15	−3.35E−3
15	13	5.66E−5	3	4	−9.99E−4	6	11	−3.02E−2	10	3	1.04E−5	13	10	2.69E−4	3	1	3.06E+0
15	14	−6.63E−4	3	5	−3.01E−5	6	12	−7.72E−5	10	4	6.57E−1	13	11	2.02E−5	3	2	5.84E−2
15	15	−9.95E−1	3	6	−1.91E−1	6	13	−3.35E−3	10	5	2.70E−2	13	12	1.09E−5	3	3	8.36E−2
16	1	3.76E−3	3	7	−9.07E−3	6	14	−8.28E−4	10	6	4.55E−5	13	13	−1.07E+0	3	4	−5.65E−3
16	2	5.57E−1	3	8	−7.85E−2	6	15	−1.26E−2	10	7	5.98E−5	13	14	−9.48E−2	3	5	−3.83E−4
16	3	9.62E−3	3	9	−1.48E−2	7	1	1.32E−1	10	8	2.78E−4	13	15	−2.93E−3	3	6	−4.32E−2
16	4	8.22E−2	3	10	−1.46E−2	7	2	4.60E−1	10	9	4.05E−3	14	1	3.46E−2	3	7	−8.44E−2
16	5	2.54E−3	3	11	−1.52E−3	7	3	1.70E−2	10	10	−1.56E−3	14	2	2.08E−2	3	8	−2.75E−1

N-like Fe (Fe^{19+})

i	i'	gf_L	i	i'	gf_L	i	i'	gf_L	i	i'	gf_L	i	i'	gf_L	i	i'	gf_L
3	9	−4.96E−2	7	1	1.28E−1	10	8	1.70E−3	13	15	−1.90E−5	3	7	−5.88E−1	11	3	8.22E−1
3	10	−5.68E−2	7	2	4.23E−1	10	9	2.20E−3	14	1	9.59E−2	4	1	1.33E−3	11	4	4.44E−4
3	11	−1.12E−3	7	3	1.01E−2	10	10	1.34E−5	14	2	1.03E−1	4	2	1.26E−2	11	5	−1.17E+0
3	12	−2.62E−5	7	4	1.52E−2	10	11	−2.52E−3	14	3	4.75E−1	4	3	−6.83E−1	11	6	−4.41E−3
3	13	−5.04E−2	7	5	3.14E−3	10	12	−9.62E−1	14	4	2.13E−2	4	4	−7.59E−3	11	7	−1.46E−4
3	14	−3.17E−2	7	6	−6.22E−3	10	13	−1.85E−5	14	5	3.13E−3	4	5	−6.56E−5	12	1	8.52E−6
3	15	−9.12E−1	7	7	−2.04E−1	10	14	−5.24E−5	14	6	2.89E−5	4	6	−1.20E−3	12	2	9.55E−4
4	1	1.51E+0	7	8	−2.38E−3	10	15	−8.71E−6	14	7	1.17E−5	4	7	−5.74E−3	12	3	1.10E−2
4	2	2.68E−1	7	9	−7.56E−3	11	1	2.61E−2	14	8	5.62E−6	5	1	1.03E−3	12	4	2.21E−1
4	3	1.16E−1	7	10	−9.99E−3	11	2	7.27E−5	14	9	2.07E−3	5	2	4.15E−3	12	5	−7.43E−6
4	4	−2.62E−3	7	11	−1.64E−1	11	3	4.22E−9	14	10	1.32E−3	5	3	1.76E−4	12	6	−7.79E−1
4	5	−1.75E−4	7	12	−1.56E−2	11	4	9.62E−4	14	11	2.07E−4	5	4	−9.32E−1	12	7	−2.46E−6
4	6	−9.58E−3	7	13	−2.26E−5	11	5	3.57E−4	14	12	6.61E−5	5	5	−6.79E−4	13	1	2.03E−1
4	7	−2.17E−2	7	14	−1.19E−3	11	6	8.23E−2	14	13	5.96E−3	5	6	−1.54E−4	13	2	6.95E−2
4	8	−4.95E−4	7	15	−8.69E−3	11	7	1.58E−4	14	14	−1.10E+0	5	7	−1.73E−2	13	3	1.36E−3
4	9	−3.19E−1	8	1	6.67E−2	11	8	4.43E−1	14	15	−6.08E−3	6	1	4.65E−3	13	4	9.70E−4
4	10	−3.95E−2	8	2	5.30E−2	11	9	1.53E−1	15	1	3.22E−1	6	2	6.71E−5	13	5	3.35E−6
4	11	−6.10E−4	8	3	4.79E−1	11	10	1.94E−5	15	2	4.18E−1	6	3	2.23E−2	13	6	1.64E−4
4	12	−1.28E−3	8	4	2.97E−1	11	11	7.66E−2	15	3	2.33E−1	6	4	−6.38E−5	13	7	−1.64E+0
4	13	−3.78E−2	8	5	6.08E−1	11	12	1.02E−1	15	4	5.20E−3	6	5	−4.93E−2	14	1	3.59E−2
4	14	−4.93E−2	8	6	5.40E−4	11	13	−9.76E−7	15	5	1.14E−3	6	6	−4.11E−5	14	2	1.52E−1
4	15	−1.93E−1	8	7	−2.01E−4	11	14	−6.52E−7	15	6	7.36E−4	6	7	−1.02E−5	14	3	3.93E−3
5	1	2.83E−1	8	8	−1.29E−1	11	15	−5.86E−6	15	7	2.19E−3	7	1	2.60E−1	14	4	9.98E−5
5	2	4.14E−3	8	9	−1.38E−1	12	1	3.60E−3	15	8	2.58E−3	7	2	2.68E−2	14	5	4.57E−5
5	3	1.80E−2	8	10	−1.82E−2	12	2	6.45E−5	15	9	2.86E−3	7	3	7.86E−2	14	6	6.92E−5
5	4	4.70E−5	8	11	−2.34E−3	12	3	3.92E−5	15	10	1.52E−3	7	4	2.40E−5	14	7	2.91E−6
5	5	−7.74E−5	8	12	−7.53E−1	12	4	2.30E−4	15	11	3.70E−4	7	5	−1.69E−2	15	1	2.23E−2
5	6	−3.99E−5	8	13	−3.91E−4	12	5	3.07E−4	15	12	0.00E+0	7	6	−2.07E−3	15	2	2.96E−2
5	7	−8.62E−3	8	14	−6.72E−3	12	6	1.79E−2	15	13	3.71E−1	7	7	−1.73E−2	15	3	6.08E−4
5	8	−2.13E−4	8	15	−3.22E−4	12	7	1.81E−2	15	14	−3.62E−4	8	1	6.14E−1	15	4	1.06E−6
5	9	−4.39E−2	9	1	3.84E−2	12	8	5.20E−6	15	15	−1.72E−2	8	2	8.16E−2	15	5	1.05E−4
5	10	−4.01E−1	9	2	1.64E−2	12	9	1.11E−1			$^2D^o$–$^2F^e$	8	3	1.79E−2	15	6	2.27E−6
5	11	−1.50E−4	9	3	8.41E−2	12	10	2.89E−1	1	1	−9.38E−1	8	4	3.28E−2	15	7	2.61E−3
5	12	−4.20E−4	9	4	3.47E−2	12	11	1.16E−1	1	2	−1.19E+1	8	5	−1.00E−1			$^2D^e$–$^2F^o$
5	13	−5.71E−3	9	5	1.60E−3	12	12	4.67E−1	1	3	−1.51E+0	8	6	−2.94E−2	1	1	−1.67E−2
5	14	−1.64E−3	9	6	7.73E−7	12	13	−1.18E−5	1	4	−4.42E−1	8	7	−3.43E−3	1	2	−4.18E+0
5	15	−7.62E−2	9	7	5.20E−4	12	14	−3.56E−5	1	5	−4.41E−2	9	1	1.42E−1	1	3	−2.74E+0
6	1	1.43E+0	9	8	−1.15E−3	12	15	−2.07E−5	1	6	−4.00E−2	9	2	2.43E−1	1	4	−7.15E+0
6	2	5.51E−2	9	9	−3.84E−2	13	1	2.92E−4	1	7	−1.08E+0	9	3	6.36E−4	1	5	−1.02E+0
6	3	1.10E−1	9	10	−1.71E−1	13	2	2.83E−5	2	1	−7.85E−1	9	4	5.99E−2	1	6	−5.72E−1
6	4	7.17E−1	9	11	−1.76E−4	13	3	5.42E−7	2	2	−2.70E−2	9	5	−3.06E−2	1	7	−6.88E−3
6	5	−2.22E−3	9	12	7.26E−2	13	4	2.01E−5	2	3	2.58E−1	9	6	1.50E−1	2	1	6.33E−1
6	6	−2.66E−1	9	13	−7.91E−5	13	5	1.00E−5	2	4	−1.30E−2	9	7	−2.68E−3	2	2	−4.24E−3
6	7	−8.27E−2	9	14	−4.46E−3	13	6	1.27E−3	2	5	−1.71E−6	10	1	4.22E−3	2	3	−3.68E−4
6	8	−6.41E−4	9	15	−2.75E−4	13	7	4.56E−3	2	6	−3.31E−4	10	2	7.03E−1	2	4	−2.20E−4
6	9	−4.78E−2	10	1	1.03E−2	13	8	3.80E−3	2	7	−3.27E+0	10	3	3.26E−3	2	5	−6.77E−4
6	10	−1.19E−2	10	2	1.54E−3	13	9	5.54E−2	3	1	−4.76E−4	10	4	9.20E−3	2	6	−9.70E−7
6	11	−1.53E−1	10	3	5.66E−3	13	10	4.43E−1	3	2	−6.67E−1	10	5	−2.45E−2	2	7	−1.57E+0
6	12	−8.88E−3	10	4	1.67E−3	13	11	4.64E−3	3	3	−3.42E−1	10	6	−5.76E−2	3	1	1.25E−2
6	13	−1.32E−3	10	5	1.29E+0	13	12	5.90E−3	3	4	−1.16E−1	10	7	−9.79E−4	3	2	−1.45E−1
6	14	−1.66E−3	10	6	4.71E−3	13	13	−1.45E−5	3	5	−1.83E−4	11	1	1.75E−3	3	3	−2.59E−2
6	15	−3.07E−3	10	7	2.91E−3	13	14	−3.29E−5	3	6	−5.25E−6	11	2	1.98E−3	3	4	−3.36E−2

N-like Fe (Fe^{19+})

i i′	gf_L	*i i′*	gf_L	*i i′*	gf_L	*i i′*	gf_L	*i i′*	gf_L	*i i′*	gf_L
3 5	−1.42E−2	11 1	8.16E−5	3 3	−2.81E−5	2 1	−4.88E−2	3 5	−1.92E−1		**$^4S^e$–$^4P^o$**
3 6	−4.53E−4	11 2	3.31E−1	3 4	−2.75E−2	2 2	−6.40E−2	3 6	−1.04E−1	1 1	2.44E−1
3 7	−1.19E−4	11 3	2.10E−1	3 5	−2.95E−3	2 3	−4.17E+0	3 7	−1.09E−1	1 2	2.40E−1
4 1	5.95E−2	11 4	7.29E−2	3 6	−1.34E−3	3 1	−1.38E+0	3 8	−1.89E−4	1 3	−2.44E−2
4 2	−1.54E−2	11 5	2.91E−3	3 7	−1.55E−4	3 2	−1.11E−2	3 9	−1.36E−3	1 4	−4.90E−1
4 3	−1.81E−2	11 6	4.13E−3	4 1	1.27E+0	3 3	−1.03E−1	3 10	−6.67E−2	1 5	−1.15E−1
4 4	−1.70E−1	11 7	−2.95E−7	4 2	−6.28E−4	4 1	5.81E−4	3 11	−4.84E−1	1 6	−1.72E−6
4 5	−8.29E−4	12 1	2.75E−4	4 3	−2.98E−3	4 2	−1.24E+0	4 1	1.06E−1		**$^4P^e$–$^4P^o$**
4 6	−7.66E−4	12 2	7.24E−2	4 4	−1.95E−1	4 3	−3.76E−5	4 2	5.91E−1	1 1	−2.92E−2
4 7	−6.43E−2	12 3	1.80E−2	4 5	−4.06E−2	5 1	1.00E+0	4 3	1.36E−3	1 2	−5.27E−1
5 1	1.82E−2	12 4	5.24E−1	4 6	−1.55E+0	5 2	4.62E−4	4 4	4.59E−4	1 3	−7.10E+0
5 2	−4.51E−3	12 5	5.78E−2	4 7	−5.28E−3	5 3	−1.54E−5	4 5	4.14E−4	1 4	−6.95E−1
5 3	−3.82E−3	12 6	5.93E−3	5 1	4.91E−5	6 1	4.74E−3	4 6	−7.85E−4	1 5	−6.42E−1
5 4	−1.03E−3	12 7	−7.99E−6	5 2	1.97E−1	6 2	5.01E−1	4 7	−6.63E−1	1 6	−7.09E−3
5 5	−2.31E−1	13 1	1.47E−4	5 3	1.70E−1	6 3	−4.94E−4	4 8	−7.31E−1	2 1	−7.42E−1
5 6	−7.13E−4	13 2	4.38E−7	5 4	6.46E−3	7 1	2.74E−3	4 9	−7.95E−4	2 2	−6.95E−1
5 7	−8.58E−3	13 3	5.25E−4	5 5	4.65E−3	7 2	3.89E−5	4 10	−4.87E−5	2 3	−3.18E−4
6 1	4.43E−4	13 4	2.36E−2	5 6	3.98E−3	7 3	−5.38E−1	4 11	−5.82E−3	2 4	−1.51E−5
6 2	−7.49E−1	13 5	1.05E+0	5 7	−4.04E−5		**$^2G^e$–$^2G^o$**	5 1	2.53E+0	2 5	−1.92E−4
6 3	−1.59E−2	13 6	2.88E−2	6 1	2.71E−4	1 1	−3.44E−3	5 2	2.16E−3	2 6	−1.24E+0
6 4	−1.42E−2	13 7	−5.11E−6	6 2	3.02E−4	1 2	−1.31E+0	5 3	1.46E−1	3 1	4.27E−1
6 5	−1.75E−4	14 1	5.98E−1	6 3	3.37E−2	1 3	−1.10E−3	5 4	6.33E−2	3 2	−4.28E−5
6 6	−6.11E−4	14 2	2.80E−4	6 4	3.63E−1	2 1	1.36E−2	5 5	2.45E−1	3 3	−3.01E−1
6 7	−1.39E−3	14 3	2.90E−6	6 5	3.00E−1	2 2	1.52E+0	5 6	3.61E−3	3 4	−6.36E−2
7 1	6.12E−4	14 4	9.03E−4	6 6	5.08E−1	2 3	−7.42E−7	5 7	6.90E−5	3 5	−3.91E−4
7 2	−1.37E−5	14 5	1.65E−3	6 7	−2.94E−4		**$^4S^o$–$^4P^e$**	5 8	−6.67E−3	3 6	−1.78E−1
7 3	−1.10E+0	14 6	4.39E−7	7 1	7.98E−2	1 1	−3.90E−1	5 9	−7.24E−2	4 1	2.70E−6
7 4	−8.19E−4	14 7	−1.37E+0	7 2	3.68E−4	1 2	−3.99E−1	5 10	−2.76E−4	4 2	−5.42E−3
7 5	−1.90E−3	15 1	4.49E−3	7 3	0.00E+0	1 3	−7.49E+0	5 11	−7.05E−3	4 3	−6.33E−2
7 6	−4.41E−4	15 2	8.36E−4	7 4	1.76E−4	1 4	−1.41E+0	6 1	1.85E−1	4 4	−5.22E−4
7 7	−2.05E−4	15 3	2.44E−3	7 5	5.67E−4	1 5	−7.21E−2	6 2	9.01E−4	4 5	−4.58E−3
8 1	2.23E−1	15 4	2.30E−4	7 6	9.96E−5	1 6	−1.11E−1	6 3	2.85E−4	4 6	−1.34E−3
8 2	4.37E−4	15 5	5.97E−5	7 7	−3.58E−3	1 7	−4.50E−1	6 4	9.57E−3	5 1	2.58E−1
8 3	−5.07E−4	15 6	6.47E−6		**$^2F^o$–$^2G^e$**	1 8	−1.52E−2	6 5	3.27E−1	5 2	1.13E−2
8 4	−5.14E−1	15 7	1.11E−4	1 1	−1.23E+0	1 9	−4.15E−2	6 6	1.10E−1	5 3	−4.79E−1
8 5	−4.95E−2		**$^2F^e$–$^2F^o$**	1 2	−9.50E−6	1 10	−6.45E−2	6 7	1.57E−1	5 4	−7.16E−2
8 6	−1.33E−1	1 1	5.58E−2	2 1	4.28E−5	1 11	−1.53E+0	6 8	4.98E−1	5 5	−3.05E−1
8 7	−1.10E−2	1 2	−1.50E−1	2 2	−5.48E−3	2 1	9.60E−2	6 9	−5.09E−1	5 6	−1.13E−2
9 1	8.91E−1	1 3	−4.01E−1	3 1	3.26E−3	2 2	2.63E−1	6 10	−2.52E−7	6 1	4.50E−2
9 2	6.89E−6	1 4	−8.86E−2	3 2	−3.68E−3	2 3	−3.03E−1	6 11	−2.77E−4	6 2	1.11E+0
9 3	1.04E−4	1 5	−2.06E−2	4 1	1.68E−1	2 4	−3.91E−1	7 1	1.70E−4	6 3	−7.33E−4
9 4	−1.20E−2	1 6	−2.09E−3	4 2	−6.86E−2	2 5	−2.30E−2	7 2	3.98E−1	6 4	−2.62E−1
9 5	−8.98E−1	1 7	−1.30E−2	5 1	1.37E+0	2 6	−7.62E−2	7 3	1.06E−1	6 5	−1.87E−1
9 6	−2.97E−1	2 1	2.19E−1	5 2	−8.96E−2	2 7	−1.57E−1	7 4	1.02E−2	6 6	−2.85E−3
9 7	−3.04E−3	2 2	−8.21E−3	6 1	6.26E−4	2 8	−6.25E−3	7 5	1.41E−4	7 1	5.41E−1
10 1	2.28E−4	2 3	−5.25E−2	6 2	−1.31E+0	2 9	−1.44E−4	7 6	3.55E−3	7 2	7.38E−3
10 2	4.06E−3	2 4	−5.32E−1	7 1	3.37E−1	2 10	−2.26E−1	7 7	1.40E−3	7 3	7.76E−4
10 3	2.96E−3	2 5	−6.01E−2	7 2	1.08E−6	2 11	−1.45E+0	7 8	4.42E−5	7 4	−9.87E−5
10 4	8.09E−3	2 6	−3.11E−3		**$^2F^e$–$^2G^o$**	3 1	2.92E−1	7 9	3.44E−5	7 5	−6.70E−1
10 5	7.31E−4	2 7	−7.99E−2	1 1	−1.40E−1	3 2	1.19E−1	7 10	5.45E−1	7 6	−4.20E−3
10 6	−9.92E−1	3 1	1.29E−4	1 2	−2.02E−2	3 3	−2.89E−2	7 11	−6.75E−1	8 1	3.33E−3
10 7	−2.17E−7	3 2	−4.46E−1	1 3	−8.30E+0	3 4	−4.33E−1			8 2	3.23E−1

N-like Fe (Fe^{19+})

i i′	gf$_L$	*i i′*	gf$_L$	*i i′*	gf$_L$	*i i′*	gf$_L$	*i i′*	gf$_L$	*i i′*	gf$_L$
8 3	4.81E−3	5 1	1.36E−4	4 8	−9.38E−4	9 8	−4.42E−5	3 7	−5.69E−1	6 2	1.44E−2
8 4	5.92E−4	5 2	9.20E−1	4 9	−2.68E−4	9 9	−4.42E−6	3 8	−1.57E−5	6 3	−3.75E−1
8 5	−6.29E−1	5 3	5.60E−2	5 1	1.39E−4	10 1	9.27E−1	3 9	−2.30E−2	6 4	−2.76E−2
8 6	−2.02E−5	5 4	−1.06E+0	5 2	1.23E+0	10 2	4.31E−3	4 1	1.40E−4	7 1	5.74E−3
9 1	3.13E−4	5 5	−8.19E−6	5 3	1.08E−3	10 3	3.80E−4	4 2	3.72E−4	7 2	1.55E+0
9 2	8.53E−4	6 1	2.99E−1	5 4	−3.71E−1	10 4	3.77E−4	4 3	3.66E−2	7 3	−2.28E+0
9 3	1.74E−1	6 2	4.16E−6	5 5	−1.40E−2	10 5	1.18E−3	4 4	3.87E−1	7 4	−7.80E−4
9 4	1.73E−1	6 3	1.88E−4	5 6	−1.36E−2	10 6	1.99E−3	4 5	4.22E−2	8 1	5.10E−1
9 5	4.91E−1	6 4	7.96E−6	5 7	−2.11E−3	10 7	1.12E−6	4 6	1.01E+0	8 2	1.51E−4
9 6	−1.97E−5	6 5	−1.71E+0	5 8	−1.44E−2	10 8	−1.97E+0	4 7	2.03E−1	8 3	3.22E−4
10 1	5.07E−1		**^{4}P^e–^{4}D^o**	5 9	−1.14E−3	10 9	−8.30E−3	4 8	−3.27E−5	8 4	−3.06E+0
10 2	2.09E−3	1 1	−3.38E−3	6 1	3.73E−1	11 1	7.97E−2	4 9	−2.33E−5	9 1	7.56E−2
10 3	3.33E−4	1 2	−1.02E+0	6 2	1.25E−3	11 2	1.42E−3	5 1	1.37E+0	9 2	6.39E−4
10 4	4.16E−4	1 3	−4.43E+0	6 3	3.34E−5	11 3	6.10E−5	5 2	1.92E−2	9 3	8.80E−5
10 5	6.32E−6	1 4	−6.28E+0	6 4	−2.96E−4	11 4	2.54E−5	5 3	1.74E−5	9 4	6.20E−3
10 6	−1.19E+0	1 5	−7.03E+0	6 5	−1.02E+0	11 5	4.59E−3	5 4	4.61E−3		**^{4}D^e–^{4}F^o**
11 1	1.36E+0	1 6	−1.25E+0	6 6	−1.16E−2	11 6	3.87E−3	5 5	2.70E−3	1 1	−1.80E−1
11 2	1.67E−2	1 7	−9.01E−1	6 7	−1.25E−1	11 7	8.08E−5	5 6	9.28E−3	1 2	−2.49E−1
11 3	7.27E−3	1 8	−6.63E−3	6 8	−7.69E−3	11 8	1.94E−2	5 7	7.51E−5	2 1	−1.22E+0
11 4	4.80E−3	1 9	−1.18E−3	6 9	−4.15E−2	11 9	−5.91E−1	5 8	6.08E−1	2 2	−5.25E−3
11 5	9.28E−6	2 1	−1.19E+0	7 1	9.95E−1		**^{4}D^e–^{4}D^o**	5 9	−1.42E−1	3 1	−3.99E−3
11 6	7.35E−1	2 2	−6.88E−1	7 2	3.04E−3	1 1	3.48E−1		**^{6}P^e–^{6}D^o**	3 2	−1.93E+0
	^{6}S^o–^{6}P^e	2 3	−7.14E−3	7 3	2.49E−3	1 2	−4.24E−5	1 1	−2.00E+0	4 1	1.21E+0
1 1	−1.32E+0	2 4	−6.63E−4	7 4	2.99E−6	1 3	−7.33E−3		**^{1}D^o–^{1}F^e**	4 2	1.07E−1
	^{4}P^o–^{4}D^e	2 5	−9.09E−5	7 5	−6.95E−4	1 4	−4.09E−1	1 1	−1.71E+0	5 1	2.43E−4
1 1	−9.68E−1	2 6	−1.74E−3	7 6	−1.14E+0	1 5	−2.47E−2	1 2	−5.26E−1	5 2	4.59E−3
1 2	−3.14E−1	2 7	−1.70E−4	7 7	−7.78E−1	1 6	−7.58E−1	1 3	−2.03E−4		**^{4}F^e–^{4}F^o**
1 3	−2.26E−1	2 8	−2.19E+0	7 8	−3.20E−3	1 7	−4.97E−3	1 4	−7.86E+0	1 1	−5.96E−1
1 4	−2.33E−5	2 9	−2.66E−2	7 9	−9.60E−2	1 8	−1.23E−1	2 1	1.68E−4	1 2	−4.66E−1
1 5	−4.18E+0	3 1	5.79E−3	8 1	2.86E−3	1 9	−2.25E+0	2 2	−2.10E+0	2 1	−5.01E−1
2 1	4.65E−5	3 2	−1.29E−4	8 2	1.01E+0	2 1	4.01E−1	2 3	−2.07E−3	2 2	−9.50E−2
2 2	−1.46E−2	3 3	−5.71E−3	8 3	6.74E−3	2 2	1.23E+0	2 4	−1.08E−1	3 1	1.58E−1
2 3	−1.44E+0	3 4	−1.33E−1	8 4	5.99E−3	2 3	−1.46E−3	3 1	8.70E−4	3 2	6.76E−1
2 4	−2.09E−3	3 5	−2.18E−1	8 5	5.61E−3	2 4	−9.49E−1	3 2	5.35E−4	4 1	1.69E−4
2 5	−5.13E−2	3 6	−2.81E−1	8 6	1.56E−3	2 5	−3.10E−3	3 3	−8.70E−4	4 2	7.19E−4
3 1	1.90E−2	3 7	−3.24E−3	8 7	−1.70E+0	2 6	−5.01E−4	3 4	−7.30E−4		**^{4}F^e–^{4}G^o**
3 2	1.85E−1	3 8	−1.72E−3	8 8	−4.31E−5	2 7	−1.35E−1	4 1	3.62E−2	1 1	−3.88E−1
3 3	4.96E−2	3 9	−1.19E+1	8 9	−9.95E−4	2 8	−1.32E−6	4 2	7.06E−2	1 2	−2.43E+1
3 4	−6.34E−2	4 1	2.00E−3	9 1	7.43E−4	2 9	−2.65E−3	4 3	−1.63E−3	2 1	−2.44E+0
3 5	−1.65E−2	4 2	−3.25E−3	9 2	7.16E−4	3 1	5.58E−1	4 4	−1.98E−2	2 2	−2.53E−1
4 1	5.40E−1	4 3	−1.59E+0	9 3	2.82E−3	3 2	1.02E−1	5 1	5.48E−1	3 1	2.07E+0
4 2	4.35E−2	4 4	−3.34E−2	9 4	2.90E−1	3 3	1.87E−3	5 2	3.92E−2	3 2	−8.40E−4
4 3	8.89E−3	4 5	−6.07E−3	9 5	3.30E−1	3 4	−2.83E−3	5 3	−7.99E−2	4 1	3.58E−3
4 4	−1.06E−1	4 6	−3.16E−3	9 6	6.12E−1	3 5	−4.01E−1	5 4	−3.01E−2	4 2	−1.91E+0
4 5	−1.61E−2	4 7	−6.93E−6	9 7	1.15E−1	3 6	−1.21E−4	6 1	1.50E+0		

O-like S (S^{8+})

Term energies relative to $2s^22p^3$ ^{4}S ionization threshold for each symmetry

i	E(Ryds)	Description
	^{1}S^e	
1	−26.7441	$2s^22p^4$
2	−18.5192	$2p^6$
3	−8.80000	$2s^22p^3$ ^{2}P $3p$
4	−5.87336	$2s2p^4$ ^{2}S $3s$
5	−4.48305	$2s2p^4$ ^{2}D $3d$
6	−4.26823	$2s^22p^3$ ^{2}P $4p$
7	−2.27714	$2s^22p^3$ ^{2}P $5p$
8	−1.20358	$2s^22p^3$ ^{2}P $6p$
9	−.74997	$2p^5$ ^{2}P $3p$
10	−.56397	$2s^22p^3$ ^{2}P $7p$
11	−.46520	$2s2p^4$ ^{2}D $4d$
12	−.36101	$2s2p^4$ ^{2}S $4s$
13	−.14773	$2s^22p^3$ ^{2}P $8p$
	^{1}S^o	
1	−8.82479	$2s^22p^3$ ^{2}D $3d$
2	−4.68364	$2s2p^4$ ^{2}P $3p$
3	−4.54686	$2s^22p^3$ ^{2}D $4d$
4	−2.61290	$2s^22p^3$ ^{2}D $5d$
5	−1.56952	$2s^22p^3$ ^{2}D $6d$
6	−.94444	$2s^22p^3$ ^{2}D $7d$
7	−.54098	$2s^22p^3$ ^{2}D $8d$
8	−.26563	$2s^22p^3$ ^{2}D $9d$
	^{1}P^e	
1	−10.0847	$2s^22p^3$ ^{2}D $3p$
2	−9.47878	$2s^22p^3$ ^{2}P $3p$
3	−5.75061	$2s2p^4$ ^{2}P $3s$
4	−4.95646	$2s^22p^3$ ^{2}D $4p$
5	−4.62250	$2s2p^4$ ^{2}D $3d$
6	−4.51256	$2s^22p^3$ ^{2}P $4p$
7	−4.31838	$2s^22p^3$ ^{2}D $4f$
8	−3.35822	$2s2p^4$ ^{2}P $3d$
9	−2.82184	$2s^22p^3$ ^{2}D $5p$
10	−2.47094	$2s^22p^3$ ^{2}D $5f$
11	−2.40063	$2s^22p^3$ ^{2}P $5p$
12	−1.68825	$2s^22p^3$ ^{2}D $6p$
13	−1.49354	$2s^22p^3$ ^{2}D $6f$
14	−1.42775	$2s^22p^3$ ^{2}P $6p$
15	−1.27184	$2p^5$ ^{2}P $3p$
16	−1.01808	$2s^22p^3$ ^{2}D $7p$
17	−.89735	$2s^22p^3$ ^{2}D $7f$
18	−.61060	$2s^22p^3$ ^{2}P $7p$
19	−.58620	$2s^22p^3$ ^{2}D $8p$
20	−.51004	$2s^22p^3$ ^{2}D $8f$
21	−.44053	$2s2p^4$ ^{2}D $4d$
22	−.30745	$2s^22p^3$ ^{2}D $9p$
23	−.24410	$2s^22p^3$ ^{2}D $9f$
24	−.21961	$2s2p^4$ ^{2}P $4s$
25	−.16478	$2s^22p^3$ ^{2}P $8p$
	^{1}P^o	
1	−22.3008	$2s2p^5$
2	−10.4322	$2s^22p^3$ ^{2}P $3s$
3	−8.66465	$2s^22p^3$ ^{2}D $3d$
4	−7.89828	$2s^22p^3$ ^{2}P $3d$
5	−5.90448	$2s2p^4$ ^{2}D $3p$
6	−5.00138	$2s2p^4$ ^{2}S $3p$
7	−4.91561	$2s^22p^3$ ^{2}P $4s$
8	−4.48229	$2s^22p^3$ ^{2}D $4d$
9	−4.25179	$2s2p^4$ ^{2}P $3p$
10	−3.94267	$2s^22p^3$ ^{2}P $4d$
11	−2.59476	$2s^22p^3$ ^{2}D $5d$
12	−2.56097	$2s^22p^3$ ^{2}P $5s$
13	−2.33015	$2p^5$ ^{2}P $3s$
14	−2.10710	$2s^22p^3$ ^{2}P $5d$
15	−1.55147	$2s^22p^3$ ^{2}D $6d$
16	−1.36938	$2s^22p^3$ ^{2}P $6s$
17	−1.10622	$2s^22p^3$ ^{2}P $6d$
18	−.93519	$2s^22p^3$ ^{2}D $7d$
19	−.86493	$2s2p^4$ ^{2}D $4p$
20	−.66404	$2s^22p^3$ ^{2}P $7s$
21	−.53460	$2s^22p^3$ ^{2}D $8d$
22	−.49974	$2s^22p^3$ ^{2}P $7d$
23	−.42085	$2s2p^4$ ^{2}D $4f$
24	−.26047	$2s^22p^3$ ^{2}D $9d$
25	−.21703	$2s^22p^3$ ^{2}P $8s$
26	−.12020	$2s2p^4$ ^{2}S $4p$
27	−.09731	$2s^22p^3$ ^{2}P $8d$
	^{1}D^e	
1	−27.3300	$2s^22p^4$
2	−9.52001	$2s^22p^3$ ^{2}D $3p$
3	−9.19828	$2s^22p^3$ ^{2}P $3p$
4	−6.66339	$2s2p^4$ ^{2}D $3s$
5	−4.77618	$2s^22p^3$ ^{2}D $4p$
6	−4.57876	$2s2p^4$ ^{2}D $3d$
7	−4.45806	$2s^22p^3$ ^{2}P $4p$
8	−4.29370	$2s^22p^3$ ^{2}D $4f$
9	−3.96617	$2s^22p^3$ ^{2}P $4f$
10	−3.68382	$2s2p^4$ ^{2}S $3d$
11	−3.34983	$2s2p^4$ ^{2}P $3d$
12	−2.72073	$2s^22p^3$ ^{2}D $5p$
13	−2.47504	$2s^22p^3$ ^{2}D $5f$
14	−2.36438	$2s^22p^3$ ^{2}P $5p$
15	−2.06849	$2s^22p^3$ ^{2}P $5f$
16	−1.63370	$2s^22p^3$ ^{2}D $6p$
17	−1.49604	$2p^5$ ^{2}P $3p$
18	−1.48219	$2s^22p^3$ ^{2}D $6f$
19	−1.25556	$2s^22p^3$ ^{2}P $6p$
20	−1.20037	$2s2p^4$ ^{2}D $4s$
21	−1.08146	$2s^22p^3$ ^{2}P $6f$
22	−.98253	$2s^22p^3$ ^{2}D $7p$
23	−.89674	$2s^22p^3$ ^{2}D $7f$
24	−.59399	$2s^22p^3$ ^{2}P $7p$
25	−.56578	$2s^22p^3$ ^{2}D $8p$
26	−.51022	$2s^22p^3$ ^{2}D $8f$
27	−.48419	$2s^22p^3$ ^{2}P $7f$
28	−.43629	$2s2p^4$ ^{2}D $4d$
29	−.28308	$2s^22p^3$ ^{2}D $9p$
30	−.24351	$2s^22p^3$ ^{2}D $9f$
31	−.16813	$2s^22p^3$ ^{2}P $8p$
32	−.09741	$2s^22p^3$ ^{2}P $8f$
	^{1}D^o	
1	−10.8517	$2s^22p^3$ ^{2}D $3s$
2	−8.57239	$2s^22p^3$ ^{2}D $3d$
3	−8.17022	$2s^22p^3$ ^{2}P $3d$
4	−5.72136	$2s2p^4$ ^{2}D $3p$
5	−5.33306	$2s^22p^3$ ^{2}D $4s$
6	−4.73686	$2s2p^4$ ^{2}P $3p$
7	−4.40771	$2s^22p^3$ ^{2}D $4d$
8	−4.08741	$2s^22p^3$ ^{2}P $4d$
9	−2.98879	$2s^22p^3$ ^{2}D $5s$
10	−2.54016	$2s^22p^3$ ^{2}D $5d$
11	−2.18058	$2s^22p^3$ ^{2}P $5d$
12	−1.78251	$2s^22p^3$ ^{2}D $6s$
13	−1.52799	$2s^22p^3$ ^{2}D $6d$
14	−1.14814	$2s^22p^3$ ^{2}P $6d$
15	−1.07739	$2s^22p^3$ ^{2}D $7s$
16	−.92177	$2s^22p^3$ ^{2}D $7d$
17	−.89732	$2s2p^4$ ^{2}D $4p$
18	−.62770	$2s^22p^3$ ^{2}D $8s$
19	−.53470	$2s^22p^3$ ^{2}D $8d$
20	−.51433	$2s^22p^3$ ^{2}P $7d$
21	−.47758	$2s2p^4$ ^{2}D $4f$
22	−.32642	$2s^22p^3$ ^{2}D $9s$
23	−.25463	$2s^22p^3$ ^{2}D $9d$
24	−.18286	$2s2p^4$ ^{2}P $4p$
25	−.12413	$2s^22p^3$ ^{2}P $8d$
	^{1}F^e	
1	−9.91885	$2s^22p^3$ ^{2}D $3p$
2	−4.95784	$2s^22p^3$ ^{2}D $4p$
3	−4.52870	$2s2p^4$ ^{2}D $3d$
4	−4.29651	$2s^22p^3$ ^{2}D $4f$
5	−3.93219	$2s^22p^3$ ^{2}P $4f$
6	−3.55826	$2s2p^4$ ^{2}P $3d$
7	−2.80525	$2s^22p^3$ ^{2}D $5p$
8	−2.47960	$2s^22p^3$ ^{2}D $5f$
9	−2.07530	$2s^22p^3$ ^{2}P $5f$
10	−1.67962	$2s^22p^3$ ^{2}D $6p$
11	−1.49324	$2s^22p^3$ ^{2}D $6f$
12	−1.08545	$2s^22p^3$ ^{2}P $6f$
13	−1.01305	$2s^22p^3$ ^{2}D $7p$
14	−.89659	$2s^22p^3$ ^{2}D $7f$
15	−.58773	$2s^22p^3$ ^{2}D $8p$
16	−.50972	$2s^22p^3$ ^{2}D $8f$
17	−.48707	$2s^22p^3$ ^{2}P $7f$
18	−.44170	$2s2p^4$ ^{2}D $4d$
19	−.29604	$2s^22p^3$ ^{2}D $9p$
20	−.24307	$2s^22p^3$ ^{2}D $9f$
21	−.09925	$2s^22p^3$ ^{2}P $8f$
	^{1}F^o	
1	−8.39265	$2s^22p^3$ ^{2}D $3d$
2	−8.11839	$2s^22p^3$ ^{2}P $3d$
3	−5.85278	$2s2p^4$ ^{2}D $3p$
4	−4.37781	$2s^22p^3$ ^{2}D $4d$
5	−4.02958	$2s^22p^3$ ^{2}P $4d$
6	−2.52747	$2s^22p^3$ ^{2}D $5d$
7	−2.14874	$2s^22p^3$ ^{2}P $5d$
8	−1.52179	$2s^22p^3$ ^{2}D $6d$
9	−1.12933	$2s^22p^3$ ^{2}P $6d$
10	−.94823	$2s2p^4$ ^{2}D $4p$
11	−.90899	$2s^22p^3$ ^{2}D $7d$
12	−.52424	$2s^22p^3$ ^{2}D $8d$
13	−.51155	$2s^22p^3$ ^{2}P $7d$
14	−.44186	$2s2p^4$ ^{2}D $4f$
15	−.25358	$2s^22p^3$ ^{2}D $9d$
16	−.11739	$2s^22p^3$ ^{2}P $8d$
	^{1}G^e	
1	−4.68061	$2s2p^4$ ^{2}D $3d$
2	−4.29121	$2s^22p^3$ ^{2}D $4f$
3	−3.89936	$2s^22p^3$ ^{2}P $4f$
4	−2.48567	$2s^22p^3$ ^{2}D $5f$
5	−2.07736	$2s^22p^3$ ^{2}P $5f$
6	−1.49499	$2s^22p^3$ ^{2}D $6f$
7	−1.08522	$2s^22p^3$ ^{2}P $6f$
8	−.89733	$2s^22p^3$ ^{2}D $7f$
9	−.51159	$2s^22p^3$ ^{2}D $8f$
10	−.49167	$2s2p^4$ ^{2}D $4d$
11	−.48634	$2s^22p^3$ ^{2}P $7f$
12	−.24328	$2s^22p^3$ ^{2}D $9f$
13	−.09840	$2s^22p^3$ ^{2}P $8f$
	^{1}G^o	
1	−8.71844	$2s^22p^3$ ^{2}D $3d$
2	−4.51385	$2s^22p^3$ ^{2}D $4d$
3	−2.59489	$2s^22p^3$ ^{2}D $5d$
4	−1.55955	$2s^22p^3$ ^{2}D $6d$
5	−.93837	$2s^22p^3$ ^{2}D $7d$
6	−.53711	$2s^22p^3$ ^{2}D $8d$
7	−.31073	$2s2p^4$ ^{2}D $4f$
8	−.26228	$2s^22p^3$ ^{2}D $9d$
	^{1}H^e	
1	−4.31943	$2s^22p^3$ ^{2}D $4f$
2	−2.49313	$2s^22p^3$ ^{2}D $5f$
3	−1.49941	$2s^22p^3$ ^{2}D $6f$
4	−.90014	$2s^22p^3$ ^{2}D $7f$
5	−.51130	$2s^22p^3$ ^{2}D $8f$

O-like S (S^{8+})

i	E(Ryds)	Description
6	−.24478	$2s^22p^3$ ^{2}D $9f$
		^{1}H^o
1	−.34035	$2s2p^4$ ^{2}D $4f$
		^{3}S^e
1	−9.63471	$2s^22p^3$ ^{2}P $3p$
2	−6.11695	$2s2p^4$ ^{2}S $3s$
3	−4.71005	$2s2p^4$ ^{2}D $3d$
4	−4.53689	$2s^22p^3$ ^{2}P $4p$
5	−2.41503	$2s^22p^3$ ^{2}P $5p$
6	−1.71560	$2p^5$ ^{2}P $3p$
7	−1.28098	$2s^22p^3$ ^{2}P $6p$
8	−.62596	$2s^22p^3$ ^{2}P $7p$
9	−.55509	$2s2p^4$ ^{2}S $4s$
10	−.46752	$2s2p^4$ ^{2}D $4d$
11	−.17416	$2s^22p^3$ ^{2}P $8p$
		^{3}S^o
1	−11.5448	$2s^22p^3$ ^{4}S $3s$
2	−8.49228	$2s^22p^3$ ^{2}D $3d$
3	−6.97617	$2s2p^4$ ^{4}P $3p$
4	−6.04532	$2s^22p^3$ ^{4}S $4s$
5	−4.65145	$2s2p^4$ ^{2}P $3p$
6	−4.41209	$2s^22p^3$ ^{2}D $4d$
7	−3.73745	$2s^22p^3$ ^{4}S $5s$
8	−2.55700	$2s^22p^3$ ^{2}D $5d$
9	−2.53205	$2s^22p^3$ ^{4}S $6s$
10	−2.14866	$2s2p^4$ ^{4}P $4p$
11	−1.82691	$2s^22p^3$ ^{4}S $7s$
12	−1.53523	$2s^22p^3$ ^{2}D $6d$
13	−1.38209	$2s^22p^3$ ^{4}S $8s$
14	−1.08158	$2s^22p^3$ ^{4}S $9s$
15	−.92359	$2s^22p^3$ ^{2}D $7d$
		^{3}P^e
1	−27.8323	$2s^22p^4$
2	−10.5571	$2s^22p^3$ ^{4}S $3p$
3	−9.64451	$2s^22p^3$ ^{2}D $3p$
4	−9.39122	$2s^22p^3$ ^{2}P $3p$
5	−7.86246	$2s2p^4$ ^{4}P $3s$
6	−5.84935	$2s2p^4$ ^{2}P $3s$
7	−5.73360	$2s2p^4$ ^{4}P $3d$
8	−5.55965	$2s^22p^3$ ^{4}S $4p$
9	−4.86503	$2s^22p^3$ ^{2}D $4p$
10	−4.67144	$2s2p^4$ ^{2}D $3d$
11	−4.48471	$2s^22p^3$ ^{2}P $4p$
12	−4.34150	$2s^22p^3$ ^{2}D $4f$
13	−3.52788	$2s2p^4$ ^{2}P $3d$
14	−3.52110	$2s^22p^3$ ^{4}S $5p$
15	−2.78676	$2s^22p^3$ ^{2}D $5p$
16	−2.52000	$2s2p^4$ ^{4}P $4s$
17	−2.47291	$2s^22p^3$ ^{2}D $5f$
18	−2.42025	$2s^22p^3$ ^{4}S $6p$
19	−2.35848	$2s^22p^3$ ^{2}P $5p$
20	−1.76606	$2s^22p^3$ ^{4}S $7p$

i	E(Ryds)	Description
21	−1.66337	$2s2p^4$ ^{4}P $4d$
22	−1.65625	$2s^22p^3$ ^{2}D $6p$
23	−1.49206	$2s^22p^3$ ^{2}D $6f$
24	−1.49206	$2p^5$ ^{2}P $3p$
25	−1.33386	$2s^22p^3$ ^{4}S $8p$
26	−1.25770	$2s^22p^3$ ^{2}P $6p$
27	−1.04996	$2s^22p^3$ ^{4}S $9p$
28	−1.00058	$2s^22p^3$ ^{2}D $7p$
29	−.89648	$2s^22p^3$ ^{2}D $7f$
		^{3}P^o
1	−23.8472	$2s2p^5$
2	−10.5642	$2s^22p^3$ ^{2}P $3s$
3	−8.56931	$2s^22p^3$ ^{2}D $3d$
4	−8.30946	$2s^22p^3$ ^{2}P $3d$
5	−6.89001	$2s2p^4$ ^{4}P $3p$
6	−5.70975	$2s2p^4$ ^{2}D $3p$
7	−5.12463	$2s2p^4$ ^{2}S $3p$
8	−4.93525	$2s2p^4$ ^{2}P $3p$
9	−4.82079	$2s^22p^3$ ^{2}P $4s$
10	−4.44843	$2s^22p^3$ ^{2}D $4d$
11	−4.10396	$2s^22p^3$ ^{2}P $4d$
12	−2.60395	$2s^22p^3$ ^{2}P $5s$
13	−2.55809	$2s^22p^3$ ^{2}D $5d$
14	−2.47975	$2p^5$ ^{2}P $3s$
15	−2.18492	$2s^22p^3$ ^{2}P $5d$
16	−2.02881	$2s2p^4$ ^{4}P $4p$
17	−1.54398	$2s^22p^3$ ^{2}D $6d$
18	−1.38094	$2s^22p^3$ ^{2}P $6s$
19	−1.14889	$2s^22p^3$ ^{2}P $6d$
20	−.92918	$2s^22p^3$ ^{2}D $7d$
21	−.90374	$2s2p^4$ ^{2}D $4p$
22	−.66928	$2s^22p^3$ ^{2}P $7s$
23	−.60863	$2p^5$ ^{2}P $3d$
24	−.53103	$2s^22p^3$ ^{2}D $8d$
25	−.52674	$2s^22p^3$ ^{2}P $7d$
26	−.26914	$2s2p^4$ ^{2}D $4f$
27	−.25755	$2s^22p^3$ ^{2}D $9d$
28	−.22312	$2s^22p^3$ ^{2}P $8s$
29	−.13936	$2s2p^4$ ^{2}S $4p$
30	−.11897	$2s^22p^3$ ^{2}P $8d$
		^{3}D^e
1	−10.0450	$2s^22p^3$ ^{2}D $3p$
2	−9.55152	$2s^22p^3$ ^{2}P $3p$
3	−6.01240	$2s2p^4$ ^{2}D $3s$
4	−5.69940	$2s2p^4$ ^{4}P $3d$
5	−4.98612	$2s^22p^3$ ^{2}D $4p$
6	−4.66776	$2s2p^4$ ^{2}D $3d$
7	−4.55331	$2s^22p^3$ ^{2}P $4p$
8	−4.33337	$2s^22p^3$ ^{2}D $4f$
9	−4.03685	$2s^22p^3$ ^{2}P $4f$
10	−3.80344	$2s2p^4$ ^{2}S $3d$
11	−3.68542	$2s2p^4$ ^{2}P $3d$

i	E(Ryds)	Description
12	−2.82412	$2s^22p^3$ ^{2}D $5p$
13	−2.47957	$2s^22p^3$ ^{2}D $5f$
14	−2.40483	$2s^22p^3$ ^{2}P $5p$
15	−2.07287	$2s^22p^3$ ^{2}P $5f$
16	−1.69451	$2s^22p^3$ ^{2}D $6p$
17	−1.65613	$2s2p^4$ ^{4}P $4d$
18	−1.54799	$2p^5$ ^{2}P $3p$
19	−1.49307	$2s^22p^3$ ^{2}D $6f$
20	−1.33085	$2s2p^4$ ^{2}D $4s$
21	−1.26605	$2s^22p^3$ ^{2}P $6p$
22	−1.08461	$2s^22p^3$ ^{2}P $6f$
23	−1.01463	$2s^22p^3$ ^{2}D $7p$
24	−.89777	$2s^22p^3$ ^{2}D $7f$
25	−.60482	$2s^22p^3$ ^{2}P $7p$
26	−.59094	$2s^22p^3$ ^{2}D $8p$
27	−.51724	$2s^22p^3$ ^{2}D $8f$
28	−.49234	$2s2p^4$ ^{2}D $4d$
29	−.48454	$2s^22p^3$ ^{2}P $7f$
30	−.29798	$2s^22p^3$ ^{2}D $9p$
31	−.24342	$2s^22p^3$ ^{2}D $9f$
32	−.17649	$2s^22p^3$ ^{2}P $8p$
33	−.09946	$2s^22p^3$ ^{2}P $8f$
		^{3}D^o
1	−10.9818	$2s^22p^3$ ^{2}D $3s$
2	−9.31638	$2s^22p^3$ ^{4}S $3d$
3	−8.65777	$2s^22p^3$ ^{2}D $3d$
4	−8.21939	$2s^22p^3$ ^{2}P $3d$
5	−7.04191	$2s2p^4$ ^{4}P $3p$
6	−5.82469	$2s2p^4$ ^{2}D $3p$
7	−5.34612	$2s^22p^3$ ^{2}D $4s$
8	−5.18300	$2s^22p^3$ ^{4}S $4d$
9	−4.80809	$2s2p^4$ ^{2}P $3p$
10	−4.49867	$2s^22p^3$ ^{2}D $4d$
11	−4.07242	$2s^22p^3$ ^{2}P $4d$
12	−3.30546	$2s^22p^3$ ^{4}S $5d$
13	−3.00819	$2s^22p^3$ ^{2}D $5s$
14	−2.59205	$2s^22p^3$ ^{2}D $5d$
15	−2.29418	$2s^22p^3$ ^{4}S $6d$
16	−2.17123	$2s2p^4$ ^{4}P $4p$
17	−2.16516	$2s^22p^3$ ^{2}P $5d$
18	−1.79353	$2s^22p^3$ ^{2}D $6s$
19	−1.68112	$2s^22p^3$ ^{4}S $7d$
20	−1.57119	$2s2p^4$ ^{4}P $4f$
21	−1.55617	$2s^22p^3$ ^{2}D $6d$
22	−1.28375	$2s^22p^3$ ^{4}S $8d$
23	−1.14064	$2s^22p^3$ ^{2}P $6d$
24	−1.08773	$2s^22p^3$ ^{2}D $7s$
25	−1.01311	$2s^22p^3$ ^{4}S $9d$
26	−.94125	$2s^22p^3$ ^{2}D $7d$
27	−.91067	$2s2p^4$ ^{2}D $4p$
		^{3}F^e
1	−9.96640	$2s^22p^3$ ^{2}D $3p$

i	E(Ryds)	Description
2	−5.86519	$2s2p^4$ ^{4}P $3d$
3	−5.05669	$2s^22p^3$ ^{4}S $4f$
4	−4.98037	$2s^22p^3$ ^{2}D $4p$
5	−4.69035	$2s2p^4$ ^{2}D $3d$
6	−4.29147	$2s^22p^3$ ^{2}D $4f$
7	−3.92702	$2s^22p^3$ ^{2}P $4f$
8	−3.59212	$2s2p^4$ ^{2}P $3d$
9	−3.24312	$2s^22p^3$ ^{4}S $5f$
10	−2.81321	$2s^22p^3$ ^{2}D $5p$
11	−2.48437	$2s^22p^3$ ^{2}D $5f$
12	−2.25303	$2s^22p^3$ ^{4}S $6f$
13	−2.07562	$2s^22p^3$ ^{2}P $5f$
14	−1.73602	$2s2p^4$ ^{4}P $4d$
15	−1.68393	$2s^22p^3$ ^{2}D $6p$
16	−1.65361	$2s^22p^3$ ^{4}S $7f$
17	−1.49649	$2s^22p^3$ ^{2}D $6f$
18	−1.26683	$2s^22p^3$ ^{4}S $8f$
19	−1.08513	$2s^22p^3$ ^{2}P $6f$
20	−1.01580	$2s^22p^3$ ^{2}D $7p$
21	−1.00090	$2s^22p^3$ ^{4}S $9f$
22	−.89938	$2s^22p^3$ ^{2}D $7f$
		^{3}F^o
1	−8.82811	$2s^22p^3$ ^{2}D $3d$
2	−8.33223	$2s^22p^3$ ^{2}P $3d$
3	−5.94774	$2s2p^4$ ^{2}D $3p$
4	−4.54417	$2s^22p^3$ ^{2}D $4d$
5	−4.11949	$2s^22p^3$ ^{2}P $4d$
6	−2.60884	$2s^22p^3$ ^{2}D $5d$
7	−2.19248	$2s^22p^3$ ^{2}P $5d$
8	−1.60232	$2s2p^4$ ^{4}P $4f$
9	−1.56715	$2s^22p^3$ ^{2}D $6d$
10	−1.15365	$2s^22p^3$ ^{2}P $6d$
11	−.97773	$2s2p^4$ ^{2}D $4p$
12	−.93876	$2s^22p^3$ ^{2}D $7d$
13	−.54168	$2s^22p^3$ ^{2}D $8d$
14	−.52788	$2s^22p^3$ ^{2}P $7d$
15	−.28429	$2s2p^4$ ^{2}D $4f$
16	−.26460	$2s^22p^3$ ^{2}D $9d$
17	−.12728	$2s^22p^3$ ^{2}P $8d$
		^{3}G^e
1	−4.86568	$2s2p^4$ ^{2}D $3d$
2	−4.29258	$2s^22p^3$ ^{2}D $4f$
3	−3.90138	$2s^22p^3$ ^{2}P $4f$
4	−2.40051	$2s^22p^3$ ^{2}D $5f$
5	−2.08064	$2s^22p^3$ ^{2}P $5f$
6	−1.49885	$2s^22p^3$ ^{2}D $6f$
7	−1.08786	$2s^22p^3$ ^{2}P $6f$
8	−.90055	$2s^22p^3$ ^{2}D $7f$
9	−.56613	$2s2p^4$ ^{2}D $4d$
10	−.50662	$2s^22p^3$ ^{2}D $8f$
11	−.48647	$2s^22p^3$ ^{2}P $7f$
12	−.24410	$2s^22p^3$ ^{2}D $9f$

O-like S (S^{8+})

i	Energy(Ryds)	Description	i	Energy(Ryds)	Description	i	Energy(Ryds)	Description
13	−.09931	$2s^22p^3\ ^2P\ 8f$	9	−1.08660	$2s^22p^3\ ^4S\ 9s$	4	−3.36912	$2s^22p^3\ ^4S\ 5d$
	$^3G^o$			$^5P^e$		5	−2.32879	$2s^22p^3\ ^4S\ 6d$
1	−8.75003	$2s^22p^3\ ^2D\ 3d$	1	−10.7924	$2s^22p^3\ ^4S\ 3p$	6	−2.22404	$2s2p^4\ ^4P\ 4p$
2	−4.52612	$2s^22p^3\ ^2D\ 4d$	2	−8.23284	$2s2p^4\ ^4P\ 3s$	7	−1.70118	$2s^22p^3\ ^4S\ 7d$
3	−2.60086	$2s^22p^3\ ^2D\ 5d$	3	−5.97819	$2s2p^4\ ^4P\ 3d$	8	−1.57769	$2s2p^4\ ^4P\ 4f$
4	−1.58010	$2s2p^4\ ^4P\ 4f$	4	−5.70578	$2s^22p^3\ ^4S\ 4p$	9	−1.29785	$2s^22p^3\ ^4S\ 8d$
5	−1.56229	$2s^22p^3\ ^2D\ 6d$	5	−3.58399	$2s^22p^3\ ^4S\ 5p$	10	−1.02267	$2s^22p^3\ ^4S\ 9d$
6	−.94034	$2s^22p^3\ ^2D\ 7d$	6	−2.63861	$2s2p^4\ ^4P\ 4s$		$^5F^e$	
	$^3H^e$		7	−2.44046	$2s^22p^3\ ^4S\ 6p$	1	−6.06208	$2s2p^4\ ^4P\ 3d$
1	−4.32002	$2s^22p^3\ ^2D\ 4f$	8	−1.80423	$2s2p^4\ ^4P\ 4d$	2	−5.06043	$2s^22p^3\ ^4S\ 4f$
2	−2.49356	$2s^22p^3\ ^2D\ 5f$	9	−1.75081	$2s^22p^3\ ^4S\ 7p$	3	−3.24828	$2s^22p^3\ ^4S\ 5f$
3	−1.49970	$2s^22p^3\ ^2D\ 6f$	10	−1.34554	$2s^22p^3\ ^4S\ 8p$	4	−2.25701	$2s^22p^3\ ^4S\ 6f$
4	−.90033	$2s^22p^3\ ^2D\ 7f$	11	−1.05623	$2s^22p^3\ ^4S\ 9p$	5	−1.81276	$2s2p^4\ ^4P\ 4d$
	$^3H^o$			$^5P^o$		6	−1.65502	$2s^22p^3\ ^4S\ 7f$
1	−.34521	$2s2p^4\ ^2D\ 4f$	1	−7.31825	$2s2p^4\ ^4P\ 3p$	7	−1.26812	$2s^22p^3\ ^4S\ 8f$
	$^5S^o$		2	−2.26418	$2s2p^4\ ^4P\ 4p$	8	−1.00193	$2s^22p^3\ ^4S\ 9f$
1	−11.8049	$2s^22p^3\ ^4S\ 3s$	3	−.10509	$2s2p^4\ ^4P\ 5p$		$^5F^o$	
2	−7.03789	$2s2p^4\ ^4P\ 3p$		$^5D^e$		1	−1.61428	$2s2p^4\ ^4P\ 4f$
3	−6.11050	$2s^22p^3\ ^4S\ 4s$	1	−6.19382	$2s2p^4\ ^4P\ 3d$		$^5G^o$	
4	−3.77768	$2s^22p^3\ ^4S\ 5s$	2	−1.85922	$2s2p^4\ ^4P\ 4d$	1	−1.58476	$2s2p^4\ ^4P\ 4f$
5	−2.56319	$2s^22p^3\ ^4S\ 6s$		$^5D^o$				
6	−2.16571	$2s2p^4\ ^4P\ 4p$	1	−9.57258	$2s^22p^3\ ^4S\ 3d$			
7	−1.83557	$2s^22p^3\ ^4S\ 7s$	2	−7.19652	$2s2p^4\ ^4P\ 3p$			
8	−1.38928	$2s^22p^3\ ^4S\ 8s$	3	−5.30507	$2s^22p^3\ ^4S\ 4d$			

Energies in ascending order from ground state for terms with effective $n \leq 4.0$, $L \leq 4$

Term	i	E(Ryds)	Term	i	E(Ryds)	Term	i	E(Ryds)	Term	i	E(Ryds)	Term	i	E(Ryds)
$^3P^e$	1	0.00000	$^3F^e$	1	17.8659	$^1P^o$	3	19.1676	$^3D^o$	5	20.7903	$^1F^o$	3	21.9795
$^1D^e$	1	0.50230	$^1F^e$	1	17.9134	$^3D^o$	3	19.1745	$^5S^o$	2	20.7944	$^3P^e$	6	21.9829
$^1S^e$	1	1.08820	$^3P^e$	3	18.1877	$^1D^o$	2	19.2599	$^3S^o$	3	20.8561	$^3D^o$	6	22.0076
$^3P^o$	1	3.98510	$^3S^e$	1	18.1975	$^3P^o$	3	19.2629	$^3D^e$	3	20.9199	$^1P^e$	3	22.0816
$^1P^o$	1	5.53150	$^5D^o$	1	18.2597	$^3S^o$	2	19.3400	$^3P^o$	5	20.9422	$^3P^e$	7	22.0987
$^1S^e$	2	9.31310	$^3D^e$	2	18.2807	$^1F^o$	1	19.4396	$^1D^e$	4	21.1689	$^1D^o$	4	22.1109
$^5S^o$	1	16.0274	$^1D^e$	2	18.3122	$^3F^o$	2	19.5000	$^5D^e$	1	21.6384	$^3P^o$	6	22.1225
$^3S^o$	1	16.2875	$^1P^e$	2	18.3535	$^3P^o$	4	19.5228	$^3S^e$	2	21.7153	$^5P^e$	4	22.1265
$^3D^o$	1	16.8505	$^3P^e$	4	18.4410	$^5P^e$	2	19.5994	$^5S^o$	3	21.7218	$^3D^e$	4	22.1329
$^1D^o$	1	16.9806	$^3D^o$	2	18.5159	$^3D^o$	4	19.6129	$^5F^e$	1	21.7702	$^3P^e$	8	22.2726
$^5P^e$	1	17.0399	$^1D^e$	3	18.6340	$^1D^o$	3	19.6620	$^3S^o$	4	21.7869	$^3D^o$	7	22.4861
$^3P^o$	2	17.2681	$^3F^o$	1	19.0041	$^1F^o$	2	19.7139	$^5P^e$	3	21.8541	$^1D^o$	5	22.4992
$^3P^e$	2	17.2752	$^1S^o$	1	19.0075	$^1P^o$	4	19.9340	$^3F^o$	3	21.8845	$^5D^o$	3	22.5272
$^1P^o$	2	17.4001	$^1S^e$	3	19.0323	$^3P^e$	5	19.9698	$^1P^o$	5	21.9278	$^3D^o$	8	22.6493
$^1P^e$	1	17.7476	$^3G^o$	1	19.0822	$^5P^o$	1	20.5140	$^1S^e$	4	21.9589	$^3P^o$	7	22.7076
$^3D^e$	1	17.7873	$^1G^o$	1	19.1138	$^5D^o$	2	20.6357	$^3F^e$	2	21.9671			

O-like S (S^{8+})

gf-values for transitions involving terms with effective $n \leq 4.0$, $L \leq 4$

$i\ i'$	gf_L	$i\ i'$	gf_L	$i\ i'$	gf_L	$i\ i'$	gf_L	$i\ i'$	gf_L	$i\ i'$	gf_L
	$^1P^e$–$^1S^o$	2 5	−2.52E−3	1 3	−6.88E−2	8 1	3.25E−1	4 8	−4.75E−5	4 5	2.05E+0
1 1	−1.56E−1	3 1	2.78E−4	1 4	−1.34E−1	8 2	3.02E−2	5 1	6.31E−1	4 6	7.75E−4
2 1	−1.03E−2	3 2	5.77E−1	1 5	−6.61E−1	8 3	8.14E−2	5 2	5.98E−1	4 7	−7.98E−5
3 1	2.69E−3	3 3	−9.37E−3		$^1F^o$–$^1D^e$	8 4	1.77E+0	5 3	8.51E−2		$^3D^o$–$^3P^e$
	$^1P^o$–$^1S^e$	3 4	−1.62E−2	1 1	2.31E+0		$^3P^o$–$^3S^e$	5 4	6.69E−2	1 1	6.18E−1
1 1	7.06E−2	3 5	−1.33E−2	1 2	7.21E−1	1 1	−1.71E−2	5 5	1.17E+0	1 2	−4.98E−2
1 2	−4.03E−1	4 1	1.66E−1	1 3	9.58E−2	1 2	−1.56E−1	5 6	−3.68E−2	1 3	−1.58E+0
1 3	−6.08E−3	4 2	4.27E−2	1 4	−1.43E−2	2 1	−3.58E−1	5 7	−4.55E−1	1 4	−9.11E−3
1 4	−4.61E−2	4 3	4.31E−3	2 1	2.94E+0	2 2	−6.16E−1	5 8	−1.71E−1	1 5	−7.38E−2
2 1	1.55E−1	4 4	2.55E−4	2 2	3.36E−1	3 1	4.78E−2	6 1	8.76E−2	1 6	−1.83E+0
2 2	5.99E−7	4 5	−2.82E−1	2 3	8.69E−1	3 2	−1.73E−4	6 2	1.13E−3	1 7	−1.39E−1
2 3	−2.73E−1		$^1D^o$–$^1P^e$	2 4	−1.96E−2	4 1	8.97E−1	6 3	2.56E−1	1 8	−5.03E−2
2 4	−1.11E−1	1 1	−2.68E−1	3 1	7.04E−1	4 2	−2.76E−3	6 4	5.72E−2	2 1	2.40E+0
3 1	1.32E−1	1 2	−7.56E−2	3 2	1.27E−3	5 1	1.19E−5	6 5	4.36E−1	2 2	2.49E+0
3 2	2.38E−5	1 3	−8.15E−1	3 3	4.32E−1	5 2	−6.73E−3	6 6	4.22E−3	2 3	2.20E−2
3 3	2.30E−3	2 1	4.19E−1	3 4	8.17E−1	6 1	2.13E−1	6 7	1.37E−5	2 4	6.69E−4
3 4	−1.90E−3	2 2	5.70E−2		$^1F^o$–$^1F^e$	6 2	3.94E−5	6 8	−7.44E−4	2 5	−9.25E−3
4 1	1.64E+0	2 3	−6.53E−3	1 1	5.12E−1	7 1	9.94E−2	7 1	1.49E−1	2 6	−2.47E−1
4 2	7.89E−5	3 1	2.35E−2	2 1	2.93E−2	7 2	9.61E−1	7 2	1.21E−2	2 7	−7.85E−1
4 3	2.17E−1	3 2	6.27E−1	3 1	6.54E−1		$^3P^o$–$^3P^e$	7 3	1.24E−1	2 8	−9.80E−2
4 4	−9.43E−3	3 3	−1.45E−5		$^1G^o$–$^1F^e$	1 1	1.12E+0	7 4	7.28E−3	3 1	4.03E+0
5 1	2.71E−3	4 1	1.76E−2	1 1	1.58E+0	1 2	−8.97E−4	7 5	3.16E−3	3 2	3.98E−2
5 2	9.91E−3	4 2	1.33E−1		$^3P^e$–$^3S^o$	1 3	−4.54E−2	7 6	1.29E−1	3 3	8.45E−1
5 3	2.25E−3	4 3	6.57E−5	1 1	−4.10E−1	1 4	−3.41E−3	7 7	4.51E−3	3 4	5.85E−3
5 4	−2.69E−4	5 1	1.94E−1	1 2	−1.33E+0	1 5	−8.44E−1	7 8	2.52E−3	3 5	−1.94E−2
	$^1P^o$–$^1P^e$	5 2	1.86E−2	1 3	−2.90E−1	1 6	−5.06E−2		$^3D^e$–$^3P^o$	3 6	−1.25E−2
1 1	−2.58E−3	5 3	6.10E−2	1 4	−4.91E−2	1 7	−1.30E+0	1 1	3.59E−3	3 7	−1.24E−2
1 2	−5.08E−3		$^1D^o$–$^1D^e$	2 1	1.12E+0	1 8	−1.42E−1	1 2	3.94E−2	3 8	−2.66E−2
1 3	−3.88E−1	1 1	5.46E−1	2 2	−1.33E−1	2 1	3.49E−1	1 3	−3.87E−1	4 1	3.63E+0
2 1	−1.90E−2	1 2	−6.28E−1	2 3	−1.96E−1	2 2	−1.84E−4	1 4	−4.94E−2	4 2	2.09E−3
2 2	−3.54E−1	1 3	−5.88E−1	2 4	−5.03E−1	2 3	−1.41E−2	1 5	−6.30E−3	4 3	3.32E−2
2 3	−2.87E−1	1 4	−2.88E−1	3 1	3.70E−1	2 4	−1.50E+0	1 6	−5.62E−1	4 4	1.94E+0
3 1	3.87E−1	2 1	8.28E−1	3 2	−4.34E−1	2 5	−5.24E−2	1 7	−5.34E−1	4 5	−1.26E−2
3 2	2.28E−2	2 2	3.00E−1	3 3	−6.57E−2	2 6	−6.26E−1	2 1	2.67E−2	4 6	−7.52E−3
3 3	−5.67E−3	2 3	1.56E−2	3 4	−1.18E−1	2 7	−3.31E−2	2 2	2.05E+0	4 7	−3.11E−3
4 1	1.03E−3	2 4	−4.71E−3	4 1	1.41E−1	2 8	−2.90E−2	2 3	−6.98E−2	4 8	−5.01E−3
4 2	3.56E−1	3 1	1.75E+0	4 2	−1.80E−3	3 1	3.94E+0	2 4	−2.01E−2	5 1	1.03E+0
4 3	−2.62E−4	3 2	3.11E−3	4 3	−4.85E−2	3 2	2.05E−1	2 5	−4.47E−3	5 2	8.07E−1
5 1	4.07E−1	3 3	2.64E−1	4 4	−4.02E−2	3 3	8.72E−1	2 6	−4.02E−2	5 3	2.62E−1
5 2	1.91E−3	3 4	−9.98E−3	5 1	3.21E−1	3 4	7.51E−3	2 7	−1.79E+0	5 4	2.26E−1
5 3	−3.47E−3	4 1	5.44E−1	5 2	4.74E−3	3 5	−2.11E−2	3 1	7.76E−1	5 5	1.02E+0
	$^1D^e$–$^1P^o$	4 2	3.57E−1	5 3	−3.76E−1	3 6	−1.14E−3	3 2	3.56E−1	5 6	−5.15E−3
1 1	−8.24E−1	4 3	2.08E−1	5 4	−2.38E−2	3 7	−5.34E−3	3 3	8.45E−4	5 7	−1.16E−1
1 2	−1.55E−1	4 4	6.66E−1	6 1	1.10E−1	3 8	−1.31E−2	3 4	4.71E−5	5 8	−4.20E−3
1 3	−6.47E−1	5 1	6.23E−2	6 2	7.48E−4	4 1	4.40E−1	3 5	−2.12E−3	6 1	1.34E−2
1 4	−8.06E−2	5 2	1.23E−1	6 3	5.34E−2	4 2	1.45E−3	3 6	−1.48E+0	6 2	5.71E−5
1 5	−3.14E−1	5 3	2.53E−1	6 4	1.38E−1	4 3	1.24E−1	3 7	−1.77E−2	6 3	7.79E−2
2 1	1.09E−2	5 4	4.95E−2	7 1	8.07E−2	4 4	5.19E−1	4 1	5.12E+0	6 4	1.86E−1
2 2	2.64E−1		$^1F^e$–$^1D^o$	7 2	6.74E−4	4 5	−2.62E−3	4 2	6.15E−3	6 5	1.36E−2
2 3	−8.13E−2	1 1	9.40E−1	7 3	9.77E−1	4 6	−1.07E−5	4 3	9.01E−5	6 6	2.21E−4
2 4	−7.72E−4	1 2	−3.61E−2	7 4	1.66E−1	4 7	−1.91E−3	4 4	1.10E−3	6 7	−1.17E−3

O-like S (S^{8+})

$i\ i'$	gf_L	$i\ i'$	gf_L	$i\ i'$	gf_L	$i\ i'$	gf_L	$i\ i'$	gf_L	$i\ i'$	gf_L
6 8	−2.62E−7	2 1	5.21E−2	7 3	5.69E−2	2 8	−7.80E−2		$^{3}\mathbf{G}^{o}$–$^{3}\mathbf{F}^{e}$	1 1	3.05E+0
7 1	2.71E−1	2 2	3.87E−3	7 4	1.34E−4		$^{3}\mathbf{F}^{o}$–$^{3}\mathbf{D}^{e}$	1 1	4.74E+0		$^{5}\mathbf{D}^{o}$–$^{5}\mathbf{P}^{e}$
7 2	7.55E−3	2 3	−6.24E−4	8 1	2.28E−2	1 1	2.39E+0	1 2	−1.85E−4	1 1	4.19E+0
7 3	1.07E+0	2 4	−8.97E−1	8 2	1.59E−2	1 2	1.27E−2		$^{5}\mathbf{P}^{e}$–$^{5}\mathbf{S}^{o}$	1 2	−4.03E−3
7 4	2.68E−2	3 1	1.59E+0	8 3	2.83E−3	1 3	−6.09E−3	1 1	2.04E+0	1 3	−1.48E+0
7 5	3.34E−2	3 2	2.50E−2	8 4	5.18E−4	1 4	−1.12E−2	1 2	−4.59E−1	1 4	−4.15E−1
7 6	5.49E−2	3 3	−2.58E−3		$^{3}\mathbf{F}^{e}$–$^{3}\mathbf{D}^{o}$	2 1	6.05E−3	1 3	−8.77E−1	2 1	1.12E+0
7 7	6.76E−3	3 4	−2.48E−2	1 1	3.03E+0	2 2	3.65E+0	2 1	1.08E+0	2 2	4.11E+0
7 8	1.32E−3	4 1	1.53E−2	1 2	−6.86E−3	2 3	−6.41E−4	2 2	−1.02E+0	2 3	−1.29E−1
8 1	1.14E+0	4 2	7.27E−1	1 3	−1.62E−1	2 4	−8.37E−6	2 3	−1.80E−3	2 4	−1.88E−2
8 2	2.22E+0	4 3	−1.17E−4	1 4	−2.09E−2	3 1	2.06E−1	3 1	8.91E−2	3 1	4.41E+0
8 3	1.91E−1	4 4	−1.44E−5	1 5	−2.88E−3	3 2	2.26E−1	3 2	1.60E+0	3 2	8.45E−2
8 4	5.15E−2	5 1	1.93E−3	1 6	−6.61E−1	3 3	3.14E+0	3 3	7.08E−2	3 3	1.19E+0
8 5	1.37E−2	5 2	1.46E−4	1 7	−1.74E+0	3 4	−1.67E−3	4 1	7.92E−1	3 4	5.68E+0
8 6	9.77E−1	5 3	−1.98E−5	1 8	−2.08E−2		$^{3}\mathbf{F}^{o}$–$^{3}\mathbf{F}^{e}$	4 2	6.99E−4		$^{5}\mathbf{D}^{o}$–$^{5}\mathbf{D}^{e}$
8 7	7.40E−1	5 4	−7.41E−1	2 1	2.14E−3	1 1	1.00E+0	4 3	3.01E+0	1 1	−1.40E+0
8 8	2.68E+0	6 1	1.15E+0	2 2	6.39E−2	1 2	−5.71E−6		$^{5}\mathbf{P}^{o}$–$^{5}\mathbf{P}^{e}$	2 1	−8.47E−1
	$^{3}\mathbf{D}^{o}$–$^{3}\mathbf{D}^{e}$	6 2	2.75E−1	2 3	1.73E−3	2 1	1.33E−1	1 1	1.06E+0	3 1	1.89E−4
1 1	−1.80E+0	6 3	2.62E+0	2 4	3.41E−4	2 2	−1.54E−4	1 2	1.98E+0		$^{5}\mathbf{F}^{e}$–$^{5}\mathbf{D}^{o}$
1 2	−1.22E−1	6 4	−3.77E−3	2 5	3.90E+0	3 1	1.83E+0	1 3	−1.14E+0	1 1	1.48E−1
1 3	−1.80E+0	7 1	4.55E−1	2 6	−9.89E−7	3 2	−8.55E−5	1 4	−1.08E−1	1 2	5.92E+0
1 4	−6.25E−4	7 2	1.20E−2	2 7	−5.66E−5				$^{5}\mathbf{D}^{e}$–$^{5}\mathbf{P}^{o}$	1 3	−1.22E−1

O-like Ar (Ar^{10+})

Term energies relative to $2s^22p^3$ ^{4}S ionization threshold for each symmetry

i	E(Ryds)	Description	i	E(Ryds)	Description	i	E(Ryds)	Description	i	E(Ryds)	Description
		^{1}S^e	26	−.59968	$2s^22p^3$ ^{2}P $8p$	20	−2.19519	$2s^22p^3$ ^{2}P $6p$	8	−3.96138	$2s^22p^3$ ^{2}D $5f$
1	−38.3063	$2s^22p^4$			**^{1}P^o**	21	−2.14004	$2s2p^4$ ^{2}D $4d$	9	−3.46064	$2s^22p^3$ ^{2}P $5f$
2	−28.6403	$2p^6$	1	−33.1089	$2s2p^5$	22	−1.97362	$2s^22p^3$ ^{2}P $6f$	10	−2.72209	$2s^22p^3$ ^{2}D $6p$
3	−13.2860	$2s^22p^3$ ^{2}P $3p$	2	−15.2932	$2s^22p^3$ ^{2}P $3s$	23	−1.70225	$2s^22p^3$ ^{2}D $7p$	11	−2.48311	$2s^22p^3$ ^{2}D $6f$
4	−9.93594	$2s2p^4$ ^{2}S $3s$	3	−13.1508	$2s^22p^3$ ^{2}D $3d$	24	−1.59220	$2s^22p^3$ ^{2}D $7f$	12	−2.14896	$2s2p^4$ ^{2}D $4d$
5	−8.19756	$2s2p^4$ ^{2}D $3d$	4	−12.1714	$2s^22p^3$ ^{2}P $3d$	25	−1.25686	$2s2p^4$ ^{2}S $4d$	13	−1.97914	$2s^22p^3$ ^{2}P $6f$
6	−6.66881	$2s^22p^3$ ^{2}P $4p$	5	−9.93087	$2s2p^4$ ^{2}D $3p$	26	−1.21060	$2s^22p^3$ ^{2}P $7p$	14	−1.73797	$2s^22p^3$ ^{2}D $7p$
7	−3.73902	$2p^5$ ^{2}P $3p$	6	−8.87011	$2s2p^4$ ^{2}S $3p$	27	−1.08042	$2s^22p^3$ ^{2}P $7f$	15	−1.59080	$2s^22p^3$ ^{2}D $7f$
8	−3.71249	$2s^22p^3$ ^{2}P $5p$	7	−8.01381	$2s2p^4$ ^{2}P $3p$	28	−1.08042	$2s^22p^3$ ^{2}D $8p$	16	−1.11715	$2s^22p^3$ ^{2}D $8p$
9	−2.25006	$2s2p^4$ ^{2}S $4s$	8	−7.40726	$2s^22p^3$ ^{2}P $4s$	29	−1.01535	$2s^22p^3$ ^{2}D $8f$	17	−1.08760	$2s^22p^3$ ^{2}P $7f$
10	−2.13143	$2s^22p^3$ ^{2}P $6p$	9	−6.92690	$2s^22p^3$ ^{2}D $4d$	30	−.86567	$2s2p^4$ ^{2}P $4d$	18	−1.01959	$2s^22p^3$ ^{2}D $8f$
11	−2.09791	$2s2p^4$ ^{2}D $4d$	10	−6.25352	$2s^22p^3$ ^{2}P $4d$	31	−.66672	$2s^22p^3$ ^{2}D $9p$	19	−.98079	$2s2p^4$ ^{2}P $4d$
12	−1.18086	$2s^22p^3$ ^{2}P $7p$	11	−5.71383	$2p^5$ ^{2}P $3s$	32	−.61218	$2s^22p^3$ ^{2}D $9f$	20	−.68081	$2s^22p^3$ ^{2}D $9p$
13	−.56880	$2s^22p^3$ ^{2}P $8p$	12	−4.11254	$2s^22p^3$ ^{2}D $5d$	33	−.58982	$2s^22p^3$ ^{2}P $8p$	21	−.61318	$2s^22p^3$ ^{2}D $9f$
14	−.15165	$2s^22p^3$ ^{2}P $9p$	13	−4.07363	$2s^22p^3$ ^{2}P $5s$	34	−.50053	$2s^22p^3$ ^{2}P $8f$	22	−.50303	$2s^22p^3$ ^{2}P $8f$
		^{1}S^o	14	−3.50806	$2s^22p^3$ ^{2}P $5d$			**^{1}D^o**			**^{1}F^o**
1	−13.3542	$2s^22p^3$ ^{2}D $3d$	15	−2.88656	$2p^5$ ^{2}P $3d$	1	−15.8113	$2s^22p^3$ ^{2}D $3s$	1	−12.7878	$2s^22p^3$ ^{2}D $3d$
2	−8.50922	$2s2p^4$ ^{2}P $3p$	16	−2.69598	$2s2p^4$ ^{2}D $4p$	2	−13.0342	$2s^22p^3$ ^{2}D $3d$	2	−12.4595	$2s^22p^3$ ^{2}P $3d$
3	−7.01108	$2s^22p^3$ ^{2}D $4d$	17	−2.55801	$2s^22p^3$ ^{2}D $6d$	3	−12.5221	$2s^22p^3$ ^{2}P $3d$	3	−9.85943	$2s2p^4$ ^{2}D $3p$
4	−4.13549	$2s^22p^3$ ^{2}D $5d$	18	−2.33878	$2s^22p^3$ ^{2}P $6s$	4	−9.69470	$2s2p^4$ ^{2}D $3p$	4	−6.78901	$2s^22p^3$ ^{2}D $4d$
5	−2.58433	$2s^22p^3$ ^{2}D $6d$	19	−2.01455	$2s^22p^3$ ^{2}P $6d$	5	−8.62738	$2s2p^4$ ^{2}P $3p$	5	−6.36438	$2s^22p^3$ ^{2}P $4d$
6	−1.65559	$2s^22p^3$ ^{2}D $7d$	20	−1.94365	$2s2p^4$ ^{2}D $4f$	6	−7.92125	$2s^22p^3$ ^{2}D $4s$	6	−4.02704	$2s^22p^3$ ^{2}D $5d$
7	−1.47523	$2s2p^4$ ^{2}P $4p$	21	−1.79547	$2s2p^4$ ^{2}S $4p$	7	−6.83299	$2s^22p^3$ ^{2}D $4d$	7	−3.56115	$2s^22p^3$ ^{2}P $5d$
8	−1.05291	$2s^22p^3$ ^{2}D $8d$	22	−1.63950	$2s^22p^3$ ^{2}D $7d$	8	−6.43754	$2s^22p^3$ ^{2}P $4d$	8	−3.40713	$2p^5$ ^{2}P $3d$
9	−.64329	$2s^22p^3$ ^{2}D $9d$	23	−1.34957	$2s2p^4$ ^{2}P $4p$	9	−4.60240	$2s^22p^3$ ^{2}D $5s$	9	−2.73680	$2s2p^4$ ^{2}D $4p$
		^{1}P^e	24	−1.26733	$2s^22p^3$ ^{2}P $7s$	10	−4.04412	$2s^22p^3$ ^{2}D $5d$	10	−2.52149	$2s^22p^3$ ^{2}D $6d$
1	−14.8765	$2s^22p^3$ ^{2}D $3p$	25	−1.10489	$2s^22p^3$ ^{2}P $7d$	11	−3.60160	$2s^22p^3$ ^{2}P $5d$	11	−2.03913	$2s^22p^3$ ^{2}P $6d$
2	−14.1307	$2s^22p^3$ ^{2}P $3p$	26	−1.04256	$2s^22p^3$ ^{2}D $8d$	12	−3.35699	$2p^5$ ^{2}P $3d$	12	−1.94759	$2s2p^4$ ^{2}D $4f$
3	−9.78975	$2s2p^4$ ^{2}P $3s$	27	−.65026	$2s^22p^3$ ^{2}P $8s$	13	−2.85444	$2s^22p^3$ ^{2}D $6s$	13	−1.61609	$2s^22p^3$ ^{2}D $7d$
4	−8.33368	$2s2p^4$ ^{2}D $3d$	28	−.63675	$2s^22p^3$ ^{2}D $9d$	14	−2.68516	$2s2p^4$ ^{2}D $4p$	14	−1.12464	$2s^22p^3$ ^{2}P $7d$
5	−7.52276	$2s^22p^3$ ^{2}D $4p$	29	−.51728	$2s^22p^3$ ^{2}P $8d$	15	−2.53022	$2s^22p^3$ ^{2}D $6d$	15	−1.03147	$2s2p^4$ ^{2}S $4f$
6	−7.07368	$2s2p^4$ ^{2}P $3d$			**^{1}D^e**	16	−2.06330	$2s^22p^3$ ^{2}P $6d$	16	−1.02843	$2s^22p^3$ ^{2}D $8d$
7	−6.96255	$2s^22p^3$ ^{2}P $4p$	1	−38.9910	$2s^22p^4$	17	−1.95385	$2s2p^4$ ^{2}D $4f$	17	−.75253	$2s2p^4$ ^{2}P $4f$
8	−6.54724	$2s^22p^3$ ^{2}D $4f$	2	−14.1818	$2s^22p^3$ ^{2}D $3p$	18	−1.82178	$2s^22p^3$ ^{2}D $7s$	18	−.62666	$2s^22p^3$ ^{2}D $9d$
9	−4.63146	$2p^5$ ^{2}P $3p$	3	−13.7817	$2s^22p^3$ ^{2}P $3p$	19	−1.62245	$2s^22p^3$ ^{2}D $7d$	19	−.52930	$2s^22p^3$ ^{2}P $8d$
10	−4.39453	$2s^22p^3$ ^{2}D $5p$	4	−10.8417	$2s2p^4$ ^{2}D $3s$	20	−1.49103	$2s2p^4$ ^{2}P $4p$			**^{1}G^e**
11	−3.96235	$2s^22p^3$ ^{2}D $5f$	5	−8.29300	$2s2p^4$ ^{2}D $3d$	21	−1.15816	$2s^22p^3$ ^{2}D $8s$	1	−8.42979	$2s2p^4$ ^{2}D $3d$
12	−3.87237	$2s^22p^3$ ^{2}P $5p$	6	−7.40328	$2s2p^4$ ^{2}S $3d$	22	−1.13719	$2s^22p^3$ ^{2}P $7d$	2	−6.68492	$2s^22p^3$ ^{2}D $4f$
13	−2.73468	$2s^22p^3$ ^{2}D $6p$	7	−7.26576	$2s^22p^3$ ^{2}D $4p$	23	−1.03078	$2s^22p^3$ ^{2}D $8d$	3	−6.18420	$2s^22p^3$ ^{2}P $4f$
14	−2.48607	$2s^22p^3$ ^{2}D $6f$	8	−7.03830	$2s2p^4$ ^{2}P $3d$	24	−.72078	$2s2p^4$ ^{2}P $4f$	4	−3.96499	$2s^22p^3$ ^{2}D $5f$
15	−2.22527	$2s^22p^3$ ^{2}P $6p$	9	−6.84055	$2s^22p^3$ ^{2}P $4p$	25	−.71775	$2s^22p^3$ ^{2}D $9s$	5	−3.46003	$2s^22p^3$ ^{2}P $5f$
16	−2.14993	$2s2p^4$ ^{2}D $4d$	10	−6.57754	$2s^22p^3$ ^{2}D $4f$	26	−.62922	$2s^22p^3$ ^{2}D $9d$	6	−2.48424	$2s^22p^3$ ^{2}D $6f$
17	−1.96652	$2s2p^4$ ^{2}P $4s$	11	−6.15065	$2s^22p^3$ ^{2}P $4f$	27	−.53828	$2s^22p^3$ ^{2}P $8d$	7	−2.21365	$2s2p^4$ ^{2}D $4d$
18	−1.73099	$2s^22p^3$ ^{2}D $7p$	12	−4.65272	$2p^5$ ^{2}P $3p$			**^{1}F^e**	8	−1.97742	$2s^22p^3$ ^{2}P $6f$
19	−1.59420	$2s^22p^3$ ^{2}D $7f$	13	−4.27095	$2s^22p^3$ ^{2}D $5p$	1	−14.6746	$2s^22p^3$ ^{2}D $3p$	9	−1.59087	$2s^22p^3$ ^{2}D $7f$
20	−1.23024	$2s^22p^3$ ^{2}P $7p$	14	−3.96184	$2s^22p^3$ ^{2}D $5f$	2	−8.23834	$2s2p^4$ ^{2}D $3d$	10	−1.08342	$2s^22p^3$ ^{2}P $7f$
21	−1.11037	$2s^22p^3$ ^{2}D $8p$	15	−3.82942	$2s^22p^3$ ^{2}P $5p$	3	−7.52730	$2s^22p^3$ ^{2}D $4p$	11	−1.01129	$2s^22p^3$ ^{2}D $8f$
22	−1.02141	$2s^22p^3$ ^{2}D $8f$	16	−3.45154	$2s^22p^3$ ^{2}P $5f$	4	−7.23915	$2s2p^4$ ^{2}P $3d$	12	−.61416	$2s^22p^3$ ^{2}D $9f$
23	−.89159	$2s2p^4$ ^{2}P $4d$	17	−3.13890	$2s2p^4$ ^{2}D $4s$	5	−6.64799	$2s^22p^3$ ^{2}D $4f$	13	−.50312	$2s^22p^3$ ^{2}P $8f$
24	−.68419	$2s^22p^3$ ^{2}D $9p$	18	−2.66268	$2s^22p^3$ ^{2}D $6p$	6	−6.16993	$2s^22p^3$ ^{2}P $4f$			**^{1}G^o**
25	−.61201	$2s^22p^3$ ^{2}D $9f$	19	−2.48448	$2s^22p^3$ ^{2}D $6f$	7	−4.37554	$2s^22p^3$ ^{2}D $5p$	1	−13.2159	$2s^22p^3$ ^{2}D $3d$

O-like Ar (Ar^{10+})

i	E(Ryds)	Description
2	−6.96415	$2s^22p^3$ ^{2}D $4d$
3	−4.11292	$2s^22p^3$ ^{2}D $5d$
4	−2.57189	$2s^22p^3$ ^{2}D $6d$
5	−1.94964	$2s2p^4$ ^{2}D $4f$
6	−1.64607	$2s^22p^3$ ^{2}D $7d$
7	−1.04843	$2s^22p^3$ ^{2}D $8d$
8	−.74543	$2s2p^4$ ^{2}P $4f$
9	−.63990	$2s^22p^3$ ^{2}D $9d$
	^{1}H^e	
1	−6.70300	$2s^22p^3$ ^{2}D $4f$
2	−3.97468	$2s^22p^3$ ^{2}D $5f$
3	−2.49008	$2s^22p^3$ ^{2}D $6f$
4	−1.59494	$2s^22p^3$ ^{2}D $7f$
5	−1.01414	$2s^22p^3$ ^{2}D $8f$
6	−.61608	$2s^22p^3$ ^{2}D $9f$
	^{1}H^o	
1	−2.01331	$2s2p^4$ ^{2}D $4f$
	^{3}S^e	
1	−14.3228	$2s^22p^3$ ^{2}P $3p$
2	−10.2299	$2s2p^4$ ^{2}S $3s$
3	−8.46912	$2s2p^4$ ^{2}D $3d$
4	−7.02144	$2s^22p^3$ ^{2}P $4p$
5	−4.99224	$2p^5$ ^{2}P $3p$
6	−3.89206	$2s^22p^3$ ^{2}P $5p$
7	−2.37277	$2s2p^4$ ^{2}S $4s$
8	−2.24985	$2s2p^4$ ^{2}D $4d$
9	−2.16025	$2s^22p^3$ ^{2}P $6p$
10	−1.23310	$2s^22p^3$ ^{2}P $7p$
11	−.60359	$2s^22p^3$ ^{2}P $8p$
12	−.17586	$2s^22p^3$ ^{2}P $9p$
	^{3}S^o	
1	−16.6146	$2s^22p^3$ ^{4}S $3s$
2	−12.9177	$2s^22p^3$ ^{2}D $3d$
3	−11.1555	$2s2p^4$ ^{4}P $3p$
4	−8.81270	$2s^22p^3$ ^{4}S $4s$
5	−8.45218	$2s2p^4$ ^{2}P $3p$
6	−6.84720	$2s^22p^3$ ^{2}D $4d$
7	−5.47357	$2s^22p^3$ ^{4}S $5s$
8	−4.16797	$2s2p^4$ ^{4}P $4p$
9	−4.05827	$2s^22p^3$ ^{2}D $5d$
10	−3.71483	$2s^22p^3$ ^{4}S $6s$
11	−2.69404	$2s^22p^3$ ^{4}S $7s$
12	−2.54129	$2s^22p^3$ ^{2}D $6d$
13	−2.04014	$2s^22p^3$ ^{4}S $8s$
14	−1.63113	$2s^22p^3$ ^{2}D $7d$
15	−1.59877	$2s^22p^3$ ^{4}S $9s$
16	−1.45895	$2s2p^4$ ^{2}P $4p$
	^{3}P^e	
1	−39.5787	$2s^22p^4$
2	−15.4154	$2s^22p^3$ ^{4}S $3p$
3	−14.3293	$2s^22p^3$ ^{2}D $3p$
4	−14.0215	$2s^22p^3$ ^{2}P $3p$
5	−12.2364	$2s2p^4$ ^{4}P $3s$
6	−9.89599	$2s2p^4$ ^{2}P $3s$
7	−9.62532	$2s2p^4$ ^{4}P $3d$
8	−8.41653	$2s2p^4$ ^{2}D $3d$
9	−8.27130	$2s^22p^3$ ^{4}S $4p$
10	−7.39047	$2s^22p^3$ ^{2}D $4p$
11	−7.26284	$2s2p^4$ ^{2}P $3d$
12	−6.92574	$2s^22p^3$ ^{2}P $4p$
13	−6.59676	$2s^22p^3$ ^{2}D $4f$
14	−5.20866	$2s^22p^3$ ^{4}S $5p$
15	−4.67152	$2p^5$ ^{2}P $3p$
16	−4.58167	$2s2p^4$ ^{4}P $4s$
17	−4.32172	$2s^22p^3$ ^{2}D $5p$
18	−3.96054	$2s^22p^3$ ^{2}D $5f$
19	−3.84424	$2s^22p^3$ ^{2}P $5p$
20	−3.61134	$2s2p^4$ ^{4}P $4d$
21	−3.53089	$2s^22p^3$ ^{4}S $6p$
22	−2.70679	$2s^22p^3$ ^{2}D $6p$
23	−2.59666	$2s^22p^3$ ^{4}S $7p$
24	−2.48406	$2s^22p^3$ ^{2}D $6f$
25	−2.23924	$2s2p^4$ ^{2}D $4d$
26	−2.18660	$2s^22p^3$ ^{2}P $6p$
27	−2.00976	$2s2p^4$ ^{2}P $4s$
28	−1.96981	$2s^22p^3$ ^{4}S $8p$
29	−1.72298	$2s^22p^3$ ^{2}D $7p$
30	−1.59196	$2s^22p^3$ ^{2}D $7f$
31	−1.55489	$2s^22p^3$ ^{4}S $9p$
32	−1.32358	$2s2p^4$ ^{4}P $5s$
	^{3}P^o	
1	−34.8957	$2s2p^5$
2	−15.4518	$2s^22p^3$ ^{2}P $3s$
3	−13.0225	$2s^22p^3$ ^{2}D $3d$
4	−12.7252	$2s^22p^3$ ^{2}P $3d$
5	−11.0652	$2s2p^4$ ^{4}P $3p$
6	−9.66839	$2s2p^4$ ^{2}D $3p$
7	−8.97114	$2s2p^4$ ^{2}S $3p$
8	−8.73380	$2s2p^4$ ^{2}P $3p$
9	−7.47295	$2s^22p^3$ ^{2}P $4s$
10	−6.88491	$2s^22p^3$ ^{2}D $4d$
11	−6.46216	$2s^22p^3$ ^{2}P $4d$
12	−5.88469	$2p^5$ ^{2}P $3s$
13	−4.12808	$2s^22p^3$ ^{2}P $5s$
14	−4.07765	$2s2p^4$ ^{4}P $4p$:
15	−4.04207	$2s^22p^3$ ^{2}D $5d$
16	−3.60793	$2p^5$ ^{2}P $3d$
17	−3.60705	$2s^22p^3$ ^{2}P $5d$
18	−2.71076	$2s2p^4$ ^{2}D $4p$
19	−2.55225	$2s^22p^3$ ^{2}D $6d$
20	−2.34896	$2s^22p^3$ ^{2}P $6s$
21	−2.06630	$2s^22p^3$ ^{2}P $6d$
22	−1.99575	$2s2p^4$ ^{2}D $4f$
23	−1.82919	$2s2p^4$ ^{2}S $4p$
24	−1.63856	$2s^22p^3$ ^{2}D $7d$
25	−1.55370	$2s2p^4$ ^{2}P $4p$
26	−1.30915	$2s^22p^3$ ^{2}P $7s$
27	−1.13749	$2s^22p^3$ ^{2}P $7d$
28	−1.04013	$2s^22p^3$ ^{2}D $8d$
29	−1.03499	$2s2p^4$ ^{4}P $5p$
30	−.65538	$2s^22p^3$ ^{2}P $8s$
31	−.63402	$2s^22p^3$ ^{2}D $9d$
32	−.53951	$2s^22p^3$ ^{2}P $8d$
	^{3}D^e	
1	−14.8298	$2s^22p^3$ ^{2}D $3p$
2	−14.2204	$2s^22p^3$ ^{2}P $3p$
3	−11.1472	$2s2p^4$ ^{2}D $3s$
4	−9.60596	$2s2p^4$ ^{4}P $3d$
5	−8.41762	$2s2p^4$ ^{2}D $3d$
6	−7.59465	$2s2p^4$ ^{2}S $3d$
7	−7.56014	$2s2p^4$ ^{2}P $3d$
8	−7.46058	$2s^22p^3$ ^{2}D $4p$
9	−6.98708	$2s^22p^3$ ^{2}P $4p$
10	−6.64130	$2s^22p^3$ ^{2}D $4f$
11	−6.15812	$2s^22p^3$ ^{2}P $4f$
12	−4.76779	$2p^5$ ^{2}P $3p$
13	−4.39880	$2s^22p^3$ ^{2}D $5p$
14	−3.96523	$2s^22p^3$ ^{2}D $5f$
15	−3.87881	$2s^22p^3$ ^{2}P $5p$
16	−3.55717	$2s2p^4$ ^{4}P $4d$
17	−3.45710	$2s^22p^3$ ^{2}P $5f$
18	−3.25554	$2s2p^4$ ^{2}D $4s$
19	−2.72648	$2s^22p^3$ ^{2}D $6p$
20	−2.48712	$2s^22p^3$ ^{2}D $6f$
21	−2.23227	$2s2p^4$ ^{2}D $4d$
22	−2.20370	$2s^22p^3$ ^{2}P $6p$
23	−1.97931	$2s^22p^3$ ^{2}P $6f$
24	−1.74234	$2s^22p^3$ ^{2}D $7p$
25	−1.59243	$2s^22p^3$ ^{2}D $7f$
26	−1.33336	$2s2p^4$ ^{2}S $4d$
27	−1.22973	$2s^22p^3$ ^{2}P $7p$
28	−1.11597	$2s^22p^3$ ^{2}D $8p$
29	−1.08911	$2s^22p^3$ ^{2}P $7f$
30	−1.05823	$2s2p^4$ ^{2}P $4d$
31	−1.00440	$2s^22p^3$ ^{2}D $8f$
32	−.78269	$2s2p^4$ ^{4}P $5d$
33	−.68492	$2s^22p^3$ ^{2}D $9p$
34	−.61372	$2s^22p^3$ ^{2}D $9f$
35	−.60079	$2s^22p^3$ ^{2}P $8p$
36	−.50264	$2s^22p^3$ ^{2}P $8f$
	^{3}D^o	
1	−15.9681	$2s^22p^3$ ^{2}D $3s$
2	−13.8971	$2s^22p^3$ ^{4}S $3d$
3	−13.1227	$2s^22p^3$ ^{2}D $3d$
4	−12.5989	$2s^22p^3$ ^{2}P $3d$
5	−11.2413	$2s2p^4$ ^{4}P $3p$
6	−9.80213	$2s2p^4$ ^{2}D $3p$
7	−8.69894	$2s2p^4$ ^{2}P $3p$
8	−7.97545	$2s^22p^3$ ^{2}D $4s$
9	−7.73388	$2s^22p^3$ ^{4}S $4d$
10	−6.94444	$2s^22p^3$ ^{2}D $4d$
11	−6.41949	$2s^22p^3$ ^{2}P $4d$
12	−4.93642	$2s^22p^3$ ^{4}S $5d$
13	−4.62650	$2s^22p^3$ ^{2}D $5s$
14	−4.19069	$2s2p^4$ ^{4}P $4p$
15	−4.10755	$2s^22p^3$ ^{2}D $5d$
16	−3.59106	$2s^22p^3$ ^{2}P $5d$
17	−3.44149	$2s2p^4$ ^{4}P $4f$
18	−3.41224	$2s^22p^3$ ^{4}S $6d$
19	−3.32998	$2p^5$ ^{2}P $3d$
20	−2.87952	$2s^22p^3$ ^{2}D $6s$
21	−2.71084	$2s2p^4$ ^{2}D $4p$
22	−2.57144	$2s^22p^3$ ^{2}D $6d$
23	−2.50170	$2s^22p^3$ ^{4}S $7d$
24	−2.05422	$2s^22p^3$ ^{2}P $6d$
25	−1.97424	$2s2p^4$ ^{2}D $4f$
26	−1.91565	$2s^22p^3$ ^{4}S $8d$
27	−1.82647	$2s^22p^3$ ^{2}D $7s$
28	−1.64554	$2s^22p^3$ ^{2}D $7d$
29	−1.51992	$2s2p^4$ ^{2}P $4p$
30	−1.51087	$2s^22p^3$ ^{4}S $9d$
	^{3}F^e	
1	−14.7336	$2s^22p^3$ ^{2}D $3p$
2	−9.81450	$2s2p^4$ ^{4}P $3d$
3	−8.46113	$2s2p^4$ ^{2}D $3d$
4	−7.57302	$2s^22p^3$ ^{4}S $4f$
5	−7.54163	$2s^22p^3$ ^{2}D $4p$
6	−7.26057	$2s2p^4$ ^{2}P $3d$
7	−6.66112	$2s^22p^3$ ^{2}D $4f$
8	−6.17254	$2s^22p^3$ ^{2}P $4f$
9	−4.84684	$2s^22p^3$ ^{4}S $5f$
10	−4.38515	$2s^22p^3$ ^{2}D $5p$
11	−3.96876	$2s^22p^3$ ^{2}D $5f$
12	−3.65971	$2s2p^4$ ^{4}P $4d$
13	−3.45917	$2s^22p^3$ ^{2}P $5f$
14	−3.36511	$2s^22p^3$ ^{4}S $6f$
15	−2.72781	$2s^22p^3$ ^{2}D $6p$
16	−2.49034	$2s^22p^3$ ^{2}D $6f$
17	−2.47221	$2s^22p^3$ ^{4}S $7f$
18	−2.22916	$2s2p^4$ ^{2}D $4d$
19	−1.97788	$2s^22p^3$ ^{2}P $6f$
20	−1.89279	$2s^22p^3$ ^{4}S $8f$
21	−1.74112	$2s^22p^3$ ^{2}D $7p$
22	−1.59298	$2s^22p^3$ ^{2}D $7f$
23	−1.49544	$2s^22p^3$ ^{4}S $9f$
	^{3}F^o	
1	−13.3589	$2s^22p^3$ ^{2}D $3d$
2	−12.7530	$2s^22p^3$ ^{2}P $3d$

O-like Ar (Ar^{10+})

i	E(Ryds)	Description
3	−9.97622	$2s2p^4$ ^{2}D $3p$
4	−7.00335	$2s^22p^3$ ^{2}D $4d$
5	−6.48036	$2s^22p^3$ ^{2}P $4d$
6	−4.13093	$2s^22p^3$ ^{2}D $5d$
7	−3.61760	$2s^22p^3$ ^{2}P $5d$
8	−3.54866	$2p^5$ ^{2}P $3d$
9	−3.45678	$2s2p^4$ ^{4}P $4f$
10	−2.77955	$2s2p^4$ ^{2}D $4p$
11	−2.57897	$2s^22p^3$ ^{2}D $6d$
12	−2.06940	$2s^22p^3$ ^{2}P $6d$
13	−1.95946	$2s2p^4$ ^{2}D $4f$
14	−1.65164	$2s^22p^3$ ^{2}D $7d$
15	−1.14217	$2s^22p^3$ ^{2}P $7d$
16	−1.05175	$2s^22p^3$ ^{2}D $8d$
17	−1.03707	$2s2p^4$ ^{2}S $4f$
18	−.76018	$2s2p^4$ ^{2}P $4f$
19	−.73557	$2s2p^4$ ^{4}P $5f$
20	−.64256	$2s^22p^3$ ^{2}D $9d$
21	−.54155	$2s^22p^3$ ^{2}P $8d$
	^{3}G^e	
1	−8.67046	$2s2p^4$ ^{2}D $3d$
2	−6.69224	$2s^22p^3$ ^{2}D $4f$
3	−6.18980	$2s^22p^3$ ^{2}P $4f$
4	−3.97335	$2s^32p^3$ ^{3}D $5f$
5	−3.46539	$2s^22p^3$ ^{2}P $5f$
6	−2.49308	$2s^22p^3$ ^{2}D $6f$
7	−2.29723	$2s2p^4$ ^{2}D $4d$
8	−1.98004	$2s^22p^3$ ^{2}P $6f$
9	−1.59398	$2s^22p^3$ ^{2}D $7f$
10	−1.08572	$2s^22p^3$ ^{2}P $7f$
11	−1.01363	$2s^22p^3$ ^{2}D $8f$
12	−.61599	$2s^22p^3$ ^{2}D $9f$
13	−.50473	$2s^22p^3$ ^{2}P $8f$
	^{3}G$^\circ$	
1	−13.2578	$2s^22p^3$ ^{2}D $3d$
2	−6.98001	$2s^22p^3$ ^{2}D $4d$
3	−4.12055	$2s^22p^3$ ^{2}D $5d$
4	−3.46624	$2s2p^4$ ^{4}P $4f$
5	−2.57590	$2s^22p^3$ ^{2}D $6d$
6	−1.98273	$2s2p^4$ ^{4}P $5f$
7	−1.95375	$2s2p^4$ ^{2}D $4f$
8	−1.64899	$2s^22p^3$ ^{2}D $7d$
	^{3}H^e	
1	−6.70393	$2s^22p^3$ ^{2}D $4f$
2	−3.97533	$2s^22p^3$ ^{2}D $5f$
3	−2.49052	$2s^22p^3$ ^{2}D $6f$
4	−1.59523	$2s^22p^3$ ^{2}D $7f$
	^{3}H$^\circ$	
1	−2.02055	$2s2p^4$ ^{2}D $4f$
	^{5}S$^\circ$	
1	−16.9283	$2s^22p^3$ ^{4}S $3s$
2	−11.2200	$2s2p^4$ ^{4}P $3p$
3	−8.89403	$2s^22p^3$ ^{4}S $4s$
4	−5.52313	$2s^22p^3$ ^{4}S $5s$
5	−4.19538	$2s2p^4$ ^{4}P $4p$
6	−3.73345	$2s^22p^3$ ^{4}S $6s$
7	−2.70734	$2s^32p^3$ ^{1}S $7s$
8	−2.04953	$2s^22p^3$ ^{4}S $8s$
9	−1.60543	$2s^22p^3$ ^{4}S $9s$
	^{5}P^e	
1	−15.6993	$2s^22p^3$ ^{4}S $3p$
2	−12.6925	$2s2p^4$ ^{4}P $3s$
3	−9.95284	$2s2p^4$ ^{4}P $3d$
4	−8.40983	$2s^22p^3$ ^{4}S $4p$
5	−5.28494	$2s^22p^3$ ^{4}S $5p$
6	−4.75530	$2s2p^4$ ^{4}P $4s$
7	−3.73442	$2s2p^4$ ^{4}P $4d$
8	−3.59441	$2s^22p^3$ ^{4}S $6p$
9	−2.62453	$2s^22p^3$ ^{4}S $7p$
10	−1.99455	$2s^22p^3$ ^{4}S $8p$
11	−1.56748	$2s^22p^3$ ^{4}S $9p$
12	−1.38504	$2s2p^4$ ^{4}P $5s$
	^{5}P$^\circ$	
1	−11.5831	$2s2p^4$ ^{4}P $3p$
2	−4.30657	$2s2p^4$ ^{4}P $4p$
3	−1.16116	$2s2p^4$ ^{4}P $5p$
	^{5}D^e	
1	−10.2568	$2s2p^4$ ^{4}P $3d$
2	−3.81802	$2s2p^4$ ^{4}P $4d$
3	−.91657	$2s2p^4$ ^{4}P $5d$
	^{5}D$^\circ$	
1	−14.2248	$2s^22p^3$ ^{4}S $3d$
2	−11.4329	$2s2p^4$ ^{4}P $3p$
3	−7.88892	$2s^22p^3$ ^{4}S $4d$
4	−5.01582	$2s^22p^3$ ^{4}S $5d$
5	−4.25910	$2s2p^4$ ^{4}P $4p$
6	−3.46584	$2s^22p^3$ ^{4}S $6d$
7	−3.43838	$2s2p^4$ ^{4}P $4f$
8	−2.53419	$2s^22p^3$ ^{4}S $7d$
9	−1.93400	$2s^32p^3$ ^{1}S $8d$
10	−1.52425	$2s^22p^3$ ^{4}S $9d$
	^{5}F^e	
1	−10.0770	$2s2p^4$ ^{4}P $3d$
2	−7.57374	$2s^22p^3$ ^{4}S $4f$
3	−4.85463	$2s^22p^3$ ^{4}S $5f$
4	−3.75833	$2s2p^4$ ^{4}P $4d$
5	−3.36881	$2s^22p^3$ ^{4}S $6f$
6	−2.47547	$2s^22p^3$ ^{4}S $7f$
7	−1.89500	$2s^22p^3$ ^{4}S $8f$
8	−1.49710	$2s^22p^3$ ^{4}S $9f$
	^{5}F$^\circ$	
1	−3.49084	$2s2p^4$ ^{4}P $4f$
2	−.75038	$2s2p^4$ ^{4}P $5f$
	^{5}G$^\circ$	
1	−3.47536	$2s2p^4$ ^{4}P $4f$

O-like Ar (Ar^{10+})

Energies in ascending order from ground state for terms with effective $n \leq 4.0$, $L \leq 4$

Term	i	E(Ryds)	Term	i	E(Ryds)	Term	i	E(Ryds)	Term	i	E(Ryds)	Term	i	E(Ryds)
$^3P^e$	1	0.00000	$^5D^o$	1	25.3539	$^5P^e$	2	26.8862	$^1P^o$	5	29.6478	$^3F^e$	3	31.1175
$^1D^e$	1	0.58770	$^3D^e$	2	25.3583	$^3D^o$	4	26.9798	$^3P^e$	6	29.6827	$^3S^o$	5	31.1265
$^1S^e$	1	1.27240	$^1D^e$	2	25.3969	$^1D^o$	3	27.0566	$^1F^o$	3	29.7192	$^1G^e$	1	31.1489
$^3P^o$	1	4.68300	$^1P^e$	2	25.4480	$^1F^o$	2	27.1192	$^3F^e$	2	29.7642	$^3D^e$	5	31.1610
$^1P^o$	1	6.46980	$^3P^e$	4	25.5572	$^3P^e$	5	27.3423	$^3D^o$	6	29.7765	$^3P^e$	8	31.1621
$^1S^e$	2	10.9384	$^3D^o$	2	25.6816	$^1P^o$	4	27.4073	$^1P^e$	3	29.7889	$^5P^e$	4	31.1688
$^5S^o$	1	22.6504	$^1D^e$	3	25.7970	$^5P^o$	1	27.9956	$^1D^o$	4	29.8840	$^1P^e$	4	31.2450
$^3S^o$	1	22.9641	$^3F^o$	1	26.2198	$^5D^o$	2	28.1458	$^3P^o$	6	29.9103	$^1D^e$	5	31.2857
$^3D^o$	1	23.6106	$^1S^o$	1	26.2245	$^3D^o$	5	28.3374	$^3P^e$	7	29.9533	$^3P^e$	9	31.3074
$^1D^o$	1	23.7674	$^1S^e$	3	26.2927	$^5S^o$	2	28.3587	$^3D^e$	4	29.9727	$^1F^e$	2	31.3403
$^5P^e$	1	23.8794	$^3G^o$	1	26.3209	$^3S^o$	3	28.4232	$^3P^o$	7	30.6075	$^1S^e$	5	31.3811
$^3P^o$	2	24.1269	$^1G^o$	1	26.3628	$^3D^e$	3	28.4315	$^5S^o$	3	30.6846	$^1P^o$	7	31.5648
$^3P^e$	2	24.1633	$^1P^o$	3	26.4279	$^3P^o$	5	28.5135	$^1P^o$	6	30.7085	$^3D^o$	8	31.6032
$^1P^o$	2	24.2855	$^3D^o$	3	26.4560	$^1D^e$	4	28.7370	$^3S^o$	4	30.7660	$^1D^o$	6	31.6574
$^1P^e$	1	24.7022	$^1D^o$	2	26.5445	$^5D^e$	1	29.3219	$^3P^o$	8	30.8449	$^5D^o$	3	31.6897
$^3D^e$	1	24.7489	$^3P^o$	3	26.5562	$^3S^e$	2	29.3488	$^3D^o$	7	30.8797	$^3D^o$	9	31.8448
$^3F^e$	1	24.8451	$^3S^o$	2	26.6610	$^5F^e$	1	29.5017	$^3G^e$	1	30.9082	$^3D^e$	6	31.9840
$^1F^e$	1	24.9041	$^1F^o$	1	26.7909	$^3F^o$	3	29.6024	$^1D^o$	5	30.9513	$^5F^e$	2	32.0049
$^3P^e$	3	25.2494	$^3F^o$	2	26.8257	$^5P^e$	3	29.6258	$^1S^o$	2	31.0694	$^3F^e$	4	32.0056
$^3S^e$	1	25.2559	$^3P^o$	4	26.8535	$^1S^e$	4	29.6427	$^3S^e$	3	31.1095			

gf-values for transitions involving terms with effective $n \leq 4.0$, $L \leq 4$

i i'	gf_L	i i'	gf_L	i i'	gf_L	i i'	gf_L	i i'	gf_L	i i'	gf_L
	$^1P^e$–$^1S^o$	4 3	1.76E−1	3 1	3.19E−1	1 7	−1.20E−2	5 6	5.85E−4		$^1D^o$–$^1D^e$
1 1	−1.28E−1	4 4	−7.32E−3	3 2	1.76E−2	2 1	1.00E−2	5 7	−1.01E−5	1 1	5.27E−1
1 2	−8.01E−2	4 5	−4.24E−3	3 3	−1.20E−3	2 2	2.30E−1		$^1D^o$–$^1P^e$	1 2	−5.47E−1
2 1	−7.79E−3	5 1	3.34E−3	3 4	−2.69E−2	2 3	−6.54E−2	1 1	−2.29E−1	1 3	−5.19E−1
2 2	−1.65E−1	5 2	8.26E−3	4 1	6.64E−4	2 4	−1.02E−3	1 2	−6.54E−2	1 4	−2.65E−1
3 1	6.44E−4	5 3	1.62E−3	4 2	3.00E−1	2 5	−2.16E−3	1 3	−6.69E−1	1 5	−5.14E−3
3 2	−1.44E−1	5 4	3.10E−5	4 3	−5.57E−5	2 6	−1.47E−1	1 4	−1.83E−3	2 1	7.92E−1
4 1	6.25E−2	5 5	−1.46E−1	4 4	−1.24E−2	2 7	−3.31E−2	2 1	3.43E−1	2 2	2.37E−1
4 2	1.26E−3	6 1	2.98E−1	5 1	3.56E−1	3 1	8.45E−5	2 2	5.91E−2	2 3	8.92E−3
	$^1P^o$–$^1S^e$	6 2	1.05E−2	5 2	2.13E−3	3 2	4.98E−1	2 3	−1.09E−3	2 4	−3.32E−3
1 1	6.45E−2	6 3	1.38E−1	5 3	−2.02E−3	3 3	−7.35E−3	2 4	−1.72E−1	2 5	−1.80E−1
1 2	−3.67E−1	6 4	3.13E−1	5 4	−3.53E−1	3 4	−1.28E−2	3 1	2.91E−2	3 1	2.05E+0
1 3	−6.03E−3	6 5	−1.18E−2	6 1	1.11E−2	3 5	−1.07E−2	3 2	5.01E−1	3 2	4.83E−3
1 4	−4.25E−2	7 1	2.56E−2	6 2	1.49E−1	3 6	−8.62E−2	3 3	−4.44E−5	3 3	2.21E−1
1 5	−4.76E−1	7 2	2.51E−3	6 3	4.46E−3	3 7	−6.24E−1	3 4	−1.80E−3	3 4	−8.31E−3
2 1	1.50E−1	7 3	1.89E−2	6 4	−1.63E−3	4 1	1.58E−1	4 1	2.25E−2	3 5	−9.01E−3
2 2	1.17E−6	7 4	1.63E−2	7 1	4.98E−2	4 2	3.94E−2	4 2	1.24E−1	4 1	5.68E−1
2 3	−2.41E−1	7 5	7.44E−4	7 2	1.43E−2	4 3	3.32E−3	4 3	3.75E−5	4 2	2.81E−1
2 4	−1.01E−1		$^1P^o$–$^1P^e$	7 3	5.84E−1	4 4	2.03E−4	4 4	−8.55E−2	4 3	1.55E−1
2 5	−2.42E−3	1 1	−1.89E−3	7 4	4.85E−4	4 5	−2.36E−1	5 1	5.51E−2	4 4	5.95E−1
3 1	1.42E−1	1 2	−4.56E−3		$^1D^e$–$^1P^o$	4 6	−2.05E−2	5 2	2.31E−2	4 5	−4.64E−1
3 2	2.67E−5	1 3	−3.96E−1	1 1	−7.32E−1	4 7	−1.08E−1	5 3	6.39E−1	5 1	6.84E−3
3 3	1.55E−3	1 4	−4.61E−1	1 2	−1.51E−1	5 1	2.31E−1	5 4	−2.47E−4	5 2	2.63E−1
3 4	−1.43E−3	2 1	−1.54E−2	1 3	−6.81E−1	5 2	1.85E−3	6 1	1.16E−1	5 3	5.26E−2
3 5	−1.25E−1	2 2	−3.06E−1	1 4	−9.37E−2	5 3	3.63E−2	6 2	3.07E−2	5 4	8.75E−3
4 1	1.82E+0	2 3	−2.40E−1	1 5	−3.47E−1	5 4	1.63E−2	6 3	1.04E−2	5 5	−1.95E−3
4 2	7.43E−5	2 4	−3.66E−3	1 6	−7.29E−3	5 5	2.85E−1	6 4	1.22E−3	6 1	8.34E−2

O-like Ar (Ar^{10+})

i i′	gf_L	*i i′*	gf_L	*i i′*	gf_L	*i i′*	gf_L	*i i′*	gf_L	*i i′*	gf_L
6 2	3.87E−1	1 3	−3.17E−1	3 3	−3.01E−1	4 9	−1.12E−3	2 7	−1.03E+0	2 9	−4.43E−1
6 3	2.02E−1	1 4	−3.90E−2	4 1	7.29E−1	5 1	7.19E−1	2 8	−3.87E−3	3 1	4.49E+0
6 4	1.41E−2	1 5	−9.47E−2	4 2	−8.90E−4	5 2	5.54E−1	3 1	7.70E−1	3 2	5.17E−2
6 5	2.15E−3	2 1	9.43E−1	4 3	−1.70E−2	5 3	7.21E−2	3 2	3.16E−1	3 3	7.06E−1
	$^1F^e$–$^1D^o$	2 2	−1.21E−1	5 1	8.53E−5	5 4	6.69E−2	3 3	3.97E−4	3 4	2.68E−3
1 1	8.09E−1	2 3	−1.51E−1	5 2	−5.39E−3	5 5	9.77E−1	3 4	1.37E−5	3 5	−1.49E−2
1 2	−2.72E−2	2 4	−4.99E−1	5 3	−7.31E−2	5 6	−3.22E−3	3 5	−6.03E−3	3 6	−2.54E−3
1 3	−6.39E−2	2 5	−5.69E−3	6 1	1.79E−1	5 7	−5.41E−1	3 6	−1.27E+0	3 7	−3.22E−3
1 4	−9.24E−2	3 1	3.38E−1	6 2	1.29E−4	5 8	−5.83E−2	3 7	−2.57E−2	3 8	−6.08E−1
1 5	−1.06E+0	3 2	−3.48E−1	6 3	−2.88E−1	5 9	−3.21E−3	3 8	−7.20E−2	3 9	−2.36E−2
1 6	−1.86E−1	3 3	−5.21E−2	7 1	4.06E−2	6 1	1.43E−1	4 1	5.28E+0	4 1	4.40E+0
2 1	1.26E−2	3 4	−4.93E−2	7 2	1.12E+0	6 2	2.49E−4	4 2	5.42E−3	4 2	4.08E−3
2 2	1.03E−4	3 5	−6.94E−1	7 3	−1.58E−2	6 3	2.09E−1	4 3	4.94E−4	4 3	2.33E−2
2 3	6.01E−2	4 1	1.36E−1	8 1	1.77E−1	6 4	2.94E−2	4 4	1.19E−3	4 4	1.58E+0
2 4	7.53E−1	4 2	−2.24E−3	8 2	1.77E−1	6 5	3.84E−1	4 5	1.68E+0	4 5	−9.70E−3
2 5	3.93E−4	4 3	−4.37E−2	8 3	−1.12E−3	6 6	8.45E−3	4 6	2.73E−3	4 6	−2.82E−3
2 6	−5.75E−3	4 4	−6.33E−2		$^3P^o$–$^3P^e$	6 7	−6.20E−6	4 7	−5.50E−4	4 7	−1.97E−3
	$^1F^o$–$^1D^e$	4 5	−1.23E−1	1 1	1.01E+0	6 8	−7.96E−1	4 8	−7.67E−3	4 8	−5.09E−2
1 1	2.20E+0	5 1	2.89E−1	1 2	−6.66E−4	6 9	−7.93E−3	5 1	4.30E+0	4 9	−9.31E−3
1 2	6.41E−1	5 2	3.38E−3	1 3	−4.40E−2	7 1	6.97E−2	5 2	2.38E−4	5 1	1.13E+0
1 3	5.41E−2	5 3	−3.28E−1	1 4	−4.37E−3	7 2	1.95E−3	5 3	8.61E−2	5 2	6.96E−1
1 4	−9.49E−3	5 4	−1.17E−2	1 5	−8.10E−1	7 3	1.14E−1	5 4	9.09E−2	5 3	2.47E−1
1 5	−1.20E−1	5 5	−3.75E−3	1 6	−2.14E−1	7 4	1.18E−1	5 5	3.64E−5	5 4	2.17E−1
2 1	3.63E+0	6 1	7.66E−3	1 7	−1.25E+0	7 5	5.25E−4	5 6	6.54E−1	5 5	1.55E+0
2 2	2.26E−1	6 2	4.46E−4	1 8	−4.32E+0	7 6	8.42E−2	5 7	1.39E−3	5 6	−2.82E−4
2 3	7.37E−1	6 3	2.49E−4	1 9	−4.26E−2	7 7	4.86E−4	5 8	5.02E−3	5 7	−9.96E−2
2 4	−1.67E−2	6 4	−4.49E−2	2 1	3.40E−1	7 8	−6.89E−4	6 1	2.00E−1	5 8	−3.58E−2
2 5	−1.34E−6	6 5	−4.57E−1	2 2	−6.98E−4	7 9	−4.50E−5	6 2	3.60E−2	5 9	−1.47E−3
3 1	7.49E−1	7 1	8.91E−3	2 3	−1.74E−2	8 1	1.30E−1	6 3	1.81E−4	6 1	3.34E−3
3 2	7.02E−4	7 2	9.60E−3	2 4	−1.29E+0	8 2	4.08E−2	6 4	1.44E−3	6 2	7.50E−4
3 3	3.75E−1	7 3	8.73E−1	2 5	−4.92E−2	8 3	3.31E−1	6 5	7.37E−4	6 3	1.71E−1
3 4	7.01E−1	7 4	−6.90E−3	2 6	−6.01E−1	8 4	4.78E−1	6 6	1.23E−5	6 4	1.45E−1
3 5	−1.97E−2	7 5	−2.53E−4	2 7	−2.91E−3	8 5	2.80E−1	6 7	1.93E+0	6 5	8.29E−3
	$^1F^o$–$^1F^e$	8 1	4.46E−3	2 8	−7.73E−4	8 6	8.26E−1	6 8	1.66E−3	6 6	1.06E−6
1 1	4.12E−1	8 2	1.16E−1	2 9	−1.48E−3	8 7	2.65E−3		$^3D^o$–$^3P^e$	6 7	−2.26E−3
1 2	−1.97E−1	8 3	1.43E−1	3 1	4.29E+0	8 8	−3.60E−3	1 1	6.00E−1	6 8	−2.60E−1
2 1	3.86E−2	8 4	8.08E−3	3 2	1.83E−1	8 9	−4.05E−4	1 2	−4.93E−2	6 9	−4.21E−3
2 2	−9.69E−3	8 5	1.09E−4	3 3	6.95E−1		$^3D^e$–$^3P^o$	1 3	−1.36E+0	7 1	5.76E−1
3 1	5.83E−1	9 1	5.96E−1	3 4	8.03E−3	1 1	2.96E−3	1 4	−1.01E−2	7 2	3.41E−3
3 2	−4.37E−1	9 2	1.55E−2	3 5	−1.59E−2	1 2	3.21E−2	1 5	−6.83E−2	7 3	6.25E−2
	$^1G^e$–$^1F^o$	9 3	1.17E−3	3 6	−8.87E−4	1 3	−3.17E−1	1 6	−1.75E+0	7 4	2.95E−1
1 1	1.03E−4	9 4	1.76E+0	3 7	−1.12E−3	1 4	−4.42E−2	1 7	−3.12E−4	7 5	2.69E−2
1 2	4.70E 2	0 5	2.27E 2	3 8	−4.43E−2	1 5	−7.44E−3	1 8	−4.08E−4	7 6	1.88E+0
1 3	1.34E+0		$^3P^o$–$^3S^e$	3 9	−2.05E−2	1 6	−5.88E−1	1 9	−1.34E−2	7 7	2.30E−4
	$^1G^o$–$^1F^e$	1 1	−1.52E−2	4 1	4.40E−1	1 7	−2.87E−1	2 1	2.29E+0	7 8	−2.63E−4
1 1	1.30E+0	1 2	−1.52E−1	4 2	8.66E−4	1 8	−9.45E−1	2 2	2.02E+0	7 9	−1.79E−3
1 2	−2.68E−1	1 3	−1.59E+0	4 3	1.12E−1	2 1	2.46E−2	2 3	1.57E−2	8 1	7.60E−2
	$^1G^o$–$^1G^e$	2 1	−3.07E−1	4 4	4.09E−1	2 2	1.77E+0	2 4	4.91E−4	8 2	1.18E−2
1 1	−5.03E−1	2 2	−5.34E 1	4 5	1.88E 3	2 3	5.85E 2	2 5	−6.82E−3	8 3	7.60E−1
	$^3P^e$–$^3S^o$	2 3	−1.54E−3	4 6	−9.33E−5	2 4	−1.55E−2	2 6	−1.04E−2	8 4	2.08E−3
1 1	−3.94E−1	3 1	4.18E−2	4 7	−2.99E−3	2 5	−2.92E−3	2 7	−5.38E−1	8 5	2.05E−2
1 2	−1.45E+0	3 2	−3.27E−5	4 8	−1.15E−1	2 6	−1.38E−2	2 8	−3.74E−3	8 6	3.90E−2

O-like Ar (Ar^{10+})

i i′	gf$_L$	*i i′*	gf$_L$	*i i′*	gf$_L$	*i i′*	gf$_L$	*i i′*	gf$_L$	*i i′*	gf$_L$
8 7	7.15E−5	4 2	6.01E−1	9 4	5.12E−4	4 2	1.01E+1	2 1	1.17E−1		**$^5P^o$–$^5P^e$**
8 8	1.04E−3	4 3	−6.69E−5	9 5	1.02E−6	4 3	1.27E+0	2 2	−5.79E−4	1 1	9.76E−1
8 9	3.26E−3	4 4	−1.33E−6	9 6	−1.74E−5	4 4	2.66E−1	2 3	−3.43E−1	1 2	1.70E+0
9 1	1.16E+0	4 5	−2.45E−1		**$^3F^e$–$^3D^o$**	4 5	1.39E−1	2 4	−1.83E−2	1 3	−1.02E+0
9 2	2.62E+0	4 6	−1.34E−1	1 1	2.62E+0	4 6	1.46E−3	3 1	1.65E+0	1 4	−1.21E−3
9 3	2.43E−1	5 1	8.86E−4	1 2	−8.08E−3	4 7	2.87E−1	3 2	−1.64E−4		**$^5D^e$–$^5P^o$**
9 4	5.87E−2	5 2	4.25E−5	1 3	−1.31E−1	4 8	5.63E−1	3 3	−1.17E+0	1 1	2.45E+0
9 5	8.12E−3	5 3	−1.76E−5	1 4	−1.62E−2	4 9	7.43E−1	3 4	−7.82E−4		**$^5D^o$–$^5P^e$**
9 6	5.66E−2	5 4	−5.87E−1	1 5	−1.61E−3		**$^3F^o$–$^3D^e$**		**$^3G^e$–$^3F^o$**	1 1	3.43E+0
9 7	7.05E−2	5 5	−7.54E−2	1 6	−3.38E−1	1 1	1.96E+0	1 1	3.78E−2	1 2	−1.86E−3
9 8	2.93E−2	5 6	−3.88E−4	1 7	−3.03E+0	1 2	1.29E−2	1 2	9.66E−2	1 3	−9.16E−1
9 9	3.54E+0	6 1	7.88E−1	1 8	−7.43E−1	1 3	−2.56E−3	1 3	3.50E+0	1 4	−7.44E−1
	$^3D^o$–$^3D^e$	6 2	2.48E−1	1 9	−1.90E−2	1 4	−1.32E−2		**$^3G^o$–$^3F^e$**	2 1	9.93E−1
1 1	−1.55E+0	6 3	2.34E+0	2 1	2.68E−3	1 5	−8.18E−1	1 1	3.89E+0	2 2	3.56E+0
1 2	−1.05E−1	6 4	−5.73E−3	2 2	2.49E−1	1 6	−6.58E−1	1 2	−1.55E−4	2 3	−1.23E−1
1 3	−1.59E+0	6 5	−1.39E+0	2 3	1.53E−2	2 1	7.39E−3	1 3	−5.92E−1	2 4	−1.60E−3
1 4	−9.03E−4	6 6	−1.92E−2	2 4	3.37E−3	2 2	2.97E+0	1 4	−5.04E−1	3 1	5.13E+0
1 5	−9.68E−3	7 1	9.88E−2	2 5	3.29E+0	2 3	−3.50E−4		**$^3G^o$–$^3G^e$**	3 2	7.50E−2
1 6	−6.07E−2	7 2	9.17E−1	2 6	−3.82E−7	2 4	−1.44E−5	1 1	−1.34E+0	3 3	1.33E−1
2 1	5.94E−2	7 3	8.30E−7	2 7	−7.16E−4	2 5	−6.13E−2		**$^5P^e$–$^5S^o$**	3 4	5.61E+0
2 2	4.49E−3	7 4	2.02E−4	2 8	−2.74E−4	2 6	−1.31E+0	1 1	1.76E+0		**$^5D^o$–$^5D^e$**
2 3	−4.68E−4	7 5	−1.64E−3	2 9	−2.70E−2	3 1	1.75E−1	1 2	−3.16E−1	1 1	−1.27E+0
2 4	−8.14E−1	7 6	−9.47E−2	3 1	8.29E−3	3 2	2.02E−1	1 3	−8.99E−1	2 1	−6.66E−1
2 5	−2.66E−4	8 1	8.97E−1	3 2	9.84E−4	3 3	2.72E+0	2 1	9.73E−1	3 1	8.48E−5
2 6	−6.06E−4	8 2	8.45E−2	3 3	5.82E−2	3 4	−2.26E−3	2 2	−8.96E−1		**$^5F^e$–$^5D^o$**
3 1	1.31E+0	8 3	7.68E−7	3 4	9.41E−2	3 5	−6.69E−2	2 3	−8.65E−4	1 1	4.62E−1
3 2	1.31E−2	8 4	9.92E−6	3 5	3.05E−3	3 6	−1.08E−2	3 1	3.21E−3	1 2	4.94E+0
3 3	−8.47E−4	8 5	6.34E−3	3 6	2.01E+0		**$^3F^o$–$^3F^e$**	3 2	1.19E+0	1 3	−4.69E−2
3 4	−2.94E−2	8 6	−1.04E−1	3 7	1.59E−7	1 1	8.12E−1	3 3	−6.78E−3	2 1	2.31E+1
3 5	−1.13E−1	9 1	2.91E−2	3 8	−9.94E−3	1 2	−2.77E−4	4 1	1.07E+0	2 2	2.99E−1
3 6	−3.64E−3	9 2	1.59E−2	3 9	−1.82E−4	1 3	−3.28E−1	4 2	3.20E−2	2 3	3.45E+0
4 1	6.48E−3	9 3	8.25E−4	4 1	2.95E−1	1 4	−1.53E−2	4 3	2.74E+0		

O-like Ca (Ca^{12+})

Term energies relative to $2s^22p^3$ ^{4}S ionization threshold for each symmetry

i	E(Ryds)	Description	i	E(Ryds)	Description	i	E(Ryds)	Description	i	E(Ryds)	Description
		^{1}S^e	25	−1.08919	$2s^22p^3$ ^{2}D $9f$	19	−3.81605	$2s^22p^3$ ^{2}D $6p$	6	−8.94241	$2s^22p^3$ ^{2}P $4f$
1	−51.8678	$2s^22p^4$	26	−1.07684	$2s^22p^3$ ^{2}P $8p$	20	−3.70718	$2s^22p^3$ ^{2}D $6f$	7	−6.15233	$2s^22p^3$ ^{2}D $5p$
2	−40.7614	$2p^6$			**^{1}P^o**	21	−3.32186	$2s2p^4$ ^{2}S $4d$	8	−5.77439	$2s^22p^3$ ^{2}D $5f$
3	−18.6407	$2s^22p^3$ ^{2}P $3p$	1	−45.9175	$2s2p^5$	22	−3.22585	$2s^22p^3$ ^{2}P $6p$	9	−5.14870	$2s^22p^3$ ^{2}P $5f$
4	−14.8717	$2s2p^4$ ^{2}S $3s$	2	−21.0270	$2s^22p^3$ ^{2}P $3s$	23	−3.06818	$2s^22p^3$ ^{2}P $6f$	10	−4.32652	$2s2p^4$ ^{2}D $4d$
5	−12.8090	$2s2p^4$ ^{2}D $3d$	3	−18.5432	$2s^22p^3$ ^{2}D $3d$	24	−2.86550	$2s2p^4$ ^{2}P $4d$	11	−3.89188	$2s^22p^3$ ^{2}D $6p$
6	−9.50948	$2s^22p^3$ ^{2}P $4p$	4	−17.3369	$2s^22p^3$ ^{2}P $3d$	25	−2.51457	$2s^22p^3$ ^{2}D $7p$	12	−3.70443	$2s^22p^3$ ^{2}D $6f$
7	−7.57580	$2p^5$ ^{2}P $3p$	5	−14.8147	$2s2p^4$ ^{2}D $3p$	26	−2.45205	$2s^22p^3$ ^{2}D $7f$	13	−3.08915	$2s^22p^3$ ^{2}P $6f$
8	−5.34180	$2s^22p^3$ ^{2}P $5p$	6	−13.6116	$2s2p^4$ ^{2}S $3p$	27	−1.90963	$2s^22p^3$ ^{2}P $7p$	14	−3.01753	$2s2p^4$ ^{2}P $4d$
9	−4.46067	$2s2p^4$ ^{2}S $4s$	7	−12.5602	$2s2p^4$ ^{2}P $3p$	28	−1.81702	$2s^22p^3$ ^{2}P $7f$	15	−2.55949	$2s^22p^3$ ^{2}D $7p$
10	−4.28356	$2s2p^4$ ^{2}D $4d$	8	−10.4094	$2s^22p^3$ ^{2}P $4s$	29	−1.68047	$2s^22p^3$ ^{2}D $8p$	16	−2.45297	$2s^22p^3$ ^{2}D $7f$
11	−3.15369	$2s^22p^3$ ^{2}P $6p$	9	−9.94763	$2p^5$ ^{2}P $3s$	30	−1.64333	$2s^22p^3$ ^{2}D $8f$	17	−1.82136	$2s^22p^3$ ^{2}P $7f$
12	−1.86258	$2s^22p^3$ ^{2}P $7p$	10	−9.86923	$2s^22p^3$ ^{2}D $4d$	31	−1.11458	$2s^22p^3$ ^{2}D $9p$	18	−1.71202	$2s^22p^3$ ^{2}D $8p$
13	−1.03318	$2s^22p^3$ ^{2}P $8p$	11	−9.03846	$2s^22p^3$ ^{2}P $4d$	32	−1.08858	$2s^22p^3$ ^{2}D $9f$	19	−1.64300	$2s^22p^3$ ^{2}D $8f$
14	−.46884	$2s^22p^3$ ^{2}P $9p$	12	−6.59612	$2p^5$ ^{2}P $3d$	33	−1.06463	$2s^22p^3$ ^{2}P $8p$	20	−1.13482	$2s^22p^3$ ^{2}D $9p$
15	−.38406	$2s2p^4$ ^{2}D $5d$	13	−5.90751	$2s^22p^3$ ^{2}D $5d$	34	−1.00620	$2s^22p^3$ ^{2}P $8f$	21	−1.08772	$2s^22p^3$ ^{2}D $9f$
		^{1}S^o	14	−5.79317	$2s^22p^3$ ^{2}P $5s$	35	−.90874	$2s2p^4$ ^{2}D $5s$	22	−1.00938	$2s^22p^3$ ^{2}P $8f$
1	−18.7897	$2s^22p^3$ ^{2}D $3d$	15	−5.17123	$2s^22p^3$ ^{2}P $5d$			**^{1}D^o**			**^{1}F^o**
2	−13.1941	$2s2p^4$ ^{2}P $3p$	16	−4.93859	$2s2p^4$ ^{2}D $4p$	1	−21.6731	$2s^22p^3$ ^{2}D $3s$	1	−18.0911	$2s^22p^3$ ^{2}D $3d$
3	−9.96973	$2s^22p^3$ ^{2}D $4d$	17	−4.11244	$2s2p^4$ ^{2}D $4f$	2	−18.3982	$2s^22p^3$ ^{2}D $3d$	2	−17.6911	$2s^22p^3$ ^{2}P $3d$
4	−5.95297	$2s^22p^3$ ^{2}D $5d$	18	−3.93064	$2s2p^4$ ^{2}S $4p$	3	−17.7714	$2s^22p^3$ ^{2}P $3d$	3	−14.7275	$2s2p^4$ ^{2}D $3p$
5	−3.80397	$2s^22p^3$ ^{2}D $6d$	19	−3.77205	$2s^22p^3$ ^{2}D $6d$	4	−14.5330	$2s2p^4$ ^{2}D $3p$	4	−9.69952	$2s^22p^3$ ^{2}D $4d$
6	−3.55319	$2s2p^4$ ^{2}P $4p$	20	−3.45865	$2s^22p^3$ ^{2}P $6s$	5	−13.3083	$2s2p^4$ ^{2}P $3p$	5	−9.17227	$2s^22p^3$ ^{2}P $4d$
7	−2.51244	$2s^22p^3$ ^{2}D $7d$	21	−3.32102	$2s2p^4$ ^{2}P $4p$	6	−11.0470	$2s^22p^3$ ^{2}D $4s$	6	−7.28790	$2p^5$ ^{2}P $3d$
8	−1.68180	$2s^22p^3$ ^{2}D $8d$	22	−3.07986	$2s^22p^3$ ^{2}P $6d$	7	−9.75367	$2s^22p^3$ ^{2}D $4d$	7	−5.81666	$2s^22p^3$ ^{2}D $5d$
9	−1.11445	$2s^22p^3$ ^{2}D $9d$	23	−2.49424	$2s^22p^3$ ^{2}D $7d$	8	−9.26220	$2s^22p^3$ ^{2}P $4d$	8	−5.23753	$2s^22p^3$ ^{2}P $5d$
		^{1}P^e	24	−2.02754	$2s^22p^3$ ^{2}P $7s$	9	−7.22744	$2p^5$ ^{2}P $3d$	9	−4.98743	$2s2p^4$ ^{2}D $4p$
1	−20.5590	$2s^22p^3$ ^{2}D $3p$	25	−1.82533	$2s^22p^3$ ^{2}P $7d$	10	−6.43655	$2s^22p^3$ ^{2}D $5s$	10	−4.08042	$2s2p^4$ ^{2}D $4f$
2	−19.6545	$2s^22p^3$ ^{2}P $3p$	26	−1.66900	$2s^22p^3$ ^{2}D $8d$	11	−5.83831	$2s^22p^3$ ^{2}D $5d$	11	−3.72099	$2s^22p^3$ ^{2}D $6d$
3	−14.7022	$2s2p^4$ ^{2}P $3s$	27	−1.14485	$2s^22p^3$ ^{2}P $8s$	12	−5.28710	$2s^22p^3$ ^{2}P $5d$	12	−3.12832	$2s^22p^3$ ^{2}P $6d$
4	−12.9677	$2s2p^4$ ^{2}D $3d$	28	−1.10654	$2s^22p^3$ ^{2}D $9d$	13	−4.92944	$2s2p^4$ ^{2}D $4p$	13	−3.10630	$2s2p^4$ ^{2}S $4f$:
5	−11.3685	$2s2p^4$ ^{2}P $3d$	29	−1.01029	$2s^22p^3$ ^{2}P $8d$	14	−4.09605	$2s2p^4$ ^{2}D $4f$	14	−3.01329	$2s^22p^3$ ^{2}P $6g$:
6	−10.5480	$2s^22p^3$ ^{2}D $4p$			**^{1}D^e**	15	−4.06391	$2s^22p^3$ ^{2}D $6s$	15	−2.72876	$2s2p^4$ ^{2}P $4f$
7	−9.87573	$2s^22p^3$ ^{2}P $4p$	1	−52.6513	$2s^22p^4$	16	−3.73612	$2s^22p^3$ ^{2}D $6d$	16	−2.46377	$2s^22p^3$ ^{2}D $7d$
8	−9.53985	$2s^22p^3$ ^{2}D $4f$	2	−19.7299	$2s^22p^3$ ^{2}D $3p$	17	−3.57144	$2s2p^4$ ^{2}P $4p$	17	−1.84828	$2s^22p^3$ ^{2}P $7d$
9	−8.67119	$2p^5$ ^{2}P $3p$	3	−19.2435	$2s^22p^3$ ^{2}P $3p$	18	−3.14761	$2s^22p^3$ ^{2}P $6d$	18	−1.64910	$2s^22p^3$ ^{2}D $8d$
10	−6.17695	$2s^22p^3$ ^{2}D $5p$	4	−15.8862	$2s2p^4$ ^{2}D $3s$	19	−2.69053	$2s2p^4$ ^{2}P $4f$	19	−1.09229	$2s^22p^3$ ^{2}D $9d$
11	−5.77882	$2s^22p^3$ ^{2}D $5f$	5	−12.9183	$2s2p^4$ ^{2}D $3d$	20	−2.66423	$2s^22p^3$ ^{2}D $7s$	20	−1.02602	$2s^22p^3$ ^{2}P $8d$
12	−5.52466	$2s^22p^3$ ^{2}P $5p$	6	−11.8926	$2s2p^4$ ^{2}S $3d$	21	−2.47017	$2s^22p^3$ ^{2}D $7d$			**^{1}G^e**
13	−4.33149	$2s2p^4$ ^{2}D $4d$	7	−11.3201	$2s2p^4$ ^{2}P $3d$	22	−1.86755	$2s^22p^3$ ^{2}P $7d$	1	−13.0823	$2s2p^4$ ^{2}D $3d$
14	−4.17007	$2s2p^4$ ^{2}P $4s$	8	−10.2495	$2s^22p^3$ ^{2}D $4p$	23	−1.78175	$2s^22p^3$ ^{2}D $8s$	2	−9.57429	$2s^22p^3$ ^{2}D $4f$
15	−3.87988	$2s^22p^3$ ^{2}D $6p$	9	−9.75853	$2s^22p^3$ ^{2}P $4p$	24	−1.65297	$2s^22p^3$ ^{2}D $8d$	3	−8.94691	$2s^22p^3$ ^{2}P $4f$
16	−3.71038	$2s^22p^3$ ^{2}D $6f$	10	−9.54956	$2s^22p^3$ ^{2}D $4f$	25	−1.18392	$2s^22p^3$ ^{2}D $9s$	4	−5.77737	$2s^22p^3$ ^{2}D $5f$
17	−3.25881	$2s^22p^3$ ^{2}P $6p$	11	−8.92225	$2s^22p^3$ ^{2}P $4f$	26	−1.09658	$2s^22p^3$ ^{2}D $9d$	5	−5.14666	$2s^22p^3$ ^{2}P $5f$
18	−2.90585	$2s2p^4$ ^{2}P $4d$	12	−8.69089	$2p^5$ ^{2}P $3p$	27	−1.03727	$2s^22p^3$ ^{2}P $8d$	6	−4.40049	$2s2p^4$ ^{2}D $4d$
19	−2.56806	$2s^22p^3$ ^{2}D $7p$	13	−6.02711	$2s^22p^3$ ^{2}D $5p$			**^{1}F^e**	7	−3.70478	$2s^22p^3$ ^{2}D $6f$
20	−2.45251	$2s^22p^3$ ^{2}D $7f$	14	−5.77677	$2s^22p^3$ ^{2}D $5f$	1	−20.3271	$2s^22p^3$ ^{2}D $3p$	8	−3.07193	$2s^22p^3$ ^{2}P $6f$
21	−1.92927	$2s^22p^3$ ^{2}P $7p$	15	−5.50720	$2s2p^4$ ^{2}D $4s$	2	−12.8454	$2s2p^4$ ^{2}D $3d$	9	−2.45472	$2s^22p^3$ ^{2}D $7f$
22	−1.71665	$2s^22p^3$ ^{2}D $8p$	16	−5.43281	$2s^22p^3$ ^{2}P $5p$	3	−11.7705	$2s2p^4$ ^{2}P $3d$	10	−1.82053	$2s^22p^3$ ^{2}P $7f$
23	−1.64436	$2s^22p^3$ ^{2}D $8f$	17	−5.13762	$2s^22p^3$ ^{2}P $5f$	4	−10.4947	$2s^22p^3$ ^{2}D $4p$	11	−1.63940	$2s^22p^3$ ^{2}D $8f$
24	−1.13951	$2s^22p^3$ ^{2}D $9p$	18	−4.32023	$2s2p^4$ ^{2}D $4d$	5	−9.55801	$2s^22p^3$ ^{2}D $4f$	12	−1.08818	$2s^22p^3$ ^{2}D $9f$

O-like Ca (Ca^{12+})

i	E(Ryds)	Description
13	−1.00861	$2s^22p^3\ ^2\mathrm{P}\ 8f$
		$^1\mathbf{G}^o$
1	−18.6201	$2s^22p^3\ ^2\mathrm{D}\ 3d$
2	−9.91268	$2s^22p^3\ ^2\mathrm{D}\ 4d$
3	−5.92409	$2s^22p^3\ ^2\mathrm{D}\ 5d$
4	−4.08510	$2s2p^4\ ^2\mathrm{D}\ 4f$
5	−3.78337	$2s^22p^3\ ^2\mathrm{D}\ 6d$
6	−2.70998	$2s2p^4\ ^2\mathrm{P}\ 4f$
7	−2.50252	$2s^22p^3\ ^2\mathrm{D}\ 7d$
8	−1.67520	$2s^22p^3\ ^2\mathrm{D}\ 8d$
9	−1.10985	$2s^22p^3\ ^2\mathrm{D}\ 9d$
		$^1\mathbf{H}^e$
1	−9.59393	$2s^22p^3\ ^2\mathrm{D}\ 4f$
2	−5.78891	$2s^22p^3\ ^2\mathrm{D}\ 5f$
3	−3.71212	$2s^22p^3\ ^2\mathrm{D}\ 6f$
4	−2.45957	$2s^22p^3\ ^2\mathrm{D}\ 7f$
5	−1.64713	$2s^22p^3\ ^2\mathrm{D}\ 8f$
6	−1.09050	$2s^22p^3\ ^2\mathrm{D}\ 9f$
		$^1\mathbf{H}^o$
1	−4.13840	$2s2p^4\ ^2\mathrm{D}\ 4f$
		$^3\mathbf{S}^e$
1	−19.8774	$2s^22p^3\ ^2\mathrm{P}\ 3p$
2	−15.2189	$2s2p^4\ ^2\mathrm{S}\ 3s$
3	−13.1379	$2s2p^4\ ^2\mathrm{D}\ 3d$
4	−9.92082	$2s^22p^3\ ^2\mathrm{P}\ 4p$
5	−9.10273	$2p^5\ ^2\mathrm{P}\ 3p$
6	−5.55162	$2s^22p^3\ ^2\mathrm{P}\ 5p$
7	−4.58597	$2s2p^4\ ^2\mathrm{S}\ 4s$
8	−4.40280	$2s2p^4\ ^2\mathrm{D}\ 4d$
9	−3.26597	$2s^22p^3\ ^2\mathrm{P}\ 6p$
10	−1.93561	$2s^22p^3\ ^2\mathrm{P}\ 7p$
11	−1.08308	$2s^22p^3\ ^2\mathrm{P}\ 8p$
12	−.50510	$2s^22p^3\ ^2\mathrm{P}\ 9p$
13	−.43923	$2s2p^4\ ^2\mathrm{D}\ 5d$
		$^3\mathbf{S}^o$
1	−22.5831	$2s^22p^3\ ^4\mathrm{S}\ 3s$
2	−18.2511	$2s^22p^3\ ^2\mathrm{D}\ 3d$
3	−16.1998	$2s2p^4\ ^4\mathrm{P}\ 3p$
4	−13.1454	$2s2p^4\ ^2\mathrm{P}\ 3p$
5	−12.0217	$2s^22p^3\ ^4\mathrm{S}\ 4s$
6	−9.77288	$2s^22p^3\ ^2\mathrm{D}\ 4d$
7	−7.42767	$2s^22p^3\ ^4\mathrm{S}\ 5s$
8	−6.60844	$2s2p^4\ ^4\mathrm{P}\ 4p$
9	−5.85649	$2s^22p^3\ ^2\mathrm{D}\ 5d$
10	−5.04668	$2s^22p^3\ ^4\mathrm{S}\ 6s$
11	−3.75202	$2s^22p^3\ ^2\mathrm{D}\ 6d$
12	−3.66456	$2s^22p^3\ ^4\mathrm{S}\ 7s$
13	−3.53230	$2s2p^4\ ^2\mathrm{P}\ 4p$
14	−2.78094	$2s^22p^3\ ^4\mathrm{S}\ 8s$
15	−2.47852	$2s^22p^3\ ^2\mathrm{D}\ 7d$
16	−2.31314	$2s2p^4\ ^4\mathrm{P}\ 5p$
17	−2.17866	$2s^22p^3\ ^4\mathrm{S}\ 9s$
		$^3\mathbf{P}^e$
1	−53.3241	$2s^22p^4$
2	−21.1669	$2s^22p^3\ ^4\mathrm{S}\ 3p$
3	−19.9100	$2s^22p^3\ ^2\mathrm{D}\ 3p$
4	−19.5193	$2s^22p^3\ ^2\mathrm{P}\ 3p$
5	−17.4780	$2s2p^4\ ^4\mathrm{P}\ 3s$
6	−14.8300	$2s2p^4\ ^2\mathrm{P}\ 3s$
7	−14.4342	$2s2p^4\ ^4\mathrm{P}\ 3d$
8	−13.0532	$2s2p^4\ ^2\mathrm{D}\ 3d$
9	−11.6954	$2s2p^4\ ^2\mathrm{P}\ 3d$
10	−11.3992	$2s^22p^3\ ^4\mathrm{S}\ 4p$
11	−10.3885	$2s^22p^3\ ^2\mathrm{D}\ 4p$
12	−9.81373	$2s^22p^3\ ^2\mathrm{P}\ 4p$
13	−9.54181	$2s^22p^3\ ^2\mathrm{D}\ 4f$
14	−8.65946	$2p^5\ ^2\mathrm{P}\ 3p$
15	−7.16141	$2s2p^4\ ^4\mathrm{P}\ 4s$
16	−7.09726	$2s^22p^3\ ^4\mathrm{S}\ 5p$
17	−6.09708	$2s^22p^3\ ^2\mathrm{D}\ 5p$
18	−5.95059	$2s2p^4\ ^4\mathrm{P}\ 4d$
19	−5.77213	$2s^22p^3\ ^2\mathrm{D}\ 5f$
20	−5.48946	$2s^22p^3\ ^2\mathrm{P}\ 5p$
21	−4.85623	$2s^22p^3\ ^4\mathrm{S}\ 6p$
22	−4.41455	$2s2p^4\ ^2\mathrm{D}\ 4d$
23	−4.19035	$2s2p^4\ ^2\mathrm{P}\ 4s$
24	−3.86138	$2s^22p^3\ ^2\mathrm{D}\ 6p$
25	−3.70610	$2s^22p^3\ ^2\mathrm{D}\ 6f$
26	−3.53915	$2s^22p^3\ ^4\mathrm{S}\ 7p$
27	−3.24336	$2s^22p^3\ ^2\mathrm{P}\ 6p$
28	−2.99899	$2s2p^4\ ^2\mathrm{P}\ 4d$
29	−2.70136	$2s^22p^3\ ^4\mathrm{S}\ 8p$
30	−2.60513	$2s2p^4\ ^4\mathrm{P}\ 5s$
31	−2.52692	$2s^22p^3\ ^2\mathrm{D}\ 7p$
32	−2.45276	$2s^22p^3\ ^2\mathrm{D}\ 7f$
33	−2.12830	$2s^22p^3\ ^4\mathrm{S}\ 9p$
34	−2.03915	$2s2p^4\ ^4\mathrm{P}\ 5d$
35	−1.91585	$2s^22p^3\ ^2\mathrm{P}\ 7p$
36	−1.72343	$2s^22p^3\ ^2\mathrm{D}\ 8p$
		$^3\mathbf{P}^o$
1	−47.9422	$2s2p^5$
2	−21.2130	$2s^22p^3\ ^2\mathrm{P}\ 3s$
3	−18.3823	$2s^22p^3\ ^2\mathrm{D}\ 3d$
4	−18.0340	$2s^22p^3\ ^2\mathrm{P}\ 3d$
5	−16.1013	$2s2p^4\ ^4\mathrm{P}\ 3p$
6	−14.4898	$2s2p^4\ ^2\mathrm{D}\ 3p$
7	−13.7257	$2s2p^4\ ^2\mathrm{S}\ 3p$
8	−13.4401	$2s2p^4\ ^2\mathrm{P}\ 3p$
9	−10.4700	$2s^22p^3\ ^2\mathrm{P}\ 4s$
10	−10.1531	$2p^5\ ^2\mathrm{P}\ 3s$
11	−9.81878	$2s^22p^3\ ^2\mathrm{D}\ 4d$
12	−9.29250	$2s^22p^3\ ^2\mathrm{P}\ 4d$
13	−7.54806	$2p^5\ ^2\mathrm{P}\ 3d$
14	−6.49627	$2s2p^4\ ^4\mathrm{P}\ 4p$
15	−5.89016	$2s^22p^3\ ^2\mathrm{D}\ 5d$
16	−5.82004	$2s^22p^3\ ^2\mathrm{P}\ 5s$
17	−5.29631	$2s^22p^3\ ^2\mathrm{P}\ 5d$
18	−4.94176	$2s2p^4\ ^2\mathrm{D}\ 4p$
19	−4.14290	$2s2p^4\ ^2\mathrm{D}\ 4f$
20	−3.97040	$2s2p^4\ ^2\mathrm{S}\ 4p$
21	−3.76704	$2s^22p^3\ ^2\mathrm{D}\ 6d$
22	−3.64712	$2s2p^4\ ^2\mathrm{P}\ 4p$
23	−3.42108	$2s^22p^3\ ^2\mathrm{P}\ 6s$
24	−3.15068	$2s^22p^3\ ^2\mathrm{P}\ 6d$
25	−2.48712	$2s^22p^3\ ^2\mathrm{D}\ 7d$
26	−2.24530	$2s2p^4\ ^4\mathrm{P}\ 5p$
27	−2.03811	$2s^22p^3\ ^2\mathrm{P}\ 7s$
28	−1.86855	$2s^22p^3\ ^2\mathrm{P}\ 7d$
29	−1.66492	$2s^22p^3\ ^2\mathrm{D}\ 8d$
30	−1.15279	$2s^22p^3\ ^2\mathrm{P}\ 8s$
31	−1.10248	$2s^22p^3\ ^2\mathrm{D}\ 9d$
32	−1.04007	$2s^22p^3\ ^2\mathrm{P}\ 8d$
		$^3\mathbf{D}^e$
1	−20.5090	$2s^22p^3\ ^2\mathrm{D}\ 3p$
2	−19.7578	$2s^22p^3\ ^2\mathrm{P}\ 3p$
3	−16.2504	$2s2p^4\ ^2\mathrm{D}\ 3s$
4	−14.4050	$2s2p^4\ ^4\mathrm{P}\ 3d$
5	−13.0585	$2s2p^4\ ^2\mathrm{D}\ 3d$
6	−12.1373	$2s2p^4\ ^2\mathrm{S}\ 3d$
7	−12.0076	$2s2p^4\ ^2\mathrm{P}\ 3d$
8	−10.5501	$2s^22p^3\ ^2\mathrm{D}\ 4p$
9	−9.89230	$2s^22p^3\ ^2\mathrm{P}\ 4p$
10	−9.55928	$2s^22p^3\ ^2\mathrm{D}\ 4f$
11	−8.93383	$2s^22p^3\ ^2\mathrm{P}\ 4f$
12	−8.83354	$2p^5\ ^2\mathrm{P}\ 3p$
13	−6.18760	$2s^22p^3\ ^2\mathrm{D}\ 5p$
14	−5.92992	$2s2p^4\ ^4\mathrm{P}\ 4d$
15	−5.77716	$2s^22p^3\ ^2\mathrm{D}\ 5f$
16	−5.61404	$2s2p^4\ ^2\mathrm{D}\ 4s$
17	−5.51190	$2s^22p^3\ ^2\mathrm{P}\ 5p$
18	−5.14711	$2s^22p^3\ ^2\mathrm{P}\ 5f$
19	−4.41193	$2s2p^4\ ^2\mathrm{D}\ 4d$
20	−3.90169	$2s^22p^3\ ^2\mathrm{D}\ 6p$
21	−3.70756	$2s^22p^3\ ^2\mathrm{D}\ 6f$
22	−3.41975	$2s2p^4\ ^2\mathrm{S}\ 4d$
23	−3.26119	$2s^22p^3\ ^2\mathrm{P}\ 6p$
24	−3.10652	$2s2p^4\ ^2\mathrm{P}\ 4d$
25	−3.06164	$2s^22p^3\ ^2\mathrm{P}\ 6f$
26	−2.57073	$2s^22p^3\ ^2\mathrm{D}\ 7p$
27	−2.45506	$2s^22p^3\ ^2\mathrm{D}\ 7f$
28	−2.02808	$2s2p^4\ ^4\mathrm{P}\ 5d$
29	−1.93076	$2s^22p^3\ ^2\mathrm{P}\ 7p$
30	−1.82076	$2s^22p^3\ ^2\mathrm{P}\ 7f$
31	−1.71898	$2s^22p^3\ ^2\mathrm{D}\ 8p$
32	−1.64427	$2s^22p^3\ ^2\mathrm{D}\ 8f$
33	−1.14227	$2s^22p^3\ ^2\mathrm{D}\ 9p$
34	−1.08955	$2s^22p^3\ ^2\mathrm{D}\ 9f$
35	−1.07980	$2s^22p^3\ ^2\mathrm{P}\ 8p$
36	−1.00917	$2s^22p^3\ ^2\mathrm{P}\ 8f$
37	−.96989	$2s2p^4\ ^2\mathrm{D}\ 5s$
		$^3\mathbf{D}^o$
1	−21.8571	$2s^22p^3\ ^2\mathrm{D}\ 3s$
2	−19.3850	$2s^22p^3\ ^4\mathrm{S}\ 3d$
3	−18.4933	$2s^22p^3\ ^2\mathrm{D}\ 3d$
4	−17.8692	$2s^22p^3\ ^2\mathrm{P}\ 3d$
5	−16.3028	$2s2p^4\ ^4\mathrm{P}\ 3p$
6	−14.6545	$2s2p^4\ ^2\mathrm{D}\ 3p$
7	−13.4048	$2s2p^4\ ^2\mathrm{P}\ 3p$
8	−11.1063	$2s^22p^3\ ^2\mathrm{D}\ 4s$
9	−10.7809	$2s^22p^3\ ^4\mathrm{S}\ 4d$
10	−9.88767	$2s^22p^3\ ^2\mathrm{D}\ 4d$
11	−9.23931	$2s^22p^3\ ^2\mathrm{P}\ 4d$
12	−7.21305	$2p^5\ ^2\mathrm{P}\ 3d$
13	−6.85775	$2s^22p^3\ ^4\mathrm{S}\ 5d$
14	−6.64178	$2s2p^4\ ^4\mathrm{P}\ 4p$
15	−6.46604	$2s^22p^3\ ^2\mathrm{D}\ 5s$
16	−5.91732	$2s^22p^3\ ^2\mathrm{D}\ 5d$
17	−5.76303	$2s^22p^3\ ^2\mathrm{D}\ 5g$:
18	−5.76055	$2s2p^4\ ^4\mathrm{P}\ 4f$:
19	−5.27114	$2s^22p^3\ ^2\mathrm{P}\ 5d$
20	−4.97470	$2s2p^4\ ^2\mathrm{D}\ 4p$
21	−4.74118	$2s^22p^3\ ^4\mathrm{S}\ 6d$
22	−4.11417	$2s2p^4\ ^2\mathrm{D}\ 4f$
23	−4.07543	$2s^22p^3\ ^2\mathrm{D}\ 6s$
24	−3.78310	$2s^22p^3\ ^2\mathrm{D}\ 6d$
25	−3.60810	$2s2p^4\ ^2\mathrm{P}\ 4p$
26	−3.47684	$2s^22p^3\ ^4\mathrm{S}\ 7d$
27	−3.13628	$2s^22p^3\ ^2\mathrm{P}\ 6d$
28	−2.69366	$2s2p^4\ ^2\mathrm{P}\ 4f$
29	−2.67606	$2s^22p^3\ ^2\mathrm{D}\ 7s$
30	−2.65921	$2s^22p^3\ ^4\mathrm{S}\ 8d$
31	−2.49921	$2s^22p^3\ ^2\mathrm{D}\ 7d$
32	−2.32370	$2s2p^4\ ^4\mathrm{P}\ 5p$
33	−2.09891	$2s^22p^3\ ^4\mathrm{S}\ 9d$
34	−1.97853	$2s2p^4\ ^4\mathrm{P}\ 5f$
35	−1.85969	$2s^22p^3\ ^2\mathrm{P}\ 7d$
36	−1.78912	$2s^22p^3\ ^2\mathrm{D}\ 8s$
		$^3\mathbf{F}^e$
1	−20.3974	$2s^22p^3\ ^2\mathrm{D}\ 3p$
2	−14.6590	$2s2p^4\ ^4\mathrm{P}\ 3d$
3	−13.1231	$2s2p^4\ ^2\mathrm{D}\ 3d$
4	−11.8106	$2s2p^4\ ^2\mathrm{P}\ 3d$
5	−10.5749	$2s^22p^3\ ^4\mathrm{S}\ 4f$
6	−10.5190	$2s^22p^3\ ^2\mathrm{D}\ 4p$
7	−9.57028	$2s^22p^3\ ^2\mathrm{D}\ 4f$
8	−8.94252	$2s^22p^3\ ^2\mathrm{P}\ 4f$
9	−6.78016	$2s^22p^3\ ^4\mathrm{S}\ 5f$
10	−6.16475	$2s^22p^3\ ^2\mathrm{D}\ 5p$

O-like Ca (Ca^{12+})

i	E(Ryds)	Description
11	−6.05042	$2s2p^4$ ^{4}P $4d$
12	−5.78326	$2s^22p^3$ ^{2}D $5f$
13	−5.14745	$2s^22p^3$ ^{2}P $5f$
14	−4.70716	$2s^22p^3$ ^{4}S $6f$
15	−4.42773	$2s2p^4$ ^{2}D $4d$
16	−3.89869	$2s^22p^3$ ^{2}D $6p$
17	−3.70876	$2s^22p^3$ ^{2}D $6f$
18	−3.45703	$2s^22p^3$ ^{4}S $7f$
19	−3.08454	$2s^22p^3$ ^{2}P $6f$
20	−3.03889	$2s2p^4$ ^{2}P $4d$
21	−2.64556	$2s^22p^3$ ^{4}S $8f$
22	−2.56425	$2s^22p^3$ ^{2}D $7p$
23	−2.45721	$2s^22p^3$ ^{2}D $7f$
24	−2.09856	$2s2p^4$ ^{4}P $5d$
25	−2.08681	$2s^22p^3$ ^{4}S $9f$
26	−1.82087	$2s^22p^3$ ^{2}P $7f$
27	−1.71515	$2s^22p^3$ ^{2}D $8p$
		$^3\mathbf{F}^{\circ}$
1	−18.7948	$2s^22p^3$ ^{2}D $3d$
2	−18.0668	$2s^22p^3$ ^{2}P $3d$
3	−14.8639	$2s2p^4$ ^{2}D $3p$
4	−9.96032	$2s^22p^3$ ^{2}D $4d$
5	−9.31393	$2s^22p^3$ ^{2}P $4d$
6	−7.45107	$2p^5$ ^{2}P $3d$
7	−5.94748	$2s^22p^3$ ^{2}D $5d$
8	−5.81311	$2s2p^4$ ^{4}P $4f$
9	−5.30746	$2s^22p^3$ ^{2}P $5d$
10	−5.03387	$2s2p^4$ ^{2}D $4p$
11	−4.09335	$2s2p^4$ ^{2}D $4f$
12	−3.79598	$2s^22p^3$ ^{2}D $6d$
13	−3.15976	$2s^22p^3$ ^{2}P $6d$
14	−3.12268	$2s2p^4$ ^{2}S $4f$
15	−2.74018	$2s2p^4$ ^{2}P $4f$
16	−2.51004	$2s^22p^3$ ^{2}D $7d$
17	−2.00020	$2s2p^4$ ^{4}P $5f$
18	−1.87287	$2s^22p^3$ ^{2}P $7d$
19	−1.67997	$2s^22p^3$ ^{2}D $8d$
20	−1.11380	$2s^22p^3$ ^{2}D $9d$
21	−1.04253	$2s^22p^3$ ^{2}P $8d$
		$^3\mathbf{G}^{e}$
1	−13.3796	$2s2p^4$ ^{2}D $3d$
2	−9.58723	$2s^22p^3$ ^{2}D $4f$
3	−8.95546	$2s^22p^3$ ^{2}P $4f$
4	−5.78854	$2s^22p^3$ ^{2}D $5f$
5	−5.15381	$2s^22p^3$ ^{2}P $5f$
6	−4.50475	$2s2p^4$ ^{2}D $4d$
7	−3.71080	$2s^22p^3$ ^{2}D $6f$
8	−3.07617	$2s^22p^3$ ^{2}P $6f$
9	−2.45925	$2s^22p^3$ ^{2}D $7f$
10	−1.82358	$2s^22p^3$ ^{2}P $7f$
11	−1.64702	$2s^22p^3$ ^{2}D $8f$
12	−1.09078	$2s^22p^3$ ^{2}D $9f$
13	−1.01075	$2s^22p^3$ ^{2}P $8f$
		$^3\mathbf{G}^{\circ}$
1	−18.6721	$2s^22p^3$ ^{2}D $3d$
2	−9.93211	$2s^22p^3$ ^{2}D $4d$
3	−5.93420	$2s^22p^3$ ^{2}D $5d$
4	−5.78767	$2s2p^4$ ^{4}P $4f$
5	−4.09416	$2s2p^4$ ^{2}D $4f$
6	−3.78982	$2s^22p^3$ ^{2}D $6d$
7	−2.71243	$2s2p^4$ ^{2}P $4f$
8	−2.50578	$2s^22p^3$ ^{2}D $7d$
9	−1.98921	$2s2p^4$ ^{4}P $5f$
10	−1.68979	$2s^22p^3$ ^{2}D $8d$
		$^3\mathbf{H}^{e}$
1	−9.59519	$2s^22p^3$ ^{2}D $4f$
2	−5.78974	$2s^22p^3$ ^{2}D $5f$
3	−3.71266	$2s^22p^3$ ^{2}D $6f$
4	−2.45993	$2s^22p^3$ ^{2}D $7f$
		$^3\mathbf{H}^{\circ}$
1	−4.14944	$2s2p^4$ ^{2}D $4f$
		$^5\mathbf{S}^{\circ}$
1	−22.9514	$2s^22p^3$ ^{4}S $3s$
2	−16.2729	$2s2p^4$ ^{4}P $3p$
3	−12.1402	$2s^22p^3$ ^{4}S $4s$
4	−7.49601	$2s^22p^3$ ^{4}S $5s$
5	−6.63117	$2s2p^4$ ^{4}P $4p$
6	−5.07891	$2s^22p^3$ ^{4}S $6s$
7	−3.68457	$2s^22p^3$ ^{4}S $7s$
8	−2.79724	$2s^22p^3$ ^{4}S $8s$
9	−2.33075	$2s2p^4$ ^{4}P $5p$
10	−2.18509	$2s^22p^3$ ^{4}S $9s$
		$^5\mathbf{P}^{e}$
1	−21.4992	$2s^22p^3$ ^{4}S $3p$
2	−18.0208	$2s2p^4$ ^{4}P $3s$
3	−14.8388	$2s2p^4$ ^{4}P $3d$
4	−11.5564	$2s^22p^3$ ^{4}S $4p$
5	−7.35422	$2s2p^4$ ^{4}P $4s$
6	−7.15446	$2s^22p^3$ ^{4}S $5p$
7	−6.12490	$2s2p^4$ ^{4}P $4d$
8	−4.90544	$2s^22p^3$ ^{4}S $6p$
9	−3.57418	$2s^22p^3$ ^{4}S $7p$
10	−2.72916	$2s^22p^3$ ^{4}S $8p$
11	−2.66476	$2s2p^4$ ^{4}P $5s$
12	−2.15537	$2s^22p^3$ ^{4}S $9p$
13	−2.11987	$2s2p^4$ ^{4}P $5d$
		$^5\mathbf{P}^{\circ}$
1	−16.7076	$2s2p^4$ ^{4}P $3p$
2	−6.77749	$2s2p^4$ ^{4}P $4p$
3	−2.38914	$2s2p^4$ ^{4}P $5p$
4	−.11847	$2s2p^4$ ^{4}P $6p$
		$^5\mathbf{D}^{e}$
1	−15.2091	$2s2p^4$ ^{4}P $3d$
2	−6.24360	$2s2p^4$ ^{4}P $4d$
3	−2.19142	$2s2p^4$ ^{4}P $5d$
4	−.02633	$2s2p^4$ ^{4}P $6d$
		$^5\mathbf{D}^{\circ}$
1	−19.7816	$2s^22p^3$ ^{4}S $3d$
2	−16.5300	$2s2p^4$ ^{4}P $3p$
3	−10.9682	$2s^22p^3$ ^{4}S $4d$
4	−6.95768	$2s^22p^3$ ^{4}S $5d$
5	−6.71831	$2s2p^4$ ^{4}P $4p$
6	−5.77730	$2s2p^4$ ^{4}P $4f$
7	−4.80048	$2s^22p^3$ ^{4}S $6d$
8	−3.51369	$2s^22p^3$ ^{4}S $7d$
9	−2.68313	$2s^22p^3$ ^{4}S $8d$
10	−2.36340	$2s2p^4$ ^{4}P $5p$
11	−2.11534	$2s^22p^3$ ^{4}S $9d$
12	−1.98835	$2s2p^4$ ^{4}P $5f$
		$^5\mathbf{F}^{e}$
1	−14.9865	$2s2p^4$ ^{4}P $3d$
2	−10.5876	$2s^22p^3$ ^{4}S $4f$
3	−6.79081	$2s^22p^3$ ^{4}S $5f$
4	−6.16875	$2s2p^4$ ^{4}P $4d$
5	−4.71290	$2s^22p^3$ ^{4}S $6f$
6	−3.46100	$2s^22p^3$ ^{4}S $7f$
7	−2.64891	$2s^22p^3$ ^{4}S $8f$
8	−2.15777	$2s2p^4$ ^{4}P $5d$
9	−2.08999	$2s^22p^3$ ^{4}S $9f$
		$^5\mathbf{F}^{\circ}$
1	−5.83657	$2s2p^4$ ^{4}P $4f$
2	−2.00906	$2s2p^4$ ^{4}P $5f$
		$^5\mathbf{G}^{\circ}$
1	−5.79697	$2s2p^4$ ^{4}P $4f$
2	−1.98894	$2s2p^4$ ^{4}P $5f$

O-like Ca (Ca^{12+})

Energies in ascending order from ground state for terms with effective $n \leq 4.0$, $L \leq 4$

Term	i	E(Ryds)	Term	i	E(Ryds)	Term	i	E(Ryds)	Term	i	E(Ryds)	Term	i	E(Ryds)
$^3P^e$	1	0.00000	$^1D^e$	2	33.5942	$^3P^e$	5	35.8461	$^1D^o$	4	38.7911	$^5S^o$	3	41.1839
$^1D^e$	1	0.67280	$^1P^e$	2	33.6696	$^1P^o$	4	35.9872	$^3P^o$	6	38.8343	$^3D^e$	6	41.1868
$^1S^e$	1	1.45630	$^3P^e$	4	33.8048	$^5P^o$	1	36.6165	$^3P^e$	7	38.8899	$^3S^o$	5	41.3024
$^3P^o$	1	5.38190	$^3D^o$	2	33.9391	$^5D^o$	2	36.7941	$^3D^e$	4	38.9191	$^3D^e$	7	41.3165
$^1P^o$	1	7.40660	$^1D^e$	3	34.0806	$^3D^o$	5	37.0213	$^3P^o$	7	39.5984	$^1D^e$	6	41.4315
$^1S^e$	2	12.5627	$^3F^o$	1	34.5293	$^5S^o$	2	37.0512	$^1P^o$	6	39.7125	$^3F^e$	4	41.5135
$^5S^o$	1	30.3727	$^1S^o$	1	34.5344	$^3D^e$	3	37.0737	$^3P^o$	8	39.8840	$^1F^e$	3	41.5536
$^3S^o$	1	30.7410	$^3G^o$	1	34.6520	$^3S^o$	3	37.1243	$^3D^o$	7	39.9193	$^3P^e$	9	41.6287
$^3D^o$	1	31.4670	$^1S^e$	3	34.6834	$^3P^o$	5	37.2228	$^3G^e$	1	39.9445	$^5P^e$	4	41.7677
$^1D^o$	1	31.6510	$^1G^o$	1	34.7040	$^1D^e$	4	37.4379	$^1D^o$	5	40.0158	$^3P^e$	10	41.9249
$^5P^e$	1	31.8249	$^1P^o$	3	34.7809	$^3S^e$	2	38.1052	$^1S^o$	2	40.1300	$^1P^e$	5	41.9556
$^3P^o$	2	32.1111	$^3D^o$	3	34.8308	$^5D^e$	1	38.1150	$^3S^o$	4	40.1787	$^1D^e$	7	42.0040
$^3P^e$	2	32.1572	$^1D^o$	2	34.9259	$^5F^e$	1	38.3376	$^3S^e$	3	40.1862	$^3D^o$	8	42.2178
$^1P^o$	2	32.2971	$^3P^o$	3	34.9418	$^1S^e$	4	38.4524	$^3F^e$	3	40.2010	$^1D^o$	6	42.2771
$^1P^e$	1	32.7651	$^3S^o$	2	35.0730	$^3F^o$	3	38.4602	$^1G^e$	1	40.2418	$^5D^o$	3	42.3559
$^3D^e$	1	32.8151	$^1F^o$	1	35.2330	$^5P^e$	3	38.4853	$^3D^e$	5	40.2656	$^3D^o$	9	42.5432
$^3F^e$	1	32.9267	$^3F^o$	2	35.2573	$^3P^e$	6	38.4941	$^3P^e$	8	40.2709	$^5F^e$	2	42.7365
$^1F^e$	1	32.9970	$^3P^o$	4	35.2901	$^1P^o$	5	38.5094	$^1P^e$	4	40.3564	$^3F^e$	5	42.7492
$^3P^e$	3	33.4141	$^5P^e$	2	35.3033	$^1F^o$	3	38.5966	$^1D^e$	5	40.4058			
$^3S^e$	1	33.4467	$^3D^o$	4	35.4549	$^1P^e$	3	38.6219	$^1F^e$	2	40.4787			
$^5D^o$	1	33.5425	$^1D^o$	3	35.5527	$^3F^e$	2	38.6651	$^1S^e$	5	40.5151			
$^3D^e$	2	33.5663	$^1F^o$	2	35.6330	$^3D^o$	6	38.6696	$^1P^o$	7	40.7639			

O-like Ca (Ca^{12+})

gf-values for transitions involving terms with effective $n \leq 4.0$, $L \leq 4$

$i\ i'$	gf$_L$	$i\ i'$	gf$_L$	$i\ i'$	gf$_L$	$i\ i'$	gf$_L$	$i\ i'$	gf$_L$	$i\ i'$	gf$_L$
	$^1\mathbf{P}^e$–$^1\mathbf{S}^o$	1 4	−4.60E−1	3 5	−9.84E−3	4 5	−2.04E−2	6 5	1.47E−3	3 1	5.28E−1
1 1	−1.10E−1	1 5	−1.61E+0	3 6	−8.22E−2	5 1	2.92E−2	6 6	2.65E−4	3 2	−3.72E−1
1 2	−7.45E−2	2 1	−1.19E−2	3 7	−4.71E−1	5 2	2.79E−2	6 7	6.15E−4	3 3	−2.74E−3
2 1	−5.59E−3	2 2	−2.74E−1	4 1	1.52E−1	5 3	5.80E−1		$^1\mathbf{F}^e$–$^1\mathbf{D}^o$		$^1\mathbf{G}^e$–$^1\mathbf{F}^o$
2 2	−1.46E−1	2 3	−2.11E−1	4 2	3.58E−2	5 4	−3.04E−4	1 1	7.20E−1	1 1	1.04E−3
3 1	3.05E−4	2 4	−2.53E−3	4 3	2.71E−3	5 5	−3.79E−2	1 2	−2.26E−2	1 2	6.40E−2
3 2	−1.27E−1	2 5	−4.82E−5	4 4	1.68E−4	6 1	1.20E−1	1 3	−5.77E−2	1 3	1.12E+0
4 1	7.74E−2	3 1	2.72E−1	4 5	−2.06E−1	6 2	1.83E−2	1 4	−8.27E−2		$^1\mathbf{G}^o$–$^1\mathbf{F}^e$
4 2	1.31E−3	3 2	1.31E−2	4 6	−1.90E−2	6 3	9.21E−5	1 5	−8.42E−1	1 1	1.10E+0
5 1	7.52E−4	3 3	−4.70E−4	4 7	−9.45E−2	6 4	5.48E−4	1 6	−2.83E−1	1 2	−2.13E−1
5 2	2.33E−1	3 4	−3.39E−2	5 1	2.29E−1	6 5	2.10E−4	2 1	4.53E−3	1 3	−9.40E−1
	$^1\mathbf{P}^o$–$^1\mathbf{S}^e$	3 5	−2.39E−2	5 2	1.17E−3		$^1\mathbf{D}^o$–$^1\mathbf{D}^e$	2 2	5.12E−4		$^1\mathbf{G}^o$–$^1\mathbf{G}^e$
1 1	5.91E−2	4 1	2.75E−4	5 3	5.96E−2	1 1	5.10E−1	2 3	8.63E−2	1 1	−5.00E−1
1 2	−3.35E−1	4 2	2.58E−1	5 4	1.39E−2	1 2	−5.18E−1	2 4	6.33E−1		$^3\mathbf{P}^e$–$^3\mathbf{S}^o$
1 3	−5.96E−3	4 3	−1.45E−5	5 5	2.46E−1	1 3	−4.40E−1	2 5	2.05E−5	1 1	−3.79E−1
1 4	−4.02E−2	4 4	−8.28E−3	5 6	7.82E−4	1 4	−2.44E−1	2 6	−3.29E−3	1 2	−1.53E+0
1 5	−5.00E−1	4 5	−7.74E−2	5 7	−8.49E−6	1 5	−3.20E−3	3 1	4.28E−3	1 3	−3.44E−1
2 1	1.46E−1	5 1	3.21E−1	6 1	6.89E−3	1 6	−2.57E−4	3 2	1.08E−2	1 4	−6.95E−2
2 2	2.12E−6	5 2	2.96E−3	6 2	2.23E−3	1 7	−5.66E−4	3 3	1.38E−2	1 5	−7.08E−2
2 3	−2.18E−1	5 3	−1.19E−3	6 3	4.33E−4	2 1	7.96E−1	3 4	7.59E−7	2 1	8.27E−1
2 4	−9.13E−2	5 4	−2.98E−1	6 4	5.79E−2	2 2	1.97E−1	3 5	8.07E−1	2 2	−1.10E−1
2 5	−1.77E−3	5 5	−5.27E−3	6 5	1.75E−3	2 3	4.72E−3	3 6	−4.35E−3	2 3	−1.32E−1
3 1	1.46E−1	6 1	8.17E−3	6 6	6.38E−1	2 4	−2.73E−3		$^1\mathbf{F}^o$–$^1\mathbf{D}^e$	2 4	−5.49E−2
3 2	2.87E−5	6 2	1.35E−1	6 7	2.02E−3	2 5	−1.97E−1	1 1	2.15E+0	2 5	−3.94E−1
3 3	7.85E−4	6 3	3.90E−3	7 1	4.33E+0	2 6	−6.61E−2	1 2	5.72E−1	3 1	3.15E−1
3 4	−1.15E−3	6 4	−1.35E−3	7 2	1.47E−4	2 7	−1.71E−2	1 3	2.82E−2	3 2	−2.90E−1
3 5	−1.13E−1	6 5	−6.44E−3	7 3	3.43E−4	3 1	2.24E+0	1 4	−7.22E−3	3 3	−4.57E−2
4 1	1.95E+0	7 1	3.59E−2	7 4	2.99E−2	3 2	6.93E−3	1 5	−8.74E−2	3 4	−5.01E−1
4 2	7.94E−5	7 2	3.37E−3	7 5	1.24E−2	3 3	1.84E−1	1 6	−1.71E−1	3 5	−1.57E−1
4 3	1.45E−1	7 3	5.47E−1	7 6	4.07E−2	3 4	−7.09E−3	1 7	−2.06E−1	4 1	1.26E−1
4 4	−6.22E−3	7 4	4.32E−4	7 7	3.13E−1	3 5	−3.91E−3	2 1	4.11E+0	4 2	−2.23E−3
4 5	−3.60E−3	7 5	−9.58E−2		$^1\mathbf{D}^o$–$^1\mathbf{P}^e$	3 6	−5.90E−2	2 2	1.48E−1	4 3	−4.03E−2
5 1	3.88E−3		$^1\mathbf{D}^e$–$^1\mathbf{P}^o$	1 1	−2.06E−1	3 7	−4.21E−2	2 3	6.48E−1	4 4	−1.42E−1
5 2	6.99E−3	1 1	−6.57E−1	1 2	−5.54E−2	4 1	6.17E−1	2 4	−1.46E−2	4 5	−2.23E−2
5 3	1.31E−3	1 2	−1.48E−1	1 3	−5.88E−1	4 2	2.59E−1	2 5	−6.88E−5	5 1	2.64E−1
5 4	2.63E−4	1 3	−7.03E−1	1 4	−9.97E−4	4 3	1.25E−1	2 6	−1.08E−1	5 2	2.67E−3
5 5	−1.20E−1	1 4	−1.01E−1	1 5	−1.27E−3	4 4	5.31E−1	2 7	−5.72E−1	5 3	−2.92E−1
6 1	3.20E−1	1 5	−3.78E−1	2 1	2.88E−1	4 5	−3.81E−1	3 1	8.06E−1	5 4	−9.68E−4
6 2	9.81E−3	1 6	−8.64E−3	2 2	5.53E−2	4 6	−3.66E−2	3 2	1.58E−3	5 5	−1.04E−2
6 3	1.19E−1	1 7	−2.48E−2	2 3	−3.33E−4	4 7	−5.92E−2	3 3	3.31E−1	6 1	3.88E−3
6 4	2.79E−1	2 1	9.33E−3	2 4	−1.40E−1	5 1	2.11E−3	3 4	6.21E−1	6 2	7.90E−4
6 5	−1.01E−2	2 2	1.88E−1	2 5	−1.44E−2	5 2	3.04E−1	3 5	−1.62E−2	6 3	6.28E−4
7 1	2.05E−2	2 3	−5.56E−2	3 1	3.19E−2	5 3	3.16E−2	3 6	−1.20E−2	6 4	−4.48E−1
7 2	3.35E−3	2 4	−1.31E−3	3 2	4.15E−1	5 4	5.54E−3	3 7	−1.13E−3	6 5	−4.73E−4
7 3	3.52E−2	2 5	−1.70E−3	3 3	−7.70E−5	5 5	−2.38E−3		$^1\mathbf{F}^o$–$^1\mathbf{F}^e$	7 1	4.88E−3
7 4	1.73E−2	2 6	−1.26E−1	3 4	−8.70E−4	5 6	−6.79E−3	1 1	3.45E−1	7 2	1.24E−2
7 5	8.31E−4	2 7	−6.32E−2	3 5	−3.02E−1	5 7	−1.27E−1	1 2	−1.95E−1	7 3	7.10E−1
	$^1\mathbf{P}^o$–$^1\mathbf{P}^e$	3 1	3.39E−5	4 1	2.22E−2	6 1	8.78E−2	1 3	2.13E−1	7 4	−2.23E−4
1 1	−1.45E−3	3 2	4.63E−1	4 2	1.10E−1	6 2	2.87E−1	2 1	4.12E−2	7 5	−3.97E−3
1 2	−4.11E−3	3 3	−5.00E−3	4 3	4.02E−5	6 3	1.85E−1	2 2	−4.63E−3	8 1	2.61E−4
1 3	−3.91E−1	3 4	−1.02E−2	4 4	−6.93E−2	6 4	1.15E−2	2 3	−1.14E−1	8 2	1.18E−1

O-like Ca (Ca^{12+})

i	i'	gf_L	i	i'	gf_L	i	i'	gf_L	i	i'	gf_L	i	i'	gf_L	i	i'	gf_L
8	3	1.38E−1	2	4	−1.15E+0	7	6	7.08E−2	5	5	2.09E−4	4	2	6.30E−3	9	4	6.25E−2
8	4	6.61E−5	2	5	−4.69E−2	7	7	5.00E−4	5	6	5.47E−1	4	3	1.57E−2	9	5	5.58E−3
8	5	−4.41E−4	2	6	−5.35E−1	7	8	−4.86E−4	5	7	1.29E−3	4	4	1.32E+0	9	6	2.62E−2
9	1	1.20E−4	2	7	−3.09E−3	7	9	−5.36E−2	5	8	4.53E−3	4	5	−8.09E−3	9	7	3.53E−2
9	2	5.20E−2	2	8	−9.18E−4	7	10	−1.93E−6	6	1	4.86E−1	4	6	−2.61E−3	9	8	1.89E−3
9	3	9.45E−4	2	9	−8.59E−4	8	1	1.17E−1	6	2	3.38E−3	4	7	−1.93E−3	9	9	2.07E−3
9	4	5.44E−1	2	10	−3.52E−3	8	2	3.57E−2	6	3	1.96E−3	4	8	−4.41E−2	9	10	3.04E+0
9	5	5.04E−6	3	1	4.49E+0	8	3	3.03E−1	6	4	3.58E−2	4	9	−4.50E−1			**^{3}D^o–^{3}D^e**
10	1	6.45E−1	3	2	1.65E−1	8	4	3.97E−1	6	5	9.92E−4	4	10	−8.99E−3	1	1	−1.38E+0
10	2	1.66E−2	3	3	5.81E−1	8	5	2.71E−1	6	6	3.81E−4	5	1	1.23E+0	1	2	−8.76E−2
10	3	4.24E−3	3	4	6.37E−3	8	6	7.40E−1	6	7	1.75E+0	5	2	6.19E−1	1	3	−1.44E+0
10	4	1.06E−2	3	5	−1.26E−2	8	7	1.30E−3	6	8	7.48E−3	5	3	2.31E−1	1	4	−9.17E−4
10	5	1.62E+0	3	6	−1.38E−3	8	8	−3.77E−3	7	1	2.98E+0	5	4	2.00E−1	1	5	−2.50E−3
		^{3}P^o–^{3}S^e	3	7	−1.37E−3	8	9	−3.59E−1	7	2	1.60E−3	5	5	1.37E+0	1	6	−1.04E−4
1	1	−1.36E−2	3	8	−4.71E−2	8	10	−5.86E−5	7	3	5.16E−2	5	6	−4.21E−4	1	7	−1.90E−3
1	2	−1.49E−1	3	9	−1.49E−1			**^{3}D^e–^{3}P^o**	7	4	1.80E−1	5	7	−8.27E−2	2	1	6.26E−2
1	3	−1.68E+0	3	10	−1.60E−2	1	1	2.53E−3	7	5	1.26E−3	5	8	−3.32E−2	2	2	4.08E−3
2	1	−2.72E−1	4	1	4.70E−1	1	2	2.44E−2	7	6	8.55E−2	5	9	−3.45E−4	2	3	−3.69E−4
2	2	−4.76E−1	4	2	6.81E−4	1	3	−2.73E−1	7	7	3.14E−3	5	10	−2.27E−3	2	4	−7.44E−1
2	3	−8.21E−4	4	3	9.45E−2	1	4	−3.69E−2	7	8	1.06E+0	6	1	2.01E−3	2	5	−3.28E−4
3	1	3.34E−2	4	4	3.35E−1	1	5	−8.14E−3			**^{3}D^o–^{3}P^e**	6	2	1.40E−3	2	6	−3.67E−3
3	2	−6.72E−6	4	5	−1.59E−3	1	6	−5.79E−1	1	1	5.83E−1	6	3	1.78E−1	2	7	−4.98E−2
3	3	−2.73E−1	4	6	−1.20E−4	1	7	−2.28E−1	1	2	−4.89E−2	6	4	1.26E−1	3	1	1.11E+0
4	1	6.11E−1	4	7	−3.94E−3	1	8	−8.13E−1	1	3	−1.22E+0	6	5	7.22E−3	3	2	5.94E−3
4	2	−3.94E−4	4	8	−1.04E−1	2	1	2.27E−2	1	4	−9.89E−3	6	6	5.01E−6	3	3	−3.32E−4
4	3	−1.39E−2	4	9	−1.83E−1	2	2	1.59E+0	1	5	−6.35E−2	6	7	−2.44E−3	3	4	−3.06E−2
5	1	3.03E−4	4	10	−4.69E−6	2	3	−4.75E−2	1	6	−1.56E+0	6	8	−2.12E−1	3	5	−1.07E−1
5	2	−4.50E−3	5	1	7.95E−1	2	4	−1.34E−2	1	7	−6.43E−4	6	9	−2.53E−4	3	6	−6.56E−2
5	3	−6.57E−2	5	2	5.13E−1	2	5	−2.18E−3	1	8	−5.41E−4	6	10	−6.55E−4	3	7	−2.70E−1
6	1	1.62E−1	5	3	6.15E−2	2	6	−6.80E−3	1	9	−1.62E−3	7	1	5.66E−1	4	1	2.00E−3
6	2	7.59E−5	5	4	6.51E−2	2	7	−8.76E−1	1	10	−2.20E−2	7	2	2.55E−3	4	2	5.08E−1
6	3	−2.27E−1	5	5	8.51E−1	2	8	−2.94E−3	2	1	2.17E+0	7	3	1.98E−2	4	3	−5.07E−5
7	1	3.20E−2	5	6	−2.16E−3	3	1	7.60E−1	2	2	1.69E+0	7	4	2.51E−1	4	4	−6.60E−5
7	2	1.00E+0	5	7	−4.59E−1	3	2	2.85E−1	2	3	1.17E−2	7	5	2.05E−2	4	5	−2.21E−1
7	3	−1.40E−2	5	8	−4.39E−2	3	3	2.22E−4	2	4	2.38E−4	7	6	1.71E+0	4	6	−2.36E−1
8	1	1.56E−1	5	9	−1.43E−2	3	4	3.80E−6	2	5	−5.39E−3	7	7	8.56E−4	4	7	−1.80E−1
8	2	1.52E−1	5	10	−1.58E−3	3	5	−8.72E−3	2	6	−4.31E−3	7	8	−3.88E−4	5	1	4.11E−4
8	3	−1.25E−3	6	1	1.87E−1	3	6	−1.12E+0	2	7	−4.69E−1	7	9	−8.65E−2	5	2	1.28E−6
		^{3}P^o–^{3}P^e	6	2	5.15E−5	3	7	−2.50E−2	2	8	−1.01E−2	7	10	−2.47E−3	5	3	−8.78E−6
1	1	9.14E−1	6	3	1.80E−1	3	8	−8.09E−2	2	9	−1.27E−2	8	1	1.13E−1	5	4	−4.78E−1
1	2	−4.61E−4	6	4	1.96E−2	4	1	5.32E+0	2	10	−3.78E−1	8	2	1.25E−2	5	5	−7.69E−2
1	3	−4.22E−2	6	5	3.45E−1	4	2	4.65E−3	3	1	4.79E+0	8	3	6.83E−1	5	6	−2.20E−3
1	4	−5.07E−3	6	6	1.20E−2	4	3	2.21E−3	3	2	5.76E−2	8	4	1.08E−2	5	7	−9.26E−3
1	5	−7.80E−1	6	7	−5.16E−7	4	4	1.12E−3	3	3	6.05E−1	8	5	2.24E−2	6	1	6.84E−1
1	6	−2.21E−1	6	8	−6.60E−1	4	5	1.40E+0	3	4	9.71E−4	8	6	4.11E−3	6	2	2.20E−1
1	7	−1.20E+0	6	9	−2.61E−2	4	6	2.31E−3	3	5	−1.22E−2	8	7	6.22E−5	6	3	2.09E+0
1	8	−4.80E+0	6	10	−3.50E−4	4	7	−7.22E−4	3	6	−2.30E−3	8	8	7.07E−4	6	4	−6.70E−3
1	9	−5.93E−1	7	1	6.72E−2	4	8	−5.69E−3	3	7	−2.39E−3	8	9	1.08E−4	6	5	−1.14E+0
1	10	−4.22E−3	7	2	1.56E−3	5	1	4.84E+0	3	8	−5.19E−1	8	10	2.31E−3	6	6	−3.09E−2
2	1	3.33E−1	7	3	1.03E−1	5	2	3.33E−4	3	9	−3.40E−1	9	1	1.14E+0	6	7	−4.72E−3
2	2	−6.56E−4	7	4	1.29E−1	5	3	1.23E−1	3	10	−4.82E−2	9	2	2.87E+0	7	1	1.49E−1
2	3	−1.75E−2	7	5	7.45E−4	5	4	9.98E−2	4	1	5.00E+0	9	3	2.66E−1	7	2	8.48E−1

O-like Ca (Ca^{12+})

$i\ i'$	gf_L	$i\ i'$	gf_L	$i\ i'$	gf_L	$i\ i'$	gf_L	$i\ i'$	gf_L	$i\ i'$	gf_L
7 3	7.49E−5	1 8	−9.75E−1	4 8	−1.65E−2	3 2	1.86E−1		$^3\mathbf{G}^o$–$^3\mathbf{F}^e$		$^5\mathbf{D}^e$–$^5\mathbf{P}^o$
7 4	1.38E−4	1 9	−1.83E−2	4 9	−1.21E−3	3 3	2.43E+0	1 1	3.30E+0	1 1	2.01E+0
7 5	−1.49E−3	2 1	2.64E−3	5 1	1.66E−2	3 4	−2.47E−3	1 2	−1.93E−4		$^5\mathbf{D}^o$–$^5\mathbf{P}^e$
7 6	−2.59E−2	2 2	3.06E−1	5 2	1.23E+1	3 5	−5.25E−2	1 3	−4.57E−1	1 1	2.89E+0
7 7	−2.65E−1	2 3	2.40E−2	5 3	1.78E+0	3 6	−1.62E−2	1 4	−3.05E+0	1 2	−9.98E−4
8 1	7.86E−1	2 4	5.97E−3	5 4	4.12E−1	3 7	−8.66E−3	1 5	−7.43E−5	1 3	−8.00E−1
8 2	3.57E−2	2 5	2.75E+0	5 5	1.23E−1		$^3\mathbf{F}^o$–$^3\mathbf{F}^e$		$^3\mathbf{G}^o$–$^3\mathbf{G}^e$	1 4	−6.58E−1
8 3	9.02E−4	2 6	−4.14E−8	5 6	5.67E−4	1 1	6.84E−1	1 1	−1.36E+0	2 1	9.08E−1
8 4	8.44E−6	2 7	−9.10E−4	5 7	1.11E−3	1 2	−5.52E−4		$^5\mathbf{P}^e$–$^5\mathbf{S}^o$	2 2	3.19E+0
8 5	2.49E−3	2 8	−2.27E−4	5 8	3.89E−2	1 3	−3.68E−1	1 1	1.57E+0	2 3	−1.04E−1
8 6	5.44E−5	2 9	−1.41E−2	5 9	9.08E−1	1 4	−1.07E−1	1 2	−2.64E−1	2 4	−7.20E−4
8 7	2.20E−3	3 1	1.03E−3		$^3\mathbf{F}^o$–$^3\mathbf{D}^e$	1 5	−1.75E−3	1 3	−8.29E−1	3 1	5.61E+0
9 1	3.26E−2	3 2	3.57E−3	1 1	1.65E+0	2 1	1.01E−1	2 1	8.86E−1	3 2	6.93E−2
9 2	1.48E−2	3 3	1.28E−1	1 2	1.23E−2	2 2	−9.28E−4	2 2	−8.04E−1	3 3	7.41E−2
9 3	5.87E−4	3 4	9.96E−2	1 3	−1.30E−3	2 3	−3.24E−1	2 3	−1.96E−3	3 4	4.70E+0
9 4	3.40E−4	3 5	2.81E−3	1 4	−1.54E−2	2 4	−8.43E−1	3 1	8.83E−4		$^5\mathbf{D}^o$–$^5\mathbf{D}^e$
9 5	7.97E−8	3 6	1.66E+0	1 5	−7.14E−1	2 5	−1.82E−3	3 2	9.61E−1	1 1	−1.17E+0
9 6	6.11E−6	3 7	8.84E−5	1 6	−2.99E−1	3 1	1.51E+0	3 3	−3.19E−3	2 1	−5.39E−1
9 7	2.00E−4	3 8	−3.61E−3	1 7	−1.24E+0	3 2	−1.86E−4	4 1	1.14E+0	3 1	2.05E−5
	$^3\mathbf{F}^e$–$^3\mathbf{D}^o$	3 9	−8.19E−5	2 1	9.23E−3	3 3	−9.93E−1	4 2	2.32E−2		$^5\mathbf{F}^e$–$^5\mathbf{D}^o$
1 1	2.34E+0	4 1	2.00E−2	2 2	2.47E+0	3 4	−2.85E−4	4 3	2.53E+0	1 1	5.76E−1
1 2	−8.58E−3	4 2	1.44E−3	2 3	−2.35E−4	3 5	−8.82E−6		$^5\mathbf{P}^o$–$^5\mathbf{P}^e$	1 2	4.11E+0
1 3	−1.09E−1	4 3	1.13E−5	2 4	−9.86E−6		$^3\mathbf{G}^e$–$^3\mathbf{F}^o$	1 1	9.01E−1	1 3	−2.54E−2
1 4	−1.33E−2	4 4	1.01E−1	2 5	−4.35E−2	1 1	9.87E−2	1 2	1.52E+0	2 1	2.32E+1
1 5	−1.11E−3	4 5	9.40E−3	2 6	−1.16E+0	1 2	1.05E−1	1 3	−8.48E−1	2 2	2.34E−1
1 6	−2.69E−1	4 6	1.34E−3	2 7	−1.02E−2	1 3	2.91E+0	1 4	−7.01E−5	2 3	2.97E+0
1 7	−2.46E+0	4 7	2.48E+0	3 1	1.54E−1						

O-like Fe (Fe^{18+})

Term energies relative to $2s^22p^3$ ^{4}S ionization threshold for each symmetry

i	E(Ryds)	Description	i	E(Ryds)	Description	i	E(Ryds)	Description	i	E(Ryds)	Description
		^{1}S^e	22	−5.17851	$2s2p^4$ ^{2}D $5d$	6	−30.6168	$2s2p^4$ ^{2}S $3d$	16	−11.1700	$2s2p^4$ ^{2}P $4f$
1	−104.587	$2s^22p^4$	23	−4.86556	$2s^22p^3$ ^{2}P $7p$	7	−29.6236	$2s2p^4$ ^{2}P $3d$	17	−9.22063	$2s^22p^3$ ^{2}D $6s$
2	−89.1793	$2p^6$	24	−4.40882	$2s^22p^3$ ^{2}D $8p$	8	−25.6444	$2p^5$ ^{2}P $3p$	18	−8.74692	$2s^22p^3$ ^{2}D $6d$
3	−39.7959	$2s^22p^3$ ^{2}P $3p$	25	−4.30511	$2s^22p^3$ ^{2}D $8f$	9	−22.2817	$2s^22p^3$ ^{2}D $4p$	19	−7.51320	$2s^22p^3$ ^{2}P $6d$
4	−34.7769	$2s2p^4$ ^{2}S $3s$	26	−3.99891	$2s2p^4$ ^{2}P $5s$	10	−21.2505	$2s^22p^3$ ^{2}D $4f$	20	−6.34790	$2s^22p^3$ ^{2}D $7s$
5	−31.8139	$2s2p^4$ ^{2}D $3d$	27	−3.23385	$2s2p^4$ ^{2}P $5d$	11	−21.1843	$2s^22p^3$ ^{2}P $4p$	21	−6.05711	$2s^22p^3$ ^{2}D $7d$
6	−23.9607	$2p^5$ ^{2}P $3p$	28	−3.18316	$2s^22p^3$ ^{2}D $9p$	12	−19.9229	$2s^22p^3$ ^{2}P $4f$	22	−5.49020	$2s2p^4$ ^{2}D $5p$
7	−20.8099	$2s^22p^3$ ^{2}P $4p$	29	−3.10999	$2s^22p^3$ ^{2}D $9f$	13	−15.2222	$2s2p^4$ ^{2}D $4s$	23	−5.28868	$2p^5$ ^{2}P $4d$
8	−13.9354	$2s2p^4$ ^{2}S $4s$	30	−3.07097	$2s^22p^3$ ^{2}P $8p$	14	−13.6007	$2s2p^4$ ^{2}D $4d$	24	−5.01768	$2s2p^4$ ^{2}D $5f$
9	−13.5444	$2s2p^4$ ^{2}D $4d$			**^{1}P^o**	15	−13.4437	$2s^22p^3$ ^{2}D $5p$	25	−4.78895	$2s^22p^3$ ^{2}P $7d$
10	−12.0814	$2s^22p^3$ ^{2}P $5p$	1	−96.3902	$2s2p^5$	16	−13.1368	$2s^22p^3$ ^{2}D $5f$	26	−4.50737	$2s^22p^3$ ^{2}D $8s$
11	−7.48949	$2s^22p^3$ ^{2}P $6p$	2	−43.2940	$2s^22p^3$ ^{2}P $3s$	17	−12.3481	$2s2p^4$ ^{2}S $4d$	27	−4.31529	$2s^22p^3$ ^{2}D $8d$
12	−5.40186	$2p^5$ ^{2}P $4p$	3	−40.0486	$2s^22p^3$ ^{2}D $3d$	18	−12.2305	$2s^22p^3$ ^{2}P $5p$	28	−3.55754	$2s2p^4$ ^{2}P $5p$
13	−5.16512	$2s2p^4$ ^{2}D $5d$	4	−38.0133	$2s^22p^3$ ^{2}P $3d$	19	−11.8161	$2s^22p^3$ ^{2}P $5f$	29	−3.25230	$2s^22p^3$ ^{2}D $9s$
14	−4.76390	$2s^22p^3$ ^{2}P $7p$	5	−34.4988	$2s2p^4$ ^{2}D $3p$	20	−11.5163	$2s2p^4$ ^{2}P $4d$	30	−3.13051	$2s^22p^3$ ^{2}D $9g$
15	−4.58705	$2s2p^4$ ^{2}S $5s$	6	−32.9368	$2s2p^4$ ^{2}S $3p$	21	−8.85331	$2s^22p^3$ ^{2}D $6p$	31	−3.12600	$2s^22p^3$ ^{2}D $9d$
16	−3.00593	$2s^22p^3$ ^{2}P $8p$	7	−31.2488	$2s2p^4$ ^{2}P $3p$	22	−8.70356	$2s^22p^3$ ^{2}D $6f$	32	−3.05186	$2s2p^4$ ^{2}P $5f$
17	−1.80800	$2s^22p^3$ ^{2}P $9p$	8	−27.4693	$2p^5$ ^{2}P $3s$	23	−7.60212	$2s^22p^3$ ^{2}P $6p$	33	−3.02569	$2s^22p^3$ ^{2}P $8d$
		^{1}S^o	9	−22.8601	$2p^5$ ^{2}P $3d$	24	−7.38187	$2s^22p^3$ ^{2}P $6f$			**^{1}F^e**
1	−40.4334	$2s^22p^3$ ^{2}D $3d$	10	−22.1596	$2s^22p^3$ ^{2}P $4s$	25	−6.12386	$2s^22p^3$ ^{2}D $7p$	1	−42.6706	$2s^22p^3$ ^{2}D $3p$
2	−32.2177	$2s2p^4$ ^{2}P $3p$	11	−21.7203	$2s^22p^3$ ^{2}D $4d$	26	−6.05397	$2p^5$ ^{2}P $4p$	2	−31.8570	$2s2p^4$ ^{2}D $3d$
3	−21.8759	$2s^22p^3$ ^{2}D $4d$	12	−20.1523	$2s^22p^3$ ^{2}P $4d$	27	−6.03361	$2s^22p^3$ ^{2}D $7f$	3	−30.4323	$2s2p^4$ ^{2}P $3d$
4	−13.4095	$2s^22p^3$ ^{2}D $5d$	13	−14.4306	$2s2p^4$ ^{2}D $4p$	28	−5.80705	$2s2p^4$ ^{2}D $5s$	4	−22.6593	$2s^22p^3$ ^{2}D $4p$
5	−12.4714	$2s2p^4$ ^{2}P $4p$	14	−13.3351	$2s^22p^3$ ^{2}D $5d$	29	−5.17649	$2s2p^4$ ^{2}D $5d$	5	−21.2363	$2s^22p^3$ ^{2}D $4f$
6	−8.84851	$2s^22p^3$ ^{2}D $6d$	15	−13.2167	$2s2p^4$ ^{2}D $4f$	30	−4.83487	$2s^22p^3$ ^{2}P $7p$	6	−19.9563	$2s^22p^3$ ^{2}P $4f$
7	−6.12068	$2s^22p^3$ ^{2}D $7d$	16	−13.1472	$2s2p^4$ ^{2}S $4p$	31	−4.71112	$2s^22p^3$ ^{2}P $7f$	7	−13.7161	$2s^22p^3$ ^{2}D $5p$
8	−4.35857	$2s^22p^3$ ^{2}D $8d$	17	−12.7949	$2s^22p^3$ ^{2}P $5s$	32	−4.67733	$2p^5$ ^{2}P $4f$	8	−13.5298	$2s2p^4$ ^{2}D $4d$
9	−3.54500	$2s2p^4$ ^{2}P $5p$	18	−12.1697	$2s2p^4$ ^{2}P $4p$	33	−4.34723	$2s^22p^3$ ^{2}D $8p$	9	−13.1288	$2s^22p^3$ ^{2}D $5f$
10	−3.15332	$2s^22p^3$ ^{2}D $9d$	19	−11.8744	$2s^22p^3$ ^{2}P $5d$	34	−4.30200	$2s^22p^3$ ^{2}D $8f$	10	−11.8573	$2s^22p^3$ ^{2}P $5f$
		^{1}P^e	20	−8.80580	$2s^22p^3$ ^{2}D $6d$	35	−3.89739	$2s2p^4$ ^{2}S $5d$	11	−11.7591	$2s2p^4$ ^{2}P $4d$
1	−42.9680	$2s^22p^3$ ^{2}D $3p$	21	−7.89474	$2s^22p^3$ ^{2}P $6s$	36	−3.20759	$2s2p^4$ ^{2}P $5d$	12	−8.96377	$2s^22p^3$ ^{2}D $6p$
2	−41.3273	$2s^22p^3$ ^{2}P $3p$	22	−7.41133	$2s^22p^3$ ^{2}P $6d$	37	−3.14540	$2s^22p^3$ ^{2}D $9p$	13	−8.70003	$2s^22p^3$ ^{2}D $6f$
3	−34.4110	$2s2p^4$ ^{2}P $3s$	23	−6.80355	$2p^5$ ^{2}P $4s$	38	−3.10898	$2s^22p^3$ ^{2}D $9f$	14	−7.39404	$2s^22p^3$ ^{2}P $6f$
4	−32.0553	$2s2p^4$ ^{2}D $3d$	24	−6.09386	$2s^22p^3$ ^{2}D $7d$	39	−3.04868	$2s^22p^3$ ^{2}P $8p$	15	−6.18411	$2s^22p^3$ ^{2}D $7p$
5	−29.7254	$2s2p^4$ ^{2}P $3d$	25	−5.44200	$2s2p^4$ ^{2}D $5p$	40	−2.97655	$2s^22p^3$ ^{2}P $8f$	16	−6.03119	$2s^22p^3$ ^{2}D $7f$
6	−25.6122	$2p^5$ ^{2}P $3p$	26	−5.10472	$2s2p^4$ ^{2}D $5f$			**^{1}D^o**	17	−5.18818	$2s2p^4$ ^{2}D $5d$
7	−22.7269	$2s^22p^3$ ^{2}D $4p$	27	−5.02180	$2s^22p^3$ ^{2}P $7s$	1	−44.6275	$2s^22p^3$ ^{2}D $3s$	18	−4.71896	$2s^22p^3$ ^{2}P $7f$
8	−21.3560	$2s^22p^3$ ^{2}P $4p$	28	−4.84592	$2p^5$ ^{2}P $4d$	2	−39.7706	$2s^22p^3$ ^{2}D $3d$	19	−4.66163	$2p^5$ ^{2}P $4f$
9	−21.2419	$2s^22p^3$ ^{2}D $4f$	29	−4.72531	$2s^22p^3$ ^{2}P $7d$	3	−38.7616	$2s^22p^3$ ^{2}P $3d$	20	−4.39676	$2s^22p^3$ ^{2}D $8p$
10	−13.7719	$2s^22p^3$ ^{2}D $5p$	30	−4.33969	$2s^22p^3$ ^{2}D $8d$	4	−34.0822	$2s2p^4$ ^{2}D $3p$	21	−4.30001	$2s^22p^3$ ^{2}D $8f$
11	−13.5952	$2s2p^4$ ^{2}D $4d$	31	−4.19453	$2s2p^4$ ^{2}S $5p$	5	−32.3707	$2s2p^4$ ^{2}P $3p$	22	−3.32236	$2s2p^4$ ^{2}P $5d$
12	−13.3069	$2s2p^4$ ^{2}P $4s$	32	−3.41817	$2s2p^4$ ^{2}P $5p$	6	−23.9619	$2p^5$ ^{2}P $3d$	23	−3.17463	$2s^22p^3$ ^{2}D $9p$
13	−13.1441	$2s^22p^3$ ^{2}D $5f$	33	−3.17267	$2s^22p^3$ ^{2}P $8s$	7	−23.4829	$2s^22p^3$ ^{2}D $4s$	24	−3.11252	$2s^22p^3$ ^{2}D $9f$
14	−12.3471	$2s^22p^3$ ^{2}P $5p$	34	−3.14184	$2s^22p^3$ ^{2}D $9d$	8	−21.5327	$2s^22p^3$ ^{2}D $4d$	25	−2.98113	$2s^22p^3$ ^{2}P $8f$
15	−11.5800	$2s2p^4$ ^{2}P $4d$	35	−2.98518	$2s^22p^3$ ^{2}P $8d$	9	−20.5056	$2s^22p^3$ ^{2}P $4d$			**^{1}F^o**
16	−8.98096	$2s^22p^3$ ^{2}D $6p$			**^{1}D^e**	10	−14.4174	$2s2p^4$ ^{2}D $4p$	1	−39.3187	$2s^22p^3$ ^{2}D $3d$
17	−8.70797	$2s^22p^3$ ^{2}D $6f$	1	−105.665	$2s^22p^4$	11	−14.0961	$2s^22p^3$ ^{2}D $5s$	2	−38.5815	$2s^22p^3$ ^{2}P $3d$
18	−7.64829	$2s^22p^3$ ^{2}P $6p$	2	−41.6857	$2s^22p^3$ ^{2}D $3p$	12	−13.2466	$2s^22p^3$ ^{2}D $5d$	3	−34.3701	$2s2p^4$ ^{2}D $3p$
19	−6.19602	$2s^22p^3$ ^{2}D $7p$	3	−40.7952	$2s^22p^3$ ^{2}P $3p$	13	−13.1836	$2s2p^4$ ^{2}D $4f$	4	−24.0471	$2p^5$ ^{2}P $3d$
20	−6.03740	$2s^22p^3$ ^{2}D $7f$	4	−36.0500	$2s2p^4$ ^{2}D $3s$	14	−12.5098	$2s2p^4$ ^{2}P $4p$	5	−21.4545	$2s^22p^3$ ^{2}D $4d$
21	−5.96314	$2p^5$ ^{2}P $4p$	5	−31.9839	$2s2p^4$ ^{2}D $3d$	15	−12.0548	$2s^22p^3$ ^{2}P $5d$	6	−20.3602	$2s^22p^3$ ^{2}P $4d$

O-like Fe (Fe^{18+})

i	E(Ryds)	Description	i	E(Ryds)	Description	i	E(Ryds)	Description	i	E(Ryds)	Description
7	−14.4898	$2s2p^4$ ^{2}D $4p$	3	−8.71461	$2s^22p^3$ ^{2}D $6f$	10	−25.5954	$2p^5$ ^{2}P $3p$	21	−12.0667	$2s^22p^3$ ^{2}P $5d$
8	−13.2223	$2s^22p^3$ ^{2}D $5d$	4	−6.04058	$2s^22p^3$ ^{2}D $7f$	11	−23.8589	$2s^22p^3$ ^{4}S $4p$	22	−8.78779	$2s^22p^3$ ^{2}D $6d$
9	−13.1670	$2s2p^4$ ^{2}D $4f$:	5	−4.30637	$2s^22p^3$ ^{2}D $8f$	12	−22.4881	$2s^22p^3$ ^{2}D $4p$	23	−7.92214	$2s^22p^3$ ^{2}P $6s$
10	−13.0277	$2s^22p^3$ ^{2}D $5g$:	6	−3.11887	$2s^22p^3$ ^{2}D $9f$	13	−21.2692	$2s^22p^3$ ^{2}P $4p$	24	−7.70287	$2s2p^4$ ^{4}P $5p$
11	−11.9928	$2s^22p^3$ ^{2}P $5d$			^{1}H^o	14	−21.2248	$2s^22p^3$ ^{2}D $4f$	25	−7.51855	$2s^22p^3$ ^{2}P $6d$
12	−11.9144	$2s2p^4$ ^{2}S $4f$:	1	−13.2604	$2s2p^4$ ^{2}D $4f$	15	−17.5132	$2s2p^4$ ^{4}P $4s$	26	−6.83997	$2p^5$ ^{2}P $4s$
13	−11.2428	$2s2p^4$ ^{2}P $4f$	2	−5.05996	$2s2p^4$ ^{2}D $5f$	16	−15.7921	$2s2p^4$ ^{4}P $4d$	27	−6.08314	$2s^22p^3$ ^{2}D $7d$
14	−8.72959	$2s^22p^3$ ^{2}D $6d$			^{3}S^e	17	−14.9252	$2s^22p^3$ ^{4}S $5p$	28	−5.50543	$2s2p^4$ ^{2}D $5p$
15	−7.46830	$2s^22p^3$ ^{2}P $6d$	1	−41.6222	$2s^22p^3$ ^{2}P $3p$	18	−13.7339	$2s2p^4$ ^{2}D $4d$	29	−5.43825	$2p^5$ ^{2}P $4d$
16	−6.04668	$2s^22p^3$ ^{2}D $7d$	2	−35.2780	$2s2p^4$ ^{2}S $3s$	19	−13.5923	$2s^22p^3$ ^{2}D $5p$	30	−5.07360	$2s2p^4$ ^{2}D $5f$
17	−5.55684	$2s2p^4$ ^{2}D $5p$	3	−32.3233	$2s2p^4$ ^{2}D $3d$	20	−13.3969	$2s2p^4$ ^{2}P $4s$	31	−5.03519	$2s^22p^3$ ^{2}P $7s$
18	−5.26237	$2p^5$ ^{2}P $4d$	4	−26.2552	$2p^5$ ^{2}P $3p$	21	−13.1321	$2s^22p^3$ ^{2}D $5f$	32	−4.79242	$2s^22p^3$ ^{2}P $7d$
19	−5.01126	$2s2p^4$ ^{2}D $5f$	5	−21.4182	$2s^22p^3$ ^{2}P $4p$	22	−12.3088	$2s^22p^3$ ^{2}P $5p$	33	−4.33341	$2s^22p^3$ ^{2}D $8d$
20	−4.76090	$2s^22p^3$ ^{2}P $7d$	6	−14.1327	$2s2p^4$ ^{2}S $4s$	23	−11.7354	$2s2p^4$ ^{2}P $4d$	34	−4.23002	$2s2p^4$ ^{2}S $5p$
21	−4.30868	$2s^22p^3$ ^{2}D $8d$	7	−13.7189	$2s2p^4$ ^{2}D $4d$	24	−10.2597	$2s^22p^3$ ^{4}S $6p$	35	−3.60616	$2s2p^4$ ^{2}P $5p$
22	−3.73285	$2s2p^4$ ^{2}S $5f$	8	−12.3684	$2s^22p^3$ ^{2}P $5p$	25	−8.92818	$2s^22p^3$ ^{2}D $6p$	36	−3.18921	$2s^22p^3$ ^{2}P $8s$
23	−3.13628	$2s^22p^3$ ^{2}D $9g$:	9	−7.66479	$2s^22p^3$ ^{2}P $6p$	26	−8.70320	$2s^22p^3$ ^{2}D $6f$	37	−3.13605	$2s^22p^3$ ^{2}D $9d$
24	−3.12002	$2s^22p^3$ ^{2}D $9d$:	10	−6.15310	$2p^5$ ^{2}P $4p$	27	−8.21593	$2s2p^4$ ^{4}P $5s$	38	−3.04576	$2s2p^4$ ^{4}P $6p$
25	−3.08128	$2s2p^4$ ^{2}P $5f$	11	−5.25995	$2s2p^4$ ^{2}D $5d$	28	−7.62172	$2s^22p^3$ ^{2}P $6p$	39	−3.02983	$2s^22p^3$ ^{2}P $8d$
26	−3.00889	$2s^22p^3$ ^{2}P $8d$	12	−4.89252	$2s^22p^3$ ^{2}P $7p$	29	−7.50710	$2s^22p^3$ ^{4}S $7p$			^{3}D^e
		^{1}G^e	13	−4.68758	$2s2p^4$ ^{2}S $5s$	30	−7.42295	$2s2p^4$ ^{4}P $5d$	1	−42.9240	$2s^22p^3$ ^{2}D $3p$
1	−32.2221	$2s2p^4$ ^{2}D $3d$	14	−3.08043	$2s^22p^3$ ^{2}P $8p$	31	−6.15938	$2s^22p^3$ ^{2}D $7p$	2	−41.4558	$2s^22p^3$ ^{2}P $3p$
2	−21.2469	$2s^22p^3$ ^{2}D $4f$	15	−1.86141	$2s^22p^3$ ^{2}P $9p$	32	−6.03357	$2s^22p^3$ ^{2}D $7f$	3	−36.5848	$2s2p^4$ ^{2}D $3s$
3	−19.9514	$2s^22p^3$ ^{2}P $4f$			^{3}S^o	33	−5.96329	$2p^5$ ^{2}P $4p$	4	−33.9665	$2s2p^4$ ^{4}P $3d$
4	−13.6881	$2s2p^4$ ^{2}D $4d$	1	−45.8448	$2s^22p^3$ ^{4}S $3s$	34	−5.71871	$2s^22p^3$ ^{4}S $8p$	5	−32.1632	$2s2p^4$ ^{2}D $3d$
5	−13.1307	$2s^22p^3$ ^{2}D $5f$	2	−39.5847	$2s^22p^3$ ^{2}D $3d$	35	−5.25320	$2s2p^4$ ^{2}D $5d$	6	−31.0274	$2s2p^4$ ^{2}S $3d$
6	−11.8258	$2s^22p^3$ ^{2}P $5f$	3	−36.3495	$2s2p^4$ ^{4}P $3p$	36	−4.84523	$2s^22p^3$ ^{2}P $7p$	7	−30.7641	$2s2p^4$ ^{2}P $3d$
7	−8.70191	$2s^22p^3$ ^{2}D $6f$	4	−32.1423	$2s2p^4$ ^{2}P $3p$	37	−4.51434	$2s^22p^3$ ^{4}S $9p$	8	−25.8622	$2p^5$ ^{2}P $3p$
8	−7.38992	$2s^22p^3$ ^{2}P $6f$	5	−24.7881	$2s^22p^3$ ^{4}S $4s$	38	−4.37756	$2s^22p^3$ ^{2}D $8p$	9	−22.7329	$2s^22p^3$ ^{2}D $4p$
9	−6.03213	$2s^22p^3$ ^{2}D $7f$	6	−21.5762	$2s^22p^3$ ^{2}D $4d$	39	−4.30189	$2s^22p^3$ ^{2}D $8f$	10	−21.3809	$2s^22p^3$ ^{2}P $4p$
10	−5.24501	$2s2p^4$ ^{2}D $5d$	7	−16.7059	$2s2p^4$ ^{4}P $4p$	40	−4.02771	$2s2p^4$ ^{2}P $5s$	11	−21.2478	$2s^22p^3$ ^{2}D $4f$
11	−4.74318	$2p^5$ ^{2}P $4f$	8	−15.4172	$2s^22p^3$ ^{4}S $5s$			^{3}P^o	12	−19.9499	$2s^22p^3$ ^{2}P $4f$
12	−4.71571	$2s^22p^3$ ^{2}P $7f$	9	−13.2639	$2s^22p^3$ ^{2}D $5d$	1	−99.1273	$2s2p^5$	13	−15.7519	$2s2p^4$ ^{4}P $4d$
13	−4.30041	$2s^22p^3$ ^{2}D $8f$	10	−12.4467	$2s2p^4$ ^{2}P $4p$	2	−43.5586	$2s^22p^3$ ^{2}P $3s$	14	−15.4146	$2s2p^4$ ^{2}D $4s$
14	−3.11472	$2s^22p^3$ ^{2}D $9f$	11	−10.5561	$2s^22p^3$ ^{4}S $6s$	3	−39.7909	$2s^22p^3$ ^{2}D $3d$	15	−13.7785	$2s2p^4$ ^{2}D $4d$
15	−2.98046	$2s^22p^3$ ^{2}P $8f$	12	−8.76699	$2s^22p^3$ ^{2}D $6d$	4	−39.1588	$2s^22p^3$ ^{2}P $3d$	16	−13.6198	$2s^22p^3$ ^{2}D $5p$
		^{1}G^o	13	−7.80254	$2s2p^4$ ^{4}P $5p$	5	−36.2357	$2s2p^4$ ^{4}P $3p$	17	−13.1372	$2s^22p^3$ ^{2}D $5f$
1	−40.1721	$2s^22p^3$ ^{2}D $3d$	14	−7.66201	$2s^22p^3$ ^{4}S $7s$	6	−33.9993	$2s2p^4$ ^{2}D $3p$	18	−12.4924	$2s2p^4$ ^{2}S $4d$
2	−21.7895	$2s^22p^3$ ^{2}D $4d$	15	−6.07097	$2s^22p^3$ ^{2}D $7d$	7	−33.0959	$2s2p^4$ ^{2}S $3p$	19	−12.3249	$2s^22p^3$ ^{2}P $5p$
3	−13.3675	$2s^22p^3$ ^{2}D $5d$	16	−5.84611	$2s^22p^3$ ^{4}S $8s$	8	−32.5571	$2s2p^4$ ^{2}P $3p$	20	−11.8940	$2s2p^4$ ^{2}P $4d$
4	−13.1974	$2s2p^4$ ^{2}D $4f$:	17	−4.60004	$2s^22p^3$ ^{4}S $9s$	9	−27.7687	$2p^5$ ^{2}P $3s$	21	−11.8203	$2s^22p^3$ ^{2}P $5f$
5	−13.0135	$2s^22p^3$ ^{2}D $5g$:	18	1.33180	$2s^22p^3$ ^{2}D $8d$	10	21.4803	$2p^5$ ^{2}P $3d$	22	−8.98605	$2s^22p^3$ ^{2}D $6p$
6	−11.2084	$2s2p^4$ ^{2}P $4f$			^{3}P^e	11	−22.2452	$2s^22p^3$ ^{2}P $4s$	23	−8.70721	$2s^22p^3$ ^{2}D $6f$
7	−8.82483	$2s^22p^3$ ^{2}D $6d$	1	−106.592	$2s^22p^4$	12	−21.6457	$2s^22p^3$ ^{2}D $4d$	24	−7.65367	$2s^22p^3$ ^{2}P $6p$
8	−6.10581	$2s^22p^3$ ^{2}D $7d$	2	−43.7893	$2s^22p^3$ ^{4}S $3p$	13	−20.5462	$2s^22p^3$ ^{2}P $4d$	25	−7.42617	$2s2p^4$ ^{4}P $5d$
9	−5.01963	$2s2p^4$ ^{2}D $5f$	3	−42.0298	$2s^22p^3$ ^{2}D $3p$	14	−16.5368	$2s2p^4$ ^{4}P $4p$	26	−7.38801	$2s^22p^3$ ^{2}P $6f$
10	−4.34832	$2s^22p^3$ ^{2}D $8d$	4	−41.1067	$2s^22p^3$ ^{2}P $3p$	15	−14.4219	$2s2p^4$ ^{2}D $4p$	27	−6.21400	$2s^22p^3$ ^{2}D $7p$
11	−3.14712	$2s^22p^3$ ^{2}D $9d$	5	−38.2234	$2s2p^4$ ^{4}P $3s$	16	−13.3054	$2s^22p^3$ ^{2}D $5d$	28	−6.10617	$2p^5$ ^{2}P $4p$
12	3.08568	$2s2p^4$ ^{2}P $5f$	6	−34.5938	$2s2p^4$ ^{2}P $3s$	17	−13.2630	$2s2p^4$ ^{2}D $4f$	29	−6.03617	$2s^22p^3$ ^{2}D $7f$
		^{1}H^e	7	−34.0359	$2s2p^4$ ^{4}P $3d$	18	−13.2070	$2s2p^4$ ^{2}S $4p$	30	−5.91867	$2s2p^4$ ^{2}D $5s$
1	−21.2782	$2s^22p^3$ ^{2}D $4f$	8	−32.1314	$2s2p^4$ ^{2}D $3d$	19	−12.8305	$2s^22p^3$ ^{2}P $5s$	31	−5.25168	$2s2p^4$ ^{2}D $5d$
2	−13.1506	$2s^22p^3$ ^{2}D $5f$	9	−30.3034	$2s2p^4$ ^{2}P $3d$	20	−12.5706	$2s2p^4$ ^{2}P $4p$	32	−4.86693	$2s^22p^3$ ^{2}P $7p$

O-like Fe (Fe^{18+})

i	E(Ryds)	Description
33	−4.72213	$2s^22p^3$ ^{2}P $7f$
34	−4.71386	$2p^5$ ^{2}P $4f$
35	−4.40427	$2s^22p^3$ ^{2}D $8p$
36	−4.30312	$2s^22p^3$ ^{2}D $8f$
37	−3.97201	$2s2p^4$ ^{2}S $5d$
38	−3.37206	$2s2p^4$ ^{2}P $5d$
39	−3.18498	$2s^22p^3$ ^{2}D $9p$
40	−3.11595	$2s^22p^3$ ^{2}D $9f$
41	−3.07485	$2s^22p^3$ ^{2}P $8p$
42	−2.98387	$2s^22p^3$ ^{2}P $8f$
43	−2.93332	$2s2p^4$ ^{4}P $6d$
	^{3}D^o	
1	−44.8896	$2s^22p^3$ ^{2}D $3s$
2	−41.1814	$2s^22p^3$ ^{4}S $3d$
3	−39.9309	$2s^22p^3$ ^{2}D $3d$
4	−38.8703	$2s^22p^3$ ^{2}P $3d$
5	−36.5079	$2s2p^4$ ^{4}P $3p$
6	−34.2571	$2s2p^4$ ^{2}D $3p$
7	−32.5200	$2s2p^4$ ^{2}P $3p$
8	−23.9464	$2p^5$ ^{2}P $3d$
9	−23.5690	$2s^22p^3$ ^{2}D $4s$
10	−22.9362	$2s^22p^3$ ^{4}S $4d$
11	−21.7455	$2s^22p^3$ ^{2}D $4d$
12	−20.4614	$2s^22p^3$ ^{2}P $4d$
13	−16.7389	$2s2p^4$ ^{4}P $4p$
14	−15.4177	$2s2p^4$ ^{4}P $4f$
15	−14.5994	$2s^22p^3$ ^{4}S $5d$
16	−14.5150	$2s2p^4$ ^{2}D $4p$
17	−14.1077	$2s^22p^3$ ^{2}D $5s$
18	−13.3463	$2s^22p^3$ ^{2}D $5d$
19	−13.2184	$2s2p^4$ ^{2}D $4f$
20	−12.5646	$2s2p^4$ ^{2}P $4p$
21	−12.0292	$2s^22p^3$ ^{2}P $5d$
22	−11.1785	$2s2p^4$ ^{2}P $4f$
23	−10.1086	$2s^22p^3$ ^{4}S $6d$
24	−9.24539	$2s^22p^3$ ^{2}D $6s$
25	−8.81854	$2s^22p^3$ ^{2}D $6d$
26	−7.80143	$2s2p^4$ ^{4}P $5p$
27	−7.50429	$2s^22p^3$ ^{2}P $6d$
28	−7.40530	$2s^22p^3$ ^{4}S $7d$
29	−7.31812	$2s2p^4$ ^{4}P $5f$
30	−6.36533	$2s^22p^3$ ^{2}D $7s$
31	−6.10368	$2s^22p^3$ ^{2}D $7d$
32	−5.66831	$2s^22p^3$ ^{4}S $8d$
33	−5.51955	$2s2p^4$ ^{2}D $5p$
34	−5.26980	$2p^5$ ^{2}P $4d$
35	−5.05028	$2s2p^4$ ^{2}D $5f$
36	−4.77856	$2s^22p^3$ ^{2}P $7d$
37	−4.51672	$2s^22p^3$ ^{2}D $8s$
38	−4.47680	$2s^22p^3$ ^{4}S $9d$
39	−4.34413	$2s^22p^3$ ^{2}D $8d$
	^{3}F^e	
1	−42.7745	$2s^22p^3$ ^{2}D $3p$
2	−34.3592	$2s2p^4$ ^{4}P $3d$
3	−32.3109	$2s2p^4$ ^{2}D $3d$
4	−30.5044	$2s2p^4$ ^{2}P $3d$
5	−22.6964	$2s^22p^3$ ^{2}D $4p$
6	−22.5959	$2s^22p^3$ ^{4}S $4f$
7	−21.2609	$2s^22p^3$ ^{2}D $4f$
8	−19.9542	$2s^22p^3$ ^{2}P $4f$
9	−15.9324	$2s2p^4$ ^{4}P $4d$
10	−14.4818	$2s^22p^3$ ^{4}S $5f$
11	−13.7962	$2s2p^4$ ^{2}D $4d$
12	−13.6266	$2s^22p^3$ ^{2}D $5p$
13	−13.1421	$2s^22p^3$ ^{2}D $5f$
14	−11.8461	$2s^22p^3$ ^{2}P $5f$
15	−11.7973	$2s2p^4$ ^{2}P $4d$
16	−10.0516	$2s^22p^3$ ^{4}S $6f$
17	−8.97478	$2s^22p^3$ ^{2}D $6p$
18	−8.71151	$2s^22p^3$ ^{2}D $6f$
19	−7.52129	$2s2p^4$ ^{4}P $5d$
20	−7.39126	$2s^22p^3$ ^{2}P $6f$
21	−7.38180	$2s^22p^3$ ^{4}S $7f$
22	−6.19137	$2s^22p^3$ ^{2}D $7p$
23	−6.03939	$2s^22p^3$ ^{2}D $7f$
24	−5.64896	$2s^22p^3$ ^{4}S $8f$
25	−5.26461	$2s2p^4$ ^{2}D $5d$
26	−4.71762	$2s^22p^3$ ^{2}P $7f$
27	−4.66185	$2p^5$ ^{2}P $4f$
28	−4.46236	$2s^22p^3$ ^{4}S $9f$
29	−4.40131	$2s^22p^3$ ^{2}D $8p$
30	−4.30489	$2s^22p^3$ ^{2}D $8f$
	^{3}F^o	
1	−40.4320	$2s^22p^3$ ^{2}D $3d$
2	−39.2125	$2s^22p^3$ ^{2}P $3d$
3	−34.5696	$2s2p^4$ ^{2}D $3p$
4	−24.3192	$2p^5$ ^{2}P $3d$
5	−21.8605	$2s^22p^3$ ^{2}D $4d$
6	−20.5790	$2s^22p^3$ ^{2}P $4d$
7	−15.5008	$2s2p^4$ ^{4}P $4f$
8	−14.5603	$2s2p^4$ ^{2}D $4p$
9	−13.3928	$2s^22p^3$ ^{2}D $5d$
10	−13.2104	$2s2p^4$ ^{2}D $4f$
11	−12.0902	$2s^22p^3$ ^{2}P $5d$
12	−11.9425	$2s2p^4$ ^{2}S $4f$
13	−11.2675	$2s2p^4$ ^{2}P $4f$
14	−8.84338	$2s^22p^3$ ^{2}D $6d$
15	−7.52909	$2s^22p^3$ ^{2}P $6d$
16	−7.35169	$2s2p^4$ ^{4}P $5f$
17	−6.11761	$2s^22p^3$ ^{2}D $7d$
18	−5.59395	$2s2p^4$ ^{2}D $5p$
19	−5.34442	$2p^5$ ^{2}P $4d$
20	−5.03270	$2s2p^4$ ^{2}D $5f$
21	−4.79801	$2s^22p^3$ ^{2}P $7d$
22	−4.35546	$2s^22p^3$ ^{2}D $8d$
23	−3.74758	$2s2p^4$ ^{2}S $5f$
24	−3.15304	$2s^22p^3$ ^{2}D $9d$
25	−3.14454	$2s2p^4$ ^{2}P $5f$
26	−3.03411	$2s^22p^3$ ^{2}P $8d$
27	−2.90981	$2s2p^4$ ^{4}P $6f$
	^{3}G^e	
1	−32.7025	$2s2p^4$ ^{2}D $3d$
2	−21.2767	$2s^22p^3$ ^{2}D $4f$
3	−19.9699	$2s^22p^3$ ^{2}P $4f$
4	−13.8498	$2s2p^4$ ^{2}D $4d$
5	−13.1488	$2s^22p^3$ ^{2}D $5f$
6	−11.8386	$2s^22p^3$ ^{2}P $5f$
7	−8.71523	$2s^22p^3$ ^{2}D $6f$
8	−7.39859	$2s^22p^3$ ^{2}P $6f$
9	−6.04170	$2s^22p^3$ ^{2}D $7f$
10	−5.32359	$2s2p^4$ ^{2}D $5d$
11	−4.75621	$2p^5$ ^{2}P $4f$
12	−4.72123	$2s^22p^3$ ^{2}P $7f$
13	−4.30622	$2s^22p^3$ ^{2}D $8f$
14	−3.67743	$2s2p^4$ ^{2}S $5g$
15	−3.11897	$2s^22p^3$ ^{2}D $9f$
16	−2.98443	$2s^22p^3$ ^{2}P $8f$
	^{3}G^o	
1	−40.2555	$2s^22p^3$ ^{2}D $3d$
2	−21.8199	$2s^22p^3$ ^{2}D $4d$
3	−15.4703	$2s2p^4$ ^{4}P $4f$
4	−13.3775	$2s^22p^3$ ^{2}D $5d$
5	−13.2179	$2s2p^4$ ^{2}D $4f$:
6	−13.0191	$2s^22p^3$ ^{2}D $5g$:
7	−11.2123	$2s2p^4$ ^{2}P $4f$
8	−8.83356	$2s^22p^3$ ^{2}D $6d$
9	−7.31895	$2s2p^4$ ^{4}P $5f$
10	−6.11112	$2s^22p^3$ ^{2}D $7d$
11	−5.03252	$2s2p^4$ ^{2}D $5f$
12	−4.35193	$2s^22p^3$ ^{2}D $8d$
	^{3}H^e	
1	−21.2807	$2s^22p^3$ ^{2}D $4f$
2	−13.1523	$2s^22p^3$ ^{2}D $5f$
3	−8.71569	$2s^22p^3$ ^{2}D $6f$
4	−6.04133	$2s^22p^3$ ^{2}D $7f$
5	−4.30684	$2s^22p^3$ ^{2}D $8f$
	^{3}H^o	
1	−13.2812	$2s2p^4$ ^{2}D $4f$
2	−5.07407	$2s2p^4$ ^{2}D $5f$
	^{5}S^o	
1	−46.3690	$2s^22p^3$ ^{4}S $3s$
2	−36.4570	$2s2p^4$ ^{4}P $3p$
3	−24.9594	$2s^22p^3$ ^{4}S $4s$
4	−16.7646	$2s2p^4$ ^{4}P $4p$
5	−15.4901	$2s^22p^3$ ^{4}S $5s$
6	−10.6081	$2s^22p^3$ ^{4}S $6s$
7	−7.84518	$2s2p^4$ ^{4}P $5p$
8	−7.84518	$2s^22p^3$ ^{4}S $7s$
9	−5.86816	$2s^22p^3$ ^{4}S $8s$
10	−4.61517	$2s^22p^3$ ^{4}S $9s$
	^{5}P^e	
1	−44.2658	$2s^22p^3$ ^{4}S $3p$
2	−39.0082	$2s2p^4$ ^{4}P $3s$
3	−34.6726	$2s2p^4$ ^{4}P $3d$
4	−24.0859	$2s^22p^3$ ^{4}S $4p$
5	−17.7627	$2s2p^4$ ^{4}P $4s$
6	−16.0639	$2s2p^4$ ^{4}P $4d$
7	−15.0410	$2s^22p^3$ ^{4}S $5p$
8	−10.3365	$2s^22p^3$ ^{4}S $6p$
9	−8.33997	$2s2p^4$ ^{4}P $5s$
10	−7.60345	$2s^22p^3$ ^{4}S $7p$
11	−7.60345	$2s2p^4$ ^{4}P $5d$
12	−5.75326	$2s^22p^3$ ^{4}S $8p$
13	−4.53398	$2s^22p^3$ ^{4}S $9p$
	^{5}P^o	
1	−37.1039	$2s2p^4$ ^{4}P $3p$
2	−16.9395	$2s2p^4$ ^{4}P $4p$
3	−7.90189	$2s2p^4$ ^{4}P $5p$
4	−3.18019	$2s2p^4$ ^{4}P $6p$
5	−.38905	$2s2p^4$ ^{4}P $7p$
	^{5}D^e	
1	−35.2404	$2s2p^4$ ^{4}P $3d$
2	−16.2306	$2s2p^4$ ^{4}P $4d$
3	−7.66475	$2s2p^4$ ^{4}P $5d$
4	−3.07651	$2s2p^4$ ^{4}P $6d$
5	−.33295	$2s2p^4$ ^{4}P $7d$
	^{5}D^o	
1	−41.7836	$2s^22p^3$ ^{4}S $3d$
2	−36.8440	$2s2p^4$ ^{4}P $3p$
3	−23.2237	$2s^22p^3$ ^{4}S $4d$
4	−16.8591	$2s2p^4$ ^{4}P $4p$
5	−15.4475	$2s2p^4$ ^{4}P $4f$
6	−14.7489	$2s^22p^3$ ^{4}S $5d$
7	−10.1963	$2s^22p^3$ ^{4}S $6d$
8	−7.86298	$2s2p^4$ ^{4}P $5p$
9	−7.46718	$2s^22p^3$ ^{4}S $7d$
10	−7.33783	$2s2p^4$ ^{4}P $5f$
11	−5.70542	$2s^22p^3$ ^{4}S $8d$
12	−4.50120	$2s^22p^3$ ^{4}S $9d$
	^{5}F^e	
1	−34.8931	$2s2p^4$ ^{4}P $3d$
2	−22.6233	$2s^22p^3$ ^{4}S $4f$
3	−16.1185	$2s2p^4$ ^{4}P $4d$
4	−14.4988	$2s^22p^3$ ^{4}S $5f$
5	−10.0637	$2s^22p^3$ ^{4}S $6f$
6	−7.61287	$2s2p^4$ ^{4}P $5d$
7	−7.38826	$2s^22p^3$ ^{4}S $7f$

O-like Fe (Fe^{18+})

i	Energy(Ryds)	Description	i	Energy(Ryds)	Description
8	−5.65441	$2s^22p^3\ ^4$S $8f$	3	−7.31704	$2s2p^4\ ^4$P $5f$:
9	−4.46629	$2s^22p^3\ ^4$S $9f$			
		$^{5}\mathbf{F}^{o}$			
1	−15.5443	$2s2p^4\ ^4$P $4f$			
2	−7.36963	$2s2p^4\ ^4$P $5f$			
3	−2.92313	$2s2p^4\ ^4$P $6f$			
4	−.24399	$2s2p^4\ ^4$P $7f$			
		$^{5}\mathbf{G}^{o}$			
1	−15.4876	$2s2p^4\ ^4$P $4f$			
2	−7.39021	$2s^22p^3\ ^4$S $7g$:			

Energies in ascending order from ground state for terms with effective $n \leq 4.0$, $L \leq 4$

Term	i	E(Ryds)	Term	i	E(Ryds)	Term	i	E(Ryds)	Term	i	E(Ryds)	Term	i	E(Ryds)
$^{3}\mathbf{P}^{e}$	1	0.00000	$^{3}\mathbf{P}^{e}$	4	65.4853	$^{5}\mathbf{S}^{o}$	2	70.1350	$^{1}\mathbf{D}^{o}$	5	74.2213	$^{1}\mathbf{D}^{e}$	8	80.9476
$^{1}\mathbf{D}^{e}$	1	0.92700	$^{1}\mathbf{D}^{e}$	3	65.7968	$^{3}\mathbf{S}^{o}$	3	70.2425	$^{3}\mathbf{S}^{e}$	3	74.2687	$^{1}\mathbf{P}^{e}$	6	80.9798
$^{1}\mathbf{S}^{e}$	1	2.00500	$^{1}\mathbf{S}^{o}$	1	66.1586	$^{3}\mathbf{P}^{o}$	5	70.3563	$^{3}\mathbf{F}^{e}$	3	74.2811	$^{3}\mathbf{P}^{e}$	10	80.9966
$^{3}\mathbf{P}^{o}$	1	7.46471	$^{3}\mathbf{F}^{o}$	1	66.1600	$^{1}\mathbf{D}^{e}$	4	70.5420	$^{1}\mathbf{G}^{e}$	1	74.3699	$^{5}\mathbf{S}^{o}$	3	81.6326
$^{1}\mathbf{P}^{o}$	1	10.2018	$^{3}\mathbf{G}^{o}$	1	66.3365	$^{3}\mathbf{S}^{e}$	2	71.3140	$^{1}\mathbf{S}^{o}$	2	74.3743	$^{3}\mathbf{S}^{o}$	5	81.8039
$^{1}\mathbf{S}^{e}$	2	17.4127	$^{1}\mathbf{G}^{o}$	1	66.4199	$^{5}\mathbf{D}^{e}$	1	71.3516	$^{3}\mathbf{D}^{e}$	5	74.4288	$^{3}\mathbf{P}^{o}$	10	82.1117
$^{5}\mathbf{S}^{o}$	1	60.2230	$^{1}\mathbf{P}^{o}$	3	66.5434	$^{5}\mathbf{F}^{e}$	1	71.6989	$^{3}\mathbf{S}^{o}$	4	74.4497	$^{3}\mathbf{F}^{o}$	4	82.2728
$^{3}\mathbf{S}^{o}$	1	60.7472	$^{3}\mathbf{D}^{o}$	3	66.6611	$^{1}\mathbf{S}^{e}$	4	71.8151	$^{3}\mathbf{P}^{e}$	8	74.4606	$^{5}\mathbf{P}^{e}$	4	82.5061
$^{3}\mathbf{D}^{o}$	1	61.7024	$^{1}\mathbf{S}^{e}$	3	66.7961	$^{5}\mathbf{P}^{e}$	3	71.9194	$^{1}\mathbf{P}^{e}$	4	74.5367	$^{1}\mathbf{F}^{o}$	4	82.5449
$^{1}\mathbf{D}^{o}$	1	61.9645	$^{3}\mathbf{P}^{o}$	3	66.8011	$^{3}\mathbf{P}^{e}$	6	71.9982	$^{1}\mathbf{D}^{e}$	5	74.6081	$^{1}\mathbf{D}^{o}$	6	82.6301
$^{5}\mathbf{P}^{e}$	1	62.3262	$^{1}\mathbf{D}^{o}$	2	66.8214	$^{3}\mathbf{F}^{o}$	3	72.0224	$^{1}\mathbf{F}^{e}$	2	74.7350	$^{1}\mathbf{S}^{e}$	6	82.6313
$^{3}\mathbf{P}^{e}$	2	62.8027	$^{3}\mathbf{S}^{o}$	2	67.0073	$^{1}\mathbf{P}^{o}$	5	72.0932	$^{1}\mathbf{S}^{e}$	5	74.7781	$^{3}\mathbf{D}^{o}$	8	82.6456
$^{3}\mathbf{P}^{o}$	2	63.0334	$^{1}\mathbf{F}^{o}$	1	67.2733	$^{1}\mathbf{P}^{e}$	3	72.1810	$^{1}\mathbf{P}^{o}$	7	75.3432	$^{3}\mathbf{P}^{e}$	11	82.7331
$^{1}\mathbf{P}^{o}$	2	63.2980	$^{3}\mathbf{F}^{o}$	2	67.3795	$^{1}\mathbf{F}^{o}$	3	72.2219	$^{3}\mathbf{D}^{e}$	6	75.5646	$^{3}\mathbf{D}^{o}$	9	83.0230
$^{1}\mathbf{P}^{e}$	1	63.6240	$^{3}\mathbf{P}^{o}$	4	67.4332	$^{3}\mathbf{F}^{e}$	2	72.2328	$^{3}\mathbf{D}^{e}$	7	75.8279	$^{1}\mathbf{D}^{o}$	7	83.1091
$^{3}\mathbf{D}^{e}$	1	63.6680	$^{5}\mathbf{P}^{e}$	2	67.5838	$^{3}\mathbf{D}^{o}$	6	72.3349	$^{1}\mathbf{D}^{e}$	6	75.9752	$^{5}\mathbf{D}^{o}$	3	83.3683
$^{3}\mathbf{F}^{e}$	1	63.8175	$^{3}\mathbf{D}^{o}$	4	67.7217	$^{1}\mathbf{D}^{o}$	4	72.5098	$^{3}\mathbf{F}^{e}$	4	76.0876	$^{3}\mathbf{D}^{o}$	10	83.6558
$^{1}\mathbf{F}^{e}$	1	63.9214	$^{1}\mathbf{D}^{o}$	3	67.8304	$^{3}\mathbf{P}^{e}$	7	72.5561	$^{1}\mathbf{F}^{e}$	3	76.1597	$^{1}\mathbf{P}^{o}$	9	83.7319
$^{3}\mathbf{P}^{e}$	3	64.5622	$^{1}\mathbf{F}^{o}$	2	68.0105	$^{3}\mathbf{P}^{o}$	6	72.5927	$^{3}\mathbf{P}^{e}$	9	76.2886	$^{3}\mathbf{D}^{e}$	9	83.8591
$^{5}\mathbf{D}^{o}$	1	64.8084	$^{3}\mathbf{P}^{e}$	5	68.3686	$^{3}\mathbf{D}^{e}$	4	72.6255	$^{1}\mathbf{P}^{e}$	5	76.8666	$^{1}\mathbf{P}^{e}$	7	83.8651
$^{1}\mathbf{D}^{e}$	2	64.9063	$^{1}\mathbf{P}^{o}$	4	68.5787	$^{3}\mathbf{P}^{o}$	7	73.4961	$^{1}\mathbf{D}^{e}$	7	76.9684	$^{3}\mathbf{F}^{e}$	5	83.8956
$^{3}\mathbf{S}^{e}$	1	64.9698	$^{5}\mathbf{P}^{o}$	1	69.4881	$^{1}\mathbf{P}^{o}$	6	73.6552	$^{3}\mathbf{P}^{o}$	9	78.8233	$^{1}\mathbf{F}^{e}$	4	83.9327
$^{3}\mathbf{D}^{e}$	2	65.1362	$^{5}\mathbf{D}^{o}$	2	69.7480	$^{3}\mathbf{G}^{e}$	1	73.8895	$^{1}\mathbf{P}^{o}$	8	79.1227	$^{5}\mathbf{F}^{e}$	2	83.9687
$^{1}\mathbf{P}^{e}$	2	65.2647	$^{3}\mathbf{D}^{e}$	3	70.0072	$^{3}\mathbf{P}^{o}$	8	74.0349	$^{3}\mathbf{S}^{e}$	4	80.3368	$^{3}\mathbf{F}^{e}$	6	83.9961
$^{3}\mathbf{D}^{o}$	2	65.4106	$^{3}\mathbf{D}^{o}$	5	70.0841	$^{3}\mathbf{D}^{o}$	7	74.0720	$^{3}\mathbf{D}^{e}$	8	80.7298			

O-like Fe (Fe^{18+})

gf-values for transitions involving terms with effective $n \leq 4.0$, $L \leq 4$

$i\ i'$	gf_L	$i\ i'$	gf_L	$i\ i'$	gf_L	$i\ i'$	gf_L	$i\ i'$	gf_L	$i\ i'$	gf_L
	$^{1}P^{e}$–$^{1}S^{o}$	6 6	−2.84E−4	5 4	−1.86E−1	3 1	2.56E−4	8 7	1.54E−1	7 6	1.60E−5
1 1	−7.87E−2	7 1	1.99E−2	5 5	−4.32E−3	3 2	4.21E−1	8 8	4.20E−1	7 7	−3.05E−1
1 2	−7.48E−2	7 2	5.03E−3	5 6	−5.43E−2	3 3	−4.06E−4	8 9	−5.72E−6		$^{1}D^{o}$–$^{1}D^{e}$
2 1	−1.49E−3	7 3	3.48E−2	5 7	−5.29E−7	3 4	−4.68E−3		$^{1}D^{o}$–$^{1}P^{e}$	1 1	4.77E−1
2 2	−1.02E−1	7 4	1.16E−2	6 1	1.74E−3	3 5	−7.58E−3	1 1	−1.61E−1	1 2	−5.30E−1
3 1	1.33E−4	7 5	9.27E−4	6 2	1.03E−1	3 6	−1.01E−1	1 2	−3.02E−2	1 3	−1.80E−1
3 2	−9.28E−2	7 6	−1.61E−1	6 3	2.04E−3	3 7	−2.74E−1	1 3	−4.60E−1	1 4	−2.13E−1
4 1	7.81E−2	8 1	5.70E−4	6 4	−8.86E−4	3 8	−1.64E−4	1 4	−8.89E−4	1 5	−2.31E−3
4 2	5.52E−4	8 2	2.50E−1	6 5	−3.62E−3	3 9	−6.72E−8	1 5	−5.16E−4	1 6	−8.59E−5
5 1	1.13E−2	8 3	1.26E−4	6 6	−3.89E−2	4 1	1.43E−1	1 6	−1.44E−4	1 7	−3.09E−5
5 2	1.55E−1	8 4	4.15E−2	6 7	−1.43E−5	4 2	2.94E−2	1 7	−2.86E−1	1 8	−1.20E−4
6 1	8.57E−5	8 5	9.91E−4	7 1	2.69E−2	4 3	1.70E−3	2 1	1.95E−1	2 1	1.00E+0
6 2	1.16E−1	8 6	−2.02E−1	7 2	2.35E−3	4 4	9.95E−5	2 2	3.46E−2	2 2	1.38E−1
7 1	2.29E−2	9 1	2.73E−3	7 3	4.12E−1	4 5	−1.47E−1	2 3	−2.12E−5	2 3	6.10E−6
7 2	1.71E−4	9 2	3.19E+0	7 4	1.10E−4	4 6	−1.45E−2	2 4	−1.02E−1	2 4	−2.20E−3
	$^{1}P^{o}$–$^{1}S^{e}$	9 3	5.64E−6	7 5	−5.73E−2	4 7	−7.06E−2	2 5	−2.84E−2	2 5	−1.90E−1
1 1	4.64E−2	9 4	1.86E−4	7 6	−9.46E−2	4 8	−3.79E−1	2 6	−2.25E−5	2 6	−5.18E−2
1 2	−2.63E−1	9 5	2.76E−2	7 7	−1.13E−4	4 9	−2.66E−4	2 7	−4.25E−2	2 7	−4.54E−2
1 3	−6.10E−3	9 6	6.30E−2	8 1	4.93E−4	5 1	2.18E−1	3 1	3.20E−2	2 8	−1.80E−4
1 4	−3.69E−2		$^{1}P^{o}$–$^{1}P^{e}$	8 2	4.14E−6	5 2	7.85E−4	3 2	2.65E−1	3 1	2.39E+0
1 5	−5.47E−1	1 1	−8.41E−4	8 3	3.16E−1	5 3	7.40E−2	3 3	−1.47E−4	3 2	1.78E−2
1 6	−1.07E−1	1 2	−3.38E−3	8 4	1.50E−3	5 4	7.73E−3	3 4	−9.32E−5	3 3	1.06E−1
2 1	1.40E−1	1 3	−3.90E−1	8 5	1.72E−3	5 5	1.58E−1	3 5	−2.48E−1	3 4	−4.41E−3
2 2	4.14E−6	1 4	−4.46E−1	8 6	−2.52E−1	5 6	9.50E−4	3 6	−1.97E−8	3 5	−2.99E−3
2 3	−1.63E−1	1 5	−1.94E+0	8 7	−1.57E−5	5 7	−1.09E−4	3 7	−2.50E−2	3 6	−5.89E−2
2 4	−7.12E−2	1 6	−2.73E−1	9 1	5.22E−4	5 8	−4.27E−4	4 1	2.44E−2	3 7	−4.92E−2
2 5	−1.02E−3	1 7	−6.11E−5	9 2	4.94E−5	5 9	−1.90E−2	4 2	7.91E−2	3 8	−1.22E−5
2 6	−1.88E−5	2 1	−2.54E−3	9 3	1.86E−4	6 1	1.26E−3	4 3	2.52E−5	4 1	7.59E−1
3 1	1.36E−1	2 2	−2.05E−1	9 4	1.96E−2	6 2	1.34E−3	4 4	−4.01E−2	4 2	2.59E−1
3 2	3.58E−5	2 3	−1.60E−1	9 5	1.40E−1	6 3	8.64E−4	4 5	−1.43E−2	4 3	4.26E−2
3 3	−8.37E−4	2 4	−1.96E−3	9 6	1.71E−1	6 4	7.35E−2	4 6	−2.25E−1	4 4	3.90E−1
3 4	−6.37E−4	2 5	−8.34E−5	9 7	−3.34E−5	6 5	1.15E−3	4 7	−1.78E−4	4 5	−2.25E−1
3 5	−8.80E−2	2 6	−4.32E−5		$^{1}D^{e}$–$^{1}P^{o}$	6 6	4.10E−1	5 1	1.54E−2	4 6	−2.75E−2
3 6	−4.32E−5	2 7	−5.48E−3	1 1	−5.02E−1	6 7	7.46E−4	5 2	2.58E−2	4 7	−4.38E−2
4 1	2.23E+0	3 1	1.95E−1	1 2	−1.46E−1	6 8	−1.39E−5	5 3	4.25E−1	4 8	−3.09E−2
4 2	7.73E−5	3 2	4.65E−3	1 3	−7.46E−1	6 9	−1.38E−3	5 4	−1.09E−4	5 1	5.46E−4
4 3	9.00E−2	3 3	−6.20E−5	1 4	−1.06E−1	7 1	5.13E+0	5 5	−2.69E−2	5 2	2.22E−1
4 4	−4.38E−3	3 4	−3.95E−2	1 5	−4.55E−1	7 2	5.98E−6	5 6	−1.55E−1	5 3	5.87E−2
4 5	−3.34E−3	3 5	−5.51E−2	1 6	−1.30E−2	7 3	1.39E−2	5 7	−1.59E−3	5 4	3.92E−3
4 6	−3.50E−6	3 6	−5.83E−6	1 7	−4.29E−2	7 4	3.32E−2	6 1	3.18E−4	5 5	−1.39E−3
5 1	6.06E−3	3 7	−5.61E−2	1 8	−1.92E−3	7 5	8.66E−3	6 2	4.48E−5	5 6	−2.90E−3
5 2	4.66E−3	4 1	2.00E−4	1 9	−4.19E−5	7 6	2.21E−2	6 3	1.31E−3	5 7	−8.30E−2
5 3	8.22E−4	4 2	1.73E−1	2 1	7.94E−3	7 7	1.91E−1	6 4	4.89E−2	5 8	−4.96E−1
5 4	7.34E−4	4 3	−2.13E−7	2 2	5.61E−2	7 8	−7.02E−3	6 5	1.09E−1	6 1	4.66E−3
5 5	−7.36E−2	4 4	−4.59E−3	2 3	−4.12E−2	7 9	−3.86E−1	6 6	2.54E−1	6 2	7.45E−5
5 6	−1.38E−2	4 5	−7.73E−2	2 4	−2.63E−3	8 1	2.75E−1	6 7	−8.27E−5	6 3	1.05E−6
6 1	3.87E−1	4 6	−3.70E−6	2 5	−4.14E−4	8 2	2.25E−5	7 1	1.12E−1	6 4	1.41E−6
6 2	9.09E−3	4 7	−1.35E−3	2 6	−6.17E−2	8 3	5.74E−5	7 2	8.23E−3	6 5	8.72E−2
6 3	8.69E−2	5 1	2.53E−1	2 7	−1.41E−1	8 4	2.84E−5	7 3	8.21E−5	6 6	4.71E−2
6 4	2.05E−1	5 2	8.14E−3	2 8	−1.40E−4	8 5	1.55E−2	7 4	1.97E−4	6 7	2.16E−2
6 5	−6.89E−3	5 3	−4.33E−4	2 9	−3.77E−5	8 6	3.54E−2	7 5	2.23E−5	6 8	9.37E−2

O-like Fe (Fe^{18+})

i i'	gf_L	i i'	gf_L	i i'	gf_L	i i'	gf_L	i i'	gf_L	i i'	gf_L
7 1	8.69E−2	2 7	−4.24E−1	1 5	−6.33E−2		**$^3P^o$–$^3S^e$**	1 11	−1.31E−4	6 8	−3.96E−1
7 2	3.03E−1	2 8	−1.28E−4	2 1	5.91E−1	1 1	−1.10E−2	2 1	3.30E−1	6 9	−1.70E−2
7 3	8.12E−2	3 1	9.83E−1	2 2	−8.69E−2	1 2	−1.45E−1	2 2	1.31E−3	6 10	−8.14E−2
7 4	6.74E−3	3 2	1.94E−2	2 3	−1.08E−1	1 3	−1.84E+0	2 3	−4.98E−3	6 11	−2.70E−4
7 5	5.37E−4	3 3	2.30E−1	2 4	−3.90E−2	1 4	−1.87E−1	2 4	−8.55E−1	7 1	6.26E−2
7 6	7.45E−5	3 4	4.52E−1	2 5	−3.51E−1	2 1	−1.96E−1	2 5	−3.67E−2	7 2	1.29E−3
7 7	3.21E−4	3 5	−9.28E−3	3 1	2.64E−1	2 2	−3.62E−1	2 6	−4.13E−1	7 3	5.27E−2
7 8	6.06E−5	3 6	−8.25E−3	3 2	−1.96E−1	2 3	−3.69E−4	2 7	−3.13E−3	7 4	1.46E−1
	$^1F^e$–$^1D^o$	3 7	−7.99E−4	3 3	−3.77E−2	2 4	−2.07E−5	2 8	−9.22E−4	7 5	4.28E−3
1 1	5.26E−1	3 8	−4.37E−1	3 4	−3.95E−1	3 1	1.20E−2	2 9	−5.31E−4	7 6	2.72E−2
1 2	−1.92E−2	4 1	3.02E−2	3 5	−1.19E−1	3 2	−3.92E−7	2 10	−1.60E−4	7 7	5.49E−4
1 3	−4.07E−2	4 2	1.60E−5	4 1	7.57E−2	3 3	−2.13E−1	2 11	−3.91E−3	7 8	−8.19E−5
1 4	−6.95E−2	4 3	3.06E−5	4 2	−5.78E−4	3 4	−5.78E−5	3 1	4.66E+0	7 9	−2.14E−2
1 5	−6.41E−1	4 4	9.36E−6	4 3	−2.68E−2	4 1	3.96E−1	3 2	1.27E−1	7 10	−1.71E−1
1 6	−6.47E−5	4 5	3.31E−2	4 4	−1.24E−1	4 2	−1.29E−4	3 3	4.13E−1	7 11	−5.41E−6
1 7	−2.89E−1	4 6	4.98E−2	4 5	−1.55E−2	4 3	−1.62E−2	3 4	1.84E−4	8 1	1.27E−1
2 1	2.38E−3	4 7	4.78E−2	5 1	2.32E−1	4 4	−1.97E−6	3 5	−7.36E−3	8 2	2.85E−2
2 2	3.29E−3	4 8	4.42E−1	5 2	1.75E−3	5 1	8.20E−4	3 6	−1.48E−3	8 3	3.07E−1
2 3	9.38E−2		**$^1F^o$–$^1F^e$**	5 3	−2.14E−1	5 2	−2.91E−3	3 7	−2.17E−3	8 4	2.43E−1
2 4	3.89E−1	1 1	2.45E−1	5 4	−6.44E−4	5 3	−4.72E−2	3 8	−3.19E−2	8 5	2.18E−1
2 5	3.46E−5	1 2	−1.69E−1	5 5	−5.78E−3	5 4	−1.57E−2	3 9	−2.72E−1	8 6	5.61E−1
2 6	−2.70E−1	1 3	−2.50E−1	6 1	1.41E−3	6 1	1.27E−1	3 10	−2.82E−4	8 7	1.79E−4
2 7	−1.37E−3	1 4	−7.54E−2	6 2	8.73E−4	6 2	1.18E−5	3 11	−1.03E−2	8 8	−2.24E−3
3 1	0.00E−5	2 1	2.81E−2	6 3	9.97E−4	6 3	−1.27E−1	4 1	6.96E−1	8 9	−2.38E−1
3 2	3.89E−2	2 2	−1.87E−3	6 4	−3.29E−1	6 4	−4.64E−1	4 2	6.80E−4	8 10	−2.37E−1
3 3	1.37E−2	2 3	−8.99E−2	6 5	−1.01E−3	7 1	2.97E−2	4 3	3.77E−2	8 11	−3.72E−5
3 4	1.66E−4	2 4	−7.86E−3	7 1	2.95E−3	7 2	7.83E−1	4 4	2.14E−1	9 1	3.98E−3
3 5	4.78E−1	3 1	4.30E−1	7 2	1.53E−2	7 3	−8.87E−3	4 5	−1.58E−3	9 2	3.39E−4
3 6	−3.07E−1	3 2	−2.35E−1	7 3	4.14E−1	7 4	−8.21E−2	4 6	−1.50E−4	9 3	2.29E−3
3 7	−5.89E−4	3 3	−2.81E−3	7 4	−5.46E−4	8 1	1.07E−1	4 7	−6.09E−3	9 4	1.05E−3
4 1	7.10E−1	3 4	−1.90E−6	7 5	−1.63E−3	8 2	6.89E−2	4 8	−8.71E−2	9 5	3.10E−3
4 2	4.63E−3	4 1	1.38E−4	8 1	2.66E−4	8 3	−8.14E−4	4 9	−1.46E−1	9 6	1.27E+0
4 3	2.07E−2	4 2	4.97E−2	8 2	9.96E−2	8 4	−1.94E−1	4 10	−6.40E−5	9 7	5.68E−3
4 4	1.99E−4	4 3	5.19E−1	8 3	1.19E−1	9 1	2.51E−4	4 11	−2.76E−5	9 8	8.19E−3
4 5	3.56E−3	4 4	−1.57E−6	8 4	7.31E−6	9 2	6.63E−2	5 1	9.79E−1	9 9	4.61E−4
4 6	1.11E−4		**$^1G^e$–$^1F^o$**	8 5	−2.74E−4	9 3	1.38E−3	5 2	4.26E−1	9 10	−9.80E−1
4 7	8.23E−1	1 1	5.44E−3	9 1	8.57E−4	9 4	−1.74E−1	5 3	4.09E−2	9 11	−9.15E−5
	$^1F^o$–$^1D^e$	1 2	7.24E−2	9 2	1.23E−1	10 1	6.95E−5	5 4	5.23E−2	10 1	1.65E−2
1 1	2.46E+0	1 3	6.90E−1	9 3	1.56E−3	10 2	1.44E−3	5 5	6.01E−1	10 2	1.64E−7
1 2	4.12E−1	1 4	−5.02E−1	9 4	3.31E−1	10 3	6.27E−2	5 6	−9.65E−4	10 3	9.15E−6
1 3	5.43E−4		**$^1G^o$–$^1F^e$**	9 5	−1.31E−4	10 4	3.23E−1	5 7	−2.90E−1	10 4	2.16E−5
1 4	1.82E−3	1 1	7.64E−1	10 1	7.22E−5		**$^3P^o$–$^3P^e$**	5 8	−2.39E−2	10 5	5.08E−4
1 5	−5.33E−2	1 2	−1.70E−1	10 2	5.88E−5	1 1	7.10E−1	5 9	−1.17E−2	10 6	1.16E−3
1 6	−1.31E−1	1 3	−7.17E−1	10 3	1.30E−3	1 2	−1.22E−4	5 10	−1.43E−2	10 7	3.58E−2
1 7	−2.06E−1	1 4	−1.72E−1	10 4	3.57E−1	1 3	−3.56E−2	5 11	−5.03E−4	10 8	2.19E−1
1 8	−2.66E−4		**$^1G^o$–$^1G^e$**	10 5	−7.59E−5	1 4	−4.10E−3	6 1	2.95E−1	10 9	2.29E−1
2 1	4.64E+0	1 1	−4.33E−1	11 1	8.26E−1	1 5	−7.47E−1	6 2	2.78E−5	10 10	1.62E−1
2 2	2.39E−2		**$^3P^e$–$^3S^o$**	11 2	1.50E−2	1 6	−2.38E−1	6 3	1.50E−1	10 11	−3.88E−6
2 3	4.86E−1	1 1	−3.53E−1	11 3	3.92E−3	1 7	−1.05E+0	6 4	7.46E−3		**$^3D^e$–$^3P^o$**
2 4	−9.60E−3	1 2	−1.68E+0	11 4	7.73E−4	1 8	−5.68E+0	6 5	2.61E−1	1 1	1.83E−3
2 5	−1.60E−6	1 3	−4.08E−1	11 5	1.19E+0	1 9	−7.94E−1	6 6	1.35E−2	1 2	5.94E−3
2 6	−9.73E−2	1 4	−1.05E−1			1 10	−4.95E−1	6 7	1.82E−5	1 3	−2.06E−1

O-like Fe (Fe^{18+})

i	i'	gf$_L$	i	i'	gf$_L$	i	i'	gf$_L$	i	i'	gf$_L$	i	i'	gf$_L$	i	i'	gf$_L$
1	4	−1.77E−2	6	6	1.81E−4	2	6	−9.32E−4	7	3	1.30E−2	2	1	6.66E−2	7	8	−1.40E+0
1	5	−9.25E−3	6	7	1.09E+0	2	7	−3.80E−1	7	4	1.81E−1	2	2	1.26E−3	7	9	−7.01E−3
1	6	−5.07E−1	6	8	2.59E−5	2	8	−8.67E−3	7	5	1.50E−2	2	3	−2.64E−4	8	1	1.03E−4
1	7	−9.53E−2	6	9	−7.32E−4	2	9	−1.76E−2	7	6	1.25E+0	2	4	−5.92E−1	8	2	2.28E−6
1	8	−6.58E−1	6	10	−2.49E−1	2	10	−2.41E−5	7	7	1.79E−3	2	5	−1.52E−3	8	3	7.91E−4
1	9	−1.45E−3	7	1	4.00E+0	2	11	−2.98E−1	7	8	−2.23E−4	2	6	−7.35E−3	8	4	1.64E−2
1	10	−1.70E−4	7	2	8.36E−4	3	1	5.49E+0	7	9	−5.59E−2	2	7	−7.41E−2	8	5	1.84E−1
2	1	1.97E−2	7	3	1.16E−1	3	2	6.25E−2	7	10	−5.57E−1	2	8	−2.66E−4	8	6	2.66E−1
2	2	1.17E+0	7	4	2.07E−1	3	3	4.35E−1	7	11	−2.18E−4	2	9	−1.94E−2	8	7	4.29E−2
2	3	−2.28E−2	7	5	3.48E−4	3	4	3.63E−4	8	1	3.77E−2	3	1	7.66E−1	8	8	3.18E−1
2	4	−1.14E−2	7	6	5.36E−2	3	5	−7.99E−3	8	2	2.91E−3	3	2	5.68E−5	8	9	−7.44E−5
2	5	−1.59E−3	7	7	1.50E−2	3	6	−1.68E−3	8	3	1.57E−5	3	3	−3.73E−5	9	1	6.98E−1
2	6	−3.64E−4	7	8	6.29E−1	3	7	−1.82E−3	8	4	5.65E−5	3	4	−3.02E−2	9	2	1.32E−2
2	7	−6.64E−1	7	9	−1.85E−3	3	8	−4.13E−1	8	5	1.03E−3	3	5	−8.96E−2	9	3	1.48E−3
2	8	−1.97E−2	7	10	−5.88E−1	3	9	−3.18E−1	8	6	1.62E−3	3	6	−7.85E−2	9	4	6.19E−6
2	9	0.00E+0	8	1	8.70E−1	3	10	−8.77E−7	8	7	1.21E−3	3	7	−3.27E−1	9	5	5.67E−4
2	10	−3.05E−5	8	2	6.62E−6	3	11	−4.49E−2	8	8	1.32E−2	3	8	−3.41E−4	9	6	5.35E−5
3	1	7.61E−1	8	3	9.18E−5	4	1	6.06E+0	8	9	4.31E−1	3	9	−2.24E−1	9	7	4.14E−4
3	2	2.23E−1	8	4	2.05E−4	4	2	1.12E−2	8	10	7.70E−1	4	1	2.73E−4	9	8	2.70E−6
3	3	5.00E−5	8	5	2.45E−4	4	3	7.82E−4	8	11	−7.88E−5	4	2	3.27E−1	9	9	−1.77E+0
3	4	1.96E−8	8	6	3.37E−6	4	4	8.41E−1	9	1	1.15E−1	4	3	−6.82E−5	10	1	4.69E−2
3	5	−1.14E−2	8	7	4.68E−2	4	5	−6.45E−3	9	2	1.58E−2	4	4	−1.17E−3	10	2	1.16E−2
3	6	−8.08E−1	8	8	4.48E−1	4	6	−1.70E−3	9	3	5.72E−1	4	5	−1.67E−1	10	3	4.17E−4
3	7	−1.55E−2	8	9	1.34E+0	4	7	−1.45E−3	9	4	5.68E−3	4	6	−2.09E−1	10	4	1.05E−4
3	8	−8.11E−2	8	10	−1.76E−3	4	8	−3.67E−2	9	5	1.56E−2	4	7	−2.04E−1	10	5	2.63E−4
3	9	−7.72E−1	9	1	4.18E−5	4	9	−3.26E−1	9	6	4.67E−4	4	8	−1.44E−4	10	6	2.72E−4
3	10	−2.78E−5	9	2	5.89E−3	4	10	−2.85E−4	9	7	2.46E−5	4	9	−7.67E−3	10	7	1.04E−5
4	1	5.20E+0	9	3	6.37E−2	4	11	−1.20E−2	9	8	3.16E−4	5	1	2.04E−5	10	8	3.00E−4
4	2	3.48E−3	9	4	4.48E−3	5	1	1.53E+0	9	9	6.13E−6	5	2	5.02E−5	10	9	−1.10E−3
4	3	8.79E−3	9	5	9.84E−6	5	2	4.95E−1	9	10	4.92E−5	5	3	6.98E−6			$^{3}F^{e}$–$^{3}D^{o}$
4	4	7.81E−4	9	6	1.43E−5	5	3	2.00E−1	9	11	2.09E−3	5	4	−2.81E−1	1	1	1.72E+0
4	5	8.67E−1	9	7	1.27E−5	5	4	1.29E−1	10	1	1.15E+0	5	5	−6.81E−2	1	2	−9.23E−3
4	6	2.87E−4	9	8	8.24E−5	5	5	1.00E+0	10	2	3.43E+0	5	6	−1.83E−3	1	3	−7.21E−2
4	7	−7.34E−4	9	9	9.38E−6	5	6	−5.39E−4	10	3	3.40E−1	5	7	−1.01E−2	1	4	−9.03E−3
4	8	−2.96E−3	9	10	3.24E−5	5	7	−5.06E−2	10	4	4.73E−2	5	8	−8.20E−4	1	5	−5.31E−4
4	9	−8.55E−3			$^{3}D^{o}$–$^{3}P^{e}$	5	8	−2.91E−2	10	5	2.54E−3	5	9	−1.46E−3	1	6	−1.97E−1
4	10	−2.11E−2	1	1	5.48E−1	5	9	−4.07E−4	10	6	1.18E−2	6	1	5.77E−1	1	7	−1.91E+0
5	1	5.95E+0	1	2	−4.41E−2	5	10	−3.88E−3	10	7	1.70E−2	6	2	1.50E−1	1	8	−2.43E−4
5	2	3.33E−4	1	3	−9.02E−1	5	11	−1.05E−3	10	8	9.50E−4	6	3	1.54E+0	1	9	−9.73E−1
5	3	1.28E−1	1	4	−1.55E−3	6	1	5.56E−4	10	9	1.01E−4	6	4	−5.80E−3	1	10	−1.88E−2
5	4	8.68E−2	1	5	−4.34E−2	6	2	2.96E−3	10	10	1.59E−5	6	5	−6.87E−1	2	1	2.30E−3
5	5	3.27E−3	1	6	−1.25E+0	6	3	1.47E−1	10	11	2.14E+0	6	6	−2.45E−2	2	2	3.31E−1
5	6	3.40E−1	1	7	−1.34E−3	6	4	1.07E−1			$^{3}D^{o}$–$^{3}D^{e}$	6	7	−1.20E−2	2	3	3.90E−2
5	7	1.22E−3	1	8	−7.69E−4	6	5	6.15E−3	1	1	−1.04E+0	6	8	−1.99E−1	2	4	9.70E−3
5	8	1.56E−3	1	9	−8.17E−4	6	6	2.77E−5	1	2	−4.38E−2	6	9	−6.07E−3	2	5	1.69E+0
5	9	−5.30E−3	1	10	−6.47E−4	6	7	−1.67E−3	1	3	−1.16E+0	7	1	1.68E−1	2	6	−6.21E−7
5	10	−3.82E−1	1	11	−3.05E−2	6	8	−1.24E−1	1	4	−9.15E−4	7	2	6.31E−1	2	7	−1.18E−3
6	1	5.64E−1	2	1	1.93E+0	6	9	−9.25E−4	1	5	−1.08E−3	7	3	2.19E−4	2	8	−4.30E−6
6	2	6.31E−4	2	2	1.14E+0	6	10	−5.98E−1	1	6	−5.19E−5	7	4	8.36E−5	2	9	−1.57E−4
6	3	8.93E−3	2	3	6.50E−3	6	11	−7.87E−4	1	7	−1.80E−4	7	5	−1.32E−3	2	10	−5.79E−3
6	4	6.44E−2	2	4	−2.90E−6	7	1	6.77E−1	1	8	−6.99E−5	7	6	−1.33E−2	3	1	4.01E−5
6	5	9.86E−4	2	5	−2.92E−3	7	2	1.91E−3	1	9	−1.45E+0	7	7	−1.57E−1	3	2	8.92E−3

O-like Fe (Fe^{18+})

i i′	gf$_L$	*i i′*	gf$_L$	*i i′*	gf$_L$	*i i′*	gf$_L$	*i i′*	gf$_L$	*i i′*	gf$_L$
3 3	1.78E−1	5 10	1.74E−4	2 7	−1.11E−2	1 6	−1.19E−4	1 3	−3.35E−1		$^5\mathbf{D}^o$–$^5\mathbf{P}^e$
3 4	8.53E−2	6 1	2.25E−3	2 8	−1.51E−6	2 1	6.09E−2	1 4	−2.35E+0	1 1	2.00E+0
3 5	2.62E−3	6 2	1.19E+1	2 9	−1.20E−2	2 2	−1.52E−3	1 5	−5.24E−1	1 2	−3.18E−4
3 6	9.97E−1	6 3	2.18E+0	3 1	1.17E−1	2 3	−2.46E−1	1 6	−6.25E−3	1 3	−6.40E−1
3 7	4.69E−5	6 4	5.21E−1	3 2	1.55E−1	2 4	−7.58E−1		$^3\mathbf{G}^o$–$^3\mathbf{G}^e$	1 4	−5.39E−1
3 8	−6.88E−1	6 5	7.82E−2	3 3	1.78E+0	2 5	−1.66E−2	1 1	−1.20E+0	2 1	7.72E−1
3 9	−1.43E−3	6 6	1.21E−3	3 4	−2.38E−3	2 6	−5.98E−5		$^5\mathbf{P}^e$–$^5\mathbf{S}^o$	2 2	2.35E+0
3 10	−9.03E−4	6 7	6.34E−4	3 5	−2.87E−2	3 1	1.25E+0	1 1	1.15E+0	2 3	−6.48E−2
4 1	2.81E−3	6 8	1.82E−3	3 6	−1.25E−2	3 2	−1.41E−4	1 2	−2.06E−1	2 4	−3.08E−4
4 2	6.07E−3	6 9	4.03E−3	3 7	−4.34E−3	3 3	−6.18E−1	1 3	−7.42E−1	3 1	6.61E+0
4 3	4.76E−3	6 10	7.13E−1	3 8	−1.21E+0	3 4	−1.32E−4	2 1	7.18E−1	3 2	5.77E−2
4 4	1.56E−1		$^3\mathbf{F}^o$–$^3\mathbf{D}^e$	3 9	−2.46E−4	3 5	−1.84E−4	2 2	−5.95E−1	3 3	3.90E−2
4 5	6.64E−3	1 1	1.14E+0	4 1	5.23E−4	3 6	−3.24E−6	2 3	−2.03E−3	3 4	3.28E+0
4 6	1.18E−4	1 2	7.26E−3	4 2	1.80E−4	4 1	1.63E−5	3 1	2.13E−4		$^5\mathbf{D}^o$–$^5\mathbf{D}^e$
4 7	1.48E+0	1 3	−3.72E−4	4 3	2.53E−3	4 2	9.82E−3	3 2	5.58E−1	1 1	−9.44E−1
4 8	−1.08E+0	1 4	−2.07E−2	4 4	6.92E−3	4 3	2.85E−1	3 3	−1.22E−3	2 1	−3.08E−1
4 9	−4.05E−3	1 5	−6.10E−1	4 5	1.24E−3	4 4	1.40E+0	4 1	1.45E+0	3 1	6.12E−5
4 10	−3.25E−4	1 6	−2.00E−1	4 6	1.02E−2	4 5	−1.80E−4	4 2	1.45E−2		$^5\mathbf{F}^e$–$^5\mathbf{D}^o$
5 1	2.08E+0	1 7	−9.62E−1	4 7	3.90E−1	4 6	−2.14E−7	4 3	1.88E+0	1 1	6.50E−1
5 2	5.08E−2	1 8	−6.01E−4	4 8	1.26E+0		$^3\mathbf{G}^e$–$^3\mathbf{F}^o$		$^5\mathbf{P}^o$–$^5\mathbf{P}^e$	1 2	2.47E+0
5 3	7.08E−2	1 9	−2.78E−1	4 9	−1.85E−4	1 1	1.50E−1	1 1	7.42E−1	1 3	−1.05E−2
5 4	2.58E−3	2 1	1.38E−2		$^3\mathbf{F}^o$–$^3\mathbf{F}^e$	1 2	1.08E−1	1 2	1.12E+0	2 1	2.37E+1
5 5	1.54E−4	2 2	1.54E+0	1 1	4.84E−1	1 3	1.74E+0	1 3	−5.26E−1	2 2	1.60E−1
5 6	1.08E−2	2 3	−1.75E−4	1 2	−1.23E−3	1 4	−1.55E+0	1 4	−5.23E−0	2 3	2.17E+0
5 7	4.70E−3	2 4	−5.74E−6	1 3	−3.84E−1		$^3\mathbf{G}^o$–$^3\mathbf{F}^e$		$^5\mathbf{D}^e$–$^5\mathbf{P}^o$		
5 8	3.61E−4	2 5	−2.01E−2	1 4	−1.72E−1	1 1	2.29E+0	1 1	1.19E+0		
5 9	2.66E+0	2 6	−9.16E−1	1 5	−1.42E−1	1 2	−2.57E−4				

F-like S (S^{7+})

Term energies relative to $2s^22p^4$ ^{3}P ionization threshold for each symmetry

i	E(Ryds)	Description
	$^2S^e$	
1	-19.6550	$2s2p^6$
2	-8.74730	$2s^22p^4$ ^{1}S $3s$
3	-6.97638	$2s^22p^4$ ^{1}D $3d$
4	-4.33441	$2s2p^5$ ^{3}P $3p$
5	-3.96740	$2s^22p^4$ ^{1}S $4s$
6	-3.64820	$2s^22p^4$ ^{1}D $4d$
7	-3.07775	$2s2p^5$ ^{1}P $3p$
8	-2.14804	$2s^22p^4$ ^{1}D $5d$
9	-1.98295	$2s^22p^4$ ^{1}S $5s$
10	-1.33095	$2s^22p^4$ ^{1}D $6d$
11	-.98255	$2p^6$ ^{1}S $3s$
12	-.84126	$2s^22p^4$ ^{1}D $7d$
13	-.52637	$2s^22p^4$ ^{1}D $8d$
14	-.47505	$2s2p^5$ ^{3}P $4p$
15	-.40035	$2s^22p^4$ ^{1}S $6s$
16	-.30844	$2s^22p^4$ ^{1}D $9d$
	$^2P^e$	
1	-9.72272	$2s^22p^4$ ^{3}P $3s$
2	-7.37659	$2s^22p^4$ ^{3}P $3d$
3	-6.91058	$2s^22p^4$ ^{1}D $3d$
4	-5.01326	$2s^22p^4$ ^{3}P $4s$
5	-4.60492	$2s2p^5$ ^{3}P $3p$
6	-4.10406	$2s^22p^4$ ^{3}P $4d$
7	-3.65620	$2s^22p^4$ ^{1}D $4d$
8	-3.12189	$2s2p^5$ ^{1}P $3p$
9	-3.04581	$2s^22p^4$ ^{3}P $5s$
10	-2.61465	$2s^22p^4$ ^{3}P $5d$
11	-2.14725	$2s^22p^4$ ^{1}D $5d$
12	-2.05691	$2s^22p^4$ ^{3}P $6s$
13	-1.80943	$2s^22p^4$ ^{3}P $6d$
14	-1.47938	$2s^22p^4$ ^{3}P $7s$
15	-1.34545	$2s^22p^4$ ^{1}D $6d$
16	-1.31326	$2s^22p^4$ ^{3}P $7d$
17	-1.11470	$2s^22p^4$ ^{3}P $8s$
18	-1.01486	$2s^22p^4$ ^{3}P $8d$
19	-.87023	$2s^22p^4$ ^{3}P $9s$
20	-.84287	$2s^22p^4$ ^{1}D $7d$
21	-.79897	$2s^22p^4$ ^{3}P $9d$
	$^2D^e$	
1	-9.34040	$2s^22p^4$ ^{1}D $3s$
2	-7.36149	$2s^22p^4$ ^{3}P $3d$
3	-6.89996	$2s^22p^4$ ^{1}D $3d$
4	-6.38903	$2s^22p^4$ ^{1}S $3d$
5	-4.65827	$2s2p^5$ ^{3}P $3p$
6	-4.54166	$2s^22p^4$ ^{1}D $4s$
7	-4.08437	$2s^22p^4$ ^{3}P $4d$
8	-3.65045	$2s^22p^4$ ^{1}D $4d$
9	-3.20710	$2s2p^5$ ^{1}P $3p$
10	-3.08295	$2s^22p^4$ ^{1}S $4d$
11	-2.60637	$2s^22p^4$ ^{3}P $5d$
12	-2.56976	$2s^22p^4$ ^{1}D $5s$
13	-2.14433	$2s^22p^4$ ^{1}D $5d$
14	-1.80597	$2s^22p^4$ ^{3}P $6d$
15	-1.57361	$2s^22p^4$ ^{1}D $6s$
16	-1.56281	$2s^22p^4$ ^{1}S $5d$
17	-1.33988	$2s^22p^4$ ^{1}D $6d$
18	-1.31400	$2s^22p^4$ ^{3}P $7d$
19	-1.01311	$2s^22p^4$ ^{3}P $8d$
20	-.98819	$2s^22p^4$ ^{1}D $7s$
21	-.84185	$2s^22p^4$ ^{1}D $7d$
22	-.79875	$2s^22p^4$ ^{3}P $9d$
23	-.74777	$2s^22p^4$ ^{1}S $6d$
	$^2F^e$	
1	-7.44870	$2s^22p^4$ ^{3}P $3d$
2	-6.93281	$2s^22p^4$ ^{1}D $3d$
3	-4.15456	$2s^22p^4$ ^{3}P $4d$
4	-3.65704	$2s^22p^4$ ^{1}D $4d$
5	-2.64305	$2s^22p^4$ ^{3}P $5d$
6	-2.14671	$2s^22p^4$ ^{1}D $5d$
7	-1.82720	$2s^22p^4$ ^{3}P $6d$
8	-1.33890	$2s^22p^4$ ^{3}P $7d$
9	-1.32988	$2s^22p^4$ ^{1}D $6d$
10	-1.02136	$2s^22p^4$ ^{3}P $8d$
11	-.84159	$2s^22p^4$ ^{1}D $7d$
12	-.80511	$2s^22p^4$ ^{3}P $9d$
	$^2G^e$	
1	-7.07210	$2s^22p^4$ ^{1}D $3d$
2	-3.70167	$2s^22p^4$ ^{1}D $4d$
3	-2.16820	$2s^22p^4$ ^{1}D $5d$
4	-1.34300	$2s^22p^4$ ^{1}D $6d$
5	-.84887	$2s^22p^4$ ^{1}D $7d$
	$^2S^o$	
1	-8.69342	$2s^22p^4$ ^{3}P $3p$
2	-4.61420	$2s^22p^4$ ^{3}P $4p$
3	-2.86465	$2s^22p^4$ ^{3}P $5p$
4	-1.95085	$2s^22p^4$ ^{3}P $6p$
5	-1.41373	$2s^22p^4$ ^{3}P $7p$
6	-1.07141	$2s^22p^4$ ^{3}P $8p$
7	-.83992	$2s^22p^4$ ^{3}P $9p$
	$^2P^o$	
1	-24.1438	$2s^22p^5$
2	-8.72599	$2s^22p^4$ ^{3}P $3p$
3	-7.99876	$2s^22p^4$ ^{1}D $3p$
4	-7.68806	$2s^22p^4$ ^{1}S $3p$
5	-5.52371	$2s2p^5$ ^{3}P $3s$
6	-4.57751	$2s^22p^4$ ^{3}P $4p$
7	-4.18425	$2s2p^5$ ^{1}P $3s$
8	-4.04827	$2s^22p^4$ ^{1}D $4p$
9	-3.56764	$2s^22p^4$ ^{1}S $4p$
10	-3.53807	$2s^22p^4$ ^{1}D $4f$
11	-3.18144	$2s2p^5$ ^{3}P $3d$
12	-2.81688	$2s^22p^4$ ^{3}P $5p$
13	-2.36872	$2s^22p^4$ ^{1}D $5p$
14	-2.15669	$2s^22p^4$ ^{1}D $5f$
15	-1.92447	$2s^22p^4$ ^{3}P $6p$
16	-1.89078	$2s2p^5$ ^{1}P $3d$
17	-1.78772	$2s^22p^4$ ^{1}S $5p$
18	-1.46626	$2s^22p^4$ ^{1}D $6p$
19	-1.38969	$2s^22p^4$ ^{3}P $7p$
20	-1.27839	$2s^22p^4$ ^{1}D $6f$
21	-1.06244	$2s^22p^4$ ^{3}P $8p$
22	-.94231	$2s^22p^4$ ^{1}D $7p$
23	-.89161	$2s2p^5$ ^{3}P $4s$
24	-.86675	$2s^22p^4$ ^{1}S $6p$
25	-.82954	$2s^22p^4$ ^{3}P $9p$
26	-.81140	$2s^22p^4$ ^{1}D $7f$
	$^2D^o$	
1	-8.75301	$2s^22p^4$ ^{3}P $3p$
2	-8.23185	$2s^22p^4$ ^{1}D $3p$
3	-4.63223	$2s^22p^4$ ^{3}P $4p$
4	-4.12956	$2s^22p^4$ ^{1}D $4p$
5	-4.00514	$2s^22p^4$ ^{3}P $4f$
6	-3.52907	$2s^22p^4$ ^{1}D $4f$
7	-3.31788	$2s2p^5$ ^{3}P $3d$
8	-2.86902	$2s^22p^4$ ^{3}P $5p$
9	-2.55608	$2s^22p^4$ ^{3}P $5f$
10	-2.37493	$2s^22p^4$ ^{1}D $5p$
11	-2.11086	$2s^22p^4$ ^{1}D $5f$
12	-1.95453	$2s^22p^4$ ^{3}P $6p$
13	-1.81908	$2s2p^5$ ^{1}P $3d$
14	-1.77654	$2s^22p^4$ ^{3}P $6f$
15	-1.45697	$2s^22p^4$ ^{1}D $6p$
16	-1.41603	$2s^22p^4$ ^{3}P $7p$
17	-1.30608	$2s^22p^4$ ^{3}P $7f$
18	-1.27711	$2s^22p^4$ ^{1}D $6f$
19	-1.07310	$2s^22p^4$ ^{3}P $8p$
20	-1.00008	$2s^22p^4$ ^{3}P $8f$
21	-.91911	$2s^22p^4$ ^{1}D $7p$
22	-.84113	$2s^22p^4$ ^{3}P $9p$
23	-.81019	$2s^22p^4$ ^{1}D $7f$
24	-.79022	$2s^22p^4$ ^{3}P $9f$
	$^2F^o$	
1	-8.36027	$2s^22p^4$ ^{1}D $3p$
2	-4.17096	$2s^22p^4$ ^{1}D $4p$
3	-4.03238	$2s^22p^4$ ^{3}P $4f$
4	-3.50955	$2s^22p^4$ ^{1}D $4f$
5	-3.37498	$2s2p^5$ ^{3}P $3d$
6	-2.92901	$2s^22p^4$ ^{1}S $4f$
7	-2.56490	$2s^22p^4$ ^{3}P $5f$
8	-2.39763	$2s^22p^4$ ^{1}D $5p$
9	-2.09558	$2s^22p^4$ ^{1}D $5f$
10	-1.92352	$2s2p^5$ ^{1}P $3d$
11	-1.78203	$2s^22p^4$ ^{3}P $6f$
12	-1.48064	$2s^22p^4$ ^{1}S $5f$
13	-1.46553	$2s^22p^4$ ^{1}D $6p$
14	-1.30931	$2s^22p^4$ ^{3}P $7f$
15	-1.27971	$2s^22p^4$ ^{1}D $6f$
16	-1.00226	$2s^22p^4$ ^{3}P $8f$
17	-.92498	$2s^22p^4$ ^{1}D $7p$
18	-.81023	$2s^22p^4$ ^{1}D $7f$:
19	-.80978	$2s^22p^4$ ^{1}D $7h$:
20	-.79175	$2s^22p^4$ ^{3}P $9f$
21	-.69889	$2s^22p^4$ ^{1}S $6f$
	$^2G^o$	
1	-4.00480	$2s^22p^4$ ^{3}P $4f$
2	-3.50060	$2s^22p^4$ ^{1}D $4f$
3	-2.56535	$2s^22p^4$ ^{3}P $5f$
4	-2.06536	$2s^22p^4$ ^{1}D $5f$
5	-1.78146	$2s^22p^4$ ^{3}P $6f$
6	-1.30857	$2s^22p^4$ ^{3}P $7f$
7	-1.28321	$2s^22p^4$ ^{1}D $6f$
8	-1.00167	$2s^22p^4$ ^{3}P $8f$
9	-.79131	$2s^22p^4$ ^{3}P $9f$
	$^2H^o$	
1	-3.52353	$2s^22p^4$ ^{1}D $4f$
2	-2.07685	$2s^22p^4$ ^{1}D $5f$
3	-1.28964	$2s^22p^4$ ^{1}D $6f$
4	-.81509	$2s^22p^4$ ^{1}D $7f$
	$^4S^e$	
1	-4.93039	$2s2p^5$ ^{3}P $3p$
2	-.65218	$2s2p^5$ ^{3}P $4p$
	$^4P^e$	
1	-9.89408	$2s^22p^4$ ^{3}P $3s$
2	-7.45862	$2s^22p^4$ ^{3}P $3d$
3	-5.08970	$2s^22p^4$ ^{3}P $4s$
4	-4.63657	$2s2p^5$ ^{3}P $3p$
5	-4.16102	$2s^22p^4$ ^{3}P $4d$
6	-3.07809	$2s^22p^4$ ^{3}P $5s$
7	-2.64758	$2s^22p^4$ ^{3}P $5d$
8	-2.07099	$2s^22p^4$ ^{3}P $6s$
9	-1.82998	$2s^22p^4$ ^{3}P $6d$
10	-1.48792	$2s^22p^4$ ^{3}P $7s$
11	-1.33951	$2s^22p^4$ ^{3}P $7d$
12	-1.12064	$2s^22p^4$ ^{3}P $8s$
13	-1.02259	$2s^22p^4$ ^{3}P $8d$
14	-.87464	$2s^22p^4$ ^{3}P $9s$
15	-.80613	$2s^22p^4$ ^{3}P $9d$
	$^4D^e$	
1	-7.66123	$2s^22p^4$ ^{3}P $3d$
2	-4.77335	$2s2p^5$ ^{3}P $3p$
3	-4.22698	$2s^22p^4$ ^{3}P $4d$
4	-2.67982	$2s^22p^4$ ^{3}P $5d$
5	-1.84796	$2s^22p^4$ ^{3}P $6d$

F-like S (S^{7+})

i	Energy(Ryds)	Description	i	Energy(Ryds)	Description	i	Energy(Ryds)	Description
6	−1.35056	$2s^22p^4\ ^3\mathrm{P}\ 7d$	2	−5.77883	$2s2p^5\ ^3\mathrm{P}\ 3s$	12	−1.00072	$2s^22p^4\ ^3\mathrm{P}\ 8f$
7	−1.02989	$2s^22p^4\ ^3\mathrm{P}\ 8d$	3	−4.68839	$2s^22p^4\ ^3\mathrm{P}\ 4p$	13	−.84356	$2s^22p^4\ ^3\mathrm{P}\ 9p$
8	−.81126	$2s^22p^4\ ^3\mathrm{P}\ 9d$	4	−3.59861	$2s2p^5\ ^3\mathrm{P}\ 3d$	14	−.79073	$2s^22p^4\ ^3\mathrm{P}\ 9f$
		$^4\mathrm{F}^e$	5	−2.90316	$2s^22p^4\ ^3\mathrm{P}\ 5p$			$^4\mathrm{F}^o$
1	−7.52396	$2s^22p^4\ ^3\mathrm{P}\ 3d$	6	−1.97404	$2s^22p^4\ ^3\mathrm{P}\ 6p$	1	−4.05353	$2s^22p^4\ ^3\mathrm{P}\ 4f$
2	−4.18578	$2s^22p^4\ ^3\mathrm{P}\ 4d$	7	−1.42952	$2s^22p^4\ ^3\mathrm{P}\ 7p$	2	−3.52272	$2s2p^5\ ^3\mathrm{P}\ 3d$
3	−2.65875	$2s^22p^4\ ^3\mathrm{P}\ 5d$	8	−1.08770	$2s^22p^4\ ^3\mathrm{P}\ 8p$	3	−2.56703	$2s^22p^4\ ^3\mathrm{P}\ 5f$
4	−1.83615	$2s^22p^4\ ^3\mathrm{P}\ 6d$	9	−.98468	$2s2p^5\ ^3\mathrm{P}\ 4s$	4	−1.78475	$2s^22p^4\ ^3\mathrm{P}\ 6f$
5	−1.34325	$2s^22p^4\ ^3\mathrm{P}\ 7d$	10	−.84254	$2s^22p^4\ ^3\mathrm{P}\ 9p$	5	−1.31115	$2s^22p^4\ ^3\mathrm{P}\ 7f$
6	−1.02499	$2s^22p^4\ ^3\mathrm{P}\ 8d$			$^4\mathrm{D}^o$	6	−1.00363	$2s^22p^4\ ^3\mathrm{P}\ 8f$
7	−.80772	$2s^22p^4\ ^3\mathrm{P}\ 9d$	1	−8.84669	$2s^22p^4\ ^3\mathrm{P}\ 3p$	7	−.79282	$2s^22p^4\ ^3\mathrm{P}\ 9f$
		$^4\mathrm{S}^o$	2	−4.66621	$2s^22p^4\ ^3\mathrm{P}\ 4p$			$^4\mathrm{G}^o$
1	−8.68684	$2s^22p^4\ ^3\mathrm{P}\ 3p$	3	−4.01452	$2s^22p^4\ ^3\mathrm{P}\ 4f$	1	−4.00669	$2s^22p^4\ ^3\mathrm{P}\ 4f$
2	−4.61181	$2s^22p^4\ ^3\mathrm{P}\ 4p$	4	−3.40303	$2s2p^5\ ^3\mathrm{P}\ 3d$	2	−2.56673	$2s^22p^4\ ^3\mathrm{P}\ 5f$
3	−2.86350	$2s^22p^4\ ^3\mathrm{P}\ 5p$	5	−2.88434	$2s^22p^4\ ^3\mathrm{P}\ 5p$	3	−1.78237	$2s^22p^4\ ^3\mathrm{P}\ 6f$
4	−1.95021	$2s^22p^4\ ^3\mathrm{P}\ 6p$	6	−2.55638	$2s^22p^4\ ^3\mathrm{P}\ 5f$	4	−1.30918	$2s^22p^4\ ^3\mathrm{P}\ 7f$
5	−1.41334	$2s^22p^4\ ^3\mathrm{P}\ 7p$	7	−1.96313	$2s^22p^4\ ^3\mathrm{P}\ 6p$	5	−1.00210	$2s^22p^4\ ^3\mathrm{P}\ 8f$
6	−1.07116	$2s^22p^4\ ^3\mathrm{P}\ 8p$	8	−1.77819	$2s^22p^4\ ^3\mathrm{P}\ 6f$	6	−.79162	$2s^22p^4\ ^3\mathrm{P}\ 9f$
7	−.83974	$2s^22p^4\ ^3\mathrm{P}\ 9p$	9	−1.42143	$2s^22p^4\ ^3\mathrm{P}\ 7p$			
		$^4\mathrm{P}^o$	10	−1.30690	$2s^22p^4\ ^3\mathrm{P}\ 7f$			
1	−8.97953	$2s^22p^4\ ^3\mathrm{P}\ 3p$	11	−1.07657	$2s^22p^4\ ^3\mathrm{P}\ 8p$			

Energies in ascending order from ground state for terms with effective $n \leq 4.0$, $L \leq 4$

Term	i	E(Ryds)	Term	i	E(Ryds)	Term	i	E(Ryds)	Term	i	E(Ryds)	Term	i	E(Ryds)
$^2\mathrm{P}^o$	1	0.00000	$^2\mathrm{D}^o$	2	15.9119	$^2\mathrm{D}^e$	3	17.2438	$^2\mathrm{S}^o$	2	19.5296	$^2\mathrm{P}^e$	6	20.0397
$^2\mathrm{S}^e$	1	4.48880	$^2\mathrm{P}^o$	3	16.1450	$^2\mathrm{D}^e$	4	17.7547	$^4\mathrm{S}^o$	2	19.5319	$^2\mathrm{D}^e$	7	20.0594
$^4\mathrm{P}^e$	1	14.2497	$^2\mathrm{P}^o$	4	16.4557	$^4\mathrm{P}^o$	2	18.3649	$^2\mathrm{P}^e$	5	19.5388	$^4\mathrm{F}^o$	1	20.0902
$^2\mathrm{P}^e$	1	14.4210	$^4\mathrm{D}^e$	1	16.4825	$^2\mathrm{P}^o$	5	18.6200	$^2\mathrm{P}^o$	6	19.5662	$^2\mathrm{P}^o$	8	20.0955
$^2\mathrm{D}^e$	1	14.8034	$^4\mathrm{F}^e$	1	16.6198	$^4\mathrm{P}^e$	3	19.0541	$^2\mathrm{D}^e$	6	19.6021	$^2\mathrm{F}^o$	3	20.1114
$^4\mathrm{P}^o$	1	15.1642	$^4\mathrm{P}^e$	2	16.6851	$^2\mathrm{P}^e$	4	19.1305	$^2\mathrm{S}^e$	4	19.8093	$^4\mathrm{D}^o$	3	20.1292
$^4\mathrm{D}^o$	1	15.2971	$^2\mathrm{F}^e$	1	16.6951	$^4\mathrm{S}^e$	1	19.2134	$^4\mathrm{D}^e$	3	19.9168	$^4\mathrm{G}^o$	1	20.1371
$^2\mathrm{D}^o$	1	15.3907	$^2\mathrm{P}^e$	2	16.7672	$^4\mathrm{D}^e$	2	19.3704	$^4\mathrm{F}^e$	2	19.9580	$^2\mathrm{D}^o$	5	20.1386
$^2\mathrm{S}^e$	2	15.3965	$^2\mathrm{D}^e$	2	16.7823	$^4\mathrm{P}^o$	3	19.4554	$^2\mathrm{P}^o$	7	19.9595	$^2\mathrm{G}^o$	1	20.1390
$^2\mathrm{P}^o$	2	15.4178	$^2\mathrm{G}^e$	1	17.0717	$^4\mathrm{D}^o$	2	19.4775	$^2\mathrm{F}^o$	2	19.9728			
$^2\mathrm{S}^o$	1	15.4503	$^2\mathrm{S}^e$	3	17.1674	$^2\mathrm{D}^e$	5	19.4855	$^4\mathrm{P}^e$	5	19.9827			
$^4\mathrm{S}^o$	1	15.4569	$^2\mathrm{F}^e$	2	17.2109	$^4\mathrm{P}^e$	4	19.5072	$^2\mathrm{F}^e$	3	19.9892			
$^2\mathrm{F}^o$	1	15.7835	$^2\mathrm{P}^e$	3	17.2332	$^2\mathrm{D}^o$	3	19.5115	$^2\mathrm{D}^o$	4	20.0142			

gf-values for transitions involving terms with effective $n \leq 4.0$, $L \leq 4$

$i\ i'$	gf_L	$i\ i'$	gf_L	$i\ i'$	gf_L	$i\ i'$	gf_L	$i\ i'$	gf_L	$i\ i'$	gf_L
	$^2\mathrm{S}^e–^2\mathrm{P}^o$	2 1	7.80E−2	3 2	1.99E−1	4 3	2.31E−3	2 1	7.32E−1	6 2	1.41E+0
1 1	4.29E−1	2 2	−3.03E−4	3 3	2.40E−1	4 4	7.10E−4	2 2	−1.92E−1		$^2\mathrm{P}^e$ $^2\mathrm{P}^o$
1 2	−2.23E−3	2 3	6.99E−3	3 4	5.46E−3	4 5	3.78E−1	3 1	2.31E−1	1 1	6.75E−1
1 3	−1.45E−2	2 4	−1.04E+0	3 5	−8.34E−3	4 6	2.21E−3	3 2	−7.45E−2	1 2	−6.21E−1
1 4	−7.08E−3	2 5	−8.03E−3	3 6	−1.88E−2	4 7	−5.84E−3	4 1	7.42E−2	1 3	−7.73E−1
1 5	−3.66E−1	2 6	−3.27E−2	3 7	−1.01E−2	4 8	−6.77E−4	4 2	−4.67E−1	1 4	−8.13E−2
1 6	−4.02E−3	2 7	−1.07E−1	3 8	−8.62E−2		$^2\mathrm{P}^e–^2\mathrm{S}^o$	5 1	3.79E−1	1 5	−5.98E−1
1 7	−1.14E−1	2 8	4.11E−2	4 1	1.82E−1	1 1	−3.45E−1	5 2	7.44E−4	1 6	−2.05E−1
1 8	−4.08E−2	3 1	8.75E−1	4 2	3.30E−1	1 2	−9.36E−2	6 1	3.01E−1	1 7	−1.04E−2

F-like S (S^{7+})

$i\ i'$	gf_L
1 8	−7.69E−4
2 1	3.95E−1
2 2	6.20E−1
2 3	4.27E−3
2 4	2.77E−3
2 5	−5.45E−4
2 6	−1.13E−1
2 7	−4.54E−4
2 8	−1.38E−2
3 1	2.80E+0
3 2	1.54E−1
3 3	7.40E−1
3 4	3.05E−3
3 5	−1.86E−2
3 6	−3.42E−3
3 7	−3.96E−2
3 8	−1.74E−1
4 1	1.89E−1
4 2	3.75E−1
4 3	4.16E−1
4 4	4.31E−2
4 5	1.83E−3
4 6	−1.47E+0
4 7	−5.01E−1
4 8	−9.22E−2
5 1	3.92E−1
5 2	8.29E−2
5 3	1.16E−1
5 4	2.20E−2
5 5	9.43E−1
5 6	−7.19E−4
5 7	−4.32E−5
5 8	−7.17E−4
6 1	3.11E−1
6 2	2.44E−1
6 3	5.45E−2
6 4	1.48E−2
6 5	1.23E−2
6 6	8.32E−1
6 7	1.36E−2
6 8	−2.29E−4
	$^2\mathbf{P}^e$–$^2\mathbf{D}^o$
1 1	−1.65E+0
1 2	−1.35E−2
1 3	−5.21E−1
1 4	−1.63E−5
1 5	−4.43E−3
2 1	5.97E−2
2 2	3.85E−2
2 3	−2.32E−2
2 4	−3.70E−3
2 5	−4.36E+0

$i\ i'$	gf_L
3 1	4.21E−2
3 2	2.80E−1
3 3	−1.64E−2
3 4	−7.66E−2
3 5	−9.43E−1
4 1	8.61E−1
4 2	5.02E−3
4 3	−2.43E+0
4 4	−5.20E−3
4 5	−1.53E−2
5 1	1.87E−1
5 2	2.19E−5
5 3	3.21E−3
5 4	−7.11E−5
5 5	−1.05E−3
6 1	2.20E−2
6 2	1.58E−2
6 3	1.11E−1
6 4	5.81E−4
6 5	−4.62E−1
	$^2\mathbf{D}^e$–$^2\mathbf{P}^o$
1 1	4.16E−1
1 2	−2.01E−1
1 3	−8.93E−1
1 4	−7.06E−4
1 5	−4.23E−2
1 6	−3.90E−1
1 7	−1.57E+0
1 8	−6.74E−2
2 1	2.56E+0
2 2	1.45E+0
2 3	1.56E−1
2 4	1.10E−3
2 5	−9.09E−3
2 6	−4.26E−1
2 7	−8.04E−3
2 8	−3.80E−3
3 1	3.30E+0
3 2	2.14E−2
3 3	7.44E−1
3 4	2.81E−2
3 5	−2.16E−2
3 6	−1.44E−2
3 7	−3.58E−2
3 8	−1.15E−1
4 1	6.96E−1
4 2	6.60E−5
4 3	6.53E−2
4 4	2.36E+0
4 5	−4.68E−3
4 6	−3.06E−3
4 7	−6.94E−4

$i\ i'$	gf_L
4 8	−8.52E−3
5 1	5.63E−1
5 2	3.79E−1
5 3	1.29E−1
5 4	4.02E−2
5 5	1.48E+0
5 6	−1.09E−2
5 7	−5.01E−3
5 8	−8.53E−2
6 1	3.26E−1
6 2	3.41E−2
6 3	7.08E−1
6 4	3.08E−2
6 5	4.24E−2
6 6	7.62E−3
6 7	−3.26E−1
6 8	−1.30E+0
7 1	1.08E+0
7 2	5.40E−1
7 3	3.16E−1
7 4	9.45E−3
7 5	5.20E−2
7 6	2.67E+0
7 7	7.26E−2
7 8	−4.39E−4
	$^2\mathbf{D}^e$–$^2\mathbf{D}^o$
1 1	−3.04E−3
1 2	−1.86E+0
1 3	−1.36E−3
1 4	−4.27E−1
1 5	−2.16E−5
2 1	5.11E−1
2 2	1.25E−1
2 3	−1.41E−1
2 4	−3.00E−2
2 5	−8.83E−1
3 1	1.72E−1
3 2	1.39E+0
3 3	−4.48E−2
3 4	−4.09E−1
3 5	−1.31E−1
4 1	5.67E−3
4 2	2.67E−2
4 3	−2.57E−3
4 4	−1.61E−3
4 5	−7.79E−3
5 1	1.04E+0
5 2	5.09E−2
5 3	−8.23E−4
5 4	−2.90E−1
5 5	−1.12E−3
6 1	9.04E−2

$i\ i'$	gf_L
6 2	6.30E−1
6 3	1.29E−4
6 4	−2.34E+0
6 5	−3.65E−5
7 1	1.40E−1
7 2	4.94E−2
7 3	1.06E+0
7 4	3.18E−3
7 5	−7.21E−2
	$^2\mathbf{D}^e$–$^2\mathbf{F}^o$
1 1	−2.29E+0
1 2	−7.54E−1
1 3	−2.55E−4
2 1	8.57E−3
2 2	−2.93E−3
2 3	−6.91E+0
3 1	8.98E−2
3 2	−1.31E−2
3 3	−1.01E+0
4 1	1.64E−2
4 2	−6.24E−3
4 3	−6.02E−2
5 1	1.37E−1
5 2	−3.83E−1
5 3	−5.90E−3
6 1	1.18E+0
6 2	−3.07E+0
6 3	−6.10E−6
7 1	6.00E−3
7 2	6.81E−4
7 3	−3.49E−1
	$^2\mathbf{F}^e$–$^2\mathbf{D}^o$
1 1	3.43E+0
1 2	6.87E−4
1 3	−8.72E−1
1 4	−9.08E−4
1 5	−6.45E−2
2 1	1.50E−3
2 2	2.31E+0
2 3	−2.83E−3
2 4	−5.53E−1
2 5	−6.03E−4
3 1	1.73E+0
3 2	2.81E−3
3 3	5.41E+0
3 4	−6.77E−6
3 5	−1.49E−3
	$^2\mathbf{F}^e$–$^2\mathbf{F}^o$
1 1	3.00E−3
1 2	−9.35E−8
1 3	−7.32E−1
2 1	1.27E+0

$i\ i'$	gf_L
2 2	−3.01E−1
2 3	−2.84E−3
3 1	1.43E−3
3 2	8.40E−6
3 3	−1.07E−1
	$^2\mathbf{F}^e$–$^2\mathbf{G}^o$
1 1	−1.16E+1
2 1	−5.40E−2
3 1	−1.59E+0
	$^2\mathbf{G}^e$–$^2\mathbf{F}^o$
1 1	4.16E+0
1 2	−1.08E+0
1 3	−6.59E−4
	$^2\mathbf{G}^e$–$^2\mathbf{G}^o$
1 1	−6.39E−6
	$^4\mathbf{S}^e$–$^4\mathbf{P}^o$
1 1	6.31E−1
1 2	5.49E−1
1 3	−5.19E−3
	$^4\mathbf{P}^e$–$^4\mathbf{S}^o$
1 1	−8.20E−1
1 2	−1.44E−1
2 1	1.59E+0
2 2	−4.03E−1
3 1	1.31E−1
3 2	−1.08E+0
4 1	7.74E−1
4 2	−8.47E−3
5 1	7.94E−1
5 2	2.46E+0
	$^4\mathbf{P}^e$–$^4\mathbf{P}^o$
1 1	−1.70E+0
1 2	−2.07E+0
1 3	−4.44E−1
2 1	1.27E+0
2 2	−4.59E−3
2 3	−2.43E−1
3 1	1.18E+0
3 2	3.31E−1
3 3	−2.82E+0
4 1	1.21E−1
4 2	2.11E+0
4 3	2.90E−2
5 1	4.74E−1
5 2	2.80E−2
5 3	2.02E+0
	$^4\mathbf{P}^e$–$^4\mathbf{D}^o$
1 1	−3.45E+0
1 2	−9.81E−1
1 3	−6.31E−3
2 1	1.47E−1
2 2	−5.67E−2

$i\ i'$	gf_L
2 3	−9.78E+0
3 1	2.04E+0
3 2	−5.06E+0
3 3	−1.97E−2
4 1	2.73E−1
4 2	2.53E−2
4 3	−1.98E−3
5 1	5.26E−2
5 2	1.86E−1
5 3	−1.40E+0
	$^4\mathbf{D}^e$–$^4\mathbf{P}^o$
1 1	3.35E+0
1 2	−1.82E−2
1 3	−8.90E−1
2 1	5.68E−1
2 2	3.66E+0
2 3	−1.41E−5
3 1	1.84E+0
3 2	1.13E−1
3 3	5.34E+0
	$^4\mathbf{D}^e$–$^4\mathbf{D}^o$
1 1	9.73E−1
1 2	−2.43E−1
1 3	−1.77E+0
2 1	2.14E+0
2 2	−3.92E−3
2 3	−6.30E−4
3 1	5.83E−1
3 2	1.66E+0
3 3	−4.20E−1
	$^4\mathbf{D}^e$–$^4\mathbf{F}^o$
1 1	−1.34E+1
2 1	−2.57E−2
3 1	−2.37E+0
	$^4\mathbf{F}^e$–$^4\mathbf{D}^o$
1 1	6.75E+0
1 2	−1.75E+0
1 3	−1.55E−1
2 1	3.53E+0
2 2	1.07E+1
2 3	−1.21E−3
	$^4\mathbf{F}^e$–$^4\mathbf{F}^o$
1 1	−1.14E+0
2 1	−2.21E−1
	$^4\mathbf{F}^e$–$^4\mathbf{G}^o$
1 1	−2.28E+1
2 1	−3.84E+0

F-like Ar (Ar^{9+})

Term energies relative to $2s^22p^4\ {}^3\mathrm{P}$ ionization threshold for each symmetry

i	E(Ryds)	Description	i	E(Ryds)	Description	i	E(Ryds)	Description	i	E(Ryds)	Description
		$^2\mathrm{S}^e$	9	−5.86746	$2s^22p^4\ {}^1\mathrm{D}\ 4d$	3	−12.3127	$2s^22p^4\ {}^1\mathrm{D}\ 3p$	24	−1.29732	$2s^22p^4\ {}^3\mathrm{P}\ 9p$
1	−29.9094	$2s2p^6$	10	−5.20654	$2s^22p^4\ {}^1\mathrm{S}\ 4d$	4	−11.9812	$2s^22p^4\ {}^1\mathrm{S}\ 3p$	25	−1.23479	$2s^22p^4\ {}^3\mathrm{P}\ 9f$
2	−13.3066	$2s^22p^4\ {}^1\mathrm{S}\ 3s$	11	−4.07491	$2s^22p^4\ {}^3\mathrm{P}\ 5d$	5	−9.44303	$2s2p^5\ {}^3\mathrm{P}\ 3s$	26	−1.06083	$2s^22p^4\ {}^1\mathrm{D}\ 8p$
3	−11.0524	$2s^22p^4\ {}^1\mathrm{D}\ 3d$	12	−4.05390	$2s^22p^4\ {}^1\mathrm{D}\ 5s$	6	−7.92030	$2s2p^5\ {}^1\mathrm{P}\ 3s$			**$^2\mathrm{F}^o$**
4	−7.97511	$2s2p^5\ {}^3\mathrm{P}\ 3p$	13	−3.52289	$2s^22p^4\ {}^1\mathrm{D}\ 5d$	7	−6.96258	$2s^22p^4\ {}^3\mathrm{P}\ 4p$	1	−12.7914	$2s^22p^4\ {}^1\mathrm{D}\ 3p$
5	−6.55424	$2s2p^5\ {}^1\mathrm{P}\ 3p$	14	−2.84935	$2s^22p^4\ {}^1\mathrm{S}\ 5d$	8	−6.48609	$2s2p^5\ {}^3\mathrm{P}\ 3d$	2	−6.81427	$2s2p^5\ {}^3\mathrm{P}\ 3d$
6	−6.24960	$2s^22p^4\ {}^1\mathrm{S}\ 4s$	15	−2.81007	$2s^22p^4\ {}^3\mathrm{P}\ 6d$	9	−6.41606	$2s^22p^4\ {}^1\mathrm{D}\ 4p$	3	−6.53104	$2s^22p^4\ {}^1\mathrm{D}\ 4p$
7	−5.87914	$2s^22p^4\ {}^1\mathrm{D}\ 4d$	16	−2.55188	$2s^22p^4\ {}^1\mathrm{D}\ 6s$	10	−5.82176	$2s^22p^4\ {}^1\mathrm{S}\ 4p$	4	−6.24340	$2s^22p^4\ {}^3\mathrm{P}\ 4f$
8	−3.53073	$2s^22p^4\ {}^1\mathrm{D}\ 5d$	17	−2.25504	$2s^22p^4\ {}^1\mathrm{D}\ 6d$	11	−5.75587	$2s^22p^4\ {}^1\mathrm{D}\ 4f$	5	−5.68537	$2s^22p^4\ {}^1\mathrm{D}\ 4f$
9	−3.35606	$2s^22p^4\ {}^1\mathrm{S}\ 5s$	18	−2.23007	$2s2p^5\ {}^3\mathrm{P}\ 4p$	12	−5.06666	$2s2p^5\ {}^1\mathrm{P}\ 3d$	6	−5.15315	$2s2p^5\ {}^1\mathrm{P}\ 3d$
10	−3.26180	$2p^6\ {}^1\mathrm{S}\ 3s$	19	−2.06412	$2s^22p^4\ {}^3\mathrm{P}\ 7d$	13	−4.33619	$2s^22p^4\ {}^3\mathrm{P}\ 5p$	7	−4.94104	$2s^22p^4\ {}^1\mathrm{S}\ 4f$
11	−2.25689	$2s^22p^4\ {}^1\mathrm{D}\ 6d$	20	−1.67526	$2s^22p^4\ {}^1\mathrm{D}\ 7s$	14	−3.79647	$2s^22p^4\ {}^1\mathrm{D}\ 5p$	8	−4.01172	$2s^22p^4\ {}^3\mathrm{P}\ 5f$
12	−2.00288	$2s2p^5\ {}^3\mathrm{P}\ 4p$	21	−1.58549	$2s^22p^4\ {}^3\mathrm{P}\ 8d$	15	−3.41478	$2s^22p^4\ {}^1\mathrm{D}\ 5f$	9	−3.82315	$2s^22p^4\ {}^1\mathrm{D}\ 5p$
13	−1.85005	$2s^22p^4\ {}^1\mathrm{S}\ 6s$	22	−1.55972	$2s^22p^4\ {}^1\mathrm{S}\ 6d$	16	−3.11700	$2s^22p^4\ {}^1\mathrm{S}\ 5p$	10	−3.40692	$2s^22p^4\ {}^1\mathrm{D}\ 5f$
14	−1.49279	$2s^22p^4\ {}^1\mathrm{D}\ 7d$	23	−1.49052	$2s^22p^4\ {}^1\mathrm{D}\ 7d$	17	−2.96269	$2s^22p^4\ {}^3\mathrm{P}\ 6p$	11	−2.78580	$2s^22p^4\ {}^3\mathrm{P}\ 6f$
15	−1.00177	$2s^22p^4\ {}^1\mathrm{D}\ 8d$	24	−1.44791	$2s2p^5\ {}^3\mathrm{P}\ 4f$	18	−2.67202	$2s2p^5\ {}^3\mathrm{P}\ 4s$	12	−2.71934	$2s^22p^4\ {}^1\mathrm{S}\ 5f$
16	−.97530	$2s^22p^4\ {}^1\mathrm{S}\ 7s$	25	−1.24699	$2s^22p^4\ {}^3\mathrm{P}\ 9d$	19	−2.40152	$2s^22p^4\ {}^1\mathrm{D}\ 6p$	13	−2.41955	$2s^22p^4\ {}^1\mathrm{D}\ 6p$
17	−.66370	$2s^22p^4\ {}^1\mathrm{D}\ 9d$	26	−1.12065	$2s^22p^4\ {}^1\mathrm{D}\ 8s$	20	−2.19483	$2s^22p^4\ {}^1\mathrm{D}\ 6f$	14	−2.18715	$2s^22p^4\ {}^1\mathrm{D}\ 6f$
18	−.47570	$2s2p^5\ {}^1\mathrm{P}\ 4p$			**$^2\mathrm{F}^e$**	21	−2.15427	$2s^22p^4\ {}^3\mathrm{P}\ 7p$	15	−2.04640	$2s^22p^4\ {}^3\mathrm{P}\ 7f$
		$^2\mathrm{P}^e$	1	−11.6143	$2s^22p^4\ {}^3\mathrm{P}\ 3d$	22	−2.07050	$2p^6\ {}^1\mathrm{S}\ 3p$	16	−1.64875	$2s2p^5\ {}^3\mathrm{P}\ 4d$
1	−14.4645	$2s^22p^4\ {}^3\mathrm{P}\ 3s$	2	−11.0018	$2s^22p^4\ {}^1\mathrm{D}\ 3d$	23	−1.71504	$2s^22p^4\ {}^1\mathrm{S}\ 6p$	17	−1.59335	$2s^22p^4\ {}^1\mathrm{D}\ 7p$
2	−11.5321	$2s^22p^4\ {}^3\mathrm{P}\ 3d$	3	−6.47819	$2s^22p^4\ {}^3\mathrm{P}\ 4d$	24	−1.64651	$2s^22p^4\ {}^3\mathrm{P}\ 8p$	18	−1.56441	$2s^22p^4\ {}^3\mathrm{P}\ 8f$
3	−10.9321	$2s^22p^4\ {}^1\mathrm{D}\ 3d$	4	−5.88520	$2s^22p^4\ {}^1\mathrm{D}\ 4d$	25	−1.57927	$2s^22p^4\ {}^1\mathrm{D}\ 7p$	19	−1.49437	$2s^22p^4\ {}^1\mathrm{S}\ 6f$
4	−8.32012	$2s2p^5\ {}^3\mathrm{P}\ 3p$	5	−4.11985	$2s^22p^4\ {}^3\mathrm{P}\ 5d$	26	−1.51945	$2s2p^5\ {}^3\mathrm{P}\ 4d$	20	−1.44928	$2s^22p^4\ {}^1\mathrm{D}\ 7f$
5	−7.54369	$2s^22p^4\ {}^3\mathrm{P}\ 4s$	6	−3.52766	$2s^22p^4\ {}^1\mathrm{D}\ 5d$	27	−1.45333	$2s^22p^4\ {}^1\mathrm{D}\ 7f$	21	−1.23684	$2s^22p^4\ {}^3\mathrm{P}\ 9f$
6	−6.58587	$2s2p^5\ {}^1\mathrm{P}\ 3p$	7	−2.84832	$2s^22p^4\ {}^3\mathrm{P}\ 6d$	28	−1.28802	$2s^22p^4\ {}^3\mathrm{P}\ 9p$	22	−1.06610	$2s^22p^4\ {}^1\mathrm{D}\ 8p$
7	−6.41356	$2s^22p^4\ {}^3\mathrm{P}\ 4d$	8	−2.25611	$2s^22p^4\ {}^1\mathrm{D}\ 6d$	29	−1.07240	$2s^22p^4\ {}^1\mathrm{D}\ 8p$			**$^2\mathrm{G}^o$**
8	−5.86727	$2s^22p^4\ {}^1\mathrm{D}\ 4d$	9	−2.08558	$2s^22p^4\ {}^3\mathrm{P}\ 7d$			**$^2\mathrm{D}^o$**	1	−6.25697	$2s^22p^4\ {}^3\mathrm{P}\ 4f$
9	−4.63138	$2s^22p^4\ {}^3\mathrm{P}\ 5s$	10	−1.59299	$2s^22p^4\ {}^3\mathrm{P}\ 8d$	1	−13.2537	$2s^22p^4\ {}^3\mathrm{P}\ 3p$	2	−5.65273	$2s^22p^4\ {}^1\mathrm{D}\ 4f$
10	−4.08327	$2s^22p^4\ {}^3\mathrm{P}\ 5d$	11	−1.49312	$2s^22p^4\ {}^1\mathrm{D}\ 7d$	2	−12.6303	$2s^22p^4\ {}^1\mathrm{D}\ 3p$	3	−4.00802	$2s^22p^4\ {}^3\mathrm{P}\ 5f$
11	−3.52566	$2s^22p^4\ {}^1\mathrm{D}\ 5d$	12	−1.42136	$2s2p^5\ {}^3\mathrm{P}\ 4f$	3	−7.09002	$2s^22p^4\ {}^3\mathrm{P}\ 4p$	4	−3.41022	$2s^22p^4\ {}^1\mathrm{D}\ 5f$
12	−3.13378	$2s^22p^4\ {}^3\mathrm{P}\ 6s$	13	−1.25548	$2s^22p^4\ {}^3\mathrm{P}\ 9d$	4	−6.71365	$2s2p^5\ {}^3\mathrm{P}\ 3d$	5	−2.78348	$2s^22p^4\ {}^3\mathrm{P}\ 6f$
13	−2.82622	$2s^22p^4\ {}^3\mathrm{P}\ 6d$			**$^2\mathrm{G}^e$**	5	−6.47562	$2s^22p^4\ {}^1\mathrm{D}\ 4p$	6	−2.18804	$2s^22p^4\ {}^1\mathrm{D}\ 6f$
14	−2.27410	$2s^22p^4\ {}^3\mathrm{P}\ 7s$	1	−11.1860	$2s^22p^4\ {}^1\mathrm{D}\ 3d$	6	−6.21926	$2s^22p^4\ {}^3\mathrm{P}\ 4f$	7	−2.04466	$2s^22p^4\ {}^3\mathrm{P}\ 7f$
15	−2.25672	$2s^22p^4\ {}^1\mathrm{D}\ 6d$	2	−5.94417	$2s^22p^4\ {}^1\mathrm{D}\ 4d$	7	−5.72279	$2s^22p^4\ {}^1\mathrm{D}\ 4f$	8	−1.56516	$2s^22p^4\ {}^3\mathrm{P}\ 8f$
16	−2.20434	$2s2p^5\ {}^3\mathrm{P}\ 4p$	3	−3.55556	$2s^22p^4\ {}^1\mathrm{D}\ 5d$	8	−4.97627	$2s2p^5\ {}^1\mathrm{P}\ 3d$	9	−1.45024	$2s^22p^4\ {}^1\mathrm{D}\ 7f$
17	−2.06876	$2s^22p^4\ {}^3\mathrm{P}\ 7d$	4	−2.27141	$2s^22p^4\ {}^1\mathrm{D}\ 6d$	9	−4.39729	$2s^22p^4\ {}^3\mathrm{P}\ 5p$	10	−1.23647	$2s^22p^4\ {}^3\mathrm{P}\ 9f$
18	−1.70685	$2s^22p^4\ {}^3\mathrm{P}\ 8s$	5	−1.50272	$2s^22p^4\ {}^1\mathrm{D}\ 7d$	10	−3.99850	$2s^22p^4\ {}^3\mathrm{P}\ 5f$			**$^2\mathrm{H}^o$**
19	−1.58525	$2s^22p^4\ {}^3\mathrm{P}\ 8d$	6	−1.44521	$2s2p^5\ {}^3\mathrm{P}\ 4f$	11	−3.80189	$2s^22p^4\ {}^1\mathrm{D}\ 5p$	1	−5.68669	$2s^22p^4\ {}^1\mathrm{D}\ 4f$
20	−1.49143	$2s^22p^4\ {}^1\mathrm{D}\ 7d$	7	−1.00618	$2s^22p^4\ {}^1\mathrm{D}\ 8d$	12	−3.40920	$2s^22p^4\ {}^1\mathrm{D}\ 5f$	2	−3.42642	$2s^22p^4\ {}^1\mathrm{D}\ 5f$
21	−1.33527	$2s^22p^4\ {}^3\mathrm{P}\ 9s$			**$^2\mathrm{S}^o$**	13	−3.00156	$2s^22p^4\ {}^3\mathrm{P}\ 6p$	3	−2.19708	$2s^22p^4\ {}^1\mathrm{D}\ 6f$
22	−1.24923	$2s^22p^4\ {}^3\mathrm{P}\ 9d$	1	−13.1791	$2s^22p^4\ {}^3\mathrm{P}\ 3p$	14	−2.77837	$2s^22p^4\ {}^3\mathrm{P}\ 6f$	4	−1.45588	$2s^22p^4\ {}^1\mathrm{D}\ 7f$
		$^2\mathrm{D}^e$	2	−7.05249	$2s^22p^4\ {}^3\mathrm{P}\ 4p$	15	−2.40707	$2s^22p^4\ {}^1\mathrm{D}\ 6p$			**$^1\mathrm{S}^o$**
1	−14.0123	$2s^22p^4\ {}^1\mathrm{D}\ 3s$	3	−4.38873	$2s^22p^4\ {}^3\mathrm{P}\ 5p$	16	−2.19051	$2s^22p^4\ {}^1\mathrm{D}\ 6f$	1	−8.71664	$2s2p^5\ {}^3\mathrm{P}\ 3p$
2	−11.4972	$2s^22p^4\ {}^3\mathrm{P}\ 3d$	4	−2.99624	$2s^22p^4\ {}^3\mathrm{P}\ 6p$	17	−2.17977	$2s^22p^4\ {}^3\mathrm{P}\ 7p$	2	−2.34462	$2s2p^5\ {}^3\mathrm{P}\ 4p$
3	−10.9276	$2s^22p^4\ {}^1\mathrm{D}\ 3d$	5	−2.17600	$2s^22p^4\ {}^3\mathrm{P}\ 7p$	18	−2.04168	$2s^22p^4\ {}^3\mathrm{P}\ 7f$			**$^4\mathrm{P}^e$**
4	−10.3585	$2s^22p^4\ {}^1\mathrm{S}\ 3d$	6	−1.65203	$2s^22p^4\ {}^3\mathrm{P}\ 8p$	19	−1.66544	$2s^22p^4\ {}^3\mathrm{P}\ 8p$	1	−14.6758	$2s^22p^4\ {}^3\mathrm{P}\ 3s$
5	−8.35627	$2s2p^5\ {}^3\mathrm{P}\ 3p$	7	−1.29694	$2s^22p^4\ {}^3\mathrm{P}\ 9p$	20	−1.61656	$2s2p^5\ {}^3\mathrm{P}\ 4d$	2	−11.6332	$2s^22p^4\ {}^3\mathrm{P}\ 3d$
6	−7.04201	$2s^22p^4\ {}^1\mathrm{D}\ 4s$			**$^2\mathrm{P}^o$**	21	−1.58542	$2s^22p^4\ {}^1\mathrm{D}\ 7p$	3	−8.40553	$2s2p^5\ {}^3\mathrm{P}\ 3p$
7	−[illegible].[illegible]5009	$2s2p^5\ {}^1\mathrm{P}\ 3p$	1	−35.2248	$2s^22p^5$	22	−1.56156	$2s^22p^4\ {}^3\mathrm{P}\ 8f$	4	−7.59053	$2s^22p^4\ {}^3\mathrm{P}\ 4s$
8	−6.38799	$2s^22p^4\ {}^3\mathrm{P}\ 4d$	2	−13.2315	$2s^22p^4\ {}^3\mathrm{P}\ 3p$	23	−1.45186	$2s^22p^4\ {}^1\mathrm{D}\ 7f$	5	−6.48948	$2s^22p^4\ {}^3\mathrm{P}\ 4d$

F-like Ar (Ar^{9+})

i	E(Ryds)	Description
6	−4.66253	$2s^22p^4\,^3$P $5s$
7	−4.12636	$2s^22p^4\,^3$P $5d$
8	−3.15248	$2s^22p^4\,^3$P $6s$
9	−2.85247	$2s^22p^4\,^3$P $6d$
10	−2.30076	$2s^22p^4\,^3$P $7s$
11	−2.21146	$2s2p^5\,^3$P $4p$
12	−2.08677	$2s^22p^4\,^3$P $7d$
13	−1.71219	$2s^22p^4\,^3$P $8s$
14	−1.59399	$2s^22p^4\,^3$P $8d$
15	−1.33940	$2s^22p^4\,^3$P $9s$
16	−1.25675	$2s^22p^4\,^3$P $9d$
	$^4\mathbf{D}^e$	
1	−11.9012	$2s^22p^4\,^3$P $3d$
2	−8.51631	$2s2p^5\,^3$P $3p$
3	−6.57690	$2s^22p^4\,^3$P $4d$
4	−4.16825	$2s^22p^4\,^3$P $5d$
5	−2.87586	$2s^22p^4\,^3$P $6d$
6	−2.28423	$2s2p^5\,^3$P $4p$
7	−2.10055	$2s^22p^4\,^3$P $7d$
8	−1.60324	$2s^22p^4\,^3$P $8d$
9	−1.45911	$2s2p^5\,^3$P $4f$
10	−1.26308	$2s^22p^4\,^3$P $9d$
	$^4\mathbf{F}^e$	
1	−11.7194	$2s^22p^4\,^3$P $3d$
2	−6.52025	$2s^22p^4\,^3$P $4d$
3	−4.14073	$2s^22p^4\,^3$P $5d$
4	−2.86013	$2s^22p^4\,^3$P $6d$
5	−2.09290	$2s^22p^4\,^3$P $7d$
6	−1.59752	$2s^22p^4\,^3$P $8d$
7	−1.42424	$2s2p^5\,^3$P $4f$
8	−1.25900	$2s^22p^4\,^3$P $9d$
	$^4\mathbf{G}^e$	
1	−1.45139	$2s2p^5\,^3$P $4f$
	$^4\mathbf{S}^o$	
1	−13.1717	$2s^22p^4\,^3$P $3p$
2	−7.04986	$2s^22p^4\,^3$P $4p$
3	−4.38749	$2s^22p^4\,^3$P $5p$
4	−2.99556	$2s^22p^4\,^3$P $6p$
5	−2.17559	$2s^22p^4\,^3$P $7p$
6	−1.65175	$2s^22p^4\,^3$P $8p$
7	−1.29674	$2s^22p^4\,^3$P $9p$
	$^4\mathbf{P}^o$	
1	−13.5392	$2s^22p^4\,^3$P $3p$
2	−9.76121	$2s2p^5\,^3$P $3s$
3	−7.19458	$2s^22p^4\,^3$P $4p$
4	−7.04299	$2s2p^5\,^3$P $3d$
5	−4.44004	$2s^22p^4\,^3$P $5p$
6	−3.03508	$2s^22p^4\,^3$P $6p$
7	−2.76293	$2s2p^5\,^3$P $4s$
8	−2.19041	$2s^22p^4\,^3$P $7p$
9	−1.76241	$2s2p^5\,^3$P $4d$
10	−1.65909	$2s^22p^4\,^3$P $8p$
11	−1.30362	$2s^22p^4\,^3$P $9p$
	$^4\mathbf{D}^o$	
1	−13.3726	$2s^22p^4\,^3$P $3p$
2	−7.14232	$2s^22p^4\,^3$P $4p$
3	−6.83634	$2s2p^5\,^3$P $3d$
4	−6.21866	$2s^22p^4\,^3$P $4f$
5	−4.41665	$2s^22p^4\,^3$P $5p$
6	−4.00111	$2s^22p^4\,^3$P $5f$
7	−3.01247	$2s^22p^4\,^3$P $6p$
8	−2.78058	$2s^22p^4\,^3$P $6f$
9	−2.18701	$2s^22p^4\,^3$P $7p$
10	−2.04380	$2s^22p^4\,^3$P $7f$
11	−1.69404	$2s2p^5\,^3$P $4d$
12	−1.64392	$2s^22p^4\,^3$P $8p$
13	−1.56145	$2s^22p^4\,^3$P $8f$
14	−1.30028	$2s^22p^4\,^3$P $9p$
15	−1.23520	$2s^22p^4\,^3$P $9f$
	$^4\mathbf{F}^o$	
1	−7.04498	$2s2p^5\,^3$P $3d$
2	−6.24920	$2s^22p^4\,^3$P $4f$
3	−4.01780	$2s^22p^4\,^3$P $5f$
4	−2.79049	$2s^22p^4\,^3$P $6f$
5	−2.05072	$2s^22p^4\,^3$P $7f$
6	−1.73557	$2s2p^5\,^3$P $4d$
7	−1.56557	$2s^22p^4\,^3$P $8f$
8	−1.23796	$2s^22p^4\,^3$P $9f$
	$^4\mathbf{G}^o$	
1	−6.26009	$2s^22p^4\,^3$P $4f$
2	−4.01026	$2s^22p^4\,^3$P $5f$
3	−2.78495	$2s^22p^4\,^3$P $6f$
4	−2.04565	$2s^22p^4\,^3$P $7f$
5	−1.56585	$2s^22p^4\,^3$P $8f$
6	−1.23697	$2s^22p^4\,^3$P $9f$

Energies in ascending order from ground state for terms with effective $n \leq 4.0$, $L \leq 4$

Term	i	E(Ryds)	Term	i	E(Ryds)	Term	i	E(Ryds)	Term	i	E(Ryds)	Term	i	E(Ryds)
$^2\mathbf{P}^o$	1	0.00000	$^2\mathbf{P}^o$	3	22.9121	$^4\mathbf{P}^o$	2	25.4635	$^2\mathbf{S}^o$	2	28.1723	$^4\mathbf{F}^e$	2	28.7045
$^2\mathbf{S}^e$	1	5.31540	$^2\mathbf{P}^o$	4	23.2436	$^2\mathbf{P}^o$	5	25.7817	$^4\mathbf{S}^o$	2	28.1749	$^4\mathbf{P}^e$	5	28.7353
$^4\mathbf{P}^e$	1	20.5490	$^4\mathbf{D}^e$	1	23.3236	$^4\mathbf{S}^e$	1	26.5081	$^4\mathbf{F}^o$	1	28.1798	$^2\mathbf{P}^o$	8	28.7387
$^2\mathbf{P}^e$	1	20.7603	$^4\mathbf{F}^e$	1	23.5054	$^4\mathbf{D}^e$	2	26.7084	$^4\mathbf{P}^o$	4	28.1818	$^2\mathbf{F}^e$	3	28.7466
$^2\mathbf{D}^e$	1	21.2125	$^4\mathbf{P}^e$	2	23.5916	$^4\mathbf{P}^e$	3	26.8192	$^2\mathbf{D}^e$	6	28.1827	$^2\mathbf{D}^o$	5	28.7491
$^4\mathbf{P}^o$	1	21.6856	$^2\mathbf{F}^e$	1	23.6105	$^2\mathbf{D}^e$	5	26.8685	$^2\mathbf{P}^o$	7	28.2622	$^2\mathbf{P}^o$	9	28.8087
$^4\mathbf{D}^o$	1	21.8522	$^2\mathbf{P}^e$	2	23.6927	$^2\mathbf{P}^e$	4	26.9046	$^4\mathbf{D}^o$	3	28.3884	$^2\mathbf{P}^e$	7	28.8112
$^2\mathbf{S}^e$	2	21.9182	$^2\mathbf{D}^e$	2	23.7276	$^2\mathbf{S}^e$	4	27.2496	$^2\mathbf{F}^o$	2	28.4105	$^2\mathbf{D}^e$	8	28.8368
$^2\mathbf{D}^o$	1	21.9711	$^2\mathbf{G}^e$	1	24.0388	$^2\mathbf{P}^o$	6	27.3045	$^2\mathbf{D}^o$	4	28.5111	$^4\mathbf{G}^o$	1	28.9647
$^2\mathbf{P}^o$	2	21.9933	$^2\mathbf{S}^e$	3	24.1724	$^4\mathbf{P}^e$	4	27.6342	$^2\mathbf{D}^e$	7	28.5681	$^2\mathbf{G}^o$	1	28.9678
$^2\mathbf{S}^o$	1	22.0457	$^2\mathbf{F}^e$	2	24.2230	$^2\mathbf{P}^e$	5	27.6811	$^2\mathbf{P}^e$	6	28.6389			
$^4\mathbf{S}^o$	1	22.0531	$^2\mathbf{P}^e$	3	24.2927	$^4\mathbf{P}^o$	3	28.0302	$^4\mathbf{D}^e$	3	28.6479			
$^2\mathbf{F}^o$	1	22.4334	$^2\mathbf{D}^e$	3	24.2972	$^4\mathbf{D}^o$	2	28.0824	$^2\mathbf{S}^e$	5	28.6705			
$^2\mathbf{D}^o$	2	22.5945	$^2\mathbf{D}^e$	4	24.8663	$^2\mathbf{D}^o$	3	28.1347	$^2\mathbf{F}^o$	3	28.6937			

F-like Ar (Ar^{9+})

gf-values for transitions involving terms with effective $n \leq 4.0$, $L \leq 4$

$i\ i'$	gf_L	$i\ i'$	gf_L	$i\ i'$	gf_L	$i\ i'$	gf_L	$i\ i'$	gf_L	$i\ i'$	gf_L
	$^2S^e$–$^2P^o$	3 1	2.43E−1	5 5	2.86E−2	6 3	2.09E−3	5 7	−7.12E−3	5 1	9.84E−1
1 1	3.88E−1	3 2	−7.43E−2	5 6	1.95E−2	6 4	4.34E−4	5 8	−1.24E−2	5 2	3.72E−5
1 2	−1.77E−3	4 1	2.38E−1	5 7	−1.41E+0	6 5	−1.09E−3	5 9	−3.61E−3	5 3	−3.47E−2
1 3	−1.49E−2	4 2	−3.29E−3	5 8	−3.01E−2	7 1	1.96E−2	6 1	2.12E−1	5 4	−4.72E−1
1 4	−5.26E−3	5 1	1.44E−1	5 9	−3.10E−1	7 2	4.40E−2	6 2	1.17E−1	5 5	−6.40E−3
1 5	−3.54E−1	5 2	−4.09E−1	6 1	1.73E−1	7 3	1.11E−1	6 3	4.26E−1	6 1	9.99E−4
1 6	−1.46E−1	6 1	4.60E−3	6 2	6.13E−2	7 4	2.12E−2	6 4	6.66E−2	6 2	3.29E−1
1 7	−6.06E−2	6 2	1.14E−3	6 3	4.24E−1	7 5	1.69E−3	6 5	1.56E−2	6 3	7.57E−6
1 8	−2.69E+0	7 1	4.42E−1	6 4	1.06E−1		**$^2D^e$–$^2P^o$**	6 6	2.06E−1	6 4	−1.70E−4
1 9	−6.48E−2	7 2	1.16E+0	6 5	2.65E−3	1 1	4.06E−1	6 7	−7.48E−3	6 5	−2.14E+0
2 1	7.61E−2		**$^2P^e$–$^2P^o$**	6 6	9.66E−1	1 2	−1.83E−1	6 8	−3.71E−2	7 1	2.56E−3
2 2	−9.19E−4	1 1	6.53E−1	6 7	2.17E−2	1 3	−7.66E−1	6 9	−1.37E+0	7 2	7.51E−1
2 3	−1.40E−2	1 2	−5.10E−1	6 8	−3.80E−4	1 4	−3.18E−3	7 1	1.54E−1	7 3	1.06E−4
2 4	−8.95E−1	1 3	−6.92E−1	6 9	−1.44E−3	1 5	−3.65E−2	7 2	5.35E−4	7 4	2.82E−5
2 5	−7.29E−3	1 4	−9.51E−2	7 1	3.14E−1	1 6	−1.69E+0	7 3	1.03E−2	7 5	−1.07E−1
2 6	−1.53E−1	1 5	−5.40E−1	7 2	3.53E−1	1 7	−8.07E−3	7 4	2.13E−1	8 1	2.72E−1
2 7	−4.15E−4	1 6	−1.10E−2	7 3	4.86E−2	1 8	−5.10E−3	7 5	4.65E−3	8 2	6.92E−2
2 8	−2.48E−3	1 7	−2.66E−1	7 4	2.77E−2	1 9	−2.03E−1	7 6	1.23E+0	8 3	8.26E−1
2 9	−2.74E−5	1 8	−4.46E−4	7 5	1.82E−3	2 1	2.43E+0	7 7	2.22E−2	8 4	5.55E−2
3 1	9.81E−1	1 9	−6.18E−3	7 6	5.99E−2	2 2	1.20E+0	7 8	−3.15E−3	8 5	5.21E−3
3 2	1.69E−1	2 1	3.18E−1	7 7	6.54E−1	2 3	1.03E−1	7 9	−6.67E−2		**$^2D^e$–$^2F^o$**
3 3	1.87E−1	2 2	5.34E−1	7 8	1.43E−3	2 4	5.40E−4	8 1	1.23E+0	1 1	−1.98E+0
3 4	3.07E−3	2 3	6.52E−4	7 9	1.08E−5	2 5	−6.61E−3	8 2	7.61E−1	1 2	−5.19E−6
3 5	−5.34E−3	2 4	2.50E−3		**$^2P^e$–$^2D^o$**	2 6	−1.77E−2	8 3	4.05E−1	1 3	−9.09E−1
3 6	−9.35E−5	2 5	3.78E−4	1 1	1.43E+0	2 7	4.02E−1	8 4	2.04E−2	2 1	1.07E−2
3 7	−1.46E−2	2 6	−6.03E−3	1 2	−1.31E−2	2 8	−4.14E−1	8 5	1.29E−2	2 2	−4.30E−2
3 8	−5.83E−3	2 7	−6.32E−2	1 3	−6.12E−1	2 9	−2.21E−2	8 6	1.04E−1	2 3	−7.46E−5
3 9	−7.23E−2	2 8	−2.36E−1	1 4	−2.07E−2	3 1	4.34E+0	8 7	2.19E+0	3 1	6.70E−2
4 1	1.98E−1	2 9	−3.03E−3	1 5	−7.23E−8	3 2	4.29E−3	8 8	2.92E−3	3 2	−9.88E−3
4 2	3.06E−1	3 1	3.40E+0	2 1	4.55E−2	3 3	6.37E−1	8 9	3.77E−3	3 3	−5.59E−3
4 3	1.02E−3	3 2	1.09E−1	2 2	4.10E−2	3 4	3.02E−2		**$^2D^e$–$^2D^o$**	4 1	1.39E−2
4 4	2.74E−4	3 3	5.89E−1	2 3	−8.01E−3	3 5	−1.64E−2	1 1	−2.94E−3	4 2	−6.05E−2
4 5	3.19E−1	3 4	6.90E−3	2 4	−1.25E−3	3 6	−2.99E−3	1 2	−1.61E+0	4 3	−3.54E−3
4 6	−1.40E−3	3 5	−1.24E−2	2 5	−6.18E−3	3 7	−2.93E−2	1 3	−1.09E−3	5 1	3.81E−3
4 7	−3.72E−3	3 6	−4.02E−4	3 1	4.51E−2	3 8	−6.17E−2	1 4	−1.56E−4	5 2	−2.52E+0
4 8	−5.47E−1	3 7	−2.23E−3	3 2	2.13E−1	3 9	−1.01E−1	1 5	−5.40E−1	5 3	−6.20E−4
4 9	−3.28E−2	3 8	−4.74E−3	3 3	−1.16E−2	4 1	9.00E−1	2 1	3.77E−1	6 1	2.03E+0
5 1	1.49E−2	3 9	−1.87E−1	3 4	−1.92E−3	4 2	1.36E−5	2 2	1.47E−1	6 2	−8.63E−4
5 2	8.43E−2	4 1	5.15E−1	3 5	−5.61E−2	4 3	7.46E−2	2 3	−1.10E−1	6 3	−2.89E+0
5 3	2.57E−1	4 2	1.37E−1	4 1	3.74E−1	4 4	1.89E+0	2 4	−6.68E−3	7 1	1.25E+0
5 4	2.66E−1	4 3	1.93E−1	4 2	4.06E−4	4 5	−3.61E−3	2 5	−3.31E−2	7 2	3.52E−5
5 5	1.70E−1	4 4	4.02E−2	4 3	−1.86E−1	4 6	−5.70E−4	3 1	1.84E−1	7 3	−6.76E−2
5 6	2.36E−1	4 5	7.91E−1	4 4	−1.27E+0	4 7	−3.22E−3	3 2	1.09E+0	8 1	9.25E−3
5 7	2.24E−4	4 6	−5.91E−5	4 5	−1.22E−3	4 8	−2.79E−5	3 3	−4.70E−2	8 2	9.01E−2
5 8	−1.42E−3	4 7	−7.63E−2	5 1	5.76E−1	4 9	−5.05E−3	3 4	−1.32E−3	8 3	2.55E−3
5 9	−5.87E−5	4 8	−6.23E−1	5 2	1.52E−3	5 1	7.05E−1	3 5	−3.14E−1		**$^2F^e$–$^2D^o$**
	$^2P^e$–$^2S^o$	4 9	−7.83E−5	5 3	−1.86E+0	5 2	2.13E−1	4 1	6.76E−3	1 1	2.80E+0
1 1	−2.98E−1	5 1	6.97E−2	5 4	−2.58E−1	5 3	2.59E−1	4 2	2.18E−2	1 2	2.43E−4
1 2	−1.16E−1	5 2	2.48E−1	5 5	−3.42E−3	5 4	6.03E−2	4 3	−2.94E−3	1 3	−9.08E−1
2 1	5.54E−1	5 3	2.46E−1	6 1	5.11E−3	5 5	1.32E+0	4 4	−3.98E−5	1 4	−3.76E−1
2 2	−1.43E−1	5 4	3.42E−2	6 2	1.22E+0	5 6	−5.88E−5	4 5	−5.18E−4	1 5	−2.89E−4

F-like Ar (Ar^{9+})

$i\ i'$	gf_L	$i\ i'$	gf_L	$i\ i'$	gf_L	$i\ i'$	gf_L	$i\ i'$	gf_L	$i\ i'$	gf_L
2 1	2.50E−3		$^{2}F^{e}–^{2}G^{o}$	3 2	−4.79E−2	4 4	−9.52E−1		$^{4}D^{e}–^{4}P^{o}$	3 2	1.23E+0
2 2	1.88E+0	1 1	−1.17E+1	4 1	4.33E−1	5 1	6.79E−1	1 1	2.72E+0	3 3	1.24E−1
2 3	−3.26E−3	2 1	−7.11E−2	4 2	−9.27E−1	5 2	2.26E−2	1 2	−4.48E−3		$^{4}D^{e}–^{4}F^{o}$
2 4	−1.04E−6	3 1	−1.50E+0	5 1	1.09E+0	5 3	1.28E+0	1 3	−1.54E+0	1 1	−2.15E−2
2 5	−4.67E−1		$^{2}G^{e}–^{2}F^{o}$	5 2	2.02E+0	5 4	3.93E−1	1 4	−2.29E−1	2 1	−4.66E+0
3 1	2.21E+0	1 1	3.37E+0		$^{4}P^{e}–^{4}P^{o}$		$^{4}P^{e}–^{4}D^{o}$	2 1	4.33E−1	3 1	1.45E−1
3 2	3.75E−3	1 2	−3.83E−3	1 1	−1.46E+0	1 1	−2.99E+0	2 2	3.14E+0		$^{4}F^{e}–^{4}D^{o}$
3 3	4.35E+0	1 3	−9.82E−1	1 2	−1.80E+0	1 2	−1.09E+0	2 3	−1.13E−2	1 1	5.50E+0
3 4	7.48E−2		$^{2}G^{e}–^{2}G^{o}$	1 3	−4.90E−1	1 3	−1.27E−1	2 4	−1.57E−2	1 2	−1.99E+0
3 5	0.00E+0	1 1	−1.03E−5	1 4	−1.93E−1	2 1	1.22E−1	3 1	2.44E+0	1 3	−5.46E−1
	$^{2}F^{e}–^{2}F^{o}$		$^{4}S^{e}–^{4}P^{o}$	2 1	1.05E+0	2 2	−1.60E−2	3 2	7.75E−2	2 1	4.50E+0
1 1	3.58E−3	1 1	5.82E−1	2 2	−9.56E−4	2 3	−1.18E−2	3 3	3.45E+0	2 2	8.33E+0
1 2	−5.28E−1	1 2	4.73E−1	2 3	−8.01E−3	3 1	1.06E+0	3 4	9.78E−1	2 3	4.08E−1
1 3	−4.33E−4	1 3	−3.72E−1	2 4	−5.76E−1	3 2	−9.83E−1		$^{4}D^{e}–^{4}D^{o}$		$^{4}F^{e}–^{4}F^{o}$
2 1	1.04E+0	1 4	−8.83E−1	3 1	4.90E−1	3 3	−1.90E+0	1 1	7.76E−1	1 1	−1.09E+0
2 2	−3.00E−4		$^{4}P^{e}–^{4}S^{o}$	3 2	2.13E+0	4 1	1.02E+0	1 2	−2.09E−1	2 1	1.78E−2
2 3	−2.20E−1	1 1	−7.18E−1	3 3	−6.16E−1	4 2	−3.28E+0	1 3	−1.11E−2		$^{4}F^{e}–^{4}G^{o}$
3 1	1.67E−3	1 2	−1.90E−1	3 4	−3.08E−1	4 3	−9.81E−1	2 1	1.87E+0	1 1	−2.30E+1
3 2	7.27E−3	2 1	1.28E+0	4 1	6.41E−1	5 1	5.19E−2	2 2	−1.05E−1	2 1	−3.56E+0
3 3	9.74E−5	2 2	−3.21E−1	4 2	1.52E−2	5 2	2.03E−1	2 3	−8.95E−1		
		3 1	3.44E−1	4 3	−1.36E+0	5 3	3.44E−2	3 1	8.56E−1		

F-like Ca (Ca^{11+})

Term energies relative to $2s^22p^4\ ^3\mathrm{P}$ ionization threshold for each symmetry

i	E(Ryds)	Description	*i*	E(Ryds)	Description	*i*	E(Ryds)	Description	*i*	E(Ryds)	Description
		$^2\mathrm{S}^e$	6	−11.0596	$2s2p^5\ ^1\mathrm{P}\ 3p$	3	−6.24279	$2s^22p^4\ ^3\mathrm{P}\ 5p$	12	−5.07074	$2s^22p^4\ ^1\mathrm{D}\ 5f$
1	−42.1745	$2s2p^6$	7	−9.92948	$2s^22p^4\ ^1\mathrm{D}\ 4s$	4	−4.27167	$2s^22p^4\ ^3\mathrm{P}\ 6p$	13	−4.28021	$2s^22p^4\ ^3\mathrm{P}\ 6p$
2	−18.7294	$2s^22p^4\ ^1\mathrm{S}\ 3s$	8	−9.19058	$2s^22p^4\ ^3\mathrm{P}\ 4d$	5	−3.10707	$2s^22p^4\ ^3\mathrm{P}\ 7p$	14	−4.00251	$2s^22p^4\ ^3\mathrm{P}\ 6f$
3	−16.0225	$2s^22p^4\ ^1\mathrm{D}\ 3d$	9	−8.58032	$2s^22p^4\ ^1\mathrm{D}\ 4d$	6	−2.36154	$2s^22p^4\ ^3\mathrm{P}\ 8p$	15	−3.64789	$2s2p^5\ ^3\mathrm{P}\ 4d$
4	−12.4791	$2s2p^5\ ^3\mathrm{P}\ 3p$	10	−7.80782	$2s^22p^4\ ^1\mathrm{S}\ 4d$	7	−1.85554	$2s^22p^4\ ^3\mathrm{P}\ 9p$	16	−3.57438	$2s^22p^4\ ^1\mathrm{D}\ 6p$
5	−10.7808	$2s2p^5\ ^1\mathrm{P}\ 3p$	11	−5.87236	$2s^22p^4\ ^1\mathrm{D}\ 5s$			**$^2\mathrm{P}^o$**	17	−3.30789	$2s^22p^4\ ^1\mathrm{D}\ 6f$
6	−9.11501	$2s^22p^4\ ^1\mathrm{S}\ 4s$	12	−5.84885	$2s^22p^4\ ^3\mathrm{P}\ 5d$	1	−48.3134	$2s^22p^5$	18	−3.10863	$2s^22p^4\ ^3\mathrm{P}\ 7p$
7	−8.60044	$2s^22p^4\ ^1\mathrm{D}\ 4d$	13	−5.21256	$2s^22p^4\ ^1\mathrm{D}\ 5d$	2	−18.6236	$2s^22p^4\ ^3\mathrm{P}\ 3p$	19	−2.94017	$2s^22p^4\ ^3\mathrm{P}\ 7f$
8	−6.93594	$2p^6\ ^1\mathrm{S}\ 3s$	14	−4.41138	$2s^22p^4\ ^1\mathrm{S}\ 5d$	3	−17.5138	$2s^22p^4\ ^1\mathrm{D}\ 3p$	20	−2.40844	$2s^22p^4\ ^1\mathrm{D}\ 7p$
9	−5.22403	$2s^22p^4\ ^1\mathrm{D}\ 5d$	15	−4.37040	$2s2p^5\ ^3\mathrm{P}\ 4p$	4	−17.1421	$2s^22p^4\ ^1\mathrm{S}\ 3p$	21	−2.36328	$2s^22p^4\ ^3\mathrm{P}\ 8p$
10	−5.03555	$2s^22p^4\ ^1\mathrm{S}\ 5s$	16	−4.05400	$2s^22p^4\ ^3\mathrm{P}\ 6d$	5	−14.2220	$2s2p^5\ ^3\mathrm{P}\ 3s$	22	−2.25160	$2s^22p^4\ ^3\mathrm{P}\ 8f$
11	−4.11349	$2s2p^5\ ^3\mathrm{P}\ 4p$	17	−3.82877	$2p^6\ ^1\mathrm{S}\ 3d$	6	−12.4871	$2s2p^5\ ^1\mathrm{P}\ 3s$	23	−2.24456	$2s^22p^4\ ^1\mathrm{D}\ 7f$
12	−3.39396	$2s^22p^4\ ^1\mathrm{D}\ 6d$	18	−3.75181	$2s^22p^4\ ^1\mathrm{D}\ 6s$	7	−10.6874	$2s2p^5\ ^3\mathrm{P}\ 3d$	24	−1.85697	$2s^22p^4\ ^3\mathrm{P}\ 9p$
13	−2.93076	$2s^22p^4\ ^1\mathrm{S}\ 6s$	19	−3.40084	$2s2p^5\ ^3\mathrm{P}\ 4f$	8	−9.88867	$2s^22p^4\ ^3\mathrm{P}\ 4p$	25	−1.77884	$2s^22p^4\ ^3\mathrm{P}\ 9f$
14	−2.35751	$2s2p^5\ ^1\mathrm{P}\ 4p$	20	−3.38987	$2s^22p^4\ ^1\mathrm{D}\ 6d$	9	−9.29437	$2s2p^5\ ^1\mathrm{P}\ 3d$	26	−1.68559	$2s2p^5\ ^1\mathrm{P}\ 4d$
15	−2.28640	$2s^22p^4\ ^1\mathrm{D}\ 7d$	21	−2.97454	$2s^22p^4\ ^3\mathrm{P}\ 7d$	10	−9.23883	$2s^22p^4\ ^1\mathrm{D}\ 4p$	27	−1.65087	$2s^22p^4\ ^1\mathrm{D}\ 8p$
16	−1.69198	$2s^22p^4\ ^1\mathrm{S}\ 7s$	22	−2.57944	$2s^22p^4\ ^1\mathrm{S}\ 6d$	11	−8.50388	$2s^22p^4\ ^1\mathrm{S}\ 4p$	28	−1.54642	$2s^22p^4\ ^1\mathrm{D}\ 8f$
17	−1.58745	$2s^22p^4\ ^1\mathrm{D}\ 8d$	23	−2.53025	$2s^22p^4\ ^1\mathrm{D}\ 7s$	12	−8.26393	$2s^22p^4\ ^1\mathrm{D}\ 4f$			**$^2\mathrm{F}^o$**
18	−1.10384	$2s^22p^4\ ^1\mathrm{D}\ 9d$	24	−2.36744	$2s2p^5\ ^1\mathrm{P}\ 4p$	13	−6.18311	$2s^22p^4\ ^3\mathrm{P}\ 5p$	1	−18.1019	$2s^22p^4\ ^1\mathrm{D}\ 3p$
19	−.91051	$2s^22p^4\ ^1\mathrm{S}\ 8s$	25	−2.29911	$2s^22p^4\ ^1\mathrm{D}\ 7d$	14	−5.54648	$2s^22p^4\ ^1\mathrm{D}\ 5p$	2	−11.0716	$2s2p^5\ ^3\mathrm{P}\ 3d$
		$^2\mathrm{P}^e$	26	−2.26703	$2s^22p^4\ ^3\mathrm{P}\ 8d$	15	−5.52610	$2p^6\ ^1\mathrm{S}\ 3p$	3	−9.45290	$2s^22p^4\ ^1\mathrm{D}\ 4p$
1	−20.0959	$2s^22p^4\ ^3\mathrm{P}\ 3s$	27	−1.79568	$2s^22p^4\ ^3\mathrm{P}\ 9d$	16	−5.08033	$2s^22p^4\ ^1\mathrm{D}\ 5f$	4	−9.18067	$2s2p^5\ ^1\mathrm{P}\ 3d$
2	−16.5911	$2s^22p^4\ ^3\mathrm{P}\ 3d$	28	−1.73131	$2s^22p^4\ ^1\mathrm{D}\ 8s$	17	−4.89870	$2s2p^5\ ^3\mathrm{P}\ 4s$	5	−9.01911	$2s^22p^4\ ^3\mathrm{P}\ 4f$
3	−15.8441	$2s^22p^4\ ^1\mathrm{D}\ 3d$	29	−1.58725	$2s^22p^4\ ^1\mathrm{D}\ 8d$	18	−4.73189	$2s^22p^4\ ^1\mathrm{S}\ 5p$	6	−8.27456	$2s^22p^4\ ^1\mathrm{D}\ 4f$
4	−12.8780	$2s2p^5\ ^3\mathrm{P}\ 3p$	30	−1.47989	$2s^22p^4\ ^1\mathrm{S}\ 7d$	19	−4.23566	$2s^22p^4\ ^3\mathrm{P}\ 6p$	7	−7.49073	$2s^22p^4\ ^1\mathrm{S}\ 4f$
5	−10.9126	$2s2p^5\ ^1\mathrm{P}\ 3p$			**$^2\mathrm{F}^e$**	20	−3.57539	$2s^22p^4\ ^1\mathrm{D}\ 6p$	8	−5.77839	$2s^22p^4\ ^3\mathrm{P}\ 5f$
6	−10.5865	$2s^22p^4\ ^3\mathrm{P}\ 4s$	1	−16.6824	$2s^22p^4\ ^3\mathrm{P}\ 3d$	21	−3.51403	$2s2p^5\ ^3\mathrm{P}\ 4d$	9	−5.57638	$2s^22p^4\ ^1\mathrm{D}\ 5p$
7	−9.22478	$2s^22p^4\ ^3\mathrm{P}\ 4d$	2	−15.9662	$2s^22p^4\ ^1\mathrm{D}\ 3d$	22	−3.31319	$2s^22p^4\ ^1\mathrm{D}\ 6f$	10	−5.06323	$2s^22p^4\ ^1\mathrm{D}\ 5f$
8	−8.58122	$2s^22p^4\ ^1\mathrm{D}\ 4d$	3	−9.30387	$2s^22p^4\ ^3\mathrm{P}\ 4d$	23	−3.08739	$2s^22p^4\ ^3\mathrm{P}\ 7p$	11	−4.25280	$2s^22p^4\ ^1\mathrm{S}\ 5f$
9	−6.54077	$2s^22p^4\ ^3\mathrm{P}\ 5s$	4	−8.60450	$2s^22p^4\ ^1\mathrm{D}\ 4d$	24	−2.95055	$2s2p^5\ ^1\mathrm{P}\ 4s$	12	−4.01227	$2s^22p^4\ ^3\mathrm{P}\ 6f$
10	−5.87459	$2s^22p^4\ ^3\mathrm{P}\ 5d$	5	−5.91964	$2s^22p^4\ ^3\mathrm{P}\ 5d$	25	−2.75732	$2s^22p^4\ ^1\mathrm{S}\ 6p$	13	−3.67240	$2s2p^5\ ^3\mathrm{P}\ 4d$
11	−5.21535	$2s^22p^4\ ^1\mathrm{D}\ 5d$	6	−5.21965	$2s^22p^4\ ^1\mathrm{D}\ 5d$	26	−2.40985	$2s^22p^4\ ^1\mathrm{D}\ 7p$	14	−3.59033	$2s^22p^4\ ^1\mathrm{D}\ 6p$
12	−4.45858	$2s^22p^4\ ^3\mathrm{P}\ 6s$	7	−4.09404	$2s^22p^4\ ^3\mathrm{P}\ 6d$	27	−2.33795	$2s^22p^4\ ^3\mathrm{P}\ 8p$	15	−3.30296	$2s^22p^4\ ^1\mathrm{D}\ 6f$
13	−4.33499	$2s2p^5\ ^3\mathrm{P}\ 4p$	8	−3.39354	$2s^22p^4\ ^1\mathrm{D}\ 6d$	28	−2.24968	$2s^22p^4\ ^1\mathrm{D}\ 7f$	16	−2.94593	$2s^22p^4\ ^3\mathrm{P}\ 7f$
14	−4.06514	$2s^22p^4\ ^3\mathrm{P}\ 6d$	9	−3.37254	$2s2p^5\ ^3\mathrm{P}\ 4f$	29	−1.84537	$2s^22p^4\ ^3\mathrm{P}\ 9p$	17	−2.48764	$2s^22p^4\ ^1\mathrm{S}\ 6f$
15	−3.39360	$2s^22p^4\ ^1\mathrm{D}\ 6d$	10	−2.99769	$2s^22p^4\ ^3\mathrm{P}\ 7d$	30	−1.71427	$2s2p^5\ ^1\mathrm{P}\ 4d$	18	−2.41915	$2s^22p^4\ ^1\mathrm{D}\ 7p$
16	−3.20986	$2s^22p^4\ ^3\mathrm{P}\ 7s$	11	−2.29825	$2s^22p^4\ ^1\mathrm{D}\ 7d$	31	−1.65941	$2s^22p^4\ ^1\mathrm{D}\ 8p$	19	−2.25507	$2s^22p^4\ ^3\mathrm{P}\ 8f$
17	−2.97994	$2s^22p^4\ ^3\mathrm{P}\ 7d$	12	−2.28857	$2s^22p^4\ ^3\mathrm{P}\ 8d$	32	−1.58519	$2s^22p^4\ ^1\mathrm{S}\ 7p$	20	−2.24075	$2s^22p^4\ ^1\mathrm{D}\ 7f$
18	−2.43102	$2s^22p^4\ ^3\mathrm{P}\ 8s$	13	−1.80572	$2s^22p^4\ ^3\mathrm{P}\ 9d$	33	−1.54775	$2s^22p^4\ ^1\mathrm{D}\ 8f$	21	−1.78215	$2s^22p^4\ ^3\mathrm{P}\ 9f$
19	−2.35719	$2s2p^5\ ^1\mathrm{P}\ 4p$	14	−1.58856	$2s^22p^4\ ^1\mathrm{D}\ 8d$			**$^2\mathrm{D}^o$**	22	−1.73890	$2s2p^5\ ^1\mathrm{P}\ 4d$
20	−2.29918	$2s^22p^4\ ^1\mathrm{D}\ 7d$			**$^2\mathrm{G}^e$**	1	−18.6465	$2s^22p^4\ ^3\mathrm{P}\ 3p$	23	−1.65296	$2s^22p^4\ ^1\mathrm{D}\ 8p$
21	−2.26538	$2s^22p^4\ ^3\mathrm{P}\ 8d$	1	−16.1947	$2s^22p^4\ ^1\mathrm{D}\ 3d$	2	−17.9083	$2s^22p^4\ ^1\mathrm{D}\ 3p$	24	−1.54884	$2s^22p^4\ ^1\mathrm{D}\ 8f$
22	−1.90312	$2s^22p^4\ ^3\mathrm{P}\ 9s$	2	−8.67746	$2s^22p^4\ ^1\mathrm{D}\ 4d$	3	−10.9878	$2s2p^5\ ^3\mathrm{P}\ 3d$			**$^2\mathrm{G}^o$**
23	−1.70830	$2s^22p^4\ ^3\mathrm{P}\ 9d$	3	−5.25303	$2s^22p^4\ ^1\mathrm{D}\ 5d$	4	−10.0096	$2s^22p^4\ ^3\mathrm{P}\ 4p$	1	−9.01145	$2s^22p^4\ ^3\mathrm{P}\ 4f$
24	−1.58767	$2s^22p^4\ ^1\mathrm{D}\ 8d$	4	−3.41499	$2s^22p^4\ ^1\mathrm{D}\ 6d$	5	−9.32961	$2s^22p^4\ ^1\mathrm{D}\ 4p$	2	−8.29497	$2s^22p^4\ ^1\mathrm{D}\ 4f$
		$^2\mathrm{D}^e$	5	−3.40323	$2s2p^5\ ^3\mathrm{P}\ 4f$	6	−9.11009	$2s2p^5\ ^1\mathrm{P}\ 3d$	3	−5.77250	$2s^22p^4\ ^3\mathrm{P}\ 5f$
1	−19.5603	$2s^22p^4\ ^1\mathrm{D}\ 3s$	6	−2.30855	$2s^22p^4\ ^1\mathrm{D}\ 7d$	7	−8.98202	$2s^22p^4\ ^3\mathrm{P}\ 4f$	4	−5.06453	$2s^22p^4\ ^1\mathrm{D}\ 5f$
2	−16.5366	$2s^22p^4\ ^3\mathrm{P}\ 3d$	7	−1.59631	$2s^22p^4\ ^1\mathrm{D}\ 8d$	8	−8.25705	$2s^22p^4\ ^1\mathrm{D}\ 4f$	5	−4.00858	$2s^22p^4\ ^3\mathrm{P}\ 6f$
3	−15.8446	$2s^22p^4\ ^1\mathrm{D}\ 3d$			**$^2\mathrm{S}^o$**	9	−6.25424	$2s^22p^4\ ^3\mathrm{P}\ 5p$	6	−3.30347	$2s^22p^4\ ^1\mathrm{D}\ 6f$
4	−15.2148	$2s^22p^4\ ^1\mathrm{S}\ 3d$	1	−18.5567	$2s^22p^4\ ^3\mathrm{P}\ 3p$	10	−5.76106	$2s^22p^4\ ^3\mathrm{P}\ 5f$	7	−2.94447	$2s^22p^4\ ^3\mathrm{P}\ 7f$
5	−12.9248	$2s2p^5\ ^3\mathrm{P}\ 3p$	2	−9.99529	$2s^22p^4\ ^3\mathrm{P}\ 4p$	11	−5.54883	$2s^22p^4\ ^1\mathrm{D}\ 5p$	8	−2.25393	$2s^22p^4\ ^3\mathrm{P}\ 8f$

F-like Ca (Ca^{11+})

i	E(Ryds)	Description
9	−2.24055	$2s^22p^4\ ^1\mathrm{D}\ 7f$
10	−1.78056	$2s^22p^4\ ^3\mathrm{P}\ 9f$
11	−1.55065	$2s^22p^4\ ^1\mathrm{D}\ 8f$
	$^2\mathrm{H^o}$	
1	−8.34005	$2s^22p^4\ ^1\mathrm{D}\ 4f$
2	−5.08558	$2s^22p^4\ ^1\mathrm{D}\ 5f$
3	−3.31523	$2s^22p^4\ ^1\mathrm{D}\ 6f$
4	−2.24793	$2s^22p^4\ ^1\mathrm{D}\ 7f$
5	−1.55557	$2s^22p^4\ ^1\mathrm{D}\ 8f$
	$^4\mathrm{S^e}$	
1	−13.3614	$2s2p^5\ ^3\mathrm{P}\ 3p$
2	−4.50875	$2s2p^5\ ^3\mathrm{P}\ 4p$
3	−.68224	$2s2p^5\ ^3\mathrm{P}\ 5p$
	$^4\mathrm{P^e}$	
1	−20.3459	$2s^22p^4\ ^3\mathrm{P}\ 3s$
2	−16.7110	$2s^22p^4\ ^3\mathrm{P}\ 3d$
3	−12.9671	$2s2p^5\ ^3\mathrm{P}\ 3p$
4	−10.6656	$2s^22p^4\ ^3\mathrm{P}\ 4s$
5	−9.31909	$2s^22p^4\ ^3\mathrm{P}\ 4d$
6	−6.57903	$2s^22p^4\ ^3\mathrm{P}\ 5s$
7	−5.92821	$2s^22p^4\ ^3\mathrm{P}\ 5d$
8	−4.50014	$2s^22p^4\ ^3\mathrm{P}\ 6s$
9	−4.34428	$2s2p^5\ ^3\mathrm{P}\ 4p$
10	−4.09690	$2s^22p^4\ ^3\mathrm{P}\ 6d$
11	−3.22100	$2s^22p^4\ ^3\mathrm{P}\ 7s$
12	−3.00098	$2s^22p^4\ ^3\mathrm{P}\ 7d$
13	−2.43771	$2s^22p^4\ ^3\mathrm{P}\ 8s$
14	−2.29179	$2s^22p^4\ ^3\mathrm{P}\ 8d$
15	−1.90878	$2s^22p^4\ ^3\mathrm{P}\ 9s$
16	−1.80715	$2s^22p^4\ ^3\mathrm{P}\ 9d$
	$^4\mathrm{D^e}$	
1	−17.0432	$2s^22p^4\ ^3\mathrm{P}\ 3d$
2	−13.1194	$2s2p^5\ ^3\mathrm{P}\ 3p$
3	−9.42720	$2s^22p^4\ ^3\mathrm{P}\ 4d$
4	−5.97974	$2s^22p^4\ ^3\mathrm{P}\ 5d$
5	−4.43603	$2s2p^5\ ^3\mathrm{P}\ 4p$
6	−4.12436	$2s^22p^4\ ^3\mathrm{P}\ 6d$
7	−3.42140	$2s2p^5\ ^3\mathrm{P}\ 4f$
8	−3.01800	$2s^22p^4\ ^3\mathrm{P}\ 7d$
9	−2.30302	$2s^22p^4\ ^3\mathrm{P}\ 8d$
10	−1.81494	$2s^22p^4\ ^3\mathrm{P}\ 9d$
	$^4\mathrm{F^e}$	
1	−16.8178	$2s^22p^4\ ^3\mathrm{P}\ 3d$
2	−9.35686	$2s^22p^4\ ^3\mathrm{P}\ 4d$
3	−5.94570	$2s^22p^4\ ^3\mathrm{P}\ 5d$
4	−4.10856	$2s^22p^4\ ^3\mathrm{P}\ 6d$
5	−3.37742	$2s2p^5\ ^3\mathrm{P}\ 4f$
6	−3.00710	$2s^22p^4\ ^3\mathrm{P}\ 7d$
7	−2.29580	$2s^22p^4\ ^3\mathrm{P}\ 8d$
8	−1.80991	$2s^22p^4\ ^3\mathrm{P}\ 9d$
	$^4\mathrm{G^e}$	
1	−3.41597	$2s2p^5\ ^3\mathrm{P}\ 4f$
	$^4\mathrm{S^o}$	
1	−18.5488	$2s^22p^4\ ^3\mathrm{P}\ 3p$
2	−9.99244	$2s^22p^4\ ^3\mathrm{P}\ 4p$
3	−6.24146	$2s^22p^4\ ^3\mathrm{P}\ 5p$
4	−4.27093	$2s^22p^4\ ^3\mathrm{P}\ 6p$
5	−3.10661	$2s^22p^4\ ^3\mathrm{P}\ 7p$
6	−2.36124	$2s^22p^4\ ^3\mathrm{P}\ 8p$
7	−1.85533	$2s^22p^4\ ^3\mathrm{P}\ 9p$
	$^4\mathrm{P^o}$	
1	−18.9902	$2s^22p^4\ ^3\mathrm{P}\ 3p$
2	−14.6038	$2s2p^5\ ^3\mathrm{P}\ 3s$
3	−11.4580	$2s2p^5\ ^3\mathrm{P}\ 3d$
4	−10.1157	$2s^22p^4\ ^3\mathrm{P}\ 4p$
5	−6.30885	$2s^22p^4\ ^3\mathrm{P}\ 5p$
6	−5.02861	$2s2p^5\ ^3\mathrm{P}\ 4s$
7	−4.30076	$2s^22p^4\ ^3\mathrm{P}\ 6p$
8	−3.81458	$2s2p^5\ ^3\mathrm{P}\ 4d$
9	−3.12517	$2s^22p^4\ ^3\mathrm{P}\ 7p$
10	−2.37433	$2s^22p^4\ ^3\mathrm{P}\ 8p$
11	−1.86479	$2s^22p^4\ ^3\mathrm{P}\ 9p$
	$^4\mathrm{D^o}$	
1	−18.7904	$2s^22p^4\ ^3\mathrm{P}\ 3p$
2	−11.1623	$2s2p^5\ ^3\mathrm{P}\ 3d$
3	−10.0590	$2s^22p^4\ ^3\mathrm{P}\ 4p$
4	−8.98684	$2s^22p^4\ ^3\mathrm{P}\ 4f$
5	−6.27797	$2s^22p^4\ ^3\mathrm{P}\ 5p$
6	−5.76555	$2s^22p^4\ ^3\mathrm{P}\ 5f$
7	−4.29414	$2s^22p^4\ ^3\mathrm{P}\ 6p$
8	−4.00750	$2s^22p^4\ ^3\mathrm{P}\ 6f$
9	−3.71089	$2s2p^5\ ^3\mathrm{P}\ 4d$
10	−3.11658	$2s^22p^4\ ^3\mathrm{P}\ 7p$
11	−2.94171	$2s^22p^4\ ^3\mathrm{P}\ 7f$
12	−2.36864	$2s^22p^4\ ^3\mathrm{P}\ 8p$
13	−2.25237	$2s^22p^4\ ^3\mathrm{P}\ 8f$
14	−1.86062	$2s^22p^4\ ^3\mathrm{P}\ 9p$
15	−1.77957	$2s^22p^4\ ^3\mathrm{P}\ 9f$
	$^4\mathrm{F^o}$	
1	−11.3721	$2s2p^5\ ^3\mathrm{P}\ 3d$
2	−9.03114	$2s^22p^4\ ^3\mathrm{P}\ 4f$
3	−5.78806	$2s^22p^4\ ^3\mathrm{P}\ 5f$
4	−4.02258	$2s^22p^4\ ^3\mathrm{P}\ 6f$
5	−3.77903	$2s2p^5\ ^3\mathrm{P}\ 4d$
6	−2.94955	$2s^22p^4\ ^3\mathrm{P}\ 7f$
7	−2.25772	$2s^22p^4\ ^3\mathrm{P}\ 8f$
8	−1.78338	$2s^22p^4\ ^3\mathrm{P}\ 9f$
	$^4\mathrm{G^o}$	
1	−9.01602	$2s^22p^4\ ^3\mathrm{P}\ 4f$
2	−5.77573	$2s^22p^4\ ^3\mathrm{P}\ 5f$
3	−4.01069	$2s^22p^4\ ^3\mathrm{P}\ 6f$
4	−2.94589	$2s^22p^4\ ^3\mathrm{P}\ 7f$
5	−2.25489	$2s^22p^4\ ^3\mathrm{P}\ 8f$
6	−1.78127	$2s^22p^4\ ^3\mathrm{P}\ 9f$

Energies in ascending order from ground state for terms with effective $n \leq 4.0$, $L \leq 4$

Term	i	E(Ryds)	Term	i	E(Ryds)	Term	i	E(Ryds)	Term	i	E(Ryds)	Term	i	E(Ryds)
$^2\mathrm{P^o}$	1	0.00000	$^2\mathrm{P^o}$	4	31.1713	$^4\mathrm{S^e}$	1	34.9520	$^2\mathrm{P^o}$	7	37.6260	$^2\mathrm{F^e}$	3	39.0095
$^2\mathrm{S^e}$	1	6.13890	$^4\mathrm{D^e}$	1	31.2702	$^4\mathrm{D^e}$	2	35.1940	$^4\mathrm{P^e}$	4	37.6478	$^2\mathrm{P^o}$	9	39.0190
$^4\mathrm{P^e}$	1	27.9675	$^4\mathrm{F^e}$	1	31.4956	$^4\mathrm{P^e}$	3	35.3463	$^2\mathrm{P^e}$	6	37.7269	$^2\mathrm{P^o}$	10	39.0745
$^2\mathrm{P^e}$	1	28.2175	$^4\mathrm{P^e}$	2	31.6024	$^2\mathrm{D^e}$	5	35.3886	$^4\mathrm{P^o}$	4	38.1977	$^2\mathrm{P^e}$	7	39.0886
$^2\mathrm{D^e}$	1	28.7531	$^2\mathrm{F^e}$	1	31.6310	$^2\mathrm{P^e}$	4	35.4354	$^4\mathrm{D^o}$	3	38.2544	$^2\mathrm{D^e}$	8	39.1228
$^4\mathrm{P^o}$	1	29.3232	$^2\mathrm{P^e}$	2	31.7223	$^2\mathrm{P^o}$	6	35.8263	$^2\mathrm{D^o}$	4	38.3038	$^2\mathrm{F^o}$	4	39.1327
$^4\mathrm{D^o}$	1	29.5230	$^2\mathrm{D^e}$	2	31.7768	$^2\mathrm{S^e}$	4	35.8343	$^2\mathrm{S^o}$	2	38.3181	$^2\mathrm{S^e}$	6	39.1983
$^2\mathrm{S^e}$	2	29.5840	$^2\mathrm{G^e}$	1	32.1187	$^4\mathrm{P^o}$	3	36.8554	$^4\mathrm{S^o}$	2	38.3209	$^2\mathrm{D^o}$	6	39.2033
$^2\mathrm{D^o}$	1	29.6669	$^2\mathrm{S^e}$	3	32.2909	$^4\mathrm{F^o}$	1	36.9413	$^2\mathrm{D^e}$	7	38.3839	$^4\mathrm{F^o}$	2	39.2822
$^2\mathrm{P^o}$	2	29.6898	$^2\mathrm{F^e}$	2	32.3472	$^4\mathrm{D^o}$	2	37.1511	$^2\mathrm{P^o}$	8	38.4247	$^2\mathrm{F^o}$	5	39.2942
$^2\mathrm{S^o}$	1	29.7567	$^2\mathrm{D^e}$	3	32.4688	$^2\mathrm{F^o}$	2	37.2418	$^2\mathrm{F^o}$	3	38.8605	$^4\mathrm{G^o}$	1	39.2973
$^4\mathrm{S^o}$	1	29.7646	$^2\mathrm{P^e}$	3	32.4693	$^2\mathrm{D^e}$	6	37.2538	$^4\mathrm{D^e}$	3	38.8862	$^2\mathrm{G^o}$	1	39.3019
$^2\mathrm{F^o}$	1	30.2115	$^2\mathrm{D^e}$	4	33.0986	$^2\mathrm{D^o}$	3	37.3256	$^4\mathrm{F^e}$	2	38.9565			
$^2\mathrm{D^o}$	2	30.4051	$^4\mathrm{P^o}$	2	33.7096	$^2\mathrm{P^e}$	5	37.4008	$^2\mathrm{D^o}$	5	38.9837			
$^2\mathrm{P^o}$	3	30.7996	$^2\mathrm{P^o}$	5	34.0914	$^2\mathrm{S^e}$	5	37.5326	$^4\mathrm{P^e}$	5	38.9943			

F-like Ca (Ca^{11+})

gf-values for transitions involving terms with effective $n \leq 4.0$, $L \leq 4$

i	i'	gf_L	i	i'	gf_L	i	i'	gf_L	i	i'	gf_L	i	i'	gf_L	i	i'	gf_L
		^{2}**S**^{e}–^{2}**P**^{o}	6	1	1.75E−2	3	6	−1.27E−3	1	6	−8.06E−4	2	4	1.53E−4	7	5	7.13E−3
1	1	3.54E−1	6	2	8.60E−3	3	7	−1.53E−2	2	1	3.67E−2	2	5	−5.08E−3	7	6	1.25E−2
1	2	−1.37E−3	6	3	6.09E−3	3	8	−2.69E−4	2	2	3.89E−2	2	6	−5.67E−3	7	7	1.43E−4
1	3	−1.49E−2	6	4	3.53E−1	3	9	−2.88E−3	2	3	−1.09E−1	2	7	−5.45E−1	7	8	−4.98E−3
1	4	−4.28E−3	6	5	6.32E−4	3	10	−1.82E−1	2	4	−2.19E−2	2	8	−2.02E−1	7	9	−1.49E−1
1	5	−3.45E−1	6	6	7.25E−4	4	1	5.38E−1	2	5	−5.88E−4	2	9	−3.12E−2	7	10	−1.10E+0
1	6	−1.51E−1	6	7	7.74E−4	4	2	1.05E−1	2	6	−1.61E−1	2	10	−5.38E−5	8	1	1.27E+0
1	7	−2.95E+0	6	8	5.59E−3	4	3	1.51E−1	3	1	4.43E−2	3	1	5.11E+0	8	2	9.17E−1
1	8	−1.51E−2	6	9	1.42E−3	4	4	3.69E−2	3	2	1.71E−1	3	2	2.11E−4	8	3	4.42E−1
1	9	−1.78E+0	6	10	1.55E−4	4	5	7.03E−1	3	3	−2.93E−2	3	3	5.53E−1	8	4	2.85E−2
1	10	−2.94E−1			^{2}**P**^{e}–^{2}**S**^{o}	4	6	−1.25E−5	3	4	−2.20E−2	3	4	2.61E−2	8	5	7.04E−3
2	1	7.46E−2	1	1	−2.62E−1	4	7	−5.57E−1	3	5	−3.71E−2	3	5	−1.37E−2	8	6	3.24E−2
2	2	−9.51E−4	1	2	−1.35E−1	4	8	−1.31E−3	3	6	−1.66E−2	3	6	−2.67E−3	8	7	3.60E−2
2	3	−1.71E−2	2	1	4.43E−1	4	9	−4.95E−2	4	1	3.03E−1	3	7	−5.59E−2	8	8	1.89E+0
2	4	−7.82E−1	2	2	−1.15E−1	4	10	−1.39E−6	4	2	2.13E−4	3	8	−2.04E−2	8	9	1.15E−3
2	5	−6.72E−3	3	1	2.33E−1	5	1	9.44E−2	4	3	−1.32E+0	3	9	−3.22E−1	8	10	3.15E−3
2	6	−1.36E−1	3	2	−7.00E−2	5	2	1.13E−1	4	4	−1.11E−3	3	10	−1.49E−2			^{2}**D**^{e}–^{2}**D**^{o}
2	7	−2.67E−3	4	1	2.31E−1	5	3	2.24E−1	4	5	−1.16E−3	4	1	1.09E+0	1	1	−2.61E−3
2	8	−1.47E−3	4	2	−5.60E−4	5	4	1.17E−1	4	6	−8.24E−3	4	2	1.67E−4	1	2	−1.42E+0
2	9	−1.07E−3	5	1	9.49E−3	5	5	2.33E−4	5	1	4.25E−2	4	3	6.90E−2	1	3	−1.55E−4
2	10	−2.59E−4	5	2	−1.62E−2	5	6	8.88E−1	5	2	1.03E+0	4	4	1.57E+0	1	4	−1.39E−3
3	1	1.05E+0	6	1	1.05E−1	5	7	−8.33E−4	5	3	1.25E−4	4	5	−3.04E−3	1	5	−5.89E−1
3	2	1.44E−1	6	2	−3.46E−1	5	8	−3.22E−2	5	4	−7.09E−2	4	6	−4.76E−4	1	6	−5.34E−2
3	3	1.53E−1	7	1	5.42E−1	5	9	−2.78E−1	5	5	−8.92E−2	4	7	−8.53E−4	2	1	2.94E−1
3	4	2.05E−3	7	2	9.80E−1	5	10	−1.88E−2	5	6	−9.41E−1	4	8	−3.40E−3	2	2	1.49E−1
3	5	−4.00E−3			^{2}**P**^{e}–^{2}**P**^{o}	6	1	1.33E−1	6	1	5.37E−1	4	9	−1.07E−1	2	3	−3.04E−2
3	6	−3.48E−5	1	1	6.35E−1	6	2	1.88E−1	6	2	6.11E−2	4	10	−3.80E−2	2	4	−7.53E−2
3	7	−1.20E−4	1	2	−4.39E−1	6	3	3.54E−1	6	3	8.04E−3	5	1	7.57E−1	2	5	−2.92E−2
3	8	−1.32E−2	1	3	−6.15E−1	6	4	1.63E−2	6	4	−1.80E+0	5	2	1.80E−1	2	6	−3.68E−5
3	9	−5.08E−3	1	4	−9.62E−2	6	5	1.38E−2	6	5	−1.16E−2	5	3	2.37E−1	3	1	1.78E−1
3	10	−6.50E−2	1	5	−4.96E−1	6	6	4.52E−3	6	6	−1.64E−2	5	4	6.04E−2	3	2	8.87E−1
4	1	2.23E−1	1	6	−3.93E−3	6	7	1.31E−3	7	1	2.42E−2	5	5	1.15E+0	3	3	−5.22E−3
4	2	2.84E−1	1	7	−6.07E−3	6	8	−1.25E+0	7	2	2.71E−2	5	6	−1.34E−4	3	4	−4.19E−2
4	3	9.38E−4	1	8	−3.27E−1	6	9	−1.21E−2	7	3	2.12E−2	5	7	−7.19E−3	3	5	−1.62E−1
4	4	7.82E−4	1	9	−6.42E−4	6	10	−2.59E−1	7	4	6.70E−2	5	8	−6.68E−3	3	6	−1.68E−1
4	5	2.71E−1	1	10	−9.84E−3	7	1	3.40E−1	7	5	1.13E−3	5	9	−1.83E−4	4	1	7.58E−3
4	6	1.63E−4	2	1	2.73E−1	7	2	3.99E−1	7	6	−7.26E−3	5	10	−2.22E−3	4	2	1.68E−2
4	7	−4.83E−1	2	2	4.62E−1	7	3	6.84E−2			^{2}**D**^{e}–^{2}**P**^{o}	6	1	3.00E−1	4	3	−1.28E−3
4	8	−6.34E−3	2	3	1.77E−5	7	4	2.60E−2	1	1	3.98E−1	6	2	1.57E−2	4	4	−2.59E−3
4	9	−3.11E−2	2	4	1.95E−3	7	5	8.77E−4	1	2	−1.62E−1	6	3	2.51E−2	4	5	−1.26E−2
4	10	−1.19E−5	2	5	−2.71E−4	7	6	1.42E−2	1	3	−6.71E−1	6	4	2.44E−1	4	6	−2.07E−1
5	1	2.62E−3	2	6	−2.03E−3	7	7	8.57E−3	1	4	−4.64E−3	6	5	1.26E−2	5	1	8.96E−1
5	2	6.01E−2	2	7	−1.61E−1	7	8	5.81E−1	1	5	−3.31E−2	6	6	1.27E+0	5	2	5.08E−5
5	3	2.68E−1	2	8	−1.06E−1	7	9	5.00E−5	1	6	−1.48E+0	6	7	−2.06E−4	5	3	−4.59E−1
5	4	1.01E−1	2	9	−5.66E−5	7	10	3.30E−5	1	7	6.18E−4	6	8	1.08E−4	5	4	2.08E−3
5	5	1.75E−1	2	10	−1.77E−2			^{2}**P**^{e}–^{2}**D**^{o}	1	8	−3.92E−2	6	9	−1.43E−2	5	5	−1.57E−3
5	6	2.42E−1	3	1	3.83E+0	1	1	−1.25E+0	1	9	−3.23E−2	6	10	−3.77E−2	5	6	−9.11E−3
5	7	−1.47E−3	3	2	8.21E−2	1	2	−1.21E−2	1	10	−2.60E−1	7	1	5.06E−2	6	1	2.40E−3
5	8	−1.99E−5	3	3	4.88E−1	1	3	−1.48E−2	2	1	2.30E+0	7	2	9.72E−2	6	2	2.21E−1
5	9	−2.85E−1	3	4	8.62E−3	1	4	7.16E−1	2	2	1.02E+0	7	3	3.56E−1	6	3	1.30E−5
5	10	−4.75E−2	3	5	−9.64E−3	1	5	−1.67E−4	2	3	7.13E−2	7	4	7.91E−4	6	4	−8.98E−6

F-like Ca (Ca^{11+})

i i'	gf_L	i i'	gf_L	i i'	gf_L	i i'	gf_L	i i'	gf_L	i i'	gf_L
6 5	−1.85E−1	8 3	6.19E−3	2 1	1.12E−2	4 1	1.19E−2	6 1	2.43E+0	8 1	9.22E−3
6 6	−2.69E−1	8 4	7.40E−1	2 2	−4.99E−2	4 2	−2.95E−4	6 2	8.26E−6	8 2	2.62E−2
7 1	1.80E−5	8 5	7.20E−3	2 3	−6.90E−3	4 3	−1.20E−2	6 3	−1.07E+0	8 3	2.95E−4
7 2	7.47E−1	8 6	−1.89E−4	2 4	−1.29E−1	4 4	−9.81E−2	6 4	−9.90E−1	8 4	−8.19E−4
7 3	3.91E−5		^{2}**D**^{e}–^{2}**F**^{o}	2 5	−6.63E+0	4 5	−9.44E−2	6 5	−2.02E−2	8 5	−5.10E−1
7 4	1.53E−5	1 1	−1.73E+0	3 1	5.24E−2	5 1	1.82E−3	7 1	5.45E−1		
7 5	−1.67E+0	1 2	−1.11E−3	3 2	−9.02E−3	5 2	−2.32E+0	7 2	7.85E−5		
7 6	−2.76E−1	1 3	−7.43E−1	3 3	−1.27E−3	5 3	−1.02E−4	7 3	−1.34E+0		
8 1	3.48E−1	1 4	−3.08E−1	3 4	−5.37E−3	5 4	−1.57E−2	7 4	−1.22E+0		
8 2	7.75E−2	1 5	−4.50E−3	3 5	−1.93E+0	5 5	−1.32E−1	7 5	−1.48E−2		

F-like Fe (Fe^{17+})

Term energies relative to $2s^22p^4\,^3\mathrm{P}$ ionization threshold for each symmetry

i	E(Ryds)	Description	i	E(Ryds)	Description	i	E(Ryds)	Description	i	E(Ryds)	Description
		$^2\mathrm{S}^e$	2	−37.0036	$2s^22p^4\,^3\mathrm{P}\,3d$	2	−19.7405	$2s^22p^4\,^1\mathrm{D}\,4d$	35	−3.87860	$2s^22p^4\,^1\mathrm{D}\,8f$
1	−90.9997	$2s2p^6$	3	−35.8734	$2s^22p^4\,^1\mathrm{D}\,3d$	3	−12.1166	$2s^22p^4\,^1\mathrm{D}\,5d$			$^2\mathrm{D}^o$
2	−39.9474	$2s^22p^4\,^1\mathrm{S}\,3s$	4	−34.9295	$2s^22p^4\,^1\mathrm{S}\,3d$	4	−11.9587	$2s2p^5\,^3\mathrm{P}\,4f$	1	−40.1755	$2s^22p^4\,^3\mathrm{P}\,3p$
3	−36.2007	$2s^22p^4\,^1\mathrm{D}\,3d$	5	−31.6599	$2s2p^5\,^3\mathrm{P}\,3p$	5	−9.13114	$2s2p^5\,^1\mathrm{P}\,4f$	2	−38.9249	$2s^22p^4\,^1\mathrm{D}\,3p$
4	−31.0119	$2s2p^5\,^3\mathrm{P}\,3p$	6	−29.0184	$2s2p^5\,^1\mathrm{P}\,3p$	6	−8.00438	$2s^22p^4\,^1\mathrm{D}\,6d$	3	−28.9606	$2s2p^5\,^3\mathrm{P}\,3d$
5	−28.5396	$2s2p^5\,^1\mathrm{P}\,3p$	7	−21.6280	$2s^22p^4\,^1\mathrm{D}\,4s$	7	−5.54022	$2s^22p^4\,^1\mathrm{D}\,7d$	4	−26.3553	$2s2p^5\,^1\mathrm{P}\,3d$
6	−22.9862	$2p^6\,^1\mathrm{S}\,3s$	8	−20.6060	$2s^22p^4\,^3\mathrm{P}\,4d$	8	−4.65352	$2s2p^5\,^3\mathrm{P}\,5f$	5	−21.8652	$2s^22p^4\,^3\mathrm{P}\,4p$
7	−20.2055	$2s^22p^4\,^1\mathrm{S}\,4s$	9	−19.5836	$2s^22p^4\,^1\mathrm{D}\,4d$	9	−3.94762	$2s^22p^4\,^1\mathrm{D}\,8d$	6	−20.6553	$2s^22p^4\,^1\mathrm{D}\,4p$
8	−19.6234	$2s^22p^4\,^1\mathrm{D}\,4d$	10	−18.7773	$2p^6\,^1\mathrm{S}\,3d$			$^2\mathrm{S}^o$	7	−20.2396	$2s^22p^4\,^3\mathrm{P}\,4f$
9	−13.1124	$2s2p^5\,^3\mathrm{P}\,4p$	11	−18.2718	$2s^22p^4\,^1\mathrm{S}\,4d$	1	−40.0416	$2s^22p^4\,^3\mathrm{P}\,3p$	8	−19.0945	$2s^22p^4\,^1\mathrm{D}\,4f$
10	−12.0708	$2s^22p^4\,^1\mathrm{D}\,5d$	12	−13.4758	$2s2p^5\,^3\mathrm{P}\,4p$	2	−21.8283	$2s^22p^4\,^3\mathrm{P}\,4p$	9	−13.7493	$2s^22p^4\,^3\mathrm{P}\,5p$
11	−11.6042	$2s^22p^4\,^1\mathrm{S}\,5s$	13	−13.1433	$2s^22p^4\,^3\mathrm{P}\,5d$	3	−13.7258	$2s^22p^4\,^3\mathrm{P}\,5p$	10	−12.9704	$2s^22p^4\,^3\mathrm{P}\,5f$
12	−10.6000	$2s2p^5\,^1\mathrm{P}\,4p$	14	−13.0238	$2s^22p^4\,^1\mathrm{D}\,5s$	4	−9.43203	$2s^22p^4\,^3\mathrm{P}\,6p$	11	−12.5402	$2s^22p^4\,^1\mathrm{D}\,5p$
13	−7.97539	$2s^22p^4\,^1\mathrm{D}\,6d$	15	−12.0506	$2s^22p^4\,^1\mathrm{D}\,5d$	5	−6.88061	$2s^22p^4\,^3\mathrm{P}\,7p$	12	−12.4351	$2s2p^5\,^3\mathrm{P}\,4d$
14	−7.08685	$2s^22p^4\,^1\mathrm{S}\,6s$	16	−11.9606	$2s2p^5\,^3\mathrm{P}\,4f$	6	−5.24079	$2s^22p^4\,^3\mathrm{P}\,8p$	13	−11.8010	$2s^22p^4\,^1\mathrm{D}\,5f$
15	−5.52270	$2s^22p^4\,^1\mathrm{D}\,7d$	17	−10.6995	$2s2p^5\,^1\mathrm{P}\,4p$:	7	−4.12452	$2s^22p^4\,^3\mathrm{P}\,9p$	14	−9.66996	$2s2p^5\,^1\mathrm{P}\,4d$
16	−5.20069	$2s2p^5\,^3\mathrm{P}\,5p$	18	−10.6434	$2s^22p^4\,^1\mathrm{S}\,5d$:			$^2\mathrm{P}^o$	15	−9.43989	$2s^22p^4\,^3\mathrm{P}\,6p$
17	−4.43066	$2s^22p^4\,^1\mathrm{S}\,7s$	19	−9.13373	$2s2p^5\,^1\mathrm{P}\,4f$	1	−99.6024	$2s^22p^5$	16	−9.00674	$2s^22p^4\,^3\mathrm{P}\,6f$
18	−3.93635	$2s^22p^4\,^1\mathrm{D}\,8d$	20	−9.10623	$2s^22p^4\,^3\mathrm{P}\,6d$	2	−40.0767	$2s^22p^4\,^3\mathrm{P}\,3p$	17	−8.23845	$2s^22p^4\,^1\mathrm{D}\,6p$
19	−3.61883	$2p^6\,^1\mathrm{S}\,4s$	21	−8.51063	$2s^22p^4\,^1\mathrm{D}\,6s$	3	−38.3602	$2s^22p^4\,^1\mathrm{D}\,3p$	18	−7.82460	$2s^22p^4\,^1\mathrm{D}\,6f$
20	−2.85256	$2s^22p^4\,^1\mathrm{D}\,9d$	22	−7.96971	$2s^22p^4\,^1\mathrm{D}\,6d$	4	−37.5923	$2s^22p^4\,^1\mathrm{S}\,3p$	19	−6.88614	$2s^22p^4\,^3\mathrm{P}\,7p$
21	−2.73554	$2s^22p^4\,^1\mathrm{S}\,8s$	23	−6.68444	$2s^22p^4\,^3\mathrm{P}\,7d$	5	−33.5880	$2s2p^5\,^3\mathrm{P}\,3s$	20	−6.61686	$2s^22p^4\,^3\mathrm{P}\,7f$
22	−2.49131	$2s2p^5\,^1\mathrm{P}\,5p$	24	−6.56524	$2s^22p^4\,^1\mathrm{S}\,6d$	6	−31.1592	$2s2p^5\,^1\mathrm{P}\,3s$	21	−5.68527	$2s^22p^4\,^1\mathrm{D}\,7p$
		$^2\mathrm{P}^e$	25	−5.85633	$2s^22p^4\,^1\mathrm{D}\,7s$	7	−28.4425	$2s2p^5\,^3\mathrm{P}\,3d$	22	−5.42928	$2s^22p^4\,^1\mathrm{D}\,7f$
1	−42.3391	$2s^22p^4\,^3\mathrm{P}\,3s$	26	−5.52016	$2s^22p^4\,^1\mathrm{D}\,7d$	8	−26.4883	$2s2p^5\,^1\mathrm{P}\,3d$	23	−5.24597	$2s^22p^4\,^3\mathrm{P}\,8p$
2	−37.1177	$2s^22p^4\,^3\mathrm{P}\,3d$	27	−5.36391	$2s2p^5\,^3\mathrm{P}\,5p$	9	−21.6817	$2s^22p^4\,^3\mathrm{P}\,4p$	24	−5.06586	$2s^22p^4\,^3\mathrm{P}\,8f$
3	−35.8639	$2s^22p^4\,^1\mathrm{D}\,3d$	28	−5.10710	$2s^22p^4\,^3\mathrm{P}\,8d$	10	−20.8947	$2p^6\,^1\mathrm{S}\,3p$	25	−4.88458	$2s2p^5\,^3\mathrm{P}\,5d$
4	−31.5857	$2s2p^5\,^3\mathrm{P}\,3p$	29	−4.64811	$2s2p^5\,^3\mathrm{P}\,5f$	11	−20.5809	$2s^22p^4\,^1\mathrm{D}\,4p$	26	−4.12642	$2s^22p^4\,^3\mathrm{P}\,9p$
5	−28.8120	$2s2p^5\,^1\mathrm{P}\,3p$	30	−4.15849	$2s^22p^4\,^1\mathrm{D}\,8s$	12	−19.2696	$2s^22p^4\,^1\mathrm{S}\,4p$	27	−4.04404	$2s^22p^4\,^1\mathrm{D}\,8p$
6	−22.7509	$2s^22p^4\,^3\mathrm{P}\,4s$	31	−4.10984	$2s^22p^4\,^1\mathrm{S}\,7d$	13	−19.1335	$2s^22p^4\,^1\mathrm{D}\,4f$	28	−4.00230	$2s^22p^4\,^3\mathrm{P}\,9f$
7	−20.6628	$2s^22p^4\,^3\mathrm{P}\,4d$	32	−4.03262	$2s^22p^4\,^3\mathrm{P}\,9d$	14	−14.2868	$2s2p^5\,^3\mathrm{P}\,4s$	29	−3.87416	$2s^22p^4\,^1\mathrm{D}\,8f$
8	−19.5863	$2s^22p^4\,^1\mathrm{D}\,4d$	33	−3.93198	$2s^22p^4\,^1\mathrm{D}\,8d$	15	−13.6439	$2s^22p^4\,^3\mathrm{P}\,5p$			$^2\mathrm{F}^o$
9	−14.2015	$2s^22p^4\,^3\mathrm{P}\,5s$			$^2\mathrm{F}^e$	16	−12.5309	$2s^22p^4\,^1\mathrm{D}\,5p$	1	−39.2137	$2s^22p^4\,^1\mathrm{D}\,3p$
10	−13.4324	$2s2p^5\,^3\mathrm{P}\,4p$	1	−37.2568	$2s^22p^4\,^3\mathrm{P}\,3d$	17	−12.2209	$2s2p^5\,^3\mathrm{P}\,4d$	2	−29.0928	$2s2p^5\,^3\mathrm{P}\,3d$
11	−13.1684	$2s^22p^4\,^3\mathrm{P}\,5d$	2	−36.1358	$2s^22p^4\,^1\mathrm{D}\,3d$	18	−11.8194	$2s^22p^4\,^1\mathrm{D}\,5f$	3	−26.6241	$2s2p^5\,^1\mathrm{P}\,3d$
12	−12.0559	$2s^22p^4\,^1\mathrm{D}\,5d$	3	−20.7930	$2s^22p^4\,^3\mathrm{P}\,4d$	19	−11.5440	$2s2p^5\,^1\mathrm{P}\,4s$	4	−20.7451	$2s^22p^4\,^1\mathrm{D}\,4p$
13	−10.6122	$2s2p^5\,^1\mathrm{P}\,4p$	4	−19.6246	$2s^22p^4\,^1\mathrm{D}\,4d$	20	−11.1309	$2s^22p^4\,^1\mathrm{S}\,5p$	5	−20.3102	$2s^22p^4\,^3\mathrm{P}\,4f$
14	−9.69281	$2s^22p^4\,^3\mathrm{P}\,6s$	5	−13.2476	$2s^22p^4\,^3\mathrm{P}\,5d$	21	−9.71156	$2s2p^5\,^1\mathrm{P}\,4d$	6	−19.0621	$2s^22p^4\,^1\mathrm{D}\,4f$
15	−9.12471	$2s^22p^4\,^3\mathrm{P}\,6d$	6	−12.0627	$2s^22p^4\,^1\mathrm{D}\,5d$	22	−9.37572	$2s^22p^4\,^3\mathrm{P}\,6p$	7	−17.6833	$2s^22p^4\,^1\mathrm{S}\,4f$
16	−7.97186	$2s^22p^4\,^1\mathrm{D}\,6d$	7	−11.9012	$2s2p^5\,^3\mathrm{P}\,4f$	23	−8.23424	$2s^22p^4\,^1\mathrm{D}\,6p$	8	−12.9997	$2s^22p^4\,^3\mathrm{P}\,5f$
17	−7.04404	$2s^22p^4\,^3\mathrm{P}\,7s$	8	−9.16746	$2s^22p^4\,^3\mathrm{P}\,6d$	24	−7.83470	$2s^22p^4\,^1\mathrm{D}\,6f$	9	−12.5860	$2s^22p^4\,^1\mathrm{D}\,5p$
18	−6.69026	$2s^22p^4\,^3\mathrm{P}\,7d$	9	−9.06992	$2s2p^5\,^1\mathrm{P}\,4f$	25	−6.85602	$2s^22p^4\,^3\mathrm{P}\,7p$	10	−12.4772	$2s2p^5\,^3\mathrm{P}\,4d$
19	−5.52260	$2s^22p^4\,^1\mathrm{D}\,7d$	10	7.07580	$2s^22p^4\,^1\mathrm{D}\,6d$	26	−6.81391	$2s^22p^4\,^1\mathrm{S}\,6p$	11	−11.7846	$2s^22p^4\,^1\mathrm{D}\,5f$
20	−5.38174	$2s2p^5\,^3\mathrm{P}\,5p$	11	−6.71709	$2s^22p^4\,^3\mathrm{P}\,7d$	27	−5.79221	$2s2p^5\,^3\mathrm{P}\,5s$	12	−10.3838	$2s^22p^4\,^1\mathrm{S}\,5f$
21	5.31028	$2s^22p^4\,^3\mathrm{P}\,8s$	12	−5.52270	$2s^22p^4\,^1\mathrm{D}\,7d$	28	5.67001	$2s^22p^4\,^1\mathrm{D}\,7p$	13	9.75845	$2s2p^5\,^1\mathrm{P}\,4d$
22	−5.11326	$2s^22p^4\,^3\mathrm{P}\,8d$	13	−5.13243	$2s^22p^4\,^3\mathrm{P}\,8d$	29	−5.43603	$2s^22p^4\,^1\mathrm{D}\,7f$	14	−9.02238	$2s^22p^4\,^3\mathrm{P}\,6f$
23	−4.19958	$2s^22p^4\,^3\mathrm{P}\,9s$	14	−4.62883	$2s2p^5\,^3\mathrm{P}\,5f$	30	−5.21692	$2s^22p^4\,^3\mathrm{P}\,8p$	15	−8.25858	$2s^22p^4\,^1\mathrm{D}\,6p$
24	−4.03969	$2s^22p^4\,^3\mathrm{P}\,9d$	15	−4.04899	$2s^22p^4\,^3\mathrm{P}\,9d$	31	−4.77546	$2s2p^5\,^3\mathrm{P}\,5d$	16	−7.81604	$2s^22p^4\,^1\mathrm{D}\,6f$
25	−3.93234	$2s^22p^4\,^1\mathrm{D}\,8d$	16	−3.93613	$2s^22p^4\,^1\mathrm{D}\,8d$	32	−4.26700	$2s^22p^4\,^1\mathrm{S}\,7p$	17	−6.62662	$2s^22p^4\,^3\mathrm{P}\,7f$
		$^2\mathrm{D}^e$			$^2\mathrm{G}^e$	33	−4.11586	$2s^22p^4\,^3\mathrm{P}\,9p$	18	−6.40567	$2s^22p^4\,^1\mathrm{S}\,6f$
1	−41.3826	$2s^22p^4\,^1\mathrm{D}\,3s$	1	−36.4956	$2s^22p^4\,^1\mathrm{D}\,3d$	34	−4.03618	$2s^22p^4\,^1\mathrm{D}\,8p$	19	−5.69905	$2s^22p^4\,^1\mathrm{D}\,7p$

F-like Fe (Fe^{17+})

i	E(Ryds)	Description	i	E(Ryds)	Description	i	E(Ryds)	Description	i	E(Ryds)	Description
20	−5.42323	$2s^22p^4$ ^{1}D $7f$	9	−9.72272	$2s^22p^4$ ^{3}P $6s$	4	−9.43115	$2s^22p^4$ ^{3}P $6p$	4	−12.6463	$2s2p^5$ ^{3}P $4d$
21	−5.07255	$2s^22p^4$ ^{3}P $8f$	10	−9.17488	$2s^22p^4$ ^{3}P $6d$	5	−6.88006	$2s^22p^4$ ^{3}P $7p$	5	−9.03609	$2s^22p^4$ ^{3}P $6f$
22	−4.90298	$2s2p^5$ ^{3}P $5d$	11	−7.06338	$2s^22p^4$ ^{3}P $7s$	6	−5.24043	$2s^22p^4$ ^{3}P $8p$	6	−6.63595	$2s^22p^4$ ^{3}P $7f$
23	−4.60672	$2s2p^5$ ^{3}P $5g$	12	−6.72221	$2s^22p^4$ ^{3}P $7d$	7	−4.12427	$2s^22p^4$ ^{3}P $9p$	7	−5.08368	$2s^22p^4$ ^{3}P $8f$
24	−4.05384	$2s^22p^4$ ^{1}D $8p$	13	−5.41338	$2s2p^5$ ^{3}P $5p$			^{4}**P**°	8	−4.98139	$2s2p^5$ ^{3}P $5d$
25	−4.00817	$2s^22p^4$ ^{1}S $7f$	14	−5.32255	$2s^22p^4$ ^{3}P $8s$	1	−40.6880	$2s^22p^4$ ^{3}P $3p$	9	−4.01091	$2s^22p^4$ ^{3}P $9f$
26	−4.00595	$2s^22p^4$ ^{3}P $9f$	15	−5.13505	$2s^22p^4$ ^{3}P $8d$	2	−34.1609	$2s2p^5$ ^{3}P $3s$			^{4}**G**°
		^{2}**G**°	16	−4.20732	$2s^22p^4$ ^{3}P $9s$	3	−29.7612	$2s2p^5$ ^{3}P $3d$	1	−20.2888	$2s^22p^4$ ^{3}P $4f$
1	−20.2791	$2s^22p^4$ ^{3}P $4f$	17	−4.05112	$2s^22p^4$ ^{3}P $9d$	4	−22.0262	$2s^22p^4$ ^{3}P $4p$	2	−12.9956	$2s^22p^4$ ^{3}P $5f$
2	−19.0609	$2s^22p^4$ ^{1}D $4f$			^{4}**D**^{e}	5	−14.5018	$2s2p^5$ ^{3}P $4s$	3	−9.02257	$2s^22p^4$ ^{3}P $6f$
3	−12.9890	$2s^22p^4$ ^{3}P $5f$	1	−37.8368	$2s^22p^4$ ^{3}P $3d$	6	−13.7929	$2s^22p^4$ ^{3}P $5p$	4	−6.62689	$2s^22p^4$ ^{3}P $7f$
4	−11.7844	$2s^22p^4$ ^{1}D $5f$	2	−31.9536	$2s2p^5$ ^{3}P $3p$	7	−12.7043	$2s2p^5$ ^{3}P $4d$	5	−5.07262	$2s^22p^4$ ^{3}P $8f$
5	−9.01831	$2s^22p^4$ ^{3}P $6f$	3	−20.9903	$2s^22p^4$ ^{3}P $4d$	8	−9.48103	$2s^22p^4$ ^{3}P $6p$	6	−4.00704	$2s^22p^4$ ^{3}P $9f$
6	−7.81667	$2s^22p^4$ ^{1}D $6f$	4	−13.5817	$2s2p^5$ ^{3}P $4p$	9	−6.91365	$2s^22p^4$ ^{3}P $7p$			
7	−6.62405	$2s^22p^4$ ^{3}P $7f$	5	−13.3227	$2s^22p^4$ ^{3}P $5d$	10	−5.86610	$2s2p^5$ ^{3}P $5s$			
8	−5.42339	$2s^22p^4$ ^{1}D $7f$	6	−12.0014	$2s2p^5$ ^{3}P $4f$	11	−5.25954	$2s^22p^4$ ^{3}P $8p$			
9	−5.07062	$2s^22p^4$ ^{3}P $8f$	7	−9.21802	$2s^22p^4$ ^{3}P $6d$	12	−5.01029	$2s2p^5$ ^{3}P $5d$			
10	−4.00564	$2s^22p^4$ ^{3}P $9f$	8	−6.74864	$2s^22p^4$ ^{3}P $7d$	13	−4.13766	$2s^22p^4$ ^{3}P $9p$			
11	−3.87031	$2s^22p^4$ ^{1}D $8f$	9	−5.41148	$2s2p^5$ ^{3}P $5p$			^{4}**D**°			
		^{2}**H**°	10	−5.15211	$2s^22p^4$ ^{3}P $8d$	1	−40.3940	$2s^22p^4$ ^{3}P $3p$			
1	−19.1420	$2s^22p^4$ ^{1}D $4f$	11	−4.67478	$2s2p^5$ ^{3}P $5f$	2	−29.2618	$2s2p^5$ ^{3}P $3d$			
2	−11.8205	$2s^22p^4$ ^{1}D $5f$	12	−4.06309	$2s^22p^4$ ^{3}P $9d$	3	−21.9414	$2s^22p^4$ ^{3}P $4p$			
3	−7.83701	$2s^22p^4$ ^{1}D $6f$			^{4}**F**^{e}	4	−20.2524	$2s^22p^4$ ^{3}P $4f$			
4	−5.43616	$2s^22p^4$ ^{1}D $7f$	1	−37.4843	$2s^22p^4$ ^{3}P $3d$	5	−13.7869	$2s^22p^4$ ^{3}P $5p$			
5	−3.87888	$2s^22p^4$ ^{1}D $8f$	2	−20.8786	$2s^22p^4$ ^{3}P $4d$	6	−12.9842	$2s^22p^4$ ^{3}P $5f$			
		^{4}**S**^{e}	3	−13.2885	$2s^22p^4$ ^{3}P $5d$	7	−12.5355	$2s2p^5$ ^{3}P $4d$			
1	−32.3174	$2s2p^5$ ^{3}P $3p$	4	−11.9129	$2s2p^5$ ^{3}P $4f$	8	−9.46010	$2s^22p^4$ ^{3}P $6p$			
2	−13.6809	$2s2p^5$ ^{3}P $4p$	5	−9.19040	$2s^22p^4$ ^{3}P $6d$	9	−9.01351	$2s^22p^4$ ^{3}P $6f$			
3	−5.46420	$2s2p^5$ ^{3}P $5p$	6	−6.73152	$2s^22p^4$ ^{3}P $7d$	10	−6.89871	$2s^22p^4$ ^{3}P $7p$			
4	−1.12667	$2s2p^5$ ^{3}P $6p$	7	−5.14193	$2s^22p^4$ ^{3}P $8d$	11	−6.62150	$2s^22p^4$ ^{3}P $7f$			
		^{4}**P**^{e}	8	−4.63652	$2s2p^5$ ^{3}P $5f$	12	−5.25485	$2s^22p^4$ ^{3}P $8p$			
1	−42.7029	$2s^22p^4$ ^{3}P $3s$	9	−4.05550	$2s^22p^4$ ^{3}P $9d$	13	−5.07117	$2s^22p^4$ ^{3}P $8f$			
2	−37.3160	$2s^22p^4$ ^{3}P $3d$			^{4}**G**^{e}	14	−4.93251	$2s2p^5$ ^{3}P $5d$			
3	−31.7113	$2s2p^5$ ^{3}P $3p$	1	−11.9807	$2s2p^5$ ^{3}P $4f$	15	−4.13204	$2s^22p^4$ ^{3}P $9p$			
4	−22.8661	$2s^22p^4$ ^{3}P $4s$	2	−4.66795	$2s2p^5$ ^{3}P $5f$	16	−4.00427	$2s^22p^4$ ^{3}P $9f$			
5	−20.8191	$2s^22p^4$ ^{3}P $4d$			^{4}**S**°			^{4}**F**°			
6	−14.2705	$2s^22p^4$ ^{3}P $5s$	1	−40.0325	$2s^22p^4$ ^{3}P $3p$	1	−29.6042	$2s2p^5$ ^{3}P $3d$			
7	−13.4716	$2s2p^5$ ^{3}P $4p$	2	−21.8250	$2s^22p^4$ ^{3}P $4p$	2	−20.3383	$2s^22p^4$ ^{3}P $4f$			
8	−13.2492	$2s^22p^4$ ^{3}P $5d$	3	−13.7242	$2s^22p^4$ ^{3}P $5p$	3	−13.0273	$2s^22p^4$ ^{3}P $5f$			

F-like Fe (Fe^{17+})

Energies in ascending order from ground state for terms with effective $n \leq 4.0$, $L \leq 4$

Term	i	E(Ryds)	Term	i	E(Ryds)	Term	i	E(Ryds)	Term	i	E(Ryds)	Term	i	E(Ryds)
$^2\mathbf{P}^o$	1	0.00000	$^4\mathbf{D}^e$	1	61.7656	$^4\mathbf{S}^e$	1	67.2850	$^2\mathbf{P}^o$	7	71.1599	$^2\mathbf{P}^o$	10	78.7077
$^2\mathbf{S}^e$	1	8.60270	$^2\mathbf{P}^o$	4	62.0101	$^4\mathbf{D}^e$	2	67.6488	$^2\mathbf{F}^o$	3	72.9783	$^4\mathbf{F}^e$	2	78.7238
$^4\mathbf{P}^e$	1	56.8995	$^4\mathbf{F}^e$	1	62.1181	$^4\mathbf{P}^e$	3	67.8911	$^2\mathbf{P}^o$	8	73.1141	$^4\mathbf{P}^e$	5	78.7833
$^2\mathbf{P}^e$	1	57.2633	$^4\mathbf{P}^e$	2	62.2864	$^2\mathbf{D}^e$	5	67.9425	$^2\mathbf{D}^o$	4	73.2471	$^2\mathbf{F}^e$	3	78.8094
$^2\mathbf{D}^e$	1	58.2198	$^2\mathbf{F}^e$	1	62.3456	$^2\mathbf{P}^e$	4	68.0167	$^2\mathbf{S}^e$	6	76.6162	$^2\mathbf{F}^o$	4	78.8573
$^4\mathbf{P}^o$	1	58.9144	$^2\mathbf{P}^e$	2	62.4847	$^2\mathbf{P}^o$	6	68.4432	$^4\mathbf{P}^e$	4	76.7363	$^2\mathbf{P}^e$	7	78.9396
$^4\mathbf{D}^o$	1	59.2084	$^2\mathbf{D}^e$	2	62.5988	$^2\mathbf{S}^e$	4	68.5905	$^2\mathbf{P}^e$	6	76.8515	$^2\mathbf{D}^o$	6	78.9471
$^2\mathbf{D}^o$	1	59.4269	$^2\mathbf{G}^e$	1	63.1068	$^4\mathbf{P}^o$	3	69.8412	$^4\mathbf{P}^o$	4	77.5762	$^2\mathbf{D}^e$	8	78.9964
$^2\mathbf{P}^o$	2	59.5257	$^2\mathbf{S}^e$	3	63.4017	$^4\mathbf{F}^o$	1	69.9982	$^4\mathbf{D}^o$	3	77.6610	$^2\mathbf{P}^o$	11	79.0215
$^2\mathbf{S}^o$	1	59.5608	$^2\mathbf{F}^e$	2	63.4666	$^4\mathbf{D}^o$	2	70.3406	$^2\mathbf{D}^o$	5	77.7372	$^4\mathbf{F}^o$	2	79.2641
$^4\mathbf{S}^o$	1	59.5699	$^2\mathbf{D}^e$	3	63.7290	$^2\mathbf{F}^o$	2	70.5096	$^2\mathbf{S}^o$	2	77.7741	$^2\mathbf{F}^o$	5	79.2922
$^2\mathbf{S}^e$	2	59.6550	$^2\mathbf{P}^e$	3	63.7385	$^2\mathbf{D}^e$	6	70.5840	$^4\mathbf{S}^o$	2	77.7774	$^4\mathbf{G}^o$	1	79.3136
$^2\mathbf{F}^o$	1	60.3887	$^2\mathbf{D}^e$	4	64.6729	$^2\mathbf{D}^o$	3	70.6418	$^2\mathbf{P}^o$	9	77.9207	$^2\mathbf{G}^o$	1	79.3233
$^2\mathbf{D}^o$	2	60.6775	$^4\mathbf{P}^o$	2	65.4415	$^2\mathbf{P}^e$	5	70.7904	$^2\mathbf{D}^e$	7	77.9744	$^4\mathbf{D}^o$	4	79.3500
$^2\mathbf{P}^o$	3	61.2422	$^2\mathbf{P}^o$	5	66.0144	$^2\mathbf{S}^e$	5	71.0628	$^4\mathbf{D}^e$	3	78.6121			

F-like Fe (Fe^{17+})

gf-values for transitions involving terms with effective $n \leq 4.0$, $L \leq 4$

i	i'	gf_L	i	i'	gf_L	i	i'	gf_L	i	i'	gf_L	i	i'	gf_L	i	i'	gf_L
		$^2S^e$–$^2P^o$	5	7	−4.40E−4	2	9	−6.04E−2	7	5	3.39E−4			$^2D^e$–$^2P^o$	5	7	−3.90E−3
1	1	2.77E−1	5	8	−2.30E−1	2	10	−3.36E−5	7	6	5.29E−3	1	1	3.83E−1	5	8	−1.75E−5
1	2	−4.79E−4	5	9	−1.44E−4	2	11	−1.51E−2	7	7	1.03E−3	1	2	−9.73E−2	5	9	−2.26E−3
1	3	−1.33E−2	5	10	−1.64E−1	3	1	4.53E+0	7	8	1.06E−6	1	3	−5.18E−1	5	10	−4.92E−4
1	4	−3.85E−3	5	11	−5.83E−4	3	2	4.04E−2	7	9	4.05E−1	1	4	−5.60E−4	5	11	−1.17E−3
1	5	−3.31E−1	6	1	3.40E−4	3	3	3.31E−1	7	10	5.01E−5	1	5	−2.54E−2	6	1	3.25E−1
1	6	−1.56E−1	6	2	1.98E−4	3	4	2.56E−3	7	11	−1.16E−6	1	6	−1.15E+0	6	2	5.07E−3
1	7	−3.14E+0	6	3	5.82E−4	3	5	−5.92E−3			$^2P^e$–$^2D^o$	1	7	−2.90E−4	6	3	9.78E−3
1	8	−3.12E+0	6	4	2.50E−5	3	6	−1.61E−3	1	1	−9.17E−1	1	8	−1.52E−4	6	4	1.75E−1
1	9	−1.38E−3	6	5	7.74E−4	3	7	−6.61E−3	1	2	−7.77E−3	1	9	−4.89E−2	6	5	9.15E−3
1	10	−3.20E−1	6	6	6.36E−1	3	8	−2.22E−1	1	3	−3.02E−3	1	10	−8.32E−4	6	6	9.43E−1
1	11	−8.54E−6	6	7	4.60E−3	3	9	−2.73E−5	1	4	−6.86E−4	1	11	−4.54E−1	6	7	−3.39E−4
2	1	7.47E−2	6	8	2.03E−3	3	10	−1.51E−4	1	5	−9.33E−1	2	1	2.31E+0	6	8	−1.05E−3
2	2	4.52E−4	6	9	−2.28E−4	3	11	−1.05E−1	1	6	−9.84E−5	2	2	6.98E−1	6	9	−1.46E−3
2	3	−4.80E−3	6	10	−6.32E−1	4	1	6.59E−1	2	1	2.53E−2	2	3	2.72E−2	6	10	−1.09E+0
2	4	−5.82E−1	6	11	−1.97E−4	4	2	8.90E−2	2	2	2.52E−2	2	4	1.68E−4	6	11	−3.20E−3
2	5	−5.54E−3			$^2P^e$–$^2S^o$	4	3	1.08E−1	2	3	−1.67E−1	2	5	−3.25E−3	7	1	7.63E−2
2	6	−1.05E−1	1	1	−1.92E−1	4	4	2.13E−2	2	4	−4.98E−3	2	6	−1.74E−3	7	2	7.28E−2
2	7	−1.79E−3	1	2	−1.76E−1	4	5	5.13E−1	2	5	−1.30E−2	2	7	−4.13E−1	7	3	3.25E−1
2	8	−6.68E−5	2	1	2.89E−1	4	6	−3.20E−5	2	6	−4.50E−3	2	8	−3.46E−2	7	4	1.75E−3
2	9	−2.14E−3	2	2	−8.05E−2	4	7	−3.43E−1	3	1	3.42E−2	2	9	−1.79E−1	7	5	6.79E−3
2	10	−1.43E−5	3	1	1.75E−1	4	8	−4.87E−2	3	2	1.08E−1	2	10	−2.30E−4	7	6	1.72E−4
2	11	−1.21E−4	3	2	−5.29E−2	4	9	−2.50E−3	3	3	−4.66E−2	2	11	−4.25E−3	7	7	3.05E−7
3	1	1.18E+0	4	1	1.89E−1	4	10	−2.52E−3	3	4	−3.69E−2	3	1	6.22E+0	7	8	4.04E−5
3	2	8.80E−2	4	2	−2.44E−4	4	11	−2.34E−3	3	5	−1.68E−2	3	2	3.35E−3	7	9	2.01E−3
3	3	1.02E−1	5	1	1.34E−3	5	1	1.60E−1	3	6	−3.43E−2	3	3	3.89E−1	7	10	−2.22E−4
3	4	2.33E−3	5	2	−6.16E−4	5	2	5.08E−2	4	1	2.41E−1	3	4	4.58E−3	7	11	−9.48E−1
3	5	−2.25E−3	6	1	8.67E−2	5	3	2.22E−1	4	2	2.29E−5	3	5	−8.91E−3	8	1	1.35E+0
3	6	−1.74E−4	6	2	−2.65E−1	5	4	9.50E−2	4	3	−8.25E−1	3	6	−1.92E−3	8	2	1.28E+0
3	7	−1.81E−3	7	1	7.31E−1	5	5	7.77E−4	4	4	−4.42E−3	3	7	−3.72E−2	8	3	4.73E−1
3	8	−8.69E−2	7	2	6.76E−1	5	6	6.55E−1	4	5	−7.05E−4	3	8	−3.10E−1	8	4	1.81E−2
3	9	−7.26E−3			$^2P^e$–$^2P^o$	5	7	−1.53E−3	4	6	−1.30E−5	3	9	−2.66E−2	8	5	3.20E−3
3	10	−1.13E−5	1	1	5.94E−1	5	8	−2.20E−1	5	1	5.61E−3	3	10	−9.59E−5	8	6	1.20E−2
3	11	−4.73E−2	1	2	−3.54E−1	5	9	−3.23E−4	5	2	8.48E−1	3	11	−6.68E−2	8	7	4.27E−3
4	1	2.89E−1	1	3	−4.51E−1	5	10	−5.64E−1	5	3	6.11E−5	4	1	1.52E+0	8	8	4.68E−4
4	2	2.31E−1	1	4	−4.95E−2	5	11	−2.81E−4	5	4	−7.47E−1	4	2	1.24E−3	8	9	1.35E+0
4	3	1.98E−3	1	5	−4.15E−1	6	1	1.04E−1	5	5	−1.63E−3	4	3	1.69E−2	8	10	4.90E−5
4	4	1.88E−3	1	6	−1.15E−3	6	2	2.17E−1	5	6	−2.36E−4	4	4	9.93E−1	8	11	−4.30E−4
4	5	1.90E−1	1	7	−1.45E−3	6	3	2.10E−1	6	1	5.02E−1	4	5	−2.25E−3			$^2D^e$–$^2D^o$
4	6	1.59E−3	1	8	−8.40E−5	6	4	1.46E−2	6	2	7.20E−3	4	6	−3.66E−4	1	1	−1.14E−3
4	7	−3.15E−1	1	9	−4.71E−1	6	5	6.20E−3	6	3	2.08E−3	4	7	−3.00E−4	1	2	−1.05E+0
4	8	−1.95E−2	1	10	−1.33E−4	6	6	1.53E−3	6	4	1.93E−4	4	8	−9.37E−2	1	3	−2.59E−4
4	9	−5.46E−4	1	11	−1.63E−2	6	7	8.86E−4	6	5	−1.36E+0	4	9	−3.44E−3	1	4	−5.97E−4
4	10	−7.60E−3	2	1	2.71E−1	6	8	4.24E−8	6	6	−2.97E−3	4	10	−1.57E−6	1	5	−9.50E−4
4	11	−1.06E−3	2	2	3.12E−1	6	9	−9.65E−1	7	1	3.43E−2	4	11	−1.69E−3	1	6	−8.58E−1
5	1	2.07E−4	2	3	3.13E−4	6	10	−5.36E−4	7	2	2.50E−2	5	1	9.43E−1	2	1	1.87E−1
5	2	2.99E−2	2	4	4.70E−4	6	11	−1.69E−1	7	3	4.47E−3	5	2	1.51E−1	2	2	1.05E−1
5	3	2.35E−1	2	5	−1.75E−4	7	1	3.64E−1	7	4	7.76E−5	5	3	1.84E−1	2	3	−3.54E−2
5	4	5.01E−2	2	6	−6.76E−4	7	2	5.20E−1	7	5	5.31E−2	5	4	3.44E−2	2	4	−4.56E−2
5	5	1.43E−1	2	7	−1.55E−1	7	3	8.02E−2	7	6	−5.63E−5	5	5	8.38E−1	2	5	−5.30E−2
5	6	1.79E−1	2	8	−3.75E−2	7	4	1.92E−2				5	6	−1.19E−4	2	6	−3.26E−2

F-like Fe (Fe^{17+})

i i′	gf$_L$	*i i′*	gf$_L$	*i i′*	gf$_L$	*i i′*	gf$_L$	*i i′*	gf$_L$	*i i′*	gf$_L$
3 1	1.30E−1	1 3	−1.98E−3		$^2F^e$–$^2D^o$		$^2G^e$–$^2F^o$	4 3	5.16E−4	3 4	2.57E+0
3 2	5.76E−1	1 4	−1.34E+0	1 1	1.58E+0	1 1	1.81E+0	4 4	−1.54E+0		$^4D^e$–$^4D^o$
3 3	−3.61E−3	1 5	−1.27E−3	1 2	5.42E−5	1 2	−1.65E−3	5 1	1.11E+0	1 1	4.24E−1
3 4	−2.85E−1	2 1	9.08E−3	1 3	−4.92E−1	1 3	−2.09E+0	5 2	1.47E−2	1 2	−8.89E−2
3 5	−3.30E−2	2 2	−1.17E−1	1 4	−1.31E−3	1 4	−3.89E−1	5 3	4.21E−3	1 3	−1.15E−1
3 6	−1.94E−1	2 3	−8.21E−3	1 5	−3.69E−1	1 5	−7.90E−4	5 4	9.96E−1	1 4	−2.11E+0
4 1	7.83E−3	2 4	−1.93E−5	1 6	−8.24E−4		$^2G^e$–$^2G^o$		$^4P^e$–$^4D^o$	2 1	1.41E+0
4 2	6.83E−3	2 5	−6.49E+0	2 1	4.10E−3	1 1	−1.91E−5	1 1	−1.93E+0	2 2	−5.70E−1
4 3	−2.16E−3	3 1	3.14E−2	2 2	1.01E+0		$^4S^e$–$^4P^o$	1 2	−2.41E−4	2 3	−2.45E−4
4 4	−2.26E−1	3 2	−2.70E−2	2 3	−1.07E−7	1 1	4.44E−1	1 3	−1.84E+0	2 4	−1.72E−2
4 5	−2.27E−3	3 3	−1.10E−2	2 4	−9.70E−1	1 2	3.00E−1	1 4	−3.14E−3	3 1	1.27E+0
4 6	−4.55E−4	3 4	−5.17E−3	2 5	−2.94E−3	1 3	−6.37E−1	2 1	7.00E−2	3 2	2.38E−3
5 1	7.27E−1	3 5	−2.07E+0	2 6	−2.27E−1	1 4	−5.33E−9	2 2	−4.16E−1	3 3	7.75E−1
5 2	1.63E−4	4 1	7.80E−3	3 1	3.32E+0		$^4P^e$–$^4S^o$	2 3	−3.49E−2	3 4	−2.75E−1
5 3	−2.92E−1	4 2	−1.14E−3	3 2	3.96E−3	1 1	−4.73E−1	2 4	−1.15E+1		$^4D^e$–$^4F^o$
5 4	−3.81E−3	4 3	−1.22E−1	3 3	1.59E−2	1 2	−3.22E−1	3 1	5.71E−1	1 1	−3.18E−1
5 5	−5.79E−8	4 4	−4.75E−3	3 4	1.60E−6	2 1	7.08E−1	3 2	−1.52E+0	1 2	−1.66E+1
5 6	−1.60E−3	4 5	−1.56E−1	3 5	2.55E+0	2 2	−1.95E−1	3 3	−1.29E−2	2 1	−2.58E+0
6 1	1.20E−3	5 1	7.22E−4	3 6	−1.81E−6	3 1	3.42E−1	3 4	−5.61E−2	2 2	−1.32E−1
6 2	2.39E−1	5 2	−1.47E+0		$^2F^e$–$^2F^o$	3 2	−3.52E−3	4 1	1.04E+0	3 1	1.31E−2
6 3	−3.07E−5	5 3	−4.40E−3	1 1	2.75E−3	4 1	2.49E−1	4 2	1.67E−3	3 2	−1.87E+0
6 4	−2.89E−1	5 4	−1.02E−3	1 2	−7.24E−1	4 2	−6.43E−1	4 3	−2.84E+0		$^4F^e$–$^4D^o$
6 5	−8.41E−6	5 5	−6.26E−2	1 3	−9.79E−3	5 1	1.64E+0	4 4	−1.78E−3	1 1	3.10E+0
6 6	−6.39E−3	6 1	1.73E+0	1 4	−3.57E−5	5 2	1.15E+0	5 1	7.74E−2	1 2	−9.74E−1
7 1	9.49E−5	6 2	4.47E−5	1 5	−1.12E+0		$^4P^e$–$^4P^o$	5 2	1.09E−2	1 3	−7.28E−1
7 2	5.58E−1	6 3	−1.30E+0	2 1	5.74E−1	1 1	−9.30E−1	5 3	8.05E−2	1 4	−4.74E−2
7 3	4.58E−5	6 4	−4.78E−3	2 2	−6.48E−3	1 2	−1.30E+0	5 4	−1.10E+0	2 1	6.71E+0
7 4	5.23E−4	6 5	−3.29E−4	2 3	−3.33E−1	1 3	−1.23E−4		$^4D^e$–$^4P^o$	2 2	3.99E−2
7 5	9.31E−7	7 1	6.82E−1	2 4	−1.65E−1	1 4	−1.12E+0	1 1	1.53E+0	2 3	5.01E+0
7 6	−1.48E+0	7 2	1.04E−4	2 5	−1.14E−2	2 1	6.07E−1	1 2	−4.49E−4	2 4	−4.33E−3
8 1	4.98E−1	7 3	2.85E−3	3 1	1.76E−3	2 2	−6.33E−5	1 3	−8.60E−1		$^4F^e$–$^4F^o$
8 2	7.35E−2	7 4	−1.91E+0	3 2	5.27E−4	2 3	−2.58E−1	1 4	−3.84E−1	1 1	−1.38E+0
8 3	5.80E−4	7 5	−6.79E−4	3 3	5.16E−6	2 4	−1.45E−1	2 1	3.16E−1	1 2	−2.29E+0
8 4	2.46E−5	8 1	1.10E−2	3 4	−1.63E−5	3 1	2.41E−1	2 2	2.02E+0	2 1	2.09E−3
8 5	5.15E−1	8 2	6.49E−3	3 5	−8.54E−2	3 2	1.41E+0	2 3	−1.14E−2	2 2	−1.92E−1
8 6	7.80E−4	8 3	4.82E−4		$^2F^e$–$^2G^o$	3 3	−3.85E−1	2 4	−4.38E−4		$^4F^e$–$^4G^o$
	$^2D^e$–$^2F^o$	8 4	1.05E−3	1 1	−1.20E+1	3 4	−7.29E−3	3 1	3.63E+0	1 1	−2.39E+1
1 1	−1.27E+0	8 5	−4.01E−1	2 1	−9.47E−2	4 1	6.33E−1	3 2	4.75E−2	2 1	−2.45E+0
1 2	−1.37E−3			3 1	−1.06E+0	4 2	1.64E−3	3 3	1.08E−2		

Ne-like S (S^{6+})

Term energies relative to $2s^22p^5\ ^2P$ ionization threshold for each symmetry

i	E(Ryds)	Description
	$^1S^e$	
1	−20.7455	$2s^22p^6$
2	−6.45659	$2s^22p^53p$
3	−3.49891	$2s^22p^54p$
4	−3.26698	$2s2p^63s$
5	−2.15235	$2s^22p^55p$
6	−1.47164	$2s^22p^56p$
7	−1.06912	$2s^22p^57p$
8	−.81168	$2s^22p^58p$
9	−.63715	$2s^22p^59p$
	$^3S^e$	
1	−7.26746	$2s^22p^53p$
2	−3.79494	$2s^22p^54p$
3	−3.43892	$2s2p^63s$
4	−2.28162	$2s^22p^55p$
5	−1.54353	$2s^22p^56p$
6	−1.11309	$2s^22p^57p$
7	−.84053	$2s^22p^58p$
8	−.65711	$2s^22p^59p$
	$^1P^e$	
1	−7.00961	$2s^22p^53p$
2	−3.66089	$2s^22p^54p$
3	−2.25226	$2s^22p^55p$
4	−1.52570	$2s^22p^56p$
5	−1.10182	$2s^22p^57p$
6	−.83299	$2s^22p^58p$
7	−.65182	$2s^22p^59p$
	$^3P^e$	
1	−7.00637	$2s^22p^53p$
2	−3.65974	$2s^22p^54p$
3	−2.25172	$2s^22p^55p$
4	−1.52540	$2s^22p^56p$
5	−1.10164	$2s^22p^57p$
6	−.83288	$2s^22p^58p$
7	−.65173	$2s^22p^59p$
	$^1P^o$	
1	−7.99207	$2s^22p^53s$
2	−5.50512	$2s^22p^53d$
3	−4.02643	$2s^22p^54s$
4	−3.09667	$2s^22p^54d$
5	−2.43425	$2s^22p^55s$
6	−2.41680	$2s2p^63p$
7	−1.97298	$2s^22p^55d$
8	−1.62283	$2s^22p^56s$
9	−1.37060	$2s^22p^56d$
10	−1.16118	$2s^22p^57s$
11	−1.00640	$2s^22p^57d$
12	−.87187	$2s^22p^58s$
13	−.77008	$2s^22p^58d$
14	−.67865	$2s^22p^59s$
15	−.60814	$2s^22p^59d$
	$^3P^o$	
1	−8.09298	$2s^22p^53s$
2	−5.82458	$2s^22p^53d$
3	−4.06111	$2s^22p^54s$
4	−3.23300	$2s^22p^54d$
5	−2.51865	$2s2p^63p$
6	−2.41881	$2s^22p^55s$
7	−2.04693	$2s^22p^55d$
8	−1.62968	$2s^22p^56s$
9	−1.41218	$2s^22p^56d$
10	−1.16554	$2s^22p^57s$
11	−1.03230	$2s^22p^57d$
12	−.87476	$2s^22p^58s$
13	−.78730	$2s^22p^58d$
14	−.68065	$2s^22p^59s$
15	−.62017	$2s^22p^59d$
	$^1D^e$	
1	−7.02623	$2s^22p^53p$
2	−3.66312	$2s^22p^54p$
3	−3.07261	$2s^22p^54f$
4	−2.25361	$2s^22p^55p$
5	−1.96761	$2s^22p^55f$
6	−1.52687	$2s^22p^56p$
7	−1.36962	$2s^22p^56f$
8	−1.11902	$2s2p^63d$
9	−1.09427	$2s^22p^57p$
10	−.99352	$2s^22p^57f$
11	−.83288	$2s^22p^58p$
12	−.76505	$2s^22p^58f$
13	−.65184	$2s^22p^59p$
14	−.60496	$2s^22p^59f$
	$^3D^e$	
1	−7.09965	$2s^22p^53p$
2	−3.68895	$2s^22p^54p$
3	−3.07818	$2s^22p^54f$
4	−2.26587	$2s^22p^55p$
5	−1.97357	$2s^22p^55f$
6	−1.53408	$2s^22p^56p$
7	−1.38110	$2s^22p^56f$
8	−1.19080	$2s2p^63d$
9	−1.10371	$2s^22p^57p$
10	−.99400	$2s^22p^57f$
11	−.83540	$2s^22p^58p$
12	−.76495	$2s^22p^58f$
13	−.65358	$2s^22p^59p$
14	−.60505	$2s^22p^59f$
	$^1D^o$	
1	−5.66887	$2s^22p^53d$
2	−3.17894	$2s^22p^54d$
3	−2.02255	$2s^22p^55d$
4	−1.39806	$2s^22p^56d$
5	−1.02351	$2s^22p^57d$
6	−.78147	$2s^22p^58d$
7	−.61611	$2s^22p^59d$
	$^3D^o$	
1	−5.66720	$2s^22p^53d$
2	−3.17819	$2s^22p^54d$
3	−2.02217	$2s^22p^55d$
4	−1.39784	$2s^22p^56d$
5	−1.02337	$2s^22p^57d$
6	−.78138	$2s^22p^58d$
7	−.61605	$2s^22p^59d$
	$^1F^e$	
1	−3.05875	$2s^22p^54f$
2	−1.95891	$2s^22p^55f$
3	−1.36083	$2s^22p^56f$
4	−.99996	$2s^22p^57f$
5	−.76566	$2s^22p^58f$
6	−.60500	$2s^22p^59f$
	$^3F^e$	
1	−3.05867	$2s^22p^54f$
2	−1.95884	$2s^22p^55f$
3	−1.36078	$2s^22p^56f$
4	−.99993	$2s^22p^57f$
5	−.76564	$2s^22p^58f$
6	−.60498	$2s^22p^59f$
	$^1F^o$	
1	−5.70702	$2s^22p^53d$
2	−3.18672	$2s^22p^54d$
3	−2.02594	$2s^22p^55d$
4	−1.96064	$2s^22p^55g$
5	−1.39988	$2s^22p^56d$
6	−1.36169	$2s^22p^56g$
7	−1.02461	$2s^22p^57d$
8	−1.00044	$2s^22p^57g$
9	−.78219	$2s^22p^58d$
10	−.76596	$2s^22p^58g$
11	−.61660	$2s^22p^59d$
12	−.60519	$2s^22p^59g$
	$^3F^o$	
1	−5.76303	$2s^22p^53d$
2	−3.21099	$2s^22p^54d$
3	−2.03822	$2s^22p^55d$
4	−1.96068	$2s^22p^55g$
5	−1.40690	$2s^22p^56d$
6	−1.36174	$2s^22p^56g$
7	−1.02899	$2s^22p^57d$
8	−1.00049	$2s^22p^57g$
9	−.78510	$2s^22p^58d$
10	−.76599	$2s^22p^58g$
11	−.61864	$2s^22p^59d$
12	−.60521	$2s^22p^59g$
	$^1G^e$	
1	−3.06952	$2s^22p^54f$
2	−1.96471	$2s^22p^55f$
3	−1.36419	$2s^22p^56f$
4	−1.36111	$2s^22p^56h$
5	−1.00207	$2s^22p^57f$
6	−1.00000	$2s^22p^57h$
7	−.76707	$2s^22p^58f$
8	−.76563	$2s^22p^58h$
9	−.60598	$2s^22p^59f$
10	−.60494	$2s^22p^59h$
	$^3G^e$	
1	−3.07066	$2s^22p^54f$
2	−1.96557	$2s^22p^55f$
3	−1.36476	$2s^22p^56f$
4	−1.36111	$2s^22p^56h$
5	−1.00246	$2s^22p^57f$
6	−1.00000	$2s^22p^57h$
7	−.76734	$2s^22p^58f$
8	−.76563	$2s^22p^58h$
9	−.60618	$2s^22p^59f$
10	−.60494	$2s^22p^59h$
	$^1G^o$	
1	−1.95932	$2s^22p^55g$
2	−1.36053	$2s^22p^56g$
3	−.99957	$2s^22p^57g$
4	−.76531	$2s^22p^58g$
5	−.60471	$2s^22p^59g$
	$^3G^o$	
1	−1.95932	$2s^22p^55g$
2	−1.36053	$2s^22p^56g$
3	−.99957	$2s^22p^57g$
4	−.76531	$2s^22p^58g$
5	−.60471	$2s^22p^59g$

Ne-like S (S^{6+})

Energies in ascending order from ground state for terms with effective $n \leq 4.0$, $L \leq 4$

Term	i	E(Ryds)	Term	i	E(Ryds)	Term	i	E(Ryds)	Term	i	E(Ryds)	Term	i	E(Ryds)
$^1S^e$	1	0.00000	$^3P^e$	1	13.7391	$^1P^o$	2	15.2403	$^3P^e$	2	17.0857	$^1D^o$	2	17.5665
$^3P^o$	1	12.6525	$^1S^e$	2	14.2889	$^3P^o$	3	16.6843	$^1S^e$	3	17.2465	$^3D^o$	2	17.5673
$^1P^o$	1	12.7534	$^3P^o$	2	14.9209	$^1P^o$	3	16.7190	$^3S^e$	3	17.3065	$^1P^o$	4	17.6488
$^3S^e$	1	13.4780	$^3F^o$	1	14.9824	$^3S^e$	2	16.9505	$^1S^e$	4	17.4785	$^3D^e$	3	17.6673
$^3D^e$	1	13.6458	$^1F^o$	1	15.0384	$^3D^e$	2	17.0565	$^3P^o$	4	17.5125	$^1D^e$	3	17.6728
$^1D^e$	1	13.7192	$^1D^o$	1	15.0766	$^1D^e$	2	17.0823	$^3F^o$	2	17.5345	$^3G^e$	1	17.6748
$^1P^e$	1	13.7358	$^3D^o$	1	15.0783	$^1P^e$	2	17.0846	$^1F^o$	2	17.5587	$^1G^e$	1	17.6759

gf-values for transitions involving terms with effective $n \leq 4.0$, $L \leq 4$

i i'	gf_L	i i'	gf_L	i i'	gf_L	i i'	gf_L	i i'	gf_L	i i'	gf_L
	$^1S^e$–$^1P^o$	1 3	−2.20E−1	3 3	−6.99E−3	2 1	−8.12E−1	2 2	3.03E−1	4 3	−1.50E+0
1 1	−2.45E−1	1 4	−6.92E−2	4 1	1.47E−2		$^3S^e$–$^3P^o$	2 3	2.74E+0		$^3D^e$–$^3D^o$
1 2	−1.61E+0	2 1	1.12E−1	4 2	2.39E−2	1 1	4.63E−1	2 4	−1.72E+0	1 1	−1.30E+0
1 3	−4.44E−2	2 2	1.69E−1	4 3	−7.33E−2	1 2	−1.53E+0		$^3P^e$–$^3D^o$	1 2	−2.36E−1
1 4	−6.10E−1	2 3	8.27E−1		$^1D^e$–$^1D^o$	1 3	−2.80E−1	1 1	−3.68E+0	2 1	3.26E−1
2 1	3.52E−1	2 4	−8.01E−1	1 1	−4.18E−1	1 4	−3.84E−1	1 2	−8.22E−1	2 2	−1.96E+0
2 2	−3.94E−1		$^1P^e$–$^1D^o$	1 2	−8.85E−2	2 1	4.98E−1	2 1	8.96E−1	3 1	1.45E+0
2 3	−1.88E−1	1 1	−1.23E+0	2 1	9.93E−2	2 2	3.13E−1	2 2	−5.54E+0	3 2	1.68E−1
2 4	−2.25E−1	1 2	−2.74E−1	2 2	−6.23E−1	2 3	5.48E−1		$^3P^o$–$^3D^e$		$^3D^e$–$^3F^o$
3 1	9.03E−3	2 1	2.99E−1	3 1	4.91E−1	2 4	−2.23E+0	1 1	−2.93E+0	1 1	−6.54E+0
3 2	9.88E−2	2 2	−1.85E+0	3 2	6.15E−2	3 1	3.97E−1	1 2	−5.80E−1	1 2	−1.66E+0
3 3	3.63E−1		$^1P^o$–$^1D^e$		$^1D^e$–$^1F^o$	3 2	4.80E−2	1 3	−2.77E−3	2 1	1.63E+0
3 4	−5.31E−1	1 1	−9.84E−1	1 1	−2.22E+0	3 3	2.23E−1	2 1	3.75E−2	2 2	−1.01E+1
4 1	1.57E−1	1 2	−1.93E−1	1 2	−5.37E−1	3 4	−1.59E−1	2 2	−1.22E−3	3 1	2.83E−2
4 2	2.25E−3	1 3	−1.49E−3	2 1	5.53E−1		$^3P^e$–$^3P^o$	2 3	−7.77E+0	3 2	1.21E−2
4 3	1.25E−1	2 1	1.35E−2	2 2	−3.40E+0	1 1	1.94E+0	3 1	1.40E+0		$^3F^o$–$^3G^e$
4 4	−7.42E−2	2 2	−5.41E−4	3 1	1.04E−2	1 2	−1.10E+0	3 2	−4.33E+0	1 1	−1.71E+1
	$^1P^e$–$^1P^o$	2 3	−2.95E+0	3 2	2.68E−3	1 3	−7.29E−1	3 3	−1.51E−2	2 1	−2.94E+0
1 1	5.84E−1	3 1	4.47E−1		$^1F^o$–$^1G^e$	1 4	−4.24E−1	4 1	4.71E−2		
1 2	−5.23E−1	3 2	−1.44E+0	1 1	−5.85E+0	2 1	2.76E−1	4 2	6.55E−2		

Ne-like Ar (Ar^{8+})

Term energies relative to $2s^22p^5$ ^{2}P ionization threshold for each symmetry

i	E(Ryds)	Description	i	E(Ryds)	Description	i	E(Ryds)	Description	i	E(Ryds)	Description
	$^{1}\mathbf{S}^{e}$		14	−1.10125	$2s^22p^59s$	3	−3.34135	$2s^22p^55d$	11	−1.02132	$2s^22p^59d$
1	−31.1600	$2s^22p^6$	15	−1.00636	$2s^22p^59d$	4	−2.30948	$2s^22p^56d$	12	−1.00058	$2s^22p^59g$
2	−10.3762	$2s^22p^53p$		$^{3}\mathbf{P}^{o}$		5	−1.69079	$2s^22p^57d$		$^{1}\mathbf{G}^{e}$	
3	−6.82823	$2s2p^63s$	1	−12.5078	$2s^22p^53s$	6	−1.29100	$2s^22p^58d$	1	−5.07539	$2s^22p^54f$
4	−5.59634	$2s^22p^54p$	2	−9.60965	$2s^22p^53d$	7	−1.01787	$2s^22p^59d$	2	−3.24863	$2s^22p^55f$
5	−3.51228	$2s^22p^55p$	3	−6.41464	$2s^22p^54s$		$^{3}\mathbf{D}^{o}$		3	−2.25563	$2s^22p^56f$
6	−2.40654	$2s^22p^56p$	4	−5.78090	$2s2p^63p$	1	−9.38518	$2s^22p^53d$	4	−2.25001	$2s^22p^56h$
7	−1.75135	$2s^22p^57p$	5	−5.32015	$2s^22p^54d$	2	−5.25297	$2s^22p^54d$	5	−1.65684	$2s^22p^57f$
8	−1.33168	$2s^22p^58p$	6	−3.89110	$2s^22p^55s$	3	−3.34085	$2s^22p^55d$	6	−1.65308	$2s^22p^57h$
9	−1.04793	$2s^22p^59p$	7	−3.37587	$2s^22p^55d$	4	−2.30920	$2s^22p^56d$	7	−1.26826	$2s^22p^58f$
10	−.92969	$2s2p^64s$	8	−2.61656	$2s^22p^56s$	5	−1.69061	$2s^22p^57d$	8	−1.26565	$2s^22p^58h$
	$^{3}\mathbf{S}^{e}$		9	−2.32897	$2s^22p^56d$	6	−1.29089	$2s^22p^58d$	9	−1.00190	$2s^22p^59f$
1	−11.4576	$2s^22p^53p$	10	−1.87962	$2s^22p^57s$	7	−1.01779	$2s^22p^59d$	10	−1.00002	$2s^22p^59h$
2	−7.08654	$2s2p^63s$	11	−1.70288	$2s^22p^57d$		$^{1}\mathbf{F}^{e}$			$^{3}\mathbf{G}^{e}$	
3	−5.95020	$2s^22p^54p$	12	−1.41544	$2s^22p^58s$	1	−5.05845	$2s^22p^54f$	1	−5.07759	$2s^22p^54f$
4	−3.68467	$2s^22p^55p$	13	−1.29904	$2s^22p^58d$	2	−3.23992	$2s^22p^55f$	2	−3.25024	$2s^22p^55f$
5	−2.50225	$2s^22p^56p$	14	−1.10426	$2s^22p^59s$	3	−2.25068	$2s^22p^56f$	3	−2.25670	$2s^22p^56f$
6	−1.81048	$2s^22p^57p$	15	−1.02351	$2s^22p^59d$	4	−1.65377	$2s^22p^57f$	4	−2.25001	$2s^22p^56h$
7	−1.37177	$2s^22p^58p$		$^{1}\mathbf{D}^{e}$		5	−1.26622	$2s^22p^58f$	5	−1.65757	$2s^22p^57f$
8	−1.08386	$2s^22p^59p$	1	−11.1430	$2s^22p^53p$	6	−1.00048	$2s^22p^59f$	6	−1.65308	$2s^22p^57h$
9	−1.00638	$2s2p^64s$	2	−5.88374	$2s^22p^54p$		$^{3}\mathbf{F}^{e}$		7	−1.26877	$2s^22p^58f$
	$^{1}\mathbf{P}^{e}$		3	−5.08401	$2s^22p^54f$	1	−5.05829	$2s^22p^54f$	8	−1.26565	$2s^22p^58h$
1	−11.1226	$2s^22p^53p$	4	−3.97717	$2s2p^63d$	2	−3.23980	$2s^22p^55f$	9	−1.00227	$2s^22p^59f$
2	−5.88006	$2s^22p^54p$	5	−3.63876	$2s^22p^55p$	3	−2.25060	$2s^22p^56f$	10	−1.00002	$2s^22p^59h$
3	−3.64093	$2s^22p^55p$	6	−3.24091	$2s^22p^55f$	4	−1.65372	$2s^22p^57f$		$^{1}\mathbf{G}^{o}$	
4	−2.47639	$2s^22p^56p$	7	−2.47680	$2s^22p^56p$	5	−1.26618	$2s^22p^58f$	1	−3.23882	$2s^22p^55g$
5	−1.79338	$2s^22p^57p$	8	−2.25310	$2s^22p^56f$	6	−1.00045	$2s^22p^59f$	2	−2.24901	$2s^22p^56g$
6	−1.35859	$2s^22p^58p$	9	−1.79377	$2s^22p^57p$		$^{1}\mathbf{F}^{o}$		3	−1.65234	$2s^22p^57g$
7	−1.06476	$2s^22p^59p$	10	−1.65541	$2s^22p^57f$	1	−9.43537	$2s^22p^53d$	4	−1.26511	$2s^22p^58g$
	$^{3}\mathbf{P}^{e}$		11	−1.35890	$2s^22p^58p$	2	−5.26280	$2s^22p^54d$	5	−.99962	$2s^22p^59g$
1	−11.1188	$2s^22p^53p$	12	−1.26733	$2s^22p^58f$	3	−3.34527	$2s^22p^55d$		$^{3}\mathbf{G}^{o}$	
2	−5.87869	$2s^22p^54p$	13	−1.06500	$2s^22p^59p$	4	−3.24119	$2s^22p^55g$	1	−3.23882	$2s^22p^55g$
3	−3.64029	$2s^22p^55p$	14	−1.00126	$2s^22p^59f$	5	−2.31160	$2s^22p^56d$	2	−2.24901	$2s^22p^56g$
4	−2.47604	$2s^22p^56p$		$^{3}\mathbf{D}^{e}$		6	−2.25109	$2s^22p^56g$	3	−1.65234	$2s^22p^57g$
5	−1.79317	$2s^22p^57p$	1	−11.2397	$2s^22p^53p$	7	−1.69207	$2s^22p^57d$	4	−1.26511	$2s^22p^58g$
6	−1.35845	$2s^22p^58p$	2	−5.91843	$2s^22p^54p$	8	−1.65392	$2s^22p^57g$	5	−.99962	$2s^22p^59g$
7	−1.06466	$2s^22p^59p$	3	−5.10042	$2s^22p^54f$	9	−1.29184	$2s^22p^58d$			
	$^{1}\mathbf{P}^{o}$		4	−4.11538	$2s2p^63d$	10	−1.26628	$2s^22p^58g$			
1	−12.3806	$2s^22p^53s$	5	−3.65438	$2s^22p^55p$	11	−1.01845	$2s^22p^59d$			
2	−9.10719	$2s^22p^53d$	6	−3.24151	$2s^22p^55f$	12	−1.00051	$2s^22p^59g$			
3	−6.36092	$2s^22p^54s$	7	−2.48540	$2s^22p^56p$		$^{3}\mathbf{F}^{o}$				
4	−5.70242	$2s2p^63p$	8	−2.25520	$2s^22p^56f$	1	−9.52191	$2s^22p^53d$			
5	−5.10566	$2s^22p^54d$	9	−1.79905	$2s^22p^57p$	2	−5.29844	$2s^22p^54d$			
6	−3.87383	$2s^22p^55s$	10	−1.65719	$2s^22p^57f$	3	−3.36297	$2s^22p^55d$			
7	−3.27043	$2s^22p^55d$	11	−1.36237	$2s^22p^58p$	4	−3.24131	$2s^22p^55g$			
8	−2.60664	$2s^22p^56s$	12	−1.26872	$2s^22p^58f$	5	−2.32164	$2s^22p^56d$			
9	−2.26917	$2s^22p^56d$	13	−1.06741	$2s^22p^59p$	6	−2.25121	$2s^22p^56g$			
10	−1.87343	$2s^22p^57s$	14	−1.00233	$2s^22p^59f$	7	−1.69830	$2s^22p^57d$			
11	−1.66573	$2s^22p^57d$		$^{1}\mathbf{D}^{o}$		8	−1.65402	$2s^22p^57g$			
12	−1.41126	$2s^22p^58s$	1	−9.38756	$2s^22p^53d$	9	−1.29597	$2s^22p^58d$			
13	−1.27441	$2s^22p^58d$	2	−5.25397	$2s^22p^54d$	10	−1.26637	$2s^22p^58g$			

Ne-like Ar (Ar^{8+})

Energies in ascending order from ground state for terms with effective $n \leq 4.0$, $L \leq 4$

Term	i	E(Ryds)	Term	i	E(Ryds)	Term	i	E(Ryds)	Term	i	E(Ryds)	Term	i	E(Ryds)
$^1S^e$	1	0.00000	$^1S^e$	2	20.7838	$^1S^e$	3	24.3317	$^3P^o$	4	25.3791	$^1P^o$	5	26.0543
$^3P^o$	1	18.6522	$^3P^o$	2	21.5503	$^3P^o$	3	24.7453	$^1P^o$	4	25.4575	$^3D^e$	3	26.0595
$^1P^o$	1	18.7794	$^3F^o$	1	21.6380	$^1P^o$	3	24.7990	$^1S^e$	4	25.5636	$^1D^e$	3	26.0759
$^3S^e$	1	19.7024	$^1F^o$	1	21.7246	$^3S^e$	3	25.2098	$^3P^o$	5	25.8398	$^3G^e$	1	26.0824
$^3D^e$	1	19.9203	$^1D^o$	1	21.7724	$^3D^e$	2	25.2415	$^3F^o$	2	25.8615	$^1G^e$	1	26.0846
$^1D^e$	1	20.0170	$^3D^o$	1	21.7748	$^1D^e$	2	25.2762	$^1F^o$	2	25.8972			
$^1P^e$	1	20.0374	$^1P^o$	2	22.0528	$^1P^e$	2	25.2799	$^1D^o$	2	25.9060			
$^3P^e$	1	20.0412	$^3S^e$	2	24.0734	$^3P^e$	2	25.2813	$^3D^o$	2	25.9070			

gf-values for transitions involving terms with effective $n \leq 4.0$, $L \leq 4$

$i\ i'$	gf_L	$i\ i'$	gf_L	$i\ i'$	gf_L	$i\ i'$	gf_L	$i\ i'$	gf_L	$i\ i'$	gf_L
	$^1S^e$–$^1P^o$	1 2	−4.37E−1	3 3	−7.25E−3	2 1	−7.85E−1	2 1	3.92E−1	5 1	7.77E−3
1 1	−2.39E−1	1 3	−1.80E−1	4 1	2.97E−1		$^3S^e$–$^3P^o$	2 2	2.31E−1	5 2	6.91E−2
1 2	−2.02E+0	1 4	−2.82E−1	4 2	1.56E−4	1 1	3.91E−1	2 3	2.35E+0	5 3	−1.27E+0
1 3	−5.14E−2	1 5	−5.35E−2	4 3	−1.15E−2	1 2	−1.23E+0	2 4	−2.56E−2		$^3D^e$–$^3D^o$
1 4	−3.33E−1	2 1	1.51E−1	5 1	1.36E−4	1 3	−2.85E−1	2 5	−1.32E+0	1 1	−1.06E+0
1 5	−5.25E−1	2 2	1.38E−1	5 2	2.07E−2	1 4	−1.35E−1		$^3P^e$–$^3D^o$	1 2	−3.88E−1
2 1	3.14E−1	2 3	7.10E−1	5 3	−3.96E−2	1 5	−5.07E−1	1 1	−2.96E+0	2 1	2.41E−1
2 2	−3.14E−1	2 4	−6.14E−3		$^1D^e$–$^1D^o$	2 1	7.62E−1	1 2	−1.29E+0	2 2	−1.62E+0
2 3	−1.73E−1	2 5	−6.81E−1	1 1	−3.38E−1	2 2	9.12E−3	2 1	6.89E−1	3 1	1.26E+0
2 4	−1.69E−1		$^1P^e$–$^1D^o$	1 2	−1.40E−1	2 3	−8.59E−2	2 2	−4.52E+0	3 2	1.47E−1
2 5	−2.06E−1	1 1	−9.87E−1	2 1	7.29E−2	2 4	−1.59E+0		$^3P^o$–$^3D^e$		$^3D^e$–$^3F^o$
3 1	1.33E−1	1 2	−4.29E−1	2 2	−5.10E−1	2 5	−3.78E−2	1 1	−2.52E+0	1 1	−5.23E+0
3 2	1.25E−2	2 1	2.30E−1	3 1	4.50E−1	3 1	9.11E−2	1 2	−7.76E−1	1 2	−2.53E+0
3 3	−1.04E−3	2 2	−1.51E+0	3 2	5.82E−2	3 2	2.73E−1	1 3	−3.80E−3	2 1	1.30E+0
3 4	−4.41E−1		$^1P^o$–$^1D^e$		$^1D^e$–$^1F^o$	3 3	6.74E−1	2 1	2.86E−2	2 2	−8.18E+0
3 5	−5.73E−2	1 1	−8.47E−1	1 1	−1.79E+0	3 4	−5.57E−3	2 2	−9.65E−4	3 1	5.63E−3
4 1	2.86E−2	1 2	−2.58E−1	1 2	−8.22E−1	3 5	−1.92E+0	2 3	−7.49E+0	3 2	1.37E−2
4 2	6.90E−2	1 3	−2.21E−3	2 1	4.40E−1		$^3P^e$–$^3P^o$	3 1	1.47E+0		$^3F^o$–$^3G^e$
4 3	4.35E−1	2 1	9.38E−3	2 2	−2.77E+0	1 1	1.68E+0	3 2	−3.73E+0	1 1	−1.72E+1
4 4	2.45E−3	2 2	−2.65E−4	3 1	4.60E−3	1 2	−8.55E−1	3 3	−1.31E−2	2 1	−2.79E+0
4 5	−5.14E−1	2 3	−2.88E+0	3 2	2.86E−3	1 3	−5.17E−1	4 1	7.80E−1		
	$^1P^e$–$^1P^o$	3 1	4.22E−1		$^1F^o$–$^1G^e$	1 4	−8.61E−1	4 2	1.72E−2		
1 1	5.03E−1	3 2	−1.24E+0	1 1	−5.86E+0	1 5	−4.04E−1	4 3	−6.26E−3		

Ne-like Ca (Ca^{10+})

Term energies relative to $2s^22p^5$ ^{2}P ionization threshold for each symmetry

i	E(Ryds)	Description	i	E(Ryds)	Description	i	E(Ryds)	Description	i	E(Ryds)	Description
	^{1}**S**^{e}		14	−1.90395	$2s^22p^58d$	15	−1.49553	$2s^22p^59f$	6	−3.36343	$2s^22p^56g$
1	−43.5856	$2s^22p^6$	15	−1.62312	$2s^22p^59s$		^{1}**D**^{o}		7	−2.53241	$2s^22p^57d$
2	−15.2029	$2s^22p^53p$	16	−1.50351	$2s^22p^59d$	1	−14.0220	$2s^22p^53d$	8	−2.47145	$2s^22p^57g$
3	−11.2355	$2s2p^63s$		^{3}**P**^{o}		2	−7.83746	$2s^22p^54d$	9	−1.93287	$2s^22p^58d$
4	−8.26365	$2s^22p^54p$	1	−17.8257	$2s^22p^53s$	3	−4.98427	$2s^22p^55d$	10	−1.89274	$2s^22p^58g$
5	−5.19689	$2s^22p^55p$	2	−14.3114	$2s^22p^53d$	4	−3.44550	$2s^22p^56d$	11	−1.52354	$2s^22p^59d$
6	−3.56759	$2s^22p^56p$	3	−9.99872	$2s2p^63p$	5	−2.52282	$2s^22p^57d$	12	−1.49836	$2s^22p^59g$
7	−2.92895	$2s2p^64s$	4	−9.23237	$2s^22p^54s$	6	−1.92653	$2s^22p^58d$	13	−1.38494	$2s2p^64f$
8	−2.59536	$2s^22p^57p$	5	−7.92836	$2s^22p^54d$	7	−1.51909	$2s^22p^59d$		^{1}**G**^{e}	
9	−1.97565	$2s^22p^58p$	6	−5.66696	$2s^22p^55s$		^{3}**D**^{o}		1	−7.58328	$2s^22p^54f$
10	−1.55347	$2s^22p^59p$	7	−5.02886	$2s^22p^55d$	1	−14.0190	$2s^22p^53d$	2	−4.85371	$2s^22p^55f$
	^{3}**S**^{e}		8	−3.82837	$2s^22p^56s$	2	−7.83625	$2s^22p^54d$	3	−3.37002	$2s^22p^56f$
1	−16.5547	$2s^22p^53p$	9	−3.47071	$2s^22p^56d$	3	−4.98368	$2s^22p^55d$	4	−3.36116	$2s^22p^56h$
2	−11.5358	$2s^22p^54p$	10	−2.76100	$2s^22p^57s$	4	−3.44517	$2s^22p^56d$	5	−2.47536	$2s^22p^57f$
3	−8.71308	$2s2p^63s$	11	−2.54078	$2s^22p^57d$	5	−2.52261	$2s^22p^57d$	6	−2.46946	$2s^22p^57h$
4	−5.41142	$2s^22p^55p$	12	−2.42179	$2s2p^64p$	6	−1.92639	$2s^22p^58d$	7	−1.89478	$2s^22p^58f$
5	−3.68992	$2s^22p^56p$	13	−2.07954	$2s^22p^58s$	7	−1.51899	$2s^22p^59d$	8	−1.89070	$2s^22p^58h$
6	−3.03601	$2s2p^64s$	14	−1.93611	$2s^22p^58d$		^{1}**F**^{e}		9	−1.49682	$2s^22p^59f$
7	−2.66039	$2s^22p^57p$	15	−1.62607	$2s^22p^59s$	1	−7.55865	$2s^22p^54f$	10	−1.49389	$2s^22p^59h$
8	−2.02077	$2s^22p^58p$	16	−1.52588	$2s^22p^59d$	2	−4.84190	$2s^22p^55f$		^{3}**G**^{e}	
9	−1.58499	$2s^22p^59p$		^{1}**D**^{e}		3	−3.36345	$2s^22p^56f$	1	−7.58681	$2s^22p^54f$
	^{1}**P**^{e}		1	−16.1660	$2s^22p^53p$	4	−2.47131	$2s^22p^57f$	2	−4.85626	$2s^22p^55f$
1	−16.1418	$2s^22p^53p$	2	−8.61556	$2s^22p^54p$	5	−1.89211	$2s^22p^58f$	3	−3.37170	$2s^22p^56f$
2	−8.60639	$2s^22p^54p$	3	−7.81160	$2s2p^63d$	6	−1.49496	$2s^22p^59f$	4	−3.36117	$2s^22p^56h$
3	−5.35345	$2s^22p^55p$	4	−7.51289	$2s^22p^54f$		^{3}**F**^{e}		5	−2.47650	$2s^22p^57f$
4	−3.65161	$2s^22p^56p$	5	−5.35468	$2s^22p^55p$	1	−7.55841	$2s^22p^54f$	6	−2.46947	$2s^22p^57h$
5	−2.64970	$2s^22p^57p$	6	−4.84886	$2s^22p^55f$	2	−4.84172	$2s^22p^55f$	7	−1.89557	$2s^22p^58f$
6	−2.01020	$2s^22p^58p$	7	−3.65274	$2s^22p^56p$	3	−3.36333	$2s^22p^56f$	8	−1.89070	$2s^22p^58h$
7	−1.57719	$2s^22p^59p$	8	−3.36726	$2s^22p^56f$	4	−2.47123	$2s^22p^57f$	9	−1.49739	$2s^22p^59f$
	^{3}**P**^{e}		9	−2.65065	$2s^22p^57p$	5	−1.89205	$2s^22p^58f$	10	−1.49390	$2s^22p^59h$
1	−16.1374	$2s^22p^53p$	10	−2.47361	$2s^22p^57f$	6	−1.49492	$2s^22p^59f$		^{1}**G**^{o}	
2	−8.60484	$2s^22p^54p$	11	−2.01130	$2s^22p^58p$		^{1}**F**^{o}		1	−4.83796	$2s^22p^55g$
3	−5.35273	$2s^22p^55p$	12	−1.89373	$2s^22p^58f$	1	−14.0785	$2s^22p^53d$	2	−3.35950	$2s^22p^56g$
4	−3.65121	$2s^22p^56p$	13	−1.61031	$2s2p^64d$	2	−7.84806	$2s^22p^54d$	3	−2.46826	$2s^22p^57g$
5	−2.64945	$2s^22p^57p$	14	−1.57128	$2s^22p^59p$	3	−4.98907	$2s^22p^55d$	4	−1.88983	$2s^22p^58g$
6	−2.01005	$2s^22p^58p$	15	−1.49541	$2s^22p^59f$	4	−4.84219	$2s^22p^55g$	5	−1.49326	$2s^22p^59g$
7	−1.57708	$2s^22p^59p$		^{3}**D**^{e}		5	−3.44809	$2s^22p^56d$		^{3}**G**^{o}	
	^{1}**P**^{o}		1	−16.2864	$2s^22p^53p$	6	−3.36318	$2s^22p^56g$	1	−4.83796	$2s^22p^55g$
1	−17.6663	$2s^22p^53s$	2	−8.66220	$2s^22p^54p$	7	−2.52440	$2s^22p^57d$	2	−3.35950	$2s^22p^56g$
2	−13.6129	$2s^22p^53d$	3	−8.01585	$2s2p^63d$	8	−2.47122	$2s^22p^57g$	3	−2.46826	$2s^22p^57g$
3	−9.86758	$2s2p^63p$	4	−7.51970	$2s^22p^54f$	9	−1.92761	$2s^22p^58d$	4	−1.88983	$2s^22p^58g$
4	−9.19688	$2s^22p^54s$	5	−5.37414	$2s^22p^55p$	10	−1.89251	$2s^22p^58g$	5	−1.49326	$2s^22p^59g$
5	−7.64668	$2s^22p^54d$	6	−4.85444	$2s^22p^55f$	11	−1.52009	$2s^22p^59d$			
6	−5.64413	$2s^22p^55s$	7	−3.66360	$2s^22p^56p$	12	−1.49778	$2s^22p^59g$			
7	−4.89053	$2s^22p^55d$	8	−3.37197	$2s^22p^56f$	13	−1.38027	$2s2p^64f$			
8	−3.81517	$2s^22p^56s$	9	−2.65740	$2s^22p^57p$		^{3}**F**^{o}				
9	−3.39271	$2s^22p^56d$	10	−2.47741	$2s^22p^57f$	1	−14.1970	$2s^22p^53d$			
10	−2.75106	$2s^22p^57s$	11	−2.01608	$2s^22p^58p$	2	−7.89510	$2s^22p^54d$			
11	−2.49320	$2s^22p^57d$	12	−1.89800	$2s^22p^58f$	3	−5.01214	$2s^22p^55d$			
12	−2.38202	$2s2p^64p$	13	−1.68531	$2s2p^64d$	4	−4.84242	$2s^22p^55g$			
13	−2.07610	$2s^22p^58s$	14	−1.57793	$2s^22p^59p$	5	−3.46108	$2s^22p^56d$			

Ne-like Ca (Ca^{10+})

Energies in ascending order from ground state for terms with effective $n \leq 4.0$, $L \leq 4$

Term	i	E(Ryds)	Term	i	E(Ryds)	Term	i	E(Ryds)	Term	i	E(Ryds)	Term	i	E(Ryds)
$^1\mathbf{S}^e$	1	0.00000	$^1\mathbf{S}^e$	2	28.3827	$^1\mathbf{S}^e$	3	32.3501	$^1\mathbf{P}^e$	2	34.9792	$^3\mathbf{D}^o$	2	35.7493
$^3\mathbf{P}^o$	1	25.7599	$^3\mathbf{P}^o$	2	29.2742	$^3\mathbf{P}^o$	3	33.5868	$^3\mathbf{P}^e$	2	34.9807	$^1\mathbf{D}^e$	3	35.7740
$^1\mathbf{P}^o$	1	25.9193	$^3\mathbf{F}^o$	1	29.3886	$^1\mathbf{P}^o$	3	33.7180	$^1\mathbf{S}^e$	4	35.3219	$^1\mathbf{P}^o$	5	35.9389
$^3\mathbf{S}^e$	1	27.0309	$^1\mathbf{F}^o$	1	29.5071	$^3\mathbf{P}^o$	4	34.3532	$^3\mathbf{D}^e$	3	35.5697	$^3\mathbf{G}^e$	1	35.9987
$^3\mathbf{D}^e$	1	27.2992	$^1\mathbf{D}^o$	1	29.5636	$^1\mathbf{P}^o$	4	34.3887	$^3\mathbf{P}^o$	5	35.6572	$^1\mathbf{G}^e$	1	36.0023
$^1\mathbf{D}^e$	1	27.4196	$^3\mathbf{D}^o$	1	29.5666	$^3\mathbf{S}^e$	3	34.8725	$^3\mathbf{F}^o$	2	35.6905			
$^1\mathbf{P}^e$	1	27.4438	$^1\mathbf{P}^o$	2	29.9727	$^3\mathbf{D}^e$	2	34.9234	$^1\mathbf{F}^o$	2	35.7375			
$^3\mathbf{P}^e$	1	27.4482	$^3\mathbf{S}^e$	2	32.0498	$^1\mathbf{D}^e$	2	34.9700	$^1\mathbf{D}^o$	2	35.7481			

gf-values for transitions involving terms with effective $n \leq 4.0$, $L \leq 4$

$i\ i'$	gf_L	$i\ i'$	gf_L	$i\ i'$	gf_L	$i\ i'$	gf_L	$i\ i'$	gf_L	$i\ i'$	gf_L
	$^1\mathbf{S}^e$–$^1\mathbf{P}^o$	1 2	−3.72E−1	3 3	−7.37E−1	2 1	−7.40E−1	2 1	4.93E−1	5 1	3.44E−2
1 1	−2.32E−1	1 3	−1.95E−1	4 1	3.17E−1		$^3\mathbf{S}^e$–$^3\mathbf{P}^o$	2 2	1.88E−1	5 2	4.37E−2
1 2	−2.34E+0	1 4	−2.05E−1	4 2	−1.05E+0	1 1	3.35E−1	2 3	7.42E−2	5 3	4.87E−2
1 3	−3.03E−1	1 5	−1.16E−1	4 3	−3.88E−2	1 2	−1.02E+0	2 4	2.01E+0		$^3\mathbf{D}^e$–$^3\mathbf{D}^o$
1 4	−2.85E−2	2 1	1.84E−1	5 1	6.06E−3	1 3	−1.85E−1	2 5	−1.12E+0	1 1	−8.83E−1
1 5	−6.51E−1	2 2	1.17E−1	5 2	1.52E−2	1 4	−1.56E−1		$^3\mathbf{P}^e$–$^3\mathbf{D}^o$	1 2	−5.13E−1
2 1	2.79E−1	2 3	8.23E−4	5 3	4.84E−2	1 5	−7.13E−1	1 1	−2.45E+0	2 1	1.47E−1
2 2	−2.60E−1	2 4	6.21E−1		$^1\mathbf{D}^e$–$^1\mathbf{D}^o$	2 1	6.62E−1	1 2	−1.66E+0	2 2	−1.38E+0
2 3	−1.50E−1	2 5	−5.95E−1	1 1	−2.81E−1	2 2	1.86E−3	2 1	5.61E−1	3 1	1.25E−1
2 4	−1.24E−1		$^1\mathbf{P}^e$–$^1\mathbf{D}^o$	1 2	−1.82E−1	2 3	−1.44E+0	2 2	−3.79E+0	3 2	−2.73E−3
2 5	−2.88E−1	1 1	−8.18E−1	2 1	4.80E−2	2 4	−1.16E−2		$^3\mathbf{P}^o$–$^3\mathbf{D}^e$		$^3\mathbf{D}^e$–$^3\mathbf{F}^o$
3 1	1.25E−1	1 2	−5.53E−1	2 2	−4.30E−1	2 5	−3.05E−2	1 1	−2.20E+0	1 1	−4.33E+0
3 2	9.05E−3	2 1	1.87E−1	3 1	1.31E−2	3 1	1.68E−1	1 2	−9.30E−1	1 2	−3.22E+0
3 3	−4.00E−1	2 2	−1.26E+0	3 2	6.54E−4	3 2	2.31E−1	1 3	−2.12E−2	2 1	1.26E+0
3 4	−8.69E−3		$^1\mathbf{P}^o$–$^1\mathbf{D}^e$		$^1\mathbf{D}^e$–$^1\mathbf{F}^o$	3 3	2.12E−2	2 1	2.24E−2	2 2	−6.82E+0
3 5	−1.47E−2	1 1	−7.37E−1	1 1	−1.48E+0	3 4	5.53E−1	2 2	−1.03E−3	3 1	6.93E−1
4 1	3.54E−2	1 2	−3.15E−1	1 2	−1.05E+0	3 5	−1.64E+0	2 3	−4.02E−1	3 2	−2.27E−2
4 2	6.01E−2	1 3	−1.76E−3	2 1	4.02E−1		$^3\mathbf{P}^e$–$^3\mathbf{P}^o$	3 1	1.30E+0		$^3\mathbf{F}^o$–$^3\mathbf{G}^e$
4 3	1.33E−2	2 1	6.79E−3	2 2	−2.33E+0	1 1	1.48E+0	3 2	−3.39E−1	1 1	−1.74E+1
4 4	3.71E−1	2 2	−2.24E−4	3 1	2.24E−1	1 2	−6.94E−1	3 3	−2.18E+0	2 1	−2.60E+0
4 5	−4.47E−1	2 3	−3.57E−1	3 2	1.16E−3	1 3	−3.64E−1	4 1	7.33E−1		
	$^1\mathbf{P}^e$–$^1\mathbf{P}^o$	3 1	3.39E−1		$^1\mathbf{F}^o$–$^1\mathbf{G}^e$	1 4	−8.33E−1	4 2	−2.91E+0		
1 1	4.37E−1	3 2	−2.59E−2	1 1	−5.89E+0	1 5	−6.15E−1	4 3	−2.82E−1		

Ne-like Fe (Fe^{16+})

Term energies relative to $2s^22p^5$ ^{2}P ionization threshold for each symmetry

i	E(Ryds)	Description	*i*	E(Ryds)	Description	*i*	E(Ryds)	Description	*i*	E(Ryds)	Description
	$^1S^e$		12	−5.94425	$2s^22p^57d$	10	−6.21142	$2s^22p^57p$	14	−2.93757	$2s2p^65f$
1	−92.8855	$2s^22p^6$	13	−4.82591	$2s^22p^58s$	11	−5.91542	$2s^22p^57f$		**$^3F^o$**	
2	−35.0457	$2s^22p^53p$	14	−4.54739	$2s^22p^58d$	12	−4.72408	$2s^22p^58p$	1	−33.5910	$2s^22p^53d$
3	−29.8312	$2s2p^63s$	15	−3.78850	$2s^22p^59s$	13	−4.52805	$2s^22p^58f$	2	−18.6954	$2s^22p^54d$
4	−19.2733	$2s^22p^54p$	16	−3.73859	$2s2p^65p$	14	−3.71386	$2s^22p^59p$	3	−11.8845	$2s^22p^55d$
5	−12.1929	$2s^22p^55p$	17	−3.58944	$2s^22p^59d$	15	−3.57760	$2s^22p^59f$	4	−11.5689	$2s^22p^55g$
6	−11.9388	$2s2p^64s$		**$^3P^o$**		16	−3.21426	$2s2p^65d$	5	−9.45400	$2s2p^64f$
7	−8.38360	$2s^22p^56p$	1	−39.1124	$2s^22p^53s$		**$^1D^o$**		6	−8.21526	$2s^22p^56d$
8	−6.12162	$2s^22p^57p$	2	−33.7828	$2s^22p^53d$	1	−33.2972	$2s^22p^53d$	7	−8.02829	$2s^22p^56g$
9	−4.66569	$2s^22p^58p$	3	−27.9232	$2s2p^63p$	2	−18.5976	$2s^22p^54d$	8	−6.01590	$2s^22p^57d$
10	−4.17412	$2s2p^65s$	4	−20.8037	$2s^22p^54s$	3	−11.8383	$2s^22p^55d$	9	−5.90018	$2s^22p^57g$
11	−3.67203	$2s^22p^59p$	5	−18.7529	$2s^22p^54d$	4	−8.18988	$2s^22p^56d$	10	−4.59453	$2s^22p^58d$
	$^3S^e$		6	−12.9229	$2s^22p^55s$	5	−6.00032	$2s^22p^57d$	11	−4.51767	$2s^22p^58g$
1	−37.1759	$2s^22p^53p$	7	−11.9159	$2s^22p^55d$	6	−4.58428	$2s^22p^58d$	12	−3.62326	$2s^22p^59d$
2	−30.2881	$2s2p^63s$	8	−11.1495	$2s2p^64p$	7	−3.61614	$2s^22p^59d$	13	−3.56999	$2s^22p^59g$
3	−19.9790	$2s^22p^54p$	9	−8.79600	$2s^22p^56s$		**$^3D^o$**		14	−2.94503	$2s2p^65f$
4	−12.5631	$2s^22p^55p$	10	−8.22936	$2s^22p^56d$	1	−33.2929	$2s^22p^53d$		**$^1G^e$**	
5	−12.0625	$2s2p^64s$	11	−6.37637	$2s^22p^57s$	2	−18.5960	$2s^22p^54d$	1	−18.1117	$2s^22p^54f$
6	−8.55980	$2s^22p^56p$	12	−6.02476	$2s^22p^57d$	3	−11.8375	$2s^22p^55d$	2	−11.5919	$2s^22p^55f$
7	−6.23072	$2s^22p^57p$	13	−4.83375	$2s^22p^58s$	4	−8.18945	$2s^22p^56d$	3	−8.04840	$2s^22p^56f$
8	−4.74050	$2s^22p^58p$	14	−4.60060	$2s^22p^58d$	5	−6.00006	$2s^22p^57d$	4	−8.02812	$2s^22p^56h$
9	−4.24714	$2s2p^65s$	15	−3.80699	$2s^22p^59s$	6	−4.58410	$2s^22p^58d$	5	−5.91176	$2s^22p^57f$
10	−3.71839	$2s^22p^59p$	16	−3.75472	$2s2p^65p$	7	−3.61602	$2s^22p^59d$	6	−5.89840	$2s^22p^57h$
	$^1P^e$		17	−3.62597	$2s^22p^59d$		**$^1F^e$**		7	−4.52523	$2s^22p^58f$
1	−36.5372	$2s^22p^53p$		**$^1D^e$**		1	−18.0628	$2s^22p^54f$	8	−4.51606	$2s^22p^58h$
2	−19.7907	$2s^22p^54p$	1	−36.5735	$2s^22p^53p$	2	−11.5709	$2s^22p^55f$	9	−3.57484	$2s^22p^59f$
3	−12.4151	$2s^22p^55p$	2	−24.4649	$2s2p^63d$	3	−8.03695	$2s^22p^56f$	10	−3.56836	$2s^22p^59h$
4	−8.51344	$2s^22p^56p$	3	−19.7956	$2s^22p^54p$	4	−5.90472	$2s^22p^57f$	11	−2.91364	$2s2p^65g$
5	−6.20018	$2s^22p^57p$	4	−18.1008	$2s^22p^54f$	5	−4.52056	$2s^22p^58f$		**$^3G^e$**	
6	−4.71643	$2s^22p^58p$	5	−12.4200	$2s^22p^55p$	6	−3.57155	$2s^22p^59f$	1	−18.1202	$2s^22p^54f$
7	−3.70809	$2s^22p^59p$	6	−11.5836	$2s^22p^55f$		**$^3F^e$**		2	−11.5978	$2s^22p^55f$
	$^3P^e$		7	−9.87025	$2s2p^64d$	1	−18.0623	$2s^22p^54f$	3	−8.05225	$2s^22p^56f$
1	−36.5318	$2s^22p^53p$	8	−8.51347	$2s^22p^56p$	2	−11.5706	$2s^22p^55f$	4	−8.02813	$2s^22p^56h$
2	−19.7888	$2s^22p^54p$	9	−8.04260	$2s^22p^56f$	3	−8.03673	$2s^22p^56f$	5	−5.91434	$2s^22p^57f$
3	−12.4143	$2s^22p^55p$	10	−6.20099	$2s^22p^57p$	4	−5.90457	$2s^22p^57f$	6	−5.89840	$2s^22p^57h$
4	−8.51296	$2s^22p^56p$	11	−5.90782	$2s^22p^57f$	5	−4.52045	$2s^22p^58f$	7	−4.52702	$2s^22p^58f$
5	−6.19989	$2s^22p^57p$	12	−4.71718	$2s^22p^58p$	6	−3.57148	$2s^22p^59f$	8	−4.51606	$2s^22p^58h$
6	−4.71624	$2s^22p^58p$	13	−4.52247	$2s^22p^58f$		**$^1F^o$**		9	−3.57614	$2s^22p^59f$
7	−3.70796	$2s^22p^59p$	14	−3.70894	$2s^22p^59p$	1	−33.3764	$2s^22p^53d$	10	−3.56836	$2s^22p^59h$
	$^1P^o$		15	−3.57284	$2s^22p^59f$	2	−18.6148	$2s^22p^54d$	11	−2.91412	$2s2p^65g$
1	−38.8841	$2s^22p^53s$	16	−3.15017	$2s2p^65d$	3	−11.8459	$2s^22p^55d$		**$^1G^o$**	
2	−32.4870	$2s^22p^53d$		**$^3D^e$**		4	−11.5680	$2s^22p^55g$	1	−11.5546	$2s^22p^55g$
3	−27.7420	$2s2p^63p$	1	−36.7614	$2s^22p^53p$	5	−9.44121	$2s2p^64f$	2	−8.02403	$2s^22p^56g$
4	−20.7321	$2s^22p^54s$	2	−24.8443	$2s2p^63d$	6	−8.19312	$2s^22p^56d$	3	−5.89552	$2s^22p^57g$
5	−18.2658	$2s^22p^54d$	3	−19.8608	$2s^22p^54p$	7	−8.02793	$2s^22p^56g$	4	−4.51400	$2s^22p^58g$
6	−12.8853	$2s^22p^55s$	4	−18.1208	$2s^22p^54f$	8	−6.00239	$2s^22p^57d$	5	−3.56677	$2s^22p^59g$
7	−11.6821	$2s^22p^55d$	5	−12.4513	$2s^22p^55p$	9	−5.89974	$2s^22p^57g$		**$^3G^o$**	
8	−11.0887	$2s2p^64p$	6	−11.6023	$2s^22p^55f$	10	−4.58563	$2s^22p^58d$	1	−11.5546	$2s^22p^55g$
9	−8.77856	$2s^22p^56s$	7	−10.0058	$2s2p^64d$	11	−4.51729	$2s^22p^58g$	2	−8.02402	$2s^22p^56g$
10	−8.09877	$2s^22p^56d$	8	−8.53037	$2s^22p^56p$	12	−3.61709	$2s^22p^59d$	3	−5.89552	$2s^22p^57g$
11	−6.36508	$2s^22p^57s$	9	−8.05304	$2s^22p^56f$	13	−3.56961	$2s^22p^59g$	4	−4.51399	$2s^22p^58g$

Ne-like Fe (Fe^{16+})

Energies in ascending order from ground state for terms with effective $n \leq 4.0$, $L \leq 4$

Term	i	E(Ryds)	Term	i	E(Ryds)	Term	i	E(Ryds)	Term	i	E(Ryds)	Term	i	E(Ryds)
$^1S^e$	1	0.00000	$^1S^e$	2	57.8398	$^1S^e$	3	63.0543	$^3D^e$	3	73.0247	$^1D^o$	2	74.2879
$^3P^o$	1	53.7731	$^3P^o$	2	59.1027	$^3P^o$	3	64.9623	$^1D^e$	3	73.0899	$^3D^o$	2	74.2895
$^1P^o$	1	54.0014	$^3F^o$	1	59.2945	$^1P^o$	3	65.1434	$^1P^e$	2	73.0948	$^1P^o$	5	74.6197
$^3S^e$	1	55.7096	$^1F^o$	1	59.5091	$^3D^e$	2	68.0412	$^3P^e$	2	73.0966	$^3D^e$	4	74.7646
$^3D^e$	1	56.1241	$^1D^o$	1	59.5883	$^1D^e$	2	68.4206	$^1S^e$	4	73.6122	$^3G^e$	1	74.7653
$^1D^e$	1	56.3120	$^3D^o$	1	59.5926	$^3P^o$	4	72.0818	$^3P^o$	5	74.1326	$^1G^e$	1	74.7738
$^1P^e$	1	56.3483	$^1P^o$	2	60.3985	$^1P^o$	4	72.1534	$^3F^o$	2	74.1900	$^1D^e$	4	74.7847
$^3P^e$	1	56.3537	$^3S^e$	2	62.5974	$^3S^e$	3	72.9064	$^1F^o$	2	74.2707	$^1F^e$	1	74.8227

gf-values for transitions involving terms with effective $n \leq 4.0$, $L \leq 4$

i i'	gf_L	i i'	gf_L	i i'	gf_L	i i'	gf_L	i i'	gf_L	i i'	gf_L
	$^1S^e$–$^1P^o$	2 1	2.51E−1	5 1	1.29E−2	1 2	5.59E−2	2 2	1.29E−1	5 2	6.15E−3
1 1	−2.20E−1	2 2	8.17E−2	5 2	2.09E−3		$^1F^o$–$^1G^e$	2 3	4.60E−3	5 3	4.03E−2
1 2	−2.94E+0	2 3	1.61E−5	5 3	1.19E−2	1 1	−6.01E+0	2 4	1.53E+0	5 4	−1.04E+0
1 3	−3.27E−1	2 4	4.55E−1	5 4	−8.65E−2	2 1	−5.84E−1	2 5	−7.37E−1		$^3D^e$–$^3D^o$
1 4	−3.69E−2	2 5	−4.14E−1		$^1D^e$–$^1D^o$		$^3S^e$–$^3P^o$		$^3P^e$–$^3D^o$	1 1	−5.88E−1
1 5	−7.37E−1		$^1P^e$–$^1D^o$	1 1	−1.86E−1	1 1	2.38E−1	1 1	−1.61E+0	1 2	−7.61E−1
2 1	2.11E−1	1 1	−5.37E−1	1 2	−2.62E−1	1 2	−6.75E−1	1 2	−2.38E+0	2 1	5.06E−1
2 2	−1.70E−1	1 2	−7.94E−1	2 1	1.88E−1	1 3	−9.71E−2	2 1	3.77E−1	2 2	−1.01E−2
2 3	−1.03E−1	2 1	1.26E−1	2 2	−1.98E−3	1 4	−1.68E−1	2 2	−2.54E+0	3 1	1.60E−1
2 4	−1.06E−1	2 2	−8.47E−1	3 1	4.75E−2	1 5	−1.04E+0		$^3P^o$–$^3D^e$	3 2	−9.06E−1
2 5	−3.86E−1		$^1P^o$–$^1D^e$	3 2	−2.85E−1	2 1	4.98E−1	1 1	−1.60E+0	4 1	1.70E+0
3 1	9.79E−2	1 1	−5.40E−1	4 1	5.55E−1	2 2	1.74E−4	1 2	−1.88E−4	4 2	1.39E−1
3 2	5.38E−3	1 2	−1.65E−3	4 2	4.98E−2	2 3	−1.07E+0	1 3	−1.29E+0		$^3D^e$–$^3F^o$
3 3	−2.92E−1	1 3	−4.28E−1		$^1D^e$–$^1F^o$	2 4	−4.85E−4	1 4	−2.19E−3	1 1	−2.83E+0
3 4	−2.59E−3	1 4	−4.10E−3	1 1	−9.72E−1	2 5	−2.02E−2	2 1	1.32E−2	1 2	−4.54E+0
3 5	−6.71E−3	2 1	3.03E−3	1 2	−1.49E+0	3 1	2.62E−1	2 2	−1.32E−1	2 1	1.05E+0
4 1	6.08E−2	2 2	−3.77E−2	2 1	3.60E−1	3 2	1.59E−1	2 3	−5.08E−4	2 2	−2.38E−2
4 2	4.28E−2	2 3	−2.62E−9	2 2	−6.44E−3	3 3	7.01E−4	2 4	−8.44E+0	3 1	6.23E−1
4 3	3.09E−3	2 4	−2.99E+0	3 1	2.16E−1	3 4	4.10E−1	3 1	8.12E−1	3 2	−4.56E+0
4 4	2.75E−1	3 1	2.40E−1	3 2	−1.56E+0	3 5	−1.12E+0	3 2	−2.11E+0	4 1	7.08E−2
4 5	−3.07E−1	3 2	−7.75E−1	4 1	1.98E−2		$^3P^e$–$^3P^o$	3 3	−3.56E−3	4 2	6.80E−3
	$^1P^e$–$^1P^o$	3 3	−2.04E−5	4 2	1.45E−3	1 1	1.09E+0	3 4	−8.69E−2		$^3F^o$–$^3G^e$
1 1	3.21E−1	3 4	2.72E−2		$^1D^o$–$^1F^e$	1 2	−4.39E−1	4 1	8.21E−1	1 1	−1.78E+1
1 2	−2.53E−1	4 1	2.91E−1	1 1	−4.16E+0	1 3	−3.91E−1	4 2	1.41E−3	2 1	−2.01E+0
1 3	−1.62E−1	4 2	1.22E−3	2 1	−4.33E−1	1 4	−5.65E−1	4 3	−2.36E+0		
1 4	−1.52E−1	4 3	−7.94E−1		$^1F^e$–$^1F^o$	1 5	−8.65E−1	4 4	−1.23E−3		
1 5	−2.12E−1	4 4	−2.45E−3	1 1	5.29E−1	2 1	7.07E−1	5 1	5.53E−2		

Na-like S (S^{5+})

Term energies relative to $2s^22p^6$ ^{1}S ionization threshold for each symmetry

i	E(Ryds)	Description	i	E(Ryds)	Description	i	E(Ryds)	Description	i	E(Ryds)	Description
	^{2}S^e			^{2}P^o			^{2}D^e			^{2}F^o	
1	−6.43692	$2s^22p^63s$	1	−5.47147	$2s^22p^63p$	1	−4.22309	$2s^22p^63d$	1	−2.25467	$2s^22p^64f$
2	−3.15189	$2s^22p^64s$	2	−2.80300	$2s^22p^64p$	2	−2.35500	$2s^22p^64d$	2	−1.44303	$2s^22p^65f$
3	−1.87278	$2s^22p^65s$	3	−1.70886	$2s^22p^65p$	3	−1.49404	$2s^22p^65d$	3	−1.00198	$2s^22p^66f$
4	−1.24045	$2s^22p^66s$	4	−1.15074	$2s^22p^66p$	4	−1.03151	$2s^22p^66d$	4	−.73603	$2s^22p^67f$
5	−.88198	$2s^22p^67s$	5	−.82764	$2s^22p^67p$	5	−.75465	$2s^22p^67d$	5	−.56344	$2s^22p^68f$
6	−.65922	$2s^22p^68s$	6	−.62384	$2s^22p^68p$	6	−.57592	$2s^22p^68d$	6	−.44512	$2s^22p^69f$
7	−.51135	$2s^22p^69s$	7	−.48705	$2s^22p^69p$	7	−.45390	$2s^22p^69d$			

Energies in ascending order from ground state for terms with effective $n \leq 5.0$, $L \leq 3$

Term	i	E(Ryds)	Term	i	E(Ryds)	Term	i	E(Ryds)	Term	i	E(Ryds)
^{2}S^e	1	0.00000	^{2}S^e	2	3.28503	^{2}F^o	1	4.18225	^{2}D^e	3	4.94288
^{2}P^o	1	0.96545	^{2}P^o	2	3.63392	^{2}S^e	3	4.56414	^{2}F^o	2	4.99389
^{2}D^e	1	2.21383	^{2}D^e	2	4.08192	^{2}P^o	3	4.72806			

gf-values for transitions involving terms with effective $n \leq 5.0$, $L \leq 3$

i i'	gf$_L$	i i'	gf$_L$	i i'	gf$_L$	i i'	gf$_L$	i i'	gf$_L$	i i'	gf$_L$
	^{2}S^e–^{2}P^o	2 2	−1.98E+0		^{2}P^o–^{2}D^e	2 2	−5.77E+0		^{2}D^e–^{2}F^o	3 1	3.28E−1
1 1	−1.32E+0	2 3	−1.41E−1	1 1	−3.71E+0	2 3	−2.08E−1	1 1	−8.82E+0	3 2	−2.65E+0
1 2	−1.63E−1	3 1	1.03E−1	1 2	−4.25E−1	3 1	1.22E−1	1 2	−1.69E+0		
1 3	−6.24E−2	3 2	1.04E+0	1 3	−2.17E−1	3 2	2.12E+0	2 1	−1.48E+0		
2 1	6.04E−1	3 3	−2.59E+0	2 1	1.01E+0	3 3	−7.58E+0	2 2	−6.72E+0		

Na-like Ar (Ar^{7+})

Term energies relative to $2s^22p^6$ ^{1}S ionization threshold for each symmetry

i	E(Ryds)	Description	i	E(Ryds)	Description	i	E(Ryds)	Description	i	E(Ryds)	Description
	^{2}S^e			^{2}P^o			^{2}D^e			^{2}F^o	
1	−10.4893	$2s^22p^63s$	1	−9.21374	$2s^22p^63p$	1	−7.51967	$2s^22p^63d$	1	−4.00949	$2s^22p^64f$
2	−5.27539	$2s^22p^64s$	2	−4.79865	$2s^22p^64p$	2	−4.18653	$2s^22p^64d$	2	−2.56632	$2s^22p^65f$
3	−3.17825	$2s^22p^65s$	3	−2.95036	$2s^22p^65p$	3	−2.65548	$2s^22p^65d$	3	−1.78195	$2s^22p^66f$
4	−2.12341	$2s^22p^66s$	4	−1.99735	$2s^22p^66p$	4	−1.83339	$2s^22p^66d$	4	−1.30895	$2s^22p^67f$
5	−1.51874	$2s^22p^67s$	5	−1.44181	$2s^22p^67p$	5	−1.34132	$2s^22p^67d$	5	−1.00198	$2s^22p^68f$
6	−1.14005	$2s^22p^68s$	6	−1.08970	$2s^22p^68p$	6	−1.02365	$2s^22p^68d$	6	−.79155	$2s^22p^69f$
7	−.88724	$2s^22p^69s$	7	−.85251	$2s^22p^69p$	7	−.80677	$2s^22p^69d$			

Energies in ascending order from ground state for terms with effective $n \leq 5.0$, $L \leq 3$

Term	i	E(Ryds)	Term	i	E(Ryds)	Term	i	E(Ryds)	Term	i	E(Ryds)
^{2}S^e	1	0.00000	^{2}S^e	2	5.21391	^{2}F^o	1	6.47981	^{2}D^e	3	7.83382
^{2}P^o	1	1.27556	^{2}P^o	2	5.69065	^{2}S^e	3	7.31105	^{2}F^o	2	7.92298
^{2}D^e	1	2.96963	^{2}D^e	2	6.30277	^{2}P^o	3	7.53894			

gf-values for transitions involving terms with effective $n \leq 5.0$, $L \leq 3$

i i'	gf$_L$	i i'	gf$_L$	i i'	gf$_L$	i i'	gf$_L$	i i'	gf$_L$	i i'	gf$_L$
	^{2}S^e–^{2}P^o	2 2	−1.70E+0		^{2}P^o–^{2}D^e	2 2	−4.72E+0		^{2}D^e–^{2}F^o	3 1	3.24E−1
1 1	−1.13E+0	2 3	−2.34E−1	1 1	−2.97E+0	2 3	−5.67E−1	1 1	−8.80E+0	3 2	−2.60E+0
1 2	−2.47E−1	3 1	9.63E−2	1 2	−8.45E−1	3 1	1.07E−1	1 2	−1.69E+0		
1 3	−8.69E−2	3 2	9.03E−1	1 3	−3.48E−1	3 2	1.63E+0	2 1	−1.47E+0		
2 1	5.30E−1	3 3	−2.22E+0	2 1	7.60E−1	3 3	−6.26E+0	2 2	−6.74E+0		

Na-like Ca (Ca^{9+})

Term energies relative to $2s^2 2p^6$ ^{1}S ionization threshold for each symmetry

i	E(Ryds)	Description	i	E(Ryds)	Description	i	E(Ryds)	Description	i	E(Ryds)	Description
	^{2}**S**^{e}			^{2}**P**^{o}			^{2}**D**^{e}			^{2}**F**^{o}	
1	−15.4433	$2s^2 2p^6 3s$	1	−13.8604	$2s^2 2p^6 3p$	1	−11.7329	$2s^2 2p^6 3d$	1	−6.26584	$2s^2 2p^6 4f$
2	−7.90726	$2s^2 2p^6 4s$	2	−7.30205	$2s^2 2p^6 4p$	2	−6.52885	$2s^2 2p^6 4d$	2	−4.01071	$2s^2 2p^6 5f$
3	−4.80891	$2s^2 2p^6 5s$	3	−4.51615	$2s^2 2p^6 5p$	3	−4.14166	$2s^2 2p^6 5d$	3	−2.78487	$2s^2 2p^6 6f$
4	−3.23197	$2s^2 2p^6 6s$	4	−3.06880	$2s^2 2p^6 6p$	4	−2.86012	$2s^2 2p^6 6d$	4	−2.04562	$2s^2 2p^6 7f$
5	−2.32104	$2s^2 2p^6 7s$	5	−2.22097	$2s^2 2p^6 7p$	5	−2.09288	$2s^2 2p^6 7d$	5	−1.56586	$2s^2 2p^6 8f$
6	−1.74748	$2s^2 2p^6 8s$	6	−1.68175	$2s^2 2p^6 8p$	6	−1.59747	$2s^2 2p^6 8d$	6	−1.23700	$2s^2 2p^6 9f$
7	−1.36305	$2s^2 2p^6 9s$	7	−1.31758	$2s^2 2p^6 9p$	7	−1.25917	$2s^2 2p^6 9d$			

Energies in ascending order from ground state for terms with effective $n \leq 5.0$, $L \leq 3$

Term	i	E(Ryds)	Term	i	E(Ryds)	Term	i	E(Ryds)	Term	i	E(Ryds)
^{2}**S**^{e}	1	0.00000	^{2}**S**^{e}	2	7.53604	^{2}**F**^{o}	1	9.17746	^{2}**D**^{e}	3	11.3016
^{2}**P**^{o}	1	1.58290	^{2}**P**^{o}	2	8.14125	^{2}**S**^{e}	3	10.6343	^{2}**F**^{o}	2	11.4325
^{2}**D**^{e}	1	3.71040	^{2}**D**^{e}	2	8.91445	^{2}**P**^{o}	3	10.9271			

gf-values for transitions involving terms with effective $n \leq 5.0$, $L \leq 3$

i i'	gf_L	i i'	gf_L	i i'	gf_L	i i'	gf_L	i i'	gf_L	i i'	gf_L
	^{2}**S**^{e}–^{2}**P**^{o}	2 2	−1.49E+0		^{2}**P**^{o}–^{2}**D**^{e}	2 2	−3.94E+0		^{2}**D**^{e}–^{2}**F**^{o}	3 1	3.11E−1
1 1	−9.96E−1	2 3	−3.14E−1	1 1	−2.46E+0	2 3	−9.09E−1	1 1	−8.85E+0	3 2	−2.44E+0
1 2	−3.18E−1	3 1	9.05E−2	1 2	−1.20E+0	3 1	9.36E−2	1 2	−1.68E+0		
1 3	−1.06E−1	3 2	8.08E−1	1 3	−4.41E−1	3 2	1.33E+0	2 1	−1.39E+0		
2 1	4.79E−1	3 3	−1.94E+0	2 1	6.08E−1	3 3	−5.27E+0	2 2	−6.86E+0		

Na-like Fe (Fe^{15+})

Term energies relative to $2s^22p^6$ ^{1}S ionization threshold for each symmetry

i	E(Ryds)	Description	i	E(Ryds)	Description	i	E(Ryds)	Description	i	E(Ryds)	Description
	^{2}**S**^{e}			^{2}**P**^{o}			^{2}**D**^{e}			^{2}**F**^{o}	
1	−35.6756	$2s^22p^63s$	1	−33.1795	$2s^22p^63p$	1	−29.7949	$2s^22p^63d$	1	−16.0418	$2s^22p^64f$
2	−18.8281	$2s^22p^64s$	2	−17.8361	$2s^22p^64p$	2	−16.5843	$2s^22p^64d$	2	−10.2688	$2s^22p^65f$
3	−11.6365	$2s^22p^65s$	3	−11.1467	$2s^22p^65p$	3	−10.5358	$2s^22p^65d$	3	−7.13013	$2s^22p^66f$
4	−7.90079	$2s^22p^66s$	4	−7.62432	$2s^22p^66p$	4	−7.28288	$2s^22p^66d$	4	−5.23735	$2s^22p^67f$
5	−5.71411	$2s^22p^67s$	5	−5.54305	$2s^22p^67p$	5	−5.33298	$2s^22p^67d$	5	−4.00898	$2s^22p^68f$
6	−4.32440	$2s^22p^68s$	6	−4.21130	$2s^22p^68p$	6	−4.07281	$2s^22p^68d$	6	−3.16698	$2s^22p^69f$
7	−3.38645	$2s^22p^69s$	7	−3.30782	$2s^22p^69p$	7	−3.21168	$2s^22p^69d$			

Energies in ascending order from ground state for terms with effective $n \leq 5.0$, $L \leq 3$

Term	i	E(Ryds)	Term	i	E(Ryds)	Term	i	E(Ryds)	Term	i	E(Ryds)
^{2}**S**^{e}	1	0.00000	^{2}**S**^{e}	2	16.8475	^{2}**F**^{o}	1	19.6338	^{2}**D**^{e}	3	25.1398
^{2}**P**^{o}	1	2.49610	^{2}**P**^{o}	2	17.8395	^{2}**S**^{e}	3	24.0391	^{2}**F**^{o}	2	25.4068
^{2}**D**^{e}	1	5.88070	^{2}**D**^{e}	2	19.0913	^{2}**P**^{o}	3	24.5289			

gf-values for transitions involving terms with effective $n \leq 5.0$, $L \leq 3$

i i'	gf_L	i i'	gf_L	i i'	gf_L	i i'	gf_L	i i'	gf_L	i i'	gf_L
	^{2}**S**^{e}–^{2}**P**^{o}	2 2	−1.08E+0		^{2}**P**^{o}–^{2}**D**^{e}	2 2	−2.61E+0		^{2}**D**^{e}–^{2}**F**^{o}	3 1	2.67E−1
1 1	−7.28E−1	2 3	−4.92E−1	1 1	−1.60E+0	2 3	−1.65E+0	1 1	−9.06E+0	3 2	−1.93E+0
1 2	−4.71E−1	3 1	7.85E−2	1 2	−1.92E+0	3 1	6.83E−2	1 2	−1.66E+0		
1 3	−1.44E−1	3 2	6.47E−1	1 3	−5.94E−1	3 2	8.85E−1	2 1	−1.11E+0		
2 1	3.88E−1	3 3	−1.41E+0	2 1	3.94E−1	3 3	−3.52E+0	2 2	−7.25E+0		

Mg-like S (S^{4+})

Term energies relative to $3s\ ^2S$ ionization threshold for each symmetry

i	Energy(Ryds)	Description
		$^1\mathbf{S}^e$
1	−5.34104	$3s3s$
2	−3.17548	$3p3p$
3	−2.41673	$3s4s$
4	−1.40478	$3s5s$
5	−1.02228	$3p4p$
6	−.90664	$3s6s$
7	−.64298	$3s7s$
8	−.58218	$3d3d$
9	−.47807	$3s8s$
10	−.36947	$3s9s$
		$^1\mathbf{P}^o$
1	−4.16720	$3s3p$
2	−2.15119	$3s4p$
3	−1.86069	$3p3d$
4	−1.46411	$3p4s$
5	−1.24822	$3s5p$
6	−.84290	$3s6p$
7	−.67954	$3p4d$
8	−.59974	$3s7p$
9	−.45880	$3s8p$
10	−.42091	$3p5s$
11	−.34722	$3s9p$
		$^1\mathbf{P}^e$
1	−1.23326	$3p4p$
2	−.31875	$3p5p$
		$^1\mathbf{D}^o$
1	−2.34897	$3p3d$
2	−.77564	$3p4d$
3	−.11610	$3p5d$
		$^1\mathbf{D}^e$
1	−3.58416	$3p3p$
2	−2.85113	$3s3d$
3	−1.71179	$3s4d$
4	−1.14168	$3p4p$:
5	−1.04671	$3s5d$:
6	−.90856	$3d3d$
7	−.73332	$3s6d$
8	−.53987	$3p4f$
9	−.53452	$3s7d$
10	−.40962	$3s8d$
11	−.32348	$3s9d$
12	−.29944	$3p5p$
		$^1\mathbf{F}^o$
1	−1.98652	$3p3d$
2	−1.51970	$3s4f$
3	−.99782	$3s5f$
4	−.74589	$3p4d$
5	−.68464	$3s6f$
6	−.50930	$3s7f$
7	−.39071	$3s8f$
8	−.30898	$3s9f$
		$^1\mathbf{F}^e$
1	−.65679	$3p4f$
2	−.05907	$3p5f$
		$^1\mathbf{G}^o$
1	−.04159	$3p5g$
		$^1\mathbf{G}^e$
1	−1.01144	$3s5g$
2	−.98830	$3d3d$
3	−.70607	$3s6g$
4	−.54132	$3s7g$
5	−.47854	$3p4f$
6	−.38610	$3s8g$
7	−.30752	$3s9g$
		$^3\mathbf{S}^e$
1	−2.49566	$3s4s$
2	−1.42460	$3s5s$
3	−1.14328	$3p4p$
4	−.92317	$3s6s$
5	−.64907	$3s7s$
6	−.48148	$3s8s$
7	−.37151	$3s9s$
		$^3\mathbf{P}^o$
1	−4.58655	$3s3p$
2	−2.18621	$3p3d$
3	−2.15214	$3s4p$
4	−1.49331	$3p4s$
5	−1.27207	$3s5p$
6	−.84535	$3s6p$
7	−.71256	$3p4d$
8	−.60122	$3s7p$
9	−.45370	$3s8p$
10	−.43785	$3p5s$
11	−.34988	$3s9p$
		$^3\mathbf{P}^e$
1	−3.50841	$3p3p$
2	−1.15818	$3p4p$
3	−.86598	$3d3d$
4	−.29611	$3p5p$
		$^3\mathbf{D}^o$
1	−2.16033	$3p3d$
2	−.76733	$3p4d$
3	−.12129	$3p5d$
		$^3\mathbf{D}^e$
1	−3.19659	$3s3d$
2	−1.72613	$3s4d$
3	−1.21046	$3p4p$
4	−1.06758	$3s5d$
5	−.73817	$3s6d$
6	−.59188	$3p4f$
7	−.53615	$3s7d$
8	−.40973	$3s8d$
9	−.32809	$3s9d$
10	−.30680	$3p5p$
		$^3\mathbf{F}^o$
1	−2.39827	$3p3d$
2	−1.59023	$3s4f$
3	−1.01616	$3s5f$
4	−.75674	$3p4d$
5	−.70202	$3s6f$
6	−.51584	$3s7f$
7	−.39450	$3s8f$
8	−.31142	$3s9f$
		$^3\mathbf{F}^e$
1	−1.01269	$3d3d$
2	−.62403	$3p4f$
3	−.05015	$3p5f$
		$^3\mathbf{G}^o$
1	−.04153	$3p5g$
		$^3\mathbf{G}^e$
1	−1.00970	$3s5g$
2	−.71475	$3s6g$
3	−.62033	$3p4f$
4	−.50465	$3s7g$
5	−.38955	$3s8g$
6	−.30855	$3s9g$

Mg-like S (S^{4+})

Energies in ascending order from ground state for terms with effective $n \leq 4.0$, $L \leq 3$

Term	i	E(Ryds)	Term	i	E(Ryds)	Term	i	E(Ryds)	Term	i	E(Ryds)	Term	i	E(Ryds)
$^{1}\mathbf{S}^{e}$	1	0.00000	$^{3}\mathbf{D}^{e}$	1	2.14445	$^{3}\mathbf{F}^{o}$	1	2.94277	$^{1}\mathbf{P}^{o}$	2	3.18985	$^{3}\mathbf{F}^{o}$	2	3.75081
$^{3}\mathbf{P}^{o}$	1	0.75449	$^{1}\mathbf{S}^{e}$	2	2.16556	$^{1}\mathbf{D}^{o}$	1	2.99207	$^{1}\mathbf{F}^{o}$	1	3.35452			
$^{1}\mathbf{P}^{o}$	1	1.17384	$^{1}\mathbf{D}^{e}$	2	2.48991	$^{3}\mathbf{P}^{o}$	2	3.15483	$^{1}\mathbf{P}^{o}$	3	3.48035			
$^{1}\mathbf{D}^{e}$	1	1.75688	$^{3}\mathbf{S}^{e}$	1	2.84538	$^{3}\mathbf{D}^{o}$	1	3.18071	$^{3}\mathbf{D}^{e}$	2	3.61491			
$^{3}\mathbf{P}^{e}$	1	1.83263	$^{1}\mathbf{S}^{e}$	3	2.92431	$^{3}\mathbf{P}^{o}$	3	3.18890	$^{1}\mathbf{D}^{e}$	3	3.62925			

gf-values for transitions involving terms with effective $n \leq 4.0$, $L \leq 3$

$i\ i'$	gf_L	$i\ i'$	gf_L	$i\ i'$	gf_L	$i\ i'$	gf_L	$i\ i'$	gf_L	$i\ i'$	gf_L
	$^{1}\mathbf{S}^{e}$–$^{1}\mathbf{P}^{o}$	3 3	−4.72E−2	3 2	7.16E−1	3 1	3.00E−1		$^{3}\mathbf{P}^{o}$–$^{3}\mathbf{D}^{e}$		$^{3}\mathbf{D}^{o}$–$^{3}\mathbf{D}^{e}$
1 1	−1.44E+0		$^{1}\mathbf{P}^{o}$–$^{1}\mathbf{D}^{e}$	3 3	−3.22E−2		$^{3}\mathbf{S}^{e}$–$^{3}\mathbf{P}^{o}$	1 1	−6.49E+0	1 1	3.67E+0
1 2	−9.84E−2	1 1	−3.13E−1		$^{1}\mathbf{D}^{o}$–$^{1}\mathbf{D}^{e}$	1 1	9.37E−1	1 2	−2.85E−1	1 2	−1.24E−3
1 3	−2.19E−3	1 2	−4.31E+0	1 1	2.10E+0	1 2	−3.39E−2	2 1	2.84E+0		$^{3}\mathbf{D}^{e}$–$^{3}\mathbf{F}^{o}$
2 1	5.59E−1	1 3	−1.56E−2	1 2	1.60E−1	1 3	−2.86E+0	2 2	−1.16E−1	1 1	−2.90E+0
2 2	−4.10E−2	2 1	3.32E−1	1 3	−1.86E−3		$^{3}\mathbf{P}^{o}$–$^{3}\mathbf{P}^{e}$	3 1	1.20E+0	1 2	−1.32E+1
2 3	−1.55E+0	2 2	5.07E−1		$^{1}\mathbf{D}^{e}$–$^{1}\mathbf{F}^{o}$	1 1	−4.36E+0	3 2	−8.05E+0	2 1	2.48E−1
3 1	2.42E−1	2 3	−2.59E+0	1 1	−3.43E+0	2 1	3.39E+0		$^{3}\mathbf{P}^{e}$–$^{3}\mathbf{D}^{o}$	2 2	−3.84E+0
3 2	−6.37E−1	3 1	5.20E−4	2 1	−1.73E+0	3 1	3.36E−2	1 1	−9.75E+0		

Mg-like Ar (Ar^{6+})

Term energies relative to 3s ^{2}S ionization threshold for each symmetry

i	E(Ryds)	Description	i	E(Ryds)	Description	i	E(Ryds)	Description	i	E(Ryds)	Description
		$^1S^e$	13	−.65085	3p5f	10	−.87408	3d4p	2	−.12181	3p6g :
1	−9.07903	3s3s	14	−.62296	3s9d	11	−.85920	3s8p	3	−.04210	3d4f :
2	−6.20795	3p3p			$^1F^o$	12	−.77240	3p5d			$^3G^e$
3	−4.32928	3s4s	1	−4.48276	3p3d	13	−.66905	3s9p	1	−2.01481	3s5g
4	−2.67185	3d3d	2	−3.04946	3s4f			$^3P^e$	2	−1.82388	3p4f
5	−2.58225	3s5s	3	−2.05271	3p4d	1	−6.63784	3p3p	3	−1.35766	3s6g
6	−2.35801	3p4p	4	−1.95987	3s5f	2	−2.99840	3d3d	4	−1.00177	3s7g
7	−1.69661	3s6s	5	−1.36730	3s6f	3	−2.63730	3p4p	5	−.77303	3s8g
8	−1.20934	3s7s	6	−1.00816	3s7f	4	−1.07918	3p5p	6	−.70329	3p5f
9	−1.00207	3p5p	7	−.91414	3p5d	5	−.30267	3p6p	7	−.60256	3s9g
10	−.89661	3s8s	8	−.77324	3s8f	6	−.19275	3d4d			
11	−.69814	3s9s	9	−.71482	3d4p			$^3D^o$			
		$^1P^o$	10	−.65121	3p5g	1	−4.78273	3p3d			
1	−7.52590	3s3p	11	−.60462	3s9f	2	−2.08854	3p4d			
2	−4.39576	3p3d			$^1F^e$	3	−.96024	3d4p			
3	−3.92885	3s4p	1	−1.89701	3p4f	4	−.79321	3p5d			
4	−3.05117	3p4s	2	−.71949	3p5f	5	−.15236	3p6d			
5	−2.36740	3s5p	3	−.41165	3d4d	6	−.05827	3d4f			
6	−1.96749	3p4d	4	−.09704	3p6f			$^3D^e$			
7	−1.59081	3s6p			$^1G^o$	1	−6.15016	3s3d			
8	−1.28160	3p5s	1	−.69360	3p5g	2	−3.34342	3s4d			
9	−1.13535	3s7p	2	−.21276	3d4f	3	−2.69410	3p4p			
10	−.87310	3d4p	3	−.07289	3p6g	4	−2.09399	3s5d			
11	−.85986	3s8p			$^1G^e$	5	−1.79525	3p4f			
12	−.69621	3p5d	1	−3.07743	3d3d	6	−1.44081	3s6d			
13	−.65774	3s9p	2	−2.00073	3s5g	7	−1.36534	3d4s			
		$^1P^e$	3	−1.70329	3p4f	8	−1.09223	3p5p			
1	−2.74218	3p4p	4	−1.34942	3s6g	9	−1.03909	3s7d			
2	−1.11070	3p5p	5	−1.00024	3s7g	10	−.79733	3s8d			
3	−.36996	3d4d	6	−.77247	3s8g	11	−.67847	3p5f			
4	−.30962	3p6p	7	−.66076	3p5f	12	−.62579	3s9d			
		$^1D^o$	8	−.59397	3s9g			$^3F^o$			
1	−5.02558	3p3d			$^3S^e$	1	−5.07922	3p3d			
2	−2.09224	3p4d	1	−4.43221	3s4s	2	−3.11896	3s4f			
3	−.96914	3d4p	2	−2.65267	3s5s :	3	−2.05731	3p4d			
4	−.81638	3p5d	3	−2.56775	3p4p :	4	−1.98853	3s5f			
5	−.15838	3p6d	4	−1.71245	3s6s	5	−1.37862	3s6f			
6	−.12128	3d4f	5	−1.21606	3s7s	6	−1.01129	3s7f			
		$^1D^e$	6	−1.05841	3p5p	7	−.92826	3d4p			
1	−6.71082	3p3p	7	−.90482	3s8s	8	−.79726	3p5d			
2	−5.70925	3s3d	8	−.70186	3s9s	9	−.77253	3s8f			
3	−3.33765	3s4d			$^3P^o$	10	−.65463	3p5g			
4	−3.03538	3d3d	1	−8.06164	3s3p	11	−.61010	3s9f			
5	−2.56094	3p4p	2	−4.81098	3p3d			$^3F^e$			
6	−2.08791	3s5d	3	−3.94751	3s4p	1	−3.17091	3d3d			
7	−1.72558	3p4f	4	−3.11321	3p4s	2	−1.86314	3p4f			
8	−1.44756	3s6d	5	−2.38066	3s5p	3	−.70845	3p5f			
9	−1.32863	3d4s	6	−1.99891	3p4d	4	−.28682	3d4d			
10	−1.06806	3s7d	7	−1.59255	3s6p	5	−.08956	3p6f			
11	−1.00328	3p5p	8	−1.29359	3p5s			$^3G^o$			
12	−.79468	3s8d	9	−1.14154	3s7p	1	−.69342	3p5g			

Mg-like Ar (Ar^{6+})

Energies in ascending order from ground state for terms with effective $n \leq 4.0$, $L \leq 4$

Term	i	E(Ryds)	Term	i	E(Ryds)	Term	i	E(Ryds)	Term	i	E(Ryds)	Term	i	E(Ryds)
$^1S^e$	1	0.00000	$^1S^e$	2	2.87108	$^3P^o$	2	4.26805	$^1S^e$	3	4.74975	$^3F^e$	1	5.90812
$^3P^o$	1	1.01739	$^3D^e$	1	2.92887	$^3D^o$	1	4.29630	$^3P^o$	3	5.13152	$^3F^o$	2	5.96007
$^1P^o$	1	1.55313	$^1D^e$	2	3.36978	$^1F^o$	1	4.59627	$^1P^o$	3	5.15018	$^3P^o$	4	5.96582
$^1D^e$	1	2.36821	$^3F^o$	1	3.99981	$^3S^e$	1	4.64682	$^3D^e$	2	5.73561	$^1G^e$	1	6.00160
$^3P^e$	1	2.44119	$^1D^o$	1	4.05345	$^1P^o$	2	4.68327	$^1D^e$	3	5.74138			

gf-values for transitions involving terms with effective $n \leq 4.0$, $L \leq 4$

i i'	gf_L	i i'	gf_L	i i'	gf_L	i i'	gf_L	i i'	gf_L	i i'	gf_L
	$^1S^e$–$^1P^o$		$^1P^o$–$^1D^e$		$^1D^o$–$^1D^e$	1 1	8.42E−1	1 2	−9.81E−1	1 2	−6.94E−4
1 1	−1.24E+0	1 1	−3.72E−1	1 1	1.72E+0	1 2	1.62E−4	2 1	2.13E+0		$^3D^o$–$^3F^e$
1 2	−6.25E−3	1 2	−3.45E+0	1 2	1.93E−1	1 3	−2.49E+0	2 2	−3.83E−5	1 1	−7.19E+0
1 3	−1.79E−1	1 3	−1.67E−1	1 3	−3.12E−2	1 4	−2.09E+0	3 1	1.22E+0		$^3D^e$–$^3F^o$
2 1	4.83E−1	2 1	2.62E−2		$^1D^e$–$^1F^o$		$^3P^o$–$^3P^e$	3 2	−6.45E+0	1 1	−3.07E+0
2 2	−1.25E+0	2 2	9.57E−1	1 1	−2.55E+0	1 1	−3.79E+0	4 1	2.20E−2	1 2	−1.31E+1
2 3	−4.26E−2	2 3	−1.40E−1	2 1	−2.60E+0	2 1	2.77E+0	4 2	1.23E−2	2 1	1.79E−1
3 1	2.11E−1	3 1	2.30E−1	3 1	1.13E−1	3 1	1.23E−4		$^3P^e$–$^3D^o$	2 2	−3.15E+0
3 2	2.76E−3	3 2	1.02E−1		$^1F^o$–$^1G^e$	4 1	1.59E+0	1 1	−7.86E+0		$^3F^o$–$^3F^e$
3 3	−5.70E−1	3 3	−1.99E+0	1 1	−3.38E+0		$^3P^o$–$^3D^e$		$^3D^o$–$^3D^e$	1 1	−5.70E+0
					$^3S^e$–$^3P^o$	1 1	−5.23E+0	1 1	3.22E+0	2 1	3.86E−4

Mg-like Ca (Ca^{8+})

Term energies relative to *3s* ^{2}S ionization threshold for each symmetry

i	E(Ryds)	Description	*i*	E(Ryds)	Description	*i*	E(Ryds)	Description	*i*	E(Ryds)	Description
	$^1S^e$		7	−3.37926	*3p4f*	7	−1.68233	*3d4d*	14	−.95043	*3p6p*
1	−13.7687	*3s3s*	8	−2.99569	*3d4s*	8	−1.45836	*3s8s*		$^3F^o$	
2	−10.1804	*3p3p*	9	−2.36772	*3s6d*	9	−1.13462	*3s9s*	1	−8.70373	*3p3d*
3	−6.76942	*3s4s*	10	−2.13238	*3p5p*	10	−.93337	*3p6p*	2	−5.15444	*3s4f*
4	−5.60391	*3d3d*	11	−1.72696	*3s7d*		$^3P^o$		3	−3.87494	*3p4d*
5	−4.39034	*3p4p*	12	−1.60348	*3p5f*	1	−12.4814	*3s3p*	4	−3.28769	*3s5f*
6	−4.02909	*3s5s*	13	−1.54923	*3d4d*	2	−8.38307	*3p3d*	5	−2.48081	*3d4p*
7	−2.71249	*3s6s*	14	−1.31081	*3s8d*	3	−6.26661	*3s4p*	6	−2.27746	*3s6f*
8	−2.10403	*3p5p*	15	−1.03473	*3s9d*	4	−5.25412	*3p4s*	7	−1.81840	*3p5d*
9	−1.92827	*3s7s*	16	−.92223	*3p6p*	5	−3.86576	*3s5p*	8	−1.67079	*3s7f*
10	−1.45464	*3s8s*		$^1F^o$		6	−3.75486	*3p4d*	9	−1.61764	*3p5g*
11	−1.32223	*3d4d*	1	−7.95736	*3p3d*	7	−2.57326	*3s6p*	10	−1.44552	*3d4f*
12	−1.13038	*3s9s*	2	−5.07148	*3s4f*	8	−2.49435	*3p5s*	11	−1.27721	*3s8f*
	$^1P^o$		3	−3.86929	*3p4d*	9	−2.38293	*3d4p*	12	−1.00822	*3s9f*
1	−11.8318	*3s3p*	4	−3.25885	*3s5f*	10	−1.85740	*3s7p*		$^3F^e$	
2	−7.86379	*3p3d*	5	−2.39812	*3d4p*	11	−1.78248	*3p5d*	1	−6.28836	*3d3d*
3	−6.25631	*3s4p*	6	−2.25200	*3s6f*	12	−1.39489	*3s8p*	2	−3.59577	*3p4f*
4	−5.16655	*3p4s*	7	−1.78642	*3p5d*	13	−1.29148	*3d4f*	3	−1.69277	*3p5f*
5	−3.86262	*3s5p*	8	−1.66149	*3s7f*	14	−1.10553	*3p6s*	4	−1.63965	*3d4d*
6	−3.71397	*3p4d*	9	−1.61211	*3p5g*	15	−1.08447	*3s9p*	5	−.66565	*3p6f*
7	−2.58389	*3s6p*	10	−1.29813	*3d4f*		$^3P^e$		6	−.05594	*3p7f*
8	−2.47489	*3p5s*	11	−1.25564	*3s8f*	1	−10.7056	*3p3p*		$^3G^o$	
9	−2.29018	*3d4p*	12	−1.00302	*3s9f*	2	−6.07765	*3d3d*	1	−1.67044	*3p5g*
10	−1.85586	*3s7p*		$^1F^e$		3	−4.63816	*3p4p*	2	−1.33463	*3d4f*
11	−1.73935	*3p5d*	1	−3.63691	*3p4f*	4	−2.18742	*3p5p*	3	−.64518	*3p6g*
12	−1.39247	*3s8p*	2	−1.81336	*3d4d*	5	−1.53890	*3d4d*	4	−.04470	*3p7g*
13	−1.17917	*3d4f*	3	−1.69580	*3p5f*	6	−.94409	*3p6p*		$^3G^e$	
14	−1.10863	*3s9p*	4	−.67317	*3p6f*	7	−.22794	*3p7p*	1	−3.63522	*3p4f*
15	−1.06642	*3p6s*	5	−.06012	*3p7f*		$^3D^o$		2	−3.21192	*3s5g*
	$^1P^e$			$^1G^o$		1	−8.35639	*3p3d*	3	−2.25257	*3s6g*
1	−4.76878	*3p4p*	1	−1.67741	*3p5g*	2	−3.92759	*3p4d*	4	−1.77062	*3d4d*
2	−2.22699	*3p5p*	2	−1.48965	*3d4f*	3	−2.50242	*3d4p*	5	−1.69513	*3p5f*
3	−1.75490	*3d4d*	3	−.64803	*3p6g*	4	−1.83392	*3p5d*	6	−1.63189	*3s7g*
4	−.96517	*3p6p*	4	−.04553	*3p7g*	5	−1.31679	*3d4f*	7	−1.26503	*3s8g*
5	−.23891	*3p7p*		$^1G^e$		6	−.75387	*3p6d*	8	−1.00070	*3s9g*
	$^1D^o$		1	−6.13396	*3d3d*	7	−.11520	*3p7d*			
1	−8.64726	*3p3d*	2	−3.50241	*3p4f*	8	−.00219	*3d5p*			
2	−3.92855	*3p4d*	3	−3.16510	*3s5g*		$^3D^e$				
3	−2.53486	*3d4p*	4	−2.24733	*3s6g*	1	−10.0595	*3s3d*			
4	−1.84306	*3p5d*	5	−1.69330	*3s7g*	2	−5.48688	*3s4d*			
5	−1.40655	*3d4f*	6	−1.59277	*3d4d* :	3	−4.70135	*3p4p*			
6	−.75516	*3p6d*	7	−1.53629	*3p5f* :	4	−3.52688	*3p4f*			
7	−.11141	*3p7d*	8	−1.26037	*3s8g*	5	−3.41824	*3s5d*			
8	−.01703	*3d5p*	9	−.99872	*3s9g*	6	−3.05101	*3d4s*			
	$^1D^e$			$^3S^e$		7	−2.37081	*3s6d*			
1	−10.7777	*3p3p*	1	−6.89631	*3s4s*	8	−2.19243	*3p5p*			
2	−9.53284	*3s3d*	2	−4.62405	*3p4p*	9	−1.77816	*3d4d*			
3	−6.11571	*3d3d*	3	−4.09833	*3s5s*	10	−1.72568	*3s7d*			
4	−5.47500	*3s4d*	4	−2.73073	*3s6s*	11	−1.63759	*3p5f*			
5	−4.53617	*3p4p*	5	−2.16650	*3p5p*	12	−1.31304	*3s8d*			
6	−3.47337	*3s5d*	6	−1.94355	*3s7s*	13	−1.03573	*3s9d*			

Mg-like Ca (Ca^{8+})

Energies in ascending order from ground state for terms with effective $n \leq 4.0$, $L \leq 4$

Term	i	E(Ryds)	Term	i	E(Ryds)	Term	i	E(Ryds)	Term	i	E(Ryds)	Term	i	E(Ryds)
$^{1}\mathbf{S}^{e}$	1	0.00000	$^{3}\mathbf{D}^{e}$	1	3.70920	$^{1}\mathbf{F}^{o}$	1	5.81134	$^{1}\mathbf{P}^{o}$	3	7.51239	$^{1}\mathbf{D}^{e}$	4	8.29370
$^{3}\mathbf{P}^{o}$	1	1.28730	$^{1}\mathbf{D}^{e}$	2	4.23586	$^{1}\mathbf{P}^{o}$	2	5.90491	$^{1}\mathbf{G}^{e}$	1	7.63474	$^{3}\mathbf{P}^{o}$	4	8.51458
$^{1}\mathbf{P}^{o}$	1	1.93690	$^{3}\mathbf{F}^{o}$	1	5.06497	$^{3}\mathbf{S}^{e}$	1	6.87239	$^{1}\mathbf{D}^{e}$	3	7.65299	$^{1}\mathbf{P}^{o}$	4	8.60215
$^{1}\mathbf{D}^{e}$	1	2.99100	$^{1}\mathbf{D}^{o}$	1	5.12144	$^{1}\mathbf{S}^{e}$	3	6.99928	$^{3}\mathbf{P}^{e}$	2	7.69105	$^{3}\mathbf{F}^{o}$	2	8.61426
$^{3}\mathbf{P}^{e}$	1	3.06310	$^{3}\mathbf{P}^{o}$	2	5.38563	$^{3}\mathbf{F}^{e}$	1	7.48034	$^{1}\mathbf{S}^{e}$	4	8.16479	$^{1}\mathbf{F}^{o}$	2	8.69722
$^{1}\mathbf{S}^{e}$	2	3.58830	$^{3}\mathbf{D}^{o}$	1	5.41231	$^{3}\mathbf{P}^{o}$	3	7.50209	$^{3}\mathbf{D}^{e}$	2	8.28182			

gf-values for transitions involving terms with effective $n \leq 4.0$, $L \leq 4$

i i'	gf$_{L}$	i i'	gf$_{L}$	i i'	gf$_{L}$	i i'	gf$_{L}$	i i'	gf$_{L}$	i i'	gf$_{L}$
	$^{1}\mathbf{S}^{e}$–$^{1}\mathbf{P}^{o}$	4 3	1.55E−3	4 1	4.96E−1	3 2	−1.47E−2	2 2	−4.29E+0	1 1	−6.43E+0
1 1	−1.08E+0	4 4	−7.87E−4	4 2	1.91E−1	4 1	1.35E−1	3 1	5.52E−5	2 1	1.60E+0
1 2	−2.02E−3		$^{1}\mathbf{P}^{o}$–$^{1}\mathbf{D}^{e}$	4 3	5.58E−3	4 2	−1.16E+0	3 2	−5.44E−5		$^{3}\mathbf{D}^{o}$–$^{3}\mathbf{D}^{e}$
1 3	−2.56E−1	1 1	−3.81E−1	4 4	2.27E−5		$^{1}\mathbf{F}^{o}$–$^{1}\mathbf{G}^{e}$	4 1	1.47E+0	1 1	2.83E+0
1 4	−2.03E−4	1 2	−2.82E+0		$^{1}\mathbf{D}^{o}$–$^{1}\mathbf{D}^{e}$	1 1	−3.36E+0	4 2	3.38E−3	1 2	−1.31E−4
2 1	4.24E−1	1 3	−1.27E−2	1 1	1.43E+0	2 1	1.66E−5		$^{3}\mathbf{P}^{o}$–$^{3}\mathbf{D}^{e}$		$^{3}\mathbf{D}^{o}$–$^{3}\mathbf{F}^{e}$
2 2	−1.04E+0	1 4	−3.09E−1	1 2	2.00E−1		$^{3}\mathbf{S}^{e}$–$^{3}\mathbf{P}^{o}$	1 1	−4.28E+0	1 1	−6.55E+0
2 3	−8.13E−3	2 1	1.02E−2	1 3	−1.92E+0	1 1	7.72E−1	1 2	−1.69E+0		$^{3}\mathbf{D}^{e}$–$^{3}\mathbf{F}^{o}$
2 4	−1.02E−1	2 2	7.87E−1	1 4	−8.85E−3	1 2	2.81E−4	2 1	1.88E+0	1 1	−2.90E+0
3 1	1.88E−1	2 3	−7.70E−1		$^{1}\mathbf{D}^{e}$–$^{1}\mathbf{F}^{o}$	1 3	−2.17E+0	2 2	−1.69E−4	1 2	−1.32E+1
3 2	8.41E−3	2 4	−2.72E−3	1 1	−1.98E+0	1 4	−1.82E+0	3 1	9.71E−1	2 1	1.38E−1
3 3	−4.93E−1	3 1	2.03E−1	1 2	−7.22E−1		$^{3}\mathbf{P}^{o}$–$^{3}\mathbf{P}^{e}$	3 2	−5.23E+0	2 2	−2.77E+0
3 4	−8.59E−1	3 2	1.24E−1	2 1	−2.50E+0	1 1	−3.33E+0	4 1	2.21E−2		$^{3}\mathbf{F}^{o}$–$^{3}\mathbf{F}^{e}$
4 1	1.04E−4	3 3	−1.20E−3	2 2	−3.78E+0	1 2	−1.33E−3	4 2	1.43E−2	1 1	−4.55E+0
4 2	7.98E−1	3 4	−1.74E+0	3 1	3.07E−1	2 1	2.28E+0		$^{3}\mathbf{P}^{e}$–$^{3}\mathbf{D}^{o}$	2 1	2.31E−3

Mg-like Fe (Fe^{14+})

Term energies relative to 3s ^2S ionization threshold for each symmetry

i	E(Ryds)	Description	i	E(Ryds)	Description	i	E(Ryds)	Description	i	E(Ryds)	Description
		$^1S^e$	6	−3.89072	3d5p	5	−3.29726	3d5d	10	−6.87403	3s6p
1	−33.5246	3s3s	7	−3.85669	3p6d	6	−2.84005	3d5g	11	−6.67354	3p5d
2	−27.5067	3p3p	8	−2.92326	3d5f	7	−1.99054	3p7f	12	−4.95936	3s7p
3	−19.9012	3d3d	9	−2.09814	3p7d	8	−.89723	3p8f	13	−4.46046	3p6s
4	−17.1950	3s4s	10	−.96940	3p8d	9	−.34054	3d6d	14	−3.83379	3p6d
5	−13.1430	3p4p	11	−.67360	3d6p	10	−.14934	3p9f	15	−3.80020	3d5p
6	−10.4905	3s5s	12	−.20208	3p9d	11	−.08782	3d6g	16	−3.73720	3s8p
7	−8.19026	3d4d	13	−.13171	3d6f			$^1G^o$	17	−2.94740	3s9p
8	−7.24608	3p5p			$^1D^e$	1	−8.27498	3d4f	18	−2.84445	3d5f
9	−6.97603	3s6s	1	−28.3941	3p3p	2	−6.40623	3p5g	19	−2.46972	3p7s
10	−5.11424	3s7s	2	−26.5397	3s3d	3	−3.63487	3p6g			$^3P^e$
11	−4.10499	3p6p	3	−20.7058	3d3d	4	−2.99172	3d5f	1	−28.3063	3p3p
12	−3.84918	3s8s	4	−14.9839	3s4d	5	−1.96006	3p7g	2	−20.6701	3d3d
13	−3.03254	3s9s	5	−13.4095	3p4p	6	−.87738	3p8g	3	−13.5769	3p4p
14	−2.97094	3d5d	6	−11.3283	3p4f	7	−.17327	3d6f	4	−8.48911	3d4d
		$^1P^o$	7	−10.9743	3d4s	8	−.13078	3p9g	5	−7.37799	3p5p
1	−30.2989	3s3p	8	−9.41050	3s5d			$^1G^e$	6	−4.18437	3p6p
2	−23.6964	3p3d	9	−8.51845	3d4d	1	−20.6789	3d3d	7	−3.13578	3d5d
3	−16.3049	3s4p	10	−7.30315	3p5p	2	−11.3866	3p4f	8	−2.29816	3p7p
4	−14.4964	3p4s	11	−6.50119	3s6d	3	−8.98619	3s5g	9	−1.10298	3p8p
5	−12.1398	3p4d	12	−6.30965	3p5f	4	−8.45724	3d4d	10	−.30397	3p9p
6	−10.0678	3s5p	13	−4.74272	3s7d	5	−6.40898	3p5f	11	−.24969	3d6d
7	−9.77629	3d4p	14	−4.34090	3d5s	6	−6.19204	3s6g			$^3D^o$
8	−7.83559	3p5s	15	−4.11336	3p6p	7	−4.58842	3s7g	1	−24.4625	3p3d
9	−7.69166	3d4f	16	−3.63608	3s8d	8	−3.62603	3p6f	2	−12.3790	3p4d
10	−6.86567	3s6p	17	−3.57209	3p6f	9	−3.50221	3s8g	3	−10.0825	3d4p
11	−6.62146	3p5d	18	−3.15030	3d5d	10	−3.12827	3d5d	4	−7.93583	3d4f
12	−4.95662	3s7p	19	−2.84602	3s9d	11	−2.84264	3d5g	5	−6.79425	3p5d
13	−4.44228	3p6s	20	−2.75982	3d5g	12	−2.76909	3s9g	6	−3.91445	3d5p
14	−3.85937	3p6d			$^1F^o$			$^3S^e$	7	−3.81300	3p6d
15	−3.77852	3s8p	1	−23.8094	3p3d	1	−17.4087	3s4s	8	−2.86026	3d5f
16	−3.66119	3d5p	2	−14.1565	3s4f	2	−13.5513	3p4p	9	−2.09531	3p7d
17	−2.94422	3s9p	3	−12.2583	3p4d	3	−10.5664	3s5s	10	−.96920	3p8d
18	−2.74961	3d5f	4	−9.87251	3d4p	4	−8.72554	3d4d	11	−.66524	3d6p
19	−2.46293	3p7s	5	−9.06108	3s5f	5	−7.36261	3p5p	12	−.19972	3p9d
		$^1P^e$	6	−7.89900	3d4f	6	−7.09626	3s6s	13	−.10071	3d6f
1	−13.7891	3p4p	7	−6.72388	3p5d	7	−5.13373	3s7s			$^3D^e$
2	−8.83162	3d4d	8	−6.28898	3s6f	8	−4.16555	3p6p	1	−27.3150	3s3d
3	−7.45303	3p5p	9	−6.27614	3p5g	9	−3.86775	3s8s	2	−15.0030	3s4d
4	−4.21727	3p6p	10	−4.61834	3s7f	10	−3.20071	3d5d	3	−13.6749	3p4p
5	−3.25175	3d5d	11	−3.90219	3p6d	11	−3.02583	3s9s	4	−11.5126	3p4f
6	−2.31778	3p7p	12	−3.71240	3d5p			$^3P^o$	5	−11.0507	3d4s
7	−1.11459	3p8p	13	−3.56742	3p6g	1	−31.2827	3s3p	6	−9.42953	3s5d
8	−.32896	3d6d	14	−3.53292	3s8f	2	−24.4766	3p3d	7	−8.86728	3d4d
9	−.28749	3p9p	15	−2.85322	3d5f	3	−16.3136	3s4p	8	−7.40994	3p5p
		$^1D^o$	16	−2.78075	3s9f	4	−14.6494	3p4s	9	−6.51263	3s6d
1	−24.8818	3p3d			$^1F^e$	5	−12.1810	3p4d	10	−6.37921	3p5f
2	−12.3803	3p4d	1	−11.7379	3p4f	6	−10.0554	3s5p	11	−4.74645	3s7d
3	−10.1418	3d4p	2	−8.92075	3d4d	7	−9.95896	3d4p	12	−4.34920	3d5s
4	−8.09652	3d4f	3	−6.49806	3p5f	8	−7.91437	3d4f	13	−4.17625	3p6p
5	−6.80334	3p5d	4	−3.68209	3p6f	9	−7.86757	3p5s	14	−3.65136	3p6f

Mg-like Fe (Fe^{14+})

i	E(Ryds)	Description	i	E(Ryds)	Description	i	E(Ryds)	Description	i	E(Ryds)	Description
15	−3.59627	$3s8d$	12	−3.81302	$3p6d$	12	−.08273	$3d6g$	7	−3.66763	$3p6f$
16	−3.27135	$3d5d$	13	−3.58494	$3p6g$			$^3G^o$	8	−3.51661	$3s8g$
17	−2.84888	$3s9d$	14	−3.54005	$3s8f$	1	−7.99136	$3d4f$	9	−3.25894	$3d5d$
18	−2.77875	$3d5g$	15	−2.96163	$3d5f$	2	−6.39677	$3p5g$	10	−2.86074	$3d5g$
19	−2.32144	$3p7p$	16	−2.79641	$3s9f$	3	−3.63265	$3p6g$	11	−2.77935	$3s9g$
		$^3F^o$			$^3F^e$	4	−2.88496	$3d5f$			
1	−24.9422	$3p3d$	1	−20.9944	$3d3d$	5	−1.95679	$3p7g$			
2	−14.2937	$3s4f$	2	−11.6693	$3p4f$	6	−.87589	$3p8g$			
3	−12.2798	$3p4d$	3	−8.65393	$3d4d$	7	−.14645	$3p9g$			
4	−10.0557	$3d4p$	4	−6.47489	$3p5f$	8	−.10455	$3d6f$			
5	−9.10940	$3s5f$	5	−3.67117	$3p6f$			$^3G^e$			
6	−8.19177	$3d4f$	6	−3.21064	$3d5d$	1	−11.6862	$3p4f$			
7	−6.75917	$3p5d$	7	−2.83241	$3d5g$	2	−9.01775	$3s5g$			
8	−6.32607	$3s6f$	8	−1.98472	$3p7f$	3	−8.82147	$3d4d$			
9	−6.29571	$3p5g$	9	−.89360	$3p8f$	4	−6.48054	$3p5f$			
10	−4.63150	$3s7f$	10	−.29988	$3d6d$	5	−6.24174	$3s6g$			
11	−3.86997	$3d5p$	11	−.14668	$3p9f$	6	−4.59732	$3s7g$			

Energies in ascending order from ground state for terms with effective $n \leq 4.0$, $L \leq 4$

Term	i	E(Ryds)	Term	i	E(Ryds)	Term	i	E(Ryds)	Term	i	E(Ryds)	Term	i	E(Ryds)
$^1S^e$	1	0.00000	$^3D^e$	1	6.20960	$^1F^o$	1	9.71520	$^1S^e$	3	13.6234	$^1D^e$	4	18.5407
$^3P^o$	1	2.24190	$^1D^e$	2	6.98490	$^1P^o$	2	9.82820	$^3S^e$	1	16.1159	$^3P^o$	4	18.8752
$^1P^o$	1	3.22570	$^3F^o$	1	8.58240	$^3F^e$	1	12.5302	$^1S^e$	4	16.3296	$^1P^o$	4	19.0282
$^1D^e$	1	5.13050	$^1D^o$	1	8.64280	$^1D^e$	3	12.8188	$^3P^o$	3	17.2110	$^3F^o$	2	19.2309
$^3P^e$	1	5.21830	$^3P^o$	2	9.04800	$^1G^e$	1	12.8457	$^1P^o$	3	17.2197	$^1F^o$	2	19.3681
$^1S^e$	2	6.01790	$^3D^o$	1	9.06210	$^3P^e$	2	12.8545	$^3D^e$	2	18.5216			

gf-values for transitions involving terms with effective $n \leq 4.0$, $L \leq 4$

$i\ i'$	gf_L	$i\ i'$	gf_L	$i\ i'$	gf_L	$i\ i'$	gf_L	$i\ i'$	gf_L	$i\ i'$	gf_L
	$^1S^e$–$^1P^o$	4 3	−3.62E−1	4 1	4.24E−1	3 2	−8.00E−4	2 2	−2.81E+0	1 1	−4.18E+0
1 1	−8.29E−1	4 4	−6.30E−1	4 2	1.45E−1	4 1	4.78E−2	3 1	1.29E−4	2 1	1.05E+0
1 2	−1.28E−3		$^1P^o$–$^1D^e$	4 3	2.95E−3	4 2	−8.11E−1	3 2	9.81E−6		$^3D^o$–$^3D^e$
1 3	−4.08E−1	1 1	−3.40E−1	4 4	7.79E−5		$^1F^o$–$^1G^e$	4 1	1.26E+0	1 1	2.18E+0
1 4	−4.24E−3	1 2	−1.92E+0		$^1D^o$–$^1D^e$	1 1	−2.61E+0	4 2	2.44E−3	1 2	−4.58E−4
2 1	3.23E−1	1 3	−3.55E−3	1 1	9.91E−1	2 1	1.46E−5		$^3P^o$–$^3D^e$		$^3D^o$–$^3F^e$
2 2	−6.72E−1	1 4	−6.51E−1	1 2	1.84E−1		$^3S^e$–$^3P^o$	1 1	−2.78E+0	1 1	−4.70E+0
2 3	−2.03E−3	2 1	9.46E−3	1 3	−1.22E+0	1 1	6.58E−1	1 2	−3.12E+0		$^3D^e$–$^3F^o$
2 4	−8.27E−2	2 2	5.80E−1	1 4	−8.83E−6	1 2	4.29E−4	2 1	1.44E+0	1 1	−2.44E+0
3 1	2.08E−3	2 3	−5.76E−1		$^1D^e$–$^1F^o$	1 3	−1.61E+0	2 2	−6.07E−4	1 2	−1.37E+1
3 2	5.16E−1	2 4	−1.49E−3	1 1	−1.12E+0	1 4	−1.38E+0	3 1	6.23E−1	2 1	7.00E−2
3 3	−2.54E−3	3 1	1.44E−1	1 2	−1.09E+0		$^3P^o$–$^3P^e$	3 2	−3.39E+0	2 2	−2.09E+0
3 4	8.14E−3	3 2	8.07E−2	2 1	−2.04E+0	1 1	−2.56E+0	4 1	1.77E−2		$^3F^o$–$^3F^e$
4 1	1.51E−1	3 3	2.68E−4	2 2	−3.46E+0	1 2	−9.91E−4	4 2	1.09E−2	1 1	−2.89E+0
4 2	4.73E−6	3 4	−1.16E+0	3 1	2.15E−1	2 1	1.50E+0		$^3P^e$–$^3D^o$	2 1	1.88E−3

Al-like S (S^{3+})

Term energies relative to $3s^2$ ^{1}S ionization threshold for each symmetry

i	Energy(Ryds)	Description	i	Energy(Ryds)	Description	i	Energy(Ryds)	Description
			11	−.23511	$3s3p$ ^{3}P $4f$	2	−.52262	$3p^2$ ^{3}P $3d$
	^{2}S^e		12	−.20847	$3s^2$ ^{1}S $9d$	3	−.23399	$3s3p$ ^{3}P $4f$
1	−2.32639	$3s3p^2$		^{2}F^o		4	−.12636	$3s3p$ ^{3}P $5p$
2	−1.80742	$3s^2$ ^{1}S $4s$	1	−1.26706	$3s3p$ ^{3}P $3d$		^{4}F^o	
3	−1.00354	$3s^2$ ^{1}S $5s$	2	−1.12462	$3s3p$ ^{1}P $3d$	1	−1.63932	$3s3p$ ^{3}P $3d$
4	−.64860	$3s3p$ ^{3}P $4p$	3	−.93757	$3s^2$ ^{1}S $4f$	2	−.39833	$3s3p$ ^{3}P $4d$
5	−.63543	$3s^2$ ^{1}S $6s$	4	−.64177	$3s^2$ ^{1}S $5f$		^{4}F^e	
6	−.44438	$3s^2$ ^{1}S $7s$	5	−.44936	$3s^2$ ^{1}S $6f$	1	−.60207	$3p^2$ ^{3}P $3d$
7	−.33540	$3s3p$ ^{1}P $4p$	6	−.34992	$3s3p$ ^{3}P $4d$	2	−.29489	$3s3p$ ^{3}P $4f$
8	−.31761	$3s^2$ ^{1}S $8s$	7	−.32648	$3s^2$ ^{1}S $7f$	3	−.13946	$3s3d^2$
9	−.24858	$3s^2$ ^{1}S $9s$	8	−.25208	$3s^2$ ^{1}S $8f$			
10	−.19784	$3p^2$ ^{1}D $3d$:	9	−.19930	$3s^2$ ^{1}S $9f$		^{4}G^e	
	^{2}P^o			^{2}F^e		1	−.26784	$3s3p$ ^{3}P $4f$
1	−3.46776	$3s^2$ ^{1}S $3p$	1	−.67370	$3p^2$ ^{1}D $3d$		^{4}H^o	
2	−1.54807	$3p^3$	2	−.33707	$3p^2$ ^{3}P $3d$			
3	−1.51989	$3s^2$ ^{1}S $4p$	3	−.11963	$3s3p$ ^{3}P $4f$			
4	−1.03712	$3s3p$ ^{3}P $3d$						
5	−.98558	$3s3p$ ^{3}P $4s$		^{2}G^e				
6	−.90867	$3s3p$ ^{1}P $3d$	1	−.64623	$3s^2$ ^{1}S $5g$			
7	−.87270	$3s^2$ ^{1}S $5p$	2	−.53386	$3p^2$ ^{1}D $3d$			
8	−.64176	$3s3p$ ^{1}P $4s$	3	−.44755	$3s^2$ ^{1}S $6g$			
9	−.56979	$3s^2$ ^{1}S $6p$	4	−.32925	$3s^2$ ^{1}S $7g$			
10	−.40531	$3s^2$ ^{1}S $7p$	5	−.25285	$3s^2$ ^{1}S $8g$			
11	−.31845	$3s3p$ ^{3}P $4d$	6	−.23817	$3s3p$ ^{3}P $4f$			
12	−.30122	$3s^2$ ^{1}S $8p$	7	−.19883	$3s^2$ ^{1}S $9g$			
13	−.23466	$3s^2$ ^{1}S $9p$		^{2}H^o				
14	−.21827	$3s3p$ ^{3}P $5s$	1	−.44489	$3s^2$ ^{1}S $6h$			
	^{2}P^e		2	−.32695	$3s^2$ ^{1}S $7h$			
1	−2.23113	$3s3p^2$	3	−.25034	$3s^2$ ^{1}S $8h$			
2	−.79600	$3s3p$ ^{3}P $4p$	4	−.19780	$3s^2$ ^{1}S $9h$			
3	−.54477	$3p^2$ ^{1}D $3d$		^{4}S^o				
4	−.36188	$3s3p$ ^{1}P $4p$	1	−1.69745	$3p^3$			
5	−.21278	$3p^2$ ^{3}P $3d$		^{4}S^e				
6	−.14011	$3s3p$ ^{3}P $5p$	1	−.73454	$3s3p$ ^{3}P $4p$			
	^{2}D^o		2	−.10579	$3s3p$ ^{3}P $5p$			
1	−1.81555	$3p^3$		^{4}P^o				
2	−1.34437	$3s3p$ ^{3}P $3d$	1	−1.45953	$3s3p$ ^{3}P $3d$			
3	−.88401	$3s3p$ ^{1}P $3d$	2	−1.05739	$3s3p$ ^{3}P $4s$			
4	−.39755	$3s3p$ ^{3}P $4d$	3	−.36968	$3s3p$ ^{3}P $4d$			
5	−.00205	$3s3p$ ^{1}P $4d$	4	−.24352	$3s3p$ ^{3}P $5s$			
	^{2}D^e			^{4}P^e				
1	−2.62841	$3s3p^2$	1	−2.84355	$3s3p^2$			
2	−2.07591	$3s^2$ ^{1}S $3d$	2	−.75289	$3s3p$ ^{3}P $4p$			
3	−1.14700	$3s^2$ ^{1}S $4d$	3	−.30846	$3p^2$ ^{3}P $3d$			
4	−.72277	$3s^2$ ^{1}S $5d$	4	−.11737	$3s3p$ ^{3}P $5p$			
5	−.70573	$3s3p$ ^{3}P $4p$	5	−.03832	$3s3d^2$			
6	−.49187	$3s^2$ ^{1}S $6d$		^{4}D^o				
7	−.40215	$3p^2$ ^{1}D $3d$	1	−1.43948	$3s3p$ ^{3}P $3d$			
8	−.38020	$3s3p$ ^{1}P $4p$	2	−.40756	$3s3p$ ^{3}P $4d$			
9	−.33856	$3s^2$ ^{1}S $7d$		^{4}D^e				
10	−.26490	$3s^2$ ^{1}S $8d$	1	−.78689	$3s3p$ ^{3}P $4p$			

Al-like S (S^{3+})

Energies in ascending order from ground state for terms with effective $n \leq 4.0$, $L \leq 3$

Term	i	E(Ryds)	Term	i	E(Ryds)	Term	i	E(Ryds)	Term	i	E(Ryds)	Term	i	E(Ryds)
$^2\mathbf{P}^o$	1	0.00000	$^2\mathbf{D}^e$	2	1.39185	$^2\mathbf{P}^o$	2	1.91969	$^2\mathbf{F}^o$	1	2.20070	$^2\mathbf{S}^e$	3	2.46422
$^4\mathbf{P}^e$	1	0.62421	$^2\mathbf{D}^o$	1	1.65221	$^2\mathbf{P}^o$	3	1.94787	$^2\mathbf{D}^e$	3	2.32076			
$^2\mathbf{D}^e$	1	0.83935	$^2\mathbf{S}^e$	2	1.66034	$^4\mathbf{P}^o$	1	2.00823	$^2\mathbf{F}^o$	2	2.34314			
$^2\mathbf{S}^e$	1	1.14137	$^4\mathbf{S}^o$	1	1.77031	$^4\mathbf{D}^o$	1	2.02828	$^4\mathbf{P}^o$	2	2.41037			
$^2\mathbf{P}^e$	1	1.23663	$^4\mathbf{F}^o$	1	1.82844	$^2\mathbf{D}^o$	2	2.12339	$^2\mathbf{P}^o$	4	2.43064			

gf-values for transitions involving terms with effective $n \leq 4.0$, $L \leq 3$

$i\ i'$	gf_L	$i\ i'$	gf_L	$i\ i'$	gf_L	$i\ i'$	gf_L	$i\ i'$	gf_L	$i\ i'$	gf_L
	$^2\mathbf{S}^e$–$^2\mathbf{P}^o$	3 2	1.29E−1	1 2	−7.10E+0	4 3	5.08E−2	2 3	−9.11E−6		$^4\mathbf{P}^o$–$^4\mathbf{P}^e$
1 1	6.31E−1	3 3	1.09E+0	1 3	−3.08E−4		$^2\mathbf{P}^e$–$^2\mathbf{D}^o$		$^2\mathbf{D}^e$–$^2\mathbf{F}^o$	1 1	5.21E+0
1 2	−4.81E−2	3 4	1.60E−2	2 1	9.48E−1	1 1	−1.64E−1	1 1	−4.22E+0	2 1	2.12E+0
1 3	−6.36E−2		$^2\mathbf{P}^o$–$^2\mathbf{P}^e$	2 2	1.54E−1	1 2	−2.59E−1	1 2	−5.71E+0		$^4\mathbf{P}^e$–$^4\mathbf{D}^o$
1 4	−2.89E+0	1 1	−4.57E+0	2 3	−7.15E−1		$^2\mathbf{D}^o$–$^2\mathbf{D}^e$	2 1	−1.02E−1	1 1	−1.43E+1
2 1	5.24E−1	2 1	2.93E−1	3 1	1.01E+0	1 1	4.41E−1	2 2	−1.18E−1		
2 2	−1.96E−1	3 1	4.57E−2	3 2	8.77E−1	1 2	1.24E−2	3 1	1.52E−2		
2 3	−1.69E+0	4 1	1.74E+0	3 3	−5.40E+0	1 3	−1.10E−3	3 2	−2.77E−1		
2 4	−6.77E−2		$^2\mathbf{P}^o$–$^2\mathbf{D}^e$	4 1	2.27E−1	2 1	6.86E+0		$^4\mathbf{S}^o$–$^4\mathbf{P}^e$		
3 1	9.57E−2	1 1	−2.86E−1	4 2	1.09E−3	2 2	4.93E−2	1 1	3.05E+0		

Al-like Ar (Ar^{5+})

Term energies relative to $3s^2$ ^{1}S ionization threshold for each symmetry

i	E(Ryds)	Description
		^{2}S^o
1	−.76172	$3p3d^2$
2	−.70188	$3p^2$ ^{3}P $4p$:
		^{2}S^e
1	−5.12980	$3s3p^2$
2	−3.56269	$3s^2$ ^{1}S $4s$
3	−2.12700	$3p^2$ ^{1}D $3d$
4	−2.05141	$3s^2$ ^{1}S $5s$
5	−1.87456	$3s3p$ ^{3}P $4p$
6	−1.52338	$3s3p$ ^{1}P $4p$
7	−1.32601	$3s^2$ ^{1}S $6s$
8	−1.18814	$3s3d^2$
9	−.93463	$3s^2$ ^{1}S $7s$
10	−.74011	$3s3p$ ^{3}P $5p$
11	−.69276	$3s^2$ ^{1}S $8s$
12	−.60366	$3p^2$ ^{1}S $4s$
13	−.53435	$3s^2$ ^{1}S $9s$
		^{2}P^o
1	−6.68097	$3s^2$ ^{1}S $3p$
2	−4.03060	$3p^3$
3	−3.24752	$3s3p$ ^{3}P $3d$
4	−3.13965	$3s^2$ ^{1}S $4p$
5	−3.07740	$3s3p$ ^{1}P $3d$
6	−2.42450	$3s3p$ ^{3}P $4s$
7	−1.99606	$3s3p$ ^{1}P $4s$
8	−1.83622	$3s^2$ ^{1}S $5p$
9	−1.42710	$3s3p$ ^{3}P $4d$
10	−1.22817	$3s^2$ ^{1}S $6p$
11	−.97860	$3s3p$ ^{3}P $5s$
12	−.96219	$3s3p$ ^{1}P $4d$
13	−.86864	$3s^2$ ^{1}S $7p$
14	−.66359	$3p^2$ ^{1}D $4p$
15	−.65226	$3s^2$ ^{1}S $8p$
16	−.53598	$3p^2$ ^{3}P $4p$
17	−.51305	$3s3p$ ^{3}P $5d$
18	−.50968	$3s^2$ ^{1}S $9p$
19	−.48024	$3s3p$ ^{1}P $5s$
		^{2}P^e
1	−5.00684	$3s3p^2$
2	−2.58635	$3p^2$ ^{1}D $3d$
3	−2.16778	$3s3p$ ^{3}P $4p$
4	−2.08204	$3p^2$ ^{3}P $3d$
5	−1.56714	$3s3p$ ^{1}P $4p$
6	−1.11217	$3s3d^2$
7	−.98142	$3p^2$ ^{3}P $4s$
8	−.77262	$3s3p$ ^{3}P $5p$
9	−.29545	$3s3p$ ^{1}P $5p$
10	−.18970	$3s3p$ ^{3}P $6p$
11	−.08865	$3p^2$ ^{1}D $4d$:
12	−.03122	$3p^2$ ^{3}P $4d$:

i	E(Ryds)	Description
		^{2}D^o
1	−4.36910	$3p^3$
2	−3.70479	$3s3p$ ^{3}P $3d$
3	−3.05673	$3s3p$ ^{1}P $3d$
4	−1.54429	$3s3p$ ^{3}P $4d$
5	−1.01882	$3s3p$ ^{1}P $4d$
6	−.85219	$3p3d^2$
7	−.73272	$3p^2$ ^{1}D $4p$
8	−.60008	$3s3p$ ^{3}P $5d$
9	−.47959	$3p^2$ ^{3}P $4p$
10	−.24359	$3p3d^2$
11	−.11514	$3s3d$ ^{3}D $4p$
12	−.04357	$3s3p$ ^{3}P $6d$
13	−.03744	$3s3p$ ^{1}P $5d$
		^{2}D^e
1	−5.50715	$3s3p^2$
2	−4.68890	$3s^2$ ^{1}S $3d$
3	−2.55678	$3s^2$ ^{1}S $4d$
4	−2.35374	$3p^2$ ^{1}D $3d$
5	−2.09583	$3p^2$ ^{3}P $3d$:
6	−1.97053	$3s3p$ ^{3}P $4p$
7	−1.77856	$3p^2$ ^{1}S $3d$:
8	−1.64108	$3s^2$ ^{1}S $5d$
9	−1.59128	$3s3d^2$
10	−1.50408	$3s3p$ ^{1}P $4p$
11	−1.13962	$3s3p$ ^{3}P $4f$
12	−1.08326	$3s^2$ ^{1}S $6d$
13	−1.08149	$3p^2$ ^{1}D $4s$
14	−.79000	$3s^2$ ^{1}S $7d$
15	−.77917	$3s3p$ ^{3}P $5p$
16	−.69840	$3s3p$ ^{1}P $4f$
17	−.59889	$3s^2$ ^{1}S $8d$
18	−.48032	$3s^2$ ^{1}S $9d$
19	−.45676	$3s3p$ ^{1}P $5p$
		^{2}F^o
1	−3.55068	$3s3p$ ^{3}P $3d$
2	−3.25176	$3s3p$ ^{1}P $3d$
3	−2.28412	$3s^2$ ^{1}S $4f$
4	−1.47851	$3s3p$ ^{3}P $4d$
5	−1.45349	$3s^2$ ^{1}S $5f$
6	−1.03902	$3s3p$ ^{1}P $4d$
7	−1.00728	$3s^2$ ^{1}S $6f$
8	−.88313	$3p3d^2$
9	−.74544	$3s^2$ ^{1}S $7f$
10	−.67059	$3p^2$ ^{1}D $4p$
11	−.56967	$3s^2$ ^{1}S $8f$
12	−.52776	$3s3p$ ^{3}P $5d$
13	−.44979	$3s^2$ ^{1}S $9f$
14	−.40337	$3s3p$ ^{3}P $5g$
15	−.38566	$3s3d$ ^{3}D $4p$

i	E(Ryds)	Description
		^{2}F^e
1	−2.73903	$3p^2$ ^{1}D $3d$
2	−2.05056	$3p^2$ ^{3}P $3d$
3	−1.36252	$3d^3$
4	−1.19878	$3s3p$ ^{3}P $4f$
5	−.75111	$3s3p$ ^{1}P $4f$
6	−.42435	$3s3p$ ^{3}P $5f$
7	−.15348	$3p^2$ ^{1}D $4d$
8	−.04510	$3p^2$ ^{3}P $4d$
		^{2}G^o
1	−.73481	$3p3d^2$
2	−.42625	$3s3p$ ^{3}P $5g$
3	−.20231	$3p3d^2$
		^{2}G^e
1	−2.51558	$3p^2$ ^{1}D $3d$
2	−1.67554	$3s3d^2$
3	−1.45623	$3s^2$ ^{1}S $5g$
4	−1.16223	$3s3p$ ^{3}P $4f$
5	−1.01789	$3s^2$ ^{1}S $6g$
6	−.77630	$3s^2$ ^{1}S $7g$
7	−.68707	$3s3p$ ^{1}P $4f$
8	−.55678	$3s^2$ ^{1}S $8g$
9	−.44351	$3s^2$ ^{1}S $9g$
10	−.38896	$3s3p$ ^{3}P $5f$
		^{2}H^o
1	−1.00141	$3s^2$ ^{1}S $6h$
2	−.73614	$3s^2$ ^{1}S $7h$
3	−.67978	$3p3d^2$
4	−.56376	$3s^2$ ^{1}S $8h$
5	−.44552	$3s^2$ ^{1}S $9h$
6	−.39496	$3s3p$ ^{3}P $5g$
		^{4}S^o
1	−4.25942	$3p^3$
2	−.60430	$3p^2$ ^{3}P $4p$
3	−.28644	$3p3d^2$
		^{4}S^e
1	−2.04423	$3s3p$ ^{3}P $4p$
2	−.78695	$3s3p$ ^{3}P $5p$
3	−.17353	$3s3p$ ^{3}P $6p$
		^{4}P^o
1	−3.82893	$3s3p$ ^{3}P $3d$
2	−2.53027	$3s3p$ ^{3}P $4s$
3	−1.47003	$3s3p$ ^{3}P $4d$
4	−1.00585	$3s3p$ ^{3}P $5s$
5	−.67992	$3p3d^2$
6	−.61195	$3p^2$ ^{3}P $4p$
7	−.51112	$3s3p$ ^{3}P $5d$
8	−.28714	$3s3p$ ^{3}P $6s$
9	−.10338	$3s3d$ ^{3}D $4p$
10	−.02662	$3s3p$ ^{3}P $6d$

i	E(Ryds)	Description
		^{4}P^e
1	−5.80453	$3s3p^2$
2	−2.27981	$3p^2$ ^{3}P $3d$
3	−2.07096	$3s3p$ ^{3}P $4p$
4	−1.81837	$3s3d^2$
5	−1.04056	$3p^2$ ^{3}P $4s$
6	−.79949	$3s3p$ ^{3}P $5p$
7	−.18151	$3s3p$ ^{3}P $6p$
8	−.06906	$3p^2$ ^{3}P $4d$
		^{4}D^o
1	−3.80178	$3s3p$ ^{3}P $3d$
2	−1.54607	$3s3p$ ^{3}P $4d$
3	−.76890	$3p3d^2$
4	−.67427	$3p^2$ ^{3}P $4p$
5	−.54135	$3s3p$ ^{3}P $5d$
6	−.51765	$3p3d^2$
7	−.17262	$3s3d$ ^{3}D $4p$
8	−.03676	$3s3p$ ^{3}P $6d$
		^{4}D^e
1	−2.55668	$3p^2$ ^{3}P $3d$
2	−2.11916	$3s3p$ ^{3}P $4p$
3	−1.22920	$3s3p$ ^{3}P $4f$
4	−.82190	$3s3p$ ^{3}P $5p$
5	−.58215	$3s3d$ ^{3}D $4s$
6	−.41474	$3s3p$ ^{3}P $5f$
7	−.18702	$3s3p$ ^{3}P $6p$
8	−.03211	$3p^2$ ^{3}P $4d$
		^{4}F^o
1	−4.07084	$3s3p$ ^{3}P $3d$
2	−1.50663	$3s3p$ ^{3}P $4d$
3	−.68995	$3p3d^2$
4	−.54099	$3s3p$ ^{3}P $5d$
5	−.39580	$3s3p$ ^{3}P $5g$
6	−.15253	$3s3d$ ^{3}D $4p$
7	−.03837	$3s3p$ ^{3}P $6d$
		^{4}F^e
1	−2.65546	$3p^2$ ^{3}P $3d$
2	−1.99463	$3s3d^2$
3	−1.28364	$3s3p$ ^{3}P $4f$
4	−.43849	$3s3p$ ^{3}P $5f$
5	−.08520	$3p^2$ ^{3}P $4d$
		^{4}G^o
1	−1.01566	$3p3d^2$
2	−.42467	$3s3p$ ^{3}P $5g$
		^{4}G^e
1	−1.28004	$3s3p$ ^{3}P $4f$
2	−.43611	$3s3p$ ^{3}P $5f$
		^{4}H^o
1	−.39955	$3s3p$ ^{3}P $5g$

Al-like Ar (Ar^{5+})

Energies in ascending order from ground state for terms with effective $n \leq 4.0$, $L \leq 4$

Term	i	E(Ryds)	Term	i	E(Ryds)	Term	i	E(Ryds)	Term	i	E(Ryds)	Term	i	E(Ryds)
$^2\mathbf{P}^o$	1	0.00000	$^4\mathbf{S}^o$	1	2.42155	$^2\mathbf{F}^o$	1	3.13029	$^4\mathbf{F}^e$	1	4.02551	$^2\mathbf{D}^e$	4	4.32723
$^4\mathbf{P}^e$	1	0.87644	$^4\mathbf{F}^o$	1	2.61013	$^2\mathbf{F}^o$	2	3.42921	$^2\mathbf{P}^e$	2	4.09462	$^2\mathbf{F}^o$	3	4.39685
$^2\mathbf{D}^e$	1	1.17382	$^2\mathbf{P}^o$	2	2.65037	$^2\mathbf{P}^o$	3	3.43345	$^2\mathbf{D}^e$	3	4.12419	$^4\mathbf{P}^e$	2	4.40116
$^2\mathbf{S}^e$	1	1.55117	$^4\mathbf{P}^o$	1	2.85204	$^2\mathbf{P}^o$	4	3.54132	$^4\mathbf{D}^e$	1	4.12429			
$^2\mathbf{P}^e$	1	1.67413	$^4\mathbf{D}^o$	1	2.87919	$^2\mathbf{P}^o$	5	3.60357	$^4\mathbf{P}^o$	2	4.15070			
$^2\mathbf{D}^e$	2	1.99207	$^2\mathbf{D}^o$	2	2.97618	$^2\mathbf{D}^o$	3	3.62424	$^2\mathbf{G}^e$	1	4.16539			
$^2\mathbf{D}^o$	1	2.31187	$^2\mathbf{S}^e$	2	3.11828	$^2\mathbf{F}^e$	1	3.94194	$^2\mathbf{P}^o$	6	4.25647			

gf-values for transitions involving terms with effective $n \leq 4.0$, $L \leq 4$

i i'	gf$_L$	i i'	gf$_L$	i i'	gf$_L$	i i'	gf$_L$	i i'	gf$_L$	i i'	gf$_L$
	$^2\mathbf{S}^e$–$^2\mathbf{P}^o$	4 1	1.62E+0	4 2	2.65E+0	1 2	1.22E−2	2 2	−5.09E+0		$^4\mathbf{P}^o$–$^4\mathbf{P}^e$
1 1	5.25E−1	4 2	−4.03E−2	4 3	−2.56E+0	1 3	−2.30E−2	2 3	−8.65E+0	1 1	4.08E+0
1 2	−1.57E−1	5 1	1.07E+0	4 4	−1.72E−1	1 4	−4.28E+0	3 1	1.95E−2	1 2	−1.43E+0
1 3	−2.93E+0	5 2	−2.89E−2	5 1	5.32E−2	2 1	5.44E+0	3 2	2.22E−1	2 1	2.16E+0
1 4	−1.19E−2	6 1	1.46E−1	5 2	7.76E−1	2 2	1.12E−1	3 3	−3.43E+0	2 2	−2.34E−4
1 5	−5.88E−1	6 2	9.22E−4	5 3	−1.64E+0	2 3	−8.70E−5	4 1	9.49E−5		$^4\mathbf{P}^o$–$^4\mathbf{D}^e$
1 6	−2.08E−1		$^2\mathbf{P}^o$–$^2\mathbf{D}^e$	5 4	−3.85E−3	2 4	−8.57E−1	4 2	1.21E−4	1 1	−2.64E+0
2 1	4.96E−1	1 1	−3.90E−1	6 1	9.99E−1	3 1	1.70E−3	4 3	−4.00E−3	2 1	1.98E−5
2 2	3.55E−4	1 2	−5.65E+0	6 2	1.90E−1	3 2	3.38E+0		$^2\mathbf{F}^o$–$^2\mathbf{F}^e$		$^4\mathbf{P}^e$–$^4\mathbf{D}^o$
2 3	−1.16E−1	1 3	−2.68E−1	6 3	1.02E−2	3 3	−8.74E−4	1 1	−4.61E−1	1 1	−1.15E+1
2 4	−8.30E−1	1 4	−1.09E−3	6 4	−2.10E−3	3 4	−2.06E−1	2 1	−6.80E−3	2 1	3.94E+0
2 5	−6.47E−1	2 1	1.35E+0		$^2\mathbf{P}^e$–$^2\mathbf{D}^o$		$^2\mathbf{D}^o$–$^2\mathbf{F}^e$	3 1	2.66E−4		$^4\mathbf{D}^o$–$^4\mathbf{D}^e$
2 6	−6.54E−2	2 2	1.80E−3	1 1	−2.68E−1	1 1	−6.96E−1		$^2\mathbf{F}^o$–$^2\mathbf{G}^e$	1 1	−5.51E−1
	$^2\mathbf{P}^o$–$^2\mathbf{P}^e$	2 3	−9.58E−3	1 2	−2.11E−1	2 1	−6.13E−1	1 1	−1.74E+0		$^4\mathbf{D}^o$–$^4\mathbf{F}^e$
1 1	−3.93E+0	2 4	−8.69E−1	1 3	−8.58E+0	3 1	−4.76E−2	2 1	−5.38E−1	1 1	−2.47E+0
1 2	−6.89E−4	3 1	5.94E−2	2 1	1.90E+0		$^2\mathbf{D}^e$–$^2\mathbf{F}^o$	3 1	4.82E−3		$^4\mathbf{D}^e$–$^4\mathbf{F}^o$
2 1	4.61E−1	3 2	2.41E−1	2 2	3.85E−1	1 1	−3.20E+0		$^4\mathbf{S}^o$–$^4\mathbf{P}^e$	1 1	4.58E+0
2 2	−1.25E−1	3 3	−5.96E−1	2 3	1.94E−3	1 2	−5.10E+0	1 1	2.65E+0		$^4\mathbf{F}^o$–$^4\mathbf{F}^e$
3 1	1.04E+0	3 4	−6.21E−2		$^2\mathbf{D}^o$–$^2\mathbf{D}^e$	1 3	−4.21E−1	1 2	−8.02E+0	1 1	−6.72E+0
3 2	−1.85E−2	4 1	1.99E−1	1 1	6.38E−1	2 1	−4.74E−1				

Al-like Ca (Ca^{7+})

Term energies relative to $3s^2$ ^{1}S ionization threshold for each symmetry

i E(Ryds) Description	*i* E(Ryds) Description	*i* E(Ryds) Description	*i* E(Ryds) Description
^{2}S^o	**^{2}P^e**	11 −2.67912 $3p^2$ ^{1}D $4s$	13 −.16581 $3s3d$ ^{1}D $4d$
1 −3.19629 $3p3d^2$	1 −8.71199 $3s3p^2$	12 −2.62432 $3s3p$ ^{3}P $4f$	14 −.00946 $3s3p$ ^{3}P $7f$
2 −2.23267 $3p^2$ ^{3}P $4p$	2 −5.57357 $3p^2$ ^{3}P $3d$:	13 −2.08688 $3s3p$ ^{1}P $4f$	**^{2}G^o**
3 −.07347 $3p^2$ ^{3}P $5p$	3 −4.98370 $3p^2$ ^{1}D $3d$:	14 −1.95711 $3s3d$ ^{3}D $4s$	1 −3.16443 $3p3d^2$
^{2}S^e	4 −4.00855 $3s3p$ ^{3}P $4p$	15 −1.89061 $3s^2$ ^{1}S $6d$	2 −2.45805 $3p3d^2$
1 −8.86000 $3s3p^2$	5 −3.71850 $3s3d^2$	16 −1.76460 $3s3p$ ^{3}P $5p$	3 −1.31186 $3s3p$ ^{3}P $5g$
2 −5.83845 $3s^2$ ^{1}S $4s$	6 −3.28363 $3s3p$ ^{1}P $4p$	17 −1.48905 $3s3d$ ^{1}D $4s$	4 −1.09516 $3p^2$ ^{1}D $4f$
3 −4.97033 $3p^2$ ^{1}D $3d$	7 −2.53083 $3p^2$ ^{3}P $4s$	18 −1.40598 $3p^2$ ^{1}D $4d$	5 −.87234 $3p^2$ ^{3}P $4f$
4 −3.84552 $3s3d^2$	8 −1.84122 $3s3p$ ^{3}P $5p$	19 −1.38097 $3s^2$ ^{1}S $7d$	6 −.63051 $3s3p$ ^{1}P $5g$
5 −3.66100 $3s3p$ ^{3}P $4p$	9 −1.35721 $3p^2$ ^{3}P $4d$	20 −1.23665 $3s3p$ ^{3}P $5f$	7 −.47022 $3s3p$ ^{3}P $6g$
6 −3.44782 $3s^2$ ^{1}S $5s$	10 −1.30102 $3p^2$ ^{1}D $4d$	21 −1.20764 $3p^2$ ^{3}P $4d$	8 −.35895 $3s3d$ ^{3}D $4f$
7 −3.16724 $3s3p$ ^{1}P $4p$	11 −1.17839 $3s3p$ ^{1}P $5p$	22 −1.16704 $3s3p$ ^{1}P $5p$	**^{2}G^e**
8 −2.25207 $3s^2$ ^{1}S $6s$	12 −.80293 $3s3p$ ^{3}P $6p$	23 −1.04876 $3s^2$ ^{1}S $8d$	1 −5.45683 $3p^2$ ^{1}D $3d$
9 −2.11126 $3p^2$ ^{1}S $4s$	13 −.71860 $3s3d$ ^{3}D $4d$	24 −.82817 $3s^2$ ^{1}S $9d$	2 −4.33917 $3s3d^2$
10 −1.73546 $3s3p$ ^{3}P $5p$	14 −.45892 $3d^3$	25 −.79202 $3p^2$ ^{1}S $4d$	3 −2.66197 $3s3p$ ^{3}P $4f$
11 −1.59799 $3s^2$ ^{1}S $7s$	15 −.26816 $3p^2$ ^{3}P $5s$	26 −.77063 $3s3p$ ^{3}P $6p$	4 −2.60596 $3s^2$ ^{1}S $5g$
12 −1.27023 $3p^2$ ^{1}D $4d$	16 −.17654 $3s3p$ ^{3}P $7p$	**^{2}F^o**	5 −2.13227 $3s3p$ ^{1}P $4f$
13 −1.19840 $3s^2$ ^{1}S $8s$	17 −.15203 $3s3p$ ^{1}P $6p$	1 −6.79616 $3s3p$ ^{3}P $3d$	6 −1.76242 $3s^2$ ^{1}S $6g$
14 −1.13660 $3s3p$ ^{1}P $5p$	18 −.05195 $3s3d$ ^{1}D $4d$	2 −6.40958 $3s3p$ ^{1}P $3d$	7 −1.36710 $3p^2$ ^{1}D $4d$
15 −.91856 $3s^2$ ^{1}S $9s$	**^{2}D^o**	3 −4.08573 $3s^2$ ^{1}S $4f$	8 −1.30719 $3s^2$ ^{1}S $7g$
16 −.75195 $3s3p$ ^{3}P $6p$	1 −7.85124 ?	4 −3.36118 $3p3d^2$	9 −1.24290 $3s3p$ ^{3}P $5f$
^{2}P^o	2 −7.00912 $3s3p$ ^{3}P $3d$	5 −3.10471 $3s3p$ ^{3}P $4d$	10 −1.00371 $3s^2$ ^{1}S $8g$
1 −10.8124 $3s^2$ ^{1}S $3p$	3 −6.20678 $3s3p$ ^{1}P $3d$	6 −2.73748 $3p3d^2$	11 −.79732 $3s^2$ ^{1}S $9g$
2 −7.44770 $3p^3$	4 −3.32654 $3p3d^2$	7 −2.61325 $3s^2$ ^{1}S $5f$	12 −.68389 $3s3d$ ^{3}D $4d$
3 −6.41755 $3s3p$ ^{3}P $3d$	5 −3.20867 $3s3p$ ^{3}P $4d$	8 −2.56177 $3s3p$ ^{1}P $4d$	13 −.67625 $3s3p$ ^{1}P $5f$
4 −6.25300 $3s3p$ ^{1}P $3d$	6 −2.57995 $3p3d^2$	9 −2.45683 $3p3d^2$	**^{2}H^o**
5 −5.25108 $3s^2$ ^{1}S $4p$	7 −2.53188 $3s3p$ ^{1}P $4d$:	10 −2.10777 $3p^2$ ^{1}D $4p$	1 −3.06822 $3p3d^2$
6 −4.39269 $3s3p$ ^{3}P $4s$	8 −2.20538 $3p^2$ ^{1}D $4p$	11 −1.80963 $3s^2$ ^{1}S $6f$	2 −1.78097 $3s^2$ ^{1}S $6h$
7 −3.85330 $3s3p$ ^{1}P $4s$	9 −2.16855 $3p3d^2$	12 −1.49858 $3s3p$ ^{3}P $5d$	3 −1.31049 $3s^2$ ^{1}S $7h$
8 −3.15160 $3s^2$ ^{1}S $5p$	10 −1.93464 $3p^2$ ^{3}P $4p$	13 −1.33437 $3s^2$ ^{1}S $7f$	4 −1.24982 $3s3p$ ^{3}P $5g$
9 −3.06621 $3s3p$ ^{3}P $4d$	11 −1.47915 $3s3p$ ^{3}P $5d$	14 −1.29077 $3s3d$ ^{3}D $4p$	5 −1.03235 $3p^2$ ^{1}D $4f$
10 −2.85782 $3p3d^2$	12 −1.45536 $3s3d$ ^{3}D $4p$	15 −1.24434 $3s3p$ ^{3}P $5g$	6 −.99409 $3s^2$ ^{1}S $8h$
11 −2.54582 $3p3d^2$	13 −1.01037 $3p^2$ ^{1}D $4f$:	16 −1.03517 $3p^2$ ^{1}D $4f$	7 −.79372 $3s^2$ ^{1}S $9h$
12 −2.47458 $3s3p$ ^{1}P $4d$	14 −.96394 $3s3d$ ^{1}D $4p$	17 −1.01077 $3s^2$ ^{1}S $8f$	8 −.65255 $3s3p$ ^{1}P $5g$:
13 −2.12287 $3s^2$ ^{1}S $6p$	15 −.91875 $3p^2$ ^{3}P $4f$	18 −.94946 $3s3d$ ^{1}D $4p$	**^{4}S^o**
14 −2.09705 $3p^2$ ^{1}D $4p$	16 −.81206 $3s3p$ ^{1}P $5d$	19 −.80476 $3s3p$ ^{1}P $5d$	1 −7.74564 $3p^3$
15 −2.08546 $3p3d^2$	17 −.59971 $3s3p$ ^{3}P $6d$	20 −.79574 $3s^2$ ^{1}S $9f$	2 −2.69080 $3p3d^2$
16 −2.02854 $3s3p$ ^{3}P $5s$:	18 −.28733 $3s3d$ ^{3}D $4f$	21 −.78487 $3p^2$ ^{3}P $4f$	3 −2.04219 $3p^2$ ^{3}P $4p$
17 −1.76831 $3p^2$ ^{3}P $4p$	19 −.10434 $3p^2$ ^{1}D $5p$	**^{2}F^e**	4 −.00947 $3p^2$ ^{3}P $5p$
18 −1.55396 $3p^2$ ^{1}S $4p$	20 −.07564 $3s3p$ ^{3}P $7d$	1 −5.75053 $3p^2$ ^{1}D $3d$	**^{4}S^e**
19 −1.51547 $3s^2$ ^{1}S $7p$	**^{2}D^e**	2 −4.87261 $3p^2$ ^{3}P $3d$	1 −3.88314 $3s3p$ ^{3}P $4p$
20 −1.44825 $3s3p$ ^{1}P $5s$	1 −9.30947 $3s3p^2$	3 −3.94726 $3s3d^2$	2 −1.80933 $3s3p$ ^{3}P $5p$
21 −1.43546 $3s3p$ ^{3}P $5d$	2 −8.25631 $3s^2$ ^{1}S $3d$	4 −2.78838 $3s3p$ ^{3}P $4f$	3 −.78578 $3s3p$ ^{3}P $6p$
22 −1.26366 $3s3d$ ^{3}D $4p$	3 −5.27673 $3p^2$ ^{1}D $3d$	5 −2.16488 $3s3p$ ^{1}P $4f$	4 −.69231 $3s3d$ ^{3}D $4d$
23 −1.13253 $3s^2$ ^{1}S $8p$	4 −4.92849 $3p^2$ ^{3}P $3d$:	6 −1.42611 $3p^2$ ^{1}D $4d$	5 −.18158 $3s3p$ ^{3}P $7p$
24 −.91107 $3s3p$ ^{3}P $6s$	5 −4.54611 $3p^2$ ^{1}S $3d$:	7 −1.32567 $3p^2$ ^{3}P $4d$	**^{4}P^o**
25 −.90031 $3p^2$ ^{1}D $4f$	6 −4.48776 $3s^2$ ^{1}S $4d$	8 −1.27607 $3s3p$ ^{3}P $5f$	1 −7.14440 $3s3p$ ^{3}P $3d$
26 −.87744 $3s^2$ ^{1}S $9p$	7 −4.27904 $3s3d^2$	9 −.76085 $3s3d$ ^{3}D $4d$	2 −4.53199 $3s3p$ ^{3}P $4s$
27 −.84065 $3s3d$ ^{1}D $4p$	8 −3.82730 $3s3p$ ^{3}P $4p$	10 −.67202 $3s3p$ ^{1}P $5f$	3 −3.10896 $3p3d^2$
28 −.78325 $3s3p$ ^{1}P $5d$	9 −3.29578 $3s3p$ ^{1}P $4p$	11 −.49273 $3s3p$ ^{3}P $6f$	4 −3.09765 $3s3p$ ^{3}P $4d$
	10 −2.78569 $3s^2$ ^{1}S $5d$	12 −.37497 $3d^3$	5 −2.13742 $3s3p$ ^{3}P $5s$

Al-like Ca (Ca^{7+})

i	Energy(Ryds)	Description
6	−2.04937	$3p^2\ {}^3\mathrm{P}\ 4p$
7	−1.45794	$3s3p\ {}^3\mathrm{P}\ 5d$
8	−1.39678	$3s3d\ {}^3\mathrm{D}\ 4p$
9	−.93380	$3s3p\ {}^3\mathrm{P}\ 6s$
10	−.58082	$3s3p\ {}^3\mathrm{P}\ 6d$
11	−.27869	$3s3p\ {}^3\mathrm{P}\ 7s$
12	−.26571	$3s3d\ {}^3\mathrm{D}\ 4f$
13	−.06935	$3s3p\ {}^3\mathrm{P}\ 7d$
14	−.00055	$3p^2\ {}^3\mathrm{P}\ 5p$
	$^4\mathrm{P}^e$	
1	−9.68447	$3s3p^2$
2	−5.21119	$3p^2\ {}^3\mathrm{P}\ 3d$
3	−4.57441	$3s3d^2$
4	−3.90845	$3s3p\ {}^3\mathrm{P}\ 4p$
5	−2.65219	$3p^2\ {}^3\mathrm{P}\ 4s$
6	−1.83001	$3s3p\ {}^3\mathrm{P}\ 5p$
7	−1.35006	$3p^2\ {}^3\mathrm{P}\ 4d$
8	−.78979	$3s3p\ {}^3\mathrm{P}\ 6p$
9	−.63353	$3d^3$
10	−.56743	$3s3d\ {}^3\mathrm{D}\ 4d$
11	−.28215	$3p^2\ {}^3\mathrm{P}\ 5s$
12	−.18082	$3s3p\ {}^3\mathrm{P}\ 7p$
	$^4\mathrm{D}^o$	
1	−7.11633	$3s3p\ {}^3\mathrm{P}\ 3d$
2	−3.22438	$3p3d^2$
3	−3.20700	$3s3p\ {}^3\mathrm{P}\ 4d$
4	−2.93410	$3p3d^2$
5	−2.13984	$3p^2\ {}^3\mathrm{P}\ 4p$
6	−1.55350	$3s3d\ {}^3\mathrm{D}\ 4p$
7	−1.42708	$3s3p\ {}^3\mathrm{P}\ 5d$
8	−.97569	$3p^2\ {}^3\mathrm{P}\ 4f$
9	−.60110	$3s3p\ {}^3\mathrm{P}\ 6d$
10	−.28778	$3s3d\ {}^3\mathrm{D}\ 4f$
11	−.09295	$3s3p\ {}^3\mathrm{P}\ 7d$
12	−.01735	$3p^2\ {}^3\mathrm{P}\ 5p$
	$^4\mathrm{D}^e$	
1	−5.53955	$3p^2\ {}^3\mathrm{P}\ 3d$
2	−3.97927	$3s3p\ {}^3\mathrm{P}\ 4p$
3	−2.74120	$3s3p\ {}^3\mathrm{P}\ 4f$
4	−2.06262	$3s3d\ {}^3\mathrm{D}\ 4s$
5	−1.84086	$3s3p\ {}^3\mathrm{P}\ 5p$
6	−1.28899	$3p^2\ {}^3\mathrm{P}\ 4d$
7	−1.27571	$3s3p\ {}^3\mathrm{P}\ 5f$
8	−.80354	$3s3p\ {}^3\mathrm{P}\ 6p$
9	−.77296	$3s3d\ {}^3\mathrm{D}\ 4d$
10	−.48492	$3s3p\ {}^3\mathrm{P}\ 6f$
11	−.19343	$3s3p\ {}^3\mathrm{P}\ 7p$
12	−.00430	$3s3p\ {}^3\mathrm{P}\ 7f$

i	Energy(Ryds)	Description
	$^4\mathrm{F}^o$	
1	−7.43931	$3s3p\ {}^3\mathrm{P}\ 3d$
2	−3.14815	$3s3p\ {}^3\mathrm{P}\ 4d$
3	−3.14666	$3p3d^2$
4	−1.48533	$3s3d\ {}^3\mathrm{D}\ 4p$
5	−1.45823	$3s3p\ {}^3\mathrm{P}\ 5d$
6	−1.25916	$3s3p\ {}^3\mathrm{P}\ 5g$
7	−.93172	$3p^2\ {}^3\mathrm{P}\ 4f$
8	−.58788	$3s3p\ {}^3\mathrm{P}\ 6d$
9	−.46494	$3s3p\ {}^3\mathrm{P}\ 6g$
10	−.39868	$3s3d\ {}^3\mathrm{D}\ 4f$
11	−.06799	$3s3p\ {}^3\mathrm{P}\ 7d$
	$^4\mathrm{F}^e$	
1	−5.65912	$3p^2\ {}^3\mathrm{P}\ 3d$
2	−4.77966	$3s3d^2$
3	−2.82721	$3s3p\ {}^3\mathrm{P}\ 4f$
4	−1.35837	$3p^2\ {}^3\mathrm{P}\ 4d$
5	−1.31479	$3s3p\ {}^3\mathrm{P}\ 5f$
6	−.83550	$3d^3$
7	−.66675	$3s3d\ {}^3\mathrm{D}\ 4d$
8	−.49952	$3s3p\ {}^3\mathrm{P}\ 6f$
9	−.01367	$3s3p\ {}^3\mathrm{P}\ 7f$
	$^4\mathrm{G}^o$	
1	−3.52656	$3p3d^2$
2	−1.31638	$3s3p\ {}^3\mathrm{P}\ 5g$
3	−1.00338	$3p^2\ {}^3\mathrm{P}\ 4f$
4	−.47706	$3s3p\ {}^3\mathrm{P}\ 6g$
5	−.30785	$3s3d\ {}^3\mathrm{D}\ 4f$
	$^4\mathrm{G}^e$	
1	−2.81343	$3s3p\ {}^3\mathrm{P}\ 4f$
2	−1.31117	$3s3p\ {}^3\mathrm{P}\ 5f$
3	−.76689	$3s3d\ {}^3\mathrm{D}\ 4d$
4	−.49981	$3s3p\ {}^3\mathrm{P}\ 6f$
5	−.45646	$3s3p\ {}^3\mathrm{P}\ 6h$
6	−.01310	$3s3p\ {}^3\mathrm{P}\ 7f$
	$^4\mathrm{H}^o$	
1	−1.25400	$3s3p\ {}^3\mathrm{P}\ 5g$
2	−.51682	$3s3p\ {}^3\mathrm{P}\ 6g$
3	−.39182	$3s3d\ {}^3\mathrm{D}\ 4f$

Al-like Ca (Ca^{7+})

Energies in ascending order from ground state for terms with effective $n \leq 4.0$, $L \leq 4$

Term	i	E(Ryds)	Term	i	E(Ryds)	Term	i	E(Ryds)	Term	i	E(Ryds)	Term	i	E(Ryds)
$^2P^o$	1	0.00000	$^4F^o$	1	3.37309	$^2S^e$	2	4.97395	$^2P^e$	3	5.82870	$^2P^o$	6	6.41971
$^4P^e$	1	1.12793	$^4P^o$	1	3.66800	$^2F^e$	1	5.06187	$^2S^e$	3	5.84207	$^2G^e$	2	6.47323
$^2D^e$	1	1.50293	$^4D^o$	1	3.69607	$^4F^e$	1	5.15328	$^2D^e$	4	5.88391	$^2D^e$	7	6.53336
$^2S^e$	1	1.95240	$^2D^o$	2	3.80328	$^2P^e$	2	5.23883	$^2F^e$	2	5.93979	$^2F^o$	3	6.72667
$^2P^e$	1	2.10041	$^2F^o$	1	4.01624	$^4D^e$	1	5.27285	$^4F^e$	2	6.03274	$^2P^e$	4	6.80385
$^2D^e$	2	2.55609	$^2P^o$	3	4.39485	$^2G^e$	1	5.35557	$^4P^e$	3	6.23799			
$^2D^o$	1	2.96116	$^2F^o$	2	4.40282	$^2D^e$	3	5.53567	$^2D^e$	5	6.26629			
$^4S^o$	1	3.06676	$^2P^o$	4	4.55940	$^2P^o$	5	5.56132	$^4P^o$	2	6.28041			
$^2P^o$	2	3.36470	$^2D^o$	3	4.60562	$^4P^e$	2	5.60121	$^2D^e$	6	6.32464			

gf-values for transitions involving terms with effective $n \leq 4.0$, $L \leq 4$

i i'	gf_L	i i'	gf_L	i i'	gf_L	i i'	gf_L	i i'	gf_L	i i'	gf_L
	$^2S^e$–$^2P^o$	4 2	−6.62E−2	4 1	1.38E−2	4 3	1.15E−1	1 3	−7.06E−1		$^4S^o$–$^4P^e$
1 1	4.55E−1	4 3	−2.07E−1	4 2	2.74E+0		$^2D^o$–$^2D^e$	2 1	−5.63E−1	1 1	2.33E+0
1 2	−1.77E−1	4 4	−3.42E−2	4 3	−7.75E−2	1 1	6.93E−1	2 2	−5.11E+0	1 2	−6.53E+0
1 3	−2.44E+0	5 1	9.53E−4	4 4	−5.11E−1	1 2	1.09E−2	2 3	−7.93E+0	1 3	−5.26E−3
1 4	−4.02E−1	5 2	1.03E−5	4 5	−3.14E−1	1 3	−3.44E+0	3 1	3.63E−3		$^4P^o$–$^4P^e$
1 5	−9.63E−3	5 3	−1.54E−4	4 6	−6.28E−2	1 4	−2.68E−2	3 2	4.82E−3	1 1	3.32E+0
1 6	−1.70E−1	5 4	−6.28E−2	4 7	−1.28E+0	1 5	−1.41E−1	3 3	−4.50E−3	1 2	−1.25E+0
2 1	4.62E−1	6 1	1.53E−1	5 1	2.35E−1	1 6	−4.16E−4	4 1	2.32E+0	1 3	−6.05E+0
2 2	6.77E−4	6 2	9.99E−4	5 2	4.10E−1	1 7	−1.55E−2	4 2	2.20E−1	2 1	2.06E+0
2 3	3.05E−3	6 3	7.07E−3	5 3	4.87E−5	2 1	4.47E+0	4 3	−5.72E−3	2 2	2.67E−6
2 4	2.06E−3	6 4	−8.79E−1	5 4	−1.41E−3	2 2	1.31E−1	5 1	3.58E−1	2 3	7.97E−4
2 5	−1.37E+0		$^2P^o$–$^2D^e$	5 5	−1.26E−3	2 3	−9.18E−1	5 2	1.43E+0		$^4P^o$–$^4D^e$
2 6	−5.63E−2	1 1	−4.18E−1	5 6	−3.80E+0	2 4	−1.00E+0	5 3	−2.64E−3	1 1	−2.61E+0
3 1	3.02E−4	1 2	−4.58E+0	5 7	−5.34E−2	2 5	−7.42E−3	6 1	1.26E−3	2 1	7.74E−5
3 2	8.59E−1	1 3	−3.08E−3	6 1	9.16E−1	2 6	−4.32E−2	6 2	7.26E−2		$^4P^e$–$^4D^o$
3 3	1.19E−2	1 4	−8.35E−3	6 2	1.65E−1	2 7	−5.03E+0	6 3	−2.75E+0	1 1	−9.33E+0
3 4	1.85E−1	1 5	−6.49E−3	6 3	2.78E−3	3 1	1.28E−3	7 1	1.30E−2	2 1	3.50E+0
3 5	4.36E−4	1 6	−6.82E−1	6 4	1.53E−2	3 2	2.97E+0	7 2	1.39E+0	3 1	2.54E+0
3 6	−2.89E−3	1 7	−3.63E−4	6 5	2.62E−3	3 3	−2.50E−1	7 3	−1.21E−2		$^4D^o$–$^4D^e$
	$^2P^o$–$^2P^e$	2 1	1.19E+0	6 6	2.75E−3	3 4	−5.78E−2		$^2F^o$–$^2F^e$	1 1	−5.38E−1
1 1	−3.43E+0	2 2	8.42E−4	6 7	−1.36E−3	3 5	−1.17E+0	1 1	−5.00E−1		$^4D^o$–$^4F^e$
1 2	−3.53E−4	2 3	−8.28E−1		$^2P^e$–$^2D^o$	3 6	−2.21E−3	1 2	−1.99E+0	1 1	−2.49E+0
1 3	−3.54E−3	2 4	−1.17E−2	1 1	−3.07E−1	3 7	−3.35E−2	2 1	−6.64E−3	1 2	−9.06E+0
1 4	−3.35E−1	2 5	−3.81E+0	1 2	−1.94E−1		$^2D^o$–$^2F^e$	2 2	−1.03E+0		$^4D^e$–$^4F^o$
2 1	4.88E−1	2 6	−2.41E−3	1 3	−6.96E+0	1 1	−6.06E−1	3 1	1.70E−4	1 1	4.03E+0
2 2	−1.39E−1	2 7	−9.48E−3	2 1	1.54E+0	1 2	−5.44E+0	3 2	2.21E−4		$^4F^o$–$^4F^e$
2 3	−2.65E+0	3 1	2.80E−3	2 2	4.15E−1	2 1	−6.77E−1		$^2F^o$–$^2G^e$	1 1	−6.33E+0
2 4	−8.50E−2	3 2	7.95E−3	2 3	1.76E−3	2 2	−2.35E+0	1 1	−1.87E+0	1 2	−6.85E+0
3 1	1.23E+0	3 3	−1.37E−1	3 1	2.12E−4	3 1	−6.05E−2	1 2	−3.50E−1		
3 2	−4.01E−2	3 4	−1.20E−1	3 2	4.06E−1	3 2	−9.96E−1	2 1	−6.17E−1		
3 3	−5.05E−2	3 5	−2.81E+0	3 3	2.45E−1		$^2D^e$–$^2F^o$	2 2	−8.37E+0		
3 4	−1.70E−1	3 6	−6.54E−2	4 1	3.05E−1	1 1	−2.54E+0	3 1	4.21E−3		
4 1	1.79E+0	3 7	−7.79E−1	4 2	1.83E−1	1 2	−4.04E+0	3 2	2.79E−5		

Al-like Fe (Fe^{13+})

Term energies relative to $3s^2$ ^{1}S ionization threshold for each symmetry

i	E(Ryds)	Description	i	E(Ryds)	Description	i	E(Ryds)	Description	i	E(Ryds)	Description
		^{2}S$^\circ$	16	−8.56884	$3p^2$ ^{1}S $4p$	19	−4.20582	$3p^2$ ^{3}P $5s$	27	−3.55585	$3p3d$ ^{1}P $4d$:
1	−16.1943	$3p3d^2$	17	−8.23224	$3s3d$ ^{3}D $4p$	20	−3.83015	$3p3d$ ^{1}D $4f$	28	−3.45497	$3s3p$ ^{3}P $6d$
2	−9.63625	$3p^2$ ^{3}P $4p$	18	−7.59145	$3s3d$ ^{1}D $4p$	21	−3.81584	$3s3p$ ^{3}P $6p$	29	−3.36994	$3p3d$ ^{1}F $4d$:
3	−4.59591	$3p3d$ ^{1}D $4d$	19	−7.15882	$3p^2$ ^{1}D $4f$	22	−3.53123	$3p3d$ ^{3}F $4f$	30	−2.85652	$3p^2$ ^{1}D $5f$
4	−4.25603	$3p3d$ ^{3}D $4d$	20	−7.14707	$3s3p$ ^{3}P $5s$	23	−3.43828	$3p3d$ ^{3}D $4f$	31	−2.79567	$3s3d$ ^{3}D $5p$
5	−3.83189	$3p^2$ ^{3}P $5p$	21	−6.40874	$3p3d$ ^{3}P $4s$	24	−3.20530	$3p^2$ ^{1}D $5d$	32	−2.75215	$3p^2$ ^{3}P $5f$
6	−1.80727	$3d^2$ ^{3}P $4p$	22	−6.28734	$3s3p$ ^{1}P $5s$	25	−3.10497	$3p^2$ ^{3}P $5d$	33	−2.53496	$3s3p$ ^{1}P $6d$
7	−.91293	$3p^2$ ^{3}P $6p$	23	−6.09616	$3s3p$ ^{3}P $5d$:	26	−2.87066	$3s3p$ ^{1}P $6p$	34	−2.04865	$3s3d$ ^{1}D $5p$
		^{2}S^e	24	−6.05460	$3s^2$ ^{1}S $6p$	27	−2.58098	$3d^2$ ^{3}P $4s$	35	−1.94108	$3d^2$ ^{3}F $4p$
1	−25.5054	$3s3p^2$	25	−5.87460	$3s3d$ ^{3}D $4f$	28	−2.45634	$3p3d$ ^{1}F $4f$	36	−1.93800	$3s3p$ ^{3}P $7d$
2	−19.1591	$3p^2$ ^{1}D $3d$	26	−5.72555	$3p3d$ ^{1}P $4s$	29	−2.17707	$3s3p$ ^{3}P $7p$	37	−1.75865	$3s3d$ ^{3}D $5f$
3	−17.3185	$3s3d^2$	27	−5.41515	$3s3d$ ^{1}D $4f$	30	−2.09037	$3s3d$ ^{3}D $5d$	38	−1.56191	$3d^2$ ^{1}D $4p$
4	−15.7072	$3s^2$ ^{1}S $4s$	28	−5.05759	$3s3p$ ^{1}P $5d$	31	−1.35475	$3s3d$ ^{1}D $5d$	39	−1.42507	$3d^2$ ^{3}P $4p$
5	−12.0282	$3s3p$ ^{3}P $4p$	29	−4.68602	$3p3d$ ^{1}D $4d$	32	−1.17137	$3s3p$ ^{1}P $7p$	40	−1.03810	$3s3d$ ^{1}D $5f$
6	−11.3108	$3s3p$ ^{1}P $4p$	30	−4.45153	$3p3d$ ^{3}F $4d$	33	−1.14584	$3p^2$ ^{3}P $6s$	41	−1.01884	$3p^2$ ^{1}D $6p$
7	−9.57594	$3p^2$ ^{1}S $4s$	31	−4.38991	$3s^2$ ^{1}S $7p$	34	−1.04581	$3s3p$ ^{3}P $8p$	42	−.94725	$3s3p$ ^{3}P $8d$
8	−9.48021	$3s^2$ ^{1}S $5s$	32	−4.13545	$3p3d$ ^{3}D $4d$:	35	−.72234	$3d^2$ ^{3}F $4d$	43	−.93858	$3s3p$ ^{1}P $7d$
9	−7.96523	$3p^2$ ^{1}D $4d$	33	−4.09140	$3s3p$ ^{3}P $6s$	36	−.61132	$3p^2$ ^{1}D $6d$	44	−.85678	$3p^2$ ^{3}P $6p$
10	−6.86860	$3s3d$ ^{3}D $4d$	34	−4.01836	$3p3d$ ^{3}P $4d$:	37	−.53377	$3p^2$ ^{3}P $6d$	45	−.75305	$3p3d$ ^{1}D $5s$
11	−6.47963	$3s3p$ ^{3}P $5p$	35	−3.85302	$3p^2$ ^{1}D $5p$	38	−.46143	$3d^2$ ^{3}P $4d$	46	−.41216	$3p^2$ ^{1}D $6f$
12	−6.39112	$3s^2$ ^{1}S $6s$	36	−3.57417	$3p^2$ ^{3}P $5p$	39	−.35516	$3s3p$ ^{3}P $9p$	47	−.34401	$3p3d$ ^{3}D $5s$
13	−6.21461	$3s3d$ ^{1}D $4d$:	37	−3.42175	$3s3p$ ^{3}P $6d$	40	−.29892	$3p3d$ ^{1}D $5p$	48	−.30402	$3p^2$ ^{3}P $6f$
14	−5.65416	$3s3p$ ^{1}P $5p$	38	−3.37881	$3p3d$ ^{1}P $4d$	41	−.11499	$3s3p$ ^{1}P $8p$	49	−.24936	$3s3p$ ^{3}P $9d$
15	−5.21582	$3p3d$ ^{3}P $4p$	39	−3.33113	$3s^2$ ^{1}S $8p$	42	−.07021	$3d^2$ ^{1}D $4d$			**^{2}D^e**
16	−4.66368	$3p3d$ ^{1}P $4p$	40	−3.28289	$3p3d$ ^{1}F $4d$			**^{2}D$^\circ$**	1	−26.1544	$3s3p^2$
17	−4.55974	$3s^2$ ^{1}S $7s$	41	−3.15825	$3s3p$ ^{1}P $6s$	1	−23.7466	$3p^3$	2	−24.4795	$3s^2$ ^{1}S $3d$
18	−3.76834	$3s3p$ ^{3}P $6p$	42	−2.92012	$3p^2$ ^{1}S $5p$	2	−22.4368	$3s3p$ ^{3}P $3d$	3	−19.6210	$3p^2$ ^{1}D $3d$
19	−3.45334	$3p^2$ ^{1}S $5s$	43	−2.78970	$3p^2$ ^{1}D $5f$	3	−21.2497	$3s3p$ ^{1}P $3d$	4	−19.1392	$3p^2$ ^{1}S $3d$
20	−3.43993	$3s^2$ ^{1}S $8s$	44	−2.69543	$3s3d$ ^{3}D $5p$	4	−16.3575	$3p3d^2$	5	−18.5550	$3p^2$ ^{3}P $3d$
21	−3.41318	$3p3d$ ^{3}F $4f$	45	−2.61038	$3s^2$ ^{1}S $9p$	5	−15.3215	$3p3d^2$	6	−18.0754	$3s3d^2$
22	−3.16029	$3p^2$ ^{1}D $5d$	46	−2.47083	$3s3p$ ^{1}P $6d$	6	−14.7138	$3p3d^2$	7	−13.3371	$3s^2$ ^{1}S $4d$
23	−2.84282	$3s3p$ ^{1}P $6p$	47	−2.28343	$3s3p$ ^{3}P $7s$	7	−11.1672	$3s3p$ ^{3}P $4d$	8	−12.2646	$3s3p$ ^{3}P $4p$
24	−2.66753	$3s^2$ ^{1}S $9s$			**^{2}P^e**	8	−10.1669	$3s3p$ ^{1}P $4d$	9	−11.8922	$3d^3$
25	−2.51227	$3p3d$ ^{1}F $4f$	1	−25.2925	$3s3p^2$	9	−9.61953	$3p^2$ ^{1}D $4p$	10	−11.4367	$3s3p$ ^{1}P $4p$
26	−2.15318	$3s3p$ ^{3}P $7p$	2	−20.0757	$3p^2$ ^{3}P $3d$:	10	−9.21077	$3p^2$ ^{3}P $4p$	11	−11.2408	$3d^3$
		^{2}P$^\circ$	3	−19.2184	$3p^2$ ^{1}D $3d$:	11	−8.48273	$3s3d$ ^{3}D $4p$	12	−10.4673	$3p^2$ ^{1}D $4s$
1	−28.6280	$3s^2$ ^{1}S $3p$	4	−17.3574	$3s3d^2$	12	−7.71878	$3s3d$ ^{1}D $4p$	13	−9.98405	$3s3p$ ^{3}P $4f$
2	−23.1610	$3p^3$	5	−12.5469	$3s3p$ ^{3}P $4p$	13	−7.38822	$3p^2$ ^{1}D $4f$:	14	−9.28960	$3s3d$ ^{3}D $4s$
3	−21.5408	$3s3p$ ^{3}P $3d$	6	−11.8817	$3d^3$	14	−7.21646	$3p^2$ ^{3}P $4f$	15	−9.19291	$3s3p$ ^{1}P $4f$
4	−21.2565	$3s3p$ ^{1}P $3d$	7	−11.4496	$3s3p$ ^{1}P $4p$	15	−6.85369	$3p3d$ ^{1}D $4s$	16	−8.66001	$3s3d$ ^{1}D $4s$
5	−15.7043	$3p3d^2$	8	−10.2029	$3p^2$ ^{3}P $4s$	16	−6.36616	$3p3d$ ^{3}D $4s$	17	−8.34111	$3s^2$ ^{1}S $5d$
6	−15.1098	$3p3d^2$	9	−8.12585	$3p^2$ ^{3}P $4d$	17	−6.25475	$3s3d$ ^{3}D $4f$	18	−8.22030	$3p^2$ ^{1}D $4d$
7	−14.6757	$3s^2$ ^{1}S $4p$	10	−8.05383	$3p^2$ ^{1}D $4d$	18	−6.10021	$3s3p$ ^{3}P $5d$	19	−7.93677	$3p^2$ ^{3}P $4d$
8	−14.5052	$3p3d^2$	11	−7.11173	$3s3d$ ^{3}D $4d$	19	−5.48398	$3s3d$ ^{1}D $4f$	20	−7.26512	$3p^2$ ^{1}S $4d$
9	−13.2353	$3s3p$ ^{3}P $4s$	12	−6.77135	$3s3p$ ^{3}P $5p$	20	−5.14286	$3s3p$ ^{1}P $5d$	21	−7.01002	$3s3d$ ^{3}D $4d$
10	−12.4424	$3s3p$ ^{1}P $4s$	13	−6.27368	$3s3d$ ^{1}D $4d$	21	−4.71155	$3p3d$ ^{3}F $4d$	22	−6.64618	$3s3p$ ^{3}P $5p$
11	−10.9085	$3s3p$ ^{3}P $4d$	14	−5.90108	$3p3d$ ^{1}D $4p$	22	−4.49643	$3p3d$ ^{1}D $4d$	23	−6.39853	$3s3d$ ^{1}D $4d$
12	−10.0527	$3s3p$ ^{1}P $4d$	15	−5.75740	$3s3p$ ^{1}P $5p$	23	−4.30504	$3p3d$ ^{3}P $4d$	24	−5.96203	$3p3d$ ^{1}D $4p$
13	−9.44712	$3p^2$ ^{1}D $4p$	16	−5.67627	$3p3d$ ^{3}P $4p$:	24	−4.14088	$3p3d$ ^{3}D $4d$	25	−5.86085	$3s3p$ ^{1}P $5p$
14	−9.05841	$3p^2$ ^{3}P $4p$	17	−5.31273	$3p3d$ ^{3}D $4p$	25	−3.91900	$3p^2$ ^{1}D $5p$	26	−5.76151	$3p3d$ ^{3}F $4p$
15	−8.97918	$3s^2$ ^{1}S $5p$	18	−4.77712	$3p3d$ ^{1}P $4p$	26	−3.71788	$3p^2$ ^{3}P $5p$	27	−5.65406	$3s^2$ ^{1}S $6d$

Al-like Fe (Fe^{13+})

i	E(Ryds)	Description	i	E(Ryds)	Description	i	E(Ryds)	Description	i	E(Ryds)	Description
28	−5.61859	$3s3p$ ^{3}P $5f$	23	−5.13111	$3s3p$ ^{1}P $5d$	32	−1.62183	$3s3d$ ^{3}D $5g$	9	−6.38570	$3s3d$ ^{1}D $4d$
29	−5.37607	$3p3d$ ^{3}D $4p$	24	−4.75973	$3p3d$ ^{3}F $4d$	33	−1.39411	$3s3d$ ^{1}D $5d$	10	−5.72851	$3p3d$ ^{3}F $4p$
30	−5.27330	$3p3d$ ^{3}P $4p$	25	−4.58146	$3s3p$ ^{1}P $5g$	34	−.89313	$3s3d$ ^{1}D $5g$	11	−5.63988	$3s3p$ ^{3}P $5f$
31	−4.80510	$3p3d$ ^{1}F $4p$:	26	−4.48502	$3p3d$ ^{1}D $4d$	35	−.85679	$3s3p$ ^{3}P $8f$	12	−5.44934	$3s^2$ ^{1}S $6g$
32	−4.72968	$3s3p$ ^{1}P $5f$	27	−4.20979	$3p3d$ ^{3}D $4d$	36	−.84741	$3d^2$ ^{3}F $4d$	13	−4.86922	$3p3d$ ^{1}F $4p$
33	−4.66716	$3p3d$ ^{1}P $4p$:	28	−4.05086	$3s^2$ ^{1}S $7f$	37	−.84404	$3s3p$ ^{1}P $7f$	14	−4.77906	$3s3p$ ^{1}P $5f$
34	−4.33342	$3p^2$ ^{1}D $5s$	29	−3.99912	$3p3d$ ^{3}P $4d$	38	−.65667	$3p^2$ ^{1}D $6d$	15	−4.04376	$3p3d$ ^{3}F $4f$
35	−4.17438	$3s^2$ ^{1}S $7d$	30	−3.87701	$3p^2$ ^{1}D $5p$	39	−.52944	$3p^2$ ^{3}P $6d$	16	−4.00069	$3s^2$ ^{1}S $7g$
36	−3.87828	$3p3d$ ^{1}D $4f$	31	−3.61992	$3p3d$ ^{1}F $4d$	40	−.43725	$3p3d$ ^{3}F $5p$	17	−3.81469	$3p3d$ ^{1}D $4f$
37	−3.82440	$3s3p$ ^{3}P $6p$	32	−3.47753	$3s3p$ ^{3}P $6d$	41	−.42346	$3d^2$ ^{1}D $4d$	18	−3.39628	$3p3d$ ^{3}D $4f$
38	−3.74924	$3p3d$ ^{3}F $4f$	33	−3.39677	$3p3d$ ^{1}P $4d$	42	−.36690	$3d^2$ ^{3}P $4d$	19	−3.23936	$3p3d$ ^{3}P $4f$
39	−3.36049	$3p3d$ ^{3}D $4f$	34	−3.17465	$3s3p$ ^{3}P $6g$	43	−.34464	$3p^2$ ^{1}D $6g$	20	−3.22336	$3s3p$ ^{3}P $6f$
40	−3.25659	$3p^2$ ^{1}D $5d$	35	−3.09806	$3s^2$ ^{1}S $8f$	44	−.25530	$3p3d$ ^{1}D $5p$	21	−3.21403	$3p^2$ ^{1}D $5d$
41	−3.21546	$3s3p$ ^{3}P $6f$	36	−2.89653	$3p^2$ ^{1}D $5f$	45	−.23700	$3p^2$ ^{3}P $6g$	22	−3.06557	$3s^2$ ^{1}S $8g$
42	−3.20864	$3s3d$ ^{3}D $5s$	37	−2.71277	$3s3d$ ^{3}D $5p$	46	−.19797	$3s3p$ ^{3}P $9f$	23	−2.94684	$3p3d$ ^{1}F $4f$
43	−3.17513	$3p3d$ ^{3}P $4f$	38	−2.63204	$3p^2$ ^{3}P $5f$			**^{2}G^o**	24	−2.75615	$3p3d$ ^{1}P $4f$
44	−3.16303	$3s^2$ ^{1}S $8d$	39	−2.53389	$3s3p$ ^{1}P $6d$	1	−16.1149	$3p3d^2$	25	−2.70899	$3p^2$ ^{1}D $5g$
45	−2.96279	$3p^2$ ^{3}P $5d$	40	−2.44396	$3s^2$ ^{1}S $9f$	2	−15.0739	$3p3d^2$	26	−2.51399	$3d^2$ ^{1}G $4s$
46	−2.89862	$3s3p$ ^{1}P $6p$	41	−2.23190	$3s3p$ ^{1}P $6g$	3	−7.55024	$3p^2$ ^{1}D $4f$	27	−2.43973	$3p^2$ ^{3}P $5g$
47	−2.71959	$3p3d$ ^{1}F $4f$	42	−2.02369	$3s3d$ ^{1}D $5p$	4	−7.17860	$3p^2$ ^{3}P $4f$	28	−2.42735	$3s^2$ ^{1}S $9g$
48	−2.65009	$3p^2$ ^{1}D $5g$			**^{2}F^e**	5	−6.39778	$3s3d$ ^{3}D $4f$	29	−2.31737	$3s3p$ ^{1}P $6f$
49	−2.62510	$3p3d$ ^{1}P $4f$	1	−20.3109	$3p^2$ ^{1}D $3d$	6	−5.62892	$3s3p$ ^{3}P $5g$	30	−2.04676	$3s3d$ ^{3}D $5d$
50	−2.52299	$3d^2$ ^{1}D $4s$	2	−19.0182	$3p^2$ ^{3}P $3d$	7	−5.53915	$3s3d$ ^{1}D $4f$			**^{4}S^o**
51	−2.50525	$3s^2$ ^{1}S $9d$	3	−17.6618	$3s3d^2$	8	−4.84059	$3p3d$ ^{3}F $4d$	1	−23.6463	$3p^3$
52	−2.49310	$3s3d$ ^{1}D $5s$	4	−11.7199	$3d^3$	9	−4.66711	$3s3p$ ^{1}P $5g$	2	−15.5367	$3p3d^2$
53	−2.31929	$3s3p$ ^{1}P $6f$	5	−10.3267	$3s3p$ ^{3}P $4f$	10	−4.58123	$3p3d$ ^{1}D $4d$	3	−9.33418	$3p^2$ ^{3}P $4p$
54	−2.27316	$3p^2$ ^{1}S $5d$	6	−9.36309	$3s3p$ ^{1}P $4f$	11	−4.11171	$3p3d$ ^{3}D $4d$	4	−4.12076	$3p3d$ ^{3}D $4d$
55	−2.15073	$3s3p$ ^{3}P $7p$	7	−8.25836	$3p^2$ ^{1}D $4d$	12	−3.61271	$3p3d$ ^{1}F $4d$	5	−3.70891	$3p^2$ ^{3}P $5p$
56	−2.04121	$3s3d$ ^{3}D $5d$	8	−8.07686	$3p^2$ ^{3}P $4d$	13	−3.22336	$3s3p$ ^{3}P $6g$	6	−1.72130	$3d^2$ ^{3}P $4p$
		^{2}F^o	9	−7.20757	$3s3d$ ^{3}D $4d$	14	−2.92455	$3p^2$ ^{1}D $5f$	7	−.86335	$3p^2$ ^{3}P $6p$
1	−22.0881	$3s3p$ ^{3}P $3d$	10	−6.40515	$3s3d$ ^{1}D $4d$	15	−2.71109	$3p^2$ ^{3}P $5f$			**^{4}S^e**
2	−21.4699	$3s3p$ ^{1}P $3d$	11	−6.15221	$3p3d$ ^{3}F $4p$	16	−2.25979	$3s3p$ ^{1}P $6g$	1	−12.3414	$3s3p$ ^{3}P $4p$
3	−16.4092	$3p3d^2$	12	−5.82765	$3p3d$ ^{1}D $4p$	17	−1.89621	$3d^2$ ^{3}F $4p$	2	−7.12004	$3s3d$ ^{3}D $4d$
4	−15.4890	$3p3d^2$	13	−5.74798	$3s3p$ ^{3}P $5f$	18	−1.79327	$3s3d$ ^{3}D $5f$	3	−6.67694	$3s3p$ ^{3}P $5p$
5	−15.1190	$3p3d^2$	14	−5.42715	$3p3d$ ^{3}D $4p$	19	−1.75123	$3s3p$ ^{3}P $7g$	4	−5.54318	$3p3d$ ^{3}P $4p$
6	−12.5076	$3s^2$ ^{1}S $4f$	15	−4.90676	$3p3d$ ^{1}F $4p$	20	−1.60678	$3d^2$ ^{1}G $4p$	5	−3.83205	$3s3p$ ^{3}P $6p$
7	−10.9968	$3s3p$ ^{3}P $4d$	16	−4.79615	$3s3p$ ^{1}P $5f$	21	−1.03700	$3s3d$ ^{1}D $5f$	6	−3.67977	$3p3d$ ^{3}F $4f$
8	−10.1735	$3s3p$ ^{1}P $4d$	17	−3.92197	$3p3d$ ^{3}F $4f$	22	−.82520	$3s3p$ ^{3}P $8g$	7	−2.19628	$3s3p$ ^{3}P $7p$
9	−9.46261	$3p^2$ ^{1}D $4p$	18	−3.80848	$3p3d$ ^{1}D $4f$	23	−.80062	$3s3p$ ^{1}P $7g$	8	−2.06534	$3s3d$ ^{3}D $5d$
10	−8.30718	$3s3d$ ^{3}D $4p$	19	−3.53604	$3p3d$ ^{3}P $4f$	24	−.44567	$3p^2$ ^{1}D $6f$	9	−1.06801	$3s3p$ ^{3}P $8p$
11	−7.97792	$3s^2$ ^{1}S $5f$	20	−3.31489	$3p3d$ ^{3}D $4f$	25	−.29062	$3p^2$ ^{3}P $6f$	10	−.34886	$3s3p$ ^{3}P $9p$
12	−7.66609	$3s3d$ ^{1}D $4p$	21	−3.28343	$3s3p$ ^{3}P $6f$	26	−.17223	$3s3p$ ^{3}P $9g$			**^{4}P^o**
13	−7.44717	$3p^2$ ^{1}D $4f$	22	−3.27232	$3p^2$ ^{1}D $5d$	27	−.13334	$3d^2$ ^{3}F $4f$	1	−22.5986	$3s3p$ ^{3}P $3d$
14	−7.00794	$3p^2$ ^{3}P $4f$	23	−3.08134	$3p^2$ ^{3}P $5d$			**^{2}G^e**	2	−16.0287	$3p3d^2$
15	−6.83656	$3p3d$ ^{3}F $4s$	24	−2.91095	$3p3d$ ^{1}F $4f$	1	−19.8391	$3p^2$ ^{1}D $3d$	3	−13.4804	$3s3p$ ^{3}P $4s$
16	−6.49246	$3p^2$ ^{1}S $4f$	25	−2.86491	$3d^2$ ^{3}F $4s$	2	−18.0767	$3s3d^2$	4	−10.9661	$3s3p$ ^{3}P $4d$
17	−6.19770	$3s3d$ ^{3}D $4f$	26	−2.70166	$3p^2$ ^{1}D $5g$	3	−12.1131	$3d^3$	5	−9.40136	$3p^2$ ^{3}P $4p$
18	−6.03601	$3s3p$ ^{3}P $5d$	27	−2.64935	$3p^2$ ^{3}P $5g$	4	−10.0399	$3s3p$ ^{3}P $4f$	6	−8.34280	$3s3d$ ^{3}D $4p$
19	−5.81938	$3p3d$ ^{1}F $4s$	28	−2.57589	$3p3d$ ^{1}P $4f$	5	−9.36378	$3s3p$ ^{1}P $4f$	7	−7.24295	$3s3p$ ^{3}P $5s$
20	−5.62822	$3s3d$ ^{1}D $4f$	29	−2.33055	$3s3p$ ^{1}P $6f$	6	−8.13974	$3p^2$ ^{1}D $4d$	8	−6.55904	$3p3d$ ^{3}P $4s$
21	−5.53173	$3s^2$ ^{1}S $6f$	30	−2.16111	$3s3d$ ^{3}D $5d$	7	−7.81569	$3s^2$ ^{1}S $5g$	9	−6.22438	$3s3d$ ^{3}D $4f$
22	−5.49503	$3s3p$ ^{3}P $5g$	31	−1.80894	$3s3p$ ^{3}P $7f$	8	−6.95218	$3s3d$ ^{3}D $4d$	10	−6.01419	$3s3p$ ^{3}P $5d$

Al-like Fe (Fe^{13+})

i	E(Ryds)	Description
11	−4.53868	$3p3d$ ^{3}F $4d$
12	−4.37396	$3p3d$ ^{3}D $4d$
13	−4.14213	$3s3p$ ^{3}P $6s$
14	−4.08161	$3p3d$ ^{3}P $4d$
15	−3.71455	$3p^2$ ^{3}P $5p$
16	−3.45445	$3s3p$ ^{3}P $6d$
17	−2.74055	$3s3d$ ^{3}D $5p$
18	−2.31283	$3s3p$ ^{3}P $7s$
19	−1.93010	$3s3p$ ^{3}P $7d$
20	−1.71528	$3s3d$ ^{3}D $5f$
21	−1.58463	$3d^2$ ^{3}P $4p$
22	−1.19021	$3s3p$ ^{3}P $8s$
23	−.93260	$3s3p$ ^{3}P $8d$
24	−.86554	$3p^2$ ^{3}P $6p$
25	−.42945	$3s3p$ ^{3}P $9s$
26	−.42159	$3p3d$ ^{3}P $5s$
27	−.24796	$3s3p$ ^{3}P $9d$
	$^4P^e$	
1	−26.7468	$3s3p^2$
2	−19.5924	$3p^2$ ^{3}P $3d$
3	−18.4393	$3s3d^2$
4	−12.3769	$3s3p$ ^{3}P $4p$
5	−12.0440	$3d^3$
6	−10.4138	$3p^2$ ^{3}P $4s$
7	−8.13687	$3p^2$ ^{3}P $4d$
8	−6.91997	$3s3d$ ^{3}D $4d$
9	−6.69836	$3s3p$ ^{3}P $5p$
10	−5.60059	$3p3d$ ^{3}P $4p$
11	−5.47096	$3p3d$ ^{3}D $4p$
12	−4.24748	$3p^2$ ^{3}P $5s$
13	−3.83906	$3s3p$ ^{3}P $6p$
14	−3.71204	$3p3d$ ^{3}F $4f$
15	−3.38892	$3p3d$ ^{3}D $4f$
16	−3.11195	$3p^2$ ^{3}P $5d$
17	−2.61439	$3d^2$ ^{3}P $4s$
18	−2.16482	$3s3p$ ^{3}P $7p$
19	−2.02746	$3s3d$ ^{3}D $5d$
20	−1.15088	$3p^2$ ^{3}P $6s$
21	−1.06793	$3s3p$ ^{3}P $8p$
22	−.67640	$3d^2$ ^{3}F $4d$
23	−.53370	$3p^2$ ^{3}P $6d$
24	−.35031	$3s3p$ ^{3}P $9p$
25	.10066	$3d^2$ ^{3}P $4d$
	$^4D^o$	
1	−22.5807	$3s3p$ ^{3}P $3d$
2	−16.1960	$3p3d^2$
3	−15.8519	$3p3d^2$
4	−11.1727	$3s3p$ ^{3}P $4d$
5	−9.48591	$3p^2$ ^{3}P $4p$
6	−8.45954	$3s3d$ ^{3}D $4p$
7	−7.32630	$3p^2$ ^{3}P $4f$
8	−6.51822	$3p3d$ ^{3}D $4s$
9	−6.23600	$3s3d$ ^{3}D $4f$
10	−6.10802	$3s3p$ ^{3}P $5d$
11	−4.65175	$3p3d$ ^{3}F $4d$
12	−4.17759	$3p3d$ ^{3}P $4d$
13	−4.11886	$3p3d$ ^{3}D $4d$
14	−3.79017	$3p^2$ ^{3}P $5p$
15	−3.48822	$3s3p$ ^{3}P $6d$
16	−2.79434	$3s3d$ ^{3}D $5p$
17	−2.77466	$3p^2$ ^{3}P $5f$
18	−1.97177	$3d^2$ ^{3}F $4p$
19	−1.94251	$3s3p$ ^{3}P $7d$
20	−1.73451	$3s3d$ ^{3}D $5f$
21	−1.65076	$3d^2$ ^{3}P $4p$
22	−.95493	$3s3p$ ^{3}P $8d$
23	−.87935	$3p^2$ ^{3}P $6p$
24	−.39140	$3p3d$ ^{3}D $5s$
25	−.31713	$3p^2$ ^{3}P $6f$
26	−.25448	$3s3p$ ^{3}P $9d$
27	−.04637	$3p3d$ ^{3}F $5d$
	$^4D^e$	
1	−20.0264	$3p^2$ ^{3}P $3d$
2	−12.4963	$3s3p$ ^{3}P $4p$
3	−10.2064	$3s3p$ ^{3}P $4f$
4	−9.40689	$3s3d$ ^{3}D $4s$
5	−7.98466	$3p^2$ ^{3}P $4d$
6	−7.23239	$3s3d$ ^{3}D $4d$
7	−6.75783	$3s3p$ ^{3}P $5p$
8	−5.89597	$3p3d$ ^{3}F $4p$
9	−5.70757	$3s3p$ ^{3}P $5f$
10	−5.62920	$3p3d$ ^{3}P $4p$:
11	−5.50799	$3p3d$ ^{3}D $4p$
12	−3.87053	$3s3p$ ^{3}P $6p$
13	−3.79953	$3p3d$ ^{3}F $4f$
14	−3.57402	$3p3d$ ^{3}D $4f$:
15	−3.42823	$3p3d$ ^{3}P $4f$:
16	−3.28187	$3s3d$ ^{3}D $5s$
17	−3.25779	$3s3p$ ^{3}P $6f$
18	−3.05891	$3p^2$ ^{3}P $5d$
19	−2.20652	$3s3d$ ^{3}D $5d$
20	−2.13068	$3s3p$ ^{3}P $7p$
21	−1.79716	$3s3p$ ^{3}P $7f$
22	−1.58781	$3s3d$ ^{3}D $5g$
23	−1.08150	$3s3p$ ^{3}P $8p$
24	−.84376	$3s3p$ ^{3}P $8f$
25	−.80057	$3d^2$ ^{3}F $4d$
26	−.50660	$3p^2$ ^{3}P $6d$
27	−.36270	$3p3d$ ^{3}F $5p$
28	−.35612	$3d^2$ ^{3}P $4d$
29	−.34586	$3s3p$ ^{3}P $9p$
30	−.19218	$3s3p$ ^{3}P $9f$
31	−.15532	$3s3d$ ^{3}D $6s$
	$^4F^o$	
1	−23.0334	$3s3p$ ^{3}P $3d$
2	−16.1579	$3p3d^2$
3	−11.0367	$3s3p$ ^{3}P $4d$
4	−8.43125	$3s3d$ ^{3}D $4p$
5	−7.30734	$3p^2$ ^{3}P $4f$
6	−6.98534	$3p3d$ ^{3}F $4s$
7	−6.46265	$3s3d$ ^{3}D $4f$
8	−6.08726	$3s3p$ ^{3}P $5d$
9	−5.55270	$3s3p$ ^{3}P $5g$
10	−4.79098	$3p3d$ ^{3}F $4d$
11	−4.29163	$3p3d$ ^{3}P $4d$
12	−4.15676	$3p3d$ ^{3}D $4d$
13	−3.48993	$3s3p$ ^{3}P $6d$
14	−3.19108	$3s3p$ ^{3}P $6g$
15	−2.77354	$3s3d$ ^{3}D $5p$
16	−2.75047	$3p^2$ ^{3}P $5f$
17	−2.03523	$3d^2$ ^{3}F $4p$
18	−1.93389	$3s3p$ ^{3}P $7d$
19	−1.83071	$3s3d$ ^{3}D $5f$
20	−1.74417	$3s3p$ ^{3}P $7g$
21	−.93308	$3s3p$ ^{3}P $8d$
22	−.86168	$3p3d$ ^{3}F $5s$
23	−.81284	$3s3p$ ^{3}P $8g$
24	−.31116	$3p^2$ ^{3}P $6f$
25	−.24798	$3s3p$ ^{3}P $9d$
26	−.16922	$3s3p$ ^{3}P $9g$
	$^4F^e$	
1	−20.2009	$3p^2$ ^{3}P $3d$
2	−18.7612	$3s3d^2$
3	−12.3598	$3d^3$
4	−10.3675	$3s3p$ ^{3}P $4f$
5	−8.10885	$3p^2$ ^{3}P $4d$
6	−7.03493	$3s3d$ ^{3}D $4d$
7	−6.04427	$3p3d$ ^{3}F $4p$
8	−5.77346	$3s3p$ ^{3}P $5f$
9	−5.56776	$3p3d$ ^{3}D $4p$
10	−3.95567	$3p3d$ ^{3}F $4f$
11	−3.54524	$3p3d$ ^{3}P $4f$
12	−3.37847	$3p3d$ ^{3}D $4f$
13	−3.30112	$3s3p$ ^{3}P $6f$
14	−3.12774	$3p^2$ ^{3}P $5d$
15	−2.97355	$3d^2$ ^{3}F $4s$
16	−2.62952	$3p^2$ ^{3}P $5g$
17	−2.11472	$3s3d$ ^{3}D $5d$
18	−1.81510	$3s3p$ ^{3}P $7f$
19	−1.63629	$3s3d$ ^{3}D $5g$
20	−.85622	$3s3p$ ^{3}P $8f$
21	.55632	$3d^2$ ^{3}F $4d$
22	−.53259	$3p^2$ ^{3}P $6d$
23	−.39987	$3p3d$ ^{3}F $5p$
24	−.34508	$3d^2$ ^{3}P $4d$
25	−.24498	$3p^2$ ^{3}P $6g$
26	−.19984	$3s3p$ ^{3}P $9f$
	$^4G^o$	
1	−16.6676	$3p3d^2$
2	−7.44657	$3p^2$ ^{3}P $4f$
3	−6.27799	$3s3d$ ^{3}D $4f$
4	−5.61560	$3s3p$ ^{3}P $5g$
5	−4.68473	$3p3d$ ^{3}F $4d$
6	−4.28057	$3p3d$ ^{3}D $4d$
7	−3.22676	$3s3p$ ^{3}P $6g$
8	−2.80913	$3p^2$ ^{3}P $5f$
9	−2.04226	$3d^2$ ^{3}F $4p$
10	−1.79719	$3s3p$ ^{3}P $7g$
11	−1.73041	$3s3d$ ^{3}D $5f$
12	−.82635	$3s3p$ ^{3}P $8g$
13	−.33239	$3p^2$ ^{3}P $6f$
14	−.17834	$3s3p$ ^{3}P $9g$
	$^4G^e$	
1	−10.3421	$3s3p$ ^{3}P $4f$
2	−7.20623	$3s3d$ ^{3}D $4d$
3	−6.05285	$3p3d$ ^{3}F $4p$
4	−5.75813	$3s3p$ ^{3}P $5f$
5	−4.04051	$3p3d$ ^{3}F $4f$
6	−3.62798	$3p3d$ ^{3}P $4f$
7	−3.50469	$3p3d$ ^{3}D $4f$
8	−3.29205	$3s3p$ ^{3}P $6f$
9	−2.55985	$3p^2$ ^{3}P $5g$
10	−2.17245	$3s3d$ ^{3}D $5d$
11	−1.81121	$3s3p$ ^{3}P $7f$
12	−1.66055	$3s3d$ ^{3}D $5g$
13	−.85299	$3s3p$ ^{3}P $8f$
14	−.84217	$3d^2$ ^{3}F $4d$
15	−.40375	$3p3d$ ^{3}F $5p$
16	−.21661	$3p^2$ ^{3}P $6g$
17	−.19625	$3s3p$ ^{3}P $9f$

Al-like Fe (Fe^{13+})

Energies in ascending order from ground state for terms with effective $n \leq 4.0$, $L \leq 4$

Term	i	E(Ryds)	Term	i	E(Ryds)	Term	i	E(Ryds)	Term	i	E(Ryds)	Term	i	E(Ryds)
^{2}P^o	1	0.00000	^{2}P^o	3	7.08720	^{2}F^e	2	9.60980	^{4}F^o	2	12.4701	^{2}P^o	8	14.1228
^{4}P^e	1	1.88120	^{2}F^o	2	7.15810	^{4}F^e	2	9.86680	^{2}G^o	1	12.5131	^{4}P^o	3	15.1476
^{2}D^e	1	2.47360	^{2}P^o	4	7.37150	^{2}D^e	5	10.0730	^{4}P^o	2	12.5993	^{2}D^e	7	15.2909
^{2}S^e	1	3.12260	^{2}D^o	3	7.37830	^{4}P^e	3	10.1887	^{4}D^o	3	12.7761	^{2}P^o	9	15.3927
^{2}P^e	1	3.33550	^{2}F^e	1	8.31710	^{2}G^e	2	10.5513	^{2}S^e	4	12.9208	^{2}P^e	5	16.0811
^{2}D^e	2	4.14850	^{4}F^e	1	8.42710	^{2}D^e	6	10.5526	^{2}P^o	5	12.9237	^{2}F^o	6	16.1204
^{2}D^o	1	4.88140	^{2}P^e	2	8.55230	^{2}F^e	3	10.9662	^{4}S^o	2	13.0913	^{4}D^e	2	16.1317
^{4}S^o	1	4.98170	^{4}D^e	1	8.60160	^{2}P^e	4	11.2706	^{2}F^o	4	13.1390	^{2}P^o	10	16.1856
^{2}P^o	2	5.46700	^{2}G^e	1	8.78890	^{2}S^e	3	11.3095	^{2}D^o	5	13.3065	^{4}P^e	4	16.2511
^{4}F^o	1	5.59460	^{2}D^e	3	9.00700	^{4}G^o	1	11.9604	^{2}P^o	6	13.4582	^{4}F^e	3	16.2682
^{4}P^o	1	6.02940	^{4}P^e	2	9.03560	^{2}F^o	3	12.2188	^{2}F^o	5	13.5090	^{4}S^e	1	16.2866
^{4}D^o	1	6.04730	^{2}P^e	3	9.40960	^{2}D^o	4	12.2705	^{2}G^o	2	13.5541	^{2}D^e	8	16.3634
^{2}D^o	2	6.19120	^{2}S^e	2	9.46890	^{4}D^o	2	12.4320	^{2}D^o	6	13.9142			
^{2}F^o	1	6.53990	^{2}D^e	4	9.48880	^{2}S^o	1	12.4337	^{2}P^o	7	13.9523			

gf-values for transitions involving terms with effective $n \leq 4.0$, $L \leq 4$

i i'	gf$_L$	i i'	gf$_L$	i i'	gf$_L$	i i'	gf$_L$	i i'	gf$_L$	i i'	gf$_L$
	^{2}S^o–^{2}P^e	3 7	−5.32E−3	4 4	−2.52E+0		^{2}P^o–^{2}D^e	5 1	1.91E−6	9 2	1.09E−1
1 1	1.22E−4	3 8	−4.53E−1	4 5	−3.91E−3	1 1	−3.86E−1	5 2	1.48E−5	9 3	3.60E−4
1 2	8.14E−1	3 9	−1.68E−2	5 1	3.33E−3	1 2	−2.86E+0	5 3	2.73E−1	9 4	1.31E−4
1 3	6.86E−2	3 10	−7.46E−3	5 2	1.55E+0	1 3	−7.67E−4	5 4	9.07E−2	9 5	6.45E−4
1 4	4.90E−2	4 1	3.88E−1	5 3	8.24E−1	1 4	−1.35E−3	5 5	2.82E−2	9 6	4.48E−3
1 5	−7.10E−6	4 2	9.46E−5	5 4	2.24E−2	1 5	−7.90E−4	5 6	4.90E−1	9 7	8.70E−4
	^{2}S^e–^{2}P^o	4 3	4.77E−4	5 5	−8.20E−4	1 6	−1.75E−3	5 7	−2.72E−5	9 8	−1.55E+0
1 1	3.26E−1	4 4	6.39E−6	6 1	2.34E−3	1 7	−1.69E+0	5 8	−1.99E−6	10 1	2.62E−1
1 2	−1.66E−1	4 5	−6.27E−9	6 2	2.67E−2	1 8	−1.07E+0	6 1	4.36E−3	10 2	6.72E−2
1 3	−1.40E+0	4 6	−2.83E−5	6 3	2.56E−2	2 1	8.68E−1	6 2	1.10E−4	10 3	1.21E−5
1 4	−3.44E−1	4 7	−9.76E−1	6 4	1.45E−1	2 2	3.02E−4	6 3	3.41E−2	10 4	9.54E−5
1 5	−4.83E−5	4 8	−3.48E−4	6 5	−3.24E−3	2 3	−6.08E−1	6 4	1.35E+0	10 5	1.53E−5
1 6	−1.94E−3	4 9	−3.92E−2	7 1	8.44E−4	2 4	−3.54E−2	6 5	9.79E−3	10 6	1.86E−3
1 7	−6.33E−4	4 10	−1.85E+0	7 2	5.51E−6	2 5	−2.05E+0	6 6	3.63E−1	10 7	4.57E−3
1 8	−5.61E−7		^{2}P^o–^{2}P^e	7 3	3.31E−5	2 6	−1.04E−5	6 7	−1.13E−5	10 8	−5.43E−3
1 9	−1.31E−1	1 1	−2.48E+0	7 4	6.75E−5	2 7	−1.88E−5	6 8	−1.11E−2		^{2}P^e–^{2}D^o
1 10	−4.69E−2	1 2	−1.72E−4	7 5	−5.48E−2	2 8	−9.56E−5	7 1	1.62E−1	1 1	−3.09E−1
2 1	7.27E−4	1 3	−5.97E−4	8 1	4.28E−3	3 1	7.33E−4	7 2	2.89E−1	1 2	−1.67E−1
2 2	5.31E−1	1 4	−1.62E−3	8 2	8.53E−3	3 2	3.27E−3	7 3	1.16E−4	1 3	−4.35E+0
2 3	6.62E−3	1 5	−5.76E−1	8 3	6.65E−1	3 3	−1.13E−1	7 4	3.03E−4	1 4	−3.47E−3
2 4	1.97E−1	2 1	4.51E−1	8 4	1.50E+0	3 4	−1.23E−1	7 5	8.68E−5	1 5	−3.22E−3
2 5	−4.26E−1	2 2	−1.43E−1	8 5	−2.62E−3	3 5	−2.13E+0	7 6	1.22E−4	1 6	−1.59E−3
2 6	−1.45E−2	2 3	−1.56E+0	9 1	1.15E−1	3 6	−6.25E−1	7 7	−2.41E+0	2 1	9.98E−1
2 7	−3.04E−4	2 4	−1.83E−3	9 2	5.57E−4	3 7	−1.95E−3	7 8	−1.42E−1	2 2	4.14E−1
2 8	−6.09E−1	2 5	−2.75E−2	9 3	8.01E−4	3 8	−6.20E−3	8 1	2.98E−3	2 3	1.89E−3
2 9	−1.96E−4	3 1	9.28E−1	9 4	6.48E−5	4 1	6.72E−3	8 2	2.00E−3	2 4	−1.16E−2
2 10	−3.68E−5	3 2	−6.12E−2	9 5	−6.23E−1	4 2	1.94E+0	8 3	3.97E−2	2 5	−3.55E−1
3 1	2.72E−3	3 3	−2.97E−2	10 1	5.91E−1	4 3	−6.54E−2	8 4	3.35E−2	2 6	−2.01E−1
3 2	4.81E−5	3 4	−6.40E−1	10 2	7.41E−6	4 4	−5.03E−1	8 5	1.76E+0	3 1	2.99E−3
3 3	6.35E−1	3 5	−3.86E−2	10 3	2.12E−4	4 5	−2.36E−1	8 6	1.79E−2	3 2	2.42E−1
3 4	6.28E−1	4 1	9.84E−1	10 4	1.61E−3	4 6	−9.98E−1	8 7	−6.45E−5	3 3	2.32E−1
3 5	−3.06E−3	4 2	−5.71E−2	10 5	3.73E−4	4 7	−1.26E−3	8 8	−5.76E−3	3 4	−7.84E−1
3 6	−4.50E−1	4 3	−2.57E−1			4 8	−6.63E−4	9 1	7.80E−1	3 5	−9.01E−2

Al-like Fe (Fe^{13+})

$i\ i'$	gf_L	$i\ i'$	gf_L	$i\ i'$	gf_L	$i\ i'$	gf_L	$i\ i'$	gf_L	$i\ i'$	gf_L
3 6	−6.55E−1	4 6	1.53E−1	1 5	−2.47E−4	8 6	3.63E−5	2 1	1.12E+0	3 2	−1.71E+0
4 1	2.22E−3	4 7	−2.92E−7	1 6	−1.04E+0		$^{2}\mathbf{F}^{o}-{}^{2}\mathbf{F}^{e}$	2 2	1.35E+0	3 3	−6.02E−1
4 2	3.46E−4	4 8	−3.87E−5	2 1	−5.13E−1	1 1	−5.00E−1		$^{4}\mathbf{S}^{o}-{}^{4}\mathbf{P}^{e}$	4 1	5.73E−1
4 3	1.32E+0	5 1	1.18E−3	2 2	−3.86E+0	1 2	−1.55E+0	1 1	1.71E+0	4 2	8.58E−5
4 4	−5.51E−2	5 2	6.84E−4	2 3	−3.09E−4	1 3	−1.59E+0	1 2	−4.01E+0	4 3	5.08E−5
4 5	−3.77E−2	5 3	1.27E+0	2 4	−4.92E−4	2 1	−8.67E−3	1 3	−2.57E−4		$^{4}\mathbf{D}^{o}-{}^{4}\mathbf{D}^{e}$
4 6	−2.00E+0	5 4	2.31E+0	2 5	−2.61E−3	2 2	−8.69E−1	1 4	−1.68E−4	1 1	−4.35E−1
5 1	1.98E−1	5 5	1.10E−2	2 6	−7.81E+0	2 3	−1.17E+0	2 1	6.82E−4	1 2	−2.23E−1
5 2	1.39E−1	5 6	1.05E+0	3 1	1.36E−2	3 1	6.12E−1	2 2	2.71E+0	2 1	9.94E−1
5 3	9.11E−3	5 7	−4.85E−5	3 2	4.20E−3	3 2	3.70E−1	2 3	6.19E−1	2 2	−2.60E−7
5 4	4.03E−4	5 8	−1.57E−3	3 3	−1.19E+0	3 3	1.87E−1	2 4	−1.86E−5	3 1	3.81E+0
5 5	1.16E−4	6 1	5.76E−4	3 4	−1.24E+0	4 1	1.20E+0		$^{4}\mathbf{S}^{e}-{}^{4}\mathbf{P}^{o}$	3 2	−4.83E−7
5 6	1.86E−6	6 2	2.62E−3	3 5	−3.44E−1	4 2	1.52E+0	1 1	3.27E−1		$^{4}\mathbf{D}^{o}-{}^{4}\mathbf{F}^{e}$
	$^{2}\mathbf{D}^{o}-{}^{2}\mathbf{D}^{e}$	6 3	6.80E−1	3 6	−1.77E−4	4 3	2.85E−1	1 2	1.55E−5	1 1	−2.07E+0
1 1	6.40E−1	6 4	6.85E−1	4 1	1.68E+0	5 1	3.10E−1	1 3	8.70E−1	1 2	−6.44E+0
1 2	8.27E−3	6 5	9.50E−1	4 2	1.58E−1	5 2	1.61E−1		$^{4}\mathbf{P}^{o}-{}^{4}\mathbf{P}^{e}$	1 3	−1.33E−3
1 3	−2.10E+0	6 6	5.66E−1	4 3	−1.24E−1	5 3	9.63E−1	1 1	2.08E+0	2 1	2.08E+0
1 4	−2.66E−2	6 7	−3.05E−5	4 4	−1.63E−1	6 1	5.80E−4	1 2	−8.62E−1	2 2	7.92E−1
1 5	−8.86E−2	6 8	−4.14E−5	4 5	−7.26E−2	6 2	6.79E−4	1 3	−3.83E+0	2 3	−2.79E−1
1 6	−2.14E−2		$^{2}\mathbf{D}^{o}-{}^{2}\mathbf{F}^{e}$	4 6	−2.69E−4	6 3	6.73E−4	1 4	−2.13E−1	3 1	1.29E−2
1 7	−1.03E−4	1 1	−4.04E−1	5 1	1.62E−1		$^{2}\mathbf{F}^{o}-{}^{2}\mathbf{G}^{e}$	2 1	6.01E−4	3 2	2.43E+0
1 8	−1.03E−1	1 2	−2.99E+0	5 2	1.26E+0	1 1	−1.66E+0	2 2	9.31E−7	3 3	−5.82E+0
2 1	2.88E+0	1 3	−5.54E−2	5 3	−4.68E−2	1 2	−3.22E−1	2 3	1.57E+0		$^{4}\mathbf{D}^{e}-{}^{4}\mathbf{F}^{o}$
2 2	1.23E−1	2 1	−6.38E−1	5 4	−7.37E−2	2 1	−7.21E−1	2 4	−2.78E−4	1 1	3.01E+0
2 3	−8.38E−1	2 2	−1.68E+0	5 5	−5.21E+0	2 2	−6.43E+0	3 1	1.82E+0	1 2	−2.26E+0
2 4	−5.26E−1	2 3	−7.60E−5	5 6	−2.47E−5	3 1	6.87E−1	3 2	2.63E−4	2 1	1.50E+0
2 5	−3.54E−3	3 1	−8.72E−2	6 1	5.70E−3	3 2	1.37E−2	3 3	3.46E−3	2 2	4.74E−6
2 6	−3.03E+0	3 2	−8.57E−1	6 2	6.77E−1	4 1	1.68E−2	3 4	−2.30E+0		$^{4}\mathbf{F}^{o}-{}^{4}\mathbf{F}^{e}$
2 7	−6.99E−6	3 3	−6.00E+0	6 3	−3.89E−1	4 2	9.63E−1		$^{4}\mathbf{P}^{o}-{}^{4}\mathbf{D}^{e}$	1 1	−5.00E+0
2 8	−3.28E−2	4 1	1.25E+0	6 4	−1.64E+0	5 1	5.71E−3	1 1	−2.13E+0	1 2	−4.06E+0
3 1	1.39E−3	4 2	4.34E−4	6 5	−2.82E−1	5 2	1.48E+0	1 2	−3.72E−3	1 3	−1.48E−3
3 2	2.15E+0	4 3	1.15E−1	6 6	−7.19E−4	6 1	1.55E−3	2 1	2.81E+0	2 1	8.05E+0
3 3	−3.11E−1	5 1	1.64E−3	7 1	4.85E−3	6 2	1.81E−5	2 2	−3.30E−5	2 2	3.88E+0
3 4	−2.77E−2	5 2	9.55E−1	7 2	6.12E−2		$^{2}\mathbf{F}^{e}-{}^{2}\mathbf{G}^{o}$	3 1	9.93E−4	2 3	−8.67E+0
3 5	−8.89E−1	5 3	3.30E−1	7 3	2.63E−6	1 1	−8.53E−1	3 2	−3.50E+0		$^{4}\mathbf{F}^{e}-{}^{4}\mathbf{G}^{o}$
3 6	−5.76E−2	6 1	8.19E−5	7 4	6.56E−6	1 2	−5.85E−1		$^{4}\mathbf{P}^{e}-{}^{4}\mathbf{D}^{o}$	1 1	−3.00E+0
3 7	−4.00E−4	6 2	4.78E−1	7 5	2.54E−5	2 1	−2.02E+0	1 1	−5.80E+0	2 1	−3.60E+0
3 8	−1.32E−2	6 3	1.15E+0	7 6	−1.77E+0	2 2	−4.19E+0	1 2	−4.14E−4	3 1	1.61E+0
4 1	1.33E−5		$^{2}\mathbf{D}^{e}-{}^{2}\mathbf{F}^{o}$	8 1	3.71E−1	3 1	−2.66E−1	1 3	−4.71E−4		
4 2	1.12E−4	1 1	−1.58E+0	8 2	7.28E−2	3 2	−4.94E+0	2 1	2.70E+0		
4 3	6.63E−1	1 2	−2.38E+0	8 3	2.98E−3		$^{2}\mathbf{G}^{o}-{}^{2}\mathbf{G}^{e}$	2 2	−4.19E−1		
4 4	1.06E−1	1 3	−3.29E−3	8 4	0.10E−4	1 1	2.90E+0	2 3	−0.17E+0		
4 5	1.80E−2	1 4	8.45E−4	8 5	1.20E−5	1 2	6.96E−1	3 1	1.47E+0		

Si-like S (S^{2+})

Term energies relative to $3s^23p\ ^2\mathrm{P}$ ionization threshold for each symmetry

i	E(Ryds)	Description
		$^1\mathrm{S}^e$
1	−2.30910	$3s^23p^2$
2	−.90074	$3s^23p\ ^2\mathrm{P}\ 4p$
3	−.56670	$3s3p^2\ ^2\mathrm{D}\ 3d$
4	−.51005	$3s^23p\ ^2\mathrm{P}\ 5p$
5	−.33279	$3s^23p\ ^2\mathrm{P}\ 6p$
6	−.23402	$3s^23p\ ^2\mathrm{P}\ 7p$
7	−.17366	$3s^23p\ ^2\mathrm{P}\ 8p$
8	−.13425	$3s^23p\ ^2\mathrm{P}\ 9p$
9	−.11302	$3s3p^2\ ^2\mathrm{S}\ 4s$
		$^3\mathrm{S}^e$
1	−.98672	$3s^23p\ ^2\mathrm{P}\ 4p$
2	−.54638	$3s^23p\ ^2\mathrm{P}\ 5p$
3	−.35170	$3s^23p\ ^2\mathrm{P}\ 6p$
4	−.32939	$3s3p^2\ ^2\mathrm{D}\ 3d$
5	−.24359	$3s^23p\ ^2\mathrm{P}\ 7p$
6	−.18007	$3s^23p\ ^2\mathrm{P}\ 8p$
7	−.13898	$3s^23p\ ^2\mathrm{P}\ 9p$
		$^3\mathrm{S}^o$
1	−1.28132	$3s3p^3$
2	−.42202	$3s3p^2\ ^4\mathrm{P}\ 4p$
		$^5\mathrm{S}^o$
1	−2.07097	$3s3p^3$
2	−.36682	$3s3p^2\ ^4\mathrm{P}\ 4p$
		$^1\mathrm{P}^e$
1	−1.04599	$3s^23p\ ^2\mathrm{P}\ 4p$
2	−.56745	$3s^23p\ ^2\mathrm{P}\ 5p$
3	−.36035	$3s^23p\ ^2\mathrm{P}\ 6p$
4	−.34781	$3s3p^2\ ^2\mathrm{D}\ 3d$
5	−.24954	$3s^23p\ ^2\mathrm{P}\ 7p$
6	−.18356	$3s^23p\ ^2\mathrm{P}\ 8p$
7	−.14072	$3s^23p\ ^2\mathrm{P}\ 9p$
		$^1\mathrm{P}^o$
1	−1.29341	$3s3p^3$
2	−1.21324	$3s^23p\ ^2\mathrm{P}\ 4s$
3	−1.03140	$3s^23p\ ^2\mathrm{P}\ 3d$
4	−.63760	$3s^23p\ ^2\mathrm{P}\ 5s$
5	−.61903	$3s^23p\ ^2\mathrm{P}\ 4d$
6	−.40193	$3s^23p\ ^2\mathrm{P}\ 5d$
7	−.39397	$3s^23p\ ^2\mathrm{P}\ 6s$
8	−.27602	$3s^23p\ ^2\mathrm{P}\ 6d$
9	−.26946	$3s^23p\ ^2\mathrm{P}\ 7s$
10	−.20087	$3s^23p\ ^2\mathrm{P}\ 7d$
11	−.19590	$3s^23p\ ^2\mathrm{P}\ 8s$
12	−.15300	$3s^23p\ ^2\mathrm{P}\ 8d$
13	−.14887	$3s^23p\ ^2\mathrm{P}\ 9s$
14	−.12205	$3s^23p\ ^2\mathrm{P}\ 9d$
		$^3\mathrm{P}^e$
1	−2.57449	$3s^23p^2$
2	−.99692	$3s^23p\ ^2\mathrm{P}\ 4p$
3	−.92782	$3p^4$
4	−.58876	$3s3p^2\ ^4\mathrm{P}\ 3d$
5	−.55427	$3s3p^2\ ^4\mathrm{P}\ 4s$
6	−.51841	$3s^23p\ ^2\mathrm{P}\ 5p$
7	−.40129	$3s3p^2\ ^2\mathrm{D}\ 3d$
8	−.34759	$3s^23p\ ^2\mathrm{P}\ 6p$
9	−.24317	$3s^23p\ ^2\mathrm{P}\ 7p$
10	−.17956	$3s^23p\ ^2\mathrm{P}\ 8p$
11	−.13801	$3s^23p\ ^2\mathrm{P}\ 9p$
		$^3\mathrm{P}^o$
1	−1.67487	$3s3p^3$
2	−1.26183	$3s^23p\ ^2\mathrm{P}\ 3d$
3	−1.21914	$3s^23p\ ^2\mathrm{P}\ 4s$
4	−.66373	$3s^23p\ ^2\mathrm{P}\ 4d$
5	−.64507	$3s^23p\ ^2\mathrm{P}\ 5s$
6	−.42002	$3s^23p\ ^2\mathrm{P}\ 5d$
7	−.39935	$3s^23p\ ^2\mathrm{P}\ 6s$
8	−.31195	$3s3p^2\ ^4\mathrm{P}\ 4p$
9	−.28118	$3s^23p\ ^2\mathrm{P}\ 6d$
10	−.27160	$3s^23p\ ^2\mathrm{P}\ 7s$
11	−.20376	$3s^23p\ ^2\mathrm{P}\ 7d$
12	−.19729	$3s^23p\ ^2\mathrm{P}\ 8s$
13	−.15391	$3s^23p\ ^2\mathrm{P}\ 8d$
14	−.14976	$3s^23p\ ^2\mathrm{P}\ 9s$
15	−.12120	$3s^23p\ ^2\mathrm{P}\ 9d$
		$^5\mathrm{P}^e$
1	−.69377	$3s3p^2\ ^4\mathrm{P}\ 3d$
2	−.62281	$3s3p^2\ ^4\mathrm{P}\ 4s$
3	−.05660	$3s3p^2\ ^4\mathrm{P}\ 4d$
		$^5\mathrm{P}^o$
1	−.37767	$3s3p^2\ ^4\mathrm{P}\ 4p$
		$^1\mathrm{D}^e$
1	−2.46964	$3s^23p^2$
2	−.95817	$3s^23p\ ^2\mathrm{P}\ 4p$
3	−.85182	$3p^4$
4	−.56852	$3s^23p\ ^2\mathrm{P}\ 4f$
5	−.53943	$3s^23p\ ^2\mathrm{P}\ 5p$
6	−.36809	$3s3p^2\ ^2\mathrm{D}\ 4s$
7	−.36725	$3s^23p\ ^2\mathrm{P}\ 5f$
8	−.33012	$3s^23p\ ^2\mathrm{P}\ 6p$
9	−.26843	$3s3p^2\ ^2\mathrm{D}\ 3d$
10	−.24734	$3s^23p\ ^2\mathrm{P}\ 6f$
11	−.23854	$3s^23p\ ^2\mathrm{P}\ 7p$
12	−.18718	$3s^23p\ ^2\mathrm{P}\ 7f$
13	−.17683	$3s^23p\ ^2\mathrm{P}\ 8p$
14	−.14433	$3s^23p\ ^2\mathrm{P}\ 8f$
15	−.13616	$3s^23p\ ^2\mathrm{P}\ 9p$
16	−.11489	$3s^23p\ ^2\mathrm{P}\ 9f$
		$^1\mathrm{D}^o$
1	−1.62946	$3s3p^3$
2	−1.15628	$3s^23p\ ^2\mathrm{P}\ 3d$
3	−.68868	$3s^23p\ ^2\mathrm{P}\ 4d$
4	−.43226	$3s^23p\ ^2\mathrm{P}\ 5d$
5	−.28982	$3s^23p\ ^2\mathrm{P}\ 6d$
6	−.20791	$3s^23p\ ^2\mathrm{P}\ 7d$
7	−.16254	$3s3p^2\ ^2\mathrm{D}\ 4p$
8	−.15499	$3s^23p\ ^2\mathrm{P}\ 8d$
9	−.12166	$3s^23p\ ^2\mathrm{P}\ 9d$
		$^3\mathrm{D}^e$
1	−1.02100	$3s^23p\ ^2\mathrm{P}\ 4p$
2	−.57539	$3s^23p\ ^2\mathrm{P}\ 4f$
3	−.55843	$3s^23p\ ^2\mathrm{P}\ 5p$
4	−.50428	$3s3p^2\ ^4\mathrm{P}\ 3d$
5	−.41572	$3s3p^2\ ^2\mathrm{D}\ 3d$
6	−.38770	$3s3p^2\ ^2\mathrm{D}\ 4s$
7	−.36875	$3s^23p\ ^2\mathrm{P}\ 5f$
8	−.35143	$3s^23p\ ^2\mathrm{P}\ 6p$
9	−.25997	$3s^23p\ ^2\mathrm{P}\ 6f$
10	−.24551	$3s^23p\ ^2\mathrm{P}\ 7p$
11	−.21349	$3s3p^2\ ^2\mathrm{S}\ 3d$
12	−.19389	$3s^23p\ ^2\mathrm{P}\ 7f$
13	−.18082	$3s^23p\ ^2\mathrm{P}\ 8p$
14	−.15283	$3s^23p\ ^2\mathrm{P}\ 8f$
15	−.13871	$3s^23p\ ^2\mathrm{P}\ 9p$
16	−.12576	$3s3p^2\ ^2\mathrm{P}\ 3d$
		$^3\mathrm{D}^o$
1	−1.82419	$3s3p^3$
2	−1.20958	$3s^23p\ ^2\mathrm{P}\ 3d$
3	−.67142	$3s^23p\ ^2\mathrm{P}\ 4d$
4	−.42786	$3s^23p\ ^2\mathrm{P}\ 5d$
5	−.34734	$3s3p^2\ ^4\mathrm{P}\ 4p$
6	−.28452	$3s^23p\ ^2\mathrm{P}\ 6d$
7	−.20556	$3s^23p\ ^2\mathrm{P}\ 7d$
8	−.16637	$3s3p^2\ ^2\mathrm{D}\ 4p$
9	−.15500	$3s^23p\ ^2\mathrm{P}\ 8d$
10	−.12106	$3s^23p\ ^2\mathrm{P}\ 9d$
		$^5\mathrm{D}^e$
1	−.86360	$3s3p^2\ ^4\mathrm{P}\ 3d$
2	−.05632	$3s3p^2\ ^4\mathrm{P}\ 4d$
		$^5\mathrm{D}^o$
1	−.40652	$3s3p^2\ ^4\mathrm{P}\ 4p$
		$^1\mathrm{F}^e$
1	−.60456	$3s^23p\ ^2\mathrm{P}\ 4f$
2	−.46209	$3s3p^2\ ^2\mathrm{D}\ 3d$
3	−.37123	$3s^23p\ ^2\mathrm{P}\ 5f$
4	−.26029	$3s^23p\ ^2\mathrm{P}\ 6f$
5	−.19122	$3s^23p\ ^2\mathrm{P}\ 7f$
6	−.14624	$3s^23p\ ^2\mathrm{P}\ 8f$
7	−.11540	$3s^23p\ ^2\mathrm{P}\ 9f$
		$^1\mathrm{F}^o$
1	−1.11630	$3s^23p\ ^2\mathrm{P}\ 3d$
2	−.63142	$3s^23p\ ^2\mathrm{P}\ 4d$
3	−.40655	$3s^23p\ ^2\mathrm{P}\ 5d$
4	−.35991	$3s^23p\ ^2\mathrm{P}\ 5g$
5	−.27858	$3s^23p\ ^2\mathrm{P}\ 6d$
6	−.25020	$3s^23p\ ^2\mathrm{P}\ 6g$
7	−.20254	$3s^23p\ ^2\mathrm{P}\ 7d$
8	−.18389	$3s^23p\ ^2\mathrm{P}\ 7g$
9	−.15552	$3s^23p\ ^2\mathrm{P}\ 8d$
10	−.14092	$3s^23p\ ^2\mathrm{P}\ 8g$
11	−.13656	$3s3p^2\ ^2\mathrm{D}\ 4p$
12	−.11737	$3s^23p\ ^2\mathrm{P}\ 9d$
13	−.11121	$3s^23p\ ^2\mathrm{P}\ 9g$
		$^3\mathrm{F}^e$
1	−.78705	$3s3p^2\ ^4\mathrm{P}\ 3d$
2	−.59711	$3s^23p\ ^2\mathrm{P}\ 4f$
3	−.46244	$3s3p^2\ ^2\mathrm{D}\ 3d$
4	−.38343	$3s^23p\ ^2\mathrm{P}\ 5f$
5	−.27228	$3s^23p\ ^2\mathrm{P}\ 6f$
6	−.21515	$3s3p^2\ ^2\mathrm{P}\ 3d$
7	−.17693	$3s^23p\ ^2\mathrm{P}\ 7f$
8	−.14032	$3s^23p\ ^2\mathrm{P}\ 8f$
9	−.11200	$3s^23p\ ^2\mathrm{P}\ 9f$
		$^3\mathrm{F}^o$
1	−1.45443	$3s^23p\ ^2\mathrm{P}\ 3d$
2	−.69152	$3s^23p\ ^2\mathrm{P}\ 4d$
3	−.43040	$3s^23p\ ^2\mathrm{P}\ 5d$
4	−.35994	$3s^23p\ ^2\mathrm{P}\ 5g$
5	−.28899	$3s^23p\ ^2\mathrm{P}\ 6d$
6	−.25025	$3s^23p\ ^2\mathrm{P}\ 6g$
7	−.20864	$3s^23p\ ^2\mathrm{P}\ 7d$
8	−.18396	$3s^23p\ ^2\mathrm{P}\ 7g$
9	−.16880	$3s3p^2\ ^2\mathrm{D}\ 4p$
10	−.15118	$3s^23p\ ^2\mathrm{P}\ 8d$
11	−.14078	$3s^23p\ ^2\mathrm{P}\ 8g$
12	−.12023	$3s^23p\ ^2\mathrm{P}\ 9d$
13	−.11125	$3s^23p\ ^2\mathrm{P}\ 9g$
		$^5\mathrm{F}^e$
1	−.91191	$3s3p^2\ ^4\mathrm{P}\ 3d$
2	−.07767	$3s3p^2\ ^4\mathrm{P}\ 4d$
		$^1\mathrm{G}^e$
1	−.58435	$3s^23p\ ^2\mathrm{P}\ 4f$
2	−.52890	$3s3p^2\ ^2\mathrm{D}\ 3d$
3	−.36555	$3s^23p\ ^2\mathrm{P}\ 5f$
4	−.25461	$3s^23p\ ^2\mathrm{P}\ 6f$
5	−.24990	$3s^23p\ ^2\mathrm{P}\ 6h$
6	−.18719	$3s^23p\ ^2\mathrm{P}\ 7f$
7	−.18361	$3s^23p\ ^2\mathrm{P}\ 7h$
8	−.14330	$3s^23p\ ^2\mathrm{P}\ 8f$
9	−.14059	$3s^23p\ ^2\mathrm{P}\ 8h$
10	−.11317	$3s^23p\ ^2\mathrm{P}\ 9f$
11	−.11109	$3s^23p\ ^2\mathrm{P}\ 9h$
		$^1\mathrm{G}^o$
1	−.36493	$3s^23p\ ^2\mathrm{P}\ 5g$

Si-like S (S^{2+})

i	Energy(Ryds)	Description	i	Energy(Ryds)	Description	i	Energy(Ryds)	Description
2	−.25295	$3s^23p\ ^2\mathrm{P}\ 6g$			$^1\mathbf{H}^o$	5	−.11118	$3s^23p\ ^2\mathrm{P}\ 9h$
3	−.18561	$3s^23p\ ^2\mathrm{P}\ 7g$	1	−.36098	$3s^23p\ ^2\mathrm{P}\ 5g$	6	−.11109	$3s^23p\ ^2\mathrm{P}\ 9j$
4	−.14196	$3s^23p\ ^2\mathrm{P}\ 8g$	2	−.25078	$3s^23p\ ^2\mathrm{P}\ 6g$			$^1\mathbf{I}^o$
5	−.11208	$3s^23p\ ^2\mathrm{P}\ 9g$	3	−.18426	$3s^23p\ ^2\mathrm{P}\ 7g$	1	−.18375	$3s^23p\ ^2\mathrm{P}\ 7i$
		$^3\mathbf{G}^e$	4	−.18364	$3s^23p\ ^2\mathrm{P}\ 7i$	2	−.14072	$3s^23p\ ^2\mathrm{P}\ 8i$
1	−.61189	$3s^23p\ ^2\mathrm{P}\ 4f$	5	−.14105	$3s^23p\ ^2\mathrm{P}\ 8g$	3	−.11120	$3s^23p\ ^2\mathrm{P}\ 9i$
2	−.51984	$3s3p^2\ ^2\mathrm{D}\ 3d$	6	−.14059	$3s^23p\ ^2\mathrm{P}\ 8i$			$^3\mathbf{I}^e$
3	−.36172	$3s^23p\ ^2\mathrm{P}\ 5f$	7	−.11142	$3s^23p\ ^2\mathrm{P}\ 9g$	1	−.25006	$3s^23p\ ^2\mathrm{P}\ 6h$
4	−.25284	$3s^23p\ ^2\mathrm{P}\ 6f$	8	−.11109	$3s^23p\ ^2\mathrm{P}\ 9i$	2	−.18375	$3s^23p\ ^2\mathrm{P}\ 7h$
5	−.24989	$3s^23p\ ^2\mathrm{P}\ 6h$			$^3\mathbf{H}^e$	3	−.14071	$3s^23p\ ^2\mathrm{P}\ 8h$
6	−.18593	$3s^23p\ ^2\mathrm{P}\ 7f$	1	−.25077	$3s^23p\ ^2\mathrm{P}\ 6h$	4	−.14061	$3s^23p\ ^2\mathrm{P}\ 8j$
7	−.18361	$3s^23p\ ^2\mathrm{P}\ 7h$	2	−.18434	$3s^23p\ ^2\mathrm{P}\ 7h$	5	−.11118	$3s^23p\ ^2\mathrm{P}\ 9h$
8	−.14231	$3s^23p\ ^2\mathrm{P}\ 8f$	3	−.14112	$3s^23p\ ^2\mathrm{P}\ 8h$	6	−.11109	$3s^23p\ ^2\mathrm{P}\ 9j$
9	−.14059	$3s^23p\ ^2\mathrm{P}\ 8h$	4	−.11148	$3s^23p\ ^2\mathrm{P}\ 9h$			$^3\mathbf{I}^o$
10	−.11237	$3s^23p\ ^2\mathrm{P}\ 9f$			$^3\mathbf{H}^o$	1	−.18375	$3s^23p\ ^2\mathrm{P}\ 7i$
11	−.11109	$3s^23p\ ^2\mathrm{P}\ 9h$	1	−.36104	$3s^23p\ ^2\mathrm{P}\ 5g$	2	−.14072	$3s^23p\ ^2\mathrm{P}\ 8i$
		$^3\mathbf{G}^o$	2	−.25084	$3s^23p\ ^2\mathrm{P}\ 6g$	3	−.11120	$3s^23p\ ^2\mathrm{P}\ 9i$
1	−.36492	$3s^23p\ ^2\mathrm{P}\ 5g$	3	−.18430	$3s^23p\ ^2\mathrm{P}\ 7g$			
2	−.25295	$3s^23p\ ^2\mathrm{P}\ 6g$	4	−.18364	$3s^23p\ ^2\mathrm{P}\ 7i$			
3	−.18560	$3s^23p\ ^2\mathrm{P}\ 7g$	5	−.14108	$3s^23p\ ^2\mathrm{P}\ 8g$			
4	−.14196	$3s^23p\ ^2\mathrm{P}\ 8g$	6	−.14059	$3s^23p\ ^2\mathrm{P}\ 8i$			
5	−.11208	$3s^23p\ ^2\mathrm{P}\ 9g$	7	−.11145	$3s^23p\ ^2\mathrm{P}\ 9g$			
		$^1\mathbf{H}^e$	8	−.11109	$3s^23p\ ^2\mathrm{P}\ 9i$			
1	−.25077	$3s^23p\ ^2\mathrm{P}\ 6h$			$^1\mathbf{I}^e$			
2	−.18434	$3s^23p\ ^2\mathrm{P}\ 7h$	1	−.25006	$3s^23p\ ^2\mathrm{P}\ 6h$			
3	−.14112	$3s^23p\ ^2\mathrm{P}\ 8h$	2	−.18374	$3s^23p\ ^2\mathrm{P}\ 7h$			
4	−.11148	$3s^23p\ ^2\mathrm{P}\ 9h$	3	−.14071	$3s^23p\ ^2\mathrm{P}\ 8h$			
			4	−.14071	$3s^23p\ ^2\mathrm{P}\ 8j$			

Energies in ascending order from ground state for terms with effective $n \leq 4.0$, $L \leq 4$

Term	i	E(Ryds)	Term	i	E(Ryds)	Term	i	E(Ryds)	Term	i	E(Ryds)	Term	i	E(Ryds)
$^3\mathbf{P}^e$	1	0.00000	$^3\mathbf{P}^o$	2	1.31266	$^3\mathbf{S}^e$	1	1.58777	$^1\mathbf{D}^o$	3	1.88581	$^3\mathbf{F}^e$	2	1.97738
$^1\mathbf{D}^e$	1	0.10485	$^3\mathbf{P}^o$	3	1.35535	$^1\mathbf{D}^e$	2	1.61632	$^3\mathbf{D}^o$	3	1.90307	$^3\mathbf{P}^e$	4	1.98573
$^1\mathbf{S}^e$	1	0.26539	$^1\mathbf{P}^o$	2	1.36125	$^3\mathbf{P}^e$	3	1.64667	$^3\mathbf{P}^o$	4	1.91076	$^1\mathbf{G}^e$	1	1.99014
$^5\mathbf{S}^o$	1	0.50352	$^3\mathbf{D}^o$	2	1.36491	$^5\mathbf{F}^e$	1	1.66258	$^3\mathbf{P}^o$	5	1.92942	$^3\mathbf{D}^e$	2	1.99910
$^3\mathbf{D}^o$	1	0.75030	$^1\mathbf{D}^o$	2	1.41821	$^1\mathbf{S}^e$	2	1.67375	$^1\mathbf{P}^o$	4	1.93689	$^1\mathbf{D}^e$	4	2.00597
$^3\mathbf{P}^o$	1	0.89962	$^1\mathbf{F}^o$	1	1.45819	$^5\mathbf{D}^e$	1	1.71090	$^1\mathbf{F}^o$	2	1.94307	$^1\mathbf{P}^e$	2	2.00704
$^1\mathbf{D}^o$	1	0.94503	$^1\mathbf{P}^e$	1	1.52850	$^1\mathbf{D}^e$	3	1.72267	$^5\mathbf{P}^e$	2	1.95168	$^1\mathbf{S}^e$	3	2.00779
$^3\mathbf{F}^o$	1	1.12006	$^1\mathbf{P}^o$	3	1.54309	$^3\mathbf{F}^e$	1	1.78744	$^1\mathbf{P}^o$	5	1.95546			
$^1\mathbf{P}^o$	1	1.28108	$^3\mathbf{D}^e$	1	1.55349	$^5\mathbf{P}^e$	1	1.88072	$^3\mathbf{G}^e$	1	1.96260			
$^3\mathbf{S}^o$	1	1.29317	$^3\mathbf{P}^e$	2	1.57757	$^3\mathbf{F}^o$	2	1.88298	$^1\mathbf{F}^e$	1	1.96993			

Si-like S (S^{2+})

gf-values for transitions involving terms with effective $n \leq 4.0$, $L \leq 4$

$i\ i'$	gf_L	$i\ i'$	gf_L	$i\ i'$	gf_L	$i\ i'$	gf_L	$i\ i'$	gf_L	$i\ i'$	gf_L
	^{1}S^e–^{1}P^o	2 1	3.78E−2	3 1	3.13E−1	1 4	−2.97E+0	1 3	−5.37E−1	2 1	2.95E−2
1 1	−2.01E−3	2 2	3.16E−3	3 2	1.83E−2	1 5	−9.82E−1	2 1	2.89E−1	2 2	6.18E−1
1 2	−6.52E−2	2 3	1.11E+0	3 3	−5.41E−3		**^{3}S^o–^{3}P^e**	2 2	1.01E+0		**^{3}D^o–^{3}F^e**
1 3	−2.71E+0		**^{1}P^o–^{1}D^e**	4 1	2.31E−1	1 1	3.23E+0	2 3	−7.08E+0	1 1	−7.77E−2
1 4	−5.82E−2	1 1	1.90E+0	4 2	9.44E−2	1 2	−2.40E−4	3 1	6.78E−1	1 2	−1.45E+0
1 5	−1.70E−1	1 2	−1.63E−1	4 3	3.14E−1	1 3	−3.97E−2	3 2	1.68E−2	2 1	−1.76E−2
2 1	2.01E−1	1 3	−3.53E−2		**^{1}D^e–^{1}F^o**	1 4	−2.33E−1	3 3	−1.92E−1	2 2	−4.79E+0
2 2	2.40E−1	1 4	−3.39E−1	1 1	−6.81E+0		**^{3}P^e–^{3}P^o**	4 1	4.23E+0	3 1	2.27E−2
2 3	1.82E−1	2 1	4.72E−1	1 2	−4.49E−1	1 1	−3.87E−1	4 2	2.68E−1	3 2	−4.83E+0
2 4	−2.89E−1	2 2	−1.66E+0	2 1	7.20E−1	1 2	−7.00E+0	4 3	6.73E−1		**^{3}F^e–^{3}F^o**
2 5	−1.03E+0	2 3	−1.55E−2	2 2	−4.45E+0	1 3	−8.03E−1		**^{3}P^o–^{3}D^e**	1 1	8.71E−2
3 1	6.63E−2	2 4	−1.31E−1	3 1	1.25E−2	1 4	−1.99E−1	1 1	−3.83E−2	1 2	−8.85E−4
3 2	6.69E−4	3 1	1.20E−1	3 2	−4.56E−2	1 5	−1.92E−1	1 2	−7.07E−1	2 1	4.77E−1
3 3	0.00E+0	3 2	−2.20E−2	4 1	1.26E−2	2 1	7.96E−2	2 1	−2.32E+0	2 2	8.17E−1
3 4	4.03E−2	3 3	−3.80E−5	4 2	2.07E−2	2 2	2.85E+0	2 2	−1.05E+0		**^{3}F^o–^{3}G^e**
3 5	3.76E−2	3 4	−1.27E+0		**^{1}D^o–^{1}F^e**	2 3	6.10E−1	3 1	−2.27E+0	1 1	−1.79E+0
	^{1}P^e–^{1}P^o	4 1	1.31E−1	1 1	−6.95E−1	2 4	−2.96E+0	3 2	−1.72E+0	2 1	−4.76E+0
1 1	3.70E−3	4 2	1.13E+0	2 1	−7.83E−1	2 5	−1.33E+0	4 1	2.23E−2		**^{3}S^o–^{3}P^e**
1 2	8.25E−1	4 3	1.68E−2	3 1	−1.69E+0	3 1	1.89E−3	4 2	−3.94E+0	1 1	0.00E+0
1 3	−4.34E−3	4 4	−1.06E−2		**^{1}F^e–^{1}F^o**	3 2	1.31E−1	5 1	2.88E+0		**^{5}S^o–^{5}P^e**
1 4	−4.33E−1	5 1	3.80E−3	1 1	2.39E−1	3 3	1.29E−2	5 2	−6.87E−1	1 1	−1.42E+1
1 5	−7.29E−1	5 2	4.34E−2	1 2	6.71E−2	3 4	−8.34E−2		**^{3}D^e–^{3}D^o**	1 2	−1.97E−2
2 1	7.10E−3	5 3	8.59E−3		**^{1}F^o–^{1}G^e**	3 5	−4.54E−2	1 1	1.27E−1		**^{5}S^e–^{5}S^o**
2 2	9.49E−2	5 4	−8.47E−1	1 1	−2.86E+0	4 1	7.52E−2	1 2	3.45E−1	1 1	0.00E+0
2 3	2.76E−4		**^{1}D^e–^{1}D^o**	2 1	−1.28E+0	4 2	2.24E−1	1 3	−2.49E+0		**^{5}S^e–^{5}P^o**
2 4	1.18E+0	1 1	−1.08E−1		**^{1}S^o–^{1}P^e**	4 3	5.77E−6	2 1	3.69E−1	1 1	−2.71E+2
2 5	3.02E−1	1 2	−5.09E+0	1 1	−2.29E−1	4 4	3.65E−1	2 2	5.55E−1		
	^{1}P^e–^{1}D^o	1 3	−1.26E−1		**^{3}S^e–^{3}P^o**	4 5	7.39E−1	2 3	3.74E−1		
1 1	3.55E−1	2 1	1.31E−1	1 1	2.29E−1		**^{3}P^e–^{3}D^o**		**^{3}D^e–^{3}F^o**		
1 2	1.08E−1	2 2	7.41E−2	1 2	6.52E−2	1 1	−2.20E−1	1 1	2.96E+0		
1 3	−2.49E+0	2 3	−7.51E−1	1 3	1.40E+0	1 2	−1.50E+1	1 2	−1.42E+1		

Si-like Ar (Ar^{4+})

Term energies relative to $3s^23p\ ^2$P ionization threshold for each symmetry

i	E(Ryds)	Description
	$^1S^e$	
1	-5.12761	$3s^23p^2$
2	-2.64105	$3p^4$
3	-2.18165	$3s^23p$ ^{2}P $4p$
4	-1.83410	$3s3p^2$ ^{2}D $3d$
5	-1.29616	$3s^23p$ ^{2}P $5p$
6	-1.07322	$3s3p^2$ ^{2}S $4s$
7	-1.00658	$3s^23d$ ^{2}D $3d$:
8	-.88093	$3s^23p$ ^{2}P $6p$
9	-.63138	$3s^23p$ ^{2}P $7p$
10	-.54860	$3s3p^2$ ^{2}D $4d$
11	-.46945	$3s^23p$ ^{2}P $8p$
12	-.36354	$3s^23p$ ^{2}P $9p$
	$^1S^o$	
1	-1.60403	$3p^3$ ^{2}D $3d$
2	-.79342	$3s3p^2$ ^{2}P $4p$
3	-.19833	$3s3p3d^2$
	$^3S^e$	
1	-2.37763	$3s^23p$ ^{2}P $4p$
2	-2.18470	$3s3p^2$ ^{2}D $3d$
3	-1.39478	$3s^23p$ ^{2}P $5p$
4	-1.17712	$3s3p^2$ ^{2}S $4s$
5	-.91699	$3s^23p$ ^{2}P $6p$
6	-.64844	$3s^23p$ ^{2}P $7p$
7	-.60551	$3s3p^2$ ^{2}D $4d$
8	-.47905	$3s^23p$ ^{2}P $8p$
9	-.36912	$3s^23p$ ^{2}P $9p$
	$^3S^o$	
1	-3.72134	$3s3p^3$
2	-1.57250	$3s3p^2$ ^{4}P $4p$
3	-1.12629	$3p^3$ ^{2}D $3d$
4	-.62706	$3s3p^2$ ^{2}P $4p$
5	-.58360	$3s3p^2$ ^{4}P $5p$
6	-.54356	$3s3p3d$ ^{4}F $4s$
7	-.20654	$3s3p3d$ ^{4}F $5s$
8	-.04467	$3s3p^2$ ^{4}P $6p$
	$^5S^o$	
1	-4.77332	$3s3p^3$
2	-1.39171	$3s3p^2$ ^{4}P $4p$
3	-.60946	$3p^3$ ^{4}S $3d$
4	-.50337	$3s3p^2$ ^{4}P $5p$
5	-.31975	$3s3p3d$ ^{4}F $4s$
6	-.04321	$3s3p^2$ ^{4}P $6p$
	$^1P^e$	
1	-2.46745	$3s^23p$ ^{2}P $4p$
2	-2.23654	$3s3p^2$ ^{2}D $3d$
3	-1.69697	$3s3p^2$ ^{2}P $3d$
4	-1.44455	$3s^23p$ ^{2}P $5p$
5	-1.05872	$3s3p^2$ ^{2}P $4s$
6	-.92144	$3s^23p$ ^{2}P $6p$
7	-.65447	$3s^23p$ ^{2}P $7p$
8	-.61914	$3s3p^2$ ^{2}D $4d$
9	-.48390	$3s^23p$ ^{2}P $8p$
10	-.37223	$3s^23p$ ^{2}P $9p$
	$^1P^o$	
1	-3.68596	$3s3p^3$
2	-3.14915	$3s^23p$ ^{2}P $3d$
3	-2.70476	$3s^23p$ ^{2}P $4s$
4	-1.78545	$3s^23p$ ^{2}P $4d$
5	-1.53564	$3s^23p$ ^{2}P $5s$
6	-1.15468	$3s3p^2$ ^{2}D $4p$
7	-1.09053	$3s^23p$ ^{2}P $5d$
8	-1.03480	$3p^3$ ^{2}D $3d$
9	-.98644	$3s^23p$ ^{2}P $6s$
10	-.77168	$3s^23p$ ^{2}P $6d$
11	-.73984	$3s3p^2$ ^{2}S $4p$
12	-.68619	$3s^23p$ ^{2}P $7s$
13	-.55950	$3s^23p$ ^{2}P $7d$
14	-.53130	$3s3p^2$ ^{2}P $4p$
15	-.50088	$3s^23p$ ^{2}P $8s$
16	-.46472	$3s3p^2$ ^{2}D $4f$
17	-.42065	$3s^23p$ ^{2}P $8d$
18	-.38598	$3s^23p$ ^{2}P $9s$
19	-.33231	$3s^23p$ ^{2}P $9d$
	$^3P^e$	
1	-5.49937	$3s^23p^2$
2	-3.12575	$3s3p^2$ ^{4}P $3d$
3	-2.58677	$3s3p^2$ ^{2}D $3d$
4	-2.34290	$3s^23p$ ^{2}P $4p$
5	-2.23647	$3s3p^2$ ^{2}P $3d$
6	-1.76705	$3s3p^2$ ^{4}P $4s$
7	-1.69694	$3p^4$
8	-1.38043	$3s^23p$ ^{2}P $5p$
9	-1.33282	$3s^23d^2$
10	-1.06376	$3s3p^2$ ^{2}P $4s$
11	-.97818	$3s3p^2$ ^{4}P $4d$
12	-.91557	$3s^23p$ ^{2}P $6p$
13	-.65766	$3s^23p$ ^{2}P $7p$
14	-.63992	$3s3p^2$ ^{2}D $4d$
15	-.60698	$3s3p^2$ ^{4}P $5s$
16	-.47959	$3s^23p$ ^{2}P $8p$
17	-.36980	$3s^23p$ ^{2}P $9p$
	$^3P^o$	
1	-4.22300	$3s3p^3$
2	-3.49194	$3s^23p$ ^{2}P $3d$
3	-2.75054	$3s^23p$ ^{2}P $4s$
4	-1.82653	$3s^23p$ ^{2}P $4d$
5	-1.55324	$3s^23p$ ^{2}P $5s$
6	-1.39262	$3s3p^2$ ^{4}P $4p$
7	-1.13975	$3s^23p$ ^{2}P $5d$
8	-1.10099	$3s3p^2$ ^{2}D $4p$
9	-1.07907	$3p^3$ ^{2}D $3d$
10	-1.00517	$3p^3$ ^{4}S $3d$
11	-.99323	$3s^23p$ ^{2}P $6s$
12	-.78516	$3s^23p$ ^{2}P $6d$
13	-.75614	$3s3p^2$ ^{2}S $4p$
14	-.68998	$3s^23p$ ^{2}P $7s$
15	-.66175	$3s3p^2$ ^{2}P $4p$
16	-.55250	$3s^23p$ ^{2}P $7d$
17	-.50628	$3s^23p$ ^{2}P $8s$
18	-.46166	$3s3p^2$ ^{4}P $5p$
19	-.42549	$3p^3$ ^{2}P $3d$
20	-.41515	$3s^23p$ ^{2}P $8d$
21	-.38783	$3s^23p$ ^{2}P $9s$
22	-.36130	$3s3p^2$ ^{2}D $4f$
23	-.32957	$3s^23p$ ^{2}P $9d$
	$^5P^e$	
1	-2.70507	$3s3p^2$ ^{4}P $3d$
2	-1.92723	$3s3p^2$ ^{4}P $4s$
3	-1.00104	$3s3p^2$ ^{4}P $4d$
4	-.68504	$3s3p^2$ ^{4}P $5s$
5	-.27932	$3s3p^2$ ^{4}P $5d$
6	-.12083	$3s3p^2$ ^{4}P $6s$
	$^5P^o$	
1	-1.50200	$3s3p^2$ ^{4}P $4p$
2	-.94164	$3p^3$ ^{4}S $3d$
3	-.52438	$3s3p^2$ ^{4}P $5p$
4	-.05330	$3s3p^2$ ^{4}P $6p$
	$^1D^e$	
1	-5.35498	$3s^23p^2$
2	-3.04032	$3p^4$
3	-2.28122	$3s^23p$ ^{2}P $4p$
4	-2.08082	$3s3p^2$ ^{2}D $3d$
5	-1.87037	$3s3p^2$ ^{2}S $3d$
6	-1.58460	$3s^23p$ ^{2}P $4f$
7	-1.46983	$3s3p^2$ ^{2}D $4s$
8	-1.39985	$3s3p^2$ ^{2}P $3d$
9	-1.32730	$3s^23p$ ^{2}P $5p$
10	-1.29584	$3s^23d^2$
11	-.98681	$3s^23p$ ^{2}P $5f$
12	-.90942	$3s^23p$ ^{2}P $6p$
13	-.69260	$3s^23p$ ^{2}P $6f$
14	-.65951	$3s^23p$ ^{2}P $7p$
15	-.64025	$3s3p^2$ ^{2}D $4d$
16	-.61190	$3s^23d$ ^{2}D $4s$
17	-.50924	$3s^23p$ ^{2}P $7f$
18	-.47551	$3s^23p$ ^{2}P $8p$
19	-.39065	$3s^23p$ ^{2}P $8f$
20	-.36809	$3s^23p$ ^{2}P $9p$
21	-.33618	$3s3p^2$ ^{2}D $5s$
22	-.30894	$3s^23p$ ^{2}P $9f$
	$^1D^o$	
1	-4.10887	$3s3p^3$
2	-3.43888	$3s^23p$ ^{2}P $3d$
3	-1.88513	$3s^23p$ ^{2}P $4d$
4	-1.28174	$3s3p^2$ ^{2}D $4p$
5	-1.18706	$3p^3$ ^{2}D $3d$
6	-1.14068	$3s^23p$ ^{2}P $5d$
7	-.78796	$3p^3$ ^{2}D $4s$
8	-.78327	$3s^23p$ ^{2}P $6d$
9	-.61148	$3s3p^2$ ^{2}P $4p$
10	-.55208	$3s^23p$ ^{2}P $7d$
11	-.42441	$3s^23p$ ^{2}P $8d$
12	-.41844	$3s3p^2$ ^{2}D $4f$
13	-.38196	$3s3p^2$ ^{2}D $5p$
14	-.35649	$3s^23d$ ^{2}D $4p$
15	-.33216	$3s^23p$ ^{2}P $9d$
	$^3D^e$	
1	-2.44157	$3s^23p$ ^{2}P $4p$
2	-2.41024	$3s3p^2$ ^{4}P $3d$
3	-2.28645	$3s3p^2$ ^{2}D $3d$
4	-2.08074	$3s3p^2$ ^{2}S $3d$
5	-1.94381	$3s^23p$ ^{2}P $3d$
6	-1.59562	$3s3p^2$ ^{2}D $4s$
7	-1.55004	$3s^23p$ ^{2}P $4f$
8	-1.40631	$3s^23p$ ^{2}P $5p$
9	-1.01135	$3s^23p$ ^{2}P $5f$
10	-.93397	$3s^23p$ ^{2}P $6p$
11	-.89733	$3s3p^2$ ^{4}P $4d$
12	-.70502	$3s^23p$ ^{2}P $6f$
13	-.69606	$3s^23d$ ^{2}D $4s$
14	-.68410	$3s3p^2$ ^{2}D $4d$
15	-.64510	$3s^23p$ ^{2}P $7p$
16	-.51601	$3s^23p$ ^{2}P $7f$
17	-.48178	$3s^23p$ ^{2}P $8p$
18	-.39510	$3s^23p$ ^{2}P $8f$
19	-.37558	$3s^23p$ ^{2}P $9p$
20	-.36546	$3s3p^2$ ^{2}D $5s$
21	-.31217	$3s^23p$ ^{2}P $9f$
	$^3D^o$	
1	-4.42404	$3s3p^3$
2	-3.42974	$3s^23p$ ^{2}P $3d$
3	-1.86751	$3s^23p$ ^{2}P $4d$
4	-1.51432	$3p^3$ ^{2}D $3d$
5	-1.44770	$3s3p^2$ ^{4}P $4p$
6	-1.15632	$3s^23p$ ^{2}P $5d$
7	-1.14942	$3s3p^2$ ^{2}D $4p$
8	-1.08646	$3p^3$ ^{4}S $3d$
9	-.83133	$3s3p^2$ ^{4}P $4f$
10	-.79036	$3s^23p$ ^{2}P $6d$
11	-.77371	$3p^3$ ^{2}P $3d$
12	-.70061	$3s3p^2$ ^{2}P $4p$
13	-.58335	$3p^3$ ^{2}D $4s$
14	-.55880	$3s^23p$ ^{2}P $7d$

Si-like Ar (Ar^{4+})

i	E(Ryds)	Description
15	−.51590	$3s3p^2$ ^{4}P $5p$
16	−.42810	$3s^23p$ ^{2}P $8d$
17	−.40451	$3p^3$ ^{4}S $4s$
18	−.37782	$3s3p^2$ ^{2}D $4f$
19	−.35619	$3s^23d$ ^{2}D $4p$
20	−.32594	$3s^23p$ ^{2}P $9d$
21	−.28909	$3s3p^2$ ^{2}D $5p$
		^{5}D^e
1	−2.96899	$3s3p^2$ ^{4}P $3d$
2	−.96352	$3s3p^2$ ^{4}P $4d$
3	−.25751	$3s3p^2$ ^{4}P $5d$
		^{5}D^o
1	−1.54294	$3s3p^2$ ^{4}P $4p$
2	−1.42524	$3p^3$ ^{4}S $3d$
3	−1.03375	$3s3p3d^2$:
4	−.84912	$3s3p3d^2$:
5	−.68708	$3s3p^2$ ^{4}P $4f$
6	−.54505	$3s3p^2$ ^{4}P $5p$
7	−.15131	$3s3p^2$ ^{4}P $5f$
8	−.11456	$3p^3$ ^{4}S $4s$
9	−.05159	$3s3p^2$ ^{4}P $6p$
		^{1}F^e
1	−2.39084	$3s3p^2$ ^{2}D $3d$
2	−1.75473	$3s3p^2$ ^{2}P $3d$
3	−1.50264	$3s^23p$ ^{2}P $4f$
4	−1.02299	$3s^23p$ ^{2}P $5f$
5	−.71479	$3s^23p$ ^{2}P $6f$
6	−.68870	$3s3p^2$ ^{2}D $4d$
7	−.52228	$3s^23p$ ^{2}P $7f$
8	−.39948	$3s^23p$ ^{2}P $8f$
9	−.31544	$3s^23p$ ^{2}P $9f$
		^{1}F^o
1	−3.23010	$3s^23p$ ^{2}P $3d$
2	−1.80558	$3s^23p$ ^{2}P $4d$
3	−1.16300	$3s3p^2$ ^{2}D $4p$
4	−1.12225	$3s^23p$ ^{2}P $5d$
5	−1.00116	$3s^23p$ ^{2}P $5g$
6	−.86715	$3p^3$ ^{2}D $3d$
7	−.79659	$3p^3$ ^{2}P $3d$:
8	−.76921	$3s^23p$ ^{2}P $6d$
9	−.69564	$3s^23p$ ^{2}P $6g$
10	−.56080	$3s^23p$ ^{2}P $7d$
11	−.51808	$3s^23p$ ^{2}P $7g$
12	−.50482	$3s3p^2$ ^{2}D $4f$
13	−.42550	$3s^23p$ ^{2}P $8d$
14	−.39201	$3s^23p$ ^{2}P $8g$
15	−.33833	$3s^23p$ ^{2}P $9d$
16	−.31210	$3s^23p$ ^{2}P $9g$
17	−.30177	$3s3p^2$ ^{2}D $5p$
		^{3}F^e
1	−2.85069	$3s3p^2$ ^{4}P $3d$
2	−2.31075	$3s3p^2$ ^{2}D $3d$

i	E(Ryds)	Description
3	−2.10394	$3s3p^2$ ^{2}P $3d$
4	−1.67519	$3s^23p$ ^{2}P $4f$
5	−1.47474	$3s^23d^2$
6	−1.02188	$3s^23p$ ^{2}P $5f$
7	−.95037	$3s3p^2$ ^{4}P $4d$
8	−.70785	$3s^23p$ ^{2}P $6f$
9	−.68931	$3s3p^2$ ^{2}D $4d$
10	−.51988	$3s^23p$ ^{2}P $7f$
11	−.39767	$3s^23p$ ^{2}P $8f$
12	−.31403	$3s^23p$ ^{2}P $9f$
13	−.26365	$3s3p^2$ ^{4}P $5d$
		^{3}F^o
1	−3.79250	$3s^23p$ ^{2}P $3d$
2	−1.86025	$3s^23p$ ^{2}P $4d$
3	−1.53767	$3p^3$ ^{2}D $3d$
4	−1.21632	$3s3p^2$ ^{2}D $4p$
5	−1.12737	$3s^23p$ ^{2}P $5d$
6	−1.07546	$3p^3$ ^{4}S $3d$
7	−1.00045	$3s^23p$ ^{2}P $5g$
8	−.78779	$3p^3$ ^{2}P $3d$
9	−.76629	$3s^23p$ ^{2}P $6d$
10	−.69718	$3s^23p$ ^{2}P $6g$
11	−.67424	$3s3p^2$ ^{4}P $4f$
12	−.55998	$3s^23p$ ^{2}P $7d$
13	−.54371	$3p^3$ ^{2}P $4s$
14	−.51229	$3s^23p$ ^{2}P $7g$
15	−.42457	$3s^23p$ ^{2}P $8d$
16	−.39413	$3s^23p$ ^{2}P $8g$
17	−.38641	$3s3p^2$ ^{2}D $4f$
18	−.33948	$3s^23d$ ^{2}D $4p$
19	−.32683	$3s^23p$ ^{2}P $9d$
20	−.31012	$3s^23p$ ^{2}P $9g$
21	−.28765	$3s3p^2$ ^{2}D $5p$
		^{5}F^e
1	−3.03567	$3s3p^2$ ^{4}P $3d$
2	−1.00294	$3s3p^2$ ^{4}P $4d$
3	−.27389	$3s3p^2$ ^{4}P $5d$
4	−.13650	$3s3p^2$ ^{4}P $5g$
		^{5}F^o
1	−.95377	$3p^3$ ^{4}S $3d$
2	−.71889	$3s3p^2$ ^{4}P $4f$
3	−.16486	$3p^3$ ^{2}P $4s$
4	−.14733	$3s3p^2$ ^{4}P $5f$
		^{1}G^e
1	−2.46383	$3s3p^2$ ^{2}D $3d$
2	−1.65394	$3s^23p$ ^{2}P $4f$
3	−1.35385	$3s^23d^2$
4	−.98302	$3s^23p$ ^{2}P $5f$
5	−.69448	$3s^23p$ ^{2}P $6f$
6	−.69353	$3s^23p$ ^{2}P $6h$
7	−.62728	$3s3p^2$ ^{2}D $4d$
8	−.50991	$3s^23p$ ^{2}P $7h$

i	E(Ryds)	Description
9	−.50860	$3s^23p$ ^{2}P $7f$
10	−.39063	$3s^23p$ ^{2}P $8f$
11	−.39035	$3s^23p$ ^{2}P $8h$
12	−.30897	$3s^23p$ ^{2}P $9f$
13	−.30850	$3s^23p$ ^{2}P $9h$
		^{1}G^o
1	−1.26033	$3p^3$ ^{2}D $3d$
2	−1.01429	$3s^23p$ ^{2}P $5g$
3	−.70439	$3s^23p$ ^{2}P $6g$
4	−.51755	$3s^23p$ ^{2}P $7g$
5	−.42297	$3s3p^2$ ^{2}D $4f$
6	−.39496	$3s^23p$ ^{2}P $8g$
7	−.35842	$3s3p^2$ ^{2}P $4f$
8	−.31216	$3s^23p$ ^{2}P $9g$
		^{3}G^e
1	−2.52555	$3s3p^2$ ^{2}D $3d$
2	−1.60221	$3s^23p$ ^{2}P $4f$
3	−1.02660	$3s^23p$ ^{2}P $5f$
4	−.71340	$3s^23p$ ^{2}P $6f$
5	−.69409	$3s^23p$ ^{2}P $6h$
6	−.67225	$3s3p^2$ ^{2}D $4d$
7	−.52086	$3s^23p$ ^{2}P $7f$
8	−.50992	$3s^23p$ ^{2}P $7h$
9	−.39808	$3s^23p$ ^{2}P $8f$
10	−.39046	$3s^23p$ ^{2}P $8h$
11	−.31400	$3s^23p$ ^{2}P $9f$
12	−.30855	$3s^23p$ ^{2}P $9h$
		^{3}G^o
1	−1.43342	$3p^3$ ^{2}D $3d$
2	−1.01446	$3s^23p$ ^{2}P $5g$
3	−.78439	$3s3p^2$ ^{4}P $4f$
4	−.70451	$3s^23p$ ^{2}P $6g$
5	−.60690	$3p^3$ ^{4}S $3d$
6	−.51838	$3s^23p$ ^{2}P $7g$
7	−.48017	$3p^3$ ^{2}P $3d$
8	−.42373	$3s3p^2$ ^{2}D $4f$
9	−.39301	$3s^23p$ ^{2}P $8g$
10	−.31218	$3s^23p$ ^{2}P $9g$
		^{5}G^e
1	−.12275	$3s3p^2$ ^{4}P $5g$
		^{5}G^o
1	−1.25074	$3p^3$ ^{4}S $3d$
2	−.75540	$3s3p^2$ ^{4}P $4f$
3	−.16349	$3s3p^2$ ^{4}P $5f$
		^{1}H^e
1	−.69607	$3s^23p$ ^{2}P $6h$
2	−.51173	$3s^23p$ ^{2}P $7h$
3	−.39183	$3s^23p$ ^{2}P $8h$
4	−.30957	$3s^23p$ ^{2}P $9h$
		^{1}H^o
1	−1.00338	$3s^23p$ ^{2}P $5g$
2	−.69805	$3s^23p$ ^{2}P $6g$

i	E(Ryds)	Description
3	−.51400	$3s^23p$ ^{2}P $7g$
4	−.40917	$3s3p^2$ ^{2}D $4f$
5	−.38744	$3s^23p$ ^{2}P $8g$
6	−.31025	$3s^23p$ ^{2}P $9g$
		^{3}H^e
1	−.69606	$3s^23p$ ^{2}P $6h$
2	−.51173	$3s^23p$ ^{2}P $7h$
3	−.39183	$3s^23p$ ^{2}P $8h$
4	−.30956	$3s^23p$ ^{2}P $9h$
		^{3}H^o
1	−1.00408	$3s^23p$ ^{2}P $5g$
2	−.80494	$3p^3$ ^{2}P $3d$
3	−.69846	$3s^23p$ ^{2}P $6g$
4	−.51415	$3s^23p$ ^{2}P $7g$
5	−.40068	$3s^23p$ ^{2}P $8g$
6	−.37658	$3s3p^2$ ^{2}D $4f$
7	−.30877	$3s^23p$ ^{2}P $9g$
		^{5}H^e
1	−.13470	$3s3p^2$ ^{4}P $5g$
		^{1}I^e
1	−.69435	$3s^23p$ ^{2}P $6h$
2	−.51017	$3s^23p$ ^{2}P $7h$
3	−.39064	$3s^23p$ ^{2}P $8h$
4	−.30867	$3s^23p$ ^{2}P $9h$
		^{3}I^e
1	−.69437	$3s^23p$ ^{2}P $6h$
2	−.51020	$3s^23p$ ^{2}P $7h$
3	−.39066	$3s^23p$ ^{2}P $8h$
4	−.30869	$3s^23p$ ^{2}P $9h$

Si-like Ar (Ar^{4+})

Energies in ascending order from ground state for terms with effective $n \leq 4.0$, $L \leq 4$

Term	i	E(Ryds)	Term	i	E(Ryds)	Term	i	E(Ryds)	Term	i	E(Ryds)	Term	i	E(Ryds)
$^3P^e$	1	0.00000	$^1F^o$	1	2.26927	$^1P^e$	1	3.03192	$^1S^e$	3	3.31772	$^1P^o$	4	3.71392
$^1D^e$	1	0.14439	$^1P^o$	2	2.35022	$^1G^e$	1	3.03554	$^3F^e$	3	3.39543	$^3P^e$	6	3.73232
$^1S^e$	1	0.37176	$^3P^e$	2	2.37362	$^3D^e$	1	3.05780	$^1D^e$	4	3.41855	$^1F^e$	2	3.74464
$^5S^o$	1	0.72605	$^1D^e$	2	2.45905	$^3D^e$	2	3.08913	$^3D^e$	4	3.41863	$^1P^e$	3	3.80240
$^3D^o$	1	1.07533	$^5F^e$	1	2.46370	$^1F^e$	1	3.10853	$^3D^e$	5	3.55556	$^3P^e$	7	3.80243
$^3P^o$	1	1.27637	$^5D^e$	1	2.53038	$^3S^e$	1	3.12174	$^5P^e$	2	3.57214	$^3F^e$	4	3.82418
$^1D^o$	1	1.39050	$^3F^e$	1	2.64868	$^3P^e$	4	3.15647	$^1D^o$	3	3.61424	$^1G^e$	2	3.84543
$^3F^o$	1	1.70687	$^3P^o$	3	2.74883	$^3F^e$	2	3.18862	$^1D^e$	5	3.62900	$^1S^o$	1	3.89534
$^3S^o$	1	1.77803	$^5P^e$	1	2.79430	$^3D^e$	3	3.21292	$^3D^o$	3	3.63186	$^3G^e$	2	3.89716
$^1P^o$	1	1.81341	$^1P^o$	3	2.79461	$^1D^e$	3	3.21815	$^3F^o$	2	3.63912	$^3D^e$	6	3.90375
$^3P^o$	2	2.00743	$^1S^e$	2	2.85832	$^1P^e$	2	3.26283	$^1S^e$	4	3.66527	$^1D^e$	6	3.91477
$^1D^o$	2	2.06049	$^3P^e$	3	2.91260	$^3P^e$	5	3.26290	$^3P^o$	4	3.67284	$^3S^o$	2	3.92687
$^3D^o$	2	2.06963	$^3G^e$	1	2.97382	$^3S^e$	2	3.31467	$^1F^o$	2	3.69379			

Si-like Ar (Ar^{4+})

gf-values for transitions involving terms with effective $n \leq 4.0$, $L \leq 4$

i i'	gf_L	i i'	gf_L	i i'	gf_L	i i'	gf_L	i i'	gf_L	i i'	gf_L
	$^1S^e$–$^1P^o$	1 3	−2.71E−2	3 2	−3.90E+0	2 7	6.98E−4	6 2	7.25E−1	6 2	2.38E−4
1 1	−1.14E−1	1 4	−1.39E−1	4 1	5.29E−2		$^3P^e$–$^3P^o$	6 3	3.17E−2	6 3	2.39E−2
1 2	−2.27E+0	1 5	−4.91E−1	4 2	−5.45E−2	1 1	−5.47E−1	7 1	5.40E−2		$^3D^e$–$^3F^o$
1 3	−1.56E−1	1 6	−2.07E+0	5 1	2.68E−3	1 2	−5.23E+0	7 2	3.30E+0	1 1	2.10E+0
1 4	−6.43E−3	2 1	1.73E−2	5 2	−1.52E−4	1 3	−1.32E+0	7 3	2.19E−3	1 2	−9.16E+0
2 1	1.59E−1	2 2	−1.53E−3	6 1	3.64E−1	1 4	−2.90E−2		$^3P^o$–$^3D^e$	2 1	6.36E−1
2 2	5.68E−3	2 3	−7.26E−5	6 2	1.30E−2	2 1	1.06E−1	1 1	−3.31E−1	2 2	−3.67E+0
2 3	3.05E−4	2 4	−4.58E−2		$^1D^o$–$^1F^e$	2 2	1.06E−2	1 2	−1.03E−3	3 1	3.91E−3
2 4	−8.73E−3	2 5	−2.09E−1	1 1	−1.42E+0	2 3	−1.01E−3	1 3	−4.02E+0	3 2	−5.44E−1
3 1	1.23E−1	2 6	−1.87E−1	1 2	−4.40E+0	2 4	−7.38E−3	1 4	−1.25E+0	4 1	6.77E+0
3 2	1.55E−1	3 1	7.29E−1	2 1	−2.39E−1	3 1	9.52E−1	1 5	−6.71E+0	4 2	−2.21E−2
3 3	3.58E−1	3 2	6.02E−4	2 2	−2.54E−1	3 2	3.52E−2	1 6	−5.47E−1	5 1	2.70E−1
3 4	−9.82E−1	3 3	−1.61E+0	3 1	4.11E−2	3 3	6.36E−4	2 1	−6.05E−7	5 2	−1.18E−4
4 1	9.48E−1	3 4	−4.08E−2	3 2	−4.79E−1	3 4	−1.37E−3	2 2	−2.21E−2	6 1	1.62E−1
4 2	1.01E−1	3 5	−1.08E−2		$^1F^e$–$^1F^o$	4 1	3.09E−2	2 3	−3.99E−1	6 2	7.65E−2
4 3	8.16E−3	3 6	−1.84E−2	1 1	1.35E−2	4 2	5.22E−1	2 4	−2.33E−1		$^3D^o$–$^3F^e$
4 4	−2.64E−3	4 1	6.80E−5	1 2	−7.05E−3	4 3	2.49E+0	2 5	−2.90E+0	1 1	−1.27E−2
	$^1S^o$–$^1P^e$	4 2	5.33E−3	2 1	5.52E−1	4 4	−2.03E+0	2 6	−3.99E−2	1 2	−1.37E+1
1 1	8.38E−5	4 3	1.77E−1	2 2	2.31E−2	5 1	5.65E+0	3 1	−2.28E+0	1 3	−4.34E+0
1 2	4.75E−2	4 4	1.78E−2		$^1F^o$–$^1G^e$	5 2	3.07E−2	3 2	−1.06E+0	1 4	−3.90E−1
1 3	4.05E−5	4 5	3.27E−2	1 1	−3.59E−2	5 3	1.58E−1	3 3	−2.69E−1	2 1	−3.61E−2
	$^1P^e$–$^1P^o$	4 6	−8.43E−1	1 2	−1.15E−1	5 4	−7.27E−2	3 4	−3.26E−2	2 2	−1.07E−2
1 1	1.46E−3		$^1D^e$–$^1D^o$	2 1	2.19E−2	6 1	6.51E−1	3 5	−2.97E−3	2 3	−2.32E+0
1 2	6.46E−2	1 1	−3.24E−1	2 2	−1.34E+0	6 2	7.13E−2	3 6	−1.08E+0	2 4	−6.70E−2
1 3	5.53E−1	1 2	−4.18E+0		$^3S^e$–$^3P^o$	6 3	2.52E−1	4 1	1.47E−1	3 1	6.64E−3
1 4	−6.36E−1	1 3	−1.70E−2	1 1	3.18E−1	6 4	6.89E−3	4 2	3.62E−2	3 2	3.36E−4
2 1	5.62E−3	2 1	4.77E−1	1 2	4.37E−1	7 1	2.86E−3	4 3	4.26E−2	3 3	2.52E−1
2 2	9.95E−4	2 2	3.41E−2	1 3	8.36E−1	7 2	2.14E+0	4 4	1.01E−3	3 4	−2.78E+0
2 3	2.70E−3	2 3	−7.67E−5	1 4	−3.21E+0	7 3	1.33E−2	4 5	1.53E−1		$^3F^e$–$^3F^o$
2 4	−1.36E−3	3 1	2.69E−1	2 1	1.43E+0	7 4	1.14E−3	4 6	−7.26E−3	1 1	2.10E−1
3 1	2.74E+0	3 2	6.81E−2	2 2	9.07E−2		$^3P^e$–$^3D^o$		$^3D^e$–$^3D^o$	1 2	−2.95E−5
3 2	1.50E−1	3 3	−5.45E−1	2 3	4.85E−2	1 1	−3.78E−1	1 1	9.40E−1	2 1	8.01E−1
3 3	7.95E−3	4 1	2.76E+0	2 4	−8.50E−2	1 2	−1.27E+1	1 2	9.84E−2	2 2	−1.72E−3
3 4	5.98E−4	4 2	3.36E−1		$^3S^o$–$^3P^e$	1 3	−1.22E−1	1 3	−1.56E+0	3 1	6.71E+0
	$^1P^e$–$^1D^o$	4 3	−2.74E−3	1 1	2.75E+0	2 1	8.95E−1	2 1	7.07E+0	3 2	−3.02E−2
1 1	2.93E−1	5 1	3.53E−1	1 2	−1.43E−1	2 2	4.75E−3	2 2	4.34E−1	4 1	8.88E+0
1 2	1.83E−1	5 2	9.02E−1	1 3	−2.36E−1	2 3	−4.68E−3	2 3	−6.12E−1	4 2	3.26E−1
1 3	−2.33E+0	5 3	1.37E−3	1 4	−1.56E−3	3 1	4.49E+0	3 1	1.93E+0		$^3F^o$–$^3G^e$
2 1	1.72E+0	6 1	3.61E−2	1 5	−1.07E−1	3 2	6.29E−5	3 2	2.03E−2	1 1	−2.80E+0
2 2	1.42E−1	6 2	8.67E−1	1 6	−4.15E−1	3 3	−6.68E−3	3 3	−7.32E−2	1 2	−1.30E+1
2 3	−4.47E−3	6 3	2.22E−1	1 7	−8.66E+0	4 1	2.38E−1	4 1	6.55E−3	2 1	2.66E−1
3 1	2.68E−4		$^1D^e$–$^1F^o$	2 1	7.81E−2	4 2	1.09E+0	4 2	1.04E−3	2 2	−9.39E+0
3 2	4.90E−1	1 1	−6.06E+0	2 2	2.35E−1	4 3	−5.65E+0	4 3	−7.21E−3		$^5S^o$–$^5P^e$
3 3	5.60E−3	1 2	−3.56E−2	2 3	1.31E−1	5 1	7.59E−2	5 1	1.74E−1	1 1	−1.09E+1
	$^1P^o$–$^1D^e$	2 1	1.16E−5	2 4	5.79E−2	5 2	2.56E−2	5 2	1.32E+0	1 2	−1.15E+0
1 1	1.43E+0	2 2	−2.42E−3	2 5	4.25E−2	5 3	−2.22E−1	5 3	−2.18E−2		
1 2	−5.29E−2	3 1	7.17E−1	2 6	4.71E−1	6 1	1.11E+0	6 1	1.60E+0		

Si-like Ca (Ca^{6+})

Term energies relative to $3s^23p$ ^{2}P ionization threshold for each symmetry

i	E(Ryds)	Description	i	E(Ryds)	Description	i	E(Ryds)	Description	i	E(Ryds)	Description
		^{1}S^e			**^{5}S^e**	21	−1.07283	$3s^23p$ ^{2}P $7d$	13	−2.24776	$3s3p^2$ ^{2}S $4p$
1	−8.92697	$3s^23p^2$	1	−1.72178	$3p^23d^2$:	22	−.94440	$3s^23p$ ^{2}P $8s$	14	−2.22025	$3s^23p$ ^{2}P $5d$
2	−5.70774	$3p^4$	2	−.64051	$3s3p3d$ ^{4}P $4p$	23	−.92679	$3s3p^2$ ^{2}D $5p$	15	−2.10494	$3s3p^2$ ^{2}P $4p$
3	−4.63243	$3s3p^2$ ^{2}D $3d$			**^{5}S^o**	24	−.81485	$3s^23p$ ^{2}P $8d$	16	−1.96843	$3s3p3d^2$
4	−4.00684	$3s^23p$ ^{2}P $4p$	1	−8.40250	$3s3p^3$	25	−.72810	$3s^23p$ ^{2}P $9s$	17	−1.82666	$3s^23p$ ^{2}P $6s$
5	−3.65280	$3s^23d$ ^{2}D $3d$	2	−2.98801	$3s3p3d^2$	26	−.64196	$3s^23p$ ^{2}P $9d$	18	−1.66137	$3s^23d$ ^{2}D $4p$
6	−2.65513	$3s3p^2$ ^{2}S $4s$	3	−2.91071	$3s3p^2$ ^{4}P $4p$			**^{3}P^e**	19	−1.54830	$3s3p^2$ ^{2}D $4f$
7	−2.40010	$3s^23p$ ^{2}P $5p$	4	−1.75513	$3p^3$ ^{4}S $4s$	1	−9.36449	$3s^23p^2$	20	−1.48641	$3s^23p$ ^{2}P $6d$
8	−1.89528	$3s3p^2$ ^{2}D $4d$	5	−1.33846	$3s3p^2$ ^{4}P $5p$	2	−6.27605	$3p^4$	21	−1.37913	$3s3p^2$ ^{4}P $5p$
9	−1.65807	$3s^23p$ ^{2}P $6p$	6	−.57588	$3s3p^2$ ^{4}P $6p$	3	−5.59407	$3s3p^2$ ^{4}P $3d$	22	−1.28323	$3s^23p$ ^{2}P $7s$
10	−1.19815	$3s^23p$ ^{2}P $7p$	7	−.09253	$3s3p^2$ ^{4}P $7p$	4	−5.13092	$3s3p^2$ ^{2}D $3d$	23	−1.26140	$3p^3$ ^{2}P $4s$
11	−1.00289	$3p^3$ ^{2}P $4p$			**^{1}P^e**	5	−4.50202	$3s3p^2$ ^{2}P $3d$	24	−1.08438	$3s^23p$ ^{2}P $7d$
12	−.90063	$3s^23p$ ^{2}P $8p$	1	−5.16798	$3s3p^2$ ^{2}D $3d$	6	−4.22057	$3s^23p$ ^{2}P $4p$	25	−.97705	$3s3p^2$ ^{2}D $5p$
13	−.72651	$3s^23d$ ^{2}D $4d$	2	−4.53984	$3s3p^2$ ^{2}P $3d$	7	−4.02313	$3s^23d^2$	26	−.96188	$3s3p3d$ ^{4}P $4s$
14	−.69998	$3s^23p$ ^{2}P $9p$	3	−4.42013	$3s^23p$ ^{2}P $4p$	8	−3.55111	$3s3p^2$ ^{4}P $4s$	27	−.94196	$3s^23p$ ^{2}P $8s$
15	−.66383	$3s3p^2$ ^{2}S $5s$	4	−2.77601	$3s3p^2$ ^{2}P $4s$	9	−2.74167	$3s3p^2$ ^{2}P $4s$	28	−.81729	$3s^23p$ ^{2}P $8d$
16	−.60612	$3s3p^2$ ^{2}D $5d$	5	−2.52659	$3s^23p$ ^{2}P $5p$	10	−2.49617	$3s^23p$ ^{2}P $5p$	29	−.72992	$3s^23p$ ^{2}P $9s$
		^{1}S^o	6	−2.00192	$3s3p^2$ ^{2}D $4d$	11	−2.44994	$3s3p^2$ ^{4}P $4d$	30	−.64388	$3s^23p$ ^{2}P $9d$
1	−4.27018	$3p^3$ ^{2}D $3d$	7	−1.74117	$3s^23p$ ^{2}P $6p$	12	−1.98350	$3s3p^2$ ^{2}D $4d$			**^{5}P^e**
2	−2.49295	$3s3p3d^2$	8	−1.48548	$3s3p^2$ ^{2}P $4d$	13	−1.78168	$3p^23d^2$	1	−5.69738	$3s3p^2$ ^{4}P $3d$
3	−2.32072	$3s3p^2$ ^{2}P $4p$	9	−1.24104	$3p^3$ ^{2}D $4p$	14	−1.72164	$3s^23p$ ^{2}P $6p$	2	−3.76477	$3s3p^2$ ^{4}P $4s$
4	−.51806	$3s3p^2$ ^{2}P $5p$	10	−1.23215	$3s^23p$ ^{2}P $7p$	15	−1.60843	$3s3p^2$ ^{4}P $5s$	3	−2.46896	$3s3p^2$ ^{4}P $4d$
5	−.40023	$3p^3$ ^{2}D $4d$	11	−.97820	$3s^23d$ ^{2}D $4d$	16	−1.47903	$3s3p^2$ ^{4}P $4d$	4	−1.68145	$3s3p^2$ ^{4}P $5s$
		^{3}S^e	12	−.95529	$3p^3$ ^{2}P $4p$	17	−1.33909	$3p^23d^2$	5	−1.50291	$3p^23d^2$:
1	−5.08247	$3s3p^2$ ^{2}D $3d$	13	−.91856	$3s^23p$ ^{2}P $8p$	18	−1.22074	$3s^23p$ ^{2}P $7p$	6	−1.16681	$3p^3$ ^{4}S $4p$
2	−4.30971	$3s^23p$ ^{2}P $4p$	14	−.86892	$3s3p3d$ ^{2}D $4p$	19	−1.18792	$3p^3$ ^{2}D $4p$	7	−1.05232	$3s3p^2$ ^{4}P $5d$
3	−2.86371	$3s3p^2$ ^{2}S $4s$	15	−.76280	$3s3p3d$ ^{2}P $4p$	20	−1.10385	$3p^23d^2$	8	−.70514	$3s3p^2$ ^{4}P $6s$
4	−2.52031	$3s^23p$ ^{2}P $5p$	16	−.71328	$3s^23p$ ^{2}P $9p$	21	−1.03487	$3s3p^2$ ^{4}P $5d$	9	−.63730	$3s3p3d$ ^{4}P $4p$
5	−1.96043	$3s3p^2$ ^{2}D $4d$	17	−.66739	$3s3p^2$ ^{2}P $5s$	22	−.91456	$3s^23p$ ^{2}P $8p$	10	−.38718	$3s3p^2$ ^{4}P $6d$
6	−1.71185	$3s^23p$ ^{2}P $6p$	18	−.58141	$3s3p^2$ ^{2}D $5d$	23	−.89359	$3p^3$ ^{4}S $4p$	11	−.29033	$3s3p3d$ ^{4}D $4p$
7	−1.22079	$3s^23p$ ^{2}P $7p$			**^{1}P^o**	24	−.83925	$3s^23d$ ^{2}D $4d$	12	−.15187	$3s3p^2$ ^{4}P $7s$
8	−.93698	$3s^23d$ ^{2}D $4d$	1	−7.06371	$3s3p^3$	25	−.75823	$3p^23d^2$			**^{5}P^o**
9	−.90526	$3s^23p$ ^{2}P $8p$	2	−6.30338	$3s^23p$ ^{2}P $3d$	26	−.71889	$3s^23p$ ^{2}P $9p$	1	−3.38003	$3s3p3d^2$:
10	−.78433	$3p^3$ ^{2}P $4p$	3	−4.77987	$3s^23p$ ^{2}P $4s$	27	−.68605	$3s3p^2$ ^{2}P $5s$	2	−3.15415	$3s3p^2$ ^{4}P $4p$
11	−.72090	$3s3p^2$ ^{2}S $5s$	4	−3.61400	$3p^3$ ^{2}D $3d$	28	−.68044	$3s3p^2$ ^{2}D $5d$	3	−1.41929	$3s3p^2$ ^{4}P $5p$
12	−.69539	$3s^23p$ ^{2}P $9p$	5	−3.48097	$3s^23p$ ^{2}P $4d$	29	−.66477	$3s3p^2$ ^{4}P $6s$	4	−1.16965	$3s3p3d$ ^{4}P $4s$
13	−.64158	$3s3p^2$ ^{2}D $5d$	6	−2.78069	$3s^23p$ ^{2}P $5s$	30	−.61031	$3p^3$ ^{2}P $4p$	5	−.59063	$3s3p^2$ ^{4}P $6p$
14	−.61318	$3s3p3d$ ^{4}P $4p$	7	−2.75035	$3p^3$ ^{2}P $3d$	31	−.59784	$3s3p3d$ ^{4}P $4p$	6	−.09532	$3s3p^2$ ^{4}P $7p$
		^{3}S^o	8	−2.66363	$3s3p^2$ ^{2}D $4p$			**^{3}P^o**			**^{1}D^e**
1	−7.12273	$3s3p^3$	9	−2.27144	$3s3p^2$ ^{2}S $4p$	1	−7.71822	$3s3p^3$	1	−9.18749	$3s^23p^2$
2	−3.65485	$3p^3$ ^{2}D $3d$	10	−2.18922	$3s3p3d^2$	2	−6.73841	$3s^23p$ ^{2}P $3d$	2	−6.18542	$3p^4$
3	−3.25267	$3s3p^2$ ^{4}P $4p$	11	−2.15723	$3s^23p$ ^{2}P $5d$	3	4.83875	$3s^23p$ ^{2}P $4s$	3	−4.91647	$3s3p^2$ ^{2}D $3d$
4	−2.92695	$3s3p3d^2$	12	−2.05204	$3s3p3d^2$	4	−3.64585	$3p^3$ ^{2}D $3d$	4	−4.69703	$3s3p^2$ ^{2}S $3d$
5	−2.03672	$3s3p^2$ ^{2}P $4p$	13	−1.93925	$3s3p^2$ ^{2}P $4p$	5	−3.54634	$3s3p3d^2$:	5	−4.20771	$3s3p^2$ ^{2}P $3d$
6	−1.69657	$3p^3$ ^{4}S $4s$	14	−1.81090	$3s^23p$ ^{2}P $6s$	6	−3.50956	$3s^23p$ ^{2}P $4d$	6	−4.11282	$3s^23p$ ^{2}P $4p$
7	−1.53438	$3s3p3d^2$	15	1.50718	$3s^23d$ ^{2}D $4p$	7	−2.99582	$3s3p^2$ ^{4}P $4p$	7	−4.00183	$3s^23d^2$
8	−1.34745	$3s3p^2$ ^{4}P $5p$	16	−1.48534	$3s3p^2$ ^{2}D $4f$	8	−2.80499	$3s^23p$ ^{2}P $5s$	8	−3.18007	$3s3p^2$ ^{2}D $4s$
9	−.61401	$3s3p^2$ ^{4}P $6p$	17	−1.44861	$3s^23p$ ^{2}P $6d$	9	−2.72584	$3p^3$ ^{2}P $3d$	9	−3.04499	$3s^23p$ ^{2}P $4f$
10	−.43955	$3p^3$ ^{2}D $4d$	18	−1.38107	$3s3p3d^2$	10	−2.64777	$3s3p3d^2$	10	−2.48090	$3s^23p$ ^{2}P $5p$
11	−.35492	$3s3p^2$ ^{2}P $5p$	19	−1.27470	$3s^23p$ ^{2}P $7s$	11	−2.57370	$3s3p^2$ ^{2}D $4p$	11	−2.14281	$3s^23d$ ^{2}D $4s$
12	−.10868	$3s3p^2$ ^{4}P $7p$	20	−1.26788	$3p^3$ ^{2}P $4s$	12	−2.35694	$3s3p3d^2$	12	−2.02846	$3s3p^2$ ^{2}D $4d$

Si-like Ca (Ca^{6+})

i	E(Ryds)	Description	i	E(Ryds)	Description	i	E(Ryds)	Description	i	E(Ryds)	Description
13	−1.96695	$3s^23p$ ^{2}P $5f$	29	−.73891	$3s3p^2$ ^{2}D $5d$	5	−2.04377	$3s3p^2$ ^{4}P $4f$	3	−4.95620	$3s3p^2$ ^{2}P $3d$
14	−1.70638	$3p^23d^2$	30	−.70550	$3s^23p$ ^{2}P $9p$	6	−1.44643	$3s3p^2$ ^{4}P $5p$	4	−4.26628	$3s^23d^2$
15	−1.69074	$3s^23p$ ^{2}P $6p$	31	−.69190	$3s3p3d$ ^{4}F $4p$	7	−.99994	$3s3p3d$ ^{4}D $4s$	5	−3.15938	$3s^23p$ ^{2}P $4f$
16	−1.46802	$3s3p^2$ ^{2}S $4d$	32	−.66551	$3s3p3d$ ^{4}P $4p$	8	−.89350	$3s3p^2$ ^{4}P $5f$	6	−2.41400	$3s3p^2$ ^{4}P $4d$
17	−1.39878	$3s3p^2$ ^{2}P $4d$	33	−.64215	$3p^3$ ^{2}P $4p$	9	−.60240	$3s3p^2$ ^{4}P $6p$	7	−2.08145	$3s3p^2$ ^{2}D $4d$
18	−1.36583	$3s^23p$ ^{2}P $6f$	34	−.61269	$3s^23p$ ^{2}P $9f$	10	−.44031	$3p^3$ ^{4}S $4d$	8	−2.01617	$3s^23p$ ^{2}P $5f$
19	−1.24635	$3s3p^2$ ^{2}D $5s$			^{3}D^o	11	−.27138	$3s3p^2$ ^{4}P $6f$	9	−1.78013	$3p^23d^2$
20	−1.20309	$3s^23p$ ^{2}P $7p$	1	−7.96270	$3s3p^3$	12	−.10020	$3s3p^2$ ^{4}P $7p$	10	−1.59937	$3p^23d^2$
21	−1.11274	$3p^23d^2$	2	−6.66041	$3s^23p$ ^{2}P $3d$			^{1}F^e	11	−1.51049	$3s3p^2$ ^{2}P $4d$
22	−1.00694	$3s^23p$ ^{2}P $7f$	3	−4.15877	$3s3p3d^2$:	1	−5.32907	$3s3p^2$ ^{2}D $3d$	12	−1.39608	$3s^23p$ ^{2}P $6f$
23	−.94374	$3p^3$ ^{2}D $4p$	4	−3.64834	$3p^3$ ^{2}D $3d$	2	−4.39680	$3s3p^2$ ^{2}P $3d$	13	−1.24352	$3p^23d^2$
24	−.90910	$3s^23p$ ^{2}P $8p$	5	−3.58634	$3s^23p$ ^{2}P $4d$	3	−3.18140	$3s^23p$ ^{2}P $4f$	14	−1.19394	$3p^3$ ^{2}D $4p$
25	−.85577	$3s^23d$ ^{2}D $4d$	6	−3.20523	$3p^3$ ^{4}S $3d$	4	−2.10100	$3s3p^2$ ^{2}D $4d$	15	−1.06833	$3s3p^2$ ^{4}P $5d$
26	−.76871	$3s^23p$ ^{2}P $8f$	7	−3.09134	$3p^3$ ^{2}P $3d$	5	−2.02684	$3s^23p$ ^{2}P $5f$	16	−1.02475	$3s^23p$ ^{2}P $7f$
27	−.71994	$3p^23d^2$	8	−3.06961	$3s3p^2$ ^{4}P $4p$	6	−1.48777	$3s3p^2$ ^{2}P $4d$	17	−.99386	$3p^23d^2$
28	−.70744	$3p^23d^2$	9	−2.68805	$3s3p^2$ ^{2}D $4p$	7	−1.39899	$3s^23p$ ^{2}P $6f$	18	−.93064	$3s^23d$ ^{2}D $4d$
29	−.69940	$3s^23p$ ^{2}P $9p$	10	−2.57681	$3s3p3d^2$	8	−1.27007	$3p^23d^2$	19	−.86872	$3s3p^2$ ^{4}P $5g$
30	−.67949	$3s3p^2$ ^{2}D $5d$	11	−2.47047	$3s3p3d^2$	9	−1.15046	$3p^3$ ^{2}D $4p$	20	−.84462	$3p^23d^2$:
31	−.60806	$3s^23p$ ^{2}P $9f$	12	−2.27641	$3s^23p$ ^{2}P $5d$	10	−1.02687	$3s^23p$ ^{2}P $7f$	21	−.78138	$3s^23p$ ^{2}P $8f$
32	−.57047	$3p^3$ ^{2}P $4p$	13	−2.15461	$3s3p^2$ ^{2}P $4p$	11	−1.01865	$3s^23d$ ^{2}D $4d$	22	−.77780	$3s3p3d$ ^{4}F $4p$
		^{1}D^o	14	−2.12111	$3s3p3d^2$	12	−.96174	$3s3p3d$ ^{2}D $4p$	23	−.69902	$3s3p^2$ ^{2}D $5d$
1	−7.54499	$3s3p^3$	15	−2.00585	$3s3p^2$ ^{4}P $4f$	13	−.78300	$3s^23p$ ^{2}P $8f$	24	−.61543	$3s^23p$ ^{2}P $9f$
2	−6.72186	$3s^23p$ ^{2}P $3d$	16	−1.87961	$3s3p3d^2$:	14	−.70999	$3s3p^2$ ^{2}D $5d$	25	−.53813	$3s3p^2$ ^{4}P $6g$
		^{3}D^e	17	−1.79602	$3p^3$ ^{2}D $4s$	15	−.61734	$3s^23p$ ^{2}P $9f$			^{3}F^o
1	−5.34524	$3s3p^2$ ^{4}P $3d$	18	−1.71567	$3s^23d$ ^{2}D $4p$			^{1}F^o	1	−7.10138	$3s^23p$ ^{2}P $3d$
2	−5.19297	$3s3p^2$ ^{2}D $3d$	19	−1.60266	$3s3p^2$ ^{2}D $4f$	1	−6.38309	$3s^23p$ ^{2}P $3d$	2	−4.18712	$3s3p3d^2$
3	−4.98123	$3s3p^2$ ^{2}S $3d$	20	−1.49704	$3s^23p$ ^{2}P $6d$	2	−3.51626	$3s^23p$ ^{2}P $4d$	3	−3.59143	$3p^3$ ^{2}D $3d$
4	−4.73230	$3s3p^2$ ^{2}P $3d$	21	−1.36506	$3s3p^2$ ^{4}P $5p$	3	−3.34418	$3p^3$ ^{2}D $3d$	4	−3.56139	$3s^23p$ ^{2}P $4d$
5	−4.37700	$3s^23p$ ^{2}P $4p$	22	−1.09295	$3s^23p$ ^{2}P $7d$	4	−3.25718	$3p^3$ ^{2}P $3d$	5	−3.20782	$3p^3$ ^{2}P $3d$
6	−3.34807	$3s3p^2$ ^{2}D $4s$	23	−1.04775	$3s3p^2$ ^{2}P $4f$	5	−2.74218	$3s3p^2$ ^{2}D $4p$	6	−2.80918	$3s3p^2$ ^{2}D $4p$
7	−3.10777	$3s^23p$ ^{2}P $4f$	24	−1.04031	$3s3p^2$ ^{2}D $5p$	6	−2.70596	$3s3p3d^2$	7	−2.77624	$3s3p3d^2$
8	−2.57436	$3s^23p$ ^{2}P $5p$	25	−.86871	$3s3p^2$ ^{4}P $5f$	7	−2.20068	$3s^23p$ ^{2}P $5d$	8	−2.59098	$3s3p3d^2$
9	−2.34036	$3s3p^2$ ^{4}P $4d$	26	−.85731	$3s3p3d$ ^{2}D $4s$	8	−1.99520	$3s3p3d^2$	9	−2.20466	$3s3p3d^2$
10	−2.20914	$3s^23d$ ^{2}D $4s$	27	−.83134	$3s^23p$ ^{2}P $8d$	9	−1.96475	$3s^23p$ ^{2}P $5g$	10	−2.19688	$3s^23p$ ^{2}P $5d$
11	−2.07475	$3s3p^2$ ^{2}D $4d$	28	−.78058	$3s3p3d$ ^{4}D $4s$	10	−1.78912	$3s3p3d^2$:	11	−1.96767	$3s^23p$ ^{2}P $5g$
12	−1.99554	$3s^23p$ ^{2}P $5f$	29	−.64833	$3s^23p$ ^{2}P $9d$	11	−1.64229	$3s^23d$ ^{2}D $4p$	12	−1.87244	$3s3p^2$ ^{4}P $4f$
13	−1.73047	$3s^23p$ ^{2}P $6p$	30	−.56004	$3s3p^2$ ^{4}P $6p$	12	−1.54238	$3s3p^2$ ^{2}D $4f$	13	−1.73115	$3s^23d$ ^{2}D $4p$
14	−1.70330	$3p^23d^2$			^{5}D^e	13	−1.45662	$3s^23p$ ^{2}P $6d$	14	−1.65446	$3s3p^2$ ^{2}D $4f$
15	−1.47961	$3s3p^2$ ^{2}S $4d$	1	−6.02507	$3s3p^2$ ^{4}P $3d$	14	−1.37209	$3s^23p$ ^{2}P $6g$	15	−1.49769	$3s^23p$ ^{2}P $6d$
16	−1.44378	$3s3p^2$ ^{2}P $4d$	2	−2.39281	$3s3p^2$ ^{4}P $4d$	15	−1.07722	$3s^23p$ ^{2}P $7d$	16	−1.37661	$3s^23p$ ^{2}P $6g$
17	−1.38394	$3s^23p$ ^{2}P $6f$	3	−1.82208	$3p^23d^2$:	16	−1.05857	$3s3p^2$ ^{2}S $4f$	17	−1.17940	$3s3p3d$ ^{4}F $4s$
18	−1.37146	$3p^23d^2$	4	−1.44891	$3p^23d^2$:	17	−1.00959	$3s3p^2$ ^{2}D $5p$	18	−1.09714	$3s^23p$ ^{2}P $7d$
19	−1.29285	$3s3p^2$ ^{2}D $5s$	5	−1.05834	$3s3p^2$ ^{4}P $5d$	18	−.97864	$3s^23p$ ^{2}P $7g$	19	−1.07328	$3s3p^2$ ^{2}S $4f$
20	−1.23332	$3p^3$ ^{2}D $4p$	6	−.72295	$3s3p3d$ ^{4}F $4p$	19	−.90171	$3s3p^2$ ^{2}P $4f$	20	−1.05439	$3s3p^2$ ^{2}P $4f$
21	−1.22184	$3s^23p$ ^{2}P $7p$	7	−.60500	$3s3p3d$ ^{4}P $4p$	20	−.81718	$3s^23p$ ^{2}P $8d$	21	−1.00928	$3s3p^2$ ^{2}D $5p$
22	−1.05237	$3s3p^2$ ^{4}P $5d$	8	−.43687	$3s3p3d$ ^{4}D $4p$	21	−.75885	$3s^23p$ ^{2}P $8g$	22	−.96089	$3s^23p$ ^{2}P $7g$
23	−1.01615	$3s^23p$ ^{2}P $7f$	9	−.35745	$3s3p^2$ ^{4}P $6d$	22	−.67749	$3s3p3d$ ^{2}F $4s$	23	−.82013	$3s3p^2$ ^{4}P $5f$
24	−1.00158	$3s^23d$ ^{2}D $4d$			^{5}D^o	23	−.64172	$3s^23p$ ^{2}P $9d$	24	−.81625	$3s^23p$ ^{2}P $8d$
25	−.92162	$3p^23d^2$	1	−4.07785	$3p^3$ ^{4}S $3d$	24	−.60429	$3s^23p$ ^{2}P $9g$	25	−.76125	$3s^23p$ ^{2}P $8g$
26	−.91192	$3s^23p$ ^{2}P $8p$	2	−3.48691	$3s3p3d^2$			^{3}F^e	26	−.70806	$3s3p3d$ ^{2}F $4s$
27	−.85791	$3p^23d^2$:	3	−3.21721	$3s3p^2$ ^{4}P $4p$	1	−5.87890	$3s3p^2$ ^{4}P $3d$	27	−.64246	$3s^23p$ ^{2}P $9d$
28	−.77705	$3s^23p$ ^{2}P $8f$	4	−3.20305	$3s3p3d^2$	2	−5.20045	$3s3p^2$ ^{2}D $3d$	28	−.61139	$3s^23p$ ^{2}P $9g$

Si-like Ca (Ca^{6+})

i	Energy(Ryds)	Description	
29	−.57731	$3s^23d$	^{2}D $4f$
	$^5\mathbf{F}^e$		
1	−6.11128	$3s3p^2$	^{4}P $3d$
2	−2.45328	$3s3p^2$	^{4}P $4d$
3	−1.82441	$3p^23d^2$	:
4	−1.08645	$3s3p^2$	^{4}P $5d$
5	−.85981	$3s3p^2$	^{4}P $5g$
6	−.77505	$3s3p3d$	^{4}F $4p$
7	−.41635	$3s3p3d$	^{4}D $4p$
8	−.36849	$3s3p^2$	^{4}P $6d$
9	−.26342	$3s3p^2$	^{4}P $6g$
10	−.05299	$3p^3$	^{4}S $4f$
	$^5\mathbf{F}^o$		
1	−3.40411	$3s3p3d^2$	
2	−2.01540	$3s3p^2$	^{4}P $4f$
3	−1.36583	$3s3p3d$	^{4}F $4s$
4	−.88535	$3s3p^2$	^{4}P $5f$
5	−.26663	$3s3p^2$	^{4}P $6f$
	$^1\mathbf{G}^e$		
1	−5.37833	$3s3p^2$	^{2}D $3d$
2	−4.14225	$3s^23d^2$	
3	−3.04836	$3s^23p$	^{2}P $4f$
4	−2.00798	$3s3p^2$	^{2}D $4d$
5	−1.95716	$3s^23p$	^{2}P $5f$
6	−1.72065	$3p^23d^2$	
7	−1.37114	$3s^23p$	^{2}P $6f$
8	−1.36046	$3s^23p$	^{2}P $6h$
9	−1.21083	$3p^23d^2$	
10	−1.01079	$3s^23p$	^{2}P $7f$
11	−.99984	$3s^23p$	^{2}P $7h$
12	−.86767	$3s^23d$	^{2}D $4d$
13	−.76754	$3s^23p$	^{2}P $8f$
14	−.76544	$3s^23p$	^{2}P $8h$
15	−.73663	$3s3p3d$	^{2}F $4p$
16	−.66393	$3s3p^2$	^{2}D $5d$
17	−.60646	$3s^23p$	^{2}P $9f$
18	−.60501	$3s^23p$	^{2}P $9h$
	$^1\mathbf{G}^o$		
1	−3.82673	$3p^3$	^{2}D $3d$
2	−2.68173	$3s3p3d^2$	
3	−1.99699	$3s^23p$	^{2}P $5g$
4	−1.96286	$3s3p^2$	^{2}D $4f$
5	−1.62262	$3s3p^2$	^{2}D $5f$
6	−1.38576	$3s^23p$	^{2}P $6g$
7	−1.05516	$3s^23p$	^{2}P $7g$
8	−.95060	$3s3p^2$	^{2}P $4f$
9	−.76643	$3s^23p$	^{2}P $8g$
10	−.62646	$3s^23d$	^{2}D $4f$
11	−.59110	$3s^23p$	^{2}P $9g$
12	−.50574	$3s3p^2$	^{2}D $6f$
	$^3\mathbf{G}^e$		
1	−5.46815	$3s3p^2$	^{2}D $3d$
2	−3.16435	$3s^23p$	^{2}P $4f$
3	−2.04782	$3s3p^2$	^{2}D $4d$
4	−2.01696	$3s^23p$	^{2}P $5f$
5	−1.65237	$3p^23d^2$	
6	−1.39768	$3s^23p$	^{2}P $6f$
7	−1.36049	$3s^23p$	^{2}P $6h$
8	−1.16292	$3p^23d^2$	
9	−1.02624	$3s^23p$	^{2}P $7f$
10	−1.00552	$3p^23d^2$	:
11	−1.00096	$3s^23p$	^{2}P $7h$
12	−.99305	$3s^23d$	^{2}D $4d$
13	−.82687	$3s3p^2$	^{4}P $5g$
14	−.78177	$3s^23p$	^{2}P $8f$
15	−.76572	$3s^23p$	^{2}P $8h$
16	−.71139	$3s3p^2$	^{2}D $5d$
17	−.61754	$3s^23p$	^{2}P $9f$
18	−.60519	$3s^23p$	^{2}P $9h$
19	−.56338	$3s3p3d$	^{4}F $4p$
	$^3\mathbf{G}^o$		
1	−4.03706	$3p^3$	^{2}D $3d$
2	−3.01842	$3s3p3d^2$	
3	−2.84469	$3s3p3d^2$	
4	−2.39469	$3s3p3d^2$	
5	−1.99330	$3s^23p$	^{2}P $5g$
6	−1.95614	$3s3p^2$	^{4}P $4f$
7	−1.67493	$3s3p^2$	^{2}D $4f$
8	−1.39008	$3s^23p$	^{2}P $6g$
9	−1.10350	$3s3p^2$	^{2}P $4f$
10	−.98272	$3s^23p$	^{2}P $7g$
11	−.85896	$3s3p^2$	^{4}P $5f$
12	−.76834	$3s^23p$	^{2}P $8g$
13	−.61163	$3s^23p$	^{2}P $9g$
14	−.52941	$3s3p^2$	^{2}D $5f$
	$^5\mathbf{G}^e$		
1	−2.05441	$3p^23d^2$	
2	−.87223	$3s3p3d$	^{4}F $4p$
3	−.82813	$3s3p^2$	^{4}P $5g$
4	−.23651	$3s3p^2$	^{4}P $6g$
	$^5\mathbf{G}^o$		
1	−3.76149	$3s3p3d^2$	
2	−2.07834	$3s3p^2$	^{4}P $4f$
3	−.90978	$3s3p^2$	^{4}P $5f$
4	−.27996	$3s3p^2$	^{4}P $6f$
5	−.23704	$3s3p^2$	^{4}P $6h$
	$^1\mathbf{H}^e$		
1	−1.36502	$3s^23p$	^{2}P $6h$
2	−1.30020	$3p^23d^2$	
3	−1.00367	$3s^23p$	^{2}P $7h$
4	−.76865	$3s^23p$	^{2}P $8h$
5	−.60750	$3s^23p$	^{2}P $9h$
	$^1\mathbf{H}^o$		
1	−2.47008	$3s3p3d^2$	
2	−1.96534	$3s^23p$	^{2}P $5g$
3	−1.58908	$3s3p^2$	^{2}D $4f$
4	−1.36674	$3s^23p$	^{2}P $6g$
5	−1.00629	$3s^23p$	^{2}P $7g$
6	−.77214	$3s^23p$	^{2}P $8g$
7	−.61232	$3s^23p$	^{2}P $9g$
	$^3\mathbf{H}^e$		
1	−1.75618	$3p^23d^2$	
2	−1.36501	$3s^23p$	^{2}P $6h$
3	−1.32752	$3p^23d^2$	:
4	−1.00365	$3s^23p$	^{2}P $7h$
5	−.84743	$3s3p^2$	^{4}P $5g$
6	−.76863	$3s^23p$	^{2}P $8h$
7	−.60748	$3s^23p$	^{2}P $9h$
	$^3\mathbf{H}^o$		
1	−3.16434	$3s3p3d^2$	
2	−1.97406	$3s^23p$	^{2}P $5g$
3	−1.63024	$3s3p^2$	^{2}D $4f$
4	−1.36711	$3s^23p$	^{2}P $6g$
5	−1.00883	$3s^23p$	^{2}P $7g$
6	−.77738	$3s^23p$	^{2}P $8g$
7	−.63646	$3s^23p$	^{2}P $9g$
	$^5\mathbf{H}^e$		
1	−.85086	$3s3p^2$	^{4}P $5g$
2	−.24697	$3s3p^2$	^{4}P $6g$
	$^5\mathbf{H}^o$		
1	−.23213	$3s3p^2$	^{4}P $6h$
	$^1\mathbf{I}^e$		
1	−1.38146	$3s^23p$	^{2}P $6h$
2	−1.36096	$3s^23p$	^{2}P $7h$
3	−1.00026	$3s^23p$	^{2}P $8h$
4	−.76604	$3s^23p$	^{2}P $9h$
	$^3\mathbf{I}^e$		
1	−1.36118	$3s^23p$	^{2}P $6h$
2	−1.00038	$3s^23p$	^{2}P $7h$
3	−.76616	$3s^23p$	^{2}P $8h$
4	−.60560	$3s^23p$	^{2}P $9h$
	$^3\mathbf{I}^o$		
1	−.23596	$3s3p^2$	^{4}P $6h$
	$^5\mathbf{I}^o$		
1	−.23604	$3s3p^2$	^{4}P $6h$

Si-like Ca (Ca^{6+})

Energies in ascending order from ground state for terms with effective $n \leq 4.0$, $L \leq 5$

Term	i	E(Ryds)	Term	i	E(Ryds)	Term	i	E(Ryds)	Term	i	E(Ryds)	Term	i	E(Ryds)
$^3P^e$	1	0.00000	$^1S^e$	2	3.65675	$^1P^e$	2	4.82465	$^1G^o$	1	5.53776	$^3D^e$	6	6.01642
$^1D^e$	1	0.17700	$^5P^e$	1	3.66711	$^3P^e$	5	4.86247	$^5P^e$	2	5.59972	$^1F^o$	3	6.02031
$^1S^e$	1	0.43752	$^3P^e$	3	3.77042	$^1P^e$	3	4.94436	$^5G^o$	1	5.60300	$^1F^o$	4	6.10731
$^5S^o$	1	0.96199	$^3G^e$	1	3.89634	$^1F^e$	2	4.96769	$^3S^o$	2	5.70964	$^3S^o$	3	6.11182
$^3D^o$	1	1.40179	$^1G^e$	1	3.98616	$^3D^e$	5	4.98749	$^1S^e$	5	5.71169	$^5D^o$	3	6.14728
$^3P^o$	1	1.64627	$^3D^e$	1	4.01925	$^3S^e$	2	5.05478	$^3D^o$	4	5.71615	$^3F^o$	5	6.15667
$^1D^o$	1	1.81950	$^1F^e$	1	4.03542	$^1S^o$	1	5.09431	$^3P^o$	4	5.71864	$^3D^o$	6	6.15926
$^3S^o$	1	2.24176	$^3F^e$	2	4.16404	$^3F^e$	4	5.09821	$^1P^o$	4	5.75049	$^5D^o$	4	6.16144
$^3F^o$	1	2.26311	$^3D^e$	2	4.17152	$^3P^e$	6	5.14392	$^1D^o$	4	5.75510	$^1D^o$	5	6.16401
$^1P^o$	1	2.30078	$^1P^e$	1	4.19651	$^1D^e$	5	5.15678	$^3F^o$	3	5.77306	$^1F^e$	3	6.18309
$^3P^o$	2	2.62608	$^3P^e$	4	4.23357	$^3F^o$	2	5.17737	$^3D^o$	5	5.77815	$^1D^e$	8	6.18442
$^1D^o$	2	2.64263	$^3S^e$	1	4.28202	$^3D^o$	3	5.20572	$^3F^o$	4	5.80310	$^3G^e$	2	6.20014
$^3D^o$	2	2.70408	$^3D^e$	3	4.38326	$^1G^e$	2	5.22224	$^3P^e$	8	5.81338	$^3H^o$	1	6.20015
$^1F^o$	1	2.98140	$^3F^e$	3	4.40829	$^1D^e$	6	5.25167	$^3P^o$	5	5.81815	$^3F^e$	5	6.20511
$^1P^o$	2	3.06111	$^1D^e$	3	4.44802	$^5D^o$	1	5.28664	$^1F^o$	2	5.84823	$^5P^o$	2	6.21034
$^3P^e$	2	3.08844	$^3P^o$	3	4.52574	$^3G^o$	1	5.32743	$^3P^o$	6	5.85493	$^3D^e$	7	6.25672
$^1D^e$	2	3.17907	$^1P^o$	3	4.58462	$^3P^e$	7	5.34136	$^5D^o$	2	5.87758	$^3D^o$	7	6.27315
$^5F^e$	1	3.25321	$^3D^e$	4	4.63219	$^1S^e$	4	5.35765	$^1P^o$	5	5.88352	$^3D^o$	8	6.29488
$^5D^e$	1	3.33942	$^1D^e$	4	4.66746	$^1D^e$	7	5.36266	$^5F^o$	1	5.96038			
$^3F^e$	1	3.48559	$^1S^e$	3	4.73206	$^1D^o$	3	5.50765	$^5P^o$	1	5.98446			

gf-values for transitions involving terms with effective $n \leq 4.0$, $L \leq 5$

i i'	gf_L	i i'	gf_L	i i'	gf_L	i i'	gf_L	i i'	gf_L	i i'	gf_L
	$^1S^e$–$^1P^o$		$^1S^o$–$^1P^e$	2 2	5.44E−1	3 2	3.97E−4	1 4	−1.32E−1	7 1	3.32E−2
1 1	−1.75E−1	1 1	6.59E−2	2 3	−1.97E−2	3 3	7.11E−4	1 5	−2.43E−5	7 2	3.85E+0
1 2	−1.80E+0	1 2	2.59E−6	2 4	−3.51E−1	3 4	−9.56E−5	2 1	5.58E−1	7 3	−8.43E−6
1 3	−1.15E−1	1 3	9.46E−6	2 5	−7.32E−2	3 5	−2.84E−1	2 2	3.22E−2	7 4	−4.93E−3
1 4	−5.23E−3		$^1P^e$–$^1P^o$	3 1	1.33E−1	3 6	−1.22E+0	2 3	−4.62E−1	7 5	−2.05E−3
1 5	−1.34E−1	1 1	6.16E−3	3 2	2.55E−3	3 7	−8.57E−2	2 4	−3.22E−3	8 1	3.73E−1
2 1	2.05E−1	1 2	3.30E−4	3 3	−8.84E−3	3 8	−4.82E−2	2 5	−1.91E+0	8 2	5.72E−2
2 2	7.85E−4	1 3	−2.27E−4	3 4	−1.60E+0	4 1	3.69E−5	3 1	2.22E+0	8 3	5.54E−3
2 3	−2.46E−4	1 4	−2.87E−1	3 5	−1.72E−2	4 2	1.57E+0	3 2	1.82E−1	8 4	2.41E−2
2 4	−1.91E−1	1 5	−9.39E−3		$^1P^o$–$^1D^e$	4 3	6.20E−2	3 3	−1.96E−2	8 5	3.83E−6
2 5	−1.52E−4	2 1	1.75E+0	1 1	1.25E+0	4 4	6.15E−2	3 4	−3.40E−3		$^1D^e$–$^1F^o$
3 1	8.38E−1	2 2	1.13E−1	1 2	−7.09E−2	4 5	9.50E−4	3 5	−5.39E−1	1 1	−4.87E+0
3 2	4.79E−2	2 3	5.30E−2	1 3	−1.19E−1	4 6	2.51E−3	4 1	3.92E−1	1 2	−5.09E−1
3 3	4.34E−4	2 4	−1.23E−3	1 4	−6.80E−1	4 7	2.84E−3	4 2	1.25E+0	1 3	−4.50E−4
3 4	−3.03E−2	2 5	−1.06E−1	1 5	−2.20E+0	4 8	−3.59E−3	4 3	−1.69E−2	1 4	−7.23E−5
3 5	−5.11E−3	3 1	4.41E−1	1 6	−5.79E−1	5 1	4.14E−3	4 4	−1.58E−3	2 1	6.14E−5
4 1	9.25E−3	3 2	1.44E−1	1 7	−3.51E−2	5 2	9.82E−3	4 5	−1.99E−2	2 2	−1.83E−3
4 2	3.28E−1	3 3	4.34E−1	1 8	−1.01E−1	5 3	1.76E−5	5 1	5.35E−3	2 3	−1.54E+0
4 3	3.24E−1	3 4	−7.58E−3	2 1	3.00E−3	5 4	2.60E−2	5 2	3.06E−1	2 4	−2.38E+0
4 4	−4.52E−3	3 5	−4.76E−1	2 2	−9.72E−4	5 5	6.55E−2	5 3	−8.86E−5	3 1	9.84E−3
4 5	−6.63E−1		$^1P^e$–$^1D^o$	2 3	−5.24E−2	5 6	4.58E−2	5 4	−7.13E−2	3 2	−4.30E−3
5 1	4.42E−4	1 1	1.47E+0	2 4	−4.33E−1	5 7	3.92E−2	5 5	−7.14E−2	3 3	−3.55E−1
5 2	9.54E−1	1 2	2.25E−1	2 5	−7.07E−1	5 8	−1.27E−2	6 1	1.29E−1	3 4	−3.74E−2
5 3	2.36E−2	1 3	−1.38E−1	2 6	−6.80E−1		$^1D^e$–$^1D^o$	6 2	1.87E−1	4 1	9.93E−4
5 4	−2.97E−4	1 4	−1.05E−3	2 7	−9.82E−1	1 1	−4.41E−1	6 3	−7.07E−5	4 2	−5.39E−3
5 5	−1.03E−2	1 5	−6.39E−1	2 8	−1.19E−2	1 2	−3.39E+0	6 4	−3.26E−1	4 3	−4.10E−1
		2 1	1.86E−2	3 1	7.07E−1	1 3	−1.55E−3	6 5	−7.39E−3	4 4	−2.12E−1

Si-like Ca (Ca^{6+})

$i\ i'$	gf_L	$i\ i'$	gf_L	$i\ i'$	gf_L	$i\ i'$	gf_L	$i\ i'$	gf_L	$i\ i'$	gf_L
5 1	2.84E+0	2 1	1.86E−3	1 5	−8.28E−2	1 8	−4.23E−1	8 4	1.19E−2	1 4	−1.77E−3
5 2	−6.66E−1	2 2	1.48E−1	1 6	−1.76E−1	2 1	1.06E+0	8 5	2.48E−3	1 5	−8.06E−3
5 3	−6.87E−2	3 1	8.36E−1	2 1	1.74E−1	2 2	3.04E−3	8 6	−2.35E−2	1 6	−2.03E+0
5 4	−3.45E−5	3 2	5.48E−4	2 2	1.28E−2	2 3	−9.97E−1	8 7	−3.71E−1	1 7	−5.75E−1
6 1	5.31E−4	4 1	1.52E+0	2 3	−9.14E−4	2 4	−3.06E−2	8 8	−2.47E+0	1 8	−5.36E−1
6 2	−2.09E+0	4 2	5.69E−5	2 4	−2.52E+0	2 5	−3.33E−3		$^{3}\mathbf{P}^{o}$–$^{3}\mathbf{D}^{e}$	2 1	1.46E+0
6 3	−4.14E−3		$^{1}\mathbf{G}^{e}$–$^{1}\mathbf{G}^{o}$	2 5	−2.13E+0	2 6	−3.30E+0	1 1	−4.49E−4	2 2	5.99E−2
6 4	−9.81E−3	1 1	−1.19E+0	2 6	−1.14E+0	2 7	−2.81E+0	1 2	−3.22E+0	2 3	−9.64E−3
7 1	2.43E−1	2 1	−5.68E−3	3 1	8.63E−1	2 8	−9.34E−1	1 3	−1.04E+0	2 4	−3.32E−1
7 2	−1.30E−1		$^{3}\mathbf{S}^{e}$–$^{3}\mathbf{P}^{o}$	3 2	7.57E−2	3 1	3.48E+0	1 4	−5.53E+0	2 5	−1.47E−1
7 3	−7.33E−3	1 1	1.33E+0	3 3	−2.21E−4	3 2	3.63E−4	1 5	−1.62E−2	2 6	−2.22E−1
7 4	−8.00E−3	1 2	2.29E−1	3 4	−5.92E−2	3 3	−6.03E−1	1 6	−5.30E−1	2 7	−8.29E−1
8 1	5.63E−2	1 3	−1.49E−3	3 5	−6.51E−1	3 4	−4.13E−2	1 7	−3.71E−1	2 8	−6.47E−1
8 2	1.02E−2	1 4	−2.58E−1	3 6	−2.55E−1	3 5	−3.35E−3	2 1	−5.90E−3	3 1	2.52E−2
8 3	6.15E−3	1 5	−1.59E−1	4 1	4.45E+0	3 6	−1.64E+0	2 2	−5.93E−1	3 2	2.37E−2
8 4	4.72E−4	1 6	−2.37E−4	4 2	1.33E−1	3 7	−4.23E−1	2 3	−3.72E−1	3 3	−1.72E−1
	$^{1}\mathbf{D}^{o}$–$^{1}\mathbf{F}^{e}$	2 1	4.88E−2	4 3	−1.47E−3	3 8	−1.15E−1	2 4	−4.24E+0	3 4	−2.29E−1
1 1	−1.32E+0	2 2	2.62E−1	4 4	−1.09E−1	4 1	2.78E−2	2 5	−1.61E−2	3 5	−7.74E−2
1 2	−2.92E+0	2 3	7.51E−1	4 5	−2.33E−1	4 2	2.35E−1	2 6	−1.41E−2	3 6	−4.21E−1
1 3	−9.37E−1	2 4	−4.25E−2	4 6	−1.20E−2	4 3	−8.29E−3	2 7	−6.69E+0	3 7	−5.85E−2
2 1	−3.70E−1	2 5	−6.64E−1	5 1	1.73E−5	4 4	−1.84E−1	3 1	1.17E−3	3 8	−1.58E−3
2 2	−2.81E+0	2 6	−2.05E+0	5 2	1.29E+0	4 5	−9.86E−2	3 2	8.11E−3	4 1	1.98E−1
2 3	−2.65E+0		$^{3}\mathbf{S}^{o}$–$^{3}\mathbf{P}^{e}$	5 3	1.17E−3	4 6	−4.09E−1	3 3	1.94E−3	4 2	1.81E+0
3 1	4.73E−1	1 1	2.41E+0	5 4	−8.67E−2	4 7	−2.56E+0	3 4	−1.82E−3	4 3	−8.61E−3
3 2	1.45E−3	1 2	−2.01E−1	5 5	−1.01E−1	4 8	−2.10E−1	3 5	−3.16E+0	4 4	−1.85E−2
3 3	−4.57E−3	1 3	−1.93E−1	5 6	−8.52E−2	5 1	2.25E−3	3 6	−1.02E+0	4 5	−7.45E−5
4 1	1.64E−2	1 4	−7.11E−2	6 1	1.34E−1	5 2	3.74E+0	3 7	−3.69E−5	4 6	−3.28E−2
4 2	5.71E−2	1 5	−7.38E+0	6 2	1.04E+0	5 3	−2.75E−2	4 1	5.75E−2	4 7	−3.95E−1
4 3	−1.66E+0	1 6	−5.32E−3	6 3	2.37E+0	5 4	−2.79E−2	4 2	3.44E−3	4 8	−2.47E−1
5 1	4.03E−1	1 7	−3.93E−3	6 4	−9.54E−3	5 5	−1.90E−2	4 3	4.92E−1	5 1	7.93E−2
5 2	3.19E−2	1 8	−2.71E−2	6 5	−3.92E−1	5 6	−1.98E−1	4 4	3.06E−2	5 2	1.57E−1
5 3	−3.06E−5	2 1	8.74E−4	6 6	−1.17E+0	5 7	−3.13E−1	4 5	3.39E−3	5 3	−4.15E−4
	$^{1}\mathbf{F}^{e}$–$^{1}\mathbf{F}^{o}$	2 2	1.43E+0	7 1	1.37E−2	5 8	−3.41E−2	4 6	−1.21E−4	5 4	−2.12E−1
1 1	2.01E−2	2 3	5.74E−2	7 2	6.07E+0	6 1	2.94E−1	4 7	−7.15E−2	5 5	−1.58E+0
1 2	−1.23E−2	2 4	9.04E−2	7 3	1.19E−1	6 2	1.05E+0	5 1	4.41E−1	5 6	−2.28E−4
1 3	−4.94E−1	2 5	4.56E−2	7 4	−9.12E−4	6 3	−5.12E−5	5 2	1.99E−1	5 7	−1.83E−2
1 4	−1.62E−3	2 6	1.62E−6	7 5	−4.66E−3	6 4	−3.97E−1	5 3	2.13E−1	5 8	−3.33E−2
2 1	2.55E+0	2 7	2.47E−3	7 6	−4.64E−2	6 5	−3.88E+0	5 4	6.92E−3	6 1	1.67E+0
2 2	−8.07E−3	2 8	−2.02E−4	8 1	5.74E−1	6 6	−2.08E−2	5 5	5.11E−2	6 2	1.97E−3
2 3	−4.79E−2	3 1	1.27E−1	8 2	4.69E−2	6 7	−1.26E−2	5 6	−2.30E−4	6 3	5.80E−7
2 4	−2.31E−1	3 2	2.26E−1	8 3	2.63E−1	6 8	−2.35E−1	5 7	−9.68E−1	6 4	3.90E−6
3 1	8.50E−1	3 3	1.73E−1	8 4	3.77E−3	7 1	1.31E−2	6 1	2.56E−1	6 5	4.40E−3
3 2	1.60E−1	3 4	3.42E−2	8 5	−9.88E−5	7 2	2.43E+0	6 2	1.34E−1	6 6	−3.06E−3
3 3	8.12E−6	3 5	1.77E−2	8 6	−8.01E−4	7 3	5.26E−5	6 3	7.33E−2	6 7	−3.60E−2
3 4	2.84E−4	3 6	4.85E−2		$^{3}\mathbf{P}^{e}$–$^{3}\mathbf{D}^{o}$	7 4	−1.93E−2	6 4	1.10E−1	6 8	−2.87E−1
	$^{1}\mathbf{F}^{o}$–$^{1}\mathbf{G}^{o}$	3 7	2.79E−3	1 1	−4.50E−1	7 5	−1.05E−1	6 5	1.41E−1	7 1	4.23E−2
1 1	−1.27E+0	3 8	4.30E−1	1 2	−1.02E+1	7 6	−3.93E−3	6 6	−1.74E−6	7 2	1.30E+0
2 1	−4.85E−2		$^{3}\mathbf{P}^{e}$–$^{3}\mathbf{P}^{o}$	1 3	−1.59E−3	7 7	−2.84E−3	6 7	−2.57E+0	7 3	8.38E−5
3 1	2.33E−3	1 1	−6.06E−1	1 4	−1.41E−1	7 8	−5.13E−2		$^{3}\mathbf{D}^{e}$–$^{3}\mathbf{D}^{o}$	7 4	9.10E−2
	$^{1}\mathbf{F}^{o}$–$^{1}\mathbf{G}^{e}$	1 2	−4.14E+0	1 5	−8.10E−1	8 1	1.13E+0	1 1	6.40E+0	7 5	6.07E−1
1 1	−6.52E−2	1 3	−1.35E+0	1 6	−5.90E−3	8 2	1.33E−1	1 2	2.48E−1	7 6	1.61E−4
1 2	−4.49E+0	1 4	−8.24E−3	1 7	−7.00E−2	8 3	2.88E−4	1 3	−3.07E−1	7 7	−1.93E−5

Si-like Ca (Ca^{6+})

$i\ i'$	gf_L	$i\ i'$	gf_L	$i\ i'$	gf_L	$i\ i'$	gf_L	$i\ i'$	gf_L	$i\ i'$	gf_L
7 8	−5.83E−5	6 2	6.75E−4	4 4	3.88E−3	2 1	1.33E+0	1 2	−1.44E+1	2 1	8.89E−5
	$^3\mathbf{D}^e$–$^3\mathbf{F}^o$	6 3	5.24E−5	4 5	−6.67E−1	2 2	−3.46E−1	2 1	2.30E−2	2 2	−1.64E−2
1 1	3.37E−2	6 4	1.01E−4	5 1	3.94E−1	2 3	−1.10E+0	2 2	−9.06E−3	2 3	−2.43E+0
1 2	−3.62E−2	6 5	−9.20E−3	5 2	3.03E−2	2 4	−5.48E−2	3 1	2.61E+0	2 4	−4.33E+0
1 3	−1.71E+0	7 1	2.23E−2	5 3	1.17E−2	2 5	−3.76E−1	3 2	−3.78E−1		$^5\mathbf{P}^o$–$^5\mathbf{D}^e$
1 4	−3.10E−2	7 2	1.29E−3	5 4	1.34E−1	3 1	6.70E+0	4 1	5.41E−4	1 1	6.28E+0
1 5	−9.47E−1	7 3	1.27E−2	5 5	−4.48E+0	3 2	−1.25E−1	4 2	−6.66E+0	2 1	1.10E+0
2 1	2.01E−1	7 4	6.66E−2	6 1	1.93E+0	3 3	−3.43E−1	5 1	4.15E+0		$^5\mathbf{D}^e$–$^5\mathbf{D}^o$
2 2	−3.84E−1	7 5	2.19E−4	6 2	7.85E−1	3 4	−2.33E−4	5 2	−2.43E−3	1 1	−4.30E+0
2 3	−3.61E−1		$^3\mathbf{D}^o$–$^3\mathbf{F}^e$	6 3	6.98E−1	3 5	−1.00E−1		$^3\mathbf{G}^e$–$^3\mathbf{G}^o$	1 2	−1.76E+0
2 4	−1.11E−2	1 1	−3.45E−3	6 4	5.69E−2	4 1	6.48E+0	1 1	−1.37E+0	1 3	−7.80E+0
2 5	−4.50E−2	1 2	−9.86E+0	6 5	−3.61E−5	4 2	−2.04E−3	2 1	6.12E−4	1 4	−1.88E+0
3 1	6.56E+0	1 3	−4.39E+0	7 1	3.99E−1	4 3	−7.91E−3		$^3\mathbf{G}^e$–$^3\mathbf{H}^o$		$^5\mathbf{D}^e$–$^5\mathbf{F}^o$
3 2	−1.39E−1	1 4	−2.82E−1	7 2	1.67E+0	4 4	−9.01E−3	1 1	−6.93E+0	1 1	−4.64E+0
3 3	−2.50E−1	1 5	−5.50E−1	7 3	2.94E−2	4 5	−7.61E−2	2 1	−1.85E−8		$^5\mathbf{D}^o$–$^5\mathbf{F}^e$
3 4	−1.41E−1	2 1	−6.35E−2	7 4	1.45E−3	5 1	9.39E−1		$^5\mathbf{S}^o$–$^5\mathbf{P}^e$	1 1	7.45E+0
3 5	−9.19E−1	2 2	−6.55E−3	7 5	2.08E−3	5 2	1.05E−3	1 1	−8.80E+0	2 1	4.73E+0
4 1	2.69E−1	2 3	−3.24E+0	8 1	1.12E+0	5 3	2.63E−2	1 2	−1.24E+0	3 1	3.96E−1
4 2	−7.71E−3	2 4	−1.08E+1	8 2	1.09E−2	5 4	5.50E−1		$^5\mathbf{P}^e$–$^5\mathbf{P}^o$	4 1	2.14E+0
4 3	−3.48E−1	2 5	−1.21E+1	8 3	1.97E−3	5 5	2.44E−4	1 1	−6.21E−3		$^5\mathbf{F}^e$–$^5\mathbf{F}^o$
4 4	−1.57E−1	3 1	1.09E+0	8 4	8.48E−3		$^3\mathbf{F}^e$–$^3\mathbf{G}^o$	1 2	−3.97E−1	1 1	−1.94E+1
4 5	−1.32E−1	3 2	2.09E−3	8 5	3.76E−3	1 1	−1.24E+0	2 1	−2.14E−2		$^5\mathbf{F}^e$–$^5\mathbf{G}^o$
5 1	1.24E+0	3 3	2.48E−3		$^3\mathbf{F}^e$–$^3\mathbf{F}^o$	2 1	−4.23E−1	2 2	−4.52E+0	1 1	−5.64E+0
5 2	−1.31E−3	3 4	1.29E−3	1 1	2.61E−1	3 1	−8.48E−1		$^5\mathbf{P}^e$–$^5\mathbf{D}^o$		
5 3	−5.54E−1	3 5	−5.80E−3	1 2	−1.53E+0	4 1	−2.00E−3	1 1	−2.20E+0		
5 4	−1.06E+1	4 1	2.86E+0	1 3	−7.75E−1	5 1	6.53E−3	1 2	−9.64E−1		
5 5	−5.29E−2	4 2	3.30E−1	1 4	−1.48E−3		$^3\mathbf{F}^o$–$^3\mathbf{G}^e$	1 3	−8.56E+0		
6 1	2.16E−1	4 3	4.66E−1	1 5	−3.74E+0	1 1	−3.19E+0	1 4	−3.25E+0		

Si-like Fe (Fe^{12+})

Term energies relative to $3s^23p$ ^{2}P ionization threshold for each symmetry

i	E(Ryds)	Description	i	E(Ryds)	Description	i	E(Ryds)	Description	i	E(Ryds)	Description
		$^1\mathbf{S}^e$	17	−4.64119	$3s3p^2\ ^2\mathrm{D}\ 5d$	41	−2.13304	$3s3p^2\ ^2\mathrm{S}\ 6p$	49	−2.04674	$3s3p^2\ ^2\mathrm{D}\ 6g$
1	−25.6432	$3s^23p^2$	18	−3.92221	$3s^23p\ ^2\mathrm{P}\ 7p$	42	−2.09168	$3s3p^2\ ^2\mathrm{D}\ 6f$			$^1\mathbf{D}^o$
2	−20.2901	$3p^4$	19	−3.82209	$3s3p^2\ ^2\mathrm{P}\ 5d$			$^1\mathbf{D}^e$	1	−23.3014	$3s3p^3$
3	−18.5543	$3s3p^2\ ^2\mathrm{D}\ 3d$	20	−3.11860	$3s^23d\ ^2\mathrm{D}\ 5d$	1	−26.0637	$3s^23p^2$	2	−22.0144	$3s^23p\ ^2\mathrm{P}\ 3d$
4	−16.7643	$3s^23d^2$	21	−2.92475	$3s^23p\ ^2\mathrm{P}\ 8p$	2	−21.0503	$3p^4$	3	−17.1805	$3p^3\ ^2\mathrm{P}\ 3d$:
5	−12.6899	$3s^23p\ ^2\mathrm{P}\ 4p$	22	−2.81750	$3p^3\ ^2\mathrm{D}\ 5p$	3	−18.9514	$3s3p^2\ ^2\mathrm{D}\ 3d$	4	−16.1817	$3p^3\ ^2\mathrm{D}\ 3d$:
6	−12.5239	$3p^23d^2$	23	−2.39579	$3s3p^2\ ^2\mathrm{D}\ 6d$	4	−18.6674	$3s3p^2\ ^2\mathrm{S}\ 3d$	5	−15.3270	$3s3p3d^2$
7	−11.7121	$3p^23d^2$	24	−2.28991	$3s^23p\ ^2\mathrm{P}\ 9p$	5	−17.7951	$3s3p^2\ ^2\mathrm{P}\ 3d$	6	−14.3477	$3s3p3d^2$
8	−10.7243	$3s3p^2\ ^2\mathrm{S}\ 4s$	25	−2.16029	$3p^3\ ^2\mathrm{P}\ 5p$	6	−17.4190	$3s^23d^2$	7	−13.8494	$3s3p3d^2$
9	−10.0492	$3p^23d^2$	26	−2.09755	$3s3p^2\ ^2\mathrm{P}\ 6s$	7	−13.7306	$3p^23d^2$	8	−11.8749	$3s^23p\ ^2\mathrm{P}\ 4d$
10	−8.82492	$3s3p^2\ ^2\mathrm{D}\ 4d$			$^1\mathbf{P}^o$	8	−12.9600	$3s^23p\ ^2\mathrm{P}\ 4p$	9	−10.5239	$3s3p^2\ ^2\mathrm{D}\ 4p$
11	−7.86338	$3s^23p\ ^2\mathrm{P}\ 5p$	1	−22.6118	$3s3p^3$	9	−12.7594	$3p^23d^2$	10	−9.44710	$3s3p^2\ ^2\mathrm{P}\ 4p$
12	−6.93955	$3s^23d\ ^2\mathrm{D}\ 4d$	2	−21.2858	$3s^23p\ ^2\mathrm{P}\ 3d$	10	−12.2290	$3p^23d^2$	11	−9.37857	$3p3d^3$
13	−6.78570	$3p^3\ ^2\mathrm{P}\ 4p$	3	−16.9047	$3p^3\ ^2\mathrm{D}\ 3d$	11	−11.9634	$3p^23d^2$	12	−8.89957	$3s^23d\ ^2\mathrm{D}\ 4p$
14	−5.28837	$3s^23p\ ^2\mathrm{P}\ 6p$	4	−15.4925	$3p^3\ ^2\mathrm{P}\ 3d$	12	−11.8588	$3p^23d^2$	13	−8.78530	$3p3d^3$
15	−5.16349	$3s3p^2\ ^2\mathrm{S}\ 5s$	5	−14.6433	$3s3p3d^2$:	13	−11.2943	$3s3p^2\ ^2\mathrm{D}\ 4s$	14	−8.69138	$3p^3\ ^2\mathrm{D}\ 4s$
16	−4.57851	$3s3p^2\ ^2\mathrm{D}\ 5d$	6	−14.3064	$3s3p3d^2$	14	−10.9197	$3p^23d^2$	15	−8.06508	$3s3p^2\ ^2\mathrm{D}\ 4f$
17	−3.82763	$3s^23p\ ^2\mathrm{P}\ 7p$	7	−14.0842	$3s^23p\ ^2\mathrm{P}\ 4s$	15	−10.6413	$3s^23p\ ^2\mathrm{P}\ 4f$	16	−7.89734	$3p3d^3$
18	−2.96464	$3s^23d\ ^2\mathrm{D}\ 5d$	8	−13.5243	$3s3p3d^2$	16	−10.1127	$3s3d^3$	17	−7.44638	$3s3p^2\ ^2\mathrm{P}\ 4f$
19	−2.81479	$3s^23p\ ^2\mathrm{P}\ 8p$	9	−11.6303	$3s^23p\ ^2\mathrm{P}\ 4d$	17	−9.82556	$3s^23d\ ^2\mathrm{D}\ 4s$	18	−7.30538	$3s^23p\ ^2\mathrm{P}\ 5d$
20	−2.42535	$3s3p^2\ ^2\mathrm{S}\ 6s$	10	−10.3504	$3s3p^2\ ^2\mathrm{D}\ 4p$	18	−9.63905	$3s3d^3$	19	−7.13812	$3p3d^3$
21	−2.31405	$3s3p^2\ ^2\mathrm{D}\ 6d$	11	−9.81799	$3s3p^2\ ^2\mathrm{S}\ 4p$	19	−9.08465	$3s3p^2\ ^2\mathrm{D}\ 4d$	20	−6.60359	$3s^23d\ ^2\mathrm{D}\ 4f$
22	−2.25193	$3s^23p\ ^2\mathrm{P}\ 9p$	12	−9.30312	$3s3p^2\ ^2\mathrm{P}\ 4p$	20	−8.46028	$3s3p^2\ ^2\mathrm{S}\ 4d$	21	−6.43248	$3p^3\ ^2\mathrm{D}\ 4d$
23	−1.96714	$3p^3\ ^2\mathrm{P}\ 5p$	13	−8.66469	$3s^23d\ ^2\mathrm{D}\ 4p$	21	−8.13397	$3s3p^2\ ^2\mathrm{P}\ 4d$	22	−5.87589	$3p^3\ ^2\mathrm{P}\ 4d$
		$^1\mathbf{S}^o$	14	−8.51833	$3p3d^3$	22	−7.86602	$3s^23p\ ^2\mathrm{P}\ 5p$	23	−5.34329	$3s3p^2\ ^2\mathrm{D}\ 5p$
1	−17.7814	$3p^3\ ^2\mathrm{D}\ 3d$	15	−8.37219	$3s^23p\ ^2\mathrm{P}\ 5s$	23	−7.55132	$3p^3\ ^2\mathrm{D}\ 4p$	24	−5.04762	$3s^23p\ ^2\mathrm{P}\ 6d$
2	−15.1845	$3s3p3d^2$	16	−8.09975	$3p^3\ ^2\mathrm{P}\ 4s$	24	−7.27499	$3s^23d\ ^2\mathrm{D}\ 4d$	25	−4.43222	$3s3p^2\ ^2\mathrm{P}\ 5p$
3	−9.85214	$3s3p^2\ ^2\mathrm{P}\ 4p$	17	−7.91567	$3p3d^3$	25	−7.10472	$3p^3\ ^2\mathrm{P}\ 4p$	26	−4.22527	$3s3p^2\ ^2\mathrm{D}\ 5f$
4	−9.58160	$3p3d^3$	18	−7.90165	$3s3p^2\ ^2\mathrm{D}\ 4f$	26	−6.83179	$3s^23p\ ^2\mathrm{P}\ 5f$	27	−3.76902	$3s^23d\ ^2\mathrm{D}\ 5p$
5	−6.50569	$3p^3\ ^2\mathrm{D}\ 4d$	19	−7.26161	$3s^23p\ ^2\mathrm{P}\ 5d$	27	−5.74924	$3s3p^2\ ^2\mathrm{D}\ 5s$	28	−3.66387	$3s^23p\ ^2\mathrm{P}\ 7d$
6	−4.58242	$3s3p^2\ ^2\mathrm{P}\ 5p$	20	−6.68109	$3p3d^3$	28	−5.60575	$3p^3\ ^2\mathrm{D}\ 4f$	29	−3.44684	$3s3p^2\ ^2\mathrm{P}\ 5f$
7	−2.15531	$3p^3\ ^2\mathrm{D}\ 5d$	21	−6.54226	$3p^3\ ^2\mathrm{D}\ 4d$	29	−5.34536	$3s^23p\ ^2\mathrm{P}\ 6p$	30	−3.22337	$3p^3\ ^2\mathrm{D}\ 5s$
8	−1.92158	$3s3p^2\ ^2\mathrm{P}\ 6p$	22	−6.27049	$3s^23d\ ^2\mathrm{D}\ 4f$	30	−4.81260	$3p^3\ ^2\mathrm{P}\ 4f$	31	−2.78670	$3s^23p\ ^2\mathrm{P}\ 8d$
9	−.41351	$3s3p^2\ ^2\mathrm{P}\ 7p$	23	−5.74755	$3p^3\ ^2\mathrm{P}\ 4d$	31	−4.74498	$3s^23p\ ^2\mathrm{P}\ 6f$	32	−2.72787	$3s3p^2\ ^2\mathrm{D}\ 6p$
		$^1\mathbf{P}^e$	24	−5.63152	$3s^23p\ ^2\mathrm{P}\ 6s$	32	−4.68238	$3s3p^2\ ^2\mathrm{D}\ 5d$	33	−2.69284	$3s^23d\ ^2\mathrm{D}\ 5f$
1	−19.3288	$3s3p^2\ ^2\mathrm{P}\ 3d$:	25	−5.26494	$3s3p^2\ ^2\mathrm{D}\ 5p$	33	−4.19971	$3s^23d\ ^2\mathrm{D}\ 5s$	34	−2.18802	$3s^23p\ ^2\mathrm{P}\ 9d$
2	−18.3637	$3s3p^2\ ^2\mathrm{D}\ 3d$:	26	−4.99330	$3s^23p\ ^2\mathrm{P}\ 6d$	34	−4.08498	$3s3p^2\ ^2\mathrm{S}\ 5d$	35	−2.13484	$3s3p^2\ ^2\mathrm{D}\ 6f$
3	−13.3355	$3s^23p\ ^2\mathrm{P}\ 4p$	27	−4.71946	$3s3p^2\ ^2\mathrm{S}\ 5p$	35	−4.04172	$3s3p^2\ ^2\mathrm{D}\ 5g$	36	−2.10110	$3p^3\ ^2\mathrm{D}\ 5d$
4	−12.8551	$3p^23d^2$	28	−4.37962	$3s3p^2\ ^2\mathrm{P}\ 5p$	36	−3.87599	$3s^23p\ ^2\mathrm{P}\ 7p$	37	−1.87498	$3s3p^2\ ^2\mathrm{P}\ 6p$
5	−12.2701	$3p^23d^2$	29	−4.16357	$3s3p^2\ ^2\mathrm{D}\ 5f$	37	−3.73930	$3s3p^2\ ^2\mathrm{P}\ 5d$			$^1\mathbf{F}^e$
6	−10.5203	$3s3p^2\ ^2\mathrm{P}\ 4s$	30	−4.01698	$3s^23p\ ^2\mathrm{P}\ 7s$	38	−3.48250	$3s^23p\ ^2\mathrm{P}\ 7f$	1	−19.5985	$3s3p^2\ ^2\mathrm{D}\ 3d$
7	−10.4712	$3s3d^3$	31	−3.69147	$3s^23d\ ^2\mathrm{D}\ 5p$	39	−3.04055	$3s^23d\ ^2\mathrm{D}\ 5d$	2	−18.1069	$3s3p^2\ ^2\mathrm{P}\ 3d$
8	−9.02953	$3s3p^2\ ^2\mathrm{D}\ 4d$	32	−3.58343	$3s^23p\ ^2\mathrm{P}\ 7d$	40	−2.98973	$3s3p^2\ ^2\mathrm{D}\ 6s$	3	−12.9026	$3p^23d^2$
9	−8.31304	$3s3p^2\ ^2\mathrm{P}\ 4d$	33	−3.01437	$3s^23p\ ^2\mathrm{P}\ 8s$	41	−2.89153	$3s^23p\ ^2\mathrm{P}\ 8p$	4	−12.5022	$3p^23d^2$
10	−8.03154	$3s^23p\ ^2\mathrm{P}\ 5p$	34	−2.77328	$3s^23p\ ^2\mathrm{P}\ 8d$	42	−2.66534	$3s^23p\ ^2\mathrm{P}\ 8f$	5	−11.6988	$3p^23d^2$
11	−7.97396	$3p^3\ ^2\mathrm{D}\ 4p$	35	−2.74966	$3s3p^2\ ^2\mathrm{D}\ 6p$	43	−2.66064	$3p^3\ ^2\mathrm{D}\ 5p$	6	−10.9847	$3s^23p\ ^2\mathrm{P}\ 4f$
12	−7.52807	$3s^23d\ ^2\mathrm{D}\ 4d$	36	−2.55912	$3p^3\ ^2\mathrm{P}\ 5s$	44	−2.51014	$3s^23d\ ^2\mathrm{D}\ 5g$	7	−10.1089	$3s3d^3$
13	−7.18721	$3p^3\ ^2\mathrm{P}\ 4p$	37	−2.54496	$3s^23d\ ^2\mathrm{D}\ 5f$	45	−2.38086	$3s3p^2\ ^2\mathrm{D}\ 6d$	8	−9.20686	$3s3p^2\ ^2\mathrm{D}\ 4d$
14	−5.66007	$3p^3\ ^2\mathrm{D}\ 4f$	38	−2.34966	$3s^23p\ ^2\mathrm{P}\ 9s$	46	−2.26949	$3s^23p\ ^2\mathrm{P}\ 9p$	9	−8.23476	$3s3p^2\ ^2\mathrm{P}\ 4d$
15	−5.43078	$3s^23p\ ^2\mathrm{P}\ 6p$	39	−2.18008	$3s^23p\ ^2\mathrm{P}\ 9d$	47	−2.12092	$3p^3\ ^2\mathrm{P}\ 5p$	10	−7.81591	$3p^3\ ^2\mathrm{D}\ 4p$
16	−4.91490	$3s3p^2\ ^2\mathrm{P}\ 5s$	40	−2.14749	$3p^3\ ^2\mathrm{D}\ 5d$	48	−2.10474	$3s^23p\ ^2\mathrm{P}\ 9f$	11	−7.61322	$3s^23d\ ^2\mathrm{D}\ 4d$

Si-like Fe (Fe^{12+})

i	E(Ryds)	Description	i	E(Ryds)	Description	i	E(Ryds)	Description	i	E(Ryds)	Description
12	−6.97334	$3s^23p$ ^{2}P $5f$	36	−2.15763	$3s3p^2$ ^{2}D $6f$	2	−13.1056	$3s^23p$ ^{2}P $4p$	13	−11.7800	$3p^23d^2$
13	−5.49037	$3p^3$ ^{2}D $4f$	37	−2.09495	$3p^3$ ^{2}D $5d$	3	−11.7521	$3p^23d^2$	14	−11.5322	$3p^23d^2$
14	−5.04262	$3p^3$ ^{2}P $4f$	38	−2.08043	$3s^23p$ ^{2}P $9g$	4	−10.9114	$3s3p^2$ ^{2}S $4s$	15	−11.1064	$3s3d^3$
15	−4.81550	$3s^23p$ ^{2}P $6f$			**^{1}G^e**	5	−8.93260	$3s3p^2$ ^{2}D $4d$	16	−10.5912	$3s3p^2$ ^{2}P $4s$
16	−4.72043	$3s3p^2$ ^{2}D $5d$	1	−19.6101	$3s3p^2$ ^{2}D $3d$	6	−7.97903	$3s^23p$ ^{2}P $5p$	17	−10.0765	$3s3d^3$
17	−4.10489	$3s3p^2$ ^{2}D $5g$	2	−17.4878	$3s^23d^2$	7	−7.42936	$3s^23d$ ^{2}D $4d$	18	−9.80818	$3s3p^2$ ^{4}P $4d$
18	−3.83685	$3s3p^2$ ^{2}P $5d$	3	−13.8270	$3p^23d^2$	8	−7.29632	$3p^3$ ^{2}P $4p$	19	−8.98006	$3s3p^2$ ^{2}D $4d$
19	−3.52476	$3s^23p$ ^{2}P $7f$	4	−12.7315	$3p^23d^2$	9	−5.39264	$3s^23p$ ^{2}P $6p$	20	−8.29890	$3s3p^2$ ^{2}P $4d$
20	−3.30988	$3s3p^2$ ^{2}P $5g$	5	−12.0095	$3p^23d^2$	10	−5.22178	$3s3p^2$ ^{2}S $5s$	21	−7.97506	$3s^23p$ ^{2}P $5p$
21	−3.15682	$3s^23d$ ^{2}D $5d$	6	−11.4846	$3p^23d^2$	11	−4.61260	$3s3p^2$ ^{2}D $5d$	22	−7.89650	$3p^3$ ^{2}D $4p$
22	−2.78925	$3p^3$ ^{2}D $5p$	7	−10.6531	$3s^23p$ ^{2}P $4f$	12	−3.86284	$3s^23p$ ^{2}P $7p$	23	−7.53108	$3p^3$ ^{4}S $4p$
23	−2.69190	$3s^23p$ ^{2}P $8f$	8	−10.4879	$3s3d^3$	13	−3.08899	$3s^23d$ ^{2}D $5d$	24	−7.25456	$3s^23d$ ^{2}D $4d$
24	−2.56431	$3s^23d$ ^{2}D $5g$	9	−8.98981	$3s3p^2$ ^{2}D $4d$	14	−2.90092	$3s^23p$ ^{2}P $8p$	25	−7.19342	$3p^3$ ^{2}P $4p$
25	−2.39034	$3s3p^2$ ^{2}D $6d$	10	−7.25327	$3s^23d$ ^{2}D $4d$	15	−2.45605	$3s3p^2$ ^{2}S $6s$	26	−6.36929	$3s3p^2$ ^{4}P $5s$
26	−2.12240	$3s^23p$ ^{2}P $9f$	11	−6.83704	$3s^23p$ ^{2}P $5f$	16	−2.35123	$3s3p^2$ ^{2}D $6d$	27	−5.57887	$3p^3$ ^{2}D $4f$
27	−2.07285	$3s3p^2$ ^{2}D $6g$	12	−5.47956	$3p^3$ ^{2}D $4f$	17	−2.26557	$3s^23p$ ^{2}P $9p$	28	−5.38627	$3s^23p$ ^{2}P $6p$
		^{1}F$^\circ$	13	−4.95774	$3p^3$ ^{2}P $4f$	18	−2.15132	$3p^3$ ^{2}P $5p$	29	−5.35021	$3s3p^2$ ^{4}P $5d$
1	−21.4174	$3s^23p$ ^{2}P $3d$	14	−4.74927	$3s^23p$ ^{2}P $6f$			**^{3}S$^\circ$**	30	−4.94441	$3s3p^2$ ^{2}P $5s$
2	−16.3419	$3p^3$ ^{2}D $3d$	15	−4.64204	$3s3p^2$ ^{2}D $5d$	1	−22.7353	$3s3p^3$	31	−4.64598	$3s3p^2$ ^{2}D $5d$
3	−16.2501	$3p^3$ ^{2}P $3d$	16	−4.15320	$3s3p^2$ ^{2}D $5g$	2	−16.8267	$3p^3$ ^{2}D $3d$	32	−3.90554	$3s^23p$ ^{2}P $7p$
4	−15.3753	$3s3p3d^2$	17	−3.54716	$3s3p^2$ ^{2}S $5g$	3	−15.8047	$3s3p3d^2$	33	−3.82202	$3s3p^2$ ^{2}P $5d$
5	−14.3294	$3s3p3d^2$	18	−3.48454	$3s^23p$ ^{2}P $7f$	4	−13.9470	$3s3p3d^2$	34	−3.60352	$3s3p^2$ ^{4}P $6s$
6	−13.9636	$3s3p3d^2$	19	−3.24237	$3s3p^2$ ^{2}P $5g$	5	−11.3602	$3s3p^2$ ^{4}P $4p$	35	−3.05412	$3s3p^2$ ^{4}P $6d$
7	−11.6946	$3s^23p$ ^{2}P $4d$	20	−3.01489	$3s^23d$ ^{2}D $5d$	6	−9.59405	$3s3p^2$ ^{2}P $4p$	36	−3.03061	$3s^23d$ ^{2}D $5d$
8	−10.4153	$3s3p^2$ ^{2}D $4p$	21	−2.66630	$3s^23p$ ^{2}P $8f$	7	−8.56450	$3p^3$ ^{4}S $4s$	37	−2.90398	$3s^23p$ ^{2}P $8p$
9	−9.25525	$3p3d^3$	22	−2.56745	$3s^23d$ ^{2}D $5g$	8	−8.52582	$3p3d^3$	38	−2.77724	$3p^3$ ^{2}D $5p$
10	−8.69094	$3s^23d$ ^{2}D $4p$	23	−2.36901	$3s3p^2$ ^{2}D $6d$	9	−7.30700	$3p3d^3$	39	−2.62413	$3p^3$ ^{4}S $5p$
11	−8.17060	$3s3p^2$ ^{2}D $4f$	24	−2.10670	$3s^23p$ ^{2}P $9f$	10	−6.47155	$3p^3$ ^{2}D $4d$	40	−2.37649	$3s3p^2$ ^{2}D $6d$
12	−7.99703	$3p3d^3$	25	−2.09050	$3s3p^2$ ^{2}D $6g$	11	−6.05355	$3s3p^2$ ^{4}P $5p$	41	−2.27989	$3s^23p$ ^{2}P $9p$
13	−7.70379	$3p3d^3$			**^{1}G$^\circ$**	12	−4.48147	$3s3p^2$ ^{2}P $5p$	42	−2.19049	$3s3p^2$ ^{2}P $6s$
14	−7.54480	$3s3p^2$ ^{2}S $4f$	1	−17.1236	$3p^3$ ^{2}D $3d$	13	−3.46495	$3s3p^2$ ^{4}P $6p$	43	−2.09101	$3p^3$ ^{2}P $5p$
15	−7.33246	$3s^23p$ ^{2}P $5d$	2	−15.2869	$3s3p3d^2$	14	−3.07510	$3p^3$ ^{4}S $5s$	44	−2.00665	$3s3p^2$ ^{4}P $7s$
16	−7.12624	$3s3p^2$ ^{2}P $4f$	3	−14.1466	$3s3p3d^2$	15	−2.14155	$3p^3$ ^{2}D $5d$			**^{3}P$^\circ$**
17	−7.11691	$3p3d^3$	4	−9.22938	$3p3d^3$	16	−1.90153	$3s3p^2$ ^{4}P $7p$	1	−23.6079	$3s3p^3$
18	−6.68473	$3s^23p$ ^{2}P $5g$	5	−8.63298	$3p3d^3$	17	−1.86003	$3s3p^2$ ^{2}P $6p$	2	−21.9758	$3s^23p$ ^{2}P $3d$
19	−6.43587	$3p^3$ ^{2}D $4d$	6	−8.25737	$3s3p^2$ ^{2}D $4f$	18	−.94088	$3s3p^2$ ^{4}P $8p$	3	−16.8351	$3p^3$ ^{2}D $3d$
20	−6.39935	$3s^23d$ ^{2}D $4f$	7	−7.82919	$3p3d^3$	19	−.40506	$3s3p^2$ ^{2}P $7p$	4	−16.7881	$3p^3$ ^{2}P $3d$
21	−5.83671	$3p^3$ ^{2}P $4d$	8	−7.39070	$3s3p^2$ ^{2}P $4f$	20	−.36496	$3p^3$ ^{4}S $6s$	5	−15.4479	$3s3p3d^2$
22	−5.30222	$3s3p^2$ ^{2}D $5p$	9	−6.81435	$3s^23d$ ^{2}D $4f$	21	−.26931	$3s3p^2$ ^{4}P $9p$	6	−15.0799	$3s3p3d^2$
23	−5.00675	$3s^23p$ ^{2}P $6d$	10	−6.66288	$3s^23p$ ^{2}P $5g$			**^{3}P^e**	7	−14.8820	$3s3p3d^2$
24	−4.67562	$3s^23p$ ^{2}P $6g$	11	−6.52666	$3p^3$ ^{2}D $4d$	1	−26.3446	$3s^23p^2$	8	−14.2280	$3s3p3d^2$
25	−4.26996	$3s3p^2$ ^{2}D $5f$	12	−4.71633	$3s^23p$ ^{2}P $6g$	2	−21.1719	$3p^4$	9	−14.1722	$3s^23p$ ^{2}P $4s$
26	−3.73965	$3s^23d$ ^{2}D $5p$	13	−4.30126	$3s3p^2$ ^{2}D $5f$	3	−20.0536	$3s3p^2$ ^{4}P $3d$	10	−11.6960	$3s^23p$ ^{2}P $4d$
27	−3.67275	$3s3p^2$ ^{2}S $5f$	14	−3.51423	$3s^23p$ ^{2}P $7g$	4	−19.2576	$3s3p^2$ ^{2}D $3d$	11	−10.8809	$3s3p^2$ ^{4}P $4p$
28	−3.59321	$3s^23p$ ^{2}P $7d$	15	−3.37608	$3s3p^2$ ^{2}P $5f$	5	−18.3612	$3s3p^2$ ^{2}P $3d$	12	−10.2956	$3s3p^2$ ^{2}D $4p$
29	−3.45496	$3s^23p$ ^{2}P $7g$	16	−2.76485	$3s^23d$ ^{2}D $5f$	6	−17.4515	$3s^23d^2$	13	−9.86396	$3s3p^2$ ^{2}S $4p$
30	−3.31020	$3s3p^2$ ^{2}P $5f$	17	−2.64586	$3s^23p$ ^{2}P $8g$	7	−13.8183	$3p^23d^2$	14	−9.57930	$3s3p^2$ ^{2}P $4p$
31	−2.76603	$3s^23p$ ^{2}P $8d$	18	−2.17794	$3s3p^2$ ^{2}D $6f$	8	−13.1413	$3p^23d^2$	15	−9.53912	$3p3d^3$
32	−2.73550	$3s3p^2$ ^{2}D $6p$	19	−2.13773	$3p^3$ ^{2}D $5d$	9	−13.1334	$3s^23p$ ^{2}P $4p$	16	−9.02440	$3p3d^3$
33	−2.66478	$3s^23d$ ^{2}D $5f$	20	−2.09352	$3s^23p$ ^{2}P $9g$	10	−12.5775	$3p^23d^2$	17	−8.77869	$3s^23d$ ^{2}D $4p$
34	−2.60394	$3s^23p$ ^{2}P $8g$			**^{3}S^e**	11	−12.5164	$3p^23d^2$	18	−8.42589	$3s^23p$ ^{2}P $5s$
35	−2.17421	$3s^23p$ ^{2}P $9d$	1	−19.1407	$3s3p^2$ ^{2}D $3d$	12	−11.9177	$3s3p^2$ ^{4}P $4s$	19	−8.25947	$3p3d^3$

Si-like Fe (Fe^{12+})

i	E(Ryds)	Description	i	E(Ryds)	Description	i	E(Ryds)	Description	i	E(Ryds)	Description
20	−8.21986	$3p^3$ ^{2}P 4s	24	−6.89802	$3s^23p$ ^{2}P 5f	26	−7.29982	$3s^23p$ ^{2}P 5d	26	−4.72562	$3s3p^2$ ^{2}D 5d
21	−8.02617	$3s3p^2$ ^{2}D 4f	25	−5.82772	$3s3p^2$ ^{2}D 5s	27	−6.53408	$3p^3$ ^{2}D 4d	27	−4.12734	$3s3p^2$ ^{2}D 5g
22	−7.91875	$3p3d^3$	26	−5.64792	$3p^3$ ^{2}D 4f	28	−6.48798	$3s^23d$ ^{2}D 4f	28	−3.86128	$3s3p^2$ ^{2}P 5d
23	−7.30549	$3s^23p$ ^{2}P 5d	27	−5.39077	$3s^23p$ ^{2}P 6p	29	−6.38242	$3p^3$ ^{4}S 4d	29	−3.51808	$3s^23p$ ^{2}P 7f
24	−6.51444	$3p^3$ ^{2}D 4d	28	−5.24707	$3s3p^2$ ^{4}P 5d	30	−5.92381	$3s3p^2$ ^{4}P 5p	30	−3.30933	$3s3p^2$ ^{2}P 5g
25	−6.45881	$3s^23d$ ^{2}D 4f	29	−5.05619	$3p^3$ ^{2}P 4f	31	−5.86789	$3p^3$ ^{2}P 4d	31	−3.08442	$3s^23d$ ^{2}D 5d
26	−5.87631	$3s3p^2$ ^{4}P 5p	30	−4.77909	$3s^23p$ ^{2}P 6f	32	−5.36109	$3s3p^2$ ^{2}D 5p	32	−3.01593	$3s3p^2$ ^{4}P 6d
27	−5.84911	$3p^3$ ^{2}P 4d	31	−4.71168	$3s3p^2$ ^{2}D 5d	33	−5.04651	$3s^23p$ ^{2}P 6d	33	−2.80285	$3p^3$ ^{2}D 5p
28	−5.65395	$3s^23p$ ^{2}P 6s	32	−4.22171	$3s^23d$ ^{2}D 5s	34	−4.91696	$3s3p^2$ ^{4}P 5f	34	−2.71906	$3s3p^2$ ^{4}P 6g
29	−5.27808	$3s3p^2$ ^{2}D 5p	33	−4.10872	$3s3p^2$ ^{2}S 5d	35	−4.49790	$3s3p^2$ ^{2}P 5p	35	−2.68662	$3s^23p$ ^{2}P 8f
30	−5.01759	$3s^23p$ ^{2}P 6d	34	−4.06903	$3s3p^2$ ^{2}D 5g	36	−4.26228	$3s3p^2$ ^{2}D 5f	36	−2.56103	$3s^23d$ ^{2}D 5g
31	−4.74656	$3s3p^2$ ^{2}S 5p	35	−3.90234	$3s^23p$ ^{2}P 7p	37	−3.76931	$3s^23d$ ^{2}D 5p	37	−2.40133	$3s3p^2$ ^{2}D 6d
32	−4.46776	$3s3p^2$ ^{2}P 5p	36	−3.77879	$3s3p^2$ ^{2}P 5d	38	−3.64688	$3s^23p$ ^{2}P 7d	38	−2.11912	$3s^23p$ ^{2}P 9f
33	−4.21400	$3s3p^2$ ^{2}D 5f	37	−3.50321	$3s^23p$ ^{2}P 7f	39	−3.44121	$3s3p^2$ ^{2}P 5f	39	−2.07777	$3s3p^2$ ^{2}D 6g
34	−4.03147	$3s^23p$ ^{2}P 7s	38	−3.13810	$3s^23d$ ^{2}D 5d	40	−3.36561	$3s3p^2$ ^{4}P 6p	40	−1.77257	$3p^3$ ^{2}D 5f
35	−3.69502	$3s^23d$ ^{2}D 5p	39	−3.02208	$3s3p^2$ ^{2}D 6s	41	−3.24563	$3p^3$ ^{2}D 5s			$^3\mathrm{F}^o$
36	−3.63780	$3s^23p$ ^{2}P 7d	40	−2.97733	$3s3p^2$ ^{4}P 6d	42	−2.79283	$3s3p^2$ ^{4}P 6f	1	−22.5163	$3s^23p$ ^{2}P 3d
37	−3.32256	$3s3p^2$ ^{4}P 6p	41	−2.91119	$3s^23p$ ^{2}P 8p	43	−2.78308	$3s^23p$ ^{2}P 8d	2	−17.6656	$3p^3$ ^{2}D 3d
38	−3.02255	$3s^23p$ ^{2}P 8s	42	−2.81280	$3p^3$ ^{2}D 5p	44	−2.75401	$3s3p^2$ ^{2}D 6p	3	−16.8098	$3p^3$ ^{2}P 3d
39	−2.77812	$3s^23p$ ^{2}P 8d	43	−2.67807	$3s^23p$ ^{2}P 8f	45	−2.64684	$3s^23d$ ^{2}D 5f	4	−16.1007	$3s3p3d^2$
40	−2.73396	$3s3p^2$ ^{2}D 6p	44	−2.52244	$3s^23d$ ^{2}D 5g	46	−2.18974	$3s^23p$ ^{2}P 9d	5	−15.4205	$3s3p3d^2$
41	−2.63370	$3s^23d$ ^{2}D 5f	45	−2.39821	$3s3p^2$ ^{2}D 6d	47	−2.15278	$3p^3$ ^{2}D 5d	6	−15.0573	$3s3p3d^2$
42	−2.61554	$3p^3$ ^{2}P 5s	46	−2.28020	$3s^23p$ ^{2}P 9p	48	−2.14146	$3s3p^2$ ^{2}D 6f	7	−14.6939	$3s3p3d^2$
43	−2.35351	$3s^23p$ ^{2}P 9s	47	−2.17364	$3p^3$ ^{2}P 5p	49	−2.06003	$3p^3$ ^{4}S 5d	8	−11.7667	$3s^23p$ ^{2}P 4d
44	−2.18841	$3s^23p$ ^{2}P 9d	48	−2.11270	$3s^23p$ ^{2}P 9f	50	−1.90224	$3s3p^2$ ^{2}P 6p	9	−10.4830	$3s3p^2$ ^{2}D 4p
45	−2.15817	$3s3p^2$ ^{2}S 6p	49	−2.05316	$3s3p^2$ ^{2}D 6g	51	−1.84806	$3s3p^2$ ^{4}P 7p	10	−9.57318	$3p3d^3$
46	−2.13785	$3p^3$ ^{2}D 5d			$^3\mathrm{D}^o$			$^3\mathrm{F}^e$	11	−9.09421	$3p3d^3$
47	−2.11684	$3s3p^2$ ^{2}D 6f	1	−23.9846	$3s3p^3$	1	−20.4311	$3s3p^2$ ^{2}D 3d:	12	−8.84318	$3s^23d$ ^{2}D 4p
		$^3\mathrm{D}^e$	2	−21.8811	$3s^23p$ ^{2}P 3d	2	−19.4020	$3s3p^2$ ^{4}P 3d:	13	−8.70013	$3p3d^3$
1	−19.6847	$3s3p^2$ ^{4}P 3d:	3	−17.6326	$3p^3$ ^{2}D 3d	3	−18.9969	$3s3p^2$ ^{2}P 3d	14	−8.52850	$3s3p^2$ ^{4}P 4f
2	−19.3465	$3s3p^2$ ^{2}D 3d:	4	−16.8302	$3p^3$ ^{2}P 3d	4	−17.7686	$3s^23d^2$	15	−8.40286	$3p3d^3$
3	−19.0666	$3s3p^2$ ^{2}S 3d	5	−16.1924	$3p^3$ ^{4}S 3d	5	−13.9001	$3p^23d^2$	16	−8.30625	$3s3p^2$ ^{2}D 4f
4	−18.6932	$3s3p^2$ ^{2}P 3d	6	−15.9762	$3s3p3d^2$	6	−13.4029	$3p^23d^2$	17	−7.97983	$3p3d^3$
5	−13.5762	$3p^23d^2$	7	−15.1935	$3s3p3d^2$	7	−12.8853	$3p^23d^2$	18	−7.64139	$3s3p^2$ ^{2}S 4f
6	−13.2299	$3s^23p$ ^{2}P 4p	8	−15.0062	$3s3p3d^2$	8	−12.6221	$3p^23d^2$	19	−7.43865	$3s3p^2$ ^{2}P 4f
7	−13.1539	$3p^23d^2$	9	−14.4098	$3s3p3d^2$	9	−12.3833	$3p^23d^2$	20	−7.27386	$3s^23p$ ^{2}P 5d
8	−12.4506	$3p^23d^2$	10	−14.3457	$3s3p3d^2$	10	−11.6340	$3p^23d^2$	21	−6.76693	$3s^23d$ ^{2}D 4f
9	−12.2621	$3p^23d^2$	11	−11.8520	$3s^23p$ ^{2}P 4d	11	−10.9831	$3s3d^3$	22	−6.61650	$3s^23p$ ^{2}P 5g
10	−11.8866	$3p^23d^2$	12	−11.0287	$3s3p^2$ ^{4}P 4p	12	−10.9205	$3s^23p$ ^{2}P 4f	23	−6.54018	$3p^3$ ^{2}D 4d
11	−11.4933	$3s3p^2$ ^{2}D 4s	13	−10.4685	$3s3p^2$ ^{2}D 4p	13	−10.5407	$3s3d^3$	24	−5.89492	$3p^3$ ^{2}P 4d
12	−11.1701	$3s3d^3$	14	−9.68398	$3p3d^3$	14	−9.69235	$3s3p^2$ ^{4}P 4d	25	−5.35011	$3s3p^2$ ^{2}D 5p
13	−10.7968	$3s^23p$ ^{2}P 4f	15	−9.65911	$3s3p^2$ ^{2}P 4p	15	−9.16653	$3s3p^2$ ^{2}D 4d	26	−5.01220	$3s^23p$ ^{2}P 6d
14	−10.5712	$3s3d^3$	16	−9.05485	$3p3d^3$	16	−8.32897	$3s3p^2$ ^{2}P 4d	27	−4.79139	$3s3p^2$ ^{4}P 5f
15	−9.84083	$3s^23d$ ^{2}D 4s	17	−8.89750	$3p^3$ ^{2}D 4s	17	−7.86815	$3p^3$ ^{2}D 4p	28	−4.68523	$3s^23p$ ^{2}P 6g
16	−9.53121	$3s3p^2$ ^{4}P 4d	18	−8.83710	$3s3p^2$ ^{4}P 4f	18	−7.41012	$3s^23d$ ^{2}D 4d	29	−4.32475	$3s3p^2$ ^{2}D 5f
17	−9.15507	$3s3p^2$ ^{2}D 4d	19	−8.82564	$3s^23d$ ^{2}D 4p	19	−6.94832	$3s^23p$ ^{2}P 5f	30	−3.75032	$3s^23d$ ^{2}D 5p
18	−8.48305	$3s3p^2$ ^{2}S 4d	20	−8.66072	$3p3d^3$	20	−5.74101	$3p^3$ ^{2}D 4f	31	−3.72245	$3s3p^2$ ^{2}S 5f
19	−8.25190	$3s3p^2$ ^{2}P 4d	21	−8.48277	$3p3d^3$	21	−5.49435	$3p^3$ ^{4}S 4f	32	−3.64406	$3s^23p$ ^{2}P 7d
20	−7.97253	$3s^23p$ ^{2}P 5p	22	−8.18708	$3p3d^3$	22	−5.28745	$3s3p^2$ ^{4}P 5d	33	−3.47798	$3s^23p$ ^{2}P 7g
21	−7.95779	$3p^3$ ^{2}D 4p	23	−8.13735	$3s3p^2$ ^{2}D 4f	23	−5.05343	$3p^3$ ^{2}P 4f	34	−3.38793	$3s3p^2$ ^{2}P 5f
22	−7.56122	$3s^23d$ ^{2}D 4d	24	−7.42781	$3s3p^2$ ^{2}P 4f	24	−4.80471	$3s^23p$ ^{2}P 6f	35	−2.78863	$3s^23p$ ^{2}P 8d
23	−7.25725	$3p^3$ ^{2}P 4p	25	−7.40034	$3p3d^3$	25	−4.73955	$3s3p^2$ ^{4}P 5g	36	−2.74782	$3s3p^2$ ^{2}D 6p

Si-like Fe (Fe^{12+})

i	E(Ryds)	Description
37	−2.73086	$3s3p^2$ ^{4}P $6f$
38	−2.72380	$3s^23d$ ^{2}D $5f$
39	−2.63447	$3s^23p$ ^{2}P $8g$
40	−2.18606	$3s3p^2$ ^{2}D $6f$
41	−2.17824	$3s^23p$ ^{2}P $9d$
42	−2.14499	$3p^3$ ^{2}D $5d$
43	−2.08597	$3s^23p$ ^{2}P $9g$
	^{3}G^e	
1	−19.7746	$3s3p^2$ ^{2}D $3d$
2	−13.5255	$3p^23d^2$
3	−12.6001	$3p^23d^2$
4	−12.4864	$3p^23d^2$
5	−11.3839	$3s3d^3$
6	−10.9140	$3s^23p$ ^{2}P $4f$
7	−9.06507	$3s3p^2$ ^{2}D $4d$
8	−7.55800	$3s^23d$ ^{2}D $4d$
9	−6.94354	$3s^23p$ ^{2}P $5f$
10	−5.76342	$3p^3$ ^{2}D $4f$
11	−5.09585	$3p^3$ ^{2}P $4f$
12	−4.80350	$3s^23p$ ^{2}P $6f$
13	−4.72658	$3s3p^2$ ^{4}P $5g$
14	−4.67776	$3s3p^2$ ^{2}D $5d$
15	−4.17429	$3s3p^2$ ^{2}D $5g$
16	−3.56084	$3s3p^2$ ^{2}S $5g$
17	−3.51635	$3s^23p$ ^{2}P $7f$
18	−3.25496	$3s3p^2$ ^{2}P $5g$
19	−3.12764	$3s^23d$ ^{2}D $5d$
20	−2.69017	$3s3p^2$ ^{4}P $6g$
21	−2.68262	$3s^23p$ ^{2}P $8f$
22	−2.58029	$3s^23d$ ^{2}D $5g$
23	−2.38385	$3s3p^2$ ^{2}D $6d$
24	−2.11810	$3s^23p$ ^{2}P $9f$
25	−2.09799	$3s3p^2$ ^{2}D $6g$
26	−1.77811	$3p^3$ ^{2}D $5f$
	^{3}G^o	
1	−17.4014	$3p^3$ ^{2}D $3d$
2	−15.7996	$3s3p3d^2$
3	−15.5300	$3s3p3d^2$
4	−14.7996	$3s3p3d^2$
5	−9.60339	$3p3d^3$
6	−8.85853	$3p3d^3$
7	−8.76140	$3s3p^2$ ^{4}P $4f$
8	−8.56686	$3p3d^3$
9	−8.29632	$3s3p^2$ ^{2}D $4f$
10	−8.02368	$3p3d^3$
11	−7.53281	$3s3p^2$ ^{2}P $4f$
12	−6.79694	$3s^23p$ ^{2}P $5g$
13	−6.53341	$3p^3$ ^{2}D $4d$
14	−6.48265	$3s^23d$ ^{2}D $4f$
15	−4.88214	$3s3p^2$ ^{4}P $5f$
16	−4.71140	$3s^23p$ ^{2}P $6g$
17	−4.33212	$3s3p^2$ ^{2}D $5f$
18	−3.53181	$3s3p^2$ ^{2}P $5f$
19	−3.41889	$3s^23p$ ^{2}P $7g$
20	−2.77472	$3s3p^2$ ^{4}P $6f$
21	−2.69759	$3s^23d$ ^{2}D $5f$
22	−2.62534	$3s^23p$ ^{2}P $8g$
23	−2.18785	$3s3p^2$ ^{2}D $6f$
24	−2.14995	$3p^3$ ^{2}D $5d$
25	−2.09142	$3s^23p$ ^{2}P $9g$
	^{5}S^o	
1	−24.6918	$3s3p^3$
2	−15.7388	$3s3p3d^2$
3	−11.0706	$3s3p^2$ ^{4}P $4p$
4	−9.50151	$3p3d^3$
5	−8.85096	$3p^3$ ^{4}S $4s$
6	−5.98080	$3s3p^2$ ^{4}P $5p$
7	−3.40334	$3s3p^2$ ^{4}P $6p$
8	−3.18518	$3p^3$ ^{4}S $5s$
9	−1.87280	$3s3p^2$ ^{4}P $7p$
10	−.92558	$3s3p^2$ ^{4}P $8p$
11	−.40494	$3p^3$ ^{4}S $6s$
12	−.28197	$3s3p^2$ ^{4}P $9p$
	^{5}P^e	
1	−20.2285	$3s3p^2$ ^{4}P $3d$
2	−13.2630	$3p^23d^2$
3	−12.1883	$3s3p^2$ ^{4}P $4s$
4	−11.8231	$3s3d^3$
5	−9.82503	$3s3p^2$ ^{4}P $4d$
6	−7.90477	$3p^3$ ^{4}S $4p$
7	−6.47684	$3s3p^2$ ^{4}P $5s$
8	−5.36583	$3s3p^2$ ^{4}P $5d$
9	−3.65353	$3s3p^2$ ^{4}P $6s$
10	−3.05665	$3s3p^2$ ^{4}P $6d$
11	−2.74167	$3p^3$ ^{4}S $5p$
12	−2.03779	$3s3p^2$ ^{4}P $7s$
13	−1.66424	$3s3p^2$ ^{4}P $7d$
14	−1.03209	$3s3p^2$ ^{4}P $8s$
15	−.78842	$3s3p^2$ ^{4}P $8d$
16	−.36170	$3s3p^2$ ^{4}P $9s$
17	−.20249	$3s3p^2$ ^{4}P $9d$
18	−.13961	$3p^3$ ^{4}S $6p$
	^{5}P^o	
1	−16.2586	$3s3p3d^2$
2	−11.1323	$3s3p^2$ ^{4}P $4p$
3	−9.26490	$3p3d^3$
4	−5.98024	$3s3p^2$ ^{4}P $5p$
5	−3.37909	$3s3p^2$ ^{4}P $6p$
6	−1.87362	$3s3p^2$ ^{4}P $7p$
7	−.92370	$3s3p^2$ ^{4}P $8p$
8	−.28576	$3s3p^2$ ^{4}P $9p$
	^{5}D^e	
1	−20.6606	$3s3p^2$ ^{4}P $3d$
2	−13.7068	$3p^23d^2$
3	−13.2283	$3p^23d^2$
4	−9.66327	$3s3p^2$ ^{4}P $4d$
5	−5.30612	$3s3p^2$ ^{4}P $5d$
6	−3.01019	$3s3p^2$ ^{4}P $6d$
7	−1.64920	$3s3p^2$ ^{4}P $7d$
8	−.77682	$3s3p^2$ ^{4}P $8d$
9	−.18435	$3s3p^2$ ^{4}P $9d$
	^{5}D^o	
1	−17.5963	$3p^3$ ^{4}S $3d$
2	−16.4136	$3s3p3d^2$
3	−16.0773	$3s3p3d^2$
4	−11.2332	$3s3p^2$ ^{4}P $4p$
5	−9.66981	$3p3d^3$
6	−9.19444	$3p3d^3$
7	−8.92694	$3s3p^2$ ^{4}P $4f$
8	−6.49749	$3p^3$ ^{4}S $4d$
9	−6.01291	$3s3p^2$ ^{4}P $5p$
10	−4.95891	$3s3p^2$ ^{4}P $5f$
11	−3.40009	$3s3p^2$ ^{4}P $6p$
12	−2.81394	$3s3p^2$ ^{4}P $6f$
13	−2.10395	$3p^3$ ^{4}S $5d$
14	−1.87812	$3s3p^2$ ^{4}P $7p$
15	−1.52784	$3s3p^2$ ^{4}P $7f$
16	−.93072	$3s3p^2$ ^{4}P $8p$
17	−.69690	$3s3p^2$ ^{4}P $8f$
18	−.29130	$3s3p^2$ ^{4}P $9p$
19	−.12890	$3s3p^2$ ^{4}P $9f$
	^{5}F^e	
1	−20.8249	$3s3p^2$ ^{4}P $3d$
2	−13.7376	$3p^23d^2$
3	−12.1284	$3s3d^3$
4	−9.78139	$3s3p^2$ ^{4}P $4d$
5	−5.74835	$3p^3$ ^{4}S $4f$
6	−5.34698	$3s3p^2$ ^{4}P $5d$
7	−4.78145	$3s3p^2$ ^{4}P $5g$
8	−3.03509	$3s3p^2$ ^{4}P $6d$
9	−2.72804	$3s3p^2$ ^{4}P $6g$
10	−1.74911	$3p^3$ ^{4}S $5f$
11	−1.65091	$3s3p^2$ ^{4}P $7d$
12	−1.46878	$3s3p^2$ ^{4}P $7g$
13	−.78542	$3s3p^2$ ^{4}P $8d$
14	−.66130	$3s3p^2$ ^{4}P $8g$
15	−.19071	$3s3p^2$ ^{4}P $9d$
16	−.10467	$3s3p^2$ ^{4}P $9g$
	^{5}F^o	
1	−16.3699	$3s3p3d^2$
2	−9.70802	$3p3d^3$
3	−8.85810	$3s3p^2$ ^{4}P $4f$
4	−4.93026	$3s3p^2$ ^{4}P $5f$
5	−2.80090	$3s3p^2$ ^{4}P $6f$
6	−1.52007	$3s3p^2$ ^{4}P $7f$
7	−.69147	$3s3p^2$ ^{4}P $8f$
8	−.12497	$3s3p^2$ ^{4}P $9f$
	^{5}G^e	
1	−14.0867	$3p^23d^2$
2	−4.75304	$3s3p^2$ ^{4}P $5g$
3	−2.70517	$3s3p^2$ ^{4}P $6g$
4	−1.46150	$3s3p^2$ ^{4}P $7g$
5	−.65304	$3s3p^2$ ^{4}P $8g$
6	−.09838	$3s3p^2$ ^{4}P $9g$
	^{5}G^o	
1	−16.8522	$3s3p3d^2$
2	−10.0785	$3p3d^3$
3	−9.00067	$3s3p^2$ ^{4}P $4f$
4	−4.98720	$3s3p^2$ ^{4}P $5f$
5	−2.82813	$3s3p^2$ ^{4}P $6f$
6	−1.55815	$3p^3$ ^{4}S $5g$
7	−1.53063	$3s3p^2$ ^{4}P $7f$
8	−.70184	$3s3p^2$ ^{4}P $8f$
9	−.13220	$3s3p^2$ ^{4}P $9f$

Si-like Fe (Fe^{12+})

Energies in ascending order from ground state for terms with effective $n \leq 4.0$, $L \leq 4$

Term	i	E(Ryds)	Term	i	E(Ryds)	Term	i	E(Ryds)	Term	i	E(Ryds)	Term	i	E(Ryds)
$^3P^e$	1	0.00000	$^1S^e$	3	7.79030	$^3G^o$	3	10.8146	$^1P^o$	8	12.8203	$^3D^o$	11	14.4926
$^1D^e$	1	0.28090	$^1P^e$	2	7.98090	$^1P^o$	4	10.8521	$^3F^e$	6	12.9417	$^5P^e$	4	14.5215
$^1S^e$	1	0.70140	$^3P^e$	5	7.98340	$^3P^o$	5	10.8967	$^1P^e$	3	13.0091	$^3P^e$	13	14.5646
$^5S^o$	1	1.65280	$^1F^e$	2	8.23770	$^3F^o$	5	10.9241	$^5P^e$	2	13.0816	$^3F^o$	8	14.5779
$^3D^o$	1	2.36000	$^1D^e$	5	8.54950	$^1F^o$	4	10.9693	$^3D^e$	6	13.1147	$^3S^e$	3	14.5925
$^3P^o$	1	2.73670	$^1S^o$	1	8.56320	$^1D^o$	5	11.0176	$^5D^e$	3	13.1163	$^1S^e$	7	14.6325
$^1D^o$	1	3.04320	$^3F^e$	4	8.57600	$^1G^o$	2	11.0577	$^3D^e$	7	13.1907	$^1F^e$	5	14.6458
$^3S^o$	1	3.60930	$^3F^o$	2	8.67900	$^3D^o$	7	11.1511	$^3P^e$	8	13.2033	$^3P^o$	10	14.6486
$^1P^o$	1	3.73280	$^3D^o$	3	8.71200	$^1S^o$	2	11.1601	$^3P^e$	9	13.2112	$^1F^o$	7	14.6500
$^3F^o$	1	3.82830	$^5D^o$	1	8.74830	$^3P^o$	6	11.2647	$^3S^e$	2	13.2390	$^3F^e$	10	14.7106
$^1D^o$	2	4.33020	$^1G^e$	2	8.85680	$^3F^o$	6	11.2873	$^1F^e$	3	13.3820	$^1P^o$	9	14.7143
$^3P^o$	2	4.36880	$^3P^e$	6	8.89310	$^3D^o$	8	11.3384	$^1D^e$	8	13.3846	$^3P^e$	14	14.8124
$^3D^o$	2	4.46350	$^1D^e$	6	8.92560	$^3P^o$	7	11.4626	$^3F^e$	7	13.4593	$^3D^e$	11	14.8513
$^1F^o$	1	4.92720	$^3G^o$	1	8.94320	$^3G^o$	4	11.5450	$^1P^e$	4	13.4895	$^1G^e$	6	14.8600
$^1P^o$	2	5.05880	$^1D^o$	3	9.16410	$^3F^o$	7	11.6507	$^1D^e$	9	13.5852	$^3G^e$	5	14.9607
$^3P^e$	2	5.17270	$^1G^o$	1	9.22100	$^1P^o$	5	11.7013	$^1G^e$	4	13.6131	$^3S^o$	5	14.9844
$^1D^e$	2	5.29430	$^1P^o$	3	9.43990	$^3D^o$	9	11.9348	$^1S^e$	5	13.6547	$^1D^e$	13	15.0503
$^5F^e$	1	5.51970	$^5G^o$	1	9.49240	$^1D^o$	6	11.9969	$^3F^e$	8	13.7225	$^5D^o$	4	15.1114
$^5D^e$	1	5.68400	$^3P^o$	3	9.50950	$^3D^o$	10	11.9989	$^3G^e$	3	13.7445	$^3D^e$	12	15.1745
$^3F^e$	1	5.91350	$^3D^o$	4	9.51440	$^1F^o$	5	12.0152	$^3P^e$	10	13.7671	$^5P^o$	2	15.2123
$^1S^e$	2	6.05450	$^3S^o$	2	9.51790	$^1P^o$	6	12.0382	$^1S^e$	6	13.8207	$^3P^e$	15	15.2382
$^5P^e$	1	6.11610	$^3F^o$	3	9.53480	$^3P^o$	8	12.1166	$^3P^e$	11	13.8282	$^5S^o$	3	15.2740
$^3P^e$	3	6.29100	$^3P^o$	4	9.55650	$^3P^o$	9	12.1724	$^1F^e$	4	13.8424	$^3D^o$	12	15.3159
$^3G^e$	1	6.57000	$^1S^e$	4	9.58030	$^1G^o$	3	12.1980	$^3G^e$	4	13.8582	$^1F^e$	6	15.3599
$^3D^e$	1	6.65990	$^5D^o$	2	9.93100	$^5G^e$	1	12.2579	$^3D^e$	8	13.8940	$^3F^e$	11	15.3615
$^1G^e$	1	6.73450	$^5F^o$	1	9.97470	$^1P^o$	7	12.2604	$^3F^e$	9	13.9613	$^3F^e$	12	15.4241
$^1F^e$	1	6.74610	$^1F^o$	2	10.0027	$^1F^o$	6	12.3810	$^1P^e$	5	14.0745	$^1D^e$	14	15.4249
$^3F^e$	2	6.94260	$^5P^o$	1	10.0860	$^3S^o$	4	12.3976	$^3D^e$	9	14.0825	$^3G^e$	6	15.4306
$^3D^e$	2	6.99810	$^1F^o$	3	10.0945	$^3F^e$	5	12.4445	$^1D^e$	10	14.1156	$^3S^e$	4	15.4332
$^1P^e$	1	7.01580	$^3D^o$	5	10.1522	$^1D^o$	7	12.4952	$^5P^e$	3	14.1563	$^3P^o$	11	15.4637
$^3P^e$	4	7.08700	$^1D^o$	4	10.1629	$^1G^e$	3	12.5176	$^5F^e$	3	14.2162	$^3D^e$	13	15.5478
$^3S^e$	1	7.20390	$^3F^o$	4	10.2439	$^3P^e$	7	12.5263	$^1G^e$	5	14.3351	$^1S^e$	8	15.6203
$^3D^e$	3	7.27800	$^5D^o$	3	10.2673	$^5F^e$	2	12.6070	$^1D^e$	11	14.3812	$^1G^e$	7	15.6915
$^3F^e$	3	7.34770	$^3D^o$	6	10.3684	$^1D^e$	7	12.6140	$^3P^e$	12	14.4269	$^1D^e$	15	15.7033
$^1D^e$	3	7.39320	$^3S^o$	3	10.5399	$^5D^e$	2	12.6378	$^3D^e$	10	14.4580	$^3P^e$	16	15.7534
$^3D^e$	4	7.65140	$^3G^o$	2	10.5450	$^3D^e$	5	12.7684	$^1D^o$	8	14.4697	$^3D^e$	14	15.7734
$^1D^e$	4	7.67720	$^5S^o$	2	10.6058	$^3G^e$	2	12.8191	$^1D^e$	12	14.4858			

Si-like Fe (Fe^{12+})

gf-values for transitions involving terms with effective $n \leq 4.0$, $L \leq 4$

i i'	gf_L	i i'	gf_L	i i'	gf_L	i i'	gf_L	i i'	gf_L	i i'	gf_L
	$^1S^o$–$^1P^e$	5 4	1.74E−4	4 3	−1.30E−4	2 8	−8.87E−3	5 14	−1.18E+0	9 5	7.02E−3
1 1	6.67E−2	5 5	9.21E−4	4 4	−6.76E−2	2 9	−4.68E−4	5 15	−2.64E−3	9 6	1.55E−3
1 2	1.27E−4	5 6	2.99E−2	4 5	−1.25E−1	2 10	−1.73E−5	6 1	6.64E−2	9 7	8.07E−7
1 3	−2.73E−9	5 7	2.21E−1	5 1	8.24E−1	2 11	−1.56E−7	6 2	1.35E−3	9 8	8.24E−2
1 4	−3.01E−1	5 8	1.53E−3	5 2	5.04E−1	2 12	−2.29E−4	6 3	4.87E−2	9 9	1.10E−4
1 5	−2.38E−1	5 9	−4.08E−1	5 3	−2.47E−3	2 13	−3.50E−4	6 4	5.91E−1	9 10	2.12E−6
2 1	7.37E−1	6 1	4.44E−4	5 4	−1.04E−1	2 14	−8.76E−3	6 5	6.33E−3	9 11	1.51E−5
2 2	1.05E−3	6 2	1.44E−4	5 5	−8.47E−2	2 15	−2.24E+0	6 6	4.88E−1	9 12	8.61E−6
2 3	−8.05E−6	6 3	7.01E−1	6 1	1.41E−1	3 1	4.67E−4	6 7	−2.05E−3	9 13	−1.15E−3
2 4	−2.58E−3	6 4	7.44E−2	6 2	1.42E−1	3 2	9.59E−1	6 8	−1.08E−1	9 14	−7.39E−4
2 5	−6.34E−1	6 5	9.09E−2	6 3	−3.58E−2	3 3	7.29E−2	6 9	−4.07E−2	9 15	−7.85E−1
	$^1S^e$–$^1P^o$	6 6	7.24E−3	6 4	−3.93E−2	3 4	4.51E−2	6 10	−2.33E−2		**$^1P^e$–$^1D^o$**
1 1	−1.75E−1	6 7	3.35E−4	6 5	−1.02E−2	3 5	2.06E−5	6 11	−6.85E−1	1 1	9.74E−1
1 2	−1.12E+0	6 8	3.43E−4	7 1	6.12E−3	3 6	1.97E−3	6 12	−8.23E−5	1 2	2.28E−1
1 3	−2.87E−4	6 9	−1.68E−7	7 2	3.17E−2	3 7	−1.73E−1	6 13	−1.41E−5	1 3	−1.29E−1
1 4	−5.29E−4	7 1	1.35E−3	7 3	−3.48E−1	3 8	−1.56E−4	6 14	−2.73E−4	1 4	−4.49E−1
1 5	−1.79E−3	7 2	7.53E−4	7 4	−2.39E−3	3 9	−4.52E−1	6 15	−1.03E−3	1 5	−2.95E−2
1 6	−1.97E−2	7 3	3.72E−3	7 5	−3.53E−3	3 10	−5.44E−2	7 1	5.41E−1	1 6	−1.24E−2
1 7	−7.12E−2	7 4	2.68E−4	8 1	2.61E−3	3 11	−3.15E−1	7 2	1.25E−4	1 7	−3.39E−1
1 8	−1.72E−3	7 5	4.57E−2	8 2	6.67E−1	3 12	−3.17E−1	7 3	2.63E−3	1 8	−3.66E−4
1 9	−5.24E−1	7 6	3.02E−1	8 3	−1.16E−3	3 13	−2.48E−4	7 4	6.46E−2	2 1	8.63E−5
2 1	2.03E−1	7 7	1.79E−2	8 4	−1.21E−3	3 14	−4.53E−2	7 5	4.82E−2	2 2	3.11E−1
2 2	7.31E−4	7 8	4.51E−3	8 5	−1.88E−2	3 15	−6.05E−4	7 6	5.30E−2	2 3	−5.10E−2
2 3	−1.74E−1	7 9	−1.08E−5	9 1	2.09E−4	4 1	3.31E−4	7 7	−7.23E−5	2 4	−8.10E−2
2 4	−9.06E−1	8 1	1.14E−1	9 2	8.07E−4	4 2	1.91E−2	7 8	−9.30E−1	2 5	−5.17E−1
2 5	−1.45E−2	8 2	7.14E−3	9 3	3.56E−1	4 3	3.43E−2	7 9	−4.55E−3	2 6	−9.56E−1
2 6	−4.71E−4	8 3	8.25E−5	9 4	3.90E−6	4 4	4.22E−2	7 10	−2.13E−4	2 7	−1.98E−1
2 7	−3.01E−6	8 4	5.84E−3	9 5	7.10E−7	4 5	3.30E−1	7 11	−3.76E−2	2 8	−1.60E−3
2 8	−4.74E−5	8 5	4.35E−3		**$^1P^o$–$^1D^e$**	4 6	3.43E−3	7 12	−7.76E−4	3 1	1.08E−1
2 9	−3.10E−5	8 6	1.56E−4	1 1	8.96E−1	4 7	−9.75E−3	7 13	−9.97E−2	3 2	6.94E−2
3 1	5.03E−1	8 7	1.73E−1	1 2	−7.50E−2	4 8	−4.97E−4	7 14	−1.72E−3	3 3	1.14E−4
3 2	5.07E−2	8 8	4.15E−3	1 3	−3.79E−2	4 9	−8.87E−2	7 15	−9.98E−3	3 4	7.38E−6
3 3	−2.61E−2	8 9	3.34E−3	1 4	−7.12E−1	4 10	−2.33E−2	8 1	1.14E−3	3 5	1.09E−4
3 4	−2.15E−1		**$^1P^o$–$^1P^e$**	1 5	−1.44E+0	4 11	−1.10E−3	8 2	5.86E−4	3 6	1.63E−6
3 5	−5.41E−1	1 1	−4.15E−3	1 6	−9.99E−5	4 12	−1.39E+0	8 3	1.39E−2	3 7	2.50E−5
3 6	−1.28E−2	1 2	−1.36E+0	1 7	−2.70E−5	4 13	−1.37E−2	8 4	1.03E−2	3 8	−1.09E+0
3 7	−3.02E−2	1 3	−7.53E−3	1 8	−2.52E−4	4 14	−5.09E−1	8 5	1.72E+0	4 1	9.83E−4
3 8	−2.93E−1	1 4	−1.51E−3	1 9	−2.33E−6	4 15	−1.06E−4	8 6	7.41E−3	4 2	4.38E−6
3 9	−2.14E−3	1 5	−1.84E−4	1 10	−9.21E−3	5 1	2.51E−3	8 7	9.52E−6	4 3	6.96E−1
4 1	1.45E−3	2 1	−5.48E−4	1 11	−5.38E−3	5 2	2.78E−4	8 8	−4.76E−3	4 4	4.28E−2
4 2	7.44E−1	2 2	−1.99E−1	1 12	−4.61E−4	5 3	3.19E−1	8 9	−4.32E−3	4 5	1.07E−1
4 3	1.42E−4	2 3	−2.53E−2	1 13	−9.82E−2	5 4	3.78E−2	8 10	−7.63E−2	4 6	2.45E−2
4 4	−1.83E−4	2 4	−7.24E−5	1 14	−8.62E−4	5 5	3.36E−2	8 11	−1.30E−2	4 7	2.75E−3
4 5	−2.20E−2	2 5	−1.37E−4	1 15	−2.39E−1	5 6	2.25E−1	8 12	−4.46E−2	4 8	−4.58E−6
4 6	−2.14E−1	3 1	2.18E−1	2 1	3.42E−4	5 7	−3.21E−2	8 13	−1.20E−2	5 1	1.86E−4
4 7	−1.08E−2	3 2	1.57E−2	2 2	−7.68E−4	5 8	−5.45E−3	8 14	−5.02E−2	5 2	6.73E−4
4 8	−6.55E−1	3 3	−2.04E−4	2 3	−5.70E−2	5 9	−2.00E−2	8 15	−1.97E−3	5 3	1.05E−1
4 9	−8.07E−3	3 4	−1.63E−1	2 4	−1.94E−1	5 10	−2.97E−1	9 1	1.77E−2	5 4	5.46E−1
5 1	1.66E−2	3 5	−5.01E−1	2 5	−1.21E+0	5 11	−1.31E−3	9 2	8.49E−4	5 5	1.67E−1
5 2	5.18E−2	4 1	2.09E−1	2 6	−8.97E−1	5 12	−5.45E−2	9 3	3.01E−4	5 6	3.04E−3
5 3	9.42E−6	4 2	3.11E−1	2 7	−9.56E−5	5 13	−6.65E−3	9 4	4.90E−3	5 7	5.80E−2

Si-like Fe (Fe^{12+})

i i'	gf_L	i i'	gf_L	i i'	gf_L	i i'	gf_L	i i'	gf_L	i i'	gf_L
5 8	−2.05E−6	4 6	1.18E−2	7 13	−1.30E−2	6 4	−1.56E−2	6 2	−6.60E−3	13 5	2.66E−4
	$^1\mathbf{D}^o$–$^1\mathbf{D}^e$	4 7	−1.45E−1	7 14	−5.52E−1	6 5	−1.60E+0	6 3	−1.35E−2	13 6	3.63E−3
1 1	4.59E−1	4 8	−7.49E−6	7 15	−4.56E−3	6 6	0.00E+0	6 4	−2.40E−1	13 7	3.65E−3
1 2	−5.39E−1	4 9	−7.78E−2	8 1	4.19E−1	7 1	1.03E−3	6 5	−1.60E+0	14 1	4.71E−4
1 3	−1.26E+0	4 10	−9.11E−2	8 2	3.63E−4	7 2	1.39E+0	6 6	−3.53E−1	14 2	1.21E−1
1 4	−3.20E−1	4 11	−4.18E−2	8 3	2.95E−4	7 3	−3.30E−2	6 7	−5.00E−3	14 3	1.54E−1
1 5	−3.69E−2	4 12	−4.08E−1	8 4	6.84E−4	7 4	−3.32E−2	7 1	3.20E−4	14 4	5.63E−3
1 6	−4.00E−2	4 13	−4.42E−4	8 5	3.96E−4	7 5	−3.85E−4	7 2	5.46E−4	14 5	1.82E−1
1 7	−2.21E−3	4 14	−9.82E−1	8 6	7.28E−4	7 6	−1.37E−4	7 3	1.08E−1	14 6	1.72E−1
1 8	−5.54E−2	4 15	−3.81E−3	8 7	2.41E−9	8 1	1.07E−2	7 4	7.18E−2	14 7	5.05E−9
1 9	−2.69E−3	5 1	4.43E−4	8 8	2.28E−1	8 2	1.87E−2	7 5	3.88E−3	15 1	9.87E−3
1 10	−1.21E−2	5 2	3.22E−2	8 9	1.71E−4	8 3	0.00E+0	7 6	1.81E−5	15 2	1.07E−5
1 11	−2.20E−3	5 3	8.75E−1	8 10	4.66E−5	8 4	2.01E−6	7 7	−1.15E−4	15 3	5.40E−4
1 12	−5.34E−3	5 4	2.63E−1	8 11	7.63E−6	8 5	−5.90E−6	8 1	2.11E−1	15 4	3.58E−4
1 13	−3.21E−1	5 5	2.98E−4	8 12	−6.90E−7	8 6	−9.48E−1	8 2	1.97E−4	15 5	8.04E−4
1 14	−4.60E−4	5 6	9.00E−2	8 13	−7.48E−3		$^1\mathbf{D}^e$–$^1\mathbf{F}^o$	8 3	3.79E−4	15 6	1.42E−4
1 15	−1.85E−1	5 7	−1.40E−1	8 14	−3.68E−4	1 1	−2.91E+0	8 4	1.33E−3	15 7	1.74E−2
2 1	2.21E+0	5 8	−1.19E−4	8 15	−1.70E−1	1 2	−7.76E−4	8 5	2.76E−4		$^1\mathbf{F}^o$–$^1\mathbf{F}^e$
2 2	−2.38E−2	5 9	−1.35E−1		$^1\mathbf{D}^o$–$^1\mathbf{F}^e$	1 3	−9.99E−5	8 6	1.75E−6	1 1	−1.71E−2
2 3	−1.06E−1	5 10	−4.90E−1	1 1	−9.47E−1	1 4	−4.05E−3	8 7	−1.87E+0	1 2	−2.11E+0
2 4	−1.12E+0	5 11	−9.82E−2	1 2	−1.56E+0	1 5	−9.86E−4	9 1	1.63E−3	1 3	−1.08E−5
2 5	−8.53E−2	5 12	−4.70E−1	1 3	−4.01E−3	1 6	−3.71E−5	9 2	5.64E−2	1 4	−1.06E−4
2 6	−2.26E+0	5 13	−1.64E−5	1 4	−1.78E−3	1 7	−1.91E+0	9 3	6.85E−1	1 5	−4.25E−5
2 7	−3.49E−4	5 14	−1.35E−1	1 5	−1.86E−4	2 1	2.77E−5	9 4	2.43E−1	1 6	−5.72E−1
2 8	−1.84E−2	5 15	−1.28E−4	1 6	−1.47E+0	2 2	−6.61E−1	9 5	3.54E−2	2 1	3.50E−1
2 9	−4.07E−4	6 1	7.79E−6	2 1	−3.66E−1	2 3	−1.75E+0	9 6	8.93E−3	2 2	2.20E−2
2 10	−5.34E−4	6 2	3.85E−3	2 2	−2.03E+0	2 4	−2.78E−1	9 7	−6.44E−4	2 3	−9.86E−3
2 11	−1.96E−3	6 3	5.26E−1	2 3	−1.23E−5	2 5	−2.28E−2	10 1	1.18E−3	2 4	−1.15E+0
2 12	−2.40E−3	6 4	1.47E+0	2 4	−1.18E−3	2 6	−7.34E−3	10 2	9.53E−1	2 5	−6.84E−1
2 13	−4.34E−2	6 5	3.43E−3	2 5	−2.10E−5	2 7	−2.48E−6	10 3	1.90E−1	2 6	−1.85E−4
2 14	−3.13E−4	6 6	2.40E−1	2 6	−2.31E+0	3 1	2.98E−2	10 4	2.14E−1	3 1	3.02E−2
2 15	−2.40E−1	6 7	−6.96E−5	3 1	4.17E−1	3 2	−3.21E−1	10 5	1.31E−1	3 2	3.82E−1
3 1	3.11E−4	6 8	−1.27E−6	3 2	1.11E−3	3 3	−8.44E−3	10 6	2.32E−2	3 3	−8.20E−2
3 2	2.41E−1	6 9	−8.15E−2	3 3	−4.40E−2	3 4	−4.73E−1	10 7	−8.88E−5	3 4	−2.37E−2
3 3	4.10E−2	6 10	−8.05E−2	3 4	−7.26E−2	3 5	−4.91E−1	11 1	7.75E−4	3 5	−9.52E−1
3 4	2.99E−2	6 11	−2.71E−3	3 5	−5.56E−1	3 6	−4.20E−2	11 2	2.00E−1	3 6	−1.62E−4
3 5	3.95E−5	6 12	−1.88E−2	3 6	−9.83E−6	3 7	−4.90E−3	11 3	6.43E−1	4 1	4.22E−1
3 6	1.88E−4	6 13	−6.43E−4	4 1	2.54E−1	4 1	1.14E−2	11 4	2.08E−1	4 2	6.19E−1
3 7	−6.81E−1	6 14	−1.01E+0	4 2	7.14E−2	4 2	−2.54E−1	11 5	3.90E−1	4 3	−1.56E−1
3 8	−6.33E−5	6 15	−9.82E−4	4 3	−2.05E−1	4 3	−2.89E−1	11 6	4.52E−3	4 4	−5.46E−1
3 9	−1.18E−1	7 1	3.52E−3	4 4	−1.37E+0	4 4	−1.36E+0	11 7	−6.93E−5	4 5	−1.88E−1
3 10	−4.52E−1	7 2	1.49E−3	4 5	−2.14E−1	4 5	−2.40E−1	12 1	8.59E−4	4 6	−5.01E−4
3 11	−1.65E−1	7 3	5.68E−1	4 6	−1.82E−6	4 6	−1.03E−1	12 2	2.42E−1	5 1	6.78E−1
3 12	−1.40E−1	7 4	4.29E−1	5 1	9.85E−1	4 7	−5.23E−4	12 3	2.83E−1	5 2	1.52E+0
3 13	−2.42E−3	7 5	8.53E−1	5 2	9.14E−4	5 1	1.47E+0	12 4	9.17E−3	5 3	−7.12E−2
3 14	−4.44E−2	7 6	1.73E+0	5 3	−6.18E−1	5 2	−1.58E−1	12 5	4.60E−3	5 4	−2.25E−2
3 15	−2.96E−4	7 7	−3.53E−5	5 4	−3.58E−1	5 3	−3.54E−3	12 6	3.30E−1	5 5	−1.26E−1
4 1	8.10E−6	7 8	−7.95E−5	5 5	−5.55E−1	5 4	−5.70E−5	12 7	−5.49E−5	5 6	−1.90E−3
4 2	1.14E+0	7 9	−3.28E−3	5 6	−9.99E−4	5 5	−7.14E−4	13 1	4.00E−2	6 1	1.79E−1
4 3	4.80E−1	7 10	−2.41E−2	6 1	8.12E−4	5 6	−3.67E+0	13 2	5.06E−3	6 2	9.67E−2
4 4	2.88E−2	7 11	−2.75E−2	6 2	1.97E−3	5 7	−7.06E−3	13 3	3.30E−4	6 3	−1.85E−2
4 5	9.63E−2	7 12	−2.16E−2	6 3	−6.66E−4	6 1	3.75E−1	13 4	4.94E−3	6 4	−1.06E−1

Si-like Fe (Fe^{12+})

i i′	gf$_L$	*i i′*	gf$_L$	*i i′*	gf$_L$	*i i′*	gf$_L$	*i i′*	gf$_L$	*i i′*	gf$_L$
6 5	−7.85E−3	7 2	3.55E−2	1 5	−4.53E+0	4 9	−1.13E−2	3 6	1.68E−1	3 3	6.69E−2
6 6	−3.47E−7	7 3	1.38E−5	1 6	−5.53E−3	4 10	−1.72E−3	3 7	1.47E+0	3 4	1.60E−1
7 1	1.13E−7	7 4	7.97E−5	1 7	−1.01E−3	4 11	−1.68E−4	3 8	7.28E−2	3 5	7.35E−2
7 2	1.25E−3	7 5	3.65E−5	1 8	−9.16E−3	4 12	−6.28E−3	3 9	1.46E−2	3 6	5.94E−4
7 3	3.93E−5	7 6	−2.35E−5	1 9	−8.57E−5	4 13	−1.79E−4	3 10	−1.39E−7	3 7	−2.31E−1
7 4	5.61E−6	7 7	−1.79E+0	1 10	−4.47E−3	4 14	−4.90E−1	3 11	−5.63E−5	3 8	−3.54E−1
7 5	1.68E−8		**$^1F^e$–$^1G^o$**	1 11	−1.09E−4	4 15	−6.27E−2	4 1	4.10E−1	3 9	−7.21E−2
7 6	−8.66E−2	1 1	−1.10E+0	1 12	−4.19E−2	4 16	−5.75E−3	4 2	6.12E−2	3 10	−7.07E−2
	$^1F^o$–$^1G^e$	1 2	−1.19E+0	1 13	−1.78E−3	5 1	2.52E−1	4 3	2.00E−4	3 11	−1.16E−1
1 1	−9.52E−2	1 3	−1.98E−1	1 14	−4.88E−3	5 2	1.11E−1	4 4	5.71E−5	3 12	−4.98E−1
1 2	−4.02E+0	2 1	−6.19E−2	1 15	−1.86E−3	5 3	5.18E−2	4 5	1.63E−3	3 13	−8.00E−1
1 3	−6.71E−5	2 2	−1.06E−1	1 16	−5.05E−1	5 4	2.03E−2	4 6	2.89E−2	3 14	−1.07E−1
1 4	−4.30E−4	2 3	−5.04E+0	2 1	5.51E−5	5 5	4.53E−4	4 7	6.99E−4	3 15	−1.79E−2
1 5	−3.24E−4	3 1	1.19E+0	2 2	9.35E−1	5 6	2.17E−4	4 8	1.07E−1	3 16	−1.32E−3
1 6	−8.39E−4	3 2	2.68E−1	2 3	3.67E−2	5 7	7.11E−4	4 9	5.00E−1	4 1	9.37E−4
1 7	−5.04E+0	3 3	2.46E−4	2 4	1.38E−1	5 8	1.16E−2	4 10	5.01E−4	4 2	2.31E+0
2 1	7.57E−1	4 1	2.43E−1	2 5	7.77E−2	5 9	2.50E−2	4 11	−7.28E−4	4 3	8.01E−1
2 2	1.13E−3	4 2	3.13E−1	2 6	7.21E−3	5 10	4.51E−3		**$^3P^o$–$^3P^e$**	4 4	7.96E−2
2 3	−3.27E−1	4 3	2.60E−2	2 7	−5.54E−1	5 11	3.87E−5	1 1	5.32E−1	4 5	2.27E−1
2 4	−1.15E−1	5 1	2.08E−2	2 8	−9.82E−2	5 12	2.24E−1	1 2	−2.14E−1	4 6	6.34E−4
2 5	−9.24E−1	5 2	6.27E−1	2 9	−2.14E−2	5 13	6.56E−2	1 3	−7.46E−1	4 7	−1.29E−2
2 6	−3.73E−1	5 3	2.32E−1	2 10	−7.30E−2	5 14	1.72E−4	1 4	−2.64E+0	4 8	−5.70E−1
2 7	−7.94E−4	6 1	6.99E−4	2 11	−5.75E−1	5 15	−3.12E−5	1 5	−3.40E−3	4 9	−1.24E−1
3 1	7.50E−1	6 2	5.07E−7	2 12	−5.16E−2	5 16	−4.01E−3	1 6	−8.78E−6	4 10	−2.78E−1
3 2	1.75E−3	6 3	4.55E−5	2 13	−1.06E−1		**$^3S^e$–$^3P^o$**	1 7	−9.89E−7	4 11	−1.86E+0
3 3	−3.51E−1		**$^1G^o$–$^1G^e$**	2 14	−4.50E−1	1 1	8.28E−1	1 8	−2.08E−2	4 12	−4.68E−2
3 4	−4.05E−1	1 1	1.04E+0	2 15	−9.51E−2	1 2	1.86E−1	1 9	−5.32E−2	4 13	−1.98E−1
3 5	−3.14E−3	1 2	2.45E−3	2 16	−1.43E−3	1 3	−3.04E−1	1 10	−9.40E−3	4 14	−1.43E+0
3 6	−1.32E+0	1 3	−2.24E−1	3 1	4.94E−4	1 4	−3.75E−2	1 11	−6.99E−4	4 15	−5.76E−3
3 7	−5.82E−5	1 4	−2.29E+0	3 2	1.64E−3	1 5	−1.02E+0	1 12	−3.48E−1	4 16	−3.15E−5
4 1	2.17E−1	1 5	−2.17E−2	3 3	1.47E+0	1 6	−1.26E−3	1 13	−9.37E−2	5 1	8.09E−4
4 2	5.61E−5	1 6	−5.07E−1	3 4	3.66E−1	1 7	−4.84E−1	1 14	−5.60E−3	5 2	3.31E−2
4 3	−3.29E−1	1 7	−8.71E−4	3 5	6.10E−2	1 8	−7.92E−1	1 15	−4.66E−5	5 3	3.16E+0
4 4	−9.81E−1	2 1	2.36E+0	3 6	2.95E−2	1 9	−1.92E−1	1 16	−5.76E−2	5 4	1.03E+0
4 5	−9.92E−2	2 2	3.02E−1	3 7	−2.29E−1	1 10	−2.24E−3	2 1	2.57E+0	5 5	2.07E−1
4 6	−1.88E+0	2 3	−2.49E−1	3 8	−2.83E−1	1 11	−2.94E−2	2 2	−8.38E−3	5 6	5.48E−3
4 7	−8.60E−3	2 4	−8.13E−1	3 9	−6.53E−2	2 1	5.81E−2	2 3	−6.97E−2	5 7	−1.85E−2
5 1	1.85E−2	2 5	−9.08E−1	3 10	−2.34E−1	2 2	1.83E−1	2 4	−1.50E−1	5 8	−1.19E−1
5 2	9.61E−1	2 6	−7.26E−1	3 11	−8.31E−2	2 3	8.73E−6	2 5	−9.99E−1	5 9	−2.14E−2
5 3	−2.19E−3	2 7	−8.11E−3	3 12	−2.47E−1	2 4	6.29E−5	2 6	−4.18E+0	5 10	−3.98E−2
5 4	−2.07E−1	3 1	2.78E−1	3 13	−4.20E−1	2 5	2.75E−4	2 7	−3.73E−5	5 11	−8.83E−2
5 5	−4.47E−1	3 2	1.82E+0	3 14	−4.02E−1	2 6	8.01E−3	2 8	−2.69E−2	5 12	−3.94E−2
5 6	−9.34E−1	3 3	−3.06E−4	3 15	−6.06E−1	2 7	3.40E−4	2 9	−8.24E−2	5 13	−9.61E−2
5 7	−1.65E−3	3 4	−7.28E−2	3 16	−8.24E−6	2 8	1.23E−1	2 10	−3.59E−3	5 14	−5.67E−1
6 1	1.84E−3	3 5	−3.43E−3	4 1	1.44E−3	2 9	4.35E−1	2 11	−5.26E−4	5 15	−1.76E+0
6 2	1.76E+0	3 6	−1.72E+0	4 2	8.57E−4	2 10	−1.55E+0	2 12	−4.17E−2	5 16	−2.94E−4
6 3	−1.32E−3	3 7	−3.91E−3	4 3	7.48E−5	2 11	−1.26E−2	2 13	−1.37E−2	6 1	2.56E−2
6 4	−1.65E−2		**$^3S^o$–$^3P^e$**	4 4	4.39E−3	3 1	9.03E−4	2 14	−3.99E−5	6 2	4.69E−6
6 5	−5.32E−1	1 1	1.75E+0	4 5	3.13E+0	3 2	2.12E−4	2 15	−3.35E−3	6 3	3.57E−3
6 6	−4.67E−3	1 2	−2.28E−1	4 6	7.52E−1	3 3	4.04E−1	2 16	−1.30E−2	6 4	5.29E−1
6 7	−1.40E−2	1 3	−1.68E−1	4 7	−3.07E−4	3 4	2.61E−1	3 1	2.56E−3	6 5	1.71E−2
7 1	2.99E−3	1 4	−6.28E−2	4 8	−4.91E−2	3 5	8.24E−2	3 2	1.31E+0	6 6	1.28E−1

Si-like Fe (Fe^{12+})

i	*i'*	gf$_L$	*i*	*i'*	gf$_L$	*i*	*i'*	gf$_L$	*i*	*i'*	gf$_L$	*i*	*i'*	gf$_L$	*i*	*i'*	gf$_L$
6	7	−1.32E−3	9	11	−2.33E−4	1	14	−7.50E−4	5	10	−2.48E−2	9	6	−1.88E+0	2	3	−6.46E−1
6	8	−4.60E−2	9	12	−1.72E−1	2	1	−4.07E−3	5	11	−1.08E−2	9	7	−1.29E−4	2	4	−1.11E−2
6	9	−5.33E−3	9	13	−4.80E−2	2	2	−6.47E−1	5	12	−3.00E+0	9	8	−7.92E−2	2	5	−3.20E+0
6	10	−1.61E+0	9	14	−2.71E−4	2	3	−1.77E−2	5	13	−8.89E−3	9	9	−1.78E−3	2	6	−7.50E−1
6	11	−8.39E−2	9	15	−2.95E−1	2	4	−3.48E+0	5	14	−3.73E−3	9	10	−1.89E−2	2	7	−1.38E−3
6	12	−1.07E−1	9	16	−3.43E+0	2	5	−1.05E−3	6	1	4.16E−1	9	11	−6.53E−1	2	8	−3.49E−2
6	13	−1.05E+0	10	1	9.50E−1	2	6	−1.80E−3	6	2	1.69E+0	9	12	−5.19E−2	2	9	−2.55E−2
6	14	−7.39E−2	10	2	4.41E−4	2	7	−3.83E−5	6	3	1.08E+0	9	13	−7.33E−3	2	10	−2.03E−2
6	15	−9.63E−2	10	3	1.12E−4	2	8	−9.04E−4	6	4	3.47E−2	9	14	−5.93E−1	2	11	−5.45E−4
6	16	−2.60E−2	10	4	1.06E−3	2	9	−2.54E−6	6	5	−1.32E−2	10	1	1.87E−4	2	12	−9.99E−3
7	1	4.97E−3	10	5	4.52E−5	2	10	−5.20E−4	6	6	−3.57E−2	10	2	2.98E−3	3	1	2.13E+0
7	2	1.71E−3	10	6	1.31E−3	2	11	−1.27E−2	6	7	−2.15E−3	10	3	5.30E−4	3	2	1.39E−3
7	3	9.28E−3	10	7	1.10E−5	2	12	−6.71E−3	6	8	−3.46E−1	10	4	4.34E−2	3	3	−6.38E−1
7	4	4.25E−1	10	8	1.93E−1	2	13	−6.96E+0	6	9	−1.90E−3	10	5	1.72E−5	3	4	−2.37E−2
7	5	1.50E−3	10	9	8.27E−1	2	14	−2.61E−2	6	10	−6.23E−2	10	6	1.42E−1	3	5	−1.52E+0
7	6	2.95E−1	10	10	4.15E−4	3	1	4.25E−2	6	11	−1.42E−2	10	7	3.92E−5	3	6	−2.76E−1
7	7	−7.36E−2	10	11	4.34E−5	3	2	3.74E−4	6	12	−9.27E−1	10	8	1.25E−5	3	7	−1.02E+0
7	8	−6.24E−2	10	12	1.29E−3	3	3	4.42E−1	6	13	−1.09E−4	10	9	3.01E−6	3	8	−2.04E−2
7	9	−7.69E−3	10	13	1.25E−4	3	4	7.83E−2	6	14	−2.19E−1	10	10	1.72E−7	3	9	−2.25E−1
7	10	−8.61E−3	10	14	−1.10E−6	3	5	−1.40E−1	7	1	1.74E−2	10	11	−1.30E−5	3	10	−3.09E−2
7	11	−1.34E−1	10	15	−4.89E−7	3	6	−7.35E−4	7	2	6.43E−1	10	12	−4.75E−5	3	11	−1.74E−3
7	12	−1.27E−1	10	16	−4.38E−3	3	7	−7.28E−1	7	3	3.25E+0	10	13	−2.08E+0	3	12	−4.60E−4
7	13	−2.99E−1	11	1	6.34E−1	3	8	−3.36E−2	7	4	1.26E−1	10	14	−1.01E−5	4	1	5.89E−2
7	14	−1.10E+0	11	2	1.16E−1	3	9	−1.10E+0	7	5	−9.70E−2	11	1	2.50E−1	4	2	1.72E−1
7	15	−3.39E−2	11	3	6.48E−2	3	10	−1.09E+0	7	6	−8.95E−4	11	2	5.95E−2	4	3	−8.24E−5
7	16	−3.18E−3	11	4	8.60E−3	3	11	−1.68E−3	7	7	−8.20E−2	11	3	1.48E−2	4	4	−2.87E−1
8	1	2.77E−1	11	5	6.43E−3	3	12	−4.41E−3	7	8	−7.96E−1	11	4	2.82E−2	4	5	−8.80E−4
8	2	2.67E−4	11	6	6.12E−4	3	13	−4.77E−4	7	9	−4.73E−2	11	5	2.88E−3	4	6	−2.37E+0
8	3	5.74E−4	11	7	1.15E−6	3	14	−1.74E−3	7	10	−1.24E+0	11	6	5.74E−3	4	7	−1.28E−5
8	4	7.77E−1	11	8	4.57E−2	4	1	8.58E−1	7	11	−3.23E−3	11	7	8.02E−4	4	8	−6.41E−1
8	5	4.40E−3	11	9	1.58E−1	4	2	3.45E−1	7	12	−2.88E−2	11	8	9.60E−5	4	9	−7.88E−1
8	6	2.52E+0	11	10	2.92E−2	4	3	1.90E−1	7	13	−6.96E−5	11	9	5.69E−4	4	10	−1.82E−1
8	7	−2.24E−3	11	11	1.19E−3	4	4	1.36E−2	7	14	−1.51E−1	11	10	2.95E−4	4	11	−1.61E−3
8	8	−6.46E−2	11	12	1.23E+0	4	5	−9.98E−1	8	1	2.58E−2	11	11	1.18E−1	4	12	−1.63E−4
8	9	−2.78E−1	11	13	3.70E−1	4	6	−6.23E−5	8	2	6.44E−2	11	12	4.04E−4	5	1	2.30E−3
8	10	−1.26E−2	11	14	2.87E−3	4	7	−3.03E−1	8	3	1.14E−2	11	13	−1.84E−4	5	2	2.62E+0
8	11	−5.85E−4	11	15	6.55E−6	4	8	−4.90E−2	8	4	2.50E+0	11	14	−1.01E−3	5	3	−4.54E−2
8	12	−7.07E−2	11	16	−2.88E−3	4	9	−1.89E+0	8	5	−2.23E−3			**$^{3}\mathrm{P}^{e}$–$^{3}\mathrm{D}^{o}$**	5	4	−6.63E−2
8	13	−5.72E−1			**$^{3}\mathrm{P}^{o}$–$^{3}\mathrm{D}^{e}$**	4	10	−1.50E−1	8	6	−5.23E−1	1	1	−4.33E−1	5	5	−5.19E−1
8	14	−6.85E−4	1	1	−8.79E−2	4	11	−9.93E−6	8	7	−4.08E−4	1	2	−6.34E+0	5	6	−1.38E−1
8	15	−1.69E+0	1	2	−2.24E+0	4	12	−3.23E−2	8	8	−4.25E−1	1	3	−2.96E−4	5	7	−7.96E−2
8	16	−8.50E−1	1	3	−9.04E−1	4	13	−2.84E−4	8	9	−9.48E−3	1	4	−2.42E−3	5	8	−6.21E−1
9	1	9.08E−1	1	4	−2.69E+0	4	14	−1.07E−4	8	10	−7.63E−2	1	5	−1.56E−3	5	9	−1.05E+0
9	2	3.56E−4	1	5	−1.15E−3	5	1	4.82E−1	8	11	−2.07E−1	1	6	−1.36E−3	5	10	−6.60E+0
9	3	4.45E−4	1	6	−1.16E−4	5	2	1.33E−1	8	12	−1.41E−1	1	7	−7.75E−3	5	11	−1.59E−4
9	4	2.62E−1	1	7	−1.90E−4	5	3	1.05E−2	8	13	−5.88E−4	1	8	−4.81E−4	5	12	−3.17E−3
9	5	1.06E−3	1	8	−2.21E−3	5	4	1.12E−1	8	14	−3.95E+0	1	9	−4.93E−5	6	1	1.51E−3
9	6	4.90E−1	1	9	−7.70E−6	5	5	−3.63E−1	9	1	8.11E−4	1	10	−7.57E−4	6	2	1.99E+0
9	7	−3.24E−4	1	10	−1.03E−3	5	6	−6.29E−6	9	2	4.46E−2	1	11	−3.11E+0	6	3	4.67E−4
9	8	−2.23E−1	1	11	−4.62E−1	5	7	−1.84E−1	9	3	1.95E−3	1	12	−1.24E+0	6	4	−2.36E−3
9	9	−1.01E+0	1	12	−2.50E−3	5	8	−5.04E−2	9	4	6.10E−1	2	1	1.00E+0	6	5	−1.49E−2
9	10	−2.06E−3	1	13	−7.38E−1	5	9	−2.03E+0	9	5	−1.82E−4	2	2	2.85E−3	6	6	−3.41E−2

Si-like Fe (Fe^{12+})

i	i'	gf_L
6	7	−2.04E−2
6	8	−3.68E−1
6	9	−3.88E+0
6	10	−5.91E−2
6	11	−9.78E−3
6	12	−1.46E−3
7	1	3.77E−4
7	2	5.40E−6
7	3	5.88E−1
7	4	3.27E−1
7	5	5.03E−2
7	6	1.46E−1
7	7	2.34E−2
7	8	7.74E−2
7	9	9.05E−4
7	10	8.55E−3
7	11	−4.28E−7
7	12	−7.28E−5
8	1	4.52E−2
8	2	6.03E−2
8	3	1.80E+0
8	4	8.26E−1
8	5	8.59E−2
8	6	3.39E−1
8	7	6.84E−2
8	8	4.60E−3
8	9	9.42E−3
8	10	1.59E−2
8	11	−5.33E−1
8	12	−6.83E−2
9	1	1.90E−1
9	2	2.27E−1
9	3	4.13E−1
9	4	1.87E−1
9	5	2.30E−2
9	6	1.04E−1
9	7	2.20E−2
9	8	1.27E−3
9	9	2.70E−3
9	10	3.67E−3
9	11	−2.27E+0
9	12	−2.04E−1
10	1	1.22E−2
10	2	2.85E−3
10	3	2.24E−2
10	4	7.53E−1
10	5	1.16E−1
10	6	4.37E−1
10	7	1.20E−1
10	8	3.88E−2
10	9	5.57E−2
10	10	5.14E−2
10	11	−9.97E−4
10	12	−3.71E−2
11	1	1.75E−4
11	2	6.65E−6
11	3	2.44E−2
11	4	9.43E−1
11	5	1.15E−1
11	6	8.78E−2
11	7	6.80E−1
11	8	4.90E−3
11	9	4.38E−4
11	10	3.16E−1
11	11	−6.56E−5
11	12	−6.87E−4
12	1	7.26E−1
12	2	9.71E−2
12	3	1.21E−2
12	4	1.92E−1
12	5	1.94E−1
12	6	3.26E−3
12	7	8.72E−2
12	8	3.14E−1
12	9	5.28E−2
12	10	1.91E−2
12	11	−1.07E−3
12	12	−1.81E+0
13	1	2.15E−1
13	2	3.51E−2
13	3	3.33E−2
13	4	5.16E−1
13	5	2.65E−1
13	6	2.29E−2
13	7	1.14E−2
13	8	8.34E−1
13	9	1.75E−1
13	10	2.76E−2
13	11	3.15E−4
13	12	−5.49E−1
14	1	9.00E−4
14	2	4.18E−5
14	3	3.40E−3
14	4	4.34E−2
14	5	2.93E+0
14	6	4.23E−1
14	7	1.25E+0
14	8	9.33E−2
14	9	8.50E−1
14	10	8.19E−2
14	11	1.36E−5
14	12	−2.56E−3
15	1	6.20E−6
15	2	2.21E−3
15	3	9.18E−4
15	4	2.05E−3
15	5	8.38E−2
15	6	2.17E+0
15	7	2.63E−1
15	8	3.02E−3
15	9	7.54E−1
15	10	4.04E−1
15	11	6.02E−6
15	12	−1.03E−6
16	1	1.39E−1
16	2	4.02E−2
16	3	2.39E−4
16	4	2.58E−4
16	5	9.92E−4
16	6	1.75E−3
16	7	1.27E−3
16	8	4.51E−3
16	9	1.36E−4
16	10	3.12E−3
16	11	1.29E−2
16	12	9.76E−4
		$^3D^o$–$^3D^e$
1	1	−4.23E+0
1	2	−5.34E−1
1	3	−1.66E−1
1	4	−1.49E−1
1	5	−3.38E−7
1	6	−6.56E−2
1	7	−1.65E−3
1	8	−4.94E−6
1	9	−2.49E−4
1	10	−2.36E−4
1	11	−1.48E+0
1	12	−5.00E−3
1	13	−1.09E−1
1	14	−5.46E−5
2	1	−3.11E−1
2	2	−2.03E−2
2	3	−6.96E−4
2	4	−1.33E+0
2	5	−2.55E−4
2	6	−1.28E−1
2	7	−2.95E−4
2	8	−8.22E−4
2	9	−4.09E−4
2	10	−1.32E−3
2	11	−3.34E−3
2	12	−6.30E−6
2	13	−1.31E+0
2	14	−3.59E−2
3	1	2.76E−1
3	2	3.60E−2
3	3	1.39E−1
3	4	2.35E−2
3	5	−1.73E−1
3	6	−2.98E−6
3	7	−1.31E+0
3	8	−2.60E−2
3	9	−1.54E+0
3	10	−8.85E−2
3	11	−1.42E−4
3	12	−1.40E−2
3	13	−5.65E−5
3	14	−8.79E−5
4	1	1.74E−3
4	2	4.27E−1
4	3	4.62E−1
4	4	4.58E−2
4	5	−1.31E−2
4	6	−1.24E−3
4	7	−1.51E+0
4	8	−1.04E−2
4	9	−3.32E−1
4	10	−2.26E+0
4	11	−2.00E−3
4	12	−6.40E−2
4	13	−2.66E−4
4	14	−1.22E−4
5	1	1.41E+0
5	2	3.08E−2
5	3	5.92E−1
5	4	1.89E−1
5	5	−4.36E−1
5	6	−6.99E−6
5	7	−1.05E−4
5	8	−2.68E−2
5	9	−3.10E+0
5	10	−3.52E−1
5	11	−8.86E−6
5	12	−6.01E−4
5	13	−5.22E−4
5	14	−3.31E−2
6	1	3.36E−3
6	2	1.17E+0
6	3	6.37E−2
6	4	4.77E−1
6	5	−4.23E−1
6	6	−5.24E−4
6	7	−3.50E−1
6	8	−1.13E−4
6	9	−3.12E−1
6	10	−7.25E−1
6	11	−3.66E−3
6	12	−4.62E−2
6	13	−1.28E−3
6	14	−1.65E−3
7	1	5.38E+0
7	2	5.48E−2
7	3	2.10E−2
7	4	6.18E−3
7	5	−6.31E−3
7	6	−2.87E−5
7	7	−3.88E−1
7	8	−4.02E−1
7	9	−7.75E−1
7	10	−6.72E−1
7	11	−2.59E−2
7	12	−4.19E+0
7	13	−7.10E−4
7	14	−1.00E−1
8	1	1.57E−1
8	2	1.71E−1
8	3	3.53E+0
8	4	1.42E−2
8	5	−6.17E−3
8	6	−3.19E−4
8	7	−1.66E−1
8	8	−3.14E−1
8	9	−1.73E−1
8	10	−2.66E+0
8	11	−4.83E−3
8	12	−1.79E+0
8	13	−2.54E−4
8	14	−2.76E−1
9	1	2.20E−1
9	2	1.74E+0
9	3	3.96E−2
9	4	2.69E+0
9	5	−5.97E−5
9	6	−2.96E−5
9	7	−2.54E−2
9	8	−2.67E−1
9	9	−1.49E−2
9	10	−1.55E−1
9	11	−7.96E−4
9	12	−4.52E−2
9	13	−2.78E−4
9	14	−1.65E+0
10	1	1.68E−1
10	2	1.71E−2
10	3	2.29E+0
10	4	4.23E−1
10	5	−1.19E−2
10	6	−1.25E−5
10	7	−1.11E−1
10	8	−5.59E−4
10	9	−4.57E−1
10	10	−6.55E−1
10	11	−3.11E−3
10	12	−7.57E−1
10	13	−7.58E−5
10	14	−7.02E−1
11	1	1.94E−5
11	2	2.71E−3
11	3	8.35E−4
11	4	3.10E−3
11	5	1.22E−5
11	6	9.61E−1
11	7	7.46E−5
11	8	2.14E−8
11	9	2.22E−7
11	10	1.33E−6
11	11	−5.08E−4
11	12	−1.68E−5
11	13	−4.25E−1
11	14	−5.42E−8
12	1	5.87E−5
12	2	1.40E−1
12	3	6.73E−4
12	4	2.63E−2
12	5	9.68E−4
12	6	2.95E−2
12	7	1.20E−5
12	8	2.09E−6
12	9	3.33E−4
12	10	3.94E−4
12	11	1.37E−1
12	12	2.64E−5
12	13	−4.29E−5
12	14	−5.41E−4
		$^3D^o$–$^3F^e$
1	1	−9.24E−3
1	2	−5.72E+0
1	3	−2.78E+0
1	4	−6.20E−2
1	5	−4.53E−4
1	6	−2.18E−3
1	7	−1.57E−3
1	8	−4.60E−5
1	9	−3.86E−3
1	10	−1.12E−4
1	11	−7.54E−3
1	12	−1.21E+0
2	1	−7.48E−2
2	2	−8.11E−3
2	3	−2.94E+0
2	4	−8.28E+0

Si-like Fe (Fe^{12+})

i	*i′*	gf_L	*i*	*i′*	gf_L	*i*	*i′*	gf_L	*i*	*i′*	gf_L	*i*	*i′*	gf_L	*i*	*i′*	gf_L
2	5	−1.06E−5	6	9	−2.62E+0	11	1	3.96E−3	4	4	−3.94E−1	10	8	−8.30E−7	2	7	−1.80E+0
2	6	−6.49E−4	6	10	−1.17E+0	11	2	5.64E−5	4	5	−4.71E−1	11	1	2.15E−1	2	8	−1.34E−2
2	7	−7.65E−6	6	11	−4.74E−2	11	3	5.17E−2	4	6	−4.36E+0	11	2	1.61E−4	2	9	−1.78E+0
2	8	−1.47E−4	6	12	−2.34E−3	11	4	3.03E−2	4	7	−3.56E+0	11	3	7.36E−4	2	10	−1.96E−2
2	9	−4.45E−4	7	1	1.75E−3	11	5	4.98E−5	4	8	−4.14E−3	11	4	3.19E−2	2	11	−6.37E−4
2	10	−3.27E−4	7	2	2.33E+0	11	6	1.53E−5	5	1	6.50E−6	11	5	3.38E−4	2	12	−1.07E−3
2	11	−6.40E−2	7	3	4.98E−1	11	7	2.73E−5	5	2	2.10E+0	11	6	1.75E−3	3	1	3.56E−1
2	12	−1.00E+1	7	4	2.84E−1	11	8	3.85E−5	5	3	6.14E−4	11	7	3.89E−3	3	2	8.02E−1
3	1	1.09E+0	7	5	−1.98E−2	11	9	1.80E−5	5	4	1.00E+0	11	8	3.98E−3	3	3	3.75E−1
3	2	8.66E−4	7	6	−1.35E−2	11	10	−3.65E−5	5	5	1.44E−2	12	1	4.66E−3	3	4	9.06E−3
3	3	7.75E−5	7	7	−1.12E+0	11	11	−2.47E−2	5	6	4.87E−2	12	2	1.43E−4	3	5	−5.51E−2
3	4	7.56E−4	7	8	−4.79E−1	11	12	−3.00E+0	5	7	6.17E−2	12	3	1.74E−1	3	6	−4.75E−1
3	5	−1.05E+0	7	9	−2.39E−1	12	1	4.96E−1	5	8	−3.83E−7	12	4	2.11E+0	3	7	−8.36E−3
3	6	−5.81E−1	7	10	−3.97E−1	12	2	2.12E−1	6	1	9.96E−1	12	5	1.07E+0	3	8	−4.92E+0
3	7	−1.20E+0	7	11	−8.67E−1	12	3	1.97E−2	6	2	1.12E−3	12	6	6.50E−3	3	9	−5.94E−2
3	8	−2.10E−3	7	12	−2.98E−2	12	4	5.95E−4	6	3	1.52E−5	12	7	3.08E−1	3	10	−2.21E+0
3	9	−5.74E−1	8	1	3.22E−4	12	5	3.30E−3	6	4	8.58E−4	12	8	4.40E−7	3	11	−1.13E−3
3	10	−2.76E−1	8	2	7.75E−1	12	6	3.42E−7	6	5	4.22E−4	13	1	6.14E−2	3	12	−1.01E−4
3	11	−5.71E−3	8	3	1.71E+0	12	7	3.18E−3	6	6	2.42E−5	13	2	4.21E−4	4	1	1.88E+0
3	12	−6.26E−4	8	4	1.80E−2	12	8	3.65E−5	6	7	8.70E−5	13	3	1.43E−3	4	2	4.26E−1
4	1	1.79E+0	8	5	−3.46E−2	12	9	3.73E−4	6	8	−6.42E+0	13	4	3.52E−7	4	3	1.60E−1
4	2	2.84E−1	8	6	−4.97E−1	12	10	8.52E−5	7	1	2.49E−4	13	5	3.01E−4	4	4	1.19E−1
4	3	3.34E−1	8	7	−8.70E−2	12	11	−5.21E−5	7	2	1.49E+0	13	6	2.79E−7	4	5	−7.77E−1
4	4	3.39E−3	8	8	−4.80E−1	12	12	−8.86E−4	7	3	2.12E−2	13	7	1.37E−6	4	6	−6.52E−1
4	5	−2.64E−2	8	9	−3.57E−1			$^3D^e$–$^3F^o$	7	4	7.81E−1	13	8	4.53E−2	4	7	−2.61E+0
4	6	−7.16E−2	8	10	−2.19E+0	1	1	3.76E−2	7	5	8.76E−3	14	1	2.66E−2	4	8	−3.53E−1
4	7	−6.45E−2	8	11	−2.73E−4	1	2	−1.73E−1	7	6	3.68E−2	14	2	1.93E−3	4	9	−4.40E−1
4	8	−2.00E−1	8	12	−5.10E−4	1	3	−9.62E−1	7	7	1.24E−1	14	3	6.32E−3	4	10	−2.49E−1
4	9	−9.34E−1	9	1	5.67E−4	1	4	−9.32E−1	7	8	−1.06E−3	14	4	6.71E−2	4	11	−8.69E−2
4	10	−3.77E−1	9	2	9.62E−2	1	5	−1.95E+0	8	1	5.00E−3	14	5	4.95E−1	4	12	−2.87E−3
4	11	−2.01E−3	9	3	9.32E−1	1	6	−5.49E−2	8	2	5.96E−4	14	6	2.68E+0	5	1	2.48E+0
4	12	−6.02E−4	9	4	2.97E−1	1	7	−1.95E−1	8	3	4.43E+0	14	7	9.96E−1	5	2	1.08E+0
5	1	6.30E−1	9	5	−1.44E−2	1	8	−1.33E−3	8	4	8.93E−1	14	8	3.15E−5	5	3	4.94E−1
5	2	1.00E+0	9	6	−1.59E−2	2	1	9.14E−3	8	5	4.61E−4			$^3F^o$–$^3F^e$	5	4	9.84E−2
5	3	8.25E−1	9	7	−8.60E−3	2	2	−3.53E−1	8	6	5.07E−2	1	1	−2.17E−1	5	5	−4.21E−2
5	4	3.34E−2	9	8	−3.57E−1	2	3	−6.59E−1	8	7	1.09E−1	1	2	−1.38E+0	5	6	−1.47E−1
5	5	−7.42E−2	9	9	−8.27E−1	2	4	−7.28E−2	8	8	−1.06E−4	1	3	−4.64E+0	5	7	−6.98E−1
5	6	−1.48E+0	9	10	−6.49E−1	2	5	−2.84E+0	9	1	9.94E−6	1	4	−3.68E+0	5	8	−1.63E−1
5	7	−1.12E−1	9	11	−6.83E+0	2	6	−1.91E+0	9	2	5.72E−2	1	5	−6.48E−6	5	9	−6.98E−1
5	8	−9.92E−1	9	12	−6.02E−2	2	7	−1.64E−2	9	3	5.68E−3	1	6	−2.98E−5	5	10	−4.86E−1
5	9	−1.61E−1	10	1	1.41E−3	2	8	−7.35E−3	9	4	1.49E−1	1	7	−4.10E−4	5	11	−3.81E+0
5	10	−3.00E+0	10	2	1.52E 3	3	1	4.16E+0	9	5	1.15E+0	1	8	−1.78E−3	5	12	−3.11E−2
5	11	−8.07E−3	10	3	1.61E−1	3	2	−8.59E−2	9	6	6.01E−1	1	9	−2.68E−5	6	1	4.72E−1
5	12	−2.32E−3	10	4	4.63E+0	3	3	−7.53E−2	9	7	5.54E−2	1	10	−2.01E−3	6	2	8.74E−1
6	1	1.23E+0	10	5	−2.68E−2	3	4	−2.99E−1	9	8	−1.80E−6	1	11	−1.73E−2	6	3	1.63E+0
6	2	8.36E−1	10	6	−1.28E−2	3	5	−1.22E−1	10	1	3.04E−5	1	12	−1.38E+0	6	4	4.03E−5
6	3	1.04E−2	10	7	−2.37E−1	3	6	−1.75E+0	10	2	1.00E−1	2	1	1.41E+0	6	5	−7.21E−2
6	4	4.04E−2	10	8	−7.35E−4	3	7	−4.01E−1	10	3	2.41E−2	2	2	5.37E−1	6	6	−5.91E−2
6	5	−1.03E+0	10	9	−3.84E−2	3	8	−3.28E−3	10	4	2.69E−1	2	3	1.41E−1	6	7	−5.78E−3
6	6	−2.22E−1	10	10	−2.16E+0	4	1	7.02E−1	10	5	6.23E−1	2	4	1.08E−3	6	8	−7.84E−2
6	7	−9.26E−2	10	11	−6.19E−1	4	2	−1.89E−2	10	6	9.30E−1	2	5	−1.23E+0	6	9	−1.00E 1
6	8	−8.96E−1	10	12	−8.11E−3	4	3	−7.98E−1	10	7	3.62E−1	2	6	−5.44E−2	6	10	−1.98E+0

Si-like Fe (Fe^{12+})

i i′	gf_L	*i i′*	gf_L	*i i′*	gf_L	*i i′*	gf_L	*i i′*	gf_L	*i i′*	gf_L
6 11	−3.32E−1	3 2	−1.13E+0	2 1	−3.65E−1	12 1	1.11E−3	3 3	1.11E+0	2 2	−4.66E+0
6 12	−2.74E−3	3 3	−2.48E+0	2 2	−1.04E+0	12 2	1.49E−2	3 4	1.05E−3	2 3	−2.19E−3
7 1	1.20E−1	3 4	−1.93E−2	2 3	−4.88E+0	12 3	1.05E−4		**$^{5}P^{o}$–$^{5}P^{e}$**	3 1	5.24E+0
7 2	1.62E+0	3 5	−3.52E−3	2 4	−2.19E+0	12 4	3.30E−2	1 1	1.82E−3	3 2	−3.80E−1
7 3	5.73E+0	3 6	−7.12E−4	3 1	−1.04E+0		**$^{3}G^{o}$–$^{3}G^{e}$**	1 2	−1.14E+0	3 3	−2.88E+0
7 4	4.82E+0	4 1	2.67E+0	3 2	−3.33E+0	1 1	1.68E+0	1 3	−8.25E−3	4 1	2.38E−1
7 5	−7.15E−2	4 2	−1.31E+0	3 3	−2.81E−1	1 2	−2.58E+0	1 4	−2.18E+0	4 2	1.01E−3
7 6	−5.31E−2	4 3	−3.17E−1	3 4	−5.09E+0	1 3	−2.76E−2	2 1	2.35E−1	4 3	1.47E−5
7 7	−1.14E−2	4 4	−8.14E−1	4 1	−3.74E−3	1 4	−5.13E+0	2 2	6.55E−5		**$^{5}D^{o}$–$^{5}F^{e}$**
7 8	−8.28E−1	4 5	−1.57E−1	4 2	−2.74E−1	1 5	−6.95E−3	2 3	3.45E+0	1 1	5.38E+0
7 9	−1.28E+0	4 6	−1.21E−2	4 3	−7.60E−1	1 6	−1.98E−3	2 4	2.81E−4	1 2	−4.29E+0
7 10	−1.33E+0	5 1	6.02E−2	4 4	−7.94E+0	2 1	5.24E+0		**$^{5}P^{o}$–$^{5}D^{e}$**	1 3	−4.12E−4
7 11	−2.14E−4	5 2	−4.36E−1	5 1	1.19E−1	2 2	−1.04E−1	1 1	3.87E+0	2 1	3.06E+0
7 12	−9.04E−4	5 3	−1.79E+0	5 2	3.78E−4	2 3	−3.50E−1	1 2	−1.37E+0	2 2	−9.98E−1
8 1	2.56E−3	5 4	−1.61E+0	5 3	1.16E−1	2 4	−1.38E+0	1 3	−2.33E−1	2 3	−3.44E−1
8 2	4.76E−4	5 5	−1.73E+0	5 4	1.89E−3	2 5	−7.73E+0	2 1	8.59E−1	3 1	4.74E−3
8 3	9.94E−3	5 6	−5.75E−3	6 1	3.92E+0	2 6	−2.53E−4	2 2	2.99E−4	3 2	−3.00E+0
8 4	1.37E−3	6 1	3.94E−4	6 2	6.70E−1	3 1	1.60E+0	2 3	1.72E−5	3 3	−8.52E+0
8 5	9.18E−7	6 2	−4.16E−2	6 3	3.16E−1	3 2	−1.80E+0		**$^{5}P^{e}$–$^{5}D^{o}$**	4 1	1.59E+0
8 6	6.70E−9	6 3	−2.18E+0	6 4	1.16E−2	3 3	−3.47E−1	1 1	−1.75E+0	4 2	8.16E−5
8 7	5.16E−6	6 4	−3.03E+0	7 1	1.62E−4	3 4	−2.70E+0	1 2	−5.71E−1	4 3	9.78E−5
8 8	8.36E−6	6 5	−3.18E+0	7 2	1.47E+0	3 5	−3.26E−1	1 3	−8.05E+0		**$^{5}D^{e}$–$^{5}F^{o}$**
8 9	8.77E−8	6 6	−1.66E−3	7 3	5.88E−1	3 6	−6.71E−4	1 4	−9.40E−2	1 1	−3.03E+0
8 10	−8.88E−8	7 1	8.00E−4	7 4	3.56E−2	4 1	2.09E+0	2 1	5.91E+0	2 1	8.74E−1
8 11	−2.73E−3	7 2	−2.61E−1	8 1	3.68E−1	4 2	−2.23E−2	2 2	1.66E+0	3 1	4.93E+0
8 12	−3.10E−1	7 3	−6.58E−1	8 2	1.34E+0	4 3	−1.14E−1	2 3	5.01E−1		**$^{5}F^{o}$–$^{5}F^{e}$**
	$^{3}F^{o}$–$^{3}G^{e}$	7 4	−1.19E+0	8 3	2.83E−1	4 4	−8.32E−1	2 4	−1.91E−5	1 1	1.09E+1
1 1	−2.66E+0	7 5	−4.88E+0	8 4	4.88E−2	4 5	−1.39E+0	3 1	3.43E−4	1 2	−3.87E−1
1 2	−4.78E−4	7 6	−7.07E−4	9 1	1.35E−1	4 6	−3.30E−5	3 2	7.90E−3	1 3	−1.23E+1
1 3	−8.35E−4	8 1	9.12E−2	9 2	3.86E−1		**$^{5}S^{o}$–$^{5}P^{e}$**	3 3	2.64E−3		**$^{5}F^{o}$–$^{5}G^{e}$**
1 4	−3.97E−4	8 2	3.17E−6	9 3	2.66E+0	1 1	−5.45E+0	3 4	−5.01E+0	1 1	−3.80E+0
1 5	−4.30E−3	8 3	4.19E−7	9 4	9.39E−2	1 2	−2.84E−3	4 1	6.59E−3		**$^{5}F^{e}$–$^{5}G^{o}$**
1 6	−1.63E+1	8 4	5.48E−6	10 1	2.81E−2	1 3	−1.18E+0	4 2	2.32E+0	1 1	−3.93E+0
2 1	1.49E−2	8 5	−4.70E−5	10 2	1.06E−2	1 4	−2.51E−4	4 3	3.41E+0	2 1	6.16E+0
2 2	−6.23E−1	8 6	−4.10E+0	10 3	4.32E−1	2 1	3.84E+0	4 4	−1.18E−4	3 1	2.14E+0
2 3	−1.41E−2		**$^{3}F^{e}$–$^{3}G^{o}$**	10 4	3.25E+0	2 2	−1.27E+0		**$^{5}D^{o}$–$^{5}D^{e}$**		**$^{5}G^{o}$–$^{5}G^{e}$**
2 4	−2.73E+0	1 1	−9.33E−1	11 1	5.89E−3	2 3	−9.05E−3	1 1	3.40E+0	1 1	−8.20E+0
2 5	−3.76E−2	1 2	−1.24E+0	11 2	1.54E+0	2 4	−3.15E+0	1 2	−3.56E−5		
2 6	−3.50E−4	1 3	−3.17E−1	11 3	1.94E−2	3 1	3.50E−1	1 3	−1.21E+1		
3 1	2.03E+0	1 4	−5.17E−1	11 4	1.88E+0	3 2	8.61E−4	2 1	1.20E+0		

P-like S (S^+)

Term energies relative to $3p^2\ ^3P$ ionization threshold for each symmetry

i	E(Ryds)	Description	i	E(Ryds)	Description	i	E(Ryds)	Description	i	E(Ryds)	Description
	$^2S^e$		5	−.34695	$3p^2\ ^1S\ 3d$:	3	−.18232	$3p^2\ ^1D\ 5p$		**$^4P^o$**	
1	−.61266	$3s3p^4$	6	−.26123	$3p^2\ ^3P\ 4d$	4	−.16040	$3p^2\ ^1D\ 4f$	1	−.52856	$3p^2\ ^3P\ 4p$
2	−.46951	$3p^2\ ^1S\ 4s$	7	−.24317	$3p^2\ ^1D\ 5s$	5	−.16017	$3p^2\ ^3P\ 5f$	2	−.27919	$3p^2\ ^3P\ 5p$
3	−.31426	$3p^2\ ^1D\ 3d$	8	−.18808	$3p^2\ ^1D\ 4d$	6	−.11135	$3p^2\ ^3P\ 6f$	3	−.17401	$3p^2\ ^3P\ 6p$
4	−.17090	$3p^2\ ^1D\ 4d$	9	−.16938	$3p^2\ ^3P\ 5d$	7	−.08183	$3p^2\ ^3P\ 7f$	4	−.11889	$3p^2\ ^3P\ 7p$
5	−.09871	$3p^2\ ^1S\ 5s$	10	−.11868	$3p^2\ ^3P\ 6d$	8	−.07651	$3p^2\ ^1D\ 6p$	5	−.08641	$3p^2\ ^3P\ 8p$
6	−.07510	$3p^2\ ^1D\ 5d$	11	−.10516	$3p^2\ ^1D\ 6s$	9	−.06666	$3p^2\ ^1D\ 5f$	6	−.06564	$3p^2\ ^3P\ 9p$
7	−.02221	$3p^2\ ^1D\ 6d$	12	−.08840	$3p^2\ ^3P\ 7d$	10	−.06265	$3p^2\ ^3P\ 8f$		**$^4D^e$**	
	$^2S^o$		13	−.08280	$3p^2\ ^1D\ 5d$	11	−.04950	$3p^2\ ^3P\ 9f$	1	−.68035	$3p^2\ ^3P\ 3d$
1	−.57054	$3p^2\ ^3P\ 4p$	14	−.07201	$3p^2\ ^1S\ 4d$		**$^2G^e$**		2	−.32190	$3p^2\ ^3P\ 4d$
2	−.29285	$3p^2\ ^3P\ 5p$	15	−.06284	$3p^2\ ^3P\ 8d$	1	−.55400	$3p^2\ ^1D\ 3d$	3	−.19428	$3p^2\ ^3P\ 5d$
3	−.17982	$3p^2\ ^3P\ 6p$	16	−.06019	$3p^2\ ^1D\ 5g$	2	−.22104	$3p^2\ ^1D\ 4d$	4	−.13023	$3p^2\ ^3P\ 6d$
4	−.12203	$3p^2\ ^3P\ 7p$	17	−.05061	$3p^2\ ^3P\ 9d$	3	−.15943	$3p^2\ ^3P\ 5g$	5	−.09340	$3p^2\ ^3P\ 7d$
5	−.08831	$3p^2\ ^3P\ 8p$		**$^2D^o$**		4	−.11077	$3p^2\ ^3P\ 6g$	6	−.07026	$3p^2\ ^3P\ 8d$
6	−.06689	$3p^2\ ^3P\ 9p$	1	−1.59561	$3s^23p^3$	5	−.09510	$3p^2\ ^1D\ 5d$	7	−.05477	$3p^2\ ^3P\ 9d$
	$^2P^e$		2	−.51927	$3p^2\ ^3P\ 4p$	6	−.08142	$3p^2\ ^3P\ 7g$		**$^4D^o$**	
1	−.75811	$3p^2\ ^3P\ 3d$	3	−.42968	$3p^2\ ^1D\ 4p$	7	−.06289	$3p^2\ ^1D\ 5g$	1	−.54416	$3p^2\ ^3P\ 4p$
2	−.67718	$3p^2\ ^3P\ 4s$	4	−.27264	$3p^2\ ^3P\ 5p$	8	−.06236	$3p^2\ ^3P\ 8g$	2	−.28438	$3p^2\ ^3P\ 5p$
3	−.43536	$3p^2\ ^1D\ 3d$	5	−.25793	$3p^2\ ^3P\ 4f$	9	−.04929	$3p^2\ ^3P\ 9g$	3	−.25802	$3p^2\ ^3P\ 4f$
4	−.37921	$3s3p^4$	6	−.18613	$3p^2\ ^1D\ 5p$		**$^2G^o$**		4	−.17636	$3p^2\ ^3P\ 6p$
5	−.33394	$3p^2\ ^3P\ 5s$	7	−.16843	$3p^2\ ^3P\ 6p$	1	−.25568	$3p^2\ ^3P\ 4f$	5	−.16437	$3p^2\ ^3P\ 5f$
6	−.25705	$3p^2\ ^3P\ 4d$	8	−.16419	$3p^2\ ^3P\ 5f$	2	−.16378	$3p^2\ ^3P\ 5f$	6	−.12020	$3p^2\ ^3P\ 7p$
7	−.20185	$3p^2\ ^1D\ 4d$	9	−.15401	$3p^2\ ^1D\ 4f$	3	−.16163	$3p^2\ ^1D\ 4f$	7	.11400	$3p^2\ ^3P\ 6f$
8	−.19989	$3p^2\ ^3P\ 6s$	10	−.11722	$3p^2\ ^3P\ 7p$	4	−.11311	$3p^2\ ^3P\ 6f$	8	−.08725	$3p^2\ ^3P\ 8p$
9	−.17007	$3p^2\ ^3P\ 5d$	11	−.11370	$3p^2\ ^3P\ 6f$	5	−.08297	$3p^2\ ^3P\ 7f$	9	−.08433	$3p^2\ ^3P\ 7f$
10	−.13331	$3p^2\ ^3P\ 7s$	12	−.08604	$3p^2\ ^3P\ 8p$	6	−.06770	$3p^2\ ^1D\ 5f$	10	−.07565	$3s3p^3\ ^5S\ 3d$
11	−.12255	$3p^2\ ^3P\ 6d$	13	−.08334	$3p^2\ ^3P\ 7f$	7	−.06337	$3p^2\ ^3P\ 8f$	11	−.06615	$3p^2\ ^3P\ 9p$
12	−.09526	$3p^2\ ^3P\ 8s$	14	−.07685	$3p^2\ ^1D\ 6p$	8	−.05004	$3p^2\ ^3P\ 9f$	12	−.06320	$3p^2\ ^3P\ 8f$
13	−.09267	$3p^2\ ^3P\ 7d$	15	−.06473	$3p^2\ ^3P\ 9p$		**$^4S^o$**			**$^4F^e$**	
14	−.08235	$3p^2\ ^1D\ 5d$	16	−.06402	$3p^2\ ^1D\ 5f$	1	−1.73660	$3s^23p^3$	1	−.71410	$3p^2\ ^3P\ 3d$
15	−.07146	$3p^2\ ^3P\ 9s$	17	−.06351	$3p^2\ ^3P\ 8f$	2	−.51734	$3p^2\ ^3P\ 4p$	2	−.33071	$3p^2\ ^3P\ 4d$
16	−.06712	$3p^2\ ^3P\ 8d$		**$^2F^e$**		3	−.27730	$3p^2\ ^3P\ 5p$	3	−.19774	$3p^2\ ^3P\ 5d$
	$^2P^o$		1	−.67077	$3p^2\ ^3P\ 3d$:	4	−.17506	$3p^2\ ^3P\ 6p$	4	−.16075	$3p^2\ ^3P\ 5g$
1	−1.49467	$3s^23p^3$	2	−.44916	$3p^2\ ^1D\ 3d$:	5	−.12411	$3p^2\ ^3P\ 7p$	5	−.13210	$3p^2\ ^3P\ 6d$
2	−.49630	$3p^2\ ^3P\ 4p$	3	−.29175	$3p^2\ ^3P\ 4d$	6	−.10103	$3s3p^3\ ^5S\ 4s$	6	−.11162	$3p^2\ ^3P\ 6g$
3	−.40290	$3p^2\ ^1D\ 4p$	4	−.22190	$3p^2\ ^1D\ 4d$	7	−.08162	$3p^2\ ^3P\ 8p$:	7	−.09452	$3p^2\ ^3P\ 7d$
4	−.28881	$3p^2\ ^1S\ 4p$	5	−.17282	$3p^2\ ^3P\ 5d$	8	−.06356	$3p^2\ ^3P\ 9p$	8	−.08197	$3p^2\ ^3P\ 7g$
5	−.26197	$3p^2\ ^3P\ 5p$	6	−.16075	$3p^2\ ^3P\ 5g$		**$^4P^e$**		9	−.07098	$3p^2\ ^3P\ 8d$
6	−.17700	$3p^2\ ^1D\ 5p$	7	−.12399	$3p^2\ ^3P\ 6d$	1	−1.01195	$3s3p^4$	10	−.06273	$3p^2\ ^3P\ 8g$
7	−.16301	$3p^2\ ^3P\ 6p$	8	−.11162	$3p^2\ ^3P\ 6g$	2	−.71035	$3p^2\ ^3P\ 4s$	11	−.05526	$3p^2\ ^3P\ 9d$
8	−.14874	$3p^2\ ^1D\ 4f$	9	−.09832	$3p^2\ ^1D\ 5d$	3	−.52144	$3p^2\ ^3P\ 3d$	12	−.04955	$3p^2\ ^3P\ 9g$
9	−.11498	$3p^2\ ^3P\ 7p$	10	.08453	$3p^2\ ^3P\ 7d$	4	−.34398	$3p^2\ ^3P\ 5s$		**$^4F^o$**	
10	−.08465	$3p^2\ ^3P\ 8p$	11	−.08196	$3p^2\ ^3P\ 7g$	5	−.29556	$3p^2\ ^3P\ 4d$	1	−.25111	$3p^2\ ^3P\ 4f$
11	−.07280	$3p^2\ ^1D\ 6p$	12	−.06663	$3p^2\ ^3P\ 8d$	6	−.20440	$3p^2\ ^3P\ 6s$	2	−.16150	$3p^2\ ^3P\ 5f$
12	−.06370	$3p^2\ ^3P\ 9p$	13	−.06273	$3p^2\ ^3P\ 8g$	7	−.18493	$3p^2\ ^3P\ 5d$	3	−.11223	$3p^2\ ^3P\ 6f$
13	−.06205	$3p^2\ ^1D\ 5f$	14	−.06203	$3p^2\ ^1D\ 5g$	8	−.13567	$3p^2\ ^3P\ 7s$	4	−.08243	$3p^2\ ^3P\ 7f$
	$^2D^e$		15	−.05289	$3p^2\ ^3P\ 9d$	9	−.12609	$3p^2\ ^3P\ 6d$	5	−.06308	$3p^2\ ^3P\ 8f$
1	−.83276	$3s3p^4$	16	−.04954	$3p^2\ ^3P\ 9g$	10	.09666	$3p^2\ ^3P\ 8s$	6	−.04981	$3p^2\ ^3P\ 9f$
2	−.60405	$3p^2\ ^1D\ 4s$		**$^2F^o$**		11	−.09127	$3p^2\ ^3P\ 7d$		**$^4G^e$**	
3	−.48753	$3p^2\ ^3P\ 3d$:	1	−.43408	$3p^2\ ^1D\ 4p$	12	−.07236	$3p^2\ ^3P\ 9s$	1	−.15945	$3p^2\ ^3P\ 5g$
4	−.39237	$3p^2\ ^1D\ 3d$	2	−.24945	$3p^2\ ^3P\ 4f$	13	−.06904	$3p^2\ ^3P\ 8d$	2	−.11079	$3p^2\ ^3P\ 6g$

P-like S (S^+)

i	Energy(Ryds)	Description
3	−.08144	$3p^2$ ^{3}P $7g$
4	−.06237	$3p^2$ ^{3}P $8g$
5	−.04930	$3p^2$ ^{3}P $9g$
	^{4}G^o	
1	−.25726	$3p^2$ ^{3}P $4f$
2	−.16442	$3p^2$ ^{3}P $5f$
3	−.11399	$3p^2$ ^{3}P $6f$
4	−.08359	$3p^2$ ^{3}P $7f$
5	−.06388	$3p^2$ ^{3}P $8f$
6	−.05039	$3p^2$ ^{3}P $9f$
	^{6}S^o	
1	−.17832	$3s3p^3$ ^{5}S $4s$
	^{6}D^o	
1	−.23573	$3s3p^3$ ^{5}S $3d$

Energies in ascending order from ground state for terms with effective $n \leq 4.0$, $L \leq 4$

Term	i	E(Ryds)	Term	i	E(Ryds)	Term	i	E(Ryds)	Term	i	E(Ryds)	Term	i	E(Ryds)
^{4}S^o	1	0.00000	^{2}S^e	1	1.12394	^{2}S^e	2	1.26709	^{4}F^e	2	1.40589	^{2}P^o	5	1.47463
^{2}D^o	1	0.14100	^{2}D^e	2	1.13254	^{2}F^e	2	1.28745	^{4}D^e	2	1.41470	^{2}D^e	6	1.47537
^{2}P^o	1	0.24192	^{2}S^o	1	1.16611	^{2}P^e	3	1.30125	^{2}S^e	3	1.42233	^{4}D^o	3	1.47859
^{4}P^e	1	0.72465	^{2}G^e	1	1.18259	^{2}F^o	1	1.30252	^{4}P^e	5	1.44104	^{2}D^o	5	1.47868
^{2}D^e	1	0.90383	^{4}D^o	1	1.19245	^{2}D^o	3	1.30692	^{2}S^o	2	1.44377	^{4}G^o	1	1.47934
^{2}P^e	1	0.97849	^{4}P^o	1	1.20804	^{2}P^o	3	1.33368	^{2}F^e	3	1.44486	^{2}P^e	6	1.47955
^{4}F^e	1	1.02252	^{4}P^e	3	1.21517	^{2}D^e	4	1.34423	^{2}P^o	4	1.44779	^{2}G^o	1	1.48093
^{4}P^e	2	1.02626	^{2}D^o	2	1.21734	^{2}P^e	4	1.35739	^{4}D^o	2	1.45222	^{4}F^o	1	1.48549
^{4}D^e	1	1.05625	^{4}S^o	2	1.21927	^{2}D^e	5	1.38964	^{4}P^o	2	1.45741			
^{2}P^e	2	1.05944	^{2}P^o	2	1.24029	^{4}P^e	4	1.39263	^{4}S^o	3	1.45930			
^{2}F^e	1	1.06583	^{2}D^e	3	1.24907	^{2}P^e	5	1.40266	^{2}D^o	4	1.46397			

P-like S (S^+)

gf-values for transitions involving terms with effective $n \leq 4.0$, $L \leq 4$

i i'	gf$_L$	i i'	gf$_L$	i i'	gf$_L$	i i'	gf$_L$	i i'	gf$_L$	i i'	gf$_L$
	$^2S^o$–$^2P^e$	3 3	2.88E−1	5 4	2.44E−2	2 4	−6.11E−3		$^2F^o$–$^2F^e$		$^4P^e$–$^4D^o$
1 1	3.21E−1	3 4	−8.13E−3	5 5	9.37E−2	2 5	−5.23E−1	1 1	4.10E−1	1 1	−1.57E−3
1 2	5.25E−1	3 5	−1.34E−1	5 6	−6.29E−2	2 6	−3.74E−1	1 2	4.85E−2	1 2	−7.79E−3
1 3	−1.24E−1	3 6	−5.47E−3		$^2P^e$–$^2D^o$	3 1	2.17E−1	1 3	−5.92E−1	1 3	−3.52E−1
1 4	−9.61E−1	4 1	1.65E−3	1 1	2.92E−2	3 2	3.45E+0		$^2F^o$–$^2G^e$	2 1	−6.85E+0
1 5	−6.43E−1	4 2	1.25E−2	1 2	−3.48E−2	3 3	6.78E−2	1 1	1.98E+0	2 2	−2.64E−2
1 6	−2.43E−1	4 3	1.20E−2	1 3	−4.41E−2	3 4	−2.34E−1		$^2F^e$–$^2G^o$	2 3	−1.06E−2
2 1	1.22E−2	4 4	1.01E−1	1 4	−4.44E−2	3 5	−5.82E−2	1 1	−1.99E+0	3 1	3.57E−3
2 2	4.71E−2	4 5	1.64E−1	1 5	−1.06E+0	3 6	−2.52E−1	2 1	−7.31E+0	3 2	−3.11E−2
2 3	1.93E−2	4 6	−1.27E−1	2 1	1.48E+0	4 1	3.68E−3	3 1	−4.60E+0	3 3	−6.93E+0
2 4	3.05E−1	5 1	3.74E−2	2 2	−3.17E+0	4 2	8.82E−2		$^2G^o$–$^2G^e$	4 1	4.06E+0
2 5	1.10E+0	5 2	2.35E−2	2 3	−5.49E−1	4 3	5.49E−2	1 1	2.29E−4	4 2	−1.02E+1
2 6	−7.43E−1	5 3	4.80E−2	2 4	−1.46E−5	4 4	1.91E−3		$^4S^o$–$^4P^e$	4 3	−1.95E−1
	$^2S^e$–$^2P^o$	5 4	5.66E−1	2 5	−8.72E−3	4 5	2.14E−1	1 1	−1.24E−1	5 1	1.05E−1
1 1	3.64E−2	5 5	2.47E+0	3 1	1.17E+0	4 6	−3.36E−1	1 2	−1.53E+0	5 2	−1.35E−4
1 2	−5.93E−3	5 6	−1.80E−1	3 2	4.77E−3	5 1	3.78E−3	1 3	−9.72E+0	5 3	−5.40E+0
1 3	−9.93E−2		$^2P^o$–$^2D^e$	3 3	−1.17E−2	5 2	2.77E−3	1 4	−2.24E−1		$^4D^o$–$^4D^e$
1 4	−3.09E−3	1 1	−6.62E−3	3 4	−4.00E−2	5 3	7.01E−1	1 5	−2.11E+0	1 1	5.23E−1
1 5	−5.63E−4	1 2	−2.52E−1	3 5	−8.08E−1	5 4	4.74E−2	2 1	1.66E−1	1 2	−2.93E+0
2 1	6.92E−1	1 3	−1.69E+0	4 1	2.16E+0	5 5	1.93E−1	2 2	1.80E+0	2 1	3.78E−3
2 2	5.63E−3	1 4	−7.06E−2	4 2	4.29E−3	5 6	2.76E−2	2 3	4.02E−2	2 2	1.32E+0
2 3	3.78E−2	1 5	−4.08E+0	4 3	4.19E−3		$^2D^o$–$^2F^e$	2 4	−1.30E+0	3 1	7.24E−1
2 4	−1.77E+0	1 6	−1.94E+0	4 4	−1.55E−1	1 1	−6.85E−2	2 5	−3.03E+0	3 2	1.61E+0
2 5	−2.28E−1	2 1	6.72E−2	4 5	−3.93E+0	1 2	−1.04E+1	3 1	8.80E−2		$^4D^o$–$^4F^e$
3 1	8.27E−1	2 2	1.69E−1	5 1	2.58E−1	1 3	−1.94E+0	3 2	3.05E−3	1 1	3.57E+0
3 2	4.41E−2	2 3	−1.15E−1	5 2	1.94E+0	2 1	1.23E+0	3 3	4.59E−2	1 2	−1.74E+1
3 3	3.28E−1	2 4	−2.93E−1	5 3	4.14E−1	2 2	−1.13E+0	3 4	2.52E+0	2 1	1.54E−1
3 4	−3.04E−2	2 5	−4.68E−1	5 4	−5.06E+0	2 3	−6.24E+0	3 5	8.63E−1	2 2	7.52E+0
3 5	−3.46E−5	2 6	−2.06E+0	5 5	−4.86E−1	3 1	1.91E−1		$^4P^o$–$^4P^e$	3 1	2.07E−3
	$^2P^o$–$^2P^e$	3 1	1.79E−1	6 1	1.27E+0	3 2	3.32E−1	1 1	2.05E−1	3 2	1.73E−1
1 1	−1.26E−2	3 2	1.79E+0	6 2	2.78E−3	3 3	−2.43E−3	1 2	4.11E+0		$^4D^e$–$^4F^o$
1 2	−5.59E−1	3 3	2.21E−1	6 3	4.46E−2	4 1	2.45E−2	1 3	−8.14E−2	1 1	−4.94E+0
1 3	−1.66E+0	3 4	−4.62E−2	6 4	2.32E−2	4 2	2.99E−1	1 4	−2.12E+0	2 1	−1.60E+1
1 4	−7.47E−1	3 5	−3.91E−1	6 5	3.82E−2	4 3	1.98E+0	1 5	−2.45E+0		$^4F^o$–$^4F^e$
1 5	−1.71E−1	3 6	−1.67E−1		$^2D^o$–$^2D^e$	5 1	8.28E−4	2 1	4.16E−2	1 1	5.72E−1
1 6	−1.68E+0	4 1	4.85E−2	1 1	−1.07E−1	5 2	3.44E−2	2 2	1.08E−3	1 2	2.11E+0
2 1	2.86E−1	4 2	4.35E−2	1 2	−1.56E+0	5 3	7.36E−2	2 3	8.36E−2		$^4F^e$–$^4G^o$
2 2	2.04E+0	4 3	4.97E−2	1 3	−1.64E+0		$^2D^e$–$^2F^o$	2 4	5.94E+0	1 1	−7.28E+0
2 3	−3.32E−2	4 4	3.19E−1	1 4	−4.44E+0	1 1	−5.47E−2	2 5	9.25E−1	2 1	−2.25E+1
2 4	−1.23E+0	4 5	5.50E−1	1 5	1.10E+0	2 1	1.60E+0		$^4P^o$–$^4D^e$		
2 5	−7.33E−1	4 6	−5.98E−2	1 6	−2.67E−1	3 1	−1.82E−2	1 1	1.86E+0		
2 6	−5.68E−1	5 1	2.81E−3	2 1	6.17E−3	4 1	9.36E−2	1 2	−9.26E+0		
3 1	1.22E−2	5 2	2.24E−2	2 2	2.53E−1	5 1	4.76E−2	2 1	7.69E−2		
3 2	2.25E−1	5 3	6.46E−2	2 3	−1.19E−1	6 1	3.47E−2	2 2	4.02E+0		

P-like Ar (Ar^{3+})

Term energies relative to $3p^2$ ^{3}P ionization threshold for each symmetry

i	E(Ryds)	Description	i	E(Ryds)	Description	i	E(Ryds)	Description	i	E(Ryds)	Description
		^{2}S^e	2	−1.82360	$3p^5$	26	−.20541	$3p^4 3d$	17	−.27135	$3p^2$ ^{3}P $8d$
1	−2.71992	$3s3p^4$	3	−1.64238	$3p^2$ ^{3}P $4p$	27	−.20413	$3p^2$ ^{1}D $8s$	18	−.25158	$3p^2$ ^{3}P $8g$
2	−2.04598	$3p^2$ ^{1}D $3d$	4	−1.49078	$3p^2$ ^{1}D $4p$	28	−.18978	$3p^2$ ^{1}D $7g$	19	−.23100	$3p^2$ ^{1}D $7d$
3	−1.69401	$3p^2$ ^{1}S $4s$	5	−1.35719	$3p^2$ ^{1}S $4p$	29	−.17659	$3p^2$ ^{1}S $6d$	20	−.21473	$3p^2$ ^{3}P $9d$
4	−1.04770	$3p^2$ ^{1}D $4d$	6	−1.06946	$3s3p^3$ ^{3}D $3d$			**^{2}D^o**	21	−.19872	$3p^2$ ^{3}P $9g$
5	−.75901	$3p^2$ ^{1}S $5s$	7	−.92878	$3p^2$ ^{3}P $5p$	1	−4.17128	$3s^2 3p^3$	22	−.19150	$3p^2$ ^{1}D $7g$
6	−.59770	$3p^2$ ^{1}D $5d$	8	−.88963	$3p^2$ ^{1}D $4f$:	2	−1.68796	$3p^2$ ^{3}P $4p$			**^{2}F^o**
7	−.36633	$3p^2$ ^{1}D $6d$	9	−.83187	$3s^2 3p3d^2$:	3	−1.54922	$3p^2$ ^{1}D $4p$	1	−1.56800	$3p^2$ ^{1}D $4p$
8	−.34925	$3p^2$ ^{1}S $6s$	10	−.79971	$3p^2$ ^{1}D $5p$	4	−1.34010	$3s3p^3$ ^{3}P $3d$:	2	−1.27742	$3s3p^3$ ^{3}D $3d$:
9	−.27362	$3s3p^3$ ^{3}P $4p$	11	−.73154	$3s3p^3$ ^{3}P $3d$:	5	−1.18182	$3s3p^3$ ^{3}D $3d$:	3	−1.08428	$3s3p^3$ ^{3}P $3d$
10	−.22434	$3p^2$ ^{1}D $7d$	12	−.66219	$3s3p^3$ ^{3}P $4s$	6	−1.03111	$3p^2$ ^{3}P $4f$	4	−1.01111	$3p^2$ ^{3}P $4f$
11	−.14018	$3p^2$ ^{1}D $8d$	13	−.61334	$3p^2$ ^{1}S $5p$	7	−.95777	$3s^2 3p3d^2$:	5	−.93287	$3p^2$ ^{1}D $4f$
12	−.13213	$3p^2$ ^{1}S $7s$	14	−.58991	$3p^2$ ^{3}P $6p$	8	−.94041	$3p^2$ ^{3}P $5p$	6	−.91067	$3s^2 3p3d^2$:
13	−.07784	$3p^2$ ^{1}D $9d$	15	−.56371	$3p^2$ ^{1}D $5f$:	9	−.88638	$3p^2$ ^{1}D $4f$	7	−.81818	$3p^2$ ^{1}D $5p$
		^{2}S^o	16	−.46943	$3p^2$ ^{1}D $6p$	10	−.85993	$3s3p^3$ ^{3}D $4s$	8	−.74660	$3s3p^3$ ^{1}P $3d$
1	−1.78026	$3p^2$ ^{3}P $4p$	17	−.46121	$3s3p^3$ ^{1}P $3d$	11	−.78502	$3p^2$ ^{1}D $5p$	9	−.66560	$3p^2$ ^{1}S $4f$:
2	−1.57281	$3s3p^3$ ^{3}D $3d$	18	−.41794	$3p^2$ ^{3}P $7p$	12	−.74459	$3s3p^3$ ^{1}P $3d$:	10	−.61977	$3p^2$ ^{3}P $5f$:
3	−.98490	$3p^2$ ^{3}P $5p$	19	−.31346	$3p^2$ ^{3}P $8p$	13	−.66741	$3p^2$ ^{3}P $5f$	11	−.56657	$3s^2 3p3d^2$
4	−.80346	$3s^2 3p3d^2$	20	−.30719	$3p^2$ ^{1}D $6f$	14	−.62230	$3p3d$ ^{1}D $4s$:	12	−.52447	$3p^2$ ^{1}D $5f$
5	−.62888	$3p^2$ ^{3}P $6p$	21	−.28864	$3p^2$ ^{1}D $7p$	15	−.58802	$3p^2$ ^{3}P $6p$:	13	−.47778	$3p^2$ ^{1}D $6p$
6	−.43838	$3p^2$ ^{3}P $7p$	22	−.27753	$3p^2$ ^{1}S $6p$	16	−.54745	$3p^2$ ^{1}D $5f$	14	−.45212	$3p^2$ ^{3}P $6f$
7	−.32424	$3p^2$ ^{3}P $8p$	23	−.27375	$3s^2 3p3d^2$	17	−.47759	$3p^2$ ^{1}D $6p$	15	−.35049	$3p^2$ ^{3}P $7f$:
8	−.29596	$3s3p^3$ ^{1}D $3d$:	24	−.23997	$3p^2$ ^{3}P $9p$	18	−.47308	$3p^2$ ^{3}P $6f$	16	−.33790	$3p^2$ ^{1}D $6f$
9	−.25214	$3p^2$ ^{3}P $9p$	25	−.21828	$3s3p^3$ ^{1}P $4s$:	19	−.42958	$3s3p^3$ ^{1}D $3d$:	17	−.31930	$3p^2$ ^{1}S $5f$
10	−.21169	$3s3p^3$ ^{3}D $4d$:	26	−.20094	$3p^2$ ^{1}D $7f$	20	−.42140	$3p^2$ ^{3}P $7p$	18	−.29303	$3p^2$ ^{1}D $7p$
		^{2}P^e			**^{2}D^e**	21	−.37103	$3p^2$ ^{3}P $7f$:	19	−.27345	$3s^2 3p3d^2$
1	−2.84329	$3p^2$ ^{3}P $3d$	1	−3.04087	$3s3p^4$	22	−.31546	$3p^2$ ^{3}P $8p$	20	−.24517	$3p^2$ ^{3}P $8f$:
2	−2.28687	$3s3p^4$	2	−2.27057	$3p^2$ ^{3}P $3d$:	23	−.31051	$3s3p^3$ ^{3}S $3d$	21	−.22946	$3p3d$ ^{3}F $4s$
3	−2.07251	$3p^2$ ^{1}D $3d$	3	−2.14937	$3p^2$ ^{1}D $3d$	24	−.30056	$3p^2$ ^{1}D $6f$	22	−.20448	$3p^2$ ^{3}P $9f$
4	−1.98903	$3p^2$ ^{3}P $4s$	4	−1.97752	$3p^2$ ^{1}S $3d$:	25	−.29197	$3p^2$ ^{1}D $7p$	23	−.18637	$3p^2$ ^{1}D $7f$
5	−1.24839	$3p^2$ ^{3}P $4d$	5	−1.89209	$3p^2$ ^{1}D $4s$	26	−.24954	$3p^2$ ^{3}P $8f$	24	−.17977	$3p^2$ ^{1}D $8p$
6	−1.08861	$3p^2$ ^{3}P $5s$	6	−1.17339	$3p^2$ ^{3}P $4d$	27	−.24059	$3p^2$ ^{3}P $9p$			**^{2}G^e**
7	−1.05939	$3p^2$ ^{1}D $4d$	7	−1.09690	$3p^2$ ^{1}D $4d$	28	−.20013	$3p^2$ ^{3}P $9f$	1	−2.45322	$3p^2$ ^{1}D $3d$
8	−.76193	$3p^2$ ^{3}P $5d$	8	−.95405	$3p^2$ ^{1}D $5s$	29	−.19193	$3p^2$ ^{1}D $7f$	2	−1.10115	$3p^2$ ^{1}D $4d$
9	−.67732	$3p^2$ ^{3}P $6s$	9	−.90342	$3p^2$ ^{1}S $4d$			**^{2}F^e**	3	−.63758	$3p^2$ ^{3}P $5g$
10	−.61378	$3p^2$ ^{1}D $5d$	10	−.73075	$3p^2$ ^{3}P $5d$	1	−2.64614	$3p^2$ ^{1}D $3d$:	4	−.61758	$3p^2$ ^{1}D $5d$
11	−.58046	$3s3p^3$ ^{3}D $4p$	11	−.61930	$3p^2$ ^{1}D $5d$	2	−2.07920	$3p^2$ ^{3}P $3d$:	5	−.51029	$3p^2$ ^{1}D $5g$
12	−.51299	$3p^2$ ^{3}P $6d$	12	−.54824	$3p^2$ ^{1}D $6s$	3	−1.22366	$3p^2$ ^{3}P $4d$	6	−.44348	$3p^2$ ^{3}P $6g$
13	−.46477	$3p^2$ ^{3}P $7s$	13	−.50221	$3p^2$ ^{3}P $6d$	4	−1.11291	$3p^2$ ^{1}D $4d$	7	−.37276	$3p^2$ ^{1}D $6d$
14	−.38844	$3s3p^3$ ^{3}P $4p$	14	−.49976	$3p^2$ ^{1}D $5g$	5	−.74870	$3p^2$ ^{3}P $5d$	8	−.32608	$3p^2$ ^{3}P $7g$
15	−.37112	$3p^2$ ^{3}P $7d$	15	−.48737	$3s3p^3$ ^{3}D $4p$	6	−.64515	$3p^2$ ^{3}P $5g$	9	−.31586	$3p^2$ ^{1}D $6g$
16	−.35124	$3p^2$ ^{1}D $6d$	16	−.42778	$3p^2$ ^{1}S $5d$	7	−.63674	$3p^2$ ^{1}D $5d$	10	−.30565	$3p^2$ ^{1}S $5g$
17	−.33815	$3p^2$ ^{3}P $8s$	17	−.37433	$3p^2$ ^{1}D $6d$	8	−.55294	$3s3p^3$ ^{3}D $4p$	11	−.24977	$3p^2$ ^{3}P $8g$
18	−.28444	$3p^2$ ^{3}P $8d$	18	−.36142	$3p^2$ ^{3}P $7d$	9	−.50609	$3p^2$ ^{3}P $6d$	12	−.23061	$3p^2$ ^{1}D $7d$
19	−.26229	$3p3d$ ^{1}D $4p$:	19	−.35471	$3s3p^3$ ^{3}P $4p$	10	−.50536	$3p^2$ ^{1}D $5g$	13	−.19747	$3p^2$ ^{3}P $9g$
20	−.25615	$3p^2$ ^{3}P $9s$:	20	−.33102	$3p^2$ ^{1}D $7s$	11	−.44763	$3p^2$ ^{3}P $6g$	14	−.19526	$3p^2$ ^{1}D $7g$
21	−.22701	$3p^2$ ^{1}D $7d$	21	−.30667	$3p^2$ ^{1}D $6g$	12	−.38101	$3p^2$ ^{1}D $6d$	15	−.18163	$3p^4 3d$
22	−.21602	$3p^2$ ^{3}P $9d$	22	−.27756	$3p^2$ ^{3}P $8d$	13	−.36421	$3p^2$ ^{3}P $7d$			**^{2}G^o**
23	−.20492	$3p^4 3d$	23	−.26751	$3p3d$ ^{1}D $4p$	14	−.32873	$3p^2$ ^{3}P $7g$	1	−1.42216	$3s3p^3$ ^{3}D $3d$
		^{2}P^o	24	−.23192	$3p^2$ ^{1}D $7d$	15	−.31010	$3p^2$ ^{1}D $6g$	2	−1.14491	$3s3p^3$ ^{1}D $3d$:
1	−4.03133	$3s^2 3p^3$	25	−.21344	$3p^2$ ^{3}P $9d$	16	−.29462	$3p3d$ ^{1}D $4p$	3	−1.00575	$3p^2$ ^{3}P $4f$

P-like Ar (Ar^{3+})

i	E(Ryds)	Description
4	−.88763	$3p^2$ ^{1}D $4f$
5	−.66002	$3p^2$ ^{3}P $5f$
6	−.53378	$3p^2$ ^{1}D $5f$
7	−.47087	$3p^2$ ^{3}P $6f$
8	−.42555	$3s^23p3d^2$
9	−.33385	$3p^2$ ^{3}P $7f$
10	−.32114	$3p^2$ ^{1}D $6f$
11	−.25808	$3p^2$ ^{3}P $8f$
12	−.20801	$3p^2$ ^{3}P $9f$:
13	−.20020	$3p^2$ ^{1}D $7f$
		^{4}S^e
1	−.42235	$3s3p^3$ ^{3}P $4p$
		^{4}S^o
1	−4.37133	$3s^23p^3$
2	−1.67927	$3p^2$ ^{3}P $4p$
3	−1.27727	$3s3p^3$ ^{3}D $3d$
4	−1.15149	$3s3p^3$ ^{5}S $4s$
5	−.93391	$3p^2$ ^{3}P $5p$
6	−.60857	$3p^2$ ^{3}P $6p$
7	−.42747	$3p^2$ ^{3}P $7p$
8	−.32244	$3p^2$ ^{3}P $8p$
9	−.29340	$3s3p^3$ ^{5}S $5s$
10	−.25633	$3s3p^3$ ^{3}S $4s$
11	−.23746	$3p^2$ ^{3}P $9p$
		^{4}P^e
1	−3.31029	$3s3p^4$
2	−2.32095	$3p^2$ ^{3}P $3d$
3	−2.04543	$3p^2$ ^{3}P $4s$
4	−1.24395	$3p^2$ ^{3}P $4d$
5	−1.09499	$3p^2$ ^{3}P $5s$
6	−.86854	$3s3p^3$ ^{5}S $4p$
7	−.75537	$3p^2$ ^{3}P $5d$
8	−.68512	$3p^2$ ^{3}P $6s$
9	−.54237	$3s3p^3$ ^{3}D $4p$
10	−.51191	$3p^2$ ^{3}P $6d$
11	−.46905	$3p^2$ ^{3}P $7s$
12	−.38662	$3s3p^3$ ^{3}P $4p$
13	−.36784	$3p^2$ ^{3}P $7d$
14	−.34097	$3p^2$ ^{3}P $8s$
15	−.27840	$3p^2$ ^{3}P $8d$
16	−.25976	$3p^2$ ^{3}P $9s$
17	−.23538	$3s3p^3$ ^{5}S $5p$
18	−.21687	$3p^2$ ^{3}P $9d$
		^{4}P^o
1	−1.70074	$3p^2$ ^{3}P $4p$
2	−1.35249	$3s3p^3$ ^{3}P $3d$
3	−1.22885	$3s3p^3$ ^{3}D $3d$
4	−.95416	$3p^2$ ^{3}P $5p$
5	−.74151	$3s3p^3$ ^{3}P $4s$
6	−.61251	$3p^2$ ^{3}P $6p$
7	−.57976	$3s^23p3d^2$
8	−.42788	$3p^2$ ^{3}P $7p$
9	−.31594	$3p^2$ ^{3}P $8p$
10	−.24287	$3p^2$ ^{3}P $9p$
		^{4}D^e
1	−2.65229	$3p^2$ ^{3}P $3d$
2	−1.23767	$3p^2$ ^{3}P $4d$
3	−.75485	$3p^2$ ^{3}P $5d$
4	−.61242	$3s3p^3$ ^{3}D $4p$
5	−.51161	$3p^2$ ^{3}P $6d$
6	−.41838	$3s3p^3$ ^{3}P $4p$
7	−.36559	$3p^2$ ^{3}P $7d$
8	−.34335	$3p^43d$
9	−.27421	$3p^2$ ^{3}P $8d$
10	−.21481	$3p^2$ ^{3}P $9d$
		^{4}D^o
1	−1.76188	$3s3p^3$ ^{3}D $3d$:
2	−1.72722	$3p^2$ ^{3}P $4p$
3	−1.34802	$3s3p^3$ ^{3}P $3d$
4	−1.18140	$3s3p^3$ ^{5}S $3d$:
5	−1.06757	$3p^2$ ^{3}P $4f$
6	−.97165	$3p^2$ ^{3}P $5p$
7	−.91623	$3s3p^3$ ^{3}D $4s$
8	−.84518	$3s3p^3$ ^{3}S $3d$
9	−.67741	$3s^23p3d^2$
10	−.62835	$3p^2$ ^{3}P $5f$
11	−.61700	$3p^2$ ^{3}P $6p$
12	−.46585	$3p^2$ ^{3}P $6f$
13	−.43862	$3s3p^3$ ^{5}S $4d$
14	−.42944	$3p^2$ ^{3}P $7p$
15	−.36114	$3p^2$ ^{3}P $7f$:
16	−.31761	$3p^2$ ^{3}P $8p$
17	−.30673	$3s^23p3d^2$
18	−.24662	$3p^2$ ^{3}P $8f$
19	−.24377	$3p^2$ ^{3}P $9p$
20	−.10700	$3p^2$ ^{3}P $9f$
		^{4}F^e
1	−2.70567	$3p^2$ ^{3}P $3d$
2	−1.26225	$3p^2$ ^{3}P $4d$
3	−.76636	$3p^2$ ^{3}P $5d$
4	−.64521	$3p^2$ ^{3}P $5g$
5	−.60107	$3s3p^3$ ^{3}D $4p$
6	−.50934	$3p^2$ ^{3}P $6d$
7	−.44784	$3p^2$ ^{3}P $6g$
8	−.36815	$3p^2$ ^{3}P $7d$
9	−.32937	$3p^2$ ^{3}P $7g$
10	−.31564	$3s3p^3$ ^{5}S $4f$
11	−.27773	$3p^2$ ^{3}P $8d$
12	−.25168	$3p^2$ ^{3}P $8g$
13	−.21694	$3p^2$ ^{3}P $9d$
14	−.19933	$3p^2$ ^{3}P $9g$
15	−.18463	$3p^43d$
		^{4}F^o
1	−1.69308	$3s3p^3$ ^{3}D $3d$
2	−1.31798	$3s3p^3$ ^{3}P $3d$
3	−1.00215	$3p^2$ ^{3}P $4f$
4	−.65021	$3p^2$ ^{3}P $5f$
5	−.56110	$3s^23p3d^2$
6	−.45071	$3p^2$ ^{3}P $6f$
7	−.33131	$3p^2$ ^{3}P $7f$
8	−.28607	$3p3d$ ^{3}F $4s$
9	−.25341	$3p^2$ ^{3}P $8f$
10	−.20012	$3p^2$ ^{3}P $9f$
		^{4}G^e
1	−.63799	$3p^2$ ^{3}P $5g$
2	−.44387	$3p^2$ ^{3}P $6g$
3	−.32638	$3p^2$ ^{3}P $7g$
4	−.24998	$3p^2$ ^{3}P $8g$
5	−.19756	$3p^2$ ^{3}P $9g$
		^{4}G^o
1	−1.53701	$3s3p^3$ ^{3}D $3d$
2	−1.05530	$3p^2$ ^{3}P $4f$
3	−.86969	$3s^23p3d^2$
4	−.65186	$3p^2$ ^{3}P $5f$
5	−.45487	$3p^2$ ^{3}P $6f$
6	−.33392	$3p^2$ ^{3}P $7f$
7	−.25532	$3p^2$ ^{3}P $8f$
8	−.20155	$3p^2$ ^{3}P $9f$
		^{6}S^o
1	−1.28727	$3s3p^3$ ^{5}S $4s$
2	−.33686	$3s3p^3$ ^{5}S $5s$
		^{6}P^e
1	−.95554	$3s3p^3$ ^{5}S $4p$
2	−.20194	$3s3p^3$ ^{5}S $5p$
		^{6}D^o
1	−2.01302	$3s3p^3$ ^{5}S $3d$
2	−.49551	$3s3p^3$ ^{5}S $4d$
3	−.00333	$3s3p^3$ ^{5}S $5d$
		^{6}F^e
1	−.29640	$3s3p^3$ ^{5}S $4f$
2	−.15582	$3s3p^23d^2$

P-like Ar (Ar^{3+})

Energies in ascending order from ground state for terms with effective $n \leq 4.0$, $L \leq 4$

Term	i	E(Ryds)	Term	i	E(Ryds)	Term	i	E(Ryds)	Term	i	E(Ryds)	Term	i	E(Ryds)
$^4S^o$	1	0.00000	$^2F^e$	2	2.29213	$^2D^o$	2	2.68337	$^2F^o$	2	3.09391	$^2D^e$	7	3.27443
$^2D^o$	1	0.20005	$^2P^e$	3	2.29882	$^4S^o$	2	2.69206	$^4S^o$	3	3.09406	$^4P^e$	5	3.27634
$^2P^o$	1	0.34000	$^2S^e$	2	2.32535	$^2P^o$	3	2.72895	$^4F^e$	2	3.10908	$^2P^e$	6	3.28272
$^4P^e$	1	1.06104	$^4P^e$	3	2.32590	$^2S^o$	2	2.79852	$^2P^e$	5	3.12294	$^2F^o$	3	3.28705
$^2D^e$	1	1.33046	$^6D^o$	1	2.35831	$^2F^o$	1	2.80333	$^4P^e$	4	3.12738	$^2P^o$	6	3.30187
$^2P^e$	1	1.52804	$^2P^e$	4	2.38230	$^2D^o$	3	2.82211	$^4D^e$	2	3.13366	$^4D^o$	5	3.30376
$^2S^e$	1	1.65141	$^2D^e$	4	2.39381	$^4G^o$	1	2.83432	$^4P^o$	3	3.14248	$^2P^e$	7	3.31194
$^4F^e$	1	1.66566	$^2D^e$	5	2.47924	$^2P^o$	4	2.88055	$^2F^e$	3	3.14767	$^4G^o$	2	3.31603
$^4D^e$	1	1.71904	$^2P^o$	2	2.54773	$^2G^o$	1	2.94917	$^2D^o$	5	3.18951	$^2S^e$	4	3.32363
$^2F^e$	1	1.72519	$^2S^o$	1	2.59107	$^2P^o$	5	3.01414	$^4D^o$	4	3.18993	$^2D^o$	6	3.34022
$^2G^e$	1	1.91811	$^4D^o$	1	2.60945	$^4P^o$	2	3.01884	$^2D^e$	6	3.19794	$^2F^o$	4	3.36022
$^4P^e$	2	2.05038	$^4D^o$	2	2.64411	$^4D^o$	3	3.02331	$^4S^o$	4	3.21984	$^2G^o$	3	3.36558
$^2P^e$	2	2.08446	$^4P^o$	1	2.67059	$^2D^o$	4	3.03123	$^2G^o$	2	3.22642	$^4F^o$	3	3.36918
$^2D^e$	2	2.10076	$^2S^e$	3	2.67732	$^4F^o$	2	3.05335	$^2F^e$	4	3.25842			
$^2D^e$	3	2.22196	$^4F^o$	1	2.67825	$^6S^o$	1	3.08406	$^2G^e$	2	3.27018			

gf-values for transitions involving terms with effective $n \leq 4.0$, $L \leq 4$

i i'	gf_L	i i'	gf_L	i i'	gf_L	i i'	gf_L	i i'	gf_L	i i'	gf_L
	$^2S^o$–$^2P^e$	3 5	−1.76E+0	4 3	3.86E−1	2 6	−2.75E−2	1 2	−1.56E−4	6 4	3.21E−2
1 1	1.98E−1	3 6	−2.90E−3	4 4	4.39E−1	2 7	−9.15E−3	1 3	−4.18E−2	6 5	6.62E−4
1 2	6.23E−2	4 1	9.90E−3	4 5	−1.12E−1	3 1	6.43E−2	1 4	−5.60E−2	6 6	−4.05E−4
1 3	6.76E−2	4 2	7.70E−3	4 6	−1.88E+0	3 2	3.75E−1	1 5	−4.91E−1	7 1	7.37E−2
1 4	4.19E−1	4 3	2.72E−1	4 7	−8.39E−1	3 3	9.98E−2	1 6	−2.32E+0	7 2	1.14E+0
1 5	−1.90E+0	4 4	9.65E−1	5 1	2.11E−2	3 4	1.43E−1	2 1	5.16E+0	7 3	4.64E−2
1 6	−1.70E−1	4 5	4.20E−2	5 2	1.05E−2	3 5	3.21E−1	2 2	−2.17E−2	7 4	4.43E−3
1 7	−2.76E−1	4 6	9.06E−6	5 3	8.64E−3	3 6	−4.12E+0	2 3	−1.50E−1	7 5	4.47E−4
2 1	3.81E−2		$^2P^o$–$^2P^e$	5 4	4.16E−2	3 7	−1.52E−1	2 4	−1.14E−3	7 6	−9.51E−3
2 2	1.03E−2	1 1	−2.40E−2	5 5	−1.04E−2	4 1	2.55E−1	2 5	−1.32E−2		$^2D^o$–$^2D^e$
2 3	8.10E−5	1 2	−1.32E−1	5 6	−3.88E−3	4 2	1.16E−1	2 6	−1.86E−1	1 1	−3.81E−1
2 4	1.22E−3	1 3	−5.78E+0	5 7	−8.58E−2	4 3	4.16E−2	3 1	5.14E−2	1 2	−1.22E+0
2 5	−2.96E−3	1 4	−6.83E−1	6 1	1.44E−1	4 4	3.42E−1	3 2	−1.62E−2	1 3	−8.76E+0
2 6	−1.34E−3	1 5	−2.12E−2	6 2	1.30E−1	4 5	1.65E+0	3 3	−3.62E−2	1 4	−1.20E−4
2 7	−4.02E−4	1 6	−1.19E−1	6 3	1.08E−1	4 6	−5.64E−1	3 4	−2.36E−2	1 5	−6.47E−1
	$^2S^e$–$^2P^o$	1 7	−1.73E−2	6 4	4.74E−3	4 7	−2.17E+0	3 5	−1.99E−2	1 6	−1.30E−2
1 1	2.09E−1	2 1	2.09E−1	6 5	3.80E−3	5 1	5.88E−2	3 6	−6.59E−1	1 7	−1.08E−2
1 2	−3.23E−3	2 2	2.10E−2	6 6	5.19E−3	5 2	5.26E−1	4 1	7.38E−1	2 1	1.06E−2
1 3	−7.13E−3	2 3	1.02E−2	6 7	−1.75E−4	5 3	1.04E−1	4 2	−2.31E+0	2 2	8.72E−2
1 4	−5.31E−2	2 4	6.18E−3		$^2P^o$–$^2D^e$	5 4	4.58E−1	4 3	−8.37E−1	2 3	4.91E−2
1 5	−2.30E−2	2 5	−6.55E−3	1 1	−4.89E−2	5 5	9.16E−2	4 4	−1.20E−2	2 4	1.06E−1
1 6	−8.32E−2	2 6	−6.15E−3	1 2	−6.10E−2	5 6	−5.98E−4	4 5	−1.02E−4	2 5	4.03E−1
2 1	1.78E+0	2 7	−5.72E−6	1 3	−1.36E−1	5 7	−8.63E−4	4 6	−2.20E−2	2 6	−8.57E−1
2 2	−6.77E−5	3 1	3.85E−1	1 4	−1.09E+1	6 1	4.95E+0	5 1	2.72E−3	2 7	−1.16E+0
2 3	−2.51E−2	3 2	2.44E−2	1 5	−3.37E−1	6 2	8.66E−3	5 2	4.47E−3	3 1	1.51E−1
2 4	−1.57E−1	3 3	2.22E−2	1 6	−8.72E−2	6 3	1.12E−2	5 3	2.10E−2	3 2	2.79E−2
2 5	−3.06E−2	3 4	1.54E+0	1 7	−2.30E−2	6 4	1.11E−2	5 4	9.89E−5	3 3	7.31E−1
2 6	−9.20E−5	3 5	−1.58E+0	2 1	2.98E−1	6 5	1.46E−2	5 5	−1.45E−1	3 4	6.68E−4
3 1	1.54E−1	3 6	−1.19E−1	2 2	1.55E−4	6 6	1.01E−2	5 6	−2.93E+0	3 5	2.32E+0
3 2	3.60E−4	3 7	−8.83E−1	2 3	2.03E−3	6 7	1.34E−3	6 1	1.30E−1	3 6	−7.92E−1
3 3	−6.14E−3	4 1	8.28E−3	2 4	4.41E−3		$^2P^e$–$^2D^o$	6 2	4.89E−1	3 7	−2.13E+0
3 4	−1.10E−2	4 2	2.30E−1	2 5	3.83E−4	1 1	3.73E−1	6 3	1.23E+0	4 1	2.11E−1

P-like Ar (Ar^{3+})

$i\ i'$	gf_L	$i\ i'$	gf_L	$i\ i'$	gf_L	$i\ i'$	gf_L	$i\ i'$	gf_L	$i\ i'$	gf_L
4 2	2.39E−3	5 3	1.81E−3	1 4	−2.58E+0	2 2	−1.11E−2	3 5	−1.15E−2	3 1	1.86E−1
4 3	4.99E−3	5 4	−1.71E−3	2 1	2.82E−1	3 1	1.58E−1		$^4\mathbf{P}^o$–$^4\mathbf{D}^e$	3 2	−6.32E−3
4 4	1.82E−3	6 1	4.41E−3	2 2	1.91E−2	3 2	2.92E−2	1 1	1.95E+0	4 1	9.53E−3
4 5	5.27E−2	6 2	8.27E−3	2 3	−1.57E−3		$^4\mathbf{S}^o$–$^4\mathbf{P}^e$	1 2	−9.68E+0	4 2	1.60E−2
4 6	−6.61E−3	6 3	1.99E−2	2 4	−3.19E−3	1 1	−3.05E−1	2 1	3.68E−1	5 1	7.99E−2
4 7	−6.17E−3	6 4	6.03E−4	3 1	7.52E−1	1 2	−1.27E+1	2 2	−6.86E−3	5 2	1.05E+0
5 1	5.82E−1		$^2\mathbf{D}^e$–$^2\mathbf{F}^o$	3 2	5.52E−3	1 3	−4.97E−1	3 1	1.73E−4		$^4\mathbf{D}^o$–$^4\mathbf{F}^e$
5 2	2.28E−2	1 1	−6.27E−2	3 3	1.36E−2	1 4	−2.27E−2	3 2	9.84E−4	1 1	4.22E−1
5 3	4.34E−3	1 2	−1.18E−1	3 4	6.49E−3	1 5	−1.57E−1		$^4\mathbf{P}^e$–$^4\mathbf{D}^o$	1 2	−1.77E+0
5 4	1.68E−2	1 3	−7.45E−2	4 1	4.08E−1	2 1	1.60E−1	1 1	−3.63E−2	2 1	3.17E+0
5 5	1.56E−2	1 4	−4.77E−2	4 2	2.81E−1	2 2	8.31E−1	1 2	−2.17E−3	2 2	−1.59E+1
5 6	−4.14E−3	2 1	−6.83E−4	4 3	5.33E−1	2 3	1.22E+0	1 3	−1.85E+0	3 1	1.71E−1
5 7	−2.25E−3	2 2	−1.63E−2	4 4	6.59E−2	2 4	−3.94E+0	1 4	−1.84E+1	3 2	−3.17E−2
6 1	1.05E+0	2 3	−6.92E−2		$^2\mathbf{F}^o$–$^2\mathbf{G}^e$	2 5	−8.06E−1	1 5	−5.35E−2	4 1	3.26E−1
6 2	4.94E−2	2 4	−1.10E+0	1 1	2.20E+0	3 1	3.55E+0	2 1	−1.20E−3	4 2	1.73E−4
6 3	1.26E−2	3 1	−2.17E−3	1 2	−1.17E+1	3 2	1.49E−1	2 2	−2.16E−1	5 1	4.13E−1
6 4	1.24E−1	3 2	−4.06E−2	2 1	3.16E−2	3 3	1.80E−4	2 3	−7.92E−4	5 2	6.70E−4
6 5	1.03E−3	3 3	−6.73E−4	2 2	−2.57E−2	3 4	−7.73E−5	2 4	−1.82E−1		$^4\mathbf{D}^e$–$^4\mathbf{F}^o$
6 6	3.64E−1	3 4	−2.19E−1	3 1	1.27E−2	3 5	−5.61E−3	2 5	−1.91E+0	1 1	−6.09E−2
6 7	9.37E−4	4 1	−1.99E−2	3 2	9.37E−5	4 1	1.95E−1	3 1	−4.71E−1	1 2	−1.10E+0
	$^2\mathbf{D}^o$–$^2\mathbf{F}^e$	4 2	−1.70E−2	4 1	1.66E−2	4 2	2.14E−1	3 2	−5.04E+0	1 3	−8.97E+0
1 1	−1.13E−3	4 3	−6.70E−4	4 2	2.30E−3	4 3	5.39E−2	3 3	−9.97E−2	2 1	3.20E−4
1 2	−1.64E+1	4 4	−1.50E+0		$^2\mathbf{F}^e$–$^2\mathbf{G}^o$	4 4	4.37E−2	3 4	−8.48E−6	2 2	1.65E−1
1 3	−7.14E−2	5 1	−4.13E+0	1 1	−2.11E−1	4 5	−7.39E−2	3 5	−4.29E−2	2 3	−1.27E+1
1 4	−6.96E−2	5 2	−1.43E−2	1 2	−8.06E−2		$^4\mathbf{P}^o$–$^4\mathbf{P}^e$	4 1	1.55E−4		$^4\mathbf{F}^o$–$^4\mathbf{F}^e$
2 1	1.27E+0	5 3	−1.90E−2	1 3	−2.20E+0	1 1	2.29E−1	4 2	2.56E−1	1 1	3.68E−1
2 2	3.17E−1	5 4	−8.03E−2	2 1	−2.66E−2	1 2	2.52E−1	4 3	1.06E−3	1 2	−1.99E−3
2 3	−8.15E+0	6 1	2.75E−3	2 2	−2.86E−1	1 3	3.84E+0	4 4	−8.26E−2	2 1	6.61E−1
2 4	−3.34E−2	6 2	3.19E−2	2 3	−3.03E+0	1 4	−3.07E+0	4 5	−4.72E+0	2 2	−1.43E−2
3 1	1.36E−1	6 3	−1.07E−2	3 1	2.56E−4	1 5	−2.20E+0	5 1	3.52E−1	3 1	1.76E+0
3 2	1.06E+0	6 4	−4.10E+0	3 2	−6.75E−1	2 1	7.93E−1	5 2	3.17E+0	3 2	1.68E+0
3 3	−3.04E−1	7 1	5.02E−1	3 3	−5.03E+0	2 2	2.23E−2	5 3	3.13E−2		$^4\mathbf{F}^e$–$^4\mathbf{G}^o$
3 4	−5.83E+0	7 2	2.05E−2	4 1	8.73E−2	2 3	2.21E−2	5 4	1.55E−3	1 1	−1.10E+0
4 1	2.70E−1	7 3	−5.59E−3	4 2	3.69E−3	2 4	−2.58E−3	5 5	−7.92E−3	1 2	−6.03E+0
4 2	9.18E−3	7 4	−1.07E−2	4 3	−9.85E−1	2 5	−7.11E−3		$^4\mathbf{D}^o$–$^4\mathbf{D}^e$	2 1	2.03E−1
4 3	−2.22E−3		$^2\mathbf{F}^o$–$^2\mathbf{F}^e$		$^2\mathbf{G}^o$–$^2\mathbf{G}^e$	3 1	1.14E+1	1 1	2.32E−4	2 2	−1.33E+1
4 4	−5.96E−2	1 1	4.20E−1	1 1	5.22E−3	3 2	2.23E−1	1 2	−3.38E−1		
5 1	1.14E−1	1 2	2.00E−1	1 2	−1.24E−2	3 3	2.73E−2	2 1	6.80E−1		
5 2	7.15E−5	1 3	−2.41E−1	2 1	5.49E−1	3 4	7.65E−4	2 2	−2.90E+0		

P-like Ca (Ca^{5+})

Term energies relative to $3p^2$ ^{3}P ionization threshold for each symmetry

^{2}S^e

i	E(Ryds)	Description
1	−5.79649	$3s3p^4$
2	−4.83781	$3p^2$ ^{1}D $3d$
3	−3.49217	$3p^2$ ^{1}S $4s$
4	−2.45954	$3p^2$ ^{1}D $4d$
5	−1.91685	$3p^43d$
6	−1.77614	$3p^2$ ^{1}S $5s$
7	−1.52948	$3s3p^3$ ^{3}P $4p$
8	−1.45576	$3p^2$ ^{1}D $5d$
9	−1.07312	$3s3p^3$ ^{1}P $4p$
10	−.98251	$3p^2$ ^{1}S $6s$
11	−.93221	$3p^2$ ^{1}D $6d$
12	−.84962	$3s3p^23d^2$
13	−.74041	$3s3p^23d^2$
14	−.65713	$3p3d$ ^{3}P $4p$
15	−.60195	$3p^2$ ^{1}D $7d$
16	−.55701	$3p^2$ ^{1}S $7s$
17	−.42983	$3p^2$ ^{1}D $8d$
18	−.31112	$3p^2$ ^{1}S $8s$
19	−.29924	$3p^2$ ^{1}D $9d$
20	−.29050	$3s3p^23d^2$
21	−.25716	$3p3d$ ^{1}P $4p$

^{2}S^o

i	E(Ryds)	Description
1	−4.20132	$3s3p^3$ ^{3}D $3d$
2	−3.53764	$3p^2$ ^{3}P $4p$
3	−3.15581	$3s3p^3$ ^{1}D $3d$:
4	−2.50374	$3s^23p3d^2$:
5	−2.03342	$3p^2$ ^{3}P $5p$
6	−1.56770	$3s3p^3$ ^{3}S $4s$
7	−1.29870	$3p^2$ ^{3}P $6p$
8	−1.24460	$3s3p^3$ ^{3}D $4d$
9	−.92038	$3p^2$ ^{3}P $7p$
10	−.82359	$3s3p^3$ ^{1}D $4d$
11	−.68412	$3p^2$ ^{3}P $8p$
12	−.52841	$3p^2$ ^{3}P $9p$

^{2}P^e

i	E(Ryds)	Description
1	−5.90673	$3p^2$ ^{3}P $3d$
2	−5.20493	$3s3p^4$
3	−4.85598	$3p^2$ ^{1}D $3d$
4	−3.85445	$3p^2$ ^{3}P $4s$
5	−2.74455	$3p^2$ ^{3}P $4d$
6	−2.50287	$3p^2$ ^{1}D $4d$
7	−2.42873	$3p^43d$
8	−2.17095	$3p^2$ ^{3}P $5s$
9	−1.96099	$3s3p^3$ ^{3}D $4p$
10	−1.71854	$3s3p^3$ ^{3}P $4p$
11	−1.68755	$3p^2$ ^{3}P $5d$
12	−1.63822	$3p^43d$:
13	−1.51032	$3s3p^3$ ^{1}D $4p$
14	−1.47386	$3p^2$ ^{1}D $5d$
15	−1.39547	$3p^2$ ^{3}P $6s$
16	−1.35679	$3s3p^23d^2$:
17	−1.17659	$3s3p^23d^2$
18	−1.14425	$3p^2$ ^{3}P $6d$
19	−1.09511	$3s3p^3$ ^{3}S $4p$:
20	−1.03582	$3s3p^3$ ^{1}P $4p$
21	−.97303	$3p^2$ ^{3}P $7s$
22	−.94185	$3p^2$ ^{1}D $6d$
23	−.92397	$3s3p^23d^2$
24	−.84580	$3p3d$ ^{3}P $4p$
25	−.82569	$3s3p^23d^2$
26	−.81414	$3p^2$ ^{3}P $7d$
27	−.78362	$3s3p^3$ ^{3}D $4f$
28	−.72286	$3p^2$ ^{3}P $8s$
29	−.71149	$3p3d$ ^{1}D $4p$
30	−.69454	$3s3p^23d^2$
31	−.63660	$3p^2$ ^{1}D $7d$
32	−.61664	$3p^2$ ^{3}P $8d$
33	−.56606	$3p3d$ ^{3}D $4p$:
34	−.54056	$3p^2$ ^{3}P $9s$:
35	−.51721	$3s3p^3$ ^{3}D $5p$
36	−.47948	$3p^2$ ^{3}P $9d$
37	−.46050	$3s3p^23d^2$

^{2}P^o

i	E(Ryds)	Description
1	−7.51415	$3s^23p^3$
2	−4.60520	$3p^5$
3	−3.58560	$3s3p^3$ ^{3}D $3d$
4	−3.37065	$3p^2$ ^{3}P $4p$
5	−3.25017	$3s3p^3$ ^{3}P $3d$:
6	−3.12203	$3p^2$ ^{1}D $4p$
7	−3.07871	$3s^23p3d^2$
8	−2.97931	$3p^2$ ^{1}S $4p$
9	−2.83545	$3s3p^3$ ^{1}P $3d$
10	−2.42307	$3s^23p3d^2$
11	−2.33156	$3s3p^3$ ^{1}D $3d$
12	−2.11513	$3s3p^3$ ^{3}P $4s$
13	−2.04407	$3p^2$ ^{1}D $4f$
14	−1.93818	$3p^2$ ^{3}P $5p$
15	−1.88556	$3s^23p3d^2$
16	−1.76514	$3p^2$ ^{1}D $5p$
17	−1.56897	$3s3p^3$ ^{1}P $4s$
18	−1.54196	$3p^2$ ^{1}S $5p$
19	−1.28047	$3p^2$ ^{3}P $6p$
20	−1.27197	$3p^2$ ^{1}D $5f$
21	−1.22747	$3s3p^3$ ^{3}D $4d$
22	−1.20281	$3p3d$ ^{3}P $4s$
23	−1.10115	$3p^2$ ^{1}D $6p$
24	−.94429	$3s3p^3$ ^{3}P $4d$
25	−.90162	$3p^2$ ^{3}P $7p$
26	−.88696	$3p^2$ ^{1}S $6p$
27	−.84656	$3s3p^3$ ^{1}D $4d$
28	−.83888	$3p^2$ ^{1}D $6f$
29	−.80317	$3p3d$ ^{1}P $4s$
30	−.73010	$3p^2$ ^{1}D $7p$
31	−.66808	$3p^2$ ^{3}P $8p$
32	−.56723	$3p^2$ ^{1}D $7f$
33	−.51925	$3p^2$ ^{3}P $9p$
34	−.50493	$3p^2$ ^{1}D $8p$
35	−.49031	$3p^2$ ^{1}S $7p$
36	−.47439	$3s3p^3$ ^{3}P $5s$

^{2}D^e

i	E(Ryds)	Description
1	−6.19572	$3s3p^4$
2	−5.11970	$3p^2$ ^{1}S $3d$:
3	−4.97190	$3p^2$ ^{1}D $3d$
4	−4.70456	$3p^2$ ^{3}P $3d$:
5	−3.73611	$3p^2$ ^{1}D $4s$
6	−2.63940	$3p^2$ ^{3}P $4d$
7	−2.55543	$3p^2$ ^{1}D $4d$
8	−2.42436	$3p^43d$
9	−2.28699	$3p^2$ ^{1}S $4d$
10	−2.01892	$3p^2$ ^{1}D $5s$
11	−1.81915	$3s3p^3$ ^{3}D $4p$
12	−1.78099	$3s3p^23d^2$:
13	−1.76065	$3p^43d$
14	−1.67263	$3s3p^3$ ^{3}P $4p$
15	−1.64146	$3p^2$ ^{3}P $5d$
16	−1.54776	$3p^43d$:
17	−1.51872	$3p^2$ ^{1}D $5d$:
18	−1.47869	$3s3p^3$ ^{1}D $4p$:
19	−1.26342	$3p^2$ ^{1}D $5g$
20	−1.25298	$3p^2$ ^{1}S $5d$
21	−1.23137	$3p^2$ ^{1}D $6s$
22	−1.20673	$3s3p^23d^2$:
23	−1.16916	$3s3p^23d^2$:
24	−1.12841	$3p^2$ ^{3}P $6d$
25	−1.09487	$3s3p^3$ ^{1}P $4p$
26	−1.01737	$3p3d$ ^{3}F $4p$
27	−.95633	$3p^2$ ^{1}D $6d$
28	−.94195	$3s3p^23d^2$
29	−.85949	$3s3p^3$ ^{3}D $4f$:
30	−.82954	$3s3p^23d^2$:
31	−.82524	$3p^2$ ^{1}D $6g$
32	−.80972	$3p^2$ ^{1}D $7s$
33	−.80318	$3p^2$ ^{3}P $7d$
34	−.77321	$3p3d$ ^{1}D $4p$
35	−.72422	$3s3p^23d^2$:
36	−.70655	$3p^2$ ^{1}S $6d$
37	−.69589	$3p3d$ ^{3}P $4p$
38	−.66778	$3s3p^3$ ^{3}P $4f$:
39	−.63291	$3p^2$ ^{1}D $7d$
40	−.61042	$3p^2$ ^{3}P $8d$
41	−.59131	$3p3d$ ^{3}D $4p$
42	−.56128	$3p^2$ ^{1}D $7g$
43	−.54644	$3p^2$ ^{1}D $8s$
44	−.53521	$3s3p^23d^2$:
45	−.50522	$3s3p^3$ ^{1}D $4f$
46	−.47842	$3p^2$ ^{3}P $9d$
47	−.46336	$3s3p^3$ ^{3}D $5p$
48	−.44080	$3p^2$ ^{1}D $8d$
49	−.42838	$3p3d$ ^{1}P $4p$:
50	−.39689	$3p^2$ ^{1}S $7d$
51	−.39362	$3p^2$ ^{1}D $8g$

^{2}D^o

i	E(Ryds)	Description
1	−7.68868	$3s^23p^3$
2	−3.90074	$3s3p^3$ ^{3}P $3d$:
3	−3.67000	$3s3p^3$ ^{1}D $3d$:
4	−3.42061	$3p^2$ ^{3}P $4p$
5	−3.33173	$3s3p^3$ ^{3}D $3d$
6	−3.21327	$3p^2$ ^{1}D $4p$
7	−3.14711	$3s3p^3$ ^{1}P $3d$
8	−2.77414	$3s^23p3d^2$:
9	−2.65697	$3s3p^3$ ^{3}S $3d$
10	−2.34430	$3s3p^3$ ^{3}D $4s$
11	−2.32416	$3s^23p3d^2$
12	−2.22223	$3p^2$ ^{3}P $4f$:
13	−2.08375	$3p^2$ ^{1}D $4f$
14	−2.03755	$3s^23p3d^2$
15	−2.03488	$3s3p^3$ ^{1}D $4s$
16	−1.93692	$3p^2$ ^{3}P $5p$
17	−1.79575	$3p^2$ ^{1}D $5p$
18	−1.47376	$3p^2$ ^{3}P $5f$
19	−1.29229	$3p^2$ ^{3}P $6p$
20	−1.28703	$3p^2$ ^{1}D $5f$
21	−1.26086	$3p3d$ ^{1}D $4s$
22	−1.20360	$3s3p^3$ ^{3}D $4d$
23	−1.11979	$3p3d$ ^{3}D $4s$
24	−1.10835	$3p^2$ ^{1}D $6p$
25	−1.02380	$3p^2$ ^{3}P $6f$
26	−1.00156	$3s3p^3$ ^{3}P $4d$
27	−.90185	$3p^2$ ^{3}P $7p$
28	−.84616	$3s3p^3$ ^{1}D $4d$
29	−.84095	$3p^2$ ^{1}D $6f$
30	−.75063	$3p^2$ ^{3}P $7f$
31	−.74447	$3p^2$ ^{1}D $7p$
32	−.69551	$3s3p^3$ ^{3}D $5s$
33	−.67112	$3p^2$ ^{3}P $8p$
34	−.57745	$3p^2$ ^{3}P $8f$
35	−.56954	$3p^2$ ^{1}D $7f$
36	−.52278	$3p^2$ ^{3}P $9p$
37	−.50295	$3p^2$ ^{1}D $8p$
38	−.45894	$3p^2$ ^{3}P $9f$
39	−.43971	$3s3p^3$ ^{3}S $4d$

^{2}F^e

i	E(Ryds)	Description
1	−5.61013	$3p^2$ ^{1}D $3d$:

P-like Ca (Ca^{5+})

i	E(Ryds)	Description
2	−4.82897	$3p^2$ ^{3}P $3d$:
3	−2.70543	$3p^2$ ^{3}P $4d$
4	−2.57621	$3p^2$ ^{1}D $4d$
5	−2.20464	$3p^4 3d$
6	−2.05742	$3p^4 3d$
7	−1.90592	$3s3p^3$ ^{3}D $4p$
8	−1.67199	$3p^2$ ^{3}P $5d$
9	−1.56363	$3s3p^3$ ^{1}D $4p$
10	−1.49162	$3p^2$ ^{1}D $5d$
11	−1.45472	$3p^2$ ^{3}P $5g$
12	−1.39725	$3s3p^2 3d^2$
13	−1.33267	$3s3p^2 3d^2$
14	−1.27180	$3p^2$ ^{1}D $5g$
15	−1.16982	$3p3d$ ^{3}F $4p$
16	−1.12409	$3p^2$ ^{3}P $6d$
17	−1.08826	$3s3p^2 3d^2$:
18	−1.01022	$3p^2$ ^{3}P $6g$
19	−.98856	$3s3p^2 3d^2$:
20	−.95832	$3p^2$ ^{1}D $6d$
21	−.91787	$3s3p^2 3d^2$:
22	−.83251	$3p^2$ ^{1}D $6g$
23	−.82053	$3p^2$ ^{3}P $7d$
24	−.81243	$3s3p^3$ ^{3}D $4f$
25	−.77982	$3p3d$ ^{1}D $4p$
26	−.74210	$3p^2$ ^{3}P $7g$
27	−.68319	$3p3d$ ^{3}D $4p$
28	−.65035	$3p^2$ ^{1}D $7d$
29	−.63735	$3s3p^2 3d^2$:
30	−.60451	$3p^2$ ^{3}P $8d$
31	−.60236	$3s3p^3$ ^{3}P $4f$
32	−.57002	$3p^2$ ^{1}D $7g$
33	−.56654	$3p^2$ ^{3}P $8g$
34	−.51770	$3s3p^2 3d^2$:
35	−.51235	$3s3p^3$ ^{3}D $5p$
36	−.49022	$3p^2$ ^{3}P $9d$
37	−.47033	$3s3p^3$ ^{1}D $4f$
38	−.44994	$3p^2$ ^{1}D $8d$
39	−.44854	$3p^2$ ^{3}P $9g$
40	−.42635	$3s3p^2 3d^2$:
41	−.39958	$3p3d$ ^{1}F $4p$
42	−.38752	$3p^2$ ^{1}D $8g$
		^{2}F^o
1	−3.79417	$3s3p^3$ ^{1}D $3d$:
2	−3.55955	$3s3p^3$ ^{3}P $3d$
3	−3.31716	$3s3p^3$ ^{3}D $3d$:
4	−3.24055	$3p^2$ ^{1}D $4p$
5	−3.09728	$3s3p^3$ ^{1}P $3d$
6	−2.85731	$3s^2 3p3d^2$
7	−2.53429	$3s^2 3p3d^2$
8	−2.37218	$3s^2 3p3d^2$
9	−2.15434	$3p^2$ ^{3}P $4f$
10	−2.06185	$3p^2$ ^{1}D $4f$
11	−1.85969	$3p^2$ ^{1}S $4f$
12	−1.80308	$3p^2$ ^{1}D $5p$
13	−1.54538	$3p3d$ ^{3}F $4s$
14	−1.42978	$3p^2$ ^{3}P $5f$
15	−1.30578	$3p^2$ ^{1}D $5f$
16	−1.19166	$3s3p^3$ ^{3}D $4d$
17	−1.12386	$3p^2$ ^{1}D $6p$
18	−1.05636	$3p^2$ ^{1}S $5f$
19	−1.00159	$3p^2$ ^{3}P $6f$
20	−.97736	$3s3p^3$ ^{3}P $4d$
21	−.91303	$3p3d$ ^{1}F $4s$
22	−.85572	$3s3p^3$ ^{1}D $4d$
23	−.85017	$3p^2$ ^{1}D $6f$
24	−.73605	$3p^2$ ^{3}P $7f$
25	−.73473	$3p^2$ ^{1}D $7p$
26	−.60434	$3p^2$ ^{1}S $6f$
27	−.57613	$3p^2$ ^{1}D $7f$
28	−.56383	$3p^2$ ^{3}P $8f$
29	−.50410	$3p^2$ ^{1}D $8p$
30	−.44661	$3p^2$ ^{3}P $9f$
31	−.41728	$3s3p^3$ ^{1}P $4d$:
32	−.41312	$3p3d$ ^{3}F $4d$:
33	−.39944	$3p^2$ ^{1}D $8f$
		^{2}G^e
1	−5.35519	$3p^2$ ^{1}D $3d$
2	−2.52927	$3p^2$ ^{1}D $4d$
3	−2.36867	$3p^4 3d$
4	−1.82205	$3s3p^2 3d^2$
5	−1.48701	$3p^2$ ^{1}D $5d$
6	−1.43678	$3p^2$ ^{3}P $5g$
7	−1.32259	$3s3p^2 3d^2$
8	−1.28240	$3p^2$ ^{1}D $5g$
9	−1.09268	$3s3p^2 3d^2$:
10	−1.02691	$3p^2$ ^{1}S $5g$
11	−1.02129	$3p3d$ ^{3}F $4p$
12	−.99965	$3p^2$ ^{3}P $6g$
13	−.94653	$3s3p^2 3d^2$:
14	−.93017	$3p^2$ ^{1}D $6d$
15	−.84085	$3p^2$ ^{1}D $6g$
16	−.81337	$3s3p^2 3d^2$:
17	−.73954	$3p^2$ ^{3}P $7g$
18	−.70994	$3s3p^3$ ^{3}D $4f$:
19	−.63667	$3p^2$ ^{1}D $7d$
20	−.60770	$3s3p^2 3d^2$
21	−.58869	$3p^2$ ^{1}S $6g$
22	−.57155	$3p^2$ ^{1}D $7g$
23	−.55842	$3p^2$ ^{3}P $8g$
24	−.54977	$3s3p^3$ ^{3}P $4f$
25	−.47716	$3s3p^3$ ^{1}D $4f$
26	−.44515	$3p^2$ ^{1}D $8d$
27	−.44334	$3p^2$ ^{3}P $9g$
28	−.41146	$3p3d$ ^{1}F $4p$
29	−.38992	$3p^2$ ^{1}D $8g$
		^{2}G^o
1	−3.98785	$3s3p^3$ ^{3}D $3d$
2	−3.57486	$3s3p^3$ ^{1}D $3d$
3	−2.72760	$3s^2 3p3d^2$:
4	−2.39922	$3s^2 3p3d^2$:
5	−2.23354	$3p^2$ ^{3}P $4f$:
6	−2.06230	$3p^2$ ^{1}D $4f$
7	−1.45947	$3p^2$ ^{3}P $5f$
8	−1.31492	$3p^2$ ^{1}D $5f$
9	−1.23486	$3s3p^3$ ^{3}D $4d$
10	−1.01501	$3p^2$ ^{3}P $6f$
11	−.87550	$3s3p^3$ ^{1}D $4d$
12	−.85399	$3p^2$ ^{1}D $6f$
13	−.74393	$3p^2$ ^{3}P $7f$
14	−.58280	$3p^2$ ^{1}D $7f$
15	−.56759	$3p^2$ ^{3}P $8f$
16	−.45772	$3p3d$ ^{3}F $4d$
17	−.44869	$3p^2$ ^{3}P $9f$
18	−.40218	$3p^2$ ^{1}D $8f$
		^{4}S^e
1	−1.75731	$3s3p^3$ ^{3}P $4p$
2	−1.21665	$3s3p^2 3d^2$
3	−.86985	$3s3p^2 3d^2$
4	−.73414	$3p3d$ ^{3}P $4p$:
5	−.30920	$3s3p^3$ ^{3}P $5p$
		^{4}S^o
1	−7.94384	$3s^2 3p^3$
2	−3.79112	$3s3p^3$ ^{3}D $3d$
3	−3.39261	$3p^2$ ^{3}P $4p$
4	−2.73598	$3s3p^3$ ^{5}S $4s$
5	−2.29692	$3s^2 3p3d^2$
6	−1.96156	$3p^2$ ^{3}P $5p$
7	−1.63913	$3s3p^3$ ^{3}S $4s$
8	−1.28827	$3p^2$ ^{3}P $6p$
9	−1.23228	$3s3p^3$ ^{3}D $4d$
10	−1.11629	$3s3p^3$ ^{5}S $5s$
11	−.90529	$3p^2$ ^{3}P $7p$
12	−.67545	$3p^2$ ^{3}P $8p$
13	−.52298	$3p^2$ ^{3}P $9p$
		^{4}P^e
1	−6.54597	$3s3p^4$
2	−5.18419	$3p^2$ ^{3}P $3d$
3	−3.93281	$3p^2$ ^{3}P $4s$
4	−2.74184	$3p^2$ ^{3}P $4d$
5	−2.44028	$3p^4 3d$
6	−2.30130	$3s3p^3$ ^{5}S $4p$
7	−2.19784	$3p^2$ ^{3}P $5s$
8	−1.89352	$3s3p^2 3d^2$
9	−1.88207	$3s3p^3$ ^{3}D $4p$
10	−1.71401	$3s3p^3$ ^{3}P $4p$
11	−1.68104	$3p^2$ ^{3}P $5d$
12	−1.62304	$3s3p^2 3d^2$
13	−1.41547	$3p^2$ ^{3}P $6s$
14	−1.39468	$3s3p^2 3d^2$
15	−1.19434	$3s3p^3$ ^{3}S $4p$
16	−1.15295	$3s3p^2 3d^2$
17	−1.10998	$3p^2$ ^{3}P $6d$
18	−.98071	$3p^2$ ^{3}P $7s$
19	−.94021	$3s3p^3$ ^{5}S $5p$
20	−.89170	$3s3p^2 3d^2$
21	−.83291	$3p^2$ ^{3}P $7d$
22	−.82016	$3s3p^3$ ^{3}D $4f$
23	−.77344	$3s3p^2 3d^2$
24	−.75367	$3p3d$ ^{3}P $4p$
25	−.72227	$3p^2$ ^{3}P $8s$
26	−.65760	$3p3d$ ^{3}D $4p$
27	−.60972	$3p^2$ ^{3}P $8d$
28	−.56546	$3s3p^2 3d^2$
29	−.55252	$3p^2$ ^{3}P $9s$
30	−.49692	$3s3p^3$ ^{3}D $5p$
31	−.47909	$3p^2$ ^{3}P $9d$
		^{4}P^o
1	−3.92978	$3s3p^3$ ^{3}P $3d$
2	−3.77123	$3s3p^3$ ^{3}D $3d$
3	−3.41705	$3p^2$ ^{3}P $4p$
4	−2.85272	$3s^2 3p3d^2$
5	−2.23373	$3s3p^3$ ^{3}P $4s$
6	−1.96977	$3p^2$ ^{3}P $5p$
7	−1.31476	$3p^2$ ^{3}P $6p$
8	−1.27547	$3p3d$ ^{3}P $4s$
9	−1.23937	$3s3p^3$ ^{3}D $4d$
10	−1.02089	$3s3p^3$ ^{3}P $4d$
11	−.91000	$3p^2$ ^{3}P $7p$
12	−.67786	$3p^2$ ^{3}P $8p$
13	−.52948	$3p^2$ ^{3}P $9p$
14	−.51269	$3s3p^3$ ^{3}P $5s$
		^{4}D^e
1	−5.61338	$3p^2$ ^{3}P $3d$
2	−2.70337	$3p^2$ ^{3}P $4d$
3	−2.60773	$3p^4 3d$
4	−1.99833	$3s3p^3$ ^{3}D $4p$
5	−1.87296	$3s3p^2 3d^2$
6	−1.75554	$3s3p^3$ ^{3}P $4p$
7	−1.64456	$3s3p^2 3d^2$
8	−1.62410	$3p^2$ ^{3}P $5d$
9	−1.27787	$3s3p^2 3d^2$
10	−1.18775	$3s3p^2 3d^2$
11	−1.15062	$3s3p^2 3d^2$
12	−1.12120	$3p^2$ ^{3}P $6d$
13	−1.01883	$3p3d$ ^{3}F $4p$
14	−.90344	$3s3p^2 3d^2$
15	−.82551	$3s3p^3$ ^{3}D $4f$
16	−.80948	$3p^2$ ^{3}P $7d$

P-like Ca (Ca^{5+})

i	Energy(Ryds)	Description	
17	−.77697	$3p3d$	^{3}P $4p$
18	−.69567	$3p3d$	^{3}D $4p$
19	−.63051	$3s3p^3$	^{3}P $4f$
20	−.61050	$3p^2$	^{3}P $8d$
21	−.58960	$3s3p^23d^2$	
22	−.53865	$3s3p^3$	^{3}D $5p$
23	−.47840	$3p^2$	^{3}P $9d$
		$^4\mathbf{D}^o$	
1	−4.44094	$3s3p^3$	^{3}D $3d$
2	−3.90993	$3s3p^3$	^{3}P $3d$
3	−3.64430	$3s3p^3$	^{5}S $3d$
4	−3.46076	$3p^2$	^{3}P $4p$
5	−3.33317	$3s3p^3$	^{3}S $3d$
6	−2.95249	$3p3d^2$	
7	−2.62256	$3p3d^2$	
8	−2.45705	$3s3p^3$	^{3}D $4s$
9	−2.27883	$3p^2$	^{3}P $4f$
10	−1.98917	$3p^2$	^{3}P $5p$
11	−1.63971	$3s3p^3$	^{5}S $4d$
12	−1.47670	$3p^2$	^{3}P $5f$
13	−1.30834	$3p^2$	^{3}P $6p$
14	−1.25574	$3s3p^3$	^{3}D $4d$
15	−1.19081	$3p3d$	^{3}D $4s$
16	−1.02694	$3p^2$	^{3}P $6f$
17	−1.01973	$3s3p^3$	^{3}P $4d$
18	−.91362	$3p^2$	^{3}P $7p$
19	−.75214	$3p^2$	^{3}P $7f$
20	−.74540	$3s3p^3$	^{3}D $5s$
21	−.67896	$3p^2$	^{3}P $8p$
22	−.63455	$3s3p^3$	^{5}S $5d$
23	−.57454	$3p^2$	^{3}P $8f$
24	−.52584	$3p^2$	^{3}P $9p$
25	−.48960	$3s3p^3$	^{3}S $4d$
26	−.45175	$3p^2$	^{3}P $9f$
		$^4\mathbf{F}^e$	
1	−5.68450	$3p^2$	^{3}P $3d$
2	−2.74348	$3p^2$	^{3}P $4d$
3	−2.44229	$3p^43d$	
4	−2.13885	$3s3p^23d^2$	
5	−1.97996	$3s3p^3$	^{3}D $4p$
6	−1.68848	$3s3p^23d^2$	
7	−1.66229	$3p^2$	^{3}P $5d$
8	−1.58467	$3s3p^23d^2$	
9	−1.45485	$3p^2$	^{3}P $5g$
10	−1.36488	$3s3p^23d^2$	
11	−1.24304	$3s3p^23d^2$	
12	−1.18673	$3s3p^3$	^{5}S $4f$
13	−1.14150	$3p^2$	^{3}P $6d$
14	−1.09244	$3p3d$	^{3}F $4p$
15	−1.04481	$3s3p^23d^2$	
16	−1.01060	$3p^2$	^{3}P $6g$
17	−.87288	$3s3p^3$	^{3}D $4f$
18	−.81539	$3p^2$	^{3}P $7d$
19	−.79335	$3s3p^23d^2$	
20	−.73947	$3p^2$	^{3}P $7g$
21	−.71498	$3p3d$	^{3}D $4p$
22	−.61784	$3s3p^3$	^{3}P $4f$:
23	−.61302	$3p^2$	^{3}P $8d$:
24	−.56596	$3p^2$	^{3}P $8g$
25	−.53159	$3s3p^3$	^{3}D $5p$
26	−.47809	$3p^2$	^{3}P $9d$
27	−.44882	$3p^2$	^{3}P $9g$
28	−.43931	$3s3p^3$	^{5}S $5f$
		$^4\mathbf{F}^o$	
1	−4.36510	$3s3p^3$	^{3}D $3d$
2	−3.87077	$3s3p^3$	^{3}P $3d$
3	−2.83459	$3p3d^2$	
4	−2.30151	$3p^2$	^{3}P $4f$
5	−1.60796	$3p3d$	^{3}F $4s$
6	−1.47728	$3p^2$	^{3}P $5f$
7	−1.27976	$3s3p^3$	^{3}D $4d$
8	−1.04431	$3s3p^3$	^{3}P $4d$
9	−1.02084	$3p^2$	^{3}P $6f$
10	−.75073	$3p^2$	^{3}P $7f$
11	−.57363	$3p^2$	^{3}P $8f$
12	−.46248	$3p3d$	^{3}F $4d$
13	−.45221	$3p^2$	^{3}P $9f$
14	−.37843	$3p^33d^2$	
		$^4\mathbf{G}^e$	
1	−1.89792	$3s3p^23d^2$	
2	−1.54992	$3s3p^23d^2$	
3	−1.43816	$3p^2$	^{3}P $5g$
4	−1.37059	$3s3p^23d^2$	
5	−1.15989	$3s3p^23d^2$	
6	−1.13310	$3p3d$	^{3}F $4p$
7	−1.00223	$3p^2$	^{3}P $6g$
8	−.89270	$3s3p^3$	^{3}D $4f$
9	−.73901	$3p^2$	^{3}P $7g$
10	−.63324	$3s3p^3$	^{3}P $4f$
11	−.56006	$3p^2$	^{3}P $8g$
12	−.44448	$3p^2$	^{3}P $9g$
		$^4\mathbf{G}^o$	
1	−4.15335	$3s3p^3$	^{3}D $3d$
2	−3.24547	$3p3d^2$	
3	−2.33966	$3p^2$	^{3}P $4f$
4	−1.49442	$3p^2$	^{3}P $5f$
5	−1.27427	$3s3p^3$	^{3}D $4d$
6	−1.03317	$3p^2$	^{3}P $6f$
7	−.75628	$3p^2$	^{3}P $7f$
8	−.57731	$3p^2$	^{3}P $8f$
9	−.45545	$3p^2$	^{3}P $9f$
10	−.44633	$3s3p^3$	^{5}S $5g$
11	−.41313	$3p3d$	^{3}F $4d$
12	−.39766	$3p^33d^2$	
		$^6\mathbf{S}^o$	
1	−2.92763	$3s3p^3$	^{5}S $4s$
2	−1.19496	$3s3p^3$	^{5}S $5s$
3	−.40956	$3s3p^3$	^{5}S $6s$
		$^6\mathbf{P}^e$	
1	−2.44017	$3s3p^3$	^{5}S $4p$
2	−2.00285	$3s3p^23d^2$	
3	−.98253	$3s3p^3$	^{5}S $5p$
4	−.29847	$3s3p^3$	^{5}S $6p$
		$^6\mathbf{D}^o$	
1	−4.76974	$3s3p^3$	^{5}S $3d$
2	−1.71908	$3s3p^3$	^{5}S $4d$
3	−.66597	$3s3p^3$	^{5}S $5d$
4	−.12917	$3s3p^3$	^{5}S $6d$
		$^6\mathbf{F}^e$	
1	−2.31723	$3s3p^23d^2$	
2	−1.33326	$3s3p^3$	^{5}S $4f$
3	−.49652	$3s3p^3$	^{5}S $5f$
4	−.03731	$3s3p^3$	^{5}S $6f$
		$^6\mathbf{G}^o$	
1	−.44811	$3s3p^3$	^{5}S $5g$
2	−.00792	$3s3p^3$	^{5}S $6g$

P-like Ca (Ca^{5+})

Energies in ascending order from ground state for terms with effective $n \leq 4.0$, $L \leq 4$

Term	i	E(Ryds)	Term	i	E(Ryds)	Term	i	E(Ryds)	Term	i	E(Ryds)	Term	i	E(Ryds)
$^{4}\mathbf{S}^{o}$	1	0.00000	$^{4}\mathbf{F}^{o}$	1	3.57874	$^{2}\mathbf{D}^{o}$	4	4.52323	$^{4}\mathbf{F}^{o}$	3	5.10925	$^{4}\mathbf{P}^{e}$	5	5.50356
$^{2}\mathbf{D}^{o}$	1	0.25516	$^{2}\mathbf{S}^{o}$	1	3.74252	$^{4}\mathbf{P}^{o}$	3	4.52679	$^{2}\mathbf{D}^{o}$	8	5.16970	$^{6}\mathbf{P}^{e}$	1	5.50367
$^{2}\mathbf{P}^{o}$	1	0.42969	$^{4}\mathbf{G}^{o}$	1	3.79049	$^{4}\mathbf{S}^{o}$	3	4.55123	$^{2}\mathbf{P}^{e}$	5	5.19929	$^{2}\mathbf{P}^{e}$	7	5.51511
$^{4}\mathbf{P}^{e}$	1	1.39787	$^{2}\mathbf{G}^{o}$	1	3.95599	$^{2}\mathbf{P}^{o}$	4	4.57319	$^{4}\mathbf{F}^{e}$	2	5.20036	$^{2}\mathbf{D}^{e}$	8	5.51948
$^{2}\mathbf{D}^{e}$	1	1.74812	$^{4}\mathbf{P}^{e}$	3	4.01103	$^{4}\mathbf{D}^{o}$	5	4.61067	$^{4}\mathbf{P}^{e}$	4	5.20200	$^{2}\mathbf{P}^{o}$	10	5.52077
$^{2}\mathbf{P}^{e}$	1	2.03711	$^{4}\mathbf{P}^{o}$	1	4.01406	$^{2}\mathbf{D}^{o}$	5	4.61211	$^{4}\mathbf{S}^{o}$	4	5.20786	$^{2}\mathbf{G}^{o}$	4	5.54462
$^{2}\mathbf{S}^{e}$	1	2.14735	$^{4}\mathbf{D}^{o}$	2	4.03391	$^{2}\mathbf{F}^{o}$	3	4.62668	$^{2}\mathbf{G}^{o}$	3	5.21624	$^{2}\mathbf{F}^{o}$	8	5.57166
$^{4}\mathbf{F}^{e}$	1	2.25934	$^{2}\mathbf{D}^{o}$	2	4.04310	$^{2}\mathbf{P}^{o}$	5	4.69367	$^{2}\mathbf{F}^{e}$	3	5.23841	$^{2}\mathbf{G}^{e}$	3	5.57517
$^{4}\mathbf{D}^{e}$	1	2.33046	$^{4}\mathbf{F}^{o}$	2	4.07307	$^{4}\mathbf{G}^{o}$	2	4.69837	$^{4}\mathbf{D}^{e}$	2	5.24047	$^{2}\mathbf{D}^{o}$	10	5.59954
$^{2}\mathbf{F}^{e}$	1	2.33371	$^{2}\mathbf{P}^{e}$	4	4.08939	$^{2}\mathbf{F}^{o}$	4	4.70329	$^{2}\mathbf{D}^{o}$	9	5.28687	$^{4}\mathbf{G}^{o}$	3	5.60418
$^{2}\mathbf{G}^{e}$	1	2.58865	$^{2}\mathbf{F}^{o}$	1	4.14967	$^{2}\mathbf{D}^{o}$	6	4.73057	$^{2}\mathbf{D}^{e}$	6	5.30444	$^{2}\mathbf{P}^{o}$	11	5.61228
$^{2}\mathbf{P}^{e}$	2	2.73891	$^{4}\mathbf{S}^{o}$	2	4.15272	$^{2}\mathbf{S}^{o}$	3	4.78803	$^{4}\mathbf{D}^{o}$	7	5.32128	$^{2}\mathbf{D}^{o}$	11	5.61968
$^{4}\mathbf{P}^{e}$	2	2.75965	$^{4}\mathbf{P}^{o}$	2	4.17261	$^{2}\mathbf{D}^{o}$	7	4.79673	$^{4}\mathbf{D}^{e}$	3	5.33611	$^{6}\mathbf{F}^{e}$	1	5.62661
$^{2}\mathbf{D}^{e}$	2	2.82414	$^{2}\mathbf{D}^{e}$	5	4.20773	$^{2}\mathbf{P}^{o}$	6	4.82181	$^{2}\mathbf{F}^{e}$	4	5.36763	$^{4}\mathbf{F}^{o}$	4	5.64233
$^{2}\mathbf{D}^{e}$	3	2.97194	$^{2}\mathbf{D}^{o}$	3	4.27384	$^{2}\mathbf{F}^{o}$	5	4.84656	$^{2}\mathbf{D}^{e}$	7	5.38841	$^{4}\mathbf{P}^{e}$	6	5.64254
$^{2}\mathbf{P}^{e}$	3	3.08786	$^{4}\mathbf{D}^{o}$	3	4.29954	$^{2}\mathbf{P}^{o}$	7	4.86513	$^{2}\mathbf{F}^{o}$	7	5.40955	$^{4}\mathbf{S}^{o}$	5	5.64692
$^{2}\mathbf{S}^{e}$	2	3.10603	$^{2}\mathbf{P}^{o}$	3	4.35824	$^{2}\mathbf{P}^{o}$	8	4.96453	$^{2}\mathbf{G}^{e}$	2	5.41457	$^{2}\mathbf{D}^{e}$	9	5.65685
$^{2}\mathbf{F}^{e}$	2	3.11487	$^{2}\mathbf{G}^{o}$	2	4.36898	$^{4}\mathbf{D}^{o}$	6	4.99135	$^{2}\mathbf{S}^{o}$	4	5.44010	$^{4}\mathbf{D}^{o}$	9	5.66501
$^{6}\mathbf{D}^{o}$	1	3.17410	$^{2}\mathbf{F}^{o}$	2	4.38429	$^{6}\mathbf{S}^{o}$	1	5.01621	$^{2}\mathbf{P}^{e}$	6	5.44097			
$^{2}\mathbf{D}^{e}$	4	3.23928	$^{2}\mathbf{S}^{o}$	2	4.40620	$^{2}\mathbf{F}^{o}$	6	5.08653	$^{2}\mathbf{S}^{e}$	4	5.48430			
$^{2}\mathbf{P}^{o}$	2	3.33864	$^{2}\mathbf{S}^{e}$	3	4.45167	$^{4}\mathbf{P}^{o}$	4	5.09112	$^{4}\mathbf{D}^{o}$	8	5.48679			
$^{4}\mathbf{D}^{o}$	1	3.50290	$^{4}\mathbf{D}^{o}$	4	4.48308	$^{2}\mathbf{P}^{o}$	9	5.10839	$^{4}\mathbf{F}^{e}$	3	5.50155			

P-like Ca (Ca^{5+})

gf-values for transitions involving terms with effective $n \leq 4.0$, $L \leq 4$

i i'	gf_L	i i'	gf_L	i i'	gf_L	i i'	gf_L	i i'	gf_L	i i'	gf_L
	$^{2}\mathbf{S}^{o}$–$^{2}\mathbf{P}^{e}$	2 11	−1.13E+0	4 7	−1.50E−3		$^{2}\mathbf{P}^{o}$–$^{2}\mathbf{D}^{e}$	6 6	−5.56E−1	1 2	−9.98E−2
1 1	3.84E−2	3 1	2.72E−1	5 1	1.32E−1	1 1	−7.18E−2	6 7	−1.49E+0	1 3	−1.48E+0
1 2	1.99E−2	3 2	5.42E−4	5 2	7.89E−2	1 2	−7.90E−7	6 8	−2.68E−2	1 4	−1.38E−1
1 3	4.06E−4	3 3	1.91E−4	5 3	5.07E−3	1 3	−3.10E−1	6 9	−3.30E−3	1 5	−5.50E−1
1 4	−1.75E−4	3 4	−1.11E−2	5 4	2.00E−1	1 4	−8.84E+0	7 1	9.06E−3	1 6	−3.56E−3
1 5	−1.87E−3	3 5	−1.04E−3	5 5	−1.56E−1	1 5	−3.05E−1	7 2	1.33E−3	1 7	−1.62E−2
1 6	−6.26E−3	3 6	−2.98E−2	5 6	−3.42E−1	1 6	−1.00E−1	7 3	5.83E−1	1 8	−3.57E−1
1 7	−3.94E−1	3 7	−7.71E−2	5 7	−5.60E−4	1 7	−1.41E−1	7 4	4.59E−4	1 9	−5.32E+0
2 1	1.42E−1	3 8	−1.38E+0	6 1	7.34E−2	1 8	−1.88E−3	7 5	3.09E−2	1 10	−6.29E−3
2 2	7.12E−2	3 9	−5.92E−2	6 2	2.16E−1	1 9	−8.36E−2	7 6	−9.63E−2	1 11	−2.85E−1
2 3	3.81E−2	3 10	−1.69E−2	6 3	4.34E−1	2 1	4.33E−1	7 7	−1.38E−2	2 1	4.03E+0
2 4	3.68E−1	3 11	−2.58E−3	6 4	5.45E−1	2 2	7.03E−4	7 8	−4.55E−3	2 2	−1.26E−2
2 5	−1.66E+0	4 1	1.95E−2	6 5	−1.54E−1	2 3	3.18E−3	7 9	−3.64E−1	2 3	−2.52E−1
2 6	−1.76E−3	4 2	1.18E−2	6 6	−1.66E+0	2 4	2.63E−4	8 1	1.09E−2	2 4	−4.21E−2
2 7	−6.09E−3	4 3	1.43E−2	6 7	−5.73E−2	2 5	−5.15E−4	8 2	7.60E−1	2 5	−9.30E−2
3 1	1.50E+0	4 4	1.94E−1	7 1	3.67E+0	2 6	−4.52E−3	8 3	2.02E−3	2 6	−5.70E−2
3 2	4.74E−2	4 5	1.35E−1	7 2	3.49E−1	2 7	−2.30E−3	8 4	4.65E−1	2 7	−4.09E−2
3 3	4.59E−2	4 6	6.79E−1	7 3	9.60E−2	2 8	−1.54E−3	8 5	6.83E−2	2 8	−9.19E−1
3 4	2.81E−4	4 7	8.04E−3	7 4	6.44E−2	2 9	−3.98E−4	8 6	−1.06E−2	2 9	−1.44E+0
3 5	−1.78E−4	4 8	8.57E−2	7 5	−2.47E−2	3 1	4.27E+0	8 7	−1.26E−2	2 10	−6.02E−1
3 6	−4.16E−4	4 9	8.32E−3	7 6	−7.91E−2	3 2	1.84E−2	8 8	−1.82E−2	2 11	−2.07E−1
3 7	−7.70E−3	4 10	−1.41E−3	7 7	−1.18E−2	3 3	4.69E−2	8 9	−4.69E+0	3 1	4.12E−3
4 1	4.84E−4	4 11	−1.46E−2	8 1	1.72E−1	3 4	2.64E−3	9 1	7.10E−2	3 2	−2.13E−2
4 2	2.64E+0		$^{2}\mathbf{P}^{o}$–$^{2}\mathbf{P}^{e}$	8 2	2.76E−1	3 5	3.41E−3	9 2	7.95E−4	3 3	−1.22E−2
4 3	2.26E−1	1 1	−7.27E−2	8 3	1.95E−2	3 6	−2.56E−3	9 3	3.62E−1	3 4	−2.19E−2
4 4	1.31E−3	1 2	−4.18E−1	8 4	5.21E−2	3 7	−2.26E−2	9 4	3.65E−1	3 5	−9.72E−3
4 5	1.79E−3	1 3	−4.60E+0	8 5	−1.18E−2	3 8	−5.52E−2	9 5	4.36E−3	3 6	−1.97E−2
4 6	−6.40E−7	1 4	−5.29E−1	8 6	−2.73E−2	3 9	−7.64E−7	9 6	−1.79E−4	3 7	−2.04E−1
4 7	−9.67E−4	1 5	−6.87E−2	8 7	−4.03E−3	4 1	1.36E−1	9 7	−2.12E−3	3 8	−9.08E−1
	$^{2}\mathbf{S}^{e}$–$^{2}\mathbf{P}^{o}$	1 6	−5.63E−2	9 1	1.49E−1	4 2	1.53E−1	9 8	−1.59E−4	3 9	−7.23E−1
1 1	2.55E−1	1 7	−2.31E−5	9 2	6.72E−1	4 3	1.14E−1	9 9	−9.60E−2	3 10	−5.70E−2
1 2	−9.76E−3	2 1	3.51E−1	9 3	4.47E−1	4 4	7.62E−2	10 1	1.04E−4	3 11	−2.91E−4
1 3	−1.90E−1	2 2	3.84E−2	9 4	1.66E−3	4 5	1.79E−1	10 2	2.91E+0	4 1	9.64E−1
1 4	−9.12E−2	2 3	3.34E−3	9 5	−1.00E−4	4 6	−2.80E+0	10 3	1.12E−3	4 2	5.04E−5
1 5	−1.03E+0	2 4	−1.03E−3	9 6	−3.51E−2	4 7	−1.16E−1	10 4	6.69E−2	4 3	−4.67E−3
1 6	−3.38E−3	2 5	−5.02E−4	9 7	−1.84E−4	4 8	−5.14E−2	10 5	1.56E−2	4 4	−1.59E+0
1 7	−5.98E−1	2 6	−5.16E−3	10 1	2.12E−2	4 9	−3.65E−2	10 6	1.56E−3	4 5	−1.62E−1
1 8	−8.11E−3	2 7	−2.39E−1	10 2	1.70E+0	5 1	4.02E−2	10 7	3.13E−3	4 6	−8.36E−1
1 9	−3.40E+0	3 1	4.67E−2	10 3	1.45E−1	5 2	1.93E−1	10 8	4.43E−6	4 7	−2.05E−1
1 10	−1.48E−3	3 2	1.89E−1	10 4	1.50E−3	5 3	1.16E−2	10 9	−9.26E−3	4 8	−9.47E−5
1 11	−3.45E−2	3 3	2.55E−2	10 5	1.94E−3	5 4	4.71E−2	11 1	2.71E−3	4 9	−1.07E−2
2 1	1.34E+0	3 4	1.10E−3	10 6	3.26E−4	5 5	2.36E−1	11 2	3.59E−1	4 10	−1.27E−1
2 2	−7.41E−4	3 5	−1.21E−3	10 7	4.14E−5	5 6	−4.34E−1	11 3	1.59E+0	4 11	−1.48E−2
2 3	−5.06E−5	3 6	−2.86E−2	11 1	3.67E−2	5 7	−1.83E−1	11 4	3.83E−3	5 1	8.53E−3
2 4	−2.73E−3	3 7	−1.10E−1	11 2	3.30E+0	5 8	−5.38E−2	11 5	2.62E−3	5 2	3.04E−4
2 5	−2.23E−1	4 1	9.72E−1	11 3	1.31E+0	5 9	−4.98E−3	11 6	3.78E−3	5 3	4.06E−2
2 6	−7.62E−2	4 2	1.36E−1	11 4	3.95E−3	6 1	2.50E−2	11 7	3.14E−3	5 4	1.68E−5
2 7	−1.50E−3	4 3	2.19E−3	11 5	1.08E−4	6 2	2.20E−1	11 8	3.14E−4	5 5	5.51E−3
2 8	−2.53E−2	4 4	9.32E−1	11 6	2.53E−2	6 3	2.19E−2	11 9	−1.48E−5	5 6	8.66E−3
2 9	−3.94E−1	4 5	−9.59E−1	11 7	6.22E−4	6 4	3.48E−1		$^{2}\mathbf{P}^{e}$–$^{2}\mathbf{D}^{o}$	5 7	2.16E−3
2 10	−7.84E−2	4 6	−1.58E−1			6 5	1.21E+0	1 1	5.50E−1	5 8	1.29E−4

P-like Ca (Ca^{5+})

i i′	gf_L	i i′	gf_L	i i′	gf_L	i i′	gf_L	i i′	gf_L	i i′	gf_L
5 9	−9.16E−2	3 9	−6.47E−5	9 7	−7.74E−5	8 3	−1.81E−4	5 6	−9.39E−2	5 1	7.06E−1
5 10	−9.19E−2	4 1	1.34E−1	9 8	−1.15E−5	8 4	−7.54E−4	5 7	−8.06E−3	5 2	5.78E−1
5 11	−1.29E+0	4 2	2.71E−4	9 9	−5.21E−5	9 1	5.73E−2	5 8	−9.20E−3	5 3	−5.03E−3
6 1	2.27E−2	4 3	4.65E−2	10 1	1.10E+0	9 2	1.53E+0	6 1	2.09E−2	5 4	−6.84E−3
6 2	1.63E−3	4 4	4.12E−2	10 2	2.57E−1	9 3	1.37E−4	6 2	1.59E−3	6 1	3.65E+0
6 3	1.37E−2	4 5	3.48E−1	10 3	3.86E−1	9 4	−2.07E−5	6 3	1.31E−2	6 2	3.49E−1
6 4	9.19E−2	4 6	−6.91E−1	10 4	2.66E−2	10 1	4.08E−2	6 4	4.39E−4	6 3	−4.41E−5
6 5	5.93E−3	4 7	−8.99E−1	10 5	7.72E−2	10 2	7.52E−1	6 5	3.38E−2	6 4	−4.41E−2
6 6	7.70E−1	4 8	−8.08E−2	10 6	5.62E−4	10 3	1.38E−2	6 6	1.26E−2	7 1	1.86E+0
6 7	2.80E−3	4 9	−1.08E−2	10 7	1.62E−2	10 4	8.46E−3	6 7	−1.14E−1	7 2	1.29E+0
6 8	4.20E−2	5 1	5.23E+0	10 8	2.74E−3	11 1	1.49E−2	6 8	−7.39E−1	7 3	7.85E−3
6 9	3.38E−4	5 2	2.26E−1	10 9	−5.50E−5	11 2	2.94E−2	7 1	1.71E−2	7 4	1.00E−2
6 10	−1.37E−4	5 3	1.87E−2	11 1	7.42E−5	11 3	1.78E−3	7 2	3.78E−2	8 1	3.08E+0
6 11	−9.69E−3	5 4	8.52E−2	11 2	6.64E+0	11 4	3.63E−4	7 3	8.95E−2	8 2	3.35E−1
7 1	3.92E−4	5 5	1.03E−1	11 3	1.60E−2		$^2D^e$–$^2F^o$	7 4	3.27E−1	8 3	2.26E−1
7 2	1.80E−1	5 6	−3.21E−2	11 4	3.82E−3	1 1	−1.62E−1	7 5	4.94E−2	8 4	4.79E−2
7 3	2.89E−1	5 7	−2.15E−1	11 5	7.87E−3	1 2	−2.24E−1	7 6	4.41E−3		$^2F^o$–$^2G^e$
7 4	1.05E−2	5 8	−2.20E−2	11 6	1.85E−1	1 3	−7.24E+0	7 7	−6.03E−3	1 1	3.63E−3
7 5	5.45E−4	5 9	−5.06E−3	11 7	1.50E−2	1 4	−1.49E+0	7 8	−3.81E−1	1 2	−1.62E−2
7 6	5.51E−3	6 1	1.95E−2	11 8	5.21E−4	1 5	−3.24E+0	8 1	5.07E−1	1 3	−4.25E−1
7 7	1.50E−2	6 2	2.35E−2	11 9	−1.56E−5	1 6	−6.13E−1	8 2	8.45E−2	2 1	5.06E−4
7 8	1.08E−2	6 3	6.60E−1		$^2D^o$–$^2F^e$	1 7	−3.12E−1	8 3	2.75E−2	2 2	−1.22E−1
7 9	1.85E−3	6 4	7.94E−3	1 1	−4.66E−5	1 8	−2.48E−1	8 4	1.36E−3	2 3	−1.86E−1
7 10	−5.59E−4	6 5	2.05E+0	1 2	1.34E+1	2 1	−1.20E−2	8 5	6.50E−4	3 1	1.25E+0
7 11	−1.43E−3	6 6	−6.15E−1	1 3	−1.79E−1	2 2	−1.39E−1	8 6	4.06E−3	3 2	−1.87E+0
	$^2D^o$–$^2D^e$	6 7	−1.41E+0	1 4	−3.45E−1	2 3	−2.79E−1	8 7	2.68E−4	3 3	−1.41E−2
1 1	−4.87E−1	6 8	−1.24E−2	2 1	4.69E−1	2 4	−4.45E−2	8 8	−2.18E−3	4 1	1.88E+0
1 2	−1.23E+0	6 9	−8.45E−2	2 2	6.09E−3	2 5	−1.67E+0	9 1	5.81E−3	4 2	−7.77E+0
1 3	−6.61E+0	7 1	8.85E−1	2 3	−1.75E−5	2 6	−4.30E−1	9 2	3.65E−2	4 3	−7.88E−3
1 4	−5.15E−2	7 2	4.35E−1	2 4	−1.63E−2	2 7	−2.45E−6	9 3	4.92E−2	5 1	3.13E+0
1 5	−9.87E−1	7 3	3.27E−1	3 1	1.78E−1	2 8	−2.57E−1	9 4	4.54E−2	5 2	−1.70E−1
1 6	−8.78E−2	7 4	5.11E−3	3 2	3.78E−5	3 1	−1.05E−1	9 5	2.34E−2	5 3	−1.51E−1
1 7	−1.26E−1	7 5	5.15E−3	3 3	−4.88E−2	3 2	−5.78E−2	9 6	1.96E−2	6 1	2.81E+0
1 8	−7.49E−6	7 6	−5.35E−2	3 4	−1.00E−2	3 3	−2.34E−2	9 7	3.31E−2	6 2	−6.35E−2
1 9	−1.70E−3	7 7	−2.71E−3	4 1	1.90E+0	3 4	−3.07E−2	9 8	1.08E−2	6 3	−1.74E−2
2 1	1.56E−1	7 8	−9.55E−2	4 2	1.38E−1	3 5	−3.80E−2		$^2F^o$–$^2F^e$	7 1	4.24E−1
2 2	1.46E−2	7 9	−2.19E−3	4 3	−5.81E+0	3 6	−3.71E−1	1 1	4.92E−1	7 2	−2.67E−4
2 3	5.82E−4	8 1	3.63E−1	4 4	−1.78E−1	3 7	−2.93E+0	1 2	1.88E−2	7 3	−7.28E−5
2 4	2.52E−3	8 2	6.09E−2	5 1	7.24E−3	3 8	−5.90E−1	1 3	−6.29E−3	8 1	3.85E−2
2 5	−2.03E−3	8 3	2.15E+0	5 2	1.67E−1	4 1	−4.76E−4	1 4	−1.79E−2	8 2	6.97E−3
2 6	−3.56E−3	8 4	1.20E−4	5 3	−5.07E−1	4 2	−7.75E−2	2 1	1.47E+0	8 3	−5.47E−6
2 7	−1.45E−2	8 5	3.13E−3	5 4	−7.10E−2	4 3	−4.00E−3	2 2	4.83E 5		$^2F^e$–$^2G^o$
2 8	2.23E 1	8 6	−4.36E−5	6 1	1.92E−5	4 4	−4.78E−2	2 3	−1.27E−4	1 1	−3.49E−1
2 9	−1.83E−3	8 7	−3.83E−2	6 2	9.25E−1	4 5	−1.88E−1	2 4	−5.11E−3	1 2	−2.08E+0
3 1	5.30E−1	8 8	−2.03E−5	6 3	−4.94E−1	4 6	−2.09E−2	3 1	1.96E−2	1 3	−1.63E+0
3 2	9.58E−4	8 9	−1.75E−3	6 4	−4.33E+0	4 7	−3.06E−1	3 2	2.73E−1	1 4	−3.48E+0
3 3	4.26E−2	9 1	4.42E−3	7 1	3.37E+0	4 8	−1.21E+0	3 3	−5.65E−3	2 1	−2.97E−2
3 4	5.37E−3	9 2	1.38E+0	7 2	3.18E−2	5 1	9.67E−4	3 4	−3.41E−1	2 2	−1.59E−2
3 5	1.11E 3	9 3	1.83E−1	7 3	−2.22E−1	5 2	−7.31E−3	4 1	3.73E−1	2 3	−1.07E+0
3 6	−1.53E−3	9 4	8.02E−1	7 4	−4.10E−2	5 3	−4.97E−1	4 2	1.75E−2	2 4	−6.21E−1
3 7	−2.15E−2	9 5	1.37E−4	8 1	2.00E+0	5 4	−2.91E+0	4 3	−9.14E−2	3 1	2.91E−6
3 8	−4.30E−2	9 6	−3.23E−3	8 2	2.22E−1	5 5	−1.01E−1	4 4	−1.76E+0	3 2	1.13E−1

P-like Ca (Ca^{5+})

$i\ i'$	gf_L	$i\ i'$	gf_L	$i\ i'$	gf_L	$i\ i'$	gf_L	$i\ i'$	gf_L	$i\ i'$	gf_L
3 3	1.15E−2	4 4	3.82E−4	3 1	1.66E+0	4 8	−3.57E−3	7 3	−3.95E−6	2 2	1.27E−1
3 4	−2.03E+0	4 5	−2.51E−1	3 2	−7.79E+0	4 9	−6.08E+0	8 1	1.27E−2	2 3	3.78E−3
4 1	5.29E−2	4 6	−2.98E+0	3 3	−3.38E−1	5 1	6.41E−1	8 2	1.46E−3	2 4	−9.04E+0
4 2	4.37E−2	5 1	6.87E−3	4 1	6.97E+0	5 2	4.56E−2	8 3	4.80E−4	3 1	2.30E+0
4 3	4.11E−2	5 2	6.04E+0	4 2	−1.22E−2	5 3	2.90E−1	9 1	2.32E−1	3 2	1.68E−2
4 4	−2.52E−1	5 3	2.22E−2	4 3	−4.86E−4	5 4	7.76E−3	9 2	1.16E+0	3 3	1.17E−2
	$^2G^o$–$^2G^e$	5 4	2.21E−2		**$^4P^e$–$^4D^o$**	5 5	1.78E−1	9 3	3.03E−2	3 4	−2.32E−1
1 1	2.09E−2	5 5	2.83E−3	1 1	−9.46E−3	5 6	2.94E−2		**$^4D^o$–$^4F^e$**		**$^4F^o$–$^4F^e$**
1 2	−8.86E−3	5 6	5.92E−4	1 2	−7.64E−1	5 7	2.07E−3	1 1	1.19E−2	1 1	5.29E−1
1 3	−5.28E−2		**$^4P^o$–$^4P^e$**	1 3	−1.50E+1	5 8	3.54E−4	1 2	−1.79E−2	1 2	−1.24E−3
2 1	2.02E+0	1 1	9.38E−1	1 4	−2.72E−2	5 9	−1.17E−4	1 3	−1.24E+0	1 3	−3.04E−1
2 2	−3.50E−3	1 2	2.39E−2	1 5	−1.55E+0	6 1	1.02E+0	2 1	6.93E−1	2 1	1.05E+0
2 3	−1.57E+0	1 3	3.51E−5	1 6	−8.40E−2	6 2	4.81E−1	2 2	−2.57E−1	2 2	−1.45E−2
3 1	8.48E+0	1 4	−7.56E−3	1 7	−2.52E−1	6 3	4.38E−1	2 3	−1.40E−1	2 3	−9.28E−1
3 2	−3.67E−3	1 5	−1.22E−1	1 8	−1.28E+0	6 4	1.03E−1	3 1	6.09E−1	3 1	2.01E+1
3 3	−2.17E−2	1 6	−2.53E−3	1 9	−1.17E−1	6 5	1.23E−2	3 2	−8.35E−3	3 2	−3.20E−4
4 1	3.66E+0	2 1	9.12E+0	2 1	−3.18E−2	6 6	4.11E−3	3 3	−7.41E−3	3 3	−6.14E−3
4 2	6.32E−3	2 2	3.68E−1	2 2	−1.48E−2	6 7	6.77E−4	4 1	3.98E+0	4 1	1.02E+0
4 3	−1.43E−5	2 3	9.91E−3	2 3	−2.99E−1	6 8	6.88E−2	4 2	−1.44E+1	4 2	1.27E+0
	$^4S^o$–$^4P^e$	2 4	−2.73E−2	2 4	−1.85E−1	6 9	−6.77E−3	4 3	−7.30E−2	4 3	9.04E−6
1 1	−3.71E−1	2 5	−1.16E−1	2 5	−2.20E+0		**$^4D^o$–$^4D^e$**	5 1	5.66E+0		**$^4F^e$–$^4G^o$**
1 2	−1.01E+1	2 6	−9.21E−2	2 6	−6.90E−1	1 1	1.17E−1	5 2	−1.77E−1	1 1	−1.93E+0
1 3	−8.06E−1	3 1	6.59E−3	2 7	−7.69E+0	1 2	−1.32E−2	5 3	−7.08E−1	1 2	−5.77E+0
1 4	−3.29E−1	3 2	2.21E−1	2 8	−1.19E−2	1 3	−1.21E+0	6 1	6.07E+0	1 3	−1.81E+1
1 5	−1.95E−2	3 3	3.19E+0	2 9	−1.64E+1	2 1	4.25E−1	6 2	−2.92E−2	2 1	1.50E−1
1 6	−1.71E−1	3 4	−2.57E+0	3 1	1.55E−3	2 2	−7.91E−2	6 3	−5.35E−3	2 2	1.62E−1
2 1	2.44E+0	3 5	−1.87E−3	3 2	−2.25E−3	2 3	−9.48E−2	7 1	4.25E−2	2 3	−1.28E+1
2 2	1.21E−1	3 6	−4.09E−5	3 3	−2.08E−3	3 1	4.18E−2	7 2	1.79E−3	3 1	2.30E+0
2 3	1.69E−4	4 1	2.35E−3	3 4	−4.73E+0	3 2	−3.76E−2	7 3	−2.80E−3	3 2	1.37E−2
2 4	−4.71E−3	4 2	4.99E−4	3 5	−8.56E−2	3 3	−1.76E−1	8 1	1.43E−1	3 3	−3.85E−3
2 5	−3.11E−2	4 3	3.19E−2	3 6	−5.87E−2	4 1	2.12E−1	8 2	2.70E−3		**$^6S^o$–$^6P^e$**
2 6	−5.61E−2	4 4	−2.76E−3	3 7	−4.62E−4	4 2	−2.62E+0	8 3	−3.47E−5	1 1	−5.59E+0
3 1	2.15E−1	4 5	−1.23E−2	3 8	−9.40E−1	4 3	−1.79E−1	9 1	1.30E−2		**$^6P^e$–$^6D^o$**
3 2	5.95E−1	4 6	−3.65E−3	3 9	−3.83E−3	5 1	5.28E+0	9 2	9.64E−4	1 1	3.77E+0
3 3	1.12E+0		**$^4P^o$–$^4D^e$**	4 1	4.29E−3	5 2	−1.58E−2	9 3	5.53E−3		**$^6D^o$–$^6F^e$**
3 4	−3.07E+0	1 1	6.48E−1	4 2	2.18E−3	5 3	−2.96E−1		**$^4D^e$–$^4F^o$**	1 1	−6.19E+0
3 5	−6.39E−2	1 2	−2.01E−2	4 3	1.10E−2	6 1	1.88E+0	1 1	−9.69E−2		
3 6	−5.05E−2	1 3	−7.80E−2	4 4	1.13E−1	6 2	−2.83E−2	1 2	−2.51E+0		
4 1	6.60E−1	2 1	6.93E−2	4 5	1.51E−1	6 3	−3.46E−5	1 3	−4.46E+0		
4 2	4.79E−1	2 2	−1.83E−1	4 6	1.98E−2	7 1	1.03E+1	1 4	−1.18E+1		
4 3	6.51E−2	2 3	−2.00E−1	4 7	−1.89E−1	7 2	2.32E−2	2 1	5.28E−2		

P-like Fe (Fe^{11+})

Term energies relative to $3p^2$ ^{3}P ionization threshold for each symmetry

i	E(Ryds)	Description	i	E(Ryds)	Description	i	E(Ryds)	Description	i	E(Ryds)	Description
		^{2}S^e			^{2}P^e	51	−2.39158	$3p^2$ ^{3}P $8d$	45	−3.41550	$3s3p^3$ ^{3}P $5d$
1	−20.5302	$3s3p^4$	1	−20.6227	$3s3p^4$	52	−2.17447	$3s3p^3$ ^{3}D $6p$	46	−3.33116	$3p^2$ ^{3}P $7p$
2	−18.8441	$3p^2$ ^{1}D $3d$	2	−19.5283	$3p^2$ ^{3}P $3d$	53	−2.06500	$3p^2$ ^{1}D $8d$	47	−3.15748	$3s3p^3$ ^{1}D $5d$
3	−14.0177	$3p^43d$	3	−18.8311	$3p^2$ ^{1}D $3d$	54	−2.03320	$3p^2$ ^{3}P $9s$	48	−3.02426	$3p^2$ ^{1}D $7p$
4	−12.2726	$3s3p^23d^2$	4	−14.7782	$3p^43d$	55	−1.87781	$3p^2$ ^{3}P $9d$	49	−2.67116	$3p^2$ ^{1}D $7f$
5	−11.9489	$3p^2$ ^{1}S $4s$	5	−13.5475	$3p^43d$	56	−1.79413	$3s3p^3$ ^{3}P $6p$	50	−2.64631	$3p^2$ ^{1}S $7p$
6	−11.8736	$3s3p^23d^2$	6	−12.9642	$3s3p^23d^2$:			^{2}P^o	51	−2.50915	$3p^2$ ^{3}P $8p$
7	−11.1972	$3s3p^23d^2$:	7	−12.6167	$3s3p^23d^2$	1	−23.4411	$3s^23p^3$	52	−2.46409	$3s3p^3$ ^{1}P $5d$
8	−10.1659	$3s3p^23d^2$:	8	−12.5252	$3p^2$ ^{3}P $4s$	2	−18.5218	$3p^5$	53	−2.20266	$3p^2$ ^{1}D $8p$
9	−9.75849	$3p^2$ ^{1}D $4d$	9	−12.1929	$3s3p^23d^2$	3	−16.8171	$3s3p^3$ ^{3}D $3d$	54	−2.02319	$3s3p^3$ ^{3}P $6s$
10	−8.38286	$3s3p^3$ ^{3}P $4p$	10	−12.0766	$3s3p^23d^2$:	4	−16.2597	$3s3p^3$ ^{3}P $3d$	55	−1.96273	$3p^2$ ^{1}D $8f$
11	−7.61325	$3s3p^3$ ^{1}P $4p$	11	−11.9110	$3s3p^23d^2$	5	−15.9527	$3s3p^3$ ^{1}D $3d$	56	−1.95795	$3p^2$ ^{3}P $9p$
12	−6.80082	$3p^23d^3$:	12	−11.5189	$3s3p^23d^2$	6	−15.5477	$3s3p^3$ ^{1}P $3d$	57	−1.81536	$3p^2$ ^{1}S $8p$
13	−6.74420	$3p^2$ ^{1}S $5s$	13	−11.0638	$3s3p^23d^2$	7	−14.7650	$3s^23p3d^2$	58	−1.81106	$3s3p^3$ ^{3}D $6d$
14	−5.96135	$3p^2$ ^{1}D $5d$	14	−10.5167	$3s3p^23d^2$:	8	−14.6358	$3s^23p3d^2$	59	−1.64859	$3p^2$ ^{1}D $9p$
15	−5.13730	$3p^23d^3$:	15	−10.3294	$3p^2$ ^{3}P $4d$	9	−13.8961	$3s^23p3d^2$			^{2}D^e
16	−4.33090	$3p^23d^3$:	16	−10.2643	$3s^23d^3$	10	−11.5047	$3p^2$ ^{3}P $4p$	1	−21.1491	$3s3p^4$
17	−4.23730	$3p^2$ ^{1}S $6s$	17	−9.84901	$3p^2$ ^{1}D $4d$	11	−11.0753	$3p^2$ ^{1}D $4p$	2	−19.2710	$3p^2$ ^{1}S $3d$
18	−4.11162	$3p^23d^3$:	18	−9.15195	$3s3p^3$ ^{3}D $4p$	12	−10.8868	$3p^2$ ^{1}S $4p$	3	−19.0534	$3p^2$ ^{1}D $3d$
19	−3.99756	$3p^2$ ^{1}D $6d$	19	−8.64337	$3s3p^3$ ^{3}P $4p$	13	−10.7205	$3p^33d^2$	4	−18.5964	$3p^2$ ^{3}P $3d$
20	−3.96035	$3s3p^3$ ^{3}P $5p$	20	−8.36681	$3s3p^3$ ^{1}D $4p$	14	−10.2257	$3p^33d^2$	5	−14.7488	$3p^43d$
21	−3.10449	$3s3p^3$ ^{1}P $5p$	21	−7.78790	$3s3p^3$ ^{3}S $4p$	15	−9.68747	$3p^33d^2$:	6	−13.6884	$3s3p^23d^2$:
22	−2.81327	$3p^2$ ^{1}D $7d$	22	−7.62582	$3s3p^3$ ^{1}P $4p$	16	−9.53733	$3s3p^3$ ^{3}P $4s$	7	−13.6804	$3p^43d$
23	−2.78201	$3p^2$ ^{1}S $7s$	23	−7.39977	$3p^2$ ^{3}P $5s$	17	−9.25176	$3p^33d^2$	8	−13.2794	$3p^43d$:
24	−2.05979	$3p^2$ ^{1}D $8d$	24	−6.74289	$3p^23d^3$:	18	−9.14499	$3p^33d^2$:	9	−12.7133	$3s3p^23d^2$:
25	−1.91398	$3p^2$ ^{1}S $8s$	25	−6.72099	$3s3p^3$ ^{3}D $4f$	19	−8.78079	$3p^2$ ^{1}D $4f$	10	−12.6447	$3s3p^23d^2$:
26	−1.73203	$3s3p^3$ ^{3}P $6p$	26	−6.37427	$3p^2$ ^{3}P $5d$	20	−8.67638	$3s3p^3$ ^{1}P $4s$	11	−12.3346	$3p^2$ ^{1}D $4s$
27	−1.54986	$3p^2$ ^{1}D $9d$	27	−6.09628	$3p^23d^3$:	21	−8.48808	$3p^33d^2$	12	−12.2252	$3s3p^23d^2$:
28	−1.32673	$3p^2$ ^{1}S $9s$	28	−6.01876	$3s3p^3$ ^{1}D $4f$	22	−8.33693	$3p^33d^2$:	13	−12.1629	$3s3p^23d^2$:
		^{2}S^o	29	−5.99406	$3p^2$ ^{1}D $5d$	23	−8.05733	$3s3p3d^3$:	14	−11.9768	$3s3p^23d^2$:
1	−17.7073	$3s3p^3$ ^{3}D $3d$	30	−5.85004	$3p^23d^3$	24	−7.82347	$3s3p3d^3$	15	−11.7222	$3s3p^23d^2$:
2	−16.0322	$3s3p^3$ ^{1}D $3d$:	31	−5.57620	$3p^23d^3$	25	−7.69258	$3s3p^3$ ^{3}D $4d$	16	−11.5800	$3s3p^23d^2$:
3	−15.0439	$3s^23p3d^2$:	32	−5.29756	$3p^23d^3$	26	−7.61378	$3s3p3d^3$:	17	−11.4700	$3s3p^23d^2$:
4	−11.9031	$3p^2$ ^{3}P $4p$	33	−4.92248	$3p^2$ ^{3}P $6s$	27	−7.47418	$3s3p3d^3$	18	−11.1397	$3s3p^23d^2$:
5	−10.6346	$3p^33d^2$	34	−4.78379	$3p^23d^3$:	28	−7.19311	$3s3p^3$ ^{3}P $4d$	19	−10.6612	$3s3p^23d^2$
6	−9.52143	$3p^33d^2$	35	−4.68993	$3p^23d^3$	29	−7.16125	$3s3p3d^3$	20	−10.2148	$3s^23d^3$
7	−8.93812	$3s3p3d^3$:	36	−4.50328	$3p^23d^3$:	30	−7.03926	$3s3p^3$ ^{1}D $4d$	21	−10.0748	$3p^2$ ^{3}P $4d$
8	−8.78044	$3s3p^3$ ^{3}S $4s$	37	−4.49384	$3s3p^3$ ^{3}D $5p$	31	−6.89337	$3p^2$ ^{3}P $5p$	22	−9.96362	$3p^2$ ^{1}D $4d$
9	−7.71495	$3s3p^3$ ^{3}D $4d$	38	−4.35306	$3p^2$ ^{3}P $6d$	32	−6.61726	$3p^2$ ^{1}D $5p$	23	−9.79550	$3s^23d^3$
10	−7.23902	$3s3p3d^3$:	39	−4.09597	$3s3p^3$ ^{3}P $5p$	33	−6.34742	$3s3p^3$ ^{1}P $4d$	24	−9.48987	$3p^2$ ^{1}S $4d$
11	−7.08201	$3p^2$ ^{3}P $5p$	40	−4.00217	$3p^2$ ^{1}D $6d$	34	−6.23555	$3p^2$ ^{1}S $5p$	25	−8.85778	$3s3p^3$ ^{3}D $4p$
12	−6.97555	$3s3p^3$ ^{1}D $4d$	41	−3.78740	$3s3p^3$ ^{1}D $5p$	35	−6.21576	$3s3p3d^3$	26	−8.61175	$3s3p^3$ ^{3}P $4p$
13	−6.52900	$3s3p3d^3$	42	−3.50910	$3p^2$ ^{3}P $7s$	36	−6.07311	$3s3p3d^3$	27	−8.34173	$3s3p^3$ ^{1}D $4p$
14	−4.74388	$3p^2$ ^{3}P $6p$	43	−3.39424	$3s3p^3$ ^{3}D $5f$	37	−5.53050	$3p^2$ ^{1}D $5f$	28	−7.72280	$3s3p^3$ ^{1}P $4p$
15	−3.85263	$3s3p^3$ ^{3}D $5d$	44	−3.27920	$3s3p^3$ ^{3}S $5p$	38	−4.64193	$3p^2$ ^{3}P $6p$	29	−7.13435	$3p^2$ ^{1}D $5s$
16	−3.73869	$3s3p^3$ ^{3}S $5s$	45	−3.16681	$3p^23d^3$:	39	−4.51386	$3s3p^3$ ^{3}P $5s$	30	−6.80422	$3p^23d^3$:
17	−3.36878	$3p^2$ ^{3}P $7p$	46	−3.13467	$3p^2$ ^{3}P $7d$	40	−4.34470	$3p^2$ ^{1}D $6p$	31	−6.68305	$3s3p^3$ ^{3}D $4f$
18	−3.14804	$3s3p^3$ ^{1}D $5d$	47	−3.10438	$3s3p^3$ ^{1}P $5p$	41	−3.97515	$3p^2$ ^{1}S $6p$	32	−6.28989	$3p^2$ ^{3}P $5d$
19	−2.54248	$3p^2$ ^{3}P $8p$	48	−2.81314	$3p^2$ ^{1}D $7d$	42	−3.82730	$3s3p^3$ ^{3}D $5d$	33	−6.27294	$3p^23d^3$
20	−1.98305	$3p^2$ ^{3}P $9p$	49	−2.71015	$3s3p^3$ ^{1}D $5f$	43	−3.74420	$3p^2$ ^{1}D $6f$	34	−6.14509	$3s3p^3$ ^{3}P $4f$
21	−1.82622	$3s3p^3$ ^{3}D $6d$	50	−2.61882	$3p^2$ ^{3}P $8s$	44	−3.57271	$3s3p^3$ ^{1}P $5s$	35	−6.08100	$3s3p^3$ ^{1}D $4f$

P-like Fe (Fe^{11+})

i	E(Ryds)	Description	i	E(Ryds)	Description	i	E(Ryds)	Description	i	E(Ryds)	Description
36	−6.02218	$3p^2$ ^{1}D $5d$	8	−14.5686	$3s^23p3d^2$	60	−2.47342	$3s3p^3$ ^{1}P $5d$	42	−4.04436	$3p^2$ ^{1}D $6d$
37	−5.96211	$3p^23d^3$:	9	−14.1420	$3s^23p3d^2$	61	−2.39624	$3s3p^3$ ^{3}D $6s$	43	−4.00748	$3p^2$ ^{3}P $6g$
38	−5.68324	$3p^23d^3$:	10	−11.6230	$3p^2$ ^{3}P $4p$	62	−2.29384	$3p^2$ ^{3}P $8f$	44	−3.80600	$3s3p^3$ ^{1}D $5p$
39	−5.65337	$3p^2$ ^{1}S $5d$	11	−11.2516	$3p^2$ ^{1}D $4p$	63	−2.21231	$3p^2$ ^{1}D $8p$	45	−3.68698	$3p^2$ ^{1}D $6g$
40	−5.64625	$3p^23d^3$:	12	−11.1472	$3p^33d^2$:	64	−1.97083	$3p^2$ ^{1}D $8f$	46	−3.34348	$3s3p^3$ ^{3}D $5f$
41	−5.55210	$3s3p^3$ ^{1}P $4f$:	13	−10.4675	$3p^33d^2$:	65	−1.96508	$3p^2$ ^{3}P $9p$	47	−3.14496	$3p^2$ ^{3}P $7d$
42	−5.31972	$3p^23d^3$	14	−9.97995	$3p^33d^2$:	66	−1.81278	$3s3p^3$ ^{3}D $6d$	48	−3.01684	$3s3p^3$ ^{3}P $5f$
43	−5.20556	$3p^2$ ^{1}D $5g$:	15	−9.89123	$3s3p^3$ ^{3}D $4s$	67	−1.80813	$3p^2$ ^{3}P $9f$	49	−2.94973	$3p^2$ ^{3}P $7g$
44	−5.04553	$3p^23d^3$:	16	−9.74315	$3p^33d^2$:	68	−1.74848	$3s3p^3$ ^{1}D $6s$	50	−2.84452	$3p^2$ ^{1}D $7d$
45	−4.88335	$3p^23d^3$	17	−9.54411	$3p^33d^2$	69	−1.65462	$3p^2$ ^{1}D $9p$	51	−2.72951	$3s3p^3$ ^{1}D $5f$
46	−4.84628	$3p^23d^3$:	18	−9.49768	$3p^33d^2$			**^{2}F^e**	52	−2.62563	$3p^2$ ^{1}D $7g$
47	−4.62731	$3p^2$ ^{1}D $6s$	19	−9.36082	$3s3p^3$ ^{1}D $4s$	1	−20.0547	$3p^2$ ^{1}D $3d$	53	−2.39046	$3p^2$ ^{3}P $8d$
48	−4.52147	$3p^23d^3$:	20	−9.31410	$3p^2$ ^{3}P $4f$	2	−18.7990	$3p^2$ ^{3}P $3d$	54	−2.28423	$3p^2$ ^{3}P $8g$
49	−4.38799	$3s3p^3$ ^{3}D $5p$	21	−9.07451	$3p^33d^2$:	3	−14.3831	$3p^43d$	55	−2.18376	$3s3p^3$ ^{3}S $5f$
50	−4.30256	$3p^2$ ^{3}P $6d$	22	−8.87756	$3p^33d^2$:	4	−14.1408	$3p^43d$	56	−2.17052	$3s3p^3$ ^{3}D $6p$
51	−4.12891	$3p^23d^3$:	23	−8.86389	$3p^2$ ^{1}D $4f$	5	−13.0283	$3s3p^23d^2$:	57	−2.08067	$3p^2$ ^{1}D $8d$
52	−4.06841	$3s3p^3$ ^{3}P $5p$	24	−8.53170	$3s3p3d^3$:	6	−12.8939	$3s3p^23d^2$	58	−2.05006	$3s3p^3$ ^{1}P $5f$
53	−4.03089	$3p^2$ ^{1}D $6d$	25	−8.31537	$3s3p3d^3$:	7	−12.4854	$3s3p^23d^2$:	59	−1.92921	$3p^2$ ^{1}D $8g$
54	−3.95185	$3p^23d^3$:	26	−8.19857	$3s3p3d^3$:	8	−12.3273	$3s3p^23d^2$:	60	−1.87092	$3p^2$ ^{3}P $9d$
55	−3.79360	$3s3p^3$ ^{1}D $5p$	27	−7.92221	$3s3p3d^3$:	9	−12.2625	$3s3p^23d^2$:	61	−1.77970	$3p^2$ ^{3}P $9g$
56	−3.66647	$3p^2$ ^{1}D $6g$	28	−7.66403	$3s3p^3$ ^{3}D $4d$	10	−11.8427	$3s3p^23d^2$:	62	−1.56636	$3p^2$ ^{1}D $9d$
57	−3.62177	$3p^2$ ^{1}S $6d$	29	−7.51618	$3s3p3d^3$:	11	−11.5457	$3s3p^23d^2$	63	−1.55060	$3s3p^3$ ^{3}D $6f$
58	−3.36776	$3s3p^3$ ^{3}D $5f$	30	−7.46442	$3s3p3d^3$:	12	−11.4984	$3s3p^23d^2$:			**^{2}F^o**
59	−3.20873	$3p^2$ ^{1}D $7s$	31	−7.29114	$3s3p^3$ ^{3}P $4d$	13	−10.7986	$3s3p^23d^2$:	1	−17.0394	$3s3p^3$ ^{3}D $3d$:
60	−3.19698	$3p^23d^3$:	32	−7.16079	$3s3p3d^3$:	14	−10.2249	$3p^2$ ^{3}P $4d$	2	−16.7130	$3s3p^3$ ^{3}P $3d$
61	−3.12721	$3p^2$ ^{3}P $7d$	33	−7.01559	$3s3p^3$ ^{1}D $4d$	15	−10.1593	$3s^23d^3$	3	−16.2575	$3s3p^3$ ^{1}D $3d$:
62	−3.10693	$3s3p^3$ ^{1}P $5p$	34	−6.97989	$3p^2$ ^{3}P $5p$	16	−9.99601	$3p^2$ ^{1}D $4d$	4	−15.9948	$3s3p^3$ ^{1}P $3d$
63	−2.93919	$3s3p^3$ ^{3}P $5f$	35	−6.88169	$3s3p3d^3$:	17	−9.05831	$3s3p^3$ ^{3}D $4p$	5	−15.4917	$3s^23p3d^2$
64	−2.83186	$3p^2$ ^{1}D $7d$	36	−6.68223	$3p^2$ ^{1}D $5p$	18	−8.42085	$3s3p^3$ ^{1}D $4p$	6	−14.8159	$3s^23p3d^2$
65	−2.72093	$3s3p^3$ ^{1}D $5f$	37	−6.50517	$3s3p3d^3$:	19	−6.85879	$3p^23d^3$:	7	−14.4204	$3s^23p3d^2$:
66	−2.61730	$3p^2$ ^{1}D $7g$	38	−6.46584	$3s3p^3$ ^{3}S $4d$	20	−6.64234	$3s3p^3$ ^{3}D $4f$	8	−11.3355	$3p^2$ ^{1}D $4p$
67	−2.43488	$3p^2$ ^{1}S $7d$	39	−6.31716	$3s3p^3$ ^{1}P $4d$	21	−6.36246	$3p^2$ ^{3}P $5d$	9	−10.9270	$3p^33d^2$:
68	−2.37468	$3p^2$ ^{3}P $8d$	40	−6.07239	$3s3p3d^3$:	22	−6.31457	$3s3p^3$ ^{3}P $4f$	10	−10.3680	$3p^33d^2$:
69	−2.30975	$3p^2$ ^{1}D $8s$	41	−5.93299	$3p^2$ ^{3}P $5f$	23	−6.26793	$3p^23d^3$	11	−10.2706	$3p^33d^2$
70	−2.12797	$3s3p^3$ ^{3}D $6p$	42	−5.57071	$3p^2$ ^{1}D $5f$	24	−6.11803	$3p^2$ ^{1}D $5d$	12	−9.72413	$3p^33d^2$:
71	−2.07413	$3p^2$ ^{1}D $8d$	43	−4.88297	$3s3p^3$ ^{3}D $5s$	25	−6.07829	$3p^23d^3$:	13	−9.53426	$3p^33d^2$:
72	−2.05476	$3s3p^3$ ^{1}P $5f$	44	−4.67193	$3p^2$ ^{3}P $6p$	26	−6.03187	$3s3p^3$ ^{1}D $4f$	14	−9.29765	$3p^33d^2$
73	−1.91602	$3p^2$ ^{1}D $8g$	45	−4.38378	$3p^2$ ^{1}D $6p$	27	−5.86529	$3p^23d^3$:	15	−9.13241	$3p^33d^2$:
74	−1.86535	$3p^2$ ^{3}P $9d$	46	−4.24816	$3s3p^3$ ^{1}D $5s$	28	−5.86303	$3p^2$ ^{3}P $5g$	16	−9.04710	$3p^2$ ^{1}D $4f$:
75	−1.78022	$3s3p^3$ ^{3}P $6p$	47	−4.10250	$3p^2$ ^{3}P $6f$	29	−5.71822	$3p^23d^3$:	17	−9.00092	$3p^2$ ^{3}P $4f$:
76	−1.72334	$3p^2$ ^{1}D $9s$	48	−3.81750	$3s3p^3$ ^{3}D $5d$	30	−5.61859	$3p^23d^3$	18	−8.81632	$3p^33d^2$:
77	−1.67410	$3p^2$ ^{1}S $8d$	49	−3.76528	$3p^2$ ^{1}D $6f$	31	−5.56236	$3p^2$ ^{1}D $5g$	19	−8.51940	$3p^2$ ^{1}S $4f$
78	−1.56434	$3s3p^3$ ^{3}D $6f$	50	−3.45958	$3s3p^3$ ^{3}P $5d$	32	−5.39770	$3s3p^3$ ^{3}S $4f$	20	−8.37684	$3s3p3d^3$:
79	−1.55939	$3p^2$ ^{1}D $9d$	51	−3.34848	$3p^2$ ^{3}P $7p$	33	−5.26629	$3s3p^3$ ^{1}P $4f$	21	−8.10036	$3s3p3d^3$
		^{2}D^o	52	−3.26354	$3s3p^3$ ^{3}D $5g$	34	−5.26302	$3p^23d^3$	22	−7.98507	$3s3p3d^3$:
1	−23.7113	$3s^23p^3$	53	−3.16515	$3s3p^3$ ^{1}D $5d$	35	−5.08675	$3p^23d^3$:	23	−7.66763	$3s3p3d^3$
2	−17.2391	$3s3p^3$ ^{3}P $3d$	54	−3.04247	$3p^2$ ^{1}D $7p$	36	−5.01869	$3p^23d^3$:	24	−7.63439	$3s3p^3$ ^{3}D $4d$
3	−16.8836	$3s3p^3$ ^{1}D $3d$:	55	−3.00374	$3p^2$ ^{3}P $7f$	37	−4.62480	$3p^23d^3$:	25	−7.58303	$3s3p3d^3$
4	−16.3477	$3s3p^3$ ^{3}D $3d$:	56	−2.67729	$3p^2$ ^{1}D $7f$	38	−4.47564	$3s3p^3$ ^{3}D $5p$	26	−7.25360	$3s3p^3$ ^{3}P $4d$
5	−16.0611	$3s3p^3$ ^{1}P $3d$	57	−2.66340	$3s3p^3$ ^{3}S $5d$	39	−4.37244	$3p^23d^3$:	27	−7.17589	$3s3p3d^3$
6	−15.3885	$3s^23p3d^2$:	58	−2.58930	$3s3p^3$ ^{1}D $5g$	40	−4.32885	$3p^2$ ^{3}P $6d$	28	−7.02946	$3s3p^3$ ^{1}D $4d$
7	−15.1773	$3s3p^3$ ^{3}S $3d$	59	−2.51424	$3p^2$ ^{3}P $8p$	41	−4.13163	$3p^23d^3$:	29	−6.98261	$3s3p3d^3$

P-like Fe (Fe^{11+})

i	E(Ryds)	Description
30	−6.70045	$3p^2$ ^{1}D $5p$
31	−6.68936	$3s3p3d^3$:
32	−6.58234	$3s3p3d^3$:
33	−6.31450	$3s3p^3$ ^{1}P $4d$
34	−5.81046	$3p^2$ ^{3}P $5f$
35	−5.62561	$3p^2$ ^{1}D $5f$
36	−5.20237	$3p^2$ ^{1}S $5f$
37	−4.37542	$3p^2$ ^{1}D $6p$
38	−4.04069	$3p^2$ ^{3}P $6f$
39	−3.80860	$3s3p^3$ ^{3}D $5d$
40	−3.79022	$3p^2$ ^{1}D $6f$
41	−3.43943	$3s3p^3$ ^{3}P $5d$
42	−3.38468	$3p^2$ ^{1}S $6f$
43	−3.26479	$3s3p^3$ ^{3}D $5g$
44	−3.17057	$3s3p^3$ ^{1}D $5d$
45	−3.04381	$3p^2$ ^{1}D $7p$
46	−2.96776	$3p^2$ ^{3}P $7f$
47	−2.86890	$3s3p^3$ ^{3}P $5g$
48	−2.69037	$3p^2$ ^{1}D $7f$
49	−2.59294	$3s3p^3$ ^{1}D $5g$
50	−2.47733	$3s3p^3$ ^{1}P $5d$
51	−2.29119	$3p^2$ ^{1}S $7f$
52	−2.26908	$3p^2$ ^{3}P $8f$
53	−2.20921	$3p^2$ ^{1}D $8p$
54	−1.97942	$3p^2$ ^{1}D $8f$
55	−1.90049	$3s3p^3$ ^{1}P $5g$
56	−1.80604	$3s3p^3$ ^{3}D $6d$
57	−1.79303	$3p^2$ ^{3}P $9f$
58	−1.65474	$3p^2$ ^{1}D $9p$
59	−1.58120	$3p^2$ ^{1}S $8f$
60	−1.50179	$3s3p^3$ ^{3}D $6g$
61	−1.49303	$3p^2$ ^{1}D $9f$
	^{2}G^e	
1	−19.6086	$3p^2$ ^{1}D $3d$
2	−14.6412	$3p^43d$
3	−13.6524	$3s3p^23d^2$
4	−12.8533	$3s3p^23d^2$
5	−12.5015	$3s3p^23d^2$:
6	−12.1391	$3s3p^23d^2$:
7	−11.8844	$3s3p^23d^2$
8	−11.6522	$3s3p^23d^2$
9	−11.2065	$3s3p^23d^2$
10	−10.5259	$3s^23d^3$
11	−9.87982	$3p^2$ ^{1}D $4d$
12	−6.84872	$3p^23d^3$:
13	−6.61038	$3s3p^3$ ^{3}D $4f$
14	−6.55213	$3p^23d^3$:
15	−6.27199	$3s3p^3$ ^{3}P $4f$
16	−6.14378	$3p^23d^3$
17	−6.12417	$3s3p^3$ ^{1}D $4f$
18	−6.02442	$3p^2$ ^{1}D $5d$
19	−5.78555	$3p^23d^3$

i	E(Ryds)	Description
20	−5.72675	$3p^2$ ^{3}P $5g$
21	−5.62025	$3p^23d^3$
22	−5.53463	$3p^2$ ^{1}D $5g$
23	−5.44850	$3p^23d^3$
24	−5.35351	$3s3p^3$ ^{1}P $4f$
25	−5.22546	$3p^23d^3$:
26	−5.03349	$3p^2$ ^{1}S $5g$
27	−4.67555	$3p^23d^3$
28	−4.48341	$3p^23d^3$:
29	−4.01374	$3p^2$ ^{1}D $6d$
30	−3.98806	$3p^2$ ^{3}P $6g$
31	−3.96345	$3p^23d^3$
32	−3.70496	$3p^2$ ^{1}D $6g$
33	−3.33846	$3s3p^3$ ^{3}D $5f$
34	−3.29840	$3p^2$ ^{1}S $6g$
35	−2.98540	$3s3p^3$ ^{3}P $5f$
36	−2.93082	$3p^2$ ^{3}P $7g$
37	−2.83142	$3p^2$ ^{1}D $7d$
38	−2.74784	$3s3p^3$ ^{1}D $5f$
39	−2.63146	$3p^2$ ^{1}D $7g$
40	−2.24699	$3p^2$ ^{3}P $8g$
41	−2.23885	$3p^2$ ^{1}S $7g$
42	−2.08698	$3p^2$ ^{1}D $8d$
43	−2.04099	$3s3p^3$ ^{1}P $5f$
44	1.93990	$3p^2$ ^{1}D $8g$
45	−1.77538	$3p^2$ ^{3}P $9g$
46	−1.55669	$3p^2$ ^{1}D $9d$
47	−1.54923	$3s3p^3$ ^{3}D $6f$
48	−1.54390	$3p^2$ ^{1}S $8g$
49	−1.46813	$3p^2$ ^{1}D $9g$
	^{2}G^o	
1	−17.3675	$3s3p^3$ ^{3}D $3d$
2	−16.6827	$3s3p^3$ ^{1}D $3d$
3	−15.2461	$3s^23p3d^2$
4	−14.4270	$3s^23p3d^2$
5	−10.6469	$3p^33d^2$
6	−10.3145	$3p^33d^2$
7	−10.1566	$3p^33d^2$:
8	−9.52063	$3p^33d^2$
9	−9.42436	$3p^33d^2$
10	−9.26532	$3p^2$ ^{3}P $4f$
11	−8.97591	$3p^2$ ^{1}D $4f$
12	−8.51591	$3s3p3d^3$
13	−8.38869	$3s3p3d^3$
14	−8.02771	$3s3p3d^3$
15	−7.80480	$3s3p3d^3$
16	−7.68209	$3s3p^3$ ^{3}D $4d$
17	−7.34971	$3s3p3d^3$
18	−7.11592	$3s3p3d^3$
19	−7.04378	$3s3p^3$ ^{1}D $4d$
20	−6.76663	$3s3p3d^3$
21	−5.89749	$3p^2$ ^{3}P $5f$

i	E(Ryds)	Description
22	−5.62991	$3p^2$ ^{1}D $5f$
23	−4.08251	$3p^2$ ^{3}P $6f$
24	−3.83264	$3s3p^3$ ^{3}D $5d$
25	−3.79669	$3p^2$ ^{1}D $6f$
26	−3.25665	$3s3p^3$ ^{3}D $5g$
27	−3.17293	$3s3p^3$ ^{1}D $5d$
28	−2.99215	$3p^2$ ^{3}P $7f$
29	−2.88809	$3s3p^3$ ^{3}P $5g$
30	−2.69547	$3p^2$ ^{1}D $7f$
31	−2.59459	$3s3p^3$ ^{1}D $5g$
32	−2.28605	$3p^2$ ^{3}P $8f$
33	−2.09998	$3s3p^3$ ^{3}S $5g$
34	−1.98350	$3p^2$ ^{1}D $8f$
35	−1.90091	$3s3p^3$ ^{1}P $5g$
36	−1.81978	$3s3p^3$ ^{3}D $6d$
37	−1.80310	$3p^2$ ^{3}P $9f$
38	−1.49936	$3p^2$ ^{1}D $9f$
39	−1.49431	$3s3p^3$ ^{3}D $6g$
	^{4}S^e	
1	−12.7354	$3s3p^23d^2$
2	−11.9879	$3s3p^23d^2$
3	−8.82178	$3s3p^3$ ^{3}P $4p$
4	−6.84673	$3p^23d^3$
5	−5.69594	$3p^23d^3$
6	−4.14135	$3s3p^3$ ^{3}P $5p$
7	−1.81546	$3s3p^3$ ^{3}P $6p$
8	−.48286	$3s3p^3$ ^{3}P $7p$
	^{4}S^o	
1	−24.1307	$3s^23p^3$
2	−17.0684	$3s3p^3$ ^{3}D $3d$
3	−14.6505	$3s^23p3d^2$
4	−11.6146	$3p^2$ ^{3}P $4p$
5	−10.5211	$3s3p^3$ ^{5}S $4s$
6	−10.2123	$3p^33d^2$
7	−9.39711	$3p^33d^2$
8	−8.89592	$3s3p^3$ ^{3}S $4s$
9	−8.59682	$3s3p3d^3$
10	−7.96586	$3s3p3d^3$
11	−7.66725	$3s3p^3$ ^{3}D $4d$
12	−7.11215	$3s3p3d^3$
13	−6.98609	$3p^2$ ^{3}P $5p$
14	−5.55646	$3s3p^3$ ^{5}S $5s$
15	−4.08093	$3p^2$ ^{3}P $6p$
16	−3.82947	$3s3p^3$ ^{3}D $5d$
17	−3.78165	$3s3p^3$ ^{3}S $5s$
18	−3.35306	$3p^2$ ^{3}P $7p$
19	−3.07913	$3s3p^3$ ^{5}S $6s$
20	−2.52531	$3p^2$ ^{3}P $8p$
21	−1.97170	$3p^2$ ^{3}P $9p$
22	−1.81796	$3s3p^3$ ^{3}D $6d$
23	−1.67497	$3s3p^3$ ^{5}S $7s$

i	E(Ryds)	Description
	^{4}P^e	
1	−21.7284	$3s3p^4$
2	−19.4172	$3p^2$ ^{3}P $3d$
3	−14.7137	$3p^43d$
4	−13.8206	$3s3p^23d^2$:
5	−13.3934	$3s3p^23d^2$
6	−13.0245	$3s3p^23d^2$
7	−12.6638	$3p^2$ ^{3}P $4s$
8	−12.6321	$3s3p^23d^2$
9	−12.1746	$3s3p^23d^2$:
10	−12.0681	$3s3p^23d^2$
11	−11.7311	$3s3p^23d^2$
12	−10.4701	$3s^23d^3$
13	−10.3180	$3p^2$ ^{3}P $4d$
14	−9.67468	$3s3p^3$ ^{5}S $4p$
15	−8.96197	$3s3p^3$ ^{3}D $4p$
16	−8.68086	$3s3p^3$ ^{3}P $4p$
17	−7.97029	$3s3p^3$ ^{3}S $4p$
18	−7.46499	$3p^2$ ^{3}P $5s$
19	−6.82697	$3p^23d^3$
20	−6.71087	$3s3p^3$ ^{3}D $4f$
21	−6.52471	$3p^23d^3$
22	−6.37416	$3p^2$ ^{3}P $5d$
23	−5.91949	$3p^23d^3$:
24	−5.66883	$3p^23d^3$:
25	−5.47508	$3p^23d^3$
26	−5.12466	$3s3p^3$ ^{5}S $5p$
27	−5.01145	$3p^23d^3$
28	−4.95165	$3p^2$ ^{3}P $6s$
29	−4.70826	$3p^23d^3$
30	−4.44967	$3s3p^3$ ^{3}D $5p$
31	−4.35572	$3p^2$ ^{3}P $6d$
32	−4.09726	$3s3p^3$ ^{3}P $5p$
33	−3.52173	$3p^2$ ^{3}P $7s$
34	−3.39242	$3s3p^3$ ^{3}D $5f$
35	−3.34416	$3s3p^3$ ^{3}S $5p$
36	−3.13908	$3p^2$ ^{3}P $7d$
37	−2.84095	$3s3p^3$ ^{5}S $6p$
38	−2.62924	$3p^2$ ^{3}P $8s$
39	−2.39165	$3p^2$ ^{3}P $8d$
40	−2.16288	$3s3p^3$ ^{3}D $6p$
41	−2.03926	$3p^2$ ^{3}P $9s$
42	−1.87830	$3p^2$ ^{3}P $9d$
43	−1.79654	$3s3p^3$ ^{3}P $6p$
	^{4}P^o	
1	−17.2997	$3s3p^3$ ^{3}P $3d$
2	−17.0808	$3s3p^3$ ^{3}D $3d$
3	−15.4270	$3s^23p3d^2$
4	−11.6570	$3p^2$ ^{3}P $4p$
5	−11.2333	$3p^33d^2$
6	−10.4009	$3p^33d^2$
7	−10.0181	$3p^33d^2$

P-like Fe (Fe^{11+})

i	E(Ryds)	Description	i	E(Ryds)	Description	i	E(Ryds)	Description	i	E(Ryds)	Description
8	−9.75365	$3s3p^3$ ^{3}P $4s$	33	−1.80857	$3s3p^3$ ^{3}P $6p$	4	−13.5120	$3s3p^23d^2$:	13	−8.28643	$3s3p3d^3$
9	−9.56022	$3p^33d^2$	34	−1.59190	$3s3p^3$ ^{3}D $6f$	5	−13.2907	$3s3p^23d^2$:	14	−7.75951	$3s3p^3$ ^{3}D $4d$
10	−9.18643	$3s3p3d^3$			**^{4}D^o**	6	−12.8138	$3s3p^23d^2$:	15	−7.72172	$3s3p3d^3$
11	−8.76982	$3s3p3d^3$	1	−18.1069	$3s3p^3$ ^{3}D $3d$	7	−12.6942	$3s3p^23d^2$	16	−7.34491	$3s3p^3$ ^{3}P $4d$
12	−8.30688	$3s3p3d^3$	2	−17.2526	$3s3p^3$ ^{3}P $3d$	8	−12.4921	$3s3p^23d^2$:	17	−5.92318	$3p^2$ ^{3}P $5f$
13	−8.03027	$3s3p3d^3$	3	−16.8444	$3s3p^3$ ^{5}S $3d$	9	−11.9567	$3s3p^23d^2$:	18	−4.09990	$3p^2$ ^{3}P $6f$
14	−7.73815	$3s3p^3$ ^{3}D $4d$	4	−16.4032	$3s3p^3$ ^{3}S $3d$	10	−10.8817	$3s^23d^3$	19	−3.85883	$3s3p^3$ ^{3}D $5d$
15	−7.46516	$3s3p3d^3$	5	−15.5773	$3s^23p3d^2$	11	−10.3006	$3p^2$ ^{3}P $4d$	20	−3.47587	$3s3p^3$ ^{3}P $5d$
16	−7.31170	$3s3p^3$ ^{3}P $4d$	6	−15.0563	$3s^23p3d^2$	12	−9.15496	$3s3p^3$ ^{3}D $4p$	21	−3.27296	$3s3p^3$ ^{3}D $5g$
17	−6.99214	$3p^2$ ^{3}P $5p$	7	−11.7386	$3p^2$ ^{3}P $4p$	13	−7.31659	$3s3p^3$ ^{5}S $4f$	22	−3.00259	$3p^2$ ^{3}P $7f$
18	−4.70636	$3p^2$ ^{3}P $6p$	8	−11.1714	$3p^33d^2$	14	−7.07796	$3p^23d^3$:	23	−2.88293	$3s3p^3$ ^{3}P $5g$
19	−4.57182	$3s3p^3$ ^{3}P $5s$	9	−10.4735	$3p^33d^2$	15	−6.84817	$3p^23d^3$:	24	−2.29312	$3p^2$ ^{3}P $8f$
20	−3.84378	$3s3p^3$ ^{3}D $5d$	10	−10.1966	$3p^33d^2$	16	−6.80396	$3s3p^3$ ^{3}D $4f$	25	−1.83232	$3s3p^3$ ^{3}D $6d$
21	−3.46785	$3s3p^3$ ^{3}P $5d$	11	−10.1254	$3s3p^3$ ^{3}D $4s$	17	−6.42793	$3s3p^3$ ^{3}P $4f$	26	−1.80812	$3p^2$ ^{3}P $9f$
22	−3.35579	$3p^2$ ^{3}P $7p$	12	−10.1004	$3p^33d^2$:	18	−6.38010	$3p^23d^3$	27	−1.50811	$3s3p^3$ ^{3}D $6g$
23	−2.52831	$3p^2$ ^{3}P $8p$	13	−9.70641	$3p^33d^2$:	19	−6.35966	$3p^2$ ^{3}P $5d$			**^{4}G^e**
24	−2.06267	$3s3p^3$ ^{3}P $6s$	14	−9.35257	$3s3p3d^3$	20	−6.14543	$3p^23d^3$	1	−13.7675	$3s3p^23d^2$:
25	−1.96953	$3p^2$ ^{3}P $9p$	15	−9.31575	$3p^2$ ^{3}P $4f$	21	−5.90682	$3p^2$ ^{3}P $5g$	2	−13.2033	$3s3p^23d^2$
26	−1.82094	$3s3p^3$ ^{3}D $6d$	16	−8.89065	$3s3p3d^3$	22	−5.90392	$3p^23d^3$	3	−12.9789	$3s3p^23d^2$:
		^{4}D^e	17	−8.68266	$3s3p3d^3$:	23	−5.63689	$3p^23d^3$	4	−12.5502	$3s3p^23d^2$
1	−20.0508	$3p^2$ ^{3}P $3d$	18	−8.45736	$3s3p3d^3$	24	−5.54338	$3s3p^3$ ^{3}S $4f$	5	−6.89103	$3s3p^3$ ^{3}D $4f$
2	−15.0716	$3p^43d$	19	−8.33076	$3s3p^3$ ^{5}S $4d$	25	−5.33243	$3p^23d^3$:	6	−6.88411	$3p^23d^3$
3	−13.7844	$3s3p^23d^2$	20	−8.25110	$3s3p3d^3$	26	−4.88557	$3p^23d^3$:	7	−6.58463	$3p^23d^3$
4	−13.4047	$3s3p^23d^2$	21	−8.17548	$3s3p3d^3$:	27	−4.52011	$3s3p^3$ ^{3}D $5p$	8	−6.45047	$3s3p^3$ ^{3}P $4f$
5	−12.8204	$3s3p^23d^2$	22	−7.72192	$3s3p^3$ ^{3}D $4d$	28	−4.33530	$3p^2$ ^{3}P $6d$	9	−6.20252	$3p^23d^3$
6	−12.6335	$3s3p^23d^2$	23	−7.32950	$3s3p^3$ ^{3}P $4d$	29	−4.04410	$3s3p^3$ ^{5}S $5f$	10	−5.98870	$3p^23d^3$
7	−12.5474	$3s3p^23d^2$	24	−7.29334	$3s3p3d^3$	30	−4.00611	$3p^2$ ^{3}P $6g$	11	−5.73359	$3p^2$ ^{3}P $5g$
8	−12.0781	$3s3p^23d^2$	25	−7.25707	$3s3p3d^3$	31	−3.43391	$3s3p^3$ ^{3}D $5f$	12	−5.61977	$3p^23d^3$
9	−11.8407	$3s3p^23d^2$	26	−7.03385	$3p^2$ ^{3}P $5p$	32	−3.15558	$3p^2$ ^{3}P $7d$	13	−4.00045	$3p^2$ ^{3}P $6g$
10	−10.2062	$3p^2$ ^{3}P $4d$	27	−6.54904	$3s3p^3$ ^{3}S $4d$	33	−3.04648	$3s3p^3$ ^{3}P $5f$	14	−3.45728	$3s3p^3$ ^{3}D $5f$
11	−9.21396	$3s3p^3$ ^{3}D $4p$	28	−5.93810	$3p^2$ ^{3}P $5f$	34	−2.94874	$3p^2$ ^{3}P $7g$	15	−3.07276	$3s3p^3$ ^{3}P $5f$
12	−8.76818	$3s3p^3$ ^{3}P $4p$	29	−4.96528	$3s3p^3$ ^{3}D $5s$	35	−2.39846	$3p^2$ ^{3}P $8d$	16	−2.93547	$3p^2$ ^{3}P $7g$
13	−7.18767	$3p^23d^3$	30	−4.70147	$3p^2$ ^{3}P $6p$	36	−2.30919	$3s3p^3$ ^{3}S $5f$	17	−2.25185	$3p^2$ ^{3}P $8g$
14	−6.89955	$3p^23d^3$:	31	−4.50611	$3s3p^3$ ^{5}S $5d$	37	−2.24826	$3s3p^3$ ^{5}S $6f$	18	−1.77997	$3p^2$ ^{3}P $9g$
15	−6.80177	$3s3p^3$ ^{3}D $4f$	32	−4.10603	$3p^2$ ^{3}P $6f$	38	−2.22714	$3p^2$ ^{3}P $8g$	19	−1.61017	$3s3p^3$ ^{3}D $6f$
16	−6.51424	$3p^23d^3$	33	−3.84713	$3s3p^3$ ^{3}D $5d$	39	−2.18677	$3s3p^3$ ^{3}D $6p$			**^{4}G^o**
17	−6.40893	$3s3p^3$ ^{3}P $4f$	34	−3.47241	$3s3p^3$ ^{3}P $5d$	40	−1.87511	$3p^2$ ^{3}P $9d$	1	−17.6393	$3s3p^3$ ^{3}D $3d$
18	−6.31969	$3p^2$ ^{3}P $5d$	35	−3.36485	$3p^2$ ^{3}P $7p$	41	−1.78223	$3p^2$ ^{3}P $9g$	2	−16.0288	$3s^23p3d^2$
19	−6.00260	$3p^23d^3$	36	−3.26361	$3s3p^3$ ^{3}D $5g$	42	−1.60110	$3s3p^3$ ^{3}D $6f$	3	−11.2468	$3p^33d^2$
20	−5.75412	$3p^23d^3$:	37	−3.00650	$3p^2$ ^{3}P $7f$			**^{4}F^o**	4	−10.5125	$3p^33d^2$
21	−5.71962	$3p^23d^3$:	38	−2.69409	$3s3p^3$ ^{3}S $5d$	1	−17.9833	$3s3p^3$ ^{3}P $3d$	5	−9.93946	$3p^33d^2$
22	−5.08104	$3p^23d^3$	39	−2.53178	$3p^2$ ^{3}P $8p$	2	−17.2023	$3s3p^3$ ^{3}D $3d$	6	−9.38870	$3p^2$ ^{3}P $4f$
23	−4.84622	$3p^23d^3$:	40	−2.50460	$3s3p^3$ ^{5}S $6d$	3	−15.4433	$3s^23p3d^2$	7	−9.38547	$3s3p3d^3$
24	−4.52344	$3s3p^3$ ^{3}D $5p$	41	−2.42996	$3s3p^3$ ^{3}D $6s$	4	−11.2146	$3p^33d^2$	8	−8.99715	$3s3p3d^3$
25	−4.33620	$3p^2$ ^{3}P $6d$	42	−2.29570	$3p^2$ ^{3}P $8f$	5	−10.6817	$3p^33d^2$	9	−8.69970	$3s3p3d^3$
26	−4.12238	$3s3p^3$ ^{3}P $5p$	43	−1.97424	$3p^2$ ^{3}P $9p$	6	−10.2977	$3p^33d^2$	10	−8.57691	$3s3p3d^3$
27	−3.42120	$3s3p^3$ ^{3}D $5f$	44	−1.82661	$3s3p^3$ ^{3}D $6d$	7	−9.60193	$3p^33d^2$	11	−7.86008	$3s3p3d^3$
28	−3.14287	$3p^2$ ^{3}P $7d$	45	−1.81008	$3p^2$ ^{3}P $9f$	8	−9.40016	$3s3p3d^3$	12	−7.74056	$3s3p^3$ ^{3}D $4d$
29	−3.04734	$3s3p^3$ ^{3}P $5f$			**^{4}F^e**	9	−9.27443	$3p^2$ ^{3}P $4f$	13	−5.96845	$3p^2$ ^{3}P $5f$
30	−2.38477	$3p^2$ ^{3}P $8d$	1	−20.1839	$3p^2$ ^{3}P $3d$	10	−8.92634	$3s3p3d^3$	14	−4.12230	$3p^2$ ^{3}P $6f$
31	−2.19285	$3s3p^3$ ^{3}D $6p$	2	−14.7455	$3p^43d$	11	−8.69276	$3s3p3d^3$	15	−3.95180	$3s3p^3$ ^{5}S $5g$
32	−1.87497	$3p^2$ ^{3}P $9d$	3	−14.1499	$3s3p^23d^2$	12	−8.49549	$3s3p3d^3$	16	−3.85458	$3s3p^3$ ^{3}D $5d$

P-like Fe (Fe^{11+})

i	Energy(Ryds)	Description		i	Energy(Ryds)	Description	
17	−3.26957	$3s3p^3$	^{3}D $5g$			**^{6}D^o**	
18	−3.01556	$3p^2$	^{3}P $7f$	1	−18.6338	$3s3p^3$	^{5}S $3d$
19	−2.89180	$3s3p^3$	^{3}P $5g$	2	−10.0111	$3s3p3d^3$	
20	−2.30181	$3p^2$	^{3}P $8f$	3	−9.51778	$3s3p3d^3$	
21	−2.19160	$3s3p^3$	^{5}S $6g$	4	−8.44869	$3s3p^3$	^{5}S $4d$
22	−2.11065	$3s3p^3$	^{3}S $5g$	5	−4.55604	$3s3p^3$	^{5}S $5d$
23	−1.83215	$3s3p^3$	^{3}D $6d$	6	−2.52938	$3s3p^3$	^{5}S $6d$
24	−1.81323	$3p^2$	^{3}P $9f$	7	−1.33769	$3s3p^3$	^{5}S $7d$
25	−1.50720	$3s3p^3$	^{3}D $6g$	8	−.57729	$3s3p^3$	^{5}S $8d$
		^{6}S^o		9	−.06243	$3s3p^3$	^{5}S $9d$
1	−10.8808	$3s3p^3$	^{5}S $4s$			**^{6}F^e**	
2	−9.86545	$3s3p3d^3$		1	−14.4439	$3s3p^23d^2$	
3	−5.67303	$3s3p^3$	^{5}S $5s$	2	−7.57383	$3s3p^3$	^{5}S $4f$
4	−3.13813	$3s3p^3$	^{5}S $6s$	3	−7.23666	$3p^23d^3$	
5	−1.70654	$3s3p^3$	^{5}S $7s$	4	−4.15604	$3s3p^3$	^{5}S $5f$
6	−.81772	$3s3p^3$	^{5}S $8s$	5	−2.30903	$3s3p^3$	^{5}S $6f$
7	−.22785	$3s3p^3$	^{5}S $9s$	6	−1.20271	$3s3p^3$	^{5}S $7f$
		^{6}P^e		7	−.48848	$3s3p^3$	^{5}S $8f$
1	−13.9784	$3s3p^23d^2$		8	−.00084	$3s3p^3$	^{5}S $9f$
2	−9.91163	$3s3p^3$	^{5}S $4p$			**^{6}G^o**	
3	−7.16420	$3p^23d^3$		1	−10.3819	$3s3p3d^3$	
4	−5.22135	$3s3p^3$	^{5}S $5p$	2	−3.96824	$3s3p^3$	^{5}S $5g$
5	−2.89103	$3s3p^3$	^{5}S $6p$	3	−2.20512	$3s3p^3$	^{5}S $6g$
6	−1.55680	$3s3p^3$	^{5}S $7p$	4	−1.13915	$3s3p^3$	^{5}S $7g$
7	−.72022	$3s3p^3$	^{5}S $8p$	5	−.44669	$3s3p^3$	^{5}S $8g$
8	−.16087	$3s3p^3$	^{5}S $9p$				

P-like Fe (Fe^{11+})

Energies in ascending order from ground state for terms with effective $n \leq 4.0$, $L \leq 4$

Term	i	E(Ryds)	Term	i	E(Ryds)	Term	i	E(Ryds)	Term	i	E(Ryds)	Term	i	E(Ryds)
$^{4}S^{o}$	1	0.00000	$^{4}P^{o}$	3	8.70370	$^{4}P^{e}$	7	11.4669	$^{2}G^{e}$	9	12.9242	$^{4}G^{o}$	5	14.1912
$^{2}D^{o}$	1	0.41940	$^{2}D^{o}$	6	8.74220	$^{2}D^{e}$	10	11.4860	$^{2}S^{e}$	7	12.9335	$^{6}P^{e}$	2	14.2190
$^{2}P^{o}$	1	0.68960	$^{2}G^{o}$	3	8.88460	$^{4}D^{e}$	6	11.4972	$^{4}D^{o}$	8	12.9593	$^{2}D^{o}$	15	14.2394
$^{4}P^{e}$	1	2.40230	$^{2}D^{o}$	7	8.95340	$^{4}P^{e}$	8	11.4986	$^{2}D^{o}$	12	12.9835	$^{2}G^{e}$	11	14.2508
$^{2}D^{e}$	1	2.98160	$^{4}D^{e}$	2	9.05910	$^{2}P^{e}$	7	11.5140	$^{2}D^{e}$	18	12.9910	$^{6}S^{o}$	2	14.2652
$^{2}P^{e}$	1	3.50800	$^{4}D^{o}$	6	9.07440	$^{4}G^{e}$	4	11.5805	$^{2}P^{o}$	11	13.0554	$^{2}P^{e}$	17	14.2816
$^{2}S^{e}$	1	3.60050	$^{2}S^{o}$	3	9.08680	$^{4}D^{e}$	7	11.5833	$^{2}P^{e}$	13	13.0669	$^{2}D^{e}$	23	14.3352
$^{4}F^{e}$	1	3.94680	$^{2}F^{o}$	6	9.31480	$^{2}P^{e}$	8	11.6055	$^{2}F^{o}$	9	13.2037	$^{2}S^{e}$	9	14.3722
$^{2}F^{e}$	1	4.07600	$^{2}P^{e}$	4	9.35250	$^{2}G^{e}$	5	11.6292	$^{2}P^{o}$	12	13.2439	$^{4}P^{o}$	8	14.3770
$^{4}D^{e}$	1	4.07990	$^{2}P^{o}$	7	9.36570	$^{4}F^{e}$	8	11.6386	$^{4}F^{e}$	10	13.2490	$^{2}D^{o}$	16	14.3875
$^{2}G^{e}$	1	4.52210	$^{2}D^{e}$	5	9.38190	$^{2}F^{e}$	7	11.6453	$^{6}S^{o}$	1	13.2499	$^{2}F^{o}$	12	14.4065
$^{2}P^{e}$	2	4.60240	$^{4}F^{e}$	2	9.38520	$^{2}D^{e}$	11	11.7961	$^{2}F^{e}$	13	13.3321	$^{4}D^{o}$	13	14.4242
$^{4}P^{e}$	2	4.71350	$^{4}P^{e}$	3	9.41700	$^{2}F^{e}$	8	11.8034	$^{2}P^{o}$	13	13.4102	$^{2}P^{o}$	15	14.4432
$^{2}D^{e}$	2	4.85970	$^{4}S^{o}$	3	9.48020	$^{2}S^{e}$	4	11.8581	$^{4}F^{o}$	5	13.4490	$^{4}P^{e}$	14	14.4560
$^{2}D^{e}$	3	5.07730	$^{2}G^{e}$	2	9.48950	$^{2}F^{e}$	9	11.8682	$^{2}D^{e}$	19	13.4695	$^{4}F^{o}$	7	14.5287
$^{2}S^{e}$	2	5.28660	$^{2}P^{o}$	8	9.49490	$^{2}D^{e}$	12	11.9055	$^{2}G^{o}$	5	13.4838	$^{4}P^{o}$	9	14.5704
$^{2}P^{e}$	3	5.29960	$^{2}D^{o}$	8	9.56210	$^{2}P^{e}$	9	11.9378	$^{2}S^{o}$	5	13.4961	$^{2}D^{o}$	17	14.5865
$^{2}F^{e}$	2	5.33170	$^{6}F^{e}$	1	9.68680	$^{4}P^{e}$	9	11.9561	$^{2}G^{e}$	10	13.6048	$^{2}P^{o}$	16	14.5933
$^{6}D^{o}$	1	5.49690	$^{2}G^{o}$	4	9.70370	$^{2}D^{e}$	13	11.9678	$^{4}S^{o}$	5	13.6096	$^{2}F^{o}$	13	14.5964
$^{2}D^{e}$	4	5.53430	$^{2}F^{o}$	7	9.71030	$^{2}G^{e}$	6	11.9916	$^{2}P^{e}$	14	13.6140	$^{2}S^{o}$	6	14.6092
$^{2}P^{o}$	2	5.60890	$^{2}F^{e}$	3	9.74760	$^{4}D^{e}$	8	12.0526	$^{4}G^{o}$	4	13.6182	$^{2}G^{o}$	8	14.6100
$^{4}D^{o}$	1	6.02380	$^{4}F^{e}$	3	9.98080	$^{2}P^{e}$	10	12.0541	$^{4}D^{o}$	9	13.6572	$^{6}D^{o}$	3	14.6129
$^{4}F^{o}$	1	6.14740	$^{2}D^{o}$	9	9.98870	$^{4}P^{e}$	10	12.0626	$^{4}P^{e}$	12	13.6606	$^{2}D^{o}$	18	14.6330
$^{2}S^{o}$	1	6.42340	$^{2}F^{e}$	4	9.98990	$^{4}S^{e}$	2	12.1428	$^{2}D^{o}$	13	13.6632	$^{2}D^{e}$	24	14.6408
$^{4}G^{o}$	1	6.49140	$^{2}S^{e}$	3	10.1130	$^{2}D^{e}$	14	12.1539	$^{4}P^{o}$	6	13.7298	$^{2}G^{o}$	9	14.7063
$^{2}G^{o}$	1	6.76320	$^{6}P^{e}$	1	10.1523	$^{4}F^{e}$	9	12.1740	$^{6}G^{o}$	1	13.7488	$^{4}F^{o}$	8	14.7305
$^{4}P^{o}$	1	6.83100	$^{2}P^{o}$	9	10.2346	$^{2}S^{e}$	5	12.1818	$^{2}F^{o}$	10	13.7627	$^{4}S^{o}$	7	14.7335
$^{4}D^{o}$	2	6.87810	$^{4}P^{e}$	4	10.3101	$^{2}P^{e}$	11	12.2197	$^{2}P^{e}$	15	13.8013	$^{4}G^{o}$	6	14.7420
$^{2}D^{o}$	2	6.89160	$^{4}D^{e}$	3	10.3463	$^{2}S^{o}$	4	12.2276	$^{4}P^{e}$	13	13.8127	$^{4}G^{o}$	7	14.7452
$^{4}F^{o}$	2	6.92840	$^{4}G^{e}$	1	10.3632	$^{2}G^{e}$	7	12.2463	$^{2}G^{o}$	6	13.8162	$^{2}D^{o}$	19	14.7698
$^{4}P^{o}$	2	7.04990	$^{2}D^{e}$	6	10.4423	$^{2}S^{e}$	6	12.2571	$^{4}F^{e}$	11	13.8301	$^{4}D^{o}$	14	14.7781
$^{4}S^{o}$	2	7.06230	$^{2}D^{e}$	7	10.4503	$^{2}F^{e}$	10	12.2880	$^{4}F^{o}$	6	13.8330	$^{4}D^{o}$	15	14.8149
$^{2}F^{o}$	1	7.09130	$^{2}G^{e}$	3	10.4783	$^{4}D^{e}$	9	12.2900	$^{2}F^{o}$	11	13.8601	$^{2}D^{o}$	20	14.8166
$^{2}D^{o}$	3	7.24710	$^{2}P^{e}$	5	10.5832	$^{4}D^{o}$	7	12.3921	$^{2}P^{e}$	16	13.8664	$^{2}F^{o}$	14	14.8330
$^{4}D^{o}$	3	7.28630	$^{4}F^{e}$	4	10.6187	$^{4}P^{e}$	11	12.3996	$^{2}P^{o}$	14	13.9050	$^{4}F^{o}$	9	14.8562
$^{2}P^{o}$	3	7.31360	$^{4}D^{e}$	4	10.7260	$^{2}D^{e}$	15	12.4085	$^{2}F^{e}$	14	13.9058	$^{2}G^{o}$	10	14.8653
$^{2}F^{o}$	2	7.41770	$^{4}P^{e}$	5	10.7373	$^{4}P^{o}$	4	12.4737	$^{2}D^{e}$	20	13.9159	$^{2}P^{o}$	17	14.8789
$^{2}G^{o}$	2	7.44800	$^{4}F^{e}$	5	10.8400	$^{2}G^{e}$	8	12.4785	$^{4}S^{o}$	6	13.9184	$^{4}D^{e}$	11	14.9167
$^{4}D^{o}$	4	7.72750	$^{2}D^{e}$	8	10.8513	$^{2}D^{o}$	10	12.5077	$^{4}D^{e}$	10	13.9245	$^{4}P^{o}$	10	14.9442
$^{2}D^{o}$	4	7.78300	$^{4}G^{e}$	2	10.9274	$^{4}S^{o}$	4	12.5161	$^{4}D^{o}$	10	13.9341	$^{4}F^{e}$	12	14.9757
$^{2}P^{o}$	4	7.87100	$^{2}F^{e}$	5	11.1024	$^{2}D^{e}$	16	12.5507	$^{2}S^{e}$	8	13.9648	$^{2}P^{e}$	18	14.9787
$^{2}F^{o}$	3	7.87320	$^{4}P^{e}$	6	11.1062	$^{2}F^{e}$	11	12.5850	$^{2}F^{e}$	15	13.9714	$^{2}P^{o}$	18	14.9857
$^{2}D^{o}$	5	8.06960	$^{4}G^{e}$	3	11.1518	$^{2}P^{e}$	12	12.6118	$^{2}G^{o}$	7	13.9741	$^{2}F^{o}$	15	14.9982
$^{2}S^{o}$	2	8.09850	$^{2}P^{e}$	6	11.1665	$^{2}P^{o}$	10	12.6260	$^{4}D^{o}$	11	14.0053	$^{2}D^{o}$	21	15.0561
$^{4}G^{o}$	2	8.10190	$^{2}F^{e}$	6	11.2368	$^{2}F^{e}$	12	12.6323	$^{4}D^{o}$	12	14.0303	$^{2}F^{e}$	17	15.0723
$^{2}F^{o}$	4	8.13590	$^{2}G^{e}$	4	11.2774	$^{2}D^{e}$	17	12.6607	$^{2}D^{e}$	21	14.0559	$^{2}F^{o}$	16	15.0836
$^{2}P^{o}$	5	8.17800	$^{4}D^{e}$	5	11.3103	$^{2}F^{o}$	8	12.7952	$^{4}P^{o}$	7	14.1126	$^{2}F^{o}$	17	15.1297
$^{4}D^{o}$	5	8.55340	$^{4}F^{e}$	6	11.3169	$^{2}D^{o}$	11	12.8791	$^{6}D^{o}$	2	14.1196			
$^{2}P^{o}$	6	8.58300	$^{4}S^{e}$	1	11.3953	$^{4}G^{o}$	3	12.8839	$^{2}F^{e}$	16	14.1346			
$^{2}F^{o}$	5	8.63900	$^{2}D^{e}$	9	11.4174	$^{4}P^{o}$	5	12.8974	$^{2}D^{o}$	14	14.1507			
$^{4}F^{o}$	3	8.68740	$^{4}F^{e}$	7	11.4365	$^{4}F^{o}$	4	12.9161	$^{2}D^{e}$	22	14.1670			

P-like Fe (Fe^{11+})

gf-values for transitions involving terms with effective $n \leq 4.0$, $L \leq 4$

i i'	gf_L	i i'	gf_L	i i'	gf_L	i i'	gf_L	i i'	gf_L	i i'	gf_L
	$^2S^o$–$^2P^e$	3 15	−5.15E−3	6 12	1.94E−3	3 8	5.37E−4	6 5	1.72E−2	9 2	6.12E−4
1 1	4.53E−2	3 16	−8.83E−2	6 13	2.48E−2	3 9	−6.55E−5	6 6	1.64E−2	9 3	5.41E−4
1 2	3.38E−2	3 17	−1.67E−4	6 14	3.29E−3	3 10	−2.41E−4	6 7	7.17E−2	9 4	9.78E−4
1 3	1.26E−3	3 18	−3.26E−4	6 15	1.49E−5	3 11	−1.57E−4	6 8	5.04E−2	9 5	5.85E−5
1 4	−3.59E−1	4 1	7.83E−2	6 16	0.00E+0	3 12	−4.70E−4	6 9	2.85E−4	9 6	4.03E−3
1 5	−3.53E−1	4 2	2.43E−2	6 17	5.13E−7	3 13	−1.23E−1	6 10	−8.95E−3	9 7	1.92E−4
1 6	−3.09E−6	4 3	1.53E−2	6 18	−4.73E−7	3 14	−2.64E−1	6 11	−1.40E−2	9 8	2.67E−3
1 7	−5.57E−1	4 4	1.35E−4		$^2S^e$–$^2P^o$	3 15	−8.05E−2	6 12	−4.95E−1	9 9	9.04E−3
1 8	−1.69E−6	4 5	1.59E−4	1 1	2.61E−1	3 16	−3.90E−3	6 13	−3.15E−5	9 10	2.24E−1
1 9	−6.44E−1	4 6	1.04E−3	1 2	−1.90E−2	3 17	−4.28E−1	6 14	−3.81E−3	9 11	3.92E−1
1 10	−7.93E−2	4 7	7.33E−5	1 3	−2.16E−1	3 18	−1.77E+0	6 15	−1.25E−2	9 12	9.89E−2
1 11	−2.45E−2	4 8	2.66E−1	1 4	−6.64E−1	4 1	1.39E−3	6 16	−3.62E−2	9 13	3.66E−5
1 12	−2.01E−2	4 9	1.46E−4	1 5	−2.25E−1	4 2	1.94E−3	6 17	−1.09E−1	9 14	1.17E−5
1 13	−3.54E−3	4 10	2.79E−5	1 6	−1.98E+0	4 3	1.70E+0	6 18	−1.93E−2	9 15	−2.60E−6
1 14	−2.21E−3	4 11	4.34E−6	1 7	−8.80E−3	4 4	1.80E−2	7 1	5.45E−4	9 16	−9.05E−4
1 15	−8.98E−5	4 12	−1.16E−4	1 8	−6.93E−3	4 5	3.06E−3	7 2	1.88E−3	9 17	−1.41E−5
1 16	−3.27E−4	4 13	−1.89E−3	1 9	−4.12E−6	4 6	5.80E−2	7 3	7.75E−3	9 18	−1.39E−4
1 17	−1.26E−3	4 14	−8.91E−5	1 10	−1.12E−2	4 7	4.91E−2	7 4	7.06E−1		$^2P^o$–$^2P^e$
1 18	−5.48E−2	4 15	−1.11E+0	1 11	−2.40E−2	4 8	3.01E−2	7 5	1.17E−1	1 1	−1.41E−1
2 1	9.31E−1	4 16	−2.84E−4	1 12	−1.14E−4	4 9	2.32E−3	7 6	1.72E−1	1 2	−5.72E−1
2 2	3.22E−2	4 17	−5.78E−3	1 13	−9.06E−5	4 10	−3.78E−4	7 7	6.28E−3	1 3	−2.52E+0
2 3	1.04E−1	4 18	−2.68E−1	1 14	−8.29E−5	4 11	−2.69E−4	7 8	6.07E−1	1 4	−1.91E−4
2 4	−1.49E−2	5 1	4.65E−4	1 15	−2.49E−3	4 12	−8.01E−3	7 9	7.70E−2	1 5	−1.50E−4
2 5	−1.03E−1	5 2	4.33E−7	1 16	−2.61E−1	4 13	−3.63E−2	7 10	8.23E−5	1 6	−9.36E−4
2 6	−4.94E−1	5 3	2.17E−6	1 17	−6.74E−3	4 14	−4.79E−2	7 11	−2.11E−6	1 7	−6.70E−4
2 7	−2.49E−1	5 4	1.73E−1	1 18	−4.81E−3	4 15	−1.69E−2	7 12	−8.07E−4	1 8	−4.21E−1
2 8	−1.80E−3	5 5	7.15E−2	2 1	7.79E−1	4 16	−9.50E−4	7 13	−8.81E−8	1 9	−4.09E−4
2 9	−3.16E−1	5 6	1.04E−1	2 2	−6.25E−4	4 17	−1.52E−2	7 14	−1.14E−4	1 10	−4.95E−7
2 10	−1.99E−1	5 7	5.73E−2	2 3	−3.37E−3	4 18	−1.81E−1	7 15	−6.06E−2	1 11	−2.84E−5
2 11	−6.19E−2	5 8	1.24E−5	2 4	−2.00E−1	5 1	1.60E−1	7 16	−8.96E−4	1 12	−9.16E−6
2 12	−4.18E−1	5 9	6.08E−4	2 5	−8.29E−3	5 2	6.03E−4	7 17	−7.80E−2	1 13	−4.46E−3
2 13	−7.80E−2	5 10	1.02E−3	2 6	−3.44E−1	5 3	5.67E−2	7 18	−6.69E−3	1 14	−2.58E−3
2 14	−2.02E−1	5 11	3.19E−2	2 7	−7.50E−1	5 4	8.06E−2	8 1	3.24E−8	1 15	−4.25E−1
2 15	−3.53E−4	5 12	6.90E−3	2 8	−4.37E−1	5 5	5.99E−3	8 2	7.33E−4	1 16	−1.03E−4
2 16	−3.13E−1	5 13	1.25E−5	2 9	−1.15E+0	5 6	3.58E−3	8 3	2.69E−5	1 17	−4.90E−1
2 17	−2.50E−4	5 14	−9.01E−5	2 10	−3.67E−2	5 7	4.05E−2	8 4	7.20E−3	1 18	−2.78E−2
2 18	−2.10E−4	5 15	−2.24E−6	2 11	−2.56E−2	5 8	1.66E−2	8 5	3.03E−2	2 1	4.19E−1
3 1	3.56E−3	5 16	−3.48E−7	2 12	−7.58E−4	5 9	3.61E−3	8 6	1.46E+0	2 2	3.79E−2
3 2	1.52E+0	5 17	−1.16E−6	2 13	−3.39E−5	5 10	−1.51E−2	8 7	7.82E−2	2 3	5.72E−4
3 3	3.00E−1	5 18	−1.16E−4	2 14	−1.11E−6	5 11	−1.58E−2	8 8	2.00E−1	2 4	−2.09E−1
3 4	−1.78E−3	6 1	7.50E−4	2 15	−1.07E−5	5 12	−7.70E−1	8 9	1.28E+0	2 5	−2.06E+0
3 5	−2.67E−2	6 2	1.65E−4	2 16	−0.75E−3	5 13	−3.26E−3	8 10	1.41E 4	2 6	1.74E 1
3 6	3.34E−2	6 3	7.98E−5	2 17	−5.54E−5	5 14	−4.97E−3	8 11	9.62E−4	2 7	−2.31E−2
3 7	−1.75E−2	6 4	6.35E−1	2 18	−1.01E−3	5 15	−8.43E−3	8 12	6.12E−4	2 8	−8.56E−5
3 8	−2.46E−3	6 5	2.28E−2	3 1	7.49E−4	5 16	−1.78E−3	8 13	3.15E−5	2 9	−8.27E−4
3 9	−3.56E−1	6 6	4.84E 1	3 2	1.08E+0	5 17	−1.36E−1	8 14	8.66E−6	2 10	−6.29E−3
3 10	−7.13E−2	6 7	4.32E−3	3 3	1.25E−1	5 18	−2.60E−2	8 15	−3.52E−4	2 11	−1.35E−4
3 11	−1.83E−1	6 8	1.16E−3	3 4	3.27E−2	6 1	1.17E−1	8 16	−4.22E−4	2 12	−6.90E−3
3 12	−5.45E−1	6 9	3.55E−1	3 5	1.65E−1	6 2	8.10E−4	8 17	−1.33E−3	2 13	−1.83E−4
3 13	−2.60E−1	6 10	9.22E−2	3 6	7.64E−3	6 3	2.56E−2	8 18	−4.47E−4	2 14	−5.81E−3
3 14	−1.68E+0	6 11	5.11E−3	3 7	3.33E−3	6 4	1.74E−1	9 1	1.88E−1	2 15	−1.02E−4

P-like Fe (Fe^{11+})

i	i'	gf_L	i	i'	gf_L	i	i'	gf_L	i	i'	gf_L	i	i'	gf_L	i	i'	gf_L
2	16	−9.29E−4	5	14	−8.09E−1	8	12	−8.79E−1	11	10	1.23E−4	14	8	1.40E−4	17	6	1.45E−3
2	17	−6.96E−5	5	15	−5.07E−3	8	13	−8.98E−1	11	11	4.07E−6	14	9	3.60E−5	17	7	1.27E−1
2	18	−5.33E−2	5	16	−1.71E−1	8	14	−1.92E−1	11	12	1.56E−4	14	10	1.60E−2	17	8	1.12E−3
3	1	1.65E−2	5	17	−7.76E−5	8	15	−9.12E−3	11	13	−6.33E−6	14	11	3.65E−3	17	9	7.87E−2
3	2	1.93E−1	5	18	−3.30E−3	8	16	−4.65E−2	11	14	−6.07E−4	14	12	3.55E−2	17	10	1.61E−1
3	3	2.79E−2	6	1	3.19E−2	8	17	−6.27E−3	11	15	−1.48E−1	14	13	8.01E−4	17	11	4.88E−1
3	4	−1.45E−1	6	2	5.02E−1	8	18	−4.63E−4	11	16	−1.41E−3	14	14	3.11E−4	17	12	2.34E−2
3	5	−1.67E−1	6	3	5.40E−1	9	1	3.65E−3	11	17	−1.19E+0	14	15	1.67E−6	17	13	9.49E−2
3	6	−2.43E−1	6	4	−1.08E−3	9	2	2.24E−3	11	18	−2.20E−2	14	16	4.56E−7	17	14	2.07E−3
3	7	−3.91E−1	6	5	−6.28E−2	9	3	1.89E+0	12	1	7.49E−3	14	17	−1.09E−5	17	15	8.65E−6
3	8	−1.13E−3	6	6	−9.89E−3	9	4	5.07E−6	12	2	5.66E−3	14	18	−3.23E−5	17	16	1.35E−3
3	9	−1.67E+0	6	7	−5.31E−2	9	5	−1.09E−3	12	3	1.27E−2	15	1	1.39E−3	17	17	1.71E−5
3	10	−9.51E−2	6	8	−3.08E−5	9	6	−2.18E−4	12	4	3.07E−4	15	2	2.92E−5	17	18	−1.05E−3
3	11	−1.38E−2	6	9	−1.09E−1	9	7	−8.25E−3	12	5	2.90E−4	15	3	4.33E−4	18	1	4.55E−5
3	12	−6.57E−3	6	10	−7.18E−4	9	8	−1.26E−3	12	6	6.48E−5	15	4	1.88E−1	18	2	7.65E−6
3	13	−3.94E−2	6	11	−7.62E−2	9	9	−1.37E−2	12	7	1.48E−4	15	5	4.36E−1	18	3	7.18E−6
3	14	−2.00E−4	6	12	−5.07E−1	9	10	−4.05E−4	12	8	2.73E−2	15	6	5.28E−1	18	4	6.86E−1
3	15	−3.40E−5	6	13	−6.10E−1	9	11	−9.17E−3	12	9	1.24E−5	15	7	2.22E−1	18	5	1.04E−1
3	16	−1.93E−4	6	14	−3.14E−2	9	12	−4.36E−5	12	10	3.13E−7	15	8	4.02E−3	18	6	1.05E−1
3	17	−3.78E−4	6	15	−1.04E−3	9	13	−5.52E−1	12	11	3.76E−4	15	9	1.77E−2	18	7	1.21E−1
3	18	−1.15E−1	6	16	−4.31E−2	9	14	−6.14E−2	12	12	4.95E−5	15	10	1.27E−1	18	8	3.52E−3
4	1	2.99E−1	6	17	−3.53E−3	9	15	−9.15E−3	12	13	1.56E−5	15	11	3.50E−4	18	9	3.65E−1
4	2	2.44E−1	6	18	−5.06E−7	9	16	−1.93E+0	12	14	−2.68E−5	15	12	4.81E−2	18	10	1.74E−1
4	3	5.61E−2	7	1	5.93E−2	9	17	−2.75E−2	12	15	−5.47E−3	15	13	1.09E−3	18	11	4.72E−2
4	4	−1.49E−2	7	2	2.50E+0	9	18	−4.08E−3	12	16	−1.78E−5	15	14	3.22E−3	18	12	7.82E−2
4	5	−2.91E−2	7	3	6.63E−1	10	1	7.38E−2	12	17	−6.15E−2	15	15	4.11E−5	18	13	9.66E−3
4	6	−1.23E−1	7	4	2.44E−5	10	2	3.68E−2	12	18	−1.81E−3	15	16	1.77E−3	18	14	1.85E−2
4	7	−3.71E−5	7	5	−3.98E−3	10	3	1.81E−5	13	1	1.17E−3	15	17	1.79E−6	18	15	4.63E−5
4	8	−4.09E−3	7	6	−1.25E−2	10	4	5.73E−6	13	2	1.67E−3	15	18	−6.74E−4	18	16	3.03E−2
4	9	−3.59E−2	7	7	−9.02E−2	10	5	1.78E−4	13	3	3.90E−4	16	1	8.58E−2	18	17	7.80E−6
4	10	−2.88E−1	7	8	−3.60E−4	10	6	7.34E−4	13	4	2.37E−1	16	2	9.40E−4	18	18	2.81E−5
4	11	−8.39E−2	7	9	−2.11E−2	10	7	2.93E−4	13	5	2.50E−1	16	3	3.19E−2			$^2P^o–^2D^e$
4	12	−3.36E−1	7	10	−3.72E−2	10	8	8.88E−1	13	6	7.81E−2	16	4	7.53E−3	1	1	−9.03E−2
4	13	−1.14E+0	7	11	−4.75E−2	10	9	2.47E−4	13	7	1.67E−3	16	5	4.49E−3	1	2	−6.16E−2
4	14	−6.59E−2	7	12	−1.71E−3	10	10	4.48E−4	13	8	1.30E−4	16	6	2.22E−3	1	3	−3.46E−1
4	15	−5.57E−4	7	13	−8.08E−1	10	11	6.38E−5	13	9	6.73E−3	16	7	5.55E−4	1	4	−5.02E+0
4	16	−5.74E−3	7	14	−1.49E−2	10	12	1.05E−5	13	10	7.07E−2	16	8	2.92E−1	1	5	−2.61E−6
4	17	−9.49E−4	7	15	−1.10E−3	10	13	−1.22E−5	13	11	2.95E−2	16	9	1.55E−3	1	6	−3.35E−4
4	18	−4.11E−5	7	16	−2.38E+0	10	14	−6.75E−4	13	12	1.96E−3	16	10	1.80E−3	1	7	−2.29E−4
5	1	2.81E+0	7	17	−1.64E−3	10	15	−6.76E−1	13	13	3.61E−4	16	11	2.03E−2	1	8	−1.64E−4
5	2	4.94E−1	7	18	−1.21E−4	10	16	−8.49E−4	13	14	−2.10E−4	16	12	7.89E−4	1	9	−8.90E−5
5	3	3.97E−3	8	1	9.31E−3	10	17	−3.54E−1	13	15	−5.10E−5	16	13	1.35E−3	1	10	−2.22E−4
5	4	−2.45E−2	8	2	3.46E−1	10	18	−9.39E−3	13	16	−1.17E−3	16	14	1.57E−3	1	11	−2.39E−1
5	5	−4.08E−2	8	3	4.17E−1	11	1	6.04E−3	13	17	−2.87E−5	16	15	4.57E−3	1	12	−3.28E−2
5	6	−1.40E−1	8	4	5.34E−5	11	2	1.01E−2	13	18	−2.02E−3	16	16	6.42E−4	1	13	−7.58E−5
5	7	−2.71E−2	8	5	−3.44E−5	11	3	1.37E−1	14	1	1.25E−5	16	17	5.05E−5	1	14	−7.12E−4
5	8	−1.19E−3	8	6	−1.60E−3	11	4	2.04E−5	14	2	4.62E−4	16	18	−9.12E−2	1	15	−1.75E−3
5	9	−1.28E−3	8	7	−1.41E−4	11	5	2.02E−4	14	3	1.63E−4	17	1	3.67E−4	1	16	−4.78E−4
5	10	−1.52E+0	8	8	−4.31E−4	11	6	6.45E−4	14	4	6.72E−1	17	2	7.14E−4	1	17	−4.41E−3
5	11	−3.29E−2	8	9	−1.07E−3	11	7	1.67E−4	14	5	1.02E−1	17	3	3.47E−5	1	18	−5.15E−4
5	12	−7.23E−1	8	10	−1.33E−1	11	8	6.77E−1	14	6	1.15E−2	17	4	4.32E−2	1	19	−3.08E−4
5	13	−2.86E−1	8	11	−2.78E−1	11	9	1.67E−5	14	7	1.30E−1	17	5	3.22E−2	1	20	−2.82E−4

P-like Fe (Fe^{11+})

i	i'	gf_L	i	i'	gf_L	i	i'	gf_L	i	i'	gf_L	i	i'	gf_L	i	i'	gf_L
1	21	−6.93E−1	4	1	3.35E−2	6	5	−6.35E−5	8	9	−2.78E−2	10	13	5.31E−3	12	17	5.32E−5
1	22	−1.22E+0	4	2	1.37E−1	6	6	−1.88E−1	8	10	−1.36E−3	10	14	4.65E−4	12	18	2.64E−5
1	23	−3.80E−3	4	3	4.61E−2	6	7	−7.64E−5	8	11	−5.00E−2	10	15	3.96E−4	12	19	−4.16E−8
1	24	−4.20E−1	4	4	1.43E−2	6	8	−4.08E−2	8	12	−1.31E−1	10	16	3.52E−6	12	20	−4.75E−8
2	1	4.65E−1	4	5	−1.20E−1	6	9	−8.49E−3	8	13	−1.68E−3	10	17	−1.17E−6	12	21	−5.26E−4
2	2	1.43E−3	4	6	−5.51E−1	6	10	−1.29E−1	8	14	−4.65E−2	10	18	−1.18E−4	12	22	−1.95E−2
2	3	1.87E−3	4	7	−1.92E−2	6	11	−4.65E−4	8	15	−5.91E−1	10	19	−1.20E−7	12	23	−5.04E−4
2	4	1.13E−5	4	8	−2.04E−2	6	12	−4.50E−2	8	16	−1.10E−2	10	20	−2.17E−3	12	24	−3.77E+0
2	5	−7.92E−4	4	9	−1.71E−1	6	13	−2.41E−4	8	17	−9.74E−2	10	21	−2.38E+0	13	1	1.48E−3
2	6	−7.43E−3	4	10	−6.14E−1	6	14	−2.18E−2	8	18	−1.36E−1	10	22	−1.12E−1	13	2	4.72E−4
2	7	−2.69E+0	4	11	−1.08E−2	6	15	−1.10E+0	8	19	−1.31E+0	10	23	−3.53E−5	13	3	5.40E−5
2	8	−1.65E+0	4	12	−2.79E−1	6	16	−8.40E−1	8	20	−1.28E+0	10	24	−1.88E−2	13	4	1.02E−4
2	9	−2.51E−4	4	13	−2.83E−1	6	17	−2.05E+0	8	21	−1.90E−2	11	1	5.17E−2	13	5	1.12E+0
2	10	−3.13E−1	4	14	−3.92E−3	6	18	−2.72E−1	8	22	−3.25E−4	11	2	1.82E−2	13	6	1.09E−1
2	11	−4.45E−3	4	15	−1.60E−2	6	19	−9.82E−1	8	23	−2.49E−2	11	3	3.78E−2	13	7	1.06E−1
2	12	−1.32E−2	4	16	−3.50E−2	6	20	−1.61E−1	8	24	−1.89E−2	11	4	9.59E−2	13	8	2.47E−1
2	13	−5.53E−3	4	17	−1.47E+0	6	21	−5.41E−5	9	1	1.53E−3	11	5	1.82E−6	13	9	2.02E−3
2	14	−2.30E−2	4	18	−3.85E−2	6	22	−2.09E−3	9	2	1.31E−3	11	6	7.65E−4	13	10	4.15E−2
2	15	−3.95E−3	4	19	−1.23E−1	6	23	−1.49E−2	9	3	3.70E−2	11	7	3.32E−5	13	11	1.21E−3
2	16	−5.77E−4	4	20	−1.48E−2	6	24	−2.82E−3	9	4	3.25E+0	11	8	1.09E−4	13	12	4.92E−3
2	17	−1.72E−4	4	21	−8.57E−3	7	1	3.41E−3	9	5	1.17E−5	11	9	2.00E−5	13	13	7.10E−4
2	18	−2.72E−2	4	22	−7.91E−5	7	2	7.80E−1	9	6	−1.81E−4	11	10	2.05E−3	13	14	2.53E−3
2	19	−6.90E−3	4	23	−8.43E−4	7	3	8.77E−2	9	7	−2.08E−4	11	11	7.17E−1	13	15	1.05E−4
2	20	−1.06E−3	4	24	−1.14E−3	7	4	4.13E−2	9	8	−4.19E−5	11	12	1.51E−1	13	16	2.07E−3
2	21	−3.93E−4	5	1	1.09E−2	7	5	−2.98E−7	9	9	−1.17E−4	11	13	1.51E−2	13	17	4.84E−4
2	22	−1.07E−3	5	2	3.19E−1	7	6	−2.22E−4	9	10	−3.65E−4	11	14	1.23E−4	13	18	6.42E−5
2	23	−2.15E−4	5	3	5.19E−1	7	7	−3.88E−2	9	11	−4.95E−3	11	15	2.20E−3	13	19	−1.81E−4
2	24	−1.60E−5	5	4	3.82E−4	7	8	−1.58E−3	9	12	−2.55E−2	11	16	1.88E−5	13	20	−7.26E−3
3	1	2.65E+0	5	5	−1.33E−2	7	9	−2.36E−2	9	13	−9.02E−2	11	17	3.04E−4	13	21	−2.93E−6
3	2	4.50E−2	5	6	−5.96E−3	7	10	−1.24E−1	9	14	−1.34E−3	11	18	4.45E−5	13	22	−5.41E−8
3	3	4.99E−2	5	7	−3.57E−2	7	11	−1.56E−2	9	15	−1.40E−3	11	19	−2.78E−6	13	23	−3.87E−3
3	4	1.45E−4	5	8	−1.83E−1	7	12	−6.55E−2	9	16	−3.98E−2	11	20	−5.00E−6	13	24	−3.06E−3
3	5	−6.26E−2	5	9	−1.33E−1	7	13	−8.13E−3	9	17	−1.01E−1	11	21	−4.38E−1	14	1	1.75E−5
3	6	−6.88E−2	5	10	−8.20E−1	7	14	−1.43E−1	9	18	−1.28E−2	11	22	−1.28E+0	14	2	8.98E−5
3	7	−8.23E−2	5	11	−2.32E−1	7	15	−6.40E−2	9	19	−5.73E−1	11	23	−3.13E−4	14	3	6.21E−5
3	8	−4.92E−1	5	12	−9.20E−1	7	16	−3.41E−1	9	20	−1.00E−1	11	24	−3.40E−3	14	4	2.50E−4
3	9	−6.03E−2	5	13	−9.32E−2	7	17	−2.30E−1	9	21	−1.28E−2	12	1	1.52E−2	14	5	1.98E−3
3	10	−5.39E−1	5	14	−1.23E−1	7	18	−2.21E+0	9	22	−9.33E−3	12	2	1.67E−1	14	6	1.18E−1
3	11	−4.85E−2	5	15	−2.84E−1	7	19	−6.35E−1	9	23	−4.07E+0	12	3	7.33E−2	14	7	9.30E−2
3	12	−3.06E−1	5	16	−1.91E−1	7	20	−1.45E+0	9	24	−3.63E−3	12	4	1.23E−1	14	8	1.99E−3
3	13	−1.23E−1	5	17	−3.13E−2	7	21	−5.25E−3	10	1	4.37E−2	12	5	1.73E−3	14	9	4.89E−1
3	14	0.26E 1	5	18	1.55E+0	7	22	−2.24E−4	10	2	1.22E−1	12	6	1.07E−4	14	10	2.78E−3
3	15	−1.11E−1	5	19	−1.02E−1	7	23	−7.97E−2	10	3	2.62E−2	12	7	2.25E−4	14	11	1.03E−2
3	16	−1.06E+0	5	20	−1.59E−2	7	24	−1.05E−2	10	4	3.55E−2	12	8	4.24E−5	14	12	8.72E−2
3	17	−1.16E−2	5	21	−7.96E−3	8	1	1.58E−4	10	5	1.40E−8	12	9	2.00E−4	14	13	1.06E−1
3	18	−5.70E−3	5	22	−1.59E−4	8	2	1.29E+0	10	6	1.53E−4	12	10	1.95E−3	14	14	3.51E−2
3	19	−9.77E−2	5	23	−1.34E−3	8	3	1.04E+0	10	7	4.87E−5	12	11	8.69E−2	14	15	2.70E−3
3	20	−4.11E−2	5	24	−1.67E−3	8	4	5.73E−3	10	8	9.87E−6	12	12	2.25E−2	14	16	7.73E−3
3	21	−1.22E−4	6	1	4.39E−2	8	5	3.12E−4	10	9	6.83E−4	12	13	2.27E−3	14	17	1.37E−3
3	22	−7.37E−3	6	2	9.50E−2	8	6	−3.87E−3	10	10	2.32E−3	12	14	3.40E−4	14	18	3.94E−4
3	23	−2.50E−4	6	3	1.86E−1	8	7	−9.54E−3	10	11	2.32E−1	12	15	6.43E−5	14	19	2.96E−4
3	24	−1.12E−4	6	4	4.83E−1	8	8	−1.04E−3	10	12	6.27E−2	12	16	1.31E−5	14	20	−1.63E−5

P-like Fe (Fe^{11+})

i	*i′*	gf_L	*i*	*i′*	gf_L	*i*	*i′*	gf_L	*i*	*i′*	gf_L	*i*	*i′*	gf_L	*i*	*i′*	gf_L
14	21	−2.67E−9	17	1	1.32E−2	1	4	−4.14E−1	3	14	−3.00E−3	6	3	7.26E−1	8	13	−6.26E−5
14	22	−9.70E−6	17	2	5.23E−3	1	5	−2.06E−4	3	15	−4.56E−3	6	4	1.18E−1	8	14	−3.94E−2
14	23	−2.26E−4	17	3	1.50E−3	1	6	−6.46E−1	3	16	−4.81E−7	6	5	1.97E−4	8	15	−1.30E−1
14	24	−1.75E−5	17	4	1.49E−3	1	7	−1.87E+0	3	17	−8.34E−5	6	6	2.43E−2	8	16	−3.89E−3
15	1	6.88E−4	17	5	5.42E−2	1	8	−1.01E−2	3	18	−6.26E−4	6	7	4.86E−2	8	17	−2.20E−3
15	2	3.09E−4	17	6	4.99E−4	1	9	−4.09E−2	3	19	−4.85E−4	6	8	5.87E−3	8	18	−2.64E−3
15	3	1.11E−5	17	7	3.38E−1	1	10	−2.03E−3	3	20	−8.57E−1	6	9	5.79E−4	8	19	−6.35E−3
15	4	7.75E−4	17	8	6.24E−2	1	11	−1.32E−2	3	21	−1.67E−3	6	10	−1.40E−3	8	20	−1.21E−2
15	5	1.26E+0	17	9	2.50E−3	1	12	−5.55E−4	4	1	1.32E−5	6	11	−1.02E−3	8	21	−6.30E−4
15	6	1.51E−7	17	10	3.21E−2	1	13	−3.65E−3	4	2	1.47E−1	6	12	−8.97E−2	9	1	2.27E−3
15	7	3.28E−1	17	11	1.46E−1	1	14	−1.61E−2	4	3	3.62E−1	6	13	−5.47E−1	9	2	1.75E−1
15	8	2.45E−1	17	12	7.70E−2	1	15	−4.25E−2	4	4	6.43E−4	6	14	−3.97E−1	9	3	2.03E−1
15	9	4.42E−1	17	13	4.37E−1	1	16	−1.19E−3	4	5	1.12E−2	6	15	−8.32E−2	9	4	9.08E−1
15	10	4.75E−3	17	14	7.52E−2	1	17	−4.43E−2	4	6	3.57E−3	6	16	−1.41E−1	9	5	3.65E−1
15	11	1.09E−2	17	15	5.01E−3	1	18	−2.54E−2	4	7	8.36E−4	6	17	−7.40E−3	9	6	8.26E−2
15	12	2.73E−2	17	16	3.56E−3	1	19	−6.02E−1	4	8	−3.14E−4	6	18	−9.91E−2	9	7	1.88E−2
15	13	1.26E−3	17	17	2.38E−3	1	20	−2.02E+0	4	9	−5.18E−5	6	19	−1.23E−2	9	8	4.97E−3
15	14	4.09E−5	17	18	7.11E−2	1	21	−4.11E−3	4	10	−2.61E−4	6	20	−6.30E−8	9	9	4.50E−2
15	15	4.63E−3	17	19	1.28E−3	2	1	2.43E+0	4	11	−1.66E−5	6	21	−2.93E−1	9	10	−2.62E−4
15	16	1.19E−2	17	20	9.00E−3	2	2	−4.51E−2	4	12	−6.42E−1	7	1	1.93E−5	9	11	−5.29E−4
15	17	2.07E−2	17	21	5.08E−6	2	3	−3.71E−1	4	13	−4.67E−2	7	2	1.60E+0	9	12	−9.93E−3
15	18	3.77E−2	17	22	8.91E−5	2	4	−1.05E−1	4	14	−1.21E+0	7	3	2.00E−1	9	13	−3.38E−2
15	19	2.45E−4	17	23	4.36E−3	2	5	−2.93E−2	4	15	−2.39E−1	7	4	2.31E−2	9	14	−6.76E−2
15	20	2.35E−6	17	24	4.62E−7	2	6	−9.56E−1	4	16	−1.73E−1	7	5	1.11E−1	9	15	−4.67E−6
15	21	9.38E−6	18	1	6.31E−3	2	7	−1.50E+0	4	17	−1.57E−1	7	6	5.43E−2	9	16	−9.72E−2
15	22	1.99E−5	18	2	6.20E−5	2	8	−7.56E−1	4	18	−3.64E−1	7	7	7.96E−2	9	17	−1.04E−1
15	23	1.91E−4	18	3	3.72E−4	2	9	−5.77E−1	4	19	−4.65E−5	7	8	1.84E−3	9	18	−5.99E−2
15	24	−6.23E−6	18	4	3.49E−3	2	10	−4.91E−6	4	20	−2.62E−3	7	9	8.60E−4	9	19	−7.97E−3
16	1	4.16E−1	18	5	1.47E−2	2	11	−5.39E−3	4	21	−5.11E−2	7	10	−2.41E−4	9	20	−7.48E−5
16	2	3.73E−2	18	6	7.36E−3	2	12	−4.78E−4	5	1	1.42E−3	7	11	−1.20E−4	9	21	−3.87E−3
16	3	2.37E−4	18	7	1.04E+0	2	13	−2.91E−4	5	2	4.61E−2	7	12	−2.82E−2	10	1	9.19E−4
16	4	4.43E−2	18	8	3.25E−1	2	14	−8.92E−3	5	3	1.46E−2	7	13	−1.56E−1	10	2	3.52E−2
16	5	1.40E−2	18	9	1.52E−4	2	15	−4.83E−2	5	4	2.80E−1	7	14	−6.24E−3	10	3	1.33E+0
16	6	1.42E−2	18	10	8.16E−2	2	16	−7.18E−6	5	5	1.61E−1	7	15	−1.19E−2	10	4	1.56E−1
16	7	4.74E−3	18	11	1.75E−1	2	17	−1.69E−4	5	6	3.40E−2	7	16	−3.14E−1	10	5	5.40E−1
16	8	5.81E−3	18	12	1.61E−1	2	18	−1.41E−3	5	7	7.38E−2	7	17	−2.95E−1	10	6	2.83E−1
16	9	4.98E−3	18	13	2.61E−1	2	19	−1.23E−2	5	8	1.61E−2	7	18	−5.85E−3	10	7	7.20E−2
16	10	3.43E−3	18	14	2.54E−2	2	20	−1.79E+0	5	9	2.09E−3	7	19	−8.32E−4	10	8	1.17E−4
16	11	6.66E−3	18	15	1.30E−3	2	21	−5.73E−4	5	10	−8.66E−4	7	20	−8.57E−6	10	9	2.61E−5
16	12	1.09E−2	18	16	1.51E−2	3	1	6.72E−2	5	11	−6.34E−4	7	21	−3.96E−1	10	10	−6.82E−4
16	13	4.20E−3	18	17	3.41E−2	3	2	−3.94E−2	5	12	−6.29E−2	8	1	8.59E−1	10	11	−2.64E−4
16	14	1.57E−3	18	18	9.63E−2	3	3	−1.16E−3	5	13	−3.33E−1	8	2	1.10E−4	10	12	−7.37E−3
16	15	4.40E−3	18	19	2.10E−2	3	4	−7.89E−5	5	14	−1.71E−1	8	3	1.87E−3	10	13	−5.79E−2
16	16	5.29E−4	18	20	1.10E−2	3	5	−2.86E−1	5	15	−1.01E−1	8	4	5.56E−3	10	14	−3.03E−3
16	17	2.66E−4	18	21	2.85E−5	3	6	−6.98E−1	5	16	−5.34E−1	8	5	1.24E−2	10	15	−1.08E−3
16	18	9.33E−8	18	22	3.03E−6	3	7	−1.31E+0	5	17	−1.02E−1	8	6	3.75E−6	10	16	−8.13E−2
16	19	8.90E−4	18	23	1.31E−2	3	8	−9.46E−1	5	18	−5.44E−1	8	7	1.19E−3	10	17	−1.59E−1
16	20	7.88E−4	18	24	8.24E−7	3	9	−1.43E+0	5	19	−4.72E−2	8	8	3.86E−3	10	18	−7.58E−2
16	21	2.44E−3			**$^2P^e$–$^2D^o$**	3	10	−2.30E−4	5	20	−8.69E−4	8	9	2.82E−3	10	19	−1.57E−2
16	22	5.29E−3	1	1	6.55E−1	3	11	−4.45E−2	5	21	−1.40E+0	8	10	−1.45E+0	10	20	−7.42E−5
16	23	6.43E−5	1	2	−1.53E−1	3	12	−1.12E−7	6	1	2.06E−3	8	11	−9.08E−1	10	21	−2.20E−1
16	24	−1.94E−5	1	3	−1.07E+0	3	13	−3.08E−4	6	2	7.03E−1	8	12	−3.99E−4	11	1	9.74E−6

P-like Fe (Fe^{11+})

i i'	gf_L	i i'	gf_L	i i'	gf_L	i i'	gf_L	i i'	gf_L	i i'	gf_L
11 2	2.05E−3	13 12	1.70E−5	16 1	2.51E−3	18 11	6.60E−2	2 17	−3.89E−1	4 21	−2.01E−3
11 3	9.21E−2	13 13	−2.56E−5	16 2	1.04E−2	18 12	1.65E−5	2 18	−1.63E−3	4 22	−7.24E−4
11 4	4.07E−1	13 14	−4.03E−2	16 3	4.28E−3	18 13	4.28E−3	2 19	−2.40E−2	4 23	−5.74E−3
11 5	7.74E−1	13 15	−1.86E−3	16 4	9.06E−3	18 14	1.53E−1	2 20	−4.03E−7	4 24	−1.60E−6
11 6	3.80E−1	13 16	−8.12E−3	16 5	4.07E−2	18 15	7.97E−1	2 21	−8.83E−4	5 1	9.65E−1
11 7	2.19E−1	13 17	−6.15E−2	16 6	1.69E−1	18 16	4.79E−3	2 22	−9.06E−4	5 2	1.94E−1
11 8	1.73E−1	13 18	−1.01E−4	16 7	2.03E+0	18 17	2.41E−2	2 23	−6.26E−3	5 3	6.96E−1
11 9	7.11E−3	13 19	−8.59E−3	16 8	1.29E+0	18 18	1.65E−3	2 24	−1.36E−3	5 4	7.78E−3
11 10	−1.90E−4	13 20	−2.25E−3	16 9	2.83E−3	18 19	1.62E−3	3 1	3.21E−1	5 5	−6.21E−2
11 11	−2.00E−4	13 21	−5.32E−2	16 10	1.09E−5	18 20	1.58E−3	3 2	8.32E−3	5 6	−3.38E−1
11 12	−6.60E−5	14 1	5.93E−4	16 11	1.09E−4	18 21	−1.71E−5	3 3	6.93E−2	5 7	−2.82E−4
11 13	−1.78E−2	14 2	3.44E−3	16 12	3.53E−4	**$^2D^o$–$^2D^e$**		3 4	1.98E−3	5 8	−5.84E−2
11 14	−2.71E−2	14 3	1.49E−3	16 13	1.48E−5	1 1	−5.23E−1	3 5	−6.80E−2	5 9	−1.71E−1
11 15	−5.96E−3	14 4	1.91E−2	16 14	−1.75E−4	1 2	−1.42E+0	3 6	−2.32E−1	5 10	−5.25E−2
11 16	−4.13E−2	14 5	5.18E−2	16 15	−4.06E−9	1 3	−3.27E+0	3 7	−2.00E−1	5 11	−5.99E−3
11 17	−1.43E−1	14 6	2.30E+0	16 16	−2.02E−3	1 4	−1.36E−1	3 8	−3.66E−2	5 12	−2.68E−2
11 18	−5.69E−1	14 7	9.69E−1	16 17	−7.41E−3	1 5	−4.12E−4	3 9	−1.06E−2	5 13	−1.21E+0
11 19	−1.00E−2	14 8	2.38E−1	16 18	−2.81E−3	1 6	−1.06E−4	3 10	−4.30E−1	5 14	−1.25E+0
11 20	−4.54E−5	14 9	1.97E+0	16 19	−1.09E−3	1 7	−2.60E−7	3 11	−1.05E−1	5 15	−3.69E−2
11 21	−3.65E−1	14 10	2.29E−4	16 20	−4.09E−3	1 8	−2.39E−3	3 12	−4.73E−1	5 16	−1.65E−1
12 1	7.60E−5	14 11	6.39E−4	16 21	−1.04E−3	1 9	−7.99E−4	3 13	−1.56E−1	5 17	−2.70E−2
12 2	5.28E−4	14 12	2.73E−5	17 1	1.85E−1	1 10	−3.66E−3	3 14	−5.59E−2	5 18	−5.47E−1
12 3	8.99E−2	14 13	−5.80E−6	17 2	5.93E−4	1 11	−7.49E−1	3 15	−1.83E−4	5 19	−1.65E+0
12 4	2.42E−1	14 14	−1.46E−3	17 3	7.28E−6	1 12	−1.33E−1	3 16	−4.97E−1	5 20	−9.18E−2
12 5	1.41E+0	14 15	−3.57E−4	17 4	2.54E−3	1 13	−1.92E−2	3 17	−1.67E−1	5 21	−7.36E−5
12 6	7.90E−3	14 16	−7.65E−4	17 5	9.78E−3	1 14	−2.89E−4	3 18	−7.14E−1	5 22	−1.43E−2
12 7	1.05E−2	14 17	−6.91E−3	17 6	4.46E−3	1 15	−1.15E−3	3 19	−7.26E−2	5 23	−4.35E−2
12 8	1.41E−1	14 18	−3.14E−2	17 7	2.44E−5	1 16	−4.15E−3	3 20	−2.07E−2	5 24	−3.26E−4
12 9	7.80E−4	14 19	−6.48E−3	17 8	3.22E−3	1 17	−1.79E−3	3 21	−4.83E−4	6 1	1.45E−1
12 10	2.05E−4	14 20	−8.07E−4	17 9	4.53E−3	1 18	−8.44E−5	3 22	−8.41E−4	6 2	7.21E−2
12 11	−2.62E−4	14 21	−5.68E−3	17 10	8.85E−2	1 19	−1.53E−4	3 23	−4.47E−4	6 3	2.08E+0
12 12	−9.17E−4	15 1	6.30E−2	17 11	4.20E−1	1 20	−3.08E−3	3 24	−6.98E−8	6 4	1.28E−1
12 13	−1.89E−2	15 2	2.56E−4	17 12	3.44E−5	1 21	−7.34E−1	4 1	2.52E+0	6 5	−3.89E−4
12 14	−2.20E−5	15 3	1.08E−2	17 13	2.75E−6	1 22	−5.96E−1	4 2	3.51E−1	6 6	−1.56E−2
12 15	−6.95E−3	15 4	7.34E−4	17 14	1.67E−7	1 23	−2.24E−4	4 3	5.16E−4	6 7	−1.27E−1
12 16	−1.53E−2	15 5	5.32E−5	17 15	1.27E−5	1 24	−5.08E−2	4 4	7.16E−2	6 8	−5.00E−3
12 17	−1.93E−1	15 6	3.57E−3	17 16	−6.15E−6	2 1	1.30E−1	4 5	−1.44E−1	6 9	−1.47E−1
12 18	−6.59E−3	15 7	2.39E−2	17 17	−3.52E−5	2 2	1.74E−2	4 6	−1.13E−1	6 10	−6.54E−1
12 19	−2.16E−2	15 8	4.51E−3	17 18	−5.10E−7	2 3	1.51E−3	4 7	−5.99E−2	6 11	−4.43E−2
12 20	−3.12E−5	15 9	9.22E−4	17 19	−1.33E−3	2 4	4.16E−3	4 8	−3.56E−1	6 12	−1.34E−3
12 21	−1.62E−1	15 10	4.62E−4	17 20	−1.50E−3	2 5	−2.50E−1	4 9	−1.42E−1	6 13	−2.66E−1
13 1	2.73E−4	15 11	3.98E−3	17 21	−7.95E−4	2 6	−1.09E−2	4 10	−9.58E−2	6 14	−7.05E−3
13 2	1.27E−2	15 12	5.11E−6	18 1	3.12E−1	2 7	−2.71E−1	4 11	−9.68E−3	6 15	−2.11E−1
13 3	3.34E−2	15 13	3.98E−6	18 2	7.28E−2	2 8	−1.00E+0	4 12	−1.92E−1	6 16	−2.27E−3
13 4	1.14E−1	15 14	−2.30E−4	18 3	2.47E−2	2 9	−1.39E−1	4 13	−9.76E−2	6 17	−9.50E−1
13 5	2.68E−1	15 15	4.41E−3	18 4	1.06E−2	2 10	−4.61E−2	4 14	−4.37E−2	6 18	−7.76E−1
13 6	1.76E−1	15 16	−1.47E−5	18 5	4.45E−3	2 11	−1.82E−1	4 15	−5.74E−1	6 19	−1.35E+0
13 7	1.21E−1	15 17	−9.37E−4	18 6	4.36E−5	2 12	−6.84E−1	4 16	−1.45E+0	6 20	−1.27E−1
13 8	4.16E−1	15 18	−3.02E−3	18 7	7.43E−4	2 13	−1.11E−2	4 17	−1.52E+0	6 21	−2.28E−3
13 9	1.94E−1	15 19	−1.00E−3	18 8	1.85E−5	2 14	−8.22E−1	4 18	−5.12E−1	6 22	−5.80E−6
13 10	5.71E−4	15 20	1.00E+0	18 9	8.71E−4	2 15	2.02E−1	4 19	−3.39E−1	6 23	−1.82E−2
13 11	6.01E−5	15 21	−2.32E−6	18 10	7.03E−3	2 16	−3.43E−2	4 20	−8.91E−2	6 24	−2.69E−4

P-like Fe (Fe^{11+})

i i'	gf_L	i i'	gf_L	i i'	gf_L	i i'	gf_L	i i'	gf_L	i i'	gf_L
7 1	2.40E−3	9 5	3.68E−4	11 9	8.46E−5	13 13	2.76E−2	15 17	6.00E−3	17 21	2.08E−5
7 2	1.08E+0	9 6	−1.95E−8	11 10	6.75E−3	13 14	2.69E−2	15 18	8.66E−3	17 22	5.44E−5
7 3	6.56E−2	9 7	−8.12E−5	11 11	1.40E+0	13 15	3.67E−2	15 19	6.16E−5	17 23	3.84E−4
7 4	1.32E+0	9 8	−7.91E−6	11 12	3.06E−1	13 16	1.39E−2	15 20	1.26E−4	17 24	−6.44E−6
7 5	−6.54E−5	9 9	−6.13E−3	11 13	3.98E−2	13 17	2.82E−5	15 21	1.18E−3	18 1	1.84E−2
7 6	−3.31E−4	9 10	−3.68E−4	11 14	1.63E−3	13 18	5.50E−4	15 22	1.19E−3	18 2	2.05E−3
7 7	−8.14E−2	9 11	−2.28E−4	11 15	4.99E−3	13 19	1.70E−7	15 23	−7.94E−6	18 3	1.48E−4
7 8	−3.04E−1	9 12	−5.29E−2	11 16	9.21E−5	13 20	−3.08E−4	15 24	−1.39E−8	18 4	9.01E−5
7 9	−6.46E−4	9 13	−5.78E−5	11 17	1.60E−7	13 21	−5.39E−5	16 1	5.63E−3	18 5	8.97E−1
7 10	−2.06E−2	9 14	−4.33E−2	11 18	−3.79E−5	13 22	−1.00E−4	16 2	7.65E−4	18 6	8.34E−2
7 11	−5.00E−3	9 15	−2.33E−2	11 19	−7.68E−4	13 23	−2.87E−4	16 3	1.08E−4	18 7	1.31E−2
7 12	−4.97E−3	9 16	−3.47E−2	11 20	−9.93E−4	13 24	−2.86E−6	16 4	2.83E−3	18 8	4.61E−1
7 13	−1.06E−1	9 17	−9.62E−2	11 21	−6.97E−1	14 1	1.73E−1	16 5	7.16E−1	18 9	9.13E−2
7 14	−2.30E−1	9 18	−8.66E−3	11 22	−8.02E−1	14 2	3.11E−3	16 6	4.45E−1	18 10	7.63E−2
7 15	−2.28E−1	9 19	−3.70E+0	11 23	−2.61E−4	14 3	8.63E−3	16 7	9.22E−2	18 11	3.73E−2
7 16	−8.22E−2	9 20	−1.16E+0	11 24	−1.33E−1	14 4	3.61E−3	16 8	1.72E−1	18 12	8.08E−3
7 17	−1.17E−1	9 21	−2.36E−3	12 1	2.05E−4	14 5	9.75E−1	16 9	5.91E−1	18 13	7.26E−4
7 18	−2.85E+0	9 22	−1.07E−2	12 2	2.63E−5	14 6	5.31E−1	16 10	2.55E−3	18 14	1.45E−1
7 19	−3.88E−2	9 23	−8.30E−1	12 3	8.52E−7	14 7	1.23E−2	16 11	1.55E−2	18 15	1.78E−2
7 20	−2.25E−2	9 24	−2.26E−3	12 4	1.15E−6	14 8	6.15E−2	16 12	1.41E−1	18 16	1.37E−2
7 21	−1.51E−3	10 1	2.32E−2	12 5	6.15E−2	14 9	1.17E−5	16 13	6.79E−3	18 17	3.66E−2
7 22	−2.81E−5	10 2	4.72E−3	12 6	8.18E−2	14 10	4.15E−1	16 14	4.12E−2	18 18	1.42E−2
7 23	−2.60E−2	10 3	8.50E−2	12 7	1.09E−1	14 11	7.00E−3	16 15	1.65E−1	18 19	1.05E−3
7 24	−5.13E−6	10 4	5.45E−2	12 8	2.10E−1	14 12	1.14E−1	16 16	3.90E−3	18 20	3.76E−3
8 1	6.54E−4	10 5	6.95E−4	12 9	2.67E−1	14 13	4.48E−2	16 17	1.10E−2	18 21	3.42E−4
8 2	4.67E+0	10 6	2.52E−5	12 10	4.64E−2	14 14	1.15E−4	16 18	5.30E−2	18 22	1.97E−6
8 3	6.02E−2	10 7	2.31E−5	12 11	2.70E−3	14 15	3.43E−3	16 19	1.45E−4	18 23	2.56E−4
8 4	1.74E−1	10 8	5.78E−4	12 12	2.18E−4	14 16	5.25E−3	16 20	5.41E−3	18 24	−8.05E−7
8 5	8.48E−4	10 9	1.22E−4	12 13	3.53E−3	14 17	2.57E−3	16 21	3.22E−5	19 1	2.69E−1
8 6	−7.02E−4	10 10	3.71E−3	12 14	1.88E−4	14 18	3.16E−2	16 22	2.76E−5	19 2	4.56E−2
8 7	−3.86E−3	10 11	3.69E−1	12 15	1.10E−4	14 19	2.14E−3	16 23	1.53E−4	19 3	3.80E−3
8 8	−3.72E−3	10 12	8.17E−2	12 16	2.53E−4	14 20	2.98E−4	16 24	−4.39E−7	19 4	1.14E−2
8 9	−6.59E−2	10 13	1.05E−2	12 17	7.98E−5	14 21	1.80E−4	17 1	1.08E−1	19 5	1.26E−2
8 10	−6.11E−2	10 14	2.85E−4	12 18	−3.89E−7	14 22	−6.16E−5	17 2	2.07E−3	19 6	3.69E−2
8 11	−1.39E−3	10 15	5.30E−4	12 19	−1.12E−5	14 23	−5.96E−4	17 3	5.56E−5	19 7	2.37E−3
8 12	−2.46E−2	10 16	−3.28E−7	12 20	−4.22E−4	14 24	−1.47E−6	17 4	3.32E−3	19 8	2.40E−2
8 13	−1.32E−2	10 17	−1.26E−5	12 21	−1.92E−4	15 1	8.87E−1	17 5	1.10E−2	19 9	9.54E−3
8 14	−3.87E−2	10 18	−3.37E−8	12 22	−9.96E−5	15 2	9.75E−3	17 6	1.15E−1	19 10	2.74E−4
8 15	−1.66E−2	10 19	−1.85E−5	12 23	−7.39E−6	15 3	3.71E−2	17 7	1.01E−3	19 11	5.80E−1
8 16	−1.96E−1	10 20	−1.58E−4	12 24	−5.70E−6	15 4	5.04E−5	17 8	5.80E−2	19 12	4.15E−1
8 17	−7.68E−4	10 21	−3.69E−1	13 1	2.54E−3	15 5	4.13E−2	17 9	1.57E−2	19 13	7.87E−2
8 18	−7.56E−1	10 22	−1.01E+0	13 2	6.16E−4	15 6	2.01E−1	17 10	1.38E−1	19 14	4.61E−2
8 19	−4.78E−1	10 23	−1.02E−3	13 3	6.98E−4	15 7	6.05E−3	17 11	4.55E−1	19 15	1.22E−2
8 20	−5.69E+0	10 24	−2.55E−2	13 4	4.87E−6	15 8	1.48E−3	17 12	1.59E−1	19 16	4.59E−3
8 21	−4.74E−3	11 1	1.22E−1	13 5	2.16E−2	15 9	1.42E−2	17 13	1.78E−1	19 17	1.94E−3
8 22	−2.86E−3	11 2	3.36E−2	13 6	1.98E−3	15 10	8.64E−2	17 14	3.55E−1	19 18	1.00E−5
8 23	−3.73E−1	11 3	9.47E−2	13 7	1.42E−1	15 11	8.10E−2	17 15	7.06E−2	19 19	3.39E−3
8 24	−2.53E−6	11 4	1.15E−2	13 8	8.01E−2	15 12	3.66E−4	17 16	7.95E−3	19 20	2.33E−3
9 1	3.22E−3	11 5	4.88E−5	13 9	6.45E−3	15 13	7.91E−3	17 17	1.14E−2	19 21	2.19E−3
9 2	5.95E−2	11 6	2.37E−4	13 10	2.39E−4	15 14	1.07E−2	17 18	2.26E−2	19 22	4.96E−3
9 3	3.74E+0	11 7	1.16E−4	13 11	7.73E−3	15 15	1.03E−2	17 19	4.04E−3	19 23	7.55E−5
9 4	1.01E+0	11 8	4.34E−3	13 12	4.09E−2	15 16	8.72E−3	17 20	2.74E−3	19 24	5.71E−4

P-like Fe (Fe^{11+})

i	i'	gf_L	i	i'	gf_L	i	i'	gf_L	i	i'	gf_L	i	i'	gf_L	i	i'	gf_L
20	1	9.01E−3	1	4	−1.99E−4	4	5	−3.05E−1	7	6	−3.41E−1	10	7	1.02E−5	13	8	5.96E−2
20	2	4.51E−1	1	5	−3.05E−3	4	6	−7.17E−2	7	7	−4.16E−4	10	8	7.62E−6	13	9	6.37E−2
20	3	1.48E−1	1	6	−3.07E−3	4	7	−1.09E+0	7	8	−8.35E−2	10	9	9.10E−5	13	10	1.15E−2
20	4	7.03E−1	1	7	−1.05E−3	4	8	−1.18E+0	7	9	−5.82E−3	10	10	1.59E−4	13	11	2.04E−3
20	5	9.45E−4	1	8	−1.88E−3	4	9	−4.84E−1	7	10	−9.12E−3	10	11	−1.27E−4	13	12	5.07E−3
20	6	3.87E−4	1	9	−8.24E−4	4	10	−4.81E−2	7	11	−9.81E−1	10	12	−6.12E−7	13	13	2.50E−4
20	7	6.64E−4	1	10	−5.20E−4	4	11	−6.83E−3	7	12	−1.73E+0	10	13	−4.63E−5	13	14	−2.81E−5
20	8	2.80E−3	1	11	−3.23E−3	4	12	−6.71E−1	7	13	−5.74E+0	10	14	−4.59E+0	13	15	−4.15E−5
20	9	6.44E−5	1	12	−1.35E−3	4	13	−2.60E−1	7	14	−6.58E−3	10	15	−1.11E−2	13	16	−1.01E−4
20	10	5.87E−6	1	13	−1.18E−3	4	14	−6.99E−3	7	15	−2.53E−2	10	16	−1.25E−1	13	17	−1.11E−2
20	11	2.71E−3	1	14	−1.19E+0	4	15	−5.96E−3	7	16	−2.40E−5	10	17	−4.39E−1	14	1	5.72E−3
20	12	4.52E−4	1	15	−1.03E−3	4	16	−1.28E−4	7	17	−2.50E−3	11	1	1.43E−2	14	2	1.75E−2
20	13	2.41E−5	1	16	−2.46E+0	4	17	−2.09E−4	8	1	5.33E−4	11	2	3.14E−1	14	3	2.44E−1
20	14	1.46E−4	1	17	−1.06E+0	5	1	2.35E+0	8	2	1.42E+0	11	3	1.54E−4	14	4	9.23E−2
20	15	2.65E−6	2	1	4.06E−1	5	2	1.91E−2	8	3	−3.34E−4	11	4	2.19E−4	14	5	8.15E−3
20	16	6.45E−7	2	2	2.34E−3	5	3	−9.43E−2	8	4	−1.46E−4	11	5	1.93E−5	14	6	2.65E−1
20	17	2.68E−4	2	3	−2.81E−2	5	4	−2.89E−1	8	5	−2.38E−2	11	6	4.41E−4	14	7	1.43E−1
20	18	2.49E−3	2	4	−8.50E−1	5	5	−8.07E−2	8	6	−2.73E−3	11	7	2.25E−5	14	8	7.10E−3
20	19	9.81E−4	2	5	−2.36E−1	5	6	−1.03E−1	8	7	−1.38E−1	11	8	5.20E−4	14	9	5.76E−3
20	20	1.70E−4	2	6	−1.38E−6	5	7	−7.17E−1	8	8	−9.37E−2	11	9	1.17E−5	14	10	1.35E−1
20	21	2.93E−1	2	7	−8.81E−2	5	8	−8.86E−2	8	9	−7.89E−2	11	10	1.43E−4	14	11	8.94E−3
20	22	7.09E−5	2	8	−3.63E−1	5	9	−2.34E−1	8	10	−7.38E−1	11	11	2.21E−4	14	12	1.42E−2
20	23	1.88E−4	2	9	−4.64E−1	5	10	−1.08E+0	8	11	−1.15E−2	11	12	1.31E−4	14	13	6.15E−5
20	24	5.49E−4	2	10	−1.02E−2	5	11	−8.19E−1	8	12	−1.10E+0	11	13	−8.17E−5	14	14	9.30E−4
21	1	1.90E−3	2	11	−2.49E−1	5	12	−2.44E−3	8	13	−2.73E+0	11	14	−4.24E−1	14	15	5.33E−4
21	2	7.48E−4	2	12	−4.68E−1	5	13	−3.75E−2	8	14	−1.97E−3	11	15	−1.76E−2	14	16	2.23E−5
21	3	1.95E−3	2	13	−4.38E−2	5	14	−8.70E−3	8	15	−4.24E−1	11	16	−3.34E+0	14	17	−4.72E−1
21	4	1.65E−6	2	14	−4.42E−5	5	15	−2.81E−5	8	16	−2.87E−2	11	17	−2.35E−1	15	1	3.72E−2
21	5	6.86E−2	2	15	−4.81E−3	5	16	−5.32E−5	8	17	−8.11E−4	12	1	2.60E−4	15	2	5.63E−2
21	6	1.11E−1	2	16	−2.87E−3	5	17	−5.53E−4	9	1	1.24E−3	12	2	8.73E−7	15	3	6.14E−2
21	7	4.74E−2	2	17	−1.53E−4	6	1	1.30E+0	9	2	2.32E+0	12	3	3.23E−3	15	4	3.42E−1
21	8	6.49E−1	3	1	5.02E−1	6	2	5.08E−1	9	3	6.83E−4	12	4	1.12E−2	15	5	5.14E−2
21	9	1.18E+0	3	2	1.25E−4	6	3	−5.10E−2	9	4	−1.01E−7	12	5	3.76E−2	15	6	3.65E−2
21	10	2.97E−2	3	3	−1.04E+0	6	4	−4.89E−6	9	5	−1.30E−2	12	6	1.35E−1	15	7	7.78E−2
21	11	3.04E−2	3	4	−1.18E−1	6	5	−3.72E−2	9	6	−1.09E−2	12	7	8.14E−2	15	8	1.65E−3
21	12	1.28E−1	3	5	−1.40E−2	6	6	−4.34E−1	9	7	−4.40E−5	12	8	1.73E−2	15	9	1.93E−2
21	13	1.27E−1	3	6	−1.66E+0	6	7	−4.09E−1	9	8	−9.57E−5	12	9	9.48E−4	15	10	3.48E−2
21	14	1.36E−1	3	7	−2.65E−1	6	8	−7.32E−1	9	9	−5.15E−2	12	10	3.16E−4	15	11	5.12E−5
21	15	3.80E−2	3	8	−5.08E−1	6	9	−1.77E−4	9	10	−7.19E−3	12	11	9.85E−4	15	12	1.29E−2
21	16	1.92E−1	3	9	−3.16E−3	6	10	−3.39E−1	9	11	−4.05E−1	12	12	1.20E−4	15	13	4.07E−5
21	17	2.27E−2	3	10	−7.08E−1	6	11	−2.65E+0	9	12	−7.35E−2	12	13	−2.41E−5	15	14	5.23E−3
21	18	3.22E−2	3	11	−4.64E−1	6	12	−7.18E−1	9	13	−3.37E−1	12	14	−4.23E−4	15	15	3.88E−5
21	19	5.18E−2	3	12	−1.20E−1	6	13	−8.87E−1	9	14	−1.15E−4	12	15	−8.16E−6	15	16	5.00E−4
21	20	3.77E−4	3	13	−1.87E−1	6	14	−1.39E−2	9	15	−5.08E+0	12	16	−4.37E−4	15	17	−2.21E+0
21	21	3.07E−4	3	14	−8.67E−4	6	15	−2.17E−2	9	16	−9.96E−2	12	17	−6.03E−8	16	1	8.52E−4
21	22	1.06E−3	3	15	−1.37E−3	6	16	−2.85E−3	9	17	−1.10E−3	13	1	2.65E−4	16	2	1.31E−3
21	23	3.21E−3	3	16	−2.87E−6	6	17	−1.23E−5	10	1	3.68E−1	13	2	6.96E−4	16	3	2.10E+0
21	24	6.56E−5	3	17	−2.77E−3	7	1	1.98E−1	10	2	7.07E−2	13	3	3.13E−1	16	4	3.07E−1
		$^2D^o$–$^2F^e$	4	1	3.13E−1	7	2	1.54E+0	10	3	2.12E−4	13	4	3.68E−1	16	5	7.40E−2
1	1	−2.29E−4	4	2	1.04E−1	7	3	−2.40E−3	10	4	1.71E−5	13	5	1.12E−1	16	6	1.78E−1
1	2	−7.90E+0	4	3	−1.22E−2	7	4	−7.80E−2	10	5	3.04E−4	13	6	3.02E−1	16	7	1.03E−1
1	3	−1.66E−5	4	4	−1.40E−1	7	5	−2.45E−2	10	6	3.24E−4	13	7	1.60E−1	16	8	6.91E−2

P-like Fe (Fe^{11+})

i	i'	gf_L	i	i'	gf_L	i	i'	gf_L	i	i'	gf_L	i	i'	gf_L	i	i'	gf_L
16	9	1.68E−2	19	10	1.97E−2	1	10	−1.94E−4	4	11	−1.70E−3	7	12	−6.67E−2	10	13	−1.58E−1
16	10	7.51E−2	19	11	2.96E−4	1	11	−7.57E−4	4	12	−2.41E−4	7	13	−2.01E−1	10	14	−7.10E−3
16	11	5.62E−2	19	12	3.64E−3	1	12	−4.16E−7	4	13	−3.28E−4	7	14	−9.70E−1	10	15	−3.51E−1
16	12	2.58E−2	19	13	1.79E−3	1	13	−5.55E−4	4	14	−1.34E−5	7	15	−5.51E−3	10	16	−1.65E−3
16	13	1.90E−3	19	14	2.60E−6	1	14	−1.24E−3	4	15	−2.36E−3	7	16	−4.68E−3	10	17	−5.81E−4
16	14	4.42E−5	19	15	6.67E−5	1	15	−4.42E−3	4	16	−4.79E−1	7	17	−8.90E−4	11	1	1.05E−1
16	15	1.45E−3	19	16	5.00E−3	1	16	−7.77E−1	4	17	−4.44E+0	8	1	1.27E−1	11	2	3.14E−1
16	16	1.02E−7	19	17	−4.91E−3	1	17	−3.21E−1	5	1	4.65E−1	8	2	4.88E−2	11	3	2.18E−1
16	17	−9.05E−3	20	1	8.57E−3	2	1	−7.96E−2	5	2	8.03E−2	8	3	5.08E−1	11	4	7.66E−3
17	1	3.59E−2	20	2	3.80E−2	2	2	−4.86E−2	5	3	2.35E−3	8	4	1.09E−1	11	5	1.14E−2
17	2	3.42E−5	20	3	8.30E−5	2	3	−3.14E−1	5	4	1.00E−4	8	5	4.87E−3	11	6	3.23E−2
17	3	1.26E−2	20	4	4.17E−4	2	4	−1.46E+0	5	5	3.67E−3	8	6	3.93E−4	11	7	6.47E−4
17	4	1.93E+0	20	5	2.40E−4	2	5	−9.05E−1	5	6	2.07E−5	8	7	1.01E−4	11	8	−2.35E+0
17	5	2.27E−2	20	6	3.46E−3	2	6	−3.70E−1	5	7	−3.39E−5	8	8	−1.61E−3	11	9	−1.54E−2
17	6	1.37E−1	20	7	7.81E−5	2	7	−4.43E−2	5	8	−2.71E−4	8	9	−8.71E−3	11	10	−3.99E−2
17	7	1.23E−1	20	8	1.08E−3	2	8	−7.08E−3	5	9	−1.03E−2	8	10	−3.47E−1	11	11	−3.33E−4
17	8	1.68E−2	20	9	1.39E−4	2	9	−1.92E−5	5	10	−4.26E−1	8	11	−8.53E−1	11	12	−3.89E−2
17	9	4.29E−2	20	10	7.99E−4	2	10	−3.11E−5	5	11	−7.22E−2	8	12	−2.28E−1	11	13	−1.61E−1
17	10	1.24E−1	20	11	1.35E−3	2	11	−1.45E−3	5	12	−6.40E−1	8	13	−8.67E−3	11	14	−3.07E−2
17	11	6.59E−3	20	12	1.56E−5	2	12	−8.94E−4	5	13	−2.10E−2	8	14	−4.04E−1	11	15	−6.53E−3
17	12	1.00E−2	20	13	3.54E−3	2	13	−5.23E−4	5	14	−1.50E+0	8	15	−1.09E−1	11	16	−2.78E−3
17	13	4.99E−5	20	14	8.05E−4	2	14	−4.02E−4	5	15	−4.65E−5	8	16	−2.50E−5	11	17	−7.42E−3
17	14	3.81E−4	20	15	8.33E−4	2	15	−1.51E−2	5	16	−8.05E−7	8	17	−1.63E−4	12	1	4.51E−1
17	15	1.12E−5	20	16	1.82E−3	2	16	−3.63E+0	5	17	−3.93E−4	9	1	2.60E+0	12	2	1.09E−1
17	16	1.87E−4	20	17	−1.28E−4	2	17	−5.00E−1	6	1	4.83E−3	9	2	1.20E−2	12	3	5.72E−1
17	17	−5.73E−2	21	1	1.09E−3	3	1	−1.34E−1	6	2	1.14E−1	9	3	2.22E−1	12	4	5.10E−2
18	1	1.88E−3	21	2	1.89E−4	3	2	−7.43E−2	6	3	3.13E−1	9	4	5.41E−2	12	5	2.62E−1
18	2	2.14E−3	21	3	6.45E−1	3	3	−1.64E−1	6	4	6.63E−1	9	5	6.05E−1	12	6	1.61E−1
18	3	3.48E−1	21	4	3.56E−1	3	4	−3.47E−3	6	5	2.01E−2	9	6	7.03E−3	12	7	1.19E−3
18	4	1.11E+0	21	5	4.25E−1	3	5	−3.37E−1	6	6	1.10E−3	9	7	3.19E−4	12	8	−5.18E−1
18	5	1.34E+0	21	6	3.19E−2	3	6	−3.86E+0	6	7	1.88E−6	9	8	−1.04E−5	12	9	−3.89E−2
18	6	5.71E−1	21	7	1.28E−2	3	7	−1.05E+0	6	8	−1.73E−3	9	9	−2.38E−2	12	10	−6.19E−2
18	7	2.23E−2	21	8	3.27E−2	3	8	−8.97E−3	6	9	−2.09E−3	9	10	−4.90E−1	12	11	−5.95E−3
18	8	2.32E−1	21	9	1.95E−1	3	9	−4.81E−4	6	10	−3.59E−1	9	11	−3.65E−1	12	12	−2.59E−3
18	9	7.72E−2	21	10	1.33E−1	3	10	−2.36E−4	6	11	−4.76E−2	9	12	−3.78E−4	12	13	−7.87E−1
18	10	1.19E−1	21	11	8.41E−3	3	11	−1.10E−4	6	12	−2.06E−1	9	13	−1.32E−1	12	14	−4.00E−1
18	11	1.04E−1	21	12	2.51E−4	3	12	−1.60E−6	6	13	−5.64E−1	9	14	−8.63E−2	12	15	−2.75E−2
18	12	2.02E−2	21	13	9.74E−2	3	13	−1.06E−4	6	14	−1.00E−1	9	15	−2.81E−2	12	16	−9.23E−3
18	13	8.30E−3	21	14	4.99E−9	3	14	−2.52E−5	6	15	−1.48E+0	9	16	−1.10E−3	12	17	−6.17E−3
18	14	9.89E−5	21	15	2.08E−4	3	15	−1.77E−3	6	16	−8.64E−3	9	17	−7.76E−8	13	1	5.49E−2
18	15	8.77E−5	21	16	2.05E−5	3	16	−1.83E+0	6	17	−2.47E−4	10	1	7.74E−2	13	2	3.07E+0
18	16	1.82E−5	21	17	−4.26E−5	3	17	−1.89E+0	7	1	6.30E−4	10	2	3.07E−2	13	3	1.18E−1
18	17	−9.25E−3			$^2D^e$–$^2F^o$	4	1	−7.33E−4	7	2	4.59E−2	10	3	1.03E−1	13	4	3.80E−2
19	1	2.85E−1	1	1	−2.15E−1	4	2	−1.37E−1	7	3	8.06E−2	10	4	4.38E−1	13	5	4.09E−2
19	2	3.93E−3	1	2	−2.43E−1	4	3	−2.94E−2	7	4	8.78E−2	10	5	3.43E−2	13	6	4.67E−4
19	3	2.18E−2	1	3	−4.69E+0	4	4	−1.94E−1	7	5	3.11E−2	10	6	1.70E−4	13	7	1.38E−2
19	4	1.31E−1	1	4	−1.92E+0	4	5	−2.90E−4	7	6	1.92E−2	10	7	5.91E−3	13	8	−5.96E−2
19	5	1.70E−2	1	5	−3.60E−1	4	6	−3.72E−1	7	7	1.66E−3	10	8	−1.35E−2	13	9	−6.04E−3
19	6	1.20E−5	1	6	−7.11E−2	4	7	−7.90E+0	7	8	−1.61E−4	10	9	−1.51E−1	13	10	−1.35E−1
19	7	1.36E−3	1	7	−2.83E−2	4	8	−2.28E−2	7	9	−1.43E+0	10	10	−2.34E−2	13	11	−2.64E−2
19	8	2.71E−3	1	8	−1.18E−2	4	9	−1.34E−4	7	10	−8.39E−2	10	11	−1.77E−2	13	12	−3.28E−1
19	9	1.03E−5	1	9	−1.66E−6	4	10	−2.78E−4	7	11	−1.32E−4	10	12	−1.23E+0	13	13	−2.22E−2

P-like Fe (Fe^{11+})

i	i'	gf_L	i	i'	gf_L	i	i'	gf_L	i	i'	gf_L	i	i'	gf_L	i	i'	gf_L
13	14	−5.66E−1	16	15	−1.36E−1	19	16	−3.62E−4	22	17	−1.75E+0	1	17	−8.21E−2	5	1	2.51E+0
13	15	−5.17E−1	16	16	−3.91E−3	19	17	−2.11E−4	23	1	3.25E−3	2	1	1.15E+0	5	2	6.52E−1
13	16	−5.02E−4	16	17	−1.82E−5	20	1	1.97E−2	23	2	7.30E−3	2	2	5.14E−3	5	3	−2.69E−2
13	17	−1.48E−3	17	1	4.23E−2	20	2	8.25E−4	23	3	5.24E−9	2	3	−2.93E−2	5	4	−1.16E−2
14	1	2.70E−1	17	2	3.05E−2	20	3	9.61E−3	23	4	5.77E−3	2	4	−6.46E−1	5	5	−2.34E−1
14	2	2.99E−1	17	3	3.55E−1	20	4	1.02E−1	23	5	6.23E−2	2	5	−2.93E−1	5	6	−2.26E−1
14	3	2.67E−1	17	4	1.05E−1	20	5	1.59E+0	23	6	8.81E−1	2	6	−1.84E−1	5	7	−3.42E−1
14	4	1.22E−1	17	5	1.14E−5	20	6	1.74E+0	23	7	3.11E+0	2	7	−4.00E−3	5	8	−3.85E−1
14	5	1.37E+0	17	6	1.53E−3	20	7	4.16E−1	23	8	1.38E−3	2	8	−2.20E+0	5	9	−1.84E−2
14	6	3.12E−1	17	7	1.08E−1	20	8	1.33E−3	23	9	2.56E−3	2	9	−2.01E+0	5	10	−4.31E+0
14	7	8.96E−4	17	8	−1.19E−5	20	9	3.28E−4	23	10	4.66E−5	2	10	−1.05E−1	5	11	−1.11E+0
14	8	−2.19E−3	17	9	−3.45E−6	20	10	1.16E−4	23	11	3.31E−3	2	11	−1.78E+0	5	12	−2.22E−2
14	9	−1.56E−2	17	10	−2.57E−2	20	11	1.36E−3	23	12	−2.97E−4	2	12	−4.21E−2	5	13	−8.70E−2
14	10	−3.38E−2	17	11	−1.46E−2	20	12	−4.37E−3	23	13	−2.87E−4	2	13	−7.87E−2	5	14	−2.18E−4
14	11	−9.16E−6	17	12	−3.10E−2	20	13	−3.40E−3	23	14	−5.68E−4	2	14	−7.67E−4	5	15	−4.64E−4
14	12	−1.03E−6	17	13	−1.12E−1	20	14	−4.21E−4	23	15	−1.84E−3	2	15	−2.53E−3	5	16	−4.62E−3
14	13	−1.90E−1	17	14	−4.18E−1	20	15	−9.07E−3	23	16	−1.35E−3	2	16	−7.26E−5	5	17	−5.65E−3
14	14	−8.68E−3	17	15	−9.06E−3	20	16	−9.19E−3	23	17	−1.04E−2	2	17	−7.99E−3	6	1	1.34E+0
14	15	−1.57E−1	17	16	−6.24E−5	20	17	−2.39E−3	24	1	5.33E−4	3	1	7.33E−3	6	2	1.96E+0
14	16	−7.98E−4	17	17	−5.48E−6	21	1	3.07E−4	24	2	9.31E−3	3	2	2.27E−1	6	3	−4.84E−3
14	17	−2.14E−3	18	1	7.14E−2	21	2	6.31E−6	24	3	2.00E−3	3	3	−9.11E−2	6	4	−9.64E−4
15	1	1.16E−2	18	2	1.06E−2	21	3	6.52E−4	24	4	1.32E−2	3	4	−1.50E−1	6	5	−5.80E−2
15	2	1.86E−2	18	3	4.80E−3	21	4	1.07E−2	24	5	1.17E−5	3	5	−7.31E−3	6	6	−1.79E−2
15	3	1.84E+0	18	4	7.51E−2	21	5	1.57E−3	24	6	1.31E−3	3	6	−2.84E−1	6	7	−8.82E−4
15	4	2.03E+0	18	5	6.92E−2	21	6	1.93E−2	24	7	2.91E−3	3	7	−6.99E−1	6	8	−2.23E−1
15	5	5.19E−1	18	6	7.19E−1	21	7	1.33E−2	24	8	9.81E−2	3	8	−4.61E−1	6	9	−1.30E−4
15	6	7.32E−2	18	7	1.26E+0	21	8	9.86E−5	24	9	9.46E−9	3	9	−5.65E−2	6	10	−2.22E−2
15	7	1.18E−1	18	8	3.62E−5	21	9	1.17E−6	24	10	1.66E−9	3	10	−5.60E−1	6	11	−1.46E+0
15	8	−6.57E−3	18	9	−2.56E−3	21	10	2.03E−6	24	11	1.62E−5	3	11	−1.02E−1	6	12	−3.20E−1
15	9	−8.99E−5	18	10	−6.64E−3	21	11	4.53E−7	24	12	2.26E−6	3	12	−3.75E−1	6	13	−9.19E−1
15	10	−1.86E−2	18	11	−3.79E−4	21	12	−9.90E−5	24	13	1.15E−7	3	13	−1.69E−1	6	14	−1.25E−2
15	11	−1.77E−2	18	12	−9.06E−2	21	13	−1.09E−4	24	14	−4.65E−5	3	14	−1.47E−3	6	15	−4.34E+0
15	12	−7.42E−5	18	13	−3.20E−2	21	14	−1.69E−6	24	15	−9.11E−5	3	15	−4.70E−2	6	16	−1.58E−3
15	13	−6.38E−2	18	14	−9.82E−2	21	15	−1.06E−2	24	16	−3.09E−3	3	16	−3.89E−3	6	17	−8.59E−5
15	14	−5.59E−3	18	15	−1.39E−2	21	16	−2.21E+0	24	17	−3.26E−3	3	17	−1.14E−1	7	1	9.51E−1
15	15	−2.21E−1	18	16	−1.06E−3	21	17	−1.08E+0			$^2F^o$–$^2F^e$	4	1	1.78E−1	7	2	1.31E−1
15	16	−2.29E−5	18	17	−3.84E−4	22	1	2.83E−3	1	1	7.20E−1	4	2	4.11E−1	7	3	−1.94E−5
15	17	−2.17E−4	19	1	4.40E−3	22	2	1.10E−3	1	2	3.40E−2	4	3	−9.19E−2	7	4	−5.00E−4
16	1	3.78E−3	19	2	4.06E−3	22	3	7.79E−3	1	3	−1.28E+0	4	4	−3.05E−2	7	5	−1.21E−3
16	2	4.03E−4	19	3	5.11E−1	22	4	8.28E−3	1	4	−3.88E−2	4	5	−1.04E−6	7	6	−1.17E−2
16	3	5.20E−1	19	4	2.18E−1	22	5	1.51E−2	1	5	−6.55E−1	4	6	−1.43E−3	7	7	−3.01E−2
16	4	2.93E+0	19	5	4.22E−3	22	6	1.63E−3	1	6	−8.25E−1	4	7	−1.11E−1	7	8	−9.92E−2
16	5	5.31E−1	19	6	1.25E−1	22	7	2.55E−2	1	7	−8.23E−3	4	8	−4.49E−4	7	9	−1.31E−2
16	6	9.93E−3	19	7	2.04E−1	22	8	2.27E−1	1	8	−4.65E−1	4	9	−2.83E−2	7	10	−1.08E−3
16	7	9.89E−2	19	8	7.42E−5	22	9	1.65E−5	1	9	−2.43E−1	4	10	−2.20E−1	7	11	−7.00E−1
16	8	−1.82E−7	19	9	6.54E−7	22	10	1.99E−5	1	10	−9.29E−1	4	11	1.22E 1	7	12	3.01E 1
16	9	−3.47E−3	19	10	−2.22E−4	22	11	2.66E−5	1	11	−9.02E−1	4	12	−4.66E+0	7	13	−8.22E−1
16	10	−2.09E−2	19	11	−4.27E−3	22	12	−4.26E−8	1	12	−2.45E−1	4	13	−1.14E+0	7	14	−6.13E−4
16	11	−8.61E−3	19	12	−9.33E−5	22	13	−1.13E−6	1	13	−5.82E−2	4	14	−2.07E−7	7	15	−3.34E−3
16	12	−3.72E−2	19	13	−9.26E−3	22	14	−3.45E−5	1	14	−3.69E−3	4	15	−2.04E−3	7	16	−2.41E−3
16	13	−4.63E−2	19	14	3.45E 2	22	15	2.01E 3	1	15	5.28E−4	4	16	−3.30E−3	7	17	−1.93E−5
16	14	−2.22E−3	19	15	−1.81E−1	22	16	−5.17E−1	1	16	−8.50E−4	4	17	−1.23E−2	8	1	1.05E−1

P-like Fe (Fe^{11+})

i i′	gf$_L$	*i i′*	gf$_L$	*i i′*	gf$_L$	*i i′*	gf$_L$	*i i′*	gf$_L$	*i i′*	gf$_L$
8 2	9.48E−2	11 3	6.40E−1	14 4	3.42E+0	17 5	2.14E−3	4 6	−4.30E−2	9 3	1.36E+0
8 3	4.07E−4	11 4	5.32E−1	14 5	8.13E−1	17 6	5.85E−4	4 7	−3.31E+0	9 4	1.52E−3
8 4	1.58E−4	11 5	1.12E+0	14 6	3.92E−1	17 7	1.75E−3	4 8	−1.05E+0	9 5	1.93E−2
8 5	3.55E−5	11 6	3.12E−1	14 7	2.06E−1	17 8	1.64E−3	4 9	−1.29E+0	9 6	9.66E−5
8 6	3.78E−4	11 7	1.76E−1	14 8	1.12E−1	17 9	5.41E−4	4 10	−9.46E−2	9 7	4.30E−7
8 7	7.88E−5	11 8	3.08E−1	14 9	3.09E−1	17 10	2.11E−3	4 11	−1.15E−3	9 8	1.51E−3
8 8	6.76E−6	11 9	8.43E−2	14 10	2.41E−2	17 11	2.96E−4	5 1	1.95E+0	9 9	1.03E−5
8 9	9.95E−5	11 10	5.75E−2	14 11	1.29E−1	17 12	6.23E−5	5 2	−2.00E−2	9 10	−3.03E−3
8 10	2.01E−4	11 11	2.43E−2	14 12	1.08E−4	17 13	6.23E−4	5 3	−1.41E−1	9 11	−5.44E−6
8 11	1.43E−5	11 12	8.48E−3	14 13	1.64E−1	17 14	5.04E−1	5 4	−3.59E−2	10 1	6.29E−5
8 12	3.98E−6	11 13	8.76E−5	14 14	1.70E−5	17 15	4.42E−3	5 5	−1.96E+0	10 2	2.05E+0
8 13	−6.03E−6	11 14	−2.24E−7	14 15	2.20E−3	17 16	2.21E−1	5 6	−1.43E−1	10 3	1.55E−1
8 14	−9.68E−2	11 15	−1.37E−7	14 16	4.02E−7	17 17	1.99E−4	5 7	−1.71E−2	10 4	5.83E−1
8 15	−6.93E−3	11 16	−7.16E−7	14 17	−3.60E−5		^{2}F^o–^{2}G^e	5 8	−1.33E−1	10 5	3.71E−1
8 16	−1.56E+0	11 17	−1.02E−3	15 1	7.27E−4	1 1	1.43E−2	5 9	−2.80E+0	10 6	9.49E−3
8 17	−2.64E−2	12 1	3.89E−4	15 2	9.01E−3	1 2	−5.82E−1	5 10	−3.32E−2	10 7	9.96E−3
9 1	6.68E−5	12 2	8.32E−6	15 3	1.22E+0	1 3	−4.14E−1	5 11	−4.52E−2	10 8	5.07E−3
9 2	2.24E−5	12 3	1.02E+0	15 4	5.41E−4	1 4	−8.72E−3	6 1	8.75E−2	10 9	1.40E−3
9 3	4.48E−1	12 4	5.84E−1	15 5	1.07E−1	1 5	−6.04E−1	6 2	−1.47E−5	10 10	2.27E−4
9 4	1.09E−1	12 5	3.80E−1	15 6	3.40E−1	1 6	−1.10E+0	6 3	−1.78E−3	10 11	−3.26E−6
9 5	3.77E−1	12 6	3.57E−2	15 7	2.93E−1	1 7	−6.70E−1	6 4	−2.84E−1	11 1	1.02E−4
9 6	8.89E−2	12 7	9.25E−3	15 8	2.31E−3	1 8	−8.19E−2	6 5	−1.47E−2	11 2	1.53E+0
9 7	5.88E−2	12 8	5.49E−1	15 9	1.97E+0	1 9	−2.99E−1	6 6	−7.21E−1	11 3	8.11E−2
9 8	4.23E−2	12 9	6.47E−1	15 10	8.05E−2	1 10	−4.29E−3	6 7	−4.00E−1	11 4	1.95E−2
9 9	4.98E−5	12 10	3.72E−2	15 11	2.53E−1	1 11	−3.80E−3	6 8	−2.73E+0	11 5	1.77E−2
9 10	2.36E−2	12 11	6.64E−2	15 12	8.54E−2	2 1	1.12E−2	6 9	−1.83E+0	11 6	1.38E−2
9 11	2.36E−4	12 12	5.35E−2	15 13	1.84E−5	2 2	−2.33E−1	6 10	−1.34E+0	11 7	1.03E−2
9 12	2.33E−3	12 13	2.00E−5	15 14	2.72E−4	2 3	−4.06E−1	6 11	−9.29E−4	11 8	2.18E−3
9 13	−7.30E−5	12 14	4.58E−6	15 15	2.94E−3	2 4	−1.09E+0	7 1	1.15E−2	11 9	8.26E−8
9 14	−3.66E−6	12 15	6.93E−6	15 16	2.27E−3	2 5	−1.65E−2	7 2	9.18E−4	11 10	4.12E−8
9 15	−9.34E−4	12 16	7.50E−6	15 17	−1.88E−7	2 6	−1.42E−1	7 3	−7.65E−3	11 11	−3.64E−6
9 16	−1.68E−6	12 17	−5.43E−6	16 1	1.88E−1	2 7	−8.13E−1	7 4	−6.83E−2	12 1	1.40E−5
9 17	−8.42E−4	13 1	1.97E−3	16 2	1.73E+0	2 8	−1.74E−1	7 5	−5.15E−3	12 2	6.10E−1
10 1	8.99E−5	13 2	1.49E−3	16 3	1.46E−2	2 9	−7.05E−2	7 6	−9.17E−2	12 3	2.85E−1
10 2	1.12E−4	13 3	1.55E+0	16 4	8.65E−7	2 10	−1.19E−3	7 7	−3.05E−1	12 4	1.57E+0
10 3	1.42E−1	13 4	9.41E−3	16 5	2.33E−4	2 11	−5.53E−3	7 8	−1.85E+0	12 5	1.81E−1
10 4	4.37E−1	13 5	1.96E−1	16 6	1.76E−3	3 1	4.21E−1	7 9	−1.13E+0	12 6	2.77E−3
10 5	4.50E−1	13 6	8.97E−2	16 7	4.01E−6	3 2	−1.20E−4	7 10	−9.63E+0	12 7	9.03E−2
10 6	1.88E−1	13 7	1.75E−4	16 8	6.29E−4	3 3	−4.04E−1	7 11	−1.86E−3	12 8	2.55E−3
10 7	2.37E−1	13 8	7.76E−2	16 9	1.25E−2	3 4	−9.41E−1	8 1	7.40E−1	12 9	1.47E−2
10 8	2.15E−2	13 9	8.50E−2	16 10	5.87E−4	3 5	−9.98E−1	8 2	6.14E−5	12 10	1.31E−3
10 9	2.34E−3	13 10	2.50E−1	16 11	3.01E−3	3 6	−2.36E+0	8 3	2.22E−3	12 11	2.38E−6
10 10	4.60E−2	13 11	6.55E−2	16 12	6.29E−4	3 7	−5.54E−1	8 4	2.04E−4	13 1	6.37E−4
10 11	4.57E−3	13 12	2.06E−3	16 13	2.31E−3	3 8	−3.32E−1	8 5	1.38E−3	13 2	3.20E−1
10 12	9.42E−3	13 13	1.19E−2	16 14	5.58E−2	3 9	−2.55E+0	8 6	2.91E−4	13 3	5.80E−1
10 13	4.47E−4	13 14	2.64E−7	16 15	9.49E−5	3 10	−1.85E−1	8 7	1.28E−5	13 4	3.57E−2
10 14	−5.94E−7	13 15	1.96E−3	16 16	5.51E−1	3 11	−4.59E−3	8 8	1.80E−5	13 5	2.21E−1
10 15	−8.01E−6	13 16	7.17E−5	16 17	6.41E−5	4 1	3.38E+0	8 9	−1.88E−6	13 6	1.21E−1
10 16	−7.31E−6	13 17	−7.71E−5	17 1	1.40E+0	4 2	−1.62E−1	8 10	−6.15E−4	13 7	3.22E−2
10 17	−1.57E−7	14 1	2.73E−4	17 2	4.15E−2	4 3	−2.87E−2	8 11	−7.26E+0	13 8	6.52E−1
11 1	4.22E−4	14 2	5.85E−7	17 3	3.49E−5	4 4	−1.19E−1	9 1	5.12E−5	13 9	1.13E−3
11 2	7.81E−5	14 3	8.55E−2	17 4	2.84E−5	4 5	−6.64E−1	9 2	3.34E−1	13 10	6.36E−3

P-like Fe (Fe^{11+})

i i'	gf_L	i i'	gf_L	i i'	gf_L	i i'	gf_L	i i'	gf_L	i i'	gf_L
13 11	4.58E−6	1 7	−1.29E−4	6 9	−7.19E−1	12 1	4.54E−3	17 3	1.44E−3	4 11	−2.35E−3
14 1	2.58E−4	1 8	−3.22E−3	6 10	−3.43E−6	12 2	8.60E−5	17 4	5.06E−6	5 1	4.32E−4
14 2	8.49E−2	1 9	−5.78E−5	7 1	2.54E−2	12 3	1.98E+0	17 5	4.73E−3	5 2	5.18E−2
14 3	1.80E−1	1 10	−8.17E+0	7 2	2.56E+0	12 4	4.15E−2	17 6	1.20E−6	5 3	2.81E+0
14 4	3.12E−1	2 1	−7.13E−2	7 3	6.40E−1	12 5	−9.37E−3	17 7	9.88E−5	5 4	1.55E−1
14 5	3.87E−1	2 2	−4.20E−2	7 4	5.71E−3	12 6	−1.17E−1	17 8	5.31E−4	5 5	1.45E−1
14 6	4.46E−2	2 3	−1.18E+0	7 5	−1.37E−1	12 7	−3.43E−2	17 9	1.20E−5	5 6	1.80E−2
14 7	1.13E−1	2 4	−1.10E+1	7 6	−1.47E−1	12 8	−1.69E−2	17 10	3.47E−4	5 7	1.44E−2
14 8	1.21E−3	2 5	−4.46E−4	7 7	−1.42E+0	12 9	−1.43E−1		$^2G^o$–$^2G^e$	5 8	1.96E−4
14 9	5.25E−2	2 6	−3.36E−4	7 8	−5.80E−2	12 10	−9.84E−6	1 1	1.97E−2	5 9	2.32E−3
14 10	1.31E−3	2 7	−4.08E−6	7 9	−1.03E+0	13 1	5.62E−4	1 2	−3.36E−2	5 10	−2.79E−4
14 11	1.07E−6	2 8	−3.08E−4	7 10	−4.24E−3	13 2	1.40E−2	1 3	−4.01E−1	5 11	−3.91E−5
15 1	8.50E−5	2 9	−1.04E−3	8 1	9.93E−2	13 3	9.82E−3	1 4	−3.18E+0	6 1	1.74E−5
15 2	3.25E−1	2 10	−1.84E+0	8 2	1.43E+0	13 4	4.06E+0	1 5	−2.92E+0	6 2	1.26E+0
15 3	9.35E−2	3 1	4.87E−1	8 3	1.53E−1	13 5	−7.50E−4	1 6	−2.81E−1	6 3	2.01E−1
15 4	7.70E−2	3 2	2.26E−1	8 4	3.01E−3	13 6	−3.36E−3	1 7	−7.68E−1	6 4	6.02E−2
15 5	1.31E+0	3 3	4.03E−3	8 5	−1.27E−1	13 7	−1.72E−2	1 8	−3.85E−5	6 5	4.28E−2
15 6	7.64E−2	3 4	3.70E−5	8 6	−3.64E−1	13 8	−1.16E−1	1 9	−2.27E−1	6 6	1.95E−2
15 7	3.26E−3	3 5	−5.10E−5	8 7	−2.03E−1	13 9	−1.35E−2	1 10	−4.69E−2	6 7	1.50E−2
15 8	3.59E−1	3 6	−1.94E+0	8 8	−2.99E−1	13 10	−2.71E−4	1 11	−2.81E−5	6 8	1.35E−1
15 9	5.67E−3	3 7	−9.40E−1	8 9	−5.11E−3	14 1	1.01E−3	2 1	2.50E+0	6 9	1.16E−3
15 10	4.07E−3	3 8	−1.15E−2	8 10	−2.59E−3	14 2	2.85E−2	2 2	−1.71E+0	6 10	3.34E−5
15 11	3.92E−5	3 9	−1.92E+0	9 1	4.77E−2	14 3	1.44E−2	2 3	−2.63E−1	6 11	−3.94E−7
16 1	6.80E−2	3 10	−5.62E−4	9 2	8.65E−1	14 4	4.88E−2	2 4	−6.54E−1	7 1	1.92E−4
16 2	2.31E−4	4 1	1.85E+0	9 3	7.71E−2	14 5	5.94E−6	2 5	−2.27E+0	7 2	1.53E−1
16 3	6.71E−3	4 2	1.34E−1	9 4	2.59E−2	14 6	6.32E−6	2 6	−1.34E+0	7 3	5.12E−2
16 4	4.48E−3	4 3	5.21E−3	9 5	−1.24E−3	14 7	−5.09E−7	2 7	−1.66E+0	7 4	2.20E−1
16 5	7.65E−4	4 4	4.23E−5	9 6	−2.28E−1	14 8	−9.87E−4	2 8	−1.08E−3	7 5	4.70E−2
16 6	2.29E−6	4 5	−6.19E−1	9 7	−1.42E−1	14 9	−2.03E−4	2 9	−1.76E+0	7 6	7.89E−2
16 7	4.59E−3	4 6	−6.93E−1	9 8	−6.84E−1	14 10	−3.77E+0	2 10	−1.04E−1	7 7	2.70E−2
16 8	1.22E−2	4 7	−2.07E−2	9 9	−1.17E−2	15 1	1.41E−2	2 11	−1.43E−3	7 8	2.17E−1
16 9	9.31E−3	4 8	−2.26E+0	9 10	−1.43E−3	15 2	7.35E−3	3 1	6.24E+0	7 9	5.37E−3
16 10	1.01E−3	4 9	−5.33E−1	10 1	1.16E−2	15 3	1.10E+0	3 2	−1.32E−2	7 10	5.85E−4
16 11	1.84E−2	4 10	−8.16E−5	10 2	3.58E−1	15 4	2.93E+0	3 3	−1.55E−1	7 11	−5.78E−6
17 1	2.06E−2	5 1	2.71E+0	10 3	1.51E+0	15 5	6.60E−4	3 4	−1.24E−1	8 1	4.12E−4
17 2	1.47E−3	5 2	8.36E−3	10 4	4.43E−2	15 6	2.65E−8	3 5	−2.67E−1	8 2	4.08E−1
17 3	4.52E−5	5 3	2.92E−4	10 5	−4.54E−2	15 7	−7.63E−6	3 6	−5.89E−2	8 3	8.26E−2
17 4	5.09E−4	5 4	1.59E−4	10 6	−3.46E−3	15 8	−7.86E−4	3 7	−2.28E+0	8 4	1.69E−2
17 5	2.90E−4	5 5	−5.48E−2	10 7	−2.40E−3	15 9	−7.68E−4	3 8	−4.17E−1	8 5	3.61E−1
17 6	1.16E−5	5 6	−4.83E−1	10 8	−1.09E+0	15 10	−1.87E−2	3 9	−3.58E+0	8 6	6.52E−1
17 7	2.52E−3	5 7	−7.83E−2	10 9	−1.30E−2	16 1	2.03E−2	3 10	−5.70E+0	8 7	1.23E−1
17 8	5.99E−3	5 8	−4.39E−1	10 10	−1.33E−3	16 2	1.40E−2	3 11	−4.80E−4	8 8	3.38E−2
17 9	4.94E−3	5 9	1.10E−2	11 1	2.28E−2	16 3	6.34E−5	4 1	1.85E+0	8 9	2.13E−3
17 10	1.54E−2	5 10	−2.47E−3	11 2	4.74E−3	16 4	2.68E−4	4 2	1.95E−7	8 10	6.75E−4
17 11	2.22E−2	6 1	1.06E+0	11 3	6.89E−3	16 5	1.39E−5	4 3	−1.23E−3	8 11	5.91E−5
	$^2F^e$–$^2G^o$	6 2	8.45E−3	11 4	1.24E−2	16 6	2.61E−5	4 4	−4.39E−4	9 1	2.61E−4
1 1	−3.97E−1	6 3	9.36E−3	11 5	−5.40E−2	16 7	4.19E−6	4 5	−8.74E−2	9 2	4.63E+0
1 2	−1.71E+0	6 4	5.59E−3	11 6	−4.46E−3	16 8	−3.01E−4	4 6	−3.61E−2	9 3	1.26E−2
1 3	−1.09E+0	6 5	−1.47E+0	11 7	3.53E−2	16 9	−1.27E−4	4 7	−1.17E−3	9 4	1.18E−1
1 4	−3.57E−1	6 6	−1.14E−2	11 8	−8.05E−3	16 10	−6.51E−2	4 8	−6.84E−2	9 5	1.77E−3
1 5	−1.58E−4	6 7	−9.33E−1	11 9	−4.51E−1	17 1	7.25E−1	4 9	4.56E+0	9 6	6.20E−1
1 6	−6.43E−4	6 8	−1.27E−3	11 10	−4.45E−4	17 2	1.74E−3	4 10	−1.27E−3	9 7	6.66E−1

P-like Fe (Fe^{11+})

i	i'	gf_L	i	i'	gf_L	i	i'	gf_L	i	i'	gf_L	i	i'	gf_L	i	i'	gf_L
9	8	2.23E−3	3	9	−2.01E−2	7	5	6.88E−2	2	7	−8.24E−2	6	3	1.31E+0	9	13	4.52E−6
9	9	4.42E−1	3	10	−7.98E−6	7	6	1.09E+0	2	8	−4.93E−1	6	4	1.54E−1	9	14	9.13E−5
9	10	1.47E−4	3	11	−1.62E+0	7	7	1.40E−2	2	9	−2.25E+0	6	5	3.10E−1	10	1	1.41E−3
9	11	8.69E−6	3	12	−5.01E+0	7	8	7.33E−3	2	10	−1.80E−1	6	6	3.29E−2	10	2	7.48E−5
10	1	6.15E−1	3	13	−8.84E−2	7	9	1.59E−1	2	11	−1.18E+0	6	7	2.06E−2	10	3	1.16E+0
10	2	1.15E−6	3	14	−8.92E−5	7	10	2.60E−1	2	12	−6.81E−2	6	8	1.25E−1	10	4	7.66E−4
10	3	2.19E−6	4	1	1.49E−1	7	11	2.96E−1	2	13	−4.60E−4	6	9	1.76E−1	10	5	1.53E+0
10	4	5.97E−6	4	2	1.88E−1	7	12	2.42E−4	2	14	−7.99E−2	6	10	3.62E−2	10	6	1.74E−1
10	5	8.14E−4	4	3	4.01E−4	7	13	4.85E−7	3	1	3.29E−3	6	11	2.01E−2	10	7	2.06E−6
10	6	2.52E−4	4	4	1.08E−5	7	14	2.31E−3	3	2	2.76E−3	6	12	4.82E−5	10	8	3.70E−1
10	7	1.04E−3	4	5	1.08E−3			**$^4S^e$–$^4P^o$**	3	3	−9.58E−3	6	13	−7.78E−7	10	9	4.64E−1
10	8	8.18E−4	4	6	3.43E−2	1	1	8.13E−1	3	4	−5.15E−2	6	14	−1.12E−4	10	10	5.49E−1
10	9	1.61E−3	4	7	7.83E−1	1	2	2.06E−1	3	5	−8.04E−3	7	1	5.44E−4	10	11	5.06E−4
10	10	3.25E−4	4	8	2.01E−2	1	3	1.90E−1	3	6	−4.62E−2	7	2	3.82E−4	10	12	1.06E−2
10	11	1.00E−1	4	9	2.41E−3	1	4	−2.94E−5	3	7	−5.73E−2	7	3	3.94E−2	10	13	9.86E−6
		$^4S^o$–$^4P^e$	4	10	3.38E−4	1	5	−1.41E−1	3	8	−1.34E−2	7	4	4.14E−1	10	14	2.18E−7
1	1	−3.90E−1	4	11	1.34E−4	1	6	−4.89E−3	3	9	−1.28E+0	7	5	6.33E−1			**$^4P^o$–$^4D^e$**
1	2	−6.12E+0	4	12	−1.61E−2	1	7	−6.51E−1	3	10	−5.22E−1	7	6	2.25E−1	1	1	5.61E−1
1	3	−1.72E−3	4	13	−2.18E+0	1	8	−7.07E−4	3	11	−1.64E+0	7	7	1.26E−2	1	2	−1.52E−1
1	4	−2.71E−4	4	14	−1.74E−1	1	9	−6.66E−1	3	12	−1.95E+0	7	8	4.05E−5	1	3	−1.70E+0
1	5	−7.21E−3	5	1	6.32E−1	1	10	−2.15E−1	3	13	−2.87E−4	7	9	3.51E−3	1	4	−1.65E−2
1	6	−4.26E−2	5	2	9.77E−2	2	1	5.65E−1	3	14	−2.56E−3	7	10	1.39E−1	1	5	−1.75E+0
1	7	−7.53E−1	5	3	1.33E−2	2	2	1.06E−1	4	1	6.44E−2	7	11	6.18E−4	1	6	−1.87E−1
1	8	−1.90E−2	5	4	4.04E−3	2	3	2.25E+0	4	2	1.46E−1	7	12	2.01E−3	1	7	−8.86E−2
1	9	−2.45E−4	5	5	7.10E−4	2	4	−1.02E−4	4	3	7.96E−5	7	13	8.38E−8	1	8	−5.62E−1
1	10	−4.54E−3	5	6	9.54E−3	2	5	−7.30E−3	4	4	8.15E−4	7	14	−1.71E−5	1	9	−3.14E−2
1	11	−9.48E−7	5	7	7.37E−2	2	6	−2.78E−4	4	5	1.01E−5	8	1	8.54E−1	1	10	−7.86E−4
1	12	−2.85E−3	5	8	7.72E−4	2	7	−3.14E−1	4	6	1.19E−1	8	2	1.04E−2	1	11	−5.00E−4
1	13	−2.10E+0	5	9	7.01E−4	2	8	−1.06E−4	4	7	2.44E+0	8	3	6.06E−4	2	1	1.06E−1
1	14	−8.95E−1	5	10	2.27E−3	2	9	−3.50E−2	4	8	7.69E−2	8	4	1.04E−4	2	2	−2.41E−1
2	1	1.44E+0	5	11	5.50E−4	2	10	−6.93E−1	4	9	8.34E−3	8	5	2.85E−4	2	3	−1.73E−2
2	2	1.12E−1	5	12	−1.26E−5			**$^4P^o$–$^4P^e$**	4	10	9.75E−4	8	6	9.13E−2	2	4	−4.08E−2
2	3	−1.10E−1	5	13	−3.90E−3	1	1	1.24E+0	4	11	6.60E−5	8	7	8.95E−1	2	5	−4.55E−1
2	4	−4.44E−1	5	14	−2.13E+0	1	2	7.33E−2	4	12	−1.59E−3	8	8	5.16E−2	2	6	−6.42E+0
2	5	−8.33E−1	6	1	1.15E−2	1	3	−1.02E−1	4	13	−1.72E+0	8	9	7.93E−3	2	7	−6.04E−1
2	6	−5.62E−2	6	2	1.12E−4	1	4	−2.93E−1	4	14	−2.21E−2	8	10	3.74E−3	2	8	−1.43E+0
2	7	−3.06E−2	6	3	7.81E−2	1	5	−2.11E−2	5	1	3.80E−6	8	11	2.12E−3	2	9	−5.47E−2
2	8	−1.39E+0	6	4	8.07E−2	1	6	−4.09E−2	5	2	1.06E−4	8	12	1.09E−5	2	10	−4.16E−3
2	9	−1.05E+0	6	5	2.48E−1	1	7	−9.93E−2	5	3	8.30E−1	8	13	5.51E−4	2	11	−1.44E−1
2	10	−2.64E−1	6	6	9.89E−1	1	8	−4.39E+0	5	4	1.07E+0	8	14	−3.97E−3	3	1	4.87E+0
2	11	−6.55E−1	6	7	3.03E−3	1	9	−2.48E−1	5	5	1.30E−1	9	1	2.56E−3	3	2	−1.41E−3
2	12	−7.10E−2	6	8	2.23E−2	1	10	−1.74E+0	5	6	3.91E−5	9	2	5.90E−5	3	3	−3.34E−2
2	13	−7.69E−5	6	9	2.08E−2	1	11	−1.77E−1	5	7	9.14E−4	9	3	1.73E+0	3	4	−3.59E−2
2	14	−3.72E−5	6	10	2.10E−2	1	12	−1.77E−3	5	8	1.82E−2	9	4	1.12E+0	3	5	−3.85E−1
3	1	1.05E−3	6	11	4.65E−2	1	13	−5.23E−5	5	9	2.40E−2	9	5	3.85E−1	3	6	−4.43E−2
3	2	4.41E+0	6	12	9.29E−5	1	14	−3.34E−3	5	10	5.37E−3	9	6	4.54E−4	3	7	−1.79E+0
3	3	2.41E−5	6	13	4.95E−6	2	1	4.85E+0	5	11	1.50E−2	9	7	1.78E−3	3	8	−1.87E+0
3	4	−2.78E−3	6	14	−2.80E−2	2	2	3.32E−1	5	12	−1.62E−3	9	8	5.12E−3	3	9	−1.19E−1
3	5	−5.00E−2	7	1	1.09E−3	2	3	−2.63E−1	5	13	−2.21E−7	9	9	1.69E−1	3	10	−2.34E−2
3	6	−5.62E−3	7	2	2.36E−4	2	4	−7.87E−1	5	14	−9.80E−7	9	10	2.25E−1	3	11	−4.55E−6
3	7	−1.29E−2	7	3	1.03E+0	2	5	−7.73E−1	6	1	5.21E−4	9	11	8.44E−3	4	1	5.86E−1
3	8	−1.06E−3	7	4	5.45E−1	2	6	−3.04E−4	6	2	2.68E−5	9	12	2.15E−2	4	2	8.53E−4

P-like Fe (Fe^{11+})

i	i'	gf_L	i	i'	gf_L	i	i'	gf_L	i	i'	gf_L	i	i'	gf_L	i	i'	gf_L
4	3	1.40E−3	8	11	−1.29E−1	2	14	−2.11E−3	6	6	3.38E−3	9	13	−1.77E−1	13	5	4.26E−5
4	4	5.08E−4	9	1	1.82E−3	2	15	−9.79E+0	6	7	−1.75E−1	9	14	−3.33E−2	13	6	1.27E−3
4	5	9.83E−5	9	2	2.50E−1	3	1	2.04E−1	6	8	−1.86E−2	9	15	−4.20E−4	13	7	3.83E−2
4	6	5.44E−5	9	3	6.57E−1	3	2	2.83E−1	6	9	−4.43E−5	10	1	1.95E−2	13	8	1.72E−5
4	7	7.18E−5	9	4	2.20E+0	3	3	1.18E−1	6	10	−2.39E−1	10	2	5.33E−1	13	9	1.15E−5
4	8	2.29E−4	9	5	9.20E−3	3	4	2.76E−1	6	11	−8.49E−2	10	3	2.82E+0	13	10	−1.68E−5
4	9	2.17E−6	9	6	2.22E−1	3	5	1.77E−2	6	12	−8.02E−3	10	4	4.88E−1	13	11	−2.85E−4
4	10	−5.86E+0	9	7	3.45E−1	3	6	4.40E−4	6	13	−5.61E−4	10	5	2.65E+0	13	12	−1.07E−4
4	11	−4.56E−1	9	8	3.59E−1	3	7	−2.63E−3	6	14	−5.97E−3	10	6	1.52E−1	13	13	−3.38E−4
5	1	9.16E−4	9	9	1.09E−1	3	8	−1.82E−4	6	15	−9.27E−4	10	7	−1.30E−3	13	14	−2.85E−3
5	2	1.92E−1	9	10	2.03E−6	3	9	−5.00E−1	7	1	7.08E−3	10	8	−1.73E−2	13	15	−3.83E+0
5	3	8.16E−2	9	11	−3.81E−4	3	10	−2.88E−1	7	2	1.32E−2	10	9	−1.86E−3	14	1	4.38E−1
5	4	3.09E−1	10	1	1.98E−3	3	11	−4.31E−4	7	3	2.71E−2	10	10	−1.38E−1	14	2	9.14E−2
5	5	8.62E−2	10	2	7.42E−2	3	12	−7.35E−1	7	4	3.17E−3	10	11	−4.82E−3	14	3	1.19E−1
5	6	1.12E−2	10	3	1.28E+0	3	13	−1.24E+0	7	5	6.41E−2	10	12	−3.63E−2	14	4	1.12E−3
5	7	2.28E−4	10	4	9.18E−2	3	14	−4.86E−2	7	6	2.94E−3	10	13	−1.26E−2	14	5	1.28E−3
5	8	1.44E−2	10	5	6.97E−1	3	15	−6.35E−3	7	7	−3.64E+0	10	14	−1.12E−4	14	6	3.60E−4
5	9	1.03E−2	10	6	3.58E−2	4	1	1.17E+0	7	8	−1.11E−3	10	15	−1.19E−3	14	7	1.47E−2
5	10	−3.93E−5	10	7	1.51E+0	4	2	7.35E−1	7	9	−2.06E−2	11	1	2.51E−3	14	8	2.70E−3
5	11	−5.94E−5	10	8	9.85E−2	4	3	1.10E−1	7	10	−4.98E−2	11	2	3.09E−4	14	9	1.03E−3
6	1	1.03E−3	10	9	9.30E−2	4	4	2.36E−1	7	11	−8.29E−1	11	3	6.66E−1	14	10	3.44E−3
6	2	5.02E+0	10	10	6.59E−6	4	5	5.01E−2	7	12	−1.49E−3	11	4	6.18E+0	14	11	1.19E−1
6	3	2.24E−1	10	11	6.10E−7	4	6	2.68E−3	7	13	−1.43E−4	11	5	1.24E+0	14	12	7.92E−4
6	4	9.00E−1			**$^4P^e$–$^4D^o$**	4	7	−8.87E−4	7	14	−4.97E−2	11	6	2.62E−1	14	13	5.14E−5
6	5	3.20E−2	1	1	−6.64E−3	4	8	−3.57E−1	7	15	−1.05E−3	11	7	8.33E−6	14	14	−4.51E−7
6	6	4.45E−2	1	2	−4.72E−1	4	9	−5.21E−1	8	1	1.72E−2	11	8	−1.55E−2	14	15	−1.10E−2
6	7	1.48E−2	1	3	−8.27E+0	4	10	−3.73E−1	8	2	1.23E+0	11	9	−1.35E−3			**$^4D^o$–$^4D^e$**
6	8	3.19E−5	1	4	−1.16E+0	4	11	−7.23E−3	8	3	5.26E−3	11	10	−1.25E−1	1	1	1.47E−1
6	9	2.64E−1	1	5	−3.14E−2	4	12	−3.65E−1	8	4	1.06E−1	11	11	−7.84E−5	1	2	−1.13E+0
6	10	−1.15E−5	1	6	−4.38E−2	4	13	−8.97E−1	8	5	1.59E−2	11	12	−1.74E−1	1	3	−1.09E+0
6	11	−6.11E−4	1	7	−1.56E−2	4	14	−4.53E−1	8	6	3.99E−1	11	13	−6.84E−1	1	4	−4.14E+0
7	1	2.32E−3	1	8	−2.30E−4	4	15	−7.57E−4	8	7	−1.09E−1	11	14	−2.97E−2	1	5	−1.05E+0
7	2	1.43E−2	1	9	−2.33E−4	5	1	3.74E+0	8	8	−6.13E−3	11	15	−2.21E−3	1	6	−1.44E+0
7	3	3.99E−1	1	10	−1.18E−2	5	2	3.07E−1	8	9	−7.65E−1	12	1	5.76E−3	1	7	−1.85E−2
7	4	2.58E−1	1	11	−1.22E+0	5	3	3.68E−3	8	10	−3.52E−3	12	2	1.84E−3	1	8	−9.34E−2
7	5	4.52E−1	1	12	−2.15E−3	5	4	8.47E−2	8	11	−1.52E−2	12	3	2.05E−2	1	9	−1.55E−2
7	6	1.26E−1	1	13	−5.67E−3	5	5	2.10E−2	8	12	−5.17E−2	12	4	6.07E−2	1	10	−9.10E−4
7	7	8.59E−1	1	14	−7.78E−4	5	6	2.17E−2	8	13	−3.63E−3	12	5	1.94E+0	1	11	−2.42E−1
7	8	7.22E−2	1	15	−8.66E−1	5	7	−5.71E−4	8	14	−1.64E+0	12	6	5.16E+0	2	1	4.43E−1
7	9	1.37E−4	2	1	−6.06E−2	5	8	−5.90E−1	8	15	−4.08E−6	12	7	4.86E−5	2	2	−1.35E−1
7	10	7.10E−6	2	2	−3.29E−2	5	9	−1.43E−1	9	1	4.44E−3	12	8	2.17E−5	2	3	−1.29E−2
7	11	−1.24E−4	2	3	1.60E−1	5	10	1.07E−2	9	2	1.19E+0	12	9	2.09E−5	2	4	−1.20E−1
8	1	1.39E−1	2	4	−2.33E+0	5	11	−8.22E−3	9	3	3.27E+0	12	10	−6.13E−4	2	5	−3.83E+0
8	2	9.91E−4	2	5	−6.18E−1	5	12	−1.93E+0	9	4	3.14E−1	12	11	−8.47E−6	2	6	−9.71E−2
8	3	1.53E−2	2	6	−1.04E+1	5	13	−3.25E−1	9	5	3.64E−2	12	12	−7.43E−4	2	7	−8.53E−2
8	4	5.30E−3	2	7	−5.88E−2	5	14	−4.28E−1	9	6	8.41E−3	12	13	−6.03E−4	2	8	−2.95E+0
8	5	4.20E−3	2	8	−1.05E−5	5	15	−1.60E−2	9	7	−1.10E−2	12	14	−9.14E−3	2	9	−3.66E−3
8	6	2.38E−6	2	9	−3.64E−4	6	1	6.24E−2	9	8	−8.20E−3	12	15	−1.59E−3	2	10	−7.17E−4
8	7	1.15E−2	2	10	−1.73E−3	6	2	2.07E+0	9	9	−1.37E−2	13	1	6.74E−3	2	11	−1.65E−1
8	8	1.10E−3	2	11	−9.91E−4	6	3	1.59E−1	9	10	−1.94E−2	13	2	2.27E−4	3	1	9.67E−2
8	9	2.06E−4	2	12	−1.19E−3	6	4	6.83E−2	9	11	−7.80E−4	13	3	9.80E−3	3	2	−2.99E−1
8	10	2.05E−3	2	13	−2.41E−3	6	5	6.92E−2	9	12	−3.03E−1	13	4	3.12E−2	3	3	−4.67E−1

P-like Fe (Fe^{11+})

i	*i'*	gf$_L$	*i*	*i'*	gf$_L$	*i*	*i'*	gf$_L$	*i*	*i'*	gf$_L$	*i*	*i'*	gf$_L$	*i*	*i'*	gf$_L$
3	4	−5.91E−2	8	1	1.37E−3	12	9	2.87E−1	2	4	−7.23E−1	6	8	−5.03E−1	10	12	−9.71E−2
3	5	−6.03E−2	8	2	2.52E+0	12	10	5.63E−7	2	5	−7.09E−2	6	9	−6.28E+0	11	1	1.69E−1
3	6	−2.69E+0	8	3	7.99E−1	12	11	−4.10E−3	2	6	−3.57E+0	6	10	−1.19E+1	11	2	7.27E−3
3	7	−2.66E−1	8	4	9.01E−3	13	1	5.16E−4	2	7	−2.28E−1	6	11	−3.66E−3	11	3	3.41E−3
3	8	−4.59E−1	8	5	1.18E−1	13	2	3.97E+0	2	8	−8.31E−1	6	12	−3.45E−4	11	4	2.41E−2
3	9	−9.87E−1	8	6	3.95E−4	13	3	2.02E+0	2	9	−2.95E−2	7	1	1.10E+0	11	5	1.27E−2
3	10	−3.12E−3	8	7	6.18E−4	13	4	3.59E−3	2	10	−4.85E−3	7	2	5.08E−4	11	6	1.39E−3
3	11	−1.62E−2	8	8	8.97E−5	13	5	1.05E+0	2	11	−8.09E−3	7	3	2.60E−3	11	7	1.28E−2
4	1	3.82E+0	8	9	6.63E−2	13	6	2.30E−1	2	12	−7.01E−2	7	4	1.45E−4	11	8	9.99E−4
4	2	−3.27E−1	8	10	−1.95E−5	13	7	1.96E−3	3	1	9.59E−1	7	5	1.17E−3	11	9	2.09E−5
4	3	−9.95E−2	8	11	−4.27E−5	13	8	1.34E−1	3	2	−3.44E−2	7	6	1.77E−4	11	10	3.46E−4
4	4	−1.82E−2	9	1	2.50E−5	13	9	3.41E−1	3	3	−6.48E−3	7	7	1.06E−4	11	11	2.60E−3
4	5	−2.02E−4	9	2	1.32E+0	13	10	1.35E−5	3	4	−2.67E−1	7	8	9.40E−5	11	12	−6.48E+0
4	6	−2.34E−2	9	3	4.98E−1	13	11	−6.54E−4	3	5	−1.04E+0	7	9	2.04E−5	12	1	5.41E−4
4	7	−1.57E−2	9	4	6.17E−1	14	1	5.06E−3	3	6	−4.42E+0	7	10	−6.55E−6	12	2	4.88E+0
4	8	−1.54E−1	9	5	6.88E−2	14	2	2.90E−2	3	7	−6.79E−1	7	11	−1.04E+1	12	3	9.40E−2
4	9	−1.29E+1	9	6	7.34E−2	14	3	2.23E+0	3	8	−2.41E+0	7	12	−1.44E+0	12	4	9.61E−1
4	10	−1.09E−2	9	7	9.92E−2	14	4	3.94E−3	3	9	−4.72E+0	8	1	2.03E−4	12	5	2.71E−2
4	11	−3.94E−3	9	8	3.05E−3	14	5	7.90E−3	3	10	−5.01E−2	8	2	6.24E−2	12	6	5.37E−1
5	1	1.94E+0	9	9	3.15E−5	14	6	1.92E−1	3	11	−1.26E−6	8	3	2.56E+0	12	7	6.60E−1
5	2	−3.68E−4	9	10	−1.38E−6	14	7	1.14E+0	3	12	−5.53E−2	8	4	9.60E−2	12	8	6.12E−2
5	3	−5.48E−3	9	11	−8.99E−6	14	8	9.55E−1	4	1	5.17E+0	8	5	2.40E−3	12	9	4.14E−5
5	4	−7.80E−2	10	1	2.49E−3	14	9	4.16E−2	4	2	−9.94E−1	8	6	1.73E−2	12	10	1.32E−2
5	5	−4.33E−1	10	2	1.54E−3	14	10	3.03E−4	4	3	−7.51E−2	8	7	3.19E−2	12	11	1.63E−5
5	6	−1.58E+0	10	3	1.20E−2	14	11	−2.13E−6	4	4	−1.26E+0	8	8	3.32E−2	12	12	−7.92E−3
5	7	−6.97E−1	10	4	7.30E−1	15	1	1.79E+0	4	5	−4.07E−1	8	9	2.74E−3	13	1	1.08E−5
5	8	−5.07E+0	10	5	3.53E−1	15	2	3.51E−3	4	6	−4.32E−1	8	10	−2.66E−3	13	2	8.26E−1
5	9	−1.96E−2	10	6	7.75E−1	15	3	3.50E−4	4	7	−8.80E−1	8	11	−1.47E−5	13	3	5.64E−2
5	10	−3.02E−3	10	7	7.01E−1	15	4	6.43E−4	4	8	−1.53E+0	8	12	−3.01E−4	13	4	1.33E+0
5	11	−8.83E−4	10	8	3.90E−1	15	5	9.55E−4	4	9	−4.02E+0	9	1	7.22E−4	13	5	1.60E+0
6	1	4.94E+0	10	9	8.03E−2	15	6	2.33E−4	4	10	−5.59E−3	9	2	1.45E−1	13	6	5.65E−1
6	2	5.81E−5	10	10	6.58E−7	15	7	7.91E−4	4	11	−5.50E−4	9	3	6.58E−1	13	7	3.50E−1
6	3	−1.89E−2	10	11	−7.01E−2	15	8	1.47E−4	4	12	−1.40E−2	9	4	9.02E−1	13	8	9.13E−1
6	4	−5.81E−2	11	1	3.24E−2	15	9	1.52E−3	5	1	4.02E+0	9	5	1.21E−1	13	9	4.17E−2
6	5	−4.39E−2	11	2	9.67E−4	15	10	6.57E−1	5	2	−7.60E−3	9	6	9.62E−2	13	10	9.19E−4
6	6	−1.15E−1	11	3	1.13E−3	15	11	−2.13E−5	5	3	−4.56E−2	9	7	2.07E−1	13	11	2.37E−6
6	7	−9.43E−2	11	4	4.33E−2			$^4D^o$–$^4F^e$	5	4	−1.80E−1	9	8	6.61E−1	13	12	−2.89E−3
6	8	−1.25E+0	11	5	8.35E−3	1	1	7.66E−3	5	5	−2.78E−2	9	9	7.89E−2	14	1	6.69E−4
6	9	−2.74E+0	11	6	2.23E−2	1	2	−1.14E+0	5	6	−2.05E−4	9	10	3.66E−3	14	2	8.63E−3
6	10	−9.82E−3	11	7	1.88E−2	1	3	−2.42E−1	5	7	−3.16E+0	9	11	−1.87E−7	14	3	1.75E+0
6	11	−1.33E−3	11	8	3.22E−3	1	4	−4.05E−1	5	8	−2.78E−1	9	12	−1.04E−3	14	4	2.67E−2
7	1	1.63E−1	11	9	5.23E−4	1	5	−2.49E+0	5	9	−1.67E+0	10	1	1.12E−2	14	5	7.13E−1
7	2	4.82E−4	11	10	5.21E−5	1	6	−5.00E−1	5	10	−3.88E−1	10	2	2.60E+0	14	6	9.50E−1
7	3	5.96E−4	11	11	−4.00E+0	1	7	−4.06E−1	5	11	−3.54E−2	10	3	1.18E+0	14	7	1.00E+0
7	4	5.10E−6	12	1	1.29E−5	1	8	−5.14E−5	5	12	−1.78E−3	10	4	2.40E−2	14	8	6.46E−1
7	5	1.11E−4	12	2	7.60E−1	1	9	−3.62E−1	6	1	8.45E−3	10	5	6.50E−3	14	9	9.83E−2
7	6	5.11E−7	12	3	2.46E−2	1	10	−4.21E−2	6	2	−3.38E−3	10	6	2.76E−2	14	10	7.72E−2
7	7	1.92E−5	12	4	9.14E−2	1	11	−4.56E−3	6	3	−2.85E−2	10	7	2.29E−3	14	11	9.52E−6
7	8	3.71E−4	12	5	2.25E−1	1	12	−1.91E−2	6	4	−2.44E−4	10	8	1.87E−3	14	12	−5.29E−5
7	9	7.20E−6	12	6	6.79E−1	2	1	4.73E−1	6	5	−2.05E−1	10	9	1.10E−2	15	1	4.26E−2
7	10	−2.05E+0	12	7	8.07E−3	2	2	−1.96E−1	6	6	−5.89E−2	10	10	6.55E−3	15	2	7.29E−6
7	11	−4.79E−1	12	8	3.49E−2	2	3	−4.37E−1	6	7	−9.50E−3	10	11	2.33E−5	15	3	3.93E−3

P-like Fe (Fe^{11+})

i	i'	gf$_L$	i	i'	gf$_L$	i	i'	gf$_L$	i	i'	gf$_L$	i	i'	gf$_L$	i	i'	gf$_L$
15	4	3.40E−4	5	7	−2.73E+0	11	5	7.28E−8	4	11	−1.70E−5	9	3	1.20E−3	1	5	−8.92E−4
15	5	9.93E−4	5	8	−1.09E−1	11	6	2.56E−4	4	12	−1.60E−4	9	4	1.20E−4	1	6	−1.66E+1
15	6	2.75E−3	5	9	−3.77E−3	11	7	4.03E−5	5	1	9.35E−5	9	5	9.78E−4	1	7	−4.18E+0
15	7	4.73E−4	6	1	3.15E−1	11	8	5.07E−5	5	2	4.74E+0	9	6	6.71E−4	2	1	2.55E+0
15	8	6.93E−4	6	2	9.18E−2	11	9	2.94E−4	5	3	4.49E−1	9	7	7.15E−4	2	2	8.21E−3
15	9	1.96E−2	6	3	1.39E−4			**$^4F^o$–$^4F^e$**	5	4	6.42E−1	9	8	1.01E−3	2	3	−9.07E−1
15	10	8.03E−3	6	4	−2.55E−1	1	1	6.41E−1	5	5	2.46E−1	9	9	8.16E−3	2	4	−5.09E−2
15	11	2.00E−4	6	5	−1.05E−1	1	2	−2.77E−1	5	6	2.48E−1	9	10	1.51E−3	2	5	−5.15E+0
15	12	−9.51E−6	6	6	−1.57E−1	1	3	−2.99E−1	5	7	1.50E−1	9	11	7.93E−1	2	6	−6.78E−2
		$^4D^e$–$^4F^o$	6	7	−1.38E+0	1	4	−5.72E−1	5	8	5.88E−1	9	12	−1.08E−3	2	7	−1.62E−1
1	1	−1.26E−1	6	8	−2.97E−1	1	5	−3.93E+0	5	9	3.16E−2			**$^4F^o$–$^4G^e$**	3	1	4.59E−1
1	2	−2.31E+0	6	9	−1.42E−3	1	6	−3.31E+0	5	10	5.60E−6	1	1	−3.96E+0	3	2	7.23E−2
1	3	−3.32E+0	7	1	6.09E−4	1	7	−3.87E−1	5	11	−9.50E−7	1	2	−1.15E+0	3	3	−1.29E+0
1	4	−1.19E−5	7	2	6.80E+0	1	8	−9.29E−1	5	12	−5.87E−5	1	3	−7.31E−1	3	4	−1.04E−1
1	5	−4.70E−4	7	3	8.46E−2	1	9	−3.11E−2	6	1	1.78E−6	1	4	−2.47E+0	3	5	−6.63E−2
1	6	−6.07E−6	7	4	−2.14E−3	1	10	−1.34E−2	6	2	2.09E−2	2	1	−7.50E−4	3	6	−2.46E−1
1	7	−6.40E−5	7	5	−5.87E−3	1	11	−5.53E−3	6	3	9.58E−3	2	2	−3.41E+0	3	7	−1.64E+0
1	8	−2.02E−2	7	6	−1.45E−1	1	12	−3.12E−1	6	4	3.05E−2	2	3	−4.00E+0	4	1	7.42E+0
1	9	−1.44E+1	7	7	−5.61E−1	2	1	8.84E−1	6	5	8.83E−1	2	4	−5.29E−1	4	2	2.54E−1
2	1	2.21E+0	7	8	−1.53E−2	2	2	−1.08E+0	6	6	1.60E−1	3	1	−1.66E−1	4	3	−5.07E−1
2	2	9.82E−5	7	9	−3.25E−3	2	3	−2.28E−3	6	7	7.61E−1	3	2	−8.54E−2	4	4	−2.21E−1
2	3	4.58E−3	8	1	6.32E−2	2	4	−1.13E+0	6	8	5.34E−1	3	3	−3.97E−4	4	5	−2.92E+0
2	4	−4.89E−1	8	2	1.75E−3	2	5	−8.13E−1	6	9	8.71E−3	3	4	−8.95E+0	4	6	−9.82E−3
2	5	−1.90E+0	8	3	1.61E+0	2	6	−1.78E+0	6	10	3.25E−3	4	1	9.81E−3	4	7	−7.93E−2
2	6	−8.08E−3	8	4	−1.53E−2	2	7	−6.08E+0	6	11	1.72E−7	4	2	1.38E−1	5	1	1.94E−2
2	7	−2.44E+0	8	5	−3.73E−2	2	8	−5.77E−1	6	12	−2.43E−5	4	3	2.25E−1	5	2	1.96E−1
2	8	−4.16E−2	8	6	−6.06E−1	2	9	−3.55E+0	7	1	5.56E−4	4	4	5.22E−2	5	3	−6.58E−1
2	9	−6.84E−3	8	7	−1.50E−2	2	10	−1.12E−2	7	2	7.37E+0	5	1	2.40E+0	5	4	−2.02E+0
3	1	3.17E+0	8	8	−8.89E−2	2	11	−4.59E−3	7	3	2.38E−1	5	2	1.99E−1	5	5	−1.50E−1
3	2	2.42E−1	8	9	−1.62E−4	2	12	−5.08E−2	7	4	9.79E−1	5	3	6.67E−2	5	6	−1.00E−2
3	3	5.55E−2	9	1	1.68E−6	3	1	1.23E+1	7	5	3.54E−1	5	4	1.72E−3	5	7	−1.15E−1
3	4	−2.53E−1	9	2	4.40E−3	3	2	−4.75E−3	7	6	8.47E−1	6	1	2.05E+0	6	1	1.57E−1
3	5	−1.38E+0	9	3	5.73E+0	3	3	−9.09E−2	7	7	5.36E−1	6	2	7.31E−1	6	2	7.35E−2
3	6	−1.71E−1	9	4	−6.22E−3	3	4	−7.20E−2	7	8	3.61E−1	6	3	3.78E−1	6	3	−1.16E−2
3	7	−1.65E+0	9	5	−2.30E−1	3	5	−4.11E−4	7	9	4.23E−1	6	4	6.35E−1	6	4	−2.09E+0
3	8	−1.56E+0	9	6	−3.71E−2	3	6	−1.14E+0	7	10	1.39E−3	7	1	7.38E−1	6	5	−1.87E−1
3	9	−1.31E−2	9	7	−1.14E+0	3	7	−1.23E−2	7	11	2.68E−6	7	2	1.76E+0	6	6	−1.37E+0
4	1	3.84E+0	9	8	−3.31E−1	3	8	−1.20E−1	7	12	−4.51E−7	7	3	9.43E−1	6	7	−5.78E+0
4	2	3.35E−2	9	9	−3.46E−4	3	9	−2.74E+0	8	1	1.28E−2	7	4	1.63E+0	7	1	1.10E+0
4	3	1.69E−1	10	1	6.17E−5	3	10	−1.37E+1	8	2	5.06E−1	8	1	2.31E+0	7	2	5.09E−3
4	4	−3.53E−1	10	2	5.99E−2	3	11	−3.56E−3	8	3	2.68E−2	8	2	1.87E−1	7	3	−1.32E−1
4	5	−2.10E−1	10	3	3.17E−4	3	12	−5.58E−4	8	4	5.82E+0	8	3	1.04E+0	7	4	−2.05E−2
4	6	−1.28E+0	10	4	1.01E−5	4	1	3.37E−3	8	5	2.50E+0	8	4	3.83E−1	7	5	−1.04E+0
4	7	−1.53E−2	10	5	1.57E−6	4	2	6.11E−3	8	6	5.18E−1	9	1	6.79E−3	7	6	−7.23E−2
4	8	−1.03E+0	10	6	3.60E−7	4	3	2.43E+0	8	7	2.21E−1	9	2	3.68E−3	7	7	−2.10E−1
4	9	−3.72E−3	10	7	−5.07E−5	4	4	3.99E−1	8	8	4.46E−2	9	3	2.31E−4	8	1	1.56E−1
5	1	1.47E−1	10	8	−1.37E−3	4	5	1.23E−1	8	9	4.98E−2	9	4	4.57E−3	8	2	8.90E+0
5	2	8.05E−2	10	9	−5.57E+0	4	6	5.97E−2	8	10	7.35E−2			**$^4F^e$–$^4G^o$**	8	3	−2.66E−1
5	3	2.16E−2	11	1	8.45E−1	4	7	1.55E−1	8	11	8.54E−5	1	1	−2.07E+0	8	4	−1.75E−1
5	4	−5.92E−1	11	2	2.55E−2	4	8	4.87E−2	8	12	−2.03E−4	1	2	−5.02E+0	8	5	−4.96E−1
5	5	−1.71E−1	11	3	3.98E−3	4	9	7.98E−3	9	1	1.87E+0	1	3	−1.83E−4	8	6	−2.02E−1
5	6	−6.77E−1	11	4	1.19E−5	4	10	−3.73E−3	9	2	2.62E−3	1	4	−6.56E−5	8	7	−1.17E+0

P-like Fe (Fe^{11+})

i i'	gf$_L$	i i'	gf$_L$	i i'	gf$_L$	i i'	gf$_L$	i i'	gf$_L$	i i'	gf$_L$
9 1	4.72E−2	10 7	−1.07E−1	12 6	1.62E−2	3 3	1.02E+0	6 4	6.07E−2	1 3	−9.85E−1
9 2	1.11E+0	11 1	8.38E−2	12 7	2.02E−3	3 4	1.13E−1	7 1	2.57E+0	2 1	1.32E+0
9 3	−1.69E−2	11 2	1.12E−2		**$^4G^o$–$^4G^e$**	4 1	1.12E−3	7 2	8.29E−1	2 2	5.59E−5
9 4	−1.40E−1	11 3	1.01E−5	1 1	−1.03E+0	4 2	5.73E−2	7 3	9.32E−1	2 3	−1.03E−6
9 5	−1.10E+0	11 4	1.03E−5	1 2	−2.95E+0	4 3	3.14E+0	7 4	3.31E−2		**$^6D^o$–$^6F^e$**
9 6	−1.91E−1	11 5	−1.14E−5	1 3	−5.22E+0	4 4	4.37E−1		**$^6S^o$–$^6P^e$**	1 1	−5.49E+0
9 7	−6.60E−1	11 6	−5.97E+0	1 4	−5.98E+0	5 1	2.35E+0	1 1	1.48E−2	2 1	6.44E+0
10 1	5.43E−3	11 7	−1.51E+0	2 1	−6.76E−1	5 2	4.74E+0	1 2	−4.07E+0	3 1	4.37E+0
10 2	1.93E+0	12 1	1.37E+0	2 2	−1.91E−1	5 3	3.09E−2	2 1	2.16E+0		
10 3	2.96E−3	12 2	3.14E−5	2 3	−4.52E+0	5 4	1.28E−1	2 2	2.82E−5		
10 4	−5.78E−4	12 3	2.05E−4	2 4	−6.22E+0	6 1	6.68E−1		**$^6P^e$–$^6D^o$**		
10 5	−2.64E−2	12 4	1.44E−4	3 1	1.17E+0	6 2	1.78E−1	1 1	7.78E+0		
10 6	−2.68E−2	12 5	1.62E−5	3 2	5.26E−1	6 3	2.14E−1	1 2	−3.44E+0		

S-like S

Term energies relative to $3s^23p^3\ ^4$S ionization threshold for each symmetry

i E(Ryds) Description	i E(Ryds) Description	i E(Ryds) Description	i E(Ryds) Description
$^1\mathbf{S}^e$	2 −.16389 $3s^23p^3\ ^4$S $4p$	4 −.02779 $3s^23p^3\ ^4$S $6f$	5 −.02359 $3s^23p^3\ ^4$S $7d$
1 −.57333 $3s^23p^4$	3 −.08284 $3s^23p^3\ ^4$S $5p$	5 −.02042 $3s^23p^3\ ^4$S $7f$	6 −.01773 $3s^23p^3\ ^4$S $8d$
$^1\mathbf{S}^o$	4 −.05069 $3s^23p^3\ ^4$S $6p$	6 −.01563 $3s^23p^3\ ^4$S $8f$	7 −.01381 $3s^23p^3\ ^4$S $9d$
1 −.02253 $3s^23p^3\ ^2$D $3d$	5 −.03545 $3s^23p^3\ ^4$S $7p$:	7 −.01235 $3s^23p^3\ ^4$S $9f$	$^5\mathbf{F}^e$
$^1\mathbf{P}^e$	6 −.02835 $3s^23p^3\ ^2$D $4p$:	$^3\mathbf{F}^o$	1 −.06256 $3s^23p^3\ ^4$S $4f$
1 −.04801 $3s^23p^3\ ^2$D $4p$	7 −.02237 $3s^23p^3\ ^4$S $8p$	1 −.01586 $3s^23p^3\ ^2$D $3d$	2 −.04005 $3s^23p^3\ ^4$S $5f$
$^1\mathbf{P}^o$	8 −.01723 $3s^23p^3\ ^4$S $9p$	$^5\mathbf{S}^o$	3 −.02782 $3s^23p^3\ ^4$S $6f$
1 −.04308 $3s^23p^3\ ^2$D $3d$	$^3\mathbf{P}^o$	1 −.27663 $3s^23p^3\ ^4$S $4s$	4 −.02044 $3s^23p^3\ ^4$S $7f$
2 −.01838 $3s^23p^3\ ^2$P $4s$	1 −.14710 $3s3p^5$	2 −.11558 $3s^23p^3\ ^4$S $5s$	5 −.01564 $3s^23p^3\ ^4$S $8f$
$^1\mathbf{D}^e$	2 −.04696 $3s^23p^3\ ^2$P $4s$	3 −.06401 $3s^23p^3\ ^4$S $6s$	6 −.01236 $3s^23p^3\ ^4$S $9f$
1 −.69844 $3s^23p^4$	$^3\mathbf{D}^e$	4 −.04069 $3s^23p^3\ ^4$S $7s$	
2 −.01606 $3s^23p^3\ ^2$D $4p$	1 −.04519 $3s^23p^3\ ^2$D $4p$	5 −.02815 $3s^23p^3\ ^4$S $8s$	
$^1\mathbf{D}^o$	$^3\mathbf{D}^o$	6 −.02064 $3s^23p^3\ ^4$S $9s$	
1 −.11567 $3s^23p^3\ ^2$D $4s$	1 −.13517 $3s^23p^3\ ^2$D $4s$	$^5\mathbf{P}^e$	
$^1\mathbf{F}^e$	2 −.11769 $3s^23p^3\ ^4$S $3d$	1 −.18024 $3s^23p^3\ ^4$S $4p$	
1 −.03802 $3s^23p^3\ ^2$D $4p$	3 −.06928 $3s^23p^3\ ^4$S $4d$	2 −.08698 $3s^23p^3\ ^4$S $5p$	
$^3\mathbf{S}^o$	4 −.04489 $3s^23p^3\ ^4$S $5d$	3 −.05160 $3s^23p^3\ ^4$S $6p$	
1 −.24151 $3s^23p^3\ ^4$S $4s$	5 −.03129 $3s^23p^3\ ^4$S $6d$	4 −.03419 $3s^23p^3\ ^4$S $7p$	
2 −.10756 $3s^23p^3\ ^4$S $5s$	6 −.02298 $3s^23p^3\ ^4$S $7d$	5 −.02433 $3s^23p^3\ ^4$S $8p$	
3 −.06089 $3s^23p^3\ ^4$S $6s$	7 −.01755 $3s^23p^3\ ^4$S $8d$	6 −.01820 $3s^23p^3\ ^4$S $9p$	
4 −.03916 $3s^23p^3\ ^4$S $7s$	8 −.01381 $3s^23p^3\ ^4$S $9d$	$^5\mathbf{D}^o$	
5 −.02729 $3s^23p^3\ ^4$S $8s$	$^3\mathbf{F}^e$	1 −.14927 $3s^23p^3\ ^4$S $3d$	
6 −.02010 $3s^23p^3\ ^4$S $9s$	1 −.06252 $3s^23p^3\ ^4$S $4f$	2 −.08036 $3s^23p^3\ ^4$S $4d$	
$^3\mathbf{P}^e$	2 −.04128 $3s^23p^3\ ^2$D $4p$	3 −.04901 $3s^23p^3\ ^4$S $5d$	
1 −.79005 $3s^23p^4$	3 −.03999 $3s^23p^3\ ^4$S $5f$	4 −.03290 $3s^23p^3\ ^4$S $6d$	

Energies in ascending order from ground state for terms with effective $n \leq 4.0$, $L \leq 3$

Term	i	E(Ryds)	Term	i	E(Ryds)	Term	i	E(Ryds)	Term	i	E(Ryds)	Term	i	E(Ryds)
$^3\mathbf{P}^e$	1	0.00000	$^5\mathbf{P}^e$	1	0.60982	$^3\mathbf{D}^o$	2	0.67237	$^3\mathbf{P}^e$	3	0.70722	$^3\mathbf{F}^e$	1	0.72753
$^1\mathbf{D}^e$	1	0.09161	$^3\mathbf{P}^e$	2	0.62616	$^1\mathbf{D}^o$	1	0.67439	$^5\mathbf{D}^o$	2	0.70969			
$^1\mathbf{S}^e$	1	0.21672	$^5\mathbf{D}^o$	1	0.64078	$^5\mathbf{S}^o$	2	0.67448	$^3\mathbf{D}^o$	3	0.72077			
$^5\mathbf{S}^o$	1	0.51342	$^3\mathbf{P}^o$	1	0.64295	$^3\mathbf{S}^o$	2	0.68250	$^5\mathbf{S}^o$	3	0.72604			
$^3\mathbf{S}^o$	1	0.54854	$^3\mathbf{D}^o$	1	0.65488	$^5\mathbf{P}^e$	2	0.70307	$^5\mathbf{F}^e$	1	0.72749			

gf-values for transitions involving terms with effective $n \leq 4.0$, $L \leq 3$

i i'	gf_L	i i'	gf_L	i i'	gf_L	i i'	gf_L	i i'	gf_L	i i'	gf_L
	$^1\mathbf{D}^e$–$^1\mathbf{D}^o$	3 1	1.78E−2	1 1	−3.90E−1	3 2	1.69E+0	1 1	5.66E+0	1 1	−4.07E+0
1 1	−9.54E−1	3 2	4.61E+0	1 2	−1.36E+0	3 3	−7.52E+0	1 2	−2.97E+0	1 2	−5.01E+0
	$^3\mathbf{P}^e$–$^3\mathbf{S}^o$		$^3\mathbf{P}^e$–$^3\mathbf{P}^o$	1 3	−2.52E−1		$^3\mathbf{F}^e$–$^3\mathbf{D}^o$	1 3	−3.07E−1	2 1	6.74E−1
1 1	−8.27E−1	1 1	−4.47E−3	2 1	−6.11E−1	1 1	1.16E+0	2 1	6.32E−2	2 2	−5.32E+0
1 2	−1.34E−1	2 1	−4.97E−4	2 2	−5.40E+0	1 2	1.06E+1	2 2	7.88E+0		$^5\mathbf{F}^e$–$^5\mathbf{D}^o$
2 1	3.42E+0	3 1	3.48E−3	2 3	−6.04E−1	1 3	5.20E+0	2 3	−4.99E+0	1 1	8.69E+0
2 2	−1.81E+0		$^3\mathbf{P}^e$–$^3\mathbf{D}^o$	3 1	8.10E−3		$^5\mathbf{P}^e$–$^5\mathbf{S}^o$		$^5\mathbf{P}^e$–$^5\mathbf{D}^o$	1 2	2.26E+1

S-like Ar (Ar^{2+})

Term energies relative to $3s^23p^3$ ^{4}S ionization threshold for each symmetry

i	E(Ryds)	Description	i	E(Ryds)	Description	i	E(Ryds)	Description	i	E(Ryds)	Description
		^{1}S^e	14	-.04188	$3s^23p^3$ ^{2}P $5f$	2	-.53930	$3s^23p^3$ ^{2}D $4d$	4	-1.01015	$3s^23p^3$ ^{2}P $3d$
1	-2.71474	$3s^23p^4$			**^{1}D^o**	3	-.24531	$3s^23p^3$ ^{2}D $5d$	5	-.50532	$3s^23p^3$ ^{2}D $4d$
2	-.71570	$3s3p^4$ ^{2}S $4s$	1	-1.35484	$3s^23p^3$ ^{2}D $4s$	4	-.16716	$3s^23p^3$ ^{2}D $5g$	6	-.41387	$3s^23p^3$ ^{2}P $4d$
3	-.63892	$3s^23p^3$ ^{2}P $4p$	2	-1.12977	$3s^23p^3$ ^{2}D $3d$	5	-.10041	$3s^23p^3$ ^{2}D $6d$	7	-.37400	$3s^23p^3$ ^{2}P $5s$
4	-.22551	$3s^23p^3$ ^{2}P $5p$	3	-.96052	$3s^23p^3$ ^{2}P $3d$	6	-.05719	$3s^23p^3$ ^{2}D $6g$	8	-.23422	$3s^23p^3$ ^{2}D $5d$
5	-.03140	$3s^23p^3$ ^{2}P $6p$	4	-.50311	$3s^23p^3$ ^{2}D $4d$	7	-.04140	$3s^23p^3$ ^{2}P $5g$	9	-.12148	$3s^23p^3$ ^{2}P $5d$
		^{1}S^o	5	-.48114	$3s^23p^3$ ^{2}D $5s$	8	-.01738	$3s^23p^3$ ^{2}D $7d$	10	-.10460	$3s^23p^3$ ^{2}P $6s$
1	-1.53082	$3s^23p^3$ ^{2}D $3d$	6	-.40544	$3s^23p^3$ ^{2}P $4d$			**^{1}H^e**	11	-.09447	$3s^23p^3$ ^{2}D $6d$
2	-.55844	$3s^23p^3$ ^{2}D $4d$	7	-.23111	$3s^23p^3$ ^{2}D $5d$	1	-.37835	$3s^23p^3$ ^{2}D $4f$	12	-.01487	$3s^23p^3$ ^{2}D $7d$
3	-.25373	$3s^23p^3$ ^{2}D $5d$	8	-.21990	$3s^23p^3$ ^{2}D $6s$	2	-.17442	$3s^23p^3$ ^{2}D $5f$			**^{3}D^e**
4	-.10514	$3s^23p^3$ ^{2}D $6d$	9	-.16721	$3s^23p^3$ ^{2}D $5g$	3	-.06236	$3s^23p^3$ ^{2}D $6f$	1	-.91844	$3s^23p^3$ ^{2}D $4p$
5	-.02034	$3s^23p^3$ ^{2}D $7d$	10	-.12251	$3s^23p^3$ ^{2}P $5d$	4	-.05710	$3s^23p^3$ ^{2}D $6h$	2	-.77792	$3s^23p^3$ ^{2}P $4p$
		^{1}P^e	11	-.09180	$3s^23p^3$ ^{2}D $7s$			**^{1}H^o**	3	-.40731	$3s^23p^3$ ^{2}D $5p$
1	-.93084	$3s^23p^3$ ^{2}D $4p$	12	-.08666	$3s^23p^3$ ^{2}D $6d$	1	-.16728	$3s^23p^3$ ^{2}D $5g$	4	-.39279	$3s3p^4$ ^{2}D $4s$
2	-.76122	$3s^23p^3$ ^{2}P $4p$	13	-.05718	$3s^23p^3$ ^{2}D $6g$	2	-.05742	$3s^23p^3$ ^{2}D $6g$	5	-.37807	$3s^23p^3$ ^{2}D $4f$
3	-.40280	$3s^23p^3$ ^{2}D $5p$	14	-.01401	$3s^23p^3$ ^{2}D $7d$	3	-.04130	$3s^23p^3$ ^{2}P $5g$	6	-.27425	$3s^23p^3$ ^{2}P $5p$
4	-.37914	$3s^23p^3$ ^{2}D $4f$	15	-.00947	$3s^23p^3$ ^{2}D $8s$			**^{3}S^e**	7	-.25394	$3s^23p^3$ ^{2}P $4f$
5	-.27272	$3s^23p^3$ ^{2}P $5p$			**^{1}F^e**	1	-.79194	$3s^23p^3$ ^{2}P $4p$	8	-.18178	$3s^23p^3$ ^{2}D $6p$
6	-.18005	$3s^23p^3$ ^{2}D $6p$	1	-.89219	$3s^23p^3$ ^{2}D $4p$	2	-.27783	$3s^23p^3$ ^{2}P $5p$	9	-.17911	$3s^23p^3$ ^{2}D $5f$
7	-.17856	$3s^23p^3$ ^{2}D $5f$	2	-.39534	$3s^23p^3$ ^{2}D $5p$	3	-.05662	$3s^23p^3$ ^{2}P $6p$	10	-.14218	$3s3p^4$ ^{4}P $3d$
8	-.07895	$3s3p^4$ ^{2}P $4s$	3	-.37468	$3s^23p^3$ ^{2}D $4f$			**^{3}S^o**	11	-.06871	$3s^23p^3$ ^{2}D $7p$
9	-.06661	$3s^23p^3$ ^{2}D $7p$	4	-.25520	$3s^23p^3$ ^{2}P $4f$	1	-1.30276	$3s^23p^3$ ^{4}S $4s$	12	-.06220	$3s^23p^3$ ^{2}D $6f$
10	-.05066	$3s^23p^3$ ^{2}P $6p$	5	-.19109	$3s^23p^3$ ^{2}D $5f$	2	-1.10240	$3s^23p^3$ ^{2}D $3d$	13	-.05395	$3s^23p^3$ ^{2}P $6p$
11	-.03387	$3s^23p^3$ ^{2}D $6f$	6	-.17654	$3s^23p^3$ ^{2}D $6p$	3	-.67698	$3s^23p^3$ ^{4}S $5s$	14	-.05049	$3s^23p^3$ ^{2}P $5f$
		^{1}P^o	7	-.15291	$3s3p^4$ ^{2}D $3d$	4	-.50465	$3s^23p^3$ ^{2}D $4d$	15	-.03062	$3s3p^4$ ^{2}D $3d$
1	-1.71261	$3s3p^5$	8	-.06344	$3s^23p^3$ ^{2}D $7p$	5	-.41475	$3s^23p^3$ ^{4}S $6s$			**^{3}D^o**
2	-1.03203	$3s^23p^3$ ^{2}P $4s$	9	-.05976	$3s^23p^3$ ^{2}D $6f$	6	-.28132	$3s^23p^3$ ^{4}S $7s$	1	-1.57303	$3s^23p^3$ ^{4}S $3d$
3	-.95152	$3s^23p^3$ ^{2}D $3d$	10	-.05711	$3s^23p^3$ ^{2}D $6h$	7	-.23503	$3s^23p^3$ ^{2}D $5d$	2	-1.26960	$3s^23p^3$ ^{2}D $3d$
4	-.79154	$3s^23p^3$ ^{2}P $3d$	11	-.04949	$3s^23p^3$ ^{2}P $5f$	8	-.20255	$3s^23p^3$ ^{4}S $8s$	3	-1.16651	$3s^23p^3$ ^{2}D $4s$
5	-.51108	$3s^23p^3$ ^{2}D $4d$			**^{1}F^o**	9	-.15330	$3s^23p^3$ ^{4}S $9s$	4	-1.04125	$3s^23p^3$ ^{2}P $3d$
6	-.36434	$3s^23p^3$ ^{2}P $5s$	1	-1.14898	$3s^23p^3$ ^{2}D $3d$			**^{3}P^e**	5	-.68444	$3s^23p^3$ ^{4}S $4d$
7	-.32952	$3s^23p^3$ ^{2}P $4d$	2	-.90835	$3s^23p^3$ ^{2}P $3d$	1	-3.03140	$3s^23p^4$	6	-.53412	$3s^23p^3$ ^{2}D $4d$
8	-.23660	$3s^23p^3$ ^{2}D $5d$	3	-.47054	$3s^23p^3$ ^{2}D $4d$	2	-1.05900	$3s^23p^3$ ^{4}S $4p$	7	-.50034	$3s^23p^3$ ^{2}D $5s$
9	-.10343	$3s^23p^3$ ^{2}P $6s$	4	-.38113	$3s^23p^3$ ^{2}P $4d$	3	-.85202	$3s^23p^3$ ^{2}D $4p$	8	-.42449	$3s^23p^3$ ^{4}S $5d$
10	-.09594	$3s^23p^3$ ^{2}D $6d$	5	-.21937	$3s^23p^3$ ^{2}D $5d$	4	-.74604	$3s^23p^3$ ^{2}P $4p$	9	-.39615	$3s^23p^3$ ^{2}P $4d$
11	-.08636	$3s^23p^3$ ^{2}P $5d$	6	-.16732	$3s^23p^3$ ^{2}D $5g$	5	-.57200	$3s^23p^3$ ^{4}S $5p$	10	-.28581	$3s^23p^3$ ^{4}S $6d$
12	-.01539	$3s^23p^3$ ^{2}D $7d$	7	-.10935	$3s^23p^3$ ^{2}P $5d$	6	-.49169	$3s3p^4$ ^{4}P $4s$	11	-.24503	$3s^23p^3$ ^{2}D $5d$
		^{1}D^e	8	-.08728	$3s^23p^3$ ^{2}D $6d$	7	-.39166	$3s^23p^3$ ^{2}D $5p$	12	-.22986	$3s^23p^3$ ^{2}D $6s$
1	-2.89974	$3s^23p^4$	9	-.05753	$3s^23p^3$ ^{2}D $6g$	8	-.37154	$3s^23p^3$ ^{2}D $4f$	13	-.20431	$3s^23p^3$ ^{4}S $7d$
2	-.80398	$3s^23p^3$ ^{2}D $4p$	10	-.04113	$3s^23p^3$ ^{2}P $5g$	9	-.35895	$3s^23p^3$ ^{4}S $6p$	14	-.16716	$3s^23p^3$ ^{2}D $5g$
3	-.73270	$3s^23p^3$ ^{2}P $4p$	11	-.01008	$3s^23p^3$ ^{2}D $7d$	10	-.27184	$3s^23p^3$ ^{2}P $5p$	15	-.15527	$3s^23p^3$ ^{4}S $8d$
4	-.37558	$3s^23p^3$ ^{2}D $4f$			**^{1}G^e**	11	-.25268	$3s^23p^3$ ^{4}S $7p$	16	-.12305	$3s^23p^3$ ^{4}S $9d$
5	-.36340	$3s^23p^3$ ^{2}D $5p$	1	-.37266	$3s^23p^3$ ^{2}D $4f$	12	-.21149	$3s3p^4$ ^{4}P $3d$	17	-.11689	$3s^23p^3$ ^{2}P $5d$
6	-.25968	$3s^23p^3$ ^{2}P $5p$	2	-.26898	$3s^23p^3$ ^{2}P $4f$	13	-.18635	$3s^23p^3$ ^{4}S $8p$	18	-.10186	$3s^23p^3$ ^{2}D $6d$
7	-.24892	$3s^23p^3$ ^{2}P $4f$	3	-.23519	$3s3p^4$ ^{2}D $3d$	14	-.17042	$3s^23p^3$ ^{2}D $5f$			**^{3}F^e**
8	-.19059	$3s3p^4$ ^{2}D $4s$	4	-.16785	$3s^23p^3$ ^{2}D $5f$	15	-.16414	$3s^23p^3$ ^{2}D $6p$	1	-.90307	$3s^23p^3$ ^{2}D $4p$
9	-.17048	$3s^23p^3$ ^{2}D $5f$	5	-.05872	$3s^23p^3$ ^{2}D $6f$	16	-.14006	$3s^23p^3$ ^{4}S $9p$	2	-.56860	$3s^23p^3$ ^{4}S $4f$
10	-.16026	$3s^23p^3$ ^{2}D $6p$	6	-.05712	$3s^23p^3$ ^{2}D $6h$			**^{3}P^o**	3	-.43178	$3s3p^4$ ^{4}P $3d$
11	-.06231	$3s^23p^3$ ^{2}D $6f$	7	-.04483	$3s^23p^3$ ^{2}P $5f$	1	-2.01253	$3s3p^5$	4	-.39798	$3s^23p^3$ ^{2}D $5p$
12	-.05573	$3s^23p^3$ ^{2}D $7p$			**^{1}G^o**	2	-1.26928	$3s^23p^3$ ^{2}D $3d$	5	-.37574	$3s^23p^3$ ^{2}D $4f$
13	-.04943	$3s^23p^3$ ^{2}P $6p$	1	-1.38719	$3s^23p^3$ ^{2}D $3d$	3	-1.06921	$3s^23p^3$ ^{2}P $4s$	6	-.36259	$3s^23p^3$ ^{4}S $5f$

S-like Ar (Ar^{2+})

i	E(Ryds)	Description	i	E(Ryds)	Description	i	E(Ryds)	Description	i	E(Ryds)	Description
7	−.25612	$3s^23p^3\ ^2$P $4f$		$^3\mathbf{G}^o$		6	−.15533	$3s^23p^3\ ^4$S $9s$	6	−.14308	$3s^23p^3\ ^4$S $8f$
8	−.25183	$3s^23p^3\ ^4$S $6f$	1	−1.42350	$3s^23p^3\ ^2$D $3d$		$^5\mathbf{P}^e$		7	−.11295	$3s^23p^3\ ^4$S $9f$
9	−.20258	$3s3p^4\ ^2$D $3d$	2	−.54325	$3s^23p^3\ ^2$D $4d$	1	−1.10769	$3s^23p^3\ ^4$S $4p$		$^5\mathbf{G}^o$	
10	−.18641	$3s^23p^3\ ^4$S $7f$	3	−.36010	$3s^23p^3\ ^4$S $5g$	2	−.59618	$3s^23p^3\ ^4$S $5p$	1	−.36021	$3s^23p^3\ ^4$S $5g$
11	−.17921	$3s^23p^3\ ^2$D $6p$	4	−.25009	$3s^23p^3\ ^4$S $6g$	3	−.52956	$3s3p^4\ ^4$P $4s$	2	−.25020	$3s^23p^3\ ^4$S $6g$
12	−.16191	$3s^23p^3\ ^2$D $5f$	5	−.24682	$3s^23p^3\ ^2$D $5d$	4	−.37562	$3s^23p^3\ ^4$S $6p$	3	−.18383	$3s^23p^3\ ^4$S $7g$
13	−.14301	$3s^23p^3\ ^4$S $8f$	6	−.18374	$3s^23p^3\ ^4$S $7g$	5	−.30676	$3s3p^4\ ^4$P $3d$	4	−.14074	$3s^23p^3\ ^4$S $8g$
14	−.11329	$3s^23p^3\ ^4$S $9f$	7	−.16727	$3s^23p^3\ ^2$D $5g$	6	−.25583	$3s^23p^3\ ^4$S $7p$	5	−.11120	$3s^23p^3\ ^4$S $9g$
	$^3\mathbf{F}^o$		8	−.14068	$3s^23p^3\ ^4$S $8g$	7	−.18766	$3s^23p^3\ ^4$S $8p$			
1	−1.51391	$3s^23p^3\ ^2$D $3d$	9	−.11115	$3s^23p^3\ ^4$S $9g$	8	−.14328	$3s^23p^3\ ^4$S $9p$			
2	−1.28977	$3s^23p^3\ ^2$P $3d$	10	−.10120	$3s^23p^3\ ^2$D $6d$		$^5\mathbf{P}^o$				
3	−.55306	$3s^23p^3\ ^2$P $4d$		$^3\mathbf{H}^e$		1	−.06226	$3s3p^4\ ^4$P $4p$			
4	−.41919	$3s^23p^3\ ^2$D $4d$	1	−.37845	$3s^23p^3\ ^2$D $4f$		$^5\mathbf{D}^e$				
5	−.25007	$3s^23p^3\ ^2$D $5d$	2	−.25000	$3s^23p^3\ ^4$S $6h$	1	−.74843	$3s3p^4\ ^4$P $3d$			
6	−.16737	$3s^23p^3\ ^2$D $5g$	3	−.18368	$3s^23p^3\ ^4$S $7h$		$^5\mathbf{D}^o$				
7	−.12408	$3s^23p^3\ ^2$P $5d$	4	−.17447	$3s^23p^3\ ^2$D $5f$	1	−1.68977	$3s^23p^3\ ^4$S $3d$			
8	−.10178	$3s^23p^3\ ^2$D $6d$	5	−.14063	$3s^23p^3\ ^4$S $8h$	2	−.74453	$3s^23p^3\ ^4$S $4d$			
9	−.05758	$3s^23p^3\ ^2$D $6g$	6	−.11112	$3s^23p^3\ ^4$S $9h$	3	−.44353	$3s^23p^3\ ^4$S $5d$			
10	−.04123	$3s^23p^3\ ^2$P $5g$		$^3\mathbf{H}^o$		4	−.29639	$3s^23p^3\ ^4$S $6d$			
11	−.01935	$3s^23p^3\ ^2$D $7d$	1	−.16741	$3s^23p^3\ ^2$D $5g$	5	−.21234	$3s^23p^3\ ^4$S $7d$			
	$^3\mathbf{G}^e$		2	−.05756	$3s^23p^3\ ^2$D $6g$	6	−.15967	$3s^23p^3\ ^4$S $8d$			
1	−.39136	$3s^23p^3\ ^2$D $4f$	3	−.04137	$3s^23p^3\ ^2$P $5g$	7	−.12446	$3s^23p^3\ ^4$S $9d$			
2	−.30604	$3s3p^4\ ^2$D $3d$		$^5\mathbf{S}^o$			$^5\mathbf{F}^e$				
3	−.24677	$3s^23p^3\ ^2$P $4f$	1	−1.37695	$3s^23p^3\ ^4$S $4s$	1	−.59834	$3s3p^4\ ^4$P $3d$			
4	−.17004	$3s^23p^3\ ^2$D $5f$	2	−.69763	$3s^23p^3\ ^4$S $5s$	2	−.53883	$3s^23p^3\ ^4$S $4f$			
5	−.06149	$3s^23p^3\ ^2$D $6f$	3	−.42459	$3s^23p^3\ ^4$S $6s$	3	−.36511	$3s^23p^3\ ^4$S $5f$			
6	−.05712	$3s^23p^3\ ^2$D $6h$	4	−.28610	$3s^23p^3\ ^4$S $7s$	4	−.25443	$3s^23p^3\ ^4$S $6f$			
7	−.04775	$3s^23p^3\ ^2$P $5f$	5	−.20594	$3s^23p^3\ ^4$S $8s$	5	−.18698	$3s^23p^3\ ^4$S $7f$			

Energies in ascending order from ground state for terms with effective $n \leq 4.0$, $L \leq 4$

Term	i	E(Ryds)	Term	i	E(Ryds)	Term	i	E(Ryds)	Term	i	E(Ryds)	Term	i	E(Ryds)
$^3\mathbf{P}^e$	1	0.00000	$^5\mathbf{S}^o$	1	1.65445	$^3\mathbf{P}^o$	3	1.96219	$^1\mathbf{F}^e$	1	2.13921	$^1\mathbf{S}^e$	2	2.31570
$^1\mathbf{D}^e$	1	0.13166	$^1\mathbf{D}^o$	1	1.67656	$^3\mathbf{P}^e$	2	1.97240	$^3\mathbf{P}^e$	3	2.17938	$^5\mathbf{S}^o$	2	2.33377
$^1\mathbf{S}^e$	1	0.31666	$^3\mathbf{S}^o$	1	1.72864	$^3\mathbf{D}^o$	4	1.99015	$^1\mathbf{D}^e$	2	2.22742	$^3\mathbf{D}^o$	5	2.34696
$^3\mathbf{P}^o$	1	1.01887	$^3\mathbf{F}^o$	2	1.74163	$^1\mathbf{P}^o$	2	1.99937	$^3\mathbf{S}^e$	1	2.23946	$^3\mathbf{S}^o$	3	2.35442
$^1\mathbf{P}^o$	1	1.31879	$^3\mathbf{D}^o$	2	1.76180	$^3\mathbf{P}^o$	4	2.02125	$^1\mathbf{P}^o$	4	2.23986	$^1\mathbf{S}^e$	3	2.39248
$^5\mathbf{D}^o$	1	1.34163	$^3\mathbf{P}^o$	2	1.76212	$^1\mathbf{D}^o$	3	2.07088	$^3\mathbf{D}^e$	2	2.25348	$^5\mathbf{F}^e$	1	2.43306
$^3\mathbf{D}^o$	1	1.45837	$^3\mathbf{D}^o$	3	1.86489	$^1\mathbf{P}^o$	3	2.07987	$^1\mathbf{P}^e$	2	2.27018	$^5\mathbf{P}^e$	2	2.43522
$^1\mathbf{S}^o$	1	1.50058	$^1\mathbf{F}^o$	1	1.88242	$^1\mathbf{P}^e$	1	2.10056	$^5\mathbf{D}^e$	1	2.28297	$^3\mathbf{P}^e$	5	2.45940
$^3\mathbf{F}^o$	1	1.51749	$^1\mathbf{D}^o$	2	1.90163	$^3\mathbf{D}^e$	1	2.11296	$^3\mathbf{P}^e$	4	2.28536	$^3\mathbf{F}^e$	2	2.46280
$^3\mathbf{G}^o$	1	1.60790	$^5\mathbf{P}^e$	1	1.92371	$^1\mathbf{F}^o$	2	2.12305	$^5\mathbf{D}^o$	2	2.28688			
$^1\mathbf{G}^o$	1	1.64421	$^3\mathbf{S}^o$	2	1.92900	$^3\mathbf{F}^e$	1	2.12833	$^1\mathbf{D}^e$	3	2.29870			

S-like Ar (Ar^{2+})

gf-values for transitions involving terms with effective $n \leq 4.0$, $L \leq 4$

i i′	gf$_L$	*i i′*	gf$_L$	*i i′*	gf$_L$	*i i′*	gf$_L$	*i i′*	gf$_L$	*i i′*	gf$_L$
	$^1S^e$–$^1P^o$	1 2	−2.45E+0	1 2	2.30E−2	3 2	1.17E−1	5 3	8.98E−2	2 1	1.85E+0
1 1	−6.46E−4	1 3	−1.26E+0		$^1F^e$–$^1G^o$	3 3	5.58E−1	5 4	1.15E−2	2 2	8.91E−1
1 2	−1.01E−1	1 4	−9.78E−3	1 1	1.15E+0	3 4	8.09E−2	5 5	4.11E+0	2 3	4.57E−3
1 3	−1.86E−1	2 1	5.94E−2		$^3S^e$–$^3P^o$	4 1	7.68E−3		$^3D^e$–$^3P^o$	2 4	4.93E+0
1 4	−3.31E+0	2 2	1.01E−3	1 1	8.18E−2	4 2	3.77E−1	1 1	3.83E−2	2 5	9.14E+0
2 1	4.70E−2	2 3	7.45E−2	1 2	4.60E−1	4 3	2.29E+0	1 2	5.65E−2		$^3F^e$–$^3F^o$
2 2	2.59E−2	2 4	−2.16E−4	1 3	2.78E−1	4 4	1.25E+0	1 3	2.51E−1	1 1	7.48E−1
2 3	2.31E−2	3 1	1.25E−6	1 4	7.75E−1	5 1	6.59E−3	1 4	3.45E−4	1 2	1.21E−1
2 4	1.99E−2	3 2	1.47E+0		$^3P^e$–$^3S^o$	5 2	4.15E−4	2 1	2.89E−2	2 1	7.40E−6
3 1	5.99E−2	3 3	4.53E−1	1 1	−1.15E+0	5 3	5.24E−2	2 2	3.53E−2	2 2	1.21E−6
3 2	3.04E−1	3 4	1.06E−2	1 2	−2.40E+0	5 4	1.51E−3	2 3	2.68E+0		$^3F^e$–$^3G^o$
3 3	8.49E−2		$^1D^e$–$^1D^o$	1 3	−2.19E−1		$^3P^e$–$^3D^o$	2 4	2.46E+0	1 1	3.37E+0
3 4	2.07E−1	1 1	−6.09E−2	2 1	3.01E+0	1 1	−2.45E−2		$^3D^e$–$^3D^o$	2 1	7.29E−4
	$^1P^e$–$^1S^o$	1 2	−1.43E+0	2 2	7.50E−3	1 2	−2.92E−1	1 1	4.40E−1		$^5P^e$–$^5S^o$
1 1	9.19E−2	1 3	−4.47E+0	2 3	−1.41E+0	1 3	−1.76E+0	1 2	6.50E−1	1 1	5.37E+0
2 1	2.56E−2	2 1	1.16E−1	3 1	4.80E−1	1 4	−1.67E+1	1 3	4.72E+0	1 2	−2.66E+0
	$^1P^e$–$^1P^o$	2 2	2.17E+0	3 2	2.97E−1	1 5	−1.52E+0	1 4	8.24E−2	2 1	5.49E−2
1 1	1.18E−1	2 3	2.29E−1	3 3	−3.75E−1	2 1	1.10E+0	1 5	−3.51E−1	2 2	7.27E+0
1 2	1.42E−1	3 1	1.09E−1	4 1	3.38E−2	2 2	2.12E−1	2 1	1.07E−1		$^5P^e$–$^5D^o$
1 3	1.21E−3	3 2	1.46E−1	4 2	3.81E−2	2 3	8.74E−2	2 2	7.50E−2	1 1	3.00E+0
1 4	−1.35E−2	3 3	5.82E−2	4 3	−1.22E−2	2 4	−9.70E−2	2 3	2.33E−1	1 2	−1.78E+1
2 1	6.75E−3		$^1D^e$–$^1F^o$	5 1	4.01E−3	2 5	−9.21E+0	2 4	2.77E−1	2 1	3.32E−1
2 2	5.19E−1	1 1	−1.53E−2	5 2	5.20E−3	3 1	8.65E−2	2 5	−4.53E−2	2 2	7.27E+0
2 3	4.27E−1	1 2	−9.01E+0	5 3	4.55E+0	3 2	1.36E−1		$^3D^e$–$^3F^o$		$^5D^e$–$^5D^o$
2 4	1.99E−2	2 1	3.98E−1		$^3P^e$–$^3P^o$	3 3	3.58E+0	1 1	1.60E+0	1 1	2.66E−1
	$^1P^e$–$^1D^o$	2 2	1.95E−1	1 1	−3.40E−1	3 4	4.79E−1	1 2	7.73E−2	1 2	−5.74E−7
1 1	1.35E−1	3 1	3.89E−1	1 2	−3.33E−2	3 5	−2.61E−1	2 1	1.90E−1		$^5F^e$–$^5D^o$
1 2	8.01E−1	3 2	4.60E−1	1 3	−8.41E+0	4 1	5.41E−3	2 2	2.36E+0	1 1	1.64E+0
1 3	1.79E−2		$^1F^e$–$^1D^o$	1 4	−3.32E+0	4 2	8.01E−1		$^3F^e$–$^3D^o$	1 2	1.10E+1
2 1	2.29E−1	1 1	2.37E−2	2 1	7.48E−3	4 3	1.20E−1	1 1	8.34E−2		
2 2	1.53E−1	1 2	2.43E+0	2 2	1.61E−3	4 4	5.83E−1	1 2	9.28E−3		
2 3	1.78E−1	1 3	1.52E−2	2 3	4.12E−3	4 5	−1.15E−3	1 3	7.58E+0		
	$^1D^e$–$^1P^o$		$^1F^e$–$^1F^o$	2 4	−4.91E−4	5 1	6.39E−2	1 4	7.88E−3		
1 1	−1.82E−1	1 1	8.36E−2	3 1	1.68E−1	5 2	2.16E−3	1 5	−1.48E−2		

S-like Ca (Ca^{4+})

Term energies relative to $3s^23p^3$ ^{4}S ionization threshold for each symmetry

i	E(Ryds)	Description
	^{1}S^e	
1	−5.83178	$3s^23p^4$
2	−3.11635	$3p^6$
3	−1.96922	$3s^23p^3$ ^{2}P $4p$
4	−1.63732	$3s3p^4$ ^{2}D $3d$:
5	−.97825	$3s^23p^3$ ^{2}P $5p$
6	−.83021	$3s^23p^23d^2$:
7	−.80446	$3s^23p^23d^2$:
8	−.49296	$3s^23p^3$ ^{2}P $6p$
9	−.43739	$3s^23p^23d^2$
10	−.22925	$3s^23p^3$ ^{2}P $7p$
11	−.16163	$3s3p^4$ ^{2}D $4d$
12	−.06526	$3s^23p^3$ ^{2}P $8p$
	^{1}S^o	
1	−4.12024	$3s^23p^3$ ^{2}D $3d$
2	−1.73811	$3s^23p^3$ ^{2}D $4d$
3	−.95913	$3s^23p^3$ ^{2}D $5d$
4	−.57349	$3s^23p^3$ ^{2}D $6d$
5	−.43630	$3s3p^4$ ^{2}P $4p$
6	−.32280	$3s^23p^3$ ^{2}D $7d$
7	−.18638	$3s^23p^3$ ^{2}D $8d$
8	−.09192	$3s^23p^3$ ^{2}D $9d$
	^{1}P^e	
1	−2.39930	$3s^23p^3$ ^{2}D $4p$
2	−2.16546	$3s^23p^3$ ^{2}P $4p$
3	−2.12716	$3s3p^4$ ^{2}D $3d$
4	−1.44665	$3s^23p^23d^2$
5	−1.30825	$3s^23p^3$ ^{2}D $4f$
6	−1.22899	$3s^23p^3$ ^{2}D $5p$
7	−1.07993	$3s^23p^23d^2$
8	−1.03988	$3s^23p^3$ ^{2}P $5p$
9	−.91523	$3s3p^4$ ^{2}P $3d$:
10	−.78598	$3s^23p^3$ ^{2}D $5f$
11	−.69613	$3s^23p^3$ ^{2}D $6p$
12	−.56130	$3p^43d^2$:
13	−.53113	$3s^23p^3$ ^{2}P $6p$
14	−.44309	$3s^23p^3$ ^{2}D $6f$
15	−.41271	$3s^23p^3$ ^{2}D $7p$
16	−.27002	$3s^23p^3$ ^{2}D $7f$
17	−.25208	$3s^23p^3$ ^{2}P $7p$
18	−.23835	$3s^23p^3$ ^{2}D $8p$
19	−.16830	$3s3p^4$ ^{2}D $4d$
20	−.14517	$3s^23p^3$ ^{2}D $8f$
21	−.12912	$3s^23p^3$ ^{2}D $9p$
22	−.07814	$3s^23p^3$ ^{2}P $8p$
23	−.06473	$3s^23p^3$ ^{2}D $9f$
	^{1}P^o	
1	−4.44212	$3s3p^5$
2	−3.40891	$3s^23p^3$ ^{2}D $3d$
3	−3.00342	$3s^23p^3$ ^{2}P $3d$
4	−2.56079	$3s^23p^3$ ^{2}P $4s$
5	−1.69931	$3s^23p^3$ ^{2}D $4d$
6	−1.46146	$3s^23p^3$ ^{2}P $4d$
7	−1.21635	$3s^23p^3$ ^{2}P $5s$
8	−.93961	$3s^23p^3$ ^{2}D $5d$
9	−.81041	$3s3p^4$ ^{2}D $4p$
10	−.74401	$3s^23p^3$ ^{2}P $5d$
11	−.61995	$3s^23p^3$ ^{2}P $6s$
12	−.54884	$3s^23p^3$ ^{2}D $6d$
13	−.45031	$3s3p^4$ ^{2}P $4p$
14	−.39755	$3s3p^4$ ^{2}S $4p$
15	−.37486	$3s^23p^3$ ^{2}P $6d$
16	−.32644	$3s^23p^3$ ^{2}D $7d$
17	−.29820	$3s^23p^3$ ^{2}P $7s$
18	−.18655	$3s^23p^3$ ^{2}D $8d$
19	−.15727	$3s^23p^3$ ^{2}P $7d$
20	−.10994	$3s^23p^3$ ^{2}P $8s$
21	−.09369	$3s3p^33d^2$:
22	−.09053	$3s^23p^3$ ^{2}D $9d$
	^{1}D^e	
1	−6.06963	$3s^23p^4$
2	−2.30795	$3s3p^4$ ^{2}D $3d$:
3	−2.17673	$3s^23p^3$ ^{2}D $4p$
4	−2.10930	$3s^23p^3$ ^{2}P $4p$
5	−1.87448	$3s3p^4$ ^{2}P $3d$:
6	−1.58432	$3s3p^4$ ^{2}S $3d$:
7	−1.40079	$3s^23p^3$ ^{2}D $4f$:
8	−1.26376	$3s^23p^23d^2$:
9	−1.20767	$3s3p^4$ ^{2}D $4s$:
10	−1.15974	$3s^23p^3$ ^{2}P $4f$
11	−1.12938	$3s^23p^3$ ^{2}D $5p$
12	−1.09774	$3s^23p^23d^2$
13	−1.01698	$3s^23p^3$ ^{2}P $5p$
14	−.85601	$3s^23p^23d^2$
15	−.74269	$3s^23p^3$ ^{2}D $5f$
16	−.66355	$3s^23p^3$ ^{2}D $6p$
17	−.62689	$3s^23p^3$ ^{2}P $5f$
18	−.54560	$3s^23p^23d^2$
19	−.51916	$3s^23p^3$ ^{2}P $6p$
20	−.43592	$3s^23p^3$ ^{2}D $6f$
21	−.39801	$3s^23p^3$ ^{2}D $7p$
22	−.37898	$3s^23p^23d^2$
23	−.26500	$3s^23p^3$ ^{2}D $7f$
24	−.24930	$3s^23p^3$ ^{2}P $6f$
25	−.24322	$3s^23p^3$ ^{2}P $7p$
26	−.23014	$3s^23p^3$ ^{2}D $8p$
27	−.15560	$3s^23p^23d^2$
28	−.14122	$3s^23p^3$ ^{2}D $8f$
29	−.12893	$3s^23p^3$ ^{2}D $9p$
30	−.10778	$3p^43d^2$
31	−.08594	$3s^23p^3$ ^{2}P $7f$
32	−.07341	$3s^23p^3$ ^{2}P $8p$
33	−.06128	$3s^23p^3$ ^{2}D $9f$
	^{1}D^o	
1	−3.88501	$3s^23p^3$ ^{2}D $3d$
2	−3.25368	$3s^23p^3$ ^{2}P $3d$
3	−2.72351	$3s^23p^3$ ^{2}D $4s$
4	−1.67598	$3s^23p^3$ ^{2}D $4d$
5	−1.53163	$3s^23p^3$ ^{2}P $4d$
6	−1.37649	$3s^23p^3$ ^{2}D $5s$
7	−.92971	$3s^23p^3$ ^{2}D $5d$
8	−.79275	$3s^23p^3$ ^{2}P $5d$
9	−.77274	$3s^23p^3$ ^{2}D $6s$
10	−.75443	$3s^23p^3$ ^{2}D $5g$
11	−.74289	$3s3p^4$ ^{2}D $4p$
12	−.66023	$3p^53d$
13	−.55154	$3s^23p^3$ ^{2}D $6d$
14	−.49994	$3s3p^4$ ^{2}P $4p$
15	−.46123	$3s^23p^3$ ^{2}D $7s$
16	−.44887	$3s^23p^3$ ^{2}D $6g$
17	−.38410	$3s^23p^3$ ^{2}P $6d$
18	−.32139	$3s^23p^3$ ^{2}D $7d$
19	−.27226	$3s^23p^3$ ^{2}D $8s$
20	−.26456	$3s^23p^3$ ^{2}D $7g$
21	−.18369	$3s^23p^3$ ^{2}D $8d$
22	−.16443	$3s^23p^3$ ^{2}P $7d$
23	−.14943	$3s^23p^3$ ^{2}D $9s$
24	−.14501	$3s^23p^3$ ^{2}D $8g$
25	−.12142	$3s3p^33d^2$
26	−.08882	$3s^23p^3$ ^{2}D $9d$
	^{1}F^e	
1	−2.33868	$3s3p^4$ ^{2}D $3d$
2	−2.25993	$3s^23p^3$ ^{2}D $4p$
3	−1.80868	$3s3p^4$ ^{2}P $3d$
4	−1.33101	$3s^23p^3$ ^{2}D $4f$
5	−1.28086	$3s^23p^23d^2$
6	−1.19594	$3s^23p^3$ ^{2}D $5p$
7	−1.16035	$3s^23p^3$ ^{2}P $4f$
8	−.81716	$3s^23p^23d^2$
9	−.69316	$3s^23p^3$ ^{2}D $6p$
10	−.68426	$3s^23p^3$ ^{2}D $5f$
11	−.62465	$3s^23p^3$ ^{2}P $5f$
12	−.55120	$3s^23p^23d^2$
13	−.44772	$3s^23p^3$ ^{2}D $6h$
14	−.42865	$3s^23p^3$ ^{2}D $6f$
15	−.41102	$3s^23p^3$ ^{2}D $7p$
16	−.29686	$3s^23p^3$ ^{2}P $6f$
17	−.26348	$3s^23p^3$ ^{2}D $7h$
18	−.25722	$3s^23p^3$ ^{2}D $7f$
19	−.23973	$3s^23p^3$ ^{2}D $8p$
20	−.15013	$3s3p^4$ ^{2}D $4d$
21	−.14388	$3s^23p^3$ ^{2}D $8h$
22	−.13538	$3s^23p^3$ ^{2}D $8f$
23	−.12798	$3s^23p^3$ ^{2}D $9p$
24	−.10664	$3s^23p^3$ ^{2}P $7f$
25	−.06191	$3s^23p^3$ ^{2}D $9h$
26	−.06005	$3s^23p^3$ ^{2}D $9f$
	^{1}F^o	
1	−3.57732	$3s^23p^3$ ^{2}D $3d$
2	−3.14871	$3s^23p^3$ ^{2}P $3d$
3	−1.63625	$3s^23p^3$ ^{2}D $4d$
4	−1.51175	$3s^23p^3$ ^{2}P $4d$
5	−.91539	$3s^23p^3$ ^{2}D $5d$
6	−.85681	$3s3p^4$ ^{2}D $4p$
7	−.75888	$3s^23p^3$ ^{2}D $5g$
8	−.75218	$3s^23p^3$ ^{2}P $5d$
9	−.66649	$3p^53d$
10	−.59074	$3s^23p^3$ ^{2}P $5g$
11	−.53926	$3s^23p^3$ ^{2}D $6d$
12	−.44882	$3s^23p^3$ ^{2}D $6g$
13	−.38343	$3s^23p^3$ ^{2}P $6d$
14	−.32075	$3s^23p^3$ ^{2}D $7d$
15	−.28618	$3s^23p^3$ ^{2}P $6g$
16	−.26366	$3s^23p^3$ ^{2}D $7g$
17	−.18201	$3s^23p^3$ ^{2}D $8d$
18	−.10188	$3s^23p^3$ ^{2}P $7d$
19	−.14465	$3s^23p^3$ ^{2}D $8g$
20	−.10112	$3s^23p^3$ ^{2}P $7g$
21	−.08838	$3s^23p^3$ ^{2}D $9d$
22	−.06242	$3s^23p^3$ ^{2}D $9g$
	^{1}G^e	
1	−2.38056	$3s3p^4$ ^{2}D $3d$
2	−1.73352	$3s^23p^23d^2$
3	−1.33460	$3s^23p^3$ ^{2}D $4f$
4	−1.21411	$3s^23p^3$ ^{2}P $4f$
5	−.96220	$3s^23p^23d^2$
6	−.94124	$3s^23p^23d^2$
7	−.76886	$3s^23p^3$ ^{2}D $5f$
8	−.59893	$3s^23p^3$ ^{2}P $5f$
9	−.47731	$3s^23p^3$ ^{2}D $6f$
10	−.44776	$3s^23p^3$ ^{2}D $6h$
11	−.35740	$3s^23p^23d^2$
12	−.28500	$3s^23p^3$ ^{2}P $6h$
13	−.28006	$3s^23p^3$ ^{2}P $6f$
14	−.26341	$3s^23p^3$ ^{2}D $7h$
15	−.24474	$3s^23p^3$ ^{2}D $7f$
16	−.18333	$3s3p^4$ ^{2}D $4d$
17	−.14394	$3s^23p^3$ ^{2}D $8h$
18	−.13547	$3s^23p^3$ ^{2}D $8f$
19	−.10059	$3s^23p^3$ ^{2}P $7h$
20	−.09848	$3s^23p^3$ ^{2}P $7f$
21	−.06190	$3s^23p^3$ ^{2}D $9h$
22	−.05800	$3s^23p^3$ ^{2}D $9f$

S-like Ca (Ca^{4+})

i	E(Ryds)	Description
		^{1}G^o
1	−3.91417	$3s^23p^3$ ^{2}D $3d$
2	−1.70579	$3s^23p^3$ ^{2}D $4d$
3	−.93997	$3s^23p^3$ ^{2}D $5d$
4	−.75399	$3s^23p^3$ ^{2}D $5g$
5	−.59164	$3s^23p^3$ ^{2}P $5g$
6	−.55286	$3s^23p^3$ ^{2}D $6d$
7	−.44831	$3s^23p^3$ ^{2}D $6g$
8	−.32835	$3s^23p^3$ ^{2}D $7d$
9	−.28616	$3s^23p^3$ ^{2}P $6g$
10	−.26387	$3s^23p^3$ ^{2}D $7g$
11	−.26339	$3s^23p^3$ ^{2}D $7i$
12	−.18661	$3s^23p^3$ ^{2}D $8d$
13	−.14431	$3s^23p^3$ ^{2}D $8g$
14	−.14384	$3s^23p^3$ ^{2}D $8i$
15	−.10167	$3s^23p^3$ ^{2}P $7g$
16	−.09144	$3s^23p^3$ ^{2}D $9d$
17	−.06221	$3s^23p^3$ ^{2}D $9g$
18	−.06185	$3s^23p^3$ ^{2}D $9i$
		^{1}H^e
1	−1.47462	$3s^23p^23d^2$
2	−1.23179	$3s^23p^3$ ^{2}D $4f$
3	−.76238	$3s^23p^3$ ^{2}D $5f$
4	−.45781	$3s^23p^3$ ^{2}D $6f$
5	−.44768	$3s^23p^3$ ^{2}D $6h$
6	−.28483	$3s^23p^3$ ^{2}P $6h$
7	−.27108	$3s^23p^3$ ^{2}D $7f$
8	−.26343	$3s^23p^3$ ^{2}D $7h$
9	−.14946	$3s^23p^3$ ^{2}D $8f$
10	−.14386	$3s^23p^3$ ^{2}D $8h$
11	−.14383	$3s^23p^3$ ^{2}D $8j$
12	−.10056	$3s^23p^3$ ^{2}P $7h$
13	−.06600	$3s^23p^3$ ^{2}D $9f$
14	−.06187	$3s^23p^3$ ^{2}D $9h$
15	−.06185	$3s^23p^3$ ^{2}D $9j$
		^{1}H^o
1	−.75459	$3s^23p^3$ ^{2}D $5g$
2	−.59135	$3s^23p^3$ ^{2}P $5g$
3	−.44850	$3s^23p^3$ ^{2}D $6g$
4	−.28619	$3s^23p^3$ ^{2}P $6g$
5	−.26375	$3s^23p^3$ ^{2}D $7g$
6	−.26339	$3s^23p^3$ ^{2}D $7i$
7	−.14457	$3s^23p^3$ ^{2}D $8g$
8	−.14384	$3s^23p^3$ ^{2}D $8i$
9	−.10131	$3s^23p^3$ ^{2}P $7g$
10	−.10052	$3s^23p^3$ ^{2}P $7i$
11	−.06247	$3s^23p^3$ ^{2}D $9g$
12	−.06185	$3s^23p^3$ ^{2}D $9i$
13	−.03002	$3s3p^33d^2$
		^{3}S^e
1	−2.21547	$3s^23p^3$ ^{2}P $4p$
2	−1.96756	$3s3p^4$ ^{2}D $3d$
3	−1.06872	$3s^23p^3$ ^{2}P $5p$
4	−.89022	$3s3p^4$ ^{2}S $4s$
5	−.77551	$3s^23p^23d^2$
6	−.53483	$3s^23p^3$ ^{2}P $6p$
7	−.25485	$3s^23p^3$ ^{2}P $7p$
8	−.22203	$3s3p^4$ ^{2}D $4d$
9	−.07884	$3s^23p^3$ ^{2}P $8p$
		^{3}S^o
1	−3.49306	$3s^23p^3$ ^{2}D $3d$
2	−2.93805	$3s^23p^3$ ^{4}S $4s$
3	−1.69267	$3s^23p^3$ ^{2}D $4d$
4	−1.60180	$3s^23p^3$ ^{4}S $5s$
5	−1.07382	$3s3p^4$ ^{4}P $4p$
6	−1.02658	$3s^23p^3$ ^{4}S $6s$
7	−.91484	$3s^23p^3$ ^{2}D $5d$
8	−.70785	$3s^23p^3$ ^{4}S $7s$
9	−.55054	$3s^23p^3$ ^{2}D $6d$
10	−.51952	$3s^23p^3$ ^{4}S $8s$
11	−.50371	$3s3p^4$ ^{2}P $4p$
12	−.39537	$3s^23p^3$ ^{4}S $9s$
13	−.32552	$3s^23p^3$ ^{2}D $7d$
		^{3}P^e
1	−6.23865	$3s^23p^4$
2	−2.72145	$3s3p^4$ ^{2}D $3d$:
3	−2.54442	$3s^23p^3$ ^{4}S $4p$
4	−2.25561	$3s^23p^3$ ^{2}D $4p$
5	−2.10386	$3s^23p^3$ ^{2}P $4p$
6	−2.10386	$3s^23p^23d^2$:
7	−1.92475	$3s3p^4$ ^{4}P $3d$:
8	−1.46662	$3s3p^4$ ^{4}P $4s$
9	−1.46662	$3s3p^4$ ^{2}P $3d$
10	−1.43159	$3s^23p^3$ ^{4}S $5p$
11	−1.31875	$3s^23p^3$ ^{2}D $4f$
12	−1.24333	$3s^23p^23d^2$
13	−1.18285	$3s^23p^3$ ^{2}D $5p$
14	−1.11151	$3s^23p^23d^2$
15	−1.03367	$3s^23p^3$ ^{2}P $5p$
16	−.97735	$3s3p^4$ ^{2}P $4s$
17	−.91554	$3s^23p^3$ ^{4}S $6p$
18	−.83900	$3s^23p^23d^2$
19	−.76957	$3s^23p^23d^2$
20	−.70829	$3s^23p^3$ ^{2}D $5f$
21	−.68488	$3s^23p^3$ ^{2}D $6p$
22	−.64661	$3s^23p^3$ ^{4}S $7p$
23	−.52518	$3s^23p^3$ ^{2}P $6p$
24	−.48216	$3s^23p^3$ ^{4}S $8p$
25	−.46605	$3s3p^4$ ^{4}P $4d$
26	−.43604	$3s^23p^3$ ^{2}D $6f$
27	−.40666	$3s^23p^3$ ^{2}D $7p$
28	−.37148	$3s^23p^3$ ^{4}S $9p$
		^{3}P^o
1	−4.85909	$3s3p^5$
2	−3.76500	$3s^23p^3$ ^{2}P $3d$:
3	−3.43675	$3s^23p^3$ ^{2}D $3d$:
4	−2.61558	$3s^23p^3$ ^{2}P $4s$
5	−1.68770	$3s^23p^3$ ^{2}D $4d$
6	−1.54412	$3s^23p^3$ ^{2}P $4d$
7	−1.23579	$3s^23p^3$ ^{2}P $5s$
8	−1.09046	$3s3p^4$ ^{4}P $4p$
9	−1.01420	$3p^53d$
10	−.93212	$3s^23p^3$ ^{2}D $5d$
11	−.79084	$3s^23p^3$ ^{2}P $5d$
12	−.77783	$3s3p^4$ ^{2}D $4p$
13	−.62770	$3s^23p^3$ ^{2}P $6s$
14	−.57731	$3s3p^4$ ^{2}P $4p$
15	−.53675	$3s^23p^3$ ^{2}D $6d$
16	−.45690	$3s3p^4$ ^{2}S $4p$
17	−.38267	$3s^23p^3$ ^{2}P $6d$
18	−.32570	$3s^23p^3$ ^{2}D $7d$
19	−.30315	$3s^23p^3$ ^{2}P $7s$
20	−.25622	$3s3p^33d^2$
21	−.18522	$3s^23p^3$ ^{2}D $8d$
22	−.16433	$3s^23p^3$ ^{2}P $7d$
23	−.11292	$3s^23p^3$ ^{2}P $8s$
24	−.09001	$3s^23p^3$ ^{2}D $9d$
25	−.02608	$3s^23p^3$ ^{2}P $8d$
		^{3}D^e
1	−2.60332	$3s3p^4$ ^{2}D $3d$:
2	−2.37257	$3s^23p^3$ ^{2}D $4p$
3	−2.24998	$3s3p^4$ ^{2}P $3d$:
4	−2.18145	$3s^23p^3$ ^{2}P $4p$
5	−1.96000	$3s3p^4$ ^{2}S $3d$:
6	−1.83424	$3s3p^4$ ^{4}P $3d$:
7	−1.44131	$3s^23p^23d^2$
8	−1.34164	$3s^23p^3$ ^{2}D $4f$
9	−1.28120	$3s3p^4$ ^{2}D $4s$:
10	−1.23417	$3s^23p^23d^2$
11	−1.18253	$3s^23p^3$ ^{2}D $5p$
12	−1.15334	$3s^23p^3$ ^{2}P $4f$
13	−1.05452	$3s^23p^3$ ^{2}P $5p$
14	−.96051	$3s^23p^23d^2$
15	−.82183	$3s^23p^23d^2$
16	−.74259	$3s^23p^3$ ^{2}D $5f$:
17	−.69748	$3s^23p^3$ ^{2}D $6p$
18	−.62298	$3s^23p^3$ ^{2}P $5f$
19	−.57124	$3s3p^4$ ^{4}P $4d$:
20	−.53400	$3s^23p^3$ ^{2}P $6p$
21	−.47664	$3s^23p^23d^2$
22	−.42531	$3s^23p^3$ ^{2}D $6f$
23	−.41405	$3s^23p^3$ ^{2}D $7p$
24	−.29497	$3s^23p^3$ ^{2}P $6f$
25	−.26547	$3s^23p^3$ ^{2}D $7f$
26	−.24983	$3s^23p^3$ ^{2}P $7p$
27	−.24142	$3s^23p^3$ ^{2}D $8p$
28	−.17513	$3s3p^4$ ^{2}D $4d$
29	−.14613	$3s^23p^3$ ^{2}D $8f$
30	−.12902	$3s^23p^3$ ^{2}D $9p$
31	−.10882	$3s^23p^3$ ^{2}P $7f$
32	−.07866	$3s^23p^3$ ^{2}P $8p$
33	−.06359	$3s^23p^3$ ^{2}D $9f$
		^{3}D^o
1	−4.16878	$3s^23p^3$ ^{2}D $3d$:
2	−3.74979	$3s^23p^3$ ^{2}P $3d$:
3	−3.33987	$3s^23p^3$ ^{4}S $3d$:
4	−2.77933	$3s^23p^3$ ^{2}D $4s$
5	−1.91334	$3s^23p^3$ ^{4}S $4d$
6	−1.70755	$3s^23p^3$ ^{2}D $4d$
7	−1.53298	$3s^23p^3$ ^{2}P $4d$
8	−1.39631	$3s^23p^3$ ^{2}D $5s$
9	−1.17068	$3s^23p^3$ ^{4}S $5d$
10	−1.10509	$3s3p^4$ ^{4}P $4p$
11	−.94520	$3s^23p^3$ ^{2}D $5d$
12	−.80839	$3s^23p^3$ ^{2}D $6s$
13	−.79069	$3s^23p^3$ ^{4}S $6d$
14	−.77486	$3s^23p^3$ ^{2}P $5d$
15	−.75478	$3s3p^4$ ^{2}D $4p$
16	−.75478	$3s^23p^3$ ^{2}D $5g$
17	−.73776	$3p^53d$
18	−.58100	$3s^23p^3$ ^{4}S $7d$
19	−.55399	$3s^23p^3$ ^{2}D $6d$
20	−.52139	$3s3p^4$ ^{2}P $4p$
21	−.46688	$3s^23p^3$ ^{2}D $7s$
22	−.45582	$3s3p^33d^2$
23	−.44842	$3s^23p^3$ ^{2}D $6g$
24	−.42839	$3s^23p^3$ ^{4}S $8d$
25	−.38712	$3s^23p^3$ ^{2}P $6d$
26	−.33591	$3s^23p^3$ ^{4}S $9d$
27	−.32643	$3s^23p^3$ ^{2}D $7d$
28	−.27562	$3s^23p^3$ ^{2}D $8s$
		^{3}F^e
1	−2.59923	$3s3p^4$ ^{4}P $3d$
2	−2.36023	$3s^23p^3$ ^{2}D $4p$
3	−2.30528	$3s3p^4$ ^{2}D $3d$:
4	−2.04846	$3s3p^4$ ^{2}P $3d$:
5	−1.80986	$3s^23p^23d^2$
6	−1.57682	$3s^23p^3$ ^{4}S $4f$
7	−1.37860	$3s^23p^23d^2$
8	−1.34297	$3s^23p^3$ ^{2}D $4f$
9	−1.23203	$3s^23p^23d^2$
10	−1.22216	$3s^23p^3$ ^{2}D $5p$
11	−1.17727	$3s^23p^3$ ^{2}P $4f$
12	−1.04639	$3s^23p^23d^2$
13	−1.00738	$3s^23p^3$ ^{4}S $5f$:
14	−.94249	$3s^23p^23d^2$
15	−.77365	$3s^23p^3$ ^{2}D $5f$
16	−.71641	$3s^23p^3$ ^{4}S $6f$

S-like Ca (Ca^{4+})

i	E(Ryds)	Description
17	−.69619	$3s^23p^3$ ^{2}D $6p$
18	−.62950	$3s^23p^3$ ^{2}P $5f$
19	−.56949	$3s^23p^23d^2$
20	−.50960	$3s^23p^3$ ^{4}S $7f$
21	−.47990	$3s3p^4$ ^{4}P $4d$
22	−.44956	$3s^23p^3$ ^{2}D $6f$
23	−.44762	$3s^23p^3$ ^{2}D $6h$
24	−.41265	$3s^23p^3$ ^{2}D $7p$
25	−.38592	$3s^23p^3$ ^{4}S $8f$
26	−.30959	$3s^23p^3$ ^{4}S $9f$
27	−.28984	$3s^23p^3$ ^{2}P $6f$
28	−.27169	$3s^23p^3$ ^{2}D $7f$
29	−.26346	$3s^23p^3$ ^{2}D $7h$
		^{3}F^o
1	−4.10040	$3s^23p^3$ ^{2}P $3d$:
2	−3.79321	$3s^23p^3$ ^{2}D $3d$:
3	−1.73189	$3s^23p^3$ ^{2}D $4d$
4	−1.54700	$3s^23p^3$ ^{2}P $4d$
5	−.96097	$3s^23p^3$ ^{2}D $5d$
6	−.91796	$3p^53d$
7	−.84343	$3s3p^4$ ^{2}D $4p$
8	−.77440	$3s^23p^3$ ^{2}P $5d$
9	−.75502	$3s^23p^3$ ^{2}D $5g$
10	−.59184	$3s^23p^3$ ^{2}P $5g$
11	−.55546	$3s^23p^3$ ^{2}D $6d$
12	−.48016	$3s3p^33d^2$
13	−.44917	$3s^23p^3$ ^{2}D $6g$
14	−.39208	$3s^23p^3$ ^{2}P $6d$
15	−.32881	$3s^23p^3$ ^{2}D $7d$
16	−.28702	$3s^23p^3$ ^{2}P $6g$
17	−.26401	$3s^23p^3$ ^{2}D $7g$
18	−.22078	$3s3p^33d^2$
19	−.18801	$3s^23p^3$ ^{2}D $8d$
20	−.16621	$3s^23p^3$ ^{2}P $7d$
21	−.14491	$3s^23p^3$ ^{2}D $8g$
22	−.12773	$3s3p^4$ ^{4}P $4f$
23	−.10177	$3s^23p^3$ ^{2}P $7g$
24	−.09211	$3s^23p^3$ ^{2}D $9d$
25	−.06259	$3s^23p^3$ ^{2}D $9g$
		^{3}G^e
1	−2.47993	$3s3p^4$ ^{2}D $3d$
2	−1.58333	$3s^23p^23d^2$
3	−1.34197	$3s^23p^23d^2$
4	−1.31310	$3s^23p^3$ ^{2}D $4f$
5	−1.12397	$3s^23p^3$ ^{2}P $4f$
6	−.95522	$3s^23p^23d^2$
7	−.77422	$3s^23p^3$ ^{2}D $5f$
8	−.60965	$3s^23p^3$ ^{2}P $5f$
9	−.46255	$3s^23p^3$ ^{2}D $6f$
10	−.44776	$3s^23p^3$ ^{2}D $6h$
11	−.30146	$3s^23p^3$ ^{2}P $6f$
12	−.28497	$3s^23p^3$ ^{2}P $6h$
13	−.27266	$3s^23p^3$ ^{2}D $7f$
14	−.26341	$3s^23p^3$ ^{2}D $7h$
15	−.22252	$3s3p^4$ ^{2}D $4d$
16	−.15138	$3s^23p^3$ ^{2}D $8f$
17	−.14395	$3s^23p^3$ ^{2}D $8h$
18	−.11082	$3s^23p^3$ ^{2}P $7f$
19	−.10059	$3s^23p^3$ ^{2}P $7h$
20	−.06699	$3s^23p^3$ ^{2}D $9f$
21	−.06190	$3s^23p^3$ ^{2}D $9h$
		^{3}G^o
1	−3.96752	$3s^23p^3$ ^{2}D $3d$
2	−1.71113	$3s^23p^3$ ^{2}D $4d$
3	−1.00095	$3s^23p^3$ ^{4}S $5g$
4	−.94241	$3s^23p^3$ ^{2}D $5d$
5	−.75490	$3s^23p^3$ ^{2}D $5g$
6	−.69535	$3s^23p^3$ ^{4}S $6g$
7	−.59153	$3s^23p^3$ ^{2}P $5g$
8	−.55470	$3s^23p^3$ ^{2}D $6d$
9	−.51100	$3s^23p^3$ ^{4}S $7g$
10	−.44919	$3s^23p^3$ ^{2}D $6g$
11	−.41818	$3s3p^33d^2$
12	−.39106	$3s^23p^3$ ^{4}S $8g$
13	−.32825	$3s^23p^3$ ^{2}D $7d$
14	−.30913	$3s^23p^3$ ^{4}S $9g$
15	−.28606	$3s^23p^3$ ^{2}P $6g$
16	−.26460	$3s^23p^3$ ^{2}D $7g$
17	−.26339	$3s^23p^3$ ^{2}D $7i$
		^{3}H^e
1	−1.75399	$3s^23p^23d^2$
2	−1.45652	$3s^23p^3$ ^{2}D $4f$
3	−1.24285	$3s^23p^23d^2$
4	−.76706	$3s^23p^3$ ^{2}D $5f$
5	−.69449	$3s^23p^3$ ^{4}S $6h$
6	−.51026	$3s^23p^3$ ^{4}S $7h$
7	−.46004	$3s^23p^3$ ^{2}D $6f$
8	−.44771	$3s^23p^3$ ^{2}D $6h$
9	−.39067	$3s^23p^3$ ^{4}S $8h$
10	−.30868	$3s^23p^3$ ^{4}S $9h$
11	−.28483	$3s^23p^3$ ^{2}P $6h$
12	−.27237	$3s^23p^3$ ^{2}D $7f$
13	−.26346	$3s^23p^3$ ^{2}D $7h$
		^{3}H^o
1	−.75567	$3s^23p^3$ ^{2}D $5g$
2	−.59209	$3s^23p^3$ ^{2}P $5g$
3	−.46742	$3s3p^33d^2$
4	−.44886	$3s^23p^3$ ^{2}D $6g$
5	−.28696	$3s^23p^3$ ^{2}P $6g$
6	−.26439	$3s^23p^3$ ^{2}D $7g$
7	−.26340	$3s^23p^3$ ^{2}D $7i$
8	−.14513	$3s^23p^3$ ^{2}D $8g$
9	−.14384	$3s^23p^3$ ^{2}D $8i$
10	−.10195	$3s^23p^3$ ^{2}P $7g$
11	−.10052	$3s^23p^3$ ^{2}P $7i$
12	−.06282	$3s^23p^3$ ^{2}D $9g$
13	−.06185	$3s^23p^3$ ^{2}D $9i$
14	−.03773	$3s3p^33d^2$
		^{5}S^o
1	−3.05000	$3s^23p^3$ ^{4}S $4s$
2	−1.65086	$3s^23p^3$ ^{4}S $5s$
3	−1.11065	$3s3p^4$ ^{4}P $4p$
4	−1.03451	$3s^23p^3$ ^{4}S $6s$
5	−.71565	$3s^23p^3$ ^{4}S $7s$
6	−.52333	$3s^23p^3$ ^{4}S $8s$
7	−.39941	$3s^23p^3$ ^{4}S $9s$
8	−.37933	$3s3p^33d^2$
		^{5}P^e
1	−2.77719	$3s3p^4$ ^{4}P $3d$
2	−2.61044	$3s^23p^3$ ^{4}S $4p$
3	−1.62661	$3s3p^4$ ^{4}P $4s$
4	−1.47304	$3s^23p^3$ ^{4}S $5p$
5	−1.41393	$3s^23p^23d^2$
6	−.94486	$3s^23p^3$ ^{4}S $6p$
7	−.66162	$3s^23p^3$ ^{4}S $7p$
8	−.51016	$3p^43d^2$
9	−.48688	$3s^23p^3$ ^{4}S $8p$
10	−.37600	$3s^23p^3$ ^{4}S $9p$
		^{5}P^o
1	−1.22653	$3s3p^4$ ^{4}P $4p$
2	−.80691	$3s3p^33d^2$
3	−.18716	$3s3p^33d^2$
4	−.05815	$3s3p^4$ ^{4}P $5p$:
5	−.04020	$3s3p^33d^2$
		^{5}D^e
1	−3.07367	$3s3p^4$ ^{4}P $3d$
2	−1.82292	$3s^23p^23d^2$
3	−1.34879	$3s^23p^23d^2$
4	−.56881	$3s3p^4$ ^{4}P $4d$
		^{5}D^o
1	−4.32148	$3s^23p^3$ ^{4}S $3d$
2	−1.97219	$3s^23p^3$ ^{4}S $4d$
3	−1.22116	$3s^23p^3$ ^{4}S $5d$
4	−1.15176	$3s3p^4$ ^{4}P $4p$
5	−.86923	$3s3p^33d^2$
6	−.79902	$3s^23p^3$ ^{4}S $6d$
7	−.57637	$3s^23p^3$ ^{4}S $7d$
8	−.43442	$3s^23p^3$ ^{4}S $8d$
9	−.35614	$3s3p^33d^2$
10	−.33906	$3s^23p^3$ ^{4}S $9d$
		^{5}F^e
1	−2.81670	$3s3p^4$ ^{4}P $3d$
2	−1.80716	$3s^23p^23d^2$
3	−1.57163	$3s^23p^3$ ^{4}S $4f$
4	−1.02566	$3s^23p^3$ ^{4}S $5f$
5	−.71229	$3s^23p^3$ ^{4}S $6f$
6	−.53871	$3s3p^4$ ^{4}P $4d$
7	−.52147	$3s^23p^3$ ^{4}S $7f$
8	−.39894	$3s^23p^3$ ^{4}S $8f$
9	−.31463	$3s^23p^3$ ^{4}S $9f$
		^{5}F^o
1	−.76652	$3s3p^33d^2$
2	−.50244	$3s3p^33d^2$
3	−.19072	$3s3p^33d^2$
4	−.15585	$3s3p^4$ ^{4}P $4f$
		^{5}G^o
1	−1.00194	$3s^23p^3$ ^{4}S $5g$
2	−.79523	$3s^23p3d^3$
3	−.69605	$3s^23p^3$ ^{4}S $6g$
4	−.52721	$3s^23p3d^3$
5	−.51151	$3s^23p^3$ ^{4}S $7g$
6	−.39184	$3s^23p^3$ ^{4}S $8g$
7	−.30981	$3s^23p^3$ ^{4}S $9g$

S-like Ca (Ca^{4+})

Energies in ascending order from ground state for terms with effective $n \leq 4.0$, $L \leq 5$

Term	i	E(Ryds)	Term	i	E(Ryds)	Term	i	E(Ryds)	Term	i	E(Ryds)	Term	i	E(Ryds)
$^{3}\mathbf{P}^{e}$	1	0.00000	$^{1}\mathbf{P}^{o}$	2	2.82974	$^{1}\mathbf{P}^{o}$	4	3.67786	$^{1}\mathbf{D}^{e}$	4	4.12935	$^{1}\mathbf{G}^{e}$	2	4.50513
$^{1}\mathbf{D}^{e}$	1	0.16902	$^{3}\mathbf{D}^{o}$	3	2.89878	$^{3}\mathbf{P}^{e}$	3	3.69423	$^{3}\mathbf{P}^{e}$	6	4.13479	$^{3}\mathbf{F}^{o}$	3	4.50676
$^{1}\mathbf{S}^{e}$	1	0.40687	$^{1}\mathbf{D}^{o}$	2	2.98497	$^{3}\mathbf{G}^{e}$	1	3.75872	$^{3}\mathbf{P}^{e}$	5	4.13479	$^{3}\mathbf{G}^{o}$	2	4.52752
$^{3}\mathbf{P}^{o}$	1	1.37956	$^{1}\mathbf{F}^{o}$	2	3.08994	$^{1}\mathbf{P}^{e}$	1	3.83935	$^{3}\mathbf{F}^{e}$	4	4.19019	$^{3}\mathbf{D}^{o}$	6	4.53110
$^{1}\mathbf{P}^{o}$	1	1.79653	$^{1}\mathbf{S}^{e}$	2	3.12230	$^{1}\mathbf{G}^{e}$	1	3.85809	$^{5}\mathbf{D}^{o}$	2	4.26646	$^{1}\mathbf{G}^{o}$	2	4.53286
$^{5}\mathbf{D}^{o}$	1	1.91717	$^{5}\mathbf{D}^{e}$	1	3.16498	$^{3}\mathbf{D}^{e}$	2	3.86608	$^{1}\mathbf{S}^{e}$	3	4.26943	$^{1}\mathbf{P}^{o}$	5	4.53934
$^{3}\mathbf{D}^{o}$	1	2.06987	$^{5}\mathbf{S}^{o}$	1	3.18865	$^{3}\mathbf{F}^{e}$	2	3.87842	$^{3}\mathbf{S}^{e}$	2	4.27109	$^{3}\mathbf{S}^{o}$	3	4.54598
$^{1}\mathbf{S}^{o}$	1	2.11841	$^{1}\mathbf{P}^{o}$	3	3.23523	$^{1}\mathbf{F}^{e}$	1	3.89997	$^{3}\mathbf{D}^{e}$	5	4.27865	$^{3}\mathbf{P}^{o}$	5	4.55095
$^{3}\mathbf{F}^{o}$	1	2.13825	$^{3}\mathbf{S}^{o}$	2	3.30060	$^{1}\mathbf{D}^{e}$	2	3.93070	$^{3}\mathbf{P}^{e}$	7	4.31390	$^{1}\mathbf{D}^{o}$	4	4.56267
$^{3}\mathbf{G}^{o}$	1	2.27113	$^{5}\mathbf{F}^{e}$	1	3.42195	$^{3}\mathbf{F}^{e}$	3	3.93337	$^{3}\mathbf{D}^{o}$	5	4.32531	$^{5}\mathbf{S}^{o}$	2	4.58779
$^{1}\mathbf{G}^{o}$	1	2.32448	$^{3}\mathbf{D}^{o}$	4	3.45932	$^{1}\mathbf{F}^{e}$	2	3.97872	$^{1}\mathbf{D}^{e}$	5	4.36417	$^{1}\mathbf{S}^{e}$	4	4.60133
$^{1}\mathbf{D}^{o}$	1	2.35364	$^{5}\mathbf{P}^{e}$	1	3.46146	$^{3}\mathbf{P}^{e}$	4	3.98304	$^{3}\mathbf{D}^{e}$	6	4.40441	$^{1}\mathbf{F}^{o}$	3	4.60240
$^{3}\mathbf{F}^{o}$	2	2.44544	$^{1}\mathbf{D}^{o}$	3	3.51514	$^{3}\mathbf{D}^{e}$	3	3.98867	$^{5}\mathbf{D}^{e}$	2	4.41573	$^{5}\mathbf{P}^{e}$	3	4.61204
$^{3}\mathbf{P}^{o}$	2	2.47365	$^{3}\mathbf{P}^{e}$	2	3.51720	$^{3}\mathbf{S}^{e}$	1	4.02318	$^{3}\mathbf{F}^{e}$	5	4.42879	$^{3}\mathbf{S}^{o}$	4	4.63685
$^{3}\mathbf{D}^{o}$	2	2.48886	$^{3}\mathbf{P}^{o}$	4	3.62307	$^{3}\mathbf{D}^{e}$	4	4.05720	$^{1}\mathbf{F}^{e}$	3	4.42997	$^{1}\mathbf{D}^{e}$	6	4.65433
$^{1}\mathbf{F}^{o}$	1	2.66133	$^{5}\mathbf{P}^{e}$	2	3.62821	$^{1}\mathbf{D}^{e}$	3	4.06192	$^{5}\mathbf{F}^{e}$	2	4.43149	$^{3}\mathbf{G}^{e}$	2	4.65532
$^{3}\mathbf{S}^{o}$	1	2.74559	$^{3}\mathbf{D}^{e}$	1	3.63533	$^{1}\mathbf{P}^{e}$	2	4.07319	$^{3}\mathbf{H}^{e}$	1	4.48466	$^{3}\mathbf{F}^{e}$	6	4.66183
$^{3}\mathbf{P}^{o}$	3	2.80190	$^{3}\mathbf{F}^{e}$	1	3.63942	$^{1}\mathbf{P}^{e}$	3	4.11149	$^{1}\mathbf{S}^{o}$	2	4.50054	$^{5}\mathbf{F}^{e}$	3	4.66702

S-like Ca (Ca^{4+})

gf-values for transitions involving terms with effective $n \leq 4.0$, $L \leq 5$

$i\ i'$	gf_L	$i\ i'$	gf_L	$i\ i'$	gf_L	$i\ i'$	gf_L	$i\ i'$	gf_L	$i\ i'$	gf_L
	$^{1}S^{e}$–$^{1}P^{o}$	2 3	1.68E−1	4 2	1.77E−1	3 2	4.57E−3	1 2	−1.32E−1	3 5	−9.13E+0
1 1	−6.29E−3	2 4	−4.06E−3	4 3	4.01E−1	3 3	−2.58E−3	1 3	−1.01E+1	3 6	−4.17E−3
1 2	−1.00E−1	3 1	1.97E−1	4 4	−7.32E−2		$^{1}F^{e}$–$^{1}G^{o}$	1 4	−4.75E−1	4 1	1.49E−2
1 3	−3.32E+0	3 2	5.03E−2	5 1	9.96E−2	1 1	8.18E−1	1 5	−1.29E−2	4 2	1.78E−1
1 4	−2.05E−1	3 3	3.19E−2	5 2	2.00E−3	1 2	−5.70E+0	2 1	5.50E−3	4 3	7.11E−1
1 5	−1.97E−3	3 4	−2.09E−2	5 3	1.97E−4	2 1	4.22E−1	2 2	1.53E−2	4 4	2.77E+0
2 1	1.42E−1		$^{1}D^{e}$–$^{1}P^{o}$	5 4	−8.91E−4	2 2	−3.76E−1	2 3	7.95E−3	4 5	−4.94E−1
2 2	4.89E−3	1 1	−3.16E−1	6 1	1.58E−1	3 1	9.26E−1	2 4	−7.93E−7	4 6	−3.79E+0
2 3	−2.01E−4	1 2	−3.12E+0	6 2	5.09E−3	3 2	−2.89E−3	2 5	−1.02E−2	5 1	8.82E−2
2 4	−7.13E−4	1 3	−5.72E−3	6 3	4.27E−4		$^{3}S^{e}$–$^{3}P^{o}$	3 1	3.09E−3	5 2	6.18E−1
2 5	−2.41E−5	1 4	−2.60E−1	6 4	3.38E−3	1 1	2.39E−1	3 2	5.14E−5	5 3	6.91E−3
3 1	7.68E−2	1 5	−2.65E−4		$^{1}D^{e}$–$^{1}F^{o}$	1 2	3.48E−1	3 3	6.02E−2	5 4	3.93E−2
3 2	1.13E−2	2 1	8.67E−2	1 1	−3.48E−4	1 3	5.69E−2	3 4	9.72E−3	5 5	−3.60E−2
3 3	2.47E−1	2 2	4.95E−2	1 2	−8.98E+0	1 4	8.82E−1	3 5	−5.92E−1	5 6	−3.65E−2
3 4	3.75E−1	2 3	5.19E−4	1 3	−2.21E−3	1 5	−1.95E−1	4 1	2.73E−1	6 1	8.82E−2
3 5	−3.70E−2	2 4	3.99E−3	2 1	1.34E−2	2 1	3.16E+0	4 2	9.54E−2	6 2	6.18E−1
4 1	1.09E+0	2 5	−3.18E−2	2 2	2.70E−2	2 2	1.94E−3	4 3	5.63E−1	6 3	6.91E−3
4 2	2.00E−1	3 1	3.88E−2	2 3	−2.44E−1	2 3	1.24E−1	4 4	9.20E−2	6 4	3.93E−2
4 3	2.48E−2	3 2	9.81E−3	3 1	4.50E−1	2 4	3.38E−2	4 5	−3.55E+0	6 5	−3.60E−2
4 4	1.45E−2	3 3	2.56E−3	3 2	1.04E−1	2 5	−2.71E−3	5 1	3.21E−1	6 6	−3.65E−2
4 5	2.16E−4	3 4	1.96E−1	3 3	−2.19E+0		$^{3}P^{e}$–$^{3}S^{o}$	5 2	7.81E−2	7 1	1.04E−2
	$^{1}P^{e}$–$^{1}S^{o}$	3 5	−3.77E−1	4 1	1.95E−1	1 1	−2.61E+0	5 3	2.43E−2	7 2	5.09E−3
1 1	9.01E−2	4 1	1.02E−4	4 2	5.51E−1	1 2	−6.11E−1	5 4	1.04E−1	7 3	1.22E−1
1 2	−5.37E−1	4 2	5.32E−2	4 3	−5.67E−1	1 3	−4.12E−3	5 5	−2.64E−2	7 4	2.79E−3
2 1	2.04E−2	4 3	3.38E−2	5 1	2.63E−1	1 4	−1.56E−1	6 1	3.21E−1	7 5	−6.54E−5
2 2	−1.01E−1	4 4	1.47E+0	5 2	9.09E−4	2 1	5.31E−3	6 2	7.81E−2	7 6	−3.68E−3
3 1	3.42E−1	4 5	−2.23E−2	5 3	−6.00E−4	2 2	2.92E−3	6 3	2.43E−2		$^{3}D^{e}$–$^{3}P^{o}$
3 2	−2.76E−3	5 1	9.30E−3	6 1	1.23E−5	2 3	−7.05E−5	6 4	1.04E−1	1 1	8.27E−2
	$^{1}P^{e}$–$^{1}P^{o}$	5 2	6.86E−4	6 2	1.38E−1	2 4	−1.44E−3	6 5	−2.64E−2	1 2	3.63E−2
1 1	6.53E−2	5 3	3.33E−2	6 3	8.84E−4	3 1	1.19E−2	7 1	8.18E+0	1 3	6.66E−3
1 2	1.31E−1	5 4	2.70E−2		$^{1}F^{e}$–$^{1}D^{o}$	3 2	2.47E+0	7 2	9.41E−3	1 4	3.16E−4
1 3	1.52E−3	5 5	−3.71E−5	1 1	8.22E−4	3 3	−1.30E−2	7 3	7.11E−2	1 5	−2.04E−2
1 4	7.28E−2	6 1	2.45E+0	1 2	7.56E−8	3 4	−1.30E+0	7 4	5.22E−2	2 1	5.38E−2
1 5	−1.36E+0	6 2	1.43E−3	1 3	1.98E+0	4 1	3.88E−1	7 5	−1.16E−3	2 2	4.32E−2
2 1	2.13E−1	6 3	2.53E−5	1 4	−5.19E−2	4 2	4.81E−1		$^{3}P^{e}$–$^{3}D^{o}$	2 3	2.02E−1
2 2	2.83E−2	6 4	4.41E−3	2 1	4.37E−2	4 3	−1.96E+0	1 1	−1.77E−3	2 4	1.32E−1
2 3	4.59E−2	6 5	1.55E−2	2 2	2.34E−2	4 4	−5.30E−3	1 2	−9.24E−2	2 5	−1.28E+0
2 4	5.53E−1		$^{1}D^{e}$–$^{1}D^{o}$	2 3	1.71E−1	5 1	5.21E−2	1 3	−1.76E+1	3 1	5.21E−1
2 5	−7.94E−2	1 1	−6.04E−2	2 4	−1.99E−3	5 2	3.46E−2	1 4	−8.13E−1	3 2	3.13E−2
3 1	2.47E−1	1 2	−4.82E+0	3 1	7.19E−1	5 3	−3.71E−2	1 5	−2.82E−4	3 3	5.18E−2
3 2	2.80E−2	1 3	−7.75E−1	3 2	4.46E−4	5 4	−7.32E−3	1 6	−9.63E−3	3 4	3.41E−1
3 3	1.60E−2	1 4	−2.30E−3	3 3	9.60E−3	6 1	5.21E−2	2 1	2.81E−1	3 5	−2.32E−4
3 4	2.76E−1	2 1	6.64E−2	3 4	−2.54E−2	6 2	3.46E−2	2 2	6.10E−3	4 1	1.61E−2
3 5	−3.61E−2	2 2	5.12E−3		$^{1}F^{e}$–$^{1}F^{o}$	6 3	−3.71E−2	2 3	2.59E−3	4 2	7.45E−3
	$^{1}P^{e}$–$^{1}D^{o}$	2 3	1.11E−1	1 1	3.11E−2	6 4	−7.32E−3	2 4	2.34E−4	4 3	3.12E−2
1 1	7.95E−2	2 4	−2.18E−1	1 2	1.36E−1	7 1	1.16E−1	2 5	−1.96E−2	4 4	4.10E+0
1 2	1.37E−2	3 1	3.13E−1	1 3	−1.53E+0	7 2	2.38E−4	2 6	−4.93E−3	4 5	−1.67E−1
1 3	7.10E−1	3 2	3.58E−1	2 1	6.98E−2	7 3	−1.51E−2	3 1	9.91E−1	5 1	1.11E+0
1 4	−1.05E+0	3 3	1.39E+0	2 2	3.43E−2	7 4	−6.62E−4	3 2	1.03E−1	5 2	8.08E−1
2 1	7.48E−2	3 4	−1.85E+0	2 3	−1.55E−1		$^{3}P^{e}$–$^{3}P^{o}$	3 3	3.49E−1	5 3	2.27E−2
2 2	9.81E−2	4 1	4.86E−2	3 1	5.90E−1	1 1	−4.81E−1	3 4	2.22E−1	5 4	7.47E−2

S-like Ca (Ca^{4+})

$i\ i'$	gf_L	$i\ i'$	gf_L	$i\ i'$	gf_L	$i\ i'$	gf_L	$i\ i'$	gf_L	$i\ i'$	gf_L
5 5	−4.38E−3	4 5	−2.18E−2	5 3	−2.31E−3	5 1	5.67E−2	6 2	1.75E−1	1 2	−9.21E−3
6 1	1.47E+1	4 6	−3.37E−1	6 1	6.68E−2	5 2	1.32E−1	6 3	3.14E−5		**$^5P^e$–$^5S^o$**
6 2	5.31E−2	5 1	1.29E−1	6 2	1.99E−1	5 3	6.75E−3		**$^3F^e$–$^3G^o$**	1 1	8.47E−2
6 3	7.37E−2	5 2	4.61E−1	6 3	−5.25E−4	5 4	4.06E−3	1 1	2.41E−3	1 2	−6.62E−2
6 4	1.51E−2	5 3	4.74E−3		**$^3F^e$–$^3D^o$**	5 5	4.88E−2	1 2	−1.20E−1	2 1	4.62E+0
6 5	−5.64E−3	5 4	3.65E−3	1 1	4.51E−1	5 6	−9.59E−3	2 1	3.16E+0	2 2	−2.30E+0
	$^3D^e$–$^3D^o$	5 5	−1.99E−4	1 2	7.84E−3	6 1	3.00E+0	2 2	−1.46E+1	3 1	6.43E−1
1 1	3.77E−1	5 6	−2.12E−2	1 3	1.94E−2	6 2	5.62E−1	3 1	1.37E−1	3 2	3.29E−2
1 2	2.11E−2	6 1	4.23E−3	1 4	1.59E−2	6 3	1.81E+0	3 2	−3.59E+0		**$^5P^e$–$^5D^o$**
1 3	3.03E−3	6 2	1.20E−2	1 5	−1.35E−6	6 4	3.84E−2	4 1	1.87E+0	1 1	1.01E+0
1 4	2.40E−2	6 3	3.91E−1	1 6	−1.11E−2	6 5	9.63E+0	4 2	−1.25E−2	1 2	−5.57E−1
1 5	−3.05E−3	6 4	1.57E−5	2 1	9.63E−4	6 6	1.74E−1	5 1	2.95E−1	2 1	2.06E+0
1 6	−6.25E−2	6 5	2.21E−4	2 2	2.46E−2		**$^3F^e$–$^3F^o$**	5 2	−5.81E−3	2 2	−1.61E+1
2 1	3.14E−1	6 6	−4.76E−3	2 3	1.03E−1	1 1	5.59E−3	6 1	7.90E−2	3 1	2.97E−1
2 2	2.81E−1		**$^3D^e$–$^3F^o$**	2 4	5.01E+0	1 2	1.52E−2	6 2	2.37E−4	3 2	1.17E−1
2 3	2.74E−2	1 1	5.42E−1	2 5	−2.40E−4	1 3	−6.66E−2		**$^3G^e$–$^3F^o$**		**$^5D^e$–$^5D^o$**
2 4	4.34E+0	1 2	4.31E−4	2 6	−7.58E−1	2 1	1.14E+0	1 1	9.68E−1	1 1	4.83E−1
2 5	−2.07E−1	1 3	−1.68E−1	3 1	1.64E−1	2 2	2.56E−1	1 2	7.53E−2	1 2	−1.78E−4
2 6	−4.68E+0	2 1	1.20E+0	3 2	4.66E−1	2 3	−3.18E+0	1 3	−9.40E−2	2 1	3.82E−3
3 1	5.15E−1	2 2	6.10E−2	3 3	2.76E−2	3 1	1.13E−1	2 1	1.59E+0	2 2	9.69E−8
3 2	4.58E−2	2 3	−9.31E+0	3 4	1.55E+0	3 2	7.28E−2	2 2	3.82E−2		**$^5F^e$–$^5D^o$**
3 3	2.43E−3	3 1	8.54E−1	3 5	−2.14E−3	3 3	−1.07E+0	2 3	5.37E−1	1 1	1.61E+0
3 4	1.11E−2	3 2	4.23E−1	3 6	−7.83E−2	4 1	3.78E−1		**$^3G^e$–$^3G^o$**	1 2	−1.18E−1
3 5	−3.66E−3	3 3	−1.21E−2	4 1	1.49E+0	4 2	9.70E−1	1 1	1.12E+0	2 1	2.32E+0
3 6	−5.79E−2	4 1	5.75E−1	4 2	5.00E−2	4 3	−5.39E−3	1 2	−1.26E−2	2 2	1.01E+0
4 1	4.50E−1	4 2	1.83E+0	4 3	5.36E−2	5 1	3.21E−1	2 1	4.80E−1	3 1	1.82E+1
4 2	5.47E−2	4 3	−5.35E−3	4 4	1.40E−2	5 2	6.56E−3	2 2	8.81E−2	3 2	1.80E+1
4 3	2.36E−1	5 1	2.05E−1	4 5	−9.96E−2	5 3	−1.92E−4		**$^3H^e$–$^3G^o$**		
4 4	2.65E−1	5 2	4.17E−1	4 6	−2.59E−2	6 1	1.40E−2	1 1	4.82E−2		

S-like Fe (Fe^{10+})

Term energies relative to $3s^23p^3$ ^{4}S ionization threshold for each symmetry

i	E(Ryds)	Description	i	E(Ryds)	Description	i	E(Ryds)	Description	i	E(Ryds)	Description
		^{1}S^e	10	−6.20090	$3p^43d^2$	21	−5.77173	$3s^23p^3$ ^{2}P $5s$	25	−5.33694	$3s^23p^3$ ^{2}P $5p$
1	−20.6392	$3s^23p^4$	11	−5.97842	$3s3p^23d^3$	22	−5.04042	$3s^23p^3$ ^{2}D $5d$	26	−5.26541	$3s3p^23d^3$
2	−15.9450	$3p^6$	12	−5.74511	$3s3p^4$ ^{2}D $4d$	23	−4.68415	$3s^23p^3$ ^{2}P $5d$	27	−5.19348	$3s3p^4$ ^{2}P $4d$
3	−13.4288	$3s3p^4$ ^{2}D $3d$	13	−5.69985	$3s^23p^3$ ^{2}D $5p$	24	−4.68224	$3s3p^4$ ^{2}D $4f$	28	−5.16053	$3s3p^23d^3$
4	−11.9149	$3s^23p^23d^2$	14	−5.47816	$3s3p^23d^3$	25	−3.55671	$3s^23p^3$ ^{2}P $6s$	29	−5.14590	$3s3p^4$ ^{2}S $4d$
5	−11.7111	$3s^23p^23d^2$	15	−5.36646	$3s^23p^3$ ^{2}P $5p$	26	−3.29678	$3s^23p^3$ ^{2}D $6d$	30	−4.77710	$3s3p^23d^3$
6	−10.3586	$3s^23p^23d^2$	16	−5.15373	$3s3p^4$ ^{2}P $4d$	27	−3.03207	$3p^33d^3$	31	−4.67989	$3s3p^23d^3$
7	−9.03022	$3s^23p^3$ ^{2}P $4p$	17	−5.09325	$3s3p^23d^3$	28	−2.97680	$3s^23p^3$ ^{2}P $6d$	32	−4.56845	$3s^23p^3$ ^{2}D $5f$
8	−8.15585	$3p^43d^2$	18	−4.81643	$3s3p^23d^3$	29	−2.94644	$3s3p^4$ ^{2}D $5p$	33	−4.44247	$3s3p^23d^3$
9	−7.40855	$3s3p^4$ ^{2}S $4s$	19	−4.60429	$3s^23p^3$ ^{2}D $5f$	30	−2.79205	$3p^33d^3$	34	−4.37201	$3s3p^23d^3$
10	−5.92973	$3p^43d^2$	20	−4.49948	$3s3p^23d^3$	31	−2.47391	$3s3p^4$ ^{2}S $5p$	35	−4.19706	$3s3p^23d^3$
11	−5.87512	$3s3p^23d^3$	21	−4.28489	$3s3p^23d^3$	32	−2.43753	$3s3p^4$ ^{2}P $5p$	36	−4.19269	$3s^23p^3$ ^{2}P $5f$
12	−5.72248	$3s3p^4$ ^{2}D $4d$	22	−4.14970	$3s3p^23d^3$	33	−2.30904	$3s^23p^3$ ^{2}P $7s$	37	−4.06365	$3s3p^23d^3$
13	−5.19968	$3s^23p^3$ ^{2}P $5p$	23	−3.80670	$3s3p^23d^3$	34	−2.26694	$3s^23p^3$ ^{2}D $7d$	38	−3.72295	$3s3p^23d^3$
14	−4.95831	$3p^43d^2$	24	−3.65797	$3s^23p^3$ ^{2}D $6p$	35	−2.22445	$3p^33d^3$	39	−3.58410	$3s^23p^3$ ^{2}D $6p$
15	−3.92743	$3s3p^23d^3$	25	−3.33584	$3s^23p^3$ ^{2}P $6p$	36	−1.95235	$3s^23p^3$ ^{2}P $7d$	40	−3.35745	$3s3p^4$ ^{2}D $5s$
16	−3.69801	$3s3p^23d^3$	26	−3.05996	$3s^23p^3$ ^{2}D $6f$	37	−1.92037	$3p^33d^3$	41	−3.30141	$3s^23p^3$ ^{2}P $6p$
17	−3.25179	$3s^23p^3$ ^{2}P $6p$	27	−2.92336	$3s3p^4$ ^{2}P $5s$	38	−1.87402	$3s3p^4$ ^{2}D $5f$	42	−3.18692	$3s3p^23d^3$
18	−2.87099	$3s3p^4$ ^{2}S $5s$	28	−2.52269	$3s3p^23d^3$	39	−1.62623	$3s^23p^3$ ^{2}D $8d$	43	−3.03883	$3s^23p^3$ ^{2}D $6f$
19	−2.34540	$3s3p^4$ ^{2}D $5d$	29	−2.48841	$3s^23p^3$ ^{2}D $7p$	40	−1.58805	$3p^33d^3$	44	−3.03510	$3s3p^23d^3$
20	−2.21163	$3s3p^23d^3$	30	−2.34567	$3s3p^4$ ^{2}D $5d$	41	−1.53675	$3s^23p^3$ ^{2}P $8s$	45	−2.69878	$3s^23p^3$ ^{2}P $6f$
21	−2.12817	$3s^23p^3$ ^{2}P $7p$	31	−2.17613	$3s^23p^3$ ^{2}P $7p$	42	−1.32331	$3p^33d^3$	46	−2.45046	$3s^23p^3$ ^{2}D $7p$
22	−1.42275	$3s^23p^3$ ^{2}P $8p$	32	−2.13081	$3s^23p^3$ ^{2}D $7f$	43	1.30789	$3s^23p^3$ ^{2}P $8d$	47	−2.30851	$3s3p^4$ ^{2}D $5d$
23	−.95074	$3s^23p^3$ ^{2}P $9p$	33	−1.85055	$3s3p^4$ ^{2}P $5d$	44	−1.18659	$3s^23p^3$ ^{2}D $9d$	48	−2.22143	$3s3p^23d^3$
24	−.67445	$3s3p^4$ ^{2}S $6s$	34	−1.76463	$3s^23p^3$ ^{2}D $8p$	45	−1.02940	$3s^23p^3$ ^{2}P $9s$	49	−2.16479	$3s^23p^3$ ^{2}P $7p$
		^{1}S^o	35	−1.53192	$3s^23p^3$ ^{2}D $8f$	46	−.92847	$3s3p^4$ ^{2}D $6p$	50	−2.11630	$3s^23p^3$ ^{2}D $7f$
1	−17.4927	$3s^23p^3$ ^{2}D $3d$	36	−1.45253	$3s^23p^3$ ^{2}P $8p$			^{1}D^e	51	−1.87179	$3s3p^4$ ^{2}P $5d$
2	−10.4462	$3s^23p3d^3$:	37	−1.28255	$3s^23p^3$ ^{2}D $9p$	1	−21.0341	$3s^23p^4$	52	−1.83326	$3s3p^4$ ^{2}S $5d$
3	−9.05245	$3s3p^33d^2$:	38	−1.12197	$3s^23p^3$ ^{2}D $9f$	2	−14.3884	$3s3p^4$ ^{2}D $3d$:	53	−1.79205	$3s^23p^3$ ^{2}P $7f$
4	−8.49308	$3s^23p^3$ ^{2}D $4d$	39	−.97311	$3s^23p^3$ ^{2}P $9p$	3	−13.7147	$3s3p^4$ ^{2}S $3d$:	54	−1.76844	$3s3p^4$ ^{2}D $5g$
5	−8.29420	$3s3p^33d^2$			^{1}P^o	4	−13.2978	$3s3p^4$ ^{2}P $3d$:	55	−1.74020	$3s^23p^3$ ^{2}D $8p$
6	−6.43719	$3s3p^4$ ^{2}P $4p$	1	−18.1501	$3s3p^5$	5	−12.7299	$3s^23p^23d^2$	56	−1.52203	$3s^23p^3$ ^{2}D $8f$
7	−5.05008	$3s^23p^3$ ^{2}D $5d$	2	−16.4709	$3s^23p^3$ ^{2}D $3d$	6	−12.3958	$3s^23p^23d^2$	57	−1.44346	$3s^23p^3$ ^{2}P $8p$
8	−3.73985	$3p^33d^3$	3	−15.6187	$3s^23p^3$ ^{2}P $3d$	7	−11.8649	$3s^23p^23d^2$	58	−1.26666	$3s^23p^3$ ^{2}D $9p$
9	−3.31020	$3s^23p^3$ ^{2}D $6d$	4	−10.7645	$3p^53d$:	8	−11.7247	$3s^23p^23d^2$	59	−1.20321	$3s^23p^3$ ^{2}P $8f$
10	−2.66092	$3p^33d^3$	5	−10.3814	$3s^23p^3$ ^{2}P $4s$	9	−11.2521	$3s^23p^23d^2$	60	−1.14408	$3s3p^4$ ^{2}D $6s$
11	−2.46166	$3s3p^4$ ^{2}P $5p$	6	−9.97419	$3s3p^33d^2$	10	−10.7859	$3s^23p^23d^2$	61	−1.11434	$3s^23p^3$ ^{2}D $9f$
12	−2.26688	$3s^23p^3$ ^{2}D $7d$	7	−9.62012	$3s3p^33d^2$	11	−9.49057	$3s^23p^3$ ^{2}D $4p$	62	−.96464	$3s^23p^3$ ^{2}P $9p$
13	−1.62809	$3s^23p^3$ ^{2}D $8d$	8	−9.10333	$3s3p^33d^2$	12	−9.35924	$3s^23p^3$ ^{2}P $4p$			^{1}D^o
14	−1.46066	$3p^33d^3$	9	−8.76235	$3s3p^33d^2$:	13	−8.05664	$3p^43d^2$	1	−17.1102	$3s^23p^3$ ^{2}P $3d$:
15	−1.18932	$3s^23p^3$ ^{2}D $9d$	10	−8.48594	$3s3p^33d^2$:	14	−7.88940	$3s3p^4$ ^{2}D $4s$	2	−10.0053	$3s^23p^3$ ^{2}D $3d$:
		^{1}P^e	11	8.43268	$3s^23p^3$ ^{2}D $4d$	15	−7.38807	$3s^23p^3$ ^{2}D $4f$	3	−11.6264	$3p^53d$
1	−14.1761	$3s3p^4$ ^{2}D $3d$	12	−8.30703	$3s3p^33d^2$:	16	−7.22257	$3p^43d^2$	4	−10.6880	$3s^23p^3$ ^{2}D $4s$
2	−12.9111	$3s3p^4$ ^{2}P $3d$	13	−7.96872	$3s^23p^3$ ^{2}P $4d$	17	−6.89679	$3p^43d^2$	5	−10.5948	$3s3p^33d^2$
3	−12.3182	$3s^23p^23d^2$	14	−7.44721	$3s^23p3d^3$:	18	−6.88815	$3s^23p^3$ ^{2}P $4f$	6	−10.2307	$3s3p^33d^2$
4	−11.7937	$3s^23p^23d^2$	15	−7.23367	$3s^23p3d^3$	19	−6.51340	$3p^43d^2$	7	−9.51143	$3s3p^33d^2$
5	−9.96071	$3s^23p^3$ ^{2}D $4p$	16	−7.09776	$3s3p^4$ ^{2}D $4p$	20	−6.29315	$3p^43d^2$	8	−9.20020	$3s3p^33d^2$
6	−9.46024	$3s^23p^3$ ^{2}P $4p$	17	−6.72002	$3s3p^33d^2$:	21	6.18803	$3p^43d^2$	9	−9.00145	$3s3p^33d^2$
7	−7.61804	$3p^43d^2$	18	−6.52853	$3s3p^4$ ^{2}S $4p$	22	−5.65355	$3s3p^4$ ^{2}D $4d$	10	−8.72031	$3s3p^33d^2$
8	−7.50208	$3s3p^4$ ^{2}P $4s$	19	−6.36169	$3s3p^4$ ^{2}P $4p$	23	−5.60840	$3s3p^23d^3$	11	−8.59649	$3s3p^33d^2$
9	−7.42697	$3s^23p^3$ ^{2}D $4f$	20	−6.06066	$3s^23p3d^3$:	24	−5.55292	$3s^23p^3$ ^{2}D $5p$	12	−8.44954	$3s3p^33d^2$

S-like Fe (Fe^{10+})

i	E(Ryds)	Description
13	−8.39033	$3s^23p^3$ ^{2}D $4d$
14	−8.07976	$3s^23p^3$ ^{2}P $4d$
15	−8.04366	$3s^23p3d^3$
16	−7.64915	$3s^23p3d^3$
17	−6.95548	$3s3p^4$ ^{2}D $4p$
18	−6.90805	$3s^23p3d^3$
19	−6.56084	$3s3p^4$ ^{2}P $4p$
20	−6.43213	$3s^23p3d^3$
21	−6.08001	$3s^23p^3$ ^{2}D $5s$
22	−5.01642	$3s^23p^3$ ^{2}D $5d$
23	−4.72927	$3s^23p^3$ ^{2}P $5d$
24	−4.67592	$3s3p^4$ ^{2}D $4f$
25	−4.45636	$3s^23p^3$ ^{2}D $5g$
26	−4.08358	$3s3p^4$ ^{2}P $4f$
27	−3.86403	$3s^23p^3$ ^{2}D $6s$
28	−3.56006	$3p^33d^3$
29	−3.28494	$3s^23p^3$ ^{2}D $6d$
30	−3.16368	$3p^33d^3$
31	−2.99130	$3s^23p^3$ ^{2}P $6d$
32	−2.96177	$3s^23p^3$ ^{2}D $6g$
33	−2.91394	$3s3p^4$ ^{2}D $5p$
34	−2.62133	$3s^23p^3$ ^{2}D $7s$
35	−2.61286	$3p^33d^3$
36	−2.49408	$3s3p^4$ ^{2}P $5p$
37	−2.42875	$3p^33d^3$
38	−2.34430	$3p^33d^3$
39	−2.26472	$3s^23p^3$ ^{2}D $7d$
40	−2.07439	$3s^23p^3$ ^{2}D $7g$
41	−1.96460	$3s^23p^3$ ^{2}P $7d$
42	−1.86327	$3s3p^4$ ^{2}D $5f$
43	−1.84791	$3s^23p^3$ ^{2}D $8s$
44	−1.83882	$3p^33d^3$
45	−1.66650	$3p^33d^3$
46	−1.62012	$3s^23p^3$ ^{2}D $8d$
47	−1.49732	$3s^23p^3$ ^{2}D $8g$
48	−1.37526	$3s3p^4$ ^{2}P $5f$
49	−1.34170	$3s^23p^3$ ^{2}D $9s$
50	−1.31293	$3s^23p^3$ ^{2}P $8d$
51	−1.30978	$3p^33d^3$
52	−1.18306	$3s^23p^3$ ^{2}D $9d$
53	−1.09309	$3s^23p^3$ ^{2}D $9g$
54	−1.03506	$3p^33d^3$
	^{1}F^e	
1	−14.2833	$3s3p^4$ ^{2}D $3d$
2	−13.5217	$3s3p^4$ ^{2}P $3d$
3	−12.5841	$3s^23p^23d^2$
4	−11.9403	$3s^23p^23d^2$
5	−11.4766	$3s^23p^23d^2$
6	−9.81426	$3s^23p^3$ ^{2}D $4p$
7	−7.78262	$3p^43d^2$
8	−7.32027	$3s^23p^3$ ^{2}D $4f$
9	−7.21693	$3p^43d^2$
10	−7.04917	$3s^23p^3$ ^{2}P $4f$
11	−6.48430	$3s3p^23d^3$:
12	−5.93324	$3p^43d^2$:
13	−5.68355	$3s^23p^3$ ^{2}D $5p$
14	−5.58687	$3s3p^4$ ^{2}D $4d$
15	−5.43985	$3s3p^23d^3$
16	−5.27649	$3s3p^23d^3$
17	−5.23663	$3s3p^4$ ^{2}P $4d$
18	−5.12744	$3s3p^23d^3$
19	−4.95407	$3s3p^23d^3$
20	−4.94706	$3s3p^23d^3$
21	−4.61255	$3s3p^23d^3$
22	−4.52973	$3s^23p^3$ ^{2}D $5f$
23	−4.40368	$3s3p^23d^3$
24	−4.27605	$3s^23p^3$ ^{2}P $5f$
25	−4.00286	$3s3p^23d^3$
26	−3.83572	$3s3p^23d^3$
27	−3.64171	$3s^23p^3$ ^{2}D $6p$
28	−3.40919	$3s3p^23d^3$
29	−3.27557	$3s3p^23d^3$
30	−3.01670	$3s^23p^3$ ^{2}D $6f$
31	−2.74261	$3s^23p^3$ ^{2}P $6f$
32	−2.48519	$3s^23p^3$ ^{2}D $7p$
33	−2.28782	$3s3p^4$ ^{2}D $5d$
34	−2.10520	$3s^23p^3$ ^{2}D $7f$
35	−1.88312	$3s3p^4$ ^{2}P $5d$
36	−1.81803	$3s^23p^3$ ^{2}P $7f$
37	−1.75881	$3s^23p^3$ ^{2}D $8p$
38	−1.74322	$3s3p^4$ ^{2}D $5g$
39	−1.51407	$3s^23p^3$ ^{2}D $8f$
40	−1.28450	$3s3p^4$ ^{2}P $5g$
41	−1.28034	$3s^23p^3$ ^{2}D $9p$
42	−1.21909	$3s^23p^3$ ^{2}P $8f$
43	−1.10952	$3s^23p^3$ ^{2}D $9f$
	^{1}F^o	
1	−16.5555	$3s^23p^3$ ^{2}P $3d$:
2	−15.8937	$3s^23p^3$ ^{2}D $3d$:
3	−11.6361	$3p^53d$
4	−10.2569	$3s3p^33d^2$
5	−10.0136	$3s3p^33d^2$
6	−9.67099	$3s3p^33d^2$
7	−9.36007	$3s3p^33d^2$
8	−9.03413	$3s3p^33d^2$
9	−8.80464	$3s3p^33d^2$
10	−8.63469	$3s3p^33d^2$
11	−8.32701	$3s^23p^3$ ^{2}D $4d$
12	−8.06686	$3s^23p3d^3$:
13	−8.04243	$3s^23p^3$ ^{2}P $4d$
14	−7.58226	$3s3p^33d^2$:
15	−7.31059	$3s^23p3d^3$:
16	−7.06861	$3s3p^4$ ^{2}D $4p$
17	−6.89074	$3s^23p3d^3$:
18	−6.37290	$3s^23p3d^3$
19	−5.00522	$3s^23p^3$ ^{2}D $5d$
20	−4.71113	$3s^23p^3$ ^{2}P $5d$
21	−4.63417	$3s3p^4$ ^{2}D $4f$
22	−4.45412	$3s^23p^3$ ^{2}D $5g$
23	−4.21573	$3s3p^4$ ^{2}S $4f$
24	−4.11898	$3s3p^4$ ^{2}P $4f$
25	−4.07291	$3s^23p^3$ ^{2}P $5g$
26	−3.91488	$3p^33d^3$
27	−3.31914	$3p^33d^3$
28	−3.27769	$3s^23p^3$ ^{2}D $6d$
29	−2.98012	$3s^23p^3$ ^{2}P $6d$
30	−2.96390	$3s3p^4$ ^{2}D $5p$
31	−2.95601	$3s^23p^3$ ^{2}D $6g$
32	−2.84789	$3p^33d^3$
33	−2.64916	$3s^23p^3$ ^{2}P $6g$
34	−2.42043	$3p^33d^3$
35	−2.34258	$3p^33d^3$
36	−2.26426	$3s^23p^3$ ^{2}D $7d$
37	−2.07172	$3s^23p^3$ ^{2}D $7g$
38	−2.05765	$3p^33d^3$
39	−1.95921	$3s^23p^3$ ^{2}P $7d$
40	−1.83767	$3s3p^4$ ^{2}D $5f$
41	−1.75877	$3s^23p^3$ ^{2}P $7g$
42	−1.62016	$3p^33d^3$
43	−1.61863	$3s^23p^3$ ^{2}D $8d$
44	−1.49461	$3s^23p^3$ ^{2}D $8g$
45	−1.40722	$3s3p^4$ ^{2}P $5f$
46	−1.36456	$3s3p^4$ ^{2}S $5f$
47	−1.30858	$3s^23p^3$ ^{2}P $8d$
48	−1.18202	$3s^23p^3$ ^{2}D $9d$
49	−1.17800	$3s^23p^3$ ^{2}P $8g$
50	−1.14578	$3p^33d^3$
51	−1.10423	$3p^33d^3$
52	−1.09192	$3s^23p^3$ ^{2}D $9g$
53	−.93182	$3s3p^4$ ^{2}D $6p$
54	−.87588	$3s^23p^3$ ^{2}P $9d$
	^{1}G^e	
1	−14.4969	$3s3p^4$ ^{2}D $3d$
2	−13.3771	$3s^23p^23d^2$
3	−12.2990	$3s^23p^23d^2$
4	−12.0309	$3s^23p^23d^2$
5	−11.2243	$3s^23p^23d^2$
6	−7.81966	$3p^43d^2$
7	−7.44262	$3p^43d^2$
8	−7.25980	$3s^23p^3$ ^{2}D $4f$
9	−7.00724	$3s^23p^3$ ^{2}P $4f$
10	−6.67284	$3p^43d^2$
11	−6.46287	$3p^43d^2$
12	−5.94495	$3s3p^23d^3$
13	−5.72671	$3s3p^23d^3$
14	−5.70016	$3s3p^4$ ^{2}D $4d$
15	−5.43011	$3s3p^23d^3$
16	−5.10531	$3s3p^23d^3$
17	−4.99192	$3s3p^23d^3$
18	−4.54177	$3s3p^23d^3$
19	−4.51892	$3s^23p^3$ ^{2}D $5f$
20	−4.38844	$3s3p^23d^3$
21	−4.23967	$3s^23p^3$ ^{2}P $5f$
22	−3.84403	$3s3p^23d^3$
23	−3.70683	$3s3p^23d^3$
24	−3.03462	$3s3p^23d^3$
25	−3.01393	$3s^23p^3$ ^{2}D $6f$
26	−2.72039	$3s^23p^3$ ^{2}P $6f$
27	−2.33579	$3s3p^4$ ^{2}D $5d$
28	−2.10423	$3s^23p^3$ ^{2}D $7f$
29	−1.80405	$3s^23p^3$ ^{2}P $7f$
30	−1.72268	$3s3p^4$ ^{2}D $5g$
31	−1.51379	$3s^23p^3$ ^{2}D $8f$
32	−1.30901	$3s3p^4$ ^{2}P $5g$
33	−1.25479	$3s3p^4$ ^{2}S $5g$
34	−1.20931	$3s^23p^3$ ^{2}P $8f$
35	−1.10967	$3s^23p^3$ ^{2}D $9f$
	^{1}G^o	
1	−17.1315	$3s^23p^3$ ^{2}D $3d$
2	−10.1885	$3s3p^33d^2$
3	−9.93367	$3s3p^33d^2$
4	−9.57077	$3s3p^33d^2$
5	−9.27086	$3s3p^33d^2$
6	−8.85255	$3s3p^33d^2$
7	−8.41280	$3s^23p^3$ ^{2}D $4d$
8	−7.94948	$3s^23p3d^3$
9	−7.44771	$3s^23p3d^3$
10	−7.24783	$3s^23p3d^3$
11	−5.04265	$3s^23p^3$ ^{2}D $5d$
12	−4.59587	$3s3p^4$ ^{2}D $4f$
13	−4.45253	$3s^23p^3$ ^{2}D $5g$
14	−4.24567	$3s3p^4$ ^{2}P $4f$
15	−4.10903	$3s^23p^3$ ^{2}P $5g$
16	−3.69712	$3p^33d^3$
17	−3.29654	$3s^23p^3$ ^{2}D $6d$
18	−3.19417	$3p^33d^3$
19	−2.96094	$3s^23p^3$ ^{2}D $6g$
20	−2.93313	$3p^33d^3$
21	−2.67652	$3p^33d^3$
22	−2.65590	$3s^23p^3$ ^{2}P $6g$
23	−2.52596	$3p^33d^3$
24	−2.27613	$3s^23p^3$ ^{2}D $7d$
25	−2.06950	$3s^23p^3$ ^{2}D $7g$
26	−1.94656	$3p^33d^3$
27	−1.83430	$3s3p^4$ ^{2}D $5f$
28	−1.76381	$3s^23p^3$ ^{2}P $7g$
29	−1.62668	$3s^23p^3$ ^{2}D $8d$
30	−1.49272	$3s^23p^3$ ^{2}D $8g$

S-like Fe (Fe^{10+})

i	E(Ryds)	Description
31	−1.43050	$3s3p^4\ ^2\mathrm{P}\ 5f$
32	−1.38150	$3p^33d^3$
33	−1.28651	$3p^33d^3$
34	−1.18714	$3s^23p^3\ ^2\mathrm{D}\ 9d$
35	−1.18023	$3s^23p^3\ ^2\mathrm{P}\ 8g$
36	−1.09194	$3s^23p^3\ ^2\mathrm{D}\ 9g$
	$^3\mathrm{S}^e$	
1	−14.1271	$3s3p^4\ ^2\mathrm{D}\ 3d$
2	−11.8672	$3s^23p^23d^2$
3	−9.60564	$3s^23p^3\ ^2\mathrm{P}\ 4p$
4	−7.57398	$3s3p^4\ ^2\mathrm{S}\ 4s$
5	−7.18935	$3s3p^23d^3$:
6	−6.28706	$3p^43d^2$:
7	−5.82625	$3s3p^4\ ^2\mathrm{D}\ 4d$
8	−5.77481	$3s3p^23d^3$
9	−5.39070	$3s^23p^3\ ^2\mathrm{P}\ 5p$
10	−5.29050	$3s3p^23d^3$
11	−4.56964	$3s3p^23d^3$
12	−4.26738	$3s3p^23d^3$
13	−3.97516	$3s3p^23d^3$
14	−3.35401	$3s^23p^3\ ^2\mathrm{P}\ 6p$
15	−2.92663	$3s3p^4\ ^2\mathrm{S}\ 5s$
16	−2.38192	$3s3p^4\ ^2\mathrm{D}\ 5d$
17	−2.17773	$3s^23p^3\ ^2\mathrm{P}\ 7p$
18	−1.45618	$3s^23p^3\ ^2\mathrm{P}\ 8p$
19	−.97431	$3s^23p^3\ ^2\mathrm{P}\ 9p$
20	−.70433	$3s3p^4\ ^2\mathrm{S}\ 6s$
	$^3\mathrm{S}^o$	
1	−16.4183	$3s^23p^3\ ^2\mathrm{D}\ 3d$
2	−11.0335	$3s^23p^3\ ^4\mathrm{S}\ 4s$
3	−10.5572	$3s3p^33d^2$
4	−9.88012	$3s3p^33d^2$
5	−9.18348	$3s3p^33d^2$
6	−8.38867	$3s^23p^3\ ^2\mathrm{D}\ 4d$
7	−8.33438	$3s3p^33d^2$
8	−7.58888	$3s^23p3d^3$
9	−7.47721	$3s3p^4\ ^4\mathrm{P}\ 4p$
10	−6.81979	$3s^23p3d^3$
11	−6.59026	$3s3p^4\ ^2\mathrm{P}\ 4p$
12	−6.44615	$3s^23p^3\ ^4\mathrm{S}\ 5s$
13	−5.02353	$3s^23p^3\ ^2\mathrm{D}\ 5d$
14	−4.25782	$3s^23p^3\ ^4\mathrm{S}\ 6s$
15	−3.46631	$3s3p^4\ ^4\mathrm{P}\ 5p$
16	−3.28631	$3s^23p^3\ ^2\mathrm{D}\ 6d$
17	−3.01606	$3s^23p^3\ ^4\mathrm{S}\ 7s$
18	−2.96611	$3p^33d^3$
19	−2.49616	$3s3p^4\ ^2\mathrm{P}\ 5p$
20	−2.35474	$3p^33d^3$
21	−2.27140	$3s^23p^3\ ^2\mathrm{D}\ 7d$
22	−2.24514	$3s^23p^3\ ^4\mathrm{S}\ 8s$
23	−2.04071	$3p^33d^3$
24	−1.74183	$3s^23p^3\ ^4\mathrm{S}\ 9s$
25	−1.62297	$3s^23p^3\ ^2\mathrm{D}\ 8d$
26	−1.46285	$3s3p^4\ ^4\mathrm{P}\ 6p$
	$^3\mathrm{P}^e$	
1	−21.3073	$3s^23p^4$
2	−15.0947	$3s3p^4\ ^2\mathrm{D}\ 3d$:
3	−14.0575	$3s3p^4\ ^2\mathrm{P}\ 3d$:
4	−13.7955	$3s3p^4\ ^4\mathrm{P}\ 3d$:
5	−13.0340	$3s^23p^23d^2$:
6	−12.6682	$3s^23p^23d^2$
7	−12.4694	$3s^23p^23d^2$
8	−11.9106	$3s^23p^23d^2$
9	−11.7784	$3s^23p^23d^2$
10	−11.1792	$3s^23p^23d^2$
11	−10.1597	$3s^23p^3\ ^4\mathrm{S}\ 4p$
12	−9.62128	$3s^23p^3\ ^2\mathrm{D}\ 4p$
13	−9.42297	$3s^23p^3\ ^2\mathrm{P}\ 4p$
14	−8.38774	$3s3p^4\ ^4\mathrm{P}\ 4s$
15	−8.31515	$3p^43d^2$
16	−7.73284	$3p^43d^2$
17	−7.58758	$3s3p^4\ ^2\mathrm{P}\ 4s$
18	−7.33317	$3s^23p^3\ ^2\mathrm{D}\ 4f$
19	−7.08192	$3p^43d^2$
20	−7.02573	$3p^43d^2$:
21	−6.64617	$3p^43d^2$:
22	−6.56641	$3s3p^23d^3$:
23	−6.41522	$3p^43d^2$:
24	−6.14798	$3s3p^4\ ^4\mathrm{P}\ 4d$
25	−6.03060	$3s^23p^3\ ^4\mathrm{S}\ 5p$
26	−6.01319	$3s3p^23d^3$
27	−5.81283	$3s3p^23d^3$
28	−5.74060	$3s3p^4\ ^2\mathrm{D}\ 4d$
29	−5.64415	$3s3p^23d^3$
30	−5.62144	$3s^23p^3\ ^2\mathrm{D}\ 5p$
31	−5.43956	$3s3p^23d^3$
32	−5.33906	$3s^23p^3\ ^2\mathrm{P}\ 5p$
33	−5.28223	$3s3p^23d^3$
34	−5.20456	$3s3p^4\ ^2\mathrm{P}\ 4d$
35	−5.07045	$3s3p^23d^3$
36	−4.90438	$3s3p^23d^3$
37	−4.79876	$3s3p^23d^3$
38	−4.73506	$3s3p^23d^3$
39	−4.55700	$3s^23p^3\ ^2\mathrm{D}\ 5f$
40	−4.40994	$3s3p^23d^3$
41	−4.26532	$3s3p^23d^3$
42	−4.07632	$3s3p^23d^3$
43	−4.01979	$3s^23p^3\ ^4\mathrm{S}\ 6p$
44	−3.89865	$3s3p^4\ ^4\mathrm{P}\ 5s$
45	−3.80678	$3s3p^23d^3$
46	−3.61628	$3s^23p^3\ ^2\mathrm{D}\ 6p$
47	−3.61376	$3s3p^23d^3$
48	−3.31940	$3s^23p^3\ ^2\mathrm{P}\ 6p$
49	−3.03493	$3s^23p^3\ ^2\mathrm{D}\ 6f$
50	−2.98783	$3s3p^23d^3$
51	−2.95878	$3s3p^4\ ^2\mathrm{P}\ 5s$
52	−2.87303	$3s^23p^3\ ^4\mathrm{S}\ 7p$
53	−2.83690	$3s3p^4\ ^4\mathrm{P}\ 5d$
54	−2.47279	$3s^23p^3\ ^2\mathrm{D}\ 7p$
55	−2.35106	$3s3p^4\ ^2\mathrm{D}\ 5d$
56	−2.17016	$3s^23p^3\ ^2\mathrm{P}\ 7p$
57	−2.15211	$3s^23p^3\ ^4\mathrm{S}\ 8p$
58	−2.11627	$3s^23p^3\ ^2\mathrm{D}\ 7f$
59	−1.86917	$3s3p^4\ ^2\mathrm{P}\ 5d$
60	−1.75914	$3s^23p^3\ ^2\mathrm{D}\ 8p$
61	−1.69294	$3s3p^4\ ^4\mathrm{P}\ 6s$
62	−1.67525	$3s^23p^3\ ^4\mathrm{S}\ 9p$
63	−1.52241	$3s^23p^3\ ^2\mathrm{D}\ 8f$
64	−1.44636	$3s^23p^3\ ^2\mathrm{P}\ 8p$
	$^3\mathrm{P}^o$	
1	−18.8741	$3s3p^5$
2	−16.8989	$3s^23p^3\ ^2\mathrm{P}\ 3d$:
3	−16.4621	$3s^23p^3\ ^2\mathrm{D}\ 3d$:
4	−12.2477	$3p^53d$
5	−10.9162	$3s3p^33d^2$
6	−10.4852	$3s^23p^3\ ^2\mathrm{P}\ 4s$
7	−10.4455	$3s3p^33d^2$
8	−10.2638	$3s3p^33d^2$
9	−9.86541	$3s3p^33d^2$
10	−9.78830	$3s3p^33d^2$
11	−9.45236	$3s3p^33d^2$
12	−9.29573	$3s3p^33d^2$
13	−9.21509	$3s3p^33d^2$
14	−8.64389	$3s3p^33d^2$
15	−8.60704	$3s3p^33d^2$
16	−8.43047	$3s^23p^3\ ^2\mathrm{D}\ 4d$
17	−8.35981	$3s3p^33d^2$
18	−8.15846	$3s^23p3d^3$
19	−8.09896	$3s^23p^3\ ^2\mathrm{P}\ 4d$
20	−7.83223	$3s^23p3d^3$
21	−7.54590	$3s3p^4\ ^4\mathrm{P}\ 4p$
22	−7.49664	$3s^23p3d^3$
23	−7.00465	$3s3p^4\ ^2\mathrm{D}\ 4p$
24	−6.69903	$3s^23p3d^3$
25	−6.66838	$3s3p^4\ ^2\mathrm{P}\ 4p$
26	−6.58725	$3s3p^4\ ^2\mathrm{S}\ 4p$
27	−5.80693	$3s^23p^3\ ^2\mathrm{P}\ 5s$
28	−5.03496	$3s^23p^3\ ^2\mathrm{D}\ 5d$
29	−4.79213	$3s3p^4\ ^2\mathrm{D}\ 4f$
30	−4.72381	$3s^23p^3\ ^2\mathrm{P}\ 5d$
31	−4.03792	$3p^33d^3$
32	−3.91699	$3p^33d^3$
33	−3.57643	$3s^23p^3\ ^2\mathrm{P}\ 6s$
34	−3.46664	$3p^33d^3$
35	−3.45760	$3s3p^4\ ^4\mathrm{P}\ 5p$
36	−3.29317	$3s^23p^3\ ^2\mathrm{D}\ 6d$
37	−3.06820	$3p^33d^3$
38	−2.99401	$3s^23p^3\ ^2\mathrm{P}\ 6d$
39	−2.96432	$3s3p^4\ ^2\mathrm{D}\ 5p$
40	−2.92712	$3p^33d^3$
41	−2.75763	$3p^33d^3$
42	−2.54706	$3s3p^4\ ^2\mathrm{P}\ 5p$
43	−2.48813	$3p^33d^3$
44	−2.48462	$3s3p^4\ ^2\mathrm{S}\ 5p$
45	−2.31780	$3s^23p^3\ ^2\mathrm{P}\ 7s$
46	−2.26526	$3s^23p^3\ ^2\mathrm{D}\ 7d$
47	−2.12597	$3p^33d^3$
48	−1.96769	$3s^23p^3\ ^2\mathrm{P}\ 7d$
49	−1.91845	$3s3p^4\ ^2\mathrm{D}\ 5f$
50	−1.85152	$3p^33d^3$
51	−1.62506	$3s^23p^3\ ^2\mathrm{D}\ 8d$
52	−1.54326	$3s^23p^3\ ^2\mathrm{P}\ 8s$
53	−1.45506	$3s3p^4\ ^4\mathrm{P}\ 6p$
54	−1.44529	$3p^33d^3$
55	−1.31649	$3s^23p^3\ ^2\mathrm{P}\ 8d$
56	−1.21713	$3p^33d^3$
57	−1.18593	$3s^23p^3\ ^2\mathrm{D}\ 9d$
58	−1.03397	$3s^23p^3\ ^2\mathrm{P}\ 9s$
59	−.93702	$3s3p^4\ ^2\mathrm{D}\ 6p$
60	−.88170	$3s^23p^3\ ^2\mathrm{P}\ 9d$
	$^3\mathrm{D}^e$	
1	−14.9109	$3s3p^4\ ^2\mathrm{D}\ 3d$:
2	−14.2944	$3s3p^4\ ^2\mathrm{P}\ 3d$:
3	−13.9022	$3s3p^4\ ^2\mathrm{S}\ 3d$:
4	−13.7110	$3s3p^4\ ^4\mathrm{P}\ 3d$:
5	−13.0469	$3s^23p^23d^2$
6	−12.6922	$3s^23p^23d^2$
7	−12.1614	$3s^23p^23d^2$
8	−11.9710	$3s^23p^23d^2$
9	−11.7371	$3s^23p^23d^2$
10	−9.91377	$3s^23p^3\ ^2\mathrm{D}\ 4p$
11	−9.53426	$3s^23p^3\ ^2\mathrm{P}\ 4p$
12	−8.05184	$3s3p^4\ ^2\mathrm{D}\ 4s$
13	−7.98813	$3p^43d^2$
14	−7.74640	$3p^43d^2$
15	−7.45037	$3p^43d^2$
16	−7.42458	$3s^23p^3\ ^2\mathrm{D}\ 4f$
17	−7.00270	$3s^23p^3\ ^2\mathrm{P}\ 4f$
18	−6.89312	$3p^43d^2$
19	−6.79571	$3p^43d^2$
20	−6.43436	$3s3p^23d^3$
21	−6.23391	$3s3p^4\ ^4\mathrm{P}\ 4d$
22	−6.11728	$3s3p^23d^3$
23	−5.99794	$3s3p^23d^3$
24	−5.92752	$3s3p^23d^3$
25	−5.81441	$3s3p^23d^3$
26	−5.79587	$3s3p^23d^3$
27	−5.73661	$3s^23p^3\ ^2\mathrm{D}\ 5p$

S-like Fe (Fe^{10+})

i	E(Ryds)	Description
28	−5.64882	$3s3p^4$ ^{2}D $4d$
29	−5.53296	$3s3p^23d^3$
30	−5.39440	$3s^23p^3$ ^{2}P $5p$
31	−5.31473	$3s3p^23d^3$
32	−5.29187	$3s3p^4$ ^{2}P $4d$
33	−5.25043	$3s3p^23d^3$
34	−5.19426	$3s3p^4$ ^{2}S $4d$
35	−5.16184	$3s3p^23d^3$
36	−5.13339	$3s3p^23d^3$
37	−4.88303	$3s3p^23d^3$
38	−4.74330	$3s3p^23d^3$
39	−4.63837	$3s3p^23d^3$
40	−4.58717	$3s^23p^3$ ^{2}D $5f$
41	−4.53138	$3s3p^23d^3$
42	−4.40561	$3s3p^23d^3$
43	−4.28787	$3s^23p^3$ ^{2}P $5f$
44	−4.15516	$3s3p^23d^3$
45	−4.10306	$3s3p^23d^3$
46	−3.93060	$3s3p^23d^3$
47	−3.66245	$3s^23p^3$ ^{2}D $6p$
48	−3.58474	$3s3p^23d^3$
49	−3.40868	$3s3p^4$ ^{2}D $5s$
50	−3.33319	$3s^23p^3$ ^{2}P $6p$
51	−3.10284	$3s3p^23d^3$
52	−3.04859	$3s^23p^3$ ^{2}D $6f$
53	−2.88058	$3s3p^4$ ^{4}P $5d$
54	−2.74758	$3s^23p^3$ ^{2}P $6f$
55	−2.49303	$3s^23p^3$ ^{2}D $7p$
56	−2.32019	$3s3p^4$ ^{2}D $5d$
57	−2.18001	$3s^23p^3$ ^{2}P $7p$
58	−2.12381	$3s^23p^3$ ^{2}D $7f$
59	−1.90686	$3s3p^4$ ^{2}P $5d$
60	−1.85279	$3s3p^4$ ^{2}S $5d$
61	−1.81818	$3s^23p^3$ ^{2}P $7f$
62	−1.77562	$3s3p^4$ ^{2}D $5g$
63	−1.76527	$3s^23p^3$ ^{2}D $8p$
64	−1.52722	$3s^23p^3$ ^{2}D $8f$
65	−1.45309	$3s^23p^3$ ^{2}P $8p$
66	−1.28493	$3s^23p^3$ ^{2}D $9p$
67	−1.22093	$3s^23p^3$ ^{2}P $8f$
68	−1.17580	$3s3p^4$ ^{2}D $6s$
69	−1.13905	$3s3p^4$ ^{4}P $6d$
70	−1.11837	$3s^23p^3$ ^{2}D $9f$
71	−.97120	$3s^23p^3$ ^{2}P $9p$
	^{3}D^o	
1	−17.5841	$3s^23p^3$ ^{2}D $3d$:
2	−16.8606	$3s^23p^3$ ^{2}P $3d$:
3	−16.1988	$3s^23p^3$ ^{4}S $3d$:
4	−11.7217	$3p^53d$
5	−11.1808	$3s3p^33d^2$
6	−10.7946	$3s^23p^3$ ^{2}D $4s$
7	−10.5471	$3s3p^33d^2$
8	−10.2205	$3s3p^33d^2$
9	−10.1195	$3s3p^33d^2$
10	−10.0109	$3s3p^33d^2$
11	−9.84888	$3s3p^33d^2$
12	−9.70908	$3s3p^33d^2$
13	−9.55266	$3s3p^33d^2$
14	−9.42599	$3s3p^33d^2$
15	−9.10972	$3s3p^33d^2$
16	−9.05320	$3s3p^33d^2$
17	−8.77080	$3s^23p^3$ ^{4}S $4d$
18	−8.62117	$3s3p^33d^2$
19	−8.54138	$3s3p^33d^2$
20	−8.44543	$3s^23p^3$ ^{2}D $4d$
21	−8.41405	$3s^23p3d^3$
22	−8.08185	$3s^23p^3$ ^{2}P $4d$
23	−7.98266	$3s^23p3d^3$
24	−7.78465	$3s^23p3d^3$
25	−7.66646	$3s^23p3d^3$
26	−7.55191	$3s3p^4$ ^{4}P $4p$
27	−7.21330	$3s^23p3d^3$
28	−7.01232	$3s3p^4$ ^{2}D $4p$
29	−6.63361	$3s3p^4$ ^{2}P $4p$
30	−6.58853	$3s^23p3d^3$
31	−6.11386	$3s^23p^3$ ^{2}D $5s$
32	−5.41825	$3s^23p^3$ ^{4}S $5d$
33	−5.11036	$3s3p^4$ ^{4}P $4f$
34	−5.04425	$3s^23p^3$ ^{2}D $5d$
35	−4.73491	$3s^23p^3$ ^{2}P $5d$
36	−4.69145	$3s3p^4$ ^{2}D $4f$
37	−4.47074	$3s^23p^3$ ^{2}D $5g$
38	−4.20930	$3s3p^4$ ^{2}P $4f$
39	−3.93960	$3p^33d^3$
40	−3.88293	$3s^23p^3$ ^{2}D $6s$
41	−3.68792	$3s^23p^3$ ^{4}S $6d$
42	−3.66147	$3p^33d^3$
43	−3.49495	$3s3p^4$ ^{4}P $5p$
44	−3.30224	$3s^23p^3$ ^{2}D $6d$
45	−3.23078	$3p^33d^3$
46	−3.21496	$3p^33d^3$
47	−2.99032	$3s^23p^3$ ^{2}P $6d$
48	−2.96023	$3s^23p^3$ ^{2}D $6g$
49	−2.94995	$3p^33d^3$
50	−2.93265	$3s3p^4$ ^{2}D $5p$
51	−2.87407	$3p^33d^3$
52	−2.68555	$3p^33d^3$
53	−2.67466	$3s^23p^3$ ^{4}S $7d$
54	−2.62707	$3s^23p^3$ ^{2}D $7s$
55	−2.51629	$3s3p^4$ ^{2}P $5p$
56	−2.46418	$3p^33d^3$
57	−2.36898	$3s3p^4$ ^{4}P $5f$
58	−2.27397	$3s^23p^3$ ^{2}D $7d$
59	−2.23579	$3p^33d^3$
60	−2.10754	$3p^33d^3$
61	−2.07240	$3s^23p^3$ ^{2}D $7g$
62	−2.02368	$3s^23p^3$ ^{4}S $8d$
63	−1.96478	$3s^23p^3$ ^{2}P $7d$
64	−1.93473	$3p^33d^3$
65	−1.88740	$3s3p^4$ ^{2}D $5f$
66	−1.88168	$3p^33d^3$
67	−1.85381	$3s^23p^3$ ^{2}D $8s$
68	−1.62650	$3s^23p^3$ ^{2}D $8d$
69	−1.58687	$3s^23p^3$ ^{4}S $9d$
70	−1.49710	$3s^23p^3$ ^{2}D $8g$
71	−1.47789	$3s3p^4$ ^{4}P $6p$
72	−1.43446	$3p^33d^3$
73	−1.43031	$3s3p^4$ ^{2}P $5f$
74	−1.34556	$3s^23p^3$ ^{2}D $9s$
75	−1.31226	$3s^23p^3$ ^{2}P $8d$
	^{3}F^e	
1	−14.8657	$3s3p^4$ ^{4}P $3d$
2	−14.3900	$3s3p^4$ ^{2}D $3d$
3	−13.9313	$3s3p^4$ ^{2}P $3d$
4	−13.5095	$3s^23p^23d^2$
5	−12.8494	$3s^23p^23d^2$
6	−12.6638	$3s^23p^23d^2$
7	−12.2820	$3s^23p^23d^2$
8	−12.1330	$3s^23p^23d^2$
9	−11.5568	$3s^23p^23d^2$
10	−9.84995	$3s^23p^3$ ^{2}D $4p$
11	−8.18653	$3p^43d^2$
12	−7.79172	$3p^43d^2$
13	−7.73175	$3s^23p^3$ ^{4}S $4f$
14	−7.61508	$3p^43d^2$
15	−7.41803	$3s^23p^3$ ^{2}D $4f$
16	−7.14381	$3p^43d^2$
17	−7.07397	$3s^23p^3$ ^{2}P $4f$
18	−6.96409	$3p^43d^2$
19	−6.78708	$3p^43d^2$:
20	−6.57029	$3s3p^23d^3$:
21	−6.20318	$3s3p^4$ ^{4}P $4d$
22	−6.15086	$3s3p^23d^3$
23	−6.02516	$3s3p^23d^3$
24	−6.01154	$3s3p^23d^3$
25	−5.73364	$3s3p^23d^3$
26	−5.71867	$3s^23p^3$ ^{2}D $5p$
27	−5.66291	$3s3p^23d^3$
28	−5.61421	$3s3p^4$ ^{2}D $4d$
29	−5.53832	$3s3p^23d^3$
30	−5.46678	$3s3p^23d^3$
31	−5.33555	$3s3p^23d^3$
32	−5.24675	$3s3p^4$ ^{2}P $4d$
33	−5.22187	$3s3p^23d^3$
34	−5.08850	$3s3p^23d^3$
35	−4.97576	$3s3p^23d^3$
36	−4.94458	$3s^23p^3$ ^{4}S $5f$
37	−4.86101	$3s3p^23d^3$
38	−4.72395	$3s3p^23d^3$
39	−4.69386	$3s3p^23d^3$
40	−4.60120	$3s^23p^3$ ^{2}D $5f$
41	−4.44157	$3s3p^23d^3$
42	−4.35323	$3s3p^23d^3$
43	−4.27209	$3s^23p^3$ ^{2}P $5f$
44	−4.24519	$3s3p^23d^3$
45	−3.95055	$3s3p^23d^3$
46	−3.64813	$3s^23p^3$ ^{2}D $6p$
47	−3.48813	$3s3p^23d^3$
48	−3.42753	$3s^23p^3$ ^{4}S $6f$
49	−3.05928	$3s^23p^3$ ^{2}D $6f$
50	−2.86771	$3s3p^4$ ^{4}P $5d$
51	−2.73850	$3s^23p^3$ ^{2}P $6f$
52	−2.51288	$3s^23p^3$ ^{4}S $7f$
53	−2.49057	$3s^23p^3$ ^{2}D $7p$
54	−2.30585	$3s3p^4$ ^{2}D $5d$
55	−2.27761	$3s3p^4$ ^{4}P $5g$
56	−2.13154	$3s^23p^3$ ^{2}D $7f$
57	−1.92087	$3s^23p^3$ ^{4}S $8f$
58	−1.88803	$3s3p^4$ ^{2}P $5d$
59	−1.81412	$3s^23p^3$ ^{2}P $7f$
60	−1.76100	$3s^23p^3$ ^{2}D $8p$
61	−1.74627	$3s3p^4$ ^{2}D $5g$
62	−1.53406	$3s^23p^3$ ^{2}D $8f$
63	−1.51311	$3s^23p^3$ ^{4}S $9f$
64	−1.29024	$3s3p^4$ ^{2}P $5g$
65	−1.28197	$3s^23p^3$ ^{2}D $9p$
	^{3}F^o	
1	−17.4846	$3s^23p^3$ ^{2}D $3d$:
2	−16.9433	$3s^23p^3$ ^{2}P $3d$:
3	−12.0333	$3p^53d$
4	−11.2004	$3s3p^33d^2$
5	−10.6650	$3s3p^33d^2$
6	−10.4535	$3s3p^33d^2$
7	−10.2303	$3s3p^33d^2$
8	−10.0915	$3s3p^33d^2$
9	−9.92890	$3s3p^33d^2$
10	−9.71878	$3s3p^33d^2$
11	−9.58724	$3s3p^33d^2$
12	−9.38420	$3s3p^33d^2$
13	−9.17068	$3s3p^33d^2$
14	−9.03871	$3s3p^33d^2$:
15	−8.48778	$3s^23p^3$ ^{2}D $4d$
16	−8.43320	$3s^23p3d^3$:
17	−8.35409	$3s3p^33d^2$:
18	−8.10255	$3s^23p^3$ ^{2}P $4d$
19	−8.08601	$3s^23p3d^3$
20	−7.79460	$3s^23p3d^3$
21	−7.51643	$3s^23p3d^3$

S-like Fe (Fe^{10+})

i	E(Ryds)	Description
22	−7.15187	$3s^23p3d^3$
23	−7.13959	$3s3p^4$ ^{2}D $4p$
24	−5.17143	$3s3p^4$ ^{4}P $4f$
25	−5.05740	$3s^23p^3$ ^{2}D $5d$
26	−4.73852	$3s^23p^3$ ^{2}P $5d$
27	−4.68796	$3s3p^4$ ^{2}D $4f$
28	−4.49257	$3s^23p^3$ ^{2}D $5g$
29	−4.28686	$3s3p^4$ ^{2}P $4f$
30	−4.27722	$3s3p^4$ ^{2}S $4f$
31	−4.07547	$3s^23p^3$ ^{2}P $5g$
32	−4.01141	$3p^33d^3$
33	−3.84594	$3p^33d^3$
34	−3.56095	$3p^33d^3$
35	−3.31354	$3p^33d^3$
36	−3.30903	$3s^23p^3$ ^{2}D $6d$
37	−3.12123	$3p^33d^3$
38	−3.01172	$3s3p^4$ ^{2}D $5p$
39	−2.96955	$3s^23p^3$ ^{2}D $6g$:
40	−2.95962	$3s^23p^3$ ^{2}P $6d$:
41	−2.93390	$3p^33d^3$
42	−2.85801	$3p^33d^3$
43	−2.77575	$3p^33d^3$
44	−2.68357	$3p^33d^3$
45	−2.65545	$3s^23p^3$ ^{2}P $6g$
46	−2.44978	$3p^33d^3$
47	−2.39359	$3s3p^4$ ^{4}P $5f$
48	−2.34776	$3p^33d^3$
49	−2.27877	$3s^23p^3$ ^{2}D $7d$
50	−2.07491	$3s^23p^3$ ^{2}D $7g$
51	−2.01692	$3p^33d^3$
52	−1.96743	$3s^23p^3$ ^{2}P $7d$
53	−1.87440	$3p^33d^3$
54	−1.87106	$3s3p^4$ ^{2}D $5f$
55	−1.76484	$3s^23p^3$ ^{2}P $7g$
56	−1.62815	$3s^23p^3$ ^{2}D $8d$
57	−1.50613	$3s^23p^3$ ^{2}D $8g$
58	−1.46180	$3s3p^4$ ^{2}P $5f$
59	−1.41494	$3s3p^4$ ^{2}S $5f$
60	−1.32891	$3p^33d^3$
61	−1.31508	$3s^23p^3$ ^{2}P $8d$
62	−1.18883	$3s^23p^3$ ^{2}D $9d$
63	−1.10057	$3s^23p^3$ ^{2}P $8g$
64	−1.09290	$3s^23p^3$ ^{2}D $9g$
65	−.94843	$3s3p^4$ ^{2}D $6p$
66	−.87961	$3s^23p^3$ ^{2}P $9d$:
67	−.87581	$3s3p^4$ ^{4}P $6f$:
		^{3}G^e
1	−14.7177	$3s3p^4$ ^{2}D $3d$
2	−13.1482	$3s^23p^23d^2$
3	−12.5863	$3s^23p^23d^2$
4	−12.2237	$3s^23p^23d^2$
5	−8.08031	$3p^43d^2$
6	−7.65970	$3p^43d^2$
7	−7.51149	$3s^23p^3$ ^{2}D $4f$
8	−7.17068	$3p^43d^2$
9	−7.11508	$3s^23p^3$ ^{2}P $4f$
10	−6.75582	$3s3p^23d^3$
11	−6.35063	$3s3p^23d^3$
12	−6.23040	$3s3p^23d^3$
13	−5.88451	$3s3p^23d^3$
14	−5.78864	$3s3p^23d^3$
15	−5.76501	$3s3p^4$ ^{2}D $4d$
16	−5.58008	$3s3p^23d^3$
17	−5.48880	$3s3p^23d^3$
18	−5.40123	$3s3p^23d^3$
19	−5.28874	$3s3p^23d^3$
20	−5.07834	$3s3p^23d^3$
21	−4.86784	$3s3p^23d^3$
22	−4.68881	$3s3p^23d^3$
23	−4.62525	$3s^23p^3$ ^{2}D $5f$
24	−4.48626	$3s3p^23d^3$
25	−4.29688	$3s^23p^3$ ^{2}P $5f$
26	−4.24844	$3s3p^23d^3$
27	−3.78727	$3s3p^23d^3$
28	−3.06845	$3s^23p^3$ ^{2}D $6f$
29	−2.75232	$3s^23p^3$ ^{2}P $6f$
30	−2.35974	$3s3p^4$ ^{2}D $5d$
31	−2.31626	$3s3p^4$ ^{4}P $5g$
32	−2.13583	$3s^23p^3$ ^{2}D $7f$
33	−1.82324	$3s^23p^3$ ^{2}P $7f$
34	−1.72640	$3s3p^4$ ^{2}D $5g$
35	−1.53422	$3s^23p^3$ ^{2}D $8f$
36	−1.31696	$3s3p^4$ ^{2}P $5g$
37	−1.25847	$3s3p^4$ ^{2}S $5g$
38	−1.22272	$3s^23p^3$ ^{2}P $8f$
39	−1.12322	$3s^23p^3$ ^{2}D $9f$
		^{3}G^o
1	−17.2342	$3s^23p^3$ ^{2}D $3d$
2	−10.9931	$3s3p^33d^2$
3	−10.5375	$3s3p^33d^2$
4	−10.3318	$3s3p^33d^2$
5	−10.2698	$3s3p^33d^2$
6	−9.90041	$3s3p^33d^2$
7	−9.61383	$3s3p^33d^2$
8	−9.29007	$3s3p^33d^2$
9	−8.94204	$3s3p^33d^2$
10	−8.44226	$3s^23p3d^3$
11	−8.41897	$3s^23p^3$ ^{2}D $4d$
12	−8.13251	$3s^23p3d^3$
13	−7.66979	$3s^23p3d^3$
14	−7.21963	$3s^23p3d^3$
15	−5.19549	$3s3p^4$ ^{4}P $4f$
16	−5.04820	$3s^23p^3$ ^{2}D $5d$
17	−4.83901	$3s^23p^3$ ^{4}S $5g$
18	−4.68894	$3s3p^4$ ^{2}D $4f$
19	−4.44244	$3s^23p^3$ ^{2}D $5g$
20	−4.26941	$3s3p^4$ ^{2}P $4f$
21	−4.12228	$3s^23p^3$ ^{2}P $5g$
22	−4.02406	$3p^33d^3$
23	−3.59698	$3p^33d^3$
24	−3.53028	$3p^33d^3$
25	−3.36607	$3s^23p^3$ ^{4}S $6g$
26	−3.29883	$3s^23p^3$ ^{2}D $6d$
27	−3.06756	$3p^33d^3$
28	−2.96790	$3s^23p^3$ ^{2}D $6g$
29	−2.89614	$3p^33d^3$
30	−2.75958	$3p^33d^3$
31	−2.67072	$3p^33d^3$
32	−2.65581	$3s^23p^3$ ^{2}P $6g$
33	−2.54788	$3p^33d^3$
34	−2.47493	$3s^23p^3$ ^{4}S $7g$
35	−2.40681	$3s3p^4$ ^{4}P $5f$
36	−2.27772	$3s^23p^3$ ^{2}D $7d$
37	−2.23256	$3p^33d^3$
38	−2.16209	$3p^33d^3$
39	−2.07655	$3s^23p^3$ ^{2}D $7g$
40	−1.89403	$3s^23p^3$ ^{4}S $8g$
41	−1.87379	$3s3p^4$ ^{2}D $5f$
42	−1.76169	$3s^23p^3$ ^{2}P $7g$
43	−1.62720	$3s^23p^3$ ^{2}D $8d$
44	−1.61802	$3p^33d^3$
45	−1.49977	$3s^23p^3$ ^{4}S $9g$
46	−1.49302	$3s^23p^3$ ^{2}D $8g$
47	−1.44476	$3s3p^4$ ^{2}P $5f$
		^{5}S^o
1	−11.2524	$3s^23p^3$ ^{4}S $4s$
2	−10.9600	$3s3p^33d^2$
3	−10.0104	$3s3p^33d^2$
4	−8.55320	$3s^23p3d^3$
5	−7.54303	$3s3p^4$ ^{4}P $4p$
6	−6.53300	$3s^23p^3$ ^{4}S $5s$
7	−4.29502	$3s^23p^3$ ^{4}S $6s$
8	−3.48185	$3s3p^4$ ^{4}P $5p$
9	−3.47331	$3p^33d^3$
10	−3.03309	$3s^23p^3$ ^{4}S $7s$
11	−2.26113	$3s^23p^3$ ^{4}S $8s$
12	−1.75040	$3s^23p^3$ ^{4}S $9s$
13	−1.47478	$3s3p^4$ ^{4}P $6p$
		^{5}P^e
1	−15.1852	$3s3p^4$ ^{4}P $3d$
2	−12.9466	$3s^23p^23d^2$
3	−10.2990	$3s^23p^3$ ^{4}S $4p$
4	−8.66500	$3s3p^4$ ^{4}P $4s$
5	−8.00114	$3p^43d^2$
6	−6.93800	$3s3p^23d^3$
7	−6.87476	$3s3p^23d^3$
8	−6.30477	$3s3p^23d^3$
9	−6.23436	$3s3p^4$ ^{4}P $4d$
10	−6.20611	$3s3p^23d^3$
11	−6.07916	$3s^23p^3$ ^{4}S $5p$
12	−5.80578	$3s3p^23d^3$
13	−5.68610	$3s3p^23d^3$
14	−5.40367	$3s3p^23d^3$
15	−4.96716	$3s3p^23d^3$
16	−4.07324	$3s^23p^3$ ^{4}S $6p$
17	−3.97007	$3s3p^4$ ^{4}P $5s$
18	−2.91309	$3s^23p^3$ ^{4}S $7p$
19	−2.85075	$3s3p^4$ ^{4}P $5d$
20	−2.16972	$3s^23p^3$ ^{4}S $8p$
21	−1.74184	$3s3p^4$ ^{4}P $6s$
22	−1.68450	$3s^23p^3$ ^{4}S $9p$
		^{5}P^o
1	−11.7418	$3s3p^33d^2$
2	−10.7993	$3s3p^33d^2$
3	−10.4418	$3s3p^33d^2$
4	−10.2277	$3s3p^33d^2$
5	−9.70598	$3s3p^33d^2$
6	−8.27355	$3s^23p3d^3$
7	−7.81475	$3s3p^4$ ^{4}P $4p$
8	−4.39664	$3p^33d^3$
9	−3.92598	$3p^33d^3$
10	−3.58579	$3s3p^4$ ^{4}P $5p$
11	−3.07665	$3p^33d^3$
12	−2.77723	$3p^33d^3$
13	−2.47574	$3p^33d^3$
14	−1.52111	$3s3p^4$ ^{4}P $6p$
15	−.35212	$3s3p^4$ ^{4}P $7p$
		^{5}D^e
1	−15.7288	$3s3p^4$ ^{4}P $3d$
2	−13.5815	$3s^23p^23d^2$
3	−12.9034	$3s^23p^23d^2$
4	−8.62105	$3p^43d^2$
5	−8.01481	$3p^43d^2$
6	−7.35959	$3s3p^23d^3$
7	−7.17983	$3s3p^23d^3$
8	−6.67854	$3s3p^23d^3$
9	−6.41148	$3s3p^23d^3$
10	−6.36350	$3s3p^4$ ^{4}P $4d$
11	−6.32594	$3s3p^23d^3$
12	−6.23419	$3s3p^23d^3$
13	−5.86835	$3s3p^23d^3$
14	−5.42888	$3s3p^23d^3$
15	−5.25561	$3s3p^23d^3$
16	−4.96459	$3s3p^23d^3$
17	−2.93387	$3s3p^4$ ^{4}P $5d$
18	−1.16984	$3s3p^4$ ^{4}P $6d$
19	−.14030	$3s3p^4$ ^{4}P $7d$

S-like Fe (Fe^{10+})

i	Energy(Ryds)	Description	i	Energy(Ryds)	Description	i	Energy(Ryds)	Description
	$^5\mathrm{D}^o$		9	−6.45781	$3s3p^23d^3$	6	−6.48956	$3s3p^23d^3$
1	−17.8493	$3s^23p^3$ $^4\mathrm{S}\ 3d$	10	−6.27596	$3s3p^4$ $^4\mathrm{P}\ 4d$	7	−6.19273	$3s3p^23d^3$
2	−11.8182	$3s3p^33d^2$	11	−5.95077	$3s3p^23d^3$	8	−5.83960	$3s3p^23d^3$
3	−10.9675	$3s3p^33d^2$	12	−5.74935	$3s3p^23d^3$	9	−2.33035	$3s3p^4$ $^4\mathrm{P}\ 5g$
4	−10.5842	$3s3p^33d^2$	13	−5.68504	$3s3p^23d^3$	10	−.83691	$3s3p^4$ $^4\mathrm{P}\ 6g$
5	−10.4208	$3s3p^33d^2$	14	−5.13605	$3s3p^23d^3$		$^5\mathrm{G}^o$	
6	−10.2705	$3s3p^33d^2$	15	−5.02769	$3s^23p^3$ $^4\mathrm{S}\ 5f$	1	−11.6877	$3s^23p3d^3$
7	−8.86617	$3s^23p^3$ $^4\mathrm{S}\ 4d$	16	−3.47347	$3s^23p^3$ $^4\mathrm{S}\ 6f$	2	−11.1912	$3s^23p3d^3$
8	−8.75633	$3s^23p3d^3$	17	−2.89749	$3s3p^4$ $^4\mathrm{P}\ 5d$	3	−10.5391	$3s^23p3d^3$
9	−8.08316	$3s^23p3d^3$	18	−2.54078	$3s^23p^3$ $^4\mathrm{S}\ 7f$	4	−9.21511	$3s^23p3d^3$
10	−7.69248	$3s3p^4$ $^4\mathrm{P}\ 4p$	19	−2.28528	$3s3p^4$ $^4\mathrm{P}\ 5g$	5	−5.32369	$3s3p^4$ $^4\mathrm{P}\ 4f$
11	−5.46230	$3s^23p^3$ $^4\mathrm{S}\ 5d$	20	−1.93885	$3s^23p^3$ $^4\mathrm{S}\ 8f$	6	−4.84187	$3s^23p^3$ $^4\mathrm{S}\ 5g$
12	−5.24886	$3s3p^4$ $^4\mathrm{P}\ 4f$	21	−1.52786	$3s^23p^3$ $^4\mathrm{S}\ 9f$	7	−4.39543	$3p^33d^3$
13	−4.40687	$3p^33d^3$		$^5\mathrm{F}^o$		8	−3.88331	$3p^33d^3$
14	−3.87599	$3p^33d^3$	1	−11.6645	$3s3p^33d^2$	9	−3.37330	$3s^23p^3$ $^4\mathrm{S}\ 6g$
15	−3.71635	$3s^23p^3$ $^4\mathrm{S}\ 6d$	2	−11.2096	$3s3p^33d^2$	10	−3.21041	$3p^33d^3$
16	−3.66180	$3p^33d^3$	3	−10.6658	$3s3p^33d^2$	11	−2.49611	$3s^23p^3$ $^4\mathrm{S}\ 7g$
17	−3.61637	$3p^33d^3$	4	−10.3434	$3s3p^33d^2$	12	−2.45049	$3s3p^4$ $^4\mathrm{P}\ 5f$
18	−3.53505	$3s3p^4$ $^4\mathrm{P}\ 5p$	5	−9.93040	$3s3p^33d^2$	13	−1.89721	$3s^23p^3$ $^4\mathrm{S}\ 8g$
19	−2.88796	$3p^33d^3$	6	−8.78500	$3s^23p3d^3$	14	−1.49909	$3s^23p^3$ $^4\mathrm{S}\ 9g$
20	−2.71510	$3p^33d^3$	7	−5.43577	$3s3p^4$ $^4\mathrm{P}\ 4f$			
21	−2.68372	$3s^23p^3$ $^4\mathrm{S}\ 7d$	8	−4.43007	$3p^33d^3$			
22	−2.44206	$3s3p^4$ $^4\mathrm{P}\ 5f$	9	−3.91799	$3p^33d^3$			
23	−2.03164	$3s^23p^3$ $^4\mathrm{S}\ 8d$	10	−3.34856	$3p^33d^3$			
24	−1.59391	$3s^23p^3$ $^4\mathrm{S}\ 9d$	11	−3.10961	$3p^33d^3$			
25	−1.50014	$3s3p^4$ $^4\mathrm{P}\ 6p$	12	−2.95362	$3p^33d^3$			
	$^5\mathrm{F}^e$		13	−2.51676	$3s3p^4$ $^4\mathrm{P}\ 5f$			
1	−15.2959	$3s3p^4$ $^4\mathrm{P}\ 3d$	14	−.94234	$3s3p^4$ $^4\mathrm{P}\ 6f$			
2	−13.4902	$3s^23p^23d^2$	15	−.00184	$3s3p^4$ $^4\mathrm{P}\ 7f$			
3	−8.53925	$3p^43d^2$		$^5\mathrm{G}^e$				
4	−7.88987	$3s^23p^3$ $^4\mathrm{S}\ 4f$	1	−13.8666	$3s^23p^23d^2$			
5	−7.29648	$3s3p^23d^3$	2	−8.28493	$3p^43d^2$			
6	−7.14459	$3s3p^23d^3$	3	−7.16287	$3s3p^23d^3$			
7	−6.82133	$3s3p^23d^3$	4	−6.92476	$3s3p^23d^3$			
8	−6.57059	$3s3p^23d^3$	5	−6.70914	$3s3p^23d^3$			

Energies in ascending order from ground state for terms with effective $n \leq 4.0$, $L \leq 4$

Term	i	E(Ryds)	Term	i	E(Ryds)	Term	i	E(Ryds)	Term	i	E(Ryds)	Term	i	E(Ryds)
$^3\mathrm{P}^e$	1	0.00000	$^3\mathrm{P}^o$	2	4.40840	$^5\mathrm{P}^e$	1	6.12210	$^3\mathrm{F}^e$	3	7.37600	$^3\mathrm{G}^e$	2	8.15910
$^1\mathrm{D}^e$	1	0.27320	$^3\mathrm{D}^o$	2	4.44670	$^3\mathrm{P}^e$	2	6.21260	$^3\mathrm{D}^e$	3	7.40510	$^3\mathrm{D}^e$	5	8.26040
$^1\mathrm{S}^e$	1	0.66810	$^1\mathrm{F}^o$	1	4.75180	$^3\mathrm{D}^e$	1	6.39640	$^5\mathrm{G}^e$	1	7.44070	$^3\mathrm{P}^e$	5	8.27330
$^3\mathrm{P}^o$	1	2.43320	$^1\mathrm{P}^o$	2	4.83640	$^3\mathrm{F}^e$	1	6.44160	$^3\mathrm{P}^e$	4	7.51180	$^5\mathrm{P}^e$	2	8.36070
$^1\mathrm{P}^o$	1	3.15720	$^3\mathrm{P}^o$	3	4.84520	$^3\mathrm{G}^e$	1	6.58960	$^1\mathrm{D}^e$	3	7.59260	$^1\mathrm{P}^e$	2	8.39620
$^5\mathrm{D}^o$	1	3.45800	$^3\mathrm{S}^o$	1	4.88900	$^1\mathrm{G}^e$	1	6.81040	$^3\mathrm{D}^e$	4	7.59630	$^5\mathrm{D}^e$	3	8.40390
$^3\mathrm{D}^o$	1	3.72320	$^3\mathrm{D}^o$	3	5.10850	$^3\mathrm{F}^e$	2	6.91730	$^5\mathrm{D}^e$	2	7.72580	$^3\mathrm{F}^e$	5	8.45790
$^1\mathrm{S}^o$	1	3.81460	$^1\mathrm{D}^o$	2	5.30200	$^1\mathrm{D}^e$	2	6.91890	$^1\mathrm{F}^e$	2	7.78560	$^1\mathrm{D}^e$	5	8.57740
$^3\mathrm{F}^o$	1	3.82270	$^1\mathrm{S}^e$	2	5.36230	$^3\mathrm{D}^e$	2	7.01290	$^3\mathrm{F}^e$	4	7.79780	$^3\mathrm{D}^e$	6	8.61510
$^3\mathrm{G}^o$	1	4.07310	$^1\mathrm{F}^o$	2	5.41360	$^1\mathrm{F}^e$	1	7.02400	$^5\mathrm{F}^e$	2	7.81710	$^3\mathrm{P}^e$	6	8.63910
$^1\mathrm{G}^o$	1	4.17580	$^5\mathrm{D}^e$	1	5.57850	$^1\mathrm{P}^e$	1	7.13120	$^1\mathrm{S}^e$	3	7.87850	$^3\mathrm{F}^e$	6	8.64350
$^1\mathrm{D}^o$	1	4.19710	$^1\mathrm{P}^o$	3	5.68860	$^3\mathrm{S}^e$	1	7.18020	$^1\mathrm{G}^e$	2	7.93020	$^3\mathrm{G}^e$	3	8.72100
$^3\mathrm{F}^o$	2	4.36400	$^5\mathrm{F}^e$	1	6.01140	$^3\mathrm{P}^e$	3	7.24980	$^1\mathrm{D}^e$	4	8.00950	$^1\mathrm{F}^e$	3	8.72320

S-like Fe (Fe^{10+})

Term	i	E(Ryds)	Term	i	E(Ryds)	Term	i	E(Ryds)	Term	i	E(Ryds)	Term	i	E(Ryds)
$^3P^e$	7	8.83790	$^3D^o$	6	10.5127	$^3S^o$	4	11.4271	$^1G^o$	6	12.4547	$^3F^o$	18	13.2047
$^1D^e$	6	8.91150	$^1D^e$	10	10.5214	$^3P^o$	9	11.4418	$^1F^o$	9	12.5026	$^3P^o$	19	13.2083
$^1P^e$	3	8.98910	$^1P^o$	4	10.5428	$^3F^e$	10	11.4573	$^5F^o$	6	12.5223	$^3F^o$	19	13.2212
$^1G^e$	3	9.00830	$^1D^o$	4	10.6193	$^3D^o$	11	11.4584	$^3D^o$	17	12.5365	$^5D^o$	9	13.2241
$^3F^e$	7	9.02530	$^5F^o$	3	10.6415	$^1F^e$	6	11.4930	$^1P^o$	9	12.5449	$^3D^o$	22	13.2254
$^3P^o$	4	9.05960	$^3F^o$	5	10.6423	$^3P^o$	10	11.5190	$^5D^o$	8	12.5509	$^3G^e$	5	13.2269
$^3G^e$	4	9.08360	$^1D^o$	5	10.7125	$^3F^o$	10	11.5885	$^1D^o$	10	12.5869	$^1D^o$	14	13.2275
$^3D^e$	7	9.14590	$^5D^o$	4	10.7231	$^3D^o$	12	11.5982	$^5P^e$	4	12.6423	$^1F^o$	12	13.2404
$^3F^e$	8	9.17430	$^3S^o$	3	10.7501	$^5P^o$	5	11.6013	$^3P^o$	14	12.6634	$^1D^e$	13	13.2506
$^3F^o$	3	9.27400	$^3D^o$	7	10.7602	$^1F^o$	6	11.6363	$^1F^o$	10	12.6726	$^3D^e$	12	13.2554
$^1G^e$	4	9.27640	$^5G^o$	3	10.7682	$^3P^e$	12	11.6860	$^3D^o$	18	12.6861	$^1D^o$	15	13.2636
$^3D^e$	8	9.33630	$^3G^o$	3	10.7698	$^1P^o$	7	11.6871	$^5D^e$	4	12.6862	$^1F^o$	13	13.2648
$^1F^e$	4	9.36700	$^3P^o$	6	10.8221	$^3G^o$	7	11.6934	$^3P^o$	15	12.7002	$^5D^e$	5	13.2924
$^1S^e$	4	9.39240	$^3F^o$	6	10.8538	$^3S^e$	3	11.7016	$^1D^o$	11	12.7108	$^5P^e$	5	13.3061
$^3P^e$	8	9.39670	$^1S^o$	2	10.8611	$^3F^o$	11	11.7200	$^5S^o$	4	12.7541	$^3D^e$	13	13.3191
$^3S^e$	2	9.44010	$^3P^o$	7	10.8618	$^1G^o$	4	11.7365	$^3D^o$	19	12.7659	$^3D^o$	23	13.3246
$^1D^e$	7	9.44240	$^5P^o$	3	10.8655	$^3D^o$	13	11.7546	$^5F^e$	3	12.7680	$^1P^o$	13	13.3385
$^5D^o$	2	9.48910	$^5D^o$	5	10.8865	$^3D^e$	11	11.7730	$^1S^o$	4	12.8142	$^1G^o$	8	13.3578
$^1P^e$	4	9.51360	$^1P^o$	5	10.9259	$^1D^o$	7	11.7958	$^3F^o$	15	12.8195	$^5F^e$	4	13.4174
$^3P^e$	9	9.52890	$^1S^e$	6	10.9487	$^1D^e$	11	11.8167	$^1P^o$	10	12.8213	$^1D^e$	14	13.4178
$^5P^o$	1	9.56550	$^5F^o$	4	10.9639	$^1P^e$	6	11.8470	$^1D^o$	12	12.8577	$^3P^o$	20	13.4750
$^3D^e$	9	9.57020	$^3G^o$	4	10.9755	$^3P^o$	11	11.8549	$^3D^o$	20	12.8618	$^1G^e$	6	13.4876
$^1D^e$	8	9.58260	$^5P^e$	3	11.0083	$^3D^o$	14	11.8813	$^3G^o$	10	12.8650	$^5P^o$	7	13.4925
$^3D^o$	4	9.58560	$^5D^o$	6	11.0368	$^3P^e$	13	11.8843	$^3F^o$	16	12.8741	$^3F^o$	20	13.5127
$^1S^e$	5	9.59620	$^3G^o$	5	11.0375	$^3F^o$	12	11.9231	$^1P^o$	11	12.8746	$^3F^e$	12	13.5155
$^5G^o$	1	9.61960	$^3P^o$	8	11.0435	$^1F^o$	7	11.9472	$^3P^o$	16	12.8768	$^3D^o$	24	13.5226
$^5F^o$	1	9.64280	$^1F^o$	4	11.0504	$^1D^e$	12	11.9480	$^3G^o$	11	12.8883	$^1F^e$	7	13.5246
$^1F^o$	3	9.67120	$^1D^o$	6	11.0766	$^3P^o$	12	12.0115	$^3D^o$	21	12.8932	$^3D^e$	14	13.5609
$^1D^o$	3	9.68090	$^3F^o$	7	11.0770	$^3G^o$	8	12.0172	$^1G^o$	7	12.8945	$^3P^e$	16	13.5744
$^3F^e$	9	9.75050	$^5P^o$	4	11.0796	$^1G^o$	5	12.0364	$^1D^o$	13	12.9169	$^3F^e$	13	13.5755
$^1F^e$	5	9.83070	$^3D^o$	8	11.0868	$^5G^o$	4	12.0921	$^3S^o$	6	12.9186	$^5D^o$	10	13.6148
$^5S^o$	1	10.0549	$^1G^o$	2	11.1188	$^3P^o$	13	12.0922	$^3P^e$	14	12.9195	$^3G^o$	13	13.6375
$^1D^e$	9	10.0552	$^3P^e$	11	11.1476	$^1D^o$	8	12.1071	$^3P^o$	17	12.9474	$^3D^o$	25	13.6408
$^1G^e$	5	10.0830	$^3D^o$	9	11.1878	$^3S^o$	5	12.1238	$^3F^o$	17	12.9532	$^3G^e$	6	13.6476
$^5F^o$	2	10.0977	$^3F^o$	8	11.2158	$^3F^o$	13	12.1366	$^3S^o$	7	12.9729	$^1D^o$	16	13.6581
$^3F^o$	4	10.1069	$^1F^o$	5	11.2937	$^3D^o$	15	12.1975	$^1F^o$	11	12.9802	$^1P^e$	7	13.6892
$^5G^o$	2	10.1161	$^3D^o$	10	11.2964	$^1P^o$	8	12.2039	$^3P^e$	15	12.9921	$^3F^e$	14	13.6922
$^3D^o$	5	10.1265	$^5S^o$	3	11.2969	$^3D^o$	16	12.2541	$^1P^o$	12	13.0002	$^3S^o$	8	13.7184
$^3P^e$	10	10.1281	$^1P^o$	6	11.3331	$^1S^o$	3	12.2548	$^1S^o$	5	13.0131	$^3P^e$	17	13.7197
$^3S^o$	2	10.2738	$^1P^e$	5	11.3465	$^3F^o$	14	12.2685	$^5G^e$	2	13.0223	$^1F^o$	14	13.7250
$^3G^o$	2	10.3142	$^1G^o$	3	11.3736	$^1F^o$	8	12.2731	$^5P^o$	6	13.0337	$^3S^e$	4	13.7333
$^5D^o$	3	10.3398	$^5F^o$	5	11.3760	$^1S^e$	7	12.2770	$^3F^e$	11	13.1207			
$^5S^o$	2	10.3473	$^3F^o$	9	11.3784	$^1D^o$	9	12.3058	$^3P^o$	18	13.1488			
$^3P^o$	5	10.3911	$^3D^e$	10	11.3935	$^3G^o$	9	12.3652	$^1S^e$	8	13.1514			
$^5P^o$	2	10.5080	$^3G^o$	6	11.4068	$^5D^o$	7	12.4411	$^3G^o$	12	13.1747			

S-like Fe (Fe^{10+})

gf-values for transitions involving terms with effective $n \leq 4.0$, $L \leq 4$

i i'	gf_L	i i'	gf_L	i i'	gf_L	i i'	gf_L	i i'	gf_L	i i'	gf_L
	$^1S^o$–$^1P^e$	2 2	3.38E−3	6 1	2.45E−3	2 5	−4.79E−2	9 7	−1.11E−3	2 8	−6.17E−1
1 1	−3.52E−1	2 3	−4.11E−5	6 2	3.99E−4	2 6	−3.80E−3	10 1	2.88E−1	2 9	−6.74E−1
1 2	−4.53E−1	2 4	−1.62E+0	6 3	1.43E+0	2 7	−2.25E−6	10 2	3.74E−6	2 10	−4.28E−1
1 3	−1.31E−1	2 5	−4.93E−7	6 4	8.59E−5	3 1	−7.71E−4	10 3	2.38E−3	2 11	−3.95E−4
1 4	−5.99E−1	2 6	−9.80E−2	6 5	8.88E−5	3 2	−5.14E−2	10 4	2.11E−1	2 12	−9.69E−5
1 5	−1.61E−2	2 7	−5.67E−2	6 6	−1.17E−6	3 3	−3.16E−3	10 5	1.48E−1	2 13	−3.07E−6
1 6	−8.03E−3	2 8	−7.76E−3	6 7	−4.54E−5	3 4	−1.81E−1	10 6	7.92E−3	2 14	−2.04E−2
1 7	−4.63E−4	2 9	−4.26E−3	6 8	−4.13E−4	3 5	−4.44E−4	10 7	−2.42E−5	3 1	3.27E−2
2 1	3.76E−4	2 10	−3.27E−3	6 9	−3.85E−3	3 6	−3.71E−2	11 1	5.09E−2	3 2	−4.82E−4
2 2	9.04E−4	2 11	−4.04E−4	6 10	−6.30E−3	3 7	−6.44E−5	11 2	4.17E−3	3 3	−5.19E−2
2 3	1.93E−2	2 12	−1.65E−3	6 11	−9.39E−3	4 1	3.63E−2	11 3	1.77E−4	3 4	−1.60E−4
2 4	4.63E−5	2 13	−2.96E−4	6 12	−9.43E−3	4 2	9.45E−2	11 4	7.09E−2	3 5	−1.11E−1
2 5	−8.22E−6	3 1	7.11E−1	6 13	−3.00E−2	4 3	4.44E−5	11 5	7.04E−1	3 6	−4.61E−3
2 6	−9.74E−8	3 2	1.50E−1	7 1	7.18E−3	4 4	1.36E−2	11 6	4.63E−2	3 7	−4.77E−4
2 7	−2.43E−1	3 3	1.03E−3	7 2	1.95E−7	4 5	−4.53E−5	11 7	−7.23E−7	3 8	−7.43E−2
3 1	4.90E−1	3 4	−2.15E−2	7 3	1.81E−2	4 6	−2.97E−4	12 1	2.53E−1	3 9	−2.83E+0
3 2	2.07E−1	3 5	−1.21E−7	7 4	6.58E−5	4 7	−2.03E−3	12 2	4.34E−2	3 10	−9.11E−1
3 3	1.40E−3	3 6	−2.47E−1	7 5	3.18E−1	5 1	7.03E−4	12 3	6.69E−2	3 11	−6.23E−3
3 4	7.89E−2	3 7	−3.04E−1	7 6	4.19E−4	5 2	1.18E−4	12 4	2.58E−1	3 12	−1.47E−2
3 5	2.12E−8	3 8	−2.55E−2	7 7	3.29E−4	5 3	3.83E−3	12 5	1.14E−3	3 13	−3.78E−5
3 6	1.98E−5	3 9	−6.77E−2	7 8	1.59E−5	5 4	1.67E−4	12 6	3.83E−3	3 14	−8.39E−5
3 7	−8.18E−3	3 10	−7.76E−1	7 9	−9.22E−5	5 5	−6.71E−2	12 7	−4.42E−5	4 1	8.76E−4
4 1	1.45E−4	3 11	−1.93E−1	7 10	−9.06E−3	5 6	−5.85E−1	13 1	1.23E−3	4 2	4.03E−3
4 2	1.29E−4	3 12	−2.82E−1	7 11	−3.10E−2	5 7	−4.01E−4	13 2	2.13E−3	4 3	1.52E−2
4 3	1.50E−2	3 13	−2.05E−3	7 12	−1.43E−4	6 1	1.12E−1	13 3	1.25E−2	4 4	1.70E−1
4 4	4.98E−3	4 1	3.25E−4	7 13	−5.66E−1	6 2	1.19E−1	13 4	1.09E−3	4 5	5.95E−2
4 5	3.48E−1	4 2	1.78E−1	8 1	1.60E−3	6 3	8.32E−4	13 5	6.62E−4	4 6	2.55E−2
4 6	5.04E−2	4 3	2.49E−4	8 2	1.25E−4	6 4	3.20E−4	13 6	5.86E−1	4 7	1.12E−2
4 7	−1.51E−4	4 4	−1.65E−3	8 3	8.12E−4	6 5	−1.36E−7	13 7	−5.40E−6	4 8	3.69E−3
5 1	6.02E−2	4 5	−4.32E−6	8 4	7.33E−3	6 6	−1.23E−3		$^1P^o$–$^1D^e$	4 9	3.34E−3
5 2	2.49E−2	4 6	−3.26E−3	8 5	5.47E−4	6 7	−6.50E−2	1 1	3.72E−1	4 10	5.82E−6
5 3	8.73E−1	4 7	−1.16E−3	8 6	7.12E−2	7 1	3.94E−1	1 2	−7.03E−2	4 11	−1.10E−4
5 4	1.52E−4	4 8	−3.50E−2	8 7	1.20E−2	7 2	1.41E−1	1 3	−5.39E−3	4 12	−1.15E−3
5 5	3.71E−3	4 9	−2.00E−1	8 8	9.17E−5	7 3	6.85E−2	1 4	−1.53E+0	4 13	−1.89E−1
5 6	1.62E−3	4 10	−5.93E−4	8 9	6.25E−5	7 4	2.88E−3	1 5	−1.02E+0	4 14	−2.12E−4
5 7	−1.31E−3	4 11	−1.31E−4	8 10	1.11E−5	7 5	9.86E−7	1 6	−4.51E−1	5 1	3.13E−1
	$^1S^e$–$^1P^o$	4 12	−2.70E−1	8 11	2.06E−5	7 6	−7.09E−5	1 7	−5.29E−2	5 2	1.43E−3
1 1	−1.50E−2	4 13	−3.17E−3	8 12	3.99E−5	7 7	−6.04E−2	1 8	−1.05E−2	5 3	4.28E−3
1 2	−1.72E−1	5 1	5.11E−3	8 13	−1.05E−5	8 1	2.47E−1	1 9	−2.40E−3	5 4	2.60E−4
1 3	−1.85E+0	5 2	1.03E+0		$^1P^o$–$^1P^e$	8 2	1.86E−1	1 10	−1.56E−2	5 5	1.79E−5
1 4	−1.60E−4	5 3	3.20E−2	1 1	−3.59E−1	8 3	1.07E−1	1 11	−1.24E−2	5 6	1.25E−3
1 5	−1.94E−1	5 4	−6.86E−3	1 2	−1.34E+0	8 4	4.30E−2	1 12	−4.86E−3	5 7	1.77E−3
1 6	−1.56E−4	5 5	−9.70E−7	1 3	−2.49E−4	8 5	2.85E−4	1 13	−2.16E−3	5 8	1.57E−3
1 7	−2.49E−4	5 6	−1.21E−2	1 4	−1.87E−2	8 6	3.84E−5	1 14	−2.25E−1	5 9	8.97E−4
1 8	−3.98E−5	5 7	−1.37E−2	1 5	−5.01E−2	8 7	−2.66E−2	2 1	1.87E+0	5 10	8.63E−4
1 9	−7.91E−4	5 8	−4.73E−4	1 6	−1.08E−3	9 1	5.55E−3	2 2	−5.68E−2	5 11	−2.38E−1
1 10	−3.24E−2	5 9	−1.09E−1	1 7	−3.72E−4	9 2	1.36E−1	2 3	−3.00E−4	5 12	−1.03E+0
1 11	−5.57E−2	5 10	−4.57E−2	2 1	−6.65E−2	9 3	5.99E−1	2 4	−5.36E−2	5 13	−1.08E−3
1 12	−3.68E−3	5 11	−1.19E−2	2 2	−1.49E−1	9 4	1.60E−2	2 5	−1.01E−1	5 14	−1.06E−2
1 13	−5.47E−1	5 12	−7.01E−2	2 3	−8.45E−3	9 5	4.67E−4	2 6	−4.52E−1	6 1	1.33E−4
2 1	1.58E−1	5 13	−6.22E−4	2 4	−1.74E+0	9 6	1.41E−4	2 7	−6.53E−1	6 2	9.09E−1

S-like Fe (Fe^{10+})

i	*i′*	gf$_L$	*i*	*i′*	gf$_L$	*i*	*i′*	gf$_L$	*i*	*i′*	gf$_L$	*i*	*i′*	gf$_L$	*i*	*i′*	gf$_L$
6	3	1.06E−1	9	13	−1.10E−3	13	9	1.72E−2	3	14	−1.71E−3	7	2	3.19E−4	3	9	−2.48E−5
6	4	7.78E−2	9	14	−5.14E−6	13	10	5.58E−4	3	15	−2.40E−1	7	3	2.69E−1	3	10	−2.99E−7
6	5	2.15E−1	10	1	4.42E−2	13	11	1.44E−3	3	16	−4.85E−1	7	4	5.25E−4	3	11	−2.26E−4
6	6	1.99E−2	10	2	5.45E−3	13	12	8.14E−2	4	1	5.08E−2	7	5	2.50E−1	3	12	−1.86E−4
6	7	9.17E−5	10	3	4.41E−1	13	13	8.47E−7	4	2	7.89E−1	7	6	6.98E−3	3	13	−3.09E−1
6	8	1.42E−4	10	4	8.86E−1	13	14	−1.51E−5	4	3	−5.32E−4	7	7	4.79E−2	3	14	−8.82E−6
6	9	3.18E−4	10	5	1.56E−1			$^1P^e$–$^1D^o$	4	4	−9.20E−4	7	8	2.74E−3	4	1	7.83E−1
6	10	2.61E−5	10	6	5.02E−2	1	1	7.26E−2	4	5	−2.90E−2	7	9	2.86E−3	4	2	4.01E−3
6	11	−4.77E−4	10	7	8.19E−3	1	2	7.90E−4	4	6	−8.57E−3	7	10	2.16E−2	4	3	2.86E−5
6	12	−1.02E−3	10	8	1.04E−1	1	3	−3.12E−1	4	7	−7.51E−3	7	11	1.68E−4	4	4	5.00E−4
6	13	−7.07E−2	10	9	4.38E−3	1	4	−2.42E−4	4	8	−3.04E−5	7	12	9.70E−5	4	5	3.46E−5
6	14	−9.60E−5	10	10	1.37E−3	1	5	−5.47E−1	4	9	−1.46E−2	7	13	6.17E−9	4	6	6.64E−5
7	1	3.97E−5	10	11	4.28E−2	1	6	−5.97E−2	4	10	−2.08E−2	7	14	7.87E−5	4	7	1.57E−3
7	2	1.67E−2	10	12	3.20E−3	1	7	−3.59E−1	4	11	−5.87E−2	7	15	7.32E−4	4	8	1.40E−5
7	3	1.63E−2	10	13	−7.46E−4	1	8	−7.94E−1	4	12	−5.01E−2	7	16	6.10E−5	4	9	6.39E−4
7	4	8.47E−2	10	14	−4.96E−3	1	9	−1.19E−1	4	13	−4.12E−3			$^1D^o$–$^1D^e$	4	10	1.03E−3
7	5	8.07E−2	11	1	1.38E−1	1	10	−1.26E−2	4	14	−3.43E−2	1	1	4.20E−2	4	11	−1.07E+0
7	6	1.64E−1	11	2	3.45E−3	1	11	−2.36E−1	4	15	−2.70E−1	1	2	−1.94E−1	4	12	−4.55E−1
7	7	9.66E−4	11	3	4.94E−2	1	12	−2.37E−4	4	16	−5.67E−1	1	3	−1.59E−1	4	13	−2.39E−3
7	8	1.64E−2	11	4	2.03E−1	1	13	−4.72E−4	5	1	3.47E−2	1	4	−2.55E−1	4	14	−1.51E−1
7	9	3.30E−3	11	5	7.55E−2	1	14	−2.13E−2	5	2	1.30E−2	1	5	−7.07E−1	5	1	2.52E−4
7	10	1.43E−3	11	6	3.10E−2	1	15	−1.47E−1	5	3	4.51E−4	1	6	−6.67E−1	5	2	5.08E−2
7	11	−1.28E−7	11	7	5.24E−4	1	16	−1.02E−2	5	4	4.88E−1	1	7	−9.11E−1	5	3	1.20E−1
7	12	−2.64E−5	11	8	3.11E−2	2	1	2.01E−2	5	5	3.99E−6	1	8	−5.08E−2	5	4	1.69E−1
7	13	−3.48E−2	11	9	1.38E−4	2	2	3.40E−1	5	6	1.39E−4	1	9	−3.78E−1	5	5	1.62E−2
7	14	−2.33E−3	11	10	1.07E−2	2	3	−2.95E−2	5	7	−1.09E−4	1	10	−5.09E−2	5	6	9.77E−3
8	1	1.40E−3	11	11	1.87E−1	2	4	−1.29E−3	5	8	−9.43E−4	1	11	−5.63E−2	5	7	2.21E−3
8	2	5.24E−1	11	12	9.62E−3	2	5	−6.42E−4	5	9	−6.33E−4	1	12	−4.17E−3	5	8	1.15E−3
8	3	3.74E−1	11	13	−3.15E−4	2	6	−5.77E−3	5	10	−1.90E−4	1	13	−4.63E−5	5	9	1.46E−4
8	4	2.39E−1	11	14	−5.03E−3	2	7	−1.18E−1	5	11	−1.85E−3	1	14	−5.27E−5	5	10	2.86E−5
8	5	1.09E−1	12	1	4.05E−3	2	8	−6.58E−2	5	12	−1.53E−2	2	1	2.81E+0	5	11	−4.11E−4
8	6	3.26E−3	12	2	2.50E−3	2	9	−1.18E−1	5	13	−7.71E−1	2	2	−1.44E−2	5	12	−1.98E−4
8	7	1.31E−2	12	3	5.25E−1	2	10	−3.34E−3	5	14	−3.16E−2	2	3	−1.31E−2	5	13	−2.72E−1
8	8	4.29E−3	12	4	3.84E−2	2	11	−5.44E−1	5	15	−9.91E−4	2	4	−1.65E−5	5	14	−2.07E−3
8	9	5.37E−3	12	5	1.96E−1	2	12	−1.69E+0	5	16	−1.73E−5	2	5	−9.86E−2	6	1	5.44E−6
8	10	1.04E−2	12	6	7.85E−2	2	13	−5.71E−2	6	1	6.71E−2	2	6	−4.25E−1	6	2	2.00E−1
8	11	2.19E−4	12	7	7.01E−1	2	14	−6.73E−2	6	2	5.91E−2	2	7	−3.07E−1	6	3	3.13E−2
8	12	1.83E−5	12	8	4.59E−1	2	15	−2.75E−1	6	3	2.98E−6	2	8	−3.01E−1	6	4	1.27E−1
8	13	−1.39E−4	12	9	5.35E−3	2	16	−2.64E−2	6	4	1.44E−1	2	9	−3.81E−1	6	5	4.37E−2
8	14	−2.14E−3	12	10	5.92E−4	3	1	1.69E+0	6	5	3.49E−5	2	10	−3.18E+0	6	6	2.71E−2
9	1	3.75E−5	12	11	3.75E−4	3	2	2.16E−2	6	6	4.57E−4	2	11	−3.82E−2	6	7	1.73E−2
9	2	1.99E−1	12	12	3.15E−5	3	3	−1.31E−3	6	7	5.73E−9	2	12	−1.89E−2	6	8	1.69E−4
9	3	2.47E−1	12	13	−8.48E−4	3	4	−2.86E−5	6	8	−1.48E−5	2	13	−3.22E−3	6	9	1.24E−3
9	4	4.30E−2	12	14	−6.74E−4	3	5	−9.24E−3	6	9	−1.30E−4	2	14	−1.76E−2	6	10	4.36E 6
9	5	2.52E−1	13	1	3.08E−2	3	6	−5.47E−2	6	10	−2.17E−4	3	1	9.50E−5	6	11	−1.76E−4
9	6	1.77E−2	13	2	2.64E 4	3	7	9.78E 3	6	11	2.40E 3	3	2	3.04E−1	6	12	−7.61E−5
9	7	1.13E−2	13	3	2.99E−3	3	8	−6.94E−2	6	12	−1.36E−3	3	3	6.43E−2	6	13	−3.37E−2
9	8	6.74E−2	13	4	1.73E−4	3	9	−7.19E−3	6	13	−8.35E−3	3	4	4.17E−4	6	14	−3.68E−4
9	9	1.18E−2	13	5	1.43E−5	3	10	−5.67E−1	6	14	−1.35E+0	3	5	6.80E−5	7	1	3.88E−4
9	10	3.52E−3	13	6	4.27E−5	3	11	−5.85E−2	6	15	−2.08E−1	3	6	1.55E−3	7	2	2.25E−2
9	11	7.01E−5	13	7	4.51E 3	3	12	2.62E 1	6	16	2.73E 4	3	7	4.12E 5	7	3	2.70E 1
9	12	5.37E−5	13	8	3.37E−4	3	13	−1.85E−3	7	1	3.22E−4	3	8	3.34E−4	7	4	5.00E−3

S-like Fe (Fe^{10+})

i	i'	gf_L	i	i'	gf_L	i	i'	gf_L	i	i'	gf_L	i	i'	gf_L	i	i'	gf_L
7	5	6.02E-2	11	1	1.25E-3	14	11	2.74E-4	3	6	-1.01E-3	11	2	4.92E-2	1	11	-1.56E+0
7	6	5.04E-2	11	2	2.87E-1	14	12	4.94E-1	3	7	-1.84E-2	11	3	1.20E+0	1	12	-1.17E-2
7	7	2.64E-2	11	3	6.80E-1	14	13	-2.44E-7	4	1	2.20E-3	11	4	2.60E-1	1	13	-2.61E-1
7	8	4.04E-4	11	4	1.01E-5	14	14	-1.04E-3	4	2	7.88E-3	11	5	3.06E-3	1	14	-9.08E-5
7	9	4.26E-3	11	5	5.04E-1	15	1	4.09E-2	4	3	1.10E-2	11	6	6.32E-4	2	1	8.61E-4
7	10	2.48E-4	11	6	4.45E-1	15	2	4.67E-2	4	4	3.07E-4	11	7	-2.88E-3	2	2	1.72E-3
7	11	-3.03E-6	11	7	2.91E-4	15	3	1.69E-1	4	5	6.76E-4	12	1	2.72E-2	2	3	-3.83E-1
7	12	-2.19E-6	11	8	5.57E-2	15	4	1.55E-1	4	6	-1.56E+0	12	2	5.44E-4	2	4	-2.40E-1
7	13	-2.02E-2	11	9	1.46E-1	15	5	2.89E-1	4	7	-2.05E-4	12	3	2.77E-2	2	5	-3.05E-2
7	14	-8.40E-4	11	10	2.89E-3	15	6	1.44E+0	5	1	1.55E-1	12	4	1.36E-2	2	6	-1.98E-1
8	1	2.81E-3	11	11	1.27E-3	15	7	2.30E-1	5	2	3.93E-2	12	5	1.95E-1	2	7	-2.23E-1
8	2	1.38E+0	11	12	2.10E-4	15	8	7.19E-2	5	3	1.42E-2	12	6	7.21E-4	2	8	-3.94E-1
8	3	8.57E-2	11	13	-3.55E-7	15	9	1.48E-1	5	4	1.85E-5	12	7	-3.94E-3	2	9	-7.26E-1
8	4	2.33E-2	11	14	-1.16E-4	15	10	8.19E-2	5	5	1.41E-3	13	1	1.05E-2	2	10	-1.16E-1
8	5	2.21E-1	12	1	1.27E-2	15	11	1.12E-3	5	6	-7.17E-5	13	2	1.37E-2	2	11	-8.88E-3
8	6	5.59E-2	12	2	3.55E-2	15	12	9.13E-2	5	7	-6.02E-1	13	3	2.05E-5	2	12	-1.99E-4
8	7	6.91E-2	12	3	9.29E-1	15	13	5.45E-8	6	1	1.82E-1	13	4	5.40E-4	2	13	-2.95E-4
8	8	4.45E-4	12	4	6.78E-1	15	14	-2.88E-9	6	2	4.76E-4	13	5	6.56E-6	2	14	-9.44E-3
8	9	1.59E-2	12	5	8.45E-3	16	1	1.17E-3	6	3	3.45E-2	13	6	3.88E-2	3	1	2.07E-1
8	10	1.41E-2	12	6	5.63E-2	16	2	2.00E-3	6	4	5.54E-3	13	7	-5.85E-7	3	2	3.22E-4
8	11	1.54E-5	12	7	1.26E-1	16	3	1.22E-3	6	5	7.71E-5	14	1	2.12E-3	3	3	-6.23E-2
8	12	3.02E-8	12	8	4.90E-2	16	4	8.41E-2	6	6	-8.79E-4	14	2	8.75E-4	3	4	-6.06E-2
8	13	-1.16E-3	12	9	1.70E-2	16	5	8.05E-2	6	7	-3.05E-1	14	3	1.34E-1	3	5	-4.05E-1
8	14	-2.94E-3	12	10	2.24E-4	16	6	5.60E-2	7	1	1.65E+0	14	4	3.16E-2	3	6	-1.35E-1
9	1	1.39E-4	12	11	1.97E-2	16	7	1.92E+0	7	2	7.78E-4	14	5	4.44E-3	3	7	-2.16E-2
9	2	1.86E-2	12	12	2.83E-3	16	8	1.39E-1	7	3	1.60E-2	14	6	3.09E-2	3	8	-1.11E-2
9	3	8.28E-1	12	13	-1.29E-5	16	9	1.45E-1	7	4	9.65E-2	14	7	-5.06E-4	3	9	-4.44E-1
9	4	2.39E-3	12	14	-1.89E-6	16	10	5.50E-3	7	5	1.73E-3	15	1	1.31E-2	3	10	-1.85E-1
9	5	3.42E-1	13	1	4.92E-1	16	11	1.81E-3	7	6	8.66E-6	15	2	1.89E-2	3	11	-4.13E-5
9	6	3.28E-1	13	2	2.91E-3	16	12	1.27E-3	7	7	-1.23E-3	15	3	7.28E-1	3	12	-1.00E-3
9	7	1.22E-2	13	3	1.31E-2	16	13	8.86E-4	8	1	3.62E-1	15	4	4.72E-2	3	13	-5.38E-3
9	8	3.07E-1	13	4	5.76E-3	16	14	2.42E-6	8	2	4.18E-2	15	5	1.38E-1	3	14	-5.16E-3
9	9	1.23E-2	13	5	1.63E-4			**$^1D^o$–$^1F^e$**	8	3	5.34E-2	15	6	9.80E-3	4	1	5.89E-4
9	10	1.15E-3	13	6	1.27E-2	1	1	-4.84E-2	8	4	9.82E-2	15	7	-3.84E-3	4	2	1.05E-1
9	11	1.45E-4	13	7	3.98E-2	1	2	-9.24E-1	8	5	1.41E-3	16	1	1.57E-2	4	3	-4.44E-5
9	12	2.19E-4	13	8	2.13E-4	1	3	-2.05E-1	8	6	3.77E-4	16	2	1.48E-3	4	4	-4.01E-1
9	13	-3.85E-3	13	9	3.42E-3	1	4	-5.93E-1	8	7	-6.07E-4	16	3	9.21E-1	4	5	-3.46E-1
9	14	-1.64E-3	13	10	1.34E-2	1	5	-4.51E-1	9	1	1.11E-2	16	4	6.09E-1	4	6	-8.87E-2
10	1	6.23E-5	13	11	1.05E+0	1	6	-2.05E-4	9	2	9.26E-3	16	5	7.57E-3	4	7	-1.23E+0
10	2	1.25E-2	13	12	7.96E-2	1	7	-1.43E-3	9	3	3.75E-1	16	6	2.69E-3	4	8	-1.22E-3
10	3	3.06E-2	13	13	-1.36E-5	2	1	-4.30E-2	9	4	1.55E-2	16	7	1.53E-3	4	9	-3.95E-1
10	4	1.70E-2	13	14	-3.30E-4	2	2	-3.83E-3	9	5	1.43E-2			**$^1D^e$–$^1F^o$**	4	10	-1.08E-1
10	5	6.96E-3	14	1	2.99E-1	2	3	-2.04E-3	9	6	1.66E-4	1	1	-6.74E-3	4	11	-4.48E-4
10	6	4.65E-2	14	2	2.04E-3	2	4	-1.02E+0	9	7	-6.43E-3	1	2	-5.09E+0	4	12	-7.53E-1
10	7	2.44E-1	14	3	3.28E-2	2	5	-2.47E+0	10	1	6.25E-3	1	3	-6.45E-8	4	13	-4.39E-2
10	8	2.93E-1	14	4	1.88E-2	2	6	-8.70E-4	10	2	3.25E+0	1	4	-1.46E-5	4	14	-3.52E-2
10	9	1.63E-4	14	5	3.49E-2	2	7	-4.55E-5	10	3	4.86E-1	1	5	-1.95E-3	5	1	6.29E-2
10	10	6.60E-3	14	6	2.21E-1	3	1	4.62E-1	10	4	1.24E-1	1	6	-5.73E-4	5	2	2.78E-1
10	11	2.95E-4	14	7	2.96E-2	3	2	8.36E-2	10	5	1.04E-2	1	7	-4.59E-3	5	3	-2.17E-2
10	12	4.49E-4	14	8	4.15E-2	3	3	9.53E-3	10	6	1.57E-3	1	8	-3.07E-3	5	4	-4.53E-3
10	13	-4.58E-3	14	9	1.82E-2	3	4	2.10E-4	10	7	-5.25E-3	1	9	-1.00E-2	5	5	-2.15E-3
10	14	-2.34E-5	14	10	8.42E-5	3	5	-1.79E-4	11	1	3.07E-1	1	10	-1.08E-3	5	6	-6.78E-1

S-like Fe (Fe^{10+})

i i'	gf_L	i i'	gf_L	i i'	gf_L	i i'	gf_L	i i'	gf_L	i i'	gf_L
5 7	−5.64E−3	9 3	9.27E−4	12 13	−1.70E+0	4 1	1.08E−1	11 4	1.74E−3	5 3	3.91E−2
5 8	−4.99E−2	9 4	−6.58E−4	12 14	−7.01E−4	4 2	2.88E−3	11 5	7.05E−3	5 4	1.40E−5
5 9	−7.54E−1	9 5	−1.73E−2	13 1	1.01E−5	4 3	1.07E−1	11 6	1.08E+0	5 5	3.91E−4
5 10	−2.22E−1	9 6	−5.42E−3	13 2	1.15E−4	4 4	2.31E−6	11 7	−2.79E−7	5 6	−1.07E−1
5 11	−1.74E−2	9 7	−3.02E−5	13 3	1.52E−2	4 5	7.52E−4	12 1	1.64E−1	6 1	1.96E−2
5 12	−3.82E−2	9 8	−4.64E−2	13 4	5.16E−1	4 6	−1.39E−4	12 2	8.84E−2	6 2	1.25E−1
5 13	−6.89E−4	9 9	−5.55E−2	13 5	9.56E−4	4 7	−4.03E−1	12 3	1.19E+0	6 3	9.41E−4
5 14	−1.94E+0	9 10	−2.53E−1	13 6	1.01E−2	5 1	8.64E−4	12 4	6.44E−1	6 4	3.72E−3
6 1	8.38E−2	9 11	−3.69E−3	13 7	2.40E−3	5 2	2.78E−1	12 5	1.56E−1	6 5	2.44E−3
6 2	3.75E−1	9 12	−5.55E−1	13 8	2.23E−4	5 3	2.39E−2	12 6	5.20E−3	6 6	−2.56E−1
6 3	−6.08E−3	9 13	−5.51E−2	13 9	1.30E−5	5 4	6.69E−4	12 7	−8.33E−4	7 1	1.18E+0
6 4	−1.08E−2	9 14	−9.94E−1	13 10	3.15E−3	5 5	1.14E−5	13 1	2.06E−2	7 2	1.39E−3
6 5	−1.54E−2	10 1	8.71E−3	13 11	3.65E−5	5 6	−6.51E−6	13 2	9.32E−4	7 3	2.01E−2
6 6	−3.95E−2	10 2	1.89E+0	13 12	5.88E−6	5 7	−1.40E−1	13 3	9.60E−2	7 4	3.61E−2
6 7	−1.62E−1	10 3	5.16E−6	13 13	−1.36E−7	6 1	1.09E+0	13 4	5.39E−2	7 5	1.21E−5
6 8	−7.34E−1	10 4	−7.70E−5	13 14	−1.33E−3	6 2	1.60E−1	13 5	2.67E−3	7 6	−1.24E−3
6 9	−1.47E−2	10 5	−6.39E−4	14 1	3.45E−3	6 3	1.99E−1	13 6	3.86E−2	8 1	3.54E−2
6 10	−3.93E−1	10 6	−1.59E−4	14 2	2.97E−2	6 4	1.16E−1	13 7	−2.64E−5	8 2	1.17E+0
6 11	−2.54E−4	10 7	−1.45E−2	14 3	2.84E−2	6 5	7.34E−4	14 1	7.49E−2	8 3	6.02E−1
6 12	−5.24E−1	10 8	−8.76E−3	14 4	3.31E−3	6 6	1.32E−4	14 2	6.69E−1	8 4	1.76E−2
6 13	−3.16E−3	10 9	−1.55E−2	14 5	7.03E−3	6 7	−4.29E−2	14 3	2.60E−3	8 5	8.09E−3
6 14	−1.15E+0	10 10	−1.46E−2	14 6	2.00E−5	7 1	3.19E−2	14 4	9.90E−1	8 6	−6.41E−3
7 1	8.18E−1	10 11	−4.84E−2	14 7	3.78E−3	7 2	4.68E−1	14 5	2.49E+0	9 1	1.99E−2
7 2	4.83E−1	10 12	−1.65E−2	14 8	2.64E−3	7 3	2.74E−1	14 6	5.76E−5	9 2	1.14E−2
7 3	−6.54E−5	10 13	−6.24E−3	14 9	1.69E−3	7 4	7.04E−2	14 7	1.68E−4	9 3	1.70E−3
7 4	−2.11E−3	10 14	−9.55E−2	14 10	9.43E−5	7 5	4.21E−2		^{1}**F**^{o}–^{1}**G**^{e}	9 4	1.40E−1
7 5	−3.99E−2	11 1	1.47E−1	14 11	8.87E−3	7 6	7.19E−5	1 1	−2.77E−3	9 5	1.97E−3
7 6	−4.96E−3	11 2	3.89E−3	14 12	2.56E−4	7 7	−6.14E−2	1 2	−8.06E−2	9 6	−2.81E−2
7 7	−4.14E−2	11 3	3.75E−5	14 13	2.62E−4	8 1	1.16E+0	1 3	−4.66E−1	10 1	1.06E−2
7 8	−1.94E−1	11 4	5.01E−4	14 14	−1.29E−4	8 2	1.51E−1	1 4	−1.62E+0	10 2	1.96E−1
7 9	−1.84E−1	11 5	2.01E−4		^{1}**F**^{o}–^{1}**F**^{e}	8 3	8.56E−4	1 5	−2.51E−2	10 3	1.47E+0
7 10	−8.31E−2	11 6	4.36E−6	1 1	−2.47E−2	8 4	5.29E−2	1 6	−7.02E−5	10 4	1.45E−1
7 11	−1.05E−2	11 7	−5.01E−5	1 2	−7.81E−1	8 5	4.12E−2	2 1	−3.66E−2	10 5	1.04E−3
7 12	−6.83E−1	11 8	−4.68E−4	1 3	−7.85E−3	8 6	3.09E−5	2 2	−2.87E−2	10 6	−1.74E−3
7 13	−1.33E−1	11 9	−1.58E−3	1 4	−2.42E+0	8 7	−3.66E−2	2 3	−3.25E−1	11 1	3.88E−3
7 14	−9.72E−2	11 10	−1.97E−3	1 5	−1.57E+0	9 1	5.17E−1	2 4	−2.40E−3	11 2	8.93E−3
8 1	1.70E+0	11 11	−9.65E−1	1 6	−3.16E−2	9 2	1.07E+0	2 5	−7.80E+0	11 3	1.07E−2
8 2	3.07E−1	11 12	−1.09E−1	1 7	−5.57E−5	9 3	1.65E−1	2 6	−1.50E−5	11 4	1.76E−3
8 3	−1.26E−5	11 13	−1.17E+0	2 1	−1.06E−2	9 4	6.77E−2	3 1	6.45E−1	11 5	2.95E−2
8 4	−1.70E−3	11 14	−3.61E−3	2 2	−1.84E−2	9 5	2.01E−1	3 2	1.85E−3	11 6	−9.50E−7
8 5	−1.68E−2	12 1	6.63E−2	2 3	−1.03E−2	9 6	7.07E−4	3 3	3.65E−3	12 1	7.24E−4
8 6	−2.55E−2	12 2	1.02E−1	2 4	4.26E−1	9 7	1.85E−4	3 4	5.53E−6	12 2	6.82E−1
8 7	−2.24E−2	12 3	8.53E−6	2 5	−9.43E−1	10 1	6.82E−3	3 5	−7.40E−5	12 3	4.97E−1
8 8	−1.25E−3	12 4	1.00E−5	2 6	−8.86E−2	10 2	1.59E+0	3 6	−4.86E−1	12 4	1.77E−2
8 9	−2.68E−1	12 5	9.00E−6	2 7	−8.89E−4	10 3	1.43E+0	4 1	2.48E−1	12 5	2.08E−1
8 10	−9.28E−2	12 6	4.12E−4	3 1	9.53E−2	10 4	5.16E−1	4 2	1.20E+0	12 6	−5.15E−4
8 11	−1.26E−4	12 7	6.81E−8	3 2	4.24E−1	10 5	9.24E−3	4 3	2.25E−2	13 1	1.81E−3
8 12	−1.76E−1	12 8	−8.20E−4	3 3	5.94E−3	10 6	2.28E−5	4 4	1.44E−4	13 2	5.30E−2
8 13	1.61E−3	12 9	7.73E−4	3 4	1.24E−3	10 7	−3.47E−5	4 5	1.15E−3	13 3	2.85E−2
8 14	−2.92E−1	12 10	−1.09E−6	3 5	−9.51E−5	11 1	1.85E−5	4 6	−1.55E−1	13 4	1.84E−2
9 1	2.11E−2	12 11	7.76E−1	3 6	1.64E−4	11 2	1.48E−4	5 1	2.37E+0	13 5	3.08E−2
9 2	6.97E−1	12 12	−9.85E−2	3 7	−1.22E−1	11 3	3.49E−3	5 2	9.08E−2	13 6	−1.16E−5

S-like Fe (Fe^{10+})

i i′	gf_L	*i i′*	gf_L	*i i′*	gf_L	*i i′*	gf_L	*i i′*	gf_L	*i i′*	gf_L
14 1	2.74E−4	6 6	−1.52E−3	7 5	3.59E−3	3 10	1.73E−4	6 11	1.33E−1	1 11	−1.10E+0
14 2	2.55E−1	6 7	−4.01E+0	7 6	−5.60E−6	3 11	−6.23E−5	6 12	1.03E+0	1 12	−4.77E−1
14 3	2.76E−2	6 8	−9.85E−5	8 1	7.42E−3	3 12	−2.80E−4	6 13	6.44E−2	1 13	−2.54E+0
14 4	1.07E−1	7 1	6.81E−4	8 2	4.14E−2	3 13	−2.11E−4	6 14	−5.58E−8	1 14	−7.10E−1
14 5	1.15E+0	7 2	8.55E−4	8 3	2.90E−1	3 14	−9.52E−3	6 15	−1.91E−7	1 15	−3.47E−4
14 6	1.00E−4	7 3	2.40E−1	8 4	6.69E−1	3 15	−8.36E−2	6 16	−1.84E−7	1 16	−1.13E−5
	$^1F^e$–$^1G^o$	7 4	3.52E−3	8 5	3.51E−1	3 16	−2.13E−1	6 17	−4.88E−3	1 17	−2.16E−1
1 1	2.01E−1	7 5	2.62E−3	8 6	−9.43E−6	3 17	−3.23E−2	7 1	1.97E−6	1 18	−2.01E−1
1 2	−2.83E−1	7 6	3.60E−4		**$^3S^o$–$^3P^e$**	4 1	1.01E−3	7 2	5.25E−4	1 19	−9.81E−5
1 3	−9.64E−2	7 7	1.58E−9	1 1	1.55E+0	4 2	1.43E+0	7 3	7.96E−1	1 20	−3.99E−3
1 4	−7.81E−1	7 8	3.97E−5	1 2	−1.75E−2	4 3	1.53E−1	7 4	2.59E−2	2 1	1.37E−3
1 5	−1.35E+0		**$^1G^o$–$^1G^e$**	1 3	−1.46E−2	4 4	5.26E−3	7 5	8.30E−1	2 2	1.75E+0
1 6	−4.18E−4	1 1	−3.43E−1	1 4	−1.52E−1	4 5	1.04E−1	7 6	3.72E−1	2 3	5.70E−2
1 7	−3.42E−3	1 2	−3.40E−1	1 5	−4.85E−1	4 6	5.79E−2	7 7	4.29E−1	2 4	2.67E−8
1 8	−4.48E−2	1 3	−4.15E+0	1 6	−3.34E−1	4 7	1.16E−3	7 8	3.51E−1	2 5	−1.06E−3
2 1	1.23E+0	1 4	−1.54E−4	1 7	−3.65E−1	4 8	4.74E−4	7 9	6.08E−1	2 6	−1.92E−3
2 2	−1.19E−2	1 5	−8.80E−1	1 8	−1.35E+0	4 9	1.88E−2	7 10	1.23E−1	2 7	−3.31E−4
2 3	−3.54E−1	1 6	−3.51E−4	1 9	−6.40E−1	4 10	2.63E−3	7 11	1.44E−4	2 8	−6.10E−3
2 4	−8.77E−1	2 1	5.68E−1	1 10	−2.15E+0	4 11	4.79E−6	7 12	5.16E−3	2 9	−3.78E−4
2 5	−4.12E−1	2 2	2.74E−5	1 11	−3.66E−3	4 12	−1.97E−6	7 13	2.42E−5	2 10	−8.68E−2
2 6	−1.55E+0	2 3	6.23E−3	1 12	−7.05E−2	4 13	−2.06E−7	7 14	5.12E−6	2 11	−1.14E−2
2 7	−4.06E−3	2 4	1.40E−2	1 13	−1.40E−2	4 14	−7.60E−3	7 15	−1.12E−6	2 12	−6.42E−2
2 8	−1.64E−2	2 5	4.68E−5	1 14	−2.14E−2	4 15	−1.50E−2	7 16	−1.16E−3	2 13	−3.19E−2
3 1	2.92E+0	2 6	−4.43E−1	1 15	−4.86E−4	4 16	−8.63E−3	7 17	−3.33E−4	2 14	−1.13E−1
3 2	−3.94E−2	3 1	5.40E−2	1 16	−2.03E−7	4 17	−7.47E−4	8 1	5.00E−3	2 15	−3.47E−1
3 3	−1.17E−1	3 2	2.75E+0	1 17	−2.07E−2	5 1	1.15E−3	8 2	3.16E−4	2 16	−8.69E−3
3 4	−3.78E−1	3 3	1.77E−1	2 1	6.14E−1	5 2	1.02E−1	8 3	5.14E−3	2 17	−9.32E−2
3 5	−8.91E−2	3 4	1.75E−3	2 2	8.61E−4	5 3	6.99E−2	8 4	4.05E−2	2 18	−6.65E−1
3 6	−5.45E−1	3 5	3.17E−3	2 3	5.19E−3	5 4	1.90E−3	8 5	7.46E−2	2 19	−1.62E−2
3 7	−3.43E−2	3 6	−3.75E−1	2 4	1.12E−5	5 5	4.28E−1	8 6	2.30E+0	2 20	−2.85E−1
3 8	−6.51E−1	4 1	2.12E−3	2 5	4.68E−3	5 6	6.76E−4	8 7	5.19E−2	3 1	3.12E−2
4 1	1.65E−1	4 2	3.24E−1	2 6	2.28E−3	5 7	1.19E+0	8 8	1.35E−1	3 2	1.35E−1
4 2	−5.67E−3	4 3	6.61E−2	2 7	1.16E−3	5 8	2.96E−2	8 9	4.16E−1	3 3	1.34E−2
4 3	−7.88E−2	4 4	2.71E−2	2 8	9.06E−4	5 9	1.33E−3	8 10	3.21E−1	3 4	9.05E−5
4 4	−4.83E−1	4 5	5.06E−3	2 9	2.03E−5	5 10	1.36E−3	8 11	1.25E−4	3 5	2.60E−4
4 5	−5.51E−3	4 6	−2.32E−3	2 10	1.68E−3	5 11	2.67E−4	8 12	1.70E−4	3 6	6.02E−1
4 6	−7.06E−2	5 1	3.87E+0	2 11	−1.63E+0	5 12	2.27E−5	8 13	7.21E−4	3 7	5.56E−3
4 7	−3.39E−3	5 2	1.06E−3	2 12	−6.71E−1	5 13	1.14E−4	8 14	1.17E−2	3 8	1.82E−3
4 8	−2.06E+0	5 3	1.53E−1	2 13	−3.25E−2	5 14	−1.62E−4	8 15	7.90E−4	3 9	3.91E−5
5 1	1.36E−2	5 4	1.56E−2	2 14	−1.84E−1	5 15	−1.59E−3	8 16	1.29E−4	3 10	3.88E−4
5 2	−1.60E−2	5 5	5.36E−2	2 15	−1.04E−2	5 16	−7.07E−2	8 17	−3.88E−8	3 11	−3.05E−5
5 3	−6.97E−4	5 6	−7.13E−2	2 16	−4.53E−4	5 17	−1.53E−7		**$^3S^e$–$^3P^o$**	3 12	−7.77E−5
5 4	−1.26E−2	6 1	1.98E−1	2 17	−1.25E−2	6 1	3.87E−1	1 1	2.03E+0	3 13	−2.60E−9
5 5	−3.89E−3	6 2	2.59E−2	3 1	3.80E−4	6 2	2.44E−3	1 2	1.65E−3	3 14	−6.31E−4
5 6	−1.74E−1	6 3	3.98E−1	3 2	2.98E−1	6 3	8.15E−3	1 3	1.84E−1	3 15	−3.10E−3
5 7	−6.86E−4	6 4	1.09E+0	3 3	2.25E−1	6 4	5.64E−3	1 4	−9.42E−2	3 16	−2.56E−1
5 8	−1.50E+0	6 5	3.93E−3	3 4	7.26E−2	6 5	1.74E−2	1 5	−9.39E−2	3 17	−3.08E−3
6 1	3.22E−1	6 6	−1.52E−2	3 5	6.25E−2	6 6	2.41E−3	1 6	−7.93E−4	3 18	−4.54E−3
6 2	7.76E−6	7 1	6.14E−4	3 6	2.29E−2	6 7	1.55E−3	1 7	−1.38E−1	3 19	−1.88E+0
6 3	3.57E−4	7 2	1.01E−3	3 7	1.90E−5	6 8	6.69E−3	1 8	−1.32E−2	3 20	−6.94E−4
6 4	−9.95E−7	7 3	7.28E−4	3 8	1.15E−2	6 9	2.99E−4	1 9	−2.09E−3	4 1	2.03E−1
6 5	−3.39E−4	7 4	1.12E−3	3 9	4.56E−4	6 10	5.63E−3	1 10	−6.96E−2	4 2	6.31E−2

S-like Fe (Fe^{10+})

i	*i'*	gf_L	*i*	*i'*	gf_L	*i*	*i'*	gf_L	*i*	*i'*	gf_L	*i*	*i'*	gf_L	*i*	*i'*	gf_L
4	3	1.22E−3	2	17	−4.61E−3	6	1	5.90E−1	9	2	6.87E−1	12	3	4.10E−2	15	4	1.38E+0
4	4	1.02E−3	3	1	5.88E+0	6	2	8.76E−4	9	3	2.51E−1	12	4	2.05E−2	15	5	3.78E−1
4	5	1.93E−4	3	2	−9.39E−3	6	3	2.42E−3	9	4	4.78E−4	12	5	1.24E−1	15	6	5.16E−1
4	6	5.42E−1	3	3	−3.52E−2	6	4	4.01E−3	9	5	1.28E−1	12	6	9.45E−1	15	7	1.06E−1
4	7	6.67E−4	3	4	−1.16E−1	6	5	2.89E−3	9	6	1.26E−1	12	7	2.35E−1	15	8	8.83E−2
4	8	3.63E−3	3	5	−5.27E−1	6	6	1.25E−3	9	7	3.40E−2	12	8	1.07E−3	15	9	5.61E−1
4	9	1.06E−3	3	6	−3.55E−1	6	7	3.11E−4	9	8	1.14E−2	12	9	4.28E−4	15	10	4.74E−4
4	10	1.21E−7	3	7	−4.11E−1	6	8	1.55E−4	9	9	7.93E−3	12	10	4.27E−3	15	11	2.57E−4
4	11	2.04E−4	3	8	−1.76E+0	6	9	4.14E−3	9	10	1.99E−2	12	11	4.82E−4	15	12	1.12E−4
4	12	1.94E−3	3	9	−2.64E−1	6	10	2.47E−3	9	11	1.75E−6	12	12	5.50E−4	15	13	1.43E−4
4	13	1.80E−5	3	10	−3.61E+0	6	11	−2.27E−2	9	12	−2.06E−4	12	13	6.78E−5	15	14	−2.07E−3
4	14	4.36E−4	3	11	−1.06E−1	6	12	−1.01E−1	9	13	−2.82E−5	12	14	−4.40E−4	15	15	−5.90E−5
4	15	1.50E−3	3	12	−6.58E−2	6	13	−2.12E+0	9	14	−1.57E−3	12	15	−4.45E−3	15	16	−2.96E−4
4	16	6.93E−4	3	13	−2.26E−4	6	14	−5.43E−2	9	15	−1.50E−2	12	16	−3.18E−2	15	17	−1.79E−4
4	17	2.41E−3	3	14	−4.46E−2	6	15	−1.03E−2	9	16	−2.40E−1	12	17	−1.07E−4	16	1	1.32E+0
4	18	6.63E−5	3	15	−2.43E−3	6	16	−2.37E−3	9	17	−2.12E−2	13	1	1.72E−3	16	2	1.80E−4
4	19	6.39E−4	3	16	−6.23E−4	6	17	−2.34E−1	10	1	1.35E−5	13	2	8.67E−1	16	3	6.24E−4
4	20	2.38E−5	3	17	−1.01E−2	7	1	5.43E−3	10	2	1.71E−1	13	3	5.19E−3	16	4	4.80E−4
		$^3P^o$–$^3P^e$	4	1	2.97E−4	7	2	6.65E−1	10	3	1.71E−1	13	4	1.75E+0	16	5	1.36E−4
1	1	5.25E−1	4	2	3.02E−1	7	3	5.43E−2	10	4	1.67E−1	13	5	1.43E−1	16	6	2.43E−3
1	2	−8.89E−3	4	3	1.80E−1	7	4	2.86E−1	10	5	2.86E−1	13	6	4.96E−2	16	7	1.68E−2
1	3	−5.31E−1	4	4	7.17E−3	7	5	1.79E−1	10	6	4.31E−3	13	7	1.19E−1	16	8	6.72E−7
1	4	−4.39E+0	4	5	1.46E−2	7	6	1.05E−3	10	7	2.58E−1	13	8	2.47E−1	16	9	8.55E−4
1	5	1.70E 1	4	6	2.67E−4	7	7	4.37E−4	10	8	1.73E−2	13	9	1.39E−2	16	10	4.61E−2
1	6	−1.43E−2	4	7	2.49E−4	7	8	3.27E−4	10	9	1.91E−2	13	10	5.65E−2	16	11	6.20E−1
1	7	−1.62E−2	4	8	−1.13E−3	7	9	5.45E−3	10	10	7.18E−3	13	11	1.32E−6	16	12	1.79E+0
1	8	−5.84E−3	4	9	−2.63E−4	7	10	6.06E−5	10	11	1.03E−4	13	12	2.98E−5	16	13	3.74E−3
1	9	−7.79E−5	4	10	−4.44E−6	7	11	−2.42E−4	10	12	−1.37E−4	13	13	4.86E−5	16	14	−1.67E−3
1	10	−7.85E−3	4	11	−1.85E−4	7	12	−6.83E−4	10	13	−1.78E−3	13	14	−8.05E−5	16	15	−1.55E−4
1	11	−2.42E−2	4	12	−2.30E−4	7	13	−3.25E−2	10	14	−4.34E−4	13	15	−4.43E−3	16	16	−1.60E−6
1	12	−1.30E−1	4	13	−7.96E−6	7	14	−1.47E−2	10	15	−9.88E−4	13	16	−5.47E−3	16	17	−1.49E−2
1	13	−2.03E−2	4	14	−6.89E−4	7	15	−1.78E−1	10	16	−1.27E−3	13	17	−2.01E−3	17	1	8.88E−3
1	14	−1.06E+0	4	15	−1.56E−1	7	16	−5.11E−1	10	17	−8.92E−5	14	1	1.51E−2	17	2	2.68E−2
1	15	−4.28E−2	4	16	−1.35E−1	7	17	−2.21E−3	11	1	2.68E−5	14	2	3.61E−2	17	3	1.14E−2
1	16	−4.25E−3	4	17	−8.16E−4	8	1	1.34E−4	11	2	6.04E−1	14	3	2.02E+0	17	4	7.76E−1
1	17	−1.22E−1	5	1	3.17E−3	8	2	8.65E−1	11	3	1.36E+0	14	4	8.46E−1	17	5	4.93E−1
2	1	2.30E−1	5	2	7.63E−1	8	3	9.53E−1	11	4	5.73E−2	14	5	1.68E+0	17	6	3.05E−1
2	2	−2.46E−2	5	3	8.93E−3	8	4	9.91E−2	11	5	1.46E−1	14	6	5.33E−1	17	7	2.01E−1
2	3	−1.77E−1	5	4	1.15E−1	8	5	2.99E−1	11	6	1.71E−2	14	7	3.46E−3	17	8	3.49E−2
2	4	−4.14E−5	5	5	2.69E−5	8	6	4.84E−3	11	7	2.16E−1	14	8	2.00E−2	17	9	2.45E−1
2	5	−6.68E−1	5	6	6.99E−2	8	7	1.98E−3	11	8	1.54E−2	14	9	1.24E−2	17	10	4.07E−2
2	6	−1.20E 1	5	7	4.03E 2	8	8	3.05E−2	11	9	1.00E−3	14	10	4.33E−4	17	11	8.50E−3
2	7	−1.98E−1	5	8	8.71E−6	8	9	3.13E−4	11	10	1.41E−3	14	11	1.11E−3	17	12	3.39E−2
2	8	−5.76E−1	5	9	3.05E−4	8	10	6.19E−4	11	11	1.77E−5	14	12	2.37E−3	17	13	1.27E−4
2	9	−4.69E+0	5	10	7.90E−5	8	11	−1.09E−5	11	12	3.08E−5	14	13	2.39E−4	17	14	7.87E−5
2	10	−5.59E−3	5	11	−4.05E−4	8	12	−7.45E−5	11	13	−5.34E−5	14	14	−3.04E−3	17	15	−7.45E−6
2	11	−1.54E−3	5	12	−1.23E−3	8	13	−1.18E−3	11	14	−7.98E−4	14	15	−1.93E−5	17	16	−2.63E−3
2	12	−8.02E−3	5	13	−4.09E−3	8	14	−3.66E−5	11	15	−4.71E−2	14	16	−7.51E−3	17	17	−2.99E−4
2	13	−1.40E−1	5	14	−4.14E−3	8	15	−5.04E−3	11	16	−3.43E−2	14	17	−2.42E−5	18	1	5.03E−4
2	14	−8.43E−3	5	15	−1.20E−1	8	16	−4.20E−1	11	17	−2.74E−3	15	1	9.45E−3	18	2	6.15E−4
2	15	6.86E−4	5	16	−4.00E−1	8	17	−3.25E−2	12	1	1.20E−3	15	2	3.95E−2	18	3	3.73E−1
2	16	−2.10E−3	5	17	−6.70E−4	9	1	3.18E−3	12	2	5.63E−3	15	3	2.15E−1	18	4	1.59E+0

S-like Fe (Fe^{10+})

i	i'	gf_L
18	5	6.54E−2
18	6	5.80E−1
18	7	3.51E−1
18	8	1.56E−1
18	9	2.90E−2
18	10	7.70E−1
18	11	4.32E−6
18	12	2.16E−4
18	13	8.86E−3
18	14	1.03E−3
18	15	1.42E−4
18	16	−2.25E−4
18	17	−5.46E−4
19	1	3.71E−2
19	2	1.07E−3
19	3	1.27E−3
19	4	3.45E−3
19	5	2.26E−3
19	6	3.01E−3
19	7	2.78E−2
19	8	7.22E−3
19	9	2.82E−2
19	10	2.68E−3
19	11	1.37E−3
19	12	8.26E−2
19	13	1.72E+0
19	14	1.92E−4
19	15	9.74E−6
19	16	−3.80E−6
19	17	−1.89E−3
20	1	1.40E−4
20	2	3.27E−4
20	3	2.32E−3
20	4	1.84E−2
20	5	1.63E+0
20	6	8.61E−1
20	7	2.01E+0
20	8	1.05E+0
20	9	1.10E+0
20	10	2.41E−1
20	11	1.39E−5
20	12	3.69E−4
20	13	5.04E−4
20	14	3.73E−4
20	15	1.56E−4
20	16	−1.87E−5
20	17	−2.08E−4
		$^3P^o$–$^3D^e$
1	1	−1.01E−1
1	2	−4.74E−1
1	3	−4.60E−2
1	4	−8.75E+0
1	5	−4.18E−1
1	6	−2.93E−2
1	7	−1.66E−3
1	8	−3.61E−2
1	9	−2.86E−2
1	10	−3.21E−2
1	11	−2.52E−3
1	12	−8.59E−1
1	13	−5.28E−3
1	14	−5.72E−4
2	1	−2.87E−2
2	2	−5.51E−2
2	3	−9.35E−1
2	4	−9.26E−3
2	5	−1.55E+0
2	6	−3.92E−2
2	7	−1.18E−2
2	8	−1.42E+0
2	9	−9.29E−1
2	10	−9.94E−3
2	11	−2.14E−6
2	12	−6.90E−3
2	13	−3.48E−4
2	14	−9.27E−5
3	1	−3.79E−2
3	2	−1.33E−1
3	3	−2.96E−2
3	4	−2.21E−1
3	5	−4.52E−1
3	6	−8.30E−2
3	7	−1.46E−1
3	8	−2.68E+0
3	9	−6.58E+0
3	10	−7.35E−2
3	11	−3.09E−4
3	12	−1.06E−4
3	13	−1.40E−4
3	14	−2.39E−5
4	1	3.09E−1
4	2	6.16E−1
4	3	4.70E−3
4	4	2.77E−3
4	5	2.19E−2
4	6	4.38E−3
4	7	−8.10E−5
4	8	−7.41E−5
4	9	−4.66E−4
4	10	−5.22E−4
4	11	−4.17E−4
4	12	−5.71E−5
4	13	−1.51E−1
4	14	−1.32E+0
5	1	3.57E−1
5	2	1.10E−1
5	3	1.01E−2
5	4	4.69E−3
5	5	9.99E−3
5	6	8.80E−4
5	7	3.08E−7
5	8	2.97E−3
5	9	3.06E−5
5	10	−5.85E−4
5	11	−3.64E−3
5	12	−5.85E−4
5	13	−6.10E−1
5	14	−7.41E−1
6	1	4.39E−2
6	2	6.31E−6
6	3	1.48E−2
6	4	2.31E−5
6	5	7.68E−3
6	6	3.47E−4
6	7	1.43E−2
6	8	4.27E−4
6	9	1.74E−3
6	10	−1.23E−1
6	11	−3.21E+0
6	12	−2.54E−1
6	13	−5.85E−4
6	14	−3.63E−4
7	1	1.38E+0
7	2	4.28E−2
7	3	1.02E−1
7	4	9.99E−2
7	5	1.67E−1
7	6	1.46E−1
7	7	2.46E−3
7	8	6.60E−4
7	9	5.64E−3
7	10	−1.10E−3
7	11	−4.20E−2
7	12	−1.75E−4
7	13	−1.61E−1
7	14	−1.23E−1
8	1	2.26E−1
8	2	4.26E−1
8	3	3.31E−2
8	4	5.35E−2
8	5	2.00E−1
8	6	1.45E−2
8	7	1.04E−3
8	8	5.30E−6
8	9	8.52E−4
8	10	−2.04E−5
8	11	−3.45E−3
8	12	−6.48E−5
8	13	−7.59E−2
8	14	−6.15E−2
9	1	2.33E+0
9	2	2.45E−1
9	3	6.18E−2
9	4	3.33E−2
9	5	1.25E−1
9	6	4.84E−2
9	7	1.40E−3
9	8	5.98E−2
9	9	9.74E−4
9	10	1.55E−7
9	11	−2.23E−5
9	12	−1.84E−3
9	13	−8.11E−3
9	14	−1.18E−5
10	1	8.13E−1
10	2	3.80E−2
10	3	1.28E−2
10	4	7.25E−3
10	5	4.80E−3
10	6	9.40E−2
10	7	1.46E−2
10	8	5.20E−4
10	9	8.83E−3
10	10	1.58E−4
10	11	−2.65E−3
10	12	−3.90E−3
10	13	−9.46E−2
10	14	−6.04E−2
11	1	5.73E−2
11	2	2.90E+0
11	3	1.39E−2
11	4	6.29E−2
11	5	9.70E−2
11	6	6.41E−1
11	7	4.00E−3
11	8	2.46E−2
11	9	3.55E−3
11	10	1.13E−6
11	11	1.47E−4
11	12	−1.02E−3
11	13	−8.94E−3
11	14	−2.64E−2
12	1	7.15E−2
12	2	3.77E−1
12	3	1.15E+0
12	4	3.79E−1
12	5	3.51E−1
12	6	2.10E−2
12	7	5.25E−1
12	8	8.84E−3
12	9	7.48E−3
12	10	6.50E−4
12	11	9.34E−5
12	12	−5.10E−4
12	13	−6.21E−2
12	14	−2.92E−4
13	1	1.71E−1
13	2	1.16E+0
13	3	2.83E−1
13	4	5.69E−1
13	5	2.96E−1
13	6	3.35E−1
13	7	7.71E−4
13	8	2.25E−1
13	9	1.13E−1
13	10	8.07E−6
13	11	1.10E−6
13	12	−7.03E−4
13	13	−3.01E−4
13	14	−9.60E−2
14	1	2.76E−3
14	2	1.98E−1
14	3	2.40E−3
14	4	2.37E+0
14	5	2.07E−2
14	6	1.32E+0
14	7	1.73E−3
14	8	5.20E−2
14	9	6.91E−1
14	10	2.08E−3
14	11	4.19E−4
14	12	−2.09E−3
14	13	−9.00E−3
14	14	−2.94E−3
15	1	2.13E−2
15	2	1.03E−1
15	3	2.39E+0
15	4	9.52E−1
15	5	4.77E−1
15	6	3.93E−1
15	7	3.92E−1
15	8	1.66E−5
15	9	7.64E−2
15	10	1.22E−4
15	11	1.54E−3
15	12	−5.68E−4
15	13	−2.05E−3
15	14	−2.13E−4
16	1	3.86E−5
16	2	2.44E−4
16	3	3.24E−2
16	4	6.68E−4
16	5	6.66E−5
16	6	4.35E−2
16	7	8.22E−2
16	8	1.83E−2
16	9	7.25E−3
16	10	7.13E−1
16	11	1.32E−1
16	12	−1.54E−4
16	13	−1.37E−5
16	14	−4.25E−5
17	1	1.29E−2
17	2	2.89E−3
17	3	1.05E+0
17	4	1.65E+0
17	5	3.80E−2
17	6	3.22E−1
17	7	3.67E+0
17	8	1.17E−1
17	9	1.87E−2
17	10	1.93E−2
17	11	6.22E−3
17	12	−7.55E−6
17	13	−8.36E−5
17	14	−5.13E−4
18	1	1.86E−5
18	2	1.09E−2
18	3	9.22E−2
18	4	9.65E−1
18	5	2.76E+0
18	6	1.34E+0
18	7	7.83E−1
18	8	1.60E−1
18	9	3.73E−1
18	10	1.33E−4
18	11	2.19E−3
18	12	−2.54E−4
18	13	−2.17E−3
18	14	−3.37E−3
19	1	1.05E−4
19	2	4.74E−4
19	3	2.02E−2
19	4	7.03E−4
19	5	4.12E−4
19	6	2.27E−4
19	7	1.05E−3
19	8	3.44E−3
19	9	3.27E−4
19	10	1.64E−1
19	11	4.97E−3
19	12	−4.42E−5

S-like Fe (Fe^{10+})

i	i'	gf$_L$	i	i'	gf$_L$	i	i'	gf$_L$	i	i'	gf$_L$	i	i'	gf$_L$	i	i'	gf$_L$
19	13	0.00E+0	2	11	−1.85E+0	4	13	−6.32E−1	6	15	−1.02E+0	8	17	−1.61E−2	10	19	−1.18E−1
19	14	−2.15E−6	2	12	−4.89E−2	4	14	−8.69E−1	6	16	−7.69E−1	8	18	−9.58E−2	10	20	−5.01E−2
20	1	1.92E−3	2	13	−5.16E−1	4	15	−1.89E+0	6	17	−1.35E−2	8	19	−1.82E−1	10	21	−4.78E−2
20	2	7.10E−2	2	14	−3.30E−1	4	16	−1.96E−1	6	18	−2.30E−1	8	20	−2.72E−2	10	22	−2.86E−3
20	3	2.52E−2	2	15	−2.67E−2	4	17	−5.37E−3	6	19	−1.83E−1	8	21	−1.76E+0	10	23	−5.06E−4
20	4	2.73E−1	2	16	−1.16E−4	4	18	−3.68E−3	6	20	−1.63E−4	8	22	−1.59E−4	10	24	−3.48E−1
20	5	1.62E+0	2	17	−1.91E−3	4	19	−8.65E−1	6	21	−2.83E−1	8	23	−3.33E−2	10	25	−2.44E−1
20	6	1.19E+0	2	18	−3.07E−3	4	20	−1.65E−4	6	22	−2.61E−3	8	24	−3.73E+0	11	1	2.82E−1
20	7	1.89E−1	2	19	−1.11E−1	4	21	−2.30E−1	6	23	−6.37E−2	8	25	−9.13E−3	11	2	3.13E−2
20	8	4.97E−1	2	20	−3.03E−3	4	22	−2.45E−4	6	24	−8.87E−1	9	1	4.45E−4	11	3	4.99E−2
20	9	8.87E−1	2	21	−7.26E−2	4	23	−7.90E−2	6	25	−4.34E−2	9	2	1.05E+0	11	4	1.72E−4
20	10	2.19E−3	2	22	−1.44E−3	4	24	−7.85E−2	7	1	1.09E+0	9	3	3.75E−2	11	5	7.59E−5
20	11	1.31E−4	2	23	−7.41E−5	4	25	−5.31E−2	7	2	1.36E+0	9	4	−3.18E−5	11	6	2.61E−1
20	12	6.54E−5	2	24	−2.44E−2	5	1	1.76E+0	7	3	1.48E−4	9	5	−6.15E−4	11	7	8.33E−4
20	13	1.24E−3	2	25	−1.24E−4	5	2	8.71E−1	7	4	−8.29E−3	9	6	−7.79E−4	11	8	1.71E−6
20	14	−7.69E−5	3	1	9.94E−2	5	3	7.31E−3	7	5	−2.48E−4	9	7	−1.54E−3	11	9	−9.23E−5
		$^3P^e$–$^3D^o$	3	2	3.18E−1	5	4	−1.00E−1	7	6	−2.43E−3	9	8	−2.66E−2	11	10	−2.57E−4
1	1	−6.67E−5	3	3	1.59E−2	5	5	−1.36E−2	7	7	−5.56E−2	9	9	−1.55E−2	11	11	−2.31E−4
1	2	−3.12E−2	3	4	−5.85E−1	5	6	−9.80E−6	7	8	−1.94E−2	9	10	−1.17E−2	11	12	−4.47E−4
1	3	−1.01E+1	3	5	−2.06E−2	5	7	−2.77E−1	7	9	−1.10E−1	9	11	−1.50E−3	11	13	−2.18E−3
1	4	−7.29E−4	3	6	−1.11E−3	5	8	−3.42E−1	7	10	−1.13E−1	9	12	−2.31E−3	11	14	−3.13E−4
1	5	−1.83E−4	3	7	−6.46E−1	5	9	−4.47E−2	7	11	−1.42E−1	9	13	−1.13E−2	11	15	−8.20E−3
1	6	−8.30E−1	3	8	−1.21E−2	5	10	−1.32E−1	7	12	−3.45E−1	9	14	−3.04E−2	11	16	−1.17E−3
1	7	−1.32E−3	3	9	−2.92E−2	5	11	5.56E−2	7	13	−7.90E−3	9	15	−5.35E−2	11	17	−5.46E+0
1	8	−7.44E−4	3	10	−5.59E−3	5	12	−6.51E−2	7	14	−3.64E−2	9	16	−1.13E−2	11	18	−7.10E−2
1	9	−4.07E−3	3	11	−6.95E−2	5	13	−8.37E−1	7	15	−1.97E−2	9	17	−1.98E−2	11	19	−1.29E−2
1	10	−7.71E−3	3	12	−8.32E−3	5	14	−7.77E−1	7	16	−5.43E−1	9	18	−7.63E−1	11	20	−2.57E−2
1	11	−2.12E−3	3	13	−1.16E+0	5	15	−2.54E−2	7	17	−6.74E−3	9	19	−3.22E−1	11	21	−1.41E−3
1	12	−1.76E−3	3	14	−2.48E−1	5	16	−6.58E−3	7	18	−5.36E−1	9	20	−6.95E−4	11	22	−9.16E−3
1	13	−1.70E−3	3	15	−3.89E−1	5	17	−6.06E−2	7	19	−1.05E−3	9	21	−1.08E+0	11	23	−4.65E−4
1	14	−3.66E−3	3	16	−2.79E−1	5	18	−9.76E−1	7	20	−2.05E−5	9	22	−5.10E−2	11	24	−4.16E−4
1	15	−4.97E−3	3	17	−3.92E−2	5	19	−7.68E−1	7	21	−5.24E−1	9	23	−3.30E−1	11	25	−6.08E−5
1	16	−2.41E−5	3	18	−4.49E−1	5	20	−2.15E−3	7	22	−3.49E−2	9	24	−4.05E−1	12	1	5.30E−3
1	17	−1.20E+0	3	19	−7.52E−2	5	21	−1.31E+0	7	23	−2.04E+0	9	25	−5.94E−1	12	2	5.01E−2
1	18	−1.63E−3	3	20	−2.70E−4	5	22	−2.57E−7	7	24	−1.09E−4	10	1	3.20E−3	12	3	1.12E−1
1	19	−2.78E−5	3	21	−1.65E−1	5	23	−1.19E−1	7	25	−1.97E+0	10	2	4.81E−4	12	4	1.56E−3
1	20	−1.57E+0	3	22	−7.12E−4	5	24	−3.72E−1	8	1	4.71E−2	10	3	7.20E+0	12	5	3.02E−4
1	21	−8.27E−3	3	23	−5.66E−3	5	25	−5.60E−1	8	2	2.86E+0	10	4	5.82E−4	12	6	1.90E+0
1	22	−5.04E−1	3	24	−8.20E−3	6	1	2.45E+0	8	3	1.97E−2	10	5	1.19E−7	12	7	7.04E−3
1	23	−1.66E−4	3	25	−2.13E−2	6	2	7.98E−2	8	4	−7.39E−6	10	6	−2.80E−3	12	8	3.18E−3
1	24	−1.69E−6	4	1	1.68E−2	6	3	6.85E−3	8	5	−2.14E−3	10	7	−2.04E−4	12	9	4.34E−4
1	25	−5.16E−3	4	2	1.20E−3	6	4	−4.50E−3	8	6	−5.17E−4	10	8	1.61E−3	12	10	2.55E−4
2	1	3.20E−1	4	3	1.99E−1	6	5	−1.21E−3	8	7	−5.85E−4	10	9	−4.76E−3	12	11	2.38E−5
2	2	4.74E−3	4	4	−1.13E−1	6	6	−1.72E−3	8	8	−1.22E−3	10	10	−1.95E−2	12	12	2.14E−4
2	3	1.36E−4	4	5	−3.98E−2	6	7	−4.56E−4	8	9	−3.08E−2	10	11	−2.73E−3	12	13	−1.80E−5
2	4	−2.40E−1	4	6	−1.54E−3	6	8	−3.14E−1	8	10	−1.19E−2	10	12	−4.94E−3	12	14	−2.22E−4
2	5	−5.92E−1	4	7	−1.77E−1	6	9	−2.97E−2	8	11	−3.08E−2	10	13	−7.37E−3	12	15	−8.19E−4
2	6	−1.58E−2	4	8	−1.46E−3	6	10	−1.89E−1	8	12	−5.65E−2	10	14	−1.66E−2	12	16	−3.75E−5
2	7	−2.47E−1	4	9	4.15E−1	6	11	−2.38E−3	8	13	−2.07E−1	10	15	−1.61E−4	12	17	−5.21E−1
2	8	−7.13E−1	4	10	−2.77E−1	6	12	−5.27E−2	8	14	−1.32E−3	10	16	−2.30E−2	12	18	−4.44E−4
2	9	−4.49E−1	4	11	−1.22E−3	6	13	8.27E−2	8	15	−4.90E−1	10	17	−3.91E−2	12	19	−3.47E−4
2	10	−9.17E−2	4	12	−1.20E+0	6	14	−9.34E−1	8	16	−1.07E−2	10	18	−1.41E−1	12	20	−2.28E+0

S-like Fe (Fe^{10+})

i i′	gf_L	i i′	gf_L	i i′	gf_L	i i′	gf_L	i i′	gf_L	i i′	gf_L
12 21	−1.16E−2	14 23	−2.56E−4	16 25	−5.50E−5	2 12	−3.50E−2	6 8	8.52E−4	10 4	8.41E−5
12 22	−1.87E−2	14 24	−5.32E−4	17 1	3.43E−1	2 13	−2.71E−4	6 9	1.51E−3	10 5	1.31E−1
12 23	−2.86E−4	14 25	−1.33E−2	17 2	1.70E−3	2 14	−1.36E−3	6 10	−2.99E+0	10 6	7.60E−2
12 24	−2.57E−6	15 1	1.56E−3	17 3	7.85E−3	3 1	−3.53E−3	6 11	−2.37E−1	10 7	2.08E−2
12 25	−2.28E−4	15 2	4.85E−4	17 4	2.40E−2	3 2	−8.70E−3	6 12	−1.23E+0	10 8	2.41E−3
13 1	1.59E−4	15 3	2.23E−3	17 5	3.91E−3	3 3	−2.51E−4	6 13	−2.67E−2	10 9	8.61E−4
13 2	2.91E−1	15 4	1.83E−1	17 6	1.16E+0	3 4	−6.29E−1	6 14	−1.57E−3	10 10	−2.68E−4
13 3	9.25E−2	15 5	8.47E−1	17 7	2.19E−3	3 5	−4.35E−1	7 1	2.03E−3	10 11	−2.32E−4
13 4	2.05E−5	15 6	7.45E−3	17 8	3.81E−4	3 6	−4.21E−2	7 2	4.05E−4	10 12	−8.52E−4
13 5	3.03E−4	15 7	1.48E−2	17 9	5.20E−3	3 7	−3.18E−2	7 3	3.19E−2	10 13	−1.13E−5
13 6	1.32E−1	15 8	2.57E−1	17 10	4.18E−3	3 8	−1.02E+0	7 4	7.31E−4	10 14	−5.15E−3
13 7	2.05E−3	15 9	1.28E−1	17 11	1.23E−3	3 9	−2.20E+0	7 5	6.53E−4	11 1	2.27E+0
13 8	1.78E−5	15 10	6.49E−3	17 12	7.96E−4	3 10	−2.13E−2	7 6	8.02E−3	11 2	6.93E−4
13 9	1.73E−6	15 11	1.80E−2	17 13	2.81E−5	3 11	−8.23E−2	7 7	3.60E−2	11 3	2.83E−1
13 10	1.11E−5	15 12	2.30E−3	17 14	5.44E−6	3 12	−1.79E−3	7 8	3.43E−5	11 4	2.41E−2
13 11	3.62E−4	15 13	3.92E−2	17 15	2.96E−5	3 13	−2.63E−3	7 9	6.72E−5	11 5	1.75E−2
13 12	1.11E−5	15 14	2.31E−3	17 16	1.51E−2	3 14	−8.07E−4	7 10	−1.18E−2	11 6	8.75E−2
13 13	2.93E−8	15 15	7.17E−5	17 17	1.37E−3	4 1	3.26E−1	7 11	−2.41E−3	11 7	8.03E−4
13 14	2.37E−6	15 16	4.11E−3	17 18	5.59E−3	4 2	9.46E−2	7 12	−8.48E−3	11 8	8.67E−4
13 15	−3.80E−7	15 17	5.81E−4	17 19	3.69E−4	4 3	3.23E−1	7 13	−4.04E−1	11 9	6.69E−2
13 16	−6.52E−5	15 18	1.75E−6	17 20	5.72E−4	4 4	1.84E−2	7 14	−1.34E−4	11 10	1.70E−4
13 17	−4.69E−3	15 19	2.37E−4	17 21	3.36E−4	4 5	3.86E−3	8 1	5.87E−1	11 11	−7.00E−6
13 18	−6.09E−4	15 20	2.73E−5	17 22	2.74E−3	4 6	4.08E−6	8 2	1.17E+0	11 12	−4.00E−3
13 19	−1.08E−3	15 21	1.98E−6	17 23	1.01E−4	4 7	3.38E−3	8 3	1.30E−1	11 13	−1.50E−2
13 20	−3.79E−3	15 22	−8.64E−8	17 24	1.03E−6	4 8	9.13E−4	8 4	2.77E−1	11 14	−1.30E−4
13 21	−2.65E−3	15 23	−2.52E−3	17 25	2.70E−6	4 9	6.09E−6	8 5	1.63E−1	12 1	1.51E+0
13 22	−4.74E+0	15 24	−1.84E−5		$^3D^o$–$^3D^e$	4 10	−1.07E−3	8 6	2.64E−2	12 2	2.21E−3
13 23	−1.56E−2	15 25	−2.32E−3	1 1	−3.68E−1	4 11	−4.56E−5	8 7	2.33E−5	12 3	6.43E−1
13 24	−1.63E−4	16 1	9.71E−3	1 2	−9.22E−1	4 12	−1.95E−7	8 8	2.89E−3	12 4	1.35E−1
13 25	−8.83E−4	16 2	7.57E−6	1 3	−1.07E−1	4 13	−6.48E−4	8 9	5.48E−3	12 5	9.04E−2
14 1	3.38E−2	16 3	2.75E−5	1 4	−3.56E−2	4 14	−4.79E−1	8 10	−1.56E−3	12 6	9.98E−2
14 2	5.26E−3	16 4	1.62E−1	1 5	−1.64E+0	5 1	2.74E−1	8 11	−1.14E−5	12 7	7.86E−3
14 3	6.20E−2	16 5	3.00E−1	1 6	−3.98E+0	5 2	2.03E−1	8 12	−1.59E−3	12 8	1.12E−3
14 4	4.29E−2	16 6	3.62E−2	1 7	−4.30E−1	5 3	3.57E−2	8 13	−1.70E−1	12 9	5.13E−3
14 5	3.38E−2	16 7	1.72E−1	1 8	−1.10E+0	5 4	2.98E−2	8 14	−1.18E−1	12 10	1.41E−3
14 6	5.76E−2	16 8	2.27E−2	1 9	−9.11E−1	5 5	1.59E−1	9 1	2.50E−1	12 11	−8.59E−8
14 7	4.35E−4	16 9	3.95E−4	1 10	−1.25E−1	5 6	1.51E−2	9 2	1.33E+0	12 12	−2.72E−3
14 8	7.30E−3	16 10	3.18E−2	1 11	−4.08E−2	5 7	3.01E−4	9 3	7.72E−3	12 13	−4.11E−3
14 9	3.04E−4	16 11	7.98E−3	1 12	−4.71E−2	5 8	2.49E−4	9 4	5.61E−3	12 14	−2.24E−2
14 10	2.09E−2	16 12	1.14E−1	1 13	−3.27E−6	5 9	5.66E−5	9 5	2.19E−1	13 1	3.79E−1
14 11	7.74E−4	16 13	1.75E−4	1 14	−3.65E−4	5 10	−4.80E−3	9 6	4.88E−3	13 2	1.87E−1
14 12	2.91E−4	16 14	1.80E−2	2 1	−6.66E−2	5 11	−5.67E−5	9 7	2.04E−2	13 3	7.24E−1
14 13	3.80E−3	16 15	2.98E−4	2 2	−1.78E−2	5 12	−1.22E−2	9 8	6.33E−2	13 4	3.80E−1
14 14	4.12E−4	16 16	4.59E−2	2 3	−6.09E−1	5 13	−1.08E−1	9 9	2.12E−4	13 5	4.65E−1
14 15	1.83E−3	16 17	4.65E−6	2 4	−3.88E−3	5 14	−9.65E−1	9 10	−5.71E−5	13 6	4.81E−1
14 16	4.44E−5	16 18	5.03E−5	2 5	−3.19E−2	6 1	9.91E−3	9 11	−1.43E−6	13 7	1.46E−1
14 17	9.03E−3	16 19	4.29E−3	2 6	−9.87E−2	6 2	1.51E−4	9 12	−7.73E−3	13 8	3.43E−2
14 18	1.68E−5	16 20	7.82E−5	2 7	−9.35E−2	6 3	5.65E−5	9 13	−2.46E−1	13 9	2.79E−3
14 19	3.64E−6	16 21	1.87E−5	2 8	−4.17E+0	6 4	5.44E−4	9 14	−5.68E−1	13 10	1.13E−4
14 20	2.84E−4	16 22	2.56E−5	2 9	−3.06E+0	6 5	1.14E−3	10 1	9.83E−2	13 11	−2.57E−8
14 21	1.20E−7	16 23	2.16E−5	2 10	−9.30E−2	6 6	1.13E−2	10 2	9.04E−1	13 12	−2.14E−3
14 22	−9.90E−6	16 24	7.25E−6	2 11	−4.64E−2	6 7	2.43E−3	10 3	1.36E−1	13 13	−1.99E−1

S-like Fe (Fe^{10+})

i i'	gf_L	i i'	gf_L	i i'	gf_L	i i'	gf_L	i i'	gf_L	i i'	gf_L
13 14	−4.39E−2	17 10	1.30E−1	21 6	3.12E−2	25 2	1.19E−2	3 11	−2.75E−4	7 7	7.02E−3
14 1	1.10E+0	17 11	1.46E−2	21 7	8.08E−1	25 3	2.03E−2	3 12	−1.46E−4	7 8	5.95E−3
14 2	1.75E−2	17 12	−3.81E−5	21 8	6.77E−2	25 4	7.79E−3	3 13	−3.34E+0	7 9	1.51E−5
14 3	1.93E+0	17 13	−1.58E−4	21 9	9.32E−1	25 5	3.45E−2	3 14	−1.45E−3	7 10	−1.71E−2
14 4	2.23E−1	17 14	−4.50E−5	21 10	1.12E−2	25 6	3.40E−2	4 1	1.43E−1	7 11	−8.17E−3
14 5	5.29E−2	18 1	7.57E−3	21 11	7.96E−3	25 7	4.60E+0	4 2	6.56E−1	7 12	−4.37E−2
14 6	3.00E−2	18 2	2.75E+0	21 12	−8.12E−6	25 8	9.99E−3	4 3	5.30E−1	7 13	−4.33E−4
14 7	5.39E−1	18 3	4.98E−1	21 13	−4.56E−4	25 9	1.34E−1	4 4	3.59E−4	7 14	−4.97E−1
14 8	1.05E−1	18 4	2.12E−2	21 14	−2.07E−3	25 10	1.09E−3	4 5	6.42E−2	8 1	1.38E−1
14 9	1.84E−2	18 5	2.30E+0	22 1	6.88E−4	25 11	6.17E−3	4 6	4.22E−3	8 2	2.81E−1
14 10	4.85E−5	18 6	1.31E−1	22 2	1.05E−4	25 12	9.62E−5	4 7	3.05E−5	8 3	5.59E−2
14 11	3.86E−5	18 7	1.73E−2	22 3	9.29E−4	25 13	2.80E−3	4 8	1.11E−5	8 4	4.02E−1
14 12	−5.70E−6	18 8	8.30E−1	22 4	5.81E−5	25 14	3.80E−4	4 9	−4.05E−6	8 5	1.84E−3
14 13	−3.19E−4	18 9	9.92E−2	22 5	3.77E−4		**$^3D^o$–$^3F^e$**	4 10	−5.23E−3	8 6	7.85E−3
14 14	−1.73E−3	18 10	3.15E−3	22 6	2.66E−2	1 1	−5.33E−1	4 11	−8.85E−2	8 7	6.50E−3
15 1	2.34E−1	18 11	1.21E−3	22 7	1.01E−2	1 2	−1.27E−1	4 12	−5.75E−1	8 8	2.90E−3
15 2	1.06E+0	18 12	−9.72E−4	22 8	1.98E−4	1 3	−1.81E+0	4 13	−5.85E−4	8 9	1.43E−3
15 3	3.20E−3	18 13	−6.59E−4	22 9	6.55E−3	1 4	−2.23E−1	4 14	−9.63E−3	8 10	−3.52E−3
15 4	1.19E+0	18 14	−3.74E−3	22 10	3.89E−1	1 5	−1.32E−1	5 1	5.90E−3	8 11	−6.89E−2
15 5	1.83E−2	19 1	2.54E−2	22 11	1.41E+0	1 6	−2.63E+0	5 2	2.63E−2	8 12	−5.97E−1
15 6	9.86E−1	19 2	4.67E−1	22 12	−4.96E−5	1 7	−7.56E−1	5 3	6.30E−2	8 13	−5.28E−4
15 7	1.52E−1	19 3	4.78E+0	22 13	−4.99E−7	1 8	−2.74E−1	5 4	1.67E−1	8 14	−7.62E−1
15 8	1.39E−5	19 4	4.33E−1	22 14	−8.44E−6	1 9	−5.38E−1	5 5	3.44E−2	9 1	2.19E+0
15 9	5.21E−2	19 5	7.57E−1	23 1	5.07E−2	1 10	−7.61E−3	5 6	1.46E−2	9 2	6.95E−2
15 10	8.89E−4	19 6	6.02E−2	23 2	1.64E−2	1 11	−1.35E−4	5 7	4.99E−5	9 3	5.17E−2
15 11	1.74E−5	19 7	1.38E+0	23 3	3.58E−2	1 12	−4.17E−3	5 8	8.06E−3	9 4	1.24E+0
15 12	−1.56E−3	19 8	2.39E−2	23 4	4.85E−3	1 13	−7.55E+0	5 9	3.04E−5	9 5	5.32E−2
15 13	−4.97E−3	19 9	4.70E−1	23 5	1.87E+0	1 14	−2.26E−3	5 10	−1.60E−3	9 6	3.84E−2
15 14	−1.16E−2	19 10	7.26E−3	23 6	3.40E+0	2 1	−3.66E−3	5 11	−1.39E+0	9 7	5.07E−3
16 1	5.63E−4	19 11	1.38E−3	23 7	5.62E−2	2 2	−6.23E−1	5 12	−1.36E−1	9 8	8.65E−4
16 2	2.14E−1	19 12	−5.58E−4	23 8	1.26E−1	2 3	−1.25E−1	5 13	−4.15E−5	9 9	1.01E−4
16 3	1.76E−1	19 13	−7.59E−5	23 9	3.11E−1	2 4	−4.00E−1	5 14	−3.41E−2	9 10	−2.56E−4
16 4	3.95E−3	19 14	−1.46E−5	23 10	4.37E−3	2 5	−1.11E+0	6 1	1.60E−2	9 11	−2.49E−1
16 5	3.62E−3	20 1	6.41E−6	23 11	5.04E−3	2 6	−1.55E−2	6 2	3.85E−2	9 12	−1.68E−3
16 6	9.67E−1	20 2	3.28E−3	23 12	1.61E−6	2 7	−6.74E−1	6 3	1.16E−2	9 13	−3.45E−3
16 7	7.59E−1	20 3	2.79E−3	23 13	9.05E−6	2 8	−3.83E+0	6 4	1.70E−3	9 14	−3.78E−1
16 8	1.89E−1	20 4	4.97E−3	23 14	−2.05E−3	2 9	−2.11E−2	6 5	1.42E−2	10 1	5.54E+0
16 9	5.16E−2	20 5	1.80E−2	24 1	1.86E−4	2 10	−2.67E−2	6 6	1.56E−3	10 2	8.50E−2
16 10	1.41E−6	20 6	9.45E−3	24 2	1.34E−3	2 11	−4.18E−4	6 7	1.28E−2	10 3	6.94E−4
16 11	2.26E−6	20 7	5.02E−2	24 3	1.38E−3	2 12	−7.14E−5	6 8	1.90E−4	10 4	2.90E−1
16 12	−2.69E−5	20 8	1.26E−2	24 4	2.33E−3	2 13	−6.71E−1	6 9	1.28E−5	10 5	7.19E−3
16 13	−3.92E−2	20 9	6.86E−3	24 5	3.47E−1	2 14	−9.49E−4	6 10	−4.79E+0	10 6	1.48E−1
16 14	−7.43E−3	20 10	2.93E+0	24 6	1.20E+0	3 1	−7.02E−2	6 11	−2.41E−3	10 7	4.92E−2
17 1	2.75E−3	20 11	3.44E−1	24 7	3.35E−2	3 2	−8.45E−3	6 12	−1.46E−4	10 8	9.98E−3
17 2	2.83E−2	20 12	−3.40E−4	24 8	1.10E−1	3 3	2.82E−2	6 13	1.62E−2	10 9	1.57E−3
17 3	7.29E−4	20 13	−2.07E−5	24 9	2.93E−1	3 4	−3.03E−2	6 14	−2.10E−3	10 10	−5.72E−4
17 4	2.83E−3	20 14	−6.71E−5	24 10	1.91E−3	3 5	−2.32E−1	7 1	9.19E−3	10 11	−5.94E−2
17 5	3.08E−2	21 1	2.48E−3	24 11	4.02E−4	3 6	−5.49E−1	7 2	6.23E−1	10 12	−6.67E−2
17 6	4.58E−3	21 2	2.02E−2	24 12	2.42E−5	3 7	−4.05E−1	7 3	2.47E−1	10 13	−2.30E−3
17 7	4.36E−3	21 3	5.07E−1	24 13	1.83E−4	3 8	−1.08E+0	7 4	4.65E−1	10 14	−7.17E−1
17 8	3.78E−3	21 4	9.05E−2	24 14	−4.66E−4	3 9	−1.50E+1	7 5	3.97E−2	11 1	4.11E−1
17 9	4.41E−3	21 5	1.91E+0	25 1	4.46E−3	3 10	−5.42E−2	7 6	7.18E−2	11 2	2.63E−1

S-like Fe (Fe^{10+})

i	i′	gf_L	i	i′	gf_L	i	i′	gf_L	i	i′	gf_L	i	i′	gf_L	i	i′	gf_L
11	3	1.64E−2	14	13	−2.42E−4	18	9	1.13E−1	22	5	5.14E−4			^{3}D^e–^{3}F^o	3	12	−3.51E−1
11	4	4.84E−1	14	14	−5.41E−2	18	10	5.34E−3	22	6	1.36E−7	1	1	3.51E−1	3	13	−1.87E+0
11	5	5.06E−2	15	1	1.45E−2	18	11	−9.14E−4	22	7	1.10E−2	1	2	1.34E−5	3	14	−4.51E−3
11	6	2.53E−1	15	2	1.41E+0	18	12	−1.19E−2	22	8	4.57E−5	1	3	−1.39E−1	3	15	−9.25E−8
11	7	2.92E−2	15	3	2.18E−1	18	13	−5.82E−2	22	9	1.49E−2	1	4	−6.33E−1	3	16	−2.70E−2
11	8	2.60E−2	15	4	1.48E+0	18	14	−4.98E−2	22	10	8.99E−2	1	5	−4.38E−1	3	17	−2.63E−1
11	9	4.43E−4	15	5	1.74E−2	19	1	7.58E−5	22	11	4.10E−6	1	6	−1.01E+0	3	18	−9.82E−3
11	10	1.34E−6	15	6	5.31E−2	19	2	1.80E−1	22	12	−1.74E−5	1	7	−2.38E−1	3	19	−1.91E−2
11	11	−2.46E−2	15	7	3.83E−2	19	3	7.03E−3	22	13	−4.58E−4	1	8	−7.29E−1	3	20	−6.28E−5
11	12	−2.39E−1	15	8	1.63E−1	19	4	3.15E−2	22	14	−6.89E−8	1	9	−2.35E−1	4	1	7.69E−3
11	13	−2.72E−5	15	9	1.33E−1	19	5	5.70E−4	23	1	1.28E−2	1	10	−4.45E−1	4	2	7.19E−2
11	14	−4.22E−2	15	10	2.20E−5	19	6	1.24E+0	23	2	8.54E−5	1	11	−5.33E−2	4	3	−5.25E−2
12	1	1.70E−1	15	11	−7.54E−5	19	7	3.45E−1	23	3	1.43E−1	1	12	−2.33E+0	4	4	−1.64E−1
12	2	4.05E+0	15	12	−4.51E−2	19	8	2.79E+0	23	4	9.07E−1	1	13	−3.57E−1	4	5	−3.41E−1
12	3	4.49E−2	15	13	−8.80E−3	19	9	1.17E−1	23	5	4.52E+0	1	14	−1.54E−1	4	6	−1.32E−1
12	4	2.97E−3	15	14	−2.11E−2	19	10	6.28E−5	23	6	3.03E−2	1	15	−8.73E−4	4	7	−4.79E−2
12	5	8.66E−2	16	1	8.36E−3	19	11	−8.46E−5	23	7	8.18E−2	1	16	−1.35E−1	4	8	−7.00E−1
12	6	4.76E−2	16	2	3.02E−1	19	12	−5.72E−3	23	8	1.30E−1	1	17	−1.59E−2	4	9	−2.85E−1
12	7	6.34E−2	16	3	7.86E+0	19	13	−9.98E−3	23	9	6.71E−2	1	18	−1.09E−4	4	10	−4.43E−2
12	8	1.55E−1	16	4	1.57E−1	19	14	−2.67E−3	23	10	2.33E−3	1	19	−2.29E−2	4	11	−3.38E−1
12	9	6.36E−4	16	5	1.09E+0	20	1	2.34E−2	23	11	6.69E−7	1	20	−1.13E−3	4	12	−4.13E−1
12	10	1.38E−3	16	6	1.49E−1	20	2	5.82E−3	23	12	−3.14E−4	2	1	1.58E+0	4	13	−1.08E+0
12	11	−1.35E−2	16	7	4.08E−1	20	3	4.15E−3	23	13	−5.79E−5	2	2	2.95E−2	4	14	−8.19E+0
12	12	−2.11E−2	16	8	1.03E−1	20	4	5.50E−3	23	14	−4.06E−4	2	3	−7.54E−1	4	15	−2.53E−2
12	13	−4.56E−4	16	9	1.60E−2	20	5	3.08E−2	24	1	4.61E−3	2	4	−4.08E−1	4	16	−3.48E−1
12	14	−3.42E−3	16	10	2.09E−4	20	6	2.34E−2	24	2	3.85E−4	2	5	−5.29E−1	4	17	−2.15E+0
13	1	1.06E−1	16	11	−7.29E−3	20	7	3.60E−3	24	3	7.03E−2	2	6	−1.59E−2	4	18	−4.50E−2
13	2	2.12E−1	16	12	−5.88E−3	20	8	1.02E−4	24	4	5.68E−2	2	7	−4.12E−1	4	19	−3.64E−1
13	3	8.83E−2	16	13	−1.14E−3	20	9	1.19E−2	24	5	2.77E+0	2	8	−1.23E−2	4	20	−1.24E−1
13	4	7.88E−1	16	14	−5.98E−3	20	10	4.58E−1	24	6	4.96E−1	2	9	−5.61E−1	5	1	3.41E+0
13	5	9.55E−1	17	1	1.76E−3	20	11	−3.43E−6	24	7	7.58E−1	2	10	−7.06E−1	5	2	2.29E−2
13	6	2.16E−3	17	2	9.59E−3	20	12	−5.46E−5	24	8	1.80E+0	2	11	−6.21E−1	5	3	−7.32E−2
13	7	4.29E−3	17	3	1.16E−3	20	13	−9.80E−3	24	9	5.00E−2	2	12	−7.38E−2	5	4	−3.57E−2
13	8	4.31E−2	17	4	7.21E−4	20	14	−1.24E−4	24	10	5.07E−4	2	13	−1.47E−1	5	5	−8.01E−2
13	9	1.13E−2	17	5	2.75E−2	21	1	7.98E−3	24	11	1.26E−4	2	14	−1.06E+0	5	6	−2.17E−1
13	10	5.44E−8	17	6	3.62E−3	21	2	3.55E−2	24	12	7.99E−6	2	15	−1.33E−2	5	7	−4.26E−1
13	11	−2.02E−2	17	7	2.22E−2	21	3	4.16E−2	24	13	−1.51E−5	2	16	−5.29E−1	5	8	−3.00E−1
13	12	−1.65E−1	17	8	4.76E−2	21	4	1.31E+0	24	14	−7.83E−7	2	17	−2.68E−1	5	9	−8.15E−2
13	13	−1.23E−3	17	9	3.75E−2	21	5	8.93E−1	25	1	7.26E−4	2	18	−3.94E−5	5	10	−7.34E−1
13	14	−2.28E−2	17	10	2.99E−4	21	6	7.19E−1	25	2	4.84E−2	2	19	−4.96E−2	5	11	−9.80E−1
14	1	4.52E−2	17	11	−2.10E−5	21	7	4.65E−1	25	3	5.47E−3	2	20	−4.84E−3	5	12	−1.00E+0
14	2	1.34E−1	17	12	−3.25E−3	21	8	1.48E−3	25	4	3.53E−1	3	1	3.30E−1	5	13	−1.45E−1
14	3	1.41E−1	17	13	−6.15E+0	21	9	9.50E−2	25	5	2.97E−1	3	2	9.67E−1	5	14	−6.47E−2
14	4	4.70E−1	17	14	−1.36E−3	21	10	2.41E−3	25	6	4.81E+0	3	3	−3.11E−4	5	15	−6.86E−7
14	5	2.19E−1	18	1	2.95E−2	21	11	−1.63E−3	25	7	3.32E−2	3	4	−1.37E−2	5	16	−5.49E−2
14	6	3.97E−3	18	2	4.02E−1	21	12	−3.21E−4	25	8	7.56E−1	3	5	−1.44E−1	5	17	−5.40E+0
14	7	1.58E−1	18	3	4.50E−1	21	13	−3.76E−5	25	9	1.21E−1	3	6	−7.36E−2	5	18	−7.07E−3
14	8	4.06E−2	18	4	8.77E−2	21	14	−1.61E−3	25	10	5.71E−5	3	7	−8.96E−4	5	19	−1.41E−2
14	9	2.04E−2	18	5	1.98E+0	22	1	1.25E−2	25	11	1.04E−3	3	8	−2.80E−1	5	20	−7.24E−2
14	10	2.26E−6	18	6	2.63E−3	22	2	1.49E−2	25	12	7.54E−8	3	9	−3.27E−3	6	1	4.38E+0
14	11	−8.66E−2	18	7	3.09E+0	22	3	4.47E−3	25	13	1.11E−4	3	10	−2.29E+0	6	2	3.29E−3
14	12	−2.08E−1	18	8	3.30E−1	22	4	9.27E−3	25	14	−3.98E−4	3	11	−1.11E+0	6	3	−1.68E−3

S-like Fe (Fe^{10+})

i	i'	gf_L	i	i'	gf_L	i	i'	gf_L	i	i'	gf_L	i	i'	gf_L	i	i'	gf_L
6	4	−5.35E−2	8	16	−5.04E−1	11	8	6.23E−4	13	20	−1.24E−4	3	3	1.19E+0	6	13	−3.90E−4
6	5	−8.03E−2	8	17	−1.42E+0	11	9	2.78E−4	14	1	2.89E−3	3	4	5.68E−3	6	14	−1.83E−1
6	6	−4.96E−5	8	18	−2.75E−1	11	10	9.83E−5	14	2	3.07E−3	3	5	2.01E−2	7	1	1.21E+0
6	7	−2.89E−1	8	19	−2.75E+0	11	11	6.96E−5	14	3	5.03E−1	3	6	6.86E−3	7	2	3.96E−1
6	8	−3.54E−4	8	20	−1.86E+0	11	12	−2.89E−5	14	4	4.92E−1	3	7	1.30E−3	7	3	3.50E−2
6	9	−6.80E−2	9	1	8.04E−3	11	13	−1.07E−5	14	5	1.10E+0	3	8	4.03E−4	7	4	2.84E+0
6	10	−2.72E−2	9	2	1.42E−1	11	14	−2.20E−6	14	6	2.69E−3	3	9	−8.18E−5	7	5	7.96E−3
6	11	−8.53E−1	9	3	9.07E−4	11	15	−3.44E−2	14	7	2.37E−3	3	10	−1.65E−4	7	6	2.83E−2
6	12	−7.18E−1	9	4	−2.62E−3	11	16	−1.73E−3	14	8	1.49E−2	3	11	−4.03E−1	7	7	7.67E−2
6	13	−6.92E−1	9	5	−3.14E−3	11	17	−1.19E−3	14	9	1.87E−2	3	12	−4.82E−3	7	8	1.03E−3
6	14	−8.15E−1	9	6	−1.10E−2	11	18	−8.34E+0	14	10	6.76E−2	3	13	−1.80E−3	7	9	3.34E−4
6	15	−5.14E−4	9	7	−6.40E−2	11	19	−8.67E−1	14	11	1.64E−1	3	14	−5.75E−2	7	10	−2.11E−4
6	16	−9.70E−1	9	8	−1.07E−3	11	20	−1.65E−4	14	12	3.13E−2	4	1	1.12E−1	7	11	−1.86E−1
6	17	−4.02E−3	9	9	−1.21E−1	12	1	1.89E−1	14	13	2.91E−3	4	2	4.46E−1	7	12	−3.00E−2
6	18	−2.62E−1	9	10	−1.98E−2	12	2	2.65E−3	14	14	6.55E−3	4	3	5.83E−2	7	13	−2.42E−4
6	19	−2.38E+0	9	11	−6.22E−2	12	3	3.48E−3	14	15	1.08E−4	4	4	1.42E−2	7	14	−1.10E+0
6	20	−1.45E−2	9	12	−2.12E−1	12	4	1.83E−3	14	16	1.60E−2	4	5	6.80E−2	8	1	2.04E+0
7	1	4.90E−2	9	13	−1.12E−1	12	5	3.05E−3	14	17	1.27E−2	4	6	6.72E−2	8	2	3.80E−1
7	2	6.97E+0	9	14	−4.42E−2	12	6	8.42E−3	14	18	1.38E−4	4	7	9.00E−3	8	3	1.58E−2
7	3	−1.09E−4	9	15	−1.09E−2	12	7	3.77E−4	14	19	1.01E−3	4	8	9.01E−4	8	4	4.69E−1
7	4	−4.02E−3	9	16	−1.57E−1	12	8	1.71E−4	14	20	1.78E−5	4	9	1.03E−4	8	5	7.15E−1
7	5	−1.04E−2	9	17	−9.24E−1	12	9	2.33E−3			$^3F^o$–$^3F^e$	4	10	−7.36E−4	8	6	1.35E−1
7	6	−9.28E−3	9	18	−1.48E−3	12	10	5.44E−4	1	1	−3.82E−2	4	11	−2.63E−2	8	7	9.32E−3
7	7	−7.52E−4	9	19	1.27E−1	12	11	1.27E−3	1	2	−6.09E−1	4	12	−2.37E+0	8	8	3.62E−3
7	8	−6.34E−2	9	20	−5.95E−2	12	12	5.12E−4	1	3	−3.96E−1	4	13	−6.86E−4	8	9	1.57E−3
7	9	−1.88E−2	10	1	4.99E−1	12	13	4.92E−3	1	4	−3.19E−1	4	14	−2.10E−1	8	10	−3.07E−5
7	10	−3.25E−2	10	2	2.08E−2	12	14	5.14E−4	1	5	−8.60E−1	5	1	6.72E−1	8	11	−3.53E−1
7	11	−1.70E−2	10	3	2.57E−4	12	15	2.83E−3	1	6	−1.83E+0	5	2	8.61E−2	8	12	−7.82E−2
7	12	−3.98E−2	10	4	1.08E−3	12	16	1.00E−3	1	7	−1.64E+0	5	3	8.90E−4	8	13	−1.14E−6
7	13	−2.08E−1	10	5	3.07E−4	12	17	8.29E−5	1	8	−4.97E+0	5	4	2.45E−1	8	14	−8.39E−2
7	14	−1.08E−1	10	6	1.44E−5	12	18	5.74E−5	1	9	−2.55E−2	5	5	1.49E−1	9	1	2.59E+0
7	15	−2.97E−3	10	7	2.13E−4	12	19	1.85E−5	1	10	−1.80E−1	5	6	4.99E−4	9	2	1.05E−1
7	16	−1.13E−1	10	8	1.73E−5	12	20	−3.34E−5	1	11	−8.62E−4	5	7	3.33E−3	9	3	8.13E−1
7	17	−1.34E−4	10	9	3.75E−8	13	1	6.89E−4	1	12	−1.94E−4	5	8	5.02E−2	9	4	8.20E−1
7	18	−1.75E−1	10	10	−2.63E−5	13	2	9.24E−6	1	13	−2.56E−1	5	9	1.91E−4	9	5	2.02E−3
7	19	−4.64E−1	10	11	−7.03E−5	13	3	2.43E−2	1	14	−1.88E−3	5	10	−1.39E−6	9	6	9.96E−2
7	20	−4.34E+0	10	12	−3.60E−4	13	4	1.23E−4	2	1	−3.83E−2	5	11	−6.66E−2	9	7	2.23E−2
8	1	3.42E−1	10	13	−1.08E−4	13	5	9.85E−3	2	2	−3.40E−1	5	12	−1.40E−1	9	8	5.70E−4
8	2	4.98E−1	10	14	−3.99E−4	13	6	1.27E−3	2	3	−1.38E+0	5	13	−1.98E−5	9	9	5.51E−3
8	3	4.15E−5	10	15	−6.05E+0	13	7	5.40E−1	2	4	−3.43E−4	5	14	−6.47E−1	9	10	−1.27E−5
8	4	−3.86E−3	10	16	−1.23E−1	13	8	4.74E−2	2	5	−7.21E−1	6	1	6.96E−1	9	11	−7.04E−2
8	5	−3.01E−4	10	17	2.02E−3	13	9	4.36E−2	2	6	−8.43E−1	6	2	4.55E−1	9	12	−8.88E−2
8	6	−4.13E−2	10	18	−5.90E−4	13	10	1.20E−2	2	7	−4.42E+0	6	3	9.15E−2	9	13	−3.75E−4
8	7	−8.01E−3	10	19	−9.33E−4	13	11	4.44E−2	2	8	−3.93E+0	6	4	3.13E−1	9	14	1.13E−2
8	8	−1.09E−1	10	20	−3.15E−5	13	12	2.68E−2	2	9	−3.36E+0	6	5	2.48E−1	10	1	1.39E+0
8	9	−2.47E−2	11	1	5.52E−2	13	13	3.44E−3	2	10	−4.16E−2	6	6	8.83E−2	10	2	1.19E+0
8	10	−2.49E−2	11	2	7.28E−1	13	14	2.36E−3	2	11	−6.76E−6	6	7	3.27E−2	10	3	7.42E−1
8	11	−8.34E−4	11	3	2.05E−7	13	15	4.38E−5	2	12	−1.68E−3	6	8	8.18E−4	10	4	1.70E−1
8	12	−4.54E−1	11	4	9.33E−4	13	16	1.73E−4	2	13	−2.26E−1	6	9	7.11E−3	10	5	3.74E−3
8	13	−1.74E−1	11	5	1.67E−3	13	17	1.15E−4	2	14	−5.55E−4	6	10	−8.16E−6	10	6	1.28E−1
8	14	−1.18E−1	11	6	1.44E−5	13	18	1.94E−5	3	1	6.23E−3	6	11	7.15E−1	10	7	2.19E−2
8	15	−2.77E−2	11	7	3.73E−5	13	19	1.45E−4	3	2	2.64E−1	6	12	−2.05E−2	10	8	1.03E−2

S-like Fe (Fe^{10+})

i i'	gf_L	i i'	gf_L	i i'	gf_L	i i'	gf_L	i i'	gf_L	i i'	gf_L
10 9	3.43E−2	14 5	2.32E−2	18 1	2.52E−3	2 4	−2.00E+0	11 2	7.24E−1	19 6	−1.33E−3
10 10	9.21E−9	14 6	8.63E−2	18 2	1.35E−3	2 5	−2.26E−3	11 3	9.90E−1	20 1	3.13E−3
10 11	−3.40E−2	14 7	1.20E+0	18 3	1.83E−2	2 6	−4.89E−5	11 4	8.59E−2	20 2	2.34E−1
10 12	−1.47E−1	14 8	1.07E−1	18 4	1.70E−2	3 1	1.34E+0	11 5	−3.34E−2	20 3	3.77E+0
10 13	−4.04E−4	14 9	5.68E−2	18 5	3.72E−2	3 2	1.47E−2	11 6	−1.32E−2	20 4	1.66E−2
10 14	−8.17E−2	14 10	6.67E−5	18 6	3.10E−4	3 3	2.91E−4	12 1	8.53E−2	20 5	1.11E−3
11 1	1.99E+0	14 11	−4.07E−3	18 7	5.56E−1	3 4	7.44E−4	12 2	1.56E+0	20 6	−1.49E−5
11 2	1.18E−1	14 12	−1.52E−2	18 8	3.04E−4	3 5	−7.50E−2	12 3	6.40E−2		$^3F^e$–$^3G^o$
11 3	3.13E+0	14 13	−2.36E−3	18 9	4.65E−2	3 6	−2.85E−1	12 4	4.00E−3	1 1	7.37E−3
11 4	1.08E−3	14 14	−2.29E−3	18 10	3.16E−1	4 1	2.12E−1	12 5	−3.14E−2	1 2	−4.18E−1
11 5	5.02E−1	15 1	3.42E−3	18 11	7.02E−7	4 2	8.78E−3	12 6	−4.07E−1	1 3	−2.82E+0
11 6	7.18E−2	15 2	2.55E−5	18 12	−7.83E−6	4 3	4.61E−4	13 1	5.85E−1	1 4	−9.77E−1
11 7	2.57E−1	15 3	7.40E−2	18 13	−1.11E−2	4 4	1.44E−3	13 2	9.75E−1	1 5	−5.70E−1
11 8	1.26E−2	15 4	9.37E−4	18 14	−1.61E−6	4 5	−1.19E+0	13 3	2.64E−1	1 6	−1.05E−1
11 9	2.86E−2	15 5	2.81E−1	19 1	3.37E−2	4 6	−3.13E−2	13 4	8.40E−2	1 7	−6.53E−1
11 10	4.67E−5	15 6	3.57E−4	19 2	7.47E−2	5 1	8.73E−2	13 5	−1.17E−2	1 8	−4.85E−2
11 11	−1.35E−1	15 7	6.17E−2	19 3	1.51E−1	5 2	7.43E−2	13 6	−5.33E−3	1 9	−1.75E−1
11 12	−2.57E−3	15 8	2.44E−2	19 4	1.95E−1	5 3	6.89E−4	14 1	3.68E−1	1 10	−1.88E−3
11 13	−1.55E−3	15 9	2.85E−3	19 5	2.25E−1	5 4	9.78E−3	14 2	7.33E−1	1 11	−1.58E−2
11 14	−1.16E−1	15 10	2.41E+0	19 6	7.02E−2	5 5	−9.75E−2	14 3	4.18E−1	1 12	−1.96E−3
12 1	9.52E−1	15 11	−7.32E−9	19 7	5.72E+0	5 6	−9.36E−1	14 4	4.88E−1	1 13	−5.54E−4
12 2	2.95E+0	15 12	−6.81E−5	19 8	7.02E−3	6 1	2.27E+0	14 5	−9.90E−4	2 1	2.54E−1
12 3	8.16E−1	15 13	−9.31E−2	19 9	8.53E−1	6 2	4.63E−1	14 6	−1.66E−1	2 2	−3.05E−1
12 4	4.92E−1	15 14	−6.67E−4	19 10	1.17E−2	6 3	1.20E−2	15 1	3.73E−2	2 3	−4.18E−1
12 5	4.03E−2	16 1	2.05E−3	19 11	1.25E−5	6 4	1.44E−3	15 2	6.92E−5	2 4	−1.15E−3
12 6	2.20E+0	16 2	2.49E−1	19 12	−4.65E−4	6 5	−2.34E−1	15 3	1.79E−3	2 5	−2.14E−2
12 7	5.14E−2	16 3	1.55E+0	19 13	−1.82E−3	6 6	−4.38E−1	15 4	1.34E−1	2 6	−3.28E+0
12 8	9.52E−2	16 4	4.19E−3	19 14	−8.21E−5	7 1	1.44E+0	15 5	−2.18E−5	2 7	−5.52E−2
12 9	1.44E−2	16 5	9.45E+0	20 1	1.15E−2	7 2	1.05E−1	15 6	−1.72E−3	2 8	−3.91E+0
12 10	7.80E−6	16 6	1.79E−1	20 2	1.93E−2	7 3	5.07E−3	16 1	1.19E−1	2 9	−6.34E−1
12 11	−3.82E−2	16 7	5.44E−1	20 3	4.67E−3	7 4	2.03E−2	16 2	1.46E+0	2 10	−1.64E−1
12 12	−3.17E−2	16 8	4.84E−1	20 4	1.60E−2	7 5	−5.79E−1	16 3	1.06E−1	2 11	−1.62E−2
12 13	−6.80E−4	16 9	2.16E−1	20 5	2.60E−1	7 6	−1.01E+0	16 4	4.25E+0	2 12	−1.15E−2
12 14	−8.65E−3	16 10	8.73E−2	20 6	1.93E+0	8 1	4.03E+0	16 5	−6.13E−3	2 13	−1.16E−2
13 1	3.23E−1	16 11	−2.47E−4	20 7	3.66E−2	8 2	1.47E−1	16 6	−3.36E−2	3 1	2.39E+0
13 2	3.57E+0	16 12	−9.01E−5	20 8	2.10E+0	8 3	4.86E−5	17 1	6.90E−4	3 2	−9.40E−2
13 3	3.48E−3	16 13	−4.19E−3	20 9	9.47E−3	8 4	5.50E−2	17 2	2.85E+0	3 3	−6.58E−2
13 4	2.87E−1	16 14	−1.79E−2	20 10	8.60E−6	8 5	−1.63E−1	17 3	4.21E−1	3 4	−5.12E−1
13 5	3.06E−2	17 1	1.96E−1	20 11	3.20E−5	8 6	−8.19E−1	17 4	2.76E+0	3 5	−2.58E+0
13 6	8.59E−1	17 2	1.25E−1	20 12	−6.01E−6	9 1	1.40E+0	17 5	−8.60E−3	3 6	−8.57E−2
13 7	2.04E+0	17 3	2.65E+0	20 13	−2.60E−5	9 2	4.66E−1	17 6	−1.62E−3	3 7	−9.39E−1
13 8	7.53E−2	17 4	1.80E−1	20 14	−3.54E−7	9 3	4.47E−3	18 1	6.37E−3	3 8	−6.01E−1
13 9	2.10E−2	17 5	9.08E−5		$^3F^o$–$^3G^e$	9 4	5.40E−2	18 2	1.64E−1	3 9	−2.78E+0
13 10	2.54E−4	17 6	3.47E+0	1 1	−1.13E+0	9 5	−2.76E−1	18 3	1.84E−2	3 10	−7.29E−2
13 11	−2.70E−3	17 7	9.10E−2	1 2	−3.98E+0	9 6	−7.44E−4	18 4	4.05E−2	3 11	−2.42E−2
13 12	−3.55E−3	17 8	2.89E−1	1 3	−5.52E−1	10 1	2.35E−2	18 5	−2.87E−6	3 12	−2.09E−2
13 13	−7.53E−5	17 9	2.42E+0	1 4	−3.83E+0	10 2	7.22E−1	18 6	−6.15E−4	3 13	−4.92E−4
13 14	−1.80E−2	17 10	6.08E−4	1 5	−2.39E−3	10 3	3.08E−1	19 1	5.39E−3	4 1	1.81E−1
14 1	7.85E−1	17 11	−1.04E−3	1 6	−4.20E−6	10 4	2.53E−3	19 2	1.41E+0	4 2	−6.94E−1
14 2	6.89E−2	17 12	−6.18E−3	2 1	−8.37E−2	10 5	−2.72E−4	19 3	7.05E−2	4 3	−4.74E−3
14 3	2.70E+0	17 13	−1.31E−3	2 2	−1.48E−2	10 6	−1.72E−1	19 4	4.72E−1	4 4	−1.75E+0
14 4	2.61E−3	17 14	−3.88E−5	2 3	−4.79E+0	11 1	6.23E−1	19 5	−9.80E−6	4 5	−9.36E−1

S-like Fe (Fe^{10+})

i	i'	gf_L	i	i'	gf_L	i	i'	gf_L	i	i'	gf_L	i	i'	gf_L	i	i'	gf_L
4	6	−3.64E−1	8	6	−1.61E−1	12	6	4.67E−2	3	6	−1.02E+0	12	4	3.93E−1	5	2	2.96E+0
4	7	−1.55E−1	8	7	−5.10E−2	12	7	4.53E−2	4	1	1.34E−1	12	5	−1.10E−6	5	3	8.24E−4
4	8	−5.83E−1	8	8	−1.62E+0	12	8	1.75E−2	4	2	4.96E−1	12	6	−9.76E−5	5	4	−2.14E−3
4	9	−1.18E−2	8	9	−1.72E−1	12	9	1.03E−2	4	3	1.55E−2	13	1	3.88E−4	5	5	−4.18E−1
4	10	−1.30E+0	8	10	−6.28E+0	12	10	4.01E−3	4	4	3.30E−2	13	2	1.24E+0	6	1	1.46E−2
4	11	−4.76E−1	8	11	−2.00E+0	12	11	1.72E−3	4	5	−4.08E−1	13	3	6.53E+0	6	2	7.94E+0
4	12	−4.37E−1	8	12	−2.75E−1	12	12	1.19E−3	4	6	−7.19E−1	13	4	8.55E+0	6	3	6.06E−3
4	13	−4.47E−2	8	13	−1.28E+0	12	13	−2.65E−6	5	1	2.36E+0	13	5	1.23E−4	6	4	1.03E−6
5	1	7.23E+0	9	1	2.00E−2	13	1	4.82E−2	5	2	6.34E−1	13	6	−1.82E−4	6	5	−2.09E−4
5	2	−3.35E−2	9	2	−3.39E−3	13	2	6.35E−4	5	3	1.19E−1			$^{5}\mathbf{S}^{o}$–$^{5}\mathbf{P}^{e}$	7	1	2.29E−1
5	3	−2.40E−1	9	3	−1.57E−2	13	3	2.08E−3	5	4	1.12E−1	1	1	3.56E−2	7	2	6.78E−6
5	4	−3.38E−1	9	4	−4.28E−3	13	4	2.30E−5	5	5	−6.97E−2	1	2	1.02E−2	7	3	7.49E−1
5	5	−6.40E−1	9	5	−1.48E−2	13	5	6.82E−3	5	6	−8.92E−2	1	3	−3.23E+0	7	4	3.28E+0
5	6	−1.35E−1	9	6	−3.56E−2	13	6	2.60E−5	6	1	1.28E−1	1	4	−7.76E−1	7	5	8.14E−5
5	7	−2.20E−1	9	7	−1.95E−2	13	7	5.29E−3	6	2	1.26E+0	1	5	−8.74E−3			$^{5}\mathbf{P}^{o}$–$^{5}\mathbf{D}^{e}$
5	8	−9.01E−2	9	8	−1.32E−2	13	8	1.34E−5	6	3	4.06E−3	2	1	1.24E−1	1	1	1.13E−3
5	9	−4.25E+0	9	9	−7.91E−1	13	9	5.87E−4	6	4	1.30E−4	2	2	5.78E−2	1	2	1.35E−1
5	10	−1.60E−2	9	10	−1.87E−1	13	10	5.29E−3	6	5	−1.96E−2	2	3	−1.06E−1	1	3	1.06E−2
5	11	−8.23E−2	9	11	−7.15E−2	13	11	1.09E−3	6	6	−1.58E−1	2	4	−1.55E−2	1	4	−3.53E−1
5	12	−9.23E−1	9	12	−4.10E−1	13	12	3.50E−4	7	1	4.43E+0	2	5	−1.52E−1	1	5	−8.27E−1
5	13	−9.97E−2	9	13	−5.45E+0	13	13	−2.57E−4	7	2	7.38E−2	3	1	2.87E+0	2	1	9.19E+0
6	1	1.32E+0	10	1	1.05E+0	14	1	1.12E−5	7	3	4.89E−1	3	2	3.24E−1	2	2	1.28E−2
6	2	−2.81E−2	10	2	4.52E−3	14	2	8.34E−1	7	4	1.50E−1	3	3	1.44E−3	2	3	2.69E−1
6	3	−1.35E−2	10	3	1.73E−4	14	3	7.06E−3	7	5	−3.10E−1	3	4	−3.04E−3	2	4	3.25E 1
6	4	−9.36E−9	10	4	2.60E 4	14	4	3.54E−1	7	6	−1.17E−1	3	5	−2.23E−2	2	5	−1.06E−1
6	5	−3.36E−1	10	5	4.91E−4	14	5	2.69E−1	8	1	5.75E+0	4	1	8.34E−2	3	1	9.34E−3
6	6	−6.74E−2	10	6	3.20E−6	14	6	2.84E−1	8	2	2.72E−1	4	2	3.30E+0	3	2	3.51E−1
6	7	−3.79E+0	10	7	−3.62E−4	14	7	2.26E−1	8	3	7.13E−1	4	3	2.67E−3	3	3	3.72E−7
6	8	−3.18E−1	10	8	−1.09E−4	14	8	6.33E−2	8	4	1.84E+0	4	4	5.58E−5	3	4	−9.91E−3
6	9	−6.23E−1	10	9	−1.16E−4	14	9	1.27E−2	8	5	−6.70E−2	4	5	−5.68E−3	3	5	−3.69E−1
6	10	−3.33E−2	10	10	−2.69E+0	14	10	1.33E−3	8	6	−3.23E−2			$^{5}\mathbf{P}^{o}$–$^{5}\mathbf{P}^{e}$	4	1	1.41E−2
6	11	−1.55E−3	10	11	−9.43E+0	14	11	1.10E−4	9	1	8.98E−1	1	1	3.17E−4	4	2	1.48E+0
6	12	−4.17E−2	10	12	−5.48E−4	14	12	3.16E−3	9	2	1.82E+0	1	2	3.45E−2	4	3	3.93E−2
6	13	−1.87E+0	10	13	−1.03E−3	14	13	1.60E−5	9	3	3.75E+0	1	3	−5.16E−5	4	4	−3.32E−1
7	1	1.81E+0	11	1	4.71E−6			$^{3}\mathbf{G}^{o}$–$^{3}\mathbf{G}^{e}$	9	4	4.18E−1	1	4	−2.25E−4	4	5	−3.71E−1
7	2	−1.36E−2	11	2	4.11E−1	1	1	−1.51E+0	9	5	−9.52E−3	1	5	−1.23E+0	5	1	1.38E−1
7	3	−3.85E−3	11	3	7.65E−2	1	2	−1.40E+0	9	6	−8.70E−2	2	1	2.81E−1	5	2	4.05E+0
7	4	−2.11E−1	11	4	1.96E−1	1	3	−7.15E−1	10	1	2.99E−1	2	2	1.71E−1	5	3	1.33E+0
7	5	−3.18E−2	11	5	6.62E−1	1	4	−1.32E+1	10	2	3.95E+0	2	3	−2.23E−5	5	4	−3.10E−2
7	6	−6.63E−2	11	6	2.91E−4	1	5	−2.13E−6	10	3	1.17E+0	2	4	−9.93E−3	5	5	−4.65E−1
7	7	−4.09E−2	11	7	2.21E−3	1	6	−4.89E−4	10	4	1.09E+0	2	5	−1.16E−2	6	1	1.03E−2
7	8	−2.96E−2	11	8	2.97E−3	2	1	1.59E−1	10	5	−5.66E−5	3	1	4.56E−1	6	2	3.47E+0
7	9	−8.78E 1	11	9	3.02E−4	2	2	3.31E−2	10	6	−6.96E−3	3	2	1.88E 2	6	3	1.08E+0
7	10	−7.20E−3	11	10	1.02E 5	2	3	2.58E−3	11	1	7.75E−2	3	3	−3.34E−7	6	4	4.71E−4
7	11	−4.07E−2	11	11	2.69E−6	2	4	2.00E−2	11	2	9.56E−1	3	4	−1.75E−4	6	5	−2.30E−3
7	12	−5.63E+0	11	12	−3.74E−6	2	5	−1.12E−1	11	3	3.51E−1	3	5	−8.42E−2	7	1	8.07E−1
7	13	−1.55E+0	11	13	−3.20E−5	2	6	6.14E−2	11	4	3.29E−1	4	1	6.47E+0	7	2	1.08E−3
8	1	4.61E−2	12	1	1.31E−7	3	1	4.19E−2	11	5	−5.64E−6	4	2	1.31E−1	7	3	2.56E−3
8	2	−1.89E−2	12	2	2.79E+0	3	2	1.96E−1	11	6	−3.07E−3	4	3	1.40E−8	7	4	1.40E 5
8	3	−1.77E−2	12	3	6.21E−2	3	3	2.49E−1	12	1	3.89E−2	4	4	−1.00E−3	7	5	7.05E−5
8	4	−4.19E−2	12	4	9.99E−4	3	4	3.32E−3	12	2	1.20E+0	4	5	−3.74E−1			$^{5}\mathbf{P}^{e}$–$^{5}\mathbf{D}^{o}$
8	5	−3.56E−4	12	5	3.99E−2	3	5	−1.71E+0	12	3	1.37E+0	5	1	2.60E+0	1	1	6.04E−1

S-like Fe (Fe^{10+})

i i'	gf_L	i i'	gf_L	i i'	gf_L	i i'	gf_L	i i'	gf_L	i i'	gf_L
1 2	−4.13E−1	5 2	7.42E−2	7 1	6.88E−3	5 4	−4.49E−4	4 1	1.99E+0	2 1	4.03E−3
1 3	−1.06E+0	5 3	3.68E−2	7 2	1.19E−4	6 1	4.18E+0	4 2	3.34E−1	2 2	−2.35E+0
1 4	−9.93E−1	5 4	3.32E−1	7 3	1.90E−4	6 2	4.83E−1	4 3	5.12E−2	3 1	1.21E+0
1 5	−2.17E+0	5 5	5.46E−1	7 4	−6.06E−6	6 3	−1.29E−4	4 4	3.64E−1	3 2	−1.66E−1
1 6	−1.84E+0	5 6	6.45E−4	7 5	−1.09E−4	6 4	−7.32E−3	4 5	2.10E−1	4 1	4.18E+0
1 7	−2.58E−3	5 7	1.10E−6	8 1	6.41E−2	7 1	7.17E−2	4 6	3.29E−3	4 2	−7.12E−1
1 8	−1.50E−2	5 8	1.51E−3	8 2	7.50E−1	7 2	3.79E−2	5 1	1.54E+0	5 1	8.30E+0
1 9	−8.25E−3	5 9	5.68E−4	8 3	3.06E−1	7 3	−9.25E−5	5 2	4.50E−2	5 2	−1.61E+0
1 10	−3.68E−2	5 10	−7.81E−5	8 4	−5.52E−5	7 4	−9.62E+0	5 3	9.27E−1	6 1	8.36E+0
2 1	9.13E+0		**$^5D^o$–$^5D^e$**	8 5	−1.20E−2	8 1	3.91E−2	5 4	8.89E−2	6 2	−4.41E−3
2 2	−3.37E−2	1 1	−6.71E−1	9 1	8.61E−3	8 2	8.97E+0	5 5	3.03E−1		**$^5F^e$–$^5G^o$**
2 3	−1.57E−2	1 2	−4.94E−3	9 2	1.50E−1	8 3	−1.13E−4	5 6	1.28E−2	1 1	−2.53E−1
2 4	−1.37E−2	1 3	−1.74E+1	9 3	2.69E+1	8 4	−8.62E−3		**$^5F^o$–$^5F^e$**	1 2	−2.31E+0
2 5	−5.77E−1	1 4	−2.72E−3	9 4	5.81E−4	9 1	1.26E−3	1 1	6.90E−6	1 3	−9.17E+0
2 6	−1.13E+0	1 5	−3.01E−4	9 5	−1.74E−3	9 2	4.18E+0	1 2	4.38E−2	1 4	−7.85E−3
2 7	−6.24E−2	2 1	2.21E−1	10 1	2.46E−1	9 3	2.23E−3	1 3	−3.43E−1	2 1	−4.28E−1
2 8	−5.01E+0	2 2	1.40E−2	10 2	1.12E−5	9 4	−5.87E−4	1 4	−8.18E−4	2 2	−1.82E−1
2 9	−1.14E+0	2 3	6.43E−2	10 3	6.09E−4	10 1	1.27E+0	2 1	8.58E−1	2 3	−5.47E+0
2 10	−1.54E−3	2 4	−1.19E+0	10 4	6.13E−5	10 2	2.71E−4	2 2	7.56E−1	2 4	−7.63E+0
3 1	9.15E−1	2 5	−7.77E−1	10 5	7.29E−5	10 3	3.62E−4	2 3	−9.01E−1	3 1	2.35E+0
3 2	4.41E−3	3 1	3.35E+0		**$^5D^o$–$^5F^e$**	10 4	1.69E−2	2 4	−4.29E−4	3 2	9.98E−1
3 3	3.29E−4	3 2	2.01E−1	1 1	−1.98E+0		**$^5D^e$–$^5F^o$**	3 1	8.33E−1	3 3	1.40E−2
3 4	2.30E−4	3 3	1.84E−4	1 2	−6.70E+0	1 1	−8.82E−2	3 2	5.94E−1	3 4	1.37E−2
3 5	6.05E−5	3 4	−4.82E−1	1 3	−1.59E−3	1 2	−6.41E+0	3 3	−1.53E−1	4 1	9.67E−3
3 6	−7.00E−8	3 5	−4.87E−2	1 4	−1.96E+1	1 3	−3.97E−2	3 4	−1.36E−5	4 2	3.69E−3
3 7	−1.07E+1	4 1	4.50E+0	2 1	1.33E−1	1 4	−1.78E+0	4 1	1.33E+1	4 3	6.12E−3
3 8	−1.57E−9	4 2	2.28E+0	2 2	1.52E−2	1 5	−1.55E+0	4 2	1.97E+0	4 4	3.25E−3
3 9	−2.58E−3	4 3	7.63E−2	2 3	−1.74E+0	1 6	−6.08E−2	4 3	−1.35E+0		**$^5G^o$–$^5G^e$**
3 10	−1.43E+0	4 4	−1.19E−1	2 4	−2.41E−2	2 1	−3.41E−1	4 4	−9.87E−4	1 1	6.49E−1
4 1	1.25E−1	4 5	−1.57E−1	3 1	4.31E−1	2 2	−9.58E−2	5 1	6.31E+0	1 2	−2.59E+0
4 2	6.48E−3	5 1	2.20E+0	3 2	9.61E−1	2 3	−3.19E+0	5 2	4.63E+0	2 1	2.28E+0
4 3	6.78E−3	5 2	3.60E−1	3 3	−1.22E−3	2 4	−1.91E−2	5 3	−3.48E−1	2 2	−1.76E−1
4 4	3.65E−3	5 3	5.84E−1	3 4	−1.78E−3	2 5	−9.31E−1	5 4	−1.31E−2	3 1	1.39E+0
4 5	5.01E−3	5 4	−1.80E−2	4 1	4.05E−2	2 6	−4.98E+0	6 1	1.08E−1	3 2	−7.65E−1
4 6	1.21E−2	5 5	−1.11E−1	4 2	6.16E−3	3 1	−3.69E−2	6 2	6.87E+0	4 1	1.31E+1
4 7	1.41E−3	6 1	3.76E+0	4 3	−1.19E−1	3 2	−1.18E−1	6 3	−5.56E−5	4 2	−1.75E−4
4 8	3.71E−5	6 2	1.21E+0	4 4	−3.98E−3	3 3	−2.08E−3	6 4	−2.55E−3		
4 9	−1.56E−5	6 3	9.89E−2	5 1	8.90E+0	3 4	−8.59E−2		**$^5F^o$–$^5G^e$**		
4 10	−6.31E+0	6 4	−1.12E−2	5 2	1.35E−1	3 5	−6.55E+0	1 1	4.23E−1		
5 1	1.83E−3	6 5	−3.72E−1	5 3	−1.84E−1	3 6	−5.12E+0	1 2	−9.84E−1		

Cl-like Ar (Ar^+)

Term energies relative to $3s^23p^4$ ^{3}P ionization threshold for each symmetry

i	E(Ryds)	Description
		^{2}S^e
1	−1.06932	$3s3p^6$
2	−.47647	$3s^23p^4$ ^{1}S $4s$
3	−.32777	$3s^23p^4$ ^{1}D $3d$
4	−.14356	$3s^23p^4$ ^{1}D $4d$
5	−.06655	$3s^23p^4$ ^{1}S $5s$
6	−.04694	$3s^23p^4$ ^{1}D $5d$
		^{2}S^o
1	−.55434	$3s^23p^4$ ^{3}P $4p$
2	−.29158	$3s^23p^4$ ^{3}P $5p$
3	−.17994	$3s^23p^4$ ^{3}P $6p$
4	−.12223	$3s^23p^4$ ^{3}P $7p$
5	−.08847	$3s^23p^4$ ^{3}P $8p$
6	−.06701	$3s^23p^4$ ^{3}P $9p$
		^{2}P^e
1	−.73741	$3s^23p^4$ ^{3}P $4s$
2	−.66221	$3s^23p^4$ ^{3}P $3d$
3	−.40065	$3s^23p^4$ ^{1}D $3d$
4	−.35616	$3s^23p^4$ ^{3}P $5s$
5	−.28183	$3s^23p^4$ ^{3}P $4d$
6	−.21087	$3s^23p^4$ ^{3}P $6s$
7	−.20842	$3s^23p^4$ ^{1}D $4d$
8	−.15587	$3s^23p^4$ ^{3}P $5d$
9	−.13905	$3s^23p^4$ ^{3}P $7s$
10	−.11874	$3s^23p^4$ ^{3}P $6d$
11	−.09874	$3s^23p^4$ ^{3}P $8s$
12	−.09252	$3s^23p^4$ ^{3}P $7d$
13	−.07428	$3s^23p^4$ ^{1}D $5d$
14	−.07354	$3s^23p^4$ ^{3}P $9s$
15	−.06042	$3s^23p^4$ ^{3}P $8d$
		^{2}P^o
1	−2.08339	$3s^23p^5$
2	−.55894	$3s^23p^4$ ^{3}P $4p$
3	−.44639	$3s^23p^4$ ^{1}D $4p$
4	−.28780	$3s^23p^4$ ^{3}P $5p$
5	−.27324	$3s^23p^4$ ^{1}S $4p$
6	−.18470	$3s^23p^4$ ^{3}P $6p$
7	−.16600	$3s^23p^4$ ^{1}D $5p$
8	−.12980	$3s^23p^4$ ^{1}D $4f$
9	−.11961	$3s^23p^4$ ^{3}P $7p$
10	−.08753	$3s^23p^4$ ^{3}P $8p$
11	−.06751	$3s^23p^4$ ^{3}P $9p$
12	−.05947	$3s^23p^4$ ^{1}D $6p$
		^{2}D^e
1	−.64497	$3s^23p^4$ ^{1}D $4s$
2	−.60542	$3s^23p^4$ ^{3}P $3d$
3	−.42144	$3s^23p^4$ ^{1}D $3d$
4	−.31193	$3s^23p^4$ ^{1}S $3d$
5	−.25106	$3s^23p^4$ ^{3}P $4d$
6	−.24124	$3s^23p^4$ ^{1}D $5s$
7	−.20564	$3s^23p^4$ ^{1}D $4d$
8	−.15914	$3s^23p^4$ ^{3}P $5d$
9	−.11576	$3s^23p^4$ ^{3}P $6d$
10	−.09150	$3s^23p^4$ ^{1}D $6s$
11	−.08893	$3s^23p^4$ ^{3}P $7d$
12	−.07453	$3s^23p^4$ ^{1}D $5d$
13	−.06192	$3s^23p^4$ ^{3}P $8d$
14	−.05041	$3s^23p^4$ ^{3}P $9d$
		^{2}D^o
1	−.56798	$3s^23p^4$ ^{3}P $4p$
2	−.43777	$3s^23p^4$ ^{1}D $4p$
3	−.29459	$3s^23p^4$ ^{3}P $5p$
4	−.24826	$3s^23p^4$ ^{3}P $4f$
5	−.18144	$3s^23p^4$ ^{3}P $6p$
6	−.17045	$3s^23p^4$ ^{1}D $5p$
7	−.15896	$3s^23p^4$ ^{3}P $5f$
8	−.12782	$3s^23p^4$ ^{1}D $4f$
9	−.12295	$3s^23p^4$ ^{3}P $7p$
10	−.11049	$3s^23p^4$ ^{3}P $6f$
11	−.08893	$3s^23p^4$ ^{3}P $8p$
12	−.08124	$3s^23p^4$ ^{3}P $7f$
13	−.06732	$3s^23p^4$ ^{3}P $9p$
14	−.06223	$3s^23p^4$ ^{3}P $8f$
15	−.05802	$3s^23p^4$ ^{1}D $6p$
		^{2}F^e
1	−.61961	$3s^23p^4$ ^{3}P $3d$
2	−.48689	$3s^23p^4$ ^{1}D $3d$
3	−.32002	$3s^23p^4$ ^{3}P $4d$
4	−.20188	$3s^23p^4$ ^{1}D $4d$
5	−.19203	$3s^23p^4$ ^{3}P $5d$
6	−.15991	$3s^23p^4$ ^{3}P $5g$
7	−.13081	$3s^23p^4$ ^{3}P $6d$
8	−.11103	$3s^23p^4$ ^{3}P $6g$
9	−.09395	$3s^23p^4$ ^{3}P $7d$
10	−.08157	$3s^23p^4$ ^{3}P $7g$
11	−.07320	$3s^23p^4$ ^{1}D $5d$
12	−.07000	$3s^23p^4$ ^{3}P $8d$
13	−.06245	$3s^23p^4$ ^{3}P $8g$
14	−.05495	$3s^23p^4$ ^{3}P $9d$
15	−.04935	$3s^23p^4$ ^{3}P $9g$
		^{2}F^o
1	−.46318	$3s^23p^4$ ^{1}D $4p$
2	−.25141	$3s^23p^4$ ^{3}P $4f$
3	−.17758	$3s^23p^4$ ^{1}D $5p$
4	−.16076	$3s^23p^4$ ^{3}P $5f$
5	−.12589	$3s^23p^4$ ^{1}D $4f$
6	−.11155	$3s^23p^4$ ^{3}P $6f$
7	−.08191	$3s^23p^4$ ^{3}P $7f$
8	−.06270	$3s^23p^4$ ^{3}P $8f$
9	−.06121	$3s^23p^4$ ^{1}D $6p$
10	−.04951	$3s^23p^4$ ^{3}P $9f$
		^{2}G^e
1	−.55870	$3s^23p^4$ ^{1}D $3d$
2	−.21314	$3s^23p^4$ ^{1}D $4d$
3	−.16014	$3s^23p^4$ ^{3}P $5g$
4	−.11124	$3s^23p^4$ ^{3}P $6g$
5	−.08173	$3s^23p^4$ ^{3}P $7g$
6	−.07862	$3s^23p^4$ ^{1}D $5d$
7	−.06257	$3s^23p^4$ ^{3}P $8g$
8	−.04944	$3s^23p^4$ ^{3}P $9g$
		^{2}G^o
1	−.24936	$3s^23p^4$ ^{3}P $4f$
2	−.15964	$3s^23p^4$ ^{3}P $5f$
3	−.12546	$3s^23p^4$ ^{1}D $4f$
4	−.11112	$3s^23p^4$ ^{3}P $6h$
5	−.11089	$3s^23p^4$ ^{3}P $6f$
6	−.08165	$3s^23p^4$ ^{3}P $7h$
7	−.08149	$3s^23p^4$ ^{3}P $7f$
8	−.06251	$3s^23p^4$ ^{3}P $8h$
9	−.06240	$3s^23p^4$ ^{3}P $8f$
10	−.04939	$3s^23p^4$ ^{3}P $9h$
11	−.04931	$3s^23p^4$ ^{3}P $9f$
		^{2}H^e
1	−.15996	$3s^23p^4$ ^{3}P $5g$
2	−.11108	$3s^23p^4$ ^{3}P $6g$
3	−.08163	$3s^23p^4$ ^{3}P $7i$
4	−.08161	$3s^23p^4$ ^{3}P $7g$
5	−.06250	$3s^23p^4$ ^{3}P $8i$
6	−.06248	$3s^23p^4$ ^{3}P $8g$
7	−.04938	$3s^23p^4$ ^{3}P $9i$
8	−.04937	$3s^23p^4$ ^{3}P $9g$
		^{2}H^o
1	−.12866	$3s^23p^4$ ^{1}D $4f$
2	−.11112	$3s^23p^4$ ^{3}P $6h$
3	−.08164	$3s^23p^4$ ^{3}P $7h$
4	−.06251	$3s^23p^4$ ^{3}P $8h$
5	−.04939	$3s^23p^4$ ^{3}P $9h$
		^{4}S^o
1	−.55200	$3s^23p^4$ ^{3}P $4p$
2	−.29071	$3s^23p^4$ ^{3}P $5p$
3	−.17953	$3s^23p^4$ ^{3}P $6p$
4	−.12200	$3s^23p^4$ ^{3}P $7p$
5	.08833	$3s^23p^4$ ^{3}P $8p$
6	−.06692	$3s^23p^4$ ^{3}P $9p$
		^{4}P^e
1	−.77929	$3s^23p^4$ ^{3}P $4s$
2	−.64275	$3s^23p^4$ ^{3}P $3d$
3	−.36749	$3s^23p^4$ ^{3}P $5s$
4	−.32731	$3s^23p^4$ ^{3}P $4d$
5	−.21534	$3s^23p^4$ ^{3}P $6s$
6	−.19755	$3s^23p^4$ ^{3}P $5d$
7	−.14160	$3s^23p^4$ ^{3}P $7s$
8	−.13217	$3s^23p^4$ ^{3}P $6d$
9	−.10022	$3s^23p^4$ ^{3}P $8s$
10	−.09461	$3s^23p^4$ ^{3}P $7d$
11	−.07467	$3s^23p^4$ ^{3}P $9s$
12	−.07106	$3s^23p^4$ ^{3}P $8d$
		^{4}P^o
1	−.60434	$3s^23p^4$ ^{3}P $4p$
2	−.30586	$3s^23p^4$ ^{3}P $5p$
3	−.18656	$3s^23p^4$ ^{3}P $6p$
4	−.12588	$3s^23p^4$ ^{3}P $7p$
5	−.09071	$3s^23p^4$ ^{3}P $8p$
6	−.06848	$3s^23p^4$ ^{3}P $9p$
		^{4}D^e
1	−.77298	$3s^23p^4$ ^{3}P $3d$
2	−.34986	$3s^23p^4$ ^{3}P $4d$
3	−.20688	$3s^23p^4$ ^{3}P $5d$
4	−.13700	$3s^23p^4$ ^{3}P $6d$
5	−.09745	$3s^23p^4$ ^{3}P $7d$
6	−.07288	$3s^23p^4$ ^{3}P $8d$
7	−.05656	$3s^23p^4$ ^{3}P $9d$
		^{4}D^o
1	−.58181	$3s^23p^4$ ^{3}P $4p$
2	−.29948	$3s^23p^4$ ^{3}P $5p$
3	−.24859	$3s^23p^4$ ^{3}P $4f$
4	−.18357	$3s^23p^4$ ^{3}P $6p$
5	−.15923	$3s^23p^4$ ^{3}P $5f$
6	−.12422	$3s^23p^4$ ^{3}P $7p$
7	−.11068	$3s^23p^4$ ^{3}P $6f$
8	−.08968	$3s^23p^4$ ^{3}P $8p$
9	−.08137	$3s^23p^4$ ^{3}P $7f$
10	−.06780	$3s^23p^4$ ^{3}P $9p$
11	−.06233	$3s^23p^4$ ^{3}P $8f$
		^{4}F^e
1	−.67749	$3s^23p^4$ ^{3}P $3d$
2	−.33408	$3s^23p^4$ ^{3}P $4d$
3	−.20042	$3s^23p^4$ ^{3}P $5d$
4	−.15992	$3s^23p^4$ ^{3}P $5g$
5	−.13366	$3s^23p^4$ ^{3}P $6d$
6	−.11103	$3s^23p^4$ ^{3}P $6g$
7	−.09549	$3s^23p^4$ ^{3}P $7d$
8	−.08157	$3s^23p^4$ ^{3}P $7g$
9	−.07162	$3s^23p^4$ ^{3}P $8d$
10	−.06246	$3s^23p^4$ ^{3}P $8g$
11	−.05571	$3s^23p^4$ ^{3}P $9d$
12	−.04935	$3s^23p^4$ ^{3}P $9g$
		^{4}F^o
1	−.25237	$3s^23p^4$ ^{3}P $4f$
2	.16152	$3s^23p^4$ ^{3}P $5f$
3	−.11208	$3s^23p^4$ ^{3}P $6f$
4	−.08227	$3s^23p^4$ ^{3}P $7f$
5	−.06295	$3s^23p^4$ ^{3}P $8f$

Cl-like Ar (Ar^+)

i	Energy(Ryds)	Description	i	Energy(Ryds)	Description
6	−.04970	$3s^23p^4\ ^3$P $9f$	10	−.04935	$3s^23p^4\ ^3$P $9f$
	^{4}**G**^{e}			^{4}**H**^{e}	
1	−.16015	$3s^23p^4\ ^3$P $5g$	1	−.15996	$3s^23p^4\ ^3$P $5g$
2	−.11125	$3s^23p^4\ ^3$P $6g$	2	−.11108	$3s^23p^4\ ^3$P $6g$
3	−.08174	$3s^23p^4\ ^3$P $7g$	3	−.08163	$3s^23p^4\ ^3$P $7i$
4	−.06258	$3s^23p^4\ ^3$P $8g$	4	−.08161	$3s^23p^4\ ^3$P $7g$
5	−.04944	$3s^23p^4\ ^3$P $9g$	5	−.06250	$3s^23p^4\ ^3$P $8i$
	^{4}**G**^{o}		6	−.06248	$3s^23p^4\ ^3$P $8g$
1	−.24960	$3s^23p^4\ ^3$P $4f$	7	−.04938	$3s^23p^4\ ^3$P $9i$
2	−.15982	$3s^23p^4\ ^3$P $5f$	8	−.04937	$3s^23p^4\ ^3$P $9g$
3	−.11114	$3s^23p^4\ ^3$P $6h$		^{4}**H**^{o}	
4	−.11100	$3s^23p^4\ ^3$P $6f$	1	−.11112	$3s^23p^4\ ^3$P $6h$
5	−.08166	$3s^23p^4\ ^3$P $7h$	2	−.08164	$3s^23p^4\ ^3$P $7h$
6	−.08156	$3s^23p^4\ ^3$P $7f$	3	−.06251	$3s^23p^4\ ^3$P $8h$
7	−.06253	$3s^23p^4\ ^3$P $8h$	4	−.04939	$3s^23p^4\ ^3$P $9h$
8	−.06245	$3s^23p^4\ ^3$P $8f$			
9	−.04941	$3s^23p^4\ ^3$P $9h$			

Energies in ascending order from ground state for terms with effective $n \leq 4.0$, $L \leq 4$

Term	i	E(Ryds)	Term	i	E(Ryds)	Term	i	E(Ryds)	Term	i	E(Ryds)	Term	i	E(Ryds)
^{2}**P**^{o}	1	0.00000	^{2}**F**^{e}	1	1.46378	^{2}**F**^{e}	2	1.59650	^{4}**D**^{e}	2	1.73353	^{2}**S**^{o}	2	1.79182
^{2}**S**^{e}	1	1.01407	^{2}**D**^{e}	2	1.47797	^{2}**S**^{e}	2	1.60692	^{4}**F**^{e}	2	1.74931	^{4}**S**^{o}	2	1.79268
^{4}**P**^{e}	1	1.30410	^{4}**P**^{o}	1	1.47905	^{2}**F**^{o}	1	1.62021	^{2}**S**^{e}	3	1.75562	^{2}**P**^{o}	4	1.79559
^{4}**D**^{e}	1	1.31041	^{4}**D**^{o}	1	1.50158	^{2}**P**^{o}	3	1.63700	^{4}**P**^{e}	4	1.75608	^{2}**P**^{e}	5	1.80156
^{2}**P**^{e}	1	1.34598	^{2}**D**^{o}	1	1.51540	^{2}**D**^{o}	2	1.64562	^{2}**F**^{e}	3	1.76337	^{2}**P**^{o}	5	1.81015
^{4}**F**^{e}	1	1.40590	^{2}**P**^{o}	2	1.52445	^{2}**D**^{e}	3	1.66195	^{2}**D**^{e}	4	1.77146	^{4}**F**^{o}	1	1.83102
^{2}**P**^{e}	2	1.42118	^{2}**G**^{e}	1	1.52469	^{2}**P**^{e}	3	1.68274	^{4}**P**^{o}	2	1.77753	^{2}**F**^{o}	2	1.83198
^{2}**D**^{e}	1	1.43842	^{2}**S**^{o}	1	1.52905	^{4}**P**^{e}	3	1.71590	^{4}**D**^{o}	2	1.78391	^{2}**D**^{e}	5	1.83233
^{4}**P**^{e}	2	1.44064	^{4}**S**^{o}	1	1.53139	^{2}**P**^{e}	4	1.72723	^{2}**D**^{o}	3	1.78880			

Cl-like Ar (Ar^+)

gf-values for transitions involving terms with effective $n \leq 4.0$, $L \leq 4$

$i\ i'$	gf_L	$i\ i'$	gf_L	$i\ i'$	gf_L	$i\ i'$	gf_L	$i\ i'$	gf_L	$i\ i'$	gf_L
	$^{2}S^{e}-^{2}P^{o}$	1 5	−2.75E−2	4 2	9.22E−3		$^{2}D^{e}-^{2}D^{o}$	2 2	−5.30E−1	1 2	−3.48E−4
1 1	1.11E−1	2 1	2.88E−3	4 3	−5.04E+0	1 1	−1.64E−2	2 3	−2.98E−2	2 1	−6.75E−2
1 2	−8.89E−3	2 2	−1.66E−1	5 1	1.10E−2	1 2	−3.01E+0	3 1	7.23E+0	2 2	−2.65E−2
1 3	−6.06E−2	2 3	−8.25E−2	5 2	3.83E−1	1 3	−1.35E−3	3 2	3.84E−3	3 1	3.70E+0
1 4	−3.70E−3	2 4	−1.95E−3	5 3	1.04E−2	2 1	−4.19E−2	3 3	−2.56E+0	3 2	−1.04E+1
1 5	−1.42E−3	2 5	−1.25E−2		$^{2}D^{e}-^{2}P^{o}$	2 2	−1.13E+0		$^{2}F^{e}-^{2}F^{o}$	4 1	1.49E−1
2 1	4.34E−1	3 1	1.96E+0	1 1	5.67E−1	2 3	−2.02E−3	1 1	−2.84E−2	4 2	−9.54E−2
2 2	1.67E−2	3 2	4.35E−1	1 2	−3.13E−1	3 1	9.37E−1	1 2	−3.35E−1		$^{4}D^{e}-^{4}P^{o}$
2 3	−1.56E−2	3 3	3.63E−1	1 3	−1.97E+0	3 2	4.71E−2	2 1	−1.13E−1	1 1	−1.52E+0
2 4	−5.64E−1	3 4	−1.03E−1	1 4	−7.64E−2	3 3	−2.49E−1	2 2	−6.30E−2	1 2	−6.41E−2
2 5	−1.57E+0	3 5	−9.58E−2	1 5	−1.01E−2	4 1	2.55E−1	3 1	2.43E−2	2 1	9.20E+0
3 1	8.16E−1	4 1	3.01E−1	2 1	5.67E−1	4 2	3.48E−1	3 2	−9.77E−1	2 2	−3.54E+0
3 2	1.33E−1	4 2	9.09E−1	2 2	−1.62E−1	4 3	−7.63E−2		$^{4}P^{e}-^{4}S^{o}$		$^{4}D^{e}-^{4}D^{o}$
3 3	5.06E−1	4 3	3.72E−1	2 3	−5.60E−2	5 1	5.28E−2	1 1	−1.40E+0	1 1	−6.06E−1
3 4	−3.98E−2	4 4	−2.07E+0	2 4	−3.14E−2	5 2	6.68E−1	1 2	−7.93E−3	1 2	−5.35E−2
3 5	−2.17E−2	4 5	−1.11E+0	2 5	−2.30E−2	5 3	8.31E−1	2 1	−7.28E−1	2 1	3.23E+0
	$^{2}P^{e}-^{2}S^{o}$	5 1	9.77E−1	3 1	3.48E+0		$^{2}D^{e}-^{2}F^{o}$	2 2	−9.96E−3	2 2	−1.19E+0
1 1	−7.07E−1	5 2	1.02E+0	3 2	1.57E+0	1 1	−5.05E+0	3 1	8.53E−1		$^{4}D^{e}-^{4}F^{o}$
1 2	−5.01E−6	5 3	2.54E−1	3 3	2.71E−1	1 2	−8.49E−2	3 2	−2.22E+0	1 1	−2.91E+0
2 1	−2.06E−1	5 4	1.03E−1	3 4	−3.15E−1	2 1	−1.08E−1	4 1	3.91E+0	2 1	−1.97E+1
2 2	−5.40E−3	5 5	−4.21E−2	3 5	−1.53E−1	2 2	−1.26E+0	4 2	−1.55E+0		$^{4}F^{e}-^{4}D^{o}$
3 1	1.29E+0		$^{2}P^{e}-^{2}D^{o}$	4 1	1.13E−4	3 1	1.61E−2		$^{4}P^{e}-^{4}P^{o}$	1 1	−2.31E+0
3 2	−4.01E−1	1 1	−3.59E+0	4 2	9.80E−1	3 2	−6.72E+0	1 1	−4.37E+0	1 2	−3.93E−2
4 1	4.00E−1	1 2	−6.05E−3	4 3	1.23E−2	4 1	4.75E−2	1 2	−1.78E−2	2 1	1.61E+1
4 2	−1.07E+0	1 3	−3.70E−3	4 4	−5.48E−1	4 2	−1.70E+0	2 1	−1.16E−1	2 2	−6.10E+0
5 1	5.00E−1	2 1	−1.68E−2	4 5	−1.15E−3	5 1	1.09E−3	2 2	−2.25E−3		$^{4}F^{e}-^{4}F^{o}$
5 2	1.97E−1	2 2	−1.01E−1	5 1	2.40E+0	5 2	2.41E−2	3 1	1.95E+0	1 1	−5.35E−1
	$^{2}P^{e}-^{2}P^{o}$	2 3	−1.41E−3	5 2	4.79E−1		$^{2}F^{e}-^{2}D^{o}$	3 2	−5.78E+0	2 1	−2.25E+0
1 1	1.39E+0	3 1	5.76E−2	5 3	5.34E−2	1 1	−7.96E−1	4 1	2.53E+0		
1 2	−2.03E+0	3 2	3.89E−2	5 4	1.06E+0	1 2	−3.65E−2	4 2	−7.71E−1		
1 3	−5.06E−1	3 3	−3.61E−2	5 5	9.14E−1	1 3	−8.32E−4		$^{4}P^{e}-^{4}D^{o}$		
1 4	−9.63E−4	4 1	1.73E+0			2 1	1.74E−1	1 1	−7.42E+0		

Cl-like Ca (Ca^{3+})

Term energies relative to $3s^23p^4\ ^3\mathrm{P}$ ionization threshold for each symmetry

i	E(Ryds)	Description	i	E(Ryds)	Description	i	E(Ryds)	Description	i	E(Ryds)	Description
		$^2S^e$	11	−.62990	$3s^23p^4\ ^3\mathrm{P}\ 6p$	16	−.32528	$3s^23p^4\ ^3\mathrm{P}\ 8p$	5	−.45083	$3s^23p^4\ ^3\mathrm{P}\ 6f$
1	−3.59304	$3s3p^6$	12	−.60543	$3s^23p^4\ ^1\mathrm{S}\ 5p$	17	−.31998	$3s^23p^4\ ^3\mathrm{P}\ 7f$	6	−.44432	$3s^23p^4\ ^3\mathrm{P}\ 6h$
2	−2.13431	$3s^23p^4\ ^1\mathrm{D}\ 3d$	13	−.49186	$3s^23p^4\ ^1\mathrm{D}\ 5f$	18	−.28089	$3s^23p^4\ ^1\mathrm{D}\ 6f$	7	−.33430	$3s^23p^4\ ^3\mathrm{P}\ 7f$
3	−1.80783	$3s^23p^4\ ^1\mathrm{S}\ 4s$	14	−.48207	$3s^23p^4\ ^1\mathrm{D}\ 6p$	19	−.28024	$3s^23p^4\ ^1\mathrm{D}\ 7p$	8	−.32641	$3s^23p^4\ ^3\mathrm{P}\ 7h$
4	−1.13062	$3s^23p^4\ ^1\mathrm{D}\ 4d$	15	−.43413	$3s^23p^4\ ^3\mathrm{P}\ 7p$	20	−.24955	$3s^23p^4\ ^3\mathrm{P}\ 9p$	9	−.29167	$3s^23p^4\ ^1\mathrm{D}\ 6f$
5	−.76799	$3s^23p^4\ ^1\mathrm{S}\ 5s$	16	−.39300	$3s3p^5\ ^1\mathrm{P}\ 4s$	21	−.24746	$3s^23p^4\ ^3\mathrm{P}\ 8f$	10	−.28347	$3s^23p^4\ ^1\mathrm{D}\ 6h$
6	−.62062	$3s^23p^4\ ^1\mathrm{D}\ 5d$	17	−.31919	$3s^23p^4\ ^3\mathrm{P}\ 8p$	22	−.19753	$3s^23p^4\ ^3\mathrm{P}\ 9f$	11	−.26469	$3s^23p^33d^2$
7	−.36454	$3s^23p^4\ ^1\mathrm{D}\ 6d$	18	−.29559	$3s^23p^4\ ^1\mathrm{D}\ 6f$			**$^2F^o$**	12	−.24990	$3s^23p^4\ ^3\mathrm{P}\ 8h$
8	−.33577	$3s^23p^4\ ^1\mathrm{S}\ 6s$	19	−.28106	$3s^23p^4\ ^1\mathrm{D}\ 7p$	1	−1.69261	$3s^23p^4\ ^1\mathrm{D}\ 4p$	13	−.23681	$3s^23p^4\ ^3\mathrm{P}\ 8f$
9	−.31331	$3s3p^5\ ^3\mathrm{P}\ 4p$	20	−.25159	$3s^23p^4\ ^1\mathrm{S}\ 6p$	2	−1.35793	$3s3p^5\ ^3\mathrm{P}\ 3d$	14	−.19746	$3s^23p^4\ ^3\mathrm{P}\ 9h$
10	−.21505	$3s^23p^4\ ^1\mathrm{D}\ 7d$	21	−.24381	$3s^23p^4\ ^3\mathrm{P}\ 9p$	3	−1.06512	$3s3p^5\ ^1\mathrm{P}\ 3d$	15	−.19674	$3s^23p^4\ ^3\mathrm{P}\ 9f$
11	−.12259	$3s^23p^4\ ^1\mathrm{D}\ 8d$			**$^2D^e$**	4	−.99571	$3s^23p^4\ ^3\mathrm{P}\ 4f$	16	−.16741	$3s^23p^4\ ^1\mathrm{D}\ 7f$
12	−.09757	$3s^23p^4\ ^1\mathrm{S}\ 7s$	1	−2.76918	$3s^23p^4\ ^3\mathrm{P}\ 3d$	5	−.85483	$3s^23p^4\ ^1\mathrm{D}\ 5p$	17	−.16555	$3s^23p^4\ ^1\mathrm{D}\ 7h$
13	−.06063	$3s^23p^4\ ^1\mathrm{D}\ 9d$	2	−2.33194	$3s^23p^4\ ^1\mathrm{D}\ 3d$	6	−.82921	$3s^23p^4\ ^1\mathrm{D}\ 4f$			**$^2H^e$**
14	−.03891	$3s3p^5\ ^1\mathrm{P}\ 4p$	3	−2.05420	$3s^23p^4\ ^1\mathrm{S}\ 3d$	7	−.64491	$3s^23p^4\ ^3\mathrm{P}\ 5f$	1	−.63962	$3s^23p^4\ ^3\mathrm{P}\ 5g$
		$^2S^o$	4	−1.99143	$3s^23p^4\ ^1\mathrm{D}\ 4s$	8	−.60947	$3s^23p^4\ ^1\mathrm{S}\ 4f$	2	−.47761	$3s^23p^4\ ^1\mathrm{D}\ 5g$
1	−1.78727	$3s^23p^4\ ^3\mathrm{P}\ 4p$	5	−1.24562	$3s^23p^4\ ^3\mathrm{P}\ 4d$	9	−.48925	$3s^23p^4\ ^1\mathrm{D}\ 5f$	3	−.44425	$3s^23p^4\ ^3\mathrm{P}\ 6g$
2	−.99312	$3s^23p^4\ ^3\mathrm{P}\ 5p$	6	−1.11138	$3s^23p^4\ ^1\mathrm{D}\ 4d$	10	−.47980	$3s^23p^4\ ^1\mathrm{D}\ 6p$	4	−.32659	$3s^23p^4\ ^3\mathrm{P}\ 7i$
3	−.63406	$3s^23p^4\ ^3\mathrm{P}\ 6p$	7	−.99489	$3s^23p^4\ ^1\mathrm{D}\ 5s$	11	−.44863	$3s^23p^4\ ^3\mathrm{P}\ 6f$	5	−.32637	$3s^23p^4\ ^3\mathrm{P}\ 7g$
4	−.44038	$3s^23p^4\ ^3\mathrm{P}\ 7p$	8	−.90616	$3s^23p^4\ ^1\mathrm{S}\ 4d$	12	−.38694	$3s^23p^33d^2$	6	−.28253	$3s^23p^4\ ^1\mathrm{D}\ 6g$
5	−.32377	$3s^23p^4\ ^3\mathrm{P}\ 8p$	9	−.75893	$3s^23p^4\ ^3\mathrm{P}\ 5d$	13	−.33024	$3s^23p^4\ ^3\mathrm{P}\ 7f$	7	−.25007	$3s^23p^4\ ^3\mathrm{P}\ 8i$
6	−.24809	$3s^23p^4\ ^3\mathrm{P}\ 9p$	10	−.61630	$3s^23p^4\ ^1\mathrm{D}\ 5d$	14	−.28574	$3s^23p^4\ ^1\mathrm{D}\ 7p$	8	−.24988	$3s^23p^4\ ^3\mathrm{P}\ 8g$
		$^2P^e$	11	−.55527	$3s^23p^4\ ^1\mathrm{D}\ 6s$	15	−.28375	$3s^23p^4\ ^1\mathrm{D}\ 6h$	9	−.19759	$3s^23p^4\ ^3\mathrm{P}\ 9i$
1	−2.85852	$3s^23p^4\ ^3\mathrm{P}\ 4s$	12	−.51199	$3s^23p^4\ ^3\mathrm{P}\ 6d$	16	−.27918	$3s^23p^4\ ^1\mathrm{D}\ 6f$	10	−.19744	$3s^23p^4\ ^3\mathrm{P}\ 9g$
2	−2.17262	$3s^23p^4\ ^3\mathrm{P}\ 3d$	13	−.48237	$3s^23p^4\ ^1\mathrm{D}\ 5g$	17	−.25551	$3s^23p^4\ ^1\mathrm{S}\ 5f$	11	−.16557	$3s^23p^4\ ^1\mathrm{D}\ 7i$
3	−2.03018	$3s^23p^4\ ^1\mathrm{D}\ 3d$	14	−.41229	$3s3p^5\ ^3\mathrm{P}\ 4p$	18	−.25140	$3s^23p^4\ ^3\mathrm{P}\ 8f$	12	−.16496	$3s^23p^4\ ^1\mathrm{D}\ 7g$
4	−1.25982	$3s^23p^4\ ^3\mathrm{P}\ 5s$	15	−.37896	$3s^23p^4\ ^1\mathrm{S}\ 5d$	19	−.20036	$3s^23p^4\ ^3\mathrm{P}\ 9f$			**$^2H^o$**
5	−1.15382	$3s^23p^4\ ^3\mathrm{P}\ 4d$	16	−.36969	$3s^23p^4\ ^3\mathrm{P}\ 7d$	20	−.16712	$3s^23p^4\ ^1\mathrm{D}\ 8p$	1	−.85496	$3s^23p^4\ ^1\mathrm{D}\ 4f$
6	−1.10338	$3s^23p^4\ ^1\mathrm{D}\ 4d$	17	−.35313	$3s^23p^4\ ^1\mathrm{D}\ 6d$	21	−.16582	$3s^23p^4\ ^1\mathrm{D}\ 7h$	2	−.49210	$3s^23p^4\ ^1\mathrm{D}\ 5f$
7	−.76218	$3s^23p^4\ ^3\mathrm{P}\ 6s$	18	−.32501	$3s^23p^4\ ^1\mathrm{D}\ 7s$	22	−.16454	$3s^23p^4\ ^1\mathrm{D}\ 7f$	3	−.44463	$3s^23p^4\ ^3\mathrm{P}\ 6h$
8	−.70916	$3s^23p^4\ ^3\mathrm{P}\ 5d$	19	−.28588	$3s^23p^4\ ^1\mathrm{D}\ 6g$			**$^2G^e$**	4	−.33681	$3s^23p^33d^2$
9	−.62217	$3s^23p^4\ ^1\mathrm{D}\ 5d$	20	−.27813	$3s^23p^4\ ^3\mathrm{P}\ 8d$	1	−2.70418	$3s^23p^4\ ^1\mathrm{D}\ 3d$	5	−.32673	$3s^23p^4\ ^3\mathrm{P}\ 7h$
10	−.51247	$3s^23p^4\ ^3\mathrm{P}\ 7s$	21	−.21913	$3s^23p^4\ ^3\mathrm{P}\ 9d$	2	−1.15546	$3s^23p^4\ ^1\mathrm{D}\ 4d$	6	−.28844	$3s^23p^4\ ^1\mathrm{D}\ 6f$
11	−.48280	$3s^23p^4\ ^3\mathrm{P}\ 6d$	22	−.21090	$3s^23p^4\ ^1\mathrm{D}\ 7d$	3	−.64223	$3s^23p^4\ ^3\mathrm{P}\ 5g$	7	−.28331	$3s^23p^4\ ^1\mathrm{D}\ 6h$
12	−.39077	$3s^23p^4\ ^1\mathrm{D}\ 6d$	23	−.19154	$3s^23p^4\ ^1\mathrm{D}\ 8s$	4	−.62634	$3s^23p^4\ ^1\mathrm{D}\ 5d$	8	−.25017	$3s^23p^4\ ^3\mathrm{P}\ 8h$
13	−.37082	$3s^23p^4\ ^3\mathrm{P}\ 8s$			**$^2D^o$**	5	−.47779	$3s^23p^4\ ^1\mathrm{D}\ 5g$	9	−.19767	$3s^23p^4\ ^3\mathrm{P}\ 9h$
14	−.35324	$3s^23p^4\ ^3\mathrm{P}\ 7d$	1	−1.81612	$3s^23p^4\ ^3\mathrm{P}\ 4p$	6	−.44611	$3s^23p^4\ ^3\mathrm{P}\ 6g$	10	−.17145	$3s^23p^4\ ^1\mathrm{D}\ 7f$
15	−.34192	$3s3p^5\ ^3\mathrm{P}\ 4p$	2	−1.64009	$3s^23p^4\ ^1\mathrm{D}\ 4p$	7	−.36490	$3s^23p^4\ ^1\mathrm{D}\ 6d$	11	−.16540	$3s^23p^4\ ^1\mathrm{D}\ 7h$
16	−.27813	$3s^23p^4\ ^3\mathrm{P}\ 9s$	3	−1.34500	$3s3p^5\ ^3\mathrm{P}\ 3d$	8	−.32767	$3s^23p^4\ ^3\mathrm{P}\ 7g$			**$^4S^e$**
17	−.26573	$3s^23p^4\ ^3\mathrm{P}\ 8d$	4	−1.00226	$3s^23p^4\ ^3\mathrm{P}\ 4f$	9	−.28286	$3s^23p^4\ ^1\mathrm{D}\ 6g$	1	−.50954	$3s3p^5\ ^3\mathrm{P}\ 4p$
		$^2P^o$	5	−1.00070	$3s^23p^4\ ^3\mathrm{P}\ 5p$	10	−.25128	$3s^23p^4\ ^1\mathrm{S}\ 5g$			**$^4S^o$**
1	−4.99190	$3s^23p^5$	6	−.90549	$3s3p^5\ ^1\mathrm{P}\ 3d$	11	−.25080	$3s^23p^4\ ^3\mathrm{P}\ 8g$	1	−1.78337	$3s^23p^4\ ^3\mathrm{P}\ 4p$
2	−1.81606	$3s^23p^4\ ^3\mathrm{P}\ 4p$	7	−.83920	$3s^23p^4\ ^1\mathrm{D}\ 5p$	12	−.21542	$3s^23p^4\ ^1\mathrm{D}\ 7d$	2	−.99203	$3s^23p^4\ ^3\mathrm{P}\ 5p$
3	−1.61932	$3s^23p^4\ ^1\mathrm{D}\ 4p$	8	−.82491	$3s^23p^4\ ^1\mathrm{D}\ 4f$	13	−.19811	$3s^23p^4\ ^3\mathrm{P}\ 9g$	3	−.63371	$3s^23p^4\ ^3\mathrm{P}\ 6p$
4	−1.44058	$3s^23p^4\ ^1\mathrm{S}\ 4p$	9	−.64124	$3s^23p^4\ ^3\mathrm{P}\ 5f$	14	−.16558	$3s^23p^4\ ^1\mathrm{D}\ 7i$	4	−.44030	$3s^23p^4\ ^3\mathrm{P}\ 7p$
5	−1.17600	$3s3p^5\ ^3\mathrm{P}\ 3d$	10	−.63433	$3s^23p^4\ ^3\mathrm{P}\ 6p$	15	−.16505	$3s^23p^4\ ^1\mathrm{D}\ 7g$	5	−.32380	$3s^23p^4\ ^3\mathrm{P}\ 8p$
6	−.98083	$3s^23p^4\ ^3\mathrm{P}\ 5p$	11	−.48844	$3s^23p^4\ ^1\mathrm{D}\ 5f$			**$^2G^o$**	6	−.24817	$3s^23p^4\ ^3\mathrm{P}\ 9p$
7	−.86486	$3s^23p^4\ ^1\mathrm{D}\ 4f$	12	−.47416	$3s^23p^4\ ^1\mathrm{D}\ 6p$	1	−1.00151	$3s^23p^4\ ^3\mathrm{P}\ 4f$			**$^4P^e$**
8	−.85167	$3s^23p^4\ ^1\mathrm{D}\ 5p$	13	−.44821	$3s^23p^4\ ^3\mathrm{P}\ 6f$	2	−.83260	$3s^23p^4\ ^1\mathrm{D}\ 4f$	1	−2.79462	$3s^23p^4\ ^3\mathrm{P}\ 3d$
9	−.74355	$3s3p^5\ ^3\mathrm{P}\ 4s$	14	−.44182	$3s^23p^4\ ^3\mathrm{P}\ 7p$	3	−.64549	$3s^23p^4\ ^3\mathrm{P}\ 5f$	2	−2.22740	$3s^23p^4\ ^3\mathrm{P}\ 4s$
10	−.68469	$3s3p^5\ ^1\mathrm{P}\ 3d$	15	−.36734	$3s^23p^33d^2$	4	−.48107	$3s^23p^4\ ^1\mathrm{D}\ 5f$	3	−1.28905	$3s^23p^4\ ^3\mathrm{P}\ 4d$

Cl-like Ca (Ca^{3+})

i	E(Ryds)	Description
4	−1.16224	$3s^23p^4\ ^3\mathrm{P}\ 5s$
5	−.77843	$3s^23p^4\ ^3\mathrm{P}\ 5d$
6	−.71832	$3s^23p^4\ ^3\mathrm{P}\ 6s$
7	−.52302	$3s^23p^4\ ^3\mathrm{P}\ 6d$
8	−.48865	$3s^23p^4\ ^3\mathrm{P}\ 7s$
9	−.40828	$3s3p^5\ ^3\mathrm{P}\ 4p$
10	−.36865	$3s^23p^4\ ^3\mathrm{P}\ 7d$
11	−.35274	$3s^23p^4\ ^3\mathrm{P}\ 8s$
12	−.27985	$3s^23p^4\ ^3\mathrm{P}\ 8d$
13	−.26761	$3s^23p^4\ ^3\mathrm{P}\ 9s$
14	−.21845	$3s^23p^4\ ^3\mathrm{P}\ 9d$
	$^4\mathrm{P}^o$	
1	−1.89224	$3s^23p^4\ ^3\mathrm{P}\ 4p$
2	−1.73119	$3s3p^5\ ^3\mathrm{P}\ 3d$
3	−1.03049	$3s^23p^4\ ^3\mathrm{P}\ 5p$
4	−.81160	$3s3p^5\ ^3\mathrm{P}\ 4s$
5	−.64647	$3s^23p^4\ ^3\mathrm{P}\ 6p$
6	−.56548	$3s^23p^33d^2$
7	−.44843	$3s^23p^4\ ^3\mathrm{P}\ 7p$
8	−.32890	$3s^23p^4\ ^3\mathrm{P}\ 8p$
9	−.25152	$3s^23p^4\ ^3\mathrm{P}\ 9p$
	$^4\mathrm{D}^e$	
1	−3.02915	$3s^23p^4\ ^3\mathrm{P}\ 3d$
2	−1.34052	$3s^23p^4\ ^3\mathrm{P}\ 4d$
3	−.79865	$3s^23p^4\ ^3\mathrm{P}\ 5d$
4	−.53650	$3s^23p^4\ ^3\mathrm{P}\ 6d$
5	−.44474	$3s3p^5\ ^3\mathrm{P}\ 4p$
6	−.37435	$3s^23p^4\ ^3\mathrm{P}\ 7d$
7	−.28320	$3s^23p^4\ ^3\mathrm{P}\ 8d$
8	−.22079	$3s^23p^4\ ^3\mathrm{P}\ 9d$
	$^4\mathrm{D}^o$	
1	−1.84534	$3s^23p^4\ ^3\mathrm{P}\ 4p$
2	−1.43155	$3s3p^5\ ^3\mathrm{P}\ 3d$
3	−1.01430	$3s^23p^4\ ^3\mathrm{P}\ 5p$
4	−.99232	$3s^23p^4\ ^3\mathrm{P}\ 4f$
5	−.65385	$3s^23p^4\ ^3\mathrm{P}\ 5f$
6	−.64017	$3s^23p^4\ ^3\mathrm{P}\ 6p$
7	−.57375	$3s^23p^33d^2$
8	−.44588	$3s^23p^4\ ^3\mathrm{P}\ 7p$
9	−.44234	$3s^23p^4\ ^3\mathrm{P}\ 6f$
10	−.32826	$3s^23p^4\ ^3\mathrm{P}\ 7f$
11	−.32606	$3s^23p^4\ ^3\mathrm{P}\ 8p$
12	−.25181	$3s^23p^4\ ^3\mathrm{P}\ 8f$
13	−.24983	$3s^23p^4\ ^3\mathrm{P}\ 9p$
14	−.19935	$3s^23p^4\ ^3\mathrm{P}\ 9f$
	$^4\mathrm{F}^e$	
1	−2.74512	$3s^23p^4\ ^3\mathrm{P}\ 3d$
2	−2.54895	$3s^23p^4\ ^1\mathrm{D}\ 3d$
3	−1.28732	$3s^23p^4\ ^3\mathrm{P}\ 4d$
4	−1.12006	$3s^23p^4\ ^1\mathrm{D}\ 4d$
5	−.77597	$3s^23p^4\ ^3\mathrm{P}\ 5d$
6	−.63882	$3s^23p^4\ ^3\mathrm{P}\ 5g$
7	−.61321	$3s^23p^4\ ^1\mathrm{D}\ 5d$
8	−.51988	$3s^23p^4\ ^3\mathrm{P}\ 6d$
9	−.47958	$3s^23p^4\ ^1\mathrm{D}\ 5g$
10	−.44367	$3s^23p^4\ ^3\mathrm{P}\ 6g$
11	−.37300	$3s^23p^4\ ^3\mathrm{P}\ 7d$
12	−.35786	$3s^23p^4\ ^1\mathrm{D}\ 6d$
13	−.32605	$3s^23p^4\ ^3\mathrm{P}\ 7g$
14	−.28397	$3s^23p^4\ ^1\mathrm{D}\ 6g$
15	−.28051	$3s^23p^4\ ^3\mathrm{P}\ 8d$
16	−.24969	$3s^23p^4\ ^3\mathrm{P}\ 8g$
17	−.21876	$3s^23p^4\ ^3\mathrm{P}\ 9d$
18	−.21123	$3s^23p^4\ ^1\mathrm{D}\ 7d$
19	−.19732	$3s^23p^4\ ^3\mathrm{P}\ 9g$
	$^4\mathrm{F}^o$	
1	−1.57961	$3s3p^5\ ^3\mathrm{P}\ 3d$
2	−1.01917	$3s^23p^4\ ^3\mathrm{P}\ 4f$
3	−.65501	$3s^23p^4\ ^3\mathrm{P}\ 5f$
4	−.51411	$3s^23p^33d^2$
5	−.45650	$3s^23p^4\ ^3\mathrm{P}\ 6f$
6	−.33768	$3s^23p^4\ ^3\mathrm{P}\ 7f$
7	−.26195	$3s^23p^4\ ^3\mathrm{P}\ 8f$
8	−.21320	$3s^23p^4\ ^3\mathrm{P}\ 9f$
	$^4\mathrm{G}^e$	
1	−.64255	$3s^23p^4\ ^3\mathrm{P}\ 5g$
2	−.44642	$3s^23p^4\ ^3\mathrm{P}\ 6g$
3	−.32793	$3s^23p^4\ ^3\mathrm{P}\ 7g$
4	−.25100	$3s^23p^4\ ^3\mathrm{P}\ 8g$
5	−.19826	$3s^23p^4\ ^3\mathrm{P}\ 9g$
	$^4\mathrm{G}^o$	
1	−1.00518	$3s^23p^4\ ^3\mathrm{P}\ 4f$
2	−.64832	$3s^23p^4\ ^3\mathrm{P}\ 5f$
3	−.53251	$3s^23p^33d^2$
4	−.45037	$3s^23p^4\ ^3\mathrm{P}\ 6f$
5	−.44432	$3s^23p^4\ ^3\mathrm{P}\ 6h$
6	−.33296	$3s^23p^4\ ^3\mathrm{P}\ 7f$
7	−.32641	$3s^23p^4\ ^3\mathrm{P}\ 7h$
8	−.27811	$3s^23p^33d^2$
9	−.25093	$3s^23p^4\ ^3\mathrm{P}\ 8f$
10	−.24987	$3s^23p^4\ ^3\mathrm{P}\ 8h$
11	−.19988	$3s^23p^4\ ^3\mathrm{P}\ 9f$
12	−.19745	$3s^23p^4\ ^3\mathrm{P}\ 9h$
	$^4\mathrm{H}^e$	
1	−.63970	$3s^23p^4\ ^3\mathrm{P}\ 5g$
2	−.44433	$3s^23p^4\ ^3\mathrm{P}\ 6g$
3	−.32661	$3s^23p^4\ ^3\mathrm{P}\ 7i$
4	−.32642	$3s^23p^4\ ^3\mathrm{P}\ 7g$
5	−.25010	$3s^23p^4\ ^3\mathrm{P}\ 8g$
6	−.24991	$3s^23p^4\ ^3\mathrm{P}\ 8i$
7	−.19762	$3s^23p^4\ ^3\mathrm{P}\ 9g$
8	−.19746	$3s^23p^4\ ^3\mathrm{P}\ 9i$
	$^4\mathrm{H}^o$	
1	−.44463	$3s^23p^4\ ^3\mathrm{P}\ 6h$
2	−.38911	$3s^23p^33d^2$
3	−.32674	$3s^23p^4\ ^3\mathrm{P}\ 7h$
4	−.25018	$3s^23p^4\ ^3\mathrm{P}\ 8h$
5	−.19767	$3s^23p^4\ ^3\mathrm{P}\ 9h$

Energies in ascending order from ground state for terms with effective $n \leq 4.0$, $L \leq 4$

Term	i	E(Ryds)	Term	i	E(Ryds)	Term	i	E(Ryds)	Term	i	E(Ryds)	Term	i	E(Ryds)
$^2\mathrm{P}^o$	1	0.00000	$^4\mathrm{P}^e$	2	2.76450	$^2\mathrm{S}^o$	1	3.20463	$^4\mathrm{D}^e$	2	3.65138	$^4\mathrm{F}^e$	4	3.87184
$^2\mathrm{S}^e$	1	1.39886	$^2\mathrm{P}^e$	2	2.81928	$^4\mathrm{S}^o$	1	3.20853	$^4\mathrm{F}^e$	2	3.68711	$^2\mathrm{D}^e$	6	3.88052
$^4\mathrm{D}^e$	1	1.96275	$^2\mathrm{S}^e$	2	2.85759	$^4\mathrm{P}^o$	2	3.26071	$^4\mathrm{P}^e$	3	3.70285	$^2\mathrm{P}^e$	6	3.88852
$^4\mathrm{F}^e$	1	2.13215	$^2\mathrm{D}^e$	3	2.93770	$^2\mathrm{F}^o$	1	3.29929	$^4\mathrm{F}^e$	3	3.70458	$^2\mathrm{F}^o$	3	3.92678
$^2\mathrm{P}^e$	1	2.13338	$^2\mathrm{P}^e$	3	2.96172	$^2\mathrm{D}^o$	2	3.35181	$^2\mathrm{P}^e$	4	3.73208	$^4\mathrm{P}^o$	3	3.96141
$^4\mathrm{P}^e$	1	2.19728	$^2\mathrm{D}^e$	4	3.00047	$^2\mathrm{P}^o$	3	3.37258	$^2\mathrm{D}^e$	5	3.74628	$^4\mathrm{F}^o$	2	3.97273
$^2\mathrm{D}^e$	1	2.22272	$^4\mathrm{P}^o$	1	3.09966	$^4\mathrm{F}^o$	1	3.41229	$^2\mathrm{P}^o$	5	3.81590	$^4\mathrm{D}^o$	3	3.97760
$^4\mathrm{F}^e$	1	2.24678	$^4\mathrm{D}^o$	1	3.14656	$^2\mathrm{P}^o$	4	3.55132	$^4\mathrm{P}^e$	4	3.82966	$^4\mathrm{G}^o$	1	3.98672
$^2\mathrm{G}^e$	1	2.28772	$^2\mathrm{D}^o$	1	3.17578	$^4\mathrm{D}^o$	2	3.56035	$^2\mathrm{G}^e$	2	3.83644	$^2\mathrm{D}^o$	4	3.98964
$^4\mathrm{F}^e$	2	2.44295	$^2\mathrm{P}^o$	2	3.17584	$^2\mathrm{F}^o$	2	3.63397	$^2\mathrm{P}^e$	5	3.83808	$^2\mathrm{G}^o$	1	3.99039
$^2\mathrm{D}^o$	2	2.65990	$^2\mathrm{S}^e$	3	3.18407	$^2\mathrm{D}^o$	3	3.64600	$^2\mathrm{S}^e$	4	3.86128	$^2\mathrm{D}^o$	5	3.99120

Cl-like Ca (Ca^{3+})

gf-values for transitions involving terms with effective $n \leq 4.0$, $L \leq 4$

i i'	gf_L
	$\mathbf{^2S^e–^2P^o}$
1 1	1.78E−1
1 2	−1.61E−2
1 3	−7.58E−2
1 4	−2.42E−2
1 5	−2.25E−1
2 1	2.89E+0
2 2	−4.98E−2
2 3	−1.40E−1
2 4	−6.35E−2
2 5	−7.53E−3
3 1	9.92E−3
3 2	5.53E−4
3 3	−2.27E−3
3 4	−1.94E+0
3 5	−1.54E−2
4 1	4.29E−2
4 2	4.33E−1
4 3	8.73E−1
4 4	2.07E−2
4 5	8.51E−4
	$\mathbf{^2P^e–^2S^o}$
1 1	−2.28E−1
2 1	−4.24E−1
3 1	−3.52E−1
4 1	2.01E+0
5 1	4.65E−2
6 1	5.33E−1
	$\mathbf{^2P^e–^2P^o}$
1 1	5.56E−3
1 2	−3.08E−1
1 3	−4.21E−2
1 4	−1.10E−2
1 5	−2.68E−1
2 1	3.96E+0
2 2	−1.32E+0
2 3	−1.35E+0
2 4	−1.95E−2
2 5	−1.46E−2
3 1	4.74E+0
3 2	−7.01E−2
3 3	−7.99E−2
3 4	−1.49E−2
3 5	−1.52E−6
4 1	8.78E−2
4 2	1.88E+0
4 3	2.70E−2
4 4	1.82E−2
4 5	−5.25E−3
5 1	3.55E−1
5 2	2.09E−1
5 3	2.09E+0
5 4	4.96E−4
5 5	4.53E−3
6 1	1.37E−2
6 2	6.43E−1
6 3	1.06E+0
6 4	3.86E−2
6 5	1.53E−3
	$\mathbf{^2P^e–^2D^o}$
1 1	−1.35E−2
1 2	−1.32E−1
1 3	−1.90E−1
1 4	−1.29E−1
1 5	−3.07E−1
2 1	−2.90E+0
2 2	−3.92E−2
2 3	−2.80E−2
2 4	−2.45E−3
2 5	−1.86E−1
3 1	−3.25E−1
3 2	−5.36E−2
3 3	−5.22E−3
3 4	−4.81E−1
3 5	−6.13E−1
4 1	7.85E−2
4 2	1.36E−1
4 3	1.33E−3
4 4	−1.38E+0
4 5	−2.95E+0
5 1	1.05E+0
5 2	2.99E−1
5 3	3.18E−2
5 4	−2.12E+0
5 5	−1.12E+0
6 1	4.66E−1
6 2	5.77E−1
6 3	5.67E−3
6 4	−1.25E+0
6 5	−2.56E−2
	$\mathbf{^2D^e–^2P^o}$
1 1	4.30E−2
1 2	−4.63E−1
1 3	−6.93E−3
1 4	−4.67E−2
1 5	−3.25E−1
2 1	1.39E−1
2 2	−5.78E−3
2 3	−1.31E−1
2 4	−8.73E−1
2 5	−8.73E−3
3 1	5.22E+0
3 2	−2.31E−1
3 3	−2.09E+0
3 4	−4.08E−3
3 5	−1.09E−3
4 1	9.95E+0
4 2	−3.11E−1
4 3	−3.01E−2
4 4	−2.42E−1
4 5	−6.56E−4
5 1	4.13E−1
5 2	3.95E+0
5 3	7.16E−1
5 4	2.50E−4
5 5	−1.20E−2
6 1	2.78E−1
6 2	4.93E−2
6 3	2.61E+0
6 4	4.23E−3
6 5	3.62E−3
	$\mathbf{^2D^e–^2D^o}$
1 1	−8.35E−2
1 2	−5.21E−1
1 3	−6.29E−2
1 4	−6.37E−2
1 5	−7.71E−4
2 1	−1.58E−2
2 2	−1.10E−1
2 3	−7.49E−5
2 4	−5.60E−3
2 5	−3.34E−2
3 1	−2.65E−3
3 2	−2.62E+0
3 3	−2.02E−2
3 4	−1.58E−2
3 5	−1.55E−2
4 1	−1.35E−1
4 2	−8.99E−1
4 3	−6.52E−5
4 4	−5.61E−2
4 5	−2.22E−1
5 1	1.39E+0
5 2	7.42E−1
5 3	2.42E−3
5 4	−6.86E−2
5 5	−1.26E+0
6 1	3.08E−1
6 2	3.54E+0
6 3	1.88E−3
6 4	−3.55E−2
6 5	−5.45E−2
	$\mathbf{^2D^e–^2F^o}$
1 1	−5.26E−2
1 2	−8.80E−2
1 3	−1.13E−2
2 1	−2.18E−2
2 2	−1.66E−2
2 3	−3.03E−4
3 1	−4.16E+0
3 2	−5.49E−2
3 3	−2.79E−2
4 1	−4.50E−1
4 2	−1.26E−3
4 3	−5.43E−1
5 1	5.65E−4
5 2	7.95E−4
5 3	−9.98E−1
6 1	3.55E−1
6 2	1.91E−2
6 3	−2.19E−3
	$\mathbf{^4P^e–^4S^o}$
1 1	−8.92E−1
2 1	−1.44E+0
3 1	5.03E+0
4 1	4.17E−1
	$\mathbf{^4P^e–^4P^o}$
1 1	−4.47E−1
1 2	−1.18E−1
1 3	−3.47E−2
2 1	−3.67E+0
2 2	−6.53E−2
2 3	−2.97E−1
3 1	2.87E+0
3 2	3.71E−2
3 3	−8.46E−1
4 1	2.21E+0
4 2	3.46E−2
4 3	−5.32E+0
	$\mathbf{^4P^e–^4D^o}$
1 1	−2.77E−2
1 2	−5.83E−1
1 3	−3.21E−1
2 1	−6.80E+0
2 2	−3.77E−2
2 3	−9.21E−2
3 1	4.90E−1
3 2	2.29E−2
3 3	−3.10E+0
4 1	3.11E+0
4 2	1.71E−2
4 3	−7.27E+0
	$\mathbf{^4D^e–^4P^o}$
1 1	−1.83E+0
1 2	−2.98E−1
1 3	−1.63E−1
2 1	1.02E+1
2 2	9.64E−2
2 3	−3.94E+0
	$\mathbf{^4D^e–^4D^o}$
1 1	−6.77E−1
1 2	−2.01E−1
1 3	−3.00E−1
2 1	3.34E+0
2 2	1.30E−2
2 3	−2.96E−1
	$\mathbf{^4D^e–^4F^o}$
1 1	−9.21E−1
1 2	−4.31E+0
2 1	8.99E−2
2 2	−1.73E+1
	$\mathbf{^4F^e–^4D^o}$
1 1	−3.40E+0
1 2	−5.05E−1
1 3	−3.28E−1
2 1	1.97E+1
2 2	1.44E−2
2 3	−7.31E+0
3 1	2.98E−1
3 2	9.33E−3
3 3	2.25E+1
4 1	1.05E−3
4 2	8.95E−2
4 3	3.05E+0
	$\mathbf{^4F^e–^4F^o}$
1 1	−9.57E−1
1 2	−6.98E−1
2 1	9.46E−3
2 2	−2.03E+0
3 1	1.81E−3
3 2	3.97E−1
4 1	7.33E−3
4 2	2.22E+0
	$\mathbf{^4F^e–^4G^o}$
1 1	−5.83E+0
2 1	−2.49E+1
3 1	5.36E+0
4 1	2.74E−2

Cl-like Fe (Fe^{9+})

Term energies relative to $3s^23p^4$ ^{3}P ionization threshold for each symmetry

i	E(Ryds)	Description	i	E(Ryds)	Description	i	E(Ryds)	Description	i	E(Ryds)	Description
		$^2\mathbf{S}^e$	5	−8.28666	$3s3p^43d^2$	14	−7.01324	$3s3p^5$ ^{3}P $4s$	3	−13.9364	$3s^23p^4$ ^{3}P $3d$
1	−16.7783	$3s3p^6$	6	−7.85328	$3s3p^43d^2$	15	−6.48712	$3s3p^5$ ^{1}P $4s$	4	−9.84197	$3p^63d$
2	−14.3091	$3s^23p^4$ ^{1}D $3d$	7	−7.41065	$3s^23p^4$ ^{3}P $4d$	16	−6.22668	$3s^23p^4$ ^{1}D $4f$	5	−9.48583	$3s^23p^4$ ^{1}D $4s$
3	−9.09651	$3s^23p^4$ ^{1}S $4s$	8	−7.32851	$3s3p^43d^2$	17	−5.67583	$3p^53d^2$	6	−8.82315	$3s3p^43d^2$
4	−8.76409	$3s3p^43d^2$	9	−7.25921	$3s3p^43d^2$	18	−5.15478	$3s^23p^4$ ^{3}P $5p$	7	−8.32450	$3s3p^43d^2$
5	−7.29532	$3s^23p^4$ ^{1}D $4d$	10	−7.23536	$3s^23p^4$ ^{1}D $4d$	19	−4.88857	$3s^23p^4$ ^{1}D $5p$	8	−7.91169	$3s3p^43d^2$
6	−7.19011	$3s3p^43d^2$:	11	−7.06341	$3s^23p^23d^3$:	20	−4.87310	$3s3p^5$ ^{3}P $4d$	9	−7.77569	$3s3p^43d^2$
7	−6.85861	$3s3p^43d^2$	12	−6.96542	$3s3p^43d^2$	21	−4.78462	$3p^53d^2$	10	−7.56550	$3s3p^43d^2$
8	−6.22260	$3s3p^43d^2$:	13	−6.61778	$3s3p^43d^2$	22	−4.51914	$3s^23p^4$ ^{1}S $5p$	11	−7.50377	$3s3p^43d^2$
9	−6.04379	$3s3p^5$ ^{3}P $4p$	14	−6.27916	$3s^23p^23d^3$	23	−4.41223	$3p^53d^2$:	12	−7.47162	$3s^23p^4$ ^{3}P $4d$
10	−5.76140	$3s^23p^23d^3$:	15	−6.16234	$3s3p^5$ ^{3}P $4p$	24	−4.26602	$3s3p^5$ ^{1}P $4d$	13	−7.31721	$3s3p^43d^2$
11	−5.45169	$3s3p^5$ ^{1}P $4p$	16	−6.11165	$3s3p^43d^2$	25	−4.11200	$3s3p^33d^3$:	14	−7.25582	$3s3p^43d^2$
12	−5.26030	$3s^23p^23d^3$	17	−5.95410	$3s^23p^23d^3$	26	−3.97529	$3s3p^33d^3$	15	−7.17844	$3s^23p^4$ ^{1}D $4d$
13	−4.93623	$3s^23p^4$ ^{1}S $5s$	18	−5.59679	$3s^23p^4$ ^{3}P $5s$	27	−3.86573	$3s^23p^4$ ^{1}D $5f$	16	−6.94328	$3s3p^43d^2$:
14	−4.88701	$3s^23p^23d^3$	19	−5.55079	$3s^23p^23d^3$	28	−3.66762	$3s3p^33d^3$	17	−6.78203	$3s^23p^4$ ^{1}S $4d$
15	−4.33550	$3s^23p^4$ ^{1}D $5d$	20	−5.49491	$3s3p^5$ ^{1}P $4p$	29	−3.49934	$3s3p^33d^3$	18	−6.70738	$3s3p^43d^2$:
16	−3.72643	$3s^23p^23d^3$	21	−5.40018	$3s^23p^23d^3$	30	−3.40752	$3s^23p^4$ ^{3}P $6p$	19	−6.45159	$3s3p^43d^2$:
17	−2.98376	$3s^23p^4$ ^{1}S $6s$	22	−5.08742	$3s^23p^23d^3$	31	−3.18726	$3s3p^33d^3$	20	−6.40741	$3s^23p^23d^3$
18	−2.83477	$3s^23p^4$ ^{1}D $6d$	23	−4.82764	$3s^23p^23d^3$	32	−3.15415	$3s^23p^4$ ^{1}D $6p$	21	−6.22970	$3s^23p^23d^3$
19	−2.52982	$3s3p^5$ ^{3}P $5p$	24	−4.71081	$3s^23p^23d^3$	33	−3.14051	$3s3p^33d^3$	22	−6.18882	$3s3p^5$ ^{3}P $4p$
20	−2.00282	$3s^23p^4$ ^{1}D $7d$	25	−4.55104	$3s^23p^4$ ^{3}P $5d$	34	−3.05956	$3s3p^33d^3$	23	−5.88382	$3s^23p^23d^3$
21	−1.93729	$3s3p^5$ ^{1}P $5p$	26	−4.31562	$3s^23p^4$ ^{1}D $5d$	35	−2.97847	$3s3p^5$ ^{3}P $5s$	24	−5.75101	$3s^23p^23d^3$
22	−1.88768	$3s^23p^4$ ^{1}S $7s$	27	−3.71064	$3s^23p^23d^3$	36	−2.83630	$3s3p^33d^3$	25	−5.59833	$3s3p^5$ ^{1}P $4p$
23	−1.41386	$3s^23p^4$ ^{1}D $8d$	28	−3.64382	$3s^23p^4$ ^{3}P $6s$	37	−2.75465	$3s^23p^4$ ^{1}S $6p$	26	−5.56896	$3s^23p^23d^3$
24	−1.24848	$3p^43d^3$	29	−3.08887	$3s^23p^4$ ^{3}P $6d$	38	−2.68311	$3s3p^33d^3$	27	−5.40482	$3s^23p^23d^3$
25	−1.22796	$3s^23p^4$ ^{1}S $8s$	30	−2.83120	$3s^23p^4$ ^{1}D $6d$	39	−2.58210	$3s^23p^4$ ^{1}D $6f$	28	−5.32201	$3s^23p^4$ ^{1}D $5s$
26	−1.04247	$3s^23p^4$ ^{1}D $9d$	31	−2.59186	$3s3p^5$ ^{3}P $5p$	40	−2.57056	$3s3p^33d^3$	29	−5.29132	$3s^23p^23d^3$
27	−.81440	$3s3p^5$ ^{3}P $6p$	32	−2.56537	$3s^23p^4$ ^{3}P $7s$	41	−2.49977	$3s3p^33d^3$	30	−5.24963	$3s^23p^23d^3$
28	−.79264	$3s^23p^4$ ^{1}S $9s$	33	−2.23339	$3s^23p^4$ ^{3}P $7d$	42	−2.43140	$3s^23p^4$ ^{3}P $7p$	31	−5.00513	$3s^23p^23d^3$
		$^2\mathbf{S}^o$	34	−1.98066	$3s^23p^4$ ^{1}D $7d$	43	−2.35709	$3s3p^33d^3$	32	−4.78450	$3s^23p^23d^3$
1	−11.0168	$3s^23p^33d^2$	35	−1.91246	$3s3p^5$ ^{1}P $5p$	44	−2.35473	$3s3p^5$ ^{1}P $5s$	33	−4.58203	$3s^23p^4$ ^{3}P $5d$
2	−10.0767	$3s^23p^33d^2$	36	−1.90357	$3s^23p^4$ ^{3}P $8s$	45	−2.14875	$3s^23p^4$ ^{1}D $7p$	34	−4.53682	$3s^23p^23d^3$
3	−8.73448	$3s^23p^4$ ^{3}P $4p$	37	−1.68943	$3s^23p^4$ ^{3}P $8d$	46	−2.00100	$3s3p^5$ ^{3}P $5d$	35	−4.34534	$3s^23p^23d^3$
4	−5.16175	$3s^23p^4$ ^{3}P $5p$	38	−1.47151	$3s^23p^4$ ^{3}P $9s$	47	−1.95851	$3s3p^33d^3$	36	−4.28662	$3s^23p^4$ ^{1}D $5d$
5	−5.02613	$3p^53d^2$	39	−1.40905	$3s^23p^4$ ^{1}D $8d$	48	−1.81374	$3s^23p^4$ ^{1}D $7f$	37	−3.90538	$3s^23p^4$ ^{1}S $5d$
6	−4.46655	$3s3p^33d^3$	40	−1.35033	$3s^23p^23d^3$	49	−1.81234	$3s^23p^4$ ^{3}P $8p$	38	−3.78146	$3s3p^5$ ^{3}P $4f$
7	−3.41913	$3s^23p^4$ ^{3}P $6p$	41	−1.32304	$3s^23p^4$ ^{3}P $9d$	50	−1.75843	$3s^23p^4$ ^{1}S $7p$	39	−3.76258	$3s^23p^4$ ^{1}D $5g$
8	−3.38685	$3s3p^33d^3$			$^2\mathbf{P}^o$	51	−1.75678	$3s3p^33d^3$	40	−3.58575	$3s^23p^23d^3$
9	−3.04796	$3s3p^33d^3$	1	−19.2443	$3s^23p^5$	52	−1.68960	$3s3p^33d^3$	41	−3.37029	$3s^23p^4$ ^{1}D $6s$
10	−2.43246	$3s^23p^4$ ^{3}P $7p$	2	−12.3944	$3s3p^5$ ^{1}P $3d$:	53	−1.60981	$3s3p^33d^3$	42	−3.23601	$3s3p^5$ ^{1}P $4f$
11	−2.40048	$3s3p^33d^3$	3	−11.4572	$3s3p^5$ ^{3}P $3d$:	54	−1.53639	$3s^23p^4$ ^{1}D $8p$	43	−3.10411	$3s^23p^4$ ^{3}P $6d$
12	−2.29772	$3s3p^33d^3$	4	−11.0474	$3s^23p^33d^2$	55	−1.45265	$3s3p^33d^3$	44	−2.81247	$3s^23p^4$ ^{1}D $6d$
13	−1.95255	$3s3p^33d^3$	5	−10.5988	$3s^23p^33d^2$	56	−1.40731	$3s^23p^4$ ^{3}P $9p$	45	−2.60945	$3s3p^5$ ^{3}P $5p$
14	−1.81919	$3s^23p^4$ ^{3}P $8p$	6	−10.0937	$3s^23p^33d^2$	57	−1.36223	$3s3p^5$ ^{1}P $5d$	46	−2.50160	$3s^23p^4$ ^{1}D $6g$
15	−1.41195	$3s^23p^4$ ^{3}P $9p$	7	−10.0758	$3s^23p^33d^2$	58	−1.31209	$3s^23p^4$ ^{1}D $8f$	47	−2.42300	$3s^23p^4$ ^{1}S $6d$
16	−1.22729	$3s3p^33d^3$	8	−9.41243	$3s^23p^33d^2$	59	−1.14623	$3s^23p^4$ ^{1}S $8p$	48	−2.28678	$3s^23p^4$ ^{1}D $7s$
		$^2\mathbf{P}^e$	9	−9.24198	$3s^23p^33d^2$	60	−1.14431	$3s3p^33d^3$	49	−2.24135	$3s^23p^4$ ^{3}P $7d$
1	−15.3882	$3s^23p^4$ ^{1}D $3d$:	10	−8.86723	$3s^23p^4$ ^{3}P $4p$	61	−1.13172	$3s^23p^4$ ^{1}D $9p$	50	−1.98152	$3s3p^5$ ^{1}P $5p$
2	−14.0119	$3s^23p^4$ ^{3}P $3d$:	11	−8.55274	$3s^23p^33d^2$			$^2\mathbf{D}^e$	51	−1.92984	$3s^23p^4$ ^{1}D $7d$
3	−9.66945	$3s^23p^4$ ^{3}P $4s$	12	−8.22393	$3s^23p^4$ ^{1}D $4p$	1	−15.2659	$3s^23p^4$ ^{1}D $3d$	52	−1.76284	$3s^23p^4$ ^{1}D $7g$
4	−8.55310	$3s3p^43d^2$	13	−8.14486	$3s^23p^4$ ^{1}S $4p$	2	−14.5764	$3s^23p^4$ ^{1}S $3d$	53	−1.69564	$3s^23p^4$ ^{3}P $8d$

Cl-like Fe (Fe^{9+})

i	E(Ryds)	Description	i	E(Ryds)	Description	i	E(Ryds)	Description	i	E(Ryds)	Description
54	−1.62042	$3s^23p^4$ ^{1}D $8s$	43	−2.27958	$3s3p^33d^3$	33	−2.80311	$3s^23p^4$ ^{1}D $6d$	36	−2.88845	$3s3p^33d^3$
55	−1.56110	$3s^23p^4$ ^{1}S $7d$	44	−2.15161	$3s3p^33d^3$	34	−2.77482	$3s^23p^4$ ^{3}P $6g$	37	−2.84015	$3s^23p^4$ ^{3}P $6f$
56	−1.50617	$3s3p^5$ ^{3}P $5f$	45	−2.14520	$3s^23p^4$ ^{1}D $7p$	35	−2.49163	$3s^23p^4$ ^{1}D $6g$	38	−2.77945	$3s3p^33d^3$
57	−1.40422	$3p^43d^3$	46	−2.07711	$3s^23p^4$ ^{3}P $7f$	36	−2.24037	$3s^23p^4$ ^{3}P $7d$	39	−2.64866	$3s3p^33d^3$
58	−1.40305	$3s^23p^4$ ^{1}D $8d$	47	−2.00980	$3s3p^33d^3$	37	−2.04043	$3s^23p^4$ ^{3}P $7g$	40	−2.55926	$3s^23p^4$ ^{1}D $6f$
59	−1.32698	$3s^23p^4$ ^{3}P $9d$	48	−1.98113	$3s3p^5$ ^{3}P $5d$	38	−1.94579	$3s^23p^4$ ^{1}D $7d$	41	−2.55231	$3s3p^33d^3$
60	−1.28163	$3s^23p^4$ ^{1}D $8g$	49	−1.91244	$3s3p^33d^3$	39	−1.75593	$3s^23p^4$ ^{1}D $7g$	42	−2.48272	$3s3p^33d^3$
61	−1.18406	$3s^23p^4$ ^{1}D $9s$	50	−1.90864	$3s3p^33d^3$	40	−1.69476	$3s^23p^4$ ^{3}P $8d$	43	−2.40116	$3s3p^33d^3$
62	−1.09209	$3p^43d^3$	51	−1.83070	$3s3p^33d^3$	41	−1.56366	$3s^23p^4$ ^{3}P $8g$	44	−2.26838	$3s3p^33d^3$
		^{2}D^o	52	−1.82341	$3s^23p^4$ ^{3}P $8p$	42	−1.52084	$3s3p^5$ ^{3}P $5f$	45	−2.18143	$3s^23p^4$ ^{1}S $6f$
1	−12.5883	$3s3p^5$ ^{3}P $3d$	53	−1.80426	$3s^23p^4$ ^{1}D $7f$	43	−1.47836	$3p^43d^3$	46	−2.15715	$3s^23p^4$ ^{1}D $7p$
2	−11.8845	$3s3p^5$ ^{1}P $3d$	54	−1.67911	$3s3p^33d^3$	44	−1.40125	$3s^23p^4$ ^{1}D $8d$	47	−2.08256	$3s^23p^4$ ^{3}P $7f$
3	−11.3888	$3s^23p^33d^2$	55	−1.58702	$3s^23p^4$ ^{3}P $8f$	45	−1.32608	$3s^23p^4$ ^{3}P $9d$	48	−2.03006	$3s3p^33d^3$
4	−10.9835	$3s^23p^33d^2$	56	−1.53071	$3s^23p^4$ ^{1}D $8p$	46	−1.27633	$3s^23p^4$ ^{1}D $8g$	49	−2.00384	$3s3p^5$ ^{3}P $5d$
5	−10.4011	$3s^23p^33d^2$	57	−1.41518	$3s^23p^4$ ^{3}P $9p$	47	−1.23475	$3s^23p^4$ ^{3}P $9g$	50	−1.89868	$3s3p^33d^3$
6	−10.1655	$3s^23p^33d^2$	58	−1.35497	$3s3p^33d^3$	48	−1.13739	$3p^43d^3$	51	−1.85664	$3s3p^33d^3$
7	−10.0294	$3s^23p^33d^2$	59	−1.32201	$3s3p^5$ ^{1}P $5d$			**^{2}F^o**	52	−1.79757	$3s^23p^4$ ^{1}D $7f$
8	−9.84944	$3s^23p^33d^2$	60	−1.30744	$3s^23p^4$ ^{1}D $8f$	1	−12.6205	$3s3p^5$ ^{3}P $3d$	53	−1.78092	$3s3p^33d^3$
9	−9.42206	$3s^23p^33d^2$	61	−1.25150	$3s^23p^4$ ^{3}P $9f$	2	−12.1503	$3s3p^5$ ^{1}P $3d$	54	−1.59096	$3s^23p^4$ ^{3}P $8f$
10	−9.08579	$3s^23p^33d^2$			**^{2}F^e**	3	−11.4326	$3s^23p^33d^2$	55	−1.54053	$3s^23p^4$ ^{1}D $8p$
11	−8.79827	$3s^23p^4$ ^{3}P $4p$	1	−15.1654	$3s^23p^4$ ^{3}P $3d$	4	−10.9325	$3s^23p^33d^2$	56	−1.50576	$3s3p^33d^3$
12	−8.46720	$3s^23p^4$ ^{1}D $4p$	2	−14.8258	$3s^23p^4$ ^{1}D $3d$	5	−10.6878	$3s^23p^33d^2$	57	−1.44028	$3s3p^5$ ^{3}P $5g$
13	−6.36908	$3s^23p^4$ ^{3}P $4f$	3	−8.65638	$3s3p^43d^2$	6	−10.2796	$3s^23p^33d^2$	58	−1.41280	$3s^23p^4$ ^{1}S $7f$
14	−6.13298	$3s^23p^4$ ^{1}D $4f$	4	−8.21148	$3s3p^43d^2$	7	−10.0392	$3s^23p^33d^2$	59	−1.35088	$3s3p^5$ ^{1}P $5d$
15	−5.65423	$3p^53d^2$	5	−8.06485	$3s3p^43d^2$	8	−9.95830	$3s^23p^33d^2$	60	−1.30388	$3s^23p^4$ ^{1}D $8f$
16	−5.19074	$3s^23p^4$ ^{3}P $5p$	6	−7.82056	$3s3p^43d^2$	9	−9.59907	$3s^23p^33d^2$	61	−1.29039	$3s3p^33d^3$
17	−5.09916	$3p^53d^2$	7	−7.68236	$3s3p^43d^2$	10	−8.98694	$3s^23p^33d^2$	62	−1.25434	$3s^23p^4$ ^{3}P $9f$
18	−4.88683	$3s^23p^4$ ^{1}D $5p$	8	−7.53652	$3s3p^43d^2$	11	−8.60957	$3s^23p^4$ ^{1}D $4p$	63	−1.12749	$3s^23p^4$ ^{1}D $9p$
19	−4.82049	$3s3p^5$ ^{3}P $4d$	9	−7.44405	$3s^23p^4$ ^{3}P $4d$	12	−6.40664	$3s^23p^4$ ^{3}P $4f$			**^{2}G^e**
20	−4.39391	$3p^53d^2$:	10	−7.20360	$3s3p^43d^2$:	13	−6.11970	$3s^23p^4$ ^{1}D $4f$	1	−15.1735	$3s^23p^4$ ^{1}D $3d$
21	−4.19565	$3s3p^33d^3$:	11	−7.15419	$3s3p^43d^2$:	14	−5.78605	$3p^53d^2$	2	−8.66179	$3s3p^43d^2$
22	−4.16889	$3s3p^5$ ^{1}P $4d$	12	−7.11721	$3s^23p^4$ ^{1}D $4d$	15	−5.75541	$3s^23p^4$ ^{1}S $4f$	3	−8.30678	$3s3p^43d^2$
23	−4.08149	$3s^23p^4$ ^{3}P $5f$	13	−6.80524	$3s3p^43d^2$:	16	−5.48052	$3p^53d^2$	4	−8.25171	$3s3p^43d^2$
24	−3.98374	$3s3p^33d^3$:	14	−6.61610	$3s^23p^23d^3$:	17	−4.94509	$3s^23p^4$ ^{1}D $5p$	5	−7.88576	$3s3p^43d^2$
25	−3.83638	$3s^23p^4$ ^{1}D $5f$	15	−6.21713	$3s^23p^23d^3$:	18	−4.87293	$3s3p^5$ ^{3}P $4d$	6	−7.51561	$3s3p^43d^2$
26	−3.74050	$3s3p^33d^3$	16	−6.08233	$3s^23p^23d^3$	19	−4.77925	$3p^53d^2$	7	−7.35177	$3s3p^43d^2$
27	−3.64409	$3s3p^33d^3$	17	−5.87441	$3s^23p^23d^3$	20	−4.43815	$3s3p^33d^3$	8	−7.24010	$3s^23p^4$ ^{1}D $4d$
28	−3.47845	$3s3p^33d^3$	18	−5.75097	$3s^23p^23d^3$	21	−4.25098	$3s3p^5$ ^{1}P $4d$	9	−7.08173	$3s3p^43d^2$
29	−3.43158	$3s^23p^4$ ^{3}P $6p$	19	−5.60948	$3s^23p^23d^3$	22	−4.12866	$3s3p^33d^3$	10	−6.80730	$3s^23p^23d^3$
30	−3.34703	$3s3p^33d^3$	20	−5.34827	$3s^23p^23d^3$	23	−4.09774	$3s^23p^4$ ^{3}P $5f$	11	−6.40706	$3s^23p^23d^3$
31	−3.20831	$3s3p^33d^3$	21	−5.31216	$3s^23p^23d^3$	24	−3.97161	$3s3p^33d^3$	12	−6.34449	$3s^23p^23d^3$
32	−3.16310	$3s3p^33d^3$	22	−5.05057	$3s^23p^23d^3$	25	−3.84165	$3s3p^33d^3$	13	−6.06933	$3s^23p^23d^3$
33	−3.13425	$3s^23p^4$ ^{1}D $6p$	23	−4.94163	$3s^23p^23d^3$	26	−3.82169	$3s^23p^4$ ^{1}D $5f$	14	−5.77550	$3s^23p^23d^3$
34	−3.06136	$3s3p^33d^3$	24	−4.76154	$3s^23p^23d^3$	27	−3.55199	$3s3p^33d^3$	15	−5.56743	$3s^23p^23d^3$
35	−2.94760	$3s3p^33d^3$	25	−4.57603	$3s^23p^4$ ^{3}P $5d$	28	−3.45033	$3s^23p^4$ ^{1}S $5f$	16	−5.52875	$3s^23p^23d^3$
36	−2.85191	$3s3p^33d^3$	26	−4.48561	$3s^23p^23d^3$	29	−3.42576	$3s3p^33d^3$	17	−5.05670	$3s^23p^23d^3$
37	−2.83114	$3s^23p^4$ ^{3}P $6f$	27	−4.27177	$3s^23p^4$ ^{1}D $5d$	30	−3.26401	$3s3p^33d^3$	18	−4.80366	$3s^23p^23d^3$
38	−2.69276	$3s3p^33d^3$	28	−4.00044	$3s^23p^4$ ^{3}P $5g$	31	−3.18647	$3s3p^33d^3$	19	−4.32012	$3s^23p^4$ ^{1}D $5d$
39	−2.66815	$3s3p^33d^3$	29	−3.79354	$3s3p^5$ ^{3}P $4f$	32	−3.15726	$3s^23p^4$ ^{1}D $6p$	20	−4.29612	$3s^23p^23d^3$
40	−2.56904	$3s^23p^4$ ^{1}D $6f$	30	−3.72391	$3s^23p^4$ ^{1}D $5g$	33	−3.14686	$3s3p^33d^3$	21	−4.03792	$3s^23p^4$ ^{3}P $5g$
41	−2.50312	$3s3p^33d^3$	31	−3.15596	$3s3p^5$ ^{1}P $4f$	34	−3.05316	$3s3p^33d^3$	22	−3.84599	$3s3p^5$ ^{3}P $4f$
42	−2.43977	$3s^23p^4$ ^{3}P $7p$	32	−3.10006	$3s^23p^4$ ^{3}P $6d$	35	−2.97483	$3s3p^33d^3$	23	−3.70840	$3s^23p^4$ ^{1}D $5g$

Cl-like Fe (Fe^{9+})

i	E(Ryds)	Description
24	−3.32968	$3s^23p^4\ {}^1\mathrm{S}\ 5g$
25	−3.24044	$3s3p^5\ {}^1\mathrm{P}\ 4f$
26	−2.82639	$3s^23p^4\ {}^1\mathrm{D}\ 6d$
27	−2.78987	$3s^23p^4\ {}^3\mathrm{P}\ 6g$
28	−2.48444	$3s^23p^4\ {}^1\mathrm{D}\ 6g$
29	−2.10194	$3s^23p^4\ {}^1\mathrm{S}\ 6g$
30	−2.04970	$3s^23p^4\ {}^3\mathrm{P}\ 7g$
31	−1.95963	$3s^23p^4\ {}^1\mathrm{D}\ 7d$
32	−1.75025	$3s^23p^4\ {}^1\mathrm{D}\ 7g$
33	−1.57108	$3s^23p^4\ {}^3\mathrm{P}\ 8g$
34	−1.54297	$3s3p^5\ {}^3\mathrm{P}\ 5f$
35	−1.41045	$3s^23p^4\ {}^1\mathrm{D}\ 8d$
36	−1.36355	$3s^23p^4\ {}^1\mathrm{S}\ 7g$
37	−1.27219	$3s^23p^4\ {}^1\mathrm{D}\ 8g$
38	−1.25037	$3p^43d^3$
39	−1.23904	$3s^23p^4\ {}^3\mathrm{P}\ 9g$
40	−1.14623	$3p^43d^3$
41	−1.04094	$3s^23p^4\ {}^1\mathrm{D}\ 9d$
	^{2}G$^\circ$	
1	−11.2022	$3s^23p^33d^2$
2	−10.8644	$3s^23p^33d^2$
3	−10.5750	$3s^23p^33d^2$
4	−10.1456	$3s^23p^33d^2$
5	−9.71244	$3s^23p^33d^2$
6	−6.41851	$3s^23p^4\ {}^3\mathrm{P}\ 4f$
7	−6.11091	$3s^23p^4\ {}^1\mathrm{D}\ 4f$
8	−5.43165	$3p^53d^2$
9	−5.16490	$3p^53d^2$
10	−4.35240	$3s3p^33d^3$
11	−4.11636	$3s^23p^4\ {}^3\mathrm{P}\ 5f$
12	−3.98195	$3s3p^33d^3$
13	−3.93298	$3s3p^33d^3$
14	−3.82191	$3s^23p^4\ {}^1\mathrm{D}\ 5f$
15	−3.70523	$3s3p^33d^3$
16	−3.67078	$3s3p^33d^3$
17	−3.58992	$3s3p^33d^3$
18	−3.41443	$3s3p^33d^3$
19	−3.30507	$3s3p^33d^3$
20	−3.04214	$3s3p^33d^3$
21	−2.85288	$3s^23p^4\ {}^3\mathrm{P}\ 6f$
22	−2.84894	$3s3p^33d^3$
23	−2.70400	$3s3p^33d^3$
24	−2.61656	$3s3p^33d^3$
25	−2.55979	$3s^23p^4\ {}^1\mathrm{D}\ 6f$
26	−2.50842	$3s3p^33d^3$
27	−2.41409	$3s3p^33d^3$
28	−2.25786	$3s3p^33d^3$
29	−2.13305	$3s3p^33d^3$
30	2.08044	$3s^23p^4\ {}^3\mathrm{P}\ 7f$
31	−2.02724	$3s3p^33d^3$
32	−1.94923	$3s3p^33d^3$
33	−1.79821	$3s^23p^4\ {}^1\mathrm{D}\ 7f$
34	−1.59602	$3s^23p^4\ {}^3\mathrm{P}\ 8f$
35	−1.41298	$3s3p^5\ {}^3\mathrm{P}\ 5g$
36	−1.40262	$3s3p^33d^3$
37	−1.30491	$3s^23p^4\ {}^1\mathrm{D}\ 8f$
38	−1.25818	$3s^23p^4\ {}^3\mathrm{P}\ 9f$
	^{4}S^e	
1	−8.38541	$3s3p^43d^2$
2	−7.83490	$3s3p^43d^2$
3	−6.91131	$3s^23p^23d^3$
4	−6.48152	$3s3p^5\ {}^3\mathrm{P}\ 4p$
5	−5.61521	$3s^23p^23d^3$
6	−2.70709	$3s3p^5\ {}^3\mathrm{P}\ 5p$
7	−1.50414	$3p^43d^3$
8	−.89959	$3s3p^5\ {}^3\mathrm{P}\ 6p$
9	−.33221	$3p^43d^3$
	^{4}S$^\circ$	
1	−10.9500	$3s^23p^33d^2$
2	−9.93154	$3s^23p^33d^2$
3	−8.70902	$3s^23p^4\ {}^3\mathrm{P}\ 4p$
4	−5.25683	$3p^53d^2$
5	−5.15657	$3s^23p^4\ {}^3\mathrm{P}\ 5p$
6	−3.92110	$3s3p^33d^3$
7	−3.56017	$3s3p^33d^3$
8	−3.41676	$3s^23p^4\ {}^3\mathrm{P}\ 6p$
9	−3.14236	$3s3p^33d^3$
10	−2.75136	$3s3p^33d^3$
11	−2.43114	$3s^23p^4\ {}^3\mathrm{P}\ 7p$
12	−1.88635	$3s3p^33d^3$
13	−1.81839	$3s^23p^4\ {}^3\mathrm{P}\ 8p$
14	−1.41142	$3s^23p^4\ {}^3\mathrm{P}\ 9p$
	^{4}P^e	
1	−15.2562	$3s^23p^4\ {}^3\mathrm{P}\ 3d$
2	−9.82465	$3s^23p^4\ {}^3\mathrm{P}\ 4s$
3	−9.38972	$3s3p^43d^2$
4	−8.73914	$3s3p^43d^2$
5	−8.37032	$3s3p^43d^2$
6	−7.88500	$3s3p^43d^2$
7	−7.74464	$3s3p^43d^2$
8	−7.47381	$3s3p^43d^2$
9	−7.43523	$3s^23p^4\ {}^3\mathrm{P}\ 4d$
10	−7.34124	$3s3p^43d^2$
11	6.62335	$3s^23p^23d^3$
12	−6.52334	$3s^23p^23d^3$
13	−6.21428	$3s3p^5\ {}^3\mathrm{P}\ 4p$
14	−6.07992	$3s^23p^23d^3$
15	−5.72939	$3s^23p^23d^3$
16	−5.62786	$3s^23p^4\ {}^3\mathrm{P}\ 5s$
17	−5.60225	$3s^23p^23d^3$
18	−5.33571	$3s^23p^23d^3$
19	−4.90792	$3s^23p^23d^3$
20	−4.56685	$3s^23p^4\ {}^3\mathrm{P}\ 5d$
21	−3.66905	$3s^23p^4\ {}^3\mathrm{P}\ 6s$
22	−3.09896	$3s^23p^4\ {}^3\mathrm{P}\ 6d$
23	−2.62641	$3s3p^5\ {}^3\mathrm{P}\ 5p$
24	−2.56454	$3s^23p^4\ {}^3\mathrm{P}\ 7s$
25	−2.23691	$3s^23p^4\ {}^3\mathrm{P}\ 7d$
26	−1.91396	$3s^23p^4\ {}^3\mathrm{P}\ 8s$
27	−1.78148	$3p^43d^3$
28	−1.69302	$3s^23p^4\ {}^3\mathrm{P}\ 8d$
29	−1.47683	$3s^23p^4\ {}^3\mathrm{P}\ 9s$
30	−1.34116	$3p^43d^3$
31	−1.32561	$3s^23p^4\ {}^3\mathrm{P}\ 9d$
	^{4}P$^\circ$	
1	−13.3163	$3s3p^5\ {}^3\mathrm{P}\ 3d$
2	−11.7028	$3s^23p^33d^2$
3	−10.7223	$3s^23p^33d^2$
4	−10.6229	$3s^23p^33d^2$
5	−9.99906	$3s^23p^33d^2$
6	−9.00655	$3s^23p^4\ {}^3\mathrm{P}\ 4p$
7	−7.21711	$3s3p^5\ {}^3\mathrm{P}\ 4s$
8	−5.98572	$3p^53d^2$
9	−5.26316	$3s^23p^4\ {}^3\mathrm{P}\ 5p$
10	−5.15975	$3s3p^33d^3$
11	−4.98022	$3s3p^5\ {}^3\mathrm{P}\ 4d$
12	−4.87920	$3s3p^33d^3$
13	−4.62478	$3s3p^33d^3$
14	−4.21864	$3s3p^33d^3$
15	−4.03961	$3s3p^33d^3$
16	−3.91785	$3s3p^33d^3$
17	−3.85446	$3s3p^33d^3$
18	−3.65252	$3s3p^33d^3$
19	−3.46955	$3s^23p^4\ {}^3\mathrm{P}\ 6p$
20	−3.38153	$3s3p^33d^3$
21	−3.16302	$3s3p^33d^3$
22	−3.05049	$3s3p^5\ {}^3\mathrm{P}\ 5s$
23	−3.04108	$3s3p^33d^3$
24	−2.91869	$3s3p^33d^3$
25	−2.63842	$3s3p^33d^3$
26	−2.53837	$3s3p^33d^3$
27	−2.45740	$3s^23p^4\ {}^3\mathrm{P}\ 7p$
28	−2.25198	$3s3p^33d^3$
29	−2.08032	$3s3p^33d^3$
30	−2.04648	$3s3p^5\ {}^3\mathrm{P}\ 5d$
31	−1.83463	$3s^23p^4\ {}^3\mathrm{P}\ 8p$
32	−1.55484	$3s3p^33d^3$
33	−1.42338	$3s^23p^4\ {}^3\mathrm{P}\ 9p$
	^{4}D^e	
1	−15.7421	$3s^23p^4\ {}^3\mathrm{P}\ 3d$
2	−9.08538	$3s3p^43d^2$
3	−8.82606	$3s3p^43d^2$
4	−8.51399	$3s3p^43d^2$
5	−8.29391	$3s3p^43d^2$
6	−7.87451	$3s3p^43d^2$
7	−7.66741	$3s3p^43d^2$
8	−7.60325	$3s^23p^4\ {}^3\mathrm{P}\ 4d$
9	−7.38606	$3s3p^43d^2$
10	−7.18822	$3s^23p^23d^3$
11	−6.82700	$3s^23p^23d^3$
12	−6.46279	$3s^23p^23d^3$
13	−6.31012	$3s3p^5\ {}^3\mathrm{P}\ 4p$
14	−6.14883	$3s^23p^23d^3$
15	−6.00680	$3s^23p^23d^3$
16	−5.83160	$3s^23p^23d^3$
17	−5.36565	$3s^23p^23d^3$
18	−5.14892	$3s^23p^23d^3$
19	−4.63016	$3s^23p^4\ {}^3\mathrm{P}\ 5d$
20	−3.98927	$3s3p^5\ {}^3\mathrm{P}\ 4f$
21	−3.13246	$3s^23p^4\ {}^3\mathrm{P}\ 6d$
22	−2.64847	$3s3p^5\ {}^3\mathrm{P}\ 5p$
23	−2.25428	$3s^23p^4\ {}^3\mathrm{P}\ 7d$
24	−1.70569	$3s^23p^4\ {}^3\mathrm{P}\ 8d$
25	−1.70512	$3p^43d^3$
26	−1.60558	$3s3p^5\ {}^3\mathrm{P}\ 5f$
27	−1.51171	$3p^43d^3$
28	−1.33347	$3s^23p^4\ {}^3\mathrm{P}\ 9d$
29	−1.14696	$3p^43d^3$
	^{4}D$^\circ$	
1	−12.7562	$3s3p^5\ {}^3\mathrm{P}\ 3d$
2	−11.7288	$3s^23p^33d^2$
3	−10.9405	$3s^23p^33d^2$
4	−10.7512	$3s^23p^33d^2$
5	−10.3218	$3s^23p^33d^2$
6	−10.2464	$3s^23p^33d^2$
7	−8.88356	$3s^23p^4\ {}^3\mathrm{P}\ 4p$
8	−6.43231	$3s^23p^4\ {}^3\mathrm{P}\ 4f$
9	−6.26654	$3p^53d^2$
10	−5.51222	$3p^53d^2$
11	−5.22490	$3s^23p^4\ {}^3\mathrm{P}\ 5p$
12	−4.91033	$3s3p^33d^3$
13	−4.85001	$3s3p^5\ {}^3\mathrm{P}\ 4d$
14	−4.67120	$3s3p^33d^3$
15	−4.43023	$3s3p^33d^3$
16	−4.23179	$3s3p^33d^3$
17	−4.12563	$3s^23p^4\ {}^3\mathrm{P}\ 5f$
18	−4.09600	$3s3p^33d^3$
19	−4.05924	$3s3p^33d^3$
20	−3.95588	$3s3p^33d^3$
21	−3.64043	$3s3p^33d^3$
22	−3.61201	$3s3p^33d^3$
23	−3.44566	$3s^23p^4\ {}^3\mathrm{P}\ 6p$
24	−3.41423	$3s3p^33d^3$
25	−3.37890	$3s3p^33d^3$
26	−3.22379	$3s3p^33d^3$
27	−3.13521	$3s3p^33d^3$
28	3.00000	$3s3p^33d^3$
29	−3.04910	$3s3p^33d^3$

Cl-like Fe (Fe^{9+})

i	E(Ryds)	Description
30	−2.85768	$3s^23p^4\ ^3\mathrm{P}\ 6f$
31	−2.80647	$3s3p^33d^3$
32	−2.68967	$3s3p^33d^3$
33	−2.44860	$3s^23p^4\ ^3\mathrm{P}\ 7p$
34	−2.30150	$3s3p^33d^3$
35	−2.20827	$3s3p^33d^3$
36	−2.09415	$3s^23p^4\ ^3\mathrm{P}\ 7f$
37	−2.09207	$3s3p^33d^3$
38	−1.99428	$3s3p^5\ ^3\mathrm{P}\ 5d$
39	−1.92940	$3s3p^33d^3$
40	−1.82696	$3s^23p^4\ ^3\mathrm{P}\ 8p$
41	−1.59821	$3s^23p^4\ ^3\mathrm{P}\ 8f$
42	−1.41819	$3s^23p^4\ ^3\mathrm{P}\ 9p$
43	−1.26002	$3s^23p^4\ ^3\mathrm{P}\ 9f$
	$^4\mathrm{F}^e$	
1	−15.4128	$3s^23p^4\ ^3\mathrm{P}\ 3d$
2	−9.28908	$3s3p^43d^2$
3	−9.01911	$3s3p^43d^2$
4	−8.58547	$3s3p^43d^2$
5	−8.20483	$3s3p^43d^2$
6	−8.02593	$3s3p^43d^2$
7	−7.79390	$3s3p^43d^2$
8	−7.49656	$3s3p^43d^2$
9	−7.47425	$3s^23p^4\ ^3\mathrm{P}\ 4d$
10	−7.00086	$3s^23p^23d^3$
11	−6.82713	$3s^23p^23d^3$
12	−6.53163	$3s^23p^23d^3$
13	−6.32715	$3s^23p^23d^3$
14	−6.19120	$3s^23p^23d^3$
15	−5.81247	$3s^23p^23d^3$
16	−5.63236	$3s^23p^23d^3$
17	−5.05962	$3s^23p^23d^3$
18	−4.59244	$3s^23p^4\ ^3\mathrm{P}\ 5d$
19	−4.01677	$3s^23p^4\ ^3\mathrm{P}\ 5g$
20	−3.82507	$3s3p^5\ ^3\mathrm{P}\ 4f$
21	−3.10861	$3s^23p^4\ ^3\mathrm{P}\ 6d$
22	−2.77722	$3s^23p^4\ ^3\mathrm{P}\ 6g$
23	−2.24551	$3s^23p^4\ ^3\mathrm{P}\ 7d$
24	−2.04281	$3s^23p^4\ ^3\mathrm{P}\ 7g$
25	−1.69873	$3s^23p^4\ ^3\mathrm{P}\ 8d$
26	−1.63587	$3p^43d^3$
27	−1.57287	$3s^23p^4\ ^3\mathrm{P}\ 8g$
28	−1.53786	$3s3p^5\ ^3\mathrm{P}\ 5f$
29	−1.48474	$3p^43d^3$
30	−1.38265	$3p^43d^3$
31	−1.32800	$3s^23p^4\ ^3\mathrm{P}\ 9d$
32	−1.23544	$3s^23p^4\ ^3\mathrm{P}\ 9g$
	$^4\mathrm{F}^o$	
1	−13.0344	$3s3p^5\ ^3\mathrm{P}\ 3d$
2	−11.6310	$3s^23p^33d^2$
3	−11.1379	$3s^23p^33d^2$
4	−10.8883	$3s^23p^33d^2$
5	−10.1683	$3s^23p^33d^2$
6	−6.59595	$3s^23p^4\ ^3\mathrm{P}\ 4f$
7	−5.86105	$3p^53d^2$
8	−5.08644	$3s3p^33d^3$
9	−4.95152	$3s3p^5\ ^3\mathrm{P}\ 4d$
10	−4.85362	$3s3p^33d^3$
11	−4.68763	$3s3p^33d^3$
12	−4.43980	$3s3p^33d^3$
13	−4.27719	$3s3p^33d^3$
14	−4.19409	$3s^23p^4\ ^3\mathrm{P}\ 5f$
15	−4.08051	$3s3p^33d^3$
16	−4.04478	$3s3p^33d^3$
17	−4.02004	$3s3p^33d^3$
18	−3.85122	$3s3p^33d^3$
19	−3.73692	$3s3p^33d^3$
20	−3.62458	$3s3p^33d^3$
21	−3.50073	$3s3p^33d^3$
22	−3.44296	$3s3p^33d^3$
23	−3.14862	$3s3p^33d^3$
24	−3.09197	$3s3p^33d^3$
25	−2.90891	$3s3p^33d^3$
26	−2.89281	$3s^23p^4\ ^3\mathrm{P}\ 6f$
27	−2.69998	$3s3p^33d^3$
28	−2.51866	$3s3p^33d^3$
29	−2.31901	$3s3p^33d^3$
30	−2.11479	$3s^23p^4\ ^3\mathrm{P}\ 7f$
31	−2.03159	$3s3p^5\ ^3\mathrm{P}\ 5d$
32	−1.77331	$3s3p^33d^3$
33	−1.61175	$3s^23p^4\ ^3\mathrm{P}\ 8f$
34	−1.44717	$3s3p^5\ ^3\mathrm{P}\ 5g$
35	−1.26930	$3s^23p^4\ ^3\mathrm{P}\ 9f$
	$^4\mathrm{G}^e$	
1	−9.25315	$3s3p^43d^2$
2	−8.71436	$3s3p^43d^2$
3	−8.21266	$3s3p^43d^2$
4	−7.78865	$3s3p^43d^2$
5	−6.94374	$3s^23p^23d^3$
6	−6.67197	$3s^23p^23d^3$
7	−6.35861	$3s^23p^23d^3$
8	−6.24612	$3s^23p^23d^3$
9	−5.76703	$3s^23p^23d^3$
10	−4.07883	$3s^23p^4\ ^3\mathrm{P}\ 5g$
11	−3.91344	$3s3p^5\ ^3\mathrm{P}\ 4f$
12	−2.79569	$3s^23p^4\ ^3\mathrm{P}\ 6g$
13	−2.05487	$3s^23p^4\ ^3\mathrm{P}\ 7g$
14	−1.71417	$3p^43d^3$
15	−1.60558	$3s3p^5\ ^3\mathrm{P}\ 5f$
16	−1.58678	$3p^43d^3$
17	−1.56176	$3s^23p^4\ ^3\mathrm{P}\ 8g$
18	−1.24925	$3p^43d^3$
19	−1.24094	$3s^23p^4\ ^3\mathrm{P}\ 9g$
	$^4\mathrm{G}^o$	
1	−11.6473	$3s^23p^33d^2$
2	−11.2635	$3s^23p^33d^2$
3	−10.6484	$3s^23p^33d^2$
4	−6.47302	$3s^23p^4\ ^3\mathrm{P}\ 4f$
5	−5.94635	$3p^53d^2$
6	−5.10523	$3s3p^33d^3$
7	−4.66357	$3s3p^33d^3$
8	−4.41383	$3s3p^33d^3$
9	−4.14367	$3s^23p^4\ ^3\mathrm{P}\ 5f$
10	−4.10791	$3s3p^33d^3$
11	−4.05318	$3s3p^33d^3$
12	−3.87431	$3s3p^33d^3$
13	−3.83160	$3s3p^33d^3$
14	−3.81096	$3s3p^33d^3$
15	−3.69444	$3s3p^33d^3$
16	−3.66670	$3s3p^33d^3$
17	−3.32375	$3s3p^33d^3$
18	−3.15767	$3s3p^33d^3$
19	−2.86689	$3s^23p^4\ ^3\mathrm{P}\ 6f$
20	−2.65083	$3s3p^33d^3$
21	−2.63786	$3s3p^33d^3$
22	−2.09874	$3s^23p^4\ ^3\mathrm{P}\ 7f$
23	−1.60197	$3s^23p^4\ ^3\mathrm{P}\ 8f$
24	−1.41478	$3s3p^5\ ^3\mathrm{P}\ 5g$
25	−1.26254	$3s^23p^4\ ^3\mathrm{P}\ 9f$

Cl-like Fe (Fe^{9+})

Energies in ascending order from ground state for terms with effective $n \leq 4.0$, $L \leq 4$

Term	i	E(Ryds)	Term	i	E(Ryds)	Term	i	E(Ryds)	Term	i	E(Ryds)	Term	i	E(Ryds)
$^2P^o$	1	0.00000	$^2G^o$	2	8.37990	$^4F^e$	3	10.2251	$^2P^e$	6	11.3910	$^2P^e$	11	12.1808
$^2S^e$	1	2.46600	$^4D^o$	4	8.49310	$^4P^o$	6	10.2377	$^4S^e$	2	11.4094	$^2P^o$	14	12.2310
$^4D^e$	1	3.50220	$^4P^o$	3	8.52200	$^2F^o$	10	10.2573	$^2F^e$	6	11.4237	$^4F^e$	10	12.2434
$^4F^e$	1	3.83150	$^2F^o$	5	8.55650	$^4D^o$	7	10.3607	$^4F^e$	7	11.4504	$^2P^e$	12	12.2788
$^2P^e$	1	3.85610	$^4G^o$	3	8.59590	$^2P^o$	10	10.3770	$^4G^e$	4	11.4556	$^4G^e$	5	12.3005
$^2D^e$	1	3.97840	$^4P^o$	4	8.62140	$^4D^e$	3	10.4182	$^2D^e$	9	11.4686	$^2D^e$	16	12.3010
$^4P^e$	1	3.98810	$^2P^o$	5	8.64550	$^2D^e$	6	10.4211	$^4P^e$	7	11.4996	$^4S^e$	3	12.3329
$^2G^e$	1	4.07080	$^2G^o$	3	8.66930	$^2D^o$	11	10.4460	$^2F^e$	7	11.5619	$^2S^e$	7	12.3856
$^2F^e$	1	4.07890	$^2D^o$	5	8.84320	$^2S^e$	4	10.4802	$^4D^e$	7	11.5768	$^4F^e$	11	12.4171
$^2F^e$	2	4.41850	$^4D^o$	5	8.92250	$^4P^e$	4	10.5051	$^4D^e$	8	11.6410	$^4D^e$	11	12.4173
$^2D^e$	2	4.66790	$^2F^o$	6	8.96470	$^2S^o$	3	10.5098	$^2D^e$	10	11.6788	$^2G^e$	10	12.4370
$^2S^e$	2	4.93520	$^4D^o$	6	8.99790	$^4G^e$	2	10.5299	$^2F^e$	8	11.7077	$^2F^e$	13	12.4390
$^2P^e$	2	5.23240	$^4F^o$	5	9.07600	$^4S^o$	3	10.5352	$^2G^e$	6	11.7286	$^2D^e$	17	12.4622
$^2D^e$	3	5.30790	$^2D^o$	6	9.07880	$^2G^e$	2	10.5825	$^2D^e$	11	11.7405	$^2D^e$	18	12.5369
$^4P^o$	1	5.92800	$^2G^o$	4	9.09870	$^2F^e$	3	10.5879	$^4F^e$	8	11.7477	$^4G^e$	6	12.5723
$^4F^o$	1	6.20990	$^2P^o$	6	9.15060	$^2F^o$	11	10.6347	$^4F^e$	9	11.7700	$^4P^e$	11	12.6209
$^4D^o$	1	6.48810	$^2S^o$	2	9.16760	$^4F^e$	4	10.6588	$^4P^e$	8	11.7704	$^2P^e$	13	12.6265
$^2F^o$	1	6.62380	$^2P^o$	7	9.16850	$^2P^e$	4	10.6912	$^2D^e$	12	11.7726	$^2F^e$	14	12.6282
$^2D^o$	1	6.65600	$^2F^o$	7	9.20510	$^2P^o$	11	10.6915	$^2F^e$	9	11.8002	$^4F^o$	6	12.6483
$^2P^o$	2	6.84990	$^2D^o$	7	9.21490	$^4D^e$	4	10.7303	$^4P^e$	9	11.8090	$^4F^e$	12	12.7126
$^2F^o$	2	7.09400	$^4P^o$	5	9.24524	$^2D^o$	12	10.7771	$^2P^e$	7	11.8336	$^4P^e$	12	12.7209
$^2D^o$	2	7.35980	$^2F^o$	8	9.28600	$^4S^e$	1	10.8588	$^4D^e$	9	11.8582	$^2P^o$	15	12.7571
$^4D^o$	2	7.51550	$^4S^o$	2	9.31276	$^4P^e$	5	10.8739	$^2G^e$	7	11.8925	$^4S^e$	4	12.7627
$^4P^o$	2	7.54150	$^2D^o$	8	9.39486	$^2D^e$	7	10.9198	$^4P^e$	10	11.9030	$^4G^o$	4	12.7712
$^4G^o$	1	7.59700	$^2D^e$	4	9.40233	$^2G^e$	3	10.9375	$^2P^e$	8	11.9157	$^4D^e$	12	12.7815
$^4F^o$	2	7.61330	$^4P^e$	2	9.41965	$^4D^e$	5	10.9503	$^2D^e$	13	11.9270	$^2D^e$	19	12.7927
$^2P^o$	3	7.78710	$^2G^o$	5	9.53186	$^2P^e$	5	10.9576	$^2S^e$	5	11.9489	$^4D^o$	8	12.8119
$^2F^o$	3	7.81170	$^2P^e$	3	9.57485	$^2G^e$	4	10.9925	$^2P^e$	9	11.9850	$^2G^o$	6	12.8257
$^2D^o$	3	7.85550	$^2F^o$	9	9.64523	$^2P^o$	12	11.0203	$^2D^e$	14	11.9884	$^2D^e$	20	12.8368
$^4G^o$	2	7.98080	$^2D^e$	5	9.75847	$^4G^e$	3	11.0316	$^2G^e$	8	12.0042	$^2G^e$	11	12.8372
$^2G^o$	1	8.04210	$^2D^o$	9	9.82224	$^2F^e$	4	11.0328	$^2P^e$	10	12.0089	$^2F^o$	12	12.8376
$^4F^o$	3	8.10640	$^2P^o$	8	9.83187	$^4F^e$	5	11.0394	$^4P^o$	7	12.0271	$^2D^o$	13	12.8752
$^2P^o$	4	8.19690	$^4P^e$	3	9.85458	$^2P^o$	13	11.0994	$^2F^e$	10	12.0407	$^4G^e$	7	12.8856
$^2S^o$	1	8.22750	$^4F^e$	2	9.95522	$^2F^e$	5	11.1794	$^2S^e$	6	12.0541	$^2G^e$	12	12.8998
$^2D^o$	4	8.26080	$^4G^e$	1	9.99115	$^4F^e$	6	11.2183	$^4D^e$	10	12.0560	$^4F^e$	13	12.9171
$^4S^o$	1	8.29430	$^2P^o$	9	10.0023	$^2D^e$	8	11.3326	$^2D^e$	15	12.0658	$^4D^e$	13	12.9341
$^4D^o$	3	8.30380	$^2S^e$	3	10.1477	$^2G^e$	5	11.3585	$^2F^e$	11	12.0901	$^2P^e$	14	12.9651
$^2F^o$	4	8.31180	$^2D^o$	10	10.1585	$^4P^e$	6	11.3593	$^2F^e$	12	12.1270	$^4D^o$	9	12.9777
$^4F^o$	4	8.35600	$^4D^e$	2	10.1589	$^4D^e$	6	11.3697	$^2G^e$	9	12.1625			

Cl-like Fe (Fe^{9+})

gf-values for transitions involving terms with effective $n \leq 4.0$, $L \leq 4$

i i'	gf_L	i i'	gf_L	i i'	gf_L	i i'	gf_L	i i'	gf_L	i i'	gf_L
	$^2S^o$–$^2P^e$	1 8	−9.86E−3	4 14	−1.18E−4	1 4	−2.79E−4	4 13	−3.71E−2	8 8	−4.76E−4
1 1	4.31E−2	1 9	−2.35E−3	4 15	−1.57E−3	1 5	−4.42E−4	4 14	−3.62E−1	8 9	−2.15E−2
1 2	1.05E−3	1 10	−1.98E−2	5 1	2.58E−1	1 6	−1.36E−3	5 1	1.46E+0	8 10	−4.44E−2
1 3	−4.99E−4	1 11	−3.83E−2	5 2	2.90E−2	1 7	−1.41E−1	5 2	1.82E−1	8 11	−4.21E−2
1 4	−1.46E−2	1 12	−9.74E−3	5 3	3.58E−4	1 8	−1.97E−2	5 3	−2.24E−4	8 12	−1.57E−2
1 5	−7.46E−2	1 13	−2.55E−4	5 4	1.97E−3	1 9	−2.91E−4	5 4	−1.77E−4	8 13	−4.04E−2
1 6	−2.93E−1	1 14	−4.34E−1	5 5	4.32E−3	1 10	−7.94E−1	5 5	−1.04E−1	8 14	−2.83E−2
1 7	−3.28E−4	1 15	−1.50E−1	5 6	5.25E−4	1 11	−3.18E−3	5 6	−1.18E−2	9 1	1.51E−1
1 8	−1.21E−1	2 1	1.74E+0	5 7	6.05E−4	1 12	−2.20E−3	5 7	−4.79E−5	9 2	4.88E−1
1 9	−6.95E−2	2 2	−4.30E−2	5 8	1.17E−4	1 13	−1.76E−5	5 8	−1.45E−2	9 3	4.98E−4
1 10	−6.10E−3	2 3	−3.87E−2	5 9	3.48E−3	1 14	−2.18E−3	5 9	−8.91E−3	9 4	−5.42E−5
1 11	−4.22E−1	2 4	−1.00E−1	5 10	3.62E−1	2 1	3.10E−1	5 10	−2.74E−3	9 5	−2.59E−4
1 12	−1.66E−1	2 5	−2.50E−1	5 11	2.61E−1	2 2	9.05E−3	5 11	−2.03E−3	9 6	−1.78E−6
1 13	−1.37E−1	2 6	−1.33E−2	5 12	1.86E−1	2 3	−2.81E−3	5 12	−3.50E−2	9 7	−1.38E−3
1 14	−4.66E−2	2 7	−5.11E−3	5 13	2.52E−2	2 4	−5.20E−2	5 13	−3.93E−1	9 8	−1.65E−2
2 1	1.53E+0	2 8	−2.71E+0	5 14	−4.30E−3	2 5	−3.28E−1	5 14	−6.30E−2	9 9	−3.18E−2
2 2	2.03E−2	2 9	−1.26E+0	5 15	−5.25E−3	2 6	−5.05E−3	6 1	8.82E−5	9 10	−4.94E−4
2 3	−2.11E−3	2 10	−8.84E−4	6 1	3.86E−4	2 7	−6.53E−3	6 2	2.10E−2	9 11	−3.12E−3
2 4	−1.51E−2	2 11	−4.60E−1	6 2	1.17E+0	2 8	−6.32E−3	6 3	−2.92E−4	9 12	−4.87E−3
2 5	−2.35E−3	2 12	−1.62E−1	6 3	3.40E−5	2 9	−4.58E−2	6 4	−1.40E−3	9 13	−1.58E−2
2 6	−4.84E−3	2 13	−8.40E−3	6 4	3.06E−2	2 10	−2.21E−2	6 5	−4.18E−4	9 14	−2.30E−1
2 7	−1.71E−2	2 14	−7.45E−3	6 5	4.95E−3	2 11	−1.43E+0	6 6	−1.21E−3	10 1	4.80E−2
2 8	−9.79E−3	2 15	−5.22E−3	6 6	4.25E−3	2 12	−1.66E+0	6 7	−1.96E−3	10 2	6.57E−2
2 9	−1.96E−2	3 1	1.38E−1	6 7	4.37E−2	2 13	−9.40E−1	6 8	−1.65E−1	10 3	7.95E−1
2 10	−2.35E−3	3 2	1.21E−4	6 8	2.73E−2	2 14	−7.78E−2	6 9	−2.77E−3	10 4	−7.23E−5
2 11	−3.79E−4	3 3	4.33E−4	6 9	3.56E−5	3 1	2.79E−3	6 10	−3.30E−5	10 5	−2.41E−4
2 12	−3.45E−2	3 4	2.90E−4	6 10	6.98E−3	3 2	1.21E−1	6 11	−1.46E−1	10 6	−1.12E−3
2 13	−3.19E−1	3 5	1.22E−3	6 11	1.74E−2	3 3	−4.52E−6	6 12	−1.53E−1	10 7	−1.21E+0
2 14	−4.19E−1	3 6	2.95E−4	6 12	6.25E−5	3 4	−2.54E−2	6 13	−1.22E−1	10 8	−7.78E−3
3 1	3.66E−2	3 7	8.78E−4	6 13	7.39E−4	3 5	−3.85E−2	6 14	−3.69E−1	10 9	−1.07E−4
3 2	5.58E−2	3 8	9.20E−4	6 14	−1.34E−4	3 6	−6.91E−4	7 1	1.33E+0	10 10	−2.19E−1
3 3	5.21E−1	3 9	2.34E−3	6 15	−1.34E−3	3 7	−9.24E−5	7 2	1.72E−3	10 11	−2.54E−4
3 4	−3.95E−5	3 10	−1.04E−2	7 1	1.92E−3	3 8	−2.04E−3	7 3	−2.83E−3	10 12	−5.32E−4
3 5	−8.50E−4	3 11	−1.37E−2	7 2	6.79E−2	3 9	−1.22E−2	7 4	−2.90E−3	10 13	−1.69E−3
3 6	−3.21E−5	3 12	−5.25E−3	7 3	3.36E−2	3 10	−2.69E−3	7 5	−1.57E−2	10 14	−1.39E−3
3 7	−1.41E+0	3 13	−1.47E+0	7 4	3.69E−2	3 11	−1.65E−2	7 6	−1.28E−2	11 1	3.49E−6
3 8	−1.44E−4	3 14	−1.63E−2	7 5	1.28E−2	3 12	−9.36E−3	7 7	−5.35E−3	11 2	3.79E+0
3 9	−1.57E−3	3 15	−4.72E−2	7 6	5.03E−1	3 13	−5.36E−3	7 8	−3.35E−2	11 3	2.65E−1
3 10	−2.34E−1	4 1	1.42E−5	7 7	1.33E−2	3 14	−5.00E−2	7 9	−5.24E−4	11 4	5.14E−8
3 11	−6.50E−4	4 2	3.16E−2	7 8	4.22E−4	4 1	3.37E−1	7 10	−2.85E−6	11 5	−2.68E−5
3 12	−4.14E−5	4 3	2.15E−2	7 9	9.56E−3	4 2	8.38E−2	7 11	−3.77E−2	11 6	−3.55E−5
3 13	−4.94E−4	4 4	6.64E−2	7 10	5.08E−4	4 3	−1.82E−4	7 12	−1.59E−2	11 7	−6.06E−3
3 14	−3.57E−4	4 5	1.64E−3	7 11	8.72E−5	4 4	−1.21E−1	7 13	−2.56E−1	11 8	−5.41E−3
	$^2S^e$–$^2P^o$	4 6	2.32E−4	7 12	6.98E−4	4 5	−4.38E−2	7 14	−2.57E−1	11 9	−4.15E−3
1 1	1.61E−1	4 7	6.32E−3	7 13	6.21E−4	4 6	−3.97E−2	8 1	1.01E+0	11 10	−8.13E−1
1 2	−1.20E−1	4 8	5.22E−5	7 14	1.03E−4	4 7	−1.11E−3	8 2	9.45E−1	11 11	−4.16E−4
1 3	−4.19E+0	4 9	1.08E−4	7 15	−1.13E−4	4 8	−2.59E−2	8 3	2.03E−5	11 12	−1.12E−2
1 4	−3.33E−1	4 10	8.37E−6		$^2P^o$–$^2P^e$	4 9	−1.94E−1	8 4	−2.01E−4	11 13	−5.15E−3
1 5	−1.68E−1	4 11	−7.38E−6	1 1	−1.92E−2	4 10	−1.06E−5	8 5	−7.11E−4	11 14	−7.45E−4
1 6	−8.66E−4	4 12	−3.74E−4	1 2	−4.56E+0	4 11	−5.44E−2	8 6	−2.08E−6	12 1	1.24E−2
1 7	−8.38E−4	4 13	−3.34E−3	1 3	−9.12E−1	4 12	−3.88E−1	8 7	−3.89E−3	12 2	1.08E+0

Cl-like Fe (Fe^{9+})

i	i'	gf_L	i	i'	gf_L	i	i'	gf_L	i	i'	gf_L	i	i'	gf_L	i	i'	gf_L
12	3	7.28E−1	15	13	7.19E−5	3	10	−2.15E−2	6	2	6.69E−6	8	14	−5.33E−4	11	6	3.22E−5
12	4	1.28E−4	15	14	−1.76E−3	3	11	−2.36E−2	6	3	3.29E−5	8	15	−2.38E−3	11	7	−9.78E−5
12	5	1.27E−5			$^2P^o$–$^2D^e$	3	12	−1.08E−3	6	4	−1.20E−5	8	16	−1.88E−2	11	8	−4.49E−6
12	6	−5.24E−4	1	1	−1.22E−2	3	13	−2.67E−2	6	5	−4.87E−5	8	17	−2.00E−2	11	9	−9.70E−4
12	7	−3.47E−2	1	2	−3.10E−3	3	14	−2.68E−1	6	6	−4.69E−4	8	18	−5.19E−3	11	10	−2.94E−3
12	8	−8.53E−3	1	3	−8.67E+0	3	15	−2.86E−4	6	7	−1.45E−3	8	19	−5.80E−2	11	11	−5.65E−2
12	9	−5.64E−4	1	4	−4.61E−4	3	16	−2.42E−1	6	8	−1.59E−2	8	20	−6.89E−4	11	12	−2.49E−1
12	10	−7.87E−1	1	5	−5.30E−1	3	17	−7.98E−2	6	9	−1.07E−7	9	1	1.35E−2	11	13	−2.47E−2
12	11	−3.19E−4	1	6	−1.12E−4	3	18	−2.20E+0	6	10	−4.63E−2	9	2	3.54E+0	11	14	−2.77E−3
12	12	−6.73E−4	1	7	−2.71E−7	3	19	−2.38E−1	6	11	−5.43E−2	9	3	2.81E−1	11	15	−7.26E−1
12	13	−2.57E−5	1	8	−4.62E−3	3	20	−3.95E+0	6	12	−1.95E−4	9	4	1.16E−3	11	16	−1.04E−4
12	14	−4.52E−5	1	9	−1.86E−5	4	1	2.38E−1	6	13	−1.76E−2	9	5	1.05E−5	11	17	−2.88E−4
13	1	3.62E−3	1	10	−3.59E−3	4	2	5.36E−3	6	14	−6.99E−2	9	6	−8.73E−5	11	18	−3.20E−3
13	2	4.97E−2	1	11	−1.35E−1	4	3	5.16E−2	6	15	−6.54E−4	9	7	−8.84E−6	11	19	−2.06E−2
13	3	5.97E−2	1	12	−1.12E+0	4	4	−1.26E−2	6	16	−3.50E−3	9	8	−8.53E−3	11	20	−2.46E−3
13	4	1.34E−5	1	13	−3.86E−3	4	5	−6.00E−6	6	17	−1.00E−2	9	9	−9.04E−3	12	1	2.23E−3
13	5	3.99E−7	1	14	−9.71E−3	4	6	−9.34E−4	6	18	−6.16E−3	9	10	−2.69E−4	12	2	8.00E−3
13	6	−2.14E−5	1	15	−7.90E−1	4	7	−3.40E−2	6	19	−1.01E−2	9	11	−3.64E−3	12	3	1.28E+0
13	7	−1.86E−2	1	16	−4.03E−6	4	8	−5.59E−1	6	20	−7.23E−5	9	12	−3.54E−3	12	4	1.55E−3
13	8	−5.05E−4	1	17	−8.28E−2	4	9	−2.99E−1	7	1	3.35E+0	9	13	−1.01E−2	12	5	5.95E−1
13	9	−1.33E−4	1	18	−2.41E−4	4	10	−2.70E−1	7	2	5.21E−4	9	14	−1.38E−3	12	6	1.28E−4
13	10	−5.04E−3	1	19	−2.36E−3	4	11	−3.65E−1	7	3	1.00E−4	9	15	−7.72E−4	12	7	1.99E−5
13	11	−1.36E−4	1	20	−2.52E−4	4	12	−5.67E−2	7	4	−3.92E−5	9	16	−6.39E−5	12	8	−4.35E−4
13	12	−4.90E−5	2	1	3.36E−1	4	13	−1.04E−1	7	5	−3.20E−3	9	17	−7.31E−2	12	9	1.24E−4
13	13	−2.43E−5	2	2	3.37E−3	4	14	−1.41E−1	7	6	−1.98E−3	9	18	−6.66E−3	12	10	−3.38E−4
13	14	−3.85E−5	2	3	4.93E−4	4	15	−1.81E−2	7	7	−3.18E−2	9	19	−6.99E−2	12	11	−7.86E−2
14	1	1.53E−2	2	4	−4.87E−1	4	16	−1.06E−1	7	8	−2.80E−3	9	20	−4.59E−3	12	12	−4.29E−1
14	2	1.01E−2	2	5	−5.86E−4	4	17	−1.30E−3	7	9	−3.16E−3	10	1	1.24E−1	12	13	−5.62E−3
14	3	2.64E−1	2	6	−2.42E−1	4	18	−1.67E−2	7	10	−2.59E−2	10	2	1.72E−3	12	14	−7.84E−4
14	4	1.30E−3	2	7	−2.08E−1	4	19	−2.45E−1	7	11	−3.64E−3	10	3	2.17E−1	12	15	−4.96E−1
14	5	6.08E−5	2	8	−1.78E−1	4	20	−2.06E−2	7	12	−1.02E−2	10	4	1.12E−3	12	16	−1.32E−6
14	6	8.06E−4	2	9	−4.30E−1	5	1	9.40E−1	7	13	−5.75E−3	10	5	4.26E−1	12	17	−1.14E−1
14	7	2.68E−3	2	10	−3.90E−1	5	2	5.89E−2	7	14	−5.32E−3	10	6	−1.01E−6	12	18	−3.18E−3
14	8	3.36E−3	2	11	−5.37E−2	5	3	6.29E−2	7	15	−2.55E−4	10	7	−9.81E−6	12	19	−5.28E−3
14	9	2.16E−3	2	12	−3.04E−2	5	4	−1.16E−2	7	16	−1.74E−2	10	8	−2.76E−3	12	20	−4.01E−3
14	10	6.35E−4	2	13	−8.46E−2	5	5	−1.02E−4	7	17	−1.66E−2	10	9	−9.28E−3	13	1	2.06E−2
14	11	3.96E−5	2	14	−5.83E−2	5	6	−3.62E−2	7	18	−5.21E−1	10	10	−4.17E−3	13	2	1.75E−1
14	12	−2.52E−4	2	15	−8.99E−6	5	7	−5.66E−3	7	19	−2.25E−2	10	11	−2.23E−1	13	3	2.37E−4
14	13	−1.06E−3	2	16	−1.13E+0	5	8	−5.39E−2	7	20	−1.87E−1	10	12	−2.03E+0	13	4	7.31E−8
14	14	−2.04E−2	2	17	−1.11E−2	5	9	−1.09E−2	8	1	1.60E−1	10	13	−5.09E−2	13	5	1.20E−2
15	1	9.24E−2	2	18	−2.16E−1	5	10	−2.25E−1	8	2	9.72E−1	10	14	−2.76E−2	13	6	9.21E−5
15	2	1.34E−3	2	19	−2.49E−1	5	11	−1.11E−2	8	3	7.00E−1	10	15	−3.09E−1	13	7	1.30E−4
15	3	1.27E−2	2	20	−7.38E−2	5	12	−5.30E−3	8	4	4.69E−4	10	16	3.73E−4	13	8	−4.63E−4
15	4	1.22E−3	3	1	3.40E−2	5	13	−1.96E−1	8	5	5.91E−4	10	17	−2.83E−3	13	9	−2.28E−4
15	5	4.49E−4	3	2	7.59E−3	5	14	−4.21E−1	8	6	−1.67E−3	10	18	−9.64E−4	13	10	−8.75E−4
15	6	1.78E−3	3	3	2.60E−1	5	15	−4.72E−4	8	7	−1.16E−4	10	19	−9.38E−4	13	11	−8.82E−4
15	7	1.34E−3	3	4	−1.58E−2	5	16	0.55E−2	8	8	−2.85E−3	10	20	−7.67E−4	13	12	−6.93E−3
15	8	2.19E−5	3	5	−1.87E−4	5	17	−4.68E−3	8	9	−8.09E−2	11	1	5.64E−3	13	13	−9.99E−3
15	9	2.25E−4	3	6	−7.61E−2	5	18	−5.52E−1	8	10	−1.19E−2	11	2	1.08E−3	13	14	−5.81E−3
15	10	6.50E−5	3	7	−2.03E−2	5	19	−4.99E−1	8	11	−1.30E−2	11	3	4.07E+0	13	15	−1.23E−1
15	11	5.30E−4	3	8	−7.90E−4	5	20	−1.70E−1	8	12	−1.29E−4	11	4	2.89E−4	13	16	−4.03E−4
15	12	1.51E−3	3	9	−5.28E−1	6	1	3.77E−2	8	13	−7.81E−3	11	5	4.65E−1	13	17	−4.49E+0

Cl-like Fe (Fe^{9+})

i i'	gf$_L$	i i'	gf$_L$	i i'	gf$_L$	i i'	gf$_L$	i i'	gf$_L$	i i'	gf$_L$
13 18	−7.46E−2	1 9	−8.61E−2	5 9	8.90E−5	9 9	1.59E−2	13 9	2.41E−2	2 14	−4.72E−2
13 19	−1.93E−3	1 10	−3.73E−1	5 10	1.36E−7	9 10	1.41E−4	13 10	4.21E−3	2 15	−2.02E−2
13 20	−3.28E−4	1 11	−1.95E−4	5 11	1.73E−4	9 11	1.53E−4	13 11	3.43E−4	2 16	−3.17E+0
14 1	1.46E−2	1 12	−1.26E−2	5 12	2.89E−6	9 12	9.26E−4	13 12	1.29E−4	2 17	−1.37E−3
14 2	2.84E−3	1 13	−3.06E+0	5 13	−3.60E−3	9 13	−1.56E−4	13 13	−1.18E−3	2 18	−1.97E−1
14 3	1.35E−2	2 1	−3.98E−2	6 1	1.87E−1	10 1	4.41E−2	14 1	6.87E−2	2 19	−1.88E+0
14 4	6.29E−3	2 2	−2.16E−2	6 2	9.24E−1	10 2	5.85E−3	14 2	1.14E+0	2 20	−3.76E−1
14 5	4.20E−2	2 3	−2.28E−3	6 3	3.57E−1	10 3	5.46E−4	14 3	2.71E−1	3 1	2.57E−1
14 6	1.46E−3	2 4	−1.88E−3	6 4	1.83E−1	10 4	6.45E−4	14 4	2.72E−1	3 2	1.42E−1
14 7	1.12E−5	2 5	−1.04E−1	6 5	3.26E−2	10 5	5.03E−3	14 5	4.55E−2	3 3	2.23E−3
14 8	1.12E−3	2 6	−6.95E−2	6 6	5.69E−2	10 6	3.26E−7	14 6	2.61E−1	3 4	−2.41E−2
14 9	1.39E−3	2 7	−2.15E−1	6 7	3.58E−3	10 7	1.96E−3	14 7	4.53E−4	3 5	−5.80E−4
14 10	7.03E−4	2 8	−4.13E−1	6 8	2.12E−2	10 8	1.75E−3	14 8	3.39E−3	3 6	−1.88E−2
14 11	3.33E−4	2 9	−1.31E+0	6 9	6.94E−3	10 9	2.31E−2	14 9	7.50E−2	3 7	−5.92E−2
14 12	1.76E−2	2 10	−4.97E+0	6 10	9.92E−5	10 10	8.86E−3	14 10	1.90E−8	3 8	−1.10E−1
14 13	3.14E−3	2 11	−3.42E−4	6 11	2.14E−5	10 11	4.69E−3	14 11	2.97E−4	3 9	−5.09E−1
14 14	4.75E−3	2 12	−2.10E−2	6 12	1.10E−3	10 12	5.40E−1	14 12	6.00E−4	3 10	−2.48E−1
14 15	9.37E−4	2 13	−1.03E+0	6 13	−1.48E−3	10 13	−5.48E−3	14 13	2.52E−4	3 11	−2.45E−1
14 16	−5.35E−5	3 1	2.24E−3	7 1	5.28E−3	11 1	3.85E−1		**^{2}D^o–^{2}D^e**	3 12	−1.44E−2
14 17	−5.02E−5	3 2	4.26E−3	7 2	1.49E−2	11 2	1.35E−2	1 1	3.97E−2	3 13	−4.79E−1
14 18	−3.79E−3	3 3	2.67E−5	7 3	7.42E−3	11 3	1.99E−1	1 2	4.74E−3	3 14	−2.92E−3
14 19	−1.93E−4	3 4	2.34E−4	7 4	7.31E−4	11 4	1.54E−1	1 3	5.33E−3	3 15	−1.48E−3
14 20	−1.22E−3	3 5	8.32E−5	7 5	1.07E−3	11 5	7.05E−2	1 4	−7.52E−5	3 16	−5.43E−2
15 1	7.91E−2	3 6	1.11E−2	7 6	4.36E−4	11 6	8.86E−3	1 5	−1.30E−3	3 17	−7.22E−5
15 2	4.01E−7	3 7	2.40E−3	7 7	7.01E−4	11 7	2.46E−1	1 6	−8.39E−2	3 18	−1.18E−1
15 3	7.39E−3	3 8	2.58E−3	7 8	1.13E−2	11 8	1.19E−3	1 7	−1.35E−1	3 19	−2.64E−1
15 4	1.19E−3	3 9	−5.18E−4	7 9	1.42E−3	11 9	1.68E−3	1 8	−4.12E−1	3 20	−9.33E−1
15 5	9.98E−1	3 10	−1.89E−4	7 10	1.64E−2	11 10	5.13E−3	1 9	−6.51E−1	4 1	1.23E−1
15 6	1.12E−3	3 11	−2.38E+0	7 11	6.83E−2	11 11	8.01E−5	1 10	−5.86E−1	4 2	6.67E−2
15 7	7.82E−4	3 12	−3.55E−2	7 12	7.97E−2	11 12	3.67E−4	1 11	−1.02E+0	4 3	3.82E−3
15 8	8.17E−4	3 13	−6.82E−4	7 13	−2.98E+0	11 13	−4.30E−4	1 12	−2.16E−1	4 4	−1.38E−2
15 9	8.27E−4	4 1	1.03E−2	8 1	2.04E+0	12 1	1.85E−1	1 13	−3.92E−2	4 5	−1.58E−4
15 10	6.87E−4	4 2	8.72E−5	8 2	4.56E−3	12 2	4.93E−2	1 14	−2.23E+0	4 6	−1.25E−2
15 11	5.31E−3	4 3	3.29E−4	8 3	9.27E−4	12 3	8.66E−3	1 15	−8.90E−3	4 7	−6.54E−2
15 12	3.35E−4	4 4	2.68E−5	8 4	4.86E−3	12 4	5.41E−3	1 16	−4.21E−1	4 8	−4.36E−1
15 13	5.66E−5	4 5	1.93E−2	8 5	4.24E−3	12 5	6.97E−1	1 17	−1.49E−2	4 9	−1.65E−1
15 14	3.18E−3	4 6	9.86E−3	8 6	6.71E−3	12 6	8.00E−2	1 18	−1.33E+0	4 10	−2.91E−1
15 15	5.89E−3	4 7	1.06E−2	8 7	2.36E−2	12 7	1.17E−1	1 19	−1.22E−1	4 11	−5.46E−1
15 16	3.50E−4	4 8	5.52E−4	8 8	2.56E−4	12 8	1.71E−1	1 20	−2.50E−1	4 12	−1.28E−2
15 17	4.68E−5	4 9	6.07E−5	8 9	1.60E−2	12 9	2.21E−4	2 1	3.24E−1	4 13	−1.36E−1
15 18	8.45E−4	4 10	2.93E−6	8 10	3.70E−3	12 10	2.12E−3	2 2	3.21E−1	4 14	−6.14E−2
15 19	−7.16E−4	4 11	4.50E−7	8 11	2.99E−4	12 11	4.50E−5	2 3	1.34E−2	4 15	−7.89E−3
15 20	−8.66E−5	4 12	−2.55E−5	8 12	6.17E−3	12 12	4.65E−4	2 4	−5.49E−1	4 16	−1.34E−1
	^{2}P^e–^{2}D^o	4 13	−4.43E−5	8 13	−1.37E−3	12 13	−2.11E−5	2 5	−4.21E−6	4 17	−3.78E−3
1 1	−2.35E−1	5 1	2.20E−3	9 1	1.67E+0	13 1	3.03E−3	2 6	−9.74E−2	4 18	−3.94E−2
1 2	−2.23E−1	5 2	1.85E−1	9 2	2.35E−1	13 2	6.97E−3	2 7	−2.31E−1	4 19	−1.88E+0
1 3	−7.33E−1	5 3	4.39E−1	9 3	3.37E−2	13 3	1.87E+0	2 8	−1.99E−1	4 20	−8.24E−3
1 4	−2.58E−1	5 4	1.09E−2	9 4	1.37E−1	13 4	3.91E−1	2 9	−1.29E−2	5 1	1.45E−1
1 5	−4.19E−1	5 5	5.87E−4	9 5	1.69E−1	13 5	4.41E−2	2 10	−4.79E−2	5 2	7.11E−2
1 6	−5.05E−1	5 6	6.93E−5	9 6	4.52E−2	13 6	8.12E−2	2 11	−2.45E−2	5 3	3.53E−2
1 7	−1.78E+0	5 7	6.89E−4	9 7	7.57E−3	13 7	1.84E−1	2 12	−6.84E−6	5 4	−5.87E−4
1 8	−4.88E−1	5 8	6.75E−3	9 8	5.51E−3	13 8	5.56E−2	2 13	−1.31E−1	5 5	−1.21E−3

Cl-like Fe (Fe^{9+})

i	i′	gf$_L$	i	i′	gf$_L$	i	i′	gf$_L$	i	i′	gf$_L$	i	i′	gf$_L$	i	i′	gf$_L$
5	6	−1.88E−3	7	18	−8.00E−3	10	10	−3.11E−3	13	2	1.06E−1	3	5	−1.87E−3	7	1	1.72E+0
5	7	−2.40E−3	7	19	−1.87E−1	10	11	−1.37E−3	13	3	1.93E−1	3	6	−1.05E+0	7	2	1.16E+0
5	8	−2.77E−4	7	20	−1.57E−2	10	12	−3.82E−5	13	4	6.51E−4	3	7	−6.77E−2	7	3	−2.72E−2
5	9	−7.27E−5	8	1	3.91E+0	10	13	−7.25E−3	13	5	3.43E−3	3	8	−4.10E−2	7	4	−3.42E−3
5	10	−3.78E−1	8	2	4.89E−1	10	14	−2.56E−3	13	6	1.48E−5	3	9	−4.40E−3	7	5	−1.13E−3
5	11	−3.69E−1	8	3	9.86E−2	10	15	−1.91E−2	13	7	8.66E−8	3	10	−6.72E−5	7	6	−5.93E−2
5	12	−2.89E−2	8	4	−4.02E−5	10	16	−1.06E−2	13	8	1.25E−3	3	11	−3.38E−2	7	7	−7.94E−2
5	13	−1.18E−1	8	5	−5.39E−3	10	17	−1.61E−4	13	9	3.53E−3	3	12	−5.12E−3	7	8	−4.65E−2
5	14	−1.20E−1	8	6	−1.12E−5	10	18	−3.75E−3	13	10	9.62E−4	3	13	−6.99E−2	7	9	−3.23E−2
5	15	−2.12E−2	8	7	−1.98E−2	10	19	−1.81E−3	13	11	6.18E−2	3	14	−2.19E+0	7	10	−1.00E−3
5	16	−8.06E−3	8	8	−4.40E−2	10	20	−8.29E−2	13	12	5.13E−1	4	1	2.33E−2	7	11	−1.16E−1
5	17	−1.30E−3	8	9	−5.71E−7	11	1	1.55E−3	13	13	9.25E−3	4	2	3.80E−1	7	12	−3.30E−2
5	18	−1.12E−1	8	10	−9.60E−3	11	2	1.34E−2	13	14	5.02E−3	4	3	−3.51E−2	7	13	−1.27E−1
5	19	−1.53E−1	8	11	−1.58E−2	11	3	8.82E−2	13	15	9.81E−2	4	4	−1.79E−2	7	14	−1.45E−1
5	20	−6.47E−2	8	12	−2.95E−3	11	4	3.24E−7	13	16	4.29E−4	4	5	−2.52E−2	8	1	8.59E−2
6	1	1.24E−1	8	13	−1.03E−1	11	5	3.44E−2	13	17	6.91E−3	4	6	−2.35E−1	8	2	3.45E+0
6	2	4.76E−1	8	14	−5.61E−3	11	6	4.05E−6	13	18	7.41E−4	4	7	−4.63E−2	8	3	−7.95E−3
6	3	6.50E−3	8	15	−3.62E−2	11	7	−1.55E−5	13	19	1.14E−4	4	8	−7.91E−1	8	4	−1.19E−2
6	4	−3.45E−3	8	16	−1.54E−2	11	8	−2.75E−3	13	20	3.40E−6	4	9	−2.17E−2	8	5	−1.13E−2
6	5	−7.10E−4	8	17	−1.00E−4	11	9	−2.96E−3			^{2}D^o–^{2}F^e	4	10	−2.83E−1	8	6	−3.18E−2
6	6	−2.59E−2	8	18	−2.78E−1	11	10	−1.78E−4	1	1	2.64E−1	4	11	−7.98E−3	8	7	−3.57E−2
6	7	−1.37E−2	8	19	−3.27E−2	11	11	−1.40E−1	1	2	6.74E−3	4	12	−7.27E−4	8	8	−3.18E−2
6	8	−2.13E−3	8	20	−7.37E−2	11	12	−8.64E−1	1	3	−3.79E−1	4	13	−4.52E−1	8	9	−7.31E−2
6	9	−2.34E−4	9	1	1.78E−1	11	13	−1.76E 2	1	4	2.35E−1	4	14	−3.55E−2	8	10	−5.77E−3
6	10	−4.60E−4	9	2	3.07E+0	11	14	−1.64E−3	1	5	−9.41E−1	5	1	3.77E−2	8	11	−2.41E−3
6	11	−2.48E−2	9	3	4.67E−1	11	15	−2.26E−1	1	6	−8.06E−1	5	2	1.62E−1	8	12	−8.14E−2
6	12	−1.55E−2	9	4	1.64E−4	11	16	−1.70E−3	1	7	−5.61E−2	5	3	−5.25E−3	8	13	−7.46E−2
6	13	−6.32E−2	9	5	8.54E−4	11	17	−2.54E−5	1	8	−3.58E−1	5	4	−3.48E−3	8	14	−8.00E−2
6	14	−4.07E−1	9	6	−4.32E−6	11	18	−1.10E−3	1	9	−6.19E−2	5	5	−1.13E−2	9	1	1.27E−1
6	15	−2.42E−2	9	7	−3.07E−4	11	19	−3.62E−3	1	10	−1.07E−1	5	6	−9.42E−2	9	2	2.12E+0
6	16	−3.14E−2	9	8	−1.13E−3	11	20	−8.38E−6	1	11	−1.47E+0	5	7	−5.33E−2	9	3	−3.83E−4
6	17	−1.26E−4	9	9	−1.03E−4	12	1	9.26E−2	1	12	−2.19E−1	5	8	−1.33E−2	9	4	−2.18E−3
6	18	−2.09E−3	9	10	−8.03E−3	12	2	9.14E−3	1	13	−7.75E−3	5	9	−5.05E−4	9	5	−6.21E−3
6	19	−1.37E−1	9	11	−6.24E−3	12	3	2.66E−2	1	14	−1.07E−1	5	10	−5.57E−2	9	6	−3.94E−4
6	20	−1.04E−3	9	12	−3.96E−3	12	4	1.69E−3	2	1	2.00E−1	5	11	−3.78E−2	9	7	−1.82E−2
7	1	1.51E+0	9	13	−4.80E−3	12	5	2.57E+0	2	2	6.72E−1	5	12	−3.30E−2	9	8	−4.08E−2
7	2	4.85E−1	9	14	−2.88E−8	12	6	6.10E−4	2	3	−6.14E−4	5	13	−1.24E−1	9	9	−2.41E−3
7	3	9.52E−2	9	15	−5.93E−3	12	7	−1.31E−5	2	4	−2.43E−2	5	14	−1.37E−1	9	10	−5.12E−3
7	4	−6.64E−4	9	16	−4.28E−3	12	8	−2.22E−4	2	5	−1.19E−2	6	1	5.56E+0	9	11	−2.11E−2
7	5	−1.28E−3	9	17	−3.74E−4	12	9	−1.27E−3	2	6	−2.21E−1	6	2	2.47E−1	9	12	−2.32E−2
7	6	−8.83E−4	9	18	−6.63E−2	12	10	−9.88E−3	2	7	−1.32E−1	6	3	−3.02E−4	9	13	−1.30E−1
7	7	−1.09E−6	9	19	−0.30E−2	12	11	−1.00E−2	2	8	2.85E 1	6	4	1.83E−4	9	14	−1.73E−1
7	8	7.00E 2	9	20	−1.52E−1	12	12	−1.68E−1	2	9	−1.30E−3	6	5	−1.10E−3	10	1	1.16E−2
7	9	−1.29E−3	10	1	1.31E−4	12	13	−1.08E−2	2	10	−2.06E+0	6	6	−1.29E−1	10	2	4.23E−3
7	10	−3.06E−2	10	2	3.53E+0	12	14	−2.14E−2	2	11	−9.49E−1	6	7	−2.26E−3	10	3	−6.03E−4
7	11	−9.18E−3	10	3	1.37E+0	12	15	−2.46E+0	2	12	−9.71E−2	6	8	−1.53E−1	10	4	−2.11E−4
7	12	−5.63E−4	10	4	6.63E−4	12	16	−6.30E−4	2	13	−5.72E−4	6	9	−1.13E−1	10	5	−1.40E−3
7	13	−2.83E−1	10	5	2.07E−3	12	17	−1.66E−1	2	14	−8.43E−1	6	10	−4.75E−2	10	6	−1.59E−2
7	14	−1.20E−2	10	6	−1.75E−4	12	18	0.15E 3	3	1	3.59E−2	6	11	−9.62E−4	10	7	−1.15E−4
7	15	−1.45E−3	10	7	−1.15E−4	12	19	−7.15E−4	3	2	7.98E−3	6	12	−2.57E−3	10	8	−6.31E−3
7	16	−3.25E−3	10	8	−2.79E−3	12	20	−5.15E−4	3	3	−9.28E−2	6	13	−6.96E−1	10	9	−1.10E−2
7	17	−8.46E−4	10	9	−3.43E−4	13	1	8.46E−1	3	4	−1.28E+0	6	14	−5.13E−3	10	10	−2.58E−2

Cl-like Fe (Fe^{9+})

i	i'	gf_L	i	i'	gf_L	i	i'	gf_L	i	i'	gf_L	i	i'	gf_L	i	i'	gf_L
10	11	−2.03E−2	1	6	−9.42E−1	5	10	−1.12E−3	10	2	1.48E+0	14	6	1.99E−1	18	10	2.90E−2
10	12	−7.98E−3	1	7	−1.05E−1	5	11	−3.33E+0	10	3	9.55E−3	14	7	4.08E−2	18	11	1.13E−3
10	13	−2.54E−2	1	8	−7.92E−1	5	12	−2.72E−2	10	4	8.98E−2	14	8	1.57E−2	18	12	−9.46E−4
10	14	−2.06E−3	1	9	−2.18E+0	6	1	7.29E−3	10	5	1.69E−1	14	9	3.10E−5	19	1	1.11E−2
11	1	2.73E−1	1	10	−9.10E−2	6	2	1.40E−2	10	6	4.23E−2	14	10	8.73E−4	19	2	1.40E+0
11	2	2.50E−2	1	11	−4.81E−7	6	3	3.89E−1	10	7	6.46E−3	14	11	4.62E−4	19	3	8.05E−1
11	3	−1.02E−6	1	12	−3.27E+0	6	4	4.01E−3	10	8	1.81E−2	14	12	−9.24E−3	19	4	1.11E−1
11	4	−6.83E−5	2	1	−2.45E−3	6	5	3.50E−5	10	9	1.09E−2	15	1	5.14E−2	19	5	1.60E−3
11	5	−1.38E−6	2	2	−3.25E−2	6	6	1.62E−2	10	10	3.13E−3	15	2	1.98E−3	19	6	2.74E−1
11	6	−2.72E−4	2	3	−1.20E−3	6	7	5.85E−4	10	11	7.74E−4	15	3	8.00E−3	19	7	1.90E−1
11	7	−1.14E−1	2	4	−1.68E−1	6	8	1.14E−5	10	12	−2.43E−3	15	4	6.20E−4	19	8	1.30E−2
11	8	−9.06E−2	2	5	−5.40E−2	6	9	4.63E−4	11	1	2.77E−1	15	5	1.91E−6	19	9	6.15E−1
11	9	−6.48E+0	2	6	−4.17E−1	6	10	1.20E−5	11	2	4.37E−1	15	6	1.09E−3	19	10	3.58E−4
11	10	−1.32E−2	2	7	−4.22E−1	6	11	−2.14E−5	11	3	9.49E−2	15	7	9.05E−3	19	11	1.18E−4
11	11	−4.06E−3	2	8	−2.05E+0	6	12	−4.93E−5	11	4	5.33E−1	15	8	6.45E−7	19	12	−3.37E−8
11	12	−2.05E−2	2	9	−1.63E+0	7	1	4.27E−3	11	5	1.61E−1	15	9	8.96E−4	20	1	5.28E−2
11	13	−6.81E−4	2	10	−5.82E−2	7	2	1.20E−2	11	6	3.14E−2	15	10	2.47E−2	20	2	1.71E−1
11	14	−4.66E−4	2	11	−1.64E−3	7	3	9.56E−2	11	7	1.88E−3	15	11	1.93E−1	20	3	1.10E+0
12	1	3.23E−2	2	12	−3.20E−2	7	4	4.22E−2	11	8	9.53E−3	15	12	−1.84E−3	20	4	1.54E−1
12	2	1.94E−1	3	1	−5.05E−2	7	5	9.50E−2	11	9	4.91E−4	16	1	4.61E−2	20	5	2.21E−3
12	3	6.41E−4	3	2	−4.89E−3	7	6	4.06E−3	11	10	7.61E−3	16	2	1.67E+0	20	6	1.61E−2
12	4	−4.24E−4	3	3	−1.71E−2	7	7	1.53E−2	11	11	3.56E−3	16	3	1.58E−1	20	7	6.67E−2
12	5	−8.10E−5	3	4	−4.59E−2	7	8	4.30E−3	11	12	−5.11E−1	16	4	5.81E−3	20	8	1.17E−2
12	6	−8.88E−4	3	5	−1.05E−2	7	9	7.64E−4	12	1	3.53E−1	16	5	2.85E−1	20	9	4.39E−3
12	7	−1.07E−3	3	6	−4.86E−1	7	10	3.56E−4	12	2	7.75E−2	16	6	4.24E−1	20	10	1.01E−1
12	8	−1.84E−3	3	7	−1.37E−1	7	11	2.18E−4	12	3	2.83E−4	16	7	7.04E−2	20	11	1.65E−4
12	9	−2.36E−2	3	8	−2.40E−2	7	12	−2.52E−5	12	4	2.41E−3	16	8	3.95E−2	20	12	−2.08E−7
12	10	−7.94E−3	3	9	−3.81E−2	8	1	1.76E−1	12	5	1.41E−2	16	9	5.42E−2			**^{2}F^o–^{2}F^e**
12	11	−5.28E−1	3	10	−1.39E+1	8	2	1.97E−1	12	6	2.12E−2	16	10	4.08E−3	1	1	4.56E−1
12	12	−4.02E+0	3	11	−1.34E−2	8	3	1.52E−3	12	7	2.18E−4	16	11	7.99E−4	1	2	7.59E−2
12	13	−1.32E−3	3	12	−3.86E+0	8	4	5.48E−2	12	8	1.58E−3	16	12	−1.09E−4	1	3	−5.73E−3
12	14	−7.88E−4	4	1	3.18E−2	8	5	2.96E−2	12	9	8.44E−4	17	1	8.58E−4	1	4	−4.22E−2
13	1	5.55E−6	4	2	8.69E−1	8	6	1.98E−2	12	10	3.62E−2	17	2	3.43E−3	1	5	−5.57E−1
13	2	9.12E−2	4	3	1.18E−2	8	7	1.39E−2	12	11	2.12E−3	17	3	4.21E−2	1	6	−4.86E−1
13	3	2.05E−5	4	4	3.09E−3	8	8	1.28E−3	12	12	−4.07E+0	17	4	1.42E−5	1	7	−4.14E−1
13	4	1.46E−3	4	5	1.63E−2	8	9	7.51E−3	13	1	2.76E+0	17	5	7.71E−3	1	8	−3.15E−1
13	5	4.76E−4	4	6	6.34E−3	8	10	8.24E−4	13	2	4.35E−2	17	6	7.07E−4	1	9	−8.92E−2
13	6	2.29E−3	4	7	9.09E−5	8	11	1.57E−4	13	3	2.09E−1	17	7	2.09E−4	1	10	−9.52E−1
13	7	1.14E−3	4	8	1.68E−7	8	12	−6.26E−3	13	4	8.75E−1	17	8	3.51E−2	1	11	−6.06E+0
13	8	3.01E−6	4	9	−3.28E−4	9	1	2.18E+0	13	5	1.66E−1	17	9	4.58E−3	1	12	−7.54E−1
13	9	2.96E−2	4	10	−7.73E−5	9	2	4.36E−1	13	6	7.20E−2	17	10	1.29E−3	1	13	−3.57E−1
13	10	3.16E−4	4	11	−6.28E−3	9	3	1.62E−1	13	7	5.73E−3	17	11	3.78E−2	1	14	−1.63E−1
13	11	2.40E−3	4	12	−2.89E−7	9	4	4.38E−1	13	8	4.80E−3	17	12	−2.16E−3	2	1	2.25E−2
13	12	1.31E−2	5	1	6.49E−3	9	5	6.45E−4	13	9	6.08E−5	18	1	1.48E−2	2	2	4.15E−1
13	13	1.07E−3	5	2	7.84E−3	9	6	1.14E−1	13	10	6.56E−3	18	2	6.45E−1	2	3	−5.44E−2
13	14	3.78E−4	5	3	1.59E−4	9	7	3.50E−2	13	11	1.86E−3	18	3	1.66E+0	2	4	−5.03E−1
		^{2}D^e–^{2}F^o	5	4	1.77E−5	9	8	9.90E−3	13	12	−6.78E−2	18	4	5.99E−2	2	5	−4.12E−4
1	1	−9.93E−2	5	5	8.12E−3	9	9	1.28E−3	14	1	1.28E+0	18	5	1.47E−1	2	6	−1.17E+0
1	2	−3.69E−1	5	6	3.20E−3	9	10	1.23E−4	14	2	1.12E+0	18	6	7.05E−3	2	7	−7.64E−3
1	3	−2.81E−1	5	7	6.95E−3	9	11	7.63E−5	14	3	6.19E−3	18	7	1.37E−1	2	8	−1.27E−1
1	4	−7.99E−1	5	8	3.32E−4	9	12	−1.33E−2	14	4	1.14E−4	18	8	1.83E−1	2	9	−8.68E−3
1	5	−8.92E−1	5	9	2.38E−3	10	1	9.01E−1	14	5	2.59E−2	18	9	2.75E−2	2	10	−6.89E−2

Cl-like Fe (Fe^{9+})

i	i'	gf_L	i	i'	gf_L	i	i'	gf_L	i	i'	gf_L	i	i'	gf_L	i	i'	gf_L
2	11	−5.07E−3	6	7	−2.52E−2	10	3	−8.17E−5	1	12	−3.57E−3	6	4	−1.55E−1	10	8	−8.81E−3
2	12	−2.23E−2	6	8	−1.55E−1	10	4	−4.97E−4	2	1	1.55E+0	6	5	−4.44E−4	10	9	−7.46E−4
2	13	−1.94E+0	6	9	−7.91E−3	10	5	−4.25E−4	2	2	−2.91E−1	6	6	−2.11E−1	10	10	−4.62E−2
2	14	−4.78E+0	6	10	−7.33E−2	10	6	−7.19E−4	2	3	−1.04E−1	6	7	−5.56E−4	10	11	−8.15E−2
3	1	1.88E−1	6	11	−1.38E−2	10	7	−1.99E−3	2	4	−3.74E−1	6	8	−6.24E−3	10	12	−3.49E−2
3	2	1.91E−1	6	12	−1.76E−2	10	8	−8.47E−4	2	5	−8.04E−1	6	9	−4.07E−1	11	1	5.43E−1
3	3	−4.32E−1	6	13	−1.55E−3	10	9	−6.07E−3	2	6	−3.26E−1	6	10	−1.01E−2	11	2	4.06E−5
3	4	−6.01E−2	6	14	−2.60E−1	10	10	−2.84E−4	2	7	−4.02E−3	6	11	−9.30E−1	11	3	−4.07E−3
3	5	−2.54E−2	7	1	4.73E+0	10	11	−5.58E−2	2	8	−8.38E−3	6	12	−2.44E−1	11	4	−1.53E−7
3	6	−6.14E−3	7	2	7.01E−2	10	12	−3.75E−4	2	9	−3.08E+0	7	1	2.01E+0	11	5	−5.03E−4
3	7	−2.42E+0	7	3	−2.47E−3	10	13	−5.75E−3	2	10	−3.51E−1	7	2	−1.60E−2	11	6	−7.98E−3
3	8	−3.72E−3	7	4	−3.07E−2	10	14	−4.09E−3	2	11	−1.15E−1	7	3	−2.44E−2	11	7	−4.01E−3
3	9	−3.58E−2	7	5	−4.08E−2	11	1	6.89E−3	2	12	−5.19E−2	7	4	−1.31E−2	11	8	−8.58E+0
3	10	−2.71E−1	7	6	−1.28E−3	11	2	7.14E−2	3	1	1.78E−1	7	5	−3.22E−3	11	9	−2.66E−2
3	11	−2.17E−1	7	7	−1.68E−2	11	3	5.09E−5	3	2	−1.56E−1	7	6	−1.34E−1	11	10	−3.18E−3
3	12	−2.08E−2	7	8	−8.72E−3	11	4	−8.94E−4	3	3	−1.40E−1	7	7	−4.82E−1	11	11	−6.66E−4
3	13	−5.58E−1	7	9	−8.46E−4	11	5	−1.88E−7	3	4	−8.44E−1	7	8	−4.30E−2	11	12	−1.30E−4
3	14	−3.51E−1	7	10	−1.95E−1	11	6	−1.97E−4	3	5	−4.68E−4	7	9	−6.70E−2	12	1	2.82E−2
4	1	3.77E−1	7	11	−9.72E−3	11	7	−4.51E−5	3	6	−1.14E+0	7	10	−3.02E−2	12	2	1.73E−5
4	2	9.76E−2	7	12	−5.73E−4	11	8	−2.32E−5	3	7	−4.52E−2	7	11	−8.21E−2	12	3	3.73E−4
4	3	−1.97E−2	7	13	−3.32E−1	11	9	−5.60E−3	3	8	−5.53E−3	7	12	−7.53E−5	12	4	1.32E−3
4	4	−1.64E−1	7	14	−9.95E−2	11	10	−3.19E−3	3	9	−2.52E−1	8	1	3.44E−1	12	5	5.64E−4
4	5	−9.30E−3	8	1	1.56E+0	11	11	−2.52E−1	3	10	−1.48E+0	8	2	−9.78E−3	12	6	4.93E−4
4	6	−5.20E−4	8	2	9.69E−1	11	12	−2.14E+0	3	11	−7.63E−2	8	3	−2.72E−3	12	7	7.51E−5
4	7	−6.17E−2	8	3	−6.49E−4	11	13	−1.13E−4	3	12	−2.71E−1	8	4	−3.49E−3	12	8	1.11E−3
4	8	−6.39E−2	8	4	−2.97E−3	11	14	−6.96E−4	4	1	7.99E−2	8	5	−4.66E−2	12	9	2.07E−3
4	9	−2.91E−4	8	5	−1.82E−2	12	1	3.03E−1	4	2	−6.94E−3	8	6	−8.61E−3	12	10	9.26E−6
4	10	−6.79E−1	8	6	−1.33E−2	12	2	5.58E−1	4	3	−3.35E−1	8	7	−6.29E−1	12	11	7.36E−7
4	11	−1.78E−3	8	7	−7.72E−2	12	3	7.18E−5	4	4	−3.32E−2	8	8	−5.68E−3	12	12	−1.91E−4
4	12	−1.21E−3	8	8	−4.47E−2	12	4	6.87E−7	4	5	−2.22E−1	8	9	−1.52E−3			$^2F^e$–$^2G^o$
4	13	−1.81E+0	8	9	−8.58E−4	12	5	2.00E−4	4	6	−6.29E−2	8	10	−4.77E−1	1	1	−2.17E−1
4	14	−3.53E−1	8	10	−3.33E−2	12	6	6.24E−5	4	7	−9.16E−2	8	11	−3.03E−1	1	2	−7.34E−4
5	1	4.50E−3	8	11	−1.13E−1	12	7	1.32E−2	4	8	−4.11E−4	8	12	−5.24E−1	1	3	−4.59E+0
5	2	2.90E−1	8	12	−1.53E−3	12	8	1.16E−2	4	9	−9.72E−1	9	1	3.58E−1	1	4	−6.37E−1
5	3	−1.63E−3	8	13	−1.14E−1	12	9	4.87E−1	4	10	−3.38E−3	9	2	−6.77E−3	1	5	−7.80E−1
5	4	−2.03E−1	8	14	−5.09E−2	12	10	2.19E−3	4	11	−4.82E−2	9	3	−9.33E−4	1	6	−8.09E+0
5	5	−7.82E−2	9	1	1.60E−2	12	11	1.63E−2	4	12	−5.51E+0	9	4	−1.07E−2	2	1	−2.38E−1
5	6	−1.96E−1	9	2	7.34E+0	12	12	6.01E−2	5	1	2.74E+0	9	5	−2.76E−2	2	2	−9.41E−3
5	7	−1.94E−3	9	3	−5.24E−4	12	13	8.57E−6	5	2	−7.42E−2	9	6	−8.27E−5	2	3	−7.76E−2
5	8	−2.95E−1	9	4	−1.30E−3	12	14	2.76E−4	5	3	−9.24E−2	9	7	−6.68E−2	2	4	−2.55E+0
5	9	−4.27E−3	9	5	−8.79E−3			$^2F^o$–$^2G^e$	5	4	−4.38E−2	9	8	−6.90E−2	2	5	−4.52E+0
5	10	1.68E−2	9	6	−3.11E−2	1	1	4.21E−3	5	5	−1.02E−1	9	9	−6.41E−3	2	6	−1.25E+0
5	11	−4.71E−2	9	7	−1.24E−3	1	2	−1.99E−3	5	6	−2.79E−1	9	10	−1.04E−1	3	1	1.84E−1
5	12	−9.28E−3	9	8	−3.34E−2	1	3	−5.81E−1	5	7	−1.05E−1	9	11	3.75E−1	3	2	8.31E−7
5	13	−1.56E−1	9	9	−6.90E−4	1	4	−5.86E−1	5	8	−5.29E−2	9	12	−3.50E−4	3	3	7.80E−4
5	14	−1.48E−1	9	10	−9.28E−2	1	5	−1.42E+0	5	9	1.66E+0	10	1	2.43E−2	3	4	4.54E−4
6	1	2.45E+0	9	11	−3.73E−2	1	6	−2.46E+0	5	10	−1.81E−1	10	2	−2.25E−4	3	5	9.06E−6
6	2	6.15E−3	9	12	−2.74E−2	1	7	−6.65E−3	5	11	−1.15E−1	10	3	−2.14E−3	3	6	−2.00E−7
6	3	−2.45E−2	9	13	−6.49E−2	1	8	−1.81E−2	5	12	−1.37E−1	10	4	−5.08E−4	4	1	9.33E−2
6	4	−7.85E−2	9	14	−2.36E−1	1	9	−1.31E−1	6	1	4.59E+0	10	5	−1.34E−3	4	2	1.44E−3
6	5	−2.75E−4	10	1	1.73E+0	1	10	3.35E−3	6	2	−1.43E−3	10	6	−6.46E−2	4	3	6.20E−2
6	6	−3.15E−2	10	2	1.19E−1	1	11	−3.30E−2	6	3	−6.04E−3	10	7	−5.93E−5	4	4	1.73E−3

Cl-like Fe (Fe^{9+})

$i\ i'$	gf_L	$i\ i'$	gf_L	$i\ i'$	gf_L	$i\ i'$	gf_L	$i\ i'$	gf_L	$i\ i'$	gf_L
4 5	1.93E−2	13 3	1.11E−1	4 6	−2.11E−1	2 9	−7.17E−2	1 7	−2.57E−3	5 11	−6.77E−2
4 6	−1.75E−4	13 4	2.23E−1	4 7	−4.08E−2	2 10	−3.10E−1	1 8	−1.95E−1	5 12	−3.85E−2
5 1	9.00E−4	13 5	8.33E−2	4 8	−1.17E−2	2 11	−1.14E−2	1 9	−4.89E−2	6 1	1.01E−1
5 2	2.14E−1	13 6	−1.40E−4	4 9	−9.70E−1	2 12	−5.06E−1	1 10	−7.35E−2	6 2	2.62E+0
5 3	2.07E−1	14 1	3.13E−1	4 10	−2.17E−1	3 1	1.46E−1	1 11	−1.34E−2	6 3	1.11E−4
5 4	5.40E−3	14 2	1.04E−2	4 11	−1.07E−1	3 2	1.08E+0	1 12	−2.80E−2	6 4	−3.66E−6
5 5	3.24E−3	14 3	1.15E−2	4 12	−6.56E−2	3 3	2.77E−3	2 1	6.27E−4	6 5	−3.79E−3
5 6	−3.35E−6	14 4	1.22E−2	5 1	9.69E+0	3 4	4.18E−5	2 2	−1.33E−5	6 6	−1.55E−3
6 1	7.23E−4	14 5	7.16E−1	5 2	−6.88E−4	3 5	−1.63E−5	2 3	−5.12E−2	6 7	−3.26E−3
6 2	1.85E−2	14 6	−2.93E−6	5 3	−1.20E−2	3 6	−4.77E−3	2 4	−9.43E−1	6 8	−5.40E−2
6 3	1.71E−1		$^2G^o$–$^2G^e$	5 4	−3.09E−7	3 7	−2.21E−4	2 5	−1.63E−2	6 9	−2.17E+0
6 4	3.36E−2	1 1	1.04E−1	5 5	−2.43E−4	3 8	−7.41E−2	2 6	−3.09E−2	6 10	−2.92E−2
6 5	4.87E−3	1 2	−6.41E−2	5 6	−2.73E−3	3 9	−3.06E+0	2 7	−1.48E+0	6 11	−1.41E−3
6 6	−1.81E−4	1 3	−1.11E+0	5 7	−3.89E−5	3 10	−5.77E−2	2 8	−1.72E−1	6 12	−1.82E−4
7 1	3.25E+0	1 4	−3.78E−2	5 8	−2.19E−4	3 11	−2.27E−4	2 9	−1.62E−4	7 1	2.97E−2
7 2	3.59E−2	1 5	−2.95E−2	5 9	−3.25E−2	3 12	−2.50E−4	2 10	−1.87E−1	7 2	1.25E+0
7 3	3.61E−4	1 6	−1.42E−1	5 10	−1.11E−7		$^4S^e$–$^4P^o$	2 11	−5.39E−1	7 3	1.51E−3
7 4	2.99E−5	1 7	−1.10E−2	5 11	−8.41E−2	1 1	5.03E−1	2 12	−1.32E−1	7 4	4.66E−4
7 5	2.91E−3	1 8	−2.73E−3	5 12	−1.83E−1	1 2	6.73E−1	3 1	5.38E−2	7 5	1.53E−3
7 6	−1.37E−1	1 9	−6.62E−1	6 1	2.17E−2	1 3	8.77E−3	3 2	−2.50E−4	7 6	4.70E−3
8 1	1.74E−1	1 10	−4.46E+0	6 2	2.53E−4	1 4	1.14E−1	3 3	−1.79E−3	7 7	1.54E−3
8 2	3.80E−1	1 11	−7.57E−1	6 3	5.94E−7	1 5	8.88E−4	3 4	−4.10E−2	7 8	6.43E−4
8 3	8.71E−3	1 12	−1.83E−4	6 4	9.73E−4	1 6	9.33E−5	3 5	−1.48E−3	7 9	5.88E−4
8 4	3.16E−3	2 1	1.01E−1	6 5	2.93E−4	1 7	−2.95E−4	3 6	−1.46E−2	7 10	2.59E−3
8 5	2.91E−3	2 2	−3.78E−2	6 6	2.92E−4	2 1	2.54E+0	3 7	−6.07E−1	7 11	−1.06E−3
8 6	−9.12E−2	2 3	−3.95E−3	6 7	1.04E−4	2 2	1.61E−1	3 8	−1.07E+0	7 12	−1.24E−2
9 1	2.44E−2	2 4	−1.58E−1	6 8	6.27E−3	2 3	2.86E−2	3 9	−7.04E−2		$^4P^o$–$^4D^e$
9 2	2.68E−3	2 5	−8.33E−1	6 9	9.42E−5	2 4	2.37E−1	3 10	−7.24E−1	1 1	5.97E−1
9 3	5.03E−2	2 6	−2.77E−1	6 10	3.36E−4	2 5	6.31E−2	3 11	−2.09E−1	1 2	−5.96E−1
9 4	8.10E−3	2 7	−1.23E+0	6 11	−2.63E−6	2 6	8.17E−5	3 12	−7.04E−1	1 3	−1.83E−1
9 5	9.38E−4	2 8	−7.53E−4	6 12	−8.58E−6	2 7	−3.21E−3	4 1	2.66E−1	1 4	−3.93E+0
9 6	−6.51E+0	2 9	−7.69E−1		$^4S^o$–$^4P^e$	3 1	9.70E−2	4 2	−9.75E−3	1 5	−3.54E−2
10 1	2.44E−1	2 10	−4.00E−1	1 1	2.96E−2	3 2	1.28E+0	4 3	−2.56E−2	1 6	−9.78E−2
10 2	1.16E−2	2 11	−2.34E+0	1 2	−4.36E−4	3 3	6.59E−1	4 4	−1.03E−2	1 7	−1.07E+0
10 3	2.82E−1	2 12	−1.20E−1	1 3	−3.10E−3	3 4	4.47E−1	4 5	−1.21E−2	1 8	−3.88E−2
10 4	2.92E−1	3 1	7.80E−1	1 4	−3.11E−2	3 5	3.80E−2	4 6	−1.10E−1	1 9	−1.17E+0
10 5	7.29E−3	3 2	−2.99E−2	1 5	−4.33E−1	3 6	7.39E−4	4 7	−7.16E−2	1 10	−4.47E−1
10 6	−1.61E−2	3 3	−2.25E−5	1 6	−3.71E−2	3 7	3.39E−4	4 8	−7.21E−3	1 11	−1.11E−1
11 1	1.58E−1	3 4	−2.05E−1	1 7	−2.73E−1	4 1	2.23E−1	4 9	−1.11E−2	1 12	−2.30E−4
11 2	4.50E−2	3 5	−2.83E−2	1 8	−9.16E−3	4 2	8.80E−4	4 10	−6.74E−2	1 13	−1.15E−3
11 3	9.49E−5	3 6	−1.18E+0	1 9	−2.86E−2	4 3	5.74E−4	4 11	−3.03E−1	2 1	1.19E−3
11 4	6.17E−2	3 7	−2.36E−1	1 10	−3.06E−3	4 4	4.90E−3	4 12	−3.41E−1	2 2	−3.64E−6
11 5	1.06E−2	3 8	−8.30E−4	1 11	−8.25E−1	4 5	4.54E−4	5 1	9.95E+0	2 3	−1.14E+0
11 6	−4.37E−3	3 9	−2.26E−2	1 12	−6.64E−1	4 6	3.93E−1	5 2	−2.59E−3	2 4	−2.26E−3
12 1	4.24E−2	3 10	−1.94E−1	2 1	3.41E+0	4 7	8.54E−1	5 3	−6.15E−4	2 5	−4.77E−1
12 2	6.58E−3	3 11	−8.73E−1	2 2	−1.72E−3		$^4P^o$–$^4P^e$	5 4	−4.85E−3	2 6	−2.04E−1
12 3	7.89E−3	3 12	−3.82E−1	2 3	−1.28E−5	1 1	1.35E−1	5 5	−3.12E−2	2 7	−1.18E+0
12 4	3.25E−2	4 1	1.69E+0	2 4	−1.04E−2	1 2	−5.23E−3	5 6	−5.27E−4	2 8	−2.55E−2
12 5	2.45E−2	4 2	−1.47E−2	2 5	−1.49E−2	1 3	−4.86E−3	5 7	−3.76E−2	2 9	−1.16E−1
12 6	−3.70E−3	4 3	−8.25E−3	2 6	−3.00E−2	1 4	−1.11E−1	5 8	−7.61E−3	2 10	−9.86E−2
13 1	1.10E−1	4 4	−2.37E−2	2 7	−1.76E−3	1 5	−3.32E−1	5 9	−1.18E−3	2 11	−1.41E−1
13 2	1.01E−1	4 5	−4.49E−4	2 8	−5.61E−3	1 6	−9.15E+0	5 10	−1.44E−1	2 12	−3.72E+0

Cl-like Fe (Fe^{9+})

i	i'	gf_L	i	i'	gf_L	i	i'	gf_L	i	i'	gf_L	i	i'	gf_L	i	i'	gf_L
2	13	−3.00E−2	6	13	−5.80E−1	5	2	1.75E−1	10	9	−4.02E−4	3	7	−6.12E−2	7	7	−4.72E−2
3	1	2.98E−1	7	1	1.29E−1	5	3	9.69E−3	11	1	1.98E−1	3	8	−1.79E−3	7	8	−1.97E+0
3	2	−1.40E−3	7	2	6.33E−3	5	4	8.15E−4	11	2	8.63E−1	3	9	−1.77E−1	7	9	−3.85E−4
3	3	−6.31E−2	7	3	1.71E−4	5	5	2.55E−3	11	3	8.15E−1	3	10	−5.59E−1	7	10	−3.76E−4
3	4	−9.02E−3	7	4	3.28E−3	5	6	6.17E−4	11	4	5.38E−1	3	11	−3.28E+0	7	11	−1.18E−3
3	5	−1.04E−3	7	5	1.75E−5	5	7	1.98E−3	11	5	1.98E+0	3	12	−5.93E−2	7	12	−4.73E−5
3	6	−1.30E−1	7	6	1.29E−2	5	8	−4.16E−3	11	6	7.68E−2	3	13	−2.25E−3	7	13	−1.56E+0
3	7	−2.03E−1	7	7	1.40E−3	5	9	−7.31E−2	11	7	4.12E−3	4	1	1.72E+0	8	1	1.39E+0
3	8	−1.80E−3	7	8	1.24E−3	6	1	1.39E−3	11	8	−7.18E−5	4	2	−5.35E−3	8	2	3.94E−3
3	9	−7.77E−3	7	9	3.59E−3	6	2	4.19E−3	11	9	−3.51E−4	4	3	−1.20E−2	8	3	2.69E−5
3	10	−1.49E−1	7	10	−3.30E−6	6	3	1.91E−2	12	1	3.12E−1	4	4	−4.07E−2	8	4	2.31E−4
3	11	−3.31E−3	7	11	−1.91E−4	6	4	1.01E−1	12	2	6.37E+0	4	5	−3.34E−1	8	5	7.10E−5
3	12	−1.19E+0	7	12	−3.78E−2	6	5	2.61E−2	12	3	1.29E−1	4	6	−1.61E−1	8	6	8.51E−4
3	13	−1.81E−2	7	13	−5.17E+0	6	6	2.99E−1	12	4	3.76E−1	4	7	−3.06E−1	8	7	3.30E−2
4	1	9.27E+0			$^4P^e$–$^4D^o$	6	7	7.08E−4	12	5	1.68E−1	4	8	−2.96E−3	8	8	1.16E+0
4	2	−3.40E−2	1	1	−8.32E−1	6	8	−1.79E−2	12	6	8.49E−3	4	9	−1.37E−2	8	9	1.14E−3
4	3	−1.08E−2	1	2	−3.09E−1	6	9	−8.85E−2	12	7	5.60E−4	4	10	−1.10E−1	8	10	2.03E−3
4	4	−1.27E−1	1	3	−1.69E+0	7	1	1.04E+0	12	8	−4.20E−4	4	11	−9.32E−1	8	11	1.19E−4
4	5	−1.15E−1	1	4	−2.67E−1	7	2	1.21E−2	12	9	−9.93E−4	4	12	−9.89E−1	8	12	2.50E−6
4	6	−3.42E−1	1	5	−8.89E−1	7	3	1.49E−1			$^4D^o$–$^4D^e$	4	13	−1.93E−2	8	13	−1.24E−3
4	7	−3.87E−1	1	6	−3.91E+0	7	4	3.59E−1	1	1	2.18E−1	5	1	1.22E+0	9	1	3.24E−3
4	8	−7.98E−2	1	7	−1.61E−2	7	5	1.73E−1	1	2	−1.28E−1	5	2	−1.16E−2	9	2	1.28E−2
4	9	−2.37E−2	1	8	−8.70E+0	7	6	3.22E−3	1	3	−2.94E−3	5	3	−2.28E−2	9	3	5.85E−2
4	10	−3.58E−1	1	9	−3.63E−3	7	7	3.15E−5	1	4	−2.02E−1	5	4	−1.17E−2	9	4	1.08E−1
4	11	−1.02E+0	2	1	9.79E−3	7	8	−1.82E−3	1	5	−1.01E+0	5	5	−4.35E−2	9	5	1.14E+0
4	12	−3.92E−1	2	2	2.22E−3	7	9	−9.37E−2	1	6	−1.51E−5	5	6	−1.02E−1	9	6	9.48E−5
4	13	−9.52E−4	2	3	1.97E−3	8	1	1.22E+0	1	7	−2.12E+0	5	7	−5.35E−1	9	7	3.31E−2
5	1	2.31E−1	2	4	2.22E−3	8	2	4.30E−2	1	8	−7.67E−2	5	8	−5.63E−2	9	8	1.71E−3
5	2	−2.94E−3	2	5	1.06E−2	8	3	1.01E+0	1	9	−9.37E+0	5	9	−1.35E−1	9	9	5.85E−4
5	3	−3.72E−4	2	6	9.95E−3	8	4	9.26E−1	1	10	−1.13E+0	5	10	−1.51E−1	9	10	5.69E−4
5	4	−3.14E−2	2	7	−4.97E+0	8	5	1.05E−1	1	11	−3.40E−1	5	11	−4.11E−1	9	11	1.53E−2
5	5	−3.15E−2	2	8	−9.38E−3	8	6	4.86E−2	1	12	−1.31E−1	5	12	−2.24E+0	9	12	2.32E−4
5	6	−2.11E−2	2	9	−2.79E−4	8	7	6.00E−3	1	13	−1.11E−1	5	13	−8.88E−3	9	13	1.01E−5
5	7	−6.40E−2	3	1	1.31E−1	8	8	−1.44E−1	2	1	1.93E−1	6	1	9.80E+0			$^4D^o$–$^4F^e$
5	8	−2.41E−2	3	2	4.15E−1	8	9	−1.76E−3	2	2	−5.75E−1	6	2	−5.29E−3	1	1	6.82E−1
5	9	−3.47E−1	3	3	4.13E−3	9	1	9.69E−2	2	3	−1.77E−1	6	3	−2.56E−2	1	2	−6.43E−2
5	10	−5.71E−1	3	4	1.16E−6	9	2	1.19E−3	2	4	−1.90E−2	6	4	−4.41E−4	1	3	−3.04E−1
5	11	−2.43E−1	3	5	4.67E−3	9	3	1.01E−1	2	5	−1.05E+0	6	5	−1.48E−2	1	4	−1.13E+0
5	12	−2.73E−1	3	6	4.59E−3	9	4	1.33E−2	2	6	−3.33E−1	6	6	−7.33E−3	1	5	−1.92E−3
5	13	−1.90E−2	3	7	−2.68E−3	9	5	7.91E−3	2	7	−5.09E−1	6	7	−1.16E−1	1	6	−2.13E+0
6	1	4.27E−1	3	8	−9.17E−3	9	6	1.68E−3	2	8	−4.24E−3	6	8	−4.76E−2	1	7	−1.71E+0
6	2	4.20E−4	3	9	−6.88E−1	9	7	3.18E−1	2	9	−1.34E−1	6	9	−6.81E−1	1	8	2.83E+0
6	3	−2.05E−5	4	1	2.46E−2	9	8	5.82E+0	2	10	−4.46E+0	6	10	−1.52E−1	1	9	−1.82E+0
6	4	−6.15E−5	4	2	5.87E−2	9	9	−1.85E−4	2	11	−8.54E−1	6	11	−8.76E−2	1	10	−3.64E−1
6	5	−2.13E−4	4	3	2.31E−2	10	1	6.02E+0	2	12	−7.98E−1	6	12	−1.15E+0	1	11	−2.13E−1
6	6	−2.04E−3	4	4	2.38E−2	10	2	3.92E−3	2	13	−1.11E−2	6	13	−3.21E−3	1	12	−6.04E−2
6	7	1.54E−1	4	5	2.98E−2	10	3	8.13E−2	3	1	1.54E+0	7	1	1.18E−1	1	13	−9.04E−4
6	8	−6.79E+0	4	6	6.55E−4	10	4	1.74E−1	3	2	−1.52E−4	7	2	3.36E−4	2	1	5.80E−2
6	9	−5.69E−6	4	7	3.83E−5	10	5	2.24E−1	3	3	−4.69E−2	7	3	−2.44E−5	2	2	−1.04E−1
6	10	−6.07E−4	4	8	−1.45E−3	10	6	2.93E−2	3	4	−9.50E−2	7	4	−5.64E−6	2	3	−1.27E+0
6	11	−1.90E−6	4	9	−1.99E−4	10	7	1.19E−2	3	5	−2.97E−1	7	5	−3.29E−4	2	4	−2.09E−4
6	12	−4.54E−3	5	1	1.14E−2	10	8	−1.13E−1	3	6	−2.36E−3	7	6	−2.15E−3	2	5	−2.34E+0

Cl-like Fe (Fe^{9+})

i	i'	gf_L	i	i'	gf_L	i	i'	gf_L	i	i'	gf_L	i	i'	gf_L	i	i'	gf_L
2	6	−9.98E−2	6	6	−1.12E−3	1	5	−3.66E+0	10	3	1.90E−3	3	4	−7.08E−6	1	3	−7.88E+0
2	7	−1.77E−1	6	7	−1.83E−4	1	6	−1.31E+1	10	4	6.85E−1	3	5	−1.35E+0	1	4	−4.04E+0
2	8	−2.92E−1	6	8	−8.10E−2	2	1	2.83E−3	10	5	1.50E−2	3	6	−5.69E−1	1	5	−5.02E−2
2	9	−1.06E−1	6	9	−4.49E−1	2	2	2.63E−2	10	6	−7.39E−5	3	7	−5.74E−2	1	6	−8.92E−3
2	10	−1.80E+0	6	10	−3.16E−1	2	3	6.34E−3	11	1	2.53E−3	3	8	−1.53E−2	1	7	−1.10E−2
2	11	−1.03E+0	6	11	−1.07E+0	2	4	4.21E−3	11	2	5.22E+0	3	9	−6.36E−2	2	1	−2.13E−2
2	12	−4.14E−1	6	12	−4.24E−1	2	5	1.45E−4	11	3	2.92E−1	3	10	−3.21E−1	2	2	−2.19E+0
2	13	−7.57E−1	6	13	−1.85E+0	2	6	−1.71E−2	11	4	1.55E+0	3	11	−3.92E−2	2	3	−7.00E−1
3	1	2.34E−1	7	1	8.27E−1	3	1	1.07E−2	11	5	1.05E−1	3	12	−2.02E+0	2	4	−9.84E−1
3	2	−7.55E−2	7	2	1.18E−3	3	2	2.89E−1	11	6	−3.07E−4	3	13	−4.73E+0	2	5	−1.00E+0
3	3	−2.79E−2	7	3	1.36E−3	3	3	1.28E−1	12	1	2.19E−1	4	1	2.10E−1	2	6	−2.62E+0
3	4	−7.23E−2	7	4	−6.77E−5	3	4	2.31E−2	12	2	1.46E+0	4	2	−2.13E−3	2	7	−2.02E+0
3	5	−1.41E−1	7	5	−4.37E−3	3	5	7.81E−3	12	3	4.66E−1	4	3	−2.87E−3	3	1	−2.36E−1
3	6	−5.04E−2	7	6	−4.25E−4	3	6	−5.38E−4	12	4	1.15E+0	4	4	−2.59E−2	3	2	−2.13E−3
3	7	−2.59E+0	7	7	−7.93E−4	4	1	8.27E−1	12	5	6.70E−1	4	5	−1.10E−3	3	3	−1.04E−1
3	8	−4.48E−2	7	8	−6.01E+0	4	2	1.46E−1	12	6	3.84E−4	4	6	−1.15E+0	3	4	−3.99E+0
3	9	−6.23E−4	7	9	−7.79E+0	4	3	6.04E−1	13	1	7.25E−1	4	7	−1.66E+0	3	5	−3.00E+0
3	10	−5.96E−2	7	10	−3.98E−4	4	4	2.01E−2	13	2	1.91E−2	4	8	−7.45E−2	3	6	−3.49E+0
3	11	−3.78E+0	7	11	−5.20E−4	4	5	3.23E−4	13	3	4.66E−3	4	9	−7.08E−2	3	7	−7.36E+0
3	12	−3.82E−2	7	12	−3.91E−5	4	6	−1.09E−3	13	4	1.33E−2	4	10	−5.37E−2	4	1	−4.30E−3
3	13	−1.23E−1	7	13	−1.05E−3	5	1	2.56E−1	13	5	1.76E−2	4	11	−6.99E−2	4	2	−5.26E−1
4	1	1.76E−1	8	1	1.79E−2	5	2	1.71E+0	13	6	2.44E−2	4	12	−7.18E−3	4	3	−3.48E−1
4	2	−5.35E−3	8	2	8.50E−4	5	3	4.56E−1			**$^4F^o$–$^4F^e$**	4	13	−9.67E−1	4	4	−1.18E−3
4	3	−1.34E−2	8	3	6.26E−4	5	4	2.27E−4	1	1	1.41E+0	5	1	2.05E+1	4	5	−3.93E−1
4	4	−2.33E−1	8	4	2.91E−3	5	5	2.89E−3	1	2	−6.77E−2	5	2	−3.93E−4	4	6	−1.88E+0
4	5	−3.51E−2	8	5	8.63E−4	5	6	−1.45E−3	1	3	−1.41E−1	5	3	−2.92E−2	4	7	−4.95E−1
4	6	−9.43E−1	8	6	3.90E−6	6	1	9.45E+0	1	4	−5.81E−3	5	4	−3.41E−3	5	1	−4.89E−3
4	7	−2.22E−1	8	7	1.33E−3	6	2	7.26E−1	1	5	−1.39E+0	5	5	−6.74E−2	5	2	−3.50E−2
4	8	−9.42E−3	8	8	2.94E−2	6	3	1.31E−3	1	6	−5.48E−1	5	6	−4.18E−5	5	3	−1.64E−1
4	9	−5.78E−2	8	9	4.94E−2	6	4	7.11E−1	1	7	−3.04E+0	5	7	−1.95E−2	5	4	−2.48E−2
4	10	−4.09E−3	8	10	4.76E−6	6	5	2.16E−1	1	8	−8.12E+0	5	8	−6.35E−1	5	5	−3.01E−2
4	11	−2.93E−1	8	11	1.78E−3	6	6	−1.59E−2	1	9	−6.08E+0	5	9	−3.99E−1	5	6	−1.56E+0
4	12	−3.26E+0	8	12	3.22E−5	7	1	3.17E+0	1	10	−3.13E−1	5	10	−2.30E−1	5	7	−1.40E−2
4	13	−6.37E−1	8	13	−3.35E−5	7	2	9.24E−1	1	11	−3.38E−1	5	11	−2.90E−1	6	1	5.65E−3
5	1	1.28E+1	9	1	9.83E−5	7	3	2.53E−3	1	12	−2.32E−1	5	12	−2.61E−1	6	2	2.02E−4
5	2	−1.59E−3	9	2	2.24E−1	7	4	5.45E−1	1	13	−1.25E−1	5	13	−9.77E−2	6	3	4.82E−4
5	3	−1.02E−2	9	3	6.43E−1	7	5	9.73E−2	2	1	4.46E−3	6	1	1.90E+0	6	4	4.80E−3
5	4	−2.33E−2	9	4	9.93E−3	7	6	−1.69E−1	2	2	−4.82E−1	6	2	8.99E−5	6	5	4.34E−4
5	5	−1.06E−1	9	5	1.13E+0	8	1	1.75E−3	2	3	−2.25E−1	6	3	5.30E−3	6	6	2.21E−4
5	6	−2.04E−2	9	6	1.15E−3	8	2	2.88E−2	2	4	−1.10E+0	6	4	5.60E−7	6	7	−7.46E−4
5	7	−1.58E−1	9	7	1.10E−2	8	3	3.46E−2	2	5	−6.15E−1	6	5	2.46E−3			**$^4F^e$–$^4G^o$**
5	8	−3.35E−1	9	8	3.79E−4	8	4	6.63E−3	2	6	−9.93E−1	6	6	7.07E−5	1	1	−2.56E−2
5	9	−1.38E−2	9	9	3.77E−4	8	5	3.14E−4	2	7	−1.65E−1	6	7	9.95E−4	1	2	−3.67E−1
5	10	−4.62E−1	9	10	1.41E−2	8	6	−8.20E+0	2	8	−3.34E−2	6	8	3.66E−1	1	3	−1.17E+1
5	11	−1.63E−1	9	11	3.41E−3	9	1	1.12E+0	2	9	−2.90E−2	6	9	5.52E−1	1	4	−1.87E+1
5	12	−7.95E−1	9	12	4.51E−4	9	2	5.20E−1	2	10	−4.12E+0	6	10	3.42E−4	2	1	3.12E−1
5	13	−6.29E−1	9	13	7.55E−6	9	3	1.29E−3	2	11	−1.49E+0	6	11	6.36E−5	2	2	4.15E−3
6	1	1.72E+0			**$^4D^e$–$^4F^o$**	9	4	1.02E+0	2	12	−1.06E+0	6	12	−1.18E−4	2	3	3.39E−4
6	2	−8.89E−3	1	1	−1.19E+0	9	5	2.47E−1	2	13	−1.10E+0	6	13	−8.25E−4	2	4	−8.44E−3
6	3	−1.31E−2	1	2	−6.24E−2	9	6	−4.76E−4	3	1	1.19E+0			**$^4F^o$–$^4G^e$**	3	1	5.76E−1
6	4	−3.03E−2	1	3	−6.08E+0	10	1	7.01E−2	3	2	−6.30E−2	1	1	−5.77E−1	3	2	6.33E−3
6	5	−1.09E−2	1	4	−4.88E−1	10	2	1.10E−2	3	3	−1.82E−2	1	2	−5.49E−1	3	3	4.33E−4

Cl-like Fe (Fe^{9+})

$i\ i'$	gf_L	$i\ i'$	gf_L	$i\ i'$	gf_L	$i\ i'$	gf_L	$i\ i'$	gf_L	$i\ i'$	gf_L
3 4	−2.41E−2	6 4	−4.72E−4	9 4	−7.17E+0	12 4	−2.07E−5	1 7	−1.38E+1	3 5	−2.00E−1
4 1	5.89E−1	7 1	1.03E+0	10 1	1.94E+0	13 1	1.54E+0	2 1	−2.25E−1	3 6	−1.24E+0
4 2	6.38E−1	7 2	1.38E+0	10 2	1.62E−1	13 2	7.93E+0	2 2	−8.10E−3	3 7	−1.03E+0
4 3	2.37E−4	7 3	5.56E−1	10 3	2.17E−1	13 3	5.40E−1	2 3	−4.14E−1	4 1	4.28E−4
4 4	−3.77E−4	7 4	−8.32E−4	10 4	−1.30E−5	13 4	4.49E−6	2 4	−4.58E−1	4 2	4.29E−4
5 1	1.47E+0	8 1	8.45E−3	11 1	8.86E+0		**$^4G^o$–$^4G^e$**	2 5	−5.19E−1	4 3	4.27E−4
5 2	8.58E−1	8 2	1.57E−1	11 2	1.39E+0	1 1	−5.57E−1	2 6	−7.36E+0	4 4	2.59E−4
5 3	1.77E−1	8 3	1.70E−2	11 3	6.26E−2	1 2	−1.19E+0	2 7	−7.80E−1	4 5	1.07E−6
5 4	−6.50E−3	8 4	−5.80E+0	11 4	−3.32E−4	1 3	−8.98E−2	3 1	−3.28E−3	4 6	5.35E−5
6 1	1.27E−1	9 1	3.03E−2	12 1	1.48E−4	1 4	−3.51E+0	3 2	−3.22E−2	4 7	−1.28E−4
6 2	5.50E−1	9 2	8.32E−2	12 2	9.59E−1	1 5	−6.74E−1	3 3	−2.33E−1		
6 3	1.84E−1	9 3	2.08E−1	12 3	8.57E−2	1 6	−3.38E−1	3 4	−2.73E+0		

Ar-like Ar

Term energies relative to $3s^23p^5$ ^{2}P ionization threshold for each symmetry

i	E(Ryds)	Description
	^{1}S^e	
1	−1.18999	$3s^23p^6$
2	−.17122	$3s^23p^5$ ^{2}P $4p$
3	−.08341	$3s^23p^5$ ^{2}P $5p$
4	−.04990	$3s^23p^5$ ^{2}P $6p$
5	−.03326	$3s^23p^5$ ^{2}P $7p$
6	−.02376	$3s^23p^5$ ^{2}P $8p$
7	−.01783	$3s^23p^5$ ^{2}P $9p$
	^{3}S^e	
1	−.21093	$3s^23p^5$ ^{2}P $4p$
2	−.09574	$3s^23p^5$ ^{2}P $5p$
3	−.05549	$3s^23p^5$ ^{2}P $6p$
4	−.03627	$3s^23p^5$ ^{2}P $7p$
5	−.02557	$3s^23p^5$ ^{2}P $8p$
6	−.01899	$3s^23p^5$ ^{2}P $9p$
	^{1}P^e	
1	−.18566	$3s^23p^5$ ^{2}P $4p$
2	−.08949	$3s^23p^5$ ^{2}P $5p$
3	−.05281	$3s^23p^5$ ^{2}P $6p$
4	−.03486	$3s^23p^5$ ^{2}P $7p$
5	−.02474	$3s^23p^5$ ^{2}P $8p$
6	−.01846	$3s^23p^5$ ^{2}P $9p$
	^{1}P^o	
1	−.28651	$3s^23p^5$ ^{2}P $4s$
2	−.11926	$3s^23p^5$ ^{2}P $5s$
3	−.10934	$3s^23p^5$ ^{2}P $3d$
4	−.06564	$3s^23p^5$ ^{2}P $6s$
5	−.06139	$3s^23p^5$ ^{2}P $4d$
6	−.04154	$3s^23p^5$ ^{2}P $7s$
7	−.03936	$3s^23p^5$ ^{2}P $5d$
8	−.02865	$3s^23p^5$ ^{2}P $8s$
9	−.02738	$3s^23p^5$ ^{2}P $6d$
10	−.02095	$3s^23p^5$ ^{2}P $9s$
11	−.02015	$3s^23p^5$ ^{2}P $7d$
	^{3}P^e	
1	−.18507	$3s^23p^5$ ^{2}P $4p$
2	−.08928	$3s^23p^5$ ^{2}P $5p$
3	−.05271	$3s^23p^5$ ^{2}P $6p$
4	−.03481	$3s^23p^5$ ^{2}P $7p$
5	−.02470	$3s^23p^5$ ^{2}P $8p$
6	−.01844	$3s^23p^5$ ^{2}P $9p$
	^{3}P^o	
1	−.30050	$3s^23p^5$ ^{2}P $4s$
2	−.12476	$3s^23p^5$ ^{2}P $3d$
3	−.11971	$3s^23p^5$ ^{2}P $5s$
4	−.06952	$3s^23p^5$ ^{2}P $4d$
5	−.06646	$3s^23p^5$ ^{2}P $6s$
6	−.04406	$3s^23p^5$ ^{2}P $5d$
7	−.04201	$3s^23p^5$ ^{2}P $7s$
8	−.03028	$3s^23p^5$ ^{2}P $6d$
9	−.02893	$3s^23p^5$ ^{2}P $8s$
10	−.02204	$3s^23p^5$ ^{2}P $7d$
11	−.02113	$3s^23p^5$ ^{2}P $9s$
12	−.01674	$3s^23p^5$ ^{2}P $8d$
	^{1}D^e	
1	−.19060	$3s^23p^5$ ^{2}P $4p$
2	−.09051	$3s^23p^5$ ^{2}P $5p$
3	−.06269	$3s^23p^5$ ^{2}P $4f$
4	−.05322	$3s^23p^5$ ^{2}P $6p$
5	−.04014	$3s^23p^5$ ^{2}P $5f$
6	−.03507	$3s^23p^5$ ^{2}P $7p$
7	−.02787	$3s^23p^5$ ^{2}P $6f$
8	−.02486	$3s^23p^5$ ^{2}P $8p$
9	−.02047	$3s^23p^5$ ^{2}P $7f$
10	−.01854	$3s^23p^5$ ^{2}P $9p$
11	−.01567	$3s^23p^5$ ^{2}P $8f$
	^{1}D^o	
1	−.11342	$3s^23p^5$ ^{2}P $3d$
2	−.06438	$3s^23p^5$ ^{2}P $4d$
3	−.04121	$3s^23p^5$ ^{2}P $5d$
4	−.02856	$3s^23p^5$ ^{2}P $6d$
5	−.02093	$3s^23p^5$ ^{2}P $7d$
6	−.01599	$3s^23p^5$ ^{2}P $8d$
7	−.01261	$3s^23p^5$ ^{2}P $9d$
	^{3}D^e	
1	−.19470	$3s^23p^5$ ^{2}P $4p$
2	−.09182	$3s^23p^5$ ^{2}P $5p$
3	−.06271	$3s^23p^5$ ^{2}P $4f$
4	−.05381	$3s^23p^5$ ^{2}P $6p$
5	−.04015	$3s^23p^5$ ^{2}P $5f$
6	−.03539	$3s^23p^5$ ^{2}P $7p$
7	−.02788	$3s^23p^5$ ^{2}P $6f$
8	−.02505	$3s^23p^5$ ^{2}P $8p$
9	−.02048	$3s^23p^5$ ^{2}P $7f$
10	−.01866	$3s^23p^5$ ^{2}P $9p$
11	−.01567	$3s^23p^5$ ^{2}P $8f$
	^{3}D^o	
1	−.11252	$3s^23p^5$ ^{2}P $3d$
2	−.06384	$3s^23p^5$ ^{2}P $4d$
3	−.04090	$3s^23p^5$ ^{2}P $5d$
4	−.02837	$3s^23p^5$ ^{2}P $6d$
5	−.02081	$3s^23p^5$ ^{2}P $7d$
6	−.01591	$3s^23p^5$ ^{2}P $8d$
7	−.01255	$3s^23p^5$ ^{2}P $9d$
	^{1}F^e	
1	−.06244	$3s^23p^5$ ^{2}P $4f$
2	−.03996	$3s^23p^5$ ^{2}P $5f$
3	−.02775	$3s^23p^5$ ^{2}P $6f$
4	−.02039	$3s^23p^5$ ^{2}P $7f$
5	−.01561	$3s^23p^5$ ^{2}P $8f$
6	−.01234	$3s^23p^5$ ^{2}P $9f$
	^{1}F^o	
1	−.11529	$3s^23p^5$ ^{2}P $3d$
2	−.06493	$3s^23p^5$ ^{2}P $4d$
3	−.04142	$3s^23p^5$ ^{2}P $5d$
4	−.04000	$3s^23p^5$ ^{2}P $5g$
5	−.02866	$3s^23p^5$ ^{2}P $6d$
6	−.02778	$3s^23p^5$ ^{2}P $6g$
7	−.02099	$3s^23p^5$ ^{2}P $7d$
8	−.02041	$3s^23p^5$ ^{2}P $7g$
9	−.01602	$3s^23p^5$ ^{2}P $8d$
10	−.01563	$3s^23p^5$ ^{2}P $8g$
11	−.01263	$3s^23p^5$ ^{2}P $9d$
12	−.01235	$3s^23p^5$ ^{2}P $9g$
	^{3}F^e	
1	−.06244	$3s^23p^5$ ^{2}P $4f$
2	−.03996	$3s^23p^5$ ^{2}P $5f$
3	−.02775	$3s^23p^5$ ^{2}P $6f$
4	−.02039	$3s^23p^5$ ^{2}P $7f$
5	−.01561	$3s^23p^5$ ^{2}P $8f$
6	−.01234	$3s^23p^5$ ^{2}P $9f$
	^{3}F^o	
1	−.11731	$3s^23p^5$ ^{2}P $3d$
2	−.06635	$3s^23p^5$ ^{2}P $4d$
3	−.04232	$3s^23p^5$ ^{2}P $5d$
4	−.04000	$3s^23p^5$ ^{2}P $5g$
5	−.02923	$3s^23p^5$ ^{2}P $6d$
6	−.02778	$3s^23p^5$ ^{2}P $6g$
7	−.02137	$3s^23p^5$ ^{2}P $7d$
8	−.02041	$3s^23p^5$ ^{2}P $7g$
9	−.01629	$3s^23p^5$ ^{2}P $8d$
10	−.01563	$3s^23p^5$ ^{2}P $8g$
11	−.01282	$3s^23p^5$ ^{2}P $9d$
12	−.01235	$3s^23p^5$ ^{2}P $9g$
	^{1}G^e	
1	−.06264	$3s^23p^5$ ^{2}P $4f$
2	−.04010	$3s^23p^5$ ^{2}P $5f$
3	−.02785	$3s^23p^5$ ^{2}P $6f$
4	−.02778	$3s^23p^5$ ^{2}P $6h$
5	−.02046	$3s^23p^5$ ^{2}P $7f$
6	−.02041	$3s^23p^5$ ^{2}P $7h$
7	−.01566	$3s^23p^5$ ^{2}P $8f$
8	−.01563	$3s^23p^5$ ^{2}P $8h$
9	−.01237	$3s^23p^5$ ^{2}P $9f$
10	−.01235	$3s^23p^5$ ^{2}P $9h$
	^{1}G^o	
1	−.04000	$3s^23p^5$ ^{2}P $5g$
2	−.02778	$3s^23p^5$ ^{2}P $6g$
3	−.02041	$3s^23p^5$ ^{2}P $7g$
4	−.01562	$3s^23p^5$ ^{2}P $8g$
5	−.01234	$3s^23p^5$ ^{2}P $9g$
	^{3}G^e	
1	−.06265	$3s^23p^5$ ^{2}P $4f$
2	−.04011	$3s^23p^5$ ^{2}P $5f$
3	−.02785	$3s^23p^5$ ^{2}P $6f$
4	−.02778	$3s^23p^5$ ^{2}P $6h$
5	−.02046	$3s^23p^5$ ^{2}P $7f$
6	−.02041	$3s^23p^5$ ^{2}P $7h$
7	−.01566	$3s^23p^5$ ^{2}P $8f$
8	−.01563	$3s^23p^5$ ^{2}P $8h$
9	−.01237	$3s^23p^5$ ^{2}P $9f$
10	−.01235	$3s^23p^5$ ^{2}P $9h$
	^{3}G^o	
1	−.04000	$3s^23p^5$ ^{2}P $5g$
2	−.02778	$3s^23p^5$ ^{2}P $6g$
3	−.02041	$3s^23p^5$ ^{2}P $7g$
4	−.01562	$3s^23p^5$ ^{2}P $8g$
5	−.01234	$3s^23p^5$ ^{2}P $9g$
	^{1}H^e	
1	−.02778	$3s^23p^5$ ^{2}P $6h$
2	−.02041	$3s^23p^5$ ^{2}P $7h$
3	−.01563	$3s^23p^5$ ^{2}P $8h$
4	−.01235	$3s^23p^5$ ^{2}P $9h$
	^{1}H^o	
1	−.04000	$3s^23p^5$ ^{2}P $5g$
2	−.02778	$3s^23p^5$ ^{2}P $6g$
3	−.02041	$3s^23p^5$ ^{2}P $7g$
4	−.01563	$3s^23p^5$ ^{2}P $8g$
5	−.01235	$3s^23p^5$ ^{2}P $9g$
	^{3}H^e	
1	−.02778	$3s^23p^5$ ^{2}P $6h$
2	−.02041	$3s^23p^5$ ^{2}P $7h$
3	−.01563	$3s^23p^5$ ^{2}P $8h$
4	−.01235	$3s^23p^5$ ^{2}P $9h$
	^{3}H^o	
1	−.04000	$3s^23p^5$ ^{2}P $5g$
2	−.02778	$3s^23p^5$ ^{2}P $6g$
3	−.02041	$3s^23p^5$ ^{2}P $7g$
4	−.01563	$3s^23p^5$ ^{2}P $8g$
5	−.01235	$3s^23p^5$ ^{2}P $9g$

Ar-like Ar

Energies in ascending order from ground state for terms with effective $n \leq 4.0$, $L \leq 4$

Term	i	E(Ryds)	Term	i	E(Ryds)	Term	i	E(Ryds)	Term	i	E(Ryds)	Term	i	E(Ryds)
$^1\mathbf{S}^e$	1	0.00000	$^3\mathbf{P}^e$	1	1.00492	$^1\mathbf{D}^o$	1	1.07657	$^3\mathbf{P}^e$	2	1.10071	$^1\mathbf{D}^o$	2	1.12561
$^3\mathbf{P}^o$	1	0.88949	$^1\mathbf{S}^e$	2	1.01877	$^3\mathbf{D}^o$	1	1.07747	$^1\mathbf{S}^e$	3	1.10658	$^3\mathbf{D}^o$	2	1.12615
$^1\mathbf{P}^o$	1	0.90349	$^3\mathbf{P}^o$	2	1.06523	$^1\mathbf{P}^o$	3	1.08065	$^3\mathbf{P}^o$	4	1.12047	$^3\mathbf{D}^e$	3	1.12728
$^3\mathbf{S}^e$	1	0.97906	$^3\mathbf{P}^o$	3	1.07028	$^3\mathbf{S}^e$	2	1.09425	$^3\mathbf{P}^o$	5	1.12353	$^1\mathbf{D}^e$	3	1.12730
$^3\mathbf{D}^e$	1	0.99529	$^1\mathbf{P}^o$	2	1.07073	$^3\mathbf{D}^e$	2	1.09817	$^3\mathbf{F}^o$	2	1.12364	$^3\mathbf{G}^e$	1	1.12734
$^1\mathbf{D}^e$	1	0.99939	$^3\mathbf{F}^o$	1	1.07268	$^1\mathbf{D}^e$	2	1.09948	$^1\mathbf{P}^o$	4	1.12435	$^1\mathbf{G}^e$	1	1.12735
$^1\mathbf{P}^e$	1	1.00433	$^1\mathbf{F}^o$	1	1.07470	$^1\mathbf{P}^e$	2	1.10050	$^1\mathbf{F}^o$	2	1.12506			

gf-values for transitions involving terms with effective $n \leq 4.0$, $L \leq 4$

i i'	gf$_L$	i i'	gf$_L$	i i'	gf$_L$	i i'	gf$_L$	i i'	gf$_L$	i i'	gf$_L$
	$^1\mathbf{S}^e$–$^1\mathbf{P}^o$	2 2	1.41E−1	1 2	−3.57E+0	2 2	2.52E+0	3 2	3.93E−3	2 1	6.93E−1
1 1	−3.20E−1	2 3	2.04E+0	1 3	−4.46E−2	2 3	1.31E−2		$^3\mathbf{D}^e$–$^3\mathbf{P}^o$	2 2	−2.82E+0
1 2	−5.99E−2	2 4	−1.73E+0	1 4	−5.75E−4	2 4	−1.57E+0	1 1	5.16E+0	3 1	1.61E+0
1 3	−7.79E−2	2 5	−2.55E+0	1 5	−1.19E−1	3 1	1.28E−5	1 2	−1.13E+0	3 2	8.72E−2
1 4	−2.16E−2		$^1\mathbf{P}^e$–$^1\mathbf{P}^o$	2 1	5.46E−2	3 2	5.63E−5	1 3	−1.76E+0		$^3\mathbf{D}^e$–$^3\mathbf{F}^o$
2 1	4.03E−1	1 1	1.03E+0	2 2	4.38E+0	3 3	3.20E+0	1 4	−4.73E−2	1 1	−1.14E+1
2 2	−2.40E−1	1 2	−5.73E−1	2 3	6.54E−1	3 4	7.09E−4	1 5	−2.58E−1	1 2	−1.08E−1
2 3	−1.01E+0	1 3	−7.42E−1	2 4	−4.80E+0		$^1\mathbf{D}^e$–$^1\mathbf{D}^o$	2 1	3.09E−2	2 1	3.49E+0
2 4	−1.65E−2	1 4	−5.62E−2	2 5	−8.10E−1	1 1	−6.87E−1	2 2	4.13E+0	2 2	−1.56E+1
3 1	1.63E−2	2 1	7.60E−3		$^3\mathbf{P}^e$–$^3\mathbf{D}^o$	1 2	−1.43E−2	2 3	3.50E+0	3 1	5.04E−2
3 2	5.24E−1	2 2	1.54E+0	1 1	−6.35E+0	2 1	2.30E−1	2 4	−3.45E−1	3 2	1.93E−2
3 3	3.29E−1	2 3	2.40E−1	1 2	−1.19E−1	2 2	−9.52E−1	2 5	−4.56E+0		$^1\mathbf{G}^e$–$^1\mathbf{F}^o$
3 4	−3.93E−1	2 4	−9.66E−1	2 1	2.02E+0	3 1	5.21E−1	3 1	1.33E−3	1 1	5.93E+0
	$^3\mathbf{S}^e$–$^3\mathbf{P}^o$		$^1\mathbf{P}^e$–$^1\mathbf{D}^o$	2 2	−8.58E+0	3 2	4.40E−2	3 2	2.57E+0	1 2	7.65E−1
1 1	9.88E−1	1 1	−2.09E+0		$^1\mathbf{D}^e$–$^1\mathbf{P}^o$		$^1\mathbf{D}^e$–$^1\mathbf{F}^o$	3 3	4.39E+0		$^3\mathbf{G}^e$–$^3\mathbf{F}^o$
1 2	−2.28E−1	1 2	−3.04E−2	1 1	1.70E+0	1 1	−3.91E+0	3 4	2.63E+0	1 1	1.67E+1
1 3	−2.73E+0	2 1	6.60E−1	1 2	−8.96E−1	1 2	−6.02E−2	3 5	3.57E−1	1 2	3.76E+0
1 4	−1.29E−4	2 2	−2.85E+0	1 3	−5.96E−2	2 1	1.22E+0		$^3\mathbf{D}^e$–$^3\mathbf{D}^o$		
1 5	−1.14E−1		$^3\mathbf{P}^e$–$^3\mathbf{P}^o$	1 4	−9.79E−2	2 2	−5.33E+0	1 1	−2.07E+0		
2 1	5.46E−4	1 1	3.06E+0	2 1	5.68E−3	3 1	1.78E−2	1 2	−7.24E−2		

Ar-like Ca (Ca^{2+})

Term energies relative to $3s^23p^5$ ^{2}P ionization threshold for each symmetry

i	E(Ryds)	Description
	^{1}S^e	
1	−3.74762	$3s^23p^6$
2	−1.08426	$3s^23p^5$ ^{2}P $4p$
3	−.58796	$3s^23p^5$ ^{2}P $5p$
4	−.37246	$3s^23p^5$ ^{2}P $6p$
5	−.25750	$3s^23p^5$ ^{2}P $7p$
6	−.18887	$3s^23p^5$ ^{2}P $8p$
7	−.14467	$3s^23p^5$ ^{2}P $9p$
	^{3}S^e	
1	−1.26258	$3s^23p^5$ ^{2}P $4p$
2	−.65122	$3s^23p^5$ ^{2}P $5p$
3	−.40249	$3s^23p^5$ ^{2}P $6p$
4	−.27410	$3s^23p^5$ ^{2}P $7p$
5	−.19911	$3s^23p^5$ ^{2}P $8p$
6	−.15183	$3s^23p^5$ ^{2}P $9p$
	^{1}P^e	
1	−1.17655	$3s^23p^5$ ^{2}P $4p$
2	−.62469	$3s^23p^5$ ^{2}P $5p$
3	−.38938	$3s^23p^5$ ^{2}P $6p$
4	−.26630	$3s^23p^5$ ^{2}P $7p$
5	−.19368	$3s^23p^5$ ^{2}P $8p$
6	−.14722	$3s^23p^5$ ^{2}P $9p$
	^{1}P^o	
1	−1.47571	$3s^23p^5$ ^{2}P $4s$
2	−1.03053	$3s^23p^5$ ^{2}P $3d$
3	−.73607	$3s^23p^5$ ^{2}P $5s$
4	−.62259	$3s^23p^5$ ^{2}P $4d$
5	−.44326	$3s^23p^5$ ^{2}P $6s$
6	−.40478	$3s^23p^5$ ^{2}P $5d$
7	−.29640	$3s^23p^5$ ^{2}P $7s$
8	−.27951	$3s^23p^5$ ^{2}P $6d$
9	−.21218	$3s^23p^5$ ^{2}P $8s$
10	−.20335	$3s^23p^5$ ^{2}P $7d$
11	−.15939	$3s^23p^5$ ^{2}P $9s$
12	−.15420	$3s^23p^5$ ^{2}P $8d$
	^{3}P^e	
1	−1.17377	$3s^23p^5$ ^{2}P $4p$
2	−.62357	$3s^23p^5$ ^{2}P $5p$
3	−.38881	$3s^23p^5$ ^{2}P $6p$
4	−.26597	$3s^23p^5$ ^{2}P $7p$
5	−.19347	$3s^23p^5$ ^{2}P $8p$
6	−.14708	$3s^23p^5$ ^{2}P $9p$
	^{3}P^o	
1	−1.75891	$3s^23p^5$ ^{2}P $3d$
2	−1.51037	$3s^23p^5$ ^{2}P $4s$
3	−.79403	$3s^23p^5$ ^{2}P $4d$
4	−.74543	$3s^23p^5$ ^{2}P $5s$
5	−.46787	$3s^23p^5$ ^{2}P $5d$
6	−.44709	$3s^23p^5$ ^{2}P $6s$
7	−.30962	$3s^23p^5$ ^{2}P $6d$
8	−.29835	$3s^23p^5$ ^{2}P $7s$
9	−.22017	$3s^23p^5$ ^{2}P $7d$
10	−.21332	$3s^23p^5$ ^{2}P $8s$
11	−.16460	$3s^23p^5$ ^{2}P $8d$
12	−.16012	$3s^23p^5$ ^{2}P $9s$
13	−.12771	$3s^23p^5$ ^{2}P $9d$
	^{1}D^e	
1	−1.19002	$3s^23p^5$ ^{2}P $4p$
2	−.62804	$3s^23p^5$ ^{2}P $5p$
3	−.57250	$3s^23p^5$ ^{2}P $4f$
4	−.39105	$3s^23p^5$ ^{2}P $6p$
5	−.36586	$3s^23p^5$ ^{2}P $5f$
6	−.26730	$3s^23p^5$ ^{2}P $7p$
7	−.25381	$3s^23p^5$ ^{2}P $6f$
8	−.19439	$3s^23p^5$ ^{2}P $8p$
9	−.18649	$3s^23p^5$ ^{2}P $7f$
10	−.14795	$3s^23p^5$ ^{2}P $9p$
11	−.14337	$3s^23p^5$ ^{2}P $8f$
12	−.12250	$3s3p^6$ ^{2}S $3d$
	^{1}D^o	
1	−1.57055	$3s^23p^5$ ^{2}P $3d$
2	−.75718	$3s^23p^5$ ^{2}P $4d$
3	−.45334	$3s^23p^5$ ^{2}P $5d$
4	−.30198	$3s^23p^5$ ^{2}P $6d$
5	−.21558	$3s^23p^5$ ^{2}P $7d$
6	−.16162	$3s^23p^5$ ^{2}P $8d$
7	−.12566	$3s^23p^5$ ^{2}P $9d$
	^{3}D^e	
1	−1.20798	$3s^23p^5$ ^{2}P $4p$
2	−.63425	$3s^23p^5$ ^{2}P $5p$
3	−.57770	$3s^23p^5$ ^{2}P $4f$
4	−.39395	$3s^23p^5$ ^{2}P $6p$
5	−.37081	$3s^23p^5$ ^{2}P $5f$
6	−.26884	$3s^23p^5$ ^{2}P $7p$
7	−.25941	$3s^23p^5$ ^{2}P $6f$
8	−.20078	$3s3p^6$ ^{2}S $3d$
9	−.19524	$3s^23p^5$ ^{2}P $8p$
10	−.17553	$3s^23p^5$ ^{2}P $7f$
11	−.14825	$3s^23p^5$ ^{2}P $9p$
12	−.14036	$3s^23p^5$ ^{2}P $8f$
	^{3}D^o	
1	−1.55162	$3s^23p^5$ ^{2}P $3d$
2	−.75544	$3s^23p^5$ ^{2}P $4d$
3	−.45267	$3s^23p^5$ ^{2}P $5d$
4	−.30164	$3s^23p^5$ ^{2}P $6d$
5	−.21539	$3s^23p^5$ ^{2}P $7d$
6	−.16150	$3s^23p^5$ ^{2}P $8d$
7	−.12558	$3s^23p^5$ ^{2}P $9d$
	^{1}F^e	
1	−.56487	$3s^23p^5$ ^{2}P $4f$
2	−.36275	$3s^23p^5$ ^{2}P $5f$
3	−.25209	$3s^23p^5$ ^{2}P $6f$
4	−.18518	$3s^23p^5$ ^{2}P $7f$
5	−.14172	$3s^23p^5$ ^{2}P $8f$
6	−.11192	$3s^23p^5$ ^{2}P $9f$
	^{1}F^o	
1	−1.53725	$3s^23p^5$ ^{2}P $3d$
2	−.76124	$3s^23p^5$ ^{2}P $4d$
3	−.45517	$3s^23p^5$ ^{2}P $5d$
4	−.36086	$3s^23p^5$ ^{2}P $5g$
5	−.30302	$3s^23p^5$ ^{2}P $6d$
6	−.25072	$3s^23p^5$ ^{2}P $6g$
7	−.21624	$3s^23p^5$ ^{2}P $7d$
8	−.18420	$3s^23p^5$ ^{2}P $7g$
9	−.16206	$3s^23p^5$ ^{2}P $8d$
10	−.14100	$3s^23p^5$ ^{2}P $8g$
11	−.12597	$3s^23p^5$ ^{2}P $9d$
12	−.11139	$3s^23p^5$ ^{2}P $9g$
	^{3}F^e	
1	−.56463	$3s^23p^5$ ^{2}P $4f$
2	−.36256	$3s^23p^5$ ^{2}P $5f$
3	−.25196	$3s^23p^5$ ^{2}P $6f$
4	−.18509	$3s^23p^5$ ^{2}P $7f$
5	−.14166	$3s^23p^5$ ^{2}P $8f$
6	−.11188	$3s^23p^5$ ^{2}P $9f$
	^{3}F^o	
1	−1.66835	$3s^23p^5$ ^{2}P $3d$
2	−.77912	$3s^23p^5$ ^{2}P $4d$
3	−.46170	$3s^23p^5$ ^{2}P $5d$
4	−.36090	$3s^23p^5$ ^{2}P $5g$
5	−.30623	$3s^23p^5$ ^{2}P $6d$
6	−.25076	$3s^23p^5$ ^{2}P $6g$
7	−.21808	$3s^23p^5$ ^{2}P $7d$
8	−.18423	$3s^23p^5$ ^{2}P $7g$
9	−.16321	$3s^23p^5$ ^{2}P $8d$
10	−.14103	$3s^23p^5$ ^{2}P $8g$
11	−.12674	$3s^23p^5$ ^{2}P $9d$
12	−.11141	$3s^23p^5$ ^{2}P $9g$
	^{1}G^e	
1	−.57250	$3s^23p^5$ ^{2}P $4f$
2	−.36637	$3s^23p^5$ ^{2}P $5f$
3	−.25409	$3s^23p^5$ ^{2}P $6f$
4	−.25004	$3s^23p^5$ ^{2}P $6h$
5	−.18641	$3s^23p^5$ ^{2}P $7f$
6	−.18372	$3s^23p^5$ ^{2}P $7h$
7	−.14253	$3s^23p^5$ ^{2}P $8f$
8	−.14067	$3s^23p^5$ ^{2}P $8h$
9	−.11248	$3s^23p^5$ ^{2}P $9f$
10	−.11115	$3s^23p^5$ ^{2}P $9h$
	^{1}G^o	
1	−.35970	$3s^23p^5$ ^{2}P $5g$
2	−.24979	$3s^23p^5$ ^{2}P $6g$
3	−.18354	$3s^23p^5$ ^{2}P $7g$
4	−.14054	$3s^23p^5$ ^{2}P $8g$
5	−.11105	$3s^23p^5$ ^{2}P $9g$
	^{3}G^e	
1	−.57369	$3s^23p^5$ ^{2}P $4f$
2	−.36727	$3s^23p^5$ ^{2}P $5f$
3	−.25470	$3s^23p^5$ ^{2}P $6f$
4	−.25004	$3s^23p^5$ ^{2}P $6h$
5	−.18683	$3s^23p^5$ ^{2}P $7f$
6	−.18372	$3s^23p^5$ ^{2}P $7h$
7	−.14283	$3s^23p^5$ ^{2}P $8f$
8	−.14067	$3s^23p^5$ ^{2}P $8h$
9	−.11270	$3s^23p^5$ ^{2}P $9f$
10	−.11115	$3s^23p^5$ ^{2}P $9h$
	^{3}G^o	
1	−.35970	$3s^23p^5$ ^{2}P $5g$
2	−.24979	$3s^23p^5$ ^{2}P $6g$
3	−.18354	$3s^23p^5$ ^{2}P $7g$
4	−.14053	$3s^23p^5$ ^{2}P $8g$
5	−.11105	$3s^23p^5$ ^{2}P $9g$
	^{1}H^e	
1	−.24997	$3s^23p^5$ ^{2}P $6h$
2	−.18364	$3s^23p^5$ ^{2}P $7h$
3	−.14060	$3s^23p^5$ ^{2}P $8h$
4	−.11109	$3s^23p^5$ ^{2}P $9h$
	^{1}H^o	
1	−.36065	$3s^23p^5$ ^{2}P $5g$
2	−.25056	$3s^23p^5$ ^{2}P $6g$
3	−.18409	$3s^23p^5$ ^{2}P $7g$
4	−.18367	$3s^23p^5$ ^{2}P $7i$
5	−.14093	$3s^23p^5$ ^{2}P $8g$
6	−.14062	$3s^23p^5$ ^{2}P $8i$
7	−.11133	$3s^23p^5$ ^{2}P $9g$
8	−.11111	$3s^23p^5$ ^{2}P $9i$
	^{3}H^e	
1	−.24997	$3s^23p^5$ ^{2}P $6h$
2	−.18364	$3s^23p^5$ ^{2}P $7h$
3	−.14060	$3s^23p^5$ ^{2}P $8h$
4	−.11109	$3s^23p^5$ ^{2}P $9h$
	^{3}H^o	
1	−.36067	$3s^23p^5$ ^{2}P $5g$
2	−.25057	$3s^23p^5$ ^{2}P $6g$
3	−.18410	$3s^23p^5$ ^{2}P $7g$
4	−.18367	$3s^23p^5$ ^{2}P $7i$
5	−.14094	$3s^23p^5$ ^{2}P $8g$
6	−.14062	$3s^23p^5$ ^{2}P $8i$
7	−.11134	$3s^23p^5$ ^{2}P $9g$
8	−.11111	$3s^23p^5$ ^{2}P $9i$

Ar-like Ca (Ca^{2+})

Energies in ascending order from ground state for terms with effective $n \leq 4.0$, $L \leq 4$

Term	i	E(Ryds)	Term	i	E(Ryds)	Term	i	E(Ryds)	Term	i	E(Ryds)	Term	i	E(Ryds)
$^{1}S^{e}$	1	0.00000	$^{1}P^{o}$	1	2.27191	$^{1}P^{o}$	2	2.71709	$^{1}P^{o}$	3	3.01155	$^{1}S^{e}$	3	3.15966
$^{3}P^{o}$	1	1.98871	$^{3}S^{e}$	1	2.48504	$^{3}P^{o}$	3	2.95359	$^{3}S^{e}$	2	3.09640	$^{3}D^{e}$	3	3.16992
$^{3}F^{o}$	1	2.07927	$^{3}D^{e}$	1	2.53964	$^{3}F^{o}$	2	2.96850	$^{3}D^{e}$	2	3.11337	$^{3}G^{e}$	1	3.17393
$^{1}D^{o}$	1	2.17707	$^{1}D^{e}$	1	2.55760	$^{1}F^{o}$	2	2.98638	$^{1}D^{e}$	2	3.11959	$^{1}D^{e}$	3	3.17512
$^{3}D^{o}$	1	2.19600	$^{1}P^{e}$	1	2.57107	$^{1}D^{o}$	2	2.99044	$^{1}P^{e}$	2	3.12293	$^{1}G^{e}$	1	3.17512
$^{1}F^{o}$	1	2.21037	$^{3}P^{e}$	1	2.57385	$^{3}D^{o}$	2	2.99218	$^{3}P^{e}$	2	3.12405	$^{1}F^{e}$	1	3.18275
$^{3}P^{o}$	2	2.23725	$^{1}S^{e}$	2	2.66336	$^{3}P^{o}$	4	3.00219	$^{1}P^{o}$	4	3.12503	$^{3}F^{e}$	1	3.18299

gf-values for transitions involving terms with effective $n \leq 4.0$, $L \leq 4$

$i\ i'$	gf_L	$i\ i'$	gf_L	$i\ i'$	gf_L	$i\ i'$	gf_L	$i\ i'$	gf_L	$i\ i'$	gf_L
	$^{1}S^{e}-{}^{1}P^{o}$	2 4	1.54E+0	2 2	1.60E−2		$^{1}D^{e}-{}^{1}D^{o}$	2 3	7.72E−2		$^{1}F^{e}-{}^{1}D^{o}$
1 1	−5.55E−1		$^{1}P^{e}-{}^{1}P^{o}$	2 3	1.72E+0	1 1	1.14E−1	2 4	6.95E+0	1 1	8.42E−1
1 2	−2.95E+0	1 1	9.22E−1	2 4	3.86E+0	1 2	−7.94E−1	3 1	1.36E+0	1 2	4.53E+0
1 3	−1.20E−1	1 2	−2.44E−1		$^{3}P^{e}-{}^{3}D^{o}$	2 1	3.59E−3	3 2	6.43E−3		$^{1}F^{e}-{}^{1}F^{o}$
1 4	−1.16E+0	1 3	−4.24E−1	1 1	1.07E+0	2 2	3.82E−1	3 3	9.00E+0	1 1	1.15E−1
2 1	4.49E−1	1 4	−4.08E−1	1 2	−7.30E+0	3 1	1.20E−1	3 4	3.34E−1	1 2	5.58E−1
2 2	−9.60E−2	2 1	1.67E−2	2 1	5.91E−2	3 2	4.98E−1		$^{3}D^{e}-{}^{3}D^{o}$		$^{3}F^{e}-{}^{3}D^{o}$
2 3	−3.40E−1	2 2	7.83E−2	2 2	3.16E+0		$^{1}D^{e}-{}^{1}F^{o}$	1 1	3.10E−1	1 1	2.59E+0
2 4	−6.64E−1	2 3	1.32E+0		$^{1}D^{e}-{}^{1}P^{o}$	1 1	6.53E−1	1 2	−2.37E+0	1 2	1.35E+1
3 1	2.60E−3	2 4	−1.52E−2	1 1	1.54E+0	1 2	−4.50E+0	2 1	3.88E−3		$^{3}F^{e}-{}^{3}F^{o}$
3 2	2.40E−2		$^{1}P^{e}-{}^{1}D^{o}$	1 2	−6.16E−3	2 1	2.75E−2	2 2	1.14E+0	1 1	2.65E−1
3 3	6.65E−1	1 1	3.61E−1	1 3	−8.62E−1	2 2	1.94E+0	3 1	3.61E−1	1 2	1.78E+0
3 4	2.03E−1	1 2	−2.43E+0	1 4	−2.47E−2	3 1	9.52E−4	3 2	1.38E+0		$^{1}G^{e}-{}^{1}F^{o}$
	$^{3}S^{e}-{}^{3}P^{o}$	2 1	2.05E−2	2 1	3.52E−2	3 2	4.20E−2		$^{3}D^{e}-{}^{3}F^{o}$	1 1	1.42E+0
1 1	4.43E−1	2 2	1.05E+0	2 2	4.22E−3		$^{3}D^{e}-{}^{3}P^{o}$	1 1	1.99E+0	1 2	6.24E+0
1 2	7.93E−1		$^{3}P^{e}-{}^{3}P^{o}$	2 3	2.32E+0	1 1	3.99E−2	1 2	−1.36E+1		$^{3}G^{e}-{}^{3}F^{o}$
1 3	−2.78E+0	1 1	3.98E−1	2 4	−3.94E−7	1 2	4.66E+0	2 1	1.39E−1	1 1	3.35E+0
1 4	−7.82E−1	1 2	2.99E+0	3 1	8.99E−4	1 3	−2.47E−2	2 2	5.78E+0	1 2	1.97E+1
2 1	2.47E−2	1 3	−3.01E+0	3 2	2.34E+0	1 4	−2.67E+0	3 1	1.13E−3		
2 2	6.91E−2	1 4	−1.13E+0	3 3	4.37E−2	2 1	2.53E−2	3 2	1.38E−1		
2 3	9.71E−1	2 1	4.36E−2	3 4	7.61E−1	2 2	9.23E−2				

Ar-like Fe (Fe^{8+})

Term energies relative to $3s^23p^5$ ^{2}P ionization threshold for each symmetry

i	Energy(Ryds)	Description
	^{1}S^e	
1	−17.2533	$3s^23p^6$
2	−9.14347	$3s^23p^43d^2$
3	−7.64888	$3s^23p^43d^2$
4	−7.23155	$3s^23p^5$ ^{2}P $4p$
5	−5.82448	$3s^23p^43d^2$
6	−5.65930	$3s3p^6$ ^{2}S $4s$
7	−4.26165	$3s^23p^5$ ^{2}P $5p$
8	−3.41254	$3p^4(^1D)3d$ ^{2}S $4s$
9	−2.81525	$3s^23p^5$ ^{2}P $6p$
10	−2.64313	$3p^63d^2$
11	−2.21730	$3p^4(^1D)3d$ ^{2}D $4d$
12	−2.12807	$3s3p^6$ ^{2}S $5s$
13	−1.99564	$3s^23p^5$ ^{2}P $7p$
14	−1.75884	$3s3p^43d^3$
15	−1.49429	$3s^23p^5$ ^{2}P $8p$
16	−1.26146	$3p^4(^1S)3d$ ^{2}D $4d$
17	−1.15935	$3s^23p^5$ ^{2}P $9p$
	^{1}S^o	
1	−5.97132	$3s3p^53d^2$
2	−5.04202	$3s^23p^33d^3$
3	−3.92767	$3s^23p^33d^3$
4	−3.55171	$3p^4(^1D)3d$ ^{2}P $4p$
5	−2.81840	$3s^23p^33d^3$
6	−2.36562	$3p^4(^3P)3d$ ^{2}P $4p$
7	−1.00161	$3p^4(^3P)3d$ ^{2}F $4f$
8	−.87515	$3p^4(^1D)3d$ ^{2}F $4f$
9	−.47482	$3p^4(^1D)3d$ ^{2}P $5p$
	^{1}P^e	
1	−8.85255	$3s^23p^43d^2$
2	−7.62983	$3s^23p^5$ ^{2}P $4p$
3	−7.42086	$3s^23p^43d^2$
4	−4.47997	$3p^4(^1D)3d$ ^{2}P $4s$
5	−4.36471	$3s^23p^5$ ^{2}P $5p$
6	−3.21203	$3p^4(^3P)3d$ ^{2}P $4s$
7	−2.86530	$3s^23p^5$ ^{2}P $6p$
8	−2.47969	$3p^4(^1D)3d$ ^{2}D $4d$
9	−2.24860	$3p^4(^1D)3d$ ^{2}P $4d$
10	−2.08370	$3p^4(^3P)3d$ ^{2}F $4d$
11	−2.03313	$3s^23p^5$ ^{2}P $7p$
12	−1.84498	$3p^4(^1D)3d$ ^{2}F $4d$
13	−1.66199	$3p^4(^1S)3d$ ^{2}D $4d$
14	−1.62816	$3s3p^43d^3$
15	−1.51482	$3s^23p^5$ ^{2}P $8p$
16	−1.44174	$3s3p^43d^3$
17	−1.17378	$3s^23p^5$ ^{2}P $9p$
18	−1.13494	$3p^4(^3P)3d$ ^{2}P $4d$
19	−1.00566	$3s3p^43d^3$
20	−.95868	$3p^4(^3P)3d$ ^{2}D $4d$
	^{1}P^o	
1	−11.8803	$3s^23p^5$ ^{2}P $3d$
2	−8.40253	$3s^23p^5$ ^{2}P $4s$
3	−6.26078	$3s^23p^5$ ^{2}P $4d$
4	−5.75689	$3s3p^53d^2$
5	−5.01370	$3s3p^6$ ^{2}S $4p$
6	−4.97483	$3s3p^53d^2$
7	−4.77449	$3s^23p^5$ ^{2}P $5s$
8	−4.71689	$3s3p^53d^2$
9	−4.08385	$3s^23p^33d^3$
10	−4.05636	$3s^23p^33d^3$
11	−3.84273	$3s^23p^5$ ^{2}P $5d$
12	−3.73318	$3p^4(^1D)3d$ ^{2}D $4p$:
13	−3.53882	$3p^4(^1D)3d$ ^{2}P $4p$
14	−3.26721	$3s^23p^33d^3$:
15	−3.22767	$3s^23p^33d^3$:
16	−3.08125	$3s^23p^5$ ^{2}P $6s$
17	−2.91928	$3s^23p^33d^3$
18	−2.77091	$3p^4(^1S)3d$ ^{2}D $4p$
19	−2.65874	$3p^4(^1D)3d$ ^{2}S $4p$:
20	−2.58229	$3s^23p^5$ ^{2}P $6d$
21	−2.49924	$3s^23p^33d^3$
22	−2.42328	$3s^23p^33d^3$:
23	−2.22301	$3p^4(^3P)3d$ ^{2}P $4p$
24	−2.14383	$3s^23p^5$ ^{2}P $7s$
25	−2.03134	$3p^4(^3P)3d$ ^{2}D $4p$
26	−1.82205	$3s^23p^5$ ^{2}P $7d$
27	−1.76697	$3s3p^6$ ^{2}S $5p$
28	−1.58755	$3s^23p^5$ ^{2}P $8s$
29	−1.38996	$3s^23p^5$ ^{2}P $8d$
30	−1.31072	$3p^4(^1D)3d$ ^{2}D $4f$
31	−1.22228	$3s^23p^5$ ^{2}P $9s$
32	−1.12944	$3p^4(^1D)3d$ ^{2}G $4f$
33	−1.08883	$3s^23p^5$ ^{2}P $9d$
34	−1.01796	$3s^23p^33d^3$
	^{1}D^e	
1	−10.3871	$3s3p^6$ ^{2}S $3d$
2	−9.24805	$3s^23p^43d^2$:
3	−8.60619	$3s^23p^43d^2$
4	−8.05815	$3s^23p^43d^2$
5	−7.84515	$3s^23p^43d^2$
6	−7.66861	$3s^23p^5$ ^{2}P $4p$:
7	−7.41356	$3s^23p^43d^2$
8	−6.57932	$3s^23p^43d^2$
9	−5.13662	$3s^23p^5$ ^{2}P $4f$
10	−4.45638	$3p^4(^1D)3d$ ^{2}D $4s$
11	−4.38074	$3s^23p^5$ ^{2}P $5p$
12	−3.76621	$3p^4(^1S)3d$ ^{2}D $4s$
13	−3.60695	$3s3p^6$ ^{2}S $4d$
14	−3.30571	$3s^23p^5$ ^{2}P $5f$
15	−3.25997	$3p^63d^2$
16	−3.09691	$3p^4(^3P)3d$ ^{2}D $4s$
17	−2.86630	$3s^23p^5$ ^{2}P $6p$
18	−2.35687	$3p^4(^1D)3d$ ^{2}P $4d$
19	−2.29467	$3s^23p^5$ ^{2}P $6f$
20	−2.27444	$3p^4(^1D)3d$ ^{2}D $4d$
21	−2.14217	$3p^4(^1D)3d$ ^{2}G $4d$
22	−2.06262	$3p^4(^3P)3d$ ^{2}F $4d$
23	−2.02729	$3s^23p^5$ ^{2}P $7p$
24	−1.83183	$3s3p^43d^3$
25	−1.73911	$3p^4(^1D)3d$ ^{2}F $4d$
26	−1.68426	$3s^23p^5$ ^{2}P $7f$
27	−1.65466	$3s3p^43d^3$
28	−1.51736	$3s^23p^5$ ^{2}P $8p$
29	−1.46100	$3p^4(^1S)3d$ ^{2}D $4d$
30	−1.29153	$3s^23p^5$ ^{2}P $8f$
31	−1.26738	$3p^4(^1D)3d$ ^{2}S $4d$
32	−1.19786	$3s3p^43d^3$
33	−1.18916	$3s3p^6$ ^{2}S $5d$
34	−1.17339	$3s^23p^5$ ^{2}P $9p$
35	−1.02623	$3s^23p^5$ ^{2}P $9f$
36	−1.00144	$3p^4(^3P)3d$ ^{2}P $4d$
	^{1}D^o	
1	−13.1139	$3s^23p^5$ ^{2}P $3d$
2	−6.23686	$3s^23p^5$ ^{2}P $4d$
3	−6.03114	$3s3p^53d^2$
4	−5.78154	$3s3p^53d^2$
5	−4.77587	$3s3p^53d^2$
6	−4.53130	$3s^23p^33d^3$
7	−4.34192	$3s^23p^33d^3$
8	−3.97984	$3s^23p^33d^3$
9	−3.82467	$3s^23p^5$ ^{2}P $5d$
10	−3.80315	$3s^23p^33d^3$
11	−3.66605	$3p^4(^1D)3d$ ^{2}P $4p$:
12	−3.61424	$3s^23p^33d^3$
13	−3.44646	$3p^4(^1D)3d$ ^{2}D $4p$
14	−3.28385	$3p^4(^3P)3d$ ^{2}F $4p$
15	−3.26305	$3s^23p^33d^3$
16	−3.14220	$3p^4(^1D)3d$ ^{2}F $4p$
17	−2.94890	$3p^4(^1S)3d$ ^{2}D $4p$
18	−2.84882	$3s^23p^33d^3$
19	−2.74395	$3s^23p^33d^3$
20	−2.55683	$3s^23p^5$ ^{2}P $6d$
21	−2.32740	$3p^4(^3P)3d$ ^{2}P $4p$
22	−2.25196	$3p^4(^3P)3d$ ^{2}D $4p$
23	−2.19189	$3s^23p^33d^3$
24	−1.83981	$3s^23p^5$ ^{2}P $7d$
25	−1.40831	$3p^4(^1D)3d$ ^{2}D $4f$:
26	−1.39029	$3s^23p^5$ ^{2}P $8d$
27	−1.23118	$3p^4(^1D)3d$ ^{2}P $4f$:
28	−1.15572	$3p^4(^1D)3d$ ^{2}G $4f$
29	−1.08773	$3s^23p^5$ ^{2}P $9d$
30	−1.00920	$3p^4(^3P)3d$ ^{2}F $4f$

Ar-like Fe (Fe^{8+})

i	Energy(Ryds)	Description	
		^{1}**F**^{e}	
1	−9.05304	$3s^23p^43d^2$	
2	−8.21764	$3s^23p^43d^2$	
3	−7.36559	$3s^23p^43d^2$	
4	−5.19720	$3s^23p^5$	^{2}P $4f$
5	−4.31525	$3p^4(^3P)3d$	^{2}F $4s$
6	−3.97789	$3p^4(^1D)3d$	^{2}F $4s$
7	−3.33732	$3s^23p^5$	^{2}P $5f$
8	−2.40355	$3p^4(^1D)3d$	^{2}P $4d$
9	−2.31716	$3s^23p^5$	^{2}P $6f$
10	−2.22330	$3p^4(^3P)3d$	^{2}F $4d$
11	−2.21827	$3p^4(^1D)3d$	^{2}D $4d$
12	−2.10445	$3p^4(^1D)3d$	^{2}G $4d$
13	−1.95440	$3s3p^43d^3$	
14	−1.87751	$3p^4(^1D)3d$	^{2}F $4d$
15	−1.72405	$3p^4(^1S)3d$	^{2}D $4d$
16	−1.69395	$3s^23p^5$	^{2}P $7f$
17	−1.42328	$3s3p^43d^3$	
18	−1.30749	$3s3p^43d^3$	
19	−1.29514	$3s^23p^5$	^{2}P $8f$
20	−1.15398	$3s3p^43d^3$	
21	−1.10421	$3p^4(^3P)3d$	^{2}P $4d$
22	−1.02094	$3s^23p^5$	^{2}P $9f$
23	−.96707	$3p^4(^3P)3d$	^{2}D $4d$
24	−.85626	$3p^4(^3P)3d$	^{2}F $5s$:
		^{1}**F**^{o}	
1	−13.0642	$3s^23p^5$	^{2}P $3d$
2	−6.28866	$3s^23p^5$	^{2}P $4d$
3	−6.04217	$3s3p^53d^2$	
4	−5.70734	$3s3p^53d^2$	
5	−5.25881	$3s3p^53d^2$	:
6	−4.70001	$3s^23p^33d^3$	:
7	−4.57445	$3s^23p^33d^3$	:
8	−3.93809	$3s^23p^33d^3$	
9	−3.85044	$3s^23p^5$	^{2}P $5d$
10	−3.73122	$3s^23p^33d^3$	:
11	−3.62630	$3p^4(^3P)3d$	^{2}F $4p$:
12	−3.57300	$3p^4(^1D)3d$	^{2}D $4p$:
13	−3.46282	$3s^23p^33d^3$	
14	−3.30473	$3s^23p^33d^3$	
15	−3.25803	$3s^23p^5$	^{2}P $5g$
16	3.21672	$3p^4(^1D)3d$	^{2}G $4p$
17	−3.10831	$3s^23p^33d^3$	
18	−3.00659	$3p^4(^1D)3d$	^{2}F $4p$
19	−2.90484	$3p^4(^1S)3d$	^{2}D $4p$
20	−2.71315	$3s^23p^33d^3$	:
21	−2.60050	$3s3p^6$	^{2}S $4f$
22	−2.58479	$3s^23p^33d^3$	
23	−2.56043	$3s^23p^5$	^{2}P $6d$
24	−2.29230	$3s^23p^33d^3$	:
25	−2.25301	$3s^23p^5$	^{2}P $6g$
26	−2.20916	$3p^4(^3P)3d$	^{2}D $4p$
27	−1.89531	$3s^23p^33d^3$	
28	−1.84455	$3s^23p^5$	^{2}P $7d$
29	−1.66146	$3s^23p^5$	^{2}P $7g$
30	−1.39467	$3s^23p^5$	^{2}P $8d$
31	−1.31038	$3p^4(^1D)3d$	^{2}P $4f$
32	−1.27401	$3s^23p^5$	^{2}P $8g$
33	−1.23354	$3p^4(^1D)3d$	^{2}D $4f$
34	−1.19284	$3p^4(^1D)3d$	^{2}G $4f$:
35	−1.09025	$3s^23p^5$	^{2}P $9d$
36	−1.03791	$3p^4(^3P)3d$	^{2}F $4f$:
37	−1.00598	$3s^23p^5$	^{2}P $9g$
		^{1}**G**^{e}	
1	−9.05306	$3s^23p^43d^2$	
2	−8.78084	$3s^23p^43d^2$	
3	−8.05230	$3s^23p^43d^2$	
4	−7.94434	$3s^23p^43d^2$	:
5	−5.24215	$3s^23p^5$	^{2}P $4f$
6	−4.30137	$3p^4(^1D)3d$	^{2}G $4s$
7	−3.35371	$3s^23p^5$	^{2}P $5f$
8	−3.14950	$3p^63d^2$	
9	−2.34458	$3p^4(^1D)3d$	^{2}D $4d$
10	−2.31752	$3s^23p^5$	^{2}P $6f$
11	−2.14873	$3p^4(^3P)3d$	^{2}F $4d$
12	−2.09083	$3p^4(^1D)3d$	^{2}G $4d$
13	−1.80205	$3s3p^43d^3$	
14	−1.74651	$3p^4(^1D)3d$	^{2}F $4d$
15	−1.69865	$3s^23p^5$	^{2}P $7f$
16	−1.41904	$3p^4(^1S)3d$	^{2}D $4d$
17	−1.39636	$3s3p^43d^3$	
18	−1.29789	$3s^23p^5$	^{2}P $8f$
19	−1.14596	$3s3p^43d^3$	
20	−1.07469	$3s3p^43d^3$	
21	−1.02863	$3s^23p^5$	^{2}P $9f$
22	−.99452	$3p^4(^3P)3d$	^{2}D $4d$
		^{1}**G**^{o}	
1	−6.24572	$3s3p^53d^2$	
2	−5.33383	$3s3p^53d^2$	
3	−4.80391	$3s^23p^33d^3$	:
4	−4.24603	$3s^23p^33d^3$	:
5	−4.14037	$3s^23p^33d^3$	:
6	−4.05691	$3s^23p^33d^3$	
7	−3.80467	$3s^23p^33d^3$	
8	−3.53483	$3p^4(^3P)3d$	^{2}F $4p$
9	−3.43233	$3p^4(^1D)3d$	^{2}G $4p$
10	−3.31972	$3s^23p^33d^3$	:
11	−3.23750	$3s^23p^5$	^{2}P $5g$
12	−3.07277	$3p^4(^1D)3d$	^{2}F $4p$
13	−2.83454	$3s^23p^33d^3$	:
14	−2.62511	$3s^23p^33d^3$	:
15	−2.25379	$3s^23p^5$	^{2}P $6g$
16	1.65800	$3s^23p^5$	^{2}P $7g$
17	−1.36052	$3p^4(^1D)3d$	^{2}P $4f$
18	−1.26990	$3s^23p^5$	^{2}P $8g$
19	−1.21543	$3p^4(^3P)3d$	^{2}F $4f$
20	−1.20227	$3p^4(^1D)3d$	^{2}D $4f$
21	−1.10302	$3p^4(^1D)3d$	^{2}G $4f$
22	−1.00325	$3s^23p^5$	^{2}P $9g$
		^{3}**S**^{e}	
1	−7.94445	$3s^23p^43d^2$	
2	−7.75003	$3s^23p^5$	^{2}P $4p$
3	−5.81786	$3s3p^6$	^{2}S $4s$
4	−4.47423	$3s^23p^5$	^{2}P $5p$
5	−3.49800	$3p^4(^1D)3d$	^{2}S $4s$
6	−2.91319	$3s^23p^5$	^{2}P $6p$
7	−2.75379	$3p^4(^3P)3d$	^{4}D $4d$
8	−2.37046	$3p^4(^1D)3d$	^{2}D $4d$
9	−2.22518	$3s3p^43d^3$	
10	−2.18406	$3s3p^6$	^{2}S $5s$
11	−2.04554	$3s^23p^5$	^{2}P $7p$
12	−1.58204	$3p^4(^1S)3d$	^{2}D $4d$
13	−1.54329	$3s3p^43d^3$	
14	−1.52799	$3s^23p^5$	^{2}P $8p$
15	−1.18313	$3s^23p^5$	^{2}P $9p$
16	−1.16387	$3s3p^43d^3$	:
17	−1.03008	$3p^4(^3P)3d$	^{2}D $4d$
		^{3}**S**^{o}	
1	−6.02319	$3s3p^53d^2$	
2	−5.22196	$3s3p^53d^2$	
3	−3.98703	$3s^23p^33d^3$	
4	−3.79217	$3s^23p^33d^3$	
5	−3.58563	$3p^4(^1D)3d$	^{2}P $4p$
6	−3.40819	$3p^4(^3P)3d$	^{4}P $4p$
7	−3.14513	$3s^23p^33d^3$	
8	−2.53816	$3s^23p^33d^3$	
9	−2.30041	$3p^4(^3P)3d$	^{2}P $4p$
10	−1.22032	$3p^4(^3P)3d$	^{4}F $4f$
11	−1.10628	$3p^4(^3P)3d$	^{2}F $4f$
12	−.75992	$3p^4(^1D)3d$	^{2}F $4f$
13	−.48698	$3p^4(^1D)3d$	^{2}P $5p$
14	−.36878	$3p^4(^3P)3d$	^{4}P $5p$
		^{3}**P**^{e}	
1	−9.48538	$3s^23p^43d^2$	
2	−9.01118	$3s^23p^43d^2$	
3	−8.50338	$3s^23p^43d^2$	
4	−8.13191	$3s^23p^43d^2$	
5	−7.86208	$3s^23p^43d^2$	
6	−7.59051	$3s^23p^5$	^{2}P $4p$
7	−7.38734	$3s^23p^43d^2$	
8	−4.53169	$3p^4(^1D)3d$	^{2}P $4s$
9	−4.39002	$3s^23p^5$	^{2}P $5p$
10	−4.32819	$3p^4(^3P)3d$	^{4}P $4s$
11	−3.29622	$3p^4(^3P)3d$	^{2}P $4s$
12	−3.26029	$3p^63d^2$	
13	−2.87325	$3s^23p^5$	^{2}P $6p$

Ar-like Fe (Fe^{8+})

i	Energy(Ryds)	Description	i	Energy(Ryds)	Description	i	Energy(Ryds)	Description
14	−2.61380	$3p^4(^3P)3d$ ^{4}D $4d$	34	−1.86874	$3s^23p^5$ ^{2}P $7d$	41	−1.21004	$3s3p^6$ ^{2}S $5d$
15	−2.49687	$3s3p^43d^3$	35	−1.77979	$3s3p^6$ ^{2}S $5p$	42	−1.17606	$3s^23p^5$ ^{2}P $9p$
16	−2.37106	$3p^4(^1D)3d$ ^{2}P $4d$:	36	−1.63960	$3p^4(^3P)3d$ ^{4}D $4f$	43	−1.13929	$3p^4(^3P)3d$ ^{2}P $4d$
17	−2.31276	$3p^4(^1D)3d$ ^{2}D $4d$	37	−1.59204	$3s^23p^5$ ^{2}P $8s$	44	−1.10771	$3s3p^43d^3$
18	−2.28135	$3p^4(^3P)3d$ ^{4}P $4d$	38	−1.39991	$3s^23p^5$ ^{2}P $8d$	45	−1.02760	$3s^23p^5$ ^{2}P $9f$
19	−2.15202	$3p^4(^3P)3d$ ^{4}F $4d$:	39	−1.31419	$3p^4(^1D)3d$ ^{2}D $4f$	46	−.97337	$3p^4(^3P)3d$ ^{2}D $4d$
20	−2.03826	$3s^23p^5$ ^{2}P $7p$	40	−1.24240	$3p^4(^3P)3d$ ^{4}F $4f$	47	−.94995	$3s3p^43d^3$
21	−2.01504	$3p^4(^3P)3d$ ^{2}F $4d$	41	−1.22458	$3s^23p^5$ ^{2}P $9s$			**$^3D^o$**
22	−1.91064	$3s3p^43d^3$:	42	−1.14130	$3p^4(^3P)3d$ ^{2}F $4f$	1	−13.0927	$3s^23p^5$ ^{2}P $3d$
23	−1.84434	$3s3p^43d^3$:	43	−1.12368	$3p^4(^1D)3d$ ^{2}G $4f$	2	−6.68789	$3s3p^53d^2$
24	−1.77894	$3p^4(^1D)3d$ ^{2}F $4d$	44	−1.09429	$3s^23p^5$ ^{2}P $9d$	3	−6.23617	$3s^23p^5$ ^{2}P $4d$
25	−1.64988	$3s3p^43d^3$:			**$^3D^e$**	4	−6.21551	$3s3p^53d^2$
26	−1.51492	$3s^23p^5$ ^{2}P $8p$	1	−10.6125	$3s3p^6$ ^{2}S $3d$	5	−6.09783	$3s3p^53d^2$:
27	−1.44918	$3p^4(^1S)3d$ ^{2}D $4d$:	2	−9.16830	$3s^23p^43d^2$	6	−5.44408	$3s3p^53d^2$
28	−1.39591	$3s3p^43d^3$:	3	−9.05272	$3s^23p^43d^2$	7	−5.03518	$3s^23p^33d^3$:
29	−1.17381	$3s^23p^5$ ^{2}P $9p$	4	−8.59747	$3s^23p^43d^2$	8	−4.94184	$3s3p^53d^2$:
30	−1.12905	$3p^4(^3P)3d$ ^{2}P $4d$	5	−7.91376	$3s^23p^43d^2$	9	−4.68613	$3s^23p^33d^3$:
31	−1.06990	$3s3p^43d^3$:	6	−7.63723	$3s^23p^5$ ^{2}P $4p$	10	−4.51790	$3s^23p^33d^3$
		$^3P^o$	7	−7.47910	$3s^23p^43d^2$	11	−4.35624	$3s^23p^33d^3$:
1	−13.5394	$3s^23p^5$ ^{2}P $3d$	8	−5.32458	$3s^23p^5$ ^{2}P $4f$	12	−4.24740	$3s^23p^33d^3$:
2	−8.49721	$3s^23p^5$ ^{2}P $4s$	9	−4.80254	$3p^4(^3P)3d$ ^{4}D $4s$	13	−4.06624	$3s^23p^33d^3$:
3	−6.67862	$3s3p^53d^2$	10	−4.50226	$3p^4(^1D)3d$ ^{2}D $4s$	14	−3.97276	$3p^4(^3P)3d$ ^{4}D $4p$
4	−6.36865	$3s^23p^5$ ^{2}P $4d$	11	−4.41647	$3s^23p^5$ ^{2}P $5p$	15	−3.83678	$3s^23p^5$ ^{2}P $5d$
5	−6.01892	$3s3p^53d^2$	12	−3.80399	$3p^4(^1S)3d$ ^{2}D $4s$	16	−3.78033	$3s^23p^33d^3$:
6	−5.78049	$3s3p^53d^2$	13	−3.64650	$3s3p^6$ ^{2}S $4d$	17	−3.69733	$3p^4(^1D)3d$ ^{2}P $4p$
7	−5.27068	$3s^23p^33d^3$	14	−3.39105	$3s^23p^5$ ^{2}P $5f$	18	−3.66914	$3p^4(^1D)3d$ ^{2}D $4p$
8	−5.12457	$3s3p^53d^2$:	15	−3.18285	$3p^4(^3P)3d$ ^{2}D $4s$	19	−3.64914	$3s^23p^33d^3$:
9	−4.94352	$3s3p^6$ ^{2}S $4p$:	16	−2.88141	$3s^23p^5$ ^{2}P $6p$	20	−3.51860	$3p^4(^3P)3d$ ^{4}P $4p$:
10	−4.83706	$3s^23p^33d^3$:	17	−2.73413	$3p^4(^3P)3d$ ^{4}D $4d$	21	−3.50839	$3p^4(^3P)3d$ ^{4}F $4p$
11	−4.77668	$3s^23p^5$ ^{2}P $5s$	18	−2.45086	$3p^4(^1D)3d$ ^{2}P $4d$	22	−3.48894	$3s^23p^33d^3$:
12	−4.63504	$3s^23p^33d^3$:	19	−2.42573	$3s3p^43d^3$	23	−3.41548	$3p^4(^3P)3d$ ^{2}F $4p$
13	−4.35602	$3s^23p^33d^3$:	20	−2.36725	$3p^4(^1D)3d$ ^{2}D $4d$:	24	−3.35844	$3s^23p^33d^3$:
14	−4.02660	$3s^23p^33d^3$:	21	−2.33832	$3s^23p^5$ ^{2}P $6f$	25	−3.31535	$3s^23p^33d^3$:
15	−3.95110	$3p^4(^3P)3d$ ^{4}D $4p$	22	−2.31558	$3p^4(^3P)3d$ ^{4}P $4d$	26	−3.17532	$3s^23p^33d^3$:
16	−3.92636	$3s^23p^33d^3$:	23	−2.25908	$3p^4(^3P)3d$ ^{4}F $4d$:	27	−3.05821	$3p^4(^1D)3d$ ^{2}F $4p$
17	−3.84489	$3s^23p^5$ ^{2}P $5d$	24	−2.20108	$3s3p^43d^3$	28	−2.92919	$3p^4(^1S)3d$ ^{2}D $4p$
18	−3.63854	$3p^4(^1D)3d$ ^{2}P $4p$:	25	−2.16235	$3p^4(^1D)3d$ ^{2}G $4d$	29	−2.87089	$3s^23p^33d^3$:
19	−3.59318	$3p^4(^1D)3d$ ^{2}D $4p$:	26	−2.10237	$3p^4(^3P)3d$ ^{2}F $4d$	30	−2.61545	$3s^23p^33d^3$:
20	−3.53346	$3p^4(^3P)3d$ ^{4}P $4p$	27	−2.04141	$3s^23p^5$ ^{2}P $7p$	31	−2.55616	$3s^23p^5$ ^{2}P $6d$
21	−3.42024	$3s^23p^33d^3$	28	−2.02126	$3s3p^43d^3$	32	−2.40713	$3p^4(^3P)3d$ ^{2}P $4p$
22	−3.33605	$3s^23p^33d^3$	29	−1.87670	$3p^4(^1D)3d$ ^{2}F $4d$:	33	−2.23637	$3p^4(^3P)3d$ ^{2}D $4p$
23	−3.09987	$3s^23p^5$ ^{2}P $6s$	30	−1.83779	$3s3p^43d^3$	34	−2.19810	$3s^23p^33d^3$:
24	−3.09247	$3s^23p^33d^3$:	31	−1.74326	$3s3p^43d^3$	35	−1.83999	$3s^23p^5$ ^{2}P $7d$
25	−2.99915	$3s^23p^33d^3$:	32	−1.71252	$3s^23p^5$ ^{2}P $7f$	36	−1.60958	$3p^4(^3P)3d$ ^{4}D $4f$
26	−2.83012	$3p^4(^1S)3d$ ^{2}D $4p$	33	−1.69079	$3p^4(^1S)3d$ ^{2}D $4d$	37	−1.39098	$3s^23p^5$ ^{2}P $8d$
27	−2.79683	$3s^23p^33d^3$:	34	−1.52180	$3s^23p^5$ ^{2}P $8p$	38	−1.35446	$3p^4(^1D)3d$ ^{2}P $4f$:
28	−2.63104	$3p^4(^1D)3d$ ^{2}S $4p$	35	−1.47198	$3s3p^43d^3$	39	−1.29700	$3p^4(^3P)3d$ ^{4}F $4f$:
29	−2.55857	$3s^23p^5$ ^{2}P $6d$	36	−1.44420	$3s3p^43d^3$	40	−1.26195	$3p^4(^1D)3d$ ^{2}D $4f$
30	−2.50729	$3p^4(^3P)3d$ ^{2}P $4p$	37	−1.34739	$3p^4(^1D)3d$ ^{2}S $4d$	41	−1.18315	$3p^4(^3P)3d$ ^{2}F $4f$:
31	−2.39082	$3s^23p^33d^3$	38	−1.31163	$3s3p^43d^3$	42	−1.15861	$3p^4(^3P)3d$ ^{4}P $4f$:
32	−2.22500	$3p^4(^3P)3d$ ^{2}D $4p$	39	−1.30297	$3s^23p^5$ ^{2}P $8f$	43	−1.11456	$3p^4(^1D)3d$ ^{2}G $4f$
33	−2.13228	$3s^23p^5$ ^{2}P $7s$	40	−1.22462	$3p^4(^3P)3d$ ^{4}D $5s$	44	−1.08760	$3s^23p^5$ ^{2}P $9d$

Ar-like Fe (Fe^{8+})

i	Energy(Ryds)	Description
		$^3F^e$
1	−9.37643	$3s^23p^43d^2$:
2	−9.18389	$3s^23p^43d^2$
3	−8.85228	$3s^23p^43d^2$:
4	−8.46983	$3s^23p^43d^2$:
5	−8.09806	$3s^23p^43d^2$
6	−7.60362	$3s^23p^43d^2$
7	−5.19685	$3s^23p^5$ $^2P\ 4f$
8	−4.50508	$3p^4(^3P)3d\ ^4F\ 4s$
9	−4.40132	$3p^4(^3P)3d\ ^2F\ 4s$
10	−4.03714	$3p^4(^1D)3d\ ^2F\ 4s$
11	−3.48704	$3p^63d^2$
12	−3.33672	$3s^23p^5$ $^2P\ 5f$
13	−2.71611	$3p^4(^3P)3d\ ^4D\ 4d$
14	−2.52218	$3s3p^43d^3$:
15	−2.42764	$3p^4(^3P)3d\ ^4F\ 4d$
16	−2.38683	$3p^4(^1D)3d\ ^2P\ 4d$
17	−2.31463	$3s^23p^5$ $^2P\ 6f$
18	−2.29531	$3p^4(^1D)3d\ ^2D\ 4d$
19	−2.25483	$3p^4(^3P)3d\ ^2F\ 4d$
20	−2.20541	$3s3p^43d^3$:
21	−2.14957	$3p^4(^1D)3d\ ^2G\ 4d$
22	−2.09502	$3p^4(^3P)3d\ ^4P\ 4d$
23	−2.00596	$3s3p^43d^3$
24	−1.78880	$3s3p^43d^3$:
25	−1.76352	$3p^4(^1D)3d\ ^2F\ 4d$
26	−1.69422	$3s^23p^5$ $^2P\ 7f$
27	−1.64064	$3s3p^43d^3$
28	−1.59301	$3p^4(^1S)3d\ ^2D\ 4d$
29	−1.53964	$3s3p^43d^3$:
30	−1.41743	$3s3p^43d^3$
31	−1.31058	$3s3p^43d^3$:
32	−1.29480	$3s^23p^5$ $^2P\ 8f$
33	−1.13403	$3p^4(^3P)3d\ ^2P\ 4d$
34	−1.07369	$3s3p^43d^3$:
35	−1.02163	$3s^23p^5$ $^2P\ 9f$
36	−1.01213	$3s3p^43d^3$:
37	−.97622	$3s3p^43d^3$
38	−.96109	$3p^4(^3P)3d\ ^2D\ 4d$:
39	−.91092	$3p^4(^3P)3d\ ^4F\ 5s$
40	−.90520	$3s3p^43d^3$:
		$^3F^o$
1	−13.3652	$3s^23p^5$ $^2P\ 3d$
2	−6.72147	$3s3p^53d^2$
3	−6.43332	$3s3p^53d^2$
4	−6.29544	$3s^23p^5$ $^2P\ 4d$
5	−5.96527	$3s3p^53d^2$
6	−5.48372	$3s3p^53d^2$:
7	−5.19926	$3s^23p^33d^3$
8	−4.86720	$3s^23p^33d^3$:
9	−4.69468	$3s^23p^33d^3$:
10	−4.48988	$3s^23p^33d^3$:
11	−4.26791	$3s^23p^33d^3$:
12	−4.24477	$3s^23p^33d^3$:
13	−4.09970	$3s^23p^33d^3$:
14	−4.06149	$3s^23p^33d^3$:
15	−3.96376	$3p^4(^3P)3d\ ^4D\ 4p$
16	−3.85445	$3s^23p^5$ $^2P\ 5d$
17	−3.80678	$3s^23p^33d^3$:
18	−3.77674	$3s^23p^33d^3$:
19	−3.75800	$3p^4(^3P)3d\ ^4F\ 4p$:
20	−3.65064	$3p^4(^1D)3d\ ^2D\ 4p$
21	−3.61010	$3s^23p^33d^3$:
22	−3.54458	$3p^4(^3P)3d\ ^2F\ 4p$
23	−3.42346	$3s^23p^33d^3$:
24	−3.35575	$3p^4(^1D)3d\ ^2G\ 4p$
25	−3.26992	$3s^23p^5$ $^2P\ 5g$
26	−3.17562	$3s^23p^33d^3$:
27	−3.13988	$3p^4(^1D)3d\ ^2F\ 4p$
28	−2.90850	$3p^4(^1S)3d\ ^2D\ 4p$
29	−2.85310	$3s^23p^33d^3$:
30	−2.64663	$3s3p^6$ $^2S\ 4f$
31	−2.57662	$3s^23p^5$ $^2P\ 6d$
32	−2.32719	$3s^23p^33d^3$:
33	−2.28524	$3p^4(^3P)3d\ ^2D\ 4p$
34	−2.26196	$3s^23p^5$ $^2P\ 6g$
35	−1.85035	$3s^23p^5$ $^2P\ 7d$
36	−1.68780	$3p^4(^3P)3d\ ^4D\ 4f$
37	−1.66554	$3s^23p^5$ $^2P\ 7g$
38	−1.40664	$3p^4(^1D)3d\ ^2P\ 4f$
39	−1.39786	$3s^23p^5$ $^2P\ 8d$
40	−1.35367	$3p^4(^3P)3d\ ^4F\ 4f$
41	−1.29272	$3p^4(^1D)3d\ ^2D\ 4f$
42	−1.27498	$3s^23p^5$ $^2P\ 8g$
43	−1.25095	$3p^4(^3P)3d\ ^4P\ 4f$
44	−1.17548	$3p^4(^3P)3d\ ^2F\ 4f$
45	−1.12990	$3p^4(^1D)3d\ ^2G\ 4f$
46	−1.09249	$3s^23p^5$ $^2P\ 9d$
47	−1.00902	$3s^23p^5$ $^2P\ 9g$
		$^3G^e$
1	−9.33102	$3s^23p^43d^2$
2	−8.98906	$3s^23p^43d^2$
3	−8.26027	$3p^63d^2$:
4	−5.28833	$3s^23p^5$ $^2P\ 4f$
5	−4.35060	$3p^4(^1D)3d\ ^2G\ 4s$
6	−3.37744	$3s^23p^5$ $^2P\ 5f$
7	−2.67654	$3p^4(^3P)3d\ ^4D\ 4d$
8	−2.48907	$3p^4(^3P)3d\ ^4F\ 4d$
9	−2.39196	$3p^4(^1D)3d\ ^2D\ 4d$
10	−2.33820	$3s^23p^5$ $^2P\ 6f$
11	−2.32149	$3s3p^43d^3$:
12	−2.22263	$3p^4(^3P)3d\ ^2F\ 4d$
13	−2.10994	$3s3p^43d^3$
14	−2.12601	$3p^4(^1D)3d\ ^2G\ 4d$
15	−1.91869	$3s3p^43d^3$:
16	−1.87276	$3p^4(^1D)3d\ ^2F\ 4d$:
17	−1.79027	$3s3p^43d^3$:
18	−1.70781	$3s^23p^5$ $^2P\ 7f$
19	−1.68844	$3s3p^43d^3$
20	−1.64964	$3p^4(^1S)3d\ ^2D\ 4d$
21	−1.53806	$3s3p^43d^3$:
22	−1.31374	$3s3p^43d^3$
23	−1.30351	$3s^23p^5$ $^2P\ 8f$
24	−1.13763	$3s3p^43d^3$
25	−1.10976	$3s3p^43d^3$:
26	−1.09606	$3s3p^43d^3$
27	−1.04183	$3p^4(^3P)3d\ ^2D\ 4d$
28	−1.01531	$3s^23p^5$ $^2P\ 9f$
		$^3G^o$
1	−6.45828	$3s3p^53d^2$
2	−6.10425	$3s3p^53d^2$
3	−5.88573	$3s3p^53d^2$
4	−5.14231	$3s^23p^33d^3$
5	−4.81864	$3s^23p^33d^3$:
6	−4.47246	$3s^23p^33d^3$:
7	−4.25147	$3s^23p^33d^3$
8	−4.14451	$3s^23p^33d^3$
9	−4.05384	$3s^23p^33d^3$
10	−4.01851	$3s^23p^33d^3$:
11	−3.80365	$3s^23p^33d^3$
12	−3.64808	$3s^23p^33d^3$:
13	−3.62559	$3p^4(^3P)3d\ ^4F\ 4p$:
14	−3.54739	$3p^4(^3P)3d\ ^2F\ 4p$
15	−3.48463	$3s^23p^33d^3$:
16	−3.37082	$3p^4(^1D)3d\ ^2G\ 4p$:
17	−3.23862	$3s^23p^5$ $^2P\ 5g$
18	−3.16616	$3p^4(^1D)3d\ ^2F\ 4p$
19	−3.07053	$3s^23p^33d^3$:
20	−2.25354	$3s^23p^5$ $^2P\ 6g$
21	−1.69340	$3p^4(^3P)3d\ ^4D\ 4f$
22	−1.65723	$3s^23p^5$ $^2P\ 7g$
23	−1.42378	$3p^4(^3P)3d\ ^4F\ 4f$
24	−1.38170	$3p^4(^1D)3d\ ^2P\ 4f$
25	−1.33218	$3p^4(^1D)3d\ ^2D\ 4f$
26	−1.27224	$3s^23p^5$ $^2P\ 8g$
27	−1.24288	$3p^4(^3P)3d\ ^2F\ 4f$
28	−1.13920	$3p^4(^1D)3d\ ^2G\ 4f$
29	−1.12586	$3p^4(^3P)3d\ ^4P\ 4f$
30	−1.00317	$3s^23p^5$ $^2P\ 9g$
31	−.82454	$3p^4(^1D)3d\ ^2F\ 4f$
		$^5S^e$
1	−9.82672	$3s^23p^43d^2$
2	−2.96135	$3s3p^43d^3$
3	−2.65441	$3p^4(^3P)3d\ ^4D\ 4d$
4	−1.02112	$3s3p^43d^3$
5	−.75349	$3s3p^43d^3$

Ar-like Fe (Fe^{8+})

i	Energy(Ryds)	Description
6	−.27196	$3p^4(^3P)3d\ ^4\mathrm{D}\ 5d$
		$^5\mathbf{S}^o$
1	−6.24845	$3s3p^53d^2$
2	−4.57525	$3s^23p^33d^3$
3	−3.54138	$3p^4(^3P)3d\ ^4\mathrm{P}\ 4p$
4	−1.32558	$3p^4(^3P)3d\ ^4\mathrm{F}\ 4f$
5	−.41076	$3p^4(^3P)3d\ ^4\mathrm{P}\ 5p$
		$^5\mathbf{P}^e$
1	−9.13683	$3s^23p^43d^2$
2	−4.49691	$3p^4(^3P)3d\ ^4\mathrm{P}\ 4s$
3	−3.14075	$3s3p^43d^3$
4	−2.85152	$3p^4(^3P)3d\ ^4\mathrm{D}\ 4d$
5	−2.73747	$3s3p^43d^3$
6	−2.38134	$3p^4(^3P)3d\ ^4\mathrm{F}\ 4d$
7	−2.31861	$3s3p^43d^3$:
8	−2.18332	$3p^4(^3P)3d\ ^4\mathrm{P}\ 4d$
9	−1.89203	$3s3p^43d^3$
10	−1.66592	$3s3p^43d^3$
11	−1.21943	$3s3p^43d^3$
12	−1.14198	$3s3p^43d^3$
13	−.82089	$3p^4(^3P)3d\ ^4\mathrm{P}\ 5s$
14	−.66028	$3s3p^43d^3$:
15	−.33591	$3p^4(^3P)3d\ ^4\mathrm{D}\ 5d$
		$^5\mathbf{P}^o$
1	−6.96591	$3s3p^53d^2$
2	−5.66239	$3s^23p^33d^3$
3	−4.95292	$3s^23p^33d^3$
4	−4.39193	$3s^23p^33d^3$
5	−4.14671	$3s^23p^33d^3$
6	−4.09448	$3p^4(^3P)3d\ ^4\mathrm{D}\ 4p$
7	−3.64268	$3p^4(^3P)3d\ ^4\mathrm{P}\ 4p$:
8	−3.56657	$3s^23p^33d^3$:
9	−1.76239	$3p^4(^3P)3d\ ^4\mathrm{D}\ 4f$
10	−1.35448	$3p^4(^3P)3d\ ^4\mathrm{F}\ 4f$
11	−.90373	$3p^4(^3P)3d\ ^4\mathrm{D}\ 5p$
12	−.44575	$3p^4(^3P)3d\ ^4\mathrm{P}\ 5p$
		$^5\mathbf{D}^e$
1	−9.75573	$3s^23p^43d^2$
2	−9.16028	$3s^23p^43d^2$
3	−4.95186	$3p^4(^3P)3d\ ^4\mathrm{D}\ 4s$
4	−3.09681	$3s3p^43d^3$:
5	−2.90381	$3s3p^43d^3$
6	−2.68674	$3p^4(^3P)3d\ ^4\mathrm{D}\ 4d$
7	−2.59840	$3s3p^43d^3$
8	−2.44926	$3p^4(^3P)3d\ ^4\mathrm{F}\ 4d$
9	−2.30451	$3s3p^43d^3$:
10	−2.19825	$3p^4(^3P)3d\ ^4\mathrm{P}\ 4d$
11	−2.14289	$3s3p^43d^3$
12	−1.92451	$3s3p^43d^3$
13	−1.56251	$3s3p^43d^3$:
14	−1.28139	$3p^4(^3P)3d\ ^4\mathrm{D}\ 5s$
15	−1.26483	$3s3p^43d^3$:
16	−1.05850	$3s3p^43d^3$
17	−.87471	$3s3p^43d^3$
18	−.27773	$3p^4(^3P)3d\ ^4\mathrm{D}\ 5d$
		$^5\mathbf{D}^o$
1	−7.28942	$3s3p^53d^2$
2	−6.45077	$3s3p^53d^2$
3	−5.54716	$3s^23p^33d^3$
4	−4.88640	$3s^23p^33d^3$
5	−4.82589	$3s^23p^33d^3$
6	−4.64281	$3s^23p^33d^3$:
7	−4.15145	$3s^23p^33d^3$:
8	−4.10274	$3p^4(^3P)3d\ ^4\mathrm{D}\ 4p$:
9	−3.94747	$3s^23p^33d^3$
10	−3.74357	$3p^4(^3P)3d\ ^4\mathrm{F}\ 4p$
11	−3.54310	$3p^4(^3P)3d\ ^4\mathrm{P}\ 4p$
12	−1.78562	$3p^4(^3P)3d\ ^4\mathrm{D}\ 4f$
13	−1.39175	$3p^4(^3P)3d\ ^4\mathrm{F}\ 4f$
14	−1.24652	$3p^4(^3P)3d\ ^4\mathrm{P}\ 4f$
15	−.87857	$3p^4(^3P)3d\ ^4\mathrm{D}\ 5p$
16	−.55736	$3p^4(^3P)3d\ ^4\mathrm{F}\ 5p$
17	−.42872	$3p^4(^3P)3d\ ^4\mathrm{P}\ 5p$
		$^5\mathbf{F}^e$
1	−9.66969	$3s^23p^43d^2$
2	−4.65286	$3p^4(^3P)3d\ ^4\mathrm{F}\ 4s$
3	−3.04854	$3s3p^43d^3$
4	−2.81208	$3s3p^43d^3$
5	−2.75552	$3p^4(^3P)3d\ ^4\mathrm{D}\ 4d$
6	−2.71208	$3s3p^43d^3$:
7	−2.48737	$3p^4(^3P)3d\ ^4\mathrm{F}\ 4d$
8	−2.33063	$3p^4(^3P)3d\ ^4\mathrm{P}\ 4d$
9	−2.22719	$3s3p^43d^3$
10	−2.10065	$3s3p^43d^3$
11	−1.74395	$3s3p^43d^3$:
12	−1.55883	$3s3p^43d^3$:
13	−1.50324	$3s3p^43d^3$
14	−.96365	$3p^4(^3P)3d\ ^4\mathrm{F}\ 5s$
15	−.93937	$3s3p^43d^3$
16	−.29531	$3p^4(^3P)3d\ ^4\mathrm{D}\ 5d$
17	−.00455	$3p^4(^3P)3d\ ^4\mathrm{F}\ 5d$
		$^5\mathbf{F}^o$
1	−6.83689	$3s3p^53d^2$
2	−5.56482	$3s^23p^33d^3$
3	−5.11186	$3s^23p^33d^3$
4	−4.63805	$3s^23p^33d^3$
5	−4.39690	$3s^23p^33d^3$:
6	−4.13787	$3s^23p^33d^3$:
7	−4.10103	$3p^4(^3P)3d\ ^4\mathrm{D}\ 4p$
8	−3.81831	$3p^4(^3P)3d\ ^4\mathrm{F}\ 4p$
9	−1.70751	$3p^4(^3P)3d\ ^4\mathrm{D}\ 4f$
10	−1.51635	$3p^4(^3P)3d\ ^4\mathrm{F}\ 4f$
11	−1.26284	$3p^4(^3P)3d\ ^4\mathrm{P}\ 4f$
12	−.90219	$3p^4(^3P)3d\ ^4\mathrm{D}\ 5p$
13	−.60531	$3p^4(^3P)3d\ ^4\mathrm{F}\ 5p$
		$^5\mathbf{G}^e$
1	−9.50646	$3s^23p^43d^2$
2	−3.11424	$3s3p^43d^3$
3	−2.90381	$3s3p^43d^3$:
4	−2.83953	$3p^4(^3P)3d\ ^4\mathrm{D}\ 4d$
5	−2.65254	$3s3p^43d^3$
6	−2.41374	$3p^4(^3P)3d\ ^4\mathrm{F}\ 4d$
7	−2.15438	$3s3p^43d^3$
8	−1.75341	$3s3p^43d^3$:
9	−1.28985	$3s3p^43d^3$
10	−.33294	$3p^4(^3P)3d\ ^4\mathrm{D}\ 5d$
		$^5\mathbf{G}^o$
1	−6.89289	$3s3p^53d^2$
2	−5.48780	$3s^23p^33d^3$
3	−5.09038	$3s^23p^33d^3$
4	−4.42936	$3s^23p^33d^3$
5	−3.76538	$3p^4(^3P)3d\ ^4\mathrm{F}\ 4p$
6	−1.74972	$3p^4(^3P)3d\ ^4\mathrm{D}\ 4f$
7	−1.50239	$3p^4(^3P)3d\ ^4\mathrm{F}\ 4f$
8	−1.34773	$3p^4(^3P)3d\ ^4\mathrm{P}\ 4f$
9	−.57169	$3p^4(^3P)3d\ ^4\mathrm{F}\ 5p$

Ar-like Fe (Fe^{8+})

Energies in ascending order from ground state for terms with effective $n \leq 4.0$, $L \leq 4$

Term	i	E(Ryds)	Term	i	E(Ryds)	Term	i	E(Ryds)	Term	i	E(Ryds)	Term	i	E(Ryds)
$^{1}S^{e}$	1	0.00000	$^{3}D^{e}$	3	8.20058	$^{1}D^{e}$	6	9.58469	$^{3}F^{o}$	4	10.9578	$^{1}S^{e}$	6	11.5940
$^{3}P^{o}$	1	3.71390	$^{3}P^{e}$	2	8.24212	$^{1}S^{e}$	3	9.60442	$^{1}F^{o}$	2	10.9646	$^{5}F^{o}$	2	11.6884
$^{3}F^{o}$	1	3.88810	$^{3}G^{e}$	2	8.26424	$^{3}D^{e}$	6	9.61607	$^{1}P^{o}$	3	10.9925	$^{5}D^{o}$	3	11.7061
$^{1}D^{o}$	1	4.13940	$^{1}P^{e}$	1	8.40075	$^{1}P^{e}$	2	9.62347	$^{5}S^{o}$	1	11.0048	$^{5}G^{o}$	2	11.7655
$^{3}D^{o}$	1	4.16060	$^{3}F^{e}$	3	8.40102	$^{3}F^{e}$	6	9.64968	$^{1}G^{o}$	1	11.0075	$^{3}F^{o}$	6	11.7695
$^{1}F^{o}$	1	4.18910	$^{1}G^{e}$	2	8.47246	$^{3}P^{e}$	6	9.66279	$^{1}D^{o}$	2	11.0164	$^{3}D^{o}$	6	11.8092
$^{1}P^{o}$	1	5.37300	$^{1}D^{e}$	3	8.64711	$^{3}D^{e}$	7	9.77420	$^{3}D^{o}$	3	11.0171	$^{1}G^{o}$	2	11.9194
$^{3}D^{e}$	1	6.64080	$^{3}D^{e}$	4	8.65583	$^{1}P^{e}$	3	9.83244	$^{3}D^{o}$	4	11.0377	$^{3}D^{e}$	8	11.9287
$^{1}D^{e}$	1	6.86620	$^{3}P^{e}$	3	8.74992	$^{1}D^{e}$	7	9.83974	$^{3}G^{o}$	2	11.1490	$^{3}G^{e}$	4	11.9649
$^{5}S^{e}$	1	7.42658	$^{3}P^{o}$	2	8.75609	$^{3}P^{e}$	7	9.86596	$^{3}D^{o}$	5	11.1554	$^{3}P^{o}$	7	11.9826
$^{5}D^{e}$	1	7.49757	$^{3}F^{e}$	4	8.78347	$^{1}F^{e}$	3	9.88771	$^{1}F^{o}$	3	11.2111	$^{1}F^{o}$	5	11.9944
$^{5}F^{e}$	1	7.58361	$^{1}P^{o}$	2	8.85077	$^{5}D^{o}$	1	9.96388	$^{1}D^{o}$	3	11.2221	$^{1}G^{e}$	5	12.0111
$^{5}G^{e}$	1	7.74684	$^{3}G^{e}$	3	8.99303	$^{1}S^{e}$	4	10.0217	$^{3}S^{o}$	1	11.2301	$^{3}S^{o}$	2	12.0313
$^{3}P^{e}$	1	7.76792	$^{1}F^{e}$	2	9.03566	$^{5}P^{o}$	1	10.2873	$^{3}P^{o}$	5	11.2343	$^{3}F^{o}$	7	12.0540
$^{3}F^{e}$	1	7.87687	$^{3}P^{e}$	4	9.12139	$^{5}G^{o}$	1	10.3604	$^{1}S^{o}$	1	11.2819	$^{1}F^{e}$	4	12.0561
$^{3}G^{e}$	1	7.92228	$^{3}F^{e}$	5	9.15524	$^{5}F^{o}$	1	10.4164	$^{3}F^{o}$	5	11.2880	$^{3}F^{e}$	7	12.0564
$^{1}D^{e}$	2	8.00525	$^{1}D^{e}$	4	9.19515	$^{3}F^{o}$	2	10.5318	$^{3}G^{o}$	3	11.3675	$^{3}G^{o}$	4	12.1109
$^{3}F^{e}$	2	8.06941	$^{1}G^{e}$	3	9.20100	$^{3}D^{o}$	2	10.5654	$^{1}S^{e}$	5	11.4288	$^{1}D^{e}$	9	12.1166
$^{3}D^{e}$	2	8.08500	$^{3}S^{e}$	1	9.30885	$^{3}P^{o}$	3	10.5746	$^{3}S^{e}$	3	11.4354	$^{3}P^{o}$	8	12.1287
$^{5}D^{e}$	2	8.09302	$^{1}G^{e}$	4	9.30896	$^{1}D^{e}$	8	10.6739	$^{1}D^{o}$	4	11.4717	$^{5}F^{o}$	3	12.1414
$^{1}S^{e}$	2	8.10983	$^{3}D^{e}$	5	9.33954	$^{3}G^{o}$	1	10.7950	$^{3}P^{o}$	6	11.4728	$^{5}G^{o}$	3	12.1629
$^{5}P^{e}$	1	8.11647	$^{3}P^{e}$	5	9.39122	$^{5}D^{o}$	2	10.8025	$^{1}P^{o}$	4	11.4964			
$^{1}G^{e}$	1	8.20024	$^{1}D^{e}$	5	9.40815	$^{3}F^{o}$	3	10.8199	$^{1}F^{o}$	4	11.5459			
$^{1}F^{e}$	1	8.20026	$^{3}S^{e}$	2	9.50327	$^{3}P^{o}$	4	10.8846	$^{5}P^{o}$	2	11.5909			

gf-values for transitions involving terms with effective $n \leq 4.0$, $L \leq 4$

i i'	gf_L	i i'	gf_L	i i'	gf_L	i i'	gf_L	i i'	gf_L	i i'	gf_L
	$^{1}S^{e}$–$^{1}P^{o}$	6 3	1.52E−2	1 8	−5.23E+0	4 4	1.72E−2	1 4	−5.69E−1	3 9	−1.72E−3
1 1	−3.11E+0	6 4	2.04E−5	1 9	−3.12E+0	4 5	1.21E−2	1 5	−7.89E−1	4 1	9.40E−2
1 2	−2.91E−1		$^{1}P^{o}$–$^{1}P^{e}$	2 1	2.15E−3	4 6	5.05E−3	1 6	−1.86E+0	4 2	1.95E−1
1 3	−4.64E−1	1 1	−3.74E−4	2 2	8.64E−4	4 7	3.92E−3	1 7	−6.50E−1	4 3	1.04E−1
1 4	−1.71E−3	1 2	−3.02E−2	2 3	4.53E−4	4 8	1.38E−4	1 8	−9.66E−1	4 4	2.24E−2
2 1	1.16E−5	1 3	−1.14E−2	2 4	−1.29E−4	4 9	−4.54E−3	1 9	−2.17E−1	4 5	3.94E−2
2 2	−2.01E−4	2 1	6.99E−4	2 5	−4.78E−2		$^{1}P^{e}$–$^{1}D^{o}$	2 1	9.06E−4	4 6	1.59E−3
2 3	−2.94E−6	2 2	−4.00E−1	2 6	−5.92E−1	1 1	2.35E−2	2 2	2.01E−3	4 7	3.05E−4
2 4	−3.72E−1	2 3	−3.99E−1	2 7	−6.21E−1	1 2	−1.98E−3	2 3	4.83E−3	4 8	3.29E−4
3 1	2.37E−3	3 1	2.90E−5	2 8	−3.23E−3	1 3	−1.11E−4	2 4	1.25E−6	4 9	−9.79E−4
3 2	2.07E−3	3 2	3.82E−1	2 9	−1.26E−2	1 4	−1.62E−1	2 5	5.00E−2		$^{1}D^{o}$–$^{1}F^{e}$
3 3	−7.43E−3	3 3	2.66E−1	3 1	7.42E−5	2 1	2.10E+0	2 6	3.62E−1	1 1	−1.14E−1
3 4	−1.54E−3	4 1	2.13E−1	3 2	2.40E−4	2 2	−1.19E+0	2 7	2.59E−1	1 2	1.02E+0
4 1	2.73E 1	4 2	6.00E−3	3 3	6.20E−4	2 3	−2.07E−2	2 8	2.36E−3	1 3	−1.28E+0
4 2	3.68E−1	4 3	6.14E−4	3 4	5.61E−3	2 4	−5.45E−3	2 9	−3.95E−1	1 4	−2.91E+0
4 3	−7.43E−1		$^{1}P^{o}$–$^{1}D^{e}$	3 5	1.82E−4	3 1	8.75E−1	3 1	1.56E−2	2 1	1.20E−2
4 4	−4.64E−3	1 1	−2.31E−2	3 6	5.23E−2	3 2	−8.46E−1	3 2	1.67E−1	2 2	2.22E−2
5 1	1.42E+0	1 2	−4.12E−3	3 7	2.03E−2	3 3	−2.36E−4	3 3	3.01E−2	2 3	9.36E−5
5 2	6.63E−2	1 3	−8.92E−3	3 8	2.39E−2	3 4	−2.97E−3	3 4	7.96E−6	2 4	−2.85E+0
5 3	1.05E−3	1 4	−1.03E−1	3 9	−1.97E+0		$^{1}D^{o}$ $^{1}D^{e}$	3 5	1.24E 2	3 1	1.54E−1
5 4	−1.44E−6	1 5	−2.13E−1	4 1	2.78E−2	1 1	−1.45E−1	3 6	3.54E−3	3 2	1.45E−4
6 1	6.79E−1	1 6	−5.46E−4	4 2	2.57E−2	1 2	−5.64E−2	3 7	1.04E−2	3 3	1.42E 3
6 2	1.21E−2	1 7	−4.79E−2	4 3	1.04E−1	1 3	−1.22E−1	3 8	9.28E−5	3 4	−1.42E−2

Ar-like Fe (Fe^{8+})

i i′	gf_L	*i i′*	gf_L	*i i′*	gf_L	*i i′*	gf_L	*i i′*	gf_L	*i i′*	gf_L
4 1	2.15E−1	1 2	−4.95E−1	4 1	5.39E−5	2 7	−1.07E+0	2 1	2.91E−3	8 5	1.06E−1
4 2	2.31E−3	1 3	−5.80E+0	4 2	4.05E−4	3 1	1.22E−1	2 2	1.36E−3	8 6	1.80E−1
4 3	4.65E−3	1 4	−3.75E−1		$^1\mathbf{G}^o$–$^1\mathbf{G}^e$	3 2	4.21E−3	2 3	1.27E−4	8 7	1.07E−2
4 4	−7.39E−3	2 1	6.45E−4	1 1	2.09E−1	3 3	1.50E−2	2 4	1.07E−3	8 8	2.62E−3
	$^1\mathbf{D}^e$–$^1\mathbf{F}^o$	2 2	2.12E−3	1 2	4.34E−5	3 4	1.71E−3	2 5	−2.20E−1		$^3\mathbf{P}^e$–$^3\mathbf{D}^o$
1 1	1.80E−1	2 3	1.55E−4	1 3	1.24E−2	3 5	7.32E−3	2 6	−2.65E+0	1 1	7.70E−2
1 2	−7.39E−3	2 4	−3.71E−1	1 4	7.77E−3	3 6	3.34E−3	2 7	−9.99E−1	1 2	−2.81E−1
1 3	−1.41E−1	3 1	5.58E−3	1 5	−8.80E−6	3 7	3.41E−3	2 8	−7.95E−3	1 3	−8.80E−1
1 4	−6.17E−1	3 2	7.11E−2	2 1	7.74E−2	4 1	1.87E−3	3 1	4.77E−2	1 4	−1.13E−1
1 5	−1.61E+0	3 3	8.28E−5	2 2	5.19E−2	4 2	2.19E−3	3 2	1.71E−1	1 5	−3.57E−1
2 1	4.52E−3	3 4	−2.23E−5	2 3	3.43E−1	4 3	5.92E−3	3 3	7.60E−2	1 6	−5.48E−1
2 2	−5.63E−3	4 1	2.62E−1	2 4	5.66E−2	4 4	1.04E−2	3 4	6.67E−3	2 1	2.08E−3
2 3	−4.33E−1	4 2	2.18E−4	2 5	−1.54E−5	4 5	1.30E−1	3 5	5.50E−4	2 2	−2.30E−2
2 4	−1.40E−1	4 3	1.25E−3		$^3\mathbf{S}^e$–$^3\mathbf{P}^o$	4 6	1.16E+0	3 6	3.87E−5	2 3	−1.64E−2
2 5	−4.70E−1	4 4	−8.25E−4	1 1	3.10E+0	4 7	6.43E−1	3 7	6.08E−7	2 4	−5.21E−4
3 1	8.39E−2	5 1	1.69E−1	1 2	1.14E−1	5 1	6.18E−2	3 8	−3.60E−2	2 5	−1.09E−1
3 2	−5.50E−3	5 2	2.52E−1	1 3	−1.38E−3	5 2	2.69E−1	4 1	2.75E−2	2 6	−2.17E−1
3 3	−4.12E−2	5 3	8.99E−6	1 4	−5.84E−1	5 3	2.13E−3	4 2	9.68E−3	3 1	2.35E−4
3 4	−7.37E−2	5 4	−8.58E−6	1 5	−1.97E−3	5 4	1.43E−4	4 3	6.67E−4	3 2	−1.68E−3
3 5	−6.77E−2		$^1\mathbf{F}^o$–$^1\mathbf{G}^e$	1 6	−1.63E−3	5 5	2.46E−2	4 4	1.61E−2	3 3	−3.96E−3
4 1	4.89E−1	1 1	−4.46E−3	1 7	−3.42E−3	5 6	9.19E−3	4 5	7.90E−3	3 4	−4.21E−2
4 2	−3.65E−3	1 2	−2.52E−1	1 8	−1.64E−1	5 7	2.19E−3	4 6	2.89E−2	3 5	−2.09E−2
4 3	−1.99E−2	1 3	−1.24E+0	2 1	1.34E−1	6 1	7.09E−2	4 7	4.29E−3	3 6	−1.35E−2
4 4	−3.77E−3	1 4	−2.29E+0	2 2	5.27E−1	6 2	4.98E−2	4 8	−5.39E+0	4 1	6.24E−2
4 5	−3.42E−3	1 5	−4.39E+0	2 3	−1.35E−2	6 3	7.62E−1	5 1	3.35E−1	4 2	−1.52E−3
5 1	2.49E+0	2 1	4.48E−3	2 4	−1.89E+0	6 4	2.63E−4	5 2	4.64E−1	4 3	−1.71E−2
5 2	−2.54E−1	2 2	1.85E−2	2 5	−1.28E−2	6 5	3.27E−4	5 3	1.20E−1	4 4	−1.68E−2
5 3	−2.80E−2	2 3	9.29E−3	2 6	−3.20E−3	6 6	1.33E−2	5 4	1.61E−2	4 5	−2.77E−2
5 4	−4.60E−3	2 4	2.08E−2	2 7	−6.14E−4	6 7	7.37E−6	5 5	4.96E−2	4 6	−3.22E−1
5 5	−2.44E−2	2 5	−4.07E+0	2 8	−6.39E−4	7 1	4.37E−2	5 6	3.40E−3	5 1	8.47E−2
6 1	1.31E+0	3 1	2.83E−1	3 1	4.24E−2	7 2	5.41E−1	5 7	2.77E−4	5 2	−4.08E−3
6 2	−2.09E+0	3 2	6.60E−3	3 2	4.71E−1	7 3	1.94E−4	5 8	−9.46E−3	5 3	−8.89E−2
6 3	−7.30E−4	3 3	3.10E−3	3 3	4.59E−3	7 4	2.79E−2	6 1	2.10E−2	5 4	−3.61E−1
6 4	−2.88E−2	3 4	8.73E−3	3 4	2.62E−4	7 5	1.39E−3	6 2	9.27E−2	5 5	−1.84E−4
6 5	−2.79E−2	3 5	−7.05E−3	3 5	4.67E−4	7 6	9.26E−4	6 3	8.40E−2	5 6	−2.88E−2
7 1	8.03E−1	4 1	3.92E−2	3 6	−2.40E−4	7 7	2.81E−5	6 4	1.74E−2	6 1	5.73E+0
7 2	−1.38E+0	4 2	5.79E−1	3 7	−7.35E−5	8 1	1.67E−2	6 5	7.93E−3	6 2	−1.21E−2
7 3	−2.07E−3	4 3	7.97E−3	3 8	−4.75E−1	8 2	8.43E−3	6 6	9.65E−4	6 3	−2.93E−1
7 4	−1.82E−5	4 4	4.14E−4		$^3\mathbf{P}^o$–$^3\mathbf{P}^e$	8 3	3.13E−3	6 7	6.28E−4	6 4	−3.20E+0
7 5	−1.41E−2	4 5	−5.50E−3	1 1	−1.94E−4	8 4	3.01E−2	6 8	−3.42E−3	6 5	−9.25E−2
8 1	4.97E−2	5 1	1.05E−2	1 2	−2.81E−2	8 5	8.13E−2	7 1	1.37E−3	6 6	−8.33E−3
8 2	−5.86E−4	5 2	1.35E−1	1 3	−7.05E−2	8 6	1.37E−1	7 2	6.31E−2	7 1	3.10E+0
8 3	−2.85E−4	5 3	1.12E−1	1 4	−7.70E−1	8 7	1.58E−3	7 3	9.65E−2	7 2	−9.64E−3
8 4	−4.26E−3	5 4	9.97E−2	1 5	−8.45E+0		$^3\mathbf{P}^o$–$^3\mathbf{D}^e$	7 4	1.17E−4	7 3	−1.54E−1
8 5	−9.80E−3	5 5	−3.18E−6	1 6	−6.13E−2	1 1	−5.48E−1	7 5	2.13E−5	7 4	−1.91E+0
9 1	1.94E−3		$^1\mathbf{F}^e$–$^1\mathbf{G}^o$	1 7	−1.86E−2	1 2	−2.49E−1	7 6	1.67E−4	7 5	−1.30E−2
9 2	2.59E−3	1 1	−5.00E−2	2 1	1.27E−3	1 3	−3.12E−1	7 7	1.42E−6	7 6	−1.20E−2
9 3	1.19E−3	1 2	−5.12E−2	2 2	2.24E−4	1 4	−3.56E+0	7 8	−3.18E−8		$^3\mathbf{D}^o$–$^3\mathbf{D}^e$
9 4	9.91E−6	2 1	−1.06E−3	2 3	2.48E−4	1 5	−7.39E−1	8 1	6.74E+0	1 1	−5.36E−1
9 5	2.82E−5	2 2	−3.58E−1	2 4	−4.07E−3	1 6	−1.03E+0	8 2	4.57E−2	1 2	−9.86E−2
	$^1\mathbf{F}^o$–$^1\mathbf{F}^e$	3 1	−8.33E−3	2 5	−9.29E−2	1 7	−1.37E+0	8 3	1.14E−2	1 3	−4.80E−2
1 1	−2.53E−5	3 2	−4.12E−4	2 6	−1.30E+0	1 8	−5.89E+0	8 4	1.26E−3	1 4	−5.41E−1

Ar-like Fe (Fe^{8+})

$i\ i'$	gf_L	$i\ i'$	gf_L	$i\ i'$	gf_L	$i\ i'$	gf_L	$i\ i'$	gf_L	$i\ i'$	gf_L
1 5	−9.90E−2		$^3\mathbf{D}^o$–$^3\mathbf{F}^e$	1 2	−3.94E−1	7 5	−1.70E−2	5 7	−9.98E−5	1 1	−6.04E−1
1 6	−5.33E+0	1 1	−3.80E−4	1 3	−1.85E−1	7 6	−2.79E−3	6 1	7.38E−2	1 2	−4.88E−1
1 7	−7.87E+0	1 2	−2.50E−1	1 4	−8.94E−4	7 7	−9.19E−3	6 2	1.60E−2	1 3	−1.70E+0
1 8	−1.33E+0	1 3	−1.61E−1	1 5	−1.40E+0	8 1	5.68E−2	6 3	4.51E−2	1 4	−2.33E−2
2 1	3.54E−2	1 4	−2.52E−1	1 6	−6.38E+0	8 2	1.28E−3	6 4	4.18E−2	2 1	−5.55E−6
2 2	9.71E−2	1 5	−4.34E+0	1 7	−2.36E−2	8 3	2.10E−3	6 5	5.74E−1	2 2	−8.38E−2
2 3	2.13E−3	1 6	−4.86E+0	2 1	2.63E−3	8 4	4.57E−3	6 6	7.44E−3	2 3	−3.93E−1
2 4	8.54E−3	1 7	−8.79E+0	2 2	−7.00E−3	8 5	1.44E−4	6 7	−9.79E−6	2 4	−9.11E−1
2 5	4.68E−5	2 1	3.63E−1	2 3	−5.14E−1	8 6	1.48E−4	7 1	1.89E−1	3 1	−1.40E−3
2 6	7.25E−3	2 2	1.94E−1	2 4	−7.06E−2	8 7	−4.79E−6	7 2	5.37E−2	3 2	−7.41E−1
2 7	4.25E−4	2 3	2.01E−5	2 5	−1.46E−2		$^3\mathbf{F}^o$–$^3\mathbf{F}^e$	7 3	4.04E−1	3 3	−5.33E−2
2 8	−6.36E−3	2 4	5.02E−6	2 6	−3.50E−2	1 1	−3.63E−3	7 4	2.15E−1	3 4	−4.81E−1
3 1	1.82E−1	2 5	1.40E−4	2 7	−4.74E−3	1 2	−1.29E−1	7 5	9.64E−3	4 1	−3.84E−2
3 2	2.21E−2	2 6	5.95E−4	3 1	5.26E−3	1 3	−1.92E−3	7 6	6.35E−3	4 2	−1.06E−1
3 3	3.15E−1	2 7	−5.78E−2	3 2	−2.17E−3	1 4	−2.32E−4	7 7	−5.14E−6	4 3	−4.78E−1
3 4	4.51E−2	3 1	6.48E−2	3 3	−9.07E−2	1 5	−3.09E+0		$^3\mathbf{F}^o$–$^3\mathbf{G}^e$	4 4	−7.88E−2
3 5	6.29E−3	3 2	2.29E−1	3 4	−1.19E−2	1 6	−1.60E+1	1 1	−1.22E−1	5 1	−8.41E−3
3 6	1.44E−1	3 3	2.97E−2	3 5	−6.89E−2	1 7	−9.06E−1	1 2	−4.68E−2	5 2	−2.23E−1
3 7	6.60E−2	3 4	6.46E−3	3 6	−3.92E−2	2 1	1.75E−1	1 3	−1.18E+1	5 3	−1.37E−2
3 8	−8.79E−2	3 5	1.11E−2	3 7	−8.55E−1	2 2	2.80E−3	1 4	−1.32E+1	5 4	−2.38E−2
4 1	2.11E−3	3 6	1.68E−2	4 1	1.24E+0	2 3	1.09E−1	2 1	7.31E−1	6 1	−2.20E−2
4 2	3.30E−3	3 7	−7.62E−1	4 2	−9.21E−2	2 4	6.99E−3	2 2	2.12E−2	6 2	−7.16E−3
4 3	6.24E−2	4 1	5.59E−5	4 3	−3.38E−3	2 5	1.32E−3	2 3	2.53E−3	6 3	−3.89E−2
4 4	1.82E−2	4 2	2.64E−3	4 4	−2.74E−2	2 6	1.21E−4	2 4	−7.35E−3	6 4	−1.10E−2
4 5	2.59E−1	4 3	2.49E−3	4 5	−3.57E−2	2 7	2.02E 3	3 1	2.30E−1	7 1	1.06E−4
4 6	1.26E+0	4 4	6.76E−4	4 6	−4.76E−1	3 1	1.16E−1	3 2	1.03E−3	7 2	5.42E−3
4 7	4.05E−1	4 5	3.13E−2	4 7	−3.65E−3	3 2	9.22E−1	3 3	4.90E−3	7 3	2.26E−7
4 8	−8.43E−1	4 6	1.91E−2	5 1	1.35E+1	3 3	2.69E−3	3 4	−7.26E−1	7 4	−2.76E−6
5 1	5.94E−2	4 7	−7.68E+0	5 2	−3.45E−3	3 4	5.51E−3	4 1	1.33E−3		$^3\mathbf{G}^o$–$^3\mathbf{G}^e$
5 2	2.61E−2	5 1	3.65E−1	5 3	−1.07E−2	3 5	1.38E−4	4 2	2.01E−4	1 1	3.20E−1
5 3	2.83E−1	5 2	9.57E−2	5 4	−1.15E+0	3 6	4.57E−3	4 3	7.91E−2	1 2	4.27E−1
5 4	2.30E−1	5 3	1.17E−1	5 5	−5.04E−2	3 7	−6.87E−2	4 4	−1.11E+1	1 3	6.26E−2
5 5	4.12E−3	5 4	3.91E−4	5 6	−7.66E−2	4 1	4.71E−3	5 1	2.65E−1	1 4	−2.21E−4
5 6	2.47E−3	5 5	8.38E−6	5 7	−1.00E−3	4 2	3.60E−2	5 2	8.95E−1	2 1	3.08E−1
5 7	4.75E−2	5 6	1.59E−3	6 1	8.39E−2	4 3	1.97E−5	5 3	3.37E−3	2 2	2.81E−2
5 8	−1.09E−2	5 7	−1.14E−1	6 2	−5.37E−4	4 4	1.05E−3	5 4	−1.91E−3	2 3	6.68E−3
6 1	1.07E−1	6 1	4.07E−1	6 3	−3.77E−1	4 5	5.37E−3	6 1	8.17E−3	2 4	−3.91E−4
6 2	6.52E−1	6 2	4.87E−2	6 4	−7.44E+0	4 6	8.89E−5	6 2	9.66E−2	3 1	1.77E−1
6 3	1.30E+0	6 3	8.97E−1	6 5	−8.63E−3	4 7	−1.08E+0	6 3	6.08E−1	3 2	1.75E+0
6 4	3.47E−2	6 4	1.14E+0	6 6	−5.73E−3	5 1	2.41E−1	6 4	−1.18E−5	3 3	4.18E−2
6 5	1.34E−2	6 5	9.21E−4	6 7	−6.21E−3	5 2	7.19E−2	7 1	1.02E−3	3 4	−8.21E−6
6 6	7.15E−4	6 6	3.94E−3	7 1	3.31E−1	5 3	4.40E−1	7 2	2.64E−2	4 1	1.35E−2
6 7	9.51E−3	6 7	−4.70E−6	7 2	−3.36E−3	5 4	2.43E−2	7 3	4.28E−4	4 2	4.40E−1
6 8	4.71E 7		$^3\mathbf{D}^e$–$^3\mathbf{F}^o$	7 3	−1.60E−1	5 5	4.90E−2	7 4	4.32E−6	4 3	7.65E−2
		1 1	7.77E−1	7 4	−2.14E+0	5 6	5.15E−3		$^3\mathbf{F}^e$–$^3\mathbf{G}^o$	4 4	2.47E−5

K-like Ca (Ca^+)

Term energies relative to $3s^23p^6$ ^{1}S ionization threshold for each symmetry

i E(Ryds) Description	i E(Ryds) Description	i E(Ryds) Description	i E(Ryds) Description
^{2}**S**^{e}	5 −.09281 $3s^23p^6$ ^{1}S $8p$	2 −.16140 $3s^23p^6$ ^{1}S $5f$	^{2}**H**^{o}
1 −.88825 $3s^23p^6$ ^{1}S $4s$	6 −.06985 $3s^23p^6$ ^{1}S $9p$	3 −.11199 $3s^23p^6$ ^{1}S $6f$	1 −.11114 $3s^23p^6$ ^{1}S $6h$
2 −.39404 $3s^23p^6$ ^{1}S $5s$	^{2}**D**^{e}	4 −.08222 $3s^23p^6$ ^{1}S $7f$	2 −.08166 $3s^23p^6$ ^{1}S $7h$
3 −.22683 $3s^23p^6$ ^{1}S $6s$	1 −.74634 $3s^23p^6$ ^{1}S $3d$	5 −.06290 $3s^23p^6$ ^{1}S $8f$	3 −.06252 $3s^23p^6$ ^{1}S $8h$
4 −.14763 $3s^23p^6$ ^{1}S $7s$	2 −.34490 $3s^23p^6$ ^{1}S $4d$	6 −.04967 $3s^23p^6$ ^{1}S $9f$	4 −.04940 $3s^23p^6$ ^{1}S $9h$
5 −.10378 $3s^23p^6$ ^{1}S $8s$	3 −.20627 $3s^23p^6$ ^{1}S $5d$	^{2}**G**^{e}	
6 −.07694 $3s^23p^6$ ^{1}S $9s$	4 −.13700 $3s^23p^6$ ^{1}S $6d$	1 −.16023 $3s^23p^6$ ^{1}S $5g$	
^{2}**P**^{o}	5 −.09755 $3s^23p^6$ ^{1}S $7d$	2 −.11127 $3s^23p^6$ ^{1}S $6g$	
1 −.65536 $3s^23p^6$ ^{1}S $4p$	6 −.07298 $3s^23p^6$ ^{1}S $8d$	3 −.08174 $3s^23p^6$ ^{1}S $7g$	
2 −.31961 $3s^23p^6$ ^{1}S $5p$	7 −.05664 $3s^23p^6$ ^{1}S $9d$	4 −.06257 $3s^23p^6$ ^{1}S $8g$	
3 −.19290 $3s^23p^6$ ^{1}S $6p$	^{2}**F**^{o}	5 −.04943 $3s^23p^6$ ^{1}S $9g$	
4 −.12935 $3s^23p^6$ ^{1}S $7p$	1 −.25235 $3s^23p^6$ ^{1}S $4f$		

Energies in ascending order from ground state for terms with effective $n \leq 5.0$, $L \leq 4$

Term	i	E(Ryds)	Term	i	E(Ryds)	Term	i	E(Ryds)	Term	i	E(Ryds)
^{2}**S**^{e}	1	0.00000	^{2}**S**^{e}	2	0.49421	^{2}**F**^{o}	1	0.63591	^{2}**P**^{o}	3	0.69535
^{2}**D**^{e}	1	0.14191	^{2}**D**^{e}	2	0.54335	^{2}**S**^{e}	3	0.66143	^{2}**F**^{o}	2	0.72685
^{2}**P**^{o}	1	0.23289	^{2}**P**^{o}	2	0.56864	^{2}**D**^{e}	3	0.68198	^{2}**G**^{e}	1	0.72802

gf-values for transitions involving terms with effective $n \leq 5.0$, $L \leq 4$

i i'	gf_L	i i'	gf_L	i i'	gf_L	i i'	gf_L	i i'	gf_L	i i'	gf_L
	^{2}**S**^{e}–^{2}**P**^{o}	2 2	−2.97E+0		^{2}**D**^{e}–^{2}**P**^{o}	2 2	−1.56E+0		^{2}**D**^{e}–^{2}**F**^{o}	3 1	2.49E+0
1 1	−1.91E+0	2 3	−9.44E−4	1 1	−5.63E−1	2 3	−3.73E−3	1 1	−1.63E+0	3 2	−1.47E+1
1 2	−5.01E−3	3 1	1.58E−1	1 2	−1.96E−3	3 1	7.89E−1	1 2	−7.33E−1		^{2}**G**^{e}–^{2}**F**^{o}
1 3	−6.25E−3	3 2	1.74E+0	1 3	−8.89E−4	3 2	5.78E+0	2 1	−1.06E+1	1 1	1.84E+1
2 1	1.09E+0	3 3	−3.82E+0	2 1	5.25E+0	3 3	−2.74E+0	2 2	−3.98E−3	1 2	3.99E−1

K-like Fe (Fe^{7+})

Term energies relative to $3p^6$ ^{1}S ionization threshold for each symmetry

i	E(Ryds)	Description
	^{2}S^e	
1	−7.19765	$3p^6$ ^{1}S $4s$
2	−3.99877	$3p^6$ ^{1}S $5s$
3	−3.94364	$3s3p^63d^2$
4	−3.11411	$3p^43d^3$
5	−2.55577	$3p^6$ ^{1}S $6s$
6	−2.47012	$3p^53d$ ^{3}P $4p$
7	−1.98625	$3p^43d^3$
8	−1.76960	$3p^6$ ^{1}S $7s$
9	−1.30290	$3p^6$ ^{1}S $8s$
10	−1.19672	$3p^43d^3$:
11	−1.00833	$3p^6$ ^{1}S $9s$
12	−.93816	$3p^53d$ ^{1}P $4p$
	^{2}S^o	
1	−6.80886	$3p^53d^2$
2	−.92822	$3p^53d$ ^{3}D $4d$
3	−.71818	$3p^53d$ ^{1}D $4d$
	^{2}P^e	
1	−4.26989	$3s3p^63d^2$
2	−3.28479	$3p^43d^3$:
3	−2.80265	$3p^43d^3$:
4	−2.59256	$3p^53d$ ^{3}P $4p$
5	−2.44765	$3p^43d^3$:
6	−2.18705	$3p^43d^3$:
7	−2.11729	$3p^53d$ ^{1}D $4p$
8	−2.05537	$3p^53d$ ^{3}D $4p$
9	−1.87056	$3p^43d^3$:
10	−1.57946	$3p^43d^3$
11	−1.50017	$3p^43d^3$
12	−1.29486	$3p^43d^3$
13	−1.03181	$3p^53d$ ^{1}P $4p$
14	−.53786	$3p^53d$ ^{3}F $4f$
15	−.07028	$3p^53d$ ^{3}F $4f$
	^{2}P^o	
1	−7.25278	$3p^53d^2$
2	−6.60282	$3p^53d^2$:
3	−6.39410	$3p^6$ ^{1}S $4p$
4	−5.80704	$3p^53d^2$:
5	−3.64527	$3p^6$ ^{1}S $5p$
6	−3.31336	$3p^53d$ ^{3}P $4s$
7	−2.36880	$3p^6$ ^{1}S $6p$
8	−1.80469	$3p^53d$ ^{1}P $4s$
9	−1.64759	$3p^6$ ^{1}S $7p$
10	−1.31455	$3p^53d$ ^{3}P $4d$
11	−1.22983	$3p^6$ ^{1}S $8p$
12	−1.07196	$3p^53d$ ^{3}F $4d$
13	−.94656	$3p^6$ ^{1}S $9p$
14	−.91360	$3p^53d$ ^{1}D $4d$
15	−.84649	$3p^53d$ ^{1}F $4d$
	^{2}D^e	
1	−11.4926	$3p^6$ ^{1}S $3d$
2	−5.13342	$3p^6$ ^{1}S $4d$
3	−4.50037	$3s3p^63d^2$:
4	−3.27350	$3p^43d^3$
5	−3.10598	$3p^6$ ^{1}S $5d$
6	−2.98235	$3p^43d^3$:
7	−2.89075	$3p^43d^3$:
8	−2.53498	$3p^53d$ ^{3}P $4p$
9	−2.45346	$3p^43d^3$:
10	−2.37902	$3p^43d^3$:
11	−2.28974	$3p^43d^3$:
12	−2.23006	$3p^43d^3$:
13	−2.19118	$3p^53d$ ^{3}F $4p$
14	−2.12250	$3p^53d$ ^{1}D $4p$:
15	−2.05224	$3p^6$ ^{1}S $6d$:
16	−1.91262	$3p^53d$ ^{3}D $4p$
17	−1.85913	$3p^53d$ ^{1}F $4p$
18	−1.65192	$3p^43d^3$
19	−1.61537	$3p^43d^3$
20	−1.48370	$3p^6$ ^{1}S $7d$
21	−1.46552	$3p^43d^3$
22	−1.20786	$3p^43d^3$
23	−1.12421	$3p^6$ ^{1}S $8d$
24	−1.00807	$3p^53d$ ^{1}P $4p$
25	−.86922	$3p^6$ ^{1}S $9d$
26	−.71117	$3p^43d^3$:
	^{2}D^o	
1	−7.37725	$3p^53d^2$:
2	−6.85807	$3p^53d^2$:
3	−5.82136	$3p^53d^2$:
4	−2.97395	$3p^53d$ ^{1}D $4s$
5	−2.89849	$3p^53d$ ^{3}D $4s$
6	−1.37163	$3p^53d$ ^{3}P $4d$
7	−1.12772	$3p^53d$ ^{3}F $4d$
8	−.96436	$3p^53d$ ^{1}D $4d$:
9	−.91562	$3p^53d$ ^{3}D $4d$
10	−.72973	$3p^53d$ ^{1}F $4d$
	^{2}F^e	
1	−4.48646	$3s3p^63d^2$:
2	−3.34156	$3p^43d^3$:
3	−3.01782	$3p^43d^3$:
4	−2.88391	$3p^43d^3$:
5	−2.49450	$3p^43d^3$:
6	−2.46523	$3p^53d$ ^{3}F $4p$
7	−2.30082	$3p^43d^3$:
8	2.22144	$3p^53d$ ^{1}D $4p$:
9	−2.17397	$3p^43d^3$
10	−2.10110	$3p^53d$ ^{3}D $4p$
11	−2.06564	$3p^53d$ ^{1}F $4p$
12	−1.93503	$3p^43d^3$
13	1.86053	$3p^43d^3$
14	−1.73456	$3p^43d^3$
15	−1.46208	$3p^43d^3$:
16	−1.13793	$3p^43d^3$:
17	−.91153	$3p^43d^3$:
18	−.35182	$3p^53d$ ^{3}P $4f$
19	−.14900	$3p^53d$ ^{3}F $4f$
20	−.00494	$3p^53d$ ^{1}D $4f$
	^{2}F^o	
1	−7.32772	$3p^53d^2$:
2	−7.20589	$3p^53d^2$
3	−6.33747	$3p^53d^2$
4	−4.13476	$3p^6$ ^{1}S $4f$
5	−3.16267	$3p^53d$ ^{3}F $4s$
6	−2.89350	$3p^53d$ ^{1}F $4s$
7	−2.64921	$3p^6$ ^{1}S $5f$
8	−1.83592	$3p^6$ ^{1}S $6f$
9	−1.34609	$3p^6$ ^{1}S $7f$
10	−1.26805	$3p^53d$ ^{3}P $4d$
11	−1.14740	$3p^53d$ ^{3}F $4d$
12	−1.02592	$3p^6$ ^{1}S $8f$
13	−.95368	$3p^53d$ ^{3}D $4d$
14	−.91649	$3p^53d$ ^{1}D $4d$
15	−.81064	$3p^6$ ^{1}S $9f$
16	−.79462	$3p^53d$ ^{1}F $4d$
	^{2}G^e	
1	−4.39608	$3s3p^63d^2$:
2	−3.09033	$3p^43d^3$:
3	−3.00652	$3p^43d^3$:
4	−2.73468	$3p^43d^3$:
5	−2.62064	$3p^43d^3$
6	−2.56933	$3p^6$ ^{1}S $5g$
7	−2.37235	$3p^53d$ ^{3}F $4p$
8	−2.17127	$3p^43d^3$
9	−2.09944	$3p^53d$ ^{1}F $4p$
10	−2.07701	$3p^43d^3$
11	−1.96160	$3p^43d^3$
12	−1.78432	$3p^6$ ^{1}S $6g$
13	−1.65260	$3p^43d^3$:
14	−1.48396	$3p^43d^3$:
15	−1.31091	$3p^6$ ^{1}S $7g$
16	−1.22629	$3p^43d^3$:
17	−1.00377	$3p^6$ ^{1}S $8g$
18	−.79308	$3p^6$ ^{1}S $9g$
	^{2}G^o	
1	−7.17309	$3p^53d^2$
2	−6.77076	$3p^53d^2$
3	−1.22871	$3p^53d$ ^{3}F $4d$
4	−.93406	$3p^53d$ ^{1}D $4d$
5	−.84175	$3p^53d$ ^{1}F $4d$
6	−.72333	$3p^53d$ ^{3}D $4d$
	^{2}H^e	
1	−3.28957	$3p^43d^3$:
2	−3.07701	$3p^43d^3$:
3	−2.60169	$3p^43d^3$:
4	−2.29253	$3p^43d^3$
5	−2.11837	$3p^43d^3$
6	−1.58573	$3p^43d^3$
7	−.22931	$3p^53d$ ^{3}F $4f$
	^{2}H^o	
1	−7.21962	$3p^53d^2$:
2	−1.10530	$3p^53d$ ^{3}F $4d$
3	−.84363	$3p^53d$ ^{1}F $4d$
	^{2}I^e	
1	−3.00275	$3p^43d^3$
2	−2.83405	$3p^43d^3$:
3	−2.37738	$3p^43d^3$:
4	−.16227	$3p^53d$ ^{3}F $4f$
	^{4}S^e	
1	−3.41155	$3p^43d^3$:
2	−2.57888	$3p^53d$ ^{3}P $4p$
3	−2.31622	$3p^43d^3$:
4	−.21543	$3p^53d$ ^{3}F $4f$
	^{4}S^o	
1	−6.80561	$3p^53d^2$:
2	−.73300	$3p^53d$ ^{3}D $4d$
	^{4}P^e	
1	−4.56485	$3s3p^63d^2$:
2	−3.63488	$3p^43d^3$:
3	−3.17931	$3p^43d^3$:
4	−2.73726	$3p^43d^3$
5	−2.54959	$3p^53d$ ^{3}P $4p$
6	−2.43274	$3p^43d^3$
7	−2.13392	$3p^53d$ ^{3}D $4p$
8	−1.88720	$3p^43d^3$
9	−1.77835	$3p^43d^3$
10	−1.29696	$3p^43d^3$:
11	−.19284	$3p^53d$ ^{3}F $4f$
	^{4}P^o	
1	−7.53361	$3p^53d^2$:
2	−3.42535	$3p^53d$ ^{3}P $4s$
3	−1.38361	$3p^53d$ ^{3}P $4d$
4	−1.13981	$3p^53d$ ^{3}F $4d$
5	−.94288	$3p^53d$ ^{3}D $4d$
6	−.16478	$3p^53d$ ^{3}P $5s$
	^{4}D^e	
1	−3.54818	$3p^43d^3$:
2	−3.42231	$3p^43d^3$:
3	−3.02437	$3p^43d^3$:
4	−2.84452	$3p^43d^3$:
5	−2.68235	$3p^53d$ ^{3}P $4p$
6	−2.51935	$3p^53d$ ^{3}F $4p$
7	−2.41104	$3p^43d^3$:
8	−2.37163	$3p^43d^3$

K-like Fe (Fe^{7+})

i	E(Ryds)	Description	i	E(Ryds)	Description	i	E(Ryds)	Description	i	E(Ryds)	Description
9	−2.23192	$3p^5 3d$ ^{3}D $4p$	3	−3.25588	$3p^4 3d^3$:	5	−.97600	$3p^5 3d$ ^{3}D $4d$			$^4\mathbf{H}^e$
10	−1.86650	$3p^4 3d^3$	4	−3.19489	$3p^4 3d^3$:	6	−.04327	$3p^5 3d$ ^{3}F $5s$	1	−3.43752	$3p^4 3d^3$:
11	−1.69993	$3p^4 3d^3$	5	−2.73194	$3p^4 3d^3$:			$^4\mathbf{G}^e$	2	−3.20796	$3p^4 3d^3$:
12	−.49194	$3s3p^6 3d$ ^{3}D $4s$	6	−2.56025	$3p^4 3d^3$:	1	−3.53870	$3p^4 3d^3$	3	−2.80653	$3p^4 3d^3$:
13	−.35899	$3p^5 3d$ ^{3}P $4f$	7	−2.47038	$3p^5 3d$ ^{3}F $4p$	2	−3.40013	$3p^4 3d^3$	4	−.18730	$3p^5 3d$ ^{3}F $4f$
14	−.22183	$3p^5 3d$ ^{3}F $4f$	8	−2.24814	$3p^4 3d^3$:	3	−3.14627	$3p^4 3d^3$			$^4\mathbf{H}^o$
		$^4\mathbf{D}^o$	9	−2.20245	$3p^4 3d^3$:	4	−2.61667	$3p^4 3d^3$:	1	−1.26786	$3p^5 3d$ ^{3}F $4d$
1	−7.74914	$3p^5 3d^2$:	10	−2.09998	$3p^5 3d$ ^{3}D $4p$:	5	−2.47672	$3p^5 3d$ ^{3}F $4p$			$^4\mathbf{I}^e$
2	−7.07490	$3p^5 3d^2$:	11	−1.70804	$3p^4 3d^3$	6	−2.11968	$3p^4 3d^3$:	1	−3.23995	$3p^4 3d^3$:
3	−3.02040	$3p^5 3d$ ^{3}D $4s$	12	−.36738	$3p^5 3d$ ^{3}P $4f$	7	−.44879	$3p^5 3d$ ^{3}P $4f$	2	−.30858	$3p^5 3d$ ^{3}F $4f$
4	−1.33903	$3p^5 3d$ ^{3}P $4d$	13	−.21043	$3p^5 3d$ ^{3}F $4f$	8	−.20541	$3p^5 3d$ ^{3}F $4f$			
5	−1.20795	$3p^5 3d$ ^{3}F $4d$			$^4\mathbf{F}^o$	9	−.01897	$3p^5 3d$ ^{3}D $4f$			
6	−.84745	$3p^5 3d$ ^{3}D $4d$	1	−7.40645	$3p^5 3d^2$:			$^4\mathbf{G}^o$			
		$^4\mathbf{F}^e$	2	−3.26683	$3p^5 3d$ ^{3}F $4s$	1	−7.58306	$3p^5 3d^2$			
1	−4.79576	$3s3p^6 3d^2$:	3	−1.44077	$3p^5 3d$ ^{3}P $4d$	2	−1.12056	$3p^5 3d$ ^{3}F $4d$			
2	−3.45912	$3p^4 3d^3$	4	−1.18829	$3p^5 3d$ ^{3}F $4d$	3	−.96052	$3p^5 3d$ ^{3}D $4d$			

Energies in ascending order from ground state for terms with effective $n \leq 5.0$, $L \leq 6$

Term	i	E(Ryds)	Term	i	E(Ryds)	Term	i	E(Ryds)	Term	i	E(Ryds)	Term	i	E(Ryds)
$^2\mathbf{D}^e$	1	0.00000	$^2\mathbf{P}^o$	3	5.09850	$^4\mathbf{F}^e$	2	8.03348	$^4\mathbf{G}^e$	3	8.34633	$^4\mathbf{H}^e$	3	8.68607
$^4\mathbf{D}^o$	1	3.74346	$^2\mathbf{F}^o$	3	5.15513	$^4\mathbf{H}^e$	1	8.05508	$^2\mathbf{S}^e$	4	8.37849	$^2\mathbf{P}^e$	3	8.68995
$^4\mathbf{G}^o$	1	3.90954	$^2\mathbf{D}^o$	3	5.67124	$^4\mathbf{P}^o$	2	8.06725	$^2\mathbf{D}^e$	5	8.38662	$^4\mathbf{P}^e$	4	8.75534
$^4\mathbf{P}^o$	1	3.95899	$^2\mathbf{P}^o$	4	5.68556	$^4\mathbf{D}^e$	2	8.07029	$^2\mathbf{G}^e$	2	8.40227	$^2\mathbf{G}^e$	4	8.75792
$^4\mathbf{F}^o$	1	4.08615	$^2\mathbf{D}^e$	2	6.35918	$^4\mathbf{S}^e$	1	8.08105	$^2\mathbf{H}^e$	2	8.41559	$^4\mathbf{F}^e$	5	8.76066
$^2\mathbf{D}^o$	1	4.11535	$^4\mathbf{F}^e$	1	6.69684	$^4\mathbf{G}^e$	2	8.09247	$^4\mathbf{D}^e$	3	8.46823	$^4\mathbf{D}^e$	5	8.81025
$^2\mathbf{F}^o$	1	4.16488	$^4\mathbf{P}^e$	1	6.92775	$^2\mathbf{F}^e$	2	8.15104	$^4\mathbf{D}^o$	3	8.47220	$^2\mathbf{F}^o$	7	8.84339
$^2\mathbf{P}^o$	1	4.23982	$^2\mathbf{D}^e$	3	6.99223	$^2\mathbf{P}^o$	6	8.17924	$^2\mathbf{F}^e$	3	8.47478	$^2\mathbf{G}^e$	5	8.87196
$^2\mathbf{H}^o$	1	4.27298	$^2\mathbf{F}^e$	1	7.00614	$^2\mathbf{H}^e$	1	8.20303	$^2\mathbf{G}^e$	3	8.48608	$^4\mathbf{G}^e$	4	8.87593
$^2\mathbf{F}^o$	2	4.28671	$^2\mathbf{G}^e$	1	7.09652	$^2\mathbf{P}^e$	2	8.20781	$^2\mathbf{I}^e$	1	8.48985	$^2\mathbf{H}^e$	3	8.89091
$^2\mathbf{S}^e$	1	4.29495	$^2\mathbf{P}^e$	1	7.22271	$^2\mathbf{D}^e$	4	8.21910	$^2\mathbf{D}^e$	6	8.51025	$^2\mathbf{P}^e$	4	8.90004
$^2\mathbf{G}^o$	1	4.31951	$^2\mathbf{F}^o$	4	7.35784	$^4\mathbf{F}^o$	2	8.22577	$^2\mathbf{D}^o$	4	8.51865	$^4\mathbf{S}^e$	2	8.91372
$^4\mathbf{D}^o$	2	4.41770	$^2\mathbf{S}^e$	2	7.49383	$^4\mathbf{F}^e$	3	8.23672	$^2\mathbf{D}^o$	5	8.59411	$^2\mathbf{G}^e$	6	8.92327
$^2\mathbf{D}^o$	2	4.63453	$^2\mathbf{S}^e$	3	7.54896	$^4\mathbf{I}^e$	1	8.25265	$^2\mathbf{F}^o$	6	8.59910	$^4\mathbf{F}^e$	6	8.93235
$^2\mathbf{S}^o$	1	4.68374	$^2\mathbf{P}^o$	5	7.84733	$^4\mathbf{H}^e$	2	8.28464	$^2\mathbf{D}^e$	7	8.60185			
$^4\mathbf{S}^o$	1	4.68699	$^4\mathbf{P}^e$	2	7.85772	$^4\mathbf{F}^e$	4	8.29771	$^2\mathbf{F}^e$	4	8.60869			
$^2\mathbf{G}^o$	2	4.72184	$^4\mathbf{D}^e$	1	7.94442	$^4\mathbf{P}^e$	3	8.31329	$^4\mathbf{D}^e$	4	8.64808			
$^2\mathbf{P}^o$	2	4.88978	$^4\mathbf{G}^e$	1	7.95390	$^2\mathbf{F}^o$	5	8.32993	$^2\mathbf{I}^e$	2	8.65855			

K-like Fe (Fe^{7+})

gf-values for transitions involving terms with effective $n \leq 5.0$, $L \leq 6$

$i\ i'$	gf_L	$i\ i'$	gf_L	$i\ i'$	gf_L	$i\ i'$	gf_L	$i\ i'$	gf_L	$i\ i'$	gf_L
	$^2S^e$–$^2P^o$	4 3	3.43E−4	5 2	5.31E−3	7 4	1.79E−4	7 7	−2.07E−3		$^2F^e$–$^2G^o$
1 1	2.60E−4	4 4	4.39E−3	5 3	2.53E−1	7 5	4.95E−5		$^2F^e$–$^2D^o$	1 1	4.52E−1
1 2	−1.49E−2	4 5	2.41E−2	5 4	1.15E−2		$^2D^e$–$^2F^o$	1 1	1.39E−1	1 2	4.62E−2
1 3	−1.29E+0	4 6	1.30E+0	5 5	7.09E+0	1 1	−1.45E−1	1 2	2.81E−2	2 1	3.63E−4
1 4	−9.15E−2		$^2P^e$–$^2D^o$	5 6	2.90E−2	1 2	−3.34E−1	1 3	7.07E−2	2 2	1.87E−2
1 5	−2.95E−1	1 1	1.77E−3	6 1	2.71E−2	1 3	−7.44E+0	1 4	−1.16E−4	3 1	6.50E−2
1 6	−4.51E−5	1 2	2.83E−1	6 2	7.87E−5	1 4	−6.01E+0	1 5	−6.82E−3	3 2	2.21E−3
2 1	2.37E−2	1 3	4.04E−3	6 3	2.56E−4	1 5	−1.19E+0	2 1	1.94E−1	4 1	3.20E−2
2 2	7.37E−2	1 4	−5.24E−5	6 4	2.72E−3	1 6	−5.83E−1	2 2	3.57E−3	4 2	1.64E−3
2 3	7.19E−1	1 5	−7.41E−4	6 5	1.07E−2	1 7	−2.07E+0	2 3	4.94E−3		$^2G^e$–$^2F^o$
2 4	2.87E−2	2 1	1.52E−3	6 6	1.61E−5	2 1	1.28E−2	2 4	−5.98E−5	1 1	8.52E−1
2 5	−1.89E+0	2 2	3.36E−2	7 1	1.97E−2	2 2	2.03E−2	2 5	−5.34E−4	1 2	2.32E−2
2 6	−4.89E−2	2 3	2.39E−6	7 2	4.76E−3	2 3	6.18E−2	3 1	4.47E−2	1 3	1.79E−1
3 1	7.61E−2	2 4	−6.49E−4	7 3	8.53E−4	2 4	−6.91E+0	3 2	8.10E−2	1 4	−8.64E−4
3 2	1.77E−1	2 5	−2.33E−5	7 4	4.70E−3	2 5	−6.53E−3	3 3	1.78E−3	1 5	−2.35E−3
3 3	1.07E−1	3 1	6.80E−2	7 5	4.73E−3	2 6	−6.13E−2	3 4	−6.04E−7	1 6	−2.38E−3
3 4	1.29E−2	3 2	8.32E−2	7 6	5.78E−3	2 7	−1.26E+0	3 5	−2.15E−5	1 7	−2.80E−4
3 5	−2.06E−1	3 3	1.34E−2		$^2D^e$–$^2D^o$	3 1	5.52E−2	4 1	9.04E−2	2 1	9.07E−1
3 6	−6.32E−3	3 4	3.43E−4	1 1	−4.04E−3	3 2	4.67E−1	4 2	3.48E−1	2 2	7.20E−3
4 1	9.67E−3	3 5	3.95E−4	1 2	−7.48E−3	3 3	2.23E−2	4 3	1.18E−3	2 3	3.47E−2
4 2	2.16E−3	4 1	7.59E−1	1 3	−1.31E+1	3 4	−1.01E−2	4 4	1.30E−4	2 4	1.45E−3
4 3	8.86E−5	4 2	3.34E−2	1 4	−1.35E−2	3 5	−4.86E−4	4 5	3.33E−6	2 5	7.32E−5
4 4	5.00E−4	4 3	2.10E−2	1 5	−9.87E−1	3 6	−1.09E−3		$^2F^e$–$^2F^o$	2 6	−1.00E−4
4 5	2.54E−5	4 4	1.44E−3	2 1	1.70E−3	3 7	−1.45E−2	1 1	3.12E−1	2 7	−1.81E−4
4 6	3.39E−4	4 5	7.08E−3	2 2	1.19E−4	4 1	7.42E−3	1 2	1.72E−2	3 1	3.70E−1
	$^2P^e$–$^2S^o$		$^2D^e$–$^2P^o$	2 3	1.67E−6	4 2	2.87E−3	1 3	1.14E−1	3 2	1.46E−1
1 1	8.32E−2	1 1	−4.14E−2	2 4	−5.31E−4	4 3	1.28E−4	1 4	−1.01E−4	3 3	4.44E−3
2 1	4.04E−2	1 2	−4.71E−2	2 5	−4.90E−3	4 4	1.38E−4	1 5	−3.00E−3	3 4	1.10E−3
3 1	1.01E−1	1 3	−1.58E+0	3 1	4.81E−1	4 5	−3.17E−4	1 6	−7.58E−4	3 5	4.13E−5
4 1	2.12E−6	1 4	−6.00E+0	3 2	2.37E−2	4 6	−3.00E−4	1 7	−1.74E−4	3 6	−2.24E−5
	$^2P^e$–$^2P^o$	1 5	−3.49E−2	3 3	2.55E−2	4 7	−7.54E−4	2 1	5.04E−3	3 7	−1.46E−4
1 1	8.15E−2	1 6	−4.54E−1	3 4	−1.35E−5	5 1	1.77E−2	2 2	1.17E−3	4 1	6.04E−1
1 2	2.06E−3	2 1	1.48E−2	3 5	−6.54E−4	5 2	9.36E−3	2 3	1.46E−2	4 2	2.87E−1
1 3	3.43E−3	2 2	9.59E−2	4 1	5.73E−2	5 3	3.57E−2	2 4	2.24E−6	4 3	7.47E−3
1 4	3.78E−2	2 3	4.70E+0	4 2	1.82E−2	5 4	1.69E+0	2 5	−8.15E−5	4 4	5.75E−5
1 5	−2.95E−6	2 4	1.13E−1	4 3	1.10E−2	5 5	7.63E−4	2 6	−1.07E−3	4 5	9.97E−3
1 6	−2.65E−3	2 5	−1.93E+0	4 4	−1.64E−5	5 6	−3.24E−2	2 7	−1.55E−5	4 6	7.01E−4
2 1	3.38E−2	2 6	−3.33E−2	4 5	−2.62E−5	5 7	−1.09E+1	3 1	3.07E−2	4 7	−1.41E−5
2 2	3.26E−3	3 1	3.34E−1	5 1	5.72E−5	6 1	4.46E−2	3 2	1.29E−1	5 1	2.70E−1
2 3	1.44E−7	3 2	3.81E−2	5 2	1.83E−7	6 2	3.34E−3	3 3	2.26E−3	5 2	4.52E−2
2 4	1.62E−4	3 3	3.19E−2	5 3	2.77E−3	6 3	2.42E−3	3 4	2.03E−5	5 3	3.14E−3
2 5	1.83E−5	3 4	1.10E−3	5 4	−4.45E−8	6 4	1.46E−3	3 5	4.32E−3	5 4	2.81E−3
2 6	9.03E−5	3 5	−9.98E−3	5 5	−8.38E−4	6 5	4.70E−4	3 6	−2.77E−5	5 5	3.73E−2
3 1	2.04E−1	3 6	−6.66E−4	6 1	9.31E−3	6 6	−2.91E−5	3 7	−2.42E−6	5 6	5.93E−3
3 2	2.79E−3	4 1	1.94E−1	6 2	5.20E−2	6 7	−1.07E−2	4 1	3.69E−2	5 7	7.68E−5
3 3	8.14E−4	4 2	1.34E−2	6 3	3.70E−4	7 1	3.80E−2	4 2	2.01E−3	6 1	1.23E−2
3 4	1.75E−2	4 3	6.35E−7	6 4	−1.84E−5	7 2	8.48E−6	4 3	5.10E−3	6 2	3.92E−3
3 5	8.76E−4	4 4	8.02E−5	6 5	−1.84E−4	7 3	1.01E−3	4 4	1.48E−7	6 3	7.17E−2
3 6	3.91E−2	4 5	4.48E−4	7 1	5.28E−3	7 4	3.33E−4	4 5	4.45E−5	6 4	1.62E+1
4 1	4.25E−1	4 6	1.25E−4	7 2	7.11E−2	7 5	1.83E−3	4 6	1.78E−5	6 5	1.50E−3
4 2	4.27E−4	5 1	1.12E−3	7 3	1.04E−3	7 6	1.02E−5	4 7	−3.57E−6	6 6	4.38E−2

K-like Fe (Fe^{7+})

i i′	gf$_L$	*i i′*	gf$_L$	*i i′*	gf$_L$	*i i′*	gf$_L$	*i i′*	gf$_L$	*i i′*	gf$_L$
6 7	1.84E+0	3 2	6.09E−1		$^4P^e$–$^4D^o$	1 3	−2.14E−4	1 3	−6.26E−5	6 1	3.82E−1
	$^2G^e$–$^2G^o$		$^2H^e$–$^2H^o$	1 1	4.94E−1	2 1	7.75E−2	2 1	3.21E−1	6 2	4.74E−1
1 1	9.16E−2	1 1	5.42E−2	1 2	6.06E−1	2 2	2.90E−1	2 2	1.32E−2		$^4F^e$–$^4G^o$
1 2	7.28E−1	2 1	1.13E−1	1 3	−1.67E−6	2 3	−1.33E−3	2 3	−6.70E−5	1 1	1.54E+0
2 1	2.41E−2	3 1	9.91E−4	2 1	1.61E−3	3 1	1.23E−1	3 1	3.53E−1	2 1	1.42E−1
2 2	6.48E−2		$^2I^e$–$^2H^o$	2 2	5.23E−2	3 2	1.08E−3	3 2	3.65E−2	3 1	2.61E−2
3 1	2.37E−4	1 1	3.76E−1	2 3	−1.05E−3	3 3	−1.07E−5	3 3	−3.68E−4	4 1	4.82E−1
3 2	2.62E−2	2 1	1.27E−1	3 1	7.50E−2	4 1	1.65E−1	4 1	3.25E+0	5 1	1.56E+0
4 1	4.79E−1		$^4S^e$–$^4P^o$	3 2	7.16E−5	4 2	4.39E−2	4 2	2.73E−1	6 1	3.44E−1
4 2	9.88E−3	1 1	2.18E−1	3 3	−1.31E−4	4 3	2.19E−4	4 3	−6.35E−5		$^4G^e$–$^4F^o$
5 1	4.81E−2	1 2	9.37E−5	4 1	4.51E+0	5 1	2.54E+0	5 1	1.17E+0	1 1	2.19E−1
5 2	4.02E−2	2 1	7.88E−1	4 2	1.00E−1	5 2	6.29E−4	5 2	2.97E−1	1 2	−1.02E−3
6 1	9.07E−6	2 2	9.94E−1	4 3	7.91E−5	5 3	6.31E−2	5 3	9.36E−3	2 1	4.33E−2
6 2	3.55E−4		$^4P^e$–$^4S^o$		$^4D^e$–$^4P^o$		$^4D^e$–$^4F^o$	6 1	1.72E−2	2 2	−1.39E−4
	$^2G^e$–$^2H^o$	1 1	1.93E−1	1 1	1.44E−1	1 1	6.69E−2	6 2	1.48E−1	3 1	1.12E+0
1 1	7.78E−1	2 1	1.25E−2	1 2	−1.65E−4	1 2	−1.36E−4	6 3	1.69E−3	3 2	1.23E−3
2 1	6.23E−2	3 1	1.67E−1	2 1	5.67E−2	2 1	1.62E−3		$^4F^e$–$^4F^o$	4 1	4.96E+0
3 1	1.71E−2	4 1	7.00E−3	2 2	1.07E−7	2 2	−3.94E−4	1 1	1.68E+0	4 2	2.00E−2
4 1	7.26E−2		$^4P^e$–$^4P^o$	3 1	1.64E−1	3 1	2.21E−1	1 2	−1.81E−4		$^4G^e$–$^4G^o$
5 1	6.72E−2	1 1	1.00E+0	3 2	2.92E−2	3 2	1.08E−3	2 1	1.82E−2	1 1	2.70E−3
6 1	1.14E−3	1 2	−7.66E−4	4 1	1.86E−1	4 1	1.58E−2	2 2	−1.51E−4	2 1	1.70E−2
	$^2H^e$–$^2G^o$	2 1	1.21E−6	4 2	7.67E−4	4 2	1.21E−2	3 1	1.10E−1	3 1	4.74E−1
1 1	5.11E−2	2 2	−7.46E−4	5 1	2.83E−1	5 1	1.15E−1	3 2	2.49E−4	4 1	1.13E−1
1 2	1.85E−3	3 1	3.21E−1	5 2	3.09E+0	5 2	7.26E−1	4 1	5.36E−1		$^4H^e$–$^4G^o$
2 1	3.96E−1	3 2	1.01E−5		$^4D^e$–$^4D^o$		$^4F^e$–$^4D^o$	4 2	1.89E−4	1 1	1.27E−1
2 2	1.51E−1	4 1	2.03E+0	1 1	1.47E−2	1 1	1.66E+0	5 1	2.75E−2	2 1	9.37E−1
3 1	1.75E+0	4 2	9.67E−1	1 2	1.02E−2	1 2	3.65E−1	5 2	7.47E−2	3 1	1.14E+1

Ca-like Ca

Term energies relative to $3p^64s\ ^2$S ionization threshold for each symmetry

i E(Ryds) Description	i E(Ryds) Description	i E(Ryds) Description	i E(Ryds) Description
$^1\mathbf{S}^e$	2 −.04939 $3p^63d\ ^2$D $4p$	3 −.06542 $3p^64s\ ^2$S $6p$	4 −.01569 $3p^64s\ ^2$S $8g$
1 −.43282 $3p^64s^2$	3 −.03915 $3p^64s\ ^2$S $5f$	4 −.05587 $3p^63d\ ^2$D $4p$	5 −.01239 $3p^64s\ ^2$S $9g$
2 −.14110 $3p^64s\ ^2$S $5s$	4 −.02804 $3p^64s\ ^2$S $6f$	5 −.03897 $3p^64s\ ^2$S $7p$	$^3\mathbf{H}^o$
3 −.07402 $3p^64s\ ^2$S $6s$	5 −.02071 $3p^64s\ ^2$S $7f$	6 −.02725 $3p^64s\ ^2$S $8p$	1 −.02778 $3p^64s\ ^2$S $6h$
4 −.06122 $3p^64p^2$	6 −.01588 $3p^64s\ ^2$S $8f$	7 −.02008 $3p^64s\ ^2$S $9p$	2 −.02041 $3p^64s\ ^2$S $7h$
5 −.04516 $3p^64s\ ^2$S $7s$	7 −.01254 $3p^64s\ ^2$S $9f$	$^3\mathbf{D}^e$	3 −.01563 $3p^64s\ ^2$S $8h$
6 −.03074 $3p^64s\ ^2$S $8s$	$^1\mathbf{G}^e$	1 −.22222 $3p^64s\ ^2$S $3d$	4 −.01235 $3p^64s\ ^2$S $9h$
7 −.02225 $3p^64s\ ^2$S $9s$	1 −.04011 $3p^64s\ ^2$S $5g$	2 −.09912 $3p^64s\ ^2$S $4d$	
$^1\mathbf{P}^o$	2 −.02788 $3p^64s\ ^2$S $6g$	3 −.05753 $3p^64s\ ^2$S $5d$	
1 −.22313 $3p^64s\ ^2$S $4p$	3 −.02049 $3p^64s\ ^2$S $7g$	4 −.03771 $3p^64s\ ^2$S $6d$	
2 −.10910 $3p^63d\ ^2$D $4p$	4 −.01570 $3p^64s\ ^2$S $8g$	5 −.02672 $3p^64s\ ^2$S $7d$	
3 −.06376 $3p^64s\ ^2$S $5p$	5 −.01409 $3p^63d^2$	6 −.02003 $3p^64s\ ^2$S $8d$	
4 −.04209 $3p^64s\ ^2$S $6p$	6 −.01239 $3p^64s\ ^2$S $9g$	7 −.01569 $3p^64s\ ^2$S $9d$	
5 −.03018 $3p^64s\ ^2$S $7p$	$^1\mathbf{H}^o$	$^3\mathbf{D}^o$	
6 −.02282 $3p^64s\ ^2$S $8p$	1 −.02778 $3p^64s\ ^2$S $6h$	1 −.06659 $3p^63d\ ^2$D $4p$	
7 −.01778 $3p^64s\ ^2$S $9p$	2 −.02041 $3p^64s\ ^2$S $7h$	$^3\mathbf{F}^e$	
$^1\mathbf{D}^e$	3 −.01563 $3p^64s\ ^2$S $8h$	1 −.02330 $3p^63d^2$	
1 −.22329 $3p^64s\ ^2$S $3d$	4 −.01235 $3p^64s\ ^2$S $9h$	$^3\mathbf{F}^o$	
2 −.09998 $3p^64s\ ^2$S $4d$	$^3\mathbf{S}^e$	1 −.09796 $3p^63d\ ^2$D $4p$	
3 −.06674 $3p^64p^2$	1 −.15820 $3p^64s\ ^2$S $5s$	2 −.06425 $3p^64s\ ^2$S $4f$	
4 −.05463 $3p^64s\ ^2$S $5d$	2 −.07874 $3p^64s\ ^2$S $6s$	3 −.04123 $3p^64s\ ^2$S $5f$	
5 −.03726 $3p^64s\ ^2$S $6d$	3 −.04774 $3p^64s\ ^2$S $7s$	4 −.02856 $3p^64s\ ^2$S $6f$	
6 −.02669 $3p^64s\ ^2$S $7d$	4 −.03208 $3p^64s\ ^2$S $8s$	5 −.02093 $3p^64s\ ^2$S $7f$	
7 −.02012 $3p^64s\ ^2$S $8d$	5 −.02305 $3p^64s\ ^2$S $9s$	6 −.01598 $3p^64s\ ^2$S $8f$	
8 −.01579 $3p^64s\ ^2$S $9d$	$^3\mathbf{P}^e$	7 −.01260 $3p^64s\ ^2$S $9f$	
$^1\mathbf{D}^o$	1 −.09977 $3p^64p^2$	$^3\mathbf{G}^e$	
1 −.09233 $3p^63d\ ^2$D $4p$	$^3\mathbf{P}^o$	1 −.04011 $3p^64s\ ^2$S $5g$	
$^1\mathbf{F}^o$	1 −.30917 $3p^64s\ ^2$S $4p$	2 −.02788 $3p^64s\ ^2$S $6g$	
1 −.06672 $3p^64s\ ^2$S $4f$	2 −.11342 $3p^64s\ ^2$S $5p$	3 −.02049 $3p^64s\ ^2$S $7g$	

Energies in ascending order from ground state for terms with effective $n \leq 4.0$, $L \leq 3$

Term	i	E(Ryds)	Term	i	E(Ryds)	Term	i	E(Ryds)	Term	i	E(Ryds)	Term	i	E(Ryds)
$^1\mathbf{S}^e$	1	0.00000	$^3\mathbf{S}^e$	1	0.27461	$^3\mathbf{P}^e$	1	0.33305	$^1\mathbf{S}^e$	3	0.35880	$^3\mathbf{F}^o$	2	0.36856
$^3\mathbf{P}^o$	1	0.12365	$^1\mathbf{S}^e$	2	0.29171	$^3\mathbf{D}^e$	2	0.33370	$^1\mathbf{D}^e$	3	0.36608	$^1\mathbf{P}^o$	3	0.36906
$^1\mathbf{D}^e$	1	0.20953	$^3\mathbf{P}^o$	2	0.31940	$^3\mathbf{F}^o$	1	0.33485	$^1\mathbf{F}^o$	1	0.36610			
$^1\mathbf{P}^o$	1	0.20968	$^1\mathbf{P}^o$	2	0.32372	$^1\mathbf{D}^o$	1	0.34049	$^3\mathbf{D}^o$	1	0.36622			
$^3\mathbf{D}^e$	1	0.21060	$^1\mathbf{D}^e$	2	0.33284	$^3\mathbf{S}^e$	2	0.35407	$^3\mathbf{P}^o$	3	0.36739			

gf-values for transitions involving terms with effective $n \leq 4.0$, $L \leq 3$

i i'	gf$_L$	i i'	gf$_L$	i i'	gf$_L$	i i'	gf$_L$	i i'	gf$_L$	i i'	gf$_L$
	$^1\mathbf{S}^e$–$^1\mathbf{P}^o$		$^1\mathbf{P}^o$–$^1\mathbf{D}^e$		$^1\mathbf{D}^e$–$^1\mathbf{D}^o$	1 2	−3.89E−1		$^3\mathbf{P}^e$–$^3\mathbf{D}^o$		$^3\mathbf{D}^e$–$^3\mathbf{D}^o$
1 1	−3.90E−6	1 1	7.43E 8	1 1	1.74E−5	1 3	−4.93E−1	1 1	−5.37E−6	1 1	−1.26E−4
1 2	−3.41E−4	1 2	−2.06E−2	2 1	−1.53E−6	2 1	6.83E−3		$^3\mathbf{P}^o$–$^3\mathbf{D}^e$	2 1	−7.72E−5
1 3	−5.84E−5	1 3	−1.20E−2	3 1	1.31E−4	2 2	2.53E+0	1 1	−1.54E−5		$^3\mathbf{D}^e$–$^3\mathbf{F}^o$
2 1	4.96E−3	2 1	4.38E−3		$^1\mathbf{D}^e$–$^1\mathbf{F}^o$	2 3	−3.02E+0	1 2	−1.11E−2	1 1	−3.40E−5
2 2	−1.75E−1	2 2	−1.73E−1	1 1	−9.85E−3		$^3\mathbf{P}^e$–$^3\mathbf{P}^o$	2 1	1.31E−2	1 2	−3.71E−2
2 3	−3.29E−1	2 3	−6.60E−1	2 1	−1.42E+0	1 1	3.32E−6	2 2	−8.49E−1	2 1	−7.38E−4
3 1	1.40E 3	3 1	1.75E−3	3 1	−8.36E−4	1 2	3.03E−5	3 1	4.98E−3	2 2	−6.18E+0
3 2	8.08E−1	3 2	8.68E−1		$^3\mathbf{S}^e$–$^3\mathbf{P}^o$	1 3	−3.92E−5	3 2	2.26E+0		
3 3	−1.16E+0	3 3	1.96E−1	1 1	1.41E−3						

Ca-like Fe (Fe^{6+})

Term energies relative to $3p^63d$ ^{2}D ionization threshold for each symmetry

i	E(Ryds)	Description	i	E(Ryds)	Description	i	E(Ryds)	Description	i	E(Ryds)	Description
		$^1S^e$	7	−1.98980	$3s3p^63d^3$	14	−.76819	$3p^63d$ ^{2}D $8g$	6	−1.63640	$3p^53d^24s$
1	−8.62780	$3p^63d^2$	8	−1.95590	$3p^63d$ ^{2}D $5g$	15	−.67958	$3p^63d$ ^{2}D $9d$	7	−1.41460	$3p^63d$ ^{2}D $6f$
2	−3.82330	$3p^63d4d$	9	−1.60690	$3p^63d$ ^{2}D $6d$	16	−.60876	$3p^63d$ ^{2}D $9g$	8	−1.24300	$3p^53d^24s$
3	−2.32540	$3p^63d$ ^{2}D $5d$	10	−1.51530	$3s3p^63d^3$	17	−.60546	$3p^43d^4$:	9	−1.03500	$3p^63d$ ^{2}D $7f$
4	−1.95680	$3p^64s^2$	11	−1.40000	$3p^63d$ ^{2}D $7s$	18	−.57869	$3p^53d^24p$	10	−.78975	$3p^63d$ ^{2}D $8f$
5	−1.56730	$3p^63d$ ^{2}D $6d$	12	−1.35940	$3p^63d$ ^{2}D $6g$			$^1F^o$	11	−.62215	$3p^63d$ ^{2}D $9f$
6	−1.12960	$3p^63d$ ^{2}D $7d$	13	−1.15870	$3p^63d$ ^{2}D $7d$	1	−5.17060	$3p^63d4p$			$^1H^e$
7	−.98029	$3p^43d^4$	14	−1.07050	$3p^53d^24p$	2	−4.77840	$3p^53d^3$	1	−1.99920	$3s3p^63d^3$
8	−.88409	$3p^53d^24p$	15	−1.03250	$3p^53d^24p$:	3	−4.04480	$3p^53d^3$	2	−1.96560	$3p^63d$ ^{2}D $5g$
9	−.85040	$3p^63d$ ^{2}D $8d$	16	−1.01910	$3p^63d$ ^{2}D $8s$:	4	−3.79550	$3p^53d^3$	3	−1.36860	$3p^63d$ ^{2}D $6g$
10	−.66526	$3p^63d$ ^{2}D $9d$	17	−.99962	$3p^63d$ ^{2}D $7g$	5	−3.16600	$3p^53d^3$	4	−1.00590	$3p^63d$ ^{2}D $7g$
		$^1S^o$	18	−.93887	$3p^43d^4$	6	−3.11030	$3p^63d4f$	5	−.97338	$3p^53d^24p$
1	−4.87910	$3p^53d^3$	19	−.91125	$3p^53d^24p$	7	−2.87570	$3p^63d$ ^{2}D $5p$	6	−.86834	$3p^43d^4$:
2	−1.32030	$3p^53d^24s$	20	−.87426	$3p^63d$ ^{2}D $8d$	8	−2.00480	$3p^63d$ ^{2}D $5f$	7	−.85995	$3p^53d^24p$
		$^1P^e$	21	−.78017	$3p^63d$ ^{2}D $9s$	9	−1.86540	$3p^63d$ ^{2}D $6p$	8	−.76879	$3p^63d$ ^{2}D $8g$
1	−4.12070	$3p^63d4d$	22	−.76557	$3p^63d$ ^{2}D $8g$	10	−1.78890	$3p^53d^24s$	9	−.61179	$3p^43d^4$:
2	−2.44180	$3p^63d$ ^{2}D $5d$	23	−.72008	$3p^53d^24p$:	11	−1.59750	$3p^53d^24s$	10	−.60744	$3p^63d$ ^{2}D $9g$
3	−2.01720	$3s3p^63d^3$	24	−.64834	$3p^63d$ ^{2}D $9d$:	12	−1.39080	$3p^63d$ ^{2}D $6f$	11	−.49641	$3p^53d^24p$
4	−1.62610	$3p^63d$ ^{2}D $6d$			$^1D^o$	13	−1.30000	$3p^63d$ ^{2}D $7p$			$^1H^o$
5	−1.16330	$3p^63d$ ^{2}D $7d$	1	−5.30690	$3p^63d4p$	14	−1.02060	$3p^63d$ ^{2}D $7f$	1	−4.70080	$3p^53d^3$
6	−1.10520	$3p^53d^24p$	2	−4.73740	$3p^53d^3$	15	−.96048	$3p^63d$ ^{2}D $8p$	2	−4.01420	$3p^53d^3$
7	−.96285	$3p^53d^24p$	3	−4.60100	$3p^53d^3$	16	−.78057	$3p^63d$ ^{2}D $8f$	3	−3.07690	$3p^63d4f$
8	−.87291	$3p^63d$ ^{2}D $8d$	4	−3.98810	$3p^53d^3$	17	−.74242	$3p^63d$ ^{2}D $9p$	4	−1.97920	$3p^63d$ ^{2}D $5f$
9	−.77449	$3p^53d^24p$	5	−3.88330	$3p^53d^3$	18	−.71274	$3p^53d^24s$	5	−1.62080	$3p^53d^24s$
10	−.67775	$3p^63d$ ^{2}D $9d$	6	−3.11850	$3p^63d4f$	19	−.61507	$3p^63d$ ^{2}D $9f$	6	−1.37560	$3p^63d$ ^{2}D $6f$
		$^1P^o$	7	−2.93180	$3p^63d$ ^{2}D $5p$			$^1G^e$	7	−1.01090	$3p^63d$ ^{2}D $7f$
1	−5.13580	$3p^63d4p$	8	−2.01850	$3p^63d$ ^{2}D $5f$	1	−9.02490	$3p^63d^2$	8	−.77359	$3p^63d$ ^{2}D $8f$
2	−4.64130	$3p^53d^3$:	9	−1.88910	$3p^63d$ ^{2}D $6p$	2	−3.94690	$3p^63d4d$	9	−.61082	$3p^63d$ ^{2}D $9f$
3	−3.40960	$3p^53d^3$:	10	−1.79570	$3p^53d^24s$	3	−2.37980	$3p^63d$ ^{2}D $5d$			$^1I^e$
4	−3.08730	$3p^53d^3$:	11	−1.40230	$3p^63d$ ^{2}D $6f$	4	−2.08470	$3s3p^63d^3$	1	−1.95900	$3p^63d$ ^{2}D $5g$
5	−2.91150	$3p^63d4f$	12	−1.35290	$3p^53d^24s$	5	−1.96790	$3p^63d$ ^{2}D $5g$	2	−1.36120	$3p^63d$ ^{2}D $6g$
6	−2.85850	$3p^63d$ ^{2}D $5p$	13	−1.31290	$3p^63d$ ^{2}D $7p$	6	−1.59840	$3p^63d$ ^{2}D $6d$	3	−1.00050	$3p^63d$ ^{2}D $7g$
7	−1.97660	$3p^63d$ ^{2}D $5f$	14	−1.02670	$3p^63d$ ^{2}D $7f$	7	−1.36680	$3p^63d$ ^{2}D $6g$	4	−.88366	$3p^53d^24p$
8	−1.85600	$3p^63d$ ^{2}D $6p$	15	−.96860	$3p^63d$ ^{2}D $8p$	8	−1.15200	$3p^63d$ ^{2}D $7d$	5	−.76601	$3p^63d$ ^{2}D $8g$
9	−1.69710	$3p^53d^24s$	16	−.78433	$3p^63d$ ^{2}D $8f$	9	−1.03990	$3p^53d^24p$	6	−.60530	$3p^63d$ ^{2}D $9g$
10	−1.37620	$3p^63d$ ^{2}D $6f$	17	−.74413	$3p^63d$ ^{2}D $9p$	10	−1.00350	$3p^63d$ ^{2}D $7g$	7	−.50886	$3p^43d^4$
11	−1.29770	$3p^63d$ ^{2}D $7p$	18	−.61842	$3p^63d$ ^{2}D $9f$	11	−.93376	$3p^43d^4$:			$^1I^o$
12	−1.02150	$3p^63d$ ^{2}D $7f$			$^1F^e$	12	−.87807	$3p^63d$ ^{2}D $8d$	1	−4.75470	$3p^53d^3$
13	−.99951	$3p^53d^24s$	1	−4.15630	$3p^63d4d$	13	−.85072	$3p^53d^24p$			$^1J^e$
14	−.96579	$3p^63d$ ^{2}D $8p$	2	−2.45560	$3p^63d$ ^{2}D $5d$	14	−.82972	$3p^53d^24p$	1	−.34113	$3p^43d^4$
15	−.94957	$3p^64s4p$	3	−1.96610	$3p^63d$ ^{2}D $5g$	15	−.76813	$3p^63d$ ^{2}D $8g$			$^1K^e$
16	−.77363	$3p^63d$ ^{2}D $8f$	4	−1.80610	$3s3p^63d^3$	16	−.71387	$3p^53d^24p$	1	−.70998	$3p^43d^4$
17	−.73725	$3p^63d$ ^{2}D $9p$	5	−1.63410	$3p^63d$ ^{2}D $6d$	17	−.65887	$3p^63d$ ^{2}D $9d$			$^3S^e$
18	−.61103	$3p^63d$ ^{2}D $9f$	6	−1.36550	$3p^63d$ ^{2}D $6g$	18	−.60698	$3p^63d$ ^{2}D $9g$	1	−4.07290	$3p^63d4d$
		$^1D^e$	7	−1.17110	$3p^63d$ ^{2}D $7d$	19	−.58090	$3p^43d^4$:	2	−2.42180	$3p^63d$ ^{2}D $5d$
1	−9.12400	$3p^63d^2$	8	−1.12890	$3p^53d^24p$			$^1G^o$	3	−1.61670	$3p^63d$ ^{2}D $6d$
2	−5.96110	$3p^63d4s$	9	−1.08190	$3p^53d^24p$	1	−4.96920	$3p^53d^3$	4	−1.15820	$3p^63d$ ^{2}D $7d$
3	−3.97870	$3p^63d4d$	10	−1.00340	$3p^63d$ ^{2}D $7g$	2	−4.45640	$3p^53d^3$	5	−1.08010	$3p^53d^24p$
4	−3.21780	$3p^63d$ ^{2}D $5s$	11	−.93311	$3p^53d^24p$	3	−3.36000	$3p^53d^3$	6	−.97261	$3p^53d^24p$
5	−2.39440	$3p^63d$ ^{2}D $5d$	12	−.87555	$3p^63d$ ^{2}D $8d$	4	−3.06350	$3p^63d4f$	7	−.86665	$3p^63d$ ^{2}D $8d$
6	−2.03150	$3p^63d$ ^{2}D $6s$	13	−.83995	$3p^53d^24p$	5	−2.04330	$3p^63d$ ^{2}D $5f$	8	−.67564	$3p^63d$ ^{2}D $9d$

Ca-like Fe (Fe^{6+})

i	E(Ryds)	Description
		$^3S^o$
1	−4.75020	$3p^53d^3$
2	−3.02800	$3p^53d^3$
3	−1.35290	$3p^53d^24s$
4	−1.25950	$3p^53d^24s$
5	−.35129	$3p^53d^24d$
		$^3P^e$
1	−9.10180	$3p^63d^2$
2	−3.96960	$3p^63d4d$
3	−2.39460	$3p^63d\ ^2D\ 5d$
4	−2.20530	$3s3p^63d^3$
5	−2.00030	$3s3p^63d^3$
6	−1.61020	$3p^63d\ ^2D\ 6d$
7	−1.44900	$3p^53d^24p$
8	−1.20200	$3p^53d^24p$
9	−1.14690	$3p^63d\ ^2D\ 7d$:
10	−1.11700	$3p^53d^24p$:
11	−1.08440	$3p^43d^4$
12	−.99178	$3p^53d^24p$
13	−.88503	$3p^63d\ ^2D\ 8d$:
14	−.84375	$3p^53d^24p$:
15	−.72196	$3p^43d^4$
16	−.67940	$3p^63d\ ^2D\ 9d$
17	−.63940	$3p^43d^4$
18	−.58234	$3p^53d^24p$:
19	−.54242	$3p^53d^24p$:
		$^3P^o$
1	−5.21030	$3p^63d4p$
2	−5.09820	$3p^53d^3$:
3	−4.78060	$3p^53d^3$:
4	−4.50640	$3p^53d^3$:
5	−3.76620	$3p^53d^3$:
6	−3.08520	$3p^63d4f$
7	−2.89480	$3p^63d\ ^2D\ 5p$
8	−1.99880	$3p^63d\ ^2D\ 5f$
9	−1.93090	$3p^53d^24s$
10	−1.86170	$3p^63d\ ^2D\ 6p$
11	−1.76850	$3p^53d^24s$
12	−1.38940	$3p^63d\ ^2D\ 6f$
13	−1.33410	$3p^64s4p$
14	−1.30340	$3p^63d\ ^2D\ 7p$
15	−1.07450	$3p^53d^24s$
16	−1.01940	$3p^63d\ ^2D\ 7f$
17	−.90271	$3p^63d\ ^2D\ 8p$
18	.77967	$3p^63d\ ^2D\ 8f$
19	−.74031	$3p^63d\ ^2D\ 9p$
20	−.61519	$3p^63d\ ^2D\ 9f$
		$^3D^e$
1	−6.00490	$3p^63d4s$
2	−4.13340	$3p^63d4d$
3	−3.23200	$3p^63d\ ^2D\ 5s$
4	−2.44710	$3p^63d\ ^2D\ 5d$
5	−2.17750	$3s3p^63d^3$
6	−2.03830	$3p^63d\ ^2D\ 6s$
7	−1.95680	$3p^63d\ ^2D\ 5g$
8	−1.70040	$3s3p^63d^3$
9	−1.62890	$3p^63d\ ^2D\ 6d$
10	−1.45230	$3p^53d^24p$
11	−1.40230	$3p^63d\ ^2D\ 7s$
12	−1.36030	$3p^63d\ ^2D\ 6g$
13	−1.22280	$3p^53d^24p$
14	−1.18130	$3p^53d^24p$
15	−1.15780	$3p^63d\ ^2D\ 7d$
16	−1.06700	$3p^53d^24p$
17	−1.02730	$3p^63d\ ^2D\ 8s$
18	−1.02660	$3p^53d^24p$
19	−1.01080	$3p^53d^24p$
20	−.99908	$3p^63d\ ^2D\ 7g$
21	−.97199	$3p^43d^4$:
22	−.94031	$3p^43d^4$:
23	−.92142	$3p^53d^24p$
24	−.87178	$3p^63d\ ^2D\ 8d$
25	−.78178	$3p^63d\ ^2D\ 9s$
26	−.76589	$3p^63d\ ^2D\ 8g$
27	−.69271	$3p^53d^24p$
28	−.67732	$3p^63d\ ^2D\ 9d$
29	−.63990	$3p^43d^4$:
		$^3D^o$
1	−5.28390	$3p^63d4p$
2	−5.10450	$3p^53d^3$
3	−4.89840	$3p^53d^3$:
4	−4.71190	$3p^53d^3$
5	−4.00840	$3p^53d^3$
6	−3.88940	$3p^53d^3$
7	−3.30190	$3p^53d^3$
8	−3.09840	$3p^63d4f$
9	−2.92140	$3p^63d\ ^2D\ 5p$
10	−2.15420	$3p^53d^24s$
11	−2.00470	$3p^63d\ ^2D\ 5f$
12	−1.88000	$3p^63d\ ^2D\ 6p$
13	−1.87050	$3p^53d^24s$
14	−1.50440	$3p^53d^24s$
15	−1.39910	$3p^63d\ ^2D\ 6f$
16	−1.37710	$3p^53d^24s$
17	−1.31100	$3p^63d\ ^2D\ 7p$
18	1.02100	$3p^63d\ ^2D\ 7f$
19	−.96717	$3p^63d\ ^2D\ 8p$
20	−.78052	$3p^63d\ ^2D\ 8f$
21	−.74318	$3p^63d\ ^2D\ 9p$
22	−.61575	$3p^63d\ ^2D\ 9f$
		$^3F^o$
1	−9.28510	$3p^63d^2$
2	−4.03880	$3p^63d4d$
3	−2.42140	$3p^63d\ ^2D\ 5d$
4	−2.19890	$3s3p^63d^3$
5	−1.99680	$3s3p^63d^3$
6	−1.96170	$3p^63d\ ^2D\ 5g$
7	−1.62690	$3p^63d\ ^2D\ 6d$
8	−1.40650	$3p^53d^24p$
9	−1.36520	$3p^63d\ ^2D\ 6g$
10	−1.25390	$3p^53d^24p$
11	−1.17450	$3p^63d\ ^2D\ 7d$
12	−1.15000	$3p^53d^24p$
13	−1.11770	$3p^53d^24p$
14	−1.09900	$3p^43d^4$:
15	−1.00860	$3p^53d^24p$
16	−1.00300	$3p^63d\ ^2D\ 7g$
17	−.96238	$3p^43d^4$
18	−.89374	$3p^53d^24p$
19	−.86498	$3p^63d\ ^2D\ 8d$
20	−.84813	$3p^53d^24p$
21	−.81661	$3p^53d^24p$
22	−.76809	$3p^63d\ ^2D\ 8g$
23	−.73571	$3p^43d^4$
24	−.72070	$3p^43d^4$
25	−.66190	$3p^63d\ ^2D\ 9d$
26	−.64609	$3p^53d^24p$
27	−.60020	$3p^63d\ ^2D\ 9g$
		$^3F^o$
1	−5.25680	$3p^63d4p$
2	−5.14400	$3p^53d^3$
3	−4.90320	$3p^53d^3$
4	−4.65570	$3p^53d^3$:
5	−4.27820	$3p^53d^3$
6	−3.70740	$3p^53d^3$
7	−3.15190	$3p^63d4f$
8	−2.91200	$3p^63d\ ^2D\ 5p$
9	−2.03580	$3p^63d\ ^2D\ 5f$
10	−1.88240	$3p^63d\ ^2D\ 6p$
11	−1.85880	$3p^53d^24s$
12	−1.81510	$3p^53d^24s$
13	−1.67550	$3p^53d^24s$
14	−1.40940	$3p^63d\ ^2D\ 6f$
15	−1.30860	$3p^63d\ ^2D\ 7p$
16	−1.03230	$3p^63d\ ^2D\ 7f$
17	−.96712	$3p^63d\ ^2D\ 8p$
18	−.78908	$3p^63d\ ^2D\ 8f$
19	−.77830	$3p^64s4f$:
20	−.73180	$3p^63d\ ^2D\ 9p$
21	−.62078	$3p^63d\ ^2D\ 9f$
		$^3G^e$
1	−4.11520	$3p^63d4d$
2	−2.43970	$3p^63d\ ^2D\ 5d$
3	−2.26810	$3s3p^63d^3$
4	−1.96910	$3p^63d\ ^2D\ 5g$
5	−1.62530	$3p^63d\ ^2D\ 6d$
6	−1.36760	$3p^63d\ ^2D\ 6g$
7	−1.24840	$3p^53d^24p$
8	−1.16520	$3p^63d\ ^2D\ 7d$
9	−1.11520	$3p^53d^24p$
10	−1.07030	$3p^53d^24p$
11	−1.00500	$3p^63d\ ^2D\ 7g$
12	−.99347	$3p^43d^4$
13	−.96635	$3p^43d^4$
14	−.91028	$3p^53d^24p$
15	−.88430	$3p^53d^24p$:
16	−.86067	$3p^63d\ ^2D\ 8d$:
17	−.84729	$3p^53d^24p$:
18	−.76869	$3p^63d\ ^2D\ 8g$
19	−.67743	$3p^63d\ ^2D\ 9d$
20	−.65069	$3p^53d^24p$
21	−.60725	$3p^63d\ ^2D\ 9g$
22	−.60164	$3p^43d^4$:
		$^3G^o$
1	−5.10530	$3p^53d^3$
2	−4.79810	$3p^53d^3$
3	−4.64060	$3p^53d^3$
4	−4.26450	$3p^53d^3$
5	−3.13400	$3p^63d4f$
6	−2.01230	$3p^63d\ ^2D\ 5f$
7	−1.95260	$3p^53d^24s$
8	−1.68110	$3p^53d^24s$
9	−1.39520	$3p^63d\ ^2D\ 6f$
10	−1.32860	$3p^53d^24s$
11	−1.02290	$3p^63d\ ^2D\ 7f$
12	−.78172	$3p^63d\ ^2D\ 8f$
13	−.61660	$3p^63d\ ^2D\ 9f$
		$^3H^e$
1	−2.21060	$3s3p^63d^3$
2	−1.96960	$3p^63d\ ^2D\ 5g$
3	−1.36800	$3p^63d\ ^2D\ 6g$
4	−1.21440	$3p^53d^24p$
5	−1.05910	$3p^43d^4$:
6	−1.00560	$3p^63d\ ^2D\ 7g$
7	−1.00000	$3p^53d^24p$
8	−.93442	$3p^53d^24p$
9	−.82721	$3p^43d^4$:
10	−.70953	$3p^63d\ ^2D\ 8g$
11	−.73826	$3p^43d^4$:
12	−.60719	$3p^63d\ ^2D\ 9g$
13	−.52548	$3p^53d^24p$
		$^3H^o$
1	−4.98780	$3p^53d^3$
2	−4.73730	$3p^53d^3$
3	3.20480	$3p^63d4f$
4	−2.04860	$3p^63d\ ^2D\ 5f$
5	−1.69570	$3p^53d^24s$
6	−1.41620	$3p^63d\ ^2D\ 6f$

Ca-like Fe (Fe^{6+})

i	E(Ryds)	Description	i	E(Ryds)	Description	i	E(Ryds)	Description	i	E(Ryds)	Description
7	−1.03620	$3p^63d\,^2\mathrm{D}\,7f$		$^5\mathbf{P}^e$			$^5\mathbf{D}^o$			$^5\mathbf{G}^e$	
8	−.79039	$3p^63d\,^2\mathrm{D}\,8f$	1	−2.37250	$3s3p^63d^3$	1	−5.45030	$3p^53d^3$:	1	−1.38960	$3p^43d^4$:
9	−.62252	$3p^63d\,^2\mathrm{D}\,9f$	2	−1.51150	$3p^53d^24p$	2	−4.88750	$3p^53d^3$:	2	−1.30490	$3p^53d^24p$
	$^3\mathbf{I}^e$		3	−1.34930	$3p^43d^4$:	3	−2.27280	$3p^53d^24s$	3	−1.16950	$3p^43d^4$:
1	−1.96110	$3p^63d\,^2\mathrm{D}\,5g$	4	−1.33600	$3p^53d^24p$	4	−1.62000	$3p^53d^24s$	4	−1.15460	$3p^53d^24p$
2	−1.36330	$3p^63d\,^2\mathrm{D}\,6g$	5	−1.02890	$3p^43d^4$:	5	−.31928	$3p^53d^24d$	5	−.91249	$3p^43d^4$:
3	−1.00240	$3p^63d\,^2\mathrm{D}\,7g$	6	−.93868	$3p^53d^24p$	6	−.18948	$3p^53d^24d$	6	−.62283	$3p^43d^4$:
4	−.95890	$3p^43d^4$:	7	−.58292	$3p^53d^24p$	7	−.05249	$3p^53d^24d$	7	−.34303	$3p^43d^4$:
5	−.92453	$3p^53d^24p$	8	−.38144	$3p^43d^4$:		$^5\mathbf{F}^e$			$^5\mathbf{G}^o$	
6	−.77590	$3p^43d^4$:	9	−.11684	$3p^43d^4$:	1	−2.59640	$3s3p^63d^3$:	1	−5.27950	$3p^53d^3$
7	−.76702	$3p^63d\,^2\mathrm{D}\,8g$		$^5\mathbf{P}^o$		2	−1.55080	$3p^53d^24p$	2	−2.07070	$3p^53d^24s$
8	−.60610	$3p^63d\,^2\mathrm{D}\,9g$	1	−4.86320	$3p^53d^3$:	3	−1.41530	$3p^43d^4$:	3	−.39341	$3p^53d^24d$
	$^3\mathbf{I}^o$		2	−2.03810	$3p^53d^24s$	4	−1.34850	$3p^53d^24p$	4	−.11509	$3p^53d^24d$:
1	−4.96670	$3p^53d^3$	3	−.38391	$3p^53d^24d$	5	−1.23020	$3p^43d^4$		$^5\mathbf{H}^e$	
2	−.06674	$3p^53d^24d$	4	−.07826	$3p^53d^24d$	6	−1.18130	$3p^53d^24p$	1	−1.33710	$3p^53d^24p$
	$^3\mathbf{J}^e$			$^5\mathbf{D}^e$		7	−.90810	$3p^43d^4$	2	−1.15860	$3p^43d^4$
1	−.79175	$3p^43d^4$	1	−1.51820	$3p^53d^24p$	8	−.86559	$3p^53d^24p$	3	−.89113	$3p^43d^4$
2	−.67179	$3p^43d^4$	2	−1.47690	$3p^43d^4$	9	−.63283	$3p^43d^4$		$^5\mathbf{H}^o$	
	$^5\mathbf{S}^e$		3	−1.27880	$3p^53d^24p$	10	−.42745	$3p^43d^4$	1	−.18046	$3p^53d^24d$
1	−1.36010	$3p^43d^4$	4	−1.22140	$3p^43d^4$		$^5\mathbf{F}^o$		2	−.01885	$3p^53d^24d$
2	−1.15800	$3p^53d^24p$	5	−1.10260	$3p^53d^24p$	1	−5.38850	$3p^53d^3$:		$^5\mathbf{I}^e$	
3	−.82038	$3p^43d^4$:	6	−.97208	$3p^43d^4$	2	−1.94690	$3p^53d^24s$	1	−1.27270	$3p^43d^4$
	$^5\mathbf{S}^o$		7	−.80617	$3p^43d^4$:	3	−.35713	$3p^53d^24d$		$^5\mathbf{I}^o$	
1	−5.50760	$3p^53d^3$	8	−.76003	$3p^53d^24p$	4	−.15046	$3p^53d^24d$	1	−.19069	$3p^53d^24d$
2	−1.38650	$3p^53d^24s$	9	−.43407	$3p^43d^4$:	5	−.08653	$3p^53d^24d$			
3	−.18854	$3p^53d^24d$	10	−.16460	$3p^43d^4$:						

Energies in ascending order from ground state for terms with effective $n \leq 4.0$, $L \leq 6$

Term	i	E(Ryds)	Term	i	E(Ryds)	Term	i	E(Ryds)	Term	i	E(Ryds)	Term	i	E(Ryds)
$^3\mathbf{F}^e$	1	0.00000	$^1\mathbf{P}^o$	1	4.14930	$^1\mathbf{D}^o$	2	4.54770	$^1\mathbf{F}^o$	3	5.24030	$^3\mathbf{D}^e$	3	6.05310
$^1\mathbf{D}^e$	1	0.16110	$^3\mathbf{G}^o$	1	4.17980	$^3\mathbf{H}^o$	2	4.54780	$^3\mathbf{F}^e$	2	5.24630	$^1\mathbf{D}^e$	4	6.06730
$^3\mathbf{P}^e$	1	0.18330	$^3\mathbf{D}^o$	2	4.18060	$^3\mathbf{D}^o$	4	4.57320	$^1\mathbf{H}^o$	2	5.27090	$^3\mathbf{H}^o$	3	6.08030
$^1\mathbf{G}^e$	1	0.26020	$^3\mathbf{P}^o$	2	4.18690	$^1\mathbf{H}^o$	1	4.58430	$^3\mathbf{D}^o$	5	5.27670	$^1\mathbf{F}^o$	5	6.11910
$^1\mathbf{S}^e$	1	0.65730	$^3\mathbf{H}^o$	1	4.29730	$^3\mathbf{F}^o$	4	4.62940	$^1\mathbf{D}^o$	4	5.29700	$^3\mathbf{F}^o$	7	6.13320
$^3\mathbf{D}^e$	1	3.28020	$^1\mathbf{G}^o$	1	4.31590	$^1\mathbf{P}^o$	2	4.64380	$^1\mathbf{D}^e$	3	5.30640	$^3\mathbf{G}^o$	5	6.15110
$^1\mathbf{D}^e$	2	3.32400	$^3\mathbf{I}^o$	1	4.31840	$^3\mathbf{G}^o$	3	4.64450	$^3\mathbf{P}^e$	2	5.31550	$^1\mathbf{D}^o$	6	6.16660
$^5\mathbf{S}^o$	1	3.77750	$^3\mathbf{F}^o$	3	4.38190	$^1\mathbf{D}^o$	3	4.68410	$^1\mathbf{G}^e$	2	5.33820	$^1\mathbf{F}^o$	6	6.17480
$^5\mathbf{D}^o$	1	3.83480	$^3\mathbf{D}^o$	3	4.38670	$^3\mathbf{P}^o$	4	4.77870	$^3\mathbf{D}^o$	6	5.39570	$^3\mathbf{D}^o$	8	6.18670
$^5\mathbf{F}^o$	1	3.89660	$^5\mathbf{D}^o$	2	4.39760	$^1\mathbf{G}^o$	2	4.82870	$^1\mathbf{D}^o$	5	5.40180	$^1\mathbf{P}^o$	4	6.19780
$^1\mathbf{D}^o$	1	3.97820	$^1\mathbf{S}^o$	1	4.40600	$^3\mathbf{F}^o$	5	5.00690	$^1\mathbf{S}^e$	2	5.46180	$^3\mathbf{P}^o$	6	6.19990
$^3\mathbf{D}^o$	1	4.00120	$^5\mathbf{P}^o$	1	4.42190	$^3\mathbf{G}^o$	4	5.02060	$^1\mathbf{F}^o$	4	5.48960	$^1\mathbf{H}^o$	3	6.20820
$^5\mathbf{G}^o$	1	4.00560	$^3\mathbf{G}^o$	2	4.48700	$^1\mathbf{F}^e$	1	5.12880	$^3\mathbf{P}^o$	5	5.51890	$^1\mathbf{G}^o$	4	6.22160
$^3\mathbf{F}^o$	1	4.02830	$^3\mathbf{P}^o$	3	4.50450	$^3\mathbf{D}^e$	2	5.15170	$^3\mathbf{F}^o$	6	5.57770			
$^3\mathbf{P}^o$	1	4.07480	$^1\mathbf{F}^o$	2	4.50670	$^1\mathbf{P}^e$	1	5.16440	$^1\mathbf{P}^o$	3	5.87550			
$^1\mathbf{F}^o$	1	4.11450	$^1\mathbf{I}^o$	1	4.53040	$^3\mathbf{G}^e$	1	5.16990	$^1\mathbf{G}^o$	3	5.92510			
$^3\mathbf{F}^o$	2	4.14110	$^3\mathbf{S}^o$	1	4.53490	$^3\mathbf{S}^e$	1	5.21220	$^3\mathbf{D}^o$	7	5.98320			

Ca-like Fe (Fe^{6+})

gf-values for transitions involving terms with effective $n \leq 4.0$, $L \leq 6$

$i\ i'$	gf_L	$i\ i'$	gf_L	$i\ i'$	gf_L	$i\ i'$	gf_L	$i\ i'$	gf_L	$i\ i'$	gf_L
	$^1S^e$–$^1P^o$	3 3	−1.27E−3	5 1	1.50E−2	1 5	−4.09E+0	1 7	−5.96E−2	2 3	2.58E−2
1 1	−1.36E−1	3 4	−1.35E−3	6 1	1.26E+0	1 6	−4.11E+0	1 8	−1.46E−5	2 4	1.63E−2
1 2	−1.45E−2	4 1	1.61E+0		**$^1F^e$–$^1G^o$**	2 1	3.54E+0	2 1	4.32E+0	2 5	−2.40E−4
1 3	−5.20E−1	4 2	4.63E−3	1 1	4.10E−3	2 2	3.17E−1	2 2	9.65E−2	2 6	−1.60E−2
1 4	−6.15E−1	4 3	−6.83E−3	1 2	1.28E−3	2 3	3.81E−2	2 3	7.43E−6	2 7	−1.02E−2
2 1	6.14E−1	4 4	−2.37E−3	1 3	−4.38E−1	2 4	5.49E−4	2 4	7.97E−3	2 8	−3.95E−1
2 2	1.66E−3	5 1	1.16E+0	1 4	−3.97E+0	2 5	−5.42E−2	2 5	−5.07E−6		**$^3F^e$–$^3F^o$**
2 3	−2.55E−2	5 2	1.01E−2		**$^1F^o$–$^1G^e$**	2 6	−2.06E+0	2 6	−1.39E−2	1 1	−1.38E+0
2 4	−2.29E−2	5 3	−3.34E−3	1 1	1.69E+0		**$^3P^e$–$^3D^o$**	2 7	−3.29E−1	1 2	−1.26E−2
	$^1S^o$–$^1P^e$	5 4	−3.30E−3	1 2	−5.23E+0	1 1	−3.87E−1	2 8	−3.44E+0	1 3	−4.04E−3
1 1	−1.92E−5	6 1	2.07E+0	2 1	2.35E−3	1 2	−2.25E−2	3 1	2.18E+0	1 4	−1.27E−1
	$^1P^o$–$^1P^e$	6 2	5.42E−4	2 2	−2.21E−2	1 3	−4.13E−1	3 2	4.46E−2	1 5	−5.97E−2
1 1	−1.09E+0	6 3	1.01E+0	3 1	9.79E−3	1 4	−3.36E−3	3 3	7.77E−3	1 6	−1.50E+1
2 1	−8.81E−3	6 4	1.51E−5	3 2	−3.37E−4	1 5	−9.09E−2	3 4	8.09E−3	1 7	−9.91E+0
3 1	2.40E−2		**$^1D^e$–$^1F^o$**	4 1	6.33E−4	1 6	−3.64E+0	3 5	2.61E−3	2 1	3.78E+0
4 1	8.40E−2	1 1	−4.18E−2	4 2	−1.42E−4	1 7	−2.09E+0	3 6	1.94E−2	2 2	2.00E−1
	$^1P^e$–$^1D^o$	1 2	−2.06E−1	5 1	9.10E+0	1 8	−8.66E+0	3 7	8.56E−3	2 3	7.20E−4
1 1	3.89E−1	1 3	−4.17E−3	5 2	6.98E−2	2 1	1.23E+0	3 8	−1.54E−3	2 4	1.88E−2
1 2	5.43E−3	1 4	−2.75E+0	6 1	8.20E−2	2 2	1.52E−2		**$^3D^e$–$^3F^o$**	2 5	6.43E−5
1 3	1.50E−4	1 5	−2.85E−1	6 2	9.16E−2	2 3	3.57E−2	1 1	−4.74E+0	2 6	−1.43E−1
1 4	−2.73E−4	1 6	−3.38E+0		**$^1G^o$–$^1G^e$**	2 4	2.50E−3	1 2	−2.70E−1	2 7	−2.98E+0
1 5	−1.62E−3	2 1	−1.76E+0	1 1	3.91E−2	2 5	3.64E−4	1 3	−5.48E−3		**$^3F^e$–$^3G^o$**
1 6	−1.76E+0	2 2	−1.49E−2	1 2	−1.60E−3	2 6	−2.34E−2	1 4	−3.69E−2	1 1	2.23E−1
	$^1P^o$–$^1D^e$	2 3	−1.05E−3	2 1	5.64E−2	2 7	−1.34E−1	1 5	−1.50E−3	1 2	−1.53E−2
1 1	3.65E−1	2 4	−3.89E−4	2 2	−4.85E−4	2 8	−4.59E+0	1 6	−2.66E−2	1 3	−2.30E+0
1 2	7.39E−1	2 5	−5.12E−2	3 1	7.15E+0		**$^3D^e$–$^3P^o$**	1 7	−5.21E−3	1 4	−6.16E+0
1 3	−1.12E+0	2 6	−3.85E−3	3 2	6.22E−2	1 1	−2.04E+0	2 1	5.55E−1	1 5	−1.41E+1
1 4	−3.68E−1	3 1	2.74E−1	4 1	5.08E+0	1 2	−1.82E−1	2 2	1.27E−2	2 1	1.23E−2
2 1	7.92E−6	3 2	1.71E−3	4 2	6.76E−1	1 3	−2.38E−2	2 3	2.73E−3	2 2	9.41E−4
2 2	9.92E−3	3 3	2.18E−6		**$^1G^e$–$^1H^o$**	1 4	−3.69E−4	2 4	1.53E−4	2 3	5.63E−2
2 3	−5.39E−3	3 4	−9.83E−4	1 1	−5.19E−1	1 5	−1.65E−2	2 5	7.76E−4	2 4	5.83E−2
2 4	−3.10E−3	3 5	−8.10E−3	1 2	−1.99E+0	1 6	−6.24E−3	2 6	−2.78E−1	2 5	−1.31E+1
3 1	7.07E−1	3 6	−2.76E+0	1 3	−5.71E+0	2 1	2.57E+0	2 7	−8.49E+0		**$^3G^e$–$^3F^o$**
3 2	2.33E−3	4 1	9.86E−1	2 1	1.91E−2	2 2	1.72E−1	3 1	2.81E+0	1 1	1.35E+1
3 3	9.72E−3	4 2	6.15E−3	2 2	1.79E−2	2 3	9.77E−3	3 2	1.42E−1	1 2	6.22E−1
3 4	−1.91E−3	4 3	9.21E−4	2 3	−6.55E+0	2 4	1.55E−4	3 3	2.53E−3	1 3	8.99E−3
4 1	3.65E+0	4 4	2.26E−4		**$^3S^e$–$^3P^o$**	2 5	−5.95E−4	3 4	1.39E−2	1 4	2.69E−2
4 2	7.65E−3	4 5	1.28E−2	1 1	1.49E+0	2 6	−4.82E−1	3 5	2.96E−4	1 5	2.58E−4
4 3	7.49E−5	4 6	−8.15E−5	1 2	8.31E−2	3 1	1.17E+0	3 6	9.90E−3	1 6	−7.78E−4
4 4	−9.86E−4		**$^1D^o$–$^1F^e$**	1 3	1.70E−3	3 2	9.55E−2	3 7	−4.49E−3	1 7	−3.36E−1
	$^1D^o$–$^1D^e$	1 1	−2.40E+0	1 4	3.07E−7	3 3	1.13E−2		**$^3F^o$–$^3D^o$**		**$^3G^e$–$^3G^o$**
1 1	4.83E−1	2 1	−2.05E−2	1 5	−3.18E−2	3 4	3.48E−4	1 1	2.56E+0	1 1	1.26E−3
1 2	1.12E+0	3 1	2.21E−3	1 6	−2.80E+0	3 5	8.78E−3	1 2	−6.24E−2	1 2	5.59E−4
1 3	1.49E+0	4 1	1.72E−8		**$^3P^e$–$^3S^o$**	3 6	−5.17E−3	1 3	−3.13E−2	1 3	1.08E−2
1 4	−7.91E−1	5 1	8.06E−5	1 1	−1.08E−2		**$^3D^e$–$^3D^o$**	1 4	−3.56E−3	1 4	7.79E−3
2 1	4.67E−3	6 1	1.55E−1	2 1	5.07E−5	1 1	−3.48E+0	1 5	−3.11E−2	1 5	−2.85E+0
2 2	2.54E−2		**$^1F^o$–$^1F^e$**		**$^3P^e$–$^3P^o$**	1 2	−9.57E−2	1 6	−2.13E+0		**$^3G^e$–$^3H^o$**
2 3	−9.98E−3	1 1	−1.13E+0	1 1	−1.37E+0	1 3	−1.40E−2	1 7	−1.86E+1	1 1	7.16E−3
2 4	−1.03E−2	2 1	−1.38E−2	1 2	−3.46E−4	1 4	−2.28E−2	1 8	−2.53E−2	1 2	9.50E−4
3 1	1.09E−3	3 1	5.45E−6	1 3	−7.13E−2	1 5	−7.35E−3	2 1	7.72E+0	1 3	−1.89E+1
3 2	3.59E−3	4 1	5.90E−3	1 4	−2.38E−4	1 6	−4.34E−2	2 2	1.37E−1		

Sc-like Fe (Fe^{5+})

Term energies relative to $3d^2$ ^{3}F ionization threshold for each symmetry

i	E(Ryds)	Description	i	E(Ryds)	Description	i	E(Ryds)	Description	i	E(Ryds)	Description
		$^2\mathbf{S}^e$	21	−.71070	$3d^2$ ^{1}D $7d$	40	−.41049	$3d^2$ ^{1}D $9p$	3	−3.82885	$3d^2$ ^{3}P $4p$
1	−4.20783	$3d^2$ ^{1}S $4s$	22	−.68369	$3d^2$ ^{3}P $7d$	41	−.38853	$3d^2$ ^{1}S $7p$	4	−2.87110	$3p^53d^4(^3P4)$:
2	−2.94577	$3d^2$ ^{1}D $4d$	23	−.64195	$3d^2$ ^{3}F $8d$	42	−.37554	$3d^2$ ^{3}P $9p$	5	−2.70908	$3p^53d^4(^3D4)$
3	−1.93138	$3d^2$ ^{1}S $5s$	24	−.60085	$3d^2$ ^{3}P $8s$			$^2\mathbf{D}^e$	6	−2.49471	$3p^53d^4(^1D4)$
4	−1.67862	$3d^2$ ^{1}D $5d$	25	−.56296	$3d^2$ ^{3}F $8g$	1	−6.92499	$3d^3(^3)$	7	−2.28853	$3d^2$ ^{3}F $4f$
5	−1.17638	$3d^2$ ^{1}G $5g$	26	−.50301	$3d^2$ ^{3}F $9d$	2	−6.50758	$3d^3(^1)$	8	−2.25440	$3d^2$ ^{1}D $4f$
6	−1.06868	$3d^2$ ^{1}D $6d$	27	−.49294	$3d^2$ ^{1}D $8d$	3	−4.64925	$3d^2$ ^{1}D $4s$	9	−2.23396	$3p^53d^4(^1D2)$
7	−.97251	$3d^2$ ^{1}S $6s$	28	−.48071	$3d^2$ ^{1}G $7g$	4	−3.02363	$3d^2$ ^{3}F $4d$	10	−2.21368	$3d^2$ ^{3}F $5p$
8	−.74319	$3d^2$ ^{1}G $6g$	29	−.46716	$3d^2$ ^{3}P $8d$	5	−2.85028	$3d^2$ ^{3}P $4d$	11	−2.15637	$3d^2$ ^{3}P $4f$
9	−.71746	$3d^2$ ^{1}D $7d$	30	−.44480	$3d^2$ ^{3}F $9g$	6	−2.80617	$3d^2$ ^{1}D $4d$	12	−2.08749	$3d^2$ ^{1}D $5p$
10	−.50188	$3d^2$ ^{1}D $8d$	31	−.41442	$3d^2$ ^{3}P $9s$	7	−2.65577	$3d^2$ ^{1}G $4d$	13	−2.06206	$3d^2$ ^{3}P $5p$
11	−.48011	$3d^2$ ^{1}G $7g$			$^2\mathbf{P}^o$	8	−2.43261	$3d^2$ ^{1}S $4d$	14	−2.02381	$3d^2$ ^{1}G $4f$
12	−.47471	$3d^2$ ^{1}S $7s$	1	−3.92777	$3d^2$ ^{1}D $4p$	9	−2.37714	$3d^2$ ^{1}D $5s$	15	−1.92081	$3p^53d^4(^1F4)$
13	−.35602	$3d^2$ ^{1}D $9d$	2	−3.79391	$3d^2$ ^{3}P $4p$	10	−1.78728	$3d^2$ ^{3}F $5d$	16	−1.62373	$3p^53d^4(^3F4)$:
14	−.30904	$3d^2$ ^{1}G $8g$	3	−3.45981	$3d^2$ ^{1}S $4p$	11	−1.74844	$3d4s^2$	17	−1.43930	$3d^2$ ^{3}F $5f$
		$^2\mathbf{S}^o$	4	−2.86784	$3p^53d^4(^1S4)$:	12	−1.64253	$3d^2$ ^{1}D $5d$	18	−1.41992	$3d^2$ ^{3}F $6p$
1	−3.97636	$3d^2$ ^{3}P $4p$	5	−2.75947	$3p^53d^4(^3D4)$:	13	−1.60464	$3d^2$ ^{3}P $5d$	19	−1.32832	$3d^2$ ^{1}D $5f$:
2	−2.47930	$3p^53d^4(^3P2)$	6	−2.30800	$3d^2$ ^{3}F $4f$	14	−1.50808	$3d^2$ ^{1}G $5d$	20	−1.30606	$3p^53d^4(^3F2)$:
3	−2.32980	$3d^2$ ^{3}F $4f$	7	−2.18002	$3d^2$ ^{1}D $4f$	15	−1.43800	$3d^2$ ^{3}F $5g$	21	−1.28563	$3d^2$ ^{3}P $5f$
4	−2.09511	$3d^2$ ^{3}P $5p$	8	−2.15761	$3p^53d^4(^1D2)$:	16	−1.42473	$3d^2$ ^{1}D $6s$	22	−1.27707	$3p^53d^4(^3P2)$:
5	−1.64428	$3p^53d^4(^3P4)$	9	−2.09846	$3d^2$ ^{1}D $5p$	17	−1.29522	$3d^2$ ^{1}D $5g$	23	−1.27508	$3d^2$ ^{1}D $6p$:
6	−1.38407	$3d^2$ ^{3}F $5f$	10	−2.03741	$3d^2$ ^{3}P $5p$	18	−1.21749	$3d^2$ ^{1}S $5d$	24	−1.23713	$3d^2$ ^{3}P $6p$
7	−1.26242	$3d^2$ ^{3}P $6p$	11	−2.00571	$3d^2$ ^{1}G $4f$	19	−1.18310	$3d^2$ ^{1}G $5g$	25	−1.10316	$3d^2$ ^{1}G $5f$
8	−1.00296	$3d^2$ ^{3}F $6f$	12	−1.72185	$3p^53d^4(^1D4)$:	20	−1.17845	$3d^2$ ^{3}F $6d$	26	−1.00817	$3d^2$ ^{3}F $6f$
9	−.81962	$3d^2$ ^{3}P $7p$	13	−1.64234	$3d^2$ ^{1}S $5p$	21	−1.05086	$3d^2$ ^{1}D $6d$	27	−1.00062	$3d^2$ ^{3}F $6h$
10	−.74124	$3d^2$ ^{3}F $7f$	14	−1.45700	$3d^2$ ^{3}F $5f$	22	−1.00867	$3d^2$ ^{3}P $6d$	28	−.98429	$3d^2$ ^{3}F $7p$
11	−.56876	$3d^2$ ^{3}F $8f$	15	−1.43224	$3p^53d^4(^1S0)$:	23	−.99926	$3d^2$ ^{3}F $6g$	29	−.89343	$3d4s$ ^{3}D $4p$
12	−.55375	$3d^2$ ^{3}P $8p$	16	−1.37199	$3p^53d^4(^3P4)$:	24	−.93457	$3d^2$ ^{1}G $6d$	30	−.87987	$3d^2$ ^{1}D $6f$:
13	−.44917	$3d^2$ ^{3}F $9f$	17	−1.29202	$3d^2$ ^{1}D $5f$	25	−.92585	$3d^2$ ^{1}D $7s$	31	−.84807	$3d^2$ ^{3}P $6f$
14	−.38248	$3d^2$ ^{3}P $9p$	18	−1.27846	$3d^2$ ^{1}D $6p$	26	−.85463	$3d^2$ ^{1}D $6g$	32	−.84006	$3d^2$ ^{1}D $7p$
		$^2\mathbf{P}^e$	19	−1.22944	$3d^2$ ^{3}P $6p$	27	−.84596	$3d^2$ ^{3}F $7d$	33	−.80947	$3d^2$ ^{3}P $7p$
1	−6.93873	$3d^3(^3)$	20	−1.17277	$3d4s$ ^{1}D $4p$	28	−.74635	$3d^2$ ^{1}G $6g$	34	−.75381	$3d^2$ ^{3}F $7f$
2	−4.57782	$3d^2$ ^{3}P $4s$	21	−1.15784	$3d^2$ ^{1}G $5f$	29	−.73438	$3d^2$ ^{3}F $7g$	35	−.74756	$3d^2$ ^{1}G $6f$
3	−3.06821	$3d^2$ ^{3}F $4d$	22	−1.02660	$3d^2$ ^{3}F $6f$	30	−.70958	$3d^2$ ^{1}D $7d$	36	−.74634	$3d^2$ ^{1}G $6h$
4	−2.94564	$3d^2$ ^{3}P $4d$	23	−.97115	$3p^53d^4(^3P2)$	31	−.67043	$3d^2$ ^{3}P $7d$	37	−.73471	$3d^2$ ^{3}F $7h$
5	−2.78338	$3d^2$ ^{1}D $4d$	24	−.86180	$3d^2$ ^{1}D $6f$	32	−.63679	$3d^2$ ^{3}F $8d$	38	−.72391	$3d^2$ ^{3}F $8p$
6	−2.33713	$3d^2$ ^{3}P $5s$	25	−.84406	$3d^2$ ^{1}D $7p$	33	−.63378	$3d^2$ ^{1}D $8s$	39	−.69773	$3d4s$ ^{1}D $4p$
7	−1.81232	$3d^2$ ^{3}F $5d$	26	−.81903	$3d^2$ ^{1}S $6p$	34	−.60638	$3d^2$ ^{1}S $6d$	40	−.59349	$3d^2$ ^{1}D $7f$
8	−1.67224	$3d^2$ ^{1}D $5d$:	27	−.81476	$3d4s$ ^{3}D $4p$	35	−.59456	$3d^2$ ^{1}G $7d$	41	−.57885	$3d^2$ ^{1}D $8p$
9	−1.61596	$3d^2$ ^{3}P $5d$	28	−.80028	$3d^2$ ^{3}P $7p$	36	−.58873	$3d^2$ ^{1}D $7g$	42	−.56669	$3d^2$ ^{3}P $7f$
10	−1.44054	$3d^2$ ^{3}F $5g$	29	−.74495	$3d^2$ ^{1}G $6h$	37	−.56235	$3d^2$ ^{3}F $8g$	43	−.56241	$3d^2$ ^{3}F $8h$
11	−1.38949	$3d^2$ ^{3}P $6s$	30	−.74209	$3d^2$ ^{3}F $7f$	38	−.50077	$3d^2$ ^{3}F $9d$	44	−.55839	$3d^2$ ^{3}F $8f$
12	−1.20244	$3d^2$ ^{3}F $6d$	31	−.72625	$3d^2$ ^{1}G $6f$	39	−.48933	$3d^2$ ^{1}D $8d$	45	−.55616	$3d^2$ ^{3}F $9p$
13	−1.17888	$3d^2$ ^{1}G $5g$	32	−.59751	$3d^2$ ^{1}D $7f$	40	−.48209	$3d^2$ ^{1}G $7g$	46	−.54435	$3d^2$ ^{3}P $8p$
14	−1.05681	$3d^2$ ^{1}D $6d$	33	−.57994	$3d^2$ ^{1}D $8p$	41	−.45802	$3d^2$ ^{3}P $8d$	47	−.48584	$3d^2$ ^{1}G $7f$
15	−1.02547	$3d^2$ ^{3}P $6d$	34	−.56640	$3d^2$ ^{3}F $8f$	42	−.44608	$3d^2$ ^{1}D $9s$	48	−.48197	$3d^2$ ^{1}G $7h$
16	−1.00030	$3d^2$ ^{3}F $6g$	35	−.54304	$3d^2$ ^{3}P $8p$	43	−.44429	$3d^2$ ^{3}F $9g$	49	−.44517	$3d^2$ ^{3}F $9f$
17	−.89516	$3d^2$ ^{3}P $7s$	36	−.48125	$3d^2$ ^{1}G $7h$	44	−.41638	$3d^2$ ^{1}D $8g$	50	−.44436	$3d^2$ ^{3}F $9h$
18	−.85603	$3d^2$ ^{3}F $7d$	37	−.48112	$3d^2$ ^{1}G $7f$			$^2\mathbf{D}^o$			$^2\mathbf{F}^e$
19	−.74430	$3d^2$ ^{1}G $6g$	38	−.44752	$3d^2$ ^{3}F $9f$	1	−4.05047	$3d^2$ ^{3}F $4p$	1	−6.73290	$3d^3(^3)$
20	−.73525	$3d^2$ ^{3}F $7g$	39	−.42388	$3d^2$ ^{1}D $8f$	2	−3.89821	$3d^2$ ^{1}D $4p$	2	−4.75569	$3d^2$ ^{3}F $4s$

Sc-like Fe (Fe^{5+})

i	E(Ryds)	Description
3	−3.11304	$3d^2$ ^{3}F $4d$
4	−2.93880	$3d^2$ ^{1}D $4d$
5	−2.87714	$3d^2$ ^{3}P $4d$
6	−2.68362	$3d^2$ ^{1}G $4d$
7	−2.51128	$3d^2$ ^{3}F $5s$
8	−1.83349	$3d^2$ ^{3}F $5d$
9	−1.67628	$3d^2$ ^{1}D $5d$
10	−1.62482	$3d^2$ ^{3}P $5d$
11	−1.56691	$3d^2$ ^{3}F $6s$
12	−1.53065	$3d^2$ ^{1}G $5d$
13	−1.43844	$3d^2$ ^{3}F $5g$
14	−1.29517	$3d^2$ ^{1}D $5g$
15	−1.26707	$3d^2$ ^{3}P $5g$
16	−1.21386	$3d^2$ ^{3}F $6d$
17	−1.18767	$3d^2$ ^{1}G $5g$
18	−1.07313	$3d^2$ ^{3}F $7s$
19	−1.06096	$3d^2$ ^{1}D $6d$
20	−1.01988	$3d^2$ ^{3}P $6d$
21	−.99976	$3d^2$ ^{3}F $6g$
22	−.94497	$3d^2$ ^{1}G $6d$
23	−.86340	$3d^2$ ^{3}F $7d$
24	−.85437	$3d^2$ ^{1}D $6g$
25	−.82508	$3d^2$ ^{3}P $6g$
26	−.77880	$3d^2$ ^{3}F $8s$
27	−.74854	$3d^2$ ^{1}G $6g$
28	−.73459	$3d^2$ ^{3}F $7g$
29	−.71432	$3d^2$ ^{1}D $7d$
30	−.67831	$3d^2$ ^{3}P $7d$
31	−.64647	$3d^2$ ^{3}F $8d$
32	−.60400	$3d^2$ ^{1}G $7d$
33	−.59145	$3d^2$ ^{3}F $9s$
34	−.58876	$3d^2$ ^{1}D $7g$
35	−.56253	$3d^2$ ^{3}F $8g$
36	−.55887	$3d^2$ ^{3}P $7g$
37	−.50381	$3d^2$ ^{3}F $9d$
38	−.49768	$3d^2$ ^{1}D $8d$
39	−.48352	$3d^2$ ^{1}G $7g$
	^{2}F$^\circ$	
1	−4.06402	$3d^2$ ^{3}F $4p$
2	−3.92601	$3d^2$ ^{1}D $4p$
3	−3.74604	$3d^2$ ^{1}G $4p$
4	−2.93311	$3p^5 3d^4(^3F2)$:
5	−2.70850	$3p^5 3d^4(^1D4)$
6	−2.62415	$3p^5 3d^4(^3D4)$
7	−2.32146	$3d^2$ ^{3}F $4f$
8	−2.27136	$3p^5 3d^4(^1F4)$
9	−2.23956	$3d^2$ ^{3}F $5p$
10	−2.20075	$3d^2$ ^{1}D $4f$
11	2.11424	$3d^2$ ^{3}P $4f$
12	−2.09694	$3d^2$ ^{1}D $5p$
13	−2.06040	$3d^2$ ^{1}G $4f$
14	−1.98467	$3p^5 3d^4(^1G2)$:
15	−1.96472	$3d^2$ ^{1}G $5p$
16	−1.80032	$3p^5 3d^4(^1D2)$:
17	−1.70673	$3d^2$ ^{1}S $4f$
18	−1.54329	$3p^5 3d^4(^3F4)$:
19	−1.48572	$3d^2$ ^{3}F $5f$
20	−1.43818	$3p^5 3d^4(^1G4)$
21	−1.41115	$3d^2$ ^{3}F $6p$:
22	−1.32051	$3d^2$ ^{1}D $5f$
23	−1.28132	$3d^2$ ^{1}D $6p$
24	−1.25212	$3d^2$ ^{3}P $5f$
25	−1.20316	$3p^5 3d^4(^3G4)$:
26	−1.17283	$3d^2$ ^{1}G $5f$
27	−1.16517	$3d^2$ ^{1}G $6p$
28	−1.05070	$3d^2$ ^{3}F $6f$
29	−.99878	$3d^2$ ^{3}F $6h$
30	−.99102	$3d^2$ ^{3}F $7p$
31	−.97319	$3d4s$ ^{3}D $4p$
32	−.87911	$3d^2$ ^{1}S $5f$
33	−.86238	$3d^2$ ^{1}D $6f$
34	−.85325	$3d^2$ ^{1}D $6h$
35	−.84634	$3d4s$ ^{1}D $4p$
36	−.84035	$3d^2$ ^{1}D $7p$
37	−.81515	$3d^2$ ^{3}P $6f$
38	−.75584	$3d^2$ ^{1}G $6f$
39	−.74787	$3d^2$ ^{1}G $6h$
40	−.73934	$3d^2$ ^{3}F $7f$
41	−.73416	$3d^2$ ^{3}F $7h$
42	−.73168	$3d^2$ ^{1}G $7p$
43	−.72497	$3d^2$ ^{3}F $8p$
44	−.59836	$3d^2$ ^{1}D $7f$
45	−.58795	$3d^2$ ^{1}D $7h$
46	−.58002	$3d^2$ ^{1}D $8p$
47	−.56938	$3d^2$ ^{3}F $8f$
48	−.56222	$3d^2$ ^{3}F $8h$
49	−.55816	$3d^2$ ^{3}P $7f$
50	−.55538	$3d^2$ ^{3}F $9p$
51	−.49103	$3d^2$ ^{1}G $7f$
52	−.48278	$3d^2$ ^{1}G $7h$
53	−.47192	$3d^2$ ^{1}G $8p$
54	−.44962	$3d^2$ ^{3}F $9f$
55	−.44426	$3d^2$ ^{3}F $9h$
	^{2}G^e	
1	−6.97631	$3d^3(^3)$
2	−4.52720	$3d^2$ ^{1}G $4s$
3	−3.04349	$3d^2$ ^{3}F $4d$
4	−2.89545	$3d^2$ ^{1}D $4d$
5	−2.78448	$3d^2$ ^{1}G $4d$
6	−2.27469	$3d^2$ ^{1}G $5s$
7	−1.79479	$3d^2$ ^{3}F $5d$
8	−1.66329	$3d^2$ ^{1}D $5d$
9	−1.56022	$3d^2$ ^{1}G $5d$
10	−1.44229	$3d^2$ ^{3}F $5g$
11	−1.32098	$3d^2$ ^{1}G $6s$
12	−1.29230	$3d^2$ ^{1}D $5g$
13	−1.25684	$3d^2$ ^{3}P $5g$
14	−1.19330	$3d^2$ ^{3}F $6d$
15	−1.19162	$3d^2$ ^{1}G $5g$
16	−1.05903	$3d^2$ ^{1}D $6d$
17	−1.00147	$3d^2$ ^{3}F $6g$
18	−.95879	$3d^2$ ^{1}G $6d$
19	−.85474	$3d^2$ ^{1}D $6g$
20	−.85141	$3d^2$ ^{3}F $7d$
21	−.84048	$3d^2$ ^{1}S $5g$
22	−.82321	$3d^2$ ^{1}G $7s$
23	−.81972	$3d^2$ ^{3}P $6g$
24	−.75059	$3d^2$ ^{1}G $6g$
25	−.73576	$3d^2$ ^{3}F $7g$
26	−.71381	$3d^2$ ^{1}D $7d$
27	−.64037	$3d^2$ ^{3}F $8d$
28	−.60967	$3d^2$ ^{1}G $7d$
29	−.58799	$3d^2$ ^{1}D $7g$
30	−.56365	$3d^2$ ^{3}F $8g$
31	−.55519	$3d^2$ ^{3}P $7g$
32	−.52957	$3d^2$ ^{1}G $8s$
33	−.50203	$3d^2$ ^{1}D $8d$
34	−.49343	$3d^2$ ^{3}F $9d$
35	−.48481	$3d^2$ ^{1}G $7g$
36	−.44508	$3d^2$ ^{3}F $9g$
37	−.41613	$3d^2$ ^{1}D $8g$
38	−.40168	$3d^2$ ^{1}S $6g$
	^{2}G$^\circ$	
1	−4.02153	$3d^2$ ^{3}F $4p$
2	−3.86448	$3d^2$ ^{1}G $4p$
3	−3.01231	$3p^5 3d^4(^3H4)$:
4	−2.65375	$3p^5 3d^4(^1F4)$
5	−2.58229	$3p^5 3d^4(^3F4)$:
6	−2.34445	$3d^2$ ^{3}F $4f$
7	−2.24408	$3p^5 3d^4(^3F2)$
8	−2.22580	$3d^2$ ^{3}F $5p$
9	−2.16967	$3d^2$ ^{1}D $4f$
10	−2.13406	$3d^2$ ^{3}P $4f$
11	−2.07336	$3d^2$ ^{1}G $4f$
12	−2.01233	$3d^2$ ^{1}G $5p$
13	−1.83731	$3p^5 3d^4(^1G2)$:
14	−1.65972	$3p^5 3d^4(^1G4)$:
15	−1.45882	$3d^2$ ^{3}F $5f$
16	−1.41835	$3d^2$ ^{3}F $6p$
17	−1.33230	$3d^2$ ^{1}D $5f$
18	−1.28907	$3d^2$ ^{3}P $5f$
19	−1.20833	$3d^2$ ^{1}G $5f$
20	−1.18557	$3d^2$ ^{1}G $6p$
21	−1.03680	$3d^2$ ^{3}F $6f$
22	−.99084	$3p^5 3d^4(^3G4)$:
23	−.99833	$3d^2$ ^{3}F $6h$
24	−.95881	$3d^2$ ^{3}F $7p$
25	−.87903	$3d^2$ ^{1}D $6f$
26	−.85368	$3d^2$ ^{1}D $6h$
27	−.83253	$3d^2$ ^{3}P $6f$
28	−.82321	$3d^2$ ^{3}P $6h$
29	−.76561	$3d^2$ ^{1}G $6f$
30	−.74903	$3d^2$ ^{1}G $6h$
31	−.74684	$3d^2$ ^{3}F $7f$
32	−.74096	$3d^2$ ^{1}G $7p$
33	−.73438	$3d^2$ ^{3}F $7h$
34	−.72157	$3d^2$ ^{3}F $8p$
35	−.60527	$3d^2$ ^{1}D $7f$
36	−.58797	$3d^2$ ^{1}D $7h$
37	−.57485	$3d^2$ ^{3}F $8f$
38	−.56383	$3d^2$ ^{3}P $7f$
39	−.56239	$3d^2$ ^{3}F $8h$
40	−.55756	$3d^2$ ^{3}P $7h$
41	−.55393	$3d^2$ ^{3}F $9p$
42	−.49654	$3d^2$ ^{1}G $7f$
43	−.48348	$3d^2$ ^{1}G $7h$
44	−.47809	$3d^2$ ^{1}G $8p$
45	−.45171	$3d^2$ ^{3}F $9f$
46	−.44437	$3d^2$ ^{3}F $9h$
	^{2}H^e	
1	−6.88971	$3d^3(^3)$
2	−3.01754	$3d^2$ ^{3}F $4d$
3	−2.74966	$3d^2$ ^{1}G $4d$
4	−1.78570	$3d^2$ ^{3}F $5d$
5	−1.55366	$3d^2$ ^{1}G $5d$
6	−1.44449	$3d^2$ ^{3}F $5g$
7	−1.29328	$3d^2$ ^{1}D $5g$
8	−1.26478	$3d^2$ ^{3}P $5g$
9	−1.19418	$3d^2$ ^{1}G $5g$
10	−1.18888	$3d^2$ ^{3}F $6d$
11	−1.00244	$3d^2$ ^{3}F $6g$
12	−.95373	$3d^2$ ^{1}G $6d$
13	−.85361	$3d^2$ ^{1}D $6g$
14	−.84832	$3d^2$ ^{3}F $7d$
15	−.82387	$3d^2$ ^{3}P $6g$
16	−.75187	$3d^2$ ^{1}G $6g$
17	−.73649	$3d^2$ ^{3}F $7g$
18	−.64081	$3d^2$ ^{3}F $8d$
19	−.60528	$3d^2$ ^{1}G $7d$
20	.58828	$3d^2$ ^{1}D $7g$
21	−.56414	$3d^2$ ^{3}F $8g$
22	−.55776	$3d^2$ ^{3}P $7g$
23	−.49671	$3d^2$ ^{3}F $9d$
24	−.48553	$3d^2$ ^{1}G $7g$
25	−.44541	$3d^2$ ^{3}F $9g$
26	−.41599	$3d^2$ ^{1}D $8g$
	^{2}H$^\circ$	
1	−3.80245	$3d^2$ ^{1}G $4p$

Sc-like Fe (Fe^{5+})

i	E(Ryds)	Description	i	E(Ryds)	Description	i	E(Ryds)	Description	i	E(Ryds)	Description
2	−2.86923	$3p^53d^4(^3G4)$	13	−.44759	$3d^2$ ^{3}F $9f$	8	−.99979	$3d^2$ ^{3}F $6g$	10	−1.26712	$3d^2$ ^{3}P $5g$
3	−2.64521	$3p^53d^4(^1G4)$	14	−.38120	$3d^2$ ^{3}P $9p$	9	−.86469	$3d^2$ ^{3}F $7d$	11	−1.18841	$3d^2$ ^{3}F $6d$
4	−2.32993	$3p^53d^4(^1G2)$			**^{4}P^e**	10	−.73469	$3d^2$ ^{3}F $7g$	12	−1.07599	$3d^2$ ^{3}F $7s$
5	−2.29043	$3d^2$ ^{3}F $4f$	1	−6.99802	$3d^3(^3)$	11	−.68455	$3d^2$ ^{3}P $7d$	13	−1.03314	$3d^2$ ^{3}P $6d$
6	−2.16504	$3d^2$ ^{1}D $4f$	2	−4.63604	$3d^2$ ^{3}P $4s$	12	−.64739	$3d^2$ ^{3}F $8d$	14	−1.00020	$3d^2$ ^{3}F $6g$
7	−2.10317	$3d^2$ ^{1}G $4f$:	3	−3.04286	$3d^2$ ^{3}F $4d$	13	−.56259	$3d^2$ ^{3}F $8g$	15	−.84995	$3d^2$ ^{3}F $7d$
8	−1.98931	$3p^53d^4(^1I4)$	4	−2.80766	$3d^2$ ^{3}P $4d$	14	−.50314	$3d^2$ ^{3}F $9d$	16	−.82513	$3d^2$ ^{3}P $6g$
9	−1.98421	$3d^2$ ^{1}G $5p$	5	−2.35427	$3d^2$ ^{3}P $5s$	15	−.46813	$3d^2$ ^{3}P $8d$	17	−.78157	$3d^2$ ^{3}F $8s$
10	−1.51751	$3p^53d^4(^3H4)$	6	−1.79699	$3d^2$ ^{3}F $5d$	16	−.44457	$3d^2$ ^{3}F $9g$	18	−.73488	$3d^2$ ^{3}F $7g$
11	−1.39722	$3d^2$ ^{3}F $5f$:	7	−1.62180	$3d^2$ ^{3}P $5d$			**^{4}D^o**	19	−.68554	$3d^2$ ^{3}P $7d$
12	−1.32628	$3d^2$ ^{1}D $5f$	8	−1.44032	$3d^2$ ^{3}F $5g$	1	−4.05667	$3d^2$ ^{3}F $4p$	20	−.63800	$3d^2$ ^{3}F $8d$
13	−1.19286	$3d^2$ ^{1}G $5f$	9	−1.39719	$3d^2$ ^{3}P $6s$	2	−3.90516	$3d^2$ ^{3}P $4p$	21	−.59345	$3d^2$ ^{3}F $9s$
14	−1.16701	$3d^2$ ^{1}G $6p$	10	−1.19477	$3d^2$ ^{3}F $6d$	3	−3.22154	$3p^53d^4(^3F4)$	22	−.56281	$3d^2$ ^{3}F $8g$
15	−1.00663	$3d^2$ ^{3}F $6f$	11	−1.02385	$3d^2$ ^{3}P $6d$	4	−2.93284	$3p^53d^4(^3P4)$	23	−.55885	$3d^2$ ^{3}P $7g$
16	−1.00078	$3d^2$ ^{3}F $6h$	12	−1.00015	$3d^2$ ^{3}F $6g$	5	−2.73953	$3p^53d^4(^3P2)$	24	−.49705	$3d^2$ ^{3}F $9d$
17	−.87678	$3d^2$ ^{1}D $6f$	13	−.89847	$3d^2$ ^{3}P $7s$	6	−2.42117	$3p^53d^4(^3D4)$	25	−.46878	$3d^2$ ^{3}P $8d$
18	−.85223	$3d^2$ ^{1}D $6h$	14	−.85250	$3d^2$ ^{3}F $7d$	7	−2.31485	$3d^2$ ^{3}F $4f$			**^{4}F^o**
19	−.82074	$3d^2$ ^{3}P $6h$	15	−.73514	$3d^2$ ^{3}F $7g$	8	−2.24359	$3d^2$ ^{3}F $5p$	1	−4.08533	$3d^2$ ^{3}F $4p$
20	−.76540	$3d^2$ ^{1}G $6f$	16	−.68179	$3d^2$ ^{3}P $7d$	9	−2.19410	$3p^53d^4(^3F2)$	2	−3.13087	$3p^53d^4(^3G4)$
21	−.74958	$3d^2$ ^{1}G $6h$	17	−.63857	$3d^2$ ^{3}F $8d$	10	−2.07103	$3d^2$ ^{3}P $5p$	3	−2.98334	$3p^53d^4(^5D4)$
22	−.74078	$3d^2$ ^{3}F $7f$	18	−.60387	$3d^2$ ^{3}P $8s$	11	−2.00664	$3d^2$ ^{3}P $4f$	4	−2.64605	$3p^53d^4(^3D4)$
23	−.73591	$3d^2$ ^{1}G $7p$	19	−.56287	$3d^2$ ^{3}F $8g$	12	−1.49941	$3d^2$ ^{3}F $5f$	5	−2.34868	$3d^2$ ^{3}F $4f$
24	−.73506	$3d^2$ ^{3}F $7h$	20	−.49910	$3d^2$ ^{3}F $9d$	13	−1.42630	$3d^2$ ^{3}F $6p$	6	−2.27761	$3p^53d^4(^3F2)$
25	−.60458	$3d^2$ ^{1}D $7f$	21	−.46485	$3d^2$ ^{3}P $8d$	14	−1.35641	$3p^53d^4(^5D4)$	7	−2.24872	$3d^2$ ^{3}F $5p$
26	−.58740	$3d^2$ ^{1}D $7h$	22	−.44472	$3d^2$ ^{3}F $9g$	15	−1.31592	$3d4s$ ^{3}D $4p$	8	−2.12065	$3d^2$ ^{3}P $4f$
27	−.56660	$3d^2$ ^{3}F $8f$	23	−.41587	$3d^2$ ^{3}P $9s$	16	−1.26790	$3d^2$ ^{3}P $5f$	9	−1.94575	$3p^53d^4(^3F4)$
28	−.56282	$3d^2$ ^{3}F $8h$			**^{4}P^o**	17	−1.23823	$3d^2$ ^{3}P $6p$	10	−1.44629	$3d^2$ ^{3}F $5f$
29	−.55586	$3d^2$ ^{3}P $7h$	1	−3.87016	$3d^2$ ^{3}P $4p$	18	−1.01429	$3d^2$ ^{3}F $6f$	11	−1.42761	$3d^2$ ^{3}F $6p$
30	−.49629	$3d^2$ ^{1}G $7f$	2	−3.07947	$3p^53d^4(^3D4)$	19	−1.00060	$3d^2$ ^{3}F $6h$	12	−1.30620	$3d4s$ ^{3}D $4p$
31	−.48396	$3d^2$ ^{1}G $7h$	3	−2.97330	$3p^53d^4(^3P4)$	20	−.98714	$3d^2$ ^{3}F $7p$	13	−1.29216	$3d^2$ ^{3}P $5f$
32	−.47450	$3d^2$ ^{1}G $8p$	4	−2.34516	$3p^53d^4(^3P2)$	21	−.83811	$3d^2$ ^{3}P $6f$	14	−1.01129	$3d^2$ ^{3}F $6f$
33	−.44875	$3d^2$ ^{3}F $9f$	5	−2.29830	$3d^2$ ^{3}F $4f$	22	−.81276	$3d^2$ ^{3}P $7p$	15	−.99879	$3d^2$ ^{3}F $6h$
34	−.44460	$3d^2$ ^{3}F $9h$	6	−2.05639	$3d^2$ ^{3}P $5p$	23	−.74621	$3d^2$ ^{3}F $7f$	16	−.99059	$3d^2$ ^{3}F $7p$
35	−.42622	$3d^2$ ^{1}D $8f$	7	−1.59737	$3p^53d^4(^5D4)$	24	−.73471	$3d^2$ ^{3}F $7h$	17	−.84770	$3d^2$ ^{3}P $6f$
36	−.41545	$3d^2$ ^{1}D $8h$	8	−1.38358	$3d^2$ ^{3}F $5f$	25	−.72577	$3d^2$ ^{3}F $8p$	18	−.74375	$3d^2$ ^{3}F $7f$
37	−.40093	$3d^2$ ^{1}S $6h$	9	−1.24778	$3d4s$ ^{3}D $4p$	26	−.57512	$3d^2$ ^{3}F $8f$	19	−.73417	$3d^2$ ^{3}F $7h$
38	−.38410	$3d^2$ ^{3}P $8h$	10	−1.24534	$3d^2$ ^{3}P $6p$	27	−.56488	$3d^2$ ^{3}P $7f$	20	−.72783	$3d^2$ ^{3}F $8p$
		^{4}S^e	11	−1.00354	$3d^2$ ^{3}F $6f$	28	−.56241	$3d^2$ ^{3}F $8h$	21	−.57424	$3d^2$ ^{3}P $7f$
1	−.12216	$3d4s$ ^{3}D $4d$	12	−.81002	$3d^2$ ^{3}P $7p$	29	−.55650	$3d^2$ ^{3}F $9p$	22	−.56924	$3d^2$ ^{3}F $8f$
		^{4}S^o	13	−.74075	$3d^2$ ^{3}F $7f$	30	−.54977	$3d^2$ ^{3}P $8p$	23	−.56222	$3d^2$ ^{3}F $8h$
1	−3.94625	$3d^2$ ^{3}P $4p$	14	−.56808	$3d^2$ ^{3}F $8f$	31	−.45104	$3d^2$ ^{3}F $9f$	24	−.55754	$3d^2$ ^{3}F $9p$
2	−3.04869	$3p^53d^4(^3P4)$	15	−.54792	$3d^2$ ^{3}P $8p$	32	−.44436	$3d^2$ ^{3}F $9h$	25	−.44947	$3d^2$ ^{3}F $9f$
3	−2.31984	$3d^2$ ^{3}F $4f$	16	−.44862	$3d^2$ ^{3}F $9f$			**^{4}F^e**	26	−.44427	$3d^2$ ^{3}F $9h$
4	−2.09966	$3d^2$ ^{3}P $5p$	17	−.37860	$3d^2$ ^{3}P $9p$	1	−7.17051	$3d^3(^3)$			**^{4}G^e**
5	−1.92985	$3p^53d^4(^3P2)$			**^{4}D^e**	2	−4.81851	$3d^2$ ^{3}F $4s$	1	−3.11380	$3d^2$ ^{3}F $4d$
6	−1.43800	$3d^2$ ^{3}F $5f$	1	−3.09922	$3d^2$ ^{3}F $4d$	3	−2.98103	$3d^2$ ^{3}F $4d$	2	−1.83684	$3d^2$ ^{3}F $5d$
7	−1.25709	$3d^2$ ^{3}P $6p$	2	−2.89126	$3d^2$ ^{3}P $4d$	4	−2.88115	$3d^2$ ^{3}P $4d$	3	−1.44228	$3d^2$ ^{3}F $5g$
8	−1.00515	$3d^2$ ^{3}F $6f$	3	−1.82907	$3d^2$ ^{3}F $5d$	5	−2.53101	$3d^2$ ^{3}F $5s$	4	−1.25778	$3d^2$ ^{3}P $5g$
9	−.81639	$3d^2$ ^{3}P $7p$	4	−1.64197	$3d^2$ ^{3}P $5d$	6	−1.78023	$3d^2$ ^{3}F $5d$	5	−1.21790	$3d^2$ ^{3}F $6d$
10	−.73969	$3d^2$ ^{3}F $7f$	5	−1.43835	$3d^2$ ^{3}F $5g$	7	−1.64439	$3d^2$ ^{3}P $5d$	6	−1.00150	$3d^2$ ^{3}F $6g$
11	−.56661	$3d^2$ ^{3}F $8f$	6	−1.21445	$3d^2$ ^{3}F $6d$	8	−1.57437	$3d^2$ ^{3}F $6s$	7	−.86691	$3d^2$ ^{3}F $7d$
12	−.55214	$3d^2$ ^{3}P $8p$	7	−1.03202	$3d^2$ ^{3}P $6d$	9	−1.43896	$3d^2$ ^{3}F $5g$	8	−.82069	$3d^2$ ^{3}P $6g$

Sc-like Fe (Fe^{5+})

i	E(Ryds)	Description
9	−.73580	$3d^2\ ^3F\ 7g$
10	−.64888	$3d^2\ ^3F\ 8d$
11	−.56361	$3d^2\ ^3F\ 8g$
12	−.55590	$3d^2\ ^3P\ 7g$
13	−.50405	$3d^2\ ^3F\ 9d$
14	−.44511	$3d^2\ ^3F\ 9g$
		$^4G^o$
1	−4.09220	$3d^2\ ^3F\ 4p$
2	−3.03985	$3p^5 3d^4(^3F4)$
3	−2.93800	$3p^5 3d^4(^3G4)$
4	−2.56457	$3p^5 3d^4(^3F2)$
5	−2.40273	$3p^5 3d^4(^3H4)$
6	−2.24808	$3d^2\ ^3F\ 5p$
7	−2.24610	$3d^2\ ^3F\ 4f$
8	−2.11185	$3d^2\ ^3P\ 4f$
9	−1.46121	$3d^2\ ^3F\ 5f$
10	−1.43133	$3d^2\ ^3F\ 6p$

i	E(Ryds)	Description
11	−1.30278	$3d^2\ ^3P\ 5f$
12	−1.01506	$3d^2\ ^3F\ 6f$
13	−.99917	$3d^2\ ^3F\ 6h$
14	−.99155	$3d^2\ ^3F\ 7p$
15	−.84965	$3d^2\ ^3P\ 6f$
16	−.82322	$3d^2\ ^3P\ 6h$
17	−.74512	$3d^2\ ^3F\ 7f$
18	−.73440	$3d^2\ ^3F\ 7h$
19	−.72835	$3d^2\ ^3F\ 8p$
20	−.57717	$3d^2\ ^3P\ 7f$
21	−.56827	$3d^2\ ^3F\ 8f$
22	−.56240	$3d^2\ ^3F\ 8h$
23	−.55783	$3d^2\ ^3F\ 9p$
24	−.55753	$3d^2\ ^3P\ 7h$
25	−.45007	$3d^2\ ^3F\ 9f$
26	−.44438	$3d^2\ ^3F\ 9h$

i	E(Ryds)	Description
		$^4H^e$
1	−3.10377	$3d^2\ ^3F\ 4d$
2	−1.83328	$3d^2\ ^3F\ 5d$
3	−1.44442	$3d^2\ ^3F\ 5g$
4	−1.26537	$3d^2\ ^3P\ 5g$
5	−1.21616	$3d^2\ ^3F\ 6d$
6	−1.00241	$3d^2\ ^3F\ 6g$
7	−.86593	$3d^2\ ^3F\ 7d$
8	−.82441	$3d^2\ ^3P\ 6g$
9	−.73648	$3d^2\ ^3F\ 7g$
10	−.64825	$3d^2\ ^3F\ 8d$
11	−.56402	$3d^2\ ^3F\ 8g$
12	−.55830	$3d^2\ ^3P\ 7g$
13	−.50362	$3d^2\ ^3F\ 9d$
14	−.44541	$3d^2\ ^3F\ 9g$
		$^4H^o$
1	−3.01439	$3p^5 3d^4(^3H4)$

i	E(Ryds)	Description
2	−2.80394	$3p^5 3d^4(^3G4)$
3	−2.33112	$3d^2\ ^3F\ 4f$
4	−1.49411	$3d^2\ ^3F\ 5f$
5	−1.03497	$3d^2\ ^3F\ 6f$
6	−1.00080	$3d^2\ ^3F\ 6h$
7	−.82076	$3d^2\ ^3P\ 6h$
8	−.75776	$3d^2\ ^3F\ 7f$
9	−.73507	$3d^2\ ^3F\ 7h$
10	−.57836	$3d^2\ ^3F\ 8f$
11	−.56282	$3d^2\ ^3F\ 8h$
12	−.55588	$3d^2\ ^3P\ 7h$
13	−.45576	$3d^2\ ^3F\ 9f$
14	−.44460	$3d^2\ ^3F\ 9h$
15	−.38412	$3d^2\ ^3P\ 8h$

Energies in ascending order from ground state for terms with effective $n \leq 4.0$, $L \leq 5$

Term	i	E(Ryds)	Term	i	E(Ryds)	Term	i	E(Ryds)	Term	i	E(Ryds)	Term	i	E(Ryds)
$^4F^e$	1	0.00000	$^2P^o$	1	3.24274	$^2D^e$	4	4.14688	$^4H^o$	2	4.36657	$^2D^e$	9	4.79337
$^4P^e$	1	0.17249	$^2F^o$	2	3.24450	$^2H^e$	2	4.15297	$^2G^e$	5	4.38603	$^4P^e$	5	4.81624
$^2G^e$	1	0.19420	$^4D^o$	2	3.26535	$^4H^o$	1	4.15612	$^2P^e$	5	4.38713	$^4F^o$	5	4.82183
$^2P^e$	1	0.23178	$^2D^o$	2	3.27230	$^2G^o$	3	4.15820	$^2P^o$	5	4.41104	$^4P^o$	4	4.82535
$^2D^e$	1	0.24552	$^4P^o$	1	3.30035	$^4F^o$	3	4.18717	$^2H^e$	3	4.42085	$^2G^o$	6	4.82606
$^2H^e$	1	0.28080	$^2G^o$	2	3.30603	$^4F^e$	3	4.18948	$^4D^o$	5	4.43098	$^2P^e$	6	4.83338
$^2F^e$	1	0.43761	$^2D^o$	3	3.34166	$^4P^o$	3	4.19721	$^2D^o$	5	4.46143	$^4H^o$	3	4.83939
$^2D^e$	2	0.66293	$^2H^o$	1	3.36806	$^2S^e$	2	4.22474	$^2F^o$	5	4.46201	$^2H^o$	4	4.84058
$^4F^e$	2	2.35200	$^2P^o$	2	3.37660	$^2P^e$	4	4.22487	$^2F^e$	6	4.48689	$^2S^o$	3	4.84071
$^2F^e$	2	2.41482	$^2F^o$	3	3.42447	$^2F^e$	4	4.23171	$^2D^e$	7	4.51474	$^2F^o$	7	4.84905
$^2D^e$	3	2.52126	$^2P^o$	3	3.71070	$^4G^o$	3	4.23251	$^2G^o$	4	4.51676	$^4S^o$	3	4.85067
$^4P^e$	2	2.53447	$^4D^o$	3	3.94897	$^2F^o$	4	4.23740	$^4F^o$	4	4.52446	$^4D^o$	7	4.85566
$^2P^e$	2	2.59269	$^4F^o$	2	4.03964	$^4D^o$	4	4.23767	$^2H^o$	3	4.52530	$^2P^o$	6	4.86251
$^2G^e$	2	2.64331	$^4G^e$	1	4.05671	$^2G^e$	4	4.27506	$^2F^o$	6	4.54636	$^4P^o$	5	4.87221
$^2S^e$	1	2.96268	$^2F^e$	3	4.05747	$^4D^e$	2	4.27925	$^2G^o$	5	4.58822	$^2H^o$	5	4.88008
$^4G^o$	1	3.07831	$^4H^e$	1	4.06674	$^4F^e$	4	4.28936	$^4G^o$	4	4.60594	$^2D^o$	7	4.88198
$^4F^o$	1	3.08518	$^4D^e$	1	4.07129	$^2F^e$	5	4.29337	$^4F^e$	5	4.63950	$^4F^o$	6	4.89290
$^2F^o$	1	3.10649	$^4P^o$	2	4.09104	$^2D^o$	4	4.29941	$^2F^e$	7	4.65923	$^2G^e$	6	4.89582
$^4D^o$	1	3.11384	$^2P^e$	3	4.10230	$^2H^o$	2	4.30128	$^2D^o$	6	4.67580	$^2F^o$	8	4.89915
$^2D^o$	1	3.12004	$^4S^o$	2	4.12182	$^2P^o$	4	4.30267	$^2S^o$	2	4.69121	$^2D^o$	8	4.91611
$^2G^o$	1	3.14898	$^2G^e$	3	4.12702	$^2D^e$	5	4.32023	$^2D^o$	8	4.73790			
$^2S^o$	1	3.19415	$^4P^e$	3	4.12765	$^4P^e$	4	4.36285	$^4D^o$	6	4.74934			
$^4S^o$	1	3.22426	$^4G^o$	2	4.13066	$^2D^e$	6	4.36434	$^4G^o$	5	4.76778			

Sc-like Fe (Fe^{5+})

gf-values for transitions involving terms with effective $n \leq 4.0$, $L \leq 5$

$i\ i'$	gf_L	$i\ i'$	gf_L	$i\ i'$	gf_L	$i\ i'$	gf_L	$i\ i'$	gf_L	$i\ i'$	gf_L
	$^2S^o$–$^2P^e$	4 1	1.60E-2	4 6	-1.82E-4	4 5	-2.78E-3	4 4	1.47E-5	2 2	2.13E-1
1 1	1.84E-1	4 2	1.41E-3	4 7	-7.23E-5	4 6	-7.02E-3	4 5	-2.75E-4	2 3	-2.15E-1
1 2	5.52E-1	4 3	2.04E-4	4 8	-5.41E-3	4 7	-2.48E-3	4 6	-8.08E-4	2 4	-5.23E+0
1 3	-8.67E-2	4 4	2.57E-4	4 9	-3.71E-4	4 8	-1.73E-3	4 7	-5.96E-4	2 5	-1.74E-2
1 4	-1.32E+0	4 5	-1.07E-3	5 1	2.66E-4	5 1	7.52E-1	4 8	-7.36E-6	2 6	-2.22E-1
1 5	-6.38E-1	4 6	-5.36E-4	5 2	1.33E-3	5 2	2.81E-1	4 9	-6.83E-4	2 7	-6.48E-2
1 6	-3.20E-1	5 1	2.41E-4	5 3	5.79E-3	5 3	1.50E-1	5 1	1.39E-2	3 1	1.89E+0
2 1	1.31E-3	5 2	3.51E-4	5 4	1.03E-3	5 4	5.25E-5	5 2	3.19E-3	3 2	4.66E-1
2 2	3.29E-3	5 3	1.36E-3	5 5	1.37E-4	5 5	-1.00E-4	5 3	1.02E-4	3 3	-5.57E-3
2 3	1.43E-2	5 4	1.63E-4	5 6	-1.97E-7	5 6	-6.25E-4	5 4	1.23E-7	3 4	-2.85E-1
2 4	3.50E-3	5 5	-4.88E-5	5 7	-1.12E-4	5 7	-6.11E-5	5 5	2.44E-6	3 5	-4.40E+0
2 5	8.05E-7	5 6	-7.46E-5	5 8	-1.01E-3	5 8	-9.44E-2	5 6	3.93E-4	3 6	-3.87E+0
2 6	-2.20E-3	6 1	1.46E-1	5 9	-2.04E-3	6 1	2.84E-1	5 7	-4.26E-4	3 7	-7.52E-2
3 1	6.63E-2	6 2	2.58E-3	6 1	1.14E-1	6 2	1.41E-2	5 8	-5.49E-5	4 1	7.08E-3
3 2	5.53E-3	6 3	2.30E+0	6 2	6.87E-2	6 3	1.11E+0	5 9	-3.77E-7	4 2	1.03E-3
3 3	1.08E+0	6 4	1.44E-4	6 3	5.41E-3	6 4	6.32E-4	6 1	5.26E-2	4 3	7.76E-4
3 4	3.92E-4	6 5	2.71E-1	6 4	7.14E-1	6 5	1.73E-3	6 2	3.17E-3	4 4	1.62E-5
3 5	1.35E-1	6 6	2.73E-6	6 5	2.23E-1	6 6	1.42E-4	6 3	2.69E-3	4 5	-1.19E-4
3 6	1.90E-4		$^2P^o$–$^2D^e$	6 6	2.40E-1	6 7	-8.70E-4	6 4	2.27E-5	4 6	-2.58E-4
	$^2S^e$–$^2P^o$	1 1	4.80E-1	6 7	3.28E-3	6 8	-1.62E-3	6 5	9.80E-3	4 7	-3.52E-4
1 1	-1.99E-2	1 2	1.94E-3	6 8	4.26E-3		$^2D^o$–$^2D^e$	6 6	1.19E-2	5 1	1.45E-5
1 2	-5.82E-4	1 3	1.73E+0	6 9	5.47E-3	1 1	4.51E-1	6 7	1.45E-3	5 2	1.07E-2
1 3	-1.77E+0	1 4	-5.95E-1		$^2P^e$–$^2D^o$	1 2	3.18E-2	6 8	-1.37E-6	5 3	6.86E-3
1 4	-4.75E-3	1 5	-5.76E-2	1 1	-4.63E-1	1 3	9.81E-2	6 9	-1.97E-3	5 4	6.92E-5
1 5	-1.76E-4	1 6	-1.55E+0	1 2	-6.77E-2	1 4	-3.25E+0	7 1	3.42E-2	5 5	4.88E-9
1 6	-7.93E-4	1 7	-8.80E-2	1 3	-6.52E-2	1 5	-2.61E-2	7 2	3.55E-2	5 6	-3.48E-7
2 1	1.18E+0	1 8	-9.89E-3	1 4	-3.43E-4	1 6	-5.71E-1	7 3	2.57E-3	5 7	-5.40E-3
2 2	1.74E-2	1 9	-8.56E-1	1 5	-2.85E-3	1 7	-8.79E-3	7 4	6.29E-1	6 1	2.26E-3
2 3	3.45E-2	2 1	2.02E-1	1 6	-1.05E-3	1 8	-1.32E-2	7 5	3.05E-1	6 2	1.80E-4
2 4	-4.93E-5	2 2	6.91E-1	1 7	-5.65E-2	1 9	-7.04E-2	7 6	4.07E-1	6 3	1.00E-3
2 5	-3.67E-4	2 3	2.41E-2	1 8	-1.82E-1	2 1	9.11E-1	7 7	3.23E-5	6 4	3.46E-4
2 6	-2.69E-2	2 4	-2.45E-1	2 1	-4.49E-1	2 2	1.86E-2	7 8	5.75E-3	6 5	2.18E-3
	$^2P^o$–$^2P^e$	2 5	-2.31E+0	2 2	-3.22E-2	2 3	2.75E+0	7 9	4.31E-3	6 6	1.12E-3
1 1	4.99E-1	2 6	-3.83E-1	2 3	-2.56E+0	2 4	-2.78E-1	8 1	4.62E-1	6 7	-2.66E-5
1 2	2.15E-2	2 7	-1.94E+0	2 4	-1.94E-3	2 5	-7.30E-1	8 2	1.51E-1	7 1	1.64E-4
1 3	-2.39E-1	2 8	-1.06E-1	2 5	-1.53E-3	2 6	-3.00E+0	8 3	1.46E-3	7 2	4.99E-3
1 4	-1.19E+0	2 9	-9.15E-3	2 6	-5.78E-4	2 7	-1.71E-3	8 4	2.53E+0	7 3	1.44E-1
1 5	-1.41E+0	3 1	9.60E-2	2 7	-1.01E-4	2 8	-3.57E-2	8 5	2.88E-1	7 4	4.65E-4
1 6	-1.42E-2	3 2	6.23E-1	2 8	-2.68E-5	2 9	-1.16E+0	8 6	6.39E-1	7 5	1.37E-3
2 1	2.48E-1	3 3	3.55E-2	3 1	2.25E+0	3 1	5.22E-3	8 7	3.07E-2	7 6	4.14E-2
2 2	1.87E+0	3 4	-6.48E-3	3 2	5.78E-1	3 2	3.76E-1	8 8	1.87E-2	7 7	4.83E-2
2 3	-5.65E-2	3 5	-1.27E-2	3 3	4.89E-1	3 3	1.53E-1	8 9	1.12E-3	8 1	5.94E-2
2 4	-8.40E-1	3 6	-3.04E-2	3 4	-7.17E-4	3 4	-3.73E-1		$^2D^o$–$^2F^e$	8 2	1.00E-2
2 5	-9.38E-1	3 7	-2.68E-1	3 5	-5.86E-3	3 5	-8.55E-1	1 1	3.55E-3	8 3	4.64E-1
2 6	-7.40E-1	3 8	-6.09E+0	3 6	-3.59E-3	3 6	-1.17E-1	1 2	2.33E+0	8 4	5.21E-4
3 1	1.90E-2	3 9	-4.25E-2	3 7	-3.16E-1	3 7	-7.11E-1	1 3	-2.64E+0	8 5	3.66E-2
3 2	6.71E-3	4 1	6.65E-3	3 8	-6.55E-1	3 8	-9.58E-2	1 4	-5.90E-1	8 6	1.18E-2
3 3	-2.57E-3	4 2	1.03E-4	4 1	9.18E-3	3 9	-8.07E-2	1 5	-3.41E-1	8 7	1.24E-1
3 4	-2.79E-2	4 3	1.17E-3	4 2	2.58E-1	4 1	3.10E-2	1 6	-1.53E-2		$^2D^e$–$^2F^o$
3 5	-2.11E-2	4 4	3.04E-4	4 3	1.13E-1	4 2	2.16E-3	1 7	-1.14E+0	1 1	-3.88E-1
3 6	-3.25E-3	4 5	-6.55E-6	4 4	-2.13E-4	4 3	7.28E-4	2 1	1.09E-1	1 2	-2.36E-2

Sc-like Fe (Fe^{5+})

i i'	gf_L	i i'	gf_L	i i'	gf_L	i i'	gf_L	i i'	gf_L	i i'	gf_L
1 3	−5.15E−2	7 7	−1.62E−3	5 6	−2.06E−3	5 5	2.32E−5	6 2	2.02E+0	2 1	2.82E+0
1 4	−4.17E−3	7 8	−1.05E−3	5 7	−3.20E−4	5 6	−5.92E−3	6 3	2.51E−4	2 2	−1.44E−1
1 5	−6.79E−2	8 1	5.88E−3	6 1	1.42E−2	6 1	3.54E−2	6 4	3.23E−4	2 3	−9.18E+0
1 6	−1.15E−1	8 2	1.09E−1	6 2	1.20E−3	6 2	1.05E−3	6 5	−5.76E−3	3 1	1.03E−2
1 7	−4.80E−2	8 3	1.53E−1	6 3	2.23E−2	6 3	5.58E−3	6 6	−3.41E−2	3 2	−2.45E−4
1 8	−1.63E−1	8 4	1.60E−4	6 4	7.02E−6	6 4	8.85E−4	7 1	2.36E+0	3 3	−9.95E−5
2 1	−3.48E−2	8 5	3.88E−6	6 5	1.99E−3	6 5	2.53E−3	7 2	6.83E−2	4 1	3.74E−3
2 2	−4.60E−2	8 6	1.64E−3	6 6	1.40E−4	6 6	−5.41E−4	7 3	5.11E−4	4 2	1.08E−3
2 3	−7.05E−1	8 7	−2.36E−3	6 7	−5.15E−4	7 1	3.09E−3	7 4	2.23E−4	4 3	8.75E−4
2 4	−2.14E−4	8 8	−1.27E−2	7 1	1.70E−1	7 2	2.89E−3	7 5	3.85E−4	5 1	3.30E−3
2 5	−2.17E−3	9 1	1.34E−1	7 2	1.71E−3	7 3	2.00E−1	7 6	−8.42E−3	5 2	1.01E−3
2 6	−3.40E−2	9 2	1.63E+0	7 3	1.05E+0	7 4	1.57E−2		$^2\mathbf{G}^o$–$^2\mathbf{G}^e$	5 3	5.66E−4
2 7	−2.26E−2	9 3	6.79E−2	7 4	9.68E−3	7 5	7.30E−2	1 1	1.24E+0	6 1	4.81E−2
2 8	−1.35E−2	9 4	3.96E−3	7 5	3.25E−2	7 6	−6.69E−4	1 2	2.14E−1	6 2	1.33E−1
3 1	−2.25E−1	9 5	6.90E−6	7 6	3.03E−3	8 1	8.22E−3	1 3	−2.57E+0	6 3	2.21E−2
3 2	−3.71E+0	9 6	5.42E−3	7 7	2.45E−2	8 2	3.89E−4	1 4	−1.28E−1		$^2\mathbf{G}^e$–$^2\mathbf{H}^o$
3 3	−1.70E−1	9 7	−4.98E−8	8 1	5.27E−2	8 3	5.16E−1	1 5	−1.99E+0	1 1	−1.25E−1
3 4	−6.04E−3	9 8	−2.71E−4	8 2	8.86E−3	8 4	5.17E−2	1 6	−1.32E−1	1 2	−3.92E−1
3 5	−1.46E−4		$^2\mathbf{F}^o$–$^2\mathbf{F}^e$	8 3	1.82E+0	8 5	1.47E−1	2 1	4.14E−1	1 3	−1.27E−1
3 6	−9.43E−3	1 1	1.73E−1	8 4	2.68E−3	8 6	−1.10E−4	2 2	5.07E+0	1 4	−6.39E−2
3 7	−4.69E−6	1 2	3.81E+0	8 5	1.48E−2		$^2\mathbf{F}^e$–$^2\mathbf{G}^o$	2 3	−1.04E+0	1 5	−2.05E+0
3 8	−9.31E−4	1 3	−4.70E+0	8 6	3.62E−2	1 1	−8.24E−4	2 4	−1.78E+0	2 1	−6.63E+0
4 1	1.12E+0	1 4	−1.68E−1	8 7	9.58E−2	1 2	−4.49E−1	2 5	−4.02E+0	2 2	−4.02E−3
4 2	4.33E−1	1 5	−4.78E−3		$^2\mathbf{F}^o$–$^2\mathbf{G}^e$	1 3	−9.95E−4	2 6	2.10E+0	2 3	−4.76E−3
4 3	5.21E−3	1 6	−1.65E−1	1 1	7.32E−1	1 4	−2.43E−2	3 1	2.32E−2	2 4	−2.75E−3
4 4	−2.15E−4	1 7	−1.92E+0	1 2	3.69E−2	1 5	−1.72E−1	3 2	5.68E−3	2 5	−3.36E−5
4 5	−4.15E−4	2 1	1.40E−1	1 3	−7.07E+0	1 6	−8.19E−2	3 3	−8.38E−5	3 1	4.46E−2
4 6	−4.38E−3	2 2	3.47E−1	1 4	−2.74E−2	2 1	−5.28E+0	3 4	−7.77E−4	3 2	−2.69E−4
4 7	−7.06E−1	2 3	−2.50E−1	1 5	−4.52E−1	2 2	−3.11E−1	3 5	−2.26E−4	3 3	−9.78E−5
4 8	−1.73E+0	2 4	−2.40E+0	1 6	−2.96E−2	2 3	−6.00E−4	3 6	−1.55E−3	3 4	−4.17E−2
5 1	4.75E−1	2 5	−4.57E−1	2 1	2.18E+0	2 4	−2.61E−5	4 1	7.39E−3	3 5	−1.09E+1
5 2	7.95E−2	2 6	−1.28E−1	2 2	1.24E−1	2 5	−1.68E−3	4 2	1.07E−2	4 1	3.17E−1
5 3	1.85E+0	2 7	−1.37E−1	2 3	−6.59E−1	2 6	−3.23E−3	4 3	5.16E−4	4 2	3.16E−4
5 4	4.33E−4	3 1	9.63E−1	2 4	−7.45E+0	3 1	7.61E−1	4 4	1.18E−4	4 3	−1.61E−2
5 5	−2.91E−4	3 2	1.13E−2	2 5	−3.09E+0	3 2	4.82E−3	4 5	3.13E−4	4 4	−1.36E−1
5 6	−1.67E−4	3 3	−6.90E−2	2 6	−6.33E−2	3 3	−1.32E−3	4 6	−8.57E−3	4 5	−2.23E−1
5 7	−2.09E−2	3 4	−2.48E−4	3 1	1.83E−1	3 4	−1.02E−3	5 1	7.09E−4	5 1	1.07E+0
5 8	−8.28E−2	3 5	−2.15E+0	3 2	4.21E+0	3 5	−9.95E−5	5 2	4.56E−6	5 2	1.04E−2
6 1	5.33E−1	3 6	−3.45E+0	3 3	−3.26E−3	3 6	−6.93E+0	5 3	3.03E−3	5 3	−6.07E−3
6 2	2.51E−2	3 7	−3.61E−3	3 4	−1.76E+0	4 1	5.41E−3	5 4	4.23E−3	5 4	−3.14E−1
6 3	6.10E−1	4 1	8.79E−3	3 5	−1.76E+0	4 2	6.60E−2	5 5	9.82E−4	5 5	−6.60E−1
6 4	5.19E−4	4 2	9.46E−3	3 6	−1.76E+0	4 3	1.02E−4	5 6	−2.91E−4	6 1	2.00E+0
6 5	−1.22E−3	4 3	3.52E−3	4 1	4.35E−4	4 4	−4.20E−5	6 1	3.56E−1	6 2	2.64E−3
6 6	−6.20E−3	4 4	−2.86E−6	4 2	5.03E−3	4 5	−8.05E−4	6 2	8.06E−3	6 3	2.47E−3
6 7	−4.50E−1	4 5	−3.21E−5	4 3	1.11E−3	4 6	−3.50E−5	6 3	2.98E+0	6 4	1.04E−3
6 8	−5.80E−1	4 6	−2.15E−3	4 4	−1.69E−4	5 1	1.78E−1	6 4	2.08E−1	6 5	6.20E−5
7 1	1.08E−3	4 7	−6.38E−3	4 5	−6.10E−4	5 2	1.31E+0	6 5	6.61E−1		$^2\mathbf{H}^o$–$^2\mathbf{H}^e$
7 2	1.24E−1	5 1	2.09E−2	4 6	−3.69E−3	5 3	7.04E−5	6 6	−2.73E−3	1 1	1.26E+0
7 3	3.75E+0	5 2	4.07E−5	5 1	4.12E−3	5 4	−1.34E−3		$^2\mathbf{G}^o$–$^2\mathbf{H}^e$	1 2	−8.69E−1
7 4	1.75E−3	5 3	9.57E−4	5 2	1.03E−2	5 5	−4.60E−3	1 1	2.02E+0	1 3	−5.00E+0
7 5	1.42E−3	5 4	8.46E−4	5 3	3.08E−4	5 6	−5.30E−3	1 2	−1.31E+1	2 1	2.56E−2
7 6	6.67E−4	5 5	6.76E−4	5 4	6.43E−6	6 1	3.10E−3	1 3	−1.15E+0	2 2	2.60E−3

Sc-like Fe (Fe^{5+})

$i\ i'$	gf_L	$i\ i'$	gf_L	$i\ i'$	gf_L	$i\ i'$	gf_L	$i\ i'$	gf_L	$i\ i'$	gf_L
2 3	−3.48E−3	4 1	3.45E−2	3 7	−2.74E+0	3 2	6.64E−4	1 5	−3.84E+0	2 4	−1.75E−2
3 1	2.06E−4	4 2	7.87E−4	4 1	1.62E+0	3 3	−1.60E−3	2 1	5.10E−3	2 5	−2.89E−3
3 2	1.62E−3	4 3	1.01E−1	4 2	5.72E−1	3 4	−9.05E−3	2 2	1.27E−3	3 1	1.40E+0
3 3	1.29E−4	4 4	3.29E−2	4 3	2.52E−3	3 5	−1.67E−5	2 3	−3.30E−5	3 2	3.48E−3
4 1	4.38E−2	4 5	−1.91E−4	4 4	6.40E−3	4 1	1.93E−2	2 4	−2.53E−4	3 3	3.88E−5
4 2	1.46E−1	5 1	8.34E−1	4 5	−9.21E−4	4 2	1.58E−2	2 5	−1.62E−4	3 4	−3.12E−1
4 3	1.67E−2	5 2	8.40E−4	4 6	−4.30E−1	4 3	7.35E−5	3 1	1.01E−1	3 5	−7.02E−2
5 1	5.41E−1	5 3	4.05E+0	4 7	−4.35E−1	4 4	−1.47E−3	3 2	3.56E−2	4 1	3.75E−1
5 2	1.94E+0	5 4	9.31E−1	5 1	1.73E−1	4 5	−1.00E−2	3 3	−4.16E−3	4 2	8.80E−3
5 3	6.97E−1	5 5	2.90E−5	5 2	2.55E+0	5 1	1.60E−2	3 4	−4.35E−3	4 3	2.22E−3
	$^4S^o$–$^4P^e$		**$^4P^o$–$^4D^e$**	5 3	6.79E−3	5 2	1.27E−3	3 5	−2.11E−2	4 4	−5.55E−1
1 1	8.93E−1	1 1	−8.59E−2	5 4	7.12E−4	5 3	3.76E−3	4 1	1.59E−1	4 5	−1.94E−1
1 2	1.21E+0	1 2	−9.50E+0	5 5	7.20E−4	5 4	5.48E−6	4 2	5.46E−3	5 1	4.76E+0
1 3	−8.83E−1	2 1	−1.77E−4	5 6	1.85E−5	5 5	−2.78E−3	4 3	3.23E−2	5 2	6.82E−3
1 4	−3.47E+0	2 2	−3.55E−3	5 7	−2.54E−3	6 1	8.25E−2	4 4	1.17E−2	5 3	4.76E−3
1 5	−5.83E−1	3 1	4.14E−4		**$^4D^o$–$^4D^e$**	6 2	1.46E−2	4 5	−2.51E−3	5 4	8.32E−3
2 1	5.87E−2	3 2	−6.87E−4	1 1	−7.99E+0	6 3	1.74E−1	5 1	6.82E−2	5 5	−2.34E−3
2 2	1.99E−4	4 1	7.69E−2	1 2	−2.22E−2	6 4	2.00E−2	5 2	2.69E−2		**$^4G^o$–$^4G^e$**
2 3	−6.46E−4	4 2	1.46E−4	2 1	−2.90E−1	6 5	1.92E−2	5 3	4.63E+0	1 1	−8.55E+0
2 4	−4.15E−3	5 1	2.71E+0	2 2	−3.23E+0	7 1	4.35E−2	5 4	1.51E+0	2 1	2.52E−4
2 5	−9.67E−4	5 2	2.78E−2	3 1	−8.03E−5	7 2	1.76E−2	5 5	1.49E−1	3 1	1.12E−2
3 1	9.28E−2		**$^4P^e$–$^4D^o$**	3 2	−8.46E−4	7 3	1.51E+0	6 1	2.83E−1	4 1	4.19E−1
3 2	5.62E−3	1 1	−1.25E+0	4 1	6.11E−7	7 4	5.72E−1	6 2	2.37E−2	5 1	2.10E−1
3 3	1.38E+0	1 2	−3.05E−1	4 2	−2.71E−4	7 5	1.99E−1	6 3	1.78E+0		**$^4G^o$–$^4H^e$**
3 4	6.36E−1	1 3	−1.64E−2	5 1	1.35E−3		**$^4D^e$–$^4F^o$**	6 4	4.76E−1	1 1	−2.77E+1
3 5	4.48E−3	1 4	−1.26E−1	5 2	1.90E−4	1 1	4.18E+0	6 5	1.66E−1	2 1	7.29E−4
	$^4P^o$–$^4P^e$	1 5	−1.03E−1	6 1	1.33E−1	1 2	3.93E−4		**$^4F^o$–$^4G^e$**	3 1	5.56E−4
1 1	7.82E−1	1 6	−5.32E−1	6 2	6.66E−2	1 3	−6.94E−5	1 1	−1.40E+1	4 1	1.26E−1
1 2	3.67E+0	1 7	−1.57E−1	7 1	4.31E+0	1 4	−2.85E−2	2 1	−8.36E−5	5 1	2.97E−4
1 3	−7.22E−1	2 1	−3.37E−1	7 2	3.47E−2	1 5	−4.00E+0	3 1	1.95E−3		**$^4G^e$–$^4H^o$**
1 4	−2.76E+0	2 2	−5.62E+0		**$^4D^o$–$^4F^e$**	1 6	−1.04E+0	4 1	4.90E−4	1 1	−5.80E−4
1 5	−1.60E+0	2 3	−1.55E−2	1 1	2.52E+0	2 1	3.69E−2	5 1	2.94E−1	1 2	−2.04E−3
2 1	2.00E−3	2 4	−2.07E−3	1 2	5.80E+0	2 2	1.00E−3	6 1	2.89E−2	1 3	−2.51E+1
2 2	8.50E−3	2 5	−9.79E−4	1 3	−2.41E+0	2 3	3.52E−8		**$^4F^e$–$^4G^o$**		**$^4H^o$–$^4H^e$**
2 3	−2.74E−4	2 6	−6.95E−4	1 4	−4.06E+0	2 4	−4.20E−3	1 1	−5.41E−1	1 1	6.37E−5
2 4	−4.04E−3	2 7	−1.11E−2	1 5	−2.64E+0	2 5	−3.72E−2	1 2	−3.16E−1	2 1	4.58E−4
2 5	−5.12E−3	3 1	5.81E+0	2 1	7.85E−1	2 6	−5.65E−3	1 3	−1.76E−1	3 1	6.26E+0
3 1	2.27E−2	3 2	1.95E−1	2 2	3.84E−1		**$^4F^o$–$^4F^e$**	1 4	−1.67E+0		
3 2	3.86E−3	3 3	4.67E−4	2 3	−5.55E+0	1 1	3.80E+0	1 5	−1.37E−2		
3 3	8.18E−5	3 4	−1.33E−3	2 4	−1.17E+1	1 2	8.40E+0	2 1	−1.05E+1		
3 4	−1.77E−3	3 5	−1.02E−3	2 5	−1.39E−1	1 3	−9.12E+0	2 2	−7.95E−3		
3 5	−2.18E−3	3 6	−3.91E−2	3 1	5.52E−3	1 4	−2.93E+0	2 3	−7.83E−3		

Ti-like Fe (Fe^{4+})

Term energies relative to $3d^3$ ^{4}F ionization threshold for each symmetry

i	E(Ryds)	Description
	^{5}P^e	
1	−3.57511	$3d^3$ ^{4}P $4s$
2	−2.36547	$3d^3$ ^{4}F $4d$
3	−2.14849	$3d^3$ ^{4}P $4d$
4	−1.78103	$3d^3$ ^{4}P $5s$
5	−1.29599	$3d^3$ ^{4}F $5d$
6	−1.11920	$3d^3$ ^{4}P $5d$
7	−1.01617	$3d^3$ ^{4}P $6s$
8	−1.00228	$3d^3$ ^{4}F $5g$
9	−.85766	$3d^3$ ^{4}F $6d$
10	−.69584	$3d^3$ ^{4}F $6g$
11	−.68708	$3d^3$ ^{4}P $6d$
12	−.62976	$3d^3$ ^{4}P $7s$
13	−.61035	$3d^3$ ^{4}F $7d$
14	−.51114	$3d^3$ ^{4}F $7g$
15	−.45647	$3d^3$ ^{4}F $8d$
16	−.44225	$3d^3$ ^{4}P $7d$
17	−.40725	$3d^3$ ^{4}P $8s$
18	−.39128	$3d^3$ ^{4}F $8g$
19	−.35395	$3d^3$ ^{4}F $9d$
20	−.30912	$3d^3$ ^{4}F $9g$
21	−.28954	$3d^3$ ^{4}P $8d$
	^{5}D^e	
1	−5.55734	$3d^4$
2	−2.25378	$3d^3$ ^{4}P $4d$:
3	−2.09347	$3d^3$ ^{4}F $4d$
4	−1.25057	$3d^3$ ^{4}F $5d$
5	−1.11574	$3d^3$ ^{4}P $5d$
6	−1.00200	$3d^3$ ^{4}F $5g$
7	−.83178	$3d^3$ ^{4}F $6d$
8	−.69571	$3d^3$ ^{4}F $6g$
9	−.68627	$3d^3$ ^{4}P $6d$
10	−.59440	$3d^3$ ^{4}F $7d$
11	−.51105	$3d^3$ ^{4}F $7g$
12	−.45039	$3d^3$ ^{4}F $8d$
13	−.43801	$3d^3$ ^{4}P $7d$
14	−.39121	$3d^3$ ^{4}F $8g$
15	−.34743	$3d^3$ ^{4}F $9d$
16	−.30907	$3d^3$ ^{4}F $9g$
17	−.28984	$3d^3$ ^{4}P $8d$
	^{5}F^e	
1	−3.74789	$3d^3$ ^{4}F $4s$
2	−2.35877	$3d^3$ ^{4}F $4d$
3	−2.17366	$3d^3$ ^{4}P $4d$
4	−1.94470	$3d^3$ ^{4}F $5s$
5	−1.29289	$3d^3$ ^{4}F $5d$
6	−1.18040	$3d^3$ ^{4}F $6s$
7	1.12710	$3d^3$ ^{4}P $5d$
8	−1.00120	$3d^3$ ^{4}F $5g$
9	−.85609	$3d^3$ ^{4}F $6d$
10	−.83331	$3d^3$ ^{4}P $5g$

i	E(Ryds)	Description
11	−.79414	$3d^3$ ^{4}F $7s$
12	−.69530	$3d^3$ ^{4}F $6g$
13	−.69126	$3d^3$ ^{4}P $6d$
14	−.60946	$3d^3$ ^{4}F $7d$
15	−.57184	$3d^3$ ^{4}F $8s$
16	−.52899	$3d^3$ ^{4}P $6g$
17	−.51077	$3d^3$ ^{4}F $7g$
18	−.51033	$3d^3$ ^{4}F $7i$
19	−.45609	$3d^3$ ^{4}F $8d$
20	−.44448	$3d^3$ ^{4}P $7d$
21	−.43142	$3d^3$ ^{4}F $9s$
22	−.39104	$3d^3$ ^{4}F $8g$
23	−.39074	$3d^3$ ^{4}F $8i$
24	−.35357	$3d^3$ ^{4}F $9d$
25	−.34507	$3d^3$ ^{4}P $7g$
	^{5}G^e	
1	−2.34902	$3d^3$ ^{4}F $4d$
2	−1.29101	$3d^3$ ^{4}F $5d$
3	−1.00010	$3d^3$ ^{4}F $5g$
4	−.85522	$3d^3$ ^{4}F $6d$
5	−.83996	$3d^3$ ^{4}P $5g$
6	−.69469	$3d^3$ ^{4}F $6g$
7	.60006	$3d^3$ ^{4}F $7d$
8	−.53248	$3d^3$ ^{4}P $6g$
9	−.51035	$3d^3$ ^{4}F $7g$:
10	−.51035	$3d^3$ ^{4}F $7i$
11	−.45547	$3d^3$ ^{4}F $8d$
12	−.39081	$3d^3$ ^{4}F $8g$
13	−.39069	$3d^3$ ^{4}F $8i$
14	−.35336	$3d^3$ ^{4}F $9d$
15	−.34724	$3d^3$ ^{4}P $7g$
16	−.30877	$3d^3$ ^{4}F $9g$
17	−.30869	$3d^3$ ^{4}F $9i$
	^{5}H^e	
1	−2.37465	$3d^3$ ^{4}F $4d$
2	−1.29818	$3d^3$ ^{4}F $5d$
3	−.99947	$3d^3$ ^{4}F $5g$
4	−.85895	$3d^3$ ^{4}F $6d$
5	−.83470	$3d^3$ ^{4}P $5g$
6	−.69439	$3d^3$ ^{4}F $6g$
7	.61125	$3d^3$ ^{4}F $7d$
8	−.52982	$3d^3$ ^{4}P $6g$
9	−.51021	$3d^3$ ^{4}F $7g$
10	−.45684	$3d^3$ ^{4}F $8d$
11	−.39063	$3d^3$ ^{4}F $7i$:
12	−.39063	$3d^3$ ^{4}F $8i$:
13	−.35428	$3d^3$ ^{4}F $9d$
14	−.34558	$3d^3$ ^{4}P $7g$
15	−.34528	$3d^3$ ^{4}P $7i$
16	−.30870	$3d^3$ ^{1}F $9g$:

i	E(Ryds)	Description
	^{5}I^e	
1	−1.00031	$3d^3$ ^{4}F $5g$
2	−.69493	$3d^3$ ^{4}F $6g$
3	−.51061	$3d^3$ ^{4}F $7g$
4	−.51015	$3d^3$ ^{4}F $7i$
5	−.39094	$3d^3$ ^{4}F $8g$
6	−.39058	$3d^3$ ^{4}F $8i$
7	−.34574	$3d^3$ ^{4}P $7i$
8	−.30861	$3d^3$ ^{4}F $9i$
	^{5}J^e	
1	−1.00185	$3d^3$ ^{4}F $5g$
2	−.69570	$3d^3$ ^{4}F $6g$
3	−.51109	$3d^3$ ^{4}F $7g$
4	−.51013	$3d^3$ ^{4}F $7i$
5	−.39126	$3d^3$ ^{4}F $8g$
6	−.39056	$3d^3$ ^{4}F $8i$
7	−.34536	$3d^3$ ^{4}P $7i$
8	−.30911	$3d^3$ ^{4}F $9g$
	^{5}S^o	
1	−2.91600	$3d^3$ ^{4}P $4p$
2	−1.60869	$3d^3$ ^{4}F $4f$
3	−1.48307	$3d^3$ ^{4}P $5p$
4	−1.13705	$3d^24s$ ^{4}P $4p$:
5	−1.03055	$3d^3$ ^{4}F $5f$
6	−.87052	$3d^3$ ^{4}P $6p$
7	−.71392	$3d^3$ ^{4}F $6f$
8	−.54742	$3d^3$ ^{4}P $7p$
9	−.52300	$3d^3$ ^{4}F $7f$
10	−.39963	$3d^3$ ^{4}F $8f$
11	−.35606	$3d^3$ ^{4}P $8p$
12	−.31505	$3d^3$ ^{4}F $9f$
	^{5}P^o	
1	−3.00075	$3d^3$ ^{4}P $4p$
2	−1.60215	$3d^3$ ^{4}F $4f$
3	−1.50946	$3d^3$ ^{4}P $5p$
4	−1.22594	$3d^24s$ ^{4}P $4p$:
5	−1.02705	$3d^3$ ^{4}F $5f$
6	−.88264	$3d^3$ ^{4}P $6p$
7	−.81174	$3d^24s$ ^{4}F $4f$:
8	−.70910	$3d^3$ ^{4}F $6f$
9	−.55339	$3d^3$ ^{4}P $7p$
10	−.52215	$3d^3$ ^{4}F $7f$
11	−.39956	$3d^3$ ^{1}F $8f$
12	−.35989	$3d^3$ ^{4}P $8p$
13	−.31541	$3d^3$ ^{1}F $9f$
	^{5}D^o	
1	−3.15178	$3d^3$ ^{4}F $4p$
2	−2.97473	$3d^3$ ^{4}P $4p$
3	−1.66622	$3d^3$ ^{4}F $5p$
4	−1.59536	$3d^3$ ^{4}F $4f$
5	−1.50153	$3d^3$ ^{4}P $5p$

i	E(Ryds)	Description
6	−1.42309	$3d^3$ ^{4}P $4f$
7	−1.13794	$3d^24s$ ^{4}F $4p$:
8	−1.04415	$3d^3$ ^{4}F $6p$
9	−1.02175	$3d^3$ ^{4}F $5f$
10	−.87924	$3d^3$ ^{4}P $6p$
11	−.85728	$3d^3$ ^{4}P $5f$
12	−.76804	$3d^24s$ ^{4}P $4p$:
13	−.71529	$3d^3$ ^{4}F $7p$
14	−.69998	$3d^3$ ^{4}F $6f$
15	−.69500	$3d^3$ ^{4}F $6h$
16	−.56083	$3d^3$ ^{4}P $6f$
17	−.55208	$3d^3$ ^{4}P $7p$
18	−.53406	$3d^3$ ^{4}F $7f$
19	−.52250	$3d^3$ ^{4}F $8p$
20	−.51054	$3d^3$ ^{4}F $7h$
21	−.45816	$3d^24s$ ^{4}F $4f$:
22	−.39903	$3d^3$ ^{4}F $9p$
23	−.39085	$3d^3$ ^{4}F $8h$
24	−.37912	$3d^3$ ^{4}F $8f$
25	−.35910	$3d^3$ ^{4}P $8p$
26	−.35018	$3d^3$ ^{4}P $7f$
	^{5}F^o	
1	−3.14121	$3d^3$ ^{4}F $4p$
2	−1.66182	$3d^3$ ^{4}F $5p$
3	−1.59343	$3d^3$ ^{4}F $4f$
4	−1.44904	$3d^3$ ^{4}P $4f$
5	−1.11458	$3d^24s$ ^{4}F $4p$:
6	−1.04096	$3d^3$ ^{4}F $6p$
7	−1.01620	$3d^3$ ^{4}F $5f$
8	−.87679	$3d^3$ ^{4}P $5f$
9	−.72618	$3d^3$ ^{4}F $6f$
10	−.71506	$3d^3$ ^{4}F $7p$
11	−.70899	$3d^24s$ ^{4}F $4f$:
12	−.69482	$3d^3$ ^{4}F $6h$
13	−.61686	$3d^24s$ ^{4}P $4f$:
14	−.53283	$3d^3$ ^{4}P $6f$
15	−.52267	$3d^3$ ^{4}F $8p$
16	−.51046	$3d^3$ ^{4}F $7h$
17	−.48831	$3d^3$ ^{4}F $7f$
18	−.39892	$3d^3$ ^{4}F $9p$
19	−.39082	$3d^3$ ^{4}F $8h$
20	−.39071	$3d^3$ ^{4}F $8f$
21	−.34800	$3d^3$ ^{4}P $7f$
	^{5}G^o	
1	−3.18143	$3d^3$ ^{4}F $4p$
2	−1.67192	$3d^3$ ^{4}F $5p$
3	−1.59797	$3d^3$ ^{4}F $4f$
4	−1.43196	$3d^3$ ^{4}P $4f$
5	−1.21506	$3d^24s$ ^{4}F $4p$:
6	−1.04598	$3d^3$ ^{4}F $6p$
7	−1.02502	$3d^3$ ^{4}F $5f$

Ti-like Fe (Fe^{4+})

i	E(Ryds)	Description
8	−.86072	$3d^3$ ^{4}P $5f$
9	−.78810	$3d^24s$ ^{4}F $4f$:
10	−.71760	$3d^3$ ^{4}F $7p$
11	−.71079	$3d^3$ ^{4}F $6f$
12	−.69455	$3d^3$ ^{4}F $6h$
13	−.54660	$3d^3$ ^{4}P $6f$
14	−.52902	$3d^3$ ^{4}P $6h$
15	−.52422	$3d^3$ ^{4}F $8p$
16	−.52091	$3d^3$ ^{4}F $7f$
17	−.51029	$3d^3$ ^{4}F $7h$
18	−.40002	$3d^3$ ^{4}F $9p$
19	−.39838	$3d^3$ ^{4}F $8f$
20	−.39070	$3d^3$ ^{4}F $8h$
21	−.35658	$3d^3$ ^{4}P $7f$
22	−.34508	$3d^3$ ^{4}P $7h$
		^{5}H^o
1	−1.59972	$3d^3$ ^{4}F $4f$
2	−1.05365	$3d^24s$ ^{4}F $4f$:
3	−1.02004	$3d^3$ ^{4}F $5f$
4	−.71102	$3d^3$ ^{4}F $6f$
5	−.69426	$3d^3$ ^{4}F $6h$
6	−.53094	$3d^3$ ^{4}P $6h$
7	−.52142	$3d^3$ ^{4}F $7f$
8	−.51010	$3d^3$ ^{4}F $7h$
9	−.39845	$3d^3$ ^{4}F $8f$
10	−.39064	$3d^3$ ^{4}F $8j$
11	−.39058	$3d^3$ ^{4}F $8h$
12	−.34619	$3d^3$ ^{4}P $7h$
13	−.31428	$3d^3$ ^{4}F $9f$
14	−.30865	$3d^3$ ^{4}F $9h$:
15	−.30865	$3d^3$ ^{4}F $9j$
		^{5}I^o
1	−1.60771	$3d^3$ ^{4}F $4f$
2	−1.03039	$3d^3$ ^{4}F $5f$
3	−.71396	$3d^3$ ^{4}F $6f$
4	−.69414	$3d^3$ ^{4}F $6h$
5	−.52935	$3d^3$ ^{4}P $6h$
6	−.52319	$3d^3$ ^{4}F $7f$
7	−.51002	$3d^3$ ^{4}F $7h$
8	−.39961	$3d^3$ ^{4}F $8f$
9	−.39052	$3d^3$ ^{4}F $8h$:
10	−.39052	$3d^3$ ^{4}F $8j$
11	−.34527	$3d^3$ ^{4}P $7h$
12	−.31509	$3d^3$ ^{4}F $9f$
13	−.30864	$3d^3$ ^{4}F $9j$
14	−.30857	$3d^3$ ^{4}F $9h$
		^{5}J^o
1	−.69436	$3d^3$ ^{4}F $6h$
2	−.51018	$3d^3$ ^{4}F $7h$
3	−.39063	$3d^3$ ^{4}F $8h$
4	−.30863	$3d^3$ ^{4}F $8j$

i	E(Ryds)	Description
		^{3}S^e
1	−2.11429	$3d^3(^3)$ ^{2}D $4d$
2	−1.61331	$3d^3(^1)$ ^{2}D $4d$
3	−1.05947	$3d^3(^3)$ ^{2}D $5d$
4	−.81838	$3d^3$ ^{2}G $5g$
5	−.64234	$3d^3(^1)$ ^{2}D $5d$
6	−.61500	$3d^3(^3)$ ^{2}D $6d$
7	−.51347	$3d^3$ ^{2}G $6g$
8	−.36347	$3d^3(^3)$ ^{2}D $7d$
9	−.32938	$3d^3$ ^{2}G $7g$
10	−.21102	$3d^3(^1)$ ^{2}D $6d$
11	−.20974	$3d^3$ ^{2}G $8g$
12	−.20709	$3d^3(^3)$ ^{2}D $8d$
13	−.12788	$3d^3$ ^{2}G $9g$
14	−.10530	$3d^3(^3)$ ^{2}D $9d$
		^{3}P^e
1	−5.34202	$3d^4$
2	−4.98047	$3d^4(^2)$
3	−3.49530	$3d^3$ ^{4}P $4s$
4	−3.48158	$3d^3$ ^{2}P $4s$
5	−2.29660	$3d^3$ ^{4}F $4d$
6	−2.09769	$3d^3$ ^{4}P $4d$
7	−2.06153	$3d^3$ ^{2}P $4d$
8	−1.94705	$3d^3(^3)$ ^{2}D $4d$
9	−1.76508	$3d^3$ ^{4}P $5s$
10	−1.75544	$3d^3$ ^{2}F $4d$
11	−1.71516	$3d^3$ ^{2}P $5s$
12	−1.56558	$3d^3(^1)$ ^{2}D $4d$:
13	−1.27333	$3d^3$ ^{4}F $5d$
14	−1.09861	$3d^3$ ^{4}P $5d$
15	−1.05286	$3d^3$ ^{2}P $5d$
16	−1.02073	$3d^3(^3)$ ^{2}D $5d$
17	−1.01051	$3d^3$ ^{4}P $6s$
18	−1.00201	$3d^3$ ^{4}F $5g$
19	−.95313	$3d^3$ ^{2}P $6s$
20	−.84884	$3d^3$ ^{2}F $5d$
21	−.84228	$3d^3$ ^{4}F $6d$
22	−.81854	$3d^3$ ^{2}G $5g$
23	−.74194	$3d^3$ ^{2}H $5g$:
24	−.69558	$3d^3$ ^{4}F $6g$
25	−.67460	$3d^3$ ^{4}P $6d$
26	−.63537	$3d^3$ ^{2}P $6d$:
27	−.62730	$3d^3$ ^{4}P $7s$
28	−.62032	$3d^3(^1)$ ^{2}D $5d$
29	−.60686	$3d^3(^3)$ ^{2}D $6d$:
30	−.58642	$3d^3(^3)$ ^{2}D $5g$:
31	−.58485	$3d^3$ ^{4}F $7d$:
32	−.56670	$3d^3$ ^{2}P $7s$
33	−.51351	$3d^3$ ^{2}G $6g$
34	−.51096	$3d^3$ ^{4}F $7g$
35	−.45035	$3d^3$ ^{4}F $8d$
36	−.44428	$3d^3$ ^{4}P $7d$

i	E(Ryds)	Description
37	−.43878	$3d^3$ ^{2}H $6g$:
38	−.41808	$3d^3$ ^{2}F $6d$
39	−.40520	$3d^3$ ^{4}P $8s$
40	−.39116	$3d^3$ ^{4}F $8g$
41	−.37544	$3d^3$ ^{2}P $7d$
42	−.35661	$3d^3(^3)$ ^{2}D $7d$:
43	−.34486	$3d^3$ ^{4}F $9d$
44	−.34284	$3d^3$ ^{2}P $8s$
45	−.32941	$3d^3$ ^{2}G $7g$
46	−.30902	$3d^3$ ^{4}F $9g$
47	−.28327	$3d^3$ ^{4}P $8d$
48	−.28173	$3d^3$ ^{2}F $6g$
		^{3}D^e
1	−5.20283	$3d^4$
2	−3.46715	$3d^3(^3)$ ^{2}D $4s$
3	−3.03804	$3d^3(^1)$ ^{2}D $4s$
4	−2.36182	$3d^3$ ^{4}F $4d$
5	−2.17461	$3d^3$ ^{4}P $4d$
6	−2.12850	$3d^3$ ^{2}G $4d$
7	−2.09767	$3d^3$ ^{2}P $4d$
8	−1.96379	$3d^3(^3)$ ^{2}D $4d$
9	−1.86038	$3d^3$ ^{2}F $4d$
10	−1.69502	$3d^3(^3)$ ^{2}D $5s$
11	−1.64781	$3d^3(^1)$ ^{2}D $4d$
12	−1.30395	$3d^3(^1)$ ^{2}D $5s$
13	−1.29724	$3d^3$ ^{4}F $5d$
14	−1.13231	$3d^3$ ^{4}P $5d$
15	−1.10040	$3d^3$ ^{2}G $5d$
16	−1.06669	$3d^3$ ^{2}P $5d$
17	−1.02164	$3d^3(^3)$ ^{2}D $5d$
18	−1.00198	$3d^3$ ^{4}F $5g$
19	−.93278	$3d^3(^3)$ ^{2}D $6s$
20	−.88017	$3d^3$ ^{2}F $5d$
21	−.85630	$3d^3$ ^{4}F $6d$
22	−.81891	$3d^3$ ^{2}G $5g$
23	−.75369	$3d^3$ ^{2}H $5g$
24	−.74314	$3d^3(^3)$ ^{2}D $5g$
25	−.69574	$3d^3$ ^{4}F $6g$
26	−.69327	$3d^3$ ^{4}P $6d$
27	−.66987	$3d^3$ ^{2}G $6d$
28	−.65388	$3d^3(^1)$ ^{2}D $5d$
29	−.62862	$3d^3$ ^{2}P $6d$
30	−.61147	$3d^3$ ^{4}F $7d$
31	−.59492	$3d^3(^3)$ ^{2}D $6d$
32	−.58752	$3d^3$ ^{2}F $5g$
33	−.54648	$3d^3(^3)$ ^{2}D $7s$
34	−.53952	$3d^3(^1)$ ^{2}D $6s$
35	−.51366	$3d^3$ ^{2}G $6g$
36	−.51107	$3d^3$ ^{4}F $7g$
37	−.45660	$3d^3$ ^{4}F $8d$
38	−.44781	$3d^3$ ^{2}F $6d$
39	−.44683	$3d^3$ ^{2}H $6g$

i	E(Ryds)	Description
40	−.43995	$3d^3$ ^{4}P $7d$
41	−.43940	$3d^3(^3)$ ^{2}D $6g$
42	−.42075	$3d^3$ ^{2}G $7d$
43	−.39124	$3d^3$ ^{4}F $8g$
44	−.37936	$3d^3$ ^{2}P $7d$
45	−.35706	$3d^3(^3)$ ^{2}D $7d$:
46	−.35192	$3d^3(^1)$ ^{2}D $5g$
47	−.35092	$3d^3$ ^{4}F $9d$
48	−.32951	$3d^3$ ^{2}G $7g$
49	−.32927	$3d^3$ ^{2}G $7i$
50	−.32283	$3d^3(^3)$ ^{2}D $8s$
51	−.30909	$3d^3$ ^{4}F $9g$
52	−.28998	$3d^3$ ^{4}P $8d$
53	−.28235	$3d^3$ ^{2}F $6g$
		^{3}F^e
1	−5.31424	$3d^4$
2	−4.98166	$3d^4(^2)$
3	−3.66285	$3d^3$ ^{4}F $4s$
4	−3.30217	$3d^3$ ^{2}F $4s$
5	−2.28227	$3d^3$ ^{4}F $4d$
6	−2.12632	$3d^3$ ^{4}P $4d$
7	−2.11588	$3d^3$ ^{2}G $4d$
8	−2.07092	$3d^3$ ^{2}P $4d$
9	−1.99411	$3d^3$ ^{2}H $4d$
10	−1.97522	$3d^24s^2$:
11	−1.92485	$3d^3$ ^{4}F $5s$
12	−1.84702	$3d^3(^3)$ ^{2}D $4d$:
13	−1.58655	$3d^3$ ^{2}F $4d$:
14	−1.53331	$3d^3$ ^{2}F $5s$
15	−1.26489	$3d^3$ ^{4}F $5d$
16	−1.17489	$3d^3$ ^{4}F $6s$
17	−1.11032	$3d^3$ ^{4}P $5d$
18	−1.09566	$3d^3$ ^{2}G $5d$
19	−1.05669	$3d^3$ ^{2}P $5d$
20	−1.03678	$3d^3$ ^{2}H $5d$
21	−1.00931	$3d^3(^3)$ ^{2}D $5d$
22	−1.00123	$3d^3$ ^{4}F $5g$
23	−.87831	$3d^3$ ^{2}F $5d$
24	−.84035	$3d^3$ ^{4}F $6d$
25	−.83295	$3d^3$ ^{4}P $5g$
26	−.81947	$3d^3$ ^{2}G $5g$
27	−.79193	$3d^3$ ^{4}F $7s$
28	−.77280	$3d^3$ ^{2}P $5g$
29	−.77093	$3d^3$ ^{2}F $6s$
30	−.75127	$3d^3$ ^{2}H $5g$
31	−.74484	$3d^3(^3)$ ^{2}D $5g$
32	−.69533	$3d^3$ ^{4}F $6g$
33	−.68344	$3d^3$ ^{4}P $6d$
34	−.66615	$3d^3$ ^{2}G $6d$
35	−.64307	$3d^3(^1)$ ^{2}D $5d$
36	−.62573	$3d^3$ ^{2}P $6d$
37	−.60730	$3d^3$ ^{2}H $6d$

Ti-like Fe (Fe^{4+})

i	E(Ryds)	Description
38	−.59544	$3d^3$ ^{4}F $7d$
39	−.58865	$3d^3$ ^{2}F $5g$
40	−.58153	$3d^3(^3)$ ^{2}D $6d$
41	−.56985	$3d^3$ ^{4}F $8s$
42	−.52866	$3d^3$ ^{4}P $6g$
43	−.51394	$3d^3$ ^{2}G $6g$
44	−.51082	$3d^3$ ^{4}F $7g$
45	−.51033	$3d^3$ ^{4}F $7i$
46	−.46648	$3d^3$ ^{2}P $6g$
47	−.45049	$3d^3$ ^{4}F $8d$
48	−.44718	$3d^3$ ^{2}F $6d$
49	−.44560	$3d^3$ ^{2}H $6g$
50	−.44029	$3d^3(^3)$ ^{2}D $6g$
51	−.43228	$3d^3$ ^{4}P $7d$
52	−.42975	$3d^3$ ^{4}F $9s$
53	−.41998	$3d^3$ ^{2}G $7d$
54	−.39110	$3d^3$ ^{4}F $8g$
55	−.39074	$3d^3$ ^{4}F $8i$
56	−.38443	$3d^3$ ^{2}F $7s$
57	−.37853	$3d^3$ ^{2}P $7d$
58	−.35802	$3d^3$ ^{2}H $7d$
59	−.35626	$3d^3(^1)$ ^{2}D $5g$
60	−.34938	$3d^3$ ^{4}F $9d$
61	−.34486	$3d^3$ ^{4}P $7g$
62	−.34477	$3d^3(^3)$ ^{2}D $7d$
	^{3}G^e	
1	−5.27045	$3d^4$
2	−3.54257	$3d^3$ ^{2}G $4s$
3	−2.29355	$3d^3$ ^{4}F $4d$
4	−2.16730	$3d^3$ ^{2}G $4d$
5	−2.08991	$3d^3$ ^{2}H $4d$
6	−2.00173	$3d^3(^3)$ ^{2}D $4d$
7	−1.86931	$3d^3$ ^{2}F $4d$
8	−1.76309	$3d^3$ ^{2}G $5s$
9	−1.63249	$3d^3(^1)$ ^{2}D $4d$
10	−1.27128	$3d^3$ ^{4}F $5d$
11	−1.11485	$3d^3$ ^{2}G $5d$
12	−1.05084	$3d^3$ ^{2}H $5d$
13	−1.02783	$3d^3(^3)$ ^{2}D $5d$
14	−1.00138	$3d^3$ ^{2}G $6s$
15	−.99964	$3d^3$ ^{4}F $5g$
16	−.88423	$3d^3$ ^{2}F $5d$
17	−.83985	$3d^3$ ^{4}P $5g$
18	−.83907	$3d^3$ ^{4}F $6d$
19	−.82017	$3d^3$ ^{2}G $5g$
20	−.77051	$3d^3$ ^{2}P $5g$
21	.74980	$3d^3$ ^{2}H $5g$
22	−.74685	$3d^3(^3)$ ^{2}D $5g$
23	−.69451	$3d^3$ ^{4}F $6g$
24	−.67682	$3d^3$ ^{2}G $6d$
25	−.64923	$3d^3(^1)$ ^{2}D $5d$
26	−.61567	$3d^3$ ^{2}G $7s$

i	E(Ryds)	Description
27	−.61195	$3d^3$ ^{2}H $6d$
28	−.60756	$3d^3(^3)$ ^{2}D $6d$
29	−.58949	$3d^3$ ^{2}F $5g$
30	−.58885	$3d^3$ ^{4}F $7d$
31	−.53241	$3d^3$ ^{4}P $6g$
32	−.51435	$3d^3$ ^{2}G $6g$
33	−.51028	$3d^3$ ^{4}F $7i$
34	−.51025	$3d^3$ ^{4}F $7g$
35	−.46537	$3d^3$ ^{2}P $6g$
36	−.45328	$3d^3$ ^{4}F $8d$
37	−.44483	$3d^3$ ^{2}H $6g$
38	−.44337	$3d^3$ ^{2}F $6d$
39	−.44131	$3d^3(^3)$ ^{2}D $6g$
40	−.42496	$3d^3$ ^{2}G $7d$
41	−.39157	$3d^3$ ^{4}F $8g$
42	−.36101	$3d^3$ ^{2}H $7d$
43	−.35997	$3d^3(^1)$ ^{2}D $5g$
44	−.35546	$3d^3(^3)$ ^{2}D $7d$
45	−.34718	$3d^3$ ^{4}P $7g$
46	−.34576	$3d^3$ ^{4}F $9d$
47	−.32989	$3d^3$ ^{2}G $7g$
48	−.32937	$3d^3$ ^{2}G $7i$
49	−.30869	$3d^3$ ^{4}F $9g$
50	−.30869	$3d^3$ ^{4}F $9i$
51	−.28380	$3d^3$ ^{2}F $6g$
52	−.28090	$3d^3$ ^{2}P $7g$
	^{3}H^e	
1	−5.31761	$3d^4$
2	−3.47108	$3d^3$ ^{2}H $4s$
3	−2.30184	$3d^3$ ^{4}F $4d$
4	−2.15468	$3d^3$ ^{2}G $4d$
5	−1.96984	$3d^3$ ^{2}H $4d$
6	−1.88654	$3d^3$ ^{2}F $4d$
7	−1.69084	$3d^3$ ^{2}H $5s$
8	−1.27531	$3d^3$ ^{4}F $5d$
9	−1.11349	$3d^3$ ^{2}G $5d$
10	−1.02045	$3d^3$ ^{2}H $5d$:
11	−.99904	$3d^3$ ^{4}F $5g$
12	−.92878	$3d^3$ ^{2}H $6s$
13	−.88282	$3d^3$ ^{2}F $5d$
14	−.84246	$3d^3$ ^{4}F $6d$
15	−.83430	$3d^3$ ^{4}F $5g$
16	.82070	$3d^3$ ^{2}G $5g$
17	−.77232	$3d^3$ ^{2}P $5g$
18	−.74964	$3d^3$ ^{2}H $5g$
19	−.74879	$3d^3(^3)$ ^{2}D $5g$
20	−.69398	$3d^3$ ^{4}F $6g$
21	−.67623	$3d^3$ ^{2}G $6d$
22	−.60960	$3d^3$ ^{4}F $7d$
23	−.58991	$3d^3$ ^{2}F $5g$
24	−.58584	$3d^3$ ^{2}H $6d$
25	−.54243	$3d^3$ ^{2}H $7s$

i	E(Ryds)	Description
26	−.52942	$3d^3$ ^{4}P $6g$
27	−.51473	$3d^3$ ^{2}G $6g$
28	−.51021	$3d^3$ ^{4}F $7i$
29	−.50985	$3d^3$ ^{4}F $7g$
30	−.46628	$3d^3$ ^{2}P $6g$
31	−.45282	$3d^3$ ^{4}F $8d$
32	−.44482	$3d^3$ ^{2}H $6g$
33	−.44312	$3d^3$ ^{2}F $6d$
34	−.44230	$3d^3(^3)$ ^{2}D $6g$
35	−.42648	$3d^3$ ^{2}G $7d$
36	−.39063	$3d^3$ ^{4}F $8i$
37	−.39048	$3d^3$ ^{4}F $8g$
38	−.36001	$3d^3(^1)$ ^{2}D $5g$
39	−.35519	$3d^3$ ^{4}F $9d$
40	−.34411	$3d^3$ ^{2}H $7d$
41	−.33011	$3d^3$ ^{2}G $7g$
42	−.32941	$3d^3$ ^{2}G $7i$
43	−.31877	$3d^3$ ^{2}H $8s$
44	−.30865	$3d^3$ ^{4}F $9i$
45	−.28396	$3d^3$ ^{2}F $6g$
46	−.28153	$3d^3$ ^{2}P $7g$
47	−.28113	$3d^3$ ^{2}P $7i$
	^{3}I^e	
1	−2.17380	$3d^3$ ^{2}G $4d$
2	−2.11219	$3d^3$ ^{2}H $4d$
3	−1.12055	$3d^3$ ^{2}G $5d$
4	−1.05467	$3d^3$ ^{2}H $5d$:
5	−.99995	$3d^3$ ^{4}F $5g$
6	−.82095	$3d^3$ ^{2}G $5g$
7	−.75264	$3d^3$ ^{2}H $5g$
8	−.75010	$3d^3(^3)$ ^{2}D $5g$
9	−.69459	$3d^3$ ^{4}F $6g$
10	−.68008	$3d^3$ ^{2}G $6d$
11	−.61107	$3d^3$ ^{2}H $6d$:
12	−.58968	$3d^3$ ^{2}F $5g$
13	−.51484	$3d^3$ ^{2}G $6g$
14	−.51035	$3d^3$ ^{4}F $7g$
15	−.51015	$3d^3$ ^{4}F $7i$
16	−.44631	$3d^3$ ^{2}H $6g$
17	−.44301	$3d^3(^3)$ ^{2}D $6g$
18	−.42976	$3d^3$ ^{2}G $7d$
19	.39076	$3d^3$ ^{4}F $8g$
20	−.39058	$3d^3$ ^{4}F $8i$
21	−.35926	$3d^3$ ^{2}H $7d$
22	−.35435	$3d^3(^1)$ ^{2}D $5g$
23	−.34574	$3d^3$ ^{4}P $7i$
24	−.33020	$3d^3$ ^{2}G $7g$
25	−.32945	$3d^3$ ^{2}G $7i$
26	−.30861	$3d^3$ ^{4}F $9i$
27	−.28360	$3d^3$ ^{2}F $6g$
28	−.28099	$3d^3$ ^{2}P $7i$
29	−.27525	$3d^3$ ^{2}G $8d$

i	E(Ryds)	Description
30	−.26161	$3d^3$ ^{2}H $7g$
31	−.26062	$3d^3$ ^{2}H $7i$
32	−.25802	$3d^3(^3)$ ^{2}D $7g$
33	−.25678	$3d^3(^3)$ ^{2}D $7i$
	^{3}J^e	
1	−2.07968	$3d^3$ ^{2}H $4d$
2	−1.04279	$3d^3$ ^{2}H $5d$
3	−1.00145	$3d^3$ ^{4}F $5g$
4	−.82035	$3d^3$ ^{2}G $5g$
5	−.75019	$3d^3$ ^{2}H $5g$:
6	−.69532	$3d^3$ ^{4}F $6g$
7	−.60497	$3d^3$ ^{2}H $6d$:
8	−.58766	$3d^3$ ^{2}F $5g$
9	−.51447	$3d^3$ ^{2}G $6g$
10	−.51080	$3d^3$ ^{4}F $7g$
11	−.51013	$3d^3$ ^{4}F $7i$
12	−.44309	$3d^3$ ^{2}H $6g$:
13	−.39105	$3d^3$ ^{4}F $8g$
14	−.39056	$3d^3$ ^{4}F $8i$
15	−.35570	$3d^3$ ^{2}H $7d$
16	−.34536	$3d^3$ ^{4}P $7i$
17	−.33003	$3d^3$ ^{2}G $7g$
18	−.32947	$3d^3$ ^{2}G $7i$
19	−.30895	$3d^3$ ^{4}F $9g$
20	−.28254	$3d^3$ ^{2}F $6g$
21	−.28111	$3d^3$ ^{2}P $7i$
22	−.26064	$3d^3$ ^{2}H $7i$
23	−.25811	$3d^3$ ^{2}H $7g$:
24	−.25686	$3d^3(^3)$ ^{2}D $7i$
	^{3}S^o	
1	−2.89972	$3d^3$ ^{4}P $4p$
2	−2.79695	$3d^3$ ^{2}P $4p$
3	−1.58199	$3d^3$ ^{4}F $4f$
4	−1.45520	$3d^3$ ^{4}P $5p$
5	−1.43906	$3d^3$ ^{2}P $5p$
6	−1.15799	$3d^3$ ^{2}F $4f$
7	−1.01364	$3d^3$ ^{4}F $5f$
8	−.85964	$3d^3$ ^{4}P $6p$:
9	−.81860	$3d^3$ ^{2}P $6p$
10	−.70384	$3d^3$ ^{4}F $6f$
11	−.59874	$3d^3$ ^{2}F $5f$
12	−.54272	$3d^3$ ^{4}P $7p$
13	−.51727	$3d^3$ ^{4}F $7f$
14	−.49108	$3d^3$ ^{2}P $7p$
15	−.43938	$3d^3$ ^{2}H $6h$:
16	−.39575	$3d^3$ ^{4}F $8f$
17	−.35348	$3d^3$ ^{4}P $8p$
18	−.31268	$3d^3$ ^{4}F $9f$
19	−.29704	$3d^3$ ^{2}P $8p$
20	−.29147	$3d^3$ ^{2}F $6f$
21	−.25581	$3d^3$ ^{2}H $7h$.

Ti-like Fe (Fe^{4+})

i	E(Ryds)	Description	i	E(Ryds)	Description	i	E(Ryds)	Description	i	E(Ryds)	Description
	^{3}P°		7	−1.66506	$3d^3$ ^{4}F $5p$	59	−.35012	$3d^3$ ^{4}P $7f$	49	−.44569	$3d^3$ ^{2}H $6h$
1	−2.98734	$3d^3$ ^{4}P $4p$	8	−1.59952	$3d^3$ ^{4}F $4f$	60	−.33286	$3d^3$ ^{2}G $6f$	50	−.44023	$3d^3(^3)$ ^{2}D $6h$
2	−2.94348	$3d^3$ ^{2}P $4p$	9	−1.48410	$3d^3$ ^{4}P $5p$	61	−.32927	$3d^3$ ^{2}G $7h$	51	−.40371	$3d^3(^1)$ ^{2}D $6p$
3	−2.85343	$3d^3(^3)$ ^{2}D $4p$	10	−1.44272	$3d^3$ ^{2}P $5p$		**^{3}F°**		52	−.39907	$3d^3$ ^{4}F $9p$
4	−2.43820	$3d^3(^1)$ ^{2}D $4p$	11	−1.41266	$3d^3(^3)$ ^{2}D $5p$:	1	−3.07143	$3d^3$ ^{4}F $4p$	53	−.39665	$3d^3$ ^{4}F $8f$
5	−1.59395	$3d^3$ ^{4}F $4f$	12	−1.40999	$3d^24s$ ^{4}F $4p$:	2	−2.94365	$3d^3$ ^{2}G $4p$	54	−.39080	$3d^3$ ^{4}F $8h$
6	−1.50897	$3d^3$ ^{4}P $5p$	13	−1.40260	$3d^3$ ^{4}P $4f$	3	−2.89148	$3d^3(^3)$ ^{2}D $4p$:	55	−.38952	$3d^3(^1)$ ^{2}D $5f$
7	−1.44417	$3d^3$ ^{2}P $5p$	14	−1.36957	$3d^3$ ^{2}P $4f$	4	−2.72816	$3d^3$ ^{2}F $4p$	56	−.35696	$3d^3$ ^{4}P $7f$
8	−1.42061	$3d^3(^3)$ ^{2}D $5p$	15	−1.34212	$3d^3$ ^{2}H $4f$	5	−2.47586	$3d^3(^1)$ ^{2}D $4p$	57	−.34392	$3d^3$ ^{2}G $8p$
9	−1.40569	$3d^3$ ^{2}G $4f$	16	−1.31518	$3d^3(^3)$ ^{2}D $4f$	6	−1.64759	$3d^3$ ^{4}F $5p$	58	−.34044	$3d^3$ ^{2}G $7f$
10	−1.35862	$3d^3$ ^{2}H $4f$	17	−1.24901	$3d^3$ ^{2}F $5p$	7	−1.59160	$3d^3$ ^{4}F $4f$	59	−.32935	$3d^3$ ^{2}G $7h$
11	−1.17291	$3d^3$ ^{2}F $4f$	18	−1.17947	$3d^3$ ^{2}F $4f$	8	−1.48310	$3d^3$ ^{2}G $5p$		**^{3}G°**	
12	−1.02401	$3d^3$ ^{4}F $5f$	19	−1.04552	$3d^3$ ^{4}F $6p$	9	−1.44106	$3d^3$ ^{4}P $4f$	1	−3.10430	$3d^3$ ^{4}F $4p$
13	−1.01286	$3d^3(^1)$ ^{2}D $5p$	20	−1.03921	$3d^3(^1)$ ^{2}D $5p$	10	−1.42409	$3d^3(^3)$ ^{2}D $5p$:	2	−2.95357	$3d^3$ ^{2}G $4p$
14	−.93334	$3d^3(^1)$ ^{2}D $4f$	21	−1.02634	$3d^3$ ^{4}F $5f$	11	−1.40663	$3d^3$ ^{2}G $4f$	3	−2.82412	$3d^3$ ^{2}H $4p$
15	−.92703	$3d^24s$ ^{4}F $4p$:	22	−.94433	$3d^3(^1)$ ^{2}D $4f$	12	−1.36812	$3d^3$ ^{2}P $4f$	4	−2.69322	$3d^3$ ^{2}F $4p$
16	−.88474	$3d^3$ ^{4}P $6p$	23	−.87446	$3d^3$ ^{4}P $6p$	13	−1.33858	$3d^3$ ^{2}H $4f$	5	−1.65562	$3d^3$ ^{4}F $5p$
17	−.84586	$3d^3$ ^{2}G $5f$	24	−.86139	$3d^3$ ^{2}G $4f$	14	−1.32887	$3d^3(^3)$ ^{2}D $4f$	6	−1.57344	$3d^3$ ^{4}F $4f$
18	−.81959	$3d^3$ ^{2}P $6p$	25	−.84581	$3d^3$ ^{4}P $5f$	15	−1.26581	$3d^3$ ^{2}F $5p$	7	−1.49164	$3d^3$ ^{2}G $5p$
19	−.81339	$3d^24s$ ^{2}D $4p$:	26	−.81982	$3d^24s$ ^{2}F $4p$:	16	−1.18861	$3d^3$ ^{2}F $4f$	8	−1.41878	$3d^3$ ^{2}G $4f$
20	−.79924	$3d^3(^3)$ ^{2}D $6p$:	27	−.81896	$3d^3$ ^{2}P $6p$	17	−1.03766	$3d^3$ ^{4}F $6p$	9	−1.41376	$3d^3$ ^{4}P $4f$
21	−.77707	$3d^3(^3)$ ^{2}D $5f$:	28	−.79767	$3d^3$ ^{2}P $5f$	18	−1.03313	$3d^3$ ^{4}F $5f$:	10	−1.39904	$3d^3$ ^{2}H $5p$:
22	−.70718	$3d^3$ ^{4}F $6f$	29	−.79385	$3d^3(^3)$ ^{2}D $6p$	19	−1.02430	$3d^3(^1)$ ^{2}D $5p$	11	−1.36982	$3d^3$ ^{2}P $4f$
23	−.60960	$3d^3$ ^{2}F $5f$	30	−.77118	$3d^3$ ^{2}H $5f$	20	−.99511	$3d^24s$ ^{4}F $4p$:	12	−1.34425	$3d^24s$ ^{4}F $4p$:
24	−.55669	$3d^3$ ^{4}P $7p$	31	−.75988	$3d^3(^3)$ ^{2}D $5f$	21	−.96646	$3d^3(^1)$ ^{2}D $4f$	13	−1.33563	$3d^24s$ ^{2}F $4p$:
25	−.52983	$3d^3$ ^{2}G $6f$	32	−.73751	$3d^24s$ ^{4}P $4p$:	22	−.92415	$3d^24s$ ^{2}F $4p$:	14	−1.25585	$3d^3$ ^{2}F $5p$
26	−.51903	$3d^3$ ^{4}F $7f$	33	−.71833	$3d^3$ ^{4}F $7p$	23	−.86926	$3d^3$ ^{4}P $5f$	15	−1.18807	$3d^3$ ^{2}F $4f$
27	−.51328	$3d^3$ ^{2}G $6h$	34	−.70850	$3d^3$ ^{4}F $6f$	24	−.86255	$3d^3$ ^{2}G $6p$	16	−1.04079	$3d^3$ ^{4}F $6p$
28	−.49153	$3d^3$ ^{2}P $7p$	35	−.69500	$3d^3$ ^{4}F $6h$	25	−.83834	$3d^3$ ^{2}G $5f$	17	−1.00955	$3d^3$ ^{4}F $5f$
29	−.47304	$3d^3(^3)$ ^{2}D $7p$	36	−.63231	$3d^3$ ^{2}F $6p$	26	−.81434	$3d^3$ ^{2}P $5f$:	18	−.96343	$3d^3(^1)$ ^{2}D $4f$
30	−.46763	$3d^3(^3)$ ^{2}D $6f$:	37	−.61170	$3d^3$ ^{2}F $5f$	27	−.80058	$3d^3(^3)$ ^{2}D $6p$:	19	−.86925	$3d^3$ ^{2}G $6p$
31	−.43953	$3d^3(^3)$ ^{2}D $6h$	38	−.56720	$3d^24s$ ^{2}P $4p$:	28	−.78451	$3d^3$ ^{2}H $5f$:	20	−.85484	$3d^3$ ^{2}G $5f$
32	−.40008	$3d^3$ ^{4}F $8f$	39	−.55180	$3d^3$ ^{4}P $7p$	29	−.76801	$3d^3(^3)$ ^{2}D $5f$	21	−.84819	$3d^3$ ^{4}P $5f$
33	−.39806	$3d^3(^1)$ ^{2}D $6p$	40	−.53541	$3d^3$ ^{4}P $6f$	30	−.75213	$3d^24s$ ^{2}D $4p$:	22	−.80708	$3d^3$ ^{2}P $5f$
34	−.38320	$3d^3(^1)$ ^{2}D $5f$	41	−.52987	$3d^3$ ^{2}G $5f$	31	−.71501	$3d^3$ ^{4}F $7p$	23	−.78967	$3d^3$ ^{2}H $6p$:
35	−.36146	$3d^3$ ^{4}P $8p$	42	−.52320	$3d^3$ ^{4}F $8p$	32	−.70515	$3d^3$ ^{4}F $6f$	24	−.78076	$3d^3(^3)$ ^{2}D $4f$:
36	−.34178	$3d^3$ ^{2}G $7f$	43	−.52044	$3d^3$ ^{4}F $7f$	33	−.69482	$3d^3$ ^{4}F $6h$	25	−.77364	$3d^3$ ^{2}H $4f$
37	−.33666	$3d^3$ ^{2}G $8f$:	44	−.51337	$3d^3$ ^{2}G $6h$	34	−.65507	$3d^24s$ ^{2}G $4p$:	26	−.74668	$3d^24s$ ^{2}G $4p$:
38	−.32921	$3d^3$ ^{2}G $7h$	45	−.51053	$3d^3$ ^{4}F $7h$	35	−.63962	$3d^3$ ^{2}F $6p$	27	−.71631	$3d^3$ ^{4}F $7p$
39	−.31348	$3d^3$ ^{4}F $9f$	46	−.49364	$3d^3$ ^{2}P $6f$	36	−.60915	$3d^3$ ^{2}F $5f$	28	−.70828	$3d^3$ ^{4}F $6f$
40	−.29875	$3d^3$ ^{2}F $6f$	47	−.49095	$3d^3$ ^{2}P $7p$	37	−.54697	$3d^3$ ^{4}P $6f$	29	−.69454	$3d^3$ ^{4}F $6h$
41	−.29640	$3d^3$ ^{2}P $8p$	48	−.46908	$3d^3(^3)$ ^{2}D $7p$	38	−.53815	$3d^3$ ^{2}G $7p$	30	−.68372	$3d^24p$ ^{4}G $4d$:
42	−.27687	$3d^3(^3)$ ^{2}D $8p$:	49	−.46603	$3d^3$ ^{2}H $6f$	39	−.52862	$3d^3$ ^{2}G $6f$	31	−.63547	$3d^3$ ^{2}F $6p$
43	−.26912	$3d^3$ ^{2}H $7f$	50	−.44695	$3d^3(^3)$ ^{2}D $6f$	40	−.52176	$3d^3$ ^{4}F $8p$	32	−.61857	$3d^3$ ^{2}F $5f$
44	−.25590	$3d^3$ ^{2}H $7h$:	51	−.43982	$3d^3(^3)$ ^{2}D $6h$	41	−.51800	$3d^3$ ^{4}F $7f$	33	−.58290	$3d^24p$ ^{4}F $4d$:
	^{3}D°		52	−.41121	$3d^3(^1)$ ^{2}D $6p$	42	−.51352	$3d^3$ ^{2}G $6h$	34	−.54052	$3d^3$ ^{2}G $7p$
1	−3.14228	$3d^3$ ^{4}F $4p$	53	−.40276	$3d^3$ ^{4}F $8f$	43	−.51045	$3d^3$ ^{4}F $7h$	35	−.53769	$3d^3$ ^{4}P $6f$
2	−2.91581	$3d^3$ ^{2}P $4p$	54	−.39955	$3d^3$ ^{4}F $9p$	44	−.50203	$3d^24p$ ^{4}G $4d$:	36	−.52901	$3d^3$ ^{4}P $6h$
3	−2.88406	$3d^3$ ^{4}P $4p$	55	−.39085	$3d^3$ ^{4}F $8h$	45	−.47956	$3d^3$ ^{2}P $6f$	37	−.52321	$3d^3$ ^{4}F $8p$
4	−2.86195	$3d^3(^3)$ ^{2}D $4p$:	56	−.37971	$3d^3(^1)$ ^{2}D $5f$	46	−.47188	$3d^3(^3)$ ^{2}D $7p$:	38	−.52187	$3d^3$ ^{2}G $6f$
5	−2.68143	$3d^3$ ^{2}F $4p$	57	−.36053	$3d^3$ ^{2}H $7f$:	47	−.45600	$3d^3$ ^{2}H $6f$	39	−.51401	$3d^3$ ^{4}F $7f$
6	−2.51233	$3d^3(^1)$ ^{2}D $4p$	58	−.35815	$3d^3$ ^{4}P $8p$	48	−.45103	$3d^3(^3)$ ^{2}D $6f$	40	−.51367	$3d^3$ ^{2}G $6h$

Ti-like Fe (Fe^{4+})

i	E(Ryds)	Description	i	E(Ryds)	Description	i	E(Ryds)	Description	i	E(Ryds)	Description
41	−.51028	$3d^3$ ^{4}F $7h$	35	−.39233	$3d^3$ ^{4}F $8f$	34	−.30856	$3d^3$ ^{4}F $9h$	6	−1.61822	$3d^3(^1)$ ^{2}D $4d$
42	−.47808	$3d^3$ ^{2}P $6f$	36	−.39057	$3d^3$ ^{4}F $8h$	35	−.29485	$3d^3$ ^{2}F $6f$	7	−1.09143	$3d^3$ ^{2}P $5d$
43	−.46566	$3d^3$ ^{2}P $6h$	37	−.39057	$3d^3$ ^{4}F $8j$	36	−.28290	$3d^3$ ^{2}F $6h$	8	−1.06944	$3d^3(^3)$ ^{2}D $5d$
44	−.46364	$3d^3$ ^{2}H $7p$	38	−.37565	$3d^3(^1)$ ^{2}D $5f$	37	−.28114	$3d^3$ ^{2}P $7h$	9	−.94865	$3d^3$ ^{2}P $6s$
45	−.45831	$3d^3$ ^{2}H $5f$:	39	−.34618	$3d^3$ ^{4}P $7h$	38	−.27213	$3d^3$ ^{2}H $8p$:	10	−.89731	$3d^3$ ^{2}F $5d$
46	−.45550	$3d^3(^3)$ ^{2}D $5f$	40	−.34524	$3d^3$ ^{2}G $8p$	39	−.26555	$3d^3$ ^{2}H $7f$:	11	−.81839	$3d^3$ ^{2}G $5g$
47	−.44502	$3d^3$ ^{2}H $6h$	41	−.34158	$3d^3$ ^{2}G $6f$	40	−.26054	$3d^3$ ^{2}H $7h$	12	−.74206	$3d^3$ ^{2}H $5g$
48	−.44072	$3d^3(^3)$ ^{2}D $6h$	42	−.32955	$3d^3$ ^{2}G $7h$	41	−.25710	$3d^3(^3)$ ^{2}D $7h$	13	−.66115	$3d^3(^1)$ ^{2}D $5d$
49	−.39917	$3d^3$ ^{4}F $9p$	43	−.31385	$3d^3$ ^{4}F $9f$	$^3\mathbf{J}^o$			14	−.63128	$3d^3$ ^{2}P $6d$
50	−.39508	$3d^3$ ^{4}F $8f$	44	−.30864	$3d^3$ ^{4}F $9j$	1	−1.41926	$3d^3$ ^{2}G $4f$	15	−.61283	$3d^3(^3)$ ^{2}D $6d$
51	−.38984	$3d^3(^1)$ ^{2}D $5f$	45	−.30860	$3d^3$ ^{4}F $9h$	2	−1.35497	$3d^3$ ^{2}H $4f$:	16	−.58665	$3d^3$ ^{2}F $5g$
52	−.35784	$3d^3$ ^{2}G $7f$	46	−.29784	$3d^3$ ^{2}F $6f$	3	−.84764	$3d^3$ ^{2}G $5f$	17	−.56425	$3d^3$ ^{2}P $7s$
53	−.35030	$3d^3$ ^{4}P $7f$	47	−.28282	$3d^3$ ^{2}F $6h$	4	−.79926	$3d^24p$ ^{2}H $4d$:	18	−.51337	$3d^3$ ^{2}G $6g$
54	−.34546	$3d^3$ ^{2}G $8p$	48	−.28083	$3d^3$ ^{2}P $7h$	5	−.77724	$3d^3$ ^{2}H $5f$:	19	−.44484	$3d^3$ ^{2}F $6d$
55	−.34508	$3d^3$ ^{4}P $7h$	49	−.27742	$3d^3(^3)$ ^{2}D $7f$	6	−.69435	$3d^3$ ^{4}F $6h$	20	−.43889	$3d^3$ ^{2}H $6g$
56	−.32945	$3d^3$ ^{2}G $7h$	50	−.27416	$3d^3$ ^{2}H $8p$:	7	−.53010	$3d^3$ ^{2}G $6f$	21	−.38235	$3d^3$ ^{2}P $7d$
57	−.31944	$3d^3$ ^{2}F $6f$	51	−.27126	$3d^3$ ^{2}H $7f$	8	−.51396	$3d^3$ ^{2}G $6h$	22	−.36307	$3d^3(^3)$ ^{2}D $7d$
$^3\mathbf{H}^o$			52	−.26052	$3d^3$ ^{2}H $7h$	9	−.51017	$3d^3$ ^{4}F $7h$	23	−.34155	$3d^3$ ^{2}P $8s$
1	−2.97737	$3d^3$ ^{2}G $4p$	$^3\mathbf{I}^o$			10	−.46063	$3d^3$ ^{2}H $6f$:	24	−.32929	$3d^3$ ^{2}G $7g$
2	−2.90386	$3d^3$ ^{2}H $4p$	1	−2.87473	$3d^3$ ^{2}H $4p$	11	−.44541	$3d^3$ ^{2}H $6h$	25	−.28196	$3d^3$ ^{2}F $6g$
3	−1.58213	$3d^3$ ^{4}F $4f$	2	−1.58561	$3d^3$ ^{4}F $4f$	12	−.44185	$3d^3(^3)$ ^{2}D $6h$	26	−.25631	$3d^3$ ^{2}H $7i$
4	−1.49197	$3d^3$ ^{2}G $5p$	3	−1.41938	$3d^3$ ^{2}G $4f$	13	−.39060	$3d^3$ ^{4}F $8h$	27	−.25548	$3d^3$ ^{2}H $7g$
5	−1.43100	$3d^3$ ^{2}H $5p$:	4	−1.41347	$3d^3$ ^{2}H $5p$:	14	−.34045	$3d^3$ ^{2}G $7f$	28	−.22843	$3d^3$ ^{2}P $8d$
6	−1.41907	$3d^24s$ ^{2}G $4p$:	5	−1.35235	$3d^3$ ^{2}H $4f$:	15	−.32961	$3d^3$ ^{2}G $7h$	29	−.21259	$3d^3(^1)$ ^{2}D $6d$
7	−1.35437	$3d^3$ ^{2}H $4f$	6	−1.18144	$3d^3$ ^{2}F $4f$	16	−.30862	$3d^3$ ^{4}F $9h$	30	−.20975	$3d^3$ ^{2}G $8g$
8	−1.35023	$3d^3(^3)$ ^{2}D $4f$	7	−1.01562	$3d^3$ ^{4}F $5f$	17	−.28271	$3d^3$ ^{2}F $6h$	31	−.20818	$3d^3(^3)$ ^{2}D $8d$
9	−1.19211	$3d^3$ ^{2}F $4f$	8	−.84679	$3d^3$ ^{2}G $5f$	18	−.26978	$3d^3$ ^{2}H $7f$:	32	−.20129	$3d^3$ ^{2}P $9s$
10	−1.01624	$3d^3$ ^{4}F $5f$	9	−.81964	$3d^24p$ ^{4}G $4d$:	19	−.26098	$3d^3$ ^{2}H $7h$	33	−.19528	$3d^3$ ^{2}F $7d$
11	−.94661	$3d^3(^1)$ ^{2}D $4f$	10	−.79374	$3d^3$ ^{2}H $6p$	20	−.25722	$3d^3(^3)$ ^{2}D $7h$	34	−.13679	$3d^3$ ^{2}H $8i$
12	−.87107	$3d^3$ ^{2}G $6p$	11	−.75956	$3d^3$ ^{2}H $5f$:	$^1\mathbf{S}^e$			35	−.13635	$3d^3$ ^{2}H $8g$
13	−.86295	$3d^24p$ ^{4}G $4d$:	12	−.70443	$3d^3$ ^{4}F $6f$	1	−5.18789	$3d^4$	36	−.12783	$3d^3$ ^{2}G $9g$
14	−.84151	$3d^3$ ^{2}G $4f$	13	−.69413	$3d^3$ ^{4}F $6h$	2	−4.44596	$3d^4(^0)$	37	−.12566	$3d^3$ ^{2}P $9d$
15	−.80562	$3d^3$ ^{2}H $6p$:	14	−.61970	$3d^3$ ^{2}F $5f$	3	−1.86842	$3d^3(^3)$ ^{2}D $4d$	$^1\mathbf{D}^e$		
16	−.79739	$3d^24p$ ^{4}F $4d$:	15	−.56734	$3d^24p$ ^{2}G $4d$:	4	−1.34771	$3d^3(^1)$ ^{2}D $4d$	1	−5.14873	$3d^4$
17	−.77786	$3d^3$ ^{2}H $5f$	16	−.52935	$3d^3$ ^{4}P $6h$	5	−.99084	$3d^3(^3)$ ^{2}D $5d$	2	−4.67335	$3d^4(^2)$
18	−.75051	$3d^3(^3)$ ^{2}D $5f$	17	−.52683	$3d^3$ ^{2}G $6f$	6	−.81807	$3d^3$ ^{2}G $5g$:	3	−3.42869	$3d^3(^3)$ ^{2}D $4s$
19	−.70138	$3d^3$ ^{4}F $6f$	18	−.51397	$3d^3$ ^{2}G $6h$	7	−.57301	$3d^3(^3)$ ^{2}D $6d$	4	−3.00228	$3d^3(^1)$ ^{2}D $4s$
20	−.69426	$3d^3$ ^{4}F $6h$	19	−.51001	$3d^3$ ^{4}F $7h$	8	−.55494	$3d^3(^1)$ ^{2}D $5d$	5	−2.12442	$3d^3$ ^{2}G $4d$
21	−.61582	$3d^3$ ^{2}F $5f$	20	−.50886	$3d^3$ ^{4}F $7f$	9	−.51309	$3d^3$ ^{2}G $6g$:	6	−2.05533	$3d^3$ ^{2}P $4d$:
22	−.54029	$3d^3$ ^{2}G $7p$	21	−.46662	$3d^3$ ^{2}H $7p$:	10	−.33841	$3d^3(^3)$ ^{2}D $7d$	7	−1.92436	$3d^3(^3)$ ^{2}D $4d$:
23	−.53279	$3d^3$ ^{2}G $5f$	22	−.46554	$3d^3$ ^{2}P $6h$	11	−.32917	$3d^3$ ^{2}G $7g$:	8	−1.76691	$3d^3$ ^{2}F $4d$
24	−.53094	$3d^3$ ^{4}P $6h$	23	−.44018	$3d^3$ ^{2}H $6f$:	12	−.20972	$3d^3$ ^{2}G $8g$	9	−1.68351	$3d^3(^3)$ ^{2}D $5s$
25	−.51979	$3d^3$ ^{4}F $7f$	24	−.44460	$3d^3$ ^{2}H $6h$	13	−.19244	$3d^3(^3)$ ^{2}D $8d$	10	−1.48352	$3d^3(^1)$ ^{2}D $4d$
26	−.51385	$3d^3$ ^{2}G $6h$	25	−.44163	$3d^3(^3)$ ^{2}D $6h$	14	−.15870	$3d^3(^1)$ ^{2}D $6d$	11	−1.29406	$3d^3(^1)$ ^{2}D $5s$:
27	−.51009	$3d^3$ ^{4}F $7h$	26	−.39324	$3d^3$ ^{4}F $8f$	15	−.12774	$3d^3$ ^{2}G $9g$.	12	−1.11171	$3d^3$ ^{2}G $5d$:
28	−.47242	$3d^3$ ^{2}H $6f$	27	−.39063	$3d^3$ ^{4}F $8j$	16	−.09489	$3d^3(^3)$ ^{2}D $9d$	13	−1.05189	$3d^3$ ^{2}P $5d$
29	−.46971	$3d^3$ ^{2}H $7p$:	28	−.39059	$3d^3$ ^{4}F $8h$	$^1\mathbf{P}^e$			14	−1.02509	$3d^3(^3)$ ^{2}D $5d$:
30	−.46502	$3d^3$ ^{2}P $6h$	29	−.34526	$3d^3$ ^{4}P $7h$	1	−3.44639	$3d^3$ ^{2}P $4s$	15	−.92826	$3d^3(^3)$ ^{2}D $6s$:
31	−.46052	$3d^3(^3)$ ^{2}D $6f$	30	−.33840	$3d^3$ ^{2}G $7f$	2	−2.12733	$3d^3$ ^{2}P $4d$	16	−.85060	$3d^3$ ^{2}F $5d$
32	−.44458	$3d^3$ ^{2}H $6h$	31	−.32961	$3d^3$ ^{2}G $7h$	3	−2.08103	$3d^3(^3)$ ^{2}D $4d$	17	−.81890	$3d^3$ ^{2}G $5g$
33	−.44122	$3d^3(^3)$ ^{2}D $6h$	32	−.31100	$3d^3$ ^{4}F $9f$	4	−1.90009	$3d^3$ ^{2}F $4d$	18	−.75349	$3d^3$ ^{2}H $5g$:
34	−.43036	$3d^24p$ ^{2}F $4d$:	33	−.30864	$3d^3$ ^{4}F $9j$	5	−1.70431	$3d^3$ ^{2}P $5s$	19	−.74311	$3d^3(^3)$ ^{2}D $5g$:

Ti-like Fe (Fe^{4+})

i	E(Ryds)	Description	i	E(Ryds)	Description	i	E(Ryds)	Description	i	E(Ryds)	Description
20	−.67107	$3d^3$ ^{2}G $6d$:	12	−1.05363	$3d^3$ ^{2}H $5d$	3	−3.50286	$3d^3$ ^{2}G $4s$	2	−2.17017	$3d^3$ ^{2}G $4d$
21	−.62303	$3d^3(^1)$ ^{2}D $5d$	13	−1.00504	$3d^3(^3)$ ^{2}D $5d$	4	−2.10816	$3d^3$ ^{2}G $4d$	3	−2.11018	$3d^3$ ^{2}H $4d$
22	−.60091	$3d^3$ ^{2}P $6d$	14	−.88386	$3d^3$ ^{2}F $5d$	5	−2.05568	$3d^3$ ^{2}H $4d$	4	−1.89769	$3d^3$ ^{2}F $4d$
23	−.59848	$3d^3(^3)$ ^{2}D $6d$:	15	−.81953	$3d^3$ ^{2}G $5g$	6	−1.91425	$3d^3(^3)$ ^{2}D $4d$	5	−1.67941	$3d^3$ ^{2}H $5s$
24	−.58744	$3d^3$ ^{2}F $5g$:	16	−.77292	$3d^3$ ^{2}P $5g$:	7	−1.75271	$3d^3$ ^{2}G $5s$	6	−1.13537	$3d^3$ ^{2}G $5d$
25	−.54406	$3d^3(^3)$ ^{2}D $7s$:	17	−.76721	$3d^3$ ^{2}F $6s$	8	−1.74434	$3d^3$ ^{2}F $4d$	7	−1.07066	$3d^3$ ^{2}H $5d$
26	−.53508	$3d^3(^1)$ ^{2}D $6s$:	18	−.75142	$3d^3$ ^{2}H $5g$	9	−1.56382	$3d^3(^1)$ ^{2}D $4d$	8	−.92502	$3d^3$ ^{2}H $6s$
27	−.51365	$3d^3$ ^{2}H $6g$:	19	−.74472	$3d^3(^3)$ ^{2}D $5g$	10	−1.09617	$3d^3$ ^{2}G $5d$:	9	−.89746	$3d^3$ ^{2}F $5d$
28	−.44664	$3d^3(^3)$ ^{2}D $6g$:	20	−.68350	$3d^3$ ^{2}G $6d$	11	−1.04887	$3d^3$ ^{2}H $5d$:	10	−.82060	$3d^3$ ^{2}G $5g$
29	−.43938	$3d^3$ ^{2}F $6g$:	21	−.67104	$3d^3(^1)$ ^{2}D $5d$	12	−1.01218	$3d^3(^3)$ ^{2}D $5d$:	11	−.77226	$3d^3$ ^{2}P $5g$
30	−.42205	$3d^3$ ^{2}F $6d$	22	−.62682	$3d^3$ ^{2}P $6d$	13	−.99637	$3d^3$ ^{2}G $6s$	12	−.74940	$3d^3$ ^{2}H $5g$
31	−.41828	$3d^3$ ^{2}G $7d$:	23	−.60378	$3d^3$ ^{2}H $6d$	14	−.85384	$3d^3$ ^{2}F $5d$:	13	−.74878	$3d^3(^3)$ ^{2}D $5g$
32	−.36713	$3d^3$ ^{2}P $7d$	24	−.58839	$3d^3$ ^{2}F $5g$:	15	−.82013	$3d^3$ ^{2}G $5g$	14	−.67897	$3d^3$ ^{2}G $6d$
33	−.35358	$3d^3(^3)$ ^{2}D $7d$	25	−.57713	$3d^3(^3)$ ^{2}D $6d$	16	−.77011	$3d^3$ ^{2}P $5g$	15	−.61037	$3d^3$ ^{2}H $6d$
34	−.35146	$3d^3(^1)$ ^{2}D $5g$:	26	−.51401	$3d^3$ ^{2}G $6g$	17	−.74970	$3d^3$ ^{2}H $5g$	16	−.58972	$3d^3$ ^{2}F $5g$
35	−.32950	$3d^3$ ^{2}G $7g$	27	−.46659	$3d^3$ ^{2}P $6g$	18	−.74675	$3d^3(^3)$ ^{2}D $5g$	17	−.54002	$3d^3$ ^{2}H $7s$
36	−.32928	$3d^3$ ^{2}G $7i$	28	−.44577	$3d^3$ ^{2}H $6g$	19	−.65939	$3d^3$ ^{2}G $6d$:	18	−.51455	$3d^3$ ^{2}G $6g$
37	−.32134	$3d^3(^3)$ ^{2}D $8s$	29	−.44148	$3d^3$ ^{2}F $6d$	20	−.64279	$3d^3(^1)$ ^{2}D $5d$	19	−.46620	$3d^3$ ^{2}P $6g$
38	−.28225	$3d^3(^1)$ ^{2}D $6g$:	30	−.44005	$3d^3(^3)$ ^{2}D $6g$	21	−.61248	$3d^3$ ^{2}G $7s$	20	−.44595	$3d^3$ ^{2}F $6d$
39	−.26980	$3d^3$ ^{2}G $8d$:	31	−.42999	$3d^3$ ^{2}G $7d$	22	−.60134	$3d^3$ ^{2}H $6d$:	21	−.44460	$3d^3$ ^{2}H $6g$
40	−.26178	$3d^3$ ^{2}H $7g$:	32	−.38352	$3d^3$ ^{2}F $7s$	23	−.58924	$3d^3$ ^{2}F $5g$	22	−.44228	$3d^3(^3)$ ^{2}D $6g$
41	−.25638	$3d^3$ ^{2}H $7i$	33	−.37683	$3d^3$ ^{2}P $7d$	24	−.57941	$3d^3(^3)$ ^{2}D $6d$:	23	−.42954	$3d^3$ ^{2}G $7d$
42	−.25576	$3d^3(^3)$ ^{2}D $7g$:	34	−.35671	$3d^3$ ^{2}H $7d$	25	−.51431	$3d^3$ ^{2}G $6g$	24	−.36018	$3d^3(^1)$ ^{2}D $5g$
43	−.21956	$3d^3$ ^{2}P $8d$:	35	−.35618	$3d^3(^1)$ ^{2}D $5g$:	26	−.46500	$3d^3$ ^{2}P $6g$	25	−.35944	$3d^3$ ^{2}H $7d$
44	−.20995	$3d^3$ ^{2}G $8g$	36	−.34085	$3d^3(^3)$ ^{2}D $7d$	27	−.44472	$3d^3$ ^{2}H $6g$	26	−.32941	$3d^3$ ^{2}G $7i$
45	−.20974	$3d^3$ ^{2}G $8i$	37	−.32970	$3d^3$ ^{2}G $7g$	28	−.44122	$3d^3(^3)$ ^{2}D $6g$	27	−.32941	$3d^3$ ^{2}F $7d$:
46	−.20753	$3d^3(^3)$ ^{2}D $8d$:	38	−.32932	$3d^3$ ^{2}G $7i$	29	−.43246	$3d^3$ ^{2}F $6d$	28	−.31729	$3d^3$ ^{2}H $8s$
47	−.20134	$3d^3(^1)$ ^{2}D $6d$	39	−.28280	$3d^3$ ^{2}F $6g$:	30	−.41009	$3d^3$ ^{2}G $7d$	29	−.28389	$3d^3$ ^{2}F $6g$
48	−.18113	$3d^3(^3)$ ^{2}D $9s$	40	−.28187	$3d^3$ ^{2}P $7g$	31	−.38994	$3d^3$ ^{2}G $8s$	30	−.28137	$3d^3(^1)$ ^{2}D $6g$:
49	−.17335	$3d^3$ ^{2}F $7d$	41	−.27627	$3d^3$ ^{2}G $8d$	32	−.35971	$3d^3(^1)$ ^{2}D $5g$	31	−.28113	$3d^3$ ^{2}P $7i$
50	−.16500	$3d^3$ ^{2}G $9d$	42	−.26127	$3d^3$ ^{2}H $7g$:	33	−.35578	$3d^3$ ^{2}H $7d$	32	−.27588	$3d^3$ ^{2}G $8d$
51	−.14973	$3d^3(^1)$ ^{2}D $7s$	43	−.25646	$3d^3$ ^{2}H $7i$	34	−.34108	$3d^3(^3)$ ^{2}D $7d$	33	−.26056	$3d^3$ ^{2}G $7g$
52	−.14193	$3d^3$ ^{2}H $8g$:	44	−.22568	$3d^3$ ^{2}P $8d$	35	−.32985	$3d^3$ ^{2}G $7g$	34	−.26056	$3d^3$ ^{2}H $7i$
53	−.13684	$3d^3$ ^{2}H $8i$	45	−.21825	$3d^3(^1)$ ^{2}D $6d$	36	−.32937	$3d^3$ ^{2}G $7i$	35	−.25754	$3d^3$ ^{2}P $7g$
54	−.13654	$3d^3(^3)$ ^{2}D $8g$:	46	−.21006	$3d^3$ ^{2}G $8g$	37	−.28357	$3d^3$ ^{2}F $6g$	36	−.25668	$3d^3(^3)$ ^{2}D $7i$
55	−.12791	$3d^3$ ^{2}G $9g$	47	−.20977	$3d^3$ ^{2}G $8i$	38	−.28061	$3d^3$ ^{2}P $7g$:	37	−.21026	$3d^3$ ^{2}G $8g$
56	−.12778	$3d^3$ ^{2}G $9i$	48	−.20428	$3d^3$ ^{2}H $8d$	39	−.26587	$3d^3$ ^{2}G $8d$	38	−.20984	$3d^3$ ^{2}G $8i$
57	−.11989	$3d^3$ ^{2}P $9d$:	49	−.19896	$3d^3(^3)$ ^{2}D $8d$	40	−.26088	$3d^3$ ^{2}H $7i$	39	−.20564	$3d^3$ ^{2}H $8d$
58	−.10253	$3d^3(^3)$ ^{2}D $9d$	50	−.18789	$3d^3$ ^{2}F $7d$	41	−.26060	$3d^3$ ^{2}H $7g$	40	−.19675	$3d^3$ ^{2}F $8d$:
59	−.09821	$3d^3$ ^{2}F $7g$:	51	−.17366	$3d^3$ ^{2}G $9d$	42	−.25689	$3d^3(^3)$ ^{2}D $7g$	41	−.17681	$3d^3$ ^{2}H $9s$
		^{1}F^e	52	−.16211	$3d^3$ ^{2}P $8g$	43	−.25657	$3d^3(^3)$ ^{2}D $7i$:	42	−.17359	$3d^3$ ^{2}G $9d$
1	−5.05856	$3d^4$	53	−.15897	$3d^3$ ^{2}F $8s$	44	−.24962	$3d^3$ ^{2}G $9s$	43	−.16193	$3d^3$ ^{2}P $8g$:
2	−3.26810	$3d^3$ ^{2}F $4s$:	54	−.14163	$3d^3$ ^{2}H $8g$	45	−.21021	$3d^3$ ^{2}G $8g$	44	−.16155	$3d^3$ ^{2}P $8i$
3	−2.17630	$3d^3$ ^{2}G $4d$	55	−.13692	$3d^3$ ^{2}H $8i$	46	−.20980	$3d^3$ ^{2}G $8i$	45	−.14108	$3d^3$ ^{2}H $7g$
4	−2.09672	$3d^3$ ^{2}P $4d$	56	−.13686	$3d^3(^3)$ ^{2}D $8g$	47	−.20723	$3d^3(^1)$ ^{2}D $6d$:	46	−.14108	$3d^3$ ^{2}H $8i$
5	−2.05607	$3d^3$ ^{2}H $4d$	57	−.12803	$3d^3$ ^{2}G $9g$:	48	−.20365	$3d^3$ ^{2}H $8d$:	47	−.13779	$3d^3(^3)$ ^{2}D $7g$
6	−1.96027	$3d^3(^3)$ ^{2}D $4d$	58	−.12780	$3d^3$ ^{2}G $9i$	49	−.19884	$3d^3$ ^{2}F $7d$:	48	−.13710	$3d^3(^3)$ ^{2}D $8i$
7	−1.80936	$3d^3$ ^{2}F $4d$	59	−.12334	$3d^3$ ^{2}P $9d$	50	−.17870	$3d^3(^3)$ ^{2}D $8d$	49	−.12817	$3d^3$ ^{2}G $9g$
8	−1.65210	$3d^3(^1)$ ^{2}D $4d$	60	−.10312	$3d^3$ ^{2}H $9d$	51	−.16532	$3d^3$ ^{2}G $9d$			**^{1}I^e**
9	−1.52339	$3d^3$ ^{2}F $5s$			**^{1}G^e**	52	−.16141	$3d^3$ ^{2}P $8g$	1	−5.19676	$3d^4$
10	−1.13707	$3d^3$ ^{2}G $5d$	1	−5.21664	$3d^4$			**^{1}H^e**	2	−2.12529	$3d^3$ ^{2}G $4d$
11	−1.08244	$3d^3$ ^{2}P $5d$	2	−4.89531	$3d^4(^2)$	1	−3.43143	$3d^3$ ^{2}H $4s$	3	−1.90120	$3d^3$ ^{2}H $4d$:

Ti-like Fe (Fe^{4+})

i	E(Ryds)	Description	i	E(Ryds)	Description	i	E(Ryds)	Description	i	E(Ryds)	Description
4	−1.11594	$3d^3$ ^{2}G $5d$:	19	−.20143	$3d^3$ ^{2}H $8d$	25	−.33507	$3d^3$ ^{2}G $7f$	35	−.33910	$3d^3$ ^{2}G $7f$
5	−1.00422	$3d^3$ ^{2}H $5d$	20	−.16153	$3d^3$ ^{2}P $8i$	26	−.32921	$3d^3$ ^{2}G $7h$	36	−.32927	$3d^3$ ^{2}G $7h$
6	−.82074	$3d^3$ ^{2}G $5g$	21	−.14107	$3d^3$ ^{2}H $8i$	27	−.29911	$3d^3$ ^{2}F $6f$:	37	−.30645	$3d^3$ ^{2}F $7p$:
7	−.75239	$3d^3$ ^{2}H $5g$:	22	−.13803	$3d^3$ ^{2}H $8g$	28	−.29344	$3d^3$ ^{2}P $8p$	38	−.29526	$3d^3(^1)$ ^{2}D $7p$:
8	−.75011	$3d^3(^3)$ ^{2}D $5g$	23	−.13726	$3d^3(^3)$ ^{2}D $8i$	29	−.27562	$3d^3(^3)$ ^{2}D $8p$	39	−.29245	$3d^3$ ^{2}F $6f$:
9	−.66755	$3d^3$ ^{2}G $6d$	24	−.12818	$3d^3$ ^{2}G $9g$	30	−.26845	$3d^3(^3)$ ^{2}D $7f$	40	−.28953	$3d^3(^1)$ ^{2}D $5f$:
10	−.58938	$3d^3$ ^{2}F $5g$	25	−.09978	$3d^3$ ^{2}H $9d$	31	−.25590	$3d^3$ ^{2}H $7h$	41	−.28188	$3d^3$ ^{2}F $6h$
11	−.57954	$3d^3$ ^{2}H $6d$	26	−.09832	$3d^3$ ^{2}F $7g$	32	−.21560	$3d^3$ ^{2}G $8f$	42	−.27375	$3d^3$ ^{2}P $8p$
12	−.51463	$3d^3$ ^{2}G $6g$	27	−.09823	$3d^3$ ^{2}F $7i$	33	−.20970	$3d^3$ ^{2}G $8h$	43	−.27118	$3d^3$ ^{2}P $7f$:
13	−.44606	$3d^3$ ^{2}H $6g$:	28	−.07954	$3d^3$ ^{2}G $9i$	34	−.16990	$3d^3$ ^{2}P $9p$	44	−.26515	$3d^3$ ^{2}H $7f$:
14	−.44300	$3d^3(^3)$ ^{2}D $6g$	29	−.07950	$3d^3$ ^{2}G $9k$	35	−.16226	$3d^3(^3)$ ^{2}D $8f$	45	−.25607	$3d^3$ ^{2}H $7h$
15	−.42202	$3d^3$ ^{2}G $7d$	30	−.06920	$3d^3$ ^{2}P $9k$:	36	−.15063	$3d^3(^3)$ ^{2}D $9p$	46	−.21697	$3d^3$ ^{2}G $8f$:
16	−.35416	$3d^3(^1)$ ^{2}D $5g$			^{1}S^o	37	−.13660	$3d^3$ ^{2}H $8h$	47	−.20974	$3d^3$ ^{2}G $8h$
17	−.34182	$3d^3$ ^{2}H $7d$	1	−2.97430	$3d^3$ ^{2}P $4p$	38	−.13373	$3d^3$ ^{2}G $9f$:	48	−.17126	$3d^3(^3)$ ^{2}D $8p$:
18	−.33003	$3d^3$ ^{2}G $7g$	2	−1.45798	$3d^3$ ^{2}P $5p$	39	−.12776	$3d^3$ ^{2}G $9h$:	49	−.16954	$3d^3(^3)$ ^{2}D $7f$:
19	−.32945	$3d^3$ ^{2}G $7i$	3	−1.18175	$3d^3$ ^{2}F $4f$	40	−.11679	$3d^3$ ^{2}F $7f$	50	−.15000	$3d^3$ ^{2}P $9p$:
20	−.28337	$3d^3$ ^{2}F $6g$	4	−.82681	$3d^3$ ^{2}P $6p$:	41	−.08823	$3d^3$ ^{2}P $9f$:	51	−.14968	$3d^3$ ^{2}F $7f$:
21	−.28099	$3d^3$ ^{2}P $7i$	5	−.61166	$3d^3$ ^{2}F $5f$			^{1}D^o	52	−.14534	$3d^3$ ^{2}P $8f$:
22	−.27032	$3d^3$ ^{2}G $8d$	6	−.49586	$3d^3$ ^{2}P $7p$	1	−2.94245	$3d^3$ ^{2}P $4p$	53	−.13704	$3d^3$ ^{2}H $8j$
23	−.26141	$3d^3$ ^{2}H $7g$	7	−.43938	$3d^3$ ^{2}H $6h$	2	−2.81378	$3d^3(^3)$ ^{2}D $4p$	54	−.13671	$3d^3$ ^{2}H $8h$
24	−.26062	$3d^3$ ^{2}H $7i$	8	−.30404	$3d^3$ ^{2}F $6f$	3	−2.69556	$3d^3$ ^{2}F $4p$	55	−.13378	$3d^3$ ^{2}G $9f$
25	−.25799	$3d^3(^3)$ ^{2}D $7g$	9	−.29879	$3d^3$ ^{2}P $8p$	4	−2.48439	$3d^3(^1)$ ^{2}D $4p$	56	−.12778	$3d^3$ ^{2}G $9h$:
26	−.25678	$3d^3(^3)$ ^{2}D $7i$	10	−.27268	$3d^24s$ ^{2}P $4p$:	5	−1.44174	$3d^3$ ^{2}P $5p$	57	−.11429	$3d^3(^3)$ ^{2}D $8p$
27	−.21031	$3d^3$ ^{2}G $8g$	11	−.25581	$3d^3$ ^{2}H $7h$	6	−1.41152	$3d^3$ ^{2}G $4f$	58	−.10796	$3d^3(^3)$ ^{2}D $9p$:
28	−.20987	$3d^3$ ^{2}G $8i$	12	−.17322	$3d^3$ ^{2}P $9p$	7	−1.40042	$3d^3(^3)$ ^{2}D $5p$:	59	−.10510	$3d^3(^1)$ ^{2}D $6f$:
29	−.19416	$3d^3$ ^{2}H $8d$	13	−.13654	$3d^3$ ^{2}H $8h$	8	−1.37809	$3d^3$ ^{2}P $4f$:	60	−.09787	$3d^3$ ^{2}F $7h$
30	−.16918	$3d^3$ ^{2}G $9d$	14	−.10802	$3d^3$ ^{2}F $7f$	9	−1.34977	$3d^24s$ ^{2}F $4p$:			^{1}F^o
31	−.16143	$3d^3$ ^{2}P $8i$			^{1}P^o	10	−1.31995	$3d^3$ ^{2}H $4f$:	1	−2.93343	$3d^3$ ^{2}G $4p$
32	−.14169	$3d^3$ ^{2}H $8g$:	1	−2.91071	$3d^3$ ^{2}P $4p$	11	−1.24677	$3d^3$ ^{2}F $5p$:	2	−2.85467	$3d^3(^3)$ ^{2}D $4p$
33	−.14106	$3d^3$ ^{2}H $8i$	2	−2.80911	$3d^3(^3)$ ^{2}D $4p$	12	−1.16738	$3d^3(^3)$ ^{2}D $4f$:	3	−2.64519	$3d^3$ ^{2}F $4p$
34	−.13810	$3d^3(^3)$ ^{2}D $8g$	3	−2.36619	$3d^3(^1)$ ^{2}D $4p$	13	−1.03594	$3d^3(^1)$ ^{2}D $5p$:	4	−2.43704	$3d^3(^1)$ ^{2}D $4p$
35	−.13719	$3d^3(^3)$ ^{2}D $8i$	4	−1.43363	$3d^3$ ^{2}P $5p$	14	−.94188	$3d^3$ ^{2}F $4f$:	5	−1.47656	$3d^3$ ^{2}G $5p$
36	−.12820	$3d^3$ ^{2}G $9g$:	5	−1.40929	$3d^3(^3)$ ^{2}D $5p$	15	−.84559	$3d^3$ ^{2}G $5f$	6	−1.41839	$3d^3(^3)$ ^{2}D $5p$:
		^{1}J^e	6	−1.38724	$3d^3$ ^{2}G $4f$:	16	−.83103	$3d^3(^1)$ ^{2}D $4f$:	7	−1.40705	$3d^3$ ^{2}G $4f$
1	−2.06183	$3d^3$ ^{2}H $4d$	7	−1.34082	$3d^3(^3)$ ^{2}D $4f$	17	−.81748	$3d^3$ ^{2}P $6p$:	8	−1.34356	$3d^3$ ^{2}P $4f$
2	−1.05282	$3d^3$ ^{2}H $5d$	8	−1.17579	$3d^3$ ^{2}F $4f$	18	−.79397	$3d^24s$ ^{2}D $4p$:	9	−1.33426	$3d^3$ ^{2}H $4f$
3	−.82034	$3d^3$ ^{2}G $5g$	9	−.99253	$3d^3(^1)$ ^{2}D $5p$	19	−.78595	$3d^3(^3)$ ^{2}D $6p$:	10	−1.32009	$3d^3(^3)$ ^{2}D $4f$
4	−.75001	$3d^3$ ^{2}H $5g$	10	−.90209	$3d^3(^1)$ ^{2}D $4f$	20	−.77299	$3d^3$ ^{2}P $5f$:	11	−1.24160	$3d^3$ ^{2}F $5p$
5	−.60175	$3d^3$ ^{2}H $6d$	11	−.83131	$3d^3$ ^{2}G $5f$	21	−.75442	$3d^3$ ^{2}H $5f$:	12	−1.17365	$3d^3$ ^{2}F $4f$
6	−.58771	$3d^3$ ^{2}F $5g$	12	−.81010	$3d^3$ ^{2}P $6p$	22	−.66519	$3d^24s$ ^{2}P $4p$:	13	−1.01408	$3d^3(^1)$ ^{2}D $5p$
7	−.51445	$3d^3$ ^{2}G $6g$	13	−.79542	$3d^3(^3)$ ^{2}D $6p$	23	−.62972	$3d^3$ ^{2}F $6p$:	14	−.94581	$3d^3(^1)$ ^{2}D $4f$:
8	−.44291	$3d^3$ ^{2}H $6g$	14	.77455	$3d^3(^3)$ ^{2}D $5f$:	24	−.59233	$3d^3(^3)$ ^{2}D $5f$:	15	−.86047	$3d^3$ ^{2}G $6p$
9	−.35434	$3d^3$ ^{2}H $7d$	15	−.60959	$3d^3$ ^{2}F $5f$	25	−.52681	$3d^3$ ^{2}G $6f$:	16	−.84043	$3d^3$ ^{2}G $5f$
10	−.33002	$3d^3$ ^{2}G $7g$	16	−.52756	$3d^3$ ^{2}G $6f$	26	−.51337	$3d^3$ ^{2}G $6h$	17	−.79452	$3d^3(^3)$ ^{2}D $6p$
11	−.32947	$3d^3$ ^{2}G $7i$:	17	−.51327	$3d^3$ ^{2}G $6h$	27	−.49014	$3d^3(^1)$ ^{2}D $6p$:	18	−.78224	$3d^3$ ^{2}P $5f$
12	−.28258	$3d^3$ ^{2}F $6g$	18	−.48803	$3d^3(^3)$ ^{2}D $6f$	28	−.48277	$3d^3$ ^{2}P $6f$:	19	.76713	$3d^3$ ^{2}H $5f$
13	−.28111	$3d^3$ ^{2}P $7i$	19	−.48610	$3d^3$ ^{2}P $7p$	29	−.46659	$3d^3$ ^{2}P $7p$	20	−.75795	$3d^3(^3)$ ^{2}D $5f$
14	−.26064	$3d^3$ ^{2}H $7i$	20	−.46816	$3d^3(^3)$ ^{2}D $7p$	30	−.46177	$3d^3$ ^{2}H $6f$:	21	−.63008	$3d^3$ ^{2}F $6p$
15	−.25796	$3d^3$ ^{2}H $7g$	21	−.43953	$3d^3$ ^{2}H $6h$	31	−.45193	$3d^3(^3)$ ^{2}D $6f$:	22	−.61066	$3d^3$ ^{2}F $5f$
16	−.25686	$3d^3(^3)$ ^{2}D $7i$	22	−.41429	$3d^24s$ ^{2}D $4p$:	32	−.43982	$3d^3$ ^{2}H $6h$	23	−.59267	$3d^24s$ ^{2}F $4p$:
17	−.21026	$3d^3$ ^{2}G $8g$	23	−.38848	$3d^3(^1)$ ^{2}D $6p$	33	−.41024	$3d^3(^3)$ ^{2}D $7p$:	24	.53687	$3d^3$ ^{2}G $7p$
18	−.20989	$3d^3$ ^{2}G $8i$	24	−.34809	$3d^3(^1)$ ^{2}D $5f$:	34	−.37647	$3d^3$ ^{2}F $5f$:	25	−.52857	$3d^3$ ^{2}G $6f$

Ti-like Fe (Fe^{4+})

i	E(Ryds)	Description	i	E(Ryds)	Description	i	E(Ryds)	Description	i	E(Ryds)	Description
26	-.51352	$3d^3$ ^{2}G $6h$	21	-.61067	$3d^3$ ^{2}F $5f$:	16	-.59872	$3d^3$ ^{2}F $5f$:	18	-.29762	$3d^3$ ^{2}F $6f$:
27	-.47279	$3d^3$ ^{2}P $6f$	22	-.54014	$3d^3$ ^{2}G $7p$	17	-.53669	$3d^3$ ^{2}G $7p$	19	-.28290	$3d^3(^3)$ ^{2}D $6h$
28	-.46855	$3d^3(^3)$ ^{2}D $7p$	23	-.53029	$3d^3$ ^{2}G $6f$	18	-.52160	$3d^3$ ^{2}G $6f$	20	-.28114	$3d^3$ ^{2}F $6h$:
29	-.45689	$3d^3$ ^{2}H $6f$	24	-.51370	$3d^3$ ^{2}G $6h$	19	-.51385	$3d^3$ ^{2}G $6h$	21	-.27131	$3d^3$ ^{2}H $7f$
30	-.44644	$3d^3(^3)$ ^{2}D $6f$	25	-.50934	$3d^24s$ ^{2}G $4p$:	20	-.46852	$3d^3$ ^{2}H $7p$	22	-.26995	$3d^3$ ^{2}H $8p$
31	-.44568	$3d^3$ ^{2}H $6h$	26	-.47764	$3d^3$ ^{2}P $6f$:	21	-.46502	$3d^3$ ^{2}P $6h$:	23	-.26053	$3d^3$ ^{2}P $7h$
32	-.44023	$3d^3(^3)$ ^{2}D $6h$	27	-.46566	$3d^3$ ^{2}P $6h$:	22	-.45824	$3d^3$ ^{2}H $6f$	24	-.25710	$3d^3$ ^{2}H $7h$
33	-.39962	$3d^3$ ^{2}F $6p$:	28	-.46202	$3d^3$ ^{2}H $7p$	23	-.45381	$3d^3(^3)$ ^{2}D $6f$:	25	-.21724	$3d^3$ ^{2}G $8f$
34	-.38556	$3d^3(^1)$ ^{2}D $5f$	29	-.45319	$3d^3(^1)$ ^{2}D $5f$:	24	-.44458	$3d^3$ ^{2}H $6h$:	26	-.20997	$3d^3$ ^{2}G $8h$:
35	-.34400	$3d^3$ ^{2}G $8p$	30	-.44502	$3d^3$ ^{2}H $6h$:	25	-.44122	$3d^3(^3)$ ^{2}D $6h$:	27	-.20997	$3d^3$ ^{2}G $8j$:
36	-.33968	$3d^3$ ^{2}G $7f$	31	-.44073	$3d^3(^3)$ ^{2}D $6h$:	26	-.37446	$3d^3(^1)$ ^{2}D $5f$:	28	-.16150	$3d^3$ ^{2}P $8j$:
37	-.33446	$3d^3$ ^{2}F $6f$:	32	-.43954	$3d^3$ ^{2}H $6f$:	27	-.34310	$3d^3$ ^{2}G $8p$	29	-.16150	$3d^3$ ^{2}G $9j$:
38	-.32934	$3d^3$ ^{2}G $7h$	33	-.38969	$3d^3(^3)$ ^{2}D $6f$:	28	-.33562	$3d^3$ ^{2}G $7f$	30	-.14861	$3d^3$ ^{2}H $8f$
39	-.30515	$3d^3(^3)$ ^{2}D $7p$	34	-.34481	$3d^3$ ^{2}G $8p$	29	-.32955	$3d^3$ ^{2}G $7h$	31	-.14572	$3d^3$ ^{2}H $9p$
40	-.28593	$3d^3$ ^{2}P $7f$	35	-.34056	$3d^3$ ^{2}G $7f$	30	-.29326	$3d^3(^3)$ ^{2}D $6f$:	32	-.14114	$3d^3$ ^{2}H $8j$:
41	-.28418	$3d^3$ ^{2}F $6f$	36	-.32945	$3d^3$ ^{2}G $7h$	31	-.28282	$3d^3(^3)$ ^{2}D $6h$	33	-.14102	$3d^3$ ^{2}P $8h$
42	-.28220	$3d^3$ ^{2}F $6h$	37	-.30774	$3d^3$ ^{2}F $7p$	32	-.28082	$3d^3$ ^{2}P $7h$:	34	-.13740	$3d^3$ ^{2}H $8h$
43	-.27458	$3d^3(^3)$ ^{2}D $8p$:	38	-.29936	$3d^3$ ^{2}F $6f$:	33	-.27671	$3d^3$ ^{2}H $7f$:	35	-.13710	$3d^3(^3)$ ^{2}D $8j$:
44	-.26534	$3d^3$ ^{2}H $7f$	39	-.28986	$3d^3$ ^{2}P $7f$:	34	-.27113	$3d^3$ ^{2}H $8p$	36	-.13326	$3d^3$ ^{2}G $9f$
45	-.26114	$3d^3$ ^{2}H $7h$	40	-.28253	$3d^3(^3)$ ^{2}D $6h$	35	-.26847	$3d^3(^3)$ ^{2}D $7f$:	37	-.12795	$3d^3$ ^{2}G $9h$
46	-.26085	$3d^3(^3)$ ^{2}D $7f$	41	-.28125	$3d^3$ ^{2}G $7h$	36	-.26052	$3d^3$ ^{2}H $7h$:	38	-.12784	$3d^3$ ^{2}P $9j$:
47	-.25630	$3d^3(^3)$ ^{2}D $7h$	42	-.26963	$3d^3$ ^{2}H $8p$	37	-.25686	$3d^3(^3)$ ^{2}D $7h$:	39	-.10814	$3d^3$ ^{2}F $7f$:
48	-.21909	$3d^3$ ^{2}G $9p$	43	-.26783	$3d^3$ ^{2}H $7f$:	38	-.21878	$3d^3$ ^{2}G $9p$	40	-.09840	$3d^3(^3)$ ^{2}D $7h$
49	-.21655	$3d^3$ ^{2}G $8f$	44	-.26355	$3d^3(^3)$ ^{2}D $7f$:	39	-.21831	$3d^3$ ^{2}G $8f$			**^{1}J°**
50	-.20980	$3d^3$ ^{2}G $8h$	45	-.26077	$3d^3$ ^{2}P $7h$	40	-.20993	$3d^3$ ^{2}G $8h$	1	-1.40115	$3d^3$ ^{2}G $4f$
51	-.16869	$3d^3$ ^{2}P $8f$	46	-.25657	$3d^3$ ^{2}H $7h$	41	-.19662	$3d^3$ ^{2}F $6f$:	2	-1.32257	$3d^3$ ^{2}H $4f$
52	-.15029	$3d^3(^3)$ ^{2}D $9p$:	47	-.22221	$3d^3$ ^{2}G $8f$	42	-.16139	$3d^3$ ^{2}P $8h$:	3	-.83507	$3d^3$ ^{2}G $5f$
53	-.14666	$3d^3$ ^{2}H $8f$	48	-.21875	$3d^3$ ^{2}G $9p$	43	-.14785	$3d^3$ ^{2}H $9p$	4	-.75816	$3d^3$ ^{2}H $5f$
54	-.14142	$3d^3(^3)$ ^{2}D $8f$	49	-.20986	$3d^3$ ^{2}G $8h$	44	-.14622	$3d^3$ ^{2}H $8f$:	5	-.56282	$3d^24s$ ^{2}G $4f$:
55	-.13706	$3d^3$ ^{2}H $8j$	50	-.17039	$3d^3(^1)$ ^{2}D $6f$:	45	-.14116	$3d^3$ ^{2}P $8j$	6	-.51396	$3d^3$ ^{2}G $6h$
56	-.13686	$3d^3$ ^{2}H $8h$	51	-.16164	$3d^3$ ^{2}G $8h$	46	-.14101	$3d^3$ ^{2}H $8h$:	7	-.51070	$3d^3$ ^{2}G $6f$
		^{1}G°	52	-.16038	$3d^3$ ^{2}P $8f$	47	-.13949	$3d^3(^3)$ ^{2}D $8f$:	8	-.44541	$3d^3$ ^{2}H $6h$:
1	-2.93391	$3d^3$ ^{2}G $4p$	53	-.14556	$3d^3$ ^{2}H $9p$	48	-.13724	$3d^3(^3)$ ^{2}D $8h$:	9	-.44188	$3d^3(^3)$ ^{2}D $6h$:
2	-2.85627	$3d^3$ ^{2}H $4p$	54	-.14386	$3d^3$ ^{2}H $8f$	49	-.13708	$3d^3$ ^{2}H $8j$	10	-.44134	$3d^3$ ^{2}H $6f$
3	-2.64954	$3d^3$ ^{2}F $4p$	55	-.14120	$3d^3$ ^{2}P $8h$			**^{1}I°**	11	-.33318	$3d^3$ ^{2}G $7f$
4	-1.48489	$3d^3$ ^{2}G $5p$:	56	-.13795	$3d^3$ ^{2}G $9f$	1	-2.83851	$3d^3$ ^{2}H $4p$	12	-.32961	$3d^3$ ^{2}G $7h$
5	-1.41243	$3d^3$ ^{2}G $4f$			**^{1}H°**	2	-1.42070	$3d^3$ ^{2}G $4f$	13	-.28270	$3d^3$ ^{2}F $6h$:
6	-1.40070	$3d^3$ ^{2}H $5p$	1	-2.93114	$3d^3$ ^{2}G $4p$	3	-1.40371	$3d^3$ ^{2}H $5p$	14	-.26098	$3d^3$ ^{2}H $7h$:
7	-1.37210	$3d^3$ ^{2}P $4f$:	2	-2.85440	$3d^3$ ^{2}H $4p$	4	-1.35082	$3d^3$ ^{2}H $4f$	15	-.25945	$3d^3$ ^{2}H $7f$
8	-1.33506	$3d^3$ ^{2}H $4f$	3	-1.47758	$3d^3$ ^{2}G $5p$	5	-1.18188	$3d^3$ ^{2}F $4f$	16	-.25722	$3d^3(^3)$ ^{2}D $7h$:
9	-1.32233	$3d^3(^3)$ ^{2}D $4f$:	4	-1.41659	$3d^3$ ^{2}H $5p$	6	-.84494	$3d^3$ ^{2}G $5f$	17	-.21261	$3d^3$ ^{2}G $8f$
10	-1.24784	$3d^3$ ^{2}F $5p$:	5	-1.40511	$3d^3$ ^{2}G $4f$	7	-.78855	$3d^3$ ^{2}H $6p$	18	-.20983	$3d^3$ ^{2}G $8j$
11	-1.17653	$3d^3$ ^{2}F $4f$:	6	-1.35336	$3d^3$ ^{2}H $4f$	8	-.77514	$3d^3$ ^{2}H $5f$	19	-.20983	$3d^3$ ^{2}G $9h$:
12	-.97139	$3d^3(^1)$ ^{2}D $4f$:	7	-1.34010	$3d^3(^3)$ ^{2}D $4f$	9	-.61165	$3d^3$ ^{2}F $5f$	20	-.16148	$3d^3$ ^{2}P $8j$:
13	-.86664	$3d^3$ ^{2}G $6p$	8	-1.16725	$3d^3$ ^{2}F $4f$	10	-.53098	$3d^3$ ^{2}G $6f$	21	-.14006	$3d^3$ ^{2}H $8f$
14	-.85162	$3d^3$ ^{2}G $5f$	9	-.93572	$3d^3(^1)$ ^{2}D $4f$	11	-.51396	$3d^3$ ^{2}G $6h$	22	-.13749	$3d^3$ ^{2}H $8h$:
15	-.79711	$3d^3$ ^{2}P $5f$:	10	-.86148	$3d^3$ ^{2}G $6p$	12	-.46125	$3d^3$ ^{2}H $7p$	23	-.13711	$3d^3$ ^{2}H $8j$
16	-.78848	$3d^3$ ^{2}H $6p$	11	-.84361	$3d^3$ ^{2}G $5f$	13	-.45925	$3d^3$ ^{2}H $6f$	24	-.12923	$3d^3$ ^{2}G $9f$
17	-.77006	$3d^3$ ^{2}H $5f$:	12	-.80221	$3d^3$ ^{2}H $5f$:	14	-.44460	$3d^3$ ^{2}P $6h$	25	-.09834	$3d^3(^3)$ ^{2}D $7h$:
18	-.76540	$3d^3(^3)$ ^{2}D $5f$:	13	-.79517	$3d^3$ ^{2}H $6p$	15	-.44164	$3d^3$ ^{2}H $6h$	26	-.07949	$3d^3$ ^{2}G $9j$
19	-.75471	$3d^24s$ ^{2}F $4p$:	14	-.77328	$3d^3(^3)$ ^{2}D $5f$:	16	-.33987	$3d^3$ ^{2}G $7f$			
20	-.63282	$3d^3$ ^{2}F $6p$	15	-.73174	$3d^24s$ ^{2}G $4p$:	17	-.32961	$3d^3$ ^{2}G $7h$			

Ti-like Fe (Fe^{4+})

Energies in ascending order from ground state for terms with effective $n \leq 4.0$, $L \leq 7$

Term	i	E(Ryds)	Term	i	E(Ryds)	Term	i	E(Ryds)	Term	i	E(Ryds)	Term	i	E(Ryds)
$^5\mathbf{D}^e$	1	0.00000	$^1\mathbf{D}^e$	4	2.55506	$^3\mathbf{D}^o$	6	3.04501	$^5\mathbf{D}^e$	3	3.46387	$^3\mathbf{P}^e$	11	3.84218
$^3\mathbf{P}^e$	1	0.21532	$^5\mathbf{P}^o$	1	2.55659	$^1\mathbf{D}^o$	4	3.07295	$^3\mathbf{G}^e$	5	3.46743	$^1\mathbf{P}^e$	5	3.85303
$^3\mathbf{H}^e$	1	0.23973	$^3\mathbf{P}^o$	1	2.57000	$^3\mathbf{F}^o$	5	3.08148	$^1\mathbf{P}^e$	3	3.47631	$^3\mathbf{D}^e$	10	3.86232
$^3\mathbf{F}^e$	1	0.24310	$^3\mathbf{H}^o$	1	2.57997	$^3\mathbf{P}^o$	4	3.11914	$^3\mathbf{J}^e$	1	3.47766	$^3\mathbf{H}^e$	7	3.86650
$^3\mathbf{G}^e$	1	0.28689	$^5\mathbf{D}^o$	2	2.58261	$^1\mathbf{F}^o$	4	3.12030	$^3\mathbf{F}^e$	8	3.48642	$^1\mathbf{D}^e$	9	3.87383
$^1\mathbf{G}^e$	1	0.34070	$^1\mathbf{S}^o$	1	2.58304	$^5\mathbf{H}^e$	1	3.18269	$^1\mathbf{J}^e$	1	3.49551	$^1\mathbf{H}^e$	5	3.87793
$^3\mathbf{D}^e$	1	0.35451	$^3\mathbf{G}^o$	2	2.60377	$^1\mathbf{P}^o$	3	3.19115	$^3\mathbf{P}^e$	7	3.49581	$^5\mathbf{G}^o$	2	3.88542
$^1\mathbf{I}^e$	1	0.36058	$^3\mathbf{F}^o$	2	2.61369	$^5\mathbf{P}^e$	2	3.19187	$^1\mathbf{F}^e$	5	3.50127	$^5\mathbf{D}^o$	3	3.89112
$^1\mathbf{S}^e$	1	0.36945	$^3\mathbf{P}^o$	2	2.61386	$^3\mathbf{D}^e$	4	3.19552	$^1\mathbf{G}^e$	5	3.50166	$^3\mathbf{D}^o$	7	3.89228
$^1\mathbf{D}^e$	1	0.40861	$^1\mathbf{D}^o$	1	2.61489	$^5\mathbf{F}^e$	2	3.19857	$^1\mathbf{D}^e$	6	3.50201	$^5\mathbf{F}^o$	2	3.89552
$^1\mathbf{F}^e$	1	0.49878	$^1\mathbf{G}^o$	1	2.62343	$^5\mathbf{G}^e$	1	3.20832	$^3\mathbf{G}^e$	6	3.55561	$^3\mathbf{G}^o$	5	3.90172
$^3\mathbf{F}^e$	2	0.57568	$^1\mathbf{F}^o$	1	2.62391	$^3\mathbf{H}^e$	3	3.25550	$^3\mathbf{F}^e$	9	3.56323	$^1\mathbf{F}^e$	8	3.90524
$^3\mathbf{P}^e$	2	0.57687	$^1\mathbf{H}^o$	1	2.62620	$^3\mathbf{P}^e$	5	3.26074	$^3\mathbf{F}^e$	10	3.58212	$^3\mathbf{D}^e$	11	3.90953
$^1\mathbf{G}^e$	2	0.66203	$^5\mathbf{S}^o$	1	2.64134	$^3\mathbf{G}^e$	3	3.26379	$^3\mathbf{H}^e$	5	3.58750	$^3\mathbf{F}^o$	6	3.90975
$^1\mathbf{D}^e$	2	0.88399	$^3\mathbf{D}^o$	2	2.64153	$^3\mathbf{F}^e$	5	3.27507	$^3\mathbf{D}^e$	8	3.59355	$^3\mathbf{G}^e$	9	3.92485
$^1\mathbf{S}^e$	2	1.11138	$^1\mathbf{P}^o$	1	2.64663	$^5\mathbf{D}^e$	2	3.30356	$^1\mathbf{F}^e$	6	3.59707	$^1\mathbf{P}^e$	6	3.93912
$^5\mathbf{F}^e$	1	1.80945	$^3\mathbf{H}^o$	2	2.65348	$^1\mathbf{F}^e$	3	3.38104	$^3\mathbf{P}^e$	8	3.61029	$^3\mathbf{S}^e$	2	3.94403
$^3\mathbf{F}^e$	3	1.89449	$^3\mathbf{S}^o$	1	2.65762	$^3\mathbf{D}^e$	5	3.38273	$^5\mathbf{F}^e$	4	3.61264	$^5\mathbf{S}^o$	2	3.94865
$^5\mathbf{P}^e$	1	1.98223	$^3\mathbf{F}^o$	3	2.66586	$^3\mathbf{I}^e$	1	3.38354	$^3\mathbf{F}^e$	11	3.63249	$^5\mathbf{I}^o$	1	3.94963
$^3\mathbf{G}^e$	2	2.01477	$^3\mathbf{D}^o$	3	2.67328	$^5\mathbf{F}^e$	3	3.38368	$^1\mathbf{D}^e$	7	3.63298	$^5\mathbf{P}^o$	2	3.95519
$^1\mathbf{G}^e$	3	2.05448	$^3\mathbf{I}^o$	1	2.68261	$^1\mathbf{H}^e$	2	3.38717	$^1\mathbf{G}^e$	6	3.64309	$^5\mathbf{H}^o$	1	3.95762
$^3\mathbf{P}^e$	3	2.06204	$^3\mathbf{D}^o$	4	2.69539	$^3\mathbf{G}^e$	4	3.39004	$^1\mathbf{I}^e$	3	3.65614	$^3\mathbf{D}^o$	8	3.95782
$^3\mathbf{P}^e$	4	2.07576	$^1\mathbf{G}^o$	2	2.70107	$^3\mathbf{H}^e$	4	3.40266	$^1\mathbf{P}^e$	4	3.65725	$^5\mathbf{G}^o$	3	3.95937
$^3\mathbf{H}^e$	2	2.08626	$^1\mathbf{F}^o$	2	2.70267	$^5\mathbf{P}^e$	3	3.40885	$^1\mathbf{H}^e$	4	3.65965	$^5\mathbf{D}^o$	4	3.96198
$^3\mathbf{D}^e$	2	2.09019	$^1\mathbf{H}^o$	2	2.70294	$^3\mathbf{D}^e$	6	3.42884	$^3\mathbf{H}^e$	6	3.67080	$^3\mathbf{P}^o$	5	3.96339
$^1\mathbf{P}^e$	1	2.11095	$^3\mathbf{P}^o$	3	2.70391	$^1\mathbf{P}^e$	2	3.43001	$^3\mathbf{G}^e$	7	3.68803	$^5\mathbf{F}^o$	3	3.96391
$^1\mathbf{H}^e$	1	2.12591	$^1\mathbf{I}^o$	1	2.71883	$^3\mathbf{F}^e$	6	3.43102	$^1\mathbf{S}^e$	3	3.68892	$^3\mathbf{F}^o$	7	3.96574
$^1\mathbf{D}^e$	3	2.12865	$^3\mathbf{G}^o$	3	2.73322	$^1\mathbf{I}^e$	2	3.43205	$^3\mathbf{D}^e$	9	3.69696	$^3\mathbf{F}^e$	13	3.97079
$^3\mathbf{F}^e$	4	2.25517	$^1\mathbf{D}^o$	2	2.74356	$^1\mathbf{D}^e$	5	3.43292	$^3\mathbf{F}^e$	12	3.71032	$^3\mathbf{I}^o$	2	3.97173
$^1\mathbf{F}^e$	2	2.28924	$^1\mathbf{P}^o$	2	2.74823	$^3\mathbf{F}^e$	7	3.44146	$^1\mathbf{F}^e$	7	3.74798	$^3\mathbf{H}^o$	3	3.97521
$^5\mathbf{G}^o$	1	2.37591	$^3\mathbf{S}^o$	2	2.76039	$^3\mathbf{S}^e$	1	3.44305	$^5\mathbf{P}^e$	4	3.77631	$^3\mathbf{S}^o$	3	3.97535
$^5\mathbf{D}^o$	1	2.40556	$^3\mathbf{F}^o$	4	2.82918	$^3\mathbf{I}^e$	2	3.44515	$^1\mathbf{D}^e$	8	3.79043	$^3\mathbf{G}^o$	6	3.98390
$^3\mathbf{D}^o$	1	2.41506	$^1\mathbf{D}^o$	3	2.86178	$^1\mathbf{H}^e$	3	3.44716	$^3\mathbf{P}^e$	9	3.79226	$^3\mathbf{P}^e$	12	3.99176
$^5\mathbf{F}^o$	1	2.41613	$^3\mathbf{G}^o$	4	2.86412	$^1\mathbf{G}^e$	4	3.44918	$^3\mathbf{G}^e$	8	3.79425	$^1\mathbf{G}^e$	9	3.99352
$^3\mathbf{G}^o$	1	2.45304	$^3\mathbf{D}^o$	5	2.87591	$^3\mathbf{P}^e$	6	3.45965	$^3\mathbf{P}^e$	10	3.80190			
$^3\mathbf{F}^o$	1	2.48591	$^1\mathbf{G}^o$	3	2.90780	$^3\mathbf{D}^e$	7	3.45967	$^1\mathbf{G}^e$	7	3.80463			
$^3\mathbf{D}^e$	3	2.51930	$^1\mathbf{F}^o$	3	2.91215	$^1\mathbf{F}^e$	4	3.46062	$^1\mathbf{G}^e$	8	3.81300			

Ti-like Fe (Fe^{4+})

gf-values for transitions involving terms with effective $n \leq 4.0$, $L \leq 7$

$i\ i'$	gf_L	$i\ i'$	gf_L	$i\ i'$	gf_L	$i\ i'$	gf_L	$i\ i'$	gf_L	$i\ i'$	gf_L
	$^5P^e$–$^5S^o$	3 1	4.66E+0	3 1	4.34E−2	6 1	6.66E−1	6 5	−2.76E−2	3 4	−1.24E+0
1 1	−1.27E+0	3 2	1.33E+0	3 2	−4.57E−2	6 2	1.58E+0	7 1	9.00E−2	3 5	−5.82E−2
1 2	−2.58E−4	3 3	−2.57E+0	3 3	−2.82E−2	6 3	−6.13E−3	7 2	3.01E−2	3 6	−8.55E−3
2 1	2.12E−2	3 4	−2.76E+0	4 1	6.79E+0	7 1	9.62E−1	7 3	1.95E+0	3 7	−3.78E−2
2 2	−4.15E+0		$^5D^e$–$^5F^o$	4 2	−2.18E+1	7 2	8.07E−1	7 4	2.30E−2	3 8	−2.70E−3
3 1	4.42E+0	1 1	−2.29E+0	4 3	−1.10E−2	7 3	−1.70E−2	7 5	−8.04E−4	4 1	−3.53E−2
3 2	−4.27E−2	1 2	−2.34E−1		$^5G^e$–$^5F^o$	8 1	9.75E−1	8 1	1.40E−1	4 2	−3.11E+0
4 1	7.87E−1	1 3	−3.88E+0	1 1	1.45E+1	8 2	4.59E−4	8 2	2.35E+0	4 3	−4.41E−1
4 2	−1.21E−3	2 1	3.34E+0	1 2	−7.58E+0	8 3	−2.70E−1	8 3	4.07E−1	4 4	−8.67E−3
	$^5P^e$–$^5P^o$	2 2	−1.80E+0	1 3	−1.76E+0	9 1	9.05E−3	8 4	5.12E−2	4 5	−2.79E−3
1 1	−3.73E+0	2 3	−6.84E+0		$^5G^e$–$^5G^o$	9 2	7.81E−1	8 5	−4.37E−1	4 6	−2.06E−2
1 2	−3.28E−3	3 1	1.84E+0	1 1	9.11E+0	9 3	−1.76E−6	9 1	8.08E−1	4 7	−2.72E−3
2 1	2.08E−2	3 2	−1.15E+0	1 2	−4.63E+0	10 1	5.21E−2	9 2	1.19E−2	4 8	−1.86E−4
2 2	−8.34E+0	3 3	−3.63E+0	1 3	−1.35E+1	10 2	1.09E−1	9 3	4.01E−2	5 1	3.76E+0
3 1	3.27E+0		$^5F^e$–$^5D^o$		$^5G^e$–$^5H^o$	10 3	−1.04E−3	9 4	2.57E−3	5 2	1.76E−1
3 2	−6.19E−2	1 1	−6.05E+0	1 1	−3.73E+1	11 1	4.58E−1	9 5	−1.20E−2	5 3	4.12E−1
4 1	2.32E+0	1 2	−2.56E−1		$^5H^e$–$^5G^o$	11 2	5.74E−6	10 1	5.64E−1	5 4	3.69E−2
4 2	−5.94E−3	1 3	−2.01E−1	1 1	2.81E+1	11 3	−2.86E−3	10 2	1.78E−2	5 5	3.14E−4
	$^5P^e$–$^5D^o$	1 4	−9.74E−4	1 2	−1.41E+1	12 1	1.28E−2	10 3	3.13E−1	5 6	3.15E−3
1 1	−2.30E−1	2 1	5.31E+0	1 3	−9.56E−1	12 2	4.64E−2	10 4	6.96E−2	5 7	−2.24E+0
1 2	−5.97E+0	2 2	2.00E−2		$^5H^e$–$^5H^o$	12 3	8.40E−7	10 5	−4.72E−3	5 8	−2.27E+0
1 3	−1.04E−2	2 3	−1.87E+0	1 1	−9.36E+0		$^3P^e$–$^3P^o$	11 1	1.01E−2	6 1	4.01E−2
1 4	−4.34E−6	2 4	−3.83E+0		$^5H^e$–$^5I^o$	1 1	−1.15E−1	11 2	8.57E−1	6 2	4.15E−2
2 1	7.64E+0	3 1	2.32E−1	1 1	−5.54E+1	1 2	−1.10E+0	11 3	5.61E−1	6 3	1.36E−2
2 2	2.85E−1	3 2	1.77E+1		$^3S^e$–$^3P^o$	1 3	−7.11E−2	11 4	3.15E−2	6 4	2.63E−1
2 3	−2.74E+0	3 3	−5.71E−4	1 1	2.26E−1	1 4	−7.23E−3	11 5	−1.76E−3	6 5	1.27E−1
2 4	−5.98E+0	3 4	−1.68E−1	1 2	5.37E−1	1 5	−7.08E−1	12 1	4.44E−3	6 6	1.23E−2
3 1	1.79E−2	4 1	3.88E+0	1 3	8.28E−1	2 1	−2.16E−1	12 2	1.08E−1	6 7	−2.20E−2
3 2	2.71E−1	4 2	1.53E−1	1 4	2.26E−3	2 2	−3.32E−2	12 3	8.75E−2	6 8	−6.12E−3
3 3	−2.56E−2	4 3	−1.21E+1	1 5	−1.03E−2	2 3	−4.06E−1	12 4	3.76E+0	7 1	3.19E−3
3 4	−1.62E−2	4 4	−2.36E−1	2 1	3.04E−6	2 4	−4.35E−1	12 5	2.38E−4	7 2	1.64E−1
4 1	1.49E−1		$^5F^e$–$^5F^o$	2 2	2.14E−2	2 5	−5.24E−4		$^3P^e$–$^3D^o$	7 3	2.50E−1
4 2	3.66E+0	1 1	−8.75E+0	2 3	6.77E−5	3 1	−1.72E+0	1 1	−9.51E−1	7 4	1.06E−1
4 3	−4.06E−2	1 2	−2.55E−1	2 4	1.68E+0	3 2	−1.82E−3	1 2	−5.51E−2	7 5	3.82E−1
4 4	−8.10E−3	1 3	−5.26E−5	2 5	−4.29E−7	3 3	−4.55E−1	1 3	−2.15E−2	7 6	1.32E−2
	$^5D^e$–$^5P^o$	2 1	1.14E+1		$^3P^e$–$^3S^o$	3 4	−2.89E−6	1 4	−4.39E−1	7 7	−5.57E−4
1 1	−2.23E+0	2 2	−4.69E+0	1 1	−5.01E−1	3 5	−3.33E−3	1 5	−1.10E−2	7 8	−2.49E−3
1 2	−1.13E+0	2 3	−1.53E+1	1 2	−1.65E−3	4 1	−1.00E−1	1 6	−4.50E−2	8 1	5.99E−1
2 1	2.92E+0	3 1	2.41E−1	1 3	−3.73E−1	4 2	−1.37E+0	1 7	−7.32E−2	8 2	7.37E−2
2 2	−2.50E+0	3 2	−1.04E−1	2 1	−4.30E−2	4 3	−6.35E−1	1 8	−4.29E−1	8 3	1.06E−1
3 1	7.30E+0	3 3	−2.35E−1	2 2	−6.55E−1	4 4	−2.18E−2	2 1	−2.07E−3	8 4	8.37E−1
3 2	−1.16E+0	4 1	5.49E+0	2 3	−1.36E−3	4 5	−3.54E−3	2 2	−1.97E−4	8 5	1.16E−3
	$^5D^e$–$^5D^o$	4 2	−1.72E+1	3 1	−8.15E−3	5 1	2.24E−2	2 3	−3.90E−6	8 6	8.15E−5
1 1	−4.81E+0	4 3	−4.29E−2	3 2	−7.76E−1	5 2	3.00E−1	2 4	−2.64E−4	8 7	−4.22E−1
1 2	−1.41E−1		$^5F^e$–$^5G^o$	3 3	−5.33E−3	5 3	7.99E−3	2 5	−1.51E+0	8 8	−1.53E−1
1 3	−3.48E−1	1 1	−1.10E+1	4 1	−7.26E−1	5 4	1.20E−3	2 6	−3.83E−3	9 1	1.73E−1
1 4	−3.77E+0	1 2	−4.18E−1	4 2	−3.96E−3	5 5	−4.18E+0	2 7	−1.65E−4	9 2	2.22E−2
2 1	5.13E+0	1 3	−8.20E−4	4 3	−6.76E−6	6 1	1.72E+0	2 8	−8.61E−3	9 3	1.05E+0
2 2	2.26E+0	2 1	1.67E+0	5 1	1.36E−1	6 2	4.51E−2	3 1	−3.15E−1	9 4	7.28E−1
2 3	−4.07E+0	2 2	−6.14E−1	5 2	1.89E−3	6 3	3.20E−1	3 2	−1.97E−1	9 5	8.96E−1
2 4	−6.37E+0	2 3	−2.36E+1	5 3	−2.08E+0	6 4	1.25E−3	3 3	−1.94E+0	9 6	4.65E−2

Ti-like Fe (Fe^{4+})

i i′	gf_L	i i′	gf_L	i i′	gf_L	i i′	gf_L	i i′	gf_L	i i′	gf_L
9 7	−6.74E−2	6 1	1.17E+0	3 6	−3.02E+0	10 2	7.02E−2	6 2	3.32E+0	2 3	−1.01E+0
9 8	−9.14E−4	6 2	2.59E−1	3 7	−5.82E−4	10 3	6.79E−1	6 3	8.40E−1	2 4	−6.41E−1
10 1	5.01E−2	6 3	5.80E−1	3 8	−1.63E−3	10 4	1.48E+0	6 4	3.10E−3	2 5	−7.37E−1
10 2	4.54E−3	6 4	6.18E−5	4 1	4.72E+0	10 5	1.06E−1	6 5	5.98E−2	2 6	−1.50E+0
10 3	3.12E−1	6 5	−2.30E−2	4 2	1.30E−1	10 6	3.71E−2	6 6	−3.91E−2	2 7	−8.97E−3
10 4	2.97E−2	7 1	1.74E−2	4 3	1.36E−2	10 7	−1.07E−5	6 7	−1.07E−1	2 8	−1.10E−1
10 5	2.96E+0	7 2	4.53E+0	4 4	9.47E−2	10 8	−8.21E−4	7 1	5.08E−3	3 1	−3.06E+0
10 6	5.92E−1	7 3	2.69E−1	4 5	7.19E−4	11 1	8.04E−3	7 2	2.55E−1	3 2	−1.79E−1
10 7	−1.14E−2	7 4	4.67E−2	4 6	2.54E−3	11 2	2.77E−7	7 3	2.15E−2	3 3	−1.81E−1
10 8	−6.26E−4	7 5	−1.29E−2	4 7	−1.64E+0	11 3	1.47E−1	7 4	1.86E−3	3 4	−1.17E−1
11 1	5.74E−2	8 1	3.47E−2	4 8	−7.53E+0	11 4	1.47E−4	7 5	5.87E−3	3 5	−1.58E−4
11 2	2.12E+0	8 2	3.44E−3	5 1	4.36E−5	11 5	3.06E−1	7 6	−5.54E−5	3 6	−3.83E−3
11 3	1.35E−1	8 3	2.76E+0	5 2	9.65E−3	11 6	4.31E+0	7 7	−6.54E−4	3 7	−1.79E−1
11 4	2.23E−3	8 4	9.75E−3	5 3	5.12E−1	11 7	8.76E−4	8 1	5.21E−2	3 8	−4.66E−6
11 5	4.96E−3	8 5	−7.04E−2	5 4	1.02E+0	11 8	−2.31E−3	8 2	3.28E−1	4 1	−3.85E−3
11 6	9.08E−3	9 1	1.77E−1	5 5	2.19E−2		$^3D^e$–$^3F^o$	8 3	1.13E+0	4 2	−3.25E−2
11 7	−4.09E−3	9 2	1.13E−2	5 6	9.28E−3	1 1	−2.68E−1	8 4	2.77E−1	4 3	−7.46E−2
11 8	−1.05E−3	9 3	6.68E−1	5 7	−8.82E−3	1 2	−4.27E−1	8 5	4.70E−2	4 4	−3.28E−2
12 1	2.42E−3	9 4	1.06E−3	5 8	−1.29E−1	1 3	−5.77E−1	8 6	−3.58E−2	4 5	−3.26E+0
12 2	5.38E−3	9 5	−3.77E−2	6 1	6.17E−2	1 4	−3.68E−1	8 7	−6.02E−2	4 6	−3.47E−1
12 3	3.88E−3	10 1	1.88E−1	6 2	3.74E−1	1 5	−1.15E−2	9 1	1.60E−1	4 7	−1.91E−3
12 4	6.70E−3	10 2	4.63E−1	6 3	1.60E−3	1 6	−2.21E−2	9 2	1.77E−1	4 8	−7.64E−4
12 5	3.26E−1	10 3	7.85E−1	6 4	1.12E+0	1 7	−1.13E−1	9 3	2.28E−1	5 1	3.09E+0
12 6	9.96E−1	10 4	3.73E−2	6 5	3.83E−2	2 1	−1.30E−1	9 4	2.45E+0	5 2	4.00E−2
12 7	1.22E−3	10 5	−2.56E−3	6 6	8.15E−5	2 2	−1.24E+0	9 5	3.42E−2	5 3	1.23E−2
12 8	3.69E−5	11 1	2.56E−2	6 7	−3.10E−2	2 3	−3.56E+0	9 6	−3.41E−2	5 4	4.66E−3
	$^3D^e$–$^3P^o$	11 2	3.88E−2	6 8	−8.71E−2	2 4	−3.54E−2	9 7	−4.26E−2	5 5	2.03E−2
1 1	−3.61E−1	11 3	1.84E−5	7 1	6.33E−2	2 5	−9.30E−2	10 1	9.73E−2	5 6	4.86E−3
1 2	−1.29E−3	11 4	2.91E+0	7 2	8.10E−1	2 6	−4.27E−3	10 2	8.27E−1	5 7	−1.54E+0
1 3	−1.27E+0	11 5	−8.66E−4	7 3	9.29E−1	2 7	−7.31E−4	10 3	2.27E+0	5 8	−1.29E+0
1 4	−1.54E−1		$^3D^e$–$^3D^o$	7 4	1.05E+0	3 1	9.24E−5	10 4	2.06E−2	6 1	5.68E−1
1 5	−9.31E−2	1 1	−1.01E−1	7 5	4.86E−2	3 2	−2.77E−2	10 5	2.14E−2	6 2	4.77E+0
2 1	−2.61E−1	1 2	−4.36E−1	7 6	1.88E−2	3 3	−8.07E−6	10 6	−2.53E−3	6 3	2.08E+0
2 2	−6.83E−1	1 3	−1.25E+0	7 7	−1.04E−2	3 4	−6.94E−2	10 7	−4.39E−5	6 4	1.31E+0
2 3	−1.19E+0	1 4	−7.94E−2	7 8	−2.46E−2	3 5	−4.65E+0	11 1	1.20E−2	6 5	6.08E−2
2 4	−3.93E−2	1 5	−1.88E−1	8 1	2.84E−2	3 6	−6.90E−6	11 2	5.45E−2	6 6	5.93E−3
2 5	−4.53E−5	1 6	−5.47E−2	8 2	6.73E−1	3 7	−3.23E−5	11 3	6.30E−6	6 7	−1.50E−4
3 1	−1.71E−5	1 7	−5.25E−2	8 3	7.96E−1	4 1	2.24E+0	11 4	1.76E−2	6 8	−4.00E−2
3 2	−1.25E−2	1 8	−3.05E−1	8 4	4.84E−1	4 2	2.12E−1	11 5	5.72E−1	7 1	5.22E−3
3 3	−6.95E−4	2 1	−4.13E−5	8 5	1.82E+0	4 3	1.08E−4	11 6	−7.74E−6	7 2	1.78E+0
3 4	−2.03E+0	2 2	−1.05E−1	8 6	5.13E−2	4 4	2.29E−2	11 7	−2.35E−3	7 3	2.32E+0
3 5	−2.72E−4	2 3	1.14E+0	8 7	4.82E−2	4 5	5.55E−5		$^3F^e$–$^3D^o$	7 4	1.23E+0
4 1	2.47E−1	2 4	−2.32E+0	8 8	−1.21E−1	4 6	−5.11E−1	1 1	−5.58E−1	7 5	3.78E−4
4 2	1.72E−2	2 5	−1.12E−1	9 1	1.45E−3	4 7	−7.98E+0	1 2	−1.40E+0	7 6	7.82E−3
4 3	6.97E−2	2 6	−1.53E−2	9 2	3.45E−1	5 1	8.87E−3	1 3	−1.66E−1	7 7	−1.17E−2
4 4	2.46E−4	2 7	−3.41E−5	9 3	9.27E−1	5 2	1.44E+0	1 4	−8.27E−1	7 8	−7.97E−2
4 5	−2.16E+0	2 8	−2.10E−3	9 4	1.22E−1	5 3	6.26E−1	1 5	−3.13E−3	8 1	1.70E−4
5 1	3.82E+0	3 1	1.69E−4	9 5	2.91E+0	5 4	1.11E−1	1 6	−2.04E−1	8 2	5.44E−1
5 2	3.75E−2	3 2	−1.12E−3	9 6	4.42E−1	5 5	9.81E−3	1 7	−5.12E−2	8 3	3.82E+0
5 3	2.06E−1	3 3	−1.16E−2	9 7	−2.40E−2	5 6	−2.60E−2	1 8	−2.51E−1	8 4	3.11E+0
5 4	3.36E−5	3 4	−3.61E−4	9 8	−6.04E−2	5 7	−1.30E−2	2 1	−7.30E−3	8 5	7.52E−2
5 5	−1.72E−1	3 5	−2.73E−1	10 1	2.06E−4	6 1	2.23E−1	2 2	−1.41E−2	8 6	1.30E−2

Ti-like Fe (Fe^{4+})

$i\ i'$	gf_L	$i\ i'$	gf_L	$i\ i'$	gf_L	$i\ i'$	gf_L	$i\ i'$	gf_L	$i\ i'$	gf_L
8 7	−7.06E−3	2 3	−2.31E−2	9 6	−1.78E−1	4 4	−6.24E+0	13 2	1.21E−2	7 5	3.71E−2
8 8	−3.48E−3	2 4	−2.76E−1	9 7	−3.87E−1	4 5	−1.63E−3	13 3	3.17E−1	7 6	−2.15E−1
9 1	8.14E−2	2 5	−1.12E+0	10 1	2.71E−1	4 6	−4.87E−3	13 4	6.72E−3	7 7	−7.58E−2
9 2	9.81E−1	2 6	−2.08E−2	10 2	3.00E+0	5 1	4.95E−1	13 5	5.13E−3	8 1	5.44E−2
9 3	4.65E−1	2 7	−3.94E−1	10 3	2.85E−2	5 2	6.78E−1	13 6	−5.79E−3	8 2	2.32E+0
9 4	2.00E−1	3 1	−4.93E+0	10 4	1.35E+0	5 3	9.10E−2		$^3G^e$–$^3F^o$	8 3	9.15E−1
9 5	4.58E−1	3 2	−2.02E−1	10 5	1.13E−3	5 4	3.68E−3	1 1	−1.55E+0	8 4	6.97E−2
9 6	4.69E−3	3 3	−9.05E−2	10 6	−9.45E−1	5 5	−2.23E−1	1 2	−2.84E−3	8 5	7.27E−2
9 7	−3.26E−1	3 4	−4.23E−2	10 7	−2.03E−1	5 6	−1.16E+1	1 3	−2.26E+0	8 6	−5.36E−3
9 8	−7.65E−2	3 5	−8.49E−5	11 1	3.14E+0	6 1	3.12E−2	1 4	−1.08E+0	8 7	−3.18E−2
10 1	5.16E−2	3 6	−1.40E−1	11 2	2.11E−2	6 2	2.57E−1	1 5	−1.35E−1	9 1	2.49E−2
10 2	2.64E+0	3 7	−3.96E−3	11 3	3.28E−2	6 3	1.10E+0	1 6	−3.67E−1	9 2	1.52E−1
10 3	4.43E−1	4 1	−2.01E−2	11 4	1.52E−3	6 4	9.26E−2	1 7	−1.38E−1	9 3	1.28E−4
10 4	1.86E+0	4 2	−3.67E−2	11 5	1.21E−3	6 5	−9.55E−2	2 1	−6.95E−2	9 4	1.23E−1
10 5	5.90E−1	4 3	−1.19E−1	11 6	−9.48E+0	6 6	−4.96E−3	2 2	−3.57E+0	9 5	1.44E+1
10 6	1.04E−1	4 4	−4.82E+0	11 7	−1.80E−1	7 1	3.24E−1	2 3	−1.46E+0	9 6	8.68E−4
10 7	−1.11E−3	4 5	−1.28E−1	12 1	9.52E−3	7 2	1.72E+0	2 4	−1.44E−1	9 7	−7.53E−4
10 8	−7.11E−2	4 6	−7.52E−3	12 2	3.25E−1	7 3	9.00E−1	2 5	−3.83E−2		$^3G^e$–$^3G^o$
11 1	2.31E+0	4 7	−3.57E−3	12 3	7.51E−2	7 4	2.09E−1	2 6	−1.71E−3	1 1	−9.31E−1
11 2	1.44E−1	5 1	6.27E+0	12 4	4.39E+0	7 5	−1.39E−2	2 7	−5.04E−3	1 2	−1.05E+0
11 3	3.98E−2	5 2	1.27E−1	12 5	2.88E−3	7 6	−3.14E−1	3 1	7.49E+0	1 3	−1.91E+0
11 4	1.09E−7	5 3	3.52E−1	12 6	−3.04E−3	8 1	3.99E−2	3 2	5.66E−1	1 4	−2.93E−1
11 5	1.87E−2	5 4	2.35E−2	12 7	−1.69E−6	8 2	1.74E−1	3 3	1.61E−1	1 5	−1.20E−1
11 6	2.67E−3	5 5	6.80E−6	13 1	4.82E−2	8 3	6.24E−2	3 4	5.22E−3	1 6	−1.18E+0
11 7	−6.91E+0	5 6	−3.79E+0	13 2	1.39E−1	8 4	7.47E−3	3 5	1.13E−3	2 1	−3.17E−1
11 8	−4.35E−2	5 7	−5.97E+0	13 3	5.17E−2	8 5	−1.10E−2	3 6	−3.23E+0	2 2	−6.10E+0
12 1	2.15E−2	6 1	4.43E−2	13 4	1.03E−2	8 6	−2.13E−2	3 7	−1.32E+0	2 3	−2.99E−1
12 2	1.15E−1	6 2	2.70E−1	13 5	3.93E+0	9 1	1.70E−1	4 1	7.93E−1	2 4	−2.27E−2
12 3	3.47E−1	6 3	3.33E−1	13 6	5.32E−2	9 2	5.27E−2	4 2	3.56E+0	2 5	−1.08E−2
12 4	1.10E+0	6 4	9.77E−2	13 7	1.11E−3	9 3	6.53E+0	4 3	5.97E−2	2 6	−3.63E−5
12 5	1.70E+0	6 5	1.06E−3		$^3F^e$–$^3G^o$	9 4	5.65E−2	4 4	1.48E−1	3 1	5.01E+0
12 6	6.30E−1	6 6	−9.40E−5	1 1	−4.65E−1	9 5	−3.06E−1	4 5	3.27E−3	3 2	7.59E−3
12 7	−2.42E−4	6 7	−2.30E−2	1 2	−7.57E−1	9 6	−8.74E−1	4 6	−1.57E−1	3 3	5.42E−1
12 8	−2.43E−4	7 1	2.09E−1	1 3	−5.97E−3	10 1	7.51E−2	4 7	−9.70E−2	3 4	1.31E−2
13 1	1.87E−3	7 2	1.59E+0	1 4	−1.08E−1	10 2	1.62E+0	5 1	1.87E−1	3 5	−2.31E+0
13 2	1.38E−2	7 3	7.53E−1	1 5	−3.53E−2	10 3	2.51E−1	5 2	1.88E+0	3 6	−6.61E+0
13 3	9.51E−2	7 4	1.30E−1	1 6	−1.43E+0	10 4	7.57E−3	5 3	6.85E+0	4 1	1.88E−1
13 4	5.24E−3	7 5	2.54E−2	2 1	−1.60E−2	10 5	−1.24E−1	5 4	8.74E−2	4 2	7.27E+0
13 5	8.22E−1	7 6	−2.76E−1	2 2	−3.29E−2	10 6	−4.67E−1	5 5	2.59E−2	4 3	1.99E−1
13 6	6.94E+0	7 7	−1.71E−1	2 3	−1.51E+0	11 1	4.15E+0	5 6	−7.82E−5	4 4	3.02E−2
13 7	1.45E−2	8 1	8.56E−2	2 4	−4.16E−1	11 2	1.47E−1	5 7	−1.21E−3	4 5	−8.68E−3
13 8	1.05E−3	8 2	2.05E−1	2 5	−1.15E−3	11 3	1.47E−2	6 1	4.09E−1	4 6	−3.31E−1
	$^3F^e$–$^3F^o$	8 3	2.79E+0	2 6	−4.43E−1	11 4	4.27E−3	6 2	3.47E−1	5 1	2.15E−2
1 1	−6.66E−1	8 4	1.59E−1	3 1	−6.30E+0	11 5	−1.27E+1	6 3	3.06E+0	5 2	2.50E−1
1 2	−2.51E+0	8 5	5.84E−2	3 2	−5.55E−1	11 6	−1.66E−2	6 4	1.62E+0	5 3	2.46E+0
1 3	−6.46E−2	8 6	−3.37E−3	3 3	−3.38E−2	12 1	1.00E−2	6 5	2.95E−4	5 4	2.69E−1
1 4	−7.78E−2	8 7	−1.46E−2	3 4	−1.44E−3	12 2	1.45E−1	6 6	−3.45E−1	5 5	−1.17E−4
1 5	−2.11E−3	9 1	3.80E−1	3 5	−2.32E−1	12 3	1.40E+0	6 7	−1.39E−1	5 6	−3.84E−3
1 6	−5.08E−2	9 2	1.18E+0	3 6	−4.93E−3	12 4	1.84E+0	7 1	2.15E−1	6 1	2.46E−1
1 7	−9.12E−1	9 3	3.85E−1	4 1	−9.75E−3	12 5	−4.51E−3	7 2	3.69E−2	6 2	1.08E+0
2 1	−3.75E−1	9 4	6.87E−1	4 2	−5.40E−2	12 6	−8.15E−5	7 3	1.41E+0	6 3	3.48E+0
2 2	−6.77E−2	9 5	2.15E−1	4 3	−4.98E−1	13 1	3.83E−3	7 4	6.97E+0	6 4	1.80E+0

Ti-like Fe (Fe^{4+})

i i'	gf_L	i i'	gf_L	i i'	gf_L	i i'	gf_L	i i'	gf_L	i i'	gf_L
6 5	−2.34E−1	1 4	−4.02E−1	5 1	1.93E−1	3 1	4.11E−1	2 1	−5.47E−3	7 4	6.17E−4
6 6	−4.96E−1	1 5	−6.16E−1	5 2	9.63E+0	4 1	2.07E−3	2 2	−2.72E−2	8 1	2.51E−2
7 1	3.36E−1	1 6	−8.60E−2	5 3	−3.41E−1	5 1	1.67E−1	2 3	−3.43E−1	8 2	1.37E−1
7 2	3.63E−1	2 1	−5.79E−2	6 1	4.94E−3	6 1	3.46E−3	3 1	−4.77E−1	8 3	1.63E+0
7 3	1.98E+0	2 2	−1.80E−1	6 2	6.56E−1		$^{1}P^{e}$–$^{1}P^{o}$	3 2	−2.17E−1	8 4	7.71E−2
7 4	3.57E+0	2 3	−6.04E+0	6 3	−1.13E−1	1 1	−2.22E−1	3 3	−1.48E−2	9 1	2.72E−1
7 5	−1.81E−1	2 4	−6.07E−1	7 1	1.02E+0	1 2	−5.13E−1	4 1	−2.40E−3	9 2	4.61E−1
7 6	−3.16E−1	2 5	−5.31E−3	7 2	4.22E+0	1 3	−2.46E−3	4 2	−3.16E−5	9 3	1.81E−2
8 1	2.40E−1	2 6	−5.51E−4	7 3	−8.59E−3	2 1	9.29E−1	4 3	−7.16E−1	9 4	1.11E−2
8 2	3.83E+0	3 1	1.60E+1		$^{3}H^{e}$–$^{3}I^{o}$	2 2	4.17E−3	5 1	7.01E−2		$^{1}D^{e}$–$^{1}F^{o}$
8 3	1.51E−1	3 2	5.08E−2	1 1	−5.75E−1	2 3	4.39E−4	5 2	5.21E−1	1 1	−5.17E−1
8 4	7.37E−3	3 3	6.94E−3	1 2	−4.10E+0	3 1	8.82E−2	5 3	6.23E−4	1 2	−2.09E−2
8 5	−2.34E−2	3 4	5.82E−3	2 1	−9.93E+0	3 2	8.95E−1	6 1	6.77E−1	1 3	−2.75E−2
8 6	−3.36E−2	3 5	−7.63E+0	2 2	−2.45E−3	3 3	4.96E−4	6 2	1.52E−1	1 4	−3.56E−2
9 1	1.10E−2	3 6	−4.74E−1	3 1	8.12E−2	4 1	3.30E−3	6 3	5.73E−3	2 1	−1.60E−2
9 2	5.20E−3	4 1	1.54E−2	3 2	−2.96E+1	4 2	8.68E−5	7 1	3.32E−1	2 2	−5.40E−2
9 3	1.41E−3	4 2	7.28E+0	4 1	4.29E−1	4 3	1.45E−2	7 2	1.08E+0	2 3	−2.42E−1
9 4	3.48E−2	4 3	1.86E+0	4 2	−2.34E−1	5 1	1.49E−1	7 3	5.27E−2	2 4	−5.43E−2
9 5	1.06E−3	4 4	2.21E−4	5 1	1.89E+0	5 2	3.08E−1	8 1	1.86E−1	3 1	−2.66E−1
9 6	−1.69E−3	4 5	−5.50E−2	5 2	−1.39E+0	5 3	7.63E−3	8 2	2.71E−2	3 2	−1.40E+0
	$^{3}G^{e}$–$^{3}H^{o}$	4 6	−5.71E−3	6 1	8.02E−2	6 1	5.08E−3	8 3	8.85E−2	3 3	−5.40E−2
1 1	−5.93E−1	5 1	7.30E−1	6 2	−3.57E−1	6 2	2.77E−5	9 1	3.67E−1	3 4	−1.45E−2
1 2	−2.85E−1	5 2	2.75E+0	7 1	5.85E+0	6 3	1.21E+0	9 2	1.29E−1	4 1	−5.68E−3
1 3	−2.78E+0	5 3	1.69E+0	7 2	−6.10E−4		$^{1}P^{e}$–$^{1}D^{o}$	9 3	3.37E−6	4 2	−9.64E−3
2 1	−6.54E+0	5 4	2.31E+0		$^{3}I^{e}$–$^{3}H^{o}$	1 1	−6.79E−1		$^{1}D^{e}$–$^{1}D^{o}$	4 3	−1.19E−2
2 2	−1.70E+0	5 5	−7.97E−1	1 1	1.81E+1	1 2	−5.13E−1	1 1	−9.74E−2	4 4	−1.56E+0
2 3	−6.19E−3	5 6	−3.41E−2	1 2	7.91E−1	1 3	−1.25E−2	1 2	−1.71E−1	5 1	1.47E+0
3 1	2.63E−1	6 1	1.39E−1	1 3	−4.67E−3	1 4	−4.61E−3	1 3	−2.90E−1	5 2	2.33E−1
3 2	2.68E−2	6 2	9.89E−1	2 1	1.47E−2	2 1	2.58E−2	1 4	−7.94E−5	5 3	6.72E−2
3 3	−1.97E+1	6 3	2.14E+0	2 2	1.22E+1	2 2	1.81E−1	2 1	−4.29E−5	5 4	2.18E−2
4 1	3.73E−1	6 4	1.37E+1	2 3	−8.24E−3	2 3	6.80E−3	2 2	−2.36E−3	6 1	3.09E−2
4 2	1.44E+0	6 5	−2.32E−1		$^{3}I^{e}$–$^{3}I^{o}$	2 4	7.62E−8	2 3	−7.78E−2	6 2	2.23E−1
4 3	−1.87E−1	6 6	−1.80E−2	1 1	8.48E−1	3 1	1.19E−1	2 4	−9.90E−1	6 3	7.85E−2
5 1	2.35E−1	7 1	4.81E−2	1 2	−2.38E−3	3 2	9.23E−2	3 1	−3.63E−1	6 4	9.81E−3
5 2	1.92E+0	7 2	1.30E−1	2 1	7.71E+0	3 3	3.78E−3	3 2	−7.55E−1	7 1	5.60E−1
5 3	−3.52E−5	7 3	3.84E+0	2 2	−6.96E−2	3 4	1.69E−3	3 3	−4.69E−2	7 2	2.11E−1
6 1	1.33E+0	7 4	3.42E−1		$^{3}J^{e}$–$^{3}I^{o}$	4 1	6.85E−2	3 4	−4.60E−2	7 3	1.08E−1
6 2	1.72E+0	7 5	−2.54E−3	1 1	2.27E+1	4 2	3.30E−2	4 1	−8.67E−5	7 4	6.67E−4
6 3	−8.40E−1	7 6	−5.28E−3	1 2	−1.15E−4	4 3	1.24E+0	4 2	−2.25E−5	8 1	1.59E−1
7 1	2.28E−1		$^{3}H^{e}$–$^{3}H^{o}$		$^{1}S^{e}$–$^{1}P^{o}$	4 4	2.55E−1	4 3	−1.99E−1	8 2	1.29E−4
7 2	1.95E−1	1 1	−6.94E−4	1 1	−2.41E−1	5 1	4.80E−1	4 4	−8.57E−1	8 3	5.51E−1
7 3	−4.98E−1	1 2	−4.32E+0	1 2	−1.23E−1	5 2	3.12E−1	5 1	2.51E−1	8 4	9.90E−2
8 1	4.05E+0	1 3	−7.17E−1	1 3	−1.00E−4	5 3	2.56E−3	5 2	4.37E−3	9 1	1.99E−1
8 2	9.58E−1	2 1	−1.50E+0	2 1	−2.35E−4	5 4	4.42E−3	5 3	2.42E−5	9 2	8.98E−1
8 3	−2.58E−4	2 2	−6.64E+0	2 2	−1.53E−4	6 1	4.47E−3	5 4	2.14E−2	9 3	2.35E−2
9 1	2.55E−3	2 3	−5.03E−3	2 3	−3.55E−1	6 2	4.11E−5	6 1	5.33E−1	9 4	6.64E−4
9 2	5.32E−3	3 1	2.86E−2	3 1	4.29E−1	6 3	3.33E−2	6 2	9.88E−1		$^{1}F^{e}$–$^{1}D^{o}$
9 3	−2.67E−3	3 2	9.05E−1	3 2	1.92E−1	6 4	4.09E−1	6 3	6.32E−3	1 1	−5.85E−3
	$^{3}H^{e}$–$^{3}G^{o}$	3 3	−4.82E+0	3 3	2.13E−3		$^{1}D^{o}$–$^{1}P^{o}$	6 4	1.05E−3	1 2	−1.11E+0
1 1	−2.38E+0	4 1	7.53E+0		$^{1}P^{e}$–$^{1}S^{o}$	1 1	−3.74E−1	7 1	2.44E−1	1 3	−1.69E−1
1 2	−3.22E+0	4 2	7.62E−3	1 1	−2.17E−1	1 2	−3.00E−1	7 2	2.00E−1	1 4	−2.26E−2
1 3	−1.54E+0	4 3	−2.64E−4	2 1	4.57E−1	1 3	−3.76E−3	7 3	1.67E−1	2 1	−6.11E−2

Ti-like Fe (Fe^{4+})

$i\ i'$	gf_L	$i\ i'$	gf_L	$i\ i'$	gf_L	$i\ i'$	gf_L	$i\ i'$	gf_L	$i\ i'$	gf_L
2 2	−2.02E−2	3 1	1.46E+0	4 3	9.38E−3	6 3	3.36E−2	8 1	3.75E−1	4 2	7.52E−1
2 3	−8.58E−1	3 2	5.54E−1	5 1	1.09E−2	6 4	6.55E−2	8 2	1.22E−2	4 3	4.82E+0
2 4	−2.93E−1	3 3	1.09E−2	5 2	1.89E+0	7 1	7.91E−1	8 3	1.84E+0	5 1	1.55E−3
3 1	2.66E−1	3 4	1.70E−2	5 3	4.63E−1	7 2	1.64E−1	9 1	1.08E−2	5 2	1.21E+0
3 2	2.29E−3	4 1	3.20E−1	6 1	1.04E−2	7 3	4.03E−1	9 2	2.39E−2	5 3	2.42E−1
3 3	3.51E−3	4 2	9.09E−2	6 2	4.46E−1	7 4	6.15E−3	9 3	1.21E−1		$^1H^e$–$^1H^o$
3 4	1.17E−3	4 3	1.12E−1	6 3	5.94E−1	8 1	8.32E−2		$^1G^e$–$^1H^o$	1 1	−4.84E−1
4 1	2.85E+0	4 4	3.63E−4	7 1	3.30E−3	8 2	4.77E−4	1 1	−2.53E−1	1 2	−2.28E+0
4 2	1.28E−1	5 1	2.14E−1	7 2	6.80E−1	8 3	2.02E+0	1 2	−2.98E−4	2 1	2.45E+0
4 3	8.08E−2	5 2	2.04E−1	7 3	1.41E−3	8 4	4.11E−1	2 1	−4.09E−2	2 2	9.02E−4
4 4	2.26E−2	5 3	1.76E−3	8 1	4.92E−3	9 1	3.20E−2	2 2	−2.95E−1	3 1	8.49E−2
5 1	3.06E−3	5 4	1.61E−3	8 2	6.49E−2	9 2	1.79E−1	3 1	−2.22E+0	3 2	3.39E+0
5 2	9.43E−1	6 1	1.01E−1	8 3	3.70E−2	9 3	2.63E−1	3 2	−6.15E−1	4 1	2.15E−2
5 3	1.28E−2	6 2	4.94E−1		$^1G^e$–$^1F^o$	9 4	4.53E+0	4 1	1.18E−1	4 2	1.99E−2
5 4	1.97E−2	6 3	6.42E−1	1 1	−4.54E−2		$^1G^e$–$^1G^o$	4 2	1.87E−1	5 1	3.41E−1
6 1	7.56E−3	6 4	1.89E−2	1 2	−1.42E+0	1 1	−1.37E+0	5 1	1.66E−1	5 2	1.43E+0
6 2	8.46E−1	7 1	1.03E−2	1 3	−1.07E−1	1 2	−8.39E−2	5 2	1.13E+0		$^1H^e$–$^1I^o$
6 3	5.69E−1	7 2	1.06E−1	1 4	−4.72E−2	1 3	−2.75E−3	6 1	6.01E−1	1 1	−3.38E+0
6 4	5.45E−2	7 3	1.63E+0	2 1	−4.09E−6	2 1	−5.89E−2	6 2	7.35E−2	2 1	1.03E−1
7 1	5.36E−5	7 4	1.32E−3	2 2	−7.03E−2	2 2	−1.30E−1	7 1	1.17E+0	3 1	5.70E−1
7 2	1.18E+0	8 1	1.90E−3	2 3	−1.15E+0	2 3	−1.08E+0	7 2	5.49E−1	4 1	5.44E−2
7 3	1.98E−1	8 2	3.01E−4	2 4	−4.91E−1	3 1	−2.22E+0	8 1	1.99E−1	5 1	2.01E+0
7 4	2.93E−1	8 3	1.92E−2	3 1	−1.30E+0	3 2	−1.37E−2	8 2	1.59E−1		$^1I^e$–$^1H^o$
8 1	1.87E−3	8 4	1.32E+0	3 2	−3.20E−1	3 3	−2.40E−3	9 1	8.63E−3	1 1	−2.27E−1
8 2	3.68E−2		$^1F^e$–$^1G^o$	3 3	−6.08E−2	4 1	1.50E+0	9 2	4.73E−2	1 2	−3.20E+0
8 3	5.69E−1	1 1	−5.47E−4	3 4	−3.53E−2	4 2	2.83E−1		$^1H^e$–$^1G^o$	2 1	6.11E+0
8 4	1.88E+0	1 2	−6.26E−1	4 1	1.81E+0	4 3	4.98E−2	1 1	−3.54E−3	2 2	6.25E−2
	$^1F^e$–$^1F^o$	1 3	−2.37E−2	4 2	7.92E−1	5 1	2.33E−3	1 2	−1.71E+0	3 1	8.33E−2
1 1	−1.81E−3	2 1	−1.19E−2	4 3	6.39E−3	5 2	2.01E+0	1 3	−5.08E−1	3 2	4.98E+0
1 2	−1.70E−1	2 2	−4.07E−1	4 4	3.07E−3	5 3	5.65E−2	2 1	2.95E+0		$^1I^e$–$^1I^o$
1 3	−3.06E−1	2 3	−1.79E+0	5 1	3.22E−2	6 1	1.71E+0	2 2	1.20E−1	1 1	−1.48E+0
1 4	−1.45E−1	3 1	1.27E+0	5 2	9.80E−1	6 2	3.86E−1	2 3	5.32E−3	2 1	5.46E−1
2 1	−6.54E−2	3 2	1.44E−2	5 3	4.25E−1	6 3	3.84E−2	3 1	3.88E−1	3 1	2.69E+0
2 2	−1.13E−2	3 3	4.52E−2	5 4	2.78E−3	7 1	1.15E+0	3 2	1.21E+0		$^1J^e$–$^1I^o$
2 3	−1.70E+0	4 1	1.21E−1	6 1	1.16E−2	7 2	4.28E−3	3 3	4.09E−2	1 1	7.60E+0
2 4	−1.36E−2	4 2	1.48E−2	6 2	2.51E+0	7 3	1.93E−1	4 1	6.13E−3		

V-like Fe (Fe^{3+})

Term energies relative to $3d^4$ ^{5}D ionization threshold for each symmetry

i	E(Ryds)	Description
		^{2}**S**^{e}
1	−3.27170	$3d^5(^5)$
2	−2.38163	$3d^4(^4)$ ^{1}S $4s$
3	−1.65252	$3d^4(^0)$ ^{1}S $4s$
4	−1.12033	$3d^4$ ^{3}D $4d$:
5	−1.00122	$3d^4(^4)$ ^{1}S $5s$
6	−.96004	$3d^4(^4)$ ^{1}D $4d$
7	−.66395	$3d^4(^2)$ ^{1}D $4d$
8	−.49519	$3d^4$ ^{3}D $5d$
9	−.46452	$3d^4(^4)$ ^{1}S $6s$
10	−.41028	$3d^4(^4)$ ^{1}D $5d$
11	−.37058	$3d^4$ ^{3}G $5g$
12	−.30319	$3d^4(^4)$ ^{1}G $5g$
13	−.25216	$3d^4(^0)$ ^{1}S $5s$
14	−.20850	$3d^4$ ^{3}D $6d$
15	−.19691	$3d^4(^4)$ ^{1}S $7s$
16	−.17497	$3d^4$ ^{3}G $6g$
17	−.13595	$3d^4(^4)$ ^{1}D $6d$
18	−.10739	$3d^4(^4)$ ^{1}G $6g$
19	−.05707	$3d^4$ ^{3}G $7g$
20	−.05094	$3d^4$ ^{3}D $7d$
21	−.04412	$3d^4(^4)$ ^{1}S $8s$
22	−.01910	$3d^4(^2)$ ^{1}D $5d$
		^{2}**S**^{o}
1	−1.99413	$3d^4(^4)$ ^{3}P $4p$
2	−1.57996	$3d^4(^2)$ ^{3}P $4p$
3	−.92773	$3d^4(^4)$ ^{3}P $5p$
4	−.77185	$3d^4(^4)$ ^{3}F $4f$
5	−.57293	$3d^4(^2)$ ^{3}P $5p$
6	−.52226	$3d^4$ ^{1}F $4f$
7	−.48568	$3d^4(^4)$ ^{3}P $6p$
8	−.43481	$3d^4(^2)$ ^{3}F $4f$
9	−.40519	$3d^4(^4)$ ^{3}F $5f$
10	−.26007	$3d^4(^4)$ ^{3}P $7p$
11	−.21962	$3d^4$ ^{3}H $6h$
12	−.21027	$3d^4(^4)$ ^{3}F $6f$
13	−.16572	$3d^4$ ^{1}F $5f$
14	−.14147	$3d^4(^2)$ ^{3}P $6p$
15	−.12464	$3d^4(^4)$ ^{3}P $8p$
16	−.10171	$3d^4$ ^{3}H $7h$
17	−.09163	$3d^4(^4)$ ^{3}F $7f$
18	−.07380	$3d^4(^2)$ ^{3}F $5f$
19	−.04137	$3d^4(^4)$ ^{3}P $9p$
20	−.02517	$3d^4$ ^{3}H $8h$
21	−.01264	$3d^4(^4)$ ^{3}F $8f$
		^{2}**P**^{e}
1	−2.94101	$3d^5(^3)$
2	−2.47727	$3d^4(^4)$ ^{3}P $4s$
3	−2.14451	$3d^4(^2)$ ^{3}P $4s$
4	−1.28586	$3d^4(^4)$ ^{3}P $4d$
5	−1.24535	$3d^4(^4)$ ^{3}F $4d$
6	−1.15040	$3d^4$ ^{3}D $4d$
7	−1.11479	$3d^4(^4)$ ^{3}P $5s$
8	−1.07108	$3d^4(^4)$ ^{1}D $4d$
9	−.99039	$3d^4$ ^{1}F $4d$
10	−.90707	$3d^4(^2)$ ^{3}F $4d$
11	−.83887	$3d^4(^2)$ ^{3}P $4d$
12	−.77384	$3d^4(^2)$ ^{3}P $5s$
13	−.64370	$3d^4(^4)$ ^{3}P $5d$
14	−.63217	$3d^4(^4)$ ^{3}F $5d$
15	−.59953	$3d^4(^2)$ ^{1}D $4d$
16	−.57665	$3d^4(^4)$ ^{3}P $6s$
17	−.51743	$3d^4$ ^{3}D $5d$
18	−.44658	$3d^4(^4)$ ^{1}D $5d$
19	−.41468	$3d^4$ ^{3}H $5g$
20	−.40028	$3d^4(^4)$ ^{3}F $5g$
21	−.37780	$3d^4$ ^{1}F $5d$
22	−.37050	$3d^4$ ^{3}G $5g$
23	−.34416	$3d^4(^4)$ ^{3}P $6d$
24	−.32771	$3d^4(^4)$ ^{3}F $6d$
25	−.30985	$3d^4(^4)$ ^{3}P $7s$
26	−.30322	$3d^4(^4)$ ^{1}G $5g$
27	−.29369	$3d^4(^2)$ ^{3}F $5d$
28	.25835	$3d^4(^2)$ ^{3}P $5d$
29	−.23807	$3d^4(^2)$ ^{3}P $6s$
30	−.22059	$3d^4$ ^{3}D $6d$
31	−.21910	$3d^4$ ^{3}H $6g$
32	−.20469	$3d^4(^4)$ ^{3}F $6g$
33	−.17631	$3d^4(^4)$ ^{3}P $7d$
34	−.17491	$3d^4$ ^{3}G $6g$
35	−.16328	$3d^4$ ^{1}F $5g$
36	−.15983	$3d^4(^4)$ ^{3}F $7d$
37	−.15710	$3d^4(^4)$ ^{3}P $8s$
38	−.14861	$3d^4(^4)$ ^{1}D $6d$
39	−.10743	$3d^4(^4)$ ^{1}G $6g$
40	−.10174	$3d^4$ ^{3}H $7i$
41	−.10143	$3d^4$ ^{3}H $7g$
42	−.08679	$3d^4(^4)$ ^{3}F $7g$
43	−.08618	$3d^4$ ^{1}F $6d$
44	−.07421	$3d^4(^2)$ ^{3}F $5g$
45	−.07386	$3d^4(^4)$ ^{3}P $8d$
40	−.00400	$3d^4$ ^{3}D $7d$
47	−.06143	$3d^4(^4)$ ^{3}P $9s$
48	−.05731	$3d^4(^4)$ ^{3}F $8d$
49	−.05681	$3d^4$ ^{3}G $7g$
50	−.02521	$3d^4$ ^{3}H $8i$
51	−.02501	$3d^4$ ^{3}H $8g$
52	−.01695	$3d^4(^2)$ ^{1}D $5d$
53	−.01018	$3d^4(^4)$ ^{3}F $8g$
54	−.00686	$3d^4(^4)$ ^{3}P $9d$
		^{2}**P**^{o}
1	−1.97786	$3d^4(^4)$ ^{3}P $4p$
2	−1.88338	$3d^4$ ^{3}D $4p$
3	−1.84645	$3d^4(^4)$ ^{1}S $4p$
4	−1.76090	$3d^4(^4)$ ^{1}D $4p$
5	−1.60679	$3d^4(^2)$ ^{3}P $4p$
6	−1.41098	$3d^4(^2)$ ^{1}D $4p$
7	−1.10187	$3d^4(^0)$ ^{1}S $4p$
8	−.92322	$3d^4(^4)$ ^{3}P $5p$
9	−.81640	$3d^4$ ^{3}D $5p$:
10	−.80500	$3d^4(^4)$ ^{1}S $5p$
11	−.76850	$3d^4(^4)$ ^{3}F $4f$
12	−.73825	$3d^4$ ^{3}G $4f$
13	−.72852	$3d^4(^4)$ ^{1}D $5p$
14	−.67640	$3d^4(^4)$ ^{1}G $4f$
15	−.66797	$3d^4$ ^{3}D $4f$
16	−.58623	$3d^4(^4)$ ^{1}D $4f$
17	−.57963	$3d^4(^2)$ ^{3}P $5p$
18	−.52421	$3d^4$ ^{1}F $4f$
19	−.48504	$3d^4(^4)$ ^{3}P $6p$
20	−.43286	$3d^4(^2)$ ^{3}F $4f$
21	−.40527	$3d^4(^4)$ ^{3}F $5f$
22	−.37807	$3d^4$ ^{3}G $5f$
23	−.37471	$3d^4(^4)$ ^{1}S $6p$
24	−.36726	$3d^4$ ^{3}D $6p$
25	−.35934	$3d^4(^2)$ ^{1}G $4f$
26	−.31705	$3d^4(^2)$ ^{1}D $5p$
27	−.30857	$3d^4(^4)$ ^{1}G $5f$
28	−.30512	$3d^4$ ^{3}D $5f$
29	−.29162	$3d^4(^4)$ ^{1}D $6p$
30	−.25853	$3d^4(^4)$ ^{3}P $7p$
31	−.22693	$3d^4(^4)$ ^{1}D $5f$
32	−.21962	$3d^4$ ^{3}H $6h$
33	−.20855	$3d^4(^4)$ ^{3}F $6f$
34	−.17913	$3d^4$ ^{3}G $6f$
35	−.17470	$3d^4$ ^{3}G $6h$
36	−.16912	$3d^4(^2)$ ^{1}D $4f$
37	−.16509	$3d^4$ ^{1}F $5f$
38	−.15115	$3d^4$ ^{3}D $7p$
39	−.14525	$3d^4(^4)$ ^{1}S $7p$
40	−.13874	$3d^4(^2)$ ^{3}P $6p$
41	−.12541	$3d^4(^4)$ ^{3}P $8p$
42	.11180	$3d^4(^4)$ ^{1}G $6f$
43	−.10858	$3d^4$ ^{3}D $6f$
44	−.10647	$3d^4(^4)$ ^{1}G $6h$
45	−.10636	$3d^4$ ^{1}I $6h$
46	−.10171	$3d^4$ ^{3}H $7h$
47	−.08948	$3d^4(^4)$ ^{3}F $7f$
48	−.07408	$3d^4(^2)$ ^{3}F $5f$
49	−.06674	$3d^4(^4)$ ^{1}D $7p$
50	−.05930	$3d^4$ ^{3}G $7f$
51	−.05679	$3d^4$ ^{3}G $7h$
52	−.05466	$3d^4(^0)$ ^{1}S $5p$
53	−.04095	$3d^4(^4)$ ^{3}P $9p$
54	−.03014	$3d^4(^4)$ ^{1}D $6f$
55	−.02517	$3d^4$ ^{3}H $8h$
56	−.01350	$3d^4$ ^{3}D $8p$:
57	−.01236	$3d^4(^4)$ ^{3}F $8f$
58	−.01102	$3d^4(^4)$ ^{1}S $8p$
		^{2}**D**^{e}
1	−3.43429	$3d^5(^5)$
2	−3.19871	$3d^5(^3)$
3	−2.85699	$3d^5(^1)$
4	−2.37606	$3d^4$ ^{3}D $4s$
5	−2.32291	$3d^4(^4)$ ^{1}D $4s$
6	−1.89868	$3d^4(^2)$ ^{1}D $4s$
7	−1.26585	$3d^4(^4)$ ^{3}P $4d$
8	−1.22806	$3d^4$ ^{3}G $4d$
9	−1.22096	$3d^4(^4)$ ^{3}F $4d$
10	−1.15644	$3d^4(^4)$ ^{1}G $4d$
11	−1.13337	$3d^4$ ^{3}D $4d$
12	−1.08642	$3d^4(^4)$ ^{1}S $4d$
13	−1.03078	$3d^4(^4)$ ^{1}D $4d$
14	−1.00306	$3d^4$ ^{3}D $5s$
15	−.97761	$3d^4$ ^{1}F $4d$
16	−.93015	$3d^4(^4)$ ^{1}D $5s$
17	−.90397	$3d^4(^2)$ ^{3}P $4d$
18	−.86807	$3d^4(^2)$ ^{3}F $4d$
19	−.74792	$3d^4(^2)$ ^{1}G $4d$
20	−.63379	$3d^4(^4)$ ^{3}P $5d$
21	−.61452	$3d^4(^2)$ ^{1}D $4d$
22	−.59558	$3d^4(^4)$ ^{3}F $5d$
23	−.58352	$3d^4$ ^{3}G $5d$
24	−.52977	$3d^4(^4)$ ^{1}G $5d$
25	−.51933	$3d^4$ ^{3}D $5d$:
26	−.49790	$3d^4(^2)$ ^{1}D $5s$
27	−.49088	$3d^4(^4)$ ^{1}S $5d$
28	−.46520	$3d^4$ ^{3}D $6s$
29	−.43749	$3d^4(^4)$ ^{1}D $5d$
30	−.41491	$3d^4$ ^{3}H $5g$
31	−.40024	$3d^4(^4)$ ^{3}F $5g$
32	−.38964	$3d^4(^4)$ ^{1}D $6s$
33	−.38693	$3d^4(^0)$ ^{1}S $4d$
34	.37548	$3d^4$ ^{1}F $5d$
35	−.37037	$3d^4$ ^{3}G $5g$
36	−.33893	$3d^4(^4)$ ^{3}P $6d$
37	−.31271	$3d^4(^4)$ ^{3}F $6d$
38	−.30327	$3d^4(^4)$ ^{1}G $5g$
39	−.30279	$3d^4$ ^{3}D $5g$
40	−.30150	$3d^4$ ^{1}I $5g$
41	−.29867	$3d^4$ ^{3}G $6d$
42	−.28127	$3d^4(^2)$ ^{3}P $5d$
43	−.25841	$3d^4(^2)$ ^{3}F $5d$
44	−.23018	$3d^4(^4)$ ^{1}G $6d$:

V-like Fe (Fe^{3+})

i	E(Ryds)	Description	i	E(Ryds)	Description	i	E(Ryds)	Description	i	E(Ryds)	Description
45	−.22603	$3d^4(^4)$ ^{1}S $6d$	15	−.74325	$3d^4(^4)$ ^{1}D $5p$	67	−.06942	$3d^4(^4)$ ^{1}D $7p$	38	−.30392	$3d^4(^2)$ ^{3}F $5d$
46	−.22252	$3d^4(^4)$ ^{1}D $5g$	16	−.73762	$3d^4$ ^{3}G $4f$	68	−.05898	$3d^4$ ^{3}G $7f$	39	−.30322	$3d^4(^4)$ ^{1}G $5g$
47	−.21939	$3d^4$ ^{3}H $6g$	17	−.67691	$3d^4(^4)$ ^{1}G $4f$	69	−.05678	$3d^4$ ^{3}G $7h$	40	−.30267	$3d^4$ ^{3}D $5g$
48	−.21201	$3d^4$ ^{3}D $6d$	18	−.67266	$3d^4$ ^{3}D $4f$	70	−.04052	$3d^4(^4)$ ^{3}P $9p$	41	−.30156	$3d^4$ ^{1}I $5g$
49	−.20462	$3d^4(^4)$ ^{3}F $6g$	19	−.66658	$3d^4$ ^{1}F $5p$	71	−.03159	$3d^4(^4)$ ^{1}D $6f$	42	−.29459	$3d^4(^4)$ ^{3}F $7s$
50	−.19723	$3d^4$ ^{3}D $7s$	20	−.59572	$3d^4(^2)$ ^{3}P $5p$	72	−.02765	$3d^4(^4)$ ^{3}P $8f$	43	−.29067	$3d^4(^2)$ ^{3}P $5d$
51	−.18756	$3d^4(^2)$ ^{1}G $5d$	21	−.59019	$3d^4(^4)$ ^{1}D $4f$	73	−.02618	$3d^4$ ^{3}H $8f$	44	−.28954	$3d^4$ ^{3}G $6d$
52	−.17481	$3d^4$ ^{3}G $6g$	22	−.55771	$3d^4(^2)$ ^{3}F $5p$	74	−.02575	$3d^4(^4)$ ^{3}F $9p$	45	−.23742	$3d^4(^2)$ ^{3}F $6s$
53	−.17048	$3d^4(^4)$ ^{3}P $7d$	23	−.52845	$3d^4$ ^{1}F $4f$	75	−.02521	$3d^4$ ^{3}H $8j$	46	−.23120	$3d^4(^4)$ ^{1}G $6d$
54	−.16370	$3d^4$ ^{1}F $5g$	24	−.48305	$3d^4(^4)$ ^{3}P $6p$	76	−.02518	$3d^4$ ^{3}H $8h$	47	−.22292	$3d^4(^4)$ ^{1}D $5g$
55	−.15801	$3d^4(^4)$ ^{3}F $7d$	25	−.47013	$3d^4(^4)$ ^{3}F $6p$	77	−.01240	$3d^4$ ^{3}D $8p$	48	−.21972	$3d^4(^4)$ ^{3}P $6g$
56	−.13579	$3d^4(^4)$ ^{1}D $6d$	26	−.43953	$3d^4(^2)$ ^{3}F $4f$:	78	−.01105	$3d^4(^4)$ ^{3}F $8f$	49	−.21960	$3d^4$ ^{3}H $6g$
57	−.12775	$3d^4$ ^{3}G $7d$	27	−.43368	$3d^4(^2)$ ^{3}P $4f$	79	−.01007	$3d^4(^4)$ ^{3}F $8h$	50	−.21771	$3d^4(^2)$ ^{1}G $5d$
58	−.11938	$3d^4(^4)$ ^{1}D $7s$	28	−.42101	$3d^4(^4)$ ^{3}P $5f$	80	−.00705	$3d^4$ ^{1}F $7p$	51	−.20712	$3d^4$ ^{3}D $6d$
59	−.10745	$3d^4(^4)$ ^{1}G $6g$	29	−.41642	$3d^4$ ^{3}H $5f$			**^{2}F^e**	52	−.20450	$3d^4(^4)$ ^{3}F $6g$
60	−.10694	$3d^4$ ^{3}D $6g$	30	−.40352	$3d^4(^4)$ ^{3}F $5f$	1	−3.42470	$3d^5(^3)$	53	−.17534	$3d^4(^4)$ ^{3}P $7d$
61	−.10598	$3d^4$ ^{1}I $6g$	31	−.37560	$3d^4$ ^{3}G $5f$	2	−3.33754	$3d^5(^5)$	54	−.17472	$3d^4$ ^{3}G $6g$
62	−.10174	$3d^4$ ^{3}H $7i$	32	−.37154	$3d^4$ ^{3}D $6p$	3	−2.46316	$3d^4(^4)$ ^{3}F $4s$	55	−.16730	$3d^4$ ^{3}H $7d$
63	−.10157	$3d^4$ ^{3}H $7g$	33	−.36009	$3d^4(^2)$ ^{1}G $4f$	4	−2.26607	$3d^4$ ^{1}F $4s$	56	−.16424	$3d^4$ ^{1}F $5g$
64	−.08715	$3d^4$ ^{1}F $6d$	34	−.31089	$3d^4(^4)$ ^{1}G $5f$	5	−2.14682	$3d^4(^2)$ ^{3}F $4s$	57	−.15557	$3d^4(^4)$ ^{3}F $7d$
65	−.08672	$3d^4(^4)$ ^{3}F $7g$	35	−.30797	$3d^4$ ^{3}D $5f$	6	−1.26453	$3d^4(^4)$ ^{3}P $4d$	58	−.14670	$3d^4(^4)$ ^{1}D $6d$
66	−.07461	$3d^4(^2)$ ^{3}F $5g$	36	−.30085	$3d^4(^4)$ ^{1}D $6p$	7	−1.24522	$3d^4(^4)$ ^{3}F $4d$	59	−.14167	$3d^4(^4)$ ^{3}F $8s$
67	−.07309	$3d^4(^4)$ ^{3}P $8d$	37	−.29052	$3d^4(^2)$ ^{1}D $5p$	8	−1.22416	$3d^4$ ^{3}H $4d$	60	−.12655	$3d^4$ ^{3}G $7d$
68	−.06384	$3d^4(^4)$ ^{1}G $7d$:	38	−.25748	$3d^4(^4)$ ^{3}P $7p$	9	−1.21149	$3d^4$ ^{3}G $4d$	61	−.10740	$3d^4(^4)$ ^{1}G $6g$
69	−.06351	$3d^4(^4)$ ^{1}S $7d$	39	−.24340	$3d^4(^4)$ ^{3}F $7p$	10	−1.15655	$3d^4$ ^{3}D $4d$	62	−.10707	$3d^4$ ^{3}D $6g$
70	−.05688	$3d^4$ ^{3}G $7g$	40	−.23262	$3d^4$ ^{1}F $6p$	11	−1.09731	$3d^4(^4)$ ^{3}F $5s$	63	−.10606	$3d^4$ ^{1}I $6g$
71	−.05677	$3d^4$ ^{3}G $7i$	41	−.22904	$3d^4(^4)$ ^{1}D $5f$	12	−1.08188	$3d^4(^4)$ ^{1}D $4d$	64	−.10181	$3d^4(^4)$ ^{3}P $7g$
72	−.05376	$3d^4$ ^{3}D $7d$	42	−.22440	$3d^4(^4)$ ^{3}P $6f$	13	−1.03574	$3d^4(^4)$ ^{1}G $4d$	65	−.10174	$3d^4$ ^{3}H $7i$
73	−.04950	$3d^4(^4)$ ^{3}F $8d$	43	−.22102	$3d^4$ ^{3}H $6f$	14	−.98887	$3d^4$ ^{1}F $4d$	66	−.10172	$3d^4$ ^{3}H $7g$
74	−.04391	$3d^4$ ^{3}D $8s$	44	−.21963	$3d^4$ ^{3}H $6h$	15	−.93694	$3d^4(^2)$ ^{3}F $4d$	67	−.08992	$3d^4$ ^{1}F $6d$
75	−.02711	$3d^4(^4)$ ^{1}D $6g$	45	−.20670	$3d^4(^4)$ ^{3}F $6f$	16	−.90045	$3d^4(^2)$ ^{3}P $4d$	68	−.08665	$3d^4(^4)$ ^{3}F $7g$
76	−.02585	$3d^4$ ^{3}G $8d$	46	−.20452	$3d^4(^4)$ ^{3}F $6h$	17	−.87304	$3d^4$ ^{1}F $5s$	69	−.08659	$3d^4(^4)$ ^{3}F $7i$
77	−.02521	$3d^4$ ^{3}H $8i$	47	−.17885	$3d^4$ ^{3}G $6f$	18	−.83719	$3d^4(^2)$ ^{1}G $4d$	70	−.07557	$3d^4(^2)$ ^{3}F $5g$
78	−.02509	$3d^4$ ^{3}H $8g$	48	−.17470	$3d^4$ ^{3}G $6h$	19	−.77491	$3d^4(^2)$ ^{3}F $5s$	71	−.07431	$3d^4(^2)$ ^{3}P $5g$
79	−.01223	$3d^4(^2)$ ^{1}D $5d$	49	−.16897	$3d^4$ ^{1}F $5f$	20	−.63911	$3d^4(^4)$ ^{3}P $5d$	72	−.07360	$3d^4(^4)$ ^{3}P $8d$
80	−.01015	$3d^4(^4)$ ^{3}F $8g$	50	−.15991	$3d^4(^2)$ ^{1}D $4f$	21	−.62853	$3d^4(^2)$ ^{1}D $4d$	73	−.06714	$3d^4$ ^{3}H $8d$
81	−.00545	$3d^4(^4)$ ^{3}P $9d$	51	−.15193	$3d^4(^2)$ ^{3}P $6p$	22	−.61539	$3d^4(^4)$ ^{3}F $5d$	74	−.06267	$3d^4(^4)$ ^{1}G $7d$
		^{2}D^o	52	−.14503	$3d^4$ ^{3}D $7p$	23	−.60651	$3d^4$ ^{3}H $5d$	75	−.05988	$3d^4$ ^{1}F $7s$
1	−1.98102	$3d^4(^4)$ ^{3}F $4p$	53	−.13361	$3d^4(^2)$ ^{3}F $6p$	24	−.58291	$3d^4$ ^{3}G $5d$	76	−.05955	$3d^4(^4)$ ^{3}F $8d$:
2	−1.95554	$3d^4(^4)$ ^{3}P $4p$	54	−.12491	$3d^4(^4)$ ^{3}P $8p$	25	−.56183	$3d^4(^4)$ ^{3}F $6s$	77	−.05678	$3d^4$ ^{3}G $7g$
3	−1.83251	$3d^4$ ^{3}D $4p$	55	−.11216	$3d^4(^4)$ ^{1}G $6f$	26	−.52732	$3d^4(^4)$ ^{1}G $5d$:	78	−.04701	$3d^4(^4)$ ^{3}F $9s$:
4	−1.81550	$3d^4(^4)$ ^{1}D $4p$	56	−.11124	$3d^4$ ^{3}D $6f$	27	−.48430	$3d^4$ ^{3}D $5d$	79	−.04573	$3d^4$ ^{3}D $7d$
5	−1.71067	$3d^4$ ^{1}F $4p$	57	−.10941	$3d^4(^4)$ ^{3}F $8p$	28	−.44529	$3d^4(^4)$ ^{1}D $5d$	80	−.02743	$3d^4(^4)$ ^{1}D $6g$
6	−1.65899	$3d^4(^2)$ ^{3}P $4p$	58	−.10648	$3d^4(^4)$ ^{1}G $6h$	29	−.41529	$3d^4(^4)$ ^{3}P $5g$	81	−.02669	$3d^4$ ^{3}G $8d$
7	−1.54289	$3d^4(^2)$ ^{3}F $4p$	59	−.10637	$3d^4$ ^{1}I $6h$	30	−.41513	$3d^4$ ^{3}H $5g$	82	−.02527	$3d^4(^4)$ ^{3}P $8g$
8	−1.32365	$3d^4(^2)$ ^{1}D $4p$	60	−.10506	$3d^4(^4)$ ^{3}P $7f$	31	−.40014	$3d^4(^4)$ ^{3}F $5g$	83	−.02521	$3d^4$ ^{3}H $8i$
9	−.91823	$3d^4(^4)$ ^{3}P $5p$	61	−.10272	$3d^4$ ^{3}H $7f$	32	−.38495	$3d^4$ ^{1}F $5d$	84	−.02518	$3d^4$ ^{3}H $8g$
10	−.90935	$3d^4(^4)$ ^{3}F $5p$	62	−.10172	$3d^4$ ^{3}H $7h$	33	−.37028	$3d^4$ ^{3}G $5g$	85	−.01136	$3d^4(^2)$ ^{1}D $5d$
11	−.80657	$3d^4$ ^{3}D $5p$	63	−.08805	$3d^4(^4)$ ^{3}F $7f$	34	−.34203	$3d^4(^4)$ ^{3}P $6d$	86	−.01014	$3d^4(^4)$ ^{3}F $8g$
12	−.78500	$3d^4(^4)$ ^{3}P $4f$	64	−.08661	$3d^4(^4)$ ^{3}F $7h$	35	−.33068	$3d^4$ ^{1}F $6s$	87	−.01006	$3d^4(^4)$ ^{3}F $8i$
13	−.77590	$3d^4$ ^{3}H $4f$	65	−.07759	$3d^4(^2)$ ^{3}F $5f$:	36	−.32426	$3d^4$ ^{3}H $6d$	88	−.00843	$3d^4(^2)$ ^{3}F $6d$
14	−.76512	$3d^4(^4)$ ^{3}F $4f$	66	−.07477	$3d^4(^2)$ ^{3}P $5f$	37	−.32133	$3d^4(^4)$ ^{3}F $6d$	89	−.00492	$3d^4(^4)$ ^{3}P $9d$

V-like Fe (Fe^{3+})

i	E(Ryds)	Description
90	−.00091	$3d^4$ ^{3}H $9d$
		^{2}F°
1	−1.96477	$3d^4(^4)$ ^{3}F $4p$:
2	−1.93565	$3d^4$ ^{3}G $4p$
3	−1.89009	$3d^4(^4)$ ^{1}G $4p$
4	−1.84039	$3d^4$ ^{3}D $4p$
5	−1.79690	$3d^4(^4)$ ^{1}D $4p$
6	−1.76243	$3d^4$ ^{1}F $4p$
7	−1.65840	$3d^4(^2)$ ^{3}F $4p$
8	−1.55320	$3d^4(^2)$ ^{1}G $4p$
9	−1.36260	$3d^4(^2)$ ^{1}D $4p$
10	−.90349	$3d^4(^4)$ ^{3}F $5p$
11	−.87894	$3d^4$ ^{3}G $5p$
12	−.82061	$3d^4(^4)$ ^{1}G $5p$
13	−.80813	$3d^4$ ^{3}D $5p$
14	−.78370	$3d^4(^4)$ ^{3}P $4f$
15	−.78102	$3d^4$ ^{3}H $4f$
16	−.76388	$3d^4(^4)$ ^{3}F $4f$
17	−.73699	$3d^4(^4)$ ^{1}D $5p$
18	−.73500	$3d^4$ ^{3}G $4f$
19	−.68784	$3d^4$ ^{1}F $5p$
20	−.67677	$3d^4(^4)$ ^{1}G $4f$
21	−.66971	$3d^4$ ^{3}D $4f$
22	−.66308	$3d^4$ ^{1}I $4f$
23	−.66068	$3d^4(^4)$ ^{1}S $4f$
24	−.59450	$3d^4(^4)$ ^{1}D $4f$
25	−.59231	$3d^4(^2)$ ^{3}F $5p$
26	−.53382	$3d^4$ ^{1}F $4f$
27	−.50532	$3d^4(^2)$ ^{1}G $5p$
28	−.46781	$3d^4(^4)$ ^{3}F $6p$
29	−.45485	$3d^4(^2)$ ^{3}F $4f$
30	−.43985	$3d^4$ ^{3}G $6p$
31	−.43323	$3d^4(^2)$ ^{3}P $4f$
32	−.42050	$3d^4(^4)$ ^{3}P $5f$
33	−.41859	$3d^4$ ^{3}H $5f$
34	−.40351	$3d^4(^4)$ ^{3}F $5f$
35	−.37810	$3d^4(^4)$ ^{1}G $6p$
36	−.37508	$3d^4$ ^{3}G $5f$
37	−.37026	$3d^4$ ^{3}D $6p$
38	−.36236	$3d^4(^2)$ ^{1}G $4f$
39	−.31136	$3d^4(^4)$ ^{1}G $5f$
40	−.30978	$3d^4(^4)$ ^{1}G $5f$:
41	−.30626	$3d^4$ ^{3}D $5f$:
42	−.30375	$3d^4$ ^{1}I $5f$:
43	−.30350	?
44	−.29566	$3d^4(^4)$ ^{1}D $6p$
45	−.24325	$3d^4(^4)$ ^{3}F $7p$
46	−.24087	$3d^4$ ^{1}F $6p$
47	−.23110	$3d^4(^4)$ ^{1}D $5f$
48	−.22368	$3d^4(^4)$ ^{3}P $6f$
49	−.22251	$3d^4$ ^{3}H $6f$
50	−.21964	$3d^4$ ^{3}H $6h$
51	−.21361	$3d^4$ ^{3}G $7p$
52	−.20656	$3d^4(^4)$ ^{3}F $6f$
53	−.20451	$3d^4(^4)$ ^{3}F $6h$
54	−.17789	$3d^4$ ^{3}G $6f$
55	−.17469	$3d^4$ ^{3}G $6h$
56	−.17101	$3d^4$ ^{1}F $5f$
57	−.15421	$3d^4(^2)$ ^{1}D $4f$
58	−.15046	$3d^4(^2)$ ^{3}F $6p$
59	−.14739	$3d^4(^4)$ ^{1}G $7p$
60	−.14511	$3d^4$ ^{3}D $7p$
61	−.11256	$3d^4$ ^{3}D $6f$
62	−.11046	$3d^4(^4)$ ^{1}G $6f$:
63	−.10981	$3d^4(^4)$ ^{3}F $8p$
64	−.10886	$3d^4(^4)$ ^{1}S $6f$
65	−.10818	$3d^4$ ^{1}I $6f$
66	−.10637	$3d^4$ ^{1}I $6h$
67	−.10443	$3d^4(^4)$ ^{3}P $7f$:
68	−.10342	$3d^4$ ^{3}H $7f$
69	−.10173	$3d^4$ ^{3}H $7h$
70	−.08858	$3d^4(^4)$ ^{3}F $7f$
71	−.08661	$3d^4(^4)$ ^{3}F $7h$
72	−.08568	$3d^4(^2)$ ^{3}F $5f$
73	−.08046	$3d^4$ ^{3}G $8p$
74	−.07496	$3d^4(^2)$ ^{3}P $5f$
75	−.06849	$3d^4(^4)$ ^{1}D $7p$
76	−.06515	$3d^4(^2)$ ^{1}G $6p$
77	−.05855	$3d^4$ ^{3}G $7f$
78	−.05677	$3d^4$ ^{3}G $7h$
79	−.03271	$3d^4(^4)$ ^{1}D $6f$
80	−.02740	$3d^4(^4)$ ^{1}D $6h$
81	−.02721	$3d^4(^4)$ ^{3}P $8f$
82	−.02664	$3d^4$ ^{3}H $8f$
83	−.02531	$3d^4(^4)$ ^{3}F $9p$
84	−.02521	$3d^4$ ^{3}H $8j$
85	−.02519	$3d^4$ ^{3}H $8h$
86	−.01438	$3d^4(^4)$ ^{1}G $8p$
87	−.01347	$3d^4$ ^{3}D $8p$
88	−.01142	$3d^4(^4)$ ^{3}F $8f$
89	−.01020	$3d^4$ ^{1}F $7p$
90	−.01007	$3d^4(^4)$ ^{3}F $8h$
91	−.00098	$3d^4(^2)$ ^{1}G $5f$
		^{2}G^e
1	−3.38135	$3d^5(^5)$
2	−3.12545	$3d^5(^3)$
3	−2.43349	$3d^4$ ^{3}G $4s$
4	−2.40883	$3d^4(^4)$ ^{1}G $4s$
5	−2.09903	$3d^4(^2)$ ^{1}G $4s$
6	−1.28881	$3d^4$ ^{3}H $4d$
7	−1.23780	$3d^4(^4)$ ^{3}F $4d$
8	−1.18349	$3d^4$ ^{3}G $4d$:
9	−1.16032	$3d^4(^4)$ ^{1}G $4d$
10	−1.13217	$3d^4$ ^{1}I $4d$
11	−1.10490	$3d^4$ ^{3}D $4d$
12	−1.06664	$3d^4$ ^{3}G $5s$
13	−1.05553	$3d^4(^4)$ ^{1}D $4d$
14	−1.01318	$3d^4(^4)$ ^{1}G $5s$
15	−.99900	$3d^4$ ^{1}F $4d$
16	−.89548	$3d^4(^2)$ ^{3}F $4d$
17	−.80074	$3d^4(^2)$ ^{1}G $4d$
18	−.70475	$3d^4(^2)$ ^{1}G $5s$
19	−.65193	$3d^4(^2)$ ^{1}D $4d$
20	−.63516	$3d^4$ ^{3}H $5d$:
21	−.61356	$3d^4(^4)$ ^{3}F $5d$
22	−.55489	$3d^4$ ^{3}G $5d$
23	−.53281	$3d^4$ ^{3}G $6s$
24	−.52836	$3d^4$ ^{3}D $5d$
25	−.51984	$3d^4(^4)$ ^{1}G $5d$
26	−.49706	$3d^4$ ^{1}I $5d$
27	−.46898	$3d^4(^4)$ ^{1}G $6s$
28	−.44401	$3d^4(^4)$ ^{1}D $5d$
29	−.41540	$3d^4(^4)$ ^{3}P $5g$
30	−.41528	$3d^4$ ^{3}H $5g$
31	−.40004	$3d^4(^4)$ ^{3}F $5g$
32	−.38967	$3d^4$ ^{1}F $5d$
33	−.37024	$3d^4$ ^{3}G $5g$
34	−.34274	$3d^4$ ^{3}H $6d$
35	−.32067	$3d^4(^4)$ ^{3}F $6d$
36	−.30306	$3d^4(^4)$ ^{1}G $5g$
37	−.30291	$3d^4$ ^{3}D $5g$
38	−.30176	$3d^4$ ^{1}I $5g$
39	−.30023	$3d^4(^4)$ ^{1}S $5g$
40	−.28420	$3d^4(^2)$ ^{3}F $5d$
41	−.27983	$3d^4$ ^{3}G $6d$
42	−.26452	$3d^4$ ^{3}G $7s$
43	−.23055	$3d^4$ ^{3}D $6d$
44	−.22992	$3d^4(^4)$ ^{1}G $6d$
45	−.22351	$3d^4(^4)$ ^{1}D $5g$
46	−.21987	$3d^4(^4)$ ^{3}P $6g$:
47	−.21976	$3d^4$ ^{3}H $6g$
48	−.21898	$3d^4$ ^{1}I $6d$
49	−.20452	$3d^4(^4)$ ^{3}F $6g$
50	−.20002	$3d^4(^2)$ ^{1}G $5d$
51	−.19898	$3d^4(^4)$ ^{1}G $7s$
52	−.17460	$3d^4$ ^{3}G $6g$
53	−.17171	$3d^4$ ^{3}H $7d$
54	−.16477	$3d^4$ ^{1}F $5g$
55	−.16170	$3d^4(^2)$ ^{1}G $6s$
56	−.15707	$3d^4(^4)$ ^{3}F $7d$
57	−.14788	$3d^4(^4)$ ^{1}D $6d$
58	−.12106	$3d^4$ ^{3}G $7d$
59	−.11191	$3d^4$ ^{3}G $8s$
60	−.10726	$3d^4(^4)$ ^{1}G $6g$
61	−.10715	$3d^4$ ^{3}D $6g$
62	−.10622	$3d^4$ ^{1}I $6g$
63	−.10512	$3d^4(^4)$ ^{1}S $6g$
64	−.10187	$3d^4(^4)$ ^{3}P $7g$
65	−.10181	$3d^4$ ^{3}H $7g$
66	−.10174	$3d^4$ ^{3}H $7i$
67	−.09077	$3d^4$ ^{1}F $6d$
68	−.08660	$3d^4(^4)$ ^{3}F $7g$
69	−.07647	$3d^4(^2)$ ^{3}F $5g$
70	−.07472	$3d^4(^2)$ ^{3}P $5g$
71	−.07278	$3d^4$ ^{3}H $8d$
72	−.06394	$3d^4$ ^{3}D $7d$
73	−.06272	$3d^4(^4)$ ^{1}G $7d$
74	−.06052	$3d^4$ ^{1}I $7d$
75	−.05678	$3d^4$ ^{3}G $7g$
76	−.04973	$3d^4(^4)$ ^{3}F $8d$
77	−.04518	$3d^4(^4)$ ^{1}G $8s$
78	−.02802	$3d^4(^4)$ ^{1}D $6g$
79	−.02555	$3d^4$ ^{3}G $8d$
80	−.02530	$3d^4(^4)$ ^{3}P $8g$
81	−.02521	$3d^4$ ^{3}H $8i$
82	−.02512	$3d^4$ ^{3}H $8g$
83	−.01659	$3d^4$ ^{3}G $9s$
84	−.01490	$3d^4(^2)$ ^{1}D $5d$
85	−.01008	$3d^4(^4)$ ^{3}F $8g$
86	−.00597	$3d^4$ ^{3}H $9d$
		^{2}G°
1	−2.00319	$3d^4$ ^{3}H $4p$
2	−1.93737	$3d^4(^4)$ ^{3}F $4p$
3	−1.89806	$3d^4$ ^{3}G $4p$
4	−1.86363	$3d^4(^4)$ ^{1}G $4p$
5	−1.73494	$3d^4$ ^{1}F $4p$
6	−1.61063	$3d^4(^2)$ ^{3}F $4p$
7	−1.56948	$3d^4(^2)$ ^{1}G $4p$
8	−.92738	$3d^4$ ^{3}H $5p$
9	−.90252	$3d^4(^4)$ ^{3}F $5p$
10	−.87199	$3d^4$ ^{3}G $5p$
11	−.81452	$3d^4(^4)$ ^{1}G $5p$
12	−.78480	$3d^4$ ^{3}H $4f$
13	−.77946	$3d^4(^4)$ ^{3}P $4f$
14	−.76482	$3d^4(^4)$ ^{3}F $4f$
15	−.73315	$3d^4$ ^{3}G $4f$
16	−.67794	$3d^4$ ^{1}F $5p$
17	−.67675	$3d^4(^4)$ ^{1}G $4f$
18	−.67046	$3d^4$ ^{3}D $4f$
19	−.66446	$3d^4$ ^{1}I $4f$
20	−.59570	$3d^4(^4)$ ^{1}D $4f$
21	−.58017	$3d^4(^2)$ ^{3}F $5p$
22	−.53856	$3d^4$ ^{1}F $4f$
23	−.51140	$3d^4(^2)$ ^{1}G $5p$
24	−.48504	$3d^4$ ^{3}H $6p$
25	−.46855	$3d^4(^4)$ ^{3}F $6p$
26	−.45747	$3d^4(^2)$ ^{3}F $4f$
27	−.43827	$3d^4$ ^{3}G $6p$

V-like Fe (Fe^{3+})

i	E(Ryds)	Description	i	E(Ryds)	Description	i	E(Ryds)	Description	i	E(Ryds)	Description
28	−.43660	$3d^4(^2)$ ^{3}P $4f$	80	−.02521	$3d^4(^4)$ ^{3}P $8h$	45	−.10707	$3d^4(^4)$ ^{1}G $6g$	26	−.37022	$3d^4$ ^{1}I $6p$
29	−.42085	$3d^4$ ^{3}H $5f$	81	−.02520	$3d^4$ ^{3}H $8h$	46	−.10648	$3d^4$ ^{1}I $6g$	27	−.36599	$3d^4(^2)$ ^{1}G $4f$
30	−.41771	$3d^4(^4)$ ^{3}P $5f$	82	−.01339	$3d^4(^4)$ ^{1}G $8p$	47	−.10186	$3d^4$ ^{3}H $7g$	28	−.31073	$3d^4(^4)$ ^{1}G $5f$
31	−.40443	$3d^4(^4)$ ^{3}F $5f$	83	−.01170	$3d^4(^4)$ ^{3}F $8f$	48	−.10177	$3d^4(^4)$ ^{3}P $7g$	29	−.30942	$3d^4$ ^{3}D $5f$
32	−.37535	$3d^4(^4)$ ^{1}G $6p$	84	−.01006	$3d^4(^4)$ ^{3}F $8j$	49	−.10174	$3d^4(^4)$ ^{3}P $7i$	30	−.30582	$3d^4$ ^{1}I $5f$
33	−.37301	$3d^4$ ^{3}G $5f$	85	−.00908	$3d^4$ ^{1}F $7p$	50	−.10174	$3d^4$ ^{3}H $7i$	31	−.25726	$3d^4$ ^{3}H $7p$
34	−.36417	$3d^4(^2)$ ^{1}G $4f$	86	−.00233	$3d^4(^2)$ ^{1}G $5f$	51	−.09129	$3d^4$ ^{1}F $6d$	32	−.23012	$3d^4(^4)$ ^{1}D $5f$
35	−.31142	$3d^4(^4)$ ^{1}G $5f$			^{2}H^e	52	−.08659	$3d^4(^4)$ ^{3}F $7g$:	33	−.22290	$3d^4$ ^{3}H $6f$
36	−.30742	$3d^4$ ^{3}D $5f$	1	−3.39455	$3d^5(^3)$	53	−.08659	$3d^4(^4)$ ^{3}F $7i$	34	−.21966	$3d^4(^4)$ ^{3}P $6h$
37	−.30392	$3d^4$ ^{1}I $5f$	2	−2.47708	$3d^4$ ^{3}H $4s$	54	−.07656	$3d^4(^2)$ ^{3}F $5g$	35	−.21965	$3d^4$ ^{3}H $6h$
38	−.25884	$3d^4$ ^{3}H $7p$	3	−1.26515	$3d^4$ ^{3}H $4d$:	55	−.07438	$3d^4(^2)$ ^{3}P $5g$	36	−.21365	$3d^4$ ^{3}G $7p$
39	−.24314	$3d^4(^4)$ ^{3}F $7p$	4	−1.23522	$3d^4(^4)$ ^{3}F $4d$	56	−.07192	$3d^4$ ^{3}H $8d$	37	−.20767	$3d^4(^4)$ ^{3}F $6f$
40	−.23723	$3d^4$ ^{1}F $6p$	5	−1.21013	$3d^4$ ^{3}G $4d$	57	−.06432	$3d^4(^4)$ ^{1}G $7d$	38	−.20450	$3d^4(^4)$ ^{3}F $6h$
41	−.23225	$3d^4(^4)$ ^{1}D $5f$	6	−1.16786	$3d^4(^4)$ ^{1}G $4d$	58	−.06210	$3d^4$ ^{3}H $9s$	39	−.17861	$3d^4$ ^{3}G $6f$
42	−.22437	$3d^4$ ^{3}H $6f$	7	−1.11270	$3d^4$ ^{3}H $5s$	59	−.06059	$3d^4$ ^{1}I $7d$	40	−.17468	$3d^4$ ^{3}G $6h$
43	−.22117	$3d^4(^4)$ ^{3}P $6f$	8	−1.07705	$3d^4$ ^{1}I $4d$	60	−.05679	$3d^4$ ^{3}G $7g$	41	−.17306	$3d^4$ ^{1}F $5f$
44	−.21965	$3d^4(^4)$ ^{3}P $6h$	9	−1.00821	$3d^4$ ^{1}F $4d$	61	−.05677	$3d^4$ ^{3}G $7i$	42	−.16367	$3d^4(^2)$ ^{1}D $4f$
45	−.21965	$3d^4$ ^{3}H $6h$	10	−.90587	$3d^4(^2)$ ^{3}F $4d$	62	−.05116	$3d^4(^4)$ ^{3}F $8d$	43	−.14893	$3d^4(^4)$ ^{1}G $7p$
46	−.21205	$3d^4$ ^{3}G $7p$	11	−.83775	$3d^4(^2)$ ^{1}G $4d$	63	−.02782	$3d^4(^4)$ ^{1}D $6g$	44	−.14507	$3d^4$ ^{1}I $7p$
47	−.20758	$3d^4(^4)$ ^{3}F $6f$	12	−.63415	$3d^4$ ^{3}H $5d$	64	−.02701	$3d^4$ ^{3}G $8d$	45	−.12492	$3d^4$ ^{3}H $8p$
48	−.20451	$3d^4(^4)$ ^{3}F $6h$	13	−.61100	$3d^4(^4)$ ^{3}F $5d$	65	−.02527	$3d^4$ ^{3}H $8g$	46	−.11218	$3d^4(^4)$ ^{1}G $6f$:
49	−.17706	$3d^4$ ^{3}G $6f$	14	−.58425	$3d^4$ ^{3}G $5d$	66	−.02523	$3d^4(^4)$ ^{3}P $8g$	47	−.11134	$3d^4$ ^{3}D $6f$
50	−.17469	$3d^4$ ^{3}G $6h$	15	−.57622	$3d^4$ ^{3}H $6s$	67	−.02521	$3d^4(^4)$ ^{3}P $8i$	48	−.10975	$3d^4$ ^{1}I $6f$
51	−.17332	$3d^4$ ^{1}F $5f$	16	−.53115	$3d^4(^4)$ ^{1}G $5d$	68	−.02521	$3d^4$ ^{3}H $8i$	49	−.10639	$3d^4$ ^{1}I $6h$
52	−.15389	$3d^4(^2)$ ^{1}D $4f$	17	−.50194	$3d^4$ ^{1}I $5d$	69	−.01006	$3d^4(^4)$ ^{3}F $8g$	50	−.10626	$3d^4(^4)$ ^{1}S $6h$
53	−.14713	$3d^4(^4)$ ^{1}G $7p$	18	−.41539	$3d^4$ ^{3}H $5g$	70	−.00354	$3d^4$ ^{3}H $9d$	51	−.10384	$3d^4$ ^{3}H $7f$
54	−.14406	$3d^4(^2)$ ^{3}F $6p$	19	−.41523	$3d^4(^4)$ ^{3}P $5g$			^{2}H^o	52	−.10175	$3d^4(^4)$ ^{3}P $7h$
55	−.12479	$3d^4$ ^{3}H $8p$	20	−.39999	$3d^4(^4)$ ^{3}F $5g$	1	−1.96218	$3d^4$ ^{3}H $4p$	53	−.10174	$3d^4$ ^{3}H $7h$
56	−.11279	$3d^4(^4)$ ^{1}G $6f$	21	−.38902	$3d^4$ ^{1}F $5d$	2	−1.93092	$3d^4$ ^{3}G $4p$	54	−.08862	$3d^4(^4)$ ^{3}F $7f$
57	−.11092	$3d^4(^4)$ ^{3}F $8p$	22	−.37025	$3d^4$ ^{3}G $5g$	3	−1.88946	$3d^4(^4)$ ^{1}G $4p$	55	−.08660	$3d^4(^4)$ ^{3}F $7h$
58	−.11028	$3d^4$ ^{1}I $6f$	23	−.33644	$3d^4$ ^{3}H $6d$	4	−1.84516	$3d^4$ ^{1}I $4p$	56	−.08171	$3d^4(^2)$ ^{3}F $5f$
59	−.10762	$3d^4$ ^{3}D $6f$	24	−.31972	$3d^4(^4)$ ^{3}F $6d$	5	−1.56588	$3d^4(^2)$ ^{1}G $4p$	57	−.07989	$3d^4$ ^{3}G $8p$
60	−.10647	$3d^4(^4)$ ^{1}G $6h$	25	−.30932	$3d^4$ ^{3}H $7s$	6	−.91783	$3d^4$ ^{3}H $5p$	58	−.06907	$3d^4(^2)$ ^{1}G $6p$
61	−.10499	$3d^4$ ^{3}H $7f$	26	−.30290	$3d^4$ ^{3}D $5g$	7	−.87840	$3d^4$ ^{3}G $5p$	59	−.05858	$3d^4$ ^{3}G $7f$
62	−.10301	$3d^4(^4)$ ^{3}P $7f$	27	−.30282	$3d^4(^4)$ ^{1}G $5g$	8	−.82252	$3d^4(^4)$ ^{1}G $5p$	60	−.05677	$3d^4$ ^{3}G $7h$
63	−.10174	$3d^4(^4)$ ^{3}P $7h$	28	−.30204	$3d^4$ ^{1}I $5g$	9	−.80731	$3d^4$ ^{1}I $5p$	61	−.04046	$3d^4$ ^{3}H $9p$
64	−.10174	$3d^4$ ^{3}H $7h$	29	−.29285	$3d^4$ ^{3}G $6d$	10	−.78330	$3d^4$ ^{3}H $4f$	62	−.03225	$3d^4(^4)$ ^{1}D $6f$
65	−.08859	$3d^4(^4)$ ^{3}F $7f$	30	−.29049	$3d^4(^2)$ ^{3}F $5d$	11	−.76531	$3d^4(^4)$ ^{3}F $4f$	63	−.02745	$3d^4(^4)$ ^{1}D $6h$
66	−.08760	$3d^4(^2)$ ^{3}F $5f$	31	−.23183	$3d^4(^4)$ ^{1}G $6d$	12	−.73491	$3d^4$ ^{3}G $4f$	64	−.02675	$3d^4$ ^{3}H $8f$
67	−.08657	$3d^4(^4)$ ^{3}F $7h$	32	−.22345	$3d^4(^4)$ ^{1}D $5g$	13	−.67541	$3d^4(^4)$ ^{1}G $4f$	65	−.02522	$3d^4(^4)$ ^{3}P $8h$
68	−.07990	$3d^4$ ^{3}G $8p$	33	−.22172	$3d^4$ ^{1}I $6d$	14	−.67315	$3d^4$ ^{3}D $4f$	66	−.02521	$3d^4$ ^{3}H $8h$
69	−.07799	$3d^4(^2)$ ^{3}P $5f$	34	−.21981	$3d^4$ ^{3}H $6g$	15	−.66770	$3d^4$ ^{1}I $4f$	67	−.02521	$3d^4$ ^{3}H $8j$
70	−.06945	$3d^4(^2)$ ^{1}G $6p$	35	−.21968	$3d^4(^4)$ ^{3}P $6g$	16	−.59145	$3d^4(^4)$ ^{1}D $4f$	68	−.01453	$3d^4(^4)$ ^{1}G $8p$
71	−.05833	$3d^4$ ^{3}G $7f$	36	−.21621	$3d^4(^2)$ ^{1}G $5d$	17	−.53904	$3d^4$ ^{1}F $4f$	69	−.01233	$3d^4$ ^{1}I $8p$
72	−.05676	$3d^4$ ^{3}G $7h$	37	−.20446	$3d^4(^4)$ ^{3}F $6g$	18	−.51081	$3d^4(^2)$ ^{1}G $5p$	70	−.01143	$3d^4(^4)$ ^{3}F $8f$
73	−.04077	$3d^4$ ^{3}H $9p$	38	−.17471	$3d^4$ ^{3}G $6g$	19	−.48301	$3d^4$ ^{3}H $6p$	71	−.01006	$3d^4(^4)$ ^{3}F $8j$
74	−.03358	$3d^4(^4)$ ^{1}D $6f$	39	−.17067	$3d^4$ ^{3}H $7d$	20	−.44614	$3d^4(^2)$ ^{3}F $4f$	72	−.00318	$3d^4(^2)$ ^{1}G $5f$
75	−.02742	$3d^4(^4)$ ^{1}D $6h$	40	−.16504	$3d^4$ ^{1}F $5g$	21	−.44019	$3d^4$ ^{3}G $6p$			^{2}I^e
76	−.02728	$3d^4$ ^{3}H $8f$	41	−.15708	$3d^4$ ^{3}H $8s$	22	−.41964	$3d^4$ ^{3}H $5f$	1	−3.50204	$3d^5(^5)$
77	−.02597	$3d^4(^4)$ ^{3}P $8f$	42	−.15419	$3d^4(^4)$ ^{3}F $7d$	23	−.40473	$3d^4(^4)$ ^{3}F $5f$	2	−2.40112	$3d^4$ ^{1}I $4s$
78	−.02574	$3d^4(^4)$ ^{3}F $9p$	43	−.12804	$3d^4$ ^{3}G $7d$	24	−.37837	$3d^4(^4)$ ^{1}G $6p$	3	−1.25016	$3d^4$ ^{3}H $4d$
79	−.02521	$3d^4$ ^{3}H $8j$	44	−.10716	$3d^4$ ^{3}D $6g$	25	−.37458	$3d^4$ ^{3}G $5f$	4	−1.21802	$3d^4$ ^{3}G $4d$

V-like Fe (Fe^{3+})

i	E(Ryds)	Description
5	−1.17129	$3d^4(^4)$ ^{1}G $4d$
6	−1.06143	$3d^4$ ^{1}I $4d$
7	−1.00937	$3d^4$ ^{1}I $5s$
8	−.85068	$3d^4(^2)$ ^{1}G $4d$
9	−.62286	$3d^4$ ^{3}H $5d$
10	−.58592	$3d^4$ ^{3}G $5d$
11	−.53279	$3d^4(^4)$ ^{1}G $5d$
12	−.50001	$3d^4$ ^{1}I $5d$
13	−.46759	$3d^4$ ^{1}I $6s$
14	−.41537	$3d^4$ ^{3}H $5g$
15	−.40009	$3d^4(^4)$ ^{3}F $5g$
16	−.37027	$3d^4$ ^{3}G $5g$
17	−.32958	$3d^4$ ^{3}H $6d$
18	−.30282	$3d^4(^4)$ ^{1}G $5g$
19	−.30237	$3d^4$ ^{1}I $5g$
20	−.29139	$3d^4$ ^{3}G $6d$
21	−.23235	$3d^4(^4)$ ^{1}G $6d$
22	−.22303	$3d^4(^4)$ ^{1}D $5g$
23	−.22243	$3d^4(^2)$ ^{1}G $5d$
24	−.22098	$3d^4$ ^{1}I $6d$
25	−.21980	$3d^4$ ^{3}H $6g$
26	−.20455	$3d^4(^4)$ ^{3}F $6g$
27	−.19808	$3d^4$ ^{1}I $7s$
28	−.17473	$3d^4$ ^{3}G $6g$
29	−.16550	$3d^4$ ^{3}H $7d$
30	−.16478	$3d^4$ ^{1}F $5g$
31	−.12701	$3d^4$ ^{3}G $7d$
32	−.10708	$3d^4(^4)$ ^{1}G $6g$
33	−.10704	$3d^4$ ^{3}D $5g$
34	−.10675	$3d^4$ ^{1}I $6g$
35	−.10186	$3d^4$ ^{3}H $7g$
36	−.10174	$3d^4(^4)$ ^{3}P $7i$
37	−.10174	$3d^4$ ^{3}H $7i$
38	−.08663	$3d^4(^4)$ ^{3}F $7g$
39	−.08659	$3d^4(^4)$ ^{3}F $7i$
40	−.07579	$3d^4(^2)$ ^{3}F $5g$
41	−.07266	$3d^4$ ^{3}H $8d$
42	−.06480	$3d^4(^4)$ ^{1}G $7d$
43	−.05680	$3d^4$ ^{3}G $7g$
44	−.05677	$3d^4$ ^{3}G $7i$
45	−.05449	$3d^4$ ^{1}I $7d$
46	.04490	$3d^4$ ^{1}I $8s$
47	−.02733	$3d^4(^4)$ ^{1}D $6g$
48	−.02609	$3d^4$ ^{3}G $8d$
49	−.02528	$3d^4$ ^{3}H $8g$
50	−.02521	$3d^4(^4)$ ^{3}P $8i$
51	−.02521	$3d^4$ ^{3}H $8i$
52	−.01010	$3d^4(^4)$ ^{3}F $8g$
53	−.01006	$3d^4(^4)$ ^{3}F $8i$
54	−.00214	$3d^4$ ^{3}H $9d$
	^{2}I^o	
1	−1.97182	$3d^4$ ^{3}H $4p$
2	−1.89257	$3d^4$ ^{1}I $4p$
3	−.92333	$3d^4$ ^{3}H $5p$
4	−.82203	$3d^4$ ^{1}I $5p$
5	−.78096	$3d^4$ ^{3}H $4f$
6	−.76691	$3d^4(^4)$ ^{3}F $4f$
7	−.73733	$3d^4$ ^{3}G $4f$
8	−.67737	$3d^4(^4)$ ^{1}G $4f$
9	−.67212	$3d^4$ ^{1}I $4f$
10	−.53136	$3d^4$ ^{1}F $4f$
11	−.48561	$3d^4$ ^{3}H $6p$
12	−.44017	$3d^4(^2)$ ^{3}F $4f$
13	−.41887	$3d^4$ ^{3}H $5f$
14	−.40578	$3d^4(^4)$ ^{3}F $5f$
15	−.37791	$3d^4$ ^{1}I $6p$
16	−.37610	$3d^4$ ^{3}G $5f$
17	−.36556	$3d^4(^2)$ ^{1}G $4f$
18	−.31151	$3d^4(^4)$ ^{1}G $5f$
19	−.30841	$3d^4$ ^{1}I $5f$
20	−.25875	$3d^4$ ^{3}H $7p$
21	−.22203	$3d^4$ ^{3}H $6f$
22	−.21966	$3d^4$ ^{3}H $6h$
23	−.21965	$3d^4(^4)$ ^{3}P $6h$
24	−.20843	$3d^4(^4)$ ^{3}F $6f$
25	−.20450	$3d^4(^4)$ ^{3}F $6h$
26	−.17870	$3d^4$ ^{3}G $6f$
27	−.17468	$3d^4$ ^{3}G $6h$
28	−.17036	$3d^4$ ^{1}F $5f$
29	−.14862	$3d^4$ ^{1}I $7p$
30	−.12579	$3d^4$ ^{3}H $8p$
31	−.11287	$3d^4(^4)$ ^{1}G $6f$
32	−.11111	$3d^4$ ^{1}I $6f$
33	−.10646	$3d^4$ ^{3}D $6h$
34	−.10301	$3d^4$ ^{3}H $7f$
35	−.10175	$3d^4$ ^{3}H $7h$
36	−.10174	$3d^4(^4)$ ^{3}P $7h$
37	−.08923	$3d^4(^4)$ ^{3}F $7f$
38	−.08659	$3d^4(^4)$ ^{3}F $7h$
39	−.07948	$3d^4(^2)$ ^{3}F $5f$
40	−.05943	$3d^4$ ^{3}G $7f$
41	−.05677	$3d^4$ ^{3}G $7h$
42	−.04105	$3d^4$ ^{3}H $9p$
43	−.02745	$3d^4(^4)$ ^{1}D $6h$
44	−.02635	$3d^4$ ^{3}H $8f$
45	−.02522	$3d^4$ ^{3}H $8h$
46	−.02521	$3d^4(^4)$ ^{3}P $8j$
47	−.02521	$3d^4(^4)$ ^{3}P $8h$
48	−.01440	$3d^4$ ^{1}I $8p$
49	−.01224	$3d^4(^4)$ ^{3}F $8f$
50	−.01006	$3d^4(^4)$ ^{3}F $8h$
51	−.00362	$3d^4(^2)$ ^{1}G $5f$
	^{2}J^e	
1	−1.26548	$3d^4$ ^{3}H $4d$
2	−1.17193	$3d^4$ ^{1}I $4d$
3	−.63953	$3d^4$ ^{3}H $5d$
4	−.53371	$3d^4$ ^{1}I $5d$
5	−.41539	$3d^4$ ^{3}H $5g$
6	−.40023	$3d^4(^4)$ ^{3}F $5g$
7	−.37023	$3d^4$ ^{3}G $5g$
8	−.34207	$3d^4$ ^{3}H $6d$
9	−.30311	$3d^4(^4)$ ^{1}G $5g$
10	−.30258	$3d^4$ ^{1}I $5g$
11	−.23268	$3d^4$ ^{1}I $6d$
12	−.21981	$3d^4$ ^{3}H $6g$
13	−.20466	$3d^4(^4)$ ^{3}F $6g$
14	−.17608	$3d^4$ ^{3}H $7d$
15	−.17466	$3d^4$ ^{3}G $6g$
16	−.16386	$3d^4$ ^{1}F $5g$
17	−.10731	$3d^4(^4)$ ^{1}G $6g$
18	−.10693	$3d^4$ ^{1}I $6g$
19	−.10186	$3d^4$ ^{3}H $7g$
20	−.10174	$3d^4(^4)$ ^{3}P $7i$
21	−.10174	$3d^4$ ^{3}H $7i$
22	−.08672	$3d^4(^4)$ ^{3}F $7g$
23	−.08659	$3d^4(^4)$ ^{3}F $7i$
24	−.07486	$3d^4(^2)$ ^{3}F $5g$
25	−.07368	$3d^4$ ^{3}H $8d$
26	−.06506	$3d^4$ ^{1}I $7d$
27	−.05677	$3d^4$ ^{3}G $7g$
28	−.02529	$3d^4$ ^{3}H $8g$
29	−.02521	$3d^4(^4)$ ^{3}P $8i$
30	−.02521	$3d^4$ ^{3}H $8i$
31	−.01015	$3d^4(^4)$ ^{3}F $8g$
32	−.01006	$3d^4(^4)$ ^{3}F $8i$
33	−.00618	$3d^4$ ^{3}H $9d$
	^{2}J^o	
1	−1.87869	$3d^4$ ^{1}I $4p$
2	−.81734	$3d^4$ ^{1}I $5p$
3	−.78172	$3d^4$ ^{3}H $4f$
4	−.73833	$3d^4$ ^{3}G $4f$
5	−.68021	$3d^4(^4)$ ^{1}G $4f$
6	−.67566	$3d^4$ ^{1}I $4f$
7	−.41988	$3d^4$ ^{3}H $5f$
8	−.37751	$3d^4$ ^{3}G $5f$
9	−.37490	$3d^4$ ^{1}I $6p$
10	−.36511	$3d^4(^2)$ ^{1}G $4f$
11	−.31301	$3d^4(^4)$ ^{1}G $5f$
12	−.31109	$3d^4$ ^{1}I $5f$
13	−.22259	$3d^4$ ^{3}H $6f$
14	−.21966	$3d^4$ ^{3}H $6h$
15	−.20451	$3d^4(^4)$ ^{3}F $6h$
16	−.17867	$3d^4$ ^{3}G $6f$
17	−.17468	$3d^4$ ^{3}G $6h$
18	−.14730	$3d^4$ ^{1}I $7p$
19	−.11387	$3d^4(^4)$ ^{1}G $6f$
20	−.11300	$3d^4$ ^{1}I $6f$
21	−.10645	$3d^4(^4)$ ^{1}G $6h$
22	−.10300	$3d^4$ ^{3}H $7f$
23	−.10175	$3d^4$ ^{3}H $7h$
24	−.08659	$3d^4(^4)$ ^{3}F $7h$
25	−.05947	$3d^4$ ^{3}G $7f$
26	−.05677	$3d^4$ ^{3}G $7h$
27	−.02741	$3d^4(^4)$ ^{1}D $6h$
28	−.02677	$3d^4$ ^{3}H $8f$
29	−.02522	$3d^4$ ^{3}H $8h$
30	−.02521	$3d^4(^4)$ ^{3}P $8j$
31	−.01375	$3d^4$ ^{1}I $8p$
32	−.01006	$3d^4(^4)$ ^{3}F $8h$
33	−.00346	$3d^4(^2)$ ^{1}G $5f$
	^{4}S^e	
1	−1.53971	$3d^4$ ^{5}D $4d$
2	−1.17409	$3d^4$ ^{3}D $4d$
3	−.88220	$3d^4$ ^{5}D $5d$
4	−.57623	$3d^4$ ^{5}D $6d$
5	−.53407	$3d^4$ ^{3}D $5d$
6	−.40624	$3d^4$ ^{5}D $7d$
7	−.37063	$3d^4$ ^{3}G $5g$
8	−.30200	$3d^4$ ^{5}D $8d$
9	−.23406	$3d^4$ ^{5}D $9d$
10	−.23223	$3d^4$ ^{3}D $6d$
	^{4}S^o	
1	−1.97819	$3d^4(^4)$ ^{3}P $4p$
2	−1.62360	$3d^4(^2)$ ^{3}P $4p$
3	−.92585	$3d^4(^4)$ ^{3}P $5p$
4	−.76499	$3d^4(^4)$ ^{3}F $4f$
5	−.58360	$3d^4(^2)$ ^{3}P $5p$
6	−.48663	$3d^4(^4)$ ^{3}P $6p$
7	−.43457	$3d^4(^2)$ ^{3}F $4f$
8	−.40295	$3d^4(^4)$ ^{3}F $5f$
9	−.25954	$3d^4(^4)$ ^{3}P $7p$
10	−.21962	$3d^4$ ^{3}H $6h$
11	−.20697	$3d^4(^4)$ ^{3}F $6f$
12	−.14582	$3d^4(^2)$ ^{3}P $6p$
13	−.12602	$3d^4(^4)$ ^{3}P $8p$
14	−.10171	$3d^4$ ^{3}H $7h$
15	−.08855	$3d^4(^4)$ ^{3}F $7f$
16	−.07651	$3d^4(^2)$ ^{3}F $5f$
17	−.04133	$3d^4(^4)$ ^{3}P $9p$
18	−.02517	$3d^4$ ^{3}H $8h$
19	−.01125	$3d^4(^4)$ ^{3}F $8f$
	^{4}P^e	
1	−3.58549	$3d^5(^3)$
2	−2.53137	$3d^4(^4)$ ^{3}P $4s$
3	−2.19440	$3d^4(^2)$ ^{3}P $4s$
4	−1.46889	$3d^4$ ^{5}D $4d$
5	−1.27131	$3d^4(^4)$ ^{3}P $4d$
6	−1.19346	$3d^4(^4)$ ^{3}F $4d$

V-like Fe (Fe^{3+})

i	E(Ryds)	Description	i	E(Ryds)	Description	i	E(Ryds)	Description	i	E(Ryds)	Description
7	−1.12789	$3d^4(^4)$ ^{3}P $5s$	26	−.26081	$3d^4(^4)$ ^{3}P $7p$	9	−.93776	$3d^4(^4)$ ^{3}P $5p$	18	−.62515	$3d^4(^4)$ ^{3}F $5d$
8	−1.12474	$3d^4$ ^{3}D $4d$	27	−.25332	$3d^4$ ^{5}D $8f$	10	−.91397	$3d^4(^4)$ ^{3}F $5p$	19	−.59096	$3d^4$ ^{3}G $5d$
9	−.93518	$3d^4(^2)$ ^{3}P $4d$	28	−.21962	$3d^4$ ^{3}H $6h$	11	−.82476	$3d^4$ ^{3}D $5p$	20	−.56735	$3d^4(^4)$ ^{3}F $6s$
10	−.89417	$3d^4(^2)$ ^{3}F $4d$			**^{4}D^e**	12	−.78265	$3d^4(^4)$ ^{3}P $4f$	21	−.54839	$3d^4$ ^{5}D $6d$
11	−.84299	$3d^4$ ^{5}D $5d$	1	−3.56341	$3d^5(^5)$	13	−.78014	$3d^4$ ^{3}H $4f$	22	−.52829	$3d^4$ ^{3}D $5d$
12	−.78696	$3d^4(^2)$ ^{3}P $5s$	2	−2.68538	$3d^4$ ^{5}D $4s$	14	−.77031	$3d^4(^4)$ ^{3}F $4f$	23	−.44446	$3d^4$ ^{5}D $6g$
13	−.63973	$3d^4(^4)$ ^{3}P $5d$	3	−2.42780	$3d^4$ ^{3}D $4s$	15	−.73686	$3d^4$ ^{3}G $4f$	24	−.41528	$3d^4(^4)$ ^{3}P $5g$
14	−.60479	$3d^4(^4)$ ^{3}F $5d$	4	−1.46054	$3d^4$ ^{5}D $4d$	16	−.70107	$3d^4$ ^{5}D $6p$	25	−.41513	$3d^4$ ^{3}H $5g$
15	−.58289	$3d^4(^4)$ ^{3}P $6s$	5	−1.33375	$3d^4$ ^{5}D $5s$	17	−.67402	$3d^4$ ^{3}D $4f$	26	−.40023	$3d^4(^4)$ ^{3}F $5g$
16	−.55450	$3d^4$ ^{5}D $6d$	6	−1.28687	$3d^4(^4)$ ^{3}P $4d$	18	−.64332	$3d^4$ ^{5}D $5f$	27	−.39434	$3d^4$ ^{5}D $7d$
17	−.51613	$3d^4$ ^{3}D $5d$	7	−1.25467	$3d^4(^4)$ ^{3}F $4d$	19	−.59772	$3d^4(^2)$ ^{3}P $5p$	28	−.37031	$3d^4$ ^{3}G $5g$
18	−.41473	$3d^4$ ^{3}H $5g$	8	−1.19859	$3d^4$ ^{3}G $4d$	20	−.57869	$3d^4(^2)$ ^{3}F $5p$	29	−.34653	$3d^4(^4)$ ^{3}P $6d$
19	−.40026	$3d^4(^4)$ ^{3}F $5g$	9	−1.11411	$3d^4$ ^{3}D $4d$	21	−.49165	$3d^4(^4)$ ^{3}P $6p$	30	−.33966	$3d^4$ ^{3}H $6d$
20	−.39616	$3d^4$ ^{5}D $7d$	10	−1.01815	$3d^4$ ^{3}D $5s$	22	−.47869	$3d^4$ ^{5}D $7p$	31	−.32984	$3d^4(^4)$ ^{3}F $6d$
21	−.37053	$3d^4$ ^{3}G $5g$	11	−.94449	$3d^4(^2)$ ^{3}P $4d$	23	−.47357	$3d^4(^4)$ ^{3}F $6p$	32	−.32654	$3d^4$ ^{5}D $7g$
22	−.34092	$3d^4(^4)$ ^{3}P $6d$	12	−.93152	$3d^4(^2)$ ^{3}F $4d$	24	−.44827	$3d^4(^2)$ ^{3}F $4f$	33	−.30370	$3d^4$ ^{3}D $5g$
23	−.31751	$3d^4(^4)$ ^{3}F $6d$	13	−.83952	$3d^4$ ^{5}D $5d$	25	−.44721	$3d^4$ ^{5}D $6f$	34	−.30119	$3d^4(^2)$ ^{3}P $5d$
24	−.31273	$3d^4(^4)$ ^{3}P $7s$	14	−.79901	$3d^4$ ^{5}D $6s$	26	−.43696	$3d^4(^2)$ ^{3}P $4f$	35	−.30000	$3d^4$ ^{5}D $8d$
25	−.30447	$3d^4(^2)$ ^{3}P $5d$	15	−.64510	$3d^4(^4)$ ^{3}P $5d$	27	−.42056	$3d^4(^4)$ ^{3}P $5f$	36	−.29894	$3d^4(^4)$ ^{3}F $7s$
26	−.29351	$3d^4$ ^{5}D $8d$	16	−.64107	$3d^4$ ^{5}D $5g$	28	−.41941	$3d^4$ ^{3}H $5f$	37	−.29459	$3d^4$ ^{3}G $6d$
27	−.28409	$3d^4(^2)$ ^{3}F $5d$	17	−.62547	$3d^4(^4)$ ^{3}F $5d$	29	−.40754	$3d^4(^4)$ ^{3}F $5f$	38	−.27130	$3d^4(^2)$ ^{3}F $5d$
28	−.24368	$3d^4(^2)$ ^{3}P $6s$	18	−.58824	$3d^4$ ^{3}G $5d$	30	−.38005	$3d^4$ ^{3}D $6p$	39	−.25001	$3d^4$ ^{5}D $8g$
29	−.23138	$3d^4$ ^{5}D $9d$	19	−.55563	$3d^4$ ^{5}D $6d$	31	−.37462	$3d^4$ ^{3}G $5f$	40	−.24335	$3d^4(^2)$ ^{3}F $6s$
30	−.22287	$3d^4$ ^{3}D $6d$	20	−.53322	$3d^4$ ^{5}D $7s$	32	−.34738	$3d^4$ ^{5}D $8p$	41	−.23267	$3d^4$ ^{3}D $6d$
31	−.21923	$3d^4$ ^{3}H $6g$	21	−.51119	$3d^4$ ^{3}D $5d$	33	−.32809	$3d^4$ ^{5}D $7f$	42	−.22512	$3d^4$ ^{5}D $9d$
32	−.20468	$3d^4(^4)$ ^{3}F $6g$	22	−.47113	$3d^4$ ^{3}D $6s$	34	−.30925	$3d^4$ ^{3}D $5f$	43	−.21972	$3d^4(^4)$ ^{3}P $6g$
		^{4}P^o	23	−.44533	$3d^4$ ^{5}D $6g$	35	−.26389	$3d^4$ ^{5}D $9p$	44	−.21959	$3d^4$ ^{3}H $6g$
1	−2.22408	$3d^4$ ^{5}D $4p$	24	−.41493	$3d^4$ ^{3}H $5g$	36	−.26209	$3d^4(^4)$ ^{3}P $7p$	45	−.20466	$3d^4(^4)$ ^{3}F $6g$
2	−1.99660	$3d^4(^4)$ ^{3}P $4p$	25	−.40028	$3d^4(^4)$ ^{3}F $5g$	37	−.25135	$3d^4$ ^{5}D $8f$	46	−.19754	$3d^4$ ^{5}D $9g$
3	−1.89478	$3d^4$ ^{3}D $4p$	26	−.39539	$3d^4$ ^{5}D $7d$	38	−.24544	$3d^4(^4)$ ^{3}F $7p$			**^{4}F^o**
4	−1.67121	$3d^4(^2)$ ^{3}P $4p$	27	−.38083	$3d^4$ ^{5}D $8s$	39	−.22366	$3d^4(^4)$ ^{3}P $6f$	1	−2.18918	$3d^4$ ^{5}D $4p$
5	−1.15509	$3d^4$ ^{5}D $5p$	28	−.37040	$3d^4$ ^{3}G $5g$	40	−.22273	$3d^4$ ^{3}H $6f$	2	−1.98676	$3d^4(^4)$ ^{3}F $4p$
6	−1.01854	$3d^4$ ^{5}D $4f$	29	−.34361	$3d^4(^4)$ ^{3}P $6d$	41	−.21963	$3d^4$ ^{3}H $6h$	3	−1.95913	$3d^4$ ^{3}G $4p$
7	−.93252	$3d^4(^4)$ ^{3}P $5p$	30	−.32790	$3d^4(^4)$ ^{3}F $6d$	42	−.20961	$3d^4(^4)$ ^{3}F $6f$	4	−1.88192	$3d^4$ ^{3}D $4p$
8	−.81972	$3d^4$ ^{3}D $5p$	31	−.32716	$3d^4$ ^{5}D $7g$			**^{4}F^e**	5	−1.68806	$3d^4(^2)$ ^{3}F $4p$
9	−.76800	$3d^4(^4)$ ^{3}F $4f$	32	−.30878	$3d^4(^2)$ ^{3}P $5d$	1	−3.42632	$3d^5(^3)$	6	−1.14440	$3d^4$ ^{5}D $5p$
10	−.73878	$3d^4$ ^{3}G $4f$	33	−.30447	$3d^4(^2)$ ^{3}F $5d$	2	−2.51606	$3d^4(^4)$ ^{3}F $4s$	7	−1.00149	$3d^4$ ^{5}D $4f$
11	−.71347	$3d^4$ ^{5}D $6p$	34	−.30243	$3d^4$ ^{3}D $5g$	3	−2.19842	$3d^4(^2)$ ^{3}F $4s$	8	−.91746	$3d^4(^4)$ ^{3}F $5p$
12	−.67617	$3d^4$ ^{3}D $4f$	35	−.29856	$3d^4$ ^{5}D $8d$	4	−1.44070	$3d^4$ ^{5}D $4d$	9	−.88631	$3d^4$ ^{3}G $5p$
13	−.65118	$3d^4$ ^{5}D $5f$	36	−.28574	$3d^4$ ^{5}D $9s$	5	−1.28912	$3d^4(^4)$ ^{3}P $4d$	10	−.82203	$3d^4$ ^{3}D $5p$
14	−.59893	$3d^4(^2)$ ^{3}P $5p$	37	−.28208	$3d^4$ ^{3}G $6d$	6	−1.26999	$3d^4$ ^{3}H $4d$	11	−.78783	$3d^4(^4)$ ^{3}P $4f$
15	−.49154	$3d^4(^4)$ ^{3}P $6p$	38	−.25047	$3d^4$ ^{5}D $8g$	7	−1.25361	$3d^4(^4)$ ^{3}F $4d$	12	−.78026	$3d^4$ ^{3}H $4f$
16	−.48374	$3d^4$ ^{5}D $7p$	39	−.23216	$3d^4$ ^{5}D $9d$	8	−1.20485	$3d^4$ ^{3}G $4d$:	13	−.77050	$3d^4(^4)$ ^{3}F $4f$
17	−.45186	$3d^4$ ^{5}D $6f$			**^{4}D^o**	9	−1.14861	$3d^4$ ^{3}D $4d$	14	−.73702	$3d^4$ ^{3}G $4f$
18	−.43854	$3d^4(^2)$ ^{3}F $4f$	1	−2.13458	$3d^4$ ^{5}D $4p$	10	−1.11150	$3d^4(^4)$ ^{3}F $5s$	15	−.70836	$3d^4$ ^{5}D $6p$
19	−.40511	$3d^4(^4)$ ^{3}F $5f$	2	−2.02440	$3d^4(^4)$ ^{3}P $4p$	11	−.92244	$3d^4(^2)$ ^{3}P $4d$	16	−.67729	$3d^4$ ^{3}D $4f$
20	−.37745	$3d^4$ ^{3}D $6p$	3	−1.96967	$3d^4(^4)$ ^{3}F $4p$	12	−.90847	$3d^4(^2)$ ^{3}F $4d$	17	−.64166	$3d^4$ ^{5}D $5f$
21	−.37538	$3d^4$ ^{3}G $5f$	4	−1.90038	$3d^4$ ^{3}D $4p$	13	−.81982	$3d^4$ ^{5}D $5d$	18	−.60267	$3d^4(^2)$ ^{3}F $5p$
22	−.35163	$3d^4$ ^{5}D $8p$	5	−1.67546	$3d^4(^2)$ ^{3}P $4p$	14	−.78912	$3d^4(^2)$ ^{3}F $5s$	19	−.48353	$3d^4$ ^{5}D $7p$
23	−.33123	$3d^4$ ^{5}D $7f$	6	−1.60549	$3d^4(^2)$ ^{3}F $4p$	15	−.64782	$3d^4(^4)$ ^{3}P $5d$	20	−.47392	$3d^4(^4)$ ^{3}F $6p$
24	−.31107	$3d^4$ ^{3}D $5f$	7	−1.12868	$3d^4$ ^{5}D $5p$	16	−.64002	$3d^4$ ^{5}D $5g$	21	−.45345	$3d^4(^2)$ ^{3}F $4f$:
25	−.26696	$3d^4$ ^{5}D $9p$	8	−1.00487	$3d^4$ ^{5}D $4f$	17	−.63639	$3d^4$ ^{3}H $5d$	22	−.44718	$3d^4$ ^{5}D $6f$

V-like Fe (Fe^{3+})

i	E(Ryds)	Description	i	E(Ryds)	Description	i	E(Ryds)	Description	i	E(Ryds)	Description
23	−.44449	$3d^4$ ^{5}D $6h$	31	−.28576	$3d^4$ ^{3}G $6d$	41	−.19752	$3d^4$ ^{5}D $9h$	14	−.44610	$3d^4$ ^{3}G $6p$
24	−.44438	$3d^4$ ^{3}G $6p$	32	−.26773	$3d^4$ ^{3}G $7s$	42	−.17961	$3d^4$ ^{3}G $6f$	15	−.44440	$3d^4$ ^{5}D $6h$
25	−.44182	$3d^4(^2)$ ^{3}P $4f$	33	−.25000	$3d^4$ ^{5}D $8i$	43	−.17469	$3d^4$ ^{3}G $6h$	16	−.42276	$3d^4$ ^{3}H $5f$
26	−.42366	$3d^4(^4)$ ^{3}P $5f$	34	−.24983	$3d^4$ ^{5}D $8g$			^{4}**H**^{e}	17	−.40799	$3d^4(^4)$ ^{3}F $5f$
27	−.41885	$3d^4$ ^{3}H $5f$	35	−.23257	$3d^4$ ^{3}D $6d$	1	−2.53129	$3d^4$ ^{3}H $4s$	18	−.37836	$3d^4$ ^{3}G $5f$
28	−.40802	$3d^4(^4)$ ^{3}F $5f$	36	−.22672	$3d^4$ ^{5}D $9d$	2	−1.28787	$3d^4$ ^{3}H $4d$	19	−.32920	$3d^4$ ^{5}D $7f$
29	−.37822	$3d^4$ ^{3}D $6p$	37	−.21990	$3d^4(^4)$ ^{3}P $6g$	3	−1.27197	$3d^4(^4)$ ^{3}F $4d$	20	−.32649	$3d^4$ ^{5}D $7h$
30	−.37576	$3d^4$ ^{3}G $5f$	38	−.21973	$3d^4$ ^{3}H $6g$	4	−1.23332	$3d^4$ ^{3}G $4d$	21	−.31163	$3d^4$ ^{3}D $5f$
31	−.34993	$3d^4$ ^{5}D $8p$	39	−.20458	$3d^4(^4)$ ^{3}F $6g$	5	−1.12638	$3d^4$ ^{3}H $5s$	22	−.26288	$3d^4$ ^{3}H $7p$
32	−.32806	$3d^4$ ^{5}D $7f$	40	−.19754	$3d^4$ ^{5}D $9i$	6	−.93354	$3d^4(^2)$ ^{3}F $4d$	23	−.25195	$3d^4$ ^{5}D $8f$
33	−.32658	$3d^4$ ^{5}D $7h$	41	−.19741	$3d^4$ ^{5}D $9g$	7	−.64820	$3d^4$ ^{3}H $5d$	24	−.25000	$3d^4$ ^{5}D $8j$
34	−.31117	$3d^4$ ^{3}D $5f$			^{4}**G**^{o}	8	−.63941	$3d^4$ ^{5}D $5g$	25	−.24996	$3d^4$ ^{5}D $8h$
35	−.26557	$3d^4$ ^{5}D $9p$	1	−2.00840	$3d^4(^4)$ ^{3}F $4p$	9	−.63266	$3d^4(^4)$ ^{3}F $5d$	26	−.22530	$3d^4$ ^{3}H $6f$
36	−.25100	$3d^4$ ^{5}D $8f$	2	−1.98801	$3d^4$ ^{3}H $4p$	10	−.60087	$3d^4$ ^{3}G $5d$	27	−.21966	$3d^4(^4)$ ^{3}P $6h$
37	−.25005	$3d^4$ ^{5}D $8h$	3	−1.92866	$3d^4$ ^{3}G $4p$	11	−.58198	$3d^4$ ^{3}H $6s$	28	−.21965	$3d^4$ ^{3}H $6h$
38	−.24594	$3d^4(^4)$ ^{3}F $7p$	4	−1.65597	$3d^4(^2)$ ^{3}F $4p$	12	−.44397	$3d^4$ ^{5}D $6g$	29	−.21712	$3d^4$ ^{3}G $7p$
39	−.22546	$3d^4(^4)$ ^{3}P $6f$	5	−.99844	$3d^4$ ^{5}D $4f$	13	−.41548	$3d^4(^4)$ ^{3}P $5g$:	30	−.21006	$3d^4(^4)$ ^{3}F $6f$
40	−.22273	$3d^4$ ^{3}H $6f$	6	−.92679	$3d^4$ ^{3}H $5p$	14	−.41530	$3d^4$ ^{3}H $5g$	31	−.20450	$3d^4(^4)$ ^{3}F $6h$
41	−.21964	$3d^4$ ^{3}H $6h$	7	−.92116	$3d^4(^4)$ ^{3}F $5p$	15	−.40005	$3d^4(^4)$ ^{3}F $5g$	32	−.19873	$3d^4$ ^{5}D $9f$
42	−.21561	$3d^4$ ^{3}G $7p$	8	−.88493	$3d^4$ ^{3}G $5p$	16	−.37034	$3d^4$ ^{3}G $5g$	33	−.19753	$3d^4$ ^{5}D $9j$
43	−.21024	$3d^4(^4)$ ^{3}F $6f$	9	−.78890	$3d^4(^4)$ ^{3}P $4f$	17	−.34705	$3d^4$ ^{3}H $6d$	34	−.19750	$3d^4$ ^{5}D $9h$
		^{4}**G**^{e}	10	−.78208	$3d^4$ ^{3}H $4f$	18	−.33151	$3d^4(^4)$ ^{3}F $6d$	35	−.18022	$3d^4$ ^{3}G $6f$
1	−3.64011	$3d^5(^5)$	11	−.76928	$3d^4(^4)$ ^{3}F $4f$	19	−.32653	$3d^4$ ^{5}D $7i$	36	−.17469	$3d^4$ ^{3}G $6h$
2	−2.48781	$3d^4$ ^{3}G $4s$	12	.74009	$3d^4$ ^{3}G $4f$	20	.32620	$3d^4$ ^{5}D $7g$			^{4}**I**^{e}
3	−1.47216	$3d^4$ ^{5}D $4d$	13	−.67879	$3d^4$ ^{3}D $4f$	21	−.31351	$3d^4$ ^{3}H $7s$	1	−1.29567	$3d^4$ ^{3}H $4d$
4	−1.26544	$3d^4(^4)$ ^{3}F $4d$	14	−.64028	$3d^4$ ^{5}D $5f$	22	−.30500	$3d^4(^2)$ ^{3}F $5d$	2	−1.23534	$3d^4$ ^{3}G $4d$
5	−1.24718	$3d^4$ ^{3}G $4d$	15	−.59364	$3d^4(^2)$ ^{3}F $5p$	23	−.30289	$3d^4$ ^{3}D $5g$	3	−.64992	$3d^4$ ^{3}H $5d$
6	−1.17666	$3d^4$ ^{3}D $4d$	16	−.48647	$3d^4$ ^{3}H $6p$	24	−.29745	$3d^4$ ^{3}G $6d$	4	−.64053	$3d^4$ ^{5}D $5g$
7	−1.14398	$3d^4$ ^{3}H $4d$:	17	−.47642	$3d^4(^4)$ ^{3}F $6p$	25	−.25000	$3d^4$ ^{5}D $8i$	5	−.60243	$3d^4$ ^{3}G $5d$
8	−1.08201	$3d^4$ ^{3}G $5s$	18	−.45488	$3d^4(^2)$ ^{3}F $4f$	26	−.24976	$3d^4$ ^{5}D $8g$	6	−.44489	$3d^4$ ^{5}D $6g$
9	−.95432	$3d^4(^2)$ ^{3}F $4d$	19	−.44471	$3d^4$ ^{3}G $6p$	27	−.21990	$3d^4(^4)$ ^{3}P $6g$:	7	−.41547	$3d^4$ ^{3}H $5g$
10	−.84616	$3d^4$ ^{5}D $5d$	20	−.44443	$3d^4$ ^{5}D $6h$	28	−.21975	$3d^4$ ^{3}H $6g$	8	−.40016	$3d^4(^4)$ ^{3}F $5g$
11	−.63957	$3d^4$ ^{5}D $5g$	21	−.44355	$3d^4$ ^{5}D $6f$	29	−.20452	$3d^4(^4)$ ^{3}F $6g$	9	−.37035	$3d^4$ ^{3}G $5g$
12	−.63443	$3d^4$ ^{3}H $5d$	22	−.44007	$3d^4(^2)$ ^{3}P $4f$	30	−.19753	$3d^4$ ^{5}D $9i$	10	−.34721	$3d^4$ ^{3}H $6d$
13	−.62031	$3d^4(^4)$ ^{3}F $5d$	23	−.42476	$3d^4(^4)$ ^{3}P $5f$	31	−.19736	$3d^4$ ^{5}D $9g$	11	−.32686	$3d^4$ ^{5}D $7g$
14	−.58931	$3d^4$ ^{3}G $5d$	24	−.42018	$3d^4$ ^{3}H $5f$	32	−.17872	$3d^4$ ^{3}H $7d$	12	−.32653	$3d^4$ ^{5}D $7i$
15	−.55101	$3d^4$ ^{5}D $6d$	25	−.40704	$3d^4(^4)$ ^{3}F $5f$	33	−.17478	$3d^4$ ^{3}G $6g$	13	−.30299	$3d^4$ ^{3}D $5g$
16	−.53785	$3d^4$ ^{3}G $6s$	26	−.37738	$3d^4$ ^{3}G $5f$	34	−.16341	$3d^4(^4)$ ^{3}F $7d$	14	−.30134	$3d^4$ ^{3}G $6d$
17	−.52548	$3d^4$ ^{3}D $5d$	27	−.32702	$3d^4$ ^{5}D $7f$			^{4}**H**^{o}	15	−.25024	$3d^4$ ^{5}D $8g$
18	−.44410	$3d^4$ ^{5}D $6g$	28	−.32650	$3d^4$ ^{5}D $7h$	1	−2.03477	$3d^4$ ^{3}H $4p$	16	−.25000	$3d^4$ ^{5}D $8i$
19	−.41548	$3d^4(^4)$ ^{3}P $5g$	29	−.31267	$3d^4$ ^{3}D $5f$	2	−1.95538	$3d^4$ ^{3}G $4p$	17	−.21990	$3d^4$ ^{3}H $6g$
20	−.41528	$3d^4$ ^{3}H $5g$	30	−.25946	$3d^4$ ^{3}H $7p$	3	1.00012	$3d^4$ ^{5}D $4f$	18	.20461	$3d^4(^4)$ ^{3}F $6g$
21	−.40012	$3d^4(^4)$ ^{3}F $5g$	31	−.25043	$3d^4$ ^{5}D $8f$	4	−.94048	$3d^4$ ^{3}H $5p$	19	−.19771	$3d^4$ ^{5}D $9g$
22	−.39702	$3d^4$ ^{5}D $7d$	32	−.24998	$3d^4$ ^{5}D $8h$	5	−.89070	$3d^4$ ^{3}G $5p$	20	−.19753	$3d^4$ ^{5}D $9k$
23	−.37030	$3d^4$ ^{3}G $5g$	33	−.24688	$3d^4(^4)$ ^{3}F $7p$	6	−.78640	$3d^4$ ^{3}H $4f$	21	−.19753	$3d^4$ ^{5}D $9i$
24	−.33646	$3d^4$ ^{3}H $6d$	34	−.22632	$3d^4(^4)$ ^{3}P $6f$	7	−.77089	$3d^4(^4)$ ^{3}F $4f$	22	−.17900	$3d^4$ ^{3}H $7d$
25	−.32653	$3d^4$ ^{5}D $7i$	35	−.22325	$3d^4$ ^{3}H $6f$	8	−.74111	$3d^4$ ^{3}G $4f$	23	−.17479	$3d^4$ ^{3}G $6g$
26	−.32629	$3d^4$ ^{5}D $7g$	36	−.21965	$3d^4(^4)$ ^{3}P $6h$	9	−.67700	$3d^4$ ^{3}D $4f$			^{4}**I**^{o}
27	.32427	$3d^4(^4)$ ^{3}F $6d$	37	−.21607	$3d^4$ ^{3}G $7p$	10	−.64555	$3d^4$ ^{5}D $5f$	1	−2.00550	$3d^4$ ^{3}H $4p$
28	−.31121	$3d^4(^2)$ ^{3}F $5d$	38	−.20970	$3d^4(^4)$ ^{3}F $6f$	11	−.49322	$3d^4$ ^{3}H $6p$	2	−.93465	$3d^4$ ^{3}H $5p$
29	−.30300	$3d^4$ ^{3}D $5g$	39	−.20451	$3d^4(^4)$ ^{3}F $6h$	12	−.45514	$3d^4(^2)$ ^{3}F $4f$	3	−.78999	$3d^4$ ^{3}H $4f$
30	−.29981	$3d^4$ ^{5}D $8d$	40	−.19798	$3d^4$ ^{5}D $9f$	13	−.44875	$3d^4$ ^{5}D $6f$	4	−.77407	$3d^4(^4)$ ^{3}F $4f$

V-like Fe (Fe^{3+})

i	E(Ryds)	Description
5	−.74172	$3d^4$ ^{3}G $4f$
6	−.49087	$3d^4$ ^{3}H $6p$
7	−.44554	$3d^4(^2)$ ^{3}F $4f$
8	−.44441	$3d^4$ ^{5}D $6h$
9	−.42488	$3d^4$ ^{3}H $5f$
10	−.40998	$3d^4(^4)$ ^{3}F $5f$
11	−.37889	$3d^4$ ^{3}G $5f$
12	−.32649	$3d^4$ ^{5}D $7h$
13	−.26166	$3d^4$ ^{3}H $7p$
14	−.25000	$3d^4$ ^{5}D $8j$
15	−.24996	$3d^4$ ^{5}D $8h$
16	−.22674	$3d^4$ ^{3}H $6f$
17	−.21966	$3d^4$ ^{3}H $6h$
18	−.21965	$3d^4(^4)$ ^{3}P $6h$
19	−.21138	$3d^4(^4)$ ^{3}F $6f$
20	−.20450	$3d^4(^4)$ ^{3}F $6h$
21	−.19753	$3d^4$ ^{5}D $9j$
22	−.19750	$3d^4$ ^{5}D $9h$
23	−.18088	$3d^4$ ^{3}G $6f$
24	−.17469	$3d^4$ ^{3}G $6h$
	^{4}J^e	
1	−1.27967	$3d^4$ ^{3}H $4d$
2	−.64586	$3d^4$ ^{3}H $5d$
3	−.41552	$3d^4$ ^{3}H $5g$
4	−.40031	$3d^4(^4)$ ^{3}F $5g$
5	−.37029	$3d^4$ ^{3}G $5g$
6	−.34541	$3d^4$ ^{3}H $6d$
7	−.32653	$3d^4$ ^{5}D $7i$
8	−.25000	$3d^4$ ^{5}D $8i$
9	−.21994	$3d^4$ ^{3}H $6g$
10	−.20474	$3d^4(^4)$ ^{3}F $6g$
11	−.19753	$3d^4$ ^{5}D $9k$
12	−.19753	$3d^4$ ^{5}D $9i$
13	−.17805	$3d^4$ ^{3}H $7d$
14	−.17472	$3d^4$ ^{3}G $6g$
	^{4}J^o	
1	−.79132	$3d^4$ ^{3}H $4f$
2	−.74355	$3d^4$ ^{3}G $4f$
3	−.44447	$3d^4$ ^{5}D $6h$
4	−.42633	$3d^4$ ^{3}H $5f$
5	−.38049	$3d^4$ ^{3}G $5f$
6	−.32657	$3d^4$ ^{5}D $7h$
7	−.25003	$3d^4$ ^{5}D $8h$
8	−.25000	$3d^4$ ^{5}D $8j$
9	−.22695	$3d^4$ ^{3}H $6f$
10	−.21966	$3d^4$ ^{3}H $6h$
11	−.20451	$3d^4(^4)$ ^{3}F $6h$
12	−.19756	$3d^4$ ^{5}D $9h$
13	−.19753	$3d^4$ ^{5}D $9j$
14	−.18167	$3d^4$ ^{3}G $6f$
15	−.17468	$3d^4$ ^{3}G $6h$
	^{6}S^e	
1	−3.96069	$3d^5(^5)$
2	−1.37454	$3d^4$ ^{5}D $4d$
3	−.81196	$3d^4$ ^{5}D $5d$
4	−.53914	$3d^4$ ^{5}D $6d$
5	−.38441	$3d^4$ ^{5}D $7d$
6	−.28800	$3d^4$ ^{5}D $8d$
7	−.22383	$3d^4$ ^{5}D $9d$
	^{6}P^e	
1	−1.52456	$3d^4$ ^{5}D $4d$
2	−.87788	$3d^4$ ^{5}D $5d$
3	−.57392	$3d^4$ ^{5}D $6d$
4	−.40499	$3d^4$ ^{5}D $7d$
5	−.30118	$3d^4$ ^{5}D $8d$
6	−.23278	$3d^4$ ^{5}D $9d$
	^{6}P^o	
1	−2.24990	$3d^4$ ^{5}D $4p$
2	−1.16488	$3d^4$ ^{5}D $5p$
3	−1.01182	$3d^4$ ^{5}D $4f$
4	−.71827	$3d^4$ ^{5}D $6p$
5	−.64670	$3d^4$ ^{5}D $5f$
6	−.48791	$3d^4$ ^{5}D $7p$
7	−.44856	$3d^4$ ^{5}D $6f$
8	−.35322	$3d^4$ ^{5}D $8p$
9	−.32922	$3d^4$ ^{5}D $7f$
10	−.26757	$3d^4$ ^{5}D $9p$
11	−.25185	$3d^4$ ^{5}D $8f$
	^{6}D^e	
1	−2.77361	$3d^4$ ^{5}D $4s$
2	−1.50145	$3d^4$ ^{5}D $4d$
3	−1.35607	$3d^4$ ^{5}D $5s$
4	−.86950	$3d^4$ ^{5}D $5d$
5	−.80939	$3d^4$ ^{5}D $6s$
6	−.64105	$3d^4$ ^{5}D $5g$
7	−.56964	$3d^4$ ^{5}D $6d$
8	−.53841	$3d^4$ ^{5}D $7s$
9	−.44532	$3d^4$ ^{5}D $6g$
10	−.40249	$3d^4$ ^{5}D $7d$
11	−.38408	$3d^4$ ^{5}D $8s$
12	−.32717	$3d^4$ ^{5}D $7g$
13	−.29959	$3d^4$ ^{5}D $8d$
14	−.28781	$3d^4$ ^{5}D $9s$
15	−.25046	$3d^4$ ^{5}D $8g$
16	−.23170	$3d^4$ ^{5}D $9d$
	^{6}D^o	
1	−2.21380	$3d^4$ ^{5}D $4p$
2	−1.15529	$3d^4$ ^{5}D $5p$
3	−1.01481	$3d^4$ ^{5}D $4f$
4	−.71368	$3d^4$ ^{5}D $6p$
5	−.65081	$3d^4$ ^{5}D $5f$
6	−.48539	$3d^4$ ^{5}D $7p$
7	−.45168	$3d^4$ ^{5}D $6f$
8	−.35168	$3d^4$ ^{5}D $8p$
9	−.33145	$3d^4$ ^{5}D $7f$
10	−.26657	$3d^4$ ^{5}D $9p$
11	−.25345	$3d^4$ ^{5}D $8f$
	^{6}F^e	
1	−1.49686	$3d^4$ ^{5}D $4d$
2	−.86844	$3d^4$ ^{5}D $5d$
3	−.64015	$3d^4$ ^{5}D $5g$
4	−.56913	$3d^4$ ^{5}D $6d$
5	−.44459	$3d^4$ ^{5}D $6g$
6	−.40219	$3d^4$ ^{5}D $7d$
7	−.32665	$3d^4$ ^{5}D $7g$
8	−.29939	$3d^4$ ^{5}D $8d$
9	−.25010	$3d^4$ ^{5}D $8g$
10	−.23157	$3d^4$ ^{5}D $9d$
11	−.19760	$3d^4$ ^{5}D $9g$
	^{6}F^o	
1	−2.25564	$3d^4$ ^{5}D $4p$
2	−1.16682	$3d^4$ ^{5}D $5p$
3	−1.00907	$3d^4$ ^{5}D $4f$
4	−.71904	$3d^4$ ^{5}D $6p$
5	−.64780	$3d^4$ ^{5}D $5f$
6	−.48832	$3d^4$ ^{5}D $7p$
7	−.44993	$3d^4$ ^{5}D $6f$
8	−.44450	$3d^4$ ^{5}D $6h$
9	−.35346	$3d^4$ ^{5}D $8p$
10	−.33034	$3d^4$ ^{5}D $7f$
11	−.32659	$3d^4$ ^{5}D $7h$
12	−.26773	$3d^4$ ^{5}D $9p$
13	−.25270	$3d^4$ ^{5}D $8f$
14	−.25005	$3d^4$ ^{5}D $8h$
	^{6}G^e	
1	−1.52212	$3d^4$ ^{5}D $4d$
2	−.87713	$3d^4$ ^{5}D $5d$
3	−.63967	$3d^4$ ^{5}D $5g$
4	−.57353	$3d^4$ ^{5}D $6d$
5	−.44421	$3d^4$ ^{5}D $6g$
6	−.40476	$3d^4$ ^{5}D $7d$
7	−.32653	$3d^4$ ^{5}D $7i$
8	−.32638	$3d^4$ ^{5}D $7g$
9	−.30104	$3d^4$ ^{5}D $8d$
10	−.25001	$3d^4$ ^{5}D $8i$
11	−.24990	$3d^4$ ^{5}D $8g$
12	−.23268	$3d^4$ ^{5}D $9d$
13	−.19754	$3d^4$ ^{5}D $9i$
14	−.19746	$3d^4$ ^{5}D $9g$
	^{6}G^o	
1	−1.00739	$3d^4$ ^{5}D $4f$
2	−.64708	$3d^4$ ^{5}D $5f$
3	−.44957	$3d^4$ ^{5}D $6f$
4	−.44444	$3d^4$ ^{5}D $6h$
5	−.33013	$3d^4$ ^{5}D $7f$
6	−.32652	$3d^4$ ^{5}D $7h$
7	−.25257	$3d^4$ ^{5}D $8f$
8	−.24999	$3d^4$ ^{5}D $8h$
9	−.19941	$3d^4$ ^{5}D $9f$
10	−.19753	$3d^4$ ^{5}D $9h$
	^{6}H^e	
1	−.63955	$3d^4$ ^{5}D $5g$
2	−.44411	$3d^4$ ^{5}D $6g$
3	−.32653	$3d^4$ ^{5}D $7i$
4	−.32630	$3d^4$ ^{5}D $7g$
5	−.25000	$3d^4$ ^{5}D $8i$
6	−.24984	$3d^4$ ^{5}D $8g$
7	−.19753	$3d^4$ ^{5}D $9i$
8	−.19742	$3d^4$ ^{5}D $9g$
	^{6}H^o	
1	−1.01820	$3d^4$ ^{5}D $4f$
2	−.65262	$3d^4$ ^{5}D $5f$
3	−.45274	$3d^4$ ^{5}D $6f$
4	−.44441	$3d^4$ ^{5}D $6h$
5	−.33212	$3d^4$ ^{5}D $7f$
6	−.32649	$3d^4$ ^{5}D $7h$
7	−.25390	$3d^4$ ^{5}D $8f$
8	−.25000	$3d^4$ ^{5}D $8j$
9	−.24997	$3d^4$ ^{5}D $8h$
10	−.20034	$3d^4$ ^{5}D $9f$
11	−.19753	$3d^4$ ^{5}D $9j$
12	−.19750	$3d^4$ ^{5}D $9h$
	^{6}I^e	
1	−.64065	$3d^4$ ^{5}D $5g$
2	−.44501	$3d^4$ ^{5}D $6g$
3	−.32695	$3d^4$ ^{5}D $7g$
4	−.32653	$3d^4$ ^{5}D $7i$
5	−.25031	$3d^4$ ^{5}D $8g$
6	−.25000	$3d^4$ ^{5}D $8i$
7	−.19776	$3d^4$ ^{5}D $9g$
8	−.19753	$3d^4$ ^{5}D $9k$
9	−.19753	$3d^4$ ^{5}D $9i$
	^{6}I^o	
1	−.44441	$3d^4$ ^{5}D $6h$
2	−.32649	$3d^4$ ^{5}D $7h$
3	−.25000	$3d^4$ ^{5}D $8j$
4	−.24997	$3d^4$ ^{5}D $8h$
5	−.19753	$3d^4$ ^{5}D $9j$
6	−.19750	$3d^4$ ^{5}D $9h$
	^{6}J^e	
1	−.32653	$3d^4$ ^{5}D $7i$
2	−.25000	$3d^4$ ^{5}D $8i$
3	−.19753	$3d^4$ ^{5}D $9k$
4	−.19753	$3d^4$ ^{5}D $9i$
	^{6}J^o	
1	−.44448	$3d^4$ ^{5}D $6h$
2	−.32657	$3d^4$ ^{5}D $7h$

V-like Fe (Fe^{3+})

Energies in ascending order from ground state for terms with effective $n \leq 4.0$, $L \leq 7$

Term	i	E(Ryds)	Term	i	E(Ryds)	Term	i	E(Ryds)	Term	i	E(Ryds)	Term	i	E(Ryds)
$^{6}S^{e}$	1	0.00000	$^{4}D^{o}$	1	1.82611	$^{2}F^{o}$	6	2.19826	$^{4}H^{e}$	3	2.68872	$^{4}F^{o}$	6	2.81629
$^{4}G^{e}$	1	0.32058	$^{2}G^{e}$	5	1.86166	$^{2}P^{o}$	4	2.19979	$^{4}P^{e}$	5	2.68938	$^{4}G^{e}$	7	2.81671
$^{4}P^{e}$	1	0.37520	$^{4}H^{o}$	1	1.92592	$^{2}G^{o}$	5	2.22575	$^{4}F^{e}$	6	2.69070	$^{2}D^{e}$	11	2.82732
$^{4}D^{e}$	1	0.39728	$^{4}D^{o}$	2	1.93629	$^{2}D^{o}$	5	2.25002	$^{2}D^{e}$	7	2.69484	$^{2}G^{e}$	10	2.82852
$^{2}I^{e}$	1	0.45865	$^{4}G^{o}$	1	1.95229	$^{4}F^{o}$	5	2.27263	$^{2}J^{e}$	1	2.69521	$^{4}D^{o}$	7	2.83201
$^{2}D^{e}$	1	0.52640	$^{4}I^{o}$	1	1.95519	$^{4}D^{o}$	5	2.28523	$^{4}G^{e}$	4	2.69525	$^{4}P^{e}$	7	2.83280
$^{4}F^{e}$	1	0.53437	$^{2}G^{o}$	1	1.95750	$^{4}P^{o}$	4	2.28948	$^{2}H^{e}$	3	2.69554	$^{4}H^{e}$	5	2.83431
$^{2}F^{e}$	1	0.53599	$^{4}P^{o}$	2	1.96409	$^{2}D^{o}$	6	2.30170	$^{2}F^{e}$	6	2.69616	$^{4}P^{e}$	8	2.83595
$^{2}H^{e}$	1	0.56614	$^{2}S^{o}$	1	1.96656	$^{2}F^{o}$	7	2.30229	$^{4}D^{e}$	7	2.70602	$^{2}S^{e}$	4	2.84036
$^{2}G^{e}$	1	0.57934	$^{4}G^{o}$	2	1.97268	$^{4}G^{o}$	4	2.30472	$^{4}F^{e}$	7	2.70708	$^{2}P^{e}$	7	2.84590
$^{2}F^{e}$	2	0.62315	$^{4}F^{o}$	2	1.97393	$^{2}S^{e}$	3	2.30817	$^{2}I^{e}$	3	2.71053	$^{4}D^{e}$	9	2.84658
$^{2}S^{e}$	1	0.68899	$^{2}D^{o}$	1	1.97967	$^{4}S^{o}$	2	2.33709	$^{4}G^{e}$	5	2.71351	$^{2}H^{e}$	7	2.84799
$^{2}D^{e}$	2	0.76198	$^{4}S^{o}$	1	1.98250	$^{2}G^{o}$	6	2.35006	$^{2}P^{e}$	5	2.71534	$^{4}F^{e}$	10	2.84919
$^{2}G^{e}$	2	0.83524	$^{2}P^{o}$	1	1.98283	$^{2}P^{o}$	5	2.35390	$^{2}F^{e}$	7	2.71547	$^{2}G^{e}$	11	2.85579
$^{2}P^{e}$	1	1.01968	$^{2}I^{o}$	1	1.98887	$^{4}D^{o}$	6	2.35520	$^{2}G^{e}$	7	2.72289	$^{2}P^{o}$	7	2.85882
$^{2}D^{e}$	3	1.10370	$^{4}D^{o}$	3	1.99102	$^{2}S^{o}$	2	2.38073	$^{4}I^{e}$	2	2.72535	$^{2}F^{e}$	11	2.86338
$^{6}D^{e}$	1	1.18708	$^{2}F^{o}$	1	1.99592	$^{2}G^{o}$	7	2.39121	$^{2}H^{e}$	4	2.72547	$^{2}D^{e}$	12	2.87427
$^{4}D^{e}$	2	1.27531	$^{2}H^{o}$	1	1.99851	$^{2}H^{o}$	5	2.39481	$^{4}H^{e}$	4	2.72737	$^{4}G^{e}$	8	2.87868
$^{4}P^{e}$	2	1.42932	$^{4}F^{o}$	3	2.00156	$^{2}F^{o}$	8	2.40749	$^{2}D^{e}$	8	2.73263	$^{2}F^{e}$	12	2.87881
$^{4}H^{e}$	1	1.42940	$^{2}D^{o}$	2	2.00515	$^{2}D^{o}$	7	2.41780	$^{2}F^{e}$	8	2.73653	$^{2}H^{e}$	8	2.88364
$^{4}F^{e}$	2	1.44463	$^{4}H^{o}$	2	2.00531	$^{4}S^{e}$	1	2.42098	$^{2}D^{e}$	9	2.73973	$^{2}P^{e}$	8	2.88961
$^{4}G^{e}$	2	1.47288	$^{2}G^{o}$	2	2.02332	$^{6}P^{e}$	1	2.43613	$^{2}I^{e}$	4	2.74267	$^{2}G^{e}$	12	2.89405
$^{2}P^{e}$	2	1.48342	$^{2}F^{o}$	2	2.02504	$^{6}G^{e}$	1	2.43857	$^{2}F^{e}$	9	2.74920	$^{2}I^{e}$	6	2.89926
$^{2}H^{e}$	2	1.48361	$^{2}H^{o}$	2	2.02977	$^{6}D^{e}$	2	2.45924	$^{2}H^{e}$	5	2.75056	$^{2}G^{e}$	13	2.90516
$^{2}F^{e}$	3	1.49753	$^{4}G^{o}$	3	2.03203	$^{6}F^{e}$	1	2.46383	$^{4}F^{e}$	8	2.75584	$^{2}F^{e}$	13	2.92495
$^{2}G^{e}$	3	1.52720	$^{4}D^{o}$	4	2.06031	$^{4}G^{e}$	3	2.48853	$^{4}D^{e}$	8	2.76210	$^{2}D^{e}$	13	2.92991
$^{4}D^{e}$	3	1.53289	$^{2}D^{e}$	6	2.06201	$^{4}P^{e}$	4	2.49180	$^{4}P^{e}$	6	2.76723	$^{4}P^{o}$	6	2.94215
$^{2}G^{e}$	4	1.55186	$^{2}G^{o}$	3	2.06263	$^{4}D^{e}$	4	2.50015	$^{2}G^{e}$	8	2.77720	$^{6}H^{o}$	1	2.94249
$^{2}I^{e}$	2	1.55957	$^{4}P^{o}$	3	2.06591	$^{4}F^{e}$	4	2.51999	$^{4}G^{e}$	6	2.78403	$^{4}D^{e}$	10	2.94254
$^{2}S^{e}$	2	1.57906	$^{2}I^{o}$	2	2.06812	$^{2}P^{o}$	6	2.54971	$^{4}S^{e}$	2	2.78660	$^{6}D^{o}$	3	2.94588
$^{2}D^{e}$	4	1.58463	$^{2}F^{o}$	3	2.07060	$^{6}S^{e}$	2	2.58615	$^{2}J^{e}$	2	2.78876	$^{2}G^{e}$	14	2.94751
$^{2}D^{e}$	5	1.63778	$^{2}H^{o}$	3	2.07123	$^{2}F^{o}$	9	2.59809	$^{2}I^{e}$	5	2.78940	$^{6}P^{o}$	3	2.94887
$^{2}F^{e}$	4	1.69462	$^{2}P^{o}$	2	2.07731	$^{6}D^{e}$	3	2.60462	$^{2}H^{e}$	6	2.79283	$^{2}I^{e}$	7	2.95132
$^{6}F^{o}$	1	1.70505	$^{4}F^{o}$	4	2.07877	$^{4}D^{e}$	5	2.62694	$^{6}F^{o}$	2	2.79387	$^{4}H^{o}$	3	2.95157
$^{6}P^{o}$	1	1.71079	$^{2}J^{o}$	1	2.08200	$^{2}D^{o}$	8	2.63704	$^{6}P^{o}$	2	2.79581	$^{6}F^{o}$	3	2.95162
$^{4}P^{o}$	1	1.73661	$^{2}G^{o}$	4	2.09706	$^{4}I^{e}$	1	2.66502	$^{2}G^{e}$	9	2.80037	$^{2}H^{e}$	9	2.95248
$^{6}D^{o}$	1	1.74689	$^{2}P^{o}$	3	2.11424	$^{4}F^{e}$	5	2.67157	$^{2}F^{e}$	10	2.80414	$^{6}G^{o}$	1	2.95330
$^{4}F^{e}$	3	1.76227	$^{2}H^{o}$	4	2.11553	$^{2}G^{e}$	6	2.67188	$^{2}D^{e}$	10	2.80425	$^{4}D^{o}$	8	2.95582
$^{4}P^{e}$	3	1.76629	$^{2}F^{o}$	4	2.12030	$^{4}H^{e}$	2	2.67282	$^{6}D^{o}$	2	2.80540	$^{2}D^{e}$	14	2.95763
$^{4}F^{o}$	1	1.77151	$^{2}D^{o}$	3	2.12818	$^{4}D^{e}$	6	2.67382	$^{4}P^{o}$	5	2.80560	$^{4}F^{o}$	7	2.95920
$^{2}F^{e}$	5	1.81387	$^{2}D^{o}$	4	2.14519	$^{2}P^{e}$	4	2.67483	$^{2}P^{e}$	6	2.81029	$^{2}S^{e}$	5	2.95947
$^{2}P^{e}$	3	1.81618	$^{2}F^{o}$	5	2.16370	$^{4}J^{e}$	1	2.68102	$^{4}F^{e}$	9	2.81208			

V-like Fe (Fe^{3+})

gf-values for transitions involving terms with effective $n \leq 4.0$, $L \leq 7$

i i'	gf_L	i i'	gf_L	i i'	gf_L	i i'	gf_L	i i'	gf_L	i i'	gf_L
	$^2S^e$–$^2P^o$	2 7	−2.46E−2	7 7	−1.14E−5	6 3	9.57E−1	3 1	6.76E−1	6 10	−6.72E−3
1 1	−1.53E−3	2 8	−4.20E−5	8 1	1.52E−1	6 4	1.47E−1	3 2	2.98E−1	6 11	−8.46E−4
1 2	−2.81E−1		$^2P^e$–$^2P^o$	8 2	2.93E−5	6 5	3.90E−3	3 3	2.36E−3	6 12	−4.13E−4
1 3	−4.90E−2	1 1	−1.73E−2	8 3	2.71E−1	6 6	9.53E−2	3 4	6.97E−1	6 13	−2.38E−2
1 4	−3.81E−1	1 2	−2.22E−2	8 4	2.33E+0	6 7	2.06E−4	3 5	2.18E−1	6 14	−3.54E−3
1 5	−2.86E−2	1 3	−4.91E−2	8 5	2.86E−2	6 8	2.95E−4	3 6	9.68E−4	7 1	1.58E−3
1 6	−3.67E−2	1 4	−1.37E−3	8 6	1.81E−3	7 1	8.73E−2	3 7	−4.37E−3	7 2	8.56E−3
1 7	−1.80E−3	1 5	−1.77E−1	8 7	6.43E−4	7 2	1.17E+0	3 8	−3.36E−1	7 3	3.08E−1
2 1	−3.19E−2	1 6	−5.11E−1		$^2P^e$–$^2D^o$	7 3	7.43E−4	3 9	−1.64E−2	7 4	6.28E−4
2 2	−7.82E−1	1 7	−3.52E−2	1 1	−2.01E−2	7 4	2.39E−1	3 10	−1.35E+0	7 5	3.04E−3
2 3	−9.65E−1	2 1	−1.81E+0	1 2	−5.16E−3	7 5	1.91E−2	3 11	−1.76E−2	7 6	1.16E−1
2 4	−1.64E−1	2 2	−5.07E−3	1 3	−7.32E−2	7 6	1.11E−3	3 12	−1.61E+0	7 7	2.13E−4
2 5	−4.37E−4	2 3	−8.75E−2	1 4	−4.22E−2	7 7	5.10E−3	3 13	−8.43E−1	7 8	3.41E−4
2 6	−4.85E−3	2 4	−1.44E−2	1 5	−1.05E−2	7 8	8.34E−5	3 14	−2.51E−1	7 9	1.98E−5
2 7	−1.39E−4	2 5	−1.21E−2	1 6	−4.65E−1	8 1	7.53E−2	4 1	5.15E−2	7 10	1.02E−5
3 1	6.15E−5	2 6	−4.87E−5	1 7	−3.05E−1	8 2	1.30E−1	4 2	2.47E−1	7 11	6.32E−6
3 2	2.00E−4	2 7	−1.46E−4	1 8	−1.79E−1	8 3	1.90E−1	4 3	6.87E−4	7 12	−2.35E−9
3 3	3.46E−4	3 1	−9.59E−3	2 1	−2.42E−1	8 4	7.11E−1	4 4	1.39E−1	7 13	−3.97E−4
3 4	4.07E−4	3 2	−4.73E−2	2 2	−2.95E+0	8 5	1.51E−2	4 5	1.70E+0	7 14	−7.05E−4
3 5	−5.28E−4	3 3	−3.07E−2	2 3	−1.61E−5	8 6	1.19E−2	4 6	2.74E−3		$^2D^e$–$^2D^o$
3 6	−7.27E−2	3 4	−6.55E−4	2 4	−1.34E−1	8 7	2.23E−6	4 7	−8.98E−2	1 1	−5.36E−1
3 7	−1.85E+0	3 5	−1.87E+0	2 5	−9.98E−3	8 8	3.53E−3	4 8	−5.20E−5	1 2	−5.86E−1
4 1	2.32E−2	3 6	−4.21E−2	2 6	−1.04E−3		$^2P^o$–$^2D^e$	4 9	−5.93E−2	1 3	−7.53E−2
4 2	4.10E−1	3 7	−5.30E−4	2 7	−2.21E−3	1 1	3.16E−1	4 10	−7.30E−2	1 4	−2.78E−1
4 3	6.14E−1	4 1	8.90E−1	2 8	−1.65E−4	1 2	1.37E−3	4 11	−3.13E−1	1 5	−4.16E−2
4 4	2.29E−1	4 2	2.70E−2	3 1	−3.82E−4	1 3	4.85E−3	4 12	−9.90E−1	1 6	−1.08E−3
4 5	6.96E−2	4 3	1.30E−1	3 2	−2.51E−4	1 4	1.84E−3	4 13	−1.09E+0	1 7	−3.59E−4
4 6	4.48E−3	4 4	1.76E−3	3 3	−2.24E−2	1 5	3.17E−2	4 14	−6.12E−3	1 8	−2.29E−3
4 7	−1.16E−4	4 5	7.07E−3	3 4	−4.21E−3	1 6	−2.72E−4	5 1	2.94E−2	2 1	−4.23E−2
5 1	1.17E−2	4 6	2.07E−5	3 5	−5.09E−1	1 7	−4.16E+0	5 2	4.22E−1	2 2	−4.81E−3
5 2	2.89E−1	4 7	−2.95E−5	3 6	−1.49E+0	1 8	−1.65E−2	5 3	4.26E−1	2 3	−2.90E−1
5 3	4.34E−1	5 1	7.30E−1	3 7	−1.16E+0	1 9	−4.34E−1	5 4	1.15E−1	2 4	−3.98E−1
5 4	1.39E−1	5 2	8.46E−2	3 8	−5.43E−4	1 10	−2.01E−1	5 5	2.62E−2	2 5	−8.21E−2
5 5	3.77E−5	5 3	4.94E−3	4 1	1.16E+0	1 11	−1.16E−1	5 6	1.73E−2	2 6	−5.73E−1
5 6	3.36E−3	5 4	6.63E−3	4 2	3.93E−1	1 12	−2.34E−2	5 7	−2.00E−3	2 7	−2.63E−2
5 7	3.95E−5	5 5	5.23E−2	4 3	8.92E−2	1 13	−5.10E−2	5 8	−2.98E−3	2 8	−9.76E−2
	$^2S^o$–$^2P^e$	5 6	1.27E−3	4 4	4.42E−2	1 14	−1.10E−3	5 9	−6.31E−3	3 1	−1.43E−4
1 1	3.56E−3	5 7	−5.87E−5	4 5	2.47E−3	2 1	5.89E−2	5 10	−1.98E−3	3 2	−7.45E−3
1 2	6.01E−1	6 1	3.47E−4	4 6	2.38E−4	2 2	5.66E−2	5 11	−5.68E−2	3 3	−6.49E−2
1 3	9.31E−3	6 2	1.56E+0	4 7	5.94E−8	2 3	2.36E−5	5 12	−6.20E−2	3 4	−1.66E−4
1 4	−1.30E+0	6 3	9.87E−1	4 8	7.57E−5	2 4	1.04E+0	5 13	−1.00E−1	3 5	−5.97E−6
1 5	−8.98E−1	6 4	1.82E−1	5 1	1.97E+0	2 5	1.95E−6	5 14	−8.87E−2	3 6	−7.67E−2
1 6	−2.69E−4	6 5	3.92E−2	5 2	7.77E−2	2 6	1.71E−4	6 1	3.06E−3	3 7	−1.09E+0
1 7	−2.67E−1	6 6	1.05E−4	5 3	1.06E−1	2 7	−3.01E−2	6 2	6.78E−2	3 8	−2.73E−1
1 8	−1.72E−2	6 7	−3.37E−4	5 4	1.07E−1	2 8	−4.36E−4	6 3	3.06E−1	4 1	−4.78E−1
2 1	3.65E−1	7 1	6.55E−1	5 5	6.76E−4	2 9	−3.71E−1	6 4	6.35E−3	4 2	−9.17E−3
2 2	2.42E−2	7 2	1.46E−2	5 6	5.39E−3	2 10	−2.36E+0	6 5	4.86E−3	4 3	−2.33E+0
2 3	6.43E−1	7 3	7.00E−2	5 7	1.88E−3	2 11	−1.24E+0	6 6	1.77E+0	4 4	−4.13E−1
2 4	−1.39E−2	7 4	2.12E−1	5 8	3.31E−7	2 12	−4.27E−2	6 7	−1.72E−5	4 5	−2.40E−3
2 5	−1.16E−2	7 5	1.06E−2	6 1	6.56E−4	2 13	−3.60E−2	6 8	−5.76E−5	4 6	−3.02E−2
2 6	−5.99E−2	7 6	2.00E−5	6 2	2.54E−2	2 14	−5.12E−1	6 9	−3.07E−3	4 7	−3.04E−3

V-like Fe (Fe^{3+})

$i\ i'$	gf$_L$	$i\ i'$	gf$_L$	$i\ i'$	gf$_L$	$i\ i'$	gf$_L$	$i\ i'$	gf$_L$	$i\ i'$	gf$_L$
4 8	−2.52E−4	11 4	1.42E−2	3 5	−6.61E−3	9 3	9.61E−1		**^{2}D^o–^{2}F^e**	4 13	−7.95E−1
5 1	−2.75E−2	11 5	1.15E−1	3 6	−1.86E−4	9 4	2.92E−2	1 1	1.09E−3	5 1	5.00E−2
5 2	−4.00E−2	11 6	3.33E−2	3 7	−2.74E−1	9 5	1.11E−1	1 2	2.55E−3	5 2	3.79E−2
5 3	−5.20E−1	11 7	7.15E−2	3 8	−7.75E−1	9 6	2.65E−3	1 3	2.32E+0	5 3	5.76E−4
5 4	−2.43E+0	11 8	2.60E−4	3 9	−1.34E−1	9 7	2.59E−2	1 4	1.13E−2	5 4	2.21E+0
5 5	−1.18E−1	12 1	2.20E−1	4 1	−8.47E−1	9 8	4.88E−4	1 5	4.01E−3	5 5	3.37E−1
5 6	−6.38E−2	12 2	8.69E−3	4 2	−1.51E−2	9 9	9.80E−4	1 6	−2.97E−1	5 6	−1.02E−2
5 7	−5.37E−4	12 3	2.47E−1	4 3	−3.85E−2	10 1	1.58E−1	1 7	−1.84E+0	5 7	−2.45E−5
5 8	−1.11E−2	12 4	9.68E−1	4 4	−3.34E+0	10 2	4.56E−5	1 8	−4.37E−1	5 8	−8.61E−3
6 1	1.26E−4	12 5	4.94E−1	4 5	−5.52E−2	10 3	1.79E+0	1 9	−8.16E−1	5 9	−1.67E−6
6 2	3.65E−6	12 6	2.25E−1	4 6	−1.88E−1	10 4	2.56E−1	1 10	−8.17E−2	5 10	−1.43E−2
6 3	−5.33E−4	12 7	1.02E−2	4 7	−2.48E−2	10 5	3.15E−2	1 11	−1.13E+0	5 11	−1.24E−3
6 4	−1.67E−3	12 8	1.32E−3	4 8	−1.06E−2	10 6	1.81E−1	1 12	−1.85E−1	5 12	−6.49E−1
6 5	−8.74E−3	13 1	1.81E−1	4 9	−3.40E−4	10 7	6.42E−4	1 13	−5.94E−3	5 13	−3.73E−1
6 6	−9.88E−5	13 2	4.79E−1	5 1	−2.55E−2	10 8	5.02E−3	2 1	1.08E+0	6 1	2.05E−2
6 7	−2.16E−3	13 3	1.65E−1	5 2	−9.80E−2	10 9	1.14E−2	2 2	2.26E−1	6 2	3.51E−1
6 8	−3.31E+0	13 4	2.05E+0	5 3	−8.19E−2	11 1	1.32E−1	2 3	2.92E−1	6 3	1.22E−3
7 1	1.17E+0	13 5	2.20E−1	5 4	−2.71E−2	11 2	6.13E−1	2 4	9.54E−3	6 4	9.48E−1
7 2	7.20E−1	13 6	1.67E−2	5 5	−3.25E+0	11 3	1.35E+0	2 5	6.51E−3	6 5	7.35E−1
7 3	2.25E−4	13 7	1.47E−2	5 6	−1.00E+0	11 4	5.88E−3	2 6	−4.34E+0	6 6	−3.56E−3
7 4	1.33E−1	13 8	9.93E−3	5 7	−2.46E−3	11 5	7.72E−2	2 7	−3.87E−1	6 7	−4.61E−3
7 5	2.97E−3	14 1	2.75E−1	5 8	−4.13E−2	11 6	3.79E−2	2 8	−1.16E+0	6 8	−2.34E−2
7 6	2.27E−2	14 2	5.12E−3	5 9	−2.56E−4	11 7	1.18E−4	2 9	−1.56E+0	6 9	−4.16E−5
7 7	7.41E−4	14 3	9.94E−1	6 1	2.24E−4	11 8	2.39E−2	2 10	−8.19E−1	6 10	1.36E−2
7 8	9.23E−7	14 4	2.64E−1	6 2	2.43E−4	11 9	1.46E−2	2 11	−1.62E−1	6 11	−1.08E−2
8 1	2.89E−1	14 5	1.42E−3	6 3	−1.01E−3	12 1	1.27E−1	2 12	−9.20E−4	6 12	−6.83E−1
8 2	3.97E−2	14 6	2.14E−2	6 4	−3.68E−5	12 2	2.00E−1	2 13	−2.38E−1	6 13	−8.73E−2
8 3	7.77E−1	14 7	3.61E−3	6 5	−9.63E−4	12 3	1.16E−1	3 1	5.74E−1	7 1	3.11E−2
8 4	2.63E−2	14 8	4.09E−5	6 6	−3.15E−3	12 4	3.72E−1	3 2	6.46E−1	7 2	2.49E−2
8 5	3.85E−5		**^{2}D^e–^{2}F^o**	6 7	−2.99E−2	12 5	1.33E−4	3 3	4.34E−1	7 3	1.62E−2
8 6	2.20E−2	1 1	−6.02E−1	6 8	−2.45E−1	12 6	1.01E+0	3 4	1.91E−2	7 4	7.81E−3
8 7	2.25E−5	1 2	−4.25E−2	6 9	−4.14E+0	12 7	2.44E−3	3 5	5.16E−7	7 5	2.25E+0
8 8	1.41E−4	1 3	−2.67E−1	7 1	5.54E−2	12 8	3.49E−3	3 6	−7.53E−5	7 6	−2.38E−3
9 1	1.95E+0	1 4	−3.43E−1	7 2	2.66E−1	12 9	3.58E−3	3 7	−1.48E−2	7 7	−1.16E−2
9 2	5.18E−1	1 5	−2.42E−4	7 3	4.86E−2	13 1	4.46E−1	3 8	−4.56E−2	7 8	−6.10E−3
9 3	7.80E−3	1 6	−1.30E−3	7 4	6.81E−2	13 2	4.12E−2	3 9	−8.66E−2	7 9	−1.45E−3
9 4	1.13E−1	1 7	−4.77E−3	7 5	8.86E−4	13 3	4.05E−1	3 10	−3.37E+0	7 10	−2.14E−5
9 5	3.18E−4	1 8	−1.21E−3	7 6	6.49E−3	13 4	1.55E−1	3 11	−1.21E−1	7 11	−2.02E−2
9 6	4.15E−2	1 9	−8.64E−3	7 7	4.72E−4	13 5	1.69E−1	3 12	−3.22E−2	7 12	−9.68E−4
9 7	8.65E−3	2 1	−4.12E−3	7 8	5.16E−3	13 6	1.08E−2	3 13	−2.17E+0	7 13	−1.10E−1
9 8	1.24E−8	2 2	−3.61E−1	7 9	2.08E−4	13 7	4.49E−2	4 1	3.63E−1	8 1	2.78E−2
10 1	8.73E−2	2 3	−5.00E−3	8 1	2.37E+0	13 8	5.84E−4	4 2	7.44E−1	8 2	3.50E−2
10 2	1.11E−2	2 4	−2.21E−1	8 2	7.85E−1	13 9	2.59E−4	4 3	2.15E−1	8 3	4.99E−4
10 3	6.39E−2	2 5	−1.96E−1	8 3	1.52E+0	14 1	4.89E−1	4 4	1.41E−1	8 4	6.72E−3
10 4	1.40E−1	2 6	−9.11E−2	8 4	4.13E−1	14 2	1.23E−2	4 5	1.10E−3	8 5	7.99E−3
10 5	3.51E−4	2 7	−1.61E−1	8 5	9.26E−2	14 3	2.20E−2	4 6	−1.31E−1	8 6	−4.76E−5
10 6	4.57E−3	2 8	−2.18E−1	8 6	1.80E−3	14 4	1.33E+0	4 7	−1.07E−1	8 7	−5.21E−4
10 7	2.07E−5	2 9	−2.20E−3	8 7	5.53E−2	14 5	6.84E−2	4 8	−5.96E−3	8 8	−3.56E−4
10 8	4.05E−3	3 1	−3.15E−3	8 8	3.13E−3	14 6	4.21E−2	4 9	−2.78E−1	8 9	6.83E−5
11 1	1.72E−2	3 2	−5.71E−3	8 9	5.66E−3	14 7	1.12E−2	4 10	−2.59E−2	8 10	−2.85E−3
11 2	5.07E−2	3 3	−1.32E−2	9 1	1.80E−1	14 8	2.81E−3	4 11	−3.86E−1	8 11	−6.16E−4
11 3	2.44E+0	3 4	−8.60E−8	9 2	1.92E+0	14 9	1.16E−3	4 12	−2.96E+0	8 12	−3.44E−3

V-like Fe (Fe^{3+})

i	i'	gf_L	i	i'	gf_L	i	i'	gf_L	i	i'	gf_L	i	i'	gf_L	i	i'	gf_L
8	13	−4.57E−3	6	6	2.00E−3	12	4	4.58E−1	6	2	1.28E+0	13	5	5.78E−3	4	7	−5.59E−2
		$^2\mathbf{F}^e$–$^2\mathbf{F}^o$	6	7	2.66E−2	12	5	2.34E+0	6	3	4.64E−1	13	6	2.53E−2	4	8	−1.06E+0
1	1	−4.65E−1	6	8	3.11E−3	12	6	6.76E−3	6	4	5.65E−3	13	7	6.83E−4	4	9	−3.55E+0
1	2	−5.11E−3	6	9	7.71E−6	12	7	9.03E−3	6	5	5.26E−2			$^2\mathbf{F}^o$–$^2\mathbf{G}^e$	4	10	−1.26E+0
1	3	−1.31E+0	7	1	1.59E+0	12	8	8.70E−3	6	6	8.83E−2	1	1	3.47E−2	4	11	−4.17E+0
1	4	−8.90E−3	7	2	3.46E+0	12	9	1.11E−2	6	7	1.15E−3	1	2	8.19E−3	4	12	−4.39E−2
1	5	−4.60E−2	7	3	5.25E−2	13	1	5.72E−1	7	1	4.67E−1	1	3	1.71E+0	4	13	−3.69E−1
1	6	−1.82E−1	7	4	6.58E−1	13	2	4.15E−2	7	2	4.31E−1	1	4	1.40E−1	4	14	−4.18E−2
1	7	−9.98E−3	7	5	1.80E−1	13	3	2.36E+0	7	3	1.20E−1	1	5	1.56E−3	5	1	1.62E+0
1	8	−1.53E−2	7	6	7.36E−2	13	4	2.12E−1	7	4	6.23E−2	1	6	−1.58E+0	5	2	1.97E−2
1	9	−1.20E−2	7	7	1.63E−2	13	5	7.76E−1	7	5	6.73E−3	1	7	−3.27E+0	5	3	1.74E−1
2	1	−2.80E−1	7	8	8.09E−3	13	6	9.03E−5	7	6	5.72E−4	1	8	−3.38E−1	5	4	1.77E−2
2	2	−2.00E−1	7	9	1.16E−3	13	7	3.50E−2	7	7	3.56E−4	1	9	−1.03E+0	5	5	4.06E−3
2	3	−3.41E−1	8	1	1.36E+0	13	8	2.23E−8	8	1	5.90E−1	1	10	−5.30E−2	5	6	−5.49E−3
2	4	−1.01E−1	8	2	8.25E−1	13	9	3.29E−3	8	2	1.87E−1	1	11	−1.42E−2	5	7	−8.90E−5
2	5	−2.90E−2	8	3	2.98E−2			$^2\mathbf{F}^e$–$^2\mathbf{G}^o$	8	3	3.08E+0	1	12	−9.15E−1	5	8	−3.18E−3
2	6	−1.12E+0	8	4	8.83E−1	1	1	−5.83E−1	8	4	5.36E−1	1	13	−1.64E−2	5	9	−1.65E−2
2	7	−7.01E−2	8	5	7.64E−2	1	2	−2.32E−1	8	5	3.36E−2	1	14	−1.13E−1	5	10	−2.05E+0
2	8	−2.16E−2	8	6	1.62E−1	1	3	−1.46E−1	8	6	5.73E−3	2	1	9.97E−1	5	11	−6.32E−1
2	9	−1.31E−3	8	7	7.52E−3	1	4	−2.38E−1	8	7	2.88E−2	2	2	2.92E−3	5	12	−1.07E−1
3	1	−1.58E+0	8	8	7.82E−4	1	5	−4.76E−2	9	1	1.39E+0	2	3	1.74E+0	5	13	−4.24E+0
3	2	−2.45E+0	8	9	7.11E−5	1	6	−2.52E−2	9	2	4.67E−1	2	4	9.62E−3	5	14	−1.52E−1
3	3	−2.05E−1	9	1	1.27E+0	1	7	−3.82E−2	9	3	1.06E−2	2	5	1.72E−5	6	1	1.06E−1
3	4	−2.36E−1	9	2	1.13E+0	2	1	−2.77E−1	9	4	1.15E+0	2	6	−6.70E−2	6	2	6.46E−2
3	5	−1.05E−1	9	3	8.12E−1	2	2	−1.15E−1	9	5	9.03E−3	2	7	−2.61E+0	6	3	1.77E−3
3	6	−1.87E−2	9	4	8.81E−2	2	3	−4.35E−1	9	6	5.61E−2	2	8	−1.32E+0	6	4	1.79E−1
3	7	−1.16E−2	9	5	6.58E−2	2	4	−1.97E−1	9	7	1.40E−3	2	9	−1.01E−1	6	5	1.84E−1
3	8	−2.53E−3	9	6	1.98E−1	2	5	−6.86E−2	10	1	1.46E−1	2	10	−1.15E+0	6	6	−1.77E−3
3	9	−5.31E−4	9	7	3.31E−2	2	6	−2.74E−2	10	2	6.11E−1	2	11	−7.97E−2	6	7	−8.05E−3
4	1	−5.46E−3	9	8	5.50E−6	2	7	−1.60E−1	10	3	1.43E−1	2	12	−1.08E+0	6	8	−8.45E−2
4	2	−8.30E−2	9	9	7.41E−5	3	1	−1.71E+0	10	4	4.02E−1	2	13	−5.67E−4	6	9	−1.07E−1
4	3	−4.45E−2	10	1	1.06E−1	3	2	−3.62E+0	10	5	1.91E−5	2	14	−7.08E−4	6	10	−2.18E−2
4	4	−1.84E−1	10	2	3.96E−2	3	3	−1.82E−1	10	6	3.76E−2	3	1	1.26E−1	6	11	−1.48E−1
4	5	−9.74E−1	10	3	1.56E+0	3	4	−3.21E−1	10	7	3.22E−2	3	2	2.91E−4	6	12	−1.15E−2
4	6	−2.93E+0	10	4	1.57E+0	3	5	−7.26E−3	11	1	8.79E−1	3	3	1.33E−1	6	13	−5.23E+0
4	7	−2.15E−2	10	5	8.63E−2	3	6	−2.37E−3	11	2	1.38E+0	3	4	4.05E+0	6	14	−2.25E−1
4	8	−1.78E−1	10	6	3.55E−1	3	7	−6.69E−3	11	3	1.80E−1	3	5	9.11E−4	7	1	1.84E−1
4	9	−1.19E−2	10	7	2.06E−2	4	1	−7.63E−3	11	4	3.56E−1	3	6	−1.08E−1	7	2	9.84E−1
5	1	−3.82E−3	10	8	1.83E−2	4	2	−3.16E−2	11	5	8.80E−5	3	7	−1.49E−1	7	3	1.40E−1
5	2	−6.78E−2	10	9	7.40E−3	4	3	−5.19E−3	11	6	1.62E−5	3	8	−1.14E+0	7	4	4.28E−4
5	3	−6.79E−3	11	1	4.72E−1	4	4	−4.78E−2	11	7	9.80E−3	3	9	−1.57E+0	7	5	1.74E−1
5	4	−3.13E−2	11	2	1.04E+0	4	5	−5.54E+0	12	1	9.75E−2	3	10	−3.43E−1	7	6	−4.41E−2
5	5	−1.73E−4	11	3	1.80E−3	4	6	−7.97E−3	12	2	2.02E−2	3	11	−3.98E−2	7	7	−5.48E−2
5	6	−1.58E−5	11	4	9.04E−2	4	7	−1.88E−1	12	3	1.56E−1	3	12	−2.94E−2	7	8	−7.61E−4
5	7	−4.06E+0	11	5	6.10E−4	5	1	−2.73E−2	12	4	2.66E−1	3	13	−8.71E−3	7	9	−1.80E−2
5	8	−2.72E−1	11	6	4.71E−2	5	2	−4.14E−2	12	5	2.20E−1	3	14	−1.57E+0	7	10	−3.01E−2
5	9	−8.60E−3	11	7	9.70E−3	5	3	−1.75E−4	12	6	1.24E−3	4	1	5.66E−2	7	11	−1.11E−1
6	1	3.82E−2	11	8	1.90E−2	5	4	−2.05E−4	12	7	2.53E−2	4	2	4.70E−1	7	12	−2.44E−1
6	2	2.60E−2	11	9	2.39E−3	5	5	−7.00E−8	13	1	1.66E−1	4	3	6.28E−1	7	13	−8.10E−2
6	3	5.64E−2	12	1	4.92E−1	5	6	−5.77E+0	13	2	1.67E−1	4	4	1.49E−1	7	14	−1.68E−2
6	4	5.73E−3	12	2	1.73E−3	5	7	−1.88E−2	13	3	3.75E−1	4	5	8.54E−3	8	1	9.41E−2
6	5	6.02E−3	12	3	6.78E−1	6	1	2.67E+0	13	4	6.19E−1	4	6	−2.90E−2	8	2	1.02E+0

V-like Fe (Fe^{3+})

i i'	gf_L	i i'	gf_L	i i'	gf_L	i i'	gf_L	i i'	gf_L	i i'	gf_L
8 3	1.53E-2	4 5	-7.42E-2	12 1	2.55E-1	7 1	8.84E-1	2 3	-2.30E+0		$^2H^e$–$^2H^o$
8 4	3.43E-2	4 6	-1.14E-2	12 2	1.91E-1	7 2	1.44E-2	2 4	-4.41E+0	1 1	-6.86E-1
8 5	3.88E+0	4 7	-1.18E-2	12 3	1.50E+0	7 3	2.80E-2	2 5	-3.15E+0	1 2	-4.81E-1
8 6	-2.68E-3	5 1	-1.19E-5	12 4	8.35E-2	7 4	5.07E-1	2 6	-4.00E-1	1 3	-2.95E-1
8 7	-1.86E-2	5 2	-7.67E-3	12 5	2.44E-2	7 5	2.89E-2	2 7	-2.27E-1	1 4	-1.71E+0
8 8	-9.88E-3	5 3	-6.39E-3	12 6	6.47E-2	8 1	8.24E-2	2 8	-1.32E+0	1 5	-1.90E-1
8 9	-3.30E-2	5 4	-1.80E-2	12 7	1.04E-2	8 2	1.23E-2	2 9	-1.38E-2	2 1	-5.63E+0
8 10	-5.01E-2	5 5	-1.38E-1	13 1	7.24E-3	8 3	1.30E+0	3 1	2.09E-2	2 2	-1.05E+0
8 11	-5.64E-2	5 6	-1.24E-2	13 2	3.04E-2	8 4	7.30E-1	3 2	1.20E+0	2 3	-6.50E-1
8 12	-2.78E-3	5 7	-5.63E+0	13 3	2.37E-1	8 5	1.07E-2	3 3	-1.43E-1	2 4	-2.10E-2
8 13	-2.39E-1	6 1	6.18E+0	13 4	5.63E-1	9 1	1.85E-6	3 4	-3.74E+0	2 5	-7.38E-5
8 14	-4.04E-2	6 2	5.70E-1	13 5	6.02E-1	9 2	3.65E-1	3 5	-3.50E+0	3 1	4.72E+0
9 1	4.35E-2	6 3	4.13E-2	13 6	3.40E-2	9 3	9.19E-2	3 6	-7.39E-1	3 2	5.32E-1
9 2	5.16E-1	6 4	2.85E-2	13 7	3.97E-2	9 4	2.59E+0	3 7	-4.31E-1	3 3	5.22E-1
9 3	1.86E-3	6 5	2.97E-2	14 1	6.50E-3	9 5	6.52E-2	3 8	-1.06E-1	3 4	1.83E-2
9 4	2.52E-2	6 6	3.32E-2	14 2	7.66E-2	10 1	6.17E-1	3 9	-2.92E-2	3 5	1.35E-5
9 5	3.67E-1	6 7	8.53E-4	14 3	1.03E-1	10 2	3.12E-1	4 1	1.42E+0	4 1	2.90E+0
9 6	-3.07E-4	7 1	2.61E-2	14 4	2.51E+0	10 3	1.01E-1	4 2	3.18E-3	4 2	1.12E-1
9 7	-5.14E-4	7 2	3.50E+0	14 5	2.54E-1	10 4	4.65E-1	4 3	-5.73E-1	4 3	1.44E-1
9 8	-1.61E-4	7 3	9.20E-1	14 6	2.57E-2	10 5	1.09E-1	4 4	-4.01E-1	4 4	1.10E+0
9 9	-2.01E-6	7 4	5.04E-2	14 7	1.64E-3	11 1	5.51E-2	4 5	-6.95E-2	4 5	7.27E-4
9 10	-1.84E-3	7 5	8.63E-2		$^2G^e$–$^2H^o$	11 2	1.62E-1	4 6	-4.99E+0	5 1	2.10E-2
9 11	-2.34E-3	7 6	1.59E-2	1 1	-2.72E-1	11 3	1.33E-1	4 7	-1.35E-2	5 2	5.50E+0
9 12	-4.29E-3	7 7	4.83E-3	1 2	-4.45E-1	11 4	5.33E+0	4 8	-2.47E+0	5 3	5.01E-1
9 13	-1.62E-6	8 1	1.77E-1	1 3	-1.37E-2	11 5	8.67E-2	4 9	-5.16E-1	5 4	4.71E-1
9 14	-1.83E-2	8 2	7.79E-1	1 4	-4.34E-1	12 1	4.45E-1	5 1	3.76E-1	5 5	6.85E-3
	$^2G^e$–$^2G^o$	8 3	4.02E+0	1 5	-2.35E-3	12 2	2.77E+0	5 2	7.71E-2	6 1	6.82E-1
1 1	-5.58E-2	8 4	9.45E-1	2 1	-2.34E-1	12 3	1.96E-1	5 3	-4.98E-6	6 2	4.29E-1
1 2	-4.69E-1	8 5	1.47E-1	2 2	-4.23E-2	12 4	1.88E-2	5 4	-1.00E-2	6 3	4.51E+0
1 3	-6.59E-1	8 6	2.71E-2	2 3	-6.49E-2	12 5	2.87E-2	5 5	-2.46E-2	6 4	1.12E+0
1 4	-7.69E-1	8 7	1.06E-2	2 4	-5.16E-1	13 1	8.08E-2	5 6	-1.36E-2	6 5	5.79E-2
1 5	-5.22E-1	9 1	1.82E-2	2 5	-2.59E-1	13 2	2.69E-2	5 7	-2.17E-1	7 1	2.60E+0
1 6	-1.79E-1	9 2	1.20E-1	3 1	-6.94E-1	13 3	8.80E-2	5 8	-1.26E+0	7 2	4.32E-1
1 7	-6.22E-2	9 3	2.88E-1	3 2	-5.20E+0	13 4	1.89E-1	5 9	-1.29E+1	7 3	3.17E-1
2 1	-1.75E-2	9 4	2.25E+0	3 3	-5.01E-1	13 5	2.18E-2	6 1	5.55E-1	7 4	7.49E-2
2 2	-2.07E-2	9 5	6.62E-4	3 4	-7.37E-1	14 1	2.57E-1	6 2	1.19E-1	7 5	1.01E-2
2 3	-6.86E-1	9 6	3.69E-3	3 5	-2.20E-3	14 2	1.53E-2	6 3	-2.28E-3	8 1	2.59E-1
2 4	-7.74E-2	9 7	2.43E-5	4 1	-6.85E-1	14 3	2.43E+0	6 4	-1.87E-2	8 2	2.19E-1
2 5	-5.12E-2	10 1	1.05E-4	4 2	-3.17E-2	14 4	2.28E-1	6 5	-9.94E-3	8 3	5.87E-1
2 6	-1.03E+0	10 2	5.22E-3	4 3	-5.94E+0	14 5	1.55E-4	6 6	-8.51E-4	8 4	4.73E+0
2 7	-8.97E-1	10 3	1.84E+0	4 4	-3.68E-1		$^2G^o$–$^2H^e$	6 7	-1.31E-1	8 5	2.82E-1
3 1	-4.91E-1	10 4	3.56E+0	4 5	-3.54E-4	1 1	3.49E-2	6 8	-3.35E-1	9 1	1.70E-1
3 2	-5.55E-1	10 5	1.18E-1	5 1	3.69E-3	1 2	3.28E+0	6 9	-1.90E-1	9 2	7.46E-2
3 3	-3.98E+0	10 6	1.19E-2	5 2	-1.28E-2	1 3	-5.42E+0	7 1	5.03E-1	9 3	2.00E-2
3 4	-9.15E-1	10 7	9.93E-5	5 3	-3.76E-3	1 4	-1.25E+0	7 2	5.81E-3	9 4	3.56E-1
3 5	-1.04E-2	11 1	3.24E-5	5 4	-1.29E-1	1 5	-4.20E-1	7 3	-5.64E-4	9 5	1.09E-3
3 6	-1.35E-2	11 2	3.29E-1	5 5	-6.88E+0	1 6	-3.70E-3	7 4	-4.13E-3		$^2H^e$–$^2I^o$
3 7	-1.48E-2	11 3	1.56E-1	6 1	2.00E+0	1 7	-1.68E+0	7 5	-2.69E-2	1 1	-4.33E-1
4 1	-5.27E-3	11 4	1.30E-1	6 2	1.18E+0	1 8	-9.84E-2	7 6	4.03E-2	1 2	4.06E-1
4 2	-4.23E-1	11 5	1.82E-1	6 3	1.04E-1	1 9	-3.81E-2	7 7	-4.95E-3	2 1	-8.02E+0
4 3	-6.15E-1	11 6	1.31E-1	6 4	1.42E-1	2 1	1.64E+0	7 8	-1.13E-2	2 2	-3.70E-1
4 4	-4.77E+0	11 7	1.10E-1	6 5	1.06E-3	2 2	1.19E+0	7 9	-2.71E-1	3 1	6.45E-1

V-like Fe (Fe^{3+})

i i′	gf_L	*i i′*	gf_L	*i i′*	gf_L	*i i′*	gf_L	*i i′*	gf_L	*i i′*	gf_L
3 2	1.07E−1	2 1	−2.82E−1	2 3	1.31E+0	8 4	8.70E−2	7 1	3.84E−2	4 6	−2.04E−3
4 1	3.45E−1	2 2	−8.08E+0	2 4	−5.77E−5	8 5	2.15E−2	7 2	2.75E+0	4 7	−3.06E−3
4 2	5.16E−1	3 1	6.67E+0	2 5	−1.31E−2	8 6	−6.64E−2	7 3	5.95E−3	4 8	−2.88E−3
5 1	9.55E−3	3 2	7.30E−1	2 6	−5.91E−2		$^{4}P^{e}$–$^{4}D^{o}$	7 4	2.47E−1	4 9	−4.90E−1
5 2	1.21E+0	4 1	1.55E−2	2 7	−8.11E−3	1 1	−3.42E−1	7 5	3.90E−2	4 10	−3.05E−1
6 1	9.78E−2	4 2	1.52E+0	2 8	−2.22E−3	1 2	−7.16E−1	7 6	3.04E−2	5 1	9.14E−2
6 2	1.86E+0	5 1	1.57E−1		$^{4}P^{e}$–$^{4}P^{o}$	1 3	−1.07E+0	7 7	2.21E−4	5 2	1.03E−1
7 1	3.36E+0	5 2	1.56E+0	1 1	−6.89E−1	1 4	−3.84E−4	7 8	−1.12E−1	5 3	4.97E−4
7 2	3.11E−1	6 1	6.18E−1	1 2	−6.55E−1	1 5	−1.28E−2	8 1	8.47E−2	5 4	1.18E+0
8 1	8.34E−1	6 2	6.76E+0	1 3	−4.24E−1	1 6	−9.33E−2	8 2	6.73E−2	5 5	5.43E+0
8 2	1.96E+0	7 1	1.23E−1	1 4	−3.90E−2	1 7	−4.80E−2	8 3	1.94E+0	5 6	3.00E−2
9 1	3.99E−2	7 2	4.07E+0	1 5	−7.45E−2	1 8	−8.12E−1	8 4	4.15E−1	5 7	3.28E−2
9 2	1.32E−1		$^{2}I^{e}$–$^{2}J^{o}$	1 6	−5.07E−1	2 1	−2.57E−3	8 5	1.91E−2	5 8	4.51E−3
	$^{2}H^{o}$–$^{2}I^{e}$	1 1	−3.85E−1	2 1	−5.29E−2	2 2	−4.64E+0	8 6	4.61E−4	5 9	−7.39E−2
1 1	2.05E+0	2 1	−9.59E+0	2 2	−3.84E+0	2 3	−1.69E+0	8 7	1.68E−3	5 10	−9.15E−4
1 2	1.01E−1	3 1	4.18E−1	2 3	−4.18E−2	2 4	−1.03E−2	8 8	−1.65E−1	6 1	7.57E−2
1 3	−7.16E+0	4 1	1.99E−1	2 4	−1.45E−2	2 5	−5.19E−3		$^{4}P^{o}$–$^{4}D^{e}$	6 2	8.74E−6
1 4	−1.95E+0	5 1	3.87E−1	2 5	−6.75E−3	2 6	−1.79E−2	1 1	5.47E−1	6 3	5.35E−4
1 5	−9.90E−1	6 1	1.38E+0	2 6	−1.04E−5	2 7	−1.58E−4	1 2	3.64E+0	6 4	6.85E−1
1 6	−1.46E+0	7 1	4.03E+0	3 1	2.92E−4	2 8	−2.26E−4	1 3	1.61E−3	6 5	2.43E−2
1 7	−7.35E−2		$^{2}I^{o}$–$^{2}J^{e}$	3 2	−2.31E−3	3 1	−1.44E−3	1 4	−4.11E+0	6 6	2.27E−3
2 1	3.67E−2	1 1	−1.98E+1	3 3	−1.67E−1	3 2	−7.62E−4	1 5	−1.82E+0	6 7	1.79E−3
2 2	7.02E−1	1 2	−3.40E−1	3 4	−3.62E+0	3 3	−1.17E−4	1 6	−1.81E−1	6 8	2.59E−3
2 3	−1.63E+0	2 1	−6.80E−1	3 5	−2.10E−3	3 4	−6.82E−2	1 7	−5.05E−2	6 9	1.81E−2
2 4	−1.34E+1	2 2	−1.12E+1	3 6	−6.91E−4	3 5	−4.24E+0	1 8	−3.63E−3	6 10	−2.90E−6
2 5	−4.15E−3		$^{2}J^{e}$–$^{2}J^{o}$	4 1	5.74E+0	3 6	−2.20E+0	1 9	−9.76E−2		$^{4}D^{e}$–$^{4}D^{o}$
2 6	−1.86E−2	1 1	2.07E−3	4 2	1.06E−2	3 7	−3.63E−3	1 10	−1.58E−3	1 1	−9.94E−1
2 7	−4.12E−1	2 1	8.80E+0	4 3	3.00E−2	3 8	−2.46E−4	2 1	7.52E−1	1 2	−1.74E−3
3 1	2.82E+0		$^{4}S^{e}$–$^{4}P^{o}$	4 4	5.04E−3	4 1	1.90E+0	2 2	1.05E−1	1 3	−1.05E+0
3 2	1.81E−1	1 1	2.64E+0	4 5	−2.32E+0	4 2	1.09E−1	2 3	4.09E−2	1 4	−5.73E−1
3 3	−1.28E−1	1 2	5.59E−2	4 6	−4.19E+0	4 3	1.93E−1	2 4	−1.67E−3	1 5	−3.74E−1
3 4	−2.33E−2	1 3	2.28E−3	5 1	3.29E−2	4 4	2.32E−2	2 5	−3.02E−1	1 6	−9.03E−3
3 5	−1.20E+1	1 4	8.85E−4	5 2	2.07E+0	4 5	1.64E−3	2 6	−6.43E+0	1 7	−1.26E−1
3 6	−4.51E+0	1 5	−7.85E−1	5 3	3.25E−2	4 6	2.73E−5	2 7	−2.15E+0	1 8	−6.48E−1
3 7	−4.91E−2	1 6	−4.91E+0	5 4	1.66E−2	4 7	−6.13E−1	2 8	−5.57E−1	2 1	−6.67E+0
4 1	9.93E−1	2 1	7.41E−3	5 5	−6.77E−4	4 8	−8.94E+0	2 9	−5.00E−1	2 2	−4.79E−3
4 2	6.15E+0	2 2	2.97E−2	5 6	−3.15E−3	5 1	5.08E−4	2 10	−3.12E−2	2 3	−9.55E−3
4 3	−8.85E−1	2 3	2.53E+0	6 1	5.00E−2	5 2	1.81E−1	3 1	8.90E−1	2 4	−9.93E−2
4 4	−5.83E−1	2 4	1.45E−1	6 2	6.22E−1	5 3	2.55E+0	3 2	1.06E−4	2 5	−1.08E−2
4 5	−2.58E+0	2 5	−3.91E−4	6 3	2.09E+0	5 4	1.17E−1	3 3	3.66E+0	2 6	−3.85E−5
4 6	−2.29E+0	2 6	−4.45E−3	6 4	7.38E−2	5 5	2.89E−3	3 4	−8.91E−2	2 7	−4.39E−2
4 7	−2.29E+0		$^{4}S^{o}$–$^{4}P^{e}$	6 5	−2.13E−2	5 6	5.95E−4	3 5	−3.20E−3	2 8	−1.26E−3
5 1	1.67E−1	1 1	6.08E−1	6 6	−5.10E−2	5 7	−2.14E−4	3 6	−4.49E−5	3 1	−4.50E−2
5 2	2.43E−1	1 2	1.33E+0	7 1	1.05E−1	5 8	−4.11E−3	3 7	−2.80E−2	3 2	−3.19E−2
5 3	−1.49E−7	1 3	7.46E−3	7 2	2.93E−1	6 1	7.85E−3	3 8	−1.64E+0	3 3	−2.89E−1
5 4	−8.26E−4	1 4	−9.47E−2	7 3	1.46E+0	6 2	9.87E−1	3 9	−2.82E+0	3 4	−5.99E+0
5 5	−1.67E−3	1 5	−3.13E+0	7 4	1.23E−1	6 3	1.83E+0	3 10	−1.31E+0	3 5	−1.23E−1
5 6	−1.38E−1	1 6	−8.17E−1	7 5	3.39E−2	6 4	1.45E+0	4 1	1.63E−1	3 6	−3.63E−3
5 7	−2.92E−1	1 7	−9.93E−1	7 6	−3.58E−2	6 5	2.26E−2	4 2	1.85E−3	3 7	−6.04E−3
	$^{2}I^{e}$–$^{2}I^{o}$	1 8	−3.90E−3	8 1	1.22E−2	6 6	3.24E−2	4 3	2.16E−1	3 8	−7.14E−6
1 1	−1.12E+0	2 1	9.38E−2	8 2	2.13E+0	6 7	−8.75E−3	4 4	−6.64E−4	4 1	8.24E+0
1 2	−2.95E+0	2 2	9.91E−3	8 3	2.14E+0	6 8	−1.55E−1	4 5	−7.71E−3	4 2	4.73E−3

V-like Fe (Fe^{3+})

i i'	gf$_L$	i i'	gf$_L$	i i'	gf$_L$	i i'	gf$_L$	i i'	gf$_L$	i i'	gf$_L$
4 3	1.49E−1	10 7	3.71E−3	8 1	4.32E−2	4 1	3.63E−1	1 2	−3.66E−1	8 5	2.53E−1
4 4	2.24E−2	10 8	−1.66E−4	8 2	1.39E−2	4 2	3.08E−1	1 3	−6.14E−1	8 6	−1.84E−2
4 5	2.94E−2		$^4D^e$–$^4F^o$	8 3	4.29E+0	4 3	1.19E−2	1 4	−4.57E−1	8 7	−1.03E−1
4 6	4.90E−4	1 1	−2.73E−1	8 4	2.58E+0	4 4	−5.61E−2	1 5	−2.02E+0	9 1	5.86E−2
4 7	−3.65E+0	1 2	−1.58E−3	8 5	8.14E−2	4 5	−1.74E−2	1 6	−7.41E−2	9 2	6.53E−2
4 8	−6.14E+0	1 3	−2.22E+0	8 6	−1.01E−3	4 6	−1.39E−1	1 7	−7.28E−1	9 3	6.66E−1
5 1	2.36E+0	1 4	−9.89E−2	8 7	−1.62E−1	4 7	−1.37E−1	2 1	−8.23E−2	9 4	6.19E+0
5 2	1.87E−3	1 5	−4.49E−4	9 1	1.36E−1	4 8	−5.67E+0	2 2	−7.63E+0	9 5	2.05E−1
5 3	1.45E−2	1 6	−3.21E−2	9 2	9.22E−2	4 9	−6.12E+0	2 3	−1.25E+0	9 6	−3.75E−3
5 4	1.23E−1	1 7	−1.38E+0	9 3	3.31E+0	4 10	−1.63E−1	2 4	−2.30E−2	9 7	−1.15E−1
5 5	8.09E−3	2 1	−8.79E+0	9 4	4.55E−3	5 1	2.03E+0	2 5	−1.28E−1	10 1	4.19E−2
5 6	4.50E−5	2 2	−2.94E−1	9 5	1.94E−1	5 2	1.33E−6	2 6	−8.10E−3	10 2	3.60E+0
5 7	−8.55E+0	2 3	−6.37E−2	9 6	1.89E−2	5 3	2.15E+0	2 7	−2.52E−4	10 3	3.74E−1
5 8	−6.05E−2	2 4	−1.66E−4	9 7	−4.18E−1	5 4	−5.34E−3	3 1	−1.57E−6	10 4	4.34E−2
6 1	7.46E−3	2 5	−3.98E−6	10 1	8.77E−3	5 5	−1.13E−2	3 2	−1.74E−2	10 5	9.17E−2
6 2	4.71E+0	2 6	−1.50E−1	10 2	1.85E−1	5 6	−1.11E−2	3 3	−1.53E−1	10 6	2.51E−3
6 3	6.09E−1	2 7	−2.93E−3	10 3	4.99E−1	5 7	−5.23E−2	3 4	−1.02E−3	10 7	−8.65E−5
6 4	7.11E−4	3 1	−1.44E−2	10 4	3.14E+0	5 8	−4.80E−3	3 5	−8.73E+0		$^4F^e$–$^4G^o$
6 5	6.71E−4	3 2	−3.74E−1	10 5	5.92E−2	5 9	−3.59E−1	3 6	−3.90E−5	1 1	−1.74E−1
6 6	1.50E−2	3 3	−1.30E+0	10 6	2.13E−3	5 10	−9.43E−4	3 7	−1.28E−3	1 2	−3.73E−1
6 7	−2.55E−2	3 4	−7.26E+0	10 7	−7.63E−8	6 1	5.57E−2	4 1	5.97E+0	1 3	−1.84E+0
6 8	−1.83E−3	3 5	−4.90E−2		$^4D^o$–$^4F^e$	6 2	2.58E−2	4 2	4.19E−2	1 4	−2.13E−1
7 1	1.16E−1	3 6	−2.21E−3	1 1	1.67E+0	6 3	4.45E+0	4 3	2.15E−2	2 1	−5.55E+0
7 2	7.09E−2	3 7	−9.59E−4	1 2	2.19E−3	6 4	1.05E−1	4 4	6.30E−2	2 2	5.35E+0
7 3	4.66E+0	4 1	5.10E−1	1 3	7.26E−4	6 5	−2.67E−2	4 5	3.86E−2	2 3	−5.64E−1
7 4	2.59E−1	4 2	5.60E−3	1 4	−1.20E+1	6 6	−4.69E−4	4 6	−2.42E+0	2 4	−1.68E−2
7 5	9.94E−3	4 3	3.80E−1	1 5	−6.63E−6	6 7	−2.28E−2	4 7	−6.42E+0	3 1	−7.99E−2
7 6	1.46E−4	4 4	9.46E−3	1 6	−4.11E−2	6 8	−1.43E−2	5 1	1.06E−1	3 2	−3.21E−2
7 7	−1.69E−1	4 5	8.54E−4	1 7	−3.30E−2	6 9	−6.08E−3	5 2	1.90E+0	3 3	−7.31E−5
7 8	−3.18E−2	4 6	−2.98E−2	1 8	−4.63E−1	6 10	−3.97E−2	5 3	2.78E−2	3 4	−1.14E+1
8 1	2.00E−1	4 7	−1.53E+1	1 9	−6.90E−2	7 1	2.57E−1	5 4	1.02E−1	4 1	8.58E−2
8 2	1.64E−2	5 1	3.91E+0	1 10	−5.87E−4	7 2	1.46E−4	5 5	2.24E−2	4 2	1.77E−1
8 3	6.22E−1	5 2	3.32E−1	2 1	8.99E−5	7 3	2.18E−3	5 6	−1.07E−2	4 3	4.97E−1
8 4	3.28E+0	5 3	1.55E−1	2 2	1.58E+0	7 4	4.37E+0	5 7	−4.76E−3	4 4	3.95E−4
8 5	9.73E−2	5 4	1.02E−2	2 3	2.21E−2	7 5	4.93E−3	6 1	1.64E−1	5 1	3.09E−1
8 6	3.87E−2	5 5	1.18E−3	2 4	−2.75E−3	7 6	1.40E−2	6 2	1.80E+0	5 2	1.58E+0
8 7	−4.51E−2	5 6	−1.25E+1	2 5	−1.32E+1	7 7	1.26E−2	6 3	1.17E+0	5 3	7.28E−3
8 8	−4.57E−2	5 7	−1.59E−1	2 6	−1.84E+0	7 8	8.98E−2	6 4	5.33E−1	5 4	2.46E−2
9 1	7.57E−2	6 1	2.03E−3	2 7	−6.14E−2	7 9	3.27E−2	6 5	2.20E−2	6 1	3.23E+0
9 2	3.04E−5	6 2	1.19E+0	2 8	−4.26E−2	7 10	−2.25E−5	6 6	−1.95E−2	6 2	7.19E+0
9 3	1.31E+0	6 3	2.88E−1	2 9	−4.98E−2	8 1	7.40E−2	6 7	−8.14E−3	6 3	1.94E−1
9 4	3.90E+0	6 4	2.06E−2	2 10	−8.15E−1	8 2	9.22E−1	7 1	1.71E−2	6 4	1.40E−1
9 5	3.89E−5	6 5	3.93E−2	3 1	1.72E−1	8 3	1.16E−9	7 2	6.21E+0	7 1	3.14E+0
9 6	1.29E−2	6 6	−2.11E−2	3 2	4.67E+0	8 4	7.01E−1	7 3	4.08E+0	7 2	6.37E−2
9 7	2.93E−2	6 7	−3.58E−5	3 3	3.68E−3	8 5	2.12E−8	7 4	3.93E−2	7 3	8.98E−2
9 8	−1.80E−1	7 1	2.88E−1	3 4	−8.60E−3	8 6	2.10E−3	7 5	6.50E−2	7 4	1.26E−1
10 1	1.65E−2	7 2	3.58E+0	3 5	−1.56E+0	8 7	9.70E−4	7 6	−1.37E−3	8 1	8.50E−2
10 2	1.40E−2	7 3	1.18E+0	3 6	−2.44E+0	8 8	1.25E−2	7 7	−1.88E−3	8 2	7.19E−1
10 3	1.80E−1	7 4	0.08E−1	3 7	−4.80E+0	8 9	7.83E−3	8 1	1.31E−3	8 3	5.18E+0
10 4	2.81E+0	7 5	1.38E−3	3 8	−9.86E−2	8 10	8.39E−3	8 2	1.10E+0	8 4	1.77E−2
10 5	2.04E−1	7 6	−1.36E−1	3 9	−3.69E−1		$^4F^e$–$^4F^o$	8 3	4.68E+0	9 1	3.69E−3
10 6	1.80E−4	7 7	−1.14E−2	3 10	−1.89E+0	1 1	−6.69E−1	8 4	4.66E−1	9 2	2.83E−1

V-like Fe (Fe^{3+})

$i\ i'$	gf_L	$i\ i'$	gf_L	$i\ i'$	gf_L	$i\ i'$	gf_L	$i\ i'$	gf_L	$i\ i'$	gf_L
9 3	4.15E+0	5 2	1.02E−1	5 1	1.07E+0	8 3	−2.30E−5	2 1	3.71E+0	3 3	1.68E−3
9 4	4.63E−2	5 3	−2.45E−3	5 2	1.57E+0		$^4G^o$–$^4H^e$	3 1	1.91E−2		$^6D^e$–$^6D^o$
10 1	2.54E+0	5 4	−1.10E−1	5 3	1.20E+1	1 1	2.81E+0	4 1	9.17E−2	1 1	−9.83E+0
10 2	2.37E+0	5 5	−5.41E−2	5 4	1.60E−2	1 2	−9.45E−1	5 1	7.04E+0	1 2	−1.11E−1
10 3	1.29E−1	5 6	−2.41E−2	6 1	7.21E−1	1 3	−1.88E+1		$^4H^o$–$^4I^e$	1 3	−7.17E−4
10 4	1.51E−2	5 7	−1.92E−2	6 2	3.48E+0	1 4	−1.15E+0	1 1	−2.21E+1	2 1	1.23E+1
	$^4F^o$–$^4G^e$	5 8	−1.32E−1	6 3	1.45E+0	1 5	−1.45E+0	1 2	−8.83E−1	2 2	−4.81E+0
1 1	2.69E+0	6 1	4.86E−1	6 4	3.13E−2	2 1	5.63E+0	2 1	−8.70E−4	2 3	−1.01E+1
1 2	7.75E−2	6 2	9.21E−3	7 1	1.28E+0	2 2	−5.51E+0	2 2	−3.27E+1	3 1	4.16E+0
1 3	−2.32E+1	6 3	8.37E+0	7 2	5.87E+0	2 3	−9.82E+0	3 1	9.74E−4	3 2	−1.34E+1
1 4	−1.24E−1	6 4	2.47E−3	7 3	2.03E−1	2 4	−2.95E+0	3 2	1.88E−3	3 3	−1.79E−2
1 5	−1.72E−1	6 5	4.43E−2	7 4	3.13E−1	2 5	−2.46E+0		$^4I^e$–$^4I^o$		$^6D^e$–$^6F^o$
1 6	−1.30E−1	6 6	3.45E−2	8 1	1.34E+0	3 1	3.54E+0	1 1	1.31E+1	1 1	−1.33E+1
1 7	−2.92E−1	6 7	−2.84E−3	8 2	3.31E−1	3 2	−1.40E+0	2 1	1.66E+0	1 2	−2.67E−1
1 8	−5.08E−2	6 8	−8.12E−3	8 3	3.19E+0	3 3	−2.38E−1		$^4I^o$–$^4J^e$	1 3	−1.26E−3
2 1	3.04E+0	7 1	3.34E−2	8 4	8.93E−2	3 4	−1.44E+1	1 1	−4.06E+1	2 1	1.26E+0
2 2	9.08E−1	7 2	1.74E−3		$^4G^e$–$^4H^o$	3 5	−1.08E+0		$^6S^e$–$^6P^o$	2 2	−2.70E−1
2 3	−4.19E−2	7 3	4.09E−1	1 1	−1.26E+0	4 1	1.19E−1	1 1	−2.44E+0	2 3	−2.48E+1
2 4	−8.99E+0	7 4	3.18E−5	1 2	−1.06E+0	4 2	−1.35E−5	1 2	−2.97E−1	3 1	5.96E+0
2 5	−9.41E−2	7 5	7.05E−3	1 3	−3.14E+0	4 3	−1.14E−3	1 3	−2.53E+0	3 2	−1.86E+1
2 6	−7.12E−1	7 6	7.99E−4	2 1	−2.24E+0	4 4	−2.94E−5	2 1	3.99E+0	3 3	−4.00E−2
2 7	−4.66E+0	7 7	1.96E−2	2 2	−1.17E+1	4 5	−1.40E−1	2 2	−1.73E+0		$^6D^o$–$^6F^e$
2 8	−6.18E−1	7 8	5.35E−3	2 3	−6.17E−4		$^4H^e$–$^4H^o$	2 3	−6.36E+0	1 1	−1.90E+1
3 1	3.67E−1		$^4G^e$–$^4G^o$	3 1	2.35E−1	1 1	−1.15E+1		$^6P^e$–$^6P^o$	2 1	6.71E+0
3 2	6.31E+0	1 1	−8.63E−1	3 2	1.13E−1	1 2	−2.58E+0	1 1	9.11E+0	3 1	1.12E+0
3 3	−6.74E−2	1 2	−4.57E+0	3 3	−3.51E+1	1 3	−1.46E−3	1 2	−3.06E+0		$^6F^e$–$^6F^o$
3 4	−1.83E+0	1 3	−5.54E−2	4 1	1.81E+0	2 1	1.46E+1	1 3	−7.44E+0	1 1	9.40E+0
3 5	−4.90E+0	1 4	−2.42E−2	4 2	8.54E−1	2 2	3.66E+0		$^6P^e$–$^6D^o$	1 2	−3.59E+0
3 6	−5.33E−1	2 1	−2.87E+0	4 3	−1.66E−3	2 3	−2.52E−2	1 1	3.07E+0	1 3	−1.06E+1
3 7	−2.65E+0	2 2	−8.50E−1	5 1	4.35E+0	3 1	1.75E−2	1 2	−9.62E−1		$^6F^e$–$^6G^o$
3 8	−3.06E+0	2 3	−7.86E+0	5 2	7.40E−2	3 2	8.56E−1	1 3	−1.45E+1	1 1	−3.88E+1
4 1	1.92E−1	2 4	−7.48E−2	5 3	−1.24E−1	3 3	−2.47E−5		$^6P^o$–$^6D^e$		$^6F^o$–$^6G^e$
4 2	1.91E+0	3 1	9.04E−2	6 1	1.76E−1	4 1	2.25E+0	1 1	5.92E+0	1 1	−3.65E+1
4 3	−2.07E−2	3 2	6.58E−1	6 2	1.43E+0	4 2	8.98E+0	1 2	−6.82E+0	2 1	1.29E+1
4 4	−1.34E−1	3 3	5.53E−4	6 3	−2.36E−1	4 3	−3.60E−3	1 3	−2.76E+0	3 1	6.12E−1
4 5	−1.45E−4	3 4	1.05E−4	7 1	1.60E+0	5 1	5.90E+0	2 1	1.01E−1		$^6G^e$–$^6G^o$
4 6	−1.56E+1	4 1	8.44E+0	7 2	1.37E+0	5 2	9.02E−1	2 2	2.37E+0	1 1	−7.69E+0
4 7	−5.43E+0	4 2	1.01E−1	7 3	−1.07E+0	5 3	−1.09E−2	2 3	8.23E+0		$^6G^e$–$^6H^o$
4 8	−4.92E−1	4 3	1.86E+0	8 1	1.07E+0		$^4H^e$–$^4I^o$	3 1	4.84E−4	1 1	−5.65E+1
5 1	3.65E−1	4 4	1.16E−1	8 2	5.19E+0	1 1	−1.66E+1	3 2	1.02E+0		

Cr-like Fe (Fe^{2+})

Term energies relative to $3d^5$ ^{6}S ionization threshold for each symmetry

i	E(Ryds)	Description	i	E(Ryds)	Description	i	E(Ryds)	Description	i	E(Ryds)	Description
	$^7S^e$			$^5S^e$		6	−.53945	$3d^5~^4P~4d$		$^5F^o$	
1	−1.92772	$3d^5~^6S~4s$	1	−1.82803	$3d^5~^6S~4s$	7	−.52578	$3d^5~^4D~5s$	1	−1.12241	$3d^5~^4G~4p$
2	−.88415	$3d^5~^6S~5s$	2	−.85982	$3d^5~^6S~5s$	8	−.50650	$3d^5~^4D~4d$	2	−1.07486	$3d^5~^4D~4p$
3	−.50965	$3d^5~^6S~6s$	3	−.50658	$3d^5~^4D~4d$	9	−.46210	$3d^5~^6S~5d$	3	−.93007	$3d^5~^4F~4p$
4	−.33194	$3d^5~^6S~7s$	4	−.49781	$3d^5~^6S~6s$	10	−.35383	$3d^5~^4F~4d$	4	−.56541	$3d^5~^6S~4f$
5	−.23339	$3d^5~^6S~8s$	5	−.32698	$3d^5~^6S~7s$	11	−.31890	$3d^5~^6S~6d$	5	−.47243	$3d^44s~^6D~4p$
6	−.17306	$3d^5~^6S~9s$	6	−.23056	$3d^5~^6S~8s$	12	−.22956	$3d^5~^6S~7d$	6	−.42346	$3d^5~^4G~5p$
	$^7P^o$		7	−.17132	$3d^5~^6S~9s$	13	−.19736	$3d^5~^4G~5d$	7	−.36933	$3d^5~^4D~5p$
1	−1.48480	$3d^5~^6S~4p$	8	−.14741	$3d^5~^4D~5d$	14	−.17865	$3d^5~^4P~5d$	8	−.36191	$3d^5~^6S~5f$
2	−.72641	$3d^5~^6S~5p$		$^5S^o$		15	−.16766	$3d^5~^6S~8d$	9	−.27352	$3d^5~^4G~4f$
3	−.56043	$3d^44s~^6D~4p$	1	−1.13651	$3d^5~^4P~4p$	16	−.15391	$3d^5~^4D~6s$	10	−.26312	$3d^44s~^4D~4p$
4	−.43685	$3d^5~^6S~6p$	2	−.40386	$3d^5~^4P~5p$	17	−.13927	$3d^5~^4D~5d$	11	−.25182	$3d^5~^6S~6f$
5	−.29270	$3d^5~^6S~7p$	3	−.23357	$3d^44s~^4P~4p$	18	−.12793	$3d^5~^6S~9d$	12	−.24406	$3d^5~^4P~4f$
6	−.20985	$3d^5~^6S~8p$	4	−.11477	$3d^5~^4P~6p$		$^5D^o$		13	−.24262	$3d^5~^4F~5p$
7	−.15784	$3d^5~^6S~9p$	5	−.08703	$3d^5~^4F~4f$	1	−1.13574	$3d^5~^4P~4p$	14	−.22738	$3d^44s~^4F~4p$
	$^7D^e$			$^5P^e$		2	−1.06012	$3d^5~^4D~4p$	15	−.21442	$3d^5~^4D~4f$
1	−.89350	$3d^5~^6S~4d$	1	−1.58189	$3d^5~^4P~4s$	3	−.91919	$3d^5~^4F~4p$	16	−.18926	$3d^44s~^4G~4p$
2	−.51046	$3d^5~^6S~5d$	2	−.55897	$3d^5~^4P~5s$	4	−.42469	$3d^44s~^6D~4p$	17	−.17971	$3d^5~^6S~7f$
3	−.33160	$3d^5~^6S~6d$	3	−.55571	$3d^5~^4P~4d$	5	−.40201	$3d^5~^4P~5p$	18	−.14933	$3d^44s~^4D~4p$
4	−.23294	$3d^5~^6S~7d$	4	−.50773	$3d^5~^4D~4d$	6	−.36591	$3d^5~^4D~5p$	19	−.14129	$3d^5~^6S~8f$
5	−.17268	$3d^5~^6S~8d$	5	−.37958	$3d^5~^4F~4d$	7	−.28738	$3d^44s~^4D~4p$	20	−.13565	$3d^5~^4G~6p$
6	−.13315	$3d^5~^6S~9d$	6	−.18655	$3d^5~^4P~6s$	8	−.27255	$3d^5~^4G~4f$	21	−.11124	$3d^5~^6S~9f$
	$^7D^o$		7	−.18483	$3d^5~^4P~5d$	9	−.24590	$3d^5~^4P~4f$		$^5G^e$	
1	−.53211	$3d^44s~^6D~4p$	8	−.14921	$3d^5~^4D~5d$	10	−.24034	$3d^5~^4F~5p$	1	−1.58639	$3d^5~^4G~4s$
	$^7F^o$		9	−.06597	$3d^5~^4G~5g$	11	−.23445	$3d^44s~^4P~4p$	2	−.58590	$3d^5~^4G~5s$
1	−.57515	$3d^5~^6S~4f$	10	−.02297	$3d^5~^4F~5d$	12	−.21314	$3d^5~^4D~4f$	3	−.56933	$3d^5~^4G~4d$
2	−.56298	$3d^44s~^6D~4p$	11	−.00954	$3d^5~^4P~7s$	13	−.17571	$3d^44s~^4F~4p$	4	−.50998	$3d^5~^4D~4d$
3	−.36352	$3d^5~^6S~5f$	12	−.00811	$3d^5~^4P~6d$	14	−.14806	$3d^44s~^4D~4p$	5	−.37757	$3d^5~^4F~4d$
4	−.25237	$3d^5~^6S~6f$		$^5P^o$		15	−.11370	$3d^5~^4P~6p$	6	−.36023	$3d^5~^6S~5g$
5	−.18529	$3d^5~^6S~7f$	1	−1.42132	$3d^5~^6S~4p$	16	−.08457	$3d^5~^4F~4f$	7	−.25016	$3d^5~^6S~6g$
6	−.14176	$3d^5~^6S~8f$	2	−1.11867	$3d^5~^4P~4p$	17	−.07970	$3d^5~^4D~6p$	8	−.21405	$3d^5~^4G~6s$
7	−.11193	$3d^5~^6S~9f$	3	−1.05570	$3d^5~^4D~4p$	18	−.06636	$3d^5~^4G~5f$	9	−.20906	$3d^5~^4G~5d$
	$^7G^e$		4	−.70816	$3d^5~^6S~5p$	19	−.04156	$3d^5~^4P~5f$	10	−.18378	$3d^5~^6S~7g$
1	−.36025	$3d^5~^6S~5g$	5	−.49241	$3d^44s~^6D~4p$	20	−.00840	$3d^5~^4D~5f$	11	−.14938	$3d^5~^4D~5d$
2	−.25019	$3d^5~^6S~6g$	6	−.43038	$3d^5~^6S~6p$		$^5F^e$		12	−.14070	$3d^5~^6S~8g$
3	−.18381	$3d^5~^6S~7g$	7	−.39891	$3d^5~^4P~5p$	1	−1.40716	$3d^5~^4F~4s$	13	−.11117	$3d^5~^6S~9g$
4	−.14072	$3d^5~^6S~8g$	8	−.36612	$3d^5~^4D~5p$	2	−.57059	$3d^5~^4G~4d$		$^5G^o$	
5	−.11118	$3d^5~^6S~9g$	9	−.28916	$3d^5~^6S~7p$	3	−.55487	$3d^5~^4P~4d$	1	−1.14840	$3d^5~^4G~4p$
	$^7H^o$		10	−.27444	$3d^5~^4G~4f$	4	−.50555	$3d^5~^4D~4d$	2	−.94020	$3d^5~^4F~4p$
1	−.25003	$3d^5~^6S~6h$	11	−.26011	$3d^44s~^4D~4p$	5	−.40038	$3d^5~^4F~5s$	3	−.43099	$3d^5~^4G~5p$
2	−.18370	$3d^5~^6S~7h$	12	−.21295	$3d^5~^4D~4f$	6	−.38129	$3d^5~^4F~4d$	4	−.28364	$3d^44s~^4H~4p$
3	−.14065	$3d^5~^6S~8h$	13	−.20914	$3d^5~^6S~8p$	7	−.20936	$3d^5~^4G~5d$	5	−.27297	$3d^5~^4G~4f$
4	−.11113	$3d^5~^6S~9h$	14	−.18741	$3d^44s~^4P~4p$	8	−.18408	$3d^5~^4P~5d$	6	−.25920	$3d^44s~^4F~4p$
	$^7I^e$		15	−.15547	$3d^5~^6S~9p$	9	−.14818	$3d^5~^4D~5d$	7	−.24465	$3d^5~^4P~4f$
1	−.18368	$3d^5~^6S~7i$	16	−.15382	$3d^44s~^4D~4p$	10	−.06596	$3d^5~^4G~5g$	8	−.24208	$3d^5~^4F~5p$
2	−.14063	$3d^5~^6S~8i$		$^5D^e$		11	−.03843	$3d^5~^4P~5g$	9	−.21269	$3d^5~^4D~4f$
3	−.11111	$3d^5~^6S~9i$	1	−2.12683	$3d^6$	12	−.03557	$3d^5~^4G~6d$	10	−.20295	$3d^44s~^4G~4p$
	$^7J^o$		2	−1.53246	$3d^5~^4D~4s$	13	−.02792	$3d^5~^4F~6s$	11	−.14194	$3d^5~^4G~6p$
1	−.14063	$3d^5~^6S~8j$	3	−.89232	$3d^44s^2$:	14	−.02278	$3d^5~^4F~5d$	12	−.08695	$3d^5~^4F~4f$
2	.11111	$3d^5~^6S~0j$	4	.85208	$3d^5~^6S~1d$	15	.00807	$3d^5~^4P~6d$	13	−.06889	$3d^5~^4G~5f$
			5	−.55733	$3d^5~^4G~4d$	16	−.00590	$3d^5~^4D~5g$	14	−.04164	$3d^5~^4P~5f$

Cr-like Fe (Fe^{2+})

i	E(Ryds)	Description
15	−.00892	$3d^5$ ^{4}D $5f$
		^{5}**H**^{e}
1	−.57357	$3d^5$ ^{4}G $4d$
2	−.37938	$3d^5$ ^{4}F $4d$
3	−.21049	$3d^5$ ^{4}G $5d$
4	−.06596	$3d^5$ ^{4}G $5g$
5	−.03843	$3d^5$ ^{4}P $5g$
6	−.03639	$3d^5$ ^{4}G $6d$
7	−.02123	$3d^5$ ^{4}F $5d$
8	−.00587	$3d^5$ ^{4}D $5g$
		^{5}**H**^{o}
1	−1.13652	$3d^5$ ^{4}G $4p$
2	−.42852	$3d^5$ ^{4}G $5p$
3	−.31575	$3d^44s$ ^{4}H $4p$
4	−.27225	$3d^5$ ^{4}G $4f$
5	−.25003	$3d^5$ ^{6}S $6h$
6	−.23226	$3d^5$ ^{4}D $4f$
7	−.21209	$3d^44s$ ^{4}G $4p$
8	−.18370	$3d^5$ ^{6}S $7h$
9	−.14139	$3d^5$ ^{4}G $6p$
10	−.14065	$3d^5$ ^{6}S $8h$
11	−.11113	$3d^5$ ^{6}S $9h$
		^{5}**I**^{e}
1	−.56977	$3d^5$ ^{4}G $4d$
2	−.20926	$3d^5$ ^{4}G $5d$
3	−.18368	$3d^5$ ^{6}S $7i$
4	−.14063	$3d^5$ ^{6}S $8i$
5	−.11111	$3d^5$ ^{6}S $9i$
		^{5}**I**^{o}
1	−.29000	$3d^5$ ^{4}G $4f$
2	−.27104	$3d^44s$ ^{4}H $4p$
3	−.08722	$3d^5$ ^{4}F $4f$
4	−.06861	$3d^5$ ^{4}G $5f$
		^{5}**J**^{e}
1	−.06597	$3d^5$ ^{4}G $5g$
		^{5}**J**^{o}
1	−.27283	$3d^5$ ^{4}G $4f$
2	−.14063	$3d^5$ ^{6}S $8j$
3	−.11111	$3d^5$ ^{6}S $9j$
		^{3}**S**^{e}
1	−1.25020	$3d^5$ ^{2}S $4s$
2	−.46616	$3d^5$ ^{4}D $4d$
3	−.36467	$3d^5(^5)$ ^{2}D $4d$
4	−.26792	$3d^5$ ^{2}S $5s$
5	−.13638	$3d^5$ ^{4}D $5d$
6	−.12501	$3d^5(^3)$ ^{2}D $4d$
7	−.06599	$3d^5$ ^{4}G $5g$
8	−.03900	$3d^5(^5)$ ^{2}D $5d$
		^{3}**S**^{o}
1	−1.06856	$3d^5$ ^{4}P $4p$
2	−.47605	$3d^5$ ^{2}P $4p$
3	−.38304	$3d^5$ ^{4}P $5p$
4	−.22104	$3d^44s$ ^{4}P $4p$
5	−.10729	$3d^5$ ^{4}P $6p$
6	−.08540	$3d^5$ ^{4}F $4f$
7	−.08312	$3d^5(^3)$ ^{2}F $4f$
8	−.00823	$3d^5(^5)$ ^{2}F $4f$
		^{3}**P**^{e}
1	−1.95203	$3d^6(^4)$
2	−1.66513	$3d^6(^2)$
3	−1.50853	$3d^5$ ^{4}P $4s$
4	−.94302	$3d^5$ ^{2}P $4s$
5	−.63085	$3d^44s$ ^{4}P $4s$
6	−.54058	$3d^5$ ^{4}P $5s$
7	−.47911	$3d^5$ ^{4}P $4d$
8	−.44387	$3d^5$ ^{4}D $4d$
9	−.35009	$3d^5(^5)$ ^{2}D $4d$
10	−.31977	$3d^5(^3)$ ^{2}F $4d$
11	−.29900	$3d^5$ ^{4}F $4d$
12	−.26951	$3d^44s$ ^{2}P $4s$
13	−.23947	$3d^5(^5)$ ^{2}F $4d$
14	−.17932	$3d^5$ ^{4}P $6s$
15	−.15474	$3d^5$ ^{4}P $5d$
16	−.12306	$3d^5$ ^{4}D $5d$
17	−.10648	$3d^5(^3)$ ^{2}D $4d$
18	−.06597	$3d^5$ ^{4}G $5g$
19	−.03054	$3d^5(^5)$ ^{2}D $5d$
20	−.00879	$3d^5$ ^{4}F $5d$
21	−.00596	$3d^5$ ^{4}P $7s$
22	−.00045	$3d^5(^3)$ ^{2}F $5d$
		^{3}**P**^{o}
1	−1.13217	$3d^5$ ^{4}P $4p$
2	−1.02835	$3d^5$ ^{4}D $4p$
3	−.98129	$3d^5(^5)$ ^{2}D $4p$
4	−.82466	$3d^5$ ^{2}S $4p$
5	−.72325	$3d^5(^3)$ ^{2}D $4p$
6	−.54435	$3d^5$ ^{2}P $4p$
7	−.45103	$3d^5(^1)$ ^{2}D $4p$
8	−.39849	$3d^44s$ ^{4}D $4p$
9	−.39461	$3d^5$ ^{4}P $5p$
10	−.35183	$3d^5$ ^{4}D $5p$
11	−.27218	$3d^5$ ^{4}G $4f$
12	−.26820	$3d^5(^5)$ ^{2}D $5p$
13	−.21610	$3d^5$ ^{2}S $5p$
14	−.20990	$3d^5$ ^{4}D $4f$
15	−.12630	$3d^44s$ ^{4}P $4p$
16	−.11525	$3d^44s$ ^{2}P $4p$
17	−.11260	$3d^5(^5)$ ^{2}D $4f$
18	−.10946	$3d^5$ ^{4}P $6p$
19	−.09034	$3d^5$ ^{4}F $4f$
20	−.08529	$3d^5(^3)$ ^{2}F $4f$
21	−.08259	$3d^44s$ ^{4}D $4p$
22	−.07582	$3d^5$ ^{4}D $6p$
23	−.06781	$3d^5$ ^{4}G $5f$
24	−.05430	$3d^5(^5)$ ^{2}G $4f$
25	−.04258	$3d^5(^3)$ ^{2}D $5p$
26	−.00973	$3d^5(^5)$ ^{2}F $4f$
27	−.00840	$3d^5$ ^{4}D $5f$
28	−.00003	$3d^44s$ ^{2}S $4p$
		^{3}**D**^{e}
1	−1.84011	$3d^6$
2	−1.45995	$3d^5$ ^{4}D $4s$
3	−1.41600	$3d^5(^5)$ ^{2}D $4s$
4	−1.18642	$3d^5(^3)$ ^{2}D $4s$
5	−.87169	$3d^5(^1)$ ^{2}D $4s$
6	−.53663	$3d^5$ ^{4}G $4d$
7	−.52058	$3d^44s$ ^{4}D $4s$
8	−.50740	$3d^5$ ^{4}D $5s$
9	−.49690	$3d^5$ ^{4}P $4d$
10	−.44296	$3d^5$ ^{4}D $4d$
11	−.42142	$3d^5(^5)$ ^{2}D $5s$
12	−.35870	$3d^5(^5)$ ^{2}D $4d$
13	−.34055	$3d^5$ ^{4}F $4d$
14	−.33279	$3d^5(^3)$ ^{2}F $4d$
15	−.29753	$3d^5(^5)$ ^{2}G $4d$
16	−.24179	$3d^5(^5)$ ^{2}F $4d$
17	−.20099	$3d^5(^3)$ ^{2}D $5s$
18	−.20002	$3d^5$ ^{4}G $5d$
19	−.18685	$3d^5$ ^{2}S $4d$
20	−.16690	$3d^5$ ^{4}P $5d$
21	−.14690	$3d^5$ ^{4}D $6s$
22	−.12834	$3d^5$ ^{4}D $5d$
23	−.11172	$3d^5(^3)$ ^{2}D $4d$
24	−.06595	$3d^5$ ^{4}G $5g$
25	−.05280	$3d^5(^5)$ ^{2}D $6s$
26	−.04077	$3d^5(^3)$ ^{2}G $4d$
27	−.03653	$3d^5(^5)$ ^{2}D $5d$
28	−.02950	$3d^5$ ^{4}G $6d$
29	−.01233	$3d^5$ ^{4}F $5d$
30	−.00730	$3d^5(^3)$ ^{2}F $5d$
31	−.00589	$3d^5$ ^{4}D $5g$
		^{3}**D**^{o}
1	−1.10821	$3d^5$ ^{4}P $4p$
2	−1.06997	$3d^5$ ^{4}D $4p$
3	−.97350	$3d^5(^5)$ ^{2}D $4p$
4	−.96271	$3d^5(^3)$ ^{2}F $4p$
5	−.91146	$3d^5$ ^{4}F $4p$
6	−.84094	$3d^5(^5)$ ^{2}F $4p$
7	−.74760	$3d^5(^3)$ ^{2}D $4p$
8	−.50052	$3d^5$ ^{2}P $4p$
9	−.41967	$3d^5(^1)$ ^{2}D $4p$
10	−.39092	$3d^5$ ^{4}P $5p$
11	−.36184	$3d^5$ ^{4}D $5p$
12	−.35419	$3d^44s$ ^{4}D $4p$
13	−.27176	$3d^5$ ^{4}G $4f$
14	−.26711	$3d^5(^5)$ ^{2}D $5p$
15	−.24445	$3d^5$ ^{4}P $4f$
16	−.23995	$3d^5$ ^{4}F $5p$
17	−.22878	$3d^5(^3)$ ^{2}F $5p$
18	−.21963	$3d^44s$ ^{4}P $4p$
19	−.21060	$3d^5$ ^{4}D $4f$
20	−.20261	$3d^44s$ ^{4}F $4p$
21	−.15979	$3d^5(^5)$ ^{2}F $5p$
22	−.11299	$3d^5(^5)$ ^{2}D $4f$
23	−.10975	$3d^5$ ^{4}P $6p$
24	−.08861	$3d^5$ ^{4}F $4f$
25	−.08590	$3d^5(^3)$ ^{2}F $4f$
26	−.08425	$3d^44s$ ^{2}P $4p$
27	−.07771	$3d^5$ ^{4}D $6p$
28	−.06735	$3d^5$ ^{4}G $5f$
29	−.06129	$3d^44s$ ^{2}F $4p$
30	−.05443	$3d^5$ ^{2}H $4f$
31	−.05052	$3d^5(^5)$ ^{2}G $4f$
32	−.04550	$3d^5(^3)$ ^{2}D $5p$
33	−.04032	$3d^5$ ^{4}P $5f$
34	−.00862	$3d^5(^5)$ ^{2}F $4f$
35	−.00766	$3d^5$ ^{4}D $5f$
		^{3}**F**^{e}
1	−1.92565	$3d^6(^4)$
2	−1.65636	$3d^6(^2)$
3	−1.40015	$3d^5(^3)$ ^{2}F $4s$
4	−1.33354	$3d^5$ ^{4}F $4s$
5	−1.29662	$3d^5(^5)$ ^{2}F $4s$
6	−.62395	$3d^44s$ ^{4}F $4s$
7	−.52008	$3d^5$ ^{4}G $4d$
8	−.48441	$3d^5$ ^{4}P $4d$
9	−.45652	$3d^5$ ^{4}D $4d$
10	−.39492	$3d^5$ ^{4}F $5s$
11	−.38291	$3d^5(^3)$ ^{2}F $5s$
12	−.35417	$3d^5(^5)$ ^{2}D $4d$
13	−.33746	$3d^5$ ^{4}F $4d$
14	−.31854	$3d^5(^5)$ ^{2}F $5s$
15	−.31626	$3d^44s$ ^{2}F $4s$
16	−.28966	$3d^5(^5)$ ^{2}G $4d$
17	−.27843	$3d^44s$ ^{2}F $4s$
18	−.27161	$3d^5(^3)$ ^{2}F $4d$
19	−.24872	$3d^5(^5)$ ^{2}F $4d$
20	−.18587	$3d^5$ ^{4}G $5d$
21	−.15846	$3d^5$ ^{4}P $5d$
22	−.13082	$3d^5$ ^{4}D $5d$
23	−.11930	$3d^5(^3)$ ^{2}D $4d$
24	−.06593	$3d^5$ ^{4}G $5g$
25	−.03840	$3d^5$ ^{4}P $5g$
26	−.03763	$3d^5(^3)$ ^{2}G $4d$
27	−.03411	$3d^5(^5)$ ^{2}D $5d$
28	−.02597	$3d^5$ ^{4}F $6s$
29	−.02424	$3d^5$ ^{4}G $6d$
30	−.02086	$3d^5(^3)$ ^{2}F $6s$

Cr-like Fe (Fe^{2+})

i	E(Ryds)	Description
31	−.00979	$3d^5$ ^{4}F $5d$
32	−.00587	$3d^5$ ^{4}D $5g$
	^{3}F$^{\circ}$	
1	−1.12692	$3d^5$ ^{4}G $4p$
2	−1.06069	$3d^5$ ^{4}D $4p$
3	−1.00686	$3d^5(^5)$ ^{2}D $4p$
4	−.96658	$3d^5(^3)$ ^{2}F $4p$
5	−.91349	$3d^5(^5)$ ^{2}G $4p$
6	−.89194	$3d^5$ ^{4}F $4p$
7	−.86062	$3d^5(^5)$ ^{2}F $4p$
8	−.74623	$3d^5(^3)$ ^{2}D $4p$
9	−.65677	$3d^5(^3)$ ^{2}G $4p$
10	−.43329	$3d^5(^1)$ ^{2}D $4p$
11	−.42327	$3d^5$ ^{4}G $5p$
12	−.40795	$3d^44s$ ^{4}D $4p$
13	−.35846	$3d^5$ ^{4}D $5p$
14	−.27116	$3d^5$ ^{4}G $4f$
15	−.27033	$3d^5(^5)$ ^{2}D $5p$
16	−.24478	$3d^5$ ^{4}P $4f$
17	−.24390	$3d^5$ ^{4}F $5p$
18	−.22913	$3d^5(^3)$ ^{2}F $5p$
19	−.21300	$3d^44s$ ^{4}G $4p$
20	−.21108	$3d^5$ ^{4}D $4f$
21	−.20776	$3d^5(^5)$ ^{2}G $5p$
22	−.18166	$3d^44s$ ^{2}F $4p$
23	−.16457	$3d^5(^5)$ ^{2}F $5p$
24	−.14233	$3d^44s$ ^{2}G $4p$
25	−.13697	$3d^5$ ^{4}G $6p$
26	−.13334	$3d^5$ ^{2}I $4f$
27	−.11327	$3d^5(^5)$ ^{2}D $4f$
28	−.08810	$3d^5$ ^{4}F $4f$
29	−.08577	$3d^5(^3)$ ^{2}F $4f$
30	−.08471	$3d^44s$ ^{4}D $4p$
31	−.07765	$3d^5$ ^{4}D $6p$
32	−.06725	$3d^5$ ^{4}G $5f$
33	−.05724	$3d^5$ ^{2}H $4f$
34	−.05084	$3d^5(^5)$ ^{2}G $4f$
35	−.04652	$3d^5(^3)$ ^{2}D $5p$
36	−.04079	$3d^5$ ^{4}P $5f$
37	−.04020	$3d^44s$ ^{2}G $4p$
38	−.00897	$3d^5(^5)$ ^{2}F $4f$
39	−.00763	$3d^5$ ^{4}D $5f$
	^{3}G^e	
1	−1.88546	$3d^6$
2	−1.51124	$3d^5$ ^{4}G $4s$
3	−1.32191	$3d^5(^5)$ ^{2}G $4s$
4	−1.08998	$3d^5(^3)$ ^{2}G $4s$
5	−.60060	$3d^44s$ ^{4}G $4s$
6	−.56767	$3d^5$ ^{4}G $5s$
7	−.52560	$3d^5$ ^{4}G $4d$
8	−.45593	$3d^5$ ^{4}D $4d$
9	−.39188	$3d^5$ ^{2}I $4d$
10	−.36264	$3d^5(^5)$ ^{2}G $5s$
11	−.36182	$3d^5(^5)$ ^{2}D $4d$
12	−.34110	$3d^5$ ^{4}F $4d$
13	−.32934	$3d^5(^3)$ ^{2}F $4d$
14	−.30192	$3d^5$ ^{2}H $4d$
15	−.28042	$3d^5(^5)$ ^{2}G $4d$
16	−.25226	$3d^5(^5)$ ^{2}F $4d$
17	−.20683	$3d^5$ ^{4}G $6s$
18	−.18951	$3d^5$ ^{4}G $5d$
19	−.13112	$3d^5$ ^{4}D $5d$
20	−.12263	$3d^5(^3)$ ^{2}G $5s$
21	−.11805	$3d^5(^3)$ ^{2}D $4d$
22	−.06592	$3d^5$ ^{4}G $5g$
23	−.06059	$3d^5$ ^{2}I $5d$
24	−.04478	$3d^5(^3)$ ^{2}G $4d$
25	−.03839	$3d^5$ ^{4}P $5g$
26	−.03672	$3d^5(^5)$ ^{2}D $5d$
27	−.03341	$3d^5$ ^{4}G $7s$
28	−.02598	$3d^5$ ^{4}G $6d$
29	−.01204	$3d^5$ ^{4}F $5d$
30	−.00588	$3d^5$ ^{4}D $5g$
31	−.00236	$3d^5(^3)$ ^{2}F $5d$
	^{3}G$^{\circ}$	
1	−1.09000	$3d^5$ ^{4}G $4p$
2	−.97793	$3d^5(^3)$ ^{2}F $4p$
3	−.94541	$3d^5$ ^{2}H $4p$
4	−.92311	$3d^5$ ^{4}F $4p$
5	−.87430	$3d^5(^5)$ ^{2}G $4p$
6	−.84598	$3d^5(^5)$ ^{2}F $4p$
7	−.65218	$3d^5(^3)$ ^{2}G $4p$
8	−.41245	$3d^5$ ^{4}G $5p$
9	−.27118	$3d^5$ ^{4}G $4f$
10	−.25323	$3d^44s$ ^{4}H $4p$
11	−.24418	$3d^5$ ^{4}P $4f$
12	−.23991	$3d^5$ ^{4}F $5p$
13	−.23489	$3d^5(^3)$ ^{2}F $5p$
14	−.21445	$3d^5$ ^{2}H $5p$
15	−.21103	$3d^5$ ^{4}D $4f$
16	−.20596	$3d^5(^5)$ ^{2}G $5p$
17	−.19381	$3d^44s$ ^{4}F $4p$
18	−.16338	$3d^5(^5)$ ^{2}F $5p$
19	−.14773	$3d^44s$ ^{4}G $4p$
20	−.13742	$3d^5$ ^{2}I $4f$
21	−.13348	$3d^5$ ^{4}G $6p$
22	−.11960	$3d^44s$ ^{2}H $4p$
23	−.11153	$3d^5(^5)$ ^{2}D $4f$
24	−.08626	$3d^5$ ^{4}F $4f$
25	−.08378	$3d^5(^3)$ ^{2}F $4f$
26	−.06773	$3d^5$ ^{4}G $5f$
27	−.05708	$3d^5$ ^{2}H $4f$
28	−.05087	$3d^5(^5)$ ^{2}G $4f$
29	−.04045	$3d^5$ ^{4}P $5f$
30	−.00981	$3d^5(^5)$ ^{2}F $4f$
31	−.00826	$3d^5$ ^{4}D $5f$
32	−.00016	$3d^5$ ^{4}G $7p$
	^{3}H^e	
1	−1.92023	$3d^6$
2	−1.33949	$3d^5$ ^{2}H $4s$
3	−.65879	$3d^44s$ ^{4}H $4s$
4	−.51329	$3d^5$ ^{4}G $4d$
5	−.38055	$3d^5$ ^{2}I $4d$
6	−.36271	$3d^5$ ^{2}H $5s$
7	−.34111	$3d^5$ ^{4}F $4d$
8	−.32780	$3d^5(^3)$ ^{2}F $4d$
9	−.29937	$3d^5(^5)$ ^{2}G $4d$
10	−.28998	$3d^5$ ^{2}H $4d$
11	−.24940	$3d^5(^5)$ ^{2}F $4d$
12	−.17968	$3d^5$ ^{4}G $5d$
13	−.06593	$3d^5$ ^{4}G $5g$
14	−.05748	$3d^5$ ^{2}I $5d$
15	−.04167	$3d^5(^3)$ ^{2}G $4d$
16	−.03840	$3d^5$ ^{4}P $5g$
17	−.02038	$3d^5$ ^{4}G $6d$
18	−.01101	$3d^5$ ^{4}F $5d$
19	−.00585	$3d^5$ ^{4}D $5g$
20	−.00227	$3d^5(^3)$ ^{2}F $5d$
	^{3}H$^{\circ}$	
1	−1.12702	$3d^5$ ^{4}G $4p$
2	−.97256	$3d^5$ ^{2}I $4p$
3	−.93778	$3d^5$ ^{2}H $4p$
4	−.88542	$3d^5(^5)$ ^{2}G $4p$
5	−.66359	$3d^5(^3)$ ^{2}G $4p$
6	−.42163	$3d^5$ ^{4}G $5p$
7	−.28932	$3d^5$ ^{2}I $5p$
8	−.27077	$3d^5$ ^{4}G $4f$
9	−.23375	$3d^44s$ ^{4}H $4p$
10	−.21783	$3d^5$ ^{2}H $5p$
11	−.21161	$3d^5$ ^{4}D $4f$
12	−.20252	$3d^5(^5)$ ^{2}G $5p$
13	−.18560	$3d^44s$ ^{4}G $4p$
14	−.14880	$3d^44s$ ^{2}H $4p$
15	−.13877	$3d^5$ ^{4}G $6p$
16	−.13651	$3d^5$ ^{2}I $4f$
17	−.11362	$3d^5(^5)$ ^{2}D $4f$
18	−.10179	$3d^44s$ ^{2}G $4p$
19	−.08605	$3d^5$ ^{4}F $4f$
20	−.08511	$3d^5(^3)$ ^{2}F $4f$
21	−.06831	$3d^5$ ^{4}G $5f$
22	−.05520	$3d^5$ ^{2}H $4f$
23	−.05260	$3d^5(^5)$ ^{2}G $4f$
24	−.01396	$3d^5$ ^{2}I $6p$
25	−.00872	$3d^5(^5)$ ^{2}F $4f$
26	−.00815	$3d^5$ ^{4}D $5f$
	^{3}I^e	
1	−1.40478	$3d^5$ ^{2}I $4s$
2	−.53732	$3d^5$ ^{4}G $4d$
3	−.44556	$3d^5$ ^{2}I $5s$
4	−.40785	$3d^5$ ^{2}I $4d$
5	−.31697	$3d^5$ ^{2}H $4d$
6	−.29424	$3d^5(^5)$ ^{2}G $4d$
7	−.19919	$3d^5$ ^{4}G $5d$
8	−.07704	$3d^5$ ^{2}I $6s$
9	−.06986	$3d^5$ ^{2}I $5d$
10	−.06593	$3d^5$ ^{4}G $5g$
11	−.04302	$3d^5(^3)$ ^{2}G $4d$
12	−.02989	$3d^5$ ^{4}G $6d$
13	−.00589	$3d^5$ ^{4}D $5g$
	^{3}I$^{\circ}$	
1	−.99319	$3d^5$ ^{2}I $4p$
2	−.92652	$3d^5$ ^{2}H $4p$
3	−.29362	$3d^5$ ^{2}I $5p$
4	−.27078	$3d^5$ ^{4}G $4f$
5	−.23419	$3d^44s$ ^{4}H $4p$
6	−.21031	$3d^5$ ^{2}H $5p$
7	−.15420	$3d^44s$ ^{2}H $4p$
8	−.13669	$3d^5$ ^{2}I $4f$
9	−.08648	$3d^5$ ^{4}F $4f$
10	−.08527	$3d^5(^3)$ ^{2}F $4f$
11	−.06795	$3d^5$ ^{4}G $5f$
12	−.05567	$3d^5$ ^{2}H $4f$
13	−.05262	$3d^5(^5)$ ^{2}G $4f$
14	−.01271	$3d^5(^5)$ ^{2}F $4f$
15	−.00777	$3d^5(^5)$ ^{2}F $5f$
	^{3}J^e	
1	−.40450	$3d^5$ ^{2}I $4d$
2	−.30937	$3d^5$ ^{2}H $4d$
3	−.06632	$3d^5$ ^{2}I $5d$
4	−.06594	$3d^5$ ^{4}G $5g$
	^{3}J$^{\circ}$	
1	−.99762	$3d^5$ ^{2}I $4p$
2	−.29377	$3d^5$ ^{2}I $5p$
3	−.27220	$3d^5$ ^{4}G $4f$
4	−.14458	$3d^5$ ^{2}I $4f$
5	−.13490	$3d^5$ ^{2}I $5f$
6	−.06863	$3d^5$ ^{4}G $5f$
7	−.05888	$3d^5$ ^{2}H $4f$
8	−.04985	$3d^5(^5)$ ^{2}G $4f$
9	−.00661	$3d^5$ ^{2}I $6p$
	^{1}S^e	
1	−1.82041	$3d^6(^4)$
2	−1.22112	$3d^6(^0)$
3	−1.20077	$3d^5$ ^{2}S $4s$
4	−.45973	$3d^5(^1)$ ^{2}D $4d$:
5	−.29281	$3d^5(^5)$ ^{2}D $4d$
6	−.25803	$3d^5$ ^{2}S $5s$

Cr-like Fe (Fe^{2+})

i	E(Ryds)	Description
7	−.09729	$3d^5(^3)$ ^{2}D $4d$
	^{1}S^o	
1	−.51410	$3d^5$ ^{2}P $4p$
2	−.20113	$3d^44s$ ^{2}P $4p$
3	−.09604	$3d^5(^3)$ ^{2}F $4f$
4	−.00819	$3d^5(^5)$ ^{2}F $4f$
	^{1}P^e	
1	−.90662	$3d^5$ ^{2}P $4s$
2	−.36456	$3d^5(^5)$ ^{2}D $4d$
3	−.34512	$3d^5(^3)$ ^{2}F $4d$
4	−.25020	$3d^5(^5)$ ^{2}F $4d$
5	−.12440	$3d^5(^3)$ ^{2}D $4d$
6	−.03795	$3d^5(^5)$ ^{2}D $5d$
7	−.02010	$3d^5(^3)$ ^{2}F $5d$
	^{1}P^o	
1	−.94728	$3d^5(^5)$ ^{2}D $4p$
2	−.79546	$3d^5$ ^{2}S $4p$
3	−.71748	$3d^5(^3)$ ^{2}D $4p$
4	−.46725	$3d^5$ ^{2}P $4p$
5	−.35282	$3d^5(^1)$ ^{2}D $4p$
6	−.25709	$3d^5(^5)$ ^{2}D $5p$
7	−.18220	$3d^44s$ ^{2}P $4p$
8	−.11063	$3d^5(^5)$ ^{2}D $4f$
9	−.10924	$3d^5$ ^{2}S $5p$
10	−.09405	$3d^5(^3)$ ^{2}F $4f$
11	−.07366	$3d^44s$ ^{2}S $4p$
12	−.05196	$3d^5(^5)$ ^{2}G $4f$
13	−.03703	$3d^5(^3)$ ^{2}D $5p$
14	−.00803	$3d^5(^5)$ ^{2}F $4f$
	^{1}D^e	
1	−1.79329	$3d^6(^4)$
2	−1.40042	$3d^5(^5)$ ^{2}D $4s$
3	−1.37609	$3d^6(^2)$
4	−1.15116	$3d^5(^3)$ ^{2}D $4s$
5	−.83608	$3d^5(^1)$ ^{2}D $4s$
6	−.43631	$3d^44s^2$:
7	−.41283	$3d^5(^5)$ ^{2}D $5s$
8	−.35227	$3d^5(^5)$ ^{2}D $4d$:
9	−.31509	$3d^5(^3)$ ^{2}F $4d$
10	−.28136	$3d^5(^5)$ ^{2}G $4d$
11	−.23912	$3d^5(^5)$ ^{2}F $4d$
12	−.19205	$3d^5(^3)$ ^{2}D $5s$
13	−.17726	$3d^5$ ^{2}S $4d$
14	−.10555	$3d^5(^3)$ ^{2}D $4d$
15	−.04939	$3d^5(^5)$ ^{2}D $6s$
16	−.03034	$3d^5(^5)$ ^{2}D $5d$
17	−.02130	$3d^5(^3)$ ^{2}G $4d$
	^{1}D^o	
1	−1.00741	$3d^5(^5)$ ^{2}D $4p$
2	−.93062	$3d^5(^3)$ ^{2}F $4p$
3	−.85708	$3d^5(^5)$ ^{2}F $4p$
4	−.71023	$3d^5(^3)$ ^{2}D $4p$
5	−.49901	$3d^5$ ^{2}P $4p$
6	−.40363	$3d^5(^1)$ ^{2}D $4p$
7	−.26848	$3d^5(^5)$ ^{2}D $5p$
8	−.24284	$3d^5(^3)$ ^{2}F $5p$
9	−.19651	$3d^44s$ ^{2}P $4p$
10	−.16237	$3d^5(^5)$ ^{2}F $5p$
11	−.15202	$3d^44s$ ^{2}F $4p$
12	−.11297	$3d^5(^5)$ ^{2}D $4f$
13	−.09288	$3d^5(^3)$ ^{2}F $4f$
14	−.05766	$3d^5$ ^{2}H $4f$
15	−.05038	$3d^5(^5)$ ^{2}G $4f$
16	−.04206	$3d^5(^3)$ ^{2}D $5p$
17	−.03094	$3d^44s$ ^{2}D $4p$
18	−.00773	$3d^5(^5)$ ^{2}F $4f$
	^{1}F^e	
1	−1.71733	$3d^6$
2	−1.36036	$3d^5(^3)$ ^{2}F $4s$
3	−1.25782	$3d^5(^5)$ ^{2}F $4s$
4	−.39553	$3d^5(^3)$ ^{2}F $5s$
5	−.36964	$3d^44s^2$:
6	−.35882	$3d^5(^5)$ ^{2}D $4d$:
7	−.33908	$3d^5(^3)$ ^{2}F $4d$
8	−.31352	$3d^5$ ^{2}H $4d$
9	−.30839	$3d^5(^5)$ ^{2}F $5s$
10	−.27442	$3d^5(^5)$ ^{2}G $4d$
11	−.21078	$3d^5(^5)$ ^{2}F $4d$
12	−.11204	$3d^5(^3)$ ^{2}D $4d$
13	−.04706	$3d^5(^3)$ ^{2}G $4d$
14	−.03431	$3d^5(^5)$ ^{2}D $5d$
15	−.03104	$3d^5(^3)$ ^{2}F $6s$
16	−.01533	$3d^5(^3)$ ^{2}F $5d$
	^{1}F^o	
1	−.97595	$3d^5(^5)$ ^{2}D $4p$
2	−.92421	$3d^5(^3)$ ^{2}F $4p$
3	−.85817	$3d^5(^5)$ ^{2}G $4p$
4	−.79990	$3d^5(^5)$ ^{2}F $4p$
5	−.73501	$3d^5(^3)$ ^{2}D $4p$
6	−.61767	$3d^5(^3)$ ^{2}G $4p$
7	−.39780	$3d^5(^1)$ ^{2}D $4p$
8	−.26278	$3d^5(^5)$ ^{2}D $5p$
9	−.23921	$3d^5(^3)$ ^{2}F $5p$
10	−.19880	$3d^5(^5)$ ^{2}G $5p$
11	−.17979	$3d^44s$ ^{2}F $4p$
12	−.15209	$3d^5(^5)$ ^{2}F $5p$
13	−.14361	$3d^44s$ ^{2}G $4p$
14	−.13331	$3d^5$ ^{2}I $4f$
15	−.11198	$3d^5(^5)$ ^{2}D $4f$
16	−.09345	$3d^5(^3)$ ^{2}F $4f$
17	−.05767	$3d^5$ ^{2}H $4f$
18	−.05407	$3d^44s$ ^{2}D $4p$
19	−.04569	$3d^5(^5)$ ^{2}G $4f$
20	−.03952	$3d^5(^3)$ ^{2}D $5p$
21	−.00686	$3d^5(^5)$ ^{2}F $4f$
	^{1}G^e	
1	−1.83602	$3d^6$
2	−1.58135	$3d^6$
3	−1.28790	$3d^5(^5)$ ^{2}G $4s$
4	−1.05307	$3d^5(^3)$ ^{2}G $4s$
5	−.52044	$3d^44s^2$:
6	−.38932	$3d^5$ ^{2}I $4d$:
7	−.35379	$3d^5(^5)$ ^{2}G $5s$
8	−.34979	$3d^5(^5)$ ^{2}D $4d$
9	−.33129	$3d^5(^3)$ ^{2}F $4d$
10	−.28829	$3d^5(^5)$ ^{2}G $4d$
11	−.25373	$3d^5$ ^{2}H $4d$
12	−.22507	$3d^5(^5)$ ^{2}F $4d$
13	−.19314	$3d^44s^2$:
14	−.11643	$3d^5(^3)$ ^{2}D $4d$:
15	−.11162	$3d^5(^3)$ ^{2}G $5s$
16	−.06053	$3d^5$ ^{2}I $5d$
17	−.02981	$3d^5(^5)$ ^{2}D $5d$
18	−.01481	$3d^5(^3)$ ^{2}G $4d$
19	−.00551	$3d^5(^3)$ ^{2}F $5d$
	^{1}G^o	
1	−.97855	$3d^5(^3)$ ^{2}F $4p$
2	−.91668	$3d^5(^5)$ ^{2}G $4p$
3	−.87035	$3d^5$ ^{2}H $4p$
4	−.82476	$3d^5(^5)$ ^{2}F $4p$
5	−.62965	$3d^5(^3)$ ^{2}G $4p$
6	−.24882	$3d^5(^3)$ ^{2}F $5p$
7	−.23258	$3d^44s$ ^{2}H $4p$
8	−.21355	$3d^5$ ^{2}H $5p$
9	−.19528	$3d^5(^5)$ ^{2}G $5p$
10	−.15910	$3d^5(^5)$ ^{2}F $5p$
11	−.14504	$3d^44s$ ^{2}F $4p$
12	−.13539	$3d^5$ ^{2}I $4f$
13	−.11234	$3d^5(^5)$ ^{2}D $4f$
14	−.11137	$3d^44s$ ^{2}G $4p$
15	−.09383	$3d^5(^3)$ ^{2}F $4f$
16	−.05640	$3d^5$ ^{2}H $4f$
17	−.04818	$3d^5(^5)$ ^{2}G $4f$
18	−.00638	$3d^5(^5)$ ^{2}F $4f$
	^{1}H^e	
1	−1.29777	$3d^5$ ^{2}H $4s$
2	−.40756	$3d^5$ ^{2}I $4d$
3	−.35404	$3d^5$ ^{2}H $5s$
4	−.34305	$3d^5(^3)$ ^{2}F $4d$
5	−.30944	$3d^5$ ^{2}H $4d$
6	−.29576	$3d^5(^5)$ ^{2}G $4d$
7	−.24884	$3d^5(^5)$ ^{2}F $4d$
8	−.06942	$3d^5$ ^{2}I $5d$
9	−.04055	$3d^5(^3)$ ^{2}G $4d$
10	−.01799	$3d^5(^3)$ ^{2}F $5d$
	^{1}H^o	
1	−.98306	$3d^5$ ^{2}I $4p$
2	−.87399	$3d^5(^5)$ ^{2}G $4p$
3	−.86238	$3d^5$ ^{2}H $4p$
4	−.63656	$3d^5(^3)$ ^{2}G $4p$
5	−.28677	$3d^5$ ^{2}I $5p$
6	−.20353	$3d^5(^5)$ ^{2}G $5p$
7	−.19969	$3d^5$ ^{2}H $5p$
8	−.17399	$3d^44s$ ^{2}H $4p$
9	−.14644	$3d^44s$ ^{2}G $4p$
10	−.13432	$3d^5$ ^{2}I $4f$
11	−.11190	$3d^5(^5)$ ^{2}D $4f$
12	−.09309	$3d^5(^3)$ ^{2}F $4f$
13	−.05499	$3d^5$ ^{2}H $4f$
14	−.05174	$3d^5(^5)$ ^{2}G $4f$
15	−.00740	$3d^5(^5)$ ^{2}F $4f$
16	−.00435	$3d^5$ ^{2}I $6p$
	^{1}I^e	
1	−1.81502	$3d^6$
2	−1.36479	$3d^5$ ^{2}I $4s$
3	−.54056	$3d^5$ ^{2}I $4d$:
4	−.43645	$3d^5$ ^{2}I $5s$
5	−.36295	$3d^5$ ^{2}H $4d$
6	−.28965	$3d^5(^5)$ ^{2}G $4d$
7	−.26745	$3d^5(^3)$ ^{2}G $4d$
8	−.07346	$3d^5$ ^{2}I $6s$
9	−.04124	$3d^5$ ^{2}I $5d$
10	−.03307	$3d^5$ ^{2}H $5d$
	^{1}I^o	
1	−.94254	$3d^5$ ^{2}I $4p$
2	−.90814	$3d^5$ ^{2}H $4p$
3	−.28099	$3d^5$ ^{2}I $5p$
4	−.20692	$3d^5$ ^{2}H $5p$
5	−.19463	$3d^44s$ ^{2}H $4p$
6	−.13489	$3d^5$ ^{2}I $4f$
7	−.09510	$3d^5(^3)$ ^{2}F $4f$
8	−.05479	$3d^5$ ^{2}H $4f$
9	−.05267	$3d^5(^5)$ ^{2}G $4f$
10	−.00835	$3d^5(^5)$ ^{2}F $4f$
11	−.00186	$3d^5$ ^{2}I $6p$
	^{1}J^e	
1	−.40028	$3d^5$ ^{2}I $4d$
2	−.30504	$3d^5$ ^{2}H $4d$
3	−.06410	$3d^5$ ^{2}I $5d$
	^{1}J^o	
1	−.98237	$3d^5$ ^{2}I $4p$
2	−.28981	$3d^5$ ^{2}I $5p$
3	−.13518	$3d^5$ ^{2}I $4f$
4	−.05527	$3d^5$ ^{2}H $4f$
5	−.05026	$3d^5(^5)$ ^{2}G $4f$
6	−.00500	$3d^5$ ^{2}I $6p$

Cr-like Fe (Fe^{2+})

Energies in ascending order from ground state for terms with effective $n \leq 4.0$, $L \leq 7$

Term	i	E(Ryds)	Term	i	E(Ryds)	Term	i	E(Ryds)	Term	i	E(Ryds)	Term	i	E(Ryds)
$^{5}D^{e}$	1	0.00000	$^{1}D^{e}$	3	0.75074	$^{3}S^{o}$	1	1.05827	$^{1}F^{o}$	2	1.20262	$^{1}P^{o}$	2	1.33137
$^{3}P^{e}$	1	0.17480	$^{1}I^{e}$	2	0.76204	$^{3}F^{o}$	2	1.06614	$^{3}G^{o}$	4	1.20372	$^{3}D^{o}$	7	1.37923
$^{7}S^{e}$	1	0.19911	$^{1}F^{e}$	2	0.76647	$^{5}D^{o}$	2	1.06671	$^{5}D^{o}$	3	1.20764	$^{3}F^{o}$	8	1.38060
$^{3}F^{e}$	1	0.20118	$^{3}H^{e}$	2	0.78734	$^{5}P^{o}$	3	1.07113	$^{1}G^{o}$	2	1.21015	$^{1}F^{o}$	5	1.39182
$^{3}H^{e}$	1	0.20660	$^{3}F^{e}$	4	0.79329	$^{1}G^{e}$	4	1.07376	$^{3}F^{o}$	5	1.21334	$^{7}P^{o}$	2	1.40042
$^{3}G^{e}$	1	0.24137	$^{3}G^{e}$	3	0.80492	$^{3}P^{o}$	2	1.09848	$^{3}D^{o}$	5	1.21537	$^{3}P^{o}$	5	1.40358
$^{3}D^{e}$	1	0.28672	$^{1}H^{e}$	1	0.82906	$^{1}D^{o}$	1	1.11942	$^{1}I^{o}$	2	1.21869	$^{1}P^{o}$	3	1.40935
$^{1}G^{e}$	1	0.29081	$^{3}F^{e}$	5	0.83021	$^{3}F^{o}$	3	1.11997	$^{1}P^{e}$	1	1.22021	$^{1}D^{o}$	4	1.41660
$^{5}S^{e}$	1	0.29880	$^{1}G^{e}$	3	0.83893	$^{3}J^{o}$	1	1.12921	$^{7}D^{e}$	1	1.23333	$^{5}P^{o}$	4	1.41867
$^{1}S^{e}$	1	0.30642	$^{1}F^{e}$	3	0.86901	$^{3}I^{o}$	1	1.13364	$^{5}D^{e}$	3	1.23451	$^{3}H^{o}$	5	1.46324
$^{1}I^{e}$	1	0.31181	$^{3}S^{e}$	1	0.87663	$^{1}H^{o}$	1	1.14377	$^{3}F^{o}$	6	1.23489	$^{3}H^{e}$	3	1.46804
$^{1}D^{e}$	1	0.33354	$^{1}S^{e}$	2	0.90571	$^{1}J^{o}$	1	1.14447	$^{3}H^{o}$	4	1.24141	$^{3}F^{o}$	9	1.47006
$^{1}F^{e}$	1	0.40950	$^{1}S^{e}$	3	0.92606	$^{3}P^{o}$	3	1.14554	$^{7}S^{e}$	2	1.24268	$^{3}G^{o}$	7	1.47465
$^{3}P^{e}$	2	0.46170	$^{3}D^{e}$	4	0.94041	$^{1}G^{o}$	1	1.14828	$^{3}G^{o}$	5	1.25253	$^{1}H^{o}$	4	1.49027
$^{3}F^{e}$	2	0.47047	$^{1}D^{e}$	4	0.97567	$^{3}G^{o}$	2	1.14890	$^{1}H^{o}$	2	1.25284	$^{3}P^{e}$	5	1.49598
$^{5}G^{e}$	1	0.54044	$^{5}G^{o}$	1	0.97843	$^{1}F^{o}$	1	1.15088	$^{3}D^{e}$	5	1.25514	$^{1}G^{o}$	5	1.49718
$^{5}P^{e}$	1	0.54494	$^{5}H^{o}$	1	0.99031	$^{3}D^{o}$	3	1.15334	$^{1}G^{o}$	3	1.25648	$^{3}F^{e}$	6	1.50288
$^{1}G^{e}$	2	0.54548	$^{5}S^{o}$	1	0.99032	$^{3}H^{o}$	2	1.15427	$^{1}H^{o}$	3	1.26445	$^{1}F^{o}$	6	1.50916
$^{5}D^{e}$	2	0.59437	$^{5}D^{o}$	1	0.99109	$^{3}F^{o}$	4	1.16025	$^{3}F^{o}$	7	1.26621	$^{3}G^{e}$	5	1.52623
$^{3}G^{e}$	2	0.61559	$^{3}P^{o}$	1	0.99466	$^{3}D^{o}$	4	1.16412	$^{5}S^{e}$	2	1.26701	$^{5}G^{e}$	2	1.54093
$^{3}P^{e}$	3	0.61830	$^{3}H^{o}$	1	0.99981	$^{1}P^{o}$	1	1.17955	$^{1}F^{o}$	3	1.26866	$^{7}F^{o}$	1	1.55168
$^{7}P^{o}$	1	0.64203	$^{3}F^{o}$	1	0.99991	$^{3}G^{o}$	3	1.18142	$^{1}D^{o}$	3	1.26975	$^{5}H^{e}$	1	1.55326
$^{3}D^{e}$	2	0.66688	$^{5}F^{o}$	1	1.00442	$^{3}P^{e}$	4	1.18381	$^{5}D^{e}$	4	1.27386	$^{5}F^{e}$	2	1.55624
$^{5}P^{o}$	1	0.70551	$^{5}P^{o}$	2	1.00816	$^{1}I^{o}$	1	1.18429	$^{3}G^{o}$	6	1.28085	$^{5}I^{e}$	1	1.55706
$^{3}D^{e}$	3	0.71083	$^{3}D^{o}$	1	1.01862	$^{5}G^{o}$	2	1.18663	$^{3}D^{o}$	6	1.28589	$^{5}G^{e}$	3	1.55750
$^{5}F^{e}$	1	0.71967	$^{3}G^{o}$	1	1.03683	$^{3}H^{o}$	3	1.18905	$^{1}D^{e}$	5	1.29075	$^{3}G^{e}$	6	1.55916
$^{3}I^{e}$	1	0.72205	$^{3}G^{e}$	4	1.03685	$^{1}D^{o}$	2	1.19621	$^{1}G^{o}$	4	1.30207	$^{5}F^{o}$	4	1.56143
$^{1}D^{e}$	2	0.72641	$^{5}F^{o}$	2	1.05197	$^{5}F^{o}$	3	1.19676	$^{3}P^{o}$	4	1.30217	$^{7}F^{o}$	2	1.56385
$^{3}F^{e}$	3	0.72668	$^{3}D^{o}$	2	1.05686	$^{3}I^{o}$	2	1.20031	$^{1}F^{o}$	4	1.32693			

Cr-like Fe (Fe^{2+})

gf-values for transitions involving terms with effective $n \leq 4.0$, $L \leq 7$

i i'	gf_L	i i'	gf_L	i i'	gf_L	i i'	gf_L	i i'	gf_L	i i'	gf_L
	$^7S^e–^7P^o$	1 3	−1.19E+0	1 3	−4.96E+0	1 3	−7.61E−1	4 7	−4.42E−3	3 3	−3.78E+0
1 1	−5.59E+0	2 1	−7.51E−1	2 1	1.20E+0	1 4	−1.55E−5	5 1	1.44E−2	3 4	−5.37E−2
1 2	−1.15E−1	2 2	−6.07E+0	2 2	−9.95E−1	1 5	−7.30E−3	5 2	2.44E−2	3 5	−3.08E−3
2 1	3.20E+0	2 3	−3.00E−1	2 3	−4.14E−1	2 1	−2.50E−4	5 3	6.46E−6	3 6	−4.75E−2
2 2	−1.00E+1	3 1	9.42E−3	3 1	2.23E−1	2 2	−2.36E−1	5 4	3.26E−2	3 7	−9.67E−3
	$^7P^o–^7D^e$	3 2	1.13E−3	3 2	−2.44E−1	2 3	−3.12E−2	5 5	3.88E−4	4 1	−1.50E−3
1 1	−2.17E+1	3 3	8.09E−3	3 3	−1.74E−1	2 4	−1.29E−1	5 6	1.80E−2	4 2	−6.01E−5
2 1	8.43E+0	4 1	2.19E−8	4 1	5.08E−3	2 5	−3.86E−1	5 7	5.48E−5	4 3	−1.11E−2
	$^7D^e–^7F^o$	4 2	5.42E−2	4 2	6.51E−4	3 1	−1.89E+0		$^3P^o–^3D^e$	4 4	−1.24E−2
1 1	−1.64E+1	4 3	2.12E−3	4 3	1.51E−6	3 2	−3.75E−1	1 1	4.96E−2	4 5	−1.36E−2
1 2	−2.55E+1		$^5D^e–^5F^o$		$^5G^e–^5G^o$	3 3	−2.02E−2	1 2	2.90E−1	4 6	−1.02E−1
	$^5S^e–^5P^o$	1 1	−2.83E+0	1 1	−1.20E+1	3 4	−3.25E−2	1 3	5.53E−2	4 7	−3.89E+0
1 1	−3.84E+0	1 2	−5.90E−1	1 2	−3.01E−1	3 5	−9.08E−3	1 4	2.23E−3	5 1	1.87E−3
1 2	−6.14E−2	1 3	−3.64E−1	2 1	4.13E+0	4 1	1.27E−3	1 5	−1.65E−4	5 2	8.94E−4
1 3	−8.86E−3	1 4	−1.73E+0	2 2	1.59E−1	4 2	4.16E−4	2 1	5.45E−1	5 3	1.01E−3
1 4	−9.34E−2	2 1	−9.11E−1	3 1	1.89E+1	4 3	1.67E−3	2 2	1.90E+0	5 4	2.92E−3
2 1	2.36E+0	2 2	−8.54E+0	3 2	6.07E−1	4 4	−1.94E−4	2 3	1.19E−1	5 5	1.09E−3
2 2	5.47E−2	2 3	−4.60E−1		$^5G^e–^5H^o$	4 5	−6.41E−4	2 4	1.04E−2	5 6	−9.60E−6
2 3	2.24E−2	2 4	−3.84E−3	1 1	−1.50E+1	5 1	4.19E−4	2 5	−1.23E−3	5 7	−1.48E−2
2 4	−7.15E+0	3 1	1.49E−2	2 1	9.34E+0	5 2	1.76E−2	3 1	4.91E−2		$^3D^e–^3F^o$
	$^5S^o–^5P^e$	3 2	4.11E−2	3 1	1.83E+0	5 3	3.20E−2	3 2	1.82E−1	1 1	−4.12E−1
1 1	1.37E+0	3 3	5.95E−3		$^5G^o–^5H^e$	5 4	2.97E−4	3 3	2.22E+0	1 2	−7.44E−3
	$^5P^e–^5P^o$	3 4	−6.49E−1	1 1	−2.00E+1	5 5	1.29E−4	3 4	2.14E−2	1 3	−1.33E−1
1 1	−7.63E−3	4 1	2.03E−1	2 1	−6.39E−2		$^3P^e–^3D^o$	3 5	−4.64E−4	1 4	−1.29E−1
1 2	−3.68E+0	4 2	4.02E−2		$^5H^e–^5H^o$	1 1	−5.03E−4	4 1	7.37E−1	1 5	−3.82E−1
1 3	−5.58E−1	4 3	7.34E−4	1 1	1.33E+1	1 2	−1.75E−1	4 2	6.92E−2	1 6	−3.22E−1
1 4	−6.30E−3	4 4	−2.84E+1		$^5H^o–^5I^e$	1 3	−1.49E−1	4 3	5.54E−4	1 7	−1.57E−1
	$^5P^e–^5D^o$		$^5D^o–^5F^e$	1 1	−3.95E+1	1 4	−1.14E+0	4 4	4.27E−1	1 8	−1.35E−1
1 1	−6.01E+0	1 1	7.74E−2		$^5S^o–^5P^e$	1 5	−7.19E−2	4 5	6.02E−4	1 9	−2.79E−1
1 2	−9.75E−1	1 2	−2.75E−1	1 1	−2.59E+2	1 6	−1.36E−1	5 1	1.01E−2	2 1	−4.38E−2
1 3	−7.80E−3	2 1	2.70E−1	1 2	−2.31E+1	1 7	−3.20E−6	5 2	7.56E−3	2 2	−4.98E+0
	$^5P^o–^5D^e$	2 2	−7.01E−3	2 1	−7.64E−4	2 1	−3.06E−2	5 3	3.33E−2	2 3	−1.33E−3
1 1	1.68E+0	3 1	6.77E+0	2 2	−2.37E+2	2 2	−1.75E−2	5 4	2.00E+0	2 4	−2.02E−1
1 2	1.27E−2	3 2	−6.62E−3		$^5P^e–^5D^o$	2 3	−5.86E−3	5 5	1.61E−3	2 5	−1.23E−1
1 3	−5.34E−1		$^5F^e–^5F^o$	1 1	−3.24E+2	2 4	−1.05E−2		$^3D^e–^3D^o$	2 6	−1.68E−1
1 4	−1.45E+1	1 1	−4.42E−1	2 1	−6.59E−5	2 5	−6.57E−1	1 1	−3.60E−2	2 7	−2.10E−1
2 1	1.49E+0	1 2	−1.21E−1		$^3S^e–^3P^o$	2 6	−4.30E−1	1 2	−7.09E−1	2 8	−3.68E−3
2 2	6.45E−1	1 3	−9.38E+0	1 1	−5.18E−5	2 7	−2.71E−1	1 3	−1.28E−1	2 9	−1.09E−4
2 3	−3.94E−3	1 4	−3.59E−3	1 2	−3.12E−2	3 1	−3.59E+0	1 4	−8.47E−2	3 1	−3.77E−2
2 4	−3.18E−2	2 1	1.12E+1	1 3	−1.06E−5	3 2	−5.15E−2	1 5	−5.31E−3	3 2	−6.67E−2
3 1	1.82E−1	2 2	1.93E+0	1 4	−1.98E+0	3 3	−1.89E−1	1 6	−1.03E+0	3 3	−3.31E+0
3 2	3.63E+0	2 3	1.18E−1	1 5	−4.64E−1	3 4	−3.17E−2	1 7	−2.49E−1	3 4	−1.99E+0
3 3	−4.86E−2	2 4	−1.13E−4		$^3S^o–^3P^e$	3 5	−3.95E−2	2 1	−3.42E−2	3 5	−1.12E−1
3 4	−1.96E−2		$^5F^e–^5G^o$	1 1	4.56E−1	3 6	−7.33E−4	2 2	−3.33E+0	3 6	−2.39E−5
4 1	2.00E−1	1 1	−2.27E−1	1 2	1.02E−1	3 7	−8.07E−3	2 3	−1.13E−1	3 7	−2.81E−3
4 2	1.02E−2	1 2	−1.23E+1	1 3	8.13E−1	4 1	1.75E−4	2 4	−9.96E−2	3 8	−1.50E−2
4 3	5.26E−2	2 1	7.49E+0	1 4	−1.91E−4	4 2	5.01E−5	2 5	−3.15E−1	3 9	−8.80E−3
4 4	5.95E+0	2 2	2.40E−1	1 5	−4.48E−2	4 3	6.02E−4	2 6	−4.48E−3	4 1	−1.16E−4
	$^5D^e–^5D^o$		$^5F^o–^5G^e$		$^3P^e–^3P^o$	4 4	3.35E−4	2 7	−6.71E−3	4 2	−1.10E−2
1 1	−2.24E−4	1 1	8.51E+0	1 1	−4.87E−1	4 5	−3.65E−3	3 1	−6.65E−2	4 3	−6.70E−3
1 2	−2.00E+0	1 2	−5.72E+0	1 2	−8.63E−2	4 6	−1.41E−2	3 2	−1.27E−1	4 4	−1.07E−3

Cr-like Fe (Fe^{2+})

i i′	gf$_L$	*i i′*	gf$_L$	*i i′*	gf$_L$	*i i′*	gf$_L$	*i i′*	gf$_L$	*i i′*	gf$_L$
4 5	−5.50E−2	7 2	1.21E−1	6 2	2.38E−3		$^{3}F^{o}$–$^{3}G^{e}$	9 4	5.36E+0	2 1	−8.11E+0
4 6	−2.87E−1	7 3	2.74E−4	6 3	2.13E−2	1 1	3.38E−1	9 5	−2.43E−4	2 2	−1.33E−1
4 7	−1.07E−2	7 4	5.08E−2	6 4	2.85E−3	1 2	4.88E+0	9 6	−5.81E−3	2 3	−8.10E−2
4 8	−5.14E+0	7 5	8.34E−2	6 5	3.77E−3	1 3	4.22E−4		$^{3}G^{e}$–$^{3}G^{o}$	2 4	−8.62E−3
4 9	−4.84E−2	7 6	−7.15E−3	6 6	6.92E−4	1 4	−4.58E−5	1 1	−6.61E−1	2 5	−1.88E−3
5 1	2.21E−3		$^{3}F^{e}$–$^{3}F^{o}$	6 7	5.54E−2	1 5	−3.87E−3	1 2	−2.27E−1	3 1	−1.81E−4
5 2	3.20E−4	1 1	−7.51E−1	6 8	2.50E−3	1 6	−2.37E+0	1 3	−9.20E−1	3 2	−3.87E−1
5 3	2.68E−3	1 2	−9.01E−2	6 9	3.61E−4	2 1	1.19E+0	1 4	−7.62E−1	3 3	−2.79E+0
5 4	1.69E−4	1 3	−1.09E−3		$^{3}F^{e}$–$^{3}G^{o}$	2 2	1.46E−2	1 5	−9.81E−1	3 4	−5.41E+0
5 5	1.19E−3	1 4	−7.36E−1	1 1	−6.51E−1	2 3	1.78E−1	1 6	−5.90E−3	3 5	−2.49E−2
5 6	8.22E−5	1 5	−6.62E−1	1 2	−2.64E−1	2 4	4.93E−5	1 7	−2.65E−1	4 1	2.79E−4
5 7	−3.54E−4	1 6	−2.48E−1	1 3	−7.98E−2	2 5	−6.24E−2	2 1	−6.88E+0	4 2	−9.43E−2
5 8	−3.70E−3	1 7	−4.16E−1	1 4	−1.10E+0	2 6	−4.24E−2	2 2	−1.89E−1	4 3	−2.40E−3
5 9	−2.11E−1	1 8	−4.04E−2	1 5	−1.17E−4	3 1	8.13E−4	2 3	−1.96E−1	4 4	−1.61E−5
	$^{3}D^{o}$–$^{3}F^{e}$	1 9	−1.84E−3	1 6	−7.56E−2	3 2	4.72E−3	2 4	−6.31E−2	4 5	−8.38E+0
1 1	7.09E−1	2 1	−6.46E−2	1 7	−3.07E−2	3 3	9.33E−3	2 5	−4.34E−3	5 1	5.72E−3
1 2	4.74E−2	2 2	−4.32E−2	2 1	−1.32E−1	3 4	3.39E−3	2 6	−4.96E−3	5 2	5.93E−2
1 3	4.79E−2	2 3	−7.48E−3	2 2	−2.55E−2	3 5	−3.67E−2	2 7	−4.78E−4	5 3	6.72E−2
1 4	3.60E−2	2 4	−6.43E−2	2 3	−1.29E−2	3 6	−1.72E−6	3 1	−2.30E−3	5 4	4.55E−3
1 5	1.12E−3	2 5	−8.37E−1	2 4	−2.22E−1	4 1	8.98E−1	3 2	−5.37E−2	5 5	1.84E−4
1 6	−1.63E−1	2 6	−5.50E−1	2 5	−2.93E−1	4 2	3.15E−1	3 3	−2.38E−3	6 1	4.35E+0
2 1	3.59E−1	2 7	−3.40E−1	2 6	−3.21E−1	4 3	4.02E−4	3 4	−2.31E+0	6 2	8.47E−2
2 2	1.69E−3	2 8	−7.27E−4	2 7	−7.45E−1	4 4	3.03E−5	3 5	−4.51E+0	6 3	4.00E−2
2 3	2.17E−2	2 9	−1.55E+0	3 1	−1.21E−1	4 5	−5.65E−4	3 6	−3.75E−1	6 4	6.73E−3
2 4	2.80E−1	3 1	−6.72E−2	3 2	−6.64E+0	4 6	−2.47E−1	3 7	−1.13E−1	6 5	4.29E−4
2 5	2.34E−4	3 2	−7.12E−2	3 3	−2.68E−3	5 1	9.86E−2	4 1	1.98E−9		$^{3}G^{o}$–$^{3}H^{e}$
2 6	−2.26E−4	3 3	−1.86E+0	3 4	−2.32E−1	5 2	1.20E−1	4 2	−8.30E−9	1 1	2.21E+0
3 1	1.50E+0	3 4	−3.10E+0	3 5	−3.41E−3	5 3	7.28E−1	4 3	−1.31E−2	1 2	3.08E−2
3 2	1.40E−2	3 5	−1.34E−1	3 6	−1.22E−3	5 4	3.54E−2	4 4	−1.70E−2	1 3	−1.53E−1
3 3	9.49E−2	3 6	−4.33E−4	3 7	−2.79E−4	5 5	−7.90E−2	4 5	−7.73E−2	2 1	4.49E−1
3 4	1.93E−3	3 7	−3.13E−1	4 1	−7.73E−2	5 6	−1.19E−1	4 6	−1.43E−3	2 2	1.49E−1
3 5	1.20E−2	3 8	−7.90E−3	4 2	−1.28E−2	6 1	1.29E+0	4 7	−7.03E+0	2 3	−4.35E−3
3 6	−1.45E−2	3 9	−4.84E−3	4 3	−4.97E+0	6 2	1.43E−3	5 1	4.73E−1	3 1	5.78E−2
4 1	2.43E−2	4 1	−1.65E−1	4 4	−1.06E+0	6 3	3.65E+0	5 2	3.48E−3	3 2	1.37E+0
4 2	9.26E−2	4 2	−7.06E−2	4 5	−7.02E−1	6 4	2.18E−2	5 3	1.55E−2	3 3	−5.00E−2
4 3	3.76E+0	4 3	−3.94E−3	4 6	−2.72E−2	6 5	−5.79E−3	5 4	2.13E−3	4 1	1.37E+0
4 4	2.54E−2	4 4	−1.35E−1	4 7	−1.04E−2	6 6	−2.86E−4	5 5	5.83E−3	4 2	3.09E+0
4 5	1.08E−1	4 5	−4.11E+0	5 1	−3.52E−2	7 1	2.16E−1	5 6	1.28E−2	4 3	−4.07E−2
4 6	−2.43E−2	4 6	−6.70E−1	5 2	−1.15E−2	7 2	7.38E−2	5 7	8.43E−3	5 1	1.58E+0
5 1	3.16E−2	4 7	−3.22E−1	5 3	−2.54E−1	7 3	9.01E−1	6 1	4.28E+0	5 2	7.34E−1
5 2	7.12E−1	4 8	−1.45E−2	5 4	−5.84E−1	7 4	2.79E−2	6 2	1.23E−1	5 3	−2.18E−3
5 3	3.58E−2	4 9	−5.19E−2	5 5	−1.39E+0	7 5	−4.41E−5	6 3	1.69E−1	6 1	7.34E−2
5 4	3.15E+0	5 1	−2.88E−2	5 6	−4.84E+0	7 6	−6.32E−2	6 4	3.78E−2	6 2	1.96E+0
5 5	3.23E−1	5 2	−3.00E−3	5 7	−1.62E−3	8 1	2.35E−1	6 5	8.70E−3	6 3	3.77E−2
5 6	−6.35E−3	5 3	−1.12E−1	6 1	4.21E−2	8 2	2.36E−3	6 6	1.19E−2	7 1	4.50E−1
6 1	2.02E−1	5 4	−6.74E−2	6 2	4.38E−3	8 3	3.14E−1	6 7	5.71E−4	7 2	1.80E−2
6 2	1.59E−1	5 5	−8.73E−2	6 3	1.05E−2	8 4	1.79E−2		$^{3}G^{e}$–$^{3}H^{o}$	7 3	6.52E−4
6 3	1.42E−1	5 6	−1.04E+0	6 4	1.10E−2	8 5	−2.04E−2	1 1	−3.45E−1		$^{3}H^{e}$–$^{3}H^{o}$
6 4	3.04E−1	5 7	−4.13E+0	6 5	3.58E−2	8 6	−2.51E−3	1 2	−1.15E+0	1 1	−1.08E+0
6 5	3.64E+0	5 8	−1.51E−1	6 6	6.40E−3	9 1	3.40E−1	1 3	−3.30E−3	1 2	−2.30E+0
6 6	−1.53E−3	5 9	−5.45E−2	6 7	8.31E−5	9 2	7.01E−3	1 4	−6.76E−1	1 3	−1.74E+0
7 1	1.80E−1	6 1	7.64E−2			9 3	8.44E−3	1 5	−6.82E−3	1 4	−1.04E−1

Cr-like Fe (Fe^{2+})

$i\ i'$	gf_L	$i\ i'$	gf_L	$i\ i'$	gf_L	$i\ i'$	gf_L	$i\ i'$	gf_L	$i\ i'$	gf_L
1 5	−1.02E−1		$^1P^e$–$^1P^o$	5 3	1.63E−4	4 2	5.83E−3	2 3	4.63E−2	2 1	−5.08E−2
2 1	−2.65E−2	1 1	6.60E−4	5 4	−4.29E−4	4 3	9.57E−2	2 4	1.15E−3	2 2	−2.92E−1
2 2	−4.84E−3	1 2	−1.75E−5		$^1D^e$–$^1F^o$		$^1F^e$–$^1F^o$	3 1	2.54E−1	2 3	−1.20E−1
2 3	−5.40E+0	1 3	−9.22E−3	1 1	−1.84E−3	1 1	−7.21E−3	3 2	7.55E−4	2 4	−1.63E−1
2 4	−3.15E+0		$^1P^e$–$^1D^o$	1 2	−2.56E−1	1 2	−1.18E−2	3 3	1.73E+0	3 1	−2.03E−1
2 5	−2.96E−3	1 1	1.94E−4	1 3	−3.25E−1	1 3	−4.29E−1	3 4	1.29E−2	3 2	−3.06E−1
3 1	1.09E−1	1 2	4.17E−5	1 4	−4.06E−2	1 4	−6.70E−1	4 1	1.59E−1	3 3	−2.39E+0
3 2	8.38E−2	1 3	−3.41E−3	1 5	−1.10E−3	1 5	−3.59E−2	4 2	1.67E−1	3 4	−4.67E−4
3 3	9.70E−3	1 4	−3.10E−4	1 6	−4.93E−3	1 6	−4.45E−3	4 3	5.03E−2	4 1	−1.48E−2
3 4	3.55E−2		$^1P^o$–$^1D^e$	2 1	−2.03E−2	2 1	−3.06E−1	4 4	9.87E−3	4 2	−1.36E−3
3 5	4.43E−4	1 1	6.39E−2	2 2	−2.73E−5	2 2	−1.46E+0	5 1	1.10E−3	4 3	−5.79E−3
	$^3H^e$–$^3I^o$	1 2	5.25E−3	2 3	−4.26E−2	2 3	−2.18E−2	5 2	1.62E−1	4 4	−2.85E+0
1 1	−1.13E+0	1 3	7.81E−1	2 4	−2.50E−3	2 4	−1.32E−1	5 3	5.78E−2		$^1G^o$–$^1H^e$
1 2	−5.00E−1	1 4	1.17E−2	2 5	−1.49E−2	2 5	−8.53E−3	5 4	1.38E−1	1 1	3.20E−1
2 1	−4.98E−1	1 5	−4.60E−6	2 6	−5.21E−1	2 6	−2.17E−4	6 1	8.41E−4	2 1	9.39E−1
2 2	−9.34E+0	2 1	3.99E−1	3 1	−1.38E+0	3 1	−4.39E−2	6 2	6.26E−1	3 1	3.27E−1
3 1	7.85E−2	2 2	1.81E−2	3 2	−3.87E−1	3 2	−5.58E−2	6 3	8.34E−2	4 1	8.10E−1
3 2	6.36E−3	2 3	3.98E−2	3 3	−5.97E−2	3 3	−4.65E−2	6 4	1.66E+0	5 1	2.53E−2
	$^3H^o$–$^3I^e$	2 4	1.30E−1	3 4	−3.91E−3	3 4	−1.81E+0		$^1G^e$–$^1G^o$		$^1H^e$–$^1H^o$
1 1	1.05E−1	2 5	1.67E−5	3 5	−9.68E−3	3 5	−9.59E−3	1 1	−1.44E−2	1 1	−2.38E−1
2 1	8.21E+0	3 1	2.61E−2	3 6	−9.46E−4	3 6	−1.20E−5	1 2	−1.05E+0	1 2	−2.47E+0
3 1	2.73E−1	3 2	2.21E−2	4 1	−3.17E−3		$^1F^e$–$^1G^o$	1 3	−1.54E−1	1 3	−1.99E−1
4 1	2.32E−1	3 3	3.02E−3	4 2	−4.07E−3	1 1	−5.06E−2	1 4	−1.11E−2	1 4	−9.28E−3
5 1	1.39E−1	3 4	6.59E−1	4 3	−8.51E−2	1 2	−8.88E−3	1 5	−9.35E−3		$^1H^e$–$^1I^o$
	$^3I^e$–$^3I^o$	3 5	7.57E−3	4 4	−5.14E−4	1 3	−4.31E−1	2 1	−5.53E−4	1 1	−2.10E−1
1 1	−9.50E+0		$^1D^e$–$^1D^o$	4 5	−1.60E+0	1 4	−1.19E−1	2 2	−2.50E−2	1 2	−3.04E+0
1 2	−6.94E−1	1 1	−6.08E−5	4 6	−1.83E−1	1 5	−1.13E−2	2 3	−6.12E−2		$^1H^o$–$^1I^e$
	$^3I^e$–$^3J^o$	1 2	−4.46E−1	5 1	1.18E−4	2 1	−1.76E+0	2 4	−4.36E−1	1 1	9.44E−2
1 1	−1.14E+1	1 3	−2.35E−1	5 2	4.47E−4	2 2	−5.07E−1	2 5	−8.87E−1	1 2	2.29E+0
	$^3S^e$–$^3S^o$	1 4	−1.23E−1	5 3	3.64E−4	2 3	−3.07E−2	3 1	−1.15E−1	2 1	4.22E−1
1 1	0.00E+0	2 1	−8.46E−4	5 4	−6.33E−7	2 4	−1.30E−3	3 2	−8.65E−1	2 2	1.41E−1
1 2	−1.04E+1	2 2	−2.10E−2	5 5	−3.63E−3	2 5	−1.25E−3	3 3	−7.66E−1	3 1	1.69E+0
1 3	0.00E+0	2 3	−1.70E−5	5 6	−6.16E−2	3 1	−3.19E−2	3 4	−6.32E−1	3 2	4.41E−1
	$^1S^e$–$^1P^o$	2 4	−1.55E−1		$^1D^o$–$^1F^e$	3 2	−2.82E−2	3 5	−1.75E−2	4 1	4.70E−1
1 1	−3.16E−1	3 1	−7.77E−1	1 1	3.93E−4	3 3	−1.25E+0	4 1	−5.46E−4	4 2	4.41E−2
1 2	−2.19E−3	3 2	−4.79E−1	1 2	4.65E−1	3 4	−1.01E+0	4 2	−1.36E−2		$^1I^e$–$^1I^o$
1 3	−6.63E−2	3 3	−5.61E−2	1 3	1.11E−2	3 5	−4.23E−2	4 3	−1.43E−2	1 1	−1.18E+0
2 1	−1.23E−2	3 4	−8.40E−3	2 1	8.34E−2		$^1F^o$–$^1G^e$	4 4	−1.69E−2	1 2	−1.02E+0
2 2	−5.45E−1	4 1	−3.09E−3	2 2	8.40E−1	1 1	2.83E−2	4 5	−2.34E+0	2 1	−3.22E+0
2 3	−9.46E−2	4 2	−2.85E−3	2 3	4.40E−2	1 2	3.48E−3		$^1G^e$–$^1H^o$	2 2	−3.01E−1
3 1	−8.12E−3	4 3	−7.54E−2	3 1	2.26E−1	1 3	1.27E−2	1 1	−5.06E−1		$^1I^e$–$^1J^o$
3 2	−1.10E−1	4 4	−1.25E+0	3 2	2.46E−2	1 4	2.27E−3	1 2	−1.73E−1	1 1	−3.16E−1
3 3	−6.22E−2	5 1	4.01E−3	3 3	1.19E+0	2 1	1.09E+0	1 3	−4.55E−2	2 1	−3.73E+0
		5 2	8.44E−4	4 1	3.91E−1	2 2	6.20E−3	1 4	−2.11E−2		

Mn-like Fe (Fe^{+})

Term energies relative to $3d^6$ ^{5}D ionization threshold for each symmetry

i	E(Ryds)	Description	i	E(Ryds)	Description	i	E(Ryds)	Description	i	E(Ryds)	Description
		$^8\mathbf{S}^e$	4	-.27132	$3d^5 4p^2$	11	-.11110	$3d^6$ ^{5}D $6h$	7	-.08164	$3d^6$ ^{5}D $7h$
1	-.24247	$3d^5 4s$ ^{7}S $5s$	5	-.25545	$3d^6$ ^{5}D $6s$	12	-.09993	$3d^6$ ^{5}D $8p$	8	-.06268	$3d^6$ ^{5}D $8f$
2	-.00097	$3d^5 4s$ ^{7}S $6s$	6	-.24220	$3d^6$ ^{5}D $5d$	13	-.08218	$3d^6$ ^{5}D $7f$	9	-.06251	$3d^6$ ^{5}D $8h$
		$^8\mathbf{P}^e$	7	-.16381	$3d^6$ ^{5}D $7s$	14	-.08162	$3d^6$ ^{5}D $7h$	10	-.04951	$3d^6$ ^{5}D $9f$
1	-.21985	$3d^5 4p^2$	8	-.15991	$3d^6$ ^{5}D $5g$	15	-.07447	$3d^6$ ^{5}D $9p$	11	-.04939	$3d^6$ ^{5}D $9h$
		$^8\mathbf{P}^o$	9	-.15583	$3d^6$ ^{5}D $6d$	16	-.06287	$3d^6$ ^{5}D $8f$	12	-.04938	$3d^6$ ^{5}D $8j$
1	-.72538	$3d^5 4s$ ^{7}S $4p$	10	-.14457	$3d^5 4s$ ^{7}S $4d$	17	-.06249	$3d^6$ ^{5}D $8h$			$^6\mathbf{I}^e$
2	-.12649	$3d^5 4s$ ^{7}S $5p$	11	-.11301	$3d^6$ ^{5}D $8s$			$^6\mathbf{G}^e$	1	-.15997	$3d^6$ ^{5}D $5g$
		$^8\mathbf{D}^e$	12	-.11104	$3d^6$ ^{5}D $6g$	1	-.42230	$3d^6$ ^{5}D $4d$	2	-.11109	$3d^6$ ^{5}D $6g$
1	-.19156	$3d^5 4s$ ^{7}S $4d$	13	-.10841	$3d^6$ ^{5}D $7d$	2	-.24168	$3d^6$ ^{5}D $5d$	3	-.08164	$3d^6$ ^{5}D $7i$
		$^6\mathbf{S}^e$	14	-.08283	$3d^6$ ^{5}D $9s$	3	-.16019	$3d^6$ ^{5}D $5g$	4	-.08162	$3d^6$ ^{5}D $7g$
1	-.99510	$3d^5 4s^2$	15	-.08160	$3d^6$ ^{5}D $7g$	4	-.15500	$3d^6$ ^{5}D $6d$	5	-.06250	$3d^6$ ^{5}D $8i$
2	-.41275	$3d^6$ ^{5}D $4d$	16	-.07989	$3d^6$ ^{5}D $8d$	5	-.11127	$3d^6$ ^{5}D $6g$	6	-.06249	$3d^6$ ^{5}D $8g$
3	-.23879	$3d^6$ ^{5}D $5d$			$^6\mathbf{D}^o$	6	-.10794	$3d^6$ ^{5}D $7d$	7	-.04939	$3d^6$ ^{5}D $9g$
4	-.21174	$3d^5 4s$ ^{7}S $5s$	1	-.84662	$3d^6$ ^{5}D $4p$	7	-.08175	$3d^6$ ^{5}D $7g$	8	-.04938	$3d^6$ ^{5}D $9i$
5	-.15392	$3d^6$ ^{5}D $6d$	2	-.41195	$3d^5 4s$ ^{5}P $4p$	8	-.08163	$3d^6$ ^{5}D $7i$			$^6\mathbf{I}^o$
6	-.13215	$3d^5 4s$ ^{5}S $5s$	3	-.37891	$3d^6$ ^{5}D $5p$	9	-.07953	$3d^6$ ^{5}D $8d$	1	-.11112	$3d^6$ ^{5}D $6h$
7	-.10758	$3d^6$ ^{5}D $7d$	4	-.34776	$3d^5 4s$ ^{5}D $4p$	10	-.06258	$3d^6$ ^{5}D $8g$	2	-.08165	$3d^6$ ^{5}D $7h$
8	-.07947	$3d^6$ ^{5}D $8d$	5	-.25116	$3d^6$ ^{5}D $4f$	11	-.06250	$3d^6$ ^{5}D $8i$	3	-.06251	$3d^6$ ^{5}D $8h$
9	-.06112	$3d^6$ ^{5}D $9d$	6	-.21967	$3d^6$ ^{5}D $6p$	12	-.06104	$3d^6$ ^{5}D $9d$	4	-.06250	$3d^6$ ^{5}D $8j$
		$^6\mathbf{S}^o$	7	-.21163	$3d^5 4s$ ^{5}F $4p$	13	-.04944	$3d^6$ ^{5}D $9g$	5	-.04939	$3d^6$ ^{5}D $9h$
1	-.40784	$3d^5 4s$ ^{5}P $4p$	8	-.16084	$3d^6$ ^{5}D $5f$	14	-.04938	$3d^6$ ^{5}D $9i$	6	-.04938	$3d^6$ ^{5}D $9j$
		$^6\mathbf{P}^e$	9	.14247	$3d^6$ ^{5}D $7p$			$^6\mathbf{G}^o$			$^6\mathbf{J}^e$
1	-.42192	$3d^6$ ^{5}D $4d$	10	-.11168	$3d^6$ ^{5}D $6f$	1	-.45889	$3d^5 4s$ ^{5}G $4p$	1	-.08163	$3d^6$ ^{5}D $7i$
2	-.24380	$3d^6$ ^{5}D $5d$	11	-.10072	$3d^6$ ^{5}D $8p$	2	-.25273	$3d^6$ ^{5}D $4f$	2	-.06250	$3d^6$ ^{5}D $8i$
3	-.18338	$3d^5 4p^2$	12	-.08202	$3d^6$ ^{5}D $7f$	3	-.23791	$3d^5 4s$ ^{5}F $4p$	3	-.04938	$3d^6$ ^{5}D $9i$
4	-.15255	$3d^6$ ^{5}D $6d$	13	-.07498	$3d^6$ ^{5}D $9p$	4	-.16147	$3d^6$ ^{5}D $5f$			$^6\mathbf{J}^o$
5	-.10735	$3d^6$ ^{5}D $7d$	14	-.06277	$3d^6$ ^{5}D $8f$	5	-.11202	$3d^6$ ^{5}D $6f$	1	-.11111	$3d^6$ ^{5}D $6h$
6	-.07925	$3d^6$ ^{5}D $8d$			$^6\mathbf{F}^e$	6	-.11111	$3d^6$ ^{5}D $6h$	2	-.08163	$3d^6$ ^{5}D $7h$
7	-.06088	$3d^6$ ^{5}D $9d$	1	-.43122	$3d^6$ ^{5}D $4d$	7	-.08223	$3d^6$ ^{5}D $7f$	3	-.06250	$3d^6$ ^{5}D $8j$
		$^6\mathbf{P}^o$	2	-.24536	$3d^6$ ^{5}D $5d$	8	-.08164	$3d^6$ ^{5}D $7h$	4	-.06250	$3d^6$ ^{5}D $8h$
1	-.81105	$3d^6$ ^{5}D $4p$	3	-.16007	$3d^6$ ^{5}D $5g$	9	-.06291	$3d^6$ ^{5}D $8f$	5	-.04938	$3d^6$ ^{5}D $9j$
2	-.65944	$3d^5 4s$ ^{7}S $4p$	4	-.15679	$3d^6$ ^{5}D $6d$	10	-.06250	$3d^6$ ^{5}D $8h$	6	-.04938	$3d^6$ ^{5}D $9h$
3	-.47851	$3d^5 4s$ ^{5}S $4p$	5	-.11118	$3d^6$ ^{5}D $6g$	11	-.04968	$3d^6$ ^{5}D $9f$			$^4\mathbf{S}^e$
4	-.39665	$3d^5 4s$ ^{5}P $4p$	6	-.10894	$3d^6$ ^{5}D $7d$	12	-.04939	$3d^6$ ^{5}D $9h$	1	-.41418	$3d^6$ ^{5}D $4d$
5	-.35809	$3d^6$ ^{5}D $5p$	7	-.08168	$3d^6$ ^{5}D $7g$			$^6\mathbf{H}^e$	2	-.23753	$3d^6$ ^{5}D $5d$
6	-.34433	$3d^5 4s$ ^{5}D $4p$	8	-.08014	$3d^6$ ^{5}D $8d$	1	-.16021	$3d^6$ ^{5}D $5g$	3	-.15293	$3d^6$ ^{5}D $6d$
7	-.24964	$3d^6$ ^{5}D $4f$	9	-.06254	$3d^6$ ^{5}D $8g$	2	-.11129	$3d^6$ ^{5}D $6g$	4	-.13151	$3d^5 4s$ ^{5}S $5s$
8	-.21260	$3d^6$ ^{5}D $6p$	10	-.06145	$3d^6$ ^{5}D $9d$	3	-.08176	$3d^6$ ^{5}D $7g$	5	-.11919	$3d^6$ ^{3}D $4d$
9	-.16013	$3d^6$ ^{5}D $5f$	11	-.04941	$3d^6$ ^{5}D $9g$	4	-.08163	$3d^6$ ^{5}D $7i$	6	-.10658	$3d^6$ ^{5}D $7d$
10	-.14043	$3d^6$ ^{5}D $7p$			$^6\mathbf{F}^o$	5	.06250	$3d^6$ ^{5}D $8g$	7	.07868	$3d^6$ ^{5}D $8d$
11	-.11131	$3d^6$ ^{5}D $6f$	1	-.81774	$3d^6$ ^{5}D $4p$	6	-.06250	$3d^6$ ^{5}D $8i$	8	-.06047	$3d^6$ ^{5}D $9d$
12	-.10431	$3d^5 4s$ ^{7}S $5p$	2	-.42712	$3d^5 4s$ ^{5}G $4p$	7	-.04945	$3d^6$ ^{5}D $9g$			$^1\mathbf{P}^o$
13	-.09768	$3d^6$ ^{5}D $8p$	3	-.37560	$3d^5 4s$ ^{5}D $4p$	8	-.04938	$3d^6$ ^{5}D $9i$	1	-.77064	$3d^6$ ^{5}D $4p$
14	-.08176	$3d^6$ ^{5}D $7f$	4	-.36746	$3d^6$ ^{5}D $5p$			$^6\mathbf{H}^o$	2	-.59969	$3d^6(^4)$ ^{3}P $4p$
15	-.07391	$3d^6$ ^{5}D $9p$	5	-.25220	$3d^6$ ^{5}D $4f$	1	-.44341	$3d^5 4s$ ^{5}G $4p$	3	-.57295	$3d^5 4s$ ^{5}S $4p$
16	-.06260	$3d^6$ ^{5}D $8f$	6	-.22754	$3d^5 4s$ ^{5}F $4p$	2	-.25057	$3d^6$ ^{5}D $4f$	4	-.50087	$3d^6$ ^{3}D $4p$
		$^6\mathbf{D}^e$	7	-.21460	$3d^6$ ^{5}D $6p$	3	-.16051	$3d^6$ ^{5}D $5f$	5	-.35749	$3d^6$ ^{5}D $5p$
1	-1.17824	$3d^6$ ^{5}D $4s$	8	-.16131	$3d^6$ ^{5}D $5f$	4	-.11147	$3d^6$ ^{5}D $6f$	6	-.33477	$3d^6(^2)$ ^{3}P $4p$
2	-.46875	$3d^6$ ^{5}D $5s$	9	-.14115	$3d^6$ ^{5}D $7p$	5	-.11112	$3d^6$ ^{5}D $6h$	7	-.28766	$3d^5 4s$ ^{5}P $4p$
3	-.42730	$3d^6$ ^{5}D $4d$	10	-.11193	$3d^6$ ^{5}D $6f$	6	-.08188	$3d^6$ ^{5}D $7f$	8	-.25577	$3d^5 4s$ ^{5}D $4p$

Mn-like Fe (Fe^+)

i	E(Ryds)	Description
9	−.24930	$3d^6$ ^{5}D $4f$
10	−.21293	$3d^54s$ ^{3}D $4p$
11	−.21009	$3d^6$ ^{5}D $6p$
12	−.18683	$3d^6(^4)$ ^{3}P $5p$
13	−.15994	$3d^6$ ^{5}D $5f$
14	−.14108	$3d^6$ ^{5}D $7p$
15	−.12119	$3d^54s$ ^{3}P $4p$
16	−.11100	$3d^6$ ^{5}D $6f$
17	−.10289	$3d^54s$ ^{3}D $4p$
18	−.09487	$3d^6$ ^{5}D $8p$
19	−.08542	$3d^6$ ^{3}D $5p$
20	−.08155	$3d^6$ ^{5}D $7f$
21	−.07275	$3d^6$ ^{5}D $9p$
22	−.06253	$3d^6$ ^{5}D $8f$
	$^4P^e$	
1	−.98458	$3d^7$
2	−.93129	$3d^6(^4)$ ^{3}P $4s$
3	−.65148	$3d^6(^2)$ ^{3}P $4s$
4	−.63315	$3d^54s^2$:
5	−.37468	$3d^6$ ^{5}D $4d$
6	−.28325	$3d^6(^4)$ ^{3}P $5s$
7	−.22670	$3d^6$ ^{5}D $5d$
8	−.22063	$3d^6(^4)$ ^{3}P $4d$
9	−.18404	$3d^6(^4)$ ^{3}F $4d$
10	−.14879	$3d^6$ ^{5}D $6d$
11	−.12310	$3d^54p^2$
12	−.10947	$3d^6$ ^{3}D $4d$
13	−.09656	$3d^6$ ^{5}D $7d$
14	−.07488	$3d^6(^4)$ ^{3}P $6s$
15	−.07440	$3d^6$ ^{5}D $8d$
16	−.05824	$3d^6$ ^{5}D $9d$
17	−.05524	$3d^6(^4)$ ^{3}P $5d$
	$^4S^o$	
1	−.60746	$3d^6(^4)$ ^{3}P $4p$
2	−.31436	$3d^6(^2)$ ^{3}P $4p$
3	−.27177	$3d^54s$ ^{5}P $4p$
4	−.19194	$3d^6(^4)$ ^{3}P $5p$
5	−.11403	$3d^54s$ ^{3}P $4p$
6	−.03071	$3d^6(^4)$ ^{3}P $6p$
7	−.02445	$3d^6(^4)$ ^{3}F $4f$
	$^4D^o$	
1	−.78896	$3d^6$ ^{5}D $4p$
2	−.58957	$3d^6(^4)$ ^{3}P $4p$
3	−.58574	$3d^6(^4)$ ^{3}F $4p$
4	−.49622	$3d^6$ ^{3}D $4p$
5	−.36311	$3d^6$ ^{5}D $5p$
6	−.34009	$3d^6(^2)$ ^{3}P $4p$
7	−.30999	$3d^6(^2)$ ^{3}F $4p$
8	−.29521	$3d^54s$ ^{5}P $4p$
9	−.27819	$3d^54s$ ^{5}D $4p$
10	−.25064	$3d^6$ ^{5}D $4f$
11	−.21335	$3d^6$ ^{5}D $6p$
12	−.20825	$3d^54s$ ^{3}D $4p$
13	−.19842	$3d^54s$ ^{3}F $4p$
14	−.18023	$3d^6(^4)$ ^{3}P $5p$
15	−.16065	$3d^6$ ^{5}D $5f$
16	−.15369	$3d^6(^4)$ ^{3}F $5p$
17	−.14067	$3d^6$ ^{5}D $7p$
18	−.13228	$3d^54s$ ^{3}F $4p$
19	−.11476	$3d^54s$ ^{3}F $4p$
20	−.11085	$3d^6$ ^{5}D $6f$
21	−.10329	$3d^6$ ^{5}D $8p$
22	−.09595	$3d^54s$ ^{3}D $4p$
23	−.08756	$3d^54p$ ^{5}P $4d$:
24	−.08179	$3d^6$ ^{5}D $7f$
25	−.07664	$3d^6$ ^{3}D $5p$
26	−.07266	$3d^6$ ^{5}D $9p$
27	−.06880	$3d^6(^4)$ ^{3}P $4f$
28	−.06257	$3d^6$ ^{5}D $8f$
	$^4D^e$	
1	−1.10602	$3d^6$ ^{5}D $4s$
2	−.84987	$3d^6$ ^{3}D $4s$
3	−.61913	$3d^54s^2$:
4	−.45550	$3d^6$ ^{5}D $5s$
5	−.42250	$3d^6$ ^{5}D $4d$
6	−.25310	$3d^6$ ^{5}D $6s$
7	−.24043	$3d^6$ ^{5}D $5d$
8	−.21909	$3d^6(^4)$ ^{3}P $4d$
9	−.19399	$3d^6(^4)$ ^{3}F $4d$
10	−.18598	$3d^6$ ^{3}D $5s$
11	−.18076	$3d^6$ ^{3}G $4d$
12	−.16139	$3d^6$ ^{5}D $7s$
13	−.15989	$3d^6$ ^{5}D $5g$
14	−.15422	$3d^6$ ^{5}D $6d$
15	−.11663	$3d^6$ ^{3}D $4d$
16	−.11193	$3d^6$ ^{5}D $8s$
17	−.11104	$3d^6$ ^{5}D $6g$
18	−.10748	$3d^6$ ^{5}D $7d$
19	−.08236	$3d^6$ ^{5}D $9s$
20	−.08161	$3d^6$ ^{5}D $7g$
21	−.07929	$3d^6$ ^{5}D $8d$
22	−.06982	$3d^54s$ ^{5}S $4d$
	$^4F^o$	
1	−.79682	$3d^6$ ^{5}D $4p$
2	−.59473	$3d^6(^4)$ ^{3}F $4p$
3	−.56448	$3d^6$ ^{3}G $4p$
4	−.50067	$3d^6$ ^{3}D $4p$
5	−.36541	$3d^6$ ^{5}D $5p$
6	−.36181	$3d^54s$ ^{5}G $4p$
7	−.29808	$3d^6(^2)$ ^{3}F $4p$
8	−.28220	$3d^54s$ ^{5}D $4p$
9	−.25207	$3d^6$ ^{5}D $4f$
10	−.23668	$3d^54s$ ^{3}D $4p$
11	−.21521	$3d^6$ ^{5}D $6p$
12	−.20222	$3d^54s$ ^{3}F $4p$
13	−.16637	$3d^6(^4)$ ^{3}F $5p$
14	−.16132	$3d^6$ ^{5}D $5f$
15	−.15749	$3d^6$ ^{3}G $5p$
16	−.14665	$3d^54s$ ^{3}G $4p$
17	−.14191	$3d^54s$ ^{3}F $4p$
18	−.13557	$3d^6$ ^{5}D $7p$
19	−.12769	$3d^54s$ ^{3}F $4p$
20	−.11370	$3d^54s$ ^{3}D $4p$:
21	−.11188	$3d^6$ ^{5}D $6f$
22	−.11110	$3d^6$ ^{5}D $6h$
23	−.09861	$3d^6$ ^{5}D $8p$
24	−.09246	$3d^54s$ ^{3}G $4p$:
25	−.08364	$3d^6$ ^{3}D $5p$
26	−.08178	$3d^6$ ^{5}D $7f$
27	−.08162	$3d^6$ ^{5}D $7h$
28	−.07362	$3d^6$ ^{5}D $9p$
29	−.06867	$3d^6(^4)$ ^{3}P $4f$
30	−.06277	$3d^6$ ^{5}D $8f$
31	−.06249	$3d^6$ ^{5}D $8h$
	$^4F^e$	
1	−1.08075	$3d^7$
2	−.93267	$3d^6(^4)$ ^{3}F $4s$
3	−.64618	$3d^6(^2)$ ^{3}F $4s$
4	−.48836	$3d^54s^2$:
5	−.39325	$3d^6$ ^{5}D $4d$
6	−.24010	$3d^6(^4)$ ^{3}F $5s$
7	−.23775	$3d^6$ ^{5}D $5d$
8	−.21878	$3d^6(^4)$ ^{3}P $4d$
9	−.20074	$3d^6$ ^{3}H $4d$
10	−.18751	$3d^6(^4)$ ^{3}F $4d$
11	−.16950	$3d^6$ ^{3}G $4d$
12	−.16007	$3d^6$ ^{5}D $5g$
13	−.13963	$3d^6$ ^{5}D $6d$
14	−.11575	$3d^6$ ^{3}D $4d$
15	−.11117	$3d^6$ ^{5}D $6g$
16	−.10203	$3d^6$ ^{5}D $7d$
17	−.08168	$3d^6$ ^{5}D $7g$
18	−.07707	$3d^6$ ^{5}D $8d$
19	−.06253	$3d^6$ ^{5}D $8g$
20	−.06007	$3d^6$ ^{5}D $9d$
21	−.05270	$3d^6(^4)$ ^{3}P $5d$
22	−.04941	$3d^6$ ^{5}D $9g$
	$^4G^o$	
1	−.61678	$3d^6$ ^{3}H $4p$
2	−.58441	$3d^6(^4)$ ^{3}F $4p$
3	−.57230	$3d^6$ ^{3}G $4p$
4	−.34751	$3d^6(^2)$ ^{3}F $4p$
5	−.30845	$3d^54s$ ^{5}G $4p$
6	−.25200	$3d^6$ ^{5}D $4f$
7	−.21149	$3d^54s$ ^{3}F $4p$
8	−.19479	$3d^54s$ ^{3}G $4p$
9	−.17765	$3d^54s$ ^{5}F $4p$
10	−.16143	$3d^6$ ^{5}D $5f$
11	−.15878	$3d^6$ ^{3}H $5p$
12	−.14975	$3d^6(^4)$ ^{3}F $5p$
13	−.14020	$3d^6$ ^{3}G $5p$
14	−.12756	$3d^54s$ ^{3}H $4p$
15	−.11342	$3d^54s$ ^{3}G $4p$
16	−.11112	$3d^6$ ^{5}D $6h$
17	−.11102	$3d^6$ ^{5}D $6f$
18	−.08993	$3d^54s$ ^{3}F $4p$
19	−.08200	$3d^6$ ^{5}D $7f$
20	−.08164	$3d^6$ ^{5}D $7h$
21	−.06872	$3d^6(^4)$ ^{3}P $4f$
22	−.06275	$3d^6$ ^{5}D $8f$
23	−.06250	$3d^6$ ^{5}D $8h$
24	−.04957	$3d^6$ ^{5}D $9f$
25	−.04939	$3d^6$ ^{5}D $9h$
	$^4G^e$	
1	−.91680	$3d^6$ ^{3}G $4s$
2	−.69133	$3d^54s^2$:
3	−.41954	$3d^6$ ^{5}D $4d$
4	−.24033	$3d^6$ ^{3}G $5s$
5	−.23887	$3d^6$ ^{5}D $5d$
6	−.21117	$3d^6$ ^{3}H $4d$
7	−.19458	$3d^6(^4)$ ^{3}F $4d$
8	−.18300	$3d^6$ ^{3}G $4d$
9	−.16019	$3d^6$ ^{5}D $5g$
10	−.15378	$3d^6$ ^{5}D $6d$
11	−.11794	$3d^6$ ^{3}D $4d$
12	−.11127	$3d^6$ ^{5}D $6g$
13	−.10722	$3d^6$ ^{5}D $7d$
14	−.08175	$3d^6$ ^{5}D $7g$
15	−.08163	$3d^6$ ^{5}D $7i$
16	−.07906	$3d^6$ ^{5}D $8d$
17	−.06258	$3d^6$ ^{5}D $8g$
18	−.06250	$3d^6$ ^{5}D $8i$
19	−.06073	$3d^6$ ^{5}D $9d$
20	−.04944	$3d^6$ ^{5}D $9g$
21	−.04938	$3d^6$ ^{5}D $9i$
	$^4H^o$	
1	−.61933	$3d^6$ ^{3}H $4p$
2	−.56693	$3d^6$ ^{3}G $4p$
3	−.36933	$3d^54s$ ^{5}G $4p$
4	−.26061	$3d^54s$ ^{3}I $4p$
5	−.25082	$3d^6$ ^{5}D $4f$
6	−.20097	$3d^54s$ ^{3}G $4p$
7	−.18207	$3d^6$ ^{3}H $5p$
8	−.16038	$3d^6$ ^{5}D $5f$
9	−.14551	$3d^6$ ^{3}G $5p$
10	−.13798	$3d^54s$ ^{3}G $4p$
11	−.13366	$3d^54s$ ^{5}G $5p$
12	−.11145	$3d^6$ ^{5}D $6f$

Mn-like Fe (Fe^+)

i	E(Ryds)	Description	i	E(Ryds)	Description	i	E(Ryds)	Description	i	E(Ryds)	Description
13	−.11112	$3d^6$ ^{5}D $6h$	3	−.06250	$3d^6$ ^{5}D $8i$	6	−.35013	$3d^6(^2)$ ^{1}D $4s$	14	−.07102	$3d^6$ ^{1}F $5s$
14	−.08188	$3d^6$ ^{5}D $7f$	4	−.04938	$3d^6$ ^{5}D $9i$	7	−.27133	$3d^54s$ ^{3}D $4s$:	15	−.05534	$3d^6(^4)$ ^{1}D $4d$
15	−.08164	$3d^6$ ^{5}D $7h$			^{2}S^e	8	−.21082	$3d^6(^4)$ ^{3}P $4d$	16	−.05238	$3d^6(^4)$ ^{3}P $5d$
16	−.06268	$3d^6$ ^{5}D $8f$	1	−.78871	$3d^6(^4)$ ^{1}S $4s$	9	−.17545	$3d^6(^4)$ ^{3}F $4d$	17	−.02771	$3d^6(^4)$ ^{3}F $6s$
17	−.04952	$3d^6$ ^{5}D $9f$	2	−.34425	$3d^6(^4)$ ^{1}S $5s$:	10	−.16464	$3d^6$ ^{3}G $4d$	18	−.01865	$3d^6$ ^{3}H $5d$
		^{4}H^e	3	−.13727	$3d^6(^4)$ ^{1}S $6s$	11	−.13524	$3d^6(^4)$ ^{1}D $5s$	19	−.01396	$3d^6(^4)$ ^{3}F $5d$
1	−.96312	$3d^6$ ^{3}H $4s$	4	−.11635	$3d^6(^0)$ ^{1}S $4s$	12	−.13158	$3d^6$ ^{3}D $5s$	20	−.00002	$3d^6$ ^{3}G $5d$
2	−.24042	$3d^6$ ^{3}H $5s$	5	−.09100	$3d^6$ ^{3}D $4d$	13	−.09044	$3d^6(^4)$ ^{1}G $4d$			^{2}F^o
3	−.21035	$3d^6$ ^{3}H $4d$	6	−.06154	$3d^6(^4)$ ^{1}D $4d$	14	−.07730	$3d^6$ ^{3}D $4d$	1	−.57940	$3d^6(^4)$ ^{3}F $4p$
4	−.19391	$3d^6(^4)$ ^{3}F $4d$			^{2}S^o	15	−.06526	$3d^6(^4)$ ^{1}S $4d$	2	−.52783	$3d^6$ ^{3}G $4p$
5	−.18298	$3d^6$ ^{3}G $4d$	1	−.54914	$3d^6(^4)$ ^{3}P $4p$	16	−.05726	$3d^6(^4)$ ^{1}D $4d$	3	−.49071	$3d^6(^4)$ ^{1}G $4p$
6	−.16021	$3d^6$ ^{5}D $5g$	2	−.32096	$3d^6(^2)$ ^{3}P $4p$	17	−.04202	$3d^6(^4)$ ^{3}P $5d$	4	−.46058	$3d^6$ ^{3}D $4p$
7	−.11128	$3d^6$ ^{5}D $6g$	3	−.21231	$3d^54s$ ^{3}P $4p$	18	−.00849	$3d^6(^4)$ ^{3}F $5d$	5	−.42123	$3d^6(^4)$ ^{1}D $4p$
8	−.08175	$3d^6$ ^{5}D $7g$	4	−.17194	$3d^6(^4)$ ^{3}P $5p$	19	−.00494	$3d^6$ ^{3}G $5d$	6	−.34108	$3d^6$ ^{1}F $4p$
9	−.08163	$3d^6$ ^{5}D $7i$	5	−.02799	$3d^6(^4)$ ^{3}P $6p$			^{2}D^o	7	−.31797	$3d^6(^2)$ ^{3}F $4p$
10	−.06259	$3d^6$ ^{5}D $8g$	6	−.02373	$3d^6(^4)$ ^{3}F $4f$	1	−.59442	$3d^6(^4)$ ^{3}P $4p$	8	−.25818	$3d^54s$ ^{3}G $4p$
11	−.06250	$3d^6$ ^{5}D $8i$			^{2}P^e	2	−.54977	$3d^6(^4)$ ^{3}F $4p$	9	−.23497	$3d^6(^2)$ ^{1}G $4p$
12	−.04945	$3d^6$ ^{5}D $9g$	1	−.91481	$3d^7$	3	−.47473	$3d^6$ ^{3}D $4p$	10	−.21007	$3d^54s$ ^{3}D $4p$
13	−.04938	$3d^6$ ^{5}D $9i$	2	−.87635	$3d^6(^4)$ ^{3}P $4s$	4	−.41343	$3d^6(^4)$ ^{1}D $4p$	11	−.17241	$3d^6(^2)$ ^{1}D $4p$
		^{4}I^o	3	−.59748	$3d^6(^2)$ ^{3}P $4s$	5	−.36300	$3d^6$ ^{1}F $4p$	12	−.14166	$3d^6(^4)$ ^{3}F $5p$
1	−.61945	$3d^6$ ^{3}H $4p$	4	−.27497	$3d^6(^4)$ ^{3}P $5s$	6	−.28505	$3d^6(^2)$ ^{3}P $4p$	13	−.13462	$3d^54s$ ^{3}D $4p$
2	−.28104	$3d^54s$ ^{3}I $4p$	5	−.20947	$3d^6(^4)$ ^{3}P $4d$	7	−.26980	$3d^6(^2)$ ^{3}F $4p$	14	−.13041	$3d^6$ ^{3}G $5p$
3	−.17512	$3d^6$ ^{3}H $5p$	6	−.16481	$3d^6(^4)$ ^{3}F $4d$	8	−.25212	$3d^54s$ ^{3}P $4p$	15	−.08667	$3d^54s$ ^{3}F $4p$
4	−.14439	$3d^54s$ ^{3}H $4p$	7	−.08924	$3d^6$ ^{3}D $4d$	9	−.23790	$3d^54s$ ^{3}D $4p$	16	−.06860	$3d^6(^4)$ ^{3}P $4f$
5	−.11112	$3d^6$ ^{5}D $6h$	8	−.07151	$3d^6(^4)$ ^{3}P $6s$	10	−.19973	$3d^6(^2)$ ^{1}D $4p$	17	−.06691	$3d^54s$ ^{3}G $4p$
6	−.08165	$3d^6$ ^{5}D $7h$	9	−.05875	$3d^6(^4)$ ^{1}D $4d$	11	−.18399	$3d^6(^4)$ ^{3}P $5p$	18	−.04619	$3d^6(^4)$ ^{1}G $5p$
7	−.06251	$3d^6$ ^{5}D $8h$	10	−.04055	$3d^6(^4)$ ^{3}P $5d$	12	−.13593	$3d^6(^4)$ ^{3}F $5p$	19	−.04072	$3d^6(^4)$ ^{1}D $5p$
8	−.06250	$3d^6$ ^{5}D $8j$	11	−.00866	$3d^6(^4)$ ^{3}F $5d$	13	−.12919	$3d^54s$ ^{3}D $4p$	20	−.03746	$3d^6$ ^{3}D $5p$
9	−.04939	$3d^6$ ^{5}D $9h$	12	−.00611	$3d^6$ ^{1}F $4d$	14	−.07531	$3d^54s$ ^{3}F $4p$	21	−.02636	$3d^6$ ^{3}H $4f$
10	−.04938	$3d^6$ ^{5}D $9j$			^{2}P^o	15	−.06851	$3d^6(^4)$ ^{3}P $4f$	22	−.02586	$3d^54s$ ^{1}D $4p$
		^{4}I^e	1	−.56325	$3d^6(^4)$ ^{3}P $4p$	16	−.06262	$3d^54s$ ^{1}D $4p$	23	−.02511	$3d^6(^4)$ ^{3}F $4f$
1	−.21063	$3d^6$ ^{3}H $4d$	2	−.48520	$3d^6$ ^{3}D $4p$	17	−.04097	$3d^6(^4)$ ^{1}D $5p$	24	−.02331	$3d^6$ ^{3}G $4f$
2	−.18120	$3d^6$ ^{3}G $4d$	3	−.44289	$3d^6(^4)$ ^{1}S $4p$	18	−.03944	$3d^6$ ^{3}D $5p$	25	−.00776	$3d^6(^2)$ ^{1}D $5p$
3	−.15997	$3d^6$ ^{5}D $5g$	4	−.41303	$3d^6(^4)$ ^{1}D $4p$	19	−.03357	$3d^6(^4)$ ^{3}P $6p$			^{2}G^e
4	−.11110	$3d^6$ ^{5}D $6g$	5	−.29024	$3d^6(^2)$ ^{3}P $4p$	20	−.02666	$3d^6$ ^{3}H $4f$	1	−.93801	$3d^7$
5	−.08164	$3d^6$ ^{5}D $7i$	6	−.25333	$3d^54s$ ^{3}P $4p$	21	−.02582	$3d^54s$ ^{1}F $4p$:	2	−.86666	$3d^6$ ^{3}G $4s$
6	−.08162	$3d^6$ ^{5}D $7g$	7	−.20010	$3d^54s$ ^{3}D $4p$	22	−.02460	$3d^6(^4)$ ^{3}F $4f$	3	−.83069	$3d^6(^4)$ ^{1}G $4s$
7	−.06250	$3d^6$ ^{5}D $8g$	8	−.17675	$3d^6(^4)$ ^{3}P $5p$	23	−.02347	$3d^6$ ^{3}G $4f$	4	−.55133	$3d^6(^2)$ ^{1}G $4s$
8	−.06249	$3d^6$ ^{5}D $8i$	9	−.14378	$3d^6(^2)$ ^{1}D $4p$:			^{2}F^e	5	−.44588	$3d^54s^2$:
9	−.04939	$3d^6$ ^{5}D $9g$	10	−.04875	$3d^6(^4)$ ^{1}S $5p$	1	−.89053	$3d^6(^4)$ ^{3}F $4s$	6	−.23096	$3d^6$ ^{3}G $5s$
10	−.04938	$3d^6$ ^{5}D $9i$	11	−.04315	$3d^6$ ^{3}D $5p$	2	−.78422	$3d^7$	7	−.20368	$3d^54s$ ^{3}G $4s$:
		^{4}J^o	12	−.03767	$3d^6(^4)$ ^{1}D $5p$	3	−.69067	$3d^6$ ^{1}F $4s$	8	−.20083	$3d^6$ ^{3}H $4d$
1	−.28522	$3d^54s$ ^{3}I $4p$	13	−.02864	$3d^6(^4)$ ^{3}P $6p$	4	−.59969	$3d^6(^2)$ ^{3}F $4s$	9	−.18489	$3d^6(^4)$ ^{3}F $4d$
2	−.11111	$3d^6$ ^{5}D $6h$	14	−.02440	$3d^6(^4)$ ^{3}F $4f$	5	−.48811	$3d^54s^2$:	10	−.16377	$3d^6$ ^{3}G $4d$
3	−.08163	$3d^6$ ^{5}D $7h$	15	−.02392	$3d^6$ ^{3}G $4f$	6	−.40460	$3d^54s$ ^{3}F $4s$:	11	−.13675	$3d^6(^4)$ ^{1}G $5s$
4	−.06250	$3d^6$ ^{5}D $8h$	16	−.01225	$3d^54s$ ^{3}D $5p$.	7	−.23136	$3d^6(^4)$ ^{3}F $5s$	12	−.10662	$3d^6$ ^{1}I $4d$
5	−.06250	$3d^6$ ^{5}D $8j$			^{2}D^e	8	−.21856	$3d^6(^4)$ ^{3}P $4d$	13	−.09547	$3d^6(^4)$ ^{1}G $4d$
6	−.04938	$3d^6$ ^{5}D $9j$	1	−.88960	$3d^7(^3)$	9	−.20660	$3d^6$ ^{3}H $4d$	14	−.07224	$3d^6$ ^{3}D $4d$
7	−.04938	$3d^6$ ^{5}D $9h$	2	−.79661	$3d^6$ ^{3}D $4s$	10	−.19197	$3d^6(^4)$ ^{3}F $4d$	15	−.05851	$3d^6(^4)$ ^{1}D $4d$
		^{4}J^e	3	−.75646	$3d^6(^4)$ ^{1}D $4s$	11	−.15192	$3d^6$ ^{3}G $4d$	16	−.02761	$3d^6$ ^{3}G $6s$
1	−.21203	$3d^6$ ^{3}H $4d$	4	−.61277	$3d^7(^1)$	12	−.09672	$3d^6(^4)$ ^{1}G $4d$	17	−.01511	$3d^6$ ^{3}H $5d$
2	−.08163	$3d^6$ ^{5}D $7i$	5	−.49670	$3d^54s$ ^{3}D $4s$:	13	−.08254	$3d^6$ ^{3}D $4d$	18	−.01111	$3d^6(^4)$ ^{3}F $5d$

Mn-like Fe (Fe^{+})

i	Energy(Ryds)	Description	i	Energy(Ryds)	Description	i	Energy(Ryds)	Description
		^{2}**G**$^\circ$	4	−.48797	$3d^6$ ^{1}I $4p$			^{2}**J**$^\circ$
1	−.60716	$3d^6$ ^{3}H $4p$	5	−.31807	$3d^54s$ ^{3}G $4p$	1	−.52106	$3d^6$ ^{1}I $4p$
2	−.57130	$3d^6(^4)$ ^{3}F $4p$	6	−.23936	$3d^6(^2)$ ^{1}G $4p$	2	−.24091	$3d^54s$ ^{3}I $4p$
3	−.52623	$3d^6$ ^{3}G $4p$	7	−.21584	$3d^54s$ ^{3}I $4p$	3	−.05970	$3d^6$ ^{1}I $5p$
4	−.48825	$3d^6(^4)$ ^{1}G $4p$	8	−.14148	$3d^6$ ^{3}G $5p$	4	−.04262	$3d^54s$ ^{1}I $4p$
5	−.37096	$3d^6$ ^{1}F $4p$	9	−.13696	$3d^6$ ^{3}H $5p$	5	−.02580	$3d^6$ ^{3}H $4f$
6	−.28862	$3d^6(^2)$ ^{3}F $4p$	10	−.11036	$3d^54s$ ^{1}I $4p$	6	−.02308	$3d^6$ ^{3}G $4f$
7	−.27405	$3d^54s$ ^{3}G $4p$	11	−.08692	$3d^54s$ ^{3}H $4p$			
8	−.20898	$3d^6(^2)$ ^{1}G $4p$	12	−.04745	$3d^6$ ^{1}I $5p$			
9	−.18227	$3d^54s$ ^{3}F $4p$	13	−.04317	$3d^6(^4)$ ^{1}G $5p$			
10	−.14523	$3d^6$ ^{3}H $5p$	14	−.02528	$3d^6$ ^{3}H $4f$			
11	−.13990	$3d^6(^4)$ ^{3}F $5p$	15	−.02498	$3d^6(^4)$ ^{3}F $4f$			
12	−.13317	$3d^54s$ ^{3}H $4p$	16	−.02414	$3d^6$ ^{3}G $4f$			
13	−.13096	$3d^6$ ^{3}G $5p$	17	−.01844	$3d^54s$ ^{3}G $4p$			
14	−.09921	$3d^54s$ ^{1}F $4p$			^{2}**I**^{e}			
15	−.07797	$3d^54s$ ^{3}F $4p$	1	−.84808	$3d^6$ ^{1}I $4s$			
16	−.06891	$3d^6(^4)$ ^{3}P $4f$	2	−.56106	$3d^54s$ ^{3}I $0s$			
17	−.04382	$3d^6(^4)$ ^{1}G $5p$	3	−.20614	$3d^6$ ^{3}H $4d$			
18	−.03856	$3d^54s$ ^{3}F $4p$	4	−.17869	$3d^6$ ^{3}G $4d$			
19	−.02574	$3d^6$ ^{3}H $4f$	5	−.13747	$3d^6$ ^{1}I $5s$			
20	−.02509	$3d^6(^4)$ ^{3}F $4f$	6	−.10571	$3d^6$ ^{1}I $4d$			
21	−.02370	$3d^6$ ^{3}G $4f$	7	−.09731	$3d^6(^4)$ ^{1}G $4d$			
		^{2}**H**^{e}	8	−.01637	$3d^6$ ^{3}H $5d$			
1	−.92221	$3d^7$	9	−.00990	$3d^6$ ^{3}G $5d$			
2	−.89585	$3d^6$ ^{3}H $4s$			^{2}**I**$^\circ$			
3	−.45881	$3d^54s^2$:	1	−.61121	$3d^6$ ^{3}H $4p$			
4	−.23235	$3d^6$ ^{3}H $5s$	2	−.49349	$3d^6$ ^{1}I $4p$			
5	−.19173	$3d^6(^4)$ ^{3}F $4d$	3	−.20576	$3d^54s$ ^{3}I $4p$			
6	−.18960	$3d^6$ ^{3}H $4d$	4	−.14378	$3d^6$ ^{3}H $5p$			
7	−.16245	$3d^6$ ^{3}G $4d$	5	−.12685	$3d^54s$ ^{1}I $4p$			
8	−.10045	$3d^6(^4)$ ^{1}G $4d$	6	−.05462	$3d^6$ ^{1}I $5p$			
9	−.08387	$3d^6$ ^{1}I $4d$	7	−.02656	$3d^6$ ^{3}H $4f$			
10	−.02807	$3d^6$ ^{3}H $6s$	8	−.02520	$3d^6(^4)$ ^{3}F $4f$			
11	−.01340	$3d^6(^4)$ ^{3}F $5d$	9	−.02445	$3d^6$ ^{3}G $4f$			
12	−.01274	$3d^6$ ^{3}G $5d$	10	−.02144	$3d^54s$ ^{1}H $4p$			
13	−.00158	$3d^6$ ^{3}H $5d$			^{2}**J**^{e}			
		^{2}**H**$^\circ$	1	−.20918	$3d^6$ ^{3}H $4d$			
1	−.57716	$3d^6$ ^{3}H $4p$	2	−.10625	$3d^6$ ^{1}I $4d$			
2	−.55812	$3d^6$ ^{3}G $4p$	3	−.01751	$3d^6$ ^{3}H $5d$			
3	−.50120	$3d^6(^4)$ ^{1}G $4p$						

Mn-like Fe (Fe^{+})

Energies in ascending order from ground state for terms with effective $n \leq 4.0$, $L \leq 7$

Term	i	E(Ryds)	Term	i	E(Ryds)	Term	i	E(Ryds)	Term	i	E(Ryds)	Term	i	E(Ryds)
$^6\mathbf{D}^e$	1	0.00000	$^4\mathbf{P}^e$	3	0.52676	$^2\mathbf{J}^o$	1	0.65718	$^6\mathbf{S}^e$	2	0.76549	$^2\mathbf{P}^o$	5	0.88800
$^4\mathbf{D}^e$	1	0.07222	$^4\mathbf{F}^e$	3	0.53206	$^2\mathbf{H}^o$	3	0.67704	$^6\mathbf{D}^o$	2	0.76629	$^2\mathbf{G}^o$	6	0.88962
$^4\mathbf{F}^e$	1	0.09749	$^4\mathbf{P}^e$	4	0.54509	$^4\mathbf{P}^o$	4	0.67737	$^6\mathbf{S}^o$	1	0.77040	$^4\mathbf{P}^o$	7	0.89058
$^6\mathbf{S}^e$	1	0.18314	$^4\mathbf{I}^o$	1	0.55879	$^4\mathbf{F}^o$	4	0.67757	$^2\mathbf{F}^e$	6	0.77364	$^4\mathbf{J}^o$	1	0.89302
$^4\mathbf{P}^e$	1	0.19366	$^4\mathbf{H}^o$	1	0.55891	$^2\mathbf{D}^e$	5	0.68154	$^6\mathbf{P}^o$	4	0.78159	$^2\mathbf{D}^o$	6	0.89319
$^4\mathbf{H}^e$	1	0.21512	$^4\mathbf{D}^e$	3	0.55911	$^4\mathbf{D}^o$	4	0.68201	$^4\mathbf{F}^e$	5	0.78498	$^4\mathbf{P}^e$	6	0.89499
$^2\mathbf{G}^e$	1	0.24023	$^4\mathbf{G}^o$	1	0.56146	$^2\mathbf{I}^o$	2	0.68475	$^6\mathbf{D}^o$	3	0.79933	$^4\mathbf{F}^o$	8	0.89604
$^4\mathbf{F}^e$	2	0.24557	$^2\mathbf{D}^e$	4	0.56547	$^2\mathbf{F}^o$	3	0.68753	$^6\mathbf{F}^o$	3	0.80264	$^4\mathbf{I}^o$	2	0.89719
$^4\mathbf{P}^e$	2	0.24695	$^2\mathbf{I}^o$	1	0.56703	$^4\mathbf{F}^e$	4	0.68988	$^4\mathbf{P}^e$	5	0.80356	$^4\mathbf{D}^o$	9	0.90005
$^2\mathbf{H}^e$	1	0.25603	$^4\mathbf{S}^o$	1	0.57078	$^2\mathbf{G}^o$	4	0.68999	$^2\mathbf{G}^o$	5	0.80728	$^2\mathbf{P}^e$	4	0.90327
$^4\mathbf{G}^e$	1	0.26144	$^2\mathbf{G}^o$	1	0.57108	$^2\mathbf{F}^e$	5	0.69013	$^4\mathbf{H}^o$	3	0.80891	$^2\mathbf{G}^o$	7	0.90419
$^2\mathbf{P}^e$	1	0.26343	$^2\mathbf{F}^e$	4	0.57855	$^2\mathbf{H}^o$	4	0.69027	$^6\mathbf{F}^o$	4	0.81078	$^4\mathbf{S}^o$	3	0.90647
$^2\mathbf{H}^e$	2	0.28239	$^4\mathbf{P}^o$	2	0.57855	$^2\mathbf{P}^o$	2	0.69304	$^4\mathbf{F}^o$	5	0.81283	$^2\mathbf{D}^e$	7	0.90691
$^2\mathbf{F}^e$	1	0.28771	$^2\mathbf{P}^e$	3	0.58076	$^6\mathbf{P}^o$	3	0.69973	$^4\mathbf{D}^o$	5	0.81513	$^6\mathbf{D}^e$	4	0.90692
$^2\mathbf{D}^e$	1	0.28864	$^4\mathbf{F}^o$	2	0.58351	$^2\mathbf{D}^o$	3	0.70351	$^2\mathbf{D}^o$	5	0.81524	$^2\mathbf{D}^o$	7	0.90844
$^2\mathbf{P}^e$	2	0.30189	$^2\mathbf{D}^o$	1	0.58382	$^6\mathbf{D}^e$	2	0.70949	$^4\mathbf{F}^o$	6	0.81643	$^4\mathbf{H}^o$	4	0.91763
$^2\mathbf{G}^e$	2	0.31158	$^4\mathbf{D}^o$	2	0.58867	$^2\mathbf{F}^o$	4	0.71766	$^6\mathbf{P}^o$	5	0.82015	$^2\mathbf{F}^o$	8	0.92006
$^4\mathbf{D}^e$	2	0.32837	$^4\mathbf{D}^o$	3	0.59250	$^6\mathbf{G}^o$	1	0.71935	$^4\mathbf{P}^o$	5	0.82075	$^4\mathbf{P}^o$	8	0.92247
$^2\mathbf{I}^e$	1	0.33016	$^4\mathbf{G}^o$	2	0.59383	$^2\mathbf{H}^e$	3	0.71943	$^2\mathbf{D}^e$	6	0.82811	$^6\mathbf{D}^e$	5	0.92279
$^6\mathbf{D}^o$	1	0.33162	$^2\mathbf{F}^o$	1	0.59884	$^4\mathbf{D}^e$	4	0.72274	$^6\mathbf{D}^o$	4	0.83048	$^2\mathbf{P}^o$	6	0.92491
$^2\mathbf{G}^e$	3	0.34755	$^2\mathbf{H}^o$	1	0.60108	$^2\mathbf{G}^e$	5	0.73236	$^4\mathbf{G}^o$	4	0.83073	$^4\mathbf{D}^e$	6	0.92514
$^6\mathbf{F}^o$	1	0.36050	$^4\mathbf{P}^o$	3	0.60529	$^6\mathbf{H}^o$	1	0.73483	$^6\mathbf{P}^o$	6	0.83391	$^6\mathbf{G}^o$	2	0.92550
$^6\mathbf{P}^o$	1	0.36719	$^4\mathbf{G}^o$	3	0.60594	$^2\mathbf{P}^o$	3	0.73535	$^2\mathbf{S}^e$	2	0.83399	$^6\mathbf{F}^o$	5	0.92604
$^4\mathbf{F}^o$	1	0.38142	$^2\mathbf{G}^o$	2	0.60694	$^6\mathbf{F}^e$	1	0.74702	$^2\mathbf{F}^o$	6	0.83716	$^2\mathbf{D}^o$	8	0.92612
$^2\mathbf{D}^e$	2	0.38163	$^4\mathbf{H}^o$	2	0.61131	$^6\mathbf{D}^e$	3	0.75094	$^4\mathbf{D}^o$	6	0.83815	$^4\mathbf{F}^o$	9	0.92617
$^4\mathbf{D}^o$	1	0.38928	$^4\mathbf{F}^o$	3	0.61376	$^6\mathbf{F}^o$	2	0.75112	$^4\mathbf{P}^o$	6	0.84347	$^4\mathbf{G}^o$	6	0.92624
$^2\mathbf{S}^e$	1	0.38953	$^2\mathbf{P}^o$	1	0.61499	$^4\mathbf{D}^e$	5	0.75574	$^2\mathbf{S}^o$	2	0.85728	$^6\mathbf{D}^o$	5	0.92708
$^2\mathbf{F}^e$	2	0.39402	$^2\mathbf{I}^e$	2	0.61718	$^6\mathbf{G}^e$	1	0.75594	$^2\mathbf{H}^o$	5	0.86017	$^4\mathbf{H}^o$	5	0.92742
$^4\mathbf{P}^o$	1	0.40760	$^2\mathbf{H}^o$	2	0.62012	$^6\mathbf{P}^e$	1	0.75632	$^2\mathbf{F}^o$	7	0.86027	$^4\mathbf{D}^o$	10	0.92760
$^2\mathbf{D}^e$	3	0.42178	$^2\mathbf{G}^e$	4	0.62691	$^2\mathbf{F}^o$	5	0.75701	$^4\mathbf{S}^o$	2	0.86388	$^6\mathbf{H}^o$	2	0.92767
$^8\mathbf{P}^o$	1	0.45286	$^2\mathbf{D}^o$	2	0.62847	$^4\mathbf{G}^e$	3	0.75870	$^4\mathbf{D}^o$	7	0.86825			
$^4\mathbf{G}^e$	2	0.48691	$^2\mathbf{S}^o$	1	0.62910	$^4\mathbf{S}^e$	1	0.76406	$^4\mathbf{G}^o$	5	0.86978			
$^2\mathbf{F}^e$	3	0.48757	$^2\mathbf{F}^o$	2	0.65041	$^2\mathbf{D}^o$	4	0.76481	$^4\mathbf{F}^o$	7	0.88016			
$^6\mathbf{P}^o$	2	0.51880	$^2\mathbf{G}^o$	3	0.65201	$^2\mathbf{P}^o$	4	0.76521	$^4\mathbf{D}^o$	8	0.88303			

Mn-like Fe (Fe^{+})

gf-values for transitions involving terms with effective $n \leq 4.0$, $L \leq 7$

i i'	gf$_L$	i i'	gf$_L$	i i'	gf$_L$	i i'	gf$_L$	i i'	gf$_L$	i i'	gf$_L$
	$^{6}S^{e}$–$^{6}P^{o}$	5 3	2.05E+0	4 2	1.05E−3	3 1	0.00E+0	1 3	−3.13E−4	1 5	−1.47E−5
1 1	−2.66E−1	5 4	−2.83E−1	4 3	2.93E−2	3 2	0.00E+0	1 4	−5.57E−1	1 6	−5.70E−1
1 2	−9.18E−2	5 5	−6.27E+0	4 4	2.21E+0	3 3	−5.99E+0	1 5	−3.24E−2	1 7	−2.87E−1
1 3	−1.04E+1	6 1	2.07E−2	4 5	−1.39E−1		$^{6}P^{e}$–$^{6}P^{o}$	1 6	−2.05E−1	1 8	−7.78E−3
1 4	−3.55E−1	6 2	2.69E−3	5 1	2.91E−1	1 1	4.11E+1	1 7	−3.03E−2	1 9	−5.50E−2
1 5	−5.34E−1	6 3	4.15E−4	5 2	2.19E−1	1 2	0.00E+0	1 8	−1.32E−3	1 10	−5.47E−1
1 6	−2.82E−2	6 4	−4.80E−3	5 3	4.86E−1	1 3	0.00E+0	2 1	−1.05E−1	2 1	−6.44E−4
2 1	2.88E+0	6 5	−3.81E−3	5 4	9.59E+0	2 1	0.00E+0	2 2	−2.33E+0	2 2	−4.29E+0
2 2	1.87E−1		$^{6}D^{e}$–$^{6}D^{o}$	5 5	−3.71E−2	2 2	4.11E+1	2 3	−3.85E−6	2 3	−1.87E−2
2 3	3.31E−2	1 1	−8.37E+0		$^{6}D^{o}$–$^{6}F^{e}$	2 3	0.00E+0	2 4	−1.50E−1	2 4	−1.89E−3
2 4	−3.46E−2	1 2	−1.03E−1	1 1	−1.53E+1	3 1	0.00E+0	2 5	−1.55E−2	2 5	−1.10E−2
2 5	−1.05E+0	1 3	−4.80E−1	2 1	4.29E−3	3 2	0.00E+0	2 6	−6.29E−1	2 6	−1.08E−1
2 6	−2.10E−3	1 4	−2.42E+0	3 1	4.32E+0	3 3	4.11E+1	2 7	−3.05E−2	2 7	−7.75E−3
	$^{6}S^{o}$–$^{6}P^{e}$	1 5	−1.05E−3	4 1	5.46E−1		$^{6}P^{o}$–$^{6}D^{e}$	2 8	−7.13E−2	2 8	−1.45E−1
1 1	5.70E−5	2 1	4.28E+0	5 1	8.62E−1	1 1	−5.51E+1	3 1	2.58E−3	2 9	−6.26E−3
	$^{6}P^{e}$–$^{6}P^{o}$	2 2	−2.72E−2		$^{6}F^{e}$–$^{6}F^{o}$	1 2	0.00E+0	3 2	−1.65E−2	2 10	−1.38E−1
1 1	7.08E+0	2 3	−1.22E+1	1 1	8.23E+0	2 1	−1.83E−9	3 3	−2.55E−4	3 1	2.17E−4
1 2	9.22E−2	2 4	−1.66E+0	1 2	−6.88E−3	2 2	−5.51E+1	3 4	−1.00E−2	3 2	−4.63E−7
1 3	2.41E−1	2 5	−2.58E−3	1 3	−5.78E−2	3 1	−1.61E−9	3 5	−5.09E−2	3 3	−2.11E−2
1 4	−1.43E−1	3 1	9.21E+0	1 4	−2.60E+0	3 2	−5.31E−9	3 6	−8.41E−5	3 4	−1.19E−3
1 5	−2.24E+0	3 2	−8.96E−3	1 5	−9.67E+0		$^{4}S^{e}$–$^{4}P^{o}$	3 7	−1.90E+0	3 5	−3.16E−3
1 6	−4.48E−3	3 3	−2.47E+0		$^{6}F^{e}$–$^{6}G^{o}$	1 1	2.15E+0	3 8	−5.80E−1	3 6	−6.97E−1
	$^{6}P^{e}$–$^{6}D^{o}$	3 4	−3.28E−1	1 1	3.75E−3	1 2	3.47E−4	4 1	4.43E−4	3 7	−6.99E−2
1 1	2.07E+0	3 5	−9.24E+0	1 2	−3.44E+1	1 3	7.49E−2	4 2	−4.19E−4	3 8	−3.88E−2
1 2	−1.38E−5	4 1	3.02E−5		$^{6}F^{o}$–$^{6}G^{e}$	1 4	3.65E−2	4 3	−3.29E−4	3 9	−2.34E+0
1 3	−6.03E−1	4 2	9.58E−3	1 1	−2.91E+1	1 5	−7.04E−1	4 4	−5.63E−2	3 10	−4.81E−3
1 4	−1.08E−1	4 3	3.01E+0	2 1	−2.56E−2	1 6	−1.09E−2	4 5	−1.97E−3	4 1	1.48E−3
1 5	−1.31E+1	4 4	4.82E−1	3 1	3.84E−1	1 7	−2.08E−3	4 6	−1.12E−1	4 2	−4.83E−4
	$^{6}P^{o}$–$^{6}D^{e}$	4 5	−5.90E−2	4 1	9.22E+0	1 8	−5.24E−2	4 7	−4.05E−1	4 3	−2.95E−3
1 1	3.58E+0	5 1	6.27E−1	5 1	4.38E−1		$^{4}P^{e}$–$^{4}S^{o}$	4 8	−5.01E−1	4 4	−4.41E−2
1 2	−2.72E+0	5 2	6.27E−3		$^{6}G^{e}$–$^{6}G^{o}$	1 1	−3.32E−1	5 1	3.87E+0	4 5	−6.16E−3
1 3	−5.71E+0	5 3	2.56E+0	1 1	6.34E−5	1 2	−4.45E−1	5 2	1.69E−4	4 6	−1.21E+0
1 4	−1.52E−1	5 4	3.31E−1	1 2	−7.22E+0	1 3	−3.33E−2	5 3	2.57E−2	4 7	−3.23E−2
1 5	−5.00E−1	5 5	−1.90E−2		$^{6}G^{e}$–$^{6}H^{o}$	2 1	−7.39E−1	5 4	2.31E−2	4 8	−6.63E−2
2 1	2.45E+0		$^{6}D^{e}$–$^{6}F^{o}$	1 1	9.35E−3	2 2	−2.20E−2	5 5	−8.04E−1	4 9	−9.06E−1
2 2	−7.19E−2	1 1	−1.13E+1	1 2	−5.16E+1	2 3	−3.85E−1	5 6	−1.55E−2	4 10	−1.33E−3
2 3	−3.32E−1	1 2	−4.40E+0		$^{6}S^{e}$–$^{6}S^{o}$	3 1	−8.90E−3	5 7	−8.11E−2	5 1	1.10E+0
2 4	−5.45E−1	1 3	−4.62E−1	1 1	−1.46E+1	3 2	−9.62E−2	5 8	−2.16E−2	5 2	1.27E−2
2 5	−8.86E−2	1 4	−9.42E−2	1 2	−1.43E+0	3 3	−2.79E−1	6 1	1.73E−4	5 3	9.56E−3
3 1	5.10E−1	1 5	−9.31E−4	1 3	−3.67E−1	4 1	−8.06E−5	6 2	1.68E+0	5 4	3.87E−2
3 2	−2.94E−2	2 1	6.59E+0		$^{6}S^{e}$–$^{6}P^{o}$	4 2	−2.64E−1	6 3	5.36E−3	5 5	−1.85E−1
3 3	−5.65E−2	2 2	−1.78E−1	1 1	−2.80E+2	4 3	−5.55E−1	6 4	7.86E−3	5 6	−1.32E−2
3 4	−6.96E−2	2 3	−7.51E−1	1 2	−2.74E+1	5 1	2.49E−4	6 5	2.84E−6	5 7	−1.01E−4
3 5	−1.50E−1	2 4	−1.88E+1	1 3	−7.03E+0	5 2	−1.29E−3	6 6	1.69E−2	5 8	−6.34E−3
4 1	2.35E+0	2 5	−4.42E−3		$^{6}S^{o}$–$^{6}P^{e}$	5 3	−1.44E−3	6 7	6.09E−4	5 9	−2.45E−2
4 2	5.40E−1	3 1	9.63E−1	1 1	−5.99E+0	6 1	5.56E−1	6 8	−1.74E−3	5 10	−9.05E+0
4 3	1.13E−1	3 2	−2.88E−5	1 2	0.00E+0	6 2	1.49E−3		$^{4}P^{e}$–$^{4}D^{o}$	6 1	1.09E−2
4 4	−5.79E−2	3 3	−4.82E−2	1 3	0.00E+0	6 3	−4.30E−3	1 1	−2.26E−1	6 2	2.94E+0
4 5	−4.71E−1	3 4	−3.88E−1	2 1	0.00E+0		$^{4}P^{e}$–$^{4}P^{o}$	1 2	−1.04E+0	6 3	1.55E−2
5 1	4.10E−1	3 5	−2.24E+1	2 2	−5.99E+0	1 1	−7.03E−1	1 3	−4.62E−1	6 4	1.55E−2
5 2	7.97E+0	4 1	1.89E+0	2 3	−1.72E−9	1 2	−6.14E−1	1 4	−3.84E−1	6 5	9.05E−3

Mn-like Fe (Fe^{+})

i	*i′*	gf$_L$	*i*	*i′*	gf$_L$	*i*	*i′*	gf$_L$	*i*	*i′*	gf$_L$	*i*	*i′*	gf$_L$	*i*	*i′*	gf$_L$
6	6	4.45E−3	8	5	7.97E−3	5	10	−6.32E+0	5	5	−2.73E−1	8	3	4.39E−2	5	3	6.35E−3
6	7	1.00E−3	8	6	−5.05E−4	6	1	5.22E−1	5	6	−8.71E−2	8	4	2.47E−2	5	4	5.25E−4
6	8	1.02E−3			**$^4D^e$–$^4D^o$**	6	2	5.95E−3	5	7	−3.17E−2	8	5	2.85E−3	5	5	−7.61E−1
6	9	−2.86E−4	1	1	−5.36E+0	6	3	6.32E−3	5	8	−2.66E−3	9	1	9.11E−2	5	6	−2.80E−1
6	10	−2.42E−4	1	2	−9.01E−2	6	4	1.42E−4	5	9	−1.53E+1	9	2	6.33E−2	5	7	−1.24E−2
		$^4P^o$–$^4D^e$	1	3	−9.96E−2	6	5	5.62E+0	6	1	7.27E−1	9	3	1.01E+0	5	8	−1.31E−2
1	1	2.69E+0	1	4	−6.09E−3	6	6	7.64E−3	6	2	5.26E−5	9	4	2.04E−1	5	9	−6.58E+0
1	2	1.24E−2	1	5	−4.68E−2	6	7	1.20E−3	6	3	4.04E−3	9	5	2.80E−4			**$^4F^e$–$^4G^o$**
1	3	−1.44E−2	1	6	−4.17E−3	6	8	6.62E−3	6	4	2.32E−4	10	1	2.55E−2	1	1	−2.15E+0
1	4	−1.73E+0	1	7	−3.58E−3	6	9	1.82E−4	6	5	5.45E+0	10	2	1.58E−3	1	2	−2.63E−1
1	5	−4.01E+0	1	8	−4.53E−1	6	10	−5.70E−4	6	6	1.81E+0	10	3	4.37E−3	1	3	−4.91E−1
1	6	−3.38E−1	1	9	−9.20E−4			**$^4D^e$–$^4F^o$**	6	7	1.97E−3	10	4	1.67E−4	1	4	−1.37E−3
2	1	3.05E−3	1	10	−3.21E−4	1	1	−7.04E+0	6	8	1.17E−2	10	5	5.73E−1	1	5	−3.08E−2
2	2	2.31E−2	2	1	−4.27E−3	1	2	−1.44E−3	6	9	−7.70E−5			**$^4F^e$–$^4F^o$**	1	6	−1.18E+0
2	3	4.18E−3	2	2	−4.24E−4	1	3	−9.34E−2			**$^4D^o$–$^4F^e$**	1	1	−8.90E−1	2	1	−1.94E+0
2	4	−5.03E−4	2	3	−4.06E−1	1	4	−1.02E−3	1	1	1.81E+0	1	2	−4.93E−1	2	2	−6.63E+0
2	5	−5.96E−4	2	4	−4.34E+0	1	5	−5.20E−1	1	2	8.19E−3	1	3	−2.41E+0	2	3	−9.90E−1
2	6	−1.02E−3	2	5	−1.19E−2	1	6	−3.12E−1	1	3	−2.43E−6	1	4	−1.77E−2	2	4	−8.87E−1
3	1	2.04E+0	2	6	−1.24E−1	1	7	−1.33E−3	1	4	−4.02E−2	1	5	−5.20E−2	2	5	−9.38E−3
3	2	7.78E−3	2	7	−3.19E−1	1	8	−9.48E−3	1	5	−9.03E+0	1	6	−6.46E−2	2	6	−1.20E−2
3	3	3.15E−3	2	8	−1.11E+0	1	9	−1.98E−4	2	1	8.78E−1	1	7	−1.85E−2	3	1	−1.23E−2
3	4	−1.64E−2	2	9	−3.91E−4	2	1	−2.09E−5	2	2	5.24E−2	1	8	−1.30E−1	3	2	−1.07E−2
3	5	−1.15E−1	2	10	−7.43E−6	2	2	−3.77E−1	2	3	1.12E−3	1	9	−3.50E−1	3	3	−2.10E−2
3	6	−6.51E−3	3	1	7.01E−2	2	3	−3.65E−1	2	4	−3.51E−3	2	1	−1.65E−2	3	4	−6.08E−2
4	1	8.06E−2	3	2	−2.14E−3	2	4	−6.31E+0	2	5	−2.19E−2	2	2	−6.10E+0	3	5	−9.28E+0
4	2	3.07E+0	3	3	−3.23E−4	2	5	−1.94E−1	3	1	1.91E−1	2	3	−5.39E−1	3	6	−3.93E−4
4	3	5.81E−3	3	4	−9.34E−2	2	6	−7.99E−1	3	2	4.86E+0	2	4	−5.73E−2	4	1	3.12E−2
4	4	−3.81E−2	3	5	−5.92E−3	2	7	−3.62E−2	3	3	3.57E−3	2	5	−2.09E−1	4	2	1.67E−3
4	5	−6.37E−2	3	6	−6.15E−3	2	8	−1.86E−2	3	4	−2.14E−4	2	6	−9.81E−1	4	3	9.24E−3
4	6	−1.81E−5	3	7	−4.54E−4	2	9	−2.85E−5	3	5	−6.81E−2	2	7	−9.73E−2	4	4	−3.27E−5
5	1	1.15E−2	3	8	−1.24E−1	3	1	3.96E−3	4	1	2.12E−1	2	8	−3.15E−2	4	5	−5.75E−2
5	2	3.29E−2	3	9	−8.06E−2	3	2	−6.50E−5	4	2	1.82E−1	2	9	−1.80E−3	4	6	−1.21E−3
5	3	5.63E−3	3	10	−9.81E−3	3	3	−4.06E−2	4	3	2.21E−3	3	1	1.92E−3	5	1	2.37E−1
5	4	5.28E+0	4	1	3.46E+0	3	4	−2.79E−3	4	4	−1.75E−4	3	2	−5.21E−3	5	2	8.14E−2
5	5	1.59E+0	4	2	8.31E−2	3	5	−5.98E−4	4	5	−3.38E−3	3	3	−2.48E−2	5	3	6.64E−2
5	6	−3.26E+0	4	3	9.42E−2	3	6	−2.57E−2	5	1	1.96E−1	3	4	−2.13E−3	5	4	−3.26E−3
6	1	3.70E−1	4	4	3.55E−3	3	7	−1.66E−1	5	2	1.02E−2	3	5	−1.24E−3	5	5	−6.21E−4
6	2	3.50E−2	4	5	−9.75E+0	3	8	−3.35E−3	5	3	2.35E−3	3	6	−4.44E−3	5	6	−2.38E+1
6	3	3.14E−2	4	6	−8.75E−3	3	9	−1.68E−2	5	4	1.18E−2	3	7	−3.68E−1			**$^4F^o$–$^4G^e$**
6	4	8.28E−2	4	7	−7.39E−4	4	1	3.98E+0	5	5	2.41E+0	3	8	−6.13E+0	1	1	3.03E−2
6	5	2.23E−2	4	8	−1.90E−2	4	2	1.54E−3	6	1	1.55E−1	3	9	−8.49E−3	1	2	−7.85E−2
6	6	−4.12E−2	4	9	−4.72E−5	4	3	8.16E−2	6	2	2.07E−1	4	1	1.52E−2	1	3	1.02E+1
7	1	3.14E−4	4	10	−2.10E−2	4	4	4.96E−6	6	3	2.35E+0	4	2	2.25E−2	2	1	3.23E−1
7	2	4.37E−2	5	1	6.05E+0	4	5	−9.10E+0	6	4	3.61E−4	4	3	4.57E−3	2	2	4.37E−2
7	3	1.87E−1	5	2	7.65E−2	4	6	−3.25E+0	6	5	3.98E−3	4	4	3.71E−7	2	3	−3.42E−3
7	4	4.08E−2	5	3	9.97E−2	4	7	−2.11E−3	7	1	1.51E−2	4	5	−6.44E−3	3	1	6.21E+0
7	5	7.68E−3	5	4	2.97E−3	4	8	−1.20E−2	7	2	9.43E−1	4	6	−2.37E−3	3	2	1.82E−2
7	6	−4.58E−2	5	5	−1.85E+0	4	9	−1.89E−2	7	3	1.44E+0	4	7	−1.97E−2	3	3	−2.06E−1
8	1	1.63E−2	5	6	−5.49E−4	5	1	8.54E−1	7	4	5.09E−2	4	8	−2.05E−1	4	1	2.63E−1
8	2	6.45E−1	5	7	−1.53E−4	5	2	8.62E−4	7	5	3.13E−3	4	9	−1.24E−3	4	2	1.88E−1
8	3	5.38E−3	5	8	−7.23E−5	5	3	2.28E−2	8	1	9.14E−3	5	1	4.70E+0	4	3	−8.04E−8
8	4	1.37E−3	5	9	−1.35E−2	5	4	1.21E−4	8	2	9.40E−2	5	2	7.17E−3	5	1	3.27E−1

Mn-like Fe (Fe^{+})

i i′	gf$_L$	*i i′*	gf$_L$	*i i′*	gf$_L$	*i i′*	gf$_L$	*i i′*	gf$_L$	*i i′*	gf$_L$
5 2	4.31E−4	3 1	1.15E−3	3 2	−8.37E−2	3 3	0.00E+0	2 2	−4.51E+1	1 2	−7.84E−5
5 3	4.77E+0	4 1	2.57E+0	3 3	−1.31E+2	3 4	0.00E+0	3 1	−1.16E−9	1 3	−1.25E−2
6 1	5.93E−1	5 1	1.75E−1	3 4	−2.59E+1		$^4D^o$–$^4D^e$	3 2	−3.94E−9	1 4	−1.47E−1
6 2	3.51E−1	6 1	4.30E−6	4 1	1.95E−2	1 1	0.00E+0		$^2S^e$–$^2P^o$	1 5	−1.95E−1
6 3	1.68E+0		$^4H^e$–$^4H^o$	4 2	7.67E−6	1 2	0.00E+0	1 1	−2.36E−2	1 6	−2.15E−6
7 1	1.19E+0	1 1	−1.02E+1	4 3	−3.81E−1	1 3	0.00E+0	1 2	−2.08E−1	1 7	−7.28E−3
7 2	8.64E−2	1 2	−8.27E−1	4 4	−7.57E−2	1 4	0.00E+0	1 3	−1.04E+0	1 8	−2.39E−2
7 3	3.96E−3	1 3	−1.25E+0	5 1	6.72E−2	2 1	0.00E+0	1 4	−2.74E−1	2 1	−5.57E−1
8 1	2.10E−1	1 4	−3.86E+0	5 2	1.42E+0	2 2	0.00E+0	1 5	−1.10E−3	2 2	−6.72E−1
8 2	8.40E−2	1 5	−1.61E−3	5 3	−1.40E−1	2 3	0.00E+0	1 6	−2.85E−3	2 3	−1.09E−1
8 3	6.48E−3		$^4H^e$–$^4I^o$	5 4	−1.25E+2	2 4	0.00E+0	2 1	1.10E−3	2 4	−9.71E−2
9 1	3.09E−4	1 1	−1.41E+1	6 1	6.15E−3	3 1	0.00E+0	2 2	1.13E−2	2 5	−4.47E−2
9 2	1.11E−5	1 2	−2.17E+0	6 2	4.63E−2	3 2	0.00E+0	2 3	8.55E−3	2 6	−1.74E−3
9 3	3.19E−1		$^4S^o$–$^4S^e$	6 3	1.38E−5	3 3	0.00E+0	2 4	5.12E−4	2 7	−3.91E−5
	$^4G^e$–$^4G^o$	1 1	−7.15E+0	6 4	−4.03E−1	3 4	0.00E+0	2 5	−2.41E−5	2 8	−3.84E−2
1 1	−3.17E−1	1 2	−5.85E−1		$^4P^o$–$^4P^e$		$^4D^e$–$^4F^o$	2 6	−2.95E−5	3 1	−6.87E−6
1 2	−5.27E−1	1 3	−1.93E−1	1 1	6.69E+0	1 1	−2.19E+2		$^2S^o$–$^2P^e$	3 2	−5.14E−4
1 3	−8.81E+0	2 1	−1.64E−3	1 2	6.94E−5	1 2	−2.14E+1	1 1	8.61E−1	3 3	−7.14E−4
1 4	−5.46E−1	2 2	−4.87E+0	1 3	−1.60E−2	1 3	−5.50E+0	1 2	1.94E−2	3 4	−1.16E−2
1 5	−1.29E−1	2 3	−9.61E−1	1 4	−1.90E−3	2 1	−8.41E−3	1 3	1.03E−3	3 5	−3.79E−2
1 6	−1.19E−5	3 1	1.55E−3	2 1	6.54E−1	2 2	−1.99E+2	1 4	−3.48E−1	3 6	−8.53E−1
2 1	−2.49E−2	3 2	−2.99E−3	2 2	6.07E+0	2 3	−3.05E+1	2 1	3.24E−2	3 7	−5.82E−1
2 2	−1.34E−1	3 3	−5.43E+0	2 3	1.97E−4	3 1	5.25E−1	2 2	8.58E−4	3 8	−6.04E−1
2 3	−1.75E−2	4 1	7.73E+0	2 4	−4.46E−2	3 2	−2.44E−2	2 3	4.06E−1	4 1	1.16E+0
2 4	−8.49E−3	4 2	−1.48E−4	3 1	1.67E−1	3 3	−1.86E+2	2 4	−3.72E−4	4 2	1.95E−1
2 5	−2.46E−1	4 3	−7.50E−3	3 2	9.27E−1	4 1	6.23E−2		$^2P^e$–$^2P^o$	4 3	2.62E−2
2 6	−1.00E−2	5 1	1.25E−5	3 3	5.66E+0	4 2	1.46E+0	1 1	−4.06E−1	4 4	6.50E−4
3 1	4.10E−3	5 2	2.87E−3	3 4	3.59E−4	4 3	−4.56E−2	1 2	−6.26E−2	4 5	1.96E−2
3 2	2.40E−3	5 3	−5.75E−3		$^4P^e$–$^4D^o$		$^4D^o$–$^4F^e$	1 3	−2.39E−3	4 6	2.06E−3
3 3	1.06E−4	6 1	1.16E+0	1 1	−2.03E+2	1 1	−3.37E+0	1 4	−1.65E−1	4 7	−2.47E−4
3 4	−2.31E−5	6 2	1.11E+1	1 2	−1.98E+1	1 2	0.00E+0	1 5	−3.50E−1	4 8	−3.15E−4
3 5	−9.65E−5	6 3	4.97E−7	1 3	−5.09E+0	1 3	0.00E+0	1 6	−1.02E−5		$^2P^o$–$^2D^e$
3 6	−4.88E+0		$^4S^e$–$^4P^o$	2 1	−2.12E−3	2 1	0.00E+0	2 1	−1.11E+0	1 1	3.32E−1
	$^4G^e$–$^4H^o$	1 1	−3.39E+0	2 2	−1.84E+2	2 2	−3.37E+0	2 2	−6.27E−2	1 2	6.18E−3
1 1	−7.85E−1	1 2	−3.28E−1	2 3	−2.82E+1	2 3	0.00E+0	2 3	−1.27E−1	1 3	1.07E−2
1 2	−1.08E+1	1 3	−8.22E−2	3 1	4.85E−1	3 1	0.00E+0	2 4	−2.92E−1	1 4	7.17E−3
1 3	−9.98E−1	2 1	2.13E−7	3 2	−6.06E−3	3 2	0.00E+0	2 5	−2.65E−1	1 5	−3.30E−3
1 4	−2.14E+0	2 2	−1.43E−2	3 3	−1.72E+2	3 3	−3.37E+0	2 6	−7.80E−2	1 6	−3.46E−3
1 5	−4.65E−7	2 3	−3.08E−3	4 1	5.76E−2		$^4F^o$–$^4F^e$	3 1	−3.96E−4	1 7	−1.89E−2
2 1	−7.57E−2	3 1	4.31E−1	4 2	1.35E+0	1 1	−3.90E+1	3 2	−1.85E−2	2 1	3.27E−1
2 2	−1.09E−2	3 2	−9.54E−3	4 3	−1.12E−2	1 2	0.00E+0	3 3	−2.35E−3	2 2	1.11E+0
2 3	−3.20E−1	3 3	−1.53E+2		$^4P^o$–$^4D^e$	1 3	0.00E+0	3 4	−4.33E−4	2 3	2.13E−2
2 4	−2.05E−1		$^4S^o$–$^4P^e$	1 1	0.00E+0	2 1	0.00E+0	3 5	−1.24E−1	2 4	9.73E−3
2 5	−7.51E−2	1 1	−1.56E+2	1 2	0.00E+0	2 2	−3.90E+1	3 6	−1.35E+0	2 5	3.50E−5
3 1	5.85E−5	1 2	−1.69E+1	1 3	0.00E+0	2 3	−1.91E−9	4 1	8.61E−1	2 6	−1.99E−3
3 2	2.31E−4	1 3	−4.66E+0	1 4	0.00E+0	3 1	0.00E+0	4 2	1.47E−3	2 7	−1.81E−5
3 3	−1.05E−2	1 4	−1.91E+0	2 1	0.00E+0	3 2	0.00E+0	4 3	6.43E−2	3 1	4.50E−1
3 4	−1.13E−3	2 1	−3.37E−2	2 2	0.00E+0	3 3	−3.90E+1	4 4	1.16E−2	3 2	5.89E−2
3 5	−3.49E+1	2 2	−1.40E+2	2 3	0.00E+0		$^4F^o$–$^4G^e$	4 5	4.42E−3	3 3	5.03E−1
	$^4G^o$–$^4H^e$	2 3	−2.32E+1	2 4	0.00E+0	1 1	−4.51E+1	4 6	−7.01E−7	3 4	3.97E−3
1 1	7.24E+0	2 4	−7.69E+0	3 1	0.00E+0	1 2	0.00E+0		$^2P^e$–$^2D^o$	3 5	2.18E−3
2 1	1.95E+0	3 1	5.16E−1	3 2	0.00E+0	2 1	−1.34E−9	1 1	−1.80E+0	3 6	−1.29E−4

Mn-like Fe (Fe^+)

i i′	gf_L	*i i′*	gf_L	*i i′*	gf_L	*i i′*	gf_L	*i i′*	gf_L	*i i′*	gf_L
3 7	−4.54E−3	4 6	−6.96E−1	4 1	−7.52E−5	4 2	1.16E−1	3 7	−3.64E−4	4 5	−8.87E−2
4 1	1.98E−1	4 7	−2.78E−1	4 2	−4.68E−2	4 3	1.77E−1	3 8	−6.63E−4	4 6	−2.85E+0
4 2	7.73E−2	4 8	−6.75E−4	4 3	−2.03E−2	4 4	2.22E−3	4 1	−4.76E−3	4 7	−1.99E+0
4 3	1.14E+0	5 1	3.47E−5	4 4	−1.04E−2	4 5	5.56E−3	4 2	−2.96E−4	5 1	2.01E−4
4 4	1.91E−3	5 2	5.26E−3	4 5	−3.39E−2	4 6	−9.32E−6	4 3	−5.37E−4	5 2	1.58E−3
4 5	1.38E−2	5 3	−1.68E−4	4 6	−6.82E−3	5 1	9.23E−5	4 4	−1.63E−3	5 3	9.93E−4
4 6	−2.20E−3	5 4	−2.83E−3	4 7	−2.16E−2	5 2	1.31E−2	4 5	−3.65E−3	5 4	1.43E−5
4 7	−2.79E−4	5 5	−5.81E−3	4 8	−3.48E−1	5 3	2.34E+0	4 6	−2.89E−2	5 5	−1.68E−3
5 1	7.18E−5	5 6	−1.00E−2	5 1	2.70E−4	5 4	2.30E−2	4 7	−2.62E−4	5 6	−3.21E−2
5 2	6.49E−2	5 7	−3.58E−4	5 2	1.65E−3	5 5	1.70E−2	4 8	−3.50E+0	5 7	−3.17E−3
5 3	2.28E−3	5 8	−3.15E−3	5 3	−3.60E−5	5 6	5.84E−3	5 1	1.48E−3	6 1	3.52E−2
5 4	2.99E−2	6 1	1.63E−3	5 4	−4.14E−3	6 1	2.36E−1	5 2	1.42E−3	6 2	1.56E−2
5 5	4.37E−3	6 2	4.67E−4	5 5	−2.11E−3	6 2	3.01E−1	5 3	6.02E−5	6 3	1.60E−3
5 6	1.37E−3	6 3	2.85E−5	5 6	−3.50E−3	6 3	1.01E−1	5 4	−2.88E−3	6 4	1.26E−3
5 7	−1.08E−3	6 4	4.39E−3	5 7	−2.18E−3	6 4	1.18E+0	5 5	−5.40E−3	6 5	−3.45E−3
6 1	3.64E−2	6 5	2.40E−4	5 8	−3.29E−3	6 5	2.27E−2	5 6	−3.76E−8	6 6	−6.70E−3
6 2	3.61E−3	6 6	−2.44E−5	6 1	7.25E−3	6 6	1.13E−4	5 7	−1.08E−2	6 7	−1.82E−3
6 3	6.77E−3	6 7	−5.60E−3	6 2	2.61E−3	7 1	2.41E−1	5 8	−6.30E−3		$^2F^o$–$^2G^e$
6 4	1.98E−1	6 8	−6.28E−4	6 3	1.51E−2	7 2	4.26E−2	6 1	9.67E−3	1 1	3.84E−1
6 5	2.26E−2	7 1	2.12E−2	6 4	4.73E−3	7 3	9.99E−3	6 2	6.56E−4	1 2	5.32E−1
6 6	4.69E−3	7 2	7.79E−3	6 5	6.84E−5	7 4	1.36E+0	6 3	6.45E−6	1 3	1.47E−1
6 7	8.36E−4	7 3	9.69E−3	6 6	−6.20E−4	7 5	6.04E−3	6 4	9.29E−3	1 4	−1.00E−5
	$^2D^e$–$^2D^o$	7 4	4.98E−3	6 7	−1.07E−4	7 6	7.11E−3	6 5	1.24E−4	1 5	−2.74E−2
1 1	−1.04E−2	7 5	8.16E−4	6 8	6.26E−1	8 1	6.40E−5	6 6	−4.38E−3	2 1	2.08E−1
1 2	−3.08E−1	7 6	7.48E−5	7 1	1.92E−2	8 2	1.11E−1	6 7	−4.93E−3	2 2	3.12E+0
1 3	−8.37E−2	7 7	−3.69E−5	7 2	1.34E−5	8 3	7.32E−3	6 8	−3.02E−3	2 3	6.61E−3
1 4	−3.56E−1	7 8	−2.50E−6	7 3	4.40E−4	8 4	9.86E−4		$^2F^e$–$^2G^o$	2 4	6.10E−4
1 5	−2.28E−1		$^2D^e$–$^2F^o$	7 4	2.31E−5	8 5	2.45E−2	1 1	−1.98E−1	2 5	−5.13E−4
1 6	−3.26E−2	1 1	−2.13E−1	7 5	9.13E−4	8 6	7.22E−3	1 2	−4.41E+0	3 1	1.69E−2
1 7	−1.58E−3	1 2	−4.44E−1	7 6	1.56E−4		$^2F^e$–$^2F^o$	1 3	−1.63E−2	3 2	3.10E−2
1 8	−1.05E−4	1 3	−8.27E−1	7 7	3.42E−6	1 1	−2.62E+0	1 4	−8.12E−2	3 3	1.65E+0
2 1	−1.14E−2	1 4	−4.66E−2	7 8	−3.84E−4	1 2	−4.19E−1	1 5	−7.51E−3	3 4	5.67E−4
2 2	−1.57E−3	1 5	−4.78E−3		$^2D^o$–$^2F^e$	1 3	−5.13E−2	1 6	−1.18E−1	3 5	−2.87E−3
2 3	−2.54E+0	1 6	−1.43E−1	1 1	2.90E−1	1 4	−2.23E−1	1 7	−4.85E−1	4 1	1.51E+0
2 4	−1.13E−2	1 7	−1.98E−2	1 2	3.68E−2	1 5	−5.27E−3	2 1	−7.43E−2	4 2	4.78E−4
2 5	−7.59E−2	1 8	−2.02E−2	1 3	3.82E−3	1 6	−1.90E−2	2 2	−4.76E−2	4 3	1.59E+0
2 6	−4.70E−3	2 1	−2.05E−3	1 4	2.11E−4	1 7	−8.05E−1	2 3	−6.81E−1	4 4	1.03E−2
2 7	−2.56E−2	2 2	−8.80E−2	1 5	−1.21E−2	1 8	−1.04E−3	2 4	−1.00E−1	4 5	−9.89E−4
2 8	−3.11E−1	2 3	−1.29E+0	1 6	−3.03E−2	2 1	−5.99E−2	2 5	−1.21E−1	5 1	2.57E−1
3 1	−1.09E−3	2 4	−1.90E+0	2 1	2.28E+0	2 2	−5.69E−1	2 6	−7.03E−2	5 2	1.57E−1
3 2	−1.50E−3	2 5	−1.49E−2	2 2	2.31E−6	2 3	−5.05E−2	2 7	−5.28E−3	5 3	9.87E−2
3 3	−2.34E−4	2 6	−5.52E−5	2 3	2.75E−3	2 4	9.07E−2	3 1	3.02E−6	5 4	5.83E−3
3 4	−2.27E+0	2 7	−5.05E−1	2 4	1.07E−2	2 5	−2.87E−1	3 2	−1.31E−3	5 5	2.16E−3
3 5	−1.06E−1	2 8	−1.89E−2	2 5	−2.45E−3	2 6	−9.62E−1	3 3	−2.27E−4	6 1	4.35E−3
3 6	−1.61E−6	3 1	−1.61E−2	2 6	−1.01E−3	2 7	−4.17E−2	3 4	−1.59E−2	6 2	5.15E−4
3 7	−1.78E−2	3 2	−7.76E−2	3 1	1.23E−1	2 8	−4.58E−1	3 5	−4.34E+0	6 3	7.94E−3
3 8	−6.27E−4	3 3	−2.36E−2	3 2	1.52E−1	3 1	−8.62E−3	3 6	−1.42E−1	6 4	9.88E−2
4 1	−5.73E−6	3 4	−2.86E−2	3 3	9.27E−3	3 2	−3.83E−2	3 7	−3.12E−2	6 5	1.80E−2
4 2	−2.79E−2	3 5	3.46E+0	3 4	8.68E−3	3 3	−4.45E−5	4 1	4.31E−4	7 1	2.91E−2
4 3	−3.12E−2	3 6	−4.31E−2	3 5	3.82E−4	3 4	−5.30E−2	4 2	−9.94E−5	7 2	5.10E−1
4 4	−2.57E−2	3 7	−1.17E−3	3 6	−1.37E−2	3 5	−8.37E−3	4 3	−3.16E−3	7 3	3.26E−4
4 5	−3.92E−2	3 8	−2.06E−2	4 1	2.48E−2	3 6	−3.11E+0	4 4	−4.77E−4	7 4	4.67E−4

Mn-like Fe (Fe^{+})

$i\ i'$	gf_L	$i\ i'$	gf_L	$i\ i'$	gf_L	$i\ i'$	gf_L	$i\ i'$	gf_L	$i\ i'$	gf_L
7 5	7.61E−5	3 2	−8.62E−2	1 2	−4.31E−4	5 4	3.66E−3	7 2	1.07E+0	2 2	−5.63E−1
8 1	2.38E−2	3 3	−5.65E−4	1 3	−6.54E−1	5 5	−1.52E−5	7 3	1.99E−2	3 1	1.17E−3
8 2	3.81E−3	3 4	−4.59E+0	1 4	−2.37E−3		$^{2}\mathbf{G}^{o}-{}^{2}\mathbf{H}^{e}$		$^{2}\mathbf{H}^{e}-{}^{2}\mathbf{H}^{o}$	3 2	2.70E−3
8 3	2.36E−2	3 5	−2.11E−2	1 5	−1.98E−2	1 1	3.39E+0	1 1	−1.57E+0		$^{2}\mathbf{H}^{o}-{}^{2}\mathbf{I}^{e}$
8 4	1.18E−1	3 6	−2.78E−6	2 1	−1.09E+0	1 2	2.36E−1	1 2	−1.04E+0	1 1	4.10E−1
8 5	1.26E−2	3 7	−1.02E−3	2 2	−3.77E+0	1 3	−2.21E−7	1 3	−3.92E−2	1 2	−2.67E−3
	$^{2}\mathbf{G}^{e}-{}^{2}\mathbf{G}^{o}$	4 1	1.95E−4	2 3	−5.35E−1	2 1	2.51E−1	1 4	−2.01E−2	2 1	1.07E−1
1 1	−8.11E−1	4 2	1.40E−4	2 4	−1.76E−1	2 2	2.32E−2	1 5	−9.01E−1	2 2	2.75E−3
1 2	−6.41E−1	4 3	−2.21E−4	2 5	−4.59E−1	2 3	−3.58E−2	2 1	−3.15E+0	3 1	3.18E+0
1 3	−8.24E−1	4 4	−2.13E−3	3 1	−5.07E−3	3 1	1.62E+0	2 2	−8.93E−1	3 2	1.04E−3
1 4	−1.10E+0	4 5	−5.70E−2	3 2	−9.63E−3	3 2	3.39E−1	2 3	−8.32E−1	4 1	2.21E+0
1 5	−1.35E−2	4 6	−3.65E−2	3 3	−2.36E+0	3 3	−3.05E−3	2 4	−1.24E+0	4 2	3.11E−2
1 6	−4.20E−2	4 7	−2.68E−2	3 4	−3.22E+0	4 1	3.39E−1	2 5	−2.39E−1	5 1	1.50E−3
1 7	−4.20E−3	5 1	2.09E−2	3 5	−1.17E−2	4 2	6.37E−3	3 1	1.20E−2	5 2	1.62E−2
2 1	−4.73E−1	5 2	1.27E−2	4 1	2.18E−3	4 3	−2.56E−3	3 2	1.37E−3		$^{2}\mathbf{I}^{e}-{}^{2}\mathbf{I}^{o}$
2 2	−1.74E−3	5 3	2.59E−2	4 2	4.33E−5	5 1	1.86E−1	3 3	1.05E−2	1 1	−1.12E−2
2 3	−3.38E+0	5 4	6.32E−3	4 3	−7.42E−3	5 2	2.59E−1	3 4	1.11E−5	1 2	−6.44E+0
2 4	−5.00E−4	5 5	−5.08E−3	4 4	−1.63E−2	5 3	8.40E−3	3 5	−5.33E−3	2 1	2.58E−2
2 5	−3.38E−2	5 6	−8.63E−4	4 5	−1.62E−6	6 1	1.29E+0		$^{2}\mathbf{H}^{e}-{}^{2}\mathbf{I}^{o}$	2 2	−3.97E−2
2 6	−1.38E−1	5 7	−1.75E−5	5 1	2.56E−3	6 2	1.78E−1	1 1	−6.48E+0		$^{2}\mathbf{I}^{e}-{}^{2}\mathbf{J}^{o}$
2 7	−4.23E−1		$^{2}\mathbf{G}^{e}-{}^{2}\mathbf{H}^{o}$	5 2	6.02E−3	6 3	3.00E−2	1 2	−3.20E−1	1 1	−7.88E+0
3 1	−1.00E−1	1 1	−8.42E−1	5 3	6.93E−3	7 1	8.14E−1	2 1	−4.05E−1	2 1	−6.72E−3

Fe-like Fe

Term energies relative to $3d^6 4s$ ^{6}D ionization threshold for each symmetry

i	E(Ryds)	Description
		^{7}S^e
1	–.11812	$3d^6 4s$ ^{6}D $4d$
2	–.06500	$3d^6 4s$ ^{6}D $5d$
3	–.04121	$3d^6 4s$ ^{6}D $6d$
4	–.02846	$3d^6 4s$ ^{6}D $7d$
5	–.02083	$3d^6 4s$ ^{6}D $8d$
6	–.01590	$3d^6 4s$ ^{6}D $9d$
		^{7}P^e
1	–.12004	$3d^6 4s$ ^{6}D $4d$
2	–.06656	$3d^6 4s$ ^{6}D $5d$
3	–.05571	$3d^6 4p^2$:
4	–.04147	$3d^6 4s$ ^{6}D $6d$
5	–.02877	$3d^6 4s$ ^{6}D $7d$
6	–.02112	$3d^6 4s$ ^{6}D $8d$
7	–.01617	$3d^6 4s$ ^{6}D $9d$
		^{7}D^e
1	–.18500	$3d^6 4s$ ^{6}D $5s$
2	–.12106	$3d^6 4s$ ^{6}D $4d$
3	–.08858	$3d^6 4s$ ^{6}D $6s$
4	–.06609	$3d^6 4s$ ^{6}D $5d$
5	–.05236	$3d^6 4s$ ^{6}D $7s$
6	–.04180	$3d^6 4s$ ^{6}D $6d$
7	–.04004	$3d^6 4s$ ^{6}D $5g$
8	–.03461	$3d^6 4s$ ^{6}D $8s$
9	–.02886	$3d^6 4s$ ^{6}D $7d$
10	–.02781	$3d^6 4s$ ^{6}D $6g$
11	–.02459	$3d^6 4s$ ^{6}D $9s$
12	–.02169	$3d^6 4s$ ^{6}D $8d$
13	–.02044	$3d^6 4s$ ^{6}D $7g$
14	–.02024	$3d^6 4p^2$:
		^{7}F^e
1	–.12240	$3d^6 4s$ ^{6}D $4d$
2	–.06653	$3d^6 4s$ ^{6}D $5d$
3	–.04199	$3d^6 4s$ ^{6}D $6d$
4	–.04005	$3d^6 4s$ ^{6}D $5g$
5	–.03404	$3d^6 4p^2$:
6	–.02885	$3d^6 4s$ ^{6}D $7d$
7	–.02783	$3d^6 4s$ ^{6}D $6g$
8	–.02108	$3d^6 4s$ ^{6}D $8d$
9	–.02044	$3d^6 4s$ ^{6}D $7g$
10	–.01007	$3d^6 4s$ ^{6}D $9d$
11	.01565	$3d^6 4s$ ^{6}D $8g$
		^{7}G^e
1	–.11971	$3d^6 4s$ ^{6}D $4d$
2	–.06556	$3d^6 4s$ ^{6}D $5d$
3	–.04148	$3d^6 4s$ ^{6}D $6d$
4	–.04006	$3d^6 4s$ ^{6}D $5g$
5	.02861	$3d^6 4s$ ^{6}D $7d$
6	–.02783	$3d^6 4s$ ^{6}D $6g$
7	–.02092	$3d^6 4s$ ^{6}D $8d$
8	–.02045	$3d^6 4s$ ^{6}D $7g$
9	–.02041	$3d^6 4s$ ^{6}D $7i$
10	–.01597	$3d^6 4s$ ^{6}D $9d$
11	–.01566	$3d^6 4s$ ^{6}D $8g$
12	–.01563	$3d^6 4s$ ^{6}D $8i$
		^{7}H^e
1	–.04006	$3d^6 4s$ ^{6}D $5g$
2	–.02784	$3d^6 4s$ ^{6}D $6g$
3	–.02045	$3d^6 4s$ ^{6}D $7g$
4	–.02041	$3d^6 4s$ ^{6}D $7i$
5	–.01566	$3d^6 4s$ ^{6}D $8g$
6	–.01563	$3d^6 4s$ ^{6}D $8i$
7	–.01237	$3d^6 4s$ ^{6}D $9g$
8	–.01235	$3d^6 4s$ ^{6}D $9i$
		^{7}I^e
1	–.04005	$3d^6 4s$ ^{6}D $5g$
2	–.02782	$3d^6 4s$ ^{6}D $6g$
3	–.02044	$3d^6 4s$ ^{6}D $7g$
4	–.02041	$3d^6 4s$ ^{6}D $7i$
5	–.01565	$3d^6 4s$ ^{6}D $8g$
6	–.01563	$3d^6 4s$ ^{6}D $8i$
7	–.01236	$3d^6 4s$ ^{6}D $9g$
		^{7}P^o
1	.37510	$3d^6 4s$ ^{6}D $4p$
2	–.21298	$3d^5 4s^2$ ^{6}S $4p$
3	–.12039	$3d^6 4s$ ^{6}D $5p$
4	–.06545	$3d^6 4s$ ^{6}D $6p$
5	–.06309	$3d^6 4s$ ^{6}D $4f$
6	–.04134	$3d^6 4s$ ^{6}D $7p$
7	–.04036	$3d^6 4s$ ^{6}D $5f$
8	–.02851	$3d^6 4s$ ^{6}D $8p$
9	–.02800	$3d^6 4s$ ^{6}D $6f$
10	–.02085	$3d^6 4s$ ^{6}D $9p$
11	–.02056	$3d^6 4s$ ^{6}D $7f$
		^{7}D^o
1	–.41312	$3d^6 4s$ ^{6}D $4p$
2	–.12610	$3d^6 4s$ ^{6}D $5p$
3	–.06746	$3d^6 4s$ ^{6}D $6p$
4	–.06325	$3d^6 4s$ ^{6}D $4f$
5	–.04232	$3d^6 4s$ ^{6}D $7p$
6	–.04046	$3d^6 4s$ ^{6}D $5f$
7	–.02900	$3d^6 4s$ ^{6}D $8p$
8	–.02806	$3d^6 4s$ ^{6}D $6f$
9	–.02119	$3d^6 4s$ ^{6}D $9p$
10	–.02059	$3d^6 4s$ ^{6}D $7f$
		^{7}F^o
1	–.38517	$3d^6 4s$ ^{6}D $4p$
2	–.12277	$3d^6 4s$ ^{6}D $5p$
3	–.06624	$3d^6 4s$ ^{6}D $6p$
4	–.06339	$3d^6 4s$ ^{6}D $4f$
5	–.04172	$3d^6 4s$ ^{6}D $7p$
6	–.04053	$3d^6 4s$ ^{6}D $5f$
7	–.02872	$3d^6 4s$ ^{6}D $8p$
8	–.02811	$3d^6 4s$ ^{6}D $6f$
9	–.02778	$3d^6 4s$ ^{6}D $6h$
10	–.02098	$3d^6 4s$ ^{6}D $9p$
11	–.02062	$3d^6 4s$ ^{6}D $7f$
12	–.02041	$3d^6 4s$ ^{6}D $7h$
		^{7}G^o
1	–.06343	$3d^6 4s$ ^{6}D $4f$
2	–.04056	$3d^6 4s$ ^{6}D $5f$
3	–.02812	$3d^6 4s$ ^{6}D $6f$
4	–.02778	$3d^6 4s$ ^{6}D $6h$
5	–.02063	$3d^6 4s$ ^{6}D $7f$
6	–.02041	$3d^6 4s$ ^{6}D $7h$
7	–.01578	$3d^6 4s$ ^{6}D $8f$
8	–.01563	$3d^6 4s$ ^{6}D $8h$
9	–.01245	$3d^6 4s$ ^{6}D $9f$
10	–.01235	$3d^6 4s$ ^{6}D $9h$
		^{7}H^o
1	–.06322	$3d^6 4s$ ^{6}D $4f$
2	–.04044	$3d^6 4s$ ^{6}D $5f$
3	–.02805	$3d^6 4s$ ^{6}D $6f$
4	–.02778	$3d^6 4s$ ^{6}D $6h$
5	.02058	$3d^6 4s$ ^{6}D $7f$
6	–.02041	$3d^6 4s$ ^{6}D $7h$
7	–.01575	$3d^6 4s$ ^{6}D $8f$
8	–.01563	$3d^6 4s$ ^{6}D $8h$
9	–.01563	$3d^6 4s$ ^{6}D $8j$
10	–.01243	$3d^6 4s$ ^{6}D $9f$
11	–.01235	$3d^6 4s$ ^{6}D $9h$
12	–.01235	$3d^6 4s$ ^{6}D $9j$
		^{5}S^e
1	–.12114	$3d^6 4s$ ^{6}D $4d$
2	–.06772	$3d^6 4s$ ^{6}D $5d$
3	–.04898	$3d^6 4s$ ^{4}D $4d$
4	–.04149	$3d^6 4s$ ^{6}D $6d$
5	–.02920	$3d^6 4s$ ^{6}D $7d$
6	–.02143	$3d^6 4s$ ^{6}D $8d$
7	–.01637	$3d^6 4s$ ^{6}D $9d$
		^{5}P^e
1	–.42880	$3d^7$ ^{4}P $4s$
2	–.11101	$3d^6 4s$ ^{6}D $4d$
3	–.07805	$3d^7$ ^{4}F $4d$
4	–.06527	$3d^6 4s$ ^{6}D $5d$
5	–.05210	$3d^6 4s$ ^{4}D $4d$
6	–.04493	$3d^7$ ^{4}P $5s$
7	–.03805	$3d^6 4s$ ^{6}D $6d$
8	–.03692	$3d^7$ ^{4}F $5d$
9	–.02740	$3d^6 4s$ ^{6}D $7d$
10	–.02178	$3d^7$ ^{4}F $5g$
11	–.02040	$3d^6 4s$ ^{6}D $8d$
12	–.01821	$3d^7$ ^{4}F $6d$
13	–.01578	$3d^6 4s$ ^{6}D $9d$
14	–.01366	$3d^7$ ^{4}F $7d$:
		^{5}D^e
1	–.56999	$3d^6 4s^2$
2	–.16610	$3d^6 4s$ ^{6}D $5s$
3	–.12670	$3d^6 4s$ ^{6}D $4d$
4	–.10512	$3d^6 4s$ ^{4}D $5s$
5	–.08080	$3d^6 4s$ ^{6}D $6s$
6	–.07796	$3d^7$ ^{4}F $4d$
7	–.07200	$3d^6 4s$ ^{6}D $5d$
8	–.05435	$3d^6 4s$ ^{4}D $4d$
9	–.04934	$3d^6 4s$ ^{6}D $7s$
10	–.04237	$3d^6 4s$ ^{6}D $6d$
11	–.04004	$3d^6 4s$ ^{6}D $5g$
12	–.03703	$3d^7$ ^{4}F $5d$
13	–.03312	$3d^6 4s$ ^{6}D $8s$
14	–.03015	$3d^6 4s$ ^{6}D $7d$
15	–.02781	$3d^6 4s$ ^{6}D $6g$
16	–.02380	$3d^6 4s$ ^{6}D $9s$
17	–.02228	$3d^6 4s$ ^{6}D $8d$
18	–.02178	$3d^7$ ^{4}F $5g$
19	–.02044	$3d^6 4s$ ^{6}D $7g$
		^{5}F^e
1	–.52951	$3d^7$ ^{4}F $4s$
2	–.14690	$3d^7$ ^{4}F $5s$
3	–.11634	$3d^6 4s$ ^{6}D $4d$
4	–.07802	$3d^7$ ^{4}F $4d$
5	–.07115	$3d^6 4s$ ^{6}D $5d$
6	–.06282	$3d^7$ ^{4}F $6s$
7	–.05598	$3d^6 4s$ ^{4}D $4d$
8	–.04051	$3d^6 4s$ ^{6}D $6d$
9	–.04005	$3d^6 4s$ ^{6}D $5g$
10	–.03699	$3d^7$ ^{4}F $5d$
11	–.03142	$3d^7$ ^{4}F $7s$
12	–.02867	$3d^6 4s$ ^{6}D $7d$
13	–.02783	$3d^6 4s$ ^{6}D $6g$
14	–.02185	$3d^6 4s$ ^{6}D $8d$
15	–.02175	$3d^7$ ^{4}F $5g$
16	–.02044	$3d^6 4s$ ^{6}D $7g$
17	–.01790	$3d^7$ ^{4}F $6d$
18	–.01714	$3d^6 4s$ ^{6}D $9d$:
19	–.01565	$3d^6 4s$ ^{6}D $8g$
20	–.01471	$3d^7$ ^{4}F $8s$
		^{5}G^e
1	–.12449	$3d^6 4s$ ^{6}D $4d$
2	–.07818	$3d^7$ ^{4}F $4d$
3	–.07025	$3d^6 4s$ ^{6}D $5d$
4	.05251	$3d^6 4s$ ^{4}D $4d$
5	–.04199	$3d^6 4s$ ^{6}D $6d$
6	–.04006	$3d^6 4s$ ^{6}D $5g$
7	–.03701	$3d^7$ ^{4}F $5d$

Fe-like Fe

i	E(Ryds)	Description
8	−.02973	$3d^64s$ ^{6}D $7d$
9	−.02783	$3d^64s$ ^{6}D $6g$
10	−.02190	$3d^64s$ ^{6}D $8d$
11	−.02178	$3d^7$ ^{4}F $5g$
12	−.02045	$3d^64s$ ^{6}D $7g$
13	−.02041	$3d^64s$ ^{6}D $7i$
14	−.01782	$3d^7$ ^{4}F $6d$
15	−.01675	$3d^64s$ ^{6}D $9d$
16	−.01566	$3d^64s$ ^{6}D $8g$
17	−.01563	$3d^64s$ ^{6}D $8i$
	^{5}H^e	
1	−.07794	$3d^7$ ^{4}F $4d$
2	−.04006	$3d^64s$ ^{6}D $5g$
3	−.03695	$3d^7$ ^{4}F $5d$
4	−.02784	$3d^64s$ ^{6}D $6g$
5	−.02179	$3d^7$ ^{4}F $5g$
6	−.02045	$3d^64s$ ^{6}D $7g$
7	−.02041	$3d^64s$ ^{6}D $7i$
8	−.01778	$3d^7$ ^{4}F $6d$
9	−.01566	$3d^64s$ ^{6}D $8g$
10	−.01563	$3d^64s$ ^{6}D $8i$
11	−.01237	$3d^64s$ ^{6}D $9g$
12	−.01235	$3d^64s$ ^{6}D $9i$
	^{5}I^e	
1	−.04005	$3d^64s$ ^{6}D $5g$
2	−.02782	$3d^64s$ ^{6}D $6g$
3	−.02179	$3d^7$ ^{4}F $5g$
4	−.02044	$3d^64s$ ^{6}D $7g$
5	−.02041	$3d^64s$ ^{6}D $7i$
6	−.01565	$3d^64s$ ^{6}D $8g$
7	−.01563	$3d^64s$ ^{6}D $8i$
8	−.01236	$3d^64s$ ^{6}D $9g$
	^{5}J^e	
1	−.02178	$3d^7$ ^{4}F $5g$
2	−.02041	$3d^64s$ ^{6}D $7i$
3	−.01563	$3d^64s$ ^{6}D $8i$
4	−.01235	$3d^64s$ ^{6}D $9i$
	^{5}S^o	
1	−.20076	$3d^64s$ ^{4}P $4p$
2	−.15718	$3d^7$ ^{4}P $4p$
3	−.04438	$3d^7$ ^{4}F $4f$
4	−.02185	$3d^7$ ^{4}F $5f$
5	−.00961	$3d^7$ ^{4}F $6f$
6	−.00222	$3d^7$ ^{4}F $7f$
	^{5}P^o	
1	−.33294	$3d^64s$ ^{6}D $4p$
2	−.24204	$3d^64s$ ^{4}D $4p$
3	−.17750	$3d^64s$ ^{4}P $4p$
4	−.15835	$3d^7$ ^{4}P $4p$
5	−.13276	$3d^64s$ ^{6}D $5p$:
6	−.10030	$3d^64s$ ^{4}D $4p$:
7	−.06310	$3d^64s$ ^{6}D $4f$
8	−.06168	$3d^64s$ ^{6}D $6p$:
9	−.04464	$3d^7$ ^{4}F $4f$
10	−.04409	$3d^64s$ ^{4}D $5p$
11	−.04036	$3d^64s$ ^{6}D $5f$
12	−.03817	$3d^64s$ ^{6}D $7p$:
13	−.02853	$3d^64s$ ^{6}D $8p$:
14	−.02800	$3d^64s$ ^{6}D $6f$
15	−.02498	$3d^7$ ^{4}P $5p$
16	−.02185	$3d^7$ ^{4}F $5f$
17	−.02056	$3d^64s$ ^{6}D $7f$
18	−.01978	$3d^64s$ ^{6}D $9p$:
19	−.01573	$3d^64s$ ^{6}D $8f$
	^{5}D^o	
1	−.36274	$3d^64s$ ^{6}D $4p$
2	−.28132	$3d^7$ ^{4}F $4p$
3	−.21984	$3d^64s$ ^{4}D $4p$
4	−.17337	$3d^64s$ ^{4}P $4p$
5	−.15461	$3d^64s$ ^{4}F $4p$
6	−.13419	$3d^7$ ^{4}P $4p$
7	−.11125	$3d^64s$ ^{6}D $5p$
8	−.09025	$3d^7$ ^{4}F $5p$
9	−.06323	$3d^64s$ ^{6}D $4f$
10	−.06252	$3d^64s$ ^{6}D $6p$
11	−.04928	$3d^64s$ ^{4}D $5p$
12	−.04440	$3d^7$ ^{4}F $4f$
13	−.04403	$3d^7$ ^{4}F $6p$
14	−.04044	$3d^64s$ ^{6}D $5f$
15	−.03821	$3d^64s$ ^{6}D $7p$
16	−.02805	$3d^64s$ ^{6}D $6f$
17	−.02755	$3d^64s$ ^{6}D $8p$
18	−.02193	$3d^7$ ^{4}F $5f$
19	−.02183	$3d^7$ ^{4}F $7p$
20	−.02060	$3d^64s$ ^{6}D $7f$
21	−.02032	$3d^64s$ ^{6}D $9p$
22	−.01828	$3d^7$ ^{4}F $8p$:
23	−.01574	$3d^64s$ ^{6}D $8f$
	^{5}F^o	
1	−.34746	$3d^64s$ ^{6}D $4p$
2	−.27784	$3d^7$ ^{4}F $4p$
3	−.21901	$3d^64s$ ^{4}D $4p$
4	−.15120	$3d^64s$ ^{4}F $4p$
5	−.11250	$3d^64s$ ^{6}D $5p$
6	−.09639	$3d^7$ ^{4}F $5p$
7	−.07909	$3d^64s$ ^{4}G $4p$
8	−.06340	$3d^64s$ ^{6}D $4f$
9	−.06249	$3d^64s$ ^{6}D $6p$
10	−.04654	$3d^64s$ ^{4}D $5p$
11	−.04453	$3d^7$ ^{4}F $6p$
12	−.04429	$3d^7$ ^{4}F $4f$
13	−.04053	$3d^64s$ ^{6}D $5f$
14	−.03863	$3d^64s$ ^{6}D $7p$
15	−.02810	$3d^64s$ ^{6}D $6f$
16	−.02778	$3d^64s$ ^{6}D $6h$
17	−.02756	$3d^64s$ ^{6}D $8p$
18	−.02191	$3d^7$ ^{4}F $5f$
19	−.02187	$3d^7$ ^{4}F $7p$
20	−.02063	$3d^64s$ ^{6}D $7f$
21	−.02041	$3d^64s$ ^{6}D $7h$
22	−.02023	$3d^64s$ ^{6}D $9p$
23	−.01990	$3d^7$ ^{4}F $8p$:
24	−.01577	$3d^64s$ ^{6}D $8f$
25	−.01563	$3d^64s$ ^{6}D $8h$
	^{5}G^o	
1	−.26659	$3d^7$ ^{4}F $4p$
2	−.15877	$3d^64s$ ^{4}H $4p$
3	−.13955	$3d^64s$ ^{4}F $4p$
4	−.09378	$3d^7$ ^{4}F $5p$
5	−.08167	$3d^64s$ ^{4}G $4p$
6	−.06340	$3d^64s$ ^{6}D $4f$
7	−.04451	$3d^7$ ^{4}F $4f$
8	−.04415	$3d^7$ ^{4}F $6p$
9	−.04050	$3d^64s$ ^{6}D $5f$
10	−.02811	$3d^64s$ ^{6}D $6f$
11	−.02778	$3d^64s$ ^{6}D $6h$
12	−.02192	$3d^7$ ^{4}F $5f$
13	−.02164	$3d^7$ ^{4}F $7p$
14	−.02059	$3d^64s$ ^{6}D $7f$
15	−.02041	$3d^64s$ ^{6}D $7h$
16	−.01577	$3d^64s$ ^{6}D $8f$
17	−.01563	$3d^64s$ ^{6}D $8h$
18	−.01245	$3d^64s$ ^{6}D $9f$
19	−.01235	$3d^64s$ ^{6}D $9h$
	^{5}H^o	
1	−.15170	$3d^64s$ ^{4}H $4p$
2	−.07541	$3d^64s$ ^{4}G $4p$
3	−.06322	$3d^64s$ ^{6}D $4f$
4	−.04448	$3d^7$ ^{4}F $4f$
5	−.04044	$3d^64s$ ^{6}D $5f$
6	−.02805	$3d^64s$ ^{6}D $6f$
7	−.02778	$3d^64s$ ^{6}D $6h$
8	−.02190	$3d^7$ ^{4}F $5f$
9	−.02058	$3d^64s$ ^{6}D $7f$
10	−.02041	$3d^64s$ ^{6}D $7h$
11	−.01575	$3d^64s$ ^{6}D $8f$
12	−.01563	$3d^64s$ ^{6}D $8h$
13	−.01563	$3d^64s$ ^{6}D $8j$
14	−.01243	$3d^64s$ ^{6}D $9f$
15	−.01235	$3d^64s$ ^{6}D $9h$
16	−.01235	$3d^64s$ ^{6}D $9j$
	^{5}I^o	
1	−.15314	$3d^64s$ ^{4}H $4p$
2	−.04439	$3d^7$ ^{4}F $4f$
3	−.02778	$3d^64s$ ^{6}D $6h$
4	−.02185	$3d^7$ ^{4}F $5f$
5	−.02041	$3d^64s$ ^{6}D $7h$
6	−.01563	$3d^64s$ ^{6}D $8h$
7	−.01563	$3d^64s$ ^{6}D $8j$
8	−.01235	$3d^64s$ ^{6}D $9h$
9	−.01235	$3d^64s$ ^{6}D $9j$
	^{5}J^o	
1	−.02778	$3d^64s$ ^{6}D $6h$
2	−.02041	$3d^64s$ ^{6}D $7h$
3	−.01563	$3d^64s$ ^{6}D $8h$
4	−.01563	$3d^64s$ ^{6}D $8j$
5	−.01235	$3d^64s$ ^{6}D $9h$
6	−.01235	$3d^64s$ ^{6}D $9j$
	^{3}S^e	
1	−.04388	$3d^64s$ ^{4}D $4d$
	^{3}S^o	
1	−.05450	$3d^7$ ^{4}P $4p$
2	−.04428	$3d^7$ ^{4}F $4f$
3	−.02188	$3d^7$ ^{4}F $5f$
4	−.01560	$3d^7$ ^{4}P $5p$:
5	−.00967	$3d^7$ ^{4}F $6f$
6	−.00231	$3d^7$ ^{4}F $7f$
	^{3}P^e	
1	−.27196	$3d^64s^2$
2	−.24803	$3d^7$ ^{4}P $4s$
3	−.18716	$3d^8$:
4	−.08150	$3d^7$ ^{4}F $4d$
5	−.04368	$3d^7$ ^{4}P $5s$
6	−.03962	$3d^7$ ^{4}F $5d$
7	−.02968	$3d^7$ ^{2}P $4s$
8	−.02423	$3d^64s$ ^{4}D $4d$
9	−.02190	$3d^7$ ^{4}F $5g$
10	−.01970	$3d^7$ ^{4}F $6d$
11	−.00968	$3d^7$ ^{4}F $6g$
12	−.00852	$3d^7$ ^{4}F $7d$
13	−.00231	$3d^7$ ^{4}F $7g$
14	−.00167	$3d^7$ ^{4}F $8d$
	^{3}P^o	
1	−.25634	$3d^64s$ ^{4}D $4p$
2	−.06462	$3d^7$ ^{4}P $4p$
3	−.04446	$3d^7$ ^{4}F $4f$
4	−.04368	$3d^64s$ ^{4}D $5p$
5	−.02191	$3d^7$ ^{4}F $5f$
6	−.01720	$3d^7$ ^{4}P $5p$:
7	−.00968	$3d^7$ ^{4}F $6f$
8	−.00231	$3d^7$ ^{4}F $7f$
	^{3}D^e	
1	−.15323	$3d^64s^2$
2	−.10801	$3d^64s$ ^{4}D $4s$
3	−.09106	$3d^7$ ^{4}F $4d$:
4	−.04677	$3d^64s$ ^{4}D $4d$:
5	−.04325	$3d^7$ ^{4}F $5d$
6	−.02190	$3d^7$ ^{4}F $5g$

Fe-like Fe

i	E(Ryds)	Description	i	E(Ryds)	Description	i	E(Ryds)	Description
7	−.02134	$3d^7$ ^{4}F $6d$	8	−.02193	$3d^7$ ^{4}F $5f$	4	−.00970	$3d^7$ ^{4}F $6f$
8	−.01486	$3d^64s$ ^{4}D $5s$	9	−.01866	$3d^7$ ^{4}F $7p$	5	−.00968	$3d^7$ ^{4}F $6h$
9	−.00968	$3d^7$ ^{4}F $6g$	10	−.00970	$3d^7$ ^{4}F $6f$	6	−.00233	$3d^7$ ^{4}F $7f$
10	−.00948	$3d^7$ ^{2}D $4s$	11	−.00968	$3d^7$ ^{4}F $6h$	7	−.00231	$3d^7$ ^{4}F $7h$
11	−.00932	$3d^7$ ^{4}F $7d$	12	−.00796	$3d^7$ ^{4}F $8p$	8	−.00066	$3d^64s$ ^{4}D $4f$:
12	−.00231	$3d^7$ ^{4}F $7g$	13	−.00233	$3d^7$ ^{4}F $7f$			**^{3}I^e**
13	−.00208	$3d^7$ ^{4}F $8d$	14	−.00231	$3d^7$ ^{4}F $7h$	1	−.02190	$3d^7$ ^{4}F $5g$
		^{3}D^o	15	−.00128	$3d^7$ ^{4}F $9p$	2	−.00968	$3d^7$ ^{4}F $6g$
1	−.28064	$3d^64s$ ^{4}D $4p$			**^{3}G^e**	3	−.00231	$3d^7$ ^{4}F $7g$
2	−.15095	$3d^7$ ^{4}F $4p$	1	−.22790	$3d^64s^2$	4	−.00231	$3d^7$ ^{4}F $7i$
3	−.07369	$3d^7$ ^{4}F $5p$	2	−.09162	$3d^7$ ^{4}F $4d$:			**^{3}I^o**
4	−.06823	$3d^7$ ^{4}P $4p$	3	−.05799	$3d^7$ ^{2}G $4s$	1	−.06053	$3d^7$ ^{2}H $4p$:
5	−.04986	$3d^64s$ ^{4}D $5p$	4	−.04586	$3d^64s$ ^{4}D $4d$:	2	−.04440	$3d^7$ ^{4}F $4f$
6	−.04442	$3d^7$ ^{4}F $4f$	5	−.04339	$3d^7$ ^{4}F $5d$	3	−.02190	$3d^7$ ^{4}F $5f$
7	−.03850	$3d^7$ ^{2}P $4p$:	6	−.02190	$3d^7$ ^{4}F $5g$	4	−.00968	$3d^7$ ^{4}F $6f$
8	−.03528	$3d^7$ ^{4}F $6p$	7	−.02140	$3d^7$ ^{4}F $6d$	5	−.00968	$3d^7$ ^{4}F $6h$
9	−.02192	$3d^7$ ^{4}F $5f$	8	−.00968	$3d^7$ ^{4}F $6g$	6	−.00231	$3d^7$ ^{4}F $7f$
10	−.01783	$3d^7$ ^{4}F $7p$	9	−.00938	$3d^7$ ^{4}F $7d$	7	−.00231	$3d^7$ ^{4}F $7h$
11	−.00969	$3d^7$ ^{4}F $6f$	10	−.00231	$3d^7$ ^{4}F $7g$			**^{3}J^e**
12	−.00968	$3d^7$ ^{4}F $6h$	11	−.00231	$3d^7$ ^{4}F $7i$	1	−.02190	$3d^7$ ^{4}F $5g$
13	−.00747	$3d^7$ ^{4}F $8p$	12	−.00211	$3d^7$ ^{4}F $8d$	2	−.00968	$3d^7$ ^{4}F $6g$
14	−.00267	$3d^7$ ^{2}D $4p$:			**^{3}G^o**	3	−.00231	$3d^7$ ^{4}F $7i$
15	−.00232	$3d^7$ ^{4}F $7f$	1	−.17425	$3d^7$ ^{4}F $4p$	4	−.00231	$3d^7$ ^{4}F $7g$
16	−.00231	$3d^7$ ^{4}F $7h$	2	−.08297	$3d^7$ ^{4}F $5p$			**^{3}J^o**
17	−.00090	$3d^7$ ^{4}F $9p$	3	−.05880	$3d^7$ ^{2}G $4p$:	1	−.00968	$3d^7$ ^{4}F $6h$
		^{3}F^e	4	−.04446	$3d^7$ ^{4}F $4f$	2	−.00231	$3d^7$ ^{4}F $7h$
1	−.34452	$3d^7$ ^{4}F $4s$	5	−.04111	$3d^7$ ^{4}F $6p$			**^{1}S^e**
2	−.27119	$3d^64s^2$:	6	−.02219	$3d^7$ ^{2}H $4p$:	1	−.11384	$3d^64s^2$
3	−.24388	$3d^64s$ ^{6}D $0s$	7	−.02189	$3d^7$ ^{4}F $5f$			**^{1}P^e**
4	−.13232	$3d^7$ ^{4}F $5s$	8	−.01983	$3d^7$ ^{4}F $7p$	1	−.02937	$3d^7$ ^{2}P $4s$
5	−.08245	$3d^8$:	9	−.00970	$3d^7$ ^{4}F $6f$			**^{1}D^e**
6	−.06110	$3d^7$ ^{4}F $6s$	10	−.00968	$3d^7$ ^{4}F $6h$	1	−.17129	$3d^64s^2$
7	−.03971	$3d^7$ ^{4}F $4d$:	11	−.00880	$3d^7$ ^{4}F $8p$	2	−.08276	$3d^8$:
8	−.03725	$3d^64s$ ^{4}D $4d$:	12	−.00233	$3d^7$ ^{4}F $7f$	3	−.00902	$3d^7$ ^{2}D $4s$
9	−.02800	$3d^7$ ^{4}F $7s$	13	−.00231	$3d^7$ ^{4}F $7h$			**^{1}D^o**
10	−.02190	$3d^7$ ^{4}F $5g$	14	−.00179	$3d^7$ ^{4}F $9p$	1	−.01768	$3d^7$ ^{2}P $4p$:
11	−.01951	$3d^7$ ^{4}F $5d$:			**^{3}H^e**			**^{1}F^e**
12	−.01380	$3d^7$ ^{4}F $8s$	1	−.28090	$3d^64s^2$	1	−.02022	$3d^64s^2$:
13	−.00968	$3d^7$ ^{4}F $6g$	2	−.09117	$3d^7$ ^{4}F $4d$:			**^{1}F^o**
14	−.00833	$3d^7$ ^{4}F $6d$:	3	−.04339	$3d^7$ ^{4}F $5d$	1	−.00055	$3d^7$ ^{2}G $4p$:
15	−.00498	$3d^7$ ^{4}F $9s$	4	−.02190	$3d^7$ ^{4}F $5g$			**^{1}G^e**
16	−.00231	$3d^7$ ^{4}F $7g$	5	−.02133	$3d^7$ ^{4}F $6d$	1	−.16966	$3d^64s^2$:
17	−.00231	$3d^7$ ^{4}F $7i$	6	−.01484	$3d^7$ ^{2}H $4s$	2	−.12468	$3d^8$:
18	−.00149	$3d^7$ ^{4}F $7d$.	7	−.00968	$3d^7$ ^{4}F $6g$	3	−.05738	$3d^7$ ^{2}G $4s$
		^{3}F^o	8	−.00933	$3d^7$ ^{4}F $7d$			**^{1}G^o**
1	−.27302	$3d^64s$ ^{4}D $4p$	9	−.00231	$3d^7$ ^{4}F $7g$	1	−.03368	$3d^7$ ^{2}G $4p$.
2	−.15883	$3d^7$ ^{4}F $4p$	10	−.00231	$3d^7$ ^{4}F $7i$			**^{1}H^e**
3	−.07612	$3d^7$ ^{4}F $5p$	11	−.00209	$3d^7$ ^{4}F $8d$	1	−.01445	$3d^7$ ^{2}H $4s$
4	−.04936	$3d^64s$ ^{4}D $5p$			**^{3}H^o**			**^{1}I^e**
5	−.04444	$3d^7$ ^{4}F $4f$	1	−.04445	$3d^7$ ^{4}F $4f$	1	−.18024	$3d^64s^2$
6	−.03812	$3d^7$ ^{4}F $6p$	2	−.03617	$3d^7$ ^{2}G $4p$:			**^{1}I^o**
7	−.02659	$3d^64s$ ^{4}D $6p$:	3	−.02194	$3d^7$ ^{4}F $5f$	1	−.03595	$3d^7$ ^{2}H $4p$:

Fe-like Fe

Energies in ascending order from ground state for terms with effective $n \leq 4.0$, $L \leq 6$

Term	i	E(Ryds)	Term	i	E(Ryds)	Term	i	E(Ryds)	Term	i	E(Ryds)	Term	i	E(Ryds)
$^5D^e$	1	0.00000	$^5F^o$	3	0.35097	$^5D^o$	6	0.43580	$^5G^o$	4	0.47621	$^3D^o$	4	0.50175
$^5F^e$	1	0.04048	$^7P^o$	2	0.35701	$^5P^o$	5	0.43722	$^3G^e$	2	0.47837	$^5S^e$	2	0.50227
$^5P^e$	1	0.14119	$^5S^o$	1	0.36923	$^3F^e$	4	0.43767	$^3H^e$	2	0.47882	$^7D^o$	3	0.50253
$^7D^o$	1	0.15687	$^3P^e$	3	0.38283	$^5D^e$	3	0.44329	$^3D^e$	3	0.47893	$^7P^e$	2	0.50343
$^7F^o$	1	0.18481	$^7D^e$	1	0.38499	$^7D^o$	2	0.44389	$^5D^o$	8	0.47974	$^7F^e$	2	0.50346
$^7P^o$	1	0.19480	$^1I^e$	1	0.38975	$^1G^e$	2	0.44531	$^7D^e$	3	0.48141	$^7F^o$	3	0.50375
$^5D^o$	1	0.20725	$^5P^o$	3	0.39249	$^5G^e$	1	0.44550	$^3G^o$	2	0.48702	$^7D^e$	4	0.50389
$^5F^o$	1	0.22253	$^3G^o$	1	0.39574	$^7F^o$	2	0.44722	$^1D^e$	2	0.48723	$^7G^e$	2	0.50443
$^3F^e$	1	0.22547	$^5D^o$	4	0.39662	$^7F^e$	1	0.44759	$^3F^e$	5	0.48754	$^7P^o$	4	0.50454
$^5P^o$	1	0.23705	$^1D^e$	1	0.39870	$^5S^e$	1	0.44885	$^5G^o$	5	0.48832	$^5P^e$	4	0.50472
$^5D^o$	2	0.28866	$^1G^e$	1	0.40033	$^7D^e$	2	0.44893	$^3P^e$	4	0.48848	$^7S^e$	2	0.50499
$^3H^e$	1	0.28909	$^5D^e$	2	0.40389	$^7P^o$	3	0.44960	$^5D^e$	5	0.48918	$^3P^o$	2	0.50537
$^3D^o$	1	0.28935	$^3F^o$	2	0.41116	$^7P^e$	1	0.44995	$^5F^o$	7	0.49089	$^7G^o$	1	0.50656
$^5F^o$	2	0.29215	$^5G^o$	2	0.41122	$^7G^e$	1	0.45028	$^5G^e$	2	0.49181	$^5F^o$	8	0.50659
$^3F^o$	1	0.29697	$^5P^o$	4	0.41164	$^7S^e$	1	0.45187	$^5P^e$	3	0.49193	$^5G^o$	6	0.50659
$^3P^e$	1	0.29803	$^5S^o$	2	0.41281	$^5F^e$	3	0.45365	$^5F^e$	4	0.49196	$^7F^o$	4	0.50660
$^3F^e$	2	0.29879	$^5D^o$	5	0.41538	$^1S^e$	1	0.45615	$^5D^e$	6	0.49203	$^7D^o$	4	0.50673
$^5G^o$	1	0.30340	$^3D^e$	1	0.41676	$^5F^o$	5	0.45748	$^5H^e$	1	0.49204	$^5D^o$	9	0.50675
$^3P^o$	1	0.31365	$^5I^o$	1	0.41685	$^5P^e$	2	0.45838	$^3F^o$	3	0.49386	$^5H^o$	3	0.50677
$^3P^e$	2	0.32195	$^5H^o$	1	0.41828	$^5D^o$	7	0.45874	$^5H^o$	2	0.49458	$^7H^o$	1	0.50677
$^3F^e$	3	0.32611	$^5F^o$	4	0.41879	$^3D^e$	2	0.46197	$^3D^o$	3	0.49629	$^5P^o$	7	0.50689
$^5P^o$	2	0.32795	$^3D^o$	2	0.41904	$^5D^e$	4	0.46487	$^5D^e$	7	0.49799	$^7P^o$	5	0.50690
$^3G^e$	1	0.34209	$^5F^e$	2	0.42309	$^5P^o$	6	0.46968	$^5F^e$	5	0.49884	$^5F^e$	6	0.50717
$^5D^o$	3	0.35015	$^5G^o$	3	0.43043	$^5F^o$	6	0.47360	$^5G^e$	3	0.49974	$^5D^o$	10	0.50746

gf-values for transitions involving terms with effective $n \leq 4.0$, $L \leq 6$

i i'	gf_L	i i'	gf_L	i i'	gf_L	i i'	gf_L	i i'	gf_L	i i'	gf_L
	$^5S^e$–$^5P^o$	2 2	1.66E−3	2 7	−4.56E+0	1 4	−1.26E+0	3 3	9.97E−1		$^5D^e$–$^5P^o$
1 1	1.10E+0	3 1	3.08E−3	3 1	1.09E−2	1 5	−5.54E−1	3 4	5.35E−1	1 1	−5.98E−1
1 2	3.05E−1	3 2	5.93E−3	3 2	6.05E−3	1 6	−4.34E+0	3 5	2.80E−2	1 2	−5.16E+0
1 3	5.13E−2	4 1	1.55E−2	3 3	2.17E−4	1 7	−3.29E−1	3 6	3.48E−2	1 3	−7.39E−2
1 4	1.16E−1	4 2	5.11E−3	3 4	1.76E−3	1 8	−2.07E−1	3 7	1.69E+0	1 4	−1.61E+0
1 5	2.79E−1		$^5P^e$–$^5P^o$	3 5	1.27E−2	1 9	−8.21E−2	3 8	3.08E+0	1 5	−8.23E−2
1 6	−4.93E−1	1 1	−2.34E−1	3 6	4.87E−4	1 10	−6.52E−2	3 9	−7.80E−4	1 6	−7.03E−1
1 7	−4.08E+0	1 2	−7.91E−2	3 7	−1.04E−3	2 1	5.12E−1	3 10	−1.21E−1	1 7	−2.28E−4
2 1	4.81E−2	1 3	−2.82E−1	4 1	9.87E−3	2 2	1.90E−2	4 1	2.10E−2	2 1	2.00E+0
2 2	3.07E−3	1 4	−9.59E−2	4 2	2.74E−1	2 3	6.99E−1	4 2	2.71E−2	2 2	3.99E−1
2 3	2.20E−4	1 5	−2.72E+0	4 3	3.59E−2	2 4	7.35E−2	4 3	1.25E−1	2 3	2.23E−2
2 4	5.17E−3	1 6	−1.31E+0	4 4	2.28E−2	2 5	2.12E−3	4 4	1.37E−1	2 4	−3.86E−2
2 5	4.18E−1	1 7	−3.33E−2	4 5	1.19E+0	2 6	1.09E−2	4 5	3.91E−2	2 5	−1.81E+0
2 6	2.35E+0	2 1	1.73E+0	4 6	3.73E+0	2 7	−1.52E−2	4 6	2.96E−1	2 6	−4.37E+0
2 7	−1.45E+0	2 2	1.14E+0	4 7	−2.77E−1	2 8	−1.81E−1	4 7	9.13E−1	2 7	−5.30E−3
	$^5P^e$–$^5S^o$	2 3	6.78E−2		$^5P^e$–$^5D^o$	2 9	−9.37E+0	4 8	3.67E−1	3 1	1.69E+0
1 1	−6.86E−1	2 4	3.53E−1	1 1	−7.34E−2	2 10	−3.85E−1	4 9	−6.79E−1	3 2	6.10E−1
1 2	−7.93E−1	2 5	9.34E−1	1 2	−1.83E−2	3 1	1.62E−1	4 10	−5.49E−1	3 3	6.30E−2
2 1	4.45E−3	2 6	−1.19E+0	1 3	−4.17E−1	3 2	1.63E+0			3 4	1.35E−1

Fe-like Fe

i	i'	gf_L	i	i'	gf_L	i	i'	gf_L	i	i'	gf_L	i	i'	gf_L	i	i'	gf_L
3	5	1.94E−1	3	1	2.71E+0	1	2	−1.57E+0	7	6	5.45E−2	5	9	−2.04E−2	5	8	−1.32E+0
3	6	−9.95E−1	3	2	2.90E−2	1	3	−1.59E+1	7	7	2.30E−2	5	10	−4.03E+0	6	1	1.88E−1
3	7	−5.02E−1	3	3	1.02E+0	1	4	−5.58E−2	7	8	−6.78E+0	6	1	1.64E−1	6	2	2.97E−1
4	1	7.12E−1	3	4	5.38E−1	1	5	−6.51E−1			$^5\mathbf{F}^e$–$^5\mathbf{D}^o$	6	2	1.31E−1	6	3	1.59E−1
4	2	1.35E+0	3	5	7.89E−3	1	6	−8.09E−2	1	1	−2.43E−1	6	3	2.47E−1	6	4	1.97E−2
4	3	1.43E−2	3	6	1.01E−1	1	7	−3.76E−2	1	2	−4.70E+0	6	4	1.09E−3	6	5	2.22E−4
4	4	4.54E−1	3	7	−1.33E+0	1	8	−5.39E−4	1	3	−2.34E+0	6	5	5.24E−6	6	6	8.47E+0
4	5	1.45E−1	3	8	−2.79E−1	2	1	4.31E+0	1	4	−2.39E+0	6	6	4.17E−2	6	7	1.48E+0
4	6	−3.31E−2	3	9	−5.00E+0	2	2	4.37E−2	1	5	−1.93E−1	6	7	2.01E−2	6	8	7.95E−3
4	7	−1.15E−2	3	10	−1.52E−1	2	3	1.81E+0	1	6	−6.21E−4	6	8	7.55E+0			$^5\mathbf{F}^e$–$^5\mathbf{G}^o$
5	1	6.90E−2	4	1	8.16E−1	2	4	−4.15E−2	1	7	−3.13E−1	6	9	1.94E−4	1	1	−1.06E+1
5	2	1.91E−1	4	2	1.67E+0	2	5	−1.47E+1	1	8	−4.82E−1	6	10	−2.28E−3	1	2	−2.78E+0
5	3	3.16E−4	4	3	1.54E+0	2	6	−3.60E−1	1	9	−5.68E−5			$^5\mathbf{F}^e$–$^5\mathbf{F}^o$	1	3	−1.58E+0
5	4	3.54E−6	4	4	1.29E−1	2	7	−2.33E−1	1	10	−1.71E−1	1	1	−2.23E−2	1	4	−2.36E−1
5	5	8.04E−1	4	5	2.03E−2	2	8	−1.62E−2	2	1	8.75E−1	1	2	−8.14E+0	1	5	−5.04E−1
5	6	3.59E+0	4	6	5.66E−2	3	1	2.42E−1	2	2	2.51E+0	1	3	−1.47E+0	1	6	−1.52E−1
5	7	−1.70E−1	4	7	2.06E−1	3	2	5.70E−2	2	3	3.77E−1	1	4	−1.55E+0	2	1	7.62E+0
6	1	2.03E−3	4	8	−7.41E−3	3	3	3.79E−1	2	4	1.27E−1	1	5	−7.67E−2	2	2	1.02E−1
6	2	3.14E−2	4	9	−1.22E−1	3	4	6.74E−3	2	5	5.93E−3	1	6	−5.90E−5	2	3	−3.34E−2
6	3	1.26E−4	4	10	−2.59E+0	3	5	−1.84E−1	2	6	−1.69E−3	1	7	−2.24E+0	2	4	−1.89E+1
6	4	2.43E−3	5	1	2.00E−1	3	6	−1.42E−3	2	7	−2.77E+0	1	8	−1.75E−2	2	5	−3.96E−1
6	5	5.63E−2	5	2	3.83E 1	3	7	3.24E−4	2	8	−7.94E+0	2	1	6.93E−1	2	6	−9.72E−1
6	6	1.85E−1	5	3	9.93E−2	3	8	−1.22E+1	2	9	−1.70E−2	2	2	4.66E+0	3	1	4.14E−2
6	7	1.35E 2	5	4	4.83E−2	4	1	1.33E+0	2	10	−6.01E−2	2	3	3.32E−1	3	2	1.29E−7
7	1	1.33E−2	5	5	1.22E−3	4	2	1.16E+0	3	1	2.61E+0	2	4	1.72E−2	3	3	5.45E 5
7	2	1.60E−4	5	6	2.62E−1	4	3	3.49E+0	3	2	2.52E−4	2	5	−1.16E+0	3	4	−1.15E+0
7	3	1.06E−2	5	7	4.17E+0	4	4	9.27E−3	3	3	3.51E+0	2	6	−1.28E+1	3	5	−8.86E−2
7	4	5.00E−2	5	8	1.64E+0	4	5	2.40E−1	3	4	1.10E+0	2	7	−1.37E+0	3	6	−2.20E+1
7	5	1.12E+0	5	9	−1.80E−1	4	6	−2.50E−3	3	5	2.13E−2	2	8	−2.30E−1	4	1	6.11E−1
7	6	3.30E+0	5	10	−1.03E+1	4	7	−1.69E−1	3	6	3.43E−1	3	1	1.19E+0	4	2	4.58E−2
7	7	−5.13E−1	6	1	7.56E−2	4	8	−7.00E−3	3	7	−7.24E−1	3	2	1.02E−1	4	3	4.06E−2
		$^5\mathbf{D}^e$–$^5\mathbf{D}^o$	6	2	1.70E+0	5	1	2.45E−1	3	8	−2.27E+0	3	3	2.02E+0	4	4	1.08E+0
1	1	−6.89E−1	6	3	1.25E+0	5	2	1.71E−1	3	9	−5.23E−1	3	4	5.14E−2	4	5	9.28E−3
1	2	−3.67E+0	6	4	6.70E−1	5	3	3.26E−1	3	10	−6.44E−2	3	5	−3.49E−1	4	6	−1.78E−4
1	3	−8.27E+0	6	5	6.53E−2	5	4	3.19E−2	4	1	9.71E−2	3	6	−9.36E−1	5	1	4.49E−1
1	4	−7.80E−1	6	6	6.24E−2	5	5	8.80E+0	4	2	1.10E+0	3	7	−6.48E−1	5	2	2.94E−3
1	5	−4.95E−2	6	7	2.44E+0	5	6	6.67E−1	4	3	6.61E−1	3	8	−5.96E+0	5	3	1.30E−4
1	6	−2.32E−2	6	8	2.85E+0	5	7	−1.86E−1	4	4	4.23E−1	4	1	2.70E−1	5	4	1.61E+0
1	7	−4.75E−1	6	9	−5.76E−4	5	8	−5.34E−1	4	5	2.61E−2	4	2	3.10E+0	5	5	2.84E−2
1	8	−2.87E−1	6	10	−7.12E−1	6	1	5.92E−2	4	6	3.94E−2	4	3	8.42E−1	5	6	−6.84E+0
1	9	−1.36E−3	7	1	9.20E−3	6	2	1.05E+0	4	7	1.06E+0	4	4	2.92E−1	6	1	4.94E−1
1	10	−7.10E−2	7	2	2.03E−3	6	3	5.31E−1	4	8	2.20E+0	4	5	7.06E−1	6	2	1.88E−2
2	1	2.84E+0	7	3	6.08E−2	6	4	6.89E−2	4	9	6.75E 1	4	6	6.80E+0	6	3	2.87E−2
2	2	5.34E−3	7	4	4.35E−2	6	5	1.10E+0	4	10	−5.97E−2	4	7	5.24E−2	6	4	1.12E+1
2	3	1.21E+0	7	5	2.67E−3	6	6	2.24E+0	5	1	1.12E−1	4	8	−4.35E−4	6	5	2.80E−1
2	4	1.22E−1	7	6	2.34E−1	6	7	6.37E−3	5	2	8.48E−2	5	1	2.13E−1	6	6	3.12E−2
2	5	−1.32E−3	7	7	6.96E+0	6	8	−9.46E−2	5	3	6.06E−4	5	2	2.88E−1			$^5\mathbf{G}^e$–$^5\mathbf{F}^o$
2	6	−5.52E−1	7	8	1.08E+0	7	1	2.89E−3	5	4	1.32E−1	5	3	6.00E−5	1	1	8.91E+0
2	7	8.66E+0	7	9	3.74E+0	7	2	5.59E−2	5	5	1.21E−4	5	4	1.47E−2	1	2	4.35E−3
2	8	−1.25E+0	7	10	−2.49E+0	7	3	1.20E−5	5	6	3.75E−1	5	5	5.10E+0	1	3	5.35E+0
2	9	−7.21E−3			$^5\mathbf{D}^e$–$^5\mathbf{F}^o$	7	4	2.20E−3	5	7	9.12E+0	5	6	3.80E−1	1	4	1.01E−1
2	10	−5.19E−1	1	1	−1.26E+0	7	5	6.15E−1	5	8	1.48E−1	5	7	2.00E−2	1	5	−4.45E+0

Fe-like Fe

i i'	gf_L	i i'	gf_L	i i'	gf_L	i i'	gf_L	i i'	gf_L	i i'	gf_L
1 6	−1.67E−1	1 3	4.66E−1	2 5	−8.64E−1	2 2	1.04E+1	3 4	−2.91E−2	4 4	−8.61E−1
1 7	−2.60E−1	1 4	1.78E+1	3 1	4.21E−1	2 3	1.59E+0	4 1	3.62E−2	5 1	5.03E−2
1 8	−2.83E−1	1 5	1.03E−1	3 2	6.06E−4	2 4	−1.57E−1	4 2	4.41E+0	5 2	3.11E+0
2 1	1.33E−1	1 6	−1.35E−2	3 3	6.39E+0		$^7\mathbf{F}^e$–$^7\mathbf{F}^o$	4 3	−8.54E−1	5 3	−6.20E−1
2 2	4.62E+0		$^5\mathbf{H}^e$–$^5\mathbf{H}^o$	3 4	−1.25E+1	1 1	6.14E+0	4 4	−2.93E−1	5 4	−3.02E−1
2 3	7.75E−1	1 1	3.72E−4	3 5	−5.20E−3	1 2	7.03E−2		$^3\mathbf{D}^e$–$^3\mathbf{P}^o$		$^3\mathbf{F}^e$–$^3\mathbf{F}^o$
2 4	3.39E−1	1 2	−2.54E−4	4 1	8.05E−1	1 3	−1.07E−1	1 1	7.67E−4	1 1	−1.26E−1
2 5	5.38E−1	1 3	−4.19E−7	4 2	1.93E−5	1 4	−8.91E+0	1 2	−5.20E−3	1 2	−6.30E+0
2 6	9.18E+0		$^5\mathbf{H}^e$–$^5\mathbf{I}^o$	4 3	4.61E+0	2 1	1.06E+0	2 1	1.21E+0	1 3	−6.43E−1
2 7	6.16E−2	1 1	8.26E−3	4 4	−4.11E−1	2 2	5.97E+0	2 2	−6.97E−1	2 1	1.20E−3
2 8	−4.35E−4		$^7\mathbf{S}^e$–$^7\mathbf{P}^o$	4 5	−1.83E−1	2 3	−2.30E−1	3 1	2.18E−2	2 2	−1.10E+0
3 1	2.67E−2	1 1	2.51E+0		$^7\mathbf{D}^e$–$^7\mathbf{D}^o$	2 4	−1.77E+0	3 2	−1.17E−2	2 3	−1.48E−1
3 2	1.37E−2	1 2	7.34E−2	1 1	5.45E+0		$^7\mathbf{G}^e$–$^7\mathbf{F}^o$		$^3\mathbf{D}^e$–$^3\mathbf{D}^o$	3 1	5.84E−3
3 3	6.02E−3	1 3	1.98E−1	1 2	−1.56E+1	1 1	2.35E+1	1 1	2.90E−3	3 2	−1.00E−1
3 4	6.09E−2	1 4	−5.17E−2	1 3	−3.50E−1	1 2	2.27E+0	1 2	−9.35E−5	3 3	−1.07E−2
3 5	2.27E+1	1 5	−6.34E+0	1 4	−1.80E−5	1 3	−7.81E−1	1 3	−6.16E−5	4 1	2.58E−1
3 6	5.22E−1	2 1	5.68E−1	2 1	7.41E+0	1 4	−4.39E−1	1 4	−2.05E−3	4 2	2.66E+0
3 7	1.12E+0	2 2	6.11E−4	2 2	1.16E+0	2 1	5.30E+0	2 1	1.97E+0	4 3	−8.35E+0
3 8	−6.38E−1	2 3	2.34E+0	2 3	−2.85E−1	2 2	2.05E+1	2 2	2.90E−1	5 1	8.49E−2
	$^5\mathbf{G}^e$–$^5\mathbf{G}^o$	2 4	1.79E−1	2 4	−8.65E+0	2 3	2.33E+0	2 3	−2.10E−1	5 2	6.07E+0
1 1	6.82E−2	2 5	−7.24E−1	3 1	6.99E−1	2 4	−4.28E−2	2 4	−5.84E−2	5 3	−1.28E+0
1 2	1.26E−3		$^7\mathbf{P}^e$–$^7\mathbf{P}^o$	3 2	9.80E+0		$^7\mathbf{G}^e$–$^7\mathbf{G}^o$	3 1	1.44E−1		$^3\mathbf{F}^e$–$^3\mathbf{G}^o$
1 3	8.93E−5	1 1	5.36E+0	3 3	−2.09E+1	1 1	−6.67E+0	3 2	4.99E+0	1 1	−7.44E+0
1 4	−2.35E−3	1 2	3.80E−1	3 4	−1.44E−3	2 1	−8.17E−1	3 3	−1.29E+0	1 2	−2.55E−1
1 5	−1.94E−4	1 3	5.83E−2	4 1	1.82E+0		$^7\mathbf{G}^e$–$^7\mathbf{H}^o$	3 4	−3.58E−1	2 1	−6.39E−2
1 6	−4.09E+0	1 4	−1.67E−1	4 2	6.19E+0	1 1	−4.86E+1		$^3\mathbf{D}^e$–$^3\mathbf{F}^o$	2 2	−1.12E−2
2 1	3.48E+0	1 5	−6.01E+0	4 3	1.44E+0	2 1	−6.70E+0	1 1	2.05E−3	3 1	−9.27E−2
2 2	3.13E−1	2 1	2.39E−3	4 4	−1.53E+0		$^7\mathbf{F}^e$–$^7\mathbf{G}^o$	1 2	1.22E−4	3 2	−1.07E−3
2 3	1.20E−1	2 2	4.81E−2		$^7\mathbf{D}^e$–$^7\mathbf{F}^o$	1 1	−3.19E+1	1 3	−5.87E−4	4 1	3.71E+0
2 4	5.41E+0	2 3	5.10E+0	1 1	8.27E+0	2 1	−6.02E+0	2 1	2.75E+0	4 2	−1.16E+1
2 5	2.42E−2	2 4	−1.19E+0	1 2	−2.22E+1		$^3\mathbf{P}^e$–$^3\mathbf{P}^o$	2 2	1.86E−1	5 1	7.31E−1
2 6	−6.43E−5	2 5	−1.18E+0	1 3	−7.84E−1	1 1	−2.81E−2	2 3	−3.77E−2	5 2	1.63E−2
3 1	5.82E−2		$^7\mathbf{P}^e$–$^7\mathbf{D}^o$	1 4	−4.78E−5	1 2	−8.88E−1	3 1	1.18E−1		$^3\mathbf{G}^e$–$^3\mathbf{F}^o$
3 2	4.94E−4	1 1	2.13E+0	2 1	1.00E+0	2 1	1.58E−2	3 2	2.40E+0	1 1	1.34E−3
3 3	1.57E−5	1 2	3.60E−1	2 2	4.63E−2	2 2	−1.41E+0	3 3	−7.65E−1	1 2	−1.18E−2
3 4	8.12E−3	1 3	−1.07E−1	2 3	−4.10E−2	3 1	2.64E−2		$^3\mathbf{F}^e$–$^3\mathbf{D}^o$	1 3	−1.32E−2
3 5	8.09E−5	1 4	−1.21E+1	2 4	−2.08E+1	3 2	−9.22E−2	1 1	−2.70E−1	2 1	4.05E−2
3 6	−1.73E+0	2 1	3.24E+0	3 1	9.62E−1	4 1	1.31E−3	1 2	−4.76E+0	2 2	8.46E+0
	$^5\mathbf{G}^e$–$^5\mathbf{H}^o$	2 2	1.38E+0	3 2	1.45E+1	4 2	−3.99E−3	1 3	−5.56E−1	2 3	−2.58E+0
1 1	3.82E−5	2 3	−4.81E−3	3 3	−2.93E+1		$^3\mathbf{P}^e$–$^3\mathbf{D}^o$	1 4	−1.34E−1		$^3\mathbf{G}^e$–$^3\mathbf{G}^o$
1 2	−2.37E−3	2 4	−2.37E+0	3 4	−3.30E−3	1 1	3.01E−3	2 1	3.33E−3	1 1	−7.28E−3
1 3	−2.99E+1		$^7\mathbf{D}^e$–$^7\mathbf{P}^o$	4 1	2.92E−1	1 2	−3.59E−2	2 2	−3.01E−1	1 2	−8.48E−3
2 1	2.24E−3	1 1	3.64E+0	4 2	7.92E−1	1 3	−4.61E−1	2 3	−3.51E−4	2 1	4.67E+0
2 2	−1.17E−3	1 2	9.03E−2	4 3	1.80E−2	1 4	−1.37E+0	2 4	−3.66E−1	2 2	−1.04E+0
2 3	−2.36E−2	1 3	−9.50E+0	4 4	−3.35E+0	2 1	3.36E−2	3 1	1.49E−3		$^3\mathbf{H}^e$–$^3\mathbf{G}^o$
3 1	4.67E−5	1 4	−4.23E−1		$^7\mathbf{F}^e$–$^7\mathbf{D}^o$	2 2	−1.65E−2	3 2	−1.93E−1	1 1	−4.40E−2
3 2	2.38E−4	1 5	−4.76E−4	1 1	1.24E+1	2 3	−2.83E−1	3 3	−1.32E−2	1 2	−4.26E−2
3 3	−1.31E+1	2 1	4.59E+0	1 2	1.34E+0	2 4	−1.68E+0	3 4	−2.19E−1	2 1	1.48E+1
	$^5\mathbf{H}^e$–$^5\mathbf{G}^o$	2 2	1.46E−1	1 3	−4.29E−1	3 1	7.70E−2	4 1	4.40E−1	2 2	−3.32E+0
1 1	1.10E+1	2 3	−8.54E−2	1 4	−8.45E−1	3 2	−3.12E−1	4 2	1.50E+0		
1 2	9.30E−1	2 4	−7.33E−2	2 1	3.03E+0	3 3	−1.45E−2	4 3	−4.47E+0		

Index to data tables

The data are grouped according to isoelectronic sequence. The following matrix gives the ionization stage (roman) and page number (arabic) corresponding to each ion.

Sequence	S		Ar		Ca		Fe	
He-like	XV	58	XVII	59	XIX	60	XXV	61
Li-like	XIV	62	XVI	63	XVIII	64	XXIV	65
Be-like	XIII	66	XV	68	XVII	70	XXIII	72
B-like	XII	73	XIV	78	XVI	82	XXII	86
C-like	XI	90	XIII	98	XV	107	XXI	116
N-like	X	126	XII	134	XIV	142	XX	151
O-like	IX	162	XI	167	XIII	173	XIX	180
F-like	VIII	188	X	191	XII	195	XVIII	199
Ne-like	VII	204	IX	206	XI	208	XVII	210
Na-like	VI	212	VIII	213	X	214	XVI	215
Mg-like	V	216	VII	218	IX	220	XV	222
Al-like	IV	224	VI	226	VIII	228	XIV	231
Si-like	III	236	V	239	VII	243	XIII	248
P-like	II	261	IV	264	VI	268	XII	274
S-like	I	297	III	299	V	301	XI	307
Cl-like			II	329	IV	332	X	335
Ar-like			I	350	III	352	IX	354
K-like					II	362	VIII	363
Ca-like					I	367	VII	368
Sc-like							VI	372
Ti-like							V	379
V-like							IV	393
Cr-like							III	409
Mn-like							II	417
Fe-like							I	427